Solar Thermal Energy Systems

Werner Platzer · Robert Stieglitz

Solar Thermal Energy Systems

Fundamentals, Technology, Applications

Volume 1

Springer

Werner Platzer
Umwelt und Natürliche Resourcen
Albert-Ludwigs-Universität Freiburg
Freiburg im Breisgau, Baden-Württemberg,
Germany

Robert Stieglitz
Institut für Neutronenphysik und
Reaktortechnik (INR)
KIT
Eggenstein-Leopoldshafen
Germany

ISBN 978-3-031-43172-2 ISBN 978-3-031-43173-9 (eBook)
https://doi.org/10.1007/978-3-031-43173-9

This Springer imprint is published by the registered company Springer Nature Switzerland AG
The registered company address is: Gewerbestrasse 11, 6330 Cham, Switzerland

Paper in this product is recyclable.

Before the final publication of this textbook, my highly esteemed co-author Prof. Robert Stieglitz, who was indispensable and tireless in the production of the book, passed away unexpectedly. He was a great person and an internationally recognized scientist who excelled in his broad knowledge of materials science and renewable energy. His loss affects not only his family, but also his colleagues and the scientific community as a whole.

Preface

Against the background of mankind's fight against climate change and its consequences, technologies and systems to provide heat, cold and electricity from the sun play a key role to support the necessary transformation of our energy systems towards a truly renewable and sustainable one. In solar thermal technologies, solar electromagnetic radiation is primarily transformed into heat, which may serve as a second step to provide us with heat and cold for thermal applications, but also electricity, chemical reactions, and fuels. The aspect of conserving energy and resources is gaining particular importance when introducing new technologies. Solar thermal systems allow the provision of a wide spectrum of temperature ranges for different purposes by means of suitable collectors. The scope of heat applications exceeds by far the use for room heating and hot water preparation in buildings. The use of solar heat for cooling and refrigeration systems, the integration into heat networks, the customized provision of industrial process heat for all kinds of industries as well as the conversion into electricity in thermal power plants are equally important applications in our current energy matrix. Although the technologies and systems are mature, future optimization using digitalization, new sustainable materials and integrated system designs will revolutionize current products on the market. Considering the technological development, the situation could be compared with that of the car industry before World War II and now. The available solutions already show great potential, but missing market capacity has not yet allowed to develop designs with minimum material content and based on real mass production. On the other hand, raw materials exploitation status and energy and climate-friendly production will force us to improve our designs soon.

This script is addressed to technically interested readers, as well as students in the Bachelor's and Master's area up to interested technically educated career changers, and is modularly structured in two parts. In the first part, the basics are worked out and the conversion of solar radiation into heat is presented, while the second part shows and discusses the applications in building technology, heat process technology, solar thermal power plants as well as solar thermal air conditioning.

Part I

Heat is intrinsically linked to energy, so at the beginning of the first part the concept of energy is highlighted, the temporal consumption profile of heat and electricity is addressed using the example of a household or the heat levels of different industrial processes. The role of heat, electricity, and fuels, which must be generated CO2-neutrally in the future as far as possible, is shown for our economies worldwide. A key lesson is that demand and generation do not necessarily coincide, and therefore, storage and load-shifting have to be developed if renewables have to cover all requirements. The vision is to develop a future decentralized and sector-coupled energy system, at a reasonable cost and with reliable and resilient supply at all times. Chapter 2 we devote to one primary renewable source of energy, our sun with its virtually infinite radiation resource. Solar irradiation on our earth has to consider the sun's extra-terrestrial radiation intensity, atmospheric attenuation, the influence of geographical location and temporal changes. The quantifying of irradiation needs suitable measuring devices for its detection. In Chap. 3 the physical basics for the optical conversion of the short-wave solar radiation into heat, as well as methods for concentrating the radiation, are presented. Measurement methods are used to quantify the optical quality of materials and devices, and mathematical description and modelling are presented for designing and optimizing the optical components. The physical and engineering description of the heat transfer by thermal conduction, thermal radiation and natural convection is included in Chap. 4, where practical problems with coupled heat transfer mechanisms are also highlighted. In general, the generation of heat and its demand occurs at different locations, so the heat must be transported by heat transfer fluids. The flow of these fluids can be either single-phase or multiphase, laminar, or turbulent. Chapter 5 examines these energy and momentum transport in solar thermal systems and discusses fundamentals, approaches to solutions, and points out engineering correlations. In Chap. 6 stationary low-temperature collectors and in Chap. 7 concentrating and tracking high-temperature collectors, solar thermal collector types with their different characteristic properties, operating modes and design alternatives are presented and discussed.

Part II

The second part of the book contains the descriptions, concepts and system designs for several application fields of the thermal energy generated by the collectors. In most solar thermal systems, the ability of storage is already conceptually included, contrary to other renewable energies like grid-integrated photovoltaics and wind, which rely on the flexibility to accept additional power of the grid. Several concepts and designs of thermal energy storage are described first in Chap. 8, which allow decoupling of the generation of heat from the consumption profiles of heat. The focus of Chap. 9 is the targeted use of solar generated low-temperature heat well below 100 °C for domestic applications, namely the production of hot water and the use in the building heating system. The chapter describes not only the essential components and basic algorithms for dimensioning and planning, but also describes the concepts for integration into heat networks. The subject of Chap. 10 is the optimal design and layout of line and point concentrating collectors in a large field consisting of mirrors, absorbers and tracking devices with the aim of efficiently generating heat. Control and regulation systems as well as the operation and maintenance of the solar field are shown. In Chap. 11, solar thermal power plants are addressed, where the heat generation is linked to a heat engine. In order to have a high conversion efficiency of the heat engine, high temperatures above 400 °C are preferable. In addition to the various concepts of solar only thermal power generation, hybrid concepts are also shown and illustrated. Industry as a sector with a huge heat demand is able to use solar thermal heat from low temperatures below 100 °C to a medium range between 100 °C and 400 °C up to more than 1000 °C. Solar heat can be integrated using very different integration concepts into supply for industrial processes. In Chap. 12, starting from the determination of the sectoral heat demand, methods for integration are shown as well as procedures for the classification of solar integration. In the final Chap. 13, methods, procedures and concepts are presented on how cooling and air conditioning can be realized with the help of thermal technologies. Finally, in two appendices, methods for the design of hydraulic components such as pumps and heat exchangers and measuring devices for the detection of thermal variables are presented, since these are often required for the design of solar thermal systems but are not specific to solar thermal energy.

The script is designed in such a way that it can also be used as a reference work. In particular, the individual chapters are written in such a way that they are largely self-contained and require little back-and-forth flipping. Nevertheless, a basic education in physics is useful for understanding the chains of argumentation and trains of thought.

Freiburg im Breisgau, Germany — Werner Platzer
Eggenstein-Leopoldshafen, Germany — Robert Stieglitz

Acknowledgements

We would like to thank our families for bearing with us, and our institutions for giving us the freedom to dedicate the necessary amount of time to the creation of this book.

Contents

About the Authors

Prof. Dr. rer.nat. Werner Platzer is a physicist and received his Ph.D. from the Albert-Ludwigs-University Freiburg in 1988. He has been working for more than 35 years for the Fraunhofer Institute for Solar Energy Systems ISE in Germany, performing research and development of solar thermal energy material, components and systems, facade technology and energy efficiency in buildings. His focus in research and teaching is on concentrating solar thermal technology for process heat and power. Until 2015, he served as Director of the Division Solar Thermal and Optics. He cofounded the Fraunhofer Centre for Solar Energy Technology in Santiago de Chile and was Executive Director there in the years 2016 and 2017. Since 2018, he coordinated the international business development of Fraunhofer ISE. He is teaching as an honorary Professor at the Faculty for Environment and Sustainability, University of Freiburg. He was appointed chairman of the IEC TC 117 on the standardization for solar thermal electricity (CSP) and has been actively involved in many European and international projects on CSP and Solar Process Heat. He has authored more than 200 articles and conference papers.

Prof. Dr.-Ing. Robert Stieglitz obtained a doctoral degree in engineering in 1994 at the University of Karlsruhe. In his professional career, he has spent some time in the power industry and joined the research centre Karlsruhe (today Karlsruhe Institute of Technology-KIT) in 1995 leading functions in several science fields associated with heat and mass transfer, instrumentation, material qualification and nuclear as well as accelerator technologies. In 2009, he was appointed as director of the Institute for Neutron Physics and Reactor Technology (INR) and he simultaneously holds a chair at the university campus as professor in Applied Thermo-Fluidics. Since 2009, he has been the Director of the CEA KIT Frederic Joliot-Otto Hahn School and served many associations as an advisor and expert such as German Nuclear Reactor Safety Commission, the Alexander von Humboldt Foundation and others. He continued the INR's more than 30-year tradition in solar thermal research through a wide range of contributions to national and international solar thermal R&D projects and industrial collaborations. His scientific record is characterized by more than 240 scientific articles/conference papers, 2 books and some book contributions in several science fields.

Part I

Fundamentals

Energy

1

Abstract

Renewable energies meet an energy market that is characterized by an energy demand or rather an exergy demand, a supply by energy carriers and disposal of end products such as carbon dioxide (CO_2). Energy supply must satisfy the worldwide population growth as well as the increasing per capita energy consumption and this simultaneously without excessive consumption of resources. The potential of regenerative energies thus depends essentially on their properties, with which they can meet the market characteristics. The profile of the energy demand, its follow-up and production costs as well as the boundary conditions determine the energy market and thus define the framework conditions for the development of regenerative energy sources, which in the future must increasingly satisfy the aspects of sustainability and resource efficiency. This chapter sheds light on these aspects, introduces key terms and discusses in particular the energy market environment in which thermal solar energy operates and finally introduces the concept of sector coupling.

Keywords

Energy demand • Energy resources • Impact on climate change • Energy forms and carriers

1.1 Energy Demand and Resources

Energy is a multifaceted term and a number of energy forms and energy carriers exist, the meaning of which will be explained later in this section. One of these forms of energy, which forms the core of this book, is heat produced by thermal solar energy systems, which can be classified as regenerative or renewable energy. Here, the term renewable or regenerative means that the energy carrier is integrally and infinitely available or can renew itself within a human generation time scale. Thus, regenerative energies differ fundamentally from fossil chemical energy

R. Stieglitz and W. Platzer, *Solar Thermal Energy Systems*,
https://doi.org/10.1007/978-3-031-43173-9_1

sources, whose availability is finite. The basically unlimited availability of the carrier does not imply, however, that a temporal and spatial availability is guaranteed at any time. Furthermore, the development of renewable sources requires the use of technical tools and raw materials, which must be available in terms of both time and resources. In a competitive market, therefore, renewable energies face other energy sources, where price is oriented on the demand-based supply of energy. This market price principle based on the match of energy production and consumption at the lowest possible price to be paid by customers, needs to modified and possibly overruled by additional considerations, public directives and regulations, in the light of climate change. These considerations include environmental damage caused by the operation, e. g. the emission of carbon dioxide (CO_2) from the use of fossil fuels, but also the consumption of resources to be used for the production of the generating facilities and for their disposal or recycling in order to ensure the long term sustainability.

A key factor in the integration of renewable energy generation technologies, especially solar sources, into a regional energy service system is therefore its geographical location. Large temporal differences between generation and consumption require either the provision of energy storage technologies or supra-regional energy balancing infrastructures where energy or energy carriers can be transported with low losses, for example electricity grids or chemical carrier networks.

Since the beginning of the industrial revolution, energy demand has been met primarily by the combustion of fossil fuels and the associated CO_2 emissions. Against the background of the climate change, a future supply system must have a footprint that is as CO_2-neutral as possible. It should be emphasized that a holistic assessment of any energy generation technology may not only consider the CO_2 emissions from the operation, but also includes corresponding CO_2 equivalents originating from production and follow-up costs, life cycle duration, resource consumption or recyclability. The environmental footprint, e. g. in terms of land use or consumption, shall also be taken into account in an assessment. Energy supply concepts, which are essentially based on volatile renewable generation sources, also require energy storage concepts as well as distribution grids and thus infrastructures that provide the energy when generation from renewable sources is not able to cover the demand. This so-called residual load, which results from the difference between consumption and demand, can be generated either fossil or largely CO_2-neutral, for example from pumped storage power plants, accumulators, or other storage technologies, but includes not only the generation facility but also corresponding distribution networks, which in turn contain a CO_2 footprint. The design of a future resilient energy supply system therefore includes the integration of energy storage and distribution in addition to a large number of generation technologies to be considered.

Due to variable geographic, meteorological and economic conditions, as well as social and cultural differences in the various regions of the world, there can be no single "*correct*" energy supply concept. Socio-political assessment variables play an dominant role and set different priorities, because they have to bear the investment burden and the consequences resulting from the implementation of the respective energy supply concept.

This chapter highlights the issues raised, introduces key concepts and discusses the energy market environment in which thermal solar energy wants to find its future role.

1.2 Energy Consumption, Economic Performance and Resource Availability

If one analyses the global per capita energy consumption of a person, it quickly becomes apparent that, in addition to historic developments, it depends in particular on the economic performance and the degree of industrialization of the respective country. If gross domestic product per person (GDP) is used as a parameter, a high value is to a significant degree associated with the provision of a large amount of energy per capita. Conversely, economic prosperity (expressed in terms of gross domestic product) allows corresponding amounts of capital to be mobilized via targeted incentive measures to enable effective energy use and thus a reduction in per capita consumption in the first place. Figure 1.1 shows the energy consumption in kilogram oil equivalents (OE; $1 kg\, OE = 41.868\, MJ/kg$) per capita of different countries of the world in relation to the gross national product per capita.

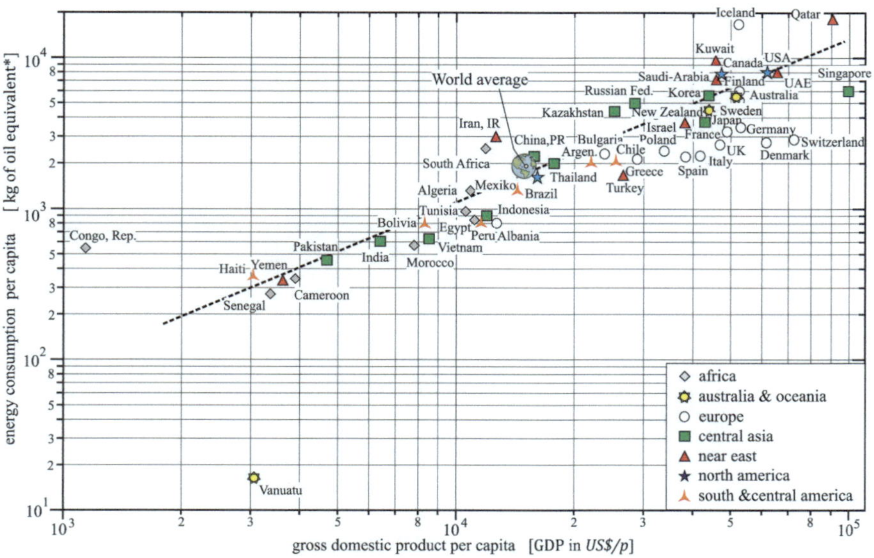

Fig. 1.1 Annual per capita energy consumption in kilograms of oil equivalents as a function of gross national product per capita per year in U.S. Dollars [US$] for various countries (World Bank, 2020). The further to the right of the centre line a country is drawn, the more effectively the energy is used

The dashed line in the graph represents an estimate for which there is no compelling functional relationship. However, it can be deduced from the figure that there are countries whose economies use energy more effectively than others that use more primary energy to achieve the same gross domestic product. For a direct comparison, the composition of the gross domestic product, the geographic structure of the countries, the weather conditions and, most importantly, the degree of industrialization or the importance of industrial production would have to be considered, which are not the focus of this book. Emerging economies often are very inefficient with their energy consumption. Most of them lack the necessary capital to improve their infrastructure. The capital required for a rational energy use is therefore in competition with the capital required to satisfy basic consumption, stabilize political power or expand production. It must be emphasized here that the size of energy consumption per capita, is not necessarily related to CO_2 emissions; for example, although Iceland has the second highest per capita energy consumption in the world, it has a renewable energy share in electricity generation of more than 99%, most of which comes from hydropower and geothermal energy.

The importance and relevance of efficient and at the same time resource-saving energy production becomes particularly obvious if one looks at the development of the world population over time as well as the average global energy consumption per capita, which is shown in Fig. 1.2. Both variables show a monotonous growth over the last twenty years. A further continuation of this trend would require the provision and generation of ever greater amounts of energy, so that measures to reverse it are imperative.

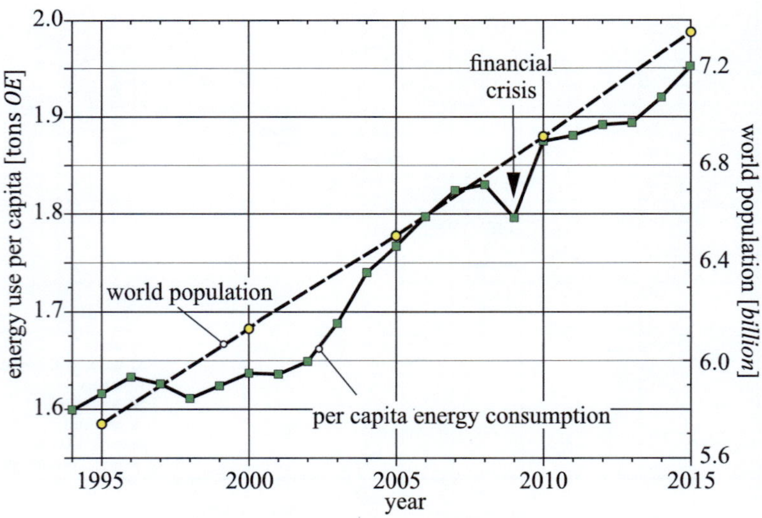

Fig. 1.2 Development of global per capita energy consumption and world population from 1994 to 2015 (World Bank, 2020), (United Nations Population Division, 2020)

Even the slowdown of one of the two trends means a steady further growth of the total energy demand. However, availability in time and place is the anchor and pivot of any competitive market economy, as it is part of all manufacturing processes of economic goods, the provision of services, communication and information as well as mobility. Until today, most of the energy production and supply in the industrialized countries is based on chemical carriers of mostly fossil nature such as oil and gas and/or coal and their established transport chains. If this annual resource consumption figure is covered solely by the available known resource reserves, then the time period in which these resources will be spent can be determined.

The global annual consumption, the secured reserves and their static range are compiled in Table 1.1. The "static range" is a hypothetical quantity and represents the ratio between the reserves and the annual world production currently determined in the survey year. The figure thus indicates how many more years the raw material will be available if consumption remains constant and reserves are fixed. The term reserve differs from that of resource. To be classified as a reserve, an energy resource must meet three conditions: It must have been confirmed by drilling/mining, it must have been accurately measured, and it must be economically recoverable with current technology at current prices. Resources, which are also listed in Table 1.1., on the other hand, describe the geologically proven reserves of an energy carrier that are not yet considered economically recoverable because the extraction requires new and/or expensive technologies. The table shows that the fossil carriers will be consumed within a few generations even at today's consumption rates, not considering their climate-damaging consequences during exploration or combustion here. At a first glance, this seems to apply to nuclear "fuels" as well; however, this is relativized by a factor of at least 60–100 if multiple recycling of the fuel in breeder reactors is considered, and by further orders of magnitude if thorium is used in nuclear reactors. If nuclear fusion is included in the consideration, range time scales result which are in comparable dimensions of renewable energies, however, they are also subject to a raw material demand as well as considerations of nuclear waste and risks..

1.3 Impact of Energy Demand and Population Growth

This continuously growing energy demand, mainly due to the thermal use of fossil fuels, is reflected in a monotonous increase of the CO_2 concentration in the atmosphere. To the natural CO_2 present in the atmosphere the anthropogenic amount is added, i.e. the amount produced by human activities. A large part of this comes from the combustion of fossil carriers, as Fig. 1.2 indicates (Fig. 1.3).

Besides the limitation of fossil fuel reserves, the release of CO_2 into the atmosphere has other serious consequences. The content of CO_2 in the atmosphere has increased from about $280ppm$ to more than $410ppm$ since the beginning of industrialization. The CO_2 present in the atmosphere absorbs the thermal radiation emitted from the earth's surface, especially in the far infrared spectral range, as

Table 1.1 Annual global consumption of different energy sources, "static" range of secured reserves and resources at today's consumption, data extracted from (British Petroleum BP p.l.c., 2021), (ENI SpA, 2021), (OECD/NEA and IAEA, 2020), reference year 2020

Energy carrier	Worldwide consumption in $[EJ = 10^{18} J]$	Static reserve $[a]$	Static resource $[a]$
Oil	173.73	95	160
Gas	152.9	85	300 – 600
Coal	156.56		
Hard coal		115	2600
Lignite		320	5500
Nuclear	23.82		
One through cycle		145	315
Multi-recycling		>8000	$\gg$10000
Renewable energy production	31.72	∞	∞
Electricity production	96.96	–	–

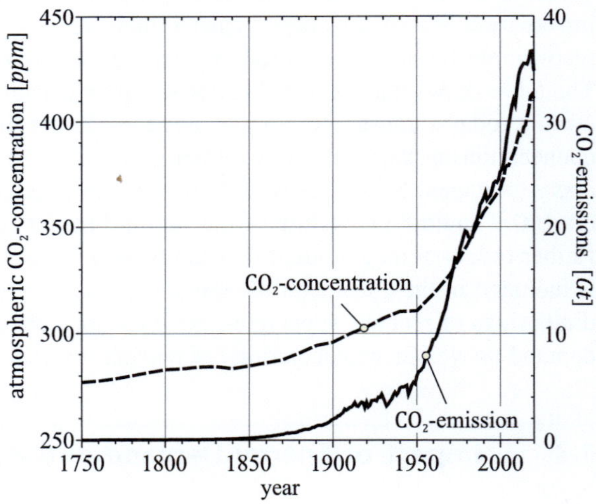

Fig. 1.3 Increase in atmospheric CO_2 concentration and CO_2 emissions from fossil fuel combustion as a function of time (Ritchie and Roser, 2020)

shown in Fig. 1.4. The Earth's atmosphere consists of about 78% nitrogen, about 21% oxygen and about 1% argon. The CO_2 portion amounts to approx. 0.04%. Due to the increased absorption caused by the increased CO_2 concentration in a previously relatively transparent waveband, the heat radiation from the earth into space is reduced and global warming occurs. Other greenhouse gas emissions like the one of methane (see Fig. 1.4) from agriculture and perma frost soil adds to the problem.

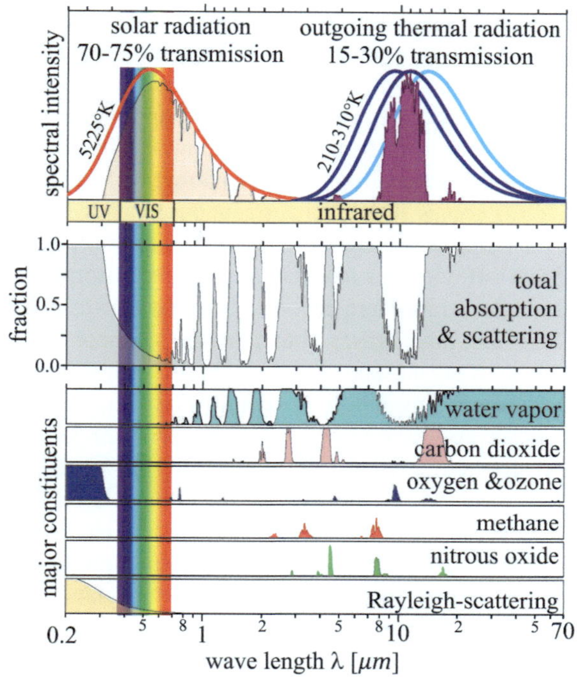

Fig. 1.4 Spectral dependence of radiation incident on and radiated from the Earth, and of the major species or mechanisms contributing to absorption/ scattering

As Fig. 1.4 shows, water vapor is the most important greenhouse gas, and its contribution to the natural greenhouse effect under clear skies is about 60%. . The origin of water vapor is mostly the natural water cycle (ocean - evaporation - precipitation - storage in the ground) and a smaller fraction from volcanism. Water vapor is thus the main source of atmospheric counter radiation and carrier of the "*natural*" greenhouse effect. The natural greenhouse effect is an important effect for the radiation budget of the earth because it leads to a global average temperature of $15°C$, which makes life on earth possible in the first place. Human activities indirectly increase the water vapor content in the atmosphere because global warming increases the air temperature and thus the evaporation rate. This is the most important feedback factor amplifying global warming. In the stratosphere, between $12-50km$ above the earth's surface, the atmosphere is always stably stratified and very dry. There, water vapor occurs only in traces, for example from air traffic and from the oxidation of methane to CO_2 and H_2O, which also contributes to the greenhouse effect. If the natural water vapor content is essential for human life, traces of water vapor present in the stratosphere are considered particularly relevant to the climate.

The earth re-radiates the absorbed heat of the sun in a wavelength range between 3 and $100\,\mu m$ (IR radiation, the blue curve, Fig. 1.4 top right). Only in a narrow range of a radiation window between $8-13\,\mu m$ an almost unhindered radiation is possible (the violet marked area, top right of the picture). The

IR weather satellites also look through the so-called radiation window in the wavelength range between $8-13\,\mu m$. Here the thermal radiation of the earth can escape almost unhindered. This allows a direct view down to the lower cloud cover and Earth's surface. Above all, the water vapor in the atmosphere is responsible for the confinement to this radiation window. It absorbs most of the radiant energy and limits the permeability of the window to $8-17\,\mu m$.

CO_2 comes second as a greenhouse gas. It limits the radiation window mainly in the range of wavelengths between 13 and 17 μm. (The maximum absorption is at 15 μm). If one goes into the range of the absorption lines/bands of the greenhouse gases, the atmosphere becomes impermeable to thermal radiation and thus opaque. There are other gases found in the atmosphere due to anthropogenic effects, such as methane and nitrogen oxides with climate-relevant absorption bands in the infrared wavelength spectrum. The increase in atmospheric concentrations of methane and nitrogen oxides also correlates with the increase in world population and the accompanying industrialization and agricultural use of the Earth's surface, as Fig. 1.5 illustrates. The climatic effectiveness of methane is more than 80 times higher than that of carbon dioxide (CO_2), because the absorption of methane has not yet reached saturation. Anthropogenic methane (CH_4) release is caused, among other things, by the use of natural gas, excessive agricultural use, and livestock farming.

The individual contributions of the trace gases to the greenhouse effect can be seen in Table 1.2. The table also shows the average residence time of the greenhouse gases in the atmosphere. Especially carbon dioxide (CO_2), nitrogen oxides (N_2O) as well as methane (CH_4) show a considerable residence time in the atmosphere, so that their effect on the climate is of a long-term nature. Their importance in terms of climatic consequences is quite of the order of chlorinated fluorochlorohydrocarbons (*CFC*'s), the use of which has been banned in many

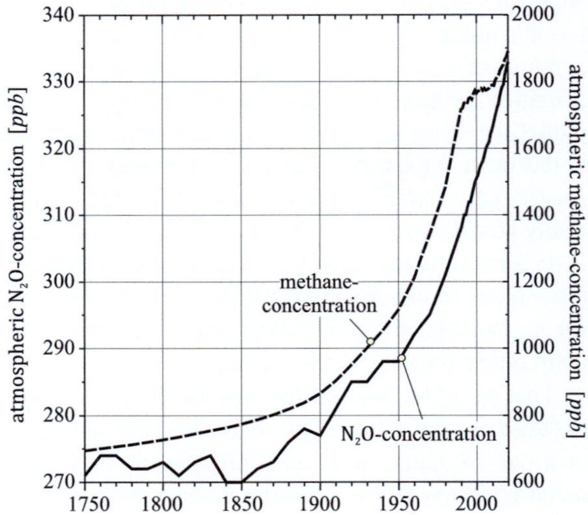

Fig. 1.5 Increase in methane and nitrogen oxide concentrations in [ppb] in the atmosphere since the beginning of industrialization from 1750, (Butler and Montzka, 2021)

Table 1.2 Contributions of climate-relevant trace gases to the greenhouse effect, their residence time in the atmosphere and the temperature effect on thermal radiation (*RF*) and the global warming potential (*GWP* reference value 20 years)

Gas	Residence time in [*a*]	Regional impact scale	Radiative forcing *RF* in $[W/m^2]$	Global warming potential*1 GWP_{20} [/]
H_2O	10^{-2}	Continental	$10^{-3} - 10^{-2}$	0
H_2	4 – 7	Local	*2	33
CO_2	~30	Global	2.16	1
CH_4	10 – 12	Global	0.54	84
N_2O	110	Global	0.21	264
CFC (chloro-fluoro-carbons)	640	Global	0.18	>5000
O_3 Stratosphere troposphere	0.1	Continental-global	–0.05 0.65	62 – 69

*1 from (Myhre et al., 2013); *2 under discussion; from (World Meteorological Organization, 2021) and the literature listed there

technical applications under international agreements. The extreme challenge is to quantify the influence of anthropogenically emitted greenhouse gases, since it is not known how each individual type of human activity affects ground-level air temperature and how the retarding effects of huge sinks like the oceans have to be considered. Since many processes influence each other, i.e. create feedbacks, the individual effects cannot simply be added together. Two numbers are often given in the assessment, and they must be clearly distinguished from each other.

One figure is a dimensionless quantity the global warming potential (*GWP*) or CO_2 equivalent of a chemical compound, which is a measure of its relative contribution to the greenhouse effect, i.e. its average warming effect on the earth's atmosphere over a certain period of time usually 100 years. It thus indicates how much a given mass of a greenhouse gas contributes to global warming compared to the same mass of CO_2. However, the global warming potential cannot be equated with the actual contribution to global warming, since the emission quantities of the various gases differ greatly. With this concept, the different contributions of individual greenhouse gases to global warming can be compared given known emission quantities.

In the Intergovernmental Panel on Climate Change (IPCC) report, another quantity used for quantification is the "stratospheric adjusted " radiative forcing (*RF*), which is the net radiative flux density at the tropopause due to the changing concentration of greenhouse gases. The term radiative flux density (*RF*) indicates the radiant energy passing through the tropopause per second and per square meter in (W/m^2). The stratospheric adjustment assumes that the temperature in the stratosphere has already adjusted to the new equilibrium temperature, but the temperature profile in the troposphere still corresponds to the old equilibrium, which is given by the stable stratification of the stratosphere. If greenhouse gas concentrations are increased, additional energy goes into the lower atmosphere and the radiative forcing and thus the *RF* factor is positive. In contrast, an example of a negative radiative forcing would be if aerosols reflect more sunlight back into space. Then less energy reaches the lower atmosphere and it cools down. The advantage of this concept is that the *RF* does not depend on complicated and poorly understood feedbacks in the climate system. The major disadvantage is that the *RF* says nothing about the changes in measurable and sensible quantities at the ground, which are the quantities of real interest for assessing climate change. Another disadvantage of this concept is that only influences that affect the temperature via radiation can be considered, so that effects such as evaporative cooling or diffusion are not taken into account, although they also affect the temperature.

The release of CO_2 into the atmosphere due to the combustion of fossil fuels can be reduced by some extent if the exhaust gas from the power plants is captured, separated and stored in natural gas reservoirs or coal mines (sequestration by carbon capture and storage technology-*CCS*). The conversion of mobility requires high energy densities, especially for heavy-duty transport by water and land, or for flight operations, which are difficult or costly to realize in terms of battery technology; here, research is still in its infancy. In the short term, the only alternatives are synthetically produced hydrocarbons or a generation and distribution network based on hydrogen as a carrier. However, the development and conversion of both electrochemical and synthetic carriers also requires a stringent analysis of the resource vector, recyclability, and an evaluation of efficiency to ensure a sustainable long-term transformation. It is obvious that any energy transformation in the sectors of households, traffic, industry as well as agriculture, trade, commerce and services consume enormous efforts in infrastructure changes, and thus initially leads to a further shortage of energy and resources. Moreover, the lower degree of availability of all intermittent renewable generation sources, such as solar and wind, forces the construction of generation facilities with installed capacities significantly higher than those of fossil power plants, in addition to the provision of new infrastructures for energy storage and distribution. However, waiting is no option, and a substantial transformation in all sectors is needed, otherwise the limitations of fossil carriers and the onset of a phase of general shortage will come sooner than expected. As the recent past shows, competition for essential resources may give rise to armed conflicts. It is therefore necessary to develop alternatives even further and, above all, to diversify all forms of renewable and far-reaching energy generation technologies to the point where a self-sustaining renewable industry can be built on the basis of renewable energies, resource efficiency and recyclability of required materials in a circular economy.

At present, using all forms of renewable energy (wind, solar, geothermal, hydro) seem to be necessary for developing a long-term solution to the energy supply. Whether a potentially sustainable use of nuclear technology, including nuclear fusion in the long term, is capable to contribute to a low risk supply of the needs of a growing global population, is an open question for the future.

1.4 Energy Forms, Carriers and Conversion Processes, Exergy-Energy

For the practical human use of energy, four primary physical forms of energy can be distinguished,

- Nuclear energy,
- Chemical energy,
- Thermal and radiant energy, and
- Mechanical (potential and kinetic) energy.

The particular form of energy requires an energy carrier either in the way of a nuclear or chemical fuel, heat, solar radiation, or mechanical energy. The type of energy carrier, in turn, determines the conversion process(es) to produce the generated form of energy (electricity, heat, or fuels). Figure 1.6 illustrates the chain from the primary energy form with the primary energy carrier used to the final energy produced, including the conversion options potentially available.

Renewable energy forms and carriers are those whose availability is either permanent or renewable up to a period of one human generation (approx. 30 – 40 years). The term renewable energy also covers a whole range of energy carriers. They also have different local and temporary availability and exergy content. A

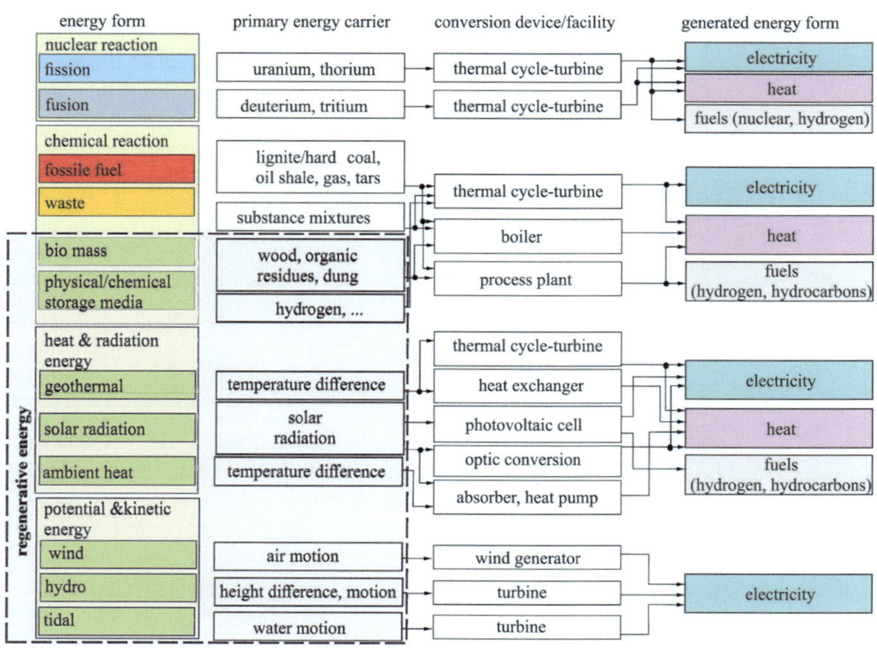

Fig. 1.6 Schematic overview of the energy conversion chain for different energy carriers by means of a conversion technology to different usable forms of energy

list of initial supply, conversion processes and forms of application is also shown in Fig. 1.6.

The conversion of the primary energy carrier into a form that can be used by the consumer requires one or more subsequent conversion process steps. In the first step, secondary energy is obtained from primary energy. In a solar thermal system this refers to the optic conversion of solar radiation in heat. In the subsequent steps, final energy or useful energy is obtained, as illustrated by the sequence of the conversion chain in Fig. 1.7. Each of these conversion processes in the chain is associated with a scalar quantity, the efficiency η_i of the respective conversion process, e.g. the absorber efficiency (η_A) or that of the thermodynamic cycle (η_{th}), where for each efficiency in the chain $\eta_i < 1$. Along an efficiency chain, the individual conversion efficiencies multiply by those of the previous step. As a result, only a fraction of the potential energy content of the primary carrier is converted into final energy directly applicable to the user, which is he described by the total efficiency along the entire chain (η_{tot}).

The general and so far, also in this book used terminology to speak of an energy problem is strictly speaking not correct. Energy is available in the world in practically unlimited quantities, if one thinks for example only of the heat content of the oceans. However, this does not initially have anything to do with local energy that can be used by the end consumer.

This approach requires a clear definition of the concept of energy. What is generally needed and partly scarce is rather exergy E. Exergy (E) is energy which is convertible into mechanical work. In the case of heat, this is the heat content that is exceeding the ambient temperature level T_{amb}. Figure 1.8 illustrates the division of heat into exergy and the fraction B that cannot be used to generate mechanical work, which is also referred to as anergy.

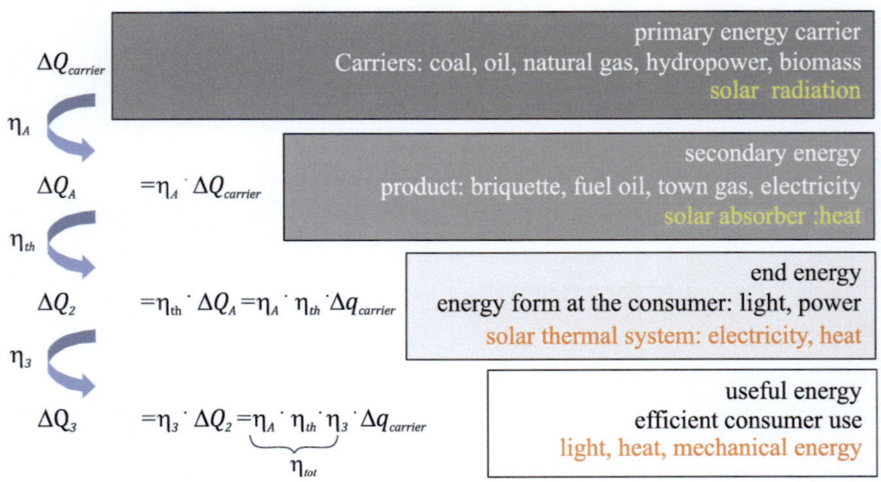

Fig. 1.7 Conversion chain of energy from primary energy source to useful energy; solar thermal accessible indicated on right in lower line

Fig. 1.8 Splitting of heat into exergy (E_{12}) and anergy (B_{12}) in the temperature-entropy diagram ($T - s$) related to the ambient temperature (T_{amb})

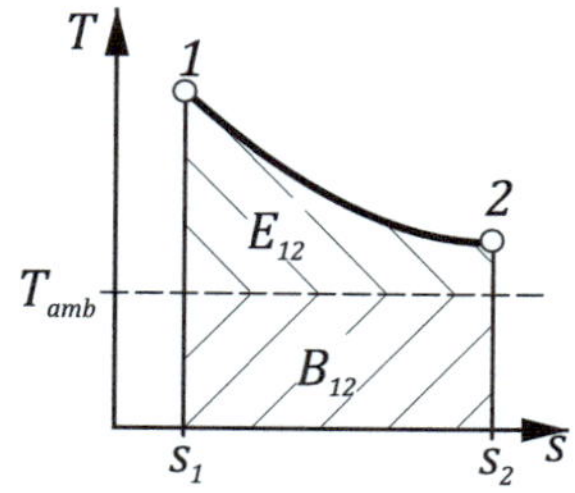

If we consider the heat Q_{12} crossing the system boundary in a process leading from point 1 to point 2, we must distinguish the energy E_{12} lying above the ambient temperature T_{amb} termed as exergy from the energy B_{12} situated below the ambient temperature called anergy.

For the use and for considering the value of energy the ambient temperature plays an important role. We value heat only, if it is at a different temperature level than the ambient. In physics this is reflected by the concepts of exergy and anergy. In the temperature-entropy diagram ($T - s$) of process, a change of the physical state, the corresponding energy quantities are proportional to the areas under the respective curves describing the physical process. When we split up the energies into two parts, the one above the ambient temperature, and the one between the absolute zero and the ambient, the following relationships result for anergy B_{12} and exergy E_{12} (Fig. 1.8):

$$Q_{12} = \int_1^2 T\,ds, \tag{1.1}$$

$$Q_{12} = E_{12} + B_{12}, \tag{1.2}$$

$$B_{12} = T_{amb} \int_1^2 ds \quad \text{with} \quad ds = \frac{dQ}{T}, \tag{1.3}$$

$$E_{12} = \int_1^2 (T - T_{amb})\,ds = \int_1^2 \left(1 - \frac{T_{amb}}{T}\right) dQ, \tag{1.4}$$

$$E_{12} = Q_{12} - T_{amb} \int_2^1 \frac{dQ}{T}, \tag{1.5}$$

where s describes the specific entropy. It can be seen from Eq. (1.3) that the fraction of the total heat which lies as exergy above the level of the ambient temperature T_{amb} is determined by a factor corresponding to Carnot's efficiency. In the terminology used here, the first law of thermodynamics has the wording:

> In a closed system, the sum of exergy and anergy remains constant.

Accordingly, the second law of thermodynamics can be formulated:

1. in irreversible processes of a closed system, exergy is converted into anergy,
2. in reversible processes the exergy remains constant,
3. anergy cannot be converted into exergy.

As soon as energy is to be converted into mechanical work, it is immediately clear that only exergy can be used. Exergy is also required for heat demand, such as for heating purposes. Of course, the required exergy input can be minimized. Figure 1.9 illustrates the problem using the example of a heat flux from a room through a wall into the environment with the temperature T_{amb}.

When a room is heated above the level of the outside temperature (ambient temperature) as indicated in Fig. 1.9, heat is added to provide temperatures above T_{amb}. The energy contained in the room is split in two fractions as exergy (E_Q) and as anergy (B_Q). Due to the heat conduction in the outer wall, as long as exergy is in the room, heat is flowing outside and is added to the infinite reservoir of ambient, thus all exergy is converted into anergy. If electric resistance heating were used for the room as illustrated in Fig. 1.10, the electric energy, which is pure exergy (E_{el}), would be converted in the resistor partly to exergy (E_Q) and partly to anergy (B_Q). In this case, the heat pumps, which will be discussed later, are more favourable, as they use only not only exergy, but also anergy for heating.

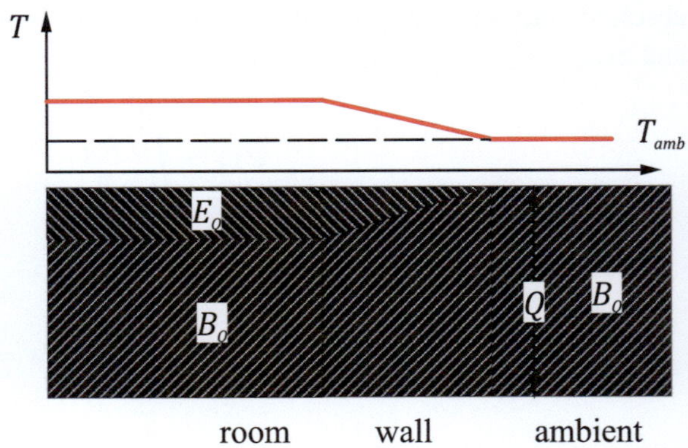

Fig. 1.9 Exergy (E) and anergy (B) using the example of a heat flow from a room through a wall into the environment with T_{amb}.

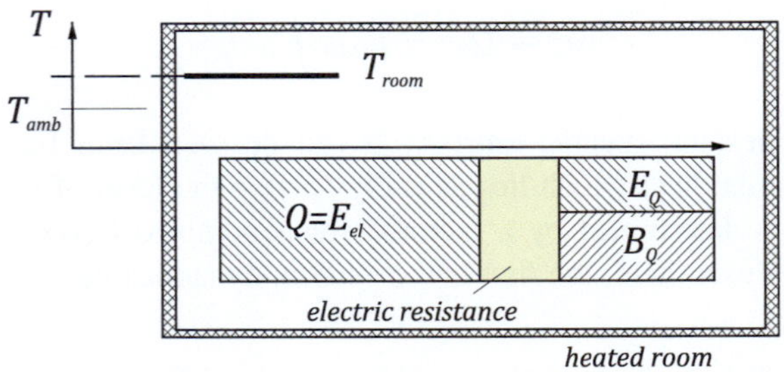

Fig. 1.10 Share of exergy and anergy for electric resistance heating of a room

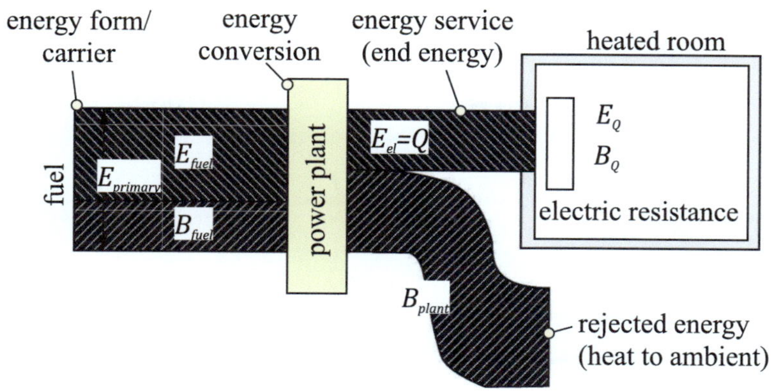

Fig. 1.11 Energy flow diagram of a coupled fuel-power plant-electric resistance heating system with division into anergy (B) and exergy (E)

This consideration can be extended to include the primary energy carrier, in the example of the heated room the fuel for the electrical energy (E_{el}). Figure 1.11 shows the heating process in a closed heated room, taking into account the power plant generating electricity. Already in the thermodynamic cycle of the power plant, part of the energy supplied by the fuel ($E_{primary}$) is lost to the environment as anergy (divided into B_{fuel} and B_{plant}). The largest fraction is the rejected heat of the thermodynamic cycle to the environment, which can be shown most clearly in the $T - s$ diagram.

The consideration of the energy flow from the fuel as energy carrier via the power plant as conversion technology to the final energy use of a single heated room can be extended to an entire national economy. The energy supply is based on different energy carriers and the final energy is broken down into different consumer sectors, such as transport, mobility, trade and residential. The result is a Sankey diagram showing an integrated national energy or energy carrier flow representation. This, however, is not an exact representation in the sense of the previously introduced anergy-exergy definition but at best a simplification. For example, waste heat flows are not identical to anergy flows, because the latter may still contain usable heat. Figure 1.12 shows such an economic Sankey diagram using the example of the United States of America.

This expresses electricity generation from renewable energy sources such as hydropower, wind power, geothermal power, and solar energy based on the heat of fossil-fuel power plants to more easily account for the differences between the conversion efficiency of renewables and the fuel utilization of combustion- and nuclear-powered systems. In this way, hydro, wind, and solar energy can be included in the energy flow on a similar basis as coal, natural gas, and oil. For this reason, the sum of the input variables for electricity differs slightly from the total electricity production shown. Distributed electricity includes only retail electricity sales and does not include self-generation. Electricity generation efficiency is calculated as the total retail electricity supplied divided by the primary energy input

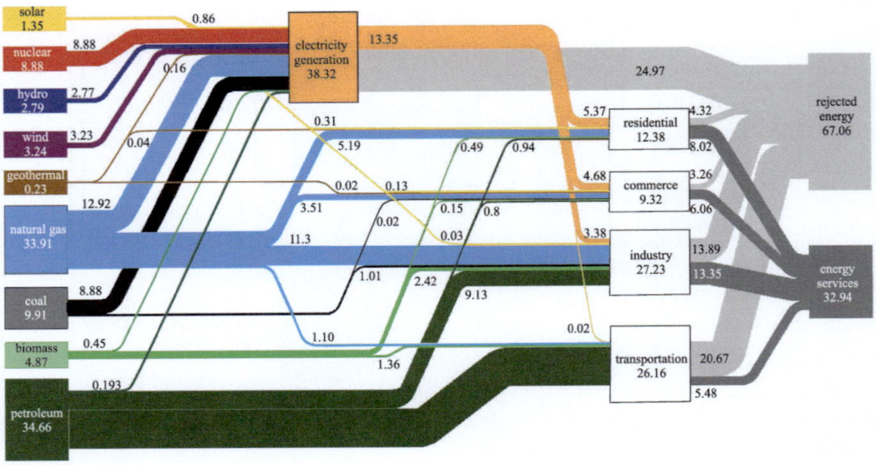

Fig. 1.12 Sankey diagram of United States energy flow in [%], where $100\% = 98EJ$, modified from (Lawrence Livermore National Laboratory, 2021)

used in electricity generation. End-use efficiency for the residential, commercial, and industrial sectors is usually estimated to be about 80%; the mobility sector is assumed to be 25% efficient.

The numbers in such a Sankey diagram indicate either the absolute amount of energy in joules, in the case of the USA $98 \cdot 10^{18} J = 98EJ$, or as a percentage of a total amount of energy to be indicated. In addition to the main energy sources, the chart also shows the breakdown of consumption in the individual sectors of the economy. The diagram also provides information on the efficiency with which primary and secondary energy is utilized. In the USA, more than 60% of the secondary energy is released into the environment in the form of excess heat. If one analyses exactly the transition from primary to secondary energy, especially for fuel carriers, one recognizes that apparently a part of the energy disappears; which of course is not the case. Rather, it is a so-called non-energy consumption, i.e. the conversion of fuels into products, such as paints, varnishes, or the like. Sankey diagrams thus provide information about the integral energy organization of a national economy and the energy-related interconnection of the sectors. However, they are only suitable to a limited extent for comparisons between countries, since these either have a different industrial development or a different geographical location. For example, the share of heat demand in Canada or the Scandinavian countries is different from that of the USA.

Likewise, Sankey as a year-integral representation does not provide information about the temporal and local availability of energy but only about the integral flow. In other words, the control and distribution structure of energy, which are the basis of the resulting integral energy flows as well as their accessibility and utilization, are not part of a Sankey diagram.

1.5 Energy Demand and Economic Considerations

The aim of this book is to show the use of thermal solar energy in the whole range from heat for domestic applications, process heat, desalination to electricity generation. The technological challenge is to analyse the solar energy supply and to transform it by means of components, processes, and systems in such a way that the solar energy is converted with high efficiency into a usable form of exergy while at the same time making rational use of resources at the end user.

Fluctuation in solar supply must be addressed in detail. If the solar supply does not match the temporal energy demand, storage must be provided in the system or substitute energy sources must be kept available. In the second case, the energy costing for renewable energy generation cannot be based on the savings of the original energy source alone (e.g., fossil carriers). Rather, a cost approach must then be considered that is based on the fuel-saving mode of use through the use of renewable energy. This is mandatory because a substitute energy carrier is still required, the provision and use of which is associated with a CO_2 footprint in the form of constructional expenditure, energy distribution or even the direct use of fossil carriers.

If the solar energy is used to generate a product such as drinking water, the storage can also be in the form of the end product. However, the cost of providing the energy production facility then shifts to the expense of a production facility that can only be partially utilized and is therefore may be oversized. In the field of industrial production, for economic reasons, the product value stream must be considerably higher than that of the energy used. To the pure energy cost calculation from expenditure (for the solar plant and maintenance) as well as energy (exergy) yield comes a further aspect. The competitiveness of an energy generation technology for service and production requires at the same time the maintenance and guarantee of a supply security.

In the light of these considerations, the statement that solar energy is free of charge, is not true, since a user-oriented supply of heat, electricity or fuel must be made available through a technical equipment, based on material resources and work. In a business sense, the use of technology turns the "free good" of solar radiation into an "economic good" that has a price. Moreover, in addition to the direct costs of operating and maintaining the system, there are also indirect costs for equipment, facilities and distribution systems that must be operated to maintain security of supply when the renewable source is not available. Currently, these additional costs in the electricity market are largely borne by consumers in the form of levies, while market deployment is supported by legal regulations such as tax depreciation or operational market regulation interventions such as the feed-in priority.

As well the electricity, the mobility and the heat market are served by different energy carriers worldwide. As we need renewable energy sources to penetrate this market, they must be either equivalent to them in terms of a 100% supply, or have advantages in at least some aspect of.

- Economic efficiency,
- Security of supply and,
- Environmental protection.

Environmental protection requirements, such as landscape consumption, can shift the market situation in the future. This aspect, which is based on social consensus, is not the subject here, but is of central importance for market penetration. If renewable energy systems do not meet the first two conditions of the general market requirements, an economic and secure supply of energy, market niches are nevertheless existing, which make a further development not only worthwhile but mandatory. Subsidies for the market introduction of renewable technologies are justified, because they offer the unique chance to reduce the long-term dependence on fossil fuels, and it is a major asset in the fight against climate change.

The treatment of solar radiation and the provision of adequate technical solar conversion systems also involves an economic analysis in terms of cost optimization. Figure 1.13 shows the diurnal variation of the energy demand broken down into electricity, heating, and hot water of an average 2.5-person household on an average cloudy working day in the transitional period of April in the course of a day in Germany in comparison with the solar area-specific irradiation. First of all, it is noticeable that the daily energy demand is subject to considerable temporal fluctuations and also does not correlate with the diurnal variation of the solar irradiation.

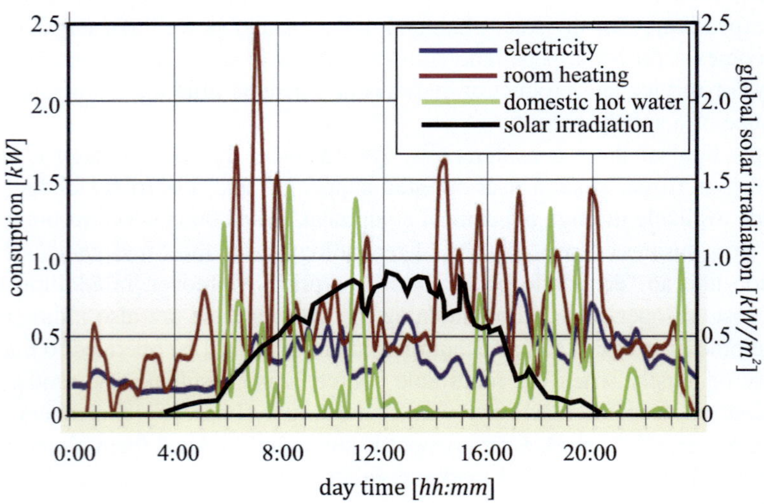

Fig. 1.13 Power demand for electricity, heating and domestic hot water of a standard household (2.5 persons) in the course of a day during the transition season (April) in comparison with the mean area-specific gross solar irradiance in Germany from (Beck,2007)

Therefore, in order to ensure the security of supply of the overall system at all times of the year and day, the integration of storage solutions into renewable energy system concepts is indispensable.

Whereas in the residential environment the temporal match of demand and generation has been discussed, for the industrial utilization of thermal energy, the level of the temperature plays a decisive role as well. With the help of solar thermal energy, a wide range of temperatures is technically accessible that would otherwise have to be provided via fossil carriers for process heat in the low-temperature range ($< 100°C$) or medium process heat range (up to approx. $400°C$). The amounts of energy used in the industrial sector are enormous, as shown by the breakdown according to process temperatures used and consumption for the production of industrial goods using Germany as an example, cf. Figure 1.14. To ensure meaningful competition, the energy to be thermally generated must correlate with the demand profile of the different industries and their time requirements for process heat. For solar thermal energy supply, the ranges up to about $400°C$ are interesting. In the upper temperature level, this simultaneously corresponds to the requirements for thermal power generation. For this purpose, systems other than those used in the domestic sector must be used, both in terms of the achievable temperature and the size of the system. In anticipation of the following chapters, concentrating collectors are required for this purpose. Leaving climate change aside, from the individual project it can be more favourable, however, to operate the low-temperature preheating with the solar plant and to reach the peak temperature with a fossil auxiliary firing in a hybrid or dual-source installation. The second source can then also compensate solar power fluctuations. Solar technology itself offers a number of possibilities for optimization, which will be examined in the course of the following chapters.

1.6 Sector Coupling—Power, Heat and Chemical Carriers

The term sector coupling does not refer to economic sectors, such as private households, commerce and services, industrial production and transport, but refers in general to the linkage of "energy sectors" electricity, heat and chemical carriers. The basic idea of the concept of sector coupling is that the sectors of the energy economy as well as the industry are to be interconnected and optimized in a common holistic approach to the energy sector, thus enabling a better and more favourable overall system. In perspective, sector coupling offers three main advantages:

- It allows decarbonizing all sectors of the economy with the help of a large/predominant share of renewable energies.
- A significant reduction in primary energy consumption through intelligent coupling of sectors using energy-efficient technologies.

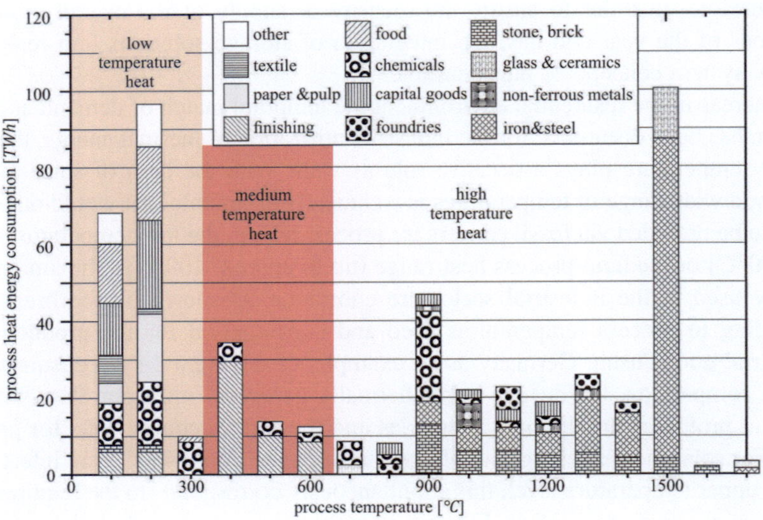

Fig. 1.14 Process heat energy consumption in [*TWh*] of Germany classified by temperature levels and different industrial goods of the manufacturing industry, modified from (Deutsche Energie-Agentur (DENA), 2016)

- Use of large and cheap energy storage outside the electricity sector through functional electricity storage (e.g. synthetic chem. carriers, heat) to satisfy demand flexibility for electrical energy during fluctuations in supply-dependent renewables.

The last point aims to reduce investments in resource-intensive, expensive electrical energy storage and to increase energy security by means of sector coupling.

But, it is also clear from the previously formulated perspective goals of sector coupling that it requires considerable challenges and material efforts that include the following aspects:

- Construction and operation of efficient conversion infrastructures (Power to Gas-$P2G$, Power to Peat-$P2H$).
- Planning, construction and maintenance of flexible transmission and distribution networks.
- Intelligent and efficient load management to meet demand-driven fluctuations while ensuring security of supply.

The concept of sector coupling is based on a diversified interaction of the different energy sources and there is no singular master solution for the different regions of the world due to different infrastructural prerequisites (degree of industrialization, orientation of the economy, etc.) but also different needs (e.g. heat demand due to geographical location or availability of renewable resources).

Figure 1.15 top outlines the current energy supply system with the generation sources participating in the energy market, the supply forms, the distribution system and the energy consumer side. Similar to the Sankey diagram, the energy transfer to the end consumer is mainly carried out via fossil energy sources in the horizontal level by means of corresponding distribution systems, for example heat or power grids, pipelines or gas stations. Vertical coupling of the consumer sectors occurs only to a very small extent on the distribution side. The functionality of this traditional energy concept is based on the fact that energy production and consumption are *balanced at all times.* Whereas this balancing of generation and consumption for electricity consumption is achieved through dedicated load management of the power grid at regional and supra-regional levels, where differences in the absolute magnitude of energy between generation and consumption at any given time t are regulated by market mechanisms (electricity trading arbitrage) and security of supply is ensured by a system of primary, secondary and ternary reserves, for example reserve power plants, hydroelectric power plants and the like. This ensures that the following equation is fulfilled within a control cycle dt.

$$\int_{production} \dot{P}(t)\,dt = E_{loss,distribution} + \int_{consumption} \dot{P}(t)\,dt, \tag{1.6}$$

The majority of the difference between generation and consumption is ensured by the use of dispatchable fossil-based power plants or quasi-loss-free storage in the form of fossil carriers. Depending on the geographic characteristics of each country, pumped storage power plants can provide base load or residual load when electricity generation is insufficient. A production characteristic on the consumer side is desired but not required, nor is an intersectoral exchange of energy with an integrated storage function. As Eq. (1.6) illustrates, there is a direct coupling of generation (left side) and consumption (right side), so that the energy flow (electricity, heat, fuels) is almost exclusively from left to right. Buffering in the form of storage is not required either on the generation side, since it is provided there by the loss-free storage of fuels, or hydropower, but especially not on the demand side. An advantage of the classical system is that the absolute of the generation capacities compared to the consumption side are small compared to a sector concept and also the hierarchically organized distribution networks of the supply side are small. However, this is bought by the fact that large amounts of energy are either released unused into the environment as thermal waste energy and large amounts of fossil carriers on the consumption side are transported to the consumer in the transport network with low efficiencies, as shown in the Sankey diagram using the example of the USA, cf. Figure 1.12.

An integral concept based on sector coupling must not only consider the volatile character of renewable sources such as wind and solar technologies, but also their conflicting goals of temporal and local availability, as well as sometimes considerable differences between generation and consumption. In order to realize this concept, which is simplified illustrated in Fig. 1.15b a massive expansion of infrastructures and systems is required on the energy generation side as well as on the

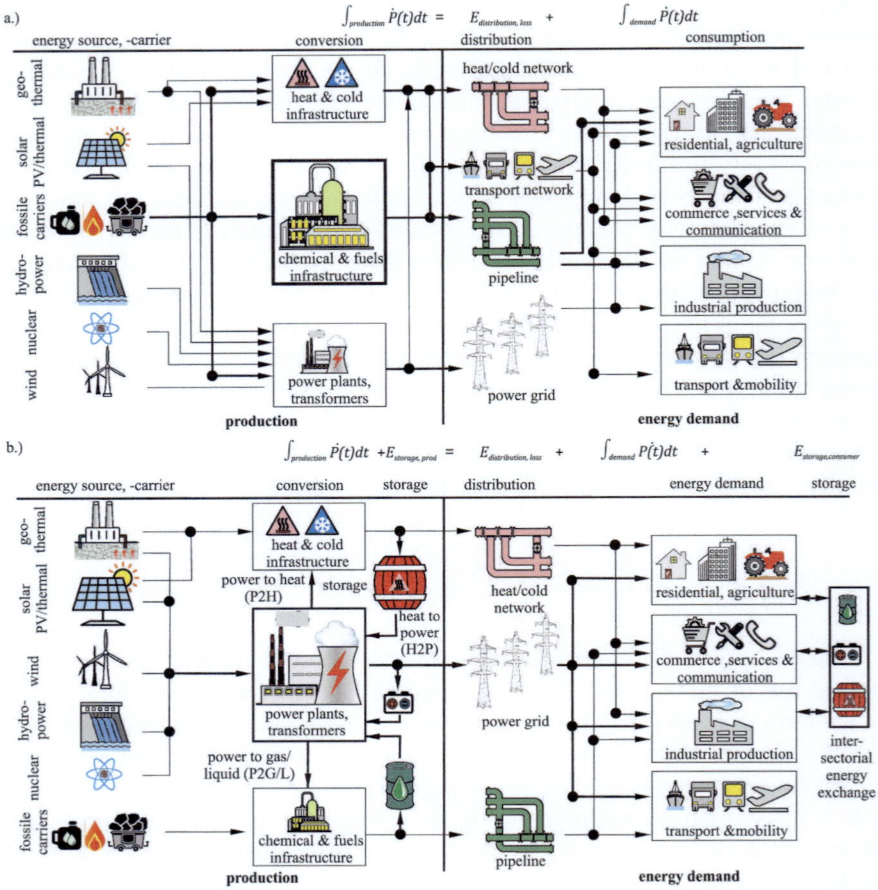

Fig. 1.15 Schematic of the generation, distribution and consumption structures of a conventional fossil based energy system **a** and a sector-coupled energy system based mainly on electricity **b** essentially fed by renewable (and nuclear) sources

distribution and consumer side. Electricity is at the centre of the generation side. In order to generate energy as CO_2-free as possible, the generation capacity of renewable sources and the associated distribution network must be massively expanded and is thus its generation capacity exceeds by far that of a conventional system. To compensate for the differences between generation and consumption at different times of the day and year, additional conversion technologies such as Power to Heat ($P2H$) and Power to Liquid/Gas ($P2G/L$) and their corresponding inversion technologies as well as additional storage facilities are required. This vertical linkage ensures not only the provision of a balancing function on the generation side, which the conventional system lacks, but also the feeding of storage to compensate for environmentally induced fluctuations in generation, thus contributing to the stabilization of the generation side, (left side of Eq. (1.7)). The distribution grid itself hardly changes its distribution function, but the preferred use of electricity as the hub of energy transport and distribution means a large-scale expansion of the power grid, since it must additionally provide energy requirements for mobility, industrial production, trade, commerce and services, as well as the private sector. Ideally, the distribution role of transport for energy products can be largely eliminated. On the consumer side, a sector-coupled concept means a mobility shift to electromobility, fuel cells or hydrogen, but also a shift to CO_2-free production technologies organized close to the consumer and thus relieve the distribution grids. Another aspect of a sector-coupled concept is that the consumer side is no longer passive, but is also assigned a producer or storage role, so that an intersectoral energy exchange is guaranteed. As a consequence of a sector-coupled energy system, both sides of Eq. (1.7) change dynamically, whereby the energy balancing quantities ($E_{storage,prod}$, $E_{storage,consumer}$) are of different dynamic sizes and can have both positive and negative signs.

$$\int_{production} \dot{P}(t)\,dt + E_{storage,prod} = E_{distribution,loss} + \int_{consumption} \dot{P}(t)\,dt + E_{storage,consumer} \quad (1.7)$$

The networking of all sectors, i.e. sector coupling in the vertical plane, should therefore contribute to the achievement of climate policy goals. This requires technical facilities, infrastructures and also market models of the various sectors to be more closely coordinated with one another in order to establish a comprehensive, intelligent energy system.

In order to drive forward the energy transition in a targeted and consistent manner, it is of great importance that the different areas and sectors of the energy system work together. So far, the focus has mostly been on the "electricity turnaround." However, in order to achieve the international climate targets, a heat and mobility transition is also imperative. Whether heat can be realized exclusively by electricity, for example via heat pumps, can be questioned, but the necessary establishment of a decentralized structure of heat networks, as well as regeneratively produced fuels with high energy density, is less so. Thus, a second important pillar of a smart energy system is a consistent expansion of generation and storage systems (especially electricity and heat) at the regional or neighbourhood level. This not only enables a consistent use of all types of waste heat and, as a consequence, a saving of fossil resources but also the relief of superregional distribution networks.

The integration of additional interconnected technologies also means a massive data flow based on modern information technologies as well as regulation and control strategies. The structure may look complex, but it also opens up the possibility of participation of new market players and the establishment of innovative business models.

What are the advantages and disadvantages of sector coupling?
Sector coupling brings with it various advantages. On the one hand, it promotes the decarbonization of all sectors through the use of renewable energies. This means that fossil fuels are gradually replaced by renewable energy sources. Sector coupling promotes flexibility in the demand for electrical energy, so that fluctuations associated with the use of renewable energies can be balanced out without having to make too large investments in energy storage. This circumstance, in turn, has a positive impact on energy security. Overall, the clear advantage of sector coupling is that it acts as a catalyst for the energy transition. By linking different sectors, synergy effects are brought about, allowing the use of renewable energy sources to be expanded further and more efficiently.

Unfortunately, compared to liquid or gaseous fuels, electricity cannot easily be stored in a vessel. Therefore, electricity must be stored electrochemically in the form of accumulators or in a modified form, such as pumped storage. However, this storage or conversion always involves energy losses. However, accumulator (often termed battery) storage is rather expensive overall and still exhibits a considerable consumption of scarce raw materials. Therefore, in some applications fossil fuels, which are expensive but can be stored at lower costs, are cheaper in business terms than technically complex electricity to heat (power to heat-$P2H$) or electricity to liquid/gas technologies (power to liquid/gas-$P2G/L$). Despite these challenges, it is essential that the energy transition is driven forward across sectors in a holistic approach with the help of intelligently deployed flexibilization and linking of power generation, power consumers and storage. It must also be clear that the realization of a sector coupling concept means a correspondingly high level of fragility with respect to attacks on the data infrastructure, since the data volumes are growing significantly and the number of active market participants is much greater. Another aspect is that the development towards a sector-coupled energy supply system must be designed to be barrier-free and technology-neutral on both the generation and consumer sides so that the most resource-conserving and energy-efficient solutions are applied and introduced into the market. This calls for research and technological development, but also for socially accepted frameworks and rules.

In this context, it is of central importance that the evaluation of a technology must be carried out on a holistic level. A simple approach to assessing a technical product or energy system in terms of CO_2 value is to allocate a CO_2 equivalent to all consumption of raw materials and inputs (i. e. self-consumption, infrastructure costs, provision of distribution networks) during the tentative projected operating life, in addition to manufacturing and recycling. This lifetime accumulated CO_2- equivalent can then be correlated with the benefit, for example in the case of an energy generation plant in CO_2 per kWh generated. If this kind of consideration is applied, no technology is really CO_2-free, including renewable energy production methods or processes, which is in accordance with the second law of thermodynamics, but it is possible to consistently search for Pareto front-optimal solutions, which represent an optimal solution in a multi-objective parameter problem.

References

British Petroleum BP p.l.c. (2021). *Statistical Review of World Energy 2021–70th edition.* London SW1Y 4PD, 1 St James's Square, UK: Whitehouse Associates, London. Retrieved from https://www.connaissancedesenergies.org/sites/default/files/pdf-actualites/bp-stats-review-2021-full-report.pdf

Butler, J. H., & Montzka, S. A. (2021). The Noaa Annual Greenhouse Gas Index (AGGI). Boulder, CO 80305–3328, Colorado, USA. Retrieved 2 28, 2022, from https://gml.noaa.gov/aggi/aggi.html

Deutsche Energie-Agentur (DENA). (2016). *Process heat in industry and commerce-Technology solutions for waste heat utilisation and renewable provision.* Chausseestraße 128 a, 10115 Berlin, Germany: Deutsche Energie-Agentur GmbH (dena). Retrieved from www.dena.de

ENI SpA. (2021). *O&GR-Volume 2: World Gas and Renewables Review 2020.* 00144 Roma, Piazzale Enrico Mattei, 1, Italy . Retrieved from https://www.eni.com/assets/documents/eng/scenari-energetici/WORLD-GAS-AND-RENEWABLES-REVIEW-2020-vol2.pdf

Lawrence Livermore National Laboratory. (2021). *Energy Flow Charts: Charting the Complex Relationships among Energy, Water, and Carbon.* 7000 East Avenue Livermore, CA 94550. Retrieved 3 1, 2022, from https://flowcharts.llnl.gov/

Myhre, G., Shindell, D., Bréon, F.-M., Collins, W., Fuglestvedt, J., Huang, J., Zhang, H. (2013). Chapter 8: Anthropogenic and Natural Radiative Forcing. In *Climate Change 2013: The Physical Science Basis. Contribution of Working Group I to the Fifth Assessment Report of the Intergovernmental Panel on Climate Change.* Cambridge University Press.

OECD/NEA &IAEA. (2020). *Uranium 2020-Resources, Production and Demand.* Organisation for Economic Co-Operation and Development. 75775 Paris Cedex 16,2 rue André-Pascal,. France: OECD/NEA Publishing.

Ritchie, H., & Roser, M. (2020). CO_2 and Greenhouse Gas Emissions. (O. W. Data, Ed.) Retrieved from https://ourworldindata.org/co2-and-other-greenhouse-gas-emissions

United Nations Population Division. (2020). *World Population Prospects.* Population Division, United Nations,, 2 United Nations Plaza, Rm. DC2–1950, New York, NY 10017 USA. Retrieved from https://population.un.org/wpp2019/Download/Files/1_Indicators%20(Standard)/EXCEL_FILES/1_Population/WPP2019_POP_F01_1_TOTAL_POPULATION_BOTH_SEXES.xlsx

World Bank. (2020). *Annual Report.* The World Bank, 1818 H Street, NW Washington, DC 20433, United States of America. Retrieved from https://data.worldbank.org/indicator

World Meteorological Organization. (2021). *WMO Greenhouse Gas Bulletin: The State of Greenhouse Gases in the Atmosphere Based on Global Observations through 2020.* Atmospheric Environment Research Division. Geneva, Switzerland: World Meteorological Organization. Retrieved from https://community.wmo.int/activity-areas/gaw

Solar Radiation 2

Abstract

In this chapter the physical processes and the composition of the sun leading to the sun as radiation source will be described. From that the properties of extra-terrestrial irradiation close to the earth are derived. Depending on the orientation of a surface different amounts of irradiation is received (horizontal, tilted, tracked). The atmosphere is interacting with this radiation through absorption and scattering depending on a number of parameters (aerosols, clouds) and modifies the incoming terrestrial radiation on the surface of the earth. For collectors not only the beam radiation directly from the sun, but also diffuse components from sky and ground reflected albedo are important. Short-wave diffuse sky radiation depends on the clearness of the sky and sun position and is in general anisotropic. Models for the angular distribution of diffuse irradiation describe this behaviour. Statistical methods exist to derive the diffuse fraction of global irradiation based on the clearness index. The basic methods for measuring solar radiation for different locations are presented, with terrestrial methods and satellite measurements. Using long-term time series representative meteorological data sets for planning purposes can be constructed.

Keywords

Sun physics • Sun geometry • Directional irradiation • Spectral solar radiation • Diffuse and global irradiation • Atmospheric attenuation • Terrestrial and satellite measurements • Statistical solar data

R. Stieglitz and W. Platzer, *Solar Thermal Energy Systems*,
https://doi.org/10.1007/978-3-031-43173-9_2

2.1 The Sun as Energy Source

2.1.1 Structure and Composition of the Sun

The sun is a star which provides the energy for life on our planet for billions of years. It draws its energy from the fusion of hydrogen nuclei, which are present as plasma in the inner core of the sun. As a result of the hot fusion process the sun has a temperature of about 10^7 °K in its interior. It is not a rigid body but is consisting of several zones as shown schematically in Fig. 2.1.

The age of our sun is estimated at about 4.5 *billion* years. During this time the hydrogen concentration has decreased by about 8% and the helium abundance therefore has reached this value 8%. However, the hydrogen supply is still sufficient to ensure that the current output can be delivered for more than additional 10 *billion* years. The sun radiates into space 4×10^{26} W of electromagnetic radiation power, of which only 2×10^{17} W reach the earth. For comparison, the current energy consumption worldwide is at about 1.6×10^{13} W. Thus, although the energy of the sun is not renewable, it may provide energy to our planet for an incredible long time and is not exhausted within a few generations of man as conventional energy sources might be.

90% of the energy is generated in the central *core* within $0.23R$, with R being the radius of the sun. The core is solid due to the high pressure and has an average density of 10^5 kg/m^3, which is about eight times the density of lead. The temperature rises to 40 *million* °K in the centre. Although the core is only 15% of the volume, it contains about 35–40% of the solar mass. From the core outwards, the energy transport takes place through radiation. However, the absorption is so strong at this high density that the radiation diffusion takes a long time, and therefore the radiation which we see now was generated near the centre of the sun about 1–2 *million* years ago.

Faster transport is achieved in the next zone, the *convection zone* extending from about $0.7R$ nearly to the edge. Convection cells exist there, which can be recognized from the outside as granules. The lifetime of such cells is a few minutes. The

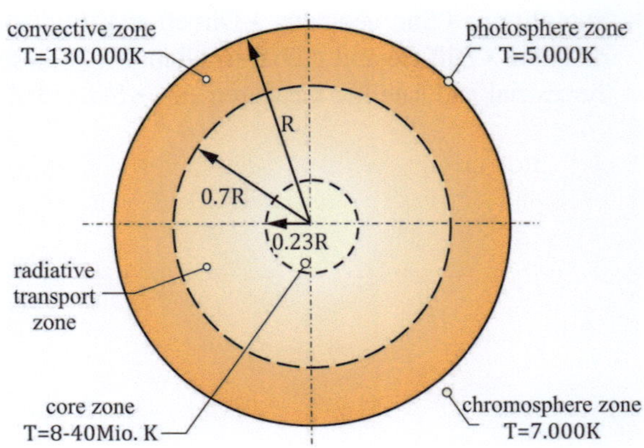

Fig. 2.1 Schematic structure of the sun

Table 2.1 Some key data related to radiation emission of the sun from McCracken and Scott (2005)

	Phys. unit	Value
Diameter	m	1.392×10^9
Surface	m^2	6.088×10^{18}
Mass distribution	H_2 He Other	0.75 0.23 0.02
Total mass	kg	1.99×10^{30}
Mean density	kg/m^3	1.41
Core temperature	K	1.6×10^7
Surface temperature	K	5776
Core pressure	MPa	4×10^{10}

Table 2.2 Some general data of the sun from McCracken and Scott (2005)

	Phys. unit	Value
Rotation of the poles	days	34
Rotation of the equator	days	25.4
Age	years	4.6×10^9
Lifetime (expected 1/2 burning time)	years	5×10^{10}
Fuel consumption H_2	Mt/s	600
Helium production	Mt/s	596
Mass difference	Mt/s	4.3

density there is about 70 kg/m^3 and the average temperature about 1.3×10^5 K. Some characteristic sun data are condensed in Tables 2.1 and 2.2.

The thin *photosphere* adjoins the convection zone. The temperature at the edge is about 5000 °K. In the low density photosphere again radiation takes over the heat transport. Here is the origin of the radiation energy emitted by the sun. In the photosphere, which is also called the reverse layer, the temperatures rise to about 7,000 K towards the outside. Its thickness is estimated to be 400–500 km and the density varies from 10^{-4} to 10^{-5} kg/m^3.

Above the photosphere lies the *chromosphere*, which is about 8,000–10,000 km thick and has an even lower density of 10^{-8} to 10^{-12} kg/m^3. The temperature in the chromosphere increases from 7,000 K inside to about 3×10^5 K outside. Taking into account the different densities and temperatures of the photo- and the chromosphere, which constitute the thin surface layer of the sun, an equivalent "*surface*" temperature of 5776 K has been derived (McCracken & Scott, 2005). Even further outside is the *corona*, where the temperature rises to 10^6 K due to magnetic heating.

2.1.2 Fusion Process Within the Sun

In order to understand the nature of the radiation generation by the sun one has to have a look at the fusion process. The sun draws its energy from the fusion of hydrogen nuclei, which due to the high temperatures are present as plasma in the inner core of the sun. In the plasma state the hydrogen atoms are present as protons (p). According to the enormous pressure and gravitational force of the sun, a multi-stage fusion process of hydrogen into helium takes place. The first step is the proton collision process ($p = {}^1H$), which leads to the generation of a deuteron ($D = {}^2H$).

$$p + p \rightarrow D + e^+ + \nu + 0.42\,\mathrm{MeV}. \tag{2.1}$$

Here, e^+ is a positron and ν an electron neutrino. Like all neutron physical processes, the collision of protons is a statistical process in which the effective cross section, the density of the reaction partners and their relative velocities play a role. The reaction cross section of a proton is comparatively small and amounts to 10^{-55} barn (1 barn $= 10^{-24}$ cm^2). The resulting positron e^+ annihilates with an electron e^- from the plasma, i.e. both are completely converted into energy. The mass of the electron–positron pair is released as kinetic energy of two gamma quanta γ

$$e^+ + e^- \rightarrow 2\gamma + 1.022\,\mathrm{MeV}. \tag{2.2}$$

The resulting deuterium (D) at some time reacts with another proton to form the light helium isotope ^{3}He:

$$D + p \rightarrow {}^3\mathrm{He} + \gamma + 5.49\,\mathrm{MeV}. \tag{2.3}$$

Two light helium nuclei ^{3}He then fuse into a single ^{4}He-nucleus (α-particle), releasing two protons. These are available for new fusion processes:

$${}^3\mathrm{He} + {}^3\mathrm{He} \rightarrow {}^4\mathrm{He} + p + p + 12.86\,\mathrm{MeV}. \tag{2.4}$$

Figure 2.2 illustrates the complete sequence of the fusion process in its individual steps. In the mass balance, the following evaluation results for hydrogen–helium fusion:

$$\begin{array}{l} 4p \equiv 4 \times 1.00727647\ u \\ {}^4\mathrm{He} - 4.0015060883\ u \\ \underline{2e^+ - 2 \times 0.00054858\,u} \\ difference\ \ 0.02650263\ u \\ => \Delta m = 4.4 \times 10^{-29}\,\mathrm{kg}, \end{array} \tag{2.5}$$

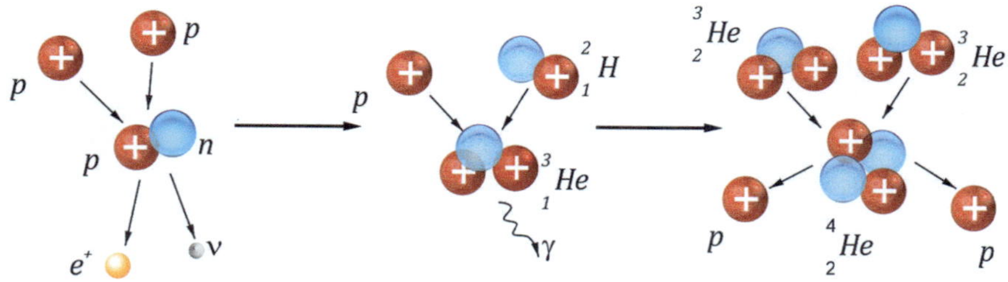

Fig. 2.2 Single steps in hydrogen fusion to Helium (Gerthsen et al., 1995)

where u is the atomic mass unit with 1.66054×10^{-27} kg. Using the Einstein formula (2.6) with c being the speed of light (=2.9979×10^8 m/s), the mass defect Δm results in an energy production per fusion process of 3.955×10^{-12} J = 24.68 MeV.

$$E = m \cdot c^2. \tag{2.6}$$

The energy from the positron–electron reaction has to be added with 1.022 MeV because the positron e^+ is annihilated thereby creating kinetic energy. This yields a value of 26.7 MeV for hydrogen–helium (H–He) nuclear fusion, equivalent to the sum of the reaction energies from Eqs. (2.2) to (2.5). However, the kinetic energy of 0.26 MeV carried away by each neutrino escaping the sun does not contribute to the sun's internal energy. Therefore, only 26.2 MeV are used for the heating process of the sun.

When use with the total mass loss rate of Table 2.2 then using Einstein's relation we can estimate solar power emitted by the sun by (2.7) and dividing it by the mass defect of a single fusion reaction we arrive at the number of fusion processes occurring per second.

$$P_{fusion,\,sun} = \Delta\dot{m} \cdot c^2. \tag{2.7}$$

The results are 3.86×10^{26} W emitted power and 9.8×10^{37} fusion reactions per second. Now, using the average distance sun-earth as the radius of a sphere around the sun we can calculate the average extra-terrestrial irradiance at this position and obtain 1374 W/m^2. This value represents the so-called solar constant (i_{SC}) being introduced in the next paragraph.

2.2 Extra-terrestrial Solar Radiation

2.2.1 Solar Constant

The solar constant i_{SC} describes a specific, long-term averaged extraterrestrial solar irradiance integrated over all wavelengths, which hits the earth perpendicularly per square meter at an average distance between earth and sun without the influence of the atmosphere. It can be calculated using the average orbit radius of the earth around the sun. From the geometry between sun and earth Fig. 2.3 the view angle θ_s of the sun disk as seen from the earth can be calculated as $\pm 0.267°$ or ± 4.65 mrad. Although the luminosity decreases from the centre of the solar disk to the outside, for most terrestrial applications, it is sufficient to assume a uniform radiation density over the entire solar disk.

The determination of the solar constant has been the subject of several investigations in the past. A measurement of the extra-terrestrial irradiance yields only a momentary picture, of course, and is dependent on solar activity. Already in the early twentieth century Abbot and colleagues from the Smithsonian institute did pioneering work using terrestrial measurements on high mountains to extrapolate to the solar constant. Abbot's value of 1322 W/m^2 was later corrected upwards by Johnson to 1395 W/m^2 (Johnson, 1954), showing the large error inherent in these methods. The availability of ballone and spacecraft permitted the determination outside of the atmosphere with a variety of instruments. This resulted in a value of 1353 W/m^2 with an estimated error of $\pm 1.5\%$ (Thekaekara, 1976). The measurements and accuracy have been re-examined by Fröhlich who published a new value of 1373 W/m^2 with an uncertainty of 1–2% (Fröhlich, 1977). This is about 1% above the previously used value of $i_{SC} = 1353\,W/m^2$. Based on additional new spacecraft measurements (Willson, 1978) in 1982 by the World Meteorological Organization in Geneva and the World Radiation Centre (WRC) have proposed

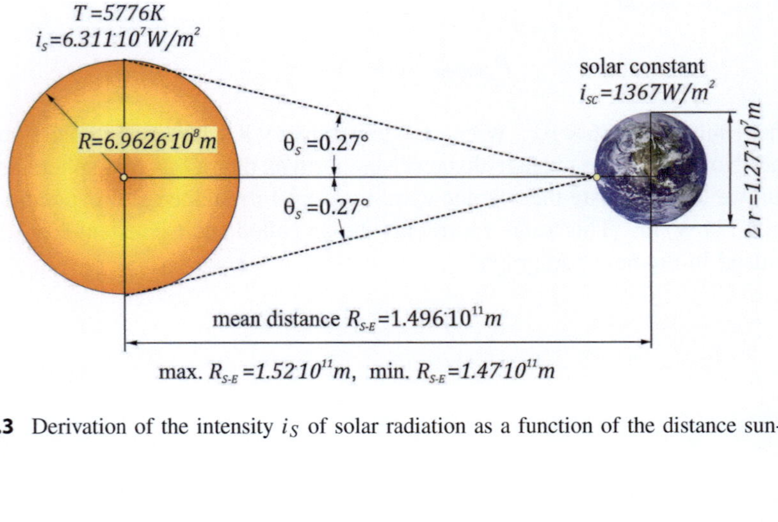

Fig. 2.3 Derivation of the intensity i_S of solar radiation as a function of the distance sun-earth R_{S-E}

$i_{SC} = 1367\,\text{W/m}^2$ as solar constant based on measurements by various authors. This will be the value which will be used in this book:

$$i_{SC} = 1367\,\text{W/m}^2 \ \pm 1\%. \tag{2.8}$$

A more detailed discussion on the history of irradiation measurement and the determination of the solar constant can be found in Duffie and Beckmann (2020). They also show how to derive the magnitude of the solar constant using the Stefan-Boltzmann law and the sun's equivalent surface temperature.

More about the periodicity and the relation of the fluctuations with the activity of the sunspots can be found at Pecker (1994) and Fröhlich et al. (1992). These reports show that in periods of about one year the solar intensity varies by less than $\pm 0.1\%$. In the UV range with wavelengths below 170 nm, the radiation can vary by a factor of 2. In the X-ray range at 0.2–3 nm the radiation power changes by up to 2 orders of magnitude. The current most accurate total spectral irradiance (TSI) values from the Total Irradiance Monitor (TIM) on NASA's Solar Radiation and Climate Experiment (SORCE) is $1360.8 \pm 0.5\ \text{W/m}^2$ during the 2008 solar minimum as compared to previous estimates of $1365.4 \pm 1.3\,\text{W/m}^2$ established in the 1990s (NASA, 2022).

Of greater importance are the fluctuations of extra-terrestrial radiation as a result of the elliptical orbit of the Earth around the Sun. As a result of the orbital eccentricity, the distance between the sun and the earth fluctuates between 1.471×10^{11} m and 1.521×10^{11} m yearly. The resulting periodically changing distance causes the irradiance on a unit area of one square meter perpendicular to the direction of irradiation to fluctuate by $\pm 3.3\%$; this results in a day-dependent extraterrestrial solar constant $i_{dir,extra}$. It can be calculated for the nth day of the year according to Eq. (2.9)

$$i_{dir,extra} = i_{sc} \cdot \left(1 + 0.033\ \cos\left(\frac{360}{365} \cdot n\right)\right), \tag{2.9}$$

in which $i_{dir,extra}$ the extraterrestrial intensity of solar radiation in watts on a surface of $1\,\text{m}^2$ is perpendicular to the direction of irradiation or to the direction of the sun/earth. Figure 2.4 shows the annual variation in extra-terrestrial solar radiation intensity per square meter as a function of the season.

As mentioned already, from the solar constant the radiant power of the Sun can be calculated by multiplying it by the surface A of that envelope around the Sun, which has the radius of the mean distance from Earth to the Sun with $R_{S-E} = 1.496 \times 10^{11}$ m. With the help of Eq. (2.10) the solar radiation power Φ is obtained:

$$\Phi = i_{SC} \cdot A = i_{SC} \cdot 4\pi \cdot R_{S-E}^2 = 3.845 \times 10^{26}\,\text{W}. \tag{2.10}$$

If that radiation power is equated with the radiation emitted by a blackbody sphere with radius of the sun, an equivalent emitting temperature of about 5800 K results—this is approximately the temperature of the sun at the inner part of the

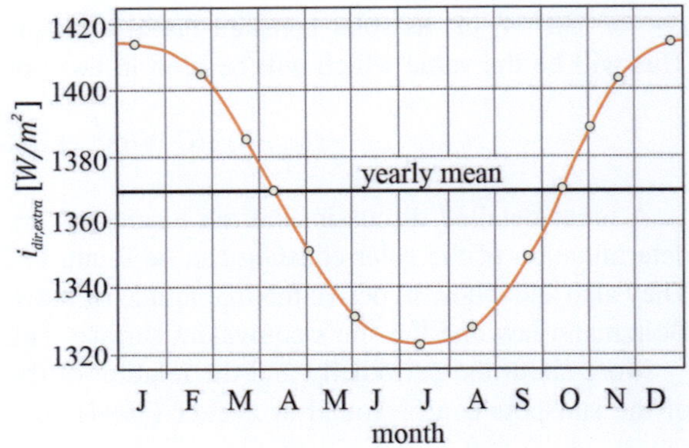

Fig. 2.4 Derivation of the extra-terrestrial intensity $i_{dir,extra}$ of solar radiation as a function of the distance sun-earth R_{S-E} along a year

relatively transparent chromosphere. With an average distance of the earth from the sun with $R_{S-E} = 1.496 \times 10^{11}$ m it results in a mean radiation density of the sun of $i_S = \Phi/(4 \cdot \pi \cdot R^2) = 6.311 \times 10^7$ W/m^2, where R is the solar radius with $R = 6.9626 \times 10^9$ m.

Using the Stefan-Boltzmann law formulated in (2.11), one can indicate the radiant power Φ emitted by a "black body" of area A and absolute temperature T. Therefore, if the radiant power Φ of the sun is known from Eq. (2.11) one can assume that the sun is a "*black radiator*" and indicate its surface temperature.

$$\Phi = \sigma \cdot A \cdot T^4, \tag{2.11}$$

where σ is the Stefan Boltzmann constant with $\sigma = 5.67.10^{-8}\,\mathrm{W}/(\mathrm{m}^2\mathrm{K}^4)$. This results in a value of the solar surface temperature $T_S = 5776\,\mathrm{K}$.

2.2.2 Spectrum of Extra-terrestrial Irradiation

The simplified picture of the structure of the sun shown above indicates that the radiation emitted by the sun into space is composed of several components and stemming from different zones close to the surface of the sun. This implies that the extra-terrestrial solar radiation has a complex spectrum that can only be approximated with a "*black body spectrum*" with a temperature of $T = 5776$K. The real extraterrestrial spectrum of solar radiation hitting the earth is shown in the Fig. 2.5 and compared to the 5776K blackbody spectrum. It covers the following wavelength ranges:

- Ultraviolet (UV) $\lambda < 380$ nm with an energy share of 6.4%,
- Visible $380 < \lambda < 780$ nm with an energy share of 48%,
- Infrared (IR) $\lambda > 780$ nm with an energy share of 45.6%.

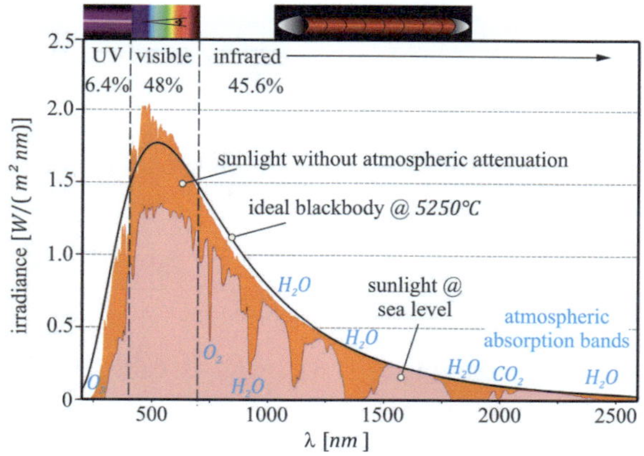

Fig. 2.5 Spectrum of solar radiation Φ as a function of wavelength λ extraterrestrial from measurements and in comparison the spectrum of a black body with 5800 °K. The intensity of the black body reduced by $((R/R_{S-E})^2 = 2.165 \times 10^{-5}$. The visible range is $380 < \lambda < 780$ nm

The visible part is determined by the spectral sensitivity of the human eye. The integration of the specific spectral irradiance over the individual wavelengths provides the energy content of the individual components. As a result, *the infrared portion of solar radiation is of the same energetic importance for solar thermal use as the visible light range.*

A real body cannot emit more thermal radiation at any wavelength than a "*blackbody*", which is therefore an ideal source of thermal radiation. Moreover, since the spectrum of the blackbody does not depend on any parameters other than temperature, and in particular on any material properties, it is a useful reference for many purposes. Its spectrum is described by Planck's radiation law (2.12).

Planck assumed in his empirically found formula that light is not absorbed and emitted continuously, but only discretely in *quanta* (today we speak of photons). According to this, the electromagnetic radiation of frequency ν can only be emitted and absorbed in discrete energy packets of the size $E = h \cdot \nu$. The Planck constant h is the quantum of electromagnetic action that relates a photon's energy to its frequency. Its magnitude was derived from the adaptation the formula law (2.12) to the experimentally determined values of the cavity radiation and assumes the value $h = 6.6261 \times 10^{-34}$ Js.

The spectral specific radiant intensity $\phi(\lambda, T)$ is dependent on wavelength (λ) and temperature (T). $\phi(\lambda, T)$ is the radiant power radiated into the entire hemisphere by the area element dA in the wavelength range between λ and $\lambda + d\lambda$. Planck's law, however, can also be formulated as a distribution in the frequency range between ν and $\nu + d\nu$. It is calculated according to Eq. (2.12):

$$\phi(\lambda, T) = \frac{2\pi\, h\, c^2}{\lambda^5} \frac{1}{exp\left(\frac{hc}{\lambda k_B T}\right) - 1}, \tag{2.12}$$

where c is the speed of light with $c = 2.9979 \times 10^8$ m/s and k_B the Boltzmann constant with $k_B = 1.36 \times 10^{-23}$ J/K. The relationship between wavelength (λ)

and frequency (ν) is given by $c = \lambda \cdot \nu$ and the unit of the spectral specific radiant intensity $\phi(\lambda, T)$ is $\mathrm{Wm^{-2}\,\mu m^{-1}}$ in the SI-system.

According to Planck's law, the thermal radiation emitted by a black body consists of a mixture of electromagnetic waves of a wide range of wavelengths. The maximum of the distribution depends on the temperature and is given by Wien's displacement law

$$\lambda_{max} = \frac{2897.8\,\mu\mathrm{m\,K}}{T}, \tag{2.13}$$

where the constant 2897.8 (μm K) is called Wien's displacement or Lummer's constant.

Figure 2.6 shows Planck's radiation spectra for different temperatures as a function of the wavelength between 300 and 1000 K in linear representation. The result is a bell shape with a distinct maximum, a steep decrease to short wavelengths and a longer decrease to long wavelengths. As required by Wien's displacement law (2.13) the position of the radiation maximum shifts to shorter wavelengths with increasing temperature. At the same time, according to the Stefan-Boltzmann law (2.11) the total specific radiation increases with the fourth power of the absolute temperature.

This disproportionate increase in radiation intensity with increasing temperature makes it difficult to display curves over a wider temperature range in a diagram. Therefore, a double-logarithmic division is chosen in the Fig. 2.7. At 300 K ($\approx$ambient temperature) the maximum of this curve is $\lambda = 10\,\mu$m. This means that the exchange of thermal radiation from objects at room temperature takes place in long-wave infrared. Typical infrared thermometers and thermographic cameras work in this area. The curve for 3000 K corresponds to the typical radiation spectrum of an incandescent lamp. Now already a part of the emitted radiation according to Fig. 2.7 is emitted in the visible spectral range. At the effective temperature of the sun of $T = 5776$ K, the radiation maximum lies in the centre of the visible spectral range.

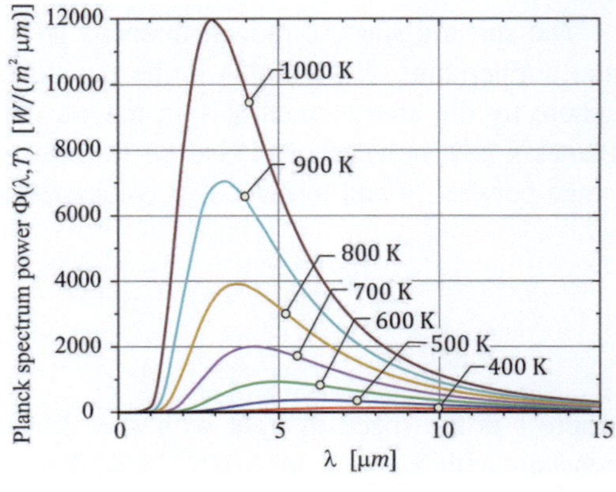

Fig. 2.6 Planckian radiation spectrum of a "*black*" radiator at different temperatures as a function of wavelength λ (μm)

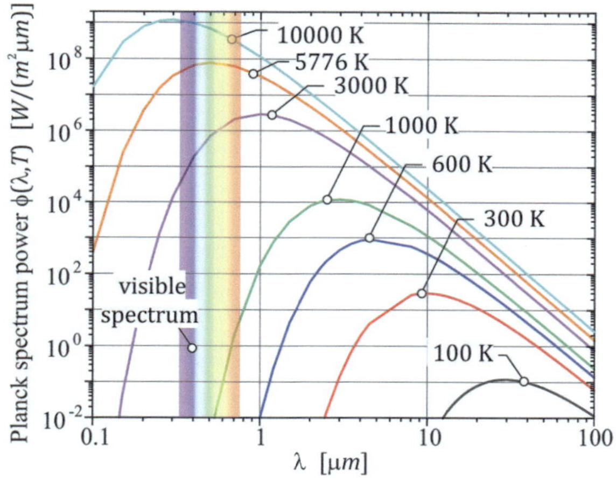

Fig. 2.7 Planck's radiation spectra as a function of wavelength λ for different temperatures in a double-logarithmic plot

2.2.3 Directional and Temporal Variation of Extra-terrestrial Irradiation

2.2.3.1 Direction of Direct Radiation

The intensity of the direct solar radiation on a surface and hence the energy that can be obtained depends mainly on the position of the earth in its orbit, the direction of the incident solar beam, and the orientation of the solar absorber (collector). Many factors play a role here, which are explained in the following section and can be expressed algebraically with the help of geometric relations. It's these:

- the elliptical orbit of the earth around the sun, the so-called ecliptic,
- the declination of the Earth, i.e. the inclination of the Earth's north–south polar axis to the ecliptic (seasonal influence),
- the rotation of the earth (influence of the time of day),
- the dependence of the position of the sun on the geographical latitude Φ and
- the influence of short-term climatic aspects (weather).

The irradiation direction of the solar direct radiation onto an arbitrarily oriented, flat surface A is determined by angular relationships. Figure 2.8 shows the different angles of such a configuration. In a first step we assume that the area of the terrestrial site is horizontal.

- θ is the incident angle of the beam radiation between the direction of irradiation and the normal of an arbitrarily oriented and tilted surface A.
- θ_z is the solar zenith angle. This is the angle between the sun and the normal on the horizontal plane where area A is located.
- Φ is the angle of the geographical latitude (e.g. Freiburg $\Phi = 48°N$) and is $\Phi = 0°$ at the equator and $\Phi = 90°$ at the poles. Latitudes on the Southern hemisphere are negative.

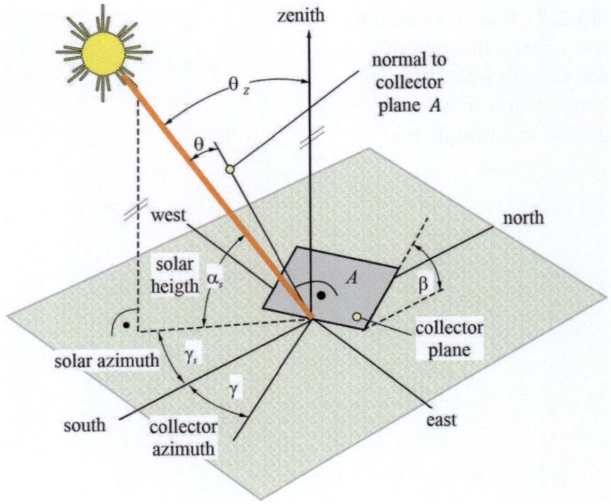

Fig. 2.8 Definition of the angles for calculating the direction of incidence of direct radiation on a surface A, which is arbitrarily oriented towards the Earth

- α_s is the sun's elevation angle or solar altitude angle. The following applies: $\alpha_s + \theta_z = 90°$.
- β is the tilt angle of the area A against the horizontal plane.
- δ is the declination of the earth and describes the angle between the sun's highest point and the equator and varies between $-23.5°$ and $+23.5°$.
- γ is the azimuth angle of the surface normal of A. It describes the angular difference of the horizontal projection of the surface normal from the south direction (West is $+90°$, East $-90°$). Observe that different systems exist to characterize the angles, e.g. in some literature South is indicated 180°.
- γ_s denotes the azimuth angle of the irradiation direction. It indicates the deviation of the horizontal projection of the irradiation direction from the south direction.
- ω is the hour angle in the plane perpendicular to the earth's rotation axis: ($1\,\text{h} = 15°$; morning $\omega < 0$ and afternoon $\omega > 0$).

Looking into the geometrical relation between different planes of earth's movement and the position of a horizontal area on the earth's surface, the declination angle, and the latitude angle, as well as the hour angle affect the angle of incidence. If we describe the direction of the solar beam by a unit vector $\vec{s}$ and the normal of the tilted plane with the vector $\vec{n}$, then the cosine of the incidence angle is given by the scalar product of the normal vector and the solar vector:

$$\theta = \cos^{-1}(\vec{s} \cdot \vec{n}) = \cos^{-1}(\cos\,\alpha_s\,\sin\,\beta\,\cos\,(\gamma_s - \gamma) + \sin\,\alpha_s\,\cos\,\beta). \quad (2.14)$$

When the expressions for the solar altitude and azimuth angles are used, this expands to the following angular relationship:

$$\cos\theta = \sin\delta\sin\Phi\cos\beta - \sin\delta\sin\Phi\cos\beta\cos\gamma + \cos\delta\cos\Phi\cos\beta\cos\omega + \cos\delta\sin\Phi\sin\beta\cos\gamma\cos\omega + \cos\delta\sin\beta\sin\gamma\sin\omega, \quad (2.15)$$

in which the hour angle ω as well as all other angles have to be used in radian. The hour angle ω here corresponds to the true solar time in degree (at 12 noon $\omega = 0°$; during the morning is $\omega < 0$ and negative, while during the afternoon is ω positive). Within a day, the hour angle can be given with an accuracy of ≈10 s with a daily correction around noon with $\omega = 0$ using Eq. (2.16)

$$\omega = \frac{360\,2\,\pi}{24}(12 - t_{loc}). \quad (2.16)$$

t_{loc} in Eq. (2.16) refers to the true local solar time, which must be corrected by the legal time difference in case of summertime. t_{loc} is counted from 0 to 24 h.

In the relation for the true solar time t_{loc} a correction Δt_{loc} for the geographical longitude L_{loc} must also be included. The longitude-dependent correction Δt_{loc} results from the difference between the longitude of the local time zone of the location L_{loc} and the standard longitude L_{st} on which the legal time is based. For Δt_{loc}, the following relationship results:

$$\Delta t_{loc} = \frac{24}{360}(L_{loc} - L_{st})\ (\text{h}). \quad (2.17)$$

The daily time differences mentioned above add up and result in deviations of up to 15 min between noon of true solar time and noon of mean solar time. The causes are the elliptical orbit around the sun and the inclination of the earth's axis to the ecliptic. After one or more rotations, the sun is seen from a different angle because of the rotation around it. As a result, the Earth needs more or less than 24 h to turn the same side to the sun again. In order to obtain the local true solar time t_{loc}, therefore, a so-called equation of time is additionally required, which describes the difference between the true and mean solar time depending on the season. This equation of time (ET) is described by Eq. (2.18).

$$ET = \sum_{k=1}^{3}[a_k\cos(k\cdot J) + b_k\sin(k\cdot J)]\ \ (\text{h});\quad J = \frac{2\pi}{365.25}\cdot n, \quad (2.18)$$

where n is the number of the day in the year starting with 1, January 1st and a_k and b_k are constants with the values $a_1 = 0.0072$, $a_2 = -0.0528$, $a_3 = -0.0012$, $b_1 = -0.1229$, $b_2 = -0.1565$ and $b_3 = -0.0041$. The maximum error of the Eq. (2.18) is 0.8 s. The result of the ET equation is shown in the Fig. 2.9 as a curve over one year.

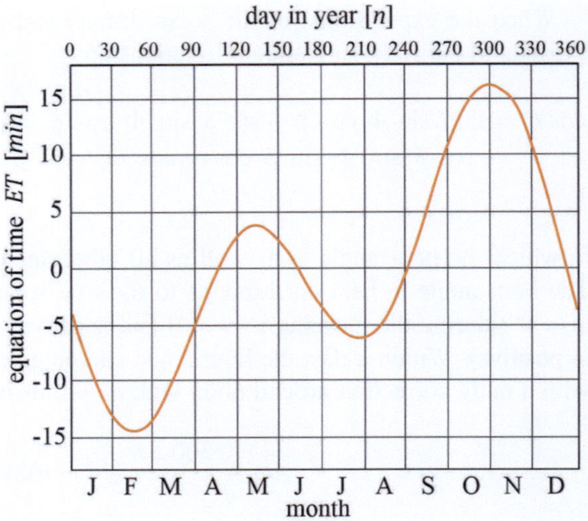

Fig. 2.9 Time correction ET from Eq. (2.18) of the true solar time due to the variation of the Earth's rotation over the year

The Eq. (2.18) ET is important if a concentrating mirror with a small acceptance angle θ_a of a few degrees is tracked with the computer. Here too, 1 h corresponds to about 15°. After all corrections over season, legal time, and longitude the local true solar time t_{loc} is given by Eq. (2.19).

$$t_{loc} = t_{st} + \frac{24}{360}(L_{loc} - L_{st}) + ET \quad \text{(h)}. \tag{2.19}$$

where t_{st} denotes the time of the local time zone (from 0 to 24 *o'clock*), L_{loc} is the longitude of the location considered and ET is the equation of time in hours according to Eq. (2.18).

The declination δ fluctuates due to the inclination of the earth's axis to the ecliptic. The declination of the Earth relative to the Sun can be determined for each day n of the year using Eq. (2.20).

$$\begin{aligned} &\delta = \arcsin\left[0.398 \cdot \sin\left(2\pi\, \tfrac{n-Eq}{365{,}25}\right)\right] + \Delta \quad (°) \\ &\Delta = 0.026 \cdot \sin\left(2\pi\, \tfrac{n-173}{365{,}25}\right) \quad \text{for} \quad 173 < n < 365; \\ &\Delta = 0 \qquad\qquad\qquad\qquad\qquad \text{for} \quad n \leq 173, \end{aligned} \tag{2.20}$$

where Eq indicates the spring equinox (lat. equinox). The value of Eq fluctuates between 79 and 79.75 (leap year: 79, leap years $+1 : 79.25$, leap years $+2 : 79.5$, leap years $+3 : 79.75$). The declination is zero at the equinox, +23.45° at the summer solstice and −23.45° at the winter solstice, as Fig. 2.10 shows.

The approximation of Eq. (2.21) gives the declination with a maximum error of 0.2°. A simplified relationship is represented by the following equation:

$$\delta = 23.45 \cdot \sin\left(360\frac{284 + n}{365}\right). \tag{2.21}$$

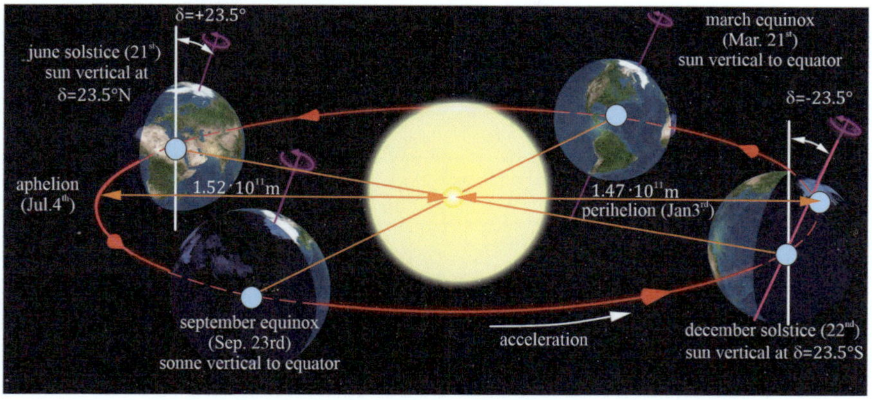

Fig. 2.10 Declination as a function of the season modified from https://allgeographynow.wordpress.com/tag/suns-declination/

The leading and trailing time compared to the mean time is called the equation of time (ET). If they are plotted against the angular distance of the Sun from the celestial equator, they take the form of an eight, a so-called analemma, which is illustrated in Fig. 2.11. In order not to have to commit to a particular location, however, the equation of time is usually shown as a function of the declination of the sun.

2.2.3.2 Direction of Irradiation on a Horizontal Surface

When irradiating a horizontal surface, the two angles are $\beta = 0°$ and $\gamma = 0°$. The angle to the surface normal changes to the zenith angle. This simplifies the Eq. (2.15):

$$\cos\theta_z = \sin\Phi\cos\delta\cos\omega. \tag{2.22}$$

For the derivation of Eq. (2.22) the cosine theorem has been used, for which the following applies: $\cos(\alpha \pm \beta) = \mathrm{co}\,s\alpha\ \cos\ \beta \mp \sin\ \alpha\ \sin\ \beta$. The sun elevation angle α_s is linked to the zenith angle θ_z in the following way $\alpha_s + \theta_z = 90°$; as also shown in the Fig. 2.12. So it follows:

$$\cos\theta_z = \sin\alpha_s. \tag{2.23}$$

If the equation of time (2.18) for ET is included in the calculation of the angle of the sun's elevation, the sun elevation is obtained as a function of time for each day of the year. The graphical representation shows that for each day there is a loop-shaped form of the solar elevation angle. Figure 2.12 shows the result with the zenith angle θ_z or the solar elevation angle α_s as a function of the azimuth angle γ_s, the declination δ or season and the time of day. An analysis of Eq. (2.15) shows that in summer in the northern hemisphere the sun rises north of the

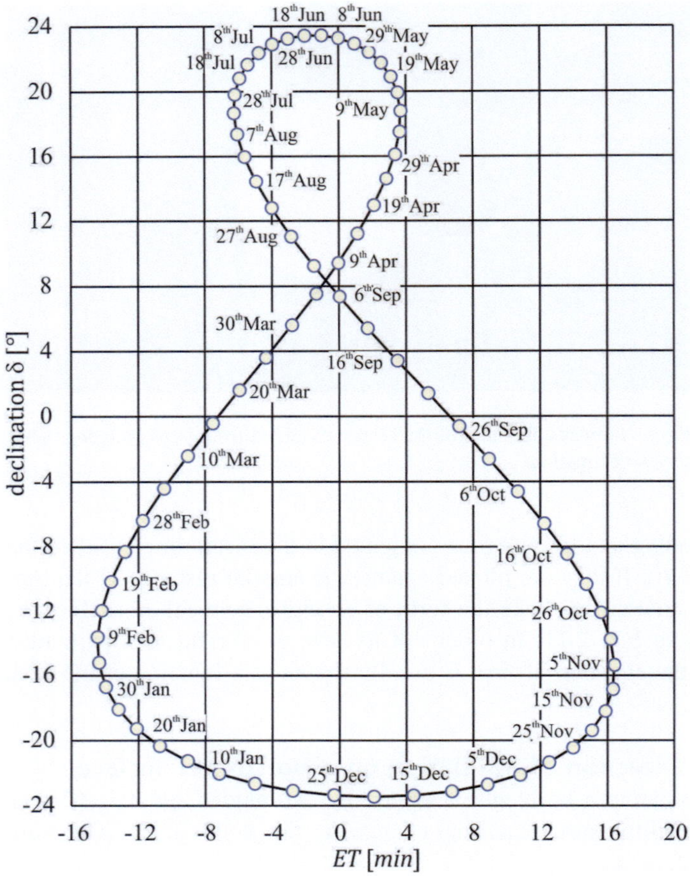

Fig. 2.11 The analemma shows the declination and equation of time (ET) for each day of the year

east–west axis. North of the polar circle depending on the latitude the sun doesn't even set at all. This means that a vertical collector at latitude 48° that is exactly oriented to the south without any tracking sees the sun on the front only between 8 and 16 o'clock (Fig. 2.12) and thus only 8 h (!!).

Figure 2.13 shows for the same location the solar elevation angle α_s as a function of the azimuth angle γ, the time of day and the declination δ (here in the form of the day in the year). In simpler representations the hourly values are entered as simple lines without consideration of the equation of time.

2.2.3.3 Direction of Irradiation on a Vertical Surface

For a vertical surface, with $\beta = 90°$ and $\gamma = 0°$ the Eq. (2.15) becomes:

$$\cos\theta_z = \sin\ \delta \cos\Phi\ \cos\gamma + \cos\ \delta\ \sin\Phi \cos\gamma \cos\omega + \cos\delta \sin\gamma \sin\ \omega. \quad (2.24)$$

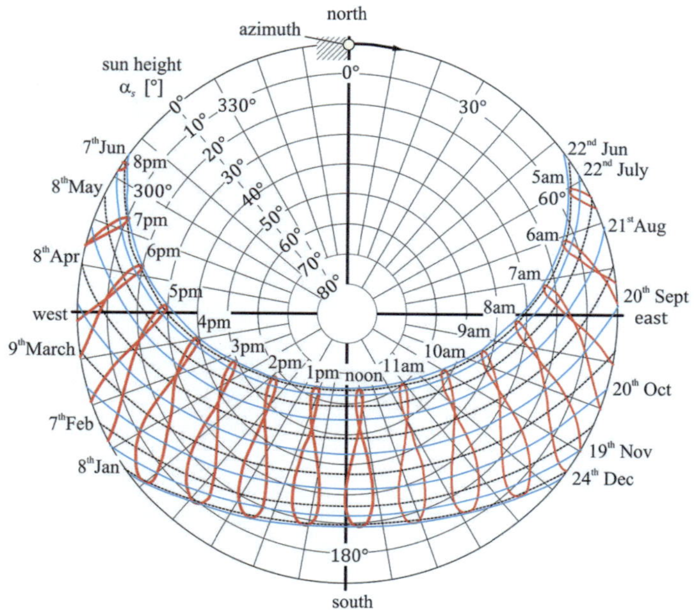

Fig. 2.12 Sun elevation angle α_s and θ_z as function of the azimuth angle γ, the time of day and the declination δ for the location Munich, latitude $\Phi = 48°\ 10'$ and $11°\ 30'$ eastern longitude; the curves to the full hours contain the time shifts according to Eq. (2.20)

Fig. 2.13 Sun elevation angle α_s as function of the azimuth angle γ, the time of day and the declination δ (here in the form of the day in the year)

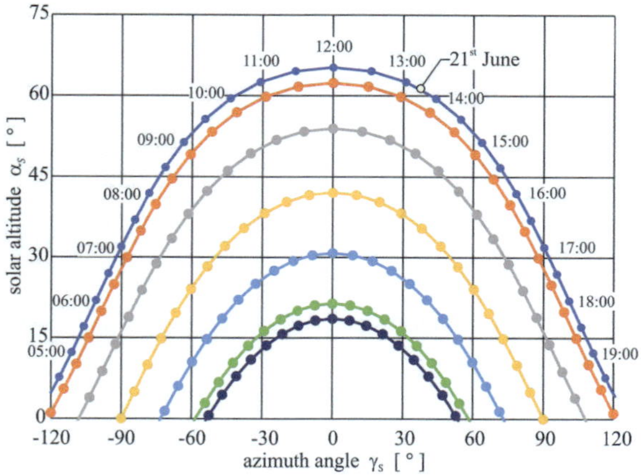

The angles of incidence of surfaces oriented to the south or north can be displayed with the difference $(\Phi - \beta)$ at an azimuth angle of $\gamma = 0°$. Figure 2.14 shows the situation schematically.

Equation (2.15) thus changes for the northern hemisphere:

$$\cos\theta_z = \sin(\Phi - \beta)\sin\delta + \cos(\Phi - \beta)\cos\omega \text{ (northern hemisphere).} \quad (2.25)$$

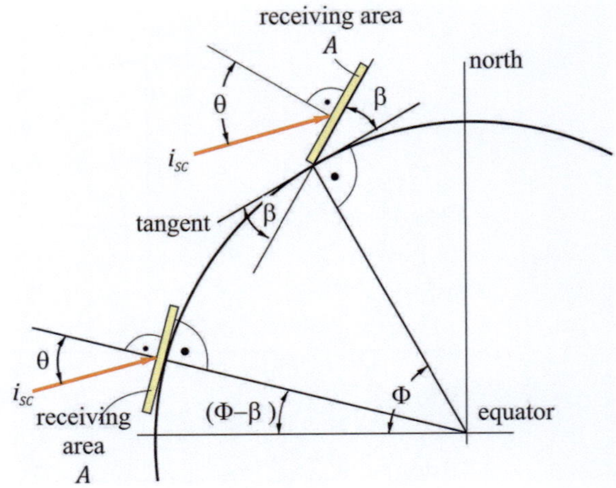

Fig. 2.14 Schematic sketch for the calculation of the radiation flux on an inclined surface A, which is aligned exactly in south-north direction ($\gamma = 0°$)

For the southern hemisphere ($\Phi + (\Phi - \beta)$) if convention of + and − is followed for Φ and δ the following relationship is obtained:

$$\cos\theta_z = \sin(\Phi + \beta)\sin\delta + \cos(\Phi + \beta)\cos\delta\ \cos\omega \text{ (southern hemisphere)} \tag{2.26}$$

2.2.3.4 Sunrise and Sunset

Equation (2.22) can also be solved according to the hour angle. At sunrise and sunset the zenith angle is $\theta_z = 90°$. If you enter the hour angle of the sunrise with ω_R and the hour angle of the sunset with ω_S and set for the hour angle of one hour $\omega = 15°$ ($\triangleq$ 1 h), you get the exact sunrise and sunset times $\omega_{R/S}$, resulting in Eq. (2.27):

$$\cos\omega_{R/S} = -\frac{\sin\Phi\ \sin\delta}{\cos\Phi\ \cos\delta} = -\tan\Phi\ \tan\delta. \tag{2.27}$$

The astronomical sunshine duration S_0 is the time difference between the rise and set times:

$$S_0 = 2 \cdot \frac{24}{360}\left|\omega_{R/S}\right|. \tag{2.28}$$

The daily solar maximum α_s is given by the latitude Φ and the declination δ according to the relation in equation

$$\theta_z = \Phi \pm \delta;\text{ respectively } \alpha_{s,max} = 90° - (\Phi \pm \delta). \tag{2.29}$$

2.2.3.5 Direct Extra-terrestrial Radiation on Horizontal or Inclined Surfaces

The larger the incidence angle of the solar irradiance on a surface, the lower the surface-related radiation density, as the Fig. 2.15 schematically shows. This is considered by the angle $\cos\theta_z$ from Eq. (2.22) for a horizontal surface A_{hor}. The radiation intensity $i_{dir,h}$ results from the product of the solar constant i_{sc} and the solid angle of irradiation θ_z

$$i_{dir,h} = i_{SC} \cdot \cos\theta_z, \tag{2.30}$$

in which the index h indicates the horizontal component.

For a surface A arbitrarily oriented towards the sun, however, the cosine of the angle from the direction of irradiation θ and the normal of an arbitrarily oriented surface A from Eq. (2.15) must be used. This results in

$$i_{dir,t} = i_{SC} \cdot \cos\theta, \tag{2.31}$$

where the subscript t is for tilted.

Often irradiation data is only available for horizontal surfaces. The conversion of irradiance on a horizontal surface A_{hor} to irradiance on an arbitrarily oriented surface is done with the help of the Eq. (2.32). The value Ψ represents the ratio of the radiation on an inclined surface compared to a horizontal one.

$$\Psi = \frac{i_{dir,t}}{i_{dir,h}} = \frac{\cos\,\theta}{\cos\,\theta_z}. \tag{2.32}$$

This will be illustrated by two examples:

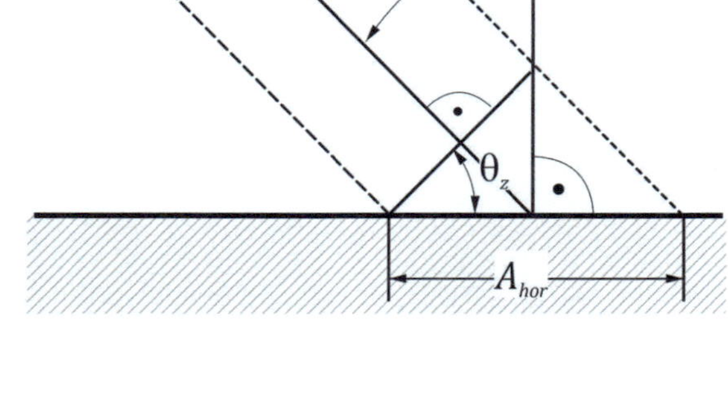

Fig. 2.15 Schematic diagram of the direct radiation of a radiation density i_{dir} on a horizontal surface A_{hor}

- On a summer day, the sun strikes a flat surface ($\beta = 0°$) at an elevation angle $\alpha_s = 60° \rightarrow \theta_z = 30°$. Direct radiation on the horizontal surface $i_{dir,h}$ with a value of $i_{dir,h} = 650\,\mathrm{W/m^2}$ is measured. If at the same location the irradiation surface is inclined by $\beta = 30°$, the angle of incidence would be $\theta = 0°$. From Eq. (2.32) a Ψ-value of $\Psi = \cos 0°/\cos 30° = 1.15$. Thus the direct radiation component for the inclined surface $i_{dir,t} = \Psi \cdot i_{dir,h} \approx 750\,\mathrm{W/m^2}$.
- On a winter day the elevation angle is $\alpha_s = 30°$ and an $i_{dir,h}$ of $150\,\mathrm{W/m^2}$ is measured. If the surface is inclined at the same angle by $\beta = 30°$, the result is $\theta = 30°$ and $\theta_z = 60°$. The Ψ-value then assumes a value of $\Psi = 1.732$ and the direct radiation component on the inclined surface is then $i_{dir,t} \approx 260\,\mathrm{W/m^2}$.

This ratio is still important for the calculation of irradiance on arbitrary surfaces from horizontal radiation values. With the help of these geometric relationships, the diurnal and annual courses of solar radiation outside the atmosphere can be calculated relatively quickly from the solar constant. They show the influence on the power density of the radiation when varying different parameters, such as orientation or location of the collector surface.

2.2.4 Daily and Monthly Sums of Extra-terrestrial Irradiation on Arbitrarily Inclined Surfaces

The sunrise and sunset angles and hence the times and the duration of sunlight were derived in Eq. (2.27) when we remember that the hour angle of 15° is equivalent to 1 h and the following relationship is assumed for the sunrise and sunset angles $\omega_R = -\omega_S$.

Under these conditions, the daily sum of the extra-terrestrial radiation E can be calculated from the integral from sunrise to sunset over time. Here it must be considered that the radiation hits the surface at an angle. The power density is thus reduced according to the Eq. (2.33) by cos θ. The corresponding functional correlation is

$$E = i_s \cdot \frac{24}{2\pi} \int_{\omega_R}^{\omega_S} \cos\theta\, dt. \tag{2.33}$$

For a horizontal surface with $\beta = 0°$ and $\gamma = 0°$ the Eq. (2.22) can be used as angle of incidence ($\cos\theta_z = \sin\Phi \sin\delta + \cos\Phi \cos\delta\,\cos\omega$). Knowing the latitude and declination, the following simplification can be applied

$$\cos\theta_z = a + b\cos\omega. \tag{2.34}$$

Using $\omega_R = -\omega_S$ the integration results in the daily sum of the extra-terrestrial irradiation E on a horizontal surface:

$$E = \frac{24 \cdot i_S}{\pi} \cdot a\omega_S + \frac{24 \cdot i_S}{\pi} b \sin\omega_S = \frac{24 \cdot i_S}{\pi}(a\omega_S + b\sin\omega_S), \tag{2.35}$$

Table 2.3 Hour angle for sunrise and sunset, the mean monthly duration of the day as well as the mean declination $\overline{\delta}$ for the respective months in the last column for a location with $\Phi = 48°$ northern latitude (Freiburg)

Month	ω_S (°)	Duration (h)	$\overline{\delta}$
January	64.2	8.6	21.4
February	74.4	9.9	13.4
March	86.8	11.6	2.9
April	93.2	12.4	9.3
May	112.2	15.0	18.8
June	118.7	15.8	23.4
July	116.1	15.5	21.6
August	105.9	14.1	13.4
September	92.6	12.3	2.3
October	79.3	10.6	9.5
November	67.1	8.9	19.3
December	61.6	8.2	23.4

where for ω the value to be entered here is in radians (1 rad $= 180°/\pi$).

Table 2.3 lists the values of the Eq. (2.35) for a location with $\Phi = 48°$ northern latitude. The maximum intensity of the extra-terrestrial irradiation is obtained around noon when $\omega = 0°$.

The following figures illustrate the effects of the angular relationships on the maximum intensity or the maximum daily sums of the extra-terrestrial radiation depending on the inclination of a surface to the horizontal (β) or the installation site (geographical latitude Φ).

Figure 2.16 shows the diurnal variation of the extra-terrestrial irradiance incident on a flat horizontal surface ($\beta = 0°$) at a latitude of $\Phi = 48° N$. Naturally, in summer the days may become almost twice as long as in winter and, additionally, due to the larger solar elevation angle α_s (Fig. 2.13) one observes also a higher direct irradiation intensity. In total the time-integrated daily irradiation (the area below the curve) in June is a factor of 8 larger than in December.

Inclined receiver surfaces oriented towards South (in the Northern hemisphere) resp. North (in the Southern hemisphere), are more favourable in winter, if they are tilted with a much larger angle of inclination β. In summer, on the contrary, less inclined surfaces (small β) are more advantageous, as the Fig. 2.17 shows. This is a combined effect of latitude and solar declination which both affect the prevailing incidence angles. Moderate inclination angles in between lead to higher intensities and daily totals, especially during the interseason periods due to their "hump curve" shape. As a rule of thumb an inclination angle of Φ should be chosen to maximise the annual irradiation sum on a surface.

Special requirements determined from the load curve of a consumer producing uniform daily totals over a certain time period or maximisation of output in a specific season may lead to an optimal collector tilt, differing from the standard orientation. Figure 2.18 shows this for the example of Cologne. With a flat horizontal orientation ($\beta = 0°$), the ratio of integrated energy per square metre from

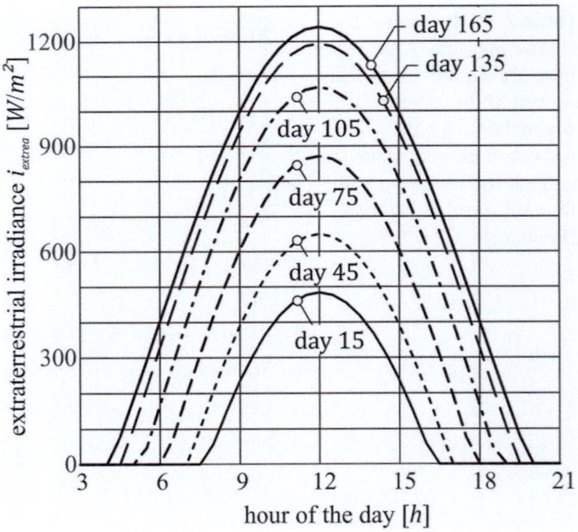

Fig. 2.16 Daily course of the intensity of extraterrestrial solar radiation i_{extra} in kW/m^2 as a function of the time of day t in hours (h) on different days of the year for the latitude of $\Phi = 48°N$ (Freiburg); angle of inclination: $\beta = 0°$ (horizontal)

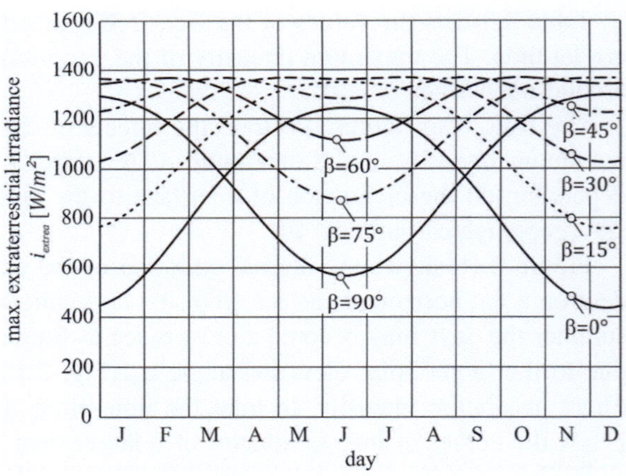

Fig. 2.17 Annual course of the daily maximum intensities of the extraterrestrial solar radiation i_{extra} in W/m^2 for the geographical latitude of $\Phi = 48°N$ (Freiburg) for different angles of inclination β; installation direction: south

summer to winter is almost a factor of four, whereas with a surface inclined at 45°, the ratio summer to winter is only 1.5. Morevover, the solar irradiation is relatively constant during the summer half year. On the other hand a tilt of 75° yields the highest values in the heating season.

Figure 2.19 shows the extra-terrestrial irradiance of a receiver surface inclined by $\beta = 45°$), for different latitudes of the northern hemisphere. At the equator ($\Phi = 0°$), maximum values of irradiation intensity occur in winter, whereas at the polar caps, solar irradiation is limited to summer only. In the polar winter of the northern hemisphere there is no irradiation due to the polar night and the irradiation power is zero.

This is also reflected in the achievable daily totals for the year as shown in the Fig. 2.20. It is noteworthy that the daily sum at an inclination of $\beta = 45°$ without

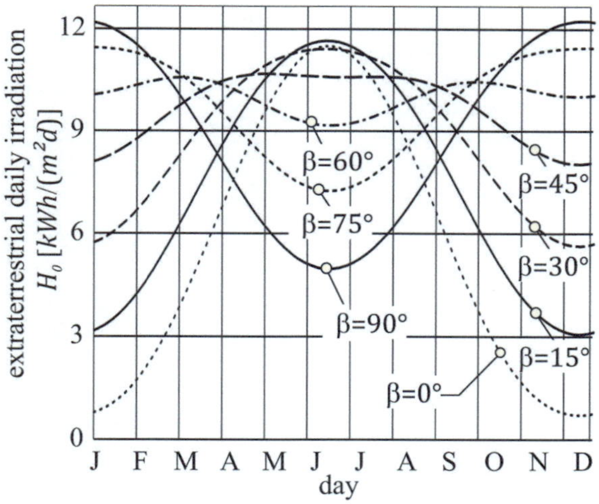

Fig. 2.18 Annual course of the daily sums of the extraterrestrial solar radiation i_{extra} in $\mathrm{kWh}/(\mathrm{m}^2\mathrm{d})$ for the geographical latitude of $\Phi = 51° N$ (Cologne) for different angles of inclination β; installation direction: south

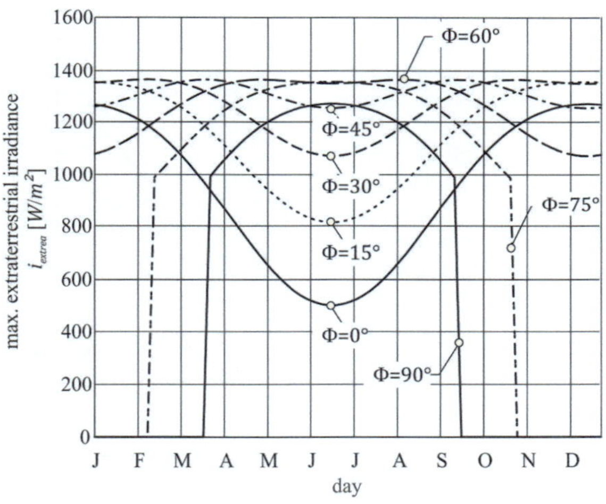

Fig. 2.19 Annual course of the daily maximum intensity of the extraterrestrial solar radiation i_{extra} in W/m^2 for different geographical latitudes Φ; installation direction: south; inclination angle: $\beta = 45°$

tracking varies only slightly at a latitude of $\Phi = 45°$. In addition, the seasonal fluctuations of the daily totals are much smaller there than, for example, near the equator.

Figure 2.21 the annual course of the daily sums of the intensity of the extra-terrestrial solar radiation for the geographical latitudes $\Phi = 34°$ (Rabat) and $\Phi = 59°$ (Stockholm) for the case that the collector surface always follows the achievable maximum intensity value both in the installation direction and the angle of inclination. A comparison of the daily sum from the Fig. 2.21 with that of a rigid collector (cf. Fig. 2.18) shows that a considerable improvement can be achieved, especially in winter. Here up to a factor of four (!) higher harvest values can be achieved. In summer the difference is smaller but still considerable, with a factor of two compared to a rigid collector. However, the Fig. 2.21 also shows the

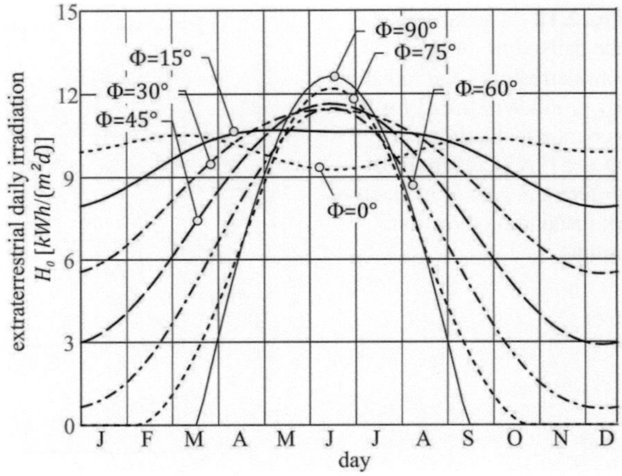

Fig. 2.20 Annual course of the daily sums (kWh/(m²d)) of the extraterrestrial solar radiation as a function of the geographical latitude Φ, installation direction: south; angle of inclination $\beta = 45°$

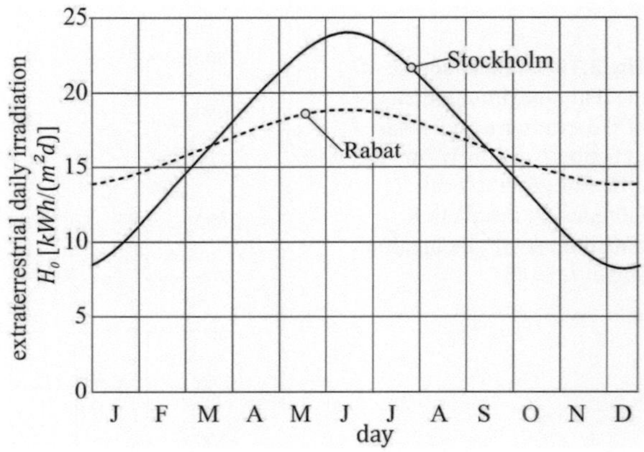

Fig. 2.21 Annual course of the daily sums (kWh/(m² d)) of the intensity of the extraterrestrial solar radiation for the geographical latitudes $\Phi = 34°$ (Rabat/Morocco) and $\Phi = 59°$ (Stockholm/ Sweden); biaxial tracking with respect to the maximum achievable irradiation power in the direction of installation

influence of the length of day. The daily sums of the irradiation on a biaxially tracked surface are calculated. In Eq. (2.33) cos = 1 is set and integrated over the length of the day. The result shows that although Stockholm is further north than Rabat, the daily sum of irradiation in summer exceeds that of Rabat. With regard to the energy yield of a collector, however, this only has an effect with a two-axis tracking system.

Due to the polar summer comparatively high maximum daily totals also occur at the North Pole, Fig. 2.22 shows. As expected, an integration of daily totals over the year shows the highest yields near the equator. The reason for this is the inclination of the earth's axis of 23.5° to the ecliptic and thus the irradiation occurring throughout the year.

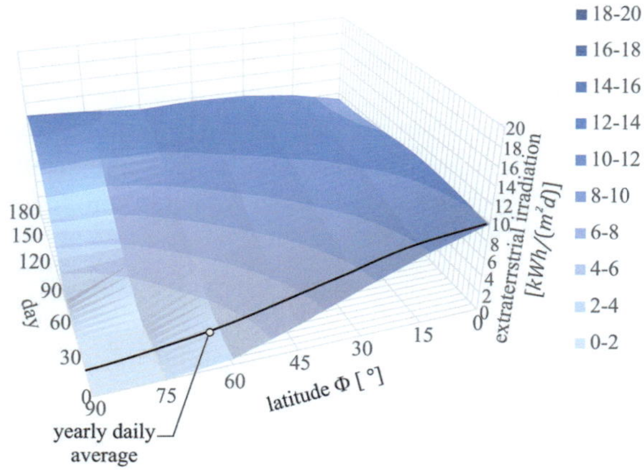

Fig. 2.22 Intensity of daily sum and time-integrated annual radiation on a flat horizontal surface as a function of latitude Φ, ecliptic skew 23° (Gerthsen et al., 1995)

2.2.5 Tracking Surfaces

According to the Fig. 2.23 the irradiation power on a surface decreases with increasing angle of incidence θ_z. For the technical use of solar energy it can therefore be advantageous to track the collector surface to the changing position of the sun. With ideal biaxial tracking, the radiation always hits the collector perpendicularly and the angle of incidence θ_z is always the same:

$$\cos\theta_z = 1. \tag{2.36}$$

However, a two-axis tracking system is technically very complex. For this reason, solar collectors are usually only tracked on one axis. The preferred tracking axes run in north/south or east/west direction. A third variant is the polar tracking with an axis parallel to the earth axis. A detailed analysis of the Fig. 2.12 allows first estimates of a reasonable control of the tracking angle.

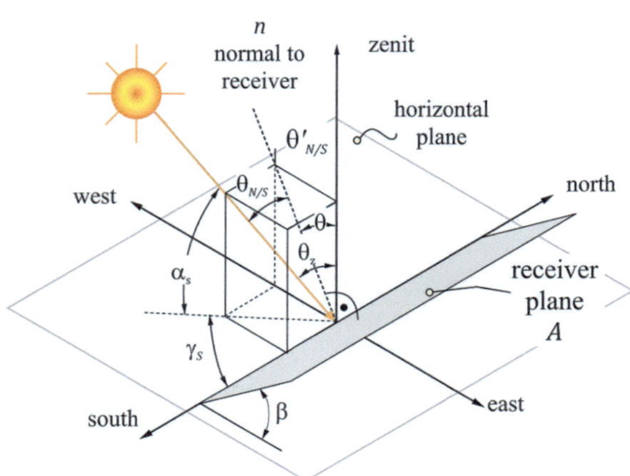

Fig. 2.23 Angular arrangement for converting the irradiation intensity to arbitrarily oriented surfaces

In this context, for the design of a solar thermal system, the demand for exergy (e.g. process heat) must be linked to the demand of the connected work process in relation to its usage time, for example the daily operating time. This requires tracking only within the period in which the demand is requested and limits the tracking effort not insignificantly. With a permanent tracking around the east/west axis, the direct radiation hits the collector with the angle $\theta_{E/W}$:

$$\cos\theta_{E/W} = \left(1 - \cos^2\delta \sin^2\omega\right)^{1/2}. \tag{2.37}$$

The inclination of the collector surface $\beta_{E/W}$ to be set is thus given by:

$$\tan\beta_{E/W} = \tan\theta_z \cdot |\cos\gamma_s|, \tag{2.38}$$

which contains the azimuth angle of the sun γ_s. For the azimuth angle γ_s applies

$$\cot\ \gamma_s = \frac{\sin\Phi\cos\omega - \cos\Phi\tan\delta}{\sin\omega} \quad \text{or} \quad \sin\gamma_s = \frac{\cos\ \delta\ \sin\omega}{\sin\theta_z}. \tag{2.39}$$

For an East/West (E/W) oriented plane, which is adjusted once a day at noon in such a way that the irradiation hits vertically, the angle of incidence depends on the hour angle. The following relationship results.

$$\cos\theta = \sin^2\delta + \cos^2\delta\cos\omega. \tag{2.40}$$

The inclination angle of the plane β corresponds to the zenith angle θ_z at noon:

$$\beta = |\theta_z| = |\Phi - \delta|, \tag{2.41}$$

where Φ is the latitude and δ the declination angle. In the case of a plane rotating around a north/south axis (N/S), the angle of incidence $\theta_{N/S}$ results from the equation

$$\cos\theta_{N/S} = \left(\cos^2\theta_z + \cos^2\delta\sin^2\omega\right)^{1/2}, \tag{2.42}$$

and the angle of inclination $\beta_{N/S}$ is calculated from the relationship

$$\tan\beta_{N/S} = \tan\theta_z|\cos(\gamma - \gamma_S)|. \tag{2.43}$$

In a constellation in which the plane is continuously polar tracked, the angle of incidence θ_{pol} varies only in the range of the declination δ:

$$\cos\theta_{pol} = \cos\delta. \tag{2.44}$$

According to the Eq. (2.31) maximum $\cos\delta = \cos 23.45° = 0.92$, i.e. only approx. 8%, the losses that occur in uniaxial polar tracking due to the deviation

Fig. 2.24 Global radiation intensity i_{glob} (W/m^2) as a function of the time of day on a horizontal surface A_{hor} compared with a tracked surface at a geographical latitude $\Phi = 50°$ on a cloudless day

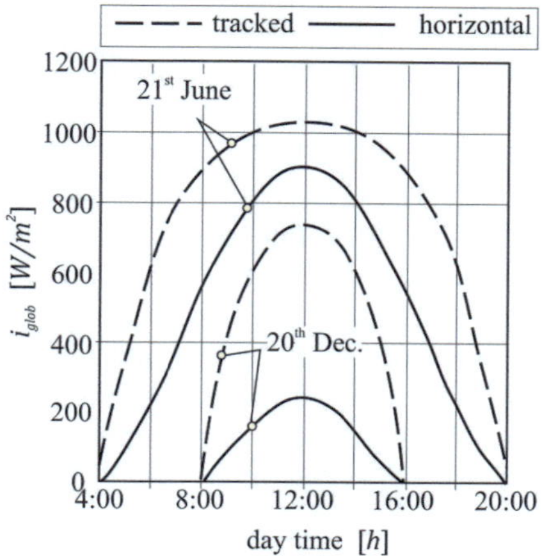

from the vertical direction of irradiation are the rotation around the polar axis corresponds to the hour angle ω. Further observations on tracking angle are described in Duffie and Beckmann (2020). However, they can also be easily derived using the trigonometric functions themselves by placing a cuboid above the angular distribution as shown in the Fig. 2.23. With the help of the projections on the surfaces of the cube, the figure allows the calculation of tracking angles using the cosine and sine theorems. For the angle of incidence of the north–south correction θ_{NSC} results:

$$\cos\theta_{NSC} = \left(\cos^2\theta_z + \cos^2\beta\sin^2\theta_z\right)^{1/2}, \tag{2.45}$$

and respectively for the tracking angle follows:

$$\tan\theta'_{NSC} = \tan\theta_z \sin\gamma_s. \tag{2.46}$$

There are large differences in the irradiation intensity between a horizontal surface and a tracked surface. On cloudless days at a geographical latitude of $\Phi = 49°$ (Karlsruhe), up to 30% higher radiant power densities in relation to the collector surface can be achieved by tracking even in summer, as the Fig. 2.24 shows. In winter this can even vary by factors. Integrated over one year, about 30% more radiant energy per square meter and year can be achieved in Germany through a targeted tracking of the collector surface.

2.2.6 Shading of Surfaces

Shading of solar surfaces occurs frequently, sometimes being a problem and in other cases a wanted regulation of the solar gain. For example the shading of collectors by neighbouring buildings restricts the positioning of these collectors in the building envelope. On the other hand, a shading of a window in summertime might avoid overheating of rooms. It is especially important, of course, for the direct beam irradiation. Therefore we treat this blocking of direct beam radiation in the chapter of extra-terrestrial radiation. The reduction of diffuse sky radiation for collector apertures is less important and can be quantified by using the view factor calculation for diffuse surfaces.

For the determination of shading we distinguish between irregular shapes like trees, mountains or the skyline of buildings, and regular geometries like collector rows or overhangs or wingwalls. In modern raytracing methods and object shapes in principle can be added and thus shading simulations for different sun positions may be performed. However, this is a very time-consuming approach and sometimes a quick assessment is needed.

For regular shapes the shading is much more amenable to analytical calculation and at least good approximations can be developed for these geometries. A second approach, however, is the qualitative use of sun path diagrams. Plots of solar altitude α_s versus solar azimuth γ_s for different days of the year may be prepared, on which shapes of obstructions (shading profiles) can be superimposed. From these diagrams one may determine when the path from the sun to the point in question is blocked.

At any point in time and at a particular latitude, φ, δ, and ω are fixed. From the equations of this see Sect. 2.2.3, the zenith angle θ_z or solar altitude angle α_s and the solar azimuth angle γ_s can be calculated. A solar position plot of θ_z and α_s versus γ_s for latitudes of $\Phi = \pm 45°$ is shown in Fig. 2.25. Lines for mean days of the months are labelled with their dates and lines of constant hour angles are labelled by hours. The angular positions of buildings, collectors, overhangs, or also natural obstructions like trees and mountains can be entered on the same plot.

The angular positions of outlines of an obstructing element may be calculated by trigonometric calculations. To illustrate the principle we give an example here.

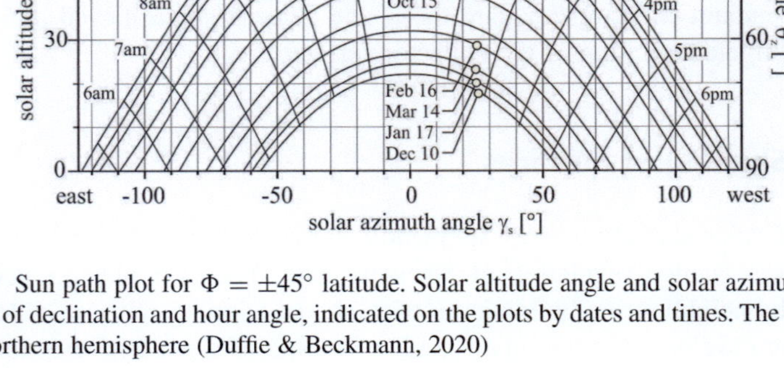

Fig. 2.25 Sun path plot for $\Phi = \pm 45°$ latitude. Solar altitude angle and solar azimuth angle are functions of declination and hour angle, indicated on the plots by dates and times. The dates shown are for northern hemisphere (Duffie & Beckmann, 2020)

Table 2.4 Calculation of angular coordinates for a building according to Eq. (2.47)

Point	D_i (m)	Δh_i (m)	Δw_i (m)	α_i (°)	γ_i (°)
1-East	45.71	26.00	−8.00	29.63	−9.93
2-South	45.00	26.00	0.00	30.02	0.00
3-West	68.77	26.00	52.00	20.71	37.10

Let us assume a collector installed on the roof of a two storey house at a height of 8 m above ground. In front of the collector at a distance of $L = 45$ m is a rectangular building wide $w = 60$ m, 34 m high parallel to the collector. The Eastern corner of the building is $\Delta w = 8$ m to the left, the Western façade $\Delta w = 52$ m to the right of the point being exactly in Southern direction. First, we calculate the angular coordinates of the 2 top corners plus the roof position right in front in direction South. Thus, we arrive using the formulae for point i

$$\alpha_i = \tan^{-1}\left(\frac{\Delta h_i}{\sqrt{L^2 + \Delta h_i^2}}\right) \text{ and } \gamma_i = \tan^{-1}\left(\frac{\Delta w_i}{\sqrt{L^2 + \Delta h_i^2}}\right). \tag{2.47}$$

Table 2.4 illustrates the computational results according to Eq. (2.47).

Vertical lines are vertical in the diagram, but horizontal lines are curved. We may calculate more points along the roof line, plot these points in the sun path plot, and shade the area of the obstructing element. The result of that example can be seen in Fig. 2.26. The precedent discussion would be exactly true only if the collecting area is an infinitesimal element. That means it is shaded or not. In reality the area is finite, and one edge of the collector might be shaded while the other is not. As a consequence a more precise treatment of shading has to take into account partial shading and the different relative angular positions of the shading objects. Then an integration over the whole collector area has to be performed. This is especially important if the shading object is close, smaller or of the same magnitude as the shaded collector aperture.

In order to illustrate this we want to consider the case of an overhang over a window. For simplicity we start with the approximation of a very long overhang so that the solar altitude angle projected onto the vertical incidence plane determines the position of the horizontal border between shadow and irradiated area. This projected angle is called the profile angle α_p. Assume a window of height H below a projection/overhang of extension L, which is a distance H_2 above the window. The plane defined by the overhand protrusion and the upper edge of the window hits the vertical plane at an angle of α_2, the plane through the lower edge at an angle $\alpha_1 > \alpha_2$. The angles are calculated as $\arctan(H_2/L)$ and $\arctan((H_1 + H_2)/L)$, respectively. For solar profile angles larger than $\alpha_p > \alpha_1$ the window is completely shaded, for profile angles $\alpha_p < \alpha_2$ it is completely irradiated. For angles in between a partial shading is observed.

A similar problem is encountered if we consider the shading of collector rows by identical rows in front of them, as illustrated in Fig. 2.27. If the solar profile

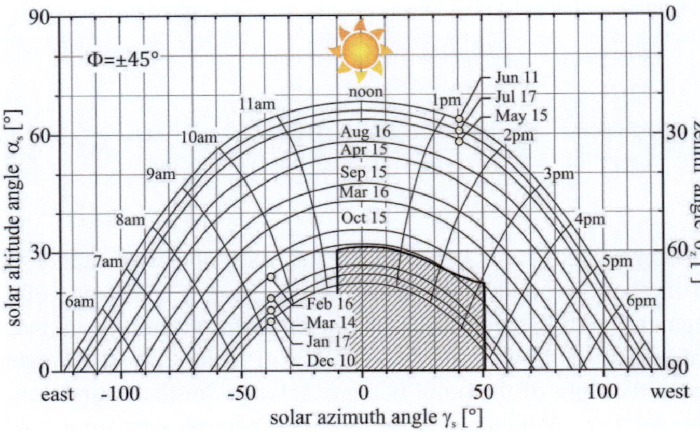

Fig. 2.26 Example of obstructing building within the sun path plot of Fig. 2.27 modified from Duffie and Beckmann (2020)

Fig. 2.27 Cross section of two collector rows

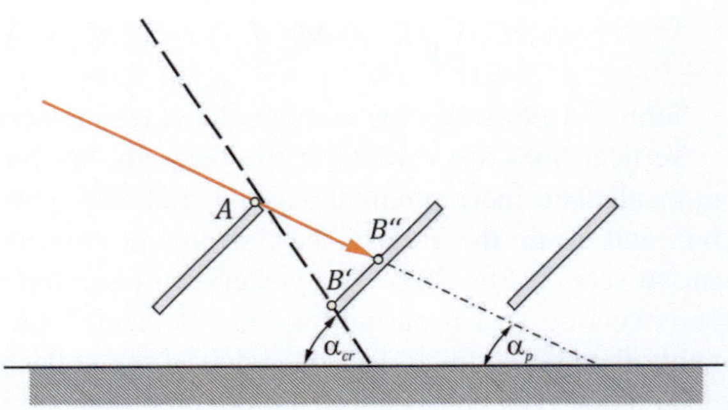

angle α_p is larger than the angle defined by the plane AB' and the horizontal plane, the collector is fully irradiated. For smaller positive angles a shading edge will appear on the second collector row in position B'' on the collector, α_p being the angle between plane AB'' and the horizontal plane. So if the collector rows are very close to each other, the times of low altitude sun with significant shading may be substantial. The optimization of collector row distance and loss of energy is a cost–benefit optimization and depends amongst other factors on the availability of ground area, natural horizon, and collector row orientation.

2.3 Terrestrial Solar Radiation

Due to the atmosphere and its interaction with the solar radiation the latter one changes intensity and spectral composition substantially before reaching the earth's surface. The direct radiation is weakened on its way through the atmosphere by

- absorption and
- scattering.

There are basically three different types of scattering mechanisms in the atmosphere, Rayleigh scattering, Mie scattering and geometrical scattering. The latter one occurs when we have large rain drops in sun light, and we can observe a colourful rainbow. The other two mechanisms are omnipresent. Rayleigh scattering occurs even in very clean atmospheres and is produced by molecular particles in the atmosphere. It causes the blue coloration of the sky and occurs on particles that are much smaller than the wavelength of light. The scattering intensity is wavelength dependent and scaled with:

$$\approx \frac{1}{\lambda^4}. \tag{2.48}$$

This means that the smaller the wavelength, the greater the scattering of the light. The direct radiation suffers a small redshift, which is also visible to the eye when travelling long distances through the atmosphere.

Mie scattering in the atmosphere is done by aerosols, dust, or salt particles with sizes of the scattering agglomerates or particles larger or approximately equal to the wavelength of the light λ. The scattering intensity of Mie scattering scales approximately with $1/\lambda$. In most cases, however, due to a distribution of particle sizes, the resulting scattered intensity is independent of the wavelength λ. Diffuse light after Mie scattering is therefore white or grey.

Several constituents of the atmosphere are responsible for absorption. CO_2, O_2 and ozone (O_3) as well as water vapour are the dominant absorbing gases in the atmosphere. The absorption of ozone in the ultraviolet range is of no energetic significance, but is of great importance for life on earth, protecting cells against high energy ultraviolet radiation. Significant losses occur in the infrared range due to absorption of water vapour at wavelengths of 930, 1120, 1400 and 1850 nm. There are also absorptions by oxygen and ozone in the infrared range. While the extra-terrestrial spectrum contains still about 5% of power at wavelengths longer than 2.5 μm, it drops to very small values due to the absorption of water vapor and CO_2. Of course, the absorption and scattering processes are dependent on the density of the atmosphere, the prevalent particles and aerosols, and the overall integrated extinction is dependent on the length of the light path through the atmosphere. Therefore, the solar spectrum on high mountains is very different from a spectrum a sea level. Industrial atmospheres have higher aerosol content. The light path through the atmosphere is changing during the course of a day. For the latter effect, assuming that other atmospheric parameters are identical, the concept of the air mass factor AM has been developed. Imagine that the atmosphere is a plane layer above the earth. Then by pure geometry depending on the solar altitude angle as, the light ray passes a different path length. The shortest path is when the sun is right above in the zenith. This is called $AM1$ for the observer's position at sea level. The length of the path increases with zenith angle θ_z. The zenith angle θ_z and the elevation angle α_s are complementary angles for which ($\theta_z = 90° - \alpha_s$)

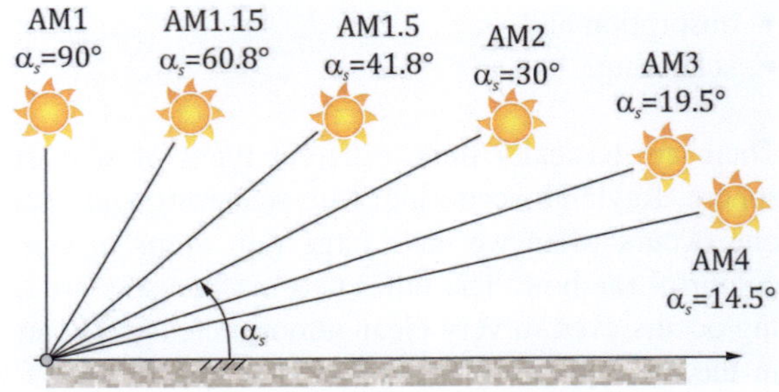

Fig. 2.28 Air mass factors AM according to Eq. (2.49)

applies. For a solar altitude of $\alpha_s = 30°$ we have twice the length of the imaginary light path through the layer. This is called $AM2$. The extraterrestrial radiation consequently is characterized as $AM0$. Generally, the optical air mass factor AM is approximated as

$$\mathrm{AM} \cong \frac{1}{\cos\theta_z} = \frac{1}{\sin\alpha_s} \rightarrow \begin{array}{ll} \mathrm{AM} = 1 & \text{for} \quad \theta_z = 0° \\ \mathrm{AM} = 2 & \text{for} \quad \theta_z = 60° \end{array}, \tag{2.49}$$

and is schematically shown in Fig. 2.28. More exactly one should consider a path length L of incident solar radiation through the atmosphere, where L_0 is the path length at zenith (i.e., normal to the Earth's surface) at sea level. The optical air mass coefficient AM is defined as:

$$\mathrm{AM} = L/L_0. \tag{2.50}$$

The approximation of Eq. (2.49) $L = L_0 \cdot \cos\theta_z$ applies to a flat atmosphere and overlooks the atmosphere's curvature and the refraction of the light beam entering the atmosphere. It predicts an infinite air mass at the horizon. It uses the direction to an apparent position of the sun, which is not the exact true position. However, it is reasonably accurate for values of θ_z up to around 70° where the error is less than 1%. A number of refinements have been proposed to model the path thickness more accurately towards the horizon, a very popular one in solar energy is that proposed by Kasten and Young (1989)

$$AM = \frac{1}{\sin(\alpha_s) + 0.50572\,(6.07995 + \alpha_s)^{-1.6364}}. \tag{2.51}$$

A more comprehensive list of such models is provided in Rapp-Arrarás and Domingo-Santos (2011) for various atmospheric models and experimental data sets. At sea level the air mass towards the horizon is approximately 38.

Figure 2.29 shows a terrestrial spectrum $AM1$ and the extra-terrestrial spectrum as a function of wavelength in (nm). For comparison, the spectrum after passing through the atmosphere is shown as a line. It can be clearly seen that a large part of the extra-terrestrial radiation does not penetrate the atmosphere in the near infrared (IR) and in the ultraviolet (UV) range. The radiation transmission is

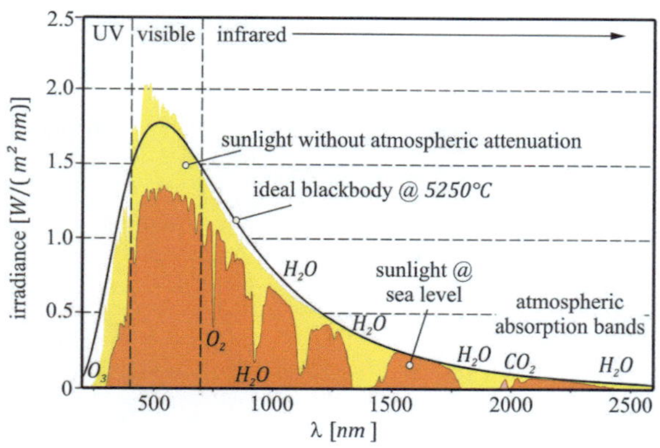

Fig. 2.29 Example of spectrum change due to Rayleigh scattering and gas absorption when passing through the atmosphere compared to extraterrestrial radiation. In particular, the absorption of water vapor in the infrared range ($\lambda > 780$ nm) should be noted. The influence of the wavelength dependence is shown in the adopted representation of Iqbal (1983). The relative air mass factor (AM) is $AM = 1$, i.e. vertical passage. The water vapour in the atmosphere was assumed to be 2 cm condensable

significantly influenced by absorption and scattering processes. Qualitatively, lower solar altitudes (larger AM) leads to a red shift of the direct solar spectrum, because of the large scattering of blue light in the long light path. This effect can be observed in an extreme case when the sun sets as a red disk when the horizon is very low.

As consequence of the atmospheric processes which affect the direct solar light beam, we have not only direct irradiance on earth. The scattering processes in the clear sky, but even more for cloudy atmospheres lead not only to a reflection of solar intensity to outer space, but also to a transmitted diffuse light coming from all over the sky hemisphere to the observer. The terrestrial irradiance coming from the sun consists of a direct beam irradiance part G_b and a diffuse irradiance part G_d. The sum is called global radiation G

$$G = G_b + G_d. \tag{2.52}$$

For the usual case that the irradiance is measured on a horizontal plane the index h is used. When a tilted plane is taken as reference (e.g. the collector plane) the index t is used. Figure 2.30 shows schematically the composition of the global irradiance.

2.3.1 Attenuation Mechanisms in the Atmosphere

In the following section we want to describe in detail and more quantitatively the attenuation mechanisms in the atmosphere.

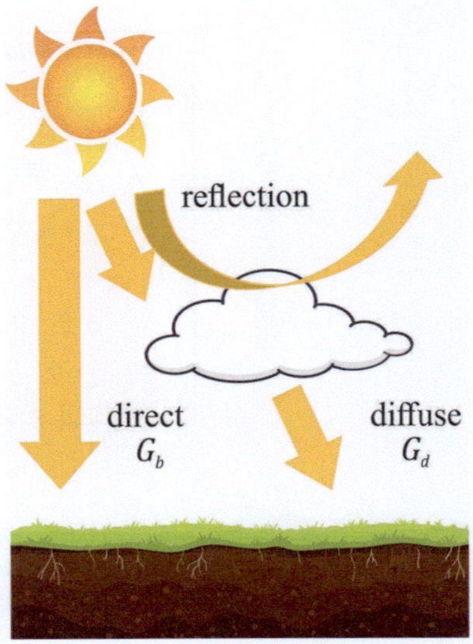

Fig. 2.30 The influence of the atmospheric layer on the solar irradiation

2.3.1.1 Absorption in the Atmosphere

Absorption is a dissipative process of a wave in a body, which means a loss of intensity. The absorption law for photon radiation has an exponential form as it is a probabilistic process. The intensity of light absorbed in a thin layer of thickness dx at penetration depth x is proportional to the intensity $I(x)$ of the radiation and proportional to an absorption coefficient (μ) dependent on the medium. It is therefore:

$$\frac{dI}{dx} = -\mu \cdot I(x) \rightarrow I(x) = I_0 \cdot \exp(-\mu \cdot x). \tag{2.53}$$

The absorption coefficient depends on the density of the material (and thus the number of particles N) and the absorption cross section (σ). If the material density is constant, this is summarised in the Lambert–Beer law, which reads to

$$I(x) = I_0 \cdot \exp(-N \cdot \sigma \cdot x) = I_0 \cdot \exp(-\delta(x)), \tag{2.54}$$

where the product $(N \cdot \sigma \cdot x)$ referred to as optical thickness $\delta(x)$. In the atmosphere, however, with increasing altitude, the density of the atmosphere and thus the number of particles in a control volume decreases. The particle density is a negative exponential function with altitude x according to the barometric height formula, as the following equation shows.

$$N(x) = N_0 \cdot \exp(-x/x_0), \tag{2.55}$$

where N is the number of molecules in the atmosphere as a function of altitude x relative to a reference altitude x_0. Inserting Eq. (2.55) into Eq. (2.54) gives a relationship for the height-dependent intensity $I(x)$, known as the Chapman function:

$$I(x) = I_0 \cdot \exp(-\delta(x)) = I_0 \cdot \exp(-N_0 \cdot \sigma \cdot x_0 \cdot \exp(-x/x_0)). \qquad (2.56)$$

If one uses selected substances such as ozone or water vapour, one can see from Fig. 2.31 that an even small ozone concentration at low atmospheric heights yields a considerable reduction of the total intensity, while for small concentrations at larger heights x almost all intensity passes through the atmosphere. Not only ozone but also water vapour can lead to large losses in intensity even in small quantities depending on the atmospheric altitude. For constant steam content the intensity attenuation effect decreases with increasing atmospheric height.

The absorption increases significantly with increasing density of the atmosphere, as illustrated in Fig. 2.32. Note that there is a dramatic attenuation for wavelengths smaller than 300 nm, which is mainly due to ozone absorption.

The light absorption on surfaces or when passing through matter is dependent on material and frequency. Depending on the possible band structure of the

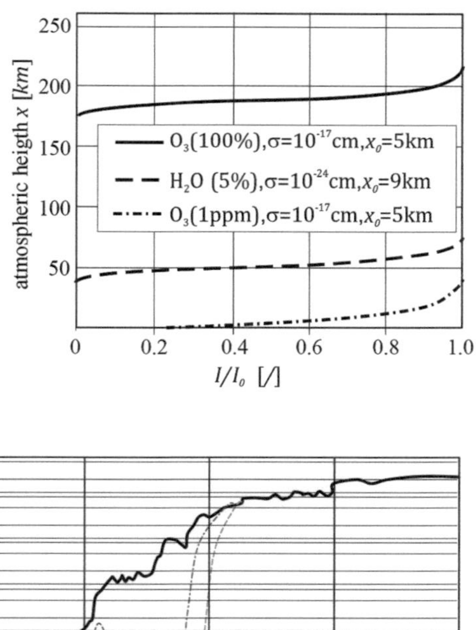

Fig. 2.31 Relative intensity change I/I_0 as function of height x for different gas concentrations and species from Hippler (2005)

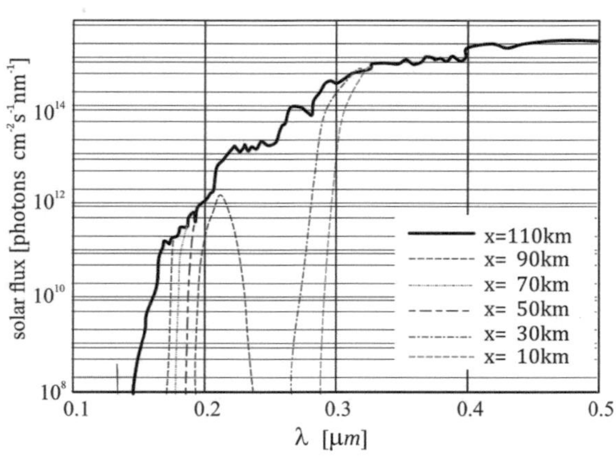

Fig. 2.32 Change in solar flux density as a function of wavelength λ at different heights from Hippler (2005)

molecules, different frequency ranges of light can be absorbed differently. This means that the absorption is different depending on the colour. Adjacent frequency components are either reflected or transmitted, depending on the material and the angle of incidence. If, for example, a surface that appears yellow is irradiated with white light, the green and red light is reflected/transmitted and blue light is absorbed. When light is absorbed, the absorbed energy is not only converted into heat, but can also be lost through other mechanisms such as fluorescence or through scattering at aerosols.

2.3.1.2 Scattering in the Atmosphere

Scatterers in the atmosphere are air molecules, aerosol particles, cloud, and rain droplets. These scatterers affect all types of electromagnetic radiation: solar radiation, thermal radiation, microwaves, and radar. Contrary to absorption scattering leads to a directional redistribution of the radiation, i.e. the radiation is deflected and usually (for elastic scattering) no energy conversion takes place as in absorption. The beam attenuation of light due to scattering can also be formulated using the Lambert–Beer's law (Eq. (2.54)) with a scattering coefficient, but additionally scattering theory yields a scattering function (scattering indicator matrix, phase function). The latter describes the distribution of the scattered radiation in the individual directions of space. As a result, the energy of the primary radiation beam decreases in the direction of the beam similarly to the absorption, but there is a gain in radiance in other directions.

All extinction due to scattering processes is described by the Lambert–Beer law. The change in wavelength-dependent intensity density $i(\lambda)$ as the radiation passes through an infinitesimal segment of thickness dx results in:

$$di(\lambda) = -\beta_\lambda \cdot i_\lambda \cdot dx, \tag{2.57}$$

where β_λ is the volume scattering coefficient $[1/m]$, which is calculated as

$$\beta_\lambda = \sigma_\lambda \cdot N. \tag{2.58}$$

In Eq. (2.58) σ_λ is the wavelength dependent scattering cross section and N the number density of particles. The total volume scattering coefficient β_λ results from the integral of all particles involved in the scattering process including multiple scattering and is evaluated by integration across the control volume:

$$\beta_\lambda = \int_{r=r_1}^{r=r_2} \sigma_\lambda(r) \cdot \frac{dN}{dr} dr. \tag{2.59}$$

Now the scattering cross section of an individual particle can be very well described with theories depending mainly on the size of the particle. For the interaction with light not only the geometric size is important. This can be understood when looking in the physics of the different scattering processes. Rayleigh scattering is an interaction of electromagnetic radiation with a dipole. When the particle reaches the size of the wavelength, the Mie scattering theory deals with the particle as a continuous medium with geometric shape and permissivity. When the particles become even bigger, the wave character of the light is less important and there is no coherent interaction of electro-magnetic (EM)-waves anymore. Then the realm of geometric optics describing the medium with a refractive index starts. Figure 2.33 shows the scattering cross section σ_λ as a function of the wave number $\kappa = (2 \cdot \pi \cdot r / \lambda)$. The range of Rayleigh scattering therefore occurs at wavenumbers $\kappa < 1$ while the laws of geometric optics apply only to wavenumbers $\kappa \gg 1$. The scattering cross-sections of Mie scattering vary considerably due to interference of waves.

In Table 2.5 the wave numbers κ, particle sizes (r) and wavelengths for some atmospheric particles are given. The particle size in particular determines the type of scattering process.

Figure 2.34 shows in double logarithmic order the areas of geometric optics, Mie- and Rayleigh scattering as a function of the mean particle radius (r) and the wavelength (λ).

Rayleigh Scattering

Rayleigh scattering describes the scattering of light in a sphere containing only scattering particles with a diameter d, whose size is much smaller than the wavelength λ. The typical particle size condition for elastic Rayleigh scattering is;

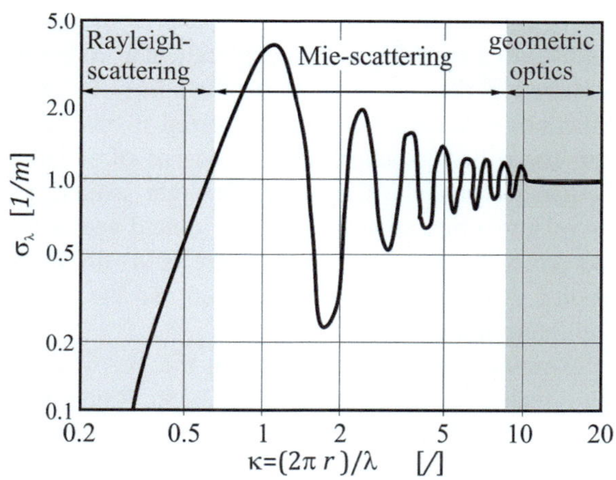

Fig. 2.33 Scattering cross section σ_λ as function of the wave number $\kappa = (2\pi r/\lambda)$ and limits of the different scattering processes

Table 2.5 Wave number κ and order of magnitude of typical air component sizes

	Typical size r	Wavelength λ (nm)	Wave number $\kappa = (2\pi r/\lambda)$
Molecule	0.1–1 nm	550	0.001–0.01
Aerosol	100 nm	550	1.2
Raindrop	1 mm	550	0.06–6

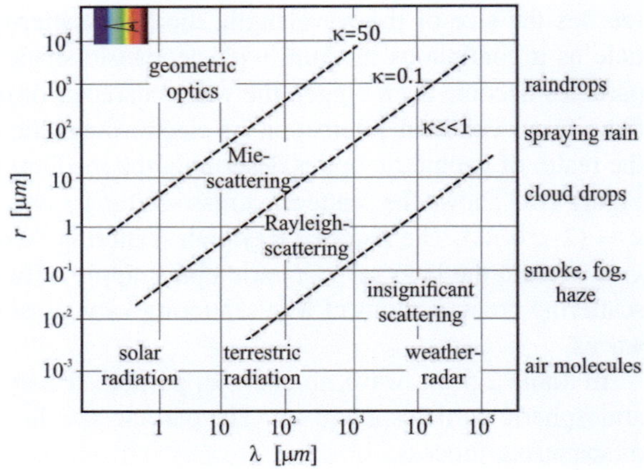

Fig. 2.34 Areas of geometric optics, Mie and Rayleigh scattering as a function of particle radius r and wavelength λ, and some typical particle representatives

$d \ll 100 \cdot \lambda$. In order to understand the physics, we first have to introduce the concept of polarization of light, i.e. of electromagnetic waves.

Light can be considered as a transverse electromagnetic wave, as shown in Fig. 2.35a, in which the wave vector $\overrightarrow{k}$ points in the direction of propagation. Perpendicular to the propagation direction are the field vectors of the electric and of the magnetic field $\overrightarrow{E}$ and $\overrightarrow{B}$, which in turn are perpendicular to each other ($\overrightarrow{E} \perp \overrightarrow{B}$). This allows a rotational degree of freedom in three-dimensional space, that of rotation around the wave vector (with time). Both vectors can be decomposed into two independent perpendicular components. Light is called linearly polarized if the field vectors oscillate only in one plane along the propagation direction. It is called circular polarized if this field amplitude rotates with time, and it is called unpolarized if it consists of a statistical mixture of both perpendicular components. Generally, the light scattering of particles is dependent on the relative orientation of the electric field vector to a particle orientation. Atoms and molecules in an electromagnetic (EM) field can be considered as oscillating dipoles as the negative electron shell and the positive nucleus interacts with the oscillating field.

Now we consider a molecule in the sky which, for the observer B, floats in the direction φ shown in the Fig. 2.35b. Rayleigh scattering describes the interaction of the EM-wave with the molecule. Two dipole moments oscillate excited by unpolarized light perpendicular to the propagation vector $\overrightarrow{k}$ with equal frequency in the plane defined by the points SAB (dipole 1, $I_{||}$) and perpendicular thereto (dipole 2, $I_{\perp}$). Any dipole radiates maximum perpendicular to its direction of oscillation, but it emits no intensity into the direction of oscillation. The angular distribution of the emitted field amplitude E_{emit} is proportional to $\sin\theta$, where θ is the angle between dipole axis $\overrightarrow{p}$ and propagation direction $\overrightarrow{k}_{emit}$. Therefore dipole 2 radiates towards B with maximum strength, while dipole 1 radiates only with the intensity $\sin^2\theta$ or $\cos^2\varphi$. The total brightness of the sky therefore changes accordingly with $\left(1 + \cos^2\varphi\right)$ as shown in Fig. 2.36b. It is darkest at 90° from the

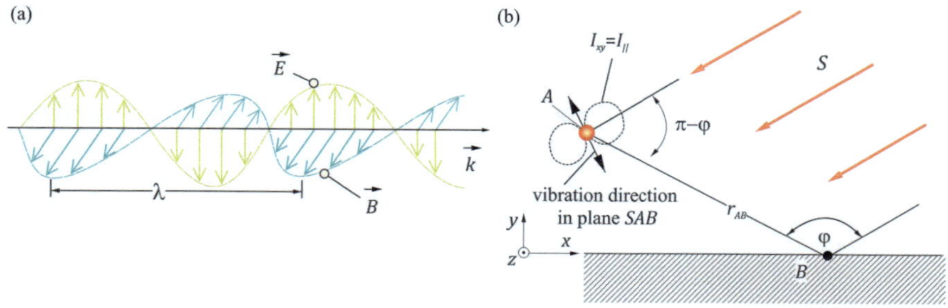

Fig. 2.35 **a** Schematic diagram of the propagation of an electromagnetic wave, wave vector $\overrightarrow{k}$, electric field $\overrightarrow{E}$ and magnetic field $\overrightarrow{B}$; **b** in an air molecule A, the portion of sunlight polarized in the plane SAB excites a dipole oscillation. The amplitude radiated to the observer B is proportional to $\cos\varphi/r_{AB}$, the intensity $I_{||} \sim \cos^2\varphi/r_{AB}^2$. For the polarization direction perpendicular to $SAB(x - y\text{-plane})$ $I_{\perp} \sim 1/r_{AB}^2$ is valid, because here the dipole oscillates perpendicular to AB into the surface

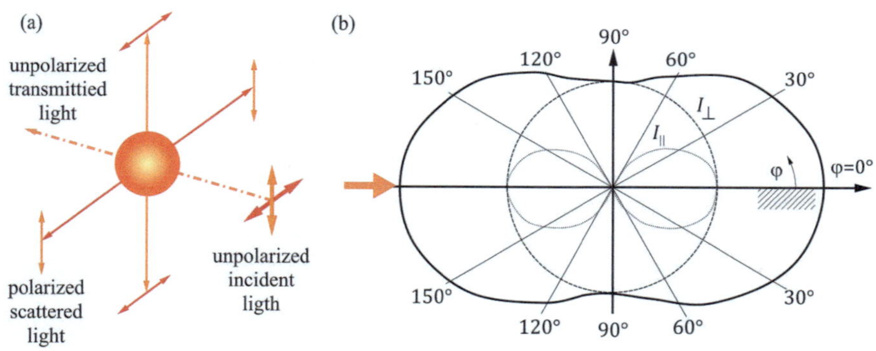

Fig. 2.36 **a** Scattering processes polarize the light. Here, dipoles emit polarized light along their axis. **b** Intensity distribution of the component $I_{\perp}$ perpendicular to the plane of vision and the plane $I_{||}$ parallel to the plane of vision of the incident unpolarized light as a function of the angle of incidence φ

sun. The eye hardly notices this slight difference, especially because the glare of the direct sun and the pale blue towards the horizon originating from mist obscure it. However, it easily can be detected using a polarization filter such as sunglasses.

As a summary the scattering molecule causes a polarization of the light, as shown in 2.36b. It emits polarized light along its axis. Thus, for Rayleigh scattering the intensity of the component perpendicular to the viewing plane $I_{\perp}$ remains constant with angle, whereas the intensity parallel to the viewing plane radiates only with $\cos^2\varphi$.

As a summary the scattering molecule causes a polarization of the light, as shown in Fig. 2.36a. It emits polarized light along its axis. Thus, for Rayleigh scattering the intensity of the component perpendicular to the viewing plane $I_{\perp}$

remains constant with angle, whereas the intensity parallel to the viewing plane radiates only with $\cos^2\varphi$.

Now let us now introduce the complete scattering function. The intensity of unpolarized light I of wavelength λ scattered by a molecule with polarizability α in Rayleigh scattering is calculated as:

$$I = I_0 \cdot \frac{1 + \cos^2\theta}{2R^2} \cdot \left(\frac{2\pi}{\lambda}\right)^4 \cdot |\alpha|^2, \tag{2.60}$$

wherein I_0 represents the intensity of the incident wave, θ the scattering angle and R the distance to the particle. The intensity loss due to Rayleigh scattering thus scales inversely proportional to the fourth power of the light wavelength. Let us assume that the particle is a small dielectric sphere with radius a and the refractive index n. Here the polarizability can be calculated as (Van de Hulst, 1981)

$$\alpha = \frac{n^2 - 1}{n^2 + 1}. \tag{2.61}$$

The scattering cross-section of such a particle is hence

$$\sigma_{\lambda R} = \frac{8\pi}{3} k^4 \cdot |\alpha|^2. \tag{2.62}$$

The Rayleigh scattering explains why the clear sky appears blue. The wavelength of blue light (450 nm) is smaller than the wavelength of red light with about 650 nm. Thus, the ratio of the effective cross-sections $\sigma_{blue}/\sigma_{red}$ is approximately 4–5.

Blue light—depending on the exact wavelength—is thus scattered about 4–5 times more strongly in the atmosphere than red light. According to Planck's radiation law, the maximum of the power radiated by the sun is green (approx. 550 nm). Rayleigh scattering shifts the power distribution of the scattered light, according to the Eq. (2.60) with λ^{-4}. As a result, the maximum moves into the UV range, as the curve of the intensity distribution of the sunlight scattered in the air shows in Fig. 2.37.

During the day, when the sun is high in the sky, the light only travels a short distance through the atmosphere. Only little blue light is scattered in other directions. Therefore, the sun appears yellow. From high-flying airplanes the sun appears "whiter" because less blue light is missing. The sum of all scattered light makes the sky appear blue from all other directions. On the moon, where a dense atmosphere is missing, the sky appears black even during the day. At low sun elevation the distance of sunlight through the earth's atmosphere is much longer. This causes a large proportion of the high frequency light components (blue) to be removed from the direct beam, leaving mainly light with long wavelengths which shifts the colour impression of the sun towards red. This effect is further may be enhanced by additional particles in the air (e.g. dust, sand).

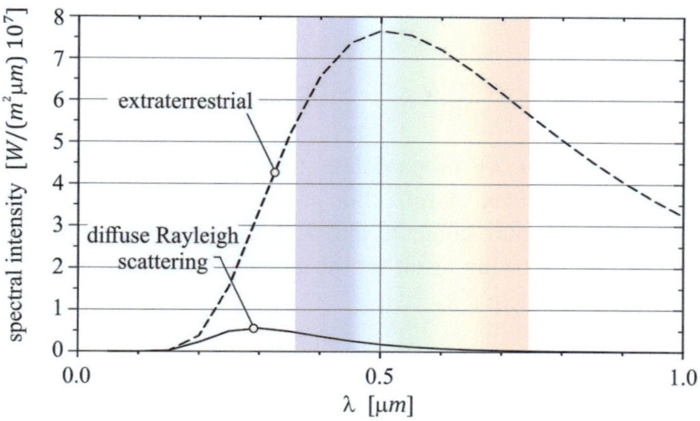

Fig. 2.37 Spectral power distribution distribution of solar radiation by Rayleigh scattering

In order to calculate how much light passes through the Earth's atmosphere due to Rayleigh scattering, the intensity formula (Eq. (2.60)) must first be integrated over the scattering angle θ and the distance from the scattering particle. The total density of scattering molecules in a volume is then considered. For an ideal gas the Avogadro number is contained for normal conditions of atmospheric pressure (1 atm) in a molar Volume of $22.4\,m^3/kmol$. From this the Loschmidt number N_0 gives the density of particles as $\approx 2.7 \times 10^{25}\,m^{-3}$. In scattering theory the extinction of a gas containing many scatters can be calculated from the cross section of the particles and the particle density. This relates also to the effective index of refraction (the real part) of the gas n_0 which is (Van de Hulst, 1981),

$$n_0 = 1 + 2\pi \cdot N_0 \cdot \alpha. \tag{2.63}$$

The optical extinction coefficient β_λ of the atmosphere has to be integrated over all particles in the vertical atmospheric profile. Since the particle density of the atmosphere decreases approximately exponentially with height (see barometric height formula, Eq. (2.55)), the true air envelope can be replaced by a homogeneous layer whose thickness H is equal to the scale height and whose particle density is equal to the particle density at the ground with an optical depth δ_λ:

$$\delta_{\lambda R} = \beta_{\lambda R} \cdot H = N_0 \cdot \sigma_{\lambda R} \cdot H = N_0 \cdot \frac{2k^4}{3\pi} \cdot \frac{(n_0 - 1)^2}{N_0^2} \cdot H, \tag{2.64}$$

which results in

$$\beta_{\lambda R} = \frac{32\pi^3}{3} \cdot \frac{(n_0 - 1)^2}{\lambda^4} \cdot \frac{1}{N_0}. \tag{2.65}$$

The equivalent layer of thickness H now yields the total transmission T_λ for the earth's atmosphere

$$T_{\lambda R} = \exp(-\beta_{\lambda R} \cdot H) = \exp(-\delta\tau_{\lambda R}). \tag{2.66}$$

N_0 here gives the density of particles at ground level $2.7 \times 10^{25}\,\text{m}^{-3}$ and $n_0 = 1.000292$ is the index of refraction of air. The extinction coefficient has the dimension m^{-1}. We may compare that with the historical results of Paetzold (1952) who gives the following result for the attenuation of light in the atmosphere for the astronomical extinction k_{ex}

$$k_{ex} = \frac{160}{3} \cdot \pi^3 \cdot \frac{(n_0 - 1)^2}{\ln 100} \cdot \frac{H}{N_0} \cdot \frac{1}{\lambda^4}, \tag{2.67}$$

where in astronomy the relation between atmospheric extinction and transmittance of an atmosphere usually is not given as an exponential law but defined as

$$T_\lambda = 10^{-0.4 \cdot k_{ex}}. \tag{2.68}$$

The scale height in this formula has been taken as $H = 7990$ m.

The optical thickness describes the attenuation of light at vertical incidence. With oblique incidence at an angle θ against the vertical, the following effective layer thickness is approximately obtained as $H(\theta)$:

$$H(\theta) = \frac{H(\theta = 0)}{\cos\theta}, \tag{2.69}$$

if plane-parallel layering is assumed. The optical thickness of the Rayleigh atmosphere must then be corrected accordingly. Stöcker (1994) specifies the value $n_0 = 1.000292$. An insertion yields:

$$\delta_{\lambda R} = \frac{0.084\,\mu\text{m}^4}{\cos\theta \cdot \lambda^4}, \tag{2.70}$$

where λ represents the wavelength to be used in micrometres. Figure 2.38 shows the transmission as a function of the angle of incidence compared to the vertical. The graph clearly shows that, for example, already at a wavelength of 0.4 μm the transmission through the atmosphere decreases significantly with a change of the incidence angle from $\theta = 0°$ to $\theta = 60°$. With flat solar radiation, i.e. large angles of incidence ($\theta \gg 0°$), the intensity drops rapidly.

Table 2.6 shows the scattering coefficients of a Rayleigh atmosphere as a function of wavelength. As can be seen they decrease considerably with increasing wavelength.

Due to the large number of small molecules in the atmosphere, multiple scattering occurs, and the resulting diffuse irradiance from the sky which behaves similarly wavelength-selectively as the extinction on the passage of light through the atmosphere, resulting in a wavelength and angle-dependent diffuse Rayleigh irradiance as shown in the Fig. 2.39.

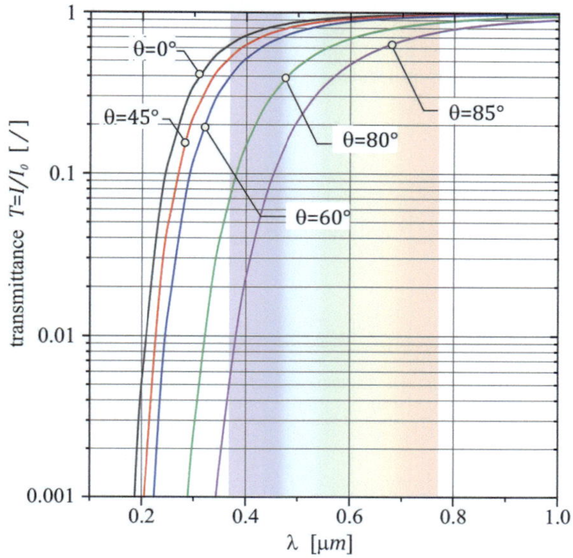

Fig. 2.38 Transmission T of Rayleigh scattering as a function of the angle of incidence θ versus the vertical incidence ($\theta = 0°$) according to Eq. (2.70)

Table 2.6 Dependence of the scattering coefficient $\beta_{\lambda R}$ on the wavelength λ according to Eq. (2.65)

Wavelength λ (nm)	Scattering coefficient $\beta_{\lambda R}$ (m^{-1})
300	147.9×10^{-6}
400	43.6×10^{-6}
800	2.6×10^{-6}
1000	1.1×10^{-6}
2000	0.07×10^{-6}

2.3.1.3 Properties of Mie-Scattering

Mie scattering occurs in particles whose dimensions d are approximately equal to the wavelength of the radiation. Strictly speaking, the historical Mie theory only applies to spherical particles, but solutions have been extended also to cylinders and ellipsoidal particles. While hygroscopic aerosols can be considered spherical, others are often non-spherical (e.g. desert aerosol). The calculation of the scattering function requires the solution of the electromagnetic wave equation for a homogeneous body, using Maxwell's equations. The solution is given by an infinite series of angular functions. On the Internet one can find freely available optical simulation tools (e.g. http://omlc.ogi.edu/software/mie/).

For Mie scattering, with increasing particle size as larger part of the radiation is scattered forward. In Mie theory, the scattering coefficient is calculated as a function of wavelength λ divided by particle diameter d. The wavelength dependence of the scattering coefficient $\beta_{\lambda M}$ is less than in the case of Rayleigh scattering; the scattering coefficient can be scaled approximately as $\beta_{\lambda M} \sim 1/\lambda$.

In principle, Mie-scattering in the atmosphere behaves similar to Rayleigh-scattering. It also shows a spectral (i.e. wavelength-dependent) behaviour and is

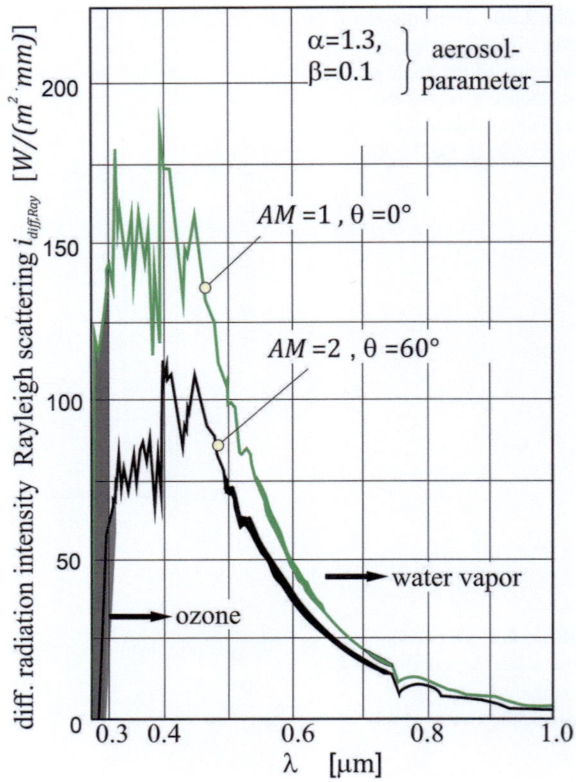

Fig. 2.39 Diffuse Rayleigh scattering caused by multiple scattering on air molecules

often referred to in the technical literature as aerosol scattering. However there are two important differences: The first is that for Rayleigh scattering backward and forward scattering effect are of the same size, whereas for Mie scattering the forward scattering dominates. Therefore, Mie scattering contributes more to the diffuse irradiance from the sky, as a comparison of the Fig. 2.39 with the Fig. 2.40 shows. The second difference is that the scattering cross section strongly depends on the parameter d/λ. As aerosols have a wide distribution of particle sizes, the resulting diffuse irradiance from Mie scattering is a mixture of different colours, as for each particle diameter a different wavelength scatters best (see Fig. 2.33). This mixture results in a white/grey diffuse irradiance $i_{diff,Mie}$ without any colour preference.

2.3.1.4 Multiple Reflections and Polarization of the Diffuse Sky Radiation

Both scattering mechanisms never occur singularly, but inevitably result in interactions in the form of multiple scattering processes of many particles and molecules, where the individual scattering mechanisms can be described by Rayleigh and Mie scattering theory. A further complication in the calculation of diffuse irradiance is the presence of the ground reflection (albedo). There are multiple reflections

Fig. 2.40 Diffuse radiation intensity due to Mie-scattering $i_{diff,Mie}$ as a function of wavelength λ at different sun elevations (air mass factors $= AM$)

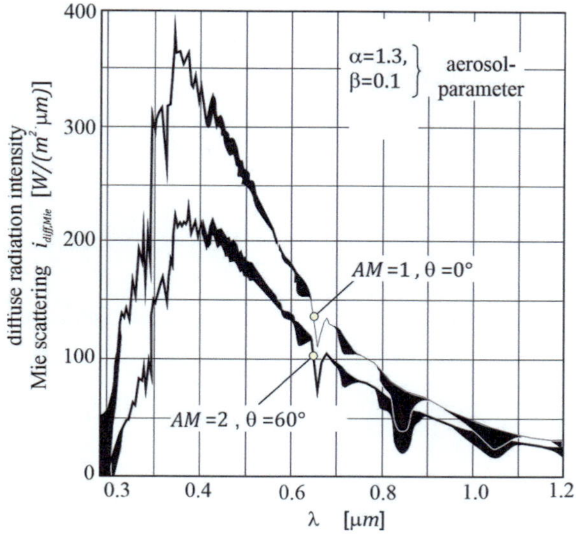

between ground and atmosphere, especially when clouds are present. This multiple reflection intensity is a third component of the spectral diffuse radiation intensity from the sky. Overall, the diffuse radiation intensity from the sky is created by many particle interactions consisting of Rayleigh, Aerosol (Mie) scattering plus diffuse light from multiple reflections between atmosphere and ground. The Fig. 2.41 shows the order of magnitude of these individual components as a function of wavelength λ. The addition of all scattered components can take up a considerable amount of the total radiation intensity and plays an important role especially when considering non-concentrating solar thermal collectors.

Fig. 2.41 Diffuse radiation intensity i_{diff} of Aerosol, Rayleigh scattering as a function of wavelength λ and by multiple reflection processes at an $AM = 2$

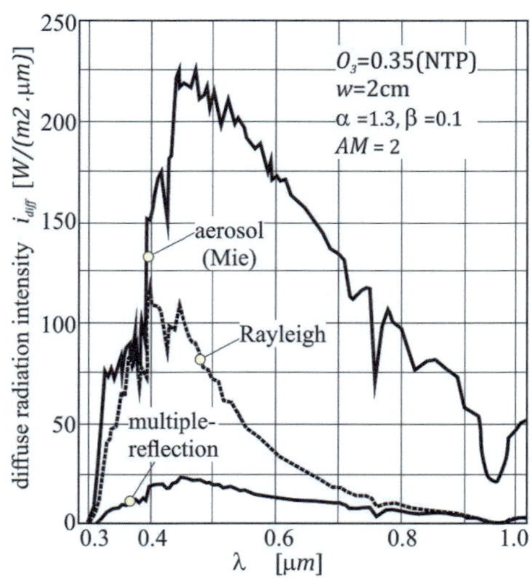

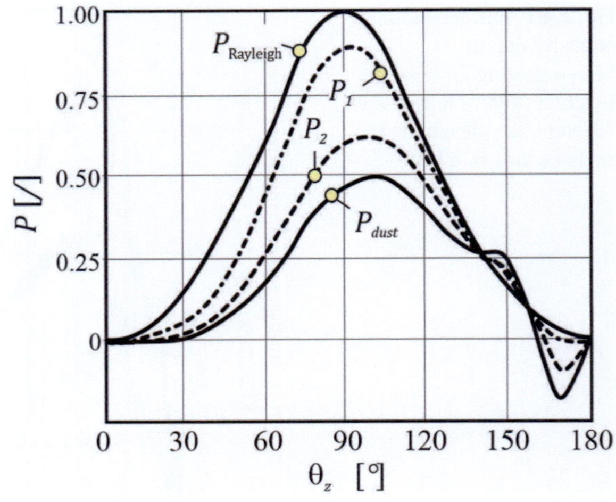

Fig. 2.42 Degree of polarization as a function of the zenith angle θ_z in a Rayleigh atmosphere and a dust atmosphere. Here is the mist component in $P_2 \gg P_1$ (Factor 60)

The polarization degree of diffuse sky light is an indicator of the accumulation of aerosols in the atmosphere. For this purpose, the degree of polarization P of the light is determined, especially in twilight. Whereas pure Rayleigh scattering exhibits full polarization P, aerosol scattering is unpolarized. The deviation from the values of the pure Rayleigh atmosphere indicates the degree of haze in the atmosphere, as shown in the Fig. 2.42. At a zenith angle of $\theta_z = 90°$ the greatest polarization occurs in the Rayleigh atmosphere. Atmospheres with a higher mist content are more turbid and exhibit a lower polarization. A high dust content initially causes a significantly reduced polarization of the light. Moreover, it leads to a shift in polarization towards larger zenith angles.

2.3.2 Technical Calculation of the Attenuation in the Atmosphere

Now we have discussed the different attenuation mechanisms of solar radiation in the atmosphere. All individual absorption and scattering processes are different and independent of each other. They therefore contribute to the overall attenuation with different spectral transmission coefficients. The direct spectral radiant power results from the product of the transmission coefficients of all components of the atmosphere $T_{i,\lambda}$ in the following way following a split-up of Iqbal (1983).

$$i_{dir,\lambda} = i_{SC,\lambda} \cdot \left(T_{Rayleigh} T_{ozone} T_{gases} T_{water\ vapour} T_{aerosol} T_{clouds}\right), \qquad (2.71)$$

which is equivalent to the addition of all individual optical thicknesses:

$$\delta_{tot,\lambda} = \delta_{Rayleigh} + \delta_{ozone} + \delta_{gases} + \delta_{water\ vapour} + \delta_{aerosol} + \delta_{clouds}. \qquad (2.72)$$

The individual transmission coefficients are again defined with an exponential function, which is described by Eq. (2.73)

$$T_{i\lambda} = \exp(-\delta_{i\lambda} \cdot AM). \tag{2.73}$$

Nowadays the spectral aerosol optical depths are becoming available and are being measured, which allows the determination of clear sky spectral irradiance.

Atmospheric turbidity characterizes the extinction effect of aerosols on solar radiation. It is well known that turbidity is the most important extinction affecting the direct and diffuse components of solar radiation under clear skies and that, as turbidity increases, the former decreases and the latter increases in about the same proportions so that the total effect of turbidity on global radiation is considerably smaller. However in many solar applications the effects have to be distinguished, for example in concentrating collectors which only accept beam radiation. Turbidity is used to characterize atmosphere in comparison to an ideal clean atmosphere without clouds.

The spectral turbidity can be defined using Ångström's law in practice, because of the complexity of Mie scattering calculation. It expresses the spectral aerosol optical depth ($SAOD$), $\delta_{a\lambda}$, at wavelength λ (in μm) through

$$\delta_{a\lambda} = \beta \cdot \lambda^{-\alpha}, \tag{2.74}$$

where the wavelength exponent α is generally observed to have values between 0 and 2.6. In this expression Ångström's turbidity coefficient, β, varies from 0 for an ideally aerosol-free atmosphere to more than 1 under extreme hazy conditions caused by sandstorms or heavy smoke plumes. The spectral aerosol transmittance is obtained from

$$T_{a\lambda} = \exp(-\delta_{a\lambda} \cdot AM). \tag{2.75}$$

where AM is the optical aerosol mass.

Instead, historically the concept of turbidity has been also used for broadband calculations. The Linke turbidity factor (T_L, defined for an air mass $AM = 2$) is a very convenient approximation to model the atmospheric absorption and scattering of the solar radiation under clear skies. It describes the optical thickness of the atmosphere due to both the absorption by the water vapor and the absorption and scattering by the aerosol particles relative to a dry and clean atmosphere. It summarizes the turbidity of the atmosphere, and hence the attenuation of the direct beam solar radiation. The larger T_L, the larger the attenuation of the radiation by the clear sky atmosphere.

The T_L denotes the transparency of the cloudless atmosphere. If the sky were perfectly dry and clean, T_L would be equal to unity (1). When the sky is deep blue, the T_L is just above 1 and still very small. In summer, in Europe, the water vapour is often large, and the blue sky is close to white. The T_L is larger than 3. In turbid atmosphere, e.g. in polluted cities, the T_L is close to 6–7.

The Rayleigh transmission factor T_{RS} is multiplied by an exponential function containing T_L and direct solar irradiance may be estimated using the solar constant i_{sc}:

$$i_{dir,h} = i_{sc} \cdot \varepsilon \cdot \cos\theta_z \ \exp\left(-0.8862 \cdot AM \cdot \frac{p}{p_0} \cdot \overline{\delta_R} \cdot T_L(AM2)\right), \qquad (2.76)$$

where ε is the intensity correction of the solar constant due to the seasonal position of the sun in the elliptic orbit. The altitude correction with respect to sea level is done by the relative density of the atmospheric pressure p/p_0. The exponential function, which is the total broadband atmospheric transmittance T_{atm} can be split up into a transmittance of the $AM1$ Rayleigh atmosphere T_{RS} and a Linke transmittance factor T_{turb}: with an exponent of air mass factor AM:

$$T_{atm} = T_R \cdot T_{turb}^{AM} = \exp\left(-\overline{\delta_R}\right) \cdot \exp\left(-0.8862 \cdot T_L(AM2) \cdot \frac{p}{p_0} \cdot AM\right). \qquad (2.77)$$

An alternative formulation can be the multiplication of broadband transmittance factors for the different attenuation processes (for the sun in the zenith $AM = 1$) absorption T_{ab}, Rayleigh scattering T_{RS} and aerosol scattering T_{ae}:

$$T_{atm}(AM) = T_{AB}^{AM} \cdot T_{RS}^{AM} \cdot T_{AE}^{AM}. \qquad (2.78)$$

Tables 2.7 and 2.8 give transmission factors and turbidity values for different terrain situations. The tables show that there are considerable differences between urban areas and the high mountains, especially at low sun elevations. The seasons also lead to a variation in turbidity; for example, the higher temperatures in summer lead to a higher water vapour content in the air and consequently to greater turbidity. Both aspects must be considered when designing solar thermal systems.

Figure 2.43, however, shows that in a clear atmosphere the attenuation is mainly caused by water vapour in the near infrared wavelength range. The attenuation of the radiation intensity at an air mass factor of $AM = 1$ corresponds to the intensity

Table 2.7 Transmission factors and annual averages of turbidity factors as a function of solar altitude α_s

		Solar elevation angle α_s (°)				
	Turbidity	90	60	30	10	5
Ideal atmosphere T_{RS}^m	1.0	0.906	0.895	0.837	0.681	0.563
Gas absorption T_{ab}^m	1.9	0.91	0.91	0.89	0.88	0.81
Aerosols in high mountains T_{ae}^m	2.75	1.0	0.99	0.96	0.85	0.75
Lowlands T_{ae}^m	3.75	0.93	0.91	0.82	0.61	0.67
Cities T_{ae}^m	5	0.84	0.81	0.69	0.42	0.27
Industrial areas T_{ae}^m	5					

Table 2.8 Monthly averages of turbidity factors in different settlement types as a function of the month

	Jan	Feb	Mar	Apr	May	Jun	Jul	Aug	Sep	Oct	Nov	Dec
City	3.1	3.2	3.5	3.9	4.1	4.2	4.3	4.2	3.9	3.6	3.3	3.1
Countryside	2.1	2.2	2.5	2.9	3.2	3.4	3.5	3.3	2.9	2.6	2.3	2.3
Mountains	1.8	1.9	2.1	2.2	2.4	2.7	2.7	2.7	2.5	2.1	1.9	1.8

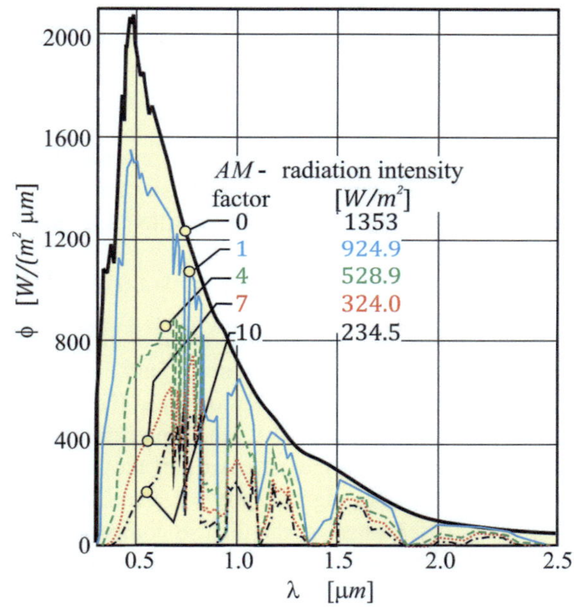

Fig. 2.43 Spectrum of direct solar radiation for different air mass bodies AM of $AM = 0-10$ as a function of wavelength λ in μm, assuming a clear atmosphere with water vapour corresponding to a precipitation of 20 and 3.4 mm ozone

attenuation of light when a layer of water of approximately 20 mm passes through. The effect of increasing air mass factors as a function of wavelength λ on the spectral radiant power is illustrated in Fig. 2.43. At the transition from an $AM = 0$ to $AM = 10$, the average radiant intensity already decreases by a factor of five (!).

2.3.3 Global, Direct, and Diffuse Irradiation

The irradiance on Earth is typically assessed on the horizontal surface. The instruments used for determining the irradiance value are able to distinguish between the direct beam radiation from the sun and the scattered diffuse irradiation from the sky. The latter one also has an angular distribution over the sky dome which in some circumstances can be quite complex. Whereas the clear sky diffuse intensity is dominated by the Rayleigh scattering and the intensity distribution due to polarization as explained already, the presence of clouds may complicate the situation.

In some cases a geometric arrangement of white clouds may even serve as a kind of concentrating element as beam radiation from the sun may be directly reflected towards the collector aperture for certain places. Other parts of the sky due to dark clouds might be contribute little to the overall diffuse irradiance. When the atmosphere is hazy with a lot of humidity, the whitish haze is especially bright around the horizon. And around the position of the sun we can detect an enhanced brightness which is called circumsolar radiation. For horizontal surfaces we look at the sum of diffuse sky and direct beam radiation. This quantity is called global irradiance. This is why in the context of solar collectors often G is used as the symbol for terrestrial irradiance instead of the more general i.

$$G_h = G_{b,h} + G_{d,h}. \tag{2.79}$$

When we have tilted surface, also reflected radiation from the ground can be accepted. But of course a part of the sky irradiance is not seen any longer by the aperture. The total irradiance on the tilted aperture is denoted by G_t:

$$G_t = G_{b,t} + G_{d,sky,t} + G_{d,refl,t}. \tag{2.80}$$

Even with clear skies, the proportion of diffuse irradiation depends on the length of the path through the atmosphere. Figure 2.44 shows this dependency in the form of the proportions of total radiation $i_{glob}(G_h)$, direct radiation $i_{dir}(G_{b,\,h})$ and diffuse radiation $i_{diff}(G_{d,\,h})$ as a function of the solar elevation angle α_s. The diffuse component hardly changes at solar elevations $\alpha_s > 40°$, while the direct radiation component increases significantly. At small solar elevation angles ($\alpha_s < 20°$) the intensity is determined by the diffuse component. Even with clear skies and vertical irradiation ($\alpha_s = 90°$), the diffuse part of the radiation is about 25% of the total irradiation intensity. Lower values are more likely to be the exception in Central Europe. They occur, for example, in mountainous areas with little aerosol concentration, environmental pollution and reduced density of air, in hours of cleaned air after heavy rains or at low temperatures where humidity is low.

Another aspect is that the total radiation i_{glob} is wavelength dependent due to scattering and absorption processes in the atmosphere. The blue sky radiation of clear skies without clouds is produced by wavelength dependent Rayleigh scattering. Therefore in this case, the largest part of the diffuse intensity i_{diff} is limited to the visible wavelength range, while at long wavelengths it has a negligible influence on the total radiation intensity i_{glob}, as the Fig. 2.45 depicts. The direct radiation intensity i_{dir}, on the other hand, is represented in the visible as well as in the near infrared wavelength range. On the other hand, heavy clouds with water droplets produce a different spectral pattern as here the larger particles result in Mie scattering. The resulting spectrum is more colour neutral (grey clouds).

The sum of both components diffuse radiation (i_{diff}) and direct radiation (i_{dir}) is called global radiation (i_{glob}). Since the distribution between diffuse and direct radiation intensity depends on the altitude, it also depends on the annual cycle. In Fig. 2.46 the typical annual cycle of global radiation in the form of the average daily values for a location in the Federal Republic of Germany. In Central

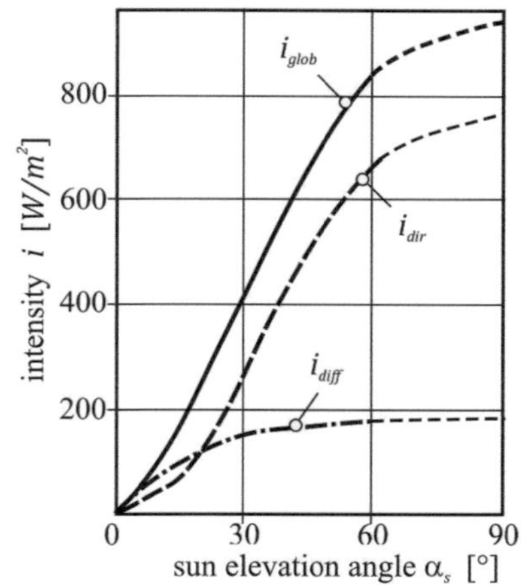

Fig. 2.44 Solar irradiance on a cloudless day (16.7.1969) at Sutton Bonington $\Phi = 53°N, 1°W$); i_{glob} = total irradiance; i_{dir} = direct irradiance on a horizontal surface; i_{diff} = diffuse irradiance; measured values are extracted curves, dashed lines are extrapolated from Monteith and Unsworth (2013)

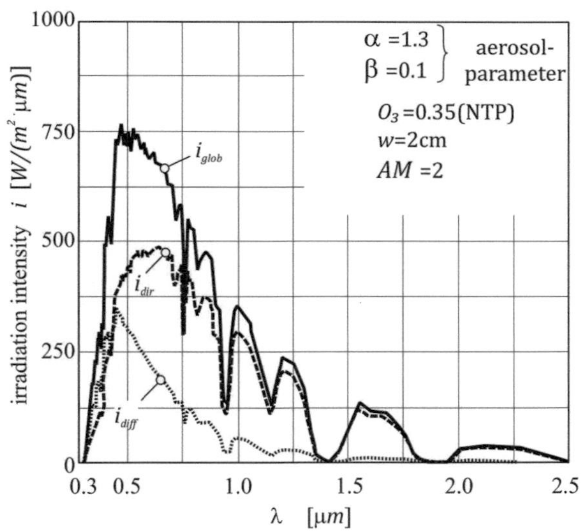

Fig. 2.45 Global, direct and diffuse radiation as a function of wavelength at the specified atmospheric parameters for a clear sky ($\Phi = 53°N, 1°W$) (Iqbal, 1983)

Europe, the diffuse part predominates. The annual average ratio of diffuse to direct radiation is around 60:40. Only in the summer months do the two components roughly balance each other out in terms of magnitude. The possible uses of both components are discussed in more detail in the following chapter.

Figure 2.47 shows the influence of different cloud types on the scattering in the atmosphere. Naturally, cumulus clouds with their high moisture content lead to a high optical thickness and reduce the proportion of radiation reaching the earth. Again there is an increased scattering with decreasing sun elevation angles and, as a consequence, the radiation intensity arriving at the ground is significantly reduced.

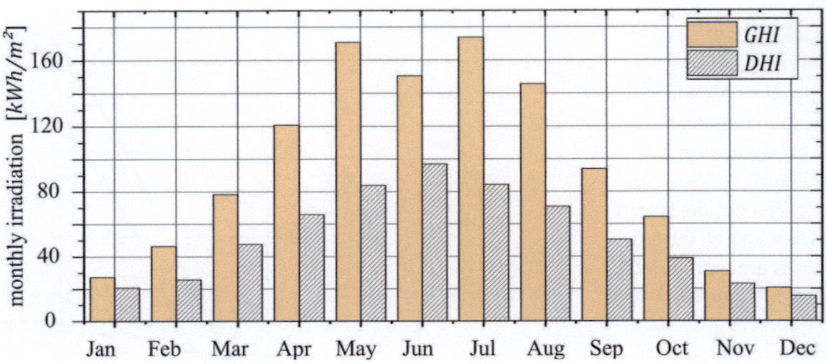

Fig. 2.46 Annual distribution of monthly global and diffuse irradiation sums on a horizontal plane for Munich/Germany

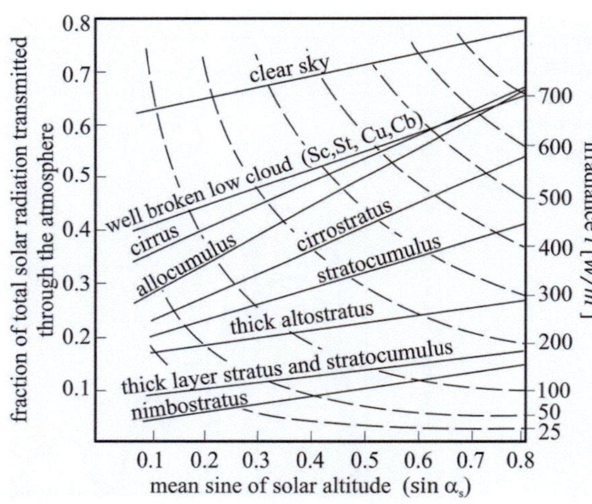

Fig. 2.47 Influence of different cloud types. The straight lines represent empirical relations between solar radiation and solar altitude for different cloud types according to measurements in the North Atlantic ($\Phi = 52° N, 20° W$). The curves represent isoplets of the illumination densities (according to Lumb). Sc = Stratocumulus; St = Stratus; Cu = Cumulus; Cb = Cumulonimbus modified from Monteith and Unsworth (2013)

2.4 Calculation of Irradiation on Arbitrarily Inclined Surfaces

Irradiation on a collector aperture consists of three main parts: direct beam radiation, diffuse radiation from the sky and the reflected radiation from the ground reaching the collector. Although the latter part also may contain directional components (e.g. reflection from a neighbouring glazed building or from a smooth water surface) very often it is assumed as diffuse. The diffuse component from the sky is produced by the scattering processes discussed previously. These lead to a generally variable distribution of radiance or sky brightness over the whole sky dome. One dominant and important part here is an enhanced brightness of the circumsolar part around the sun, the second one a typical horizon brightening at the horizon. Both effects can be observed already qualitatively with the bare eye.

Let us look onto the qualitative distribution of normalized intensity as a function of the zenith angle θ_z for clear and cloudy weather shown in the Fig. 2.48. It can be observed that within a few degrees to the sun position a strong decrease of diffuse intensity occurs. This is the so-called circumsolar radiation, which is solar beam radiation scattered in single events in the forward direction. A second part for a clear sky is the rather isotropic background intensity received rather uniformly from most of the sky hemisphere. The decrease of diffuse radiation in an angular range about 90° from the sun's position on a clear day, which is shown in the logarithmic scale of the Fig. 2.48 is explained by the dipole behaviour of the molecules at which Rayleigh scattering of matter occurs. In clear weather the Rayleigh scattering on air molecules predominates. The third part is horizon brightening from large zenith angles which originates from the fact that the number of scatterers in horizontal direction contributing to a diffuse intensity is high when we have good visibility. These three parts will be treated in an idealized way in the different diffuse sky models.

When we measure diffuse and direct components of irradiance on the horizontal plane, all these effects in summary lead to the measured value. However, when the measured irradiance shall be used to calculate the irradiance on a tilted aperture plane of a solar thermal collector, the angular view factors of the different parts

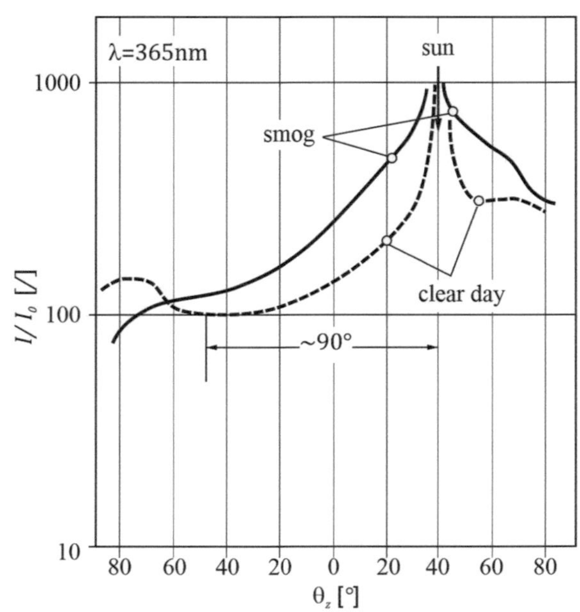

Fig. 2.48 Relative intensity I/I_0 of solar radiation (at $\lambda = 0.365\,\mu m$) as a function of the zenith angle θ_z in a plane where the sun is located, for Los Angeles on a clear day and on a day with smog (adapted from Coulson (1975))

change, and therefore a model is needed for the consideration of the different components.

2.4.1 Circumsolar Radiation

Depending on the atmospheric conditions, different directional distributions of the radiation intensity I result. When the sky is clear, about 75% of the diffuse sky radiation comes with approximately constant intensity from all directions of the firmament and about 25% comes from the direction of the sun. For simulation calculations of solar systems, this proportion is often referred to as the circumsolar part of direct radiation. This is permissible because direct radiation meters have an aperture angle of about 5° anyway.

The following Fig. 2.49 shows measured intensities of direct beam and circumsolar intensity. The measurement is taken in the angular range $[0, 30]$ mrad around the sun. The intensity drops exponentially fast outside the range of the sun disk rim at 4.65 mrad, where the intensity is nearly constant. If one integrates the circumsolar intensity Φ_{cs} from the $\delta_s = 4.65$ to $\delta_\Delta = 30$ mrad, the outer limit of the circumsolar region, θ indicates the radial angular displacement in the solar disk and $\phi(\theta)$ expresses the solar intensity profile assuming radial symmetry and divides it by the total intensity Φ_{tot} in this angular region one gets the circum-solar ratio CSR:

$$\Phi_{tot} = 2\pi \int_0^{\delta_\Delta} \phi(\theta) \sin\theta d\theta \approx 2\pi \int_0^{\delta_\Delta} \phi(\theta)\theta d\theta; \tag{2.81}$$

$$\Phi_{cs} = 2\pi \int_{\delta_s}^{\delta_\Delta} \phi(\theta) \sin\theta d\theta \approx 2\pi \int_{\delta_s}^{\delta_\Delta} \phi(\theta)\theta d\theta; \tag{2.82}$$

$$CSR = \frac{\Phi_{cs}}{\Phi_{tot}}. \tag{2.83}$$

The circumsolar profiles now can be classified by their CSR value. $CSR =$ 4–5% is a typical value for a clear sky. $CSR = 17\%$ is often taken as an average for slightly overcast sky.

Depending on the optical concentrator system, a large part of the CSR radiation moving with the sun's position can also be used in tracking collectors. As the acceptance angle (θ_a) of typical pyrheliometers anyway is about 4°–5° the CSR is recorded in meteorological measurements as beam radiation from the sun.

High circumsolar ratios (CSR) can significantly reduce the capture efficiency of high concentration collectors, due to a larger fraction of flux spillage. A changing CSR yields to a modification of the sun-shape distribution (intensity versus angle), see Fig. 2.50b. The actual sun shape is most strongly influenced by prevailing atmospheric conditions, particularly the level of particulate matter or moisture in

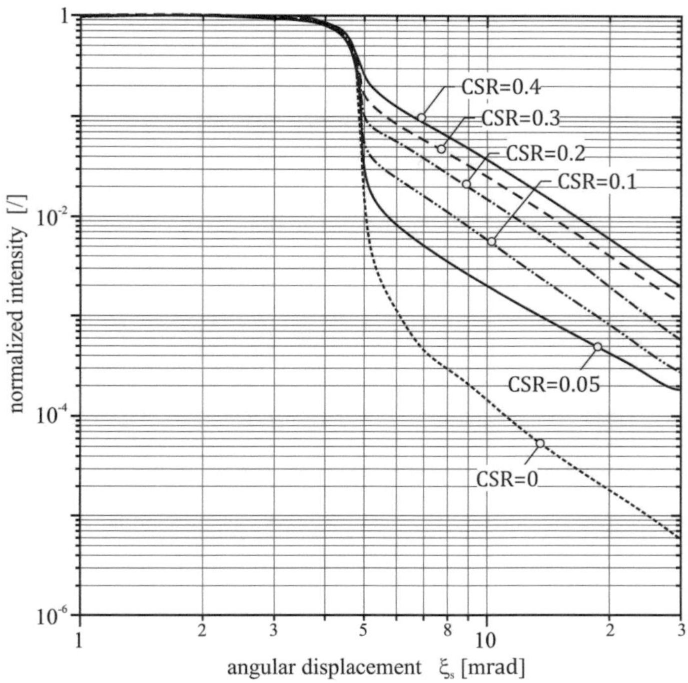

Fig. 2.49 Measurement data of different circumsolar conditions from Buie et al. (2003)

the air. A number of different sun-shape distributions can be found in literature however, for computational purposes the most commonly used is the one proposed by Buie et al. (2003). In Buie's model, the brightness at any point is normalized against the central intensity, which yields the following expressions:

$$\phi(\theta_s) = \begin{cases} \frac{\cos(0.326\,\theta)}{\cos(0.308\,\theta)} & \text{for } 0^\circ \leq \theta \leq 4.65\,\text{mrad} \\ \theta^\gamma \ \exp\kappa & \text{for } \ \theta > 4.65\,\text{mrad} \end{cases}, \tag{2.84}$$

with

$$\kappa = 0.9\ln(13.5\,CSR)\,CSR^{-0.3};\ \gamma = 2.2\ln(0.52\ CSR)\ CSR^{0.43} - 0.1. \tag{2.85}$$

The exact solar shape distribution for yield calculation only plays a relevant role for concentrators with very high optical concentration ratios ($C > 10^3$), if the optical surface is also very good there. For collectors with lower concentrations, a flat intensity distribution, often called pill-box sunshape, is sufficient and provides only a few percent deviation.

The directionality of diffuse sky radiation is complex and can be in principle detected by sky cameras using lenses taking pictures of the hole hemisphere. When calibrated correctly, the radiance L of the different angular regions may be derived.

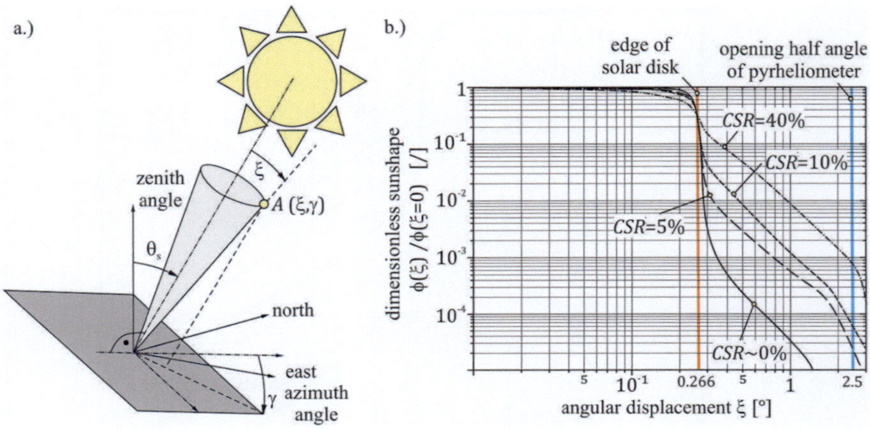

Fig. 2.50 **a** Coordinate system describing sky radiance starting from point $A(\xi, \gamma)$; **b** dimensionless solar radiance distribution as a function of angular distance ξ from the solar center ($\xi = 0$) at different atmospheric densities expressed as CSR. With sun disc limit (red line) and acceptance angle of a typical pyrheliometer indicated (blue line)

In practice however, simplified sky models serve to calculate the acceptance of solar irradiation on collector apertures. These are certainly simplifications, but for a large aperture say of a flat-plate collector it seems less important to model the exact angular distribution of the sky intensity. Also in thermal systems modelling we are mainly interested in time-averaged energy densities. And even if the time period is as short as 5 min, the changing radiances of moving clouds average out quite a lot.

The circumsolar part is also included in the typical measurement of the direct normal irradiance DNI. We consider an element of the sky whose angular position is defined by the angular distance ξ from the center of the Sun and the associated azimuth angle γ, as shown in Fig. 2.50, and from which the broadband celestial radiance $L(\xi, \gamma)$ emanates. The ISO definition of DNI is given by,

$$DNI = \int_0^{2\pi} \int_0^{\alpha_l} P(\xi, \gamma) L(\xi, \gamma) \cos\xi \, \sin\xi d\xi d\gamma, \tag{2.86}$$

in which $P(\xi, \gamma)$ is the penumbra function or often termed acceptance function. $P(\xi, \gamma)$ vanishes if ξ exceeds a limit angle α_l. The aperture solid angle precisely defined with the help of the penumbra function can be roughly described by means of an equivalent angle half-width, which is called the acceptance, aperture half-width angle. The value of the penumbra function $P(\xi, \gamma)$ results from the fraction of parallel light rays falling on the aperture from the angles (ξ, γ) and reaching the sensor element of the pyrheliometer. The penumbra function can thus be calculated from the geometric dimensions of the pyrheliometer. For angles ξ_l smaller than 5° the deviation of the $\cos\xi$ term from 1 is negligible and assuming a radial symmetry of the celestial ray density in the vicinity of the sun's position in a clear sky we get:

$$DNI = \int_0^{2\pi} P(\xi)L(\xi)\sin\xi d\xi, \tag{2.87}$$

in which $P(\xi)$ and $L(\xi)$ are the azimuthal averages of $P(\xi,\gamma)$, $L(\xi,\gamma)$. The radiance $L(\xi)$ is a function of the optical density of the atmosphere and thus the aerosol fraction as well as the Rayleigh spectral and aerosol optical depth, the aerosol phase function, and other variables. The DNI is thus calculated from the sky radiance and the penumbra function, of the sun disk and the circumsolar region. In the definition of the DNI according to Eq. (2.86) lies the actual dilemma since it contains neither a prescription of limiting angle α_l nor a penumbra function. If different values are used for both variables, different amounts of the integrated radiance near the sun are obtained. At the outer edge of the atmosphere the angular extent of the sun is equal to the extent of the solar disk, i.e. δ_s, but the scattering of particles and molecules in the atmosphere results in an extension around the circumsolar region, which can be up to 5° and from which the radiation can also fall on a collector. Figure 2.50b shows measured data of the normalized radiation distribution as a function of the angular distance ξ from the solar center at different atmospheric states.

The direct radiance thus results from the radiance emanating from the Sun and the circumsolar region and is a function of the angular position relative to the centre of the Sun. The so-called sunshape, or $s_s(\xi)$, is then a dimensionless quantity in which the azimuthal mean radiance profile is normalized by the radiance at the centre of the sun including the circumsolar region.

$$s_s(\xi) = \frac{S_s(\xi)}{S_s(\xi = 0)} = \int_0^{2\pi} L(\xi,\gamma)d\gamma. \tag{2.88}$$

Even without the influence of the Earth's atmosphere, the radiance of the solar disk decreases with increasing angular distance from the centre of the sun. This effect is called limb darkening and is a function of wavelength. For example, the radiance at the edge of the solar disk at 1000 nm is about 55% of the radiance at the solar centre, while the radiance at 370 nm drops to about 20% (Blanc et al., 2014).

2.4.2 Isotropic Sky Model

The simplest model of the sky (and ground reflected) diffuse radiation is the isotropic model. Here the sky intensity is assumed to be constant over the hemisphere. Similarly the ground reflected radiation is taken as isotropically diffuse.

Table 2.9 Ground reflection coefficients (ρ_g) of various soil types for solar radiation

Soil type	ρ_g (/)
New snow	0.75–0.95
Old snow	0.40–0.70
Sandy soil	0.15–0.40
Asphalt	0.15
Meadows and fields	0.12–0.30
Streets and houses	0.15–0.25
Forests	0.05–0.20
Dark floor	0.07–0.10
Water (depending on incidence angle of radiation, pollution, algae content, …)	0.03–0.22

This means that all diffuse radiation falling on a tilted collector aperture with tilt angle β can be quantified by the isotropic intensity and the view factor for the surface to the sky and the ground respectively. The direct irradiance has to consider the incidence angle with the cosine-factor:

$$G_t = G_{b,n} \cdot \cos\theta + G_{d,h} \cdot \frac{1+\cos\beta}{2} + \rho \cdot G_h \cdot \frac{1-\cos\beta}{2}, \tag{2.89}$$

where ρ is the ground reflectivity or albedo. $G_{b,n}$ is the direct normal beam irradiance and θ the incidence angle on the aperture area.

The isotropic sky model is easy to apply and to understand. It tends to be conservative in the sense that it underestimates G_t in typical situations. However, improved models have been developed which consider the horizon brightening or the circumsolar radiation.

The reflectivity of different ground surfaces (ρ_g) is given in the Table 2.9. This reflectivity is often referred to as albedo (Latin albus = "white") and is a measure of the reflectivity of diffusely reflecting surfaces. The albedo is the ratio of reflected light to incident light and is therefore between 0 and 1. It is particularly important in meteorology, as it allows estimations to be made about how much air warms up over different surfaces. The albedo of ice and snow in the polar regions influences considerably the radiation balance of the Earth. Even different types of snow show a considerably different but high degree of reflection, while that of water is almost negligible except for low solar altitudes. It is important that the diffuse albedo is nearly independent on the incidence angle of the solar radiation, in contrary to the direct reflectance of a smooth water surface.

2.4.3 Anisotropic Sky Models

In the situation of Fig. 2.51 let us consider the circumsolar part which is measured—depending on the sky conditions—partly as diffuse fraction (for large CSR

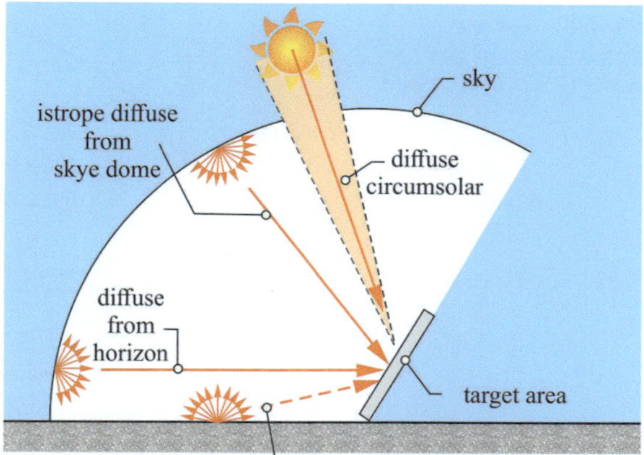

Fig. 2.51 Irradiation from an anisotropic sky on a tilted collector aperture

conditions) and partly as beam radiation (for clear skies with small CSR). Hay and Davies modified the simple isotropic in an easy way to correct the underestimation of directional radiation by introducing an anisotropy index A_i. A_i is a function of the transmittance of the atmosphere for beam radiation and given as

$$A_i = \frac{I_{bn}}{I_{0n}}, \tag{2.90}$$

where I_{bn} is the hourly terrestrial beam radiation and I_{0n} the extraterrestrial value. The fraction A_i of diffuse irradiation is now considered as circumsolar radiation and treated like beam radiation with an (approximate) incidence angle θ. Therefore, the total irradiance on the tilted surface (G_t) is modified in this model to:

$$G_t = \left(G_{b,n} + G_{d,h} \cdot A_i\right) \cdot \cos\theta + (1 - A_i) \cdot G_{d,h} \cdot \frac{1 + \cos\beta}{2} + \rho \cdot G_h \cdot \frac{1 - \cos\beta}{2}. \tag{2.91}$$

This is a slight improvement, but the horizon brightening is still not covered. An extension of the Hay and Davies model by Reindl et al. (1990) (the so-called HDKR-model) tries to also incorporate this feature.

$$\begin{aligned} G_t = {} & \left(G_{b,n} + G_{d,h} \cdot A_i\right) \cdot \cos\theta + (1 - A_i) \cdot G_{d,h} \\ & \cdot \frac{1 + \cos\beta}{2} \cdot \left(1 + f \cdot \sin^3\left(\frac{\beta}{2}\right)\right) + \rho \cdot G_h \cdot \frac{1 - \cos\beta}{2}, \end{aligned} \tag{2.92}$$

where f corrects for the horizon brightening,

$$f = \sqrt{G_{bh}/G_h}. \tag{2.93}$$

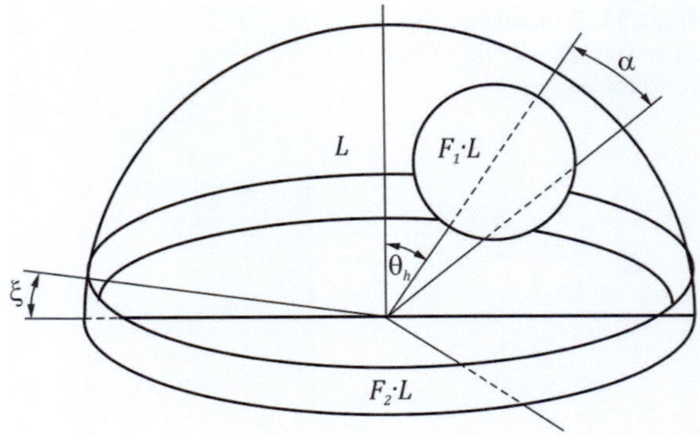

Fig. 2.52 The sky representation of Perez's model (Perez et al., 1986)

A different approach was undertaken by Perez et al. (1986) when dividing up the sky hemisphere in certain defined regions for circumsolar and horizon brightening using factors for the enhanced brightness (see Fig. 2.52). Here, L is considered as the sky radiance originating from the main portion of the dome, $(F_1 \cdot L)$ comes from the circumsolar zone, with $\alpha = 15°$ the half-angle of the circular region centered on the sun's position, and $(F_2 \cdot L)$ comes from the horizon with the horizon band angular thickness of $\xi = 6.5°$.

The diffuse part of the irradiation calculation for a tilted surface with this model can be written as

$$G_{d,t} = G_{d,h} \cdot \left[F_1 \cdot \frac{\max[\cos\theta, 0]}{\max\left[\cos\theta_z, \cos 85°\right]} + (1 - F_1) \cdot \frac{1 + \cos\beta}{2} + F_2 \cdot \sin\beta \right]. \tag{2.94}$$

The factor for the first circumsolar part does consider the incidence angle, but also the reduction of circumsolar area during sunset. Even for zenith angles larger than 90° some intensity is still available. The brightness coefficients are functions of three parameters zenith angle θ_z, a sky clearness ε and a brightness Δ. They are given by:

$$\varepsilon = \frac{G_{bn} + G_d}{G_d}, \tag{2.95}$$

$$\Delta = \frac{AM \cdot G_d}{G_0}. \tag{2.96}$$

The equations for calculating F_1 and F_2 are:

$$F_1 = |f_{11} + f_{12} \cdot \Delta + f_{13} \cdot \theta_z|^+, \tag{2.97}$$

$$F_2 = f_{21} + f_{22} \cdot \Delta + f_{23} \cdot \theta_z, \tag{2.98}$$

Table 2.10 Perez brightness coefficients f_{ij} for the anisotropic sky (Perez et al., 1988)

Range of ε	f_{11}	f_{12}	f_{13}	f_{21}	f_{22}	f_{23}
1.000–1.065	−0.008	0.588	−0.062	−0.060	0.072	−0.022
1.065–1.230	0.130	0.683	−0.151	−0.019	0.066	−0.029
1.230–1.500	0.330	0.487	−0.221	0.055	−0.064	−0.026
1.500–1.950	0.568	0.187	−0.295	0.109	−0.152	0.014
1.950–2.800	0.873	−0.392	−0.362	0.226	−0.462	0.001
2.800–4.500	1.132	−1.237	−0.412	0.288	−0.823	0.056
4.500–6200	1.060	−1.600	−0.359	0.264	−1.127	0.131
6.200 $\rightarrow \infty$	0.678	−0.327	−0.250	0.156	−1.377	0.251

where the zenith angle θ_z has to be given in radians. Table 2.10 with the statistically derived coefficients f_{ij}, which depend on the brightness range ε, has been published in Perez et al. (1988). The development of anisotropic sky models took into account the statistical contribution sky brightness conditions on the irradiance of tilted apertures of collectors. If however, for some reason the exact distribution of brightness due to clouds should be incorporated, a more detailed knowledge on the sky radiance is needed. This cannot be provided by simple irradiance measurments, but pictures of the whole clouded sky should be taken by sky cameras. However, in most applications the foregoing analysis gives sufficient accuracy to energy calculations of solar thermal systems.

2.5 Statistical Distribution of Irradiation

The frequency of periods with various radiation levels of clear, overcast, and cloudy days is important for a realistic assessment of the performance of solar systems. We know from our daily experience that the intensity of beam radiation varies substantially over "good" and "bad" days. However the categories cannot be assigned only by measuring the global irradiance level as this varies in both day categories due to solar position and time as well. A better characterizing parameter is the so-called clearness index which relates the measured terrestrial irradiation level to the incident irradiation for the same time period at the same location outside the atmosphere, to the extra-terrestrial irradiation also on a horizontal plane. We differentiate between hourly, daily, and monthly clearness indices k_T, K_T and $\overline{K}_T$:

$$k_T = \frac{I}{I_0}; \tag{2.99}$$

$$K_T = \frac{H}{H_0}; \tag{2.100}$$

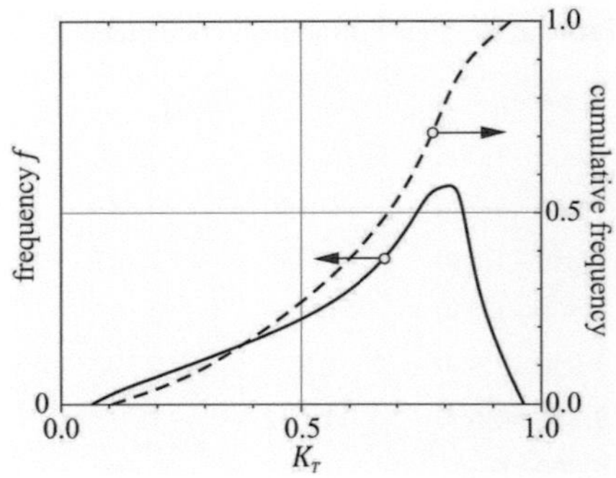

Fig. 2.53 Example of frequency of occurrence of days with various clearness index for a specific monthly average $\overline{K}_T$ (dashed: frequency distribution, solid: cumulative distribution)

$$\overline{K}_T = \frac{\overline{H}}{\overline{H}_0}; \tag{2.101}$$

in which I, H an $\overline{H}$ are the hourly, daily, or monthly sums of the irradiation on a horizontal plane, with the index 0 for extraterrestrial values. This clearness index is therefore a kind of atmospheric transmittance and quantifies the extinction in the atmosphere. For a location with a particular monthly clearness index $\overline{K}_T$ we find that the frequency distribution of days with various values of K_T is depending on the monthly average. If $\overline{K}_T$ is low we must have a large number of days also with cloudy and overcast sky and hence small K_T, if $\overline{K}_T$ is high, then there have to be much more days with clear sky and high K_T. A typical example of such a frequency distribution is shown in Fig. 2.53.

Liu and Jordan (1960) found that the cumulative distribution curves for locations having the same values $\overline{K}_T$ are very close to each other, even though the locations varied widely in latitude and elevation. Based on this they have developed generalized distribution curves of K_T versus fractional time. Bendt et al. (1981) have developed equations to represent the Liu and Jordan distributions, based on additional 20 years of data for 90 locations in the United States. These equations represent the Liu and Jordan curves very well for values of $f(K_T)$ less than 0.9–above 0.9 the correlations overpredict the frequency. The equations have the form:

$$f(K_T) = \frac{\exp(\gamma \cdot K_{T,min}) - \exp(\gamma \cdot K_T)}{\exp(\gamma \cdot K_{T,min}) - \exp(\gamma \cdot K_{T,max})}, \tag{2.102}$$

where γ is can be determined implicitly from the condition that the average of the distribution is the known $\overline{K}_T$. An explicit solution from a curve fit is given by

$$\gamma = -1.498 + \frac{1.184\,\xi - 27.182 \cdot \exp(-1.5\xi)}{K_{T,max} - K_{T,min}}, \tag{2.103}$$

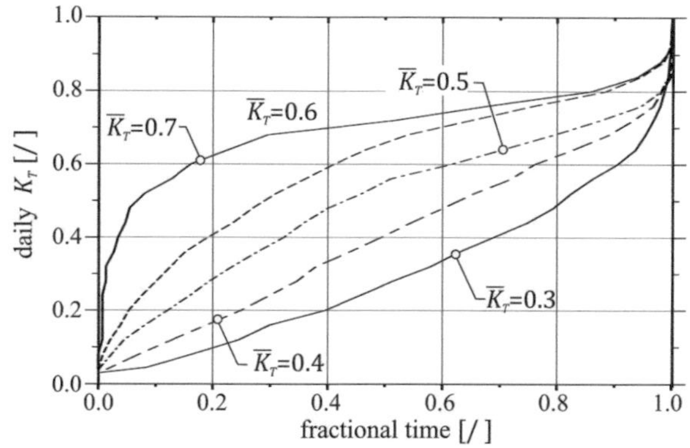

Fig. 2.54 Generalized functions of $f(K_T)$ over the monthly time fraction as a function of the monthly mean clearness (Liu & Jordan, 1960) in comparison to the functional representations (Bendt et al., 1981) (also applicable to hourly averages)

with $\xi = (K_{T,max} - K_{T,min})/(K_{T,max} - \overline{K}_T)$. For the extreme limits of the distribution the following values are recommended:

$$K_{T,max} = 0.6313 + 0.267 \cdot \overline{K}_T - 11.9(\overline{K}_T - 0.75)^8,$$
$$K_{T,min} = 0.05. \quad (2.104)$$

The universality of the Liu and Jordan distributions has been questioned, and in fact, for tropical climates an alternative formulation is suggested. And deviations were found for locations at high altitudes.

Figure 2.54 shows the K_T-values as a function of the time fraction. It has been shown that the same set of curves can also be used to derive $\overline{K}_T$ hourly values of K_T from daily values. Here, for a specific $\overline{K}_T$ a statistical distribution of K_T-values for individual days or subsets of a month can be given from the monthly mean values (e.g. for $\overline{K}_T = 0.7$, 7 days (25%) have a daily value of $K_T \approx 0.4$–0.7; 14 days (50%) have a ratio of $K_T \approx 0.7$–0.8 and the remaining 25% are above $K_T \approx 0.8$).

It has been observed however, that the generalized distribution curves for hourly and daily curves for a location are very similar (Whillier, 1965). Therefore, the distribution curves above may also be used for hourly clearness indexes K_T using the same monthly average $\overline{K}_T$.

The importance of these findings is that for many historical weather data only monthly values have been recorded and are still available. On the basis of the statistical distributions we can, however, construct artificial data for shorter time scales. Of course, for a realistic generation of weather data from long-term averages one needs not only know the statistical distribution of values but also time-correlations of time series. Stochastic modelling takes into account also the perseverance of weather situations. A good introduction into the topic is given by Aguiar et al. (1988) and more refined methods to construct with high confidence synthetic data can be found in Ngoko et al. (2014).

Similarly to the question, which temporal distribution of days with clear, overcast, and cloudy skies exist for a location, from the clearness indexes in turn it can be answered on a statistical basis, what the diffuse fraction of the corresponding hour, day or month is. The variable from which the diffuse fraction x_{dif} of the global irradiation can be derived, is the clearness index.

As the methods which present diffuse and beam fractions are derived from statistical evaluations, the individual results may differ substantially from measured data, especially when short time intervals like hours are considered. Individual cloud formations and changes in physical meteorological conditions (temperature, humidity) will certainly influence the reality whilst the statistics will only average over many similar clearness values.

The usual approach for hourly values is to correlate $x_{dif} = I_d/I$, the hourly fraction of the diffuse part in the horizontal global irradiation versus the hourly clearness index K_T, as illustrated in Fig. 2.55. Although the scatter of data is substantially, a tendency can be recognized. Erbs et al. (1982) have suggested the following correlation to be used based on US and Australian weather stations:

$$x_{dif} = \begin{cases} 1.0 - 0.249 \cdot k_T & \text{for } k_T < 0.35 \\ 1.5571 - 1.84 \cdot k_T & \text{for } 0.35 \leq k_T < 0.75 \\ 0.177 & \text{for } k_T \geq 0.75 \end{cases}, \tag{2.105}$$

A similar correlation has been developed by Reindl and Beckman (1990) based on different European and US stations. In this correlation a small influence of the solar altitude angle has been incorporated:

$$x_{dif} = \begin{cases} 1.020 - 0.254 \cdot k_T + 0.0123 \cdot \sin\alpha_s & \text{for } k_T < 0.3 \\ 1.400 - 1.749 \cdot k_T + 0.1770 \cdot \sin\alpha_s & \text{for } 0.3 \leq k_T < 0.78 \\ 0.486 \cdot k_T - 0.182 \cdot \sin\alpha_s & \text{for } k_T \geq 0.78 \end{cases}. \tag{2.106}$$

Reindl and Beckman (1990) investigated the addition of three more variables solar altitude, ambient temperature, and relative humidity in the predictor, and reduced then the numbers of variable. The single variable correlation is very close to the curve of Orgill and Hollands (1977), and the solar altitude yields the largest improvement in reduction of residuals (as given in Eq. (2.106), see Fig. 2.56). However, adding temperature and humidity even improved the result.

Similarly to correlation of hourly values, many scientific papers in the past have investigated the correlations of daily and monthly averaged diffuse fraction versus clearness index H_d/H_0 versus K_T and $\overline{H}_d/\overline{H}_0$ versus $\overline{K}_T$ (Table 2.11). The longer the averaging period, the smaller the range of observed and realized values. The correlation functions for monthly averages are typically in the range of $0.3 < \overline{K}_T < 0.7$. And the curves are usually linear or close to linear functions, with the scattering of measured values becoming closer. This is a consequence of statistics when we deal with averaged values. The outliers are singled out.

In some situations it is interesting to know how a solar system would perform on an average "typical" day. We are not interested in fluctuations and variations. Then we may construct a typical day based on a diffuse fraction from a correlation of monthly averages. How can the diffuse and global radiation be distributed of

Table 2.11 Monthly mean values of irradiation $\overline{H}$, clearness index $\overline{K}_T$, temperature $\overline{T}_a$ and degree days DD for two cities in the Federal Republic of Germany

		Jan	Feb	Mar	Apr	May	Jun	Jul	Aug	Sep	Oct	Nov	Dec
Bochum	51.17°												
$\overline{H}$	MJ/m^2	2.29	4.81	9.56	16.14	18.84	20.95	19.07	17.02	12.03	6.81	3.05	1.76
$\overline{K}_T$		0.274	0.356	0.441	0.524	0.497	0.51	0.484	0.51	0.498	0.428	0.321	0.255
$\overline{T}_a$	°C	1.7	2.5	4.9	8.3	13.1	15.7	17.2	16.5	13.9	9.6	5.1	2.7
DD	K_d	536	462	437	321	60	0	0	0	14	291	417	505
Hamburg	53.63°												
$\overline{H}$	MJ/m^2	2.29	5.99	11.56	18.66	24.18	25.35	23.59	19.83	14.96	8.04	3.23	1.94
$\overline{K}_T$		0.432	0.497	0.569	0.625	0.645	0.619	0.602	0.607	0.636	0.556	0.402	0.351
$\overline{T}_a$	°C	0.3	1.0	3.5	7.5	13.3	15.4	17.1	16.2	13.6	8.8	4.2	1.6
DD	K_d	580	504	481	345	56	0	0	0	35	316	444	539

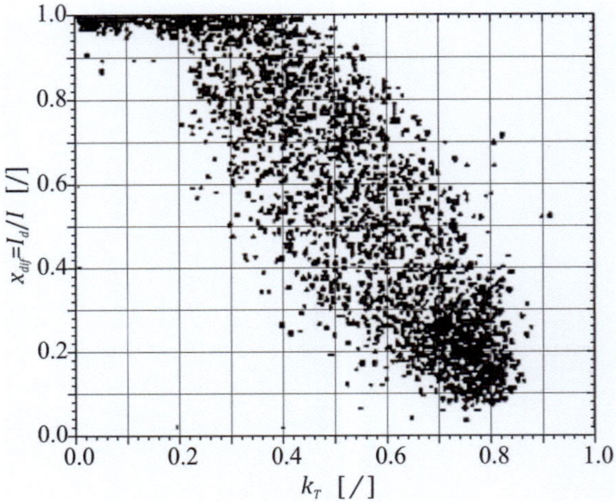

Fig. 2.55 Example of measured diffuse radiation fraction $x_{dif} = I_d/I$ as a function of global radiation in relation to extraterrestrial radiation on a horizontal surface as hourly mean value, adapted from Reindl (1988)

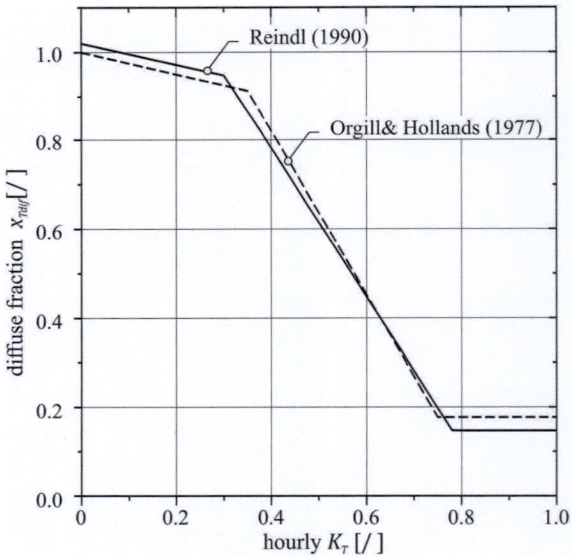

Fig. 2.56 Example of two correlations for diffuse fraction x_{dif} versus clearness index on an hourly basis (Orgill & Hollands, 1977; Reindl, 1988)

the hours of such a day? Already Liu and Jordan have given an answer to such a question. They developed from statistical studies a model how to distribute the hourly global and diffuse radiation over the day. The ratio of hourly to daily total global radiation is called r_t, the ratio of diffuse hourly to daily total diffuse r_d:

$$r_t = \frac{I}{H} \quad \text{and} \quad r_d = \frac{I_d}{H_d}. \tag{2.107}$$

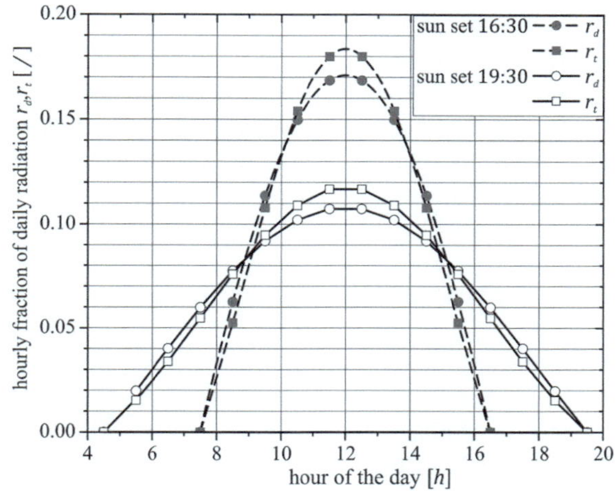

Fig. 2.57 Relationship between hourly and daily global irradiation on a horizontal surface for two days with sun set at 16:30 h and 19:30 h solar time

Collares-Pereira (1979) developed a functional form describing these ratios:

$$\begin{aligned} r_d &= \frac{\pi}{24} \cdot \frac{\cos\omega - \cos\omega_s}{\sin\omega_s - \omega_s \cdot \cos\omega_s}; \\ r_t &= (a + b\cos\omega) \cdot r_d; \\ a &= 0.409 + 0.5016 \cdot \sin\left(\omega_s - \frac{\pi}{3}\right); \\ b &= 0.6609 - 0.4767 \cdot \sin\left(\omega_s - \frac{\pi}{3}\right); \end{aligned} \tag{2.108}$$

where ω_s is the sunset hour angle. The curves of r_t of the relationship between hourly and daily global radiation is shown for hours in the afternoon as a function of the sunset hour angle in Fig. 2.57.

With the simple means described in the last paragraph we are able to model even starting from simple monthly irradiation values solar hourly irradiation values which we need at least to model the performance of a solar thermal system over the day. Modern computer programs use much shorter time steps of 1–5 min, but the input data there are usually still based on hourly data. If not measured, of course we do not know the statistical distribution over time, the correct time series which is dependent on many weather variables, the prominent ones being clouds, rain fronts and temperature drops. However, we can do better than just producing an average monthly day. This anyway would not give a representative result for a solar system performance but a conservative underestimation. But, at least using the distribution of days in a month we can produce by throwing the dices—in science terms with random Monte Carlo methods—a realistic distribution of clear, overcast, and cloudy days producing the correct monthly average. And similarly methods have been developed to create hourly or sub-hourly data of irradiation. Of course, the limits of the physical reality have to be considered, so global terrestrial irradiation might not exceed extra-terrestrial at any time. The computer program Meteonorm (Remund et al., 2010) has exactly incorporated such stochastic methods to create time series of beam and diffuse irradiation, and the values are even correlated with other variables like humidity and temperature. It is unlikely that a decrease in irradiance will lead to an immediate temperature drop.

But, although artificial weather data are in fact a big help, we need at least a good experimental basis of this, and the best solution anyway is to directly measure the real weather data in a time resolution which is needed. With this question we will deal in the next paragraph.

2.6 Quantitative Measurement of Solar Radiation

Acquiring information on the typical solar irradiation available at potential plant sites is of great importance to the design of solar power plants. The climatological average annual solar irradiance is most important in the planning process, but the characteristic frequency distribution also plays an important role. We talk here about solar resource assessment, in contrast to solar irradiation forecasting, which is the process of predicting solar radiation conditions on different time scales (from minutes to days) at a given time in the future. Ground based measurements are essential both at pre-operational and operational stage. However, satellite data provides broader temporal and special coverage, and therefore long-term estimates are commonly based on satellite-derived solar resource assessment. For a complete assessment of solar thermal plant design and operation, auxiliary meteorological parameters need to be analysed, e.g., ambient air temperature, wind speed and relative humidity. Although these quantities are important regarding plant performance and malfunction, we will not deal in detail with their measurement. Wind for example may influence heat loss, especially of uncovered collectors, wind gusts may be a risk for the safe integration of roof-mounted collectors or might harm mirror structures of concentrating mirrors. Ambient air temperature and humidity affect the necessity of climatization, have an impact on drying substances or determine the choice of heat-rejecting coolers.

The term *irradiance* is used to describe the solar power (instantaneous energy flux) falling on a unit area per unit time, e.g., in W/m^2. The term *irradiation* relates to the amount of solar energy falling on a unit area over a specified time interval such as a day, a month, or a complete year. In climatology, solar irradiation typically is given in J/m^2 summed over the period of a day, whereas for solar engineering, solar irradiation is often reported in units of kWh/m^2.

According to the standard (ISO 9488, 2021), the symbol for the irradiance is G and for the irradiation is H. In practice, often the same symbol (e.g. DNI) is used for irradiance and irradiation. In that case, they have to be differentiated by context or by the attached units.

2.6.1 Terrestrial Measurements

Measuring the Sun's energy incident to the Earth's surface is a difficult field measurement exercise. The measurement technology applied today is based on the conversion of electromagnetic radiant energy to heat or electricity. The sensors for such measurement instrumentation are e.g. photodiodes and thermopiles. The first utilizes the photoelectric effect to convert direct sunlight, and the second one the thermoelectric effect to convert generated temperature differences into an electrical current, which then can be measured.

Instruments for measuring solar radiation are of two basic types:

- A pyranometer is an instrument for measuring total hemispherical solar (beam plus diffuse) radiation with a hemispherical (180°) field of view, usually on a horizontal surface (global irradiance). If shaded from the beam radiation by a shade ring or disc, a pyranometer measures diffuse radiation.
- A pyrheliometer is an instrument using a collimated detector for measuring solar radiation at normal incidence from the sun and from a small portion of the sky (typically 5°) around the sun (i.e., beam radiation) with narrow field of view.

Pyrheliometers operate with a sun-tracking system that positions the sensor adequately towards the sun. Consequently, when considering the measurement of direct sunlight, the requirements for solar measurement stations are advanced since tracking systems bring high technical and mechanical requirements.

2.6.1.1 Thermal Sensors

With thermal solar radiation sensors, the radiant energy is initially converted into thermal energy by means of a black absorbing surface. A temperature difference is created in comparison with a white, reflecting or a shaded surface, as shown in Fig. 2.58 left. The small difference creates an electrical signal by a thermopile working with the thermo-electric effect. This electrical output can be measured with a voltmeter.

Due to the functional principle of thermopile radiometers, they are sensitive in a wide spectral range. Ideally, the black surface of thermopile instruments should absorb all incoming irradiance, like an ideal black body. However, only 97–98%

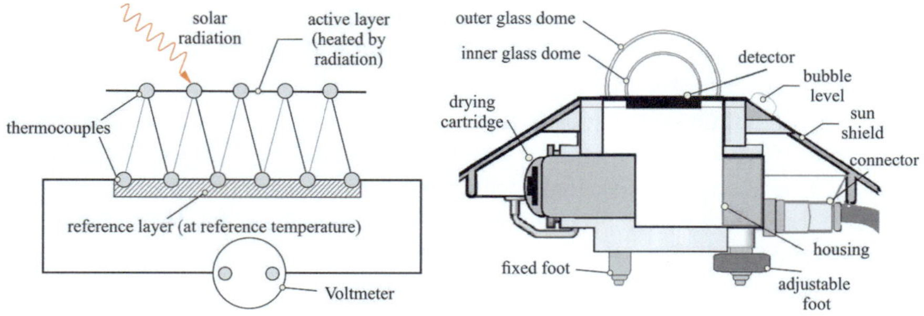

Fig. 2.58 Pyranometer principle (left) schematical functional principle: (right) set-up functional device

of the total solar energy is absorbed by the thermal detector. In addition, thermal losses by convection and radiation have to be minimized so that the temperature of the black surface is a good measure for the absorbed radiation and is not influenced by wind or sky conditions. Glass domes and glass windows are used for that purpose, as illustrated in Fig. 2.58 right. In summary, the sensors do not behave as ideal blackbodies but rather as small solar thermal collectors, so the instrumentation needs to be calibrated and compensation corrections are used to deduce the solar irradiation.

A thermopile element consists of a large number of thermocouple junction pairs, electrically connected in series. A thermocouple consists of two dissimilar metals connected together. The absorption of thermal radiation by the active (or 'hot') thermocouple junctions increases its temperature to T_1 and the reference (or 'cold') junction is kept at a fixed temperature T_2. The differential temperature between the active ('warm') and the inactive ('cold') junction produces an electromotive force directly proportional to the differential temperature created. This is called the thermoelectric effect. The magnitude and direction of the electromotive force is affected by the type of metal and the temperature difference between the hot and cold ends. The relationship between the temperature difference and the output voltage of a thermocouple is nonlinear and is approximated by a polynomial interpolation.

Thermal processes need some time to reach stationarity, therefore the instruments have response times to abrupt changes in irradiation. This means that the sampling time Δt_s of high-quality measurements should be short enough (smaller than response times) in order to track adequately the signal and to be able to correctly average the incoming radiation. The averaged measurements using a good number of sampling data then can be stored after a recording interval $\Delta t_r \gg \Delta t_s$. Recording intervals of 5 min are certainly adequate for solar thermal collector applications. When producing the irradiation data for experimental validation or for high-quality simulations of solar thermal systems, other inaccuracies like zero point accuracy or tilt errors may affect the result. For example zero point errors can be often estimated looking at recorded signals at nighttime where no solar radiation should reach the sensor (but watch out for light sources in the neighbourhood!). Negative values often can be found then which indicate that especially for clear skies with low sky temperatures the thermal radiation exchange with sky produces some uncompensated losses. Of course, these thermal losses might also exist during daytime but are obscured by the large solar irradiance. If this is the case the solar signal might be systematically underestimated in such conditions. Therefore, the quality of the instrumentation is assessed by a number of factors listed in the Table 2.12, which affects the final accuracy of the measurement.

Figure 2.59 shows examples of thermopile pyranometers designed to measure hemispherical irradiance. The hemispherical glass domes allow the capture of GHI without a bias of incidence angles, when installed on a horizontal plane. For solar energy applications often the insolation on a tilted plane is of interest. If this is the case, pyranometers are installed for example in the installation plane of solar thermal collectors with tilt angles from 15° to 60° depending on latitude and

Table 2.12 Uncertainties and pyranometer classes according to ISO 9060 (2018)

Category	Secondary standard	First class	Second class
Response time (95% final value)	<15 s	<30 s	<60 s
Zero point accuracy			
(a) response 200 W/m^2 thermal radiation	$\pm 7\ W/m^2$	$\pm 15\ W/m^2$	$\pm 30\ W/m^2$
(b) response 5 K/h temperature change	$\pm 2\ W/m^2$	$\pm 4\ W/m^2$	$\pm 8\ W/m^2$
Stability/drift per year	±0.8%	±1.5%	±3.0%
Non-linearity (related to 1000 W/M^2)	±0.5%	±1.0%	±3.0%
Tilt error (0°−90°)	±0.5%	±1.0%	±5.0%
Spectral response deviation	±3%	±5%	±10.0%
Influence of temperature range of 50 K	2%	4%	8%
Overall measurement uncertainty (daily sums)	±3%	±5%	±10.0%

Fig. 2.59 Set-up of a pyranometer

application. These instruments directly capture the global tilted irradiance (GTI) mostly including a part of the irradiation reflected from the ground.

A disadvantage of thermopile pyranometers in comparison to photoelectric pyranometers is a higher price and frequent soiling of the glass dome. Also, measurements might be biased due to weather impact. Common phenomena to be considered here are dew or frost on the glass dome. To overcome this, advanced pyranometer systems have integrated ventilation and heating. Figure 2.59 shows an example of a heated, ventilated pyranometer next to a standard pyranometer. Heavy frosting on the glass dome will falsify the data and thereby generate data loss. To avoid such losses, the instrumentation must be selected depending on meteorological conditions. A disadvantage of air-conditioned systems is the higher price and the higher power consumption. Therefore power supply must be adjusted to the needs of the measurement system.

Although the DNI and diffuse irradiance is defined in ISO-9488 (2021) it is quite vague due to the lack of specification of the angular distribution of the solar radiation. Experimental instruments include or block not only the sun disc, but also a part of the sky. This leads to the requirement of a correction factor if direct radiation only from the solar disc or diffuse radiation including circumsolar radiation is needed. The measured DNI on site e.g. by conventional thermopile pyrheliometers is not necessarily larger than the effective DNI for a concentrator system. This originates from the larger aperture angle of pyrheliometers or equivalent pyranometric systems when compared to that of concentrating collectors leading to an overestimation of measured irradiance since it includes the irradiance from the solar disc and a larger circumsolar region. Also the diffuse radiation blocked by a shadowband has to be corrected.

Measurements of diffuse radiation can be made with pyranometers by shading the instrument from beam radiation. This is usually done by means of a shading ring, as shown in Fig. 2.60. The ring is used to allow continuous recording of diffuse radiation without the necessity of continuous positioning of smaller shading devices; adjustments need to be made for changing declination only and can be made every few days. The ring shades the pyranometer from part of the diffuse radiation, and a correction for this shading must be estimated and applied to the observed diffuse radiation (LeBaron et al., 1990). The corrections are based on assumptions of the distribution of diffuse radiation over the sky and typically are factors of 1.05–1.2 depending on the exact geometry. Typical shade ring correction factors, to illustrate their trends and magnitudes, is shown in Fig. 2.61.

Examples of thermal pyrheliometers that are designed to measure DNI are shown in Fig. 2.62. The benefit of pyrheliometers is the direct measurement of DNI and therefore such instrumentation complies with ISO 9060:2018 Class A standards. In order to capture direct sunlight, pyrheliometers are mounted to a sun tracker as can be seen in Fig. 2.62. For redundant measurements, a station usually

Fig. 2.60 Pyranometer with shadowband

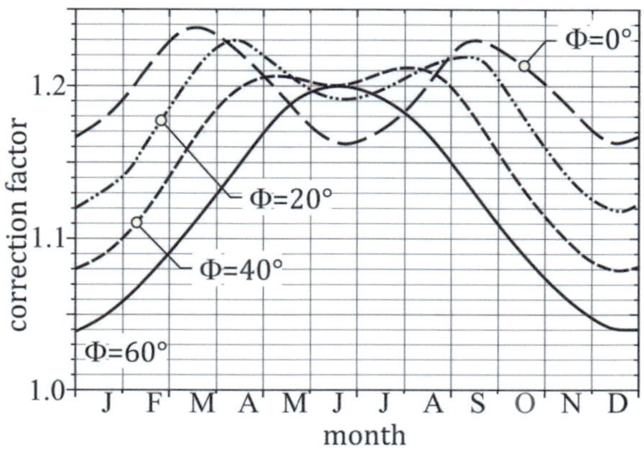

Fig. 2.61 Typical shade ring correction factors to account for shading of the detector from diffuse radiation. Adapted from Coulson (1975)

Fig. 2.62 Pyrheliometers (*Source* NREL (2009), Solar Radiation Research Laboratory)

holds two pyrheliometers and thereby reduces the risk of data loss. In order to avoid dew or frost to affect the measurements, Class A pyrheliometers have integrated window heating, such as the Hukseflux DR30-D1, shown in Fig. 2.62. For the needs of data quality monitoring, it is recommended to consider pyrheliometers with inbuilt tilt sensors, which allow the orientation of the instrument to be monitored.

The definition of the DNI according to Eq. (2.86) is an experimental definition and relates the DNI to the specific measurement device being used. The amount of measured circumsolar scattered irradiance depends on the state of the atmosphere and on the specific penumbra function of the instrument. The corresponding penumbra function can be approximated geometrically according to Fig. 2.63 by introducing three angles: the opening half-angle α, the slope angle α_s, and the limit angle α_l, defined as

$$\alpha = a\tan(R/L);\ \alpha_s = a\tan((R-r)/L);\ \alpha_l = a\tan((R+r)/L), \tag{2.109}$$

where R, r and L are characteristic dimensions of the instrument. Since the angles are small one approximate for α, so that $\alpha \approx (\alpha_s + \alpha_l)/2$. This definition results in the so-called geometric penumbra function, which describes the fraction of collected radiant flux by an optical aperture depending on the off-axis angles. For off-axis angles $0° \leq \alpha \leq \alpha_s$, the penumbra function $P(\alpha)$ is unity; if $\alpha \geq \alpha_l$ correspondingly $P(\alpha) = 0$. The penumbra function monotonically drops from 1 to 0 in the interval $[\alpha_s, \alpha_l]$. The opening half-angle α is the main characteristic for the description of the field of view of a pyrheliometer, and it corresponds approximately to the centre of the transition range $[\alpha_s, \alpha_l]$. The viewing angle Ω is defined as the solid angle of a cone with apex angle 2α:

$$\Omega = 2\pi(1 - \cos\alpha). \tag{2.110}$$

The detailed procedure, which is based on a multi-step approach, to arrive finally at a consistent DNI measurement value is described in Blanc et al. (2014).

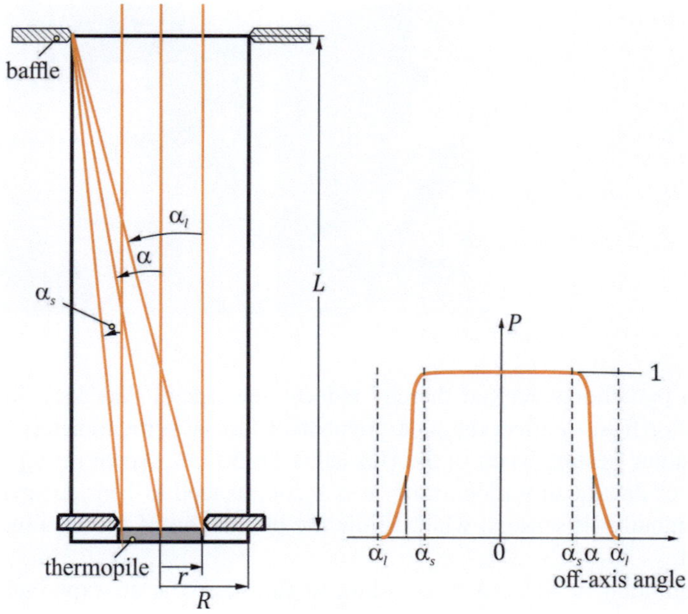

Fig. 2.63 Circular opening of a thermopile pyrheliometer to acquire DNI with definition of the slope angle α_s, opening half-angle α and limit angle α_l and corresponding penumbra function $P(\alpha)$ as afunction of the off-axis angle α (right)

2.6.1.2 Photoelectric Sensors

A photodiode sensor is usually a silicon semiconductor with $p - i - n$ structure or $p - n$ junction. These sensors convert radiant energy directly into an electrical signal. When a photon of sufficient energy strikes the diode, it creates an electron–hole pair. This mechanism is also known as the inner photoelectric effect. If the absorption occurs close to the junction, the pair is separated by the inner field: holes move toward the anode and electrons toward the cathode, and a photocurrent is produced. And in first order the photocurrent is linearly proportional to the irradiance. Photodiodes can be used under either reverse bias (photoconductive mode) or zero bias (photovoltaic mode), as depicted in Fig. 2.64.

Diodes have extremely high resistance when reverse biased. This resistance is reduced when light of an appropriate frequency shines on the junction. Hence, a reverse biased diode can be used as a light detector by monitoring the current running through it. A photodiode signal can be measured as a voltage or a current. Current measurement demonstrates far better linearity, offset, and bandwidth performance. The generated photocurrent is proportional to the incident light power.

Under zero bias, solar radiation incident on the diode causes a current across the device, leading to forward bias, which in turn induces 'dark current' in the opposite direction to the photocurrent. This is called the photovoltaic effect, which is the basis for solar cells. Figure 2.65 shows an exemplary photoelectric pyranometer.

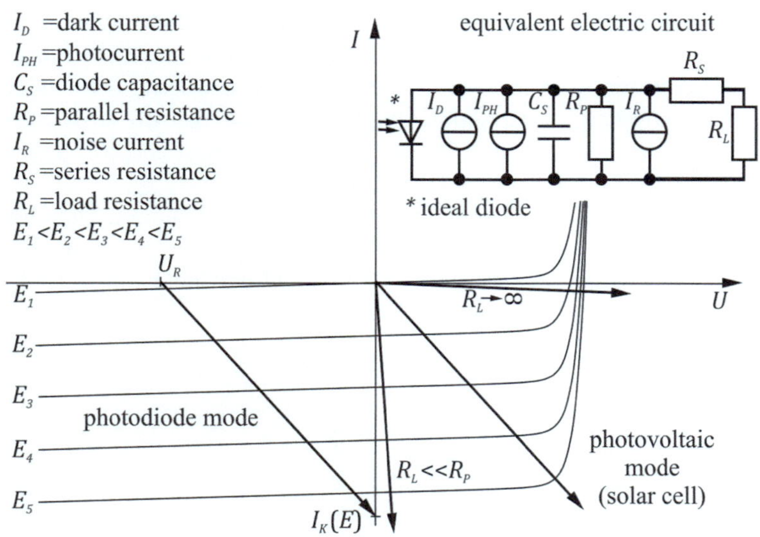

Fig. 2.64 I-V characteristic of a photodiode. The linear load lines represent the response of the external circuit: I = (Applied bias voltage − diode voltage)/total resistance. The points of intersection with the curves represent the actual current and voltage for a given bias, resistance and illumination (modified from Hess (2011))

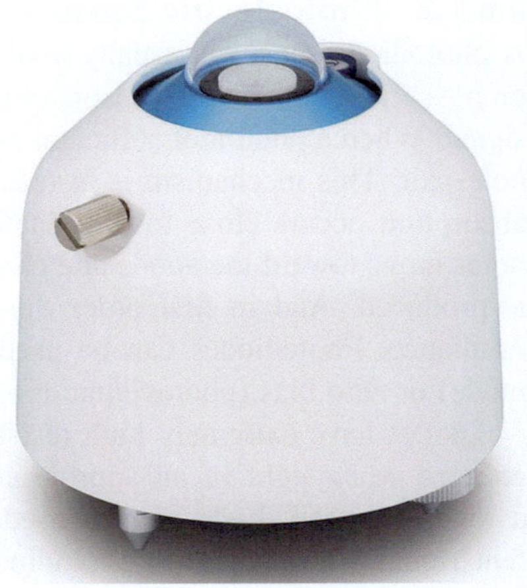

Fig. 2.65 Photoelectric pyranometers (Jrloginov, 2016)

The white diffusing layer on top of the Si diode cell shall improve the angular sensitivity.

They are less precise instruments than the thermopile instruments and have some limitations on their use. They are also less expensive than thermopile instruments and are easy to use. The main disadvantage of photovoltaic detectors is their spectrally selective response. Figure 2.66 shows a typical terrestrial solar spectrum and the spectral response of a silicon solar cell. Spectral response typically is in the range between 0.4 and 1.1 μm, and it is not uniform. If the spectral distribution of incident radiation was fixed, a calibration could be established that would remain constant; however, there are some variations in spectral distribution with clouds and atmospheric water vapor. One may estimate that the error introduced because of spectral response is ±5% maximum under most conditions of natural daylight and is ±3% under typical conditions.

Photovoltaic detectors have additional characteristics of interest. Their response to changing radiation levels is essentially instantaneous and is linear with radiation. The temperature dependence is ±0.15%/°C maximum. In order that the

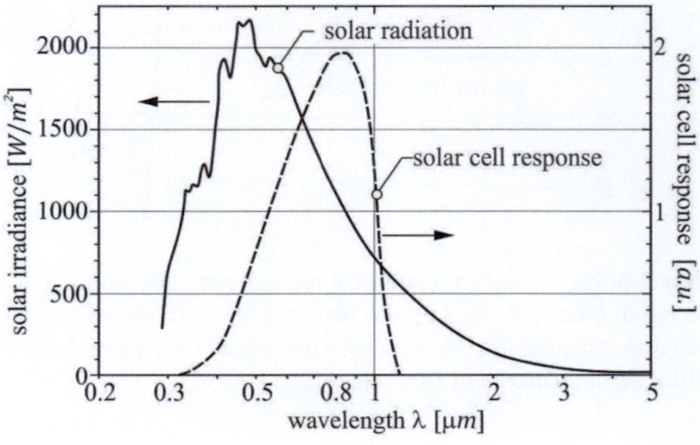

Fig. 2.66 Spectral sensitivity of photoelectric sensors

dependence of response on the angle of incidence is removed, the sensor is often fitted with a white diffusing cover that substantially removes the incidence of the radiation. The response of the detectors is independent of its orientation.

The advantages especially in the robustness and fast response led to the development of commercial Rotating shadowband irradiometers or pyranometers (RSI or RSP respectively) usually employ such a photoelectric radiometer (Fig. 2.65) to measure incident solar radiation. As shown in Fig. 2.67, the pyranometer is mounted on a 'head unit' apparatus, which permits unobstructed measurement of global horizontal irradiance. DNI can be determined from a single photoelectric pyranometer, if it is assembled in an arrangement that periodically blocks the direct beam radiation and causes it to measure global irradiance and diffuse irradiance alternately. The black shadowband rotates around the radiometer and thereby shades it off. While shaded, the radiometer is measuring horizontal diffuse irradiance (DHI). Accordingly, both GHI and DHI can be measured with one instrument. Since the shadowband is motor driven, and RSP's power consumption is demanding for stand-alone off-grid solar measurement stations sufficient power supply must be ensured.

RSP operation is realized by a program coding, which drives a motor control module. The code is usually run by a datalogger, which processes and collects the measured values. Typically, the shadowband rotates over the photodiode once every 30 s, taking approximately 1 s for this motion. During this period, the photoelectric pyranometer signal is sampled approximately 1,000 times. When the sensor is completely shaded from the sun by the shadowband, the lowest pyranometer readings occur. The software determines the DHI by finding the average of minimum values. For these instruments narrow-to-broadband corrections need to be applied to derive the full solar range. Ideally, these consider the spectral effect of lower air masses and also that of various contents of atmospheric trace gases and aerosols.

Fig. 2.67 RSP instrument operated by National Renewable Energy Laboratory-NREL (2009)

DNI measurements derived by RSPs are not reaching the exact same accuracy compared to pyrheliometer measurements, as long as the pyrheliometer tracking is accurate and the measurement device is maintained and properly cleaned. Soiling of measurement devices is often an issue with pyrheliometers, as these are easily affected by dust and dirt accumulation. Accurate pyrheliometer-based *DNI* measurements require daily and careful cleaning. In real life, the cleaning of pyrheliometers is often not practical since daily station visits can be very expensive and challenging for remote areas—the typically preferred areas for CSP application. RSP-based measurements are less susceptible to soiling since the impact is reduced for photoelectric effect. Accordingly, RSP measurements require less attendance and maintenance. In compliance with internationally recognized Tier2-standards, station maintenance and cleaning of the RSP should be performed twice a week. In case low soiling rates are given, station visits may be reduced down to once per week. Little soiling can be expected, e.g., in remote areas with sufficient ground cover by grass. For a desert environment, however, high soiling rates must be considered. Not only station maintenance but also the purchase price of RSP instruments is reduced compared to pyrheliometer systems. For maximum data reliability, solar monitoring stations with both sorts of instruments are typically configured with redundant sensors, which also allow comparison of both sensors regarding parallel measurement of the same *DNI* intensity. However, highest standards for solar measurements require pyrheliometer-based measurements of *DNI*. Such standards are classified as Tier1.

2.6.1.3 Sky Cameras and Measurement of Circumsolar Radiation

The first measurements of the solar aureole were carried out during total solar eclipses at the end of the nineteenth century. Between 1920 and 1955, the Smithsonian Institution measured the circumsolar irradiance coming from an annular region concentrically positioned around the sun. Their instruments had ring-shaped apertures and measured the irradiance coming from the angular region between 3.5° and 14.5° and later between 8.5° and 18.5° (half angles) (Hoyt, 1979). Also pyrheliometers with different collimators have been used.

Measurements of the solar radiance profile including the solar disk and the circumsolar region have been carried out by LBL. The measurements of LBL are of special importance, as approximately 180,000 measurements collected at 11 different sites between 1976–1981 were later digitally published in the LBL reduced data base (Noring et al., 1991). The instrument had a small circular aperture and measured the radiance coming from nearly point-like regions around and inside the solar disk.

Other groups used analogue photographic techniques. In the 90s CCD cameras with a highly linear response were used by Schubnell (1992) and Neumann et al. (2002). The data acquired with this approach agreed well with the LBL data base.

2.6.2 Satellite Data Evaluation

In cases where ground-based measurements are not available, satellite-derived solar radiation values are used. Satellites measure reflected radiation from the Earth's surface in several wavelength bands. Known albedo values by location and complex models and algorithms can be used to determine global, diffuse, and direct beam irradiance components. Raw data are available from various satellite operators. In turn, these are processed by several different organizations providing solar resource satellite data services. In some cases, these are commercial services and in other cases, research, or government. One of the best known and most extensive data sets is provided by NASA. Today, their data is provided through multiple platforms.

The methodology for calculating direct and global irradiance data from satellite data is differing from one institution to the other. From the early development of look-up tables already in the early years scientists have tried to model surface solar irradiance (SSI) data based on satellite observations, but after the year 2000, satellite-based estimation of SSI became increasingly mature thanks to the modern satellite sensors that have onboard calibration for both narrow and broadband channels, as well as further development of remote sensing methods. With quantitative remote sensing information regarding aerosols, water vapor, ozone etc. have become or are becoming available, which increasingly supported the accurate modelling of the radiative process. Therefore, quite sophisticated algorithms have been developed in which detailed radiative transfer processes are considered (Huang et al., 2019).

One major step is usually the determination of irradiation for clear sky conditions (see Fig. 2.68a) in one channel of the transfer calculations. Of the various atmospheric constituents under clear-sky conditions, ozone, water vapor, and aerosols are three critical components because of their influences on SSI and their variability in the atmosphere. Ozone and water vapor absorb solar radiation at certain wavelengths, which means that their loadings can be effectively retrieved by the sharp contrasts between the absorption windows. Aerosols are more challenging to retrieve, particularly over land. Factors like cloud contamination and the difficult differentiation of light reflected by aerosols and surfaces makes the determination of AOD (aerosol optical depth) over bright surfaces prone to large uncertainties.

When the sky is not clear, but clouds are partly covering shading the Earth's surface, solar irradiation is effectively altered (see Fig. 2.68b). Clouds play a critical role in solar radiation. This corresponds to our common perception that clouds obscure the direct sun but may provide a different diffuse light than the clear sky. This is dependent on cloud shape, cloud cover and cloud properties. In view of the decisive influence of clouds on SSI, to some extent, SSI satellite estimation is conducted around how to accurately account for radiative attenuation by clouds (cloud scattering and absorption) in the atmosphere. Towards this end, there are two different strategies:

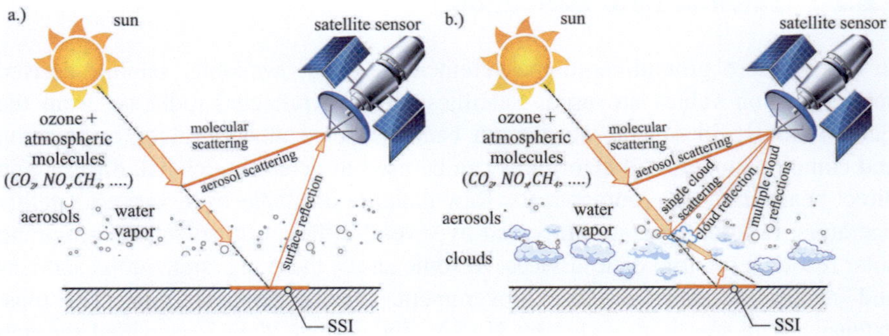

Fig. 2.68 Simplified relations between satellite observations and SSI according to one-dimensional radiative transfer theory for **a** clear sky and **b** cloudy sky modified from Huang et al. (2019)

- directly inferring the attenuation of clouds and subsequently SSI from top-of-atmosphere satellite pictures, or
- estimating the attenuation of clouds from cloud models.

Figure 2.69 and the following short description outlines one of the first modern methodologies to use satellite data for the determination of global and direct solar irradiance. It is taken from the ISIS model description, a method developed by DLR in the ISCCP (International Satellite Cloud Cover Programme). The radiative transfer through the Earth's atmosphere is calculated as 1-dimensional transport through a horizontally layered, plane-parallel atmosphere. The two-stream approximation of the diffuse und direct radiative transfer equation is used to derive the spectrally integrated diffuse und direct irradiances at the surface. Absorption of atmospheric gases is included by a 32 band calculation covering the whole solar spectrum from 250 to 4300 nm.

Fig. 2.69 Methodology for calculating direct normal and global irradiance from satellite data as developed in the ISIS project modified from German Aerospace Center (DLR) (2006)

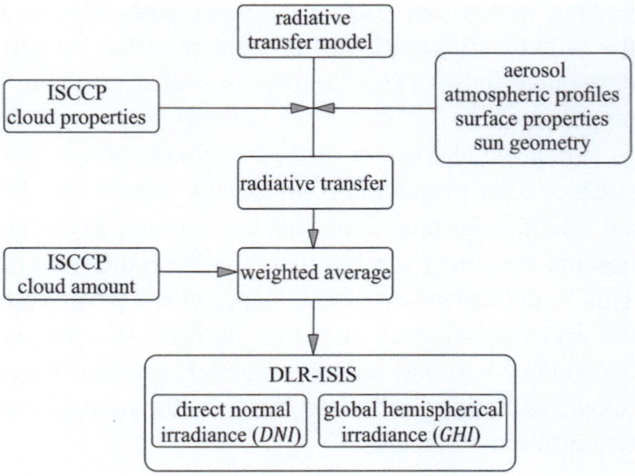

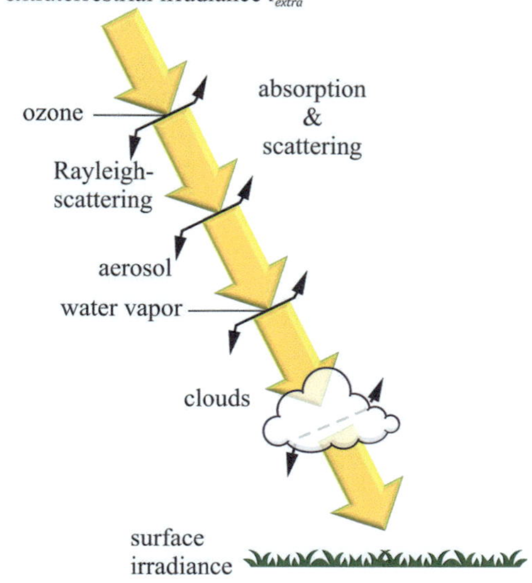

Fig. 2.70 Schematic modelling process in a radiative transfer model for clear sky

Cloud optics are included as two parametrizations for the optical properties of cloud particles. For water clouds Mie-calculations of liquid water droplets are used, for ice clouds a different parametrization is chosen. The clear sky radiative transfer algorithm and parameterizations are included in the radiative transfer library libRadtran. For the derivation of the DLR-ISIS data set, a program was built around this library to ensure the correct combination of atmospheric parameters, initiate radiative transfer calculations (Fig. 2.70), and handle the results; further details may be taken from Lohmann et al. (2006).

For clouds the models are based on the ISCCP D1 cloud data (see ISCCP homepage for more information about cloud data sets). It provides 3-hourly cloud observations on a 280 km × 280 km equal area grid, see Fig. 2.71. ISCCP differentiates between 15 cloud types. The classification includes three intervals of optical thickness in three cloud levels: low, middle, and high clouds. Low and middle cloud types are further divided into water and ice clouds, high clouds are always ice clouds.

For DLR-ISIS, optical thickness, cloud top pressure and cloud phase given in the ISCCP data set are processed to generate clouds for the radiative transfer calculations. One radiative transfer calculation is carried out for each occurring cloud type assuming 100% cloud coverage, plus one calculation for clear sky. For the final result, irradiances are weighted with the cloud amount for each cloud type and for clear sky, respectively, and summed.

The input of the aerosol optical thickness is split into two parts: tropospheric aerosol and stratospheric aerosol. The optical thickness of the tropospheric aerosol is taken from the NASA-GISS data set. Optical properties of the most common aerosol types were measured, and the distribution of aerosol particles are modelled

Fig. 2.71 Principle scheme of calculation of cloud cover index from image evaluation of different spectral channels

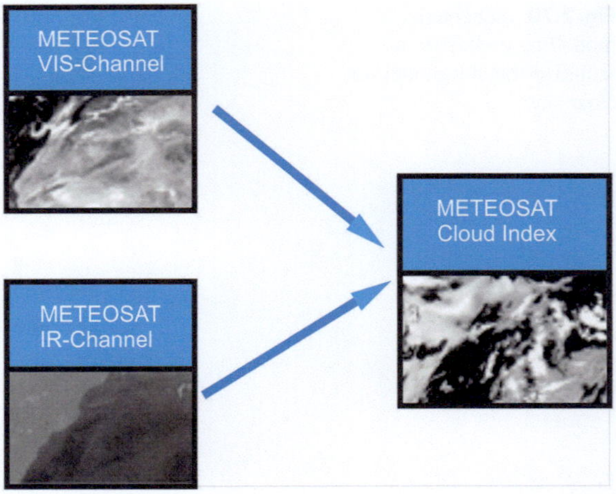

for individual months with the help of emission data and transport models. Stratospheric aerosol is mostly caused by large volcanic eruptions. The data set here also consist of monthly averages of optical thickness of the stratospheric aerosol. Ozone and water vapor profiles are also provided in the ISCCP input data set, based on the Total Ozone Mapping Spectrometer (TOMS) and the TIROS Operational Vertical Sounder (TOVS). Together with the surface albedo and the mean elevation above sea level for each grid box included in the ISCCP data set, these are used to derive irradiances for the DLR-ISIS data set.

Although the methodologies have been developed further and new versions of the semi-empirical approach with improved accuracy are being used in the engines such as SolarGIS, in the NSRBD of NREL and the Indian SUNY project, the basic principle is similar.

The derivation of solar irradiation values from satellite data is realized by radiative transfer calculations. Irradiation is calculated from infrared and visible channels of satellites with a nominal spatial resolution of 3 km $\times$ 3 km at ground level and 5 km $\times$ 5 km at ground level for the second and first generation of Meteosat, respectively. The annual sums of satellite-derived data can coincide with ground-measured data within a range of $\pm$ 5%. However, in some cases, deviations of around 20% have been observed (Gueymard, 2010). For shorter time periods like single months, days or hours, the deviation of satellite-derived and measured solar irradiance values can increase significantly (Zelenka et al., 1999). Due to insufficient accuracy of satellite data, precise ground-measured data are essential for validation and calibration of long-term satellite-derived data. The minimum measurement period should cover a duration of 1 year, so that at least one complete seasonal cycle can be evaluated (Suri & Cebecauer, 2014). Where to place the contribution of satellite data in a determination of statistical solar irradiance at a site is depicted in Fig. 2.72. Table 2.13 illustrates the input data used for the solar radiation model.

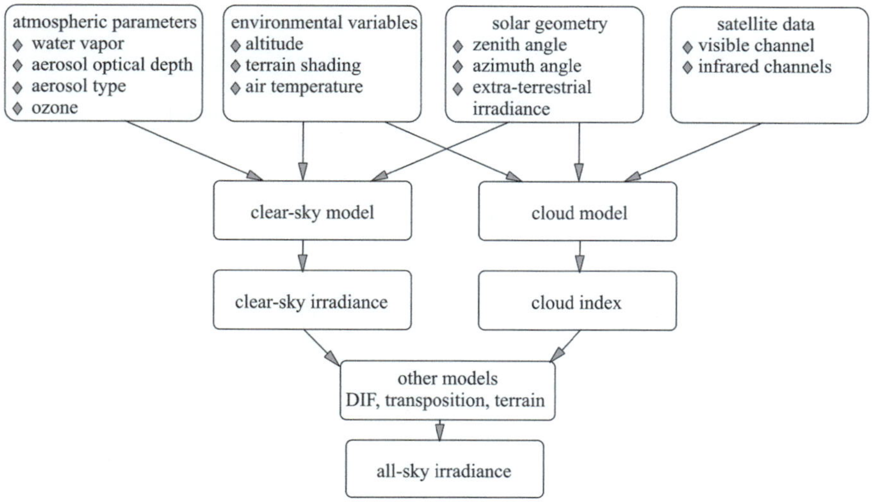

Fig. 2.72 Evaluation path to determine all-sky irradiance, with elements, data and models contributing to the result

Table 2.13 Inputs to a satellite model from Cebecauer and Suri (2016)

Inputs into the SolarGIS model	Source of input data	Time period	Original time step	Approximate grid resolution
Cloud index	Meteosat MFG (PRIME) satellites (EUMETSAT)	1994–2004	30 mins	3.0 km × 3.1 km
	Meteosat MSG satellites (EUMETSAT)	2005–present	15 mins	3.6 km × 3.8 km
Atmospheric optical depth AOD	MACC (ECMWF)	2003–present	6 h (monthly avg. before 2003)	85 and 125 km
Water vapor	CFSR/GS (NOAA)	1994–present	1 and 3 h	35 and 55 km
Elevation and horizon	SRTM-3 (SRTM)	–	–	250 m
SolarGIS primary data outputs (GHI and DNI)	–	1994–present	15 mins	250 m

At present, there is no standard procedure for processing solar radiation data or a set of procedures to be followed for solar resource assessments. As a result, and due to pressing deadlines for projects, project developers do not always follow a strict approach for solar resource assessment. The majority of the decisions for site selection of CSP plants are still based on the usually rough initial site assessments.

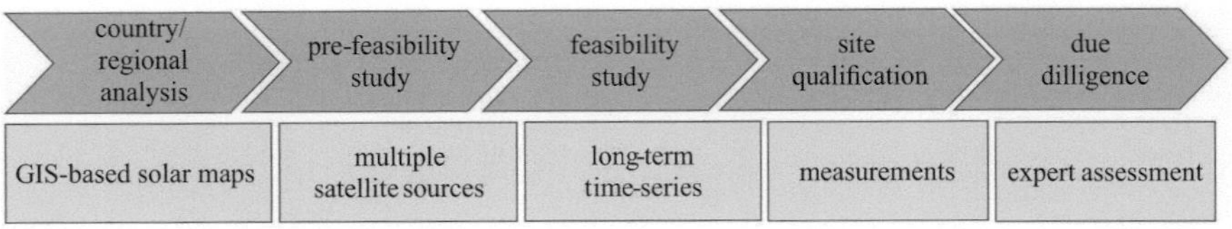

Fig. 2.73 Recommended steps for solar resource analysis during project development from Meyer (2010)

In such cases, solar radiation conditions such as annual cycle of DNI, frequency distribution of DNI, inter-annual variability, and the uncertainty have not been assessed in detail. As a result, the current knowledge of solar resource available at the foreseen location remains limited.

Based on practical experience and analysis of various data products and methods, the best practices to achieve high-quality assessments with reasonable effort are proposed. Following Meyer (2010) or Sengupta et al. (2017) the recommended procedure is illustrated in Fig. 2.73 and involves the following steps:

- When no detailed assessment of a project location has been made, multiple satellite data sources and, if available, ground measurements, should be taken into consideration. At this point, long-term average values of DNI from these sources may suffice.
- Calculate a quality-weighted best average and determine the resulting uncertainty by Gaussian error propagation.
- Once a project site has been confirmed for further development of the CSP project, a suitable measurement station should be installed at the site in order to reach the 1 year of measurement data in parallel with the project approval process.
- Determine the long-term best estimate for the specific project site (in 1 km × 1 km resolution) and based on as many years of satellite data as available. For DNI, a minimum of 10 consecutive years should be considered, because inter-annual variability is high.
- Multiple site-specific data sets from satellites and ground-based measurements, with overlapping time periods, can provide independent information and increase accuracy when combined. Only reliable data sets shall be considered. The uncertainty of the data is determined individually for each data set.

Site-specific satellite data time series are required to reflect the long-term history of irradiation and must be adapted based on overlapping time periods between different satellite and measurement data sets. Section 2.6.1 introduced the principles of solar ground-based measurements and pointed out the benefit of accurate measurements, providing less risk to power performance estimates and control. However, ground-based measurements come at great expense, preventing from long-term measurement campaigns. To derive long-term best estimates, it is necessary to consider data series exceedingly more than a decade. Thus, the lowest level of risk in solar resource assessment is achieved when long-term satellite data is recalibrated to ground-based measurements. This will reduce uncertainty. Furthermore, when simulating power performance of a CSP plant, data of sub-hourly temporal resolution should be considered (Hummon et al., 2012). Highest temporal resolution is given for ground based data, where common sampling rates are 1 min. Thus, adapting satellite data to ground data holds both advantages of long-term and high-temporal resolution. For spatial resolution of satellite data, at least a 10 km × 10 km spatial resolution is required. Preferably, it should be 1 km × 1 km resolution.

The accuracy and reliability of the satellite data provider is important as random and systematic errors can result in data of poor quality. Satellite providers often have a regional focus, so that one provider is not always recommended for all global regions. The current state of satellite-derived methods leaves room for improvement. Also, ground-based measurements can be of poor quality depending on the individual maintenance of the measurement station over the complete measurement period. If a location is seriously considered during the site qualification phase, the satellite-derived time series should be adapted to the ground-measured data by, e.g., using the procedure of Chhatbar and Meyer (2011). However, such adaptions require temporal overlap of ground- and satellite-based data.

2.6.3 Composition of Representative Time Series of Solar Irradiation

A Typical Meteorological Year (TMY) is a data set giving values in high time resolution. It shall well represent the characteristic meteorological conditions of average years at a site.

From the long-term satellite data, which are corrected with the measurement data, the average DNI can be derived, representing the so-called $P50$ value. $P50$ means that the annual solar irradiation has a probability of 50% being above and 50% below this value. It is the arithmetic average of all annual values. Often, $P50$ is referred to as best estimate and represents the statistical level of confidence to which the estimated solar resource exceeds with 50% probability. In other words, the $P50$-value is given as the solar irradiance, which is expected to be exceeded with a probability of 50%. From the corrected satellite data, the climatological average of DNI can also be derived.

Generally in statistics PXX-measures represent a value that is exceeded by $XX\%$ of the population of a data set, meaning that $P70$ or $P90$ values for annual solar irradiation represents a data set where 70% or even 90% of the years exhibit better results. $P50$ refers to 50% of the years exceeding the value and is the highest of these quantities. $P70/P90$ are pessimistic values regarding the representative irradiation. Therefore, they are a common quantity in the financial world, and widely applied for PV and wind energy projects. Thus, P-values of solar radiation (DNI) are provided to judge the reliability of the solar resource of a project, a typical example is illustrated in Fig. 2.74. Accordingly, the DNI value related

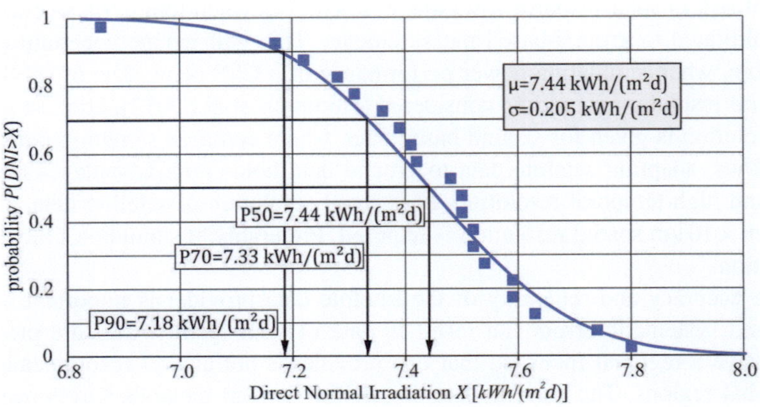

Fig. 2.74 Example of a fitted normal distribution to daily average DNI values of 20 different years. The values for P50, P70 and P90 are indicated

to $P90/P70$ for a given site are lower compared to $P50$, which represents the long-term average. Such values are derived based on the uncertainty of the data basis. A normal distribution is fitted to all available annual values, and from the determined parameter the critical value for $P70$ or $P90$ is determined.

A typical meteorological year (TMY), data set, is a compilation of extracts of data from various real years, assembled such that the annual DNI matches the long-term average for that site. Such data sets are a valuable input for modelling typical system output. For the creation of the most reliable CSP-specific typical meteorological years, site-specific measurements are of great value. According to Hoyer-Klick et al. (2009) data selection for TMY building should be done in a way that for each month only data that deviates by less than 5% from the climatological average is considered. The annual average in the TMY should be within 1% of the climatological average.

Finally, for bankable expert opinions on meteorological conditions for CSP, usually data with a lower probability of underperformance is required in order to satisfy conservative approaches from banks and lenders. This said, additional TMY s based on a more conservative approach are required to assess the risk of lower irradiance values. For this purpose, typically either $P70$ or $P90$ data sets are computed. These represent the data of a typical year with values that are expected to be exceeded in 70% or 90% of all years, respectively.

Alternatively, a risk assessment using performance simulation based on several good and bad years could give sufficient comfort to the banks. Such time series can be derived from satellites but should also be adapted to site-specific characteristics based on ground-measured data. This approach would allow a more detailed assessment of the influence of variable DNI conditions.

Also, by processing of at least 10 years of data, $P70$ or $P90$ values could be derived, which usually are used as basis to calculate the financial base case for a project. Compared to the simpler approach of only using $P70$ or $P90$ years, the advantage of using multiple years is manifold. First, the effects of meteorological variability and uncertainty are considered. Additionally, uncertainties resulting from technical parameters (describing the plant or the uncertainty of the performance simulation models) can be assessed in more detail.

2.6.4 Data Sources for Direct and Global Irradiation

Ground based solar and other meteorological data are available nowadays from many sources, however with different accuracy and temporal resolution. Measurement stations are operated by national meteorological services, airports, academic institutions, and even commercial entities, for instance when the run a large PV power plant. In addition interpolation algorithms are available to approximate meteodata for locations close to existing measurement stations.

Scientific measurement campaigns on solar irradiation in a large global scale started in the 1960s with the creation of the World Radiation Data Centre by the World Meteorological Organisation (WMO) in 1965. Since then a number of quality stations measuring also global solar irradiation have been maintained by national meteorological services around the world for many years. The archived data are still available (World Radiation Data Centre-WRDC (World Radiation Data Centre, 2022)) and the number of stations is indicated in Fig. 2.75.

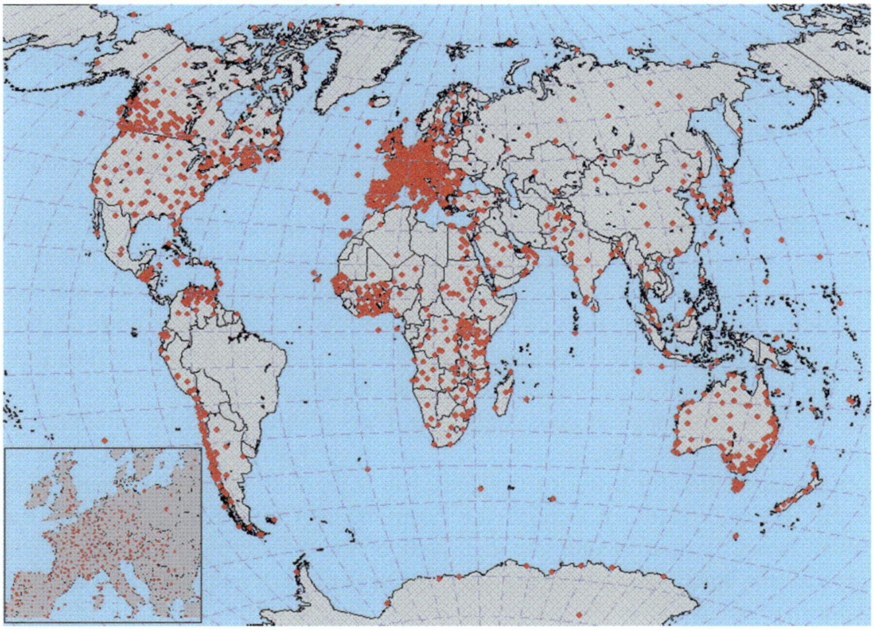

Fig. 2.75 WRDC stations worldwide from World Radiation Data Centre (2022)

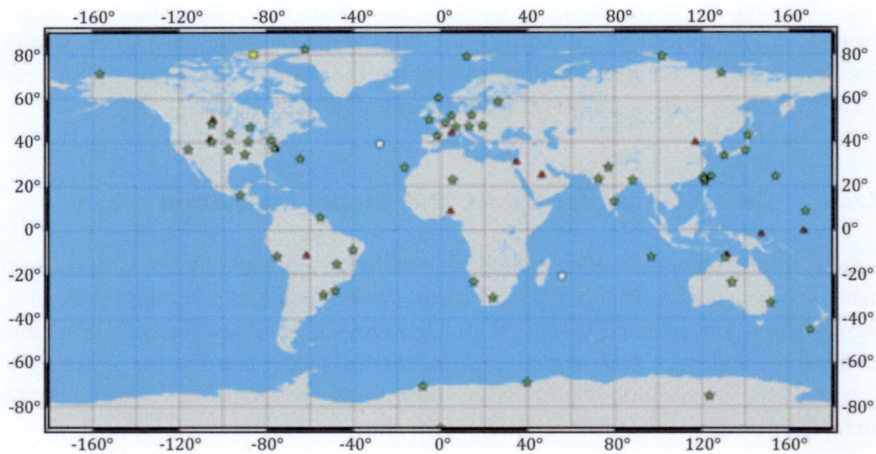

Fig. 2.76 BSRN network of currently 64 stations worldwide from Base Surface Radiation Network (BSRN) (2023)

Compared to the many years of daily, monthly and yearly sums of global irradiation from this network, systematic world-wide scientific measurements of direct normal irradiation DNI has been started only 1992 with the Baseline Surface Radiation Network (BSRN, (Base Surface Radiation Network (BSRN), 2023)), which is a project of the World Climate Research Programme (WRCP) to monitor with high precision all terrestrial radiation components (global, direct, diffuse, UV, Infrared) to detect differences due to climatic change; an indication of operating stations is provided in Fig. 2.76. These long-term data sets are mainly used for the improvement of climate models, but also e.g. for the assessment of temporal intervariability of years, and for the calibration of radiation models based on satellite data.

Weather data including solar radiation and temperature data can be divided into two major classes: historical meteodata and so–called typical meteorological years (TMY). Historical data is just "real" data: usually measured (but sometimes modeled) data from a particular location for a given period of record. Typical years are artificially generated years assembled to match the long term data from a particular location using a particular statistical measure. Typical data may also be "real" data but may not be a contiguous year—the data may be comprised of months from multiple years.

Ag good source for historical weather data is the U.S. National Climatic Data Centre (NCDC, (National Centers for Environmental Information)). NCDC can provide hourly historical data for thousands of locations around the world. This data may not always be complete; data items or periods of record may be missing. Many national weather services also provide data, e.g. the Deutsche Wetterdienst DWD for Germany, the China Meteorological Bureau, the India ISHRAE Weather Data files. Some of the data provided are typical years, others are historical records. Internationally, the Solar and Wind Energy Resource Assessment (SWERA) project, funded by the United Nations Environment Program, is developing high quality information on solar and wind energy resources in 14 developing countries. Currently typical year hourly data are available for 156 locations.

A tool used widely as meteorological input for simulation of solar applications is the global climatological database Meteonorm (www.meteonorm.com). It's a combination of a climate database, a spatial interpolation tool, and a stochastic weather generator. In this way typical years with hourly or minute time resolution can be calculated for any site. The input of Meteonorm for global radiation is the Global Energy Balance Archive (GEBA, (Global Energy Balance Archive-GEBA)). All other meteorological parameters are taken from databases of WMO and NCDC (periods 1961–90 and 1996–2005). The stochastic generation of global radiation is based on a Markov chain model for daily values and an autoregressive model for hourly and minute values (Aguiar et al., 1988 or Aguiar & Collares-Pereira, 1992). The generation of temperature is based on global radiation and measured distribution of daily temperature values of approx. 5000 sites. Meteonorm generates also additional parameters like precipitation, wind speed or radiation parameters like diffuse and direct normal irradiance.

Satellite-derived solar and meteorological data are provided for example by NASA (NASA—National Aeronautics and Space Administration). Also commercial providers present maps on a monthly or annual basis e.g. SolarGIS. PV-GIS (SOLARGIS) is a free tool to find solar data worldwide as well. Programs have been sponsored by the World Bank to provide good data for developing countries. For instance, NASA's Land Processes Distributed Active Archive Centre (LPDAAC) offers DNI data to the public without charge for any location across the globe. However, freely available data only holds monthly averages and thereby does not hold information on fluctuations throughout the day. The data is derived from 22 consecutive years of satellite data with a basic resolution of approximately 30 km, which has been processed to a $1° \times 1°$ geographical grid.

All satellite data providers have different temporal and spatial coverage, different temporal, and spatial resolution, use different algorithms and different inputs. As a result, solar radiation values derived for a given place from different satellite providers can differ significantly. There are many advantages associated with satellite-derived values:

- Satellites have high spatial resolution covering most of the sites of interest.
- Compared to ground-based measurement campaigns, the temporal coverage of meteorological satellites is extensive. Some data date back to the late 1970s and hold almost continuous coverage since the 1980s. Since satellites cover such long terms of data, the historic data basis can be applied to estimate future irradiation, and thus can serve as basis for planning and sizing of solar systems. Furthermore, inter-annual variations in energy yield can be estimated. Such estimates are vital for risk analysis of CSP plants.
- Satellite-derived solar radiation values (real-time data) can be utilized for monitoring and managing dispersed solar power in the grid.

References

Aguiar, R., & Collares-Pereira, M. (1992). Statistical properties of hourly global radiation. *Solar Energy, 48*(3), 157–167. https://doi.org/10.1016/0038-092X(92)90134-V

Aguiar, R., Collares-Pereira, M., & Conde, J. (1988). Simple procedure for generating sequences of daily radiation values using a library of Markov transition matrices. *Solar Energy, 40*(3), 269–279. https://doi.org/10.1016/0038-092X(88)90049-7

Base Surface Radiation Network (BSRN). (2023, February 16). *Global baseline network for surface radiation for the Global Climate Observing System* (J. R. European Commission, Ed., & D. C. Lanconelli, Compiler). Alfred-Wegener-Institut. https://bsrn.awi.de/contact-persons/

Bendt, P., Collares-Pereira, M., & Rabl, A. (1981). The frequency distribution of daily insolation values. *Solar Energy, 27*(1), 1–5. https://doi.org/10.1016/0038-092X(81)90013-X

Blanc, P., Espinar, B., Geuder, N., Gueymard, C., Meyer, R., Pitz-Paal, R., & Wilbert, S. (2014). Direct normal irradiance related definitions and applications: The circumsolar issue. *Solar Energy, 110*, 561–577.

Buie, D., Monger, A., & Dey, C. (2003). Sunshape distributions for terrestrial solar simulations. *Solar Energy, 74*(2), 113–122. https://doi.org/10.1016/S0038-092X(03)00125-7

Cebecauer, T., & Suri, M. (2016). Site-adaptation of satellite-based DNI and GHI time series: Overview and SolarGIS approach. *AIP Conference Proceedings, 1734*, 150002-1–150002-11. https://doi.org/10.1063/1.4949234

Chhatbar, K., & Meyer, R. (2011). The influence of meteorological parameters on the energy yield of solar thermal plants. In *Proceedings of the SolarPACES 2011 Conference* (p. 8). https://doi.org/10013/epic.37915.d001

Collares-Pereira, M. A. (1979). The average distribution of solar radiation-correlations between diffuse and hemispherical and between daily and hourly insolation values. *Solar Energy, 22*(2), 155–164. https://doi.org/10.1016/0038-092X(79)90100-2

Coulson, K. (1975). *Solar and terrestrial radiation-methods and measurements*. Academic Press.

Duffie, J. A., & Beckmann, W. A. (2020). *Solar engineering of thermal process, photovoltaics and wind*. Wiley.

Erbs, D., Klein, S., & Duffie, J. (1982). Estimation of the diffuse radiation fraction for hourly, daily and monthly-average global radiation. *Solar Energy, 28*(4), 293–302. https://doi.org/10.1016/0038-092X(82)90302-4

Fröhlich, C. (1977). The solar output and its variation. In O. R. White (Ed.), *Contemporary measures of the solar constant* (p. 93). Colorado Associated University Press.

Fröhlich, C., Foukal, P. V., Hickey, J. R., Hudson, H. S., & Willson, R. C. (1992). Solar irradiance variability from modern measurements. In *Sun in time* (3rd ed.). Univ. of Arizona Press.

German Aerospace Center (DLR). (2006). *Derivation of irradiances from ISCCP cloud data*. Linder Höhe, 51147 Cologne, Germany. Retrieved February 27, 2023, from http://www.pa.op.dlr.de/ISIS/method.html

Gerthsen, C., Kneser, H. O., & Vogel, H. (1995). *Physik. Ein Lehrbuch zum Gebrauch neben Vorlesungen* (17. Auflage ed.). Springer.

Global Energy Balance Archive-GEBA. (n.d.). *Central database for the worldwide instrumentally measured energy fluxes at the surface* (E. Z. Institute for Climate and Atmospheric Sciences, Ed.). Retrieved March 20, 2022, from https://geba.ethz.ch/

Gueymard, C. (2010). Variability in direct irradiance around the Sahara: Are the modeled datasets of bankable quality? In *Proceedings of the 16th SolarPACES Conference* (p. 9).

Hess, G. (2011, June 22). *I-V curve of a photodiode and the equivalent circuit*. Retrieved February 27, 2023, from https://commons.wikimedia.org/w/index.php?curid=15579778

Hippler, R. (2005). *Atmosphärenphysik*. Universität Greifswald.

Hoyer-Klick, C., Hustig, F., Schwandt, M., & Meyer, R. (2009, September 15–18). Characteristic meteorological years from ground and satellite data. In *Proceeding of the 15th SolarPACES Symposium*.

Hoyt, D. (1979). Pyrheliometric and circumsolar sky radiation measurements by the Smithsonian Astrophysical Observatory from 1923 to 1954. *Tellus, 31*(3), 217–239.

Huang, G., Li, Z., Li, X., Liang, S., Yang, K., Wang, D., & Zhang, Y. (2019). Estimating surface solar irradiance from satellites: Past, present, and future perspectives. *Remote Sensing of Environment, 233*, 111371. https://doi.org/10.1016/j.rse.2019.111371

Hummon, M., Ibanez, E., Brinkman, G., & Lew, D. (2012). Sub-hour solar data for power system modeling from static spatial variability analysis. In *Conference Paper, 2nd International Workshop on Integration of Solar Power in Power Systems*, Lisbon, Portugal, 12–13th November.

Iqbal, M. (1983). *An introduction to solar radiation*. Academic Press Inc.

ISO 9060. (2018). *Solar energy—Specification and classification of instruments for measuring hemispherical solar and direct solar radiation.*

ISO 9488. (2021). *Solar energy—Vocabulary.* ISO/TC 180 Solar energy.

Johnson, F. (1954, December). The solar constant. *Journal of the Atmospheric Sciences*, 431–439. https://doi.org/10.1175/1520-0469(1954)011<0431:TSC>2.0.CO;2

Jrloginov. (2016, July 16). *Class A secondary standard pyranometer*. Retrieved February 27, 2023, from https://commons.wikimedia.org/w/index.php?curid=86344332

Kasten, F., & Young, A. T. (1989). Revised optical air mass tables and approximation formula. *Applied Optics, 28*(22), 4735–4738.

LeBaron, B., Michalsky, J., & Perez, R. (1990). A simple procedure for correcting shadowband data for all sky conditions. *Solar Energy, 44*(5), 249–256. https://doi.org/10.1016/0038-092X(90)90053-F

Liu, B. Y., & Jordan, R. C. (1960). The interrelationship and characteristic distribution of direct, diffuse and total solar radiation. *Solar Energy, 4*(3), 1–19. https://doi.org/10.1016/0038-092X(60)90062-1

Lohmann, S., Schillings, C., Mayer, B., & Meyer, R. (2006). Long-term variability of solar direct and global radiation derived from ISCCP data and comparison with reanalysis data. *Solar Energy, 80,* 1390–1401. https://doi.org/10.1016/j.solener.2006.03.004

McCracken, G., & Scott, P. (2005), *Fusion: the Energy of the Universe*. Elsevier Academic Press, Burlington MA.

Meyer, R. (2010). Recommendations for bankable meteorological site assessments for solar thermal power plants. In *Proceedings of SolarPACES 2010* (p. 8). http://www.solarpaces2010.org/

Monteith, J., & Unsworth, M. (2013). *Principles of environmental physics-plants, animals, and the atmosphere* (4th ed.). Elsevier Science. ISBN 9780123869937.

NASA—National Aeronautics and Space Administration. (n.d.). *The POWER Project-Prediction of worldwide Energy Resources* (P. Stackhouse, Compiler). 300 E. Street SW, Suite 5R30, Washington, DC 20546, United States of America. Retrieved March 20, 2022, from https://power.larc.nasa.gov/

NASA. (2022, March 14). *EOS Project Science Office—Earth Observatory* (N. G. Center, Editor). https://earthobservatory.nasa.gov/features/SORCE/sorce_03.php

National Centers for Environmental Information. (n.d.). *Archives of atmospheric, coastal, geophysical, and oceanic research* (N. O. (NOAA), Ed.). 1401 Constitution Avenue NW, Room 5128, Washington, DC 20230, United States of America. Retrieved March 20, 2022, from https://www.ncei.noaa.gov/

National Renewable Energy Laboratory-NREL. (2009). *NREL pyrheliometer comparison: NPC2009.* Solar Radiation Research Laboratory. https://midcdmz.nrel.gov/apps/npcphotos.pl?year=2009

Neumann, A., Witzke, A., Jones, S., & Schmitt, G. (2002). Representative terrestrial solar brightness profiles. *Journal of Solar Energy Engineering, 124*(2), 198–204. https://doi.org/10.1115/1.1464880

Ngoko, B., Sugihara, H., & Funaki. (2014). Synthetic generation of high temporal resolution solar radiation data using Markov models. *Solar Energy, 103*, 160–170. https://doi.org/10.1016/j.solener.2014.02.026

Noring, J., Grether, D., & Hunt, A. J. (1991). *Circumsolar radiation data: The Lawrence Berkeley Laboratory reduced data base.* National Renewable Energy Lab.

Orgill, J., & Hollands, K. (1977). Correlation equation for hourly diffuse radiation on a horizontal surface. *Solar Energy, 19*(4), 357–359. https://doi.org/10.1016/0038-092X(77)90006-8

Paetzold, H. K. (1952, May). Ein Beitrag zur atmosphärischen Extinktion. *Astronomische Nachrichten, 281,* 17–22. https://doi.org/10.1002/asna.19522810106

Pecker, J.-C. (1994). The sun: A synoptic view. *Vistas in Astronomy, 38*(2), 111–164. https://doi.org/10.1016/0083-6656(94)90028-0

Perez, R., Stewart, R., Arbogast, C., Seals, R., & Scott, J. (1986). An anisotropic hourly diffuse radiation model for sloping surfaces: Description, performance validation, site dependency evaluation. *Solar Energy, 36*(6), 481–497. https://doi.org/10.1016/0038-092X(86)90013-7

Perez, R., Stewart, R., Seals, R., & Guertin, T. (1988). *The development and verification of the Perez diffuse radiation model.* Sandia National Lab.

Rapp-Arrarás, Í., & Domingo-Santos, J. M. (2011). Functional forms for approximating the relative optical air mass. *Journal of Geophysical Research, 116,* D24308-13. 0148-0227/11/2011JD016706

Reindl, D. (1988). *Estimating diffuse radiation on horizontal surfaces and total radiation on tilted surfaces.* Solar Energy Laboratory. University Wisconsin Madison: Faculty Mechanical Engineering. http://digital.library.wisc.edu/1793/47660

Reindl, D., & Beckman, D. J. (1990). Diffuse fraction correlations. *Solar Energy, 45*(1), 1–7. https://doi.org/10.1016/0038-092X(90)90060-P

Reindl, D., Beckman, W. A., & Duffie, J. (1990). Evaluation of hourly tilted surface radiation models. *Solar Energy, 45*(1), 9–17. https://doi.org/10.1016/0038-092X(90)90061-G

Remund, J., Müller, S., Schilter, C., & Rihm, B. (2010). The use of Meteonorm weather generator for climate change studies. In *Proceedings of the 10th EMS Annual Meeting; 8th European Conference on Applied Climatology* (Vol. 7, pp. EMS2010-417).

Schubnell, M. (1992). Sunshape and its influence on the flux distribution in imaging solar concentrators. *Journal of Solar Energy Engineering, 114*(4), 260–266. https://doi.org/10.1115/1.2930015

Sengupta, M., Habte, A. G., & Wilbert, S. R. (2017, December). *Best practices handbook for the collection and use of solar resource data for solar energy applications: Second edition.* National Renewable Energy Laboratory (NREL).

SOLARGIS. (n.d.). *Online data and apps.* Bottova 2A, 811 09 Bratislava, Slovakia. Retrieved March 20, 2022, from https://solargis.com/

Stöcker, H. (1994). *Taschenbuch der Physik.* Harry Deutsch.

Suri, M., & Cebecauer, T. (2014). *Satellite-based solar resource data: Model validation statistics versus user's uncertainty.* Environmental Science, ASES SOLAR 2014 Conference, San Francisco, 7–9 July 2014. https://solargis2-web-assets.s3.eu-west-1.amazonaws.com/public/publication/2014/1f0b376723/Suri-Cebecauer-ASES-Solar2014-Satellite-Based-Solar-Resource-Data-Model-Validation-Statistics-Versus-User-Uncertainty.pdf

Thekaekara, M. (1976). Solar radiation measurement: Techniques and instrumentation. *Solar Energy, 18*(4), 309–325. https://doi.org/10.1016/0038-092X(76)90058-X

Van de Hulst, H. C. (1981). *Light scattering by small particles.* Dover Publications Inc.

Whillier, A. (1965). Solar radiation graphs. *Solar Energy, 9*(3), 164–165.

Willson, R. C. (1978). Accurate solar 'constant' determinations by cavity pyrheliometers. *Journal of Geophysical Research, 83*(C8), 4003–4008. https://doi.org/10.1029/JC083iC08p04003

World Radiation Data Centre. (2022). *Meteorological Services world solar radiation data* (V. M. Observatory, Ed., & H. o. Dr. Anatoly V. Tsvetkov, Compiler). Retrieved March 20, 2022, from http://wrdc.mgo.rssi.ru/wrdc_en_new.htm

Zelenka, A., Perez, R., Seals, R., & Renné, D. (1999). Effective accuracy of satellite-derived hourly irradiances. *Theoretical and Applied Climatology, 62,* 199–207. https://doi.org/10.1007/s007040050084

Optical Conversion

3

Abstract

Solar collector systems always need to maximize the solar gain from the short-wave solar radiation of the sun and hence minimize optical losses. Many components and different materials are used in a solar collector to convert solar radiation into thermal heat. It is crucial to understand the principles and limitations of the interactions of transparent, reflecting, and absorbing components with the electromagnetic radiation. The concepts of spectral selectivity and optical concentration are introduced and several methods for achieving that are described.

Keywords

Radiation transport • Transparent media • Reflector • Absorber • Spectral selectivity • Directional optical property • Anti-reflection • Optical concentration • Imaging concentrators • Non-imaging optics • Fresnel optics

3.1 Electromagnetic Radiation

Solar radiation is electromagnetic radiation emitted from the sun. Due to the internal processes in the hot sun the electromagnetic waves emitted have energies mainly corresponding to a wavelength range between approximately $0.2\mu m$ and $2.6\mu m$ as discussed in Chap. 2. However, when dealing with solar technology, we also find electromagnetic radiation with other wavelengths. Notably, we also consider the thermal radiation transport with waves of wavelengths depending on the temperature of the emitting body. The typical range where the associated black-body or Planck spectra of thermal radiation have their energy is between 5 and $100\mu m$ for ambient temperatures, and between 2 and $40\mu m$ for $600° K$; the spectrum is depicted in Fig. 3.1. Although ultraviolet UV, visible and near infrared solar radiation and thermal radiation all is transported by electromagnetic waves, due to the different energy densities the interaction with matter is different. We will deal with that in a separate chapter. At first, however we want to introduce

R. Stieglitz and W. Platzer, *Solar Thermal Energy Systems*,
https://doi.org/10.1007/978-3-031-43173-9_3

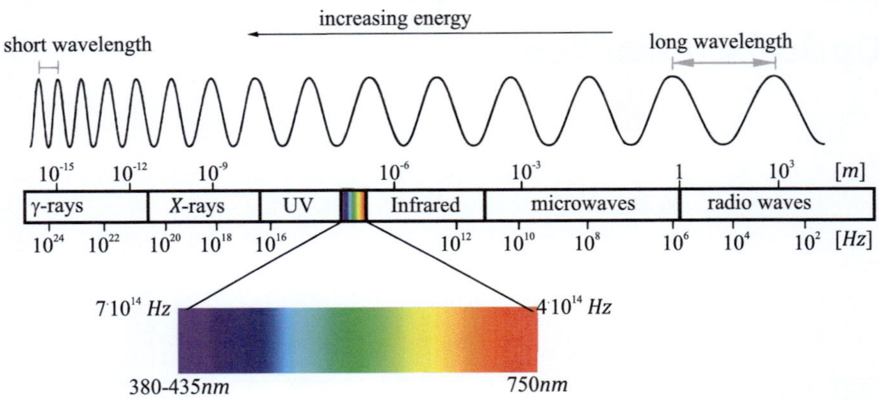

Fig. 3.1 Overview over energies and wavelength spectrum for different electromagnetic waves

the basic physical laws which govern the propagation of electromagnetic waves, the Maxwell equations.

An electromagnetic wave, such as a light wave, is described by the electric field strength vector ($\overrightarrow{E}$), the magnetic induction vector ($\overrightarrow{B}$) and the wave propagation vector ($\overrightarrow{k}$), which are perpendicular to each other, thus:

$$\vec{E}\perp\vec{B},\ \vec{E}\perp\vec{k}\ \text{ and }\ \vec{k}\perp\vec{B}. \tag{3.1}$$

In classical electromagnetic theory the Maxwell equations describe how electric and magnetic fields are related to each other and to electric charges and electric current under given boundary conditions. Together with the Lorentz force, they explain all phenomena of classical electrodynamics. They, therefore, also form the theoretical basis of optics and electrical engineering.

If we only want to discuss the propagation of electromagnetic waves, of light and thermal radiation, in vacuum and in a continuous medium, the equations for the electric and magnetic fields simplify, because we need not to consider source terms. Also, we need only solve the equations for one field as the other one is coupled directly to that. Using the so-called wave equation, we can deal with a wide range of optical phenomena which are important for solar energy technologies; may it be the transmission of solar radiation in transparent glass panes or the scattering of light due to aerosol particles. So essentially, we have to find a solution for the electric field vector ($\overrightarrow{E}$) within a given domain and for pre-scribed boundary conditions for the following equation

$$\frac{d^2}{dt^2}\vec{E} = \frac{1}{\varepsilon_0\mu_0\varepsilon_r\mu_r}\cdot\Delta\vec{E} = \frac{1}{\varepsilon\cdot\mu}\cdot\Delta\vec{E}, \tag{3.2}$$

in which ε is the absolute permittivity of a medium characterizing the polarizability of the medium, μ the absolute magnetic permeability of a medium related to the magnetisation of a medium and ε_0 the vacuum permittivity or electric constant given by $\varepsilon_0 = 8.8541878128 \cdot 10^{12}\ As/(Vm)$ and μ_0 the magnetic permeability of vacuum given by the constant $\mu_0 = 4 \cdot \pi \cdot 10^{-7}\ H/m$.

For vacuum the relative permittivity and permeability is unity (1). The product of permeability and permittivity is related to the Index of refraction n and the absolute speed of light in the medium c resp.in vacuum c_0.

$$c = \frac{1}{\sqrt{\varepsilon.\mu}}, \tag{3.3}$$

$$c_0 = \frac{1}{\sqrt{\varepsilon_0 \mu_0}} = 2.997893 \cdot 10^8 m/s, \tag{3.4}$$

$$n = \frac{c}{c_0}. \tag{3.5}$$

One solution of the wave Eq. (3.2) are plane waves with a linear propagation direction. The wave can be represented with respect to the electric field vector $\overrightarrow{E}$ in the following form (similar for the magnetic field $\overrightarrow{H}$):

$$\overrightarrow{E}(t, z) = \overrightarrow{E}_0 \cdot \cos(\omega \cdot t - k \cdot z), \tag{3.6}$$

where ω is the angular frequency, t is time and $k = 2\pi/\lambda = \omega \cdot \sqrt{\varepsilon \cdot \mu}$ is the absolute value of the wave vector $\overrightarrow{k}$. A projection of the wave in the $x - y$-plane provides a line, as illustrated in Fig. 3.2.

The polarization of light is defined by the direction of the electric field vector. If the electric field vector of a wave always points in the same direction perpendicular to the z-axis (see Fig. 3.2), the wave is called linearly polarized. If a wave propagates along the z-axis $\overrightarrow{k} = k \cdot \overrightarrow{e}_z$ and the contributions $\overrightarrow{E}_{0x}$ and $\overrightarrow{E}_{0y}$

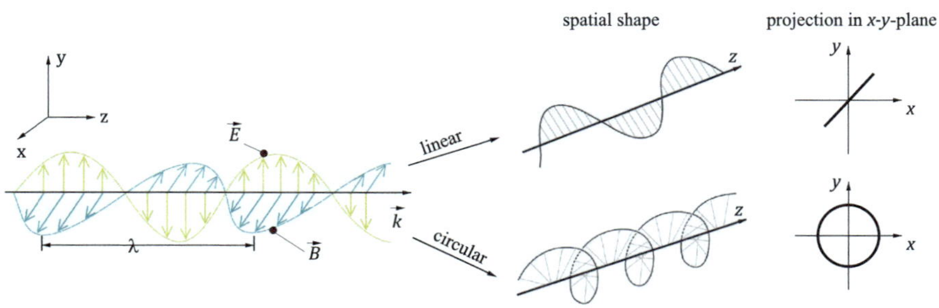

Fig. 3.2 Light as an electromagnetic wave. Polarization as a form of the spatial propagation of the electric field vector and the projection into the $x - y$-plane of a linearly polarized wave (top) and a circularly polarized wave (bottom)

of the wave vector are equal but phase shifted by $\pi/2$, it is called a circularly polarized wave. The electric field vector describes a spiral and it applies:

$$\vec{E}_x(t, z) = \vec{E}_{0x} \cdot \cos(\omega t - kz) \text{ and } \vec{E}_y(t, z) = \vec{E}_{0y} \cdot \cos\left(\omega t - kz - \frac{\pi}{2}\right). \quad (3.7)$$

Both forms of polarization are sketched in the Fig. 3.2.

There is a multitude of polarization types depending on the orientation in time of the resulting amplitude vector (E_x, E_y) perpendicular to the propagation direction.

In the following paragraph we want to deal with the physical reasons for the interaction of electromagnetic waves with matter, which in classical continuum theory results in the functions of permeability and permittivity. As we are dealing in solar mostly with non-magnetic materials, we assume in the further development $\mu_r = 1$.

More general for a continuum the electromagnetic field can be written as a complex function. This is important for a medium where due to energy dispersion the waves can be also attenuated and extinguished (absorption of energy and conversion into thermal energy). We write the solutions of the Maxwell equations in complex form:

$$\vec{E}(t, z) = \vec{E}_0 \cdot \exp\left(i \cdot \omega \cdot \left(t - \frac{n}{c_0} \cdot z\right)\right) \cdot \exp\left(-\omega \cdot \frac{k}{c_0} \cdot z\right). \quad (3.8)$$

Here, the second part describes the attenuation and damping of an electromagnetic wave due to interaction with a dispersive medium. The intensity decreases exponentially and is described by the imaginary part of the complex index of refraction. The attenuation of the electromagnetic wave depends on its complex refractive index, into which the electrical and magnetic permittivities of the media ε_r and μ_r between the interfaces enter in a complex form. The complex refractive index $\tilde{n}$ is:

$$\tilde{n} = n + i \cdot k. \quad (3.9)$$

Please note that the imaginary part of the complex index of refraction is denoted k– as the absolute value of the wave vector characterizing wavelength and direction of a plane wave. It depends on the context which quantity should be referred to. The imaginary part of the index of refraction causes absorption. It leads to an attenuation of electromagnetic radiation in the direction of propagation. The attenuation coefficient α is related to k by

$$\alpha = \frac{4\pi}{\lambda} k \text{ and is given as } \alpha = \frac{1}{I} \frac{d}{dz} I, \quad (3.10)$$

The spatial reduction of the radiant intensity in dependence of the layer thickness z

$$I(z) = I(z = 0)e^{(\alpha \cdot z)} \approx |E|^2, \tag{3.11}$$

is proportional to $|E|^2$ and falls off exponentially with increasing$z$. For transparent dielectric media the penetration depth must be large compared to its thickness$z$, whereas for metals it is exceedingly small. The energy flux of the radiation is given by the Poynting vector, and the time-average is obtained by squaring the amplitudes of the electric field.

$$I = \left|\vec{S}\right| = \left|\vec{E} \times \vec{H}\right| = \frac{n}{\mu \cdot c_0}\left|\vec{E}\right|^2 = \frac{n}{\mu \cdot c_0}\left(E_x^2 + E_y^2 + E_z^2\right). \tag{3.12}$$

For a complex formulation the complex conjugate product is needed to obtain the (real) intensity:

$$I = \frac{\tilde{n}}{\mu \cdot c_0}\vec{E} \cdot \overrightarrow{E^*} = \frac{\tilde{n}}{\mu \cdot c_0}\left(E_x \cdot E_x^* + E_y \cdot E_y^* + E_z \cdot E_z^*\right). \tag{3.13}$$

3.2 Radiation Transport in Transparent Media

3.2.1 Basic Definitions and Nomenclature

When we talk about the processes of transmission, reflection, absorption and scattering in transparent media, we are interested in the impact of components like polymeric films and glass panes on the solar radiation in different wavelength ranges, generally speaking on radiative power or energy. Radiometric terms describe the distribution of the radiation's power in space. Here, several quantities are being used, describing on the one side the radiation, and on the other side the material optical properties. In experimental measurements these properties are determined. Therefore, we use here the nomenclature of radiometry (Table 3.1).

A basic quantity is the radiant power or radiant flux Φ, defined as the amount of radiant energy dQ per unit time dt.

$$\Phi = \frac{dQ}{dt} \quad [\mathrm{W}]. \tag{3.14}$$

The radiant intensity is defined by:

$$I = \frac{d\Phi}{d\Omega} \quad [W/sr]. \tag{3.15}$$

And expresses the radiant power per unit solid angle and does not include any information about the emitting source or receiving element. To describe the radiant

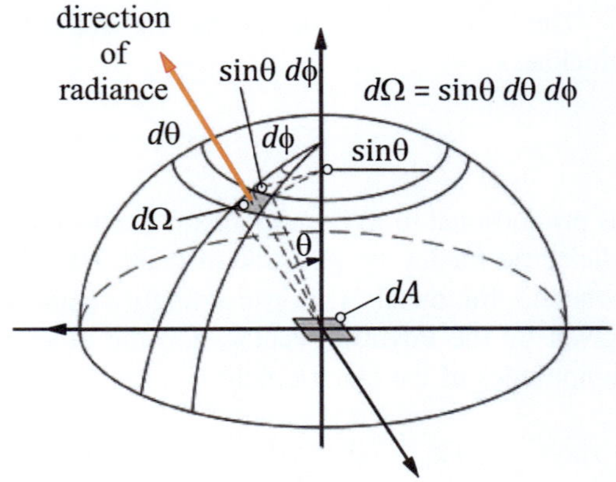

Fig. 3.3 Visualization of radiance L, the radiant flux per unit solid angle $d\Omega$ per projected area dA

flux emitted by a surface A into a specific solid angle Ω, the term radiance is defined, as depicted in Fig. 3.3. The radiance L in $W/(sr \cdot m^2)$ is the radiant power per unit solid angle per unit *projected* source area and thus results in:

$$L = \frac{d^2\Phi}{\cos\theta dAd\Omega}. \tag{3.16}$$

To distinguish between radiance leaving and incident on a surface, reflected radiance is in the following L_r or L_2 and incident radiance L_i or L_1.

The term irradiance E in W/m^2 describes the radiant power incident on a surface. It can be expressed as the integrated radiance over a defined solid angle in space as:

$$E = \frac{d\Phi}{dA} = \int_0^{2\pi} L_i d\Omega. \tag{3.17}$$

The index 'e' denotes the radiometric range. For photometry in the visual wavelength range the index 'v' is used. The optical properties of components now are defined in the Table 3.2. As an example, the mathematical definition of the reflectance is given in the following equation (the definition for transmittance and absorptance is analogous).

Reflectance R describes the ratio of radiant flux Φ_r reflected from a surface dA to the incident radiant flux Φ_i. Often it is additionally relevant to analyse radiant flux Φ_i incident from a specific solid angle Ω_i onto a surface dA and the radiant flux Φ_r reflected into a specific solid angle Ω_r. The reflectance is given by:

$$\rho = \frac{d\,\Phi_r}{d\Phi_i} = \frac{\int_{\Omega r} L_r d\Omega_r}{\int_{\Omega_i} L_i d\Omega_i}. \tag{3.18}$$

The reflection and transmission coefficients describe the ratio of magnitude of the amplitudes of the considered electromagnetic field to the incident field. When

Table 3.1 Definition of radiometric quantities (ISO 80000–7, 2019)

Quantity	Symbol	Unit	Description
Flux	Φ_e	W	Radiative flux from a source (EM radiation)
Irradiance	E_e	W/m^2	Radiative specific flux falling onto a surface (EM radiation)
Radiosity/spec. radiance	M_e	W/m^2	Radiative specific flux emitted by a surface (EM radiation)
Intensity	I_e	W/sr	Radiative flux per unit solid angle (from or into a cone) (EM radiation)
Radiance	L_e	$W/(sr \bullet m^2)$	Radiative specific intensity emitted by a surface into a solid angle (EM radiation)

Table 3.2 Definition of optical properties in radiometry according to (ISO 80000–7, 2019); all dimensionless

Property	Symbol	Description
Reflection coefficient	r	Magnitude of reflected amplitude divided by amplitude of incident radiation
Transmission coefficient	t	Magnitude of transmitted amplitude divided by amplitude of incident radiation
Transmissivity	τ	Transmitted power through a surface or a medium divided by incident power
Reflectivity	ρ	Reflected power from a surface divided by incident power
Absorptivity	α	Absorbed power at a surface divided by incident power
Emissivity	(ε)	Emitted power from a surface divided by maximum blackbody emitted power
Transmittance	τ	Transmitted power through a component divided by incident power
Reflectance	ρ	Reflected power from a component divided by incident power
Absorptance	α	Absorbed power by a component divided by incident power
Emittance	ε	Emitted power from a component divided by maximum blackbody emitted power

we speak about "-ivity" in the context of optical properties, we speak about intrinsic material or surface properties. The terms ending on "-ance", however, describe the properties of complete components, for instance a pane of glass where the reflectivity on the front and the back surface adds to the reflectance, and where the thickness of the materials has an impact on that as well. Also, inhomogeneous materials are described by transmittance and reflectance terms. Sometimes the difference is minute and mainly shows the emphasis we want to give, for example when we talk about emittance or emissivity of a solar absorber. Strictly speaking the emissivity describes the emissive power of a material, and the emittance the emissive power of the absorber material on, say the copper substrate.

We will see in many literature texts, that not always a clear distinction has been kept. This makes it sometimes complex to compare results from different authors (coming maybe from different academic background) when they write about the same things. Symbols moreover have always been interpreted in the context they are used. A symbol a for example might denote a multitude of physical quantities, from solar absorptivity to thermal diffusivity to heat transfer coefficient. In the optical realm, moreover, the conditions for in- and outgoing radiation must be understood as well. Is the incoming radiative flux directional or diffuse, is the outgoing flux considered integral or is the angular distribution of radiation considered? Is the flux measured spectrally or for a certain wavelength range? Very often indexing may give some indication what is meant, but although quantities may be mathematically better defined using multiple indexing, the comprehensibility of text usually decreases. In doubt the context is important. Therefore, we hope that the reader will not be completely confused, when sometimes we do not stick slavishly to a certain connotation.

3.2.2 Origin of Absorption and Reflection in Transparent Media

Solid matter can absorb electromagnetic radiation. The degree of absorption depends on its electronic band structure (see for a thorough treatment e.g. (Ashcroft, Mermin, Wei, and Mermin, 2016). Electromagnetic radiation (or photons, if you want to use the particle representation of light!) provide the energy to excite quantum states of the solids with higher energy. With solar and thermal radiation, the excitation of electrons and phonons is possible, whereas for the nuclei the energy is not sufficient. In solids, the outer electrons are no longer firmly bound to the atoms of the solid but can move freely in electron bands. Two bands are important for absorption:

- The valence band and,
- The conduction band.

The classification of substances into band models is described herein with the help of quantum mechanics. At temperature zero, the valence band is the highest fully occupied energy band. Bands which are energetically above, unoccupied, or partially occupied are called conduction bands. In semiconductors and insulators (dielectrics), there is a forbidden area between the two bands called the band gap. The electrical and thermal conductivity of the solid depends on the one hand on whether the conduction band at temperature zero is occupied, partially occupied or unoccupied, and on the other hand on the temperature of the crystal. A fully occupied band, just like an unoccupied band, does not contribute to charge transport. The reason for this is that electrons in fully occupied bands cannot absorb energy, for example, through an electric field, because for them there are no unoccupied

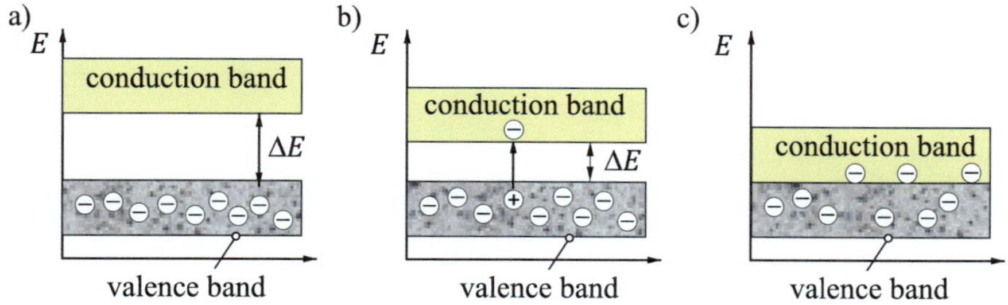

Fig. 3.4 Energy bands in an electrical non-conductor (**a**), a semiconductor (**b**) and a metal (**c**)

states with suitably higher energy. Only a partially occupied band enables a net current induced by the electric field.

Semiconductors and insulators have a band gap ΔE, semiconductors in the range of $1eV$ and for insulators well above $3eV$. Since the valence bands are fully occupied at $0K$, no electrical charge transport is possible. If sufficient energy is supplied to a semiconductor by increasing the temperature or by irradiating with light, electrons can overcome the band gap and be lifted into the conduction band. In this way, the unoccupied band can be partially occupied. These electrons and the holes remaining in the valence band both contribute to the conductivity of the electric current.

According to this idea, insulators like glasses have an unoccupied conduction band and a very large band gap ($\Delta E > 3eV$, see Fig. 3.4a). At room temperature and even at much higher temperatures, only a few electrons can be transferred from the valence band to the conduction band by thermal excitation. As a result, they have a high electrical resistance. However, the conduction band can be partially filled by electrons from the valence band overcoming the energy gap between the valence and conduction band by absorbing light. This requires light with an energy that exceeds a certain lower limit, which is given by the size of the energy gap.

Like for insulators, the conditions for a semiconductor are similar, but here the band gap is relatively small, typically ($1eV < \Delta E < 2eV$ corresponding to $0.6 - 1.2\ \mu m$ wavelength, see Fig. 3.4b), this can be overcome more easily by thermal or optical energy supply. An electron can be lifted into the conduction band and is freely movable here. At the same time, it leaves a gap in the valence band which can be filled by neighbouring electrons. Semiconductors, therefore, have pronounced absorption bands with a sharp edge at a maximum wavelength. The Fig. 3.5 illustrates this dependence of the absorption coefficient of various semiconductors on the wavelength in the visible and near infrared range. The figure exhibits a sharp cut-off edge of the absorption coefficient of all semiconductors when a certain wavelength is exceeded.

However, the following properties might make actual absorption impossible: If the semiconductor has an ideal interface with air, due to the large complex part of the index of refraction, its reflectivity is particularly high at wavelengths for which it has a large absorption coefficient (extinction coefficient). To avoid reflection, the

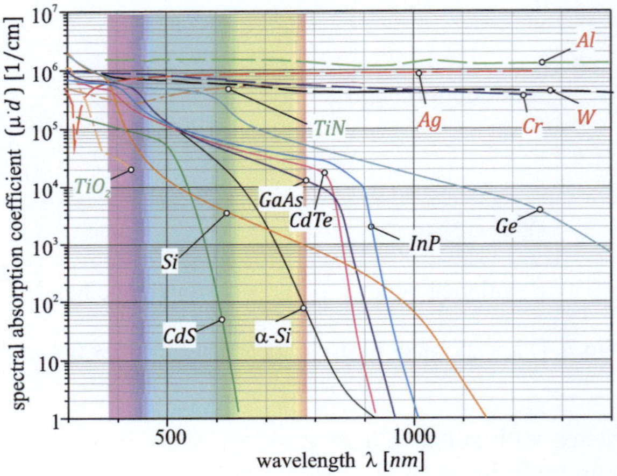

Fig. 3.5 Spectral absorption coefficients of various semiconductors in the visible light range and in the near infrared as a function of wavelength λ (McNaught and Wilkinson, 1997)

refractive index must not change abruptly at the interface. One possibility is to use a transparent coating with intermediate index of refraction, or to emboss a surface texture on the surface with structures in the nanometre range. Then, the structure cannot be resolved by light with the wavelength of visible light and the effective index of refraction is changing continuous from air to solid material. Thus, the surface of a photocell requires special treatment. If the surface looks red when viewed, then this treatment is missing or insufficient. This is because the cell reflects exactly the light that should actually be absorbed.

In a metal one usually does not speak of a conduction or valence band. For monovalent metals, the highest occupied energy band is half filled, as shown in Fig. 3.4c. In polyvalent metals, the energy bands partially overlap. Electrons can therefore change to a higher energy state when small electric field strengths are applied and can thus move more or less freely and contribute to the flow of electric current. Metals are, therefore, good electrical conductors. An increase in temperature leads to a decrease in conductivity because the increased scattering of the electrons causes a lower average velocity (Wiedemann–Franz law). Electrons can also be raised to higher energy states in the conduction band by absorbing light. Metals are therefore in principle ideal light absorbers, as indicated in Fig. 3.5 exhibiting no cut-off wavelength at all. However, similar to semiconductors, the optical constants again prohibit that for an ideal smooth surface, because light is almost completely reflected ($\rho \rightarrow 1$) exactly when the absorption coefficient is large. Metallic conductors with ideal polished surfaces therefore do barely absorb light. This contradiction is resolved by the fact that the permittivity leading to high absorption inside the body also lead to high reflectivity of the interface air-metal which prohibits an intrusion of light into the volume.

3.2.3 Polarization, Refraction, and Reflection

When light, that is electromagnetic radiation, crosses the border between one medium and another, then the boundary conditions for the EM-radiation according to the Maxwell equations determine the interaction with the interface. If the incident wave encounters the interface, it is split into two new waves, as depicted in Fig. 3.6. One is refracted while proceeding into the second medium and the other is reflected, see (Born and Wolf, 1999).

In the continuum theory the media are characterized by the permissivity and permeability. Valid solutions for the dominating wave equation must fulfil boundary conditions at the interface. And derived from this mathematical theory the laws of refraction, of reflection and transmission at the boundary result. The change in optical constants ε and μ of the medium (or$\tilde{n}$) determines the relation between frequency (the colour of light) and its propagation (the velocity of light). Whereas the frequency does not change, the velocity of light changes and hence the energy density. This is the reason for the change of direction when a wave crosses the boundary between two media. The resulting law in optics then is the law of refraction.

Although the following topics can be reformulated also in complex numbers, we will stick to the formulation in real numbers. The complex number formulation is useful for metals and the laws of Snellius and Fresnel have formally the same shape as the simpler real formulation for dielectrics, however the interpretation of a complex incidence angle is less straight-forward.

Snellius' law of refraction is used to describe the transition of an electromagnetic wave at an interface where both media have a different wave velocity. The changed phase velocity causes a deflection of the wave, i.e. a change in direction.

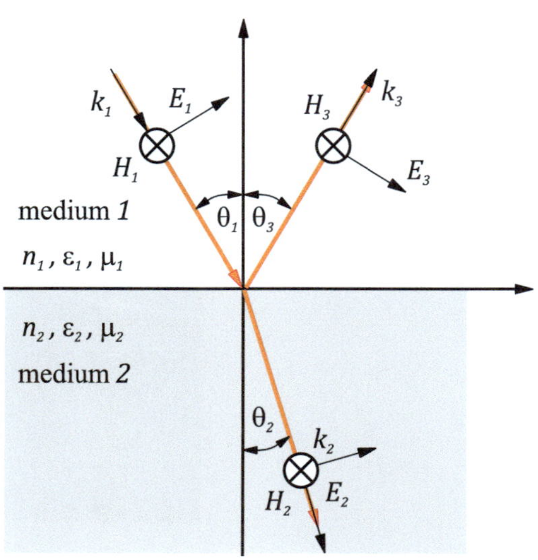

Fig. 3.6 Reflection and refraction of a plane wave with p polarization

In the field of optics, when passing from one transparent medium to another, this means that the light beam is refracted.

If you look at a light beam that is incident from medium 1 into medium 2 at an angle of θ_1, the non-reflected portion continues to travel at an angle of θ_2 in medium 2 relative to the perpendicular. Light travels in any isotropic medium at the velocity of propagation dependent on the refractive index n_i, which indicates the ratio of the velocity of light in a vacuum c_0 to that of the respective medium (c_i).

$$n_1 = \frac{c_0}{c_1} \text{ and } n_2 = \frac{c_0}{c_2}. \tag{3.19}$$

Here, the refractive index is a material constant dependent on the wavelength, which causes a splitting of the different wavelengths of the incident beam during refraction. In a dispersion prism, this wavelength dependence allows an incident light beam to be deflected from its original direction of propagation by refraction at two non- parallel surfaces. Since the strength of the deflection depends on the wavelength, it is possible to spectrally decompose polychromatic light, as shown in the Fig. 3.7.

The Table 3.3 shows typical refractive indices of some liquids and glasses as averaged over the solar spectrum. An example of the wavelength dependence, also called dispersion, is shown in Fig. 3.8. This functional form is dependent on the proximity of a resonance frequency in the medium which determines the shape of the complex permittivity function.

If two parallel incident light beams meet an ideal interface between two media of different optical density, an additional distance $\lambda_1 = (c_1 \cdot t)$ in medium 1 (Fig. 3.9) is geometrically obtained for the second beam. The first light beam must travel an additional distance $\lambda_2 = (c_2 \cdot t)$ in medium 2. At the interface, one finally obtains, via geometrical relations and using the Eq. (3.19) Snellius' law of refraction

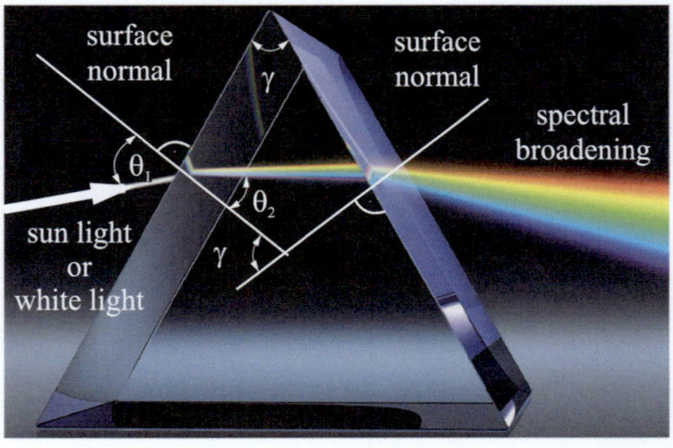

Fig. 3.7 Spectral decomposition of polychromatic light in a dispersion prism

Table 3.3 Refractive indices (n) of some selected transparent liquids and solids

Material	Refractive index n [/]	Material	Refractive index n [/]
Dry air (1 bar)	1.000278	Acrylic glass (PMMA)	1.49
Ice (−8 °C)	1.309	Optical lenses	1.51 – 1.58
Water (0 °C)	1.3346	Quartz glass	1.544 – 1.553
Ethanol	1.3614	Sapphire/Ruby	1.757 – 1.779
Ethyl glycol	1.4318	Lead glass	1.805
Carbon tetrachloride	1.46	Cubic zircon	2.17 – 2.21
Glycerine	1.47	Zinc sulphide	2.37
Benzene	1.49	Diamond	2.419
Epoxy resin	1.55 – 1.63	Silicon carbide	2.65 – 2.69

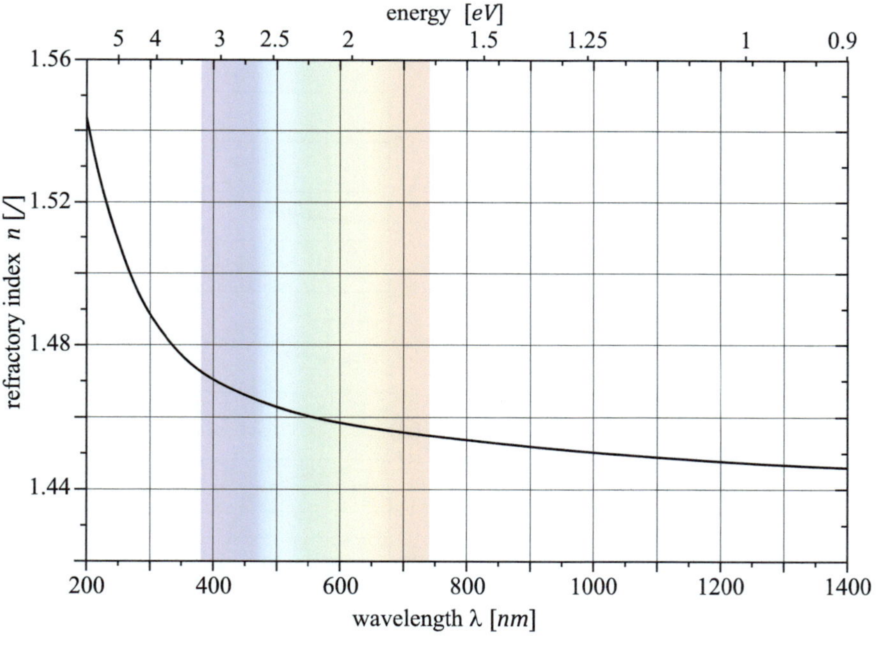

Fig. 3.8 Example of wavelength dependent dispersion in SiO_2

$$\frac{\sin\theta_1}{\sin\theta_2} = \frac{n_2}{n_1} = \frac{c_1 \cdot t}{c_2 \cdot t} = \frac{\lambda_1}{\lambda_2}. \tag{3.20}$$

For absorbing media in which the refractive index n is defined as a complex number, the law of refraction can also be applied in a similar modified form, however

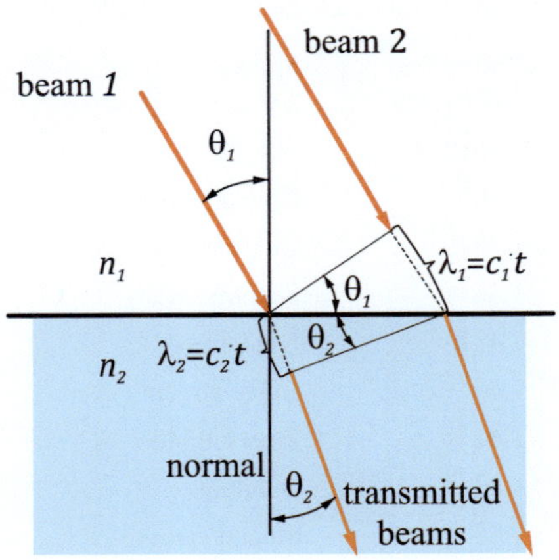

Fig. 3.9 Sketch for the derivation of Snellius' law of refraction

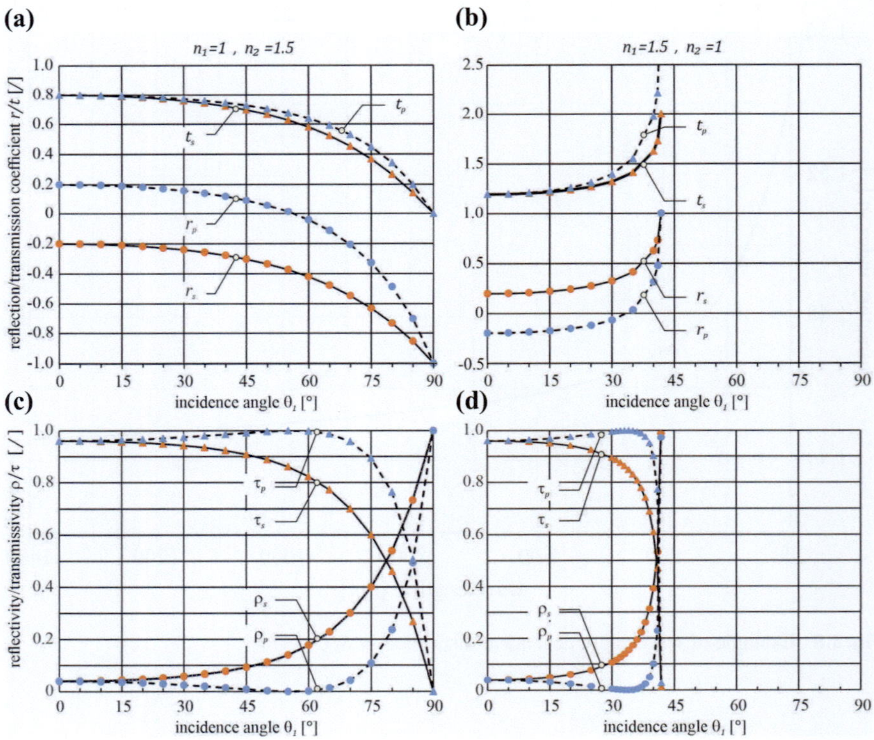

Fig. 3.10 Calculated amplitude ratios r_i, t_i (top) and reflectivity/transmittanceρ_i, τ_i (bottom) for the interface air ($n = 1$) and glass ($n = 1.5$). Light incident on interface with angle θ_1 from the air side (**a**, **c**) and the glass side (**b**, **d**)

the interpretation of the complex quantities is not straight-forward. It describes the relation between propagation direction, propagation velocity and attenuation. It should be remembered that the even the resulting "angle of refraction" is a complex quantity.

When the wave passes from a medium with smaller optical density into a medium with larger density ($n_2 > n_1$) then the refraction is towards the surface normal $\theta_2 < \theta_1$. Contrary, when the wave passes from the dense medium to the thinner medium, the refraction is away from the surface normal. The optical laws are valid in forward and backward direction equally. An example is the situation when a diver below water is looking to the water surface into the environment above. One might see the full landscape within a smaller cone of view. But there is obviously a limiting angle where the refracted beam already is travelling parallel to the surface, the water surface in air in our case. A larger incidence angle on the interface then leads to a transmitted wave with propagation direction into the source medium, i.e. the beam is reflected. The solution of the wave equation for these boundary conditions leads to the case of total reflection of an electromagnetic wave, where the wave is completely reflected back into the output medium. The angle where total reflection starts is referred to as the critical angle θ_c, for which

$$\theta_c = \arcsin\left(\frac{n_2}{n_1}\right). \tag{3.21}$$

If the critical angle ($\theta > \theta_c$) is exceeded, a total reflection is obtained in which $\theta_1 = \theta_2$ Snellius' law merely states in which direction a wave (or beam of light) is deflected, but not how much of the wave is transmitted or reflected at the transition between the two media. The total reflection can be easily observed by holding a mirror under a water surface. If the mirror is turned towards the water surface, total reflection sets in from a certain angle, the critical angle ($\theta = \theta_c$).

When we deal with the question which part of the intensity of light is transmitted or reflected, we need information also for the amplitudes of the electromagnetic waves, not only the directions. This is determined with the help of Fresnel's formulas.

Since an electromagnetic wave field vector can be represented as a linear combination of component perpendicular and a component parallel to the incident plane. Treating the boundary conditions at the interface, the individual components of the amplitudes can thus be determined. For this purpose, amplitude transmission coefficients t_s, t_p and amplitude reflection coefficients r_s, r are defined in perpendicular and parallel directions, which indicate the ratio of the amplitude of the outgoing or reflected electric field to the amplitude of the incoming component of the incident electric field $(E_{0e})_s$ or $(E_{0e})_p$. These coefficients thus describe the amplitude damping in the respective direction, i.e. in perpendicular or parallel direction.

A well-known case of the Fresnel equations is for dielectric materials in which the complex refractive index k in Eq. (3.9) is negligible and close to zero. However, the Fresnel equations can be solved in a universal form for all materials, i.e. for glasses, metals, etc., in which then the amplitude ratios t_s, t_p or r_s, r_p are given as a function of the complex refractive index $\tilde{n}$, the relative angles (θ_1, θ_2) and the magnetic permeability (μ_{r1}/μ_{r2}) between the media, compare for example (Nolting, 2009) or (Demtröder, 2004).

A dielectric material may absorb to some extent light, in fact otherwise one would not have coloured dielectrics. However, the imaginary part of the index of refraction $k \ll 1$ is so small that a simplified extinction coefficient can be derived (Beer's law). For this case of dielectrics, for example the pairing air-glass, the Fresnel formulas can be simplified considerably. The relations between electromagnetic radiation on either side of the interface can be reduced as the Fresnel laws for real numbers use only the real part of the refractive index n_i of the respective medium. First of all, the component which is polarized linearly perpendicular (index:s) to the plane of incidence shall be considered. It is also often referred to in the literature as the transversal electrical (TE) component. The following transmission amplitude coefficients t_s and amplitude reflection coefficients r_s result for the vertical polarization:

$$t_s = \left(\frac{E_{0t}}{E_{0e}}\right)_s = \frac{2n_1 \cdot \cos\theta_1}{n_1\cos\theta_1 + n_2\cos\theta_2} = \frac{2\sin\theta_2\cos\theta_1}{\sin(\theta_1+\theta_2)},$$
$$r_s = \left(\frac{E_{0r}}{E_{0e}}\right)_s = \frac{n_1 \cdot \cos\theta_1 - n_2\cos\theta_2}{n_1\cos\theta_1 + n_2\cos\theta_2} = -\frac{\sin(\theta_1-\theta_2)}{\sin(\theta_1+\theta_2)}. \tag{3.22}$$

Correspondingly, for parallel polarization (TM) the correlations are:

$$t_p = \left(\frac{E_{0t}}{E_{0e}}\right)_p = \frac{2n_1 \cdot \cos\theta_1}{n_2\cos\theta_1 + n_1\cos\theta_2} = \frac{2\sin\theta_2\cos\theta_1}{\sin(\theta_1+\theta_2)\cos(\theta_1-\theta_2)},$$
$$r_p = \left(\frac{E_{0r}}{E_{0e}}\right)_p = \frac{n_2 \cdot \cos\theta_1 - n_1\cos\theta_2}{n_2\cos\theta_1 + n_1\cos\theta_2} = -\frac{\tan(\theta_1-\theta_2)}{\tan(\theta_1+\theta_2)}. \tag{3.23}$$

The equations after the second equals sign are based on Snellius' law of refraction and the addition theorem. The calculated amplitude coefficients r_i, t_i for the substance pairing light from air in glass (a) and light from glass in air (b) are shown in Fig. 3.10a, b as a function of the angle of incidence.

Looking at the vertical incidence of light on the interface between media 1 and 2, the angle of incidence is $\theta_1 = \theta_2$, and the amplitude reflection coefficient r_s or r_p is then:

$$r_s = \left(\frac{E_{0r}}{E_{0e}}\right)_s = \frac{n_1 - n_2}{n_1 + n_2} = -r_p = \left(\frac{E_{0r}}{E_{0e}}\right)_p. \tag{3.24}$$

Using the fact that the intensity I is proportional to the square of the amplitude of the electric field E, one can calculate the intensity of the portion reflected at the interface. In the vertical transition of visible light from air ($n_1 = 1$) to quartz glass ($n_1 = 1.46$), 3.5% of the incident intensity is reflected regardless of the polarization.

Figure 3.10a and b show that there are amplitude coefficients for both transmission and reflection which have a negative sign. In this case a phase shift of the wave by 180° (180° $= \pi$) occurs physically. The amplitude reflection coefficient r_p has a zero crossing, therefore, the light polarized parallel to the plane of incidence is completely refracted and only the part polarized perpendicular to the plane of incidence is reflected. An analysis of Eq. (3.22) shows that the following must apply to this case:

$$r_p = -\frac{\tan(\theta_1 - \theta_2)}{\tan(\theta_1 + \theta_2)} = 0 \rightarrow \theta_1 + \theta_2 = 90°. \tag{3.25}$$

This means that the beam refracted at the angle (θ_2) is perpendicular to the reflected beam.

If polarized light were also reflected parallel to the plane of incidence, this would result in a longitudinally oscillating light wave. Due to the properties of the electromagnetic field, such a field cannot exist, since, for example, no radiation can occur in the direction of the dipole axis. Physically, the incoming light beam causes an excitation of the electrons of the material, which leads to an oscillation called displacement polarization. This results in a collection of atomic dipoles that oscillate in the direction of polarization and thus emit secondary waves. Inserting the relation (3.25) into Snellius' law of refraction (Eq. (3.20)) gives the following angular relation for the angle referred to as Brewster angle θ_B:

$$\theta_B = \arctan\left(\frac{n_2}{n_1}\right). \tag{3.26}$$

This means that light polarized parallel to the plane of incidence and incident at Brewster angle θ_B is not reflected (see also Fig. 3.11).

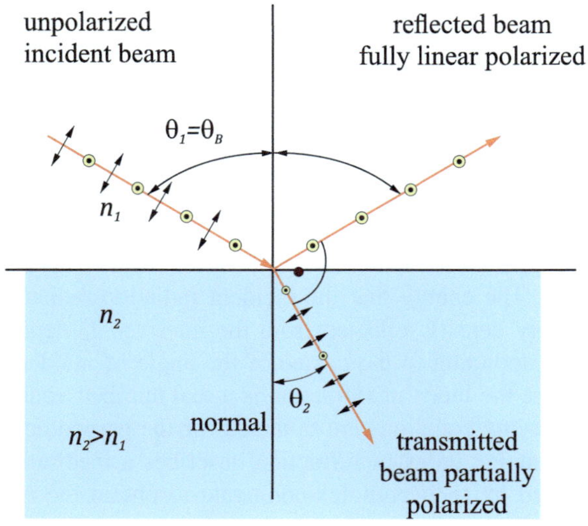

Fig. 3.11 Schematic sketch to explain the Brewster angle θ_B

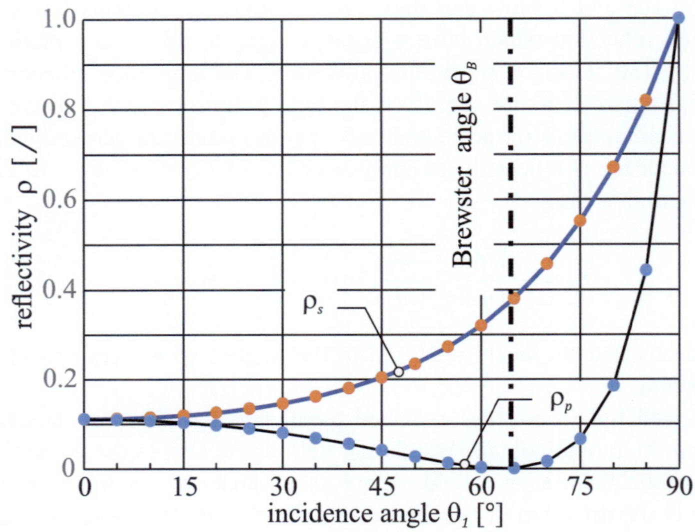

Fig. 3.12 Position of the Brewster angle at the transition from an optically thinner ($n_1 = 1$) to a denser medium ($n_2 = 2$) and calculated reflection amplitude as a function of the angle of incidence θ_1

This parallel polarized light penetrates the surface completely. Light with perpendicular polarization, on the other hand, is also partially reflected, as can be seen in the Fig. 3.12. Therefore, the reflected light is completely linearly polarized. One may observe polarisation of natural unpolarized light when looking at obliquely reflected light approximately incident under the Brewster angle e.g. from a low sun reflected by a water surface of a smooth lake, or by window shields from a car, when one uses sunglasses with polarizers. Rotating the sunglasses by 90° changes the transmitted intensity substantially, because these glasses are selectively transmitting only one linear polarization.

If light enters from an optically denser medium into an optically thinner one, i.e. $n_1 > n_2$, the amplitude ratios become complex from a certain angle onwards. From this critical limit angle θ_c, (cf. Eq. (3.21), the total reflection occurs, cf. Figure 3.13 for the pairing glass ($n_1 = 1.5\ 5$)$\rightarrow$air ($n_2 = 1$) the Brewster angle is $\theta_B = 33.7°$ and the angle for total reflection is $\theta_c = 41.8°$.

So far, only amplitude coefficients have been considered. Much more important in application is the change in relative intensity as it passes through the interface. The energy that the incident radiation radiates onto an interface I_e and the energy density reflected from the interface I_r depends on the angle of incidence corresponding to the cosine of the angle of incidence. For the case of dielectrics where the index of refraction is a real number leading to real field amplitudes, the following equations are obtained for the perpendicular or parallel components (if a complex index of refraction describes a medium, the amplitudes must be multiplied by their complex-conjugate to obtain the real reflection and transmission

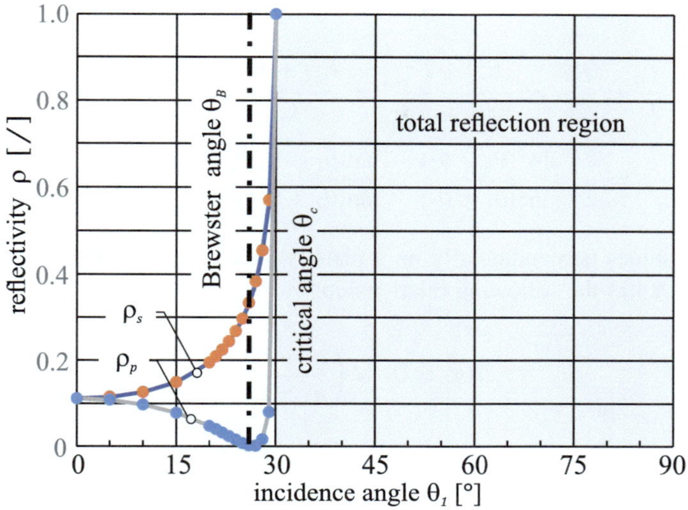

Fig. 3.13 Reflection amplitudes in both directions at the transition of light from an optically denser medium ($n_1 = 2$) to a thinner one ($n_2 = 1$), as well as the position of the Brewster angle (θ_B) and the critical angle (θ_c)

coefficients):

$$R_{s/p} = \left(\frac{E_{0r}}{E_{oe}}\right)^2_{s/p} = r^2_{s/p} \text{ and } T_{s/p} = \frac{n_2 \cdot \cos\theta_2}{n_1 \cdot \cos\theta_1} \cdot \left(\frac{E_{0t}}{E_{oe}}\right)^2_{s/p} = \frac{\tan\theta_1}{\tan\theta_2} \cdot t^2_{s/p}, \tag{3.27}$$

wherein i represents the s- or p-polarized component, respectively. Without absorption, the following energy current balance applies for the individual polarization directions:

$$T_s + R_s = 1 \text{ and } T_p + R_p = 1. \tag{3.28}$$

We denote here the reflectivity and transmissivity with capital R and T to avoid confusion when we talk about reflectance and transmittance of components like glass sheets or polymer films having two interfaces with surrounding air. However, strictly speaking also in this case the symbols ρ and τ could be used if the context is clear.

According to Fresnel's equations, unpolarized light on a smooth surface is partially reflected when passing from a material with refractive index n_1 to a material with n_2. If one assumes this unpolarized light, the individual components must be determined separately according to the proportions parallel (p) and perpendicular (s) to the surface. If one additionally postulates that the size of the perpendicular and parallel light portion of the incident beam was identical, i.e. $(E_{0e})_s = (E_{0e})_p$, the following relationship results for the total reflection coefficient R:

$$R = \frac{I_r}{I_e} = \frac{E_r^2}{E_{0e}^2} = \frac{E_{rs}^2 + E_{rp}^2}{E_{0es}^2 + E_{0ep}^2} = \frac{E_{rs}^2 + E_{rp}^2}{2E_{0es}^2} = \frac{1}{2}\left(R_s^2 + R_p^2\right) \\ = \frac{1}{2}\left(\frac{\sin^2(\theta_1 - \theta_2)}{\sin^2(\theta_1 + \theta_2)} + \frac{\tan(\theta_1 - \theta_2)}{\tan(\theta_1 + \theta_2)}\right) . \tag{3.29}$$

If light shines perpendicularly on a plate, which is $\theta_1 = 0$, then the reflection coefficient R has the following relationship:

$$R(\theta_1 = 0) = \left(\frac{n_1 - n_2}{n_1 + n_2}\right)^2 . \tag{3.30}$$

If, for example, visible unpolarized light falls perpendicularly on the interface between air ($n_1 = 1$) and quartz glass ($n_1 = 1.46$), then according to the Eq. (3.30) the portion of 3.5% of the incident intensity is reflected independently of the polarization.

3.2.4 Scattering and Absorption in Media

As a rule, when light passes through a body, partial absorption occurs in the transparent material itself, reducing the intensity of the light beam. This attenuation (or extinction) of the radiation intensity is a function of the path length of the beam traveling through the material. The intensity I in the transparent body when the beam passes through it can be described as follows:

$$\frac{dI}{dx} = -\mu \cdot I(x), \tag{3.31}$$

in which the extinction coefficient μ depends on the material. The extinction may be caused by a mixture of absorption processes, where radiation energy is converted into heat—or in the molecular picture into kinetic energy of the molecules, and scattering, where the direction of the radiation is changed. Scattering may occur by interaction of light with phonons (vibrational waves of the solid) or with small particles embedded in the medium. These particles may be much smaller than the wavelength of light and hence may not be visible. Depending on their chemical composition the particles can be also absorbing and scattering. In such a case the attenuation depends essentially on the concentration c_k of the absorbing substance or particles in the transparent medium and the molar wavelength-dependent extinction coefficient ε_λ of all elements contained in the transparent body. Thus, the extinction coefficient is a function of both quantities, and the relation is $\mu = f(c_k, \varepsilon_\lambda)$.

This extinction coefficient (which may be spectral or broadband, depending on the accuracy needed in the calculations) in general is a sum of scattering coefficient σ and absorption coefficient κ:

$$\mu = \kappa + \sigma. \tag{3.32}$$

In contrast to colour dispersion in refraction, the dependence of then extinction on wavelength cannot be described by a simple function. The dependence on wavelength is complex and determined by the transparent medium and it complex index of refraction, and by molecular properties of the dissolved substances or embedded particles. There are theories which try to derive the complex index of refraction and the extinction properties for mixtures of substances (e.g. for metallic or oxidic particles embedded in a dielectric matrix), the so-called effective medium theories (Bruggeman, Maxwell-Garnet, etc., see e.g. (Wang and Pan, 2008)). It is not only important which particles sizes and chemistry we have for a specific case, but also which grain structure and topology is found for the mixture. Depending on that, different theories have to be applied to give good estimates.

Generally, coloured substances show one or several defined absorption maxima on the scale of wavelengths. Other dyes show a continuously increasing or decreasing absorbance with the wavelength in the visible range. These properties are based on the molecular structure (especially the electron configurations in the molecules). Therefore, the spectral reflectance or absorptance in the visible and infrared spectral range can be used for chemical analysis, where many substances can be directly identified and quantitatively measured due to their specific spectral extinction patterns, may it be iron oxide, metallic particles, organic chlorophyll, or haemoglobin.

Even for a dielectric medium with a very small imaginary part k of the index of refraction for relatively thick transparent media the spectral absorption is non-negligible. For example, if we assume $k = 10^{-6}$ and want to calculate the reduction of the intensity for a wavelength $\lambda = 1\ \mu m$ and a medium thickness of $d = 4\ mm$ the spectral attenuation according to electromagnetic wave theory is:

$$I(\lambda, d) = I_0 \cdot \exp\left(-\frac{4\pi}{\lambda} \cdot k \cdot d\right). \tag{3.33}$$

Evaluating the exponential yields 95.1 or 4.9% absorption with the thickness of 4 mm. For real dielectric materials it is easier to provide a solar absorption coefficient $\kappa(\lambda)$ which includes already the full dependence on wavelength. Using the broadband value is critical. Especially for polymers with a very pronounced spectral absorption bands the broad band approximation fails when treating different film thicknesses and spectral calculation is preferable.

When we consider the absorption in a dielectric layer, where the intensity travels with a propagation direction $\theta\prime$ across a layer with thickness d, we must take into account also the longer path length. Using the extinction law of Lambert (3.33) we define an internal attenuation factor or "internal transmissivity" x_i:

$$x_i = \exp\left(\frac{-\mu \cdot d}{\cos\theta\prime}\right). \tag{3.34}$$

The transmission behaviour of conventional glasses, which consist of a multitude of individual chemical elements, is much more inconsistent. The spectral dependence of transmission is essentially determined by the composition of the glass. For solar covers, a high transmission in the visible spectrum up to the near infrared (approx. $2-3\ \mu m$) is required in particular, since the intensity of the solar radiation must be utilized in this wavelength range. In the solar range, it is mainly the iron oxide content of the glass that influences the energy transmission most. The ultraviolet part of the spectrum is nearly blocked completely but carries only a very small fraction of the solar energy. Figure 3.14 shows the spectral transmittance of glasses with different iron oxide contents. As the iron content decreases, the transparency of glass increases, especially in the near infrared part of the solar radiation spectrum. To thermal radiation, which occupies the wavelength range of $3\ \mu m$ and above, glass is practically opaque. It is therefore not possible to create a thermal image with a conventional infrared camera through a window. Figure 3.14 shows that for optimum glass coverage, glass with as little iron oxide as possible should be used. However, with iron oxide contents of less than 0.1%, the production costs of the glasses increase considerably. Table 3.4 provides for some glass types the absorptivity and transmissivity in the visible spectrum ($\lambda < 3\ \mu m$) of a 4 mm thick glass plate.

An alternative to a glass cover may be a polymeric film. Figure 3.15 shows the spectral transmission behaviour of a polyvinyl fluoride Tedlar film. Like most plastic films, this film shows considerable transparency in the long-wave infrared range. This transparency leads to large radiant heat losses, especially at high absorber temperatures. As a transparent cover for a collector, it is therefore

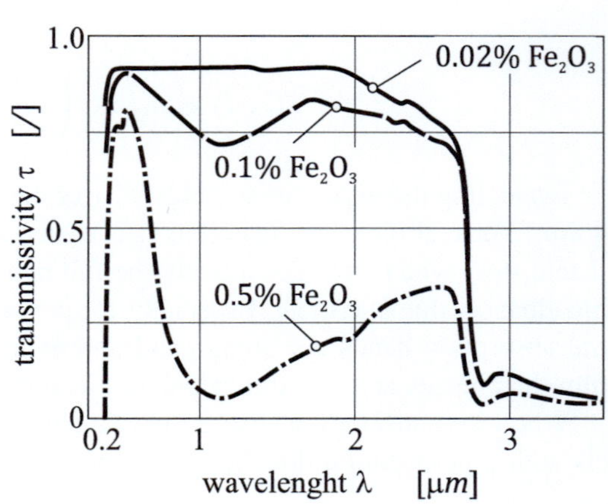

Fig. 3.14 Spectral transmittance of glasses of different iron oxide content from (Duffie and Beckmann, 2020)

Table 3.4 Absorption and transmission coefficients in the visible spectrum ($\lambda < 3\mu m$) for some transparent materials

Material	$\kappa[1/m]$	x_i $(d = 4mm)$	$\tau_{vis}[/]$
White solar glass	4	0.984	0.94
Float glass	16	0.938	0.861
Greenish glass	32	0.88	0.808
Green glass	80	0.726	0.666

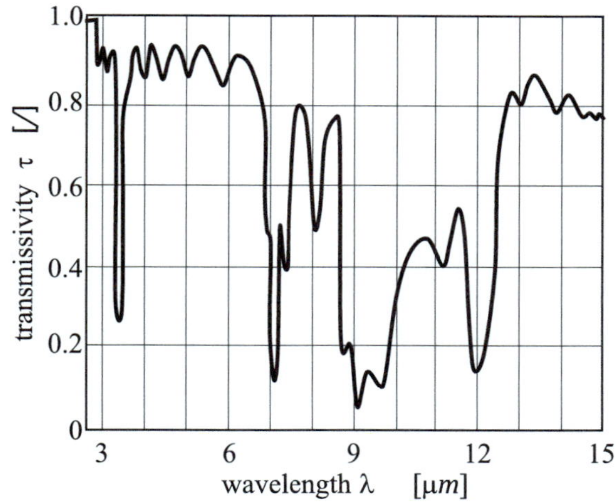

Fig. 3.15 Spectral transmission coefficient of a polyvinyl fluoride Tedlar film modified from (Duffie and Beckmann, 2020)

only slightly suitable. Films often show this only slightly wavelength-selective behaviour.

Colours and spectral absorptance patterns play a decisive role in the natural environment. Reflections from natural environment have an impact on the efficiency of some solar energy applications which also accept these radiative reflections. For example, the albedo, the reflection form of the ground has a smaller effect on the output of a tilted solar thermal collector placed in this environment. The absorption of solar in water, in swimming pools or lakes, has an impact on the temperature distribution in the water.

Let us take the first example. Especially at low sun elevations ($\alpha_s < 40°$) the diffuse scattering and thus the environment plays a larger role in the generation of solar energy. The efficiency of a solar thermal system can be significantly improved by including the immediate vicinity of the collector in an efficiency calculation for the design and alignment of the collector. Adding white or light grey gravel on the ground increases the irradiation on a collector row. In cold climates such effects may be noticed in winter when snow is covering the ground. For some solar systems a low-cost technical installation increasing ground albedo might even improve economics. Figure 3.16 depicts the spectral reflectance of several ground types.

In the case of water, the following can be said. The reflection coefficient of water decreases monotonically with increasing wavelength. Clear and deep waters

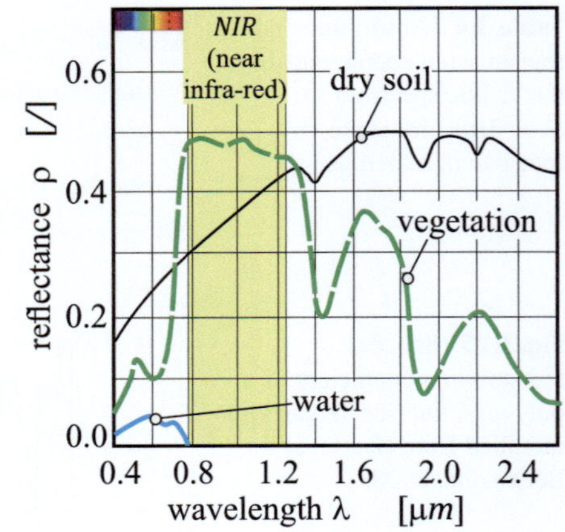

Fig. 3.16 Spectral reflectance of different soil formations according to (Thiemann, 2000)

reflect only visible light; in the infrared range their reflectance is practically zero. With increasing turbidity and decreasing depth, however, the degree of reflection rises, so that turbid and shallow waters in the near infrared range show quite measurable reflection values, as seen in Fig. 3.17.

Spectral reflection is determined in detail by the presence and concentration of dissolved and suspended organic and inorganic constituents. Suspended solids act as a scatterer, dissolved substances absorb selectively. The chlorophyll content is an important indicator of the status of water bodies. The Fig. 3.18 shows the spectral reflectance at different chlorophyll contents of the water. In waters low in chlorophyll, the reflection is large at small wavelengths and decreases monotonically with increasing wavelength. A high chlorophyll content, on the other hand, shows a homogeneous almost constant reflection up to a wavelength of $\lambda \approx 550\ nm$ (green) and then drops sharply.

Figure 3.19 shows a distorted water transmission photograph up to a depth of about 5 m with optical selective absorption. In water, the longer wavelengths are absorbed by the water body already in the uppermost 20 cm. The result is a relative

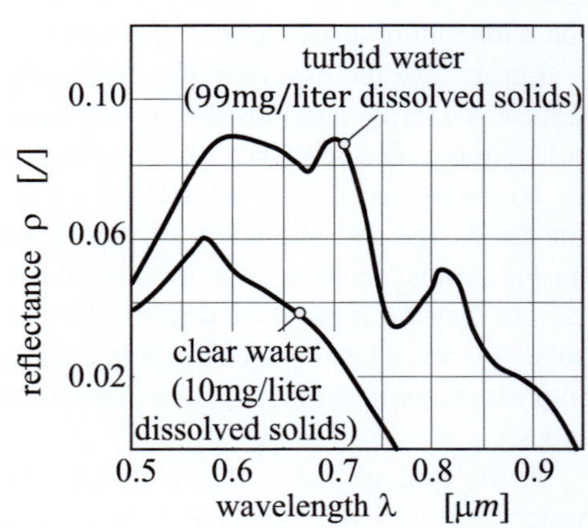

Fig. 3.17 Spectral reflection of clear and turbid water from (Jarmer et al., 2005)

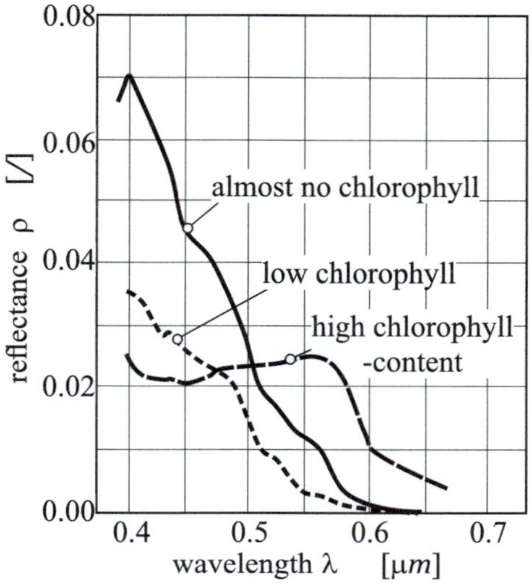

Fig. 3.18 Spectral reflectance at different chlorophyll contents of water from (Jarmer et al., 2005)

energy maximum at 0.5 m (limit blue/green). From a depth of about two meters, no radiation with a wavelength greater than 0.8 μm penetrates the water. Depths of about 20 meters are only reached by the blue and green spectral range.

However, other ground configurations also exhibit a special spectral behaviour that allows remote sensing of often difficult to access areas of the earth's surface with the help of satellites. Figure 3.16 shows the different spectral behaviour for dry soil, vegetation, and water. It can be clearly seen that in vegetation a high proportion of the incident radiation is already reflected in the near infrared, whereas in dry soil the maximum is shifted towards large wavelengths. Water differs significantly from this behaviour and shows only a low reflectivity.

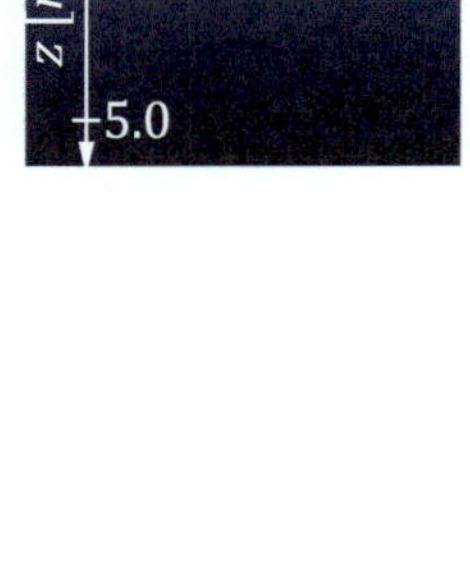

Fig. 3.19 Transmission photograph of unpolarized light entering water

3.2.5 Transmission and Reflection in (Semi-)transparent Components

In this subchapter we want to discuss the transmittance, reflectance and absorptance of common collector elements with respect to solar radiation. The most accurate treatment will consider the spectral properties of the materials, variable incidence angles and polarization effects. In practical situations however, a quick assessment using broadband properties and average conditions for polarization and incidence angle also are helpful. The formulae given can be applied to do the complex calculations, however they can be applied also in a simplified way.

The first basic physical law with respect to the optical properties, transmission, reflexion, and absorption, is energy conservation. It states that the sum of transmittance, reflectance and absorptance is unity. This is true for each wavelength and each incidence angle:

$$\tau(\lambda, \theta, \phi) + \rho(\lambda, \theta, \phi) + \alpha(\lambda, \theta, \phi) = 1. \tag{3.35}$$

This is of course also valid when all the quantities are integrated over wavelength or over a solid angle region or over both. In other words, any radiation flux is either transmitted, reflected, or absorbed.

The second basic optical law is given by Kirchhoff's radiation law which describes the relation between absorption and emission of a body. It is equivalent to the second law of thermodynamics (whereas Eq. (3.35) is equivalent to the first law of thermodynamics, and therefore, sometimes is called Kirchhoff's first radiation law):

$$\varepsilon(\lambda, \theta, \phi, T) = \alpha(\lambda, \theta, \phi, T). \tag{3.36}$$

It is important to state here that the temperature of the body in question also may play a role, not because a body at higher temperature radiates generally more emissive power (that is covered by the Planckian function of blackbody radiation), but because optical constants might change with temperature (e.g. when a phase change of the material occurs). This 2nd law will play a role in the subchapter on selective absorbers further below.

Let us now consider the case of a transparent body, for example a glass pane. How can the optical properties as defined in the Sect. 3.2.1 and as described in Eq. (3.35) be calculated form the optical constants?

For this we consider a single pane optical material, a layer with a certain thickness d. When a beam of light passes from a medium into the transparent layer, e.g. a sheet of glass, a large number of reflections occur. The incident beam is partially reflected at the upper entrance interface while another part penetrates into the layer towards the second interface. There at the exit from the transparent medium the light beam is again partially reflected at the interface. The part not reflected at the interface leaves the layer. This process repeats infinitively but the energy contained in the layer becomes smaller and smaller, as illustrated in Fig. 3.20. The energy balance should account also for absorption in the (semi-)transparent medium.

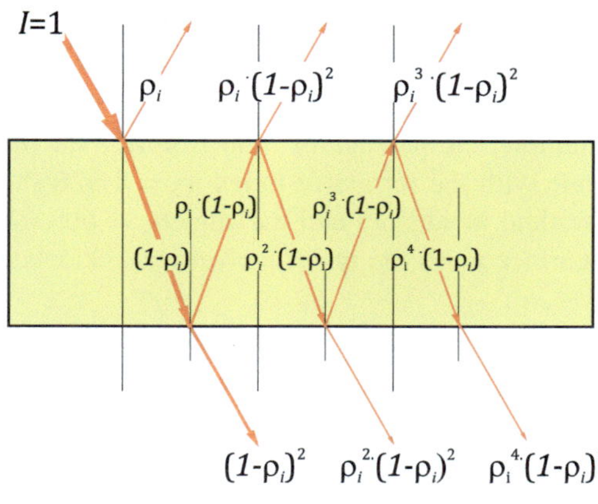

Fig. 3.20 Schematic diagram of the transmission of a light beam at the interfaces of an ideal transparent medium in which the internal attenuation factor is x_i

The reflectivity and transmissivity at the border between the two media are given by the Eqs. (3.23) and (3.22) for the parallel and perpendicular polarization states. In order not to confuse these values with the resulting transmittance and reflectance values for the complete layer, we denote reflectivity in the following paragraphs with the symbol R and transmissivity with the symbol T (instead of ρ and τ).

The transmittance of one single sheet $\tau_{s/p}$ emerging on the underside of the panel in one of the two directions is then calculated as the sum of the individual passing components. In each passage the internal extinction in the material reduces the intensity by a factor of x_i (see Eq. (3.38)). In many conditions the infinite sum can be approximated with good accuracy using the first zero and first order components. Only for large reflection values for instance for oblique incidence the multiple reflections gain in importance. Since, according to the equations, the reflectivity values $R_{s/p}$ are less than unity, the transmitted energy fraction decreases with each reflection. The following relationship results for the transmittance τ_i:

$$\tau_{s/p} = \left(1 - R_{s/p}\right)^2 \cdot x_i \sum_{j=0}^{\infty} x_i^{2j} R_{s/p}^{2j} = \frac{\left(1 - R_{s/p}\right)^2 \cdot x_i}{1 - R_{s/p}^2 x_i^2}, \tag{3.37}$$

where s/p represents the variable for the perpendicular (s) and parallel (p) component.

The reflectance ρ is similarly calculated by summing up the infinite series of reflected parts:

$$\rho_{s/p} = R_{s/p} + \left(1 - R_{s/p}\right)^2 \cdot R_{s/p} \cdot x_i^2 \sum_{j=0}^{\infty} x_i^{2j} R_{s/p}^{2j} = R_{s/p} \cdot \left[1 + \frac{\left(1 - R_{s/p}\right)^2 \cdot x_i^2}{1 - R_{s/p}^2 x_i^2}\right]. \tag{3.38}$$

For incident unpolarized light, the arithmetic average of both results for the two polarization states has to be considered.

A fundamental result of this consideration is that the passage of unpolarized light even through an ideal glass pane without any absorption leads to an angle-dependent transmittance. We want to illustrate the magnitude of this effect in an example, the passage of light from the air ($n_1 = 1$) through a conventional glass pane with the refractive index $n_2 = 1.526$ should be viewed at the angle $\theta_1 = 0°$ (vertical incidence) and the angle$\theta_1 = 60°$. The reflectance at the first air in glass interface gives the result for vertical incidence according to the Eq. (3.39):

$$R(\theta_1 = 0°) = \frac{1}{2}\left(r_s^2 + r_p^2\right) = \left(\frac{n_1 - n_2}{n_1 + n_2}\right)^2 = 0.04336. \tag{3.39}$$

For the incidence angle 60°, Eq. (3.20) yields the exit angle of $\theta_1 = 34.57°$; thus the corresponding amplitude reflection amplitudes can be determined according to Eqs. (3.23) and (3.22). With an irradiation under the angle $\theta_1 = 60°$ one obtains a reflectivity for unpolarized radiation of:

$$\begin{aligned} R(\theta_1 = 60°) &= \frac{1}{2}\left(r_s^2 + r_p^2\right) = \frac{1}{2}\left[\left(-\frac{\sin(\theta_1 - \theta_2)}{\sin(\theta_1 + \theta_2)}\right)^2 + \left(\frac{\tan\sin(\theta_1 - \theta_2)}{\tan(\theta_1 + \theta_2)}\right)^2\right]. \\ &= 0.075523 \end{aligned} \tag{3.40}$$

This means that at an angle of $\theta_1 = 60°$ a share of 7.5% of the incident radiation intensity is already reflected on the first surface of a glass pane, while only about 92% passes the pane and is available for an absorber to convert into thermal energy.

At the second surface of the pane, the calculations repeated this time with a reversed order of media yield exactly the same result. We will arrive with the Fresnel formulae at the same values, not matter whether we consider the interface where radiation enters the glass from air, or whether we consider the one where it enters air from glass. Simply summing up the two reflectivities gives a first approximation of the reflectance of the glass pane. When we want to be exact, we must also consider the multiple reflections, which becomes more important for oblique incidence. Figure 3.21 illustrates this dependence of reflectivity as a function of incident angle θ_1 for different types of glass from Table 3.4.

Calculation of optical properties for a several transparent layers

We want to consider now a second pane of glass added. How is the transmittance and reflectance of a glazing consisting of several glass sheets calculated? The following procedure is giving a very general approach which is iterative and can be applied to as many layers one may think of.

Consider a stack of N layers with transmittance τ_N and reflectance ρ_N. As the stack may not be symmetrical the reflectance on the back side which we denote by ρ_N^* might be different. This could be due to coatings on the individual layers, or due to different layers in the series. For example, a white bright layer on the back side and a black dark layer on the front side would be asymmetric in reflectance.

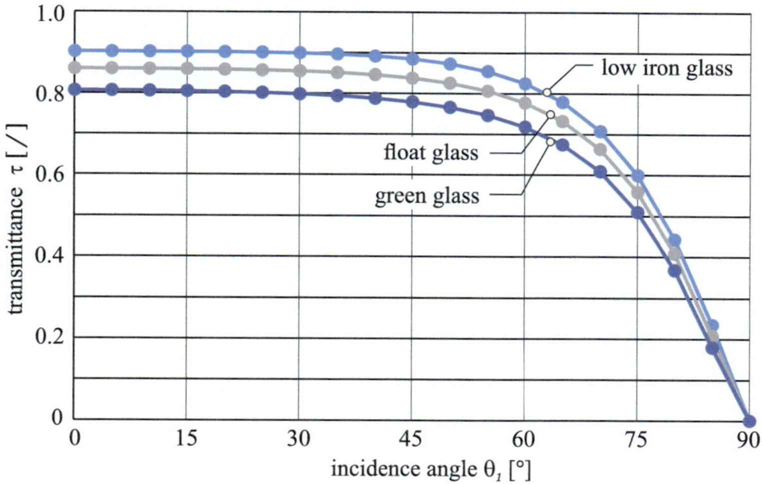

Fig. 3.21 Transmittance of different glass panes, 4 mm thickness and different iron-oxide content, i.e. low-iron glass ($\kappa = 4m^{-1}$), normal float glass ($\kappa = 16m^{-1}$) and green glass ($\kappa = 32m^{-1}$) from Table 3.4

Now we would like to put in front of this stack another set of M layers, having the optical properties transmittance τ_M, reflectance ρ_M and back reflectance ρ_M^*. The combined stack $M + N$ then has the following properties:

$$\tau_{M+N} = \frac{\tau_M . \tau_N}{1 - \rho_N \cdot \rho_M^*}; \tag{3.41}$$

$$\rho_{M+N} = \rho_M + \frac{\rho_N \cdot \tau_M^2}{1 - \rho_N \cdot \rho_M^*}; \tag{3.42}$$

$$\rho_{M+N}^* = \rho_N^* + \frac{\rho_M^* \cdot \tau_N^2}{1 - \rho_N \cdot \rho_M^*}; \tag{3.43}$$

$$\alpha_{M+N}^{(M)} = \alpha_M \cdot \left(1 + \frac{\tau_M \cdot \rho_N}{1 - \rho_N \cdot \rho_M^*}\right); \tag{3.44}$$

$$\alpha_{M+N}^{(N)} = \frac{\tau_M \cdot \alpha_N}{1 - \rho_N \cdot \rho_M^*}. \tag{3.45}$$

Simplification for no absorption

For the case where there is absolutely no absorption in the pane, we reach a simplified formula:

$$\tau = 1 - \rho = 1 - \frac{1}{2}\left(r_s^2 + r_p^2\right) = \frac{1}{2}\left[\left(1 - r_s^2\right) + \left(1 - r_p^2\right)\right] = \frac{1}{2}\left(t_s^2 + t_p^2\right), \tag{3.46}$$

where r_s and r_p describe the amplitude reflection coefficients and t_s and t_p are the amplitude transmission coefficients, respectively. Assuming an originally unpolarized irradiation on the transparent pane, the total transmission coefficient, which includes both directions, is:

$$\tau = \frac{1}{2}\left[\frac{1-R_p}{1+R_p} + \frac{1-R_s}{1+R_s}\right]. \tag{3.47}$$

An analogous procedure can also be carried out for the case of N several panes arranged parallel to each other. If there is no absorption within the layer, the following equation can be derived:

$$\tau = \frac{1}{2}\left[\frac{1-R_p}{1+(2N-1)R_p} + \frac{1-R_s}{1+(2N-1)R_s}\right]. \tag{3.48}$$

If, for example, one looks again at the vertical incidence of light ($\theta_1 = 0°$) from the air on a double pane, one obtains for the coefficient of transmission:

$$\tau(\theta = 0) = \frac{1}{2}\left[\frac{1-R_p}{1+(2N-1)R_p} + \frac{1-R_s}{1+(2N-1)R_s}\right] = \frac{1-R_s}{1+3R_s} = 0.85. \tag{3.49}$$

This already means a loss of intensity of about 15% for an ideal double glazing without absorption. An oblique incidence angle different from the vertical incidence leads to a further loss of light intensity. At an angle of incidence of $\theta_1 = 60°$, two panes without any absorption only result in a total transmittance of $\tau(\theta_1 = 60°) = 0.76$. The result of the observation of ideal glass panes shows that the intensity available after a glass pane is sensitive with respect to the refractive index and angle of incidence and even without absorption a significant loss of intensity can already occur. Since absorption is a passive damping process, glasses in solar thermal systems must have a low absorption; at the same time, care must be taken to ensure that radiation is as perpendicular as possible. The Fig. 3.22 shows the total transmission coefficient τ as a function of the angle of incidence θ_1 for a different number of ideal absorption-free panes arranged in parallel one behind the other. A value of $n_2 = 1.526$ was used as the refractive index for the glasses.

The calculation is of course simplified and does not consider the effects of polarization and multiple reflection as addressed in Sect. 3.2.3. Both effects lead to the fact that for incident angles of $\theta_1 > 45°$ both the transmission through the pane becomes slightly larger and the reflectivity of the pane slightly decreases. This effect of the real values for τ and ρ is illustrated in Fig. 3.23, showing that the simplification is conservative but relatively accurate.

In order to quantify the attenuation, we will first look at the Fig. 3.24. There, a beam of intensity I_0 hits a glass pane at an entrance angle θ_1. Assuming a homogeneous isotropic transparent body with a constant extinction coefficient, the

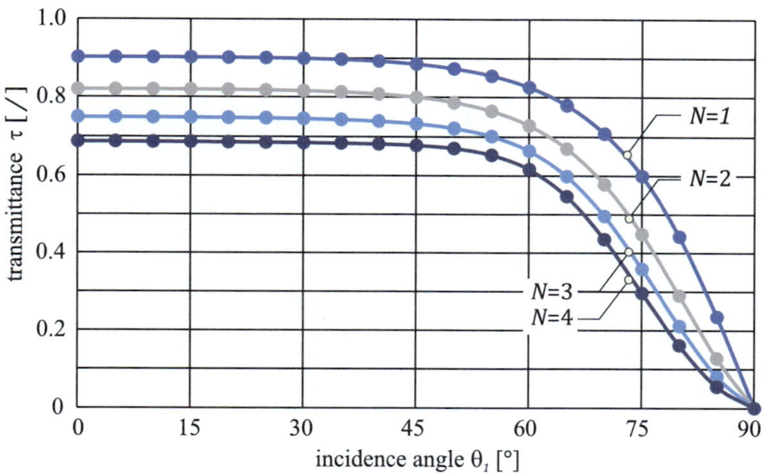

Fig. 3.22 Calculated ideal transmission coefficient τ (without absorption $\mu = 0$) as a function of the angle of incidence $\theta_1$for a different number N of parallel arranged panes for the pairing of air ($n_1 = 1$) and quartz glass ($n_2 = 1.526$)

Fig. 3.23 Transmittance and reflectance of double glazing (stack of two identical float glasses), using different approximations (exact: polarization and multiple reflections both considered, w/o polarization: multiple reflections considered, calculation of double stack with average properties of single float glass, w/o pol and multiple reflections: calculations as before but neglecting multiple reflections between 2 panes)

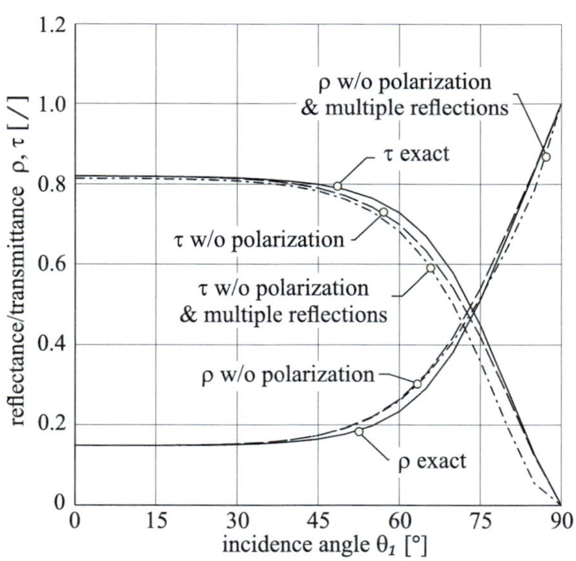

Eq. (3.31) can be integrated over the disc thickness d and one obtains the following result for the intensity for the single passage of the beam through the plate:

$$I(x = d) = I_0 \cdot (1 - R_i)^2 \cdot \exp\left(-\mu \cdot \frac{d}{\cos\theta_2}\right). \tag{3.50}$$

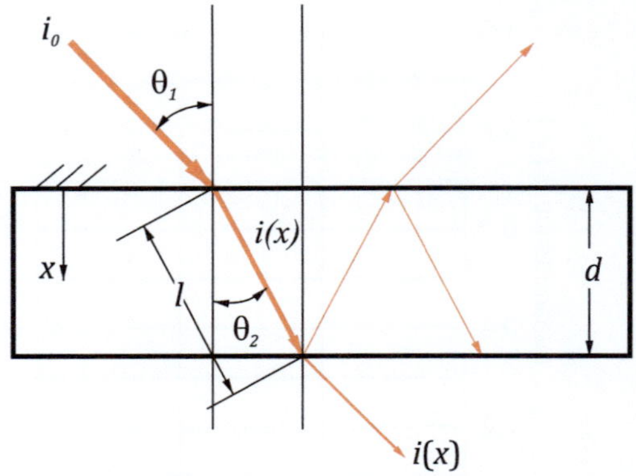

Fig. 3.24 Schematic diagram of the transmission with absorption

The factor $(1 - R_i)^2$ describes the entrance and exit from the material and the exponential function describes the light absorption along the path. The index i stands for the parallel and perpendicular reflection in the material. Since many reflections occur in the material, the total intensity i emerging from the layer is i_τ:

$$I_\tau(x = d) = I_0(1 - R_i)^2 \cdot \exp\left(-\mu \cdot \frac{d}{\cos\theta_2}\right) \cdot \sum_{j=1}^{\infty}\left[R_i^2 \exp\left(-2\mu \cdot \frac{d}{\cos\theta_2}\right)\right]^j. \tag{3.51}$$

If the transparent body reflects only slightly, i.e. i is small ($R_i \ll 1$, attention i depends on the angle θ_1) and absorbs only slightly ($\mu \to 0$), the exponential expression $\left[\exp(-2\mu d/\cos\theta_2)\right]$ approaches unity (1). If, in addition, the perpendicular and parallel reflection coefficients are approximately the same, i.e. $R_s \approx R_p$, then one obtains i_τ for the intensity of the outgoing (transmitted) beam:

$$I_\tau(x = d) = I_0 \cdot \frac{(1 - \rho)}{(1 + \rho)} \cdot \exp\left(-\mu \cdot \frac{d}{\cos\theta_2}\right). \tag{3.52}$$

The solution of the exponential attenuation of the intensity thus obeys the Lambert–Beer law, which was already discussed in Chap. 2 in connection with absorption in the atmosphere. The transmission coefficient in the case of absorption is reduced compared to the ideal case by the exponential expression$\exp(-2\mu d/\cos\theta_2)$. An analogous procedure for N layers arranged in parallel one behind the other can be carried out in the same way as for the case without absorption losses, so that the following results are obtained for τ_a andτ_{aN}:

$$\tau_a = \underbrace{\frac{(1 - \rho)}{(1 + \rho)}}_{\text{reflection}} \cdot \underbrace{\exp\left(-\mu \cdot \frac{d}{\cos\theta_2}\right)}_{\text{absorption}} \text{ and } \tau_{aN} = \left(\frac{1 - \rho}{1 + \rho}\right)^N \cdot \exp\left(-N\mu \cdot \frac{d}{\cos\theta_2}\right). \tag{3.53}$$

The transmittance defined in Eq. (3.53) has a reflection part τ_ρ and a pure absorption part τ_a and is only valid for small absorption coefficients. It is often common in literature to combine both terms multiplicatively, so that absorption is separated from reflection, i.e. $\tau = \tau_\rho \cdot \tau_a$.

The absorption in the glass leads to a significant reduction of the transmission intensity. If, for example, the material pairing air and quartz glass ($n_2 = 1.526$) is selected again and $(\mu d) = 0.05m^{-1}$ is applied for the absorption, the intensity of the light after passing through the pane is reduced from 93 to 88%, considering the absorption and with vertical irradiation ($\theta_1 = 90°$). The Fig. 3.25 shows this reduction of the transmission coefficient as a function of the angle of incidence with several panes placed one behind the other for the ideal case and taking absorption into account. The figure clearly shows that even with two panes of glass connected in series, taking absorption into account $(\mu d) = 0.064\ m^{-1}$, the intensity behind the pane assumes a value which corresponds to the order of magnitude of four ideal glass panes arranged one behind the other. This illustrates that even a small proportion of reflections and absorption can lead to a considerable loss of intensity. This loss of intensity is no longer available for use by an absorber.

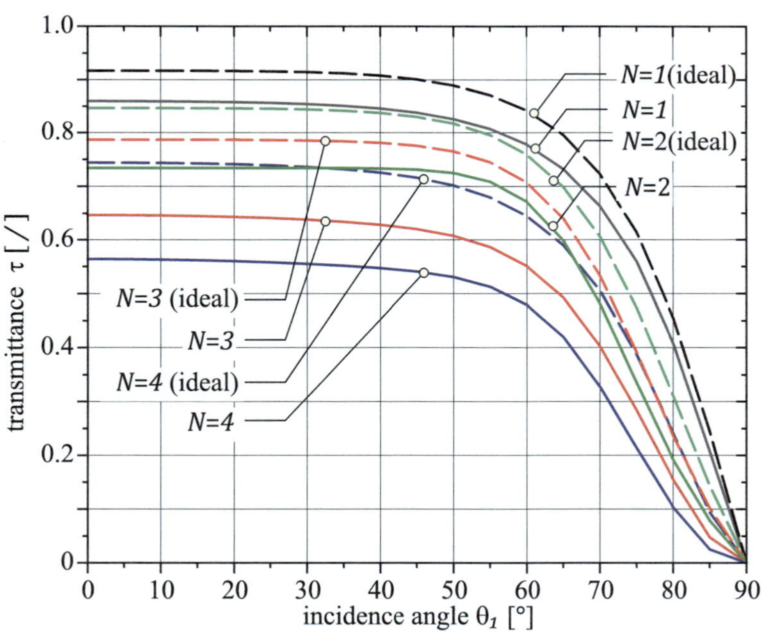

Fig. 3.25 Calculated transmission coefficient τ_a as a function of the angle of incidence θ_1 for a different number N of parallel panes of air-quartz glass ($n_1 = 1, n_2 = 1.526$) arranged one behind the other, assuming an absorption $(\mu d) = 0.05m^{-1}$. The respective curves of an ideal glass pane arrangement $(\mu d) = 0$ show the dashed lines

3.2.6 Anti-Reflection Coatings

When electromagnetic waves pass through the interface between two media with different refractive indices, a part of the radiation is reflected. For the transition $n_1 < n_2$ (for example $n_1 = 1$ as ambient medium air and n_2 as substrate material), the larger the refractive index difference and the larger the angle of incidence θ_1 the more radiation of the incident light is reflected at the interface. For the example of float glass $n_2/n_1 = 1.5$ a reflection at perpendicular incidence ($\theta_1 = 0°$) results already in 4% and at $\theta_1 = 60°$ in 9%, i.e. a reflection loss more than twice as large.

It is extremely desirable to reduce the occurring reflections for solar energy systems in general, as these reflections reduce the efficiency of the optical conversion. But reflections are also disturbing in other aspects. Reflections from solar energy systems might disturb visually in sensitive areas e.g. public places, critical traffic areas or airports. With the help of anti-reflection layers, disturbing reflections can be reduced, for example in technical optics e.g. in spectacle lenses as well as for solar components like windows and collector covers, thus increasing the functionality. Depending on the application different technical solutions for anti-reflection coating must be applied. It is important whether a certain wavelength region needs to be anti-reflected or a broadband measure for the whole solar spectrum is looked for. Also, weatherability, sensitivity against scratching etc. also impacts the choice of technology. The various technical methods will be described and discussed in the following paragraphs.

3.2.6.1 Interference Coatings

One of the most common methods is interference coatings. In these layers, the degree of reflection on the coated surface is reduced by means of destructive interference of the reflected rays.

The use of thin layers is nowadays an established process for the anti-reflection coating of surfaces. To reduce the reflected light at the interface of two optical media with different refractive indices (n_1, n_s), an optical layer is applied at this interface whose refractive index (n_s) lies between that of the two optical media, i.e. $n_1 < n_s < n_2$. The Fig. 3.26 shows schematically the course of a light wave when passing through a thin transparent layer of thickness d. At each interface (air $\rightarrow$ layer and layer $\rightarrow$ substrate), the penetration of the light beam causes part of the incident light to be transmitted and another part to be reflected. The intensity remaining in each partial beam is calculated using Fresnel's formulae. In most cases, the multiple reflections occurring within the thin layer as well as the absorption occurring within them are neglected in such processes. The two reflected partial waves R_1 and R_2 come from the same light source and therefore oscillate synchronously for a certain time, i.e. they are coherent. Since the wave reflected at the interface layer-substrate covers a longer optical path than the wave reflected at the front interface air-layer, R_1 and R_2 have a phase difference between them. Depending on the optical thickness ($n_s \bullet d$) of the layer and the direction

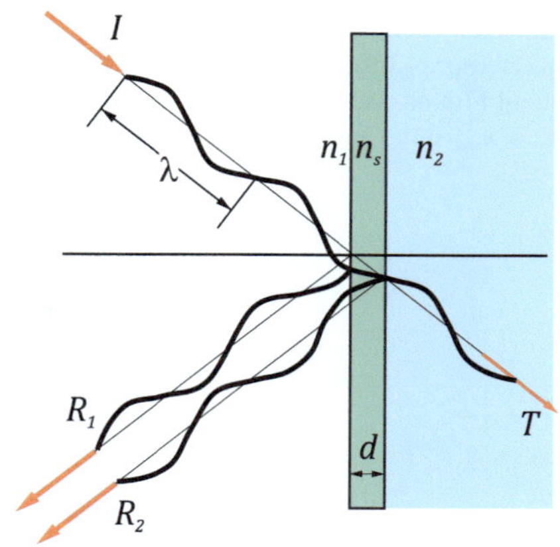

Fig. 3.26 Sketch to describe the destructive interference using $\lambda/4$ layers for $n_1 < n_s < n_2$

of the refractive index transition to the substrate ($n_s < n_2$,$n_s > n_2$), these partial reflections can now intensify or cancel each other out.

The dependence of the phase position of the partial beams on the light wavelength λ as well as the optical thickness of the layer *d* is described by the following equation assuming a perpendicular incidence of light,

$$\delta = \frac{2 \cdot \pi}{\lambda} \cdot n_s \cdot d, \tag{3.54}$$

where δ is the path difference, *d* is the thickness of the layer and n_s is its refractive index.

If the optical thickness of a low-refractive layer is a quarter of the light wavelength $\lambda/4$) or an odd multiple of $\lambda/4$, the phase difference between the two reflected partial waves after travelling back and forth through the layer or an odd multiple of π. The total reflection is the sum of the two reflected waves. With an optical layer thickness d of $\lambda/4$, the partial waves are shifted in such a way that the signs of the amplitudes are opposite. As a result, these values partially cancel each other in the sum (at $n_1 < n_s < n_2$). In the case of weakening there is a so-called destructive interference. The $\lambda/4$-thickness of a single layer is the basis for the design of most interference layer systems, whose detailed calculation and generation in (Willey, 2002) is described.

The opposite effect is achieved by using a highly refractive layer, as can be seen in oil films floating on water. The refractive indices for air, oil and water are $n_{air} = 1$, $n_{oil} = 1.5$ and $n_{water} = 1.33$. Since oil is denser than air, the light is refracted and broken down into its spectral colours when it enters the gasoline. At the water surface under the oil it is then reflected, penetrates the oil again and is broken again. At the boundary with the air it is reflected again, sent back into the oil. If the gear difference is $\lambda/2$, the waves cancel each other out with this refractive index combination ($n_{oil} > n_{water}$). If the path difference is a quarter of the light wavelength $\lambda/4$) or an odd multiple of $\lambda/4$, the amplitudes are amplified, and a colour film can be seen (see Fig. 3.27). Soap bubbles also show colourful

Fig. 3.27 Photograph of the interference (diffraction) of an oil film on water $n_s = n_{oil} > n_2 = n_{water}$

effects due to interference. If the soap film thickness is less than $5\mu m$, the light reflected on the inside interferes with that on the outside.

For a complete antireflection coating of the substrate surface ($\rho \to 0$) not only the phase condition described above must be fulfilled, but also the amplitude condition, i.e. the amplitudes of both partial waves must be equal (amplitude condition: $R_1 = R_2$). This requirement can only be met if the following condition is met for the refractive index of the layer according to the Fresnel formulae:

$$n_s = \sqrt{n_1 \cdot n_2}, \tag{3.55}$$

when light passes from glass ($n_2 = 1.5$) to air ($n_1 = 1$), this means that the layer must have a refractive index of $n_s \approx 1.22$. However, this refractive index, which is necessary according to theoretical analyses, cannot be realized industrially, since dielectric solid materials with this refractive index are not known in the visible spectral range (VIS = Visual spectral range). The substance magnesium fluoride (MgF_2) with a refractive index of $n \approx 1.38$ comes relatively close to this value. For an anti-reflection coating with a single layer, therefore, a MgF_2 layer is usually used, which brought Smakula to industrial application as early as 1935 and in 1957 (Smakula, 1957) fully published; details are available in (Anders, 1967) to find. With a $\lambda/4$-MgF_2 layer, the glass surface can only be anti-reflective from $\rho = 4\%$ to1.41%, as the Fig. 3.28 shows: the reflection coefficient ρ is a function of the wavelength λ and, therefore, the effect is limited to a limited wavelength interval. In addition to the significantly reduced reflection compared to the glass pane, a wavelength dependence of the reflection coefficient can also be observed.

If the light beam does not hit the interface perpendicularly but at an angle, the optical path length in the coating changes according to Snellius' law of refraction, so that a different optimum coating thickness d results, which is calculated

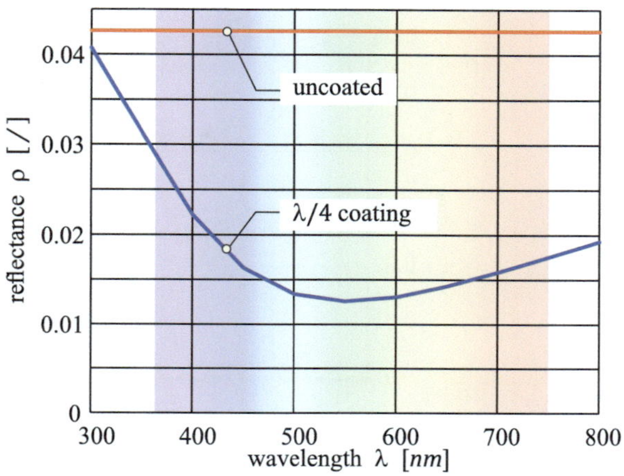

Fig. 3.28 Reflectance ρ as a function of wavelength once with $\lambda/4$ coating and once without

according to equation

$$d = \frac{\lambda}{4 \cdot n_s} \cdot \sqrt{1 + \left(\frac{n_1}{n_s}\right)^2 \cdot \sin\theta_1}, \tag{3.56}$$

The design wavelength is the wavelength for which the $\lambda/4$-layer was calculated. A zero-percent reflection is generally conceivable for any wavelength. For an antireflection coating in the visible range, the design wavelength is usually specified as 550 nm, as this is approximately the middle of the visible spectrum. This results in a layer thickness d of 137.5 nm for $\lambda/4$ at $\lambda = 550\ nm$.

Since in practice a single layer does not allow a reflection factor of 0%, the use of layer stacks is used. When using two $\lambda/4$-layers, a more effective reduction of reflection at one wavelength can be achieved. After applying a medium refractive $\lambda/4$-layer ($n_s > n_2$) as the first layer on the substrate, it is possible to achieve a reflection of 0% at the design wavelength by applying another layer with a low refractive index. These systems are also called V-reflection or $V - coat$ due to the shape of the spectral reflection curve. The Fig. 3.29 shows this using a glass coated with a medium refractive λ/4-layer Al_2O_3 ($n_s = 1.7$) and then with a low refractive λ/4-layer MgF_2 ($n_s = 1.38$). At the design wavelength, almost no reflection occurs now.

To obtain an anti-reflection effect over a wider wavelength range, anti-reflection coatings with three layers have been developed. By introducing a highly refractive $\lambda/2$-layer ($n_s = 2.3$) into the two-layer mirror coating, the reflection at two wavelengths can theoretically be reduced to zero, which is why this design is also called $W - Coat$, see (Pulker, 1999). Figure 3.30a illustrates this type of layer stacking. For very broadband anti-reflective coating systems, the use of at least 4 to 6 layers (and sometimes even more) are necessary, which results in much more complex technical designs. Thus, the reflection can be eliminated in almost the entire visible wavelength range as Fig. 3.30b represents. However, all types of

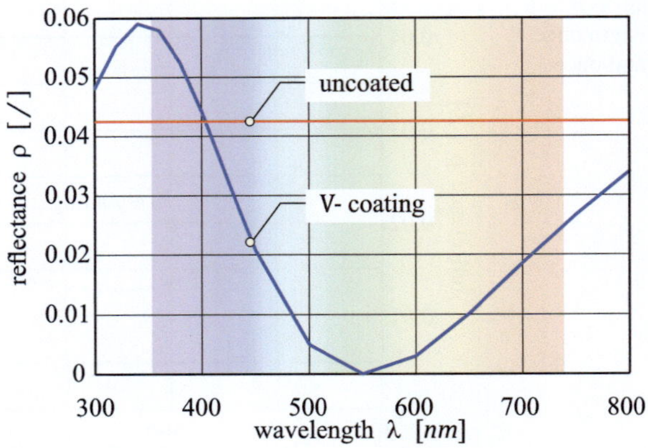

Fig. 3.29 Reflectance of glasses without and with a $V - Coat$ ($2 \bullet \lambda/4$-layers with Al_2O_3 ($n_s = 1.7$) and then with MgF_2 ($n_s = 1.38$)

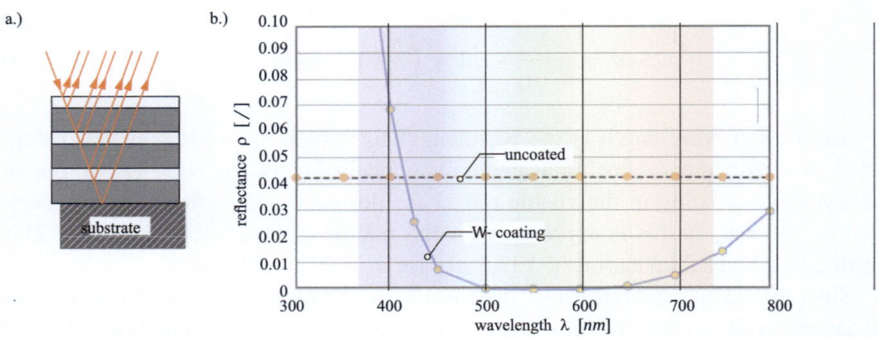

Fig. 3.30 (**a**) layer stacking of a $W - Coat$; (**b**) spectral reflectance without coating and with a $W - Coat$ according to (Pulker, 1999)

destructive interference also limit the transmission of light intensity, so that there is a certain optimum of the number of layers.

3.2.6.2 Porous Coatings

An alternative to anti-reflection coating by interference coatings is the use of porous coatings, such as the introduction of a surface structuring as used by (Gombert and Rommel, 1997) can be described. If a solid is mixed with air in a size dimension below the wavelength of light, very low refractive indices can be achieved. In fact, such media have a refractive index between that of air and that of the material. In the case of porous layers, this can be determined by the volume fractions of different media (e.g. by the size or density of pores) according to (Bläsi, 2000) can be specifically influenced.

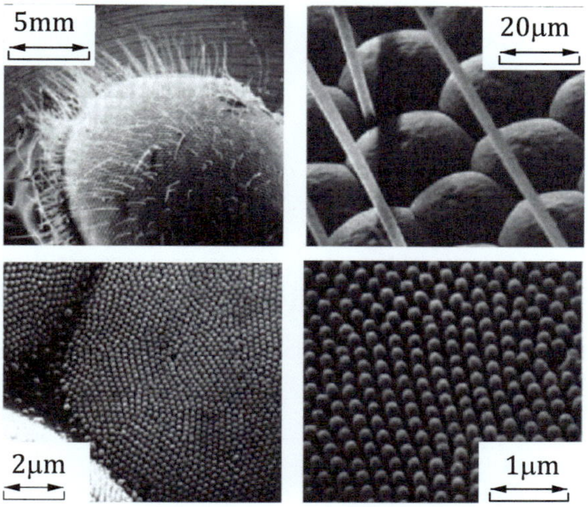

Fig. 3.31 Scanning electron microscope images of the surface of a moth eye (MacLeod, 2001)

A reflex reduction of the surface can also be achieved by introducing a suitable microstructure, which is called “moth eye structure” in reference to the surface structure of insect eyes. The model for this anti-reflection principle was the eye of nocturnal moths on their cornea (see Fig. 3.31). The individual compound eyes have structures that are very similar to a periodic hexagonal surface relief lattice with periods of $230nm$. The introduction of this structure leads to a refractive index gradient at the interface of an optically transparent medium, which increases continuously towards the substrate (Bläsi et al., 2002).

3.2.6.3 Micro-Structured Surfaces

Surfaces with “moth-eye structures” are also known as periodic surface profiles, which can be described as optical gratings. Optical gratings are locally periodic structures that act on the amplitude or phase of light radiation. The lattice properties are determined by the material properties, the structural shape and structural period. The latter can be used for a classification. It is divided into three areas:

- Structure period less than the wavelength of the electromagnetic radiation (a),
- Structural period approximately equal to the wavelength of the electromagnetic radiation (b),
- Structural period larger than the wavelength of the electromagnetic radiation (c).

Different diffraction efficiencies result, which correspond to the share of redistributed energy for the respective diffraction orders. The Fig. 3.32 shows schematically the beam path with different ratios of structure period to wavelength.

If the structure period is much larger than the wavelength (case c.), significantly more diffraction orders are capable of propagation than in the case of approximately the same order of magnitude (case b.). It can propagate for $\lambda \gg \Lambda$ in

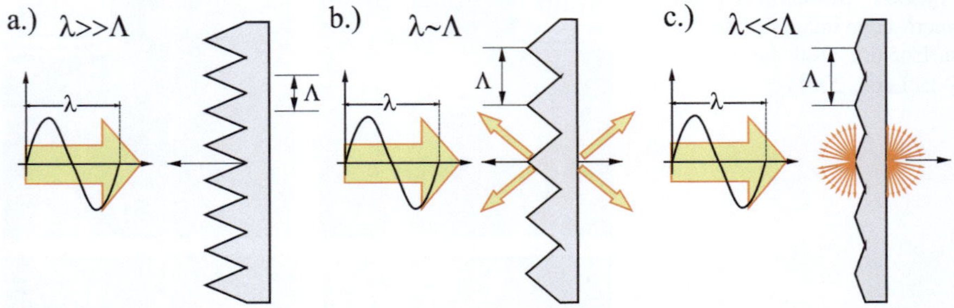

Fig. 3.32 Ratio of grating period Λ to wavelength λ of the irradiated electromagnetic wave determines the optical behaviour of the grating. (**a**) $\lambda \gg \Lambda$, (**b**) $\lambda \approx \Lambda$ und (**c**) $\lambda \ll \Lambda$

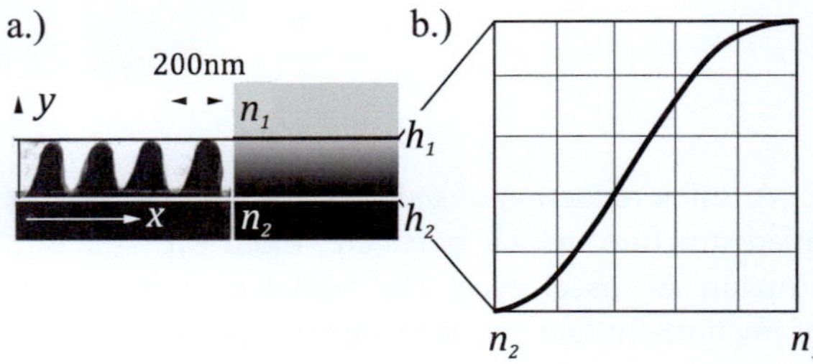

Fig. 3.33 Scanning electron microscope image of a section through the moth eye structure on the cornea of nocturnal insects after (Miller et al., 1966) and the resulting refractive index curve from Gäbler (Gäbler, 2005)

each case only the zeroth diffraction-order in reflection and transmission. The light passes through the structured interface as through a plane interface, which means that the region can be considered optically homogeneous and described with an effective refractive index (Gombert, et al., 1999). Figure 3.33a show a section through a moth's eye made with a scanning electron microscope. Above the height y of the eye there is a continuous increase in material density. The resulting refractive index is shown in Fig. 3.33b.

Thus, for $\lambda \gg \Lambda$, i.e. at long wavelengths, one obtains hardly any reflections and a high transmission. At the other extreme (case c.), small wavelengths and large structures result in very many reflection orders and less transmission.

For the description of the optical properties of material mixtures or the moth-eye structure, there are now many methods, models and simulations that seem suitable for the different aspects. For example, RCWA (Rigorous Coupled Wave Analysis) is well suited for the simulation of periodic gratings. For the description of the structures, the model of the effective media approach (EMA) is often favoured, which physically essentially uses material mixture under consideration of two constituents. Three mixing models are technically applied: the Maxwell–Garnett, the Bruggeman and the Lorentz-Lorenz mixing model.

3.3 Spectral Selectivity

3.3.1 Introduction

The concept of *spectral selectivity* is a key concept in solar thermal energy conversion. We deal with solar radiation basically in a wavelength range between 0.2 and 2.6 μm, which is responsible for our system energy gains, and with thermal radiation a different wavelength region between about 2 μm and 50 μm (depending on the temperatures considered), which is responsible for system energy losses. Optical properties of materials and components used therefore are searched for which take different values in these spectral regions. And this is called spectral selectivity:

3.3.1.1 Selective Transmittance

One spectral selectivity we discuss about is the *selective transmittance* of a transparent element. Here, we distinguish two important classes of selective transmission:

Visual selectivity: high light transmittance in the visual part of the spectrum ($0.38 - 0.78$ μm, weighted with the eyes sensitivity curve$V(\lambda)$, as defined in (ISO 2353, 2005) combined with minimum solar transmittance ($0.3 - 2.5$ μm, weighted with a standard solar spectrum (ISO 9845–1, 1992), (ISO 9050, 2003):

- $\tau_{vis} \gg \tau_e$;
- $\rho_{vis} \ll \rho_e$.

This kind of selective glazing does not transmit the near infrared invisible part of the solar spectrum and hence may reduce the cooling load of buildings. As in windows the heat transport shall be minimized also, the visual selectivity often is produced by a coating of the base glass with low thermal emittance. In principle however, visual selectivity could also be enhanced by layers transparent to the visual light and absorbing in the near infrared range (NIR) between 0.8 μm and 2.5 μm . Water or polymers could be used theoretically for such purpose.

Thermal selectivity: high solar transmittance is combined with low thermal transmittance and minimized heat transport. Ordinary float glass and many transparent polymers are already an example for this selective transmission properties, as only a very small part of the thermal radiation is transmitted directly through these media (of course depending on the thickness of the polymer film or glass). The heat transport however may be further improved by using suitable coatings with low thermal emissivity on the transparent element, which reduce further the thermal radiative transport.

- $\tau_e \gg \tau_{th}$;
- $\rho_e \ll \rho_{th}$.

The performance of a candidate high-performance selective glazing can be characterized by its visual and solar transmittance and its thermal emittance. Using Kirchhoff's law, spectral transmittance and emittance are dependent very much on reflectance $\rho(\lambda)$ for transparent materials,

$$\tau(\lambda) = 1 - \rho(\lambda) - \alpha(\lambda), \tag{3.57}$$

$$\varepsilon(\lambda) = \alpha(\lambda). \tag{3.58}$$

Using standard spectrophotometers, solar reflectance and transmittance are usually measured in the $0.3 - 2.5\mu m$ wavelength range at near-normal angle of incidence. Similarly, in the infrared range above $2\mu m$ emittance is determined by measuring near-normal hemispherical reflectance and transmittance with an integrating sphere, and by using Eqs. (3.57) and (3.58). This may lead to unrealistically low predictions of radiative heat transport because the effective emittance relevant for that is systematically underestimated. The reason is that the property determined from normal-hemispherical reflectance is the normal emittance, which is usually smaller than hemispherical emittance relevant for the overall radiative heat loss of a surface. An alternative is the calorimetric measurement of radiative heat loss of a surface in vacuum at the relevant temperature. Whereas spectral reflectance can be measured at room temperature, total emittance must be measured at elevated temperatures. An average emittance over a wavelength range is frequently reported from spectral data weighted with blackbody curves

$$\varepsilon(T) = \frac{\int_{\lambda_{\min}=0}^{\lambda_{\max}=\infty} \varepsilon(\lambda, T) \cdot e_B(\lambda, T) d\lambda}{\sigma \cdot T^4}, \tag{3.59}$$

where $\sigma = 5.67 \cdot 10^{-8} W/\left(m^2 K^4\right)$ is the Stefan-Boltzmann constant and $e_B(\lambda, T)$ is the spectral irradiance of a blackbody curve from

$$e_B(\lambda, T) = \frac{c_1}{\lambda^5 \cdot \left[\exp\left(\frac{c_2}{\lambda \cdot T}\right) - 1\right]}, \tag{3.60}$$

where $c_1 = 3.7405 \cdot 10^8 W/\left(\mu m^4 m^2\right)$ and $c_2 = 1.43879 \cdot 10^4 \mu m K$ are Planck's first and second radiation constants, respectively (Howell, Mengüc, Daun, and Siegel, 2020).The actual performance of an absorber at high temperatures may not correspond to the calculated emittance.

Using the spectral data as described in the previous section we may determine similarly the solar and visual transmittance and reflectance by integration using a standard solar spectrum as weighting factor:

$$\tau_e = \frac{\int_{\lambda_{\min}=0}^{\lambda_{\max}=2.6} \tau(\lambda) \cdot e_s(\lambda)\, d\lambda}{\int_{\lambda_{\min}=0}^{\lambda_{\max}=2.6} e_s(\lambda)\, d\lambda}, \tag{3.61}$$

$$\tau_{vis} = \frac{\int_{\lambda_{\min}=0}^{\lambda_{\max}=2.6} \tau(\lambda) \cdot V(\lambda) \cdot e_s(\lambda)\, d\lambda}{\int_{\lambda_{\min}=0}^{\lambda_{\max}=2.6} V(\lambda) \cdot e_s(\lambda)\, d\lambda}, \tag{3.62}$$

where $e_s(\lambda)$ is a suitable solar standard spectrum and $V(\lambda)$ is the eye sensitivity. For window application a global irradiation spectrum is taken depending on the standard. A recommended standard global spectrum here is ISO 9845 solar global spectrum $AM1.5$ (ISO 9845–1, 1992).

3.3.1.2 Selective Absorptance

Another spectral selectivity is the relation of solar absorption and thermal emission for *spectrally selective absorbers*. Here, the purpose is a high absorption in the solar range combined by a low emittance in the thermal range to reduce thermal radiation losses.

The surface of an ideal selective absorber is black (i.e. completely absorbing) in the solar wavelength range between $0.3\mu m$ and $2.6\mu m$ approximately. To reduce the thermal heat losses via emitted radiation from the absorber tubes, the surfaces on the other hand should ideally have a very low emittance for the wavelengths that they emit.

Absorber surfaces are governed by the basic principles of optical behaviour of materials. Key to this are two main principles:

- The absorptance of a surface at a particular wavelength (i.e. the fraction of incident radiation that it absorbs) is equal to its emittance at the same wavelength (the amount of radiation it emits expressed as a fraction of the amount that an ideal black body would emit).
- All the radiation incident on a surface must either be absorbed, reflected, or transmitted, so absorptance, transmittance and reflectance must sum to 1.

The key to the idea of a selective surface is that the wavelengths that a hot solar receiver will emit are different to the solar spectrum wavelengths. Whilst absorptance must equal emittance at each specific wavelength, they can and do vary with wavelength.

As a consequence, the ideal reflectance function of an effective so-called selective solar absorber is a step function between a minimum value of reflectance (hence high absorptance) in the solar wavelength range and a high reflectance (hence low emittance) in the infrared (IR) thermal radiation range (see Fig. 3.34). Ideally the lower reflectance value limit would approach 0% while the upper reflectance limit would be 100%, however in reality these values cannot be achieved. Perfect gold surfaces have reflectance values in the IR of nearly 98%, whilst blackened surfaces with 98% absorptance in the solar wavelength range are also possible. For lower temperatures there is no energetically significant overlap between the thermal emission spectrum of a body (with the blackbody spectrum according to Planck being the upper limit) and the solar spectrum. A critical wavelength λ_{crit} for the step function can be defined which separates the two respective

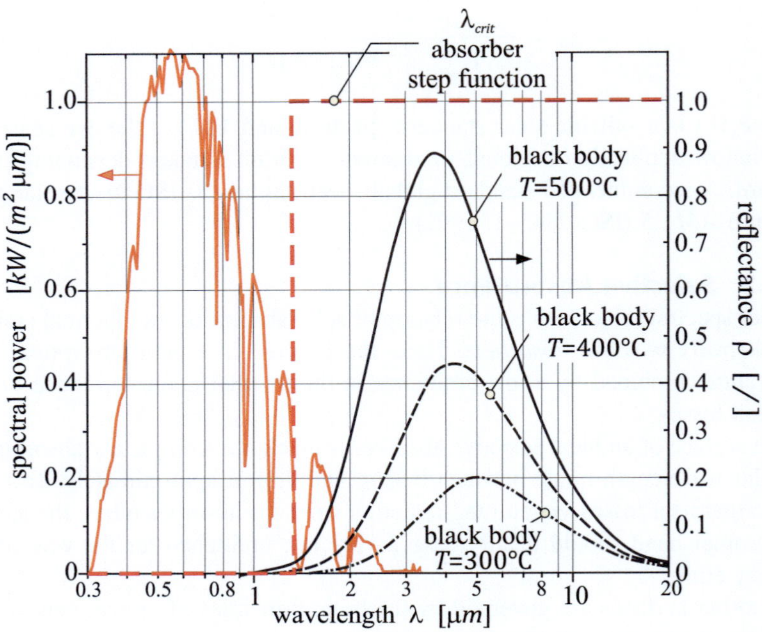

Fig. 3.34 Reflectance of an ideal selective solar absorber together with standard solar spectrum ($AM1.5$ direct, not concentrated) and blackbody spectra (300 $°C$, 400 $°C$, 500 $°C$)

wavelength regions. For higher temperatures however, the thermal emission spectrum is shifted towards shorter wavelengths which implies an increasing overlap of the spectra. Therefore, the critical wavelength λ_{crit} must be optimized with respect to solar absorptance and thermal emittance for different operation temperatures.

When the Critical wavelength λ_{crit} of the ideal reflectance step function is varied (see Fig. 3.34), the integrated values of solar absorptance α_s and of thermal emittance ε_{IR} for a specific temperature change due to a different overlap with solar and thermal spectra. For a low value of the critical wavelength a part of the solar spectrum overlaps with the high reflectance, and therefore solar reflectance increases—and due to the basic principles solar absorptance decreases. When the critical wavelength exceeds the maximum wavelength of significant solar radiation (at about $2.6\mu m$) a further increase leaves the integrated value constant, as illustrated in Fig. 3.35.

In contrast, the infrared emittance rises with increasing critical wavelength, as a part of the low reflectance/high absorptance curves enters the infrared region where the thermal spectrum is significant. Consequently, shifting the critical wavelength to higher values in order to gain a higher absorption will always result in a higher emittance for a given temperature. Similar to a transparent window coating, the performance of a candidate solar absorber can be characterized by its solar absorptance and thermal emittance, both depending on the spectral reflectance. For opaque (zero transmittance) materials, Eq. (3.57) can be simplified setting $\tau = 0$.

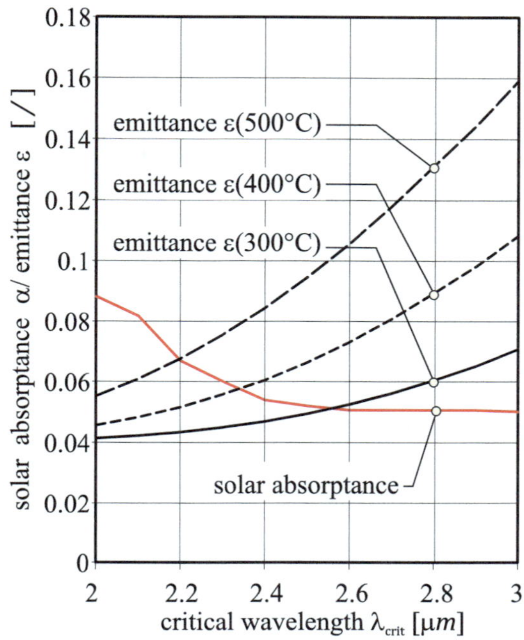

Fig. 3.35 Solar absorptance and thermal emittance for absorber temperatures 300 °C, 400 °C *and* 500 °C versus critical wavelength λ_{crit} of an ideal selective absorber (step function) with short wavelength reflectance 0.04 and long wavelength reflectance 0.96

Development of spectrally selective materials depends on reliable characterization of their optical properties. The determination of the thermal emittance is as described in the last chapter on coated glass, only setting transmittance to zero and using Eqs. (3.59) and (3.60).

The actual performance of an absorber at high temperatures may not correspond to the calculated emittance. This is because small errors in measured ρ can lead to large errors in small values of ε. Calculating the emittance from spectral data taken at room temperature assumes that the spectral characteristics do not change with increasing temperature. This is only valid if the material is invariant for example does not undergo a phase change (as do some titanium containing materials), break down or undergo oxidation (as do paints and some oxide coatings) at higher temperatures. It is important before using high-temperature emittance calculated from room temperature data, that the calculated data is verified with high-temperature emittance measurements for each selective coating. Therefore, it is preferable to also measure the total emittance at the operating temperatures and conditions (Brunotte, Lazarov, and Sizmann, 1992). In addition, measurement errors might lead to differences to the direct calorimetric measurement.

Emittance is a surface property and depends on the surface condition of the material, including the surface roughness, surface films, and oxide layers. Coatings typically replicate to some degree the surface roughness of the substrate. Therefore, to facilitate development, it is important to measure the emittance of each coating-substrate combination as well as the uncoated substrate when developing a solar selective coating. Using the spectrally measured reflectance as described before we may determine similarly the solar absorptance α_e by integration using a standard solar spectrum as weighting factor.

$$\alpha_e = \frac{\int_{\lambda_{\min}=0}^{\lambda_{\max}=2.6} [1 - \rho(\lambda, T)] \cdot e_s(\lambda) d\lambda}{\int_{\lambda_{\min}=0}^{\lambda_{\max}=2.6} e_s(\lambda) d\lambda}, \tag{3.63}$$

where $e_s(\lambda)$ is a suitable solar standard spectrum. For concentrating collectors only direct irradiation is used, hence a direct solar spectrum should be taken. A recommended standard spectrum here is ASTM G173-03 $AM1.5$ Direct Normal (ASTM G173-03, 2020).

3.3.2 Selective Glazings

In order to produce spectrally highly selective glazings with better performance than the ordinary float glass transparent coatings are applied on the glass. As for all glazings the reduction of U-value is a target, a low emissivity coating is the key element. For commercial coatings two different coating types are used: the sputtering of ultra-thin noble metal films or the spray pyrolysis of doped semiconductor oxides. In both materials the number of free electrons determine the spectral limit between transparent and reflective properties of the coating. The properties of these coating types are described in the next paragraphs.

3.3.2.1 Noble Metal Coatings (Soft Coatings)

Very thin noble-metal coatings produced by sputtering are, at least in principle, the simplest solution for reaching a significant short wavelength transmittance combined with long-wavelength reflectance. The best optical properties are obtainable with copper, silver, and gold (Valkonen et al., 1984). Alternative materials may be TiN (Claesson et al., 1990) and aluminium. Thin silver coatings stand out as the superior material on account of their low absorption of luminous and solar radiation (Valkonen and Karlsson, 1987) and are mainly used for commercial coatings.

It is necessary to improve the transmittance by additional antireflection layers. Dielectrics with high refractive indices—such as Bi_2O_3, In_2O_3, SnO_2, TiO_2, ZnO and ZnS - give the largest enhancement of the transmittance. By selecting the thicknesses of the three-layer configuration properly, one can optimize for a warm climate (minimum solar gain with high visible transmission) or a cold climate (maximum solar gain combined with low emissivity). Figure 3.36 shows the spectral optical properties of two silver-based coated glasses, one optimized for high solar gain, the other one for reasonable solar rejection with similar visible transmission.

3.3.2.2 Doped Oxide Semiconductor Coatings

Doped oxide semiconductor coatings by spray pyrolysis offer an alternative to the earlier discussed noble-metal based coatings. The materials which are known

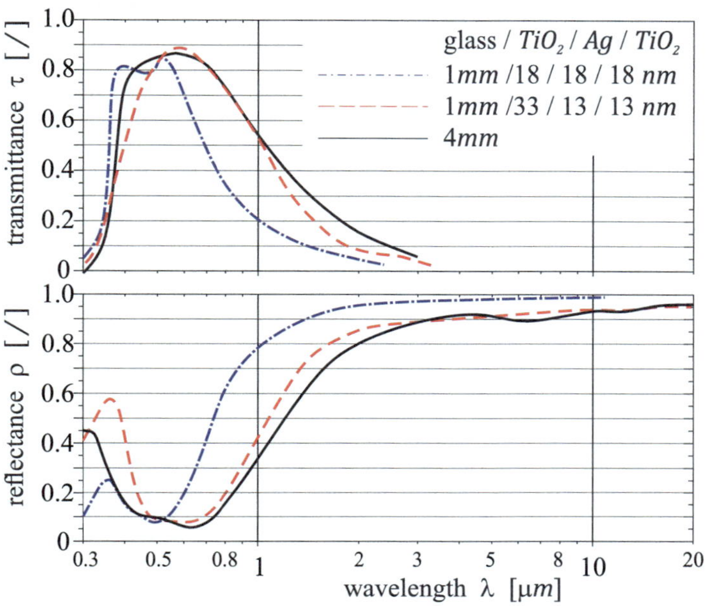

Fig. 3.36 Spectral transmittance and reflectance of glass with silver-based coatings, solar control coating: high visible transmittance with maximal solar reflectance, Solar gain coating: increased solar gain in the solar infrared (above $0.7\mu m$), (Granqvist, 1991a, 1991b)

to be useful are oxides based on zinc, cadmium, indium, tin, thallium, lead, and alloys of these. The required doping is often achieved by the addition of a foreign element; particularly good properties have been obtained with $SnO_2 : F$, $SnO_2 : Sb$, $In_2O_3 : Sn$ and $ZnO : Al$ (Granqvist, 1990), see Fig. 3.37. The industrial products are based on tin oxide, where the coating is directly done in the float process. A specific and important advantage of the doped oxide semiconductors is their excellent chemical and mechanical durability, which allows their use on glass surfaces exposed to the air.

The required coating thickness t is usually around $t \approx 1\mu m$, much thicker than the multilayer soft coatings! This is done to avoid iridescence with unpleasant coloured fringes. The human eye then senses a uniform coloration. However, the use of a much larger thickness than the one demanded for a low τ is clearly inefficient in terms of materials utilization, coating time, and cost. Further, thick oxide semiconductor coatings can display some light scattering, sometimes referred to as haze.

This phenomenon is associated with the occurrence of large crystallites as well as surface roughness, see cf. (Schade and Smith, 1985).

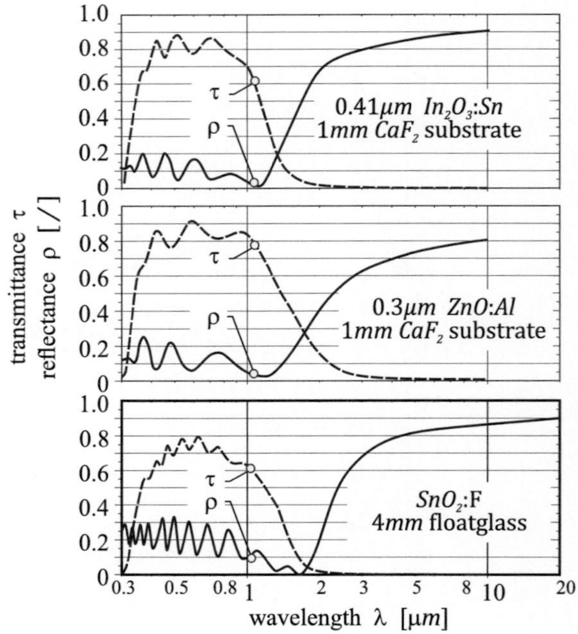

Fig. 3.37 Transmittance and reflectance for thick pyrolithic semiconductor oxide coatings, (Granqvist, 1990)

3.3.3 Design of Selective Absorbers

Selective absorber surface coatings can be categorized into several distinct types based on different physical principles:

a. Intrinsic,
b. Semiconductor–metal tandems,
c. Multilayer absorbers,
d. Multi-dielectric composite coatings,
e. Textured surfaces, and
f. Selectively solar-transmitting coating on a blackbody-like absorber.

Whereas the first two principles need only one homogeneous material, the other ones work with layers of different optical properties.

(a) Intrinsic absorber coatings use a material having intrinsic properties that result in the desired spectral selectivity.
(b) Semiconductor–metal tandems absorb short wavelength radiation where photon energy is above the semiconductor bandgap and have low thermal emittance because of the metal layer.
(c) Multilayer absorbers use multiple reflections between layers to absorb light and can be tailored to be efficient selective absorbers.

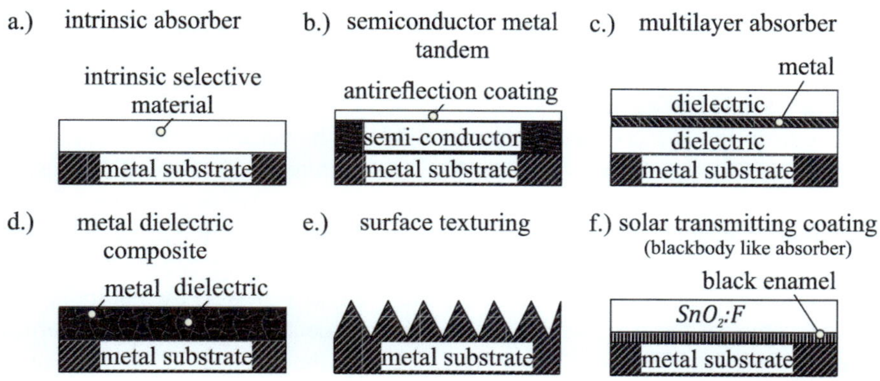

Fig. 3.38 Schematic designs of six types of coatings and surface treatments for selective absorption of energy, adapted from (Kennedy, 2002)

(d) Metal-dielectric composites—cermets—consist of fine metal particles in a dielectric or ceramic host material.
(e) Textured surfaces can produce high solar absorptance by multiple reflections among needle-like, dendritic, or porous microstructure.
(f) Additionally, selectively solar-transmitting coatings on a blackbody-like absorber are also used but are typically used in low-temperature applications.

These constructions are shown schematically in Fig. 3.38a-f, respectively, and are discussed in more detail below.

3.3.3.1 Intrinsic Absorbers

Intrinsic selective absorbers consist of a single material with intrinsic selective property of the material. No naturally occurring material exhibits intrinsically ideal solar-selective properties, but some have mild selective properties. Intrinsic properties are found in transition metals and semiconductors, but both need to be greatly modified to serve as an intrinsic absorber.

Examples include the transition metal tungsten W (Agnihotri and Gupta, 1981) and a series of semiconducting materials, for example MoO_3-doped Mo (Seraphin, 1979), Si doped with B, HfC (Seraphin and Meinel, 1976), ZrB_2 (Randich and Allred, 1981), SnO_2 and others. Hafnium carbide (HfC) could be useful as an absorbing selective surface at elevated temperatures because of its high melting point. However, the absorptance of HfC has to be increased e.g. by an antireflective (AR) layer. Using Zirconium bromide ZrB_2 one can reach the best selectivity with $\alpha = 88\%$ and $\varepsilon(100°C) = 8\%$, which may be even inproved by a $Si_3N_4 - AR$-Layer to $\alpha = 93\%$ and $\varepsilon(100°C) = 10\%$ (Randich and Allred, 1981).

Intrinsic absorber materials are usually structurally very stable but optically less effective than other types of selective absorbers. Therefore, historically, research in intrinsic absorbers has not been very productive because there are no ideal intrinsic materials; but the intrinsic materials are finding increasing use as a component in high-temperature absorber multilayers and composite coatings.

3.3.3.2 Surface Texturing

Surface texturing is a common technique to obtain higher absorption due to optical trapping of solar energy. When surfaces are textured with a lateral dimension close to the wavelength to be absorbed, they appear as a series of cavities. Thus, structures smaller than $1\mu m$ absorb solar energy whereas for longer wavelengths the texture cannot be resolved and appears highly reflective and mirror-like to thermal energy. The emittance can be adjusted (higher or lower) by modifying the microstructure of the surfaces with ion-beam treatments. The selective properties depend on the ratios of mean height deviations and the autocorrelation distance to the wavelength (Cuomo et al., 1975).

Needle-like, dendrite, or porous microstructures on the same scale as the wavelength of the incident radiation exhibit both wavelength and some directional selectivity. The surface of the microstructure must be protected from damage caused by surface contact or abrasion. Selection of a material having a high intrinsic absorption coefficient can further optimize the absorptance. A number of methods exist to prepare textured microstructures, (Seraphin and Meinel, 1976) for example:

- Liquid–solid phase separations.
- Lithography in combination with ion-etching.
- Unidirectional solidification of eutectic alloys.
- Lithography with X-rays.
- Ion-exchange reactions between metals—isothermal transport occurs between two metals where the difference in the work function (ΔE_w) is $> 0.2eV$, e.g. for $Cu - Ni$ alloy.
- Vapor–liquid-solid (VLS) mechanism grown in a controlled way, whiskers on substrates from the liquid alloy zone at the interface (e.g., Si, Ge or group $III - V$ whiskers).
- Vapor deposition—the condensation of a metal or compound from the gas phase onto a substrate by Chemical Vapour Deposition (CVD) or Physical Vapour Deposition (PVD) (e.g., $Ni - Al2O3$, Ni).
- Oxidation of metals at high temperature—the growth of whiskers on metals by the oxidation process in air or O_2 at high temperature ($400° - 850\ °C$), e.g. $Fe_2O_3 - Fe$, steel, $CuO - Cu$, phosphor bronze, $ZnO - Zn$, brass, W; Ni or Mo.

These techniques can be useful in texturing selective surfaces. Examples for absorber surfaces are dendritic tungsten or structured Copper, Nickel or stainless steel surfaces (Zhao, 2007). Typical values are $\alpha = 92\%$ and $\varepsilon(100°C) = 22\%$; using Nickel which is not stable at high temperatures the emittance can be as low as $\varepsilon(100\ °C) = 10\%$, see (Kennedy, 2002).

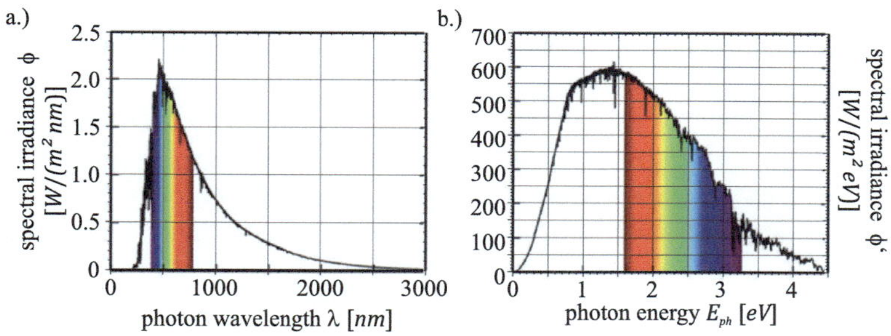

Fig. 3.39 Spectral irradiance as function of wavelength λ in nm (left) and photon energy in eV (right), visible light indicated in colour

3.3.3.3 Semiconductor–metal Tandems

Semiconductors with bandgaps from about $\sim 0.5eV$ to $1.26eV$, corresponding to wavelengths of $2.5\mu m$ to $1.0\mu m$ (cf. Figure 3.39), absorb short-wavelength radiation and are transparent to longer wavelengths. Hence, an underlying metal provides low emittance to give the desired spectral selectivity to semiconductor–metal tandems. Semiconductors of interest include Si ($1.1eV$), Ge ($0.7eV$), and PbS ($0.4eV$) (Agnihotri and Gupta, 1981). An antireflection treatment (coatings or porous surfaces) is needed because the useful semiconductors have high refractive indices resulting in large detrimental reflectance. Si-based designs produced by Chemical-Vapor Deposition (CVD) are well known that are suitable for mid-to high-temperature applications. Best values to be reached are quoted as $\alpha = 91\%$ and $\varepsilon(100°C) = 10\%$ as mentioned in (Kennedy, 2002).

3.3.3.4 Multilayer Absorbers

Multilayer absorbers or multilayer interference stacks consist of many alternating thin layers of transparent dielectric layers and semi-transparent metallic or semiconducting layers, for which the schematic design is illustrated in Fig. 3.40. The lowest opaque thick coating must be metallic to provide a low emittance. The alternating thin-layer stack is optimized for high solar absorption while being transparent to longer wavelengths. Thus, the stack can be designed to become an efficient selective absorber.

The basic physics of multilayer absorbers is well understood, and computer modelling can predict the optical properties given by an optimum multilayer design of candidate materials (Born and Wolf, 1999), provided adequate optical constants

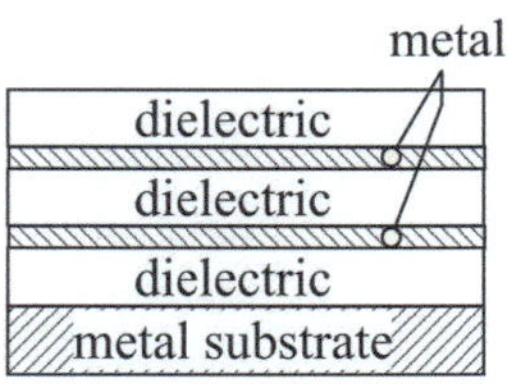

Fig. 3.40 Schematic designs of multilayer absorber film structure adapted from (Kennedy, 2002)

for the thin films are available. Multilayer interference stacks have high solar absorption, low thermal emittance, and can be stable at elevated temperatures ($\geq 400°C$) depending on the materials used. Several multilayer absorbers using different metals (e.g. Mo, Ag, Cu, Ni) and dielectric layers (e.g., Al_2O_3, SiO_2, CeO_2, ZnS) have been cited in the literature for high-temperature applications (Kennedy, 2002). A very well-known example is the AMA-coating ($Al_2O_3 - Mo - Al_2O_3$) with an absorption $\alpha = 92 - 95\%$ and $\varepsilon(100°C) = 6 - 10\%$ (Thornton et al., 1980).

3.3.3.5 Metal-Dielectric Composite Coatings (Cermets)

These absorbers consist of an Anti-Reflection (AR) layer on top of a composite layer, which is highly absorbing in the solar region (i.e., black) and transparent in the IR, deposited onto a highly IR-reflective metal substrate, for which two types are depicted in Fig. 3.41. The highly absorbing metal-dielectric composite, or Cermet, consists of fine metallic nanoparticles in a dielectric or ceramic matrix. These films are transparent in the thermal IR region, while the absorption in the solar region is either due to inter-band transitions in the metal or due to small particle resonances. The cermet layer may have uniform metal content, but graded profiles are preferable to enhance absorption especially towards the substrate. The metal-dielectric concept offers a high degree of flexibility, and the solar selectivity can be optimized by proper choice of constituents, coating thickness, particle concentration, size, shape, and orientation. The solar absorptance can be boosted with a suitable choice of substrates and AR layers, which can also provide protection (for example, from thermal oxidative degradation). A variety of techniques, such as electroplating, anodization, inorganic pigmentation of anodized aluminium, Chemical Vapour Deposition (CVD), and co-deposition of metal and insulator materials by Physical Vapor Deposition (PVD), can produce the composite coatings.

In cermets, solar absorptance is mainly determined by the response of the absorbing particles. There is a shift of the absorption and scattering cut-offs to larger wavelengths when the particle radius, r, increases. This effect is accompanied by a reduction in the scattering and absorption efficiencies roughly proportional to $1/r$ (Arancibia-Bulnes, Estrada, and Ruiz-Suarez, 2000).

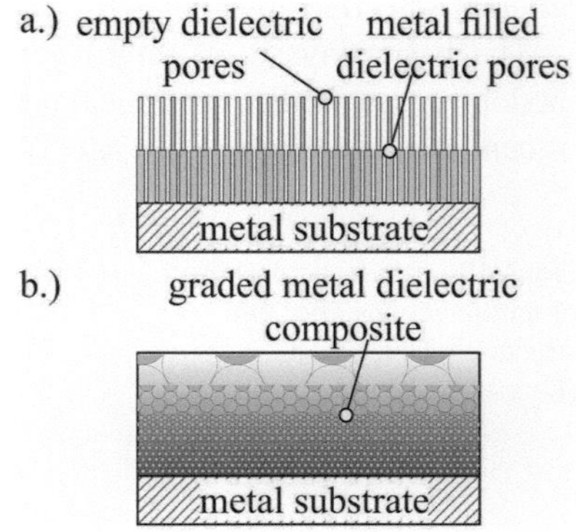

Fig. 3.41 Schematic designs of two different metal-dielectric solar selective coatings adapted from (Kennedy, 2002)

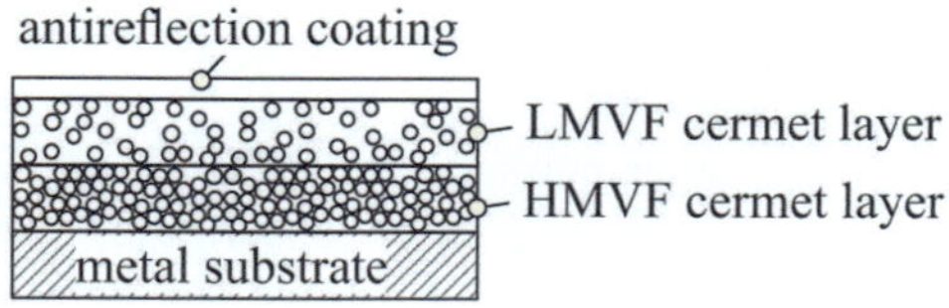

Fig. 3.42 Schematic design of double-cermet film structure adapted from (Kennedy, 2002)

In a graded cermet (Fig. 3.41b), the reflectance from the cermet is reduced by gradually increasing the metal volume fraction, hence the refractive index, as a function of depth from the surface to the base of the film. A double-cermet film structure has been developed similarly increasing the absorption of a homogeneous cermet layer like the graded cermets (Zhang and Mills, 1992). Solar radiation is effectively absorbed internally and by phase interference in double-cermet solar coatings.

Further, it is easier to deposit the double-cermet selective coating than graded-cermet layer selective surfaces. The typical double-cermet layer film structure from surface to substrate consists of the following arrangement:

- An AR layer that enhances solar absorption.
- An absorbing layer composed of two homogenous cermet layers,
- A low-metal-volume fraction (LMVF) cermet layer on
- A high-metal-volume fraction (HMVF) cermet layer; and
- A metallic infrared reflector layer to reduce substrate emittance (Zhang and Mills, 1992),

which is shown in Fig. 3.42. High selectivity can be achieved with these absorber types. Typical values are $\alpha = 95\%$ and $\varepsilon(100°C) = 5\%$, see (Kennedy, 2002).

3.3.3.6 Selectively Solar-Transmitting Coating on a Blackbody-Like Absorber

A selectively solar-transmitting coating on a blackbody-like absorber is the last concept. Here the low-emitting surface is the top layer, which transmits radiation in a wavelength selective manner. This solar-transmitting coating can be a highly doped semiconductor (e.g., $SnO_2 : F$, $SnO_2 : Sb$, $In_2SO_3 : Sn$, and $ZnO : Al$) over an absorber with a proven long-term durability. Some low-temperature flat-plate collectors have used black enamel as the absorber material (Granqvist, 1991a, 1991b). Highly doped semiconductors may be useful with high-temperature black absorber materials. The selectivity, however, is limited for this type of coatings.

3.3.4 Production Methods for Selective Layers

The typical layer thicknesses are in the micrometre and nanometre range, up to monomolecular layers. This also makes the use of expensive materials economical if the desired effect can be achieved despite small quantities. By using thin-film technology, material efficiency can be combined with energy resource conservation in a suitable way.

A thin layer applied to a substrate must not only have predetermined optical properties (refractive index, absorption), but must also meet additional technical requirements so that the product can withstand everyday stress. This includes in particular the outdoor use of solar thermal systems:

- Mechanical strength (adhesion, hardness, abrasion and scratch resistance, centring, ultrasound),
- Thermal resistance (cold, heat),
- Chemical resistance (solvents, detergents, sweat, UV radiation, humidity).

The thermal expansion coefficient of the coating material must not normally differ from that of the substrate. In most cases, the coatings already have much higher internal stresses during production than are caused by temperature differences during operation, for example. Flaking of the layer therefore occurs (especially with thicker layers) when the internal stresses are superimposed on the load or the adhesion of the layer to the substrate is too low.

The combination of all the above-mentioned requirements means that only a few substances are even considered as coating materials. In the field of optical thin films, for example, not every refractive index n is available due to the small selection of coating materials. The Table 3.5 lists the refractive index of some common optical coating materials.

In principle, there are several processes for solar thermal purposes to produce thin films. The most common methods are the following:

- Chemical Vapor Deposition (CVD) is a gas phase reaction (usually at or near the substrate surface). The reaction gases are simultaneously fed into the reaction chamber with the substrate to be coated. The mostly preheated gases are thermally activated by the heated substrate and react with each other. The desired material is separated and chemically bound (chemisorption). Besides countless CVD variants, which differ in working pressure and other process

Table 3.5 Average refractive indices n of some selected optical coating materials

Material	Molecular formula	$n[/]$
Magnesium fluoride	MgF_2	1.38
Silicon dioxide	SiO_2	1.46
Aluminium oxide	Al_2O_3	1.7
Zirconium dioxide	ZrO_2	2.05
Praseodymium	$PrTiO_3$	2.1
Titanium oxide	TiO_2	2.3
Zinc sulphide	ZnS	2.3

Table 3.6 Overview of process parameters (temperature, pressure) of common CVD processes and their application areas

CVD-process	Temperature $T[°C]$	Pressure $p[mbar]$	Activation type	Intended use
APCVD	$400 - 1300$	1000	Thermal	Poly silicon deposition for conductor tracks, gate oxide, epitaxy of silicon wafers
LPCVD	$500 - 1000$	$0.1 - 10$	Thermal	Conductor paths, gate oxide SiO_2, silicon nitride Si_3N_4, polysilicon
HFCVD	$150 - 750$	$0.01 - 200$	Thermal	Carbon-based deposition (diamond, amorphous carbon, carbon nanotubes), silicon-based layers (amorphous / crystalline Si, Si_3N_4, ...
PECVD	$200 - 500$	1	Thermal plasma	SiO_2

parameters and are listed in the Table 3.6, there are also some coating processes which represent strongly modified CVD processes.

- Plasma polymerization: In this process, gaseous monomers excited by a plasma form a highly crosslinked layer on a substrate.
- Atomic layer deposition (ALD): Atomic layer deposition is a strongly modified CVD process in which the reaction or sorption on the surface stops automatically after the surface is completely covered. This self-limiting reaction is carried out in several cycles (with intermediate rinsing steps), so very good aspect ratios and exact layer thicknesses can be achieved.

The chemical principles of the CVD process are described in Powell et al. (Powell, Oxley, Blocher, and Klerer, 1966) in detail.

In contrast to CVD processes, physical vapour deposition (PVD) is based on purely physical processes. PVD is also a vacuum-based coating process in which, in contrast to the CVD process, the layer is formed directly by condensation of a material vapour on the starting material. The procedure is characterised by three successive processes. First, a gas (vapor) generation of the layer forming particles takes place; in the second stage the vapor is transported to the substrate and finally the vapor is condensed on the substrate which then forms the layer.

- Thermal evaporation

In thermal evaporation as shown in Fig. 3.43, the evaporation material is heated until it evaporates at a suitable evaporation rate. There are three "subprocesses" depending on the type of evaporator used (inductive, resistive or electron beam evaporators are possible). To ensure the deposition of high-quality and homogeneous layers, it is necessary to keep the space between the evaporator and the substrate as free of material (i.e. vacuum) as possible. The interactions (usually impacts) of the particles with residual gas atoms can bind or scatter them in such

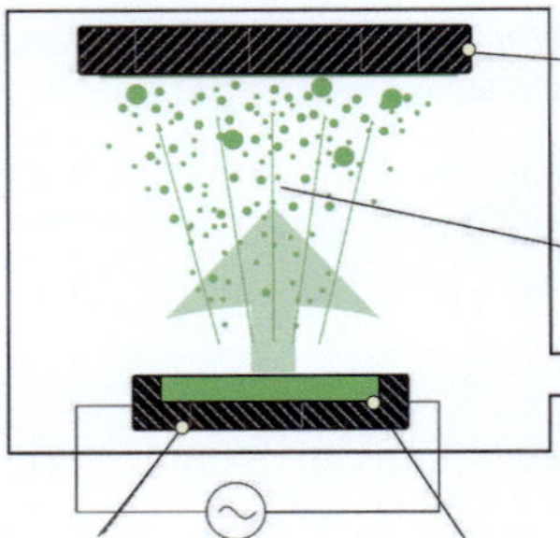

Fig. 3.43 Schematic representation of surface coating by thermal evaporation using a resistance heater

a way that the reproducibility of the coating cannot be guaranteed. Oscillating crystals are often used to measure and control the evaporation rate and layer thickness (alternatively, optical monitoring is also used).

- Sputter deposition:

During sputtering (also called cathode sputtering) particles are removed from the surface by ion bombardment. This process can be used to clean the surface, e.g. of oxides or water that have entered the material through production, processing, or storage. In thin-film technology, this physical process is also used to atomize material from the target, i.e. to transfer it into the gas phase. The resulting gaseous material is then fed onto the substrate to be coated and condenses there, as shown in the Fig. 3.44. Compared to evaporation, sputter deposition has the advantage that alloys can also be transferred to the wafer in "the same ratio". However, it must be considered that different materials have different sputtering coefficients, i.e. they can be atomized to different degrees. The layer thickness is often controlled by means of a time switch-off.

- *Ion plating.* Ion plating is a vacuum-based and plasma-assisted PVD process for metals and metal compounds. Vaporized metal (e.g. by arc discharge) is fed into a plasma. There a part of the metal vapour cloud ionises and is accelerated towards the substrate. The metal ions form a layer on the substrate surface, which is sputtered back together with the substrate material at the beginning by the constant bombardment by metal ions.
- *Ionized cluster beam deposition (ICBD) technology.* The cluster beam technique is also a modified evaporation process. The crucible used for evaporation is kept closed. The heating of the evaporation material creates an overpressure in the sealed crucible. If this steam is discharged through a nozzle, an adiabatic expansion causes sudden cooling. Neutral atomic clusters are formed, which partially dissolve when they hit the substrate surface and are deposited distributed over the surface.
- Molecular beam epitaxy (MBE).
- Electrochemical process: Galvanization.

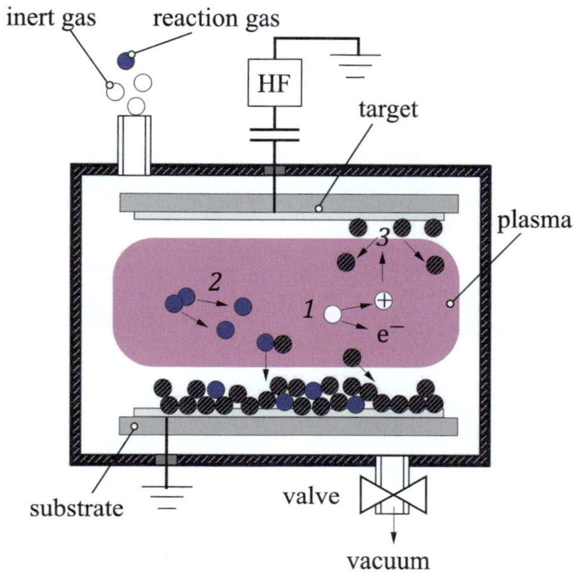

Fig. 3.44 Diagram of a sputter deposition system (reactive HF sputtering) for surface coating

- Electroplating is the electrochemical deposition of metallic deposits (coatings) on objects. In electroplating, an electric current is sent through an electrolytic bath. At the positive pole (anode) is the metal to be applied, e.g. copper or nickel, at the negative pole (cathode) is the object to be coated. The electric current detaches metal ions from the consumption electrode and deposits them on the target object by reduction. In this way, the object to be finished is evenly coated on all sides with the target metal. The longer the object is in the bath and the higher the electric current, the thicker the metal layer (e.g. copper layer) becomes.
- Paint coatings and dipping method.
- Rotation coating (spin coating).
- Self-growing oxide films.

Microlayers are often produced by combined processes.

One of the best known is the sol–gel process, which is used to produce microporous layers of homogeneous thickness. For the production of coatings, the brines used as coating solutions must first be applied to the substrate surface. In principle, all coating processes that can also be used for paint systems are suitable for this purpose. The coating technology used has a decisive influence on the desired properties such as layer thickness, homogeneity, or low defect density. In sol–gel processes, many process parameters must be considered, such as viscosity of the coating solution, flammability of the solvent, volatility of the solvent, geometry of the substrate (planar, curved, cylindrical, complex shapes), size of the surface, flexibility, and fracture behaviour of the substrate (flat glass or metal strip) and many

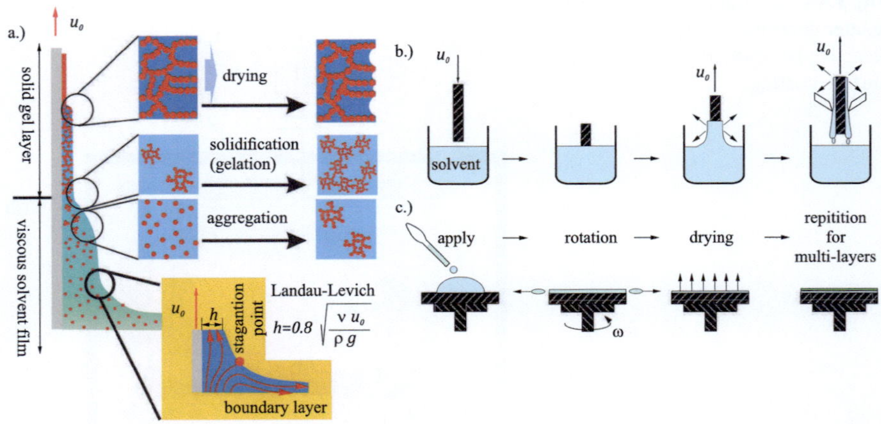

Fig. 3.45 (**a**) Schematic representation of the continuous solidification of a thin surface coating along a vertical flat plate using the sol–gel process; (**b**) dip-coating and (**c**) spin-coating

more. During the entire coating process and the drying of the sol, the hydrolysis and condensation reactions of the precursor molecules (or precursor) used will continue until the aggregation of the sol particles results in a solid gel film. The process is sketched in Fig. 3.45a.

The pre-treatment and cleaning of the substrate is of central importance. Contamination can lead to inadequate wetting with the coating solution. Particulate impurities such as dust also often lead to local defects. In order to improve the adhesion of the coating, metal surfaces are often provided with a thin oxide layer by brief heating (tempering of the surface). The simplest way to evenly wet a surface with the substrate is to immerse and pull it out at a constant speed. This process is known as *dip coating* and is shown in Fig. 3.45b. In *spin coating*, which is shown in Fig. 3.45c, the coating solution is distributed by fast rotation, which is why planar or slightly convex surfaces are particularly suitable. Compared to dip coating, the maximum substrate size is mechanically more limited.

Hybrid polymer coating materials may contain little or no solvent, while the proportion of volatile components in inorganic systems is relatively high. In general, however, the microscopic processes during drying play an important role. During film production, a gel network forms before the solvent has completely evaporated. During further drying from the gel film, liquid menisci are formed which allow capillary forces to act on the pore walls. The magnitude of these forces can be described by a modified form of the Kelvin equation. The forces increase strongly with decreasing meniscus radius and reach high values for gel films with typical pore diameters well below$100 nm$.

(Sakka, 2005) describes that the solid gel layers can only condense perpendicular to the substrate plane during drying. In addition, an unhindered shrinkage of the film in the substrate plane would lead to a "pulling back" of the film from the edges of the substrate and thus to damage. This tendency is counteracted by the adhesion of the layer to the surface. If the tensile stresses occurring in the plane exceed the internal stability of the gel film, drying cracks will occur.

For the production of micro-structured surfaces, the use of mostly multi-stage, more complex processes are required. Structure generation can be done, for example, as in semiconductor manufacturing, using masks, lithography or even electron beam processing.

In holographic lithography, two expanded laser beams are superimposed at a certain angle. The interference image of this superposition is a sinusoidal intensity modulation with which a photoresist (light-sensitive material) is exposed. During development, this photoresist is ablated, creating a surface grid that can be moulded into metals with virtually no structural loss. This impression, in turn, serves as a pattern for the structuring of polymers with which large areas of anti-reflective, transparent surfaces can be produced at low cost.

In this context, a process to produce micro-structured surfaces, the so-called *LIGA* process, which was specially developed in Karlsruhe, will be discussed. The German acronym *LIGA* (stands for the process steps: lithography, electroplating and impression taking) describes a process based on a combination of deep lithography, electroplating and micro impression taking. The *LIGA* method was developed in the early 1980s at the then Nuclear Research Centre Karlsruhe by a team led by Becker and Ehrfeld (Becker, et al., 1982) was developed as part of the development of the separation nozzle process for uranium enrichment in order to be able to produce extremely small separation nozzles. The process is carried out in five steps:

- The starting material is a flat substrate, for example a silicon wafer or a polished disk made of beryllium, copper, titanium, or other materials. The substrate, if not already electrically conductive, is provided with a metallic "seed layer", usually by sputter deposition or evaporation. A thick photo or X-ray sensitive positive resist (often PMMA) is applied to the starting layer Fig. 3.46a).
- The resist is exposed (Fig. 3.46b).
- After developing, a negative mould of the metal structure remains (Fig. 3.46c).
- In an electroplating process, a metal is deposited on the substrate in those areas where the resist has been removed during development (i.e. where the seed layer has been exposed) (Fig. 3.46d).
- After removal of the resist, the substrate, the seed layer, and the electrodeposited metal remain.

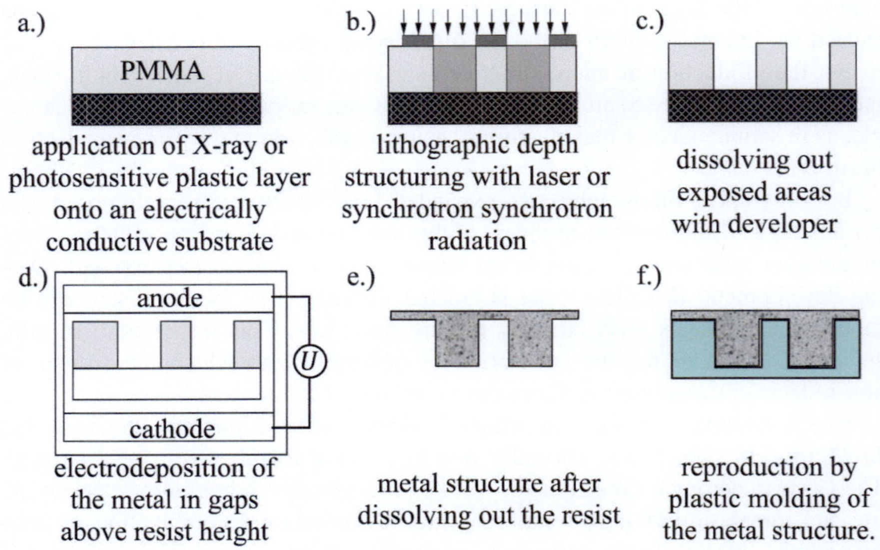

Fig. 3.46 Individual process steps of the LIGA process for the production of microstructured surfaces (Becker, et al., 1982)

Now there are different possibilities for the further procedure:

a. By etching the seed layer (which now functions as a sacrificial layer) and possibly the substrate, (small) metallic components can be produced directly.
b. By further electroplating ("overgrowth") and subsequent removal of substrate and germ layer, a mould insert can be produced from the lithographically produced microstructure, which is then built into a moulding tool, with which in turn the ultimately desired plastic component is moulded, for example by injection moulding or hot stamping (Fig. 3.46e and f).
c. Alternatively, the mould insert can be cut directly from the substrate with the lithographically produced microstructure, for example by spark erosion, and incorporated into a moulding tool.

If the steps of exposure, development and electroplating are repeated several times, more complex structures develop, but these must taper towards the substrate (the wafer), otherwise the component would not be able to be released from the mould. This limits the complexity of the structure to be produced. If the mould produced by *LIGA* is used for injection moulding, a special feature results in contrast to the production of macroscopic parts. There is no need to provide holes for the air present in the mould to escape, as the unevenness of the contact surface between the mould and its counterpart is sufficient for the air to escape when manufacturing components several hundred micrometres in size. Only the holes for feeding the material to be sprayed must be created.

3.3.5 Low and Medium Temperature Absorber Layers in Air/Vacuum

Selective absorbers for flat-plate and vacuum tube collectors were developed in the 1980's to improve the performance of these collectors. The coatings were mainly based on the Cermet principle. They had to be stable in air for stagnation temperatures up to $250 - 300\ ^{\circ}C$. Commercially the first products were produced with galvanic deposition. Until about 1995 essentially two fundamentally different types of absorber layer were prepared:

- The first absorber layer called "black chrome" proved to be very stable with good selectivity. The layer is applied by a galvanic bath process, which means inherently wastewater containing heavy metals and especially chromium is produced. The layer thickness of a black chrome layer is about $5 - 10\ \mu m$ and must be removed in the recycling of used absorbers.
- The second layer called "black nickel" is a type of "nickel-pigmented alumina": It has a more limited lifetime due to sensitivity against moisture in the collector. Moreover, the optical selectivity is modest. The production of these layers is, however, less energy-intensive and polluting. A very successful product in the first years of selective collectors was the coated strip-absorber which could be applied from a roll.

In 1995, the TiNOX GmbH was the first company with a highly selective Cermet absorber coating for solar thermal collectors on the market, produced by physical vapour deposition (PVD) in vacuum. The TiNOX absorber layer is composed of several functional layers: a metal substrate is used as a carrier material having good IR reflection. A thin titanium carbide compound serves as an adhesion-promoting layer and diffusion barrier. The absorber layer is made of a titanium compound with oxygen and nitrogen. This patented The Titanium-oxynitride gave the "TiNOX" company its name. On top of this layer there is an additional anti-reflection and protective layer of silicon oxide. This minimizes the surface reflection. Due to interference effect the optimized anti-reflection layer reflects small amounts of blue light, so that the absorber appears "blueish". The properties of these commercial products are given amongst others in Table 3.7.

Decisive for selection of a suitable selective absorber layer is the range of the absorber temperature achieved in the collector, and the maximum expected exposure, which is dependent on the location (climate, roof pitch, ...), the type of collector (flat, tubular storage collector) and the intended use (process heat or

Table 3.7 Commercial selective absorber coatings and their specific characteristics

Product	TiNOX	Sunselect	Diamanta	Sunstrip	Gibo
Functional layer	TiN_xO_y	Cermet	–	$Ni - NiO$ black nickel	$Cr - CrO$ black chrome
Substrate	Cu	Cu	$Cu(Al$, stainless steelt	Al	Cu
α_s [/]	0.95	0.95	0.9 – 0.96	0.92 – 0.96	0.96
ε_{IR} [/]	0.05	0.05	0.02 – 0.05	0.1	0.1
$\alpha_s/\varepsilon_{IR}$ [/]	19	19	12.85 – 19.2	9.2 – 9.6	9.6
Production route	PVD	Sputtering	Combined sputtering CVD	Sputtering	Sputtering or electro-plating

power use). Although there is no clear distinction between low and high temperature applications, for the absorbers, an imaginary border at about $250°C$ is considered for example in (Kennedy, 2002) for the low to intermediate temperature range and above $250°C$ for the high temperature range.

For the absorber layers, the temperature range is essential, because chemical reactions like oxidation are strongly temperature-dependent, and because due to different thermophysical properties of substrate material and coating mechanical stresses might develop.

Several physical principles and a wide range of different production methods can be used to produce solar selective absorbers. The selectivity and corresponding optical properties are very important for assessing the relevance for solar thermal applications. However, other factors like stability and cost of production certainly come into consideration. Multilayer cermet absorber coatings seem to be the most adaptable and selective class of high-temperature absorber coatings used up to now. Although the inherent properties of the coatings (materials chosen, number of barrier layers and contact layers, thicknesses, and number of different layers as well as gradients) are important for the optical properties and the stability of a material, other factors add to that. The roughness and purity of the substrate surface, the substrate material and phase transitions have large impacts on the practical applicability of an absorber coating type. Production processes must be stable and controlled to guarantee homogeneous properties and long-term performance of selective absorbers at elevated temperatures.

As a consequence, good laboratory results cannot automatically be transformed into good products. Nevertheless, laboratory research and numerical optimization and modelling of absorber coatings can stimulate development. A large number of historical approaches for selective absorbers in the medium temperature $(150 - 250°C)$ have been described in a comprehensive review (Kennedy, 2002). The various approaches are listed in the overview Table 3.8. Especially the sol-gel production technology might be interesting commercially, especially for the coating of tubular absorbers for solar process heat collectors.

Table 3.8 Different absorber layer systems for the mid temperature range when operating in a vacuum and air (Kennedy, 2002), $TSSS$=thickness sensitive spectrally selective layer, SS=stainless steel

Material	Substrate	Fabrication process	Absorptance α_s	Emittance $\varepsilon(100°C)$	Stability limit [°C] vacuum air	Comm. product
$TSSS$	Al	Painted	0.92	0.015	<135	Solaris Select
PbS	Al	Evaporation pyrolysis	$0.93 - 0.99$	$0.1 - 0.21$	~200	
$NiCrO_x$	SS Al polyamide	Sputtering	0.8 $0.92 - 0.93$	0.14 0.06	<200 <200	
coloured SS	SS	Chem. convection	$0.62 - 0.93$	0.1	<200	SEL INCO
Black Nickel NiS/ZnS $NiSn$	Ni / steel Cu / steel	El.-deposition sol–gel	0.88–0.96 0.92–0.98	0.03–0.1 0.08–0.25	<200 300	Maxorb black crystal
Graphite films $aC : H/Cr$ $A - C : H$	Cu Al	MF-pulse PVD/PEVD	0.92 0.876	0.025 0.061	250 250	
Ge in Si Si PbS	Cu, SS	Painted	0.91 0.83 0.96	0.7 0.7 0.7	300	
Ag diel	Al	Sol–gel			300	
Black copper $Cu/Cu_2O/Cu$	Cu	El.-deposition	0.97 - 0.98	0.02	370250	
Black chrome Cr/Cr_2O_3 Mo/Cr_2O_3	$Ni - Cu$ Cu / steel	El.-deposition	0.97 0.95	0.09 < 0.3	400 3 50 425	MTI, Chrome Coat, Energie Solaire
TiN_xO_y	Cu	El. beam vapor deposition / sputtering	0.92	0.06	400	TiNOX, Almeco
$CuFeMnO_4/Si$	glass,Si	Sol–gel	0.6	0.29–0.39		
$Cr, Fe, Mo,$ $SS, Ta,$ $Ti,$ Wsilicates	Cu	DC reactive sputtering	$0.76 - 0.82$ $0.81 - 0.86$	$0.02 - 0.3$	400 250	

(continued)

Table 3.8 (continued)

Material	Substrate	Fabrication process	Absorptance α_s	Emittance $\varepsilon(100°C)$	Stability limit [°C] vacuum air	Comm. product
$Cr, Fe, Mo,$ $SS, Ta,$ $Ti, WCarbide$	Cu	DC reactive sputtering	$0.76 - 0.81$ $0.84 - 0.90$	0.02 $0.035 -$ 0.06	400 250	
$Ni - NiO$	Al	Reactive sputtering	0.96	0.10	300	Sunstrip
Ni-pigmented Al_2O_3	Al	Anodization	$0.85 - 0.97$	$0.08-0.21$	400	Showa

3.3.6 Absorber Layers for High Temperature Applications

In concentrating systems higher temperatures are achieved. Here, it is very difficult to produce absorber coatings stable in air. Therefore, the development of vacuum receivers came first, which also had the advantage of reducing even further the heat losses when compared to a system filled with air. However, the tightness of the vacuum tube over many years of lifetime is another challenge. Especially for organic heat transfer media this is an issue.

Historically the first commercially successful providers of vacuum receivers and corresponding absorber coatings for parabolic trough collectors after the pioneers LUZ were Solel Solar Systems, Israel, and Schott Solar CSP GmbH, Germany. The absorber tube is covered by a borosilicate glass tube to keep the vacuum and to reduce the heat losses. Polished stainless steel tubes (mean roughness index $R_a < 0.2\mu m$) are heated in vacuum for 10 minutes at a temperature of $600°C$. Then the coating system consisting of Al_2O_3 diffusion barrier, the $Mo-$ IR-mirror, the Mo/Al_2O_3-cermet and the SiO_2 AR layer are deposited on the tube (Lanxner and Elgat, 1990). Solel and Schott for a long time dominated the market with their coatings stable up to about $500°C$.

In 2005 in Italy a DC (Direct Current) sputtered absorber system of a $500nm$ thick tungsten (W) or ZrN IR-mirror, a cermet of TiN_x, ZrN_x or HfN_x in AlN. The AR-layer is made from AlN or Al_2O_3. The absorber coating is stable in vacuum up to $580°C$ and reaches $\alpha > 95\%$ and $\varepsilon(580°C) < 12\%$. It is specially developed and used at the moment for receivers with molten salt as a heat transfer fluid (HTF) exceeding the temperature of the HTF thermo-oil at $390°C$.

The market of the suppliers of vacuum receiver has undergone major changes in recent years. While the Spanish company Rioglass now has taken over the technologies of Solel and Schott, in China other producers developed. Table 3.9 provides an overview of the properties of some commercial products.

An overview of various laboratory development of high-temperature selective coatings stable in vacuum or even in air is given in Table 3.10. Many systems use the expensive metals platinum (Pt), Molybdenum (Mo) or Tungsten (W) as

Table 3.9 Comparison of absorptance and emittance of commercial high-temperature vacuum tube receiver coatings

Product	Rioglass Solar S.A.U (Rioglass Solar SA, 2022)	Archimede Solar Energy S.r.l (Tagle-Salazar and Nigam, 2020)	Royal Tech CSP Limited	Himin Solar Co. Limited (Hi-min SOLAR, 2022)
α $AM1.5$ direct (ASTM)	0.9452	0.95	0.93	0.96
$\varepsilon(T = 400°C)$	0.0953	0.073	0.098	0.11
τ ASTM	0.96	0.965	0.961	0.96
Max. operation temperature [$°C$]	350	580	400	400

infrared mirror and Pt in Al_2O_3 as Cermet. Material and the slow high frequency sputtering process make these designs costly. A commercially successful development could be based on silver (Ag) in a DC sputtering process using polished stainless steel as a substrate. Theoretically this would lead to lower emissivity at high temperatures and therefore better solar thermal efficiencies. But a problem of silver is the agglomeration in vacuum, but that can be overcome by a special multi-layer design using barrier layers.

For use at high temperatures in air no selective absorber coatings are commercially available. There are concepts for absorbers for temperatures above $350°C$ at laboratory scale. One of the most stable material candidates for absorber systems is platinum (Pt). Many authors describe it as a stable IR-mirror, and embedded in Al_2O_3 it forms a stable cermet (Thornton et al., 1980; Vien et al., 1985; and Sella, et al., 1989).

The most extensive investigations describe the influence of different process parameters on the stability of the system $Pt-$ IR-mirror, Pt/Al_2O_3 -cermet and Al_2O_3 AR-layer on glass and on stainless steel with an Al_2O_3 diffusion barrier. Sputtering with RF (Radio Frequency) produces better results than DC. Also sputtering with higher substrate temperatures seems to be beneficial for stability and low emittance. However, platinum is an extremely expensive material, and secondly RF sputtering is slow, which makes a commercial production process expensive. Also, the substrate heating requires longer production times and additional investment costs. Highly alloyed substrates are necessary which are too expensive for receiver tubes.

Therefore, less expensive materials and processes are still required. Substrates used should be only stainless steel tubes and not extremely expensive materials, at least for linear reflectors. The suitability of many other materials Cr, Mo, Ni, and Ta have therefore been investigated as IR-Mirrors. One problem is the relatively fast oxidation in air for these materials, although embedding in the cermet layer helps. Silver (Ag) is an excellent IR-mirror and on the other hand has, as a noble metal, a relatively low tendency for oxidation. The main degradation mechanism is the formation of Ag_2S through reaction with H_2S in air, see (Graedel,

Table 3.10 Common absorber layer systems for high temperature applications in air and vacuum adapted from (Kennedy, 2002), SS=stainless steel

Material	Substrate	Fabrication process	Absorpt α_s	Emittance $\varepsilon(100°C)$	Stability limit $[°C]$ vacuum	Stability lim $[°C]$ air
$Ni - Al_2O_3$	$Mo-Ni-SS$	RF sputtering	0.94	0.07	500	350 – 400
$SiO_2 - AR$	$Ni - Al$	RF sputtering	0.94	0.04	350 – 500	600
$Co - Al_2O_3$	$steel$	RF sputtering	0.96	0.16(350)	500	600
$Mo - Al_2O_3$	Cu	CVD	0.97 – 0.98	0.1 – 0.07(400)		
$W - Al_2O_3$		RF sputtering	0.85	0.04		
$Pt - Al_2O_3$			0.90 – 0.98	0.08		
$Al_2O_3 - Pt - Al_2O_3$						
$Mo - Al_2O_3$	Cu	DC sputtering	0.96	0.06(350)	500	
$SS - AlN$			0.95	0.10(350)	350 – 500	
$Mo - AlN$			0.92 – 0.94	0.08 – 0.1(350)	500	
$W - AlN$					500	
Quasicrystals	Cu, Si		0.90	0025	500	400
Multi-layer cermet			0.86 – 0.92	0.031 – 0.05	550	
$Si_3N_4/Si - Ge/Ag$	SS, Al	CVD	0.89	0.0389 (300)0.0545 (500)		650(He)
$Ni - -SiO_2$	Al, Cu	DC reactive sputtering	0 .90 – 0.96.	0.03 – 0.14		400 – 800(Ar)
$Cr - SiO.$						
$Al - AlN_x - AlN$	SS	DC reactive sputtering	0.97	0.10	500	
CuO	Cu, SS	Electroplating	0.91	0.18	700	< 400
$Ag/CuO/Rh_2O_3/CeO_2$	SS	Organ.-metal. spray	0.90	0.10	775	500
$CeO_2/Ag/Pt/CuO,$			0.86 – 0.88	0.1	700	550
$Rh/Rh_2O_3/Ag/Pt$			0.88 – 0.92	0.06 – 0.12		500
$CeO_2/CuO/CoO, Mn_2O_3/Pt$						

(continued)

Table 3.10 (continued)

Material	Substrate	Fabrication process	Absorpt α_s	Emittance $\varepsilon(100°C)$	Stability limit $[°C]$ vacuum	Stability lim $[°C]$ air
Black cobalt Co_3O_4/Co $Ni - Co_3O_4/Co$			$0.96 - 0.92$ 0.95	$0.71 - 0.017$ 0.10		$400 - 650$
Black molybdenum $Mo - MoO_2$		CVD	0.94	0.3(500)	500	350
Black tungsten $W - WO_x$		CVD	0.83	0.15	800	
$Au - TiO_2$ $Au - MgO$	SS Mo, SS	Sol–Gel RF sputtering	0.85 $0.90 - 0.93$	0.01(400) $0.04 - 0.1$	> 500	400
ZrC_xN_y $Al_2O_3/ZrC_xN_y/Ag$ $ZrO_x/ZrC_x/Zr$	Al S S		0.85 0.91 0.90	0.074(325) 0.05(325) 0.05(20)	600 700 700	125 175
TiN $Ti_{1-x}Al_xN$	Cu, Al	DC reactive sputtering	0.80	$14 - 0.4$		500 $750 - 900$
$M_bO_c + M\prime Fe_2O_4$	$Ni - Mo$ alloy	Brushing, plasma	> 0.90	> 0.45		700 1060

(continued)

Table 3.10 (continued)

Material	Substrate	Fabrication process	Absorpt α_s	Emittance $\varepsilon(100°C)$	Stability limit $[°C]$ vacuum	Stability lim $[°C]$ air
VB_2, NbB_2, TaB_2	Glass	DC reactive	0.99	$0.95 - 0.97$	$2300 - 3040$	500
$TiB_2, ZrB_2LaB_6,$	ZrB_2	sputtering	$0.88 - 0.93$	$0.08 - 0.12$	$(Melt)$	
$WSi_2, TiSi_2,$		CVD				
$Si_3N_4, AR - ZrB2$						
C-textured Cu	Cu		0.9	0.04(20)	400	< 300
textured Ni	Ni		0.92 ± 0.2	0.09 ± 0.02	> 440	> 500
textured SS	SS		0.93 ± 0.02	0.22 ± 0.02	550	> 600
textured Cr	Cr		$0.8 - 0.9$	$0.1 - 0.3$		800
W whiskers,Mo, Rh, Pt, W, HfC, Au			0.98	>0.26		
NiO, CoO_3						

1992). Also, the thin layers tend to agglomerate for temperatures above $200°C$, cf. (Hollingsworth Smith and Gurev, 1977). Therefore, protection and adhesion layers are needed to stabilize Ag in the system. Using that approach absorber coatings on stainless steel tubes with $\alpha > 95\%$ and $\varepsilon(450°C) < 12\%$ could be produced but do not have long-term stability for testing temperatures of $500°C$ (Hildebrandt, 2009). However, based on the emittance of polished stainless steel a CrO_x cermet in combination with a $SiO_x - AR$ layer was stable 3000h at $500°C$. The optical properties reached for such a system were $\alpha = 94\%$ and $\varepsilon(450°C) = 18\%$ (Hildebrandt, 2009). For solar towers even, higher temperatures have to be reached, but the requirement on selectivity is lower due to the higher concentration. Here absorption is the key parameter.

High solar absorptance (non-selective) coatings can be easily applied by spray coating. The coating equipment is cheap and the possibility of being able to recoat the receiver on the tower after a performance loss due to degradation is a big advantage. During the coating process attention must be paid to the homogeneity of application as the rather low heat conductivity of the thick coatings (around $100\mu m$) might lead to hot spots and result in a degradation of the receiver. Currently developments are ongoing.

3.3.7 Degradation Mechanisms of Selective Absorber Layers

Absorber coatings of the various sorts that have been discussed are precisely formulated to achieve the optical properties that are desired. They must operate at elevated temperatures, subject to temperature fluctuations, and need to achieve long lifetimes to be practically useful.

Surface temperatures experienced during operation which exceed the production temperatures of the absorber systems may lead to structural reordering within the materials, to diffusion and redox reactions. The optical properties will usually change as a consequence. High temperatures above $400°C$ pose a special challenge because kinetics for most processes is much faster at higher temperatures. A short overview of degradation processes is given here based on reference (Hildebrandt, 2009).

3.3.7.1 Diffusion Processes

There are different kinds of diffusion processes that can contribute to performance degradation. Diffusion of atoms along interstitial lattice sites or lattice vacancies plays a minor role if the material in question exhibits lattice disorder like grain boundaries or dislocations. The diffusion along these defects is several orders of magnitude faster than the first process and is also called “short circuit diffusion” (Heumann and Mehrer, 1992). Sputtered layers may also have a high porosity. The diffusion of molecules, for example of oxygen from air, may take place.

Diffusion processes may take place between substrate and absorber systems or within the individual layers. In cermet layers even, the diffusion within the cermet itself changes the gradient of constituents and therefore the optical properties of the system.

To prevent or at least substantially reduce the diffusion of atoms e.g. from the substrate (Fe, C) into the IR-mirror layers, barrier layers can be prepared. Oxides, for example Al_2O_3 have been used for these layers. High density of the barrier layer and continuous layer formation is important. These layers must cope also with thermal expansion and stresses of the substrates because cracks will destroy the barrier functionality. A different approach is taking a thick IR- mirror metal layer as barrier.

3.3.7.2 Oxidation

The oxygen in air is obviously a threat for degradation due to oxidation in high temperature absorbers operated in air. Metallic layers of the IR-mirror and the metallic particles in the cermet layer are subject to oxidation. Barrier layers between surrounding air and the absorber system must prevent oxygen diffusion without destroying the desired optical properties of the absorber system.

3.3.7.3 Redox Reactions

Redox reactions are important in cermet layers even in vacuum. During the production of a reactively sputtered cermet Me_1/Me_2-oxide,[1] e.g. a Mo/Al_2O_3-cermet, the metallic constituent Me_1 (Mo) usually is also partially oxidized whereas the second metal Me_2 (Al) is not completely oxidized. During a tempering process redox reactions drive the system towards chemical equilibrium, i.e. the oxidized Mo will be reduced and give the oxygen to the metallic aluminium. The size and shape of metallic particles in a cermet might also change. All these processes provoke strong spectral changes in the reflection spectrum and hence a change in solar absorptance and thermal emittance.

3.3.7.4 Thermo-Mechanical Stresses

The temperature differences occurring within high temperature receivers and / or, unequal thermal expansion coefficients of adjacent layers may lead to thermally induced stresses on the individual layers. This in turn may induce flaking or chipping off of layers. Cracks develop and barrier layers are destroyed locally. The problems may be ameliorated by introducing special adhesion layers between problematic materials.

3.3.7.5 Other Environmental Stresses

Apart from the mechanisms described above, a number of other environmental stresses may deteriorate the optical properties of an absorber system reversibly or irreversibly. For instance, for non-hermetically sealed receivers open to air, humidity and even dust, several reactions may occur. For example, dust on a hot receiver

---

[1] Me_1 And Me_2 references general combinations of metal 1 and metal 2.

may fasten on the surface irreversibly. Condensation of humidity during cold periods (night-time) in combination with gases like sulphur or carbon dioxide may produce acids attacking surface layers of the absorber. Sulphur itself can be a problem e.g. for silver IR-mirrors, if the oxide is reduced. For plants built in the proximity of an ocean, salt may react aggressive with both steel and absorber system.

Although accelerated testing (Brunold, Frei, Carlsson, Möller, and Köhl, 2000a, 2000b) may try to incorporate all conceivable effects, there will be always a risk that the exposure in the real environment will have additional effects. The conditions at a specific location may be different from what is expected in a standard atmosphere, and interactions between different mechanisms may not be the same as in a test. Nevertheless, it is important to develop methods for estimating long-term stability and predicting service lifetime as far as possible to reduce the risks for degradation and failure.

In addition to the initial efficiency, long-term stability is an important requirement for absorber coatings. Complex designs of absorber coating systems are frequently susceptible to changes upon exposure to elevated temperature. However, these changes may even help to improve the performance of a freshly produced absorber system. Tempering processes may improve the optical properties but should stop after a short period of time. Slow continuous changes on the other hand are likely to lead to long-term performance reduction and must be quantified and eliminated.

In order to quantify "degradation" of an absorber system a so-called Performance Criterion (PC) is required. For flat-plate collector selective absorber testing (i.e., non-concentrating, $1-2X$ sunlight intensity) such a criterion has been developed (Brunold, Frei, Carlsson, Möller, and Köhl, 2000a, 2000b). The PC describes the influence in the change of solar absorption ($\Delta\alpha_s$) and emittance ($\Delta\varepsilon$) on the solar fraction:

$$PC = -\Delta\alpha_s + 0.25 \cdot \Delta\varepsilon \leq 0.05. \tag{3.64}$$

It provides a single quantifiable parameter for maximum acceptable degradation. Service lifetime testing for this criterion is performed by exposing the absorber coatings of flat-plate collectors for $200h$ at $250°C$. If the material survives, it is then exposed for $75h$ at $300°C$, followed by $600h$ at $40°C/95\%$ relative humidity (RH), then $85h$ at $60°C/95\%RH$, for more details see (Brunold, Frei, Carlsson, Möller, and Köhl, 2000a, 2000b). After exposure testing, the emittance is typically measured at 100 °C. A similar criterion has been developed for testing the service lifetime of high-temperature absorbers for CSP applications. It is obvious that for concentrating collectors due to the concentration a change in the absorptance is even more important than for non-concentrating collectors. On the other hand, emittance at higher temperatures is more important due to the Stefan-Boltzmann law. Therefore, we would suggest as a first approach to use a PC for concentrating collectors:

$$PC = -\Delta\alpha_S + \frac{0.25 \cdot f}{C} \cdot \Delta\varepsilon \leq 0.05 \text{ with } f = \frac{\sigma \cdot \left(T^4_{\text{op,CSP}} - T^4_{\text{amb}}\right)}{\sigma \cdot \left(T^4_{\text{op,FPC}} - T^4_{\text{amb}}\right)} \text{ and}$$
$$T_{op,CSP} = 273.2 + 340\ K;\ T_{op,FPC} = 273.2 + 60\ K;\ T_{amb} = 273.2 + 25\ K. \tag{3.65}$$

where C is the concentration factor, f is a factor describing the higher operating temperatures compared to flat-plate hot water collectors, T_{amb} is ambient temperature; T_{op} is the operating temperature of the solar system and the subscripts FPC denote a flat-plate collector and CSP a concentrating collector for solar thermal power plant.

Thermal stability is sometimes determined by the thermal properties of the individual materials or the processing temperature parameters. Durability or thermal stability is typically tested by heating the selective coating, typically in a vacuum oven or in air at elevated temperatures, for a relatively short duration (thousands of hours) compared to the desired lifetime ($25 - 30 years$). This procedure of accelerated indoor testing may pose problems if cascaded processes and interactions during exposure occur (Kennedy, 2002). Degradation of high-temperature absorbers usually changes the reflectance spectrum significantly and causes increasing emittance; therefore, optical properties are sensitive indicators to monitor degradation with exposure.

3.4 Optical Concentration

3.4.1 Motivation for Optical Concentration

The solar collector energy balance is strongly dependent on the absorber or fluid temperatures reached. The thermal loss as compared to an ordinary black absorber plate can be reduced by many means, but in fact they limit the maximum temperatures which can be achieved. Concentration of solar radiation is an elegant possibility to extend the temperature range of solar collectors. When the low power density of solar radiation with maximum around $1000 W/m^2$ can be focused from a large collecting area, the aperture, to a much smaller absorber area, then the losses for a specified absorber temperature can be reduced dramatically. And although the thermal losses increase non-linearly with absorber temperature, the break-even temperature where losses and gains for the small area are equal (for collectors usually called stagnation temperature), increases also very much. Absorber temperatures of more than $1000°C$ can be achieved in practical systems, and very often the limitation is rather the stability of the collector material rather than the energy balance. This is the reason why concentration of solar radiation is such a powerful tool.

The ratio between collecting aperture A_a and reradiating receiver area A_r is called the geometric concentration ratio C_g, independently of the optical efficiency of the optical concentrator system trying to achieve that:

$$C_g = \frac{A_a}{A_r}. \tag{3.66}$$

Optical efficiency is influenced by the optical properties of real materials like mirrors or lenses which do not have 100% reflectance or transmittance, respectively, even for perfect concentrator geometry. On top of that the shapes of real optical concentrators deviate to some extent from the ideal geometry, on the nano level due to surface irregularities up to the macro level due to mechanical stresses, imperfect support structures, thermal expansion, wind load sensitivity and many other factors. In some applications optical efficiency also might be compromised by design, for example if a homogeneous irradiation distribution on the receiver area is a target. Also, economic considerations play a role, and therefore often concentrator designs lead to a non-perfect concentration with some radiation not directed towards the receiver area. The optical efficiency (η_{opt}) is a measure for the irradiation collected by the aperture compared with the radiation reaching the receiver and is thus defined by.

$$\eta_{opt} = \frac{\iint^{A_a} I_0(x,y)dxdy}{\iint^{A_r} I(x,y)dxdy} = \frac{A_a \bullet I_0}{\iint^{A_r} I(x,y)dxdy}. \tag{3.67}$$

The intercept factor, which also frequently used especially for imaging concentrators (the ones which have a focal point or line as the image of the sun), is very similar to that definition, as it is a measure for the irradiation redirected towards the receiver and reaching the receiver. In the case of a flat receiver area one could write the intercept factor γ as:

$$\gamma = \frac{\iint_{-\infty}^{+\infty} I(x,y)dxdy}{\iint^{A_r} I(x,y)dxdy}. \tag{3.68}$$

The optical efficiency relates the received radiation to the input of the optical concentrator system, the intercept to the (redirected) output of the concentrator. When we look at a heliostat for example, the average reflectance of the heliostat surface should be multiplied by γ to get optical efficiency η_{opt}.

An optical concentration ratio is defined for any part of the receiver area as the intensity at the receiver $I(x, y)$ divided by the incoming solar intensity I_0 hitting the aperture. The optical concentration is often reported as a factor X or as X suns. It may be specified as peak optical concentration for the focal area or averaged over the whole receiver:

$$C_{opt} = \frac{I_r}{I_0}. \tag{3.69}$$

The average optical concentration ratio is obviously related to the intercept factor by:

$$\overline{C_{opt}} = \eta_{opt} \cdot C_g. \tag{3.70}$$

Concentrating systems only make use of direct beam radiation. They do not collect diffuse radiation, except some concentrators with rather small concentration ratio in the range of 1.3 to 5. The reason is that the acceptance angle due to fundamental physics is becoming smaller when the concentration ratio increases. This leads to a smaller and smaller view angle of the concentrator system to the sky, resulting in negligible diffuse contribution. Therefore, generally a part of the global irradiation is not used in concentrating systems. The higher temperatures possible can be offset by small energy gains in climates with little direct beam radiation. This is often a reason for selecting a concentrating or a non-concentrating system design for solar thermal applications. For very high point concentrating systems one has even to consider that a part of the circum-solar radiation (CSR) may be cut out due to a small acceptance angle. So, these limits to concentration must be understood and considered in the system design and selection.

3.4.2 Limits of Optical Concentration of Solar Radiation

The concentration ratio C_g determines the absorber temperature and thus also the long-term performance of the solar thermal system. It is easy to see, through simple considerations, that there are some limits to the concentration ratio. It's these:

- Divergence of solar radiation.
- Technical design of the imaging optics (focal point or focal surface design).
- Precision of the image (surface quality, cleanliness, temperature, …).

In addition, there are also material-technical questions concerning the mirrors or lenses as well as the type of tracking and its quality.

The sun produces direct divergent irradiation on the Earth's surface where the solar radiation comes within a cone divergence angle θ_s as illustrated in Fig. 3.47, the angle of view of the sun disk. This has a consequence for the maximum optical concentration possible. Assume a two-dimensional, perfectly concentrating system, which concentrates the incident radiation with a circular aperture of the surface A_a onto a receiver (or absorber) of the surface A_r. At the same time, both surfaces should be ideally perpendicular to the direction of irradiation. According

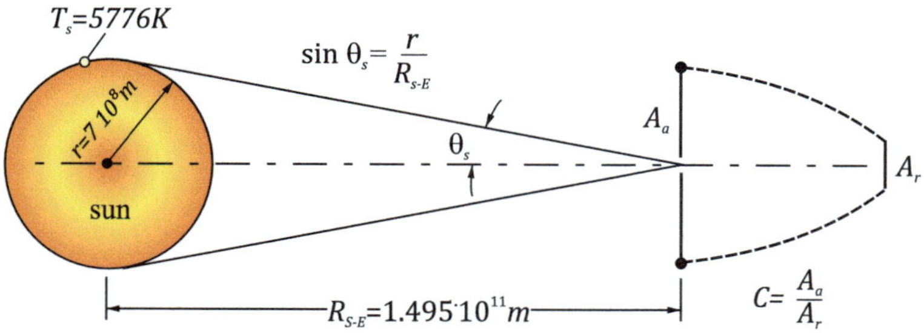

Fig. 3.47 Derivation of the angular distribution of the solar radiation

to the sketch, the solar radiation power $\dot{Q}_{S-a}$. indicated in the Eq. (3.71) falls on the aperture and further on the receiver.

$$\dot{Q}_{S-a} = A_a \frac{r^2}{R_{s-E}^2} \sigma T_s^4, \tag{3.71}$$

where A_a is the aperture area, R_{s-E} the average distance of the earth from the sun with $R_{s-E} = 1.495 \cdot 10^{11}\, km$, r is the radius of the sun with $r = 7 \cdot 10^8\, m$ and σ the Stefan-Boltzmann constant ($\sigma = 5.67 \cdot 10^{-8} W/\left(m^2 K^4\right)$). T_s describes the surface temperature of the sun. The angle of view of the sun θ_s directly results as:

$$\sin\theta_S = \frac{r}{R_{s-E}} \text{ with } \theta_s = 0.27^\circ = 16'. \tag{3.72}$$

Assuming that the receiver is a "black radiator", the following relationship results for the radiation power of the absorber (receiver) in the direction of the sun $\dot{Q}_{r-s}$:

$$\dot{Q}_{r-s} = A_r \cdot \sigma \cdot T_r^4 \cdot \varepsilon_{r-s}. \tag{3.73}$$

In Eq. (3.73) T_r describes the temperature of the absorber (receiver). ε_{r-s} is the emission coefficient between absorber (receiver) and sun reflects the radiation exchange between the receiver and the sun. The maximum emission can be the radiation of a black body, i.e. the maximum emission coefficient is $\varepsilon_{r-s} = 1$. According to the second law of thermodynamics, in a borderline case the receiver temperature (T_r) can be at most the radiation temperature of the sun, i.e. in a limiting case $T_r = T_s$. It follows directly from this:

$$\dot{Q}_{s-E} = \dot{Q}_{r-s} \rightarrow A_r \cdot T_r^4 = A_a \cdot \left(\frac{r}{R_{S-E}}\right)^2 T_S^4. \tag{3.74}$$

With the result of the viewing angle considerations resulting in Eq. (3.75) one obtains a maximum possible concentration ratio for C for the two-dimensional case of:

$$C_{\max,2-\dim} = \frac{A_a}{A_r} = \frac{R^2_{s-E}}{r^2} = \frac{1}{\sin^2\theta_s}. \tag{3.75}$$

Assuming the maximum angle on the aperture of $\theta_s = 0.27° = 16'$ the maximum concentration ratio of a two-dimensional concentrator is $C_{\max,2-\dim}$:

$$C_{\max,2-\dim} \approx 4.5613 \cdot 10^4. \tag{3.76}$$

For a linear concentrating mirror system that is one-dimensional, it can be shown with the help of similar considerations that the maximum achievable one-dimensional concentration ratio $C_{\max,1-\dim}$ is limited to the following value:

$$C_{\max,1-\dim} = \frac{1}{\sin\theta_s} \approx 213. \tag{3.77}$$

In practice, however, such values cannot be achieved due to surface defects and adjustment errors of the mirrors. Higher concentration ratios can be achieved with spherically symmetrical mirrors than with cylinder-symmetrical ones. Paraboloids with a C_g -value of $C = 5 \cdot 10^3 - 8 \cdot 10^3$ have been realized with great effort; however, the values are usually around $C_g \approx 10^3$. In line-focusing systems, technical values around 30 are achieved.

Another important parameter for concentrating solar systems is the so-called acceptance angle θ_A. In a real solar receiver, the position of the sun and thus the solar elevation angle α_s as well as the direction of irradiation and thus the azimuth angle γ_s changes due to the sun's daily motion. The acceptance angle θ_a is now defined as the angle at which the solar radiation is just about geometrically able to hit the absorber. (This assumption implies that a complete reflection occurs on the contour surface ($\rho = 1$) and the sun is a point source, which is also idealized). In fact, the solar disk is seen on Earth at an angle of $(2 \cdot \theta_s) \approx 0.54°$ (see Fig. 3.47).

If the acceptance angle θ_A now decreases below the viewing angle θ_s of the solar disk S, the radiation power emitted by the sun can no longer be fully utilized. This effect is known as radiation loss. If the angle of acceptance corresponds to the angle of view of the sun ($\theta_a = \theta_s$), maximum use is made of the irradiation power. From this, a maximum possible geometric concentration ratio C_g can be derived, as described by (Kreider, 1979). Through geometric considerations, it is easy to consider that with smaller absorber dimensions, the acceptance angle θ_a of a collector also decreases monotonically. However, a decreasing acceptance angle means an increasing concentration ratio and thus increasing absorber temperatures when the collector is ideally adjusted towards the sun.

The maximum theoretical concentration as discussed above and given in Eqs. (3.76) and (3.77) can be derived also from the optical theorem of conservation of etendue in a perfect optical system. In simple terms, the Étendue describes how a bundle of rays expands in geometrical optics. To calculate this, the cross-section and the solid angle are required, as well as the direction vector of the radiation. For an infinitesimal area immersed in a medium with refractive index n the Etendue is given by

$$dG = n^2 \cos\phi \, dA \, d\Omega, \tag{3.78}$$

in which dA is the surface, $d\Omega$ is the solid angle extended by the ray bundle with a propagation at angle ϕ with respect to the surface normal. This may be interpreted also a volume in phase space $d\phi d\theta$ in Hamiltonian optics. If a light cone has a divergence angle of θ_{in}, the integration over the solid angle and a finite area A yield

$$G_{in} = \pi n^2 \sin^2 \theta_{in} \, \mathrm{A}. \tag{3.79}$$

The quantity $(n\sin\theta_{in})$ in optics is called the numerical aperture NA. The Etendue is conserved not only in free space but also when refraction or reflection occurs. Consider now a concentrator system immersed in a medium with refractive index n_2 and focusing all radiation on a receiver area A_r and incoming radiation on an aperture A_a coming within a cone of θ_1 in a medium with refractive index n_1. Then the conservation of Etendue requires that $G_1 = G_2$ which implies that

$$\pi \, n_1^2 \sin^2 \theta_{in} \, \mathrm{A}_a = \pi \, n_2^2 \sin^2 \theta_{out} \, \mathrm{A}_r. \tag{3.80}$$

The situation is sketched in Fig. 3.48. The concentration C is given then by

$$C = \frac{\mathrm{A}_a}{\mathrm{A}_r} = \frac{{n_2}^2 \sin^2\theta_{out}}{{n_1}^2 \sin^2\theta_{in}}, \tag{3.81}$$

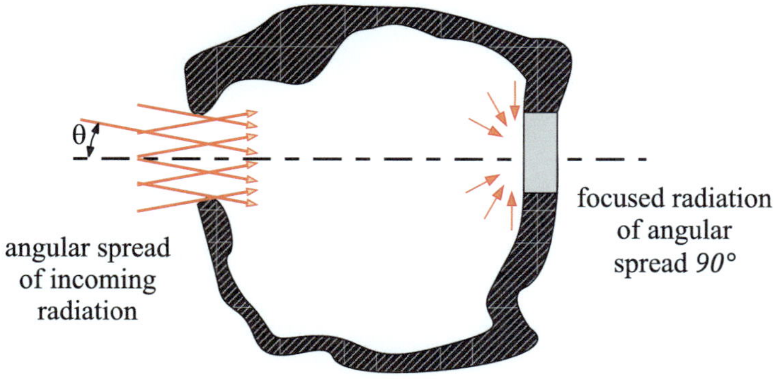

Fig. 3.48 Maximum concentration from Etendue conservation (considering media with refractive index n_i)

and the maximum concentration for incoming solar radiation within an angle θ_s in air ($n_1 = 1$) is reached with $\theta_{out} = 90°$ leading to

$$C_{\max,2-\dim} = \frac{A_a}{A_r} = \frac{n^2}{\sin^2\theta_S}. \tag{3.82}$$

The difference of Eqs. (3.82) and (3.75) is the factor n_2 which gives a larger concentration when the optical concentrator is immersed in a medium with contact to the receiver. In optical microscopy this is used when the lens is immersed in a liquid in which the object is immersed thus providing a larger numerical aperture. In concentrator optics a secondary concentrator made of glass or acrylic glass connected to the receiver surface would utilize this effect.

As can be seen from burning mirror experiments, the temperature on the imaging surface increases with increasing concentration ratio. To derive the maximum absorber temperature, therefore, the energy incident on the absorber is first determined, which results from the product of the aperture area A_a and the radiant power on the aperture ($\dot{Q}_{s-E}$). If the relation (3.75) is inserted into Eq. (3.74), the radiant power incident on the aperture is $\dot{Q}_{s-E}$:

$$Q_{s-E} = A_a \cdot \sigma \cdot T_S^4 \cdot \frac{1}{C_{\max,2-\dim}}. \tag{3.83}$$

Since the absorber temperatures also increase with increasing concentration ratio, radiation losses make up an ever larger share of the total losses and dominate the loss share at high temperatures. If only radiation losses of the absorber are considered as losses, the greatest losses are obtained when the absorber is a blackbody radiator, i.e. the emission coefficient $\varepsilon_{a-S} = 1$. Thus, the power loss $\dot{Q}_{a-s}$ is given by the Eq. (3.72). The highest absorber (receiver) temperatures $T_{r,max}$ are obtained when the gains of irradiation correspond to the losses to the outside, i.e. $\dot{Q}_{s-E} = \dot{Q}_{a-s}$. Equating the gains with the losses physically means that the collector efficiency is zero. The highest possible absorber temperature $T_{r,max}$ thus is:

$$T_{r,max} = T_s \cdot \sqrt[4]{\frac{C}{C_{\max,2-\dim}}}, \tag{3.84}$$

where $T_{r,max}$ in this sense refers to the stagnation temperature of a concentrating system.

3.4.3 Reflector Materials

Solar thermal concentrators are in most cases constructed with reflecting materials and not with glass or polymeric lenses. One of the reasons is certainly that over the wide solar wavelength range of solar thermal collectors the dispersion of dielectric materials is substantial, leading to different focal lengths when using dielectric materials. This is not the case for reflector materials as the reflection law is independent on wavelength.

The basis of reflectors are metallic layers. The ones with the highest solar reflectance are using Aluminium (Al) or Silver (Ag). Contrary to dielectric materials the reflectivity of metals is even high for very thin layers. This is cause by the substantial imaginary part of the complex index of refraction as introduced in Eq. (3.9). For calculation of the reflectivities of layers the Fresnel laws (3.22) and (3.23) as well as Snell's law (3.20) have to be used with complex mathematics. For example, a complex angle of refracted electromagnetic waves is given by

$$\sin\theta_1 = (n - ik)\sin\theta_2. \tag{3.85}$$

For the common solar reflector materials aluminium and silver Fig. 3.49 shows the reflectance R_p and R_s for thick layers exhibiting no interference effects as a function of angle of incidence θ_1.

The optical properties of media described by complex indices of refraction vary with wavelength λ due to specific excitations of the material involved. Within the solar spectral range these excitations are typically electronic excitations leading to resonant absorption in the UV, metallic excitations (e.g. plasmon excitations) in the Vis–NIR range and molecular excitations (e.g. vibration absorption bands) in the NIR and IR. With the spectral variation of the complex index of refraction, transmittance and reflectance are also a function of wavelength. In Fig. 3.50 the reflectance $R(\lambda)$ at normal incidence ($\theta = 0°$) is plotted versus the wavelength for a thick layer of pure silver and aluminium (Palik, 1985). The reflectance of silver increases steeply from UV to an almost constant high reflectance over the

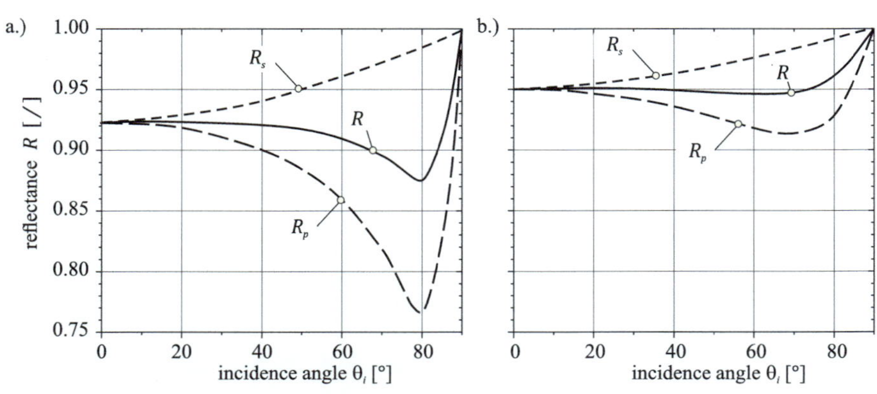

Fig. 3.49 Simulated reflectance R_p, R_s and mixed reflectance R at metallic surfaces as function of the incidence angle θ_1 at a wavelength $\lambda = 510nm$ for a thick aluminum layer **a** and a thick silver layer **b** using optical constants from (Palik, 1985)

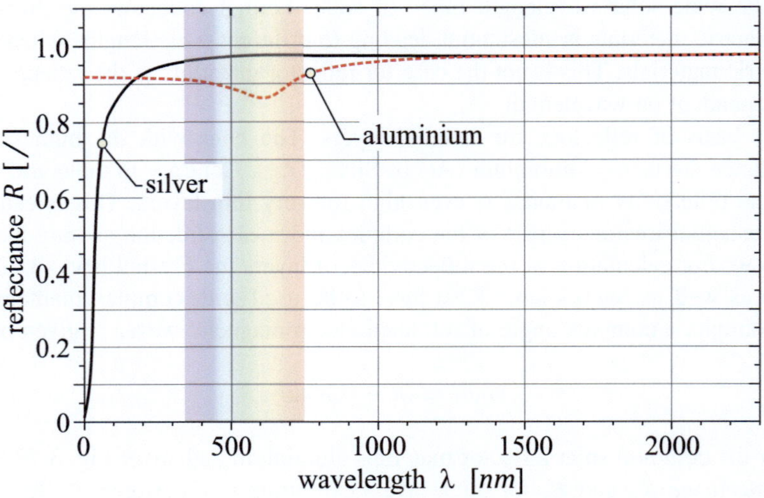

Fig. 3.50 Computed reflectance ρ of silver (Ag, straight) and aluminium (Al, dashed) versus the wavelength λ for 'unpolarized' radiation and near-normal incidence

solar wavelength range, whereas aluminium displays a material-specific minimum in the near infrared between $800nm$ and $900nm$ due to an absorption band.

Although silver is more expensive than the abundant aluminium most reflectors in concentrator devices are based on silver. We must distinguish between first surface and second surface reflectors, as illustrated in Fig. 3.51.

The classical bathroom mirror is a second surface reflector, because the front glazing here protects the thin silver layer from the humid environment. Silver

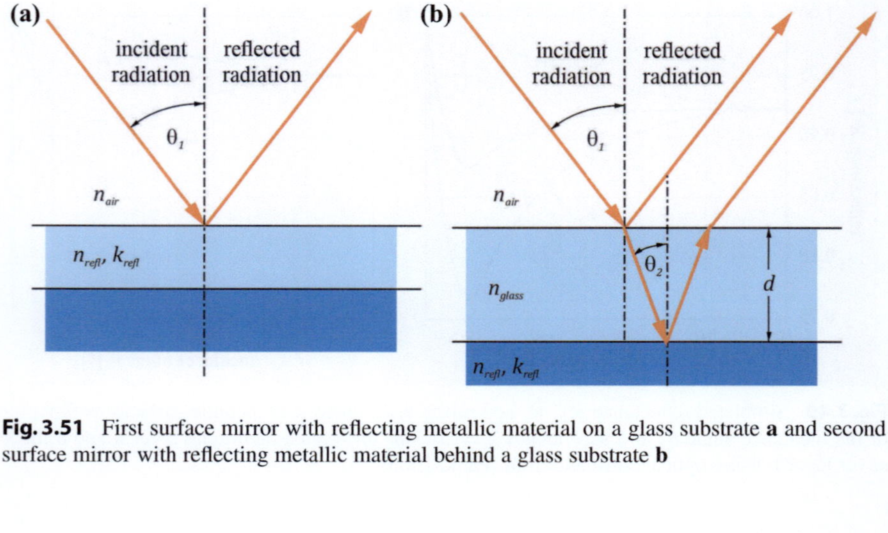

Fig. 3.51 First surface mirror with reflecting metallic material on a glass substrate **a** and second surface mirror with reflecting metallic material behind a glass substrate **b**

Table 3.11 Solar reflectance of Ag and Al refector materials (normal incidence)

Material	Solar reflectance
Silver, fresh, electroplated	0.96
Aluminium, fresh, high-purity	0.91
Aluminium, foil 99.5% purity	0.86
Silver, back surface, anti-reflected glass	0.94
Silver, back surface, white glass	0.91
Aluminium, front surface, sputtered	0.89
Aluminium, front surface, added SiO coating	0.87

is very sensitive to humidity and other components in the atmosphere (e.g. sulphides). Also, the back and edges of a silver mirror must be protected, usually some layers of protecting lacquer. First surface mirrors are not protected by a thick dielectric layer, but very often ultra-thin protecting layer of sputtered oxides are applied. Aluminium is better suitable for first surface mirrors as it is self-passivating with the formation of an aluminium-oxide layer. Also, high-reflectance reflectors can be optically improved by adding stacks of dielectric layers. The principle of constructive interference is applied to the dielectric coating of aluminium-based solar reflectors, (see Table 3.11).

The key requirement for a mirror surface is a reflectance close to the value of one in the solar wavelength spectrum, where reflectance indicates the fraction of incident radiation intensity that is reflected by the surface. All solar thermal mirrors aim at specular reflection, where the direction of the incident solar radiation according to Snellius' law of refraction and the direction of the outgoing light form the same angle with the normal of the mirror surface. However, since the reflectance of a metallic surface is different for different wavelengths, it must be specified for a particular wavelength or range of wavelengths.

The fraction of sunlight that is concentrated on the receiver only by the mirror material can be described by the "solar weighted reflectance" $\rho_r(SW, \theta_i, \theta_c)$ without considering other possible imperfections such as shape, receiver orientation, and tracking errors addressed within the context of the optical defects. An acceptance half-angle of $\theta_c = 7.5 - 12.5\ mrad$ (usually 12.5 $mrad$) as indicated in Fig. 3.52 is considered appropriate for typical parabolic trough designs. The solar weighted reflectance $\rho_r(SW, \theta_i, \theta_c)$ cannot be determined directly because spectrophotometers can measure specular reflectance at a given θ_i only in a narrow wavelength range. Therefore, $\rho_r(SW, \theta_i, \theta_c)$ must be approximated.

The hemispherical reflectance $\rho_h(\lambda, \theta_i, h)$ is measured over a wavelength range representative of the terrestrial solar spectrum ($\lambda = 250 - 2500\ nm$) using a spectrophotometer covering the solar spectrum wavelength domain and an integrating sphere (typically for an incidence angle $\theta_i = 8°$) relative to standards. The spectrum is then weighted by the standard terrestrial solar spectrum to calculate a sun-weighted hemispherical reflectance, $\rho_h(250 - 2500\ nm)$, as a meaningful

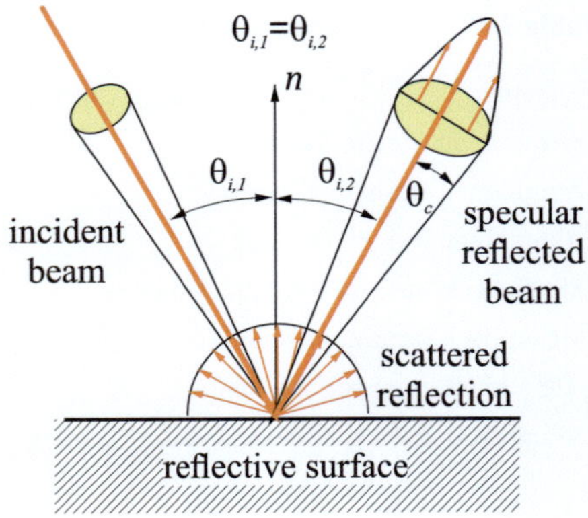

Fig. 3.52 Sketch to determine solar weighted specular reflectance using specific acceptance half angles

single measure of optical performance. The procedure and measurement are specified in standards (e.g. (ASTM G173-03, 2020), at an air mass factor $AM = 1.5$. The specular reflectance $\rho_s(\lambda, \theta_i, \theta_c)$ is measured with a specular reflectometer at $\theta_c = 3.5, 7.5, 12.5mrad$ with an incidence angle θ_i=15° for $\lambda = 660nm$.

For highly specular glass or front-surface reflectors, the specular reflectance can be described by the ratio between the specular reflectance $\rho_s(\lambda, \theta_i, \theta_c)$ and the hemispherical reflectance $\rho_h(\lambda, \theta_i, h)$ at the same wavelength, which is assumed to be constant and equal to the ratio of the solar-weighted values. The solar-weighted specular reflectance is then approximated by:

$$\rho_r(SW, \theta_i, \theta_c) = \frac{\rho_r(\lambda = 660nm, \theta_i, \theta_c)}{\rho_h(\lambda = 660nm, \theta_i, \pi)} \cdot \rho_h(SW, \theta_i, \pi). \tag{3.86}$$

The solar-weighted reflectance thus indicates the proportion of solar energy that is reflected by a mirror and is therefore the decisive quality criterion for efficient mirrors. This data provided by the manufacturer refer always to clean mirror.

3.4.4 Linear-Imaging Concentrators

The simplest case of a concentrator is a parabola of focal length f in two-dimensional space. A translational extension of the two-dimensional shape then forms a linear concentrator, a parabolic trough. The light rays incident parallel to the y-axis are focused by the parabola contour to the focal point F of the parabola, as shown in the Fig. 3.53. Provided that the lines FP and PQ are of equal length, the tangent is always correct to focus all vertical beams into the focal point F. According to geometric algebra, the following relationship is obtained for the contour shape of the parabola:

$$y + f = \sqrt{x^2 + (y - f)^2} \rightarrow y = \frac{1}{4f}x^2. \tag{3.87}$$

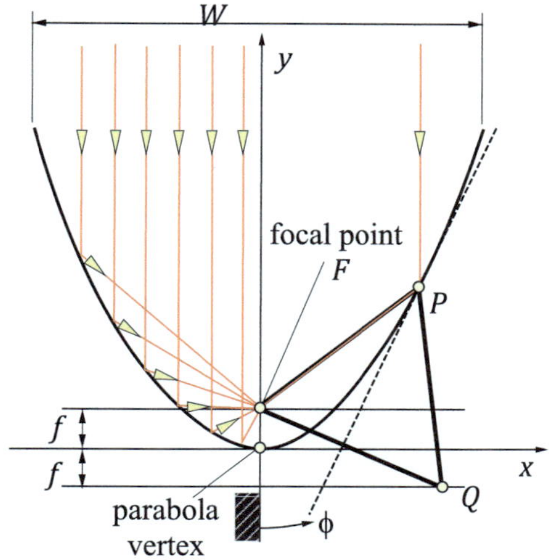

Fig. 3.53 Concentration of incident radiation in a parabolic concentrator with focal length f

The result of this simple observation is that as the focal length f of the parabola increases; a flatter parabola becomes necessary to focus the light in a point-like manner. In practice, this means a smaller contour length and thus less material requirement for the same mirror width (as will be seen later, constant concentration ratio). On the other hand, the distance to the focal point increases, so that mirror errors have a greater effect.

Although modern manufacturing machines allow the production of parabolic contours, it is technically easier to produce precise circular segments. Under certain conditions the parabola can be replaced by a segment of a circle. The following relationship results from the circular equation:

$$x^2 + (y - R)^2 = R^2 \rightarrow y = R \pm R\sqrt{1 - \left(\frac{x}{R}\right)^2}, \tag{3.88}$$

where R denotes the radius of the circle and x, y represent the coordinates from the Fig. 3.53.

If the surface segment DE is small in comparison to the radius R ($DE/R \ll 1$), it is also true for all coordinate ratios x/R that these are much smaller than one ($x^2/R^2 \ll 1$). In this case, a Taylor series expansion of the root expression in Eq. (3.88) yields:

$$\sqrt{1 - \left(\frac{x}{R}\right)^2} = 1 - \frac{1}{2} \cdot \left(\frac{x}{R}\right)^2 - \frac{1}{8}\left(\frac{x}{R}\right)^4 - \frac{1}{16}\left(\frac{x}{R}\right)^6 - \ldots \approx 1 - \frac{1}{2} \cdot \left(\frac{x}{R}\right)^2. \tag{3.89}$$

Inserting the result in Eq. (3.88) we obtain:

$$y = \frac{1}{2} \cdot \left(\frac{x}{R}\right)^2. \tag{3.90}$$

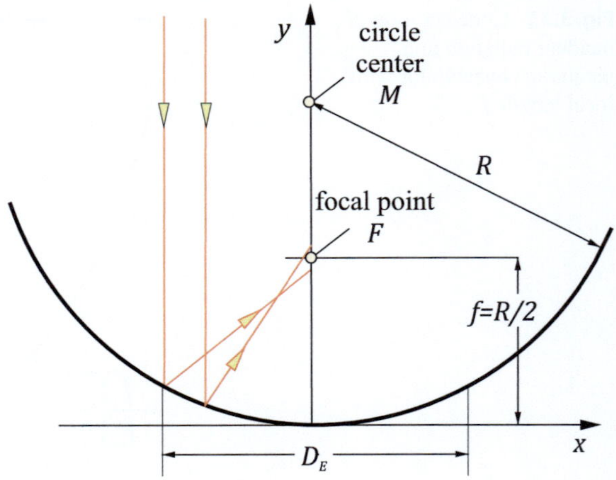

Fig. 3.54 View of the focal length of a circle segment as a replacement for a parabolic contour

Thus, the focal length f of a circle segment is $f = R/2$, as illustrated in Fig. 3.54.

Theψ_r rim angle ψ_r of a parabola is defined as the angle between the axis and a line from the focus to the physical edge of the concentrator. Together, the focal length f and rim angle ψ_r of a parabolic concentrator completely define its cross-sectional geometry. The rim angle of a parabola is given by

$$\tan\psi_r = \frac{W/2}{f-h} = \frac{W/2}{4f^2 - (W/2)^2}, \tag{3.91}$$

where W is the width and $h = y_r$ is the full depth of the parabola,

$$h = \frac{W^2}{16f}. \tag{3.92}$$

Assuming W as the width of the aperture, one can find a relation between the rim angle and the diameter to focal length ratio (d/f) and it results after some transformations:

$$\frac{f}{W} = \frac{1}{4\tan(\psi/2)}. \tag{3.93}$$

It follows directly from the relationship that a small rim angle corresponds to a large focal length (for a given aperture diameter) and vice versa. The rim angle is an important design feature of collectors, especially because it affects the concentration ratio.

The parabolic effect of focusing on a single focal line only occurs with perfectly parallel incoming radiation. As noted above, sunlight has a range of incident angle due to the physical extent of the solar source. This reduces the optical concentration achievable at the focus of parabolas, as discussed in the following sections.

3.4.4.1 Limits with Flat Receivers

Each point on a parabolic mirror will reflect a cone of rays that matches the angular distribution of the solar source (half-angle size θ_s). Consider the size of the spot formed by the cone of rays reflected from the points on the mirror when incident on a flat target placed in the focal plane (see Fig. 3.53). The rays from the rim will form the widest such spot. The distance x_r of the rim reflection point from the vertical axis is

$$x_r = r \cdot \sin\psi_r, \tag{3.94}$$

and the width of the focal spot on the focal plane due to reflection from this point will be

$$D = \frac{2r \cdot \sin\theta_S}{\cos\psi_r}. \tag{3.95}$$

The reflected cone from a single spot on the mirror will form an elliptical spot on the target with a major axis length of d. This applies equally for dishes and troughs. Clearly, the size of the focal-plane spot depends on the received incidence angle θ, so the accumulated effect of all the spots across the mirror will not be a true image of the sun formed at the focal plane. This is the reason that solar concentrator optics are referred to as non-imaging optics. If the rim angle ψ_r is small, then the distance mirror to the focus r is approximately equal to f over the whole mirror, and then, to a good approximation, it will form a true image of the sun in the focal plane with the image diameter being

$$\mathrm{D} = \frac{2f \cdot \sin\theta_s}{\cos\psi_r} \approx 2f \cdot \sin\theta_s. \tag{3.96}$$

On the other hand, for a dish or trough with a flat receiver and a rim angle of 90°, the mirror elements at the very edge will make an infinitely long image, spreading their radiation over the entire focal plane. If the receiver is large enough to accept reflected spots from the entire mirror surface, then the diameter of the receiver will be defined by the reflected spot size from points at the edge of the mirror with $x = W$ and $\phi = \psi_r$. The geometric concentration ratio for a parabolic trough with a flat receiver will then be,

$$C_g = \frac{A_a}{A_r} = \frac{W}{D}, \tag{3.97}$$

leading to

$$C_g = \frac{\sin\psi_r \cdot \cos\psi_r}{\sin\theta_S} = \frac{\sin 2\psi_r}{2\sin\theta_S}. \tag{3.98}$$

The optimal rim angle $\psi_{r,opt}$ can be found by setting the derivative of Eq. (3.98) to zero, i.e.

$$\frac{dC_g}{d\psi_r} = 0, \tag{3.99}$$

which yields the maximum concentration at $\psi_{r,opt} = 45°$ for the flat receiver and ideal shape of the parabola. When using the divergence of the sun with $\theta_s = 4.65 mrad$ we get as maximum concentration $C_{g,max,flat}$,

$$C_{g,max,flat} = \frac{1}{2\sin\theta_s} \approx 108. \tag{3.100}$$

When the beam is reflected from an imperfect mirror with imperfect shape, an additional dispersion half-angle δ must be added to θ_s in the formulae. Then the equivalent beam spread θ_s^* with

$$\theta_s^* = \theta_s + \delta \tag{3.101}$$

should be used, resulting, of course, in a smaller maximum concentration.

3.4.4.2 Limits for Cylindrical Receivers

More useful for solar energy applications is the consideration of a receiver with circular cross-section as shown in Fig. 3.55. In this case, the diameter of the target should be at least

$$\mathrm{D} = 2r \cdot \sin\theta_S. \tag{3.102}$$

For a trough with a cylindrical receiver, then

$$C_g = \frac{A_a}{A_r} = \frac{2 \cdot r \cdot \sin\psi_r}{2ßr \cdot \sin\theta_S} = \frac{\sin\psi_r}{\pi \sin\theta_S}. \tag{3.103}$$

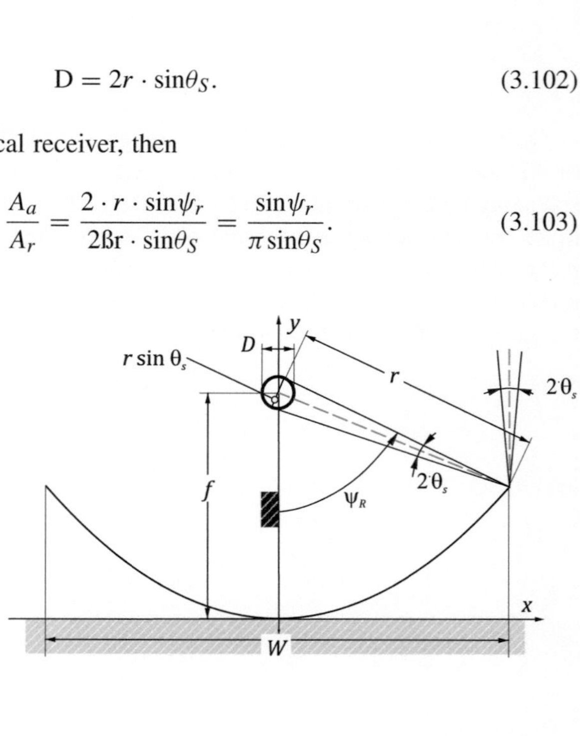

Fig. 3.55 Parabolic concentrator with circular target (cross section of parabolic trough)

Again, the shading fraction of the receiver tube on the mirror is neglected. Solving for maximum geometric concentration ratio as before, the optimal trough rim angle now is calculated as 90°, and for this angle the maximum concentration is

$$C_{g,max,tube} = \frac{1}{\pi \sin\theta_s} \approx 68. \tag{3.104}$$

Note that many companies quote however the concentration of their product as described by Eq. (3.97) for a flat receiver – arguing that the projection of the tube is the relevant receiver area. However, the area of the receiver emitting thermal radiation is undoubted ($\pi \bullet D \bullet L$) and not the projection ($D \bullet L$) when L is the receiver length oriented normal to the cross-sectional plane. Also, it should be emphasized that this analysis only considers the geometric concentration ratio, and the above derivation indicates that the local contributions to focal spot size vary as a function of reflection radius x, so the geometric concentration ratio limits are less than the optical concentration ratio limits at the centre of the focal spot.

3.4.5 Point-Imaging Concentrators

When we rotate the parabola in Fig. 3.53 around its focal axis, we get a paraboloid, a concentrator with a point focus. The width of the parabola W then is the diameter of a circular aperture. Instead of focusing on a focal line as for a trough there is a focal point for this three-dimensional concentrator. This rotational concentrator is called parabolic dish.

It follows directly from the relationship (3.93) that a small rim angle corresponds to a large focal length (for a given aperture diameter) and vice versa. The rim angle is an important design feature of parabolic dishes, especially because it affects the concentration ratio. A perfect paraboloid mirror shows the highest concentration ratio on a flat receiver (as shown in 3.4.4.1) at a rim angle of $\psi = 45°$, and we now consider how the concentration ratio behaves when deviating from this ideal rim angle. If the rim angle becomes very small, the reflector is far from the focal point for a given diameter, so that the divergence of radiation, which is unavoidable due to the spread of the solar disk, produces a larger solar image. This leads to a lower concentration ratio at small edge angles for a given collector diameter. Therefore, reflector geometries with small rim angles are unfavourable for large concentration ratios.

Assuming a receiver with a flat circular aperture in the focal plane and a large rim angle ($\psi > 45°$), as shown in Fig. 3.56 for beam 2, large rim angles lead to an expansion of the focal spot exceeding the aperture plane.

The surface area of a mirror with a large edge angle is larger than the surface area of a mirror with the same diameter and smaller edge angle. This affects the material requirements and the weight of the collector. The absolute mirror surface

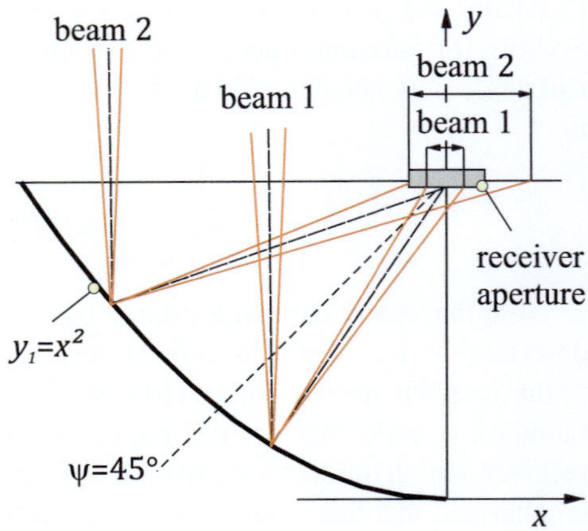

Fig. 3.56 Radiation path of an angle-variant irradiation on a parabolic mirror at different rim angles ψ

(A_{mir}) is

$$A_{mir} = 4\pi f^2 \frac{\sin^2\psi}{(1+\cos\psi)^2}. \tag{3.105}$$

Therefore, not only parabolic mirrors with very small edge angles, but also mirrors with very large edge angles are not favourable for a high concentration ratio. In addition to these considerations about a sufficiently large concentration ratio of parabolic mirrors, there are other aspects related to the rim angle. One advantage of mirrors with a very small rim angle is that as the rim angle decreases, the parabolic shape can be better approximated by a spherical shape, which is more favourable for manufacturing reasons, since many mirrors are segmented and a larger paraboloid with sufficiently large rim angles can be composed of many spherical shaped, small facets. Moreover, there is a relationship between the rim angle of the collector and the receiver type. Small rim angles mean low angular tolerance of incident radiation at the receiver, while large rim angles have the opposite effect, meaning that cavity receivers are used at rim angles ($\psi < 50°$), while external receivers allow larger rim angles. Since dish systems target high temperatures, thermal radiation losses play an excellent role, so cavity receivers are in the development focus.

The diameter of the focal point in the receiver (d_{rec}) can be calculated by using Eq. (3.105), which affects the size of the receiver. Here, the acceptance angle (θ_A) to the parabolic dish appears, which describes the angle of worst condition that collect all reflected solar radiation from the concentrator at small time for tracking,

$$d_{rec} = \frac{f\theta_A}{\cos\psi(1+\cos\psi)}, \tag{3.106}$$

with θ_A in (rad). If one considers the radiation distribution of the sun, large concentration ratios such as those of a dish receiver result in an extreme flux maximum present at the actual focal point, which exceeds the permissible temperatures of a possible absorber material. Therefore, cavity geometries are usually chosen for the receiver targeting in the design to achieve the most uniform wall radiation flux distribution possible. Here, spherical receiver shapes offer a good starting point for shape optimization with which a desirable shape (inverted bulb) with nearly uniform distribution can be achieved; if necessary, the central receiver can also be supplemented by a secondary reflector.

The geometrical concentration factor for flat circular receivers can be derived similarly to the paragraphs 3.4.4.1 and 3.4.4.2. For flat receivers the concentration is

$$C_{g,flat} = \frac{\sin^2\psi_r \cdot \cos^2(\psi_r + \theta_S)}{4 \cdot \sin^2\theta_S} - 1. \tag{3.107}$$

For a spherical bulb receiver (which is different from the inverted bulb mentioned above) the concentration is given by

$$C_{g,sphere} = \frac{\sin^2\psi_r}{4 \cdot \sin^2\theta_S} - 1. \tag{3.108}$$

The optimum rim angles here again are 45° for flat disks and 90° for the spherical receiver. As described above, the formulae are valid for perfect mirrors, otherwise the equivalent beam spread θ_s^* (3.101) has to be used.

3.4.6 Non-Imaging Concentrators

Concentration ratios corresponding to the ideal concentrator reach the levels achieved by (Winston, 1974) proposed Compound-Parabolic Concentrators, which are often abbreviated as CPC in the technical literature. Rabl investigated (Rabl A., 1976) different absorber geometries with adapted CPC contours and compares them with parabolic trough concentrators (PTC).

In combination with a flat absorber, the contour of a CPC is composed of two parabolic branches whose focal points are located at the opposite ends of the absorber strip. The parabola axes run parallel to the shadow lines that start from the contour ends and meet the opposite absorber ends. The angle between these two shadow lines gives twice the acceptance angle. The Fig. 3.57 shows schematically the geometrical structure of a Compound-Parabolic Concentrator (CPC), whose contour is composed of two parabolic branches.

In the case of a round absorber, the shadow lines from the upper contour ends touch the absorber as a tangent, as shown in the Fig. 3.58. Below the absorber between the points where the shadow lines meet, the contour has the shape of two involutes. Above the points of impact, it is composed of a sequence of intersections of parallels to the shadow lines and tangents to the absorbers. Further combinations of absorber geometries and CPC contours are described in Rabl (Rabl A., 1976) as well as in (Kreith and Kreider, 1978).

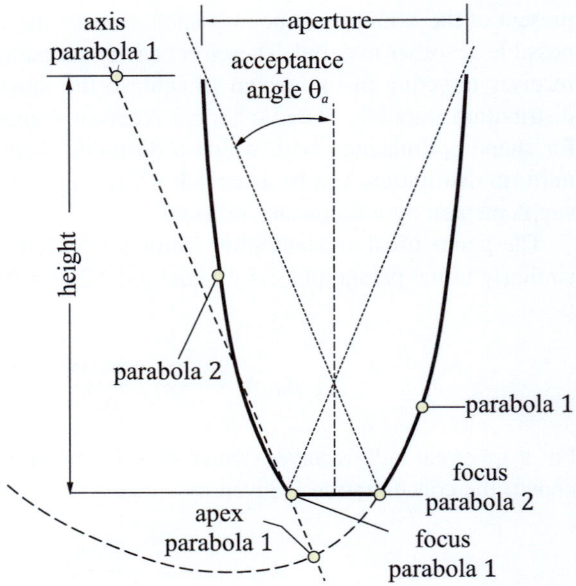

Fig. 3.57 Schematic diagram of the contour of a Compound Parabolic Concentrator (CPC) with a flat absorber strip; its mirror contour is composed of two parabolic branches

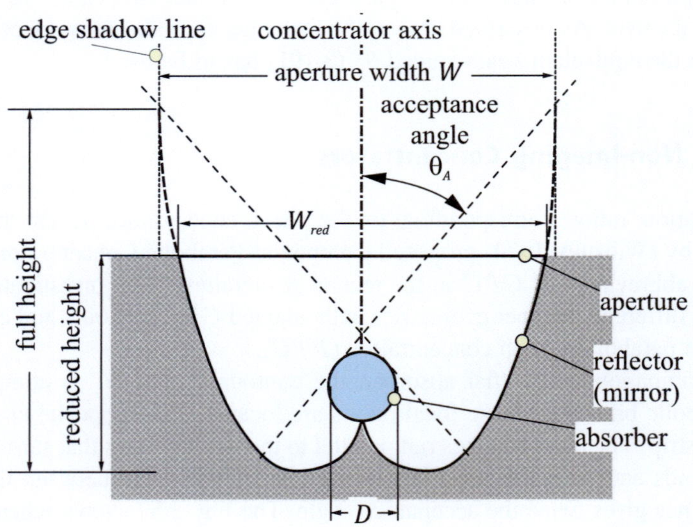

Fig. 3.58 Schematic sketch of the contour geometry of a compound parabolic concentrator (CPC) with round absorber tube at full or reduced height

Figure 3.59 shows the ratio of the reflector contour length to the aperture width W as a function of the concentration ratio C at various specified acceptance angles θ_a. The graph illustrates two disadvantages of CPC concentrators (shown here for flat absorbers) As the concentration ratio C_g increases, the contour length grows faster than the aperture width W. This would lead to very large reflector heights and therefore collectors of the CPC type are generally only built with a shortened or reduced height.

Figure 3.60 illustrates a comparison of a full height CPC and reduced height $CPC's$. In addition, at high concentration ratios C the CPC level, the acceptance angle θ_a decreases very rapidly due to the large height. This in turn requires a large number of tracking steps to ensure a high daily operating time.

The design of full-height CPC reflectors leads to complex multiple reflections on the mirror surface at high concentration ratios. Figure 3.61 shows the mean number of reflections n_ρ as a function of the concentration ratio C_g at different

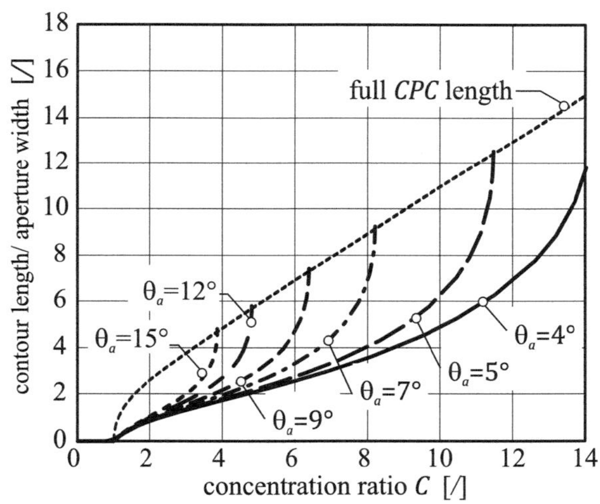

Fig. 3.59 Ratio of the reflector contour length to the aperture width W as a function of the concentration ratio C at a given acceptance angle θ_a in (°) for CPC collector types with full and reduced height

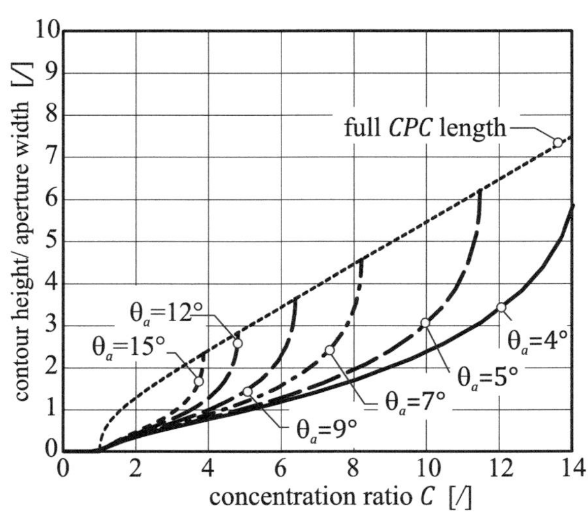

Fig. 3.60 Ratio of the concentrator height to the aperture width W as a function of the concentration ratio C at a given acceptance angle θ_a in (∘) for CPC collector types with full and reduced height

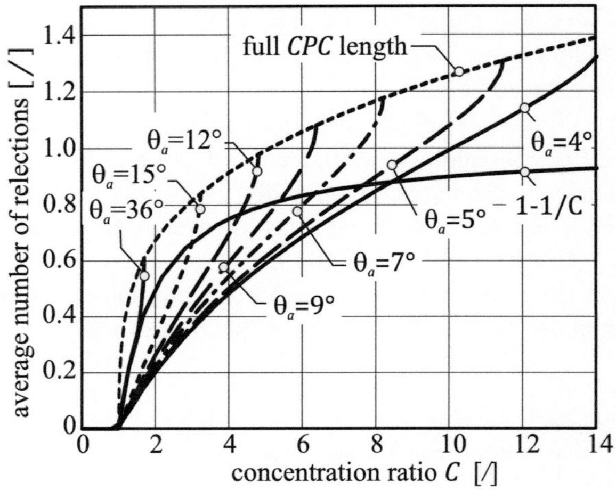

Fig. 3.61 Mean number of reflections n_ρ (right) as a function of the concentration ratio C at a given acceptance angle θ_a for CPC collector types with full and reduced height

given acceptance angles θ_a. This figure shows that the average number of reflections until the radiation from the reflector edge areas reaches the absorber increases with increasing concentration ratio C_g. Since the reflection coefficient is $\rho < 1$, the optical efficiency decreases with the number of reflections proportionally with ρ^n.

Most applications of CPC concentrators are therefore limited to small concentration ratios, typically $C_g \leq 2$. There are several collectors on the market with fixed CPC concentrators as back-up for vacuum tubes or they are built into such tubes. On the other hand, (Niemann and Schreitmüller, 1993) showed that double concentrating $CPC's$ in combination with good vacuum tubes offer efficiencies of $\eta \approx 50\%$ at operating temperatures of $200°C$ and $\Delta T/i = 0.1K/\left(Wm^{-2}\right)$.

In the concentration range of $C_g \leq 2$, CPC collectors compete economically with other non-imaging concentrators that are easier to manufacture. These include, for example, V-shaped channels with flat mirrors on both sides of the absorber, whose reflector cross section forms a V-shape. The Fig. 3.62 schematically shows the geometrical structure of a V-shaped channel. Rabl (Rabl A., 1976) shows, using equations of linear algebra, that up to an angular deviation of δ^* between the direction of irradiation and the normal, all irradiation hitting the aperture surface comes to the absorber. At angular deviations greater thanθ_C, no radiation at all reaches the absorber. The limit angle θ_C is thus the sum of the angle at the tip of the V-groove with $(2 \cdot \theta_V)$ and δ^*:

$$\theta_C = 2\theta_V + \delta^*. \tag{3.109}$$

Between θ_C and δ^*, the proportion of the incident radiation decreases approximately linearly with the angle. With this approximation the possible concentration

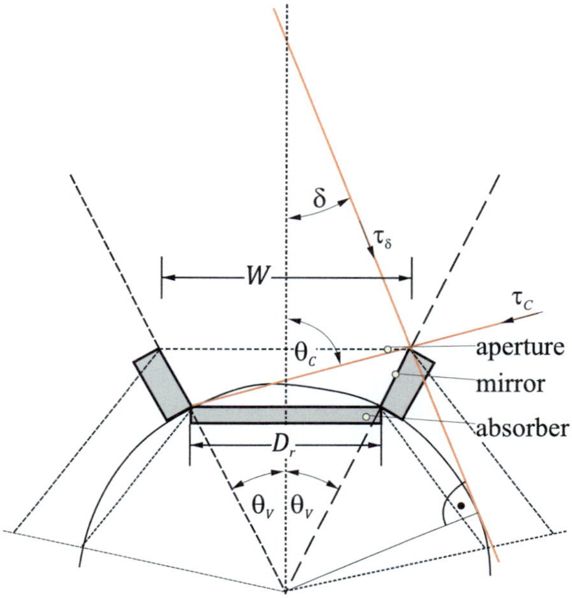

Fig. 3.62 V-shaped trough with a flat absorber and two flat lateral mirrors; up to an angular deviation δ^* between the direction of irradiation and the normal, all the irradiation hitting the aperture surface reaches the absorber; at angular deviations greater than θ_c, no more radiation at all hits the absorber

ratio $C_{g,V}$ can be given:

$$C_{g,V} = \frac{1}{\sin(\delta^* + \theta_V)}. \tag{3.110}$$

It is difficult to draw a comparison between the CPC and the V-channel, as several sizes have to be considered. (Rabl A., 1976) contrasts examples where the average number of reflections is equal. If this number is 0.75 and the acceptance angle $\theta_a = 19.5°$ a truncated CPC reaches a concentration of$C_{g,CPC} = 2.9$. In contrast, the V-trough with half the trough angle $\theta_V = 11°$ reaches a concentration of$C_{g,V} = 2$. The ratio of mirror length to receiver width is 2.8 for the CPC and 2.5 for the V-trough. Only with small C_g and trough angles of $\theta_V \leq 20°$ do both concentrator types reach comparable values. For direct evaporation tests of water, flat plate collectors with a V-trough concentrator (location Karlsruhe,$\Phi = 49°$) were used, on which the collector mirrors were mounted on the long sides so that they could be swivelled. The long sides and thus also the axes of rotation of the mirrors were in east–west direction. A tracking system was dispensed with to avoid constructively complex flexible steam lines, and a rod system connected the mirror ends. This meant that one lock was sufficient for simultaneous adjustment of both mirrors during tracking.

3.4.7 Fresnel Concentrators

3.4.7.1 Primary Mirror Field

The primary mirror or reflector array consists of multiple rows of slightly curved mirrors, rotating with half the angular speed of the sun in the transversal projection path over the rows. The basic geometry (see Fig. 3.63) has variables; as primary mirror width $w_{pr,i}$, gap Δx_i, focal length f_i and height h_i which all may be functions of distance x_i of mirror row i from the receiver unit.

There are several optical loss mechanisms related to geometry. Even for perfect reflectors, cosine losses occur due to off-normal incidence on the reflectors most of the time. Blocking considers the partial interception of reflected rays by mirror rows closer to the receiver tower. In addition, mirror rows may partially shade rows behind them as seen from the sun (shading). Additionally, any real mirror has a small but finite absorption and—if dust or a rough surface must be considered – scattering extinction both reduce the direct radiation reflected towards the receiver. Blocking and shading is dominant for low solar altitudes.

According to (Abbas et al., 2018) there are four principal classes of LFC primary mirror designs:

•

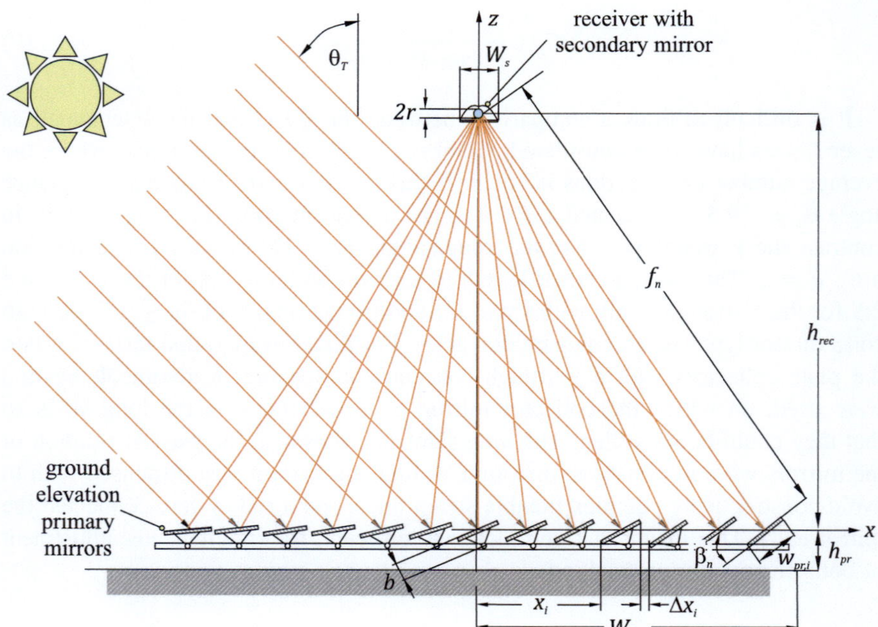

Fig. 3.63 Geometrical transversal cross-section of a primary reflector array of a Fresnel collector (w_i=mirror width, Δx_i= gap between neighbouring mirrors, b=distance between rotational axis and mirror axis, f_i=focal length)

- Constant mirror width and constant gap (CW-CG),
- Constant mirror width and variable gap (CW-VG),
- Variable mirror width and constant gap (VW-CG),
- Variable mirror width and variable gap (VW-VG).

The advantage of constant mirror width is of course that the production line does not need to deal with different mirror sizes. So, it is a question whether variable mirror width options will be chosen by a commercial development. Providing a variable gap to minimize blocking and shading losses however may be solved by mounting options—but of course at the expense of additional assembly effort and more careful design. This design however will result in a somewhat larger field area for the same reflector area.

A further design option is that the focal lengths of individual mirror rows can be constant or variable. The focal lengths should be adapted to the distance of the mirror rows to the receiver aperture. In practice, however, it is too costly to produce different focal lengths for every row, so practical designs typically utilize one to four different focal lengths. In addition to losses due to optical design there are other factors which degrade the quality of the solar image on the receiver. Rough, dusty, or scratched mirror surfaces as well as shape or contour inaccuracies widen the reflected beam. Also tracking inaccuracies must be avoided.

Optical design tries to optimize the size and efficiency of the reflector array (to increase concentration) with respect to economics. The further away a reflector row is from a receiver tower, the lower is its time-averaged optical efficiency. In financial terms every additional investment of adding a row has less and less addition to overall output. This is one reason for the CLFR approach where the mirror rows can be reoriented, according to sun position, to reflect to a receiver in better position.

3.4.7.2 Secondary Optics

For identical receiver area large primary reflector fields result in larger concentration, however, to enhance the efficient collection of the reflected radiation secondary concentrator optics may be used. This allows larger fields, as the receiver aperture will be larger than the absorbing area. It is important that radiation impinging on the receiver aperture should be directed towards the absorber tube or tubes. Non-imaging optics provides us with the means to design such secondary concentrators. Several principles have been proposed, from truncated compound parabolic concentrators (CPC) to elliptic or hyperbolic concentrators. Typically, there is slightly more than one reflection on average in a secondary concentrator, which means that reflectivity should be very high, because energy is lost with any reflection. In addition, any absorption by the reflector heats up the secondary concentrator due to the high intensity. This must be considered carefully in the choice of secondary material and shape accuracy. Deformation and degradation may be caused by high temperatures.

***CPC* Receiver**

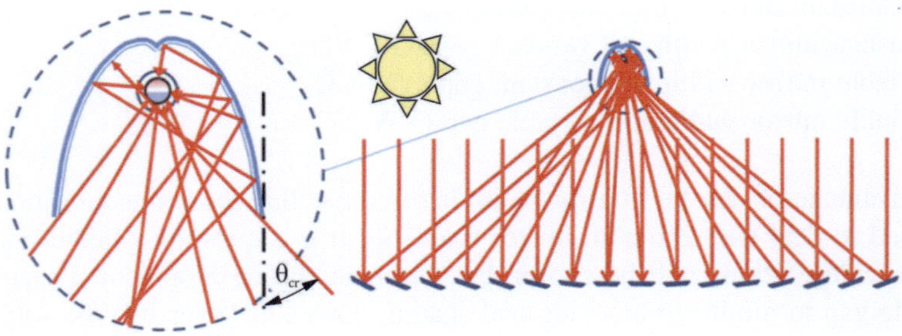

Fig. 3.64 CPC Receiver within a linear Fresnel collector (LFR)

The geometry used in most commercial approaches is a truncated CPC, as illustrated in Fig. 3.64. A CPC reflector ideally redirects all rays from a certain field of view (within the critical acceptance angle θ_{cr} which should cover the primary reflector field) towards the receiver surface. Truncation reduces the length of the CPC without much sacrifice in concentration (Rabl and Winston, 1976). The advantage of this design is a relatively shallow geometry using little reflector material. An ideal CPC with a cylindrical absorber would have a cusp nearly touching the absorber tube. To avoid that, the cusp often is flattened out, allowing some rays within the theoretical acceptance range to escape the concentrator. Experimental methods have been developed to assess the quality of the secondary concentrator early in development, see (Platzer, Heimsath, Hildebrandt, and Georg, 2008).

XX-SMS Fresnel concentrator
A similar concept from non-imaging optics using the Simultaneous Multiple Surface method yields a taller secondary concentrator but resulting in higher concentration (Canavarro et al., 2014) or (Guerreiro et al., 2015), which is depicted schematically in Fig. 3.65. It is called the LFR XX-SMS concept. This advanced optical design obviously has consequences on other issues like construction of the tower structure (due to higher wind loads) or concentrator material costs. It must be shown that the additional optical concentration leading eventually to better thermal performance at low irradiation, justifies the effort. According to (Guerreiro et al., 2015). a concentration of 73.7 can be reached using 16 mirrors 1.06 m wide within a field 20.1 m wide. The tower has to be 9.8 m high. The optical efficiency is close to 70%.

TERC concentrator
A further secondary concentrator type has been designed by Chaves and Collares-Pereira also based on etendue conservation using non-imaging optics maximum (Chaves and Collares-Pereira, 2010). However, in this case the optics is designed for a modified primary mirror field where the rotational axes of the mirrors are not on one altitude level. The primary field mirrors of one segment between receiver towers, point to different receivers alternatingly, similarly to the CFLR concept as

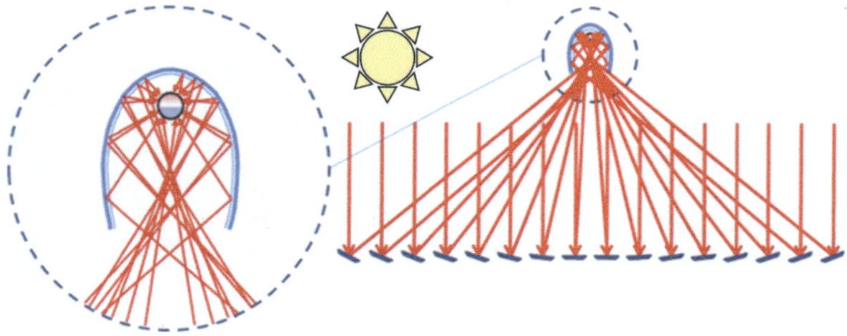

Fig. 3.65 LFR XX SMS Concept according to (Guerreiro et al., 2015)

depicted in Fig. 3.66. In this way the primary mirror field can be extremely compact minimizing the gaps. The mirrors further away from the receiver are elevated to avoid blocking. The receiver in this case is a V-shaped absorber with two wings of mirrors on the sides. These concentrators are called Tailored Edge Ray Concentrators (TERC). From a theoretical point this is an ideal optical concentrator with highest possible optical efficiency for infinitesimally small mirrors. Practically viewpoints lead to finite size mirrors and mirrors in one plane which brings the efficiency down.

Adaptive Design Concentrator (ADC)
NREL's adaptive design concentrator is not defined by a set of equations but by a repetitive procedure to determine the next segment of the reflector from a previously defined one (Zhu, 2017). The resulting secondary shape looks very similar to a truncated CPC and is also determined by the design variables as indicate in Fig. 3.67a; absorber tube diameter, aperture width and height position of the tube above the secondary aperture plane. A recent comparative optical study concludes that the optical performance is very similar to that of a CPC, but however may have some benefits at higher flux intensities (Abbas et al., 2018).

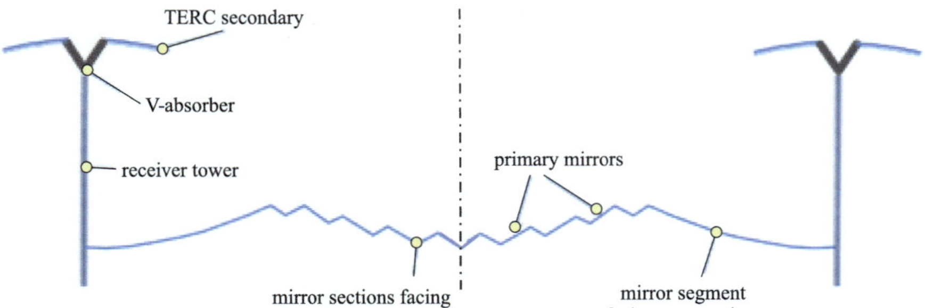

Fig. 3.66 Concentrator for several receivers with Fresnel primary and truncated secondary mirrors (TERC) after (Chaves and Collares-Pereira, 2010)

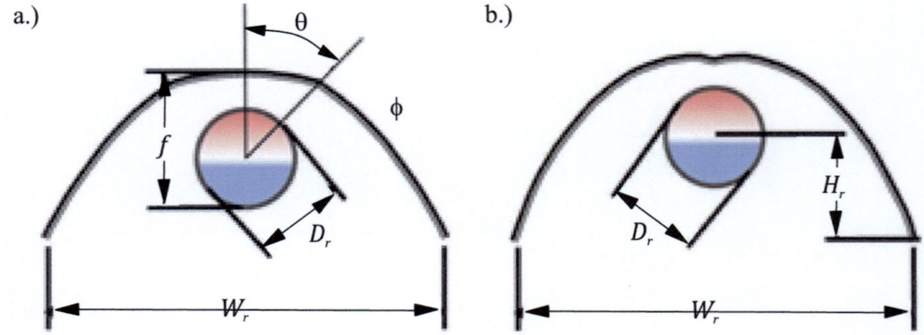

Fig. 3.67 Design parameters in an ADC (**a**) and a SPSC (**b**)

Segmented Parabolic Secondary Concentrator (SPSC)

This shape is composed of a parabolic trough broken at a given point to change its tilt. The result is a kind of reflecting cavity with 3 segments of curved walls which is depicted in Fig. 3.67b. The design parameters are the receiver diameter, the rim angle of the bottom segment where the parabola is cut, and then the tilt angle for the wing segments. In the comparison of (Abbas et al., 2018) the SPSC has a clearly lower performance which could be accepted if cost would be lower.

3.4.7.3 Comparison of Design Options

A combination of primary concentrator with a receiver and possibly a secondary concentrator results in an angle-dependent optical efficiency of the collector. The situation is depicted in Fig. 3.68. Let us define the angles transversal, the longitudinal and the solar longitudinal incidence angles θ_T, θ_L and θ_{LS} as

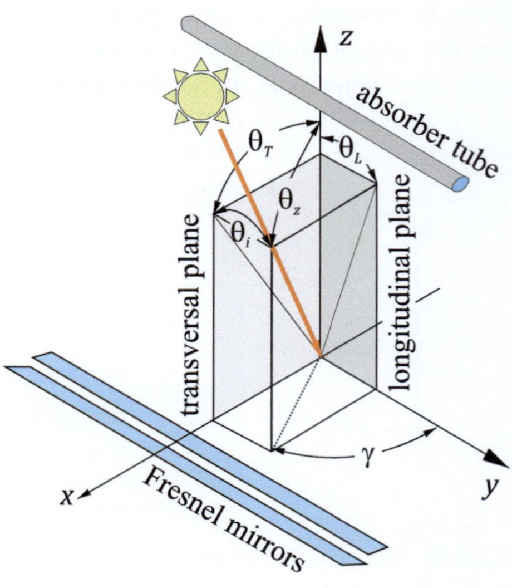

Fig. 3.68 Incidence angles for a Linear Fresnel Collector (LFC) from (Mertins, 2006)

$$\theta_T = \tan^{-1}(|\sin\gamma| \cdot \tan\theta_z); \tag{3.111}$$

$$\theta_L = \tan^{-1}(\cos\gamma \cdot \tan\theta_z); \tag{3.112}$$

$$\theta_{LS} = \sin^{-1}(\cos\gamma \cdot \sin\theta_z). \tag{3.113}$$

The incidence angle modifier function (IAM) describes the optical efficiency for off-normal incidence angles normalized by optical efficiency at normal incidence. It is usually written as the product of a transversal and a longitudinal function assuming transversal and longitudinal angle dependence to be independent, which is a reasonable assumption for most incidence angles θ and reads mostly to:

$$IAM = K(\theta_{LS}, \theta_T) = K_T(\theta_{LS} = 0, \theta_T) \cdot K_L(\theta_{LS}, \theta_T = 0), \tag{3.114}$$

where the coefficients are provided by the manufacturer of a collector assembly. A typical IAM distribution as a function of the incidence angle θ_i is shown in Fig. 3.69 for a Linear Fresnel collector (LFC) in comparison with a parabolic trough (PTC), which exhibit significant differences further discussed in Chaps. 7 and 10.

In a comparison of design options theoretical maximization of optical efficiency should not be the target. This approach certainly leads to designs with high concentration and preserving etendue, however, it must be kept in mind that at the end manufacturing, construction and operating cost will also determine the final economic performance. Complex solutions with variable component geometries, with changing altitude of mirror rows, with complicated construction manuals will produce extra costs and sources of errors. The decision which approaches to use and which to abandon due to cost certainly depends very much on experience and production technology of an individual commercial entity. So, a purely academic design recommendation will probably not meet the final expectations. The optimized design is always a dialogue between science and commercial considerations.

3.4.8 Practical Concentrators

So far, we assumed error-free concentrators. However, practical designs of mirrors will have tolerances in the contours (roughness radius waviness), tracking or surface contamination. Kreider (Kreider, 1979) propose to assume a Gaussian distribution for the radiation distribution of the sun. In fact, the sun is about 2.5 times brighter in the center of the solar disk than at the edge (Giovanelli and Beck, 1987). The defects of the concentrator fan out the beam reflected in the direction

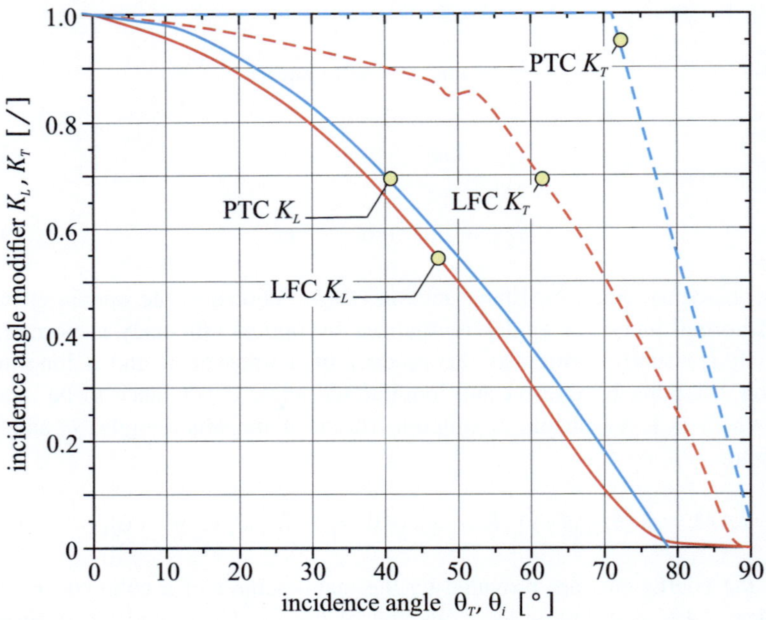

Fig. 3.69 Simulated incidence angle modifiers (IAM, relative to DNI) for the Linear Fresnel Collector (LFC) for transversal (K_T) and longitudinal (K_L) directions, in comparison to the Parabolic Trough Collector (PTC) (here transversal K_T describes the row-to-row shading) from (Morin et al., 2012)

of the absorber. With standard distributions of the individual errors (from geometric ripples, imaging, adjustment, and tracking), their sum forms the total optical error σ_{opt} of the reflector:

$$\sigma_{\text{opt}}^2 = 4\sigma_{\text{shape}}^2 + \sigma_{\text{rough}}^2 + \sigma_{adjust}^2 + \sigma_{\text{track}}^2. \tag{3.115}$$

Together with the assumed radiation distribution of the sun, this gives the standard deviation of the reflected ray, which is represented by a total standard deviation σ_{tot} in the way,

$$\sigma_{\text{tot}}^2 = \sigma_{\text{opt}}^2 + \sigma_{\text{sun}}^2. \tag{3.116}$$

This will be elaborated further in the context of the collectors in Chaps. 7 and 10. For the linear Fresnel reflector e.g. the tracking angle error has to be added as $4\sigma_{\text{track}}^2$.

3.5 Measurement Methods

3.5.1 Spectral Measurements—Reflectance and Transmittance

Optical materials should be characterized spectrally if detailed investigations are needed for optimization. Broadband measurements over the whole solar spectrum using solar simulators are possible, however, the quality of the simulator spectrum as well as other criteria like homogeneity of irradiation and temporal stability lead very often to lower accuracy. The classical characterization instrument for scientific spectral measurements are double-beam spectrometers, where a light beam is split up in to two parts, one being used as the 100% reference beam and the other with the sample in the light path for determination of the transmitted or reflected radiation. The beam path of a double-beam spectrometer is shown in Fig. 3.70.

Single beam spectrometer are useful tools in practice when spectral information has to be collected in mobile applications or outdoors. The accuracy is lower because of the inherent problem that the reference of 100% is less secure. On the one hand side, an unstable light source might lead to artefacts in the spectral curve as it takes a certain time to collect the information. Although nowadays with CCD-detector arrays and very fast data acquisition a scan of a total spectrum might need only milliseconds, nevertheless the exact determination of the reference can be a problem. If the materials are transparent and non-scattering, the direct beam transmittance or reflectance can be collected very precisely using a double-beam design using gratings and lock-in-techniques with shutters as shown in Fig. 3.70. When diffuse parts of the light also shall be collected (for determining the hemispherical reflectance or transmittance, so-called integrating spheres are used).

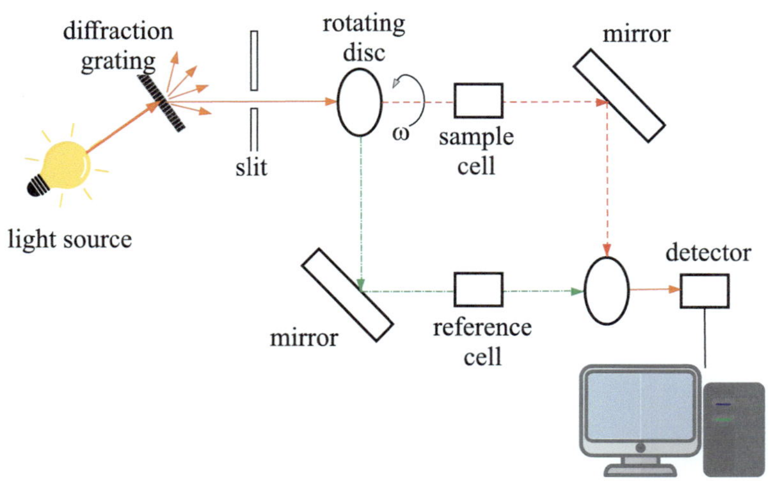

Fig. 3.70 Schematical design of double-beam spectrometer

3.5.2 Scattering and Surface Characterization

Scattering of light due to surface roughness or soiling particles reduces the amount of reflected intensity in the direction predicted by the reflection law. Therefore, a straightforward measurement technique is the determination of direct reflectance with a defined aperture acceptance angle as described in the guideline developed by an international working group of IEA SolarPACES (Fernandez-Garcia, et al., 2018). An equipment measuring the integrated total scattering signal TIS can be a so-called Coblentz sphere (with specularly reflecting walls redirected all scattering to a sensor) or an Ulbricht sphere (with diffusely reflecting walls and a sensor detecting the brightness level) (Kadkhoda, et al., 2000). However, the directionality of the scattering should be known for concentrating devices especially. The bidirectional reflectance distribution function (BRDF) is a comprehensive characterization, where narrow forward or wide-angle scattering can be quantified depending on incidence and exit angles. The geometrical definition of the $BRDF$ is given in Fig. 3.71.

In the photometric measurement set-up, a sensor with a narrow acceptance angle measuring only a cone with width $d\Omega_2$ is moved over the complete hemisphere above the reflector to measure the distribution for a specific incidence angle. The mathematical description is given in Eq. (3.117).

$$BRDF = \frac{\text{differential radiance}}{\text{differential irradiance}} = \frac{dL_2(\theta_1, \varphi_1; \theta_2, \varphi_2; E_1)}{dE_1(\theta_1, \varphi_1)}, \tag{3.117}$$

where the incident solid angle (θ_1,φ_1) and the scattered solid angle (θ_2,φ_1) are given in spherical coordinates. For practical application the $BRDF$ is often approximated by a simplified term [41]

$$BRDF|_{\theta_1,\varphi_1} = \frac{dL_2(\theta_1, \varphi_1; \theta_2, \varphi_2)}{dE_1(\theta_1, \varphi_1)} = \frac{d\Phi_2(\theta_2, \phi_2)}{d\Omega_2 dA \cos\theta_2} \frac{dA}{d\Phi_1} \approx \frac{\Phi_2}{\Omega_2 \cos\theta_2} \frac{1}{\Phi_1}, \tag{3.118}$$

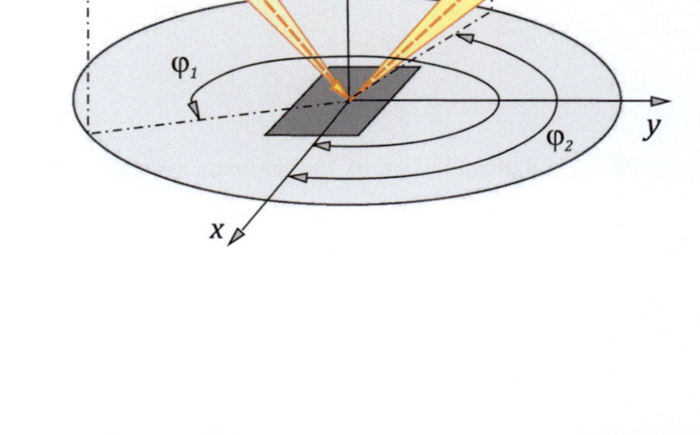

Fig. 3.71 Geometrical definition of bidirectional reflectance distribution function

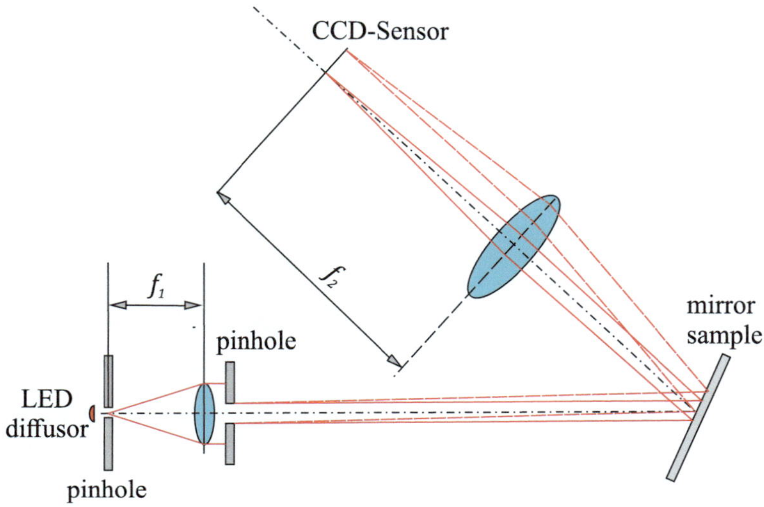

Fig. 3.72 Schematic representation of set-up for specular reflectance measurement. Sample measurement; adapted from (Heimsath et al., 2015)

valid for a nearly collimated beam of incident light reflected from a small differential surface elementdA. The $BRDF$ is relating the radiant power Φ_2 scattered into a solid angle Ω_2 to the incident power Φ_1For analysis of surface scatter it is sometimes useful to drop the cosine correction, the resulting term is then called angle resolved scatter function (ARS), which is the cosine corrected$BRDF$.

$$ARS|_{\theta_1,\varphi_1} = \frac{\Phi_2}{\Omega_2}\frac{1}{\Phi_1}. \tag{3.119}$$

Experimental design in the past used collimated light or laser light to illuminate a reflecting small sample material, and a sensor with a high sensitivity over several decades of intensity e.g. a PIN diode were used to detect the scattered light by moving it over the hemisphere over the complete hemisphere. Such an instrument is described for example in (Platzer and Bergkvist, 1993). A more modern version using a highly linear CCD camera has been developed at Fraunhofer able to detect a forward scattering range in one shot (Fig. 3.72).

As seen in Fig. 3.73b the (logarithmic) intensity plot of a scattering surface may show a pattern without rotational symmetry. Integrated over the azimuth angle the fast-declining intensity over few $mrad$ can be better seen in Fig. 3.73a.

3.5.3 Deflectometry

Other methods are needed when the geometrical shape of the reflector is not ideal. Then deviations from the ideal concentrator will result in an optical error, i.e.

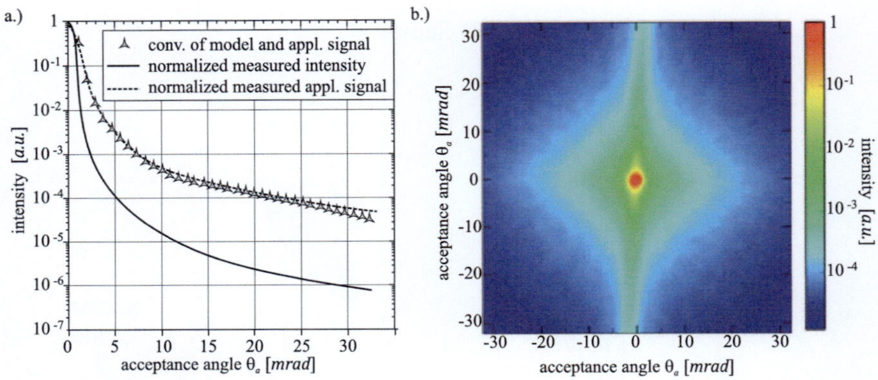

Fig. 3.73 Measured angle resolved scattering distribution of an aluminium based reflector. (left) Measurement data (blue) and convolution (red-crosses), as well as apparatus signal (green) obtained by integration (right) 2-D plot of the measured intensity. Data were obtained at $\lambda = 455nm$ and $\theta_i = 15°$ (Heimsath, 2020)

the beam reflected towards the receiver will have a wider divergence spread as compared to the ideal situation. The situation is depicted in Fig. 3.74.

The large deviations might have an origin in torsion, thermo-mechanical stress, and gravity forces, which due to limited stiffness of the reflector construction lead to macroscopic distortions. on thermal movements. Waviness or indentations on a smaller scale may be due to mechanical damages in the production process. The microscopic surface roughness of the material discussed above adds to these errors. If the curved cross section of the concentrator lies in the $x-z$-plane, the local slope deviation is defined as the difference between the measured local surface gradient

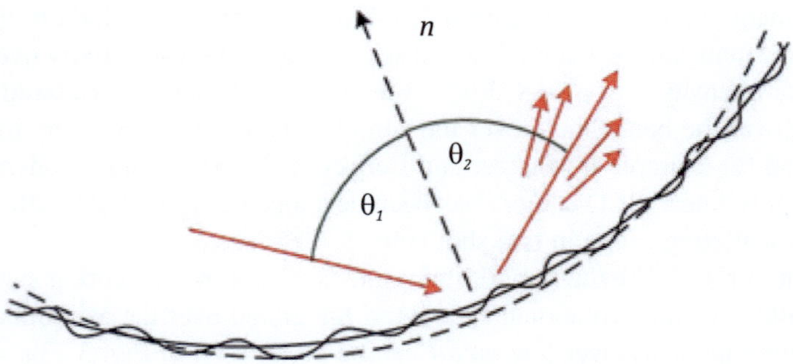

Fig. 3.74 Large and medium scale geometrical error of a reflector surface. The dashed line shows the ideal surface, the solid line a distorted average surface and the thin solid line shows the actual surface, including small surface undulations or scratches. The angular scattering of the reflected solar rays at the surface normal n, the angle of reflectance θ_2 and the indicdent solar ray at the incidence angle θ_1 are sketched

dz/dx_m and the local surface gradient of the design shape dz/dx_d:

$$sd_x = \frac{\partial z}{\partial x}_m - \frac{\partial z}{\partial x}_{id} with \overline{sd_x} = \sqrt{\frac{1}{A}\int\int_A sd_x dxdy}. \qquad (3.120)$$

Experimentally large areas have to be analysed, with only sometimes parts of a millimetre deviation from the original shape. This needs an experimental instrumentation working exact even for large samples like parabolic trough mirrors. Contour accuracy of solar reflectors is measured either indirectly, by analysis of the focal line or focal spot, or directly. For direct measurement techniques either the reflection of laser profilometry (Maccari and Montecchi, 2007) or camera-based image processing techniques such as photogrammetry (Pottler, Lüpfert, Johnston, and Shortis, 2005), video processing or use of other reflected images are used. (Xiao et al., 2012).

A very powerful technique working with the reflected image of a pattern is the fringe reflection technique (FRT) or deflectometry (Heimsath, Platzer, Bothe, and Li, 2008) (Kammel, 2005). The advantage is its fine lateral resolution and its ability to provide the desired information for the entire reflector area. The principle is based on a phase-shifting method in which multiple sinusoidal patterns are reflected on the reflector surface to determine the surface gradients of the specular reflecting surface by analysing the reflected image mathematically, see Fig. 3.75.

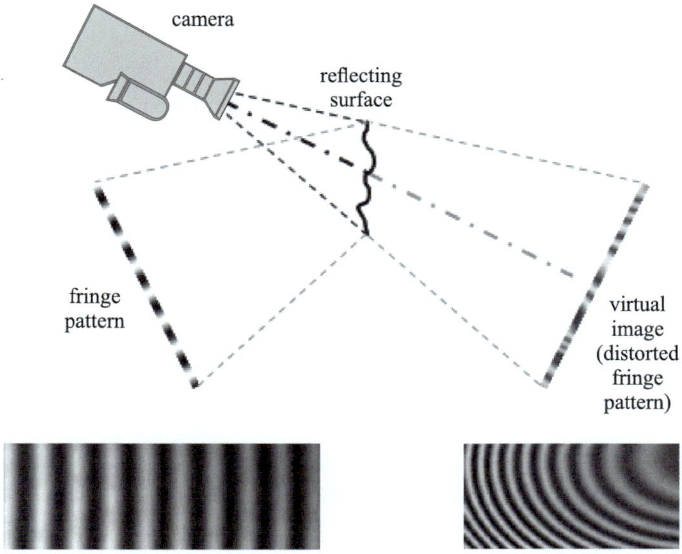

Fig. 3.75 Deflectometry measurement principle. A regular fringe pattern is displayed on a monitor (TFT) or projected onto a target. A camera records the reflected pattern. Irregularities on the specular surface result in a distorted fringe pattern which is evaluated by a phase-shifting technique, adapted from (Burke, Li, Heimsath, Kopylow, and Bergmann, 2013)

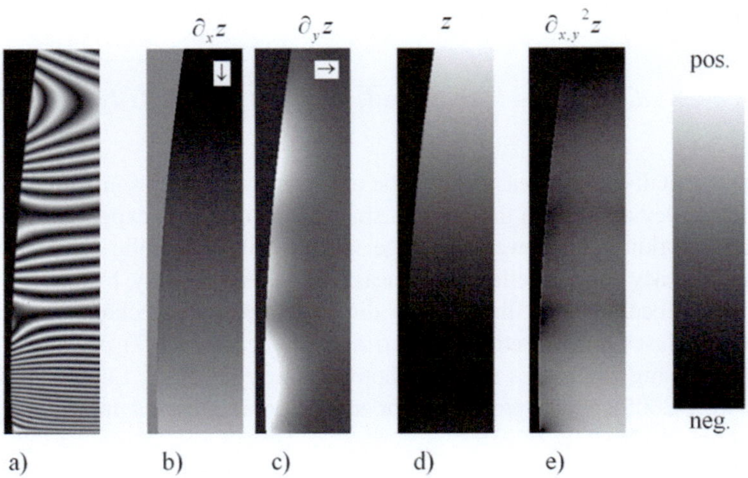

Fig. 3.76 Principal results of fringe reflection or deflectometry techniques. **a** reflected pattern image **b** gradient in x- and **c** y- direction, **d** surface height map and **e** local curvature (Heimsath, Platzer, Bothe, and Li, 2008)

Results of FRT measurements are shown as examples in the following figures. Figure 3.76a display the distorted fringe pattern on part of a reflector. Slope distortions in the range of $mrad$ are clearly visible and demonstrate the sensitivity of the method. The fringe distortions are evaluated by phase measurement and deliver surface gradients in two directions, Fig. 3.76b and c. A height map of the reflector may be calculated by integration, Fig. 3.76d. The curvature of the reflector in Fig. 3.76e shows the slope change per lateral distance and is obtained by differentiation.

In Fig. 3.76 one may see how the image of the reflected and distorted sinusoidal pattern (a) is evaluated to generate local surface gradients. The gradients are calculated in the $x-$ (b) and $y-$ (c) directions. The surface height map (d) is generated by integration and the local curvature (e) by further differentiation. If mirror concentrators are described by such position resolved data, which may be imported into raytracing models, then the effect of the localized slope deviations on the intensity distribution on an absorber geometry can be simulated. This interference-based technology (constructive for $\lambda/2$ and destructive for $\lambda/4$) provides resolution accuracy on the order of wavelength (λ), but it includes only limited directional information. These disadvantages can be compensated by multi-wavelength interferometry, light polarization, white light interferometry (low coherence length) or acousto-optical modulation.

In addition to interferometry, there are a number of other methods for measuring specular surfaces, such as the time-of-flight method (ToF) and triangulation, which are based on the use of coherent light. If unstructured light is used, intensity methods are the obvious choice. Based on the basic principles shown in Fig. 3.77, different signal coding methods can also be used via the operating mode of the

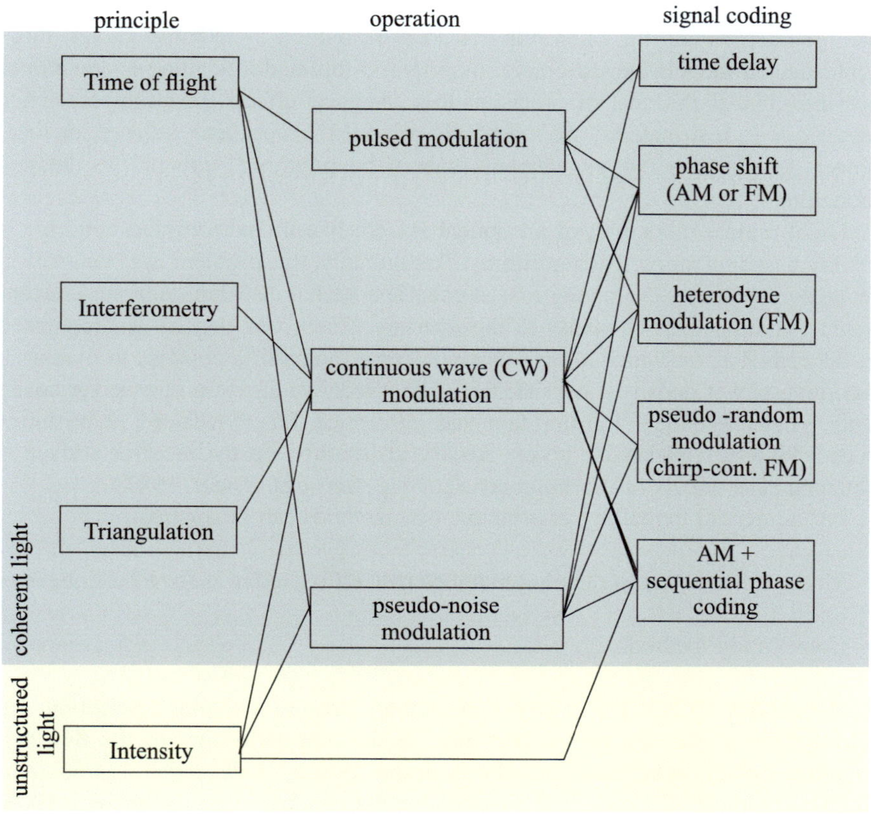

Fig. 3.77 Classification of measurement methods for measuring specular surfaces, with different modes of operation and signal coding options

source (continuous, pulsed, or pseudo-noise methods), based on either time offset or amplitude modulation (AM) or phase modulation (FM) or mixed methods. Successfully also with high time resolution were used so far, the double layer projection method (DLP- (Hillenbrand et al., 2012) and/or the time-of-flight method (ToF), see (Brenneis, 2022).

3.6 Modelling of Optical Systems

For the optical modelling of simple geometries as found in windows or non-concentrating collectors, as treated in Chap. 6 usually a simplified one-dimensional treatment of optical radiation fluxes with balancing equations is sufficient. Only in the case of local shading effects depending on the exact three-dimensional position of transparent and non-transparent components more sophisticated modelling is needed. For concentrating collectors, however, as treated in Chap. 10, the complex geometries and the often required local resolution of radiative fluxes, these sophisticated models are extremely important. Simplified modelling is sometimes possible, but at the cost of precision loss. Angular effects, resulting efficiency losses due to real material properties and non-ideal shapes and surfaces, or local power densities on certain elements cannot be predicted without the detailed modelling.

Local optical modelling of an optical system like a concentrating collector is mainly based on numerical algorithms. Traditionally, this problem has been solved by analytical means evaluating first angular and spatial distribution of the concentrated radiation density planar to the aperture. Then, this density is propagated to the reflector, on which a Gaussian scattering function is applied, to obtain an incoming power density a next element like a receiver aperture or receiver cover. Both, angle-dependent transmission and absorption functions need to be introduced. Finally, a piecewise power density distribution on the receiver surface is obtained. The details of this approach may be taken from (Jeter, 1987).

For numerical modelling in principle two methods can be applied.

- Monte Carlo based ray tracing methods (MCRT), similar as for PTC collectors (cf. 9.3.2.1).
- Convolution methods.

Both methods MCRT and convolution methods require a source description, i.e. a modelling of the sun disk accounting for the non-uniformity of the flux distribution on the solar disk and the widespread solar half-angle of $4.66 mrad$ (0.267°). Almost all simulations assume that the rays pass through air with 100% transmittance, and the reflectance of the reflectors (primary and secondary mirrors is constant; the same is true for transmittance and refractive index of the receiver glass cover. However, in MCRT modelling also the angular-dependent optical properties may be included exactly. A difference between both modelling approaches is that MCRT methods are probabilistic, and only by increasing the number of emitted photons, the resulting power density curve approaches a smooth distribution. The convolution solution, however, is based on the evaluation of a convolution with a Gaussian distribution function, and thus relatively fast. Therefore, this virtual advantage of MCRT methods is not only paid by higher computational effort but also counterbalanced by a fine spatial discretization of all mirror elements to attain a high-quality solution.

In the following two sections we would like to illustrate in detail the two approaches, each time with an example of a specific concentrating collector.

3.6.1 Monte Carlo Raytracing Techniques (MCRT)—Example PTC Collector

Monte-Carlo based Ray Tracing methods ($MCRT$) are based on the probabilistic treatment oft he interaction of a light ray with optical elements like mirrors, absorber, glasses etc.—each interaction of the ray or beam is governed by probability distribution functions. The calculation of a huge number of rays then should reproduce the optical behaviour. The process could also be imagined in a way that a large number of photons are followed on their passage through an optical system. If they hit for example a mirror, a random number and the probability of reflection decides whether the individual photon is reflected or absorbed.

$MCRT$ use randomized rays as indicated in Fig. 3.78 in the example of a PTC, which are traced along their beam path based on a model relying on geometrical optics. The MC simulation first requires a model for the angular distribution of the solar source. The calculation of the ray energy transport from the source to the absorber by means of a $MCRT$ simulation is then structured sequentially as follows:

- First, the position, direction, and energy of each ray on the aperture plane of the PTC are initialized by probability distribution functions.
- Then, based on geometrical optics, the path of each ray, considering reflection at the mirror, transmission and refraction in the glass tube, and absorption at the receiver at each interface, is determined by a series of uniformly distributed random numbers. The ray then is split up or one ray selected with an energy weight, both according to the values of the optical coefficients
- In the third step, the position of the intersection of each ray with the receiver surface with the corresponding energy content is calculated and stored.
- If the relative change in, for example, the energy content for one local target element falls below a pre-determined limit, statistical convergence is assumed

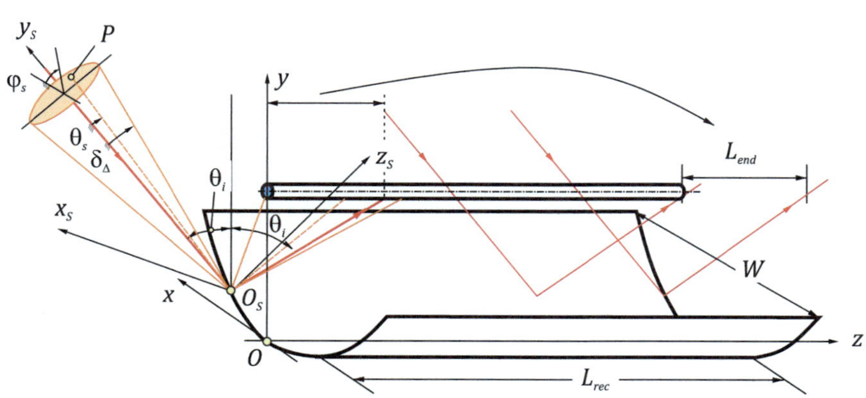

Fig. 3.78 Ray path in a PTC for an irradiation with a longitudinal incident angle of θ_i

for this element and the energy flux density on the absorber surface can be calculated. The entire simulation is finished if all tallies of all target elements have attained statistical convergence, which some time require considerable amounts of rays.

MCRT procedure formulation

In order to trace an arbitrary ray emanating from the sun in as source the world coordinate system $O_s x_s y_s z_s$ must be transformed in the local coordinate system of the PTC, given by $Oxyz$. The coordinate of hitting point (x_0, y_0, z_0) is given by

$$(x_0, y_0, z_0) = \left(W \cdot \kappa_1, x_0^2/4f, L_{rec} \cdot \kappa_2\right), \tag{3.121}$$

in which κ_1, κ_2 are random numbers. Assuming as starting point of a ray P (see Fig. 3.78) on the solar disk its world coordinates are given by,

$$(x_s, y_s, z_s) = (l\sin\theta_s\cos\varphi_s, l\cos\theta_s, l\sin\theta_s\sin\varphi_s), \tag{3.122}$$

where l is the distance $\overline{PO_S}$. θ_s and φ_s denote the radial angle of a sun point and the circumferential one. From this the unit direction vector of the incident ray in the world coordinate system is $\overrightarrow{S} = (\sin\theta_s\cos\varphi_s, -\cos\theta_s, \sin\theta_s\sin\varphi_s)$. Further, the radial ($\theta_s$) and the circumferential angle (φ_s) of the ray´s starting point in the sun are described

$$(\theta_s, \varphi_s) = \left(\arctan\sqrt{\kappa_3}\tan\delta, 2\pi\kappa_4\right), \tag{3.123}$$

withκ_3, κ_4 being again random numbers. The transformation of the world-coordinate system $O_s x_s y_s z_s$ to the PTC coordinate system by translation and rotation is expressed by a transformation matrix$M$, for details see (Zou et al., 2017). The inner normal unit vector of the parabolic reflector ($\overrightarrow{n_r}$) and the glass envelope ($\overrightarrow{n_g}$) are given by

$$\overrightarrow{n_r} = \left(-\frac{x}{\sqrt{x^2+4f^2}}, \frac{2f}{\sqrt{x^2+4f^2}}, 0\right), \overrightarrow{n_g} = \left(\frac{-2x}{d_g}, -\frac{2(y-f)}{d_g}, 0\right), \tag{3.124}$$

in which d_g is the outer diameter of the glas tube. For the unit vectors of the angle of incidence ($\overrightarrow{\lambda}$), the normal unit vector ($\overrightarrow{n}$) and the unit reflected vector ($\overrightarrow{r}$) on the surface (mirror or glass cover), using Fresnel's law, we obtain

$$\vec{r} = \vec{\lambda} - 2 \cdot \left(\vec{\lambda} \cdot \vec{n}\right) \cdot \vec{n}. \tag{3.125}$$

The optical behaviour independently whether reflection, transmission or absorption on a surface is determined as a function of a uniformly generated random number. For example, a uniformly generated random number (κ_3) is compared to the reflection coefficient (ρ) of the mirror. When a ray hits the surface of the mirror the ray is reflected if $\kappa_3 < \rho$, otherwise it is discarded. Other optical quantities along the beam path such as transmission and reflection in the glass cover and absorption at the receiver tube are determined in a similar manner.

In order to initialize the position of the ray on the Sun, a probability distribution function is required, which can be derived using a sunshape model. This describes the intensity distribution in transversal direction perpendicular to the receiver tube. It is given by

$$F(x) = \frac{\int_0^x \phi(\theta) \sin\theta \, d\theta}{\int_0^{\delta_\Delta} \phi(\theta) \sin\theta \, d\theta}, \tag{3.126}$$

in which $\phi(\theta)$ is the solar intensity profile assuming radial symmetry and δ_Δ the outer extent of the solar aureole, finally yielding a probability density function

$$f(x) = \frac{\phi(x) \sin(x)}{\int_0^{\delta_\Delta} \phi(\theta) \sin\theta \, d\theta}. \tag{3.127}$$

Since the direct sampling method is barely applicable mostly acceptance-rejection sampling methods (or hit-and-miss method) are used for random variable sampling, as in other Monte-Carlo techniques as well.

The sun is imaged using countless individual line light sources. These sources are reflected at the mirror surface and those that reach the absorber surface provide the intensity for optical conversion. The image of any line light source in the optical cone with one unit height can be derived using a sunshape model and the intensity $\psi(\theta)$ is obtained by,

$$\psi(\theta) = 2\int_0^{\sqrt{\delta_\Delta^2 - \theta^2}} \phi\left(\sqrt{\theta^2 + \tau_s^2}\right) d\tau_s, \tag{3.128}$$

where θ is the radial angular displacement of line light source. From this, the total energy in any angle span ($\Delta\theta$) in the light cone can be calculated and further the fraction of energy limited by a given radial angular displacement (θ) of a given line light source can be determined. It can now happen that only a part of the rays located around the light cone are intercepted after reflection on the absorber tube, as shown in Fig. 3.79. This radial intercept angle (θ_γ) can be determined analytically for each position of a point A, x_A,

$$\theta_\gamma = \arcsin\left(\frac{d_{rec}}{\left(2\sqrt{x_A^2 + \left(x_A^2/4f - f^2\right)}\right)}\right), \tag{3.129}$$

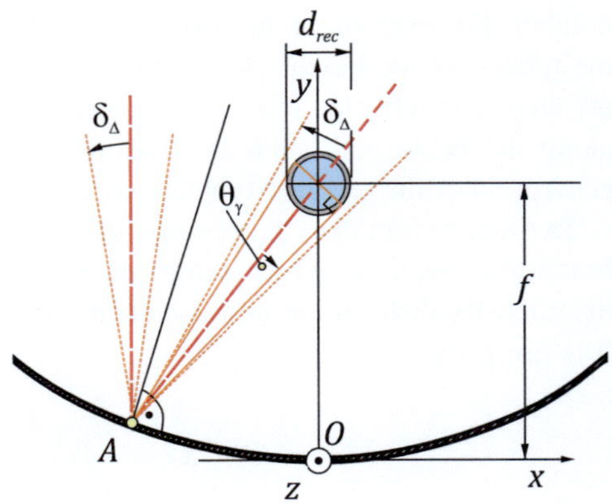

Fig. 3.79 Partially intercepted rays in a PTC collector

in which d_{rec} describes the outer receiver diameter. Again analytically, this allows to define a minimum receiver diameter (d_{min}) given by (3.130) which ensures that all reflected rays reach the receiver. However, in the probabilistic MCRT result, there will be an intensity profile decreasing strongly within the cone of θ_γ, but also outside this cone intensity will be found. Therefore, quantitatively it can be evaluated for which receiver radius how much intensity is lost and spilled.

$$d_{min} = 2\left(W^2/16\,f + f\right) \cdot \sin\delta_\Delta\,. \tag{3.130}$$

However, it is also clear from Fig. 3.78 that part of the incident rays at the angle of incidence (θ) hits the receiver tube after reflection offset on the z-axis. If the entire reflector length L is scanned, a part of the rays no longer reaches the absorber; end losses occur. If the collector row continues infinitely in direction z, the reflected rays can be intercepted by the neighboring collector, but at the end of a collector row, end losses are unavoidable. With the help of trigonometry, the final loss length (L_{end}) as shown in Fig. 3.78 can be calculated:

$$L_{end} = \left(x_A^2/4f + f\right) \cdot \tan\theta_i. \tag{3.131}$$

Due to the longitudinal incidence angle (θ_i) the effective solar radiation intensity (DNI_{eff}) falling on the aperture is reduced on the basis of Lambert's cosine law according to:

$$DNI_{eff} = DNI\,\cos\theta_i. \tag{3.132}$$

This reduction of effective intensity is called cosine-loss and unavoidable. Moreover, it constitutes the largest fraction of losses.

Using the locally evaluated radiance on the receiver tube, one can determine a local concentration factor (C_L), which is defined as the ratio of the local energy flux density ($I(\varphi_{rec})$) to the effective solar irradiance, incident on the aperture.

$$C_L = \frac{I(\varphi_{rec})}{DNI_{eff}}. \tag{3.133}$$

By sampling enough rays to match the statistic convergence criterion set in advance of the computation, which may require in complex designs up to 10^8 rays, which is not a computational challenge since the Monte-Carlo technique allows in a simple manner for massive parallelisation. Despite the fact that the MCRT method describes in a quasi-exact manner the transport along the ray path, it stills contains simplifications. A drawback is that the ray transport at the interface in many codes is considered to be governed by constants as for reflectivity of the mirror, transmissivity of the receiver glass hull and absorptivity of the receiver. Of course, this distorts the results to some extent as all quantities are dependent on the incidence angle (θ_i). It is not a fundamental problem, but a problem of calculation time, when probabilistic distribution functions for the angular optical properties are built into a MCRT code. Especially the transmittance of cover glass (τ_{cover}) depends on the beam direction of its intersection with the absorber surface, so that a simple counting of the accumulated intensity as done in most MCRT-codes yields an overestimation of the yield.

3.6.2 Modelling LFR Collectors -Convolution Methods

Shading and blocking modelling can be done as for a PTC with simple calculation methods, however, requiring in contrast to PTC also consideration of transversal angle dependence (θ_T) in LFC. Details may be taken from (Sharma et al., 2015), (Canavarro et al., 2014). On the other hand, the distribution of incident power on the receiver tube is a localized effect, which can be modelled by MCRT or convolution methods.

The radiance distribution of the sun as well as the optical errors are regarded as statistically independent distributions. The distribution of the optical errors σ_{tot} and the sunshape s_s can be combined by convolution to a resulting distribution L_2 by means of a convolution integral:

$$L_2(\alpha) = \int_{\alpha^*=-\frac{\pi}{2}}^{\frac{\pi}{2}} \sigma_{opt}\left(\alpha - \alpha^*\right) \cdot s_s\left(\alpha^*\right) \, d\alpha^*. \tag{3.134}$$

This approach is used in the analysis of heliostat fields for tower power plants (Walzl, 1977). Also, the mapping of the heliostat into the image plane of the absorber at an idealized parallel irradiation is taken as a probability density distribution. The convolution of the distributions of sun, heliostat and optical errors results again in a probability density distribution in the image plane of the absorber. The determined function corresponds to the radiant flux density distribution in this image plane. To keep the computational effort of the convolution low, the probability density distributions are approximated by Gauss-Hermite polynomials (Winter et al., 1991). Due to the orthogonality of the polynomials, the convolution can be determined from the moments of the distribution using arithmetic operators. Occlusion or shading losses are not explicitly included in the convolution and are only considered proportionally. The flux density distribution is location dependent.

For Fresnel collectors, the angular distribution of the secondary reflector must be additionally determined from the spatial relationship of heliostat and absorber.

For line-focusing systems, the use of the Gauss-Hermite polynomials can be considerably simplified, since angles of incidence deviating from the transverse plane ($\theta_i > 0°$) change only the length (l) of the reflected rays. But the angular relationships with respect to the transverse plane are preserved. The actual length ($l\prime$) then results from:

$$l' = \frac{l}{\cos \theta_i}. \tag{3.135}$$

First, the primary mirror is discretized into small mirror elements. Further, the reflected radiation fraction at a point P of the receiver aperture is determined (compare Fig. 3.80a). This depends exclusively on the angles α_1 and α_2 at which the mirror element is seen by point P deviating from the ideal reflection. If α_1 and α_2 have the same sign and at the same time the angular range of the radiance distribution L_2 is larger, the point P cannot receive radiation from the mirror element under consideration. With different angular sign but also larger amount than the angular range of L_2, P receives the total irradiance of the source. The radiance distribution $L_2(\alpha)$ includes the source i.e. the sun with its distribution (sunshape) as well as the optical error σ_{opt} of the primary mirrors and is calculated with the help of Eq. (3.135). The convolution must be determined only once because the radiance distribution does not change at the location of the mirror element. Due to the geometrical arrangement of the primary mirrors, a partial shading of the mirror element is possible, which can be described by a flat reduction factor. However, in order to reflect the distribution correctly, the shading angle β (cf. Fig. 3.80b) must be included in the convolution. This results in $L_2(\alpha)$ for the radiance distribution,

$$L_2(\alpha) = \int_{-\frac{\pi}{2}}^{\frac{\pi}{2}} \begin{cases} \alpha* < \beta \ \sigma_{opt}(\alpha - \alpha*)\, L(\alpha*)\, d\alpha* \\ \alpha* > \beta \ 0 \end{cases}, \tag{3.136}$$

which Fig. 3.81 illustrates as a function of the angle α for an unshaded and a shaded primary mirror element.

In principle, this radiance distribution does not change with the location of the mirror element, which is why it is sufficient to determine the convolution only

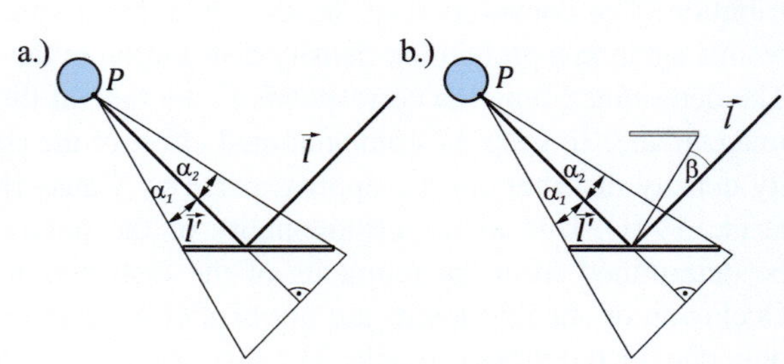

Fig. 3.80 Sketch for determining the reflected radiation fraction at a primary mirror element; (a) without shading and (b) with shading

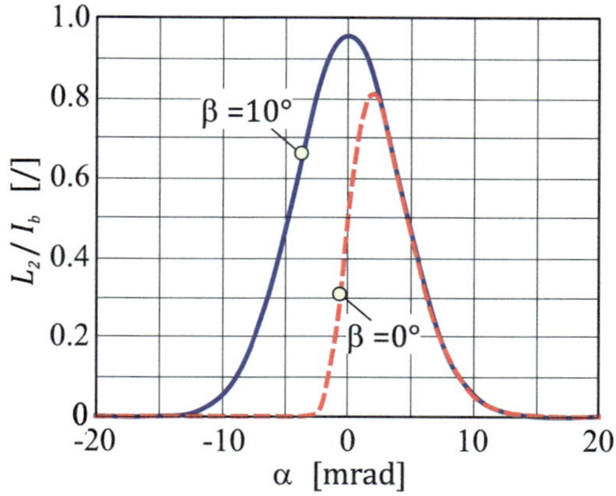

Fig. 3.81 Calculated normalised convolution $L_2(\alpha)$ for unshaded ($\beta = 10°$) and partially shaded primary mirror element ($\beta = 0°$)

once. The radiance transmitted through the mirror element to the point P can be calculated by integrating the convolution $L_2(\alpha)$ from α_1 to α_2 in the form of

$$E(\alpha_1, \alpha_2, \beta) = \int_{\alpha_1}^{\alpha_2} L_2(\alpha, \beta)\, d\alpha, \tag{3.137}$$

where the reflected radiance is the area below the normalised convolution L_2 in the boundaries of α_1 and 2, which is depicted in Fig. 3.82a. Equation (3.137) actually only applies to infinitely long primary mirrors, but this restriction is only relevant when considering the row end losses. The consideration also applies to a sun position outside the transverse plane ($\theta_i > 0$). The limiting angles $\alpha_{1,2}$ are given by the points of the primary mirror element boundaries x_i:

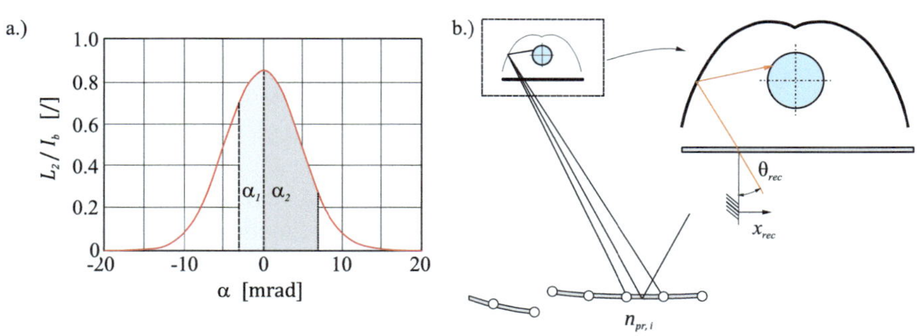

Fig. 3.82 **a** Calculated reflected radiation fraction as area below the normalised convolution in the limits from α_1 to α_2; **b** Coordinate definition for the acceptance range of the secondary reflector of a Fresnel collector

$$\alpha_{1,2} = \cos^{-1}\left(\frac{\left(\vec{P} - \vec{x}_{i,1,2}\right) \cdot \vec{l}'}{\left|\left(\vec{P} - \vec{x}_{i,1,2}\right)\right| \cos\theta_i}\right). \tag{3.138}$$

The total illuminance $E(x_{rec})$ at location P of the receiver results from the summation of the proportions of all mirror elements $n_{pr,i}$ with:

$$E(x_{rc}) = I_b \cdot \sum\nolimits_{i=1}^{n_{pr}} E(\alpha_1, \alpha_2, \beta). \tag{3.139}$$

For Fresnel systems, the procedure must be extended, since up to now only the depicted portion of the direct radiation of a mirror element at a location on the receiver is determined. For a Fresnel collector, however, the radiation reaching the absorber is important, which is not identical to the above result. The coordinates of a light beam incident on the aperture are determined by the angle θ_{rec} and the coordinate x_{rec} (cf. Fig. 3.82b). Whether the beam hits the absorber tube can be calculated using the forward ray tracing method. The result can be summarised in a matrix representing the acceptance range $\Gamma_{rec}(x_{rec}, \theta_{rec})$ of the secondary reflector. The total radiation E_{abs} reaching the absorber results from the summation over all receiver locations (x_{rec}) and each primary mirror element $(x_{pr,i})$ by:

$$E_{abs} = \rho_{pr} \cdot \sum_{n_{x_{rec}}} \sum_{n_{pr}} \cos(\theta_{rec})\ E(\alpha_1, \alpha_2, \beta)\ \Gamma_{rec}(x_{rec}, \theta_{rec})\ \tau_{rec}(\theta_{rec})\ \Delta x_{rec}. \tag{3.140}$$

A typical result of such a calculation is shown in Fig. 3.83. Plotted is the relative illuminance as the sum of the mirror elements of a primary mirror over the location x_{rec} and the influence of the utilisation of the receiver, which is also a function of θ_{rec} and x_{rec}. In the centre, the mirror directly hits the absorber tube. At the edges of this area, there is a strong attenuation due to the angle-dependent absorption, as the absorber tube is hit at increasingly unfavourable angles. This is followed by the area that reaches the target, i.e. the receiver, via one and then two reflections.

Another central result is depicted in Fig. 3.84a, in which the acceptance range $\Gamma_{rec}(x_{rec}, \theta_{rec})$ of a secondary reflector is shown as a function of the length x_{rec}. The solar radiation that reaches the input aperture of the receiver within the black bordered areas directly or indirectly hits the absorber tube. Figure 3.84b illustrates the phase diagram of the radiation incident from the primary mirror field. The area shadowed by the receiver at $\theta_{rec} = 0°$ is clearly visible. Due to their greater distance, the outer mirrors smear their image over a wider area. In order to lose as little radiation as possible, both images should overlap as well as possible.

Although the convolution method provides both temporally and spatially resolved data, its implementation is more complex and time consuming.

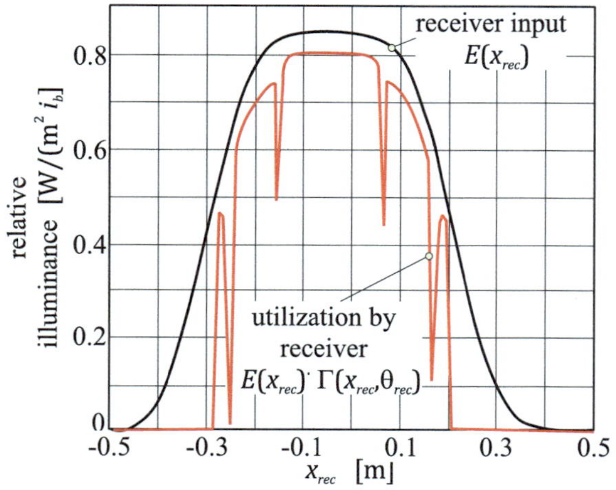

Fig. 3.83 Relative illuminance distribution at the input aperture of the receiver $E(x_{rec})$ of a linear Fresnel collector and its utilisation by the receiver $E(x_{rec})\Gamma_{rec}(x_{rec}, \theta_{rec})$

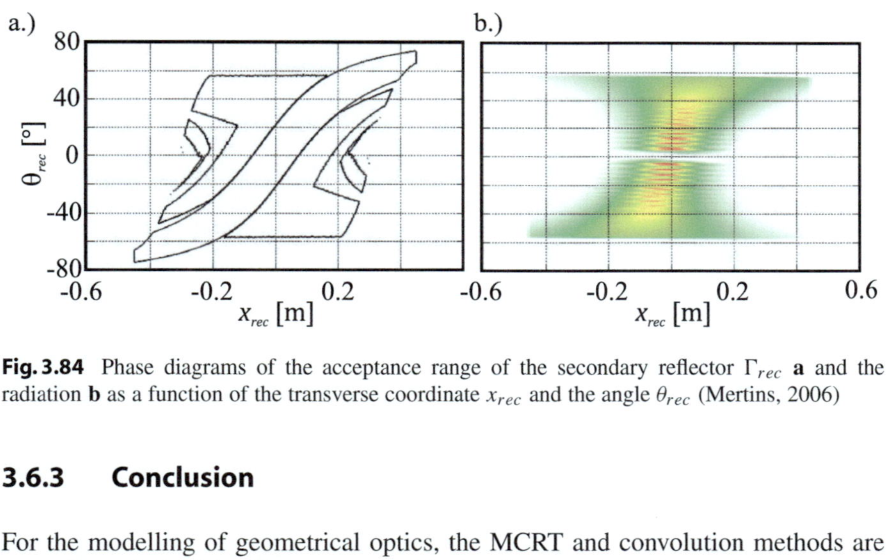

Fig. 3.84 Phase diagrams of the acceptance range of the secondary reflector Γ_{rec} **a** and the radiation **b** as a function of the transverse coordinate x_{rec} and the angle θ_{rec} (Mertins, 2006)

3.6.3 Conclusion

For the modelling of geometrical optics, the MCRT and convolution methods are versatile and flexible. They can provide detailed results with spatial resolution. Although commercial MCRT tools often use simplified assumptions to reduce simulation time, such as optical properties which are constant with incident angle, or not spectrally dependent, in principle all these properties can be included in the code by providing an addition probability distribution function. However, the more properties are included in the probabilistic code, the longer takes a calculation of a ray path. And more over, much more photons with different (spectral) properties have to be emitted and followed in order to get a decent statistic with a representative result.

Other optical simulation software than presented here, e.g. on spectral properties of transparent glasses, coatings, and thin-films, therefore are more sophisticated in the optical calculations, but deterministic and they do not take into account spatially resolved interactions but calculate only a one-dimensional radiation balance in one propagation direction.

References

Abbas, R., Sebastián, A., Montes, M., & Valdés, M. (2018). Optical features of linear Fresnel collectors with different secondary reflector technologies. *Applied Energy, 232*, 386–397. https://doi.org/10.1016/j.apenergy.2018.09.224

Agnihotri, O. P., & Gupta, B. K. (1981). *Solar selective surfaces.* Wiley-Interscience.

Anders, H. (1967). *Thin Films in Optics.* Focal Press Ltd.

Arancibia-Bulnes, E., & C., & Ruiz-Suarez, J. (2000). Solar absorptance and thermal emittance of cermets with large particles. (I. P. Ltd, Ed.). *Journal of Physics. D. Applied Physics, 33*, 2489–2496. https://doi.org/10.1088/0022-3727/33/19/320

Ashcroft, N., Mermin, N., Wei, D., & Mermin, N. (2016). *Solid State Physics* (2nd ed.). Cengage Learning.

ASTM G173–03. (2020). *Standard Tables for Reference Solar Spectral Irradiances: Direct Normal and Hemispherical on 37° Tilted Surface.* West Conshohocken, PA, 19428–2959, USA: ASTM Headquarters . doi:DOI: https://doi.org/10.1520/G0173-03R20

Becker, E. W., Ehrfeld, W., Münchmeyer , D., Betz, H., Heuberger, A., Pongratz, S., . . . von Siemens, R. (1982). Production of separation-nozzle systems for uranium enrichment by a combination of X-ray lithography and galvanoplastics. *Naturwissenschaften,* Vol 69(11), pp. 520-523.

Bläsi, B. (2000). *Holographisch hergestellte Antireflexoberflächen für solare und visuelle Anwendungen.* Dissertation, Albert-Ludwigs-Universität Freiburg i. Br., Freiburg im Breisgau.

Bläsi, B., Aufderheide, K., & Abott, S. (2002). Entspiegeln mit Mottenaugenstrukturen. *Kunststoffe, Jahrgang, 92*, 50–53.

Born, M., & Wolf, E. (1999). *Principles of Optics-Electromagnetic Theory of Propagation.* Cambridge University Press. https://doi.org/10.1017/CBO9781139644181.

Brenneis, B. (2022). Long Range Optical Distance Sensors for Liquid Metal Free Surface Detection. *Journal of Nuclear Engineering, 3*(4), 364–372. https://doi.org/10.3390/jne3040023.

Brunold, S., Frei, U., Carlsson, B., Möller, K., & Köhl, M. (2000). Accelerated life testing of solar absorber coatings: Testing procedure and results. *Solar Energy, 68(4)* , pp. 313-323. doi:https://doi.org/10.1016/S0038-092X(00)00034-7.

Brunold, S., Frei, U., Carlsson, B., Möller, K., & Köhl, M. (2000). Round robin on accelerated life testing of solar absorber surface durability. *Solar Energy Materials and Solar Cells,* 61(3), pp. 239-253. doi:https://doi.org/10.1016/S0927-0248(99)00113-0.

Brunotte, A., Lazarov, M., & Sizmann, R. (1992). Calorimetric measurements of the total hemispherical emittance of selective surfaces at high temperatures. *Proc. SPIE 1727, Optical Materials Technology for Energy Efficiency and Solar Energy Conversion XI, 1727*. doi:https://doi.org/10.1117/12.130497.

Burke, J., Li, W., Heimsath, A., Kopylow, C. v., & Bergmann, R. B. (2013). Qualifying parabolic mirrors with deflectometry. *Journal of the European Optical Society - Rapid publications; Vol 8.*

Canavarro, D., Chaves, J., & Collares-Pereira, M. (2014). Simultaneous multiple surface method for linear fresnel concentrators with tubular receiver. *Solar Energy, 110*, 105–116. https://doi.org/10.1016/j.solener.2014.09.002.

Chaves, J., & Collares-Pereira, M. (2010). Etendue-matched two-stage concentrators with multiple receivers. *Solar Energy, 84*(2), 196–207. https://doi.org/10.1016/j.solener.2009.10.022.

Claesson, Y., Georgson, M., Roos, A., & Ribbing, C.-G. (1990). Optical characterisation of titanium-nitride-based solar control coatings. *Solar Energy Materials, 20*(5–6), 455–465. https://doi.org/10.1016/0165-1633(90)90035-Y.

Cuomo, J., Ziegler, J., & Woodall, J. M. (1975). A new concept for solar energy thermal conversion. *Applied Physics Letters, 26*(10), 557–550.

Demtröder, W. (2004). *Experimentalphysik 2 (2* (Auflage). Springer Verlag.

Duffie, J. A., & Beckmann, W. A. (2020). *Solar Engineering of Thermal Process, Photovoltaics and Wind*. John Wiley & Sons Inc.

Fernandez-Garcia, A., Sutter, F., Montecchi, M., Sallaberry, F., Heimsath, A., Heras, C., & Le Baron, E. a.-G. (2018). *International Guideline Version 3.0: Parameters and method to evaluate the solar reflectance properties of reflector materials for concentrating solar power technology.* www.solarpaces.org: IEA SolarPACES.

Gäbler, D. (2005). *Entspiegelung von Silizium-Photodioden nach dem Vorbild der Nanooptik von Mottenaugen.* Diplomarbeit TU Ilmenau, Illmenau.

Giovanelli, R., & Beck, R. (1987). *Geheimnisvolle Sonne.* VCH Verlagsgesellschaft.

Gombert, A., & Rommel, M. (1997). Breitbandige Antireflexbeschichtungen. (F. Sonnenenergie, Ed.) *Solare Gebäudetechniken*, 81–86.

Gombert, A., Glaubitt, W., Rose, K., Dreibholz, J., Bläsi, B., Heinzel, A., . . . Wittwer, V. (1999). Subwavelength-structured antireflective surfaces on glass. *Thin Solid Films*, Vol 351(1), pp. 73-78.

Graedel, T. E. (1992). Corrosion Mechanisms for Silver Exposed to the Atmosphere. *J. Electrochem. Society, 139*(7), 1962–1970.

Granqvist, C. (1990). Window Coatings for the Future. *Thin Solid Films*, 193-194, pp. 730-741. doi:dx.doi.org/https://doi.org/10.1016/0040-6090(90)90225-3.

Granqvist, C. (1991). Energy-efficient windows: Present and fortcoming technology. In C. (. Granqvist, *Materials Science for Solar Energy Conversion Systems* (pp. 106–167). Oxford: Pergamon Press plc.

Granqvist, C. (1991). Solar energy materials-Overview and some examples. *Appl. Phys. A*, 52, pp. 83-93.

Guerreiro, L., Canavarro, D., & Collares Pereira, M. (2015). Efficiency improvement and potential LCOE reduction with an LFR-XX SMS plant with storage. *Energy Procedia*, 69, pp. 868-878. doi:doi: https://doi.org/10.1016/j.egypro.2015.03.112.

Heimsath, A. (2020). *Optical efficiency of concentration solar collectors - Investigation of loss mechanisms of solar reflectors.* Dissertation, Universität Bremen, Bremen.

Heimsath, A., Platzer, W., Bothe, T., & Li, W. (2008). Characterization of optical components for linear Fresnel collectors by the fringe reflection method. *Proc. 14th SolarPACES International Symposium, Las Vegas, 3–7 Mar. 2008.*

Heimsath, A., Schmid, T., & Nitz, P. (2015). Angular Resolved Specular Reflectance Measured with VLABS. *Energy Proc., 69*, 1895–1903.

Hildebrandt, C. (2009). *Hochtemperaturstabile Absorberschichten für linear konzentrierende solarthermische Kraftwerke.* Dissertation, Universität Stuttgart, Institut für Thermodynamik und Wärmetechnik, Stuttgart.

Hillenbrand, M., Stieglitz, R., & Neitzel, G. (2012). Detection of liquid–metal, free-surface flow using the DLP measurement technique. *Experiments in Fluids*, 52, pp. 179–192. doi:doi.org/https://doi.org/10.1007/s00348-011-1214-z.

Hi-min SOLAR. (2022). Parabolic Trough Receiver data sheet. Sun-Moon Mansion, Solar Valley Road, Dezhou, Shandong, PR China. Retrieved March 22, 2022, from http://himinsun.com/1-3-parabolic-trough-receiver.html.

Hollingsworth Smith, P., & Gurev, H. S. (1977). Silicon Dioxid as a High Temperature Stabilizer for Silver Films. *Thin Solid Films, 45*, 159–168.

Howell, J., Mengüc, M. P., Daun, K., & Siegel, R. (2020). *Thermal Radiation Heat Transfer* (7 ed.). CRC Press ,Taylor & Francis Group, LLC. Retrieved from https://doi.org/10.1201/9780429327308.

ISO 2353. (2005). *Photometry—The CIE system of physical photometry.*

ISO 80000–7. (2019). *Quantities and units—Part 7: Light and radiation.*

ISO 9050. (2003). *Glass in building—Determination of light transmittance, solar direct transmittance, total solar energy transmittance, ultraviolet transmittance and related glazing factors.*

ISO 9845–1. (1992). *Solar energy—Reference solar spectral irradiance at the ground at different receiving conditions — Part 1: Direct normal and hemispherical solar irradiance for air mass 1,5.* ICS:27.160 Solar energy engineering.

Jarmer, T., Serwaty, J., & Knote, C. (2005). *Grundlagen und Praxis radiometrischer; Abschnitt: Fernerkundliche Erfassung photosynthetisch aktiver Pigmente und weiterer Blattinhaltsstoffe.* Universität Trier.

Jeter, S. M. (1987). Analytical determination of the optical performance of practical parabolic trough collectors from design data. *Solar Energy, 39*(1), 11–21.

Kadkhoda, P., Müller, A., Ristau, D., Duparré, A., Gliech, S., Lauth, H., Bennett, J. M. (2000). International round-robin experiment to test the International Organization for Standardization total-scattering draft standard. *Applied optics*, 39(19), pp. 3321–3332.

Kammel, S. (2005). *Deflektometrische Untersuchung spiegelnd reflektierender Freiformflächen.* Dissertation, Universität Karlsruhe.

Kennedy, C. E. (2002). *Review of mid- to high-temperature solar selective absorber materials.* Technical Report , National Renewable Energy Laboratory, Golden, Colorado.

Kreider, J. F. (1979). *Medium and high temperature solar processes.* Academic Press Inc.

Kreith, F., & Kreider, J. F. (1978). *Principles of solar engineering.* Hemisphere Publishing Corporation.

Lanxner, M., & Elgat, Z. (1990). March 12–13). Solar selective absorber coating for high service temperatures, produced by plasma sputtering. (e. G. C., Ed.) *SPIE- Proc. Optical Materials Technology for Energy Efficiency and Solar Energy Conversion IX, 1272*, 240–249.

Maccari, A., & Montecchi, M. (2007). An optical profilometer for the characterisation of parabolic trough solar concentrators. *Solar Energy, 81*(2), 185–194.

MacLeod, A. (2001). *Thin Film Optical Filters.* Bristol: Adam Hilger Ltd.

McNaught, A. D., & Wilkinson, A. (1997). *IUPAC. Compendium of Chemical Terminology (2nd* (edition). Blackwell Scientific Publications.

Mertins, M. (2006). *Technische und wirtschaftliche Analyse von horizontalen Fresnel-Kollektoren.* Uinversität Karlsruhe: Fakultät Maschinenbau.

Miller, W. H., Möller, A. R., & Bernhard, C. G. (1966). The corneal nipple array. In C. G. Bernhard (Ed.), *Functional organization of the compound eye* (pp. 21–33). Pergamon Press.

Morin, G., Dersch, J., Platzer, W., Eck, W., & Häberle, A. (2012). Comparison of Linear Fresnel and Parabolic Trough Collector plants. *Solar Energy*, 86, pp. 1-12. doi:doi:https://doi.org/10.1016/j.solener.2011.06.020

Niemann, M., & Schreitmüller, K. R. (1993). Investigations on high efficiency collectors for industrial process heat generation up to 200°C. *Proc. of ISES World Congress Budapest, Hungary.*

Nolting, W. (2009). *Grundkurs Theoretische Physik 7: Viel-Teilchen-Theorie.* Springer Verlag.

Palik, E. (1985). *Handbook of optical constants of solids* (8th ed.). Academic Press.

Platzer, W. J., & Bergkvist, M. (1993). Bulk and Surface Light-Scattering from Transparent Silica Aerogel. *Solar Energy Materials and Solar Cells, 31*(2), 243–251.

Platzer, W., Heimsath, A., Hildebrandt, C., & Georg, A. (2008). Quality Control of concentrating collector components for the optimization of performance. *Proc. 14th SolarPACES International Symposium, Mar 4–7, 2008.* Las Vegas, USA.

Pottler, K., Lüpfert, E., Johnston, G. H., & Shortis, M. R. (2005). Photogrammetry. A Powerful Tool for Geometric Analysis of Solar Concentrators and Their Components. *J. Sol. Energy Eng. 127 (1)*, pp. 94–101.

Powell, C. F., Oxley, J. H., Blocher, J. M., & Klerer, J. (1966). Vapor Deposition. *Journal of The Electrochemical Society,* 113.

Pulker, H. K. (1999). *Coatings on Glass* (2nd ed.). Elsevier Science B.V.

Rabl, A. (1976). Comparison of solar concentrators. *Solar Energy, 18*, pp. 93-111.

Rabl, A. (1976b). Optical and thermal properties of compound parabolic concentrators. *Solar Energy, 18*(6), 497–511.

Rabl, A., & Winston, R. (1976). Ideal concentrators for finite sources and restricted exit angles. *Applied Optics, 15*(11), 2880–2883. https://doi.org/10.1364/AO.15.002880

Randich, E., & Allred, D. (1981). Chemically vapor-deposited ZrB2 as a selective solar absorber. *Thin Solid Films, 83*(4), 393–398. https://doi.org/10.1016/0040-6090(81)90646-5

Rioglass Solar SA. (2022). Receiver Tubes for Linear CSP. Poligono Industrial Villallana Pola de Lena, 33695, Spain. Retrieved March 22, 2022, from https://www.rioglass.com/our-products/hce-tubes.html

Sakka, S. (2005). Applications of Sol-Gel Technology. In S. Sakka (Ed.), *Handbook of Sol-Gel Science and Technologie (Vol (III*, pp. 599–760). Kluwer Academic Publishers.

Schade, H., & Smith, Z. (1985). Mie scattering and rough surfaces. *Applied Optics, 24*(19), 3221–3226. https://doi.org/10.1364/AO.24.003221

Sella, C., Bichri, A., Martin, J. C., Lafait, J., Driss-Khodja, K., & Berthier, S. (1989). Adjustable optical properties of coatings based on cermet thin films near the percolation threshold. *Physica a: Statistical Mechanics and Its Applications, Vo., 157*(1), 555–560.

Seraphin, B. (1979). Spectrally selective surfaces and their impact on photothermal solar energy conversion. In B. Seraphin, *Solar Energy Conversion* (pp. 5–55). Springer,. doi:https://doi.org/10.1007/3-540-09224-2_2.

Seraphin, B., & Meinel, A. (1976). Optical Properties of Solids — New Developments. *In B. Seraphin. Amsterdam: North-Holland.* https://doi.org/10.1016/0030-3992(76)90029-3.

Sharma, V., Nayak, J., & Kedare, S. B. (2015). Comparison of line focusing solar concentrator fields considering shading and blocking. *Solar Energy, 122*, 924–939. https://doi.org/10.1016/j.solener.2015.10.011.

Smakula, A. (1957). *Acta Crystallographica, Vol. 10*, 735–863.

Tagle-Salazar, P., & Nigam, K. R.-S. (2020). Parabolic trough solar collectors: A general overview of technology, industrial applications,energy market, modeling, and standards. (D. Gruyter, Ed.) *Green Processing and Synthesis, 9*, pp. 595–649. doi:https://doi.org/10.1515/gps-2020-0059.

Thiemann, S. (2000). *Das Prinzip Fernerkundung.* Scientific Technical Report 00/04, Geo-Forschungs Zentrum Potsdam, Potsdam.

Thornton, J. A., Penfold, A., & Lamb, J. L. (1980). Sputter-deposited Al2O3/Mo/ Al2O3 selective absorber coatings. *Thin Solid Films, 72*, 101–109.

Valkonen, E., & Karlsson, B. (1987). Optimization of metal-based multilayers for transparent heat mirrors. *Energy Research, 11*(3), pp. 397-403. doi:doi.org/https://doi.org/10.1002/er.4440110309.

Valkonen, E., Karlsson, B., & Ribbing, C.-G. (1984). Solar optical properties of thin films of Cu, Ag, Au, Cr, Fe Co. *Ni and Al. Solar Eneregy, 32*(2), 211–222. https://doi.org/10.1016/S0038-092X(84)80038-9

Vien, T. K., Sella, C., Lafait, J., & Berthier, S. (1985). Pt-Al2O3 selective cermet coatings on superalloy substrates for photothermal conversion up to 600°C. *Thin Solid Films, 126*, 17–22.

Walzl, M. D. (1977). A Solar Flux Density Calculation for a Solar Tower Comcentrator Using A Two-Dimensional Hermite Function Expansion. *Solar Energy, 19*, 239–253.

Wang, M., & Pan, N. (2008). Predictions of effective physical properties of complex multiphase materials. *Materials Science and Engineering* R, 63, pp. 1-30. doi:doi:https://doi.org/10.1016/j.mser.2008.07.001.

Willey, R. (2002). *Practical Design and Production of Thin Films* (2nd ed.). Marcel Dekker Inc.

Winston, R. (1974). Principles of solar concentrators of a novel design. *Solar Energy, 16*(2), 89–95.

Winter, C. J., Sizmann, R. L., & Vant-Hull, L. L. (1991). *Solar power plants: Fundamentals, technology, systems, economics.* Springer Verlag.

Xiao, J., Wei, X., Lu, Z., Yu, W., & Wu, H. (2012). A review of available methods for surface shape measurement of solar concentrator in solar thermal power applications. *Renewable and Sustainable Energy Reviews, 16*(5), 2539–2544.

Zhang, Q. C., & Mills, D. (1992). New cermet film structures with much improved selectivity for solar thermal applications. *Appl. Phys. Lett.*, 60, pp. 545-547. doi:doi.org/https://doi.org/10.1063/1.106602.

Zhao (2007). *Spectrally selective solar absorbing coatings prepared by dc magnetron sputtering.* Uppsala Schweden. Acta Universitatis Upsaliensis.

Zhu, G. (2017). New adaptive method to optimize the secondary reflector of linear Fresnel collectors. *Solar Energy, 144*(1), 117–126. https://doi.org/10.1016/j.solener.2017.01.005.

Zou, B., Yao, Y., & Jiang, Y. (2017). A detailed study on the optical performance of parabolic trough solar collectors with Monte Carlo Ray Tracing method based on theoretical analysis. *Solar Energy, 147*, 189–201. https://doi.org/10.1016/j.solener.2017.01.055.

4 Heat Transfer

Abstract

Thermal technologies and thus also solar thermal always involve heat transfer within a body or between bodies. In some cases the heat transfer has to be enhanced in order to achieve a highly efficient optical conversion of solar radiation into heat, in other cases a minimal heat loss over the system boundaries to the environment is the target. In particular, the optical conversion of short wave solar irradiance (electromagnetic waves) to heat occurs in solids. Within the solid body the heat is released mostly volumetrically in thin layers or layer stacks causing not only temperature rise but also differential thermal elongation. This heat has to be transported via passive means towards a heat sink for utilization. At the same time, however, a body that is warmer than its surroundings loses this heat in the form of thermal radiation and convection, and as the temperature rises, this proportion increases rapidly. This conflicting feature emphasizes the excellent importance of understanding passive heat transfer mechanisms in solar thermal applications. In this context we consider as passive heat transfer processes such as heat conduction and heat radiation, which are not associated with a medium transport. Convection, however, which enhances heat transport through molecular heat conduction in fluid media by a movement of fluid volumes, will be treated in the Chap. 5. The driving mechanisms of passive heat transport are scalar quantities such as pressure and temperature, which characterize the thermodynamic state in a continuum, as well as waves, which propagate in a medium. The propagation of the waves in a solid or fluid medium, in turn, depends on its molecular thermo-physical properties and in cases of solids even its lattice structure. For this reason, any temperature difference leads unavoidably to a heat transfer changing the thermal properties of a solar thermal system, but also represents a key variable in component design. An important technical parameter here is, among others, the maximum permissible wall temperature. After an introductory description of the distinction between heat transfer and transport processes, the steady and transient heat conduction in gases, fluids and solids is covered. Successively, radiative heat

R. Stieglitz and W. Platzer, *Solar Thermal Energy Systems*,
https://doi.org/10.1007/978-3-031-43173-9_4

transport in different configurations as well as methods to evaluate to transferred quantities are elaborated. Finally, as a result of a heat transport by conduction and radiation across interfaces mixed situations may occur, in which density differences in the adjacent fluid/gas are induced yielding to natural buoyant convective fluid motion increasing the heat transfer. Some selected cases of combined heat transfer problems are covered in the last section.

Keywords

Steady and transient heat conduction • Conduction in solid and gases • Heat radiation • Radiosity • Irradiance • Radiation transport

4.1 Definition of a Passive Heat Transport

In thermodynamics, energy that crosses the boundary of a system is called heat if the energy transport is caused solely by a temperature difference between the system and its environment. According to the second main theorem of thermodynamics, heat always flows in the direction of lower temperature within a body and across the body interface. Adopting this approach the term passive heat transfer describes a heat energy transport that occurs exclusively via microscopic collisions within a continuum or interactions of waves (phonons, photons, electrons), i.e. molecular or electromagnetic interactions. Thus, it is not a process necessarily associated with a matter transport such as a fluid flow, however, it describes the heat transport within a continuum and across a continuum interface. This passive type of heat transport is either influenced by the thermo-physical material properties or via the surface properties of the interface. Thus, the solution of a passive heat transfer problem in terms of thermodynamics answers the following questions:

- heat transfer though molecular conduction in a continuum and,
- heat exchange across interface by radiation between bodies.

General thermodynamics considers macroscopically a continuum in equilibrium or quasi-equilibrium states called phases, while heat transfer considers the exchange of heat. This exchange of heat changes the internal energy of the volume under consideration. Since thermodynamics considers only the initial and final states, it is not able to make a statement about how fast or intense the heat transfer occurs. In contrast to thermodynamics, which operates with homogeneous systems having the same properties, the so-called phases, heat transfer is a continuum theory in which spatially extended fields with variable properties occur, which may also depend on time. In most cases the driving temperature differences play the dominant role in heat transfer. Therefore, temperatures differences ΔT are used, where the temperature T is described in relation to a fixed reference temperature T_0:

$$\Delta T = T - T_0. \tag{4.1}$$

Using as reference temperature $T_0 = 273.15$ K, ΔT corresponds to the temperature in degree Celsius. Temperature differences are independent of the reference temperature, so that its specification can usually be omitted. The continuum theory of steady-state or transient heat transfer at an interface deals with all possible forms of heat energy exchange between and within media. In multi-body problems, for example, there is a superposition of heat conduction and radiation. Typical cases are packed beds, insulating materials or the gap between the glass cover and the absorber of a solar collector.

However, the temperature differences along the interface of a body generated by passive heat transport can also induce density gradients that trigger macroscopic fluid motion (buoyant or natural convection), so that thermal conduction, thermal radiation, and convective heat transport interact. Since the radiation and moreover convection are non-linear physical phenomena, only a few cases allow for analytical access. Some such coupled issues are addressed in a simplified manner at the end of the chapter, here especially emphasis is put on the derivation of dimensionless numbers to characterize an engineering phenomenon in order to decide in which manner it can be solved. The detailed treatment of convective heat transport by natural convection is provided in Chap. 5.

4.2 Heat Conduction in a Continuum

Heat conduction is a net transfer of kinetic energy of unordered movements (Brownian motion) between neighbouring molecules in otherwise non-moving media due to a temperature gradient. In metals, the free electrons also transfer energy. In opaque media, energy is transported by thermal conduction alone; in gases and liquids, energy transport by the flowing motion of a fluid (convection) and by thermal radiation is often superimposed on the thermal conduction process. Depending on the medium in which conductive heat transport occurs, the carriers of energy transport are atoms, molecules, electrons, or phonons. The latter are energy quanta of elastic waves, which are responsible for the transport of thermal energy in non-metals and—besides electrons—also in metals.

Let us consider the simplest case of a flat, homogeneous wall (or a gap homogeneously filled with a continuum at rest) with temperature T_1 on one side and temperature T_2 on the other side as depicted in Fig. 4.1. The energy transfer in the heat-conducting material due to the temperature difference can be described by the vector field of the heat flux density $\dot{q}$, where

$$\dot{q} = \overrightarrow{\dot{q}}\,(\vec{x}, t), \tag{4.2}$$

is a vector and indicates the strength and direction of an energy flow, which can also depend on the time t, at a location characterized by the vector $\vec{x}$. The corresponding temperature field T is in contrast to the heat flux a scalar but also a function of location $\vec{x}$ and time t and is given by

$$T = T(\vec{x}, t). \tag{4.3}$$

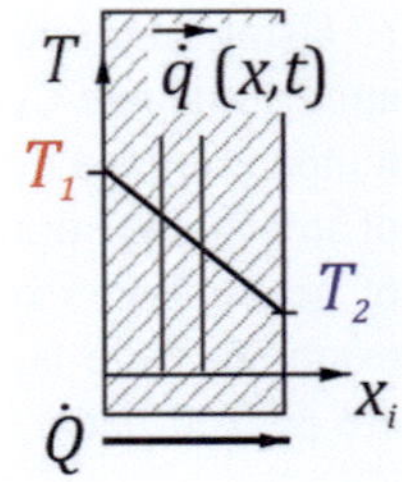

Fig. 4.1 Sketch of heat transport by heat conduction in a continuum

For a steady process ($\partial/\partial t = 0$) the time dependence of t disappears. All points of the body, which have the same temperature T at a certain time, can be thought of as connected by a surface a so-called isotherm. The strongest temperature change occurs normal to the isotherms and is defined by the temperature gradient ∇T:

$$\nabla T = \frac{\partial T}{\partial x} + \frac{\partial T}{\partial y} + \frac{\partial T}{\partial z}. \tag{4.4}$$

The heat flux density $\vec{q}$ is defined such that the heat flux $d\,\dot{Q}$ through an arbitrarily oriented element dA is given by

$$d\dot{Q} = \vec{q}\,(\vec{x},t) \cdot \vec{n} dA = \left|\vec{q}\right| \cdot \cos\beta \cdot dA, \tag{4.5}$$

where $\vec{n}$ denotes the plane normal vector forming the angle β with $\vec{q}$. The heat flux is thus a time-related energy, i.e. a power, and has the dimension J/s = W and the heat flux density accordingly an area-related reading given in W/m_2. If we now assume the temperature gradient as the cause of the occurring heat flux it is obvious to formulate a proportionality between cause and effect and the first Fourier law results:

$$\vec{q} = -k \cdot \nabla T = -k \cdot \frac{dT}{d\vec{x}}. \tag{4.6}$$

The minus sign appearing in the Eq. (4.6) considers the second law of thermodynamics, wherein heat always flows in the direction of the lower temperature. The proportionality constant k is a material property and is called specific thermal conductivity and has the SI-Units W/(mK). It depends on the temperature and pressure. Only if the material is isotropic, the specific thermal conductivity is a scalar quantity, otherwise it is a tensor of the second order.

For the heat flow $d\dot{Q}$ through an arbitrarily oriented surface element dA, substituting the Eq. (4.6) into (4.5) yields the following result:

$$d\dot{Q} = -k\;\nabla T\;(\vec{n}\;dA) = -k \cdot dA \cdot \left(\frac{\partial T}{\partial n}\right). \tag{4.7}$$

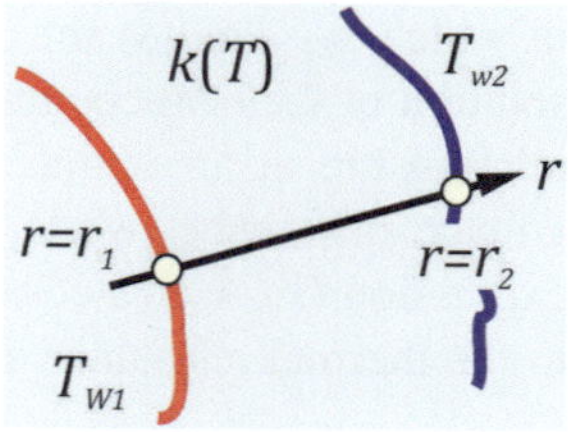

Fig. 4.2 Sketch of steady conductive heat transport in an arbitrary continuum

The basis of solving complicated heat conduction problems is the differential equation for the temperature field in a medium at rest, called the heat conduction equation. The solution of a heat conduction problem means to determine the temperature field $T(\vec{x}, t)$ in its spatial and temporal dependence. Once it is known, the associated heat flux density field can be calculated using Fourier's law and the heat fluxes transported by heat conduction can be determined at arbitrary locations of the body.

In the stationary case of a one-dimensional heat conduction, the temperature field T depends only on one spatial coordinate $\vec{r}$. The same applies to a hollow sphere, which is therefore included in the following considerations for the sake of completeness. If we consider first a stationary one-dimensional heat conduction, then the temperature is only a function of the spatial coordinate r, where at the place r the temperature $T = T(r)$ appears and for each $r = const.$ an isothermal surface is obtained. Assume that for $r = r_1$ the continuum has the temperature $T = T_{W1}$, respectively for $r = r_2$ a temperature $T = T_{W2}$, see Fig. 4.2.

If $T_{W1} > T_{W2}$, then a quantity of heat flows by heat conduction from r_1 over the surface A_1 to r_2 into the surface A_2. According to Fourier's law in Eq. (4.6), the heat flow in [W/m^2] through a surface $A(r)$ along r is given in the following way:

$$\dot{Q} = \dot{q}(r)A(r) = -k(T)\frac{dT}{dr}A(r), \tag{4.8}$$

or

$$-k(T)dT = \dot{Q}\frac{dr}{A(r)}. \tag{4.9}$$

Integration of Eq. (4.9) over the wall thickness from r_1 to r_2 ($\delta = r_2 - r_1$) leads to

$$-\int_{T=T_{W1}}^{T=T_{W2}} k(T)dT = \dot{Q}\int_{r=r_1}^{r=r_2} \frac{dr}{A(r)}, \tag{4.10}$$

and with the mean value theorem of the integral calculus we obtain:

$$-k_m(T_{W2} - T_{W1}) = \dot{Q}\frac{\delta}{A_m} \text{ or } \dot{Q} = \frac{k_m}{\delta}A_m(T_{W1} - T_{W2}), \tag{4.11}$$

in which k_m denotes the average specific thermal conductivity. In this form, the solution of the heat conduction equation corresponds to that of the current–voltage relationship at an ohmic resistor. The temperature difference is equivalent to a voltage difference, while the heat flow is equivalent to an electric current. The expression $(k_m A_m)/\delta$ corresponds to the thermal conductance G and its reciprocal to the thermal resistance R_H

$$R_H = \frac{\delta}{k_m A_m} = \frac{1}{G}\left[\frac{K}{W}\right]. \tag{4.12}$$

By analogy with Ohm's law in electricity, the following is then obtained for the heat flow $\dot{Q}$:

$$\dot{Q} = \frac{(T_{W2} - T_{W2})}{R_H}. \tag{4.13}$$

This analogy allows the use of further methods from electrical engineering. The average thermal conductivity k_m is given by:

$$k_m = \frac{1}{(T_{W2} - T_{W1})} \int\limits_{T=T_{W1}}^{T=T_{W2}} k(T)dT, \tag{4.14}$$

which can be simplified often considering small temperature differences. If the temperature dependence of thermal conductivity in the considered temperature interval is negligible a constant thermal conductivity $k_m = k$ can be used, while in case of a linear temperature dependence of the thermal conductivity, we obtain:

$$k_m = \frac{1}{2}[k(T_{W1}) + k(T_{W2})]. \tag{4.15}$$

For a flat wall, area A normal to the heat flux does not change with the spatial coordinate r, so $A \neq A(r)$ holds. In the case of a hollow cylinder or a hollow sphere this is different, and one obtains for the area A:

$$A(r) = \begin{cases} A_1 = A_2 & \text{for a planar wall} \\ 2\pi Lr & \text{for a hollow cylinder of length } L \\ 4\pi r^2 & \text{for a hollow sphere} \end{cases}. \tag{4.16}$$

Accordingly, the following result for the mean area A_m is obtained:

$$A_m = \begin{cases} A_1 = A_2 = \frac{1}{2}(A_1 + A_2) & \text{for a planar wall} \\ \frac{(A_2 - A_1)}{\ln\left(\frac{A_2}{A_1}\right)} & \text{for a hollow cylinder} \\ \sqrt{A_1 A_2} & \text{for a hollow sphere} \end{cases}. \tag{4.17}$$

If one assumes a constant temperature-independent specific thermal conductivity $k_m = const.$, then the following results:

$$-dT = \frac{\dot{Q}}{k}\frac{dr}{A(r)} \tag{4.18}$$

and integration over r yields the sought temperature distribution $T(r)$:

$$\frac{T(r) - T_{W2}}{T_{W1} - T_{W2}} = \begin{cases} \frac{r_2 - r}{r_2 - r_1} & \text{for a planar wall} \\ \frac{\ln\left(\frac{r_2}{r}\right)}{\ln\left(\frac{r_2}{r_1}\right)} & \text{for a hollow cylinder} \\ \frac{\frac{1}{r} - \frac{1}{r_2}}{\frac{1}{r_1} - \frac{1}{r_2}} & \text{for a hollow sphere} \end{cases} . \tag{4.19}$$

Thus, for $k_m = const.$ the central result for the temperature distribution given by Eq. (4.19) is:

- in a flat wall, the dimensionless wall temperature scales linearly with the thickness r,
- in a hollow cylinder, the temperature changes logarithmically with the cylinder radius r,
- in and inside the wall of a hollow sphere, temperature is hyperbolically dependent on the coordinate r.

To derive the general equation of conductive heat transfer, we first apply the first law of thermodynamics to a closed system. To do this, we consider a continuous region of any size, which we imagine to be cut out of the heat-conducting body, as shown in Fig. 4.3.

The first law of thermodynamics provides the power balance in the differential Volume $dV = dx \cdot dy \cdot dz$ and reads to:

$$\frac{dU}{dt} = \sum_i \dot{Q}_i + \dot{W} + q_V dV. \tag{4.20}$$

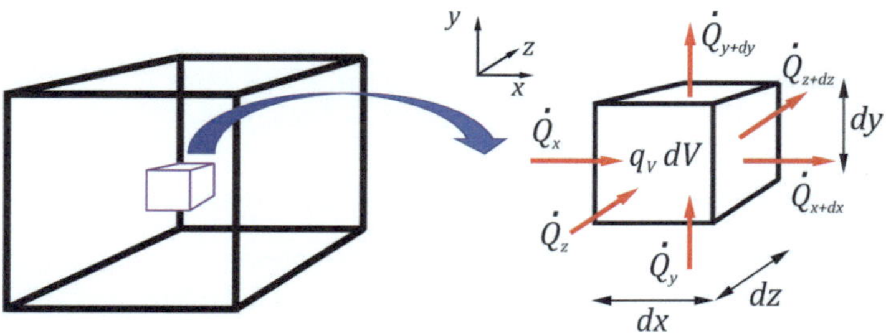

Fig. 4.3 Conductive heat transport in a differential control volume

It attributes the temporal change of the internal energy U within the volume element dV to three sources; the heat flows $\dot{Q}_i$ and the technical power flows $\dot{W}_i$ in the volume element dV over the system boundary and the released heat power $q_V dV$ present in the volume element dV with the volume specific source term q_V. In a quiescent incompressible medium the mechanical power flow is zero ($\dot{W}_i = 0$). The specific volumetric power release q_V acting as source term can be of different types (either electric, chemical reactions or nuclear radiation), but in solar thermal energy it is especially given by optical conversion of solar radiation into heat within the optically selective absorber. Thus q_V depends on the local coordinates and the time:

$$q_V = q_V(\vec{x}, t) \quad \left[W/m^3\right]. \tag{4.21}$$

If we consider heat conduction in a solid, the density changes due to temperature and pressure changes are marginal and we can apply the material model of an incompressible body with constant specific density ($\rho = const.$), resulting in the following:

$$m = \rho \cdot V_{\mathrm{a}} \text{ and } U = u \cdot m = u \cdot \rho \cdot V, \tag{4.22}$$

in which m denotes the mass und u the specific internal energy. Under this assumption, the change in internal energy can transform in the following way:

$$\frac{dU}{dt} = \frac{d}{dt}\int_V \rho u dV = \rho \int_V \frac{du}{dt} dV, \tag{4.23}$$

where the integral extends over the volume element dV. The hereby introduced mass-specific internal energy u of the incompressible body depends on its temperature. The change of internal energy with temperature is related to the specific heat capacity c_V in the following way:

$$du = c_V(T) \quad dT, \tag{4.24}$$

so that insertion in (4.24) yields

$$\frac{dU}{dt} = \rho \int_V c_V(T) \frac{\partial T}{\partial t} dV. \tag{4.25}$$

In order to calculate the heat flux $\dot{Q} = \sum_i \dot{Q}_i$ that crosses the system boundary first consider an element dA of the surface whose normal $\vec{n}$ is directed outwards. The heat flux $d\dot{Q}$ flowing into the area through dA is calculated to:

$$d\dot{Q} = -\overrightarrow{\dot{q}}\,\vec{n} dA. \tag{4.26}$$

By integrating all heat fluxes and using the integral theorem of Gauss, we obtain:

$$\dot{Q} = -\int_A \vec{q} \cdot \vec{n} dA = -\int_V \nabla \cdot \vec{q} \, dV. \tag{4.27}$$

Substituting in Eqs. (4.25) and (4.27) the power balance (Eq. 4.20) then yields:

$$\frac{dU}{dt} = \sum_i \dot{Q}_i + \underbrace{\dot{W}}_{=0} + q_V dV,$$

$$\rho \int_V c_V(T) \frac{\partial T}{\partial t} dV = -\int_V \nabla \cdot \vec{q} \, dV + \int_V q_V \quad dV. \tag{4.28}$$

This volume integral vanishes for arbitrarily chosen balance domains only if the integrand itself equals zero and it results:

$$\rho c_V(T) \frac{\partial T}{\partial t} = -\nabla \cdot \vec{q} + q_V. \tag{4.29}$$

In the last step we use Fourier's law (Eq. 4.6) and combine the heat flux density $\vec{q}$ with the temperature gradient and obtain the differential equation for the sought temperature field:

$$\rho c_V(T) \frac{\partial T}{\partial t} = \nabla \cdot (k(T) \nabla T) + q_V. \tag{4.30}$$

Equation (4.30) often called second Fourier equation is a first order differential equation in time and second order in space and a linear equation if c_V and k are invariant (constant) in the temperature interval considered. The assumption of constant thermophysical material properties also yields that the square of the Nabla operator becomes the Laplace operator Δ. The Laplace operator has a different form depending on the formulation of the geometric problem. The corresponding forms are listed in Table 4.1.

If no source terms are considered and constant material values are assumed, the Eq. (4.30) is simplified to:

$$\frac{\partial T}{\partial t} = \frac{k}{\rho \cdot c_V} \cdot \Delta T = a \cdot \Delta T. \tag{4.31}$$

Table 4.1 Mathematical forms of the Laplace operator ΔT for different coordinate systems

Cartesian	Cylindric	Spherical
$\frac{d^2T}{dx^2} + \frac{d^2T}{dy^2} + \frac{d^2T}{dz^2}$	$\frac{1}{r}\frac{\partial}{\partial r}\left(r\frac{\partial T}{\partial r}\right) + \frac{1}{r^2}\frac{\partial^2 T}{d\varphi^2} + \frac{\partial^2 T}{\partial z^2}$	$\frac{1}{r^2}\frac{\partial}{\partial r}\left(r^2\frac{\partial T}{\partial r}\right) + \frac{1}{r^2 sin^2\Theta}\frac{\partial^2 T}{d\varphi^2} +$ $\frac{1}{r^2 sin\Theta}\frac{\partial}{\partial\Theta}\left(\sin\Theta\frac{\partial T}{\partial\Theta}\right)$

Table 4.2 Typical examples of the magnitude of thermal diffusivity a of some materials

Material	a[m^2/s 10^{-6}]
Metals	5–10
Glass	0.6–0.8
Stone, ceramics	0.5–1.5
Water	0.143
Organic liquids	0.09–0.1
Air 20 °C, 0.1 MPa	21.4

The expression $a = k/(\rho c_V)$ denotes a substance size combination, which is referred to in the literature as the temperature coefficient, thermal diffusivity, or thermal diffusion coefficient. The thermal diffusivity has the unit $\left[\mathrm{m}^2/\mathrm{s}\right]$ analogous to the dynamic viscosity. Typical values of thermal diffusivity of some substances are listed in Table 4.2. Metals and gases have the high thermal diffusivities. This means that the temperature equalization in metals and gases is about the same. Liquids and non-metallic substances have lower thermal diffusivities, so temperature equalization is slower there.

The following generally valid statements can be derived directly from the second Fourier equation without presence of volumetric heat sources.

- For $(\partial^2 T/\partial x_i^2 > 0)$, the immediate result is $(\partial T/\partial t) > 0$. In other words, in areas of positive curvature of the temperature curve, there is an increase in temperature.
- If $(\partial^2 T/\partial x_i^2 < 0)$, we get $(\partial T/\partial t) < 0$. A negative curvature means a decrease of the temperature with increasing time.

All in all, it can be stated that the transient influence works towards a straightening of the temperature distribution, because for $t \to \infty$ the linear relationship must be achieved.

4.2.1 Initial and Boundary Conditions, Dimensionless Numbers

Initial conditions define the temperature distribution in the volume at a reference time, often referred to as $t = 0$, i.e. $T(x_i, t = 0) = T_0(x_i)$. The boundary conditions either define the temperature at the boundaries of the considered volume, or they make statements about its time derivative. There are essentially three types of thermal boundary conditions, whose physical meaning are depicted in Fig. 4.4:

- The Dirichlet (or first-type) boundary condition specifies the values that a solution needs to take along the boundary $\partial\Omega$ of the domain Ω and reads to:

$$T(x_i, t)|_{\partial\Omega} = T_{\partial\Omega}(x_i, t)|_{\partial\Omega}, \tag{4.32}$$

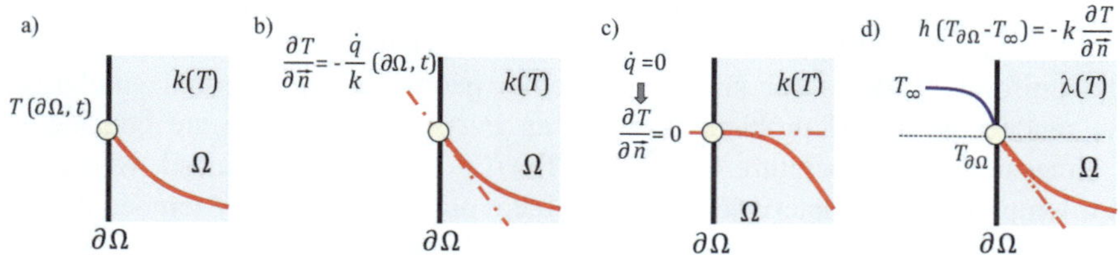

Fig. 4.4 Different thermal boundary conditions to the heat transport equation. **a** Dirichlet; **b** general Neumann condition; **c** homogeneous Neumann condition corresponding to adiabatic wall; **d** Robin or mixed boundary condition

where $T_{\partial\Omega}$ is a known value defined on the boundary $\partial\Omega$.

- The Neumann (or second-type) boundary condition specifies the values in which the derivative of a solution is applied within the boundary of the domain and is given by:

$$\left.\frac{\partial T(x_i, t)}{\partial n}\right|_{\partial\Omega} = T_{\partial\Omega}(x_i, t)|_{\partial\Omega}. \tag{4.33}$$

Regarding the thermal problem considered a Neumann condition means a prescribed heat flux from/to surface boundary $\partial\Omega$, which is in solar thermal applications given by solar radiation but could also be the electrical power provided e.g. by trace heating elements. One specific case is the homogeneous Neumann boundary condition denoting an adiabatic boundary which reads to $(\partial T/\partial n)|_{\partial\Omega} = 0$.

- The Robin boundary condition or third type boundary condition specifies a combination of the temperatures *and* the values of its derivative on the boundary of the domain. The linear Robin condition,

$$-k\left.\frac{\partial T(x_i, t)}{\partial n}\right|_{\partial\Omega} = h(T_{\partial\Omega} - T_\infty). \tag{4.34}$$

applies to problems where heat conduction in the domain is connected with a convective heat transfer at the boundary $\partial\Omega$, expressed by a heat transfer coefficient h, while the non-linear Robin condition

$$-k\left.\frac{\partial T(x_i, t)}{\partial n}\right|_{\partial\Omega} = \varepsilon\left(T_{\partial\Omega}^4 - T_{\mathrm{S}}^4\right) + h(T_{\partial\Omega} - T_\infty). \tag{4.35}$$

addresses the heat transfer with combined surface radiation and convection in which the surface radiates with heat to surrounding surface of temperature T_S with a rate ε proportional to its emissivity current temperature. In Eqs. (4.34) and (4.35) T_∞ refers to the fluid temperature far away the boundary $\partial\Omega$.

To solve the heat conduction differential equation, it is useful to introduce dimensionless quantities, with the help of which the number of variables is reduced and uniform solutions for similar problems (geometric and thermal similarity) are made possible. If problems without an internal heat source are considered a dimensionless temperature $\Theta = (T - T_C)/(T_0 - T_C)$ is introduced, where T_C is a temperature characteristic for the problem and T_0 is the body temperature at the beginning. If internal heat sources are considered, the normalization is done in such a way that the prefactor of the second spatial derivative is omitted, so that the dimensionless temperature assumes the following scaling $\Theta \sim k(T - T_0)/\left(q_V L^2\right)$. If, for example, a flat absorber strip with the homogeneous initial temperature T_0 cools down steadily in a colder environment until the plate temperature corresponds to the ambient temperature T_{amb}, the ambient temperature T_{amb} is to be selected as the characteristic temperature T_C. The length scale on the x-coordinate is obtained by normalization to a characteristic length L, for example the plate thickness, which leads to $\xi = x/L$.

Furthermore, it is useful to introduce a dimensionless time, the so-called Fourier number (Fo):

$$Fo = \frac{k}{\rho c_V} \cdot \frac{t}{L^2} = \frac{at}{L^2}, \tag{4.36}$$

which can be interpreted as the ratio of the transport rate to the storage rate. For transient heat conduction, it describes the ratio of the rate at which sensible heat is transported by conduction to the rate at which it is absorbed.

Problem specific boundary conditions of the differential equation may also be made dimensionless. In many practical cases the heat, which is transported in a body by heat conduction, is transferred at its surface by convection into a surrounding fluid of temperature T_{amb}. At the body surface-fluid interface (index W) the continuity of the heat flux must be fulfilled. If one introduces the previous scaling for the de-dimensioning, Eq. (4.34) read in dimensionless form:

$$-\frac{1}{\Theta_W} \frac{\partial \Theta}{\partial \xi}\bigg|_{\mathrm{W}} = \frac{hL}{k} = Bi, \tag{4.37}$$

in which $\Theta = (T - T_{amb})/(T_0 - T_{amb})$ and $\Theta_W = (T_W - T_{amb})/(T_0 - T_{amb})$. The boundary condition for the heat conduction problem in the body is now a function of a dimensionless quantity $Bi = (hL)/k$ called Biot number. In the case of heat transfer through the surface of a body, the Biot number indicates the ratio of the body's thermal resistance to the heat transfer resistance of the surrounding medium. Thus, the Biot number is formally formed equal to the Nusselt number, where, however, the specific thermal conductivity k_F of the fluid is used instead of the solid at rest k, and the length scale L has a different meaning. A large Biot number means that temperature differences within the solid body are larger than in the boundary layer of the fluid, so that an improvement of the external heat transfer, e.g. by forced instead of free convection, does not significantly accelerate the heat transfer process. From the solution of this heat conduction problem in dimensionless form, the dimensionless temperature of the body is calculated. This has the form:

$$\Theta = f(\xi, Fo, Bi). \tag{4.38}$$

For some practical problems with one-dimensional, unsteady heat conduction, solutions in analytical form are known. Some of them are presented in the subsequent subchapters as examples.

Normally, the Nusselt number Nu is used to describe the heat transfer of a body to flowing fluids. The Nusselt number is also a dimensionless parameter from similarity theory that measures the improvement in heat transfer from a surface. It relates the intensity of a convective heat transfer at a solid surface to one imagined at a fluid at rest when pure heat conduction is acting. However, the Nusselt number can also be understood as a dimensionless gradient of the temperature at a surface. If one uses the fluid and the dimensions of the fluid space as a reference system, Nu results in:

$$Nu = \frac{hL}{k_F}. \tag{4.39}$$

However, the Nusselt number, in addition to the geometry due to its different reference frame, depends essentially on flow parameters of the fluid such as the Reynolds number and the Prandtl number, as shown in Chap. 5. If one uses the fluid system with a normalization of the temperature $\Theta = (T - T_\infty)/(T_W - T_\infty)$. and the length normalization of the fluid system $\xi = x/L_F$ the reference system for de-dimensioning, then the dimensionless temperature gradient corresponds to that of the Nusselt number:

$$-\left.\frac{\partial \Theta}{\partial n}\right|_{\mathrm{W}} = Nu. \tag{4.40}$$

4.2.2 Stationary Heat Conduction

Stationary or steady state heat conduction treats problems where the temperature field does not change with time. A constant temperature difference between surfaces or a constant heat flux into a boundary cause such stationary conditions.

4.2.2.1 Stationary Heat Conduction in a Solar Absorber

In any solar thermal collector, heat gain occurs through optical conversion in a continuum volume being at rest. The resulting increase in temperature causes heat not only to be transported to the desired heat sink, but also to be lost by conduction through structures or radiation exchange with all collector boundaries.

Let there be an absorber strip as shown in Fig. 4.5, which absorbs the solar radiation i in [W/m^2] over the entire upper surface and conducts the heat to a pipe

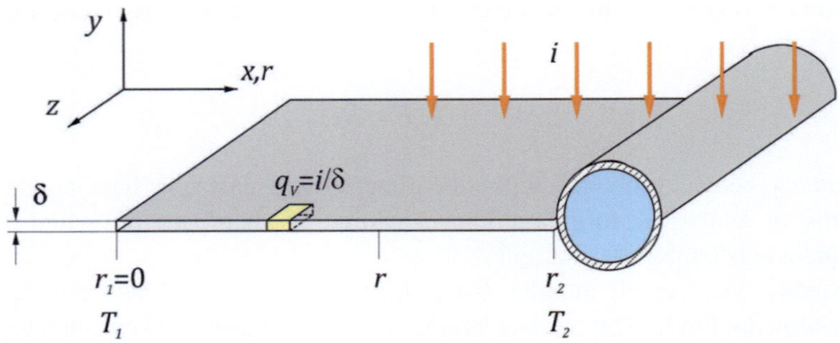

Fig. 4.5 Schematics of a "volumetric" heat release in an absorber strip and the conductive heat transport to the heat transfer medium

in which the heat transfer fluid is located. Dividing the irradiation power by the thickness of the absorber strip δ gives a volumetric power $q_V(x, y, z)$ in [W/m^3]. In most cases, the absorber plates are so thin that the temperature difference normal to the plate can be neglected ($\partial/\partial y \approx 0$). Likewise, the heat conduction parallel to the heat transfer tubes can be neglected ($\partial/\partial z \approx 0$). This results in a one-dimensional case with a volumetric source $q_V = q_V(r)$ as a first approximation. The heat conduction with an internal heat source also occurs in the glass covers, but in terms of magnitude it is only significant for concentrating systems in which large temperature differences can occur.

For the absorber plate, the irradiance would have to be converted to the cross-section of the absorber strip to equal a power density. Assuming a stationary case ($\partial/\partial t \approx 0$), Eq. (4.29) simplifies to:

$$\frac{d^2T}{dr^2} + \frac{q_V(r)}{k} = 0. \tag{4.41}$$

The general solution with a volumetric power density independent of r is thus given:

$$T = -\frac{q_V}{2\lambda} r^2 + C_1 r + C_2, \tag{4.42}$$

where C_1 and C_2 are integration constants. Their value depends on the respective boundary condition, i.e. whether, for example, the absorber strips are adjacent to a heat transfer tube on one or both sides. For a strip attached to a heat transfer tube on one side only and the maximum temperature $T_1(r_1 = 0)$ at the free end, as sketched in Fig. 4.5, the temperature curve is obtained:

$$T(r) = T_1 - \frac{q_V}{2k} \cdot r^2. \tag{4.43}$$

The temperature then decreases with increasing r_2. The average temperature of the absorber strip can be calculated if r_2 is the interface to the heat transfer tube and $r_1 = 0$:

$$\overline{T} = \frac{\int_{r=r_1}^{r=r_2} T(r)dr}{\int_{r=r_1}^{r=r_2} dr} \text{ and taking } r_1 = 0 \rightarrow \overline{T} = T_1 - \frac{1}{3}\frac{1}{2}\frac{q_V}{k} r_2^2. \tag{4.44}$$

Thus, the average temperature differs from the maximum temperature at the free end by one third (1/3) of the temperature difference between both boundaries of the strip $\Delta T = T_2 - T_1$:

$$\overline{T} = T_1 - \frac{1}{3}\Delta T. \tag{4.45}$$

In the case of a cylinder, the steady-state heat conduction equation assuming again constant thermal conductivity (k = const.) and a power density q_V independent of the coordinate r takes the following form:

$$\frac{1}{r}\frac{d}{dr}\left(r\frac{dT}{dr}\right) + \frac{q_V}{k} = 0. \tag{4.46}$$

The first integration provides the following relationship

$$r\frac{dT}{dr} = -\frac{q_V}{2k}r^2 + C_1, \tag{4.47}$$

and the solution reads to:

$$T(r) = -\frac{q_V}{4k} r^2 + C_1 \cdot \ln(r) + C_2. \tag{4.48}$$

Considering a solid cylinder with $r_1 = 0$ as an example, the boundary condition at this point is:

$$\left.\frac{dT}{dr}\right|_{r=r_1=0} = 0. \tag{4.49}$$

If $T_{1,2}$ are the temperatures in the centre of the cylinder and at the surface, respectively, the boundary condition (Eq. 4.49) leads to $C_1 = 0$. Assuming T_1 as the temperature at $r_1 = 0$, the second integration constant is obtained and as a solution for the temperature distribution $T(r)$ the result:

$$C_2 = T_1 + \frac{q_V}{4k}r_1^2 \text{ and } T(r) = \frac{q_V r_1^2}{4k}\left(1 - \frac{r^2}{r_1^2}\right) + T_1. \tag{4.50}$$

The temperature also changes here with r, however, decreasing quadratically.

For a hollow cylinder with an inner radius $r_1 \neq 0$ and an outer radius r_2, as well as the assumption that the inner side is adiabatic ($\partial T/\partial n = 0$), the following boundary condition results for the inner side:

$$\left.\frac{dT}{dr}\right|_{r=r_1} = 0. \tag{4.51}$$

The solution then leads to the following temperature distribution:

$$T(r) = T_1 - \frac{q_V}{4k}\left(r^2 - r_1^2\right) + \frac{q_V}{2k} r_1^2 \ln\frac{r}{r_1}. \tag{4.52}$$

Equations (4.45), (4.50) and (4.52) illustrate that the temperature depends essentially on q_V and thus on the power density generated volumetrically in the absorber by optical conversion, i.e., i_{glob}/δ. In addition to conductive heat conduction in the absorber strip, the temperature of the absorber is also determined by its thermal conductivity a, its ability to transfer heat in the form of radiation, which is given by emissivity ε and discussed in the following chapters of thermal radiation, and convective heat transfer to the heat sink, so that the temperature is a function of several influencing parameters in the following form:

$$\Delta T = f\left(i_{glob}, \varepsilon, a, \delta \ldots\right). \tag{4.53}$$

4.2.2.2 Stationary Heat Transfer Through a Wall

In practical heat transfer problems, heat conduction in a solid and convective heat transfer in a fluid boundary layer at the surface of the solid are often coupled. For large temperature differences, coupling with heat transfer by radiation is added. If the temperature is considered as a potential, as already mentioned, a total thermal resistance can be calculated from the individual thermal resistances in analogy to electrical engineering.

Let us consider the heat transfer through a plane wall of thickness δ surrounded on one side by fluid 1 of temperature T_1 and on the opposite side by fluid 2 of temperature T_1 with $T_1 > T_2$, as shown in Fig. 4.6. Due to the higher temperature, heat is transferred from fluid 1 to fluid 2. The heat flow is tfirst convective from fluid 1 to the wall surface, then the heat is transferred conductive in the wall to the opposite wall surface and there it is transferred again to fluid 2 by means of convective heat transfer. The total process consisting of heat conduction and convective heat transfer is called heat transfer. The resulting qualitative temperature profile from Fluid 1 to Fluid 2 is shown in Fig. 4.6.

The heat flow which is transferred due to the temperature difference between fluid 1 and fluid 2 can be calculated as

$$\dot{Q} = \frac{T_1 - T_2}{R_H} \tag{4.54}$$

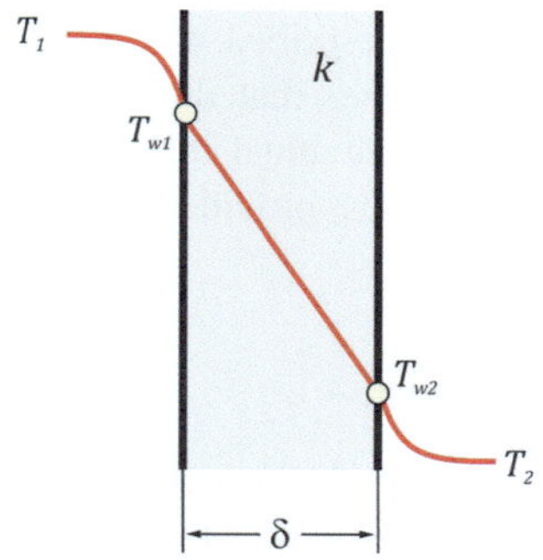

Fig. 4.6 Qualitative temperature profile during heat transfer through a planar wall. flat wall

in analogy to Ohm's law. If one assumes a one-dimensional heat transport normal to the wall surface, one obtains a total thermal resistance R which is the sum of the individual thermal resistances in the following way:

$$R_H = \frac{1}{h_1 A} + \frac{\delta}{kA} + \frac{1}{h_2 A} \quad \left[\frac{K}{W}\right], \tag{4.55}$$

where the convective thermal resistances are of the form $R_{conv,i} = 1/(h_i A)$ and the thermal resistances are given by $R_{cond,i} = \delta/(\lambda A)$. The heat flux transferred from fluid 1 to fluid 2 can then be given by a temperature difference $(T_1 - T_2)$,

$$\dot{Q} = k_{tot} A (T_1 - T_2), \tag{4.56}$$

in which k_{tot} is the total heat transfer coefficient in $\left(W/\left(m^2 K\right)\right)$, which combines in a single coefficient the specific heat conduction and heat transfer quantities and is related to the total thermal resistance R_H as follows:

$$k_{tot} = \frac{1}{R_H A}. \tag{4.57}$$

In a cylindrical tube of length L and thickness δ, through which fluid 1 flows on the inside and fluid 2 flows over it on the outside, one can also assume one-dimensional, now radial heat transport, resulting in the following total thermal resistance R_H:

$$R_H = \frac{1}{k_{tot} A} = \frac{1}{h_1 A_1} + \frac{\delta}{k A_m} + \frac{1}{h_2 A_2}, \tag{4.58}$$

where $A_1 = (2\pi r_1 L)$ denotes the inner shell area, $A_2 = (2\pi r_2 L)$ is the outer shell area, $\delta = (r_2 - r_1)$ is the pipe wall thickness, and $A_m = (A_2 - A_1)/\ln(A_2/A_1)$ is the logarithmically averaged cross-sectional area already introduced in Eq. (4.17). If we solve Eq. (4.58) for the total heat transfer coefficient k_{tot}, we see that the value for k_{tot} for a pipe must be specified in terms of the cross-sectional area A to which the value relates, which is not necessary for a flat wall because the areas are the same.

The described case of heat transfer from fluid 1 through a plane wall to fluid 2 illustrates that the total thermal resistance is composed of a series connection of the individual resistances and can be represented as the sum of the individual resistances, provided that one-dimensional heat transfer is present.

$$R_H = \sum_i R_i. \tag{4.59}$$

If the planar wall between the fluids consists of several layers j of thickness δ_j with different thermal conductivities k_j, the total thermal resistance when the individual resistances are connected in series is given by:

$$R_H = \frac{1}{k_{tot} A} = \frac{1}{h_1 A_1} + \sum_j \frac{\delta_j}{k_j A} + \frac{1}{h_2 A_2}. \tag{4.60}$$

The summation term changes for cylindrical bodies with the number n of layers of different thermal conductivity k_i or spheres of radius r in the following way:

$$\frac{1}{R_H} = \frac{1}{\sum_{i=1}^{n} \frac{1}{k_i} \ln\left(\frac{r_{i+1}}{r_i}\right)} \text{ (cylinder)}; \quad \frac{1}{R_H} = \frac{1}{\sum_{i=1}^{n} \frac{1}{k_i} \left(\frac{1}{r_i} - \frac{1}{r_{i+1}}\right)} \text{ (sphere)}. \tag{4.61}$$

Conversely, the analogy of dealing with electrical resistors can also be applied to the parallel connection of thermal resistances, provided that one-dimensional heat transport is given by the parallel thermal resistances.

4.2.2.3 Variable Thermal Conductivity and Kirchhoff Transformation

Often the thermal conductivity of the bodies analysed is temperature dependent, and a constant heat conductivity is normally present temperature only if small temperature intervals are considered. Moreover, the thermal conductivity substantially varies between the different types of continua appearing in in heat transfer problems. The Fig. 4.7 illustrates the temperature dependence of some gases, solid crystalline substances and metals appearing in solar thermal engineering and shows clearly that there are orders of magnitude differences depending on the material type, but also on the material structure. The physics is for this behaviour will be discussed in Sect. 4.2.4.

In case the difference between a temperature level T_1 and T_2 is in the range of a few percent, an average value can be determined for the thermal conductivities k_m in this range:

$$k_m = \frac{1}{(T_1 - T_2)} \int_{T=T_1}^{T=T_2} k(T) dT, \tag{4.62}$$

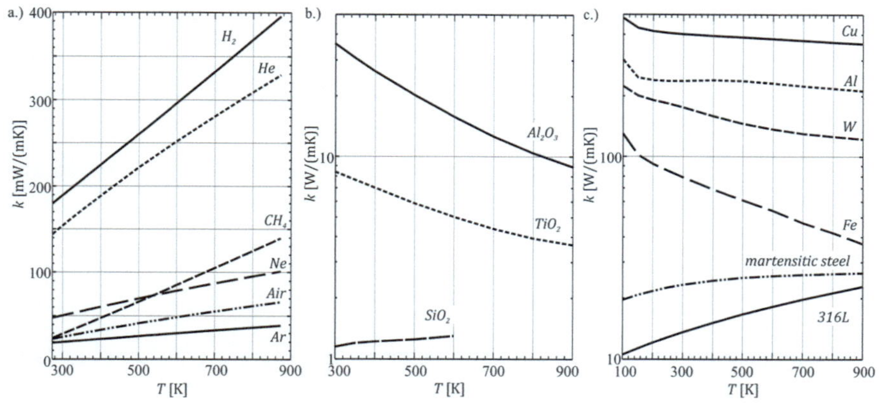

Fig. 4.7 Temperature dependent molecular heat conductivity k of **a** some gases, **b** ceramic materials based on ion bonding and **c** pure metals together construction metals

which can be reasonably used in calculations. Nonetheless, in this case, even in a flat wall, the temperature distribution is no longer linear.

The Kirchhoff transformation can be used to reduce the partial differential equation (Eq. 4.30), which becomes non-linear due to the temperature dependence of the thermal conductivity, to an equivalent linear equation that can be solved for steady-state problems. To do this, one introduces a new temperature variable T^* in the form:

$$k^* = \int_{T=T_1}^{T=T_2} k(T)dT \quad \text{and} \quad T^* = \frac{1}{k_m} \int_{T=T_1}^{T=T_2} k(T), \tag{4.63}$$

which adjusts the "normal" temperature to the variable thermal conductivity. This yield $k^* dT^* = k(T)dT$. From the definition of T^* we get for $T^*(T = T_1) = 0$ and for $T^*(T = T_2) = 1$. The transformed temperature T^* is monotonically increasing and $T(T^*)$ is the inverse function of $T^*(T)$. Rewriting the nonlinear differential equation yields for the stationary case:

$$k^* \frac{\partial^2 T^*}{\partial x_i^2} + q_V = 0, \tag{4.64}$$

and thus, a linear equation, provided that q_V does not depend on the temperature.

4.2.2.4 Two-Dimensional Heat Conduction

The heat flow from an isothermal surface A_1 to another isothermal surface A_2 in arbitrary geometrical arrangements is calculated in a general way by:

$$\dot{Q} = -\lambda \iint_{A_1} \left(\frac{\partial T}{\partial n}\right)_1 dA_1 = k \iint_{A_2} \left(\frac{\partial T}{\partial n}\right)_2 dA_2. \tag{4.65}$$

By introducing a form coefficient S of the type,

$$S = \frac{\iint_{A_1} \left(\frac{\partial T}{\partial n}\right)_1 dA_1}{(T_2 - T_1)} = \frac{\iint_{A_1} \left(\frac{\partial T}{\partial n}\right)_1 dA_1}{(T_2 - T_1)}, \tag{4.66}$$

and normalizing it to a characteristic length L, e.g. S/L, one obtains a dimensionless form factor S^* and ultimately a heat flux per unit length:

$$\frac{\dot{Q}}{L} = kS^*(T_2 - T_1). \tag{4.67}$$

In engineering data bases these dimensionless form factors (S^*) are tabulated for different geometrical arrangements.

4.2.3 Transient Heat Conduction

Time-varying temperature fields $T(\vec{x}, t)$ play a decisive role in many technical applications. Their calculation allows, for example, to answer questions about the duration of a heating or cooling process, about the heat dissipated or absorbed, or about the location inside the body where a maximum permissible temperature is reached at a certain time. For geometrically simple body shapes, a large number of analytical exact solutions are known, see e.g. (Carslaw & Jaeger, 1986) among others. Here, only some practically relevant and simple solutions shall be discussed, which are relevant for solar thermal systems. Transient heat transfer methods play an excellent role not only in evaluating the performance of collectors, but also in determining the thermal conductivity of materials, heating cooling of buildings, temporal resolution of thermocouples, etc., so an understanding of these effects is essential.

4.2.3.1 One-Dimensional Heat Flow with Abrupt Change of the Ambient State

Let us first consider a flat plate which is located in an environment of temperature T_∞. Normalizing the problem using the following scaling

$$\xi = \frac{x}{L}; \quad \Theta = \frac{T - T_\infty}{T_B - T_\infty} \quad \text{and} \quad \tau = \frac{at}{L^2}, \tag{4.68}$$

where L denotes the characteristic length, a the heat condcutance and T_B the temperature at the beginning, yields:

$$\frac{\partial \Theta}{\partial \tau} = \frac{\partial^2 \Theta}{\partial \xi^2}. \tag{4.69}$$

The direction of the heat flow vector in this case always coincides with a coordinate direction ξ and the planes $\xi = const.$ are then isothermal surfaces. The initial condition is given by $\Theta(\xi, \tau = 0) = 1$, while as boundary condition we assume the general case of a Robin boundary condition, in which the ratio of the internal thermal resistance of the plate to the external heat transfer resistance from the plate to the environment is expressed using a Biot number (Bi). This yield:

$$\left(\frac{1}{Bi}\frac{\partial\Theta}{\partial\xi} + \Theta\right)\Bigg|_{|\xi|=1} = \begin{cases} 1 & \text{for } \tau = 0 \\ 0 & \text{for } \tau > 0 \end{cases}. \tag{4.70}$$

The formulation has the advantage that both a heating ($\Theta_\infty > \Theta_B$) and a cooling ($\Theta_\infty < \Theta_B$) of a body is described and tends towards zero for $\tau \to \infty$.

A consequence of the normalization of the boundary condition is that all tangents to the temperature curves at the surface of the body ($\xi = 1$) intersect at one point. The coordinates of this reference point are ($\Theta = 0, \xi = 1 + 1/Bi$). If the heat transfer from the plate to the environment is very small (i.e. the thermal resistance is high), the reference point is far away; conversely, an excellent heat transfer means that it is close to the wall and the Robin boundary condition mutates to a boundary condition of the first kind, i.e. $\Theta(\xi = 1, \tau > 0) = 0$.

With the help of a geometrical consideration, the maximum temperature difference $\Delta\Theta_{max}$ occurring in the body can be estimated as follows:

$$\Delta\Theta_{max} < \lambda\frac{Bi}{1 + Bi}. \tag{4.71}$$

From the figure it can be seen that at small times (e.g. the time $\tau = \tau_1$ in Fig. 4.8), for which $Fo \leq 0.01$ holds, the temperature in the centre of the plate has not yet noticed the temperature change at the interface. This behaviour is called boundary layer character. Conversely, this means that the thickness with its heat capacity has no influence on the temperature field and the plate body can be regarded as semi-infinitely extended. By introducing dimensionless wall distance $\xi' = (1 - \xi)/\left(2\sqrt{\tau}\right)$ and a modified Biot number formed with the dimensionless wall distance $Bi' = Bi\sqrt{\tau}$ the following special solution of the differential Eq. (4.69) applies:

$$\Theta(\xi', Bi') = \operatorname{erf}(\xi') + \exp(\xi'^2) \cdot \exp(\xi' + Bi')^2 \cdot \left[1 - \operatorname{erf}(\xi' - Bi')\right], \tag{4.72}$$

where erf is the Gaussian error integral defined as

$$\operatorname{erf}(x) = \frac{2}{\sqrt{\pi}} \cdot \int\limits_{x=0}^{z} \exp(-x^2)dx. \tag{4.73}$$

The latter is usually solved numerically and Fig. 4.9 shows the evaluation of the Gaussian error integral. For short times, the temperature in the centre of the body corresponds to the initial temperature and the temperature of the surface $\Theta_S(\xi' = 0)$ is obtained from Eq. (4.72) in dimensionless notation by

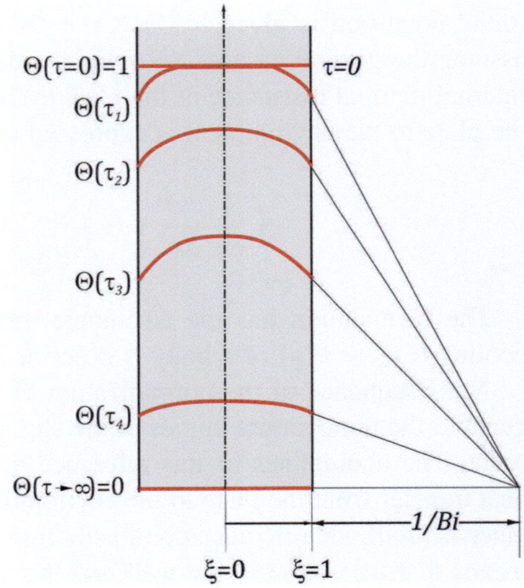

Fig. 4.8 Schematic dimensionless temperature profile $\Theta(\xi, \tau, Bi)$ for the transient heat transfer in a plate using Robin boundary conditions

$\Theta_S = \exp\left(Bi'^2\right) \cdot \left[1 - \mathrm{erf}\left(Bi'\right)\right]$. For $Bi' > 2$, Eq. 4.72 yields a dimensionless surface temperature of $\Theta_S = 1/\left(\sqrt{\pi}Bi'\right)$ with an error less than 1%. For $Bi' \to \infty$ the surface temperature gets zero ($\Theta_S = 0$), i.e. there is a boundary condition of the first kind, and from Eq. (4.72) only the first summand remains:

$$\Theta_S = erf\left(x'\right) = \mathrm{erf}\left(\frac{x}{2 \cdot \sqrt{a \cdot t}}\right). \tag{4.74}$$

The instantaneous heat flux density through the surface is then obtained by:

$$\dot{q}_S(\xi = 1, t) = \frac{k}{\sqrt{\pi \cdot a \cdot t}}(T_B - T_S) = \frac{\sqrt{\lambda \cdot \rho \cdot c_V}}{\sqrt{\pi \cdot t}}(T_B - T_S). \tag{4.75}$$

Thus, the heat flux decreases with inverse of the square root of time $(1/\sqrt{t})$ and is proportional to the quantity $b = \sqrt{k \cdot \rho \cdot c_V}$ called thermal effusivity or heat penetration coefficient. The integral mean value of the heat flux density at the

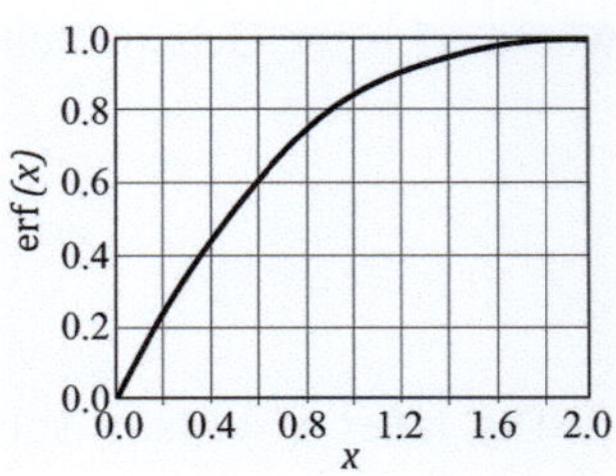

Fig. 4.9 Schematic evaluation of the Gaussian error integral

surface from $t = 0$ to t is calculated as:

$$\dot{q}_S(\xi = 1, t) = \int_{t=0}^{t} \dot{q}_S(t)\, dt = \frac{2 \cdot k \cdot t}{\sqrt{\pi \cdot a \cdot t}}(T_B - T_S). \tag{4.76}$$

The mean temperature $\overline{T}$ can be used to calculate the heat added or removed in a given time. For this purpose, the integral caloric mean temperature $\overline{T}$ is required, which is defined by:

$$\overline{T} = \frac{\int_V (\rho \cdot c_V \cdot T) dV}{\int_V (\rho \cdot c_V) dV}[K] \quad \text{or} \quad \overline{\Theta} = \frac{\left(\overline{T} - T_\infty\right)}{(T_B - T_\infty)}[/]. \tag{4.77}$$

With the initial and the mean temperature, the change of the heat content is determined. This results in:

$$Q = \rho \cdot V \cdot c_V \cdot \left(T_B - \overline{T}\right) = m \cdot c_V \cdot \left(T_B - \overline{T}\right)\left(1 - \overline{\Theta}\right)\text{in}[W], \tag{4.78}$$

where $\overline{T}$ denotes a dimensional mean temperature. Finally, for the mean temperature one obtains:

$$\left(T_B - \overline{T}\right) = \frac{2 \cdot A \cdot k \cdot t}{V \cdot \rho \cdot c_V \cdot \sqrt{\pi \cdot a \cdot t}}(T_B - T_S) = \frac{2 \cdot A}{V} \cdot \frac{\sqrt{a \cdot t}}{\pi} \cdot (T_B - T_S). \tag{4.79}$$

The ratio of the surface area A to the volume V is $1/L$ for a plate (, where it should be noted that L is half the thickness of the plate), $4/L$ for a circular cylinder and $6/L$ for a sphere. Thus, the surface temperature calculated from Eq. (4.74) can be used to determine the mean temperature of the body. Equations 4.74 and 4.79 are valid only if the Biot number Bi is very large ($Bi \gg 1$). Physically, this assumption means that the heat transfer coefficient h tends to infinity ($\rightarrow \infty$), i.e. there is perfect heat transfer. For sufficiently long times τ, the temperature profiles inside the body of finite thickness are similar to each other and can all be described by a location function $f(\xi)$, which is reduced in scale with increasing time.

Introducing a multiplicative approach to the differential Eq. (4.69) for the temperature in the following way $\Theta = g(\tau) \cdot f(\xi)$, we obtain in each case for the time scale function $g(\tau) = \exp\left(-m^2\tau\right)$. Therein, m is a constant to be determined from the boundary conditions. The form of the location function $f(\xi)$ depends on whether it is a plate, a cylinder, or a sphere. Unfortunately, these functions alone are not able to represent the temperature field for short times, especially the initial temperature distribution. Therefore, a representation in the form of an infinite series as:

$$\Theta = \sum_{i=1}^{n} C_i(m_i) \cdot f(m_i\xi), \tag{4.80}$$

is necessary, whose coefficients C_i have to be determined in such a way that a finite series describes the initial temperature distribution $\Theta(\xi, \tau = 0) = 1$ with a minimum of the error sum of squares and converges to this initial temperature distribution for $n \to \infty$.

For the one-dimensional plate of thickness L, the general solution of the differential equation for the temperature can be represented as a Fourier series of in the form:

$$(x, t) = \sum_{n=1}^{\infty} [C_1 \cos(B \cdot x) + C_2 \sin(B \cdot x)] \cdot \exp(-C_3 \cdot a \cdot t), \tag{4.81}$$

in which the constants B and C_i can be determined with the help of the boundary conditions. If the heat transfer coefficient is infinite on the outside (i.e., the external wall temperature is equal to the temperature of the surrounding fluid corresponding to $1/Bi \to 0$), the solution in the case of the infinite plate is:

$$\begin{aligned} \Theta = \frac{T(x,t) - T_\infty}{T_A - T_\infty} &= \frac{4}{\pi} \sum_{n=1}^{\infty} \frac{1}{n} \sin\left(\frac{n \cdot \pi \cdot x}{2 \cdot L}\right) \exp\left(-\frac{n^2 \pi^2 a t}{4L^2}\right) \\ &= \frac{4}{\pi} \sum_{n=1}^{\infty} \frac{1}{n} \sin\left(\frac{n \cdot \pi \cdot x}{2 \cdot L}\right) \exp\left(-\frac{n^2 \pi^2 Fo}{4}\right), \end{aligned} \tag{4.82}$$

where n is a series of odd values with $n = 1, 3, 5, \ldots, (2n - 1)$. For $\tau > 0.5$ the local and the mean temperature Θ and $\overline{\Theta}$ are:

$$\Theta = \frac{4}{\pi} \cos\left(\frac{\pi x}{2L}\right) \exp\left(-\frac{n^2 \pi^2 a t}{4L^2}\right) \text{and}\, \overline{\Theta} = \frac{8}{\pi} \exp\left(-\frac{n^2 \pi^2 a t}{4L^2}\right), \tag{4.83}$$

and the heat flux density at the surface results to:

$$\dot{q}_S = \frac{2k}{L} (T_B - T_\infty) \exp\left(-\frac{\pi^2 a t}{4L^2}\right). \tag{4.84}$$

4.2.3.2 Approximations for Transient Heat Conduction Problems in Practice

From the analytical solutions (of Eq. 4.81), the temperature field $\Theta = \Theta(\tau, \xi, Bi, a_V^*)$ can be calculated as a function of time, location, external heat transfer coefficient and body shape. Herein, a_V^* is a dimensionless ratio in the from $a_V^* = (AL)/V$, where A is the surface area, V the volume and L half the thickness of a plate. If one defines for the transient heat conduction in the solid an internal heat transfer coefficient h_i with the heat flux through the surface $\dot{Q}$ and the difference between the caloric mean temperature $\overline{T}$ and the surface temperature T_S being $h_i = \dot{Q}/\left[A\left(\overline{T} - T_S\right)\right]$ it can be expressed in asymptotic equations

Table 4.3 Constants for different body shapes of the long term asymptotic in one-dimensional transient heat transfer

	Plate	Cylinder	Sphere
a_V^*	1	2	3
m_1	$\pi/2$	2,4048	π

for long and short times using additionally the non-dimensional Nusselt number $Nu = (h_i L)/\lambda$, for more details see e.g. (von Boeckh & Wetzel, 2011). Accordingly, the following results for the long-term asymptote with the values from the Table 4.3:

$$Nu(t \to \infty) = Nu_\infty = \frac{2 + a_V^* + Bi}{1 + \left(a_V^* Bi\right)/m_1^2}, \tag{4.85}$$

whereas the mean value of the short-term asymptote gives the relationship:

$$Nu(t \to 0) = Nu_0 = \frac{1}{\sqrt{\tau}} \frac{\sqrt{\pi} + 10Bi\sqrt{\tau}}{1 + 5Bi\sqrt{\pi\tau}}. \tag{4.86}$$

By superimposing the asymptotes, a mean Nusselt number Nu:

$$Nu = \sqrt{Nu_\infty^2 - 0.04 + (Nu_0 + 0.2)^2}, \tag{4.87}$$

is obtained, which can be used to approximately calculate a caloric mean temperature $\overline{\Theta}$:

$$\overline{\Theta}(\tau) = exp\left(-\frac{a_V^* \quad \tau}{1/Bi + 1/Nu(\tau)}\right). \tag{4.88}$$

The argument of the preceding exponential function represents a dimensionless time. If one analyses the contributing dimensions from the numerator and denominator $(kAt)/(\rho c_V V)$, one recognizes that they correspond to the NTU (Number of Transfer Units) values of a heat exchanger.

To determine the surface temperature, which results from the thermal resistance ratio k/h_o, with $1/k = 1/h_o + 1/h_{i,t}$, one needs the instantaneous heat transfer coefficient instead of the time average values (Eq. 4.86), which is:

$$Nu_{0t} = \frac{1}{\sqrt{\tau}} \frac{2.3\sqrt{\pi} + 2Bi\sqrt{\tau}}{2\left(2.3 + Bi\sqrt{\pi\tau}\right)}. \tag{4.89}$$

The superposition of the asymptotes gives the instantaneous value of the Nusselt number Nu_t,

$$Nu_t = \sqrt{Nu_\infty^2 - 0.04 + (Nu_{0t} - 0.2)^2}, \tag{4.90}$$

from which the surface temperature $\Theta_0(\tau, \xi = 1)$ can be calculated:

$$\Theta_0(\tau, \xi = 1) = \frac{\overline{\Theta}}{1 + Bi/Nu_t}. \tag{4.91}$$

The temperature in the centre of the body $\Theta_m = \Theta(\tau, \xi = 0)$ exhibits a time offset with respect to the time change of the caloric mean temperature, which can be taken into account by means of the introduction of a new time scale $\tau_m = (\tau - \Delta\tau)$ with the variable dead time $\Delta\tau$:

$$\begin{gathered}\Delta\tau = \left(1/\Delta\tau_\infty^m + 1/\tau^m\right)^{-\frac{1}{m}} \text{ with } m = \frac{30+Bi}{10.8+Bi/4} + \frac{0.16}{\sqrt{\tau}} \text{ and} \\ \frac{1}{\Delta\tau_\infty} = 4 + \frac{2a_V^*(6+a_V^*+2Bi)}{6+a_V^*+Bi(2.71+0.03a_V^*)}.\end{gathered} \tag{4.92}$$

The temperature at the centre of the body now reads using $\tau_m = (\tau - \Delta\tau)$:

$$\Theta_m = \Theta(\tau, \xi = 0) = \exp\left(-\frac{a_V^* \tau_m}{1/Bi + 1/Nu(\tau_m)}\right). \tag{4.93}$$

The approximate Eqs. (4.85)–(4.93) show a maximum difference of less than 3% compared to the exact solutions, also for the caloric mean temperature and for the surface temperature.

4.2.3.3 Coupled Interface Boundaries

If two bodies 1 and 2 of different temperature are brought into contact with each other at time $t = 0$, as sketched in Fig. 4.10, the contact temperature T_C is established after an arbitrarily short time at the surface of both bodies. The two bodies can have different material properties, but since they have the same contact area, the temperature at the interface of the two bodies is the same, so that the boundary condition at the contact point is as follows:

$$T_1(x = 0) = T_2(x = 0). \tag{4.94}$$

In the case of real interfaces of solid bodies, this condition does not necessarily have to be fulfilled; for example, roughness, oxide layers can be produced by imperfect contact, so that an additional contact resistance occurs. This type of contact leads to a boundary condition that describes the continuity of the heat flux at the contact point.

$$k_1 \left.\frac{\partial T}{\partial n}\right|_{1,x=0} = k_2 \left.\frac{\partial T}{\partial n}\right|_{2,x=0}. \tag{4.95}$$

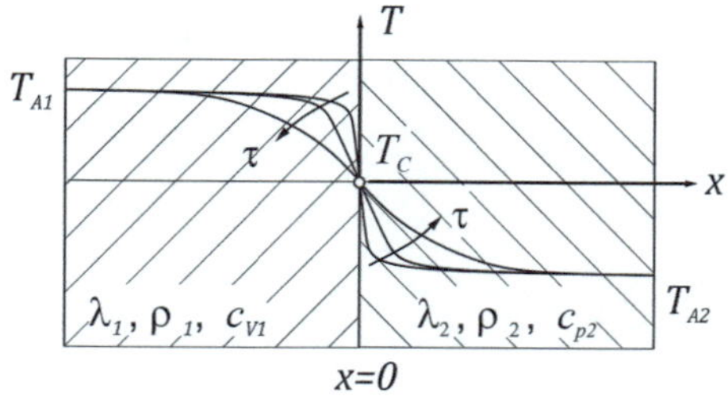

Fig. 4.10 Sketch for the temporal temperature evolution of the contact of two bodies of different temperature

Assuming an ideal contact, we have a boundary condition of the first kind at the contact point $x = 0$, the solution of which is Eq. (4.74):

$$\begin{aligned} \frac{T-T_C}{T_{A1}-T_C} &= -\mathrm{erf}\left(\frac{x}{2\cdot\sqrt{a_1\cdot t}}\right) \text{ for } x < 0 \text{ and} \\ \frac{T-T_C}{T_{A2}-T_C} &= \mathrm{erf}\left(\frac{x}{2\cdot\sqrt{a_2\cdot t}}\right) \text{ for } x > 0. \end{aligned} \tag{4.96}$$

This condition for the continuity of the temperature at the interface can be inserted into Eq. (4.94)

$$\sqrt{k_1 \cdot \rho_1 \cdot c_{p1}}(T_{A1} - T_C) = \sqrt{k_2 \cdot \rho_2 \cdot c_{p2}}(T_C - T_{A2}), \tag{4.97}$$

and can be resolved according to the contact temperature T_C so that the following is obtained:

$$T_C = \frac{\left(T_{A1} + \sqrt{\frac{k_2\cdot\rho_2\cdot c_{p2}}{k_1\cdot\rho_1\cdot c_{p1}}}\cdot T_{A2}\right)}{\left(1+\sqrt{\frac{k_2\cdot\rho_2\cdot c_{p2}}{k_1\cdot\rho_1\cdot c_{p1}}}\right)} = \frac{\left(T_{A1} + \frac{b_2}{b_1}\cdot T_{A2}\right)}{1+\frac{b_2}{b_1}}. \tag{4.98}$$

Thus, the contact temperature T_C depends on the ratio of the thermal effusivities $b_i = \sqrt{(k\rho c_V)_i}$ and it is always closer to the initial temperature of the body that has the higher thermal effusivity. Equation (4.98) can be used to explain why different bodies feel differently warm and the contact temperature is determined by the ratio of the thermal effusivity of the two bodies.

A copper plate has a much larger thermal effusivity than the human hand. Therefore, one feels approximately the temperature of the copper plate. With a Styrofoam plate it is opposite, here one feels rather the temperature of the hand. As long as the temperature change does not reach deeper layers of the body, the contact temperature remains constant. If larger layers are covered, a gradual temperature equalization occurs, which is determined by the heat content of the two bodies. This can be demonstrated, for example, on an aluminium foil. If one touches a very thin, hot aluminium foil, a contact temperature arises at the first moment which corresponds approximately to the temperature of the aluminium foil. Since the foil is very thin and thus has little mass to store the heat, it cools down very quickly. It is not able to heat the skin in such a way that the temperature increase reaches the nerves with which the temperature is registered. As a

Table 4.4 Thermal effusivities b of selected matter in $\left[J/\left(m^2 K\sqrt{s}\right)\right]$ (some take from https://thermtest.com/materials-database)

Material	$b\left[J/\left(m^2 K\sqrt{s}\right)\right]$	Material	$b\left[J/\left(m^2 K\sqrt{s}\right)\right]$
Graphite	16318	Cu (liquid)	27700
Si (solid)	7660	Fe (liquid)	16100
Fe (solid)	12600	Al (liquid)	17000
Stainless-steel (316L)	7230	$NaNO_3$	1921
Aluminium (solid)	23688	Na (liquid)	1381
Copper (solid)	36983	Si (liquid)	11600
Concrete stone	1260	Water (H_2O)	1588
MgO	5990	Oil (motor 10W30)	519
Al_2 O*3*	5890	Diesel	491
SiO_2	3610	Ethanol	589
CaO	7530	Methane (CH_4)	7
Granite	2490	He	12
SiC(SiC76Si24)	11266	Hydrogen (H_2)	272
Glass (avg. property)	1419	Paper	305
Hard wood	380	Human skin	1120

result, the foil feels only moderately warm. If you touch an aluminium plate that is equally hot, you will burn your fingers. Some typical thermal effusivities of solid matter are listed in Table 4.4.

The importance of the thermal effusivity especially appears in in the storage of sensible heat, where it characterizes the ability to exchange thermal energy with its surroundings. Since the effusivity is related to the surface heat exchange the most favourable medium at high heat loaded surfaces is that which can absorb (dissipate) more heat without raising temperature. Thermal effusivity is also of importance to measure heat fluxes by means of calorimetric methods, where the transient heat conduction plays a prominent role.

4.2.3.4 Limiting Cases with $Bi = 0$ and $Bi \to \infty$

If the Biot number approaches zero ($Bi = 0$), this means that the external heat transfer coefficient h is very small or the thermal conductivity λ of the material is very large. In this case, the temperature in the body is independent of the location and therefore only dependent on time. For very large Biot numbers ($Bi \to \infty$), i.e. for very large external heat transfer coefficients h or very small thermal conductivity of the body under consideration, the surface temperature T_S approaches the external temperature T_∞, which is depicted in Fig. 4.11a. The surface temperature is equal to the external temperature. Figure 4.11c shows the case for $Bi = 0$. Here, the temperature inside the body is independent of the location. Figure 4.11b illustrates the temperature behaviour for finite Biot numbers, where the temperature depends on both location and time.

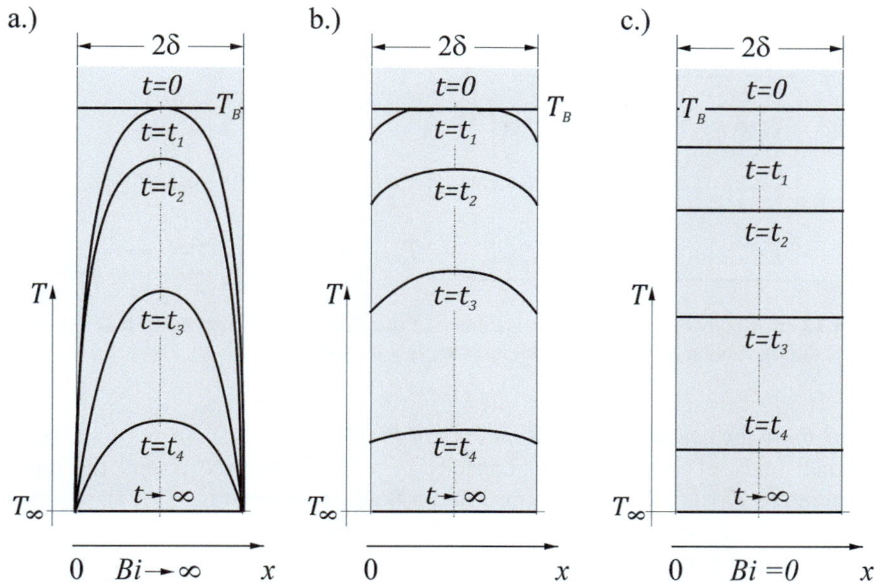

Fig. 4.11 Qualitative temperature profile of transient heat conduction at $Bi \to \infty$ (**a**), real Biot number (**b**) and $Bi = 0$ (**c**)

4.2.3.5 Solar Thermal Relevant Cases of Temperature Changes for $Bi \ll 1$

In many technical processes, it is not the temperature distribution in the body that is of interest during cooling or heating, but the mean temperature change. If the Biot numbers are small (the external heat transfer coefficient is smaller than that within the body), there is only a small temperature difference between the average temperature and the temperature in the centre of the body between the surfaces. If the Biot numbers are less than 0.5, the calculations can be done with fairly good accuracy such that the temperature in the whole body is taken as the mean temperature. For Biot numbers smaller than unity, a calculation performed in this way is still acceptable for estimations.

- A small body heats/cools transiently in a large environment with homogeneous temperature. This occurs when an absorber foil heats up quickly due to solar radiation, but the air in the collector box is still cold (danger of thermal shock!).
- Unsteady heating or cooling for bodies of approximately the same mass. The volumetric part of an absorber in a collector is relatively small, but its mass is of similar size compared to the air surrounding it (density air !).
- Heat transfer through a flowing heat transfer medium. This case occurs, for example, during collector operation, loading and unloading of thermal storage.

Subsequently, these cases and their results are discussed.

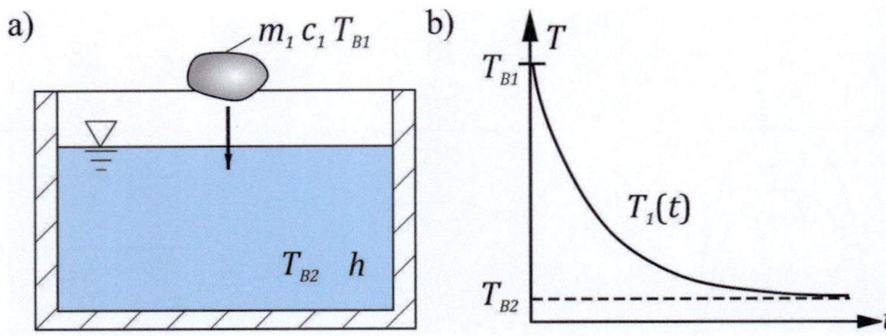

Fig. 4.12 **a** A body of small mass m_1 is immersed in a fluid pool of large mass. **b** Temperature profile during cooling of a body of small mass m_1 in a fluid pool

Small Body Immersed in Large Mass Pool of Fluid

A small body of mass m_1, heat capacity c_1 with the surface A and the initial temperature T_{B1} is suddenly dropped into a fluid of temperature T_{B2}, as sketched in Fig. 4.12a. Assume now that $T_{B1} > T_{B2}$ and the mass of the fluid is so large that no temperature change occurs in the fluid due to the immersion of the body. At the surface of the body, the heat transfer coefficient h is considered to have a given value and the walls of the vessel towards the ambient are supposed to be isolated. Then the heat flow transferred from the body to the fluid is given by the change in the heat content of the body over time as:

$$\dot{Q} = -m_1 \cdot c_1 \cdot \frac{dT_1}{dt}. \tag{4.99}$$

The transferred heat flow is determined by the heat transfer coefficient h and the temperature difference between the fluid and the body $(T_1 - T_{B2})$. If we note that when the body cools, its temperature change is negative and the temperature difference is positive, the result for the heat flow at the interface from the body to the fluid is:

$$\dot{Q} = h \cdot A \cdot (T_1 - T_{B2}). \tag{4.100}$$

Since the heat flow out of the body must equal the heat flow into the fluid, Eqs. (4.99) and (4.100) can be equated and an ordinary differential equation of the form is obtained:

$$\frac{dT_1}{dt} = -\frac{h \cdot A \cdot}{m_1 \cdot c_1}(T_1 - T_{B2}). \tag{4.101}$$

Replacing the temperature T_1 by the temperature difference $(T_1 - T_{B2})$, the following expression is obtained after separating the variables:

$$\frac{d(T_1 - T_{B2})}{(T_1 - T_{B2})} = -\frac{h \cdot A \cdot}{m_1 \cdot c_1} \cdot dt. \tag{4.102}$$

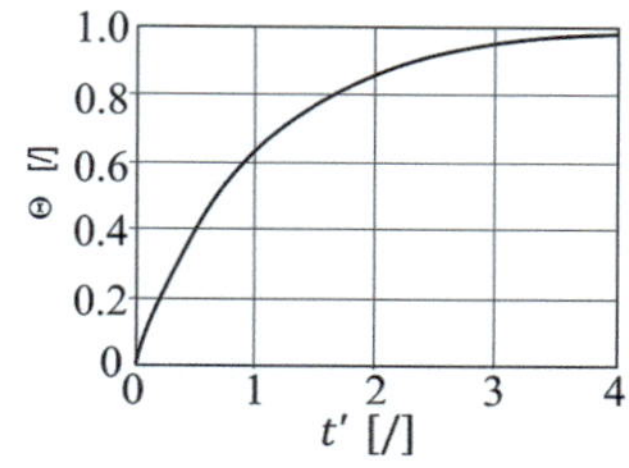

Fig. 4.13 Dimensionless temperature profile Θ as a function of dimensionless time t' when a small body cools in a large pool

If we assume a constant heat transfer coefficient h and a constant heat capacity (c_1), we can integrate this equation and obtain:

$$(T_1 - T_{B2}) = (T_{B1} - T_{B2}) \cdot exp\left(-\frac{h \cdot A}{m_1 \cdot c_1} \cdot t\right). \tag{4.103}$$

Figure 4.12b shows schematically the temporal evolution the body cooling. The temperature of the small body approaches the fluid temperature negatively exponentially asymptotically with increasing time. For the dimensionless representation, the dimensionless temperature and dimensionless time t' can be introduced

$$\Theta = \frac{(T_{A1} - T_1)}{(T_{A1} - T_{A2})} = 1 - \exp(-t') \text{ with } t_0 = \frac{h \cdot A}{m_1 \cdot c_1} \text{ and } t' = \frac{t}{t_0}. \tag{4.104}$$

Here, the time t_0 denotes the time in which the immersed body is heated by $1°K$ at $1°K$ temperature difference with respect to the surrounding fluid. The dimensionless representation allows to depict the solution for the temperature in a universal form, as it is shown in Fig. 4.13.

Body Immersed in Fluid of Similar Mass

In this limiting case, we will consider the immersion of a body in a basin thermally isolated from the outside and filled with a fluid with a relatively small fluid mass (and small dimension L) as illustrated in Fig. 4.14a. At immersion, the body has temperature T_{B1} and the fluid has initial temperature T_{B2}. The mass and the specific heat capacity of the body have index 1, respectively that of the fluid index 2. Since the fluid mass is no longer large, the temperature of the fluid is changed by the immersion of the body. The heat flux released by the body corresponds to the change over time of its heat content and one obtains:

$$-\dot{Q} = m_1 \cdot c_1 \cdot \frac{dT_1}{dt}. \tag{4.105}$$

For reasons of energy conservation, the heat flow absorbed by the fluid must be of the opposite magnitude to the change in the heat content of the fluid over time, so that:

$$-\dot{Q} = m_2 \cdot c_2 \cdot \frac{dT_2}{dt}. \tag{4.106}$$

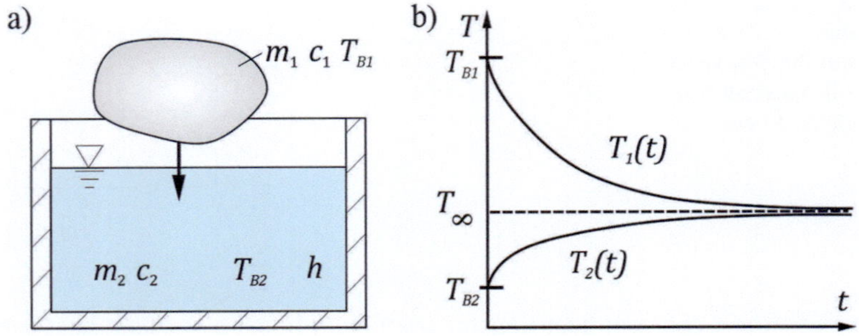

Fig. 4.14 **a** A body of small mass is immersed in a fluid of comparable mass. **b** Temporal temperature profile during cooling of a body of small mass m_1 in a fluid pool of comparable mass

The heat flux is determined by the heat transfer coefficient h and the driving temperature difference between fluid and body $(T_1(t) - T_2(t))$. The heat flow is thus given by:

$$\dot{Q} = h \cdot A \cdot (T_1 - T_2). \tag{4.107}$$

For the change of the temperature difference between fluid and body, we obtain from Eqs. (4.106) and (4.107) the relation:

$$d(T_1 - T_2) = -\dot{Q} \cdot \left[\frac{1}{m_1 \cdot c_1} + \frac{1}{m_2 \cdot c_2}\right] dt. \tag{4.108}$$

The heat flow from Eq. (4.107) can thus be substituted into Eq. (4.108), yielding the following differential equation

$$\frac{d(T_1 - T_2)}{(T_1 - T_2)} = -h \cdot A \cdot \left[\frac{1}{m_1 \cdot c_1} + \frac{1}{m_2 \cdot c_2}\right] dt. \tag{4.109}$$

Provided that the masses m_i, the heat transfer coefficient h and the specific heat capacities c_i are constant, Eq. (4.109) can be integrated. The integration is performed from time $t = 0$ to t and from initial temperatures T_{B1} and T_{B2} to temperatures $T_1(t)$ and $T_2(t)$ and results in the time dependent temperature difference $(T_1(t) - T_2(t))$:

$$(T_1(t) - T_2(t)) = (T_{B1} - T_{B2}) \cdot \exp\left[-h \cdot A \cdot \left(\frac{1}{m_1 \cdot c_{p1}} + \frac{1}{m_2 \cdot c_{p2}}\right) \cdot t\right]. \tag{4.110}$$

Equation (4.110) can be used to determine the temperature difference between the fluid and body, but not the temperature of the body or fluid. The temperature of the body and fluid can be determined from Eqs. (4.105) and (4.106) for any time.

Since in the equations the masses m_i and specific heat capacities c_i are considered as constant, the heat removed and added in time t can be calculated

$$Q(t) = m_1 \cdot c_1 \cdot (T_{B1} - T_1(t)); \quad Q(t) = m_2 \cdot c_p \cdot (T_2(t) - T_{B2}). \tag{4.111}$$

Equation (4.110) is also valid for infinite times at which temperature equilibrium is reached between the body and the fluid. The temperature after equilibration for $t \to \infty$ is calculated with the aid of equilibrium to be

$$T(t \to \infty) = T_\infty = \frac{m_1 \cdot c_1 \cdot T_{B1} + m_2 \cdot c_2 \cdot T_{B2}}{m_1 \cdot c_1 + m_2 \cdot c_2}. \tag{4.112}$$

For the temporal changes of the temperature one obtains from the two previous equations:

$$\frac{T_{B1} - T_1}{T_2 - T_{B2}} = \frac{T_{B1} - T_\infty}{T_\infty - T_{B2}} = \frac{m_2 \cdot c_2}{m_1 \cdot c_1}. \tag{4.113}$$

Equation (4.113) can be solved for T_1 or T_2 and inserted into Eq. (4.110). The result is the temperature of the body and fluid, the time course of which is shown schematically in Fig. 4.14b. Finally, the time negative exponentially changing temperatures of the bodies 1 and 2 reads to:

$$\begin{aligned} T_1 - T_\infty &= (T_{B1} - T_\infty) \cdot \exp\left[-h \cdot A \cdot \left(\frac{1}{m_1 \cdot c_1} + \frac{1}{m_2 \cdot c_2}\right) \cdot t\right]; \\ T_2 - T_\infty &= (T_{B2} - T_\infty) \cdot \exp\left[-h \cdot A \cdot \left(\frac{1}{m_1 \cdot c_1} + \frac{1}{m_2 \cdot c_2}\right) \cdot t\right]. \end{aligned} \tag{4.114}$$

Transient Heat Transport Through a Flowing Heat Transfer Medium

Figure 4.15 shows an externally insulated container filled with a fluid 2. This container is heated or cooled by a heat exchanger flowing with the mass flow $\dot{m}_1$. The fluid 2 in the container is separated from the heat transfer fluid 1 by a wall. The temperature of the fluid in the vessel is T_{B2} at the beginning of heating or cooling. The heat transfer fluid has the temperature T_{B1} at the inlet and $T_1(t)$ at the outlet. It is cooled or heated by the fluid 2 in the vessel. The heat transfer takes place with a constant heat transfer coefficient h. After an infinitely long time, the fluid in the vessel reaches the inlet temperature of the heat transfer medium T_{B1}, after which the fluid temperature of the heat transfer medium no longer changes. Assuming that the mass flow rate of the heat transfer fluid $\dot{m}_1$, the heat transfer coefficient h, the specific heat capacities c_i and local fluid temperatures are constant, the time temperature changes of the fluid 2 (T_2) and those of the heat transfer fluid 1 at the outlet T_2 can be calculated. The result is without going into the derivation:

$$T_{B1} - T_2 = (T_{B1} - T_{B2}) \cdot \exp\left[\frac{\dot{m}_1 \cdot c_1}{m_2 \cdot c_2} \cdot \left(\exp\left(\frac{-h \cdot A}{\dot{m}_1 \cdot c_1}\right) - 1\right) \cdot t\right]. \tag{4.115}$$

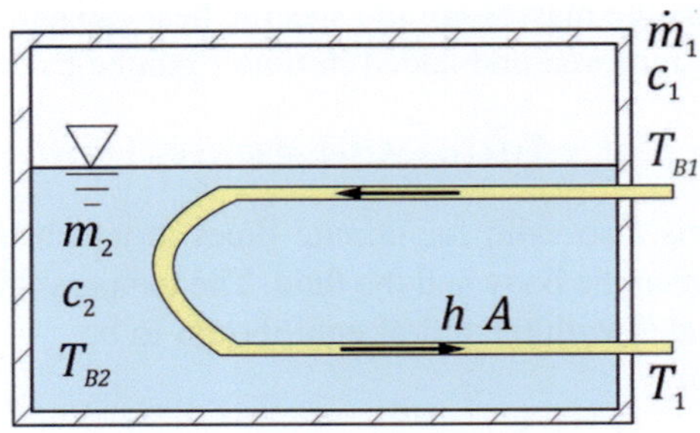

Fig. 4.15 Heating/cooling of a fluid 2 in an insulated container by means of a heat exchanger 1 through which the fluid 1 and the mass flow $\dot{m}_1$

4.2.3.6 Periodic Temperature Changes

Another boundary condition that occurs in many applications is the periodic change of the environmental condition, such as the daily change of solar radiation on a collector or a building structure.

When a periodic external temperature variation hits a plate, a heat wave propagates through the plate. Due to thermal conduction in the solid, the amplitude of the imposed external temperature variation is reduced. The damping of the wave, i.e. the attenuation of the amplitude, is determined by the thermal effusivity coefficient b. However, not only the amplitude of the temperature changes as the wave passes through the wall, but also the time response. A large specific thermal storage coefficient $s = (\rho \cdot c)$ in (J/(m^{3}K)) causes a slower progression of the temperature wave. The imposed temperature variation is therefore modulated not only in amplitude but also in phase, as shown in Fig. 4.16a, where a sinusoidal heat wave is assumed on the outside of the wall. As a rule of thumb, the higher the bulk density, the longer the phase shift and the higher the heat storage coefficient.

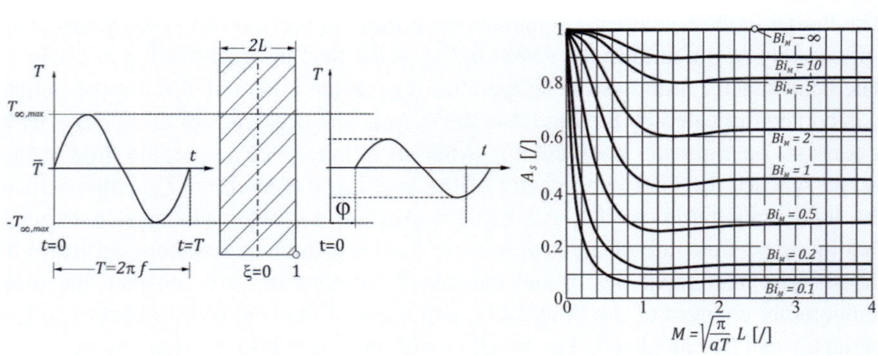

Fig. 4.16 **a** Schematic response behaviour of the establishing outer wall temperature ($T_S(\xi = 1)$)(right) of a wall with the thermal conductivity λ for a periodically changing outer temperature T_∞ (left) of period T. **b** Dimensionless amplitude of a plate surface temperature as a function of M and Bi

The simplest form of this type of an imposed external boundary condition is the specification of the ambient temperature as a harmonic oscillation around a mean value $\overline{T}$ with the amplitude $A = \left(T_{\infty,\max} - \overline{T}\right)$:

$$T_\infty - \overline{T} = \left(T_{\infty,max} - \overline{T}\right) \cdot \sin\left(2\pi \frac{t}{T}\right), \tag{4.116}$$

in which T specifies the period of the imposed temperature disturbance. Assuming that this boundary condition has existed for such a long time that any disturbances from an initial temperature distribution have decayed, the following temperature distribution results inside a body

$$T - \overline{T} = \left(T_{\infty,\max} - \overline{T}\right) \cdot A \cdot \sin\left(2\pi \frac{t}{T} + \varphi\right), \tag{4.117}$$

where A denotes the amplitude and φ the the phase shift of the temperature wave, which are functions of the normalized distance $\xi = x/L$ from the centre, the Biot number Bi and a parameter M:

$$M = L \cdot \sqrt{\pi/(at)}, \tag{4.118}$$

which can be interpreted as the reciprocal value of Fourier number. In case of slow changes of ambient temperature (with a large period duration $T \gg \left(\pi L^2\right)/a$) corresponding to a small M, the heat wave is hardly damped and the amplitude A is close to unity; at the same time there is hardly any phase shift, i.e. $\varphi \approx 0$. The temperatures inside the body therefore follow the fluctuation of the of the ambient temperature almost completely. For large values of M, which imply a rapid change of the ambient temperature, the amplitude A decreases ccontinuously from the surface ($\xi = 1$) to the centre ($\xi = 0$) of the body. Thereby, a negative phase shift occurs, increasing in the same direction. For sufficiently large values of M, only a narrow edge zone of the body follows the external temperature variation corresponding to a boundary layer character. The analytical expressions for the amplitude and phase shift can be universally represented for a plate in the following way, (details for the cylinder and sphere can be obtained from (von Boeckh & Wetzel, 2011):

$$\begin{gathered}
A(\xi, M, Bi) = \sqrt{\frac{F_R^2(M\xi)+F_I^2(M\xi)}{G_R^2(M,Bi)+G_I^2(M,Bi)}} \text{ and } \varphi(\xi, M, Bi) = \arctan\left(\frac{F_I G_R - F_R G_I}{F_R G_R - F_I G_I}\right), \\
\text{in which } F_R(M\xi) = \cosh(M\xi) \cdot \cos(M\xi), \\
F_I(M\xi) = \sinh(M\xi) \cdot \sin(M\xi), \\
\left.\frac{dF_R}{d(M\xi)}\right|_{\xi=1} = \mathrm{F}'_R = \sin M \cdot \cos M - \cosh M \cdot \mathrm{sinM}, \\
\left.\frac{dF_I}{d(M\xi)}\right|_{\xi=1} = \mathrm{F}'_I = \cosh M \cdot \sin M - \sinh M \cdot \cos M \text{ with} \\
G_R(M, Bi) = \left(F_R(M\xi) + \frac{M}{Bi} \cdot \frac{dF_R}{d(M\xi)}\right)_{\xi=1} \text{ and} \\
G_I(M, Bi) = \left(F_I(M\xi) + \frac{M}{Bi} \cdot \frac{dF_{RI}}{d(M\xi)}\right)_{\xi=1}.
\end{gathered} \tag{4.119}$$

The Fig. 4.16b illustrates the amplitude $A_S(\xi = 1)$ at the surface of the plate as a function of M and Bi_M, which is defined as $Bi_M = Bi/M = (h/\lambda)\sqrt{(aT)/\pi}$.

Due to the external heat transfer resistance 1/h, the harmonic oscillation of the ambient temperature (according to Eq. 4.116) is transmitted to the surface T_S with a smaller amplitude A_S and a time offset φ_S, which is called negative phase shift, see schematic sketch in Fig. 4.16b so that:

$$T_S - \overline{T} = \left(T_{\infty,max} - \overline{T}\right) \cdot A_S \cdot sin\left(2\pi \frac{t}{T} + \varphi_S\right). \tag{4.120}$$

The time shift φ_S of the surface temperature T_S with respect to the ambient temperature can reach for thin bodies ($M \to 0$) and simultaneously a small external heat transfer coefficient ($Bi/M \to 0$) a quarter of the period, i.e. $\varphi_S = -\pi/2$. Then, however, the amplitude of the surface temperature oscillation A_S is also almost zero as well, in other words the wall does not notice anything from the surroundings.

For large values of $M\xi$ the plate can be considered as a semi-infinite body and Eq. (4.119) simplify and a boundary layer character is obtained. By introducing a boundary layer coordinate η and a modified Biot number Bi_M in the following form,

$$\eta = M(1 - \xi) = \sqrt{\frac{\pi}{(aT)}} y \text{ and } Bi_M = \frac{h}{a}\sqrt{\frac{(aT)}{\pi}}, \tag{4.121}$$

in which y denotes the coordinate from the plate surface towards the plate center the following result is obtained for the amplitude $A(\eta, Bi_M)$ and the phase shift $\varphi(\eta, Bi_M)$:

$$A(\eta, Bi_M) = \frac{\exp(-\eta)}{\sqrt{(1 + 2/Bi_M + 2/Bi_M^2)}} \text{ and } \varphi(\eta, Bi_M) = -\left(\eta + \arctan\left(\frac{1}{1 + Bi_M}\right)\right). \tag{4.122}$$

With the imposed sinusoidal boundary condition (4.116), the direction of the heat flow is reversed after half a period, i.e. heat is stored in the plate during the first half-wave and released to the environment during the next half-wave. Thus, the instantaneous heat flux is constantly changing. If we integrate over a half period ($t = T/2$), we obtain the amount of heat that the plate stores in the structure during the first half-wave in the form of internal energy and releases again during the next half-period. Using the abbreviations F'_R and F'_I from Eq. (4.119), one obtains the amount of heat stored after one half period $Q_{T/2}$:

$$Q_{T/2} = \frac{2}{\pi} A^* \cdot k \cdot \frac{T}{2} \sqrt{\frac{\pi}{(aT)}} \cdot \left(\frac{2F'_R + 2F'_I}{G_R^2 + G_I^2}\right) \cdot \left(T_{\infty,\max} - \overline{T}\right), \tag{4.123}$$

in which A^* represents the surface area of the plate. If we define a heat transfer coefficient $k_{T/2}$ in the way

$$Q_{T/2} = A^* \cdot \frac{T}{2} \cdot k_{T/2} \cdot \left(T_{\infty,\max} - \overline{T}\right), \tag{4.124}$$

in which the time-averaged temperature difference $\Delta\overline{T}$ is obtained from Eq. (4.116) by,

$$\Delta\overline{T} = \frac{1}{T/2}\int_{t=0}^{T/2} \left(T_\infty - \overline{T}\right)dt = \frac{2}{\pi} \cdot \left(T_{\infty,\max} - \overline{T}\right), \tag{4.125}$$

and the comparison of variables of Eqs. (4.123) and (4.124) lead to:

$$k_{T/2} = k \cdot \sqrt{\frac{\pi}{(aT)} \cdot \left(\frac{2F'_R + 2F'_I}{G_R^2 + G_I^2}\right)}. \tag{4.126}$$

The right-hand side of Eq. (4.126) is a function of M and Bi_M and for $Bi_M \to 0$, meaning a negligible internal heat transfer resistance compared to the external one, one gets

$$\lim_{Bi_M \to 0}\left(\frac{k_{T/2}}{k} \cdot \sqrt{\frac{(aT)}{\pi}}\right) = Bi_M \text{ or } \lim_{Bi_M \to 0}\left(k_{T/2}\right) = h. \tag{4.127}$$

In the other limiting case $Bi_M \to \infty$, an internal heat transfer coefficient h_i is obtained for $k_{T/2}$, which is

$$h_i = k \cdot \sqrt{\frac{\pi}{(aT)} \cdot \left(\frac{2F'_R + 2F'_I}{G_R^2 + G_I^2}\right)}. \tag{4.128}$$

Knowing the internal heat transfer coefficient, one can determine an average internal Nusselt number $Nu_{i,m} = (h_i/\lambda) \cdot \sqrt{(aT)/\pi}$. For large values of M, $Nu_{i,m}$ tend towards the constant value of $Nu_{i,m} = \sqrt{2}$, while in the opposite case for small values of M has the limit value $Nu_{i,m} = 2M$. If the internal heat transfer coefficient h_i is analyzed at a fixed period T for a plate, it passes through a maximum with increasing plate thickness, which is located at $M = 1.2$.

Practically occurring periodic boundary conditions usually deviate more or less strongly from the simple harmonic oscillation; however, they can be represented by Fourier analysis as sums of harmonic oscillations.

4.2.3.7 Application of Transient Heat Conduction to Building Walls

Based on the findings of the previous section, various conclusions can be drawn regarding the application in building technology and collector development. For example, the smaller the heat penetration coefficient b (see e.g. Table 4.4) of the room boundary surfaces, the faster the heating of a room with a certain heat output of the heating device. Rapid heating over time may be desirable for reasons of comfort in winter; this requires a propagation of a temperature field in a material, i.e. a high thermal diffusivity a. In the heating period, for example, heating energy can be saved by lowering the night temperature; in this case, a high thermal diffusivity a is disadvantageous. In the building sector, the storable energy is given by a storage coefficient s, which is the product of the density ρ and the heat capacity c of the material, and thus $s = (\rho c)$ in [J/(m^{3}K)]. However, large storage masses (high heat storage coefficients s) are desired for thermal inertia in summer, this means only a small heating of the building at high solar radiation. These storage masses can be used efficiently if there is an exterior thermal insulation with a small thermal conductivity. This creates a small contact temperature T_C and the building heating is significantly delayed.

This result, obtained on the basis of heuristic considerations, can be analysed on the basis of the temporal heating and cooling behaviour of typical building walls as a function of the selected building materials. Let us therefore analyse the heating and cooling behaviour of two walls for a given outside air temperature $T_{A,o} =$ 5 °C and a fixed inner room air temperature of $T_{A,i} = 20\,°\mathrm{C}$. The wall 1 consists of $\delta = 0.24\,\mathrm{m}$ thick aerated concrete ($\rho = 500\,\mathrm{kg/m^3}$) a thermal conductivity $k =$ 0.16 W/(mK), resulting. in a thermal resistance of $R_H = 1.5$ K/W (or U-value of $\Lambda = k/\delta = 0.66$ W/(m^{2}K)) according to Eq. (4.55). Wall 2 is composed of two layers at first a $\delta = 0.06\,\mathrm{m}$ thick polystyrene layer ($\rho = 30\,\mathrm{kg/m^3}$, $k = 0.04$ W/(mK)) followed by a 10 cm wide heavy concrete layer ($\rho = 2500\,\mathrm{kg/m^3}$, $k = 2.1$ W/(mK)) corresponding to $\Lambda = 0.646$ W/(m^{2}K) being almost equal to that of wall 1. Figure 4.17 shows the heating (left) and cooling (right) behaviour of the two different walls as a function of time for the same given external air temperature $T_{A,o}$.

Both walls reveal a completely different time behaviour of the surface temperature on the inner wall side $T_{S,i}$. Wall 1 offers advantages during the heating period since the inside wall temperature $T_{S,i}$ increases faster than in wall 2. At the same time, however, this wall exhibits a higher heat loss to the outside because the wall-normal temperature gradient $\partial T/\partial n$ is larger (compare Eq. 4.7). These high heat losses are also evident in the cooling of the wall. The massive wall 1 loses heat much faster than wall 2, and an uncomfortable living feeling quickly results after a winter night. Conversely, in summer the interior air in wall 1 heats up rapidly, so that an uncomfortably warm climate can establish. In this context, the question immediately arises as to the most energy-efficient way of insulating a wall. Which type of insulation is more favourable in a house depends on the way the room is used.

Figure 4.18 shows two identically thick aerated concrete walls on which the insulation is applied either externally or internally. With the exterior insulation (left

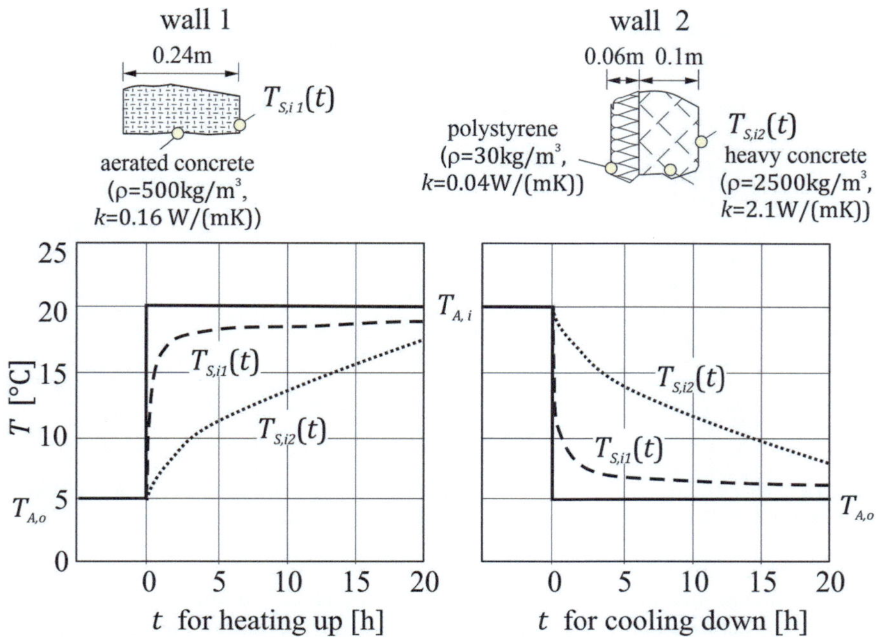

Fig. 4.17 Heating and cooling curves of the internal wall temperatures $T_{S,i}$ of two different walls as a function of time at for a fixed outside air temperature of $T_{A,o} = 5$ °C and a set internal room air temperature $T_{A,i} = 20$ °C

figure, a), the temperature gradient within the insulation layer is very high. This means that both at high temperatures in summer and low temperatures in winter, the building mass temperature hardly changes. The storage effect of the concrete walls is thus significantly increased. This type of insulation is disadvantageous in winter, if a "cold house" has to be heated. The heat loading of the solid storage wall by solar radiation is low and lasts for a relatively long time. In the case of interior insulation, rapid heating/cooling of the rooms by the direct solar radiation occurs, this is advantageous in the case of temporary use of the room or intermittent use. However, the storage effect of the relatively thick concrete wall is lost. In addition, at the interface concrete insulation in the case of internal insulation can fall below the dew point in winter, which can lead to undesirable meld growth. This type of insulation is not very useful for residential buildings.

If, for example, strong fluctuations in the outside temperature occur on the outer shell of a building on several consecutive summer days (24-h daily cycle), then a heat wave propagates through the outer components towards the interior every day. Heat conduction in the building material reduces the amplitude of the outside temperature fluctuation. The attenuator, i.e. the attenuation of the amplitude, is determined by the heat penetration coefficient b as discussed in Sect. 4.2.3.6. However, not only the amplitude of the temperature changes as it passes through the wall but also the time response. A large specific thermal storage coefficient s

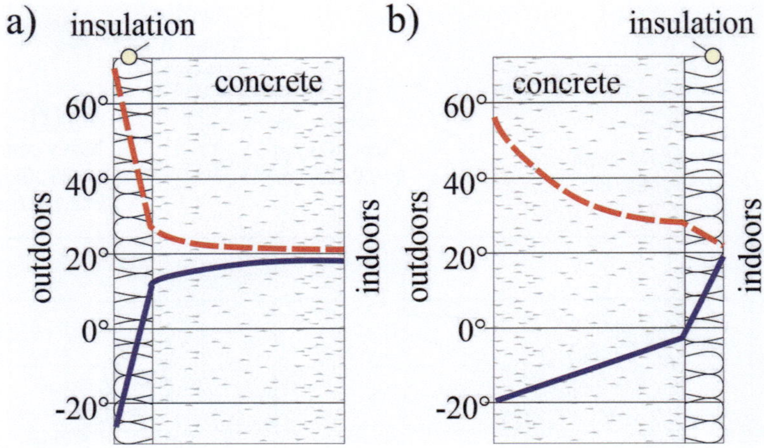

Fig. 4.18 Schematic temperature curve for exterior insulation (**a**) and interior insulation (**b**) of two identically thick building material layers of concrete and insulation material

causes a slower advance of the temperature wave. Thus, the temperature variation is modulated not only in amplitude but also in phase. Insulation materials based on renewable raw materials usually have a higher density and heat storage capacity and thus offer better thermal protection in summer. The ratio of the surface wall temperatures outside to the value inside is referred to in building technology as the temperature-amplitude ratio or TAR. The phase shift φ describes the time delay of the occurrence of the maxima on the outside and inside due to the storage of heat in the structural material. Figure 4.19 shows the effect of TAR on the indoor air temperature in a building with the same k-value but different heat storage capacities indexed as lightweight construction and heavy construction on ten consecutive summer days, determined by numerical simulation. It can be clearly seen that the TAR is significantly smaller for the heavy construction method.

A cooling time t_c is often specified in building services engineering. It indicates in hours how long it takes for a building component to cool down to the same temperature level in all layers under constant heat extraction. The longer the cooling time t_c of the outer walls, the more comfortable the indoor climate. It is defined as the ratio of the specific thermal heat storage coefficient s and the equivalent heat transfer coefficient h_{eq}, which scopes the thermal resistance of the insulation layer (k/δ), and the external heat transfer coefficient h. It has as SI unit the value in seconds, i.e. $t_c = s/h_{eq}$ [s].

$$h_{eq} = \left(\frac{1}{h} + \frac{\delta}{k}\right)^{-1}. \tag{4.129}$$

Values for t_c of more than 120 h are considered very good. In general, the greater the density of the material, the lower the specific heat (c). In the case of solid building materials, it varies so little that the bulk density provides sufficient

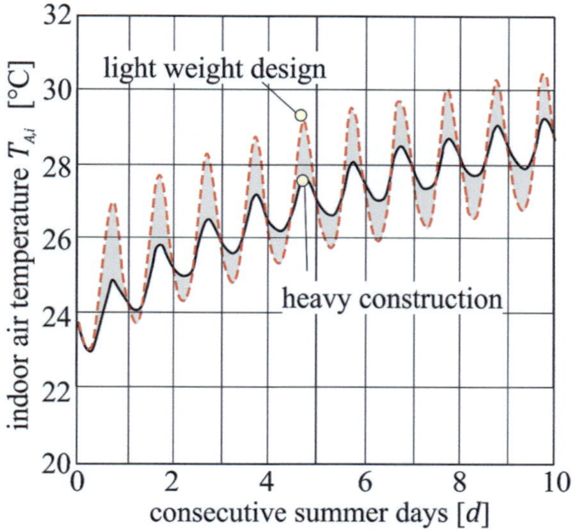

Fig. 4.19 Simulated temperature amplitude ratio (TAR) for 10 consecutive days for a light construction and a heavy construction

orientation for the heat storage capacity (thickness × bulk density = mass per unit area). The simple rule is that the more heat a material can store, the more sluggishly it reacts during heating and cooling, which is often referred to as so-called amplitude damping. Table 4.5 shows typical values of a solid wall of constant thickness 17.5 cm with different insulation thickness and the resulting h_{eq}-values. This constellation results in different cooling times, temperature-amplitude ratios (TAR) and phase shifts φ.

4.2.4 Molecular Heat Conductivity

In the context of the Kirchhoff-Transformation in the section of steady heat transfer we recognized large variations of the thermal conductivity in a continuum. Both, the order of magnitude differences in the temperature-dependent thermal conductivity of different materials, illustrated in Fig. 4.20, and their nature of temperature dependence vary considerably. Thermal conductivity is obviously a transport property of a body and a high thermal conductivity of a structural and/or functional material allows large heat fluxes; moreover, different materials exhibit a substantial temperature dependent thermal conductivity. Functional materials as the spectral selective solar thermal absorbers target at high temperatures to attain high collector efficiencies, while as thermal insulation aim at low heat losses to the ambient requiring low thermal conductivities. According to Fourier's law determining the heat flux magnitude, this flux through matter is associated is associated with the heat conductivity. Relevant for this heat transfer is obviously an exchange of energy between neighbouring particles. Apparently, in gases and liquids where the particles can freely move and are not bound to a lattice structure this energy transfer is established by collisions. In solids that is different because the atoms

Table 4.5 Solid wall of constant thickness 17.5cm made of different materials with different insulation thickness, the resulting h_{eq}-values. Resulting cooling times t_c, TAR and phase shifts φ in $[h]$

Material	Insulation thickness δ [cm]	h_{eq} $\left[W/(m^2K)\right]$	Cooling characteristic time t_c $[h]$	TAR [–]	Phase shift φ $[h]$
Clay wall	0	2.59	11.08	0.492	5.5
	2	1.13	48.58	0.082	8.2
	4	0.72	86.2	0.043	8.5
	8	0.42	161.7	0.022	8.8
	12	0.30	237.8	0.015	9.3
	16	0.23	314	0.011	9.5
Brick wall	0	3.26	4.2	0.867	2.5
	2	1.14	38.39	0.106	7.5
	4	0.73	68.33	0.057	7.8
	8	0.42	128.5	0.029	8.2
	12	0.30	189.2	0.019	8.7
	16	0.23	250.4	0.014	9.2
Aerated	0	1.23	5.7	0.672	4.1
	2	0.61	25.0	0.161	8.5
	4	0.47	37.3	0.100	9.0
	8	0.32	62.4	0.056	9.6
	12	0.24	87.6	0.039	10.2
Lightweight	0	1.59	3.14	0.861	2.5
	2	0.88	6.14	0.597	4.1
	4	0.61	9.31	0.422	5.0
	8	0.36	15.9	0.253	6.0
	12	0.27	22.9	0.176	6.7

are integrated in a lattice structure; nonetheless the atoms are elastically linked to each other by bonding forces. Thereby, they are able to transfer vibrational energy from one particle to the next in the lattice structure. For this reason, the transport of vibrational energy in non-metallic solids can also be thought of as a motion of quasi-particles. The quasi-particles are called phonons, which have a dual character as a wave and as a particle similar as photons. Since the phonon energy transfer occurs quantised, the lattice arrangement and therefore the atom distance plays an essential role. This can be fundamentally derived by solid state physics.

By this description the heat conductivity in gases, liquids as well as solids can be formulated in a general way as an energy transfer from colliding particles by an energy diffusion. An exception are metals where an additional effect

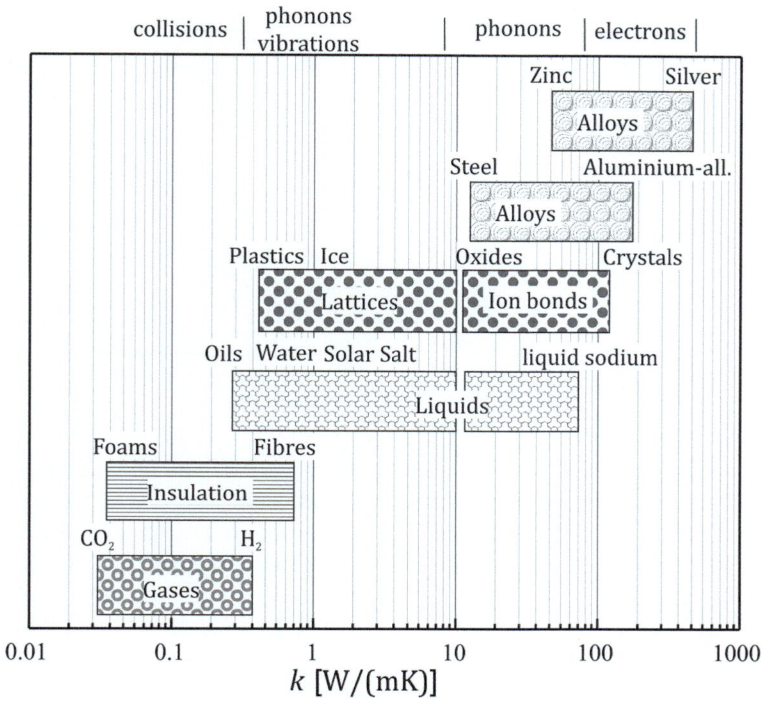

Fig. 4.20 Thermal conductivities k of various matter and leading processes for the heat conduction

is superimposed. Aside from the phonons also freely moving particles, the electrons, contribute to the energy transfer. These free conduction electrons of the electron gas outweigh the phonon contribution especially in pure metals is and is responsible for their good thermal conductivity.

4.2.4.1 Thermal Conductivity of Solids

Generally, in solids the energy propagation is described by the motion of phonons and these phonons transfer energy via collisions. This collisional energy transfer in non-electrically conducting solids makes heat conduction processes similar to that in liquids and gases. In the meantime between two phonon collisions of a phonon, the phonon travels on average a distance ℓ, which is called the mean free path as illustrated in Fig. 4.21. This mean free path refers to the motion of phonons in a three-dimensional frame of reference. But, relevant for the heat transfer in the one-dimensional case is only the motion of phonons in the direction of the heat flux and here in x-direction. The mean free path length ℓ is related to the mean free path components in x-, y- and z-direction. If we assume an isotropic substance, then the phonons can move in all spatial directions similarly and hence between two collisions the phonons travel the same distance on time average. If the phonons are assumed to be the thermal energy carriers, they cool down by the amount of

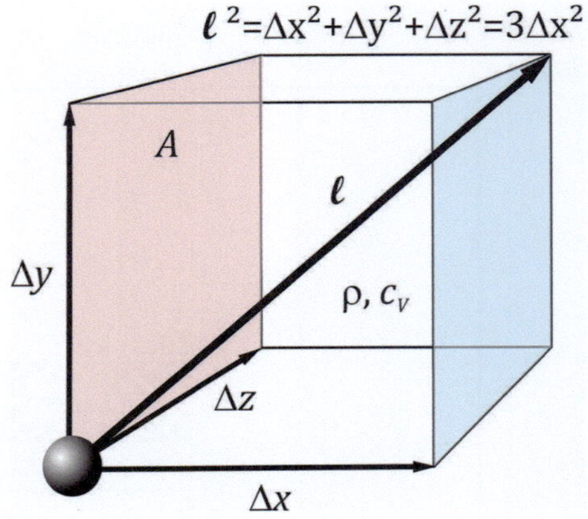

Fig. 4.21 Definition of mean free path ℓ in threedimensional space

ΔT due to the collisions when passing the distance Δx. The phonon itself losses energy through this but has transferred it to the next phonon. The magnitude of the energy transfer Q through a surface A at a temperature change ΔT results from the specific heat capacity c_V and the mass of the domain m ($m = \rho \cdot V$), which is given by:

$$Q = m \cdot c_V \cdot \Delta T = \rho \cdot V \cdot c_V \cdot \Delta T. \tag{4.130}$$

The average distance is given by the time between two collisions τ during which the heat Q is transferred. The thermal power $\dot{Q}$ through the area A is the given by $\dot{Q} = Q/\tau$. By inserting Fourier's law $\dot{Q}/A = \lambda \Delta T/\Delta x$ in Eq. (4.130) and replacing Δx by the expression for the mean free path 0 ℓ, we obtain

$$k = \frac{1}{3} c_V \cdot \rho \cdot v \cdot \ell, \tag{4.131}$$

where $v(= \ell/\tau)$ is the mean velocity of phonons which is approximately equal to the speed of sound. The existence of a certain finite value of ℓ is the result of phonon scattering on phonons, on defects of crystal lattice and, in particular on grain boundaries of the entire specimen. The temperature dependence of $k(T)$ is determined by that of ℓ and c_V on temperature. The Eq. (4.131) states that for no-metallic solids, the greater the mass specific heat capacity and the greater the density, the greater the thermal conductivity. Thus, for a high thermal conductivity, it is advantageous if the matter is composed of very many, but light particles. The specific heat capacity is not a constant value but depends on the temperature and the same holds for the mean free path, while the phonons, which are essentially sound waves, can be considered to be independent of temperature to a first approximation. In contrast to gases, atoms in solids cannot move freely but are bound to specific locations, also rotation is not possible. The movements of the atoms are thus limited to oscillation around an equilibrium position, whereby the oscillations can take place in all three spatial directions. This results in a degree of freedom for the atoms (f) in solids, which must be counted twice because of the potential and kinetic energy inherent in an oscillation. With effectively six degrees of freedom ($f = 6$), the molar heat capacity of solids is theoretically three times the value of the universal gas constant ($\mathcal{R}$), which is called the Dulong-Petit-law, which reads to:

$$c_V = \frac{f}{2}\mathcal{R} = 3 \cdot \mathcal{R} = 3 \cdot N_A \cdot k_B, \tag{4.132}$$

where $\mathcal{R}$ can be expressed by the product of the Avogadro number N_A and the Boltzmann constant k_B, with $k_B = 1.380649 \cdot 10^{-23}$J/K. At high temperatures, it is mainly the mean free path length impacting the thermal conductivity and relatively strong lattice vibration occur. The many and violent lattice oscillations then influence each other, i.e. phonons are scattered by other phonons. Then the phonons travel only short distances before they are scattered again by other phonons. As the temperature increases, there are more phonons and the mean free path declines. Since the phonon density or population is proportional to the temperature, the mean free path length decreases inversely proportional to the temperature and thus in the temperature domain relevant for solar thermal engineering the thermal conductivity scales with $k \sim 1/T$.

The deduction explains the dependence $k(T)$ in crystal dielectric in general but does not account for deviations from a constant lattice due to atomic impurities of other elements. At a high temperature the resistance offered by impurities to heat transport is independent of temperature. This allows to estimate the degree of crystal purity from variation of its thermal conductivity with temperature. The same fact enables us to determine the distribution between a glassy and crystal substance in the natural mineral if the glass is considered as an impurity in a crystal.

In amorphous solids, too, a dependence of the thermal conductivity with increasing temperature is observed. However, the mean free path length is relatively independent of the temperature, since these do not have a regular lattice structure, but consist almost entirely of lattice perturbations. Therefore, the mean free path length of these materials is of the order of the atomic distance. The fact that the thermal conductivity of amorphous materials also changes with increasing temperature must therefore originate from the change in the heat capacity. In case of amorphous materials, the heat capacity is often greater than in crystalline materials, but the mean free path length is substantially smaller. This usually results in lower overall thermal conductivity values compared to crystalline matter.

4.2.4.2 Thermal Conductivity of Metals

A high thermal conductivity inherent appears in metals as shown in Fig. 4.20. Additionally, k for nearly all pure metals declines with increasing temperature. Another effect also observed is, that the coefficient of thermal conductivity for steel is substantially lower than for pure iron (see also Fig. 4.7), an effect which is observed in nearly all alloys. Also, for similar metal alloy compositions, different thermal conductivities are obtained. Obviously, this depends on the lattice configuration, which can be modified by mechanical and thermal treatment; e.g. quenching and cold treatment of metals yield a decrease of the thermal conductivity k by introducing additional lattice defects, while reversely pre-heating up to a high temperature anneals of defects and diffusion of alloy constituent to grain boundaries and thus an increasing thermal conductivity. Therefore, the knowledge of the treatment of construction alloys is of crucial importance especially for high thermally loaded materials.

In contrast to the non-metallic solids the thermal conductivity in metals depends not only on phonons but to essentially on the motion and interaction of current carriers, i.e., conduction electrons. Thus, the thermal conductivity k of metals is a superposition of lattice thermal conductivity by phonons k_{ph}, and conduction electrons k_e, where $k_e \gg k_{ph}$. If we assume a flow of free electrons and postulate that the length of their free path does not depend on velocity, then the Wiedemann–Franz law is obtained, postulating that the ratio of specific thermal conductivity to that of the specific electrical conductivity (σ) equals the Lorenz constant (L) timest he temperature, i.e.:

$$\frac{k}{\sigma} = L \cdot T \text{ with } L = \frac{k}{\sigma\, T} = \frac{\pi^2}{3} \cdot \left(\frac{k_B}{e}\right)^2 = 2.44 \cdot 10^{-8} \frac{WV}{AK^2}, \tag{4.133}$$

in which k_B is the Boltzmann constant and e is the elementary charge with $e = 1.602175 \cdot 10^{-19} As$. The Wiedemann–Franz law is used to calculate k_e from the measured electrical conductivity. The experimental findings show that the thermal conductivity of metals at high temperatures is a slightly decreasing function of temperature, however, there are numerous exceptions; e.g. k of iron strongly decreases with temperature while niobium exhibits a positive gradient.

The heat transport in semiconductors is more complex than in dielectrics, although they are of general interest as selective absorbers for solar thermal application since the band gap corresponds to the energy of the solar spectrum. The reason for the complex derivation of the thermal conductivity is the same order of magnitude of the phonon and electron part of the thermal conductivity. Moreover, impurities, bipolar diffusion, and other little studied factors considerably affect the heat transfer.

4.2.4.3 Thermal Conductivity of Liquids

The thermal conductivity in liquids is mostly based on the Bridgman model, relying that the sound speed v in a liquid exceeds by an order of magnitude the speed of the molecules determined from the kinetic theory. Another difference is that the coefficient for thermal conductivity k is based on the heat capacity at a constant pressure c_p, so that the following relation is obtained:

$$k = \rho \cdot c_p \cdot v \cdot \mathcal{L}, \tag{4.134}$$

where $\mathcal{L}$ denotes the intermolecular spacing with $\mathcal{L} = \Delta - d$ being Δ the center-to-center distance of molecules of diameter d. The coefficients of thermal conductivity and viscosity of liquids obey more, and no simple proportionality characteristic of gases exists between λ and their viscosity μ. One reason for this is the presence of intermolecular force fields in liquids as e.g. van der Waals forces, which prevent to some extent the relative motion of neighbouring molecules. For increasing temperature, the kinetic energy of the thermal motion of the molecules rises and thus yields a decreasing the attraction forces. As a consequence, the thermal conductivity k decreases. For many liquids the temperature dependence can be approximated by a linear function $k(T)$ with the proportionality factor from -0.0005 to -0.002 for different types of liquids. The sound speed for liquids is mainly experimentally determined; however, semiempirical relations as the one suggested by Missenard (see e.g. in Grifoni & Passerini, 2006) are used; which read to:

$$k = B \cdot c_p \cdot \sqrt{\rho \cdot T_m}, \tag{4.135}$$

where B is assumed as a constant for all the liquids with the same number of atoms in the molecule N, being $B = N^{-1/4}$ and T_S the boiling point of the liquid. All liquids exhibit an increasing thermal conductivity with increasing pressure, which is based on the fact that compression increases the mutual attraction of the molecules, however, this effect is marginal for pressure below 5 MPa.

Contrary to gas mixtures in which components are mixed at the atomic and molecular level, the structure of systems with liquid components can differ essentially and reliable assessments should be based on manufacturer data. In solar thermal applications this is of particular importance, as water–glycol mixtures are used for frost protection in domestic applications or solar salt mixtures for power plant as well as sensible heat storage purposes. Based on generalized experimental data Filippov (1968) suggested an empirical relation for estimating thermal conductivity of solutions 1 and 2:

$$k = k_1(T) \cdot C_1 + k_2(T) \cdot C_2 - 0.72/k_1(T) - k_2(T)/(C_1 \cdot C_1), \tag{4.136}$$

where C_1 and C_2 are the mass concentrations of components with thermal conductivities k_1 and k_2 respectively. For the calculation of the thermal conductivity of electrolytes (salt solutions) often the relation suggested by Missenard is recommended:

$$k = 9.3 \cdot 10^{-2} \cdot N \cdot m^{-5.6} \cdot \sqrt{\rho \cdot T_m}, \tag{4.137}$$

in which m represents the molar mass of the molcules, N the number of atoms in a molecule, ρ the electrolyte density and the T_m melting temperature.

4.2.4.4 Thermal Conductivity of Insulation Materials and Gases

Table 4.6 shows some typical thermal conductivities of solids, such as metals, liquids, composites, and gases. From Table 4.6 and Fig. 4.20 it is clear that there is a significant difference between the values of solids and especially gases, the causes of which have been discussed in the previous sections. It is interesting to compare the thermal conductivity values of the gases with those of insulating materials and loose fills, which is given in Table 4.7. Both, exhibit similar values being substantially smaller than those of liquids and metals.

Insulating materials are inhomogeneous agglomerations of gas-filled structures (fibres, beads, porous bodies). In these inhomogeneous materials there is a mixture of radiation and conductive-convective heat transport occurring on a microscopic scale, which constitutes in an equivalent "heat conductivity k" on a macroscopic scale. As long as there are many interactions of radiative or conductive exchange in a layer of the materials (say within a few cm), the Fourier law of conduction holds. However, if the layers become very thin, the exact transport equations of in the structures (e.g. between the fibres in mineral wool) have to be solved. In the case of pebble bed fills, there are only a few surfaces and mostly only point contacts between particles occur. These small thermal bridges cause a large thermal conductance resistance and there are long paths of the heat transport by conduction processes compared to the layer thickness of the medium, as illustrated by Fig. 4.22. In order to calculate for this type of pebble beds as they e.g. appear

Table 4.6 Typical specific thermal conductivity k in [W/(mK)] of some materials at standard pressure (1013 m bar) and temperature (25 °C)

Metal	k [W/(mK)]	Liquids	k [W/(mK)]	Solids	k [W/(mK)]	Gases	k [W/(mK)]
Aluminium	221	Gasoline	0.14	Concrete	2.1	Ar	0.016
Bismuth	8.4	Glycerine	0.286	Glass	0.76	He	0.143
Lead	35	Fuel oil	0.14	Limestone	2.2	CO_2	0.015
Iron	80.2	Machine oil	0.124	Marble	2.8	CH_4	0.034
Gold	314	Ice at 0 °C	2.2	Sand	0.58	O_2	0.023
Potassium	135	H_2O (10 °C)	0.58	Sandstone	2.3	N_2	0.02
Copper	393	H_2O (60 °C)	0.644	Mild steel	40	H_2	0.18
Brass	120	H_2O (100 °C)	0.679	Stainless steel	15	Vacuum	0.0
Sodium	133			Brick	0.5–1.4	Xe	0.0051
Silver	429			Diamond	2300	Air	0.0245
Tungsten	167			Silicon	148		
Zinc	110			Sulphur	0.269		
Tin	67				2.1		

Table 4.7 Specific thermal conductivity k in $[W/(mK)]$ of some insulation materials and loose fill beds

Insulating materials	k $[W/(mK)]$	Insulating materials	k $[W/(mK)]$
Expanded polystyrene (EPS)	0.035–0.05	Aerated concrete	0.08–0.25
Extruded polystyrene	0.032–0.04	Poroton	0.08–0.45
Glass wool	0.035–0.05	Foam glass crushed stone	0.08
Cork	0.035–0.046	Perlite	0.04–0.07
Foam glass	0.04	Straw bales	0.038–0.07
Wool	0.035	Sand loose	0.04
Vacuum insulation panel	0.004–0.006	Wood pressed	0.1–0.2

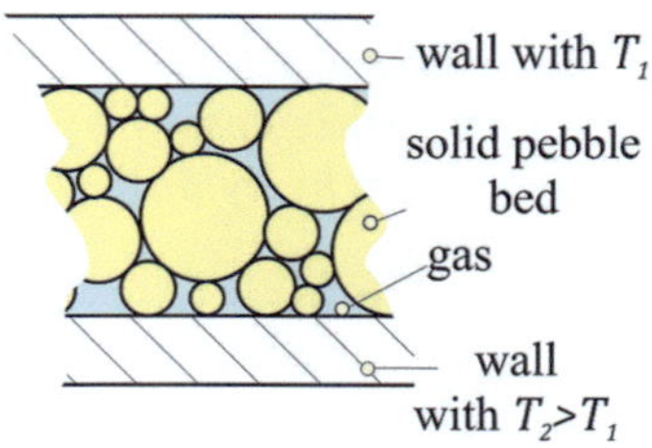

Fig. 4.22 Schematic sketch of heat conduction in a pebble

in filler beds of thermal energy storages the effective thermal conductivity, the mixed form between gas and solid conduction as well as thermal radiation must be considered, which we will discuss at the end of this chapter.

In gases, heat transport does not occur within a fixed lattice; rather, in gases, the gas atoms or their molecules encounter each other as they move randomly, transferring energy from particles with higher energy to those with lower energy. The gas molecules can move both translationally and rotationally. Thus, the energy transfer in gases is determined by the changes in translational as well as rotational energy as the gas atoms collide. Higher temperatures are equivalent to higher molecular energies.

For the gas to be conceived as a continuum, many collisions of the gas atoms with each other must occur. This is dealt within the idealized treatment of kinetic gas theory. The requirement for a high collision frequency directly implies that the mean free path length ℓ of the gas atoms or molecules between each other must be small compared to the dimension of the control space (vessel or gap in which the heat exchange is to take place). The mean free path length ℓ of the gas atoms between two collisions is given by the collision cross-section σ_{12} (in cm^2) and the particle or atom density n of the gas. A collision between two particles with particle radii r_1 and r_2 occurs when the particles approach each other to the distance of $(r_1 + r_2)$. For "point" gas atoms, a collision would occur when one of two particles fly around the second particle through a disk of cross-section $\pi(r_1 + r_1)^2 = \sigma_{12}$, as outlined in Fig. 4.23.

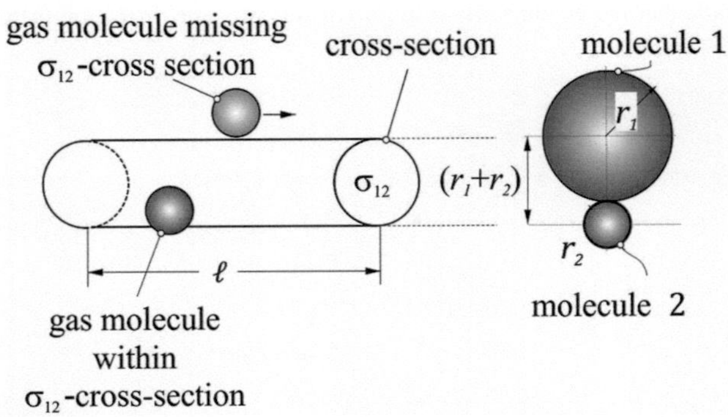

Fig. 4.23 Concept of the effective cross section σ_{12} for the collision between two gas molecules

If one imagines a particle stream with particle density n_1 flying through a volume with particle density n_2, collisions on a path element dx reduce the original particle stream n_1 by dn_1:

$$dn_1 = -n_1 \cdot \sigma_{12} \cdot n_2 \, dx. \tag{4.138}$$

Integration along path x yields a negative exponential function for the decrease in particle density n_1 along path x relative to the original particle density $n_{1,0}$ at location $x = 0$. This yield:

$$n_1(x) = n_1(x = 0) \cdot \exp(-\sigma n_2 x). \tag{4.139}$$

The mean distance ℓ between two collisions is then given by

$$\ell = \frac{\int_{x=0}^{x\to\infty} x dn_1}{\int_{x=0}^{\infty} dn_1} = \frac{\int_{x=0}^{x\to\infty} x dn_1}{n_1(x = 0)}. \tag{4.140}$$

Also important is the number of collisions, which is described with the help of the collision frequency z_{12} between the two types of particles. The collision frequency is the number of collisions per time, which the incoming particles n_1 have with the particle 2 located in the volume V. The particles 1 have an average velocity $\overline{v}_1$, so that the following relationship results for the number of collisions between the particles 1 and 2:

$$z_{12} = \frac{n_2 \cdot V}{\Delta t} = \frac{\sigma_{12} \cdot \ell \cdot n_2}{\Delta t} = \sigma_{12} \cdot \overline{v}_1 \cdot n_2. \tag{4.141}$$

If the particles of type 2 are also moving, the mean velocity is replaced by the mean relative velocity of both particles $\overline{v}_{12}$ and a modified impact number z_{12} is obtained. Finally, the impact density Z_{12} results from the multiplication of the impact number z_{12} by the particle density n_1 of the particles 1 entering the control volume V:

$$z_{12} = \sigma_{12} \cdot \overline{v}_{12} \cdot n_2 \text{ and } Z_{12} = \sigma_{12} \cdot \overline{v}_{12} \cdot n_2 \cdot n_1 \left[\frac{1}{(m^3 s)} \right]. \tag{4.142}$$

As a rule, the gas molecules do not all have the same velocity, rather there is a statistical distribution of velocities in both types of particles, which is given by the Boltzmann distribution. For the sake of simplicity, we consider further only one type of particle which has a statistical velocity distribution, and as a result we obtain the following relation for the mean free path length ℓ:

$$\ell = \frac{\overline{v}_1 \cdot \Delta t}{z_{11} \cdot \Delta t} = \frac{1}{\sqrt{2}} \cdot \frac{1}{\sigma \cdot n_1}. \tag{4.143}$$

In a plane gas-filled gap as sketched in Fig. 4.24, which is bounded by two walls of temperature T_1 and T_2, where $T_1 > T_2$, heat transport occurs from the hot to the cold side. For this purpose, at an arbitrary point x_1 in the gap, one can think of an intermediate plane with the thickness of the free path length ℓ below and above the plane x_1. If we first consider the number of particles that pass over the interfaces at $(x_1 + \ell)$ and $(x_1 - \ell)$, we obtain a net particle flux ΔN:

$$\Delta N = N(x_1 - \ell) + N(x_1 + \ell) = [n(x_1 - \ell) - n(x_1 + \ell)] \cdot A \cdot \frac{\ell}{6}, \tag{4.144}$$

where A denotes the cross-sectional area. The factor $1/6$ results from the fact that all particles in the control volume can move in the three translational directions (x, y and z) as well as the three rotational directions. The boundary transition from N to dN yields the following result using Eq. (4.139):

$$dN = -2 \cdot A \cdot \frac{\ell}{6} \cdot \ell \cdot \frac{\partial n}{\partial x} = \frac{\ell^2}{3} \cdot A \cdot \frac{\partial n}{\partial x}. \tag{4.145}$$

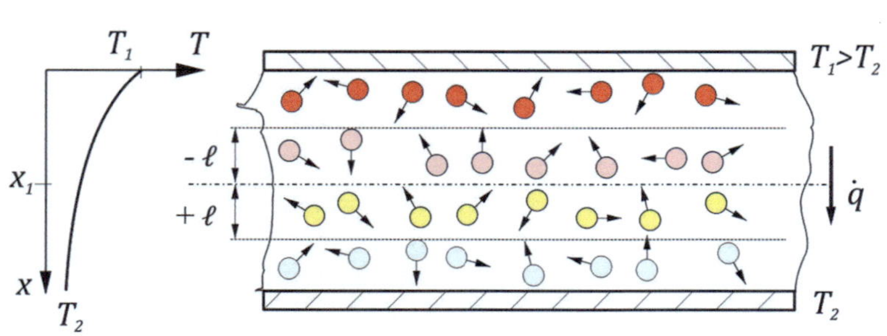

Fig. 4.24 Sketch of the derivation of the heat conduction in a gap

Thus, the number of particles exchanged over the imaginary control volume is proportional to the square of the mean free path length ℓ and the gradient of the particle density $\partial n/\partial x$. Further, one can define a particle flow in the x-direction J_x and using $\ell = \overline{v} \cdot t$ assuming a constant mean velocity yield:

$$J_x = \frac{dN}{A \cdot dt} = -\underbrace{\frac{\ell}{3} \cdot \overline{v}}_{\text{diffusion constant}} \frac{\partial n(x,t)}{\partial x}. \tag{4.146}$$

Each molecule has a certain amount of heat Q, which is given by the number of degrees of freedom f of its motion, and is accordingly calculated as follows

$$Q = \frac{1}{2} \cdot f \cdot k_B \cdot T. \tag{4.147}$$

Therein, f is the number of degrees of freedom—3 of translation and 3 of rotation, depending on the type of molecule, k_B is the Boltzmann constant, and T is the absolute temperature in Kelvin. The heat flux transported from the hot to the cold side in the gap is then given by the product of the particle flux and the heat stored in each particle to give:

$$\dot{q} = -\frac{\ell}{3} \cdot \frac{\partial}{\partial x}\left(\overline{v} \cdot n \cdot \frac{f}{2} \cdot k_B \cdot T\right). \tag{4.148}$$

If we now express the mean kinetic energy of the gas particles E_{kin}, given by $E_{kin} = m\overline{v}^2/2$, by the temperature T of the gas, given by $E_T = 3/2(k_B T)$, we obtain the ideal gas relation.

$$p = n \cdot k_B \cdot T \,\text{with}\, n = 1/V, \tag{4.149}$$

in which n denotes the particle density $\left[1/\text{m}^3\right]$. Thus, if the temperature T changes but the pressure p remains constant, according to the general gas law, the particle density n must change in the opposite sense so that the ratio $(T/V) \sim (n \cdot T)$ remains constant. Assuming the validity of the gas law, only the mean velocity $\overline{v}$ can change in the term for the derivative in Eq. (4.148). This gives the following relationship for the specific heat flux

$$\dot{q} = -\frac{\ell}{6} \cdot n \cdot f \cdot k_B \cdot T \frac{\partial \overline{v}}{\partial x}. \tag{4.150}$$

However, the mean velocity is also related to the temperature. As already shown in connection with the probability of collisions among each other (Eq. 4.142), a frequency distribution of velocity exists in a gas, which is described with the help of the Maxwell–Boltzmann distribution. The distribution F of the gas molecules with velocity v is given in the one-dimensional case by the relation:

$$F(v) = \sqrt{\frac{m}{2 \cdot \pi \cdot k_B \cdot T}} \cdot \exp\left[-\frac{m \cdot v^2}{2 \cdot k_B \cdot T}\right], \tag{4.151}$$

where m denotes the particle mass. If the mean velocity is defined as the sum of the individual velocities of the particles and divided by their total number N, the mean velocity $\overline{v}$ of the gas-molecule composite is given by the following equation

$$\overline{v} = \frac{1}{N}\sum_{i=1}^{i\to\infty} v_i = \int_{v=0}^{v\to\infty} v \cdot F(v)\,dv = \sqrt{\frac{8}{\pi}\cdot\frac{k_B \cdot T}{m}}. \tag{4.152}$$

This allows us to convert the derivative of the mean velocity in Eq. (4.150) into a differential of the temperature.

$$\frac{\partial\overline{v}}{\partial x} = \frac{\partial}{\partial x}\sqrt{\frac{8}{\pi}\cdot\frac{k_B T}{m}} = \frac{\partial}{\partial T}\sqrt{\frac{8}{\pi}\cdot\frac{k_B T}{m}}\cdot\frac{\partial T}{\partial x} = \frac{1}{2}\cdot\frac{1}{\sqrt{T}}\sqrt{\frac{8}{\pi}\cdot\frac{k_B}{m}}\cdot\frac{\partial T}{\partial x} = \frac{1}{2}\cdot\frac{\overline{v}}{T}\cdot\frac{\partial T}{\partial x}. \tag{4.153}$$

This expression for the change in mean velocity can now be substituted into Eq. (4.150) and the result obtained for the heat flux density $\dot{q}$ transferred.

$$\dot{q} = -\frac{n}{12}\cdot f\cdot k_B\cdot\ell\cdot\overline{v}\cdot\frac{\partial T}{\partial x}. \tag{4.154}$$

The equation for the heat flow corresponds to that of the 1st Fourier law (Eq. 4.6) at the beginning of this chapter. By a comparison of variables, the following relation is obtained for the thermal conductivity $k(T)$ of an ideal gas:

$$k_{Gas} = -\frac{\overline{v}}{12}\cdot\ell\cdot n\cdot f\cdot k_B\left[\frac{W}{(mK)}\right]. \tag{4.155}$$

Thus, the thermal conductivity of the gas can be analysed as a function of the individual contributions. If we consider the average velocity of the molecules $\overline{v}$, we get a higher conductivity of the gases scaling with $\sqrt{T}$ (from Eq. 4.152). The higher the temperature, the more collisions occur, and thus more heat can be transferred. Another result of the molecular velocity analysis is that the thermal conductivity of the gases decreases correspondingly with $1/\sqrt{m}$ as the molecular mass increases. Table 4.8 gives the thermal conductivity of some gases at standard pressure and temperature. The decrease of the specific thermal conductivity with increasing molecular mass can be seen particularly well for the noble gases; for example, for xenon it is only about one third of that of helium.

In Eq. (4.143), it was found that the mean free path length ℓ is inversely proportional to the particle density n, thus $k \sim 1/n$. Using the ideal gas equation, the particle density n is directly proportional to the pressure p, thus $p \sim n$.) If we insert this relationship into the conductivity relationship, we obtain the striking result that the thermal conductivity of a gas does not depend on the particle density. Thus, the thermal conductivity of a gas is independent of the pressure p; likewise, it can be shown that the viscosity of gases is independent of the pressure

Table 4.8 Measured specific thermal conductivity k in $[W/(mK)]$ of some gases at standard pressure (1013 m bar) and temperature (25 °C)

Gas	k [W/(mK)]	Gas	k [W/(mK)]
Hydrogen (H_2)	0.168	Helium (He)	0.144
Methane (CH_4)	0.030	Neon (Ne)	0.046
Air	0.0262	Argon (Ar)	0.016
Carbon dioxide (CO_2)	0.0146	Krypton (Kr)	0.0088
SO_2	0.0086	Xenon (Xe)	0.0051

as well. This statement is true as long as the mean free path length ℓ is small compared to the width of the considered gap or the vessel dimension. Therefore, for a calculation of the gas conductivity, it is mandatory to first determine the mean free path length ℓ_{gas} of the gas molecules at a certain residual gas pressure p_{gas}. As a rule of thumb, the following relationship can be used here:

$$\ell_{gas} = \frac{C_1(p_{gas})}{1 + C_2/T}[m], \tag{4.156}$$

where T is the absolute temperature in Kelvin and C_1 and C_2 represent two gas-specific constants. Table 4.9 provides some data for selected gases.

For air at normal pressure and room temperature, $\ell_{gas} \approx 0.6\mu m$. Next, the vessel dimensions δ must be checked. This can be the distance of the walls over which the driving temperature gradient occurs. For insulation materials, this can be the dimensions of the pores or the spacing of foils. For most insulation materials, $\ell_{gas} \gg \delta$. This means that the mean free path ℓ_{gas} in these materials is not so much determined by molecule–molecule collisions but is rather limited by the mean foil spacing δ. According to (Kaganer, 1969), the following relationship then applies

Table 4.9 Constants for determining the mean free path of various gases ℓ_{gas}. C_1 applies for a pressure of $p_0 = 133\ Pa$; for $p_{gas} \neq p_0$: $C_1(p_{gas}) = (p_0/p_{gas}) \cdot C_1(p_0)$, (Stephan et al., 2019)

Gas	$C_1(p_0)[\mu m]$	$C_2[K]$
Hydrogen (H_2)	105.6	76
Nitrogen (N_2)	61	112
Oxygen (O_2)	68.7	132
Helium (He)	160	79
Neon (Ne)	111.9	56
Argon (Ar)	70.3	169
Krypton (Kr)	59.6	142
Xenon (Kr)	48.7	252
Water (H_2O)	95	600
Carbon monoxide (CO)	60.2	100
Carbon dioxide (CO_2)	57	273

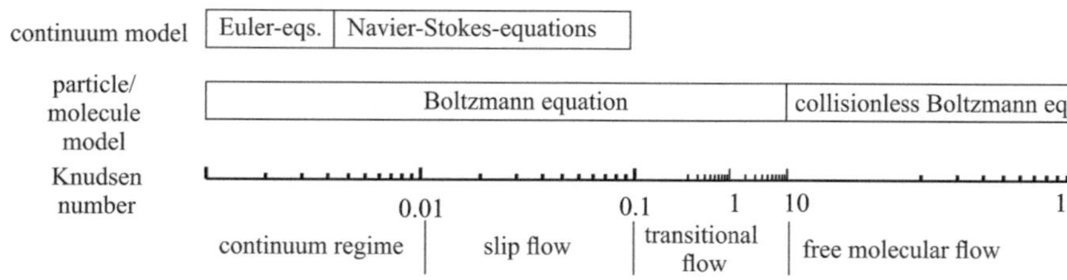

Fig. 4.25 Classification of flow regimes and associated continuum models as a function of the Knudsen number (Bird, 1994)

to the residual gas thermal conductivity k_{gas}:

$$\ell_{gas}(p_{gas}) = \frac{k_0}{1 + 2\beta Kn}\left[\frac{W}{(mK)}\right]. \tag{4.157}$$

In Eq. (4.157), k_0 denotes the thermal conductivity of the gas at standard pressure, β a weighting factor and Kn the Knudsen number. The Knudsen number Kn indicates the ratio of the mean free path length in the gas ℓ_{gas} to the typical characteristic dimension δ of the problem e.g. the pipe diameter or wall distance and is defined as follows:

$$Kn = \frac{\ell_{gas}}{\delta} \cdot . \tag{4.158}$$

The magnitude of the Knudsen number serves as a criterion to characterize the flow regime and the continuum model to be applied for computations. As Fig. 4.25 illustrates continuum models can be applied up to roughly $Kn \approx 0.1$.

The weighting factor β contains a correction quantity k_c and the adiabatic index γ, which is the ratio of the isochoric heat capacity c_V and the isobaric heat capacity c_p. The adiabatic index γ of an ideal gas is,

$$\gamma = \frac{c_p}{c_V} = \frac{f+2}{f} \quad [/], \tag{4.159}$$

where f denotes the degree of freedom of the gas. The accommodation coefficient α is also included in the weighting factor β, so that the exact formulation is as follows:

$$\beta = \frac{2 \cdot k_c}{\kappa + 1} \cdot \frac{2 - \alpha}{\alpha}. \tag{4.160}$$

Some correction values k to the specific heat capacity c_V for the calculation of β in Eq. (4.160) are listed in Table 4.10. The accommodation coefficient α is usually only imprecisely known. For the light gases (H_2, He) $\alpha \approx 0.3$, for the heavy gases for example nitrogen (N_2) α is about unity (Wakao & Vortmeyer, 1971). For air $\beta \approx 1.63$.

Table 4.10 Correction values k_c (to specific heat c_V) to calculate β from Eq. (4.160)

Gas	$T[°C]$ −182.6	−78.4	0	100	200	300	400
Hydrogen (H_2)	2.35	–	2.03	–	2.12	–	2.19
Nitrogen (N_2)	–	–	1.95	–	–	–	–
Oxygen (O_2)	–	–	1.915	–	–	–	–
Air	–	1.94	1.95	1.945	1.943	–	–
Carbon monoxide (CO)	–	–	1.865	–	–	–	–
Carbon dioxide (CO_2)	–	1.805	1.67	1.60	1.56	1.55	–
Water (H_2O)	–	–	–	1.41	1.395	1.41	1.425
Helium (He)	–	–	2.43		–	–	–
Neon (Ne)	–	–	2.50		–	–	–
Krypton (Kr)	–	–	2.49		–	–	–
Xenon (Xe)	–	–	2.49		–	–	–

According to Eq. (4.160), the gas thermal conductivity k_{gas} becomes small if k_0 is chosen to be as small as possible, as in the case of noble gases, for example, or if the Knudsen number Kn is aimed for to be as large as possible. Noble gases with their small conductivity are used in the construction industry for very high-quality multiple-pane glazing. In some cases, the cavity between the panes is filled with krypton or even xenon.

Technically, the more common method is to reduce the pressure to increase the Knudsen number. At low pressures, heat conduction then depends only on the number of particles n and then decreases proportionally with the pressure. There is a significant increase in the free path length at pressures lower than 100 Pa. However, in turn this means a rough vacuum is therefore insufficient to reach the dimensions of a macroscopic system with the free path length. Since the evacuation of a volume causes the mean free path length ℓ_{gas} to increase, the lower the pressure, the larger the Knudsen numbers (- or the smaller the $\lambda_{gas}-$ values). The mean free path of some gases ℓ_{gas} as a function of pressure p_0 in $[Pa]$ is given in Table 1.11. This method has been standard for thermos cans or, for example, Dewar vessels in cryogenics for over 100 years. In the building industry, it has been pursued as a vision for some time, but here problems often arise in the practical implementation in the area of sealing surfaces, the mechanical stability of large glass surfaces and, last but not least, the technical implementation on site at the construction site. Therefore, vacuum insulation elements for building applications have only recently become available. It also follows from Eq. (4.158) that large Knudsen numbers can be obtained by specifying very small film (or wall) spacings δ or, as described in Table 4.11, by using fine-grained fills with very fine pore/particle diameters δ.

By evacuation and by means of small δ-values, it is finally possible on a technical scale to completely eliminate the thermal conductivity component k_{gas}. Plates

Table 4.11 Mean free path length of some gases ℓ_{gas} in $[m]$ as a function of pressure p_0 in $[Pa]$ at standard temperature $T = 25\,°C$

Gas	$p_0[Pa]$ 10^{-6}	10^{-3}	1	10^3	10^5
Air	$6.8 \cdot 10^3$	6.8	$6.8 \cdot 10^{-3}$	$6.8 \cdot 10^{-6}$	$6.8 \cdot 10^{-8}$
Argon (Ar)	$7.2 \cdot 10^3$	7.2	$7.2 \cdot 10^{-3}$	$7.2 \cdot 10^{-6}$	$7.2 \cdot 10^{-8}$
Carbon dioxide (CO_2)	$4.5 \cdot 10^3$	4.5	$4.5 \cdot 10^{-3}$	$4.5 \cdot 10^{-6}$	$4.5 \cdot 10^{-8}$
Water (H_2O)	$4.2 \cdot 10^3$	4.2	$4.2 \cdot 10^{-3}$	$4.2 \cdot 10^{-6}$	$4.2 \cdot 10^{-8}$
Helium (He)	$19.6 \cdot 10^3$	19.6	$19.6 \cdot 10^{-3}$	$19.6 \cdot 10^{-6}$	$19.6 \cdot 10^{-8}$
Nitrogen (N_2)	$6.7 \cdot 10^3$	6.7	$6.7 \cdot 10^{-3}$	$6.7 \cdot 10^{-6}$	$6.7 \cdot 10^{-8}$
Neon (Ne)	$14 \cdot 10^3$	14	$14 \cdot 10^{-3}$	$14 \cdot 10^{-6}$	$14 \cdot 10^{-8}$
Oxygen (O_2)	$7.2 \cdot 10^3$	7.2	$7.2 \cdot 10^{-3}$	$7.2 \cdot 10^{-6}$	$7.2 \cdot 10^{-8}$

filled with mineral or polymer foam cores are commonly used for vacuum insulation. Figure 4.26 shows the thermal conductivity of some insulation materials as a function of the gas pressure in the fill. It is clear from the figure that as the dimensions of the voids decrease (smaller δ-values), the specific thermal conductivity is lower at the same pressure. Furthermore, a fine-grained bed reaches a saturation value already at higher gas pressures. This saturation value is given by the dimension of the voids. The drop of the thermal conductivity with decreasing gas pressure is often referred to in literature as Knudsen effect.

The better the insulation capacity is at low gas pressures, the more the local contact areas within the insulation material play a role. For example, they are smaller in spheres than in fibre-like composites. Figure 4.27a illustrates this by comparing

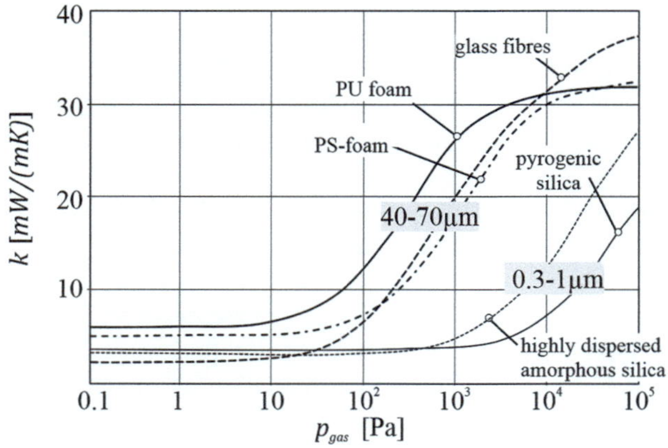

Fig. 4.26 Specific thermal conductivity k of several high insulation materials in $[\mathrm{mW/(mK)}]$ as a function of gas pressure p_{gas} in the bulk in $[Pa]$ at 20 °C (dry nitrogen) for different mean effective pore diametera. PU = polyurethane, PS = polystyrene, silica materials (specific surface, 300 m^2/g), (IfB GmbH Wolfgang Sorge, 2009)

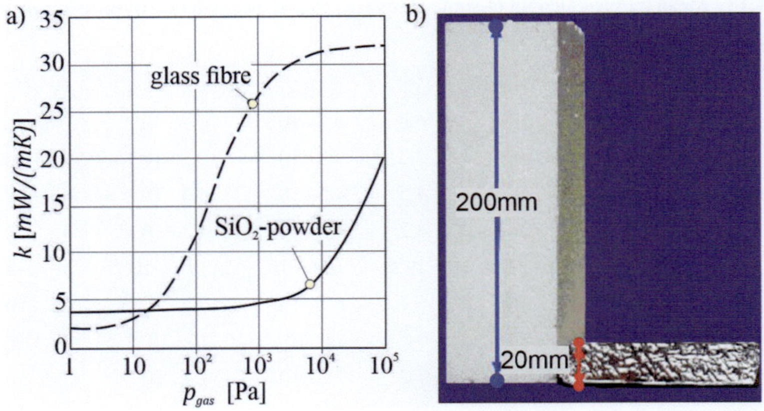

Fig. 4.27 Thermal conductivity of glass fibre and silicon oxide powder fill k in [mW/(mK)] as a function of gas pressure in nitrogen at 25 °C. **b** Thickness comparison of a solid polystyrene (PS) board and a vacuum insulated PS fill with the same heat transfer coefficient

a glass fibre with a silicon oxide spherical fill. The bulk material reaches the value of thermal conductivity of the pure material already at relatively large pressures, whereas in the case of the glass fibre composite, depending on the contact areas and gap dimensions, an optimum insulation effect is only achieved at much lower gas pressures. The performance of such insulation is illustrated by Fig. 4.27b, which shows the comparison of a solid polystyrene (PS) sheet with an amorphous vacuum-insulated pyrogenic silica fill. The solid structure builds 10 times thicker with the same heat transfer coefficient.

Note, the vacuum filled beds are rather exceptional limiting cases. For insulation material the effective heat conductivity (k) is usually a mixture of heat conduction of (residual) gas and insulation materials; moreover, for high emissivities and especially at higher temperatures also radiative effects contribute to the effective heat conductivity k, which is then constituted by a super-position of heat transfer coefficients arising from heat conduction, radiation, and convection.

4.3 Radiation Heat Transfer

From a microscopic physics point of view, there are only two mechanisms of heat transfer; either momentum transfer due to local collisions or diffusion (the transfer of heat via molecular interactions) and radiation, where the transfer of heat is realized via photons, which can be conceived as quasi-particles similar as phonons or as electromagnetic waves. Convection, being the bulk transport of a fluid, is not precisely a heat transfer mechanism. The physics of radiation transport is distinctly different than those of diffusion transport. The diffusion of particles in gases or fluids and of quasi-particles (phonons) in solids is a local phenomenon, meaning that the rate of heat transfer, at a point in space, precisely depends only on the local nature about the point, i.e., the temperature gradient and its thermo-physical properties at this point. Of course, the temperature field also depends on the boundary and initial conditions imposed on the system. Radiation, on the other hand, is not confined to a local point; the flux of radiation at a point, in general, directly, and instantaneously dependents on the radiation flux from all other visible points in a system. Unlike diffusion, radiation thus acts over a distance. Accordingly, the mathematical description of radiation transport employs an integral equation for the radiation field balancing all in- and output, as opposed to the differential equation for heat diffusion via thermal conduction.

4.3.1 Fundamental Radiation Properties

Radiation can be conceived as the transport of energy by discrete photons, and the basic relationship between the energy of a photon e, the frequency ν or wavelength $\lambda = c_0/\nu$ of the wave is provided by Planck's relation,

$$e = h \cdot \nu = \frac{h \cdot c_0}{\lambda}\cdot, \qquad (4.161)$$

where h is Planck's constant $\left(h = 6.6261 \cdot 10^{-34}\mathrm{J/s}\right)$ and c_0 the speed of ligth in vacuum $c_0 = 2.9979 \cdot 10^8\mathrm{m/s}$. Thus, radiation is quantized. Thermal radiation occurs in a range of the electromagnetic spectrum of energy emission, and the full electromagnetic spectrum includes an enormous range of energy-bearing waves, of which the thermal radiation covers only a small part parts as Fig. 4.28 illustrates. From the graph it becomes visible that thermal radiation refers to the spectrum of radiation from the ultra-violet range via the visible spectrum ($\lambda = 0.4 - 0.7\,\mu\mathrm{m}$) to the infrared (IR) or far-IR wavelengths. Comparing these wavelengths with the atomic ones it becomes clear that by solar radiation the rotational and vibrational energy levels of molecules can be excited and thus energy is transferred by the solar spectrum to the molecules in the form of heat. This has been subject of Chap. 3 (optical conversion). There it could be demonstrated that by selective absorption of thermal radiation a raise of the temperature is achieved. In turn this means, thermal radiation emission is a direct result of vibrational and rotational motions of molecules, atoms, and electrons of a substance and temperature is a measure of these activities. Thus, the rate of thermal radiation emission increases with increasing temperature. Additionally, thermal radiation is a volumetric phenomenon. However, for opaque solids such as metals, radiation is considered to be a surface phenomenon, since the radiation emitted by the interior region never reaches the surface.

A body that emits the maximum possible radiation power is called a "blackbody". The radiation emitted by a blackbody has a wavelength distribution. In addition to the extra-terrestrial formulation, se Chap. 2, in which the gas density does not play a role, the refractive index n occurs in a body and the specific

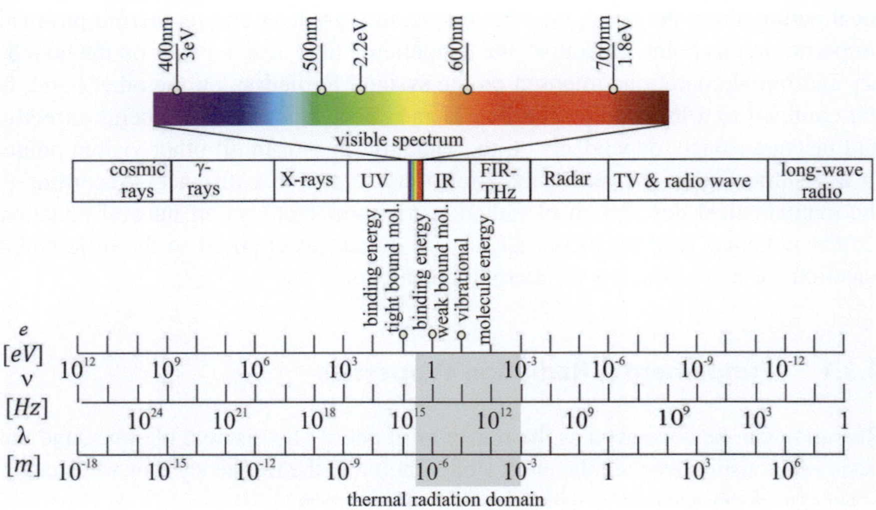

Fig. 4.28 Thermal radiation spectrum as function molecule energy e $[eV]$, wavelength λ $[m]$ and frequency ν $[Hz]$

spectral radiosity $e(\lambda, T)$ in $[\mathrm{W}/(\mathrm{m}^2\mu\mathrm{m})]$ is given in complete form by:

$$e_{b\lambda}(\lambda, T) = \frac{2\pi\ h\ c^2}{n^2 \cdot \lambda^5}\ \frac{1}{\exp\left(\frac{hc}{\lambda k_B T}\right) - 1}. \tag{4.162}$$

in which k_B is the Boltzmann constant ($k_B = 1.36 \cdot 10^{-23} J/K$).

4.3.2 Blackbody and Its Concept

In the following, a "blackbody" is considered. Physically, such a "blackbody" is an idealized body that completely absorbs electromagnetic radiation incident on it at any wavelength; it is not transparent to any radiation through and does not reflect or scatter anything. At the same time the emissivity is maximal at all wavelengths. Thus, any real body cannot emit more thermal radiation than a blackbody at any wavelength. It is the ideal thermal radiation source. Moreover, its spectrum depends *only* on its temperature. In particular, material properties have no influence. Thus, it is the reference source for numerous theoretical and practical purposes. The total radiation emitted by the surface a blackbody e_b per unit area into half-space follows from the integration of Planck's relation over all wavelengths from $\lambda = 0$ to $\lambda \to \infty$ and yields the Stefan-Boltzmann law:

$$e_b = \int_{\lambda=0}^{\lambda\to\infty} e_{b\lambda}(\lambda, T) d\lambda = \sigma T^4 \left[\frac{W}{m^2}\right], \tag{4.163}$$

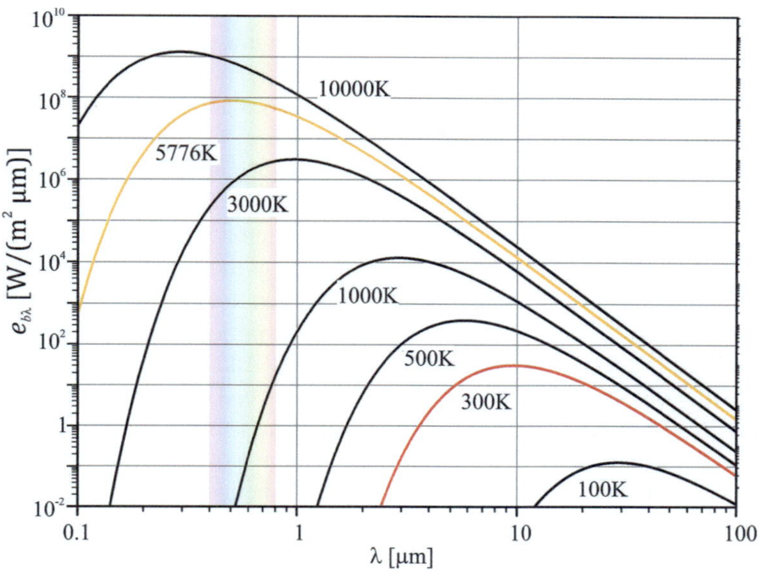

Fig. 4.29 Blackbody spectral power distribution $e_{b\lambda}$ as a function of the wavelength at different tempratures

in which $\sigma = 5.67 \cdot 10^{-8} \mathrm{W}/\left(\mathrm{m}^2 \mathrm{K}^4\right)$ is the Stefan-Boltzmann constant. A plot of $e_{b\lambda}$ as a function of the wavelength λ is illustrated in Fig. 4.29 on a double logarithmic scale and reveals to essential aspects. An increase of the temperature T yields a a rise of the spectral emissive power at all wavelengths and moreover the wavelength corresponding to the maximum power shifts towards the shorter wavelengths. The value of the wavelength at which the maximum occurs $\lambda_{\max}$ is predicted by the Wien displacement law,

$$\lambda_{\max} = \frac{2898}{T} \quad [\mu \mathrm{mK}]. \tag{4.164}$$

An ideal black body cannot be realized technically. There are no materials that completely absorb the electromagnetic waves independent of frequency. Even the production of a body that comes close to the ideal black body is difficult. A sooty surface has an absorptivity of about $\alpha \approx 0.96$ and is sufficient for simple purposes, amorphous graphite absorbs about 99% of the incoming radiation. For practical applications only, the absorption and emission properties of the radiation source are of interest, but not its shape. Therefore, instead of a surface, the aperture of a cavity radiator or simply a long blind hole is used. This allows a better representation of the ideal properties of a blackbody radiator, even if the inner surfaces have a low emissivity. Most real substances, on the other hand, reflect a portion of the incident radiation ($\rho > 0$) and absorb only a portion described by the absorptance or absorption coefficient $\alpha < 1$. Likewise, the emissivity ε differs from 1.

Although the blackbody emissive power from a surface has as quantity the units of [W/m^2] being the same as a heat flux, but it is *not* a vector quantity which can be resolved into directional components. The emissive power describes the energy leaving a surface per unit area of the surface and does not describe the directional characteristics of the radiation as it leaves the surface.

In order to describe the directional propagation of radiation through space, we have to introduce a quantity referred to as the radiance I. The radiance is defined by means of the Fig. 4.30 in which radiation is emitted from a small surface element ΔA and travels in all directions. A fraction of the radiation hit a small area element on a hemisphere enveloping ΔA, being denoted as dA. The net amount of energy $d\dot{Q}$ falling on dA is proportional to the projected area of the emitting surface which is $(\Delta A \cos\theta)$, and to the target dA. Additionally, it scales inverse square of the distance between source and target. i.e. $1/r^2$. The latter originates from the fact that the area of the hemisphere scales as r^2 and the total power emitted to the hemisphere, which is equal tot the total power emitted by the source is constant. Hence the power per unit area on the hemisphere goes as $1/r^2$. The target area dA is given by the distance r and the polar coordinates θ, ϕ:

$$dA = r^2 \sin\theta d\theta d\phi. \tag{4.165}$$

Now, $d\dot{Q}$ is proportional to dA/r^2, and

$$\frac{dA}{r^2} = \sin\theta d\theta d\phi \equiv d\Omega \tag{4.166}$$

defines a differential solid angle $d\Omega$ with steradian (sr) as unit. Because the surface area A of a sphere is $\left(4\pi r^2\right)$, the definition implies that a sphere subtends 4π steradians ($\approx 12.56637 sr$) at its centre. By means of the solid angle definition, the power emitted towards the target gets proportional to the projected area of

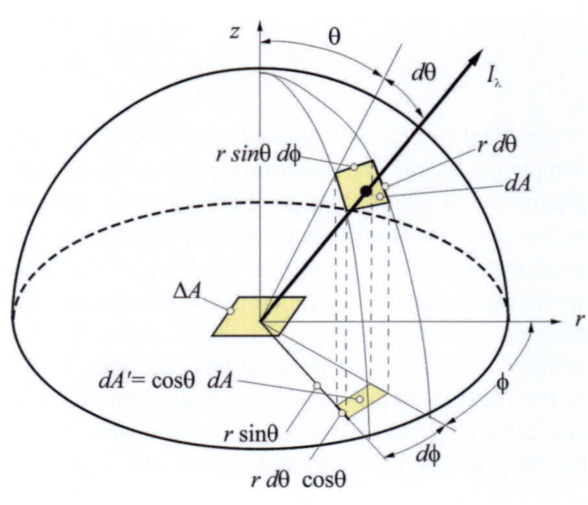

Fig. 4.30 Sketch do define the radiation intensity I

the source and the solid angle subtended by the target with respect to the source. Thereby, the power is not only be proportional to the magnitude but also the directionality of the radiation leaving the source and propagating through space. Since the power $d\dot{Q}$ is in $[W]$, and $d\dot{Q}$ is proportional to $(\Delta A \cdot \cos\theta \cdot d\Omega)$, it follows that radiation propagating through space must have units of $[W/(m^2 sr)]$. This quantity is called radiance L and is defined

$$L = \frac{d\dot{Q}}{\Delta A \cdot \cos\theta \cdot d\Omega}. \quad (4.167)$$

Although the radiance L displays the directional distribution of the propagating radiation, it is not a vector rather the intensity is a scalar that is dependent on the directional coordinates θ and ϕ together with the spatial coordinates and time.

Inherently, we assumed that the radiance L (Eq. 4.167) is a wavelength–integrated quantity. Thus, the spectral radiance L_λ is defined similarly but on a per–unit–wavelength basis in the way,

$$L_\lambda = \frac{d\dot{Q}}{\Delta A \cdot \cos\theta \cdot d\Omega \cdot d\lambda}, \quad (4.168)$$

and the integral of L_λ over all wavelength corresponds to L.

If the ΔA emits as a perfect blackbody the radiance distribution leaving the surface is independent of direction, which means that the blackbody radiance L_b landing on the hemisphere in Fig. 4.30 isn´t a function of θ. (Note that the power per unit area landing on the hemisphere $(d\dot{Q}/dA)$ is a function of θ.) If the total power leaving the black source is $\dot{Q} = e_b \Delta A$ and Eq. (4.167), then the total power landing on the target the hemisphere is,

$$\begin{aligned}\Phi = \dot{Q} = e_b \Delta A = \int d\dot{Q} = L_b \Delta A \int\limits_{\Omega=4\pi} \cos\theta d\Omega \\ = L_b \Delta A \int\limits_0^{4\pi} d\phi \int\limits_0^{\pi/2} \cos\theta \sin\theta d\theta = \pi L_b \Delta A.\end{aligned} \quad (4.169)$$

Here, we used $L = L_b = const$. Since all blackbody radiators emit diffusely, the power density of a blackbody is,

$$e_b = \pi L_b, \quad (4.170)$$

and in the case of monochromatic radiation:

$$e_{b\lambda} = \pi L_{b\lambda}. \quad (4.171)$$

Assuming an ideally diffuse surface ΔA with an isotropic distribution of radiance L in all directions in half-space, i.e. that of a blackbody, and substituting

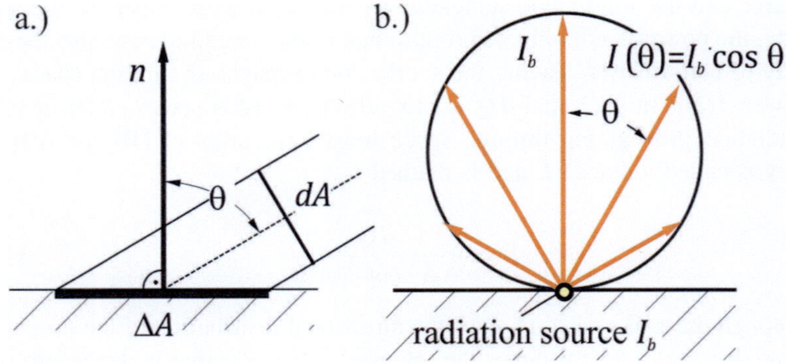

Fig. 4.31 **a** Independence of the emitted radiant intensity (radiant intensity I) from the direction of observation. **b** Radiance I over a diffusely radiating surface according to Lambert's cosine law

this into Eq. (4.167) using (4.166), we obtain the intensity $I(\theta)$ in $[W/sr]$ of the black surface ΔA in direction θ around the solid angle $d\Omega$, which is illustrated in Fig. 4.31a.

$$I(\theta) = \frac{d\dot{Q}}{d\Omega} = \Delta A \cdot L_b \cdot \cos\theta = I_b \cdot \cos\theta. \tag{4.172}$$

This law is called Lambert's cosine law. The radiation intensities I plotted as correspondingly long arrows above an area element of such a radiator come to rest with their tips on a sphere, as Fig. 4.31 shows. A good approximation for a Lambertian radiator is paper (for visible light!). Small air pockets between the fibres form scattering centres for the visible light. If they are missing, for example by impregnating the paper with water, paper loses its reflective properties and becomes transparent.

The intensity being a scalar describes the directional distribution of radiation energy transfer, while the radiative heat flux $\vec{q}_R$, is a vector representing the net direction and magnitude of radiant energy transfer in space. Consider a small surface element in space, dA, oriented so that its normal vector points in the positive z-direction, and radiation, arriving from all directions with an intensity distribution $I(\theta, \phi)$, passes through the element, then the net rate at which energy dQ is transported across the surface, is

$$d\dot{Q} = dA \int_{\phi=0}^{\phi=2\pi} \int_{\theta=0}^{\pi} I(\theta, \phi) \cos\theta \sin\theta d\theta d\phi. \tag{4.173}$$

It is important that the term $\cos\theta$ appears because $(dA\cos\theta)$ is the projected area of the element with respect to the θ-direction. The quantity $d\dot{Q}$ can be either positive, negative, or zero. The sign indicates the direction of a net transfer across dA being positive in positive z-direction. It is essential to note that a zero value does not necessarily imply that $I(\theta, \phi) = 0$, it rather means that all positive and negative contributions annihilate. This occurs if the intensity distribution is isotropic, i.e., independent of direction.

If one calculates for a black body according to Stefan-Boltzmann law (Eq. 4.163) the power of the emitted radiation and sets for the temperature $392.8^\circ K$, one gets for the solar constant a value of $i_{SC} = 1357 W/m^2$, which nearly corresponds to the value of the solar constant (see Chap. 2). A black-body on an extra-terrestrial position with one side facing perpendicular to the sun and the other surfaces ideally insulated could thus not reach a temperature above about $120^\circ C$. However, there are two ways to exceed that temperature limitation. The first is concentrating mirror systems that increase the irradiance on the target body; the second is to use selective surfaces that absorb well in the solar radiation wavelength range but emit little in the thermal radiation wavelength range at solar collector operating temperatures.

4.3.3 Radiation Balance

Most real substances, reflect part of the incident radiation expressed by coefficient named reflectance ρ being $\rho > 0$ and absorb only a portion described by the absorptance or absorption coefficient $\alpha < 1$. Similarly, the emissivity ε differs from unity. A grey radiator is a body whose surface does not completely absorb the incident radiation and therefore does not radiate the full radiation back into space at a given temperature. The numerical value of how "grey" the surface is expressed by the absorptivity α (- wich can equal the emissivity ε, see later Kirchhoff's law). An ideal gray radiator, which does not exist in reality, describes a body whose emissivity ε does not change with the wavelength l (see also in the following section).

Moreover, if a body is transparent, a part of the radiation will pass through it, as depicted in Fig. 4.32, which is characterized by the transmittance τ. The part of the radiation that is incident on the body will be the same as the part of the radiation that is not incident on the body. The fraction of the incident radiation that is reflected from the surface is given by the reflectance. Since only an equal or lesser intensity can be reflected back, the reflectance ρ is always $\rho < 1$. In transparent bodies, some electromagnetic (or thermal) radiation can pass through the body. This transparency by the transmittance τ characterizing the part of radiation passing the body. An energy balance of the surface element in Fig. 4.32 states that the energy incident on the surface element $(I \cdot dA)$ must correspond to the portions that leave the surface element by reflection and transmission and lead to heating by absorption. Assuming thermodynamic equilibrium, the first law of thermodynamics leads to the following relationship:

$$\alpha + \tau + \rho = 1. \tag{4.174}$$

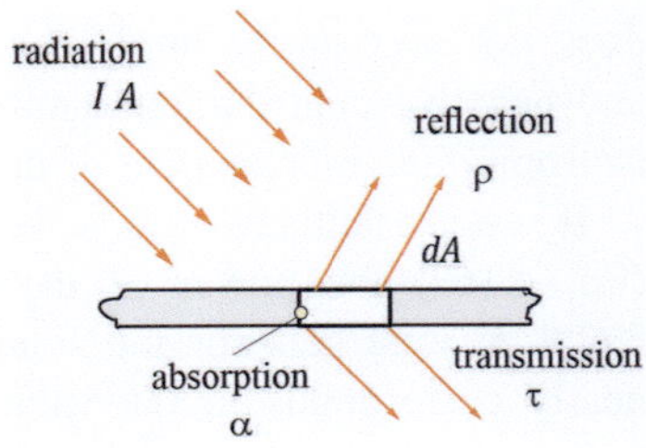

Fig. 4.32 Sketch of the radiation balance on a real body

The absorption coefficient α is similar in its behavior to the emission coefficient ε. For example, a surface with an emission or absorption coefficient of $\alpha = 0.5$ absorbs half of the incident radiation, but at a given temperature also emits only half of the thermal radiation compared to a blackbody. Heat radiation can be reduced by using bare metal surfaces. i.e. metal layers on rescue blankets and insulating bags, mirroring in thermos flasks and super insulation. To increase the thermal radiation of a metallic body that reflects light well, it can be covered with a "dark" matter coating in the relevant wavelength range of thermal radiation (infrared range). Technically, this is used, for example, in painting radiators with paint (which emits radiation well in the infrared range), in black anodizing aluminium heat sinks in computers and electrical equipment, or in enamelling stove pipes and metal furnaces.

The relation (4.174) suggests that the absorptance α, the transmittance τ and the reflectance ρ are constants, which is unfortunately not the case. Since thermal radiation, like solar radiation, has a spectral behaviour, it is also dependent on location, wavelength, and temperature.

Opaque bodies do not let any radiation through, so that for their transmittance $\tau = 0$ is valid. Depending on the body, the radiation can be reflected at the interface to different degrees. This reflection at the surface can be specular or diffuse, as outlined in Fig. 4.33. If the angle between the surface normal and the incident ray is equal to the angle of the outgoing ray and surface normal (incidence angle = emerging angle, i.e. $\theta_1 = \theta_2$), it is a specular reflection. In the case of ideal diffuse reflection, the reflected radiation is isotropically distributed in intensity in all directions in the half-space above the reflecting surface (compare Fig. 4.33b). Real surfaces reflect are neither ideally specular nor ideally diffuse, and a space-dependent reflection results, as illustrated by Fig. 4.33c. Well-polished surfaces come very close to the ideal case with a specular reflection behaviour, while very rough surfaces tend to a homogeneous diffusive reflection.

Thus, an analysis of the thermal radiation exchange requires a characterization of the surface properties of the surfaces involved in the radiation problem.

4.3.4 Surface Properties and Radiation

As recognized in the context of the radiation balance of a body surface can emit, absorb, and reflect radiation and we have assigned to the surface coefficients

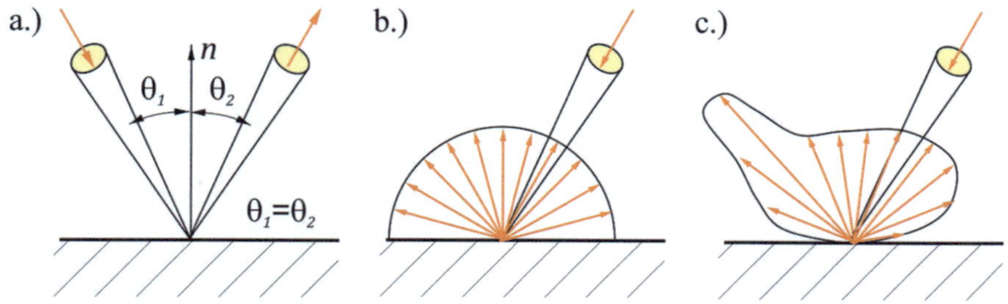

Fig. 4.33 Reflection at an ideally specular (**a**), at an ideally diffuse (**b**) and at a real, mixed reflecting surface (**c**)

describing absorptivity and reflectivity, i.e., the absorptivity represents the fraction of incident radiation on a surface that is absorbed by the surface. Nonetheless, more precise definitions need to be developed to account for the directional and spectral dependencies of the radiation falling on or emitted by a surface.

4.3.4.1 Emissivity

One of the advantages of a blackbody is that the intensity emitted is independent of direction and given by $I_b = e_b/\pi$. Likewise, the spectral (or wavelength–dependent) blackbody intensity is independent of direction and given by $I_{b\lambda} = e_{b\lambda}/\pi$. However, a real body surface emits a spectral intensity into some direction θ and ϕ and within a small wavelength interval about λ, whichs always be less than the spectral blackbody intensity for a surface at the same temperature. We can therefore define the spectral directional emissivity $\varepsilon'_\lambda = I_{e\lambda}(\theta, \phi)/I_{b\lambda}$, in which $I_{e\lambda}$ is the emitted intensity from the surface. This spectral directional emissivity represents the most irreducible information on the emission properties of a surface and a function of direction, wavelength, temperature as well as the physical and chemical properties of the surface, but much too complex for technical application. By integration of ε'_λ over the unit sphere the spectral hemispherical emissivity ε_λ is obtained which represents the directional average. It is defined by:

$$\varepsilon_\lambda = \frac{\int_{\phi=0}^{\phi=2\pi} \int_{\theta=0}^{\pi/2} I_{e\lambda}(\theta, \phi) \cos\theta \sin\theta d\theta d\phi}{\int_{\phi=0}^{\phi=2\pi} \int_{\theta=0}^{\pi/2} I_{b\lambda} \cos\theta \sin\theta d\theta d\phi}$$
$$= \frac{\int_{\phi=0}^{\phi=2\pi} \int_{\theta=0}^{\pi/2} \varepsilon_\lambda{}'(\theta, \phi) \cos\theta \sin\theta d\theta d\phi}{\pi} = \frac{e_\lambda}{e_{b\lambda}}. \tag{4.175}$$

That is,ε_λ is the ratio of the emitted spectral power from the surface, e_λ to the blackbody power $e_{b\lambda}$. Likewise, the total directional emissivity ε' is obtained by an integration over all wavelength to obtain a wavelength average by

$$\varepsilon' = \frac{\int_{\lambda=0}^{\infty} I_{e\lambda} d\lambda}{\int_{\lambda=0}^{\infty} I_{b\lambda} d\lambda} = \frac{\int_{\lambda=0}^{\infty} \varepsilon'_\lambda I_{b\lambda} d\lambda}{I_b}. \tag{4.176}$$

This total directional emissivity is of technical importance and observed in many practises. Figure 4.34 illustrates the directional emissivity as a function of the incidence angle for various electrically non-conducting materials. The deviation of the emissivity from that of a Lambertian radiator is particularly evident for the non-conductors in the region of large angles θ. Because of the comparatively low radiation power in this angular range, the technical impact remains marginal. The assumption of ideally diffuse radiating surfaces is therefore reasonably well fulfilled for electrical non-conductors.

Metallic bare surfaces are far from a Lambert radiator behaviour as Fig. 4.35 illustrates, where a clear dependence of the total directional emissivity can be seen.

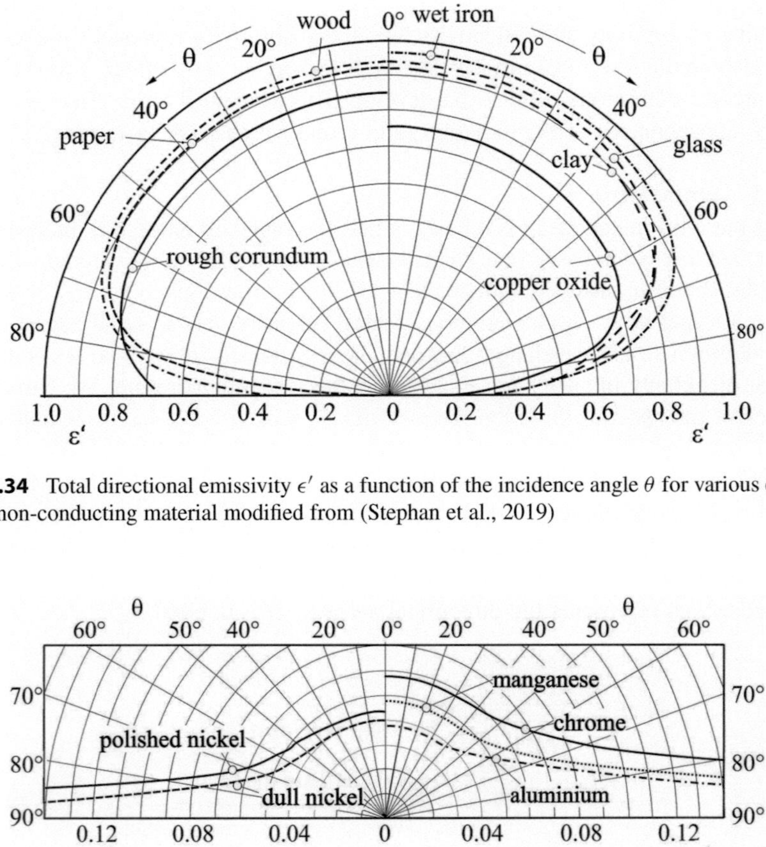

Fig. 4.34 Total directional emissivity ϵ' as a function of the incidence angle θ for various electrically non-conducting material modified from (Stephan et al., 2019)

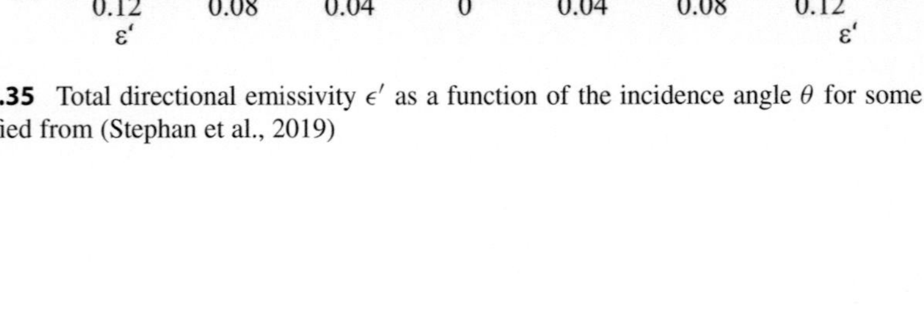

Fig. 4.35 Total directional emissivity ϵ' as a function of the incidence angle θ for some metals modified from (Stephan et al., 2019)

Finally, the total hemispherical emissivity ε is obtained from either a directional average of ε' or a wavelength average of ε_λ;both yield the same result, which is:

$$\varepsilon = \frac{\int_{\lambda=0}^{\infty}\int_{\phi=0}^{\phi=2\pi}\int_{\theta=0}^{\pi/2} I_{e\lambda}(\theta,\phi)\cos\theta\sin\theta d\theta d\phi d\lambda}{\int_{\lambda=0}^{\infty}\int_{\phi=0}^{\phi=2\pi}\int_{\theta=0}^{\pi/2} I_{b\lambda}\cos\theta\sin\theta d\theta d\phi d\lambda}$$

$$= \frac{\int_{\phi=0}^{\phi=2\pi}\int_{\theta=0}^{\pi/2}\varepsilon'(\theta,\phi)\cos\theta\sin\theta d\theta d\phi}{\pi} = \frac{\int_{\lambda=0}^{\infty}\varepsilon_\lambda I_{b\lambda}d\lambda}{I_{b\lambda}} = \frac{e}{e_b}. \tag{4.177}$$

4.3.4.2 Absorptivity

Let us assume an opaque surface ($\tau = 0$) exposed to an incident source of spectral intensity given by $I_\lambda(\theta,\phi)$ shining on its surface.When the radiation strikes the surface a fraction of it will be absorbed by the surface, the remainder will be reflected. Defining the absorbed intensity as $I_{\alpha\lambda}$ the spectral directional absorptivity α'_λ is defined by:

$$\alpha'_\lambda = \frac{I_{\alpha\lambda}}{I_\lambda}, \tag{4.178}$$

and it describes the fraction of incident spectral intensity absorbed by the surface. The spectral directional absorptivity describes the fundamental absorption properties of the surface. Further, similar with as the emissivity averages can be introduced with the direction, wavelength, or both by the corresponding integrations.

The spectral hemispherical absorptivity α_λ writes to:

$$\alpha_\lambda = \frac{\int_{\phi=0}^{\phi=2\pi}\int_{\theta=0}^{\pi/2} I_{\alpha\lambda}(\theta,\phi)\cos\theta\sin\theta d\theta d\phi}{\int_{\phi=0}^{\phi=2\pi}\int_{\theta=0}^{\pi/2} I_\lambda\cos\theta\sin\theta d\theta d\phi}$$

$$= \frac{\int_{\phi=0}^{\phi=2\pi}\int_{\theta=0}^{\pi/2}\alpha'_\lambda(\theta,\phi)\mathrm{I}_\lambda(\theta,\phi)\cos\theta\sin\theta d\theta d\phi}{q_\lambda} = \frac{q_{\alpha\lambda}}{q_\lambda}, \tag{4.179}$$

in which q_λ ist he spectral flux in direction of the surface and $q_{\alpha\lambda}$ the spectral absorbed flux. The total directional absorptivity α' is:

$$\alpha' = \frac{\int_{\lambda=0}^{\infty} I_{\alpha\lambda}d\lambda}{\int_{\lambda=0}^{\infty} I_\lambda d\lambda} = \frac{\int_{\lambda=0}^{\infty}\alpha'_\lambda I_\lambda d\lambda}{I}, \tag{4.180}$$

and, finally, the total hemispherical absorptivity α is:

$$\alpha = \frac{\int_{\lambda=0}^{\infty}\int_{\phi=0}^{\phi=2\pi}\int_{\theta=0}^{\pi/2} I_{\alpha\lambda}(\theta,\phi)\cos\theta\sin\theta d\theta d\phi d\lambda}{\int_{\lambda=0}^{\infty}\int_{\phi=0}^{\phi=2\pi}\int_{\theta=0}^{\pi/2} I_\lambda\cos\theta\sin\theta d\theta d\phi d\lambda}$$

$$= \frac{\int_{\phi=0}^{\phi=2\pi} \int_{\theta=0}^{\pi/2} \alpha'(\theta,\phi) I(\theta,\phi) \cos\theta \sin\theta d\theta d\phi}{q} = \frac{\int_{\lambda=0}^{\infty} \alpha_\lambda q_\lambda d\lambda}{q} = \frac{q_\alpha}{q}. \tag{4.181}$$

At a first glance the definitions look quite similar to those of the emissivity, however, there are especially with respect to the solar thermal application fundamental differences. In the emissivity relations for ε_λ and ε Eqs. (4.175) and (4.177) definition the blackbody intensity I_b appears which is not a function of the direction, which allowed for a removal from integral over direction. But, with respect to the absorptivity the incident intensity irrespective if it is spectral (I_λ) or total (I) is, in general, a function of direction, and it cannot be removed from the corresponding definition (Eqs. 4.179, 4.181) for the hemispherical absorptivity.

Thus, it is important to emphasize that the thermal emission by radiation is a function of the surface only, while regarding the absorption the directional intensity distribution falling on a surface as well as the wavelength dependent absorptivity α_λ play a role. As illustrated already in Chap. 3, a surface illuminated by the sun from the normal direction ($\theta = 0°$) has a different hemispherical absorptivity than the same surface illuminated by the sun at an oblique angle ($\theta > 60°$), or by the sun on a cloudy day, which is dominated by diffuse illumination. Likewise, the total absorptivity (either directional or spectral) is dependent on the spectral distribution of the incident radiation. A spectral selective of a solar thermal receiver exposed to visible light exhibits a total absorptivity and correspondingly a high temperature, whereas than the same surface illuminated by IR radiation source will barely heat up, since the absorptivity in this wavelength domain is marginal.

4.3.4.3 Reflectivity

The complexity to describe the reflection is one step higher than for the absorption, because not only the incident radiation direction plays a role but also the direction of the outgoing (reflected) radiation. We therefore introduce the superscripts „–" for the incident and „+" to assign the reflected intensity. By means of this notation one can define a spectral bi-directional reflectivity ρ''_λ:

$$\rho''_\lambda(\Omega, \Omega') = \frac{I_\lambda^+(\Omega')}{\pi I_\lambda^-(\Omega)}. \tag{4.182}$$

The spectral directional–hemispherical reflectivity ρ'_λ is obtained by integration of the reflected intensity over the entire hemisphere

$$\rho'_\lambda = \frac{\int_{\phi'=0}^{2\pi} \int_{\theta=0}^{\pi/2} I_\lambda^+(\theta', \phi') \cos\theta' \sin\theta' d\theta' d\phi'}{\pi I_\lambda^-(\theta,\phi)}$$

$$= \frac{1}{\pi} \int_{\phi'=0}^{2\pi} \int_{\theta=0}^{\pi/2} \rho''_\lambda(\theta,\phi,\theta',\phi') \cos\theta' \sin\theta' d\theta' d\phi'. \tag{4.183}$$

All of the radiation incident on the surface is either absorbed by the surface or reflected. Assuming an opaque body ($\tau = 0$), we get:

$$\alpha'_\lambda + \rho'_\lambda = 1. \tag{4.184}$$

4.3.5 Kirchhoff's Law and Grey Bodies

To derive Kirchhoff's radiation law, first assume that a body is in a cavity of temperature T. The body is supposed to be in thermal equilibrium with the cavity. If the cavity walls behave inwardly like a perfect blackbody radiator, they emit radiation e_b in the direction of the blackbody according to the Stefan-Boltzmann law (Eq. 4.163). In the state of equilibrium, a blackbody for $t \to \infty$ attains the same temperature as the cavity walls. The incident radiation at the location of the blackbody with area A is thus ($\dot{q} \cdot A$). In equilibrium, the blackbody emits radiation also according to Eq. (4.163), which is given by the product ($e_b \cdot A$). Because of the conservation of energy, the power absorbed by the body must be equal to its emitted power, so that holds:

$$e_b \cdot A = \dot{q} \cdot A. \tag{4.185}$$

If we now consider a real body in the same cavity, it will absorb a lower radiation power by the absorptivity α ($\alpha < 1$). Furthermore, the real body emits a radiation power lower by the factor ε ($\varepsilon < 1$) than a blackbody. In temperature equilibrium, the absorbed radiation must be equal to the emitted one, so that:

$$\underbrace{\varepsilon \cdot e_b \cdot A}_{\substack{\text{emitted}\\\text{power}}} = \underbrace{\alpha \cdot \dot{q}}_{\substack{\text{absorbed}\\\text{power}}} \cdot A. \tag{4.186}$$

Up to this point, no restrictions have been made with regard to the spectral, spatial and temperature behaviour of the body. As discussed in the context of the reflectivity ρ, both emissivity and absorptivity depend on the wavelength (λ), the solid angles (θ, ϕ) and the temperature (T). In order to maintain the thermal equilibrium, however, the real body must re-emit the absorbed amount of energy at the same wavelengths in the same directions in order to replace the energy taken from the black cavity ("Hohlraum"). The division of Eq. (4.179) with Eq. (4.179) thus results in,

$$\frac{e_\lambda}{e_{b\lambda}} = \alpha'_\lambda(\theta, \phi, \lambda, T) = \varepsilon'_\lambda(\theta, \phi, \lambda, T), \tag{4.187}$$

where θ, ϕ are the solid angles in sphere. The key statement of Kirchhoff's law is thus:

In thermal equilibrium, for the same frequencies and directions, the directional spectral absorptivity and the directional spectral emissivity are equal.

Technically, this means that good absorbers are also good emitters. Kirchhoff's radiation law initially applies only in thermal equilibrium; for the most part, however, it is also applicable to a good approximation for bodies that are not in thermal equilibrium with their surroundings, as long as their directional spectral absorption and emission coefficients do not change under these conditions.

The identity of absorption and emission coefficients applies *only* to the directional spectral absorptivity and the directional spectral emissivity. These quantities, which describe the explicit directional and wavelength dependence of the absorption and emission processes, are not known or available for many materials. Usually, for a specific material, the hemispherical spectral emission coefficient $\varepsilon_\lambda(\lambda, T)$ integrated over all directions of the half-space or the total hemispherical emissivity ε, which is still a function of the temperature. The equality of the integrated absorptivity with emissivity holds only in a few special cases, since the integrated absorptivities also depend on the directional and wavelength distribution of the incident radiation. This means that total absorptivities and emissivities are not pure material properties.

Kirchhoff's radiation law is valid in the following cases:

- For ideally diffusely radiating surfaces the emissivity is direction-independent and we get (Baehr & Stephan, 2008):

$$\alpha_\lambda(\lambda, T) = \varepsilon_\lambda(\lambda, T), \tag{4.188}$$

- for ideal grey bodies $\varepsilon \neq \varepsilon(\lambda)$ and thus

$$\alpha(\theta, \phi, T) = \varepsilon(\theta, \phi, T), \tag{4.189}$$

- for diffuse and grey radiating bodies

$$\alpha(T) = \varepsilon(T). \tag{4.190}$$

Real bodies are often diffuse radiators to a good approximation, while the requirement for grey radiating surfaces can be poorly met. Non-metals (i.e. electrical insulators, dielectrics) behave approximately like diffuse emitters with few exceptions. Moreover, in many cases their directional spectral emissivity is reasonably constant for wavelengths above about $1 - 3\mu m$. Thus, for radiative transfer in the long-wavelength region (especially thermal radiation at temperatures that are not too high), dielectrics can often be treated approximately as diffuse grey radiators, and Eq. (4.190) applies. In the case of metals, on the other hand, the directional dependence of the emissivity usually does not allow an approximation

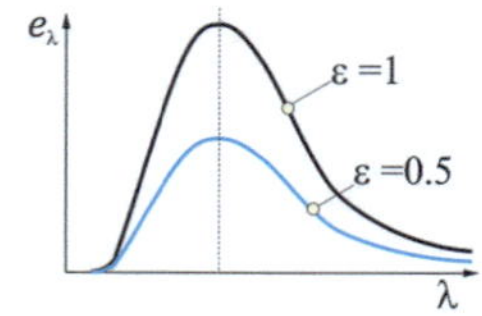

Fig. 4.36 Schematic representation of the radiation intensity ε_λ as a function of the wavelength λ for a black and a diffuse and gray radiator ($\varepsilon = 0.5$)

by a diffuse radiator. In addition, their spectral emissivity is not constant even at long wavelengths, so they are not grey radiators either; thus, Eq. (4.190) is not satisfied. Oxide layers or contamination, however, can strongly change the radiation properties of metals and make them approach those of dielectrics. Important for the validity of Kirchhoff's law is the assumption of thermodynamic equilibrium as well as the consideration of the radiation properties of the body. If it is fulfilled, the radiation emission of a diffuse grey radiator is reduced by the factor ε compared to a black radiator; it results:

$$e_{\varepsilon<1} = \varepsilon \cdot \sigma T^4. \tag{4.191}$$

For the case of a diffuse and grey emitting radiator, the spectrum is shifted downward by a factor e over all wavelengths, as schematically sketched in Fig. 4.36. However, real surfaces not only emit less than a blackbody, but their emissivity varies with wavelength. It is therefore often necessary to start from the monochromatic properties and normalize to the blackbody as shown in Eq. (4.175) and then to integrate over all wavelengths.

The grey assumption can fail miserably, however, when the incident radiation has a significantly different spectra than the emitted radiation. In the context here, as an example sunlight is falling on a solar collector. The radiation spectrum of sunlight is similar to that of a blackbody with a temperature of $T \approx 5800K$, which is concentrated to a considerable fraction in the visible wavelengths. Solar collectors, on the other hand, typically operate at a temperature of around $T \approx 350K$, and correspondingly their thermal emission at this temperature is concentrated in the mid IR wavelengths. For such conditions, a collector with a high spectral absorptivity in the visible yet a small spectral emissivity in the IR is desirable.

4.3.6 Radiosity and Irradiance

By now we considered only the radiation leaving or accepted by a surface, which is of rather academic nature. Technically relevant, however, is the knowledge of the net rate of heat transfer to each surface in the domain considered. Thus, in the next step we assume N surfaces which can exchange radiation heat, and for simplicity we consider their surfaces to be diffuse so that the emissivity and absorptivity are not a function of the direction, but to maintain generality, the properties are initially taken to be wavelength–dependent, so that $\alpha \neq \varepsilon$.

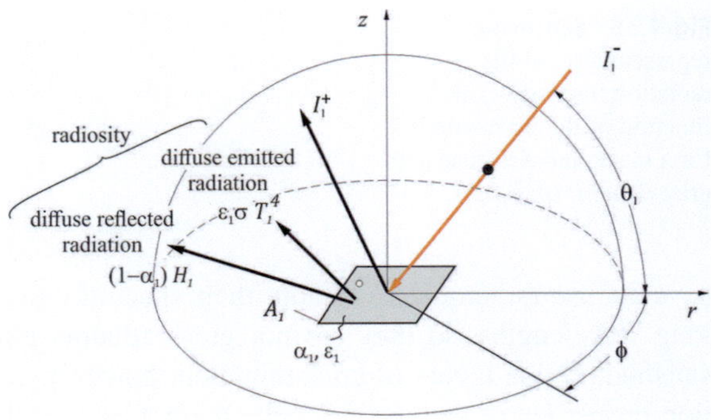

Fig. 4.37 Incoming and emerging energy for a diffuse surface A_1

In order to define the basic quantities of radiation heat exchange we consider a surface 1, with an area A_1 and a uniform constant total hemispherical emissivity ε_1 and absorptivity α_1, see Fig. 4.37.

The irradiance H_1 in $[W/m^2]$ is the flux of radiant energy reaching the surface averaged over the surface area and the total radiant energy falling on the surface is thus

$$H_1 \cdot A_1 = \int_{A_1} \int_{\Omega} I_1^- \cos\theta_1 d\Omega \, dA_1, \tag{4.192}$$

where I_1^- represents the intensity falling on the surface A_1 depending on the direction (θ, ϕ) of the incident radiation. The radiosity J_1 decribes the surface averaged radiant energy flux emerging the surface:

$$J_1 \cdot A_1 = \int_{A_1} \int_{\Omega} I_1^+ \cos\theta_1 d\Omega dA_1, \tag{4.193}$$

where I_1^+ denotes the intensity leaving the surface A_1. The radiosity is connected to the irradiance H_1 by:

$$J_1 = \varepsilon_1 \cdot e_{b1} + \rho_1 \cdot H_1 = \varepsilon_1 \cdot e_{b1} + (1 - \alpha_1) \cdot H_1. \tag{4.194}$$

Thus, the radiosity consists of two parts, the emission from its own surface, which is given by $e_1 = \varepsilon\sigma T_1^4$, complemented by the reflected $(\rho_1 = (1 - \alpha_1))$ fraction of the incident radiation. The irradiance H_1 depends on the incoming radiation field, which is incident on surface A_1 from all surrounding surfaces at a "viewing angle". Hence, radiosity J_1 represents the total radiation energy streaming from a surface, per unit area per unit time, and is the summation of the reflected and the emitted radiation.

The net average heat flux of the surface $(\dot{q}_1)$ is the difference of the flux arriving and the one emerging from the surface,

$$\dot{q}_1 = J_1 - H_1 = \varepsilon_1 \cdot e_{b1} - \alpha_1 \cdot H_1, \tag{4.195}$$

and thus, relates the emissive flux from the surface to the absorbed incident flux. The net balance ($\dot{q}_1 \cdot A_1$) corresponds to the net rate of the heat transfer from/to the surface similar as in heat conduction. If $(\dot{q}_1 \cdot A_1) = 0$ the surface is adiabatic otherwise it describes power required tobe supplied to the surface to keep it at the constant Temperature T_1. Eliminating H_1 from the previous equations one obtains

$$\dot{q}_1 = \frac{1}{(1 - \alpha_1)}(\varepsilon_1 \cdot e_{b1} - \alpha_1 \cdot J_1). \tag{4.196}$$

In case of blackbody radiation ($\alpha = \varepsilon = 1$), the radiosity J_1 corresponds to e_{b1} and thus the solution of (4.196) is trivial, however, the essential message here is that the net transfer is associated with a balance equation accounting for emission and absorption (or reflectance).

For surface which are opaque, diffuse, and grey ($\varepsilon_1 = \alpha_1$) and considering them to be isothermal Eq. (4.194) simplifies

$$J_1 = \varepsilon_1 \cdot e_{b1} + (1 - \varepsilon_1) \cdot H_1. \tag{4.197}$$

Using an energy balance, the net rate of radiation heat transfer from a surface 1 of surface area A_1 can be expressed as

$$\dot{Q}_1 = \dot{q}_1 \cdot A_1 = A_1 \cdot (J_1 - H_1) = A_1\left(J_1 - \frac{J_1 - \varepsilon_1 \cdot e_{b1}}{1 - \varepsilon_1}\right) = \frac{A_1\varepsilon_1}{1 - \varepsilon_1}(e_{b1} - J_1). \tag{4.198}$$

In electrical analogy to Ohm's law, a thermal resistance can be defined as:

$$\dot{Q}_1 = \frac{(e_{b1} - J_1)}{R_1} \text{ with } R_1 = \frac{1 - \varepsilon_1}{A_1\varepsilon_1}, \tag{4.199}$$

where R_1 can be conceived as "*surface resistance*" to radiation. Note, that the surface resistance to radiation for a blackbody is zero. For insulated or adiabatic surfaces, the net heat transfer through them is zero. In these cases, the surface is called re-radiating surface and thus there is no net heat transfer to a re-radiating surface.

4.3.7 Radiation Transport Between Bodies

Radiation transport differs from conduction and convection as heat transport mechanisms in that the observed net heat transport cannot be divided into two heat transports only mathematically. The net heat transport from surface 1 to surface 2 is given by the heat transport from 1 to 2 minus the heat transport from 2 to 1. Thus, the thermal radiation transports the heat not only in one direction, as in a diffusive energy transport in a continuum of, for example, heat conduction (from warm to cold), but at the same time the thermal radiation emitted by the colder body also hits the warmer body and can be absorbed by it. In technical heat transfer problems, usually only the effective heat exchange by thermal radiation between two or more bodies with known surface temperatures is of interest. Several methods are available to determine this heat exchange, such as the net radiation method, the zone method, Monte Carlo methods or the method of discrete ordinates. For the sake of clarity, the following derivations are limited to the net radiation method, more details of the complementary calculation methods or a more refined description of the radiative heat exchange can be found in (Modest, 2014).

4.3.7.1 Radiation Exchange Between Gray, Diffusely Radiating Infinite Surfaces

If we consider two plane, grey, equally sized parallel plates 1 and 2, both will emit thermal radiation according to their temperature according to Eq. (4.190) and according to their emissivity ε_i. The radiation emitted by plate 1 will be partly absorbed and partly reflected by plate 2. The radiation emitted by plate 1 will be partly absorbed and partly reflected by plate 2. In the simplest case, the intensity of the reflected beam is lower by the reflection coefficient $\rho_2 = (1 - \varepsilon_2)$. The reflected part of the beam again hits plate 1, where it is in turn partially absorbed and reflected. With each further reflection, another part of the radiation is absorbed. As a result, the radiation energy decreases by $(1 - \varepsilon_i)$ each time, as schematically sketched in Fig. 4.38.

The sum of all absorbed radiation fractions results in a geometric series, which yields for radiated heat flux $\dot{q}_{12}$ from plate 1 to plate 2:

$$\dot{q}_1 = \sum_{k=1}^{k\to\infty} \varepsilon_1 \cdot \varepsilon_2 \cdot (1-\varepsilon_1)^k \cdot (1-\varepsilon_2)^k \cdot \sigma \cdot T_1^4. \tag{4.200}$$

If the areas A_i of the two plates are equal and the length dimension (L) is considerably larger than the gap width, the summation of Eq. (4.200) yields the following result:

$$\dot{q}_1 = \sigma \cdot T_1^4 \cdot \frac{1}{\frac{1}{\varepsilon_1} + \frac{1}{\varepsilon_2} - 1}. \tag{4.201}$$

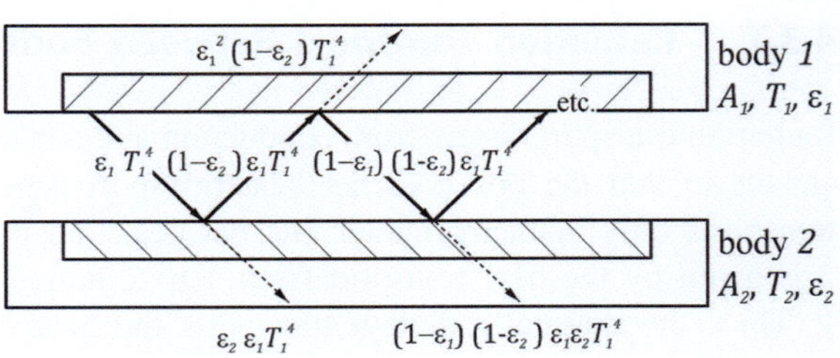

Fig. 4.38 Radiation exchange between two planes, parallel, gray radiating plates

Conversely, there is a radiation flux from the body 2 to the body 1 with the result:

$$\dot{q}_2 = \sigma \cdot T_2^4 \cdot \frac{1}{\frac{1}{\varepsilon_1} + \frac{1}{\varepsilon_2} - 1}. \tag{4.202}$$

Thus, the net radiation balance is the difference of the radiation intensities absorbed by both surfaces:

$$\dot{q}_{12} = \sigma \cdot \frac{\left(T_1^4 - T_2^4\right)}{\frac{1}{\varepsilon_1} + \frac{1}{\varepsilon_2} - 1}. \tag{4.203}$$

In the case of radiation heat exchange between two concentric tubes, the exchange areas are different in size, so that a geometry factor is added if the axial extension of the tubes is much larger than the gap between the tubes. Thus, all emitted radiation is transferred from tube 1 to tube 2 and vice versa. For this purpose, assume that the area of the inner tube is A_1 and A_2 is that of the cladding tube. In this simple case, the heat flow from the inner tube A_1 to the outer tube can be given by the equation below, where the heat flow is related to the area of the inner tube with A_1.

$$\dot{q}_{12} = \frac{\sigma\left(T_1^4 - T_2^4\right)}{\frac{1}{\varepsilon_1} + \frac{A_1}{A_2}\left(\frac{1}{\varepsilon_2} - 1\right)}. \tag{4.204}$$

Obviously, the relative orientation of the radiating surfaces plays a role, which is depicted here by an area ratio.

4.3.7.2 Radiation Exchange Between Diffuse Radiating Bodies

To determine the magnitude of the relative orientation of two radiating bodies, we consider two radiating surfaces A_i and A_j, each with homogeneously distributed surface temperatures T_i and T_j, as shown in Fig. 4.39. Let H_i be the total irradiance in $\left[W/m^2\right]$ incident on A_i and J_i the total radiosity in $\left[W/m^2\right]$ outgoing from A_i (uniformly). Since J_i includes the reflected portion of H_i in addition to the radiation emitted by A_i $\left(\varepsilon_i \sigma T_i^4\right)$, we obtain immediately according to Eq. (4.194).

$$J_i = \varepsilon_i \cdot e_{bi} + \rho_i \cdot H_i = \varepsilon_i \cdot e_{bi} + (1 - \alpha_i) \cdot H_i, \tag{4.205}$$

with $e_{bi} = \sigma T_i^4$. Now only a fraction of the emitted radiosity J_i reaches the surface A_j. The dimensionless number $F_{ij} \in [0, 1]$, which has yet to be determined, is called the view factor (alternatively often used are expression as configuration factors, form factors, angle factors or shape factors). This view factor depends exclusively on the shape and relative position of the surfaces A_i and A_j with

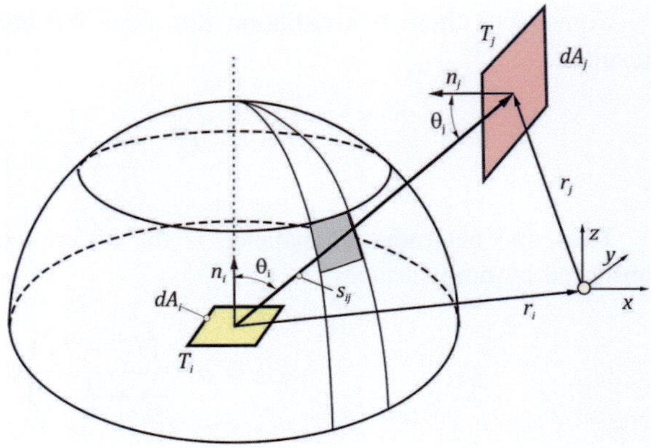

Fig. 4.39 Two surfaces with radiation exchange

respect to each other. The distance vector $\vec{s}_{ij} = \vec{r}_j - \vec{r}_j$ between the two surface elements dA_i and dA_j depends on the angles θ_i and θ_i in the following form:

$$\cos\theta_i = \frac{\vec{n}_i \cdot \vec{s}_{ij}}{\left|\vec{s}_{ij}\right|} \text{ and } \cos\theta_j = \frac{\vec{n}_j \cdot \vec{s}_{ij}}{\left|\vec{s}_{ij}\right|}, \tag{4.206}$$

in which $\vec{n}_i$ and $\vec{n}_j$ denote the area normal vectors of dA_i and dA_j.

The total radiant energy emitted from the surface element dA_i is $(J_i dA_i)$ with the radiosity according to (4.205), of which the fraction $(J_i dA_i \cos\theta_i)$ is emitted in the direction of $\vec{s}_{ij}$. The radiant energy incident on the surface element dA_j is calculated to be $\left(J_i dA_i \cos\theta_i \cos\theta_j\right)$. Thus, the radiant energy incident on the entire surface A_j and emitted from the entire surface A_i is calculated to be:

$$\mathrm{J}_{ji} = \frac{J_i}{\Gamma} \int\limits_{A_i} \int\limits_{A_j} \cos\theta_i \cos\theta_j dA_j dA_i, \tag{4.207}$$

in which Γ represents a normalization factor yet to be determined. Since for a hemisphere of the surface A_j with the radius $\left|\vec{s}_{ij}\right|$ the incident radiation energy J_{ji} must correspond to the radiation energy J_i emitted by the surface A_i for energy conservation reasons the following must apply:

$$\frac{1}{\Gamma} \int\limits_{A_i} \int\limits_{A_j} \cos\theta_i \cos\theta_j dA_j dA_i = 1. \tag{4.208}$$

Evaluating Eq. (4.208) for the hemisphere with radius $\left|\vec{s}_{ij}\right|$, it follows that $\theta_j = 0$ and we obtain:

$$\int\limits_{A_i} \cos\theta_i dA_i = \int\limits_0^{2\pi} \int\limits_0^{\pi/2} \mathrm{s}_{ij}^2 \cos\theta_i \sin\theta_i d\theta_j d\phi = \Gamma. \tag{4.209}$$

From Eq. (4.207) together with (4.209) we immediately obtain the definition equation for the view factor as the ratio of J_{ji} (radiant energy incident on the surface A_j and emitted from the surface A_i) to $(J_i A_i)$ (total radiant energy emitted from the surface A_i):

$$F_{ij} = \frac{J_{ji}}{J_i A_i} = \frac{1}{A_i} \int\limits_{A_i} \int\limits_{A_j} \frac{\cos\theta_i \cos\theta_j}{\pi \left|\vec{s}_{ij}\right|^2} dA_j dA_i. \tag{4.210}$$

Provided that no other heat processes occur, in thermodynamic equilibrium the radiative heat fluxes from i to j and from j to i must correspond to each other for energy conservation reasons and one obtains the reciprocity rule:

$$A_i \cdot F_{ij} = A_j \cdot F_{ji}. \tag{4.211}$$

Since a surface can emit at most the power it has also received, the sum of the visibility factors from surface i to all other surfaces j must equal unity. This results in the summation rule, which in its general form reads to:

$$\sum_j F_{ij} = 1. \tag{4.212}$$

Since the view F_{ij} factor is of fundamental importance for a thermal radiation exchange calculation some general aspects shall be summarized here:

- The view F_{ij} factor ranges between zero and one.
- $F_{ij} = 0$ indicates that two surfaces do not see each other directly. $F_{ij} = 1$ indicates that the surface j completely envelopes surface i.
- The radiation that strikes a surface does not need to be absorbed by that surface.
- F_{ii} is the fraction of radiation leaving surface i that strikes itself directly. $F_{ii} = 0$ for plane or convex surfaces, and $F_{ii} \neq 0$ for concave surfaces, meaning that they see each other for example the surface of a cylinder inner side.
- Radiation analysis of an enclosure consisting of N surfaces requires the calculations of N^2 view factors, which form a matrix. However, all of these calculations are not necessary. Once a sufficient number of view factors are available, the rest of them can be found using the reciprocity and summation rule.
- The summation rule can be applied to each surface of an enclosure by varying i from 1 to N (number of surfaces). Thus, the summation rule gives N equations. Also, reciprocity rule gives $(0.5N(N-1))$ additional equations. Therefore, the total number of view factors that need to be evaluated directly for an N-surface enclosure becomes $(0.5N(N-1))$.
- The superposition rule states that the view factor from a surface i to a surface j is equal to the sum of the view factors from surface i to the parts of surface j and reads to $F_{i(j+k)} = F_{ij} + F_{ik}$.

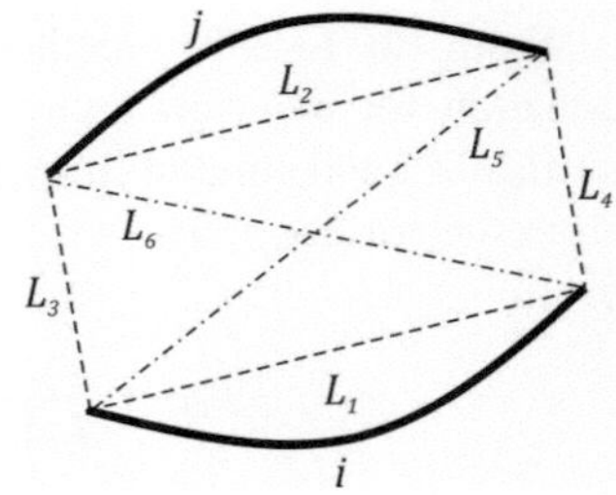

Fig. 4.40 Cross-string method to determine a view factor in 2D-geometries

- A good way to check the correctness of a thermal radiation calculation is to assume blackbody radiation everywhere in the geometry under consideration. The product $(A_i \cdot F_{ij})$, which can be understood as a radiation exchange factor matrix, must always be symmetrical, otherwise there is a calculation error.

Geometries such as channels and ducts that are very long in one direction can be considered two-dimensional (since radiation through end surfaces can be neglected). The view factor between their surfaces can be determined by the cross-string method as shown by the example in Fig. 4.40 in the following way:

$$F_{ij} = \frac{\sum(\text{crossed strings}) - \sum(\text{uncrossed strings})}{2 \cdot (\text{string on surface} i)} = \frac{(L_5 + L_6) - (L_3 + L_4)}{2L_1}. \tag{4.213}$$

What is finally missing is to determine the exchange?

For the simple problem stated with two bodies j and i we had $(H_i \cdot A_i) = (J_j \cdot F_{ji} \cdot A_j)$, expressing the irradiance emitted by i corresponds to the radiosity from surface j considering the view factor of both. This can be generalized to a N–surface system:

$$H_i \cdot A_i = \sum_{j=1}^{N} J_j \cdot F_{ji} \cdot A_j. \tag{4.214}$$

Using the reciprocity rule (Eq. 4.211) allows to eliminate A_i and inserting this into Eq. (4.194) we obtain system of equations for the radiosities J_i:

$$J_i = \varepsilon_i \cdot e_{bi} + (1 - \alpha_i) \cdot \sum_{j=1}^{N} J_j \cdot F_{ij}, \tag{4.215}$$

with $i = 1, 2, \ldots N$. Knowing the temperature of each surface i ($e_{bi} = \sigma T_i^4$) and determining the view factors the system of equations can be solved for the radiosity at each surface and using Eqs. (4.195) or (4.196) we get the heat flux $\dot{q}_i$. Unfortunately, in many technical applications the heat transfer to a surface is known, and the temperature of the surface are an unknown and of relevance. A typical example is the insulated surface leading to $q_i = 0$. If the temperatures are not known also e_{bi} becomes an unknown. Then the path follows first from $\dot{q}_i$:

$$\dot{q}_i = J_i + \sum_{j=1}^{N} J_j \cdot F_{ij}, \tag{4.216}$$

to obtain the set of radiosities. Once the radiosities are obtained, the temperature of the surface can be obtained from Eq. (4.195).This equation has a network interpretation, since multiplying J_j using the summation rule $\left(\sum F_{ij} = 1\right)$ yields:

$$\dot{Q}_i = \dot{q}_i \cdot A_i = \sum_{j=1}^{N} \left(J_i - J_j\right) \cdot F_{ij} \cdot A_i. \tag{4.217}$$

The general procedure is as follows: say our system has $M < N$ surfaces on which the temperature is prescribed, and $N - M$ surfaces on which $\dot{q}_i$ is prescribed. We apply Eq. (4.215) to the M surfaces with specified temperature, and Eq. (4.216) to the $N - M$ surfaces with specified $\dot{q}_i$. Thereby, N linear equations for the radiosities $J_1, J_2, \ldots. J_N$ are obtained. And once we have solved for these quantities, we can calculate either the heat flux or the temperature of the surface.

Similar as for heat conduction it is possible to treat the radiation problem as an electrical circuit. If we consider two diffuse, grey, and opaque surfaces of arbitrary shape maintained at uniform temperatures, then the net rate of radiation heat transfer from surface i to surface j can be expressed as:

$$\dot{Q}_{ij} = A_i \cdot J_i \cdot F_{ij} - A_j \cdot J_j \cdot F_{ji} = A_i \cdot F_{ij} \cdot \left(J_i - J_j\right), \tag{4.218}$$

where in the second part the reciprocity rule has been applied. In analogy with Ohm's law, a resistance R_{ij} can be defined as:

$$\dot{Q}_{ij} = \frac{\left(J_i - J_j\right)}{R_{ij}} \text{ with } R_{ij} = \frac{1}{\left(A_i \cdot F_{ij}\right)}, \tag{4.219}$$

in which R_{ij} is called the *space resistance* to radiation. In a cavity consisting of N-surfaces, the conservation of energy requires that the net heat transfer from surface i to be equal to the sum of the net heat transfers from i to each of the N surfaces of the cavity leading to:

$$\dot{Q}_i = \sum_{j=1}^{N} \dot{Q}_{ij} = \sum_{j=1}^{N} \frac{\left(J_i - J_j\right)}{R_{ij}}. \tag{4.220}$$

Using the derived a relationship for the net radiation from a surface (Eq. 4.199) and inserting this in previous relation, we get:

$$\frac{(e_{bi} - J_i)}{R_i} = \sum_{j=1}^{N} \frac{(J_i - J_j)}{R_{ij}}. \tag{4.221}$$

Thus, the heat transfer from surface i can therefore be interpreted as a sum of currents flowing from i to allother surfaces, with $(J_i - J_j)$ being the potential (or voltage) difference and R_{ij} the resistancebetween i and j.

To solve simple radiation problems necessitates that either the temperature or the net rate of heat transfer must be given for each of the surfaces to obtain a unique solution for the unknown surface temperature and heat transfer rates. Using the electrical network method, the following procedure is subsequently conducted:

- form an enclosure; consider fictitious surface(s) for openings, room, etc.;
- draw a surface resistance associated with each surface of the enclosure;
- connect the surface resistances with space resistances;
- solve the radiation problem (radiosities) by treating it as an electrical network problem.

Note, that this method is not practical for enclosures with many surfaces.

For very large problems in which many surfaces are involved a different approach is used. Considering N grey emitting surfaces ($\varepsilon_i = \alpha_i, \rho_i = (1 - \varepsilon_i)$, then all absorptions and reflections among the N surfaces are participating in the radiation problem, as sketched in Fig. 4.41 by the surfaces i, j and k. For this purpose, it is useful to introduce Gebhart factor which describe the ratio of radiation absorbed by any other surface versus the total emitted radiation from the given surface; thus, it is defined by:

$$B_{ij} = \frac{\dot{Q}_{ij}}{\varepsilon_i A_i \sigma T_4^4}. \tag{4.222}$$

As such, the Gebhart factor becomes the radiation exchange factor between a number of surfaces: It has to be emphasized that the Gebhart factor approach assumes that the surfaces are *grey* and are illuminated *diffusely* and *uniformly*. The total heat flux of surface i is $(\varepsilon_i \sigma T_4^4)$, but only $(\varepsilon_i \sigma T_4^4 F_{ij})$ reaches surface j with F_{ij} the dimensionless view factor. Thus, the surface j absorbs the fraction $(\alpha_i F_{ij} = \varepsilon_i F_{ij})$. This fraction is called primary exchange. From the emissions from surface i to surface k, $(\alpha_k F_{ik})$ is absorbed by k, from which surface k reflects again $(\rho_k F_{ik} = (1 - \varepsilon_k) F_{ik})$, which is absorbed by surface j $(\alpha_k F_{ik} B_{kj})$. This fraction coming from i must be traced over all surfaces and thus the Gebhart factors can be mathematically represented in the following way.

$$B_{ij} = \varepsilon_j \cdot F_{ij} + \sum_{k=1} \rho_k \cdot F_{ik} \cdot B_{kj} = \varepsilon_j \cdot F_{ij} + \sum_{k=1} (1 - \varepsilon_k) \cdot F_{ik} \cdot B_{kj}. \tag{4.223}$$

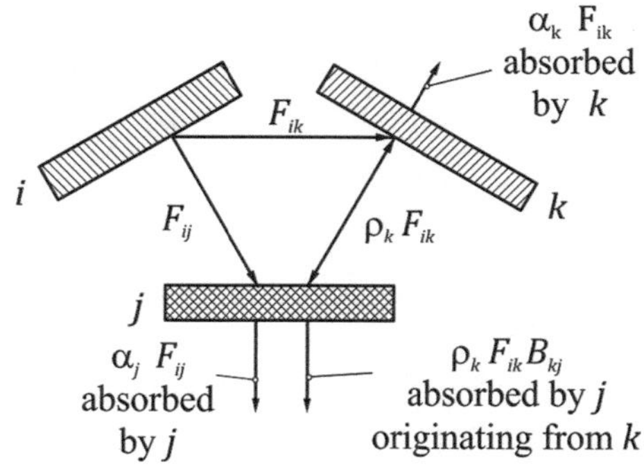

Fig. 4.41 Radiation path of thermal rays in a three-body problem

Since $B_{ij} \leq 1$, Eq. (4.223) can be solved iteratively, that is, starting with $B_{ij}^0 = 0$ and obtaining $B_{ij}^1 = (\alpha_j F_{ij})$ as a first approximation. The calculation is repeated until the relative change of the Gebhart factors falls below a given threshold. Another possibility is to set $B_{ij} = (\delta_{ik} B_{kj})$, where δ_{ik} is the Kronecker tensor. This allows to transform the Eq. (4.223) into a matrix equation which reads to

$$\sum_{k=1} (\delta_{ij} - \rho_k F_{ik}) \cdot B_{kj} = \alpha_j \cdot F_{ij}. \tag{4.224}$$

Thereby one obtains a modified space resistance $R'_{ij} = (\varepsilon_i A_i B_{ij})$. It is obvious that for all $\varepsilon_i = \alpha_i = 1$ and thus omitting reflection ($\rho_i = 0$) the Gebhart factor become identical to the view factor, i.e. $B_{ij} = F_{ij}$ and thus the results for the blackbody are obtained, which is a recommended consistency check in any radiation calculation.

4.3.8 Radiation in the Presence of the Sun

If a surface A_1 is enclosed by a very large surface A_2, the view factor $F_{12} \to 1$. The surface A_2, however, hardly sees the surface A_1 and therefore receives almost no thermal radiation from it; as a consequence, $F_{12} \to 0$ is obtained. In this special case, the radiation exchange is independent of the surface properties of the large surface A_2, and the result is:

$$\dot{Q}_{12} = \varepsilon_1 \cdot A_1 \cdot \sigma \cdot \left(T_2^4 - T_1^4\right). \tag{4.225}$$

In other words, the large surface A_2 behaves like a blackbody. This special case corresponds to the situation of a collector which sees the sky. The radiation exchange of a collector with the area $A_C = A_1$ the emissivity $\varepsilon_C = \varepsilon_1$ and the temperature T_C with the sky of the temperature T_{sky} is thus given by the following equation:

$$\dot{Q}_{12} = \varepsilon_C \cdot A_C \cdot \sigma \cdot \left(T_C^4 - T_{sky}^4\right). \tag{4.226}$$

However, the atmosphere is not a black body, because at wavelengths between 8 and 14 μm it is almost transparent (cf. Chap. 2). So how the effective sky temperature is obtained?

If we assume the atmosphere to be a grey body and then the atmospheric long-wave irradiance in the range of 3 to 100 μm (I_{sky}) can be approximated as a fraction of the blackbody emissive power, termed as effective sky emissivity ε_{sky}, at the surface level ambient temperature (T_a).

$$\varepsilon_{sky} = \frac{I_{sky}}{\sigma \cdot T_a^4}, \tag{4.227}$$

where σ ist he Stefan-Boltzmann constant. Assuming the sky as blackbody we can define a sky temperature $I_{sky} = \sigma \cdot T_{sky}^4$. If we insert this in Eq. (4.227) we obtain

$$T_{sky} = \varepsilon_{sky}^{1/4} \cdot T_a, \tag{4.228}$$

and thereby, we recognize that the effective sky temperature is lower than the surface level air temperature, since ε_{sky} cannot exceed unity. As a consequence, the apparent radiation temperature is several degrees below the sensible temperature reading of a thermocouple in air, which is not of relevance for non-concentrating systems or low temperature applications, because the thermal radiation losses are low compared to other loss mechanisms, however, for high concentrating applications ranging from solar cookers to solar towers, we require a closer look.

The sky emissivity relates the atmospheric radiation, which emanates from the gases in the air, some of which absorb and emit in the far infrared range. Although oxygen and nitrogen constitute by far the majority (about 99% by volume) both gases are transparent to such radiation and do not absorb or emit in the far infrared. Also, carbon dioxide, ozone, and other asymmetrical molecules do not have much influence on the variation of atmospheric radiation received at the ground, see (Lienhard, 2001). Thus, the variation in the water vapor in the atmosphere near the ground is a major contributor to the atmospheric radiation and about 90% of the atmospheric radiation received at the ground level originates at an altitude less than 1 km (Bliss, 1961). Although the effective sky temperature is always lower than the ambient air temperature due to a decrease in elevation, but the presence of clouds in the sky usually tend to increase the effective sky temperature, causing it to approach the clear-sky temperature. In general, the effective sky emissivity of the atmosphere can be expressed as a function of ambient temperature (T_a), relative humidity ($\phi[/]$), the partial pressure of water vapor (p_v) and dew-point temperature (T_{dp}). The partial pressure of water vapor (p_v) in $[Pa]$ and dew point temperature T_{dp} $[K]$ als well as the moisture content d $(kg/(kgdryair))$ can be expressed as a function of ambient temperature (T_a) the Magnus expression (Alduchow & Eskridge, 1996):

$$\begin{aligned} p_v &= 610.94\left(\frac{\phi}{100}\right)\exp\left(\frac{17.625(T_a-273.15)}{T_a-30.11}\right), \\ T_{dp} &= \frac{243.04\ln(p_v/610.94)}{17.625-\ln(p_v/610.94)} + 273.15, d = \frac{0.622p_v}{p_a-p_v}, \end{aligned} \tag{4.229}$$

in which p_a is the air pressure in $[Pa]$. For energy balance applications in solar engineering, the effective emissivity during the daytime is more favourable while for night-time passive cooling applications, the effective emissivity during the night-time is more favourable. Therefore, often parametric clear-sky Brunt models for both daytime and night-time are used to calculate the clear sky emissivity $\varepsilon_{sky,c}$ (Alados, 2012):

$$\varepsilon_{sky,c} = 0.598 + 0.057\sqrt{p_v}\,\text{day};\ \varepsilon_{sky,c} = 0.633 + 0.057\sqrt{p_v}\,\text{night}. \quad (4.230)$$

Apart from these primary parameters, the sky emissivity additionally also depends on the cloudiness of the sky, moisture content, and so forth, which have to be taken from statistical meteorological data on site. The presence of clouds modifies the longwave irradiance I because the radiation emitted by water vapor and other gases in the lower atmosphere is supplemented by the emission from clouds. Therefore, under cloudy conditions, the effective sky emissivity is higher compared to clear-sky value ($\varepsilon_{sky,c}$). Parametric models validated by numerous experiments are then used to estimate the all-sky condition for the longwave sky irradiance I in $[W/m^2]$ for an all-day model, see (Li & Yiang, 2017):

$$I = I_{sky,c}(1 - 0.78CF) + 0.380\sigma T_a^4 CF^{0.95}\phi^{0.17}, \quad (4.231)$$

in which CF (%) is the cloud cover fraction in the sky dome. For a simple assessment of the performance of solar thermal components the relationship of Berdahl and Fromberg (1982) is sufficient, which reads to:

$$T_{sky} = T_a\left[0.711 + 0.0056 \cdot T_{dp} + 0.00004T_{dp}^2 + 0.013\cos(15 \cdot t)\right]^{1/4}, \quad (4.232)$$

where T_{sky} is the radiation temperature of the sky in $[K]$ and t is the number of hours after midnight. The relation is derived from measured data at temperatures from −20° to + 30 °C. The deviations of the radiation temperatures of the sky from the air temperatures can be 5 °C in hot and humid climate and up to 30 °C in cold and dry conditions.

The following example is intended to demonstrate the mode of operation of a radiation shield which is used, among other things, in high-temperature receivers for solar towers as night cover. If several such shields, which are technically designed as, for example, thin aluminium or steel foils, are arranged one behind the other, this is referred to as multi-layer insulation, or MLI for short. Figure 4.42 shows a schematic diagram of this. A spacer is usually placed between two foils so that the foils do not come into contact with each other. It shall be considered in the following how the number of foils increases the insulation performance. The walls and the foils are assumed to have an emissivity ε and the spacers transfer the heat flux by means of heat conduction corresponding to its heat conduction resistivity according to $C = (k \cdot A)$. If the areas are equal, the radiation resistance is obtained for $R = (\varepsilon \cdot A)/(2 - \varepsilon) \sim (\varepsilon \cdot A)/2$ (see e.g. Eq. 4.203). In the steady

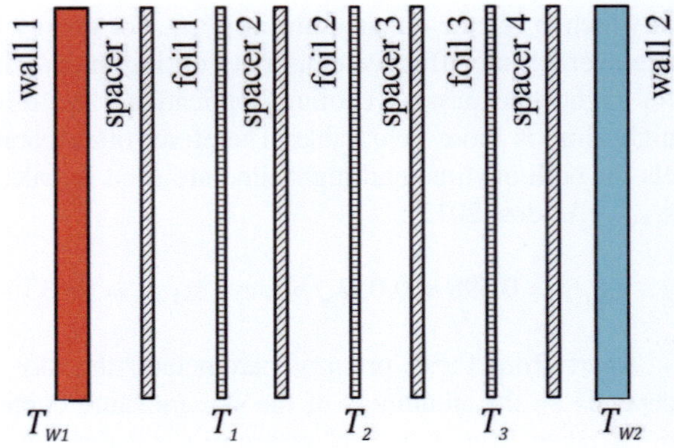

Fig. 4.42 Sketch for determining the insulation effect in a multi-layer insulation with three foils and four spacers

state, the same heat flux flows in each interstitial space, which can be calculated as follows:

$$\begin{aligned}\dot{q}_{W1\to T1} &= C \cdot (T_{W1} - T_1) + \sigma \cdot R \cdot \left(T_{W1}^4 - T_1^4\right) = \dot{q}_{1\to 2},\\ \dot{q}_{1\to 2} &= C \cdot (T_1 - T_2) + \sigma \cdot R \cdot \left(T_1^4 - T_2^4\right) = \dot{q}_{2\to 3},\\ \dot{q}_{2\to 3} &= C \cdot (T_2 - T_3) + \sigma \cdot R \cdot \left(T_2^4 - T_3^4\right) = \dot{q}_{3\to W2},\\ \dot{q}_{3\to W2} &= C \cdot (T_3 - T_{W2}) + \sigma \cdot R \cdot \left(T_3^4 - T_{W2}^4\right).\end{aligned} \tag{4.233}$$

Summing up all heat fluxes the temperatures of the foils are eliminated and we obtain:

$$\dot{q}_{W1\to W2} = \frac{C}{4} \cdot (T_{W1} - T_{W2}) + \frac{\sigma \cdot R}{4} \cdot \left(T_{W1}^4 - T_{W2}^4\right). \tag{4.234}$$

The main result of the MLI structure is that the insulation effect, which is composed of heat conduction and radiation losses, reduces by a factor of $(N + 1)$ with each foil layer.

In the case of radiation exchange outside a cavity under the influence of the sun, Kirchhoff's law with $\varepsilon_i = \alpha_i$ no longer applies. The emissivity differs from absorptivity and thus $\varepsilon_i \neq \alpha_i$. The reason lies in the definition of the absorptivity α_i, which is related to the spectrum of the sun. The radiant heat flux from the sun or irradiance is,

$$I_s = \int\limits_{\lambda=0}^{\lambda\to\infty} I_{s\lambda}(\lambda, T) d\lambda, \tag{4.235}$$

while the absorptivity of a surface is given by the following relation:

$$\dot{q}_\alpha(T) = \int\limits_{\lambda=0}^{\lambda\to\infty} \alpha(\lambda, T) \cdot I_{s\lambda}(\lambda, T)\, d\lambda. \tag{4.236}$$

The ratio of the previous two equations yields the temperature dependent absorption coefficient:

$$\alpha(T) = \frac{\int_{\lambda=0}^{\lambda\to\infty} \alpha(\lambda, T) \cdot I_{s\lambda}(\lambda, T)\, d\lambda}{\int_{\lambda=0}^{\lambda\to\infty} I_{s\lambda}(\lambda, T)\, d\lambda}. \tag{4.237}$$

Equation (4.237) is the basis of "passive" solar thermal energy. Only this validity allows to manipulate the surface temperatures by a specific choice of the surface properties. This will be shown by means of an example.

A plate with the area A floats in space and is ideally insulated on the side facing away from the sun ($\partial\dot{q}/\partial n = 0$). On the other side the sun shines with solar constant $i_{sc} = 1370 W/m^2$. The temperature of the space is the temperature of the universe $T_s = 0° K$. Searched and unknown is now the plate temperature T. The plate has the radiation exchange factor $R = (\varepsilon A)$ and absorbs thermal power from the sun according to $\dot{Q} = (\alpha \cdot A \cdot i_{sc})$. The power absorbed by the plate must equal the power received from the sun and thus:

$$\dot{Q} = \alpha \cdot A \cdot i_{SC} = \sigma \cdot R \cdot \left(T^4 - T^4_{Space}\right). \tag{4.238}$$

Solving the equation for temperature finally yields:

$$T = \left(\frac{\alpha}{\varepsilon}\frac{i_{Sc}}{\sigma}\right)^{\frac{1}{4}}. \tag{4.239}$$

This dependence on (α/ε) is used in so-called SSMs (second surface mirrors), for example in radiators, to emit heat despite incident solar radiation. For this purpose, a metal film (e.g. silver or aluminum) is vapor-deposited on the back of a carrier material (e.g. polyester film). In front of this is a Teflon or plastic film that is transparent to solar radiation, i.e. permeable. The metal layer reflects most of the solar radiation and absorbs only a little ($\alpha = 0.08 - 0.14$). The outer layer emits with a high emission coefficient ε ($\varepsilon = 0.6 - 0.8$) and prevents excessive heating. Table 4.12 shows some temperatures of different materials according to Eq. (4.239).

Table 4.12 Temperatures of a sunlit plate in space

Gas	$\alpha[-]$	$\varepsilon[-]$	$\alpha/\varepsilon[-]$	$T[K]$
Gold	0.24	0.03	8	661
Aluminium	0.13	0.05	2.6	499.08
Black paint	0.97	0.9	1.08	400.46
White paint	0.18	0.85	0.212	366.62
SSM	0.08	0.8	0.1	221.02

4.4 Practical and Coupled Radiative-Convective Heat Transfer Problems

In solar energy and building physics, the heat exchange between surfaces separated by an infrared transparent medium (usually air) is a very common problem. Examples are the heat transfer from the absorber to the collector casing, or the heat exchange of walls with different temperatures in a room or the heat exchange between heating surfaces (radiator/convector) with its surrounding, or the heat exchange between collector surface with the environment and the sky. As air has a minimal infrared absorption coefficient. When considering the radiation exchange between neighbouring buildings or walls in a room with only a few meters distance, the absorption can be neglected. This is not the case for the complete atmosphere layer of several kilometres. Here infrared radiation is effectively hold back – on of the basic pre-requisites of our life on earth.

In an engineering treatment the radiation exchange between surfaces can be described as described in Sect. 4.3.7. But there is another mode of exchanging heat: the convective heat transfer from a surface to the air. Also here a convective heat transfer coefficient to air can be defined:

$$h_{conv} = \frac{\dot{q}_{conv}}{T_s - T_{air}} \tag{4.240}$$

in which T_s denotes the surface temperature, and T_{air} the air temperature outside the thin boundary layer. If the surface temperatures in a problem are given, radiative and convective heat transfer can be separately calculated and then the two solutions add to a net outgoing heat flux for each surface. This quasi-linearization can be applied if the temperature differences are not too large, usually in the range of $\Delta T < 100K$, unless non-linear effects come into play. However, if for some surfaces not temperatures but heat fluxes ($\dot{q}$) are given, a coupled problem appears requiring non-linear equations be solved.

Additional challenges appear, if the gas itself is absorbing and emitting radiation, or within a cavity a solid structure is in radiative interaction with the boundaries creating temperature differences in the gas leading to convection. In such a context the superposition of radiation and convection collapses and the combined heat transfer problem to be treated in a fully coupled way.

In solar collector problems, a coupling between convection and the radiation transport is possible if for example polymer films as internal glazings are coming too close to a selectively emitting absorber, or if transparent insulation materials (see Chap. 6) like honeycomb structures or aerogel, which are partially transparent to infrared radiation are used to improve the heat losses in the air gap. Simplified methods to solve the coupled equations have been developed alongside these developments, and some details are discussed for example in (Pflüger & Wittwer, 1984; Platzer, 1988, 1992).

One important result is that by using an appropriate air gap in this case between the selective absorber and the absorbing-emitting structure, the heat transport modes may be effectively decoupled, with the results that a much lower combined heat transfer can be observed, which is close to the one calculated with the additive simplification. Other research where coupled heat transfer of both modes, radiation and convection occurs, is the energy loss of stellar atmospheres. Although combined heat transfer is needed in many practical problems in engineering, the treatment still seems to be a challenge in many cases (Dombrovsky, 2022). The separation of the individual effects is often not easy, because apart from the given collector or storage dimensions and the material specifications, the environmental conditions such as solar irradiation (direction and intensity) also external parameters such as wind overflow of the collector or flow through the storage or the wall panels determine which of the heat transfer mechanisms plays an excellent role. Thus, the key to an engineering assessment and the subsequent decision on how a problem can be solved is the evaluation of dimensionless number characterizing the problem, which require some engineering assessment. In the next step a simple analytical example of pure radiative heat transfer within a cylinder is addressed to illustrate how a radiation heat transfer calculation is executed Based on this we subsequently address two practical combined problems appearing in solar thermal systems such as

- the heat transfer between two glass panes in buildings.
- the effective thermal conductivity of packed beds, which are used as thermal energy storages in gas-cooled systems, among others, both in low-temperature systems such as packed beds in air collectors and in high-temperature systems.

4.4.1 Non-Dimensional Numbers Characterizing Combined Problems

Despite the complexity of conjugate radiation-convection and heat-conduction problems, it is possible to assign certain dominant phenomena by forming characteristic dimensionless parameters. This allows the selection of suitable solution approaches and methods, either described in the literature such as (Howell et al., 2020a, 2020b) or being depicted by commercially available numerical solution packages like COMSOL Multiphysics.

Thus, when a considerable heat conduction or convection occurs simultaneously with thermal radiation in the presence of an absorbing emitting medium, a more complex mathematical treatment is needed. In principle, if none of the modes is negligible, a non-linear integrodifferential equation will result for the energy equation for the general problem having combined modes of heat transfer. Sometimes the formulation can be simplified, for example when the absorbing-emitting medium can be considered as optically thick, so that emitted infrared photons may travel very short before being absorbed again. Under such conditions the so-called diffusion approximation in radiation transport is applicable, leading to differential terms only instead of the integral equations. The other extreme case, an optically very thin medium also allows the simplification of additive solutions. However, as combined mode problems in general are mathematically complex, it is usually not possible to obtain an analytical solution even for geometrically simple cases.

Nevertheless, there is the possibility to characterize the coupled problem with the help of dimensionless numbers before starting the solution procedure. The Stark number (N), also called the conduction and radiation parameter, weights the energy transport by heat conduction versus radiation and reads:

$$N = \frac{(\alpha + \sigma_s) \cdot k_s}{4 \cdot \sigma \cdot \Delta T^3}, \tag{4.241}$$

in which α is the total, linear absorption/emission coefficient, σ_s the total isotropic scattering coefficient, k_s the thermal conductivity of the solid, σ the Stefan Boltzmann-constant and $\Delta T^3 = T_a^3 - T_b^3$ is determined from two characteristic temperatures (T_a, T_b) of the problem considered. If N falls, thermal radiation effects increase. Roughly, three characteristic regimes for N can be distinguished, i.e. $N = 1, 0.1, 0.01$. The dimensionless numbers for the other heat transfer modes can also be derived as the ratio between the heat fluxes due to convection ($\dot{q}_{conv}$), radiation ($\dot{q}_{rad}$) and conduction ($\dot{q}_s$). These heat fluxes read to

$$\dot{q}_{conv} = u_0 \rho c_p \Delta T; \dot{q}_{rad} = (\alpha + \sigma_s)\sigma \Delta T^4; \dot{q}_s = k_s \Delta T / d, \tag{4.242}$$

in which u_0 is the mean characteristic velocity, ρ the specifc density, c_p the heat capacity and d a characteristic length scale. The Boltzmann number (Bo) determines the relative importance of the energy transported by forced convection to thermal radiation and is given by

$$Bo = \frac{\dot{q}_{conv}}{\dot{q}_{rad}} = \frac{u_0 \rho c_p \Delta T}{(\alpha + \sigma_s)\sigma \Delta T^4}. \tag{4.243}$$

Typically, three regimes exist at the Boltzmann number, i.e. $Bo = 0.1, 1, 10$, which implies an increasing influence of convection on the versus radiative heat transfer. The convection-conduction number (Cn) characterizes the ratio between heat flux generated by forced convection to that of molecular thermal conduction

$$Cn = \frac{\dot{q}_{conv}}{\dot{q}_s} = \frac{u_0 \rho c_p d}{k_s}. \tag{4.244}$$

where d specifies the characteristic length of the system along the direction of heat conduction, typical domains are separated by $Cn = 1, 10, 100$. In case of buoyancy induced problem, where the temperature differences in a fluid induce natural convection, the characteristic velocity u_0 has to be replaced by the buoyant mean velocity u_b given by $u_b = \sqrt{g\beta \Delta T d}$, in which g is the gravity acceleration and β the volumetric expansion coefficient.

When considering radiative heat transfer problems, two other non-dimensional quantities must be considered in advance. The linear scattering albedo (ω), defined as,

$$\omega = \frac{\sigma_s}{(\alpha + \sigma_s)}. \tag{4.245}$$

is the ratio between the scattering coefficient and the extinction coefficient and describes the relative importance of scattering compared to absorption/emission. Two scattering regimes can be distinguished. For $\omega \approx 0$, a strongly absorbing/emitting material or a non-scattering medium exists; while $\omega \approx 1$ represents a strongly scattering medium.

The optical thickness (or opacity, τ), which we described in Chap. 3 as product of refractory index (n) and homogeneous medium thickness (d), can be conceived as the ability of a medium to attenuate radiation and is defined as

$$\tau = (n \cdot d) = (\alpha + \sigma_s) \cdot d, \tag{4.246}$$

where d is the characteristic length of the medium layer. Essentially, there are four different physical states here:

- non-participating medium $\tau \approx 0$;
- optically thin medium $\tau \ll 1$, where the radiative heat transfer is determined by the interfacial emission and the radiation of the medium is limited;
- self-absorbing medium $\tau \approx 1$, in which the contributions of the interfaces and the internal radiation balance each other;
- optically thick medium $\tau \gg 1$, where radiation is essentially a local phenomenon and radiative transfer behaves like a diffusion process (similar as molecular transport).

For transient heat transfer phenomena, the characteristic diffusion time of the solid (τ_d), has to be considered, which is defined as follows;

$$\tau_d = \frac{d^2}{a} = \frac{(\rho c_p)_s}{k_s} \cdot d^2, \tag{4.247}$$

where a specifies the thermal conductivity. The diffusion time τ_d can be interpreted as a measure of the time required to establish thermal equilibrium in the solid. When heat transfer occurs through the liquid–solid interface, the thermal activity ratio (TAR) given by,

$$TAR = \frac{\sqrt{(\rho k c_p)_f}}{\sqrt{(\rho k c_p)_s}}, \tag{4.248}$$

describes the ratio between the heat transfer coefficients of the liquid to the solid media and can be used to quantify the interfacial heat flux. High values of the TAR values indicate a low heat flux, while small TAR values reflect a large heat flux.

4.4.2 Radiative Transfer Equation and Simple Example

Since the radiative transfer describes a net exchange between the different materials involved in a heat transfer problem, the determination of the heat transfer is fundamentally different from that in solids, which are linked via the thermophysical material properties, or via interfaces, which are characterized by a media transport such as flows over a wall. To solve energy conservation in a control space, the energy conservation equation must be satisfied in a control volume at the given thermal boundary conditions. For a translucent medium, the energy conservation equation in general form is given by

$$\rho c_p \frac{dT}{dt} = \beta T \frac{dp}{dt} + \nabla \cdot (k\nabla T - \dot{q}_{rad}) + \dot{q} + \Phi_{diss}. \tag{4.249}$$

In it, ρ and c_p describe the specific density and heat capacity, T the temperature, β the volumetric expansion coefficient, p the pressure, k the thermal conductivity, $\dot{q}$ a vo-lumetric heat source/sink, and Φ_{diss} the heat release by volumetric viscous dissipation. d/dt denotes the substantial derivative.

To obtain the radiation contribution $\dot{q}_{rad}$ to the energy equation, radiative transfer relations must be provided. Since some of the radiation terms such as absorption, emission, and reflection depend not only on wavelength but also on temperature, the radiative transfer equation (RTE) must be solved simultaneously with the energy conservation equation to determine the temperature distribution and radiation properties.

To obtain the temperature distribution in a control volume by solving Eq. (4.249), we need $(\nabla \cdot \dot{q}_{rad})$ as function of the temperature distribution. Unfortunately, $(\nabla \cdot \dot{q}_{rad})$ depends on the scattering of the radiation, which must be determined within the medium. There are now two options to obtain the required relations for $(\nabla \cdot \dot{q}_{rad})$.

The first is to derive $\dot{q}_{rad}$ and then differentiate to obtain $(\nabla \cdot \dot{q}_{rad})$. The second option is to directly determine $(\nabla \cdot \dot{q}_{rad})$ by the local radiation interaction within a control volume in the medium, since $(\nabla \cdot \dot{q}_{rad})$ is the difference in absorbed radiation emitted in the same volume element. This in turn depends on the absorption coefficient and the intensity incident from all directions. The radiation energy required for the energy equation (Eq. 4.249) is the total radiation over all wavelengths, so for both absorption and emission relationships are required to obtain the difference of the total radiation Ω_i incident from all solid angles, absorbed locally in the volume, and the locally emitted radiation. These two quantities (absorbed and emitted radiation) are described by the two integral terms over all wavelengths on the right-hand side of the following equation.

$$-\nabla \cdot \dot{q}_{rad} = \underbrace{\int_{\lambda=0}^{\infty} \alpha_{\lambda} \left(\int_{\Omega_i=0}^{4\pi} I_{\lambda}(\Omega_i) d\Omega_i \right) d\lambda}_{absorption} - \underbrace{4\pi \int_{\lambda=0}^{\infty} \alpha_{\lambda} I_{\lambda b} d\lambda}_{emission} . \tag{4.250}$$

To determine the radiative power density ($\dot{q}_{rad}$), we now still need the radiation intensity (I_{λ}) along a propation direction $\vec{s}$. The radiation intensity is given as balance equation containing all contributions (propagation, emission, absorption, and scattering) in the volume element, which is called the radiative transfer equation (RTE) reading in its general form to

$$\frac{dI_{\lambda}}{d\vec{s}} = \alpha_{\lambda}(\vec{s}) I_{\lambda b}(T, \vec{s}) - \beta_{\lambda} I_{\lambda}(\vec{s}) + \frac{\sigma_{\vec{s},\lambda}}{4\pi} \int_{\Omega_i=0}^{4\pi} I_{\lambda}(\vec{s}, \Omega_i) \Phi_{\lambda}(\Omega, \Omega_i) d\Omega_i , \tag{4.251}$$

in which α_{λ}, β_{λ} and $\sigma_{\vec{s},\lambda}$ are the absorption, extinction, and scattering coefficients and Φ_{λ} is the scattering phase function. The RTE expresses the general conservation of radiant energy along a line of sight in an absorbing, emitting, and scattering medium. The solution of this differential equation is by no means trivial, especially in engineering geometries, but often a number of simplifications can be made, such as the existence of an opaque medium ($\tau = 0$), assumptions of a gray beam ($\varepsilon = const$), and many others that reduce the problem to an interface problem.

There are a number of approximate solvers for the RTE equation, which are essentially based on the assumption of local thermodynamic equilibrium. Mostly, however, the solution is done numerically. Some of these solution methods are applicable to special environments and an overview is given, for example, in (Howell et al., 2020a, 2020b). The numerical methods most commonly used in radiative heat transfer and their main features can be summarized as follows:

Monte Carlo

- based on conceptually simple physics, which represents quasi-exactly the behaviour of a single ray;
- completely general; can solve time-dependent and 3D problems with arbitrary geometry;
- easy to program.
- Challenge: reduction of statistical error, which can be reduced by tracking more initial rays at the expense of computation time (requiring longer computer times)

Discrete Ordinates Systems

- Complex programmability;
- poor convergence on complex scattering phenomena with large gradients.

Invariant Embedding

- Difficult programmability;
- limited to 1D problems;
- includes all orders of multiple scattering and the computed radiances have no statistical errors;
- extremely fast.

Due to their simplicity, quasi-exact representation of the behaviour of a single ray, and most importantly, the extreme increase in computer capacity, Monte Carlo methods are now mostly used to solve $RTEs$.

Radiative heat transfer phenomena are an often underestimated or neglected aspect in the calculation of temperature distributions in technical assemblies. Using a simple example of a cylinder, the radiative heat exchange between finite surfaces will be sketched here to vividly illustrate the calculation procedure presented in Sect. 4.3.7. With the coordinate designations given in Fig. 4.43, one obtains

$$x = r/h;\, y = r/h \rightarrow z = 1 + \left(1 + y^2\right)/x^2,$$

for which we obtain as view factor F_{12} between the area A_1 and A_2:

$$F_{12} = 1/2\left(z - \sqrt{z^2 - (2y/x)^2}\right)$$

We now assume the hollow cylinder with the dimensions $r = 0.1m$ and the height $h = 0.25m$ to be insulating to the outside and there should be no heat conduction within the cylinder, and we want to determine the radiation exchange factors and temperatures.

First, the areas A_1 and the radiation exchange factors F_{ij} must be calculated.

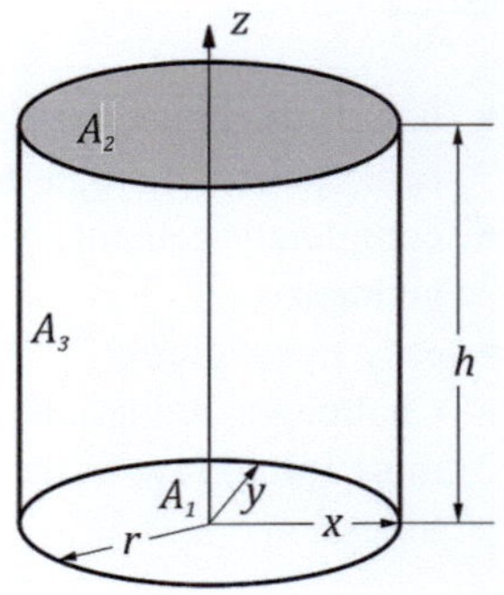

Fig. 4.43 Schematic sketches for the determination of the visibility factors in a closed cylinder

- Areas: The areas of the cylinder under consideration are

$$A_1 = A_2 = \pi \cdot r^2 = 0.031416\,\mathrm{m}^2 and A_3 = 2 \cdot \pi \cdot r \cdot h = 0.15708\,\mathrm{m}^2.$$

- View factors: For $n = 3$ surfaces, there exist are $n^2 = 9$ view factors. Since the ground and top surfaces are considered to plane parallel, they cannot see themselves. This results in $F_{11} = F_{22} = 0$. The view factor from 1 to 2 yields for $x = 0.4$, $y = 0.4$ and $z = 8.25$ to $F_{12} = 0.123$. Since symmetry applies, $F_{12} = F_{21} = 0.123$. The visibility factor F_{31} results from the summation rule (Eq. 4.212), according to which $F_{11} + F_{12} + F_{13} = 1$ must be to $F_{13} = 0.877$.
- Also now the symmetry is applicable, and it is $F_{23} = F_{13} = 0.877$.
- The symmetry also requires $F_{31} = F_{32}$. With the help of the reciprocity law, F_{31} can now be determined. It is $(A_1 \cdot F_{13}) = (A_3 \cdot F_{31})$. Thus, $F_{31} = (A_1/A_3) \cdot F_{13} = (r/2h) \cdot F_{13} = 0.1754 = F_{32}$.
- The inner shell of the cylinder sees itself, so $F_{31} \neq 0$ (!!!). Now the summation law is applied, and it yields $F_{31} + F_{32} + F_{33} = 1$ and thus $F_{33} = 0.6942$. Thus, the view factor matrix is complete, and it results in:

$$F_{ij} = \begin{bmatrix} 0 & 0.123 & 0.877 \\ 0.123 & 0 & 0.877 \\ 0.1754 & 0.1754 & 0.6942 \end{bmatrix}$$

- Accordingly, the following matrix is obtained for the radiation exchange factors.

$$R_{ij} = A_i \cdot F_{ij} = \begin{bmatrix} 0 & 0.03864 & 0.02755 \\ 0.03864 & 0 & 0.02755 \\ 0.02755 & 0.02755 & 0.101945 \end{bmatrix}$$

The radiation exchange factor matrix must always be symmetrical, otherwise there is a calculation error. This is an important checking mechanism for complex technical calculations in difficult geometries. The determination of the radiation exchange factors, and their symmetry behaviour also apply to grey bodies.

In the following it is assumed that all three surfaces are isolated from each other. Let surface 1 be at $T_1 = 550K$ and surface 2 at $T_2 = 300K$. The steady state temperature ($t \to \infty$) of the shell surface is obtained from the heat balance leading to:

$$\sigma \cdot \left[\mathrm{R}_{31} \cdot \left(\mathrm{T}_1^4 - \mathrm{T}_3^4\right) + \mathrm{R}_{32} \cdot \left(\mathrm{T}_2^4 - \mathrm{T}_3^4\right)\right] = 0.$$

From this, after solving for the shell surface temperature T_3, we get

$$T_3^4 = \frac{R_{31} \cdot T_1^4 + R_{32} \cdot T_2^4}{R_{31} + R_{32}} = \frac{T_1^4 + T_2^4}{2} \to T_3 = 472.4K$$

It is noteworthy that without the influence of convection, the temperature T_3 is approximately at the level of the hot surface A_1.

Now the same cylinder should not be "black" radiating but only one black surface A_1 with $\varepsilon_1 = 1$ and two gray radiating surfaces A_2 with $\varepsilon_2 = 0.1$, A_3 with $\varepsilon_3 = 0.75$; the temperatures $T_1 = 550K$ and $T_2 = 300K$ are assumed identical as before. The first two calculation steps for the areas (A_i) and the view factor matrix (F_{ij}) remain the same, however we have now to evaluate the Gebhart factors (B_{ij}) according to Eq. (4.224) leading to

$$B_{ij} = \begin{bmatrix} 0.0727 & 0.176 & 0.90968 \\ 0.1762 & 0.00479 & 0.8190 \\ 0.2426 & 0.0218454 & 0.7356 \end{bmatrix}.$$

This in turn results in the following radiation exchange matrix

$$R_{ij} = \begin{bmatrix} 0.00228 & 0.00554 & 0.028578 \\ 0.00554 & 0.000015 & 0.002573 \\ 0.028578 & 0.002573 & 0.08666 \end{bmatrix}.$$

If we now calculate the temperature of the shell surface A_3, we obtain from the heat balance a value of $T_3 = 539.34$ K and thus approx. 69 °C more (!!!) than in the case of the black body assumption. Even if the example is of an academic nature, it shows that it is possible to manipulate the temperatures on interfaces by controlling the radiation properties of the surfaces, which is of excellent importance for the integrity of components, especially in high-temperature applications.

4.4.3 Coupled Heat Transfer Between Two Glass Panes

Using Fig. 4.44a, let us first consider the heat transfer between two parallel vertical panes of glass, as typically occurs in windows in buildings. When calculating heat losses through a double-glazed window, the heat flow in the individual sections of the window is composed of components of radiative, convective, and conductive heat transfer on the outer and inner surfaces of the double-glazed window and in the gap between the panes. In this case, the value of each of these components at a known temperature difference ΔT in each separate section of heat transfer is calculated using the coefficients of convective h_{conv} and radiative heat transfer h_{rad}, as well as the thermal conductivity of gas (k_{gas}) and the glass panes (k_I, k_{II}), respectively. This interaction of different heat transfer mechanisms can be clearly represented in the form of an electrical equivalent circuit as shown in Fig. 4.44b.

In the first step, we will only consider the radiative heat loss through a single-chamber double-glazed window installed in the opening of the building envelope. In order to consider the problem in a one-dimensional way, we assume that the areas of all four glass surfaces are equal, and the emission coefficients take the values ε_1, ε_2, ε_3 and ε_4. Let the temperature of the inner surface of the room be

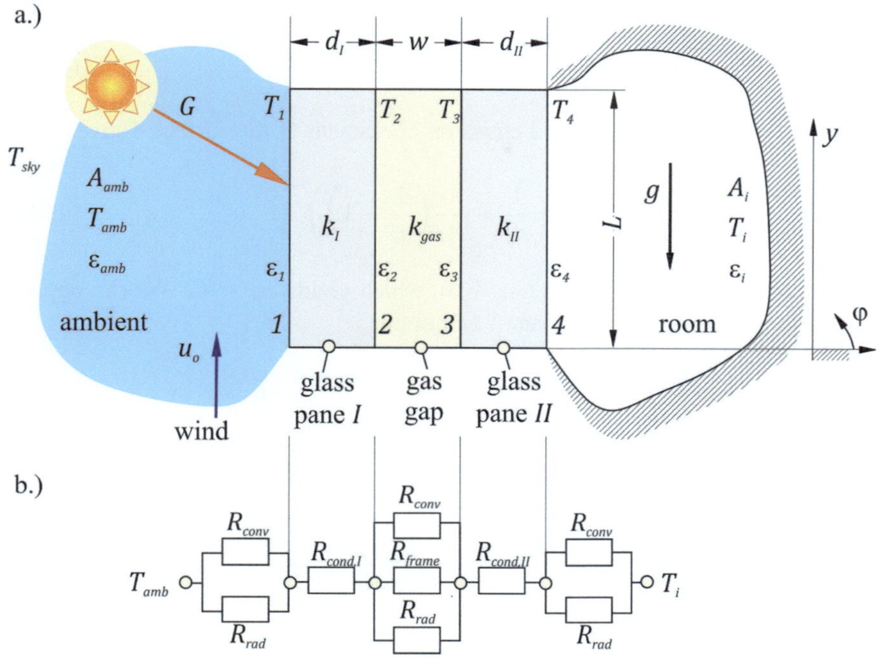

Fig. 4.44 Sketch to calculate the combined convective-radiative and conductive heat transfer across a double-glazed window(a) and corresponding equivalent electric circuit expressed by heat resistanccs (R_i) (b)

constant and be T_i, while that of the surroundings is also constant and has the value T_{amb}, where $T_{amb} < T_i$. Thus, heat is radiative transferred from the room to the surroundings. The objective is to determine the glass temperatures T_I, T_{II} and the radiant heat flux density $\dot{q}_{rad}$ through the window. For the analysis of radiant heat transfer, we assume that there is no temperature gradient across glass thickness, i.e. $T_I = T_1 = T_2$ or $T_{II} = T_3 = T_4$. This is sufficiently justified since the thermal conductivity of the glass is considerably greater than that of the gas in the gap between the panes.

The temperatures T_{amb} and T_i are the radiation temperatures of the surfaces that differ from the ambient temperature Tamb and the room air temperature T_i. In indoor surfaces the air temperature is usually equal to the radiant temperature, while the equation of the radiant temperature and the air temperature in the outdoor window surface reflects a dense cloudiness of the sky, otherwise the radiant temperature of the sky must be determined, (see cf. Sect. 4.3.8). Here, we further assume the identity of temperatures.

The surface of the space A_i and the surface of the inner glass A_4 form a closed system, and since the surface of the glass A_4 is flat does not radiate on itself. Thus, we can determine the heat flux density $\dot{q}_{rad}$ transferred by radiation from the inner surface of the room to the inner surface of the glass by thermal radiation.

$$\dot{q}_{rad,4\to i} = \varepsilon_{4i} \cdot \sigma \cdot \left(T_i^4 - T_{II}^4\right), \tag{4.252}$$

in which ε_{4i} denotes the reduced emission coefficient of the surface 4 given by

$$\varepsilon_{4i} = \left(\frac{1}{\varepsilon_4} + \frac{A_4}{A_i}\left(\frac{1}{\varepsilon_i} - 1\right)\right)^{-1}. \tag{4.253}$$

Usually $A_i \gg A_4$ so that $A_4/A_i \to 0$, which yields $\varepsilon_{4i} \approx \varepsilon_4$. We can apply a similar approach for the glass pane I and obtain:

$$\dot{q}_{rad,amb\to 1} = \varepsilon_1 \cdot \sigma \cdot \left(T_I^4 - T_{amb}^4\right). \tag{4.254}$$

At last, we calculate the radiative heat flux density of the two panes with temperatures T_I and T_{II}, whose emission coefficients of the surfaces are ε_2 and ε_3, which is given as follows:

$$\dot{q}_{rad,2\to 3} = \varepsilon_{23} \cdot \sigma \cdot \left(T_{II}^4 - T_I^4\right), \tag{4.255}$$

where the reduced emission coefficient ε_{23} is given by the following relation.

$$\varepsilon_{23} = \left(\frac{1}{\varepsilon_2} + \frac{1}{\varepsilon_3} - 1\right)^{-1}, \tag{4.256}$$

In the steady state of radiative heat transfer through a double-glazed window, the values of all heat fluxes are equal. We apply here the solution method used in the calculation of heat transfer by radiation through a system of screens, since the glasses in a double-glazed window correspond to the role of screens in heat radiation. Thus, this methodology for double-glazed windows is also applicable to any number of parallel arranged glass panes. This results in the radiative heat flux density of:

$$\dot{q}_{rad} = \frac{1}{\left(\frac{1}{\varepsilon_1} + \frac{1}{\varepsilon_2} + \frac{1}{\varepsilon_3} + \frac{1}{\varepsilon_3} - 1\right)}\left(T_i^4 - T_{amb}^4\right). \tag{4.257}$$

After determining $\dot{q}_{rad}$, T_I and T_{II}, the values of the radiative heat transfer coefficients h_{rad} can be calculated for each of the heat transfer points through the window, for which we assume $\dot{q}_{rad,ij} = h_{rad,ij} \cdot \left(T_i - T_j\right)$. This yields $h_{rad,amb1}$ for the outer glass environment; $h_{rad,23}$ for the glass interspace; $h_{rad,4i}$ for the inner pane surface facing the room. The value of the total thermal resistance ($R_{rad,tot}$) for radiant heat transfer is composed of the sum individual thermal resistance, which is given by

$$R_{rad,tot} = \frac{1}{h_{rad,amb1}} + \frac{1}{h_{rad,23}} + \frac{1}{h_{rad,4i}} = R_{rad,amb1} + R_{rad,23} + R_{rad,4i}$$

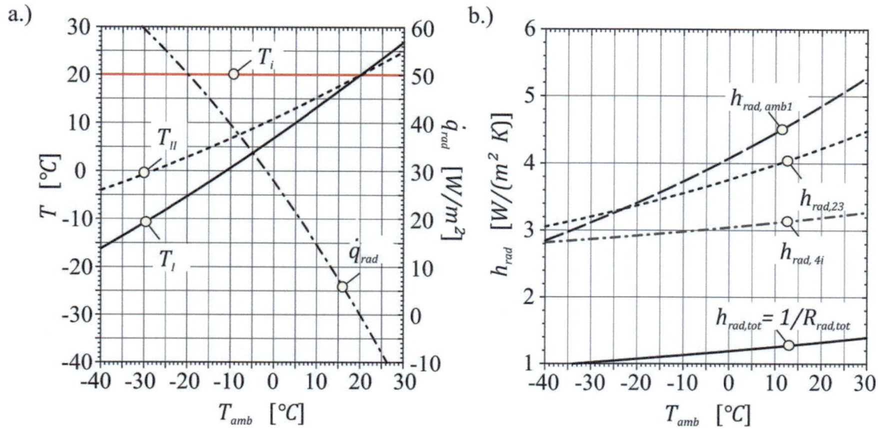

Fig. 4.45 **a** Calculated temperatures T_I and of T_{II} the window panes and radiative heat flux ($\dot{q}_{rad}$) for a given inner room temperature T_i as a function of the ambient temperature T_{amb} for a glas emissivity of $\varepsilon = 0.85$; **b** corresponding total ($h_{rad,tot}$) and individual radiative heat transfer coefficients ($h_{rad,ij}$)

$$R_{rad,tot} = \frac{1}{h_{rad,tot}}. \tag{4.258}$$

Figure 4.45a shows the temperatures of the two panes (T_I, T_{II}) as a function of the ambient temperature (T_{amb}) and the heat flux ($\dot{q}_{rad}$) resulting from radiation exchange alone, assuming an emissivity of $\varepsilon = 0.85$ on each surface of the panes. The corresponding heat transfer coefficient ($h_{rad,ij}$) due to radiation and the total heat transfer $\left(h_{rad,tot}\right)$ coefficient is shown in Fig. 4.45b.

The figure shows that the coefficient of radiative heat transfer decreases both with a decrease in the maximum temperature of two surfaces and with an increase in the temperature difference of these surfaces. The coefficient of radiative heat transfer ($h_{rad,23}$) between the pane surfaces also decreases monotonically with decreasing arithmetic mean temperature of the panes. Thus, with the increase of the temperature difference at a double-glazed window when the ambient temperature is lowered, the total radiation thermal resistance ($R_{rad,tot}$) increases, which in turn means that the radiative heat transfer coefficient decreases as the ambient temperature decreases. Thereby, a reduction of the radiative heat loss can only be realized by means of two methods.

- increase the number of glasses arranged in parallel behind each other.
- the use of glasses with one/more heat-shielding coatings (low emissivity-ε).

While the first method is already known to us from the problem of radiation screens, it makes sense to deal with the second method in particular. For this purpose, let us consider that one or both inner sides of the panes are provided with a low-emissivity coating ($\varepsilon = 0.05$) instead of the prior value of $\varepsilon = 0.05$.

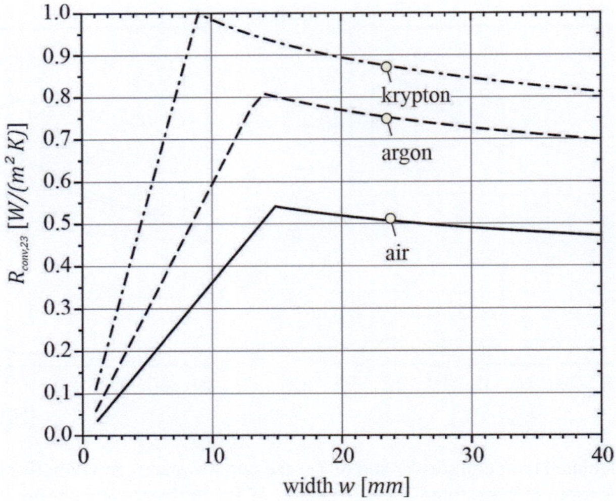

Fig. 4.46 Calculated convective heat resistance ($R_{conv,23}$) as a function of the gas gap width w for temperature difference between the inner pane surfaces of $(T_2 - T_3) = 20°C$ for three gas gap fillings

Since the reduced emission coefficient ε_{23} (see Eq. 4.256) immediately impacts the radiative heat flux exchanged between the panes (Eq. 4.257), the ratio of the radiative heat fluxes for the three cases reads to:

$$\begin{aligned}\dot{q}_{rad}(\varepsilon_2, \varepsilon_3 = 0.85)/\dot{q}_{rad}(\varepsilon_2 = 0.05, \varepsilon_3 = 0.85)/\dot{q}_{rad}(\varepsilon_2, \varepsilon_3 = 0.05) \\ \approx 11/1.84/1.\end{aligned} \tag{4.259}$$

In other words, the specific radiation-related heat loss can be substantially reduced by a low-emissivity coating. The case of emission-reduced coating of the interface 3 is particularly important for solar thermal collectors, since we have seen in the Chap. 3 on optical conversion that emission in the infrared range is a particular development goal for an optically selective absorber.

In the next step, we will consider when heat transfer by conduction and convection is superposed to the radiative heat transfer. The total heat flow ($\dot{q}$) through the glass panes is obtained as a parallel connection of the individual thermal resistances composed of conduction, radiation, and convection, as indicated in Fig. 4.44, and reads to:

$$\dot{q} = U \cdot (T_i - T_{amb}), \tag{4.260}$$

in which U denotes the thermal transmittance in $(W/(m^2K))$.

This total heat transfer coefficient consists of parts comprising the outer pane surface to into the environment (k_{amb1}), the heat conduction in panes I and II, the convective heat transfer between panes I and II as well as the heat radiation between them termed as the total thermal conductance of the glazing (k_t), and finally the combined radiative-convective heat transport from the pane into the room (k_{4i}), which can be expressed by:

$$\frac{1}{U} = \frac{1}{h_{amb1}} + \frac{1}{h_t} + \frac{1}{h_{4i}}. \tag{4.261}$$

The total thermal conductance of the glazing (h_t) is obtained by assuming continuity of the heat flux at each interface and is calculated for M parallel panes with N number of gas spaces as follows

$$\frac{1}{h_t} = \sum_1^N \frac{1}{h_s} + \sum_1^M \frac{d_j}{k_j}, \tag{4.262}$$

where N is the number of gas spaces, d_j the thickness of the of each pane j and k_j its thermal conductivity. Neglecting the contribution of the frame to the heat losses and considering only the heat flow normal to the pane, the thermal resistance from radiation and that from conduction and convection represents a parallel circuit according to the analogue of the electrical equivalent circuit. Thus, the total heat transfer coefficient of each gap ($h_{s,l}$) can be calculated as the sum of the mixed conductive and convective heat transfer coefficient ($h_{g,l}$) and the radiation heat transfer coefficient ($h_{rad,l}$) as follows.

$$h_{s,l} = h_{g,l} + h_{rad,l}, \tag{4.263}$$

In the next step, the external heat transfer from the pane to the environment (h_{amb1}) is determined, which results from a radiation component ($h_{rad,amb1}$) and the convective component ($h_{conv,amb1}$). These are calculated as

$$h_{rad,amb1} = \sigma \cdot \varepsilon_1 \cdot \left(T_{g,1} + T_{amb}\right)\left(T_{g,1}^2 + T_{amb}^2\right), \tag{4.264}$$

where σ is the Stefan-Boltzmann's constant and $T_{g,1}$ is the outer surface temperature of the glass pane to the ambient, which is not constant anymore as in the case of radiative heat transfer. The convective part is calculated by

$$h_{g,amb1} = 0.86 \cdot \frac{u_0^{3/5}}{L^{2/5}} \quad \left(W/\left(m^2K\right)\right), \tag{4.265}$$

where u_0 describes the mean wind speed along the glazing, and L the lateral dimension of the window. $h_{g,amb1}$ shall not be below $5W/\left(m^2K\right)$ even is wind conditions may yield lower values as recommended by (Kalogirou, 2013). For standard conditions and vertical arrangement the European Norm for glass in buildings (European Standard EN673, 2011) has a standard value of $25W/\left(m^2K\right)$ for the purposes of comparison of glazing U-values, which does not consider the improvement of the U-value due to the presence of externally exposed coated surfaces with an emissivity lower than $\varepsilon_1 = 0.837$ (corresponding to corrected emissivity of soda lime glass).

Similar to the external heat transfer the inter heat transfer coefficient (h_{4i}) from the inner pane to the room can be evaluated, in which the radiative part is:

$$h_{rad,4i} = \sigma \cdot \varepsilon_4 \cdot \left(T_{g,4} + T_i\right)\left(T_{g,4}^2 + T_i^2\right), \tag{4.266}$$

while the European norm here simplifies to $h_{rad,4i} = 4.898 \cdot \varepsilon_4 W/\left(m^2 K\right)$. Also for the convective part ($h_{g,4i}$) the European Norm provides a constant value of $h_{g,4i} = 3.6W/\left(m^2 K\right)$, so that $h_{4i} = h_{rad,4i} + h_{g,4i} = 7.7W/\left(m^2 K\right)$.

The thermal conductivity of the gas in the gap between the panes ($h_{gas,23}$) is given by the following equation:

$$h_{gas,23} = Nu \cdot \frac{k_{gas}}{w}, \tag{4.267}$$

in which k_{gas} is the thermal conductivity of the gas in the gap and w the width of the gap interspace. The Nusselt number is a dimensionless parameter and has the physical meaning of the ratio between the magnitude of the heat flux transferred by convection and the magnitude of the heat flux transferred by thermal conductivity through the stationary gas filling the interlayer. To determine the Nusselt number, we actually need the basic equations of convective transport of a fluid, which are the subject of Chap. 5. Here, we will limit ourselves to quantifying the natural convective heat transfer expressed by the Nusselt number using correlations. The Nusselt number correlations themselves depend on the Rayleigh number (Ra) and the inclination angle of the pane gap described by the angle φ with respect to the gravity field vector g, so that $Nu = f(Ra, \varphi)$.

The Rayleigh number itself is the product of the Grashof number (Gr) and the Prandtl number (Pr), i.e. $Ra = (Gr \cdot Pr)$. The Grashof number (Gr) describes the ratio of the buoyancy force acting on a fluid element and the counteracting viscous fluid forces caused by the temperature difference between the panes ($T_2 - T_3$) or a heat flux ($\dot{q}$) applied

$$Gr = \frac{g \cdot \beta \cdot (T_2 - T_3) \cdot L^3}{\nu^2} = \frac{g \cdot \beta \cdot \dot{q} \cdot L^4}{\nu^2 k_{gas}}, \tag{4.268}$$

with L the length of the panes, ν the kinematic gas viscosity, β its volumetric expansion coefficient (for gases $\beta = 1/T = 1/(273.15 + (T_2 + T_3)/2)$), k_{gas} is thermal conductivity, and g the acceleration due to gravity.

The Prandtl number describes the ratio of the transport coefficient of the velocity field versus the temperature field and is defined as follows:

$$Pr = \frac{\rho \nu c_p}{k_{gas}} = \frac{\nu}{a}. \tag{4.269}$$

in which ρ denotes the specific density of the gas and c_p the specific heat capacity. Following the recommendations for the Nusselt-number correlation of the European- Norm, we get for glazings

$$Nu = A \cdot (Gr \cdot Pr)^n, \tag{4.270}$$

in which for vertical glazings $A = 0.35$ and $n = 0.38$, while for horizontal glazings $A = 0.16$ and $n = 0.28$; for $\varphi = 45°$ values of $A = 0.1$ and $n = 0.31$ are stated. Note, that these values contain quite rough approximations with errors in the range of $\pm 10\%$ with respect to the Nusselt number.

Let us first consider a vertical double glazing. For small temperature differences or narrow gaps between the panes, the temperature gradient is constant, and the temperature drops linearly to the colder side. Thus, the heat flow through the gap corresponds to the heat flow transferred by the thermal conductivity of the quiescent gas filling the gap. This mode, in which the Nusselt number takes the value of unity, i.e. $Nu = 1$, is called conductive. As the Rayleigh number increases, the intensity of the gas circulation increases and gradually deviates from the linear distribution. Thus, the Nusselt grows slowly. Initially, a stable laminar state is established, which is characterized by a single convection cell. From about a Rayleigh number of $Ra = 6 \cdot 10^3$, a second cell forms and the Nusselt number continues to increase. Each further increase in the Ra number leads to a gradually more complex flow structure and both the convection velocity and thus the Nusselt number increase continuously. A higher heat transfer from the hot pane towards the cold one expressed yields also to decrease of the heat resistance ($R_{conv,23}$), which is given by

$$R_{conv,23} = \frac{1}{Nu} \cdot \frac{w}{k_{gas}} \tag{4.271}$$

The resulting thermal heat resistance depends on the Nusselt number relationship, the temperature difference between the glass surfaces ($T_2 - T_3$), the thickness of the gas layer w, and the physical properties (type) of the gas, and the impact of the inclination φ with respect to the vertical plane which we subsequently analyse.

- Effect of the thickness of the gas layer (w) on its thermal resistance

The Fig. 4.44 shows the convective heat resistance ($R_{conv,23}$) using the Nusselt number relationship as a function of the distance w of both panes for a vertical double glazing filled with air, at a mean temperature of $10°C$ and a temperature difference between the inner pane surfaces of $(T_2 - T_3) = 20°C$. The curve consists of two sections. A linear section in which the heat resistance grows linearly with the gap width. This is followed by the section in which the heat resistance decreases by the onset of convection. The transition occurs at the point where the Nusselt-number relationship takes on the value of unity and represents the optimal pane distance (or optimal gas layer thickness) with the smallest convective heat losses. This optimal width is associated with a Rayleigh number domain in the range of $3 \cdot 10^3 \leq Ra \leq 10^4$. From this, one can already derive a measure for reducing convective heat transport. If individual elements are introduced into the glass gap, which reduce the active height for the formation of a natural convective flow, the Rayleigh number, which scales with the third (or fourth) power of the length L, is reduced, and Nu keeps small. Such measures are used, for example, in transparent thermal insulation. These so-called convection suppression devices (CSD) have been investigated very early in the solar collectors in order to suppress the convection in the air gab of a flat-plate collector. Parallel films (slats) or tube / honeycomb structures have been investigated as early as in the 1960s. The aspect ratios $(AR = L/w)$ of these devices were in a range of $AR = 3 - 8$ and still showed some convection inside the cells structures. Later on with a further development of the so-called transparent insulation structures larger aspect ratios $AR > 10$ could be produced, which practically suppressed convection even for temperature differences between absorber and glazing of $80K$ and more. Cross section and total thickness of the structures proved also to be important (see Platzer, 1984).

- Impact of the thermophysical gas properties

Thermal resistance can be increased by using fill gases with lower thermal conductivity. Table 4.13 shows the physical properties of some such gases in a temperature range between -10 and $20°C$. The dependency of the heat resistance in relation to changed gas properties assuming a constant temperature difference of $20°C$ between the two glass interior surfaces is also shown in Fig. 4.46. If the heat conductivity of the gas decreases, i.e. from air via argon to krypton, the gradient of the heat resistance increases in the conductive (linear) heat transfer mode. However, the value of the optimal thickness of the gas layer decreases, since the maximum heat resistance is achieved for gases with lower heat conductivity at smaller gap widths. By reducing the optimal thickness of the gas layer, both the amount of expensive inert gas can be reduced, and also multi-chamber double-glazed windows of acceptable thickness can be produced.

- Dependence of the temperature difference of the glass surfaces $(T_2 - T_3)$

The optimum thickness of the gas gap (w_{opt}) between the panes is directly linked to the temperature difference between the disks via the Nusselt number relationship (Eq. 4.270), so that the following relation can be established,

$$w_{opt}^3 \cdot (T_2 - T_3) = Ra_{opt}\left(\frac{\nu^2}{g\beta Pr}\right) = C, \tag{4.272}$$

where the constant C depends on the type of gas. Thereby, an increase in the temperature difference between the glasses leads to a decrease in the optimum distance between them, which is caused by an enhanced convective heat transfer.

Table 4.13 Thermophysical values of some gases from (European Standard EN673, 2011)

Gas $T[°C]$	density ρ $[kg/m^3]$	dynamic viscosity μ $[kg/(ms) \cdot 10^5]$	thermal conductivity k_{gas} $[mW/(mK)]$	spec. heat conductivity c_p $[J/(kgK)]$
Air				
−10	1.326	1.661	23.36	1008
0	1.277	1.711	24.16	
10	1.232	1.761	24.96	
20	1.189	1.811	25.76	
Argon				
−10	1.829	2.038	15.84	519
0	1.762	2.101	16.34	
10	1.699	2.164	16.84	
20	1.640	2.228	17.34	
Krypton				
−10	3.832	2.260	8.42	245
0	3.690	2.330	8.70	
10	3.560	2.400	9.00	
20	3.430	2.470	9.26	

In this case, the corresponding value of $R_{conv,23,opt}$ also decreases.As a result, at $(T_2 - T_3) \geq 25°C$, the distance between the glasses will practically no longer affect the value of thermal resistance. Such undesirable effect of temperature difference on the value of thermal resistance can be reduced if the gas layer with glass is divided into two parts of equal thickness. Then half of the total temperature difference is realized on each of the halves. This is another way to increase the thermal resistance ($R_{conv,23,opt}$) of convective heat transfer through the glass.

- Effect of pane inclination φ on the convective heat transfer resistance

The main effect of inclination of the double glazing φ to the gravity vector g is the reduction of the effective height above which a convection flow is formed. In the case of a horizontally arranged double glazing, i.e. $\varphi = 0°$, the effective height for determining the Rayleigh number corresponds only to the gap width w, while in the purely vertical arrangement considered so far ($\varphi = 90°$) it corresponds to the complete pane height L. Thus, naturally, higher Rayleigh numbers result from an increasing φ according to the sine function of φ, and the convective heat transport between the panes reaches a maximum with a vertical arrangement of the double glazing. This is a complex problem of fluid mechanics, which is usually solved computationally with the help of CFD (see Chap. 5). Nevertheless, there are a number of Nusselt number correlations in the literature, which are supported by numerous experiments. Table 4.14 gives some of the most commonly used relations, more details can be found in Summ et al. (2021), Stephan et al. (2019) and others.

The solution of the whole combined radiation, heat conduction and convection problem consisting of Eqs. (4.260)–(4.263) to determine the combined heat transfer through a single-chambered glass package is iterative. First, we will determine the thermal resistance of the convective heat transfer through the intermediate gas layer $R_{conv,23}$ for which we need the temperatures of the inner surfaces of the glasses T_2 and T_3. First of all, all constant parameters have to be taken into account, such as w, T_i, T_{amb}, $k_{conv,amb1}$, $k_{konv,4i}$, d_j, k_j, ε_i, ε_{ij}, as well as the properties of the filling gas. In a first approximation, we choose arbitrary values for the temperature of the inner surfaces of the glasses T_2 and T_3 from the interval between T_i and T_{amb}. For example, for glasses we can use the condition of equality of temperature drops in all sections: $(T_2 - T_3) = (T_2 - T_{amb}) = (T_i - T_3)$, from which we obtain the average temperature of the gas layer $(T_2 - T_3)/2$. For this temperature we obtain Ra, Nu and the value $R_{conv,23}$. Next, we calculate the individual components and the total thermal resistance of the convective heat transfer and obtain the value of the heat flux from which we obtain the temperatures of all glass surfaces as a first approximation. The values of the temperature of the inner surfaces of the glasses T_2 and T_3 obtained in the first approximation are used in the second approximation for repeated calculations and the loop is repeated until in two successive approximations the values do not change within a previously determined error interval, usually the third decimal digit is used for this.

The double glazing is a similar problem as a solar thermal collector. In the solar collector also a coupling between convection and the radiation transport is possible if for example polymer films as internal glazings are coming too close to a selectively emitting absorber, or if transparent insulation materials (see Chap. 6) like honeycomb structures or aerogel, which are partially transparent to infrared radiation are used to improve the heat losses in the air gap. Simplified methods to solve the coupled equations have been developed alongside these developments, and some details are discussed for example in (Pflüger & Wittwer, 1984; Platzer, 1988, 1992).

One important result is that by using an appropriate air gap in this case between the selective absorber and the absorbing-emitting structure, the heat transport modes may be effectively decoupled, with the results that a much lower combined heat transfer can be observed, which is close to the one calculated with the simplified additive superposition. Other research where coupled heat transfer of both modes, radiation and convection occurs, is the energy loss of stellar atmospheres. Although combined heat transfer is needed in many practical problems in engineering, the treatment still seems to be a challenge in many cases (Dombrovsky, 2022). Simplified superposition is not always correct.

Table 4.14 Nusselt number correlations for inclined cavities modified from Summ et al. (2021), Stephan et al. (2019) von Boeckh & Wetzel (2011), $Ra_\varphi = Racos\varphi$; value of the terms indicated by "+"only considered if > 0, else being 0; $\varphi > 90°$ heated from above

Source	Nu correlation	$Ra/Ra_\varphi/Gr$ [/]	φ [°]	valid AR [/]
Ozoe et al. (1975)	$Nu = 0.109Ra_\varphi^{1/3}$ for $T = const$	$10^3 < Ra < 10^5$	< 10	8–16
Hollands et al. (1976)	$Nu = 1 + 1.446\left(1 - \frac{1708}{Ra_\varphi}\right) \cdot \left(1 - 1708\frac{sin(1.8\varphi)^{8/5}}{Ra_\varphi}\right)^+ + \left(\left(\frac{Ra_\varphi}{5830}\right)^{1/3} - 1\right)^+$ for $T = const.$	$Ra < 10^6$	$0 \leq \varphi \leq 60$	50
Randall et al. (1979)	$Nu = 0.118Ra_\varphi - 4520 - 29$ for $T = const.$	$2 \cdot 10^3 < Ra < 2.2 \cdot 10^5$	$45 \leq \varphi \leq 90$	9–36
Schinkel (1980)	$Nu = CRa^b$ $\varphi = 0°C = 0.08$; $\varphi = 20°C = 0.075$; $\varphi = 40°C = 0.074$; $\varphi = 60°C = 0.072$; $\varphi = 90°C = 0.062$; $b = 1/3$ for $T = const.$ $b = 1/4$ for $\dot{q} = const.$	$10^5 < Ra < 4 \cdot 10^6$	$0 \leq \varphi \leq 90$	6–27
Inaba (1984)	$Nu = 1 + 1.21\left(1 - \frac{2500}{Ra_\varphi}\right)$ $2.5 \cdot 10^3 < Ra < 6 \cdot 10^3$ $Nu = 0.199Ra_\varphi^{0.258}$ $6 \cdot 10^3 < Ra < 4 \cdot 10^5$ $Nu = 0.785Ra_\varphi^{0.33}$ $4 \cdot 10^5 < Ra < 1.2 \cdot 10^6$ $Nu = 0.271AR^{-0.21}(Rasin\varphi)^{1/4}$ $5 \cdot 10^3 < Rasin\varphi < 1.2 \cdot 10^6$ and $60 \leq \varphi \leq 120$	$2.5 \cdot 10^3 < Ra < 1.2 \cdot 10^6$	$60 \leq \varphi \leq 120$	5–83

(continued)

Table 4.14 (continued)

Source	Nu correlation	$Ra/Ra_{\varphi}/Gr$ [/]	φ [°]	valid AR [/]
ElSherbiny (1996)	$Nu_{180} = \left(1 + \left(0.212 Ra^{0.136}\right)^{11}\right)^{1/11}$ $Nu_{120} = \left(1 + \left(0.0566 Ra^{0.332}\right)^{4.76}\right)^{1/4.76}$ $Nu = Nu_{180} + \frac{180-\varphi}{60}(Nu_{120} - Nu_{180})$	$10^2 < Ra < 10^6$	$120 \leq \varphi \leq 180$	20 – 80
Matuska (2009)	$Nu = \left(0.1464 - 2.602 \cdot 10^{-4}\varphi - 2.046 \cdot 10^{-6}\varphi^2\right) Ra^{0.29}$	$10^4 < Ra < 2 \cdot 10^6$	$0 \leq \varphi \leq 90$	n.a
Eismann (2015)	$Nu = 1 + Nu_1 + Nu_2$ $Nu_1 = 1.44\left(1 - \frac{1708}{Ra+1708 R_c}\right)\left(1 - (sin 1.8\varphi)^{1.6} \frac{1708}{Ra+1708 R_c}\right)$ $Nu_2 = \left(\left(\frac{Ra+5830 R_c}{5830}\right)^{0.39} - 1\right)(1 + C R_c)$		$10^3 < Ra < 10^6$	n.a

4.4.4 Effective Thermal Conductivity of Packed Beds

Packed beds consisting of spheres, fibres, or irregular other geometric bodies find engineering applications in many places, including solar thermal issues. Excellent examples are the insulating materials already mentioned in Sect. 4.2.4.4. Other potential applications include thermal heat storage systems, where a solid filling material is used as a heat storage material, which is loaded or unloaded with the help of a heat transfer fluid, for example in a thermocline. As we have seen from the example of insulating materials, the effective thermal conductivity (ETC) does not correspond to that of either the solid or the gas. All these examples are combined effects of heat conduction, convection, and heat radiation.

The most common model for determining the effective thermal conductivity of a packed bed is based on combining all relevant heat transfer mechanisms into a single representative value, the effective thermal conductivity of a packed bed (k_{bed}), which is composed of three contributing components.

The first component is the effective thermal conductivity (k_{eff}) of the bed with a fluid at rest, which in turn includes four different heat transfer mechanisms:

- heat conduction through the solid particle;
- heat conduction through the contact area between adjacent particles, considering the surface roughness;
- heat conduction through the stagnant liquid/gas phase; and
- thermal radiation between solid surfaces.

The second component is the increased effective thermal conductivity of the gas ($k_{g,eff}$), which is due to convective fluid motion flowing through the interstices of the packing parallel to the wall while the solid phase is at rest.

The third component occurs when both the fluid and solid phases are in motion ($k_{s,eff}$), such as during particle transfer in particle receivers. This increases heat transfer in the solid phase due to particle motion of the particles.

As a result, for the effective thermal conductivity of the packed bed (k_{bed} often also referred to as ETC), one obtains a (simplified) superposition of the three components given by:

$$k_{bed} = k_{eff} + k_{g,eff} + k_{s,eff}. \tag{4.273}$$

Due to the increased computer capacities, it is possible to simulate the heat transfer properties of packed beds with the help of the discrete element method (DEM), in which the heat transport over the contact surfaces and the exchange of radiation between the individual particles i and j is determined, whereby the walls and the flow through them are considered as boundary conditions. Since the packed bed is usually a mixture of solid particles of different size and thus porosity and the flowing fluid, however, an exact heat transfer model for the bed is also numerically a very complex and computationally expensive problem.

When considering heat transfer in a packed bed at the micro level, one possible assumption is that each individual particle i has a characteristic temperature T_i. As a consequence, the heat transfer equation for particle i can be written as

$$M_i \cdot c_{p,i} \frac{dT_i}{dt} = \sum_i^{n_m} \dot{Q}_{ij} + \dot{Q}_i, \tag{4.274}$$

where M_i and $c_{p,i}$ represent the particle mass and specific heat capacity, respectively, and Q_i denotes the volumetric heat output originating from the internal heater, while $\dot{Q}_{ij}$ is the heat flux transferred through each neighboring particle j. $\dot{Q}_{ij}$ is then summed over all n_m particles adjacent to particle i. Consequently, the heat transfer of the entire granular system results from a system of linear equations for all particles as well as additional thermal equations describing the initial and boundary conditions.

Before that, however, the we´ve to determine some unknown geometrical quantities, like the arrangement (packing density-porosity), the shape of the single particles, their size distribution and contact areas, but also a model for the thermal contact to determine the unknown $\dot{Q}_{ij}$, which scopes conduction, convective and radiative heat transport.

Geometric Parameters of a Packed Bed

One of the first bed quantities is the porosity, i.e. the gas volume fraction within a total volume. This void fraction (μ) determines not only the packing density but also, in convective problems, significantly the pressure drop through the bed and is defined as follows:

$$\mu = V_g / V_T = V_g / \left(V_g + V_s\right), \tag{4.275}$$

where V_g is the volume of void-space (such as fluids/gases) and V_T is the total or bulk volume of material, including the solid and void components $\left(V_g + V_s\right)$.

From the knowledge of the void fraction, the effective heat capacity of the packed bed is automatically calculated as the weighted average between the heat capacity of the liquid and the solid filler, reading to:

$$\left(\rho \cdot c_p\right)_{eff} = \mu\left(\rho \cdot c_p\right)_g + (1 - \mu) \cdot \left(\rho \cdot c_p\right)_s, \tag{4.276}$$

in which ρ is the specific density and c_p the specific heat capacity.

Far from being constant, the porosity of a packed bed varies from the wall to the centre of a bed. In addition, it matters whether the bed is densely packed or loose random packed. One relevant quantity required for the bed type is given by the coordination number which describes the number of particles in contact with the particle under consideration. This is relevant to model the heat transport phenomena through packed beds. Typical coordination numbers for dense sphere packings are between 5 and 6 in the bed bulk with a porosity (void fraction) of

$\mu = 0.39$. Detailed correlations to coordination numbers and the corresponding porosities can be found e.g. in (van Antwerpen et al., 2010).

Heat Transfer Modelling
In general the heating power between the particles i and j, $\dot{Q}_{ij}$ can be expressed as

$$\dot{Q}_{ji} = -\dot{Q}_{ij} = -H_{ij}(T_i - T_j), \tag{4.277}$$

In which H_{ij} is the effective heat conductance $[W/K]$ between particles i and j. The total heat transfer is then described by:

- heat conduction through the solid and fluid phase;
- heat conduction through the contact area of the of solid between particle i and j;
- radiant heat transfer between solid surfaces in case of gases
- radiant heat transfer between adjacent voids in gases; and the
- convective heat transfer, including the solid–fluid convection and the fluid mixing convection.

This allows us to represent the effective thermal conductance (H_{ij}) exactly in the following way,

$$H_{ij} = H_{ij}^{t} + H_{ij}^{c} + H_{ij}^{rad}, \tag{4.278}$$

where H_{ij} is composed of the solid contact heat conduction fraction H_{ij}^{t}, the solid–fluid solid conduction fraction H_{ij}^{c} and the radiation fraction H_{ij}^{rad}. For the exact description, the consideration of thermal contacts between neighboring particles with separating gaps is relevant, since the solid–fluid-solid heat conduction and the radiation part occur, although the heat conduction through contacting solids does not contribute.

Solid–solid Contact of Pebbles
A solid-state heat conduction exists between two contacting particles with a contact area, as shown in Fig. 4.47a. Since the temperature distribution in the region near the contact is significantly different from that in the region far from the contact, the solid contact conduction must be considered. There is an analytical approximate solution for the effective thermal conductivity H_{ij}^{t} between two smooth elastic spheres with a contact area of radius r_t, which can be written as follows

$$H_{ij}^{t} = 2r_t k_s = \frac{4\sqrt{R_e \delta_n}}{1/k_{si} + 1/k_{sj}} = \frac{4}{1/k_{si} + 1/k_{sj}}\left(\frac{3F_n R_e}{4E_e}\right)^{1/3}, \tag{4.279}$$

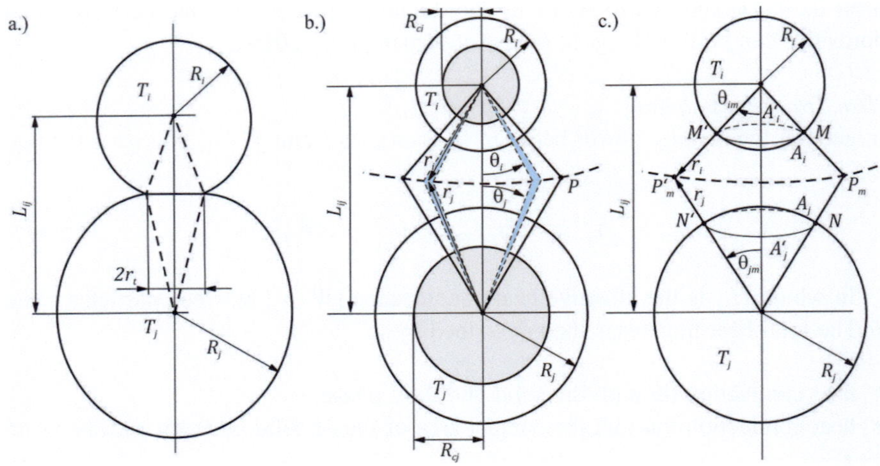

Fig. 4.47 Sketch of the heat transfer between two neighboring particles. **a** solid-state heat conduction via particle contact; **b** heat transfer via solid–fluid-solid conduction; **c** heat transfer between particles via radiation

in which E_e, R_e and δ_n are given by

$$E_e = \left(\frac{1-\nu_i^2}{E_i} + \frac{1-\nu_j^2}{E_j}\right)^{-1}; R_e = \left(\frac{1}{R_i} + \frac{1}{R_j}\right); \delta_n = R_i + R_j - L_{ij}, \quad (4.280)$$

where k_s; ν, E and R are the conductivity, Poisson's ratio, Young's modulus, and radius of the particles, respectively, and the subscript i and j indicates the corresponding particle number. F_n describes the normal contact force caused by the normal overlap dn. E_e and R_e are the equivalent modulus and radius defined by Hertzian contact theory. The solid heat conductivity k_s in Eq. (4.279) is replaced by $2k_{si}k_{sj} = (k_{si} + k_{sj})$ to account for particles with different thermal properties. Equation (4.279) shows the relationship between the conductance and the normal contact force, which allows the exact model to account for the effects of an external mechanical load on the effective thermal conductivity.

Solid–fluid-Solid Contact

In most liquid-filled packed beds, the solid phase has a greater thermal conductivity than the liquid phase, so that the temperature distribution over the solid phase is more uniform, affecting in particular the core part of the particle (grey background in Fig. 4.47b). Therefore, it can be assumed that each particle has an isothermal core part with radius R_c with a representative temperature. For bulk beds of the same particle size, the gaps are small, while for beds with different particle sizes, the gaps are larger. As an approximation to the far-field boundary condition, one assumes a fluid bed at rest in a large zone with a radius 30-times the maximum particle radius. The heat flux by solid–fluid-solid heat conduction from the particle centroid of higher temperature to the particle centroid of lower temperature passes through a point P, as shown in Fig. 4.47b, which is equidistant from both particle surfaces. Based on the above assumptions of an isothermal particle core and heat flow path, the heat flow $d\dot{Q}_{ij}^{c}$ through the infinitesimal volume inside the smaller sphere i shown in Fig. 4.47b)is given by

$$d\dot{Q}_{ij}^{c} = 2\pi r^2 sin\theta_i d\theta_i \left(-k_s \frac{dT}{dr}\right), \tag{4.281}$$

θ_i is the angle between the central axis and the investigated heat flux path and r is the distance from the corresponding particle center. By integrating from the edge of the core part (r = R_{ci}; T = T_i) to the outer boundary of the smaller particle (r = R_i; T = $T_{\theta,\,i}$), we obtain the heat flux to particle j; and doing this in a similar way for the opposite direction, we finally obtain the fraction of differential heat flux dQ_{ij}^{c} between particles i and j

$$d\dot{Q}_{ij}^{c} = \frac{2\pi k_g\left(T_i - T_j\right)}{\frac{1/R_{ci}-1/R_i}{K_{si}sin\theta_i d\theta_i} + \frac{1/R_i-1/r_i}{sin\theta_i d\theta_i} + \frac{1/R_j-1/r_j}{sin\theta_j d\theta_j} + \frac{1/R_{cj}-1/R_j}{K_{sj}sin\theta_j d\theta_j}}, \tag{4.282}$$

with the details of the calculation can be found in the article of (Chen et al., 2019). where $K_{si} = k_{si}/k_g$ and $K_{sj} = k_{sj}/k_g$ denote the dimensionless solid conductivities of particles i and j, respectively. Applying the cosine law for the triangle spanned by the points O_i, O_j and P from Fig. 4.47b yields the following equations.

$$cos\theta_i = \frac{r_i^2 + L_{ij}^2 - r_j^2}{2r_i L_{ij}} = \frac{1 - \Delta\alpha^2}{2\delta_i}; \; cos\theta_j = \frac{r_j^2 + L_{ij}^2 - r_i^2}{2r_j L_{ij}} = \frac{1 - \Delta\alpha^2}{2(\delta_i + \Delta\alpha)} + \Delta\alpha, \tag{4.283}$$

in which $\delta_i = r_i/L_{ij}$, $\delta_j = r_j/L_{ij}, \alpha_i = R_i/L_{ij}, \alpha_j = R_j/L_{ij}$ and $\Delta\alpha = \left(r_j - r_i\right)/L_{ij} = \left(R_j - R_i\right)/L_{ij}$ are the dimensionless length scales of the different radii and the particles radius difference. By this dimensionless parameter Eq. (4.282) mutates

$$dQ_{ij}^{c} = \frac{\pi k_g L_{ij}\left(T_i - T_j\right)\left(1 - \Delta\alpha^2\right)d\delta_i}{(A + B)\delta_i^2 - 2(1 - \Delta\alpha B)\delta_i - (1 - \Delta\alpha B)\Delta\alpha}, \tag{4.284}$$

in which

$$A = \frac{1}{\alpha_i K_{si}}\left(\frac{R_i}{R_{ci}} - 1\right) + \frac{1}{\alpha_i}; \; B = \frac{1}{\alpha_j K_{sj}}\left(\frac{R_j}{R_{cj}} - 1\right) + \frac{1}{\alpha_j}, \tag{4.285}$$

The integration of Eq. (4.282) yields an analytic solution in the following way:

$$H_{ij}^{c}=\frac{\dot{Q}_{ij}^{c}}{\left(T_{i}-T_{j}\right)}\begin{cases}\frac{\pi k_{g}L_{ij}(1-\Delta\alpha^{2})}{2\sqrt{|\Lambda|}}ln\left|\frac{1-Y_{1}}{1+Y_{1}}\right| & \text{for } \Lambda>0\\ \frac{\pi k_{g}L_{ij}(1-\Delta\alpha^{2})}{A+B}\left(\frac{1}{\delta_{imin}}-\frac{1}{\delta_{imax}}\right) & \text{for } \Lambda=0\ ,\\ \frac{\pi k_{g}L_{ij}(1-\Delta\alpha^{2})}{\sqrt{|\Lambda|}}arctanY_{2} & \text{for } \Lambda<0\end{cases} \tag{4.286}$$

with

$$\begin{gathered}Y_{1}=\frac{X_{max}-X_{min}}{1-X_{max}X_{min}};\ Y_{2}=\frac{X_{max}-X_{min}}{1+X_{max}X_{min}},\\ X_{max}=\frac{\Delta\alpha B-1+(A+B)\delta_{imax}}{\sqrt{|\Lambda|}};\ X_{min}=\frac{\Delta\alpha B-1+(A+B)\delta_{imin}}{\sqrt{|\Lambda|}},\end{gathered} \tag{4.287}$$

where $\Lambda=((1+\Delta\alpha A)/(1-\Delta\alpha B))$ represents the discriminant for the integration. The upper limit of the integral δ_{imax} is directly coupled with the nominal area A_n at the interface can be expressed using a geometric relationship as follows:

$$\delta_{imax}=1/2\left(\sqrt{1+\frac{4A_{n}}{\pi L_{ij}^{2}\left(1-\Delta\alpha^{2}\right)}}-\Delta\alpha\right). \tag{4.288}$$

The lower integral value δ_{imin} is given by the direct particle contact so that δ_{imin} can be calculated as follows.

$$\delta_{imin}=1/2\left(\sqrt{1+\frac{4r_{t}^{2}}{L_{ij}^{2}\left(1-\Delta\alpha^{2}\right)}}-\Delta\alpha\right). \tag{4.289}$$

where the contact radius r_t can be solved from Eq. (4.279). For spherical particles of the same size ($R_i = R_j$ and $\Delta\alpha = 0$), the solution of the Cheng-Yu -Zulli (Cheng et al., 1999) model is obtained directly.

Radiation Heat Transfer

In order to determine the radiative heat transfer through a packed bed, we must first make some assumptions to make the physical description computationally accessible. These are:

- All particles in the bed consist of opaque spheres whose size is larger than the wavelength of the radiation
- All particle surfaces should be grey radiating so that E is not a function of solid angle.
- The temperature gradient with respect to the absolute temperature should be much smaller than one in the packed bed, i.e. $\Delta T/T \ll 1$;
- Radiative heat transfer between two adjacent particles takes place in a cone spanned by an imaginary re-radiating surface spanned by the points $MP_mNN'P_mM$' as shown in Fig. 4.47c.

Given these assumptions we can immediately calculate the radiative heat flux between i and j which reads to

$$\dot{Q}_{ij}^{rad} = \frac{\sigma\left(T_i^4 - T_j^4\right)}{\frac{1-\varepsilon_i}{\varepsilon_i A_i} + \frac{1}{A_i F_{ij}+[1/(A_i F_{iR})+1/(A_j F_{jR})]} + \frac{1-\varepsilon_j}{\varepsilon_j A_j}}, \tag{4.290}$$

in which ε_i and ε_j are the emissivty of the partcles i and j and $A_i = 2\pi R_i^2(1 - cos\theta_i)$ and $A_j = 2\pi R_j^2(1 - cos\theta_j)$ describe the areas through which the radiation heat exchange occurs, i.e. MM' and NN'. The variables F_{ij}, F_{iR}, and F_{jR} describe the view factors of the surface pairs MM' and NN', MM' and $MP_mNN'P_m'M'$, and NN' and $MP_mNN'P_m'M'$. Using the reciprocity rule of the view factor we get

$$A_i F_{iR} = A_i(1 - F_{ij}); A_j F_{jR} = A_j(1 - F_{ji}) = A_j - A_i F_{ij}. \tag{4.291}$$

Inserting Eqs. (4.291) in (4.290) we obtain

$$\begin{aligned} H_{ij}^{rad} &= \frac{\dot{Q}_{ij}^{rad}}{(T_i - T_j)} = \frac{\sigma A_i\left(T_i^4 - T_j^4\right)/(T_i - T_j)}{\frac{1-\varepsilon_i}{\varepsilon_i} + \frac{1-\varepsilon_j}{\varepsilon_j A_j/A_i} + \frac{1+A_j/A_i-2F_{ij}}{A_j/A_i-F_{ij}^2}} \\ &\approx \frac{4\sigma A_i T_i^3}{\frac{1-\varepsilon_i}{\varepsilon_i} + \frac{1-\varepsilon_j}{\varepsilon_j A_j/A_i} + \frac{1+A_j/A_i-2F_{ij}'}{A_j/A_i-F_{ij}^2}} \end{aligned} \tag{4.292}$$

while the approximation is justified only if the temperature difference is small.

By summing the heat fluxes over all particles at a given externally applied heat flux density $\dot{q}$ or a temperature difference ΔT, the total heat flux through the packed particle bed is obtained, which is characterized by the geometrical parameters such as packing density (porosity-μ), the coordination parameter and the particle size distribution. The method is indispensable if one has a complex particle bed and is dependent on an accurate determination of the effective thermal conductivity, since the temperature differences lead to a thermal expansion of the particles and thus, depending on the temperature, the heat conduction through point contacts but also the other heat conduction mechanisms change. Figure 4.48 shows a result of such a DEM calculation in a vertical cube with a heat flux $\dot{q}$ $(100W/m^2)$ applied at the top and a lower constant wall temperature $T_w = 421°C$, for a ceramic bed (Li_2TiO_3 particles of a size $d_p = 1.91mm$) and a packing density $\mu = 0.385$. A comparison with the measured data shows a deviation of max. 16%, which occurs especially near the wall.

The complete numerical determination is very time-consuming and only makes sense if the safety behaviour of the system is also connected with the bed behaviour, as for example in nuclear applications.

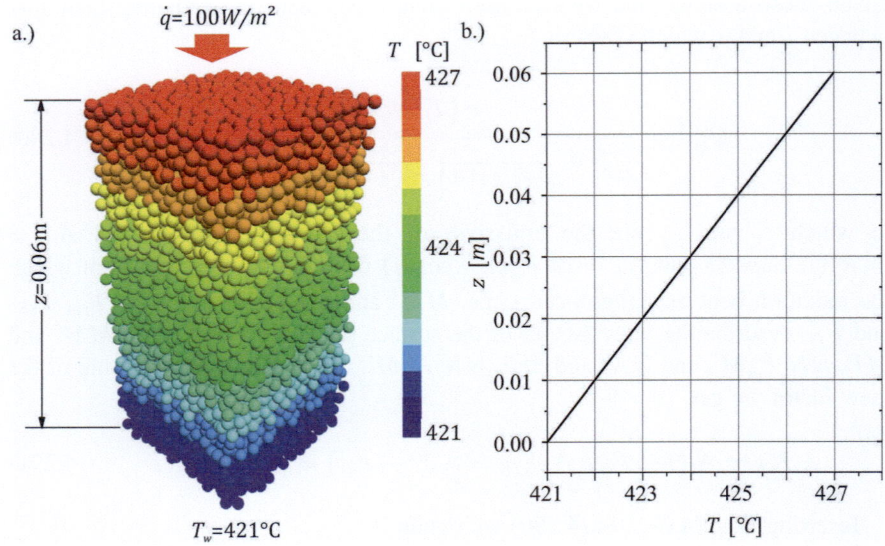

Fig. 4.48 Computed temperature distribution in a monosized spherical particle bed (Li_2TiO_3, $d_p = 1.91mm$, $\mu = 0.385$, $\varepsilon = 08$) exposed to a heat flux $\dot{q}$ ($100W/m^2$) with a constant bottom wall temperature $T_w = 421°C$ with helium as gas adapted from (Chen et al., 2019)

For pebble beds subjected to small mechanical loads, such as in thermal storage or insulation materials, however, the determination of the effective thermal conductivity (k_{eff}) can be substantially simplified by imagining the individual heat transfer mechanisms as analogous to an equivalent electrical circuit. This concept, shown in Fig. 4.49b, considers the entire heat transfer in the gas and solid phases as electrical resistors connected in parallel. The heat transfer in the solid consists of the heat conduction across the contact surface of the particles, the heat conduction through the stagnant gas layer, and the radiative transfer which is in series with the heat conduction through the spheres. In parallel to this is the heat conduction in the gas as well as the radiative component. This cell-based approach to the evaluation of the effective thermal conductivity (k_{eff}) leads to the following equation:

$$k_{eff} = \mu\left(k_g + h_{rad,v}\Delta L\right) + \frac{(1-\mu)}{\frac{1}{\frac{k_g}{l_v}+h_p+h_{rad,s}} + \frac{l_s}{k_s}}; \tag{4.293}$$

in which k_g and k_s describe the thermal conductivities of the gas and the solid; $h_{rad,v}$ and $h_{rad,s}$ reflect the heat transfer coefficients for the thermal radiation from void to void and between solid surfaces. ΔL denotes the effective length between the centres of two adjacent particles along the main direction of the heat flux, l_v specifies the thickness of a film of fluid which yields the same heat resistance as the stagnant fluid close to the contact point, l_s is the thickness of a slab of solid

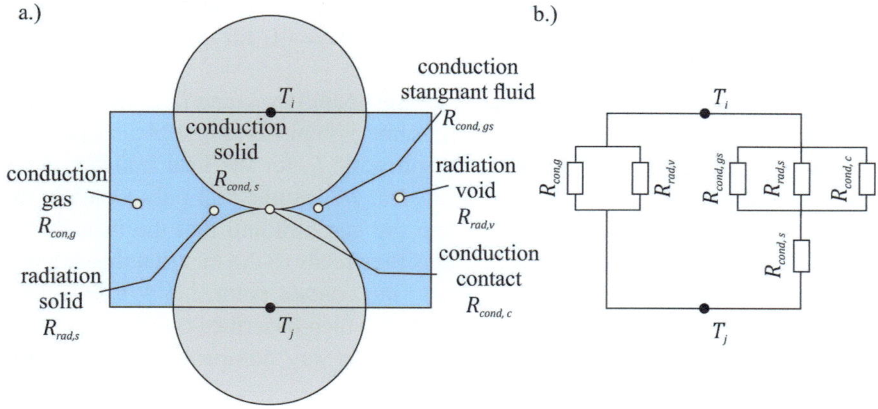

Fig. 4.49 Presentation of a unit cell with the heat transfer mechanisms involved (**a**) and transfer to an equivalent electrical circuit (**b**)

providing the same heat transfer resistance as the pebble and h_p is the heat transfer coefficient for the heat transfer through the contact surfaces between adjacent particles.

This Eq. (4.293) implies that one can estimate the effective thermal conductivity of the entire packed bed from the effective conductivity of a single contact within a uniform bed, which is reasonable because most thermal contacts in a bed of uniformly sized spherical particles with low mechanical loading have similar effective conductivities, and thus the entire bed has the same properties for all local contacts. This also explains why these cell-based theoretical models for predicting the effective thermal conductivity of pebble beds with the spherical sized particles, such as the model of (Kunii & Smith, 1960), are so successful. Ultimately, the problem can thus be reflected in form of a unit cell as illustrated in Fig. 4.49a.

The contribution of thermal radiation is taken into account by means of the heat transfer coefficients $h_{rad,v}$ and $h_{rad,s}$, which are defined as follows

$$h_{rad,v} = \left(4\sigma T^3 + \frac{\mu}{2(1-\mu)}\frac{(1-\varepsilon)}{\varepsilon}\right);\ h_{rad,s} = 4\sigma T^3\left(\frac{\varepsilon}{2-\varepsilon}\right), \tag{4.294}$$

in which ε denotes the emissivity of the particle surface, while σ is the Stefan-Boltzmann constant given by $5.67 \cdot 10^{-8} W/\left(m^2 K^4\right)$.

Using the electrical equivalent circuit (i.e., the notion of unit cell), one can represent the effective thermal conductivity in the following way,

$$\frac{k_{eff}}{k_g} = \mu\left(1 + \frac{h_{rad,v}}{k_g}\beta d_p\right) + (1-\mu)\frac{\beta}{\frac{1}{\frac{d_p}{l_v} + \frac{d_p}{k_g}(h_p + h_{rad,s})} + \frac{l_s}{d_p}\left(\frac{k_g}{k_s}\right)}, \tag{4.295}$$

where β is defined in the following way $\beta = \Delta L/d_p$, where β is in the range between 0.895 for dense sphere packing and $\beta = 1$ for loose packing, see (Trevisan et al., 2021).

What is unknown in Eq. 4.290 so far are the quantities l_s and l_v. Assuming that l_s is the length of a cylinder that has the same volume as the spherical particle, we get $l_s = 2/3d_p$. To determine the length scale l_v for the heat transfer of the gas phase near the contact point of the spheres, we assume that only a part of the total heat flux $(1/n)$ is transferred through the gas film and that the heat flux is parallel to the axis of the two particles. This then leads to the fact that this fraction corresponds to an angle θ associated with one contact point, i.e. $1/n = sin^2\theta$. Thus, a distribution function ϕ can be derived which describes the heat flux over the near-contact fuid region, given in the following way, compare (Kunii & Smith, 1960).

$$\phi = \frac{1}{2}\frac{\left(\frac{K-1}{K}\right)^2 sin^2\theta}{\ln(K-(K-1)cos\theta)-\left(\frac{K-1}{K}\right)(1-cos\theta)} - \frac{2}{3K}, \tag{4.296}$$

in wich K is $K = k_s/k_g$.

What is missing at last is an information about the packing state of the particles, which reflects the number n of contact points. For the densest possible sphere packing one obtains $n = 4\sqrt{3}$ while a loose packing has $n \sim 1.42$ only. This gives $\mu_1 = 0.476$ for a loose packing and $\mu_2 = 0.26$ for a dense ball packing. For real packed beds, it can be assumed that the actual one is a mixture of the densest and the loose ball packing. Since ϕ is proportional to the thermal resistance, the real value for such a bed should approximately correspond to an additive function of ϕ_1 (loose packing) and ϕ_2 (dense packing), so that for a spherical packed bed with a given void fraction μ, the following relation is obtained for the distribution function ϕ:

$$\phi = \phi_2 + (\phi_1 - \phi_2)\frac{\mu - \mu_2}{\mu_1 - \mu_2}. \tag{4.297}$$

The effective thermal conductivities (k_{eff}) determined using the unit cell model show a slightly larger deviation ($\pm 25\%$) than the exact DEM models when compared to experimental data but are conservative in an engineering sense. Figure 4.50 shows experimental results of a graphite pebble bed ($\varepsilon = 0.8$; $\mu = 0.385$) consisting of monodisperse spheres of a diameter of $d_p = 0.06m$ in helium ($k_s/k_g \approx 135$) with the model calculation.

This allows us to analyse qualitatively the influence of different bed parameters on the effective thermal conductivity, which is a function of the bed packing (ϕ, μ), the particle diameter (d_p), the emissivity (ε), the ratio of solid to gas thermal conductivity (k_s/k_g). Details on the influence of these parameters, as well as a set of correlations and their constraints for the determination of packed bed parameters, radiation exchange factors, mean coordination number as a function of porosity can be found in the review studies of van Antwerpen et al. (2010), Chen et al. (2019). Here is just some general information provided.

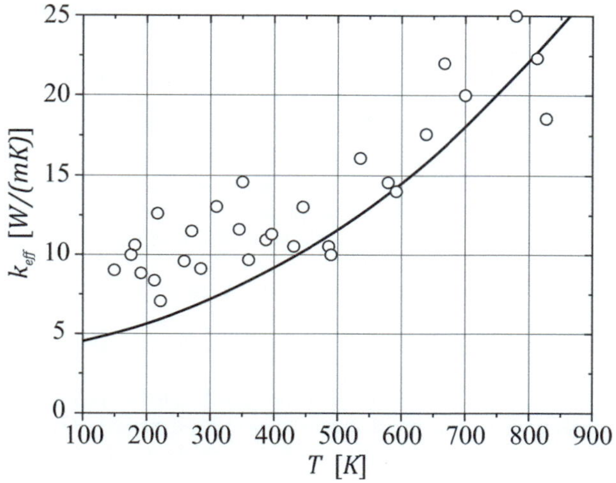

Fig. 4.50 Comparison of experimental (Niessen & Ball, 2000) and predicted (Kunii & Smith, 1960) effective thermal conductivity of a mono-sized spherical graphite particle bed ($\varepsilon = 0.8$; $\mu = 0.385, d_p = 0.06m$) for varying temperatures in helium ($k_s/k_g \approx 135$)

It is obvious that the radiation exchange increases with increasing emissivity (ε), so that a high emissivity is desirable for a high effective thermal conductivity (k_{eff}). What is not so obvious is that the interaction between thermal conduction and radiation can be neglected when the ratio between the conductivity of the solid and that of the gas k_s/k_g is less than 100. For $k_s/k_g > 100$, both the heat conduction and radiation fluxes are affected by the k_s/k_g ratio, and quantitatively in different ways.

The relative contributions of each heat transfer mechanism to the total heat transfer also depend on the particle diameter. For a small particle diameter, the contributions due to conduction are larger than the heat transfer due to radiation and are only slightly dependent on the bed temperature. Conversely, heat transfer by radiation increases rapidly with increasing particle diameter and overtakes heat transfer by conduction; for particles of $2cm$ (and $\varepsilon = 0.8$), a relevant heat radiation contribution starts at about $400°C$. The effective thermal conductivity also increases with increasing bed temperature, mainly because of the increasing contribution of radiation. In packed beds consisting of two components with the same size distribution but different materials, the effective thermal conductivity increases with increasing volume fraction of the particles with the higher thermal conductivity, because they contribute more to the k_{eff} compared to the particles with the lower thermal conductivity.

References

Alados, I., Foyo-Moreno, I., & Alados-Arboledas, L. (2012). Estimation of downwelling longwave irradiance under all-sky conditions. *International Journal of Climatology, 32*(5), 781–793.

Alduchow, O., & Eskridge, R. (1996). Imporved Magnus form approaximation of saturation vapr pressure. *Journal of Applied Meteorology, 35*(4), 601–609.

Baehr, H. D., & Stephan, K. (2008). *Wärme- und Stoffübertragung* (6th ed.). Springer.

Berdahl, P, Fromberg, R. (1982). The thermal radiance of clear skies. *Solar energy, 29*, 299–314.

Bird, G. A. (1994). *Molecular gas dynamics and the direct simulation of gas flows.* Oxford University Press.

Bliss, R. (1961). Atmospheric radiation near the surface of the ground: A summary for engineers. *Solar Energy, 5*, 103–120.

Carslaw, H., & Jaeger, J. C. (1986). *Conduction of Heat in Solids* (2nd ed.). Oxford University Press.

Chen, l., Wang, C., Moscardini, M., Kamlah, M., & Songlin, L. (2019). A DEM-based heat transfer model for the evaluation of effective thermal conductivity of packed beds filled with stagnant fluid: Thermal contact theory and numerical simulation. *International Journal of Heat and Mass Transfer, 132*, 331–346. https://doi.org/10.1016/j.ijheatmasstransfer.2018.12.005

Cheng, G. J., Yu, A. B., & Zulli, P. (1999). Evaluation of e!ective thermal conductivity from the structure of a packed bed. *Chemical Engineering Science, 54*(19), 4199–4209. https://doi.org/10.1016/S0009-2509(99)00125-6

Dombrovsky, Leonid A. (2022): Specialty Grand Challenge for Heat Transfer and Thermal Power. In: Front. Therm. Eng. 2, Artikel 862070. https://doi.org/10.3389/fther.2022.862070

Eismann, R. (2015). Accurate analytical modeling of flat plate solar collectors: Extended correlation for convective heat loss across the air gap between absorber and cover plate. *Solar Energy, 122*, 1214–1224. https://doi.org/10.1016/j.solener.2015.10.037

ElSherbiny, S. (1996). Free convection in inclined air layers heated from above. *International Journal of Heat and Mass Transfer, 39*(18), 3925–3930. https://doi.org/10.1016/0017-9310(96)00047-6

EN 673. (2011). *Glass in building—Determination of thermal transmittance (U-value)-calculation method.* ICS 81.040.20, European Committee for Standardization, Avenue Marnix 17, B-1000 Brussels.

Filippov, L. P. (1968). Liquid thermal conductivity research at Moscow University. *International Journal of Heat and Mass Transfer, 11*(2), 331–345. https://doi.org/10.1016/0017-9310(68)90161-0

Grifoni, R. C., & Passerini, G. (2006). *Transport properties of organic liquids -advances in fluid mechanics* (5th ed.). WIT Press.

Hollands, K., Unny, T., Raithby, G., & Konicek, L. (1976). Free convective heat transfer across inclined air layers. *Journal of Heat Transfer, 98*(2), 189–193. https://doi.org/10.1115/1.3450517

Howell, J. T., Mengüc, M., & Siegel, R. (2020a). *Thermal radiation heat trasnfer* (7th ed.). CRC PRess, Taylor & Francis Group, LLC. https://doi.org/10.1201/9780429327308

Howell, J., Mengüc, M. P., Daun, K., & Siegel, R. (2020b). *Thermal radiation heat transfer* (7 ed.). CRC Press, Taylor & Francis Group, LLC. https://doi.org/10.1201/9780429327308

IfB GmbH Wolfgang Sorge. (2009). Presentation "Vakuumdämmung im Bauwesen" (in German). Nürnberg. Retrieved September 20, 2022, from http://vip-bau.de/

Inaba, H. (1984). Experimental study of natural convection in an inclined air layer. *International Journal of Heat and Mass Transfer, 27*(8), 1127–1139. https://doi.org/10.1016/0017-9310(84)90040-1

Kaganer, M. G. (1969). Thermal insulation in cryogenic engineering. *Israel Program for Scientific Translations*, 212–220.

Kalogirou, S. (2013). *Solar energy engineering* (2nd ed.). Academic Press. https://doi.org/10.1016/B978-0-12-5

Kunii, D., & Smith, J. M. (1960). Heat transfer characteristics of porous rocks. *AIChE Journal, 6*(1), 71–78. https://doi.org/10.1002/aic.690060115

Li, M., & Yiang, Y. C. (2017). On the determination of atmospheric longwave irradiance under all-skyconditions. *Solar Energy, 144*, 40–48.

Lienhard, J. (2001). *A heat transfer textbook* (3rd ed.). Massachusetts Institute of Technology.

Matuska, T. Z. (2009). *A mathematical model and design tool "Kolektor 2.2"*. Czech Technical University in Prague Faculty of Mechanical Engineering, Department of Environmental Engineering.

Modest, M. (2014). *Radiative heat transfer*. Academic Press.

Niessen, H., & Ball, S. (2000). *Heat transport and afterheat removal for gas cooled reactors under accident conditions*. International Atomic Energy Agency.

Ozoe, H., Sayama, H., & Churchill, S. (1975). Natural convection in an inclined rectangular channel at various aspect ratios and angles—experimental measurements. *International Journal of Heat and Mass Transfer, 18*, 1425–1431. https://doi.org/10.1016/0017-9310(75)90256-2

Pflüger, A., & Wittwer, V. (1984). Exact Analysis Of Radiative And Conductive Heat Transfer Through Radiative Grey Films. In: *Optical Materials Technology for Energy Efficiency and Solar Energy Conversion III*. SPIE.

Platzer, W. J. (1988). *Solare transmission und wärmetransportmechanismen bei transparente wärmedämmmaterialien*. Dissertation. Albert-Ludwigs-Universität, Freiburg. Fakultät für Physik.

Platzer, W. J. (1992). Calculation procedure for collectors with a honeycomb cover of rectangular cross section. In: *Solar Energy, 48*(6), S. 381–393.

Randall, K., Mitchell, J., & El-Wakil, M. (1979). Natural convection heat transfer characteristics of flat plate enclosures. *Journal of Heat Transfer, 101*(1), 120–125. https://doi.org/10.1115/1.3450901

Schinkel, W. M. (1980). *Natural convection in inclined air-filled enclosures*. Pijnacker.

Stephan, P., Kabelac, S., & Kind, M., Mewes, D. (2019). *VDI-Wärmeatlas* (12th ed.). Springer Vieweg.

Summ, T., Abdalsalam, M., Ehrenwirth, M., Trinkl, C., & Zörner, W. (2021). A detailed ivestigation on the cnvective heat transfer inside the enclosed cavity of insulated glass solar thermal flat-plate collectors. *Proceedings of the international solar world congress 2021* (p. 11). International Solar Energy Society. https://doi.org/10.18086/swc.2021.22.09

Trevisan, S., Wang, W., & Laumert, B. (2021). Coatings utilization to modify the effective properties of high temperature packed bed thermal energy storage. *Applied Thermal Engineering, 185*, 116414–116512. https://doi.org/10.1016/j.applthermaleng.2020.116414

van Antwerpen, W., du Toit, C., & Rousseau, P. (2010). A review of correlations to model the packing structure and effective thermal conductivity in packed beds of mono-sized spherical particles. *Nuclear Engineering and Design, 240*, 1803–1818. https://doi.org/10.1016/j.nucengdes.2010.03.009

von Boeckh, P., & Wetzel, T. (2011). *Heat transfer -basics and practise*. Springer.

Wakao, N., & Vortmeyer, D. (1971). Pressure dependency of effective thermal conductivity of packed beds. *Chemical Engineering Science, 26(10)*, 1753–1765.

Momentum and Energy Transport

5

Abstract

When solar radiation is converted into heat, thermal losses occur mainly due to passive heat transport modes as there are conduction through gases and solids, heat radiation and natural convection. In the latter one, already macroscopic mass transport driven by buoyancy effects can be observed. Convective heat transport is connected to momentum transport and movement of mass. Driven by mechanical pumps or ventilators the forced convective heat transport is supporting the thermal system by transporting heat with the help of a heat transfer medium from the place of optical conversion to the consumer. Effective heat and mass transport thus plays a key role in the design of a solar thermal system. Convective momentum and heat transport covers many aspects and is a-priori a nonlinear phenomenon. For example, flows can differ in the type of momentum transport, which can be forced, mixed or buoyancy-driven. In addition, the mass flow can have different flow patterns, for example laminar or turbulent. In the engineering treatment, a stability parameter plays a role in predicting the transition between the flow regimes. These regimes are even more complex in two-phase flows. Furthermore, the mass flow can be confined by walls, e.g. in tubes or between plates, often called *bounded flows*. On the other hand mass flow may fully include bodies as *external flow* for instance when an air flow occurs around a thermal storage tank. The interaction of mass flow and momentum transport within the media and with solid bodies not only leads to heat transfer but also to friction, which determines the hydraulic performance. External drives like pumps have to be dimensioned in a way that the pressure loss due to these hydraulic forces can be overcome. Last but not least, the effects of cold/heat can lead to phase changes, such as boiling, evaporation and condensation, which do not alter only the momentum characteristics but also the heat transfer characteristics of the transport. There is an extensive literature on convective momentum and energy transport in its various forms. The objective of this section is to elaborate the basic ideas of single-phase and multiphase momentum and energy transport, to show methods for classifying

R. Stieglitz and W. Platzer, *Solar Thermal Energy Systems*,
https://doi.org/10.1007/978-3-031-43173-9_5

the flow type, and based on this, to present engineering approaches as well as relevant correlations for quantitative calculations, which can be used to carry out assessments.

Keywords

Laminar • Turbulent momentum transport • Boiling • Two-phase flow • Friction • Hydraulic resistance • Natural • Mixed • And forced convection

5.1 Single Phase Convective Momentum and Heat Transport

The aim of this basic first section is to formulate the conservation equations of mass and heat transport in fluids, which will enable us to determine characteristic parameters which can be used to classify different flow forms. For example we may distinguish between laminar and turbulent flows. Since the latter is analytically scarcely accessible, approaches and methods are discussed which allow a numerical calculation. These will be outlined with some examples. In the calculation it is crucial whether the model considers the physical effect(s) occurring in the engineering problem in an appropriate way, which will be highlighted in the last section.

In order to get a more transparent approach to convective transport, which in contrast to passive transfer mechanisms is associated with mass transport, we start with a heuristic consideration.

In a macroscopic mass flow of fluid volumes we find a superposition in the microscopic scale of random motions of individual atoms and molecules. Both scales contribute to the heat transport. At low densities and/or low flow velocities, and when the fluid has a sufficiently high thermal conductivity, heat conduction created by microscopic stochastic kinetics dominates the macroscopic convective heat transport. As the heat capacity or the velocity of the flowing fluid increases, the heat transport due to the macroscopic fluid motion exceeds the molecular processes of heat conduction until these can finally be completely neglected.

Not only the heat transport of a fluid, but also the heat transfer between a flowing fluid and a solid wall is of interest if both have different temperatures. The interaction of the fluid flowing close to the wall creates a thin layer with large velocity and temperature gradients in the fluid, the so-called boundary layer, in which the velocity increases from zero at the wall, henceforth noted as $y = 0$, to almost the value of the undisturbed flow. For flows in pipes, the velocity in the core flow u_0 plays a similar role. The thickness of this "*viscous*" boundary layer is denoted by δ_v. Similarly, the fluid temperature changes across the near-wall zone, called the "*thermal*" boundary layer, starting from the wall temperature $T = T_w$ at $y = 0$ to almost the temperature $T = T_\infty$ at a greater distance from the wall. The thickness of this thermal boundary layer is denoted by δ_{th}. For pipe flows, the mean fluid temperature T_f averaged over the pipe cross-section is used instead of T_∞. The hydrodynamic and thermal boundary layers can differ considerably in

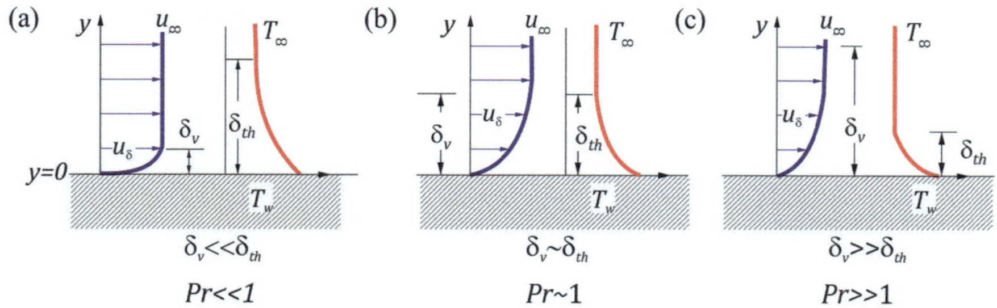

Fig. 5.1 Influence of the Prandtl number on the length scale of the thermal and viscous boundary layer in a two-dimensional flow at a constant wall temperature T_w

thickness from each other and depend largely on the thermophysical properties of the fluid, as indicated in Fig. 5.1.

The origin of the fluid motion can be a forced flow driven by an external drive such as a pump, wind, etc. Moreover, also density or concentration gradients may give rise to convection termed as free or natural convection. Here the buoyancy forces caused by density differences as a result of temperature differences, different phase distributions or different concentrations, e.g. in solutions, result in fluid motion. In practice, superpositions of several forms of convection may occur. If the buoyant flow moves the fluid perpendicular to the direction of the main forced flow, free convection can account for a considerable fraction of the total convective heat transfer. Convective heat transfer need not be limited to the transport of sensible heat. In the case of vapor formation or condensation, the transport of latent heat is added. The formation of vapor bubbles on a wall also causes a change of fluid movements and even may enhance or restrict the heat transfer. Very high heat transfers therefore occur during the condensation of steam on cold walls, for example.

Regardless of the type of convection, the relationship of the heat flux density at the fluid wall interface ($\dot{q}_w$) with the driving temperature difference at the wall and the fluid ($T_w - T_\infty$) is linked with a heat transfer coefficient h in $(\mathrm{W}/(\mathrm{m}^2\mathrm{K}))$ in the following form as already developed in Chap. 4:

$$\dot{q}_w(y=0) = h \cdot (T_w - T_f) \; or \; \dot{q}_w(y=0) = h \cdot (T_w - T_\infty). \tag{5.1}$$

The treatment of heat transfer at the interface is transformed into the functional determination of the heat transfer coefficient by Eq. (5.1). Directly at the wall, the velocity of the fluid approaches zero (non-slip condition). Thus, the heat transfer there takes place only by heat conduction and Fourier's law of heat conduction applies:

$$\dot{q} = -k\left(\frac{dT}{dy}\right)_w, \tag{5.2}$$

in which k denotes the thermal conductivity. This yields for the heat transfer coefficient:

$$h = -\frac{k}{\left(T_w - T_f\right)} \cdot \left(\frac{dT}{dy}\right)_w, \tag{5.3}$$

h is thus determined by the slope of the temperature profile at the wall and the nominal difference between the wall and fluid temperatures. The calculation of the heat transfer coefficient thus requires a knowledge of the temperature field in the fluid, which is strongly influenced by the local velocity field.

The heat transfer coefficient h has been introduced and defined in Chap. 4. The concept of boundary layers has to be filled in the following section in order to understand methods with may treat convective heat transfer quantitatively. Similarly to the heat transfer, ad- and desorption, evaporation or condensation at a wall surface will also depend on the exact properties of the boundary layer.

First, a few order-of-magnitude considerations should be made. The different fluid types, such as air, oil, water, or liquid metals, have completely different thermal conductivities. Liquid metals exhibit a high thermal conductivity k $(\mathrm{W/(mK)})$ and a comparatively low specific heat capacity c_p (J/(kgK)). Oils, on the other hand, have a high kinematic viscosity ν $(\mathrm{m^2/s})$ compared with water or air. The thermal conductivity can be combined with the density ρ $(\mathrm{kg/m^3})$ and the heat capacity c_p to give the thermal diffusivity a $\left(\mathrm{m^2/s}\right)$. This thermal diffusivity has the same SI unit as the viscosity ν. The ratio ν/a yields a dimensionless ratio called the Prandtl number Pr. The physical meaning of the Prandtl number is that it weights the transport coefficient of the velocity field versus that of the temperature field, and thus the Prandtl number describes the ratio of momentum transport versus that of diffusive thermal transport. It is defined as follows:

$$Pr = \frac{\rho \nu c_p}{k} = \frac{\nu}{a}. \tag{5.4}$$

While oils typically show Prandtl numbers of the order of $O\left(10^2 - 10^6\right)$ and "conventional fluids" such as air or water show a value around unity, however, depending on the temperature, the Prandtl number of liquid metals is $Pr = 10^{-3} - 10^{-2}$. Table 5.1 shows the Prandtl number of some Newtonian media at different temperatures. It shows that there exists no Newtonian fluid in the Prandtl number range between $0.05 < Pr < 0.5$. A Newtonian fluid is one where there is a linear relationship between stress and strain-rate. The ratio of stress to strain-rate is the viscosity of the fluid.

With this in mind, consider the flow of a Newtonian fluid of temperature T_∞ over a semi-infinite plate of constant temperature T_w, as plotted in the Fig. 5.1 for various fluids of different Prandtl numbers such as $Pr \ll 1$, $Pr \sim 1$ and $Pr \gg 1$. Case (a) corresponds to liquid metals, which have good molecular thermal conductivity and low viscosity, hence $Pr \ll 1$. The thickness of the viscous boundary layer δ_v is negligible compared to the thickness of the thermal boundary layer δ_{th}.

Table 5.1 Typical temperature dependent Prandtl numbers of some fluids

Temperature (°C)	Mercury	Air	Water	Engine oil
0	0.0288	0.72	13.6	4.7×10^4
20	0.0249	0.71	7.02	1×10^4
100	0.0162	0.70	1.74	2.6×10^2

For water or air, both have approximately the same order of magnitude, while for oils the viscous boundary layer is significantly thicker than the thermal one.

The following sections are structured as follows. First, the basic differential equations of energy and momentum transfer in a continuum as well as the boundary and initial conditions are explained. Based on this, the basic ideas of the solution are shown, where the focus is on the identification of the essential dimensionless quantities or scaling dependencies.

A balance in methodical approach is needed in the engineering applications, between a detailed calculation of local quantities, with the help of finite difference/volume methods, so called computational fluid dynamics (CFD), and an empirical integrated description of the system behaviour or the system yield. The latter make use of scale separation methods and considers mainly the central physical effects quasi one-dimensionally. For example, CFD is required for the design of thermally highly loaded receivers, while a simplified integral description is relevant for the control of mass flows, the dynamic overall system behaviour or also economic performance quantities. In order to take both aspects into account, the basic ideas for the treatment of both local and system level problem solving approaches are presented and correlations or challenges are listed, and reference is made to relevant literature.

5.1.1 Conservation Equations of Mass, Momentum, and Energy

The conservation equations are balances of mass, momentum, and energy at a control volume, in which all fluxes of these control quantities are considered over all inlet and outlet surfaces of the three spatial directions.

This basic idea shall be demonstrated by a simple two-dimensional example of an incompressible ($\rho = const.$), single-phase, single-component flow in a closed control volume to illustrate the logic. In the volume element of dimensions dx aand dy shown in the figure, the mass flow leaving the control volume must be equal to that entering it. The fluid mass flows into the element and leaves it again exclusively as a result of the flow of the fluid. A transport of this type is called convective or advective. For mass flows that go beyond the boundaries of the element, conservation of mass must apply. This states that the sum of the mass flows across the volume boundaries is equal to the change in density in the volume element, or zero for incompressible fluids. Using the mass flows given in Fig. 5.2,

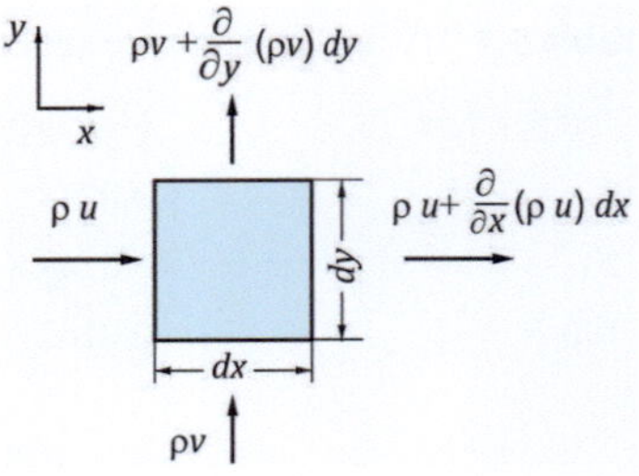

Fig. 5.2 Sketch for the derivation of the mass conservation equation

a mass balance can be established for the simplified, two-dimensional case in the following manner:

$$(\rho u)dy + (\rho v)dx - \left[\rho u + \frac{\partial(\rho u)}{\partial x}dx\right]dy - \left[\rho v + \frac{\partial(\rho v)}{\partial y}dy\right]dx = 0. \quad (5.5)$$

After rearranging the equation, some terms cancel out and after dividing by $dxdy$, we finally get:

$$\frac{\partial(\rho u)}{\partial x} + \frac{\partial(\rho v)}{\partial y} = 0. \quad (5.6)$$

For an incompressible medium, the mass conservation or continuity equation reads to:

$$\nabla \cdot \vec{u} = 0 \text{ or } \frac{\partial u}{\partial x} + \frac{\partial v}{\partial y} + \frac{\partial w}{\partial z} = 0. \quad (5.7)$$

where $\vec{u}$ is the velocity vector composed of the velocity components (u, v, w) in the x, y and z-direction.

The dynamic behaviour of a fluid motion is described by the momentum conservation equations. In analogy to the conservation of mass, a balancing of the momentum changes $(\rho \cdot u \cdot dxdydz)$ and forces acting on the infinitesimal volume element is carried out and as a result a vector equation is obtained for each of the spatial directions in the manner:

$$\begin{aligned}
\rho\left(\frac{\partial u}{\partial t} + u\frac{\partial u}{\partial x} + v\frac{\partial u}{\partial y} + w\frac{\partial u}{\partial z}\right) &= \rho\, f_x + \frac{\partial \sigma_{xx}}{\partial x} + \frac{\partial \tau_{yx}}{\partial y} + \frac{\partial \tau_{zx}}{\partial z},\\
\rho\left(\frac{\partial v}{\partial t} + u\frac{\partial v}{\partial x} + v\frac{\partial v}{\partial y} + w\frac{\partial v}{\partial z}\right) &= \rho\, f_y + \frac{\partial \tau_{xy}}{\partial x} + \frac{\partial \sigma_{yy}}{\partial y} + \frac{\partial \tau_{zy}}{\partial z},\\
\rho\left(\frac{\partial w}{\partial t} + u\frac{\partial w}{\partial x} + v\frac{\partial w}{\partial y} + w\frac{\partial w}{\partial z}\right) &= \rho f_z + \frac{\partial \tau_{xz}}{\partial x} + \frac{\partial \tau_{yz}}{\partial y} + \frac{\partial \sigma_{zz}}{\partial z}, \quad (5.8)
\end{aligned}$$

where $\vec{f} = (f_x, f_y, f_z)$ represent the volume forces occurring e.g. by gravity or imposed electric or magnetic fields. Herein, σ describes the stress normal to the surface and τ the shear stresses tangential to the sides of any control volume, with a linear relationship between stress and strain for a Newtonian medium (Lamb, 1993). On this basis, one can derive the following equations for Newtonian fluids:

$$\sigma_{xx} = -p + 2\mu\frac{\partial u}{\partial x} - \frac{2}{3}\mu\nabla\cdot\vec{u}, \quad \sigma_{yy} = -p + 2\mu\frac{\partial v}{\partial y} - \frac{2}{3}\mu\nabla\cdot\vec{u},$$
$$\sigma_{zz} = -p + 2\mu\frac{\partial w}{\partial z} - \frac{2}{3}\mu\nabla\cdot\vec{u},$$
$$\tau_{xy} = \tau_{yx} = \mu\left(\frac{\partial v}{\partial x} + \frac{\partial u}{\partial y}\right), \quad \tau_{xz} = \tau_{zx} = \mu\left(\frac{\partial w}{\partial x} + \frac{\partial u}{\partial z}\right),$$
$$\tau_{yz} = \tau_{zy} = \mu\left(\frac{\partial w}{\partial y} + \frac{\partial v}{\partial z}\right), \tag{5.9}$$

where $\mu = (\rho\ \nu)$ represents the specific dynamic viscosity (Pa s) of the fluid and p the pressure.

Substituting Eq. (5.9) into (5.8), we obtain the complete momentum conservation equation, known in the literature as the Navier–Stokes equations, which is a set of partial non-linear differential equations of elliptic character. Almost all analytical studies of viscous fluid flows are based on them. They are general in the sense that they also apply to compressible Newtonian fluids of variable viscosity, such as gases. For constant density and viscosity (incompressible fluid and small temperature differences) the Navier–Stokes equations simplify to:

$$\rho\left(\frac{\partial u}{\partial t} + u\frac{\partial u}{\partial x} + v\frac{\partial u}{\partial y} + w\frac{\partial u}{\partial z}\right) = \rho\ f_x - \frac{\partial p}{\partial x} + \mu\left(\frac{\partial^2 u}{\partial x^2} + \frac{\partial^2 u}{\partial y^2} + \frac{\partial^2 u}{\partial z^2}\right),$$
$$\rho\left(\frac{\partial v}{\partial t} + u\frac{\partial v}{\partial x} + v\frac{\partial v}{\partial y} + w\frac{\partial v}{\partial z}\right) = \rho\ f_y - \frac{\partial p}{\partial y} + \mu\left(\frac{\partial^2 v}{\partial x^2} + \frac{\partial^2 v}{\partial y^2} + \frac{\partial^2 v}{\partial z^2}\right),$$
$$\rho\left(\frac{\partial w}{\partial t} + u\frac{\partial w}{\partial x} + v\frac{\partial w}{\partial y} + w\frac{\partial w}{\partial z}\right) = \rho\ f_z - \frac{\partial p}{\partial z} + \mu\left(\frac{\partial^2 w}{\partial x^2} + \frac{\partial^2 w}{\partial y^2} + \frac{\partial^2 w}{\partial z^2}\right),$$

$$\text{or} \quad \underbrace{\frac{d\vec{u}}{dt}}_{\text{inertial term}} = \underbrace{f}_{\text{gravity, buoyancy, capillary forces,}\ldots} - \underbrace{\frac{1}{\rho}\nabla p}_{\text{pressure force}} + \underbrace{\nu\,\nabla^2\vec{u}}_{\text{viscous friction forces}}, \tag{5.10}$$

in which $\nabla^2 = \nabla\cdot\nabla$ is the Laplace operator, which is given by the relation $\nabla^2 = \left(\partial^2/\partial x^2 + \partial^2/\partial y^2 + \partial^2/\partial z^2\right)$.

Any fluid dynamic problem involving the determination of velocity and pressure distribution as a function of spatial coordinates and time requires the simultaneous solution of the continuity equation and the Navier–Stokes equations, considering the special initial and boundary conditions. Although in most cases this system of equations is not accessible to an analytical solution, in many cases it can be solved numerically. Nevertheless, there are a few specific cases where the nature of the flow is such that it can be simplified considerably and thus be submitted to an analytical solution.

A similar procedure can be applied to the derivation of the energy conservation equation. The derivation includes the first law of thermodynamics, which links the thermal energy with the work done on the system and the total internal energy within the control volume, which finally reads to:

$$\rho\frac{dU}{dt} = \nabla\cdot(k\nabla T) + \sigma_{xx}\frac{\partial u}{\partial x} + \sigma_{yy}\frac{\partial v}{\partial y} + \sigma_{zz}\frac{\partial w}{\partial z} + \tau_{xy}\left(\frac{\partial v}{\partial x} + \frac{\partial u}{\partial y}\right) + \tau_{yz}\left(\frac{\partial w}{\partial y} + \frac{\partial v}{\partial z}\right) + \tau_{xz}\left(\frac{\partial u}{\partial z} + \frac{\partial w}{\partial x}\right), \quad (5.11)$$

where k is the thermal conductivity of the fluid, U is the internal energy and T is the temperature.

The internal energy U is related to the fluid enthalpy h in the following way: $h = U + p/\rho$. The operator d/dt describes the total derivative with respect to time. If the relations for the stress and strain on a fluid element (Eq. (5.9)) are substituted into Eq. (5.11), the simplified energy conservation equation is obtained:

$$\rho\frac{dU}{dt} = \nabla\cdot(k\nabla T) - p\nabla\cdot\vec{u} + \mu\Phi \text{ with}$$

$$\Phi = 2\left[\left(\frac{\partial u}{\partial x}\right)^2 + \left(\frac{\partial v}{\partial y}\right)^2 + \left(\frac{\partial w}{\partial z}\right)^2\right] + \left(\frac{\partial v}{\partial x} + \frac{\partial u}{\partial y}\right)^2 + \left(\frac{\partial w}{\partial y} + \frac{\partial v}{\partial z}\right)^2 + \left(\frac{\partial u}{\partial z} + \frac{\partial w}{\partial x}\right)^2 - \frac{2}{3}(\nabla\cdot\vec{u})^2. \quad (5.12)$$

In this, Φ represents the dissipation function and describes the rate of dissipation of mechanical energy per unit mass due to viscosity. The first term on the right describes the net rate of heat conduction in a fluid element per control volume, the second the reversible work done on the control volume, and the last term gives the amount at which viscous forces do irreversible work in the form of viscous dissipation or viscous heating on the control volume. Assuming an incompressible fluid where the change in internal energy is described by $dU = c_p \cdot dT$, the following formulation is obtained:

$$\rho c_p\frac{dT}{dt} = \nabla(k\nabla T) + \mu\Phi \text{ with}$$

$$\Phi = 2\left[\left(\frac{\partial u}{\partial x}\right)^2 + \left(\frac{\partial v}{\partial y}\right)^2 + \left(\frac{\partial w}{\partial z}\right)^2\right] + \left(\frac{\partial v}{\partial x} + \frac{\partial u}{\partial y}\right)^2 + \left(\frac{\partial w}{\partial y} + \frac{\partial v}{\partial z}\right)^2 + \left(\frac{\partial u}{\partial z} + \frac{\partial w}{\partial x}\right)^2. \quad (5.13)$$

If the thermal conductivity λ in the considered temperature regime is constant, the equation simplifies further, and we obtain:

$$\underbrace{\frac{dT}{dt}}_{\substack{\text{convective}\\ \text{heat transport}}} = \underbrace{\frac{k}{\rho c_p}\nabla^2 T}_{\text{heat conduction}} + \underbrace{\frac{\mu}{\rho c_p}\Phi}_{\substack{\text{heat generation by}\\ \text{viscous dissipation}}} \quad . \tag{5.14}$$

The continuity equation, the Navier–Stokes equations, and the conservation of energy equation together provide a complete set for describing any flow within a considered control volume. But, they also pose an almost insurmountable difficulty in the mathematical sense, given the large number of simultaneous equations to be solved and the presence of non-linear terms such as $(u\partial u/\partial x)$.

Because of the non-linearities, the superposition principle of engineering mechanics is not applicable and therefore complex flows cannot be composed of several simple flows as for example by the theory of potential flows, which are irrotational and incompressible and thus the viscous term vanishes, see e.g. Kreith and Goswami (2005). Exact solutions exist for some simple cases where nonlinear effects are small (approximate solutions) or zero. This class of solutions occurs in creeping flows and is important in lubricant film theory. In most engineering applications, however, the nonlinear terms are often larger than the other of the Navier–Stokes equations. An important parameter in this is the hydraulic Reynolds number (Re), a dimensionless quantity that weights the ratio of inertial forces versus viscous forces in a problem and is shown in Eq. (1.16). Finally, the product of the Prandtl number Pr with the Reynolds number Re gives the Peclet number Pe, which describes the ratio of convective heat transport to that by heat conduction:

$$Re = \frac{u_0\, d}{\nu} \text{ and } Pe = RePr = \frac{u_0 d}{a}, \tag{5.15}$$

in which d denotes a characteristic dimension of the hydraulic problem considered.

According to the definition of Reynolds number, creeping flows are characterized by small Reynolds numbers, while in most engineering problems Reynolds numbers well above unity occur.

At last, two observations should be noted. First, velocity and temperature fields are always coupled when the fluid has a temperature dependent density and/or viscosity. Second, temperature fields are similar to velocity fields provided certain conditions are met. If the pressure gradient is zero ($\nabla p = 0$) and $\Phi = 0$ and $f = 0$ and if additionally, the Prandtl number is one ($Pr = 1$), then this similarity is fulfilled (however, this also requires a similarity of the boundary conditions).

5.1.2 Boundary Conditions and Similarity Laws

The system of partial differential equations requires for the solution the specification of boundary and initial conditions either at the in-/outlet and the boundaries of the considered domain. The type of the initial and boundary conditions corresponds mathematically to that specified for the passive heat transfer mechanisms to be of Dirichlet, Neumann or Robin type, however, they scope a wider set of physical quantities such as scalars like flow rate or pressure as well as vectoral or tensorial quantities like e.g. velocity fields, shear stress at interfaces at walls or gas/liquid boundaries and thus impact the flow motion type and the fluid state. Especially, the Navier–Stokes equations can show large variations in mathematical behaviour, ranging from almost completely elliptic to almost completely hyperbolic, which has implications when it comes to prescribing admissible boundary conditions. There is also a discrepancy between mathematically valid boundary conditions and practically useful boundary conditions.

It is, therefore, of paramount importance that physically realistic, well-posed boundary conditions are set either in computational fluid dynamic (CFD) computations or piping network simulations, otherwise severe difficulties are encountered in obtaining solutions. The single, most common cause of rapid divergence of thermal-hydraulics simulations is the inappropriate selection of boundary conditions. Due to the multitude of potentially applicable boundary conditions we limit here to the most common ones as representatives to cover most applications.

5.1.2.1 Hydraulic Boundary Conditions

The hydraulic boundary conditions to be applied at the boundary of a fluid in contact with another medium depends on the nature of this other medium either it is solid, liquid, or gas. It has to be distinguished between in-/outlet and domain confining boundary conditions.

A common pressure boundary condition (p) at in-/outlet of incompressible flows assuming there are no external forces is of Dirichlet type reading to:

$$p = \rho \cdot \vec{g}\left(\vec{r} - \vec{r_0}\right) + p_0 \text{ or } p = p_0, \tag{5.16}$$

in which $\vec{g}$ denotes the gravity vector, $\vec{r}$ the location vector, ρ the density of the fluid and the index "0" indicates the reference conditions. The first term provides the hydrostatic pressure, while the second one determines the system reference pressure. Whenever possible, it is recommended to locate the pressure boundary in a region where the approximate definition of the hydrostatic pressure is applicable, or to define a boundary that is perpendicular to the gravity vector.

Often also a Neumann type boundary condition is applied imposing a constant pressure gradient, which imposes intrinsically a fully developed flow at the outlet, meaning that the velocity gradient in flow direction does not change. It reads to:

$$\frac{\partial p}{\partial \vec{n}} = const. \text{ or } \vec{u} - (\vec{u} \cdot \vec{n})\vec{n} = 0, \tag{5.17}$$

where $\vec{n}$ denotes the outward unity vector on the surface.

At a *solid–fluid interface* the velocity and temperature are continuous. Contained in this boundary condition is the "no-slip" condition: namely, that the tangential velocity of the fluid in contact with the boundary of the solid is equal to that of the boundary. If the wall is at rest, the condition posed along the wall boundary $\partial\Omega$ of the domain Ω is given by:

$$\vec{v}|_{\partial\Omega} = 0. \tag{5.18}$$

For an inviscid fluid, the no-slip condition does not apply, and only the normal component of velocity is continuous. If the wall is permeable, the tangential velocity is continuous and the normal velocity is arbitrary; the temperature boundary condition for this case depends on the nature of the injection or suction at the wall. Assuming an inviscid fluid no boundary layer develops, and the condition reads to:

$$(\vec{v} \cdot \vec{n})|_{\partial\Omega} = 0; \tag{5.19}$$

expressing that there is no flow across the boundary.

At a *liquid–liquid interface* the velocity components are continuous across the interface and the same holds for the normal and tangential stress components.

At a *liquid–gas interface*s as they occur e.g. in two-phase flows the interaction of the liquid with the gas is of relevance. In most cases the tangential stresses in the gas are generally negligible and the boundary conditions imposed are, first, that the normal velocity of the liquid at the interface equals to the normal velocity of the interface; and secondly the pressure of the liquid at the interface is the same as that of the gas. If surface tension (σ) is important across the interface a jump of the pressure occurs according to

$$\nabla p = \left(p_l - p_g\right) = \sigma \cdot \left(\frac{1}{R_1} + \frac{1}{R_2}\right); \tag{5.20}$$

where R_1 and R_2 are the radii of curvature of the intercepts of the interface by two orthogonal planes containing the normal to the interface.

For hydraulic piping networks instead of a pressure often a mass flux boundary condition is imposed. It specifies the mass flux ($\dot{m}$) flowing into the domain across an inlet boundary $\partial\Omega$ and can be specified in a number of ways, while most often the following formulation is used:

$$-\int_{\partial\Omega} \rho \cdot (\vec{u} \cdot \vec{n}) \partial\Omega = \dot{m}. \tag{5.21}$$

In case of a CFD computation this condition must be complemented by a further constraint that the tangential velocity components are set to zero on the boundary $\partial\Omega$:

$$(\vec{u} \times \vec{n})|_{\partial\Omega} = 0. \tag{5.22}$$

Although many combinations of constraints are conceivable, only a few of these combinations allow a solution without over constraining the problem. Here, some permissible combinations are stated:

- walls only;
- walls and inlet and at least one outlet;
- walls and inlet and at least one constant pressure boundary;
- walls and constant pressure boundaries.

Particular care must be taken in applying the outlet boundary condition. It can only be used if all flows entering the calculation domain are given by means of inlet boundary conditions (i.e. velocity and scalars fixed at inlet) and is only recommended for flow domains with a single exit. From a hydraulic point of view the exit pressures determines the flow split between multiple outlets so it is better to specify this quantity at outlets than setting there (zero-gradient) outlet conditions. It is not permitted to combine an outlet condition with one or more constant pressure boundaries, because the zero-gradient outlet condition specifies neither the flow rate nor the pressure at the exit, thus leaving the problem underspecified.

5.1.2.2 Thermal Boundary Conditions

In addition to hydraulics, the description of energy transport requires a definition of thermal wall conditions; a systematic description can be found in Shah and London (1978). Only the most common conditions will be addressed here.

The simplest Dirichlet-type thermal boundary condition along the flow boundary $\partial\Omega$ is that of a constant wall temperature (T_w) circumferentially and axially (i.e. in the flow direction x). This boundary condition occurs in solar thermal applications in condensers and evaporators and is as follows:

$$T_w|_{\partial\Omega} = const. \tag{5.23}$$

The second boundary condition to be assigned to the Neumann type is that of constant axial and circumferential heat flux.

$$\dot{q}_w|_{\partial\Omega} = const. \tag{5.24}$$

Technically, this characterizes a case in which the wall material has low thermal conductivity, and the wall thickness is constant.

Both of the above-named boundary conditions are somehow academic or artificial boundary conditions since they are technically barely feasible. Nevertheless, they are relevant for rough estimations because they reflect the upper and lower limits of the temperatures at the fluid/wall interfaces, respectively.

A condition typical for electric resistance heaters, but also for the one in heat exchangers, is the assumption of a constant heat flow with circumferentially constant wall temperature and axially constant wall heat flux.

$$\dot{q}_w|_{\partial\Omega} = \dot{q}_w(y, z), \quad T_w = T_w(x), \tag{5.25}$$

in which x denotes the axial direction and y, z the circumferential coordinates. It must be emphasized that this condition may only be used if the wall materials exhibit a high heat conductivity compared to that of the fluid, i.e. $\lambda_w \gg \lambda$, or the wall is much thinner than the characteristic dimension of the problem.

A mixed boundary condition of the Robin type is the assumption of an axially constant wall temperature and a finite thermal wall normal resistance.

$$T_{w0} = T_{w0}(y, z) \; ; \; T_w = T_w(x, y, z); \lambda\left(\frac{\partial T}{\partial n}\right)_w = h_0(T_{w0} - T_w);$$
$$\frac{1}{h_0} = \frac{t_w}{\lambda_w} + \frac{1}{h_i}. \tag{5.26}$$

Here T_{w0} is the temperature on the outside of the heater/cooler, T_w the temperature at the fluid/wall interface, h_0 the heat transfer coefficient at the fluid/wall interface, λ_w represents the thermal conductivity of the wall, t_w is its thickness, n is the wall normal unit vector, and h_i is the heat transfer coefficient at the inlet where heat transfer begins. Physically, the finite thermal wall resistance leads to an upstream heating of the fluid.

Convective boundary condition with axially constant heat flux and finite wall-normal thermal resistance. In addition to Eq. (5.26), it is assumed here that the heat conduction at the channel periphery (lateral direction) is negligibly small, which is particularly applicable to selective absorbers. This results in:

$$\dot{q}_w = \dot{q}_w(y, z) \; ; \; k\left(\frac{\partial T}{\partial n}\right)_w = h_0(T_{w0} - T_w). \tag{5.27}$$

Heat conduction boundary condition with axially constant heat flux and finite heat conduction at the channel wall circumference, in which s represents a circumferential coordinate. This boundary condition is an extension of Eq. (5.24) and reads:

$$q_w = q_w(y, z) \; ; \; \frac{q_w}{k} - \left(\frac{\partial T}{\partial n}\right)_w + \frac{k_w t_w}{\lambda}\left(\frac{\partial^2 T}{\partial s^2}\right) = 0. \tag{5.28}$$

This list of boundary conditions can be applied to both laminar and turbulent flows and illustrates the need for a detailed description when recalculating experiments and designing a thermal system. Regardless of the correct choice of boundary conditions in the simulation, this says nothing about the quality of the experiments, which can also be influenced by buoyancy effects in the laminar case, for example.

5.1.3 Similarity Laws in Fluid Mechanics

In the context of the conservation equations at the beginning of the chapter, the Reynolds, Peclet and Prandtl numbers were introduced as dimensionless parameters. We found that the magnitude of these dimensionless ratios separates different transport phenomena of momentum and energy transport. Thus, the equality of dimensionless ratios ensures the conversion of geometrically similar problems into each other. Therein lies the most important significance of the dimensionless quantities for the application in convective problems. It is the basis of every model experiment, in which with the help of measurements on reduced/enlarged models, even with different fluids, statements can be made about the subsequent application execution on a large or small scale. Already in the context of passive heat transfer in Chap. 4 we have found the Fourier number (Fo) and the Biot number (Bi) as important dimensionless quantities to characterize time dependent transient heat conduction problems. But how does one systematically arrive at the derivation of characteristics and similarity laws?

In principle, there are four different methods, which are briefly outlined in the following. As further literature we refer to e.g. to Kreith et al. (2011).

- π-Theorem of Buckingham

In every technical system force, energy, time and one or more characteristic lengths occur. So, there is a number of n different quantities Q_k, for which there is a relation $f(Q_1, \ldots, Q_n) = 0$. Each of these quantities Q_i is dimensionally related, i.e. its physical quantity can be represented in SI units in the form of kg, m, s and degree K. These basic quantities $A_1, \ldots, A_m$ cannot be represented by power products of the others. Thus, each of the quantities Q_i can be determined as a power product of the basic quantities A_i. The question arises: how many dimensionless quantities π of the form $\pi = Q_1^{i_1} \cdot Q_2^{i_2}, \ldots\ldots, Q_n^{i_n}$ are there? Comparing the exponents of the base variables yields a homogeneous linear system of equations of m equations for n unknown i_1, i_2 to i_n. The solvability of the system of equations is thus reduced to the rank r of a dimension matrix. Buckingham's π-theorem now states: If there are n quantities Q_1, ..., Q_n and there is a relation between them, then there are exactly $(n - r)$ dimensionless ratios, where $r \leq m \leq n$ is the rank of the dimension matrix. There is then also a relation $f(\pi_1, \pi_2, \ldots, \pi_{n-r}) = 0$ which solves the problem. This is a relatively mathematical consideration of the derivation of similarity parameters, and the question often arises for the user: how are the free unknowns to be chosen so that one gets meaningful π-ratios? There is no rule here and only experience helps to filter out the physically meaningful parameters.

- Ratios from conservation equations

Ratio formation on the basis of conservation equations is a heuristic procedure in principle. The starting point is the respective conservation equations of the considered problem. For example, the conservation of momentum Eq. (5.10) provides a balance of forces at a control volume of the fluid. As shown in the vectoral notation of the equation, the individual terms describe the contribution of a force to the balance. This differential representation can be expressed in terms of characteristic quantities of the flow, geometry, and fluid by selecting characteristic quantities of the thermal–hydraulic problem as e.g. the mean flow velocity u_0, a characteristic dimension l, and a refe-rence temperature T. This results in:

$$F_i \sim u\frac{\partial u}{\partial x} \sim \frac{u_0^2}{l}\;;\; F_p \sim \frac{1}{\rho}\frac{\partial p}{\partial x} \sim \frac{p}{\rho \cdot l}\;;$$
$$F_\upsilon \sim \nu\frac{\partial^2 u}{\partial x^2} \sim \nu\frac{u}{l^2};\; F_g \sim g, \tag{5.29}$$

in which F_i represents the inertia force, F_p the pressure force, F_υ the friction force and F_g the gravity force. The ratio of inertia force to friction force, i.e. F_i/F_υ, thus gives the Reynolds number. A similar procedure can be followed for the energy conservation Eq. (5.15). Table 5.2 shows a selection of the ratios (π-quantities) to be obtained by this procedure.

Table 5.2 Dimensionless numbers determined from force and energy ratios

Force ratio		Energy ratio	
Euler number	$Eu = \frac{\text{pressure}}{\text{inertia}} = \frac{p}{\rho u_0^2}$	Peclet number	$Pe = \frac{\text{convective transp.}}{\text{heat conduction}} = \frac{u_0 l}{a}$
Reynolds number	$Re = \frac{\text{inertia}}{\text{viscosity}} = \frac{u_0 l}{\nu}$	Eckert number	$Ec = \frac{\text{viscosity}}{\text{convective transp.}} = \frac{u_0^2}{c_p \Delta T}$
Froude number	$Fr = \frac{\text{inertia}}{\text{gravity}} = \frac{u_0^2}{gl}$	Damköhler number	$Da = \frac{\text{reaction enthalpy}}{\text{convective transp.}} = \frac{\Delta H}{c_p \Delta T}$
Weber number	$Wb = \frac{\text{inertia}}{\text{capillary}} = \frac{\rho u_0^2 l}{\sigma}$	Fourier number	$Fo = \frac{l^2}{at}$
Grashof number	$Gr = \frac{\text{buoynancy}\cdot\text{inertia}}{(\text{viscosity})^2} = \frac{gl^3 \beta \Delta T}{\nu^2}$		
Strouhal number	$Str = \frac{l}{u_0 t}$		
Composite numbers			
Stokes number	$Sto = Eu \cdot Re = \frac{p}{\mu}\frac{l}{u_0}$	Prandtl number	$Pr = \frac{Pe}{Re} = \frac{\nu}{a}$
		Rayleigh number	$Ra = Pr \cdot Gr = \frac{g\beta l^3 \Delta T}{\nu a}$

In the relationships of Table 5.2, p indicates pressure, ρ density, ρ kinematic viscosity, g gravity, σ surface tension, β volume expansion coefficient, μ dynamic viscosity, and t time. l and u_0 denote a characteristic length and a velocity, respectively; ΔT is a characteristic temperature difference, a is the temperature conductivity, c_p is the heat capacity, and Q is a heat source or sink.

- Method of differential equations

This method provides the best possible similarity statement, since one starts directly from the differential equations describing the corresponding problem. The scaling laws for the individual influencing variables are obtained directly. The method consists of several successive steps:

- Formulation of the differential equations and boundary conditions.
- De-dimensioning of all occurring quantities with the help of adjusted parameters, for example a characteristic length l, velocity u_0, ...
- Summarization of the incoming parameters to dimensionless parameter groups yields the ratios and as a result the similarity laws.
- Size order approximation of the dimensionless differential equation leads to the reduction of the problem, since only terms of the leading order are considered, and supplies thereby the solution approach for the treatment of the problem.

In the presentation of the boundary layer equations, this method is explicitly applied so that it will not be further discussed here.

- Similarities by transformation of variables

The convective transport of momentum and energy is usually described by a set of partial differential equations and the associated boundary and initial conditions. With the help of the method of the transformations of the independent and dependent variables variable combinations are searched in such a way that the number of the independent variables is reduced by one. By the resulting summary of the variables a similarity statement is obtained in such a way that different fields for example temperature or flow velocities can be converted into each other. The applicability of the method of the transformation of the variables is given if the solution function is continuous thus none of the independent variables takes an excellent value.

This method was already used unconsciously in Sect. 4.2.3 (transient heat conduction) and allows the conversion of different temperature fields at different heat conductivity and times into each other. Other examples of application of the method of transformation of variables known in fluid mechanics are the 1st Stokes problem for the plate, similar solutions of the boundary layer equations, heat transfer at the heated vertical plate, and many others.

5.1.4 Vorticity, Stream Function and Total Head in Incompressible Fluids

The rotation of a fluid element at a given point in the flow domain is related to the vector cross product of the velocity, which reads to:

$$\nabla \times \vec{u} = \left(\frac{\partial w}{\partial y} - \frac{\partial v}{\partial z}\right) \cdot \vec{e}_x + \left(\frac{\partial u}{\partial z} - \frac{\partial w}{\partial x}\right) \cdot \vec{e}_y + \left(\frac{\partial v}{\partial x} - \frac{\partial u}{\partial y}\right) \cdot \vec{e}_z, \tag{5.30}$$

in which $\vec{u} = (u, v, w)$ is the velocity field and $\vec{e}_i$ the unit vector in each direction of a 3D-space. This vector product $\nabla \times \vec{u}$ is called vorticity or $\vec{\omega}$. The Navier–Stokes equation for incompressible ($\rho = const$) flow, Eq. (5.10), may also be written in the form:

$$\frac{d\vec{u}}{dt} = -\frac{1}{\rho}\nabla \mathrm{p} + \vec{g} - \mu(\nabla \times (\nabla \times \vec{u})), \tag{5.31}$$

from which can observe that if viscous forces act on a fluid, the flow intrinsically contains vorticity. Thus, if $\nabla \times \vec{u} \neq 0$ the flow is called rotational, reversely a flow in which the vorticity vanishes is termed as irrotational. Further applying the vector identity,

$$(\vec{u} \cdot \nabla)\vec{u} = \nabla\left(\frac{\vec{u}^2}{2}\right) - \vec{u} \times (\nabla \times \vec{u}) \rightarrow \Delta\vec{u} = \nabla(\nabla \cdot \vec{u}) - \nabla \times (\nabla \times \vec{u}), \tag{5.32}$$

together with the definition of the vorticity $\vec{\omega}$ and using mass conservation ($\nabla \cdot \vec{u} = 0$) the Navier–Stokes equation can be re-written:

$$\rho\frac{\partial \vec{u}}{\partial t} + \nabla\left(p + \frac{\rho}{2}u^2 + \rho g z\right) = \rho\, \vec{u} \times \vec{\omega} - \mu(\nabla \times \vec{\omega}). \tag{5.33}$$

If the flow is said to be *irrotational*, i.e. $\vec{\omega} = \nabla \times \vec{u} = 0$, a velocity potential $\Phi = \nabla\vec{u}$ can be defined and by integration with respect to all spatial variables the unsteady Bernoulli-equation for irrotational, incompressible flows is obtained.

$$\rho\frac{\partial \Phi}{\partial t} + \left(p + \frac{\rho}{2}u^2 + \rho g z\right) = F(t). \tag{5.34}$$

The arbitrary function $F(t)$ introduced by the integration can be summarized in the potential function Φ or is determined by the boundary conditions.

Hence, irrotational flows are always potential flows, even if the flow is compressible. Because the viscous term in Eq. (5.33) vanishes identically for $\vec{\omega} = 0$, it would appear that the preceding Bernoulli equation is valid even for viscous flow. Potential solutions of hydrodynamics are in fact exact solutions of the full Navier–Stokes equations. Such solutions, however, are not valid near solid boundaries or bodies because the no-slip condition generates vorticity and causes non-zero $\vec{\omega}$.

The potential flow solution often used in external flows such as in air foil theory for the determination of lift is invalid in all those parts of the flow field that have been "*contaminated*" by the spread of the vorticity by convection and diffusion.

If we take the curl ($\nabla\times$) of Eq. (5.33), being aware that the curl of any gradient is zero, we obtain that for the vorticity an equation which is independent of the pressure p:

$$\frac{d\vec{\omega}}{dt} = (\vec{\omega}\cdot\nabla)\vec{u} + \nu\Delta\vec{\omega}, \tag{5.35}$$

where $\nu = \mu/\rho$ is the kinematic viscosity in (m^2/s). Therein, first inviscid term on the right describes the rate of change of vorticity by stretching and tilting of vortex lines while the second one represents the rate of change of $\vec{\omega}$ due to diffusion of vorticity, which may be generated, for example, at solid boundaries due to the no-slip condition or by rotation such as Coriolis forces. In inviscid, two-dimensional flow both terms on the right-hand side vanish, and the equation reduces to $d\vec{\omega}/dt = 0$. Physically, this means the vorticity of a fluid particle remains constant as it moves, which is called Helmholtz's theorem.

For a two-dimensional planar, incompressible flow the mass conservation (continuity) reduces to $\partial u/\partial x + \partial v/\partial y = 0$, is identically satisfied by introducing a stream function Ψ defined by:

$$u = \frac{d\Psi}{dy}, v = -\frac{d\Psi}{dx}. \tag{5.36}$$

Physically, the stream function Ψ expresses the volume flow rate between streamlines. Stream functions can be similarly defined to satisfy identically the continuity equations for incompressible cylindrical and spherical axisymmetric flows and for these flows, as well as the preceding planar flow, as long as they are steady.

In inviscid flows the physical meaning of irrotational flow can be illustrated by the relation between the rotation or vorticity $\vec{\omega}$, and the total head, as given by $\left(p/\rho + v^2/2 + gz\right)$. If we assume a steady flow ($\partial/\partial t = 0$) the Bernoulli Eq. (5.33) shows that that the gradient of the total head depends upon the vorticity. Since the vorticity is by definition perpendicular to the velocity vector $\vec{u}$, the gradient of the total head has no component along a streamline. This in turn means that along a streamline in a compressible, inviscid, steady flow the following relation must be satisfied:

$$\frac{p}{\rho} + \frac{v^2}{2} + g\,z = const. \tag{5.37}$$

Thus, an irrotational and steady flow of an incompressible fluid has a constant total head throughout the flow field and Bernoulli's equation is valid throughout the flow field.

Although this section may seem somewhat theoretical at first, the result of considering irrotational flows is that one can introduce a superposition principle of forces along a streamline, which is actually denied by the nature of the Navier–Stokes equations. Thus, the Bernoulli approach prepares the basic pre-requisite for the stream tube (or stream filament) theory considered in a later section, which also allows the inclusion of viscous effects such as pressure loss along a current filament in an additive manner, provided certain conditions are met.

5.1.5 Newtonian and Non-Newtonian Fluid Behaviour

The knowledge of fluid rheology is important in the design of process equipment such as pipelines, pumps, heat exchangers, mixers and lots of other thermal components and the rheological behaviour of fluids can be divided into Newtonian and non-Newtonian categories. Unfortunately, most fluids show non-Newtonian characteristics over at least part of the shear rate range that can be achieved in practice. There are many different forms of non-Newtonian behaviour, and each of these forms has to be considered separately in the design process. Although fluids, organic and inorganic liquids, nearly all gases, inorganic salts, molten metals and salts, and gases exhibit Newtonian flow characteristics this mostly holds when the temperature is well above the melting point. Nonetheless, many liquids industrial significance, especially of multi-phase nature as mixtures, emulsions such as nanofluids, dispersions and suspensions, solutions (both natural and manmade) do not conform to the Newtonian postulate. For most liquids, the viscosity decreases with temperature and increases with pressure, while for gases it increases with both temperature and pressure. Broadly, higher is the viscosity of a substance, more resistance it presents to flow and thus more challenging to pump. Table 5.3 provides typical values of viscosity for scores of common fluids and we immediately recognize that viscosity spreads of several orders of magnitude. One can argue that a solid can be treated as a fluid whose viscosity tends towards infinity, $\mu \to \infty$.

A Newtonian fluid is defined in such a way that at constant temperature and pressure the response of the fluid is characterized by a linear relationship

Table 5.3 Dynamic viscosity μ in (Pa s) of some liquids at room temperature

Fluid	μ (Pa s)	Fluid	μ (Pa s)
Air	10^{-5}	100% Glycerol	1.5
Water	10^{-3}	Honey	10
Ethyl alcohol	1.2×10^{-3}	Syrup	10^{2}
Mercury	1.5×10^{-3}	Bitumen	10^{8}
Ethylene glycol	2×10^{-2}	Molten glass	10^{12}
Olive oil	0.1		

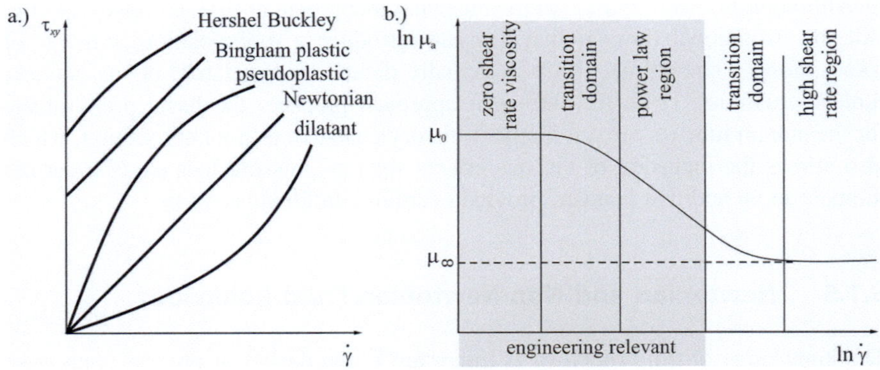

Fig. 5.3 **a** Flow curves characteristics of different viscous fluid types. **b** Flow curve of a pseudoplastic fluid with the different domains of apparent flow viscosites μ_a as function of the strain rate $\dot{\gamma}$

between the applied shear stress (τ) and the rate of shear ($\dot{\gamma}$) and the constant of proportionality is the dynamic viscosity (μ):

$$\tau_{xy} = \mu \frac{du}{dy} = \mu\, \dot{\gamma}, \tag{5.38}$$

where x refers to the direction of the shear stress y the direction of the velocity gradient. It is important to note that in a Newtonian fluid the dynamic viscosity is independent of the shear rate and thus Eq. (5.38) represents a constitutive equation specific to each fluid. Plotting a so-called flow curve, indicating the relation of shear stress as a function of strain rate as schematically shown in Fig. 5.3a is the usual procedure to visualize the viscous properties of a fluid. In such a figure, a Newtonian fluid represents a straight line of origin.

The dynamic viscosity μ has to be distinguish from the frequently used kinematic viscosity ν to which it is related by quantity $\mu = (\nu\ \rho)$. For μ the term shear viscosity is also in common use, and it varies first from substance to substance, but beyond that it is also a function of temperature. Strictly, μ even depends on the pressure (in contrast to the kinematic viscosity), but the pressure dependence can almost always be neglected. The kinematic viscosity ν can also be interpreted as a mass-specific viscosity.

Figure 5.4 contains dynamic viscosities of a larger number of substances as a function of reciprocal temperature. The plot shows that the numerical values of the gas viscosities are all within about one power of ten. In contrast, those of liquids differ by more than five orders of magnitude. Basically, viscosity is the internal resistance to motion due to molecular motion. This leads to collisions of molecules of different average velocities and consequently to an exchange of momentum. Macroscopically, this can be recognized as shear stress. Due to the low density of gases, the number of such molecular collisions is low and thus their viscosity is small. Because of the high density of liquids, the number of molecular collisions and therefore their dynamic viscosity is relatively large. While gases become more viscous with increasing temperature, the viscosity of liquids generally decreases with T. This is due to two different mechanisms. Two different mechanisms are responsible for this, one of which predominates in liquids, the other in gases. In liquids, intermolecular attractive forces must be overcome during shear. These forces generally weaken with increasing temperature because the liquid expands and the mean distance increases. In addition, the mean kinetic energy of the molecules increases, and with it the frequency of space changes. Both lead to reduced toughness with increasing temperature. Since the liquid density hardly changes with pressure, the pressure dependence of the viscosity of liquids can almost always be neglected. The average molecular distance for gases is about one order of magnitude higher than that for liquids. Therefore, intermolecular forces between them can often be neglected (ideal gases; the viscosity is therefore almost pressure-independent for ideal gases up to pressures of 10 MPa, although the molecular distances decrease sharply at this pressure change and thus the forces acting between the molecules increase by powers of ten). With increasing temperature, the molecular motion and thus the number of molecular collisions and, as a consequence, the toughness increase. The kinetic theory of gases leads to an increase in viscosity with $\sqrt{T}$. The application form chosen in Fig. 5.4 shows that for many fluids there is a proportionality $\ln \mu \sim 1/T$, at least to a first approximation, since an approximately linear progression is obtained in the diagram. The real behaviour, however, shows more or less strong deviations from the simple proportionality.

The non-Newtonian fluids are categorized in pseudoplastic (or shear thinning) and dilatant (or shear thickening). Other types as those exhibiting first an initial yield stress after which it acts as a Newtonian fluid, called Buckingham plastic, as e.g. toothpaste and those named Hershel-Buckley, which show a yield stress after which it becomes pseudoplastic. Although non-Newtonian fluids do not have a constant viscosity, it is useful to assign them an apparent viscosity (μ_a) for description using the fluid dynamic equations. This is defined as the ratio of the local shear stress to the shear rate at this point, i.e. $\mu_a = \tau_{xy}/\dot{\gamma}$.

In a next step the constitutive equations need to be defined, which relate the shear stress or apparent viscosity and the shear rate through the rheological properties of the fluid. For engineering purposes, simpler equations are mostly sufficient, and the general idea is described here only.

Since many of the non-Newtonian fluids considered in engineering applications are pseudoplastic, we use them to illustrate typical flow curves and constitutive equations. Figure 5.3b shows qualitative flow curve for a typical pseudoplastic fluid using logarithmic coordinates. The graph exhibits that at low shear rates the fluid shows a Newtonian behaviour with a constant apparent viscosity of μ_a (so-called the *zero-shear rate viscosity*). Increasing the shear rates entering the apparent viscosity begins to decrease until it becomes a straight line. This domain is called the power law region and is an important region in fluid mechanics and heat transfer. Increasing the shear rates beyond the power law region, another transition region appears until again the fluid gets Newtonian again. This behaviour is explained in

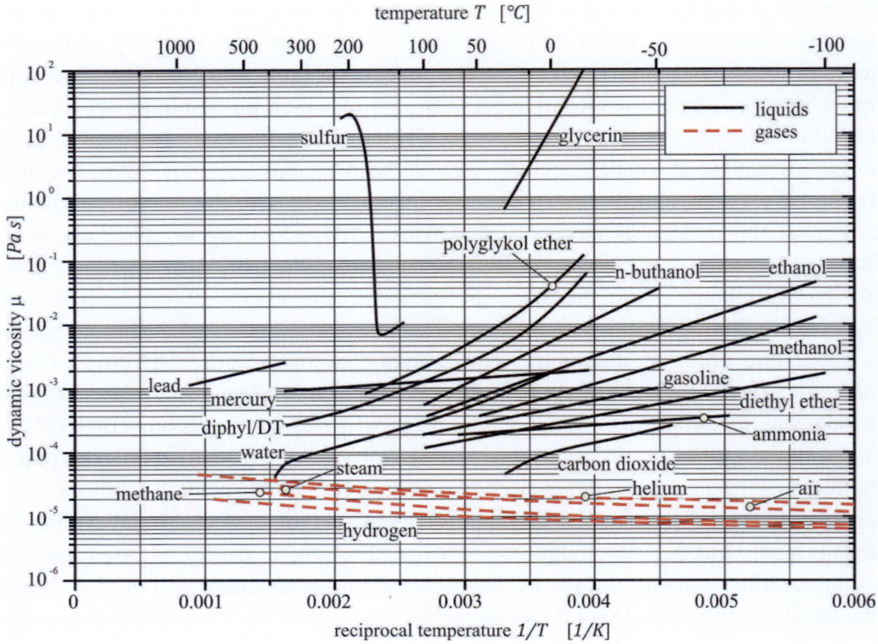

Fig. 5.4 Dynamic viscosity μ (Pa s) of some gases and liquids as a function of temperature T

polymer solutions or melts by the stretching of the molecular chains as a result of shear. The molecules become more mobile, and the viscosity decreases to a certain final value. Low shear rates do not lead to any significant change in the molecular clusters, and the viscosity initially remains unchanged. The grey shaded region indicates the domain where most of the engineering applications occur. A power law region as indicated in Fig. 5.3b can be in the simplest case defined by a constitutive equation as

$$\tau_{xy} = K \cdot \dot{\gamma}^n \text{ or } \mu_a = K \cdot \dot{\gamma}^{n-1}. \tag{5.39}$$

Here, K is termed fluid consistency and n the flow index. If $n = 1$, the fluid becomes Newtonian and K turns into the dynamic viscosity. Because of its simplicity, the power law constitutive equation is most often used in rheological studies, but at times it is inappropriate because it has several inherent flaws and anomalies. For example, if one considers the flow of a pseudoplastic fluid ($n < 1$) through a circular duct, because of symmetry at the centre of the duct the shear rate (velocity gradient) becomes zero and thus the apparent viscosity from Eq. (5.39) tends to infinity, for more details see Boger and Yeow (1988).

In order to avoid misunderstandings, it must be emphasized that a non-Newtonian media behaviour has nothing to do with a strongly variable viscosity change of a fluid, as it can be observed, for example, as a temperature-variable behaviour in water–glycol mixtures and must be considered accordingly in the pump design. Especially for fluid mixture for which the properties are not available in handbooks or other reference materials a measurement of the rheological properties is recommended and a discussion of the various instruments for rheological instruments can be found e.g. in Whorlow (1979).

5.1.6 Laminar Momentum Exchange

The laminar momentum exchange of a steady two-dimensional flow of an incompressible fluid of constant kinematic viscosity ν is described by the following equations:

$$\begin{aligned} &\frac{\partial u}{\partial x}+\frac{\partial v}{\partial y}=0; \\ &u\frac{\partial u}{\partial x}+v\frac{\partial u}{\partial y}=-\frac{1}{\rho}\frac{\partial p}{\partial x}+\nu\left(\frac{\partial^2 u}{\partial x^2}+\frac{\partial^2 u}{\partial y^2}\right); \\ &u\frac{\partial v}{\partial x}+v\frac{\partial v}{\partial y}=-\frac{1}{\rho}\frac{\partial p}{\partial y}+\mu\left(\frac{\partial^2 v}{\partial x^2}+\frac{\partial^2 v}{\partial y^2}\right). \end{aligned} \tag{5.40}$$

This is a system of nonlinear partial differential equations of elliptic type. A uniform solution does not exist and there are few exact solutions under rather restrictive boundary conditions. Fortunately, most laminar exchange processes can be simplified considerably. At large Reynolds numbers, almost all exchange processes occur in a thin layer called the viscous boundary layer. Boundary layer simplifications are not limited to flows over solid walls; free jets or wakes behind a body can also have boundary layer character.

First, let us consider a stationary, planar and a fully developed flow in a channel as shown in Fig. 5.5. In this context, fully developed means that the velocity component u does not change in the x-direction; thus $\partial u/\partial x = 0$. For continuity reasons, $\partial v/\partial y = 0$. This ultimately means that the v-component of the velocity is constant. Excluding suction or blowing across the walls, one immediately obtains that the transverse velocity is $v = 0$. Thus, the convective terms in Eq. (5.40) on the left side disappear and in the force balance only the expression $dp/dy = 0$ remains, which means that the pressure in wall-normal direction is constant and depends only on the flow direction x.

$$\frac{dp}{dx}=\mu\frac{d^2u}{dy^2}=\frac{d\tau}{dy}. \tag{5.41}$$

Since u is only a function of the wall-normal direction y, the partial derivative ∂ can be replaced by the simple derivative d. Only $\tau = \mu\partial u/\partial y$ remains from the stress tensor. An integration leads directly to a linear relation for the shear stress because of the symmetry which requires $\tau(y = H) = 0$ and we obtain:

$$\tau_w=-\frac{dp}{dx}H. \tag{5.42}$$

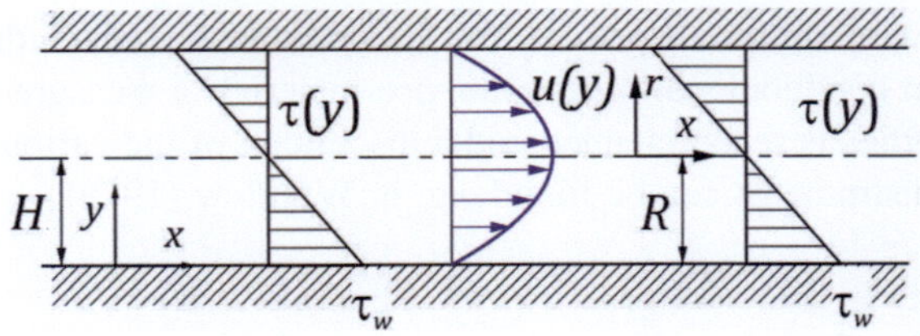

Fig. 5.5 Shear stress and velocity distribution of a fully developed flow in a planar duct or pipe

This equation links the wall shear stress with the pressure gradient and results in dimensionless form in the relationship

$$\frac{\tau(y)}{\tau_w} = 1 - \frac{y}{H}. \tag{5.43}$$

With $\tau = \mu \partial u / \partial y$ and a further integration the velocity distribution is obtained. If we set $u(y = H) = u_{max}$ we get:

$$u_{max} = \frac{\tau_w H}{2\mu} = -\frac{dp}{dx} \cdot \frac{H^2}{2\mu} = \frac{\Delta p H^2}{2\mu L}, \tag{5.44}$$

where $\Delta p / L = -dp/dx$ is the pressure drop along the channel of length L. Finally, for the dimensionless velocity distribution, we obtain:

$$\frac{u}{u_{max}} = \frac{y}{H}\left(2 - \frac{y}{H}\right), \tag{5.45}$$

which corresponds to the classical parabolic Hagen-Poiseuille profile shown in Fig. 5.5. The same result is obtained for a laminar, fully developed pipe flow leading to:

$$\frac{\tau(r)}{\tau_w} = \frac{r}{R};\ u(r) = \frac{\Delta p}{4\mu\ L}\left(R^2 - r^2\right);$$
$$u_{max} = \frac{\Delta p\ R^2}{4\mu\ L};\ \frac{u}{u_{max}} = 1 - \left(\frac{r}{R}\right)^2, \tag{5.46}$$

where R in this context represents the pipe radius. The factor ½ is explained by the fact that in a pipe flow, compared to a plate flow, the ratio of the channel cross-sectional areas on which the pressure acts are twice as large as the surface on which the viscous forces act to compensate for the pressure. The volumetric flow $\dot{V}$ in a pipe flow thus calculates to:

$$\dot{V} = \int_{r=0}^{r=R} 2\pi r\ u(r) dr = \frac{\pi\ \Delta p\ R^4}{8\ \mu\ L}. \tag{5.47}$$

It is important to emphasize the scaling of the flow rate $\dot{V}$ with $\dot{V} \sim \Delta p/L$ and $\dot{V} \sim R^4$. For the same applied pressure gradient, a 10% increase in the tube diameter leads to a 46% increase in flow rate. Assuming a mean velocity u_0 is defined using $u_0 = \dot{V}/A$ with A as cross-sectional area, $u_0 = 1/2 u_{max}$ is obtained. For engineering purposes, the pressure drop is a crucial parameter. Using $\dot{V} = \left(u_0\ \pi\ R^2\right)$ and for the diameter $D = 2R$, we get for the pressure loss over a duct length L:

$$\Delta p = \frac{\rho}{2} u_0^2 \frac{L}{D} \frac{64}{\text{Re}}. \tag{5.48}$$

Defining a dimensionless quantity in form of a friction factor f_D in the way $\Delta p = (\rho/2)(L/D)u_0^2$ results in the Darcy formulation of the friction law of a pipe flow:

$$f_D = \frac{64}{Re}. \tag{5.49}$$

In it, Re is the hydraulic Reynolds number defined as $Re = u_0 D/\nu$. Thus, the pressure drop in a laminar pipe flow depends linearly on the velocity and the Reynolds number is the only occurring variable in all momentum exchange processes. One has to be careful about the normalization of the friction factor, if e.g. the Fanning friction factor f_F is mentioned, the friction factor is defined as the ratio of the wall shear stress τ_w to the dynamic pressure $(u_0^2 \cdot \rho/2)$. Then the Fanning friction factor for the tube reads instead of Eq. (5.49) to $f_F = 16/Re$. It should be noted that the Fanning friction factor is also used in the technical literature. The Fanning friction factor is 1/4 of the Darcy friction factor.

Forming an expression for frictional head loss and normalizing instead of the dynamic pressure with the gravitational pressure ($\rho\ g$), we get the so-called frictional head loss h_L in (m):

$$h_L = \frac{\Delta p}{\rho\ g} = 32 \frac{\nu\ u_0}{g} \frac{L}{D^2}. \tag{5.50}$$

If we use the Darcy formulation for the friction factor f_D, we obtain:

$$h_L = f_D \cdot \frac{L}{D} \frac{u_0^2}{2g}. \tag{5.51}$$

5.1.7 Boundary Layer Approximation

As we have already seen, due to the wall non-slip condition, a boundary layer is formed as a result of the viscosity of the fluid, within which the velocity changes from zero to the value of the external flow, as schematically illustrated in Fig. 5.6.

According to Prandtl's hypothesis developed in 1904, the effects of fluid friction at high Reynolds numbers are limited to a thin layer near the boundary of a body, while the rest of the domain behaves as inviscid. The thickness of the boundary layer is unknown a priori. Nevertheless, an estimate of its wall-normal extension can be obtained by means of order of magnitude estimations. If one introduces the following dimensionless variables,

$$x' = \frac{x}{L} \; ; y' = \frac{y}{\delta(x)} \; ; u' = \frac{u}{u_0} \; ; p' = \frac{p}{\rho_0 u_0^2}, \tag{5.52}$$

choosing the scaling in such a way that the ratios are of $O(1)$, and in a further step takes the length L and the mean flow velocity u_0 of the external flow as a reference variable in such a way, that

$$v' = \frac{v}{u_0} \; ; \rho' = \frac{\rho}{\rho_0} = 1 \quad (\text{since } \rho = const.), \tag{5.53}$$

and inserts all into the continuity equation the following result is obtained,

$$\frac{v}{u_0} \sim \frac{\delta}{L}, \tag{5.54}$$

since u', x' and y' are all of order $O(1)$. Inserting the chosen scaling into the Navier–Stokes equation (5.40) for the x-direction, the result is that the Reynolds number appears as a determinant parameter characterizing a viscous flow. Returning to the estimation and postulating that the viscous terms are of the same order of magnitude as the inertia terms and the pressure forces, it immediately follows that the thickness of the boundary layer scales as follows:

$$\frac{\delta}{L} \sim \frac{1}{\sqrt{Re}}. \tag{5.55}$$

A similar insertion of the dimensionless scaling in momentum equation for the y-direction shows that for $Re \gg 1$ the derivative of the pressure in wall-normal direction disappears. This means that the pressure perpendicular to the wall is constant within the boundary layer and is thus imposed by the inviscid external

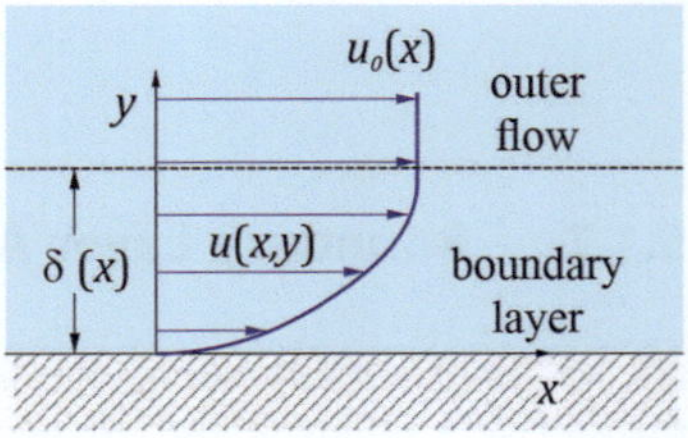

Fig. 5.6 Flow profile of a boundary layer flow

flow. Thus, the pressure p in the boundary layer is not an unknown quantity but a boundary condition. The boundary layer equations are thus:

$$\frac{\partial u}{\partial x}+\frac{\partial v}{\partial y}=0 \text{ and } u\frac{\partial u}{\partial x}+v\frac{\partial u}{\partial y}=-\frac{1}{\rho}\frac{dp(x)}{dx}+\nu\frac{\partial^2 u}{\partial y^2}. \tag{5.56}$$

There are two equations to determine the unknowns $u,\ v=f(x,y)$. The pressure and the velocity gradient of the external flow are coupled via the Bernoulli equation (5.34). The boundary conditions to the system of equations are:

$$\begin{aligned} y=0:&\quad u=v=0,\\ y=\delta:&\ u=u_0(x). \end{aligned} \tag{5.57}$$

The system of equations of the boundary layer equations is thus parabolic and is a typical initial boundary value problem. Because of the simplification of the viscous terms, viscous dissipation is actively restricted to the y-direction only. There is no upstream information, and all flow quantities can be calculated straight forward, i.e. downstream. This is a considerable simplification compared to the elliptic Navier–Stokes equations, where a simultaneous calculation in all flow regions is required.

The main objective of laminar momentum exchange is to determine the local friction loss coefficient f_D and/or the global friction coefficient f_F. The reason is that such solutions include both friction factor–Reynolds number relations and a shear rate parameter, which is of relevance if non-Newtonian liquids are considered. The latter allows the designer to determine the shear rate region in which the system considered is operating and thus to choose the appropriate solution. Thus, for a laminar fully developed flow or a Newtonian or a modified power law fluid in a circular duct, the product of the friction factor and a certain Reynolds number is a constant depending on the flow index, n, and the shear rate parameter β.

$$f_D\cdot Re_h=const.(n,\beta), \tag{5.58}$$

where f_D is the Darcy friction factor and Re_h the Reynolds number built with the hydraulic diameter, which is defined as:

$$d_h=A/P, \tag{5.59}$$

with A the cross-sectional area of the conduit and P the wetted perimeter. For Non-Newtonian fluids the modified power law Reynolds number must be used, which contains a modified dynamic viscosity μ^* given by:

$$\mu^*=\frac{\mu_0}{1+\beta} \text{ with } \beta=\frac{\mu_0}{K}\left(\frac{u}{d_h}\right)^{1-n}. \tag{5.60}$$

Therein, β is the shear rate parameter and K the fluid consistency. When the $\log_{10}\beta$ is less than approximately -2, the duct is operating in a Newtonian region as indicated in Fig. 5.3b and therefore classical Newtonian solutions can be used. Considering the laminar momentum exchange of Newtonian fluids, the Darcy factor f_D and the Faning friction factor f_f can be expressed as,

$$f_f(x) = f\left(Re_x^m, e/d_h\right) \quad \text{and} \quad f_D = f\left(Re_L^n, e/d_h\right), \tag{5.61}$$

where Re_x is the local Reynolds number and Re_L is the Reynolds number formed with the characteristic dimension of the channel or along the body contour. Herein, the ratio e/d_h accounts for the surface roughness of the channel or on the body.

For a channel flow $n = -1$ holds and there the friction factor does not change in flow direction for fully developed flows. This differs from the boundary layer flow along a plate since there the boundary layer thickness along the plate increases in flow direction x while the local friction coefficient decreases with x. In this case the exponent $m = -1$ results for the local friction coefficient. In this case, the local friction coefficient has the exponent $m = -1/2$ and the total friction coefficient has the exponent $n = -1/2$. It is crucial in this context that the only two dimensionless quantities occurring are the Reynolds number and the surface property scaled with a characteristic dimension D or L. For fully developed flows, the inertial force can be replaced by the driving pressure gradient.

The following solution approaches for the calculation of laminar flows can be distinguished:

- *Analytical solutions* of the Navier–Stokes equations existing only for a few special cases.
- *Numerical solutions* of the Navier–Stokes equations are extremely time-consuming, due to the elliptic character of the nonlinear partial differential equations (PDE). They constitute a boundary value problem and no consecutive methods as for parabolic partial differential equations can be applied. In engineering practice, however, these are of little importance. For the limiting case $Re \to \infty$, the Navier–Stokes equations mutate into the boundary layer equations. In the practical cases, the Reynolds number is usually large enough to apply the boundary layer equations. An exception is the plate leading-edge problem for which, as a pre-requisite, $\partial\delta/\partial x \ll 1$ and $\partial^2/\partial x^2 \ll \partial^2/\partial y^2$ are not satisfied.
- *Exact solutions* of the boundary layer equations exist only for a few cases and denote here an arbitrarily exact numerical solution of an ordinary differential equation. For the class of self-similar solutions, the shape of the individual velocity profile does not change in the direction of flow, and it can be regressed to a similar profile by a suitable coordinate transformation.

For arbitrary external flows, the velocity profile changes in the flow direction and there are no self-similar solutions. In such cases, the boundary layer equations must be solved numerically. This can be done in three ways:

- *Numerical solutions* of the boundary layer equations exist for a variety of problems—using difference methods, the parabolic boundary layer equations can be numerically integrated as an initial value problem. For stability reasons, implicit procedures are preferable to explicit ones; several procedures are described in detail in Cebeci and Bradshaw (1977).
- *Integral methods* for solving the boundary layer equations such as Pohlhausen and V. Karman schemes, demanding low computational and memory requirements. Integral methods are still used, especially for the calculation of turbulent boundary layers. An extensive discussion of the various numerical solution approaches can be found in Walz (1966). Despite many indisputable advantages of integral methods, they have the central disadvantage; that the quality of the results depends on the adequate assumptions of the chosen initial profile.
- *Empirical solutions* are mainly used where complex geometries do not allow a direct numerical solution.

Finally, for the sake of completeness, two problems should be mentioned.

- These are higher orders of boundary layer theory, such as perturbation approaches, the so-called coupled asymptotic expansions, which are a further development of Prandtl's boundary layer concept. An overview of the different models is given in the books of van Dyke (1975) or Cole (1968).
- Another not considered aspect are three-dimensional boundary layers. Solution approaches for this are described (Schlichting & Gersten, 2000 or White, 1991).

A summary of the pressure losses or friction factors occurring in laminar flows is given together with the representation of the laminar heat transfer.

5.1.8 Laminar Energy Exchange

In a flowing fluid, energy and momentum are exchanged if

- thermal energy is supplied to or dissipated from the system;
- the kinetic energy of the flow is of the order of magnitude of the internal energy, so that dissipation (internal friction) causes the fluid temperature to rise.

In the first case a distinction is made between forced, mixed, and buoyant convection. In forced convective systems, the flow is driven by an external source, for example, a pressure gradient generated by a pump, while in free or natural convection, density differences (resulting from temperature differences) are the driving forces of the flow.

Considering stationary, two-dimensional, and single-phase flow of a Newtonian fluid, we obtain the following system as the set of equations to be solved for laminar flows with energy transport:

$$\frac{\partial u}{\partial x}+\frac{\partial v}{\partial y}=0;\ \rho\left(u\frac{\partial u}{\partial x}+v\frac{\partial u}{\partial y}\right)=-\frac{\partial p}{\partial x}+\mu\left(\frac{\partial^2 u}{\partial x^2}+\frac{\partial^2 u}{\partial y^2}\right)+\rho g_x;$$
$$\rho\left(u\frac{\partial v}{\partial x}+v\frac{\partial v}{\partial y}\right)=-\frac{\partial p}{\partial y}+\mu\left(\frac{\partial^2 v}{\partial x^2}+\frac{\partial^2 v}{\partial y^2}\right)+\rho g_y;$$
$$c_p\left(u\frac{\partial T}{\partial x}+v\frac{\partial T}{\partial y}\right)=k\left(\frac{\partial^2 T}{\partial x^2}+\frac{\partial^2 T}{\partial y^2}\right)+u\frac{\partial p}{\partial x}+v\frac{\partial p}{\partial y}+\Phi. \tag{5.62}$$

where k is assumed to be the specific heat conductivity constant and $\vec{g}$ is the gravity field vector in the form $\vec{g}=(g_x,g_y)$. Φ is the dissipation function formulated in Eq. (5.12).

In this section, only steady-state, incompressible laminar flows of a Newtonian fluid with constant material properties are considered. All forms of bulk forces such as phase change, mass transport and chemical reactions are omitted here. A very complete presentation of laminar heat transport correlations and solution approaches can be found in Shah and London (1978).

Until now we had assumed that the hydrodynamic and thermal quantities do not change in the direction of flow, i.e. $d/dx=0$, which is an idealized idea, since for both the viscous and thermal boundary layers are already fully formed. However, when entering a channel, both must develop along the channel walls, so that actually four types of laminar channel flow domains can be distinguished:

- fully developed,
- hydrodynamically evolving,
- thermally developing and
- simultaneously developing

flows. The latter form states that the flow is hydrodynamically and thermally evolving simultaneously.

Figure 5.7a sketches the distinguishing features of the individual flow forms and regions. It represents a flow of a fluid with constant velocity u_0 and temperature T_0 entering a duct of an arbitrary cross-section at $x=0$.

If we first assume an isothermal flow and postulate that there is no heat generation or dissipation in the fluid, the fluid experiences no temperature change and a viscous boundary layer is formed by the viscosity of the fluid starting at $x=0$. The viscous boundary layer grows in wall-normal thickness downstream according to Prandtl's boundary layer theory proportionally with the duct length corresponding to $1/\sqrt{Re_x}$. The boundary layer divides the flow field into two domains; a viscous region at the walls and a frictionless region around the centre of the channel. At $x=l_{hy}$, the viscous boundary layers have merged and cover the entire channel cross-section. The length $0\leq x\leq l_{hy}$ is called the hydrodynamic inlet

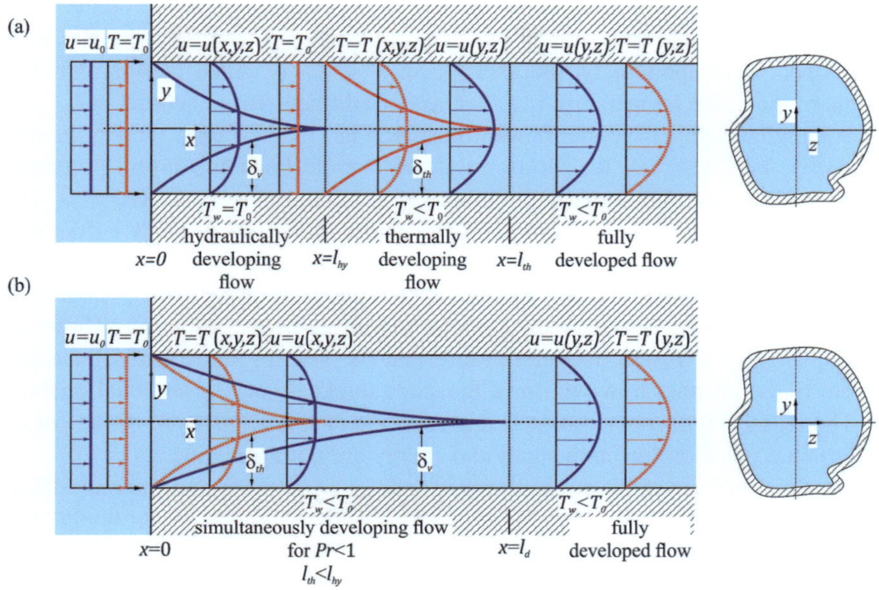

Fig. 5.7 Classification of a laminar channel flow at a boundary condition with constant wall temperature T_w into different regions. **a** Hydrodynamically developing flow followed by thermally evolving (and hydraulically developed) flow; **b** simultaneously developing flow in liquid metals ($Pr \ll 1$). Solid lines indicate velocity profiles, while dotted lines outline temperature profiles

region. In the inlet region, the velocity depends on all three spatial coordinates, while for $x > l_{hy}$ the velocity distribution is independent of the axial coordinate x.

After reaching the entrance length we assume that the channel wall temperature is lowered to a constant value with respect to the inlet temperature of the fluid, i.e. $T_w < T_0$. In this case, the thermal effects gradually diffuse from the channel wall towards the core flow starting at $x = l_{hy}$. The magnitude of the thermal diffusion is determined by the thickness of the thermal boundary layer δ_{th}, which increases along the axial coordinate x.

A similarity analysis based on the boundary layer approximation approach shows that the thermal boundary layer becomes thicker according to $\delta_{th}(x) \sim 1/\sqrt{Re \cdot Pr}$. The application of the boundary layer approximation to the thermal case also reveals that the thermal boundary layer separates the flow field into a zone near the wall influenced by the thermal effects and an uninfluenced region in the flow core. One result of the dimensional analysis is the scaling of the ratio of the thicknesses of the thermal and viscous boundary layers:

$$\frac{\delta_{th}}{\delta_v} \sim \frac{1}{\sqrt{Pr}}. \tag{5.63}$$

At $x = l_{th}$, the influence of the thermal effects extends over the entire channel cross-section and downstream of this point the flow is termed thermally fully developed. Correspondingly, the area $l_{hy} < x < l_{th}$ is called thermal entrance domain; there, the temperature is a function of all three spatial coordinates.

A simultaneous developing flow is illustrated in Fig. 5.7b for a Prandtl number $Pr \ll 1$. For this case, the viscous and the thermal boundary layers propagate from the channel wall towards the centre of the channel starting at $x = 0$. The decisive parameter here is the Prandtl number. The kinematic viscosity ν describes the propagation of the momentum flow (velocity) in the same way as the thermal diffusivity a describes the propagation of the energy transport (temperature). If $Pr = 1$, then viscous and thermal effects propagate at the same rate. However, the equality of diffusion does not ensure that the viscous and thermal boundary layers in the channel flow will have the same thickness at a given axial position x. The reason for this paradoxon is that for $Pr = 1$, the application of the partial differential equation for momentum and energy are not analogous.

As shown in Fig. 5.7b, in the range $0 < x < l_d$, viscous and thermal effects simultaneously propagate toward the center of the channel. Accordingly, this region is also called the inlet (entrance) region in this case. Obviously, the length l_d also depends here on the Prandtl number and velocity as well as temperature depend on all three spatial coordinates. Only for $x > l_d$, i.e. in the entrance region, both become axially invariant and depend only on y and z, i.e. the following applies: $u = u(x, y)$ and $T = T(x, y)$.

5.1.9 Flow and Heat Transfer Parameters

The flow characteristics of all channel flows are specified in terms of hydraulic parameters, which are universally specified as dimensionless quantities whenever possible. For a hydraulic inflow, the axial inlet length x^+ is defined as follows:

$$x^+ = \frac{1}{Re} \frac{x}{d_h}, \tag{5.64}$$

where d_h is the hydraulic diameter. The hydraulic entrance length l_{hy} is described as the length in flow direction required by the flow to reach 99% of the magnitude of the fully developed flow velocity. The dimensionless hydraulic entrance length, as it is mostly given in textbooks, is thus given by the expression $l_{hy}^+ = l_{hy}/(d_h \cdot Re)$.

The mean fluid temperature often also called bulk (-sometimes mixing) temperature T_m is defined by the following expression:

$$T_m = \frac{1}{A\, u_0} \int_A uT \, dA, \tag{5.65}$$

and thus, the circumferentially averaged but local axial heat transfer coefficient h_x is given by:

$$\dot{q}_{w,x} = h_x\left(T_{w,m} - T_m\right), \tag{5.66}$$

where $T_{w,m}$ is the mean wall temperature and T_m is the fluid bulk temperature (from Eq. (5.65)).

The heat flux $\dot{q}_w$ and the temperature difference $\left(T_{w,m} - T_m\right)$ are directional and thus vectorial quantities. In the notation of Eq. (5.66), the direction of heat flow points from the wall toward the fluid and, consistent with this, the temperature falls from the wall toward the fluid. In contrast, if $\dot{q}_w$ points the heat flow from fluid in the direction of the wall, the difference is $\left(T_m - T_{w,m}\right)$. The heat transfer coefficient hm averaged over the path of the flow x is obtained by integration from $x = 0$ to x in the following form:

$$h_m = \frac{1}{x} \int\limits_{x=0}^{x=x} h_x \, dx. \tag{5.67}$$

The ratio of the convective heat transfer h to the heat transfer by molecular heat conduction k/d_0 is described as the Nusselt number Nu. The circumferentially averaged but local axial Nusselt number Nu_x is defined by:

$$Nu_x = \frac{h_x d_h}{k} = \frac{\dot{q}_{w,x} d_h}{k\left(T_{w,m} - T_m\right)}. \tag{5.68}$$

Thus, the Nusselt number describes nothing other than the dimensionless temperature gradient at the wall. However, the Nusselt number can also be understood as the ratio of two lengths, namely that of the characteristic channel dimension to the local thickness of the thermal boundary layer. The average Nusselt number Nu_m based on h_m in the thermal inlet (entrance) region is thus given by:

$$Nu_m = \frac{1}{x} \int\limits_0^x Nu_x dx = \frac{h_m d_h}{k} = \frac{\dot{q}_{w,m} d_h}{k(\Delta T)_m}. \tag{5.69}$$

The expression for $(\Delta T)_m$ could be problematic to determine since it depends on the thermal boundary conditions. The dimensionless axial thermal inlet length x^* is defined by:

$$x^* = \frac{1}{Pe} \frac{x}{d_h} = \frac{1}{Re\ Pr} \frac{x}{d_h}. \tag{5.70}$$

The thermal entrance length l_{th} is defined as the axial length required to attain a local Nusselt number Nu_x corresponding to 1.05 times the Nusselt number of a thermal developed flow. The dimensionless thermal inlet length is thus given by the following expression $l_{th}^{*} = l_{th}/(d_h\ Pe)$. The Stanton number St is also often used. It describes the ratio of the heat flux dissipated by the wall to the enthalpy difference of the external flow. Accordingly, the Stanton number is defined as follows:

$$St = \frac{h}{\rho\ c_p\ u_0} = \frac{\dot{q}_w}{\rho\ c_p u_0\ \Delta T} = \frac{Nu}{Re\ Pr}. \tag{5.71}$$

5.1.10 Laminar Heat Transfer in Tubes

The qualitative behaviour of the laminar momentum exchange as a function of the individual initial and boundary conditions can be clearly illustrated in the technically most relevant case of a pipe flow. It is useful to distinguish the four different flow domains from Fig. 5.7, considering the respective thermal boundary conditions.

- Fully developed flow

Assuming a constant wall temperature ($T_w = const.$), the temperature distribution in circular tubes can be specified exactly, provided the flow is non-dissipative, has no energy sources or axial heat conduction. The expression for the mean fluid bulk temperature is given by the following asymptotic formula for $x^* > 0.0335$, see Bhatti (1985):

$$\frac{T_w - T_m}{T_w - T_{in}} = 0.819048\ exp\left(-2\lambda_0^2 x^*\right) \text{ with } \lambda_0 = 2.70436442, \tag{5.72}$$

where T_{in} denotes the inlet temperature. It should be noted that although the local temperature T is simultaneously a function of the radial and axial coordinates, the mean fluid temperature T_m depends only on the axial coordinate (the dimensionless temperature $(T_w - T)/(T_w - T_m)$ is a function of the radius only). The mean Nusselt number of a fully developed flow is thus:

$$Nu_m = \frac{\lambda_0^2}{2} =^* 3.6568. \tag{5.73}$$

For developed flows with constant wall temperature, the Nusselt number is thus a constant and independent of the Prandtl number. However, if the Peclet number Pe is less than 10, axial heat conduction cannot be neglected and the following asymptotic expression is obtained, cf. Michelsen and Villadsen (1974):

$$Nu_m = \begin{cases} 4.180654 \ - \ 0.18346 \cdot Pe \text{ for } & Pe < 1.5 \\ 3.656794 \ + \ \frac{4.4870}{Pe^2} & \text{for } \ Pe > 5.0 \end{cases}. \tag{5.74}$$

For the local Nusselt number Nu_x at a location x, calculated from the beginning of heating or cooling, the asymptotes are as follows:

$$Nu_x = \begin{cases} 3.6568 & \text{for} \quad Pe \cdot (d_h/x) < 1 \\ 1.077 \cdot (Pe \cdot d_h/x)^{1/3} & \text{for} \quad Pe \cdot (d_h/x) > 1 \end{cases}, \tag{5.75}$$

where Pe is the Peclet number defined as $Pe = Re \cdot Pr$.

For a constant wall heat flux ($\dot{q}_w = const.$), the temperature distribution in circular tubes for non-viscous dissipative systems, without energy sources and axial heat conduction can also be derived analytically and it is obtained:

$$\left(\frac{T_w - T}{T_w - T_m}\right) = 6\left(1 - \left(\frac{r}{R}\right)^2\right)\left(\frac{3 - (r/R)^2}{11}\right);$$
$$k\left(\frac{T_w - T}{\dot{q}_w d_h}\right) = \frac{11}{48} = \frac{1}{Nu} = \frac{\lambda}{h d_h}. \tag{5.76}$$

The Nusselt number for the case with constant heat flow is thus about 20% larger than that at constant wall temperature. The qualitative form of the mean fluid temperature distribution for both boundary conditions as a function of the axial coordinate x is shown in Fig. 5.8.

Regarding the boundary condition of a convectively heated/cooled wall, it is assumed that the temperature along the channel is constant in flow direction, but a finite wall-normal thermal resistance exists. This thermal resistance can be embedded in an external convective heat transfer coefficient h_e, which is expressed by a dimensionless Biot number. The Biot number Bi is defined by $Bi = (h_e d_h)/k$ (see Sect. 4.2).

Via the Biot number the influence of the thermal wall resistance can be considered. In this case, the Biot number is inversely proportional to the total thermal resistance R_w, i.e. $Bi = 1/R_w$. The total thermal resistance is given by

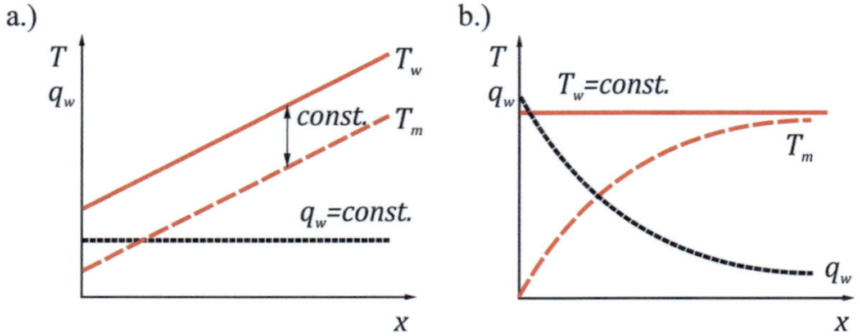

Fig. 5.8 Qualitative distribution of the mean fluid temperature T_m, the wall temperature T_w and the wall heat flux $\dot{q}_w$ as a function of the axial coordinate x for **a** constant wall heat flux $\dot{q}_w$ and **b** const. wall temperature T_w

$R_w = (kt_w)/(k_w d_h) + k/(h_e d_h)$. Thus, the limiting case of constant wall temperature corresponds to $Bi \to \infty$ and the other limiting case of constant wall heat flux implies $Bi = 0$. Therefore, the Biot number can be taken as the ratio of the wall thermal resistance to that of the fluid. For a constant h_e (Hickmann, 1974), an asymptotic solution for the Nusselt number is obtained and the global mean Nusselt number $Nu_{o,m}$ is:

$$Nu = \frac{\frac{48}{11} + Bi}{1 + \frac{59}{220} Bi}, \frac{1}{Nu_{o,m}} = \frac{1}{Nu} + \frac{1}{Bi} \quad \text{with} \quad Nu_{o,m} = \frac{h_e d_h}{k} = \frac{q_w d_h}{k(T - T_m)}. \tag{5.77}$$

$Nu_{o,m}$ is insensitive with respect to a circumferential change in the Biot number after Sparrow et al. (1978).

- Hydraulically developing flow

The hydraulic inflow problem has been studied in many ways and depending on the Reynolds number, different solutions exist, which can be categorized as follows:

- Solutions using the boundary layer equations for $Re > 400$;
- Solutions using the Navier–Stokes equations for $Re < 400$;
- Solutions for creeping flows for $Re \to 0$.
- Boundary layer type solutions are described in detail in Shah and London (1978). Technically, the solution of Horbeck is mostly used. According to this, the dimensionless hydraulic inlet length l_{hy}^+ + and the dimensionless pressure gradient Δp^* are obtained:

$$l_{hy}^+ = 0.0565, \ \Delta p^* = 13.74\sqrt{x^+} + \frac{1.25 + 64x^+ - 13.74\sqrt{x^+}}{1 + 2.1 \times 10^{-4}\left(x^+\right)^{-2}}. \tag{5.78}$$

Near the inlet of the flow into the channel, both the axial diffusion of the momentum and the radial pressure distribution are relevant. Adequate consideration of these aspects inherently leads to the inclusion of the Reynolds number and requires accurate knowledge of the velocity profile at the inlet. With the help of computational fluid dynamic codes (CFD) an arbitrarily exact solution can be obtained. Assuming a constant velocity far from the inlet, Chen (1973) obtains a relation for the hydraulic inlet length.

$$l_{hy}^+ = 0.056 + \frac{0.6}{Re(1 + 0.035Re)}. \tag{5.79}$$

- Thermally developing flows

At a constant wall temperature ($T_w = const.$) and neglecting viscous dissipation, axial fluid heat conduction, and other energy sources/sinks, the solution for the local Nusselt number Nu_x is

$$Nu_x = \begin{cases} 1.077 \cdot (x^*)^{-\frac{1}{3}} - 0.7 & \text{for } x^* \leq 0.1 \\ 3.657 + 6.874 \cdot \left(10^3 x^*\right)^{-0.488} exp(-57.2x^*) & \text{for } x^* \geq 0.1 \end{cases}, \tag{5.80}$$

and the relation for the mean Nusselt number Nu_m in the whole range of x^* is

$$Nu_m = 3.66 + \frac{0.0668}{x^{*1/3}\left(0.04 + x^{*2/3}\right)} \tag{5.81}$$

The error of the analytical relation is of the order of 2% compared to a full numerical solution. The influence of axial heat conduction is negligible for $Pe > 50$, but the thermal entrance length l^*_{th} increases considerably, from 0.033 for $Pe \to \infty$ to 0.5 for $Pe = 1$. The dimensionless temperature distribution T as a function of r/R and x^* is shown Fig. 5.9. The mean local Nusselt number Nu_x as well as the mean Nusselt number Nu_m are shown in Fig. 5.10.

Analogously, the thermal inlet length can be solved for a constant wall heat flux ($\dot{q}_w = const.$). The mean local Nusselt number Nu_x and the mean Nusselt number Nu_m can be derived by means of approximations:

$$Nu_x = \begin{cases} 1.302 \cdot \left(x^*\right)^{-1/3} - 1.0 & \text{for } x^* \leq 5 \times 10^{-5} \\ 1.302 \cdot \left(x^*\right)^{-1/3} - 0.5 & \text{for } 5 \times 10^{-5} \leq x^* \leq 1.5 \times 10^{-3} \\ 4.364 + 8.68 \cdot \left(10^3 x^*\right)^{-0.56} \exp\left(-41x^*\right) & \text{for } x^* \geq 1.5 \times 10^{-3} \end{cases}; \tag{5.82}$$

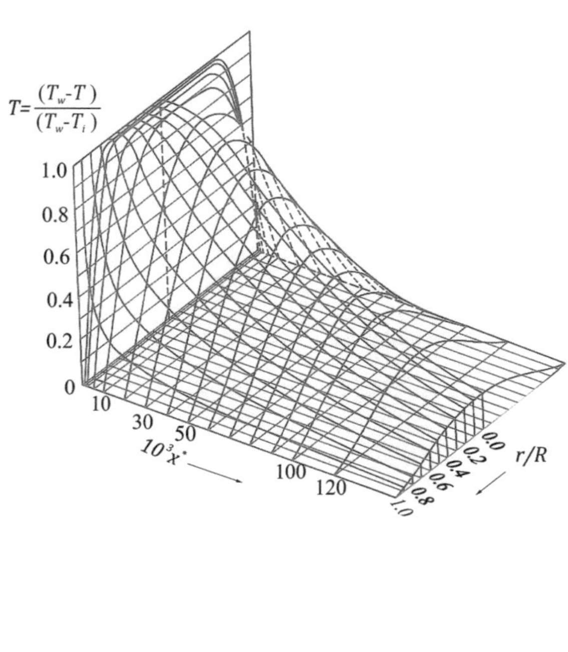

Fig. 5.9 Dimensionless temperature distribution of a laminar thermal pipe inlet flow as a function of the dimensionless length x^* for $T_w = const.$ from Grigull and Tratz (1965)

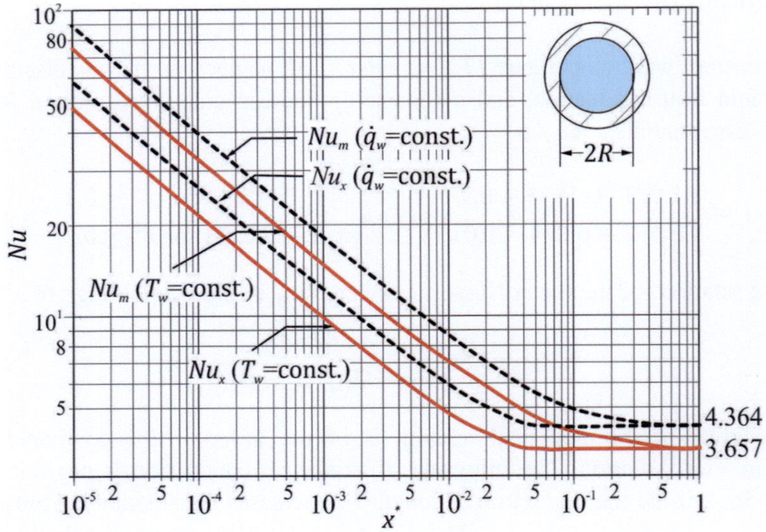

Fig. 5.10 Local Nu_x and mean Nu_m as a function of dimensionless inlet length x^* for a hydraulically inlet and thermally inlet pipe flow for $T_w = const.$ and $\dot{q}_w = const.$

$$Nu_m = \begin{cases} 1.953 \cdot (x^*)^{-1/3} & \text{for} x^* \leq 0.03 \\ 4.364 + 0.0722/x^* & \text{for} x^* \geq 0.03 \end{cases}. \tag{5.83}$$

The approximation error is about ±2% compared to the exact solution. Again, the axial heat conduction is negligible for $Pe > 10$, provided $x^* > 0.005$. This is less than for the constant wall temperature case. The thermal inlet length l^*_{th} is 0.0430527. The temperature distribution is given in Fig. 5.11 and the Nusselt numbers can be taken from Fig. 5.10.

For the convective heated/cooled wall boundary condition, a solution involving the Biot number can also be determined using a series expansion (Hsu, 1968). The local Nusselt numbers as a function of the dimensionless distance x^* for different Biot numbers are shown in Fig. 5.12.

- Simultaneous developing flow

If the flow is not hydraulically fully developed, the heat transfer solution depends directly on the Prandtl number.

Constant wall temperature ($T_w = const.$)

For $Pr \to \infty$, the flow is already fully developed hydraulically before a temperature profile form. In the case of a well molecular conducting fluid, assuming the idealized boundary case $Pr = 0$, the temperature profile forms much faster than the velocity profile, so that in this case during the formation of the temperature field the velocity field changes only marginal downstream. The unmodified

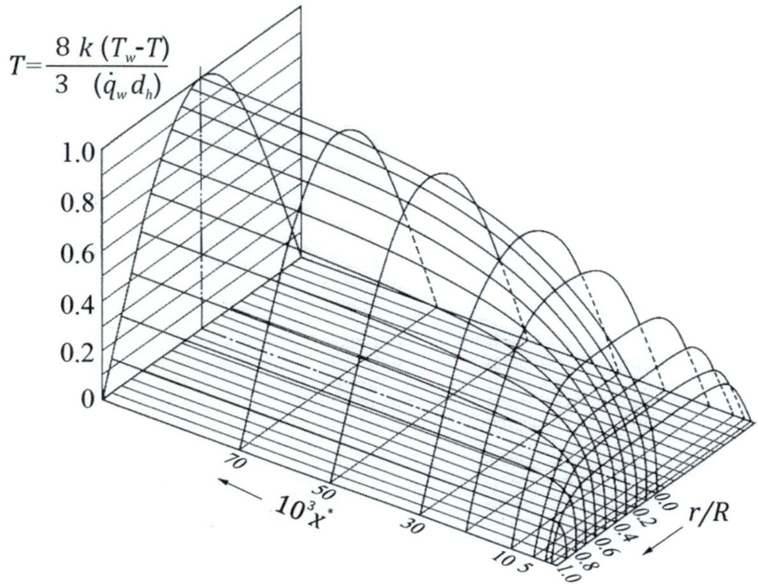

Fig. 5.11 Dimensionless temperature distribution of a laminar thermal pipe inlet flow as a function of the dimensionless length x^* for $\dot{q}_w = const.$ from Grigull and Tratz (1965)

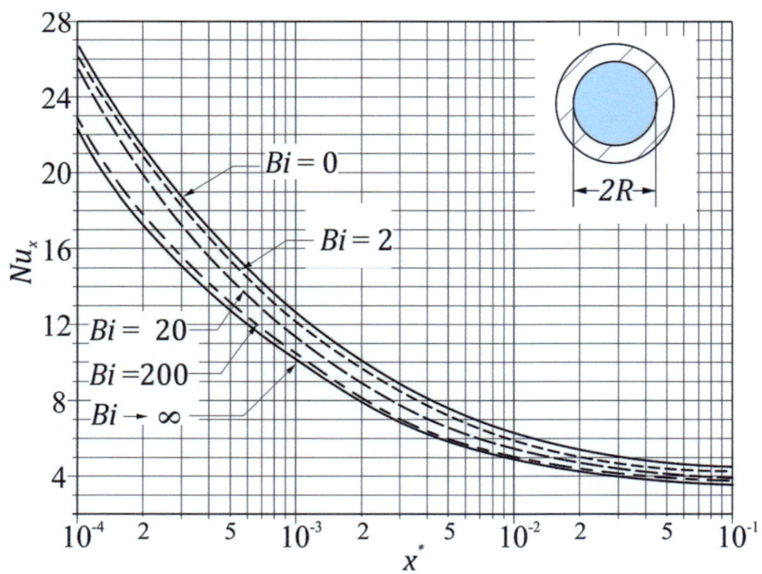

Fig. 5.12 Local Nu_x as a function of the dimensionless inlet length x^* of a hydraulically developed and thermally evolving laminar pipe flow for mixed convective thermal boundary conditions expressed by means of the Biot number Bi

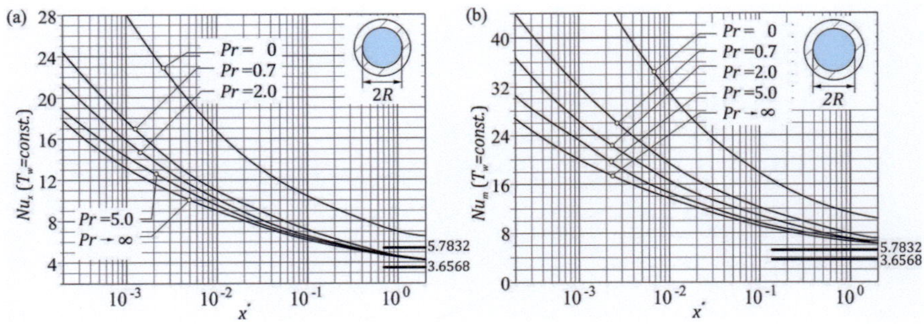

Fig. 5.13 Local and mean Nusselt numbers Nu_x (**a**) and Nu_m (**b**) as a function of the dimensionless entrance length x^* for simultaneous inlet laminar pipe flows with $T_w = const.$ and for different fluids (Pr)

velocity profile with its small boundary layer has a high velocity near the wall, which leads directly to a significantly increased heat transport and thus to larger Nusselt numbers than for the developed flow. Analytical solutions can be found for the idealized cases $Pr = 0$ and $Pr \rightarrow \infty$, while series expansions involving Bessel functions are required for the other Prandtl numbers. The mean and local Nusselt numbers Nu_x and Nu_m as a function of the dimensionless inlet length x^* for different Prandtl numbers are shown in Fig. 5.13a, b.

The dependence of the local heat transfer coefficient Nu_x is well described by the following approximate equation for media with different Prandtl numbers.

$$Nu_x = \frac{1}{2}\left(\frac{2}{1 + 22\ Pr}\right)^{\frac{1}{6}} \cdot \sqrt{Re\, \mathrm{Pr}(d_h/x)}. \tag{5.84}$$

Note that for $Pr = 0$, the local and mean Nusselt numbers have the value 5.7832 as the lowest limit, while the others have the value 3.6568 as the lower limit. The thermal entrance length for this type of flow is:

$$l_{th}^* = \begin{cases} 0.028 \text{ for } Pr = 0 \\ 0.037 \text{ for } Pr = 0.7\ . \\ 0.033 \text{ for } Pr \rightarrow \infty \end{cases} \tag{5.85}$$

Constant wall heat flux ($\dot{q}_w = const.$)

Using a similar approach shows the limit value for the local mean Nusselt number and the idealized fluid with $Pr = 0$ for large x^* of 8, which is about 40% more than for constant wall temperature. The dependence of the local Nusselt number as a function of the inlet length is shown in Fig. 5.14. Not only is the heat transfer greater at constant wall heat flux, but the thermal entrance length also increases.

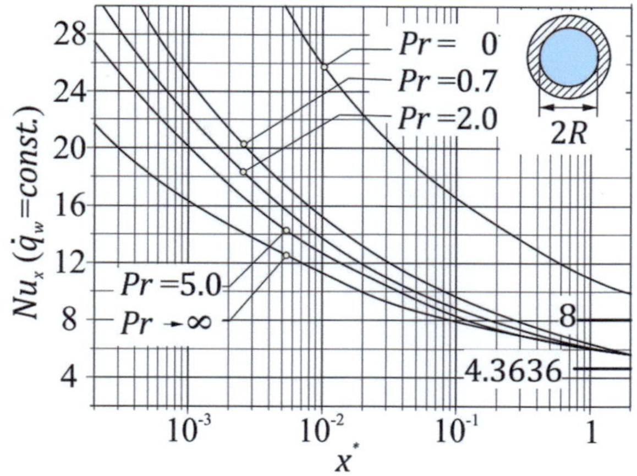

Fig. 5.14 Local and mean Nusselt numbers Nu_x (**a**) and Nu_m (**b**) as a function of the dimensionless entrance length x^* for simultaneous evolving pipe flows with $\dot{q}_w = const.$ and for different fluids (Pr)

It is about 50% more for well molecular conducting fluids than at constant wall temperature. Shah and London (1978) suggest the following values:

$$l_{th}^* = \begin{cases} 0.042 \text{ for } & Pr = 0 \\ 0.053 \text{ for } & Pr = 0.7 \\ 0.043 \text{ for } & Pr \to \infty \end{cases} . \tag{5.86}$$

The local Nusselt number Nu_x for a pipe flow reads to:

$$Nu_x = \begin{cases} 0.459 \cdot Pr^{1/3} \cdot \sqrt{(Re \cdot d_h/x)} \text{ for } & Pr \approx 1 \\ 0.464 \cdot Pr^{1/3} \cdot \sqrt{(Re \cdot d_h/x)} \text{ for } & Pr \to \infty \end{cases} . \tag{5.87}$$

- Influence of the temperature dependence of the thermo-physical properties

The dimensionless parameters relevant for laminar heat transfer, such as Nu and Re, depend on thermophysical properties of the fluid, and the solutions were derived assuming constant material values. For the practical use of the equations, one can use for the material values those at the mean temperature of the fluid flow $T_m = (T_e + T_o)/2$, where T_e is the temperature at the entrance and T_o that at the outlet, however, this refers only if the temperature differences are not too large. The direction of the heat flow (heating or cooling) influences the heat transfer for temperature-dependent material data. Laminar flow experiments of gases, such as air, nitrogen, and helium in the range $0.5 < T/T_w < 2.0$ have shown that the heat transfer coefficient changes less than 10%, where T is the mean Kelvin temperature of the gas and T_w is the Kelvin temperature of the tube wall. For liquids, a correction factor must be considered, which is obtained from the ratio of the Prandtl numbers of the liquid at the mean bulk temperature T_m and

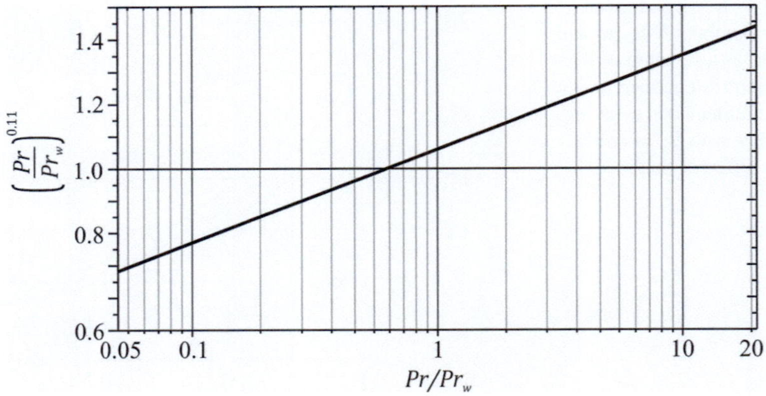

Fig. 5.15 Consideration of the temperature dependence of fluids during heat transfer in the tube

that at the wall, respectively. The effective Nusselt number Nu is then obtained multiplicatively from, see Fig. 5.15:

$$Nu = Nu_m \cdot \left(\frac{Pr}{Pr_w} \right)^{0.11} . \tag{5.88}$$

Note, that for turbulent flows also for gases a Prandtl number ratio of bulk fluid temperature to fluid wall interface temperature is required as well.

The main subject of laminar heat transfer was the different thermal boundary conditions, which lead to considerably different Nusselt numbers depending on the respective hydrodynamic inlet conditions. The influence of the fluid properties in dimensionless form thus plays a decisive role in inlet flows. The main point in laminar heat transfer is directed to the heat that can be dissipated from the wall as a function of geometry and boundary conditions. The wall heat flux, on the other hand, which is given by Fourier's law ($\dot{q}_w = -k(\partial T/\partial n)_w$, depends only on the molecular heat conduction of the fluid and the temperature gradient at the wall.

Although laminar heat transport is a purely molecular transport process, the flow field plays a central role. The flow is responsible for the convective transport of internal energy or enthalpy; it influences the temperature gradient. Obviously, heat transfer can be improved by increasing the flow velocity. The higher kinetic energy allows a larger amount of energy to be removed convectively from the fluid wall interface. For incompressible flows, this process is described only by one characteristic number. In forced convective systems this number is the Reynolds number, while in buoyant flows of natural convection it is given by the Grashof number. In both characteristic numbers, the molecular viscosity appears, which is responsible for momentum transport. In the heat exchange in a fluid, the ratio of the molecular viscosity to the molecular thermal conductivity $a = k/(\rho\, c_p)$, and thus the Prandtl number (Pr), plays the decisive role. In contrast to the Reynolds or Grashof number, Pr is a pure material constant and depends only on the thermo-physical fluid properties.

The normalization of the wall heat flux with a suitable temperature difference is called the heat transfer coefficient $h = \dot{q}_w/\Delta T$. The heat transfer coefficient can be de-dimensioned in different ways. The Stanton number is the ratio of the wall heat flux to the enthalpy difference and does not include heat conduction. More common is the use of the Nusselt number, in which the heat transfer coefficient is multiplied by a characteristic length and weighted with respect to molecular heat conduction.

For pure forced convection, the results can all be presented in universal form:

$$\frac{Nu_x}{\sqrt{Re_x}} = f\left(Pr^n\right) \text{ with } 0 < n < 1. \tag{5.89}$$

- The expression $Nu_x/\sqrt{Re_x}$ grows with increasing Prandtl number. This does not mean that the heat transfer also increases with increasing Prandtl number, because the Nusselt number itself includes the molecular heat conduction.
- For fully developed flows, the Nusselt number is constant. The constant depends only on the thermal boundary conditions.

The Prandtl number only influences the heat transfer for developing flows. There are several ways to derive the heat transfer relations:

- Analytical solutions of the conservation equations (only possible in simple cases with drastic restrictions).
- Numerical solution of the conservation equations (very computationally intensive).
- Exact solutions of the boundary layer equations. Numerical solutions of the boundary layer equations (require mathematical effort but can be easily applied with commercial CFD programs).
- Approximate solutions using integral methods with lower accuracy than numerical methods.
- Analogies between heat and momentum transport are only applicable for $Pr = 1$ and result in an exact solution.
- Empirical correlations must be used for complex geometries.

5.1.11 Turbulent Momentum Exchange

In most engineering applications, however, the flow is turbulent, which is illustrated in Fig. 5.16 by the famous colour thread experiment first conducted by Osborne Reynolds 1883. If the flow velocity is continuously increased expressed by an increasing Reynolds number a transition from laminar flow characterized by a stratified ink layer as illustrated in Fig. 5.16a change first into transitional flow and finally into a turbulent flow if a critical velocity is exceeded. He determined a

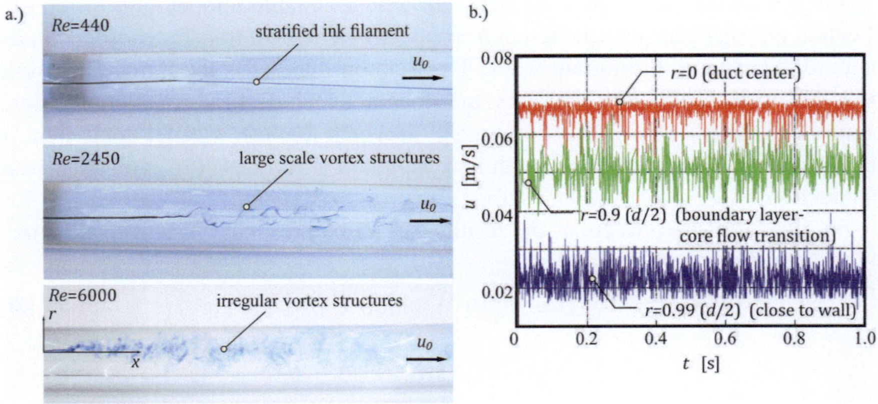

Fig. 5.16 **a** Reproduction of the ink filament Osborne Reynolds experiment by a water flow in a horizontal duct at three different Reynolds numbers; **b** measured velocities art $Re = 6000$ at three radial positions of the experiment in (**a**)

value for this laminar turbulent transition to occurs in a circular duct at a critical Reynolds number of $Re_{crit} \approx 2.300$ and assumed that this transition is a stability problem. From this simple experiment two main questions arise:

- How does turbulence evolve from laminar flow? This is indeed a stability problem, first solved by Tollmien in 1929 using a perturbation approach. There is a myriad of literature on this for example Schlichting and Gersten (2000), Betchov and Criminale (1967) and White (1991).
- What are the specific characteristics of a turbulence flow? Why does a turbulent flow remain turbulent? Here, the key to answering this question lies in the nature of turbulence production. As shown in the velocity recordings in the water ink flow in Fig. 5.16b obviously viscous shear of the boundary towards the core impacts the time dependent behaviour of fluid motion.

Turbulent flows are always unsteady, three-dimensional, and stochastic. From experiments we know that the transition from laminar to turbulent is accompanied by the following phenomena:

- increase of the pressure drop;
- increase of the boundary layer thickness.

Both are caused by turbulent diffusion. Due to the intense macroscopic fluctuations, the turbulent velocity profile becomes "fuller", and the turbulent boundary layer becomes thicker compared to the laminar flow. All variables in the turbulent flow field depend on the spatial coordinates x, y, and z and time t. The unknowns of a turbulent isothermal flow are thus all the velocity components of the velocity vector $\vec{u}$ namely u, v, w and the pressure p, which are all functions of the three spatial coordinates and the time.

There are several ways to treat turbulent flow problems and the user has to specify, which type of solution approach is required. In simplified terms, a distinction can be made between two classes of application and analysis questions.

The design and analysis of thermally highly stressed components requires detailed local two- or three-dimensional, often transient, computationally validated thermal–hydraulic modelling, which is solved by means of computational fluid dynamics (CFD). In contrast to this are system analyses, as they are relevant for regulation and control or for techno-economic analyses (yield, performance,). These are based on scale-separation methods and are therefore often one- or one-and-a-half-dimensional in nature. Therefore, the system analysis tools used there are mainly based on simplified validated correlations, which require a qualified user to adequately apply flow domain decomposition approaches and to select suitable models.

The focus of a textbook cannot be a detailed presentation of the model-technical subtleties of a three-dimensional turbulent CFD simulation as well as the details of a system-analytical calculation, so that the following presentation concentrates on the description of the essential methodical thought processes in the solution for both solution classes. The next two sections are dedicated to elaborate the modelling used for the detailed computation of turbulent flows by means of CFD approaches.

Detailed thermal hydraulic simulations can be roughly categorized into three larger classes.

- Direct numerical simulation (DNS) solve the Navier–Stokes equations are numerically without applying any turbulence model requiring that the entire range of spatial and temporal scales of the turbulence must be resolved in the computational grid from the smallest dissipative scales (Kolmogorov scales η) where the fluctuations are dissipated, up to the integral scale L, associated with the motions containing most of the kinetic energy. The Kolmogorov scale is $\eta = \left(\nu^3/\varepsilon\right)^{1/4}$, in which ν is the kinematic viscosity and ε the rate of kinetic energy dissipation, while the integral scale L is of order of the boundary layer dimensions. The challenge of a DNS is that the integration of the integral scale L requires a spatial resolution of N of points along a given mesh direction with increments h so that, $(N \cdot h) > L$, while simultaneously $h < \eta$ needs to be matched. Since the rate of the kinetic energy dissipation scales $\varepsilon \approx u'^3/L$, with u' the root mean square (RMS) of the velocity fluctuation the spatial resolution requirement of a 3D DNS scales with $N^3 \geq \left(Re'\right)^{9/4}$, where Re' is the Reynolds number of the turbulent fluctuation $Re' = \left(u' \cdot L\right)/\nu$. Additionally, the smallest vortices have to be resolved in time requiring in order to be accurate a time step (Δt) small enough such that a fluid particle moves only a fraction of the mesh spacing h in each step, i.e. $C = \left(u' \cdot \Delta t\right)/h < 1$, which is called Courant number condition and the ratio correspondingly C is the Courant number. In order to get a representative solution, the total time interval simulated must be at least proportional to the turbulence time scale $\tau = L/u'$. Merging all definitions, the number of time steps required by a DNS scales with $L/\eta \sim Re'^{3/4}$ and together with required spatial resolution required about Re'^3 operation is required. This disqualifies DNS for practical engineering applications since memory requirements exceed the capacity of the most powerful computer currently available. Nonetheless, the DNS is a powerful and indispensable complementary tool to get access to the fundamental physics of turbulence, e.g. by identifying transfer processes of eddy cascades. Moreover, DNS provide a quasi-accurate solution to develop and validate turbulence models for practical applications, such as sub-grid scale models for Large eddy simulation (LES) and to improve models for methods that solve the Reynolds-averaged Navier–Stokes equations (RANS), since the DNS serves data experimentally not accessible even by future sensing technologies.

- Large eddy simulation (LES) relies on the theory of self-similarity, which assume that the large eddies of the flow are dependent on the geometry while the smaller scales are quasi-"universal". By this approach the large eddy scales are explicitly solved in a simulation and implicitly also the smaller eddies by means of a subgrid-scale model (SGS model). Formally, one may think of filtering as the convolution of a function with a filtering kernel G, which yields $\overline{u}_i = \int G\left(\vec{x} - \vec{\xi}\right) u\left(\vec{\xi}\right) \mathrm{d}\,\vec{\xi}$. By this filtering the velocity is decomposed in a mean and a fluctuating part, $u_i = \overline{u_i} + u_i'$ similar as in a RANS simulation. Substituting in this decomposition and a similar one for the pressure and then filtering provides the filtered Navier–Stokes equations for the resolved "large scale" field.

$$\frac{\partial \overline{u_i}}{\partial t} + \rho \overline{u}_j \frac{\partial \overline{u_i}}{\partial x_j} = -\frac{1}{\rho} \frac{\partial \overline{p}}{\partial x_i} + \frac{\partial}{\partial x_j}\left(\nu \frac{\partial \overline{u_i}}{\partial x_j}\right) + \frac{1}{\rho} \frac{\partial \tau_{ij}}{\partial x_j}. \tag{5.90}$$

Thereby it is assumed that the filtering operation and the differentiation operation commute and the errors associated with this assumption are small. The additional term $\partial \tau_{ij} / \partial x_j$ originates from the non-linear advection terms and the turbulent shear stress tensor termed as Reynolds stress tensor is $\tau_{ij} = \overline{u_i} \cdot \overline{u_j} - \overline{u_i u_j}$. For the sub-grid-scale or residual field a similar equation is developed. Therein, the stress tensor is related proportionally proportional to the trace-less mean strain rate tensor S_{ij}, which is called Boussinesq approximation. Thereby, the the SGS is correlated with mean flow quantities:

$$\tau_{ij} - \frac{1}{3} \tau_{kk} \delta_{ij} = -2\, \rho\, \nu_t S_{ij} = -2\, \rho\, \nu_t \left(\frac{\partial \overline{u_i}}{\partial x_j} + \frac{\partial \overline{u_j}}{\partial x_i}\right), \tag{5.91}$$

in which ν_t is the subgrid-scale turbulent viscosity. The subscale information is the inserted into the filtered Navier–Stokes equations leading to:

$$\frac{\partial \overline{u_i}}{\partial t} + \rho \overline{u}_j \frac{\partial \overline{u_i}}{\partial x_j} = -\frac{1}{\rho} \frac{\partial \overline{p}}{\partial x_i} + \frac{\partial}{\partial x_j}\left(\nu + |\nu_t| \frac{\partial \overline{u_i}}{\partial x_j}\right). \tag{5.92}$$

In subgrid-scale models, there are a variety of approaches, ranging from isotropic dissipation of small eddy structures (Smagorinski), to algebraic models, to dynamic models that consider wall distance (Wale, RNG), or consider local stress tensors (structural modelling). These are:

- Smagorinsky model (Smagorinsky, 1963),
- Algebraic Dynamic model (Germano et al., 1991),
- Dynamic Global-Coefficient model (You & Moin, 2007),
- Localized Dynamic model (Kim & Menon, 1995),
- WALE (Wall-Adapting Local Eddy-viscosity) model (Nicoud & Duos, 1999),
- RNG-LES model,
- Structural modelling (Lu et al., 2008).

It must be mentioned that also the spatial resolution is substantially reduced compared to a DNS, the spatial resolution is substantially reduced, the Courant number requirements still applies for the temporal resolution. Also, for the spatial resolution the mesh—Reynolds number, in which as characteristic length scale the mesh dimension h is used, must be below 2. This limits the application of LES simulations considerable, but as computing power substantially increases is gaining more and more importance also for technical flows. However, the drawbacks potentially impacting the solution quality is that the dissipation and its commuting properties with the main flow field is imposed a-priori and requires a pre-emptive knowledge of the solution by the user to choose adequate SGS-models and additionally the frequency information of the subgrid scale is lost.

- Reynolds averaged Navier-Stokes based turbulence models (RANS). For engineering applications most relevant are the RANS models, since they are robust and fast posing limited computational efforts. Hence, some space is devoted to them subsequently.

Fundamental ideas of turbulence modelling

If we consider a statistically stationary flow, the flow variables can be divided into a time-averaged value and an additional fluctuation value around this average or mean value. This kind of consideration, which is called Reynolds decomposition, provides the correlation for the velocity in flow direction (x-direction):

$$u_i(x_i, t) = \overline{u_i(x_i)} + u_i'(t, x_i). \tag{5.93}$$

The time-averaged velocity is described by the following equation:

$$\overline{u_i(x_i)} = \frac{1}{\Delta t} \int_{t}^{t+\Delta t} u_i(x_i, t)\, dt, \tag{5.94}$$

in which the time interval Δt over which integration is performed must be large enough to maintain the ability to capture possible time variations of the flow in the control volume. By definition, the time average of the fluctuations is zero, but the mean of their squares is different from zero. The turbulence intensity (ϕ) or Root Mean Square (RMS) is given in the following manner.

$$\phi = \sqrt{\overline{u_i'^2}} = \sqrt{\overline{u'^2 + v'^2 + w'^2}}. \tag{5.95}$$

The relative intensity of the velocity fluctuations, defined as the degree of turbulence Tu, is therefore:

$$Tu = \frac{\sqrt{\overline{u_i'^2}}}{|\overline{u_i}|}, \tag{5.96}$$

To get a sense of the orders of magnitude of turbulence intensity, it is about 1% downstream of a fine screen, about 10% near a wall, and more than 10% in a free jet. In general, turbulent flows can be divided into three classes:

- *Isotropic turbulence*: The statistical data of the flow field in the whole flow field are independent of the direction ($\overline{u'^2} = \overline{v'^2} = \overline{w'^2}$). They are invariant with respect to the translation and rotation of the coordinate system.
- *Homogeneous turbulence*: All statistical properties depend only on the direction and not on the location. They are invariant with respect to translational motion.
- *Anisotropic or shear turbulence*: This type is of practical interest in engineering applications, as it occurs in boundary layers, free jets, and channel flows, etc.

Reynolds equations

The steady state Navier–Stokes equations are valid for all turbulent flows of a continuum, but the complexity as well as the random nature of the turbulent exchange processes prevents for most technical configurations a direct numerical access to obtain an exact solution of these equations. Thus, most current methods for describing turbulent flows are based on the time-averaged Navier–Stokes equations. These equations are referred to as the Reynolds equations. Time averaging of the equations of motion generates new terms that can be interpreted as apparent stress gradients or heat flux quantities associated with turbulence. These new quantities must be related to the main flow using appropriate turbulence models. This modelling approach involves further assumptions and approximations. Therefore, the solution of a turbulent flow problem with the help of the Reynolds equations does not completely follow a self-contained basis based on basic physics, since the additional assumptions made must be made close to the system of equations on the one hand and on the other hand establishes well-founded but simplified links. The following equations based on Reynolds averaging are valid for a flow of an incompressible Newtonian fluid:

The application of Reynolds decomposition yields for the continuity equation:

$$\frac{\partial \overline{u_i}}{\partial x_i} = 0 = \frac{\partial \overline{u}}{\partial x} + \frac{\partial \overline{v}}{\partial y} + \frac{\partial \overline{w}}{\partial z}, \tag{5.97}$$

and accordingly, for the Reynolds averaged momentum equation:

$$\rho \overline{u_j} \frac{\partial \overline{u_i}}{\partial x_j} = -\frac{\partial \overline{p}}{\partial x_i} + \frac{\partial}{\partial x_j}\left(\mu \frac{\partial \overline{u_i}}{\partial x_j} - \rho \overline{u'_i u'_j}\right). \tag{5.98}$$

In Eq. (5.95), the apparent or virtual Reynolds shear stress tensor appears $\tau_{tur} = -\rho \overline{u'_i \cdot u'_j}$, which is significantly different from the molecular stress tensor given by $\tau_{ij} = \mu \cdot \left(\partial \overline{u_j}/\partial x_i + \partial \overline{u_i}/\partial x_j\right)$, as shown in Eq. (5.9). It is important to note that the additional Reynolds stress tensor appears only with respect to the time-averaged flow. The equation of motion for the instantaneous quantities does not include any additional turbulence term. The transition from the time-dependent equations of motion to the Reynolds equations formulates the challenge of turbulence modelling. The "price" of the time averaging is the introduction of the unknown Reynolds stress tensor arising from the nonlinear convective terms of the Navier–Stokes equations. The main task of solving the turbulence problem is to relate this Reynolds stress tensor to the time-averaged quantities of the velocity field using appropriate turbulence models. This can be done by three main categories of RANS-based turbulence models:

- Linear eddy viscosity models,
  - algebraic models (Cebeci-Smith model, Baldwin-Lomax model, …),
  - one equation models (Prandtl's one equation model, Baldwin Lomax),
  - two equation models ($k - \varepsilon$ types, $k - \omega$ types).
- Nonlinear eddy viscosity models
  - Explicit algebraic Reynolds stress models (EARSM),
  - V2f models.
- Reynolds stress model (RSM)

The procedure to evaluate the unknown shear stress tensor with the help of balance equations. Like any other transport quantity, these must satisfy a general form. The types of balance equations, often referred to as transport equations, are the basis of almost all recent closure approaches regardless of whether they are algebraic or tensorial in nature. The starting point for the formulation of the transport equations is the derivation of the Navier–Stokes equation for the instantaneous values. Following a strict mathematical derivation, the transport equations for the Reynolds stress tensor are:

$$\underbrace{\overline{v_k}\frac{\partial}{\partial x_k}\left(\overline{v_i'v_j'}\right)}_{C}+\underbrace{\overline{v_i'v_k'}\frac{\partial\overline{v_j'}}{\partial x_k}+\overline{v_j'v_k'}\frac{\partial\overline{v_i'}}{\partial x_k}}_{P}+\underbrace{2\nu\frac{\partial\overline{v_i'}}{\partial x_k}\frac{\partial\overline{v_j'}}{\partial x_k}}_{DS}$$
$$\underbrace{-\overline{\frac{p'}{\rho}\left(\frac{\partial v_i'}{\partial x_j}+\frac{\partial v_j'}{\partial x_i}\right)}}_{PSC}+\underbrace{\frac{\partial}{\partial x_k}\left[\overline{v_i'v_j'v_k'}-\nu\frac{\partial}{\partial x_k}\left(\overline{v_i'v_j'}\right)+\overline{\frac{p'}{\rho}\left(\delta_{kj}v_i'+\delta_{ki}v_j'\right)}\right]}_{D}=0. \quad (5.99)$$

In this equation C describes the convective change and P the turbulence production, which is the product of the Reynolds stress tensor with the time-averaged velocity gradients. The term DS represents turbulent dissipation, which is essentially a negative production term, and PSC is the pressure-shear correlation, which contributes to flow reorganization in a similar way to diffusion D. The production term is one of the most important, as it describes that only in flows with a velocity gradient (shear flow) can preserve turbulence.

The velocity gradient of the time-averaged velocity is thus the turbulence generator and the energy required for this is extracted from the main flow. The tensor equation consists of nine components. An important case of the transport equation is obtained by the formal reduction $i = j$, for which the transport equation of the turbulent kinetic energy k is obtained. Defining k as $k = 1/2 \cdot \overline{v_i'^2} = 1/2 \cdot \left(\overline{u'^2}+\overline{v'^2}+\overline{w'^2}\right)$, the transport equation of the turbulent kinetic energy has the following appearance:

$$\underbrace{\overline{v_k}\frac{\partial k}{\partial x_k}}_{C}+\underbrace{\overline{v_j'v_k'}\frac{\partial\overline{v_j'}}{\partial x_k}}_{P}+\varepsilon+\underbrace{\frac{\partial}{\partial x_k}\left[\overline{\left(k+\frac{p'}{\rho}\right)v_k'}-\nu\frac{\partial k}{\partial x_k}-\nu\frac{\partial}{\partial x_j}\left(\overline{v_j'v_k'}\right)\right]}_{D}=0, \quad (5.100)$$

where ε is the dissipation of turbulence. The turbulent energy is not an averaged quantity and the turbulent dissipation ε is described by the following equation:

$$\varepsilon=\nu\overline{\frac{\partial v_j'}{\partial x_k}\left(\frac{\partial v_j'}{\partial x_k}+\frac{\partial v_k'}{\partial x_j}\right)}. \quad (5.101)$$

The transport equations play an essential role in the development of modern turbulent flow modelling and methods. Although the Reynolds stress tensor can be formulated using the transport equations, these equations do not solve the closure problem because they themselves again contain unknown correlations. For these correlations, new transport equations must be formulated, in which, in turn, unknown correlations appear. The order of the obtained tensors increases with each step by a factor of one.

To summarize, the closure problem cannot be solved directly. There are always more unknowns than available equations. The only possible solution is to use closure assumptions, which in the simplest case can be applied directly to the Reynolds equations, such as Prandtl's mixing length approach. Further closure concepts consist in the use of numerically determined correlations, the application of physical model concepts and last but not least semi-empirical methods. Here, only the main ideas shall be communicated.

Fundamental ideas of linear eddy viscosity RANS turbulence modelling

Even simple algebraic models show reasonable results for plane two-dimensional flows but are hardly relevant for technical applications where three-dimensional flows occur, since higher orders of turbulence models are needed. These use one or more partial differential equations consisting of the modified Navier–Stokes equations for the turbulent kinetic energy k, the dissipation of the energy ε and the components of the turbulent stress tensor τ_{ij}.

A simple turbulence model in which a partial differential equation for the turbulent kinetic energy is related to an algebraic expression for the turbulent length scale l is the Prandtl mixing length approach. This model type is called a one-equation model or $k - l$ model.

An extension is represented by the likewise linear two-equation models, in which partial differential equations for the kinetic energy and its dissipation are solved. The best-known model is the $k - \varepsilon$ model, introduced by Harlow and Nakayama (1968)and later modified by Jones and Launder (1972) and Launder and Spalding (1974). It is the most widely used model in engineering applications and is implemented in almost all commercial fluid dynamic computational programs (CFD codes). Many applications of the $k - \varepsilon$ model use analytical wall functions to handle near-wall flow. In the alternative case, additional terms to the $k-$ and $\varepsilon-$equations must be introduced to extend the applicability to the treatment of the viscous sublayer (Chien, 1982). In this context, the viscous sublayer is often referred to as the low turbulent Reynolds number region $\left(\sqrt{k}l/\nu\right)$. This inner model is essential for flows in complex geometries, flow separation, or abrupt transverse cross-sectional area changes. The uncertainties in modelling such inner region models in complex flows with large anisotropy significantly limits the applicability framework of the $k - \varepsilon$ model.

The $k - \varepsilon$ models have been modified to include ta description of viscous sublayer. The necessary extensions are called "low Reynolds number" $k - \varepsilon$ models (Rodi, 1993). These additional extensions to the $k - \varepsilon$ model to capture lift effects or streamline changes require two-equation models. The most widely used models are those of Ng and Spalding (1972) and Wilcox and Traci (1976). All such models involve a formulation of turbulent kinetic energy and differ substantially in their treatment of the gradient of diffusion. The most striking difference, however, is the choice of dependent variable for the second transport model equation from which the characteristic length scale is determined. Two-equation models often coexist, as in the "shear stress turbulence model" (SST). The SST model combines the advantages of the $k - \varepsilon$ model with those of the $k - \omega$ model, where ω is the vorticity. The main advantage of the SST model is that in the near-wall region in the $k-\omega$ model, an analytical solution for the viscous sublayer is known (Wilcox, 1986). The linkage of the near-wall $k - \omega$ model to the $k - \varepsilon$ model treating the bulk flow is done by means of blending functions. Complicated multi-equation methods require the solution of partial differential equations of all components of the turbulent shear stress tensor. A summary of the description of the turbulence models with a large number of details can be found in the works of Reynolds (1976) and Arpaci and Larsen (1984).

The non-linear eddy viscosity models use a similar concept as the linear eddy viscosity concepts by relating the mean turbulence field to the mean velocity field. The main difference, however, is to introduce a nonlinear function not only dependent on the mean strain but additionally also the vorticity fields or even other turbulence variables. Thus, they do not require blending functions; prominent representatives are the EARSM and the V2f model widely used in computations due to their robustness.

Basic thoughts on RSM models in RANS turbulence modelling

Here, only the essential ideas of the Reynolds Stress Model (RSM) are described which a-priori do not assume that the turbulent shear stress is proportional to the strain rate, which leads for a two-dimensional, incompressible flow to:

$$-\rho\overline{u'v'} \neq \mu_t\left(\frac{\partial u}{\partial y} + \frac{\partial v}{\partial x}\right). \tag{5.102}$$

μ_t is the turbulent viscosity given by $\mu_t = \left(k^2/\varepsilon\right)$. The most widely known Reynolds stress models are those of Hanjalic and Launder (1972), Launder et al. (1975) and Donaldson (1972). RSM models require the solution of at least three additional partial differential equations (PDE) compared to the $k-\varepsilon$ model. Moreover, even RSM models still require approximations and assumptions for the double and triple correlation, being barely accessible by experimental techniques. Fortunately, the progress in DNS Simulation opens here a gate by providing a data basis for a closed-form confirmation. A simplification of RSM modelling known as the algebraic stress or flow model ASM is described in detail in Rodi (1981). In the ASM model, the transport of Reynolds stresses is assumed to be proportional to the transport of kinetic energy. For boundary layer flows without buoyancy effects, the ASM approach results in the following relationship for the turbulent shear stress.

$$\overline{u'v'} = C_\mu \frac{k^2}{\varepsilon}\frac{\partial u}{\partial y}, \tag{5.103}$$

which at first sight looks identical to the result of the $k-\varepsilon$ model, but in the ASM C_μ is a function of the ratio of turbulence production to dissipation of turbulent kinetic energy and thus no longer a constant.

Turbulent boundary layer equations

Many technical flows have boundary layer character. The laminar boundary layer equations cannot be applied to the turbulent case because the fluctuation velocities v_k' are fast in space and time and thus their derivatives cannot be neglected with respect to the other terms, even though their absolute value is an order of magnitude smaller than that of the mean values. Therefore, the boundary layer equations must be derived directly within the Reynolds equations and the transport equations. However, the procedure is the same as in the laminar case; details can be found in the books by Burmeister (1993) or Anderson et al. (1984). In two-dimensional case, the turbulent boundary layer equations are:

$$\frac{\partial \overline{u}}{\partial x}+\frac{\partial \overline{v}}{\partial y}=0,\ \rho\left(\overline{u}\frac{\partial \overline{u}}{\partial x}+\overline{v}\frac{\partial \overline{u}}{\partial y}\right)=-\frac{dp_\delta}{dx}+\frac{\partial}{\partial y}\left(\mu\frac{\partial \overline{u}}{\partial y}-\rho\overline{u'v'}\right), \tag{5.104}$$

where p_δ is the pressure at the edge of the boundary layer towards the bulk flow.

A similar procedure can be used for the transport equations and the transport equations for the shear stress and the turbulent kinetic energy are obtained, which for a two-dimensional flow are as follows:

$$\underbrace{\overline{u}\frac{\partial\left(\overline{u'v'}\right)}{\partial x}+\overline{v}\frac{\partial\left(\overline{u'v'}\right)}{\partial y}}_{C}+\underbrace{\overline{v'^2}\frac{\partial \overline{u}}{\partial y}}_{P}+\underbrace{2\nu\left(\overline{\frac{\partial u'}{\partial x}\frac{\partial v'}{\partial x}}+\overline{\frac{\partial u'}{\partial y}\frac{\partial v'}{\partial y}}+\overline{\frac{\partial u'}{\partial z}\frac{\partial v'}{\partial z}}\right)}_{DS}$$

$$+\underbrace{\overline{\frac{p'}{\rho}\left(\frac{\partial u'}{\partial y}+\frac{\partial v'}{\partial x}\right)}}_{PSC}+\underbrace{\frac{\partial}{\partial y}\left(\overline{u'v'^2}-\nu\frac{\partial}{\partial y}\left(\overline{u'v'}\right)+\frac{\overline{p'u'}}{\rho}\right)}_{D}=0;$$

$$\underbrace{\overline{u}\frac{\partial k}{\partial x}+\overline{v}\frac{\partial k}{\partial y}}_{C}+\underbrace{\overline{u'v'}\frac{\partial \overline{u}}{\partial y}}_{P}+\varepsilon+\underbrace{\frac{\partial}{\partial y}\left(\overline{\left(k+\frac{p'}{\rho}\right)v'}-\nu\frac{\partial k}{\partial y}-\nu\frac{\partial \overline{v'^2}}{\partial y}\right)}_{D}=0. \tag{5.105}$$

Most terms of the two transport equations except for the local pressure fluctuations can be determined experimentally. Fortunately, pressure fluctuations do not appear in the kinetic energy equation. If channel walls limit the flow, the velocity fluctuations disappear as the wall is approached, and so do the Reynolds shear stresses; the result is

$$lim_{y\rightarrow 0}\mu\frac{\partial \overline{u}}{\partial y}=\tau_w. \tag{5.106}$$

If one assumes a fully developed, two-dimensional, turbulent flow, then in the vicinity of the wall all statistical quantities depend only on the lateral (y-) coordinate. Since the derivative of the velocity in the axial direction is zero, the continuity equation leads to the result that the mean velocity in the y-direction is also zero. In this particular case, the convective terms and the pressure in the Reynolds equation are zero, and there remains

$$\frac{\partial}{\partial y}\left(\mu\frac{\partial \overline{u}}{\partial y} - \rho\overline{u'v'}\right) = 0. \tag{5.107}$$

With the boundary condition on the wall we get:

$$\frac{\tau_w}{\rho} = \nu\frac{\partial \overline{u}}{\partial y} - \overline{u'v'}. \tag{5.108}$$

Obviously, only the parameters ν and τ_w/ρ occur and one can define a shear stress velocity u_τ which is as follows:

$$u_\tau = \sqrt{\frac{\tau_w}{\rho}} = u_0\sqrt{\frac{f}{2}}, \tag{5.109}$$

in which f is the dimensionless friction factor (normalized by with $f = 2\tau_w/\left(\rho u_0^2\right)$. If one defines the dimensionless coordinates u^+ and y^+ in the way:

$$u^+ = {}^{\overline{u}}/_{u_\tau},\ y^+ = {}^{(yu_\tau)}/_{\nu}, \tag{5.110}$$

the velocity distribution near the wall can be given in a universal representation of the form $u^+ = f\left(y^+\right)$. The solution for f in the immediate vicinity of the wall is thus immediately given, since the velocity fluctuations near the wall are zero and the adhesion condition holds at the wall, so that $u^+ = y^+$. This describes the so-called viscous sublayer, in which the molecular shear stresses are considerably higher than the Reynolds stresses. With increasing wall distance, the Reynolds stresses increase and at the same time the molecular fraction becomes negligible. This leads to a logarithmic expression for the dimensionless velocity in the form:

$$u^+ = \frac{1}{C_1} lny^+ + C_2. \tag{5.111}$$

Between the viscous sublayer and the fully turbulent boundary layer there is a transition layer, for which different but mostly logarithmic functions are used. Van Kármán suggests for the velocity distribution in the boundary layer:

- Viscous sub-layer: $u^+ = y^+$ for $y^+ < 5$;
- Transition layer: $u^+ = 5lny^+ - 3.05$ for $5 < y^+ < 30$;
- Fully turbulent boundary layer: $u^+ = 2.5lny^+ + 5.5$ for $y^+ > 30$.

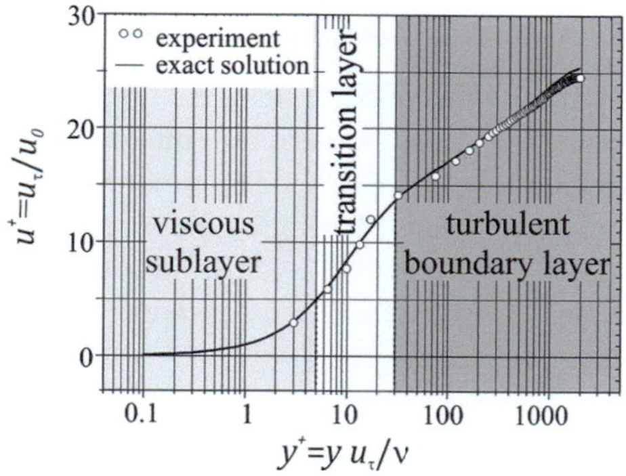

Fig. 5.17 Measured dimensionless wall velocity u^+ of a pipe flow as a function of the dimensionless wall distance y^+ in lead bismuth at $Re = 8 \times 10^4$ from Schulenberg and Stieglitz (2010) in comparison with the exact solution (Reichhardt, 1951)

At the transition from the anisotropic turbulent boundary layer to the region of isotropic turbulence of the bulk flow, the logarithmic wall law no longer applies. It changes into a wake law, which is also called the centre law in pipe or channel flows.

An overview of the different definitions between the viscous sublayers and the developed boundary layer is given by Schlichting and Gersten (2000). The existence and form of the laminar sublayer, the transition layer, and their attachment to the logarithmically scaling turbulent boundary layer has been experimentally proven even at high Reynolds numbers, as illustrated in Fig. 5.17.

The transition of the anisotropic turbulent boundary layer into an isotropic core flow causes the logarithmic functions to lose their applicability. Within this range, wake laws are introduced. There is a myriad of articles on this, most of which differ only in the choice of the constants of the wake profile. A good overview is given here by the works of Zagarola and Smits (1998), Barenblatt and Chorin (1998) and Guo and Julien (2003).

For the modelling of turbulence in a broad set of literature is available. The work of Rodi (2005) provides a detailed overview of the entire range, but also references to best practice guidelines, as well as the cross-references listed therein. The technical implementation within the framework of finite difference and volume methods is well described in Versteeg and Malalasekera (2007), among others.

5.1.12 Turbulent Energy Exchange

A similar formalism to that used for momentum exchange can also be used to derive the turbulent equations of energy exchange, which are for steady incompressible flow:

$$\rho c_p\left(v_k\frac{\partial T}{\partial x_k}\right) = \lambda\frac{\partial^2 T}{\partial x_k^2} \text{ or } \rho c_p\left(u\frac{\partial T}{\partial x} + v\frac{\partial T}{\partial y} + w\frac{\partial T}{\partial z}\right)$$
$$= \lambda\left(\frac{\partial^2 T}{\partial x^2} + \frac{\partial^2 T}{\partial y^2} + \frac{\partial^2 T}{\partial z^2}\right). \tag{5.112}$$

Reynolds equations of turbulent energy exchange

Similar to the velocity field, the temperature field in the Reynolds turbulent energy exchange splits the current temperature T into a fluctuating component T' and a time-averaged value $\overline{T}$. The time-averaged component of the temperature should not be confounded with the mean temperature T_m. The former is a local temperature at a discrete point, while the latter is a flow-averaged quantity over the channel cross-section. The Reynolds approach yields the following result for turbulent energy exchange:

$$\rho c_p\left(\overline{v_k}\frac{\partial \overline{T}}{\partial x_k}\right) = \lambda\frac{\partial^2 \overline{T}}{\partial x_k^2} - \underbrace{\rho c_p\frac{\partial}{\partial x_k}\left(\overline{v_k' T'}\right)}_{RH}, \tag{5.113}$$

in which the term RH represents a virtual turbulent heat flux vector arising from the Reynolds averaging approach. It thus describes a macroscopic energy exchange arising from the fluctuation components. In contrast to the Reynolds shear stress tensor, the turbulent heat flux is a vector quantity, since the transport quantity of the energy is a scalar (temperature), while for the momentum a vector quantity is already present. Similarly, transport equations must be derived for the turbulent heat flux. In contrast to momentum transport, turbulent energy exchange is in most cases a passive transport process; exceptions are buoyant flows. As in the case of momentum transport, boundary layer equations can also be derived for energy transport, which read for a two-dimensional, stationary, incompressible flow to:

$$\rho c_p\left(\overline{u}\frac{\partial \overline{T}}{\partial x} + \overline{v}\frac{\partial \overline{T}}{\partial y}\right) = -\frac{\partial}{\partial y}\left(-\lambda\frac{\partial \overline{T}}{\partial y} + \rho c_p\overline{v'T'}\right). \tag{5.114}$$

5.1.13 Buoyancy and Mixed Convection in Closed Cavities

Until now, we had considered energy transport only as a forced convective transport, but the energy thermal supply also changes the density of a fluid regardless of whether it is a gas, a liquid, or a mixture. With an appropriate energy input, the buoyancy velocity induced by the density gradient can be significantly greater than the mean flow velocity u_0 produced by an external pressure gradient, e.g. by body forces or a pump. If both types of flow drive appear the flow regime is called mixed convection. In the extreme case, for example, a solar thermal system can be driven only by means of buoyancy, such as a thermosiphon as shown in Fig. 5.18.

In buoyant flows, the mean velocity can thus no longer play the excellent role, since the density differences due to the temperature difference between the wall T_w and the bulk T_m generate the flow itself. These density differences, which give the buoyancy to the flow, originate from a temperature coefficient of volume expansion

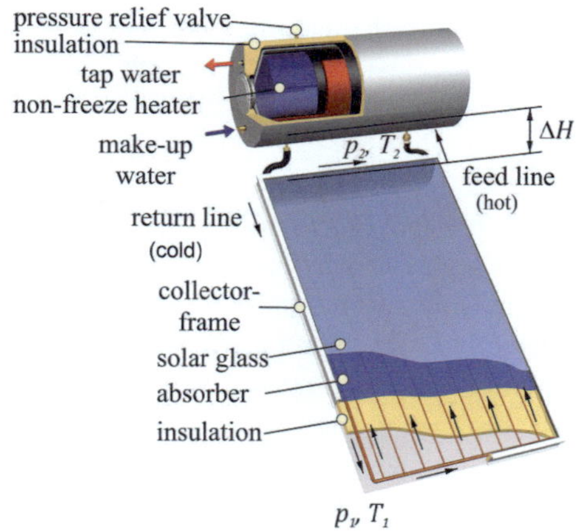

Fig. 5.18 Example of open thermosiphon system for drinking water preparation

(or thermal expansion) β, which is defined as follows:

$$\beta = \rho - \rho_m = -\frac{1}{\rho}\left(\frac{\partial \rho}{\partial T}\right), \tag{5.115}$$

where ρ_m is the density of the bulk flow and ρ the local fluid density.

In many engineering applications as e.g. the heat losses at solar receiver panels, both mechanisms occur at the same time. Therefore, heat transfer by mixed convection is the result of a superposition of buoyancy-driven natural and forced convection, which due to the non-linearity of the convective terms cannot simply be added. Thus, to account for buoyant flows, a new dimensionless quantity arises, the so-called Grashof number Gr, which can be conceived as the ratio of the buoyancy force acting on a fluid element to the counteracting viscous force. The Grashof number corresponds entirely to the Archimedes number used in civil engineering if the variation of the density is caused by a change of temperature T only as in wall bounded/internal flows. Since in most cases (heat flow by radiation on the absorber) not the wall temperature but the heat flux is provided, the wall temperature T_w can be converted into a relation containing the wall heat flux $\dot{q}_w$ by means of Fourier's law.

$$Gr = \frac{g \cdot \beta \cdot (T_W - T_m) \cdot L^3}{\nu^2} = \frac{g \cdot \beta \cdot \dot{q}_w \cdot L^4}{\lambda \, \nu^2}, \tag{5.116}$$

with ν the kinematic viscosity and g the acceleration due to gravity. Thus, for free convection, the Nusselt number no longer depends only on the Reynolds and Prandtl numbers, but also on the Grashof number, i.e. $Nu = Nu(Gr, Pr)$. With the Grashof number describing the contribution of the natural convection flow (buoyant force), and the Reynolds number Re_L describing the imposed flow (inertia), the relative effect of buoyancy on mixed convection can also be expressed though the Richardson number:

$$Ri = \frac{Gr}{Re_L^2}, \tag{5.117}$$

where the length L denotes the scale where the heat is applied to. For varying length scales and hydraulic circuits, i.e. containing a horizontal L and a vertical H, the Richardson number is considered to be more suitable to characterize mixed convection. The extrema $Ri \to 0$ and $Ri \to \infty$ represent pure forced and pure natural convection, respectively. A rough separation of the transition between the flow regimes is suggested by Gebhart (1988) for $Ri \approx 0.3$ from forced to mixed convection and $Ri \approx 16$ from mixed to pure natural convection. New, numerical based studies supported by an unfortunately rare experimental data bases (Garbrecht, 2017), however, show that the especially the onset of the transition from forced to mixed convection starts already at $Ri \approx 0.1$.

A first estimation based on analytical considerations of the transition from forced convection to mixed convection provides the following relationship suggested by Müller and Ehrhard (1999):

$$\frac{\sqrt{Gr}}{Re} < 1. \tag{5.118}$$

This ratio considers the natural convective lift influence on the flow via the Grashof number. The influence of the forced flow is characterized by the Reynolds number Re. If Re dominates in the ratio and it becomes smaller than one, buoyancy can be neglected. If the Grashof number Gr increases due to a higher heat flow or due to the given dimensions (see power of L!), the buoyancy effects caused by density differences must also be considered. According to this criterion, however, it is not possible to estimate whether mixed convection or pure natural convection is present. In order to distinguish more precisely between natural, mixed, and forced convection, the following criterion determined on the basis of experiments has been extended by Müller and Ehrhard (1999):

$$\frac{Ra^{1/3}}{Re^{4/5} \cdot Pr^{2/5}} = c^*. \tag{5.119}$$

The Rayleigh number $Ra = Gr \cdot Pr$ contained therein is formed from the product of the Grashof number Gr and the Prandtl number Pr and also represents a characteristic dimensionless parameter for describing natural convective effects. The Rayleigh number is an extension of the Grashof number and includes molecular viscous fluid properties as well as molecular thermal properties. The exponent of the Rayleigh number in the condition for c^* results from a boundary layer consideration of the thermal boundary layer of a buoyant flow whose thickness scales with $Ra^{-1/3}$. For pure forced convection, the thickness of the turbulent thermal boundary layer scales with $Re^{-4/5}/Pr^{-2/5}$ (Aicher & Martin, 1997). Thus, for given Ra, Re, and Pr, the parameter c^* provides information about the mechanism of heat transfer from the wall across the boundary layer into the flow core. Aicher and Martin (1997) define the following ranges; for $c^* < 0.2$ forced convection dominates, while for $c^* > 0.4$ heat transfer by natural convective effects is dominant. By plotting the ratio of the measured Nusselt number Nu to the Nusselt number Nu_{FT} resulting from pure forced convective heat transfer under the same boundary conditions, it is possible to classify measurement results into natural, mixed, and forced convection.

A wide range of literature exists to estimate the magnitude of buoyancy velocities caused by density differences versus forced convective mean flow velocity generated by external pressure forces (pumps). Before performing a CFD-based simulation of the flow field, it is recommended to use analytical relationships to estimate the buoyancy-induced velocities and to determine the Grashof and Rayleigh numbers, respectively. Although collector flow geometries can have complex geometric configurations, they can often be limited to the geometry of a point or line source or a vertical and horizontal plate. Analytical relationships for determining the lift velocities achievable in this context can be found, for example, in Kakac et al. (1985) or in the section on engineering relationships.

Another aspect of the methodology of analytical penetration should not be forgotten. Buoyant flows exhibit a multitude of flow patterns. At low heating powers, thermal conduction exists and only after a certain temperature difference is exceeded, a density difference occurs in which the buoyancy overcomes the viscous resistance, laminar convection sets in, which is characterized by an ordered cellular structure. As the simplest case, consider the Rayleigh–Benard cell shown in Fig. 5.19, in which a fluid is heated from below and cooled from above. If the fluid height (d) in the direction of the gravity vector g is smaller than the dimensions occurring in the $x - y$ plane ($x \gg d$, $y \gg d$), a critical Rayleigh number Ra_{crit} can be determined for the transition from the heat conduction state to laminar convection. The critical Rayleigh number is $Ra_{crit} = 1708$ for very good thermal conductive walls (Biot number $Bi \gg 1$), which usually occur in the copper absorber tubes in low temperature systems. Laminar convection is essentially two-dimensional. In high-temperature systems, where stainless steel tubes with a smaller molecular thermal conductivity are used, the Biot number effect must be considered. In this case, higher wall temperatures occur and thus the critical Rayleigh number decreases, in the adiabatic case ($Bi \rightarrow 0$) down to $Ra_{crit} = 1296$.

The laminar convection with its mainly two-dimensional cellular pattern is followed by the region of three-dimensional convection. This transition depends now in addition to the Rayleigh number also on the Prandtl number. As the Rayleigh number continues to grow, the still structured three-dimensional convection is accompanied by a stochastic three-dimensional convection in which distinct flow patterns are no longer discernible.

First flow pattern maps, in which the flow domain boundaries are illustrated as a function of the Rayleigh number and the molecular Prandtl number, were developed for the Rayleigh–Benard cell. Corresponding to these maps, Nusselt number

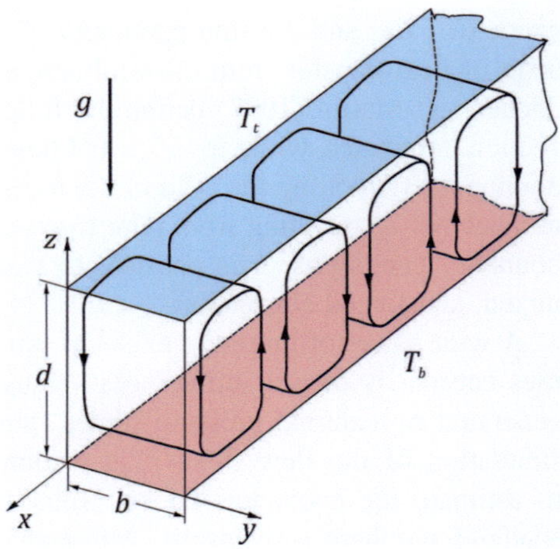

Fig. 5.19 Structure of a Rayleigh–Benard cell, schematic flow pattern of a quasi-two-dimensional buoyant flow

correlations as a function of Prandtl number and Rayleigh number were filed, based on the work of O'Toole and Silveston (1961), Castaing et al. (1989), and Kek and Müller (1993). Table 5.4 provides common Nusselt number relationships and their Rayleigh number range of validity for different fluids (Prandtl number) and Fig. 5.20 shows flow pattern map of a Rayleigh–Benard cell as a function of Pr and Ra.

For the vertical gap of height h and lateral thickness d, in which one vertical side is heated ($T = T_h$) and the other respectively cooled ($T = T_c$), a different flow dynamic results than in the case of the Rayleigh Benard cell (compare Fig. 5.21a) because even the smallest temperature differences lead to the formation of a large laminar convection cell, which has boundary layer character. In the boundary layer, there is an upward movement on the hot side and a corresponding downward movement on the cold side, while the core is almost at rest. This quiescence in the core leads to a linear temperature drop across the core characterized

Table 5.4 Ranges of flow patterns as a function of the Rayleigh number Ra and the Nusselt numbers Nu that can be calculated with it as a function of the molecular Prandtl number Pr for a Rayleigh–Benard cell from Müller and Ehrhard (1999)

$0 < Ra < 1.7 \times 10^3$	$Nu = 1$	
$Pr < 0.05$	$0.1 \leq Pr \leq 100$	$0.65 < Pr \leq 1.5$
$1.7 \times 10^3 < Ra < 7 \times 10^3$ $Nu = 0.59 \cdot Ra^{0.072}$	$1.7 \times 10^3 < Ra < 3.5 \times 10^3$ $Nu = 0.0024 \cdot Ra^{0.82}$	
$10^4 < Ra < 4 \times 10^4$ $Nu = 0.62 \cdot Ra^{0.31}$	$3.5 \times 10^3 < Ra < 10^5$ $Nu = 0.23 \cdot Ra^{1/4}$	
$4 \times 10^4 < Ra < 2.5 \times 10^5$ $Nu = 0.2 \cdot Ra^{1/5}$	$10^5 < Ra < 10^9$ $(\delta_{th}/d) \sim Ra^{-1/3}$ $Nu = 0.1 \cdot Ra^{0.31} \cdot Pr^{0.84}$	$10^7 < Ra < 10^{14}$ $(\delta_{th}/d) \sim Ra^{-2/7}$ $Nu = 0.23 \cdot Ra^{0.282}$

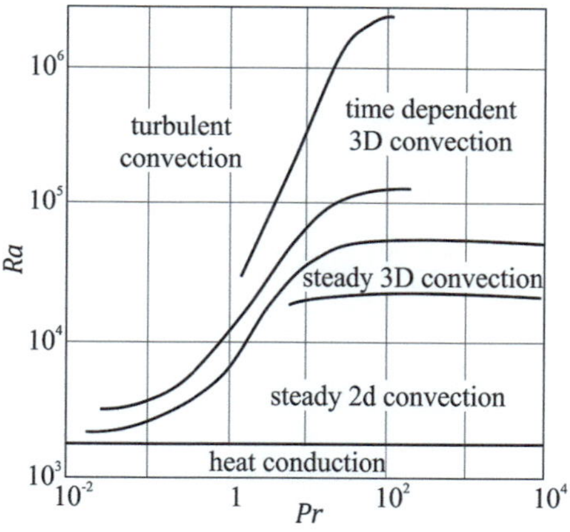

Fig. 5.20 Flow map of convection patterns establishing in a Rayleigh–Benard cell as a function of Rayleigh number *Ra* and molecular Prandtl number *Pr* according to Müller and Ehrhard (1999)

by molecular heat conduction. This region occurs between $10^5 < Ra < 10^8$. Further temperature increase also results in a cellular structure in the core. These are co-rotating vortices whose axes are horizontal in the gap and have a fixed spacing. Depending on the fluid properties, the cellular structure can be stationary or time dependent. For $Pr \ll 1$ they are mostly stationary while $Pr \gg 1$ leads to transient waves. A further increase in the temperature difference $\Delta T = (T_h - T_c)$ subsequently leads to counter-rotating secondary vortices, which are formed at the points between the large vortices where the maximum shear occurs. If the temperature difference is increased even further, a condition is created which is determined by the formation of turbulent boundary layers on both sides of the plate. The interaction between the boundary layers of both sides is only slight. In the core region of the flow, the flow is almost ideally mixed and thus almost isothermal.

In contrast to the Benard cell, two length scales occur inherently in the vertical gap, characterized by the aspect ratio $\epsilon = (h/d)$. Analogous to the Benard cell, flow maps can be generated which depend on the Rayleigh number Ra, the Prandtl number Pr, and the aspect ratio ϵ, and are thus three-dimensional. One such plot is illustrated in Fig. 5.21b, which shows the results of the transitions for two different assumptions. The solid boundary lines show the limits for the assumption that buoyancy is considered only in momentum conservation (so-called Boussinesq approximation). If the buoyancy effects are also considered in the energy equation, the other boundary lines result, which, however, require considerably more far-reaching modelling approaches. Table 5.5 gives some Nusselt number relationships for the vertical gap given in the literature.

From the consideration of flow in closed cavities, it is clear that an analytical consideration of the flow problem is mandatory prior to the creation of a numerical calculation model, so that with their help the selection of correct turbulence models and closure approaches can be made.

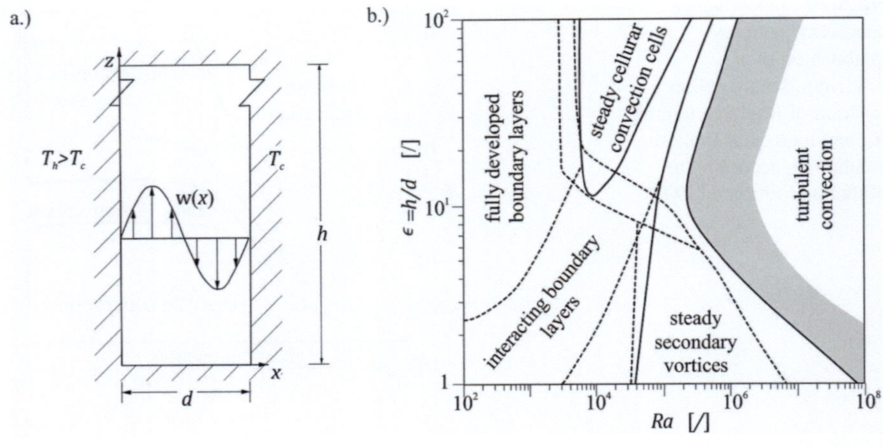

Fig. 5.21 **a** Schematic sketch of a vertical gap heated with $T = T_h$ on the left and cooled with $T = T_c$ on the right; **b** flow map of buoyant flow in a vertical gap for air ($Pr = 0.71$) from Chenoweth and Paolucci (1986), showing aspect ratio e as a function of Rayleigh number Ra. (—) ideal Boussineq approximation; (- - -) without Boussinesq approximation

Table 5.5 Nusselt number relations in heat transfer in a vertical gap as a function of Rayleigh number Ra, aspect ratio ϵ, and Prandtl number Pr from Müller and Ehrhard (1999)

$Ra < 7 \times 10^3$ $Nu = 1$	$10^3 < Ra < 10^5$ $Nu = \epsilon^{1/4} \cdot Ra^{1/4}$	$10^7 < Ra \cdot \epsilon^{-3} < 5 \times 10^9$ $Nu_m = 0.36 \cdot Pr^{0.051} \epsilon^{0.36} \cdot Ra^{1/4}$
$Nu = 1$	$10^3 Ra5 \cdot 10^5$, $Pr = 0.7$ $Nu_1 = 0.0605 \cdot Ra^{1/3}$ $Nu_2 = \left[1 + \left(\frac{0.104 \cdot Ra^{0.239}}{1+\left(\frac{6310}{Ra}\right)^{1.36}}\right)^3\right]^{\frac{1}{3}}$ $Nu_3 = 0.242 \cdot (Ra \cdot \epsilon)^{0.272}$ $Nu_m = \max(Nu_1, Nu_2, Nu_3)$	$5 \times 10^9 \leq a \cdot \epsilon^{-3} \leq 10^{11}$ $Nu_m = 0.096 Pr^{0.051} \cdot \epsilon^{0.1} \cdot Ra^{0.3}$

5.1.14 Analogy of Heat and Momentum Transport, Solution Approaches for Local Turbulent Simulations

In analogy between Newton's friction law for the laminar shear stress τ, a law for the virtual turbulent shear stress τ_{tur} can be derived:

$$\tau = -\mu \frac{\partial \overline{u}}{\partial y} \quad \text{and} \quad \tau_{tur} = -\mu_t \frac{\partial \overline{u}}{\partial y}, \tag{5.120}$$

where μ_t is a virtual viscosity caused by the turbulence, which is called eddy viscosity. It should be noted here that the eddy viscosity is not a property of the fluid like the dynamic viscosity μ, but depends on the time-averaged flow variables. In a similar way as the molecular momentum diffusion ν, the molecular eddy diffusion ε_M is often used to characterize a turbulent flow, ν and ε_M being defined as:

$$\nu = \frac{\mu}{\rho} \quad \text{and} \quad \varepsilon_M = \frac{\mu_t}{\rho}. \tag{5.121}$$

By analogy with Fourier's law of heat conduction, a law for turbulent heat conduction $\dot{q}_{tur}$ can also be formulated:

$$\dot{q} = -k\frac{\partial \overline{T}}{\partial y} \quad \text{and} \quad \dot{q}_{tur} = k_{tur}\frac{\partial \overline{T}}{\partial y}, \tag{5.122}$$

in which k_{tur} represents an apparent (eddy) thermal conductivity, which is also not a fluid property. It also depends on the time-averaged flow variables and the mean temperature. By analogy with the molecular thermal diffusivity a, an eddy diffusivity of heat ε_H is defined:

$$a = \frac{k}{\rho c_p} \quad \text{and} \quad \varepsilon_H = \frac{k_{tur}}{\rho c_p}. \tag{5.123}$$

Thus, a turbulent Prandtl number Pr_t can be introduced complementary to the molecular Prandtl number Pr:

$$Pr = \frac{\nu}{a} \quad \text{and} \quad Pr_t = \frac{\varepsilon_M}{\varepsilon_H}. \tag{5.124}$$

Experiments to measure the turbulent Prandtl number Pr_t in tubes showed that the turbulent Prandtl number is a function of the Reynolds number Re, the molecular Prandtl number Pr and the wall distance (y/R), and exactly the determination definition of the turbulent Prandtl number Pr_t is thus:

$$Pr_t = f(Re, Pr, y/R) = \frac{\overline{u'v'}}{\overline{v'T'}} \cdot \frac{\partial T/\partial y}{\partial u/\partial y}. \tag{5.125}$$

Eckert and Drake (1972) collected measurement results of turbulent Prandtl number for different fluids including liquid metals. They write, "The experimental difficulties for detecting Pr_t are apparently so great that there is no satisfactory coherence of results for small Prandtl numbers", which holds still today. Assuming a two-dimensional turbulent pipe inlet flow, the resulting wall shear stress τ and heat flux $\dot{q}$ can be expressed using the turbulent exchange quantities in the following way:

$$\frac{\tau}{\rho} = (\nu + \varepsilon_M)\frac{d\overline{u}}{dy} \quad \text{and} \quad \frac{\dot{q}}{\rho c_p} = -(a + \varepsilon_H)\frac{d\overline{T}}{dy}. \tag{5.126}$$

The Reynolds analogy is used in most technical applications. In this context, it must be explicitly pointed out that in flows with recirculation, separation, shear flows, free jets, there is a strong anisotropy of the turbulence field. There, the validity of the $k-\varepsilon$ model and the applicability of the Reynolds analogy presented here is problematic. This cannot be improved by the user by choosing a higher spatial resolution of the computational grid. Rather, the applicability of the chosen viscous and thermal wall functions and boundary conditions must be verified by empirical values and with the help of analytical considerations before starting the calculation.

In fully developed flows, $\nu \ll \varepsilon_M$ and $a \ll \varepsilon_H$ are valid except in the immediate vicinity of the wall. If one assumes that the ratio of turbulent heat transport to turbulent momentum transport is constant, in the extreme case $\varepsilon_M = \varepsilon_H$ this leads directly to the result $Pr_t = 1$, which is called Reynolds analogy:

$$c_p d\overline{T} = -\frac{\dot{q}}{\tau} d\overline{u}. \tag{5.127}$$

The equation can be easily integrated, provided that a no-slip condition at the wall, the wall temperature T_w and $\overline{T}(y = 0)$ are known. If the following assumptions are made for the connection of the flow and temperature field to the external flow, $\overline{u} = u_0$ and $\overline{T} = T_m$, which are defined in the following way:

$$c_p \int_{T_w}^{T_m} d\overline{T} = -\frac{\dot{q}}{\tau} \int_0^{u_0} d\overline{u} \quad \text{or} \quad c_p(T_w - T_m) = \frac{\dot{q}_w}{\tau_w} u_0, \tag{5.128}$$

the Reynolds analogy in a pipe flow yields the following analytical results for the friction factor f_λ and the Stanton number St:

$$f_\lambda = 8\frac{\tau_w}{\rho u_0^2} \quad \text{and} \quad St = \frac{Nu}{Re\ Pr} = \frac{f_\lambda}{8}. \tag{5.129}$$

The Stanton number measures the ratio of the heat transferred into a fluid to the thermal capacity of the fluid. By the above relation the Reynolds analogy relates heat transfer to the friction. Generally, this approach leads to relatively good results for molecular Prandtl numbers with $Pr > 0.71$, provided the anisotropy of the flow field is small. Therefore, it is usually implemented as a default setting in commercial computational codes that include the $k - \varepsilon$ model. Nevertheless, it must be emphasized that a $k - \varepsilon$ model reduces the turbulent dissipation from a tensor to a constant (question of anisotropy) and the Reynolds analogy is an additional constraint with the assumption $\varepsilon_M \sim \varepsilon_H$.

This problem of anisotropy of the momentum field as well as the use of the Reynolds analogy in turbulence heat transfer problem is known for a long time and critically discussed by several authors (Kays, 1994 or Carteciano & Grötzbach, 2003). To solve the closure problem of the turbulent heat fluxes RH in Eq. (5.113), different approaches can be taken, filling several textbooks. In the following, only the basic ideas of the solution approaches of the closure problem for the turbulent heat flows as well as the order of the methods shall be touched upon.

- *Semiempirical or empirical solution methods of zero and first order.* All such models do not allow insight into turbulent exchange mechanisms between temperature and velocity fluctuations and are only applicable after careful examination of the valid parameter range of the dimensionless numbers (Re, Pr, Ra) and for Pr_t around unity.
- *Turbulent Prandtl numbers from analytical solutions* are difficult to treat and are essentially limited to simple geometries but provide important information about the near-wall behaviour and the thermal wall functions. Unlike transport equation models, they include the temporal behaviour of the fluctuating variables and thus their spectral behaviour, which allows comparison with results from a direct numerical simulation. Unfortunately, for engineering applications, such methods require considerable effort and intervention in commercial CFD design tools.
- *Modelling of turbulent heat flux using transport equations* require an experienced CFD user to analyse the posed flow problem in terms of the potentially occurring phenomena and to choose appropriate thermal and viscous wall functions for this purpose. In highly anisotropic flows, heat transfer calculations using the $k - \varepsilon$ turbulence model and a turbulent Prandtl number $Pr_t = 0.9$ (Reynolds analogy) can lead to unphysical results such as a negative wall heat flux. In such cases, modelling the turbulent heat fluxes using transport equations of, for example, an algebraic heat flux model (AHFM) (Shams et al., 2019) or the Reynolds stress turbulence model (RSM) (Khan & Joshi, 2015) is essential. Nevertheless, such a procedure always leads to triple correlations, which are not known “a priori” and have to be determined either by experimental benchmarks or by “Direct Numerical Simulation” (DNS). Increasing computational power provided substantial progress in computing local turbulent heat transfer problems by means of AHFM, RSM and also large eddy simulations (LES); nonetheless, pro-found expert knowledge is still required to tune the computations to experimental observations. Nonetheless, the degree of uncertainty is continuously decreasing.
- *Direct numerical simulation (DNS)* of the time-dependent Navier–Stokes equations is in principle an exact solution to the problem. Although memory resources and computational speeds have increased considerably, their application has so far remained limited to a few mostly academic cases. Recent DNS studies have focused on the influence of the Prandtl number in the case of isotropic turbulence (Kasagi & Ohtsubo, 1992), near-wall flows (Kawamura et al., 1999 or Piller et al., 2002), and buoyant flows (Otic et al., 2005). All of them include results for the time-dependent as well as the mean values of the temperature and velocity field depending on the boundary conditions. Therefore, DNS is one of the few tools that allows to validate the approaches and constants chosen in the transport sliding methods. This is especially important because the experimental determination of these quantities is hardly possible

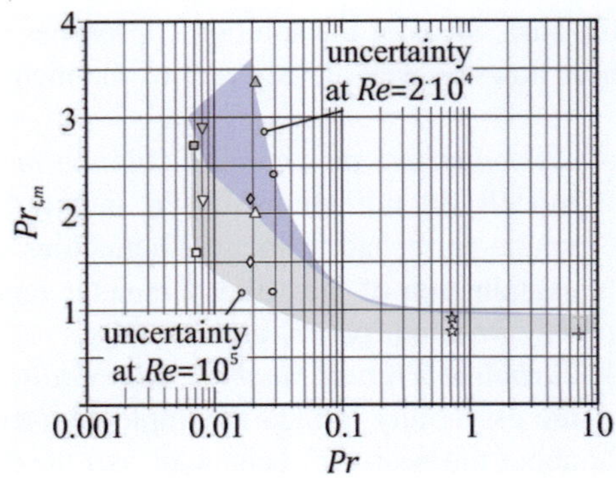

Fig. 5.22 Measured mean turbulent Prandtl number $Pr_{t,m}$ of a pipe flow as a function of molecular Prandtl number Pr (multiple fluids) for two different Reynolds numbers (Stieglitz, 2007)

in the absence of suitable measurement methods. Today's measurement methods are not capable of simultaneously measuring several field quantities with sufficiently precise time and spatial resolution.

Apart from the turbulent Prandtl number, which is difficult to determine experimentally, the assumption of a constant Pr_t value is problematic for many flows, especially for liquid metals. Figure 5.22 shows the measured mean turbulent Prandtl number $Pr_{t,m}$ of a circular pipe flow as a function of the molecular Prandtl number (several fluids) for two different Reynolds numbers (Stieglitz, 2007). It is clear from the graph that for $Pr < 1$ the uncertainties are considerable, although with the help of improved calculation methods the confidence has steadily increased in recent years. All experimentally determined observations allow the following conclusions concerning the mean turbulent Prandtl number $Pr_{t,m}$:

- For $Pr \geq 1$ (air water and oil, etc.) the turbulent Prandtl Pr_t number varies little and is almost independent of the Reynolds number and the molecular Prandtl number.
- The influence of turbulent Prandtl number and Reynolds number increases simultaneously with falling molecular Prandtl number and falling Reynolds number.
- The turbulent Prandtl number increases with falling Prandtl number and falling Reynolds number.

5.1.15 External Flows

Flow losses or heat transfer in external flows are often neglected in solar thermal textbooks. However, they play an important role in heat transfer, e.g. at plate collectors or thermal heat storages, but also along the often kilometre-long absorber rows of linear-focusing solar power plants or at the absorbers of solar towers. The

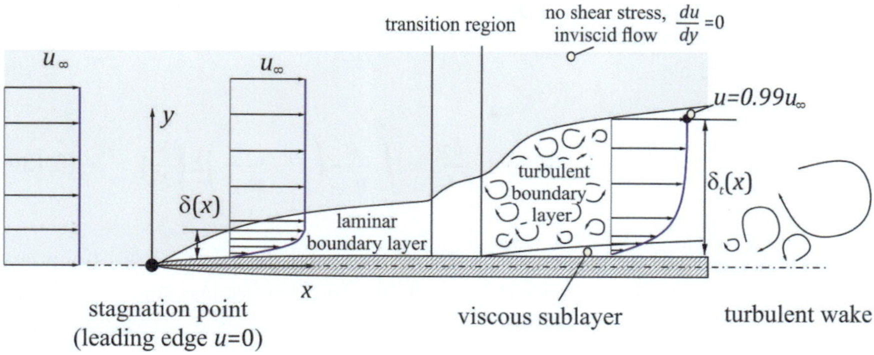

Fig. 5.23 Schematic evolution of a developing boundary layer flow over a flat plate

same applies to the flow resistance of the large mirrors, the structural integrity of solar updraft towers, which have to take environmental influences into account in an appropriate manner.

A number of phenomena that occur in external flow at high Reynolds number over a plate. Let's assume a plate as illustrated in Fig. 5.23. As the flow hits the object the free stream flow divides at the stagnation point, where due to the non-slip condition the velocity is zero. Due to the viscosity first, a laminar boundary layer develops which grows as the flow proceeds downstream in x-direction. The initially laminar boundary layer turns at a transition point into a turbulent boundary layer. Boundary layers thickening on the surfaces cause only a slight displacement of the streamlines of the external flow. Once the flow leaves the body separation may occur in the region of increasing pressure on the rear of the body, since the boundary layer fluid does not longer stay in contact with the wall surface. The fluid domain that formed the boundary layers generates a viscous wake behind the plate.

Because the boundary layer is thin, disturbance of the inviscid flow outside the boundary layer is negligible, and the pressure gradient along the surface is close to zero. The transition from laminar to turbulent boundary layer flow on a flat plate occurs if the Reynolds number based on x exceeds a critical Reynolds number of $Re_x = 5 \times 10^5$. This transition may occur at lower Re_x if the surface is rough, or the pressure increases downstream in flow direction, or separation occurs. For very low-turbulence inflow, considerably larger values of $Re_{x,crit}$ are obtained, and for high-turbulence inflow, considerably lower values. Downstream the transition, the turbulent boundary layer thickens more rapidly than the laminar boundary layer caused by increased shear stress at the body surface.

Boundary layer disturbance thickness δ is defined as the wall normal distance y, from the surface to a position at which the velocity within the boundary layer, $u(y)$ reaches $0.99u_\infty$ of the local freestream, and its velocity profile merges smoothly and asymptotically into the freestream, being challenging to measure. Thus, two integral measures of boundary layer thickness, which are easier to acquire experimentally are defined, the displacement thickness δ_1 and the momentum thickness δ_2, being defined as:

$$\frac{\delta_1}{\delta} = \int_0^\infty \left(1 - \frac{u}{u_\infty}\right) d\left(\frac{y}{\delta}\right) \quad \text{and} \quad \frac{\delta_2}{\delta} = \int_0^\infty \frac{u}{u_\infty}\left(1 - \frac{u}{u_\infty}\right) d\left(\frac{y}{\delta}\right). \tag{5.130}$$

Physically, δ_1 can be precepted as the wall normal distance a solid boundary would need to be shifted towards the free stream in a frictionless flow to generate a volumetric flow deficit caused by the boundary layer; while δ_2 corresponds to a thickness of a fluid layer with velocity u_∞, for which the momentum flux is the same as the deficit in momentum flux within the boundary layer. For a laminar flat-plate boundary layer according to Blasius similar self-similar velocity profiles in the boundary layer expressed in the form $u/u_\infty = f(y/\delta)$ can be assumed so that velocity profiles do not vary downstream with x. The result of the Blasius solution for the boundary layer thickness δ and the dimensionless wall shear stress C_f read to,

$$\frac{\delta}{x} = \frac{5}{\sqrt{Re_x}} \quad \text{and} \quad C_f = \frac{\tau_w}{(\rho/2)u_\infty^2} = \frac{0.664}{\sqrt{Re_x}}, \tag{5.131}$$

indicating that the laminar boundary layer thickness grows downstream with $\sqrt{x}$ and the shear stress accordingly with $1/\sqrt{x}$.

Regarding a turbulent boundary layer, the velocity profile can be expressed by a power law, $u/u_\infty = (y/\delta)^{1/n}$, in which n is an integer between 6 and 10, mostly 7 is taken. Both, boundary layer thickness and shear stress cannot be analytically calculated and for both empirical correlations are used:

$$\frac{\delta}{x} = \frac{0.382}{Re_x^{1/5}} \quad \text{and} \quad C_f = \frac{\tau_w}{(\rho/2)u_\infty^2} = \frac{0.0594}{Re_x^{1/5}}. \tag{5.132}$$

Hence, for a turbulent boundary layer the thickness increases with $x^{4/5}$ while the wall shear stress follows $x^{-1/5}$. A comparison of velocity distribution of laminar and turbulent flows in Fig. 5.23 exhibits that the turbulent velocity profile has much more momentum than the laminar profile.

Viscous shear always causes a net breaking force on any fluid particle within the boundary layer and even assuming zero pressure gradient ($dp/dx = 0$) shear forces only can never bring a fluid particle to rest, because the shear stress never vanishes for a finite downstream value of x (remind in laminar and boundary layers the wall shear stress scales with $x^{-1/2}$ and $x^{-1/5}$). Since the wall shear stress is defined by $\tau_w = \mu(\partial u/\partial y)_{y=0}$ the velocity gradient cannot diminish. In accelerated flows characterized by a negative pressure gradient ($dp/dx < 0$) as shown in Fig. 5.24 the pressure forces occur, conversely a positive pressure gradient ($dp/dx > 0$) yield pressure forces opposing the fluid motion and thus are a necessary condition for flow separation. Since turbulent boundary layers have a higher momentum than laminar ones, they are less fragile to flow separation in case of a positive pressure gradient.

Anybody subjected to viscous flow experiences shear stresses at the interfaces due to the viscosity of the fluid. At the same time, pressure differences result from the flow around the body. Both lead to a force acting on the body, whereby the force in the main flow direction is defined as drag and that perpendicular to it as lift. Both drag and lift forces are of importance for large-scale objects such as heliostats, solar upwind draft towers are well as parabolic troughs, which have to withstand environmental extreme environmental conditions without losing performance. For engineering purposes, the resulting drag and lift forces are calculated from experimentally derived coefficients (C_D, C_L) where both are normalized by the dynamic pressure.

$$C_D = \frac{F_D}{(\rho/2)u_\infty^2 \cdot A} \quad \text{and} \quad C_L = \frac{F_L}{(\rho/2)u_\infty^2 \cdot A}. \tag{5.133}$$

Note here that A is the area upon which the coefficient is based on. While A for drag coefficients is given by the projected frontal area in case of lift it is based on projected planform area. The total friction drag force acting on a plane plate aligned with the flow direction is obtained by integrating the shear stress distribution along the surface and the corresponding drag coefficient is defined as friction force divided by dynamic pressure and wetted area in contact with the fluid. As the shear stress is a function of the Reynolds number, so is the drag coefficient, which is depicted in Fig. 5.26a for a flat plate.

A thin flat surface exposed normal to a fluid flow has no area parallel to the flow direction, thus, there can be no friction force parallel to the flow; all drag is caused by pressure forces. Observations shows that drag coefficients C_D for objects with sharp edges are almost independent of the Reynolds number, if the

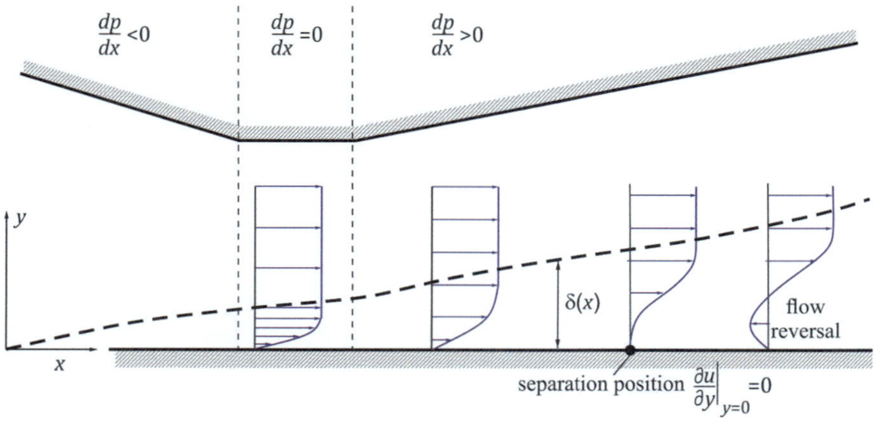

Fig. 5.24 Sketch of the boundary layer shape evolution as function of the pressure gradient

Reynolds number is large ($Re > 10^3$), because the flow separation points are pinned to the geometry discontinuity of the object. Smoothing the object shape reduces the drag considerably. A comparison of the drag coefficients for the hemisphere and the parabolic shapes as they appear for solar mirrors facing into and away from the flows is illustrated in Fig. 5.26b. Also note that the drag coefficient for a two-dimensional object (long square cylinder) is about twice that for the corresponding three-dimensional object (square cylinder with $b/h = 1$).

For smoothed objects exposed to wind loads the drag coefficients depend on the Reynolds number, the angle of attack of the wind α, and the height to diameter ratio. Moreover, for smoothed structures its own roughness but also the interaction with other adjacent structures yields Reynolds number dependent effects such as vortex shedding posing additional loads acting on the structures. Detailed drag and lift coefficient data and methods to evaluate the loads exposed to environmental effects may be taken from Veritas (2010). Exemplarily, here the values for a cylinder are shown since they are relevant for towers or chimneys of upwind power plants. For a smooth cylinder the flow domains are termed sub-critical flow if the drag-coefficient remains constant ($Re_D < 2 \times 10^5$), which is characterized by a laminar boundary layer. It is followed by a transcritical domain, at which the drag coefficients undergo at first a rapid drop reaching a local minimum before it increases again. At about $Re_D > 3 \times 10^6$ the supercritical flow domain with a fully turbulent boundary layer is reached where the drag coefficient recovers to an almost constant but lower value than in laminar flow. The dependence of the drag coefficient as function of the Reynolds number built with the diameter D of the cylinder is shown in Fig. 5.25.

If the total drag force on a slender body with the characteristic cross-sectional dimension d and the finite length l is estimated, the integrated sum of sectional force contributions is reduced due to the finite dimensions and the drag coefficient is multiplied by a reduction factor κ according to Table 5.6: Values of drag reduction factor κ for cylinders of finite length l and slenderness d. For cylinders with one end abutting on to another cylinder or a wall in such a way that free

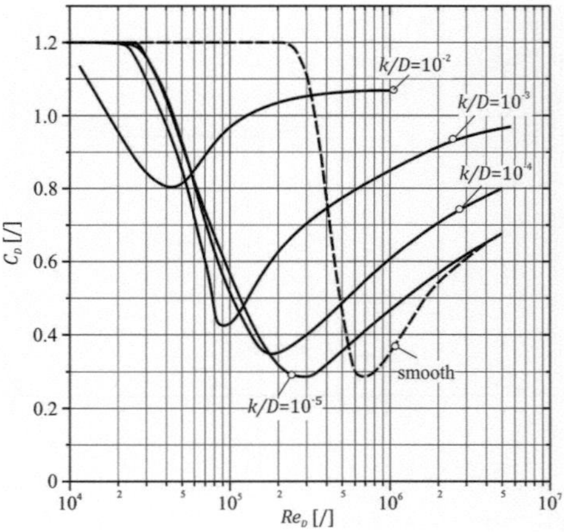

Fig. 5.25 Drag coefficient C_D for a fixed circular cylinder as a function of Re_D for steady flow in critical flow regime at various surface cylinder roughnesses

Table 5.6 Values of drag reduction factor κ for cylinders member of finite length l and slenderness d

l/d (/)	2	5	10	20	40	50	100
Cylinder sub-critical flow	0.58	0.62	0.68	0.74	0.82	0.87	0.98
Cylinder super-critical flow	0.80	0.80	0.82	0.90	0.98	0.99	1.00
Flat plate normal to flow	0.62	0.66	0.69	0.81	0.87	0.90	0.95

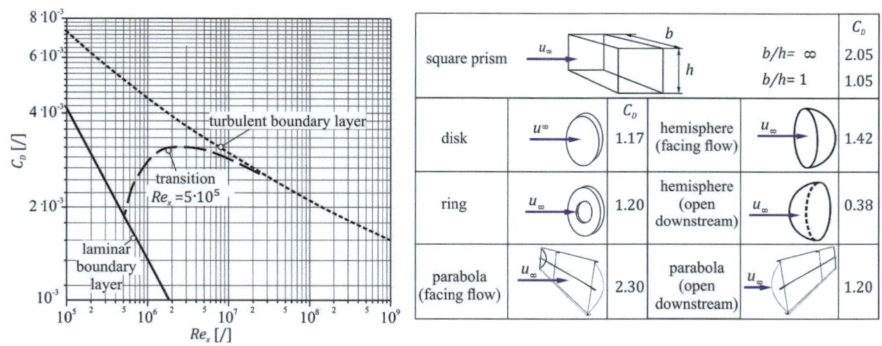

Fig. 5.26 Drag coefficient C_D as a function of the Reynolds number Re_x for a flat plate (left). Drag coefficient for selected geometries with sharp edges (for $Re > 10^3$, right)

flow around that end of the member is prevented, the ratio l/d should be doubled for the purpose of determining the drag reduction factor κ. When both ends are abutted as mentioned, the drag coefficient C_D should be taken equal to that for an infinitely long cylinder.

In the case of unbounded flow or flows around bodies, as in closed ducts, there is a distinction between forced and mixed convection and, accordingly, the heat transfer depends on the respective characteristic number and the thermal boundary conditions. As soon as the surface temperature of the body differs from that of the fluid, heat is transferred as forced convection. Conversely as soon as the bulk motion of the fluid results exclusively from the difference in temperature of the solid surface and the fluid, the mechanism is natural convection. For both type of problems, the velocity and temperature of the fluid far away from the solid surface are the free-stream velocity and free-stream temperature. Both, u_∞ and T_∞ are usually known or specified.

The task is usually to assess the heat flux from or to the surface, dependent on the given temperature or heat flux or to evaluate the temperature distribution, which requires the determination of the convective heat transfer coefficient h,

$$\dot{q} = h \cdot (T_w - T_\infty), \tag{5.134}$$

whereby in the following elaborations it is to be paid attention carefully whether in each case the local heat transfer expressed by the subscript "x" or the averaged heat transfer with the index "l" is addressed. It should also be pointed out, in the case of flows around body problems, that many correlations for determining the convective heat transfer coefficient are based on experimental data, which are subject to some uncertainty, even though the experiments are conducted under carefully controlled conditions. The resulting uncertainties are therefore often of the order of ±10%.

Similar as in a duct flow the temperature of the fluid changes from the surface temperature (T_w) to the free-stream temperature (T_∞) at the edge of the temperature boundary layer. We have seen that laminar velocity boundary layer thickness δ depends on the Reynolds number scaling with $\delta/x = 5/\sqrt{Re_x}$. Hence, also the thermal boundary layer thickness δ_{th} depends both on Re_x and Pr. Based on a similitude boundary layer analysis the thickness scaling of velocity to thermal boundary layers is:

$$\begin{aligned} Re_x < Re_{crit}: & \frac{\delta}{\delta_{th}} = \mathrm{Pr}^{1/3} \text{ for } Pr > 0.7; \\ & \frac{\delta}{\delta_{th}} = \mathrm{Pr}^{1/2} \text{ for } Pr \ll 1; \quad . \\ Re_x > Re_{crit}: & \frac{\delta}{X} = \frac{0.37}{Re_x^{0.2}} \quad \delta \approx \delta_{th} \end{aligned} \tag{5.135}$$

Viscous dissipation and high-speed effects can be neglected if $Pr^{1/2} \cdot Ec/2 \ll 1$, where the Eckert number Ec is defined by $Ec = u_\infty^2/\left(c_p \cdot (T_w - T_\infty)\right)$.

Flat plate

For a laminar flow over a plate, the local heat transfer (Nu_x) and the heat transfer averaged over the plate length l (Nu_l) are calculated at isothermal wall surface temperature ($T_w = const.$) as:

$$Nu_{x,lam} = 0.332 \cdot \sqrt{Re_x} \cdot Pr^{1/3}, \tag{5.136}$$

in which $Nu_{x,lam} = (h \cdot l)/\lambda$ and $Re_x = (u_\infty \cdot x)/\nu$. The mean Nusselt number $Nu_{l,lam}$ of a plate averaged over the length l is obtained by integrating the Eq. (5.136) to:

$$Nu_{l,lam} = 0.664 \cdot \sqrt{Re_l} \cdot Pr^{1/3}, \tag{5.137}$$

where for the Reynolds number the length l is to be used.

In case of a turbulent flow over plate the local Nusselt number $Nu_{x,turb}$ at constant wall temperature ($T_w = const.$) can be calculated with the following equation:

$$Nu_{x,turb} = \frac{0.0296 \cdot Re_x^{4/5} \cdot Pr}{1 + 2.185 \cdot Re_x^{-0.1} \cdot (Pr - 1)}, \tag{5.138}$$

for which the thermophysical material values for the mean temperature T_m of the fluid are to be used. For the length averaged turbulent Nusselt number $Nu_{l,turb}$, the average drag coefficient $C_D = 0.037 \cdot Re_l^{-0.2}$ of the turbulent boundary layer flow over the plate length l can be used, so that for $Nu_{l,turb}$ the following equation applies:

$$Nu_{l,turb} = \frac{0.037 \cdot Re_l^{0.8} \cdot Pr}{1 + 2.443 \cdot Re_l^{-0.1} \cdot \left(Pr^{2/3} - 1\right)}. \tag{5.139}$$

The range of validity of the equation is $5 \times 10^5 < Re_l < 10^7$ and $0.5 < Pr < 2 \times 10^3$.

With constant wall heat flux ($\dot{q}_w = const$), the local laminar Nusselt number at the location x away from the beginning of the plate can be calculated using the equation:

$$Nu_{x,lam} = 0.460 \cdot \sqrt{Re_x} \cdot Pr^{1/3}, \tag{5.140}$$

which is recommended for $Re_x < 10^5$ and $0.6 < Pr < 2 \times 10^3$. Churchill and Ozoe (1973) recommend the following single correlation for all Prandtl numbers:

$$Nu_{x,lam} = \frac{0.886 \cdot \sqrt{Re_x} \cdot Pr^{1/2}}{\left(1 + \left(\frac{Pr}{0.0207}\right)^{2/3}\right)^{1/4}}. \tag{5.141}$$

Note that for surfaces with uniform heat flux the local convective heat transfer coefficient is used to determine the local surface temperature. The total heat transfer rate being known, an average heat transfer coefficient is not needed and not defined.

For turbulent boundary layers with constant wall heat flux ($\dot{q}_w = const$) the following relations are recommended to evaluate the local turbulent Nusselt number $Nu_{x,turb}$:

$$\begin{aligned} Re_{cr} < Re_x < 10^7 &: \quad Nu_{x,turb} = 0.0296 \cdot Re_x^{4/5} \cdot Pr^{1/3} \\ Re_x > 10^7 &: \quad Nu_{x,turb} = 1.596 \cdot Re_x \cdot (\ln Re_x)^{-2.584} \cdot Pr^{1/3}. \end{aligned} \tag{5.142}$$

In laminar boundary layers, the convective heat transfer coefficient with uniform heat flux is approximately 36% higher than with uniform surface temperature. In case of turbulent boundary layers, the difference is negligible and the correlations for the local convective heat transfer coefficient can be used for both uniform surface temperature and uniform heat flux.

If the boundary layer is initially laminar followed by a turbulent boundary layer at $Re_x = Re_{cr}$, the for $0.7 < Pr < 60$ the following correlations are suggested for the length averaged turbulent Nusselt number $Nu_{l,turb}$:

$$Re_{cr} < Re_L < 10^7 : \quad Nu_{l,turb} = \left(0.664\sqrt{Re_l} + 0.037\left(Re_l^{4/5} - Re_{cr}^{4/5}\right)\right)Pr^{1/3},$$

$$Re_x > 10^7: \quad Nu_{l,turb} = \left(1.963 Re_l \cdot (\ln Re_l)^{-2.584} - 871\right) Pr^{1/3} \tag{5.143}$$

In most practical cases, even at intermediate Reynolds numbers, a laminar boundary layer will not form along the entire length of the plate due to blunt plate edges and the turbulence of the inflowing medium. If measurement results are compiled, they can be represented by a "mean curve" which also includes the area of the turbulent boundary layer. As Gnielinski (1975) has shown, a mean curve for a large range of the Prandtl number can be represented by the relation:

$$Nu_{l,0} = \sqrt{Nu_{l,lam}^2 + Nu_{l,turb}^2}, \tag{5.144}$$

in which the relations for the laminar and turbulent averaged Nusselt numbers according to the thermal boundary conditions are inserted. The range of application applies to mean Nusselt numbers at $10^1 < Re < 10^7$ and $0.5 < Pr < 2 \times 10^3$ and for the fluid properties the values are to be used at the mean temperature $T_m = (T_{in} + T_{out})/2$.

Flow over cylinders

The heat transfer coefficients of cross-flow over tubes and profile cylinders play a role, for example, in tower receivers, the hot tanks of sensitive heat storage systems, but also in the often kilometre-long absorber tubes of solar farms. For the calculation of the mean Nusselt numbers of cross-flowed tubes or profiled cylinders, the same heat transfer relationships apply as for the flat plate; the so-called "overflow length" l is to be used as the characteristic length when calculating the corresponding Reynolds and Nusselt numbers. The overflow length is the distance a flowing particle can travel along the entire surface and is defined as the total surface area A of a body divided by the perimeter l_c of the shadow area of the body, i.e. $l = A/l_c$, which is sketched in Fig. 5.27. For a long pipe of diameter d and length L, the overflow length is thus obtained as:

$$l = \frac{\pi\, d\, L}{2\, L} = \frac{\pi}{2} d. \tag{5.145}$$

For the average Nusselt number of cross-flowed tubes and profile cylinders, the following relation is given as an approximation equation:

$$Nu_{l,0} = 0.3 + \sqrt{Nu_{l,lam}^2 + Nu_{l,turb}^2}, \tag{5.146}$$

for which the laminar and turbulent relations at the corresponding thermal boundary conditions are to be used. The minimum value for $Nu_{l,0}$ originates from the

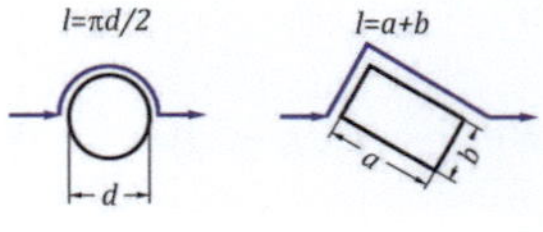

Fig. 5.27 Sketch for the definition of the overflow length l

Table 5.7 Exponents of heat transfer on a transversely flowed cylinder of constant wall temperature ($T_w = const.$) in different Reynolds number ranges

Re_d	Flow regime	C	m	n
$Re_d < 40$	Creeping	0.76	0.4	0.37
$40 < Re_d < 10^3$	Laminar	0.152	0.5	0.37
$10^3 < Re_d < 2 \times 10^5$	Transitional	0.26	0.6	0.37
$2 \times 10^5 < Re_d < 4 \times 10^7$	Turbulent	0.023	0.8	0.40

fact that cross-flowed cylinders always have a finite length, so that a minimum heat flow occurs even in a stationary environment. The mean Nusselt number of the cross-flowed cylinder therefore asymptotically approaches a minimum, which is approximately given by $Nu_{min} \approx 0.3$.

Exactly, the mean heat transfer (Nu_l) on transversely flowed cylinders with $T_w = const.$ is calculated according to the basic Eq. (5.147), where depending on the Reynolds number range Re_d related to the diameter, the exponents shown in Table 5.7 have to be used.

$$Nu_d = C \cdot Re_d^m \cdot Pr^n. \tag{5.147}$$

For the thermal boundary condition $\dot{q}_w = const.$ the following relation can be found in the literature, which is valid in the range $10^{-2} < Re_d < 10^5$:

$$Nu_d = 0.42 \cdot Pr^{0.2} + 0.57 \cdot Re_d^{0.5} \cdot Pr^{\frac{1}{3}}. \tag{5.148}$$

In the case of a cylinder with longitudinal flow, the following relations are obtained for the local Nusselt number Nu_x and the mean Nusselt number (Nu_l) in the range for $Re_{x,crit} < 2 \times 10^5$:

$$Nu_x = 0.332 \cdot Re_x^{1/2} \cdot Pr^{1/3} \cdot \left[1 + \frac{4.6}{Re_x^{1/2}}\left(\frac{x}{d}\right)\right];$$
$$Nu_l = 0.664 \cdot Re_L^{1/2} \cdot Pr^{1/3} \cdot \left[1 + \frac{2.3}{Re_x^{1/2}}\left(\frac{L}{d}\right)\right]. \tag{5.149}$$

For turbulent inflow, the correlation (5.149) expands multiplicatively by a geometry factor K_{cyl}:

$$K_{cyl} = \left[1 + 0.37 \cdot \frac{2 \cdot l}{d \cdot Re_d^{1/5}}\right]^{0.14}. \tag{5.150}$$

Free natural convection occurs in particular where temperature differences between the surfaces of two bodies occur in open cavities, for example in cavity receivers of solar thermal tower power plants, but also behind transparent heat insulation in thermal insulation. Again, the Nusselt number describes the heat transfer $\dot{Q}$ from the body to the environment. It is defined as follows and depends, as in closed systems, on the Rayleigh and Prandtl numbers, and furthermore on the geometry and the thermal boundary conditions

$$Nu^* = \frac{\overline{h} \cdot L}{k} = \frac{\dot{Q} \cdot L}{A_w \cdot k \cdot (T_w - T_\infty)} = f\big(Ra, Pr, \varepsilon_f\big), \tag{5.151}$$

In which $\overline{h}$ denotes the geometry averaged heat transfer coefficient, L a characteristic geometric length, k the heat conductivity of the fluid, A_w the area through which the heat is transferred and $\Delta T = (T_w - T_\infty)$ the temperature difference between the body wall and the ambient. The heat transfer correlation in the case of free convection can be calculated with the help of the general equation according to Michejew:

$$Nu = C_i^j(Pr) \cdot Ra^n \cdot \varepsilon_f, \tag{5.152}$$

where ε_f represents a form factor. The subscript i indactes in the following correlations, whether the flow is laminar (lam) or turbulent ($turb$), while the superscript indicates if the surface orientation is vertical (v) or horizontal (h) with respect to the gravity vector g. When calculating the Rayleigh number, it is necessary to pay attention to the correct choice of the characteristic length l. For vertical plane plates, for example receivers, the characteristic length l corresponds to the plate height. For example, the transition from laminar to turbulent heat transfer occurs for a vertical plate at $Ra_l = 10^{12}$. The fluid properties are taken as average of the wall and ambient temperatures, whereas the temperature coefficient of volume expansion β is evaluated at T_∞ for external natural convection.

For all external body flows, a similar dependence on the Prandtl number Pr is observed, provided it is larger than $Pr = 0.5$, so that according to Churchill and Usagi (1972) the following correlations average Nusselt-number correlations can be used for laminar and turbulent flows:

$$\begin{aligned} C_{lam} &= \frac{0.671}{\left(1 + (0.492/Pr)^{9/16}\right)^{4/9}}; \\ C_{turb}^v &= \frac{0.13\ Pr^{0.22}}{\left(1 + 0.61\ Pr^{0.81}\right)^{0.42}}; \\ C_{turb}^h &= 0.14\left(\frac{1 + 0.0107\ Pr}{1 + 0.01\ Pr}\right). \end{aligned} \tag{5.153}$$

Since there is often a coexistence of laminar and turbulent flows, the laminar and turbulent Nusselt numbers (Nu_{lam}, Nu_{turb}) are superimposed similar to Gnielinski's approach (compare Eq. (5.144)) but complemented by introducing a blending exponents accounting for body shape and orientation in order to compute an average Nusselt number according to.

$$Nu_l = \left((Nu_{lam})^m + (Nu_{turb})^m\right)^{1/m}. \tag{5.154}$$

The equation for Nu_{lam} in this section is usually expressed in terms of Nu_0, the Nusselt number that would be valid if the thermal boundary layer were thin. The difference between Nu_{lam} and Nu_0 accounts for the effect of the large boundary layer thicknesses encountered in natural convection.

Subsequently the correlations for the body shapes depicted in Fig. 5.28 are provided considering them to be heated ($T_w > T_\infty$) and immersed in a fluid at rest. In case $T_w < T_\infty$ the same correlations apply with good approximation, for a geometry that is rotated 180° relative to the gravitational vector; for example, the correlations for a horizontal heated upward facing flat plate applies to a cooled downward-facing flat plate of the same plane form, and for the driving temperature gradient the absolute value is taken, i.e. $|T_w - T_\infty|$.

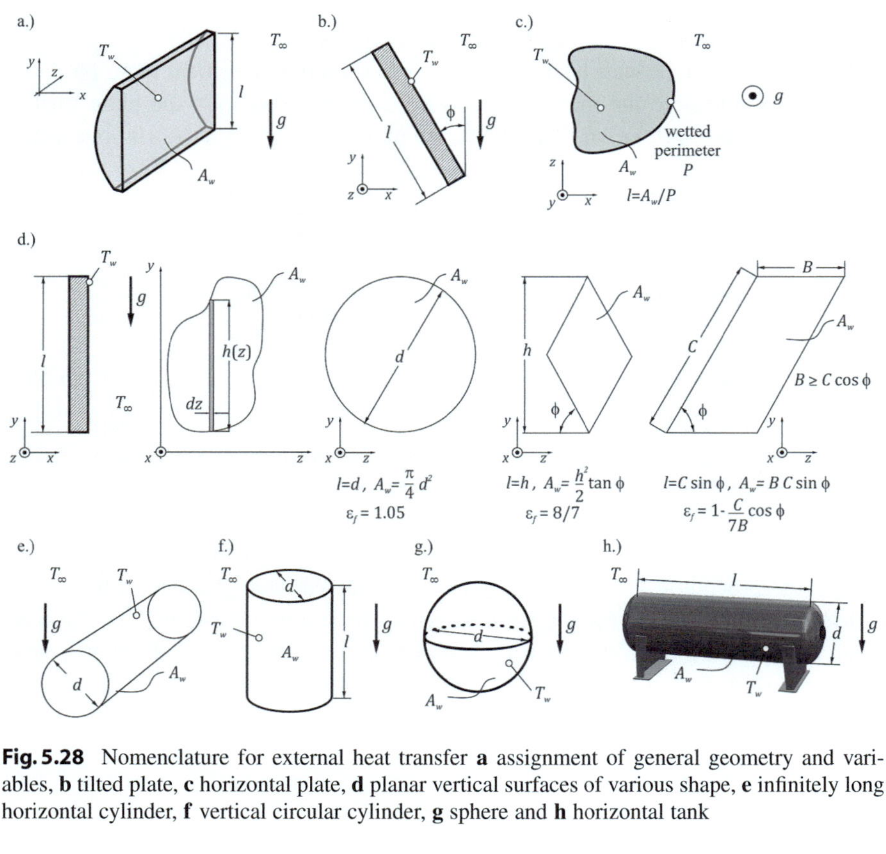

Fig. 5.28 Nomenclature for external heat transfer **a** assignment of general geometry and variables, **b** tilted plate, **c** horizontal plate, **d** planar vertical surfaces of various shape, **e** infinitely long horizontal cylinder, **f** vertical circular cylinder, **g** sphere and **h** horizontal tank

- The Nusselt number along a vertical plate ($\phi = 0°$) as shown in Fig. 5.28b for $1 < Ra < 10^{12}$ is given for a constant wall temperature ($T_w = const.$) by:

$$Nu_0 = C_{lam} \cdot Ra^{1/4};\ Nu_{lam} = \frac{2}{\ln(1 + 2/Nu_0)};$$
$$Nu_{turb} = C^v_{turb} \cdot \frac{Ra^{1/3}}{\left(1 + 1.4 \times 10^9 Pr/Ra\right)}, \tag{5.155}$$

In which C_{lam} and C^v_{turb} are to be taken from Eq. (5.153) and Nu_l is computed by inserting the expressions of Eq. (5.155) for Nu_{lam} and Nu_{turb} in relation (5.154) using as exponent $m = 6$. The average heat transfer at vertical surfaces with fully turbulent convection ($Ra > 10^{12}$) and constant wall temperature is

$$Nu_l = \left[0.825 + 0.387 \cdot (Ra_L \cdot K(Pr))^{\frac{1}{6}}\right]^2 \text{ with } K(Pr) = \left[1 + \left(\frac{0.492}{Pr}\right)^{\frac{9}{16}}\right]^{-\frac{16}{9}}. \tag{5.156}$$

For gases, $K(Pr)$ is constant and is $K(Pr) = 0.348$.

In case of constant wall heat flux ($\dot{q}_w = const$) along a vertical plate ($\phi = 0°$) the aim is to evaluate the mean temperature difference between the plate surface and the ambient fluid, i.e. $\overline{\Delta T}$. In the range of $15 < (Nu^* \cdot Ra) < 10^5$ one gets:

$$Nu_0 = G_{lam} \cdot \left(Nu^* \cdot Ra\right)^{1/5};\ Nu_{lam} = \frac{1.83}{\ln(1 + 1.83/6Nu_0)};$$
$$Nu_{turb} = C^v_{turb}{}^{3/4} \cdot \frac{(Nu^* \cdot Ra)^{1/4}}{\left(1 + 1.4 \times 10^9 Pr/Ra\right)} \text{ with}$$
$$G_{lam} = \frac{6}{5} \cdot \left(\frac{Pr}{4 + 9\sqrt{Pr} + 10Pr}\right)^{1/5}. \tag{5.157}$$

The Nusselt number Nu_l is again obtained by evaluating Nu_{lam} and Nu_{turb} (Eq. (5.153)) inserted in Eq. (5.154) and using $m = 6$.

- For tilted downward facing plates ($-90° < \phi < 0°$) at constant wall temperature the procedure is first to calculate the removed heat $\dot{Q}$ for a vertical plate using for the gravity vector instead of g the expression $g \cos\phi$. For the horizontal downward plate also, the removed heat $\dot{Q}$ is to be evaluated using $g \sin\phi$ instead of g. It is recommended to take the larger of both values.

- In case of inclined upward facing plates ($0° < \phi < 90°$) a similar procedure is applied and also the heat flux $\dot{Q}$ is calculated for a vertical plate with $g \cos \phi$ and correspondingly for a horizontal plate using $g \sin \phi$ and use the maximum of both values
- For horizontal upward isothermal surfaces ($\phi = 90°$) of various platforms, correlations are given in terms of a modified length scale l, scaled as $l = A_w/P$ according to Fig. 5.28c for $Ra > 1$:

$$Nu_0 = 0.835 C_{lam} \cdot Ra^{1/4};\ Nu_{lam} = \frac{1.4}{\ln(1 + 1.4/Nu_0)};$$
$$Nu_{turb} = C_{turb}^{h} \cdot Ra^{1/3}. \tag{5.158}$$

The procedure is similar as before to obtain, however the exponent is $m = 10$, while for constant heat flux ($\dot{q}_w = const$) ΔT is to be replaced by $\overline{\Delta T}$.

- For horizontal downward-facing plates ($\phi = -90°$) of various plane forms, the buoyant fluid motion is directed towards the plate so that only a very weak force drives the fluid along the plate and only laminar flows are reported in literature. For $Ra < 10^{10}$ and $Pr \geq 0.7$ the next equations apply:

$$Nu_0 = H_{lam} \cdot Ra^{1/5};\ H_{lam} = \frac{0.527}{\left(1 + (1.9/Pr)^{9/10}\right)^{2/9}};$$
$$Nu = \frac{2.45}{\ln(1 + 2.45/Nu_0)}. \tag{5.159}$$

- In case of vertical, tilted plates of various planform at constant temperature the line of constant z in in Fig. 5.28d represents the line of steepest gradient on the plate. Provided all such lines intersect the plate edges twice the thin-layer (Nu_0) heat transfer value can be found by subdividing the body into strips of width dz calculating the heat transfer from each strip and adding. For laminar flow from an isothermal vertical plate, this results in:

$$Nu_0 = \varepsilon_f \cdot C_{lam} \cdot Ra^{1/4};\ \varepsilon_f = \left(\frac{l^{1/4}}{A_w} \int_0^z h(z)^{3/4} dz\right). \tag{5.160}$$

The form factor of some plate shapes is provided in Fig. 5.28d. In case of a vertical plate the expression for Nu_0 in Eq. (5.155) is replaced by Eq. (5.160), while for inclined plates depending on the orientation (upwards/downwards) the procedure for an ideal plate is to be used.

- Horizontal cylinders, as they appear e.g. in parabolic trough collectors, can be considered as infinitely long cylinders and the expression for Nu_{lam} and Nu_{turb} read to

$$Nu_0 = 0.772\, C_{lam} \cdot Ra^{1/4};$$

Table 5.8 Coefficients C_{turb} for flow along a horizontal cylinder at different Prandtl numbers

Pr	0.01	0.022	0.1	0.71	2.0	6.0	50	100
Horizontal cylinder	0.077	0.81	0.90	0.103	0.108	0.109	0.100	0.097

$$Nu_{lam} = \frac{2f}{(1 + 2f/Nu_0)}; \; Nu_{turb} = C_{turb} \cdot Ra^{1/3}, \tag{5.161}$$

for which C_{turb} depends strongly on the Prandtl number Pr and values are provided in Table 5.8. For $Ra > 10^{-2}$, $f = 0.8$ can be used, while for $Ra < 10^{-2}$, the functional relationship $f = \left(1 - 0.13/Nu_0^{0.16}\right)$ is used. To find Nu, the values of Nu_{lam} and Nu_{turb} from Eq. (5.161) are substituted into Eq. (5.154) taking $m = 15$.

- Vertical cylinders are used as sensitive heat storage tanks and for high Ra values and large diameter d, the heat transfer from the wall of a vertical cylinder approaches that for a vertical flat plate. Thus, the equations for Nu_0 and Nu_{lam} for a vertical plate of length l (Eq. (5.155)) can be rewritten as $Nu_{0,p}$ and Nu_p. At smaller Ra and diameter d, transverse curvature yields lateral exchange which is has to be considered using

$$Nu_{lam} = \frac{0.9\,\xi\,Nu_p}{\ln(1 + 0.9\xi)}, \; \xi = \frac{2\,l}{d\,Nu_{0,p}}, \tag{5.162}$$

  for which only in the laminar case correlations are reported in literature. In order to get Nu_l for Nu_{lam} the Eq. (5.162) has to be used and for Nu_{turb} Eq. (5.155), which have to inserted in Eq. (5.154) using the exponent $m = 10$.
- In case of spheres also Eq. (5.154) is taken with the exponent $m = 6$ and

$$Nu_{lam} = 2 + 0.878\,C_{lam} Ra^{1/4}, \; Nu_{turb} = C_{turb} \cdot Ra^{1/3}, \tag{5.163}$$

  where for C_{turb} the data displayed in Table 5.8 can be used.
- Regarding combined shapes, such as the tank displayed in Fig. 5.28h with spherical end caps, calculate the heat transfer from the cylinder of length l, the heat transfer from a sphere of diameter d and add to obtain the total transfer. Other shapes can be treated in a similar manner.

High vertical components are found, for example, in pipelines of solar towers, as well as open gaps behind transparent thermal insulation. To prevent condensation from forming, which contributes to degradation, these are back-ventilated, and the open interfaces allow air circulation, which is associated with heat loss. In addition, ribbed structures are used as emergency capacitors to shorten stagnation periods.

- In vertical narrow cavities ($l/w << 1$), as shown in the figure Fig. 5.29a, with constant wall temperature (where wall temperatures T_1 and T_2 can be different but not between T_∞), the Nusselt number is given by:

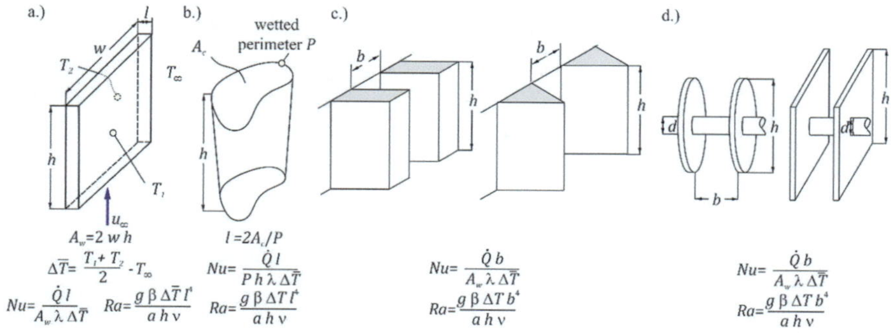

Fig. 5.29 Geometries for natural convective heat transfer in open cavities and corresponding nomenclature; **a** parallel plates, **b** arbitrary surface, **c** vertical fins and **d** horizontal fins

$$Nu = \left(\left(\frac{Ra}{f\,Re}\right)^m + \left(C_1 C_{lam} Ra^{1/4}\right)^m\right)^{1/m}, \tag{5.164}$$

for $Ra \leq 10^5$, where the product of friction factor and Reynolds number ($f\,Re$) is taken for fully developed plate flow at the corresponding aspect ratio and C_1 characterizes the increased heat transfer compared to a vertical plate according to the chimney effect. In Eq. (5.164) the Nusselt number refers to the total heat power transfer $\dot{Q}$ to the fluid form all cavity surfaces. For parallel arranged plates $f\,Re = 24$ with $m = -1.9$ using $C_1 \approx 1.2$ for gases applies. C_1 tends to unity for increasing Pr as well as for increasing l/w.

For cylindrically shaped gaps $f\,Re = 16$ holds with $m = -1.03$ and $C_1 \approx 1.17$ for gases. Other cross-sectional areas, as depicted in Fig. 5.29b, like concentric squares ($f\,Re = 14.23$), hexagonal shrouds ($f\,Re = 15.05$) Eq. (5.164) applies as well by inserting $m = -1.5$ and $C_1 \approx 1.2$ for gases (Kreith & Goswami, 2005). The heat transfer per unit cross-sectional area, $\dot{Q}/A_c$, for a given channel length h and temperature difference, passes through a maximum at approximately $Ra_{max} = \left(f\,Re C_1 C_{lam}/2^{1/m}\right)^{4/3}$ and provides the hydraulic radius $l = 2A_c/P$.

- Fins are frequently used as heat sinks, e.g. in condensers, and the heat transfer at low Ra is associated with a large uncertainty. Conservatively, for $Ra < 2 \times 10^2$ and U-shaped fins according to Fig. 5.29c, the following relation equation finds use,

$$Nu = \left(\left(\frac{24}{Ra}\right)^2 + \left(C_1 C_{lam} Ra\right)^{-2}\right)^{-1/2};\ C_1 = \left(1 + \left(\frac{h}{b}\right), 1.16\right)_{min}, \tag{5.165}$$

underestimating the heat transfer since heat losses at the fin edges and three-dimensional heat conduction within the structures are not considered. In case of triangular fins, the relation:

$$Nu = C_{lam} Ra^{1/4} \left(1 + \left(\frac{3.26}{Ra^{0.21}} \right)^3 \right)^{-1/3}, \tag{5.166}$$

can be used for $0.4 < Ra < 5 \times 10^6$.

- Horizontal fins, as illustrated in Fig. 5.29d, separate in circular or rectangular geometries for which the following relations may be taken for $5 < Ra < 10^4$ and air:

$$\begin{aligned} \text{circular fins: } Nu &= \frac{1}{8} Ra^{0.55} \left(1 - \exp\left(-\frac{137}{Ra} \right) \right)^{-0.294}, \\ \text{rectangular fins: } Nu &= \left(\left(\frac{Ra^{0.89}}{18} \right)^{-2.7} + \left(0.62 Ra^{1/4} \right)^{-2.7} \right)^{-1/2.7} \end{aligned} \tag{5.167}$$

5.1.16 Thermal-Hydraulics of Pipes and Engineering Correlations

Pipe system and pipe grids are central elements to transfer the heat generated within the solar absorber to a functional component, be it a steam generator for power generation, a thermal storage unit for intermediate storage of thermal energy, or direct heat application in processes or by the consumer. Pipelines or pipe networks play a central role here, and their optimization goals are the lowest possible pressure drop (to keep pumping work low), efficient heat transfer (to keep the components compact), and small heat losses to the environment.

The fundamentals and methods of one-dimensional stream filament theory often termed also as stream tube theory are the starting point for the design of piping networks or for system analyses. Here the flow is considered along a stream thread, which is a model concept, and which must be imagined as a thin stream tube with variable cross-sectional area formed by streamlines. The flow along the streamline is treated as a one-dimensional process. The state variables velocity, pressure, density, and temperature do not change over the cross-section of the flow filament, but only in the direction of flow. In the considerations of the streamline theory, among others, a one-dimensional form of the continuity equation is used, which considers the law of conservation of mass.

Since the streamlines are by definition the tangents of the velocity vectors, no fluid mass passes through the shell of the stream tube. This means that solid-walled channels with flow through them form stream tubes. If the changes in the flow variables across the cross-section of the flow tube are small compared to the changes along the flow tube, the approximately one-dimensional changes in the flow variables along the abstracted flow path can be calculated. The image of a flow field is represented more accurately by stream filaments the smaller

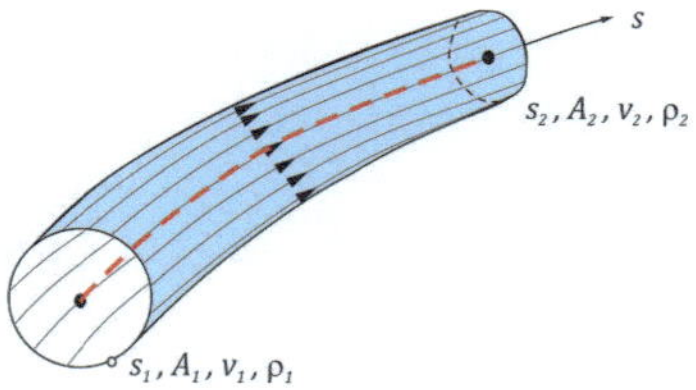

Fig. 5.30 Stream filament in a stream tube

their cross sections. Flows through pipes and in flumes can be treated as a single stream tube if values of the flow variables averaged over the cross section are used. This is the importance of streamline theory in engineering: It is often possible to treat two- or even three-dimensional flows approximately in one dimension. When applying relationships from the stream filament theory, it must be checked whether the idealizations made describe the processes correctly, e.g. whether the changes in the state variables are actually much smaller in the transverse direction than in the direction of flow.

Since the mass of a material volume remains constant, the following applies to such a material piece of a current filament, which is shown in Fig. 5.30:

$$\frac{d}{dt}\int_{s_1(t)}^{s_2(t)} \rho(s,t)A(s,t)ds = 0. \tag{5.168}$$

For the derivation of an integral with respect to time, whose integrand and limits depend on time, Leibniz's rule applies and using $ds/dt = u_i$, the following is obtained:

$$\int_{s_1(t)}^{s_2(t)} \frac{d(\rho A)}{dt} ds + \rho_2 A_2 u_2 - \rho_1 A_1 u_1 = 0. \tag{5.169}$$

For the steady-state flow of an incompressible fluid through a stream filament, it follows: $(u_1 A_1) = (u_2 A_2)$, $(u\ A) = const.$ or $((u\ dA) + (A\ du)) = 0$. In a stream filament, $(u\ A)$ just represents the volumetric flow rate through a cross section A of the stream filament, it is $\dot{V} = (u\ A)$. Trivially, this means that for a steady flow of an incompressible fluid, the volumetric flow through a stream tube is the same in any cross section, i.e., $\dot{V}_1 = \dot{V}_2$ or $d\dot{V} = 0$.

For steady-state pipe flow, the momentum equation relates the sum of forces in a given direction acting on a stream filament in a closed conduit, to the net momentum flux through the control volume in the same direction. The most common forces F are pressure and friction

$$\sum_{i=1}^{i=n} F_i = (\rho_2 A_2 u_2) \cdot u_2 - (\rho_1 A_1 u_1) \cdot u_1. \tag{5.170}$$

The energy equation in a general form for an incompressible pipe flow reads then to:

$$\frac{p_1}{\rho g} + z_1 + \frac{u_1^2}{2g} = \frac{p_2}{\rho g} + z_2 + \frac{u_2^2}{2g} - h_p + h_t + h_f, \tag{5.171}$$

in which the unit is a length scale (m), and p, ρ and g denote the pressure, the density, and the gravity constant. The first three terms of both sides denote the pressure head $(p/(\rho g))$, the elevation head with respect to a ground normal level, and the velocity head $(u_i^2/(2g))$. The last three terms on the right-hand side are the total dynamic head supplied by the pump (h_p), the head loss by a turbine (h_t) and the total friction losses from 1 to 2.

5.1.16.1 Friction Losses in Pipes and Pipe Elements

In this context pressure changes resulting from incompressible flow for pipes with circular geometry are considered, however, the results can be generalized for a pipe of noncircular geometry by replacing diameter D in any of the equations with the hydraulic diameter, D_h, defined as $D_h = A/P$, in which A is the cross-sectional area of the conduit and P denotes the wetted perimeter. To calculate the factors influencing the volumetric flow rate, which are particularly important in the design of pipelines, the minimum volume flow is usually given and the pressure difference between the inlet and outlet is sought. With the help of the energy equation of the stream tube theory, the Darcy-Weisbach relationship can be derived:

$$h_f = f\frac{L}{D}\frac{u^2}{2\,g} = f\frac{L}{g}\frac{8}{\pi^2}\frac{\dot{V}^2}{D^5}, \tag{5.172}$$

in which L denotes the duct length, $\dot{V}$ the volumetric flow rate and f the tube friction factor depending on the material properties of the pipe and the flow velocity.

The friction factor f is a measure of pipe roughness and has been evaluated experimentally to create the Moody chart. According to that laminar flow exists if $Re < 2.300$, beyond that a critical zone exists up to $Re \sim 4 \times 10^3$, there friction loss calculations yield ambiguous results because it is impossible to determine a unique value of f. For $Re > 4 \times 10^3$, the flow becomes turbulent and f is a function of Re and relative pipe roughness (e/D), where e is the equivalent roughness height and D is pipe diameter. At $Re > 10^4$, f depends only on the relative roughness height, e/D and this domain is termed fully turbulent flow. The Reynolds number at which this transition occurs depends on the relative roughness of the pipe; a Moody chart for pipe flow is illustrated in Fig. 5.31.

According to Blasius, in the range $3 \times 10^3 < Re < 10^5$ the simple formula applies:

$$f = \frac{0.3164}{Re^{1/4}}, \tag{5.173}$$

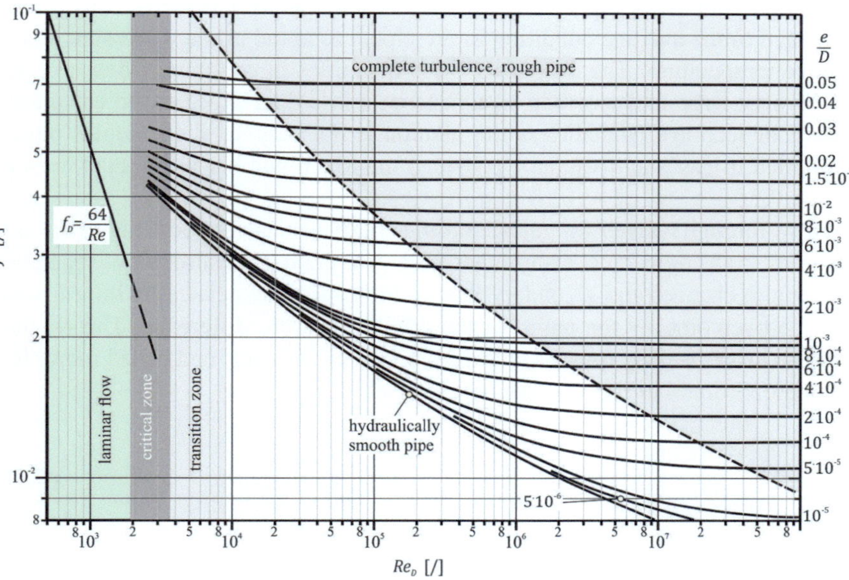

Fig. 5.31 Moody diagram for pressure losses in a pipe flow

and for higher ranges $10^4 < Re < 10^6$ the equation of Konakov is still convenient to handle:

$$f = (1.8 \cdot \lg Re - 1.5)^{-2}. \tag{5.174}$$

In the range $10^5 < Re < 5 \times 10^7$ the implicit formula of Filomenko is considered to be reliable with deviations of less than 0.5%:

$$\frac{1}{\sqrt{f}} = 1.819 \cdot \lg Re - 1.64. \tag{5.175}$$

Considering not ideally smooth tubes but also rough tubes, then, if the roughness flow is fully developed, the quadratic resistance law holds, and f is independent of Re. According to Prandtl and v. Kármán the following relation holds

$$\frac{1}{\sqrt{f}} = 2 \cdot \lg\left(\frac{D}{e}\right) + 1.14. \tag{5.176}$$

According to Nikuradze, the implicit equation of Colebrook and White applies to the turbulent flow in the entire range of smooth and rough surfaces:

$$\frac{1}{\sqrt{f}} = 1.74 - 0.869 \cdot \ln\left(\frac{2\,e}{D} + \frac{18.7}{Re\sqrt{f}}\right). \tag{5.177}$$

The material for the kinematic viscosity values ν as well as the velocity u are to be related to the average condition in the pipe, i.e. to the average pressure and the average temperature. The change of these quantities in continuous flow direction can be very large under certain circumstances; it is then expedient to calculate in individual sections. Nonetheless the relation is not only valid for liquids but can be applied to gases and vapours, provided the Mach number in the duct does not exceed 0.3.

Flow through valves or gates, throttles, expansions/contractions, elbows, T-junctions, transitions, and many other pipe grid elements causes flow separation, momentum changes and secondary flows which are all associated by losses, which must be overcome by the pump or the driving pressure difference. The resulting pressure loss is usually expressed as resulting head loss, h_m, and called minor loss. h_m is proportional to the velocity height

$$h_m = K_l \frac{1}{2g}\left(\frac{\dot{V}}{A}\right)^2 = K_l \frac{u^2}{2\,g}, \tag{5.178}$$

in which K_l is the minor loss coefficient being analogous to $f \cdot (L/D)$, g the gravity constant, A_m the cross-sectional area at the inlet of the minor loss element and u the corresponding velocity there. The summation of all friction and minor losses in a pipe system can be expressed by a head loss h_l:

$$h_l = h_f + h_m = \left(\sum\left(f \frac{L}{D} \frac{1}{2g\ A_p^2}\right) + \sum\left(K_l \frac{1}{2g\ A_m^2}\right)\right)\dot{V}^2. \tag{5.179}$$

It is mandatory to use the correct inner pipe diameter for each pipe section and minor loss. In order to simplify calculations, minor losses are often expressed also in form of an equivalent pipe length $l_{equ} = K_l(D/f)$. It represents the length of pipe that produces the same head loss as the local or minor loss. This is a simple, but potentially inaccurate, method of including minor losses. The problem is that the friction coefficient varies from pipe to pipe, so the equivalent length will not have a unique value. When minor losses are truly minor, this problem becomes academic. However, when local losses are an important part of the analysis, it is recommended that the minor loss coefficients K_l be used rather than an equivalent length.

The geometry of a pipe helix can be described, as shown in Fig. 5.32a, by the inner diameter D of the pipe wound into a helix, by the mean diameter D_h of the helix and by the pitch h of the helix. As a result of the curvature of the tube, centrifugal forces occur during the flow, which cause a secondary flow in the form of a double vortex termed as Dean vortices. The secondary flow causes an increase in pressure loss compared to the straight pipe. The pressure loss that a fluid suffers when flowing through a pipe helix is calculated analogous to a straight pipe

$$\Delta p = f \frac{L}{D} \rho \frac{u^2}{2}, \tag{5.180}$$

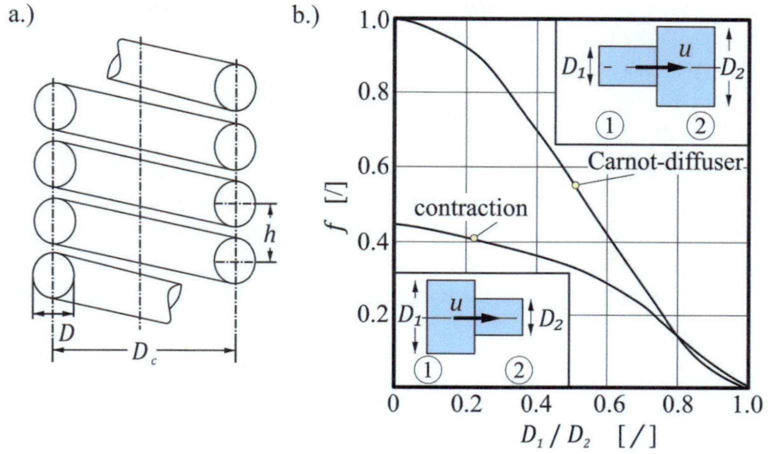

Fig. 5.32 **a** Geometry of a helical coil; **b** dimensionless pressure drop coefficients f for a discontinuous pipe cross-sectional constriction

where L is the length of the pipe wound into the helix. The drag coefficient f defined by this relation depends on the Reynolds number Re and on the ratio D/D_m, where D_m denotes the mean diameter of curvature of the pipe helix, which is determined according to

$$D_m = D_c\left(1 + \left(\frac{h}{\pi\ D_c}\right)^2\right). \tag{5.181}$$

Noticeable differences between D_m and D_c arise only for strongly curved pipes with a large pitch h of the helix, and for calculating f $D_m = D_c$ can be set if $h < D_c$. The drag coefficients for isothermal laminar and turbulent flows can be calculated using:

$$f = \frac{64}{Re}\left(1 + 0.33\left(\lg\left(Re \cdot \sqrt{\frac{D}{D_m}}\right)\right)^4\right) \text{ for } Re < \left(Re_{crit} \cdot \sqrt{D/D_m}\right)$$

$$f = \frac{0.3164}{Re^{1/4}}\left(1 + 0.095 \cdot \sqrt{\left(\frac{D}{D_m}\right)Re^{1/4}}\right) \text{ for } Re < 10^5;$$

$$\text{with } Re_{crit} = 2300\left(1 + 8.6\left(\frac{D}{D_m}\right)^{0.45}\right). \tag{5.182}$$

In solar thermal piping networks, many pipe connections are required due to the low power density, so the most common geometries are discussed in this context. A collection of formulas for the determination of head losses for pipe installations can be found in numerous standards handbooks available worldwide. Here, relations are provided from American Water Works Association (AWWA), American Society of Mechanical Engineers (ASME), American Society for Testing and Materials (ASTM), American National Standards Institute (ANSI) and Stephan et al. (2019), for details of the characteristics of pipe elements it is useful to use data sheets of the vendor/manufacturer.

Discontinuous duct cross-sectional changes occur, for example at manifolds and headers or at penetrations such as Carnot diffusers or pipe contractions, result in pressure loss coefficients f, as shown in Fig. 5.32b. Of central importance is that the pressure loss coefficients for this type of constriction are directional. Regarding contractions a certain downstream length l_d is required to equilibrate the velocity, which calculates to.

$$\begin{aligned} l_d &= 0.13 \cdot Re_2 \cdot D_2 \quad \text{(laminar flow);} \\ l_d &= 0.0575 \cdot Re_2 \cdot D_2 \ \text{(tubulent flow),} \end{aligned} \tag{5.183}$$

where the subscript 1 denotes the position upstream the discontinuity and 2 downstream, respectively. If a fluid enters a larger space from a tube, e.g. into a tube extension or from the tube bundle of a heat exchanger, an expanding jet with strong turbulence is formed which reattaches to the tube wall after 8–10 times D_2. Compared to a change of the static pressure due to the velocity change, which can be calculated according to Bernoulli's law, a momentum or shock loss occurs in the so-called Carnot diffuser, which—for the case of constant flow velocity over the tube cross-section ($Re \to \infty$)—can be calculated exactly.

$$\Delta p = (1 - A_1/A_2)^2 \cdot \frac{\rho}{2} \cdot u_1^2 \tag{5.184}$$

In the case of turbulent flow, the deviation from Eq. (5.184) is small (<5%), in the case of laminar flow, due to the more pronounced velocity profile, the loss according to Eq. (5.184) is about 33% smaller.

With continuous contraction according to Fig. 5.33a, no flow separation occurs, provided the opening angle $\alpha < 40°$ remains. A special case of discontinuous cross-section narrowing is the pipe inlet, which also depends on the shape of the inlet. The corresponding pressure loss coefficients f_i are shown in Table 5.9. For continuous cross-sectional changes of the pipe inlet geometry at angle α (Fig. 5.33), the following relationship between pressure loss coefficient f and pipe diameter ratios D_1/D_2 is obtained:

$$f = k(\alpha) \cdot \left(1 - \frac{D_1}{D_2}\right)^2 \tag{5.185}$$

in which the factor $k(\alpha)$ is obtained from experimental data and assumes the values assumed in provided in Table 5.10.

Thus, in convergent tubes almost the entire pressure is converted into flow energy. With larger angles α, one approaches the case of sudden contraction, which causes much stronger losses and the data may be taken from Fig. 5.33b. To small

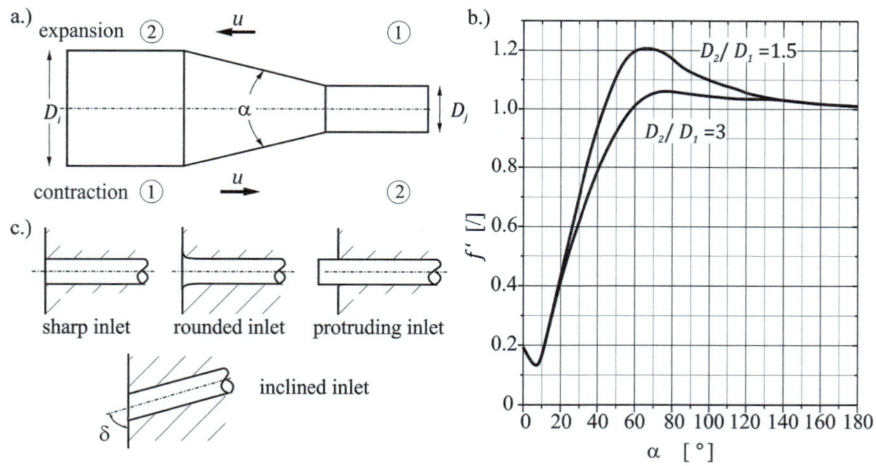

Fig. 5.33 **a** Nomenclature for continuous pipe expansion and contractions; **b** drag coefficient f' of expanded pipe sections as a function of the opening angle α and the ratio of pipe diameters D_2/D_1; **c** representation of common pipe inlet geometries

Table 5.9 Pressure loss coefficient f_i of different pipe inlet geometries

Pipe inlet type	f_i (/)
Sharp inlet	1.69 (90°) 0.45 (normal curved)
Rounded inlet	0.005 (very smooth) 0.05 (normal)
Protruding inlet	~ 3 (Borda mouth)
Inclined inlet	$= 0.5 + 0.3\cos\delta + 0.2\cos^2\delta$

Table 5.10 Experimentally determined loss factors $k(\alpha)$ of a continuous pipe cross-section contraction as a function of the opening angle α (from Fig. 5.33a)

α (°)	5	7.5	10	15	20	40	180
$k(\alpha)$	0.13	0.14	0.16	0.27	0.43	1.0	1.0

angles yield long conical pipe segment; then the additional pipe length must be calculated, typically using the mean cross-sectional area A_m and the mean velocity u_m.

Conversely, flow separation cannot be prevented in continuous pipe expansions due to wall friction and pressure build-up and there is no ideal conversion of flow energy into pressure energy (according to the Bernoulli equation). The losses are

Table 5.11 Reynolds number at which the minimum loss coefficient in an expansion is obtained at different divergence angles α (from Fig. 5.33a)

$Re\ (\times 10^5)$	0.5	1	1.5	2
α (°)	10	8.4	7.6	7.1

calculated as:

$$\Delta p = f' \cdot (1 - A_1/A_2)^2 \cdot \frac{\rho}{2} \cdot u_1^2, \tag{5.186}$$

with f' as the drag coefficient, whose dependence on the opening angle α is shown in Fig. 5.33b. The minimum of the losses occurs at low expansions and depends on the Reynolds number, as shown in Table 5.11. At even higher Re numbers, it is necessary to go down to $\alpha = 5–6°$. The roughness of the tube wall is of influence, so that, for example, $\alpha = 7°$ was found to be the preferred angle for a smooth surface ($f = 0.02$), and $\alpha = 9°$ for a rougher surface ($f = 0.04$).

When discharging from vessels, the area ratio in Eq. (5.186) $A_1/A_2 = 0$. In addition, there is an extremely strong dependence of the loss coefficient f_i on the design of the orifice, for which the values from Table 5.9 can be used. In order to calculate the outgoing volumetric flow rate $\dot{V} = (u_2 \cdot A_2)$ when the overpressure $(p_2 - p_1)$ prevails at the orifice, the acceleration work, and the resistance of a connecting pipe f (with the length L) as well as possibly further fittings must be added to the resistance coefficient of the orifice. The force balance around the outlet translates into a pressure difference:

$$\Delta p = (p_2 - p_1) = \left(1 + f_i + f \cdot \frac{L}{D_2}\right) \cdot \frac{\rho}{2} \cdot u_2^2, \tag{5.187}$$

which allows the outlet velocity and volume to be determined.

In circular pipe bends, the pressure loss coefficient f_b depends on the bend radius to diameter ratio (R/D), the angle of curvature and the surface roughness inside the pipe. In hydraulically smooth pipes, the following formula is obtained for f_b, which is shown in Fig. 5.34a for a bend angle of $\varphi = 90°$.

$$\begin{aligned} f\left(\frac{D}{R}\right) &= 0.986859 - 0.658175 \cdot \left(\frac{D}{R}\right) + 0.0207726 \cdot \left(\frac{D}{R}\right)^2 \\ &\quad - 0.027726 \cdot \left(\frac{D}{R}\right)^3 + 0.001274 \cdot \left(\frac{D}{R}\right)^4; \\ k(\varphi) &= 0.06364 + 0.01149 \cdot \varphi - 1.3167 \times 10^{-5} \cdot \varphi^2;\ f_{Kr} = k(\varphi) \cdot f(D/R) \end{aligned} \tag{5.188}$$

With rough tubes, the pressure drop increases significantly, as illustrated by the curve in Fig. 5.34a. Standard orifices, standard nozzles or Venturi tubes are often used for flow measurement, the shape of which is described in DIN EN ISO 5167

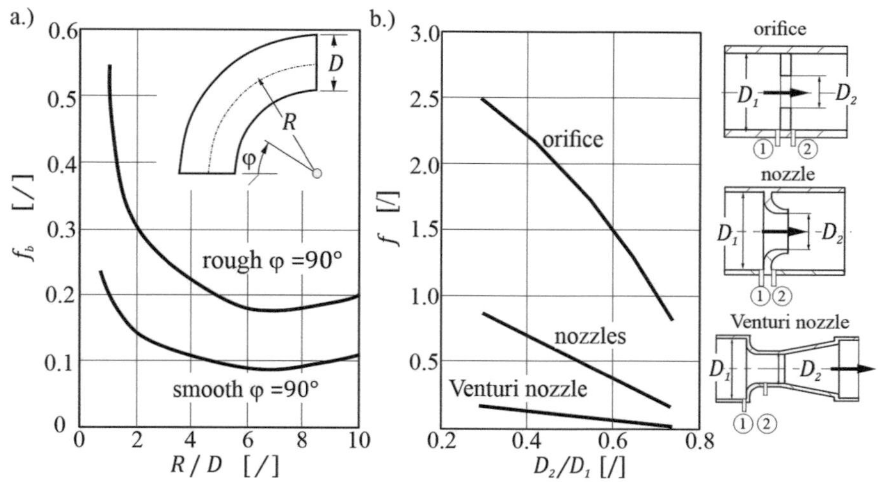

Fig. 5.34 **a** Pressure loss coefficients f_b of a circular tube elbow with smooth and rough surface as a function of the radius to diameter ratio (R/D); **b** pressure loss coefficients f of orifices, nozzles and Venturi tubes as a function of D_2/D_1

(2004). These also generate a pressure drop characterized by loss coefficient f, which is shown in Fig. 5.34b for the different designs.

For right-angled and oblique tees, bends, and elbows, the respective pressure loss coefficients can be found in the manufacturers' data sheets. The same applies to different valve types (gate or butterfly type), throttles, where the pressure loss coefficient is a function of the opening fraction.

5.1.16.2 Convective Heat Transfer in Pipes and Piping Elements

The engineering design of components often requires an estimation of the achievable heat transfer. In this section, some of the most important single-phase correlations for duct flows are compiled, similarly as before for external flows or flows through cavities. According to the previously discussed sections, the heat transfer relationships can be divided into different types of equations, which are shown in Table 5.12. The problem-specific Nusselt number correlations also include correction factors, exponents and constants resulting from the geometry, the temperature-dependent thermophysical properties of the fluid and boundary layer considerations.

Whenever heat transfer correlations are used, correction factors K_p must be considered to cover the following aspects:

- temperature influence on the material values (K_T),
- developing or fully developed flows (K_l),
- geometry influences (K_{dh}),
- convection influences (K_c),
- combined influences (K_p).

Table 5.12 Equation structure and incorporated correction factors/exponents for the different single-phase heat transfer modes

Heat transfer type	Dim.-less quantity equation type	Correction factors and exponents
Forced convection	$Nu = f(Re, Pr)$	$Nu = C \cdot Re^m \cdot Pr^n \cdot K_p$
Natural convection	$Nu = f(Gr, Pr)$	$Nu = C \cdot (Gr \cdot Pr)^n \cdot K_p$
Mixed convection	$Nu = f(Re, Gr, Pr)$	$Nu = C \cdot Re^m \cdot Pr^n \cdot Gr^k \cdot K_p$

The application of the adequate correction factors must be checked for each calculation based characteristic number equations. Additionally, care has to be taken of using the correct length scaling in the equations, the validity range of the respective correlations as well as their confidence interval must be considered in particular. Furthermore, the reference temperature T_{ref} for the material values must be considered, since this results in the correction factors K_T, which accommodates for the influence of viscosity due to temperature gradients. For gases and liquid metals, the following applies to T_{ref}:

$$T_{ref} = \frac{T_f + T_w}{2}, \tag{5.189}$$

in which T_f denotes the fluid temperature and T_w that of the wall, for all other liquids $T_f = T_w$ is assumed. Since the fluid temperatures change due to heat transfer, a nomenclature has been adopted in determining the fluid temperature value (T_f) for the individual flow problems, which Table 5.13 reflects. After calculating the fluid temperature and the reference temperature, the correction factors of the temperature effect (K_T) can be determined for the different heat transfer modes, as shown in Table 5.14. Here, the molecular Prandtl numbers Pr at the fluid temperature T_f and the value at the wall Pr_w at the wall temperature T_w are to be used. In case of missing data, a value of $K_T = 1$ is applied.

The hydraulic (l_{hy}) and thermal (l_{th}) inlet lengths of laminar flows depend on the geometry and the thermal boundary condition. Table 5.15 gives an estimate for this. The correction factor of developing or inflowing flows (K_l) results in $K_l = \left(1 + (d/l)^{2/3}\right)$ for laminar flows.

Technically more relevant are turbulent pipe flows, which constitute nearly all piping grids of thermal systems and for molecular Prandtl-numbers $Pr > 0.5$ the

Table 5.13 Fluid reference temperature T_f to be used when using heat transfer correlations

Flow configuration	Fluid reference temperature T_f
Bounded flows (pipes, ducts, bundles)	$T_f = \frac{T_{f,in} + T_{f,out}}{2}$
Boundary layers, external flows	$T_f = T_\infty$
Short pipes, inflows	$T_f = T_{in}$
Condensation, evaporation of fluids	$T_f = T_{sat}$ or $T_f = \frac{T_w + T_{sat}}{2}$

Table 5.14 Temperature correction factor calculation K_T for different heat transfer types and fluids

	Liquids	Gases
Duct lows	$K_T = (Pr/Pr_w)^{0.11}$	$K_T = 1$
Unbounded flows	$K_T = (Pr/Pr_w)^{0.25}$	$K_T = \left(T_f/T_w\right)^{0.12}$
Free convection	$K_T = (Pr/Pr_w)^{0.25}$	$K_T = 1$

Table 5.15 Dependence of hydraulic (l_{hy}) and thermal (l_{th}) entrance lengths on geometry and thermal boundary condition for laminar duct flows

Geometry	$l_{hy}/(d_h Re)$	$l_{hy}/(d_h Re Pr)$	
		$T_w = const.$	$\dot{q}_w = const.$
Circular duct	0.065	0.055	0.07
Plane channel	0.01	0.014	0.02
Annulus	0.01–0.015	0.05	0.06

Reynolds analogy can be used for technical assessments, which ties the friction factor f to the heat transfer characteristics expressed by the Nusselt number Nu, as e.g. the Colburn relation $\left(St \cdot Pr^{2/3}\right) \sim f/2$ for pipe flows, for details see Bejan (2013). Moreover, similar as for the laminar energy exchange (cf. Sect. 5.1.10) also for the turbulent pipe flow three different flow domains can be separated:

- fully developed flow (either hydraulically or thermally),
- hydraulically developing and
- thermally developing,

for which also the cases of $T_w = const.$ and $\dot{q}_w = const.$ can be separated, which fortunately do not differ as strong as in the laminar case.

- Fully developed pipe flow

For engineering assessment fully thermal and hydraulically developed pipe flow establishes due to the effective wall normal momentum transfer after a downstream entrance length of about $l/d_h \approx 10 - 40$ for $Pr > 0.5$. More exactly, Cengel and Cimbala (2006) provide a Reynolds number dependent relation $l/d_h = \left(4.4/Re^{1/6}\right)$. DNS calculations show that even after hundred hydraulic diameters upstream information still exists and, thus, regarding scientific measurements specific precautions have to be set to achieved fully developed flow. Fully developed turbulent flow is present at Reynolds numbers $Re \geq 10^4$. For both thermal boundary conditions $T_w = const.$ and $\dot{q}_w = const.$, practically the same mean Nusselt numbers result and the following equation for the heat transfer applies for $Pr > 0.5$:

$$Nu_m = \frac{(f/8) Re\ Pr}{1 + 12.7\sqrt{f/8}\left(Pr^{2/3} - 1\right)} \left(1 + \left(\frac{d}{l}\right)^{2/3}\right) \text{ with}$$

$$f = (1.8 \log Re - 1.5)^{-2}, \tag{5.190}$$

where f is the friction factor illustrating the tight connection of heat transfer and friction in higher Prandtl number fluids. This formula is valid up to $Re \leq 10^6$. For large length $l/d > 60$ the simpler Dittus-Bölter relation can be applied with an accuracy of about 20%

$$Nu_m = 0.23\, Re^{4/5} Pr^n, \tag{5.191}$$

in which the Prandtl number exponent is $n = 0.4$ when the fluid is being heated and $n = 0.3$ if the fluid is being cooled. The Dittus-Boelter relation is valid in the range $10^4 \leq Re \leq 1.2 \times 10^5$ for $Pr > 0.5$. For Reynolds number in the transition regime $2 \times 10^3 \leq Re \leq 10^4$ an intermittency factor γ is introduced blending the Nusselt number of the laminar flow ($Nu_{m,lam}$) to that of the turbulent flow ($Nu_{m,turb}$), e.g. from Eq. (5.190)

$$Nu_m = (1 - \gamma) \cdot Nu_{m,lam} + \gamma \cdot Nu_{m,turb} \text{ with } \gamma = \frac{Re - 2300}{10^4 - 2300}, \tag{5.192}$$

in which for the laminar Nusselt number the values at the corresponding thermal boundary are to be used. The reference temperature T_f is to be calculated according to Table 5.13 and to obtain the effective Nusselt number (Nu), Nu_m must be multiplied by a temperature correction factor K_T from Table 5.14. For liquid metals ($Pr < 0.1$) the accuracy of the correlations is in the range of the Dittus-Bölter relation at $Re \sim 10^4$ getting more accurate for higher Re. According to Kakac et al. (1987) the following relations exhibit high confidence

$$\begin{aligned} Nu_m &= 4.5 + 0.0156\, Re^{0.85} Pr^{0.86} \quad \text{for } T_w = const., \\ Nu_m &= 5.6 + 0.0165\, Re^{0.85} Pr^{0.86} \quad \text{for } \dot{q}_w = const., \end{aligned} \tag{5.193}$$

which are valid from $10^4 \leq Re \leq 5 \times 10^6$. Similar as indicated for high Prandtl number fluids a temperature correction factor must be introduced to account for change of the temperature dependence of the thermo-physical properties, which require a multiplication of Nu_m with a temperature ratio in the form, $Nu = Nu_m(T_m/T_i)^n$ to obtain the effective Nusselt-number (Nu), where T_m represents the bulk temperature and T_i is the inlet temperature. The exponent for the different thermal boundary conditions read to

$$\begin{aligned} T_w = const.: n &= 0.08 + 2.2 \times 10^{-4} T_m & T_m &> 600\,\text{K} \\ n &= 0.08 & T_m &< 600\,\text{K}, \\ \dot{q}_w = const.: n &= \exp\left(5.9 \times 10^{-3} T_{m-6.91}\right) & T_m &> 600\,\text{K}, \\ n &= 0 & T_m &< 600\,\text{K}. \end{aligned} \tag{5.194}$$

- hydraulically developing flow

Turbulent hydraulically developing flow exhibit with respect to the heat transfer a similar behaviour as fully developed flow and thus the correlations for turbulent flow can be used unaltered respecting the corresponding Re and Pr regimes.

- thermally developing flow

Turbulent thermally developing flows occur at the entrance e.g. of absorber pipes of solar farms and for Prandtl number fluids $Pr > 0.5$ also the correlations of fully developed turbulent can be used with high confidence. For liquid metals the thermal entrance length differs from those of higher Prandtl number fluids and depend not only on the Reynolds number but also on the thermal boundary condition. Figure 5.35a, b illustrate the thermal developing length l_{th}/d in turbulent pipe flows for various molecular Prandtl numbers as function of Re. The mean Nusselt number correlations for this condition are (Bejan, 2013):

$$Nu_m = 4.8 + 0.0156\, Re^{0.85} Pr^{0.93} \quad \text{for } T_w = const.,$$
$$Nu_m = 6.3 + 0.0167\, Re^{0.85} Pr^{0.93} \quad \text{for } \dot{q}_w = const., \tag{5.195}$$

which are close to that of fully developed flow and are reported to be valid for $10^4 \leq Re \leq 10^6$. Here, also the temperature correction applies as in fully developed turbulent flows.

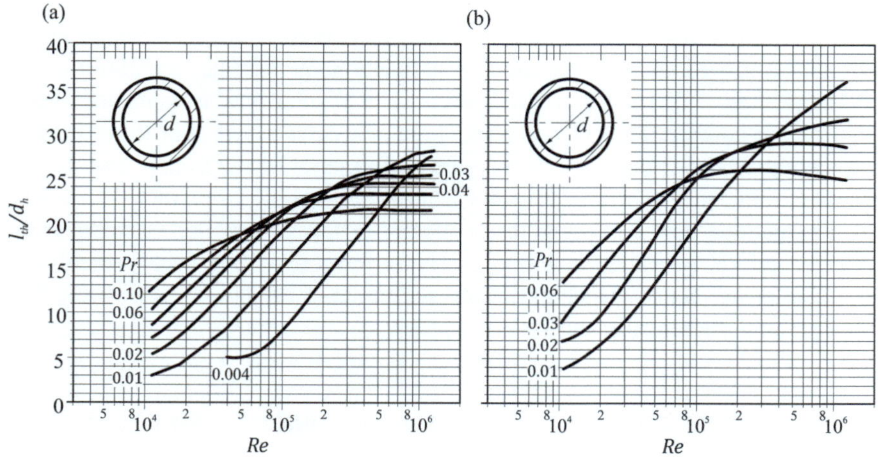

Fig. 5.35 Dimensionless thermal entrance length l_{th}/d as a function of the Reynolds number in a hydraulically fully developed flow of a smooth circular tube for different Prandtl numbers **a** $T_w = constant$ and **b** $\dot{q}_w = const$

5.2 Multiphase Momentum and Energy Exchange

If a collector exceeds the saturation temperature of the heat transfer fluid, which is observed, for example, in connection with the collector stagnation, evaporation of part of the cooling medium may occur. Due to the expansion during the phase transition, there may be a significant increase in pressure and temperature in the system. The existence of a two-phase flow substantially changes the thermal hydraulics discussed so far and therefore requires basic knowledge of two-phase flow.

The simplest and most familiar form of two-phase flow occurs during boiling. Starting from boiling, an attempt will be made to show some model ideas which are important for the further description, as they are also valid for the operating conditions which are encountered in solar collectors. Further details can be found in the books Collier and Thome (1996), Incropera and DeWitt (2001) as they are beyond the scope of this script. Empirical relations of pressure losses and heat transfer in two-phase flow can be found in the work of Streicher (1991) and the literature listed therein. Heinzel et al. (1994), in addition to providing the basics, also addresses the two-phase problem in the solar collector, focusing on the development of direct-evaporation boiling water collectors. When a collector or several collectors are at a stagnation, the same processes occur in principle, and in almost all cases the occurrence of flow instabilities is observed during the transition from single-phase to two-phase flows, which are addressed in the last section.

Two-phase flow effects must also be appropriately considered in the system analytical design and construction of collector systems in order to predict potential stall scenarios or their duration as a function of boundary conditions. The design of large-scale solar systems or even thermal solar power plants necessarily requires the consideration of potential shutdown cases since both their duration and the associated scenarios influence the performance of the system and thus its economic efficiency. For this purpose, it must be possible to predict the individual flow patterns that can occur in a collector or a segment. For example, if the collector field is connected incorrectly, reverse flows may even occur, resulting in no usable power being available. For this reason, an overview of the possible flow forms in an absorber tube will be given below.

5.2.1 Boiling Phenomenology-Homogeneous Boiling-Phase Diagrams

If a vapor bubble forms on a heated wall along which a fluid flows, the bubble will not only remove the latent heat (heat of vaporization) when it detaches from the wall, but its interaction with the fluid will cause mixing and thus agitation. This mixing disturbs the flow boundary layer but leads to an improvement in heat transfer. The formation of a bubble requires superheating of the fluid, the dynamics of which will be considered below.

Let us first consider homogeneous boiling. A vapor bubble of radius r in a liquid is subjected not only to the external pressure p_l imposed by the liquid column, but also to an additional pressure caused by the surface tension σ, so that the vapor bubble pressure p_v is given by:

$$p_v = p_l + \frac{2\,\sigma}{r}. \tag{5.196}$$

In addition, the vapor pressure in a bubble that has a concave curved surface is less than the pressure over a flat surface p_∞ of the same fluid at the same temperature T. Here one can derive the following relation:

$$p_v = p_\infty exp\left(\frac{-2\sigma\ v_l\ M}{r\ RT}\right) \approx p_\infty\left(1 - \frac{2\sigma\ v_l}{p_\infty r\ v_g}\right), \tag{5.197}$$

where R is the universal gas constant ($R = 8.314\,\text{J}/(\text{mol K})$), M is the molecular mass (g/mol), and v_g and v_l are the specific volumes of vapor and liquid, respectively. Substituting the Eq. (5.196) into the relation (5.197) yields:

$$(p_\infty - p_l) = \frac{2\sigma}{r}\left(1 + \frac{v_l}{v_g}\right). \tag{5.198}$$

To calculate the superheat due to the pressure, difference $(p_\infty - p_l)$ the Clausius–Clapeyron equation can be used. This is obtained by considering a closed adiabatic system in which there are vapor and liquid which are in thermodynamic equilibrium. According to the 2nd law of thermodynamics, the equilibrium state of a closed system is characterized by the maximum of its entropy. Under the constraints that the internal energy U and the volume V are constant:

$$dS = 0 \quad (U = const., V = const.) \tag{5.199}$$

The entropy (S) of the two-phase system is composed additively of the entropies of the two phases, whose state variables are marked with the subscript "v" and "l" for the gas phase (vapor) and the liquid phase, respectively. This results in the following for the entropy:

$$S = S_l + S_v = m_l s_l(u_l, v_l) + m_v s_v(u_v, v_v), \tag{5.200}$$

in which the small letters m, s, u and v represent the specific molar fractions, entropies, internal energies and volume fractions of the vapor phase and liquid phase, respectively. Inserting the relations for the thermodynamic quantities and forming the differential, we obtain for dS:

$$dS = m_l\left(\frac{1}{T_l} - \frac{1}{T_{v\,l}}\right)du_l + m_l\left(\frac{p_l}{T_l} - \frac{p_v}{T_v}\right)dv_l$$

$$+\frac{1}{T_v}(u_v - u_l + p_v(v_v - v_l) - T_v(s_v - s_l))dm. \tag{5.201}$$

In thermodynamic equilibrium, $dS = 0$. Since the differentials du_l, dv_l, and dm_l are independent of each other, the value of each of the three brackets must be zero, which yields the following condition for the two-phase equilibrium:

$$T_l = T_v = T, \quad p_l = p_v = p, \quad u_v + pv_v - Ts_v = u_l + pv_l - Ts_l. \tag{5.202}$$

As a result, two phases can only be in equilibrium if they have the same temperature and pressure. In addition, their specific free enthalpies h or Gibbs energies g must match, thus it applies:

$$g = u + pv - Ts = h - Ts. \tag{5.203}$$

A differentiation of the Gibbs function yields:

$$dg = -s\,dT + v\,dp, \tag{5.204}$$

to which the boundary condition $dg_l(T, p) = dg_v(T, p)$ is applied and this gives:

$$-s_l dT + v_l dp = -s_v dT + v_v dp. \tag{5.205}$$

The result is the Clapeyron relationship:

$$\frac{dp}{dT} = \frac{s_v - s_l}{v_v - v_l}. \tag{5.206}$$

If instead of the entropy difference $(s_v - s_l)$ one introduces an enthalpy difference $(h_v - h_l)$, then using $((h_v - h_l) = T(s_v - s_l)$ one obtains the Clausius-Clapeyron equation in the form:

$$T\frac{dp}{dT} = \frac{(h_v - h_l)}{(v_v - v_l)} \quad \text{or} \quad \frac{dp_{sat}}{dT} = \frac{\Delta h}{T_{sat}(v_v - v_l)}, \tag{5.207}$$

where p_{sat} is the saturated steam pressure, T_{sat} the saturated steam temperature and h the enthalpy of vaporization. The Clausius–Clapeyron equation can be used to calculate the phase boundary line between the liquid and gaseous phases, the so-called boiling point curve. Equation (5.207) can be further simplified under the following boundary conditions:

- the specific volume of the liquid is small, i.e. $v_l/v_v \ll 1$, and
- for small temperature changes the saturated vapor can be treated as an ideal gas, thus $(pv_v) = (R/M)T$ holds, and for
- the enthalpy of vaporization shall be $h(T) = const.$

Under these assumptions, the following result is initially obtained:

$$\frac{dp}{p} = \frac{M}{R\,T^2}\Delta h\, dT. \tag{5.208}$$

An integration of the pressure from p_l to p_∞ and from saturation temperature (T_{sat}) to vapor bubble temperature (T_b) leads to

$$\ln\left(\frac{p_\infty}{p_l}\right) = -\frac{M}{R}\left(\frac{1}{T_b} - \frac{1}{T_{sat}}\right)\Delta h = -\frac{M\Delta h}{RT_bT_{sat}}(T_b - T_{sat}). \tag{5.209}$$

and considering the surface tension, the following is obtained

$$(T_b - T_{sat}) = \frac{RT_bT_{sat}}{M\Delta h}\ln\left(1 + \frac{2\sigma}{rp_l}\cdot\left(1 + \frac{v'}{v''}\right)\right). \tag{5.210}$$

The assumptions $v_l/v_v \ll 1$ and $(2\sigma/(rp_l)) \ll 1$ further simplify Eq. (5.210) by using series expansion for $\ln(1 + x)$ while neglecting higher order terms:

$$(T_b - T_{sat}) = \Delta T_{sat} = \frac{RT_{sat}^2}{M\Delta h}\left(\frac{2\sigma}{r\,p_l}\right). \tag{5.211}$$

($\ln(1 + x) = x - x^2/2 + x^3/3 - x^4/4 + \cdots$. The pressure difference between the bubble and the fluid must thus be compensated by an appropriate superheat temperature to maintain a bubble with a stable size. Because of the reciprocal dependence of the pressure p_l on the bubble radius ($\sim 1/r$), bubble formation in a homogeneous fluid starts only at the temperature significantly higher than the boiling temperature, i.e. the spontaneous or homogeneous nucleation temperature.

In this context, it is useful to first take a look at phase and state diagrams, as schematically shown in Fig. 5.36. Within such a space spanned by the specific volume v, the pressure p and the temperature T, there are curves, the so-called phase boundary lines, which delimit different aggregate states from one another. In the areas enclosed by them, several phases are in thermodynamic equilibrium. The individual phase boundary lines are called

- vaporization curve, which indicates the phase boundary liquid to vapor between the tripe line and the critical point.
- the sublimation curve, which separates the solid from the vapor state between the zero point and the triple line, and
- the dew point line, which is often referred to as the condensation curve.

The vaporization and sublimation curves can be combined to form the vapor pressure curve. The critical point has no degree of freedom, while all other curves exhibit one degree of freedom of either temperature or pressure. Within the phase spaces, two degrees of freedom exist, namely pressure and temperature.

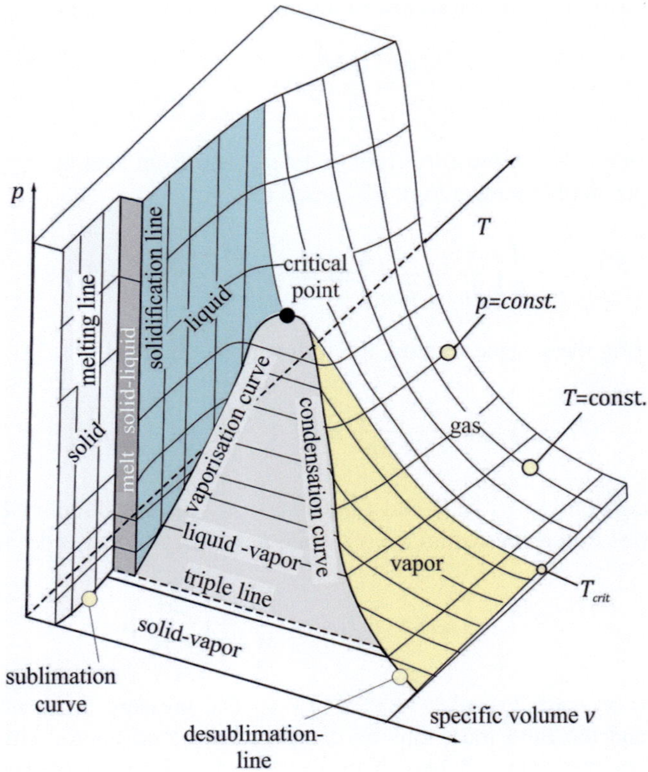

Fig. 5.36 Three-dimensional schematic representation of a phase diagram

Table 5.16 provides boiling temperatures at normal pressure, temperatures and pressures of the triple point and critical points of some selected fluids. It illustrates that as the molar mass number increases, the critical pressure and temperature increases as well. A similar increase of both quantities can be observed with increasing surface interaction of the molecules.

Figure 5.37 shows an example of a $p - v$ diagram of carbon dioxide (CO_2). To the left of the critical point, the vaporization separates the region of subcooled, saturated liquids from the two-phase domain. To the right of the critical point, the condensation (dew point) curve separates the two-phase region from the saturated vapor region. Above the critical point, there is no more phase transition. Points within the boundary line correspond to stable states with a mixture of saturated vapor and saturated liquid. The mass ratio of vapor and liquid of a state is described by the ratio of the distances on the line AE to the boundary lines of the saturated liquid and saturated vapor, respectively.

For an ideal gas, the isotherm is characterized by the stability condition $(\partial p/\partial v)_{T=const.} < 0$. In the $p - v$ diagram, the critical point denotes the inflection point of the critical isotherms. As the boundary line is approached, the state

Table 5.16 Boiling point, triple point, and critical point of some fluids

	Fluid	Boiling temp. @ 0.1 MPa (K)	Triple point T (°K)	p (MPa)	Critical T_{crit} (°K)	Point p_{crit} (MPa)
H_2	Hydrogen	20.28	13.80	7.04×10^{-3}	33.19	1.315
N_2	Nitrogen	77.34	63.14	1.252×10^{-2}	126.19	3.396
O_2	Oxygen	90.23	54.36	1.463×10^{-4}	154.60	5.043
CO_2	Carbon dioxide	194.69	216.59	0.518	304.13	7.377
NH_3	Ammonia	239.81	195.50	6.1×10^{-3}	405.40	11.359
H_2O	Water	373.15	273.16	6.117×10^{-4}	647.10	22.104
CH_4	Methane	111.55	90.67	1.169×10^{-2}	190.56	4.599
C_2H_6	Ethane	184.60	91.60	1.169×10^{-6}	305.30	4.89
C_3H_8	Propane	231.11	85.03	1.685×10^{-10}	369.82	4.248
C_4H_{10}	Iso-butane	262.12	113.55	1.948×10^{-8}	425.13	3.796
CH_3OH	Methanol	337.82	175.52	8.11	513.38	8.216
C_2H_5OH	Ethanol	351.52	150.00	6.34	489.00	6.591
$C_2H_6O_2$	Glycol	470.48	260.57	*n.a.*	720.00	8.23
He	Helium	4.222	2.177	5.04×10^{-3}	5.30	0.228
Ne	Neon	27.104	24.556	4.34×10^{-2}	44.00	2.720
Ar	Argon	87.302	83.806	6.89×10^{-2}	151.08	4.85
	Air	78.8	59.75	5.27×10^{-3}	133.00	3.95
$NaNO_3$	Sodium nitrate	653.15	*n.a.*	*n.a.*	*n.a.*	*n.a.*
KNO_3	Potassium nitrate	673.15	*n.a.*	*n.a.*	*n.a.*	*n.a.*
Na	Sodium	1156.90	370.97	*n.a.*	2503.70	35.00
Hg	Mercury	629.81	234.15	1.65×10^{-9}	234.29	*n.a.*
Pb	Lead	2016.65	*n.a.*	*n.a.*	4460.00	160.00

of a gas increasingly deviates from the ideal gas. This behaviour of the real gas is described by the van der Waals equation:

$$\left(p + \frac{a}{T^n v^2}\right)(v - b) = RT. \tag{5.212}$$

where a and b are constants and n is the Berthelot or van der Waals exponent. In addition to the external pressure p, there is the internal pressure described by $\left(a/\left(T^n v^2\right)\right)$, which is based on the attractive forces between the molecules. The co-volume b must be subtracted from the volume because the molecules with their finite radius require a certain volume which is no longer available for free movement. Within the two-phase transition region in Fig. 5.37, the van der

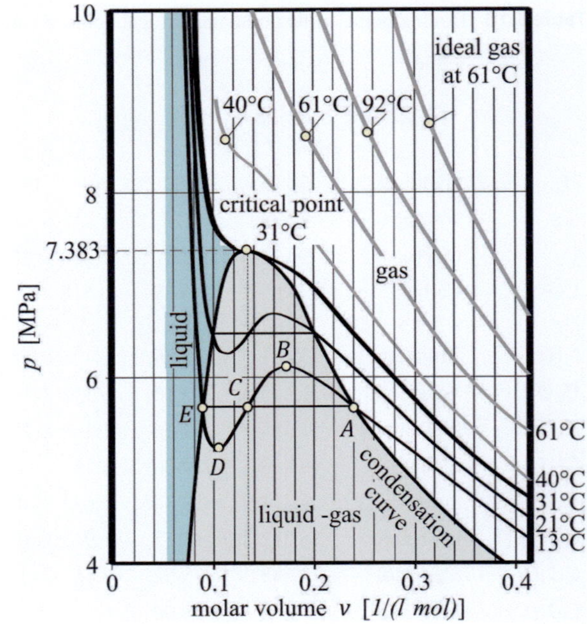

Fig. 5.37 Section of the $p - v$ diagram of carbon dioxide (CO_2)

Waals relationship assumes a curve corresponding to the line through the points $ABCDE$.

With $T < T_{crit}$, on the other hand, an initially unphysical behavior is observed. The pressure decreases with compression. This weakness of the van der Waals equation is corrected by the so-called Maxwell construction. Here, a horizontal straight line is placed in such a way that the areas enclosed by the straight line and the graph are equal. This horizontal line describes the phase transition. A further compression now leads to the transition of vapor particles into the liquid phase. The pressure remains constant and corresponds to the vapor pressure. Once the phase transition is complete, further compression leads to an extreme increase in pressure, which is also expected when liquids are compressed. The value of the intrinsic volume b, which cannot be undercut, can also be read off.

In technical systems, changes of state are mostly isobaric, e.g. imposed by a pump pressure, and therefore the $T - v$-diagram schematically shown in Fig. 5.38 shall be used for explanation. If the isobar p_1 is considered, the two-phase region is located between E and A, in which all states between E and A are additively composed of two phases. With the van der Waals equation there are now further points, or lines DE, BD and AB.

The state on line ED can also be assumed by water, although with respect to the stability condition $(\partial p/\partial v)_{T=const.} < 0$. It is called metastable and as a consequence fluids can be superheated. Coming from E, the fluid is slowly heated as a supercooled fluid and should actually boil at E. Instead, you get what is called a superheated fluid, between E and D. This is unstable. If any disturbance occurs,

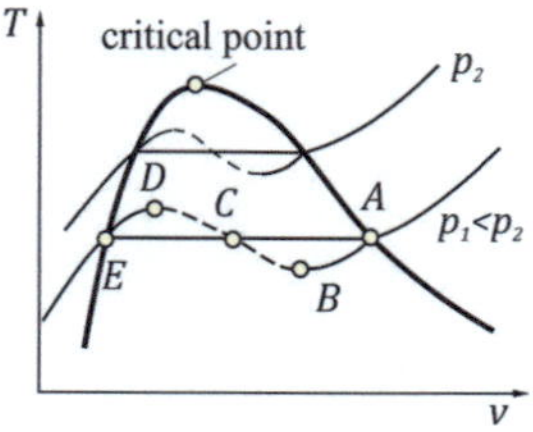

Fig. 5.38 Schematic sketch of a $T - v$ diagram to explain superheating and subcooling

the superheated liquid jumps from a point between ED to a point on the line AE. This means that the fluid suddenly changes into two phases.

The corresponding reverse phenomenon, called supercooling, is known from meteorology, and is depicted by the branch AB. Coming from the right from the superheated vapor region, the vapor should start to condense at A, e.g. in clouds. But it does not do this. It remains supercooled vapor. If there is a disturbance, for example by an airplane jet, the cloud suddenly forms droplets. This supercooling plays an important role in meteorology and can amount to several degrees Celsius.

At points B and D, the van der Waals curves exhibit a maximum and a minimum, respectively. There $(\partial p/\partial v)_{T=const.} = 0$. In between there is an unstable region where $(\partial p/\partial v)_{T=const.} > 0$. In this region, metastable states cannot exist and cannot occur. The latter can be seen again better in the $p-v$ diagram. Between B and D any arbitrary small expansion would lead to a pressure increase and thus to a further expansion.

In the $p - v$ diagram, the area, and the points CDE above the equilibrium line AE must equal the area and ABC below the equilibrium line.

With the understanding of the phase diagrams it is possible to define the limit against overheating, because at the minima the conditions must be fulfilled

$$(\partial p/\partial v)_{T=const.} = 0 \quad \text{and} \quad \left(\partial^2 p/\partial v^2\right)_{T=const.} > 0. \tag{5.213}$$

With the conditions at the critical point $(\partial p/\partial v)_{T=crit.} = 0$ and $\left(\partial^2 p/\partial v^2\right)_{T=crit.} > 0$, the constants in the van der Waals equation can be determined. The result is $a = 3$ and $b = 3$. A normalization of the state variables with the values at the critical point in the manner $\pi = (p/p_{crit})$, $\theta = (T/T_{crit})$ and $\phi = (v/v_{crit})$ transforms the Eq. (5.212) into:

$$\left(\pi + \frac{3}{\theta^n \phi^2}\right)(\phi - 3) = \frac{8}{3}\,\mathrm{T}. \tag{5.214}$$

For the exponent n, van der Waals propose $n = 0$ and Berthelot $n = 1$. Along the superheat limit of a liquid, according to Eq. (5.212) $(\partial \pi/\partial \phi)_{\theta=const.} = 0$ must hold, yielding the following expression for the reduced temperature, the so-called spontaneous nucleation "sN":

$$\theta_{sN}^{n+1} = \frac{(3\phi - 1)}{4\phi^3}. \tag{5.215}$$

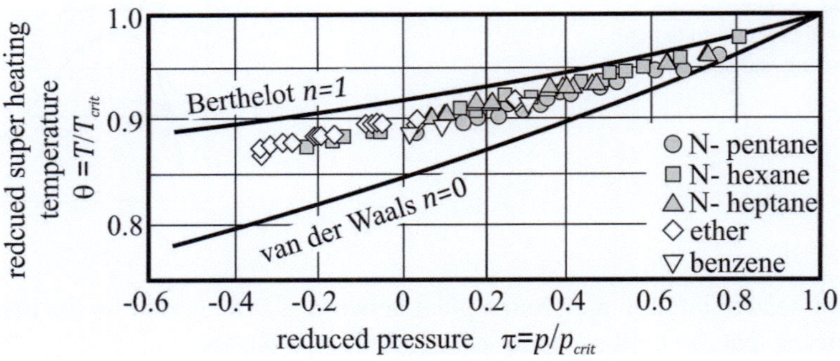

Fig. 5.39 Comparison of the superheat limit according to Eq. (5.215) with the exponent of van der Waals with $n = 0$ and Berthelot with $n = 1$ as well as measured values from Eberhart and Schnyders (1973)

Eberhart and Schnyders (1973) published a comparison of measured superheat limits with calculated values (compare Fig. 5.39). The best fit of the measured values is obtained for $n = 1/2$. Substituting in Eq. (5.215), a modified Berthelot equation is obtained. (The negative reduced pressure in Fig. 5.39 corresponds to conditions where the fluid is subjected to external forces.) The atmospheric pressure is taken to be 0.1 MPa small compared to the critical pressure of water at 22.5 MPa, so that the reduced pressure $\pi \to 0$. For $\pi = 0$, it follows from Eqs. (5.214) and (5.215) $\phi = 2/3$. Substituting this result into the relation (5.215), we obtain for the spontaneous nucleation temperature:

$$\theta_{sN} = \left(\frac{27}{32}\right)^{1/(1+n)}. \tag{5.216}$$

Thus, the reduced temperature of the superheat limit takes the value of 0.844, 0.893, and 0.919, depending on van der Waals, Berthelot, or the modified Berthelot equation.

Collier and Thome (1996) quote a value of $T_{sN} = 320.7\,°\mathrm{C}$ for water under atmospheric conditions as the temperature for homogeneous nucleation. These values considerably exceed the temperature level aimed at for boiling water collectors during operation. Such stagnation temperatures can only be reached in case of malfunctions, when the heat sink fails, and full solar irradiation is present. However, in the presence of gases, the superheating temperature decreases rapidly because the partial pressure p_p of trapped gases stabilizes the steam bubble. In such a case, Eq. (5.196) modifies to:

$$p_v - p_p = p_l + \frac{2\,\sigma}{r}. \tag{5.217}$$

The superheat temperature T_{sh} thus results in:

$$(T_{sh} - T_{sat}) = \frac{R T_{sat}^2}{M \Delta h}\left(\frac{2\sigma}{r p_l}\right) \ln\left(1 + \left(\frac{2\sigma}{r p_l}\right)\left(1 + \frac{v_l}{v_v}\right) - \frac{p_p}{p_l}\right). \tag{5.218}$$

5.2.2 Boiling at Surfaces

In technical systems, especially in solar collectors for steam generation, which are aerated and deaerated at least once a day, gases dissolved in water must always be expected. In addition to these gas bubbles carried in the water, gas-filled pores in the pipe wall provide additional boiling nuclei. In the latter case, the wetting behaviour of the water on the pipe wall and thus the contact angle plays an excellent role. The free enthalpy of formation $\Delta G(r)$ of a boiling nucleus is reduced by the factor φ at a planar interface, which depends on the wetting angle θ in the following way (Bankoff, 1957):

$$\varphi = \frac{\left(2 + 2\cos\theta + \cos\theta\sin^2\theta\right)}{4}. \tag{5.217}$$

In the limiting case $\theta = 0°$, which means complete wetting, $\varphi = 1$ and the enthalpy of formation does not change. Without any wetting $\theta \to 180°$ and with $\varphi = 0$ no superheating is required. In wall pores opening to the liquid with an angle β, the effective contact angle θ increases, as shown in Fig. 5.40. In the case of a shallow cavity or pores, this effect is marginal. However, with more acutely tapered cavities, the effective contact angle can reach 180° and the required superheat approaches zero. In the case of poor wetting and small β, the curvature of the liquid surface in the notch can reverse and the enthalpy of formation becomes negative. Thus, supercooled boiling is possible. In the later sections describing the flow forms, attention must therefore be paid to the wetting effect because bubble boiling at the duct walls significantly impacts the flow.

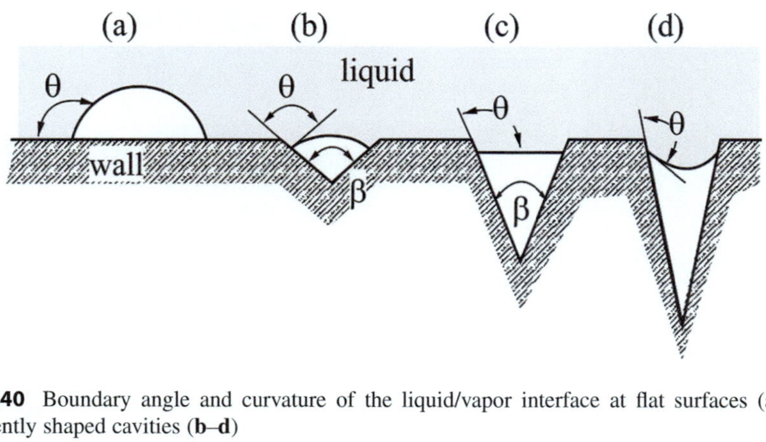

Fig. 5.40 Boundary angle and curvature of the liquid/vapor interface at flat surfaces (**a**) and differently shaped cavities (**b–d**)

5.2.3 Pool Boiling

Vaporization of liquid at the surface of a body immersed in a pool of liquid at rest is generally referred to as pool boiling. A technical distinction is made between pool boiling (boiling in free flow) and flow boiling in tubes, since the nature of the pool boiling process differs considerably depending on the conditions at which boiling occurs from that in tubes. Numerous parameters as the level of heat flux, the thermophysical properties of the liquid and vapor, the surface material and finish, and the physical shape of the heated surface affect the boiling process.

Pool boiling can be explained very well using the Nukiyama curve, which is depicted in Fig. 5.41 for water at 0.1 MPa in the form of a plot of the transferred heat flux ($\dot{q}$) versus the difference between wall (T_w) and saturated steam temperature (T_{sat}). Using the boiling curve, five different regions can be distinguished during boiling.

Since a liquid does not immediately change its phase when it reaches saturation temperature unless it is already in contact with vapor, no vapor bubbles form in region I. This is due to the difference in density between the wall and saturated vapor temperatures. Natural convection, driven only by density differences, sets in, continuously transporting colder fluid towards the wall. This transfer process leads to an increase of the heat transfer from $A - B$.

Once the fluid temperature is above the saturation temperature, small vapor bubbles form at surface pores or cavities of the wall, called vapor nuclei. The formation of the vapor bubbles indicates the transition to region II, that of nucleate

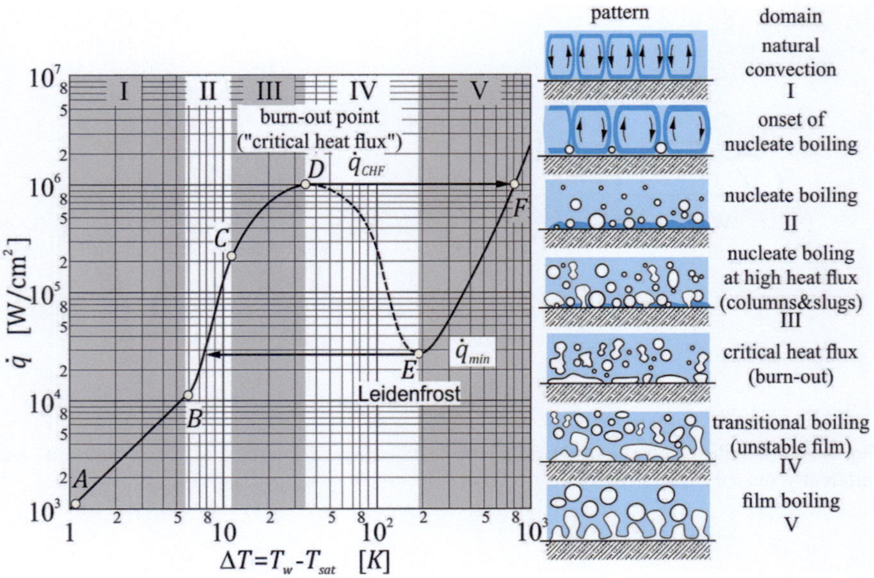

Fig. 5.41 Boiling curves during pool boiling for water at atmospheric pressure (left) and various stages of vessel boiling modified from Collier and Thome (1996)

boiling, at point B ($B = ONB$ "onset of nucleate boiling"). The onset of nucleate boiling is associated with a steep increase in the heat transfer coefficient; even a small increase in the wall heating temperature is associated with a significantly higher heat flux transferred. First, vapor bubbles appear, grow, and slowly leave the wall. There, is a sufficient distance between the bubbles, which is why region II is called individual or isolated bubble boiling.

If the temperature is increased further, the distance between the bubbles near the wall is shortened and one bubble has barely left the wall before the next one is formed. This continues until the bubbles coalesce into continuous columns or slugs of vapor departing from the wall. This marks the transition to Region III, which is called slug or column nucleate boiling ("high-heat-flux nucleate boiling regime"). This process cannot continue indefinitely.

As temperature difference rises further, vapor bubbles are generated so quickly that they prevent the fluid from entering the heated wall. The heat flux at which this boiling crisis occurs is called burn-out-point or critical heat flux (CHF). It marks the beginning of area IV, the transitional boiling. In this region, large parts of the wall are covered by a vapor film and liquid reaches the heated wall only intermittently. It is characterized by a rapid drop in the heat transfer coefficient as the wall temperature increases.

The transition boiling is extremely unstable and the dashed line in Fig. 5.41 can hardly be achieved technically, since in most cases the surface temperature is not controlled but rather the heating power, e.g. solar irradiation. Thus, in power-controlled configurations if the critical heat flux is exceeded, the temperature of the wall abruptly jumps to a higher value (F) and film boiling (area V) starts instantaneously. This is indicated by the arrow. This sudden rise in temperature often has considerable consequences for the heater or the interface material. During film boiling, there is a homogeneous vapor film on the wall and the wall temperature is so high that the heat conduction through the vapor film is large enough to vaporize liquid on the other side.

The lowest wall temperature at which a closed vapor film covers the wall is called the Leidenfrost temperature. If the wall temperature is continuously lowered after entering the film boiling region (region V), this hysteresis phenomenon becomes clearly visible. The wall superheat follows the line from E to D and the closed vapor film remains stable until the Leidenfrost temperature is reached. Then the vapor film collapses abruptly. This collapse leads to an abrupt transition from film boiling back to region II of the nucleate boiling along the arrow.

For the domain of bubble boiling (bubble boiling in free flow) in pool boiling the following relation applies:

$$\frac{\alpha}{\alpha_0} = c_w \cdot F(p^*) \cdot \left(\frac{\dot{q}}{\dot{q}_0}\right)^n, \tag{5.220}$$

α_0 is a reference value at $\dot{q} = 20\,\text{kW/m}^2$ and $p^* = p/p_{crit} = 0.1$, which is determined experimentally. The impact of roughness is considered by $c_w = (R_a/R_{a0})^{0.133}$, in which as reference roughness $R_{a0} = 0.4\,\mu m$ is used. For the pressure correction $F(p^*)$ from $p^* = 0.1$ to p^* the following relations can be used for liquids other than liquid metals, which exhibit a too high surface tension:

$$\text{fluids except water: } F(p^*) = 1.2 \cdot p^{*0.27} + \left[2.5 + \frac{1}{1-p^*}\right] \cdot p^*,$$
$$\text{water: } F(p^*) = 1.73 \cdot p^{*0.27} + \left[6.1 + \frac{0.68}{1-p^{*2}}\right] \cdot p^{*2}. \quad (5.221)$$

In solar collectors, the transition from nucleate boiling to film boiling will occur only when the liquid supply is limited and stagnant. A comparative consideration of the numerical values of the heat flux densities with surface powers of the solar irradiation below 1000 W/m^2 quickly shows that in a container, even without convective heat transfer, the advantages of the increased heat transfer during boiling are not utilized due to the lack of heat flux density. Only a significant concentration of solar irradiation with mirrors allows an improvement of the efficiency by boiling. This can be further increased by improved heat conduction in the absorber board to the heat dissipation tubes.

The following relationships then apply to the exponent n:

$$\text{fluids except water: } n(p^*) = 0.9 - 0.3 \cdot p^{*0.3},$$
$$\text{water: } n(p^*) = 0.9 - 0.3 \cdot p^{*0.15}. \quad (5.222)$$

If no experimental value is known for α_0, it can be estimated using relationships based on the bubble break-up diameter and the wetting angle. Details on this can be found in Collier and Thome (1996), among others.

The maximum attainable heat flux in the nucleate boiling regime termed as departure of nucleate boiling (DNB) or critical heat flux (CHF) is reached when a hydrodynamic instability destabilizes the vapor jets rising from the heater surface; this domain of bubble boiling ends at the critical heat flux density $\dot{q}_{CHF}$, which for water is given by the relation

$$\dot{q}_{CHF} = 0.149 \cdot \sqrt{\rho_v} \Delta h \cdot (\sigma \cdot g \cdot (\rho_l - \rho_v))^{1/4}, \quad (5.223)$$

where ρ_l represents the density of the liquid and ρ_v that of the vapor, σ is the surface tension and Δh the vaporization enthalpy (latent heat). This relation considers that Taylor and Helmholtz-instabilities govern the collapse of a liquid vapor interface if a certain velocity is exceeded, for details see Bejan and Kraus (2003). Equation (5.223) holds for strictly planar surfaces, other geometries require a correction factor which may be taken from Bejan and Kraus (2003) as well.

For an upward-facing heated surface, vapor generated at the interface during stable film boiling is released as bubbles at the nodes of a standing two-dimensional Taylor wave pattern. The Taylor instability of the interface assumes that the minimum is reached when the vapor generation rate becomes too small to sustain Taylor wave action on the free interface for stable generation of bubbles.

This formulates the relation for the minimum heat flux at the Leidenfrost-point reading to:

$$\dot{q}_{min} = 0.09 \cdot \rho_v \Delta h \cdot \left(\frac{\sigma \cdot (\rho_l - \rho_v)}{(\rho_l + \rho_v)^2} \right)^{1/4}. \qquad (5.224)$$

Regarding the film boiling the heat transport across the vapor film from the wall to the liquid interface depends on convection, conduction, and radiation. The radiation depends on the wall surface shape. However, if radiation is negligible the film boiling heat transfer is independent of the material properties and surface quality. For buoyancy-driven laminar film boiling over a vertical flat isothermal surface in a pool of saturated liquid, the local heat transfer coefficient h from the surface is given by:

$$h = \left(\frac{k_v^3 g \; \rho_v(\rho_l - \rho_v)\Delta h}{4 \; \mu_v(T_w - T_{sat})x} \right)^{1/4}, \qquad (5.225)$$

where k_v is the heat conductivty of the vapor, x the void fraction and the μ_v dynamic viscosity of the vapor.

5.2.4 Flow Boiling

In a two-phase flow, different flow patterns are observed depending on the volume fraction of the vapor and liquid phases. Their knowledge is of fundamental importance, since in most cases the various processes in two-phase flow can only be calculated more precisely if the distribution of the two phases is adequately understood. Flow boiling in tubes is perhaps the most complex convective process encountered in applications, since the heat transfer includes both a convective contribution and a nucleate boiling contribution, whose relative importance depends on the specific conditions. The local flow boiling heat transfer coefficient is depends on vapor quality, mass velocity, heat flux, duct geometry and orientation, the fluid properties and as a result of the preceding mentioned quantities certain characteristic flow patterns establish. These two-phase flow patterns have a particular identifying characteristic and thus flow pattern maps are used to predict the transition from one type of two-phase flow pattern to another. Thereby not only the flow pattern occurring at the particular local conditions under consideration is identified but also the heat transfer characteristics delineating the individual flow patterns are separated and accessible via correlations. Because the flow boiling coefficient is a function of vapor quality, these calculations are typically done locally in thermal design methods.

For a better understanding of two-phase flows, flow patterns are defined which are characterized by certain features. For an upward directed two-phase flow in a vertical pipe, the flow patterns illustrated in Fig. 5.42a can be observed and qualitatively, the individual patterns are described as follows:

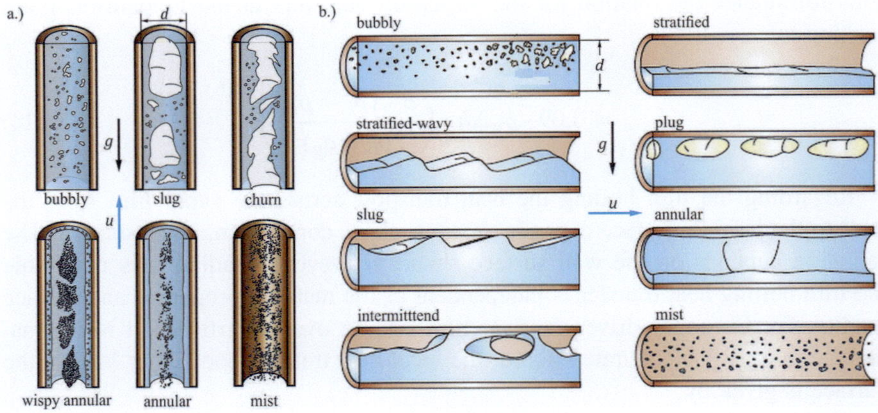

Fig. 5.42 Observed flow patterns in adiabatic tubes (increasing gas fraction from left to right, or from top to bottom, modified according to Thome and Cioncolini (2015)): **a** vertical tube, **b** horizontal (or low inclined) tube

- *Bubbly flow*: a large number of small bubbles being considerably smaller than the channel diameter d are dispersed in a continuous liquid flow. The bubbles vary in size and shape but are rather densely packed. Moreover, they interact with each other by collisions, coalescence, break-up. This flow form occurs at small gas volume fractions and is often observed in subcooled boiling before the fluid attains T_{sat}.
- *Slug flow*: The gas flows as plugs which fill almost the entire pipe cross-section and is separated by liquid slugs which may contain small bubbles. These bubbles exhibit an almost cylindrical shape with dimensions of the order of the tube diameter with a hemispherical nose and a blunt fluctuating tail and are often termed as *Taylor bubbles*.
- *Churn flow*: This transitional flow type occurs between stable slug flow to stable annular flow and features tails of the elongated bubbles beginning to break up and creating large voids in the liquid slugs, which in turn become unstable themselves. This flow type is to be avoided in technical applications since it is associated with pressure and flow rate oscillations. It is usually observed at larger pipe diameters and higher pressure.
- *Annular flow*: In annular flow, a liquid film flows on the inner wetted perimeter of the channel with a higher velocity of the gas (or vapor) core in the centre. The film gas interface is characterized by short wavelength waves, due to the shear stress by the velocity difference of liquid film and gas. Thereby a bidirectional transfer occurs, gas bubbles can be trapped by the film and vice versa liquid droplets are entrained in the gas. When the flow depicts entrained droplets traveling in periodic coherent structures as clouds or wisps of liquid in the central vapor core, some flow pattern maps refer to this as *wispy annular flow*, see Thome and Cioncolini (2015).

- *Mist flow*: In mist flow, the entire liquid is transported by the gas in the form of droplets. Mist flow is encountered only at high gas velocities and low liquid volume fractions if the annular film is thinned by the shear of the gas core on the interface until it becomes unstable. It is also observed in flow boiling if the critical heat flux is surpassed and thus appears often during evaporation under higher pressure.

In the flow of two-phase mixtures in a horizontal or only slightly inclined pipe, different flow patterns occur than in vertical pipes, since, due to gravity and density, the liquid flows predominantly in the lower half of the pipe and the gas for the most part in the upper half. The flow patterns in a horizontal pipe and in a vertical pipe differ from each other the more the influence of gravity is greater than that of frictional forces on the flow. Especially at low velocities, flow forms develop in the horizontal pipe that are not observed in the vertical pipe. The flow forms in the horizontal pipe can be classified as bubbly, stratified, slug, surge, annular, and splash flow. Figure 5.42b qualitatively outlines these flow patterns.

- Bubbly flow: Multiple small gas bubbles are fine dispersed in the continuous liquid-phase, depending on the liquid velocity, where at lower values an accumulation in the top of the duct occurs.
- Stratified flow: At low liquid and gas velocities, gas and liquid separate due to buoyancy. Since the shear stress cause by the velocity difference between the two phases is small it can be balanced by the surface tension and the interface is smooth. Typically, more than half of the perimeter of the tube is dry while the remainder is wet.
- Stratified wavy flow: If the gas velocity in the stratified flow increases the interfacial shear between the phases causes waves directed towards the top of the tube. A considerable fraction of the upper pipe part keeps dry.
- Intermittent flow. A further rise of the gas velocity, yields waves to wet the top perimeter, typically leaving behind a liquid film such that the entire perimeter of the tube is nearly always wet. The large amplitude waves are separated by smaller amplitude waves not wetting the top and entrained bubbles can be identified in the large waves, origination from being trapped by the rollover of the waves having been trapped by the rolling motion of the waves. „The liquid slugs separating such large, elongated bubbles could also be described as large periodic amplitude waves and hence gives the name of this regime": intermittent flow. Intermittent flow is sometimes subdivided into two other regimes: plug flow and slug flow, see Thome and Cioncolini (2015).
- Annular flow: Even higher gas flow rates lead to the formation of a continuous liquid annular ring like film around the perimeter of the tube, which at high gas flow rates has a uniform thickness. Rising the gas flow fraction further, the top of the tube with its thinner film becomes dries out first, so that the annular film covers only the lower tube perimeter.

- Mist flow: Finally, at very high gas velocities, the liquid is dispersed as small droplets in the almost continuous gas phase. From time to time small rivulets of liquid can be observed after the breakdown of the liquid film.

Flow patterns in non-adiabatic heated tubes

A typical example of complete evaporation of a liquid at not too high pressure in a vertical tube and the successive flow regions occurring in the process is shown in Fig. 5.43. Supercooled liquid enters the tube. When the liquid layer near the wall is sufficiently superheated as a result of the external heating, bubbles form on the wall which break off at a certain size and migrate into the flow core. This is initially subcooled, and the bubbles condense again (range of boiling in subcooled liquid). When the entire liquid has reached the boiling temperature T_{sat}, chains of small bubbles are entrained in the flow. As the number of bubbles increases, the bubbles combine, and larger bubbles are formed (so-called bubble flow). With further heat input, these can grow together to form long plugs (slug flow region) or form very large irregular bubbles, with many small bubbles also present (churn flow region). With increasing evaporation, the liquid walls between the plugs or large bubbles tear open and an annular flow develops. Droplets are torn out of the liquid film by the fast-flowing vapor and transported away from the vapor core. As a result of the evaporation, the liquid film becomes thinner and thinner. At very low film thickness, it begins to breakup and then disappears completely. A mist flow is present. When all the liquid droplets have evaporated, the vapor is superheated.

During complete condensation (heat removal from the tube), superheated or saturated steam enters the tube. The vapor condensation on the pipe wall forms a liquid film and an annular flow is created. As the film thickness increases, droplets are torn out of the film. With further condensation, slugs form on the film, which grow together and fill the entire cross-section of the tube. During the subsequent bubbly flow, the bubbles become smaller due to heat dissipation and finally disappear. If the vapor condenses in the horizontal tube, the wavy flow and stratified flow are the main phenomena that occur with larger liquid volume fractions.

In a simplified form, the transitions between the individual flow states in a horizontal pipe are shown in Fig. 5.44, which also indicates qualitatively. The heat transitions in the horizontal pipe are also shown. Correspondingly, the heat transitions are indicated qualitatively the distribution of the heat transfer coefficient along the flow path. The processes in the horizontal tube, as they occur for example in Fresnel collector or parabolic trough collector systems, differ from those in the vertical tube, as they typically occur in a low-temperature collector or a solar tower configuration.

Starting from pure single-phase convective heat transfer characterized by a heat transfer coefficient h, the heat transfer in the two-phase region increases significantly with the onset of boiling. The convective fraction also increases with the rising vapor content. In the domain of stratified and wavy flow, a considerable

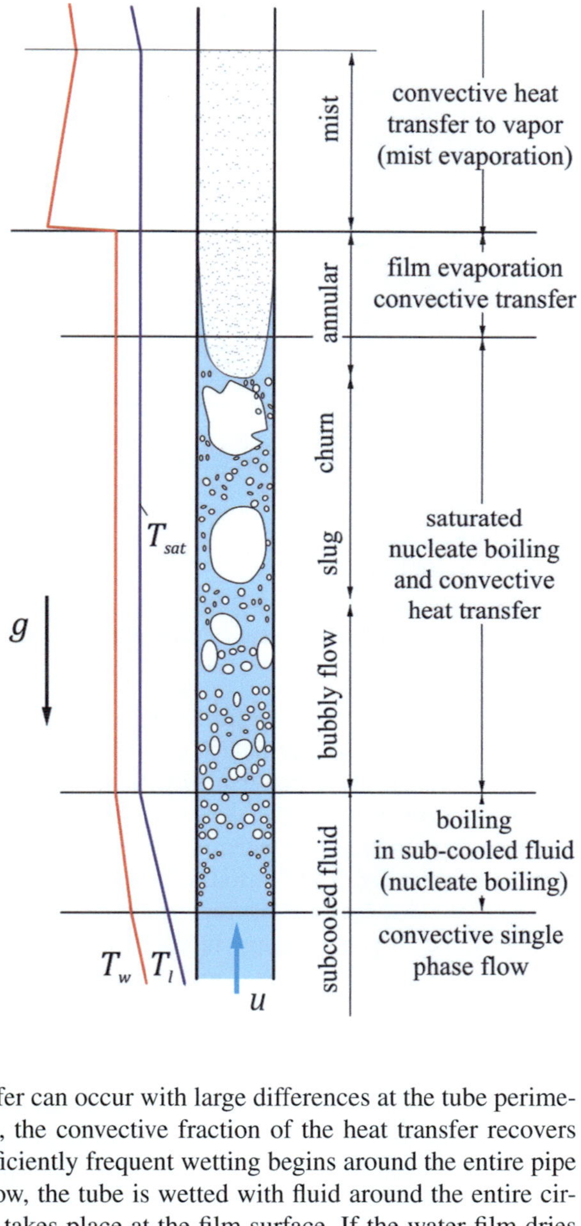

Fig. 5.43 Sketch of flow patterns, heat transfer regions and course of the wall temperature T_w and the fluid temperature T_l during evaporation in the vertical pipe

degradation of the heat transfer can occur with large differences at the tube perimeter. In slug or annular flows, the convective fraction of the heat transfer recovers again, since sufficient or sufficiently frequent wetting begins around the entire pipe circumference. In annular flow, the tube is wetted with fluid around the entire circumference and evaporation takes place at the film surface. If the water film dries out, the heat transfer changes to a state in which only steam exists. The transition from nucleate boiling dominated regime to evaporation dominated ones (slug, annular or intermittent), is termed a departure from nucleate boiling (DNB) or the CHF condition. This transition is associated with a decreasing total effective heat transfer performance causing a rapid rise of the wall temperatures. An uncontrolled

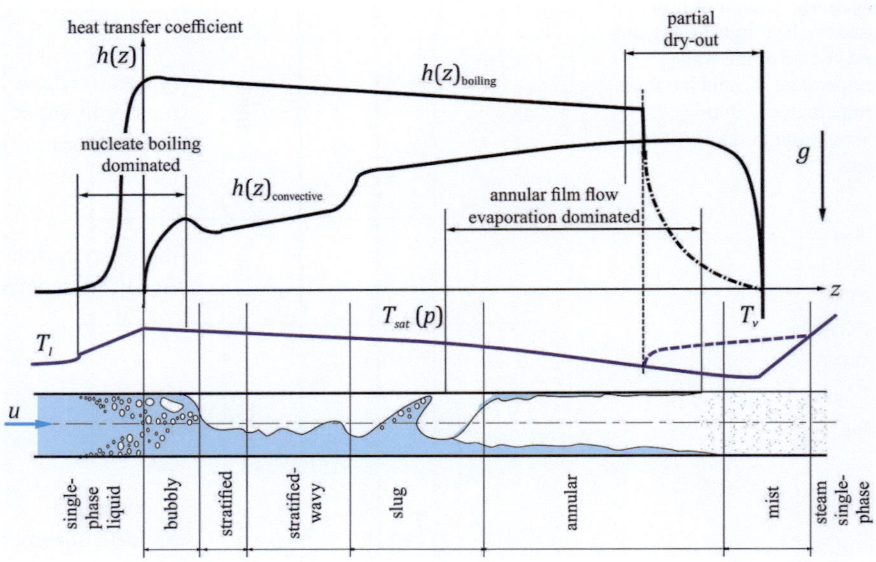

Fig. 5.44 Flow patterns and heat transfer regions with indication of the qualitative course of the fluid temperature T_l and the heat transfer coefficient (h) during evaporation in a horizontal pipe

operation in such a domain has to be prevented because the equipment is not usually designed to operate under such conditions. Another option as e.g. followed in direct evaporation linear power plants is the integration of steam dryers within a recirculation concept.

The corresponding flow patterns are of great importance for heat transfer and pressure drop, as Fig. 5.44 illustrates. In order to be able to predict their occurrence and the associated pressure loss and heat transfer, they are represented in the form of so-called flow pattern maps, in which the occurrence of the flow patterns is indicated as a function of flow variables and material parameters.

5.2.4.1 Definitions and Characteristic Numbers in Two-Phase Flows

Two-phase flows of gas and liquid are characterized by a temporally and spatially variable phase boundary and interaction forces between both phases. This results in difficult analytical accessibility for both momentum and heat transport. Reliable analytical models require a detailed investigation of the individual subprocesses that occur in two-phase flow and often depend on each other in many ways. Due to the increase in computing power in recent years, it is now possible to use numerical computational programs to solve the analytical models, which consist of complex systems of equations, some of which are multiply indeterminate. For a macroscopic description of the processes in the two-phase region, simplified part-empirical models are mostly still used today and state of the art in all system thermal–hydraulic codes. This requires the definition of the variables essential to characterize a two-phase flow. The most important of the parameters are:

- $\dot{M}_g$, $\dot{M}_l$ mass flow of the gas and liquid,
- ρ_g, ρ_l density of gas and liquid,
- μ_g, μ_l dynamic viscosity of gas and liquid,
- σ surface tension,
- d pipe diameter and β the inclination angle versus the horizontal plane and
- g the gravitational acceleration.

In addition, there are inlet effects, influences of the channel geometry, nature of the channel walls and many more. In the following derivations, the subscript "g" denotes the gas/vapor phase and "l" (liquid) the liquid phase, u_m is the mean velocity and ν the kinematic viscosity. As for single-phase flows, the main variables can be combined to form a set of independent, dimensionless numbers, which can be conceived force ratios, such as the Reynolds number Re, the Froude number Fr and the Weber number Wb, which are defined as follows:

$$Re = \frac{u_m \cdot d}{\nu}; \; Fr = \frac{u_m^2 \cdot d}{g \cdot d}; \; Wb = \frac{u_m^2 \cdot \rho_l \cdot d}{\sigma}, \tag{5.226}$$

Here, the Froude number can be understood as the ratio of inertial forces versus gravity, and the Weber number as the ratio of inertial forces versus surface tension forces. Ratios of thermo-physical properties, such as the density ratio of the two phases ρ and the viscosity ratio E of the two phases, form a further class of dimensionless ratios:

$$\rho = \frac{\rho_g}{\rho_l}; \; E = \frac{\mu_g}{\mu_l}. \tag{5.227}$$

Last, the fraction of the total mass flux $\dot{x}$ often termed as flow quality and the mean velocity u_m are defined by

$$\dot{x} = \frac{\dot{M}_g}{\dot{M}_l + \dot{M}_g} \quad \text{and} \quad u_m = \frac{\dot{M}_g}{\rho_g \cdot A_C} + \frac{\dot{M}_l}{\rho_l \cdot A_C}, \tag{5.228}$$

in which A_C represents the total flow cross-section intersected by both phases.

The formation of a kinematic viscosity from the values ν_g and ν_l of the two phases is difficult and will be discussed later. In the same way, another set of dimensionless combinations could be formed from the ten influencing variables mentioned. However, the number of variables is always reduced to seven according to Buckingham's π-theorem (cf. Sect. 5.1.3). If heat transfer processes are also considered, four further influencing variables are added. These are the specific heat capacities and the heat conduction coefficients of the two phases, from which the Prandtl numbers for the vapor and the liquid result as two further dimensionless parameters.

Additional definitions are required for further consideration of the models and the flow form maps, for which the following, all time-related quantities are indicated by a dot as superscript. In addition, specific quantities, i.e. quantities related to area or mass, are written in lower case. The balance equations known from the single-phase flow can be formed for the two-phase flow with the quantities of the individual phases as well as with average quantities of the two phases. The total flow cross-section A_C is occupied partly by the gas and partly by the liquid.

$$A_C = A_g + A_l. \tag{5.229}$$

The total mass flux or volume flux of the two-phase flow is the sum of the fluxes of the individual phases:

$$\dot{M} = \dot{M}_g + \dot{M}_l \quad \text{and} \quad \dot{V} = \dot{V}_g + \dot{V}_l. \tag{5.230}$$

The proportion of the gas mass flux is given in Eq. (5.228). Accordingly, the following applies to the share of the vapor volume flow $\dot{\alpha}$ in the total volume flow:

$$\dot{\alpha} = \frac{\dot{V}_g}{\dot{V}_g + \dot{V}_l}. \tag{5.231}$$

For the liquid phase, the fractions according to Eq. (5.230) are thus $(1 - \dot{x})$ and $(1 - \dot{\alpha})$. Analogous to Eqs. (5.230) and (5.231), the volumes and masses V, M and volume (α) or mass (x) specific vapor fractions are determined:

$$M = M_g + M_l;\ V = V_g + V_l;\ \alpha = \frac{V_g}{V_g + V} \text{ and } x = \frac{M_g}{M_g + M_l}. \tag{5.232}$$

There is a functional dependence between the quantities of phase velocities of gas and liquid u_g, u_l, their mass flows $\dot{M}_g$, $\dot{M}_l$ and densities ρ_g, ρ_l. For two-phase flows, gas–liquid, the dependence can be given in the form

$$\frac{\dot{M}_g}{\dot{M}_l} = \frac{\rho_g}{\rho_l} \cdot S \cdot \frac{\alpha}{(1-\alpha)}, \tag{5.233}$$

where for any kind of averaging and for reasons of mass conservation $\dot{x}_l = 1 - \dot{x}$, and $\alpha_l = (1 - \alpha)$ holds here. In particular, it is clear from this relationship that the volume fraction alpha depends quite generally on the velocity ratio S, the density ratio and the vapor content of a two-phase flow. In general, the values $\dot{x}$ and x or $\dot{\alpha}$ and α differ because the two phases move at different velocities in the channel. To illustrate this, imagine the limiting case where vapour bubbles rise upwards through a column of liquid at rest. There $\dot{x} = 1$ or $\dot{\alpha} = 1$ but $x < 1$ or $\alpha < 1$. The average density in a volume element is given by:

$$\rho_m = \frac{M}{V} = \alpha \cdot \rho_g + (1-\alpha) \cdot \rho_l. \tag{5.234}$$

It is determined from the respective density of the individual phases multiplied by their volume fractions, and the average flow density can thus be determined.

$$\dot{\rho}_m = \frac{\dot{M}}{\dot{V}} = \dot{\alpha} \cdot \rho_g + (1 - \dot{\alpha}) \cdot \rho_l. \tag{5.235}$$

The respective mean axial velocity u_j of phase j can be calculated from the volume flux $\dot{V}$ of the phase j considered and its occupied cross-sectional area A_j via:

$$u_v = \frac{\dot{V}_g}{A_g} \text{ and } u_l = \frac{\dot{V}_l}{A_l}. \tag{5.236}$$

The average velocity of the two-phase mixture u_m in the flow channel is thus obtained as follows

$$u_m = \frac{\dot{V}}{A_C} = \dot{\alpha} \cdot u_g + (1 - \dot{\alpha}) \cdot u_l. \tag{5.237}$$

If the mass flux or specific volume flux density of the mixture is related to the cross-sectional area, the corresponding density of mass flux $\dot{m}$ or volume flux is obtained:

$$\dot{m} = \frac{\dot{M}}{A_C} = (u_m \cdot \rho_m) \text{ and } \dot{v} = \frac{\dot{V}}{A_C} = u_m. \tag{5.238}$$

When forming the ratios of two-phase flow, a velocity v_j is often calculated, which would be obtained if phase j flows exclusively in the duct. This velocity v_j is termed superficial velocity.

$$v_g = \frac{\dot{V}_g}{A_C} = u_v \cdot \alpha \text{ and } v_l = \frac{\dot{V}_l}{A_C} = u_l \cdot (1 - \alpha). \tag{5.239}$$

Because two-phase flows are of particular importance in applications, additional terms for velocities have been introduced to describe the transport processes. Drift velocity $u_{j,u}$ is the deviation of the actual phase velocity u_j from the total related phase velocity $(u = u_g + u_l)$ and these drift velocities are linked in a simple way to the relative velocity $(u_g - u_l)$:

$$\begin{gathered} u_{g,u} = (1 - \alpha)\left(u_g - u_l\right) \text{ and } u_{l,u} = -\alpha\left(u_g - u_l\right) \\ \text{with } u_{g,u} = \left(u_g - u\right) \text{ and } u_{l,u} = (u_l - u). \end{gathered} \tag{5.240}$$

Another characteristic parameter for two-phase flows was introduced by Lockhart and Martinelli (1949). It is defined as the ratio of the frictional pressure losses for the case where gas and liquid each flow through the pipe alone. If $(dp/dx)_l$ and $(dp/dx)_g$ denote respectively the pressure loss of the liquid and gas volume flows averaged over the total cross-section, then the Martinelli parameter X^2 is as follows

$$X^2 = (dp/dx)_l/(dp/dx)_g. \tag{5.241}$$

This parameter is thus, among other things, a measure of the volume fraction of the flow, because for $X^2 \gg 1$ the two-phase flow consists predominantly of liquid while for $X^2 \ll 1$ predominantly gas occur.

The most essential parameters expressed by the j-th phase are the mass flow rate ($\dot{m}_j$), the velocity (u_j) of the phase and the volume fraction (α_j) of the phase, while subsequently $G_j = m_j/A_C$ is used as mass flux of the j-th phase. All quantities are interrelated via the following relations:

$$v_j = \frac{G_j}{\rho_j}; u_j = \frac{v_j}{\alpha_j}; x_j = \frac{\dot{m}_j}{\sum_j^n m_j} = \frac{G_j}{\sum_j^n G_j}; \alpha_j = \frac{\left(x_j/\left(\rho_j u_j\right)\right)}{\sum_j^n\left(x_j/\left(\rho_j u_j\right)\right)}. \tag{5.242}$$

In most engineering calculations, the above parameters are defined as average quantities across the whole flow area A_C. It should be noted, however, that details of the multiphase flow may incorporate local variations across the flow area. Then, G_j, v_j, and α_j are often defined on a local basis, varying with transverse position across the flow area.

5.2.4.2 Flow Maps

Flow maps have been developed with the help of experimental observations to delineate the different flow forms from each other. A considerably simplified representation of the states can be made for a given gas–liquid mixture, the most common being here experimentally studied the mixture of air/water, by means of the two control parameters related liquid and gas velocity $U_l = (1-\alpha)u_l$ and $U_g = \alpha\, u_g$. Such a map was constructed from a large experimental database first by Mandhane et al. (1974) and it depicts the individual flow patterns observed as a function of liquid and gas volume flows for a horizontal/vertical pipe configuration, as e.g. illustrated in Fig. 5.45.

The boundaries between flow patterns are not sharp and the experimental observations show hysteresis depending on the transition type. In addition, for vertical pipe flows, the pipe length (l) and the shape of the flow inlet play a role. It significantly influences the transition between slug flow and dispersed/annular flow. By linearly stretching the referenced velocities for air–water mixtures with property parameters that consider density and surface tension, Mandhane et al. (1974) were able to generalize their flow maps to other gas–liquid mixtures. Since the more detailed investigation of the two-phase flow, therefore, a large number of different maps for the prediction of the flow pattern have been developed. Essentially, flow regimes are assigned to different flow patterns in a two-dimensional way. One flow pattern map used in petrochemistry and applied to different pairs of substances is the flow pattern map outlined in Fig. 5.46, which is based on the derivation according to Baker (1954). It is based on extensive experiments with air/water mixtures

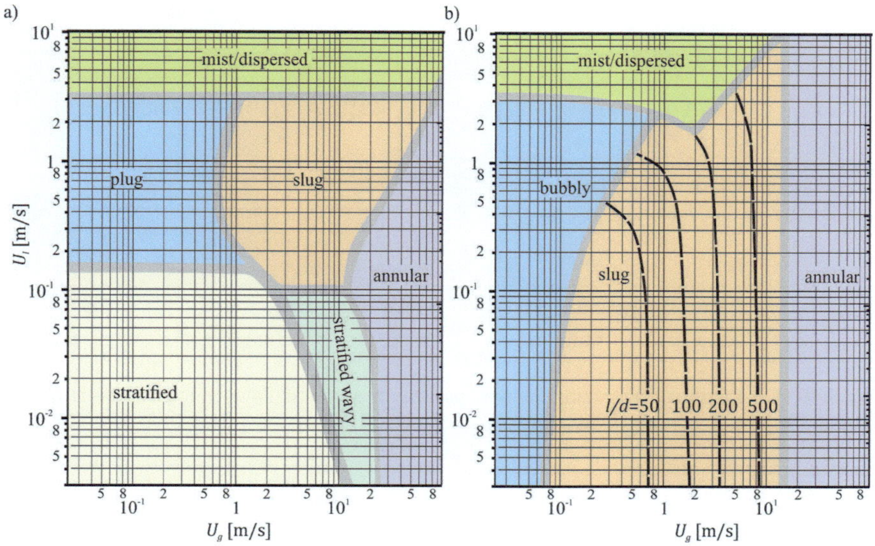

Fig. 5.45 Flow map after Mandhane et al. (1974) for air–water flows at 0.1 MPa for a horizontal tube (**a**, $d = 25$ mm) and a vertical tube (**b**, $d = 51$ mm). l/D denotes the ratio of pipe length to pipe diameter, modified according to Müller (2016)

at in pipes of diameter $25 < d < 100$ mm. For application to other gas/liquid systems and other pressure ranges the gas or liquid mass flow densities used as parameters are extended. Ratios are formed from the material properties of the fluid pair at hand with those of air and water, respectively, at atmospheric pressure and 20 °C:

$$\lambda_b = \left[\frac{\rho_g}{\rho_{Luft}} \cdot \frac{\rho_l}{\rho_{H2O}}\right]^2 \psi_b = \left(\frac{\sigma_{H2O}}{\sigma_l}\right) \cdot \left[\frac{\mu_l}{\mu_{H2O}} \cdot \left(\frac{\rho_{H2O}}{\rho_l}\right)^2\right]^{\frac{1}{3}} \tag{5.243}$$

With the flow maps, one usually obtains satisfactory results for rough calculations. At higher pressures, however, larger deviations result. Various researchers tried to improve the Baker diagram and to transfer it to a non-adiabatic flow, but so far, no satisfactory solution has been achieved. By this type of representation, one is limited to two independent variables associated with the x-axis and the y-axis. Most often, the superficial velocities of the fluid and gas, or generalized parameters containing them, are chosen for these. Hewitt and Roberts (1969) use as axis parameters the momentum current densities $(\rho_j u_j^2)$ of the two phases. However, this map can only serve as a rough estimate of the flow shape, since the plot versus the momentum current density is not adequate to reflect the influence of the physical fluid properties and the channel diameter.

Therefore, the idea is to study the transitions from one flow form to another, and to describe them for a wide range of applications if possible. In this case,

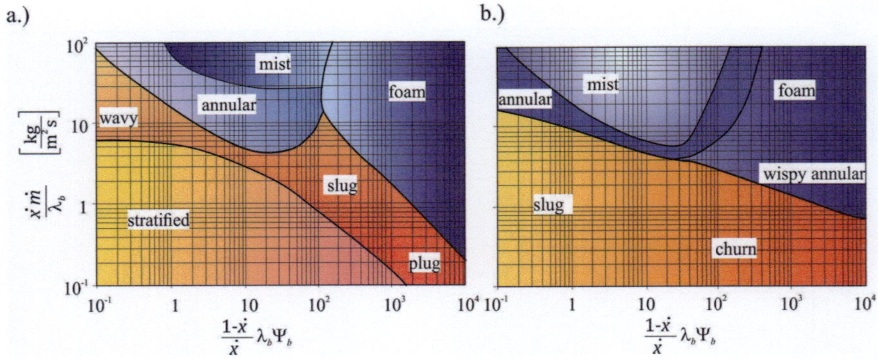

Fig. 5.46 Diagram for determining the flow pattern for adiabatic flows in the horizontal (**a**) and vertical (**b**) pipes according to Baker (1954)

the shape of the transition lines is described as a function of other important flow parameters such as the liquid and vapor mass flow densities, the fluid properties, the duct diameter, and the inclination of the duct. A flow shape map can then be generated for a set of these quantities. Taitel and Duckler (1976) developed such a model for predicting the flow shapes, which was used early with some modifications in programs for safety calculations of nuclear power plants. Here, the transition lines of the flow forms were extended by transition areas, as shown in Fig. 5.47 by using hatched areas. There exist in this formulation dependencies to the homogeneous velocity u_h, the gas volume fractions α and the steam superheat $(T_g - T_{sat})$. Furthermore, a distinction is made between stratified and unstratified flow (depending on the homogeneous velocity). Limits are the Taylor bubble velocity u_{Tb} and a minimum gas volume fraction in the considered node as well as in the nodes below and above it. If unstratified flow is present, this model distinguishes between regions of different heat flux densities (or temperature difference $\left(T_g - T_{sat}\right)$). Below the critical heat flux density (Pre-CHF), the flow patterns shown in Fig. 5.43 occur. Above the film evaporation limit (post-dry out), flow reversal can occur (e.g., inverted annular, inverted slug). The transition functions to the different flow forms are described in detail, for example, in the program manual (RELAP Code Development Team, 1995) or similar manuals of code packages. Decisive for the user of such simulation programs is the examination whether and to what extent the implemented models and in particular their limitations satisfy the respective application.

Since two-phase flows in particular play a role in solar farm concepts (parabolic trough or Fresnel power plants), this area is analysed again in more detail here. Three flow families can be observed in horizontal and slightly inclined pipes,

- stratified flows in smooth or wavy form,
- intermittent flows in the form of surge and plug flows, and
- dispersive flows in the form of bubble or ring-droplet flow.

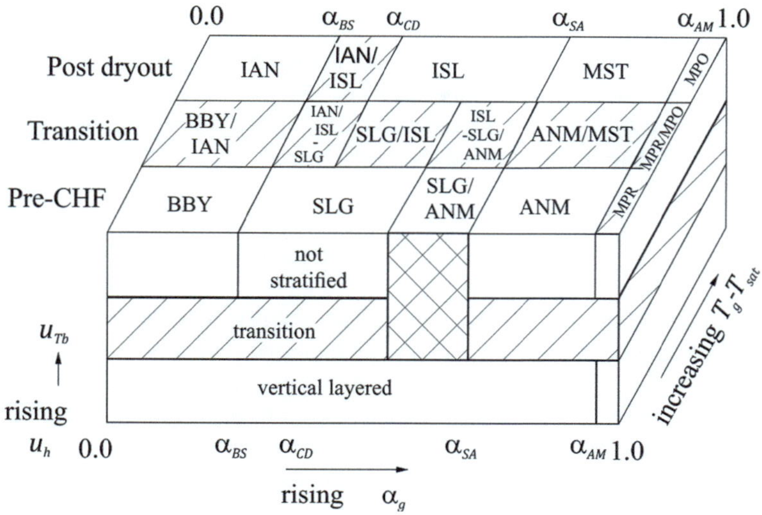

Fig. 5.47 Schematic of the 3D flow pattern map for a vertical pipe from the RE-LAP simulation program (RELAP Code Development Team, 1995) Hatched areas indicate transition regions. IAN-inverted annular flow, ISL = inverted plug flow, BBY = bubbly, SLG = slug, ANM = annular mist flow, MST = mist flow, MPR = mist flow pre-CHF, MPO = mist flow post-CHF

A condition for the transition from stratified to intermittent flow can be derived using an instability condition for a soliton wave. The boundary curve is given by a modified Froude number Fr:

$$Fr = \sqrt{\frac{\rho_g}{\rho_l - \rho_g}} \frac{U_g}{\sqrt{d\, g \cos\beta}}, \tag{5.244}$$

as a function of Martinelli parameter X (Eq. (5.241)). Here, β is the inclination angle of the tube, g is the gravity and d is the tube diameter. The transition from smooth to wavy stratified flow is determined by the Kelvin–Helmholtz instability condition

$$K = \sqrt{\frac{\rho_g}{\rho_l - \rho_g}} \frac{U_g}{\sqrt{dg \cos\beta}} \sqrt{\frac{U_l d}{\nu_l}} \geq 20\alpha\sqrt{(1-\alpha)}, \tag{5.245}$$

in which ν_l denotes the kinematic viscosity of the liquid phase. For the volume fractions, a clear relationship to the Martinelli parameter can be established, so that a curve in the form $K(X)$ can be formulated.

The transition from annular flow to intermittent flow in pipes which are not too steeply inclined is determined by the minimum possible liquid fraction in a slug interspersed with gas bubbles. Intermittent slug flow sets in according to experimental observations for $(1-\alpha) \geq 0.24$, which corresponds to a value of the

Martinelli parameter of $X \approx 1.6$. The state of dispersed bubble flow and intermittent flow is caused by turbulent agitation, by gravity, and by bubble breakup and coalescence. The turbulence intensity in the liquid phase is largely determined by the pressure drop in the liquid phase. For this purpose, Taitel and Duckler (1976) introduce the ratio of related pressure drop of the liquid phase and hydrostatic buoyancy of the bubbles as a parameter in the form:

$$T^2 = \left|\frac{dp}{dx}\right|_l \frac{1}{(\rho_l - \rho_g) g \cos\beta}. \tag{5.246}$$

The transition between the two regions can then be represented as a function T of the Martinelli parameter X. After evaluating the functional relationships, the flow map has the form shown in Fig. 5.48, where each of the boundary curves $K(X)$, $F(X)$, and $T(X)$ is individually assigned to an ordinate. Later this theory of flow pattern boundaries has been generalized such, that two-phase flows in pipes with any angle of inclination can be classified. The transition conditions are either represented graphically or can be determined point by point with the aid of a computer program. The various state variables and variables of the two-phase flow as well as their derivatives, such as the pressure gradient, the volume fraction, and the heat transfer coefficient, depend to a large extent on the flow form. Therefore, in many cases of one-dimensional and quasi-stationary processes, it is aimed to develop calculation methods for two-phase flows individually for the characteristic flow patterns, which is complex and extensive. The different calculation models are then to be linked with each other via flow maps or computer-aided transition conditions in order to determine the two-phase states sufficiently accurately for technical systems, for example for steam generators. The range-based calculation of two-phase flows has not been satisfactorily solved to date. Therefore, correlations derived on the basis of extensive experimental data are currently still used for certain technically relevant quantities such as pressure drop and heat flows. These two-phase laws are addressed in the following sections.

5.2.4.3 One-Dimensional Two-Fluid Model

The formal derivation of the one-dimensional two-fluid model is carried out in a simplified heuristic manner with the help of the notion of a layered flow in the pipe, as illustrated in Fig. 5.49. When deriving the one-dimensional two-fluid equations from the general basic equations for multiphase flows, some fundamental assumptions are made in connection with the spatial, in this case cross-sectional, and temporal averaging. In the case of spatial or temporal averaging of a product of state variables f and g, the following generally applies $\langle f \cdot g\rangle = c\langle f\rangle \cdot \langle g\rangle$. In the simplest representation of the one-dimensional two-fluid model, $c = 1$ is set.

The thermodynamic equations of state and the constitutive equations for local quantities also apply to the averaged quantities. The heat conduction and the change of shear stresses in the flow direction as well as the dissipation power of the frictional forces in the respective phases are neglected. Local thermodynamic equilibrium is assumed in each phase. Against each other, the phases can

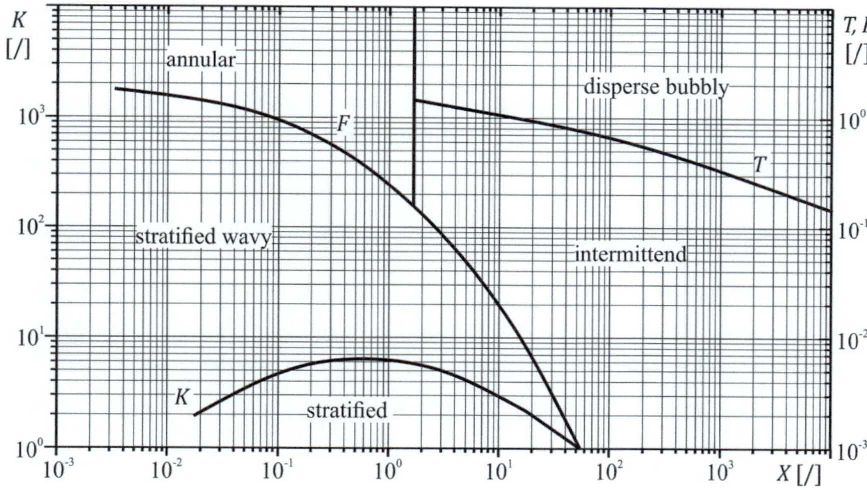

Fig. 5.48 Flow map for horizontal and slightly inclined pipes showing K, F and T as a function of the Martinelli parameter X according to Taitel and Duckler (1976)

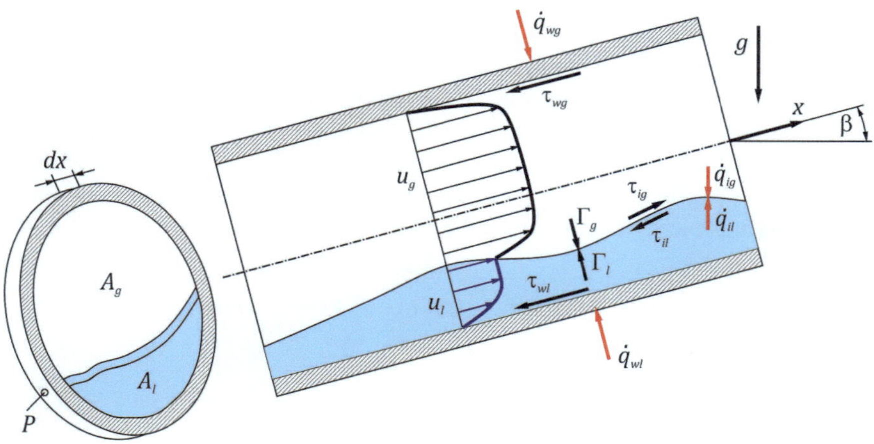

Fig. 5.49 Control volume for a separate two-phase treatment flow with designations

be in thermodynamic non-equilibrium. In vertical two-phase flow, the pressure is assumed to be constant over the cross-section of the pipe, which in many applications is also justified with good approximation for flows in horizontal pipes. The effect of interfacial tensions can be neglected for the time being. However, it reappears in a general formulation or when introducing constitutive equations to close the model equations.

To represent the one-dimensional conservation equations, the following terms are used for the averaged quantities.

- Geometry: cross-sectional area $A_g + A_l = A_C$, perimeter $P_g + P_l = P_C$, wetted perimeter of gas, liquid at position i P_{gi}, P_{li}, P_i, cross-sectional fraction of phase $\varepsilon_g = \alpha = A_g/A_c, \varepsilon_l = A_l/A_c, \alpha + \varepsilon_l = 1$.
- State variables: velocities u_g, u_l, pressures and densities p_g, p_l, ρ_g, ρ_l, mass flux densities $\dot{m}_g, \dot{m}_l$, specific inner energy and enthalpy e_g, e_l, h_g, h_l.
- Constitutive variables: wall shear stress τ_{wg}, τ_{wl}, wall heat flux $\dot{q}_{wg}, \dot{q}_{wl}$, interfacial shear stress at position i τ_{ig}, τ_{il}; interfacial heat flux densities $\dot{q}_{ig}, \dot{q}_{il}$, mass source densities at the phase interface Γ_g, Γ_l, velocity and pressure at interface at position i u_i, p_i.
- Transfer at phase transition: momentum source term due to mass exchange at interface area $M_g^{\Gamma_g}$, $M_g^{\Gamma_l}$, power source density due to wall shear stress at interfaces and walls $L_{\tau g}, L_{\tau l}$, power source density due to mass exchange at the interface $L_g^{\Gamma_g}$, $L_l^{\Gamma_l}$, power source density due to gravity and heat supply Lg, L_q.

Due to local equilibrium, the following relationships hold between the constitutive variables at the interface:

$$\Gamma_g - \Gamma_l = 0;\ \tau_{ig} - \tau_{il} = 0;\ \dot{q}_{ig} - \dot{q}_{il} = 0;\ M_g^{\Gamma_g} - M_g^{\Gamma_g} = 0;\ L_g^{\Gamma_g} - L_l^{\Gamma_l} = 0. \tag{5.247}$$

Those parts of the quantities that are based on mass transfer sum to zero, but if surface tensions play a role, other terms appear in the momentum and energy conservation equations for the respective phases that take surface tensions into account. For simplicity, surface tension effects will not be considered here. Under these simplified assumptions, the one-dimensional conservation of mass, momentum and energy for the two-fluid model are as follows:

$$\frac{\partial}{\partial t}(\rho_g \alpha A) + \frac{\partial}{\partial x}(\dot{m}_g \alpha A) = 0;\ \frac{\partial}{\partial t}(\rho_l (1-\alpha) A) + \frac{\partial}{\partial x}(\dot{m}_l (1-\alpha) A) = 0. \tag{5.248}$$

$$\begin{aligned}
&\frac{\partial}{\partial t}(\dot{m}_g \alpha A) + \frac{\partial}{\partial x}(\dot{m}_g u_g \alpha A) = -\alpha A \frac{\partial p_g}{\partial z} - \tau_{wg} P_g - \tau_{ig} P_i - \alpha A \rho_g g \sin\beta + M_g^{\Gamma_g};\\
&\frac{\partial}{\partial t}(\dot{m}_l (1-\alpha) A) + \frac{\partial}{\partial x}(\dot{m}_l (1-\alpha) u_l A)\\
&\quad = -(1-\alpha) A \frac{\partial p_l}{\partial z} - \tau_{wl} P_l - \tau_{il} P_i - (1-\alpha) A \rho_l g \sin\beta + M_l^{\Gamma_l};
\end{aligned} \tag{5.249}$$

$$\begin{aligned}
&\frac{\partial}{\partial t}(\rho_g E_g \alpha A) + \frac{\partial}{\partial x}\left(\dot{m}_g \left(h_g + \frac{u_g^2}{2}\right) \alpha A\right) = L_{\tau g} + L_{gg} + L_g^{\Gamma_g} + L_{qg};\\
&\frac{\partial}{\partial t}(\rho_l E_l (1-\alpha) A) + \frac{\partial}{\partial x}\left(\dot{m}_l \left(h_l + \frac{u_l^2}{2}\right)(1-\alpha) A\right) = L_{\tau l} + L_{gl} + L_l^{\Gamma_l} + L_{ql}
\end{aligned} \tag{5.250}$$

Herein $E_j = e_j + u_j^2/2$ represents the energy density, h_j is the specific enthalpy of the respective phase, and on the right-hand side of the equation are listed, without further specification, the power components of shear stress, gravity, momentum exchange by mass transfer between phases, and heat fluxes. Here one can also introduce the assumption $p_l = p_g = p$. For the simplest case of an incompressible flow, the equations contain the six state variables u_g, u_l, e_g, e_l, p and α. In addition, there are a large number of constitutive variables, which can be reduced to the following independent variables Γ, $(\tau_{wg} \cdot P_g)$, $(\tau_{wl} \cdot P_l)$, $(\tau_{ig} \cdot P_{wi})$, $(\dot{q}_{wg} \cdot P_g)$, $(\dot{q}_{wl} \cdot P_l)$, $(\dot{q}_{il} \cdot P_{wl})$ when the momentum and energy equations are derived in detail. For the constitutive variables, correlations to the state variables must be established on the basis of experimental findings or theoretical considerations in order to close the conservation equations and apply them to the solution of two-phase flow problems. In each case, the model correlations for the constitutive variables are to be developed for specific flow domains. Such correlations have been given in general form by Ishii and Hibiki (2011).

5.2.4.4 Homogeneous Model of Two-Phase Flow

In the homogeneous model, it is assumed that a homogeneous fluid flows in the flow channel, as indicated in Fig. 5.50. Average values are formed from the properties of the two phases. These are assumed to be constant over the entire flow cross-section.

The two-fluid model can be simplified if, for technical reasons, one is only interested in the integral states of the two-phase flow such as total mass flow, total pressure drop and total heat transfer. By adding two conservation equations for each of the individual phases, one then obtains three conservation equations for the total mass, total momentum, and total energy of the two-phase mixture. These equations can be represented in a form corresponding to the form of the basic one-dimensional fluid mechanics equations for compressible media if the densities, wall shear stress, and wall heat flux are introduced into the mixture equations as weighted quantities. Using the general definitions and with simple algebraic transformations, one obtains for the two-phase mixture the following

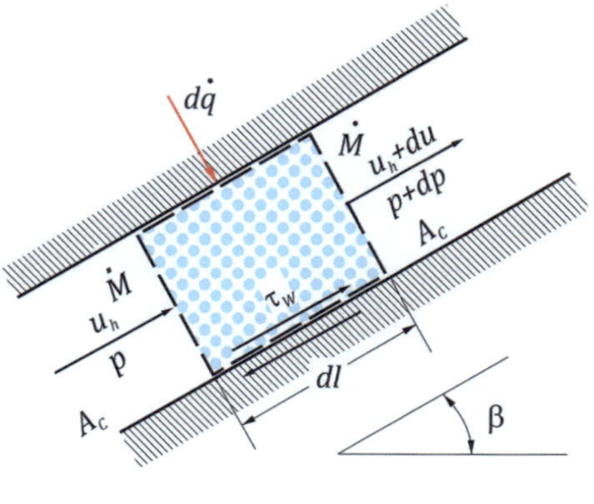

Fig. 5.50 Control volume of a non-adiabatic two-phase flow in the inclined channel assuming the homogeneous two-phase flow model

three conservation equations:

$$\frac{\partial \rho_h}{\partial t} + \frac{1}{A_c}\frac{\partial \dot{M}}{\partial x} = 0 \text{ with } \rho_h = \alpha\rho_g + (1-\alpha)\rho_l;$$
$$\frac{\partial \dot{M}}{\partial t} + \frac{\partial}{\partial x}\left(\frac{1}{\rho_M}\cdot\frac{\dot{M}^2}{A_C}\right) = -A_C\frac{\partial p}{\partial x} - \tau_w \cdot P - A_C\rho_h g \sin\beta;$$
$$\text{with } \dot{x}_g = \dot{x};\ \frac{1}{\rho_M} = \frac{\dot{x}^2}{\alpha\rho_g} + \frac{(1-\dot{x})}{(1-\alpha)\rho_l};\ \tau_w = \tau_{wg}P_g + \tau_{wl}P_l;$$
$$\frac{\partial E}{\partial t} + \frac{1}{A_C}\frac{\partial}{\partial x}\left(\dot{M}\left(h + \frac{1}{\rho_E^2}\frac{\dot{M}^2}{2A_C}\right)\right) = L_{\tau w} + L_g + L_{qw};$$
$$E = \rho_g E_g + \rho_l E_l;\ h = \dot{x}h_g + (1-\dot{x})h_l;\ \frac{1}{\rho_E^2} = \frac{\dot{x}^3}{\alpha^2\rho_g^2} + \frac{(1-\dot{x})^3}{(1-\alpha)^2\rho_l^2}. \quad (5.251)$$

It can be seen that the densities of the individual phases are weighted differently in the various conservation equations, so in addition to the homogeneous density of the mixed phase ρ_h, the momentum density ρ_M and the energy density ρ_E. They have different effects compared to single-phase flows. Although the model for the mixture flow has simplified by reducing the equations, the new problem now arises that the volume fraction α must be correlated with the vapor content $\dot{x}$ by an empirical relationship in order to use the simplified model for problem solving. In general, α is related to $\dot{x}$ by the velocity ratio $S = u_g/u_l$ of the phases (see Eq. (5.233)). Therefore, empirical links are sometimes also determined for the velocity ratio S and the relationship is inserted into the mixture equations.

By their nature, the mixture equations are suitable for application calculations when dispersed flow forms such as bubbly or droplet flows are present. They can be simplified even further if mechanical equilibrium between the phases is assumed, i.e. if the disperse phase has the same velocity as the homogeneous phase in each case. The mixture model simplified in this way is also called a "*homogeneous flow model*". It has special significance because of its simplicity and can be reliably applied to disperse flows with very small volume fractions of the disperse phase. With the assumption $S = 1$ there is an unambiguous relation between α and $\dot{x}$ according to Eq. (5.233):

$$\alpha_h = \frac{1}{1 + \frac{1-\dot{x}}{\dot{x}}\cdot\frac{\rho_g}{\rho_l}}, \quad (5.252)$$

where is the α_h is the homogeneous void fraction. With this relation, all other density definitions for momentum ρ_M and energy ρ_E reduce to the expression $\rho_h = \alpha\rho_g + (1-\alpha)\rho_l$ in the mixture balance, i.e., $\rho_h = \rho_M = \rho_E$.

There are a number of averaging methods for the homogeneous dynamic toughness μ_h. The following relationships for calculating the viscosity of the homogeneous fluid are described (Collier & Thome, 1996)

$$\frac{1}{\mu_h} = \frac{\dot{x}}{\mu_g} + \frac{(1-\dot{x})}{\mu_l}; \; \mu_h = \dot{x}\mu_g + (1-\dot{x})\mu_l; \; \mu_h = \rho_h\left[\dot{x}u_g\mu_g + (1-\dot{x})u_l\mu_l\right]. \tag{5.253}$$

These averaging calculations can lead to very different results. Since the calculation of mean viscosities already encounters great difficulties for liquid mixtures, the formation of mean values for gas/liquid mixtures is very questionable. Thus, even results for pressure drop calculated with different definitions for μ_h show sometimes considerable deviations from each other. The homogeneous model is the simplest of all two-phase flow models. It can certainly be extended to thermally non-equilibrium disperse flows with evaporation and condensation processes. Mechanical equilibrium is not possible for dispersive two-phase flows with larger of the disperse phase is generally not given. In order to account for this fact and still use the idea of a good mixing of the phases for simplification the drift flow model has been developed, which is described next.

5.2.4.5 Drift Model

The drift flow model is based on the basic idea that both phases are well mixed but move relative to each other and generally have different thermodynamic states. The scope of the model is therefore primarily in dispersed flows, i.e., bubbly, foam, and droplet flows. However, there are also attempts to extend the model to plug and ring flows. The model is based on a density-weighted mixture velocity um

$$u_m = \frac{\alpha\,\rho_g u_g + (1-\alpha)\rho_l u_l}{\alpha\,\rho_g + (1-\alpha)\rho_l}. \tag{5.254}$$

and considers the relative velocities by so-called drift velocities, which are first introduced locally as follows:

$$u_{g,u} = u_g - U; \, u_{l,u} = u_l - U, \tag{5.255}$$

where U is the local total volumetric flow density, represented by the related local velocities $U = U_g + U_l$. From these definitions, the mean values of volume fraction alpha, related total volume flow U and a mean drift velocity yet to be defined can be linked by cross-sectional averaging. This is done by multiplying the relation Eq. (5.255) with the local volume fraction α and a cross-sectional averaging. After algebraic transformations one first obtains the relation:

$$\langle\alpha\, u_{g,u}\rangle = \langle\alpha\, u_g\rangle - C\,\langle\alpha\rangle U, \tag{5.256}$$

Using the definitions $\overline{u}_{g,u} = \langle\alpha u_{g,u}\rangle/\alpha$ and $U_g = \langle\alpha u_g\rangle$ as cross-section-weighted drift and averaged related gas velocity, the cross-section-averaged gas volume fraction is given by

$$\langle\alpha\rangle = U_g/\left(CU + \overline{u}_{g,u}\right). \tag{5.257}$$

To define a mean drift velocity, the volume-weighted quantities $\overline{u}_{g,u}$ and $\overline{u}_g = \langle \alpha u_g \rangle / \langle \alpha \rangle$ are now introduced. For the representation in the conservation equations the relations $\overline{u}_{g,u} = \overline{u}_g - U$ and $\overline{u}_{l,u} = \overline{u}_l - U$ are used. To simplify the notation, the averaging symbols are again omitted in the following. The drift velocity can thus be understood as the velocity of the phase relative to a surface moving at the mixing velocity U. With the relation for the mixture velocity $U = \alpha u_g + (1-\alpha) u_l$, the relation of drift velocity and relative velocity of the phases to each other can be established directly.

$$u_{g,u} = (1-\alpha)\left(u_g - u_l\right);\ u_{l,u} = -\alpha\left(u_g - u_l\right). \tag{5.258}$$

From this relationship and the defining Eq. (5.254) follows a relationship between u_g, u_m, and $u_{g,u}$ of the form:

$$u_g = u_m + \frac{\rho_l}{\rho_h} u_{g,u};\ u_l = u_m - \frac{\rho_l}{\rho_h} \frac{\alpha}{(1-\alpha)} u_{g,u}. \tag{5.259}$$

If these expressions are inserted into the conservation equation for the mixtures and if, furthermore, the conservation equation for the gas phase is retained for the description of phase transitions, then, after some algebraic transformations, four conservation equations result for the state variables mean velocity u_m, pressure p, mean enthalpy h_m and volume fraction α.

$$\begin{aligned}
&\frac{\partial \rho_h}{\partial t} + \frac{1}{A_c} \frac{\partial (\rho_h u_m A_c)}{\partial x} = 0;\\
&\frac{\partial \left(\alpha \rho_g\right)}{\partial t} + \frac{1}{A_c} \frac{\partial \left(\alpha A_c \rho_g u_m\right)}{\partial x} + \frac{1}{A_c} \frac{\partial}{\partial x} \left(\alpha A_c \frac{\rho_g \rho_l}{\rho_h} u_{g,u}\right) = \frac{\Gamma_g}{A_c};\\
&\frac{\partial (\rho_h u_m)}{\partial t} + \frac{1}{A_c} \frac{\partial \left(A_c \rho_h u_m^2\right)}{\partial x} + \frac{1}{A_c} \frac{\partial}{\partial x} \left(A_c \frac{\alpha}{1-\alpha} \frac{\rho_g \rho_l}{\rho_h} u_{g,u}^2\right)\\
&= -\frac{\partial p}{\partial x} - \tau_w \cdot \frac{P}{A_c} - \rho_h g \sin\beta;\\
&\frac{\partial (\rho_h h_m)}{\partial t} + \frac{1}{A_C} \frac{\partial}{\partial x} (A_C \rho_g u_m h_m)\\
&= \frac{1}{A_C} \left(\dot{q}_{wg} P_g + \dot{q}_{wl} P_l\right) + \frac{\partial p}{\partial t} + u_m \frac{\partial p}{\partial x} + u_{g,u} \frac{\rho_l - \rho_g}{\rho_h} \frac{\partial p}{\partial x}\\
&- \frac{1}{A_C} \left(A_C \alpha \frac{\rho_g \rho_l}{\rho_h} u_{g,u} \Delta h_{lg}\right) + \frac{L_{diss}}{A_C};
\end{aligned} \tag{5.260}$$

in which Δh_{lg} denotes the vaporization, L_{diss} the power released by dissipation enthalpy and h_m is the density averaged enthalpy given by:

$$h_m = \frac{\alpha\, \rho_g h_g + (1-\alpha) \rho_l h_l}{\alpha\, \rho_g + (1-\alpha) \rho_l}. \tag{5.261}$$

This model is used especially for ring and plug flow. Since in ring flow the liquid droplets in the gas core flow much faster than the liquid film on the wall, large errors occur, especially at high loadings, due to the formation of an average velocity for the liquid. Therefore, in this case, the liquid transport in the film and core are often considered separately. A successful application of the simple drift flow model depends essentially on whether constitutive relations can be developed for the drift velocity $u_{g,u}$.

5.2.4.6 Pressure Loss Calculations in Multi-phase Flows

Pressure losses in two-phase flows are of great importance in power plant and process engineering. Therefore, robust empirical pressure drop relationships similar to those in single-phase flow hydraulics have been developed on the basis of measurements and simple model notions that do not distinguish between specific flow forms. Nevertheless, the notion of a dispersed flow on the one hand and that of a separate two-phase flow on the other has led to two different variants for pressure drop relationships. In general, the total pressure drop in a pipe or line section is composed additively of the pressure drop due to viscous friction (f), acceleration (a) and gravity (g) as sketched in Fig. 5.51 in the following way.

$$\left(\frac{dp}{dx}\right)_{tot} = \left(\frac{dp}{dx}\right)_f + \left(\frac{dp}{dx}\right)_a + \left(\frac{dp}{dx}\right)_g . \tag{5.262}$$

While in horizontal straight pipe sections of constant cross-section usually only the friction component (laminar/turbulent) is relevant, in contraction or expansion elements, such as valves or elbows, the acceleration component often predominates, similar to single-phase flow, but far more pronounced. This is because in two-phase flows, heat supply or pressure change not only mean a change in density, which is reflected in variable gas/liquid velocities, but can also lead to evaporation and/or condensation. Also, for two-phase flows lots of correlations are available, see e.g. Friedel (1979), Collier and Thome (1996), however, great care has to be

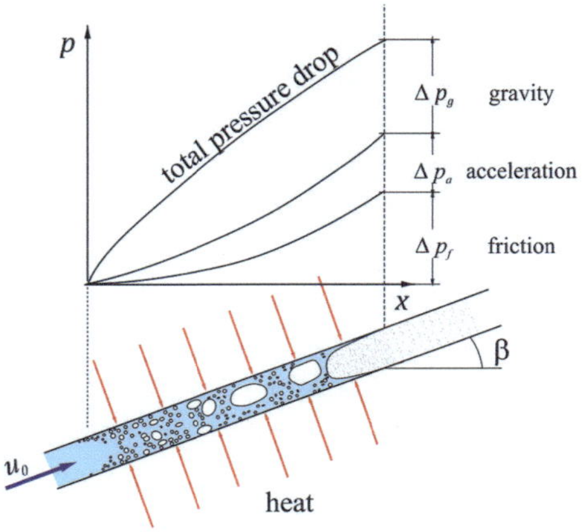

Fig. 5.51 Schematic sketch showing the components of the pressure drop in an inclined pipe with heat supply

taken, on which basis the correlations are formed. The most common approaches are the homogeneous model or the separate model:

Homogeneous model (friction)

The pressure gradient in a pipe is calculated similarly to a single-phase flow as follows:

$$\left(\frac{dp}{dx}\right)_f = \frac{4}{d}\tau_w \text{ with } \tau_w = \frac{f_{F,2ph}}{4}\frac{\rho_h}{2}u^2, \tag{5.263}$$

In which d is trhe hydraulic diameter, $f_{F,2ph}$ the Fanning friction factor of the two-phase flow, $f_{F,2ph}$ the homogeneous density. The coefficient of friction is considered as a function of a Reynolds number of the two-phase flow to be defined and for the respective velocity a corresponding coefficient of friction is to be chosen, for example for a annular-droplet flow a value of $f_{F,2ph}$ of $f_{F,2ph} \cong 0.02$ is valid for a rough pipe wall—if the volume fraction of the gas is large, $\alpha \to 1$ the f_F—value of the single-phase gas flow can be chosen, i.e. $f_F = f_{F,g}$, vice versa at $\alpha \to 0$ the value of the liquid $f_F = f_{F,l}$, where in the two-phase Reynolds number (Re_{2ph}) appears the viscosity of the gas or liquid. Often based on the classical implicit relation of Prandtl for single-phase fully turbulent flows, the following relation is also used

$$\frac{f_{F,2ph}}{4} = 0.0014 + \frac{1}{8}Re_{2ph}^{-0.32}. \tag{5.264}$$

To form the Reynolds number, the total mass flow density and a mixture viscosity are used, for example, in the following form

$$\mu_{2ph} = \frac{U_g}{U}\mu_g + \frac{U_l}{U}\mu_l = \dot{x}\mu_g + (1-\dot{x})\mu_l \text{ or } \frac{1}{\mu_{2ph}} = \frac{\dot{x}}{\mu_g} + \frac{1-\dot{x}}{\mu_l}. \tag{5.265}$$

For condensable mixtures such as water and water vapor, a so-called two-phase multiplier is also frequently used for practical calculations. This multiplier is defined as the ratio between the pressure drop in the current two-phase flow with mass flux $\dot{M}$ to the pressure drop of the condensed liquid flow with the same pipe cross-section and mass flux:

$$\Phi_{l0}^2 = \frac{(dp/dx)_{2ph}}{(dp/dx)_{l0}}, \tag{5.266}$$

where the suffix "*l0*" indicates that the pressure gradient of the pure liquid at the same mass flux form the reference. Textbooks give correlations that reflect values essentially as a function of gas–liquid properties and vapor content and take the following form $f = \left(\mu_g/\mu_l, \rho_g/\rho_l, \dot{x}\right)$.

Separate model (friction)

The separate model is based on the idea that both phases flow in two separate regions of the pipe, as in a layered or annular flow, but that pressure equilibrium nevertheless exists in both phases regardless of the flow pattern. The total pressure drop is then described by the momentum equation for the two-phase mixture given and e.g. for the horizontal pipe of constant cross-section, the following follows for steady-state flows:

$$\frac{dp}{dx} = \langle \tau_w \rangle \frac{P}{A_C} + \frac{\dot{M}^2}{A_C^2} \frac{d}{dx} \left(\frac{\dot{x}^2}{\alpha \rho_g} + \frac{(1-\dot{x})^2}{(1-\alpha)\rho_l} \right). \tag{5.267}$$

If no heat is applied across the edge of the tube, $\dot{x}$ and α do not change along the tube and the wall shear stresses are in equilibrium with the compressive force. If one introduces a Martinelli parameter analogous to Eq. (5.241) for the two-phase multiplier in such a way that they relate the pressure drop in the two-phase flow to the pressure drop of the gas or liquid phase flowing alone in the tube, one obtains and alpha do not change along the tube and the wall shear stresses are in equilibrium with the compressive force.

$$\Phi_g^2 = \frac{(dp/dx)_{2ph}}{(dp/dx)_g} \text{ and } \Phi_l^2 = \frac{(dp/dx)_{2ph}}{(dp/dx)_l}. \tag{5.268}$$

These quantities are related to the so-called Martinelli parameter, the pressure decay ratio, based on experimental data. The Martinelli can be calculated explicitly if the related velocities of the two-phase flow and its laminar or turbulent flow state are known. The turbulence states are determined by the Reynolds number of the gas or liquid flow and for $Re > 2300$ a turbulent state is assumed. Accordingly, there are four possible forms of the Martinelli parameter depending on whether the gas and liquid phases are laminar or turbulent flow.

The Martinelli pressure drop relationship is based on a relatively small in system pressure ($p < 0.4\,\text{MPa}$) and pipe diameter ($d < 3\,\text{cm}$) and have an uncertainty of up to 40%. For more accurate design calculations, empirical pressure drop correlations have been developed based on a large database and taking into account other specific dependencies on two-phase mass flow density (Reynolds number), surface tension (Weber number $We = \left(\dot{m}^2 d\right)/(\rho_h \sigma)$), and gravity (Froude number $Fr = \dot{m}^2/\left(gd\,\rho_h^2\right)$) all calculated using the homogeneous density ρ_h. They represent complex functional relationships between dimensionless ratios that capture the physical relationships. The Friedel correlation for a horizontal tube is given here as an example.

$$\Phi_{l0}^2 = E + \frac{3.24 F H}{Fr^{0.045} We^{0.035}} \text{ with } E = (1-\dot{x})^2 + \dot{x}^2 \frac{\rho_l}{\rho_g} \frac{f_{F,g0}}{f_{F,l0}};$$
$$F = \dot{x}^{0.78}(1-\dot{x})^{0.24};\ H = \left(\frac{\rho_l}{\rho_g}\right)^{0.91} \left(\frac{\mu_g}{\mu_l}\right)^{0.19} \left(1 - \frac{\mu_g}{\mu_l}\right)^{0.7}. \tag{5.269}$$

Acceleration pressure losses

Compared to single-phase flow, acceleration pressure losses occur more frequently in pipe extensions or contractions, in pipe bends and pipe junctions. For design purposes in apparatus engineering, the pressure loss relationships are given in analogy to corresponding relationships from single-phase flow, neglecting wall friction.

Since a deceleration of the flow generally occurs with cross-section expansion, a separation of the phases is to be expected with two-phase flows. Therefore, it is obvious to use the separate flow model in the form of the mixture balance equations for the calculations of a pressure change. From the conservation equations for the two-phase mixture of the separate model, a relationship for the pressure recovery in the Carnot diffuser (cf. Fig. 5.32b) can be easily given if wall friction is neglected.

$$p_2 - p_1 = \dot{m}_1^2 \frac{A_1}{A_2} \left(\frac{1}{\rho_{M,1}} - \frac{A_1}{A_2} \frac{1}{\rho_{M,2}} \right) \tag{5.270}$$

Here $\rho_{M,1}, \rho_{M,2}$ denote the momentum density (cf. Eq. (5.252)). If we switch to single-phase flow, i.e. choose $\dot{x} = 0$ or $\dot{x} = 1$, the momentum density changes to the density of single-phase flow and so does the expression for pressure recovery. By its nature, this expression can only be evaluated if volume fraction and vapor content can be linked between cross sections 1 and 2. At low system pressures, when $p \ll p_{crit}$ holds and no phase transitions occur due to evaporation, the volume fraction of the gas α and the vapor content $\dot{x}$ do not change and in this case the analogy to single-phase flow is evident, $\rho_{M,1} = \rho_{M,2}$. Observations show that downstream, however, equilibrium of two-phase flow is reached only after relatively long running distances, about 30–70 diameters. This fact requires a pressure-dependent correlation for the change of the vapor content in more accurate calculations. In the case of vapor–liquid flows with phase transitions, empirical correlations between volume fraction α and vapor content $\dot{x}$ are used.

For pipe contraction from 1 to 2, an acceleration of the two-phase flow occurs, which leads to better mixing, therefore a pressure loss calculation can be carried out here according to the homogeneous flow model with a good approximation. Since the law of single-phase flow applies to the homogeneous model, with the homogeneous density as the only characteristic feature, the known relationship of single-phase flow is obtained:

$$p_2 - p_1 = \frac{\dot{m}_2^2}{\rho_h}\left(1 - \left(\frac{A_2}{A_1}\right)^2 + \left(\frac{A_2}{A_1} - 1\right)^2\right), \tag{5.271}$$

in which the two fractions in the parenthesis can be identified as the reversible and irreversible fractions of the pressure drop.

In the pipe elbow, detachment also occurs on the inside at higher velocities, causing a phase separation due to centrifugal forces in two-phase flows. In this case, gas phase usually accumulates on the inside, while the fluid flows in the outer area of the elbow. Dispersed flows temporarily become layered flows. A new equilibrium between the phases corresponding to the inlet conditions is also established only after 30–70 pipe diameters. This means that fully developed two-phase flows can rarely be assumed for technical designs in plant engineering.

For arbitrary cross-sections, the acceleration pressure loss is determined integrally over a distance x along the duct. It results from the equation of inertia. The acceleration du/dt of the mass $dM = (A\Delta x\,\rho)$ corresponds to the force difference, i.e. the change in the product $(A\,p)$.

$$\frac{\partial(A \cdot p)}{\partial x}\Delta x = -A\Delta x \cdot \rho\frac{du}{dt} \text{ with } \frac{du}{dt} = \frac{\partial u}{\partial t} + \frac{\partial u}{\partial x} \cdot \frac{\partial x}{\partial t} = \frac{\partial u}{\partial t} + u \cdot \frac{\partial u}{\partial x}. \tag{5.272}$$

The insertion of the right side into the left provides:

$$p \cdot \frac{\partial A}{\partial x} + A \cdot \frac{\partial p}{\partial x} = -A \cdot \rho\frac{\partial u}{\partial t} - A \cdot \rho \cdot u \cdot \frac{\partial u}{\partial x}. \tag{5.273}$$

For a pipe with an invariant cross-sectional area, as is the case for an absorber tube, the term with $\partial A/\partial x$ cancels out and the acceleration pressure loss is composed of two components, the acceleration over distance and the acceleration over time.

$$\frac{\partial p}{\partial x} = \rho\frac{\partial u}{\partial t} + \rho \cdot u \cdot \frac{\partial u}{\partial x} = \Delta p_{a0} + \Delta p_{a1}. \tag{5.274}$$

The acceleration along the flow path is given by $\dot{m} = (\rho u)$ and after local discretization into the nodes i and $i + 1$:

$$\Delta p_{a1} = \frac{\left(\dot{m}_i^2 - \dot{m}_{i+1}^2\right)}{2\rho}. \tag{5.275}$$

When accelerating over time, one obtains after time discretization:

$$\Delta p_{a0} = \frac{\Delta x}{\Delta t}\left(\dot{m}_{i,t} + \dot{m}_{i,(t-1)}\right). \tag{5.276}$$

Pressure change due to height

The change in pressure due to the geodetic change in elevation is given by the following equation, where β is the angle of slope with respect to the horizontal plane.

$$\Delta p_H = g \cdot sin\beta \cdot \rho \cdot \Delta x = \rho \cdot g \cdot \Delta H. \tag{5.277}$$

5.2.5 Density Waves, Critical Mass Flux and Cavitation

When a two-phase mixture flows out of a pressure reservoir through a narrow cross-section, a limitation of the mass flow occurs above a certain critical pressure difference of container, for example a dam drum, and the consumer's extraction pressure $(p_1 - p_0)$. A further reduction of the pressure, for example by a further opening of a consumer valve, then leads to no further increase of the mass flux. An analogous phenomenon can be observed with compressible flow through a Laval nozzle. The physical cause is identical in both cases, because above the critical pressure ratio, wave-like pressure or density disturbances in the fluid can only propagate downstream because the flow velocity has become larger than the propagation speed of small disturbances. It is not possible to influence the flow region upstream by a change of state further downstream. In compressible flow, these are sonic or Mach waves, and in two-phase flow, they are small changes in the volume fraction of the gas phase. The propagation velocity a of these small disturbances is given in both cases by the isentropic change of density with pressure:

$$a^2 = \left(\frac{\partial p}{\partial \rho}\right)_s. \tag{5.278}$$

The calculation of the critical mass flow in a cross-sectional constriction is performed as in gas dynamics from the conservation equations for mass and momentum using certain equations of state for the gas and the liquid phase. The initial equations are those for the two-phase mixture Eq. (5.251) or (5.260). The critical mass flux $\dot{m}_{crit} = (A^* a^* \rho^*)$ is given by the propagation speed a^* of the density wave at the narrowest flow cross-section A^*.

The relation shows again the typical problem in a two-phase flow, since the critical mass flow depends on the definition of the two-phase density ρ_{2ph}, which has a different shape depending on the mixing model (separate or homogeneous model). Since the flow is always accelerated during outflow or overflow, it is usually assumed that there is good mixing of the phases and that the homogeneous density ρ_h describes the mixing well. Differentiation of the expression for this density at constant entropy, i.e., in the approximation that the changes of state occur adiabatically in each phase but with phase changes at the phase boundaries, then yields

$$\left(\frac{1}{a_{2ph}^2}\right)_h = \left(\frac{\partial \rho}{\partial p}\right)_s = \frac{1}{\rho_h^2}\left(\frac{1}{\rho_l^2 a_l^2} + \dot{x}\left(\frac{1}{\rho_g^2 a_g^2} - \frac{1}{\rho_l^2 a_l^2}\right) - \left(\frac{\partial \dot{x}}{\partial p}\right)_s \left(\frac{1}{\rho_g} - \frac{1}{\rho_l}\right)\right), \tag{5.279}$$

in which a_g, a_l are the sound speed of the gas and liquid, and the index "h" refers to the homogeneous model. Since the speed of sound in the liquid is much greater than that in the gas, one can further simplify the equation:

$$\left(a_{2ph}^2\right)_h = a_g^2\left(\dot{x} + \frac{\rho_g}{\rho_l}(1-\dot{x})\right)^{-2} \cdot \frac{1}{\left(\dot{x} - \left(\frac{\partial \dot{x}}{\partial p}\right)_s a_g^2 \frac{\rho_g}{\rho_l}(\rho_l - \rho_g)\right)}. \tag{5.280}$$

Here, it can be clearly seen that the propagation velocity of the density wave depends essentially on the vapor content and on the change of the vapor content in isentropic thermodynamic equilibrium. In many technically relevant flows, an evaporation process in thermodynamic equilibrium by pressure reduction does not take place in the narrowest flow cross-section because the pressure reduction occurs too quickly and is too small. This means that the relaxation time for evaporation is significantly greater than the transit time of the flow through the narrowest cross-section. This case is referred to as inhibited or frozen thermodynamic equilibrium, characterized by $(\partial \dot{x}/\partial p)_s = 0$. If the vapor content is divided by the volume fraction α, the frozen density-wave velocity in a homogeneous two-phase flow is given by the expression:

$$\left(a_{2ph}^2\right)_h = \frac{a_g^2}{\alpha\left(\alpha + \frac{\rho_l}{\rho_g}(1-\alpha)\right)^2} \approx \frac{\rho_l}{\rho_g}\frac{a_g^2}{\alpha^2(1-\alpha)}. \tag{5.281}$$

It is noteworthy here that the velocity of the density wave is much lower than the speed of sound in the gas and assumes the smallest value for $\alpha \approx 0.5$. This behavior has been well confirmed experimentally in air–water flows. Surprising here is the low value of the wave velocity compared to the sound velocity of the gas. It is in the minimum below 10% of the sound speed of the gas. This has the consequence that the mass flow limitation at geometrical constrictions of flow conduits already occurs at very low two-phase mass flux densities. This is technically extremely significant in connection with the outflow of gas–liquid mixtures from of gas–liquid mixtures from pressure vessels.

Analytical models for the calculation of critical mass flux can be roughly classified according to whether thermodynamic and mechanical equilibrium between the phases is assumed or not. Outflow experiments show that, in general, neither thermodynamic nor mechanical equilibrium is established locally in an outflow process. During flash evaporation, temperature differences develop between the phases, which are not balanced by heat transfer at the interfaces during the short overflow process. At the same time a differential velocity between the phases.

Long pipe connections between pressure reservoirs, on the other hand, lead to the reattachment of the free jet to the wall after a certain distance, and independently of this, after a certain relaxation run length, to the setting of the thermodynamic equilibrium with decisive steam formation rates. The noticeable increase of the vapor content in the flow then results in a significant two-phase friction and acceleration pressure drop.

The two situations shown can each be described in rough approximation by a one-dimensional homogeneous two-phase flow in thermodynamically fully inhibited or in full equilibrium. The two limiting cases of complete equilibrium and completely inhibited equilibrium are captured by simplifying Eqs. (5.279) and (5.280) for the critical propagation speed of a homogeneous two-phase flow. Assuming that $a_g^2 \ll a_l^2$ holds for the sound velocities in the vapor and liquid phases, the simplified expression is obtained for the critical and maximum mass flow densities, respectively:

$$\dot{m}^*_{crit} = \frac{1}{\sqrt{\frac{\dot{x}}{\rho_g a_g^2} - \left(\frac{1}{\rho_g} - \frac{1}{\rho_l}\right) \cdot \left(\frac{\partial \dot{x}}{\partial p}\right)_s}}, \tag{5.282}$$

in which the outflow at fully inhibited thermal equilibrium is characterized by $(\partial \dot{x}/\partial p)_s = 0$. The linkage of the states in the narrowest flow cross-section with the rest values in the overpressure reservoir is done via the equation of state of the corresponding gas–liquid mixture, for example water vapor state tables.

In summary, the thermodynamic and mechanical equilibrium assumptions made in the simple models are too broad for a quantitative comparison of experiment and model. In order to capture the actual conditions when two-phase mixtures flow through nozzles, orifices or pipe nozzles under strong pressure relief, the local and temporal deviation from thermodynamic equilibrium and the mechanical interaction between the phases must be considered in the model, here the devoted reader is referred to expert literature.

In fluid flows of high velocity, a reduction in pressure when flowing around bodies and corners can lead to localized vapor or gas formation. This phenomenon is called cavitation. It is localized flash evaporation followed by condensation, or the release of dissolved foreign gases from the fluid due to pressure relief. They sometimes occur in hydraulic fluid machinery such as pumps and turbines and other hydraulic components such as valves. Undesirable side effects such as deterioration of operating performance, noise generation, mechanical vibration and localized material wear can be observed. The prevention and control of cavitation is therefore of great importance for hydraulic machine and plant engineering. A detailed description of cavitation can be found in Franc and Michel (2006).

Cavitation can start in a single-component flow when the local static pressure in the flow reaches and falls below the thermodynamic saturation vapor pressure p_{sat} of the fluid. This necessary condition for cavitation onset is characterized in a frictionless incompressible flow by a dimensionless cavitation number. The cavitation number is similar in structure to the Euler number and gives a measure of when the fluid cavitates. It expresses the relationship between a local pressure difference caused by a restriction $(p - p_{sat})$ and the kinetic energy per volume of the flow $(\rho u^2/2)$ and is used to characterize energy losses in the flow, where a perfect frictionless flow corresponds to a Euler number of 0.

$$Ca = 2\frac{(p - p_{sat})}{\rho u^2}. \tag{5.283}$$

For values of $Ca \leq 0$, evaporation of the fluid may occur. Due to mechanical and thermodynamic real effects, the "ideal" critical value of the cavitation coefficient $Ca = 0$ for the onset of cavitation, however, shifts to positive or negative coefficients.

The observed manifestations of cavitation are similarly diverse as the flow forms of two-phase flows in pipes and always transient. They are classified according to increasing vapor content as cloud, bubble, layer, and super cavitation. Collectives of bubbles in the micron and submicron range, which can form in strong sheared boundary layers, are called cloud cavitation. Extended coherent bubble flows develop in the negative pressure regions of detached flows behind body edges, on the suction side of adjusted profiles in pumps, or in regions of detached flows. This is referred to as bubble cavitation. The formation of coherent vapor or gas films occurs in partial areas of the profile or body edge due to bubble coalescence when the vapor or gas volume fraction exceeds a critical level of $\alpha \sim 0.5$. This form is called layer cavitation. If the body is surrounded by the vapor film at very high flow velocities and correspondingly low local pressures on the whole or on the largest parts of its contour area, this is called super-cavitation.

In addition to the cavitation number, other parameters influence cavitation; these include geometry, fluid properties as viscosity, surface tension, parameters of the real gas equations, thermal conductivity, heat capacity, latent heat, and concentration of foreign substances in the fluid such as foreign gases or particles. Furthermore, for the onset and the dynamic behaviour of the cavitation process, the thermal micro-processes for the activation of nucleation nuclei and the degree of turbulence of the flow are also important.

5.2.6 Instabilities in Two-Phase Flows

Instabilities can occur in two-phase flows. The considerations in the following refer to heated channels—single or several parallel channels—with increasing steam content in the direction of flow. A pressure difference is imposed on the channels together from the outside. Transients may involve changes in mass flow rate, pressure or pressure drop within the duct, density or changes in flow patterns and consequently heat transfer.

From Fig. 5.44 it could be seen that two-phase flow—except for layered flow—provides better heat transfer than single-phase flow at the same mass flow rate. On the other hand, there is a higher pressure drop and the occurrence of flow instabilities. The instabilities can be aperiodic or periodic in nature. If parallel channels are operated with flow ratios near transitions to instabilities, periodic oscillation can arise from inherently aperiodic processes due to communication between the parallel channels and in interaction with the heat transfer and vapor formation rate.

Aperiodic instabilities arise primarily when several flow states are stable under the same external conditions. Instabilities are triggered by transitions from one flow form to another, resulting in, among other things, a different pressure drop. If such a transition in the boiling tube of a collector results in a reduced steam release rate and heat transfer, the steam utilization is impaired, or the solar efficiency deteriorates due to its temperature dependence. The boiling tube wall can be overheated to such an extent that, for example, the selective coating of the absorber is damaged.

Periodic oscillations can result in vibration excitations, against which the collector and its components must be protected. From the preceding sections and experiments presented later, it is clear that oscillations of the mass flow rate can lead to an improvement of the heat transfer and can be advantageous in the application. The pre-requisite for advantageous behaviour is that the oscillations do not grow undamped to unacceptably high amplitudes.

A general distinction is made between *static* and *dynamic instabilities*, depending on whether it is sufficient to describe the problem with static equations, or to predict the onset or occurrence of the instability.

The simplest flow instability at a phase interface is wave instability, which is caused by the effect of relative velocity between phases, surface tension, and acceleration forces, and is known as Kelvin–Helmholtz and Rayleigh–Taylor instability. These types of instabilities likewise play a central role in the deformation and decay of bubbles, drops, liquid lamellae, and liquid films.

Schematically, Fig. 5.52 depicts an interface in which a gas and a liquid move relative to each other with the phase velocity $u_g - u_l$. As soon as the velocity difference is greater than zero, small wave-like perturbations are fanned in time at a certain magnitude of surface tension and forces acting externally on the interface, e.g., gravity. Here, the fanning rate of the disturbance generally depends on the wavelength.

The perturbation with a wavelength m with the largest amplification rate, leads to the decay of the smooth interface and the formation of a new spatial phase distribution characterized by the intrinsic new length scale λ_m. This length scale also determines the size of droplets and bubbles after the remodelling process. From a linear stability analysis for a frictionless infinitely extended two-layer flow

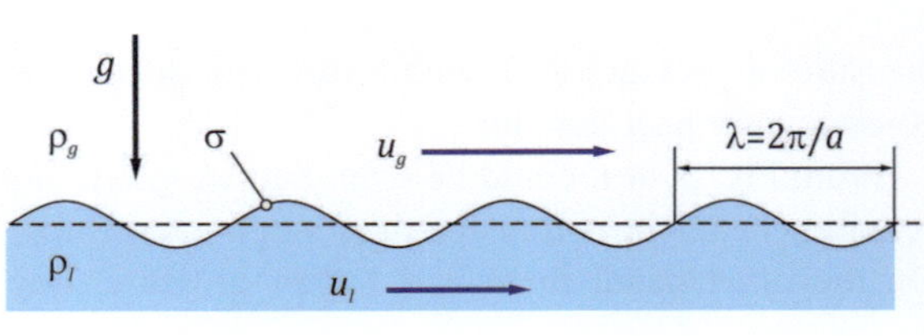

Fig. 5.52 Sketch of the stability of an interface of two stratified fluids with different velocities u_l, u_g

yields an amplification rate c_i of the form:

$$c_i = \frac{1}{a}\sqrt{\frac{\rho_l \rho_g (u_g - u_l)^2 a^2}{\rho_l + \rho_g} - \frac{\sigma\, a^3 - g(\rho_g - \rho_l)a}{\rho_l + \rho_g}}, \tag{5.284}$$

in which ρ_l, ρ_g are the densities, a is the wave number ($\lambda = 2\pi/a$), σ and the surface tension. The acceleration here is due to the gravity g. Any other acceleration with corresponding effect can take its place. The maximum of the amplification rate as a function of the wave number can be determined from $\partial c_i/\partial a = 0$. If there is no velocity difference between the phases ($u_l = u_g$), the maximum amplified perturbation has the wavelength λ_m:

$$\lambda_m = 2\pi\sqrt{\frac{3\,\sigma}{g(\rho_l - \rho_g)}}. \tag{5.285}$$

This assumes that the denser liquid is accelerated in the direction of the less dense one. Otherwise, only a damped oscillation occurs. This is referred to as the Rayleigh–Taylor instability. The cut off wavelength for amplified perturbations is given by a vanishing amplification rate c_i and is:

$$\lambda_c = 2\pi\sqrt{\frac{\sigma}{g(\rho_l - \rho_g)}}. \tag{5.286}$$

This length scale λ_c is called the Laplace length. It can be concluded that the interface is stable against small perturbations with smaller wavelengths and does not undergo any permanent change because the surface tension keeps the accelerating forces in equilibrium. For perturbations with longer wavelengths, the interface deforms permanently. An example of this type of interfacial instability is the prevention of leakage of liquids from containers with sufficiently small orifice openings in the bottom or the decay-free rise of gas bubbles of diameter $d_b \leq \lambda_c$ in a liquid.

When significant velocity differences $(u_g - u_l)$ occur between phases and interfacial tensions are present, the amplification of interference is limited downward by the cutoff wavelength:

$$\lambda_c = 2\pi\left(\frac{\sigma}{\rho_g (u_g - u_l)^2}\right). \tag{5.287}$$

Here, it is assumed that the densities of the phases are very different as in gas and liquid under normal conditions, i.e. $\rho_g \ll \rho_l$. The perturbation with the strongest amplification has the wavelength $\lambda_m = 1.5\lambda_c$. For two-phase flows, the consequence of this capillary instability is that undamped capillary waves due to shear velocities at the phase interface cause the interfaces to break up. For example, large droplets break up into smaller components when the droplet dimension is larger than the cut-off wavelength λ_c. This instability of interfaces against shear motion is called Kelvin–Helmholtz instability. This fact also explains the Weber number criterion for the decay of liquid jets and liquid drops, which states that a decay of moving liquid volumes with free surfaces occurs when the Weber number We_L formed with the characteristic length L of the volume exceeds unity, i.e.

$$We_L = \frac{\rho_g (u_g - u_l) L}{\sigma} > 1. \tag{5.288}$$

The two instabilities presented are of great influence for bubble and droplet formation in two-phase flows and explain some important phenomena in boiling and condensation processes, for example, the collapse of bubble boiling and film boiling can be explained as Rayleigh–Taylor instabilities. The formation of wavy layer flows of homogeneous bubble and droplet flows is directly related to the Kelvin–Helmholtz instability, which can be described in terms of a Weber number criterion.

Dynamic instabilities—Ledinegg-instability

In addition to the small-scale interfacial instabilities that have an important function for the phase distribution, there are other fundamental large-scale instability mechanisms that determine the timing of two-phase flow in hydraulic systems with phase transitions, such as those involving boiling, condensation, and cavitation processes. Since such instabilities can lead to uncontrolled mechanical pressure and shock loads and further to thermally triggered stresses in channel walls of the system, the stability limit for such processes is practically a design and operational criterion for such systems as steam generators and fluid machinery. Typical two-phase instability can occur in a system consisting of two pressure reservoirs and two hydraulic components connected in series in between, for example a pump and an evaporator tube with constant heat supply, as shown in Fig. 5.53.

The stability behaviour of the system is determined by the different pressure-mass flow dependencies of the pump and the evaporator tube. While the pressure build-up (Δp) of a pump typically decreases monotonically with increasing mass flow ($\dot{m}$), the two-phase pressure drop in the evaporator tube shows a non-monotonic behaviour in an average mass flow range, which is mainly caused by the different contributions of friction and acceleration components to the total pressure drop of the two-phase flow in the evaporator tube (compare Sect. 5.2.4.6 Eq. (5.262)). The monotonic branches of the pressure loss curve are determined by the high vapor fraction at low mass fluxes and by the high liquid volume fraction at high mass fluxes. In general, therefore, the system can assume three steady-state operating conditions with a change in mass flow, given by the intersections A, B, C of the two pressure-mass curves for pump and evaporator tube. Assuming a small variation of the quantities $\dot{m}$ or p in the vicinity of the steady states, a simple consideration shows that states A and C are stable, while a pressure and mass flow change in the vicinity of B due to a power change of the pump causes a pressure loss change with exactly the opposite sign. This leads to a transition to one

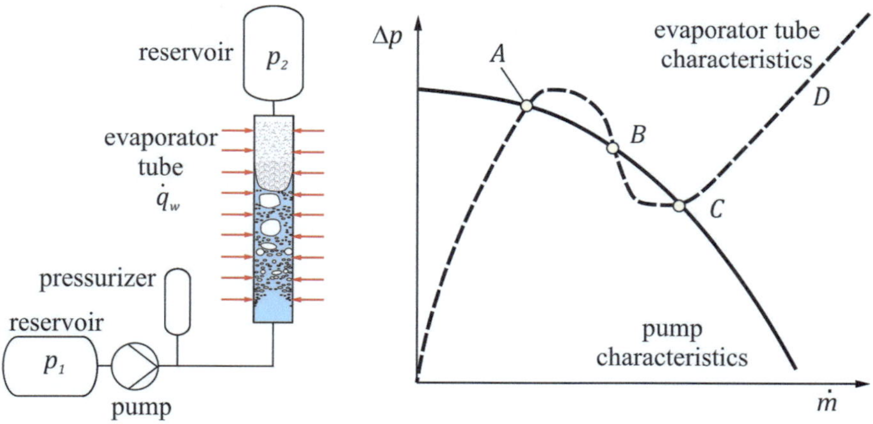

Fig. 5.53 Two-phase system with possible (Ledinegg) instability, left: Flow system, right: pressure-mass flow characteristics of the pump and the evaporator tube

of the two stable operating states A or C. The simple considerations show that for the stability of an operating point the condition $\partial \Delta p / \partial \dot{m} > 0$ is necessary, with Δp as the driving system pressure through the pump. Since this time behaviour triggered by the instability is given by a simple transient from an unstable to a stationary stable state, the system is said to be statically stable or unstable.

Ledinegg (1938) was the first to investigate this instability, which is named after him. According to Ledinegg, an aperiodic instability can occur in a flow channel with a constant heat flux ($\dot{q}_w$) when a subcooled fluid enters and is partially evaporated. In contrast to before, the occurrence of the instability requires that the static pressure drop increases with decreasing mass flow rate $\dot{m}$, that is, when $\partial \Delta p / \partial \dot{m} < 0$. The frictional pressure drop in turbulent flow is proportional to the square of the mass flow rate, i.e., $\Delta p \sim \dot{m}^2$. It dominates at large mass fluxes or small heat flux rates and accordingly affects the total pressure drop. In a certain range, the acceleration pressure drop can increase with increasing heat output or decreasing flow rate in such a way that a transition to a new stable condition with smaller total mass flux rate and large vapor volume fraction at the channel outlet occurs, resulting in the operating point A.

The influence of subcooling, given by the enthalpy difference $(h' - h_i)$, where h_i is the inlet enthalpy, is shown in Fig. 5.54. Instability arises with decreasing mass flow density and exceeding of a critical subcooling of the liquid at the channel inlet. In the case of several parallel channels and an externally imposed equal pressure drop for all channels, e.g. in steam boilers or in solar collectors, steam blockages with strongly reduced heat removal can develop.

A similar phenomenon, based on the same instability, can also be observed in nature geysers. Heat is supplied to a cavern filled with supercooled water and connected to a higher water reservoir via a slender channel. The water heats to saturation temperature and begins to boil. The steam that forms leaves the cavern in

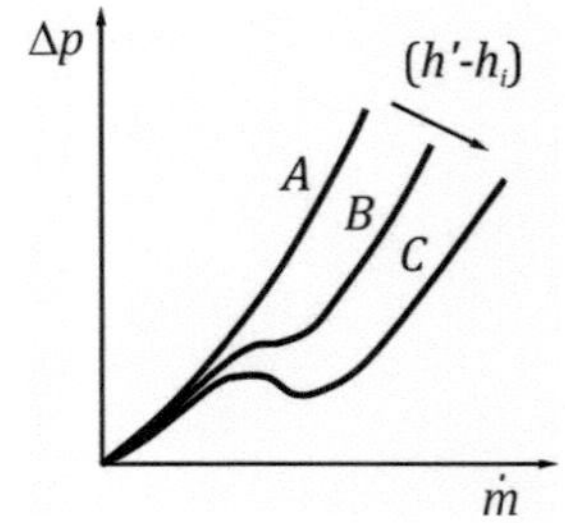

Fig. 5.54 Instability of the two-phase flow in the heated channel with increasing inlet subcooling

the form of a two-phase mixture via the connecting channel. Due to the increasing steam content, the hydrostatic pressure in the cavern is reduced. This leads to further increased evaporation and to further water discharge from the cavern. As the water deficit increases, steam generation is reduced and supercooled water can enter the cavern from the upper reservoir against a reduced steam flow and bring the boiling process to a complete standstill. As the heat source heats the cavern again the penetrated water to saturation temperature the process repeats itself and with constant external conditions (heat supply in the cavern, heat removal from the reservoir) a periodic process with typical relaxation character is established.

The Ledinegg criterion can be used to estimate whether aperiodic instabilities of the type described above occur during the evaporation of a liquid that enters the evaporator tube subcooled. The criterion considers the friction and acceleration pressure loss. The coefficient of friction $f_{F,2ph}$ is applied according to a homogeneous model of two-phase flow and heat flux density ($\dot{q}_w$) at the tube wall is assumed to be stationary as well as constant along the entire tube length. This provides the enthalpy difference ($\Delta h - h_i$) by which the liquid may be at most subcooled at the pipe inlet:

$$(\Delta h - h_i) \leq \frac{K_1 \cdot K_2}{1 + K_1 - \sqrt{\frac{3}{2}\, K_1 \cdot \left(\frac{K_1}{2} + 1\right)}} \text{with } K_1 = \frac{f_{F,2ph} L}{2d} \text{ und } K_2 = h_r \frac{\rho_g}{\rho_l}, \tag{5.288}$$

where L is the tube length, d is the tube diameter, ρ_g and ρ_l are the vapor density and fluid density, respectively, and Δh is the enthalpy of vaporization. Then, instability is avoided by decreasing subcooling or with increasing friction pressure loss, for example, by using an orifice plate at the pipe inlet. Pederson and May (1982) describe a formula that treats the friction pressure loss divided by inlet or outlet zones as well as the pipe section. They assume the simplified relationship for total pressure loss. The studies were specifically set up for solar systems with parabolic trough collectors.

Dynamic instabilities—density wave instability

Another system-wide two-phase instability, which takes the form of intense oscillatory fluctuations in the gas volume fraction, is the so-called density-wave

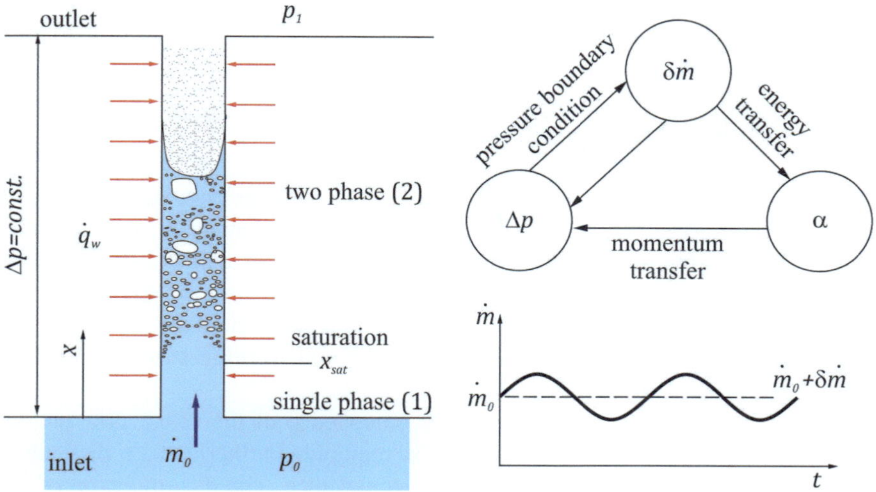

Fig. 5.55 Principle diagram for the formation of the density-wave instability (left) and feedback effects occurring there (top right) according to Müller (2016)

instability. The instability is common in systems with phase transitions and is due to a feedback of mass flow, vapor formation rate or condensation rate, and pressure drop in the boiling section or condensation section. This instability can be described analytically for a piping system with constant heat supply and removal using the one-dimensional two-fluid model and a linear stability analysis. Here, only the mechanism will be described using a simple evaporator tube tied between two pressure reservoirs with constant pressures p_0 and p_1, compare Fig. 5.55.

If the evaporator tube is subjected to a small periodic perturbation of the mass flow $\partial\dot{m}$ at the inlet, the position of the saturation temperature and thus the limit of the onset of evaporation inside the tube follows this fluctuation, because mass flow oscillations in the range of single-phase flows include enthalpy fluctuations. Mass flow variations and changes in the path of the single-phase flow in the pipe result in pressure variations $\partial\Delta p_1$ in the region of the single-phase flow. In the two-phase region, enthalpy disturbance also acts as a disturbance in the vapor volume fraction α, which propagates as a density wave in flow direction. The change in vapor content in conjunction with the mass flow and length perturbation leads to increased pressure perturbations $\partial\Delta p_2$ in the two-phase region. Since the total pressure difference is imposed on the evaporator tube by the system, the individual pressure fluctuations $\partial\Delta p_1$ and $\partial\Delta p_2$ must balance each other. This implies a feedback between the two-phase and single-phase flow domains and leads to a resonant amplification of the initial small disturbance if the disturbances are in phase. The result is massive fluctuations in the extension of the boiling domain and the vapor content of the evaporator section. This should be avoided for reasons of increased thermal wall loads in technical systems. The specification of limits for density wave instability is therefore important for the design of tubular steam generators. The tendency to unstable behaviour increases when several evaporator channels are connected in parallel.

5.2.7 Phase Change Heat Transfer

The calculation of heat transfer in evaporator tubes depends on a large number of influencing variables. These are mainly the mass flow density $\dot{m}$, the heat flux density $\dot{q}_w$, the fraction of the vapor mass flow $\dot{x}$, the boiling pressure p_{sat}, and the material properties of the liquid and gaseous phases of the heat transfer medium. Due to the occurrence of a two-phase flow, there are different flow pattern described earlier in this chapter with different wetting states of the tube wall.

To calculate the heat transfer, therefore, knowledge of the actual flow pattern in the pipe segment under consideration is required. Furthermore, a distinction must be made between a vertical and a horizontal pipe. The determination of the flow pattern by means of flow pattern maps is useful for manual calculations for estimations, but in the simulation the transitions of the flow forms must be brought into functional dependencies. In complex models, this leads to an accumulation of a large number of equations which is not very descriptive. For this reason, the description of the calculation of the heat transfer using the flow form maps is omitted here.

For heat transfer (except for subcooled boiling) in horizontal tubes, Streicher (1991) uses a subdivision of flow forms as shown in Table 5.17 to calculate heat transfer in heat pump circuits. The heat transfer coefficient for the different ranges is calculated as a mixture of the heat transfer for the saturated single-phase flow h_l and the heat transfer coefficient for bubble boiling h_b. For the bubble flow region (index I), the average heat transfer coefficient h_I, at $\dot{x} > 0$, the following relationship is assumed:

$$h_I = \sqrt{\left(h_l^2 + h_b^2\right)} \text{ with } Nu_l = \frac{h_l \cdot d_{hg}}{\lambda_l} = 0.023 \cdot \mathrm{Re}_l^{0.8} \cdot Pr_l^{0.4}, \tag{5.290}$$

where the Reynolds (Re_l) and Prandtl numbers (Pr_l) for the liquid are evaluated in the following way:

$$Re_l = \frac{u_l \cdot d_{hg} \cdot \rho_l}{\mu_l}; Pr_l = \frac{\mu_l \cdot c_{p,l}}{\lambda_l}, \tag{5.291}$$

where the actual mean flow velocity u_l and the modified hydraulic diameter d_h are determined in the following way.

$$u_l = \frac{\dot{m} \cdot (1 - \dot{x}_m)}{\rho_l \cdot (1 - \dot{\alpha}_m)}; d_{hg} = d_h \cdot \sqrt{1 - \dot{\alpha}_m}, \tag{5.292}$$

in which d_h the geometric hydraulic diameter.

Table 5.17 Limits of the fraction of the vapor volume flux $\dot{\alpha}_{lim}$ for the transition from one to another flow pattern according to Streicher (1991)

Froude number Fr	limiting value $\dot{\alpha}_{lim}$	Flow pattern transition type
Arbitrary	$\dot{\alpha}_{lim} = 0.38$	Bubbly → stratified
$0.06 \leq Fr \leq 0.1$ $Fr \leq 0.06$	$\dot{\alpha}_{lim} = 1.09 + 0.16 \ln Fr$ $\dot{\alpha}_{lim} = 0.38$	Stratified → stratified wavy
$Fr \geq 0.1$	$\dot{\alpha}_{lim} = 0.52 + 0.085 \ln Fr$	Stratified → plug-annular
$Fr \leq 0.1$	$\dot{\alpha}_{lim} = 0.60 + 0.087 \ln Fr$	Stratified wavy → annular
$Fr \geq 0.1$	$\dot{\alpha}_{lim} = 0.80$	Stratified → annular

The fraction of the steam volume flux $\dot{\alpha}$ is obtained by:

$$\dot{\alpha} = \beta - 2.55 \cdot \beta \cdot (1 - \beta)^{0.36} \cdot (Fr \cdot Ga)^{-0.11} \text{ with } \beta = \frac{1}{1 + \left(\frac{1-\dot{x}}{\dot{x}}\right) \cdot \frac{\rho_g}{\rho_l}}, \tag{5.293}$$

in which the Galilei number Ga appears, which is defined in the following way

$$Ga = \frac{g \cdot \lambda_c^3 \cdot \rho_l^2}{\mu_l^2}. \tag{5.294}$$

In the Galilei number appears the Laplace's constant λ_c (see Eq. (5.286)), which is a ratio of surface tension forces versus gravity and density differences. The heat transfer for bubble boiling h_b near ambient pressure is obtained by an empirical numerical equation of the type:

$$h_b = 1.95 \cdot \dot{q}_w^{0.72} \cdot p^{0.24}, \tag{5.295}$$

where the heat transfer coefficient h_b in $(\mathrm{W}/(\mathrm{m}^2\mathrm{K}))$, the heat flux $\dot{q}_w$ in $(\mathrm{W}/\mathrm{m}^2)$ and the pressure p in (bar) are to be used. With a range of validity of: $10^4 < \dot{q}_w\,(\mathrm{W}/\mathrm{m}^2) < 10^6$ and $0.5 < p\,(\mathrm{bar}) < 20$, the previous equation is well suited for transfer to a solar absorber tube, but other relationships are required for power plant applications. In the range of wave flow (index II) and layer flow (index III), for the calculation of the heat transfer coefficient $h_{II} = h_{II}I$ the relation is valid:

$$h_{II} = h_{III} = 0.58 \cdot \dot{x}_m^{-0.194} \cdot \sqrt{h_l^2 + h_b^2}, \tag{5.296}$$

where $\dot{x}_m$ represents the average vapor coefficient in the considered area. For the heat transfer coefficient in the area of the wavy annular flow (index IV) applies:

$$h_{IV} = \sqrt{h_l^2 + h_b^2}, \tag{5.297}$$

while in the case of annular flow (index V) the heat transfer coefficient h_V is calculated from

$$h_V = \sqrt{h_l^2 + \left(1 + 36.5 \times 10^{-9} \cdot K_D^{1.5}\right) \cdot h_b^2};\; K_D = \frac{\dot{m} \cdot \Delta h_g \cdot \dot{x}_m}{\dot{q}_w \cdot \dot{\alpha}_m \cdot \frac{\rho_g}{\rho_l}}. \tag{5.298}$$

Δh_gin Eq. (5.298) denotes the evaporation enthalpy. With the equations it is possible to determine the heat transfer coefficient from the wall to the fluid with sufficient accuracy for the boundary conditions occurring in a thermal solar collector. However, intermittent flow forms such as slug flow or plug flow are not considered. For these, separate flow stability considerations may have to be made, the fundamentals of which can be found, for example, in Heinzel (1998).

5.2.8 Condensation in Tubes and Pipes

In a solar collector operated with water or water–glycol mixtures, or in direct evaporation solar farm concepts, steam generation intrinsically occurs in the collector absorber. The potential range of this steam in the piping network essentially determines the control of the system, its temporal availability as well as the dimensioning of the components of the solar circuit. In contrast to solar thermal power plants, the formation of steam in solar thermal low-temperature systems is undesirable.

In case of a collector stagnation, the generated steam is pressed from the collector into the connecting pipes. Since the temperature there is lower, it can condense there. The steam production capacity is also determined, among other things, by the thermal losses of the piping, so that it is necessary to consider the heat transport from the inside of the steam-filled piping towards the surroundings. This requires a consideration of the condensation processes in the piping. In addition, the amount of residual liquid in the collector, the fraction of steam and liquid in the individual pipe sections must be known. A detailed consideration of condensation in inclined pipes, as occurs in real solar thermal systems, is omitted here and only the basic physical processes in vertical or horizontal pipes are outlined below. Special effects such as condensate subcooling or steam superheating are not discussed, rather reference is made to the technical literature.

Although in low-temperature thermal solar systems mostly binary mixtures of water and propylene glycol are used as heat transfer medium, most technical considerations assume a pure water vapor flow in the pipes. Experimental investigations show that fractional evaporation of the fluid occurs in early stagnation phases (Rommel et al., 2007). The phase change that takes place in the collector initially allows a water vapor flow to develop in the piping, with the maximum vapor range being reached almost exclusively in this early stagnation phase. Only when the heat transfer medium is highly concentrated in the collector field and the steam production rates are comparatively low does the glycol fraction in the steam flow increase. Another advantage of the calculations of water vapor flows is that the necessary material values for water are known very precisely, in contrast to water-propylene glycol mixtures. The description of condensation within the tube requires knowledge of the flow state. Since flow and temperature fields are hardly accessible analytically even at low complexity of the problem, mostly dimensionless numbers are determined here as well by means of the similarity theory. An illustrative introduction of the indices used in this section can be found in Baehr and Stephan (2008).

5.2.8.1 Steam Velocities at Stagnation

First, the velocity of the steam flow in the stagnation case must be determined. The direct experimental measurement requires a considerable metrological effort. However, if the steam production rate is known, the inlet velocity of the steam from the collector ug into the piping can be estimated.

$$u_g = \frac{P_g}{\Delta h_V \; A_C} = \frac{P_g^* \; A_{col}}{\Delta h_V \; A_C}. \tag{5.298}$$

Therein, the flow cross-section A_C is given by the tube inner diameter d. Both the steam production rate P_g and d depend on the collector aperture area A_{col}. Table 5.18 shows typical required pipe sizes for different collector areas A_{col} and system types (low-flow or high-flow systems). If, for example, the values of a low-flow system $15\,\mathrm{l}/\left(\mathrm{m}^2\mathrm{h}\right)$ are taken as a basis, the steam flow velocities ug and steam mass flow densities as a function of the specific steam production rate P_g^* given in Table 5.19 result at the inlet to the pipe. Table 5.19 shows that the steam velocity at the inlet of the pipeline u_g depends only on the specific steam production power. The resulting steam mass flow, however, is additionally a function of the pipe cross-section d, which in turn depends on the collector area. In addition, it must be noted that in most cases, during stagnation, steam propagation occurs through both the supply and return lines, halving the flow values listed in Table 5.18. The largest steam production rates to date have been measured on an evacuated tube collector at Fraunhofer ISE and were approximately 400 W/m^2 (Rommel et al., 2007).

5.2.8.2 Condensation in a Vertical Pipe

When saturated steam comes into contact with a surface whose temperature is below the saturation temperature of the steam, the steam liquefies. The liquid precipitates on the surface and serves as a source of further condensate. This precipitation can occur in the form of individual drops or as a closed liquid film.

The occurrence of drop condensation is decisively accelerated by impurities on the cold surface, into which the wettability of the cooling surface and the surface tension of the condensate enter. However, droplet condensation plays only a minor role in solar thermal systems.

A simple configuration for phase change is the condensation of a vapor on a cold vertical surface, which is illustrated in Fig. 5.56. The condensate forms a film on the surface driven in a vertical configuration by gravity. At first the film

Table 5.18 Typical dimensioning of the solar circuit piping, in which the dimensions of the pipes were selected as a function of the volume flow so that a comparable flow velocity of the fluid u_g of about $u_g = 0.35$ m/s results (Remmers, 2001)

A_{col} (m^2)	d (m) High flow $30\,\text{l}/(\text{m}^2\text{h})$	d (m) Low flow $15\,\text{l}/(\text{m}^2\text{h})$	d (m) High flow $15\,\text{l}/(\text{m}^2\text{h})$
20	0.025	0.016	0.016
30	0.032	0.020	0.020
50	0.039	0.025	0.025
80	0.050	0.032	0.032
130	0.072	0.039	0.032
160	0.072	0.050	0.039
200	0.085	0.050	0.050
250	0.085	0.072	0.050
300	0.103	0.072	0.072
400	0.103	0.085	0.072
u_g (m/s)	0.33	0.37	0.35
A_{col}/A_c	3.95×10^4	8.94×10^4	1.09×10^5

Table 5.19 Specific steam production rate P_g^* (reference area = collector aperture area) of a typical low-flow system $15\,\text{l}/(\text{m}^2\text{h})$ and resulting steam flow velocity and steam mass flow rate at the collector outlet

P_g^* (W/m^2)	u_g (m/s)	$\dot{M}_g$ $(\text{kg}/(\text{m}^2\text{s}))$
20	0.38	0.8
50	0.96	2.1
100	1.92	4.2
200	3.83	8.3
300	5.75	12.5
400	7.66	16.7

is laminar near the top and relatively thin. As the film proceeds downwards its thickness increases continuously vapor condenses at the vapor film interface. If the film exceeds a certain thickness first signs of a transition to a nonlaminar flow are observed referred to a *wavy flow region*, in which a regular sequence ripples are observed. This region is last followed by a turbulent film flow, provided that the cooled wall is long enough.

Figure 5.56b illustrates that even in a laminar film flow interacts with the vapor boundary layer. At the vapor film interface the temperature corresponds to the saturation temperature to the local pressure along the wall T_{sat}, while the saturation temperature is lower than that of the vapor far away the wall (T_∞) but higher than that of the wall T_w. Through the shear stress at the liquid–vapor interface, the falling jet of vapor aids the downward flow of the liquid film. The vapor in the falling jet is colder than the vapor reservoir and warmer than the liquid in the film attached to the wall. The complexity of the two-phase flow gets more

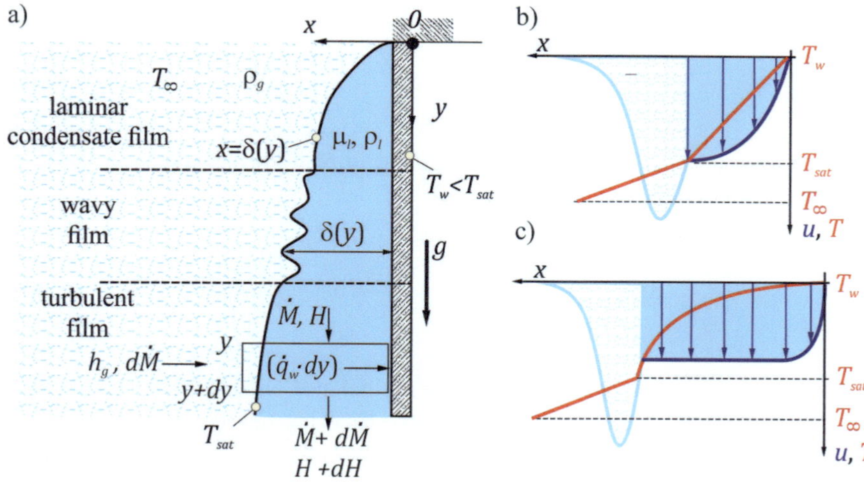

Fig. 5.56 (**a**) Flow regions of film condensation on a vertical cold surface and (**b**) qualitative wall-normal temperature and velocity distribution in laminar flow and (**c**) turbulent flow

sophisticated the wavy and turbulent flow domain (Fig. 5.56c). Moreover, in case of very large films covering all flow regimes, the overall heat transfer rate from the vapor reservoir to the wall is dominated by the contributions from the wavy and turbulent sections.

Laminar film condensation

If vapor condenses on a vertical cold surface, a liquid film forms which flows downward driven by gravity. If the vapor velocity in the vicinity of the condensate is low and the liquid film is thin, the flow in the condensate is laminar. The simplest case of laminar film flow when the wall surface is isothermal is described by Nusselt's water film theory. Heat removal is mainly by conduction from the condensate surface towards the wall and convective heat transfer within the film can be neglected. Assuming a Newtonian fluid, a constant wall temperature T_w, the saturated vapor temperature T_{sat} and a temperature-independent dynamic viscosity μ_l the motion in gravity direction reads to:

$$\underbrace{\rho_l\left(u\frac{\partial u}{\partial x}+v\frac{\partial v}{\partial y}\right)}_{\text{inertia}}=-\frac{dp}{dy}+\mu_l\frac{\partial^2 v}{\partial y^2}+\rho_l g=\underbrace{\mu_l\frac{\partial^2 v}{\partial x^2}}_{\text{friction}}+\underbrace{g(\rho_l-\rho_g)}_{\text{sinking}}, \tag{5.300}$$

where due to the slenderness of the film, the vertical pressure gradient in the liquid is the same as the hydrostatic pressure gradient in the external vapor, i.e. $(dp/dy) = \rho_g g$. Assuming inertia to be negligible compared the equation can be simply integrated twice in wall normal direction x assuming at the wall no-slip condition ($v(x = 0) = 0$) the vertical velocity profile is obtained:

$$v(x, y) = \frac{g}{\mu_l}(\rho_l - \rho_g)\delta^2\left(\frac{x}{\delta} - \frac{1}{2}\left(\frac{x}{\delta}\right)^2\right), \tag{5.301}$$

in which $\delta(y)$ is the film thickness. Integrating in wall normal direction the local mass flow ($\dot{M}$) is obtained.

$$\dot{M} = \int_0^\delta \rho_l v(x, y)dx = \frac{g\rho_l}{3\mu_l}(\rho_l - \rho_g)\delta^3. \tag{5.302}$$

The film thickness is obtained by an energy balance on an infinitesimal control volume ($dy \cdot \delta(y)$) as indicated in Fig. 5.56a. Entering this control volume from the right is the saturated vapor stream $d\dot{M}$, the enthalpy flow rate of which is $\left(hg \cdot d\dot{M}\right)$. The vertical enthalpy inflow H is given by

$$H = \int_0^\delta \rho_l v\left(h_f - c_{p,l}(T_{sat} - T)\right)dx. \tag{5.303}$$

The specific enthalpy of the film is smaller than the specific enthalpy of saturated liquid (h_f) because the liquid is subcooled ($T < T_{sat}$). Considering that the temperature across the film from the vapor film towards the wall with the temperature T_w changes linearly due to heat conduction the enthalpy H can be related to the mass flow $\dot{M}$.

$$H = \dot{M}\left(h_f - \frac{3}{8}c_{p,l}(T_{sat} - T_w)\right). \tag{5.304}$$

The formation of a condensate mass flow requires the provision of a condensation heat dissipation at the wall ($\dot{q}_w$), which is given in a simple way by heat conduction ($\dot{q}_w = (h_l/\delta)(T_{sat} - T_w)$). Balancing the thermal power in the control volume and using the previous relation one obtains

$$\begin{aligned} 0 &= H - (H + dH) + h_g d\dot{M} - \dot{q}_w dy \\ &= \underbrace{\left(h_{fg} + \frac{3}{8}c_{p,l}(T_{sat} - T_w)\right)}_{h_{fg'}} d\dot{M} - \frac{h_l}{\delta}(T_{sat} - T_w), \end{aligned} \tag{5.305}$$

where the latent heat h_{fg} and a sensible-heat contribution accounting for the cooling of the fresh condensate to temperatures below forms a modified *augmented* latent heat of condensation h'_{fg}. Combining Eq. (5.302) with Eq. (5.305) and integrating from $x = 0$ to $x = \delta$ yields the laminar film thickness $\delta(y)$ as a function y:

$$\delta(y) = \left(\frac{4\, h_l \mu_l}{\rho_l(\rho_l - \rho_g) g} \frac{(T_{sat} - T_w)}{h'_{fg}} \cdot \right)^{\frac{1}{4}}. \tag{5.306}$$

Thus, the laminar film thickness increases with the power ¼ of the vertical length L. Since the local film heat transfer coefficient is via Fourier's law coupled to the film thickness and reads to:

$$h(y) = \frac{\dot{q}_w}{(T_{sat} - T_w)} = \frac{h_l}{\delta(y)} = \left(\frac{\rho_l(\rho_l - \rho_g) g\, h_l^3}{4\mu_l} \frac{h'_{fg}}{(T_{sat} - T_w)} \cdot \frac{1}{y} \right)^{\frac{1}{4}}. \tag{5.307}$$

This allows the average heat transfer coefficient $\overline{h}_L$ in the film of length L to be determined and also the average Nusselt number

$$\overline{h}_L = \frac{4}{3} h_{y=L};\ \overline{Nu_L} = \frac{\overline{h}_L L}{h_l} = 0.943 \left(\frac{\rho_l(\rho_l - \rho_g) g}{h\, \mu_l} \frac{h'_{fg} L^3}{(T_{sat} - T_w)} \right)^{\frac{1}{4}}. \tag{5.308}$$

For engineering evaluations, the liquid properties are best evaluated at the average film temperature $(T_{sat} + T_w)/2$. The condensation heat transfer coefficient may also be written in terms of the film Reynolds number ($Re_y = 4\dot{M}/\mu_l$), with $\dot{M}$ from Eq. (5.302) which is then

$$\frac{h_y}{h_l} \left(\frac{\mu_l^2}{\rho_l(\rho_l - \rho_g) g} \right) = 1.1 Re_y^{-1/3};\ \frac{\overline{h}_L}{h_l} \left(\frac{\mu_l^2}{\rho_l(\rho_l - \rho_g) g} \right) = 1.47 Re_L^{-1/3}. \tag{5.309}$$

Rohsenow (1998) showed that if the condensate temperature profile is allowed to be nonlinear to account for convection effects in the condensate film, the augmented latent heat expression requires a modification, for which the following expression has been shown to be successful $h'_{fg} = h_{fg} + 0.68 c_{p,l}(T_{sat} - T_w)$.

Laminar film condensation with wave formation

As the running length of the condensation film increases, its thickness and flow velocity increase. Even though laminar flow is still present within the film, waves form at the film-vapor interface which enhance the heat transfer, which is then about 10–25% higher than that of Nusselt's water skin theory. There are different models that take this wave formation into account when determining the heat transfer, which can be specified using a critical Reynolds number.

$$Re_{crit} = 0.392 \left[\left(\frac{\sigma}{\rho_l \cdot g} \right)^{\frac{1}{2}} \left(\frac{g}{\nu_l^2} \right)^{\frac{1}{3}} \right]^{\frac{3}{4}}, \tag{5.310}$$

where ν_l is the kinematic viscosity of the condensate and σ is its surface tension. According to Rohsenow, surface waves can be observed in the Reynolds number range of $256\,Pr^{0.44} \leq Re \leq 400$. Where Pr is the molecular Prandtl number. Within these run lengths, heat transfer is improved over laminar flow and textbooks suggest a multiplication factor of 1.15 over Eq. (5.308) for the average heat transfer coefficient in this range. Somewhat more accurately, the influence of film flow ripple can be accounted for by a Reynolds number dependent ripple correction factor f_{rib}. Here approaches of different authors exist, which do not differ very much; mostly the following relation is found

$$f_{rib} = \begin{cases} 1 & \text{for } Re < 1 \\ Re^{0.04} & \text{for } Re \geq 1 \end{cases} \tag{5.311}$$

Turbulent film condensation

As the film length increases, the flow changes into a turbulent condensate flow. There are different specifications as to the Reynolds number from which the transition occurs. In addition to the Reynolds number, the molecular Prandtl number is also included in the heat transfer relationships. The following relationship is commonly used:

$$\begin{aligned} Nu &= \frac{h\left(\frac{\nu_l^2}{g}\right)^{\frac{1}{3}}}{\lambda_l} = 0.0325 Re^{1/4}\sqrt{Pr}\ ; \\ Re &= \left(89 + 0.024\sqrt{Pr}\left(\frac{Pr}{Pr_w}\right)^{\frac{1}{4}}(f - 2300)\right); \\ f &= \frac{c_{p,l}(T_{sat} - T_w)}{h_{fg}}\,\frac{1}{Pr}\,\frac{y}{(\nu_l^2/g)^{\frac{1}{3}}} \end{aligned} \tag{5.312}$$

where $c_{p,l}$ is the heat capacity of the condensate. The approach is valid for $1 \leq Pr \leq 25$ and $400 \leq Re \leq 7 \times 10^5$. The material values are to be used at saturation temperature and Pr_w is to be formed with the wall temperature T_w. For practical purposes, the heat transfer in each region of condensate flow on a vertical surface is summarized as follows:

$$h = \sqrt[4]{(f_{rib} \cdot h_{lam})^4 + h_{turb}^4}, \tag{5.313}$$

f_{rib} is calculated according to Eq. (5.311), the heat transfer coefficient of laminar film flow according to (5.307) and that of turbulent film flow according to (5.312).

Influence of the steam flow velocity

The previous considerations include condensation of steam with low flow velocities or steam at rest. However, during the collector stagnation significant steam flow velocities occur, this also affects the condenser of a Rankine cycle. The influence of a steam flow on the condensate film at a vertical steam-water film interface can be neglected up to a steam velocity of approximately 5 m/s. The condensate film flow then depends on the steam velocity. The condensate film flow then depends only on gravity and the heat transfer relationships mentioned for steam at rest apply.

Only at very high steam production rates do considerable flow velocities occur, in which the flowing steam exerts a shear stress τ_δ on the condensate film, which depends on the drag coefficient f:

$$\tau_\delta = f \cdot \frac{\rho_g \cdot u_g^2}{4}, \tag{5.314}$$

For a laminar steam flow of steam ($Re < 2320$), the drag coefficient f_D is independent of the pipe wall roughness and $f_D = 64/Re$ applies (compare Eq. (5.49)). In the turbulent regime, the steam flow also behaves like a conventional Newtonian fluid, so that with respect to the pressure drop, the relationships worked out in Sect. 5.1.16 apply. The shear stress τ_δ affects the mean condensate flow velocity of the film uf and the condensate mass flow rate draining through a plane perpendicular to the wall. The following relationships are obtained for both:

$$u_0 = \frac{(\rho_l - \rho_g)g}{3\,\mu_l}\,\delta^2 \pm \frac{\tau_\delta\,\delta}{2\,\eta_L},$$
$$\dot{M}_l = u_0\,\rho_l\,b\,\delta = \frac{(\rho_l - \rho_g)g\,b}{3\,\mu_l}\,\delta^3 \pm \frac{\rho_l\,\tau_\delta\,\delta^2\,b}{2\,\mu_l}, \tag{5.315}$$

in which b is the wetted depth, i.e. for a tube $b = \pi d$. The positive sign applies to downward flowing steam, the negative sign accordingly to upward flowing steam. In the laminar film flow region, the heat transfer results directly from the film thickness via the relationship $h = \lambda_l/\delta$.

5.2.8.3 Condensation in the Horizontal Pipe

Whereas in vertical pipes the condensate flows off the wall as a film, in a horizontal pipe—depending on the flow velocity of the steam and the proportion of liquid in the pipe cross-section—various flow forms can occur which are similar to those during heating (compare Sect. 5.2.4.2 Horizontal pipes). At low flow velocities, the influence of gravity predominates and, depending on the fluid content, a stratified, slug or plug flow is formed. At high flow velocities, annular or even bubbly flow establishes, which can also occur at high steam velocities in vertical pipes if the steam shear stress is large compared to the drag effect of gravity.

A stratified flow is formed if

$$u_g = \frac{x\,\dot{M}}{A\left[g\,d\,\rho_g\,(\rho_l - \rho_g)\right]} \leq 1, \tag{5.316}$$

where x represents the vapor content. During stagnation, steam velocities of about 2 m/s typically occur at the inlet to the piping. Assuming a pipe cross-section of $d = 0.016$ m, Eq. (5.316) yields a value of $u_g = 0.233$ m/s and a Reynolds number of $Re = 4610$, so that mostly a stratified flow is obtained. In the calculations, it should be noted that the flow velocity decreases more and more with the rising flow length within the tube. Therefore, the following consideration is limited only to this type of two-phase flow in horizontal pipes.

In stratified flow, the condensate flows downward from the pipe wall and collects at the bottom of the pipe. Considering the pressure drop in the flow direction of the steam, the condensate fills the entire cross-section at the end of long tubes. The heat transfer coefficient h in this type of flow is largely dependent on the thickness of the bottom condensate and is calculated according to:

$$h = F(\gamma)\left(\frac{\rho_l\,(\rho_l-\rho_g)g\,h_{fg}\,\lambda_l^3}{d\,\mu_l\,(T_{sat}-T_w)}\right)^{\frac{1}{4}}, \tag{5.317}$$

in which h_{fg} is the latent heat. The thinner the soil condensate film, the greater the factor F and the heat transfer to the tube wall improves. Table values exist for F. The wetting angle γ can hardly be determined by measurement. Baehr and Stephan (2008) propose empirical equations for the calculation of the heat transfer coefficients, which consider the influence of the steam flow on the turbulent condensate film. For the local heat transfer, they give the following correlation:

$$Nu = 0.023\,Re^{0.8}\,Pr^{0.4}\left[(1-x)^{0.8} + \frac{3.8\,(1-x)^{0.04}\,x^{0.76}}{p^{0.38}}\right];$$

$$Nu = \frac{h\,d}{\lambda_l};\ Re = \frac{u_m\,d}{\nu_L};\ u_m = \frac{4\,\dot{M}}{\rho_L\pi\,d^2};\ Pr = \frac{\nu_l}{a_l}\ \text{and}\ p^+ = \frac{p}{p_{crit}}. \tag{5.318}$$

The critical pressure p_{crit} of water is 22.12 MPa. However, there is disagreement in the literature about the range of validity of this relation. Shah and Sekulic (2003) quote a validity of the calculation method for steam velocities from 3 to 300 m/s and mass flow densities between $10.8 \le \dot{M}\left(\text{kg}/(\text{m}^2\text{s})\right) \le 210.6$, while Collier and Thome (1996) limit the validity range to mass flow densities above $\dot{M}\left(\text{kg}/(\text{m}^2\text{s})\right) > 200$. For slow vapor flows, the latter recommend a different relationship:

$$Nu = C\,Re^n\,Pr^{1/3}\ \text{with}\ Re_e = \frac{\dot{M}_e\,d}{\mu_l}\ \text{and}\ \dot{M}_e = \left[(1-x) + x\left(\frac{\rho_l}{\rho_g}\right)^{1/2}\right]. \tag{5.319}$$

The quantities C and n are calculated depending on the flow pattern. For $Re_e < 5\times10^4$, $C = 0.0265$ and $n = 0.8$, and for $Re_e > 5\times10^4$, $C = 5.03$ and $n = 1/3$. Since the steam production rates escaping at stagnation lead to steam flow velocities in the range of 2 m/s and mass flow densities of $4\left(\mathrm{kg}/\left(\mathrm{m}^2\mathrm{s}\right)\right)$ for most house plants, the determination method according to Eq. (5.319) is more suitable for these cases.

References

Aicher, T., & Martin, H. (1997, October). New correlations for mixed turbulent natural and forced convection heat transfer in vertical tubes (Elsevier, Ed.). *International Journal of Heat and Mass Transfer, 40*(15), 3617–3626.

Anderson, D. A., Tannehill, J. C., & Pletcher, R. H. (1984). *Computational fluid mechanics and heat transfer*. Hemisphere Publishing Co.

Arpaci, V. S., & Larsen, P. S. (1984). *Convection heat transfer*. Prentice Hall Inc.

Baehr, H. D., & Stephan, K. (2008). *Wärme- und Stoffübertragung* (6. Auflage ed.). Springer.

Baker, O. (1954, July). Design of pipe lines for simultaneous flow of oil and gas. *Oil and Gas Journal, 53*, 185–195.

Bankoff, S. G. (1957). Ebullition from solid surfaces in the presence of a pre-existing gaseous phase. *Transactions ASME, 79*, 735ff.

Barenblatt, G. I., & Chorin, A. J. (1998). Turbulence: An old challenge and new perspectives. *Meccanica, 33*(5), 445–468.

Bejan, A. (2013). *Convection heat transfer* (4th ed.). Wiley.

Bejan, A., & Kraus, A. (2003). *Heat transfer handbook*. Wiley.

Betchov, R., & Criminale, O. W. (1967). *Stability of parallel flows*. Academic Press.

Bhatti, M. S. (1985). *Fully developed temperature distribution in a circular duct tube with uniform wall temperature.* Owens-Corning Fiberglas Cooperation.

Boger, D., & Yeow, Y. (1988). *Ullmann's encyclopedia of industrial chemistry-chapter fluid mechanics* (Vol. 1). VCH Verlagsgesellschaft mbH.

Burmeister, L. C. (1993). *Convective heat transfer* (2nd ed.). Wiley-Interscience.

Carteciano, L. N., & Grötzbach, G. (2003). *Validation of turbulence models in the computer code FLUTAN for a free hot sodium jet in different buoyancy flow regimes.* Forschungszentrum Technik und Umwelt.

Castaing, B., Gunaratne, G., Heslot, F., Kadanoff, L., Libchaber, A., Thomae, S., & Zanetti, G. (1989). Scaling of hard thermal turbulence in Raleigh-Benard convection. *Journal of Fluid Mechanics, 204*, 1ff.

Cebeci, T., & Bradshaw, P. (1977). *Momentum transfer in boundary layers.* McGraw-Hill Inc.

Cengel, Y., & Cimbala, J. M. (2006). *Fluid mechanics fundamentals and applications.* McGraw-Hill Higher Education.

Chen, R. Y. (1973). Flow in the entrance region at low Reynolds number. *Journal of Fluids Engineering, 95*, 153–158.

Chenoweth, D. R., & Paolucci, S. (1986). Natural convection in an enclosed vertical air layer with large horizontal temperature differences. *Journal of Fluid Mechanics, 169*, 173–210.

Chien, K. Y. (1982). Prediction of channel and boundary-layer flows with a low-Reynolds-number turbulence mode. *AIAA Journal, 20*(1), 33–38.

Churchill, S., & Ozoe, H. (1973). Correlations for laminar forced convection with uniform heating in flow over a plate and in developing and fully developed flow in a tube. *Journal of Heat Transfer, 18*, 78ff.

Churchill, S., & Usagi, R. (1972). A general expression for the correlation of rates of transfer and other. *AIChE Journal, 18*, 1121–1128.

Cole, J. D. (1968). *Perturbation methods in applied mathematics.* Blaisdell Publishing Co. Ginn and Co.

Collier, J. G., & Thome, J. R. (1996). *Convective boiling and condensation* (3rd ed.). Oxford University Press.

DIN EN ISO 5167. (2004). *Durchflussmessung von Fluiden mit Drosselgeräten in voll durchströmten Leitungen mit Kreisquerschnitt.* Beuth Verlag GmbH.

Donaldson, C. (1972). Calculation of turbulent shear flows for atmospheric and vortex motions. *AIAA Journal, 10*, 4–12.

Eberhart, J. G., & Schnyders, H. (1973). Application of the mechanical stability condition to the prediction of the limit of superheat für normal alkanes, ether and water. *Journal of Physical Chemistry, 77*(23), 2730.

Eckert, E. R., & Drake, R. M. (1972). *Analysis of heat and mass transfer*. Mc Graw-Hill.

Franc, J.-P., & Michel, J.-M. (2006). *Fundamentals of cavitation.* Springer.

Friedel, L. (1979, June). Improved friction drop correlations for horizontal and vertical two-phase pipe flow. *European Two-phase Flow Group Meeting*, Paper E2.

Garbrecht, O. (2017). *Large eddy simulation of three-dimensional mixed convectionon a vertical plate.* Disseration RWTH Aachen, Fakultät Maschinenwesen.

Gebhart, B. (1988). *Buoyancy-induced flows and transport* (1st ed.). Springer.

Germano, M., Piomelli, U., Moin, P., & Cabot, W. H. (1991). A dynamic sub-grid scale eddy viscosity model. *Physics of Fluids A, 3*, 1760–1765.

Gnielinski, V. (1975). Berechnung mittlerer Wärme- und Stoffübergangskoeffizienten an laminar und turbulent überströmten Einzelkörpem mit Hilfe einer einheitlichen Gleichung. *Forsch-Ing Wes, 41*, 145–150.

Grigull, U., & Tratz, H. (1965). Thermischer Einlauf in ausgebildeter laminarer Rohrströmung. *International Journal of Heat and Mass Transfer, 8*, 669–678.

Guo, J., & Julien, P. Y. (2003). Modified log-wake law for turbulent flow in smooth pipes. *Journal of Hydraulic Research, 41*(5), 493–501.

Hanjalic, K., & Launder, B. E. (1972). A Reynolds stress model of turbulence and its application to asymmetric shear flows. *Journal of Fluid Mechanics, 52*, 609–638.

Harlow, F. H., & Nakayama, P. I. (1968). *Transport of turbulence energy decay rate.* Los Alamos Science Lab.

Heinzel, V. (1998). *Solare Prozeßwärmeerzeugung -Siedewasserkollektoren.* Universität.

Heinzel, V., Holzinger, J., & Simon, M. (1994, August). Flow patterns in horizontal and tilted tubes of direct evaporating solar (H. E. Soc., Ed.). In *Proceedings of the ISES Solar World Congress* (Vol. 4, pp. 135–140).

Hewitt, G. F., & Roberts, D. N. (1969). *Studies of two-phase flow patterns by simultaneous x-ray and flash photography.* HMSO. UKAEA.

Hickmann, H. J. (1974). An asymptotic study of the Nusselt-Graetz problem, Part 1: Large x behaviour. *International Journal of Heat and Mass Transfer, 96*, 354–358.

Hsu, C. J. (1968). Exact solution to entry region of Laminar heat transfer with axial conduction. *Chemical Engineering Science, 23*, 457–468.

Incropera, F. P., & DeWitt, D. P. (2001). *Fundamentals of heat and mass transfer* (5th ed.). Wiley.

Ishii, M., & Hibiki, T. (2011). *Thermo-fluid dynamics of two-phase flow* (2nd ed.). Springer.

Jones, W. P., & Launder, B. E. (1972, February). The prediction of laminarization with a two-equation model of turbulence. *International Journal of Heat and Mass Transfer, 15*(2), 301–314.

Kakac, S., Aung, W., & Viskanta, R. (1985). *Natural convection: Fundamentals and applications.* Hemisphere Publishing.

Kakac, S., Shah, R., & Aung, W. (1987). *Handbook of single phase convective heat transfer* (A. W. Publication, Ed.). Wiley.

Kasagi, N., & Ohtsubo, Y. (1992). Direct numerical simulation of low Prandtl number thermal field in a turbulent channel flow. In F. Durst et al. (Eds.), *Turbulent shear flows* (Vol. 8, pp. 97–119). Springer.

Kawamura, H., Abe, H., & Matsuo, Y. (1999, June). DNS of turbulent heat transfer in channel flow with respect to Reynolds and Prandtl number effects. *International Journal of Heat and Fluid Flow, 20*(3), 196–207.

Kays, W. M. (1994). Turbulent Prandtl number. Where are we? *Journal of Heat Transfer, Transactions ASME, 116*, 284–296.

Kek, V., & Müller, U. (1993, July). Low Prandtl number convection in layers heated from below. *International Journal of Heat and Mass Transfer, 36*(11), 2795–2804.

Khan, Z., & Joshi, J. (2015). Comparison of k–ε, RSM and LES models for the prediction of flow pattern in jet loop reactor. *Chemical Engineering Science, 127*, 323–333. https://doi.org/10.1016/j.ces.2015.01.054

Kim, W., & Menon, S. (1995). A new dynamic one-equation subgrid-scale model for large eddy simulation. In *33rd Aerospace Sciences Meeting and Exhibit, Reno NV-USA.*

Kreith, F., & Goswami, D. Y. (2005). *The CRC handbook of mechanical engineering* (2nd ed.). CRC-Press LLC.

Kreith, F., Manglik, R. M., & Bohn, M. S. (2011). *Principles of heat transfer* (7th ed.). Cengage Learning.

Kunii, D., & Levenspiel. (2013). *Fluidization engineering* (2nd ed.) (Brenner, Ed.). Butterworth-Heinemann-Elsevier.

Lamb, H. (1993). *Hydrodynamics* (6th ed.). Cambridge University Press.

Launder, B. E., Reece, G., & Rodi, W. (1975). Progress in the development of a Reynolds-stress turbulence closure. *Journal of Fluid Mechanics, 68*(3), 537–566.

Launder, B. E., & Spalding, D. B. (1974). The numerical computation of turbulent flows. *Computer Methods in Applied Mechanics and Engineering, 3*, 269–289.

Ledinegg, M. (1938). Instability of flow during natural and forced circulation. *Die Wärme, 61*, 8.

Lockhart, R.W., Martinelli, R.C. (1949). *Chemical engineering progress, 45*, pp. 39–48.

Lu, H., Rutland, C. J., & Smith, L. M. (2008). A posteriori tests of one-equation LES modeling of rotating turbulence. *International Journal of Modern Physics C, 19*(12), 1949–1964. https://doi.org/10.1142/S0129183108013394

Mandhane, J., Gregory, G., & Aziz, K. (1974). A flow pattern map for gas–liquid flow in horizontal. *International Journal of Multiphase Flow, 1*, 537–553.

Michelsen, M. L., & Villadsen, J. (1974). The Graetz problem with axial conduction. *International Journal of Heat and Mass Transfer, 17*, 1391–1402.

Müller, U. (2016). Strömung mit mehreren Phasen. In H. Oertel (Ed.), *Prandtl - Führer durch die Strömungslehre.* Springer Vieweg Fachmedien. https://doi.org/10.1007/978-3-658-08933-7_9

Müller, U., & Ehrhard, P. (1999). *Freie Konvektion und Wärmeübertragung.* C.F. Müller-Verlag.

Ng, K. H., & Spalding, D. B. (1972). Turbulence model for boundary layers near walls. *Physics of Fluids, 15*, 20–30.

Nicoud, F., & Duos, F. (1999). Subgrid-scale modelling based on the square of the velocity gradient tensor. *Flow, Turbulence and Combustion, 62*, 183–200.

Otic, I., Grötzbach, G., & Wörner, M. (2005). Analysis and modelling of the temperature variance equation in turbulent natural variance equation in turbulent natural (C. U. Press, Ed.). *Journal of Fluid Mechanics, 525*, 237–261.

O'Toole, J. L., & Silveston, P. L. (1961). Correlation of convective heat transfer in confined horizontal layers. *Chemical Engineering Progress Symposium Series, 67*, 81ff.

Pacio, J. W. (2013). Assessment of liquid metal technology status and research paths for their use as efficient heat transfer fluids in solar central receiver systems. *Solar Energy, 93*, 11–22. https://doi.org/10.1016/j.solener.2013.03.025

Pederson, R. J., & May, E. K. (1982). *Flow instability during direct steam generation in a line-focus solar collector system.* Solar Energy Research Institute Report.

Piller, M., Nobile, E., & Hanratty, T. J. (2002). DNS study of turbulent transport at low Prandtl numbers in a channel flow. *Journal of Fluid Mechanics, 458*, 419–441.

Reichhardt, H. (1951, Juli). Vollständige Darstellung der turbulenten Geschwindigkeitsverteilung in glatten Leitungen. *Zeitschrift für Angewandte Math Mech, 31*(7), 208ff.

RELAP Code Development Team. (1995). *RELAP5/MOD3 code manual volume I–IV: Code structure, system models, and solution methods.* Idaho National Engineering Laboratory: Lockheed Idaho Technologies Company.

Remmers, K.-H. (2001). *Große Solaranlagen: Einstieg in Planung und Praxis* (2. überarb. Auflage ed.). Solarpraxis AG.

Reynolds, W. C. (1976). Computation of turbulent flows. *Annual Review of Fluid Mechanics, 8*, 183–208.

Rodi, W. (1981). Progress in turbulence modelling for incompressible flows. *AIAA Journal, Paper 81-0045*.

Rodi, W. (1993). *Turbulence models and their application—a state of the art review.* Monograph Int.l Ass. for Hydraulic Res. (IAHR): A.A. Balkema.

Rodi, W. (2005). *Engineering turbulence modelling and experiments 6* (E. I. ETMM6, Ed.). Elsevier Science.

Rohsenow, W. M. (1998). *Handbook of heat transfer*. McGraw-Hill.

Rommel, M., Siems, T., Schüle, K., Mehnert, S., & Thoma, C. (2007). *Entwicklung von Techniken zur Beherrschung des Stillstandsbetriebs.* www.solarkombianlagen-xl.info

Schlichting, H., & Gersten, K. (2000). *Boundary-layer theory* (8th ed.). Springer.

Schulenberg, T., & Stieglitz, R. (2010). Flow measurement techniques in heavy liquid metals. *Flow Measurement Techniques in Heavy Liquid Metals, 240*, 2077–2087.

Shah, R. K., & London, A. L. (1978). *Laminar flow forced convection in ducts: A source book for compact heat exchanger analytical data* (Vols. Advances in heat transfer, Supplement 1). Academic Press.

Shah, R. K., & Sekulic, D. (2003). *Fundamentals of heat exchanger design.* Wiley.

Shams, A., De Santis, A., & Roelofs, F. (2019). An overview of the AHFM-NRG formulations for the accurate prediction of turbulent flow and heat transfer in low-Prandtl number flows. *Nuclear Engineering and Design, 355*, 110342. https://doi.org/10.1016/j.nucengdes.2019.110342

Smagorinsky, J. (1963). General circulation experiments with the primitive equations I the basic experiment. *Monthly Weather Review, 91*, 99–164.

Sparrow, E. M., Patankar, S. V., & Shahrestani, H. (1978). Laminar heat transfer in a pipe subjected to circumferentially varying external heat transfer coefficients. *Numerical Heat Transfer, 1*, 117–127.

Stephan, P., Kabelac, S., & Kind, M. M. (2019). *VDI-Wärmeatlas* (12th ed.). Springer Vieweg.

Stieglitz, R. (2007). Chapter 10: Low Prandtl number thermal hydraulics. In Nuclear Energy Agency NEA-OECD (Ed.), *Handbook on lead-bismuth eutectic alloy and lead properties, materials, compatibility, thermalhydraulics and technologies* (pp. 399–478). NEA-OECD.

Streicher, W. (1991). *Simulation des instationären Verhaltens von Kompressionswärmepumpen.* Universität Graz.

Taitel, Y., & Duckler, A. A. (1976). A model for predicting flow regime transitions in horizontal and near horizontal gas-liquid flow. *AIChE Journal, 22*(1), 47–55.

Thome, J. R., & Cioncolini, A. (2015). Chapter 2: Two-phase flow pattern maps for macrochannels. In T. J. R., *Encyclopedia of two-phase heat transfer and flow* (Vol. 3, p. 1112). World Scientific Publishing.

van Dyke, M. (1975). *Perturbation methods in fluid mechanics* (Annotated ed.). Parabolic Press.

Veritas, D. N. (2010). *Environmental conditions and environmental loads—Recommended practice.* DNV-RP-C205, October 2010, Bærum, Norway.

Versteeg, H. K., & Malalasekera, W. (2007). *An introduction to computational fluid dynamics* (2nd ed.). Pearson Education Limited.

Walz, A. (1966). *Strömungs- und Temperaturgrenzschichten.* G. Braun Verlag.

White, F. M. (1991). *Viscous fluid flow* (2nd ed.). McGraw-Hill Science/Engineering/Math.

Whorlow, R. W. (1979). *Rheological techniques.* Wiley & Sons.

Wilcox, D. C. (1986, January). Multiscale model for turbulent flows (A. I. Astronautics, Ed.). In *Proceedings of the 24th AIAA Aerospace Sciences Meeting* (pp. 1311–1320).

Wilcox, D. C., & Traci, R. M. (1976). *A complete model of turbulence.* NASA.

You, D., & Moin, P. (2007). A dynamic global-coefficient subgrid-scale eddy-viscosity model for large-eddy simulation in complex geometries. *Physics of Fluids, 19*(6), 065110.

Zagarola, M. V., & Smits, A. J. (1998). Mean flow scaling of turbulent pipe flow. *Journal of Fluid Mechanics, 373*, 33–79.

Stationary Solar Thermal Collectors

6

Abstract

Solar thermal collectors are the core components of solar thermal energy systems, converting the solar radiation into heat, which is transported to a demand location by active or passive means. Solar thermal collectors can use the whole solar spectrum, not only a specific spectral part. But converted solar radiation cannot be fully used as inevitable thermal losses from the hot absorbers to the colder environment cannot be completely avoided. Therefore, the design of solar collectors must compromise—depending on the specific application—between maximizing the optical gains and the minimization of thermal losses. Highly transparent covers and selective absorbers with small thermal emissivities are used to achieve this target. Concentration by mirror optics and consecutive reduction of receiver area is a powerful tool to obtain an even better energy balance also for very high temperatures. Stationary collector types without or with small concentration are described in their main characteristics and the internal processes described, as well as the concentrating collector types which need one- or two-axis tracking devices to be able to collect the direct solar radiation from the sun over the day. In a last part collector testing methods are described and discussed, as well as the stagnation of collectors at which the highest absorber temperatures are reached without any net energy output, as thermal losses and solar gains just cancel out.

Keywords

Flat-plate collector • Vacuum tube collector • Collector equation • Theory • Parabolic trough • Linear Fresnel reflector • Thermal performance testing

6.1 Introduction

Solar thermal collectors are special kinds of heat-exchangers which transform incoming radiation energy into heat, typically of a fluid which is used to transport the heat to a distant demand. They can be designed for a multitude of applications

R. Stieglitz and W. Platzer, *Solar Thermal Energy Systems*,
https://doi.org/10.1007/978-3-031-43173-9_6

and locations, each having specific requirements with respect to the solar radiation input and the temperatures of the heat output.

In Sect. 6.2 the energy balance of a collector and the physical characteristics will be described mathematically, mainly for the case of a simple flat-plate collector. However, the principles described, and the concepts developed are also valid for other collector times if applied accordingly.

For this reason, we have a huge variety of collector designs including non-concentrating collectors using natural global radiation with a maximum intensity of about 1100 W/m^2 and concentrating collectors which focus the incoming radiation onto small focal spots or linear focal lines with intensities of up to many kW/cm^2.

After dealing with some basic definitions and principles, in Sects. 6.3 and 6.4. different collector types in both categories will be described, including the functionality and specifications of the individual components used. Then the topic of performance testing and efficiency of solar collectors will be treated in Sect. 6.4. Section 6.5 is dealing with the specific problem of stagnation in collectors, which may occur if heat is not needed or cannot be transported away from the collectors.

6.1.1 Basic Structure and Operation of a Collector

Although designs of collectors differ substantially in geometry, choice of materials and mode of operation, very often one finds in some form or other the main components of a collector:

- absorber,
- transparent cover,
- heat transport system,
- thermal insulation.

The aperture is defined as the active optical part of the collector, which accepts solar radiation for conversion to heat. For concentrating collectors additionally, one finds a concentrator optic and an adapted aperture system.

The role of the absorber is the transformation of incoming solar radiation into thermal energy. This heat raises the temperature of the absorber above ambient. Resulting heat losses due to this temperature difference may be reduced by using a transparent cover. A heat transport system usually uses a fluid to transport the generated heat to a different place. However, it is also possible that the heat naturally is transported to a colder place by conduction, thermal radiation, or natural convection. Thermal insulation is used to suppress heat losses of the absorber parts which need not be irradiated or of the heat transport system.

We illustrate the components with a typical flat-plate collector (Fig. 6.1) as an example: A flat absorber sheet of metal is coated black on the side facing the sun. The glass cover at a distance of 2–3 cm from the absorber reduces convective and radiative heat losses of the absorber surface facing the environment. On the back side of the collector some mineral wool or insulating foam is used to reduce the

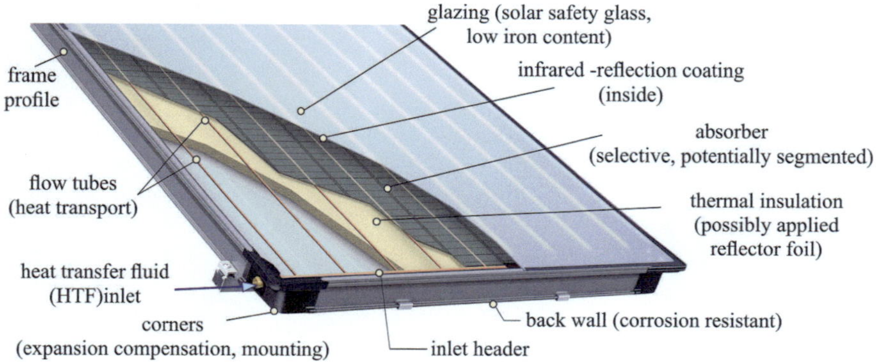

Fig. 6.1 Principal components of a solar flat-plate thermal collector

heat losses from the back side of the absorber sheet. And many pipes carrying water or a similar heat transfer fluid transport the resulting useful heat to a heat consumer or heat storage.

In a solar tower system with a central receiver and a heliostat field, as illustrated in Fig. 6.2, many mirrors of the heliostats—the optical concentrator—are oriented any time in a way to reflect the incoming direct radiation from the sun towards a limited receiver area on a tower. Here the absorber may consist of a multitude of pipes located in this area. A transparent glass cover does not exist, but for instance in the example shown in the picture an open cavity restricts the heat transport to the environment by protecting the absorber tubes from wind and reducing the view factor to sky and environment to reduce the thermal radiation exchange. Thus, this component is also "covering" the absorber but is 100% transparent for the incoming concentrated radiation from the heliostat field. The fluid in the absorber pipes must be suitable for the high temperatures of several hundred degrees reached. Therefore air, steam or molten salt are examples for such a fluid. Thermal insulation may be used in the cavity receiver to avoid peripheral heat losses.

Even the Earth, our blue planet, is a solar thermal collector. We have all the basic components present. Ground and water absorb solar radiation and convert it to heat. The atmosphere is partly withholding thermal radiation to escape to outer space and thus acts as a transparent cover during the day and as thermal insulation during the night. The air of the atmosphere transports heat to other shaded places, being responsible for compensating temperature differences on earth and having temperatures suitable for the existence of life everywhere.

So, in fact "solar collector" is a term which is very general and has many conceptual variations. Collectors may be mounted stationary or use a one-axis or two-axis tracking device to follow the sun's movement. The concentration of light varies from 1 (no concentration) to more than 10^3, the thermal insulation may be just the atmosphere (in an uncovered collector) or an evacuated space with low emissivity areas, so that convective-conductive as well as radiative heat transfer

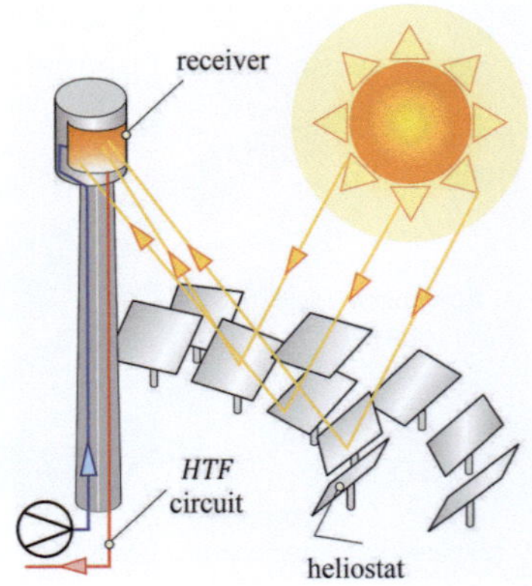

Fig. 6.2 Principal layout of a solar tower with heliostat field

is minimized. Figure 6.3 depicts selection of possible solar collectors using some fluid for transporting the heat to a user. But even this figure does not cover all variety. The heat may be transported away with different media, including solid walls for a passive transport. And in hybrid PVT collectors the absorber is combined with a photovoltaic module to produce heat and electricity simultaneously. But let us first talk about the distinction between passive and active solar.

6.1.2 Distinction Between Active and Passive Solar Thermal Energy

As mentioned already, we distinguish between passive and active solar thermal collectors depending on the existence of an active circulation of a heat transfer fluid. The system is called *active* if a mass transport induced by a mechanical system like a pump or ventilator is included. If the heat transport is taking place only based on a temperature gradient without any external means (e.g. by conduction through a massive absorber wall) the system is called *passive*. In the passive solar application, the heat converted from solar radiation should be used locally close to the absorber. Table 6.1 lists some general pros and cons of passive vs. active systems in a qualitative manner.

Popular examples for passive solar collectors are windows or transparently insulated solar walls. In fact, a complete building could be seen as a huge but inhomogeneous passive solar collector where the sun serves to partially cover the heat demand for comfort living during the heating season. With solar shading devices such a collector can be actively shut down to some extent to avoid overheating in summertime.

It is this aspect which led to the branding of "passive solar building design" or "passive-solar green architecture" which has a developed set of own design rules. Within these designs sometimes also heat transport by natural convection is considered. Hot air from a solar glass house or conservatory may be let into a house by openings or channels, thus providing passive solar heating even to rooms

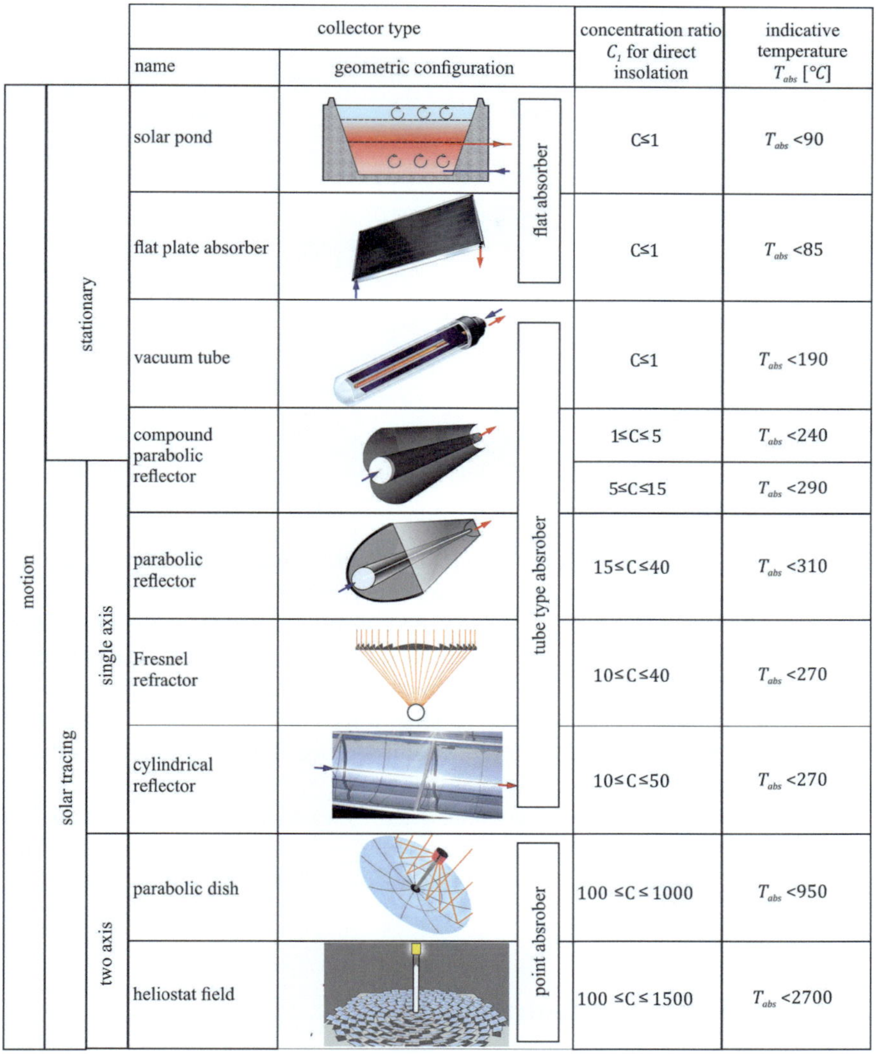

Fig. 6.3 Selection of different solar collector concepts

Table 6.1 Pros and Cons for active/passive solar systems

Area	Passive solar	Active solar
Efficiency	Low	High
Investment cost	Low	High
Operation cost	None	Noticeable
Control	Difficult	Good
Power use	Negligible	Noticeable
Maintenance	Little	Noticeable

Table 6.2 Normal solar transmittance (τ), reflectance (ρ) and absorptance (α) of selected transparent media

Material	Thickness t (mm)	Transmittance τ (–)	Reflectance ρ (–)	Absorptance α (–)
Float glass (soda lime)	4	0.836	0.075	0.089
Green glass (soda lime)	4	0.623	0.062	0.315
White glass (soda lime)	4	0.908	0.080	0.012
Borosilicate glass	3.3	0.926	0.048	0.026
Polyvinylchloride (PVC)	1	0.836	0.082	0.082
Polymethyl methacrylate (PMMA)	1	0.878	0.071	0.051
Polycarbonate (PC)	0.1	0.856	0.093	0.051
Polytetrafluoroethylene (PTFE)	0.1	0.924	0.046	0.030
Polyvinyl fluoride (PVF)	0.1	0.919	0.067	0.014
Polyethylene terephthalate (PETP)	0.1	0.882	0.097	0.022

further away from the actual collector. If ventilators would be used to increase the air transport volume, this system would be converted in an active system.

Another widespread use of a passive solar collector is the thermosyphon collector which works by a heat transport via natural convection. A solar liquid collector is connected directly to a storage tank directly above the collector. When the sun heats up the liquid inside, due to gravitational forces the less dense hot water rises to the storage and some colder water from the storage flows towards the collector inlet, thus circulating relatively slowly.

As consequence we can conclude that often components may be used both in a passive or in an active system. Nevertheless, we will see that an optimization of a collector for a passive or for an active application will result in differences of the design.

6.1.3 Impact of Concentration

Two fundamental categories of collectors are often called non-concentrating and concentrating collectors. Another way to label the different categories is whether the collectors require tracking mechanisms for their perfect operation or not. However, physics is not black and white and, therefore, one may find intermediate cases. The fundamental design parameter to classify the concentrating quality of a collector is the geometrical concentration ratio C_g. This parameter describes the relation between two areas of aperture (A_a) and receiver (A_r), as illustrated in Fig. 6.4. The receiver or absorber are receiving the concentrated radiation, where power is converted into heat and from which heat again can be lost by convection and thermal radiation (energy loss). Secondly, the aperture area of the collector which collects all the natural solar irradiation and redirects that as far as possible to the receiver (energy gain).

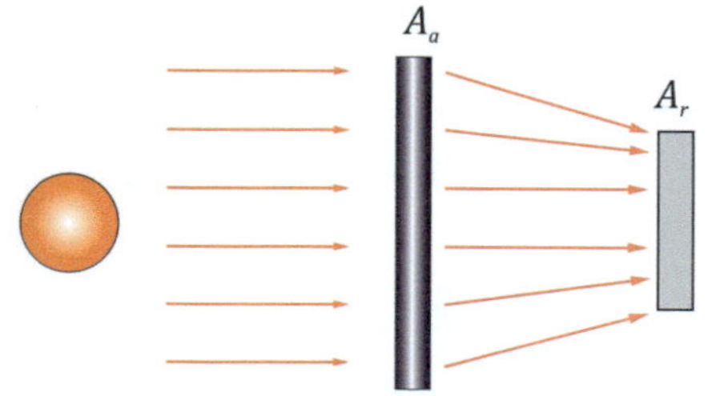

Fig. 6.4 Sketch aperture/receiver relation

For $C_g < 3$ we will find collectors using concentrators which are able to cover the whole sun-path over the day or even year without any adjustment, i.e. without tracking. This is a small concentration value. For values between 5 and 100 you have medium concentration where at least tracking along one axis is needed, and only the direct and circumsolar part of the solar irradiance can be used. Above 100 we speak about high concentration ratios, requiring two-dimensional tracking.

6.1.4 Typical Operating Temperatures and Applications

The relationship between concentration ratio and temperature is illustrated by the schematic diagram of different collector systems outlined in the Fig. 6.5, in which typical application temperatures and concentration ratios are given.

Wth increasing power density and temperature at the absorber, the demands on the collector designs increase. In case of high concentration ratios, the two-axis tracking is useful. Tracking not only increases the collection of direct solar radiation by better adjustment of the mirrors, but also ensures a constant level of power generation over a longer period.

When increasing typical operating temperatures for a collector, the heat losses usually increase more than linearly. Therefore, it is key to reduce the heat losses due to design when higher operating temperatures are to be achieved. Only then the heat losses will not reach the same magnitude as the solar gains, which would mean no useful heat can be transferred from the collector to the demand site. The stationary heat loss $\dot{Q}_{loss}$ of a collector is proportional to the heated (absorber) area, the temperature difference, and the so called U_L-value of the collector, the area-specific heat loss coefficient. One option to reduce of the value of U_L is to use low-emissive surfaces to lower thermal radiation emission, and to use convection barriers like additional glazings to reduce the convection heat transfer. Advanced methods are the utilization of noble gases or vacuum in gaps and cavities to reduce heat conduction. The other option is opened up by optics: If the solar radiation can be concentrated into a small absorber area compared to the collecting aperture area, the heat loss will be diminished accordingly for the same temperature difference—and hence it is possible to reach higher temperatures with higher concentration ratios.

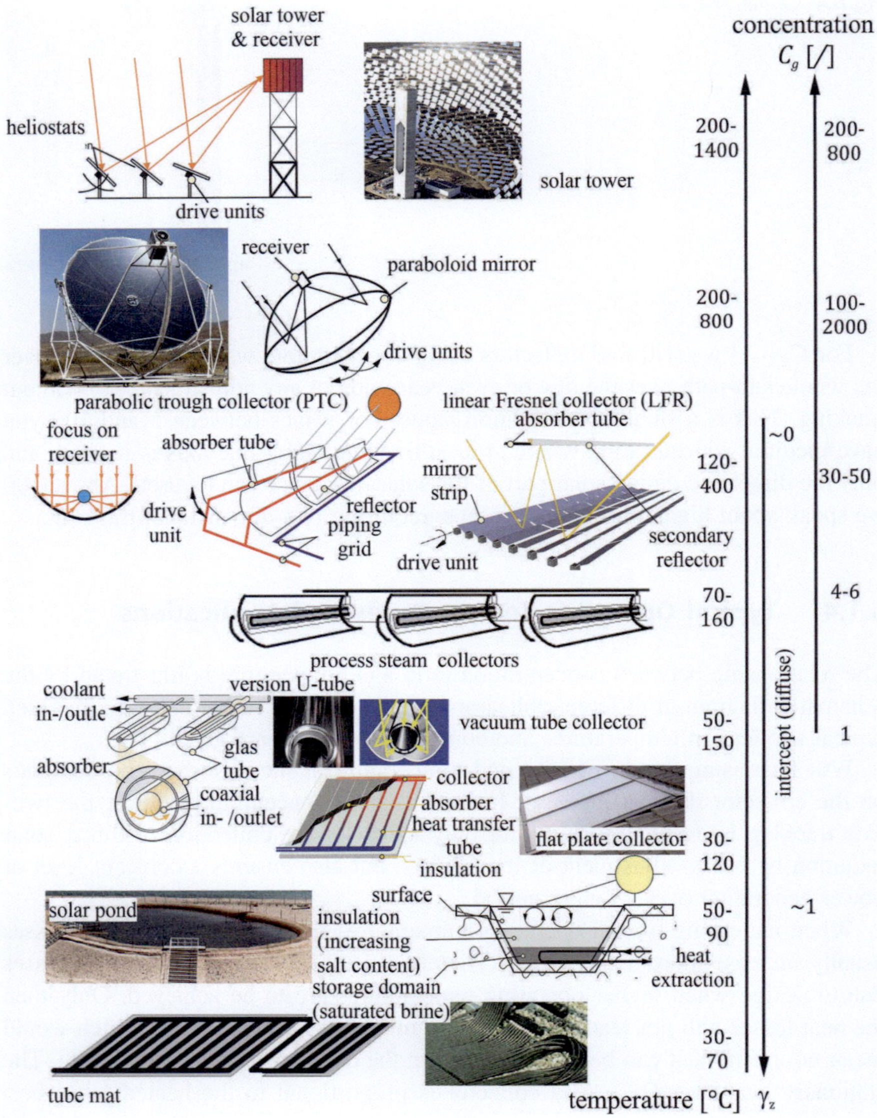

Fig. 6.5 Several collector types with indication of characteristic temperatures and geometric concentration factors (C_g)

The geometrical concentration ratio C_g is only one part of the story as the optical system has an efficiency—the optical quality of the materials used, and the perfect geometry determine the optical losses of a concentrator. Optical efficiency defined as the ratio of output of a real concentrator divided by the concentrator with ideal geometry and ideal surfaces (100% reflectivity) must be multiplied with C_g to get the real optical concentration factor C_{opt}.

The overview in Fig. 6.5 suggests that choosing a temperature level automatically leads to the choice of collector type. But that is not true in many cases. On the one hand side you have seasonal variations in the demand and the efficiency of collectors as well. When you need hot water with a certain temperature only for summertime where the outdoor temperature is high and solar irradiation is abundant, then a simpler collector may be the better and more economic situation compared to the case where the same hot water is needed in wintertime with possibly minus degrees outside. Then a collector with lower heat losses is needed.

On the other hand also, the climate and location are important for the collector choice, for example, when the production of 150 °C steam for an industrial site is perfectly possible with linearly concentrating collectors like Parabolic Trough or Linear Fresnel. The latter one might be the more economical solution for locations closer to the equator say up to $\Phi = 30°$ latitude. But the same steam is needed in a location (with similar *DNI*) further North, say at $\Phi = 40°$ latitude, due to the lower solar altitudes the Parabolic Trough might be the better solution. If in this location, however, due to meteorological conditions the *DNI* is relatively small and more diffuse irradiation is available, then a non-concentrating Vacuum Tube Collector (VTC) on an adequately tilted roof could perform better. Therefore, in many situations we have not to adhere to fixed rules and prejudices, but we should use our knowledge and adequate design rules to choose the best collector—whether that is the one producing more energy, or energy for less cost, or energy for the least investment.

6.2 Theoretical Description of Solar Collectors

6.2.1 Energy Balance

The theoretical description of a collector is based on the (quasi-)stationary heat energy balance between thermal losses and solar gains of a collector. For dynamical considerations this balance is later enhanced by thermal capacity terms for the collector materials and the heat transfer fluid (if that applies). The useful power $\dot{Q}_u$ which may cover an existing heat demand and thus the corresponding efficiency η of a solar collector is described in this so-called collector equation which in its first form has been introduced by Whillier and Hottel (Whillier, 1964), modified by and thus is called Hottel-Whillier(-Bliss) equation. Graphically, this energy balance is schematically shown in Fig. 6.6.

The area-related useful power $\dot{q}_u = \dot{Q}_u / A_c$ from a collector (in the picture it is a flat plate collector—but, the principle applies also to other collector types) is thus calculated from the balance of the solar gain, given by global irradiance G (in W/m^2)) times an optical efficiency η_{opt}, and the thermal losses from the absorber which in the stationary case may be described by a heat loss factor U_L of the collector times the temperature difference between mean absorber T_{abs} and ambient temperature T_{amb}. The optical efficiency often is also called the effective transmittance-absorptance-product $(\tau\alpha)_e$, because the main optical losses are

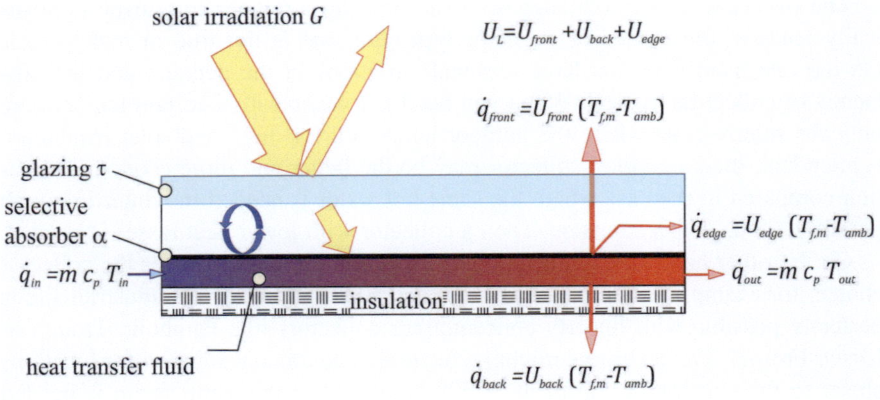

Fig. 6.6 Principal design of a solar thermal collector and energy balance

characterised by the angular-dependent transmission (τ) of the cover glass and the absorptivity of the thermal absorber (α). It is called the effective product, because multiple reflections between absorber and cover glass enhance the gain, as well as so-called secondary thermal gains induced by absorption in the cover glass(es). Of course, all optical quantities have a spectral dependence and are varying with the incidence angle of the incoming solar light. Similarly, the U_L-value is not a constant but in fact its value changes appreciably with temperatures involved, with wind velocities and even IR-radiation of the sky. The balance shown in Eq. (6.1) therefore shows a relationship for the specific net power where all quantities characterizing heat loss and optical efficiency are effective averages for the situation considered:

$$\dot{Q}_u = A_c \cdot \left[G \cdot (\tau\alpha)_e - U_\mathrm{L} \cdot (T_{abs} - T_{amb})\right]. \tag{6.1}$$

Here, the absorber temperature (T_{abs}) for which the useful power becomes zero is called the stagnation temperature. There are equivalent variants of this collector efficiency equation. There main difference is the use of different collector temperatures.

$$\dot{Q}_u = A_c \cdot F^{'} \cdot \left[G \cdot (\tau\alpha)_e - U_\mathrm{L} \cdot \left(T_{f,m} - T_{amb}\right)\right]; \tag{6.2}$$

$$\dot{Q}_u = A_c \cdot F_R \cdot \left[G \cdot (\tau\alpha)_e - U_\mathrm{L} \cdot \left(T_{f,in} - T_{amb}\right)\right]. \tag{6.3}$$

$T_{f,m}$ is the mean fluid temperature in the collector, and $T_{f,in}$ is the inlet temperature of the fluid. F' is called the collector efficiency factor and F_R is called the heat removal factor. Both factors for real collectors are smaller than 100% but should be ideally very close to that limiting value. In Hottel and Whillier (1955) could show with this collector equation Formula (6.3) that efficiency η increases with increasing solar irradiation.

In the equations $(\tau\alpha)_e$ represents the optical conversion of transmitted short wave solar radiation at the absorber into long wave heat. The factor U_{L} can be conceptually split up in the case of a flat plate collector, being the sum of the heat losses through the front cover U_{front}, the rear insulation layer U_{back} and the side walls U_{edge} of the collector box:

$$U_{\mathrm{L}} = U_{back} + U_{edge} + U_{front}. \tag{6.4}$$

The first two factors in Eq. (6.4) can be kept small by using insulation layers and making a collector slim and large. The front side, on the other hand, requires optimization regarding high optical transmission of the incident radiation and, on the other hand, high thermal insulation in order to keep convective losses to the environment low.

Similar to an electrical circuit, the front cover heat loss factor U_{front} adds up reciprocally according to the Eq. (6.5) from the heat transfer coefficients between the absorber and the glass cover $h_{abs-\mathrm{gl}}$, the heat conduction through the glass pane (with the thickness δ_{gl} and its thermal conductivity k_{gl}) and the heat transfer coefficient $h_{\mathrm{gl}-amb}$ from the glass to the environment:

$$\frac{1}{U_{front}} = \frac{1}{h_{abs-\mathrm{gl}}} + \frac{\delta_{gl}}{k_{gl}} + \frac{1}{h_{\mathrm{gl}-amb}}. \tag{6.5}$$

In Eq. (6.5) the total heat transfer coefficient $h_{abs-\mathrm{gl}}$ usually dominates the front side heat loss. It is composed of heat transfer by convection and conduction ($h_{conv,abs-\mathrm{gl}}$) as well as heat radiation ($h_{rad,abs-\mathrm{gl}}$) in the air gap between absorber and glass:

$$h_{abs-gl} = h_{\mathrm{conv},abs-gl} + h_{\mathrm{rad},abs-gl}. \tag{6.6}$$

Conductive and convective heat transfer can be described $\dot{q} = h_{conv} \cdot \Delta T = h_{conv} \cdot (T_{abs} - \mathrm{T}_{amb})$ where h_{conv} again is a complex function depending on the temperature difference between glass and absorber, the tilt angle β oft he collector, the dimensions of the gap and the fill gas involved. The physics is described in detail in Chaps. 4 and 5. Similarly, idealized as the thermal radiation exchange between two infinitely large planes, from the absorber to the glass, the radiative thermal flux is:

$$\dot{q}_{\mathrm{rad},abs-\mathrm{Gl}} = \frac{\sigma}{\frac{1}{\varepsilon_{abs}} + \frac{1}{\varepsilon_{\mathrm{gl}}} - 1} \cdot \left(T_{abs}^4 - T_{\mathrm{gl}}^4\right). \tag{6.7}$$

wherein T_{abs} represents the absorber temperature and T_{gl} represents that of the glass plate; ε_{abs} and $\varepsilon_{\mathrm{gl}}$ represent the emission coefficients and σ is the Stefan-Boltzmann constant, the value of which is $\sigma = 5.67 \cdot 10^{-8} \mathrm{W}/(\mathrm{m}^2\mathrm{K}^4)$. Using the transformation $\left(T_{abs}^4 - T_{\text{gl}}^4\right) = \left(T_{abs}^2 + T_{\text{gl}}^2\right)\left(T_{abs} + T_{\text{gl}}\right) \cdot \left(T_{abs} - T_{\text{gl}}\right)$ one gets the result for the radiative heat transfer coefficient:

$$h_{\text{rad},abs-\text{gl}} = \frac{\dot{q}_{\text{rad}}}{\left(T_{abs} - T_{\text{gl}}\right)} = \frac{\sigma}{\frac{1}{\varepsilon_{abs}} + \frac{1}{\varepsilon_{\text{gl}}} - 1} \cdot \left(T_{abs}^2 + T_{\text{gl}}^2\right)\left(T_{abs} + T_{\text{gl}}\right). \quad (6.8)$$

In all these considerations we must understand, that, the quantities given in the formulae like temperatures, irradiation, heat fluxes when investigated in detail are variable fields having complex distributions which may be modelled in so called Computational Fluid Dynamics programmes (CFD). However, for engineering purposes the simple description in Eq. (6.1) and following is extremely useful and powerful to describe the collector in a system and to determine characterizing parameters in testing procedures.

In a limited temperature range of several 10 °K both, convective and radiative heat transfer functions can be linearized based on average temperature or temperature difference. And for a very narrow temperature range, of 10–20 °K they can even be approximated by a constant value with good accuracy.

For pure heat conduction the gas conductivity k_{gas} determines the heat transfer coefficient $h_{cond,abs-gl} = \left(k_{gas}/d_{abs-gl}\right)$ through the gap of thickness d_{abs-gl} between the glass and absorber plate. Air is usually used as the gas filling here. If the temperature difference is large enough so convection sets on, the heat transfer coefficient for conduction must be multiplied by the Nusselt number Nu to get the convection coefficient $h_{conv,abs-gl}$ (in which $Nu = 1$ assumes pure heat conduction). Further details on the calculation of conductive and convective heat transport are discussed in Chaps. 4 and 5, respectively.

The heat exchange or heat loss due to radiation between the absorber and the glass plate decreases with the emission coefficients ε_i (see Eqs. (6.7) and (6.8)). For standard uncoated glass, however, $\varepsilon_{\text{gl}} \approx 0.87$ applies, so the emission coefficient of the absorber surface in the range of long-wave thermal radiation must be kept small. The same surface absorbs the solar radiation. It should therefore have in the optimal case a high solar absorptivity and a low thermal emissivity. Such a surface is called a "selective absorber surface", as the optical properties reflectivity or absorptivity change drastically when going from the short wavelength solar range to the long wavelength thermal region, as illustrated in Fig. 6.7.

When we specify the collector area, different area is used depending on the purpose of the description or sometimes preference of the author. Collector efficiency might be defined relative to absorber or to aperture area. Gross collector area (total area) A_G should be used to specify the space needed for installation. It is the area between the outer boundaries of the collector, generally the outer boundaries of the collector casing.

The aperture area A_a is the transmission aperture of the collector for vertical or angular solar radiation incidence, generally the visible glass area within the pane seal. In the case of vacuum tube collectors without reflectors, it is the product of the inner diameter, the unshaded cylindrical inner length, and the number of tubes. In the case of reflectors attached to the reverse side or in the collector, it is

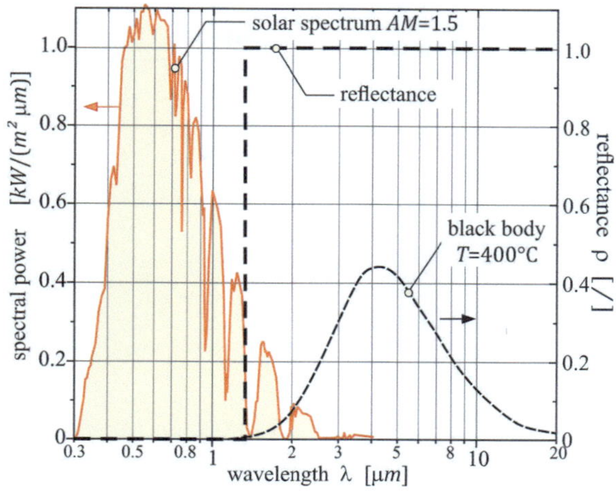

Fig. 6.7 Spectral reflectivity/absorptivity for an ideal selective absorber

determined using the projection of the area formed by the absorber and reflector in the collector level. The smallest area is the absorber area A_{abs} which equals the flat or curved absorber surface which is relevant for the thermal loss of the collector. For even more detailed explanations of these area definitions see the ISO 9806 standard specifying collector testing (ISO 9806, 2017) or the ISO 9488 on solar terminology (ISO 9488, 2021). The active absorber area generally corresponds to the absorber area A_a. Some collectors, however, are constructed so that the absorber area below the collector frame is larger than the aperture area. The active absorber area is then the absorber area not affected by shading from the frame. In this case it is equal to the aperture area A_a.

6.2.2 Efficiency Equation

By dividing the net thermal power $\dot{Q}_u$ in Eq. (6.1) with the collector area (A_a) and solar irradiance (G), the efficiency of the collector (η) is obtained directly:

$$\eta = \frac{\dot{Q}_u}{A_c \cdot G} = (\tau\alpha)_e - U_L \cdot \frac{(T_{abs} - T_{amb})}{G}. \tag{6.9}$$

The idealised Eq. (6.9) can first be analysed in more detail with the help of a diagram, which is shown in Fig. 6.8. For a temperature difference of $(T_{abs} - T_{amb}) = 0$, the collector characteristic curve corresponds exclusively to the conversion ratio $(\tau\alpha)_e$ and thus indicates which proportion of the solar radiation is converted into heat. The area above the horizontal of the conversion ratio reflects the proportion of optical losses of a collector. The area below the collector curve indicates the useful heat, while the area between the useful power and the optical losses quantifies the heat losses.

The graph can be further analysed. For small temperature differences, the efficiency is essentially determined by the conversion ratio $(\tau\alpha)_e$. This means that

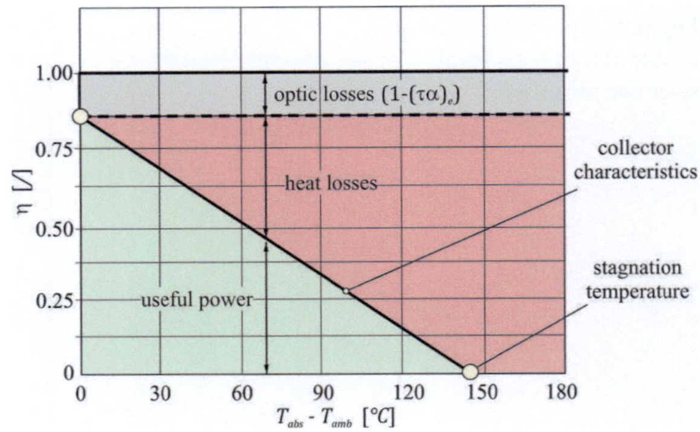

Fig. 6.8 Graphic analysis of the collector efficiency from Eq. (6.9)

even in the case of poorly thermally insulating collectors such as tube mats, a high degree of efficiency is still achieved with small temperature differences, provided that the conversion ratio is sufficiently high. However, as the temperature difference between the absorber surface and the environment ($T_{abs} - T_{amb}$) increases, all heat losses increase, and especially convective but also radiation heat losses then determine the total heat loss. For the latter, we must therefore introduce an addition that considers the non-linear character of the convective or radiant heat losses. The collector characteristic curve also shows another point, the stagnation or standstill temperature. This is the point at which the collector efficiency becomes zero, i.e. the collector does not produce any useful output. This occurs when, for example, the coolant circulation pump fails or the thermal storage tank of the collector circuit is full, so that the collector inlet temperature increases with time and thus also the heat losses to the environment. The occurrence of a stagnation temperature is a characteristic of all solar thermal collectors, regardless of whether they are concentrating or not. The stagnation temperature increases the lower the U_L value, i.e. the lower its heat transfer loss to the environment. A closer look at Eq. (6.9) also shows that the efficiency depends on the irradiation. The greater the irradiation, the lower the relative importance of the heat losses. For example, for a flat-plate collector for domestic hot water heating, this means that in winter, when irradiation is low (G), the importance of good thermal insulation is considerably more important than in summer, since G is in the denominator.

If the assumption of a narrow temperature difference with constant heat transfer coefficients and heat loss factor U_L does not apply the linearization leads to a quadratic loss term in the efficiency equation, as shown in Eq. (6.10), see c.f. (Frei, Häuselmann, Flückinger, & Frey, 1995).

$$\eta = (\tau\alpha)_e - \frac{1}{G}\left[u_0 \cdot (T_{abs} - \mathrm{T}_{amb}) + u_1 \cdot (T_{abs} - \mathrm{T}_{amb})^2\right]. \tag{6.10}$$

This is the typical approximative parametrization of collector testing for non-concentrating collectors. Other factors are also introduced in the empirical loss equation describing the dependence for example on wind speed. For concentrating collectors due to the high temperatures which often leads to a dominating radiative

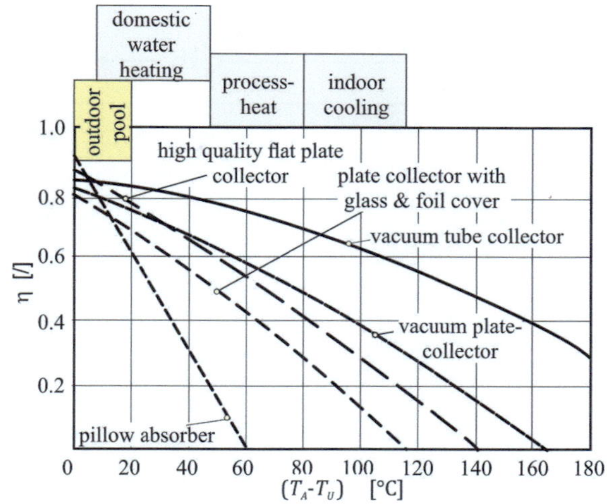

Fig. 6.9 Examples of different low temperature collector types with linear and quadratic efficiency curves with indication of a potential area of use

heat transport the polynomial expansion up to the 4th power in temperature difference is preferable. Figure 6.9 illustrates an example of a linear and a quadratic efficiency curve of different collector types and their potential utilization domain, which will be discussed subsequently.

6.2.3 Internal Flow and Temperature Distribution

To understand the different variations (6.1)–(6.3) of the collector equation, we now discuss the internal heat flows within the collector. At first, we will have a look at the heat transfer between absorber and heat transfer fluid. For simplicity we use the example of a fin-type absorber sheet of a flat-plate or vacuum tube absorber.

The thermal coupling of the absorber strip to the absorber tube influences the absorber efficiency factor F'. The thickness of the absorber plate, its thermal conductivity and the distance from the pipe is described by the so-called *fin factor* F. The absorber efficiency factor $F\prime$, however, relates the thermal resistance from the absorber fin to the ambient to the resistance from the fluid to the ambient. Therefore, also the thermal resistance of the bonding connection pipe-plate, the resistance of the pipe walls and the convective heat transfer in the pipe fluid must be considered as well. The decisive factor here is the effective heat transfer surface and thus the transfer to the fluid, which in turn depends on the flow rate. The fin factor F for a classical ideal fin absorber, as shown in a cross-sectional view in Fig. 6.10 neglecting the two-dimensional heat flux within the very thin absorber sheet can be modelled quite well discretizing the fin and solving the one-dimensional conduction heat transport within the fin for constant flux (absorbed solar power) from the fin end to the pipe section.

A simple thermal resistor model as sketched in Fig. 6.11 then is required to calculate the absorber efficiency factor F' for that situation.

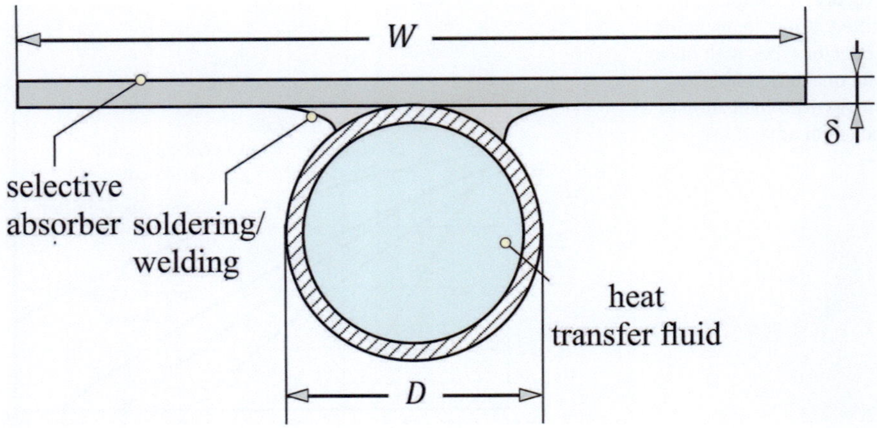

Fig. 6.10 Idealized absorber fin geometry

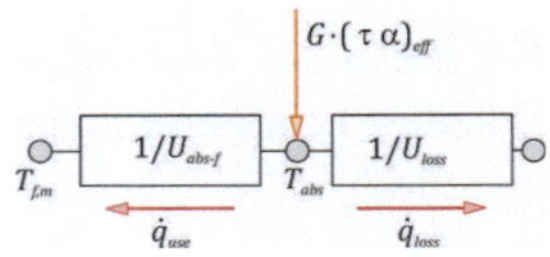

Fig. 6.11 Electrical equivalent diagram for determining the absorber efficiency factor F' of a flate plate collector

The temperature distribution in a thin absorber fin between tube and absorber sheet may be approximated with a one-dimensional heat transfer problem. The fin shall be discretized in infinitesimal elements receiving solar radiation and conducting the absorbed heat horizontally along the fin according to the temperature distribution and the conductance of the fin. Figure 6.12 shows schematically the situation with the infinitesimal discrete element of length Δx of the fin.

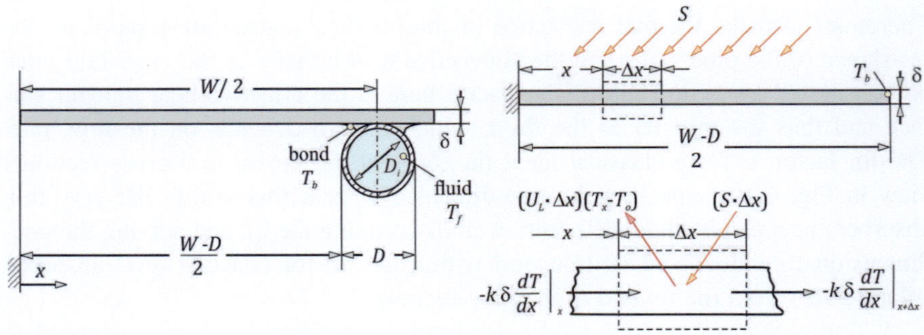

Fig. 6.12 Schematical drawing of absorber fin with tube after Duffie and Beckmann (2020)

The energy balance for the discrete element is

$$S \cdot \Delta x - U_L \cdot \Delta x \cdot (T_x - T_a) - k \cdot d \cdot \left(\left.\frac{dT}{dx}\right|_{x+\Delta x} - \left.\frac{dT}{dx}\right|_{x} \right) = 0. \tag{6.11}$$

Using $S = \left((\tau\alpha)_e \cdot G\right)$ and x = position on fin. This leads to a differential equation of 2nd order with adiabatic boundary condition on one side and a center tube temperature on the other side of the symmetrical half fin. If we define the two variables:

$$m = \sqrt{\frac{U_L}{k \cdot d}};\ \psi = (T - T_a) - \frac{S}{U_L}. \tag{6.12}$$

This leads to the simple differential equation

$$\frac{d^2\psi}{dx^2} - \mathrm{m}^2 \cdot \psi = 0, \tag{6.13}$$

which can be solved taking into account the two boundary conditions. The solution for the temperature distribution as a function of x along the fin is then

$$\frac{T_x - T_{amb} - S/U_L}{T_0 - T_{amb} - S/U_L} = \frac{\cosh(m \cdot x)}{\cosh(m \cdot (W - D)/2)}, \tag{6.14}$$

where W denotes the fin width, D the tube diameter, T_0 is the temperature of the absorber right at the center of the tube position. Summing up the heat flux per unit length perpendicular to the fluid flow direction from the fin sheet to the central region and the part directly above the tube one gets the useful gain per unit length is

$$\dot{q}'_u = [(W - D) \cdot F + D] \cdot [S - U_L \cdot (T_0 - T_{amb})]. \tag{6.15}$$

The solution for the simple model of the absorber fin yields the following result

$$F = \frac{\tanh[m \cdot (W - D)/2]}{m \cdot (W - D)/2}, \tag{6.16}$$

where m is given by Eq. (6.12), U_L in $(\mathrm{W}/(\mathrm{m^2K}))$ is absorber heat loss factor, k in (W/(mK)) the thermal conductivity of absorber plate, W the fin width in (m) and D the diameter of pipe also in (m). The fin factor F is shown as a function of fin width W, thermal conductivity k and thickness d of the absorber sheet in Fig. 6.13.

Now the heat collected in the fin must be transferred to the fluid circulating in the absorber tube. When the bonding resistance R_b and the heat transfer coefficient h_{fi} to the fluid inside the tube are considered, the resulting equation is

$$\dot{q}'_u = W \cdot F' \cdot \left[S - U_L \cdot \left(T_f - T_{amb}\right)\right]. \tag{6.17}$$

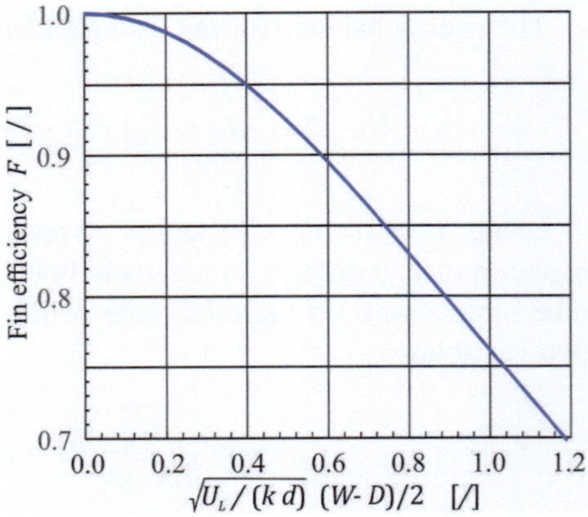

Fig. 6.13 Fin efficiency F for tube and sheet solar collectors

The fin factor F then is used to calculate F' according to Fig. 6.10 as

$$F' = \frac{1/U_L}{W \cdot \left[\frac{1}{U_L \cdot [(W-D) \cdot F + D]} + R_b + \frac{1}{\pi \cdot D_i \cdot h_{fi}} \right]}; C_b = \frac{k_b \cdot w_b}{\overline{\delta}_b}, \tag{6.18}$$

in which C_b in (W/(mK)) is bond conductance, k_b in (W/(mK)) the thermal conductivity of bond, $\overline{\delta}_b$ in (m) the average thickness of bond and w_b in (m) the width of bond, respectively.

A physical interpretation for F', the collector efficiency factor, results from examining Eq. (6.17). F' presents the ratio of actual useful energy gain to the useful gain if the collector fluid temperature would be identical to the absorber temperature. Another interpretation becomes clear when one recognises that the denominator of Eq. (6.18) is the heat resistance $1/U_0$ from fluid to ambient air. The numerator is the resistance from the absorber plate to ambient air. Therefore,

$$F\prime = \frac{U_0}{U_L}. \tag{6.19}$$

An example calculation again for typical parameters of a flat-plate liquid solar thermal collector is shown in Fig. 6.14. As can be seen there the overall heat loss to the ambient characterized by U_L is a decisive factor. Therefore, the curves are shown for simple collectors with black absorber coating and for state-of-the art collectors with selective absorbers and low emissivity values.

Fig. 6.14 Computed temperature distribution in an absorber produced from high-performance concrete (Hermann et al., 2015)

The exact form of the function F' may differ for appreciably for other collector geometries, e.g. for parabolic trough collectors. However, the fundamental principle and the considerations of designing a good collector by making the thermal resistance between absorber solar radiation and transporting fluid as small as possible is a general design law. While the approximate considerations above can be used for designing highly efficient fin-tube collectors, varying sheet thickness and tube spacing dependent mainly on the thermal conductivity of the sheet material, other more complex geometries may be simulated by 2D heat conduction programs, but the concept of F' is valid even then. When we integrate now Eq. (6.17) over the collector length L and use the average mean fluid temperature we arrive at the collector equation introduced already earlier, which reads to:

$$\dot{Q}_u = A_c \cdot F' \cdot \left[G \cdot (\tau\alpha)_e - U_{\mathrm{L}} \cdot \left(T_{f,m} - T_{amb}\right)\right], \tag{6.20}$$

with $T_{f,m}$ being the mean fluid temperature. The Fig. 6.15 illustrates the collector efficiency factor F' as a function of the tube spacing for different heat loss factors U_{L} and absorber fin conductance $k \cdot \delta$, where k is the absorber conductivity and δ the absorber fin thickness. Other assumptions taken for these calculations are: linear bond conductance $C_b = 500$ W/(m·K) and internal convection coefficient of absorber pipe $h_{fi} = 2000$ W/(m^2 · K).

It is obvious that the mass flow through a solar collector for given solar irradiation has a strong impact on the fluid temperature distribution along flow direction and hence also on the mean fluid temperature. A low mass flow would allow a much higher temperature rise during pass-through compared to a high mass flow, because a higher solar energy gain is accumulated during the longer passage time. While in reality the temperature distribution on an absorber area surface even with homogeneous irradiation has a two-dimensional distribution due to the

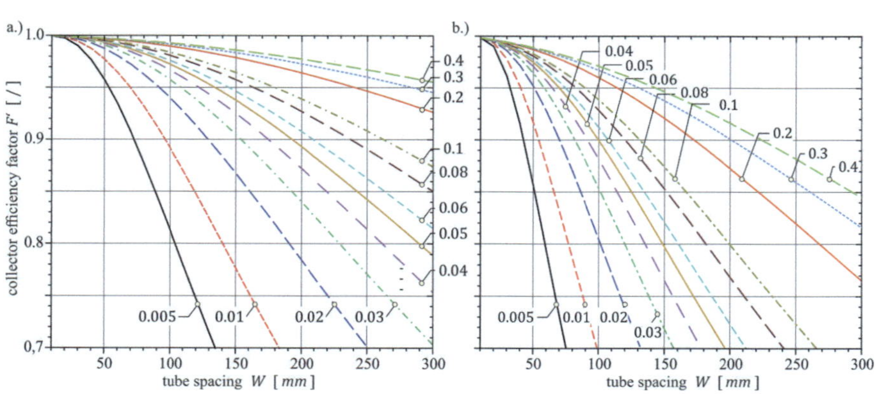

Fig. 6.15 Calculated collector efficiency η factor F' as a function of the tube spacing for different heat loss factors U_{L} and fin conductance $(k \cdot \delta)$ in (W/K) (curve indicator at right hand side of graphs); **a** $U_{\mathrm{L}} = 2\mathrm{W}/\left(\mathrm{m}^2\mathrm{K}\right)$, **b** $U_{\mathrm{L}} = 4\mathrm{W}/\left(\mathrm{m}^2\mathrm{K}\right)$

slightly inhomogeneous hydraulic flow of fluid through individual tubes, a one-dimensional temperature distribution along the main flow direction gives generally a good approximation for most geometries. Assume therefore a fin collector with identical fin elements. Here the fluid temperature along each tube is determined by a constant solar gain, but an increasing heat loss due to the rising fluid and absorber temperature. The energy balance of an infinitesimal tube element with length dy, and assuming that F' and U_L are constant over a small temperature range, one obtains a 1st order differential equation:

$$\dot{m} \cdot c_p \cdot \frac{dT_f}{dy} = n \cdot W \cdot F' \cdot \left[S - U_L \cdot \left(T_f - T_{amb}\right)\right], \tag{6.21}$$

where $\dot{m}$ is the total collector mass flow and c_p the fluid specific heat capacity. Integrating the equation over the tube length and assuming that all tubes inlet and outlet temperatures are the same, using $A_c = (n \cdot W \cdot L)$, the result is:

$$\frac{T_{f,out} - T_{amb} - S/U_L}{T_{f,in} - T_{amb} - S/U_L} = \exp\left(-\frac{A_c \cdot U_L \cdot F'}{\dot{m} \cdot c_p}\right). \tag{6.22}$$

It is convenient now to define a quantity that relates the actual useful energy gain of a collector to the useful energy gain if the absorber temperature were at fluid inlet temperature. This quantity is called the collector heat removal factor F_R:

$$F_R = \frac{\dot{m} \cdot c_p \cdot \left(T_{f,out} - T_{f,in}\right)}{A_c \cdot \left[S - U_L \cdot \left(T_{f,in} - T_{amb}\right)\right]}. \tag{6.23}$$

With some rearrangement of the terms and using Eq. (6.22) the factor F_R can be expressed as

$$F_R = F' \cdot F'' = F' \cdot \frac{\dot{m} \cdot c_p}{A_c \cdot U_L \cdot F'} \cdot \left[1 - \exp\left(-\frac{A_c \cdot U_L \cdot F'}{\dot{m} \cdot c_p}\right)\right], \tag{6.24}$$

where F'' is called the collector flow factor. Using the dimensionless quantity γ as the collector capacitance rate, being defined as

$$\gamma = \frac{\dot{m} \cdot c_p}{A_c \cdot U_L \cdot F'}, \tag{6.25}$$

we can see the importance of flow rate for the maximisation of the collector flow factor in Fig. 6.16. In solar thermal system design in fact different flow rate regimes are being used. There are so called low-flow systems where γ is in the range of 1–3 with a low F'', and high-flow systems with γ larger than 10 with a F'' close to 1. Although the transfer of heat to the fluid is reduced in low-flow systems there are other advantages which may counteract these disadavantages. We will discuss this further in the chapter on low temperature systems in Chap. 9.

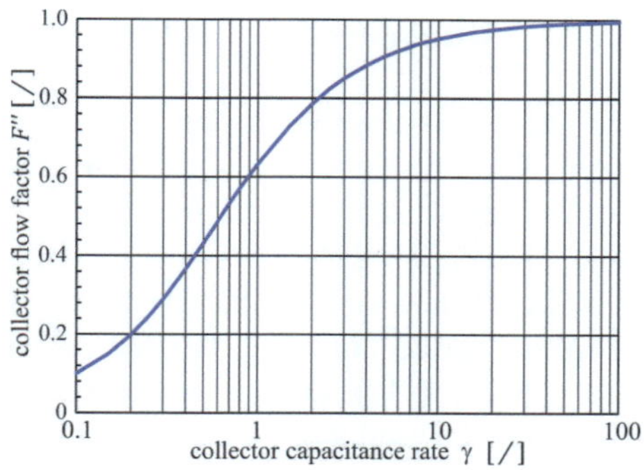

Fig. 6.16 Collector flow factor F'' versus dimensionless collector capacitance rate γ.

The quantity F_R is equivalent to the effectiveness of a conventional heat exchanger, which is defined as the ratio of the actual heat transfer to the maximum heat transfer rate. The maximum possible useful energy gain of a collector is when the whole collector is at the inlet fluid temperature—then heat losses to the ambient are minimum.

$$\dot{Q}_u = A_c \cdot F_R \cdot \left[G \cdot (\tau\alpha)_e - U_L \cdot \left(T_{f,in} - T_{amb}\right)\right], \tag{6.26}$$

with $T_{f,in}$ being inlet fluid temperature. This is the third form of the collector energy balance equation, the so-called Hottel-Whillier equation (Whillier, 1964). Whereas Eq. (6.20) serves as the basis for standardized collector testing in European standards (ISO 9806, 2017), the equation in the form (6.26) is the basis for the ASME collector test procedure (ANSI/ASHRAE Standard 93-2003, 2003).

6.3 Non-concentrating Collectors

According to the basic collector structure and geometry traditionally we distinguish three main collector groups which have established themselves in the market of solar low temperature collector systems:

- uncovered absorbers,
- flat plate collectors,
- vacuum collector,
- special collector types like air collectors, PV-T collectors.

In the following, the characteristics and performance of these three types are described. Certainly, we see numerous subtypes of collector designs that are specially adapted to specific applications and use. For example, the material choice is influenced by the targeted temperature range (polymeric or metallic absorbers), and also the heat transfer fluid used does vary (e.g. water, water–glycol, air).

6.3.1 Uncovered Collectors

The uncovered absorber design is certainly the simplest approach to produce hot water. No collector cover is used to lower heat losses to the environment, and thus the performance is very much dependent on the environmental variables like wind, ambient air and sky temperature. In the simplest case the absorbers consist of black UV-resistant flexible plastic tubes, which are placed on a suitable mostly horizontal surface. In more complex systems, the pipes are combined into absorber mats (see Fig. 6.17). Other uncovered collectors consist of metallic absorber sheets which may be used as complete facade or roof elements (Gombert, Reineke-Koch, Fenske, & Hofmann, 2007). An interesting variation is the transpired air collector, where fresh air for a building is sucked through tiny holes of an absorber sheet, heating up during the passage and in the rear cavity formed with a wall or roof.

The optical efficiency is high, not reduced by a cover glass, due to the high absorptance in the range of $0.8-0.95$. However, the thermal loss increases dramatically with fluid temperature rising above ambient temperature, as the uncovered surface due to convection and thermal radiation has on the average a heat transfer coefficient of around $20\ \mathrm{W}/\left(\mathrm{m}^2\mathrm{K}\right)$. This allows to raise the water temperature only few degrees above the ambient temperature.

In short, the characteristic features of uncovered collectors are:

- very simple construction,
- low price,
- easy installation. They can easily be laid on flat surfaces.
- Low efficiency due to high heat losses,
- low temperature applications, e.g. preheating of water and solar swimming pool heating,
- no winter operation possible, and hence no frost protection measures necessary.

Figure 6.17 shows a simple absorber mat using many tubes made of flexible, black coloured plastic with intermediate bars. The flexibility of the individual elements allows easy installation and decommissioning on any surface. Low system pressures ensure the use of conventional pipe connection systems for installation technology. The blackened water-filled pipes are made of polyethylene (PE),

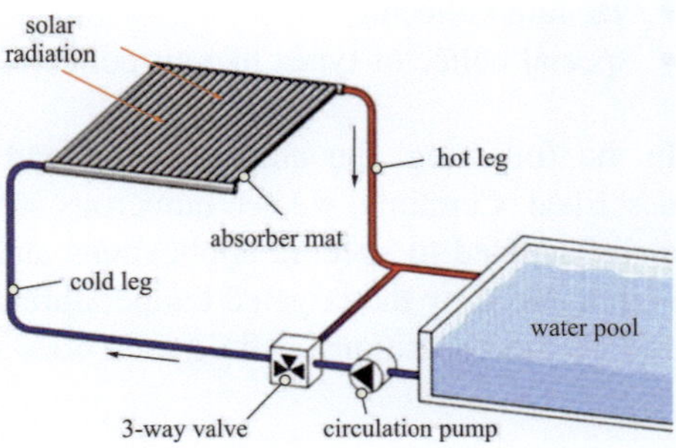

Fig. 6.17 Set-up of mat absorber collector circuit

polypropylene (PP) or more stable ethylene-propylene-diene monomer rubber (EPDM). For swimming pool applications, the absorber surface is about 50–80% of the pool surface. The simple design requires only a two-point control for efficient operation. Above a certain temperature difference between the pool and the absorber temperature, a pump is switched on, which sends the water back into the pool through the absorber.

6.3.2 Liquid Flat-Plate Collectors (FPC)

If a second basic element of collector design is added, a transparent cover for the absorber, then we can increase the working temperatures and efficiency of the solar collector substantially. In this case also a back-side insulation for the flat absorber is added thus both front and back side heat losses are reduced. Depending on the number and type of transparent collector covers and the absorber type this flat-plate collector as sketched in Fig. 6.18 may reach absorber temperatures well above 200 °C. Hence metal sheet absorbers from copper, aluminium or steel are an appropriate material choice. The absorber coating itself may be a simple black absorber e.g. using a paint which has a high thermal emissivity, or a selective absorber, which combines high absorptance α_s in the solar wavelength range with low thermal emissivity ε_{th}. The heat produced in the solar conversion process is transported in tubes connected to the backside of these metal absorbers. If only a simple absorber strip is connected to a single pipe, we call that an absorber fin. Nowadays however most collectors use full area metal sheet absorbers with a complete manifold of pipes attached. We distinguish between flat-plate liquid collectors and air collectors. Whereas in the first type the fluid is water or a water–glycol mixture and is pumped through pipes connected to the absorber sheet, in the case of air more design alternatives are common (cf. Fig. 6.33). Air may flow above, through or behind the absorber or in combinations of that. In the following paragraphs the design and effect on the collector performance of different components will be discussed.

6.3.2.1 Transparent Cover

The most popular solution for a transparent cover of a flat-plate collector is a sheet of glass. It has several functions. First of all, the cover shall reduce the thermal

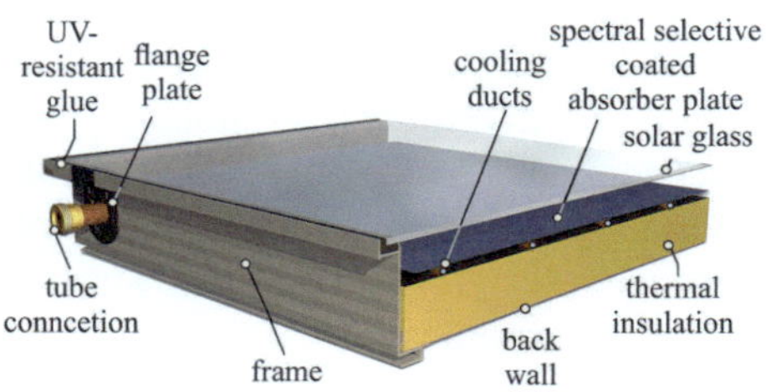

Fig. 6.18 Components of a liquid flat plate collector

front-side losses of the absorber due to convection and thermal radiation. For thermal radiation is glass a good absorber, if it has a sufficient thickness of 3 mm or more. It also creates a tight and narrow air gap between absorber and absorber, thus introducing a thermal resistance characterized by the natural convection of air. High convective heat losses due to wind are only possible from the outermost, much colder glass surface. A second requirements is, of course, that a very large amount of solar incoming radiation will be transmitted to the absorber. The solar transmittance of the glazing therefore shall be maximised. The third important task of the cover is to keep rainwater, hail, snow, ultraviolet radiation, birds remaining, organic growth of fungi or atmospheric pollution from the sensitive materials in the collector, i.e. the absorber coating and the thermal insulation. In days, when the absorber sheets had a less favourable appearance, also structured glass with micro-pyramids or other patterns obscured the view for the observers.

Depending on the glass quality and the treatment of the glazing different optical efficiencies of flat-plat collectors may be realized. Borosilicate glass and sodium lime glass are used for commercial collector covers. Soda lime glass is produced in huge thermal ovens on a fluid mercury bath with an extremely flat surface as float glass. Very often the glass is toughened in a thermal treatment process to avoid large sharp flakes when breakage occurs. As toughened glass breaks into very small fragments this type is often also called “security glass”. Polymeric glazings are sometimes used as outer or as second cover materials, as they have less weight and may have higher transmittance. Very thin polymer films, however, are too sensitive for the outer surfaces, as they might degrade fast due to environmental stresses or even, they tear when birds or cats are using them as sitting area.

Table compares the solar transmittance for normal incidence of typical solar cover glass materials. The different reflectance is due to the index of refraction of the materials which is given as averaged over the solar range in Table.

The difference between green, normal float and white glass is that they are produced from sands which contains silicate oxide, but also different contaminations of iron oxide Fe_2O_3. White sands with very small iron oxide content produce little absorption in the near infrared. Green glass is produced from sand with high iron-oxide content. Therefore, a large part of the transmitted red light is absorbed, and the light emitted from the edges appears as green, which is the complementary colour.

If further increase of transmittance is needed, then surface treatment of the pure white glass may reduce the index of refraction at the glass-air interface, resulting in a lower reflectivity. This antireflection (AR) treatment of glass can be applied on one or on both sides of the glass sheet. One possibility is the etching of the glass surface with fluoric acid, resulting in a nano porous surface layer with intermediate refraction index. Alternatively, and a more environmentally friendly production is the coating with a thick solgel dip coating process. A silicate network structure is forming on the glass surface, and when the fluid solvent evaporates then similarly an open nano porous surface layer is created. A further production method could be a PVD coating of oxides on the glass. This AR-coatings reduce the reflectivity not

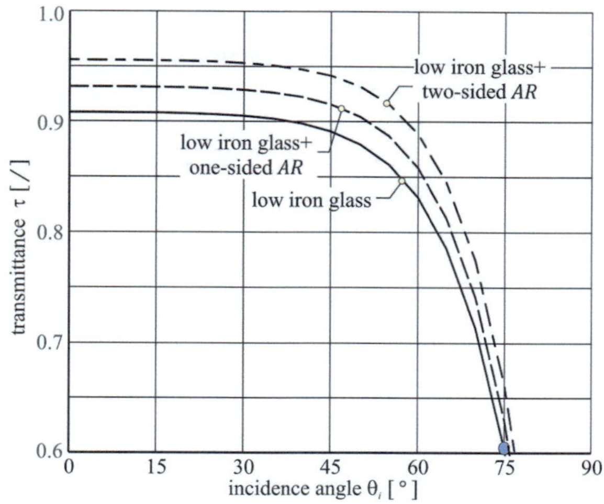

Fig. 6.19 Solar light transmittance as function of incidence angle θ_i of low iron-oxide content glass without Anti-reflection layer (AR) with one-sided and two-sided AR-coating

only for normal incidence θ_i, but have even a high effect for off-normal incidence, as illustrated in Fig. 6.19.

As a second collector cover—even with AR-coatings—reduces the solar gains and increases the weight of the collector, the benefit due to additional reduction of convective and radiative heat loss must be meticulously taken into account. The Fig. 6.20 shows the thermal network of a two-cover collector in terms of convective (c) and radiative (r) resistances. From that can be concluded that for a collector with low absorber emissivity ε_{abs} the radiative resistance between absorber and first cover glass is the dominating one. A second cover glass then may only be reasonable if the absorber temperatures are high, which implies that large temperature differences between absorber and glazings leading to a low convective resistance in the air gap between absorber and first glazing, which is "short-circuiting" the heat flux. In this case a second glazing might be appropriate.

To investigate in detail whether a second cover may be useful for increasing the useful energy output of the collector we have to investigate the physical models of the collector heat loss according to the presented network, not only the empirical parameterized heat loss of Eq. (6.10). Within the gaps between absorber, 1st glazing and 2nd glazing we must quantify radiative and convective heat transfer

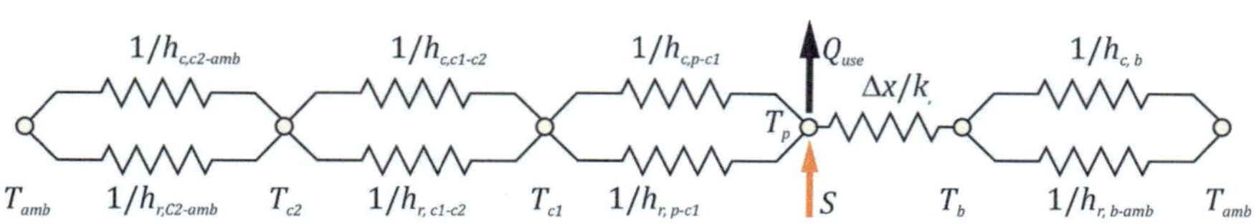

Fig. 6.20 Electrical equivalent circuit for a double-glazed collector consisting of convective and radiative resistors connected in parallel

coefficients, which in the case of convection also depend on the resulting temperatures. With some temperatures assumed we will determine the resulting heat fluxes, which in turn may change the temperatures in stationary equilibrium. Therefore, the solution of the network equations is iterative.

The radiative heat transfer in the gaps can be approximated by shallow cavities with the formula for infinite parallel plates:

$$h_{r,p-c1} = \frac{\sigma \cdot \left(T_{c1}^2 + T_p^2\right) \cdot (T_{c1} + T_p)}{\frac{1}{\varepsilon_{c1}} + \frac{1}{\varepsilon_p} - 1}, \tag{6.27}$$

$$h_{r,c1-c2} = \frac{\sigma \cdot \left(T_{c2}^2 + T_{c1}^2\right) \cdot (T_{c2} + T_{c1})}{\frac{1}{\varepsilon_{c1}} + \frac{1}{\varepsilon_{c2}} - 1}. \tag{6.28}$$

The radiative heat transfer to the environment from the outer cover "$c2$" is dominated by the radiation exchange with the sky having an equivalent sky temperature T_{sky} which for clear skies may be drastically below the ambient temperature T_{amb}. While for cloudy skies the equivalent radiative temperature of the sky dome is close to the air temperature, for clear skies due to low humidity content of the atmosphere the collector "sees" partially the cold background of outer space, resulting in equivalent temperatures 40–50° Kelvin below air temperature.

$$\begin{aligned} h_{r,c2-sky} = {} & \varepsilon_{c2} \cdot \sigma \cdot \left(T_{sky}^2 + T_{c2}^2\right) \cdot \left(T_{sky} + T_{c2}\right) \cdot (1 + \cos\beta)/2 \\ & + \varepsilon_{c2} \cdot \sigma \cdot \left(T_{amb}^2 + T_{c2}^2\right) \cdot (T_{amb} + T_{c2}) \cdot (1 - \cos\beta)/2. \end{aligned} \tag{6.29}$$

Convective heat transfer is characterized by Nusselt numbers depending mainly on the temperature differences of adjacent walls, the gap width d and the inclination β of the collector. We use for illustration the following correlation for convective heat transfer in a closed and inclined shallow cavity:

$$\begin{aligned} Nu = {} & 1 + 1.44 \cdot \left[1 - \frac{1708 \cdot (\sin 1.8\beta)^{1.6}}{Ra \cdot \cos\beta}\right] \cdot \left[1 - \frac{1708}{Ra \cdot \cos\beta}\right]^+ \\ & + \left[\left(\frac{Ra \cdot \cos\beta}{5830}\right)^{1/3} - 1\right]^+, \end{aligned} \tag{6.30}$$

where Nu ist he Nusselt number, Ra the Rayleigh number and β in (rad) the tilt angle of collector with respect to the horizontal plane. The Rayleigh number is defined by:

$$Ra = \frac{g \cdot T_m^{-1} \cdot \Delta T \cdot d^3}{\nu \cdot \alpha}, \tag{6.31}$$

in which g in (m/s^2) denotes gravitational constant, T_m in (°K) mean absolute gas temperature in the gap, d in (m) gap width, ΔT in (K) temperature difference over gap, ν in (m^2/s) the kinematic viscosity of gas, α in (m^2/s) the thermal diffusivity of gas.

The convective heat transfer coefficients in the gap then are:

$$h_{c,p-c1} = Nu_1 \cdot \frac{k}{d_1}, \tag{6.32}$$

$$h_{c,c1-c2} = Nu_2 \cdot \frac{k}{d_2}, \tag{6.33}$$

where k (W/(mK)) gas conductivity, d_1 in (m) is gap width between absorber and 1st cover and d_2 in (m) the gap width between 1st and 2nd cover.

For the convective heat transfer from outer cover to the air we take a simple correlation, e.g.

$$h_{c,c2-amb} = \left(2.8 + 3.0 \cdot \frac{u_0}{\mathrm{m/s}}\right) \mathrm{in} \left(\frac{\mathrm{W}}{\mathrm{m^2K}}\right), \tag{6.34}$$

where u_0 in (m/s) represents the wind speed, and d_1 in (m) is gap width between absorber and 1st cover and d_2 in (m) the gap width between 1st and 2nd cover. More sophisticated correlations for flat areas with different inclination angles or bodies can be found in literature for wind driven situations with forced convection (using the Reynolds number) or for wind-still situations with natural convection (using the Rayleigh number), for more details see e.g. Rohsenow (1998), Stephan et al., 2019) or Chap. 5 of this book.

From this a solution for the thermal network is obtained in terms of the different heat loss coefficients. Figure 6.21 illustrate the calculated total front heat loss coefficient (U_t) as function of absorber temeprature for different ambient temperatures for single glazed collectors (top) and for double glazed collectors (bottom).

A different possibility of suppressing convection is the use of so-called transparent insulation materials like honeycombs or capillary structures. Especially useful for the collector cover are combinations of a protecting glass-sheet with an open absorber-vertical cover behind. The narrow cavities or slits between the transparent walls of the structure are subduing convection efficiently. It is important that the structures do not touch the absorber surface. In direct contact this might on the one hand side enhance infrared emission from the absorber, on the other hand polymeric materials will not be stable enough the take the high absorber temperatures. Figure 6.22 shows a commercial collector using such a honeycomb structure (TIGI Solar, 2011) and a close up on the material.

Convection suppression by choosing gas fillings with a lower conductivity than air is a common approach and utilized e.g. in noble-gas filled window glazings. Similarly, solar thermal collectors can be built based on such fillings. However, one critical design criterion is the collector gas volume. As the collector shall be gas-tight, a warming up of the collector results due to the gas expansion in a higher pressure level on the glass pane, where a breakage must be avoided. Evacuation of the air gap between absorber and cover glass or even of the complete collector

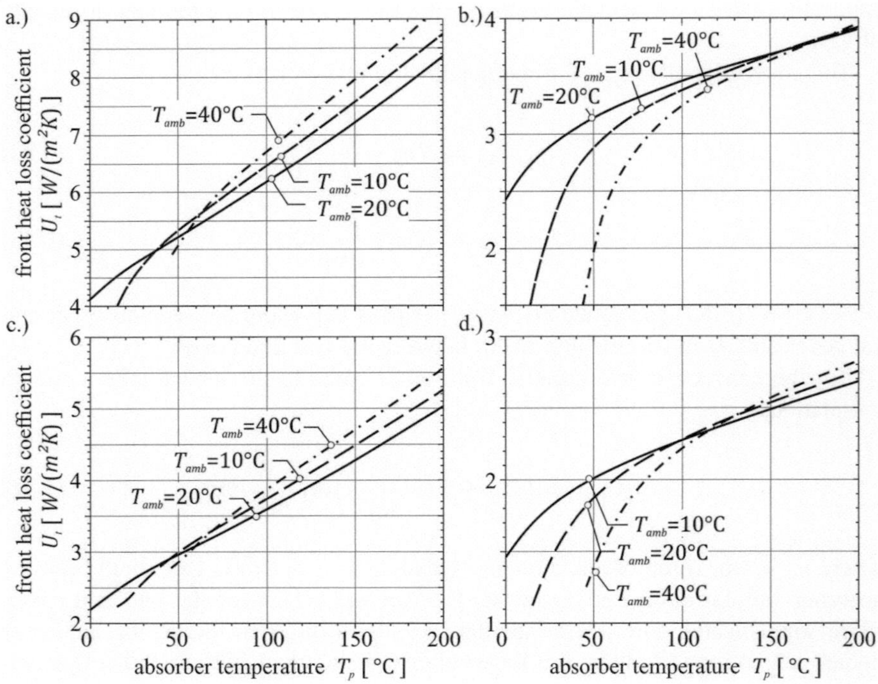

Fig. 6.21 Calculated heat loss coefficient U_t for different ambient temperatures as function of absorber temperature T_p and for a single glazed cover (**a**, **b**) and double glazed version (**c**, **d**). Left graphs black absorber $\varepsilon_p = 0.95$; right graphs selective absorber $\varepsilon_p = 0.05$

Fig. 6.22 Commecially available Honeycomb Collector from TIGI Solar (2011)

theoretically is the best suppression method of convection. The Rayleigh number Ra scales with the square of the air density and a temperature rise yields thus a pressure increase and thereby to an enhanced heat loss to the ambient. One may see the in Fig. 23 the resulting Nu-number of an inclined single-glazed collector of tilt $\beta = 45°$ and air gap 20 mm for a temperature difference of 40 °K as a function of density. At a pressure of $p = 0.2$ bar in the collector convection is suppressed effectively which is expressed by $Nu = 1$. If the conductive heat transport through the air gap shall be reduced further, even lower pressures are needed challenging the solar glass integrity. Such a low-vacuum collector thus is able to suppress air

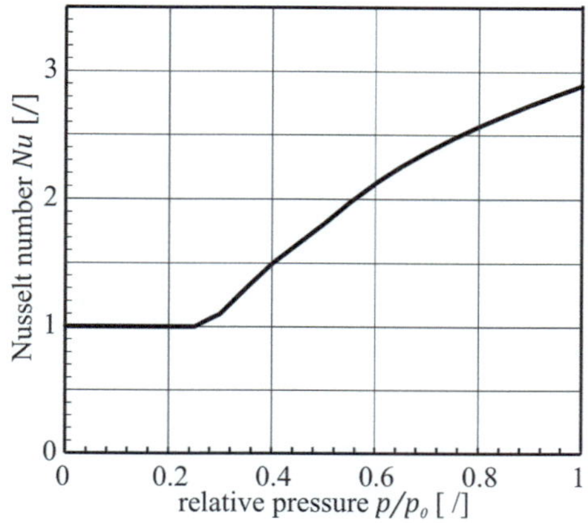

Fig. 6.23 Nusselt number for an inclined flat plate collector ($\beta = 45°$) as a function of the relative pressure (~ air density) for a collector with an air gap of 20 mm and a temperature difference of $\Delta T = 60$ °K

convection; however, the gas conductivity is still present and contributing to the heat loss.

Suppressing gas conduction needs a further evacuation to a high-vacuum collector. When the free mean path length of air molecules between mutual collisions is reaching the dimensions of the air gap (Knudsen regime), then the thermal gas conductivity is determined by the equation:

$$\frac{k_{gas}(p)}{k_{gas}(p = 1atm)} = \frac{1}{1 + 3.75 \cdot Kn} \text{ with } Kn = \frac{\delta}{L} \approx \frac{0.008313}{(1 + \frac{116}{T})L \cdot p}, \tag{6.35}$$

in which Kn is the dimensionless Knudsen number, k_{gas} in (W/(mK)) the thermal conductivity of gas, δ in (m) the mean free path of gas molecules, L in (m) a characteristic dimension of container (pore, air gap) and p in (Pa) the gas pressure.

At about $p = 10^{-5}$ Pa the final limit of conductive heat transport through the air gap in the collector is reached, if that gap dimension is $L = 1$ mm, as illustrated in Fig. 6.24, for more details see Chap. 4.

However, the practical problem is that with a pressure difference of close to an atmosphere from inside to outside, the forces on the cover glass are enormous, and in order not to break even special toughened glass must be supported about every 50–100 mm. This in turn creates a problem of heat conduction through the support pillars, which have to be extremely slim, but mechanically stable.

Figure 6.25 shows a solution of a low-vacuum collector commercially produced in the 2000 years, where holes have been punched into the absorber plates so that the stainless-steel support pillars only tough the cold back and front surfaces. The convection was well suppressed in both air gaps on the front and back side of the absorber. The collector U-value was therefore reduced considerably compared to an unevacuated selective flat-plate collector. The low pressure within this collector leads to a slow infiltration of air, as the sealings were not completely air-tight. For this reason, a possibility to reduce the pressure in the collector in a routine maintenance step using a simple pump (e.g. a waterjet pump) was foreseen by the manufacturer (see Fig. 6.25).

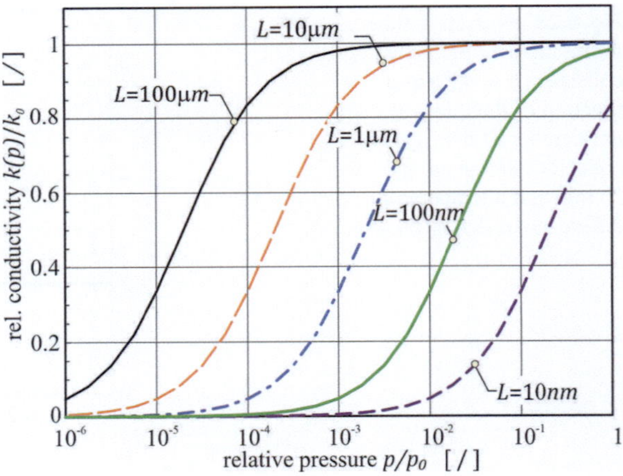

Fig. 6.24 Relative conductivity of the air gap to normal conditions as a function of the relative pressure (p/p_0) for different dimensions of the container L; free molecular regime (Knudsen number $Kn > 1$) requires small dimensions L or low relative pressures (p/p_0)

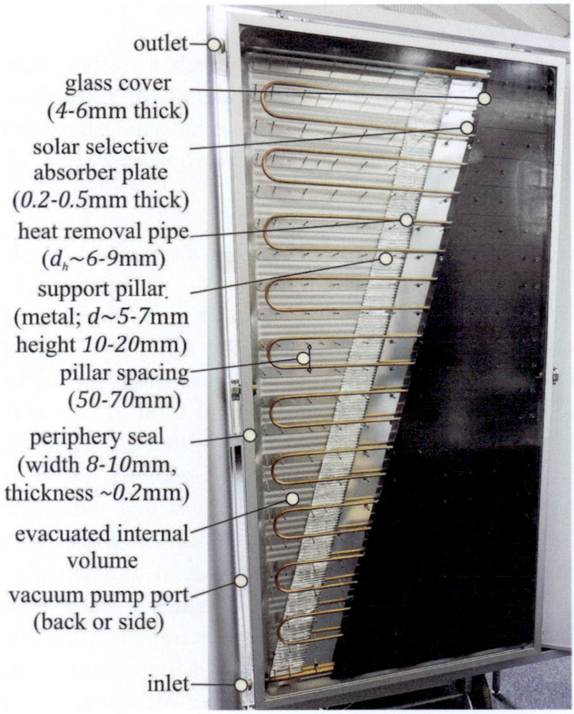

Fig. 6.25 Low-vacuum flat-plate collector width some typical component dimensions

For a real high-vacuum flat-plate collector evacuation not only requires a vacuum pump (e.g. a turbomolecular pump), it also would be very difficult to evacuate a thin layer of say 1 mm air through one valve, as this would take a long time. Therefore, such high-vacuum collectors need a much better sealing technology, so they do not need to be serviced because of vacuum loss. High-vacuum collectors have been developed some years ago utilizing glass-to-metal sealings and a getter

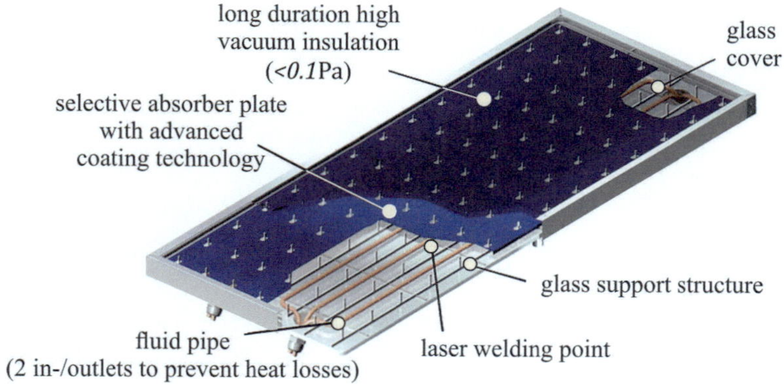

Fig. 6.26 Cross-sectional cut through a high-vacuum flat-plate collector

Fig. 6.27 Collector efficiency based on gross area for three different flat-plate collectors (unevacuated, low and high vacuum) vs temperature difference between collector and ambient $G = 850$ W/m^2

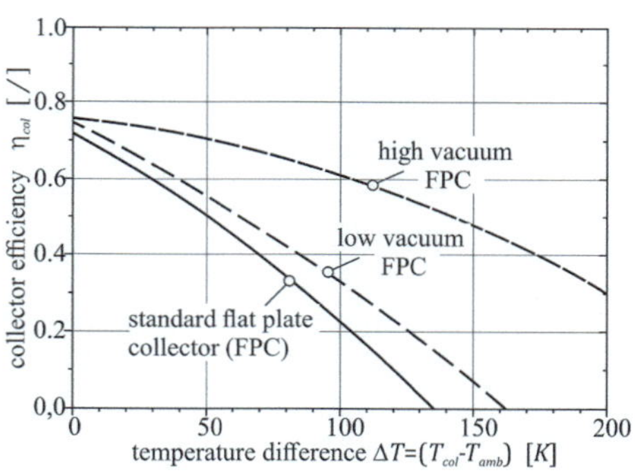

solution to keep the ultra-high vacuum (UHV) inside. No organic materials can be used inside these extremely slim collectors (see as an example Fig. 6.26). The collector efficiency of a low-vacuum and a high vacuum flat plate collector is shown in the next Fig. 6.27. Here it can be seen the convection suppression increases the efficiency noticeably in the temperature range 60–100 °C (here the comparison is with a non-evacuated collector of the same manufacturer). High vacuum, however, enables the flat-plate collector to be used also in up to 180 °C for medium temperature applications (solar cooling or solar process heat).

6.3.2.2 Absorber Plate

The connection of the absorber plate with the absorber tubes carrying the heat transfer fluid plays an important role. Several different joining techniques are available here, as the Fig. 6.28 shows. The absorber strips can either be connected in a fixed way with the absorber by welding, soldering, or gluing (left 4 options). Other options are attachment by mechanical means (roll-bond option pressing two sheets together or clamping) or extrusion as a whole structure, while the latter is mainly conducted mainly with polymer absorbers.

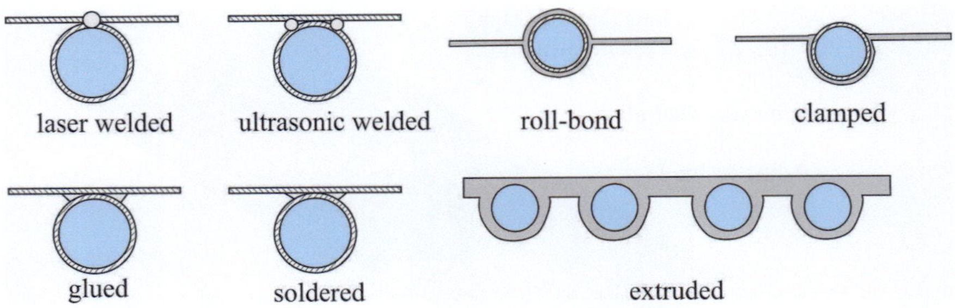

Fig. 6.28 Different possible way of connector flow channel (pipes) with the absorber sheet material

Other possibilities include pressing entire structures including cooling channels or inflating pressed sheets using gas pressure. The cooling channel cross-section can also be adapted to the individual applications, although this requires computer-aided simulation during the design phase. Figure 6.29 shows some examples.

Table 6.3 lists typical absorber efficiency factors F' as a function of the flow rate for the three joints shown. Other joining options include pressing or blowing processes in which parts are left out of the sheet metal and treated with release agents to prevent them from joining together. Afterwards these areas are inflated, so fine canals can be built there.

Another important aspect of the collector design is the internal piping. There are potentially several possibilities. The arrangement of the flow channels in the absorber can be serial (Fig. 6.30a)—serpentine collector—for low-flow systems

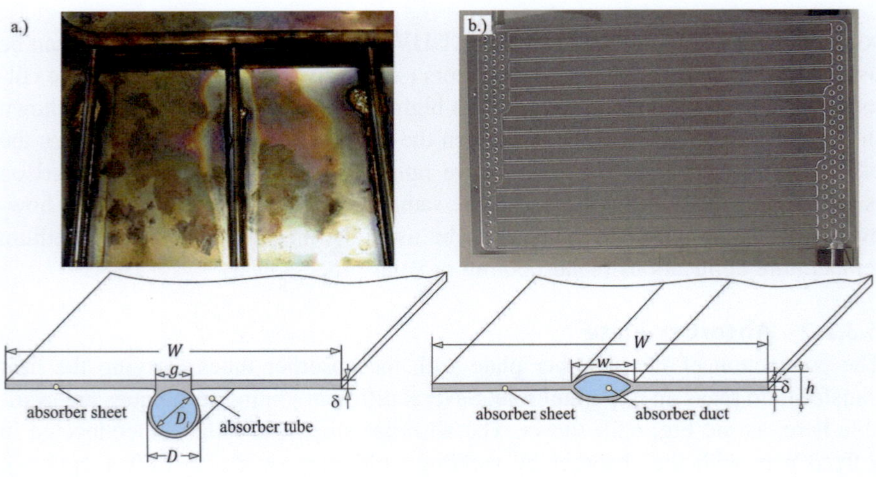

Fig. 6.29 Tube-fin absorber (**a**) and roll bond absorber **b**) in view (top row) and in cross-section (bottom row); © Fraunhofer ISE (left) © CGA Technologies S.p.A.(right)

Table 6.3 Absorber efficiency factor F' for three connection methods of absorber plate with absorber pipe for different mass flows

Joining method absorber and tube	F' at (20 L/min)	F' at (40L/min)	F' at (80 L/min)
Roll-in procedure	0.89	0.91	0.92
Soldering/welding	0.87	0.91	0.93
Rollers	0.88	0.89	0.92

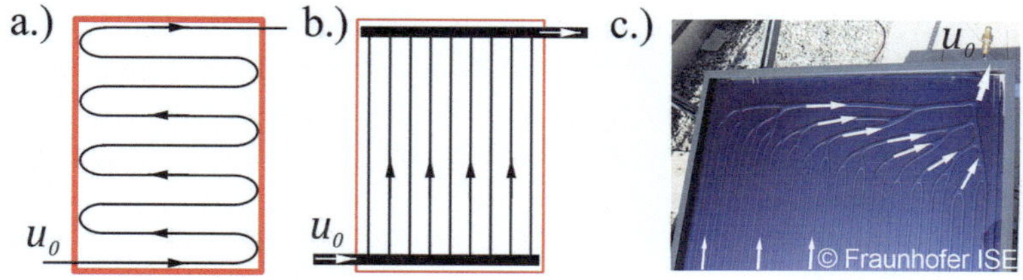

Fig. 6.30 Piping and distribution arrangement in flat plate collectors in a serial circuit (**a**), parallel (**b**) and a type taken from biological (**c**) models (Hermann, 2011)

and collectors working on natural convection, parallel (Fig. 6.30b)—harp collector—which is the standard design for high-flow systems, or in a branched structure (Fig. 6.30c)–bionic collector which uses the principles developed in nature to approach a very even temperature distribution over the collector width combined with a minimum hydraulic resistance. With serial flow control in a meander, high flow rates through the long flow path result in high pressure losses, which require a powerful pump and lead to high electricity consumption. In addition, there are large temperature differences in the collector box and between the fluid inlet and outlet. Therefore, this kind of collector is suitable for low mass flow systems, e.g. in thermosyphon systems. When the flow channels are connected in parallel, the temperature distribution in the collector is more uniform. It must be ensured that the flow rates in the subchannels are approximately equal, so that local evaporation of the heat transfer fluid does not cause the collector to come to a standstill. Also, uneven temperature distribution between different riser pipes would result in a lower resulting efficiency and exergy. In the case of flat plate collectors with their relatively small differences in useful temperature, this is ensured by suitable pressure loss ratios between the distributor—or header—and absorber tubes. For vacuum tube collectors with higher useful heat, more complex numerical calculations are required. This also applies to the complex distribution of flows and temperatures in the bionic branching systems.

6.3.2.3 Thermal Insulation

Thermal insulation is necessary in flat plate collectors to reduce heat losses. Generally, the back wall and the sides of the collector casing are lined with these materials. As well as rock and glass wool, PUR foams and melamine resin are used. In rare cases, smaller manufacturers offer collectors with sheep's wool, flax or other»natural« insulation. The insulation materials must withstand the high stagnation temperatures of the collectors. In particular, the rear side insulation, on which the absorber mostly lies, must have high thermal stability. The insulation should not evolve gas, even at very high temperatures, otherwise inside deposition on the cover pane is to be expected. Some typical values of thermal conductivity for insulation materials are shown in Fig. 6.31. Important for the protection of insulation against wetness are seals, at the casing, at the glass cover and at the tube ducts. They must prevent the penetration of rainwater over the entire operating period, however due to pressure and humidity transport (micro-climate) considerations, the collector should be able to exchange air with the environment (slight ventilation). Above all, the seals at the tube ducts must withstand very high temperatures in the case of stagnation and mechanical loading (movements of tubes due to temperature expansion).

Thermal Insulation Durability

Peuser, Remmers, and Schnauss (2011) investigated materials degradation of collectors, having served already 10–20 years of operation. In these collectors the insulation consisted almost exclusively of polyurethane (PUR) rigid expanded

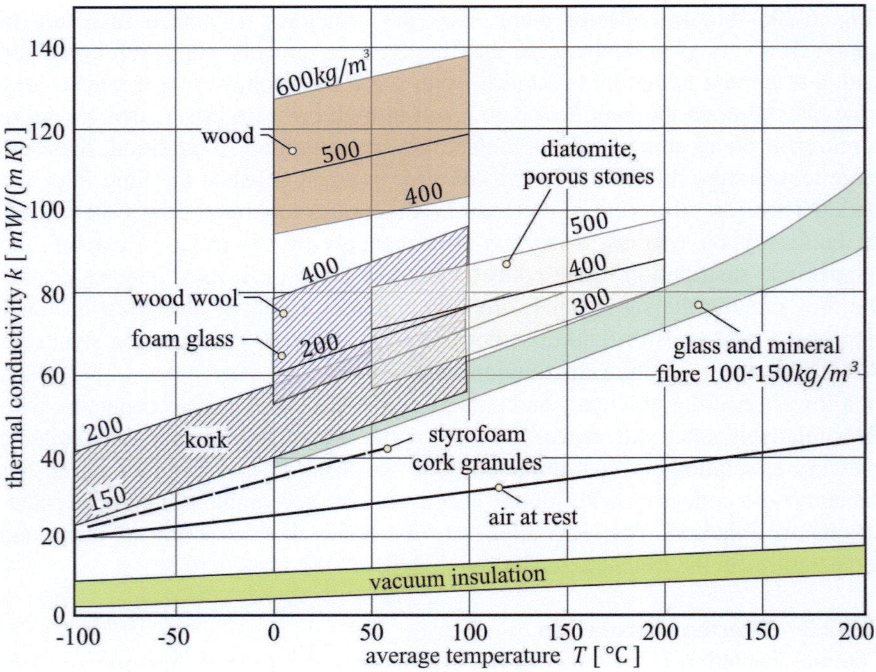

Fig. 6.31 Thermal conductivity k (W/(mk)) of selected insulating materials as function of the temperature in (°C)

Fig. 6.32 Right: rock wool insulation of a selective flat plate collector

Fig. 6.33 Seal between collector casing and collector tubes, (new-right and after more than one decade of use-left)

sheets or mineral wool. Often the insulation was having an aluminium lamination (laminated on one or both sides). Mineral wool is still used frequently today for which an examples is illustrated in Fig. 6.32. PUR sheets (not in combination with other materials) are rare. When opening the collectors, the mineral wool insulation was undamaged and seemed to be in good condition, regardless of whether it was used in selective or non-selective collectors or whether aluminium lamination was present or not. Also, the thermal conductance of insulation from a 16-year old collector resulted in a value of $U(10\ °C) = 0.034$ W/(m^{2}K), which is close to the value for new material.

The results of the investigations proved that except for some minor optical impact, the mineral wool materials had still good performance after many years of operation while the PUR mats were discoloured brown on the surface at the absorber and hardened at the top. The intensity of the discolouration follows the temperature development in the absorber. Even when the insulation mat was protected from the absorber by the mineral wool, damage to the PUR foam could be observed. In selective collectors where PUR foam was directly touching the absorber back, the polymer foam material partly crumbled away. In contrast, none of the PUR material in non-selective collectors showed discolouration which is a clear indication of lower temperature load. As the cell structures of the PUR foam were partially destroyed due to high temperatures and gas could escape, the thermal conductance of the PUR increased, however still being lower in comparison to mineral wool (PUR foam manufacturers state thermal conductivity values of around $k = 0.030$ W/(mK) for insulation mats without propellant). Laboratory tests showed that thermal conductivities are still low even after partial gas evolution of the propellant, but have deteriorated by approx. 20% and by 25–45% for the degraded 15 year old material depending on the colouration—compared with the new material $k(10\ °C) = 0.020$ W/(mK); after one week in an oven at 150 °C: $U(10\ °C) = 0.024$ W/(m^{2}K) The continuing good insulating properties of

PUR rigid expanded foams even after many years are confirmed by measurements on this material from two 13–14 year old collectors. The thermal conduction was maintained. For a non-discoloured insulation from a non-selective collector still good insulating properties of PUR rigid expanded foams can be seen after 15 years with about $k(10\ °\text{C}) = 0.025\ \text{W/(m}^2\text{K)}$, whereas it is comparable with that of new mineral wool for selective absorbers. Therefore, even if PUR rigid foam is not optimal for use in selective collectors due to their low temperature resistance the damage which occurs is not critical to the function of the collector. Problematic possibly are very efficient collectors used in large-scale installations which are frequently exposed to high stagnation temperatures. Also, the seals at the tube ducts are to be viewed more critically than the insulation. Figure 6.33 shows two of the seals made of an elastomer, after several years of operation with some stagnation periods in new and aged conditions. The seals at the collector inlet are typically facing lower temperatures, while those at the outlet get deformed and highly crumbled and therefore more likely to fail than those at the collector inlet.

Current Thermal Insulation Materials

The thickness of the rear insulation is often between 40 and 70 mm. Increasing the thickness of the insulation beyond this leads only to a small improvement in the U-value. In collectors with frames made of aluminium or stainless steel, an edge insulation can prevent a thermal bridge between the absorber and frame. This markedly improves the U-value. Furthermore, some thermal bridges to the tube ducts should be avoided by using suitable insulation. Careful construction also considering the vibration during transport, has proven to be more effective than further increases in insulation thicknesses. Panels with deposits on the inside of the cover should be avoided as interaction with other materials in the collector, such as glass and absorber coatings. Therefore, stability and compatibility should be tested before their use. It should be carried out with every material before its use since the high temperatures in the collector cause gas evolution in almost all materials. Sheep´s wool, flax, hemp, or similar natural materials are used in some collector designs, but some of these materials have proven to be unsatisfactory in practice. This may be noticed by condensation on the inside of the collector cover or by decomposition.

Requirements for Sealing Materials

Materials based on ethylene-propylene-dien-monomer (EPDM) have proven themselves for the seal profiles used between the frame and the pane of glass. The adhesives used are mostly silicon-based. In this case, care should be taken since many of the products on offer have completely different compositions despite having the same name. When using the adhesives, adhesive power, material softening, and gas evolution must be tested as well as their long-term stability. The seals in the tube ducts should be temperature resistant to the stagnation temperature (larger than 180 °C) to guarantee permanent sealing against leaks. Some silicon-based materials are suitable for this purpose.

6.3.3 Air Collectors (Absorber Design, Ventilation Energy)

Air collectors use air as a heat transfer medium, and therefore need channels with large cross-section for the air transport. Otherwise their basic construction corresponds to that of a flat plate collector, as Fig. 6.1 sketches. The surface-structured absorber is placed in a metal tray and transfers the heat energy provided by the sun to the air flow below. However, it is also possible to use other materials with large surface area, like woven textiles or matrix absorber, even pebble heaps consisting, for example, of lava stones, which are heated by solar radiation and transfer the heat to the air. Modern air collectors achieve efficiencies $\eta = 50$–70% in terms of heat transfer. Since the air ducts have a large cross-section, a solar thermal system with air collectors is more voluminous than liquid collectors. Depending on the basic air heater designs the transfer of solar heat to the air has a different collector efficiency factor F'. These are shown for a wider variety of geometries in Fig. 6.34.

Since air itself can store almost no energy (over a long period of time), there are two alternatives for using the heat:

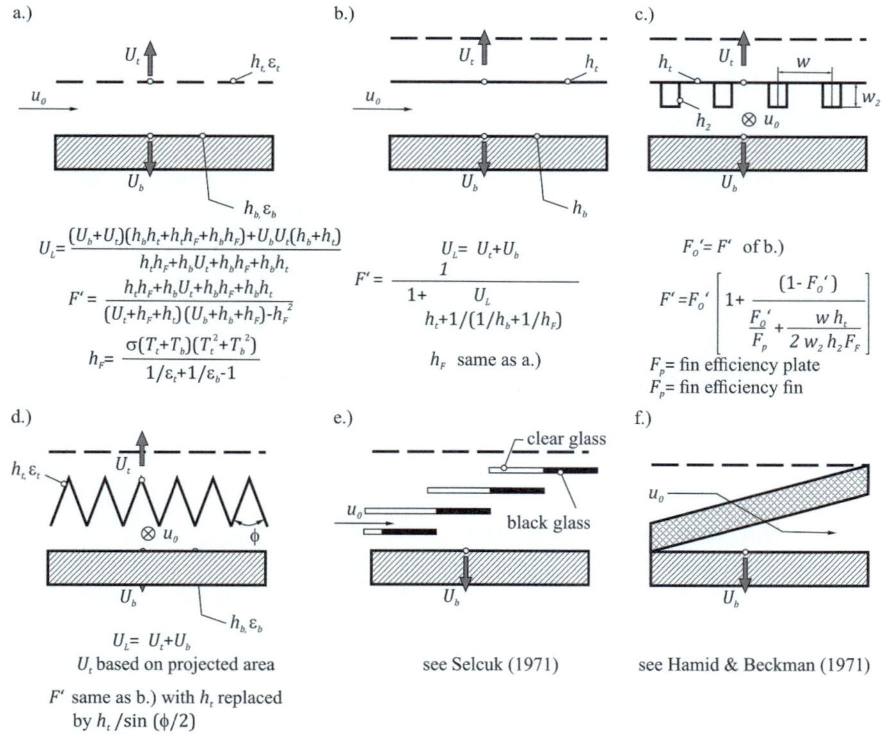

Fig. 6.34 Typical design of absorber geometries for air collectors and corresponding F' factors; Selcuk (1971) and Hamid and Beckman (1971)

Table 6.4 Pros and Cons of air-cooled collectors for active/passive solar systems

Advantages	Disadvantages
No freezing	Large duct cross-section
No steam formation	High ventilation energy
No damage of HTF by stagnation	Lower efficiency
Leaks are less problematic	Collector thickness
Lower critical start-up level	

- direct air heating with moderately heated warm air ventilated directly into a building with high fresh air demand (holiday homes, apartment building, swim halls, factory halls etc.). The combination with an air-heating system is most favourable.
- Air is heated to high temperatures and heat is transferred to a storage by an air–liquid heat exchanger (water storage) or solid storage materials (rocks, pebbles, bricks, or hollow walls etc.). Only after charging the storage, the cooler air may be used for direct heating of rooms.

In agriculture and industry, air collectors have been successfully used for several years to dry hay, grain, herbs, or other biomass. Very favourable applications are solar dryers and swim halls, as they have a high ventilation heat requirement even in summer. But also the heating of buildings with lower comfort temperatures and high ventilation requirement is a popular application. Some pros and cons of air cooled collectors are listed in Table 6.4.

A variant which is especially suitable for large industrial buildings and halls is the transpired uncovered metal sheet collector (Wang, Shukla, & Liu, 2017), which is depicted in Fig. 6.35. In this product the outside air is sucked through a perforated absorber sheet and heated up in the passage. The air flow must be adjusted to reach the desired temperature. Behind the absorber fixed to a vertical façade the air is guided through the air-channel between absorber and wall to the building interior. With this technology preheating is of fresh air is the target. The coloured and stable absorber coatings provide an architecturally interesting option of a wall cladding.

6.3.4 Vacuum Tube Collectors

A vacuum tube collector (VTC) consists of an evacuated glass tube, inside which there are either concentric cylindrical or fin-type absorbers. A cost-efficient way to produce vacuum tubes is by using two concentric glass tubes which are containing an evacuated volume (the Dewar-type Sydney collector). The required vacuum—as the characteristic distance L between absorber and glass tube is in the order of magnitude 1 cm—must be as low as 10^{-7}–10^{-8} atmospheres. In this case the thermal conduction of gas can be suppressed and only the radiative component contributes to the heat loss. We can see some cross sections of VTCs in Fig. 6.36.

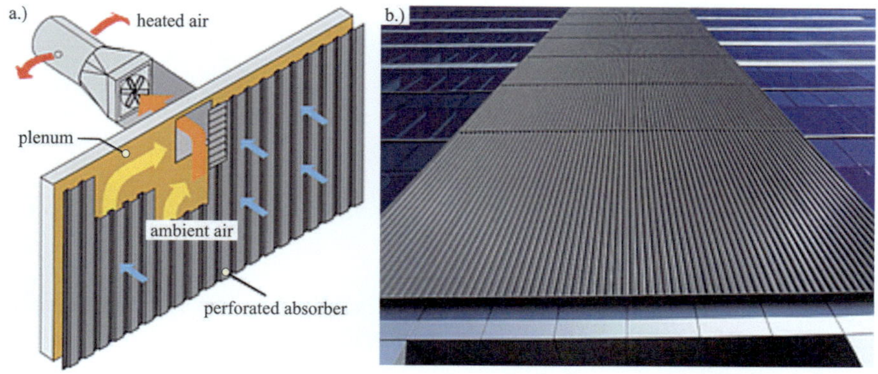

Fig. 6.35 Transpired façade air-collector—functional principle (**a**) and example building façade using SolarWall air collector (**b**) (SolarWall Europe Sarl., 2023)

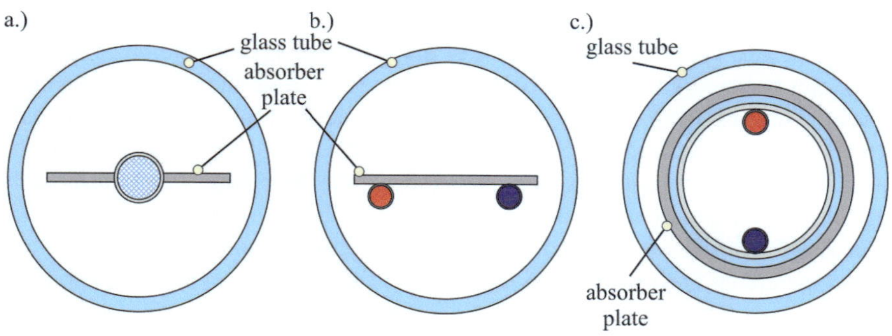

Fig. 6.36 Cross sectional cut of different vacuum tube collector (VTC) types; **a** heat pipe, **b** direct flow and **c** Dewar type

Vacuum collectors differ from flat plate collectors by an almost optimal thermal insulation within the collector box. The absorber strips are located inside an evacuated glass flask. This practically eliminates convection losses completely. Only the heat radiation occurs as a loss factor. This is minimized by a selective coating of the absorber strips. The heat transfer to the plant circuit is either direct or by means of the so-called heat-pipe system. We will discuss the three absorber basic VTC types, heat-pipe VTC, direct-flow VTC with fins and Sydney collector in the next subsections individually.

6.3.4.1 Heat-Pipe Vacuum Tube Collector

Vacuum collectors based on the heat-pipe principle use no circulation of fluid through the vacuum tube as other types of vacuum collectors with direct flow, and their mode of operation is shown in the Fig. 6.37. The heat pipe inside the glass tube is coated black and absorbs the solar radiation. It is closed on both sides

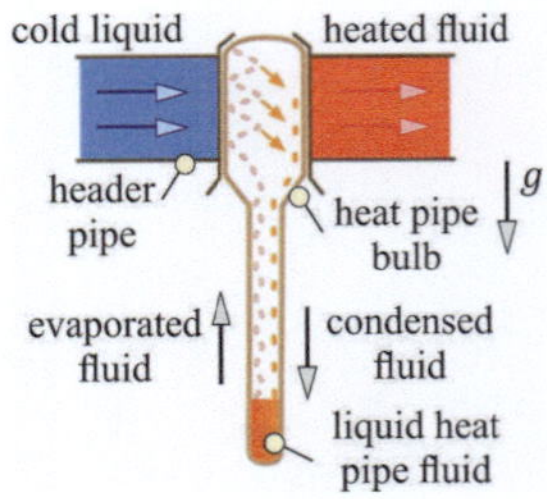

Fig. 6.37 Operation principle of a heat pipe

and is hermetically sealed itself, containing a rapidly evaporating liquid, usually methanol or other types of alcohol, at an appropriate pressure.

Due to the heat transferred from the absorber to the liquid in the heat pipe, this liquid evaporates and rises to the top due to its lower density. Therefore, usually a heat-pipe requires an installation inclined at a certain angle as it is dependent on gravity. A heat exchanger is located at the upper end. Here, the steam condenses and transfers the heat of evaporation to the secondary circuit. This secondary circuit is connected to the heat circulation system of the solar circuit. The advantages of the heat pipe system are greater ease of maintenance, as defective pipes in a collector can be replaced individually without having to drain the transport medium. Since the heat exchanger is connected to the secondary circuit "dry", a better south orientation of the individual absorber fins can be made possible in larger plants by specially rotating the absorber tubes. In addition, this type requires a considerably smaller pump capacity, since there are no pressure losses within individual tubes. Heat Pipes are inherent safe and protect the solar circuit from overheating—when the temperatures are too high no condensation can occur anymore and transfer heat. A disadvantage of the heat-pipe compared to the coaxial system is the higher price, which is mainly due to the more complex design. Such a dry connection is illustrated in Fig. 6.38.

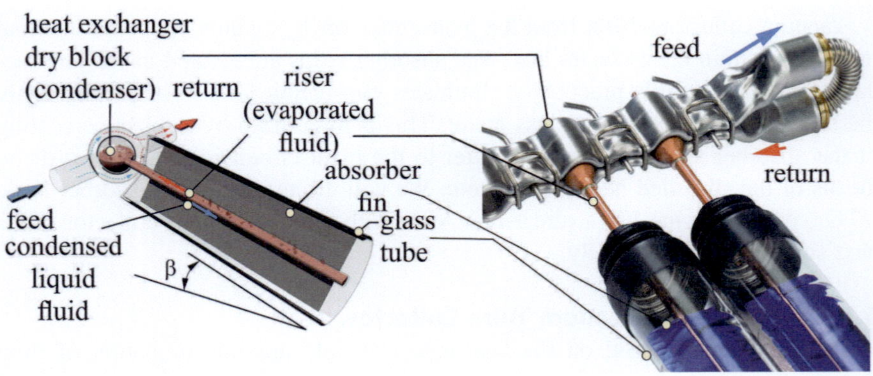

Fig. 6.38 Vacuum tube according to the heat pipe principle with dry connection

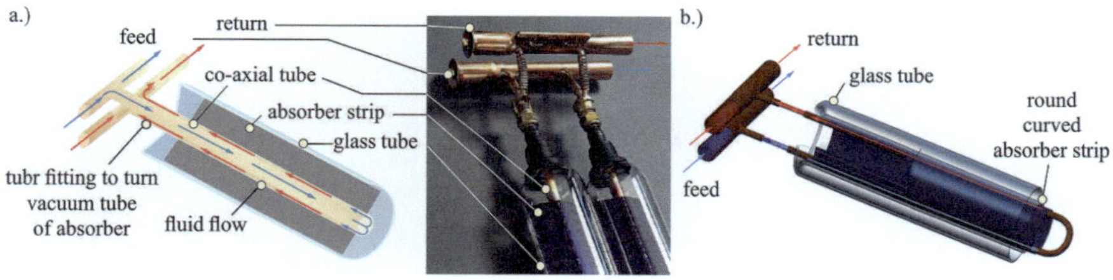

Fig. 6.39 Vacuum tube collectors (VTC) with fins; coaxial tube design **a** with a typical head design and **b** U-type tube design

The inside of the glass tube can be evacuated with a vacuum-tight connection between the metal tubes and the surrounding glass tube. The vacuum thermally insulates the absorber against the glass tube. One of the central challenges is to ensure a vacuum-tight connection between the glass tube and the metallic absorber over many years. Occasionally the glass tubes contain a getter. This can be heated inductively from the outside to restore the vacuum during the service life.

6.3.4.2 Direct-Flow Vacuum Tube Collector with Fins

There are two different types of direct flow vacuum tubes, the coaxial vacuum tube collector, and the U-tube collector. Both are collector tubes connected on one side to the circulation. However, in principle the fin-type collectors exist also as two-sided flow through collectors. In the coaxial version, the absorber fin is directly connected to a single coaxial heat exchanger tube through which the heat transfer medium flows directly. The heat transfer medium absorbs the heat from the absorber via the heat exchanger tube. The heat exchanger tube opens into the collector pipe. The fin collectors have the advantage that solar radiation may hit the absorber fin vertical when the sun is at an optimal position. Deviations from the south direction can be compensated for by axial rotation (maximum approx.25°) of the vacuum tubes. One problem for the usual oneside collectors is the temperature difference between inlet and outlet being very close to each other and connected to the same fin. The critical upper side, the so-called head, of such a collector tube is shown in the Fig. 6.39. There, relative strains occur because of the different temperatures, which must exclude failure in every potentially conceivable operating condition by suitable design measures. Typical design measures to minimize stress are metal bellows integrated.

6.3.4.3 Dewar-Type Tube Collector "Sydney Collector"

Another solution is a double glass tube with the vacuum in the intermediate area similar to the principle of a thermos flask or Dewar. The absorber is coated onto the inner glass tube. The absorber cross-section is smaller than the outer glass tube, so that the usable absorbing surface is smaller than the collector surface. This can be compensated by rear side reflector sheets.

The two glass tubes lying one inside the other are welded together and evacuated. The absorber is vapour-deposited on the inner glass tube and the absorbed

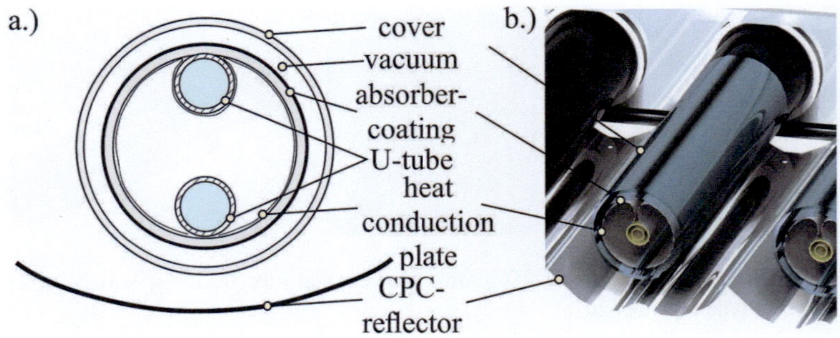

Fig. 6.40 **a** Schematic diagram of a direct flow U-tube collector. **b** Improvement of the heat generation of vacuum collectors by rearward mounting of CPC mirrors

heat is transferred to the heat transfer medium via a heat conduction sheets and absorber tubes inserted in them. This is sketched in Fig. 6.40a. The cold fluid enters at the bottom for preheating and returns in the top to reach temperatures. CPC mirrors are used at the rear of the absorber to ensure that the incident radiation can also be utilized. As an alternative also here, a coaxial tubing is possible (Fig. 6.40b), which exhibits better standstill behaviour than a U-tube.

During standstill the absorber tube and the pipes reach high temperatures. If, after a standstill, cooler fluid first flows back into one of the two U-tube legs on one side—which can hardly be avoided—then this leg is cooled down considerably within a very short time. Cooling can be more than 100 °C within a few seconds. The pipe is stiffened by fixing it to the absorber plate. The only possibility to reduce resulting thermal stresses remains in the pipe section between the absorber plate and the penetration of the vacuum pipe. In many cases, only a few such situations are sufficient until the already critical vacuum-tight connection glass tube/metal tube fails. In addition, the U-tube type has a lower square metre capacity.

The optical losses of vacuum collectors are comparable or slightly higher than those of flat collectors (single-glazed, curved surface). The conversion factor η_0 is in the range between 0.65 and 0.85. Due to the excellent thermal insulation of the vacuum in the order of 10^{-4}–10^{-5} mbar at a collector temperature of 150 °C and an ambient temperature of 20 °C, vacuum collectors achieve high efficiency even at very high operating temperatures or low irradiation. The heat loss factor U_L is in the range 0.5–1.2 W(m^{2}K) which is due to the thermal radiation loss.

Due to these values the efficiency curve of a typical vacuum tube collector as shown in Fig. 6.41 indicates efficiencies of 50% for high irradiation even at higher temperatures above 100 °C. VTC are preferably used for heating support in older buildings (high-temperature heating) and for process heat applications. They are also an option if space is limited. A disadvantage of vacuum tube collectors compared to flat plate collectors is their significantly higher price for high quality products in Europe. In China VTCs of a simpler quality with shorter lifetime, however, are very cheap being a mass product.

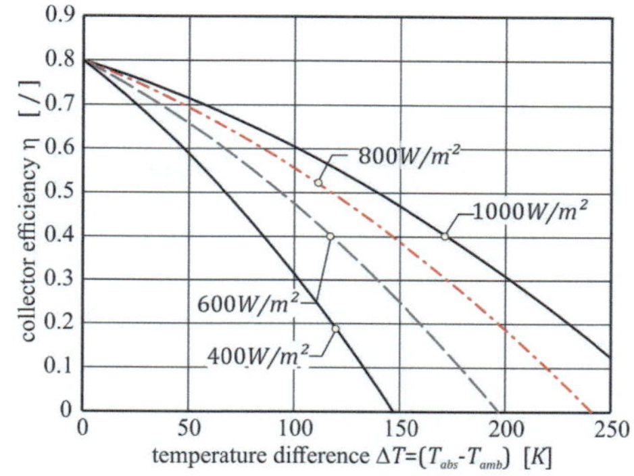

Fig. 6.41 Efficiency curve η of a vacuum tube collector as a function temperature difference for different solar irradiation intensities G in (W/m^2) for a collector with $u_0 = 1.45\ W/(m^2K)$ and $u_1 = 0.005\ W/(m^2K)$ according to Eq. (6.10)

The quality of the vacuum in the glass tube can be easily checked. At a sufficiently low internal pressure, a "good vacuum", there is neither heat conduction nor convection in the free space around the absorber. The absorber is selectively coated and behaves approximately like a Lambertian spreader in terms of thermal radiation. The temperature is the same almost everywhere around the glass tube circumference. If the pressure increases, heat conduction occurs, and the glass tube temperature is highest at the point with the smallest distance to the absorber. With a temperature peak at the zenith of the glass tube, i.e. at the top, the internal gas pressure is sufficiently high that natural convection can form. Additionally, getters are built into vacuum tubes as indicators. If the internal pressure increases due to diffusion or a leak in the glass tube, the getter absorbs the acid and nitrogen. This can be done at operating temperature, or the getter is heated by induction heating. If the leakage rate exceeds the absorption capacity of the getter, this can be seen from its colour change—usually white.

6.4 Stationary Collectors with Low Concentration

The use of solar thermal energy to supply industrial process heat is one of the most promising areas for solar market growth. Whereas low temperature ($T < 100$ °C) applications and the higher end of medium temperature applications (200 °C $< T <$ 400 °C) are well suited for stationary collectors and line-focusing tracking collectors, respectively, for the temperatures gap in between there are few developments available. Non-imaging collectors of the type flat-plate collectors with integrated CPC (Compound Parabolic Concentrator, see Chap. 3), if carefully designed, could possibly serve this temperature range. Another option is to use flat-plate collectors with reduced heat loss for higher temperatures and/or a radiation enhancement method using external mirrors redirecting part of the global radiation towards the collector.

6.4.1 CPC-Collectors

CPC reflectors may be integrated into a flat-plate collector. This leads to a larger thickness of the collector casing, and a CPC volume with an internal tubular or sheet absorber, in which due to the depth of the structure air usually is convecting. Therefore, the heat loss in such collectors is a challenge. Commercial first developments were not very successful, as described in IEA TASK 33 (Weiss & Rommel, 2008), see e.g. Fig. 6.42.

Due to the laws of optics it can be shown that collectors with $C_{max} < 2$ has a resulting half acceptance angle of $\theta_a > 30°$. As the declination of earth's orbit determining the seasonal path variation of the sun is $\delta_{max} = \pm 22.5°$, in practice, even lower values of C are preferred.

Non-imaging CPC collectors can be designed for quasi-stationary (or seasonally tilt adjusted) operation, making use of the advantages of a low concentration $C < 2$ while not depending on tracking systems. In this way a low-cost production could benefit from the dismissal of tracking devices. Many factors, from the geometrical design to materials, production, installation conditions, O&M, operation temperature and resource availability have an impact on the performance and cost, and must be accounted for to find the optimum collector design (Osório et al., 2019).

Stationary collectors with appropriate tilt and CPC trough orientation East–West could cover then the solar path for the whole year. At equinox the solar path should be just in front of the aperture of the collector, such that the normal on the aperture would be in the orbit of the sun. And the tilt of the CPC collector would be equal to the latitude:

$$\beta = \Phi. \tag{6.36}$$

For the whole year the projected transversal incidence angle on the aperture is within $\Phi \pm \delta_{max}$, and provided, the acceptance angle is large enough, the direct beam radiation from the sun should be accepted throughout the year by the optical concentrator. If a smaller acceptance angle is chosen, a seasonal adjustment is sufficient.

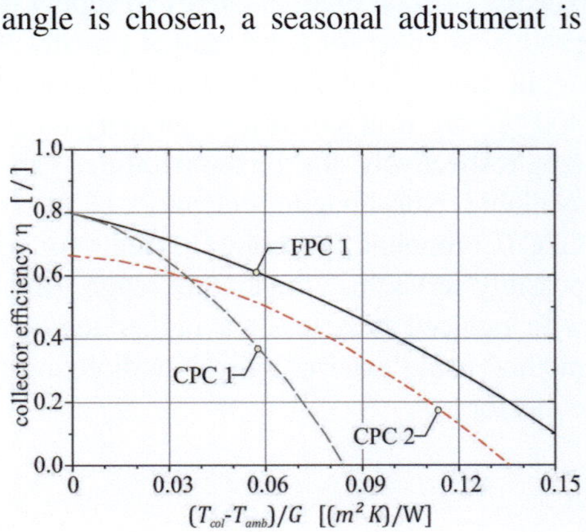

Fig. 6.42 Efficiency curve η of a two CPC collectors as a function reduced temperature difference for $G = 800$ W/m^2

One would think that for a CPC trough with an acceptance angle $\theta_a > \delta_{max}$ the condition should be always fulfilled. However, the analysis of the incoming solar radiation with direction cosines μ_E, μ_N, μ_H with the normalized radiation vector $\overrightarrow{k_s} = (\mu_E, \mu_N, \mu_H)$ shows a different situation (Spirkl et al., 1998). Rays are only accepted if and only if

$$\left|\frac{\mu_N}{\mu_H}\right| \leq \tan\theta_a. \tag{6.37}$$

Using the fact that the vector is normalized the direction cosine along the aperture normal μ_H can be eliminated, which leads to the equation

$$\mu_N^2 \leq \left(1 - \mu_N^2 - \mu_E^2\right)\tan^2\theta_a. \tag{6.38}$$

Putting this into the normalization equation where the sum of cosines squared equals unity, one obtains

$$\frac{\mu_N^2}{\sin^2\theta_a} + \mu_E^2 \leq 1. \tag{6.39}$$

The Fig. 6.43 shows the critical acceptance angles for the incidence direction cosines μ_N and μ_E for a CPC trough with critical acceptance angle 35° and 23.5°.

The solid collector just touches the sun path at the solstice and as a result the collector achieves no operational hours on the summer and winter solstice days. The dashed collector has a large enough acceptance angle (θ_a) to still capture a significant fraction of solar energy even during the solstices. The unit circle in the plot represents the "horizon" of the collector (as for a horizontal collector at the equator)—if the collector would be at another latitude and then would be tilted, a part of the ground would be included there. The horizon to be seen by the collector in this case then would appear also as a half ellipse in the lower half of the map.

The combination of a CPC type reflector with an evacuated tube collector is a commercial solution already on the market, as already shown in Sect. 6.3.4. However, here the concentration factor C is close to 1, and the advantage of the reflector is mainly to collect also global direct and diffuse radiation falling in between the tube absorbers. However, with a different shape also higher concentration can be reached—considering that then only a part of the diffuse radiation can be redirected to the absorber.

If higher temperatures shall be reached by a CPC type collector, the design must be more sophisticated. Absorber shape and concentrator shape must match each other. As already shown in Chap. 3 the CPC shape is different for tubular and flat absorbers. There are also geometries where two sides of a flat absorber are irradiated. A critical issue with the optical design is always that the hot absorber on the one hand side should not touch the cold reflector, and that rays falling into a gap between absorber and reflector shall not escape again.

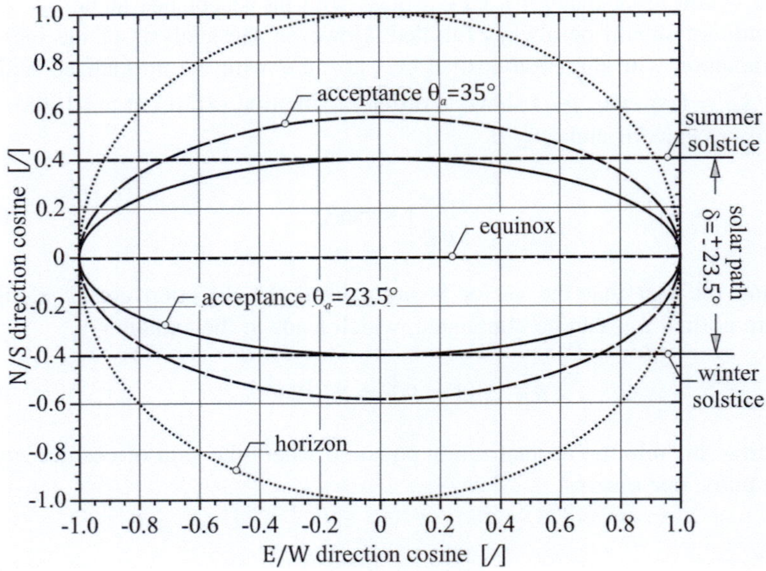

Fig. 6.43 Direction Cosine Map. The full hemisphere is shown as the unit circle. The sun travels along horizontal paths each day, bound by the upper and lower dashed black lines representing the summer and winter solstices. The sun path during the equinox is the horizontal dashed line ($y = 0$). The parts of the sky which are "seen" by a solar collector are drawn for acceptance angles $\theta_a = 23.5°$ (solid) and $\theta_a = 35°$ (dashed)

The design and development of an experimental medium-temperature solar thermal collector, the external compound parabolic concentrator (XCPC), is presented in Widyolar et al., (2018). Here, a non-imaging reflector is paired with an evacuated tube absorber for efficient and low-cost heat collection between 100 and 250 °C. The absorber geometry is optimized under the constraint of being assembled with an ultrasonic welding machine, with a final pentagon-shaped absorber selected. This approximates a circular absorber. The external CPC trough is designed with an acceptance angle of 40°. The modified absorber shape, gap loss, and truncated reflector result in a geometric efficiency of 93% compared to an ideal CPC. The simulations for the manufactured prototype predicted an optical efficiency of 71% and thermal efficiency of 50% at 200 °C. Experimental test results (optical, thermal, stagnation) have shown an optical efficiency of 62% and a thermal efficiency near 50% at 200 °C with a final stagnation temperature of $333°C$. This example shows that in principle with such a design high temperature can be produced with a stationary collector. Whether that collector is cost efficient and practical enough, must be shown, but the theoretical potential is demonstrated.

6.4.2 Radiation Enhancement with External Reflectors

Another option to reach higher temperatures, this time with flat-plate collectors, is the possibility to enhance the radiation on the collector aperture by external reflectors. When between two rows of tilted flat-plate collectors the gap can be filled with mirrors, then a part of the diffuse radiation from the sky, and for most of the time direct beam radiation may be reflected on the collector aperture. Due to reflectivity losses and spillage due to incomplete reflection matching with the collector aperture, of course not a doubling of the incident radiation can be achieved, but a factor between 1.3 and 1.6. In the German Reflec Project (Hess et al., 2010) such a collector assembly has been designed and optimized. A prototype of such a collector has been installed to provide heat for a laundry at elevated temperatures. The flat-plate collectors used were double-glazed selective collectors with low heat losses.

The concept of a collector field with reflector segments in between the rows is shown in the next picture as a CAD impression (Fig. 6.44).

After optimization of the reflector concept, where an expensive CPC shape reflector has been replaced by an assembly of three flat reflector segments, prototypes have been built and tested, which is illustrated in Fig. 6.45. The incidence angle modifier (IAM) of such a collector-reflector unit is biaxial and asymmetric for incidence of beam radiation in the vertical plane. The incidence angle modifier in the longitudinal plane is just the same as of an ordinary flat-plate collector without reflector.

The transversal incidence angle modifier for the collector in the vertical plane is depicted in Fig. 6.45. It shows a maximum enhancement of radiation at about 30° incidence. This is the angle where the direct sun is fully projected onto the flat-plate collector. The reflector itself shades in continuously radiation from the ground in the range −10° until −40°. After that some radiation enters again through the gap between reflector and collector.

However, this part is shielding ground reflected diffuse radiation, which is much lower typically than the diffuse sky irradiance.

The performance at higher temperatures compared to other flat-plate collectors can be deduced from the efficiency curves as shown in Fig. 6.46b.

Taking into account the complex optical performance of the collector, the benefit over a year can be seen in the annual solar gains. For the location of Würzburg,

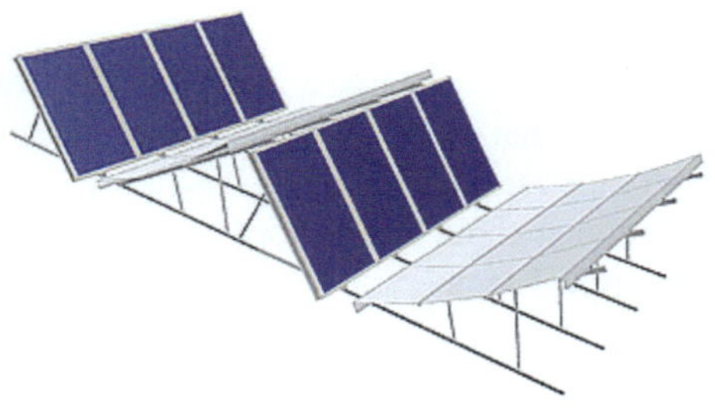

Fig. 6.44 CAD impression of RefleC collectors 4 on a flat roof. Three flat segments of 900 mm approximate the shape of a one-sided CPC with $\theta_c = 35°$ with a trough slope of $\beta = 55°$

Fig. 6.45 Prototype of the RefleC collector on the outdoor test rig of Fraunhofer ISE; the collector is tilted $\beta = 55°$

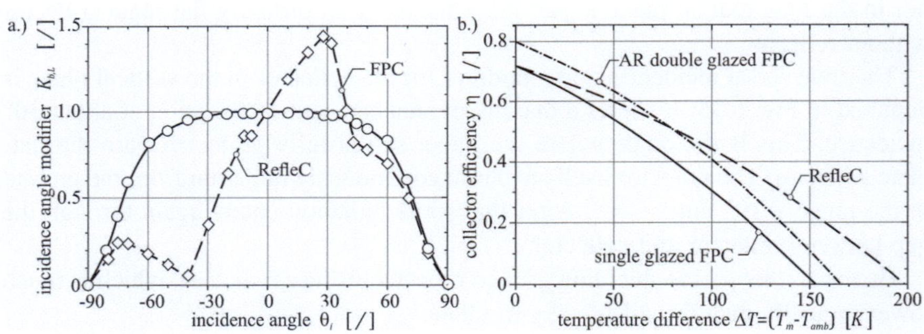

Fig. 6.46 **a** Transversal incidence angle modifier of the RefleC collector (beam irradiance in the vertical plane); the negative values between −35° and −90° correspond the view angle of the ground for a collector til of 55° (after Hess et al., 2010); **b** Efficiency curves of an improved FPC with booster reflector (Reflec), a single glazed standard FPC and an improved AR double-glazed FPC without booster as function of temperature difference for G = 1000 W/m^2

Germany, a comparison between the RefleC collector using a double-glazed improved flat-plate collector (cover glass is using antireflexive (AR) coatings on both side; second glazing is a highly transmitting polymeric film) and the same collector without a reflector shows an impressive larger yield especially for higher operational temperatures (Table 6.5).

Table 6.5 Annual gain of RefleC and AR-double glazed flat-plate at constant inlet temperatures T_{in}. The RefleC gains are related to 1 m^2 flat-plate aperture to illustrate the additional gain by the booster reflectors (location Würzburg/Germany). Extrapolated from Hess (2015), Hess, et al. (2010)

System	Annual yield of collector w/o booster reflectors $[kWh/(m^2a)]$	Annual yield of collector with booster reflectors $[kWh/(m^2a)]$
Optimal tilt β	40°	55°
Inlet $T_{in} = 40°C$	645	771
Inlet $T_{in} = 80°C$	350	427
Inlet $T_{in} = 120°C$	145	271

6.5 Advanced Collector Topics

Although solar thermal collector technologies for hot-water and heating applications are mature and highly efficient, innovative approaches exist, which improve efficiency, reduce costs and materials use, integrate new functions in a collector or improve the aesthetical appearance of the technology. In the next sections some of the more recent developments are shortly characterized.

6.5.1 Bionic Collectors

Solar flat-plate collectors with harp absorber are a state-of-the-art technology with very good performance. However, the hydraulic resistance of the different riser channels is slightly different, leading to different flow and temperature distribution over the complete absorber surface. The aim of a new development of improved absorber sheets was to develop solar collectors with improved absorbers based on bionic principles (Hermann, 2005a, 2005b). The bionic FracTherm® channel structures as shown in Fig. 6.47 are multiply branched in a fractal way to obtain a uniform flow distribution, a low pressure drops as well as a high thermal efficiency. The pattern resembles the veins of leaves or in the human body. The aluminium absorbers are produced by a roll-bond process. In the roll-bond process, the channel structure is printed with a separating ink on an aluminium sheet. A second sheet is then rolled over the first sheet, and they are cladded together except for the areas printed with the separating ink. In a further step, pressurized air is applied to inflate the flow channels through which the solar fluid is to flow. When the channels produced in this way are visible on both sides of the absorber, they are called Double-Side-Inflated panels (DSI), as opposed to One-Side-Flat panels (OSF).

Not only rectangular shaped absorbers can be design but also other shapes, for example triangles. This allows in principle a good building integration. The collectors have been experimentally and numerically investigated for a wide temperature range to cover various applications, as illustrated in Fig. 6.48. A specific FracTherm® algorithm for designing the flow channels has been developed. A

Fig. 6.47 Partial view of selectively coated FracTherm® roll-bond absorber

challenge of the development has been the coating of the absorber after the production of the channel plate. To prevent corrosion in Aluminium, it is necessary to work with appropriate heat transfer fluids. The collector has been tested and developed in a European project Bionicol (Hermann, 2011).

Through a partnership in 2012 between CGA Technologies and the Finnish company Savosolar Oy, spectrally selective roll-bond solar absorbers can meanwhile be offered commercially. In the usual production process, absorber tubes are connected to pre-coated metal sheets. In the roll bonding process, however, the coating is applied to the finished absorber.

The FracTherm® principle is not limited to the design of solar absorber but can be applied also to other applications where homogeneous flows with minimum hydraulic resistance are useful, for example in cooling structures of semiconductor elements or for heating elements in buildings.

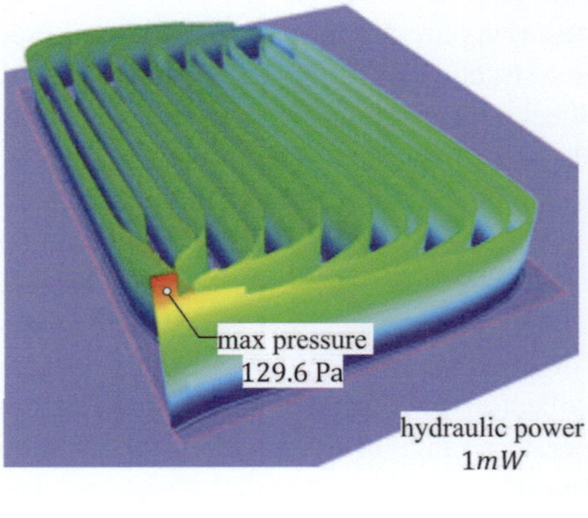

Fig. 6.48 Numerically simulated pressure distribution in a FracTherm® absorber plate (Hermann, 2005a, 2005b)

6.5.2 Collectors Using High-Performance Concrete

Usually absorbers are fabricated from metals with high thermal conductivity, because the thickness of the absorber sheet and the conductivity determine the F'-factor to a large extent (see Sect. 6.2.3). An attempt to design similarly efficient absorbers, but produced from a cheaper building material, is realized in the TABSOLAR® concept (https://www.tabsolar.de/). A special ultra-high performance concrete (UHPC) can be used to manufacture complex structures as needed. The material combines mechanical robustness with good thermal properties. The target of producing large area collectors for integration in a façade used the approach of developing prefabricated flow-through façade elements made of for different façade systems (ventilated curtain wall (VHF), external thermal insulation composite system (ETICS) and sandwich wall). A solar fluid flowing in channels through the absorber element made of UHPC transports the extracted heat to the use (Fig. 6.49). Inside the building, corresponding elements can also be used as so-called thermoactive building systems (TABS), possibly in addition to a classic concrete core activation system. Hence the name TABSOLAR®. The channels are formed directly from UHPC using the so-called membrane vacuum thermoforming process. The forming of the shape is shown in Fig. 6.50.

In the development of the concept several technological challenges had to be met. The first and basic challenge was the production of a stable UHPC absorber, which would withstand internal fluid pressure and temperature changes with leakage. In order to reach higher temperatures, the development of a selective coating on the UHPC absorber is necessary. As the substrate concrete has a high emissivity as such, a low emitting coating for the microscopically rough surface of concrete had to be developed. A prototype is depicted in Fig. 6.51.

The third challenge using the UHPC material with rather low thermal conductivity for an absorber was to reach a high F'-factor of collector (Fig. 6.52).

Three different product families have been conceived. The TABSOLAR® Premium design may achieve yields and temperature levels comparable to commercially available solar flat-plate collectors with a spectrally selective coating

Fig. 6.49 Bionic UHPC absorber cut into two pieces (modified from Hermann et al., 2020)

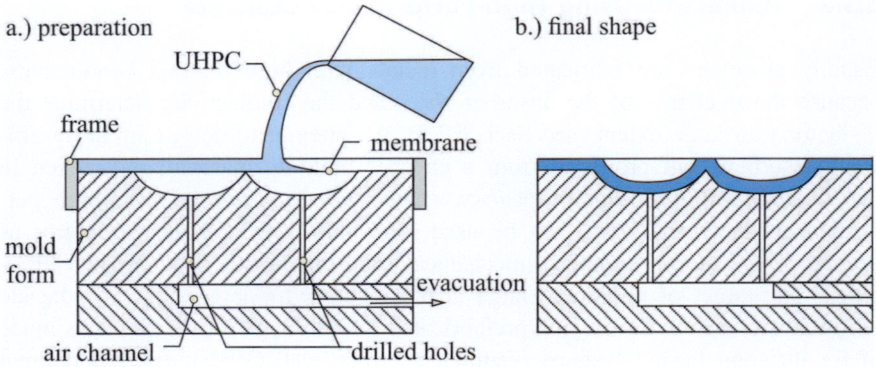

Fig. 6.50 First steps in membrane vacuum deep-drawing process **a** preparation of UHPC paste, **b** forming in mould by evacuation

Fig. 6.51 Spectrally selective coating on UHPC absorber (https://tabsolar.de/spektralselektive-beschichtung)

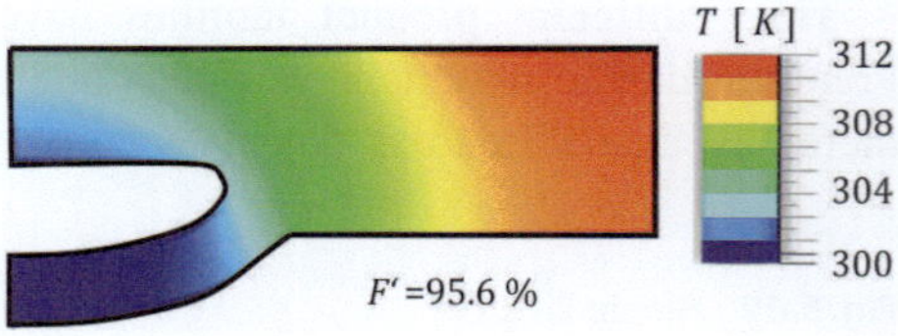

Fig. 6.52 Numerical simulation of temperature distribution and heat flux in UHPC-absorber resulting in $F' = 95.6\%$ from Hermann et al. (2015)

and anti-reflective glazing. TABSOLAR® Economy stands for painted or integrally coloured UHPC collectors with low-E glazing and slightly lower yields and temperature levels. Finally, in the TABSOLAR® Design product family, the collectors are unglazed and can be provided with a wide variety of structures and colours—see also Fig. 6.53. The focus here is on the design aspect; yields and temperature levels are even lower here (Hermann et al., 2020).

The first two product families are designed for direct solar applications for domestic hot water preparation or for use in solar combi systems. The third product family is characterized by a wide range of architectural design options. Since it

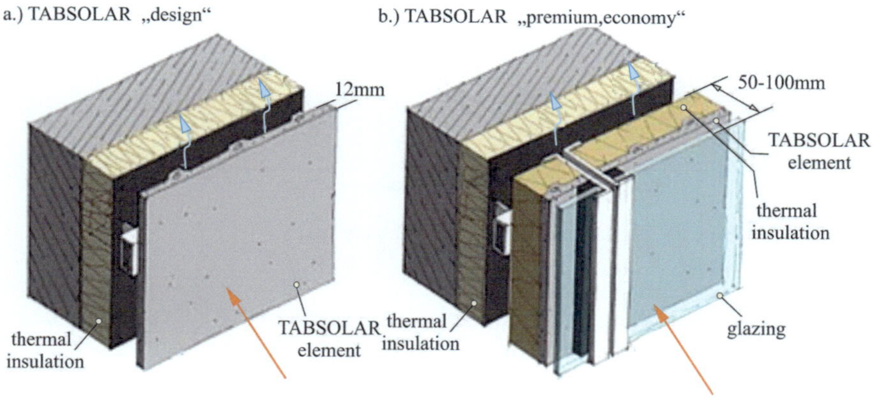

Fig. 6.53 TABSOLAR® rear-ventilated elements; **a** product family "Design" as uncovered absorber and **b** product family "Premium/economy" as glazed collector with low-ε glazing or selective absorber (modified from Hermann et al., 2020)

achieves significantly lower temperature levels and efficiencies, it is suitable, for example, as a low-temperature heat source for heat pumps, for domestic hot water preheating or for a swimming pool heating system.

6.5.3 Photovoltaic-Thermal Collectors (PVT)

The drop in photovoltaic (PV) module prices and the resulting area competition on the building roofs has stimulated interest in the development of photovoltaic/thermal (PVT) collectors which convert solar energy into both electricity and heat. At present modern PV modules typically convert about 20% of the incident solar radiation into electricity, depending on the employed cell technology. This implies that a major part of the solar spectrum remains unused and is transformed into heat. Hybrid PVT collectors on the other hand can make use of this excess heat by cooling the solar cells using a heat transfer fluid. The electrical and the thermal performance of PVT collectors is lower than that of separate PV panels and conventional thermal collectors. However, they emphasized that two PV/T collectors together produce more energy per unit surface area than one PV panel and one thermal collector next to each other. This is particularly important in applications where surface area availability is very significant. In this way, solar energy can be harvested on cascading exergy levels: first high-grade exergy is produced in the form of electricity, then low-grade exergy in the form of heat. In this way, PVT collectors make optimum use of the solar resource with high utilization of available area. PVT collectors can be categorized according to their heat transfer medium (air and liquid) and their design (unglazed, glazed, and concentrating). Each collector design has its specific temperature levels and accordingly suitable applications.

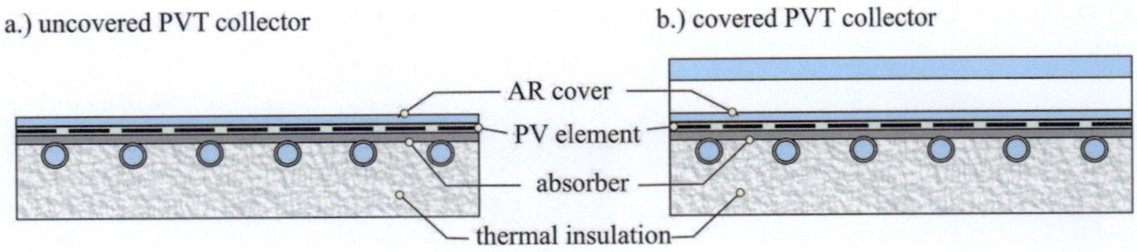

Fig. 6.54 Flat-plate PVT collectors **a** uncovered and **b** covered

At present the European market is dominated by unglazed PVT collectors delivering heat at low temperatures, however, at operating temperatures below ambient they can achieve some heat gains to be used by heat pumps. Glazed PVT collectors reach higher temperature levels. However, these collector types have often relatively high thermal losses, as they do not use selective coatings. In the next sections we will discuss the main types in practice, uncovered and covered flat-plate collectors and discuss their design and performance (Fig. 6.54).

Extensive reviews have been written on the large variety of PVT collectors (which we will not cover here), for example by Bandaru et al. (2021) who included a survey of different reviews, or Chow (2010) by looking into design variations and the importance of parameters like the impact of temperature on PV-cells, of the bonding quality of PV cells to liquid heat exchanger, of efficient cooling of the cells and thermal design options using different cover glasses with and without air gap (Zondag, 2008). Single sheet and tube absorbers were used very often in designs, but channelled plates are also interesting because of the more homogeneous temperature distribution. Uncovered collectors are widespread, as it is not easy to reduce the heat losses from a single cover collector. A selective absorber is neither possible on top nor below the PV layer. Early products have heat losses like non-selective black flat-plate collectors.

To achieve a sustainable PVT market development, performance and reliability of products is a key factor, and it can only be characterized by establishing functioning testing standards. The testing of thermal and electrical performance is complex in the case of PVT, and different situations must be considered.

Especially for uncovered PVT collectors the thermal performance has not only be determined with collector temperatures above ambient, as covered in (ISO 9806, 2017). The quasi-dynamic test method seems to be able to cover the wide range of PVT products. But as the uncovered PVT collectors are connected to a heat pump and used as an ambient energy source, also operating temperatures below ambient must be considered. For extracting heat from the ambient, the topics of condensation, rainfall and infrared radiation must be considered in the testing equation. An overview over existing approaches is given in Kramer et al. (2020).

The electrical condition of the PV cells is even important for thermal testing. The standard testing in hybrid operation where the PV module is operated in the maximum power point (MPP), yields lower thermal gains than a testing in open circuit mode (OC, disconnected with no current flowing). The optical efficiency

η_0 is higher in OC-mode because the photons excited in the PV-cells eventually thermalize and produce thermal power, whereas in the MPP-mode the electrical power is extracted. The electrical power extracted is. On the other hand, dependent on the cell temperatures. In most designs the absorber with the heat transfer fluid is coupled to the cells and should cool the PV as compared to the standard PV module operation. Especially for covered PVT-collectors, this contradicts the objective of reaching higher temperatures in the solar fluid. As a consequence, the electrical and thermal output ideally should be parameterized in a testing method simultaneously. Here a promising method has been developed (Kramer & Helmers, 2013) but not yet evaluated for all types of PVT collectors.

The reliability and durability of PVT modules are especially challenged at elevated temperatures and higher humidity loads. The test methods available from the IEC and ISO standards are covering the specifics of PV and ST module's, which often are similar for PVT modules. But the PVT specific loads and the operating conditions arising from the system integration of PVT are only partially covered in the set of existing standards (e.g. stagnation temperature approval, mechanical load resistance). Therefore, further work is needed to close this gap.

6.5.3.1 Uncovered PVT Collectors

Unglazed PVT collectors are beneficial for operation near ambient temperatures particularly because of high heat transfer to ambient air even in times of low radiation. Among PVT collector types, unglazed PVT collectors are widely popular in the northern European market mainly as an additional heat source for heat pumps. Such a collector is based on a PV module on the back of which an absorber plate (a heat extraction device) is attached. The purpose of the absorber plate is twofold. Firstly, to cool the PV module and thus improve its electrical performance and secondly to collect the thermal energy produced, which would have otherwise been lost as heat to the environment.

The thermal performance of a coverless PVT collector is reduced especially at high temperatures due to heat losses from the top. However, the uncovered PVT collectors have a better electrical performance than covered ones as already shown in a design comparison by Zondag et al., (2003).

The most common connection to couple the photovoltaic panel to the thermal absorber is the gluing technique. Here the entire PV laminate is bonded to the flat plate, by means of thermo-conductive glues or adhesives, which shall be resistant to high temperatures. However, this technique presents several problems. With the gluing technique, air bubbles may be trapped between the PV laminate and the plate decreasing the heat transfer coefficient and creating inhomogeneity in the cell's temperature distribution. Additionally, a risk of condensation between the two parts occurs (Aste et al., 2014).

Concerning the manufacturing of uncovered, liquid-based photovoltaic-thermal (PVT) collectors, design requirements or low expenses have certainly high priority, not only the overall optimization of the thermal efficiency. The specific impact of inevitable thermal resistances caused by wavy surfaces and imperfect contacts between PV modules and heat exchangers is still a topic of investigation and optimization. In their paper (Brötje et al., 2018) focus on the performances of uncovered PVT collectors with detachable mounting of PV modules on heat exchangers and the effect of macroscopic air gaps. Two prototypes with thin film modules were built without using silicon glue, other adhesives or heat conductive paste. They measured the thermal zero-loss efficiencies for two adhesive-free collectors at $\eta_{0,M} = 66.8\%$ (copper pipe meander laser-welded on aluminium sheet) and $\eta_{0,M} = 69.2\%$ (aluminium roll-bond) based on gross areas. These results are similar or even better than the efficiencies of commercially available collectors. Due to the thermal resistance of air gaps, they found temperature differences of 12 K and higher between absorbing sheets and heat carrier, which have to be considered while estimating the electrical yield by means of system simulations. As comparison they also built a prototype with a very thin adhesive layer between PV module and a roll-bond heat exchanger. This optimized heat transport led to a higher optical efficiency of $\eta_{0,M} = 76.8\%$. Simulations using numerical, two-dimensional models show good agreement with experimental results and proved to importance of the thermal coupling (Brötje et al., 2018).

A second important parameter seems to be the backside heat exchanger efficiency, especially for the use as thermal source for a heat pump system. The solar gain itself is only a fraction of the ambient energy to be provided. The heat exchange with ambient air is key. With an PVT collector optimized for heat exchange with ambient air, front and back side should be used to increase heat exchanger area Chugani (2020) investigated the performance of a PVT collector which is thermally coupled with a fin-tube heat exchanger on the backside of PV. The fin surface area is 10 times larger than the PV area in each module, which specifically increases the use of ambient energy from ambient air and hence the PVT collector works as a good environmental heat exchanger. In addition to convective heat gains in times of low irradiation particularly in winters or nights, condensation heat gains, as well as heat gains through the phase change to frost occur on larger fins surfaces, which improves the thermal collector output and makes it suitable for working as a sole heat source of a heat pump. The authors show that especially when the PVT collector is the only heat source for a heat pump, the improved heat exchange allows to use smaller collector areas compared to a standard unglazed PVT collector as a reference for reaching the same seasonal performance factor.

Using the PVT collector as an ambient heat exchanger naturally leads to a large value of the thermal loss coefficients in the collector testing (see (ISO 9806, 2017) testing of WISC collectors, WISC = wind and infrared sensitive collectors). As a solar collector therefore, the efficiency is only good for very low temperatures. The solar gain seems to be less important than the gains from ambient air.

6.5.3.2 Covered PVT Collectors

A different development route than the one just presented is followed in the design of covered PVT collectors. The cover glass certainly leads to a lower optical solar gain, due to the additional reflectance. On the other hand, its purpose is to decrease the thermal loss coefficient to the environment. A collector like that may produce heat at higher temperatures with good efficiency as long as the sun is shining, however, the heat gain through ambient air at low temperatures is also drastically reduced.

Hybrid PVT collectors generate solar electricity and heat simultaneously in one component. PV modules typically have efficiencies between 15 and 20%. So, it makes sense to dissipate the unused part of the solar radiation as heat to a fluid and use it for domestic hot water or space heating. This makes better use of limited solar surfaces and generates more electricity and heat. However, covered PVT collectors often have a low thermal efficiency for operating temperatures 40–80 °C. A selective absorber cannot be applied on top of the PV-cells in the airgap of a covered collector such as Fig. 6.54b as all solar radiation would be absorbed. On the other hand, below in direct contact with the PV cells and the protecting glass and possibly lamination film the low emissivity surface of the absorber is blocked. One idea to design a good selective PVT collector is to have a partially transparent PV front cover, as also used in building integrated PV (BIPV), combined with a selective absorber. Here, the rear contacts of the PV cells must be provided by a grid. However, this option seems not be investigated often, probably because the transparency of the solar cells is limited (Zondag, 2008). Another option is the use of transparent selective coatings, as in principle used for windows, for the PVT collector as cover glass or as IR-reflective coating on the PV cover glass, e.g. on either side of the air gap. Both approaches have their challenges. Several design aspects of such a covered PVT-collector using a low-e coating have been tackled successfully.

The combination of photovoltaics and solar thermal energy in one component represents an exciting technological challenge. Good thermal coupling of the PV cell to the fluid is of central importance to ensure the lowest possible cell temperatures. Low-E coatings must be optimized to simultaneously have a high transmittance for high PV efficiency and a low emissivity for high thermal efficiency. The high temperatures in stagnation can be addressed either by overheating protection or adapted materials (glues, laminates). Finally, the PVT collector, solar thermal application, and building technology must be coordinated to exploit the full potential of this technology. A suitable low-e coating has been demonstrated by Lämmle et al., (2016). The highly transparent low-e coating is based on silver, specifically optimized for the application in PVT collectors. The coating combines a thin silver layer with a multi-layer system of low-refractive SiO_2 and high-refractive TiO_2. It reaches a low emissivity of ε (100 °C) = 0.13 while maintaining a high solar transmittance of $\tau_{AM1.5} = 0.79$ (see Fig. 6.55).

The densely packed monocrystalline PV cells are directly laminated on an aluminium sheet absorber. Thus, providing a good contact between cells and fluid. Two layers of EVA serve as adhesive, encapsulant and vapour barrier for the PV cells. A thin electrical insulation layer is spray-coated on the aluminium sheet before lamination. The cross-section of the collector is shown in Fig. 6.56.

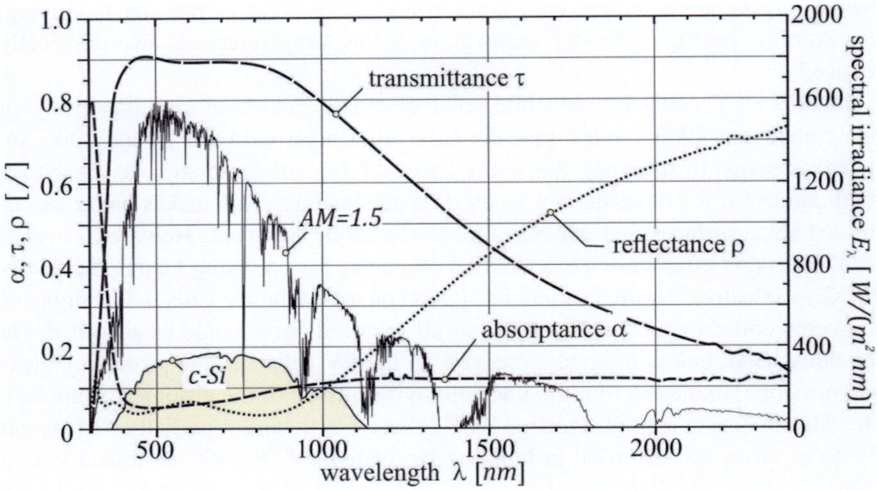

Fig. 6.55 Measured transmittance, reflectance and absorptance spectra for a PVT low-e glass cover, modified from Lämmle et al., (2016)

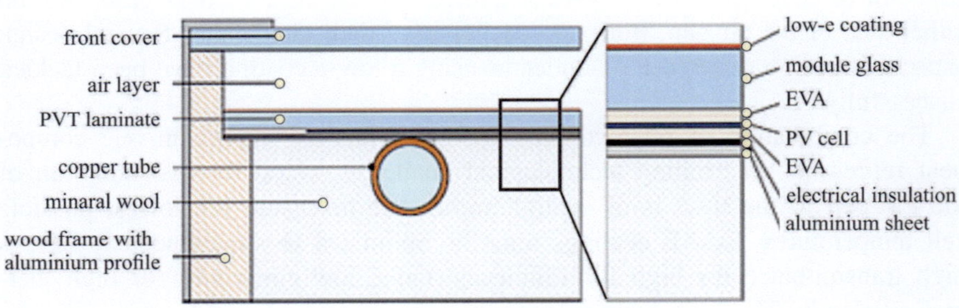

Fig. 6.56 Cross section of PVT collector with low-e glazing and directly laminated absorber from Lämmle et al., (2016)

In a comparison with a PVT collector without low-e coating the thermal performance was very much improved, as indicated in Fig. 6.57. In an optimized construction of the PV-module-absorber compound a high resistance against mechanical loads could be achieved. The improved thermal performance, however, leads to possible stagnation temperatures of about 160 °C, which either requires better temperature-resistant materials for glueing or an overheating prevention. The economic potential of such a development is large. When roof area is available without limitation, the individual thermal and electrical efficiencies of separate solar thermal collectors and PV-modules are higher than the ones of a PVT system. If, however, the area is limited, the best combined output can be produced with a PVT system. Several annual performance simulations have been performed comparing both alternatives in different applications.

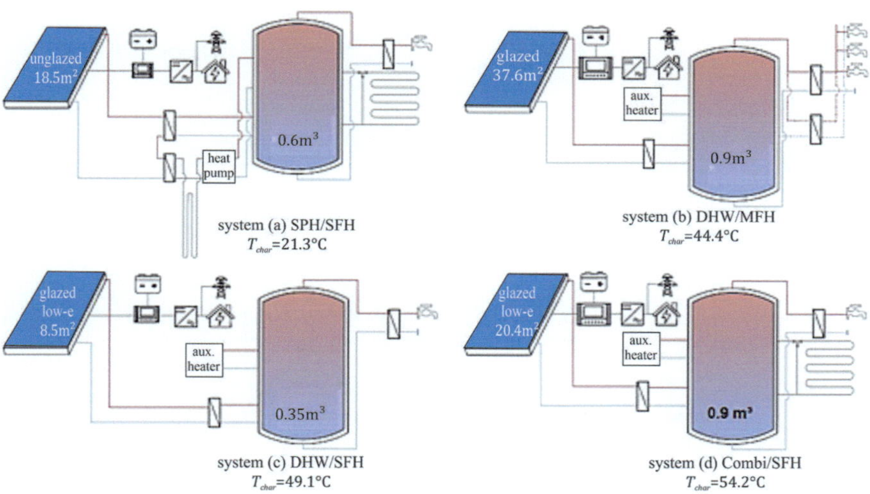

Fig. 6.57 Electrical and hydraulic layout of PVT systems **a–d** including PVT technology and required collector area (Lämmle, 2018)

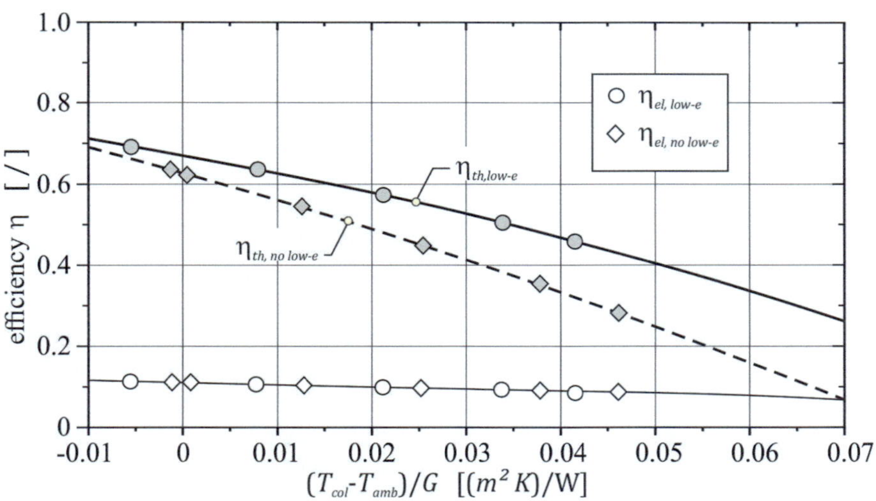

Fig. 6.58 Thermal and electrical efficiency curves (lines) with test points (markers) measured in MPP mode of PVT collectors with and without low-e from Lämmle et al., (2016)

(a) Solar-assisted heat pump system with ground heat exchanger for DHW and room heating,
(b) DHW in a multi-family house MFH,
(c) DHW in a single family house SFH,
(d) Combi system in a SFH.

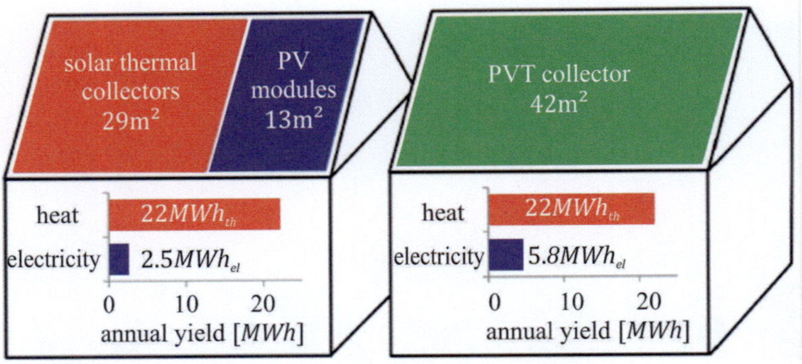

Fig. 6.59 Comparison of building with roof covered by separate solar thermal and photovoltaic panels vs. roof covered with covered PVT collectors producing the same solar thermal yield per year (Lämmle, 2018)

Table 6.6 Electricity surplus generation compared to a side-by-side PV + ST system with the same thermal yield and using the same roof area PV (Lämmle et al., 2017)

System	Covered PVT collector (%)	Covered PVT collector/low-e (%)
System a—SHP/SFH	187	318
System b—DHW/MFH	42	134
System c—DHW/SFH	16	90
System d—Combi/SFH	15	85

Figure 6.57 compares the thermal and electric efficiency of PVT collectors with and without low-e coating. Model calculations for a demonstration building show that on a given available roof area, 15–318% more electricity is generated with the same thermal yield compared to a separate installation of solar thermal and PV (Lämmle et al., 2017). The lower the solar thermal system temperature (lowest in case (a) and increasing towards (d)) the higher the electricity surplus. Using low-e coatings, covered PVT collectors can be suitable for conventional solar thermal applications such as domestic hot water and space heating. Table 6.6 shows that the collector using the low-e coating has a much higher improvement due to its better thermal performance.

6.5.4 Building Integration of Solar Thermal Collectors

In the building sector, the combination of energy efficiency measures and the use of renewable energy sources reduces the consumption of fuels and electricity and thereby reduces emissions of CO_2 and other pollutants (Aelenei et al., 2016).

The optimal use of the solar components in buildings has an influence on architectural design and building concepts. The complex architectural integration of solar components into the building envelope is of critical importance to the appearance, form, aesthetics, investment, and operating cost as well as the whole environmental impact within the life cycle of the buildings. From an architectural perspective, additional specifications for components may be required such as mounting position, appearance, dimensions, geometrical shape, physical functions, light permeability, colour, and construction details. Building integrated solar thermal systems (BISTS) can be used as well in the construction of new buildings as in the refurbishment of buildings. With true integration, not just add-on design, additional functions are fulfilled, foremost to be mentioned is the thermal performance which is strongly influenced by materials, constructional integration, and detailing. Thus, BISTS may also improve the quality of life for inhabitants (e.g. thermal comfort, daylight use). The increasing implementation of BISTS may also stimulate a transition of the building product industry towards a more open attitude with respect to innovative building designs, new functionalities, and competitiveness of conventional building claddings. The influence of BISTS on the appearance of single buildings and building groups can be significant. However, the correct application of BISTS is not without its challenges for architects and building industry.

The main function of BISTS is to produce thermal energy. In the case of hybrid systems BIPV/T electricity will also be produced. A whole range of additional functions related to building physics and constructional requirements can be addressed by BISTS (Platzer, Smyth, & Kennedy, 2014):

- Thermal insulation
- Acoustic insulation
- Humidity regulation
- Rain and wind tightness
- Solar protection
- Daylighting
- Structural functions
- Fire resistance
- Security protection

There are different levels of building integration depending on the number of functions being delivered by BISTS. While partially integrated solar thermal systems have a lower level of functionality, fully integrated systems are characterized by higher complexity. The external layers of the building envelope and of the solar thermal collectors has an influence on the aesthetic potential and design options (Krstic-Furundzic, 2007). Solar thermal components can be used to replace normal building components possibly showing a multifunctional potential in the external skin.

Clearly the multi-functionality of a collector makes it applicable to integration and can provide the advantage for the designer to use less building elements. A prominent example is the application of building integrated solar thermal façade collectors which may remove the need for conventional cladding materials. That again is reflected in investment costs.

In terms of functions, light permeability requires a new type of semi-transparent collectors (Giovannetti, 2014; Ritter, 2014). Various light transmission grades and interesting lighting effects can be produced inside a building. For example, the variation of profiles, partial use of absorber and transparent areas in the aperture, redirection of light by slats, different arrangements and distances between vacuum tubes can result in different effects achieved by the shadows and light (Fig. 6.60).

Achieving functional requirements must be accompanied by fulfilling aesthetic requirements.

This requires that the functional and aesthetic aspects are considered simultaneously, considering the various building aesthetics, building physics and solar thermal collector mounting criteria categories. The aesthetical possibilities like dimensions, form, jointing, colors, texture etc. of a solar thermal collector can all be influenced by the materials and constructional elements of the collector that are used. Even the type of fluid used (liquid for solar water heating systems or air for solar air heating systems) may affect the eventual appearance (Table 6.7), see Platzer, Smyth, and Kennedy (2014). Passive solar collectors like the Trombe wall or Translucent/transparent insulation use the thermal mass of the wall directly

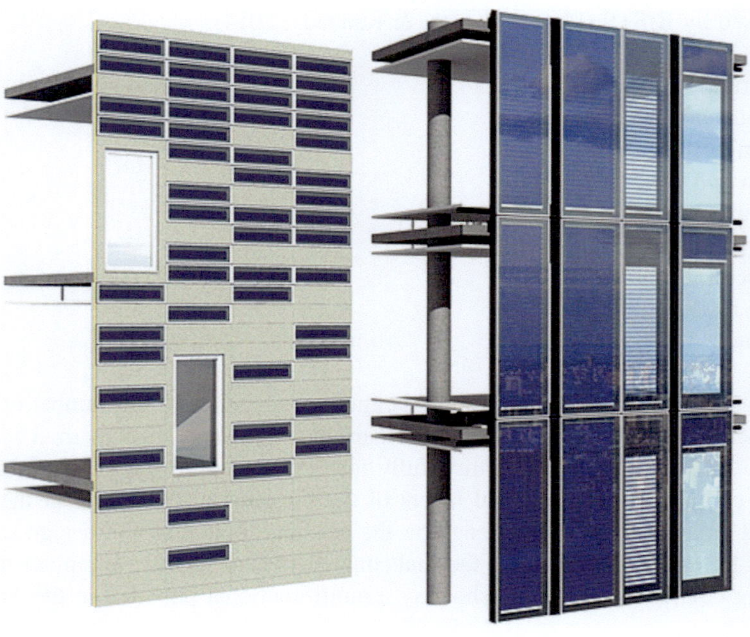

Fig. 6.60 Visualization showing building integration of a horizontal strip collector in an opaque façade (left) and of a solar thermal collectors (opaque glazed absorber semi-transparent solar thermal blinds) in a glazed transparent façade (right) (Maurer, 2018)

Table 6.7 Connection between functionality and aesthetic criteria

Aspects	Individual criteria
Building aesthetics	Dimensions
	Color
	Materials (texture, surface structure, heat capacity)
	Optical reflectivity and glossiness
	Sealing, joints, and frames
	Design harmony
	Structural integration
	Use context
Building physics	Mechanical characteristics
	Moisture and water permeability
	Thermal insulation performance
	Acoustic characteristics
Practical operation	Mounting and installation ease
	Construction stability
	Maintenance

as absorbing element and thus some aspects are already dictated by the building structure (Chan, 2010; Platzer, 2000).

6.6 Collector Performance Testing and Certification

In performance testing we must distinguish the different approaches which all have their benefits and shortcomings. First of all, there is indoor and outdoor testing.

Indoor and Outdoor Testing

In both cases we need a good control of the environmental variables which affect the thermal performance of a collector:

- Solar irradiance level and direction.
- Ambient air and radiative temperatures.
- Wind speed and direction.

Whereas in outdoor testing we need to find representative situations which are sufficiently stable and characteristic for the range of operational conditions in indoor testing the ambient conditions can be controlled by using artificial solar simulators (lamps with a spectral distribution close to a standard solar irradiation spectrum) and by controlling temperatures by thermostats, heating and cooling of surfaces and rooms. In this way very constant and reproducible conditions can be created, which lead to very exact and reproducible measurements. On the other hand, it is not easy to be close to realistic conditions, especially for the solar simulation. Not only the spectral distribution of the best class of simulators is not exactly a solar spectrum, as illustrated in Fig. 6.61a, also it is extremely difficult to design simulators with sufficient homogeneity for large areas in the collector aperture plane and with similar beam divergence. The latter problem can only be solved for measuring relatively small concentrator devices indoors by using high precision paraboloids with a good point source. Flat-plate collectors and solar photovoltaic panels due to their flat angular response between 0° and 50° incidence on the other hand have little sensitivity to the divergence of simulator light as long as the main part comes from a source in front of the sample. Therefore, for such collectors, large simulator lamp arrays have been designed which provide a reliable and quite homogeneous irradiance in the collector plane, which are exemplarily shown in Fig. 6.61b.

Stationary and Quasi-Dynamic Testing

Solar collectors have a complex dynamic response to changes in environmental conditions. Material thermal capacities as well as the properties of the heat transfer fluid circulating through the interior of the collector influences the reaction to changes. Therefore, it is much easier to use a stationary approach using a collector energy balance equation as given by Eq. (6.1). The model has been used very successfully over many years, and a physical modelling of the dynamic thermal response is not required, when the conditions for testing are kept nearly constant within a certain small range. This is the approach of stationary testing as specified for example in ISO 9806 (2017), DIN EN 12975-1 (2011). Indoor testing lends itself naturally to this approach. For outdoor testing the conditions on constant conditions have to be relaxed to some extent, and small changes in the collector response are taken into account by adding a single-knot effective capacity to the time-dependent temperature change. This quasi-dynamic approach is not able to cope with all variable situations but in many labs outdoor periods can be found where solar irradiation (around noon) and temperatures are quite stable. The

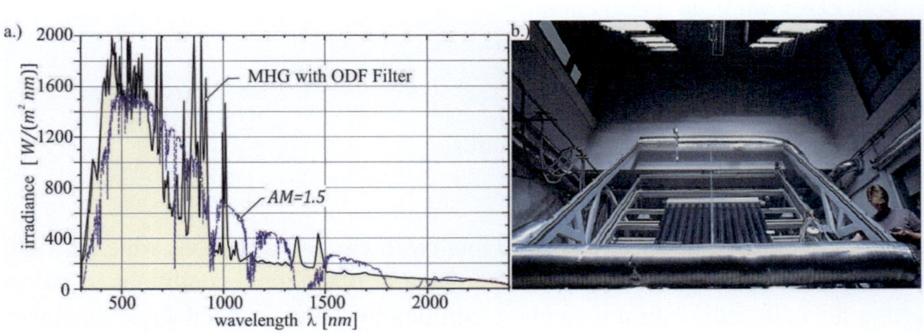

Fig. 6.61 **a** Comparison of a class I solar simulator spectrum (using HMI lamps) with the standard solar spectrum Global $AM1.5$; **b** Solar simulator in the thermal testing lab of Fraunhofer ISE

change in incidence angle can be compensated by tracking the sun with the collector at a constant incidence angle; a test stand for this type of testing is illustrated in Fig. 6.62.

Collector and Solar Field Testing

Although similar in theory the testing of large solar fields and individual collectors has different challenges. Small collectors can be tested indoors and outdoors, and the test equipment itself can be a standardized closed loop circuit where the returning fluid of the collector is cooled down by a chiller to the required and desired inlet temperature of the collector. High quality sensors can be used and included in the testing circuit, as the collector unit usually is installed *into* the existing circuit. A testing of a collector is in most cases used for a product certification that is to guarantee the product specifications of a certain type of collector which can be bought as commodity. The production of the manufacturer then must be certified to be able to guarantee the uniformity of the production.

However, for large solar fields the situation is different: First, these fields shall be tested as installed, they are unique entities and no mass product. Therefore, they should be tested at the site and in the condition as they were installed by a technology provider. A deinstallation and reinstallation at a test centre could give different results. Therefore, a temporary installation of the testing infrastructure including heat exchangers, pumps, fluid conditioning is required. Therefore, movable testing labs may be designed for smaller installations where all equipment of the circuit, sensors and data acquisition are including in a transportable container.

On the other hand, the thermal power produced by large solar fields is huge and dependent on the installation. Building a test circuit specifically laid out and designed for exactly the case considered is economically not feasible. A scientifically improved twin of the balance of plant already installed in the application (pumps, heat exchangers, expansion vessel, control, heat rejection and sensors) would drive up the project cost enormously. Therefore, field testing is for practicality relying on existing infrastructure, on the existing balance of plant. It may be possible to control mass flows and temperature levels to some extent, in order to cover a certain range of temperatures in the testing. But it is usually not possible to reach the qualification for a quasi-dynamic testing as put forward in the

Fig. 6.62 Outdoor tracker for testing thermal collectors at Fraunhofer ISE

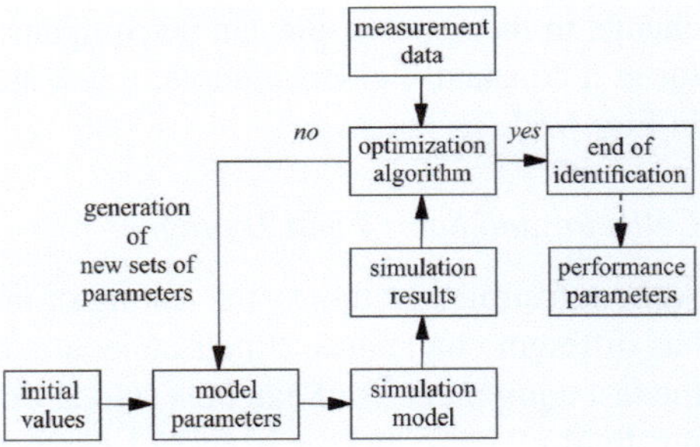

Fig. 6.63 Schematical procedure of parameter identification procedure for large scaled solar fields

standards. For these situations scientists have developed dynamic testing methods which allow to evaluate even completely dynamic environmental conditions – solar irradiation with moving clouds, temperature drops and other realistic changes. The basic idea of dynamic testing is that the collector field can be simulated with a dynamic simulation program. The underlying equations in the collector model must be parametrized (similar to the quasi-dynamic model), but the output at each time-step is dependent on the previous time step (contrary to the quasi-dynamic evaluation). It is still an area of current research how to reach reproducible and reliable detailed results with such a method, especially when industrial sensor qualities must be used which are not equivalent to scientific sensors. The principle of the dynamic testing with parameter identification is schematically illustrated in Fig. 6.63. One drawback is, that the simulation program using the identified parameters of the solar collector field may give realistic results and may predict the performance correctly, however, one may not transfer the parameters into a similar but different simulation program which might use different algorithms.

6.6.1 Standardization (ISO, ASHRAE)

As we have distinguished now the different types of performance testing, let us look into the standardization of testing. Here in international or national standards the exact conditions to be used during the testing as well as the prescribed instrumentation, the precision of sensors and the evaluation of the tests is described in detail.

As we discussed different fundamental approaches for performance testing, we did not yet focus on the kind of collector equation used for the evaluation of stationary or quasi-dynamic testing. However, although the theoretical basis is identical different standards use either the mean fluid temperature as variable or the collector inlet temperature. The European and international testing standards (ISO 9806, 2017) use the first approach while the American ASHRAE test (ANSI/ASHRAE Standard 93–2003, 2003) uses the second one. Both approaches require the mass flows to be in a representative range of collector operation and for both the focus is small collectors with a few square meters collector area which experience only a small temperature rise within the collector under normal operation.

The collector heat loss and the heat loss factor U_L is considered representative as an average for the whole collector.

The advantage of the European way to test collectors is that the parameters are independent (within the limits said before) of flow rates using in testing, whereas the advantage of the ASHRAE approach is that with these parameters it is easy to calculate the performance of collectors put into a series. In the following discussion we focus now on the international standard ISO 9806 which describes both stationary and quasi-dynamic testing. In addition, we only present the case of liquid heating collectors, although in the standard also other collector types, e.g. air heating collectors, are being treated in a similar way.

Stationary testing is based on the stationary collector equation without a heat capacity. The heat loss is parametrized using possibly a few coefficients which describe the heat loss related to wind speed, to thermal radiation and

$$\dot{Q}_u = A_c \cdot \left[\eta_{0,hem} G_{hem} - a_1(T_m - T_{amb}) - a_2(T_m - T_{amb})^2 - a_3 u'(T_m - T_{amb}) + a_4\left(E_L - \sigma T_{amb}^4\right) - a_6 u' G_{hem} - a_7 u'\left(E_L - \sigma T_{amb}^4\right) - a_8(T_m - T_{amb})^4\right], \tag{6.40}$$

with $a_4 = \eta_{0,hem}(\varepsilon/\alpha)$, $a_6 = \eta_{0,hem} \cdot b_u$ and $a_7 = \eta_{0,hem} \cdot b_u \cdot (\varepsilon/\alpha)$. In this equation u' is a reduced wind speed, E_L is the thermal irradiance in the collector plane and b_u is a coefficient describing wind dependence. Depending on the collector, some coefficients can be safely set to zero. For details refer to the standard document.

In this case a hemispherical optical efficiency and irradiance is being used, because with solar simulators it is not practically possible to separate a direct beam radiation having a similarly small divergence as the sun's radiation from the diffuse irradiation.

For the *quasi-dynamic testing* outdoors, the testing conditions cannot be stationary due to the changing weather conditions. However only testing periods satisfying a partially stable measurement are specified in the standard. Due to the change of temperature with time and irradiation levels an additional term is being introduced in the testing equation for the extracted useful power, which is an effective heat capacity of the collector a_5:

$$\dot{Q}_u = A_c \cdot \left[\eta_{0,b} K_b G_b + \eta_{0,d} K_d G_d - a_1(T_m - T_{amb}) - a_2(T_m - T_{amb})^2 - a_3 u'(T_m - T_{amb}) - a_5 \frac{dT_m}{dt} + a_4\left(E_L - \sigma T_{amb}^4\right) - a_6 u'^{G_{hem}} - a_7 u'^{(E_L - \sigma T_{amb}^4)} - a_8(T_m - T_{amb})^4\right]. \tag{6.41}$$

The terms K_b and K_d described the incidence angle modifier functions for beam and diffuse irradiation. K_b can be a function depending one incidence angle (e.g. a flat plate collector) or on two incidence angles (e.g. for vacuum tubes or linearly concentrating collectors).

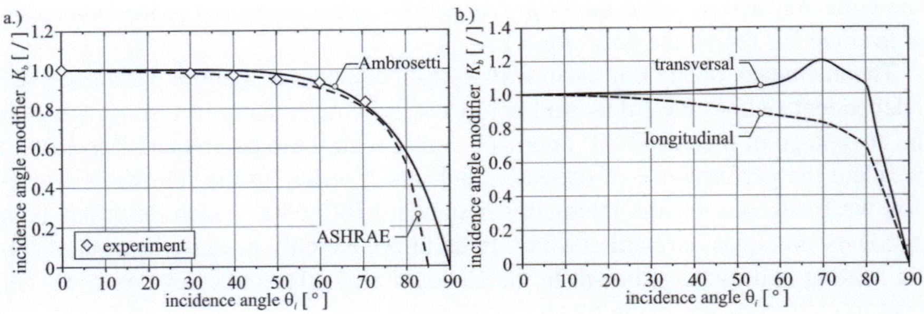

Fig. 6.64 **a** Incidence angle modifier K_b for a flat-plat collector (line: Ambrosetti function $r = 4.8$; dashed: ASHRAE function $b_0 = -0.1$); **b** Incidence angle modifiers $K_{b,T}$ and $K_{b,L}$ vacuum tube collector with interpolation (line: transversal data; dashed: longitudinal data)

Depending on the collector type the incidence angle modifier function could be one- or two-dimensional. For flat-plate collectors a radially symmetric one-dimensional incidence angle modifier is assumed which is usually described by a single fit parameter, see Fig. 6.64a. The function describing very well the optical performance of a collector covered with glazing(s) is the Ambrosetti function with the parameter r to be fitted to experimental results according to (ISO 9806, 2017):

$$K_b(\theta) = 1 - \left[\tan\frac{\theta}{2}\right]^r. \tag{6.42}$$

This equation improves the simple function specified by ASHRAE 93-77:

$$K_b(\theta) = 1 + b_0 \cdot \left(\frac{1}{\cos\theta} - 1\right), \tag{6.43}$$

where θ is the incidence angle, and r respectively b_0 are parameters for fitting.

For a collector without rotational symmetry like a vacuum tube collector, we have a longitudinal incidence angle modifier K_L for incidence angles θ_L in the plane along the absorber tube which defines the optical axis, and a transversal incidence angle modifier K_T for incidence angles θ_T in the plane perpendicular to the optical axis (ISO 9806, 2017), as e.g. illustrated in Fig. 6.64b. The incidence angles are given by

$$\theta_T = \tan^{-1}(\tan\theta \cdot \sin\gamma), \tag{6.44}$$

$$\theta_L = tan^{-1}(tan\theta \cdot \cos\gamma), \tag{6.45}$$

where θ is the incidence angle on the collector plane, and γ is the angle between the optical axis and the incidence plane of the solar radiation in the spherical coordinate system.

The bidirectional incidence angle modifier for such a collector and arbitrary incidence angles can be approximated by (ISO 9806, 2017):

$$K_b(\theta_L, \theta_T) = K_b(\theta_L, 0) \cdot K_b(0, \theta_T) = K_{b,L}(\theta_L) \cdot K_{b,T}(\theta_T). \tag{6.46}$$

When doing stationary testing in the lab or outdoors, the radiation should always come from the same direction, and a test can be performed at a certain incidence angle $\theta_{testing}$, which should be kept constant during the test. For outdoor testing this implies that collectors must be mounted on solar tracker racks, which move in a way that the incidence angle is constant during the test period. This implies small inaccuracies as collector inclination (with an impact on the convection heat transfer) and the view angle to the sky (with an impact on radiation heat transfer and on diffuse irradiance level on the collector) changes. In the lab it is often very difficult to produce directional and homogenous irradiation on a collector plane using a rack of lamps.

This is much easier in quasi-dynamical testing, because here the conditions of irradiation may change during the testing period. The collector is fixed and is irradiated by the moving sun. In this way the incidence angle modifier function can be parameterized, and fitting parameters can be extracted using the experimental time series data.

The standards mentioned up to now have mainly been developed to test individual collector products used for heating or hot-water production. However, it is also possible to test complete collector fields. Collectors like the concentrating collectors for solar thermal power already as individual modules have an aperture area which is exceeding several hundred square meters. In addition, very often the target is not the quality of an individual component but the quality of an installation. In both cases it would be costly to transport the collector to a test site, install and test it there, and then tear it down and move it to the intended installation site. Moreover, the new installation could lead to different performance. All these arguments lead to the requirement of in-situ testing at the intended installation site. Test equipment and experts must be available now at this site, which is challenging for the logistics and cost. From this background the requirement of shortening the test period as much as possible has led to the development of *dynamic test procedures* (DT), which are able to digest all different dynamically changing test conditions, where quasi-dynamic testing (QDT) would exclude these periods. In these procedures for the evaluation of the dynamic performance of a collector, a numerical simulation model is required that shall reproduce the dynamic behaviour of the collector. To this end, any dynamic simulation model can be used, mostly incorporating a discretization of the (simplified) Navier–Stokes equations. Along the fluid flow the collector is segmented therefore in several N nodes. A numerical difference between a QDT method and a DT method is that for QDT a parameterized collector equation is used for each individual time-step to minimize the difference between modelled and measured heat output of the collector. A multi-linear regression analysis is performed, which is a non-iterative, fast matrix method minimizing the error of the measured output power of the collector versus the approximated/modelled output power. Although there is a single heat capacity term with a parameter characterizing the thermal capacity of the collector and connected to the temperature derivative with time, essential it is not a dynamic method. The DT methods, on the other hand, can model a collector performance using a detailed model, in which thermal capacities of fluid and collector materials, as well as time-lags of in the fluid flow can be included. There is a real thermal simulation of the collector. The minimization may not only be done with collector output power, but also with collector temperatures. A flow scheme of the identification process comparing simulation and experimental time series of data is shown in Fig. 6.63. A validation of such a DT method, which can be used for a large collector field as well as individual collectors, has been done in Fahr, Gumbel, Zirkel-Hofer, and Kramer (2018).

Also, these methods use the comparison of the power output measured with the power output simulated in the numerical tool. The simulation model therefore needs parameters to quantify optical performance and thermal losses. Then the parameter identification method as sketched in Fig. 6.63 will be used, where an optimization method is used to find the parameters for which the agreement between the two power output time series is best (least-squares minimization). Depending on the collector, the testing period, and the location sometimes not the complete range of incidence angles can be realized (e.g. it is difficult to realize normal incidence for a Linear Fresnel collector for latitudes larger than 22.5°). If certain angles however are needed for a full simulation, interpolation or extrapolation using an optical raytracing tool is a possibility. A comparison of quasi dynamic testing (QDT) and dynamic testing (DT) for concentrating collectors including the complex incidence angle modifiers has been described by Hofer et al., (2015) and an overview over procedures and standards for testing concentrating collectors has been given in Hofer et al., (2016). Since then the IEA SolarPACES published guideline for dynamic collector testing, and a first standard is referring to this method (IEC6 2862-5-2, 2022). In the standards for collector testing also other tests are described, which are needed also for the quality assessment and certification. Apart from thermal performance testing gives several additional functional tests to be performed in a full cycle according to ISO 9806 which are addressed in Table 6.8.

6.6.2 Certification

A product certification is an action or process of providing something with an official document attesting to a status or level of achievement. With the certification document a certain quality can be attested. This helps the product in the market, as buyers/customers may trust that this product useful and has a guaranteed performance. For collectors and collector systems several certification schemes exist. They all look in the different performance criteria, from thermal performance to durability and resistance against external stresses. In Europe, the Solar Keymark scheme is implemented. The Solar Keymark was developed by the Solar Heat

Table 6.8 Tests according to ISO 9806 (2017)

Sub clause	Testing type
6	Internal pressure test for fluid channels
7	Leakage test
8	Rupture and collapse test
9	High-temperature resistance
11	Exposure test
12	External thermal shock test
13	Internal thermal shock test
14	Rain penetration test
15	Freeze resistance test
16	Mechanical load test
17	Impact resistance test
20	Thermal performance test
28	Pressure drop measurement
18	Final Inspection

Europe/ESTIF and CEN (European Committee for Standardisation) in close co-operation with leading European test labs and with the support of the European Commission. It is the main quality label for solar thermal products and is widely spread across the European market and beyond. In the USA the Solar Rating and Certification Council (SRCC) has a similar role, In the China the Golden Sun certification and in Australia the Standard Mark™ program exists.

Certification often is a requirement in many countries to receive subsidies in a support program. The certification is issued by a certification body, an independent organization watching and guaranteeing that the rules are followed. Solar Keymark has a network of such certification bodies and accredited test laboratories as well as technical inspectors. The test laboratories must perform tests according to international standards. When the products pass the physical tests, the certification can be issued. However, a regular inspection of production facilities is another part of the scheme. In this way the production uniformity and constancy can be guaranteed.

Products holding a certificate can be found in a publicly accessible data base, see for example (Solar Keymark, no date). Tools provided by Solar Keymark allow also to calculate in a simplified way the annual yield of different collectors for various European locations.

Accredited laboratories often do not only work for one certification scheme but are able to support the different global schemes. This also means that they must be able to follow different testing standards. In the area of thermal performance of collectors, the approaches are quite different. But with the following section at least an approximate conversion of results is possible between International ISO and American ASHRAE standards.

6.6.3 Conversion of Efficiency Equations

Generally, the results and parameter from the two approaches of ISO and ASHRAE can be converted into each other, using a procedure we will describe in the next paragraph. However, it should be stressed that ideally the test data should be determined at flowrates corresponding to those found in reality. The hydraulic turbulence is changing with Reynolds number (Re) and may give rise to small changes in the heat transfer between collector piping and fluid. For the conversion we have to assume in order to make analytical corrections for flow rates that the only noticeable effect of changing flow rate is to change the temperature gradient in the flow direction, and that we can neglect changes in F' due to different internal heat transfer coefficients.

The results of a collector thermal performance according to ISO 9806 is an efficiency curve as presented in Fig. 6.65. The parameters determined from the experimental data to fit the test equation are listed in Table 6.9.

The mass flow during testing in ISO 9806 is chosen in a way, that the collector is operating in a typical condition with respect to turbulent flow and heat transfer within the tubes of the collector. Now assuming that within a certain range the impact of a change of flow is negligible on the value of F', these results are valid for a range of mass flows. The mean fluid temperature in Eq. (6.2) is approximated by the arithmetic average of inlet and outlet temperature, which is equivalent in the limit of high mass flow, where the temperature profile within the collector is

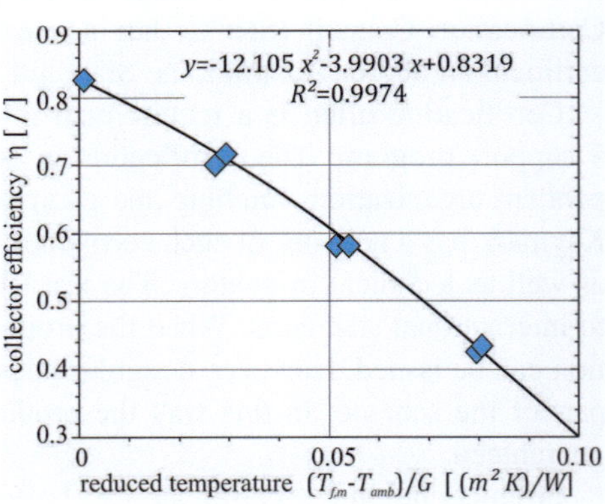

Fig. 6.65 Efficiency curve for a flat plate liquid collector tested according to ISO9806 for irradiation $G = 1000$ W/m^2

Table 6.9 Test parameter according to ISO 9806 for the collector curve of Fig. 6.65

Test parameter	Symbol	Value	Unit
Optical efficiency	$F_{av}(\tau\alpha)$	0.832	–
Thermal loss coefficient	$F_{av}(u_0)$	3.990	W/(m^{2}K)
Thermal loss coefficient (2nd order)	$F_{av}(u_1)$	0.0121	W/(m^{2}K)

linear. Therefore, the testing equation is correctly written (to distinguish it from Eq. (6.2)) as:

$$\dot{Q}_u = A_c \cdot F_{av} \cdot \left[G \cdot (\tau\alpha)_e - U_L \cdot \left(T_{f,av} - T_{amb}\right)\right]. \tag{6.47}$$

In (ANSI/ASHRAE Standard 93-2003, 2003) the results on the contrary are presented in the form of collector Eq. (6.3) using the collector inlet temperature. Here, the mass flow is extremely important for the results—a low mass flow will result in much higher outlet temperatures than a high mass flow, and therefore the overall heat loss will be higher. Given a certain mass flow within the range of validity of test results in Table 6.9 one may convert these results to parameters used by ASHRAE. The conversion formulae are:

$$F_R = F_{av} \cdot \left[1 + \frac{A_c \cdot F_{av} \cdot U_L}{2\dot{m} \cdot c_p}\right]^{-1} = F_{av} \cdot \left[1 + \frac{1}{2\gamma_{av}}\right]^{-1}, \tag{6.48}$$

where F_{av} is approximately the same as F'. A flow rate correction for F_R can be done with the ratio r:

$$r = \frac{F_R|_{use}}{F_R|_{test}} = \frac{\frac{\dot{m} \cdot c_p}{A_c \cdot F' \cdot U_L} \cdot \exp\left(-\frac{A_c \cdot F' \cdot U_L}{\dot{m} \cdot c_p}\right)\Big|_{use}}{\frac{\dot{m} \cdot c_p}{A_c \cdot F' \cdot U_L} \cdot \exp\left(-\frac{A_c \cdot F' \cdot U_L}{\dot{m} \cdot c_p}\right)\Big|_{test}} = \frac{\gamma \cdot \exp\left(-\frac{1}{\gamma}\right)\Big|_{use}}{\gamma \cdot \exp\left(-\frac{1}{\gamma}\right)\Big|_{test}}. \tag{6.49}$$

Again, here F' may be approximated by F_{av} from a test according to ISO 9806—or from F_R under test conditions directly using the equation:

$$(F' \cdot U_L) = -\frac{\dot{m} \cdot c_p}{A_c \cdot} \cdot \ln\left(1 - \frac{A_c \cdot F_R \cdot U_L}{\dot{m} \cdot c_p}\right)\Big|_{test}. \tag{6.50}$$

These equations are straightforward if U_L can be approximated by a constant. However, very often a second order approximation for the temperature dependence of U_L is used, and hence U_L must be evaluated for the temperature conditions specified in the test. Then the heat loss U_L is expressed as a linear function of temperature difference as in the collector equation, which changes the dimensionless flow rate γ, the flow factor F' and consequently also the flow factor relation r. It is advisable to check whether for the mass flow conditions considered the linear approximation of temperature profile can be utilized.

6.7 Collector Stagnation

The *stagnation of* a solar system is defined as the state in which there is no more forced circulation in the solar circuit and thus no heat can be extracted from the absorber. A stagnation occurs if, on the one hand, heat can no longer be dissipated because the storage tank has reached its maximum temperature and the control system must therefore switch off the pump, or, on the other hand, the pump or the control system fail.

The first case can be prevented by additional heat dissipation or by optimized control. In the event of pump or controller failure (e.g. due to power failure or defective sensor), stagnation cannot be avoided in forced flow systems. In general, shutdown in today's efficient collectors is associated with evaporation of the fluid in the absorber.

When the system is stopped, the temperature of the collector rises until the losses are as high as the gains, i.e. the efficiency drops to zero. In the simplest case of a linear efficiency relation with a constant heat loss factor U_L, the absorber stagnation temperature $T_{abs,stag}$ can be calculated in a simple way:

$$T_{abs,stag} = \frac{\eta_0}{U_L} \cdot \mathrm{G} + T_a. \tag{6.51}$$

In the case of a linear temperature dependence of U_L a quadratic equation must be solved instead. In Fig. 6.66 the absorber stagnation temperature for the collector efficiency curve as shown in Fig. 6.65 is plotted as a function of the radiation intensity G for the constant and the linear approximation of the heat loss factor for an ambient temperature of 20 °C.

As can be seen in Fig. 6.66, the stagnation temperatures of modern flat plate or vacuum tube solar collectors with U_L values in the range 1–4 $(\mathrm{W}/(\mathrm{m}^2\mathrm{K}))$ are well above the evaporation temperature of water at atmospheric pressure. Although the evaporation temperature of the collector fluid rises with pressure in the collector loop (typically around 145 °C for 0.6 MPa pressure), a collector shutdown during sunny days usually is associated with an evaporation of the heat transfer medium. The processes during a stagnation situation with evaporation can be divided according to Lustig (2002) into five successive phases. Since the shutdown scenario affects the entire solar thermal system, Fig. 6.67 shows a typical system diagram of a simple solar circuit system with pressurized collector loop and forced circulation.

Due to the evaporation of the heat transfer fluid caused by solar radiation, the system pressure also changes over time, if no safety valve is opened and releases

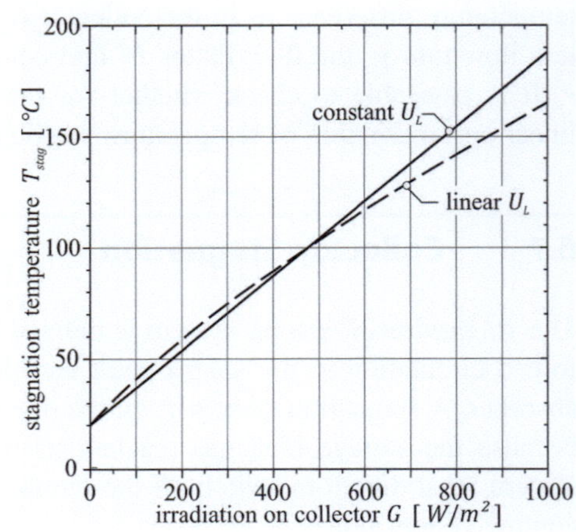

Fig. 6.66 Stagnation temperature T_{stag} in (°C) as a function of the irradiance G in $(\mathrm{W}/\mathrm{m}^2)$ on the absorber surface for constant and linear heat transfer factors U_L for ambient temperature of 20 °C for collector efficiency as in Fig. 6.65

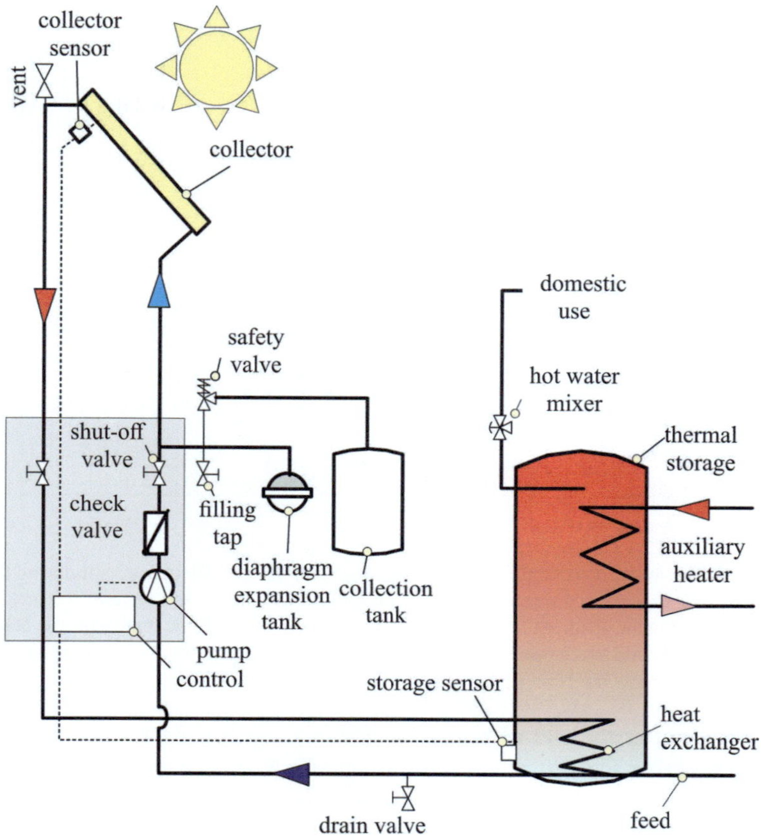

Fig. 6.67 Simplified schematic structure of a domestic solar thermal collector system

steam. This pressure change in turn affects the boiling temperature and the steam-filled pipe space. Figure 6.68 shows qualitatively the development of the system pressure and the steam/fluid fraction of the five different phases of a collector stagnation; details are described in Dimitrova (2006) or Scheuren (2008).

Phase 1: Liquid expansion

After switching off the collector circuit pump, the collector temperatures rise rapidly and evenly until the evaporation process begins at a point on the absorber material in the upper part of the collector. Up to this point, the increase in system pressure is only slight and corresponds to the temperature-related volume expansion of the fluid in the pipeline network.

Phase 2: Emptying the collector

Evaporation begins at a boiling temperature which is determined by the local pressure at this point of the collector. As the system pressure is still low at this point

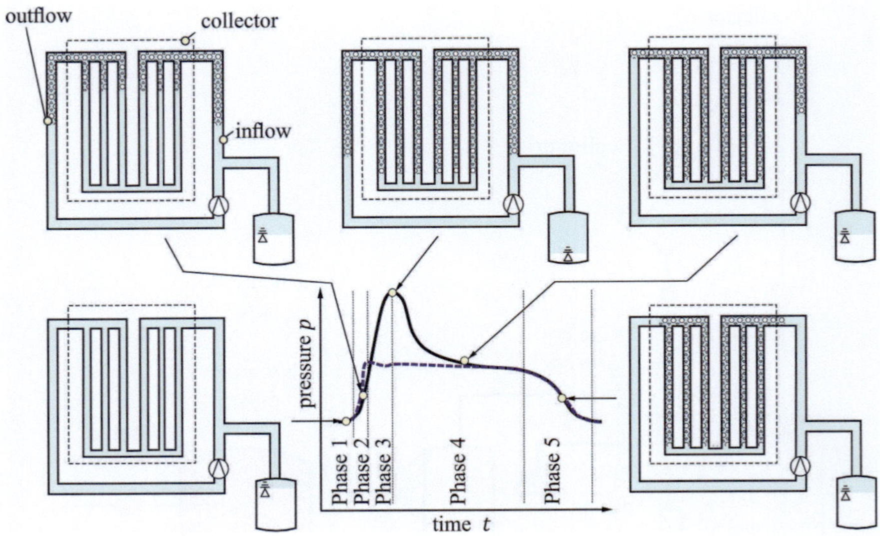

Fig. 6.68 Schematic representation of the steam distribution in a collector circuit during the five different phases of a collector shutdown. Bottom middle: Course of the system pressure p as a function of time t in the different collector stagnation phases; the solid line represents the behaviour of a system that is difficult to drain, the dashed line that of a system that is easy to drain

(immediately after the end of normal operation), the boiling temperature is also lower than during the rest of the system shutdown. A small amount of medium evaporates and pushes a larger part of the liquid collector contents out of the absorber tubes until finally the supply and return lines are also filled with saturated steam. This results in a relatively rapid increase in pressure, as large quantities of liquid are pushed into the expansion tank. During this phase, the system pressure increases rapidly and thus the boiling temperature in the system also increases. The duration of this phase is only a few minutes. The temperatures of the supply and return lines in the plant room and the supply line to the expansion tank are initially low (cooled down after pump stagnation) and later rise when the hot collector contents are inserted. However, they do not reach the values of the collector temperatures due to the cooling losses in the pipes. If the lines are short and have low cooling losses, high temperature loads can occur on the expansion tank and other system components, especially if evaporation only starts at higher temperatures at high system pressure. Long and uninsulated lines to the expansion tank significantly reduce this temperature load. Likewise, a hanging attachment of the membrane expansion vessel can prevent the vessel membrane from overheating.

Phase 3: Boiling in the collector—Phase with saturated steam

This is a phase with a slowing pressure increase. A smaller, but not insignificant part of the collector content is still present in liquid form in the absorber and collection pipes. In this phase, either liquid medium or saturated steam in thermal equilibrium with the liquid is observed at all points on the absorber. The supply and return lines are increasingly pressed downwards towards the membrane expansion vessel with approximately the same liquid level (communicating vessels) and the heat exchanger can be reached by steam. This can also lead to steam formation on the secondary side of the heat exchanger. The system pressure reaches its maximum value, which also results in the highest boiling temperature. The dissipated power is determined by the collector efficiency at high temperatures, taking into account the energy stored in the tubes and the absorber. According to the principle of the heat pipe, energy is transported very effectively at almost constant temperature from the heat source to all heat sinks, whereby the vapour condenses again and runs off downwards as a liquid. The temperature differences between heat source (collector) and heat sink (e.g. heat exchanger) are low (of the order of a few degrees Kelvin), as there are only slight pressure differences (flow pressure losses of the steam). However, concentration differences in the water–glycol system must also be considered (fractional distillation). The maximum temperature load of the plant components reached by the steam thus depends on the maximum plant pressure via the saturated steam temperature. At the beginning of this phase, the temperatures of the supply and return lines in the plant room and the supply line to the expansion tank can drop, as the steam area now increases much more slowly, and the lines filled with liquid cool down. A sudden rise in temperature up to boiling point only occurs when steam appears at these points.

Phase 4: Boiling in the collector—Phase with saturated steam and superheated steam

In the upper collector area the liquid begins to evaporate completely. These areas overheat and the absorber temperature increases. As a result, collector efficiency drops further and the amount of energy to be transported away with steam decreases, because less energy is transported into the system areas filled with steam, so that the "loss area" of the system can also decrease. The volume of steam decreases rapidly even with constant solar radiation. The system pressure decreases (and with it the saturated steam temperature) and liquid is slowly pushed from the expansion tank into the return line. The non-return valve prevents the latter on the flow side. As the superheating areas in the collector increase, the system pressure decreases further, and the liquid level of the return flow reaches the collector inlet. The flow line also fills slowly through condensate. The overheating phase can last several hours on cloudless days and only ends when the solar radiation decreases.

Phase 5: Refilling the collector

The collector is refilled via the return line, whereby the temperatures here drop rapidly. The expansion vessel presses the liquid back into the collector. Depending on the position of the non-return valve, this is done via the supply and return lines, or only via the return line. In the latter case, the flow line is filled via the collector with a time delay.

Measurements and numerical observations show that the collector stagnation depends essentially on the arrangement of the piping system and the internal piping of the collector. It is important that heat transfer fluid is driven out of the collector by the formed steam, and thus the maximum steam load is reduced when the fluid is pressed into cooler parts of the loop. If however the fluid is captured by steam pockets due to gravity in the hot collector, stagnation leads to the evaporation of nearly all fluid content of a collector.

References

Aelenei, L., Smyth, M., Platzer, W., Norton, B., Kennedy, D., Kalogirou, S., & Maurer, C. (2016). Solar thermal systems—Towards a systematic characterization of building integration. *Energy Procedia, 91*, 897–906.

ANSI/ASHRAE Standard 93-2003. (2003). *Methods of testing to determine thermal performance of solar collectors.* 1791 Tullie Circle, Ne, Atlanta, GA30329: ASHRAE, Inc.

Aste, N., Del Pero, C., & Leonforte, F. (2014). Water flat plate PV–thermal collectors: A review. *Solar Energy, 102*, 98–115.

Bandaru, S. H., Becerra, V., Khanna, S., Radulovic, J., Hutchinson, D., & Khusainov, R. (2021). A review of photovoltaic thermal (PVT) technology for residential applications: Performance indicators, progress, and opportunities. *Energies, 14*(13), 3853.

Brötje, S., Kirchner, M., & Giovannetti, F. (2018). Performance and heat transfer analysis of uncovered photovoltaic-thermal collectors with detachable compound. *Solar Energy, 170*, 406–418.

Chan, H.-Y. R. (2010). Review of passive solar heating and cooling technologies. *Renewable and Sustainable Energy Reviews, 14*(2), 781–789. https://doi.org/10.1016/j.rser.2009.10.030

Chow, T. T. (2010). A review on photovoltaic/thermal hybrid solar technology. *Applied Energy, 87*(2), 365–379.

Chugani, B. (2020). Model validation and performance assessment of unglazed photovoltaic-thermal collectors with heat pump systems. In A. G. Charalambides. International Solar Energy Society.

Dimitrova, K. G. (2006). *Thermische Effekte im Kollektorfeld solarthermischer Anlagen.* Dissertation, Techn. Universität, Fak. Maschinenbau, Chemnitz.

DIN EN 12975-1. (2011). *Thermal solar systems and components—Solar collectors—Part 1: General requirements.* Beuth Verlag GmbH. https://doi.org/10.31030/1728174

Duffie, J. A., & Beckmann, W. A. (2020). *Solar engineering of thermal process, photovoltaics and wind*. Wiley.

Fahr, S., Gumbel, U., Zirkel-Hofer, A., & Kramer, K. (2018). In situ characterization of thermal collectors in field installations. In *Conference EuroSun 2018* (pp. 1–10). International Solar Energy Society. Retrieved September 2, 2022, from http://proceedings.ises.org/paper/eurosun2018/eurosun2018-0167-Fahr.pdf

Frei, U., Häuselmann, T., Flückinger, F., & Frey, R. (1995). *Leistungsdaten Thermischer Sonnenkollektoren.* Solarenergie Prüf- und Forschungsstelle, Technikum Rapperswil. Infoenergie Brugg, Postfach 310, CH-5200 Brugg.

Giovannetti, F. K. (2014). Development of an insulated glass solar thermal collector. *Energy Procedia, 48*, 58–66. https://doi.org/10.1016/j.egypro.2014.02.008

Gombert, A., Reineke-Koch, R., Fenske, K., & Hofmann, T. (2007). Optische Beschichtungen für Solarkollektoren-Technologien und Qualitätssicherung. In F. Sonnenenergie (Ed.), *FVS-Jahrestagung 2007 zu Produktionstechnologien für Solaranlagen* (p. 23). Leibniz Universität Hannover.

Hamid, Y., & Beckman, W. A. (1971). Performance of air-cooled radiatively Heated screen matrices. *Journal of Engineering Power, 93*(2), 221–224. https://doi.org/10.1115/1.3445553

Hermann, M. (2005a). *Bionische Ansätze zur Entwicklung energieeffizienter Fluidsysteme für den Wärmetransport.* Dissertation, Universität Karlsruhe, Faculty for Mechanical Engineering, Karlsruhe.

Hermann, M. (2005b, April). Entwicklung des FracTherm-Absorbers—Simulationen und Experimente. In *Proceedings of 15. Symposium Thermische Solarenergie OTTI* (pp. 94–99).

Hermann, M. (2011). *Development of a bionic solar collector with aluminium roll-bond absorber.* BIONICOL Project Final Report, European Commission, 7th Framework Programme.

Hermann, M., et al. (2015). *TABSOLAR—Solarabsorber und andere thermisch aktive Bauteile aus Ultrahochleistungsbeton (UHPC)—Schlussbericht Förderkennzeichen: 03ET1117A/C/D.* Retrieved June 12, 2022, from https://tabsolar.de/externer-bericht/schlussbericht_TABSOLAR/at_download/file

Hermann, M., et al. (2020). TABSOLAR II – Solarabsorber und andere thermisch aktive Bauteile aus Ultrahochleistungsbeton (UHPC). *Schlussbericht Förderkennzeichen: 03ET1310A/B/C/D.*, 115.

Hess, S. (2015). *Low-concentrating, stationary solar thermal collectors for process heat generation.* Shaker Verlag.

Hess, S., Oliva, A., Di Lauro, P., Klemke, M., Hermann, M., Eisenmann, W., … Kramp, G. (2010). Solar heat for industrial processes: RefleC collector development and system design. In *Conference Proceedings/EuroSun 2010, International Conference on Solar Heating, Cooling and Buildings* (pp. 1–8). International Solar Energy Society.

Hofer, A., Büchner, D., Kramer, K., Fahr, S., Heimsath, A., Platzer, W., & Scholl, S. (2015). Comparison of two different (quasi-) dynamic testing methods for the performance evaluation of a Linear Fresnel Process Heat Collector. *Energy Procedia, 69*, 84–95. https://doi.org/10.1016/j.egypro.2015.03.011

Hofer, A., Valenzuela, L., Janotte, N., Burgaleta, J., Arraiza, J., Montecchi, M., … Scholl, S. (2016). State of the art of performance evaluation methods for concentrating solar collectors. In *AIP Conference Proceedings 1734* (pp. 020010, 9 p.). https://doi.org/10.1063/1.4949034

Hottel, H., & Whillier, A. (1955). Evaluation of flat-plate solar collector performance. *Transactions of the Conference on the Use of Solar Energy II, 3 (Part 2 Thermal Processes)*, 74–104.

IEC 62862-5-2. (2022). *Solar thermal electric plants Part 5-2. Systems and components. General requirements and test methods for large-size linear Fresnel collectors.*

ISO 9488. (2021). *Solar energy—Vocabulary.* ISO/TC 180 Solar energy.

ISO 9806. (2017). *Solar energy—Solar thermal collectors—Test methods.* ICS 27.160 Solar energy enigneering, ISO/TC 180 Solar Energy.

Kramer, K., et al. (2020). *Status Quo of PVT characterization.* Report B1, IEA SHC Task 60.

Kramer, K., & Helmers, H. (2013). The interaction of standards and innovation: Hybrid photovoltaic–thermal collectors. *Solar Energy, 98*, 434–439.

Krstic-Furundzic, A. (2007). PV integration in design of new and refurbishment of existing buildings: educational aspect. *Journal of the Association of Arab Universities for Basic and Applied Sciences, 4*(Supplement), 135–146.

Lämmle, M. (2018). *Thermal management of PVT collectors.* Technische Fakultät. Freiburg im Breisgau: Albert-Ludwigs-Universität.

Lämmle, M., Hermann, M., Kroyer, T., Mülhöfer, G., Oliva, A., Thoma, C., … Jäger, H. (2017). PVTgen2: Entwicklung und Pilotfertigung eines abgedeckten photovoltaisch-thermischen Hybrid-Kollektors der zweiten Generation mit verbesserter thermischer Leistungsfähigkeit. 113. Freiburg.

Lämmle, M., Kroyer, T., Fortuin, S., Wiese, M., & Hermann, M. (2016). Development and modelling of highly-efficient PVT collectors with low-emissivity coatings. *Solar Energy, 130*, 161–173.

Lustig, K. (2002). *Experimentelle Untersuchungen zum Stillstandsverhalten thermischer Solaranlagen.* Dissertation, Universität Karlsruhe, Fak. Maschinenbau, Karlsruhe.

Maurer, C. (2018). *Worldwide innovation: A venetian blind that supplies solar thermal energy.* Retrieved from https://blog.innovation4e.de/en/2018/09/19/worldwide-innovation-a-venetian-blind-that-supplies-solar-thermal-energy/

Osório, T., Horta, P., & Collares-Pereira, M. (2019). Method for customized design of a quasi-stationary CPC-type solar collector to minimize the energy cost. *Renewable Energy, 133*, 1086–1098.

Peuser, F. A., Remmers, K.-H., & Schnauss, M. (2011). *Solar thermal systems: Successful planning and construction* (1st ed.). Taylor & Francis Ltd.

Platzer, W. (2000). Transparent insulation materials and products. In Y. Goswami (Ed.), *Advances in solar energy* (Vol. 14, pp. 33–66). ASES.

Platzer, W., Smyth, M., & Kennedy, D. (2014). *State of the art on building integrated solar thermal systems.* TU1205—Building Integration of Solar Thermal Systems (BISTS), Deliverable.

Ritter, X. L. (2014). *Solarthermie in der Fassade.* Retrieved from http://ritter-xl-solar.com/uploads/media/Booklet_Fassadenkollektor.pdf

Rohsenow, W. M. (1998). *Handbook of heat transfer*. McGraw-Hill.

Scheuren, J. (2008). *Untersuchungen zum Stagnationsverhalten solarthermischer Kollektorfelder.* Dissertation, Universität, Fachbereich Maschinenbau, Kassel.

Selcuk, K. (1971). Thermal and economic analysis of the overlapped-glass. *Solar Energy, 13*, 165–191. https://doi.org/10.1016/0038-092X(71)90002-8

Solar Keymark. (n.d.). https://solarkeymark.eu/. Retrieved from The Origin and Composition of the Solar Keymark. https://solarkeymark.eu/

SolarWall Europe Sarl. (2023). SolarWall air heating systems-product information. 66 Avenue des Champs Elysees, 75008 Paris, France. Retrieved February 27, 2023, from https://www.solarwall.com/technology

Spirkl, W., Ries, H., Muschaweck, J., & Winston, R. (1998). Nontracking solar concentrators. *Solar Energy, 62*(2), 113–120.

Stephan, P., Kabelac, S., & Kind, M., Mewes, D. (2019). *VDI-Wärmeatlas* (12th ed.). Springer Vieweg.

TIGI Solar. (2011). Product whitepaper honeycomb collector. 3 Hamechonai St., Hod Hasharon, Israel. Retrieved from http://www.tigisolar.com/assets/files/Honeycomb%20Collector%20whitepaper%20Oct11(1).pdf

Wang, Y., Shukla, A., & Liu, S. (2017). A state of art review on methodologies for heat transfer and energy flow characteristics of the active building envelopes. *Renewable and Sustainable Energy Reviews*, 1102–1116.

Weiss, W., & Rommel, M. (2008). *Process heat collectors-state of the art.* IEA SHC-Task 33 and SolarPACES-Task IV: Solar Heat for Industrial Processes. IEA SHC. Retrieved September 20, 2022, from https://task33.iea-shc.org/publications

Whillier, A. (1964). Performance of black-painted solar air heaters of conventional design. *Solar Energy, 8*(1), 31–37. https://doi.org/10.1016/0038-092X(64)90008-8

Widyolar, B., Jiang, L., Ferry, J., & Winston, R. (2018). Non-tracking East-West XCPC solar thermal collector for 200 celsius applications. *Applied Energy, 216*, 521–533.

Zondag, H. A. (2008). Flat-plate PV-Thermal collectors and systems: A review. *Renewable and Sustainable Energy Reviews, 12*(4), 891–959.

Zondag, H. A., Vries, D. W., van Helden, W., van Zolingen, R., & van Steenhoven, A. A. (2003). The yield of different combined PV-thermal collector designs. *Solar Energy, 74*(3), 253–269.

Concentrating Collectors

7

Abstract

High-temperature collectors using optical concentration and tracking, are described in this chapter. The determination and optimization of the optical performance of line-focusing and point-focusing technologies is key for reaching high temperatures. We classify therefore the collectors by their optical concentrator principle. Although also the selected heat transfer fluid has an impact on the temperature level which can be achieved, due to fluid stability and receiver material durability issues, this aspect is only treated in Chap. 10 on high temperature collector systems and fields. Here, the focus is on the main components and designs of individual collectors. As solar radiation is transformed into heat in the receiver, the design of this element is mainly responsible for minimizing the thermal losses. However, thermal losses play a minor role, the higher the optical concentration. We present receiver designs, determine the selection of materials, and define the performance. Finally, some line- and point-focusing collector types are compared with respect to their optical performance characteristics.

Keywords

Classification of high temperature collectors • Line-focusing collectors • Parabolic trough and Fresnel collector • Point focusing collectors • Solar tower • Heliostats • Parabolic dish and Scheffler dish

7.1 Classification of High-Temperature Concentrating Collectors

We may classify concentrating tracking collectors by their optical concentration principle (Fig. 7.1). Line and point-focusing collectors have a very different range of achievable and real optical concentration factors. The receiver element converting the concentrated solar intensity into high temperature heat is a second key element. Its design is a consequence of the concentrator, and the thermal losses of

R. Stieglitz and W. Platzer, *Solar Thermal Energy Systems*,
https://doi.org/10.1007/978-3-031-43173-9_7

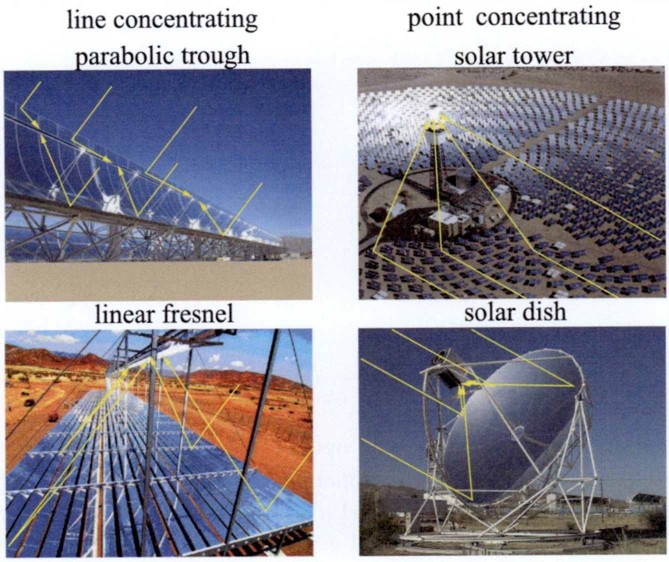

Fig. 7.1 Classification of solar thermal collector types according to the type of the solar concentration

the complete collector are mainly influenced by this design. The critical choice of materials which are able to withstand the high temperatures, is a part of the collector design. The geometrical concentration determined by design, and the optical quality which is dependent on the selection of optical materials like mirrors, their production and mounting, within the two main groups of line- and point-focusing systems determines to a large extent the maximum temperatures which can be achieved. Also, the land consumption of a collector assembly is associated with the concentration type.

However, another option of classification, is often useful with respect to the selection of the heat transfer fluid. In Chap. 6 on stationary collectors we found for example the classification air and liquid collector. In the case of temperatures well above 100 °C the choice of heat transfer fluid determines for example the maximum allowable fluid temperature, which has an impact on the capability to drive connected thermal processes, whether that is a thermodynamical cycle for producing electricity, desalination, or an industrial process. Materials in the open or closed heat transfer fluid (HTF) loop must be selected accordingly. As we will see in Chap. 10, the maximum achievable fluid temperature also has an influence on the size of the solar field (via the pressure and heat losses in the solar receiver piping system), the choice of thermal storage medium, and ultimately the type of thermodynamical cycle which needs the heat.

We focus solely on line-focusing and point-focusing collectors which track the sun and hence work only with direct solar irradiation, in contrast to stationary non-tracking collectors without concertation which also use the diffuse part of the global irradiance.

The concentrating collectors are able to reach high temperature levels exceeding by far 150–200 °C. The practical limits will be discussed. In a first step the optical concentration process of solar radiation from the source (sun) via the imaging system (mirrors) and the conversion process (receiver) is discussed. Based on these considerations, conversion efficiencies or cumulative yields can be obtained, describing the characteristic properties of the individual collector technologies.

Due to the different imaging technologies and geometric arrangements of both line- and point-concentrating mirror/receiver systems, there is also a different utilization of the beam radiation in the individual receiver technologies.

The angular distribution of the solar radiation intensity (solar disc and circumsolar radiation as discussed in Chap. 2), and the optical imaging of the concentrator determine the spatial optical conversion and distribution of heat generated in the receiver, which determines optical efficiency and heat losses of the receiver. The incoming beam radiation is subject to atmospheric influences and varying position of the sun during the day and the seasons. The conversion process therefore is dependent on the solar incidence angles and the geometrical arrangement of the mirror concentrator in relation to the solar source (e.g. for line-focusing collectors the axis might be oriented East–West, North–South or in between). Depending on the concentrator design and the tracking, the utilization of solar radiation is reduced by geometric restrictions leading to effects such as mutual shading of mirrors, or shading by structures, for example a tower construction, a linear receiver or support pylons.

7.1.1 Concentration, Exergy, and Operating Temperatures

In concentrating collectors, the solar irradiation power is directed towards the absorber with the help of mirrors and/or lenses. Mirrors are favored in solar thermal technology, as the lenses show strong achromatism due to the large wavelength range. In both cases the focal spot shall be much smaller than the collecting aperture. We distinguish between imaging concentrators where an image of the sun is formed on the absorber, or non-imaging concentrators without any image formation.

However, redirection of light towards a specific target, also imaging, is only possible with beam radiation. An immediate consequence is that focusing systems are only able to use the direct radiation component, but not the diffuse radiation component, which accounts for about 40% of the incident radiation in Germany. In arid regions around the equator, however, the proportion of direct radiation is between 70 and 80%.

Several concentrator types can be realized. In mirror geometries with rotational symmetry the focus may be a single small spot (ideally a point-focus), whereas in mirror shapes with translational symmetry the focal area will be a linear strip (ideally a line-focus). The concentration respectively the radiation intensity on the absorber with point-focus systems can be much higher than with line-focus concentrators.

The basic idea of concentrating light has been shown already in Chap. 3. Light from a source is incident on a concentrator aperture are A_a from a certain direction. Due to finite sizes of light source the light always has a small divergence and is not completely parallel. In the concentrator the directional light is redirect towards a smaller receiver area A_r. The geometrical concentration factor C_g then is defined as:

$$C_g = \frac{A_a}{A_r}. \tag{7.1}$$

An ideal concentrator is one with the maximum possible concentration. The maximum concentrator is determined by the θ divergence angle θ_s of the incoming light. From optical theory using the conservation of étendue the upper limit for concentration can be derived. We will find the same result considering the second law of thermodynamics. For the radiation coming from the sun the divergence angle due to the finite disk size of the sun is 0.27° (half-angle). The maximum concentration for line-focus systems using concentration in one dimension is:

$$C_{max,1D} \leq \frac{1}{\sin\theta_s} = 212. \tag{7.2}$$

The maximum concentration for point-focus systems focusing in two dimensions is

$$C_{max,2D} \leq \frac{1}{\sin^2\theta_s} \approx 4.5 \cdot 10^4. \tag{7.3}$$

Maximizing overall system efficiency in terms of energy is an appropriate goal. However, the restriction to purely energy terms may miss some information needed for optimization. The second law of thermodynamics addresses the inherent of irreversibility of processes and leads to the limiting conversion efficiency of a heat engine between isothermal reservoirs, given by the well-known Carnot limit on the efficiency, η_C:

$$\eta_C = 1 - \frac{T_2}{T_1}. \tag{7.4}$$

In this equation, temperatures T_1 and T_2 are absolute temperatures to be provided in Kelvin (K), where T_2 is the lower temperature of the Carnot cycle, at which heat is released, usually close to the ambient temperature. Clearly, operating at low driving temperatures T_1 to achieve high receiver efficiencies due to small heat losses does not result in the highest Carnot efficiency. The efficiency of an ideal Stirling engine is approximated by the Carnot cycle quite well. When mounted in the focus of a receiver, at steady state the Stirling engine operates between a single high-temperature source and a fixed low-temperature environment. However, for systems where a heat transfer fluid (HTF) is used between receiver and power cycle, the power cycle is no longer operating between effective isothermal reservoirs. The thermodynamic cycles approximating these situations can be the Rankine cycle for a steam turbine or the Brayton cycle for a gas turbine. Nevertheless, the Carnot efficiency give the upper limit of work that can be extracted between two isothermal reservoirs. It is the basis for the useful thermodynamic concept called exergy.

The exergy of a system is defined as the maximum amount of work that could usefully be performed when bringing the (thermodynamic) system to equilibrium with its low-temperature environment. In this sense, also a certain mass of a substance such as a heat transfer fluid has an exergy content. Thus, every energy flow can be associated with a corresponding exergy flow, being the ideal amount of work that could be extracted from it. Exergy analysis can be a powerful tool in understanding thermal systems which are designed to produce power or chemical actions, and it can be used to design systems in low-exergy applications to analyze which low temperature sources might be best used to provide a certain energy service as for example heating a building.

Table 7.1 illustrates the temperature dependence of the Carnot efficiency together with the exergy-to-energy ratio calculated for some potential HTFs at the same temperature. Considering realistically achievable temperatures yields a reduced potential to extract work. Also, when heat is transferred from an isothermal source to an HTF heated up to the same temperature the exergy content is reduced. It is an irreversible process destroying exergy. For a given temperature, the Carnot efficiency assuming an infinite isothermal source is higher than the exergy-to-energy ratio of a working fluid having the same temperature as this is cooling down when work is performed. Energy in a finite amount of a substance at a certain temperature can be considered as having a work potential according to the full range from its starting temperature down to ambient temperature according to temperature decrement. Steam on the other hand has a temperature profile that is strongly affected by the phase change from steam to water. Depending on where the condensation/evaporation temperature is relative to the maximum steam temperature, its exergy to- energy ratio is different.

The concept of regenerative feedwater heating in a steam cycle using bled steam can be understood using the exergy concept. Partially expanded steam has already given up some of its exergy to power generation, so using it to pre-heat feedwater destroys less exergy than using the highest temperature source to heat the feedwater from ambient temperature. Thinking about a concentrated solar power (CSP) system, preheated HTF in the receivers utilizes the high exergy potential of the solar radiation better than cold HTF. On the other hand, the consequence is a higher receiver average temperature, and the thermal losses are increased. Also, higher flow rates must be maintained as the energy increment per unit mass flow is smaller.

A simplified analysis of the performance of a solar thermal energy system is useful in demonstrating the competing design demands. From the theory of

Table 7.1 Thermodynamic efficiency metrics (relative to environment at 298 °K, 0.1 Mpa)

Temperature (K)	Carnot-efficiency η_C (/)	Constant c_p	Exergy-energy-ratio steam 15 Mpa	Steam 5 MPa	Typ. steam cycle efficiency (/)
5800	94.9	84.1			N.a.
1273	77.0	56.1			N.a.
973	69.9	48.3	0.469	0.435	0.4 – 0.45
673	56.5	35.9	0.421	0.388	0.35 – 0.4
573	48.9	29.8	0.294	0.376	0.25 – 0.3
473	38.1	22.0			

reversible heat engines, power cycles running from high-temperature thermal reservoirs are most efficient, but thermal losses increase as receiver temperatures increase. Using simplified expressions for the various loss mechanisms, and some typical parameter values, gives the results for receiver efficiency as a function of concentration ratio and temperature shown in Fig. 7.2. It can be seen that at low temperatures, all efficiencies reduce to the average absorptivity value. As temperature is increased, efficiencies increase with higher concentration systems at the same temperature difference. In a real system, operating temperature can be controlled through variation of HTF flow rate, for example having higher receiver efficiencies at any given temperature. The various curves intersect the x-axis at the corresponding stagnation temperature. To simulate an idealized operating regime for maximizing power production, the receiver efficiency must be multiplied by a power cycle conversion efficiency to yield an overall ideal system efficiency η_{sys}.

To a large extent, power cycle efficiency dependence on operating conditions is complex and must be established from empirical performance curves. The difference in exergy value of the HTF before and after transfer of heat to the working fluid at input to the power cycle provides a good indicator, with real power cycles extracting something over 70% of the available exergy as work. The Carnot expression provides a good qualitative illustration of the principles, however. If the receiver efficiency results in Fig. 7.2 are multiplied by the Carnot efficiency (according to Eq. (7.4)) pertaining to the receiver temperature at each point to give a semi-ideal system efficiency, the results are as shown in Fig. 7.3. There is now a clear peak efficiency at a particular temperature for each concentration ratio. Higher concentration ratios have higher peak efficiencies, and these occur at higher receiver temperatures.

Without going into the different technology options for point-focusing systems, it is useful to recall the basic idea for the development of high-concentration collectors. The development of central receivers is essentially targeting at high collector outlet temperatures. This is essentially driven by the increase of the solar-to-thermal receiver efficiency $\eta_{rec,th}$, which balances the effective useful heat extracted by the receiver to the solar radiation impinging on the receiver surface. The useful heat contains the optical conversion minus the heat losses. By doing

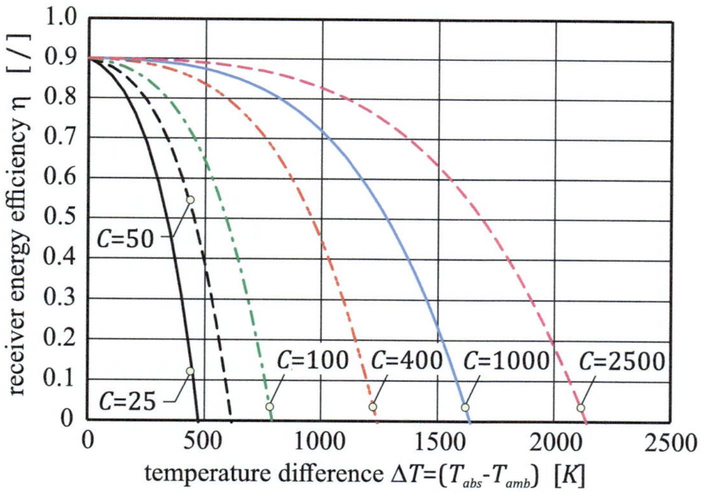

Fig. 7.2 Ideal receiver energy efficiency (η_{rec}) for $DNI = 800$ W/m^2 as function of the temperature difference between absorber and ambient for different concentration ratios C_g; optical efficiency (solar absorptance) 0.9 and emissivity 0.9, assuming $h_{conv} = 20$ W/(m^{2}K) (dashed lines)

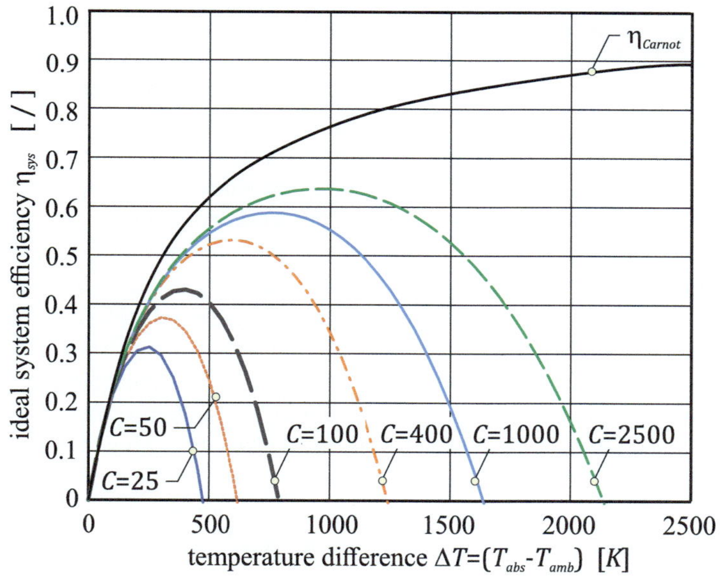

Fig. 7.3 Ideal system efficiency $\eta_{sys} = (\eta_{rec} \cdot \eta_C)$ given for $DNI = 800$ W/m^2 as function of the temperature difference between absorber and ambient for different concentration ratios C_g; optical efficiency (solar absorptance) 0.9 and emissivity 0.9, assuming no convective losses and $h_{conv} = 20$ W/(m^{2}K)

so $\eta_{rec,th}$ can be derived in a simplified manner from a thermal balance in the following way:

$$\eta_{rec,th} = \frac{\alpha \dot{q}_{rec} - \dot{q}_{loss}}{\dot{q}_{rec}} = \alpha - \frac{\varepsilon\sigma\left(T_{rec}^4 - T_{amb}^4\right) \mp h_{conv}(T_{rec} - T_{amb})}{\eta_{opt,conc} \cdot DNI \cdot C_g}, \quad (7.5)$$

in which is $\dot{q}_{rec}$ the area specifc irradiance on the receiver and $\dot{q}_{loss}$ describes the sum of the radiative and convective losses from the receiver to the ambient, both in (W/m^2). α is the effective solar absorptance of the receiver and ε its thermal emittance, while T_{rec} and T_{amb} denote the receiver surface temperature and the ambient temperature, respectively. σ is the the Stefan-Boltzmann constant and h is the heat transfer coefficient from the receiver to the ambient. The optical efficiency of the concentrator system is described by $\eta_{opt,conc}$, while DNI is the specific irradiance on the aperture area in (W/m^2) and C_g is the concentration ratio, given as collector aperture area divided by the receiver area.

In principle, $\eta_{rec,th}$ for a given temperature difference increases the higher the optical concentration ratio $\eta_{opt,conc}C_g$ gets. In this sense point-focusing collectors from a thermodynamic point of view are more efficient than line-focusing systems. On the other hand, as receiver outlet temperature increases, the thermal receiver efficiency decreases, because the area specific heat losses increase, and, as shown in a simplified example in Fig. 7.2 finally a stagnation temperature is reached where the receiver efficiency is zero. A dilemma of the efficiency relation is not immediately apparent. Both, the absorptivity α and the total emissivity ε are spectral quantities, but moreover also strongly temperature dependent. While the solar irradiance spectrum does not change substantially, and α can be considered constant, this is not true for the thermal emissivity ε which is a strongly temperature-dependent quantity and increases dramatically with absorber temperature. The spectral emissivity is constant in the considered temperature range, but the thermal emission described by a blackbody spectrum inevitably increases with increasing absorber temperature and shifts its maximum closer to the solar wavelengths according to Wien's displacement law. It even extends into the visible spectrum at very high temperatures (c.f. the red glow of hot iron). The consequence of this behavior is that, to develop appropriate high-temperature absorbers, the focus must be on maintaining high absorptivity at high temperatures, while heat emission-related losses caused by an increasing emissivity at high temperatures are increasingly difficult to be avoided.

However, if the relative temperature difference to ambient temperature is used to drive a thermodynamic cycle for a working machine, a higher receiver fluid outlet temperature causes the Carnot efficiency η_C of the working machine (compare Eq. (7.4)) to grow monotonically. As a result the system efficiency η_{sys} being the product of the thermal collector efficiency and the Carnot efficiency, i.e.$\eta_{sys} = \eta_{rec,th} \cdot \eta_C$, exhibits a local maximum, which exceeds 300–400 °C for line focusing systems as depicted in Fig. 7.3. Thus, the potential capability of point-concentrating systems to achieve exergy efficiencies than line-focusing systems originates from two causes. At high concentration ratios the area-specific heat losses decrease since the concentration factor is in the denominator. Secondly, the Carnot efficiency increases with increasing temperature which, according to the first law of thermodynamics, allows more work to be extracted. It must be emphasized that the development of vacuum receivers with much smaller thermal losses than assumed in the Fig. 7.3 lead also to higher optimum working temperatures.

7.1.2 Radiation Density Distribution of the Sun and the Circum-Solar Radiation

The direct normal irradiance (DNI) describes the irradiance on a surface perpendicular to the vector from the observer to the center of the sun. The DNI is the essential input responsible for the yield of concentrating collectors. The angular distribution of the direct and circumsolar radiances, *i.e.* the sunshape, as detailed in Chap. 2, has a significant effect on the performance of concentrating collectors especially for high geometric concentration factors C_g. Depending on atmospheric conditions, a significant fraction of solar radiation coming from the sun is forward scattered by atmospheric constituents and forms the circumsolar region.

Concentrating collectors do not only collect the solar radiation directly originating from the sun's disk, but also a fraction of the circumsolar radiation. The absolute fraction depends strongly on the collector type and the concentrator design and quality.

7.2 Line Focusing Systems

Line-focusing collector systems have the largest market penetration of all high-temperature collector systems, such that more than 80% of the world's operating CSP power plants are based on line-focusing collector technologies, of which more than 90% are parabolic trough power plant types. In the following sections, we focus first on a full description of the parabolic trough collector's (PTC) for high-temperature applications, with the aim of describing the collector performance. Here, optical efficiency, incident angle modifier, thermal losses as key factors are influenced by design and material choices. Similarly, we will describe later the more variable designs of linear Fresnel collectors (LFC). Although both, parabolic trough, and linear Fresnel, are line concentrating technologies, the optical paths in the concentrators and the spatial distribution of optical conversion and heat transfer in the receivers differ.

7.2.1 Parabolic Trough Collectors

7.2.1.1 Optics of a Parabolic Trough

A parabolic trough solar collector (PTC) consists of mirrors of parabolic cross-section and a receiver element, which is composed of a tubular receiver containing

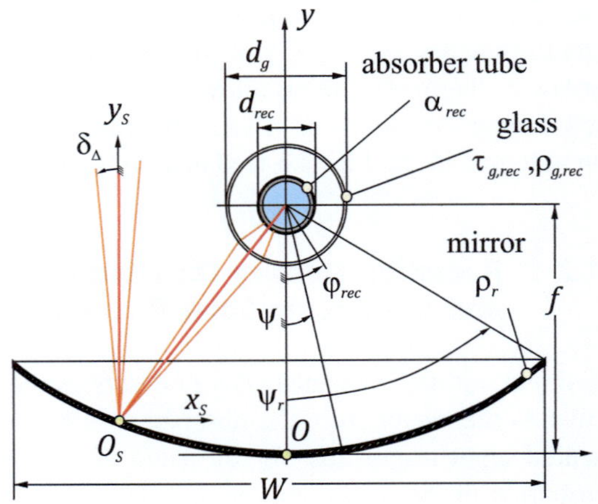

Fig. 7.4 Cross section of a parabolic trough collector (PTC) and definition of the geometry

the heat transfer fluid, which is concentrically enveloped by a glass tube. The cross-section of such a generic PTC is sketched in Fig. 7.4. In this simple case of a concentrator of a parabola in two-dimensional space, the light rays incident parallel to the *y*-axis are focused by the parabola contour to the focal point *F* of the parabola, as shown in the Eq. (7.6). According to geometric algebra, the following relationship for the contour is obtained

$$y + f = \sqrt{x^2 + (y - f)^2} \rightarrow y = \frac{1}{4f}x^2. \tag{7.6}$$

The incident beam of the solar radiation given by the direct normal irradiation including the circumsolar irradiation with an angular width δ_Δ first strikes the reflector of aperture width W, rim angle ψ_r, and focal length f. There, the incident beam is redirected towards the receiver with reflectivity ρ_r of the mirror, which is also a function of the incident angle θ. The change in direction of the beam at the reflector changes the distribution of the radiance in the direction of the receiver. At the glass envelope of the receiver, a part of the beam, ($\rho_{g,rec}$), is reflected dependent on the incidence on the glass tube into the environment, and the other part of the radiation continues in the direction of the receiver with a transmittance ($\tau_{g,rec}$) of the glass cover. A spatially distributed radiation density—the fraction which hits the receiver tube—then impinges on the absorber surface and is converted into heat by the absorptivity (α_{rec}). The process of the photothermal conversion is complex, and the distribution of the solar energy flux on the absorber is the boundary condition for heat transfer calculation. The heat is transferred through the absorber tube wall to the heat transfer fluid (HTF) circulating through the absorber tube.

For the determination of the thermal collector yield in the context of a power plant or process heat facility, the use of simple models calculating the energy balances is sufficiently accurate, since it aims at the determination of integral quantities, such as heat yields or losses per meter collector length. However, the design

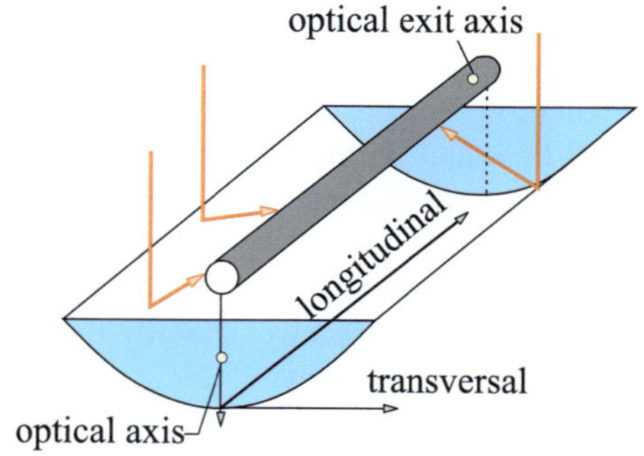

Fig. 7.5 Schematical design of a parabolic trough collector

of a combination of reflector and receiver, or the detailed modelling of heat transfer to a two-phase flow pattern requires a detailed determination of local temperatures, heat fluxes and heat losses, for example to identify thermally highly loaded spots and to adjust or optimize the component design accordingly.

When the parameter focal length f of the parabola increases, a flatter parabola is the consequence. In practice, this means a smaller contour length and thus less material requirement for the same mirror width (i.e. constant concentration ratio). On the other hand, the distance to the focal point increases, so that possible mirror errors have a larger effect. Also, the irradiated section of the absorber tube becomes smaller.

Based on this optical element collectors are designed using a long reflecting mirror ideally with the cross section of this parabola. Figure 7.5 shows this design schematically defining the optical axis along the absorber tube and the longitudinal and transversal directions.

Looking at Fig. 7.4 this shows the parameters of a cross section. All incident rays with a projection into the transversal plane which is parallel to the optical axis are reflected by the mirror to the focal point where the absorber is located. However, depending on the longitudinal incidence projection the rays end on a different position along the tube. In fact, for large longitudinal incidence angles rays might be reflected to a focus outside the finite length of a parabolic trough. This is considered as end loss.

The end loss formula for a collector trough of finite length L and focal length f is

$$\eta_{endloss}(\theta) = 1 - \frac{f}{L} \cdot \left(1 + \frac{W}{48 \cdot f^2}\right) \cdot \tan\theta. \tag{7.7}$$

where θ is the incidence angle on the tracked aperture, which is the angle in the longitudinal optical plane.

Basic Parameters of a Parabolic-Trough Collector

As solar concentrating devices, PTCs require solar tracking systems to modify their position with the changing apparent sun position in the sky from sunrise to sunset. Movement of this type of solar collector needs only one degree of freedom, an axial rotation with the axis ideally in North–South (*NS*–) or East–West (*EW*–) orientation. The position of the PTC must be such that the sun vector, the collector focal line, and the normal to the collector aperture plane are in the same plane. For PTC the resulting incidence angle is the angle between surface normal and solar beam direction. It affects the energy reaching the receiver pipe as mainly the cosine effect reduces the incident solar flux on the reflector. Diffuse solar radiation reflected onto the receiver tube by the concentrator is negligible compared to the concentrated direct part.

The most important PTC parameters are the geometric concentration ratio C_g, acceptance angle (δ_a), rim angle ψ_r, and peak optical efficiency at normal incidence $\eta_{opt,0}$. These parameters are explained in the following paragraphs. The geometric concentration ratio, C_g, is the ratio between the collector aperture area and the total absorber tube area (see Eq. (7.8)). This concentration ratio is usually about 25, although theoretically, the maximum is in the order of 70. High concentration ratios are associated with higher working temperatures. The geometric concentration ratio is given by Eq. (7.8):

$$C_g = \frac{W \cdot L}{\pi \cdot d_{rec} \cdot L} = \frac{W}{\pi \cdot d_{rec}}, \tag{7.8}$$

where d_{rec} is the outer diameter of the receiver steel pipe, L is collector length, and W is the parabola width. The acceptance angle, δ_a, is the maximum angle that can be formed by two rays on a plane transverse to the collector aperture in such a way that, when they are reflected by the parabolic mirrors, they intercept the absorber pipe. The wider the collector acceptance angle is, the less accurate the sun tracking system has to be. Small acceptance angles as explained in the optics Chap. 3 are associated with high concentration ratios and require the installation of very accurate solar-tracking systems leading often to higher costs. To capture all direct beam solar radiation the minimum acceptance half-angle should be at least the solar divergence half-angle $\delta_s = 0.27$ which contains the solar disk as seen from the Earth. It must be chosen by the designer which part of the circumsolar radiation shall be captured as well. Recommended acceptance angles for commercial PTCs are in the range 1–2°. Higher concentration with smaller acceptance angles demands very accurate and more expensive solar-tracking systems, while wider angles with lower concentration lead to lower working temperatures of the heat transfer fluid.

The relevant parameters identified for optical gain are the dimensions of the trough given by the rim angle ψ_r, and the geometric concentration ratio C_g, which are defined by:

$$\frac{f}{W} = \frac{1 + \cos \psi_r}{4 \sin \psi_r}; \; C_g = \frac{W \cdot L_{rec}}{\pi \cdot d_{rec} \cdot L_{rec}} = \frac{\sin \psi_r}{\pi \sin \theta}, \tag{7.9}$$

in which θ is the incidence angle on the aperture. The acceptance angle δ_a describes the maximum angle that can be formed by two rays in a plane normal to the collector aperture such that, when reflected from the parabolic mirrors, they just hit the absorber tube. Small acceptance angles thus imply large concentration factors that require the installation of accurate sun tracking systems. The minimum acceptance angle is ($2 \cdot \delta_S = 0.532°$), which is the average solid angle at which the solar disk is seen from Earth. Therefore, any parabolic trough collector (PTC) with an acceptance angle less than ($2 \cdot \delta_S$) always looses some direct solar irradiance. Real acceptance angles for commercial PTCs are in the range of 1–2°. The outer diameter of the absorber (d_{rec}) must be larger than the diameter of the light cone to prevent rays missing the absorber tube. When we only consider the sun disk as source (excluding circum-solar), the required minimum outer diameter of the absorber can be calculated by

$$d_{rec} = 2\left(\frac{W^2}{16f} + f\right)\sin\theta_s. \tag{7.10}$$

The rim angle (ψ_r) is the angle between the reflected rays from the reflector edge and y-axis, wich is given by

$$\psi_r = \begin{cases} \arcsin\left(\frac{8(W/f)}{(W/f)^2} + 16\right) & \text{for } 0° < \psi_r \le 90° \\ 180 - \arcsin\left(\frac{8(W/f)}{(W/f)^2} + 16\right) & \text{for } 90° < \psi_r < 180° \end{cases}. \tag{7.11}$$

The upper half of the receiver tube absorbs solar radiation directly, while at the bottom side the concentrated solar beam focused by the reflector mirrors is optically converted by the receiver.

A further consequence is that the permissible acceptance angle δ_a also depends on the distance between absorber and reflector. In the case of the parabolic contour, this is largest at the edge or rim. The average value over the entire contour segment of a parabolic trough reaches a minimum if the rim angle (ψ_r) of the parabola is $\psi_r = 120°$. However, parabolas with this rim angle have a strong curvature at the vertex, which can lead to problems during production. Furthermore, the distance at the vertex can become too narrow for the installation of the glass envelope tube or mirror cleaning. Therefore, parabolas with smaller rim angles are often chosen.

For cylindrical tube receivers, usual rim angles are in the range 70–110°. Smaller rim angles are not advisable because they reduce the aperture surface and lead to comparatively large focal lengths. Rim angles over 110° need a lot of reflector material because the surface is enlarged without significantly increasing the area of the aperture plane.

Optical and thermal losses in a parabolic-trough collector. Optical losses characterized by the peak optical efficiency are very important in a PTC because they are about 25% of the total solar flux incident on the PTC aperture plane. Optical losses are associated with the following four parameters, illustrated in Fig. 7.6.

Reflectance $\rho_r < 1$ of the collector reflective surface. Typical solar spectral reflectance values of the best clean silver-based low iron glass mirrors are around 0.93–0.94. Reflective films have usually a specular reflectance $\rho_r < 0.9$.

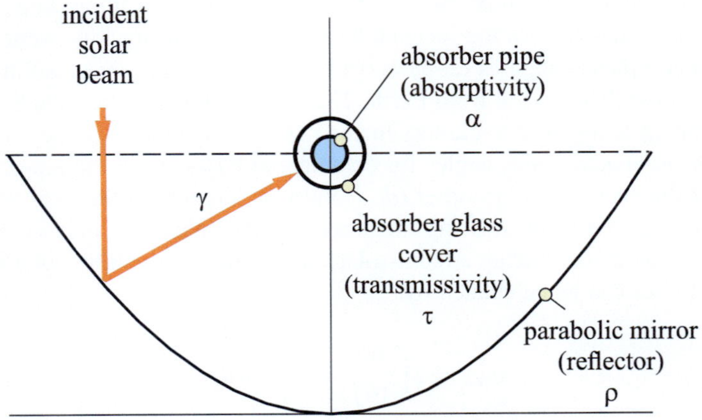

Fig. 7.6 Parameters impacting the optical efficiency ($\eta_{opt,0}$) of a PTC

Intercept factor γ. This is the fraction of the direct solar radiation reflected by the mirrors that finally reach the active surface of the receiver tube. The other part reflected by the mirrors does not reach the receiver pipe due to either microscopic imperfections of the reflectors, macroscopic errors in the parabolic-trough concentrator shape (e.g. inaccuracies during assembly), mechanical deformation of the PTC, flexible bellows, or shadowing by the receiver tube supports. Therefore the intercept factor for 0° incidence angle of the solar beam radiation can be calculated as the product of two parameters: a factor γ_1 related to optical imperfections (micro-roughness, errors in the parabolic shape, mechanical deformation (i.e. bending plus torsion) of the support structure) and a factor γ_2 related to shadowing by the receiver tube supports, the active length factor of the receiver tube, which accounts for the shadowing effect of the flexible bellows and their protective shield onto the steel absorber tube. Typical values are $\gamma_1 < 0.97$ and $\gamma_2 < 0.97$, giving an overall intercept factor γ in the range of 0.91–0.93 for high-quality PTCs.

Transmittance of the glass cover $\tau_{g,rec}$. The steel receiver tube is often covered by a glass cylinder to reduce thermal losses. A fraction of the direct solar radiation reflected by the mirrors onto the glass cover of the receiver pipe is unable to penetrate it. The transmittance $\tau_{g,rec}$ of the cylindrical glass for the relevant incidence angles on the glass surface is typically very close to the normal transmittance $\tau_{g,rec} = 0.93$ and can be increased up to 0.96 by antireflective coatings applied on both sides of the glass cover.

Absorptance of the receiver selective coating α_{rec}. This parameter quantifies the amount of energy absorbed by the steel receiver pipe divided by radiation reaching it. This parameter is typically 0.95 for receiver pipes with a cermet selective coating, and slightly lower (0.92) for pipes coated with black nickel or chrome.

Multiplication of these four parameters (intercept factor, mirror reflectance, glass transmittance and receiver absorptance) for an incidence angle $\theta_i = 0°$ gives the peak optical efficiency of the PTC $\eta_{opt,0}$:

$$\eta_{opt,0} = \rho_r \tau_{g,rec} \alpha_{rec} \gamma. \tag{7.12}$$

$\eta_{opt,0}$ is usually in the range of 0.74–0.79 for clean, good-quality PTCs. Taking the four optical parameters included in the peak optical efficiency into consideration, it represents the percentage of beam solar radiation converted to heat for normal incidence.

The effect of non-normal incidence angles is quantified by the incidence angle modifier $K(\theta)$, abbreviated *IAM*. This accounts for the fact that the change in the angle of incidence of direct solar radiation (θ_i) affects the four optical parameters mentioned previously and the collector's useful aperture area. The incidence angle modifier (*IAM*) corrects for these additional reflection and absorption losses. This parameter is usually provided by the manufacturer and results either from extensive calculations supported by validation experiments or as an empirical correlation to experimental data on a parabolic trough collector. The *IAM* $K(\theta)$ includes all optical and geometric losses in a PTC generated by an incidence angle $\theta_i = 0°$, without considering the $\cos\theta$ effect, and without accounting for the end-loss factor (7.7). Thus, we get for the incidence angle dependent optical efficiency $\eta_{opt}(\theta_i)$:

$$\eta_{opt}(\theta_i) = \eta_{opt,0} \cdot K(\theta_i)\eta_{endloss}(\theta_i). \tag{7.13}$$

This formulation contains by its factorization some simplifications, e.g. that the absorption or transmission of incident radiation on the absorber tube or on the cover glass is constant with circumferential position. However, the simple structure of Eq. (7.13) and the comparatively small deviations from the exact result make it ideally suited to determine different collector field or solar circuit arrangements with regard to the thermal yield for any location quickly and with good accuracy.

The incidence angle modifier is usually given by a polynomial equation such that it is equal to 0 for high θ_i (>85°), and to 1 for $\theta_i = 0°$. A representative incidence angle modifier for a PTC similar to a EuroTrough collector is given by:

$$K(\theta_i) = \max\left[0, 1 - \frac{7 \cdot 10^{-4} \cdot \theta_i + 36 \cdot 10^{-6} \cdot \theta_i^2}{\cos\theta_i}\right], \tag{7.14}$$

where θ_i is given in degrees.

However, this simple optical efficiency approach has some drawbacks. Although thermal collector yield calculations consider daily and annual variations of a typical meteorological year (TMY) and the associated direct normal irradiation (DNI), they do not include the influence of variable circumsolar radiation (CSR) on the intercept factor. This has to be modelled in more detail. What is more important from a technical point of view, especially for the constructive design of a collector and of the receiver, is a determination of local circumferential and longitudinal temperature distribution on the receiver (absorber) surface. This is the basic pre-requisite for the design of the receiver, the selection of suitable structural and functional materials, for example the selective absorber, and is dependent on the dimensioning and positioning of spacers, brackets, pylons carrying the support structure etc. and thus the structural design of the receiver relative to the parabolic reflector. The formula Eq. (7.13) does not produce any location-dependent power values of the radiation density at the thermally loaded components. This requires local modelling for example using ray tracing techniques as described in Chap. 3.

7.2.1.2 Thermal Performance

Regarding the thermal performance the integral approach models characterize the heat gain of the fluid along the flow path through the receiver tube using as heat source the optically converted heat minus the radiative and conductive-convective heat losses through via the glass envelope and the bracket and spacer elements. An energy balance model of heat loss in a vacuum receiver has been described for example in Burkholder & Kutscher (2008, 2009). In a similar way, even air-filled receivers many be treated, via engineering correlations outlined in Chaps. 4 and 5. Otherwise measured parameter determined in a receiver test can be used. Typically, the empirical thermal loss function is provided by the collector manufacturer in form of a polynomial function of a temperature difference between mean fluid or absorber surface temperature in the receiver and the ambient temperature. It contains linear terms representing conductive heat losses, quadratic and/or fourth order terms approximating convective and radiation heat losses. For a real local determination of heat losses to the ambient the temperatures at a discrete position z of the collector is obtained as a function of the flow rate, the flow regime and the imposed solar irradiation using computational fluid dynamic (CFD) models.

Thermal losses are very important in PTCs because they significantly affect the overall collector efficiency. The total thermal loss of a PTC $\dot{Q}_{loss}$ is due to radiative, convective, and conductive processes from the receiver tube to its surrounding. Although they all are well-known mechanisms and can be modelled explicitly, often—not knowing the precise design data—in testing they are experimentally determined under real solar conditions at several temperatures. These experimental results are then processed to determine the best-fit coefficients b_1 and b_2 *or* b_1 and b_4 depending on which fit is better. Usually, evacuated tube receivers are characterized by b_1 and b_4:

$$\frac{\dot{Q}_{loss}}{A_{ap}} = \dot{q}_{loss} = b_1 \cdot \Delta T + b_2.\Delta T^2 + b_4 \cdot \Delta T^4, \tag{7.15}$$

where b_1,b_2 and b_4 are coefficients obtained from the experimental values, and ΔT in (°K) is calculated from the difference of the average temperature of the working fluid in the collector (T_m) and the ambient temperature (T_{amb}).

Although now an international standard exists for specifying this testing previously to this the manufacturers delivered their receivers with an equation different from (7.15) to calculate the thermal losses. Today's high-temperature PTCs are provided with evacuated receiver tubes, thus avoiding convection losses between the steel pipe and its glass cover.

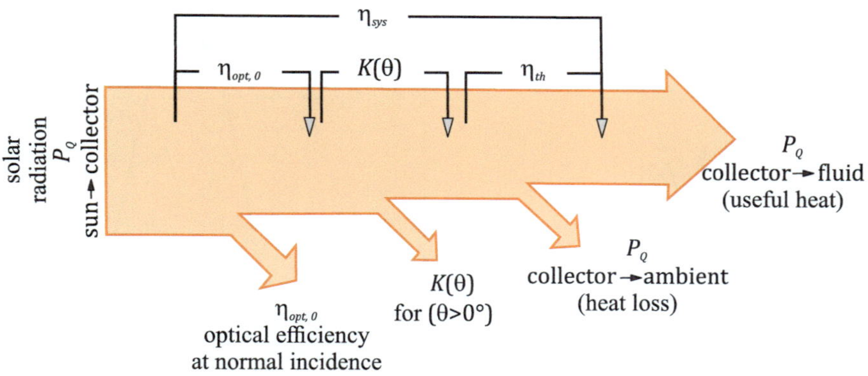

Fig. 7.7 Energy flows for PTC

With the determination of the heat losses, the entire loss chain of a collector is thus mapped, which results from the optical efficiency at perpendicular irradiation ($\eta_{opt,0}$), the consideration of the IAM ($K(\theta)$) and the inclusion of the heat losses, which is depicted in Fig. 7.7.

This allows the specific collector yield to be calculated at any time of the day depending on the collector orientation, for example north–south (NS) or east–west (EW). And from the integration over the collector area and over the day, the total energy yield results. Figure 7.8 shows the calculated specific power as a function of solar time for a Eurotrough-100 collector at the Seville site on a summer and winter day for NS and EW orientations, respectively. The yield in the NS orientation is greater in the EW direction over the year, with this being an advantage especially in the summer. Only in winter is the EW orientation superior to the NS orientation in terms of yield, but the irradiance values are significantly lower in winter.

7.2.1.3 Optical Errors and Effects on Collector Efficiency

For small concentration factors, the optical errors of the concentrator and the mirror material have only a minor effect. For C_g below 4–5 there is still a usable portion of the diffuse radiation which decreases with increasing concentration. On the other hand, the higher the concentration, the more important is the precise manufacturing of the concentrator shape and the selection of mirror materials with small optical spread. The quantification of the optical quality is given by the width σ of a Gaussian beam. The larger the value of σ the wider the beam spread. And diffusing surface or deformed surface may widen up the beam spread. Good concentrators usually achieve a total optical error σ_{opt} of a few mrad only. In the following sections we describe the methodology of coming to a total error and show some representative results of optical simulations.

Apart from optical quality of the mirrors which has a permanent effect, there may be also temporary effects due to soiling of mirrors. Soiling is the accumulation

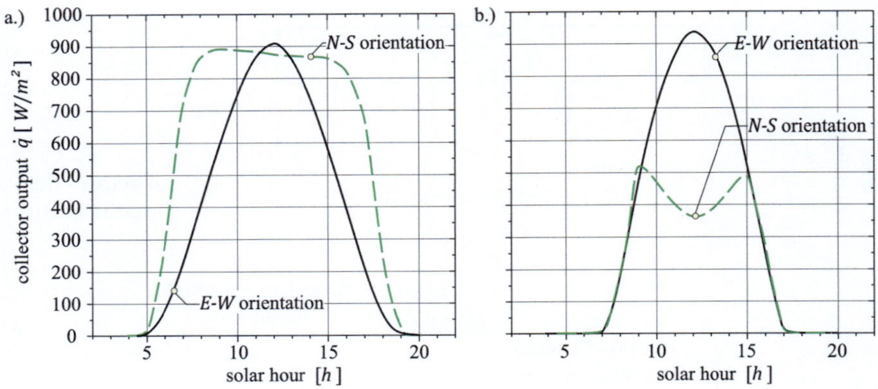

Fig. 7.8 Simulated daily thermal output of a Eurotrough-100 parabolic trough collector located in Seville/Spain ($\Phi = 37°14'$), on a summer (**a**) and winter (**b**) clear day and with two different PTC orientations

of dirt (dust particles, aerosols, salt particles, biological growth) on a mirror and may be at least partially removed by washing/cleaning processes. Especially in dry areas with little vegetation the concentration of fine dust particles in the air is high which come from wind erosion.

The accumulation of particles sticking to the mirror surface depends on the particle size and chemistry, the adhesive forces of the surface (characterized by the contact angle) and environmental parameters (humidity, wind velocities). Soiling leads to light scattering and absorption and hence to a drastical reduction of specular reflectance of a mirror, which in turn is characterized by an additional beam spread σ. From geometry considerations the impact of dirt particles on the reflectance varies with incidence angle. Different soiling processes can be parametrized very well with a model developed by Heimsath and Nitz (2019) even for different wavelengths as illustrated in Fig. 7.9.

Imperfections in material and construction, beam spread of the sun, soiling all need to a lowering of the optical efficiency. However, the sensitivity is different for large and small parabolic trough collectors and also differs to Linear Fresnel collectors as indicated in Fig. 7.10. The reason is that these collectors have different focal lengths, and the displacement of a beam is the product of an angular error and a length dimension, which is in the order of the focal length.

Optical Errors Affecting PTC Collector Performance

As introduced in Sect. 3.4.6 any deviation of from the ideal mirror receiver geometrical configuration strongly impacts the intercept factor γ. Theses geometrical inaccuracies are expressed by the term optical errors and its standard deviation. To achieve a low overall optical error σ_{opt} the entire chain from manufacturing to installation requires a stringent quality assurance posing a challenging task to ensure a high precision equipment. To the standard deviation of the overall optical

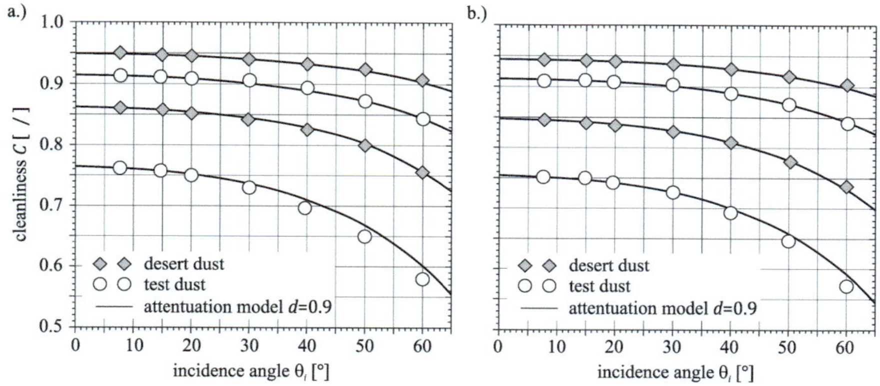

Fig. 7.9 Comparison of model (lines) and measured values (symbols) at different wavelengths as a function of incidence angle (θ_i) (left:λ = 455 nm, right λ = 533 nm). Crosses correspond to mirrors soiled with desert dust, circles correspond to samples soiled with test dust modified from Heimsath and Nitz (2019)

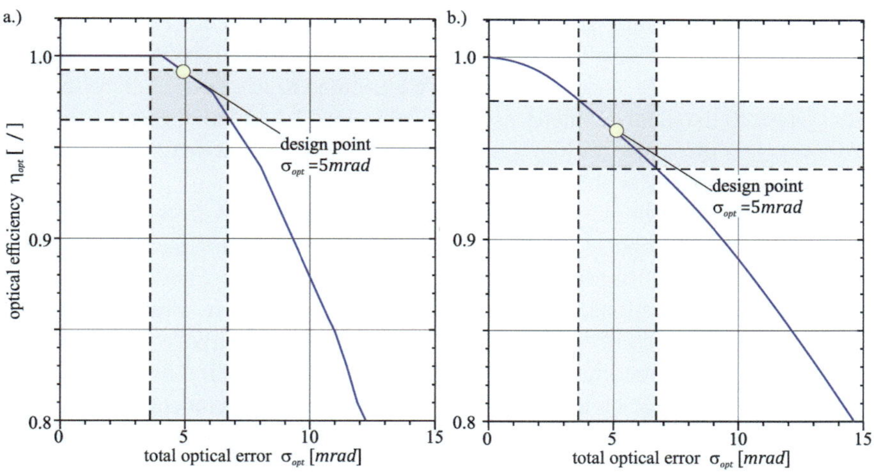

Fig. 7.10 Impact of total optical error on the optical efficiency (Heimsath, Bern, van Rooyen, & Nitz, 2014a, 2014b, 2014c); Heimsath, Cuevas, Hofer, Nitz, & Platzer, 2014a, 2014b, 2014c). **a** Large scale PTC and **b** a small scale LFC. Both collectors were designed for 5 mrad optical efficiency (Design Point). The effect of an increase or decrease of 1 mrad shape accuracy is indicated by grey lines

error σ_{opt} several factors contribute such as the standard deviations of the specularity error σ_{spec}, the slope error σ_{slope}, the misalignment or displacement error σ_{dis} and the tracking error σ_{trac}, which are depicted in Fig. 7.11. The specularity error σ_{spec} describes the glass quality or exactly to the degree of which the collected rays respect to the law of reflection. Macroscopic surface deviations in

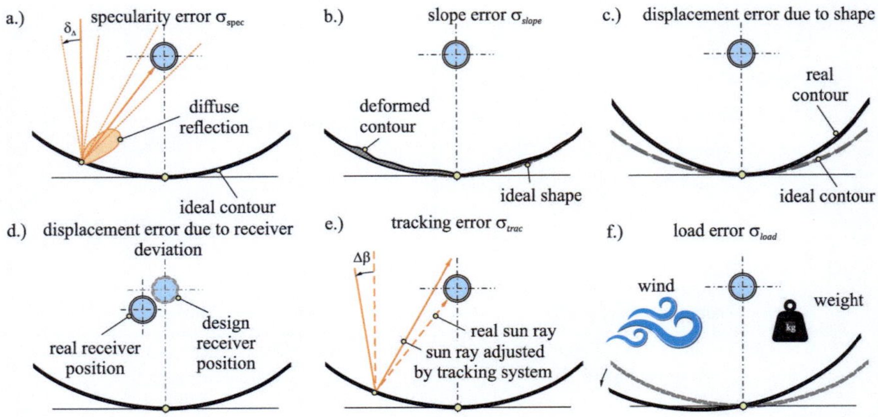

Fig. 7.11 Geometrical error types appearing in a PTC collector

position and slope are summarized as slope errors σ_{slope} are directly linked to the manufacturing quality and to the local ripples or the aberrations on the mirrors. Displacement errors σ_{dis} are caused either by a misalignment of the facets that compose the collector or a miscellaneous arrangement of the receiver with respect to the mirror. These displacement errors are attributed to geometrical deformations with respect to the ideal geometry as indicated in Fig. 7.11c and d. Finally, tracking errors σ_{trac} as illustrated in Fig. 7.11e occur either due to incorrect measurement or aging of the drives or their gearboxes.

To quantify and, moreover, to separate operational effects from quality assurance arguments mandatory to design, characterize and assess the collector performance all arguments associated with fabrication, erection and mounting are separated from operational impacts. This allows to analyse already computationally the robustness of a PTC design by executing sensitivity studies for all parameters affecting the intercept factor. Shape errors originating from operational loads such as bending or torsion of the structures due to wind or weight loads occur especially in periods with large ambient gradients (temperature, wind cloud, fields) are termed loading errors.

Thereby, averaged over time and over the entire collector array of collectors, all these errors can be assumed to be approximately linear independent and exhibiting a Gaussian distribution according to Rabl (1985). Additionally, all standard deviations of the optical errors can be summarized in the form of a sum of the contributing individual terms,

$$\sigma_{opt}^2 = \sum_i^n (a_i\sigma_i)^2 = \sigma_{spec}^2 + \left(2\sigma_{slope}\right)^2 + \sigma_{dis}^2 + \sigma_{trac}^2, \tag{7.16}$$

in which a_i represents a weighting factor. The beam spread of the sunrays, when interacting with the imperfect mirror (concentrator), is represented by a weighting factor of two and thus twice all other the standard deviations. This refers to all mirror components such as mirror shape, mirror support position and mirror support angle. Since the sun shape represents also a statistical random error independent of the collector design a so-called standard deviation of the total error σ_{tot} is used

$$\sigma_{tot}^2 = \sigma_{opt}^2 + \sigma_{sun}^2. \tag{7.17}$$

The most commonly used, straight forward and accurate method to quantify the standard deviation of the total optical error is statistical ray tracing (MCRT), in which normal distribution functions of the angular deviations of each contributing ray are evaluated by means of using perfect geometrical optics. Once the local deviations fall below a predefined threshold the solution is converged, see e.g. Pottler et al. (2014). The central advantage of this Monte-Carlo concept is that all statistical deviations of the contributing effects can be condensed in one single parameter, the so-called intercept factor γ (see Eq. (7.12)). This intercept factor γ is then a function of the parabolic trough rim angle ψ_r, the longitudinal incidence angle θ_i and the total beam width σ_{tot}, as qualitatively shown in Fig. 7.12a. An increasing beam width yields a monotonic decrease of the intercept factor. Bendt Rabl, Gaul, and Reed (1979) interpret the influence of the incidence angle on the beam pattern as the apparent broadening of the solar disk with increasing incidence angle, resulting in an increase in the path of the reflected beam from the mirror to the absorber. Larger total beam widths, have a particular effect at greater incidence angles, which becomes relevant in the winter period when the incidence angle at noon for the north–south oriented trough collector exceeds values $\theta_i >$ 30° at widths $\delta > 23.5°$, which are additionally associated with a higher CSR. Figure 7.12b depicts that the intercept drastically declines for increasing incident angles.

Robustly designed and professionally installed parabolic trough collectors today achieve an overall quality of σ_{tot} = 6–6.5 mrad. This requires extensive quality assurance steps and control measures, which already start with the single part production in series production. Proof is provided by deflectometric measurements for mirrors and modules and 3D point measurements for metal frame structures. This process continues during installation on the solar field where the focus is to achieve a balanced quality across all components and assembly steps involved. At the end, after the assembly is completed, the achieved geometric quality is measured in the field by means of photogrammetry, deflectometry, tachymetry, inclinometers providing real data. The real geometry is then analysed again using mostly by means of statistical ray tracing models applying standard deviation parameters σ_i, which is shown in Fig. 7.12c broken down by the individual components of a collector according to specifications of the manufacturer or operational experience. The Fig. 7.12c illustrates that for a real PTC collector slope, displacement and sun-shape statistical standard deviations are of the same order of magnitude, while tracking deviations contribute usually to a much smaller fraction. Finally, Fig. 7.12d depicts a computed ray-tracing result based on measurement of a real PTC surface highlighting reflected rays in a 1 m width section with strong

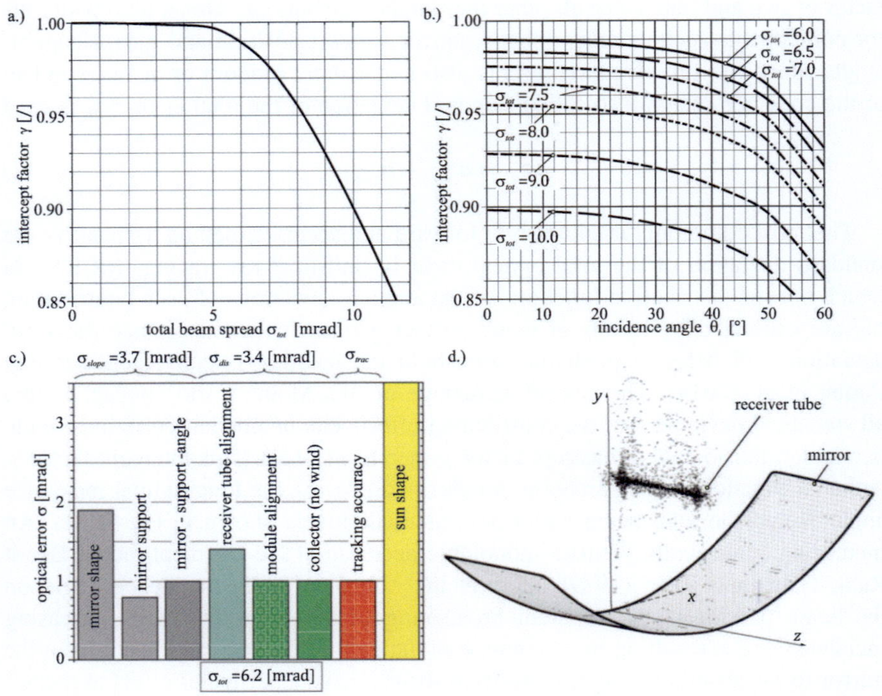

Fig. 7.12 **a** Computed intercept factor γ against total beam spread σ_{tot}; **b** Calculated intercept for different total beam spreads σ_{tot} as function of the incident angle θ_i on intercept factor; **c** share of the various optical error sources on the total beam spread σ_{tot} at $\theta_i = 0°$, **a–c** modified from Pottler et al. (2014); **d** simulated behaviour of the reflected rays in a 1 m width section with strong deformation modified according to García-Cortés et al. (2012). All graphs refer to the EuroTrough collector geometry

deformation. The graph in coincidence with operational experience shows that the positions of the greatest deviations depend strongly on the installation type of the mirror facets; many designs exhibit larger deviations occurring at the outer borders of the collector mirror (large x-values). Also, operational data for several plants reveal that larger deviations compared to an ideal optics is obtained for a collector in a horizontal position rather than in a vertical position. Stochastic effects are particularly relevant for the design of the collector, but during operation or installation of systems, systematic deviations occur, which in some cases have a more significant negative influence on the collection factor. Systematic deviations mainly include asymmetric shape deviations such as too shallow supports, deviations of the centre of gravity from the axis of rotation and misalignment of the tracking sensor. Symmetric systematic effects can also reduce the pickup factor, such as mirror fields that are too open or too closed and/or concentrator shapes.

Each collector design must be analysed with respect to the sensitivity of manufacturing, measurement deviations of the drive or inclination sensors, actuator deviations or bending of the collector support structure, which are reflected in the collection factor. Finite element methods (FEM) are typically used to analyse the strength of the support structures with the associated deformations in terms of bending and/or torsion of the structures, for more details see (Meiser, Schneider, Lüpfert, Schiricke, & Pitz-Paal, 2017). With the help of numerical calculations, it is possible to compensate for systematic alignment deviations, for example, by an offset in the tracking angle.

Simulated Optical PTC Collector Characteristics

PTC performance at Normal Incidence ($\theta_i = 0°$).

Figure 7.13a shows a simulated circumferential distribution of local concentration factors (C_L) at normal irradiance ($\theta_i = 0°$) for a PTC collector of realistic dimensions at different circumsolar irradiance fractions. First, the local concentration factor C_L shows four different ranges. A so called shelter zone (1) with a high but reduced C_L caused by the shadowing of the receiver tube, then a segment (2) with a significant increase of the concentration ratio up to a maximum, followed by a Sect. 3 characterized by a reduction of the local irradiance and finally a receiver Sect. 4, which only sees the sun ($C_L = 1$) and is called the isolation zone. If we first consider a $CSR \approx 0$, we find that significant local power density gradients over a few degrees (here for $\varphi_{rec} > 62°$) occur in circumferential direction. Furthermore, there is a significant decrease of the local concentration factor C_L in the shelter domain compared to the peak power position. For the considered receiver at $\varphi_{rec} \approx 89°$ no radiation reaches the receiver ($C_L = 0$), before with increasing φ_{rec} a value of $C_L = 1$ establishes. Analyzing the impact of a variable CSR, the local concentration factor in regions 1 and 2 falls strictly monotonically with increasing CSR. Thus, the optical efficiency of the PTC collector $\eta_{opt,0}$ decreases steadily with increasing CSR, which is immediately apparent, because the angular angle variation of the sun θ_s increases and thus the path of solar irradiation through the atmosphere (SOD). The reduction of $\eta_{opt,0}$ in the example is about 22% at an incidence angle $\theta_i = 0°$ from $CSR = 0$ (no atmospheric scattering) to a $CSR = 0.8$ (extremely hazy sky). At the same time, however, the power gradients expressed by C_L also flatten out in domains 1 and 2, and a shelter effect is hardly observed at high CSR. Conversely, the increase of CSR in region 3 leads to a minor increase of the local concentration ratio compared to $CSR \approx 0$.

The graph also shows that for $CSR > 0$ the optical efficiency $\eta_{opt,0}$ increases with increasing receiver diameter. This is because the larger the diameter, the larger the radial intercept angle (θ_γ), and thus more reflected rays are received by the absorber. It can also be seen from the figure that the larger the CSR value becomes, the smaller the optical efficiency is. The difference of optical efficiency between two CSR s decreases with the increase of receiver diameter, indicating that the optical efficiency is less sensitive to the CSR for a PTC with larger receiver diameter.

With the help of local determination of the radiation incident on the receiver, the geometry of the PTC collector (receiver diameter-d_{rec}, focal length-f, aperture

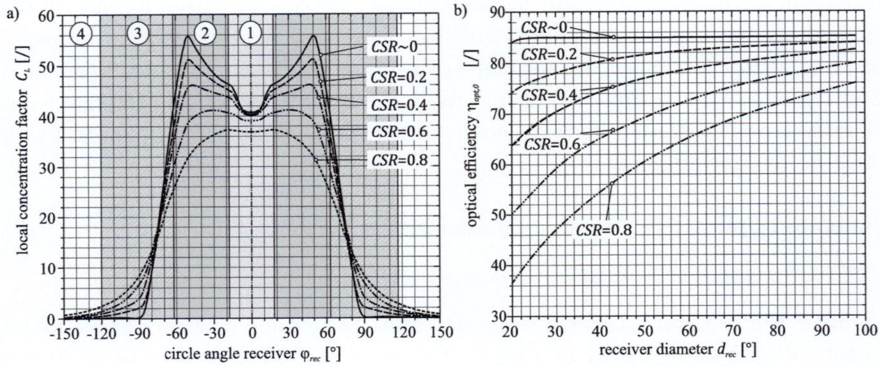

Fig. 7.13 Computed optical properties of a PTC collector for different CSR modified from Zou et al. (2017) (SEGS-LS-2PTC collector: $W = 5$ m, $f = 1.84$ m, $L_{rec} = 7.8$ m, $d_{rec} = 0.07$ m, $d_g = 0.115$ m, $\alpha_{rec} = 0.96$, $\tau_g = 0.95$, $\rho_r = 0.93$); **a** local concentration factor C_L as function of azimuthal receiver angle φ_{rec}; **b** optical efficiency $\eta_{opt,0}$ for different receiver diameters d_{rec}

width-W) can now be optimized to achieve an optimum optical efficiency. Analyzing the optical efficiency as a function of the receiver diameter, as shown in Fig. 7.13b, we see that there is a critical minimum diameter at $CSR \approx 0$. This is because at $CSR \approx 0$ all rays come from the solar disk (with negligible aureole) whose radial angle is 4.65 mrad ($\delta_s = 4.65$ mrad) and using Eq. (7.10) we can calculate this analytically. A diameter larger than the critical minimum diameter does not lead to any increase in the optical efficiency when $CSR \approx 0$. Conversely, for smaller values of the receiver diameter, some of the reflected rays escape around the absorber tube, which reduces the optical efficiency.

Figure 7.14a shows the distribution of the radial intercept angle (θ_γ) as a function with absolute abscissa value x_A. The intercept angle decreases monotonically with increasing the abscissa up to W, which means that the larger the aperture width W is, the smaller the average θ_γ becomes, whereby more rays do not reach the receiver and as consequence the optical efficiency $\eta_{opt,0}$ decreases with increasing aperture width for a given focal length f. This behavior requires a larger PTC receiver diameter d_{rec} to maintain a large optical efficiency. The effect of focal length f on the optical efficiency for $CSR = 0.4$ in Fig. 7.14b shows that the optical efficiency initially increases and reaches a local maximum and then decreases monotonically with further increase of the focal length. The cause is the dependence of the radial intercept angle (θ_γ) on the focal length f at different reflection points A shown in Fig. 7.14b. For $x_A \geq 1$, the radial intercept angle first increases with increasing focal length before it reaches a local maximum value. Beyond this local maximum θ_γ decreases monotonically with increasing f. Further, Fig. 7.14b shows that the intercept angle exhibits a peak at an increasing focal length fp as x_A rises. Therefore, f for a given PTC is determined by the combined effects of all reflection points A on the parabolic mirror. This indicates

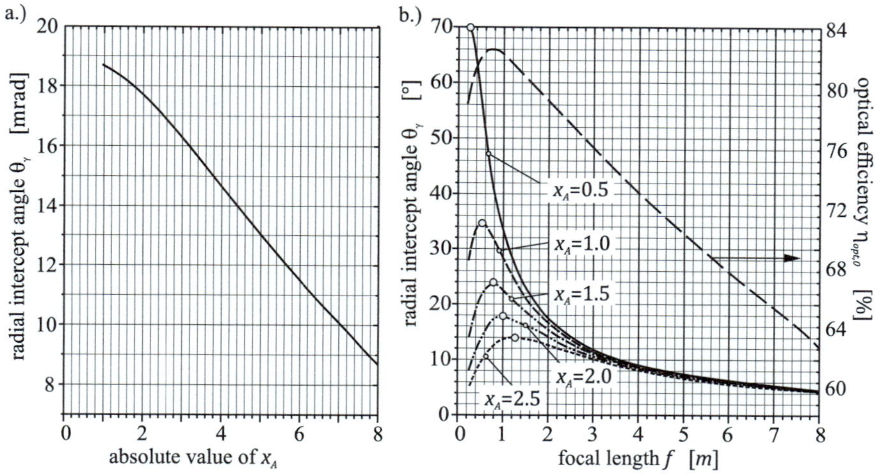

Fig. 7.14 Calculated optical properties of a PTC collector modified from Zou et al. (2017); (SEGS-LS-2PTC collector: $W = 5$ m, $L_{rec} = 7.8$ m, $d_{rec} = 0.07$ m, $d_g = 0.115$ m, $\alpha_{rec} = 0.96$, $\tau_g = 0.95$, $\rho_r = 0.93$; $CSR = 0.4$); **a** radial intercept angle θ_γ as function of x_A ($f = 1.84$ m) **b** radial intercept angle θ_γ for different x_A and optical efficiency $\eta_{opt,0}$ as function of focal length f (circles indicating maximum at lateral reflector width)

that the larger the aperture width W is, the larger the optimum f and the receiver diameter should be.

Figure 7.15a shows the results of the distribution of C_L for different aperture widths W. All C_L distribution curves as a function φ_{rec} exhibit the four regions as before. As the aperture width increases the maximum of C_L increases continuously, too, while the C_L value of the Shelter region for $\varphi_{rec} = 0°$ remains constant. Overall, the energy flux in regions 2 and 3 increases with increasing aperture width W, making region 4, which receives the direct irradiance, smaller. When the aperture width falls below a threshold value (here $W < 5$ m), the receiver area $\varphi_{rec} > 90°$ receives no concentrated radiation, while exceeding a certain aperture width (here $W > 12$ m) the entire receiver tube circumference is reached by concentrated radiation. Furthermore, it can be observed that the sheltering area becomes smaller and smaller when W falls below a threshold value (here 2 m) and disappears completely at a certain aperture width (here $W < 0.92$ m). The distribution of the intercept factor (γ), the dependence of the rim angle and the minium required receiver diameter as a function of the aperture width W are depicted in Fig. 7.15b. The intercept factor drops significantly above a critical aperture width W, since as the aperture width increases, the required absorber diameter also increases according to Eq. (7.10) and at some point, exceeds the size of the real absorber. Thus, part of the concentrating radiation is lost, which reduces the intercept factor. Increasing aperture width are associated with a rising rim angle (ψ_r) and require larger minimum receiver diameters (d_{min}). In other words, an increase of the aperture width

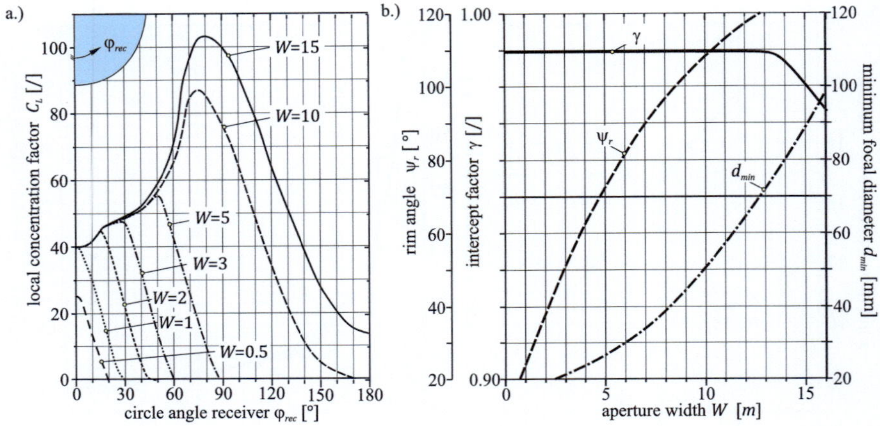

Fig. 7.15 Calculated optical properties of a PTC collector modified from Zou et al. (2017); (SEGS-LS-2PTC collector: $f = 1.84$ m, $L_{rec} = 7.8$ m, $d_{rec} = 0.07$ m, $d_g = 0.115$ m, $\alpha_{rec} = 0.96$, $\tau_g = 0.95$, $\rho_r = 0.93$); **a** circumferential local concentration factor C_L for different aperture width W; **b** intercept factor (γ), rim angle (ψ_r) and minimum receiver diameters (d_{min}) as a function of the aperture width W

W always means an increase of the rim angle (ψ_r) and an increase of the minimum receiver diameter (d_{min}). More radiation can be absorbed above the receiver surface, but with increasing aperture width the utilization of the radiation on the receiver decreases, so that above a certain aperture width, even with perpendicular irradiation, the intercept and thus the optical efficiency decreases.

PTC Performance at Non-normal Incidence Angle θ_i.

As illustrated in Chap. 6, for the integral collector modelling the impact of the incidence angle on the optical efficiency is depicted by the product of the peak optical efficiency $\eta_{opt,0}$ and the incidence angle modifiers (*IAM*) considering the temporal variations of the sun position with respect to the collector. Because PTCs are equipped with sun-tracking systems so that the transversal incidence angle θ_T of solar radiation is kept very small, ideally zero or very close to zero (<0.5°), so that the *IAM* due to transversal incidence angles does not play any significant role.

An increasing incidence angle θ_i causes part of the rays reflected by the mirror to exit at one end of the PTC and not reach the receiver, which naturally leads to end losses and thus to a reduction in optical efficiency. If the incidence angle θ_i exceeds a certain limit, the optical efficiency η_{opt} tends to zero because no beam reaches the receiver. Another effect of an increasing angle of incidence is that the image of the sun on the receiver becomes larger, because the distance of the reflected rays from the reflector to the receiver increases with growing incidence angle. This leads to a decrease in the power density and thus in the local concentration ratio C_L. Moreover, this effect smears the circumferential power distribution of regions 1 to 3 shown in Fig. 7.13a and results in a circumferential more uniform but smaller power distribution at large incident angles, similar to that at large

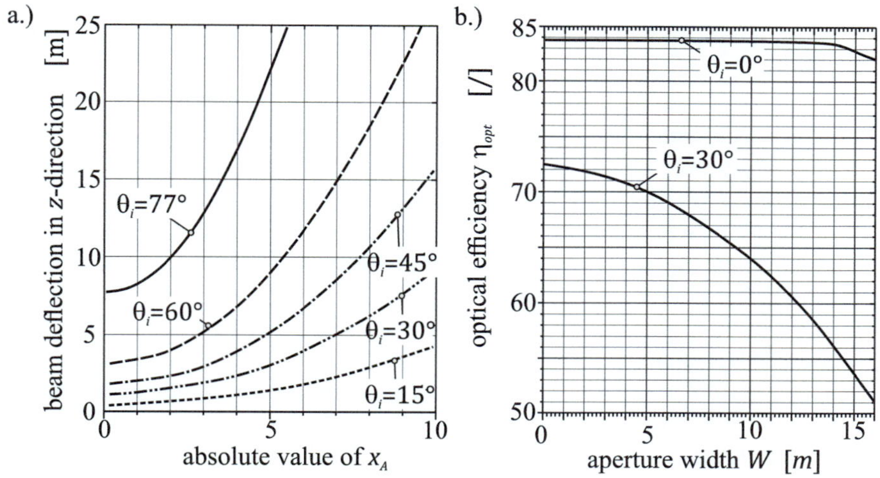

Fig. 7.16 Calculated optical properties of a PTC collector modified from Zou et al. (2017); (SEGS-LS-2PTC collector: $W = 5$ m, $f = 1.84$ m, $L_{rec} = 7.8$ m, $d_{rec} = 0.07$ m, $d_g = 0.115$ m, $\alpha_{rec} = 0.96$, $\tau_g = 0.95$, $\rho_r = 0.93$); **a** axial beam deflection (z-direction) for different incident angles θ_i; **b** optical efficiency η_{opt} as function of the aperture width W for $\theta_i = 0°$ and 30°

CSR, and characterized by a large-scale absence of a shelter effect at $\varphi_{rec} = 0°$. Considering the beam deflection in the longitudinal z-direction along the absorber tube for different incidence angles θ_i at a given point A (with the absolute value of the abscissa x_A, see Fig. 7.4 (A is the point where the incoming beam hits the mirror)), as shown in Fig. 7.16a, it is evident that the larger the aperture width (W), the larger the end losses become with increasing incidence angle θ_i. The position with a larger circumferential angle φ_{rec} can only receive reflected beams of larger values of x_A, which in turn leads to larger beam deflections in the axial z-direction. Therefore, the length of the section with very small local concentration factor C_L will be larger for larger φ_{rec} than for small φ_{rec}. For the SEGS-PTC considered, it is seen that at an angle of incidence of $\theta_i = 77°$, that the beam deflection by the reflector is larger than the length of the receiver ($L_{rec} = 7.8$ m), and no beam reflected on the mirror at any point of x_A reaches the receiver. Finally, Fig. 7.16b shows the optical efficiency η_{opt} as a function of the aperture width W at an angle of incidence of $\theta_i = 0°$ and 30°. As the incidence angle increases the optical efficiency decreases moreover; for increasing the aperture width this effect increases rapidly.

7.2.1.4 Technical Design Components of a Parabolic Trough Unit

A functional unit of a PTC receiver consists mainly of four components.

- a support structure,
- the receiver with the selective absorber enclosed in a glass envelope and the heat carrier fluid circulating inside,

- the curved mirrors mounted on the support structure, and
- a tracking system that tracks the receiver and mirror according to the position of the sun.

Each component must meet specific design criteria and ensure both functionality and availability. Any deviation due to faulty installation, poor design or tracking quality or great sensitivity to environmental influences such as wind, temperature and or humidity will result in a reduction of the optical yield. In Sect. 7.2.1.1, the basic design was discussed, while here the technical design of some realized collectors is considered.

Commercial Designs

Many commercial designs are based on a precommercial development from a European project, which is the Eurotrough design. In the years after 2000 mainly Spanish and US companies developed their own PTC design with little variations, as the aperture width of 5.8 m had been optimized for the 70 mm absorber tubes on the market. Some of them used prefabricated and preformed parts. Most structures were made of galvanized steel, but also aluminum and metal-polymer composites were used.

The implementation of the parabolic trough power plants during the first decade of the 21st century made evident that cost degression was slow using the same design. New developments investigated the possibility to achieve a larger solar field cost reduction if the size of the PTC was increased. The reduction in the number of specific components (e.g. ball joints, pylon foundations, drive units, local controllers, and loop specific piping) for solar fields with the same nominal power and the design of steel structures requiring less manpower cost for transport and on-site assembly could achieve significant cost reductions. Therefore, in 2010 the development of the Ultimate Trough design was started by the engineers of Schlaich Bergermann und Partner (SBP) and the Fraunhofer Institute for Material Flow and Logistics (IML) under the coordination of Flabeg Holding GmbH (Flabeg) (Riffelmann et al., 2014). The new aperture width of 7.5 m and receivers with 90 mm diameter were leading to an aperture area of 1689 m^2 and a claimed 25% solar field cost reduction. The main contribution to this was a 30% cost reduction of the solar field assembly cost. Another competitor design of large aperture PTC was the HelioTrough design with an aperture area of 1263 m^2 and an aperture width of 7.5 m (Janotte et al., 2014). Other industrial groups were following using optimized aperture widths from 6.9 to 8.2 m. The size difference between standard and large collector is shown in Table 7.2.

Although most of the R&D effort in the late 20th century and early in the 21st century was devoted to large aperture PTCs for large solar thermal power plants, new smaller troughs have also been developed for process heat applications with temperatures below 300 °C. These collectors can provide process heat to a wide range of applications replacing natural gas, such as crop drying and food processing. Industrial processes such as biofuel production, water purification, desalination, and absorption-chiller air conditioning for commercial and

Table 7.2 Typical technical design parameters of PTC for power plants

Parameter	ET-150 Standard	large PTC
Parabola width (m)	5.77	7.51
Collector length SCA (m)	147.8	242
Module length SCE (m)	12.27	24
Number of modules SCE	12	10
Outer diameter of steel receiver pipe (m)	0.07	0.09
Inner diameter steel receiver pipe (m)	0.065	0.085
Aperture area (m^2)	822.5	1689
Peak optical efficiency $\eta_{opt,o}(/)$	0.75	N.a

industrial buildings are similarly feasible with these small PTCs. These applications are grouped under the acronym SHIP (Solar Heat for Industrial Processes). The main difference to so-called large CSP applications is often, that the solar field sizes typically are in the range of a few hundred to a few thousand square meters, whereas for solar power plants in the 50–200 MW range the field sizes were several hundred thousand or even millions of square meters aperture area. Consequently, local production on-site is not cost-effective for those SHIP collectors and a high degree of pre-fabrication combined with standardized shipping and fast mounting is desired. This is reflected in the geometrical designs of the collectors, for which typical data are collected in Table 7.3. One of the first companies developing PTC for industrial uses was IST (USA) which installed several commercial plants in the United States. The technology was purchased by the Spanish group Abengoa, and their collectors (PT1 and RMT) were then marketed by Abengoa Solar (www.abengoasolar.com). Both collectors had non-evacuated receiver pipes with a black chrome coating and glass envelope, and maximum working temperatures of 288 and 204 °C, respectively. Another small PTC for process heat applications is the PTC-1800 collector developed by the German-Turkish company, SOLITEM (http://solitemgroup.com). The maximum working temperature of this collector is 200 °C and the collector aperture area is 36.65 m^2. The Australian company, NEP Solar Pty Ltd. (www.nep-solar.com), also developed small PTCs, the PolyTrough-1200 and PolyTrough-1800, suitable for temperatures up to 230 °C.

The 24 m long, 1.2 m wide PolyTrough-1200 collector (or 1800 mm wide PolyTrough-1800) consists of composite reflector panels with aluminum sheets which can be mounted either on roofs or at ground level. A more recent development is the PT950 solar collector from the company PROTARGET (www.protarget-ag.com) which has an aperture width of 3 m and due to the use of silicone-based oil as HTF may achieve a maximum working temperature of 425 °C. The Mexican company Inventive Power (https://inventivepower.com.mx) has been very successful economically and installed more than 100 SHIP projects in Mexico up

Table 7.3 Technical parameters of small PTC process heat collectors

Parameter	PowerTrough 110	PowerTrough 250	PTC-1100	PTC950	PolyTrough-1800
Manufacturer	Inventive power (MX)	Inventive power (MX)	Solitem (D)	Protarget (D)	NEP Pty (AUS)
Structure	Galvanized steel	Galvanized steel	Aluminum	Galvanized steel	Aluminum
Reflectors	Aluminum	Aluminum	Aluminum	Tempered glass	Composite
Module size (m)	1.21 × 3.06	2.5 × 4.16	2.5 × 1.1	3 × 12	1.8 × 24
Receiver tube	Non-evacuated, selective coating, borosilicate glass	Evacuated, selective coating, borosilicate glass	Non-evacuated, selective coating, borosilicate glass	Evacuated, selective coating, borosilicate glass	Evacuated, selective coating, borosilicate glass
Max. operating temp. T(°C)	130	260	200	425	230
HTF	Water			Silicone oil	

to 2019 using their own PTC designs. The success of Inventive Power demonstrates that there is a good market niche for SHIP applications in countries with a high level of direct solar irradiance.

Support Structure

The main functions of the support structure are to fix the collector components in a defined position and to provide structural stiffness and stability against varying environmental conditions. Steel or aluminum are mostly used as structural materials. From a structural-mechanical point of view, the PTC support structure consists of three parts: the main support (columns, piles, and box), the frame and the receiver brackets as depicted in Fig. 7.17.

- The main support serves as the structural anchor of the collector and holds it in place. The main support is structurally capable of resisting wind loads without exhibiting large torsion or bending which substantially impair the optical conversion and thus the yield.
- The frame provides rigidity to the mirrors, allowing them to always maintain their parabolic shape. In addition, it must serve torque transmission for tracking the mirrors.
- The brackets are attached to the mount of the mirrors, and the receiver is fitted via receiver supports to the frame so that it is in the focal line of the parabola.

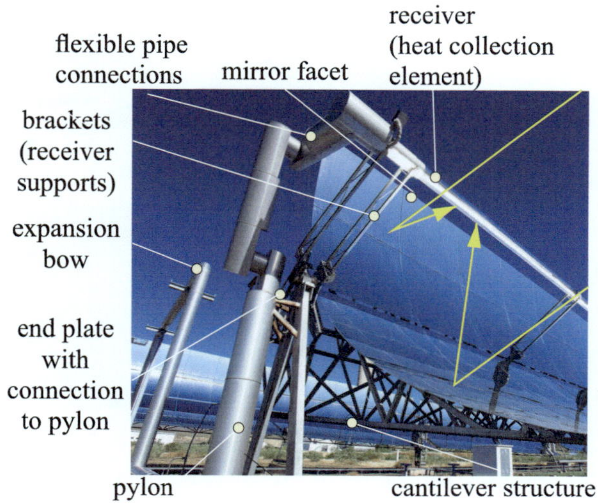

Fig. 7.17 Structural composition of a parabolic trough collector from DLR (2008)

The structural design of the connection between the frame and the receiver unit is such that there is no heat loss in the direction of the frame structure. Almost all designs use thermal insulation materials of various types.

The central task of the supporting structure of a parabolic trough is the correct mirror positioning and therefore primarily requires a high stiffness, since any deviation from the ideal contour leads to losses in the optical efficiency of the system. The stiffness refers to both the static loads, which are mainly due to the weight of the receiver elements, some of which are more than 100 m long, and the dynamic loads, which are mainly due to the aperture surfaces exposed to the wind. The entire collector structure, consisting of support structure and frame, must be designed to withstand these loads with only very slight geometric deviations. In addition, high stiffness allows for longer troughs and thus fewer poles and tracking units, and thus reduces solar field costs. High stiffness is realized with consequent use of lightweight construction while keeping material and manufacturing costs low, which not only allows for weaker foundations and fewer number of tracking mechanisms with lower power consumption but also reduces deformations due to dead weight. In particular, the power consumption of the actuators contributes not insignificantly to the parasitic energy consumption of a parabolic trough power plant, which is relatively high compared to other solar thermal power plants. It should be considered that the solar field cost contributes to more than 30–40% of the total cost of a PTC plant (IRENA—International Renewable Energy Agency, 2013). This means that reducing the cost of the solar field has a large impact on the total cost of the power plant and its economics. Usually, the supporting structure of a collector consists of a main body, which in most cases is a space frame

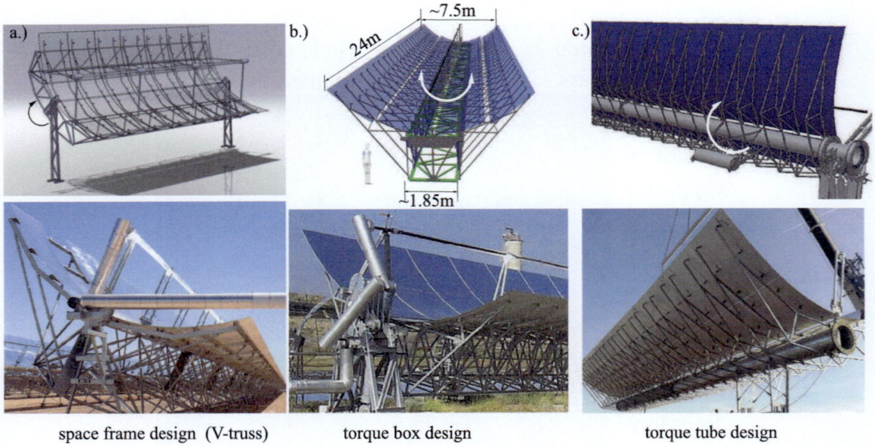

Fig. 7.18 Sketches of the support structure of PTC collectors (top) and realization in different power plants or prototypes (bottom); **a** space frame design—Solana 2 Power Station; **b** torque box design—Eurotrough—150 DLR; **c** torque tube design—Noor 2-SENER

or a tubular structure made of steel or aluminum. Other elements of the support structure are:

- mirror support points on the space frame structure or on special cantilevers,
- receiver support, also called heat collector element (*HCE*) support,
- structure for attachment to the pylons,
- pylons, including drive pylons, and foundations.

Modern structures target at a high degree of pre-fabrication rate off-site at the manufacturer and final assembling on site, where the focus is to minimize the work hour for assembling a solar collector assembly (*SCA*). Typical structural frame designs are frame space designs, torque box designs for large, scaled module of power stations of up to 200 m in length per single collector assembly composed of several collector modules which in length range of up to 20 m. Such designs do not only allow to use aluminum to reduce weigth but also allow for designs with a center of weight below the mirror and thereby minimizing stress for static loads. Torque tube designs as sketched in Fig. 7.18a mainly use steel as constructional material and are characterized by a simple and cost-effective design. Figure 7.18 show the different structural design types with support structure, cantilevers, and fixation of the *HCE*.

Table 7.4 shows that there is no uniform design of both the structural form as well as the structural material chosen; nevertheless, the trend in power stations is towards longer modules and thus also longer overall units to reduce the number of drives. At the same time, larger aperture widths are aimed at, with up to 7.5 m being achieved commercially. Nevertheless, the upward size growth is limited due to the strength of the available structural materials for the support structures, as the collectors must be in structural integrity even at wind speeds of up to 50 m/s. The verification for this is done by coupled fluid dynamic and structural mechanics calculations, which include the resistance and deflection as well as critical positions.

Substructure

A precise assembly of the steel structural profiles, overall structural rigidity and low cost is required to achieve a good parabolic trough concentrator. Therefore, most PTC designs for large collectors require assembly jigs, as depicted in Fig. 7.19 left, in order to meet the design tolerances, which are compared to the assembly size very small, typically of the order of ± 1 mm. The collector design shown in Fig. 7.19 right has a so-called central space frame, called torque box, which provides good rigidity and prevents torsion, which is important for a sufficient optical performance.

Another PTC design approach is the replacement of the torque box by a central steel tube (called the torque tube, see Fig. 7.20). The main advantage of PTC torque box designs is their usually better performance and rigidity under wind loads, while their main disadvantage is higher assembly cost due to the number of steel profiles that is required by high-precision specifications. Torque tube designs, on the other hand, are usually somewhat cheaper but may be more prone to deformation from gravity (bending) and wind loads (torsion) than those using a torque box. In both cases, rigidity in wind load situations of both designs is good enough to guarantee high optical performance.

Concentrating Mirror Surface

The most common parabolic mirrors of almost all commercially realized parabolic trough power plants consist of metal-coated glass mirrors, with the coating consisting almost exclusively of silver. The mirrors have proven to be durable, showing little decrease in mirror reflectance even after more than ten years of operation considering regular cleaning. Figure 7.22a shows the spectral reflectance of some potential mirror materials in comparison with the solar spectrum containing direct irradiance and circumsolar radiation at an air mass factor of $AM = 1.5$. The figure exhibits that for silver (Ag) the reflectance collapses in the UV spectral domain (≈ 320 nm) containg already a certain fraction of the spectral solar energy. This UV transparency of silver can cause a degradation of epoxy-based glues attaching the silver layer to the steel liner forming the mirror frame. This degradation is mainly expressed by blistering. To avoid blistering the silver layer backside is usually covered by copper. Except for this small spectral reflectance gap silver exhibits an excellent reflectance along the rest of the solar spectrum being close to unity, while all other metals except for aluminum such as gold, copper or chromium exhibit a smaller reflectance. Thus, integrated over the entire spectrum silver represents the prime candidate as mirror material.

The glass for the mirror facets is usually produced using the float glass process, in which molten glass is continuously applied to a liquid tin bath where it gradually solidifies. This process enables high planarity of the glass surface. The

Table 7.4 Structural design, dimensions and optical performance of selected parabolic trough concentrators (PTC) for solar thermal electric plants, data from Tagle-Salazar and Nigam (2020) complemented by manufacturer data sheets

Concentrator (developer/ name)	support structure design type/ material	Aperture width W/ module length L_{rec} (m)	Assembly length (m)	conc. ratio C^* (/) focal length f (m)	Optical efficiency $\eta_{opt,0}$ (/)	Mirror reflectance ρ_r (/)
Luz International						
LS -2	torque tube/ steel	5/7.8	47	71/1.49	0.737	0.94
LS 3	V-truss/ steel	5.76/	99	82/1.71	0.8	0.94
Eurotrough						
ET-100	Torque box/ steel	5.76/12	99.5	82/1.71	0.75	0.94
ET-150	Torque box/ steel	5.76/12	148.5	82/1.71	0.75	0.94
ENEA	Torque tube/ steel	5.76/12.5	100	77	0.78	0.94
Acciona SGX-2	Space frame/ alu	5.77/12	100–150	82/1.72	0.77	N.a
Sener Trough	Torque tube + stamped cantilever/ steel	5.76/12	150	80/1.70	0.76	0.935
Flagsol						
SKAL -ET 150	Torque box/ steel	5.77/12	148.5	82/1.71	0.8	N.a.
Helio Trough	Torque tube	6.77	191	82/n.a	0.82	0.95
Ultimate Trough	Torque box	7.51	246	76/n.a.	0.8	0.94
Skyfuel						
Sky trough	Space frame /alu	5.7/12	115	72/1.71	0.76	0.94
Sky trough DSP	Space frame /alu	7.6	150	75/n.a.	0.76	0.935
Abengoa						
Solucar PT-2	N.a	4.4/12	148.5	63/1.73	0.75	0.93
Astr0	Torque box/ steel	5.76	150	80/n.a	0.77	N.a
Phoenix	Space frame/ steel	5.76	150	81/n.a	0.79	0.95

(continued)

Table 7.4 (continued)

Concentrator (developer/ name)	support structure design type/ material	Aperture width W/ module length L_{rec} (m)	Assembly length (m)	conc. ratio C^* (/) focal length f (m)	Optical efficiency $\eta_{opt,0}$ (/)	Mirror reflectance ρ_r (/)
Solarlite						
SL 4600	Torque tube/ steel/alu	4.6/12	Up 120	66/1.2	0.75	0.94
SL 5770	N.a	5.77/12	Up 120	82/1.71	N.a	>0.94
Gossamer space frame						
LAT 73	V-struts/steel	7.3/12	192	103/2.56	0.77	0.95

NB: C^* is not the correct thermodynamically defined maximum concentration factor using the circular aperture $A_r = \pi.d_{rec}.L_{rec}$, but the commercially used factor based on the projection of the receiver: $A_r^* = d_{rec}.L_{rec}$.

Fig. 7.19 Assembly of Eurotrough Collector in a jig (left) and parabolic trough design using space frame (right) (Zarza Moya, 2020)

Fig. 7.20 PTC mirror support structure based on torque tube design (Zarza Moya, 2020)

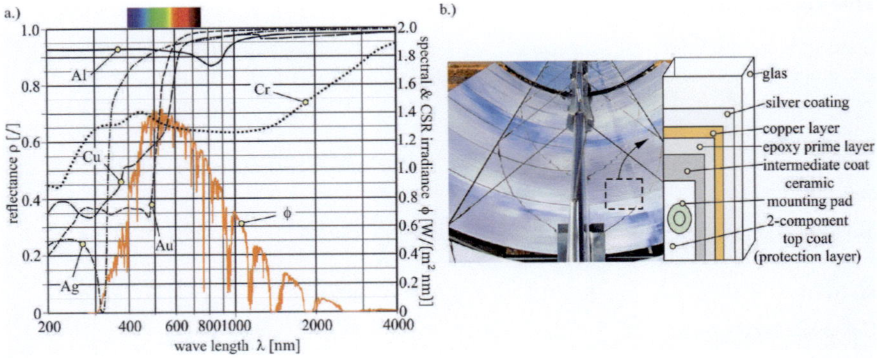

Fig. 7.21 **a** spectral reflectance of several mirror metals in comparison with the spectral irradiance of direct and circumsolar irradiation at air mass factor $AM = 1.5$ according to ISO 9845-1 (1992), ASTM G173-03 (2020); **b** Cross-section through the layered structure of a parabolic reflector mirror of a PTC

glass coating itself consists of metal oxide free glass to minimize light absorption, especially by iron oxide, which keeps transmissivity high. After the glass is cut to the right size, it is formed into the parabolic shape in a furnace. One option is to form the glass in a mold by its own weight; in other manufacturing routes, the glass is formed to the desired parabolic shape using a pressing process. Industrially, geometric mirror accuracies are achieved in which the tilt error is in the order of 0.132° from the ideal design contour shape. For Eurotrough applications, an intercept factor of 99.9% was obtained for a 70 mm diameter absorber tube (the common absorber tube diameter of most commercial receivers) and 95.5% for a 40 mm diameter absorber tube.

The mirrors have a multilayer structure as illustrated in Fig. 7.21b. The first layer below the glass is the reflective layer, is usually made of silver deposited by spray coating. Due to the transparency of silver in the UV range, there is a copper layer below the silver layer. Below this is a base layer, an intermediate layer, and a protective two component topcoat, which are epoxy coatings that encapsulate the metal–glass bond. In most solar mirrors applied to date, the base and intermediate layers contain for durability reasons some percentage of lead, but layers without copper or lead are under development. The thickness of the entire mirror assembly is about 4–5 mm. The following figure shows the multilayer structure. Below the topcoat ceramic pads are used to mount on the substrate to achieve good focusing. Typical average solar weighted direct reflectivity of the PTC mirrors is in the range of $\rho_r(SW, \theta_i, \theta_c) = 0.95$.

The solar field costs of a parabolic trough plants can reach up to 50% of the total investment cost and therein the mirror cost constitute a significant, for more details see Sect. 10.2. So, the search for alternative materials with high potential to reduce costs is a high priority. Research here is focused on mirror composites, which is directed either from silver-coated polymer films or from front-side aluminum mirrors.

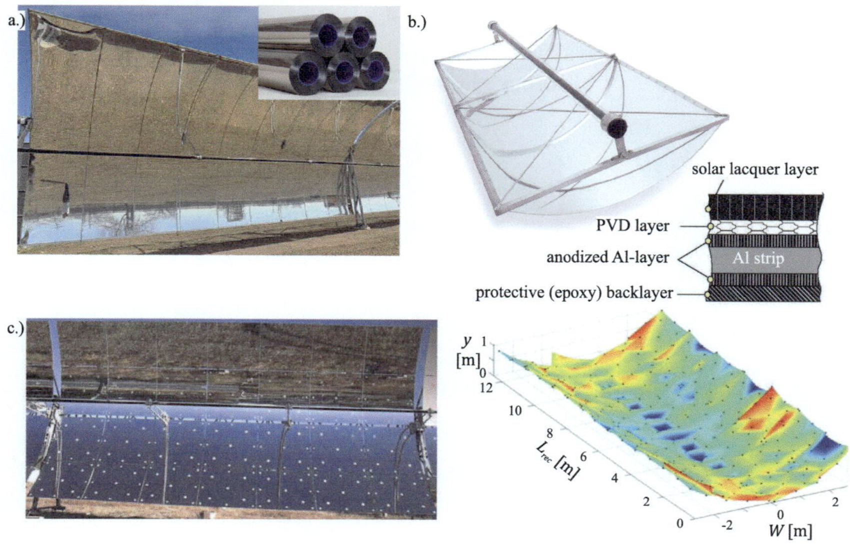

Fig. 7.22 **a** Advanced parabolic trough solar collector from SkyFuel covered with a silver coated polymer film by ReflecTech™ (Kurup & Turchi, 2015); **b** PTC collector with aluminized mirror from ALANOD GmbH & Co. KG (2021); **c** Photogrammetric 3D-analysis of Eurotrough PTC collector from measurements modified from Lüpfert et al. (2003) and reconstructed mirror shape right (red indicating 7 mrad deviation from ideal value denoted in blue)

Silver-coated polymer films as reflector materials are designed as roll-up reflective films, as shown in Fig. 7.22a (top right), which can be applied to any smooth, nonporous material and consists of multiple polymer layers with a reflective silver layer. Mirrors under the proprietary brand name ReflecTech were tested in the SEGS facilities for over seven years and showed no decrease in reflectance, although the reflectance of about 0.94 was slightly lower than that of the glass versions.

The other variant is front aluminized mirrors (Fig. 7.22b), which consist of an aluminum reflective layer, an anodized layer below and above it, and other protective layers. Due to the lower reflectance of the solar spectrum of aluminum compared to silver. The reflectance of the mirrors marketed under the brand name Alanod MIRO-SUN^{R} is lower than that of the silver-based mirror types. A value of $\rho \approx 0.9$ is given here. The mirror performance of a PTC is measured either during manufacturing or after installation on-site. In both cases the mirror is scanned spatially at discrete positions and by means of a photogrammetric analysis to gather a spatial picture of the area distribution in terms of surface quality at clean conditions as indicated in Fig. 7.22c, while during operation deflection measurements are executed at discrete mirror positions in the solar field.

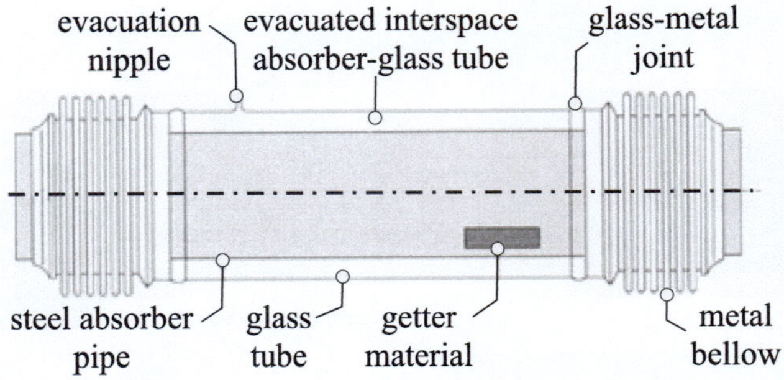

Fig. 7.23 Schematical design of an evacuated tube receiver

Receiver (Heat Collecting Element-HCE) Design

A receiver tube for PTC is composed by an inner steel pipe containing the heat transfer fluid and an outer glass tube surrounding the steel pipe. The glass tube is produced from a robust low iron borosilicate glass. The steel pipe is coated with a selective absorber surface with a high solar absorptance and low thermal emittance, for more details see in Chap. 3. The glass tube is usually coated with an antireflective layer to achieve a higher solar transmittance.

Receivers may be either evacuated or non-evacuated. The evacuated ones are still having small thermal losses even for temperatures above 300 °C because they have a high vacuum (i.e., 10^{-5} mbar) between the steel pipe and the glass cover, thus prohibiting thermal conduction and convection losses. Figure 7.23 shows the schematical design of an evacuated receiver. The glass cover of has to be connected to the steel pipe in a flexible way to avoid thermal stress, e.g. by using stainless steel expansion bellows which compensate the different thermal expansion of glass and steel and keep the annular gap hermetically sealed. The exact shape of the bellows is dependent on the manufacturer. A second means to keep the vacuum are chemical 'getters' placed in the gap. They shall absorb hydrogen molecules which might form in the HTF and diffuse through the steel tube. The diameter of the PTC receivers depends on the aperture width of the parabola. The standard outer diameter of the steel absorber tube is 70 mm for a parabola with up to 6 m. For bigger parabolas, steel tubes of 80 and 90 mm are used. For the small process heat collectors with aperture widths of 1–25 mm absorber tube of 1–2 in. are often used. Further typical technical data may be taken from Table 7.5.

The core of parabolic trough collectors are the receivers also named "heat collection elements"-HCE, which are located at the focus of the parabolic trough mirrors. They consist of a steel tube through which the heat transfer medium circulates enveloped by a glass cover tube, which is evacuated to reduce heat losses. Of crucial importance is the selective coating of the steel tube, which ensures high absorption of the incident concentrated solar radiation in the solar spectrum with

Table 7.5 Typical technical parameters of commercial vacuum tube receivers for PTC

Property	Typical value	Range of commercial products
Solar absorptance α	0.95	0.94–0.96
Thermal emittance ε (400 °C)	0.09	0.07–0.11
Steel pipe diameters (mm)	70	70–90
Pipe material	Stainless steel	–
Glass cover diameter (mm)	125	115–145
Glass cover material	Borosilicate glass + AR	
Transmittance cover glass τ	0.96	0.95–0.97
Heat transfer fluid (HTF)	Thermal oil	Thermal or silicone oil, molten salt, direct (water/steam)
Maximum temperature T(°C)	400	300–550

simultaneously high reflectance in the far IR spectrum to minimize radiation losses to the environment. The residual gas pressure in the gap is of order of 10–100 Pa.

Receiver Glass Tube

The receiver glass must exhibit a spectral selective behavior expressed by a high transmittance in the solar spectrum and a high reflectance in the heat emission spectrum, as outlined in Chap. 3 (optical conversion). The principal physics and the corresponding means to reduce either reflectance in the solar spectrum by means of anti-reflection coatings as well as absorptance by choosing low iron content glass, usually borosilicate glass, has been explained there. With the help of both measures, PTC glass tubes today achieve transmittance values of more than $\tau \approx 0.96$. Nonetheless, we focus here on specific PTC receiver collector challenges.

Since the glass envelope almost completely encloses the heat source, convective, conductive, and radiant heat losses emanating from the absorber tube must be minimized as far as possible. Convective losses can be substantially reduced by evacuating the gap between hull and receiver; typical residual gas pressures in the gap are about 1–100 Pa. This requires a minimum glass thickness, typically in the order of 3–4 mm to achieve structural hull integrity at all operations conditions. Conductive heat losses are reduced by appropriate choice of structural materials. A special challenge is the mismatch of the specific linear expansion between glass and the structural material. Therefore, materials with small specific thermal conductivities towards the glass envelope are used, for example, stainless steel. Moreover, wall thicknesses are adapted in such a manner to keep thermal stresses low and simultaneously maintain a high thermal insulation to ambient.

In PTC collectors, thermal radiation is the largest heat loss mechanism. For example, according to Wien's displacement law, the maximum of heat radiation at 600 °C is about 3.32 μm, of which already a small part is in the visible light range, irretrievably leaving the glass envelope. Fortunately, glasses exhibit a spectrally selective behavior (see Chap. 3), characterized by a low transmittance in

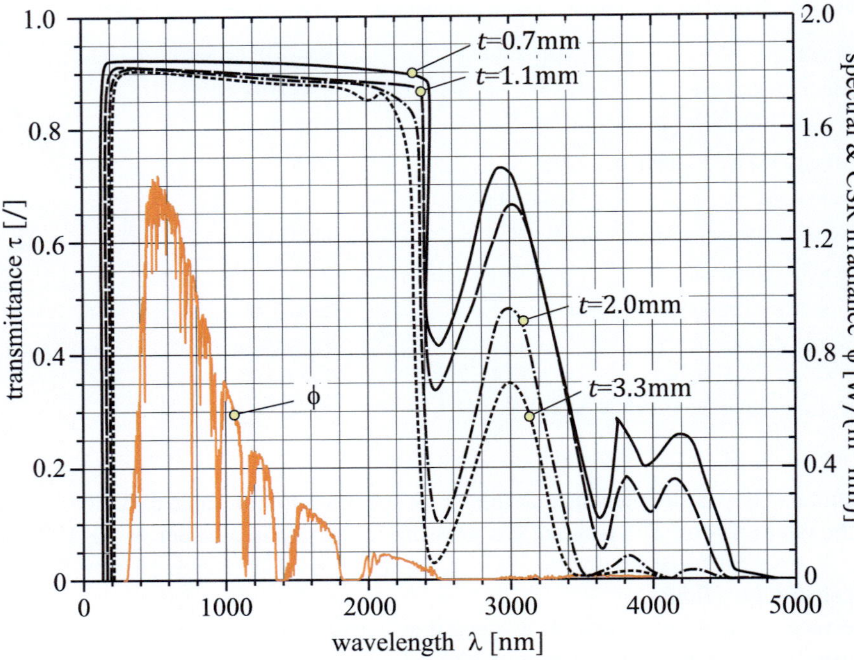

Fig. 7.24 Spectral transmittance of uncoated commercially available borosilicate glass of different thickness t (SCHOTT Technical Glass Solutions GmbH, 2018) in comparison with the spectral irradiance of direct and circumsolar irradiation at air mass factor $AM = 1.5$

the infrared region. Thus, the glass envelope acts as a thermal radiation reflector and contributes to thermal insulation, since the thermal radiation emitted by the absorber tube remains within the glass envelope. Nevertheless, the glass envelope is heated and thus generates additional thermal radiation back onto the absorber tube. Thus, glasses to be used must exhibit a high thermal resistance. To provide sufficiently high resistance to thermal shock and temperature gradients high boron content glasses are taken, compared to soda-lime glasses used in the past. These borosilicate glasses are usually coated with an anti-reflective layer enhancing transmittance to values exceeding 0.96 and allow for temperatures of about 450–500 °C.

Despite this favorable behavior the addition of boron reduces the thermal expansion coefficient of the glass considerably, posing significant design challenges at each end of the receiver assembly where a glass-metal bond is required. Care is taken to ensure that the glasses contain no iron or other mainly oxide metal impurities which cause high absorptance especially in the visible light spectrum. Even untreated glasses exhibit a high transmittance down to 300 nm UV with a value of $\tau \geq 0.92$ in the visible and near IR wavelength range, as well as low intrinsic fluorescence and low solarization, as shown in Fig. 7.24 for an uncoated Borosilicate glass in comparison with the solar spectrum.

Selective Absorber on Receiver Tube

In the selective absorber on the receiver tube the optical conversion takes places according to the principles mentioned in Chap. 3 based spectral selective behavior characterized by spectral high absorption of the incident concentrated solar radiation in the solar spectrum (250–2500 nm) over a long lifetime, typically 25 years are considered. Simultanously, the emissivity in the far infred ($\lambda > 3$ μm) should be as small as possible. Since the glass envelope of the receiver is evacuated and thus losses due to thermal conduction or convection are reduced, heat losses due to radiation dominate. With increasing temperatures, the spectra of the solar radiation and that of a blackbody radiator increasingly overlap, as illustrated in Fig. 7.25. For an average collector temperature of 350 °C, high absorption is only possible at the expense of higher emission values. Silver has one of the highest reflectances of all materials in the infrared spectrum and, therefore, it is primarily used as an IR mirror. High absorber temperatures place high demands on the stability of the absorber layers due to the mobility and solubility of metals in other metals or composites. Even when used in a vacuum, in which no degradation occurs due to oxidation processes, the development of absorbers with high temperature stability is a major challenge in terms of materials technology. Thus, the details of the design of commercial absorbers as well as their fabrication route are the intellectual property of the manufacturers. Scientific developments of selective absorbers such as spectral analyses show that most of the used absorber types in PTC collectors consist of multilayer cermets ("*cer*" = ceramic and "*met*" metal) as described in Chap. 3. This cermet layer is applied on the IR-mirror layer and consists of materials such as Cr_2O_3, Al_2O_3, AlN, SiO_2 in which metal nanoparticles are integrated. The top layer is a pure ceramic anti-reflective coating mostly made of Al_2O_3 and SiO_2, more details can be found in (Cao, McEnaney, Gang, & Ren, 2014).

Depending on the material pairing, additional adhesive layers are required to ensure the adhesion of the functional layers to the structural tube material. In case of oil as heat transfer medium, additional gas diffusion barrier layers are applied to minimize gas diffusion through the absorber tube and to maintain the insulation vacuum. If molten salt is used as the heat transfer medium, coolant temperatures of about 550 °C are possible. Such temperatures require additional diffusion barriers to prevent the migration of metal from the IR mirror into the tube substrate. The principal spectral reflectance characteristics of a selective absorber is illustrated by Fig. 7.25.

Technical Design

High power plant efficiency requires high steam turbine inlet temperatures and thus high heat transfer fluid (HTF) temperatures. Above about 390 °C, most thermal oils preferably used in PTC troughs decompose and develop hydrogen in a dissociation reaction, which may be released into the vacuum gap, diffusing through the steel, and deteriorating the vacuum there. Getter material in the vacuum tube, improved diffusion barriers for the steel tube and a hydrogen removal in the circulating liquid can reduce the problem largely. In direct steam generation PTC concepts, in which the thermal oil as HTF is replaced by water, higher steam temperatures

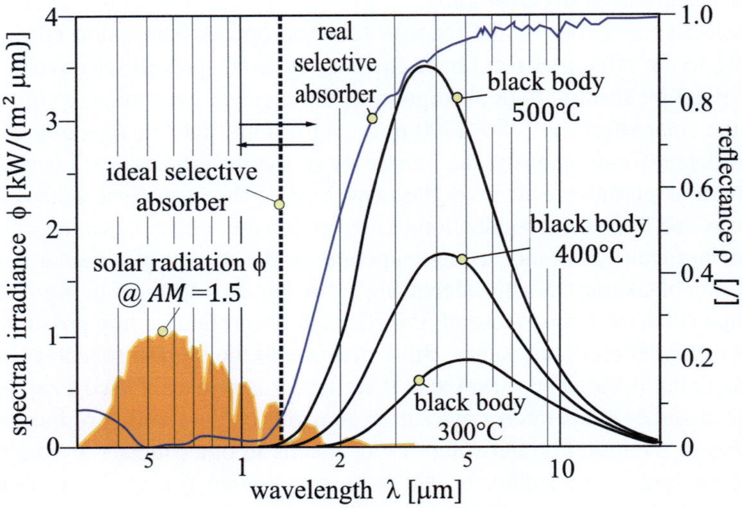

Fig. 7.25 Increasing overlap of the solar spectrum ($AM = 1.5$ direct, non-concentrated) with the spectra of a blackbody radiator for temperatures between 300 and 500 °C; a typical absorber spectrum and an optimal selective absorber are also shown for illustration modified from Hildebrandt (2009)

are achieved by superheating the steam in the collector directly. However, the use of water as HTF requires high pressures increasing the thickness of receiver tube. In alternative concepts using molten salts or liquid metals as HTF lower pressures are needed, and the degradation of the vacuum due to hydrogen diffusion through the tube can be avoided. But, due to their capability to attain substantially higher HTF temperatures, other technical measures are required to limit the enhanced heat losses through heat radiation via the glass to the ambient or heat conduction through the brackets connecting the HCE to the support structure. Additionally, using molten salts or metals as HTF demands sophisticated control and operating strategies to avoid solidification (or freezing) of the heat transfer fluid during the night.

A key technical aspect of all line-focusing systems is the piping connection of a collector module string to the manifold leading to the power plant unit. Due to the temperature differences during operation of the receivers, there is a differential elongation of the metallic absorber pipe compared to the mirror units and the support as well as the entire collector line. Since the lengths of a string can be several hundred meters, a single compensator in an HCE string would be very costly and hardly able to absorb the entire elongation. Therefore, compensation for the different thermal elongation of the collector and the structural-mechanical superstructure is achieved by means of bellows on each absorber element. From the effects described above, there are three immediate critical issues for the design of a long-life HCE:

Table 7.6 Coefficient of (linear) thermal expansion α_{th} of some metals and glasses (Schöttl et al., 2019)

Metals	$\alpha_{th}(\cdot 10^{-6}/K)$	Ceramics, glasses	$\alpha_{th}(\cdot 10^{-6}/K)$
Aluminum	25	Al_2O_3	6.5–8.8
stainless steel	17	Si	2.6
ASTM Nickel alloy	12	Soda-lime silicate glass	9.2
Titanium (Ti)	8.8	Borosilicate glass	4.6
Tungsten (W)	5	Silica (96% pure)	0.8
Kovar alloy (Fe–Ni–Co)	5	Silica (99.9% pure)	0.55

- breakage of the glass-to-metal seal,
- thermal oil degradation,
- weathering resistance of the glass tube.

Breakage of the glass-to-metal joint is the main cause of receiver module damage in power plants, mainly due to different thermal expansion coefficients of the thermal fluid tube and glass envelope. Table 7.6 shows the thermal expansion coefficients (CTE, α_{th}) of some metals and glasses. To compensate for the different thermal expansion coefficients between steel and glass, in the past the metal–glass joint was fused into the glass body by means of a continuously decreasing metal cross-section. All new designs use either tungsten or a Kovar between glass and steel, circumventing the problem of differential elongation; the details of the bond, however, are the intellectual property of the manufacturer. Figure 7.26a illustrates the principle of metal–glass bonding by the two approaches.

When thermal oils are used, but also when steam is superheated in direct evaporating collectors, hydrogen can be generated, which diffuses atomically through the steel and recombines at the interface to the vacuum space. This leads to a deterioration of the vacuum and increases heat losses by a factor of two to three. The diffusion of the hydrogen can be slowed down, but not prevented, by permeation barriers made of aluminum oxide, for example, which are arranged between the steel tube and the selective absorber layer. To maintain good vacuum over a long period of operation, a gettering or getter capsule (on the opposite side of the focal line) is preferably integrated at the coldest point of the vacuum chamber, which absorbs the hydrogen. Some products include vacuum level indicators to allow visual inspection of the receivers during operation. This is outlined in Fig. 7.26b and c.

All receivers are prefabricated at the manufacturer and only installed on site. As the size of the power plant has increased, so have the internal diameters of the tubes, and commercial systems have sizes of 70–90 mm with lengths of 2–4 m, as shown in Table 7.7 for some commercial products. To reduce the number of circuits in large solar arrays and allow for long arrays, coolant pressures of 6 MPa are not uncommon. The design of the connection between the receiver tubes aims at a compact construction to ensure a maximum active absorber length on the one

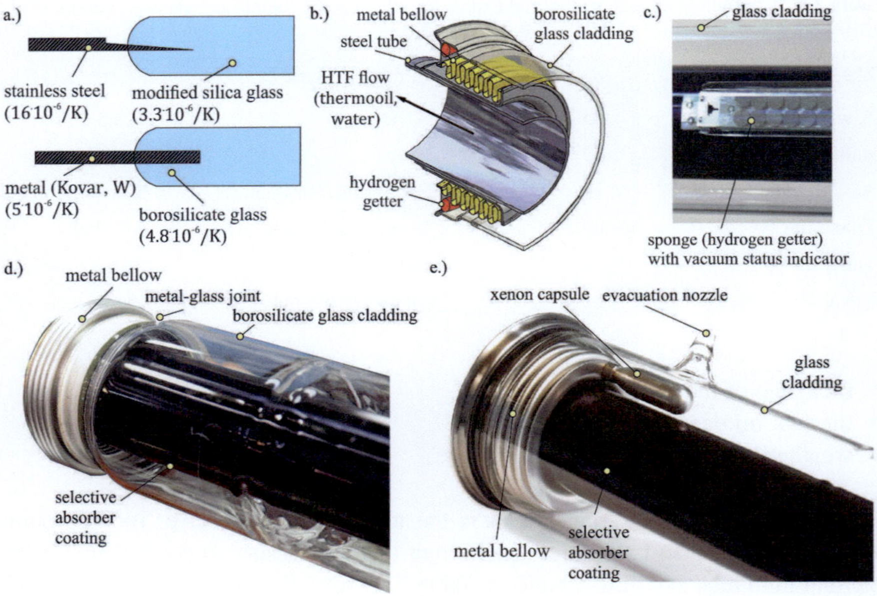

Fig. 7.26 **a** principle of metal–glass bonding to compensate different CTE by geometry and use of adapted material; **b** hydrogen getter options to maintain vacuum and allowing visual receiver integrity inspection (**c**), from Nagarajan and Barshilia (2010). Commercial receiver design external bellow (Siemens Concentrated Solar Power, Ltd., 2011 now Rioglass) and external bellow (SCHOTT Solar CSP GmbH, 2013)

hand and to keep the heat loss via the metallic tube connection elements small on the other hand, at the same time the compensation of the relative elongation of the tubes during operation must be ensured. Two development lines have become established on the market, a design variant with an external bellows and one with an internal compensation element, as shown in Fig. 7.26d and e, respectively.

Solar Tracking System

Parabolic troughs as every concentrating system have to track the sun to reach a continuous concentration of the direct solar radiation to provide maximum efficiency at the given site and environmental conditions. In contrast to linear Fresnel in a parabolic trough the whole assembly (mirror and absorber) is simultaneously single axis tracking the sun's position.

Two types of collector tracking directions are used in PTCs, namely north–south ($N - S$) and east–west ($E - W$) tracking, according to the direction of rotation of the collector aperture plane as sketched in Fig. 7.27. In north–south tracking, the receiver axis and the rotation axis are aligned in the east–west direction, which means that the collector aperture is rotated from north to south. Conversely, in east–west tracking, the aperture plane and the rotation axis are aligned in the

Table 7.7 Characteristic data of some PTC receivers from Tagle-Salazar and Nigam (2020) and manufacturer data sheets

Manufacturer/ model ID	Rioglass UVAC 90 7G	Rioglass PTR 70 4G	Archimede HCEMS11 (molten salt)	Archimede HCESHS12 (direct evap.)	Solar lite SL 4600	Sunda Seido 6-3
HTF tube						
Length (m)	4.061	4.06	4.06	4.06	4.06	4.06
Diameter *d* (mm)	88.9	70	70	70	70	70
Thickness t_w (mm)	2.5	2	3	N.a.	2–3	2.5
Material	Stainless steel	Stainless steel	Stainless steel	Stainless steel	Stainless steel	Stainless steel
Glass tube length (m)						
Length (m)	3.91	3.77	3.9	3.9	3.9	3.9
Diameter *d* (mm)	135	125	125	125	115–125	115
Thickness t_w (mm)	3	2.5	3	3	2	N.a.
Transmittance τ (/)	0.964	0.97	0.966	0.966	0.96	0.95
Selective coating						
Absorptance α (/)	0.962	0.94	0.95	0.95	0.095	0.94
Emittance ε@ 400 °C (/)	0.095	0.095	0.073	0.073	0.14	0.12
Solar field						
Max. pressure (MPa)	4	4.1	3	10.4	<4	4
Max. temp. T (°C)	450	400	580	550	393	450
Design						
Lifetime (years)	25	25	25	25	25	N.a.
Vacuum pressure (Pa)	<10	<10	<10	<10	N.a.	N.a.

north–south direction and the aperture plane rotates from east to west. The north–south tracking has the advantage of lower energy consumption during tracking, but is associated with higher end losses, while with east–west tracking systems it is the other way around: lower end effect but higher energy consumption. Without prejudice to solar field design, north–south tracking is almost exclusively used to achieve maximum thermal yield.

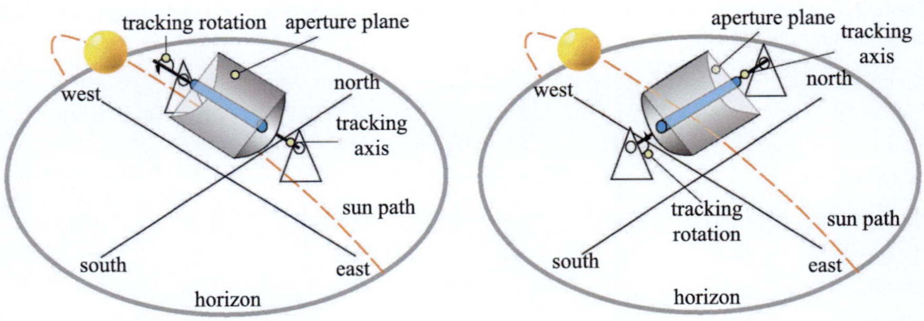

Fig. 7.27 Definition of east–west mounted collector with north–south tracking (left) and north–south mounted collector with east–west tracking (right)

The tracking angle φ_{track} as deviation from the normal of the horizontal plane is for any horizontal collector with its transversal plane azimuth angle γ_c relative to the south direction is defined by solar azimuth γ_s and altitude α_s by

$$\varphi_{track} = \sin^{-1}\left(\frac{\sin(\gamma_s - \gamma_c)}{\tan\alpha_s}\right), \tag{7.18}$$

$$\alpha_s = \sin^{-1}(\sin\lambda \cdot \sin\delta + \cos\lambda \cdot \cos\delta \cdot \cos\omega), \tag{7.19}$$

$$\gamma_s = \sin^{-1}\left(\frac{cos\delta \cdot \sin\omega}{\cos\alpha_s}\right), \tag{7.20}$$

where λ denotes the latitude, δ the declination and ω the hour angle. The incidence angles on the tracked collector aperture with tilt φ_{track} are given in the following equations—θ_{NS} for the north–south mounted collector ($\gamma_c = 0°$) and θ_{EW} for the east–west mounted collector ($\gamma_c = 90°$

$$\cos\theta_{NS} = \cos\delta\sqrt{\sin^2\omega + [\cos(\beta-\lambda)\cos\omega - \tan\delta\sin(\beta-\lambda)]^2} \tag{7.21}$$

$$\theta_{NS} = \cos^{-1}\left[cos\delta\sqrt{sin^2\omega + [cos(\beta-\lambda)cos\omega - tan\delta sin(\beta-\lambda)]^2}\right]; \tag{7.22}$$

$$\cos\theta_{EW} = \cos\delta\sqrt{\cos^2\omega + \tan^2\delta} \tag{7.23}$$

$$\theta_{EW} = \cos^{-1}\left[cos\delta\sqrt{\cos^2\omega + \tan^2\delta}\right]. \tag{7.24}$$

Here, β denotes a possible tilt of a parabolic trough along the rotation axis. This is not a realistic option for large power plant collector of several meter length, but possibly for short roof-integrated collectors or collectors mounted on a slope of a hill. When the tilt is equal to the latitude, then we speak of a polar-mounted collector, where the rotational axis is parallel to the earth's axis. In this case the incidence angle is constant over the day and equal to the declination of the sun's orbit.

Knowing exactly position and solar time of a collector, in principle the tracking angle can be calculated. However, to generally realize a tracking of the receiver, a solar field sensor, which detects the irradiation, an actuator consisting of a motor and a gearbox for the rotation of the collector, and a position sensor, which detects the angular position of the collector, are required. All these elements are part of a regulation and control chain and are themselves subject to manufacturing tolerances or inaccuracies and errors that must be considered in the regulation, regardless of whether they are sensors or drives or gears.

Active tracking systems are standard in concentrating solar power plants, with a further subdivision into closed-loop and open-loop systems, more information on details may be taken from Moga et al. (2015), Lee et al. (2009). The main difference between these types of active tracking systems is the way of signal conversion and feedback with the actuator.

Closed-loop active tracking systems consist of three components, a light sensor, a control system, and a drive system. The sensor sends a signal to the control system when the collector is not properly aligned with the sun, prompting the control system, where a comparison is made. If it is misaligned, the controller sends a signal to the actuator, which applies torque to the structure until the vector value falls below a threshold. Thus, closed loop tracking systems are pure feedback controls. Light receiving diodes or PV cells are the most commonly used sensors for closed loop tracking. The main advantage of this tracking method is the high precision tracking, but it is susceptible to the sensitivity of the light sensor and its positioning/s in the solar field or due to wanted or environmental shading of the sensor. A key disadvantage is the inability of the control system to recover or find the direction of the sun during long periods of cloud cover.

Open loop tracking systems consist of a controller and a drive. During operation, the algorithm programmed in the control system, based for example on a clock, signals the actuator to change the collector position. Open-loop tracking systems can be further subdivided into pure time-controlled tracking systems or altitude/azimuth tracking systems. Timed tracking systems use algorithms based only on time, so that the collector is controlled by incremental movements. Altitude/azimuth trackers use algorithms based on astronomical data as a function of location and time.

Hybrid tracking has been developed to overcome the disadvantages of closed-loop and open-loop tracking. The basic strategy of hybrid tracking is to use an algorithm (open-loop) to track the sun and reduce misalignment by sensors (closed-loop). A typical hybrid loop tracking consists of the algorithm calculating the position of the sun and activating the drive, and later correcting the misalignment by the light sensor if it occurs.

Increased computer capabilities such as wireless measurement signal transmission and the use of data buses have led to the widespread adoption of hybrid control systems, which not only include the position of the collector in relation

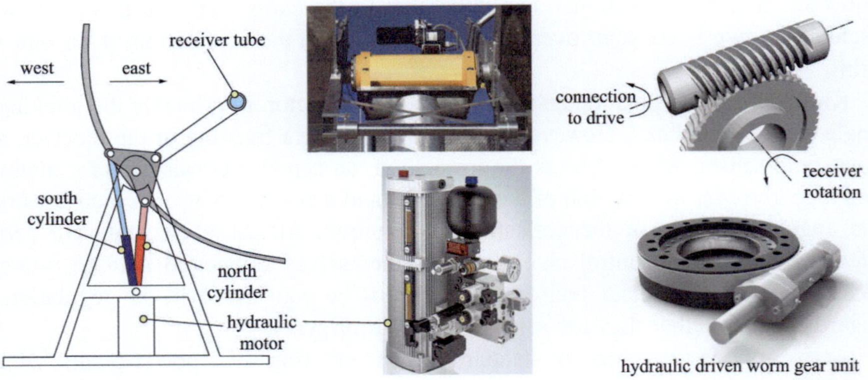

Fig. 7.28 Hydraulic drive operated receiver tracking principle (left), hydraulic motor gear box unit (top, middle); functional principle of backlash-free actuators based on a screw drive with a constructive design (right)

to the sun, but also model-based representations of the individual receiver circuit strings and the reference to the respective overall power plant operation strategy. Such systems not only take into account the plant itself and its different operating modes, such as start-up or shutdown, standby, control mode, but automatic switch to safety positions in case of extreme weather situations e.g. provided by external information provided by weather forecast systems.

Tracking, especially with parabolic receivers, places extreme demands on the drives and gearboxes, not only in terms of precision but also because of the enormous masses that must be operated without backlash under highly variable environmental conditions over long periods of time with as little maintenance as possible. Both hydraulic drives, which the figure outlines, and either hydraulic or worm gear units driven directly by electric motors have become established on the market. Figure 7.28 illustrates some options of tracking systems and drives being installed in solar fields.

7.2.2 Linear Fresnel Collectors

The focusing properties of large parabolic reflectors can be simulated by small reflector elements distributed over some suitable (ground or roof) plane surface. This allows large concentrator systems to be built up from small elements, avoiding the large structures and cleaning accessibility problems associated with very large reflectors. A single-axis linear Fresnel collector (LFC) system is composed of many long row reflectors that together focus sunlight on elevated linear tower receivers with one or more absorber tubes running parallel to the reflector rotational tracking axis. This allows a large size array to be constructed inexpensively using very similar or identical long focal length (and therefore almost flat) glass mirror elements.

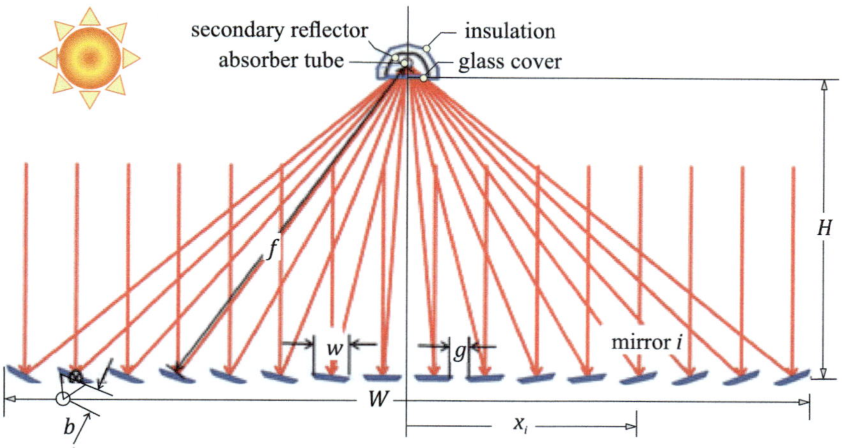

Fig. 7.29 Geometrical transversal cross-section of a primary reflector array with relevant geometric parameters (w = mirror width, g = gap between neighbouring mirrors, b = distance between rotational axis and mirror axis, f = focal length)

Today's usual single-axis tracking LFCs differ from parabolic troughs in that the reflectors are composed of many long row segments which focus collectively on an elevated long linear tower receiver running parallel to the reflector rotational axis (Fig. 7.29). Unlike for parabolic troughs, the LFC receiver is fixed in space, and the reflectors rotate to maintain focus on the fixed receiver. The reflector rows all track through the same number of degrees during the day as the sun moves but are inclined at different angles at any one time because they have different positions relative to the tower target. This allows a large size basic unit made up of identical relatively small reflectors of long focal length, which in turn leads to the ability to use almost flat, low-cost glass mirror elements.

The LFC approach to Concentrating Solar Power (CSP) today reaches a commercial maturity approaching that of trough systems. Specific commercial projects focused on different approaches have driven the development of LFC, showing the versatility of the concept. In a recent development, the first 50 MW_{el} commercial molten salt power plant using a linear concentrator is not based on the trough principle but on the LFC concept. Similarly, LFC collector designs exists for the application of solar process heat in industry. Different specifications lead to smaller collector dimensions, often lower temperature demands and thus more cost-effective material choices. The transportability of smaller collectors is key for the success in this market.

For both applications the fundamental advantages of LFC are essential:

- reflector production form low-cost flat mirrors,
- stationary long receivers,
- low wind loads due to smaller elements,

- high compactness of aperture leads to optimum area use,
- highly flexible optical design (high concentration) and easy adaptability to different heat transfer fluids.

The disadvantage of LFR compared to Parabolic Troughs is certainly the stronger sensitivity to low sun positions, especially in high latitude applications, and hence a lower optical efficiency which must be compensated by a larger solar aperture.

In contrast to a mirror of a parabolic trough, which consists of fixed segments or a single piece, the primary mirror of Fresnel collectors consists of n mostly identical primary mirror strips mounted on a steel structure fixed to the ground via a support system. In order to achieve a reduction of the solar image on the receiver, i.e. a large concentration ratio C, the primary mirrors are often marginally parabolically curved. Since the focal lengths f of the mirrors are very large in relation to their width W_{pr}, elastic curvature is possible. Depending on the focal length and mirror width, the maximum deflection at the center of the mirror is <1 mm. Several frames are connected to form a module, and the modules form a long row. Via motor drives the mirror strips (primary mirrors) are manipulated to reflect solar radiation onto the receiver depending on the position of the sun. Figure 7.30a shows the main geometrical parameters of an LFR collector primary mirror field in relation to the receiver.

The simplest example of a primary mirror strip curvature is a parabola. The parabolic contour is given by $y = ax^2$. From this approach results a focal length f of the primary mirror of $f = 1/(4a)$. A focal point (or a focal line) results with parallel incident light only for parabolic reflectors and there also only if the irradiation is parallel to the parabolic main axis, as sketched in Fig. 7.30b (top). If

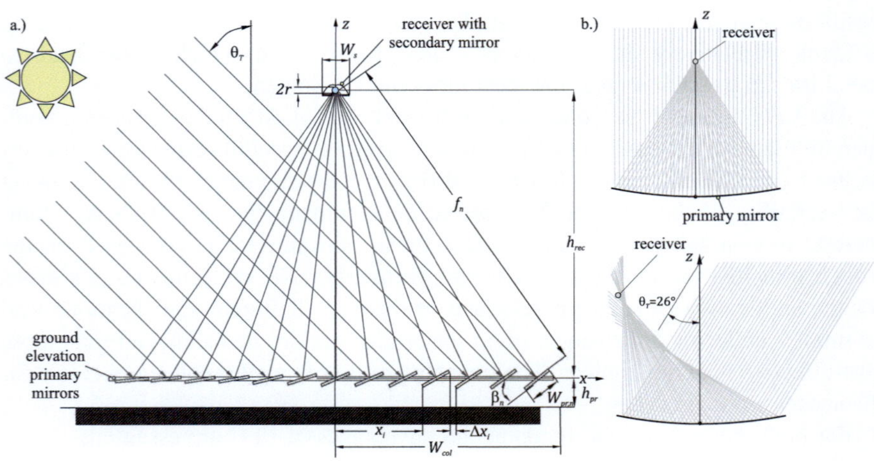

Fig. 7.30 **a** Geometry parameters and coordinate system defining the relation between primary mirrors and receiver of a LFR collector; **b** astigmatism due to deviation of incident radiation from the major axis of the parabola. Parallel irradiation $\theta_T = 0°$ (top), irradiation at $\theta_T = 26°$ (bottom)

the irradiation deviates from this, the focal point widens to a caustic. This widening leads to an aberration and is called astigmatism. Figure 7.30b (bottom) illustrates this for a transverse incidence angle of $\theta_T = 26°$. In a Fresnel collector, only the primary mirror, which is in direct alignment with the sun and the receiver, has approximately the condition for a focal line. However, exactly this mirror is shaded by the receiver. All other mirrors show a more or less strong astigmatism. The larger the ratio of primary mirror width $W_{pr,i}$ to target aperture and the smaller the ratio of primary mirror width $W_{pr,i}$ to focus length f, the stronger the spillage due to astigmatism.

The receiver unit of a linear Fresnel collector (LFR) contains a spectral solar selective absorber, often a secondary reflector, which e.g. is composed of compound parabolic concentrator (CPC), and sometimes a thermal insulation behind it. The receiver, which is permanently mounted at a height of several meters, should be protected from wind and dust by an anti-reflection coated glass pane or a cylindrical glass tube. The LFR collector design offers a vast variety design freedom. Subsequently, the model description of the integral and local LFR collector performance refers only to the standard design of the receiver composed of single tube design, which is located in the center of a housing consisting of a CPC mirror and covered at the bottom by a glass plate, which is illustrated in Fig. 7.34a.

7.2.2.1 Optics of LFC

Primary Mirror Field

The primary mirror or reflector array consists of multiple rows of slightly curved mirrors, rotating with half the angular speed of the sun in the transversal projection path over the rows. The shape ideally is a parabolic segment with a focal length determined by the distance to the receiver. The basic geometry illustrated Fig. 7.29 has variables; width of the mirror element (w_i), gap g_i between the mirror strips, the focal length (f_i) of each mirror to the receiver and the height (h_i) of a a segment to the absorber. which all may be functions of distance (x_i) of the mirror row i from the receiver tower.

There are several optical loss mechanisms related to geometry. Even for perfect reflectors, cosine losses occur due to off-normal incidence on the reflectors most of the time. Blocking considers the partial interception of reflected rays by mirror rows closer to the receiver tower. In addition, mirror rows may partially shade rows behind them as seen from the sun (shading). Additionally, any real mirror has a small but finite absorption and—if dust or a rough surface must be considered—scattering extinction both reduce the direct radiation reflected towards the receiver. Blocking and shading is dominant for low solar altitudes. Blocking and shading certainly can be minimized for a primary field if the outer mirror rows are positioned with a higher altitude above ground. The disadvantage might be that the receiver height must be increased to some extent.

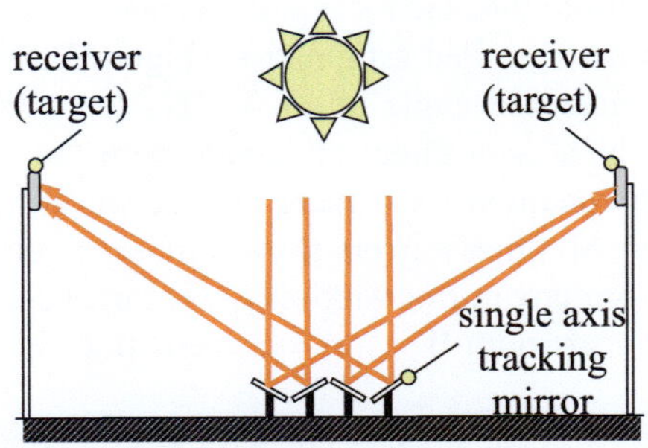

Fig. 7.31 Ray path of a compact linear Fresnel reflector (CFLR) showing interleaving of mirrors minimizing shading between mirrors based on Mills and Morrison (2000)

We may distinguish four principal classes of LFC primary mirror designs, see Abbas et al. (2018):

- Constant mirror width and constant gap (CW-CG),
- Constant mirror width and variable gap (CW-VG),
- Variable mirror width and constant gap (VW-CG),
- Variable mirror width and variable gap (VW-VG).

The advantage of constant mirror width is that the production line does not need to deal with different mirror sizes. Thus, it is a question whether variable mirror width options will be favored by a commercial development. Providing a variable gap to minimize blocking and shading losses however could be solved by mounting options at the expense of additional assembly effort and more careful design. Probably this design will result in a larger field area for the same reflector area.

A further design option is that the focal lengths of individual mirror rows can be constant or variable. The focal lengths should be adapted to the distance of the mirror rows to the receiver aperture. In practice it is too costly to produce different focal lengths for every row, so practical designs typically utilize only few (1–2) different focal lengths. This aspect is related to the question of constant width. A compromise could be to produce a small number of mirror groups, each with a different curvature and width. In addition to losses due to optical design there are other factors which degrade the quality of the solar image on the receiver. Rough, dusty, or scratched mirror surfaces as well as shape or contour inaccuracies widen the reflected beam. Also tracking inaccuracies should be avoided.

Optical design tries to optimize the size and efficiency of the reflector array (to increase concentration) with respect to economics. The further away a reflector row is from a receiver tower, the lower is its time-averaged optical efficiency. In financial terms every additional investment of adding a row has less and less addition to overall output. This is one reason for the compact linear Fresnel reflector approach, called CLFR and illustrated in Fig. 7.31, where the mirror rows can be reoriented, according to sun position, to reflect to a receiver in better position.

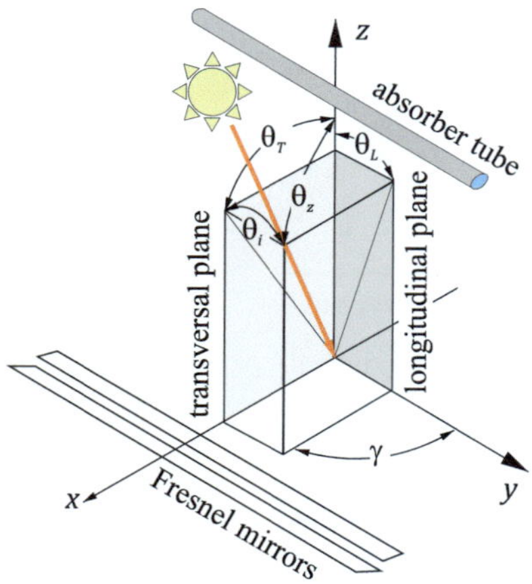

Fig. 7.32 Incidence angles for a linear Fresnel collector (LFC) according to Mertins (2006)

Secondary Optics

Larger concentration can be easily achieved for a linear Fresnel collector (LFC) by adding one more mirror row to the primary reflector field. However, to guarantee a good optical efficiency of the concentrator system a secondary concentrator often is used to collect the rays coming from the reflector field and redirect them to the absorber tube. Typically, the receiver aperture is larger than the absorbing area by a factor between 1.5 and 2. It is important that radiation impinging on the receiver aperture is directed towards the receiver tube. Non-imaging optics gives us several options to design such secondary concentrators. The principles proposed range from truncated compound parabolic concentrators (CPC) to elliptic or hyperbolic concentrators, which are discussed in the Sect. 7.2.2.2 of dealing with LFC receivers. As usually there is slightly more than one reflection on average in a secondary concentrator, the reflectivity should be chosen very high to maximize efficiency. Moreover, any absorption by the reflector heats up the secondary concentrator due to the high intensity. This must be considered carefully in the choice of the secondary reflector material. If temperatures become too high in the cavity, deformation and degradation will be the result.

Incidence Angle Modifier

A combination of primary concentrator with a receiver and possibly a secondary concentrator results in an angle-dependent optical efficiency of the collector. With two incidence angles (θ_i) the situation may be characterized as depicted in Fig. 7.32.

Let us define the angles transversal, the longitudinal and the solar longitudinal incidence angles θ_T, θ_L and θ_{LS} as

$$\theta_T = \tan^{-1}(|\sin\gamma| \cdot \tan\theta_z); \tag{7.25}$$

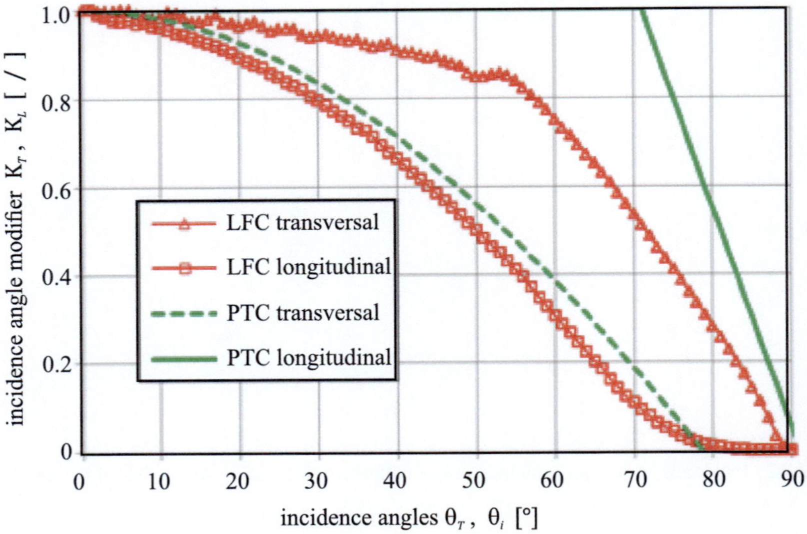

Fig. 7.33 Simulated incidence angle modifiers IAM (relative to DNI) for the Linear Fresnel Collector (LFC) for transversal (K_T) and longitudinal (K_L) directions, in comparison to the Parabolic Trough Collector (PTC) (here transversal K_T describes the row-to-row shading) from Morin et al. (2012)

$$\theta_L = \tan^{-1}(\cos\gamma \cdot \tan\theta_z); \tag{7.26}$$

$$\theta_{LS} = \sin^{-1}(\cos\gamma \cdot \sin\theta_z). \tag{7.27}$$

The incidence angle modifier function (IAM) describes the optical efficiency for off-normal incidence angles normalized by optical efficiency at normal incidence. It is usually written as the product of a transversal and a longitudinal function:

$$IAM = K(\theta_{LS}; \theta_T) = K_T(\theta_{LS} = 0; \theta_T) \cdot K_L(\theta_{LS}; \theta_T = 0). \tag{7.28}$$

The Fig. 7.33 illustrates the transversal and longitudinal IAM (K_T, K_L) of a LFC collector in comparison with a parabolic trough collector (PTC) as a function of the incidence angle ($\theta_{i,T}$).

In a comparison of design options theoretical maximization of optical efficiency should not be the target. This approach certainly leads to designs with high concentration and preserving etendue, however, it must be kept in mind that at the end manufacturing, construction and operating cost will also determine the final economic performance. Complex solutions with variable component geometries, with changing altitude of mirror rows, with complicated construction manuals will produce extra costs and sources of errors. The decision which approaches

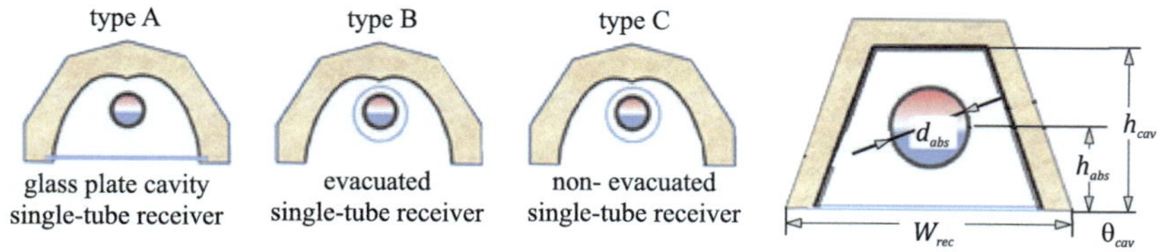

Fig. 7.34 Configurations for single-tube receivers with secondary reflector according to Montes et al. (2017) and Tsekouras et al. (2018)

to use and which to abandon due to cost certainly depends very much on experience and production technology of an individual commercial entity. Therefore, a purely academic design recommendation may fail to meet the expectations. The optimized design is always a compromise between theoretical and commercial considerations.

7.2.2.2 LFC Receivers and Thermal Performance

Thermal efficiency and heat losses of the receivers are critical to collector performance. But there is not a single optimal design, as depending on the design of the primary reflector field, its size, shape, and construction can be quite different. From a fundamental viewpoint we know two basic receiver options: Single tube receivers and multiple tube cavity receivers.

While single tube receivers usually have a secondary reflector to improve the optical concentration of the LFC and to avoid high intercept losses, multi-tube cavity receivers are without secondary optics. As already stated, the average number of reflections in a secondary is slightly above 1, and about 3–4% of the energy is lost in this process. It is a critical issue that the surface might degrade due to soiling, corrosion, and high temperatures. These risks must be carefully compared with the possible advantage of high concentration. Cavity receivers without secondary utilizing one or several tubular absorbers on the other hand have the potential disadvantages of higher thermal losses, larger optical spillage, and limited primary reflector field width.

Single Tube Receivers

Three different basic configuration for single-tube receivers with secondary reflector have been investigated by Montes et al. (2017) or Tsekouras et al. (2018) and are illustrated in Fig. 7.34. The implications are quite different for all designs. Whereas in (b) and (c) the reflectors are open to environmental impact from humidity, sand etc., the evacuated tube receiver protects the absorber effectively. In (a) both absorber and reflector may be protected against sand, but there is a risk for overheating of the reflector material. A selective absorber coating on the tube is open to oxidation in air. Traditionally secondary CPC-reflectors with acceptance angle defined by field width are considered. More sophisticated approaches have been proposed, therefore, using secondary concentrators of different shape based on non-imaging optics.

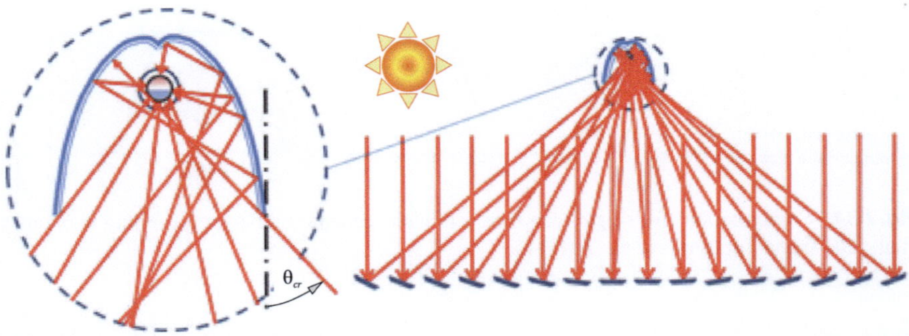

Fig. 7.35 CPC receiver with LFR

Design Options

The geometry used in most commercial approaches is a truncated CPC. A CPC reflector ideally redirects all rays from a certain field of view (within the critical acceptance angle θ_{cr}), which should cover the primary reflector field) towards the receiver surface as shown in Fig. 7.35. Truncation reduces the length of the CPC without much sacrifice in concentration (Rabl & Winston, 1976). The advantage of this design is a relatively shallow geometry using little reflector material. An ideal CPC with a cylindrical absorber would have a cusp nearly touching the absorber tube. To avoid that, the cusp often is flattened out, allowing some rays within the theoretical acceptance range to escape the concentrator. Experimental methods have been developed to assess the quality of the secondary concentrator early in development (Platzer, Heimsath, Hildebrandt, & Georg, 2008).

A similar concept from non-imaging optics using the Simultaneous Multiple Surface method yields a taller secondary concentrator but resulting in higher is called the LFR XX- SMS concept, which is depicted in Fig. 7.36a, see Canavarro et al. (2014), Guerreiro et al. (2015). This advanced optical design obviously has consequences on other issues like construction of the tower structure (due to higher wind loads) or concentrator material costs. It must be shown that the additional optical concentration leading eventually to better thermal performance at low irradiation, justifies the effort. According to (Guerreiro et al., 2015) a concentration of 73.7 can be reached using 16 mirrors 1.06 m wide within a field 20.1 m wide. The tower must be 9.8 m high. The optical efficiency is close to 70%.

A further secondary concentrator type has been designed by Chaves and Collares-Pereira also based on non-imaging optics, see Chaves and Collares-Pereira (2010). However, in this case the optics is designed for a modified primary mirror field where the rotational axes of the mirrors are not in one altitude level. The primary field mirrors of one segment between receiver towers, point to different receivers alternatingly, similarly to the CFLR concept. In this way the primary mirror field can be extremely compact minimizing the gaps. The receiver is a V-shaped absorber with two mirror wings. These concentrators are called Tailored Edge Ray Concentrators (TERC), see Fig. 7.36b. From a theoretical point this is

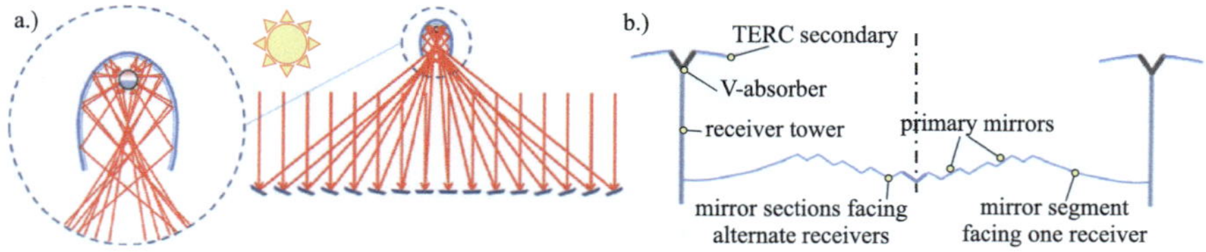

Fig. 7.36 Ray path in a LFR XX SMS collector concept (**a**) and geometrical set-up of TERC-concentrator (**b**) for several receivers according to (Chaves & Collares-Pereira, 2010)

an ideal optical concentrator with highest possible optical efficiency for infinitesimally small mirrors. Practically viewpoints lead to finite size mirrors and mirrors in one plane which brings the efficiency down.

Other secondary concentrator design principles have been put forward. The Adaptive Design Concentrator (ADC) is not defined by a set of equations but by a repetitive procedure to determine the next segment of the reflector from a previously defined one, cf. (Zhu, 2017).The resulting secondary shape looks very similar to a truncated CPC. The Segmented Parabolic Secondary Concentrator (SPSC) shape is composed of a parabolic trough broken at a given point to change its tilt. The result is a kind of reflecting cavity with 3 segments of curved walls. A recent comparative optical study concludes that the optical performance of ADC is very similar to that of a CPC, whereas the SPSC has a clearly lower performance which could be accepted if cost would be lower, see (Abbas et al., 2018).

Commercial Receivers
In commercial LFC the CPC geometry has been used similar to that shown in Fig. 7.37. It is of *Type a* described in Fig. 7.34. This receiver type is based on absorbers which should be selective and stable in air. The radiative-convective heat losses in the space between secondary walls and the aperture glazing are complex and can be modelled with CFD methods. However, for a faster assessment of heat loss dependence on temperature, incidence angles and irradiation levels a simplified method based on heat balance of different segments has been developed and used successfully (Heimsath, Cuevas, Hofer, Nitz, & Platzer, 2014a, 2014b, 2014c). The insulation level of the secondary to the ambient is certainly important as the hot air tends to stay in the top part of the cavity. Also, the air tightness level of the aperture window influences the wind dependence of heat losses. Air tightness stops not only forced convection inside the cavity but can also reduce dust deposition on the absorber and mirror surfaces. This may also be the reason that no LFC receiver seems to be of *Type c* (Fig. 7.34). That solution would be like the one of simple parabolic troughs for industrial use, where the absorber tubes are contained in a glass tube, but not evacuated. Mirror cleaning of the secondary mirror is tedious, especially for large tower heights. Therefore, the glass window for the receiver seems the better solution.

Using a vacuum tube receiver to reduce heat losses further is a viable alternative to the approach shown above. Then the geometry of *Type b* is chosen. It is a big

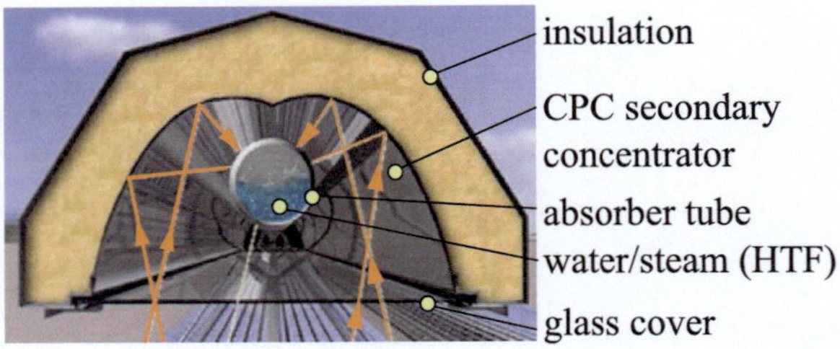

Fig. 7.37 Cross-sectional sketch of the NOVA-1 Novatec/Frenell receiver with secondary reflector (courtesy Frenell GmbH)

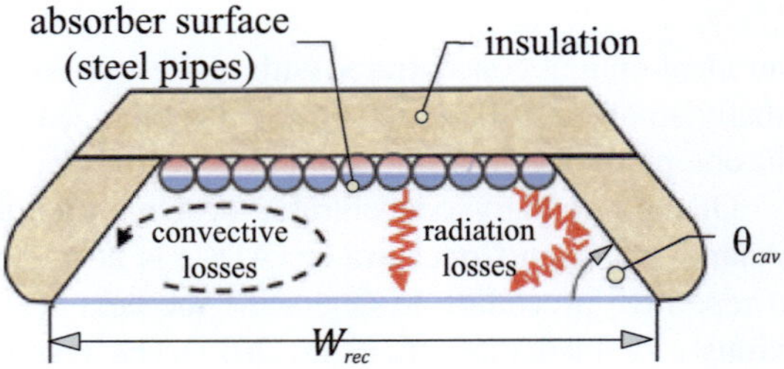

Fig. 7.38 Schematic cross-sectional sketch of a multi-tube cavity receiver

advantage that mirror surfaces are not heated anymore by convective and radiative losses from the absorber tube. Instead, they will be cooled to some effect by wind, which is important as the absorbed concentrated solar intensity still might heat up the reflector. As there is no need for thermal insulation, the receiver structure may be lighter and less voluminous, this reducing the shading of primary mirrors. Commercial companies here use standard size evacuated receiver tubes as used for Parabolic Troughs. The SuperNOVA collector of Frenell is also able to reach temperatures up to 550 °C when molten salt is used as a heat transfer fluid, see Morin, Mertins, Kirchberger, and Selig (2011).

Multi-tube Receivers

Multi-tube cavity receivers have been investigated theoretically by many authors in the past, looking into radiative and convective heat losses depending on the geometry with new studies published in recent years, see e.g. Di Canio, Tretyl, Jur, and Watson (1977–1979). A review on the newer studies is given in Bellos (2019).The designs cover usually 3–8 receiver tubes in a cavity and the parameters shown in Fig. 7.38 are varied in the different proposals. Also, here a glass or plastic cover as shown in Fig. 7.38 may protect the absorber tubes from dust and help to reduce thermal losses.

For the practical development, this receiver type has been chosen by the company Ausra which later became Areva Solar, see Conlon, Johnson, and Hanson (2011). These commercial cavity receivers in the early years used 8–16 tubes. Flow considerations, heat transfer capacity and availability of suitable tubes probably influenced this choice. It is important to separate the tubes in the receiver sufficiently to allow differential thermal expansion between the tubes which have different temperatures. This was analyzed first in a paper by Pye, Morrison, and Behnia (2003) looking into thermal convective and radiation losses (Fig. 7.38).

In later versions of the receiver the number of tubes was reduced to 8–10, with larger diameters of the tubes because larger primary fields needed sufficient flow. Areva Solar has also tested a collector technology line called SSG4 (Conlon, Johnson, & Hanson, 2011) with a superheated steam receiver pressurized to 9.2 MPa, unlike the previous saturated steam units.

Comparison of Heat Loss and Collector Efficiency

Only a few heat loss models of commercial collectors have been published and data are available. Although heat loss is not desirable, one must be cautious—higher heat losses do not necessarily imply lower efficiency, because the concentration also is very important. The main part of the collector energy balance is the solar gain where geometrical concentration and optical efficiency are multiplied by the direct solar irradiance DNI. The thermal losses are proportional to absorber area and influenced by receiver characteristics such as emissivity, evacuation, or wind protection by cover glass. For high solar fluxes the heat losses are often nearly negligible—for low solar fluxes they may play a greater role, i.e. when DNI is small or when incidence angle modifier or shading from other objects reduce the solar flux on the receiver.

Figure 7.39 presents heat loss data from company data as a function of the difference between mean absorber temperature and ambient temperature; data taken from Morin (2005), Bernhard et al. (2009), Industrial Solar GmbH (2022), Novatec Solar (2012). Depending on the solar irradiation and resulting temperature gradients in the absorber tubes, the fluid temperature during solar operation is lower than the absorber surface temperature. The heat resistance of the absorber tube and the convective heat transfer into the fluid cause a temperature drop. During solar gains the difference is typically around 10–30 °K. The F'-factor of the receiver is smaller than unity. For one-phase heat transfer fluid and high mass flow the mean fluid temperature may be estimated as the arithmetic mean between inlet and outlet temperature. For two-phase steam generation during the passage through the tubes the main part of the steam is wet, i.e. evaporation occurs at the saturation temperature. Therefore, apart from the small section for preheating and superheating before and after the evaporation section the fluid temperature is constant at the saturated steam temperature.

For the company Frenell published data exist for two collectors: The NOVA-1 collector is intended to produce saturated steam, while the SuperNOVA collector using the same concentration but having a special vacuum tube receiver for direct steam generation similar to the standard receivers used for parabolic troughs (see test data e.g. Burkholder & Kutscher, 2009). These receivers have thicker metal walls to contain the high pressure superheated steam. Similarly, the Industrial Solar LF-11 has an evacuated tube receiver but uses lower concentration. The SuperNOVA collector clearly has the lowest loss per aperture area (Fig. 7.39). This is not only due to the evacuated absorber (as the example of Industrial Solar LF-11

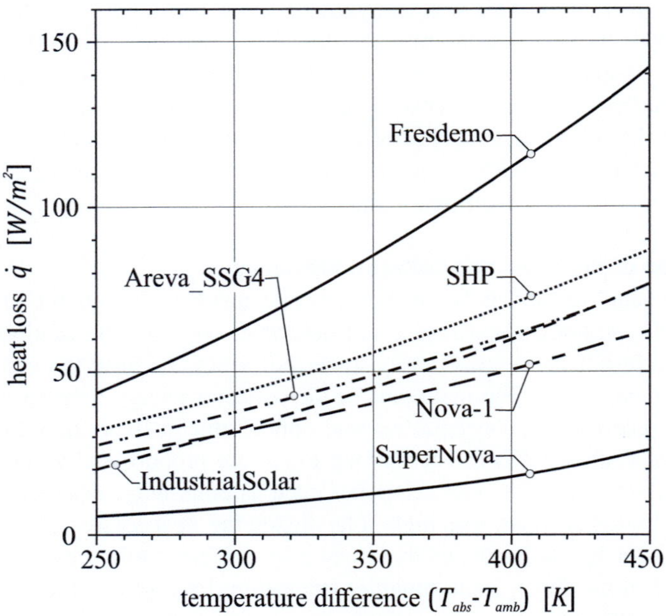

Fig. 7.39 Heat loss per aperture/reflector area for different commercial LFC developments

Table 7.8 Thermodynamical geometric (C_g) and intensity based (C_g^*) concentration factors of some commercial collectors

Collector type	Concentration C_g (thermodynamic)	Concentration C_g^* (intensity based)
SPG Fresdemo	34.1	107.1
SHP V1.0	17.6	55.2
Areva SSG4	20.2	63.6
ISG LF-11	25.0	78.6
Nova-1	54.6	171.4
SuperNova	54.6	171.4

shows), but also caused by the high geometrical concentration C_g. Thermodynamically the absorber tube circumference is relevant for the concentration, however typically in CSP industry the mirror aperture area is related to tube diameter which gives a factor π higher factor C_g^*. The reason given is that mainly the projected area (diameter × length of the tube) is illuminated by the focused sun. It may be seen as a kind of intensity-based concentration. In Table 7.8, therefore, both concentration factors are compared and stated for commercial collectors with heat losses shown in Fig. 7.39. The value of C_g^* may be compared to the concentration 78 of the Eurotrough parabolic trough collector.

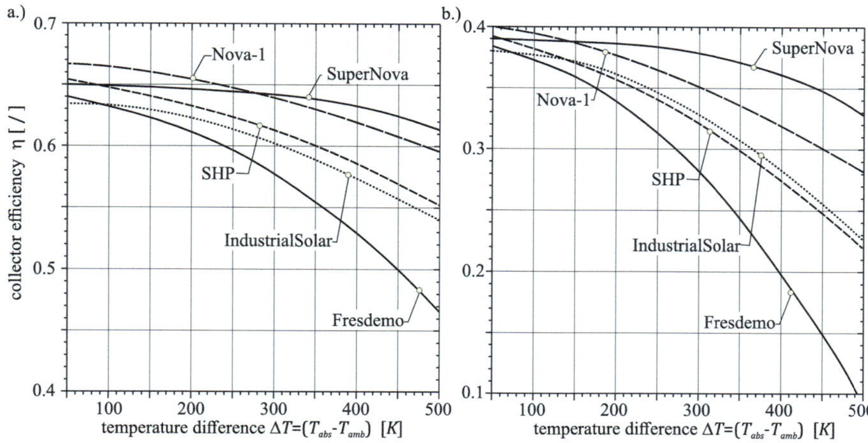

Fig. 7.40 Collector efficiency for commercial LFC as a function of the temperature difference between mean absorber and ambient temperature ΔT. **a** $DNI = 950$ W/m^2 and sun in the zenith ($IAM = 1$); **b** $DNI = 600W/m^2$ and low sun ($IAM = 0.6$)

When we consider and compare overall collector efficiency curves or collector output power, both the concentration and the optical efficiency is of utmost importance. The heat losses are less important especially for high DNI-values, as illustrated in Fig. 7.40a. The comparison with a case of lower DNI and less favourable incidence angle is shown in Fig. 7.40b. Due to off-angle incidence (sun not in the zenith resulting in an IAM of 60%) and lower DNI the solar gain is reduced, and all curves show lower efficiency. However, the curves of the different collectors show a wider spread compared to Fig. 7.40a. For the SSG4 collector no published data exist to estimate the optical efficiency, therefore the collector efficiency curve has been left out here.

Optical errors have a considerable effect on the thermal efficiency of a LFC. This means there is a large potential to improve first designs which only considered a part of the factors. In Fig. 7.41 the different optical loss effects for normal incidence are quantified. The case study shows optical loss effects for normal incidence ($\theta_T = 0°$; $\theta_i = 0°$). The first part shows losses due to the primary mirror field geometry (cosine, shading, blocking, η_{opt}) which are in more detailed outlined in Chap. 10, then receiver losses (acceptance of secondary geometry and receiver shading), and losses due to material properties (reflection, transmission, absorption). The light blue columns show a plausible range for the optical efficiency of a collector for lower quality components and high-quality components.

The geometrical losses of a standard LFC can be influenced by simple design optimizations using the parameters mirror width, gaps, receiver height and number of mirror lines. They are already very small. The receiver design still has some optimization potential, and better materials can contribute to a higher optical efficiency. For off-normal incidence the geometrical design is much more important,

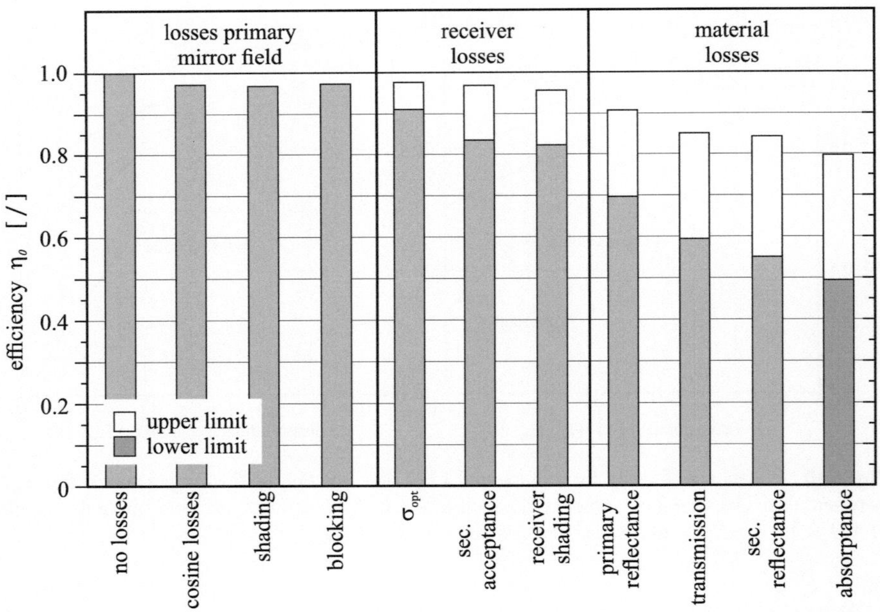

Fig. 7.41 Optical loss effects—overview on geometrical and material al losses from Heimsath et al. (2014a, 2014b, 2014c)

and the avoidance of mirrors blocking or shading each other is a key element of design.

The different optical losses of a linear Fresnel collector can be quantified for typical cases. There is a large range of the achieved optical efficiency depending on the design effort and material selection. Figure 7.41 shows the different effects indicating an upper and lower limit of the resulting optical efficiency η_0 by cumulating the effects of basic geometry, mirror shape and optical properties of the materials selected.

7.2.2.3 Evaluation Local LFR Collector Performance—Convolution Methods

Due to the complex geometrical Fresnel collector structure, the local optical modelling of an LFR collector is mostly executed numerically. In principle two methods can be applied:

- Monte Carlo based ray tracing methods (MCRT), similar as for PTC collectors.
- Convolution methods.

Both methods MCRT and convolution methods require a source description, i.e. a modelling of the sun disk accounting for the non-uniformity of the flux distribution on the solar disk and the widespread solar half-angle of 4.66 mrad (0.267°).

Almost all simulation assume that the rays pass through air with 100% transmittance, and the reflectance of the reflectors (primary and secondary mirrors is constant; the same is true for transmittance and refractive index of the receiver glass cover. The difference between both modelling approaches is that in contrast to the convolution method the MCRT solutions still contains the direction of the integral solar flux and by implementing an angle-dependent absorptance the product can be evaluated exactly and thus the solar conversion in the receiver tube is more accurately resolved. However, on the other hand, the MCRT methods need to simulate the ray paths of an enormously large number of photons to achieve a reasonable convergence of the solution. A convolution method provides a unique function as solution, while the MCRT provides rather results of a numerical experiment including uncertainties in each data point, which can be reduced by making many experiments. This is paid by higher computational effort especially when using a fine spatial discretization of all mirror elements to attain a high-quality solution. It can be shown that both approaches agree well and provide a code-to-code validation. The absolute typical computation time for MCRT exceeds that of the convolution more than an order of magnitude (Mertins, 2006), however, it can be faster, since it allows for parallelization using multi-processor computing.

A simpler modelling of just shading and blocking modelling can be provided much faster and are similar as for a PTC, however, requiring in contrast to PTC also consideration of transversal angle dependence (θ_T) in LFR collector. Details may be taken from Sharma et al. (2015), Canavarro et al. (2014) or Qiu et al. (2015).

For their simplified method (Sharma et al., 2015) have derived numerous analytical expressions to illustrate the correlations between the various geometric parameters and annual shading losses, annual blocking losses, and annual energy losses. These correlations are significant because they can be used to quickly and accurately evaluate the effects of a variety of factors on optical performance. Analyses of LFR collectors with different mirror shifts and widths show only marginal changes in annual efficiency. For example, the numerical calculations show that increasing the receiver height by 30% may increase the amount of energy by 8% and increasing the mirror length by 30% can increase the amount of energy by 24% for short collectors. Conversely, increasing the mirror width and decreasing the receiver height leads to an increase in shading.

Simulated Optical Efficiency of LFR Collectors

LFR concentrator, generally have inferior performance due to optical losses in the mirror field compared to a PTC. These include blocking of reflected radiation and shading of adjacent mirrors from incoming radiation, similar as in parabolic troughs, within the module itself, which are not present in a PTC. The reason for the reduced optical efficiency of an LFR collector arises from its geometrical set-up consisting of parallel arranged primary mirror strips with a co-planar tracking axis. By such a configuration the collector cannot conserve the solar beam etendue even if the Fresnel reflector (primary mirror) consists of an infinite number of infinitesimally small mirrors forming a quasi-straight line, as schematically shown

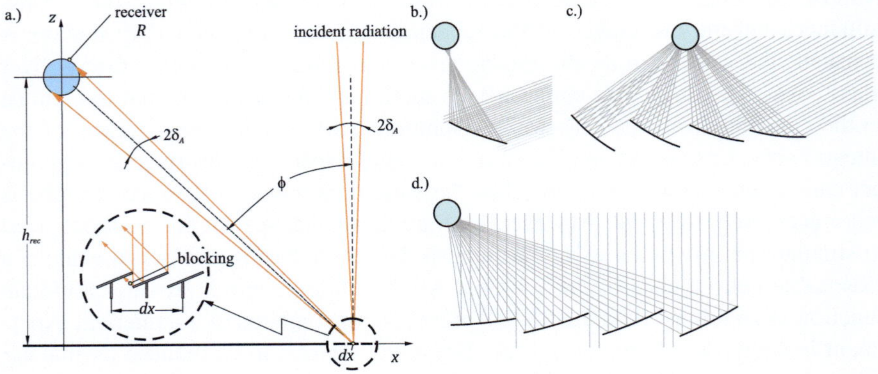

Fig. 7.42 **a** Etendue mismatch of incoming and emerging beam in a linear Fresnel collector array; **b** visualization of ray paths of cosine loss effect; **c** shadowing and **d** blocking/through losses

in Fig. 7.42a. In order to explain this essential feature, let us consider a small segment of the mirror on the *x*-axis deflecting a beam onto a circular receiver *R*. Therein, the infinitesimal incident etendue dU_i illuminating dx is given by:

$$dU_i = 2dx \sin \delta_A. \tag{7.29}$$

But the infinitesimal etendue dU_o emitted from dx within a cone $2\delta_A$ as well as inclined by an angle ϕ with respect to normal incidence to dx and pointing to the center of the receiver is given by:

$$dU_o = 2dx \sin \delta_A \cos \phi, \tag{7.30}$$

where $\cos \phi$ is the projection factor of dx in direction of the receiver. β itself depends on the position of the mirror in $x-$ direction and is given by

$$\beta(x) = \tan^{-1}(x/h_{rec}), \tag{7.31}$$

in which h_{rec} denotes the receiver height above the mirror plane. Thus, the beam width (etendue) of the incident light dU_i is greater than the etendue of the light emitted by dx dU_o, so the etendue is not preserved. The incident light that does not reach the receiver is lost due to shadowing or blocking between the heliostats or passes through the heliostats without being captured (through losses). To visualize shading, blocking or through losses they a shown in Fig. 7.42c–d, compared to cosine losses (see Fig. 7.42a) also present in PTC collectors.

In the case of cosine losses, which are shown schematically in Fig. 7.42b, only the primary mirror surface projected perpendicular to the incident solar radiation reflects. Therefore, the proportion of solar radiation related to the mirror surface corresponds to the cosine of the mirror normal and the direction of incidence. Depending on the position of the sun, the cosine losses are lowest for the primary mirror closest to the direct connection from the position of the sun and the receiver. This mirror never experiences a perpendicular incidence because it is shaded by the receiver itself. Particularly with larger transversal angles θ_T, shadowing of neighbouring primary mirrors occurs, as can be seen in Fig. 7.42c, whereby the shadowed part remains ineffective because it does not reach the receiver. It can be seen from Fig. 7.42c that the transverse angle at which shadowing begins depends mainly on the geometric arrangement of the primary mirror field and the receiver. A larger distance between the primary mirrors (called gap or shift) shifts the onset of shading to lower sun angles.

For small transversal angles especially for primary mirrors far out in x-direction, a part of their mirror surface may reflect radiation against the back of the neighbouring mirror (compare Fig. 7.42d). Increasing the receiver height h_{rec} reduces the tilt angle β_n, which means that mutual occlusion can be almost completely avoided. If the primary mirrors are pulled further apart, the masking losses also decrease and then have hardly any influence on the optical efficiency. However, this measure leads to a lower area utilization of the solar field, which is described by the fill factor FF of the solar field. The fill factor is defined as the quotient of the effective total primary mirror area A_{pr} and the solar field area A_{SF} occupation of the collector.

$$FF = A_{pr}/A_{SF}. \quad (7.32)$$

Primary Mirrors

If we assume primary reflectors of equal width, the curvature and inclination of each primary reflector must be adjusted so that the rays intersect the absorber at the focal point after reflection from all primary reflectors. The primary reflectors can have linear, parabolic, or circular profile shapes with a large radius of curvature. Furthermore, the primary mirrors can have the same focal length f or a variable radius of curvature among themselves. For a simplified description of the calculation process, we assume,

- that the primary mirrors perfectly follow the apparent motion of the sun,
- the solar radiation is vertically incident on the primary mirror, and
- the solar flux is constant within a cone with a half vertex angle $\delta_S \approx 4.65$ mrad.

For the sake of simplicity, we will consider reflectors with the same width. The angle of inclination β_n of each reflector is set so that the incident rays are reflected onto the absorber at the focal point. To prevent shadowing of a reflector by an adjacent reflector, a constant distance (called shift or gap) between two adjacent mirrors distance is provided. The $n-th$ reflector in the LFR array is thus characterised by three parameters, namely the lateral position (x_n), inclination (β_n) and shift (Δx_n). The angle that the $n-th$ primary mirror makes with the horizontal plane is the tilt angle (β_n) and is calculated according to:

$$\beta_n = \frac{1}{2}\tan^{-1}\left(\frac{x_n + (W_{pr}/2)\cos\beta_{n-1}}{f_n - (W_{pr}/2)\sin\beta_{n-1}}\right). \tag{7.33}$$

The gap (shift) between the adjacent mirrors shifts (Δx_n) is provided to enhance the performance of the LFR system by reducing the shade created by adjacent mirrors, particularly at large incidence angles and can is given by

$$\Delta x_n = W_{pr}\sin\beta_{n-1}\tan(2\beta_n + \varepsilon_0), \tag{7.34}$$

in which ε_0 solar subtense angle in mrad The lateral position of the $n-th$ mirror assuming collector symmetry can be acquired from the center of the mirror field through the relation:

$$x_n = x_{n-1} + W_{pr}\cos\beta_{n-1} + \Delta x_n \tag{7.35}$$

for $n > 1$. Equations (7.33)-(7.35) are solved iteratively to obtain the values of x_n, β_n and Δx_n, starting from the initial guess of $\beta_n = 0$, Δx_n, $x_0 = -W_{pr}/2$, $x_1 = W_{pr}/2$ and $n = 1, 2, ..., m$, with 'm' the total number of primary mirrors. The curvature radius of the primary reflectors depends on the focal length (f_n) and tilt angle (β_n) of the individual reflectors. The following Eqs. (7.36)-(7.38) are then used to determine the radius of curvature of circular/parabolic primary reflectors. Optical angle (ϕ_n) is defined as the angle between the optical axis and the line from the focus to the primary reflector being

$$\phi_n = \tan^{-1}(x_n/h_{rec}) \tag{7.36}$$

from which the focal length of the $n-th$ primary mirror is obtained by

$$f_n = \sqrt{x_n^2 + h_{rec}^2} \tag{7.37}$$

Thereby, the radius of the $n-th$ parabolic shaped ($r_{pr,n,p}$) or circular shaped (($r_{pr,n,c}$)) primary mirror is pre-scribed by

$$r_{pr,n,p} = f_n/2 \text{ or } r_{pr,n,c} = 2f_n/\cos\beta_n \tag{7.38}$$

Since the radius of curvature of the primary mirror is much larger than the receiver aperture, the profile of the primary reflector has little effect on the flux distribution arriving at the receiver, regardless of whether the primary mirror shape is circular or parabolic. The aim of the primary reflector optimization is to adjust the focal length and the radius of primary mirror curvature so that the energy absorption within the receiver cavity is maximized. Simulations show that the flux distributions to be achieved in the receiver cavity for different primary reflector profiles differ only marginally as long as there is a focus (Balaji et al., 2016).

Secondary Reflector and Receiver
With the aid of secondary reflectors, for a given combination of primary reflectors and absorbers, the target aperture can be enlarged, which allows the total heat flow on the absorber tube to be increased considerably. The receiver receives radiation from the primary mirror field from a large angular range, often more than $\pm 50°$. Due to this relatively large acceptance angle, the additional concentration achieved by the secondary reflector is rather small, so that the actual task of the secondary reflector is not to concentrate the radiation, but to redirect it in order to direct rays reaching the entrance aperture of the secondary receiver onto the absorber. Figure 7.43a shows the calculated beam path when using a planar primary mirror on a secondary receiver with a CPC profile, and Fig. 7.43b a primary mirror with a long focal length. If the divergence of the solar radiation (sunshape) is superimposed on the primary mirror in the form of an optical error (e.g. $\sigma_{opt} = 3.5$ mrad), it is clear from Fig. 7.43c that with the use of the CPC reflector all incident radiation continues to reach the absorber. The increased target aperture allows more primary mirrors to be installed relative to a target aperture, further reducing not only cost but also specific heat losses through convection. The latter is based on the fact that the secondary reflector system, which contains the absorber, is integrated into a thermally insulated structure that also serves as protection against wind. The use of a secondary reflector also facilitates the sun tracking process, the line focus of the primary reflector is not a point but a target area. One effect of using a secondary reflector is a more uniform circumferential distribution of solar irradiance on the absorber tube and thus a reduction in thermal loads and especially gradients, which is less than with parabolic troughs, which is illustrated in Fig. 7.43d for a computed solar flux distribution of on the perimeter of an LFR absorber tube. In contrast to the PTC (compare Fig. 7.13a), the absorber tube of the LFR experiences significant solar irradiation over almost the entire circumferential length. The secondary reflector geometry depends on the acceptance angle $(2\theta_{a,s})$ of the mirror shape, whereby the acceptance angle is defined as the angle over which the flux may deviate from the normal to the aperture, but still just reaches the absorber surface after the back reflection.

Figure 7.44 shows the angular dependence (θ_T) of the different loss mechanisms. The cosine losses here are defined differently from the PTC case: for the PTC the incidence cosine on the aperture plane is considered, whereas for the LFC often the incidence cosines on the individual mirror elements are taken into account. Therefore, an integrated cosine loss exists already for normal incidence $\theta_T = 0°$. The solid curve shows the cumulated cosine loss without considering the shading and masking losses on the geometric optical efficiency η_{opt} in the transverse plane. Since the mirror normal is the angle bisector of the irradiation direction and the mirror-receiver direction, the cosine losses do not approach zero even at flat sun angles.

If the sun's position deviates from the transverse plane by angle θ_i, part of the solar radiation reflected from the primary mirror field to the receiver is reflected beyond the ends of the receiver and can no longer be utilized. The usable portion can be estimated with the ratio of receiver height h_{rec} and collector length L_{col},

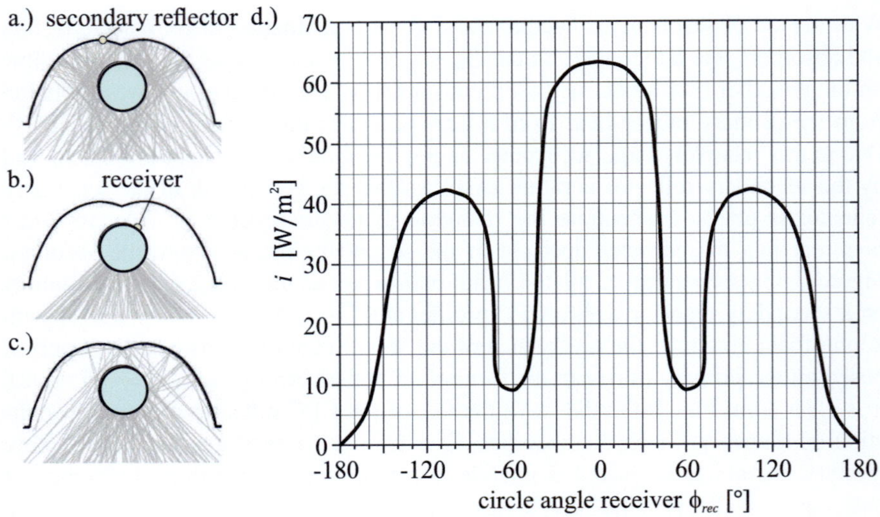

Fig. 7.43 MCRT calculated beam path in a secondary reflector (Mertins, 2006). **a** flat primary mirror ($f_{pr} \rightarrow \infty$), **b** primary mirror with large focal length f_{pr} ($f_{pr} = 17$ m); **c** primary mirror of large focal length ($f_{pr} = 17$ m) and optical error ($\sigma_{opt} = 3.5$ mrad) with additional divergence of solar radiation (sunshape); **d** computed solar flux distribution as a function of the circumferential angle ϕ_{rec} ($W_{pr} = 1.07$ m, $L_{rec} = 12$ m, $2r = 70$ mm, $h_{rec} = 8$ m, $\sigma_{opt} = 4.65$ mrad) modified from Balaji et al. (2016)

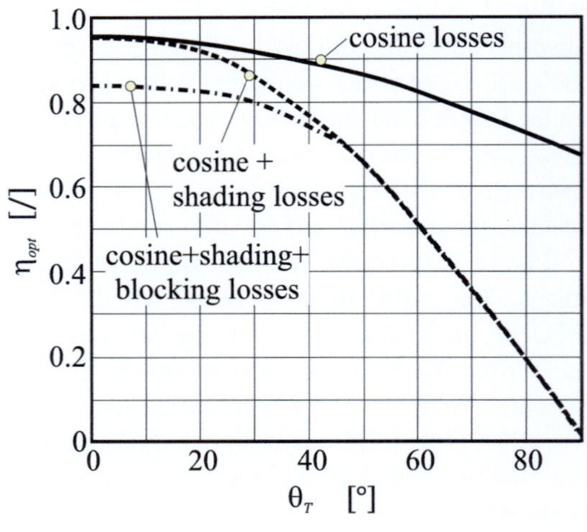

Fig. 7.44 Influence of cosine, shading and masking losses on the geometric efficiency η_{opt} as a function of the transverse angle θ_T for different collector geometries ($h_{rec} = 9$ m, $\Delta x = 0.1$ m)

collector width W_{col} and the angle of incidence θ_{LS} (IEC6 2862-5-2, 2022):

$$\eta_{end} = 1 - \tan(\theta_{LS}) \cdot \frac{h_{eff}}{L_{col}};$$

$$h_{eff} = \sqrt{\left(\frac{W_{col}}{4}\right)^2 + h_{rec}^2}. \qquad (7.39)$$

This effect is called end loss, and it is calculated for mirror rows and receiver of equal length and same position. Consequently, the shorter the absorber is in relation to the receiver height, the higher the end-of-row losses are. For a north–south orientation, row-end losses are greater at locations of higher northern latitude Φ. To avoid this, the receiver could be extended beyond the primary mirror field. Effort and benefit must be compared here. The receiver itself casts a shadow on the primary mirror field that migrates with time, so these shadowed areas do not contribute to the receiver gain. The shadow of the receiver runs moves across the primary mirror field as a function of the transverse angles (θ_T). This effect is called receiver shadowing. The portion obscured by the structure in this plane is of the order of about 2%, which is the quotient of the support span and the width of the supports. For each pass, the portion, called the reduction factor, is reduced by a flat 2%.

Geometrically, a linear Fresnel collector is defined by a multitude of geometric parameters, including:

- number of mirror rows N,
- distance between the mirrors Δx_n,
- mirror shape primary and secondary reflector (flat, parabolic, circular)
- mirror height,
- mirror width $W_{pr,n}$,
- height of the receiver assembly h_{rec},
- size of the receiver assembly,
- geometry of the receiver assembly,
- length of the collector assembly/line length (L_{rec}, L).

All the above geometric parameters can vary from one collector to another and it extremely difficult to optimize a linear Fresnel collector with respect to most or all the relevant parameters, especially in conjunction with the optical properties of the collector. In most cases, one starts from a basic empirical pattern of a linear Fresnel collector and optimizes the few selected parameters that are most important for overall optical and/or economic performance.

In this context, the question often arises as to which parameter is best suited as a comparative variable for evaluating the different collector configurations. At the collector level, the evaluation of the individual configurations is based on annual average values for a defined location. In the literature, different parameters can be found, such as the $\eta_{\text{opt, a}}$ annual optical efficiency ($\eta_{opt,a}$) and the$\eta_{\text{en,ann}}$ annual energetic efficiency ($\eta_{en,ann}$), but also $\eta_{\text{ex,ann}}$ annual exergetic efficiency ($\eta_{ex,ann}$), which are defined as follows.

$$\eta_{opt,ann} = \frac{\int_0^{365}\left[\eta_0 \cdot K(\theta_T, \theta_L) \cdot G_{b,n}\right]dt}{\int_0^{365} G_{b,n} \cdot \cos\theta dt}; \qquad (7.40)$$

$$\eta_{en,ann} = \frac{\int_0^{365}\left[\eta_0 \cdot K(\theta_T, \theta_L) \cdot G_{b,n} - \frac{U_L}{C_g} \cdot \left(T_{avg} - T_{amb}\right)\right]^+ dt}{\int_0^{365} G_{b,n} \cdot \cos\theta dt}; \tag{7.41}$$

$$\eta_{ex,ann} = \frac{\int_0^{365}\left[\eta_0 \cdot K(\theta_T, \theta_L) \cdot G_{b,n} - \frac{U_L}{C_g} \cdot \left(T_{avg} - T_{amb}\right)\right]^+ \cdot \left(1 - \frac{T_{amb}}{T_{out}}\right) dt}{\int_0^{365} G_{b,n} \cdot \cos\theta dt}; \tag{7.42}$$

in which $G_{b,n}$ is the direct normal solar irradiance incident on the collector aperture with an incidence angle θ, $K(\theta_T, \theta_L)$ is the incidence angle modifier function for direct normal irradiance (including the incidence cosine), A_a the collector aperture area,T_{avg} is the average fluid temperature in the collector, and T_{out} the outlet fluid temperature of the collector loop. The annual energy efficiency considers only the positive values of the balance between collector gains and collector losses over the year, which implies that for lower irradiance values the collector must be switched off. The average fluid temperature T_{avg} must be estimated from the operational details similar to the outlet temperature T_{out}. When mass flow through the collector is varied to hold the outlet temperature constant, the values are more or less identical with the set point temperatures, however for fixed mass flow, the estimation is more difficult and might be calculated iteratively for each time step.

Indeed, the formulations of Eqs. (7.40) to (7.42) are simplified and have to be understood in their meaning. The optical efficiency $\eta_{opt,ann}$ provides a measure of how much solar radiation is provided to the receiver over a year, including the effects of off-normal incidence. In this view it can be conceived as a measure of the optical quality of the collector-assembly for a specific location in contrast to the optical efficiency at normal incidence $\eta_{opt,0}$, which is independent of latitude and climate. The annual energy efficiency gives a measure of the total thermal energy provided by the collector for a specified temperature range. And the exergetic annual efficiency includes as metric the Carnot efficiency calculated with the outlet temperature and hence gives a measure of possible work (or electric generation) to be provided.

The result of every local collector calculation is the determination of the optical efficiency, which includes the largest losses. Here, a distinction must be made between the optical efficiency at perpendicular incidence and the annual optical efficiency. Figure 7.45a shows the distribution of the different shares of the loss mechanisms in the optical efficiency $\eta_{opt,0}$ at normal incidence ($\theta_i = 0°$) and integrated over the year $\eta_{opt,ann}$ (Fig. 7.45b). The figure refers to a specific optimized Fresnel collector absorber, but the proportional distribution of other LFR collectors is similar in magnitude.

In the optical efficiency $\eta_{opt,0}$ at normal incidence, the material-specific properties such as absorption, reflection and transmission are of decisive influence. Large values provide information about the quality of the construction and the materials used.

The optical annual efficiency $\eta_{opt,ann}$, which is much more important for estimating annual yields, is mainly influenced by the geometry-related losses such as

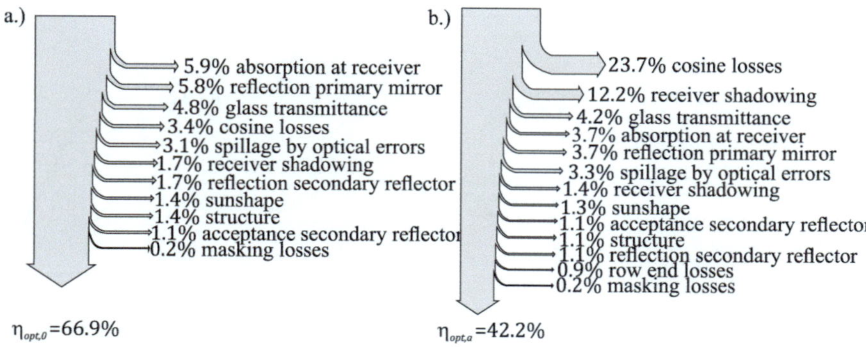

Fig. 7.45 Sankey diagrams for the optical loss fractions at normal incidence $\eta_{opt,0}$ (**a**) and annual yield $\eta_{opt,ann}$ (**b**) of an optimised Fresnel collector (Mertins, 2006)

cosine and shading losses. In typical optimisation for power plant planning, losses due to spillage and sunshape of about 4.5% are assumed. This value can be further reduced during fine optimisation of the collector geometry. However, it should be noted that a focus on reducing these losses is immediately detrimental to the levelized cost of electricity ($LCoE$) or heat, because this would require reducing the absorber or receiver height as well as the number of mirrors to reduce the length of the reflected rays and thus the spillage. A lower receiver height in turn increases the cosine losses due to the larger angle of incidence of the primary mirrors. A lower number of mirrors increases the specific heat losses and the specific costs of the collector.

Sensitivity Analysis of the Optical Losses in a Fresnel Collector

Typically, the first step of LFR collector design optimization refers to the distance between two adjacent rows of mirror strips, which requires a calculation of the intercept for different incidence angles in order to analyze the interactions between the rows in terms of cosine losses, blocking and shading, especially at high angles of incidence. The highest value of the intercept factor is usually reached when the distance between the mirror strips is zero, at which the Fresnel collector behaves similarly to a parabolic trough collector although it exhibits lower intercepts. This is not a practical solution, as it poses considerable challenges to both the mirror strips and their maintenance, installation, and cleaning, but also the installation of the tracking systems. The Fig. 7.46a illustrates the intercept γ of a LFR collector at a given site in north–south orientation at one time as a function of the receiver height h_{rec} for different mirror strip shifts Δx. Exceeding a mirror gap of $\Delta x =$ 0.1 m yields a significant loss in the intercept. It can be also noticed in the figure that for receiver heights h_{rec} below 8–10 m the intercept drops to unacceptably low values. Small gap sizes Δx correspond to large fill factors FF (see Eq. (7.32)) being used in other optimization studies as initial step.

The second step is to analyze the time-of-day response of a collector. Since an LFR collector is single-axis tracking cosine losses play the greatest role integrally

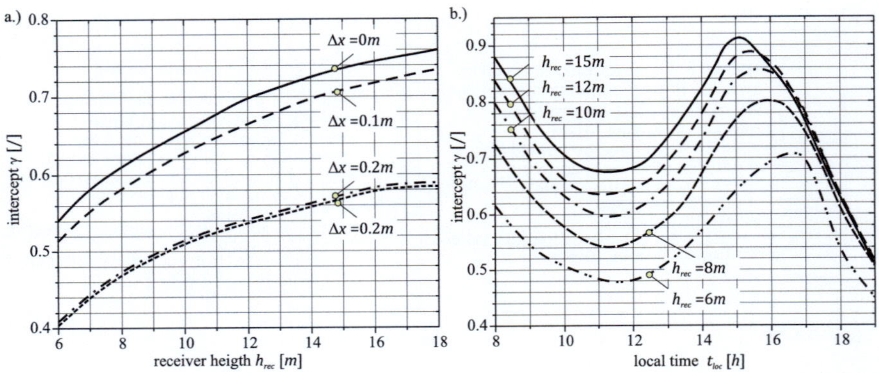

Fig. 7.46 Simulated intercept factors γ for a LFR-collector ($\Phi = 36.33°$, $\lambda = 10°21'$, $W = 1.25$ m) in north–south orientation; **a** gap width Δx for different receiver heigths h_{rec} at local time 10 am; **b** time-dependent intercept for different receiver heights h_{rec}; modified from Eddhibi et al. (2017)

over the year; these in turn depend on the orientation of the collector, the location, and the local time. In contrast to parabolic troughs, LFRs show a more complex time response of the diurnal cycle, as illustrated in Fig. 7.46b. The simulations show two peaks of the intercept factor throughout the day (from 8 am to 7 pm); the first in the morning and the second at the site shown at 3 pm. Such calculations are particularly relevant for control and regulation at the power plant level, as the thermal output of the solar field is usually selected to be greater than that of the power plant unit (solar multiple $SM > 1$) and thus feeds a thermal storage or buffer. The graphic also shows that the intercept depends essentially on the receiver height above the primary mirrors. Large intercept factors require receiver heights in the order of 7–10 m, a further increase of the receiver height is not sensible, since no significant additional gain in the intercept factor is achieved, on the contrary, the associated technical effort leads to a disproportionate increase in costs. As can also be seen from Fig. 7.46b, the intercept factor reaches its maximum when the angle of incidence is high and decreases at noon when the angle of incidence is close to zero, which is due to the orientation of the LFR collector considered ($N - S$). When LFR is oriented north–south, it follows the sun from east to west. It reaches the maximum value of the intercept factor twice, the first time in the morning and the second time in the afternoon, which corresponds to the lowest cosine losses. However, the lowest value of the intercept factor is reached at noon, resulting in the highest cosine losses. Compared to the east/west configuration, the north/south configuration has a more constant annual output; in summer it collects a larger amount of energy than the other configurations. Although it has a problem in winter, as it collects less collector yield than the East/West orientation (especially at higher geographic latitudes).

The next step is to optimize the height of the receiver above the primary mirror field together with the number of mirror rows. A time around noon is chosen

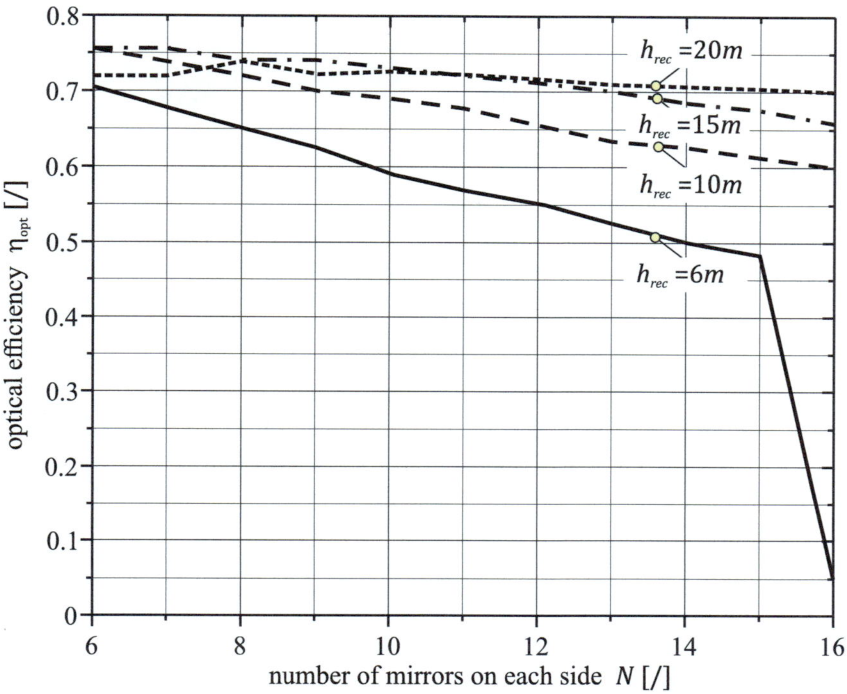

Fig. 7.47 Optical efficiency η_{opt} for a LFR-collector ($\Phi = 36.33°$, $\lambda = 10°21'$, $W = 1.25$ n) a one time instance in north–south orientation as a function number of mirrors on each side N and different receiver heights h_{rec} modified Eddhibi et al. (2017)

where the angle of incidence is close to zero to minimize blocking and shadowing effects and the line size is the optical efficiency. Figure 7.47 shows that the optical efficiency decreases monotonically as the number of mirror rows increases for a certain receiver height and a defined gap width. Therefore, the amount of blocking and shadowing produced by the rows far from the receiver has a greater impact than the amount of reflected solar radiation in the center of the primary field. The optical efficiency increases in direct proportion to the height of the receiver from the primary mirror field, as the blocking and shadowing effect is less the higher the receiver is placed. The spillage effect is defined as solar beams that miss hitting the receiver due to design specifications or errors. This effect increases for an LFR-collector that consists of a large reflective area and low receiver height as indicated in Fig. 7.47. The choice of optimal receiver height requires the consideration of blocking, shadowing, and scattering losses which all targeted to be reduced to a minimum. Thus, most commercial designs use about 9–12 mirors on each side with a receiver height form 7–10 m, allowing for a robust power up-and downscaling of the collector with a low sensitivity and a manageable tracking effort.

The diverse geometry parameters of an LFR collector also allow studies on the influence of the curvature of the primary mirrors, the shape of the secondary reflector, the number and size of the absorber tubes in the secondary receiver, on which there is a rich technical literature. Almost all concepts on the market have minimally cylindrically curved primary mirrors, which have a similar performance at typical receiver heights as parabolically curved ones but are significantly cheaper to manufacture (Abbas, Montes, Piera, & Martínez-Val, 2012). For secondary reflector shapes, truncated CPC designs have gained acceptance. As far as the number of absorber tubes is concerned, the standard is a centrally arranged single tube with diameters of 0.06–0.09 m, depending on the heat transfer medium and the pressure grading. Larger absorber tube diameters allow a reduction of the solar irradiance hitting the circumference, but at the same time the heat losses increase disproportionately (Morin et al., 2012).

Optical Errors Affecting LFR Collector Performance

In principle, the same geometric errors can occur in an LFR collector as in a parabolic trough (cf. 7.2.1.3.1), such as specularity error, displacement error due to shape or receiver deviation or load error. The segmented design and the considerably smaller component sizes of a LFR allow for considerably better manufacturing accuracies, provided all quality assurance aspects are met, so that their impact on the optical efficiency is smaller than for a PTC. On the other hand, tracking errors and slope errors are much more important for a LFR than for a PTC due to the larger focal distance. Compared to a PTC the standard deviation of the optical error in a LFR modifies with respect to the tracking error, which is doubled, to:

$$\sigma_{opt}^2 = \sum_{i}^{n} (a_i \sigma_i)^2 = \sigma_{spec}^2 + \left(2\sigma_{slope}\right)^2 + \sigma_{dis}^2 + (2 \cdot \sigma_{trac})^2. \tag{7.43}$$

The effects of the tracking error (σ_{trac}) on the optical efficiency and the local circumferential absorber concentration ratio at a tilt error $\sigma_{slope} = 2.5$ mrad are shown in Fig. 7.48a, b. Figure 7.48a shows that the optical efficiency $\eta_{i,opt}$ of the LFR decreases rapidly with increasing σ_{trac}, because the larger σ_{trac}, the more solar radiation reaches the secondary receiver aperture. Figure 7.48b depicts that the distribution of the local concentration ratio is preserved in its fundamental shape as long as $\sigma_{trac} < 2.5$ mrad; above this it varies more randomly if σ_{trac} is too large, e.g. $\sigma_{trac} = 3.5$ mrad. It should also be noted that local hot spots with high concentration ratio σ_{trac}, e.g. $\sigma_{trac} = 1.5$ mrad, may occur. In order to achieve high efficiency and to avoid a too high local solar flux, a high tracking accuracy of the order of $\sigma_{trac} = 0.5$ mrad is required, which can be realised by using a stepper motor.

Figure 7.48c, d illustrate the effects of slope σ_{slope} error at normal incidence. Again, the optical efficiency $\eta_{i,opt}$ decreases as σ_{slope} increases. More drastic is the decrease of the local concentration ratio (see Fig. 7.48d), which in the example decreases by more than one third when σ_{slope} increases from 0 to 2.5 mrad. The solar flux absorbed on the upper part of the absorber is only of the order of about 35% of the total flux for $\sigma_{slope} = 2.5$ mrad, and this allows the system to use the

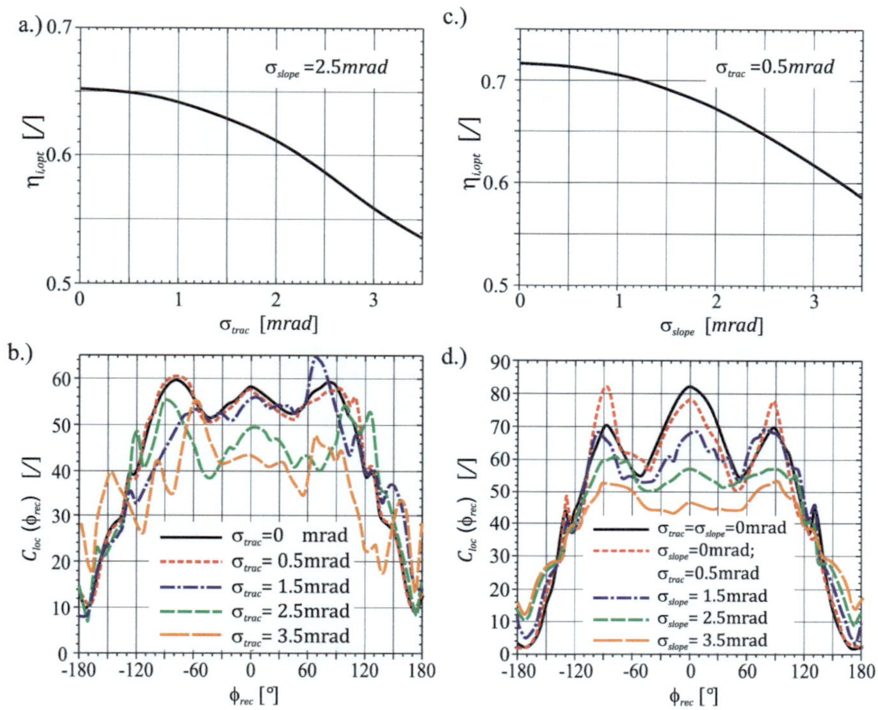

Fig. 7.48 Effects of tracking (σ_{trac}) and slope (σ_{slope}) errors in a LFR-collector at normal incidence ($W_{pr,i} = 0.6$ m, $h_{rec} = 8$ m, $\Delta x = 0.85$, CPC half acceptance angle $\theta_A = 56°$) modified from Qiu et al. (2015). Optical efficiency $\eta_{i,opt}$ (**a**) and local concentration ratio (**b**) as function of circumferrential receiver anlge ϕ_{rec} for different tracking errors σ_{trac} in (mrad at a slope error of $\sigma_{slope} = 2.5$ mrad; Optical efficiency $\eta_{i,opt}$ (**c**) and local concentration ratio (**d**) as function of of circumferrential receiver anlge ϕ_{rec} for different slope errors σ_{slope} in (mrad) at constant tracking error $\sigma_{trac} = 0.5$ mrad

entire circumferential surface of the absorber. On the one hand, this means that a more uniform radiation flux can be achieved by sacrificing a small efficiency with a reasonable slope error, e.g. $\sigma_{slope} = 2.5$ mrad, that can be achieved with existing technologies. On the other hand, the system cost can be reduced by using mirrors with slope error in this range.

7.2.2.4 Technical Designs of Linear Fresnel Systems

Mechanical Set-Up of a Linear Fresnel Collector

The design of linear Fresnel collectors consists of rows of mirrors arranged in parallel, which in turn consist of segmented primary reflectors placed near the ground at a height of about 0.5–2 m. Due to this design, the area-specific weight is very low; optimized systems have <30 kg/m^2 and thus have a much lower value than parabolic trough collectors. The design results in very low wind loads and thus changes in optical performance due to torsion are small. Moreover, as well as the

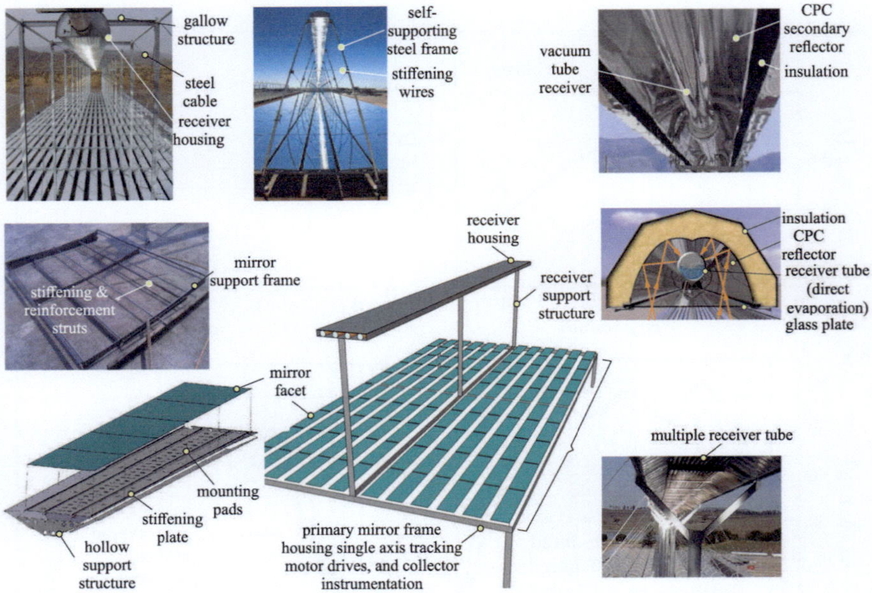

Fig. 7.49 Set-up of a linear Fresnel collector; top left gallow structure (Bernhard et al., 2009); self-support structure receiver structure (top middle); mirror facet support structure types: frame type (left middle), plate type (left bottom). Right: Vacuum type receiver in open CPC (top) former AREVA), single tube receiver (direct evaporation, FRENELL, middle); open multi-tube cavity receiver (Abbas et al., 2016), bottom)

good weight distribution places low demands on the foundations. For a parabolic trough the concrete foundations constitute about two thirds of the total weight of about 118 kg/m^2 which is not needed for a light-weight linear Fresnel construction (Viebahn, Kronshage, Trieb, & Lechon, 2008). The segmented structure does not require heavy machinery for installation. The light-weight support system has as well as modular design allows for relatively low collector costs.

The mirror rows can be erected with close spacing and allow a high land use factor with about 63% of the plant area used as aperture whereas for the Andasol PTC plant in Spain the aperture is only 26% of the total plant area. This saves valuable ground space especially in more densely populated regions, or for smaller process heat installations, the integration on the horizontal roof is more effective. The primary mirrors can be cleaned automatically, for example with small robots and almost without water consumption, and individual mirror rows are easily accessible; both aspects allow relatively low operation and maintenance costs.

The receiver of the LFR collector is fixed on a stand structure mostly made of steel and does not move, which in turn does not require flexible high-pressure and high-temperature pipe compensators as e.g. in parabolic trough systems. It is located several meters above the ground. A typical picture of the mechanical set-up of LFR collector is shown in Fig. 7.49 center.

For the elevation of the receiver support, there are concepts with a steel gallows structure fixed to the ground by steel cables like in FRESDEMO (Fig. 7.49 top left) or cantilevered steel frame structures. There are also several concepts for primary strip supports, all based on a steel frame. The uniaxially movable mirror facets are mounted on the frame. On this frame structure not only, the tracking motors/ gearboxes are installed but also the instrumentation and controls for each collector. For the facet carriers on which the primary mirrors are located, there are concepts based on a frame structure with support struts on which the mirrors are glued or screwed. Other concepts are based on a rigid aluminum or steel body in which support pads are integrated to fix the mirror facets.

The receivers are usually fixed and suspended from the steel structure at a height of several meters. This allows good accessibility and automatic cleaning. For the receiver elements, secondary receivers based either on a CPC structure (Fig. 7.49 right top and center) or on a W-trough have prevailed, which in turn are located in a thermally insulated steel housing to keep heat losses low along the often several hundred meter long piping lines. The receiver tubes themselves vary considerably depending on the choice of heat transfer medium and the design of the selective receiver tube. Some concepts are based on receiver tubes that rely on vacuum tubes, as in the parabolic trough. This allows small convective heat losses and eliminates the need for a glass plate cover on the bottom of the secondary reflector (Fig. 7.49, top right), but results in a costly cleaning of the same. Another concept, especially for direct water evaporating receivers, relies on a single tube placed in the center of a CPC secondary reflector, which is covered at the bottom by a planar glass plate. This allows easy cleaning but higher convective heat losses. Another design consists of several parallel receiver tubes placed in a trapezoidal cavity, also open at the bottom, located on the top of the secondary receiver.

7.2.2.5 Applications and Outlook

The typical fields of application of linear Fresnel collectors cover a considerably wider spectrum than that of power plant applications as compared to parabolic trough systems. Existing LFR collectors are operational in the following application areas such as

- solar power plants with and without storage,
- solar process heat such as textile processing (bleaching, dyeing, …..) or food processing (drying, distillation, evaporation),
- solar-assisted heating/comfort cooling,
- sea water desalination and wastewater treatment,

In general, the PTC shows higher optical efficiency than the LFR because the segmented primary reflectors result in optical losses due to blocking and shadowing, and the aperture is horizontal and not tilted towards the sun as for the PTC. The row end optical losses of the LFR cannot be neglected if the collector length

(L_{col}) is small, cf. Equation (7.39). Because there is a longitudinal and a transversal $IAM(\theta_T, \theta_L)$ in the LFR, which are both smaller than one (in contrast to the PTC), for a single collector there is intrinsically a smaller optical efficiency than for the PTC, especially noticeable at higher geographical latitudes. On the other hand, the row-shading effect of PTC rows also has a strong impact on the optical efficiency for low sun altitude in the mornings and evenings.

Due to the robust, simple, and cost-effective structure and the lower requirements for maintenance and operation, the smaller area requirements, and the horizontal installation of the primary field an LFR collector is often quite attractive for applications in industry where space on roofs and ground is limited. Two optical aspects seem to be less in favor of a LFR than a PTC collector.

- lower overall optical efficiency at high transversal incidence angles,
- larger variation of optical efficiency during the day as a consequence.

The first aspect may be compensated by lower collector cost per area, the second aspect plays no role when thermal storage can be used to manage the mismatch between hourly generation and demand. When no vacuum receivers are used, certainly the length-specific heat losses are higher than for the standard PTC collectors, however, that is also true for PTC collectors without vacuum receivers used in industrial applications. Moreover, higher possible concentration can compensate this.

Although the maturity of LFC technology in several aspects: optics, flexibility, and aptness to heat transfer media, thermal efficiency and system integration has been shown already, there are still not enough demonstration plants worldwide. LFC must market its specific advantages and the still large potential for technical and economic improvement. On the other hand, there is still an appreciable optimization potential.

To play on the advantages of linearly concentrating systems, several issues must be emphasized:

- Modularity and repetitiveness of reflectors and receivers.
- Short distance between concentrator and receiver.
- Piping of fields.
- Compactness of solar fields.
- Scalability of power plant size.
- Cleaning solutions for soiled mirrors.
- Possibility for maintenance during operation.
- Free choice of collector orientation.
- Minimization of system costs.
- Adaptability to variable land flatness, slope, and form.

The potential of LFC is considerable when advanced optical designs are realized. As already the use of molten salt technology with temperatures up to 550 °C have been demonstrated, and the combination with direct two-tank storage solutions is straightforward, the different daily generation profile with a peak at noon compared with the typical "Hat-curve" of parabolic troughs does not seem to matter anymore. Whether the linear Fresnel collector or the point-focus "Fresnel" collector, the solar tower using heliostat fields have advantages, is therefore not clear for this temperature range, but possibly for even higher temperatures. For the classical "solar salt" consisting of a mixture of $40\%KNO_3/60\%NaNO_3$ at 550 °C, other local conditions and of course economics will have the highest impact on selection.

A cost-effective improvement of an LFR collector is often associated with the design of the mirror-receiver configuration. In principle there are four measures to increase LFR collector efficiency:

- increasing absorber thermal performance,
- design of the primary reflector shape and arrangement,
- design of the secondary reflector, and finally,
- reduction of optical losses.

The individual measures will be briefly outlined below, details of which can be found in the literature listed.

Absorber Thermal Performance

Absorber optimization has two main goals; increasing thermal efficiency and homogenizing the circumferential radiation flux distribution.

The heat transfer side can be improved by optimizing the heat transfer fluid, for example by optimizing heat transfer using finned structures or by mixing additives into the heat transfer fluid, which show only marginal improvement potential (Bellos, 2019). Much more effective, however, is the reduction above all convective heat losses at the receiver, which are significantly greater than the heat radiation losses. Studies show that wind protection and the use of vacuum tubes increase the annual thermal efficiency by up to 10% (Cagnoli et al., 2018).

The non-uniform heat flux on the receiver allows for higher optimization potential and the literature is full of a variety of new structures designed to achieve better heat flux uniformity. A summary can be found in Bellos (Bellos, 2019). The most effective way seems to be to guide the target lines of the primary mirrors so that a part of the heat flux reaches the upper part of the absorber. With this measure, the optical efficiency is not affected. In the same direction, the use of segmented parabolic secondary mirrors (SPC) with the use of variable different target lines. With both measures, both a higher optical efficiency and a larger circumferential receiver flux distribution can be achieved, which is significantly better than the conventional TC (Trapezoidal Concentrator) and the CPC (Compound Parabolic Concentrator).

Primary Reflector Design

The primary reflector is a key component of the LFR system because it redirects the solar rays to the absorber or secondary reflector, and therefore its structure determines the final optical efficiency. Innovative primary reflector designs aim at minimizing shadowing and blocking effects, such as the Compact Linear Fresnel Reflector (CLFR) concept, where the solar rays are directed to two linear receivers, without compromising the ground use efficiency as illustrated in Fig. 7.31. However, the elimination of shading and blocking effects of the CLFR lead to losses in the concentration ratio. The use of the third dimension in positioning the mirrors, i.e. having primary mirror axis at different heights, may also reduce shading and blocking, allowing larger concentration factors.

Secondary Reflector and Receiver Design
The use of a secondary reflector in LFR systems is common practice to improve optical efficiency, and many different shapes have been proposed in the literature as presented in this chapter and Chap. 3. The different shapes show not only different optical performance, but also manufacturing cost. This is often more important than the pure chase for higher efficiency. Very critical for a good cavity receiver is the stability of the reflector and the low absorptivity for the concentrated radiation. Cleanliness and cleaning options of the optical system are important to consider in the design.

Optical End Loss Reduction
The minimization of end losses in LFR systems is possible. (Zhu et al., 2017) proposed the scalable linear Fresnel reflector (SLFR). The structure can constantly adjust to the longitudinal incidence angle of the sun, avoiding end losses, as outlined in Fig. 7.50a. The optical efficiency of a standard LFR can be raised up to 9%. However, a key advantage of the LFR collector, the low cost simple single-axis fixed mirrors are given up in this approach. Another advanced concept envisions a two-axis tracking system in which the primary reflector can slide along V-shaped rails to compensate for end losses. This design, shown in Fig. 7.50b, increases the optical efficiency compared to the single-axis tracking LFR collector by 8% to 50% depending on the normalized collector efficiency, especially at short collector lengths, moving receivers were also analysed to reduce the end losses. The effects of lengthening the receiver and moving the receiver, as shown in Fig. 7.50c, achieved an annual yield increase of about 50% when the length of the receiver is doubled, but the thermal losses also grow. Therefore, this type of measures has great importance Fresnel collectors with short length dimensions.

7.3 Point Focusing Systems

There are two main lines of development for point-focusing receivers. These are on the one hand so-called solar towers, where a heliostat field focuses the solar irradiation onto a central receiver installed close to the top in a tower. Each of the individual biaxial tracking mirrors is therefore similar to a Fresnel mirror, which redirects the solar radiation to the receiver with different focal lengths. The basic beam path of a solar tower concept is shown in Fig. 7.51a. With the help

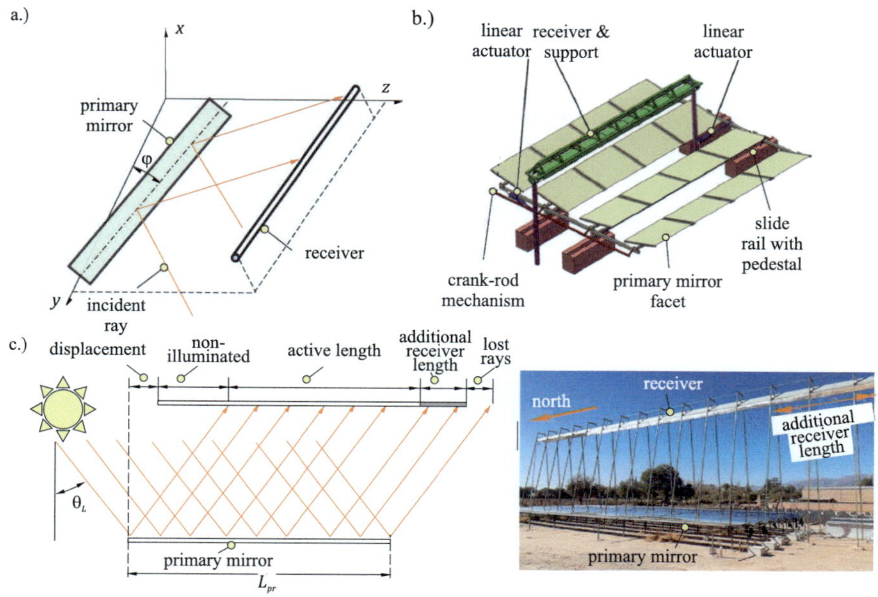

Fig. 7.50 Design measures to reduce end losses of LFR collectors; **a** scalable linear Fresnel reflector concept by Zhu et al. (2017); **b** two-axis tracking LFR system as proposed by Yang et al. (2018); **c** receiver extension along longitudinal direction as proposed and realized (modified from Bellos, 2019)

of large heliostat arrays, solar towers with more than 100 MW nominal readings have already been put into operation.

The second line of development are so called parabolic dishes, where a parabolic monolithic or segmented mirror focuses to a focal point, where the receiver is coupled to a heat engine (Stirling engine or micro gas turbine). The entire setup composed of mirror and receiver plus heat engine unit must track the position of the sun biaxially, as illustrated in Fig. 7.51b. As it is designed as full functional unit the dish systems perform power scaling by replicating the base unit. An intrinsic disadvantage is that the storage of thermally generated excess power is technically only possible to a limited extent. Therefore, dish systems are mostly used where either high temperature or electricity is to be generated independently of the grid.

7.3.1 Dish Collectors

Dish concentrating collectors use paraboloidal mirrors that track the sun and focus solar radiation into a receiver where it is absorbed and transferred to a heat engine/generator or alternatively into a heat transfer fluid that is transported to a ground-based plant. Dish concentrators due to their principle have the highest

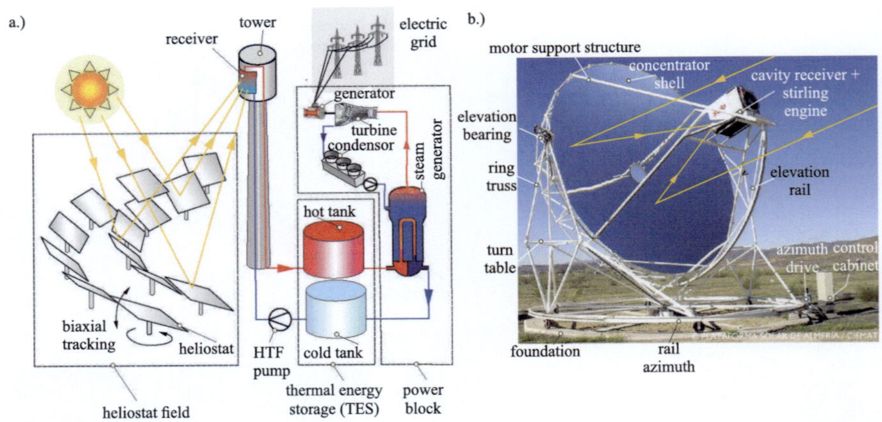

Fig. 7.51 Components and radiation solar beam path of point-focusing solar thermal systems; **a** solar tower concept **b** solar dish concept (EuroDISH © Ciemat, La Plataforma Solar de Almería (PSA))

optical efficiencies, the highest concentration ratios, and the highest overall conversion efficiencies of all the concentrating collectors. Most development with dish concentrators involved the use of receiver integrated Stirling engines for direct production of electricity. However also other concepts are possible.

The modular approach was seen in the beginning as an advantage of Dish-Stirling systems, however, the cost reduction of the directly competing photovoltaic (PV) systems, has been far more successful than expected. Technological progress, large market volume, and strong competition let PV panel cost fall to fall below 400US$/kW in many markets and therefore outperformed the CSP technologies for electric power generation without storage. Thermal energy storage (TES) is possible in principle for dish engine systems but suffers from small size leading to cost drawbacks and considerable technical challenges. The attempts to develop storage solutions for dish have not brought a breakthrough so far. Dish engine systems without storage directly compete with PV and have clearly lost this competition. Consequently, most of the companies active with dish systems have failed commercially or moved to other areas. However, as the efficiency advantages of dishes are considerable, they might have a potential in direct solar chemical processing at very high temperatures, which can only be achieved by dish technology.

7.3.1.1 Concentrator Optics

The ideal case of a concentrator focusing parallel rays onto a focal point in 3 dimensions is a rotational paraboloid or parts of that. If the focal point is on the center axis of the full parabola, then the rotationally symmetric reflector forms a rotational paraboloid with a height h, as illustrated in Fig. 7.52. Mathematically it is described as a parabola of focal length f in three-dimensional space. The light rays incident parallel to the y-axis are focused by the parabola contour to the focal point F of the parabola. Provided that the lines FP and PQ are of equal

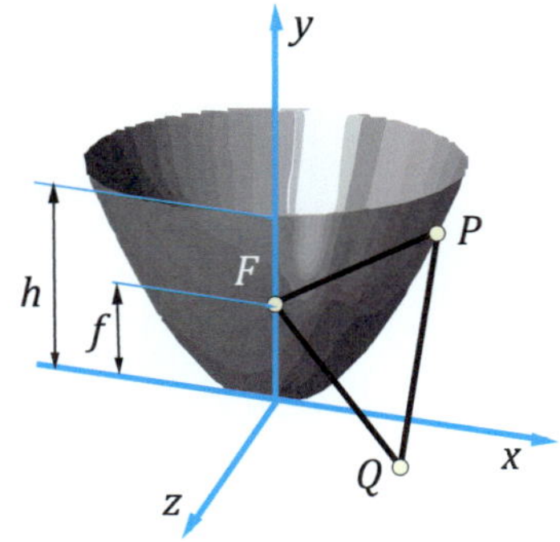

Fig. 7.52 Sketch of a rotational paraboloid of height h

length, the tangent is always correct to focus all vertical beams into the focal point F. Similar to the two-dimensional case of the trough as the focal length f of the parabola increases, a flatter paraboloid becomes necessary to focus the light in a point-like manner. The advantage of the flatter reflector area is less material requirement for the same aperture area (i.e. constant concentration ratio). The disadvantage is that when the distance to the focal point increases, possible mirror or tracking errors have a larger effect.

Using this optical element collectors using a symmetric rotational reflecting mirror ideally with the cross section of a parabola are called a parabolic dish collector (PDC). In the PDC all rays incident parallel to the parabolic axis are reflected by the mirror to a focal point where an absorbing element is located. Divergent irradiation within a cone of half-angle δ leads to a widening up of the focal point to a finite size focal spot when the absorber plane is going through the focal point. Depending on the angle formed with the axis in the incident plane the rays would intersect at a smaller or larger distance than the focal distance. Thus, the permissible acceptance angle θ_a also depends on the distance between focal point and the reflector. In the case of the parabolic contour, the ray length is maximum for the rays coming from the rim.

It should be noted that also partial sections from a paraboloid can be used as a concentrator. If the axis of the paraboloid is not included in the aperture, we call the reflector an off-axis paraboloid. In such a case the advantage is that the receiver can be positioned more flexibly with respect to the mirrors (see later in Fig. 7.53). It is also possible to move the reflecting area in such a way that the sun always is focused on a fixed spot in space. Then we talk about a fix-focus concentrator. This so-called Scheffler dish is presented in Sect. 7.3.1.1.2.

Parabolic Dish Optics

A parabolic dish requires a solar tracking system following the sun position in 3 dimensions all day long. The axis of the paraboloid must point towards the source, the sun. The movement of this dish collector thus has two degrees of freedom and is a two-dimensional tracking in angular space. The incidence angle for a dish is the angle between paraboloidal axis and solar beam direction. Due to the

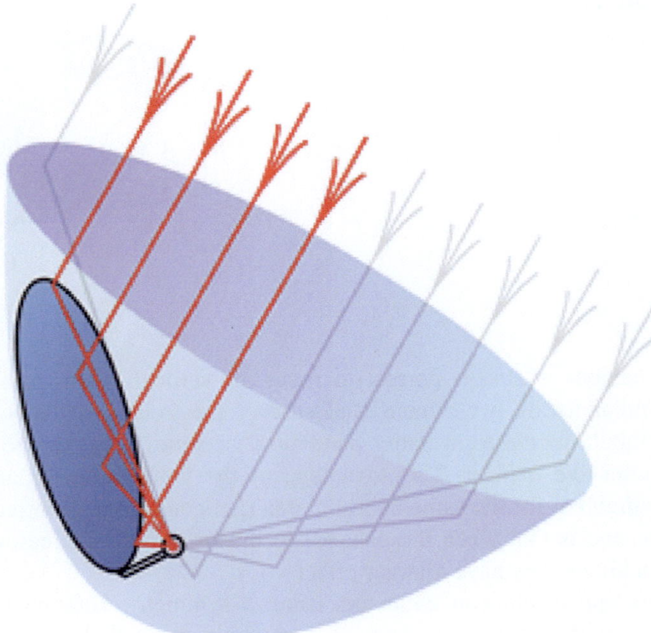

Fig. 7.53 Schematical view of an off-axis paraboloidal concentrator

two-dimensional tracking no cosine effect reduces the incident solar flux on the aperture.

The most important PDC parameters are geometric concentration ratio C_g, acceptance angle (θ_a), rim angle (ψ_r), and peak optical efficiency. The geometric concentration ratio, C_g, is the ratio between the collector aperture area and the total absorber area. For a dish collector we distinguish between flat receivers and spherical receivers. Depending on that we have different concentration ratios the geometric concentration ratio for a flat receiver is given by Eq. (7.8). Here, the shading of the reflector by the receiver itself is neglected as it depends on the coating of the receiver: If the absorber is only on one side (viewing the reflector) the area of the receiver should be not considered as aperture—then $C_{g,flat}$ has to be subtracted from Eq. (7.44); if the receiver is absorbing on both sides, the full width of the dish is considered as aperture and Eq. (7.44) is correct:

$$C_{g,flat} = \frac{\frac{\pi}{4} \cdot W^2}{\frac{\pi}{4} \cdot d^2} = \frac{W^2}{d^2} = \left(\frac{\sin 2\psi_r}{2 \cdot \sin \theta_a} \right)^2, \tag{7.44}$$

and it is maximal for a rim angle of $\psi_r = 45°$.

The geometric concentration ratio for a spherical receiver is given by Eq. (7.45)–here nothing should be subtracted as we assume absorption over the whole sphere:

$$C_{g,sphere} = \frac{\frac{\pi}{4} \cdot W^2}{\pi \cdot d^2} = \frac{W^2}{4 \cdot d^2} = \left(\frac{\sin \psi_r}{2 \cdot \sin \theta_a}\right)^2, \tag{7.45}$$

and it is maximal for a rim angle of $\psi_r = 90°$. In Eqs. (7.44) and (7.45) d describes the diameter of the receiver area (disc or sphere), W is the parabola width, θ_r the rim angle of the dish and θ_s the solar divergence half-angle. The acceptance half-angle, θ_a, is the maximum angle that can be formed by a ray with the paraboloid axis that it intercepts still the receiver area. To capture all direct beam solar radiation the minimum acceptance half-angle should be at least the solar divergence half-angle $\theta_s = 0.27°$ which contains the solar disk as seen from the Earth. It must be chosen by the designer which part of the circumsolar radiation shall be captured as well. The small acceptance angles do demand very accurate and expensive solar-tracking systems, but high concentration values can be achieved.

$$C_{g,sphere} = \frac{\frac{\pi}{4} \cdot W^2}{\pi \cdot d^2} = \frac{W^2}{4 \cdot d^2} = \left(\frac{\sin \psi_r}{2 \cdot \sin \theta_a}\right)^2. \tag{7.46}$$

The rim angle, ψ_r, which is directly related to the concentrator arc length, can be calculated as a function of the parabola focal distance (f) and aperture width (W) by

$$\tan \psi_r = \frac{8 \cdot f \cdot W}{W^2 - 16f^2} \tag{7.47}$$

For parabolic dishes often, flat receivers are used therefore usual rim angles are in the range of 45°. Smaller rim angles are not advisable because they reduce the aperture surface and lead to comparatively large focal lengths.

Optical losses are very important in a PDC due to the extremely high concentration of the total solar flux incident on the PTC aperture plane. Optical losses are associated with the following four parameters:

Reflectance $\rho < 1$ of the collector reflective surface. Typical solar spectral reflectance values of the best clean silver-based low iron glass mirrors are around 0.93–0.94. Reflective films have usually a specular reflectance $\rho < 0.9$.

Intercept factor γ. This is the fraction of the direct solar radiation reflected by the reflecting dish that finally reaches the active surface of the receiver. The intercept factor for 0° incidence angle is caused by optical imperfections (micro-roughness, errors in the parabolic shape, mechanical deformation of the support structure) and some shadowing by the receiver support or structure.

Absorptance of the receiver coating α. This parameter quantifies the amount of energy absorbed by the spherical or flat receiver area divided by radiation reaching it. This parameter is typically 0.96 for black receivers. The temperatures reached can be extremely high exceeding 1000 °C. There are no selective absorbers stable at these temperatures. On the other hand, the high concentration C_g in the range on

1500–4000 leads to good thermal efficiencies. In fact, often small cavity receivers with small open aperture windows are incorporated in designs to reduce convective and radiative losses.

Multiplication of these parameters (intercept factor, mirror reflectance and receiver absorptance) for an incidence angle $\theta_i = 0°$ gives the peak optical efficiency of the PDC $\eta_{opt,0}$:

$$\eta_{opt,0} = \rho \cdot \alpha \cdot \gamma. \tag{7.48}$$

In total the $\eta_{opt,0}$ should be in the range of 0.9 for clean, good-quality PDCs. The effect of non-normal incidence angles for parabolic dishes does not apply, because the aperture should always track the sun position.

Scheffler Dish Optics

The Scheffler reflector, named after the inventor Wolfgang Scheffler, is a solar concentrator design which maintains a fixed focus while only having a single axis tracking mechanism. It is applied to produce steam for solar cooking, industrial processes an electricity generation. Ideally the reflector is an off-axis segment of a paraboloid mirror, see Fig. 7.53. The fix focus characteristics of the Scheffler dish is achieved by adjusting seasonally the curvature of the reflector in dependence of the solar declination and tracking the sun throughout the day about an axis parallel to the earth's axis of rotation. This design makes the construction and operation of high temperature solar concentrators accessible to developing nations. Because of the fix focus feature, it is possible for example to cook food in a pot at the concentration point inside a building, hence persons can be close to the focus point without being at risk to be harmed.

For a detailed ray tracing simulation of a single Scheffler Reflector as well as an array of reflectors a flexible algorithm has been developed by van Rooyen, Heimsath, Nitz, and Platzer (2011). The results are visualized in a 3D map for varying solar declination and solar hour angle. The occurring shading of the array of Scheffler Reflectors is also studied. The modelled reflective surface of a Scheffler Reflector comprises of rectangular planar mirror tiles of variable size. The algorithm adjusts the curvature of the paraboloidal segment depending on the solar declination (Fig. 7.54).

The dish typically is produced with a cross bar structure consisting of square steel or aluminum bars and angle sections to form a shape of parabola with metal channels bolted above the square bars for fixing mirrors. Small rectangular flat mirrors of solar grade mirrors or aluminum mirror sheets are then fixed on the structure to approximate the paraboloidal shape. The shape of the flexible support structure is adapted seasonally by telescopic clamps or strings. The daily tracking can be done with a small motor drive and gears to rotate the complete structure at 15° per hour with centre of gravity in the rotational axis. A design guideline for the construction of small dishes can be obtained from Solare Brücke (2022). Expanding the technology to larger dishes was a prerequisite for commercial projects. In India several companies produce steam generation systems mainly

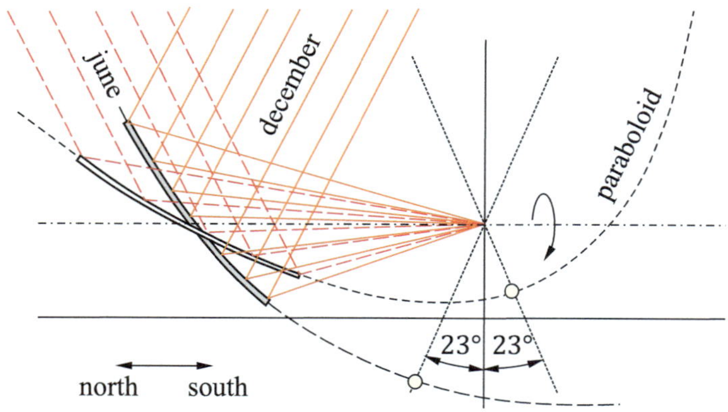

Fig. 7.54 Declination of the solar orbit for which the curvature of the dish segment is seasonally adapted. In this way optical axis of the paraboloid is tilted ±23° according to the declination

for cooking, and a 1 MW solar power plant "India One" has been constructed using a field of 770 dishes each 60 m^2 aperture and an iron cavity steam generator at the focal point (Brahma Kumaris, 2011).

7.3.1.2 Practical Considerations and Experiences Parabolic Dish Collectors

A dish system consists of a paraboloidal shaped concentrator, tracking system, a receiver, optionally an engine with a generator. All dish systems are based on the same three basic assemblies, as shown in Fig. 7.55:

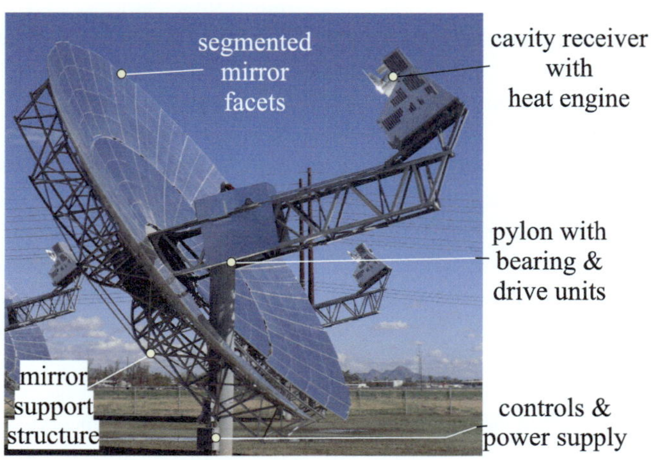

Fig. 7.55 25 kW Disk-Stirling base unit of Maricopa Solar Project CSP Project modified from United Sun Systems International Ltd. (2012)

- solar dish concentrator and support structure (reflector unit),
- receiver installed in the focal point (with/without heat engine) and
- solar tracking system (mechanical structure with bearings and drives).

and a system control unit. The concentrator tracks the sun biaxially in such a way that the optical axis of the concentrator always points to the sun. A reflective mirror surface on the paraboloidal concentrator, either metallized glass or plastic or anodized aluminum, reflects incident sunlight to a small region called the focus. The ideal shape of the reflecting surface of a solar concentrator is a paraboloid. Due to the three-dimensional structure, there is a multitude of different engineering solution options regarding the positioning of bearings and tracking systems, the connection of the receiver with the reflector system, which results from the sizes of the respective system.

Facetted reflectors (i.e., consisting of individual spherical or parabolically shaped segments) and full-surface reflectors are encountered in different designs. For facetted concentrators, several mirror segments are mounted on a supporting structure. The segments are attached and oriented individually. The segments may either consist of glass mirrors or substrates covered with reflecting foil or thin-glass mirrors. For full-surface concentrators, the entire concentrator surface, e.g. in the metallic membrane technology, is shaped parabolically by a forming process.

Practical considerations of wind load and weight lead to typical aperture sizes of 1–25 m diameter. The receiver captures the radiation as high-temperature thermal energy in a fluid that is either the working fluid for a receiver-mounted engine or is used to transport the energy to ground-based processes. A typical engine size is 3–25 kW_{el}. In the overwhelming number of cases Stirling engines were used, but also micro-turbines using the Brayton cycle were tested. Maintenance of the engines in the past has been a major cost driver. Maximum size of dishes up to now is 500 m^2 (developed by ANU—Australian National University). The bigger systems utilized rather steam receivers, high-temperature heat transfer fluid or compressed air receivers for the micro-turbines (Brayton cycle).

Tracking may be accomplished generally in two ways:

- Azimuth-elevation tracking in which the dish rotates in a plane parallel to the earth's surface (azimuth) and around an axis perpendicular to it (elevation). A rotation about both axes is necessary throughout the day.
- Polar-equatorial tracking where the dish rotates about an axis parallel to the Earth's axis of rotation. The rotational speed is constant with 15°/h. The other axis, the declination axis, is perpendicular to the polar axis and needs to be adjusted only gradually <0.5 per day.

Many pylon-based dish designs use slewing drives for azimuth and linear drives for the elevation movement. Both azimuth and elevation drives can profit from the large market for tracking PV and heliostats where similar solutions are used. One difference to PV applications is however that high stiffness, low backlash, and highly accurate positioning is necessary. As slewing drives are a considerable cost factor, new and advanced solutions are needed. Turntable-based dish mounting and tracking (e.g. EuroDish) is characterized by small and cost-effective gears but needs many structural parts to be assembled. They also need large foundations which must be equipped with rails for big systems. In the design of the tracking system always the compromise of accuracy and cost must be found.

High receiver surface temperatures of 600–800 °C or even 1000 °C and more result in considerable thermal losses by convection, infrared radiation, and heat conduction. Thermal radiation losses increase with the power of 4 in temperature (in Kelvin) and are dominant at high temperatures. This can give an unacceptably low efficiency if simple open receivers were used. Thus, efficient receivers for dish systems are cavity receivers with a small opening (aperture) through which concentrated sunlight enters.

As with the other approaches to concentrator design, the concept of using mirrored dishes to focus the sun has been around since 200 BC. One of the earliest technical implementations of a dish system was by Augustin Mouchot, who built dish-driven engine systems as early as 1864 and displayed one at the WorldExhibition in 1878 in Paris. The dish technology of today was revitalized in the early 1980s following the oil price shock in the 1970s. Many dish prototypes were designed and built by researchers and commercial companies, using different materials like stretched silvered plastic membranes, bonded thin and segmented thick glass mirrors, and even from tiny flat mirrors bonded to a paraboloid support material.

Receivers were often direct insolation receivers for Stirling engines, but also heat pipes and spherical steam receivers. Commercial developments stimulated larger technology progress for Dish-Stirling in the decade until 2010, with companies like Stirling Energy Systems (SES), Schlaich Bergermann & Partner (spb), Infinia, HelioFocus, Solar System Pty, South West Solar Technologies mainly in US, Australia and Israel. However, the rise of PV brought a rapid end to commercial project developments.

Nevertheless, new developments have appeared again, this time tackling also the question of storage integration or direct chemical reactors in the focus. An overview of these activities is given by Schiel and Keck (2012).

The achievable concentration ratios C, are higher than those of central receiver towers; geometric concentration ratios of peak $C = 1.3 \cdot 10^4$ have already been achieved, see Hafez et al. (2017). This is due, among other things, to the fact that the focal length is much smaller than for towers, and spillage, atmospheric attenuation, as well as surface and tracking errors are smaller. However, the power densities realized so far correspond only to those of solar towers. The theoretically achievable temperatures due to the high concentration ratios are far above those of available receiver materials, temperatures of more than 800 °C have been reached in experiments. This results in significantly higher thermal–mechanical energy conversion efficiencies. Dish/Stirling systems achieve more than 30% peak solar-to-grid, with an average solar-to-electric efficiency in the range of 16–25% under typical conditions. Power conversion options include Stirling engines

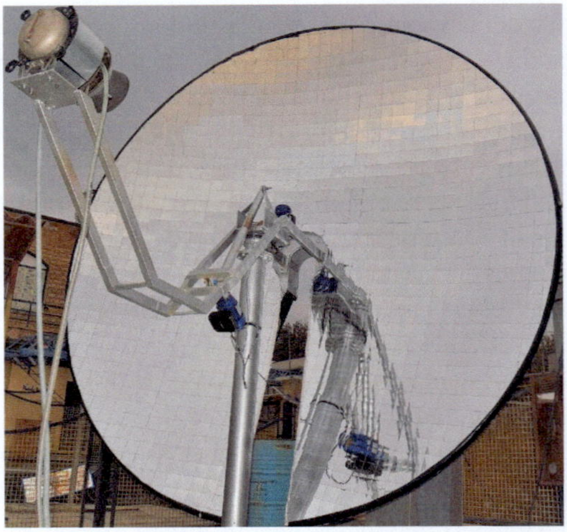

Fig. 7.56 1 kW free-piston Stirling engine installed in the Kerman pilot plant, using helium as fluid at maximum operating conditions of 800 K and 1 MPa from Gholamalizadeh and Chung (2017)

or microturbines, which require more compressor work than Rankine cycle conversion systems, but the waste heat from condensation is eliminated and water consumption is also minimal. Subsequently the individual assembly units are discussed.

Structural Concentrator Design

From previous design considerations, we know that the reflecting mirrors must be positioned on an approximately parabolic structure, and the thermal output results from the projected concentrator area. Typical solar concentrators are about 3–10 m in diameter, although large concentrators up to 17 m in diameter have also been fabricated for Stirling engines for power generation. Typical focal lengths range from 4 to 12 m. Almost all support structures and columns at these dimensions are made of steel for stiffness reasons. There are several structural designs for the reflector. The simplest type of structure is a closed reflector surface, as in the Eurodish collector (compare Fig. 7.51b), the other option is that of segmented facets mounted on a common support structure, compare Fig. 7.55. Closed reflector structures require shells exhibiting a continuous paraboloidal surface. They either consist of one large piece or are composed of precisely manufactured segments connected to each other. If they are made of one large piece, often a membrane type is chosen, see e.g. in from Schiel et al. (2004). Stretched membrane heliostats have been built with reflecting areas of more than 200 m^2 and up to 17 m in diameter. One disadvantage of a large component is the transport of the shell, so many other concepts show a simplified structure in which the tensioned membrane is replaced by sandwich panel mirrors supported at the circumference by a ring beam. The panels can either be reinforced plastic layers with a foam core and stiffened by means of a radial rib along the centerline, or consist of thin glass mirrors bonded to a plate, also stiffened radially to form the reflective surface, as shown in Fig. 7.56.

Fig. 7.57 Fresnel paraboloid collector with cavity receiver for process heat installed at a Hotel site in India modified from Brossard (2010)

The most frequent applied is a segment assembly, in which the reflecting surface is constructed of a space frame/glass mirror type.

In contrast, in a multi-facet design, each facet has its own curvature and alignment characteristics according to its position within the reflector assembly, and each facet can be aligned individually. The entire system, of course, requires an optical axis. This can be achieved by assembling the facets in the form of a large paraboloid. A multi-facet design has the advantage of not having to construct a large paraboloid surface in one step, which is time-consuming and expensive. In addition, facet constructions allow the use of very small spherical curvatures, since each individual facet has a small rim angle focal ratio. Assembling facets with "marginal" errors allows a good approximation to a paraboloid. The production of small facets is also advantageous because small components can be manufactured with very high precision and, in addition, fine adjustment to the focal point can be made during on-site installation.

A special case of multi-surface construction is a central Fresnel collector, where many single mirror facets are arranged to form a Fresnel lens, similar to a lighthouse. Figure 7.57 illustrates such a design of a reflector for solar process heat supply installed heliostat at the Maurya Sheraton Hotel in New Delhi (India).

Table 7.9 Range of reflectivity (ρ) and emissivity (ε) of some selected reflector options realized in solar dish systems modified from Hafez et al. (2017)

Reflector composite	Operation temperature T (°C)	ρ (/)	ε (/)
Polymer film/non-metal	125	0.98	0.02
Aluminum Acrylic	95	0.98	0.02
Silver Acrylic	95–100	0.92–0.97	0.02–0.08
Aluminum	100–500	0.86–0.97	0.03–0.14
Silver/copper with polymer	95	0.97	0.03
Polished stainless steel	80	0.50	0.50
Cermet	90	0.95	0.05

As with the solar towers, both azimuth-elevation and polar tracking of the mirror assembly have been realized for the dish systems, although on large systems the former has a preference that it can be rolled.

There are also a number of concepts of mirror facet shapes that have found their way into experimental implementation, ranging from flat mirrors, spherical mirrors, parabolic mirror segments, to cylinder mirrors, conical constructions to elliptical mirrors; an overview about realized mirror shapes and solar tracking types can be found in Hafez et al. (2017).

Reflector Material

The optical power incident on the receiver aperture ($\dot{q}_{rec}$) surface is given by the product of the concentrator reflectivity (ρ), the concentrator geometric concentration ratio (C), and the direct normal irradiance (DNI) in (W/m^2)

$$\dot{q}_{rec} = \rho \cdot C \cdot DNI \tag{7.49}$$

As can be seen from Eq. (7.49), in addition to the shape of the mirror, the reflectivity of the mirror material is of particular importance for achieving high power densities per unit area on the receiver. In principle, any material with good reflectivity is suitable, but in dish systems, mirrors with silver ($\rho = 0.95$) or aluminum ($\rho = 0.9$) as reflector material are in the foreground of applications due to their high reflectivity, especially when the receiver system is coupled with a heat engine.

However, in addition to applications for electricity generation, which require the highest possible receiver temperatures, dish systems are also being used as a cost-effective alternative for the decentralized provision of process heat, and so there are several demonstrators in which reflective polymer films, silver/aluminum with polymer films, polished stainless steel and cermet have been used as reflector material. It should be noted that reduced reflectivity automatically results in heating of the reflector, which should be considered in the appropriate combination of materials chosen. Table 7.9 shows the reflectivity or emissivity of various materials used in dish reflectors.

Also, the optical collector losses in dish systems result from geometric errors, the limited reflectivity of the mirror material and the radiation distribution of the sun. Similar as with solar towers, the geometric errors can be divided into several subtypes. These are composed of the relative optical errors of beam quality (σ_{bq}), astigmatism (σ_{ast}) and tracking (σ_{trac}) and the sun shape error σ_{sun} (typically assumed to $\sigma_{sun} \approx 2.5$ mrad). Theσ_{tot} total optical error (σ_{tot}) is then given by

$$\sigma_{tot} = \sqrt{S_0^2 \cdot \left(\sigma_{sun}^2 + \sigma_{bq}^2 + \sigma_{ast}^2 + \left(2 \cdot \sigma_{trac}^2\right)\right)}, \tag{7.50}$$

where S_0 is the actual distance from the dish to the focus point. In contrast to solar towers the atmospheric attenuation plays hardly any role due to the small distance between reflector and receiver, which is only a few meters.

In the beam quality factor (σ_{bq}), deviations of the mirror curvature from the ideal shape and imperfections of the reflecting surface due to waviness and roughness from manufacturing and assembly are taken into account. These are called slope errors (σ_{slope}), compare Sect. 7.2.2.3.2. Statistically, they are assumed to be uniformly distributed and described by a circular normal distribution. The slope error σ_{slope} is defined with respect to the surface normal vector; therefore, its effect is doubled in the reflected beam.

$$\sigma_{bq}^2 = \left(2 \cdot \sigma_{slope}\right)^2. \tag{7.51}$$

Receiver and Heat Engine

Conventional dish receivers for power generation are based on the use of gaseous heat transfer fluids with Stirling machines that use either hydrogen, helium or nitrogen, or Joule-Brayton based miniature gas turbines that use air as a heat carrier. The receiver is responsible for absorbing the reflected and concentrated solar radiation and transferring the heat to a heat-carrier fluid. A detailed description of concepts used in the design of receivers for Dish/Stirling Engine systems is found in Stine and Diver (1994).

A very advanced concept is the AMTEC receiver where sodium is vaporized at low pressure in the receiver and the vapor is passed through a solid electrolyte through an AMTEC assembly, compare Fig. 7.71, thereby generating electricity directly, whereby the waste heat (>600 °C) is subsequently converted further in a thermal engine. These concepts require dedicated design measures for which details may be taken from Wua, Xiao, Cao, and Li (2010) or Tanaka (2010).

The receivers are usually designed as a cavity receiver with a small opening through which the concentrated sunlight enters. The geometry of the receiver opening serves to limit the radiation emission and convection losses, with the optimum depth being one to 1.5 times the diameter of the aperture. Various methods can be used to transmit the absorbed radiation to the fluids. It is also possible to use secondary concentrators at the receiver opening to increase the intercept without increasing the size of the receiver aperture, or to reduce the opening diameter for a specific intercept. These are then trumpet-shaped in relation to the surface and

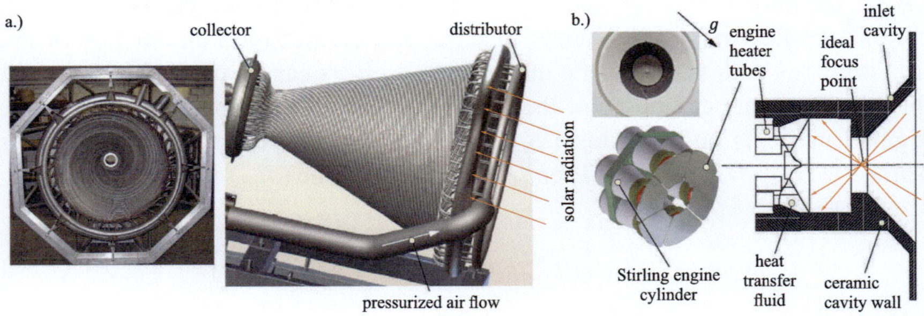

Fig. 7.58 **a** Front and side view of a direct illuminated pressurized air receiver of a gas turbine from BraytonEnergy LLC (2022), **b** cross-sectional cut to the principle of a indirect illumination cavity receiver type (left top-front view, left bottom connection to Stirling engine)

transmit the reflections from a large area through the cavity receiver opening similar to those used in solar tower concepts. The result is an increase in the intercept without the need to increase the receiver opening. A secondary concentrator generally improves the performance of a parabolic receiver and may reduce the negative impact of one or all components of the optical fault, but a secondary concentrator increases the cost of the collector, since the reflection coefficient for dish systems must be very high due to the high concentration ratio and therefore requires a well-designed cooling system. In the construction of the receivers themselves there are two different concepts:

- In the case of directly illuminated tube receivers, small parallel arranged tubes are used, through which the working fluid enters directly into the area of the concentrated solar radiation. The tube surfaces form the absorption surface. To increase the residency time in the solar radiation zone, especially for gas-cooled receivers, these are then pitched, as depicted in Fig. 7.58a for a pressurized air receiver of a gas turbine.
- Other receivers use an indirect illumination concept with a focus at the narrowest point of the cavity, cf. Fig. 7.58b. At the lower end, on a cone, is the absorber behind which the heat transfer fluid is arranged. There are two sub-species; one option transfers the heat directly to the plates of a Stirling engine, or a heat-pipe concept is used, in which a liquid metal absorber surface evaporates and condenses at the heat plates of the Stirling engine (reflux receiver).

7.3.2 Solar Towers—Central Receiver Systems

A concentrating solar tower system (STS), also called a central receiver system (CRS), essentially consists of three core components.

- A solar tower with a receiver installed at a certain height. If the conversion of the thermal receiver power does not take place in the tower itself, it contains the lines for the coolant supply and removal from the receiver.
- The receiver itself, in which the optical conversion of the solar radiation into heat takes place and which is dissipated with the help of the coolant. The concept and design of the receiver essentially determine the arrangement of the heliostat field and the height of the tower.
- The mirror field, which is decoupled from the coolant system, consists of heliostats that track the position of the sun on two axes, in which the solar radiation is focused on the receiver with the help of target point strategies based on measurement data from sensors in the solar field.

All three components must be precisely coordinated to enable a cost-optimal, competitive solar tower system and are discussed in detail in the following sections. The core component, which determines the arrangement of the heliostats around the tower, is the receiver mounted on the tower, which can be up to 250 m in height, and there are several designs. At present, it is clear that no receiver technology is economically superior to any other. Heliostat design, receiver type, material choices and control issues are intimately coupled, and will be discussed in detail in the following sections. But, first a short overview shall be given to demonstrate the complexity of the overall design.

Like the parabolic dish a solar tower is working with two-dimensional concentration on a focal point, not a line. The solar tower with the heliostat field is the corresponding equivalent of the linear Fresnel collector. Also, here the ideal paraboloidal shape is broken up according to the Fresnel principle into individual facets. These mirror facets are movable in two angular directions with their support structure and called heliostats.

7.3.2.1 Receiver Configuration Options

The receiver usually is the starting point in the design consideration. The heliostat components, control and field arrangement are very much depending on the chosen receiver which in turn is connected to the choice of heat transfer fluid used in the collector. The concentrating collector as defined here therefore consist of one or multiple receivers mounted on towers and connected to a heat transfer fluid loop transporting the energy to a power block or to a storage system, plus a suitable heliostat field for the receiver(s). The two fundamental design options of receivers are

- panel receivers mounted around a tower viewing the whole landscape around the tower (360°–receiver, Fig. 7.59a), a sub class is formed by external star (Fig. 7.59b) or rectangular receivers,
- cavity receivers with a limited view factor, to which simple cavity receivers as shown in Fig. 7.59c, but also face down receivers (Fig. 7.59d) and beam-down receivers belong (Fig. 7.59e).

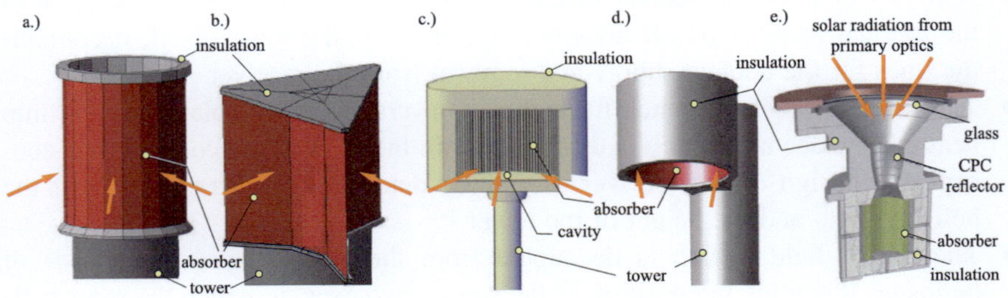

Fig. 7.59 Different receiver concepts of solar towers; external receivers with 360° solar field using external tubular receivers (**a**) or star receiver (**b**) and cavity receiver types sinple cavity (**c**), face down receiver type (**d**) and beam down receiver with secondary optics (**e**)

A further option is the use of a secondary optics which redirects the rays from the heliostat field to the ground. Such a concept referred to as beam down concept, allows the receiver to stand on the ground, open at the top, and eliminates the need for coolant routing to the turret, but at the expense of secondary optics with optical losses. Figure 7.59 schematically illustrates the different receiver type options, which will be discussed in more detail in Sect. 7.3.2.3.

The receivers with a full 360° view factor utilize a heliostat field distributed around the tower, which is more or less approximating a circular arrangement, shown in Fig. 7.60a. Depending on latitude and other factors like weather characteristics, intended load curve, available space etc. the circular form might be distorted and also the tower might not be in the center of gravity of the field. For higher latitude a larger fraction of the heliostats might be favorable on the side of the tower opposite the dominant sun position (in the Northern hemisphere on the North of the tower). Cavity receivers with a limited view factor use and receive only reflected radiation from heliostats within this view angles, which is depicted in Fig. 7.60b. Therefore, the heliostat field is asymmetric positioned on one side of the tower.

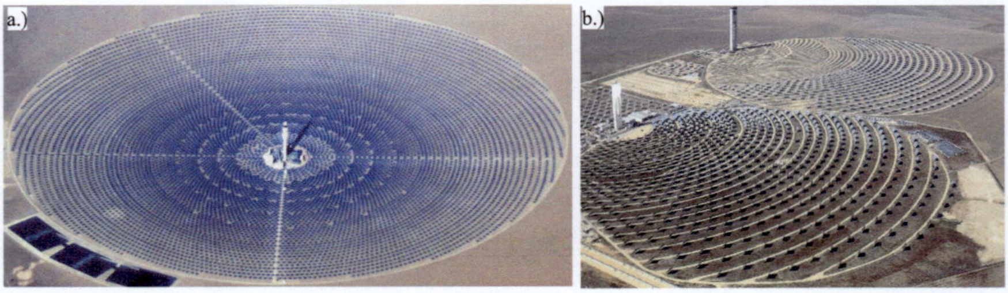

Fig. 7.60 Circular heliostat field with 360° receiver (left) from Crescent Dunes and (**a**) asymmetric heliostat field with cavity receiver at PS10/20 (right) (**b**) modified from (commons.wikimedia.org)

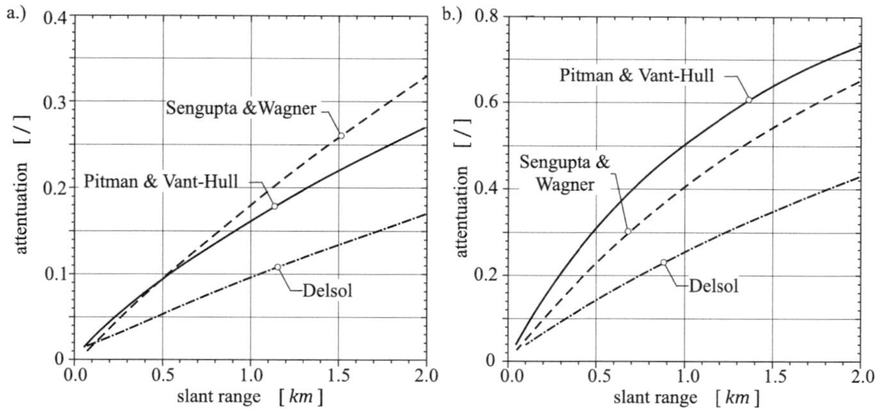

Fig. 7.61 Calculated atmospheric attenuation from the heliostat to the receiver over the slant range distance for different models. **a** clear day, **b** hazy day from Cardemil et al. (2014) for a 100 m high tower, Bom Jesus da Lapa/Bahia Brazil ($\Phi = 13.25\ S$) DNI level (990 W/m^2) and water vapor concentration ($\upsilon = 11.2$ g/m^3)

The distance of the individual heliostats from the receiver varies very much. Heliostats very close to the tower have a short overall distance, but the incidence angle for the receiver is larger than for heliostats further away. For a cavity receiver therefore, it might be advantageous to tilt it towards the ground, which improves the optical acceptance from near-by heliostats. Depending on the haze or dust in the atmosphere the beam between heliostats at the outer rim of the field (say 1–2 km away) can be attenuated appreciably. An example of attenuation depending on visibility is shown in Fig. 7.61.

Consequently, the heliostats too close and too far away contribute less to the overall thermal power in the receiver. It is important that the coverage of heliostats on the ground is optimized to reduce required area. However, several factors must be considered and respected:

- Heliostat from a row in front of the considered heliostat should not block reflected beam radiation—the heliostat needs a free view to the receiver at least most of the time.
- Heliostats in the direction of the sun should not shade the considered heliostat—a condition which cannot be fully fulfilled especially for low sun altitudes.
- In no circumstance should heliostats be able to mechanically touch each other (in some state of rotation).
- Heliostats must be cleaned and maintained—therefore some maintenance roads have to be planned inside the fields e.g. for the vehicles for spray-washing.

A very common approach to avoid most of blocking and shading is the staggered arrangement of heliostats: The heliostat behind a row of heliostats is always put

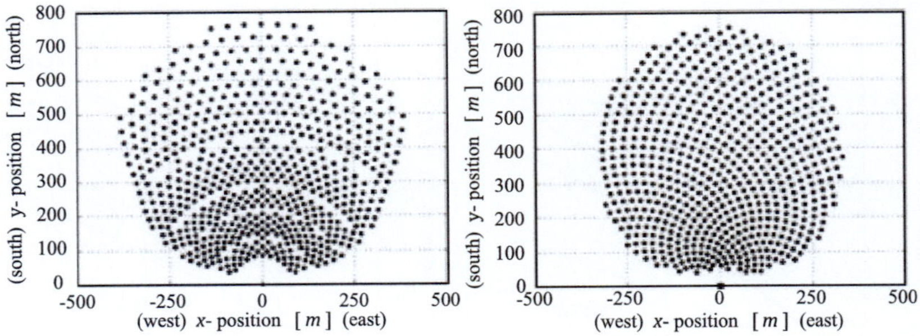

Fig. 7.62 Standard and optimized heliostat field layout, for concepts of optimization see e.g. Schöttll et al. (2019) or details in Chap. 10

into the gap between two heliostats in front. It is the same arrangement as singers in a choir have when the want to see the conductor!

The consequence is that in a row at a larger radius the distances between heliostats become larger and larger. This is often compensated by having this staggered arrangement not over the whole field but only in certain areas where positioning starts anew in each sector. In Fig. 7.62 a standard field layout and an optimized on is shown. The difference in annual optical efficiency is less than 0.6% but this may make a difference in economy.

The selection of the heliostats for a specific receiver and a specific position in a heliostat field has already several challenges:

- selection of heliostat size and shape (from flat to facetted to curved) suitable to receiver aperture,
- selection of tracking axes and control,
- heliostat support structure and stability,
- positioning of the heliostats in a field (minimization of mutual blocking and shading) depending on tower position and height, and,
- concentration factor using the optimization of heliostat area (or number).

The individual heliostats themselves have very different sizes. The reflective area of a heliostat ranges from 1 to 200 m^2. The ideas behind that are very well summarized in the review by Blackmon (2020) which also provides an overview over existing heliostat designs. There are three categories of costs for heliostats:

- Cost proportional to area (reflector material),
- size dependent costs (drive units, pedestals, motors etc.),
- fixed costs for each heliostat (sensors, encoders, limit switches etc.).

Qualitatively, it is obvious that many small heliostats need a lot of work for calibration they use many controllers, sensors, drives and motors. Also, one needs

Fig. 7.63 **a** eSolar heliostats (2 m^2) in a fixed array at the Sierra SunTower installation from Ricklin et al. (2014); **b** Segmented large scaled mirrors (118 m^2) from Crecent Dunes solar thermal facility with heliostats modified from National Renewable Energy Laboratory-NREL (2017)

more support structures, and it is important many steps in installation, maintenance and operation are automatized. Large heliostats on the contrary reduce these efforts, but they must be stable against large wind loads, the structures become quite massive, and the shape of the reflectors must be adapted whereas for very small heliostats even flat mirrors are adequate. So, simplified it is a consideration of few high-quality heliostats against a large number of mass-produced heliostats. It depends on materials and components selected, on the receiver size, on the installation and calibration routines, on the efficiency needed due to the size of the heliostat fields which approach small versus large heliostats is more cost-effective. However, using one technical solution shows also that a size optimization led to appreciable cost reductions of 30–40%. It seems possible that with reasonable assumptions heliostats might be produced well below 100USD/m^2. The Fig. 7.63 give an impression on small and large scaled heliostats. The heliostats are discussed in more detail in Sect. 7.3.2.5.

Traditionally, the heliostat field is being placed in a flat area, where only minimal ground height variations are existing. However, in principle heliostats can also be placed in a three-dimensional hilly landscape. Placing the heliostat on a slope facing the dominant sun position can even be used to improve the optical efficiency.

The optical design requires a complex determination of the optical path via a heliostat to the receiver to achieve an efficient, highly concentrating optical system. At the same time, the concentrating solar radiation hitting the receiver during optical conversion generates temperatures that reach the limits of the currently available structural and functional materials.

To ensure efficient and, above all, sufficiently quantitative predictive control of the heliostats and thus power generation in the receiver at all times, both the control and regulation of the thermal and the widely spatially distributed optical system (heliostats) with the sensors must be optimal and be coordinated at any time.

This requires many optimizations that are difficult to access analytically, which is why the development of solar towers in the past was essentially limited to experimental demonstrators. Only the massive increase in the computing power available, supplemented by innovative, high-precision production processes, advances in structural and functional material development, wireless signal transmission and refined modeling approaches make it possible today to predict and design efficient solar towers and their operational behavior. This computational design architecture has made the development of large solar towers so attractive that the growth rate in the construction of new solar tower systems, often referred to as solar power towers (SPT), has exceeded that of linear concentrator systems over the past decade.

For solar tower collectors many different receiver types and corresponding heat transfer media already have been developed and used in prototypes. The most common commercial heat transfer media are direct steam, where the steam is produced directly in the receiver tubes at a high pressure with temperature up to 480 °C, and molten salt (so-called Solar Salt, typically a 40 : 60 wt% mixture of potassium and sodium nitrate) where temperature up to 565 °C are possible. Fluids are usually circulated from and to the receiver consisting of tubes coated black. The standard here is a non-selective highly temperature-stable paint. To reach higher temperatures, the materials for the receiver and the heat transfer media must be changed. Other salt mixtures (nitrates, carbonates or even chlorides) are being investigated. The problem of carbonates and chlorides is the higher corrosivity of these hot salts. Liquid metals like NaK are also interesting candidates. It seems to be important that the fluid material is chemically stable, also against oxidation because of the fire hazard of leaks, and can be bought at reasonable cost. It is key that materials cannot harm the environment when leaking. A completely different approach is to use air as a heat carrier. The advantage of air is that it is non-corrosive, poses no environmental hazard and it has virtually no cost. It is stable even for temperatures up to 1500 °C. The disadvantage is the low heat capacity or density. The latter one can be somewhat improved when compressed air is used. But still large air channels and in operation a high electrical consumption due to ventilators are needed. In Chap. 11 the interlinks and interdependency between coolant choice and power plant concept are analyzed in more depth.

7.3.2.2 Receiver Design

There are different options described in literature to categorize different receiver types. A uniform way of classifying receiver concepts, however, is challenging, because the receiver design with the associated selection of functional materials and heat transfer fluid (HTF) routing within the receiver is irrevocably linked to the size and aperture angle of the receiver aperture and the heat dissipation capability of the selected heat transfer fluid, and thus determines the size and shape of the heliostat array. For example, from an optical point of view, it is easy to build a heliostat field that produces a peak concentration of more than 2 MW/m^2 on a receiver, but only a few receivers are able to dissipate more than 1–2 MW/m^2, such as liquid metals or molten salts, even with optimal coolant configurations, due to their heat transfer properties and capabilities.

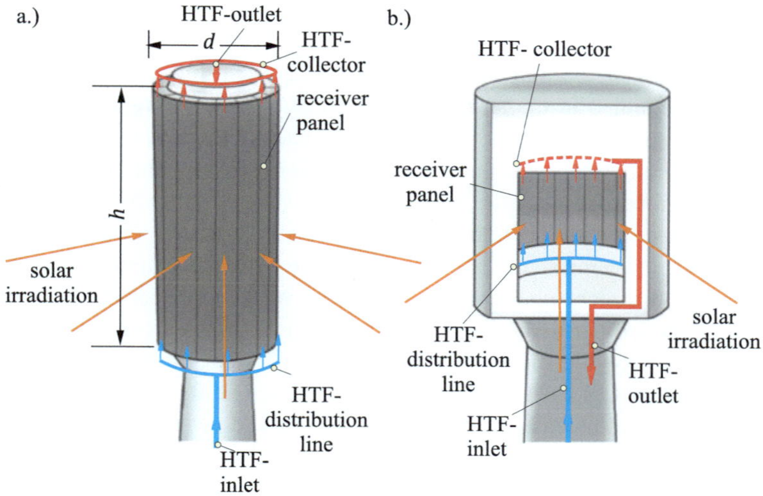

Fig. 7.64 Schematic diagram of an external **a** and a cavity **b** receiver concept used in solar tower installations

Principal Design Options

A basic distinction of receiver types is the question whether it is an external receiver, as outlined in Fig. 7.64a, in which the concentrated solar radiation is captured on the circumference of the tower, or a cavity receiver, in which the receiver is located in a cavity within the tower, as shown in Fig. 7.64b. In the case of the latter, there are further options: secondary concentrator optics can enlarge the aperture for a smaller cavity opening, beam-down or face-down cavities as shown in Fig. 7.59c–e are used as well. The orientation of the cavity can be optimized in terms of view factor and thermal losses.

It must be emphasized that the receiver design decisively depends on the heat transfer fluid (or coolant) used in the receiver, which determines both the maximum possible fluid temperature and the usable temperature difference between inlet and outlet. A holistic receiver development strategy provides a multi-dimensional optimization problem comprising the interaction of

- constraints on functional material properties for optical conversion,
- permissible heat flux density on the receiver design,
- coolant flow configuration,
- radiation and convection heat losses of the receiver,
- heliostat field design.

Therefore, before individual receiver designs are presented, various aspects in the development of receivers are discussed in a general way. This then also explains why there are many efficient solutions for central receiver types.

Constraints on Functional Material Properties for Optical Conversion

The design of the receiver directly affects via the attainable temperature the thermal efficiency of the individual receiver concept. On the other hand, the temperature level attained in the receiver poses constraints not only to the functional materials (absorptivity, emissivity, reflectivity) to achieve a high optical conversion but also to structural materials in terms of corrosion, erosion, and thermal stresses. Stress is experienced especially in transients of start-up and shut down. Central receiver concepts strive for high coolant outlet temperatures, therefore selective absorbers with low emissivity (ε) to limit radiation losses at temperatures above 500 °C are a technical challenge. The critical wavelength for the step-change from low reflectivity in the solar range to high reflectivity in the thermal range is tricky because of the increasing overlap of solar and thermal spectrum with higher temperatures. Solar absorption is more important than emissivity. The aim of the receiver is to increase the net energy captured in the receiver. For example, for a solar flux density of 1 MW/m^2 a reduction in the receiver absorptivity α by 0.02 translates in a decrease of heat gain of 20 kW/m^2. To compensate that with a reduction in radiative heat loss for an assumed surface temperature of 800 °K a reduced emissivity ε from 1.0 to 0.12 is required. This is hardly possible. A typical compromise for the critical wavelength should be between 1.5 and 2 μm. Other challenges are the operating conditions. Especially in atmospheric conditions with high oxygen content, corrosion can hardly be avoided. Receiver pipes undergoing a structural phase transition below the absorber temperature increase thermal stresses on the coatings.

Impact on Heliostat Field Design

A limited receiver view factor, e.g. of a billboard or cavity receiver rather than a cylindrical external receiver, may enhance the power density in MW/m^2, While a cylinder has a 360° field of view to the environment, a billboard type is limited to a 180° field of view, often even less, although a downward sloping front aperture can bring a larger ground area into the field of view. Using a polar aligned billboard design with a 120° field of view to the heliostats, high performance can be achieved around noon, as most heliostats have an angle of incidence of sunlight close to 0° ("cosine effect" is marginal). However, if the azimuth of the sun changes (at other times of the day), the optical power on the target aperture decreases rapidly. Conversely, with an external cylindrical receiver, only a moderate power is obtained around noon, but when the azimuth of the sun changes from its noon value, which can be more than 100° in summer, it results in only small changes in optical efficiency. This is because the nearly circular nature of the surrounding heliostat field, approximately maintains the average value of the cosine over the entire field. The effects of determining the receiver concept therefore directly involve a restriction in the design and shape of the heliostat array. The latitude of the tower location as well as the target power has an impact on the decision, whether a billboard, a cavity or an external cylindrical 360°-receiver is being used.

If, however, the selected working fluid allows high receiver outlet temperatures or low heat transfer coefficients (e.g. gases), a cavity concept is preferred, since the thermal losses with cavity concepts are considerably lower than those of an external receiver directly facing the environment.

One of the most important advantages of central receivers is the relatively short pipe lengths from the receiver to storage and demand site. Decentralized tower concepts, which are currently also considered, partially eliminate this advantage.

Effect of the Permissible Heat Flux Density on the Receiver Design

Most commercial receivers only allow surface power densities in the order of up to 1.0 MW/m^2 (e.g. for molten salt) with a certain safety margin from a heat transfer point of view, so that larger receivers are used than the minimum size of the optically achievable focal spot area. The ratio of height to diameter of an external receiver, often referred to as an aspect ratio, should be larger than unity, so that heliostats in the near range can be effective, since they see optically only a substantially shortened receiver. Since the target a certain outlet temperature, a minimum pipe length is required for a given flux density distribution. This usually leads to an aspect ratio between 1 and 2. The permissible solar flux is then achieved by defining several heliostat target planes (two to six) on the receiver, to each of which specific parts of the heliostat field are assigned. An intelligent strategy that estimates the radius of the beam and shifts the image of the heliostats associated with the outer target planes so that their beam remains just on the receiver results in a very low spillage, while the peak flux is significantly reduced (by a factor of 2–4, depending on the optimization strategy). A second requirement may be that the ratio between the north and south side of the flux density be limited to ensure a solar flux which is as uniform as possible on the receiver surfaces to minimize circumferential flux gradients and thus thermal stresses.

Coolant Flow Configuration

Depending on heat transfer fluids there are both open and closed receiver concepts.

- Open concepts are possible with the heat transfer fluid air. Ambient air is heated by sucking it through a fixed structure where the heat is generated by solar absorption. The other option is direct optical conversion on particles which are carried by an air flow are freely falling in the gravitational field. Open concepts have the advantage that there are no heat transfer losses due to heat conduction via walls or other structures. Conversely, they are susceptible to convective/radiative heat losses through wind.
- In closed receiver concepts, the optical conversion is realized pipe receiver panels. The generated heat flux is transported conductively through the pipe walls, and then transferred by forced convection to a fluid. Closed concepts inevitably imply a pressure-resistant pipe network as well as corresponding conveying machines (e.g. pumps, compressors, ...). In contrast to open concepts, closed arrangements allow both large coolant mass flow densities and a significantly larger heat transfer density to be achieved than in open arrangements. This

enables very compact receiver dimensions, which also mean low heat losses to the ambient.

There are also several experimentally validated concepts with respect to flow guidance. The flow guidance within receivers plays an important role, as it influences both the heat loss and the thermomechanical stress distribution, resulting from the temperature distribution, especially during transients. All receiver concepts presented in the remainder of this chapter aim at a homogeneous temperature distribution to keep the thermo-mechanical stress levels low. To achieve the highest possible coolant outlet temperature from the receiver over the longest possible period of the day, several options are possible for flow control. The simplest solution is the simple upward flow through a tube or a plate and the adjustment of the flow in such a way that the required coolant outlet temperature is reached at the upper end of the tube, as illustrated in Fig. 7.65a. Often, however, the vertical extension of the receiver is not sufficient to achieve the desired temperature lift along its length. An increase in the residence time then can be achieved with a meandering flow path, as shown in Fig. 7.65b by splitting a tube array into segments (top) through which flow passes in succession or a plate design with an integrated meander flow channel (bottom). If even longer residence times are required, the concept of a quasi-volumetric receiver can be used, in which cooling channels are positioned normal to the solar irradiation and flow is alternated several times to achieve the desired target outlet temperature. The concept of an open volumetric air receiver and a quasi-volumetric closed receiver is shown in Fig. 7.65c and d.

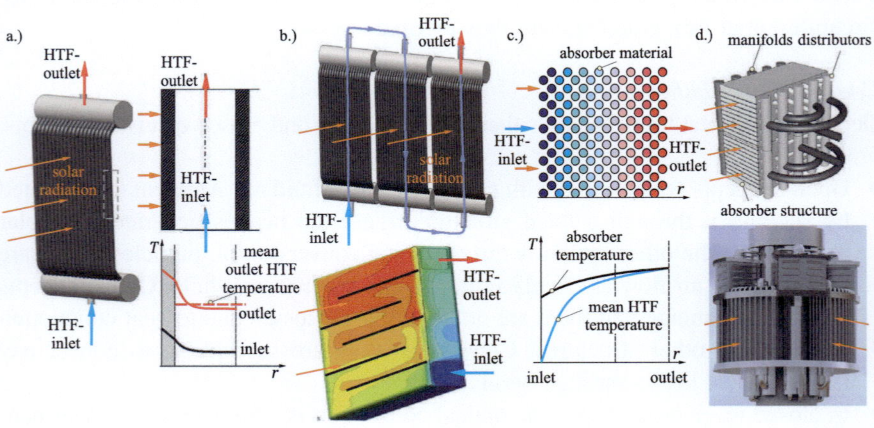

Fig. 7.65 **a** Billboard receiver and its typical HTF temperature evolution along a single tube (right); **b** Meander flow receiver to enlarge HTF residence type (top) subsequent bundle array type (bottom) monolithic billboard type receiver (for sodium/salt as HTF); **c** volumetric open air receiver; **d** closed gas/direct steam generator volumetric receiver with indirect cooling modified from Kalbhor (2014)

The basic form of the flow channels is either tube bundles or plate elements. Both designs allow a modular construction and are therefore suitable for both external and cavity receiver designs. The general structure of the tubular receiver consists of parallel tubes arranged in rows, whose outer surfaces are heated by the concentrated solar rays. A tubular receiver is simple, easy to build, inexpensive and easy to monitor and maintain. The tubular design allows for parallel but also staggered tube arrangements. Another key advantage is that heat transfer relationships for tubes are reliably known and available with high accuracy for both single-phase and two-phase flows at many Prandtl numbers (i.e. quasi for arbitrary fluids) regardless of their orientation, which substantially facilitates the predictive performance determination of the receiver.

Monolithic plate concepts are becoming increasingly important, as they allow a high degree of receiver coverage and thus have large heat flux densities and at the same time smaller temperature gradients than tube concepts. Due to their construction, they have a lower compressive strength than tube concepts; but additive manufacturing processes, especially in low-pressure receiver concepts with liquid salts or liquid metals as heat carriers, allow fluid-adapted high surface utilization rates at extremely high surface power densities.

Radiation and Heat Losses of the Receiver

In the case of the cavity receivers, the radiation and convection heat losses are lower in comparison with external receiver designs with a surrounding heliostat field. They therefore usually show a greater annual optical efficiency but require higher towers and greater tracking accuracy in order to "see" all heliostats in a northern (or southern) heliostat field in comparison to a surrounding field with the same given power. However, also the hot surfaces of a cavity receiver emit thermal radiation via the aperture. The internal volume of the cavity is filled with air, heated by the internal temperatures at the absorbing surfaces. with the gradient to the low temperatures at the aperture entrance can create turbulent buoyant flows that can carry large convective heat fluxes to leave the cavity. Depending on the aperture opening and the wind conditions prevailing at the aperture, the area specific heat dissipation from the cavity can even exceed that of external receiver concepts.

Cavity receivers are particularly suitable for systems with very high temperatures, where a small ratio between aperture and absorber area is advantageous and high fluxes can be absorbed. However, a cavity must both support and enclose the heat-absorbing surface. It is clearly larger than the support structure required for an external receiver with the same absorbed flux. It must also be well thermally insulated to reduce heat conduction losses through the large surface area. Cavity receivers therefore require large, heavy, and expensive enclosures, such that low-cost steel frame constructions as towers are usually ruled out. In addition, cavity receivers require a higher tower due to the limited cone angle.

7.3.2.3 Receiver Types

In central receivers, liquids (e.g. water, molten salt, liquid metals), gases (e.g. air,$s-CO_2$, helium) and solid media (e.g. ceramic particles) are used as heat transfer and/or storage media. Almost all fluid receivers use plate segments (billboard, staggered, monolithic) or tube bundle arrangements which have a (sometimes) selective coating on the side receiving the solar radiation and are cooled in the core by the heat transfer fluid (HTF). The tube fields can be contained in a cavity receiver or be arranged in a cylindrical or cubic configuration.

Due to the variety of commercially realized concepts but also in prototypes already proven receiver concepts, here the receiver types are divided according to the HTF type in

- gas-cooled receiver,
- liquid cooled receiver and
- particle receiver,

for which only the functional principle is explained as well as typical temperatures, pressures and thermal efficiencies are addressed in their order of magnitude. This also includes a presentation of the advantages but also challenges, otherwise reference is made to the secondary literature listed in the mentioned sources. An overview can be found in Ho (2017) or Coventry et al. (2015).

Gas-Cooled Receivers

Gas-cooled receivers allow intrinsically high receiver temperatures without any phase change of the fluid and, when using inert environmentally friendly gases, low costs for the heat transfer fluids and marginal environmental requirements for the construction of a plant. Conversely, due to their low thermal conductivity and low heat capacity, gases require high flow rates to realize a sufficiently good heat transfer so that the receiver material as well as the pipelines remain within material sustainable limits. The low thermal inertia requires good control and regulation logic to largely eliminate flow instabilities. Thus, the demands on the structural material increase and so do its costs. Conversely, the use of pressure-charged closed systems allows direct conversion into electrical energy with the aid of the Joule-Brayton process at high thermodynamic efficiencies but requires the use of expensive high-temperature and high-pressure piping as well as greater parasitic losses due to high specific work by operation of the compressors. With the gases themselves, thermal energy storage of a possible thermal surplus thermal power from the solar field is only possible with indirect thermal energy storage concepts such as brick Cowper or rock bed storage. Three sub-families of gas-cooled receivers are known in the literature:

- volumetric air receivers,
- closed (pressure-charged) receivers,
- gas-particle mixture receivers.

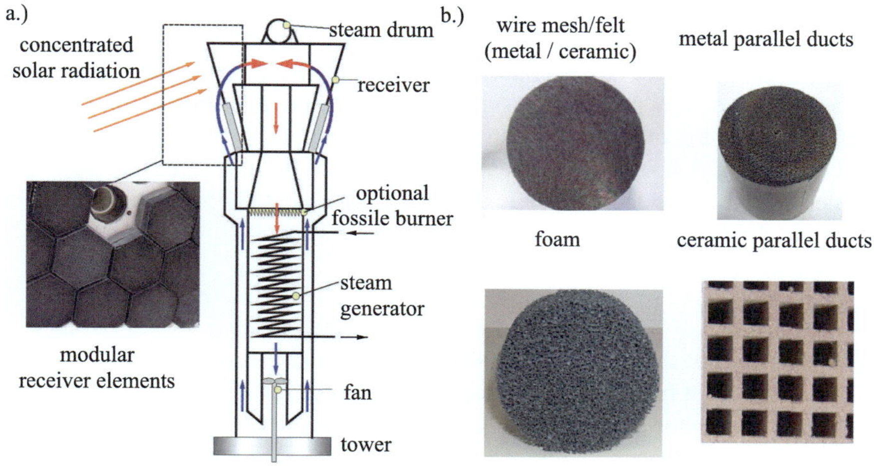

Fig. 7.66 **a** Schematic design of a solar tower with open air receiver. **b** Different designs of the absorber structures of volumetric open-air receiver

Volumetric Air Receivers

Open air receivers are a simple and safe design of a receiver system in terms of plant engineering, since even a possible failure of the fan does not immediately lead to destruction of the receiver unit. Figure 7.66a shows the schematic design of an open-air receiver system in which a steam generator is integrated into the tower, which can be supported by a fossil burner if necessary.

The receiver unit consists of a volumetric absorber that receives the solar radiation and is usually modular. The air is drawn through channels or honeycomb blocks that are irradiated by the concentrated sunlight. The air heats up as it passes through the irradiated structure, and ideally the light should penetrate deep into the receiver so that the warmest part of the receiver is inside, away from the aperture, so that the heat radiation losses to the surroundings remain low. However, most studies have shown that the opening of the volumetric receiver is the hottest part, resulting in significant radiation losses to the environment. The receiver structure consists of either a metallic or ceramic material, which is constructed either as a knitted fabric, felt, foam or ordered in parallel channel structures, as Fig. 7.66b shows. Here, the concentrated radiation penetrates the absorber material and generates heat, which has led to the term volumetric receiver. With the help of a fan, ambient air is drawn in from outside through the absorber material and the air flow absorbs the heat stored in the absorber. With open air receivers, the choice of material determines the maximum turbine inlet temperatures. With Inconel alloys up to 950 °C and with ceramic silicon carbide up to 1000 °C air temperatures have already been realised in experiments (Avila-Marin, 2011), and power densities of more than 800 kW/m^2 at receiver efficiencies of about 70% have also been demonstrated. Air as a heat transfer medium has many advantages here; it is non-toxic, not excessively corrosive, available everywhere in any quantity, and is also non-combustible and easy to handle.

One disadvantage of the open-air receiver concept is the low power density due to the fixation on low process pressure. This requires high air flow rates, which in turn only allow for smaller temperature differences. Impressive advantages should not be overlooked, however. With a modular design of the absorber unit, individual segments can be changed without lowering the pressure in the system. This allows high availability. The optimization potential of this operating concept essentially comprises two points.

- The reduction of heat losses on the front side. Analytical considerations as well as experiments show that a two-zone receiver, whose sun-facing side consists of a transparent honeycomb structure to which a ceramic structure is connected downstream, allows receiver efficiencies of more than 90%. The principle of operation corresponds to that of a transparent thermal insulation, see Pitz-Paal et al. (1991). The quartz honeycomb directs solar radiation into the receiver, so it reaches the ceramic absorber part of the receiver almost unhindered. The heat radiation losses from the ceramic receiver parts are partially absorbed in the quartz glass part and can be recovered as usable heat, since the quartz glass channels are also cooled by the gas flow. This results in low temperatures in the quartz glass structure, especially at the receiver inlet.
- The second way is to increase the air recirculation rate (the ratio between the air mass flow recirculated and the total air mass flow through the receiver). Heat recovery of the heat still present in the air after a heat exchange to the power conversion system increases the thermal efficiency of air-cooled solar thermal receivers. Typical optimal air recirculation rates are around 70%.

Other deficits that can be reduced but not eliminated by intelligent design is the soiling of the receiver structure by sand, and the heterogeneous heating of the receiver. Large temperature gradients cause strong changes in the temperature-dependent density and viscosity of the air, which lead to flow instabilities. Absorber materials with low porosity seem to eliminate this to a large extent.

Closed (Pressure-Charged) Receivers

In the case of gas-cooled pressure-charged receivers, the selective absorber, which is subject to high thermal loads, is cooled with the aid of a forced convective fluid flow. The pressure charging principle intrinsically requires a closed gas circuit in which the gas is compressed to pressures of up to several MPa with the aid of a compressor and is conveyed through corresponding cooling openings in the optically selective part of the receiver in which the optical conversion of the concentrated solar radiation takes place. As a result, heat is removed from the receiver unit either directly to the power conversion system or to a secondary circuit by means of a heat exchanger. There are several different concepts here.

Due to the higher pressure in the gas line network, higher power densities can be dissipated in the receiver and the achievable heat transfer coefficients can be higher than those of an open system by a factor of 100 ($h > 10^4$ W/(m^{2}K)) and even reach values of liquid-operated receivers. The closed (pressurized) gas receiver faces the solar field and all receivers tested successfully to date are of the cavity type, in which the concentrated solar radiation generated by the heliostat field is guided into the cavity via a secondary receiver. In the case of the absorber cavity, either the absorption can take place directly on the cooling channel, which can be either conical, cylindrical, or hemispherical. Some concepts increase the heat transfer surface, for example, by integrating silicon carbide foams, which additionally also increases the residence time of the heat transfer fluid in the receiver. Conversely, this measure shown in Fig. 7.67a means an increase in the pressure loss. Other flow guides with direct conversion on the cooling channels either place on a pipe bundle (Fig. 7.67b) or on the circumferential flow guide (Fig. 7.67c). A disadvantage of the direct conversion at the cooling channel is that the pressure-carrying tubes are exposed to extreme temperatures which correspondingly presuppose the use of high-temperature-certified, pressure-resistant materials. At the same time, the high temperatures lead to continuous growth of an oxide layer on the selective absorber material, which has a low thermal conductivity. In indirect, pressure-charged receiver concepts, this limitation is circumvented by the use of a quartz glass window, and the conversion can take place either on porous (Fig. 7.67d) or on cushion-shaped absorbers (Fig. 7.67e), which are surrounded by gas on both sides by means of cooling slots and thus do not require any strength functions. Figure 7.67f shows a further decoupling of the absorption with a simultaneously large heat transfer surface, in which a porcupine structure assumes the absorption function. Only a few demonstration projects rely on a purely solar application, the power spectrum of which is limited to the order of magnitude of a few MW, while the concepts aimed at a commercial application, some of which are listed in Table 7.10, are aimed at solar-fossil hybrids in order to substantially reduce the CO_2 footprint of fossil power plants in a sustainable manner.

One of the key challenges of a pressurized gas-cooled receiver is its low thermal inertia, which intrinsically leads to high lateral and transverse temperature gradients associated with correspondingly high thermo-mechanical stresses, and the optical loss and cost of the transparent window opening. Therefore, the operation of gas-cooled receivers requires special attention to control and regulation to uniformly achieve the desired receiver outlet temperatures necessary to operate a gas turbine or downstream power conversion system at high efficiency. Special technical precautions must also be taken to ensure receiver integrity during irradiation power peaks or fluctuations, for example, in the event of misalignment of heliostats or compressor pressure or flow rate fluctuations. Technical options include shutters or fast-acting adjustment mechanisms of the secondary receivers, which take corrective action by means of sensors installed in the solar field and the pipe network of the coolant.

Pressure-charged receivers appear to be less suitable for large power plant capacities in purely solar operation also in the future, since in addition to the limited possibilities for thermal energy storage, the design of large-scale receivers in particular requires technically complex constructions. As a thermal energy storage

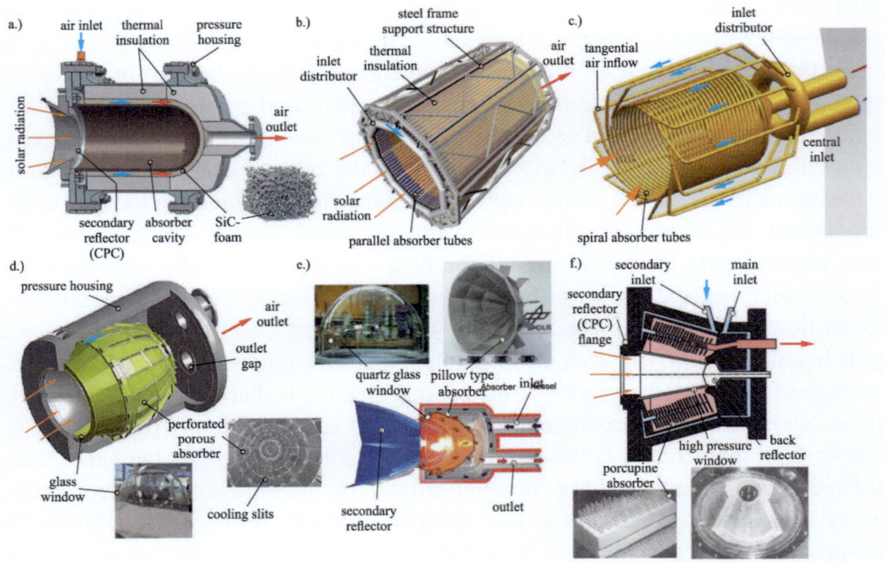

Fig. 7.67 Technical versions of pressurized air receiver; **a** Cylinder receiver with supporting foam heat transfer surface (Poživil et al., 2014); **b** Cylindrical tube bundle receiver according to Korzynietz et al. (2016); **c** tangent cooled cylinder receiver according to Wang et al. (2015); **d** indirect barrel receiver with porous absorber structure modified from del Río et al. (2015); **e** indirect hemispherical receiver with rear-cooled cushion absorber (Deutsches Zentrum für Luft- und Raumfahrt (DLR), 2008); **f** indirect receiver with porcupine-like absorber plates modified from Karni et al. (1997)

system for pressure-charged receivers, the only direct storage option is a particle bed, in which gas solids are heated and, when heat is required, this heat is fed back to the gas compare (Stein & Buck, 2017), which is used in particular during start-up to quickly reach the operating temperature. However, this requires a second compressor.

Another consideration is the choice of structural materials, because in pressurized receivers, a large part of the piping network must withstand temperatures of more than 550 °C, so the use of stainless steel is largely ruled out. In the demonstrators realized so far with temperatures up to and above 1000 °C, nickel alloys (Inconel), aluminum silicates or silicon carbide (SiSiC) are used (Avila-Marin, 2011).

A challenge for all receivers operated with air is the oxidation of metallic structural materials and as an alternative to air as a heat transfer medium, nitrogen and especially helium are suitable for closed Joule-Brayton circuits, mainly due to their inertness; for corresponding concepts see Forsberg et al. (2007), Dunham and Iverson (2014) and the literature listed therein. Although the choice of the cooling medium due to its thermo-physical properties is essential in the design of the cooling channel geometries and dimensions of the receiver, unless material limits are exceeded, the choice of the HTF is more of an argument of downstream electrical

Table 7.10 Characteristic data of some pressure charged gas cooled receiver design experimentally studied according to the design illustrated in figures

Technology orientation	Low pressure PSI	SOLUGAS	SOLGATE	PEGASE	DIAPR
Max. heat flux density (MW/m^2)	4.36	0.7–1	0.4	1	–4
Design type	Cavity	Cavity	Cavity	Cavity	Cavity
Inner cavity shape	Cylindrical with foam	Cylindrical	Conical	Cylindrical	Conical
Absorbing surface	Plane surface	Annular tubes bundle	Screen behind glass	Spiral cylindrical tube	Porcupine absorber behind glass
Peak surface temperature (°C)	900	1000	1140	1100	1200
Fluid	Air (He)	Air	Air	Air	Air
Max. fluid temperature (°C)	553 (619)	800	950	980	1000
Pressure (MPa)	0.4 (0.6)	0.9	0.65	0.4	2
Thermal efficiency (/)	0.77 (0.78)	Up to 0.81	0.79	~0.7	0.8
Conversion type	Solar only	Solar-fossil	Solar-fossil	Solar-fossil	Solar only

power conversion, so this aspect will be discussed in more detail in Chap. 11 in the context of the Joule-Brayton cycle for the different gases. In this context, the prejudice that power densities of more than 1 MW/m^2 cannot be reliably cooled with the aid of a gas-cooled wall should also be dispelled. Helium-cooled divertors have shown in pulse tests that, with appropriate design, power densities of more than 1 MW/m^2 can be removed even with large gradients and in fast transients, see You et al. (2018). Nevertheless, a dedicated mode of operation is required, which is difficult to achieve in the free field of a solar tower.

Small Particle Gas Receivers

This is a subgroup of pressure-charged receivers in which microparticles are used that are present as a suspension in the carrier gas. The solar radiation concentrated over the transparent front screen leads to a volumetric absorption at the particles and thus to a uniform heating in the fluid volume in contrast to the purely gaseous variants, which are characterized by a large temperature gradient of the wall and bulk of the fluid. These types of receivers have so far been tested more in an academic environment, as there are several challenges to overcome. For uniform mixing, it must be ensured that the fluidized gas-particle bed is homogeneously mixed at all flow rates so that there is a constant particle concentration and thus a constant volumetric optical conversion. At the same time, the design must be such that no recirculation zones can occur where particles agglomerate and change the flow configuration and thus the heat dissipation. In addition, a suitable window configuration must be found in such a way that the window transparency is maintained during the lifetime even under changing pressures.

Liquid Cooled Receiver

Liquid cooled receivers are probably the most common type of high temperature receivers in solar towers. Liquid heat transfer media have the advantage over gases that they have a high density, allowing comparatively large mass flows to be realized for cooling at low velocities. Thereby the pressure losses and the temperature rise per unit length in a receiver tube element is kept low, which reduces pumping power and minimizes the thermo-mechanical stresses in the structure. Another property is their incompressibility, which shrinks the pumping work required to circulate the heat transfer fluid. Thus, the mean flow velocities required to dissipate the receiver heat is the range of a few meters per second only, whereas with gases mean flow velocities of more than 50 m/s are not uncommon. This leads not only to lower corrosion and erosion rates on the structural material walls, but also to liquid-cooled receivers being more compact (i.e. smaller) than gas-cooled receivers at the same total thermal power. Depending on the heat transfer fluid, there are different designs.

Water/Steam Receivers

Water and direct steam central receivers are probably the longest known receiver design, in which the concentrating solar radiation is converted in the receiver and the heat is transferred to the water. In principle, external receivers can be used as receivers, as for example in the Ivanpah Solar Electric Energy Generating System using billboard type panel receivers, see Fig. 7.68a, but also cavity designs of the receivers have been realized, as in the demonstrator PS10 in Seville/Spain or Khi Solar One in South Africa, see Fig. 7.68b, and c. Direct evaporation systems have the advantage of avoiding exergy losses and costs associated with heat exchangers between the receiver and the power unit. However, to achieve high collector efficiencies, and thus high thermodynamic efficiencies in a downstream power conversion system, high pressures are required. While the first demonstrators such as PS10 operated at a pressure of 4.5 MPa using a saturated steam process, modern plants operate at pressures and temperatures comparable to fossil-fired power plants, Ivanpah and Khi for example at 12–16 MPa and 530–560 °C receiver outlet temperature. At such high temperatures, a direct transition from water to superheated steam within one vertical flow path in the receiver is no longer possible, so the flow path is divided into three segments, one of single-phase heating of the water, one of evaporation, and a final one of superheating of the steam. This is done by taking advantage of the fact that the radiance is not uniformly distributed over the receiver surface and designing the flow path within the receiver so that preheating and superheating are positioned in regions of lower irradiance.

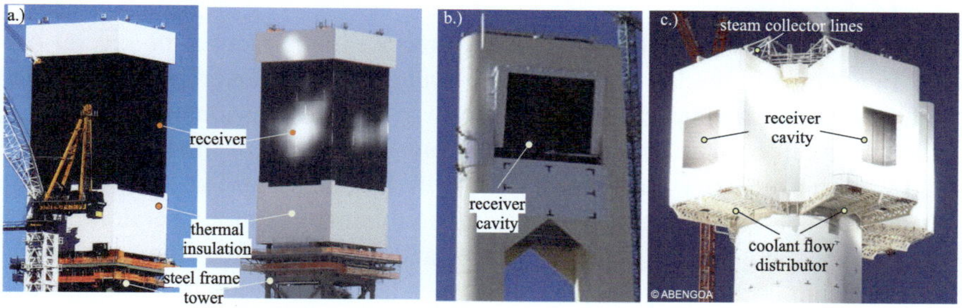

Fig. 7.68 **a** External direct superheated steam generating receiver of Ivanpah, CA Solar Electric Energy Generating System (left) no focussing and partial illumination (right); **b** single cavity saturated steam receiver of PS10 Seville Spain; **c** multiple direct superheated steam generating receiver of Khi Solar One SA (image Abengoa Solar)

In the early development stages of direct evaporation receivers, saturated steam systems were used, such as in PS10/PS 20. Due to the relatively low pressure (4.5 MPa) and thus low temperatures (receiver outlet temperatures approx. 300 °C), the efficiencies of the steam turbine are rather low, but the technical operating risk is low as well. The low pressure not only has the advantage that the tube walls for pressurization in the receiver tube can be kept relatively thin compared to high-pressure cooled systems and thus the heat transfer is greater, but it also offers the possibility of short-term storage of excess thermal power, for example in a steam accumulator or a steam drum.

In the case of high-pressure systems whose steam parameters exceed 10 MPa, the receiver design is more complex; not only must the flow path be divided into three parts because the heat transfer coefficients for water, saturated steam and superheated steam are different, but if reheating is used, an additional further flow path must be integrated. Direct thermal energy storage of excess thermal energy in a steam accumulator is not feasible, since it requires such large-scale pressure-resistant containers that they cannot be built with currently available materials. However, the high superheated steam temperatures allow the heat to be transferred by means of a heat exchanger to an indirect thermal energy storage facility, such as a liquid salt storage facility.

The advantages of using water are obvious. It is a technology known from power plant technology, so that not only the availability of the components—with the exception of the receiver—but also maintenance chains and quality assurance procedures are established and thus the capital risk is low. At the same time, the technical challenges during normal operation are high, e.g. high temperature gradients with variable solar irradiation and the associated thermal stresses can only be controlled by active flow control. The low flexibility in thermal storage (often none), means low capacity factors for the power plant and thus the availability of the power plant. Nevertheless, one of the largest tower power plants in the world, the Ivanpah power plant with 377 MW_{el}, is based on a water-cooled receiver technology.

Liquid Salt Receiver

Liquid salt as a heat transfer medium in a receiver is attractive for several reasons. In addition to the low specific price, liquid salts have a similar high thermal conductivity and a comparable heat capacity to water, but without undergoing a phase change like water. This allows a low-pressure design of the receivers using small wall thicknesses with compact construction volumes and water-like cooling structures. Another positive argument is the usable maximum temperature, which allows temperatures in the order of 600 °C and thus fits perfectly to a Clausius-Rankine steam cycle of a power conversion system. In addition, the use of liquid salts allows the use of direct thermal storage concepts, in which the heat transfer fluid of the receiver and the thermal storage fluid are identical. Compared to indirect concepts, this eliminates the need for components such as heat exchangers and pumps. A disadvantage of all salts is that they have a relatively high melting point, so that suitable measures must be taken to prevent "freezing" under all possible operating conditions. Measures range from draining strategies, recirculation concepts to auxiliary electrical heating systems. Further challenges with salts are the corrosion and dissociation behavior (compare Sect. 5.3.2.4), which require monitoring.

The first project of a molten salt receiver was deployed in Gemasolar/Spain, where on a 140 m tower in an external cylindrical receiver salt is heated from 290° to 565 °C to drive a 20 MW_{el} steam turbine. The excess thermal energy of the receiver is fed into a direct two-tank thermal energy storage system, which has a maximum storage capacity of 15 h of the downstream electrical power block. The Crescent Dunes power plant, which is more than a factor of 5 larger in electrical power output when commissioned, has a 195 m high tower on which an external receiver is mounted that is also cylindrical and about 30 m tall, as shown in Fig. 7.69a. The thermal energy extracted from the receiver is fed into a direct two-tank thermal energy storage system with a capacity of 10 h. The tower, the storage facility, and the receiver are very compact. The compactness of the tower, storage, and power block is illustrated in Fig. 7.69b.

Almost all liquid salt cooled projects realized so far rely on the external receiver, although in the case of the tower, steel frame structures for smaller power units do represent a cost-effective alternative to reinforced concrete structures of large power classes, as shown in Fig. 7.69c.

Ho and Iverson (2014) have also analyzed in detail alternative salt mixtures such as fluoride, chloride, and carbonate salts, which exhibit a higher temperature stability than the most common nitrate-based salt mixtures. But they conclude that while collector efficiencies of up to 90% are achievable, each of the salt classes considered has particular challenges. Fluoride salts are stable up to 1000 °C even under high heat flux, so that salt dissociation plays a minor role, but the melting points of the alternative salt mixtures are considerably higher than the nitrate-based salts. Chlorides have a lower melting point but their corrosion behavior towards based structural materials such as steel increases according to an Arrhenius function with increasing temperatures. Carbonate salts are cheap but can dissociate at high temperatures to form CO_2, so they require stabilizers and thus coolant

Fig. 7.69 Molten salt tower receiver installed. Crescent Dunes power station external cylindrical receiver (**a**) with the storage and power block parts at the foot of the tower (**b**), fotos modified from (BrightSource Energy Inc., 2023); external molten salt receiver of Delingha, China 10 MW tower on a steel frame tower (image (CosinSolar, 2023))

chemistry control. The trend in research here is more and more towards binary and ternary mixtures, which on the one hand exhibit lower melting point temperatures and simultaneously to increase the dissociation temperature limit to increase the usable temperature spread. Another focus is to improve the aging behavior and reduce corrosion, more details can be found in Vignarooban et al. (2015) and the literature cited therein.

Liquid Metal Receiver

The motivation for the use of liquid metal cooled receivers comes from the limitation of previously tested salts to temperatures around 600 °C. Liquid metals, such as sodium, lithium but also metals such as tin or lead or lead alloys allow here a single-phase low-pressure operation up to temperatures of 850 °C in case of sodium or over 1000 °C for other liquid metals, whereby the limitation is essential determined by the constraints of the tube structure material. The main advantage of liquid metals over salts is their outstandingly good thermal conductivity, especially in the case of sodium, which allows significantly larger heat fluxes compared to other HTF's. This not only results in smaller thermo-mechanical stresses due to the smaller temperature differences between front and back in the receiver structure, but also allows significantly smaller receiver areas with the same total power, which conversely requires greater optical accuracy of the heliostat field.

The advantages of liquid metals and especially of sodium as the heat transfer fluid compared to molten salts are particularly evident at high power densities in the optimum geometry configuration. The more compact design of the receiver (more than 45% smaller surface area) leads to a substantial reduction in convection and radiation losses, whereby the influence of the reduced surface area in particular is far more significant than the losses due to the possible significantly higher

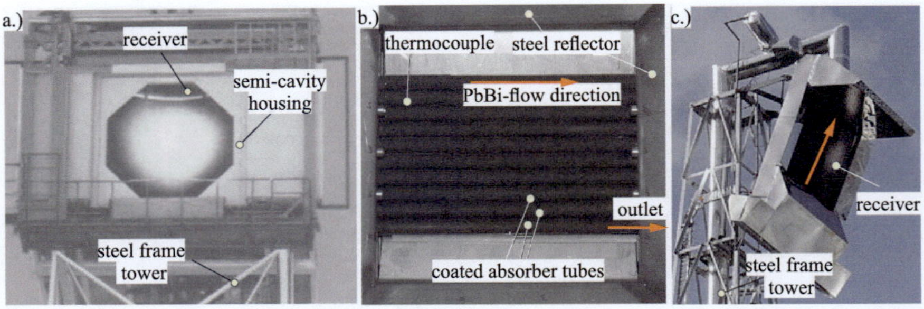

Fig. 7.70 **a** Sulzer sodium half-cavity receiver installed in Almeria/Spain (Kesselring & Selvage, 1986), **b** PbBi-eutectic cooled reflector receiver in Karlsruhe/Germany (Flesch, 2021), **c** Sodium cooled external plate receiver Jemalong/Australia (Coventry et al., 2015)

absorber temperatures, see Boerema et al. (2012). The potential for an optimized sodium system compared to molten salts is estimated to be of the order of 20%.

Although the potential of liquid metal cooled receivers was identified early on, field testing has so far only been carried out in the kW range in demonstration facilities, with the initial focus on sodium, see Fig. 7.70a; but liquid lead–bismuth has also been tested in receivers. Heat flux densities of more than one MW/m^2 could be dissipated with both types of receivers. All Fig. 7.70a–c show that both horizontal and vertical up flow configurations were analyzed, as well as cavity, half-cavity, but also shallow external receiver designs.

Despite the numerous advantages, several fluid-specific challenges to a designer have prevented market penetration by now. Common to all liquid metals is that their price is significantly higher than that of solar salts and their heat capacity is lower compared to salts, making direct thermal storage less economically attractive, which can only be compensated to a limited extent by the higher usable temperature difference.

There are also some arguments that stand in the way of widespread use. Metals such as tin and lead, as well as their alloys, are largely chemically inert in air, but have a high density that is not desirable on tall towers. In addition, their chemical compatibility with structural materials, especially steels, is difficult and they require dedicated oxygen control systems to both prevent dissolution of the steel alloy components within the HTF and simultaneously prevent oxidation of the coolant. Such oxidation problems or incompatibilities with metallic structural materials are alien to alkali metals such as lithium, sodium, or the eutectic sodium–potassium alloy, so that they have been successfully used in nuclear systems for decades. They can be cleaned unproblematically in-situ by means of cold traps. Moreover, corrosion and erosion play only a marginal role along the entire temperature domain. Their challenge lies in the design of a safe receiver system that excludes contact with oxygen or water and water vapor. Alkali metals, e.g. sodium are self-igniting in air just above the melting point if the temperature exceeds about 140 °C. The reaction of sodium with water, for example, releases a large enthalpy of 285.5 KJ/mol. Leakages of sodium-bearing tubes were also the reason for the termination of the sodium-cooled receiver project in Almeria. Nevertheless, the use of sodium has increased again in recent years in the context of successful tests on prototype solar towers.

Alkali metals, unlike all other heat transfer fluids, open up not only the use in thermodynamic cycles but also a direct conversion into electricity. As fluids with a high specific electrical conductivity, the option of direct thermal-electrical conversion exists, on the one hand, with the help of the AMTEC (Alkali Metal Thermal Electric Converters) process or using a magnetohydrodynamic (MHD) generator.

Thermo-electric converters as AMTEC transform thermal energy into electrical energy. Therein, absorbed heat is transported with a pressure difference across a solid electrolyte (in case of sodium it is a beta"-alumina solid electrolyte or BASE) to generate an electro-chemical potential. This is a direct transformation of electro-chemical energy to electricity, which allows to describe the AMTEC process through the same electro-chemistry principles as in a galvanic cell. At the interface of the electrolyte, the sodium atom separates from the electron. Driven by the pressure difference, the sodium atom diffuses through the electrolyte to the cathode and recombines there, while the electron is fed into the circuit at the anode, as illustrated in Fig. 7.71. Due to the stripping reaction and recombination, the sodium loses heat so that the temperature at the anode is lower than at the cathode side. Details of the AMTEC process can be found in Díez de los Ríos Ramos (2018). AMTEC has a theoretical efficiency close to that of the Carnot cycle about ~45% can be theoretically obtained even for liquid sodium at 800 °C, however, in practise about half of the value is achieved. The conversion efficiency even increases for vaporized sodium in a two-phase sodium receiver and has been proposed as topping cycle solar power tower (Hering et al., 2012), since the heat level of 600 °C at the cathode is suitable to operate a Rankine based steam turbine.

The other option is given by a liquid metal magnetohydrodynamic generator (LMMHD) producing electricity by accelerating the electrically conducting liquid through a nozzle with a velocity u_0 perpendicular an applied magnetic field B, according to $u_0 x B$. To accelerate the liquid metal flow, there are several possibilities. The simplest way is to use the heat of the receiver to drive a buoyant flow, however, attaining only small velocity and a poor balance while, boiling in a two-phase liquid metal receiver yields higher velocities and temperatures, which can be further enhanced by gas injection. However, the latter requires two-phase flow liquid metal receivers. All these concepts have been analyzed mostly only conceptually, and experimentally only on a small scale (Coventry et al., 2015).

Particle Receiver

Particle-based receivers use small solid particles that are heated by concentrated sunlight. The particles can be heated to temperatures above 1000 °C, and used to generate process heat or electricity. A major advantage of particle receivers is that the particles can be irradiated directly, eliminating the heat transfer resistance and flow limitations associated with indirect heating through tube or duct walls. However, indirect receiver concepts also exist in which concentrating solar radiation,

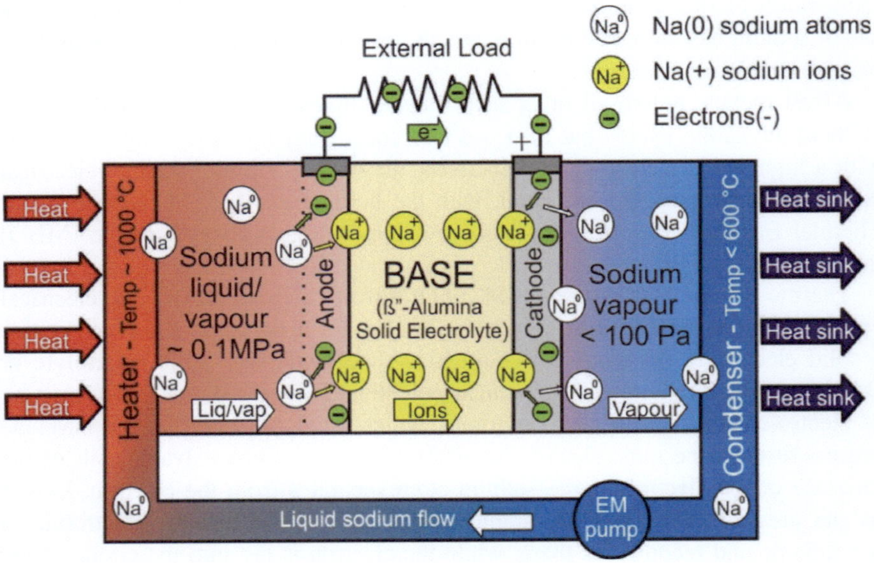

Fig. 7.71 Schematic diagram of the operation of an AMTEC generator for direct thermo-electric conversion

like gas-cooled receivers, are either heated directly through a glass wall or heat up as a fluidized bed against a wall.

A major advantage of direct particle receivers without any walls is that the heat is absorbed directly in the particles, so that solar flux limitations and as a result thermal stresses or relative thermal elongation of the absorber structure as well as pressure increases due to volumetric expansion of the heat carrier are eliminated. However, particle flows are not completely free of temperature limitations, effects particle agglomeration as well as clumping on all transport routes must be prevented or limited to a large extent. There are a variety of different designs of particle-based receivers, including free-falling or rotating cylindrical flow, tubular receivers with or without fluidization, concepts with multi-path recirculation, respectively, and downward-facing configurations; an overview is given in Ho (2017) and Ho and Iverson (2014).

As mentioned, particle receivers can be categorized as either direct or indirect particle heating receivers, where in direct particle receivers, particles are directly irradiated as they pass through the receiver, while in indirect particle receivers, either tubes or other enclosed volumes are used to transport and heat the particles. The direct particle receiver class scopes the following concepts:

- open gravity driven particle receivers,
- guided (or obstructed) falling particle receivers,
- rotating centrifugal receivers and
- fluidized particle receivers,

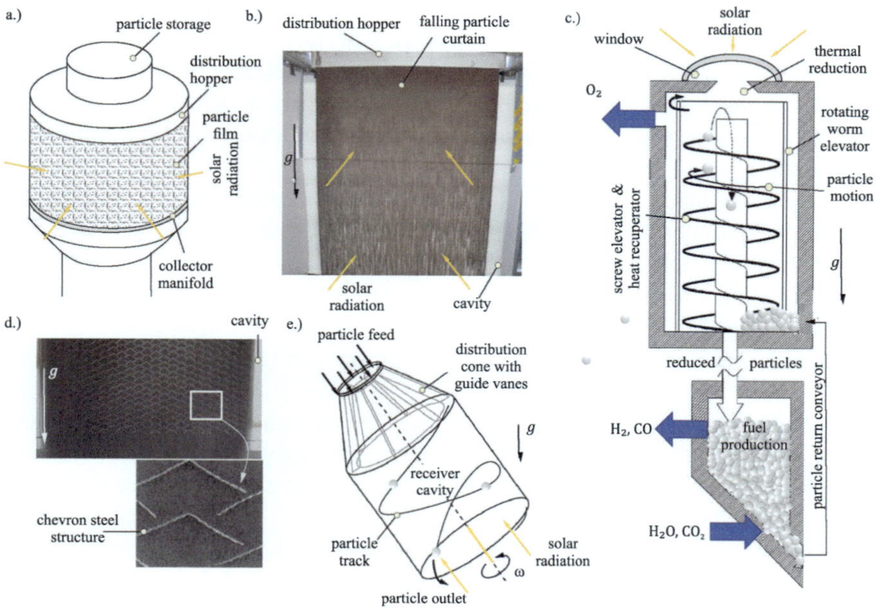

Fig. 7.72 Various direct particle receiver design concepts: **a** external receiver for a surround heliostat field; **b** Gravity driven falling particle curtain cavity receiver (**b**) and guided falling particle receiver with array of staggered chevron-shaped structure (**c**) both modified from Ho (2016); **d** thermo-chemical particle receiver from Ermanoski, Siegel, and Stechel (2013); rotating centrifugal receiver principle (**e**)

while to the indirect particle receiver relate to:

- gravity-driven particle flow through enclosures (plates/tubes),
- fluidized particle flow through tubes.

Direct Particle Receivers

- *Open-gravity driven particle receivers*

In an open gravity-driven particle receiver, particles fall in the form of a particle curtain along a wall in an external receiver type (Fig. 7.72a) or inside a cavity as a free curtain as depicted in Fig. 7.72b, where they are irradiated by incident concentrated sunlight. The curtain is created by a hopper above the receiver, which is fed with the particles by a conveyor.

To control the mass flow through the receiver to achieve the desired particle outlet temperature for a given radiant flux density, sliders or adjustable orifices are used. Although larger particle mass fluxes lead to an increase in the volume fraction of solid material in the curtain and thus a higher absorption rate of sunlight, which allows a higher thermal efficiency of the receiver to be achieved, the additional shading and blocking reduces the outlet temperature of the particles. One option to increase the outlet temperature is to increase the residence time of the particles in the zone of focused solar radiation, by multiple recirculation of the particles in the cavity, with the temperature successively increasing in each recirculation step. However, this requires additional temperature-resistant conveying equipment.

In open cavities, particles can be lost through the aperture, especially if the wind blows parallel to the aperture and the cavity depth is shallow. One option to prevent this is to circulate air or provide an air curtain in front of the particle curtain to reduce particle discharge from the receiver and thus convective losses.

- *Guided (or obstructed) falling particle receivers*

Flow-guided particle receivers aim from the start to extend the residence time of particles in concentrated sunlight by slowing the velocity of particle flow through the introduction of obstacles or porous structures, while ensuring direct absorption of concentrated solar energy.

Obstacles can be, for example, staggered mesh grids such as chevron-shaped grid structures made of stainless steel, as shown in Fig. 7.72d, in which Ho (2016) experimentally demonstrated more than 300 °C per meter of illuminated particle path temperature rise at 1 MW/m^2 irradiation. In another design, a spiral ramp is used, where particle flow is provided by a mechanically excited vibration in addition to gravity and has been successfully tested in experiments. Principally, this design requires beam-down optics, with large particle flows posing a technical challenge in this concept. In another concept study with beam-down optics, shown in Fig. 7.72c, particles are conveyed upward to the aperture by a screw elevator, where they are heated by the concentrated radiation and then fall down the hollow tube of the screw. This particle-based receiver design has been developed as part of a thermo-chemical reactor to reduce particles, which are then oxidized to produce either hydrogen or carbon monoxide.

- *Rotating centrifugal receivers*

The basic principle of operation is that on the top of the receiver, via a funnel, particles are fed into a cone provided with vanes into a rotating receiver which is open at the bottom, and into which concentrated sunlight enters from below, as shown in Fig. 7.72e. The centrifugal force of the rotating container causes the particles to spiral along the walls of the container, while being irradiated all the way by the concentrated sunlight, and finally leaving the container at the bottom, where they are trapped in a collector. Experiments with a downward tilted receiver and an irradiance of 670 kW/m^2 have determined particle exit temperatures of 900 °C and receiver efficiencies of about 75% (±4%). Challenges include maintaining a constant and sufficient mass throughput of particles at larger scales, the need for parasitic energy, and the reliability of a large rotating receiver system, especially in the presence of circumferential velocity variations.

The challenge of all more directly guided receiver designs is that they are open to ambient and thus associated to a particle loss. In addition, local overheating of structures located directly in the irradiation poses a lifetime problem, as does their erosion during operation. Other design-dependent technical challenges to be solved include non-uniform irradiance distributions on the particle curtain, variable mass flow rates, wind effects through the aperture, reliability of particle handling equipment.

Indirect Particle Receivers

- *Gravity-driven particle flow through enclosures (plates/tubes)*

Indirectly heated particle receivers consist of transparent structures (tubes or flat channels) where the particles flow downwards under gravity within the housing. The structures are irradiated by concentrated solar radiation, and heat absorption by the particles occurs both via the radiation and by heat transfer with the wall. Experiments here show that heat transfer in tubes is limited where wall contact of the particles is lost. Challenges include ensuring sufficient mass flow, ensuring homogeneous radiation flux within the cavity.

- *Fluidized particle receivers*

Fluidized bed receivers have been experimentally analyzed at Sandia, where in a vertical transparent silica tube, particles are fluidized with compressed air from below and solar irradiation occurs from above. The particle transport to achieve the required mass flow and at the same time fluidize the particle flow in the receiver tube is one of the challenges to be overcome to avoid local hot spots where the allowable material temperatures of the structures are exceeded. In addition, a suitable window for the pressurized receiver as well as the development of a solid–gas suspension system that ensures a homogeneous particle concentration and temperature inside the vessel is required.

The advantages of all indirect receiver designs are the absence of particle loss through the aperture opening and thus lower heat losses than with an open cavity receiver. However, the development of the additional interface is still a technical problem that has not yet been solved if similarly, high temperatures as in open receiver concepts are to be achieved.

7.3.2.4 High Temperature Structural and Functional Materials

Since the temperatures in the receiver take on the highest values in the whole system, a receiver design is intrinsically at the limits of the available structural materials as well as the functional materials.

The consideration of both aspects is essential for a sustainable receiver concept, which must ensure high availability under all time scales and at the same time a long lifetime. Failure mechanisms, manufacturing procedures as well as in-service operation, maintenance or replacement must be considered.

Structural Materials

The choice of structural materials is coupled to that of the heat transfer fluid which determines the maximum range of the operating temperatures.

In principle, several classes of materials are available where material temperatures above 500 °C are feasible.

- Fe—based alloys,
- Ni—based alloys,
- Co—based alloys,
- Alloys of refractory metals (W, Mo, Nb or Ta),
- Alloys based on intermetallic phases,
- Precious metal alloys (Pt),
- Ceramic Matrix Composites (*CMC*), such as Si_f–SiC composites.

The creep rupture strength which's determination is very complex, is the basis for the design of high-temperature components and is specified with the quantity $R_{m/time/temperature}$. There are two high temperature ranges. One with a time-independent material behavior, i.e. operation in the hot yield stress range, and one with a time-dependent material behavior, the so-called creep rupture range. The limit is the temperature at which the intersection of the yield strength at 0.2% elongation at a given temperature and the creep rupture strength R_m for 10^5 hours at the same temperature intersect.

High-temperature corrosion describes the reaction of a material with its environment at elevated temperature. In the absence of an electrolyte four aspects affects affect material performance.

- Oxidation. Here, a distinction must be made between external oxidation, which describes the formation of an oxide coating at the interface of the material and the HTF, and internal oxidation.
- Carburization, which reflects the internal carbic formation.
- Nitriding with internal or external nitride formation and
- sulfidation.

Regarding corrosion, its alternating effect with mechanical stress must also be analyzed. Two aspects play a role here; stress corrosion cracking, which leads to inter- and trans crystalline cracks due to chemical attack under stress, and fatigue corrosion cracking, which involves a combined effect of cyclic stress and aggressive media.

Figure 7.73a illustrates schematically the limits of creep rupture strength R_m at 10^4 h of some potentially applicable high-temperature material classes as a function of temperature. According to this figure, martensitic steels can be used up to about 600 °C with adequate strength, while the use of austenitic steels pushes the limit upwards by a maximum of 100 °C. Significantly higher temperatures are only attainable with nickel-based alloys, which are used in particular in gas turbine construction. For all metallic materials, however, the main failure mechanisms at

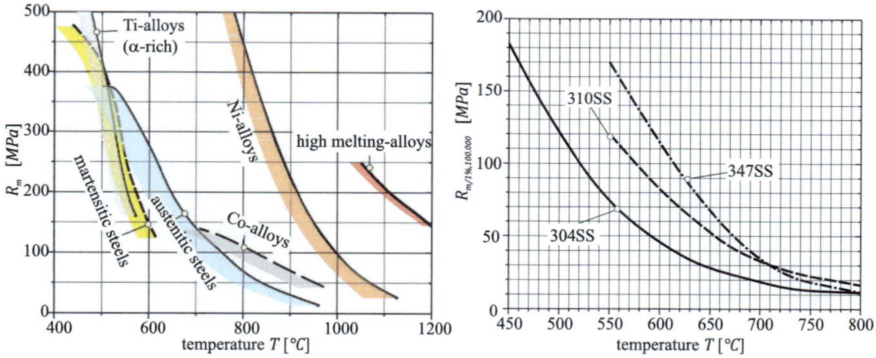

Fig. 7.73 **a** Tensile strength limits R_m in (MPa) of some structural material classes as a function of temperature; **b** Creep strength of some austenitic stainless steels at 1% creep over 10^5 h as a function of temperature modified from Davis (2000)

high temperatures, i.e. for $T > 0.4 \cdot T_m$ with T_m the melting temperature of the metal/metal alloy, are creep and high-temperature corrosion.

Before describing the properties of the individual material classes in rudimentary form, it is useful to first introduce a few definitions.

Steels

All iron-based materials use measures that impede the movement of dislocations in the material and slow diffusion to increase creep strength and high-temperature corrosion resistance. These include particle hardening, foreign phases, or the introduction of foreign particles, for example oxide dispersion hardened steels (ODS).

For temperatures up to 600–620 °C, martensitic steels (9–12% chromium content) can be used as a high-temperature material characterized by good creep rupture strength (creep resistance), high thermal conductivity reducing thermal stresses and a good scale resistance. With suitable adjustment of the water chemistry, martensite can also be used due to the formation of a corrosion-resistant magnetite layer; this also applies to liquid metals and lead alloys. Only in the case of air receivers, oxidation at temperatures above 600 °C can lead to the formation of a continuously growing iron oxide layer which tends to flake off.

The austenitic steels are high-temperature resistant and, depending on the alloy composition, can be used up to 700 °C, but with low strength there. They are characterized by good creep resistance due to partial carbide formation either by MC ($M = Ti, Nb$) or $M_{23}C_6$ (M = Cr, Fe, Mo). Due to the high chromium content (Cr > 16.5%) they exhibit a very good high temperature corrosion resistance, so that except for liquid lead alloys or tin as heat carrier a long life is guaranteed. A disadvantage of all austenite is their low specific thermal conductivity, which can lead to high thermal stresses, especially during transients, and a high sensitivity to thermal fatigue.

When selecting a material for receivers, a high number of thermal cycles must be considered, whereby the number of thermal cycles during the entire service life is greater than that of conventional thermal power plants. The lower design limit is at least $2 \cdot 10^4$ cycles for an estimated service life of 30 years, which is why 10^5 cycles are usually assumed for design purposes. This thermal cycle number applies not only to the receiver but to all components of the heat transfer network but occurs more severely in the receiver. In addition, the receiver is not isothermally loaded, so several temperature ranges and resulting thermal stresses must be considered. Important aspects include consideration of loss of ductility and tensile strength, creep and carburization rates, fatigue crack growth rates, and oxidation resistance. Figure 7.73b shows the loss of material strength at 1% creep of various autentic steels as a function of operating temperature. High thermal stresses are also the trigger for crack growth, which is observed with increasing temperatures in all metallic materials.

Nickel and Nickel-Based Alloys
Nickel and nickel-based alloys have a face-centered cubic (fcc) lattice as the basic configuration and do not exhibit allotropic modifications; they show good corrosion resistance when chromium is alloyed in. Creep strength enhancement can be achieved by alloying measures, with nickel offering a myriad of alloying element options, such as Fe, Cr, Co, Mo, W, Nb, Ta, Al, Ti, Si, C, B, Zr, Hf All nickel alloys exhibit low thermal conductivity and thus high susceptibility to thermal stress; at the same time, the alloy also shows high coefficients of thermal expansion, and thus the tendency to thermal fatigue.

In recent years, nickel-based superalloys have been developed, which are ultimately two-phase microstructures consisting of a disordered γ-matrix and an ordered γ'-phase. With the aid of targeted process control during production, different solidification states can be set, thereby reducing the number of grain boundaries, and increasing the proportion of γ'-phase. For example, single-crystalline solidified nickel superalloys for turbine blades have no grain boundaries and are extremely creep-resistant up to approx. 1100 °C. The oxidation behavior in air is good at high temperatures due to the formation of an oxidic protective layer but must be checked in individual cases. Combustion experiments show that the protective layer formation is reduced in the presence of CO_2. Another disadvantage is the limited weldability of nickel alloys. Further, Ni and its alloys exhibit a good solubility to all high mass number metals, making the choice of a lead or tin alloy as a cooling medium difficult and requiring the development of appropriate protective coatings and in-situ HTF coolant chemistry control and conditioning equipment.

The variation of crack growth rates (da/dN) for a fixed value of stress intensity factor range (ΔK) is shown in Fig. 7.74 for different Inconel alloys (600, 625, 718 and $X - 750$) and stainless steels. The observed variations could be explained by the temperature dependence of the elastic modulus and the yield strength. The measurements show that the crack growth rate increases dramatically with increasing temperature, from room temperature to about 900 °C by a factor of 25. Since

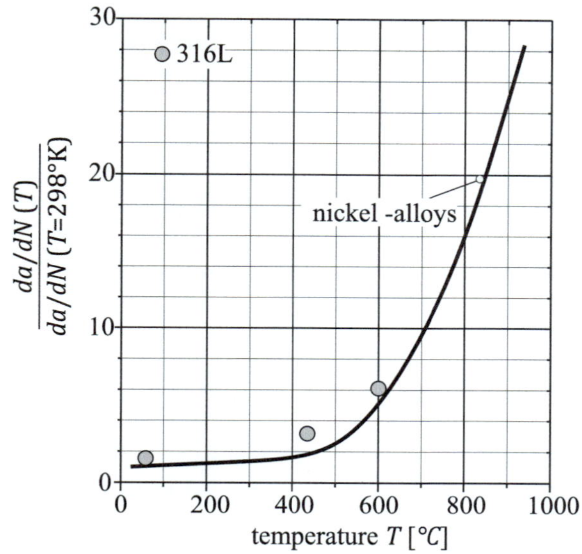

Fig. 7.74 Relative change of fatigue-crack-growth rates (da/dN) compared to room temperature of nickel alloys from Viswanathan (1993) and 316 L as a function of temperature T (°C) for a stress intensity factor range ($\Delta K = 30$ MPa$\sqrt{\text{m}}$)

the experimental data were obtained for different ΔK differences and temperature ranges, the different material classes cannot be compared directly. But, when the values are analyzed for a constant stress intensity factor range (ΔK), as shown in Fig. 7.74, a clear trend of increasing crack growth rate with increasing temperature can be seen. The figure shows that the growth rates at temperatures up to about 50% of the melting point (0.5 T_m = 550–600 °C) show little influence versus temperature; above this threshold, crack growth increases rapidly with increasing temperatures. The crack growth rates for all materials at temperatures up to 600 °C relative to the rates at room temperature can be estimated with a maximum correlation factor of 5 (2 for ferritic steels).

Taken together, both Figs. 7.73 and 7.74 show that a holistic receiver design must consider not only surface heat flux provided by the concentrated solar radiation but also thermomechanical and thermocycling effects in all expected temperature ranges to make a reliable lifetime prediction.

Cobalt-Based Alloys

These alloys are suitable for use up to 800 °C and are characterized by good weldability compared with nickel alloys. They exhibit good creep resistance based on solid solution hardening with W, Mo, Ta, Nb and particle hardening with carbides. At the same time, they show a lower temperature dependence of the creep resistance than the nickel alloys. Although nickel alloys already have a high price compared to steels, cobalt alloys are generally even more expensive. Other disadvantages are their lower strength compared to nickel alloys, since no hard phase content is possible. Also, lattice transformation of Co can occur during overheating, resulting in a decrease in creep resistance. Therefore, the fcc matrix of cobalt alloys is usually stabilized with Fe and Ni, further details on nickel, cobalt and their alloys may be taken from Davis (2000).

High Melting Alloys

Well-known from the construction of conventional power plants are niobium and molybdenum alloys, as well as tungsten in the field of nuclear fusion. There is a tendency for the yield strength $R_{m/time/temperature}$ to shift from niobium via molybdenum to tungsten to ever higher values well above 1200 °C for W. The challenge with these alloys, apart from the high specific price and the relatively small selection of possible joining processes with conventional materials, is above all the scaling resistance under gas atmospheres containing oxygen or nitrogen, where special protective coatings have already been developed for some applications. At temperatures above 1000 °C, hydrogen embrittlement also plays a role, negatively affecting this mostly bcc-oriented metal matrix alloys in terms of strength and fracture behavior.

Ceramic Matrix Composites (CMC)

In many technical high-temperature applications, such as in flame tubes, turbines or furnaces, the commonly used nickel-based superalloys are being replaced by ceramic matrix composites (*CMC*), whereby characteristic properties comparable to metals are achieved through the use of long fibers and the embedding of the ceramic. The most important high-temperature fiber-reinforced ceramics currently available on an industrial scale are C/C, C/SiC, SiC/SiC and Al_2O_3/Al_2O_3, which differ from the ceramic base material in the following respects:

- higher elongation at break of up to one percent,
- significantly higher crack resistance,
- extreme thermal shock resistance,
- better dynamic load capacity,
- anisotropic properties oriented to the fiber orientations.

Fiber-reinforced ceramics are usually manufactured in three steps:

- Placement and fixation of the fibers in the desired component shape,
- placement of the ceramic matrix material between the fibers, with five basic options; deposition of ceramics from a hot gas mixture, generation by pyrolysis of a polymer containing Si and C atoms, generation by chemical reaction or by sintering at low temperatures (1000°–1200 °C) and production by electrophoresis with ceramic powder.
- Finishing and, if required, further post-treatment steps such as the application of coatings.

Accordingly, the secret of fiber-reinforced ceramics with high crack resistance is to ensure during manufacture that bridging across the crack is possible. This presupposes that the matrix can slide along the fibers, i.e. that it is only weakly mechanically connected to the fibers and that the connection between fibers and matrix allows sliding. When used as a structural material, attention should be paid to the form in which the composite ceramic is present, i.e., either in two-dimensional form as stacked fabric plies or as crossed plies of one-dimensionally aligned fibers before. This type of *CMC* exhibits anisotropic behavior, as crack propagation between these layers is not hindered by any fiber bridging, making potential delamination a weak point of the material. Quasi-isotropic material behavior can be specifically achieved by introducing three-dimensional fiber structures.

SiC/SiC-CMC*s* reinforced with continuous fibers of silicon carbide are of particular interest for applications in high-temperature solar thermal receivers because they exhibit high temperature stability in oxidizing atmospheres due to a high melting point and the formation of a protective silicon dioxide layer, lower density (~2.6 g/cm^3), higher specific strength, and lower creep rate at high temperatures compared to nickel-based superalloys. The fiber reinforcement leads to completely different failure mechanisms compared to monoclinic, more brittle ceramics or metallic materials. Silicon carbide as a matrix has a high thermal conductivity, but this decreases with increasing temperature; conversely, the fiber has a low thermal conductivity. As a composite material, Sic–SiC reaches values on the order of steels ($\lambda \approx$ 12–22 W/(mΔK)) and has been successfully in performance tests for solar applications, see Ortona et al. (2015).

Although SiC/SiC-CMCs have excellent temperature stability, the presence of water vapor can lead to volatilization of the silica by formation of silicon hydroxide. This requires a coating to be applied to the CMC, and these environmental barrier coatings (*EBC*) are produced by thermal spraying, usually silicate-based. The coatings are multi-layered and consist of a silicon bond, one or more intermediate layers, and a silicate top layer with low silica activity. This low activity results in less formation of silicon hydroxide when exposed to water vapor, and thus improved corrosion resistance. However, these silicate-based coatings are susceptible to cracking, which leads to the formation of silica deposits on the silicon binder layer, eventually resulting in severe corrosion or spalling of the coating. Newer robust processes rely on pack cementation to produce oxides other than silica.

Selective Absorber Materials for Tower Receivers

Next-generation central tower receivers are expected to attain heat transfer temperatures exceeding 650 °C to achieve higher efficiency of the downstream power conversion system and thus better economics. Higher temperatures and long lifetimes, which imply high cycling, are particularly challenging for the optically solar selective absorber (SSA) receiver material, which must maintain high absorptivity ($\alpha \rightarrow 1$ for $\lambda \leq 2.5$ µm) at these high temperatures to maximize absorbed energy while maintaining low emissivity ($\varepsilon \rightarrow 0$) to minimize heat radiation losses. In contrast to linear focusing collectors, the volumetric heat release in high temperature receivers is not at least one order of magnitude greater, but also the temperature. Thus, ensuring the functionality of SSA, both in its structural design and composition, represents a unique development challenge, even if the physical and material-specific concepts for achieving spectral selectivity do not differ from those elaborated in Chap. 3.

In all windowless receiver concepts, such as open-air receivers but also liquid cooled receivers or the particle receivers, the selective efficiency (η_{sel}) is therefore the key parameter. It is defined as the energy retained by the absorber, considering both absorption and emission, in relation to the energy incident on the surface. Other requirements for the absorber material include:

- high thermal conductivity,
- high temperature chemical and structural stability through entire operation range,
- low thermal expansion,
- low creep during thermal cycling and long durability (>25 years),
- high resistance to oxidation,
- compliance with local regulations on use of materials for health, safety, or environmental reasons,
- inexpensive and easy to apply to the absorber tube,
- good adhesion of the absorber to the heat transferring surface.

In addition, external absorbers, i.e. those not protected by a cover, should ideally be dirt and abrasion resistant to minimize cleaning, repair, and maintenance costs. Furthermore, the oxidation resistance to air and the stability of its structure during its lifetime are of particular importance to ensure the availability of the overall system. In the case of window-based receiver concepts, such as pressure-charged gas-cooled receivers, the high-temperature behavior and the optical properties of the glass also play a special role, while some requirements for the receiver material, such as the oxidation behavior, among others, become somewhat less important.

First High Temperature Absorber Materials

For the parabolic trough collector technology, often cermets ($M - Al_2O_3$, $M = Mo, W, Ti, Ta$ et al.) are used as solar selective absorber (SSA) coatings, but although there the durability in a vacuum receiver is proven, the application in atmosphere in tower receivers with enhance oxidation and ageing accelerated by increasing temperatures poses a significant challenge.

The absorber substrates themselves are made of different materials with different melting points, either aluminum (600 °C), different steels (about 1500 °C) or ceramics (1000–3000 °C). The choice of material for the absorber tubes depends first on the temperature range for which the receiver system is designed—possibly well above 1000 °C for tower systems, but if solar radiation is absent, the temperature can drop below zero degrees overnight. Untreated metals such as aluminum,

Fig. 7.75 Development chain of an intrinsic $W-Ti$ laminate **a** Notched impact specimen of the laminate from Reiser et al. (2015), **b** plastic deformation of a $W-Ti$ laminate tube; **c** delamination of the tube under oxygen; **d** W-laminate tube for the Almeria test before (left) and after the high-temperature tests (middle) and IR image during operation (right)

copper, steel but also tungsten is ruled out because they can be highly reflective. With the help of coatings, however, this can be changed. Coatings which withstand high temperatures in atmosphere, however, also are quite rare. The emissivity for the high temperature, say at 600 °C is usually large even with an optimized optical design of the selective absorber and its critical wavelength. The different selective absorber concepts and examples of high-temperature selective coatings are presented in Sect. 3.3.4.

The main product on the market for solar towers is a proprietary non-selective coating called Pyromark 2500. However, absorbers coated with Pyromark must be repainted on an annual basis, which reduces collector availability. The manufacturer, Tempil, states that Pyromark 2500 resists temperatures up to 1093 °C and can be applied on steel, aluminum, alloys, and ceramics. Pyromark 2500 has a very high solar absorptance ($\alpha = 0.96$), but unfortunately it also has a very high thermal emittance ($\varepsilon \approx 0.9$) at the temperatures exceeding 600 °C.

Modern manufacturing and fabrication methods as well as the low requirement for low emissivity imply that intrinsic absorbing absorbers should not be excluded from general considerations. Figure 7.75 shows such a development chain for tungsten. By means of an alternating diffusion welded titanium and tungsten foils (Fig. 7.75a), a ductile tube was produced which also behaves ductile under plastic loading as illustrated in Fig. 7.75b. However, it is not stable to oxidation in air and delaminates, as depicted in Fig. 7.75c. Applying a titanium carbide coating, an oxidation-resistant high-temperature-stable selective absorber tube is obtained, as shown in Fig. 7.75d, which can easily withstand temperatures above 1000 °C with high absorption.

7.3.2.5 Heliostats

The design of a solar tower system, i.e. the size and position of the receiver on the tower, and the determination of the number, type and position of the heliostats is a problem with a high number of degrees of freedom and can only be solved with computational help, the results of which allow to analyze the efficiency and the yield of the heliostat field. This is always a two-step process.

First, a large heliostat field is generated as input and after positioning the selected heliostats (which may be of several types for near and far field positions), for this initial field the annual energy that each heliostat provides to the receiver is evaluated. To calculate this energy, yield an optical model is required to determine the optical efficiency not only of each individual mirror, but also its interaction with the neighboring mirrors, which scopes numerous loss effects occurring in the heliostat field, namely:

- cosine loss, i.e., the reduction of effective mirror size due to off-axis reflection,
- shadowing of specular surfaces by neighboring heliostats,
- blocking of reflected light by adjacent heliostats,
- attenuation of reflected light between heliostat and receiver due to absorption and scattering processes in the atmosphere,
- reflected light reaching the receiver plane but not entering the aperture ("spillage") due to incomplete heliostat alignment and/or an oversized reflection image.

If the annual yield of the initial heliostat field is known as well as its annual integral optical efficiency; however, one is still far away from the optimal heliostat arrangement for the tower and the question arises, how can an optimal arrangement look like? This question is dealt with in Chap. 10 on the high temperature solar fields. In the following we would like to present the heliostat as a component with a large degree of freedom concerning design and material selection.

Heliostat Design Options

The cost of the heliostat array accounts for about more than one-third of the total plant cost of a solar tower power plant, so a variety of efforts have been made in recent years to optimize this aspect, the most recent reports are Coventry et al. (2016) or Pfahl et al. (2017). Simultaneously, an individual heliostat must meet several criteria such as the accuracy of the device and the local conditions at the heliostat field. The list of these criteria has a major impact on design and cost. They include the required optical quality, any external loads to be sustained, the required performance of the tracking system, protective measures for the heliostat and receiver, the lifetime of individual components, and the overall heliostat system.

Before we turn to the design criteria, it is first useful to identify the individual components of a heliostat. An individual heliostat as depicted in Fig. 7.76 essentially consists of five assemblies, which have different functionalities:

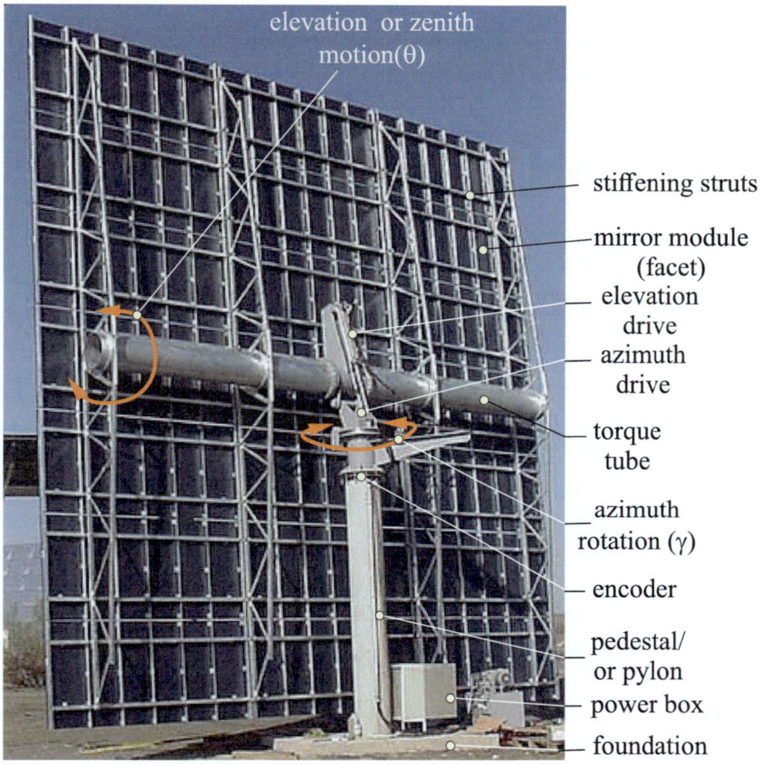

Fig. 7.76 Typical T-type heliostat with pedestal and frame assembly (modified foto from © Plataforma Solar de Almería)

- (canted) mirrors using a highly reflecting surface to project the sunlight to the receiver termed as *reflecting module.*
- structure or mirror fixation, which is realized by means of an adjustable support structure serving degree of freedom,
- pylons and *foundation* for connection to the ground,
- *mirror canting, measurement and control devices* to focus the mirror (canting) and determination of position of the mirror surface and its control,
- *drive mechanisms* for biaxial tracking of the mirror surface.

Heliostats may have different shapes and are not all rectangular. The company SBP Sonne for example designed a heliostat, the Stellio type, with pentagon shape (see https://www.sbp.de/projekt/stellio-heliostat/). This has the advantage that heliostats can be positioned closer to each other, as the risk of touching with edges is reduced.

Heliostats with membrane mirrors are a second option showing the variability of design. The concentrator consists of two membranes that are connected by a compression ring. A vacuum pump creates a slight negative pressure in the space between the two membranes. This negative pressure leads to a curved deformation of the concentrator. This enables the desired bundling of the incident radiation. The membranes are made of either plastic foils or metal foils, the latter being more durable.

The advantage of a membrane heliostat compared to a faceted heliostat is that good optical quality is achieved with low production and assembly costs. The time-consuming alignment (canting) of the individual elements is no longer necessary, as it is only a matter of a concentrator surface that has to be aligned. Disadvantages of the membrane heliostat are the higher costs compared to the faceted heliostat and a larger self-consumption of electricity caused by the vacuum pump.

Heliostat Sizing

The cost-optimal size of a heliostat is a question that cannot be answered easily. Current projects show that the focus is on two size classes: medium-sized heliostats with mirrors up to 40 m^2 and very large mirrors in the size class of approx. 120 m^2. The design is always the result of an optimization process, which is dependent on costs of individual components, on the approach to manage complexity versus replicability. The main advantage of large heliostats is that fewer components, foundations and drives are required per square meter of mirror area, but at the same time the static and dynamic wind loads increase significantly as well as the requirements for high power and especially high torque motors with high precision (Blackmon, 2020).

Conversely, the main advantage of smaller heliostats is that the wind loads are smaller and the superstructures lower, and simultaneously the requirements for drives and foundations (weight per m^2) are reduced. The reason for the lower area-specific weight is that an increase in the chord length of the mirror panel quadratically increases the mirror area and, therefore, all other dimensions of the heliostat also increase according to the chord length so that the loads caused by the wind peaks remain the same. However, with the increase in size, the volume and mass of the heliostat grow to the third power of the chord length. Higher weight means higher costs, so smaller heliostats seem to be more advantageous due to their lower specific weight, if additional costs are not eaten up by more effort in setup, adjustment, and maintenance during operation. Communication and control algorithms in the required IT are much more extensive than for large heliostats. With more advanced computing and cheaper hardware there could be a tendency towards smaller heliostats in future, but as of today there is no clear breakthrough. And as mentioned earlier, the selection of heliostats is dependent on plant size and receiver type, so that different approaches might be successful in different projects.

Reflecting Module/Mirror Facets

The reflective module of a heliostat usually consists of one or more mirrors (also called facets) arranged in a usually spherical pattern (canting) to concentrate the reflected light onto the receiver surface. In addition, each mirror is often mounted slightly concave to help concentrate the reflected beam to a focus. Flat mirrors are also occasionally used to simplify the design. Subdivision into facets is not only for technical reasons in manufacturing, but also facilitates handling during construction and allows better control of curvature for larger heliostats. Typical facet sizes are between 1 and 10 m^2. The module must perform accurately under a wide range of operating conditions such as varying wind speeds and ambient temperatures should have the following characteristics:

- high optical performance (e.g., reflectance/transmittance, specularity, geometric configuration).
- low specific weight,
- long lifetime ($\approx 25y$) and low maintenance costs.

Typical heliostat mirror facets are based on so-called glass/metal structures. This is a sandwich construction in which a second surface mirror (i.e., the reflective material is covered by a transparent material placed between the sun and the reflective surface). This double layer is bonded to the support structure using an adhesive and sealed at the surrounding facet edges. The silver reflective surface is thus protected from one side by the glass but must also be protected from the back. In monolithic mirrors, a copper layer is used with a protective layer for the copper. Mirrors made of laminated glass use glass on both the front and back sides, so the reflective film is sandwiched between two layers of glass. The reflective layer is made of silver or aluminum, with silver being advantageous over aluminum due to its high reflectivity ($\rho \approx 0.95$–0.97) over a wide range of wavelengths because its reflectivity ($\rho \approx 0.88$–0.92) is lower and alumnium has an undesirable absorption band at about 800 nm.

The transparent cover is made almost exclusively of low-iron glass because it is nearly inert, abrasion resistant, and inexpensive and available in large quantities, and has high strength and simultaneous low absorption across the wavelength spectrum. Current commercially fabricated reflectors exhibit a reflection coefficient of $\rho \approx$ ~0.93–0.94 without excessive corrosion or UV degradation. The development goals are, on the one hand, to increase the reflectance to values of $\rho \approx 0.95$–0.96 and to use low-maintenance anti-soiling coatings to reduce operating costs by cleaing the mirrors. Another disadvantage of the glass-metal composite is the relatively high area specific weight, about 7.5 kg/m^2 for a common 3 mm thick mirror, which must be supported by the mechanical structure and positioned by the drive mechanism with appropriate moments.

Mirror Support Structure

The support structure of a heliostat consists of two large assemblies:

- Base segment, which anchors the assembly to the floor.
- Frame structure, which holds the mirror facets rigidly in position.

Most often, the basic construction of medium or large heliostats is a so-called T-type, where the torque tube and the pylon together form a "*T*"; both are connected by the drive system for the azimuth and elevation movement. The base provides the necessary ground clearance for the reflection module. The frame assembly supports the arrangement of the mirror modules or facets. The modules are joined to the assembly using adhesives or other mechanical fasteners. Each mirror module typically has a slight concave curvature and is also tilted (aligned) with respect to the plane of the support structure to better focus reflected sunlight on the receiver to improve performance. The torque tube is connected to the drive system, while the mirror facets are attached to the cross beams. Trusses are preferred, especially for larger heliostats, because their depth can be varied to achieve the required stiffness with little weight loss. In some designs, the beams connected to the mirrors are replaced by perforated pressed plates. The azimuth/elevation gear is installed at the connection between the base and the torque tube. The aim of the design is to give the heliostat structure the necessary stiffness and to minimize deflections due to wind loads. For large heliostats, this means a heavy torque tube, which increases the specific weight, material, and cost of the heliostat structure. One of the great advantages of the T-type design (where there are open or closed designs), with multiple facets is the ability to set the heliostat downward (compare Fig. 7.76), where setting this safety position prevents damage to the reflective module in hailstorms or similar conditions. However, this comes at the cost of slightly reduced optical efficiency due to the smaller effective area.

Glass and steel have different thermal expansion ratios. Therefore, the mirrors are often connected to the support structure by pins that are flexible in both directions of the mirror plane. In the example shown in Fig. 7.77a, steel strips are used that are flexible in one direction and connected to the cross-members by riveted joints, allowing rotation in the other direction. Only the central pins (above the torque tube) hold the gravity load of the mirrors in the upright orientation and are attached to the cross members by two riveted joints. The pins are bonded to the mirrors by pads that increase the area of force application and rotate to avoid high stresses in the glass. The height of the pads perpendicular to the mirror plane determines the shape of the mirror panels. The number of pads is determined by the wind loads and the required accuracy of the parabolic curvature of the mirror field. For smaller heliostats, wind and other loads are significantly lower, allowing lighter frame structures that either require only a very small pedestal, or are mounted on a floor structure (compare Fig. 7.77b) shared by several adjacent heliostats to avoid a costly foundation. Small mirrors also do not require a true frame structure (especially a torque tube), so simple structures are sufficient to provide stiffness to the mirror. For small facets with sufficient adhesive thickness, the flexibility of the adhesive is able to compensate for the differences in thermal expansion between steel and glass, and no pins are required.

Foundation

The function of the foundations is static protection, which naturally depends on the size and weight of the structure. It is important to prevent rotation of the pylon. For large heliostats the bases are mounted on a reinforced concrete pier or concreted directly into the ground, lighter heliostats allow mounting on an above ground ballast frame or ballast foundations consisting of a concrete ground anchor

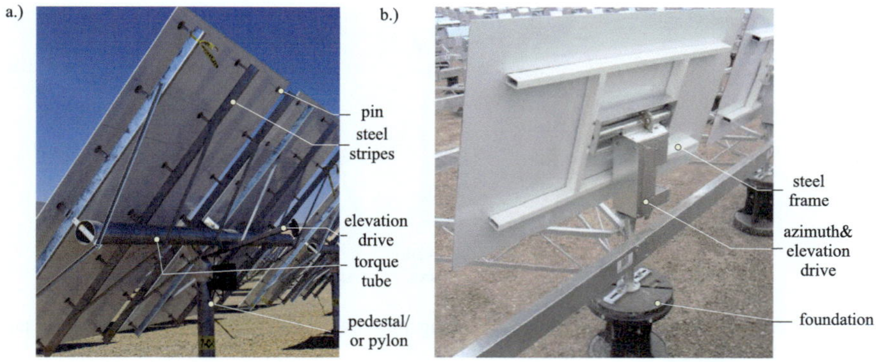

Fig. 7.77 **a** T-Type Heliostat with connection of glass mirrors and steel support structure via flexible pins (for compensation of different thermal expansion of steel and glass) and pads modified from Pfahl et al. (2017); **b** Low cost support structure for small heliostats (Téllez, Villasante, & Burisch, 2014)

weighted with natural site material, more details can be found in Coventry et al. (2016).

Mirror Canting, Measurement and Control

The heliostat's task is to reflect sunlight onto a predetermined target on the receiver. This requires a procedure that involves several aspects.

- alignment of the individual mirror facets of a heliostat to the receiver aperture.
- calibration measurement of the flux profile on the target.
- control logic to adjust the heliostats.

The term canting refers to the alignment of the individual mirror facets of a heliostat on the carrier structure with respect to the receiver and has as its goal to maximize the thermal yield on the receiver. For small heliostats consisting of a single facet, the curvature of the facet and the applied canting are identical, resulting in a continuously canting surface.

Many studies have shown that the radius of curvature of the heliostat should be different along the tangential and sagittal planes to reduce aberrations in the reflected image. Therefore, the energy gained from a particular heliostat field can be maximized using a targeted curvature technique. The applied curvature technique also depends on the tracking method of the heliostat. All heliostats have two axes of motion, most of which provide azimuth and elevation rotation as shown in Fig. 7.78a.

Regarding azimuth-elevation tracking usually a first vertical axis (azimuth) and a second horizontal axis are used, although a first horizontal axis is also possible. The methodology for this tracking type chosen for this process is, however, not unique, but can be classified into two groups: on-axis and off-axis. In case of on-axis canting the facets are aligned to give a perfect image, when the heliostat

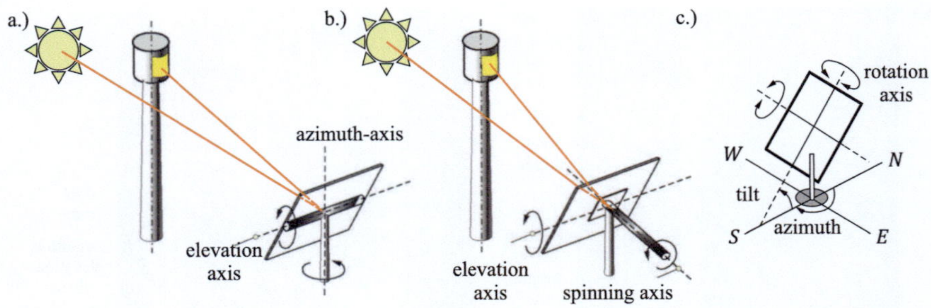

Fig. 7.78 Schematic sketch for heliostat tracking options **a** azimuth-elevation tracking; **b** target aligned tracking; **c** polar-mounted elevation tracking

center, the target and the sun are collinear, i.e. are on a line. For off-axis canting, the facets are arranged to optimize the convergence of all images on the solar receiver for a particular time and day of the year.

An alternative tracking option is the polar-mounted elevation tracking as shown in Fig. 7.78c. Here the first axis is fixed in north–south orientation parallel to the earth's axis (polar mounting). The rotation around this axis tracks the daily solar path. Due to the changing declination of the solar orbit the tilt angle must be adjusted every day so that the reflected rays hit the receiver. For this heliostat tracking the corners of the rectangular mirror stay parallel to the north–south direction. This implies that heliostats can be placed closer to each other than for the azimuth-elevation tracking. A more compact assembly is possible, which is beneficial for the near field.

A third alternative option is the target-aligned tracking, in which the target-aligned heliostats allow rotation around the fixed axis connecting the center of the heliostat to the target, so that the second axis of the surface is vertical to the plane defined by the sun, target and mirror center. A second rotation tilts the mirror perpendicular to this plane so that the reflected image is within the receiver target area (Fig. 7.78b). In principle, this procedure reduces the astigmatism of rectangular shaped heliostats but generates additional effort, since each individual heliostat must be aligned to the target.

The procedures for setting a correct heliostat canting can be divided into two broad categories mechanical and optical options, an overview can be found in Yellowhair and Ho (2010) among others.

- Mechanical methods (e.g., using gauge blocks or inclinometers) involve calculating the required facet canting angles in advance and manually measuring and adjusting each facet to achieve the desired angles. These methods are labor intensive, tedious, and prone to several errors, primarily due to structural gravity sag and local slope errors at the measurement point.
- Optical methods (such as photogrammetry, fringe reflection, parallel laser beam projection, scanning laser projection, or reflection of a known image) typically

rely on applying a correction to the mirror surfaces so that an undistorted reflection of a known image is obtained. Optical methods are not only able to correct for facet tilt, but also measure focal length errors and generally provide higher accuracies. The application of the canting process requires precision and labor and is therefore a factor that contributes to the overall cost of the heliostat during the manufacturing and installation phases.

The heliostat has the task of reflecting sunlight onto a pre-determined target. Therefore, the heliostat controller must know the orientation of the mirror to determine deviations and precisely control the actuators to minimize optical deviations. The conventional approach is calibration measurements and open-loop control. Modern control systems rely on closed-loop systems that use sensor signals for feedback. The pre-requisite for an open loop is very low statistical error and stable, predictable systematic behavior. Further, they assume that the heliostat model represents all real-world imperfections. Closed-loop systems are limited only by the backlash of the actuators, the measurement accuracy of the feedback sensor, and the motor control. Difficult influences such as deformation due to gravity, wind, and temperature can be compensated for as long as the initial sensor-heliostat calibration remains intact. Closed loop tracking accuracy is usually better, which reduces the spillage to achieve the desired overall flux distribution on the receiver.

Drive Mechanisms

The actuators control the position of the heliostat frame assembly by rotating or inclining the structure to project solar radiation onto the fixed target. The drive unit consists essentially of two assemblies, the power source, and the transmission unit. The requirements for the heliostat drive unit demand only low speed but high torque. For optical efficiency, however, high transmission precision and low backlash are required, since tracking errors in the range of mrad can already lead to a significant power loss of more than 10% of the total power. In addition, the above-mentioned requirements for the drives should be maintained during the typical service life of about 25 years, with the lowest possible maintenance expected, which means that all seals, wear, and abrasion of the drives should be designed to last as long as possible. These stringent requirements mean that the drive and gearbox are a cost factor of the same order of magnitude as the mirrors (Coventry et al., 2016). The mechanical transmission of forces and movements from the energy source (motor or hydraulic pump) to the rotary axes of the heliostat can be achieved using different technical concepts. Important for the selected transmission solution are the transmission ratio, the transmission accuracy, and the stiffness, whereby the efficiency of the transmission is of secondary importance. There are a number of different drive solutions also used in industry, which meet the requirements for required power, speed and accuracy, so that there is no one-size-fits-all solution; mostly a drive concept consisting of an electric motor combined with a mechanical transmission gear is used in commercial heliostat systems, including worm and spur gears, screw and nut, chain, harmonic drives, whose characteristics are briefly outlined below and illustrated in Fig. 7.79.

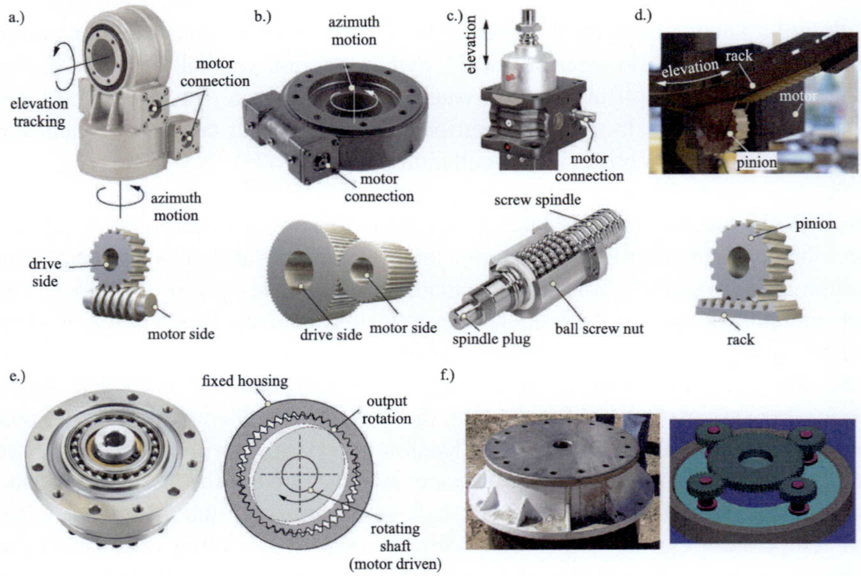

Fig. 7.79 Different types of gearboxes for tracking heliostats and their operating principles; **a** Worm gearboxes, **b** spur gear/planetary gearboxes, **c** screw-nut gearboxes, **d** chain/rack-pinion gearboxes, **e** harmonic gear drive, and **f** planocentric gearboxes adapted and modified from Coventry et al. (2016)

- Worm gears allow large gear ratios in each stage (10–30), with unfortunately low efficiencies. Backlash can be reduced by preloading the gears but increases wear and thus aging. They are wide used as drive mechanisms of Heliostats in central receiver plants.
- Spur gears provide smooth mechanical transmission at moderate gear ratios (2–5) and therefore require a large number of stages. Construction ions such as planetary gears offer larger gear ratios in each stage. The efficiencies are high, but the irreversibility of the gear cannot be guaranteed.
- Screw and nut gear drive are used when linear motion is part of the transmission chain, screw and nut solutions can be considered to achieve large gear ratios. When using ball screws, the smoothness and efficiency are considerable, but the cost is higher, and the system is reversible. Friction screw and nut solutions, on the other hand, allow irreversible step designs.
- Chain-pinion or rack-pinion solutions offer the possibility of low-cost transmission. In principle, they are based on contact forces. In case of the chain option, the chain is not firmly connected to the assembly to be moved, and thus the achievable stiffness is lower, this also applies to accuracy. However, high transmission ratios can be achieved at high efficiencies.
- Harmonic drive gears, also known as "expansion shaft gears," rely on contact and deformation to provide high gear ratios of up to 100, high accuracy, high

efficiency, and backlash-free transmission of rotary motion. They are often used in high-precision solutions, such as in machine tools or robotics, but are costly.

- Planocentric drives (Winsmith drives) offer exceptional gear reduction (over 30000 : 1) in an extremely compact package; unfortunately, heliostats seem to be the only application for this concept, and they cannot benefit from industrial system applications.

7.4 Comparison of Optical Performance of Parabolic Trough, Linear Fresnel, and Solar Towers

It is not possible to conclude from the optical performance of the individual line- or point-focusing systems alone, which of the concepts is better or worse, because for a complete solar thermal system its optical and thermal behavior, storage possibilities, the temporal and temperature suitability for the respective application includes many more parameters to be considered. For example, the different orientation of a parabolic trough or a linear Fresnel might produce different solar heat generation profiles during a day, but when storage is included to match demand and generation, the exact profile is not important at all. An optical and thermal performance can be determined for each of the presented mirror-receiver concepts. Using the optical performance, the consequences in receiver absorbed power and heat transfer to the fluid can be derived as well as thermal losses. The required quality of the optical components (mirrors, absorbers, glass) for a certain power demand can be determined, not necessarily representing a cost optimized. The overall optimum, however, is dependent on additional factors such as thermal energy storage capacity or the storage concept, spatial limitations, and many other parameters more.

Nevertheless, it is useful to compare the individual technologies in terms of their time-dependent optical performance. In the following, a linear Fresnel collector is compared with a parabolic trough and a central receiver tower qualitatively and quantitatively since these concepts offer a power (up-)scaling at least in principle. The comparison considers equal primary reflector/heliostat characteristics, as far as possible equal receiver characteristics, which are summarized in Table 7.11, and is based on the article of Kincaid et al. (2018). However, here it has to be noted that the characteristic properties may vary for different products of the same collector type and the intention here is to illustrate the characteristic differences of the technologies.

Figure 7.80 illustrate the optical efficiency (η_{opt}) as a function of transverse (θ_T) and longitudinal (θ_L) incidence angle at the same location for all three technologies, determined using a Monte Carlo ray tracing calculation, where the longitudinal angle of the tower is defined as the angle with respect to the east–west axis. The PT design exhibits the highest optical performance except at large longitudinal incidence angles, where the central receiver takes larger values. The

Table 7.11 Optical properties of a linear Fresnel (LF), a parabolic trough (PT) and a central receiver (CR) system for comparison of optical performance

Component	Linear Fresnel	Parabolic Trough	Central Receiver
Primary reflector			
Slope error σ_{slope} (mrad)	3	3	1.53
Specularity σ_{spec} (mrad)	1.5	1.5	1.5
Tracking σ_{trac} (mrad)	1	1	1
Optical error σ_{opt}(mrad)	6.5	6.3	3.95
Reflectivity ρ(/)	0.94	0.94	0.94
Cleanliness η_{cl}(/)	0.97	0.97	0.97
Concentration ratio C(/)	37	26	1089
Receiver			
Absorptance α(/)	0.96	0.96	0.94
Transmittance τ(/)	0.97	0.97	–
Cleanliness η_{cl}(/)	0.98	0.98	–
Bellow shading f_{bel}(/)	0.96	0.96	–
End loss f_{end}(/)	0.98	0.98	–
Secondary reflector			
Reflectivity ρ(/)	0.94	–	–
Slope error σ_{slope} (mrad)	2	–	–
Specularity σ_{spec} (mrad)	1.5	–	–

parabolic trough technology generally has a high optical efficiency, which is insensitive to changes in transverse incidence angle until shadowing effects from the adjacent rows become significant. The Fresnel collector has a similar high optical efficiency as the central receiver at low angles of incidence, but the CR technology shows higher performance at higher angles of incidence.

Figure 7.81 depict the stacked optical losses due to various error sources as a function of the transverse and longitudinal incidence angles of the three technologies. To get a picture of the general characteristic differences three technologies, we divide the optical losses into three categories. The cosine losses of the collector denote the effect in which the aperture is not perpendicular to the incident solar rays, while the intercept loss without cosine losses subsumes shadowing, blocking and spillage losses. The optical losses due to the component properties are represented by the reflectivity, the absorptivity of the receiver, or the transmittance of the glass envelope shielding the absorber (if present).

For uniaxial, i.e. line-focused collectors that track in the transverse plane, the cosine loss is less than for point-focused CR. Conversely, in the longitudinal direction, the cosine loss is greater for both line-focused collectors than for the CR. As the transverse incidence angle increases, the shadowing and blocking losses grow rapidly for the LF collector. In contrast, the cosine effect is responsible for most

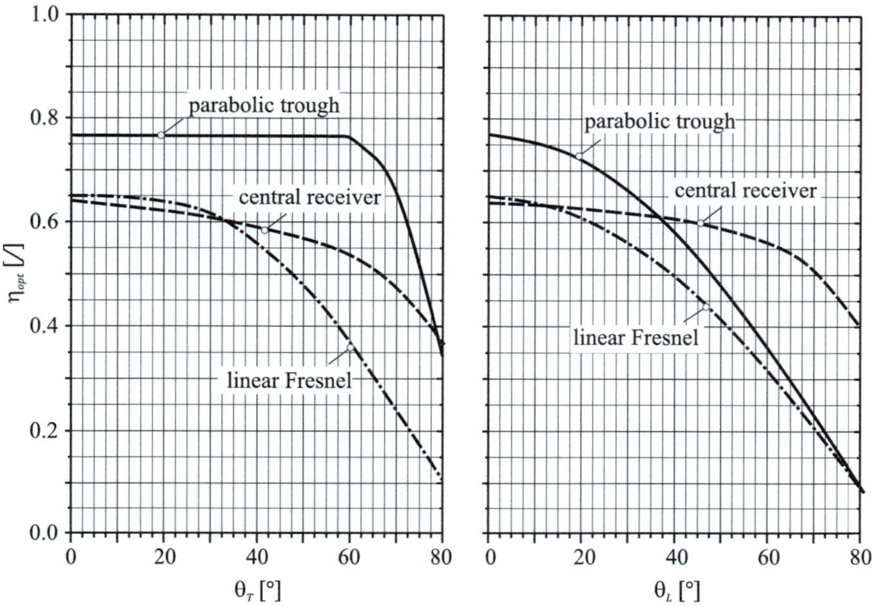

Fig. 7.80 Optical efficiency (η_{opt}) as a function of transverse (θ_T, left) and longitudinal (θ_L, right) incidence angles for parabolic trough (PT), central receiver (CR), and linear Fresnel (LF) collectors

of the optical losses in the longitudinal direction for both the LF and PT collectors since the reflectors cannot track in this direction. If the row spacings in the PT field are sufficiently large, a shadowing effect only occurs at transverse incidence angles of 70° and above, but then grows rapidly. The CR collector behaves fundamentally different due to its biaxial tracking and shows in both directions a comparable dependence of the optical loss as a function of transverse and longitudinal incidence angle (which is also due to the circular heliostat field in this example). Even at normal incidence the cosine loss is significant at the CR, because the inclination angle of a heliostat increases with the distance from the tower. Since a heliostat field increases in size for a given tower height, its cosine loss also increases even at normal incidence. The same holds for line-focusing systems as the aperture size increases; however, the contribution to optical losses is small compared to CR collectors due to the difference in scale.

Annual Efficiency

Annual optical efficiency ($\eta_{opt,a}$) is defined as the average efficiency weighted by DNI at each hour of a year:

$$\eta_{opt,a} = \frac{\sum_{t=0}^{8760h} DNI \cdot \eta_{opt,0} \cdot IAM(\theta_T, \theta_L)}{\sum_{t=0}^{8760h} DNI} \tag{7.52}$$

in which $\eta_{opt,0}$ denotes the the optical efficiency at zero degree incidence.

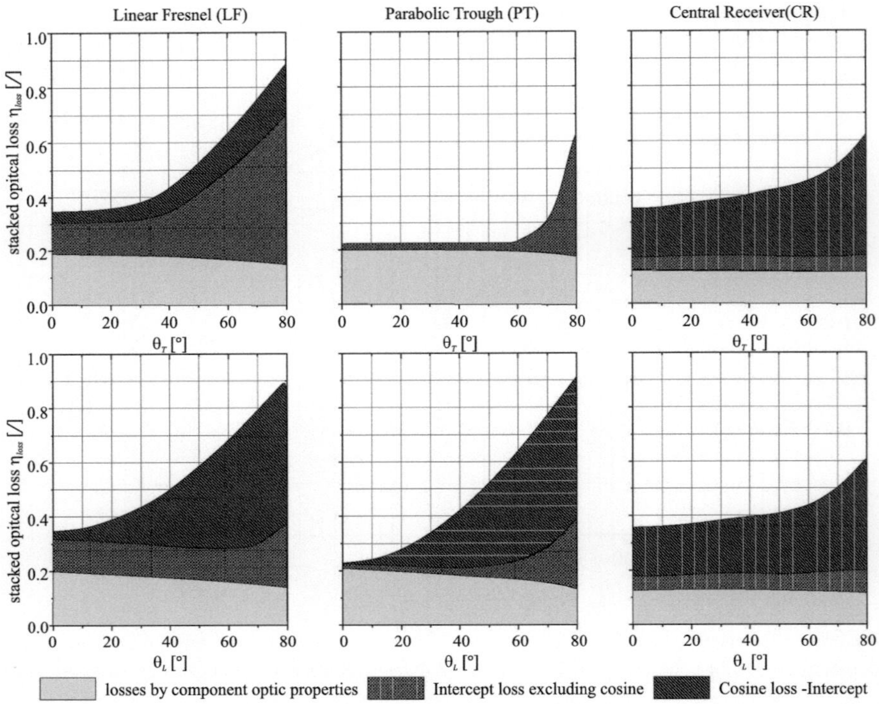

Fig. 7.81 Comparison of calculated stacked optical losses as a function of transverse (θ_T, top row) and longitudinal incidence angle (θ_L,bottom) for linear Fresnel, parabolic trough, and central revceiver at the same site, modified from Kincaid et al. (2018)

The three collector systems behave completely differently in summer and winter as shown in Fig. 7.82. In summer, the PT collector shows the highest and most balanced efficiency throughout the day. The LF system shows a comparable performance to the central receiver but drops opposite it in the morning and evening. In winter, on the other hand, the central receiver is superior to linear technologies, while in particular the linear Fresnel system suffers from a relatively low efficiency at high angles of incidence. The temporarily higher optical efficiency of the PT system compared to the central receiver in the morning and afternoon results from its lower sensitivity to transverse angles of incidence. During the day, the transverse angle of incidence approaches zero at midday, but the longitudinal angle of incidence is higher in the sky than in the morning hours due to the low sunlight. This leads to a lower optical efficiency for the PT receiver at noon.

The power generated on three consecutive days by the different collector systems per square meter of mirror area in summer and winter are shown in Fig. 7.83.

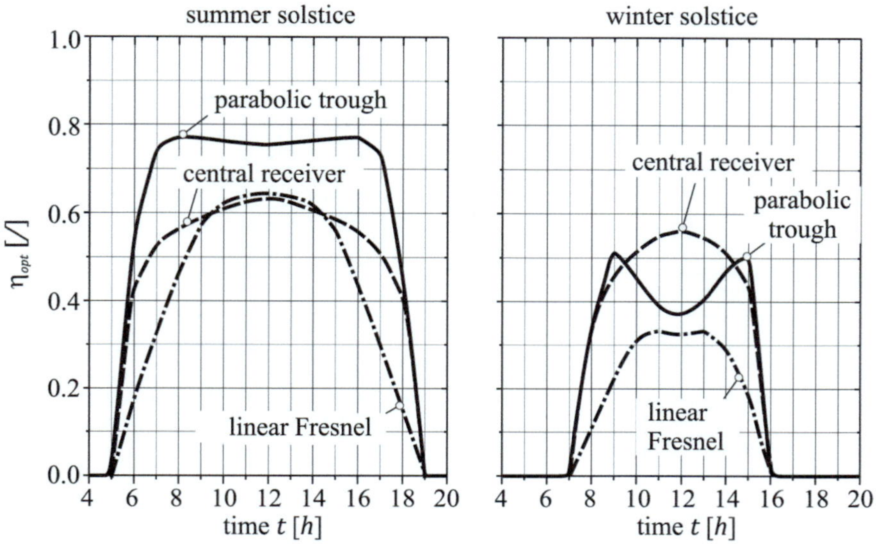

Fig. 7.82 Simulated hourly optical efficiency η_{opt} in summer (**a**) and winter (**b**) by linear Fresnel, parabolic trough, and central receiver technology, site: China Lake, California $\Phi = 35°39$ °N

The PT collector shows the highest power extraction in the summer week while the central tower receiver is superior in the winter.

The calculated monthly average optical efficiency of the respective technologies normalized by the hourly DNI is shown in Fig. 7.84a, assuming a minimum operating stow angle of 10°. The parabolic trough achieves the highest efficiency during most of the year, while the solar tower shows the most balanced values as an annual average.

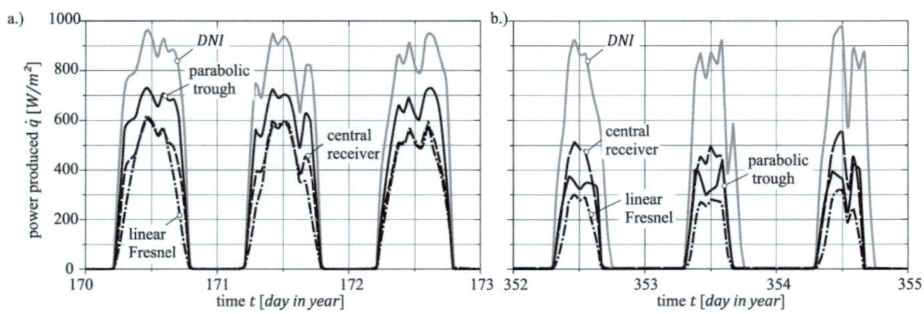

Fig. 7.83 Computed specific produced power (per m^2 mirror aperture) in summer (**a**) and winter (**b**) by linear Fresnel, parabolic trough, and central revceiver technology during three days, site: China Lake, California $\Phi = 35°39$ °N

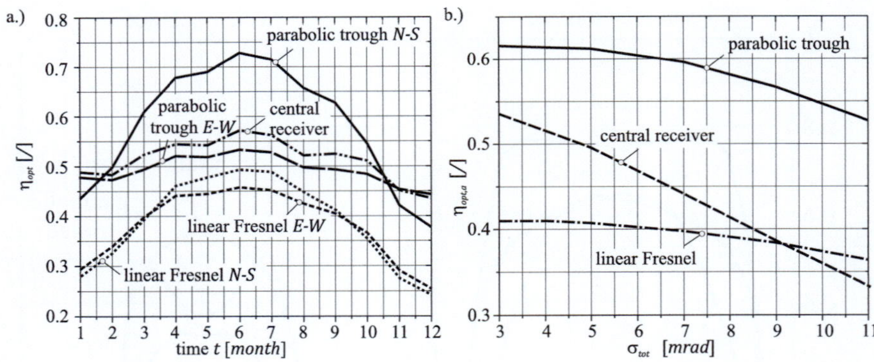

Fig. 7.84 Sensitivity of optical efficiency (η_{opt}) as a function of collector orientation over the annual cycle (**a**) and as a function of total optical error (σ_{tot}), **b** of parabolic trough, linear Fresnel, and solar tower; site: China Lake, California $\Phi = 35°39$ °N

Line-focusing collectors show different characteristics depending on their orientation north–south (N–S) or east–west (E–W) as illustrated in Fig. 7.84a. For the parabolic trough, the east–west orientation leads to a decrease in annual efficiency from 60 to 49%, but a greater uniformity in the annual mean, which may be desirable in the application. The sensitivity of the linear Fresnel system with respect to the orientation is rather low, while for central receivers, provided they have a circumferential field, the orientation does not play a role.

It has to be emphasized that the efficiency consideration is not sufficient to determine optimum installation cases. A lower efficiency might be compensated by a larger aperture area, if the area-specific costs are smaller than for the competitive solution. A conclusion whether a CRS, A PTC or LFC system is best for a certain application cannot be drawn from the consideration of the annual and diurnal optical performance.

Sensitivity of Optical Performance with Respect to Total Optical Collector Error
The optical performance of a CSP system is significantly affected by the total optical error (σ_{tot}) of the beam path, so small values imply high technical cost both in term of investment but also maintenance. Figure 7.84b illustrates the annual optical efficiency ($\eta_{opt,a}$) as a function of the total collector optical error (σ_{tot}). All technologies show a monotonic decrease with increasing total optical error, with the sensitivity being small for linear Fresnel systems and largest for the central tower, so that for $\sigma_{tot} > 9$ mrad the optical efficiency of the tower falls below that of the linear Fresnel collector. The parabolic trough is more sensitive to optical efficiency than a linear Fresnel collector, but even at large optical errors it still shows efficiencies of the order of an optimal central receiver system.

References

Abbas, A., Montes, M., Piera, M., & Martínez-Val, J. (2012). Solar radiation concentration features in Linear Fresnel Reflector arrays. *Energy Conversion and Management, 54*, 133–144. https://doi.org/10.1016/j.enconman.2011.10.010

Abbas, R., Montes, M., Rovira, A., & Martinez-Val, J. (2016). Parabolic trough collector or linear Fresnel collector? A comparison of optical features including thermal quality based on commercial solutions. *Solar Energy, 124*, 198–215. https://doi.org/10.1016/j.solener.2015.11.039

Abbas, R., Sebastián, A., Montes, M., & Valdés, M. (2018). Optical features of linear Fresnel collectors with different secondary reflector technologies. *Applied Energy, 232*, 386–397. https://doi.org/10.1016/j.apenergy.2018.09.224

ALANOD GmbH & Co. KG. (2021). *Efficiency-solar-surface.* Product catalogue, Egerstr. 12· 58256 Ennepetal. Germany. Retrieved December 10, 2022, from www.alanod-solar.com

ASTM G173-03. (2020). *Standard tables for reference solar spectral irradiances: direct normal and hemispherical on 37° tilted surface*, West Conshohocken, PA, 19428-2959. ASTM Headquarters. https://doi.org/10.1520/G0173-03R20

Avila-Marin, A. L. (2011). Volumetric receivers in solar thermal power plants with central receiver system technology: A review. *Solar Energy, 85*, 891–910. https://doi.org/10.1016/j.solener.2011.02.002

Balaji, S., Reddy, K., & Sundararajan, T. (2016). Optical modelling and performance analysis of a solar LFR receiver system with parabolic and involute secondary reflectors. *Applied Energy, 179*, 1138–1151. https://doi.org/10.1016/j.apenergy.2016.07.082

Bendt, P., Rabl, A., Gaul, H. W., & Reed, K. A. (1979). Optical analysis and optimization of line focus collectors. In: *NASA STI/Recon Technical Report N 80.*

Bellos, E. (2019). Progress in the design and the applications of linear Fresnel reflectors—A critical review. *Thermal Science and Engineering Progress, 10*, 112–137. https://doi.org/10.1016/j.tsep.2019.01.014

Bernhard, R., de LaLaing, J., Kistner, R., Eck, M., Eickhoff, M., Feldhoff, J., … Morin, G. (2009). Linear fresnel collector demonstration at the PSA—Operation and investigation. In *Proceedings of 15th SolarPACES Conference*, Berlin, 15–18 September 2009.

Blackmon, J. B. (2020). Heliostat size optimization for central receivers. In K. Lovegrove, & W. Stein (Eds.), *Concentrating solar power technology* (2nd ed., pp. 585–632). Elsevier Woodhead Publishing.

Boerema, N., Morrison, G., Taylor, R., & Rosengarten, G. (2012). Liquid sodium versus Hitec as a heat transfer fluid in solar thermal central receiver systems. *Solar Energy, 86*, 2293–2305. https://doi.org/10.1016/j.solener.2012.05.001

Brahma Kumaris. (2011). *India One.* Retrieved from https://solar.brahmakumaris.com/links/

BraytonEnergy LLC. (2022). SolarCAT Solar Compressed Air Turbine. 75B Lafayette Road, Hampton, NH 03842. Retrieved December 21, 2022, from https://www.braytonenergy.net/our-projects/solarcat/

BrightSource Energy Inc. (2023). 1999 Harrison Street Suite 2150, Oakland, CA 94612, California, USA. Retrieved February 27, 2023, from https://www.brightsourceenergy.com/breakthrough-technologies

Brossard, D. (2010, May 16). *Heliostat—Maurya Sheraton Hotel—New Delhi* (Flickr.com, Ed.) Retrieved May 3, 2023, from https://www.flickr.com/photos/string_bass_dave/4609871733/in/photostream/

Burkholder, F., & Kutscher, C. F. (2008). Heat-loss testing of Solel's UVAC3 parabolic trough receiver: Technical report. In *Heat-loss testing of Solel's UVAC3 parabolic trough receiver: Technical Report*. National Renewable Energy Laboratory.

Burkholder, F., & Kutscher, C. (2009). *Heat loss testing of Schott's 2008 PTR70 Parabolic Trough Receiver.* National Renewable Energy Lab. (NREL), Golden, CO, USA. https://doi.org/10.2172/1369635

Cagnoli, M., Mazzei, D., Procopio, M., Russo, V., Savoldi, L., & Zanino, R. (2018). Analysis of the performance of linear Fresnel collectors: Encapsulated vs. evacuated tubes. *Solar Energy, 164*, 119–138. https://doi.org/10.1016/j.solener.2018.02.037

Canavarro, D., Chaves, J., & Collares-Pereira, M. (2014). Simultaneous multiple surface method for Linear Fresnel concentrators with tubular receiver. *Solar Energy, 110*, 105–116. https://doi.org/10.1016/j.solener.2014.09.002

Cao, F., McEnaney, K., Chen, G., & Ren, Z. (2014). A review of cermet-based spectrally selective solar absorbers. *Energy & Environmental Science, 7*(5), 1615–1627.

Cardemil, J., Starke, A., Scariot, V., Grams, I., & Colle, S. (2014). Evaluating solar radiation attenuation models to assess the effects of climate and geographical location on the heliostat field efficiency in Brazil. *Energy Procedia, 49*, 1288–1297. https://doi.org/10.1016/j.egypro.2014.03.138

Chaves, J., & Collares-Pereira, M. (2010). Etendue-matched two-stage concentrators with multiple receivers. *Solar Energy, 84*(2), 196–207. https://doi.org/10.1016/j.solener.2009.10.022

Conlon, W. M., Johnson, P., & Hanson, R. (2011, July 12–14). Superheated steam from CLFR solar steam generators. In *Proceedings of ASME 2011 Power Conference* (Vol. 1, pp. 301–307). https://doi.org/10.1115/POWER2011-55174

CosinSolar. (2023). SUPCON SOLAR Delingha 10MW Tower CSP Plant. No. 307, Liuhe Road, Binjianga, District, Hangzhou, Zhejiang, China. Retrieved Febraury 27, 2023, from http://www.supconsolar.com/en/index/index.html

Coventry, J., Andraka, C., Pye, J., Blanco, M., & Fisher, J. (2015). A review of sodium receiver technologies for central receiver solar power plants. *Solar Energy, 122*, 749–762. https://doi.org/10.1016/j.solener.2015.09.023

Coventry, J., Campbell, J., Xue, Y., Hall, C., Kim, J.-S., Pye, J., … Beath, A. (2016). *Heliostat cost down scoping study.* Final Report, ASTRI—Australian Solar Thermal Research Initiative.

Davis, J. R. (2000). *Nickel, Cobalt, and their alloys-ASM specialty handbook* (Vol. 1), Materials Park, OH 44073, USA. ASM International. Retrieved from www.asminternational.org

del Río, A., Korzynietz, R., Brioso, J., Gallas, M., Ordóñez, I., Queroa, M., & Diaz, C. (2015). Soltrec–Pressurized volumetric solar air receiver technology. *Energy Procedia, 69*, 360–368. https://doi.org/10.1016/j.egypro.2015.03.042

Deutsches Zentrum für Luft- und Raumfahrt (DLR). (2008). Solarstrahlungsempfänger (Solar-receiver) auf dem Turmkraftwerk in Almería. Institut für Technische Thermodynamik, Solarforschung.

Di Canio, D., Tretyl, W., Jur, F., & Watson, C. (1977–1979). *Line-focus solar thermal central receiver research study.* Report DOE/ET/20426-1, FMC Corporation, Santa Clara, CA.

Díez de los Ríos Ramos, N. (2018). *Technology developments for the Alkali Metal Thermal to Electric Converter (AMTEC) Test Facility.* Dissertation, Faculty of Mechanical Engineering, Karlsruher Institut für Technologie (KIT), Karlsruhe Germany.

Dunham, M., & Iverson, B. (2014). High-efficiency thermodynamic power cycles for concentrated solar power systems. *Renewable and Sustainable Energy Reviews, 30*, 758–770. https://doi.org/10.1016/j.rser.2013.11.010

Eddhibi, F., Amara, M., Balghouthi, M., Qoaider, L., & Guizani, A. (2017). Analytic optical design of a Linear Fresnel solar collector with variable parameters. *Journal of Materials and Environmental Sciences, 11*(8), 4068–4084.

Ermanoski, I., Siegel, N., & Stechel, E. B. (2013). A new reactor concept for efficient solar-thermochemical fuel production. *Journal of Solar Energy Engineering, 135*(3), 031002 (10 p.). https://doi.org/10.1115/1.4023356

Flesch, J. (2021). *LBE-cooled tube receiver performance—Design aspects and high-flux operation in a solar furnace.* Dissertation, Karlsruhe Institute of Tchnology (KIT), Faculty Chemical Egnigneering and Process Technology, Karlsruhe.

Forsberg, C., Peterson, P. F., & Zhao, H. (2007). High-temperature liquid-fluoride-salt closed-brayton-cycle solar power towers. *Journal of Solar Energy Engineering, 129*(2), 141–146. https://doi.org/10.1115/1.2710245

García-Cortés, S., Bello-García, A., & Ordóñez, C. (2012). Estimating intercept factor of a parabolic solar trough collector with new supporting structure using off-the-shelf photogrammetric equipment. *Applied Energy, 92*, 815–821. https://doi.org/10.1016/j.apenergy.2011.08.032

Gholamalizadeh, E., & Chung, J. D. (2017). Exergy analysis of a pilot parabolic solar dish-stirling system. *Entropy, 19*(10), 509 (12 p.). https://doi.org/10.3390/e19100509

Guerreiro, L., Canavarro, D., & Collares Pereira, M. (2015). Efficiency improvement and potential LCOE reduction with an LFR-XX SMS plant with storage. *Energy Procedia, 69*, 868–878. https://doi.org/10.1016/j.egypro.2015.03.112

Hafez, A., Soliman, A., El-Metwally, K., & Ismail, I. (2017). Design analysis factors and specifications of solar dish technologies for different systems and applications. *Renewable and Sustainable Energy Reviews, 67*, 1019–1036. https://doi.org/10.1016/j.rser.2016.09.077

Heimsath, A., & Nitz, P. (2019). The effect of soiling on the reflectance of solar reflector materials—Model for prediction of incidence angle dependent reflectance and attenuation due to dust deposition. *Solar Energy Materials and Solar Cells, 195*, 258–268. https://doi.org/10.1016/j.solmat.2019.03.015

Heimsath, A., Bern, G., van Rooyen, D., & Nitz, P. (2014a). Quantifying optical loss factors of small linear concentrating collectors for process heat application. *Energy Procedia, 48*, 77–86. https://doi.org/10.1016/j.egypro.2014.02.010

Heimsath, A., Cuevas, F., Hofer, A., Nitz, P., & Platzer, W. J. (2014b). Linear Fresnel collector receiver. *Energy Procedia, 49*, 386–397.

Heimsath, A., Cuevas, F., Hofer, A., Nitz, P., & Platzer, W. J. (2014c). Linear Fresnel collector receiver: Heat loss and temperatures. *Energy Procedia, 49*, 386–397. https://doi.org/10.1016/j.egypro.2014.03.042

Hering, W., Stieglitz, R., & Wetzel, T. (2012). Application of liquid metals for solar energy systems. *EPJ Web of Conferences, 33*, 7. https://doi.org/10.1051/epjconf/20123303003

Hildebrandt, C. (2009). *Hochtemperaturstabile Absorberschichten für linear konzentrierende solarthermische Kraftwerke.* Dissertation, Universität Stuttgart, Institut für Thermodynamik und Wärmetechnik, Stuttgart.

Ho, C. K. (2016). A review of high-temperature particle receivers for concentrating solar power. *Applied Thermal Engineering, 109*, 958–969. https://doi.org/10.1016/j.applthermaleng.2016.04.103

Ho, C. K. (2017). Advances in central receivers for concentrating solar applications. *Solar Energy, 152*, 38–56. https://doi.org/10.1016/j.solener.2017.03.048

Ho, C., & Iverson, B. D. (2014). Review of high-temperature central receiver designs for concentrating solar power. *Renewable and Sustainable Energy Reviews, 29*, 835–846. https://doi.org/10.1016/j.rser.2013.08.099

IEC 62862-5-2. (2022). *Solar thermal electric plants Part 5-2. Systems and components. General requirements and test methods for large-size linear Fresnel collectors.*

Industrial Solar GmbH. (2022). Fresnel Collector LF-11 Datasheet. Basler Str. 115, D-79115 Freiburg, Germany. RetrievedMarch 29, 2022, from https://www.industrial-solar.de/en/technologies/fresnel-collector/

IRENA -International Renewable Energy Agency. (2013). *Renewable Power Generation Costs in 2012: An overview.* Abu Dhabi, C67 Office Building, Khalidiyah (32nd) Street, United Arab Emirates. Retrieved from www.irena.org

ISO 9845-1. (1992). *Solar energy—Reference solar spectral irradiance at the ground at different receiving conditions—Part 1: Direct normal and hemispherical solar irradiance for air mass 1,5.* ICS:27.160 Solar energy engineering.

Janotte, N., Feckler, G., Kötter, J., Decker, S., Herrmann, U., Schmitz, M., & Lüpfert, E. (2014). Dynamic performance evaluation of the HelioTrough® collector demonstration loop-towards a new benchmark in parabolic trough qualification. *Energy Procedia, 49*, 109–117. https://doi.org/10.1016/j.egypro.2014.03.012

Kalbhor, A. (2014). Modified solar central receiver in concentrated solar power systems. *International Journal of Engineering and Technical Research (IJETR), 12*(2).

Karni, J., Kribus, A., Doron, P., Rubin, R., Fiterman, A., & Sagie, D. (1997). The DIAPR: A high-pressure, high-temperature solar receiver. *Journal of Solar Energy Engineering—Transactions of the ASME, 119*, 74–78. https://doi.org/10.1115/1.2871853

Kesselring, P., & Selvage, C. (1986). *The IEA/SSPS solar thermal power plants—Facts and figures—Volume 1: Central Receiver System (CRS).* International Energy Agency (IEA).

Kincaid, N., Mungas, G., Kramer, N., Wagner, M., & Zhu, G. (2018). An optical performance comparison of three concentrating solar power collector designs in linear Fresnel, parabolic trough, and central receiver. *Applied Energy, 231*, 1109–1121. https://doi.org/10.1016/j.apenergy.2018.09.153

Korzynietz, R., Brioso, J., del Río, A., Quero, M., Gallas, M., Uhlig, R., … Teraji, D. (2016). Solugas—Comprehensive analysis of the solar hybrid Brayton plant. *Solar Energy, 135*, 578–589. https://doi.org/10.1016/j.solener.2016.06.020

Kurup, P., & Turchi, C. S. (2015). *Parabolic trough collector cost update for the system advisor model (SAM).* 15013 Denver West Parkway,Golden, CO 80401. National Renewable Energy Laboratory (NREL). Retrieved from https://www.nrel.gov/docs/fy16osti/65228.pdf

Lee, C.-Y., Chou, P.-C., Chiang, C.-M., & Lin, C.-F. (2009). Sun tracking systems: A review. *Sensors, 9*, 3875–3890. https://doi.org/10.3390/s90503875

Lüpfert, E., Zarza-Moya, E., Geyer, M., Nava, P., Langenk, J., Schiel, W., … Mandelberg, E. (2003). Eurotrough collector qualification complete—Performance test results from PSA. In *Proceedings of ISES Solar World Congress* (p. 12), June 14–19, Göteborg, Sweden

Mertins, M. (2006). *Technische und wirtschaftliche Analyse von horizontalen Fresnel-Kollektoren.* Dissertation, Universität Karlsruhe, Faculty for Mechanical Engineering, Karlsruhe.

Mills, D. R., & Morrison, G. (2000). Compact Linear Fresnel reflector solar thermal powerplants. *Solar Energy, 68*(3), 263–283. https://doi.org/10.1016/S0038-092X(99)00068-7

Moga, D., Sita, I. V., Stroia, N., Dobra, P., Moga, R., & Petreus, D. (2015). Sensing and control strategies in tracking solar systems. In *Proceedings of International Conference on Control Systems and Science* (pp. 989–995), 23–25 May 2015. https://doi.org/10.1109/CSCS.2015.143

Montes, M. J., Abbas, R., Muñoz, M., Muñoz-Antón, J., & Martínez-Val, J. (2017). Advances in the linear Fresnel single-tube receivers: Hybrid loops with non-evacuated and evacuated receivers. *Energy Conversion and Management, 149*, 318–333. https://doi.org/10.1016/j.enconman.2017.07.031

Morin, G. (2005). *Weiterführende Forschung und Entwicklung zum Fresnel-Konzept sowie Vorbereitung eines Demonstrationsvorhabens.* Report MAO1-GM-0512-E02, Fraunhofer ISE, Freiburg.

Morin, G., Mertins, M., Kirchberger, J., & Selig, M. (2011). SuperNOVA–construction, control & performance of steam superheating linear Fresnel collector. In *Proceedings of 17th SolarPACES International Symposium.* Granada, Spain.

Morin, G., Dersch, J., Platzer, W., Eck, W., & Häberle, A. (2012). Comparison of Linear Fresnel and parabolic trough collector plants. *Solar Energy, 86*, 1–12. https://doi.org/10.1016/j.solener.2011.06.020

Nagarajan, S., & Barshilia, H. (2010). *Review of sputter deposited mid- to high- temperature solar selective coatings for flat plate/evacuated tube collectors and solar thermal power generation applications.* Bangalore 560017, India. National Space Laboratories.

National Renewable Energy Laboratory-NREL. (2017, August 4). Retrieved May 3, 2023, from Mirror Field Crescent Dunes Solar Thermal Facility. https://www.flickr.com/photos/nrel/35965146980/in/photostream/

Novatec Solar. (2012). Turnkey Solar Boiler. Karlsruhe, Germany. Retrieved December 12, 2021, from https://www.energy-xprt.com/products/solar-boiler-251465

Ortona, A., Yoon, D., Fend, T., & Smirnova, O. (2015). Tubular Si-infiltrated SiCf/SiC composites for solar receiver application—Part 2: Thermal performance analysis and prediction. *Solar Energy Materials & Solar Cells, 140*, 382–387. https://doi.org/10.1016/j.solmat.2015.04.034

Pfahl, A., Coventry, J., Röger, M., Wolfertstetter, F., Vásquez-Arango, J.-F., Gross, F., … Liedke, P. (2017). Progress in heliostat development. *Solar Energy, 152*, 3–37. https://doi.org/10.1016/j.solener.2017.03.029

Pitz-Paal, R., Morhenne, J., & Fiebig, M. (1991). A new concept of a selective solar receiver for high temperature applications. *Solar Energy Materials, 24*, 293–306. https://doi.org/10.1016/0165-1633(91)90070-2

Platzer, W., Heimsath, A., Hildebrandt, C., & Georg, A. (2008). Quality control of concentrating collector components for the optimization of performance. In *Proceedings of 14th SolarPACES International Symposium*, March 4–7, 2008, Las Vegas, USA.

Pottler, K., Ulmera, S., Lüpfert, E., Landmann, M., Röger, M., & Prahl, C. (2014). Ensuring performance by geometric quality control and specifications for parabolic trough solar fields. *Energy Procedia, 49*, 2170–2179. https://doi.org/10.1016/j.egypro.2014.03.230

Poživil, P., Agab, V., Zagorskiy, A., & Steinfeld, A. (2014). A pressurized air receiver for solar-driven gas turbines. *Energy Procedia, 49*, 498–503. https://doi.org/10.1016/j.egypro.2014.03.053

Pye, J., Morrison, G., & Behnia, M. (2003). Modelling of cavity receiver heat transfer for the Compact Linear Fresnel reflector. In *Proceedings of ISES Solar World Congress-Solar Energy for a Sustainable Future* (p. 10), 26–29 November 2003. Gothenburg, Sweden.

Qiu, Y., He, Y.-L., Cheng, Z., & Wang, K. (2015). Study on optical and thermal performance of a linear Fresnel solar reflector using molten salt as HTF with MCRT and FVM methods. *Applied Energy, 146*, 162–173. https://doi.org/10.1016/j.apenergy.2015.01.135

Rabl, A. (1985). *Active solar collectors and their applications* (1st ed.). Oxford University Press.

Rabl, A., & Winston, R. (1976). Ideal concentrators for finite sources and restricted exit angles. *Applied Optics, 15*(11), 2880–2883. https://doi.org/10.1364/AO.15.002880

Reiser, J., Franke, P., Weingärtner, T., Hoffmann, J., Hoffmann, A., & Rieth, M. (2015). Tungsten laminates made of ultrafine-grained (UFG) tungsten foil—Ageing of tungsten–titanium (W–Ti) laminates. *International Journal of Refractory Metals and Hard Materials, 51*, 264–274. https://doi.org/10.1016/j.ijrmhm.2015.04.032

Ricklin, P., Slack, M., Rogers, D., & Huibregtse, R. (2014). Commercial readiness of eSolar next generation heliostat. *Energy Procedia, 49*, 201–208. https://doi.org/10.1016/j.egypro.2014.03.022

Riffelmann, K., Richert, T., Nava, P., & Schweitzer, A. (2014). Ultimate Trough®—A significant step towards cost-competitive CSP. *Energy Procedia, 49*, 1831–1839. https://doi.org/10.1016/j.egypro.2014.03.194

Schiel, W., & Keck., T. (2012). Parabolic dish concetrating solar power systems. In K. Lovegrove, & W. Stein (Eds.), *Concentrating solar power technology* (pp. 284–321). 80 High Street, Sawston, Cambridge CB22 3HJ, UK. Woodhead Publishing.

Schiel, W., Hunt, B., Keck, T., Schweitzer, A., Weinrebe, G., & Partner, S. B. (2004, May 25–27). Solar thermal power plants for central or distributed electricity generation. In *Proceedings of Power-Gen Europe* (p. 22).

SCHOTT Solar CSP GmbH. (2013). Advanced qualification methods for parabolic trough receivers. Riyadh, November 20th 2013, Kingdom Saudi Arabia. Retrieved December 20, 2022, from https://www.dlr.de/sf/Portaldata/73/Resources/dokumente/veranstaltungen/saudisch-d_ws_riad_11-2013/SCHOTT_Solar_Saudi-German_CSP-WS_Riad_20112013.pdf

SCHOTT Technical Glass Solutions GmbH. (2018). *BOROFLOAT® 33 & Functional Coatings-Product information.* Technical Information sheet, Otto-Schott-Strasse 13,07745 Jena,Germany. Retrieved December 18, 2022, from www.schott.com/borofloat/coatings

Schöttll, P., Rohani, S., Leonardi, E., Pisani, L., Les, I., Mutuberria, M., & Nitz, P. (2019). Solar field heliostat selection based on polygon optimization and boundaries. *AIP Conference Proceedings, 2126*, 8. https://doi.org/10.1063/1.5117565

Sharma, V., Nayak, J. K., & Kedare, S. (2015). Comparison of line focusing solar concentrator fields considering shading and blocking. *Solar Energy, 122*, 924–939. https://doi.org/10.1016/j.solener.2015.10.011

Siemens Concentrated Solar Power, Ltd. (2011). Siemens UVAC 2010. http://www.energy.siemens.com/hq/pool/hq/power-generation/renewables/solar-power-solutions/concentrated-solar-power/downloads/E50001-W320-A104-V2-4A00_DA_UVACreceiver_US_ohne%20Israel.pdf. Siemens AG, Energy Sector, Freyeslebenstrasse 1, 91058 Erlangen, Germany.

Solare Brücke. (2022, September 20). Retrieved from Scheffler reflector: http://www.solare-bru ecke.org/index.php/en/die-scheffler-reflektoren

Stein, H., & Buck, R. (2017). Advanced power cycles for concentrated solar power. *Solar Energy, 152*, 91–105. https://doi.org/10.1016/j.solener.2017.04.054

Stine, W., & Diver, R. (1994). *A compendium of solar dish/stirling technology.* Sandia National Laboratories.

Tagle-Salazar, P., & Nigam, K. R.-S. (2020). Parabolic trough solar collectors: A general overview of technology, industrial applications,energy market, modeling, and standards (D. Gruyter, Ed.) *Green Processing and Synthesis, 9*, 595–649. https://doi.org/10.1515/gps-2020-0059

Tanaka, K. (2010). Concept design of solar thermal receiver using alkali metal thermal to electric converter (AMTEC). *Current Applied Physics, 10*, S254–S256. https://doi.org/10.1016/j.cap. 2009.07.031

Téllez, F., Villasante, C., & Burisch, M. (2014). *State of the art in heliostats and definition of specifications-survey for a low cost heliostat development.* Deliverable 12.1, EU Project STAGE STE.

Tsekouras, P., Tzivanidis, C., & Antonopoulos, K. (2018). Optical and thermal investigation of a linear Fresnel collector with trapezoidal cavity receiver. *Applied Thermal Engineering, 135*, 379–388. https://doi.org/10.1016/j.applthermaleng.2018.02.082

United Sun Systems International Ltd. (2012, December 31). Maricopa dish-stirling plant. Retrieved March 29, 2023, from https://commons.wikimedia.org/wiki/File:Maricopa_Dish-Sti rling_plant_01.jpg

van Rooyen, D. W., Heimsath, A., Nitz, P., & Platzer, W. (2011). An optical model of the fix-focus scheffler reflector. In *Proceedings of 30th ISES Solar World Congress* (pp. 4527–4537), Kassel, 28 August–2 September 2011. https://doi.org/10.18086/swc.2011.28.30

Viebahn, P., Kronshage, S., Trieb, F., & Lechon, Y. (2008). *Final report on technical data, costs, and life cycle inventories of solar thermal power plants.* Deliverable 12.2 Project 502687 NEEDS EU 6th Framework Programme.

Vignarooban, K., Xu, X., Arvay, A., Hsu, K., & Kannan, A. (2015). Heat transfer fluids for concentrating solar power systems—A review. *Applied Energy, 146*, 383–396. https://doi.org/10. 1016/j.apenergy.2015.01.125

Viswanathan, R. (1993). *Damage mechanisms and life assessment of high-temperature components* (2nd ed.). Metals Park, Ohio 44073. ASM International.

Wang, K., Wu, H., Wang, D., Wang, Y., Tong, Z., Lin, F., & Olabi, A. (2015). Experimental study on a coiled tube solar receiver under variable solar radiation condition. *Solar Energy, 122*, 1080–1090. https://doi.org/10.1016/j.solener.2015.10.004

Wua, S.-Y., Xiao, L., Cao, Y., & Li, Y. (2010). A parabolic dish/AMTEC solar thermal power system and its performance evaluation. *Applied Energy, 87*, 452–462. https://doi.org/10.1016/j.ape nergy.2009.08.041

Yang, M., Zhu, Y., & Taylor, R. (2018). End losses minimization of linear Fresnel reflectors with a simple, two-axis mechanical tracking system. *Energy Conversion and Management, 161*, 284–293. https://doi.org/10.1016/j.enconman.2018.01.082

Yellowhair, J., & Ho, C. (2010). Heliostat canting and focusing methods: An overview and comparison. *Energy Sustainability*, 609–615. https://doi.org/10.1115/ES2010-90356

You, J. H., Visca, E., Barett, T., Böswirth, B., Creszensi, F., Dompatil, F., … Vorpahl, C. (2018). European divertor target concepts for DEMO: Design rationales and high heat flux performance. *Nuclear Materials and Energy, 16*, 1–11. https://doi.org/10.1016/j.nme.2018.05.012

Zarza Moya, E. (2020). Parabolic-trough concentrating solar power systems. In K. S. Lovegrove (Ed.), *Concentrating solar power technology* (2nd ed., p. 832). Woodhead Publishing. https:// doi.org/10.1016/C2018-0-04978-6

Zhu, G. (2017). New adaptive method to optimize the secondary reflector of linear Fresnel collectors. *Solar Energy, 144*(1), 117–126. https://doi.org/10.1016/j.solener.2017.01.005

Zhu, Y., Shi, J., Li, Y., L., W., Huang, Q., & Xu, G. (2017). Design and thermal performances of a scalable linear Fresnel reflector solar system. *Energy Conversion and Management, 146*, 174–181. https://doi.org/10.1016/j.enconman.2017.05.031

Zou, B., Yang, H., Yao, Y., & Jiang, J. (2017). A detailed study on the effects of sunshape and incident angle on the optical performance of parabolic trough solar collectors. *Applied Thermal Engineering, 126*, 81–91. https://doi.org/10.1016/j.applthermaleng.2017.07.149

MIX
Papier aus verantwortungsvollen Quellen
Paper from responsible sources
FSC® C105338

If you have any concerns about our products,
you can contact us on
ProductSafety@springernature.com

In case Publisher is established outside the EU,
the EU authorized representative is:
Springer Nature Customer Service Center GmbH
Europaplatz 3, 69115 Heidelberg, Germany

Printed by Libri Plureos GmbH
in Hamburg, Germany

Solar Thermal Energy Systems

Werner Platzer · Robert Stieglitz

Solar Thermal Energy Systems

Fundamentals, Technology, Applications

Volume 2

Springer

Werner Platzer
Umwelt und Natürliche Resourcen
Albert-Ludwigs-Universität Freiburg
Freiburg im Breisgau, Baden-Württemberg,
Germany

Robert Stieglitz
Institut für Neutronenphysik und
Reaktortechnik (INR)
KIT
Eggenstein-Leopoldshafen
Germany

ISBN 978-3-031-43172-2 ISBN 978-3-031-43173-9 (eBook)
https://doi.org/10.1007/978-3-031-43173-9

This Springer imprint is published by the registered company Springer Nature Switzerland AG
The registered company address is: Gewerbestrasse 11, 6330 Cham, Switzerland

Paper in this product is recyclable.

Before the final publication of this textbook, my highly esteemed co-author Prof. Robert Stieglitz, who was indispensable and tireless in the production of the book, passed away unexpectedly. He was a great person and an internationally recognized scientist who excelled in his broad knowledge of materials science and renewable energy. His loss affects not only his family, but also his colleagues and the scientific community as a whole.

Preface

Against the background of mankind's fight against climate change and its consequences, technologies and systems to provide heat, cold and electricity from the sun play a key role to support the necessary transformation of our energy systems towards a truly renewable and sustainable one. In solar thermal technologies, solar electromagnetic radiation is primarily transformed into heat, which may serve as a second step to provide us with heat and cold for thermal applications, but also electricity, chemical reactions, and fuels. The aspect of conserving energy and resources is gaining particular importance when introducing new technologies. Solar thermal systems allow the provision of a wide spectrum of temperature ranges for different purposes by means of suitable collectors. The scope of heat applications exceeds by far the use for room heating and hot water preparation in buildings. The use of solar heat for cooling and refrigeration systems, the integration into heat networks, the customized provision of industrial process heat for all kinds of industries as well as the conversion into electricity in thermal power plants are equally important applications in our current energy matrix. Although the technologies and systems are mature, future optimization using digitalization, new sustainable materials and integrated system designs will revolutionize current products on the market. Considering the technological development, the situation could be compared with that of the car industry before World War II and now. The available solutions already show great potential, but missing market capacity has not yet allowed to develop designs with minimum material content and based on real mass production. On the other hand, raw materials exploitation status and energy and climate-friendly production will force us to improve our designs soon.

This script is addressed to technically interested readers, as well as students in the Bachelor's and Master's area up to interested technically educated career changers, and is modularly structured in two parts. In the first part, the basics are worked out and the conversion of solar radiation into heat is presented, while the second part shows and discusses the applications in building technology, heat process technology, solar thermal power plants as well as solar thermal air conditioning.

Part I

Heat is intrinsically linked to energy, so at the beginning of the first part the concept of energy is highlighted, the temporal consumption profile of heat and electricity is addressed using the example of a household or the heat levels of different industrial processes. The role of heat, electricity, and fuels, which must be generated CO2-neutrally in the future as far as possible, is shown for our economies worldwide. A key lesson is that demand and generation do not necessarily coincide, and therefore, storage and load-shifting have to be developed if renewables have to cover all requirements. The vision is to develop a future decentralized and sector-coupled energy system, at a reasonable cost and with reliable and resilient supply at all times. Chapter 2 we devote to one primary renewable source of energy, our sun with its virtually infinite radiation resource. Solar irradiation on our earth has to consider the sun's extra-terrestrial radiation intensity, atmospheric attenuation, the influence of geographical location and temporal changes. The quantifying of irradiation needs suitable measuring devices for its detection. In Chap. 3 the physical basics for the optical conversion of the short-wave solar radiation into heat, as well as methods for concentrating the radiation, are presented. Measurement methods are used to quantify the optical quality of materials and devices, and mathematical description and modelling are presented for designing and optimizing the optical components. The physical and engineering description of the heat transfer by thermal conduction, thermal radiation and natural convection is included in Chap. 4, where practical problems with coupled heat transfer mechanisms are also highlighted. In general, the generation of heat and its demand occurs at different locations, so the heat must be transported by heat transfer fluids. The flow of these fluids can be either single-phase or multiphase, laminar, or turbulent. Chapter 5 examines these energy and momentum transport in solar thermal systems and discusses fundamentals, approaches to solutions, and points out engineering correlations. In Chap. 6 stationary low-temperature collectors and in Chap. 7 concentrating and tracking high-temperature collectors, solar thermal collector types with their different characteristic properties, operating modes and design alternatives are presented and discussed.

Part II

The second part of the book contains the descriptions, concepts and system designs for several application fields of the thermal energy generated by the collectors. In most solar thermal systems, the ability of storage is already conceptually included, contrary to other renewable energies like grid-integrated photovoltaics and wind, which rely on the flexibility to accept additional power of the grid. Several concepts and designs of thermal energy storage are described first in Chap. 8, which allow decoupling of the generation of heat from the consumption profiles of heat. The focus of Chap. 9 is the targeted use of solar generated low-temperature heat well below 100 °C for domestic applications, namely the production of hot water and the use in the building heating system. The chapter describes not only the essential components and basic algorithms for dimensioning and planning, but also describes the concepts for integration into heat networks. The subject of Chap. 10 is the optimal design and layout of line and point concentrating collectors in a large field consisting of mirrors, absorbers and tracking devices with the aim of efficiently generating heat. Control and regulation systems as well as the operation and maintenance of the solar field are shown. In Chap. 11, solar thermal power plants are addressed, where the heat generation is linked to a heat engine. In order to have a high conversion efficiency of the heat engine, high temperatures above 400 °C are preferable. In addition to the various concepts of solar only thermal power generation, hybrid concepts are also shown and illustrated. Industry as a sector with a huge heat demand is able to use solar thermal heat from low temperatures below 100 °C to a medium range between 100 °C and 400 °C up to more than 1000 °C. Solar heat can be integrated using very different integration concepts into supply for industrial processes. In Chap. 12, starting from the determination of the sectoral heat demand, methods for integration are shown as well as procedures for the classification of solar integration. In the final Chap. 13, methods, procedures and concepts are presented on how cooling and air conditioning can be realized with the help of thermal technologies. Finally, in two appendices, methods for the design of hydraulic components such as pumps and heat exchangers and measuring devices for the detection of thermal variables are presented, since these are often required for the design of solar thermal systems but are not specific to solar thermal energy.

The script is designed in such a way that it can also be used as a reference work. In particular, the individual chapters are written in such a way that they are largely self-contained and require little back-and-forth flipping. Nevertheless, a basic education in physics is useful for understanding the chains of argumentation and trains of thought.

Freiburg im Breisgau, Germany — Werner Platzer
Eggenstein-Leopoldshafen, Germany — Robert Stieglitz

Acknowledgements

We would like to thank our families for bearing with us, and our institutions for giving us the freedom to dedicate the necessary amount of time to the creation of this book.

Contents

About the Authors

Prof. Dr. rer.nat. Werner Platzer is a physicist and received his Ph.D. from the Albert-Ludwigs-University Freiburg in 1988. He has been working for more than 35 years for the Fraunhofer Institute for Solar Energy Systems ISE in Germany, performing research and development of solar thermal energy material, components and systems, facade technology and energy efficiency in buildings. His focus in research and teaching is on concentrating solar thermal technology for process heat and power. Until 2015, he served as Director of the Division Solar Thermal and Optics. He cofounded the Fraunhofer Centre for Solar Energy Technology in Santiago de Chile and was Executive Director there in the years 2016 and 2017. Since 2018, he coordinated the international business development of Fraunhofer ISE. He is teaching as an honorary Professor at the Faculty for Environment and Sustainability, University of Freiburg. He was appointed chairman of the IEC TC 117 on the standardization for solar thermal electricity (CSP) and has been actively involved in many European and international projects on CSP and Solar Process Heat. He has authored more than 200 articles and conference papers.

Prof. Dr.-Ing. Robert Stieglitz obtained a doctoral degree in engineering in 1994 at the University of Karlsruhe. In his professional career, he has spent some time in the power industry and joined the research centre Karlsruhe (today Karlsruhe Institute of Technology-KIT) in 1995 leading functions in several science fields associated with heat and mass transfer, instrumentation, material qualification and nuclear as well as accelerator technologies. In 2009, he was appointed as director of the Institute for Neutron Physics and Reactor Technology (INR) and he simultaneously holds a chair at the university campus as professor in Applied Thermo-Fluidics. Since 2009, he has been the Director of the CEA KIT Frederic Joliot-Otto Hahn School and served many associations as an advisor and expert such as German Nuclear Reactor Safety Commission, the Alexander von Humboldt Foundation and others. He continued the INR's more than 30-year tradition in solar thermal research through a wide range of contributions to national and international solar thermal R&D projects and industrial collaborations. His scientific record is characterized by more than 240 scientific articles/conference papers, 2 books and some book contributions in several science fields.

Part II

Systems and Applications

8 Thermal Energy Storage

Abstract

The storage of thermal energy is a core element of solar thermal systems, as it enables a temporal decoupling of the irradiation resource from the use of the heat in a technical system or heat network. Here, different physical operating principles are applicable, which enable the energy to be stored. Central parameters for the technical and economic evaluation of storage systems are storage capacity and length, whereby the storage materials and fluids in particular play a decisive role since they influence the charging and discharging dynamics as well as the efficiency. In the course of the chapter, various storage concepts, their basic mode of operation as well as terminology and evaluation variables are explained and presented.

Keywords

Thermal energy storage • Sensible heat • Latent heat • Storage tanks • Chemical heat storage • Sorption • Storage systems • Storage materials

8.1 Introduction

In the field of solar energy use, a large fraction of the solar energy is converted at times of day and year when it cannot be used. Accordingly, short and long-term storage facilities are needed to compensate for this. Thermal energy storage (TES) stores energy in the form of heat whereas for example electrochemical batteries store electricity. High- and medium-temperature storage systems are used for industrial process heat applications and solar thermal power plants, low-temperature heat storage systems for buildings. For the choice of storage unit, the form of energy, energy and power density, cycle efficiency and duration, self-discharge rate, system life, reliability and costs are important parameters that have a considerable influence on the economic efficiency.

R. Stieglitz and W. Platzer, *Solar Thermal Energy Systems*,
https://doi.org/10.1007/978-3-031-43173-9_8

In order to get an idea of the magnitude of the required storage capacity without focusing on an individual application purpose, it makes sense to consider the energy content of available primary energy sources as potential storage. Figure 8.1 shows the specific energy content *E* of several direct energy sources in (kWh/kg). This figure does not make any distinction with regard to the type of energy stored, whether it is chemical, mechanical, or thermodynamical in nature. Figure 8.1 shows that the specific energy contents of the primary energy carriers hydrogen as well as the fossil carriers based on hydrocarbons have the highest specific energy densities. The energy content stored in electrochemical redox systems such as batteries is low compared to the fossil carriers and is essentially limited by the electrochemical voltage series of the respective substances. Pumped storage and compressed air storage also have low energy densities, so that storing significant amounts of energy by solar generation either requires a considerable amount of equipment and/or sufficient space (and thus social acceptance), which is also reflected in the required investment costs.

The storage question is of central importance for the future use of solar thermal energy as a potential substitute for fossil primary energy sources. The storage of solar heat in thermal energy storage systems (TESS) depends very much on the application. Heat for domestic hot water needs to be stored for few days in order to bridge the gap between cloudy and sunny periods, and to have warm water available whenever it is needed. When it comes to low-temperature heating, the winter

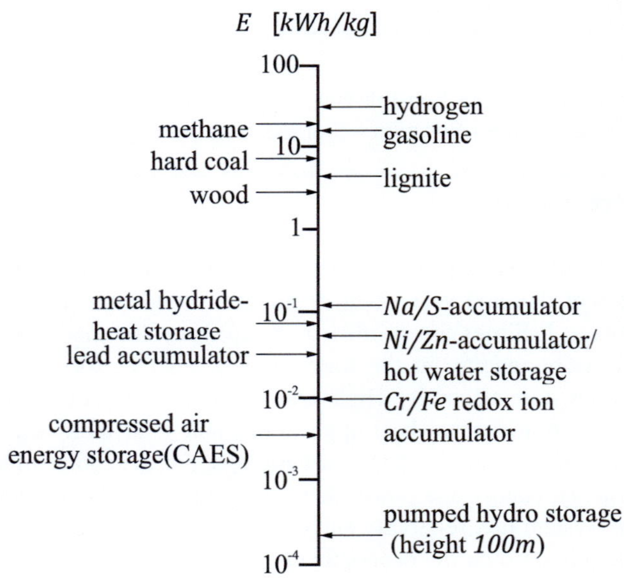

Fig. 8.1 Specific densimetric energy density E [kWh/kg] of available primary energy source/ energy storage options

period is critical. With systems storing heat only for a few days, the solar collector area required to cover the demand even in wintertime may be uneconomically large and even not feasible. Seasonal storage comes into the picture, when solar heat from the summer shall be used to cover the demand in winter. This requires generally very large storage volumes, on the one hand side because of the large amount energy to be stored, on the other hand because unavoidable heat losses are less important for huge volumes. An overview of thermal storage options and benefits is given in (Ernström, 2020).

For industrial use, a secure energy supply adapted to the work rhythm with approx. 9 hours per day and a 5—day week must be guaranteed. Otherwise, a substitute heat supply must be provided, and the solar heat supply can only be remunerated according to a fuel-saving operation, the lowest tariff.

The benefits of storage systems in various applications, where they are used on a cyclical basis, can be seen on several levels:

- Improvement of generation capacity

Neither demand nor generation of heat for heating, cooling and power applications is seldom constant over time. Excess generation during periods with lower demand can be used to charge a TES in order to avoid switching of the generation. On the other hand, during high-demand periods a generator can be dimensioned below maximum needed demand and by discharging a TES the demand can be met. This feature allows the installation of smaller heat generation capacities and results in a higher load factor for these units.

- Decoupling of load curves

If systems are combined in cogeneration of heating, cooling and/or power, the individual load curves might be rather different. For instance, in a combined heat and power plant (CHP), the operation can be optimised for efficiency of the CHP-unit or for maximum profitability. If a TES is installed the supply of necessary heat to consumers can be guaranteed with much less restrictions on the operational schedule.

- Economic benefits

With a TES the consumers may shift their demand to a period with possibly lower tariffs, while generators may decide to supply in the market rather in periods of higher tariffs (e.g. for electricity in the CHP case, or for solar power plants in the grid).

- Increase of system reliability

As storage allows to buffer energy even for interruptions in the generation or supply, the reliability of the application is increasing. In solar thermal heating systems low irradiation hours due to intermittent clouds can be bridged without problems. When we come to power-generating systems the idea is to charge a storage tank suitable for electrical energy generation during peak hours. The storage allows a temporal extension of the solar power production from daytime into the evening, he availability and the smoothing of the power plant operation (with the associated longer service life for turbine and generator).

- Increase of system efficiency

Using thermal energy storage, the operation of systems may be improved in a way that unfavourable part-load scenarios with low efficiency can be avoided. For example, turbines in a solar power plant can run at design point with better thermodynamic efficiency even when the solar radiation input is too low to provide the full power without storage.

8.2 Storage Characterisation

8.2.1 Physical Principles

Thermal energy supplied by solar thermal processes can be in principle stored directly as thermal energy and as chemical energy (Steinmann, 2020) The direct storage of heat is possible as sensible and latent heat, while the thermo-chemical storage involves reversible physical or chemical processes based on molecular forces.

- Storage as sensible heat involves a change of temperature in the storage medium.
- Storage as latent heat, also called phase-change, involves a state change in the storage medium associated with the release/expenditure of latent heat and can be carried out using any phase transition: solid/solid, solid/liquid, liquid/liquid, liquid/vapor, or solid/vapor.
- Thermo-chemical storage involves heat released or needed by reversible thermo-chemical reactions or thermo-physical absorption and adsorption processes.

Regardless of the storage principle, the respective storage concept requires a storage medium (or several media) that can be either in solid, liquid, or gaseous phase. The heat needs to be fed into and released from the media. In the simplest case, the heat transport takes place via a fluid that is used both in the solar field and in a storage tank; this is then referred to as direct storage. If a heat transfer takes place from a heat transfer medium to a separate storage medium, this is called indirect storage.

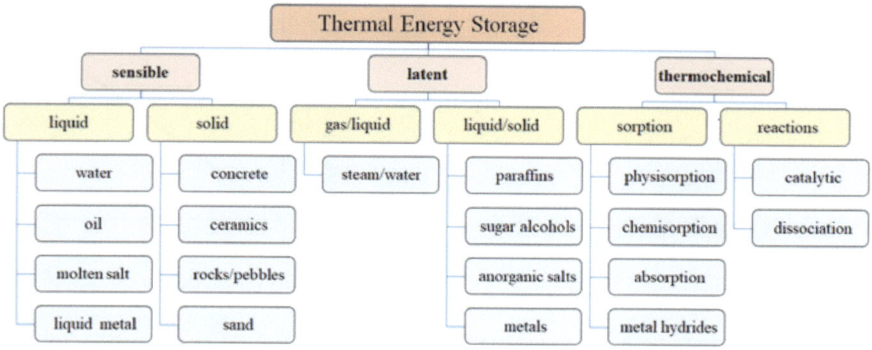

Fig. 8.2 Thermal energy storage options and classification by principle as well as media state

The different storage principles lead to different storage density and thus storage volume, but of course also the temperature range of the various materials also impacts the usability for certain applications. The materials used for thermal energy storage can lead to many different approaches and hence storage systems. On overview of the basic classification of thermal storage its classification by principles and storage media state is illustrated in Fig. 8.2.

For the sake of simplicity, we will divide thermal energy storage into the following three classes. Of, course in storage systems, combinations of individual options, even across the classes, is possible.

- sensible heat storages (SHS),
- latent heat storages (LHS), and
- thermo-chemical storages (TCS).

8.2.1.1 Sensible Heat Storage

Sensible heat storage systems achieve storage by raising the temperature of a medium; therefore, the sensible storage materials undergo no change in phase over the temperature range of the storage process. Sensible heat storage in a material depends strongly on its specific heat capacity, which determines the energy density and the thermal diffusivity, which in turn impacts the rate at which that heat can be released and extracted.

The energetic heat content Q in (J) of any incompressible continuum stored as *sensible heat* is written as:

$$Q = \int_{T=T_1}^{T=T_2} M \cdot c_p(T)\ dT = M \cdot \overline{c_p} \cdot \Delta T, \tag{8.1}$$

where M is the mass, $c_p(T)$ the specific temperature dependent heat capacity and T_2, T_1 the upper and lower temperatures considered. For simplicity often, the mean

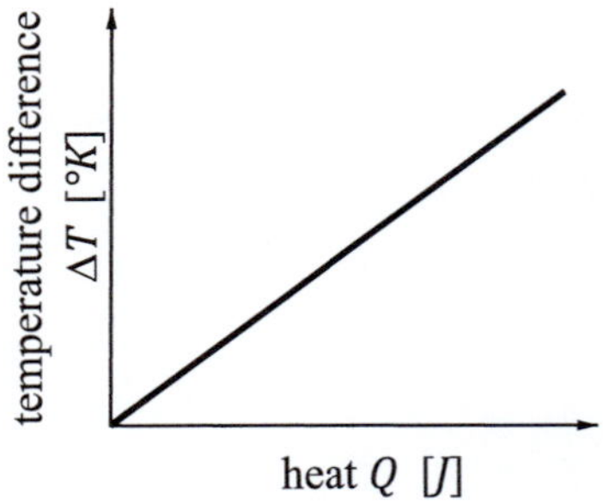

Fig. 8.3 Heat content Q as function of temperature change (ΔT) in a sensible thermal energy storage

specific heat capacity of the storage medium $\overline{c_p}$ is used. ΔT describes a characteristic temperature difference from an initially arbitrarily selectable reference temperature. But, according to the relationship in Eq. (8.1) assuming a constant mean heat capacity the stored heat energy Q depdends linearly on the temperature difference ΔT, as indicated in Fig. 8.3.

The specific heat capacity c_p for materials can vary quite substantially. It can vary from $4.2kJ/(kgK)$ for water to values much lower as shown in the Table 8.2 giving an overview of liquid and solid materials. For gases the specific heat is more complicated. For gas as a compressible substance it depends on the process. If the gas is kept in a fixed volume, the pressure will increase with temperature and the relevant property is c_v, the specific heat at constant volume. If, however, the pressure is kept constant and the volume is allowed to vary, the specific heat at constant pressure c_p is relevant. The ratio $\kappa = c_p/c_v$ describes the heat capacity ratio. In thermal physics and thermodynamics, it is also known as the adiabatic index, isotropic expansion factor or Laplace's coefficient. The magnitude of κ is related to the molecular structure and is affecting the heat which can be stored. The sensible specific heat of a material is characterizing the kinetic energy of the molecules and, therefore, the degrees of freedom and the number of atoms in the molecule are relevant for the heat capacity.

8.2.1.2 Latent Heat Storage (LHS)

The latent heat storage systems utilize materials that change phase at a temperature that falls within the upper and lower limit of the solar field. In doing so, they exploit the latent heat, or enthalpy $\Delta h_{m/b}$, associated with phase transition. Phase change phenomena vary from solid–solid, liquid–vapor and solid–liquid transitions, however, the latter is typically used due to its low volumetric expansion compared to the liquid–gas transition and due to its high latent heat compared to the solid–solid transition.

The heat content of a material within a temperature range, where a phase-change occurs is consisting of the sensible heat, and the latent heat has to be added. During the phase change the temperature in the phase change material (PCM) remains constant, as illustrated in Fig. 8.4. Exceeding the phase change temperature again the heat content increases linearly with the temperature as like

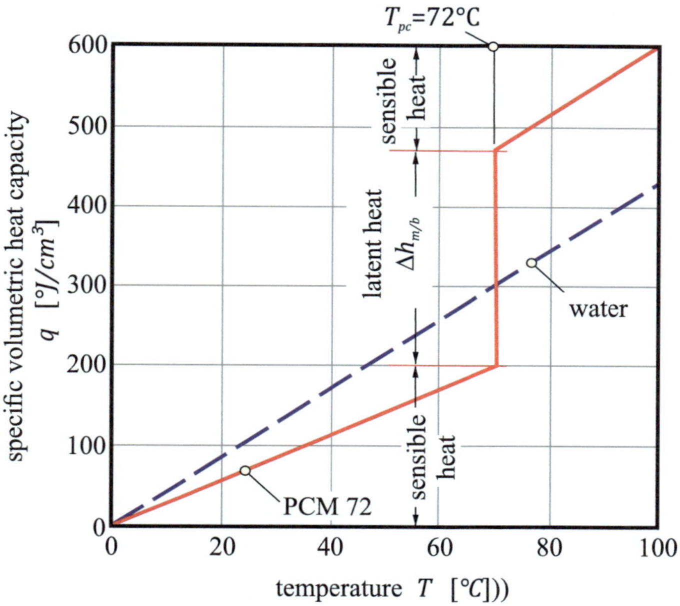

Fig. 8.4 Comparison of stored specific heat content as a function of temperature between water (sensible heat) and PCM −72 (phase change material) between 0° and 100 °C

for sensible heat storages. The heat content for a phase-change occurring within the temperature range between T_1 and T_2 is calculated as

$$Q = \int_{T=T_1}^{T=T_{pc}} M \cdot c_{p,1} \cdot dT + M \cdot \Delta h_{m/b} + \int_{T=T_{pc}}^{T=T_2} M \cdot c_{p,2} \cdot dT$$
$$= M \cdot \left[\overline{c_{p,1}} \cdot \left(T_{pc} - T_1\right) + \Delta h_{m/b} + \overline{c_{p,2}} \cdot \left(T_2 - T_{pc}\right)\right], \quad (8.2)$$

where $\overline{c_{p,1}}$ is the mean specific heat capacity for $T \leq T_{pc}$ and $\overline{c_{p,2}}$ corresponds to the mean specific heat capacity for $T \geq T_{pc}$, respectively. Note, if several phase-changes occur with the range between T_1 and T_2 several latent heats have to be included. The major advantage of heat storage by means of phase change materials are the comparatively high specific latent heat energies stored in PCM materials compared to the specific heats of liquids or gases, which a comparison of PCM −72 with water in Fig. 8.4 depicts. Typical PCM materials are paraffins, acids, salts or salt eutectics depending on the intended temperature level, see e.g. (Agyenim et al., 2010).

8.2.1.3 Thermo-Chemical Energy Storage (TCS)

The thermo-chemical storage systems rely on heat to drive reversible chemical reactions; thus, the storage medium must have the ability to completely dissociate in the temperature range provided. In this storage concept, the reaction in the

Table 8.1 Bonding strengths—strengths are typically (cited ranges vary) (https://science.jrank.org/pages/984/Bond-Energy.html)

Physical force	Bonding energy [kJ/mol]	Bonding energy [eV/mol]	Dissociation temperature [°C]
Van der Waals	$0.4 - 4$	$4 \cdot 10^{-3} - 4 \cdot 10^{-2}$	$-100 - 100$
Hydrogen bond	$10 - 40$	$0.1 - 0.4$	$0 - 100$
Dipole attractions	$40 - 400$	$0.4 - 4$	$100 - 800$
Covalent bonds	$200 - 1000$	$2 - 10$	$500 - 1500$
Ionic bond	$700 - 4000$	$7 - 40$	$1000 - 3000$

forward direction is endothermic while the reverse reaction is exothermic. The amount of heat stored in a chemical reaction depends on the heat of reaction and the extent of conversion given by:

$$Q = M \cdot a_r \cdot \Delta H, \tag{8.3}$$

where, a_r is the fraction reacted and ΔH is the heat of reaction per unit mass, and M the mass content.

This involves some kind of reversible exotherm/endotherm chemical reaction with thermo-chemical materials (TCM). Depending on the reactants, this method can allow for an even higher storage capacity than LHS. The energy is stored in a binding energy on a molecular level. Depending on the binding energy, different temperatures are needed to drive the reversible dissociation of the reactants. The weakest molecular binding is the Van der Waals force based on induced polarity of molecules. Low temperature levels below 100 °C are sufficient to dissociate these bonds, as indicated in Table 8.1. The next in strength are hydrogen bonds and then covalent chemical bonds and for dissociating they need temperatures of several hundred to thousand °K. Ionic bonds belong to the strongest binding forces of molecules. Generally, there are several components, either as fluid (gas or liquid) and possibly a solid part (e.g. a porous network for water vapor adsorption. Therefor the resulting temperature is dependent on the heat capacities of the different components and the mass fractions in the reaction volume. The reversible reaction means that there is an equilibrium depending on temperature in the reaction volume, which is responsible for the fractions of associated and dissociated substances.

8.2.2 Storage Capacity

Storage capacity Q is the useful energy content of the storage in kWh or MJ. For a known demand or design heat load D the heat content can be directly converted into ideal storage time (assuming no losses) or storage hours (N_{stor}). This is rather a measure of capacity in a defined application context than a categorization of typical time, the heat is stored in the storage. This quantity may also be called "Equivalent full load hours" ($EFLH$). It is determined by the ratio between storage capacity Q, in MWh_{th} ($1MWh = 3.6 \cdot 10^6 kJ$), and thermal power, in MW, demanded by the application system at design nominal conditions ($\dot{Q}_{demand,nom}$).

$$N_{stor} = \frac{Q}{\dot{Q}_{demand,nom}}. \tag{8.4}$$

One example would be a molten-salt storage for a solar thermal power plant with parabolic troughs as installed several places worldwide with steam turbines producing $50MW_{el}$ at the design point. Let the design efficiency of such a turbine be 36%, then the design thermal load is $138.9MW_{th}$ to be produced by the steam generating system. In many project storage hours of about 8h are chosen. This implies that the turbine could run 8h respectively at full load only supplied by the storage. The stored heat required for that is then $1111MW_{th}$. If the storage material is Nitrate Solar Salt with a heat capacity of 1.6 kJ/(kgK), and the temperature spread would be 100 °K, the required storage mass would be $2.5 \cdot 10^4 tons$.

8.2.3 Storage Duration

A slightly different concept is the storage duration, which describes the time over which a certain amount of heat is being stored typically. It is related to storage hours and capacity, but not identical. For example, it could be interesting to collect solar energy in the summer period and store a capacity of 240 s torage hours (10 days) that with little losses (say using thermo-chemical storage) for 6 months until it is needed in wintertime. Then, the storage duration exceeds the number of storage hours by large.

So, energy storage has to be designed in order to meet certain requirements also in time. Charging and de-charging time as well the capacity is related to these requirements. Whereas for electrical storage, the response can be in the subsecond range, thermal storage is characterized by thermal inertia. Typical short-term thermal energy storage therefore shall be fully charged and discharged within a few hours. We can make classification of storage types:

- Short-term storage with a capacity in the range of few hours,
- medium-term storage, with a capacity in the range of several days,
- long-term or seasonal storage, with a capacity in the range of months.

Long-term thermal storage can be operated with high efficiency in the low-temperature range for local heat supply, while in the high-temperature range the efficiency is not very appropriate due to the high storage losses.

8.2.4 Storage Heat Losses

In most cases the storage material is enclosed in a container and has an elevated temperature. Certainly, that is the case with sensible storage where the storage principle is that addition of heat to the material results in a temperature rise. However also for latent heat storages the temperature of the phase-change (evaporation–condensation for liquid–gas or melting–freezing for liquid–solid changes) may very well be above the ambient temperature around the storage. And in chemical storage reaction enthalpies may create elevated temperatures as well. Therefore, in all of these cases a temperature differential to the ambient environment is present, which results in a heat loss. Thermal insulation of storage material containers thus is an important issue. The dominant part of that can be estimated by taking the geometric shape of a container (say a cylinder), calculate the envelope area and estimate the thermal resistance of the casing. Thermal insulation materials add a thermal resistance to the pure surface heat transfer resistance.

Thermal bridges add also to the overall U-value measured in $(W/(m^2K))$ of a storage. There are geometric thermal bridges related to edges and corners. Here, exact 3D-thermal heat loss simulation may quantify more exactly the overall U-value than just geometric areas and thickness of thermal insulation. But, in addition natural convection within the storage or convective heat transfer across the storage boundaries towards the ambient may increase thermal losses. The fluid in inlet and outlet ducts usually cools down faster as the main body of the thermal storage container. Then the cooler fluid may start convecting due to natural convection i.e. density differences, if the ducts are not shaped in a way to prevent that.

The heat loss is proportional to the overall thermal conductance U times the surface area A of the storage in (m^2), which is called UA-value (W/K), and the temperature difference between the storage inside (T_{stor}) to the ambient (T_{amb}). Thus, the heat losses of a thermal storage unit can be determined from:

$$Q_{loss} = M \cdot \frac{d\left(c_p \cdot (T_{sto} - T_{amb})\right)}{\mathrm{dt}} = -U \cdot A \cdot (T_{sto} - T_{amb}), \tag{8.5}$$

in which the product $(U \cdot A)$ determines the heat flow Q_{loss} from the storage tank. The solution of Eq. (8.5) is

$$T_{sto} - T_{amb} = (T_{sto} - T_{amb})_{t=0} \cdot \exp\left(-\frac{U \cdot A}{M \cdot c_p} \cdot t\right), \tag{8.6}$$

in which $(T_{sto} - T_{amb})_{t=0}$ denotes the initial temperature difference of the storage tank compared to the ambient temperature T_{amb} which is assumed constant. The exponential drop in the temperature difference is common to all single-phase heat storage tanks and can be applied in the same way to low-temperature storage tanks for the provision of domestic hot water. The time constant $(t_{1/2})$ of this storage tank, i.e. the time in which the existing temperature difference drops to half of the initial value due to insulation losses, is given by the following relationship,

$$t_{1/2} = \frac{M \cdot c_p}{U \cdot A} \cdot \ln 2, \tag{8.7}$$

and is called storage half-life.

8.2.5 Economic Assessment

For the design of a storage system storage capacity and charging/discharging power are the most fundamental parameters, and hence a basic economic assessment is looking at the storage material costs in €/kWh, the construction of the storage volume also in €/kWh, and at the cost of the heat exchanger system in €/kW.

These are the most important cost drivers in TES, and a selection of cost-effective materials is key for a good economical design. The numbers are, however, no values which can be taken simply from literature, but have to be seen in the overall design concept and the engineering of the system. For instance, a doubling of the storage temperature spread from 100 to 200 °K using the same sensible storage material does half the material cost for a certain storage capacity. This may go along with a choice of more expensive materials in other parts of the plant.

8.2.6 Storage Materials

For all storage materials one of the most important parameters is the energy density, either volumetric in kWh/m^3, or gravimetric in kWh/kg. This relates to the heat capacity and the included phases changes within the temperature interval considered. Often thermal conductivity plays a role, especially for solid materials, because then it determines and usually limits the rate of charging and discharging. Cost is important for the economic assessment. In many cases the storage material cost is the dominant part of the overall system cost for the construction of a TES system. A generic overview of several sensible storage materials is given in Table 8.2 and for some phase-change materials in Table 8.3.

Thermal stability and other reaction process like oxidation or reduction of components are crucial, as they determine the operation temperature range and other design parameters of the thermal storage system.

Beside the storage type and the charge/discharge characteristics the storage media must satisfy several other aspects. These are these:

- High storage capacity. With economical dimensions, a storage unit should store as much energy as possible to keep space requirement low.

Table 8.2 Thermophysical data of some potential single-phase storage media as well as their volume-specific heat capacity, weight-specific costs and costs per thermally generated kWh from (Barlev et al., 2011)

Fluid	T_{min} [°C]	T_{max}[°C]	ρ [kg/m^3]	λ [W/(mK)]	$(c_p \cdot \rho)$ [kWh/(m^{3}K)]	Cost/kg (US$/kg)	Cost/kWh $(US\$/kWh_{th})$
Water	50	140	1000	0.67	1.161	10^{-4}	
Mineral oil	200	300	770	0.12	0.556	0.3	4.2
Synthetic oil	250	350	900	0.11	0.575	3.0	43.0
Silicone oil	300	400	900	0.10	0.525	5.0	80.0
Nitrite salts	250	450	1825	0.57	0.760	1.0	12.0
Nitrate salts	265	565	1870	0.52	0.831	0.5	3.7
Carbonate salts	450	850	2100	2.0	1.050	2.4	11.0
Sodium	140	550	850	71	0.307	2.0	21.0

- Fast and efficient transfer of heat during charging (storage of heat) as well as during de-charging (withdrawal), which for many storage concepts requires good thermal conductivity.
- Abundant availability and costs to ensure competitiveness.
- Charge cycle stability, compatibility to construction and heat exchanger materials. Often, the solar supply implies that the storage tank can only be partially charged, or full loads follow several times in quick succession. The transfer fluid and storage material must retain its thermophysical properties without degradation under different load scenarios and at the same time be compatible over the entire temperature spectrum with the structural materials used for the components (such as pump, heat exchanger, tank filling material). This is because corrosion or layer formation significantly change the storage capacity as well as the temporal storage characteristics.

8.2.7 TES System Configurations

Thermal energy storage technologies are classified also in terms of applied process and loading method meant to direct thermal storage and indirect thermal storage.

Table 8.3 Thermophysical data of some potential phase change storage media as well as their volume-specific heat capacity and indicative weight-specific costs and costs per thermally generated kWh

Fluid	T_{min} [°C]	T_{max} [°C]	$\rho[kg/m3]$	Δh_m Δh_b [kJ/kg]	k [W/(mK)]	$(c_p \cdot \rho)$ [kWh/(m^{3}K)]	Cost/kg (US$/kg)
Water	50	140	1000	333 2257	0.67	1.161	10^{-4}
Sodium nitrate $NaNO_3$	306	380	2260	182.3 –	0.50	0.857	0.2
Potassium nitrate KNO_3	334	750	2110	267 –	0.50	0.914	0.3
Potassium hydroxide KOH	360	1320	2040	150 –	0.50	0.482	24
Salt ceramics ($NaCO_3$-$BaCO_3/MgO$	500 – 850	–	2600	420 –	5.00	2.160	17
Table salt $NaCl$	801	1465	2170	7035 6604	5.00	0.514	1.2
Sodium carbonat Na_2CO_3	851	1600	2530	520 –	5.00	0.194	2.6

- Direct storage: heat transfer medium and storage medium are the same, implying that no heat exchanger is needed.
- Indirect storage: the heat transfer medium is different from the storage medium, and a heat exchanger is needed.

Indirect storages always have a role when there are specific reasons to justify the costs of a heat exchanger. The heat transfer medium might be more expansive due to some additional requirements (freezing properties, viscosity, pump compatibility etc.). Also, when dealing with food and beverages on the demand side, the risk for contamination is reduced by adding a heat exchanger.

Also, solid materials might be used as heat transfer medium (e.g. solid pellets, granules or powder, sand, but in most cases fluid materials are used (liquids and gases) which are easier to transport and pump.

An additional feature of storage systems is the circulation of the storage medium:

- Active systems: when the storage medium is circulated through the storage system.
- Passive systems: when the storage medium does not circulate.

When the storage medium is a fluid and is able to flow between the tanks, the systems are referred to as active type systems. If the storage medium is also used as the heat transfer fluid, the system is referred to as direct-active system. An additional heat exchanger is needed when the storage fluid and heat transfer fluid are different, and the unit is referred to as indirect-active type. In cases where the storage medium is solid, the HTF passes through the storage material only for charging and discharging. Such system is called a passive type.

In an *active storage system*, the storage material itself circulates between the heat exchangers for energy transfer. The employed heat exchangers can also be solar receivers or steam generators.

In *passive storage systems*, the HTF carries energy received from the energy source to the storage medium during charging and receives energy from the storage system when discharging (these systems are also called regenerators). The arrangement for the HTF to flow through the storage medium is a major parameter that dictates heat transfer in the unit. When the heat transfer fluid is a liquid and the heat capacity of the solid in the storage system is not negligible, the system is called a dual storage system. Passive storage systems may utilize solids such as rocks, sand or concrete for sensible heat storage materials, or phase change materials for storing thermal energy.

8.2.8 Storage Charge and Discharge

Charging is the process during which energy is transferred or supplied to the thermal storage system. This definition can be qualified depending on the acquired storage level:

- full charge: charging ends when the storage level is 100%;
- partial charge: charging ends when the storage level is below 100%.

Discharging is the process during which energy is transferred or extracted from the thermal storage system. This definition can be qualified depending on acquired storage level:

- full discharge: discharging ends when the storage level is 0%;
- partial discharge: discharging ends when the storage level is above 0%.

The thermal power for charging or discharging is also an important parameter of a storage system. It is specified in kW and may not be constant over time.

8.2.8.1 Liquid HTF

Charging and de-charging of thermal storages considers generally an active transport of a medium from and to the storage. However, in certain cases also passive discharging and charging might be possible. For example, a massive concrete wall in a greenhouse or conservatory may be charged passively by the solar radiation being absorbed at the surface of the wall, and de-charged by convection and thermal radiation loss to the ambient air and room inventory. In general, however, a heat transfer medium is pumped or ventilated to the storage, where it may release or take up heat and return back. The hydraulic loop from the solar collector may be separated from the hydraulic loop connecting the storage to the demand, but these circuits also can be coupled. For liquid HTF we distinguish between direct and indirect storage systems as described above.

8.2.8.2 Steam as HTF

In case steam is considered as HTF, then the isothermal evaporation and condensation property of the two-phase system water-steam is of essential importance. Assume that some heat shall be transferred to a storage fluid from dry steam (steam quality$\alpha = 1$, cf. see Sect. 5.2) at a pressure levelp_1. Then the temperature which can be reached is the saturated steam temperature $T_{sat}(p_1)$ minus a temperature difference related to the heat exchanger (HX), typically$\Delta T = 10 - 20°\text{K}$. If the storage is a sensible heat storage, then the storage temperature will have increased from a starting temperature T_0 to the end temperature$(T_{sat} - \Delta T)$. In the converse mode the storage fluid with temperature $(T_{sat} - \Delta T)$ will heat up water and transfer heat to the water for evaporation. When the steam is fully evaporated in the HX then the liquid should be cooled down to starting temperature T_0 again. But, as the temperature difference between liquid and steam in the isothermal

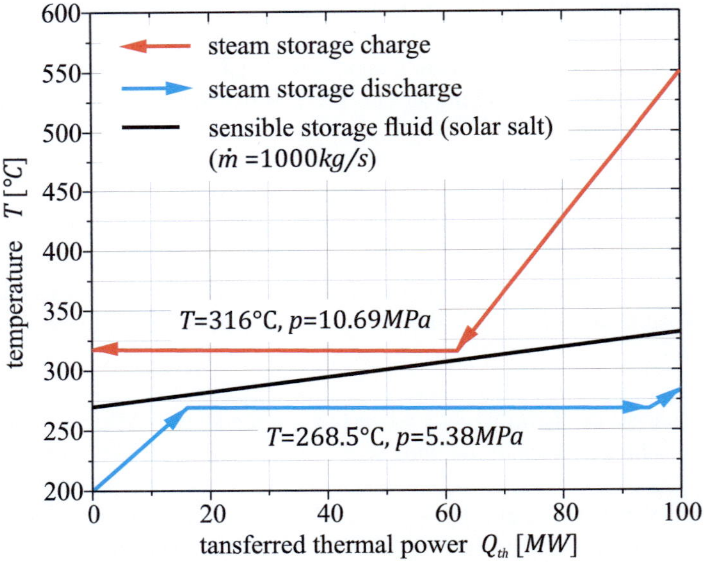

Fig. 8.5 Comparison of temperature levels for steam storage charge and discharge compared to a sensible heat liquid storage due to the pressure–temperature specifity of steam in the evaporation region

region always has to be positive with the ΔT as above, the resulting evaporation temperature will be below$T_{sat}(p_1)$, necessarily at a different pressure p_1 at $T_{sat}(p_1)$=$T_{sat}(p_1) - 2 \cdot \Delta T$. A numerical example is shown in Fig. 8.5.

However, when a phase-change material (PCM) can be found with a melting temperature just below the upper steam temperature, then both processes—evaporation and crystallizing of PCM as well as condensation and melting of PCM—may occur isothermally. This leads to a resulting steam temperature and pressure at discharge which is much higher than when using sensible storage. A corresponding numerical example is shown in Fig. 8.6.

8.2.9 Storage Efficiency

The storage efficiency is the ratio between the energy gained by the heat transfer fluid, in a full discharge process, and the energy supplied to the thermal storage system, in a full charge process. The charge and discharge processes should be consecutive, so that heat losses over time are not included.

Several factors can lead to efficiencies below 100%. First of all, thermal capacities are a source for efficiency losses. During a charge process also, solid container walls and construction elements are being heated up. During a discharge process not all this energy may be transferred back to the fluid at an adequate temperature. Secondly, in a heat exchanger of an indirect storage a certain temperature

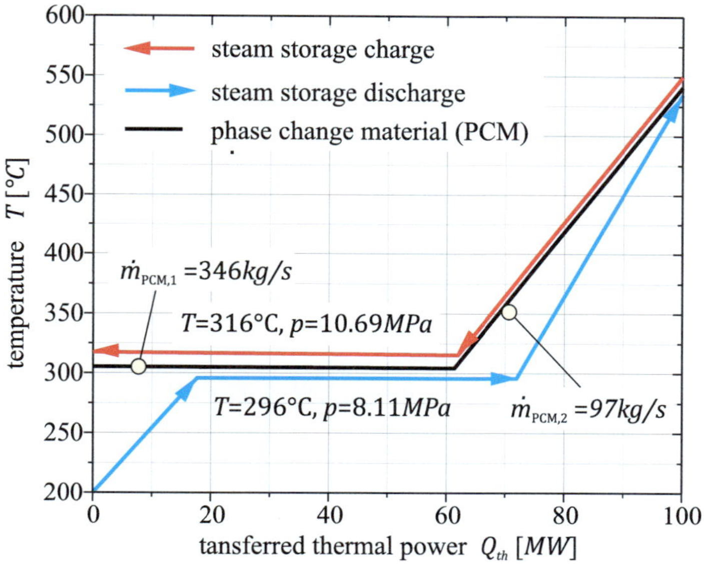

Fig. 8.6 Temperature and pressure levels for charging and de-charging of phase-change materials (PCM) by steam in the evaporation region (with a melting temperature between condensation and evaporation temperature of the steam

difference, say 10 °K is required to effectively transfer heat. That means that due to charging and discharging of this temperature difference is lost twice, resulting in less sensible discharged heat than sensible heat needed for charging. For steam charging and discharging not only the temperature, but also the pressure difference of input and output steam plays a role.

It is obvious that the storage efficiency not only depends on materials and temperature levels, but also on the exact operation of the charging and discharging process. Especially the rated thermal power of the process can influence the efficiency.

This efficiency of a thermal storage unit is expressed by the so-called round-trip efficiency (η_{stor}). Here, it is important to pay attention to what is used as the reference value for the evaluation. In a "heat-to-heat" evaluation, which is mainly used in domestic applications, the round-trip efficiency describes the ratio between the energy input to the storage (before storage) and the energy output of the storage (after storage) and is given by Eq. 7.8

$$\eta_{stor} = \frac{Q_{th,discharge}}{Q_{th,charge}}. \tag{8.8}$$

The absolute value of η_{stor} depends on how efficient the different conversion steps are; with good thermal insulation, values well above 0.9 can be achieved. In a "heat-to-electricity" evaluation, the efficiency at which electricity is generated in relation to the efficiency at nominal load must also be considered. Heat is usually less efficient at partial load than at full load.

8.3 Evaluation Criteria

For the evaluation of a thermal energy storage system (TESS) a whole range of criteria can be given.

8.3.1 Technical Criteria

Beside the storage type and the temporal appropriation ability the storage media must satisfy several other aspects. Important quality criteria for a thermal energy storage are referring to (Hadorn 2005a, b).

- capacity and density (kWh/m^3 = kWh/kg kg/m^3),
- charging and discharging rate (kW),
- efficiency of storage (Q_{out}/Q_{in}, $Q_{out} = Q_{in} - Q_{loss}$),
- stability (mechanical, chemical),
- chemical compatibility with the container and exchange loop,
- reversibility during a number of cycles,
- cost (€/kWh): material, container, room needed, access, maintenance, number of cycles achieved,
- toxicity of the material,
- recyclability, assessed through a Life Cycle Analysis (LCA).

The storage material is not the only component of a storage. The heat exchanger plays a crucial role since heat has to be injected into the store when it is available (say noon during a solar day) and withdrawn from the store at any time even in peak load conditions. The heat exchanger should therefore also satisfy some basic criteria:

- rating (max power in and out),
- hydraulic pressure losses,
- cost (€/kW).

High capacity and density will lead to more economical dimensions of the container. Good thermal conductivity of the storage material is a main criterion which supports a fast and efficient transfer during charging (loading/storing of heat) as well as during discharging (unloading/withdrawal of heat) with reasonable equipment.

The material availability and costs are the main factor to ensure competitiveness, especially for large storages, because they constitute a large part of the investment.

Reversibility or charge cycle stability is related also to material compatibility. Often, the solar supply means that the storage tank can only be partially charged, or full loads follow several times in quick succession. The transfer fluid must retain its thermophysical properties without degradation under different load scenarios and at the same time be compatible over the entire temperature spectrum with the structural materials used for the components (such as pump, heat exchanger, tank filling material). This is because corrosion or layer formation significantly changes the storage capacity as well as the temporal storage characteristics.

8.3.2 Environmental Criteria

The materials and the operational practices used for a TESS shall not have negative impacts on human or animal health or generally on the natural environment and ecology. Materials shall not be toxic or otherwise dangerous, e.g. explosive. The environment should not be adversely affected during extraction (e.g. mining), manufacture, distribution, installation, operation, and dismantling, from cradle to grave. Recycling of materials should be self-evident after dismantling a plant. A sustainable way of designing thermal systems has to balance well risks and benefits. If risks are identified, then specific prevention measures might help to mitigate them. For instance, the risk of a leak in a system containing mineral or synthetic oil which be disastrous for the potable water reserves can be minimized by constructional measures (a container collecting possible leakage oil), by operation and maintenance measures (regular inspections) and by control systems (detector for pressure loss).

8.3.3 Economic Criteria

An economic justification for a storage system normally requires that the annualized capital and operating costs for the system with TES be less than those without TES, supplying the same loads and periods. Often TES systems are designed in a way that the higher capital costs are more than compensated by higher generation (which is especially true for solar systems) and hence lower cost per energy supplied.

The key parameters involved in the evaluation of the cost effectiveness of a TES system include:

- hourly load profile of the heat demand,
- hourly generation profile of the solar system,
- rated capacity and rated power of the TES system,
- control system for charging/discharging of the TES.

The economic information needed includes:

- cost for the alternative heat supply per kWh (if necessary, with different cost levels for specific times),
- investment and operation costs for the TES system,
- financial incentives and interest rate for capital.

Economic evaluation in its simplest way calculates the payback period for a TES system installed. However, this depends very much on financial parameters as the conditions for loans. A more substantial comparison also considers the risks of the existing system without TES and compares levelized annual heat costs of the solution with and without TES. Fuel or electricity prize increase should be estimated taking into account the future policy changes with respect to non-renewable energies and climate change (CO_2 emission costs).

Well-designed TES systems may help to reduce the dimensions of heat supply systems and thus reduce investment costs there. Often the gaps between demand hours and preferential generation can be closed or mitigated. This is true for solar systems where generation is only possible during the day when the sun has sufficient power, but is may also be relevant for other energy sources. Fluctuating generation can be smoothened out and TES may serve to provide the heat at the right time with the right power. Applications and processes can operate in a chosen design point with better efficiency and possibly also with better stability and durability. Control issues are reduced substantially when using a storage.

If due to a better and smaller design of the heat transport system ducts, pipes, heat exchangers can be reduced in size, also operating costs due to pumping or ventilation might be lower.

In order to evaluate this, the complete design of the thermal processes used is sometimes necessary.

In buildings especially, the difference between active and passive storage is important. Passive storage (e.g. in thermally activated building mass) might react much slower and with less possibilities to control the release of heat. Therefore, the prediction of demand and supply is important for the correct operation of the system.

Thermal design of heat distribution is connected to the temperature levels which have to be provided by a storage. The lower the supply temperature the larger the storage capacity for a specific volume (if sensible heat storage is used). In these cases, an optimization of temperature levels is required to optimize the sizing of the storage and the distribution system in an economical way.

8.3.4 Integration

When considering the integration of a TES into an existing thermal facility or into a solar system, several options usually exist. Direct and indirect storage, active and passive storage, sensible or latent heat storage, single storage, or multi-storage solutions, central or decentral solutions, material choices can be included in the selection and design process. In most cases a deep knowledge on the specifications of the application is needed.

- Which power and energy demand profiles have to be satisfied with the TES system?
- Are there compatibility and process requirements on the storage?

For instance, in order to heat food and beverage in industry, stainless steel heat exchangers are needed to prevent the contamination of the food. The heat exchanger size has to be chosen according to the allowed temperature peaks in a process—if the temperature limits are small then large heat exchanger areas are required which adds to the cost. For the storages themselves the heat exchanger designs of internal heat exchangers often can be very specific, and that often has an impact in storage efficiency, for example due to improper stratification, i.e. mixing of temperature levels.

Another aspect of integration is not the integration of TES into a heat supply system, but the architectural integration of a TES in a building. The storage may be located inside or outside a building. Inside the building special service rooms could hold the storage, and floors should be stable enough to carry the weight of the storage. However, seasonal storages for house heating have been also incorporated in the precious living area of houses, and in this way heat losses of the TES contribute to the heating of the living area.

Outside the building positioning in open air or below in the ground are possible options. Especially for seasonal storages in heating grids pit-hole thermal energy storage provides a very economical way of constructing such a large volume for TES.

8.4 Sensible Heat Storage (SHS)

For sensible heat storage solid and liquid media are being used, because here the heat capacity is high. Gases have a low density—even when compressed—and hence the product of specific heat and density is much smaller than for liquid and solid materials. The difference in liquids and solids lies mainly in the different possibilities to transfer the heat to and from the storage media. In solid materials thermal conductivity is the key parameter characterizing the heat transfer, whereas in liquid media convection may occur and be enforced to guarantee a high heat transfer rate. Generally, liquid media can be charged and de-charged much faster than solid media, but dimensions of the solid volumes are important. If a heat transfer fluid flows through a tank filled with small solid particles the time necessary to heat up or cool down the solid storage can be minimized.

8.4.1 Integration of SHS

Integration of sensible heat storage into a system is favoured when large temperature spread is available or can be utilized. The heat content in the storage as given by Eq. (8.1) is proportional to the upper and lower temperature limit of the storage. The upper limit is usually linked to the stability of the storage material (e.g. for a pressure-less water storage we have an upper design limit of typically 90–95 °C, because steam formation should be avoided) or to the capabilities of the heat source. The lower limit is given by ambient temperature and useful temperature in the system. In the example of the hot water storage the lower temperature is given by the tap water temperature, as the tap water replaces the hot water drawn from the storage. In order to provide a decent temperature for a shower or dishwashing, the preheated water may have to be heated up to the demand design temperature, but the energy content of the preheating is reducing the heat amount. If no auxiliary heating is available in the system, then the lower temperature limit for the storage would be the lowest useful temperature needed by the consumer. For an example, in a process heat system requiring hot water circulation to a heat exchanger with say 60 °C hot water demand for cleaning, being serviced only from the storage, the energy content of a hot water tank is zero when the water is at or below 60 °C. Cooling down below that lower temperature limit implies a negative energy content, which has been added to the storage before it can be recharged again, i.e. the "charge state" of a thermal storage may also take negative levels. If, however, there is an auxiliary heating in the storage (say electric heating, then the return temperature of 45 °C from the heat exchanger would be the lower limit. The examples are made quantitatively in Table 8.4.

Table 8.4 illustrates that the maximum capacity of a hot water storage can vary quite significantly depending on the application.

A sensible heat storage can be integrated in a system requiring heat not only in one way. Within complex systems often several temperature levels exist. These levels, the heat source characteristics, the heat transfer options, and the storage material selection influence the positioning of SHS in the system. Several

Table 8.4 Specific energy content of water storage in different applications

Application	Lower temp. limit T_1 [°C]	Upper temp. limit T_1 [°C]	Spec. heat content q [kWh/m^3]
Solar hot water with auxiliary heating	12	90	91.0
Solar hot water w/o auxiliary heating	35	90	64.2
Solar process heat with auxiliary heating	45	90	52.5
Solar process heat w/o auxiliary heating	60	90	35.0

storages might be included at different locations and with different temperature requirements. Charging and de-charging power has to be also considered.

When heat exchanger units are incorporated then the temperature differences driving the heat flow into and out of the storage have to be considered. A sufficiently fast heat transfer rate in a plate-heat exchanger providing hot potable water at 45 °C from a buffer storage may need typically $10° Kelvin$ driving temperature between service temperature and storage temperature. That means that the storage should be able to provide 55 °C hot water.

8.4.2 Liquid Storage Media

Liquid storage is the most common SHS. Hot liquid is stored in tanks which usually are thermally insulated to reduce the heat losses to the environment. From the storage principles we should distinguish three idealized cases:

- fully mixed heat storage,
- fully stratified heat storage,
- variable-volume liquid storage.

In the mixed storage heat is transferred to a liquid contained in a fixed storage volume, and the liquid due to internal convection is ideally at the same temperature.

8.4.2.1 Stratification Within a Storage Tank

In the ideally stratified storage the liquid has two temperature levels. There is a hot partial volume at the upper temperature limit of the storage, and a second cold part with the lower temperature level. Heat is transferred to the fluid just below the hot part (as the hot fluid due to density differences is above the cold one) at the hot temperature level. In this way the volume fraction of the hot part increases. Obviously, this is an idealized concept. Firstly, there is a heat exchange by conduction and maybe even convection between the hot and cold part of the storage leading to a temperature equalization. In practice there is a so-called thermocline between hot and cold fraction with intermediate temperatures. This thermocline zone will increase in width with time. If convection is induced in the liquid storage, then this stratification might be even destroyed, and a mixed state of the storage is observed, see cf. (Haller et al. 2009c).

Secondly the temperature provided at charging may deviate from the ideal design temperature of the hot storage part. Still stratification with a thermocline zone is the result, as temperatures below design temperature would be fed in at a lower lever into the thermocline if possible—this requires a sophisticated charging strategy. We will discuss stratified charging in detail later in the chapter.

Stratified storage has a higher exergy content than a mixed storage. We take as an example a water storage in an environment with 20 °C ambient temperature. A solar thermal system provides 60 °C hot water to the storage. In a fully stratified

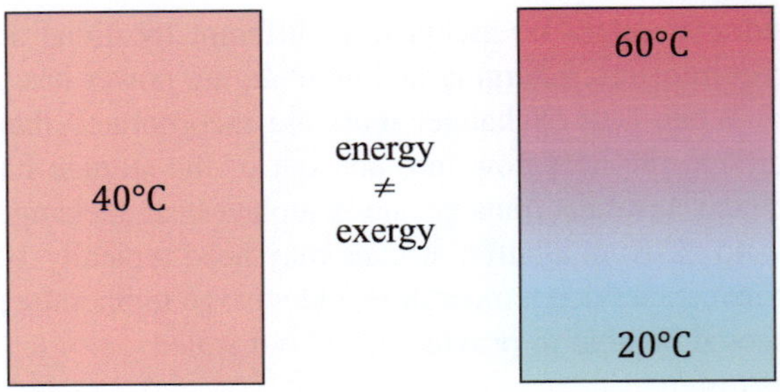

Fig. 8.7 Difference between a mixed and a stratified storage temperature distribution, having the same energy content but different exergy content

storage we consider the case where 50% of the volume is at the cold temperature 20 °C and 50% at the upper temperature of 60 °C. Then the energy content is the same as for a fully mixed storage with the temperature of 40 °C (see Fig. 8.7).

The energy and exergy calculation for the two situations is as follows:

$$Q_{strat} = M \cdot c_p \cdot 50\% \cdot \left(333\,^\circ K - 293\,^\circ K\right),$$
$$Q_{mix} = M \cdot c_p \cdot \left(313\,^\circ K - 293\,^\circ K\right). \tag{8.9}$$

$$Ex_{strat} = M \cdot c_p \cdot 50\% \cdot \left(333^\circ K - 293\,^\circ K\right) \cdot (1 - 293/333);$$
$$Ex_{mix} = M \cdot c_p \cdot \left(313\,^\circ K - 293\,^\circ K\right) \cdot \left(1 - \frac{293}{313}\right). \tag{8.10}$$

In our example although the energy content of the two storages is identical and amount to 23.3 kWh/m^3 for the temperatures indicated in Eq. (8.8) and Fig. 8.7, the exergy content is different. The stratified storage has nearly double the exergy content of the mixed storage, due to the overall low temperatures 2.8kWh/m^3 versus 1.5 kWh/m^3. The difference becomes smaller relatively when the percentage of the hot water is increased. For 100% hot water obviously both storage conditions lead to the same fully charged 60 °C storage.

A second advantage of the stratified storage is that the solar collectors are working at more convenient conditions. The return flow from the storage to the collector is at a lower temperature level and therefor the collector heat loss is lower, leading to a higher collector efficiency.

For many solar thermal systems water is the ideal material for storing heat. On the one hand, side water is cheap, and it has the highest specific heat capacity of all materials. On the other hand, energy can be added to the store or withdrawn by transporting the storage medium itself, without the need of a heat exchanger. Usually this is utilized either for the collector loop or for the demand side. Although water might be the basis of heat transfer on both sides, usually the quality of the water is different. On the demand side often drinking water is needed, whereas on the collector side additional substances like anti-freeze or anti-corrosion substances are added. But where no heat exchanger is necessary the additional temperature drop is eliminated.

In order for a storage tank to maintain its thermal stratification, a number of constructive measures are necessary. Stratification can be obtained with different storage designs and configurations of heat exchangers (Fertahi et al. 2018; Fernández et al., 2021). Most important is that as little convection as possible is induced by the heat transfer to the storage. The kinetic momentum of the water stream has to be taken out by diffusers, deflection, or baffle plates. In a potable water tank where the hot water is tapped from the hot top, the cold-water inlet pipe is at the bottom of the storage, at the lowest temperature level. In Fig. 8.8 the situations of different inlets are shown as simulated with a CFD model. In a simple vertical pipe inlet, the cold water has a momentum and intrudes far into the hot zone of the store. Such an inlet can only be used with very small velocities, otherwise stratifications is destroyed easily. A fixed plate (right hand side) forces the water to change direction from vertical to horizontal movement and leads to less mixing of cold inlet and shot storage water. For optimizing the flow other inlet nozzles for taking out the vertical momentum can be designed. One is the so-called Metro inlet in the centre part of the Figure.

A second constructional parameter is the aspect ratio height to diameter of a tank. The recommended height-diameter ratio is 2:1 to 4:1. Reclining horizontal tanks are not favorable for stratification. Tank walls are usually a few degrees colder than the hot water, and therefore induce natural ventilation loops which destroy stratification. The form of the container tank may support or reduce such movements. Thirdly, special attention must also be paid to the destruction of the stratification at the hot water withdrawal point, which may be caused by water cooled in the outlet piping circulating back into the storage tank (single-pipe circulation) when no water is drawn. To prevent these heat losses which may reach 15–20% of the total storage capacity, a good solution is that the hot water outlet

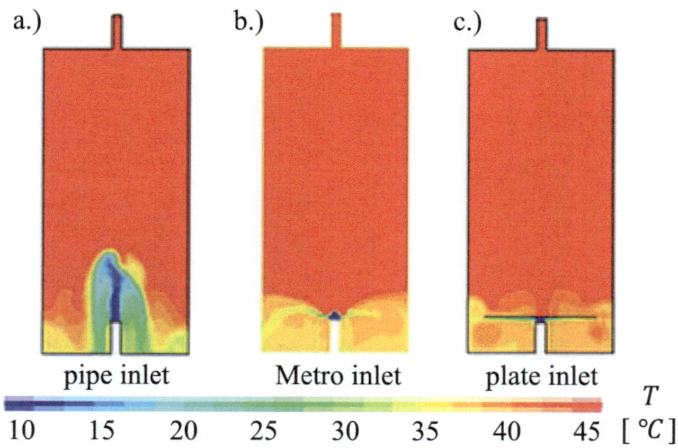

Fig. 8.8 Convection induced by different inlet configurations in a water tank; **a** pipe, **b** Metro and **c** plate

should be provided with a 180° pipe bend (siphon). Thermal bridges should also be avoided as far as possible in all other pipe connections of the storage tank. If possible, the pipe connections should run in the cold—i.e. lower—area of the storage tank.

Of course, the partial mixing of the cold water with the tank volume depends very much on the duration of the water flow and the velocities in the inlet section. Figure 8.9 shows an example where within a draw period 50% of the storage volume has been replaced by cold water. The right idealized stratified storage shows the 50% hot and cold-water sections clearly separated. The effect of partial mixing of different inlets after a 30 second period after the draw to let the water come to a rest now can be seen in the central 3 pictures. Plate inlet and Metro inlet have still a large volume fraction of hot water in the tank separated from the cold water with a thermocline region, but the pipe inlet has led to a substantial mixing of the two temperature levels in the lower 3 quarters of the tank.

Stratification might also be destroyed by charging the storage, not only in the de-charging situations. When heat is added to the storage ideally the heat should be added to the zone in the tank with the same temperature as the charging heat transfer fluid. The charging from the collector loop via a heat exchanger however also has to take place in the lower part of the storage as otherwise the heat from the collector loop would be useless for the storage volume below the heat exchanger. The heat transfer is best when the heat exchanger with the hot collector fluid is transferring heat to the cold part of the storage due to the big temperature difference. When temperature become equal, no charging can take place anymore. Heat added in this way via an internal heat exchanger to the lower part of the storage heats up storage water which slowly rises according to the density differences. Similarly, when the heat is transferred by an external heat exchanger to the storage water, the inlet and outlet positions can be chosen accordingly. This slow natural

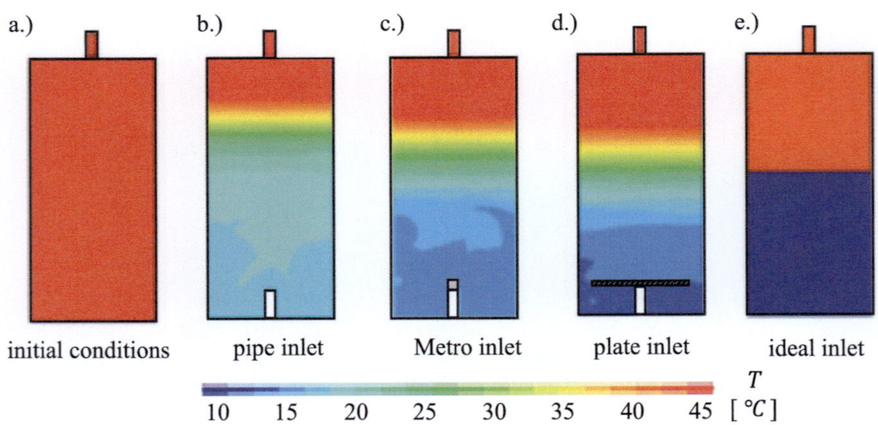

Fig. 8.9 Thermal stratification after a 50% draw-off from a potable water tank (after a 30 second standstill period), for 3 different inlet configurations (as in Fig. 8.8); **a** left: initial temperature conditions before draw, **e** ideal stratification

convection loop due to gravitational force and density differences, however, may be enforced by inappropriately high heat transfer rates. Commercial products try to avoid the natural convection by designing heat exchangers which transfer the heat ideally to the zone of equal temperature. Heated water is introduced into the layer of the storage tank where the temperature of the incoming water is equal to that of the storage tank. This takes place purely passively in a riser pipe (compare storage tank in Fig. 8.40). We will discuss this later in the section of specific storage systems.

The simplest way to describe storage stratification is a graphical representation showing a two-dimensional temperature distribution in the tank as shown in Fig. 8.9. Also, it can be expressed by several *stratification indices* describing the stratification without considering the history of charging and discharging processes. Numerical figures comparing the experimental storage process with a hypothetical storage process and, thus, including information about the history of the storage process are called *stratification efficiencies*, for more details see (Haller et al., 2009a, b).

The following indices have been discussed for quantifying the stratification:

- thermocline width Δz,
- thermocline mean temperature gradients,
- stratification index based on energy or exergy analysis.

A stratification index based on energy can be defined summing over the energy content in the cold and hot zone excluding a thermocline zone as

$$I_{strat,energy} = \frac{\int_{z_{cold}}^{z_{+15\%}} m(z') \cdot c_p \cdot (\mathrm{T}(z') - T_{cold}) dz' + \int_{z_{-15\%}}^{z_{hot}} m(z') \cdot c_p \cdot (T_{hot} - \mathrm{T}(z')) dz'}{m_{sto} \cdot c_p \cdot (T_{hot} - T_{cold})}, \quad (8.11)$$

where $z_{+15\%}$ is the height above the cold zone where the temperature $T_{cold} + 15\%(T_{hot} - T_{cold})$ is reached. Similar, $z_{-15\%}$ is the height below the hot zone where the temperature $T_{cold} + 15\%(T_{hot} - T_{cold})$ is reached, which is illustrated in Fig. 8.10.

A second law stratification index based on exergy may be defined by calculating the exergy for the actual temperature profile A_{actual} as well as the exergy for ideally mixed and stratified profiles A_{mix} and A_{step} as indicated in with the same energy content of the storage. Then the exergy-based index can be defined as

$$I_{strat,exergy} = \frac{A_{actual} - A_{mix}}{A_{step} - A_{mix}}. \quad (8.12)$$

Stratification efficiencies based on the first law of thermodynamics most often calculate a fraction of energy that is recovered with a certain charging

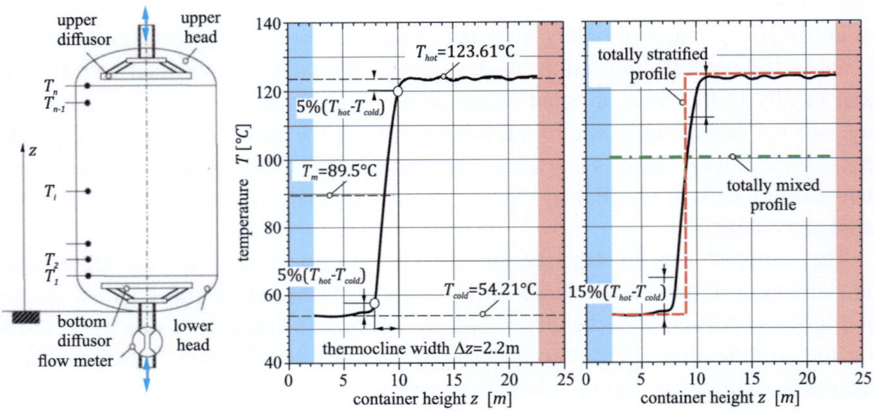

Fig. 8.10 Heights for calculation of the stratification index based on energy analysis with totally mixed, totally stratified and actual temperature distribution modified from (Fernández et al., 2021)

and discharging procedure with fixed inlet temperature and mass flow. Between the charging and discharging, a storing period may, or may not, be included. Equation (8.13) is one example for defining a charging efficiency over time.

$$\eta_{ch}(t) = \frac{m_{sto} \cdot c_p \cdot (T_m(t) - T_{ini})}{m_{sto} \cdot c_p \cdot (T_{\text{inlet}} - T_{ini})} = \frac{T_m(t) - T_{ini}}{T_{\text{inlet}} - T_{ini}}, \tag{8.13}$$

where $T_m(t)$ is the mean time dependent in the storage and T_{ini} the initial temperature.

High stratification behavior is particularly important when a lot of time is needed to circulate the storage volume. This is the case with low-flow systems as well as with large storage tanks and long-term storage systems. In principle, when loading the storage tank, care must be taken to ensure that there is no mixing with water of lower temperature. External heat exchangers provide usually good stratification behavior with small zones of mixing, provided that the mass flows are low. Domestic hot water (DHW) with a useful temperature is quickly available with the appropriate temperature spread in the collector.

In order for a storage tank to maintain its thermal stratification, a number of constructive measures are necessary. The flow at the inlet to the storage tank is calmed down by means of built-in components, so that mixing of the incoming cold water with already heated water is avoided and the stratification is not disturbed. Since the mixing area is smaller than without inlet calming, this also results in a higher stratification coefficient N. If the incoming cold water is not slowed down, water located further up cools down and the corresponding layers must be reheated. The lowest storage layers are slightly heated by the turbulence. As a result, the collector efficiency decreases, and the solar circuit switches off earlier when the irradiation decreases.

The recommended height-diameter ratio is 2 : 1 to 4 : 1. Reclining tanks should not be used in conjunction with a solar system. To maintain temperature stratification in the storage tank, the following rules should be observed for the cold-water inlet and the hot water withdrawal point: Domestic hot water storage tanks should in any case have a baffle plate at the cold-water inlet to avoid swirling of the incoming cold water with warmer water from higher layers. Special attention must also be paid to the destruction of the stratification at the hot water withdrawal point, which is caused by the cooled water sinking from the pipes back into the storage tank (single-pipe circulation). To prevent these heat losses of up to 15% of the total storage capacity, the hot water outlet should be provided with a 180° pipe bend (siphon). Thermal bridges should also be avoided as far as possible in all other pipe connections of the storage tank. If possible, the pipe connections should run in the cold—i.e. lower—area of the storage tank.

8.4.2.2 Modelling of Temperature Stratification Within the Storage Tank

We have seen that temperature stratification within the storage tank is an important question in the design and dimensioning of the storage tank, because cold, warm, and hot water are simultaneously present in the storage tank. Due to the different density of the water and the gravity force, the hot water is always on the top. A lower zone that is as cold as possible often provides the return temperature for the solar system which then can operate at a low temperature level with good efficiency, even with little solar radiation. This temperature stratification is more pronounced for slim and high storage tanks. The preservation of thermal stratification is of great importance for the efficiency of a solar system.

To simulate the thermal integration of the storage tank into the thermal solar system, mathematical models are required to describe the thermal–hydraulic effects within the storage tank. As always, such a model is based on certain assumptions that represent a simplification of the actual physical facts. Accordingly, the model may only reflect the performance of the real storage tank more or less. The requirements for models depend entirely on the application. For annual simulations of systems, it is sufficient if the model produces the correct outlet temperature at a given time. It is sufficient to determine a one-dimensional axial temperature distribution in this case. Although physically in hot water storage tanks natural convection is a dominant phenomenon, a simple one-dimensional temperature distribution in hot water storage tanks with heat losses mainly uses the equations for transient two-dimensional heat conduction with simple modifications.

In order to describe the temperature stratification in a storage tank, the simplest case of a mixed storage tank is considered first. Here the storage tank temperature is considered independent of the location. In terms of temperature distribution, it therefore differs considerably from a real storage tank, as sketched in the left part of Fig. 8.11. When loading an ideally mixed hot water tank via an internal heat exchanger, the heat balance at the tank provides the following result:

$$\frac{d\dot{Q}}{\mathrm{dt}} = \dot{m}_{\mathrm{HX}} \cdot c_p \cdot \left(T_{1'} - T_1^{''}\right) - (U \cdot A)_{\mathrm{sto}} \cdot (T_{\mathrm{sto}} - T_{amb}), \tag{8.14}$$

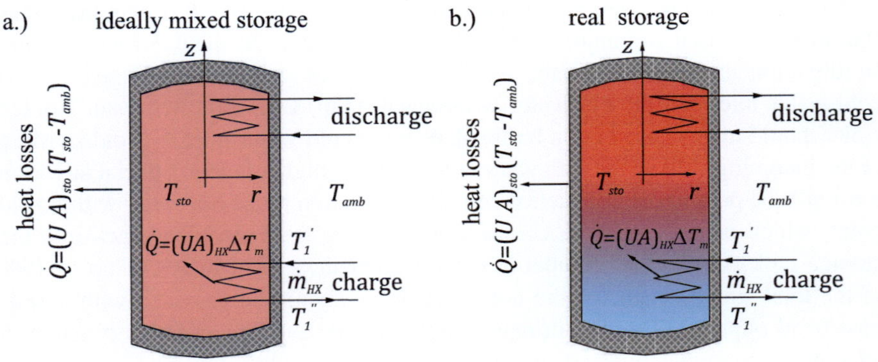

Fig. 8.11 Qualitative comparison of the temperature of an ideally mixed tank (**a**) and a real tank (**b**)

where U is the heat transfer coefficient, T_{sto} is the average storage tank temperature and T_{amb} is the ambient temperature. If the ambient temperature T_{amb} is constant, the material values and c_p and the heat transfer coefficient U are independent of the temperature and the heat capacity of the fluid in the heat exchanger can be neglected, an analytical solution exists.

The heat balance in the heat exchanger then is written as:

$$\dot{m}_{HX} \cdot c_p \cdot \left(T_1' - T_1''\right) = (U \cdot A)_{HX} \cdot \frac{\left(T_1' - T_1''\right)}{\ln\left(\frac{T_1' - T_{sto}}{T_1'' - T_{sto}}\right)}, \tag{8.15}$$

leading to:

$$\ln\left(\frac{T_1' - T_{sto}}{T_1'' - T_{sto}}\right) = \frac{(U \cdot A)_{HX}}{\dot{m}_{HX} \cdot c_p}. \tag{8.16}$$

If a heat exchanger efficiency is defined, one will get the following result:

$$\varepsilon = \left(\frac{T_1' - T_1''}{T_1' - T_{sto}}\right) = \frac{\text{actual heat transfer}}{\text{maximum heat transfer}} = 1 - \exp\left[-\frac{(U \cdot A)_{HX}}{\dot{m}_{HX} \cdot c_p}\right]. \tag{8.17}$$

The expression in brackets in the exponential function is called the number of transfer units "Number of Transfer Unit" (NTU) or "Heat Capacity Current" or "Dimensionless Transfer Area" or "Dimensionless Dwell Time", see Sect. 5.5.3.2 Eq. (5.385). Inserting Eq. (8.17) into (8.14) yields

$$\begin{aligned} & m_{sto} \cdot c_{p,sto} \cdot \frac{dT_{sto}}{\mathrm{dt}} \\ & = \dot{m}_{HX} \cdot c_p \cdot \varepsilon \cdot \left(T_1' - T_{sto}\right) - (U \cdot A)_{sto} \cdot \left(T_{sto} - T_{amb}\right). \end{aligned} \tag{8.18}$$

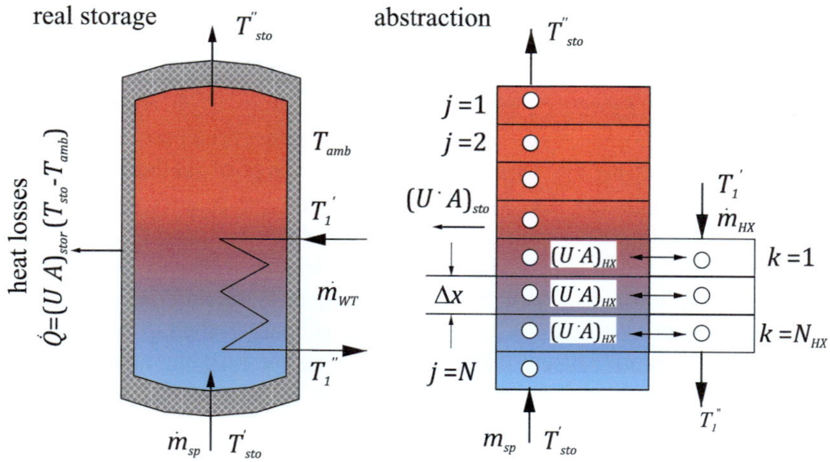

Fig. 8.12 Model of the hot water tank with heat supply via heat exchanger and direct discharge in real configuration (left) and abstraction (right)

The solution is obtained by setting the initial condition of the temperature of the storage tank $T_{sto,0}$ at time $t = 0$:

$$T_{sto}(t) = C_1 + \left(T_{\text{sto},0} - T_{amb}\right) \cdot \exp(-C_2 \cdot t) \;\; \text{with} \quad C_1 = \frac{\dot{m}_{HX} \cdot c_p \cdot \varepsilon \cdot T_1{\prime} + (U \cdot A)_{sto} \cdot T_{amb}}{\dot{m}_{HX} \cdot c_p \cdot \varepsilon + (U \cdot A)_{\text{sto}}}, \quad C_2 = \frac{\dot{m}_{\text{HX}} \cdot c_p \cdot \varepsilon + (U \cdot A)_{\text{sto}}}{m_{sto} \cdot c_{p,\text{ sto}}}. \tag{8.19}$$

For very long times ($t \to \infty$) the storage tank temperature T_{sto} tends towards the value C_1. If the heat losses of the storage tank to the environment are low (i.e. $(U \cdot A)_{sto=0}$), the storage tank assumes the inlet temperature of the loading heat exchanger, i.e. $T_{sto} = T_1{\prime}$.

For real water storages with stratification, models were developed to approximate their operation, in connection with the solar thermal use of the Solar Storage Testing Group (SSTG) (Marshall and Barragan) in which the storage is divided into N_{sto} layers, as shown in Fig. 8.12. Here, the storage is charged by an internal heat exchanger, divided in N_{HX} layers, while the hot water is drawn directly. For each layer in the storage a heat balance is calculated in the following way,

$$m_{\text{sto}} \cdot c_{\text{sto},j} \cdot \frac{dT_{sto,j}}{dt} = \underbrace{\dot{m}_{\text{sto}} \cdot c_{p,sto} \cdot \left(T_{\text{sto},j+1} - T_{\text{sto},j}\right)}_{I} + \underbrace{\xi_1 \cdot (U \cdot A)_{\text{HX},k} \cdot \left(T_{HX,k} - T_{\text{sto},j}\right)}_{II} + \underbrace{\xi_2 \cdot k_{eff,sto} \cdot A_{\text{sto,cs}} \cdot dx \cdot \frac{d^2 T_{\text{sto}}}{dx^2}}_{III}$$

$$- \underbrace{(U \cdot A)_{\text{sto},j} \cdot \left(T_{\text{sto},j} - T_{amb}\right)}_{IV} \tag{8.20}$$

In the Eq. (8.20) ξ_1 and ξ_2 are control variables, k_{eff} is an effective thermal conductivity of the storage fluid and $A_{\text{sto,cs}}$ is the cross-sectional area of the storage tank. The left side of the Eq. (8.20) describes the change of the heat content of a layer segment of thickness Δx. The *term I* represents the heat transport by direct discharge, where $T_{sto,j+1} > T_{sto,j}$ (no inversion of the temperature stratification is assumed). If no water is discharged, this term is zero. *Term II* occurs only in segments that are in contact with the heat exchanger. A logical control variable ξ_1 is introduced for this purpose. When the storage is charged then $\xi_1 = 1$ otherwise $\xi_1 = 0$. Thus, *term II* reflects the heat input from the k-th segment of the heat exchanger into the j-th segment of the storage tank which have to be in the same height level. The *term III* takes into account the heat convection and conduction between the layers within the storage tank, modelled in a simplified approach using an effective thermal conductivity $k_{eff,sto}$ is used, which is larger than the heat conductivity of the fluid. Through $k_{eff,sto}$ it is thus possible to simulate heat transport by convection during charging, if it is chosen very large. In numerical algorithms this often requires small time step sizes and long simulation times. Therefore, often during charging and discharging of the tank this term is neglected. This procedure hence avoids possible inversion layers of temperature stratification, which occur by mixing the layers after each time step. *Term IV* in Eq. (8.20) reflects the height-dependent heat losses of the storage tank to the environment. It is easy to include here also different insulation levels or effective thermal losses by thermal bridges in the respective layer.

In addition to the storage tank, the heat exchanger is also divided into *k-segments*, which feed heat into the corresponding storage layers. The balance equation for the *k-th* heat exchanger segment is

$$\begin{aligned} m_{\text{HX}} \cdot c_{p,\text{HX}} \cdot \frac{dT_{HX,k}}{dt} &= \underbrace{\dot{m}_{\text{HX}} \cdot c_{p,\text{HX}} \cdot \left(T_{\text{HX},k-1} - T_{\text{HX},k}\right)}_{I} \\ &+ \underbrace{(U \cdot A)_{\text{HX},k} \cdot \left(T_{\text{sto},j} - T_{\text{HX},k}\right)}_{II} \\ &+ \underbrace{k_{\text{eff,HX}} \cdot A_{\text{HX}} \cdot dx \cdot \frac{d^2 T_{HX,k}}{dx^2}}_{III}. \end{aligned} \tag{8.21}$$

The structure corresponds to the Eq. (8.20) of the storage tank, here for an internal heat exchanger without the term of the heat losses to the environment. In the case of shell-and-tube heat exchangers exposed to the environment, however, these heat losses must be considered. The axial heat conduction in the heat exchanger is usually negligible (term $III \approx 0$) since convective transport mainly determines the heat transfer. If one neglects the capacity of the heat transfer fluid

$(m_{\text{HX}} \cdot c_{p,HX} << m_{\text{sto}} \cdot c_{p,\text{sto}})$ and uses the average temperature difference instead of the average logarithmic temperature difference, the Eq. (8.21) changes to the Eq. (8.15). The system of $N_{sto} + N_{HX}$ coupled partial differential equations can be solved numerically with little effort.

Modelling of stratified storages and the numerical solution of the differential equations has been used with different methods like explicit Euler, implicit Crank–Nicholson, predictor–corrector or Runge–Kutta. An alternative approach assumes of a plug-flow model. Here, increments of volume (segments) of fluid from either the heat source or heat are entering the collector at an appropriate position. Shifting the position of all existing segments in the tank between inlet and outlet. The size of segment depends on the flow rate and the time step used, corresponding to the considerations in Chap. 5 related to the Courant number, for more details see (Kleinbach et al., 1993).

8.4.3 Solid Storage Media

Solid storage media offer very high densities and, therefore, the volumetric heat capacity may even exceed the one of water. Especially metals as elements with high atomic weight have very good heat storage properties. However, these are industrially produced materials with a competitive market, e.g. when steel is considered. Therefore, often price limitations prohibit a use in very large storage if other options are available. On the other hand, rock and sand materials combine reasonable heat storage capacity with good economy as cheap abundant materials may be mined from nature. An interesting new approach is to use waste materials from industrial processes like slags from steel production (Ortega et al., 2015) or glass and ceramic slags from municipal waste (Gutierrez et al., 2016) which is availbale at no or even negative cost as dangerous materials like asbestos are being contained in that way. Table 8.5 gives an overview of typical thermophysical properties of solid storage materials.

One of the key challenges of solid storage materials is the heat transfer from and towards the solid storage material and then the incorporation/extraction in or from the material. Thus, solid thermal storages belong to the class of indirect thermal storages requiring a HTF for the heat transfer/extraction. A high thermal conductivity leads to a better uptake and release of heat. The larger the thermal diffusion constant $a = k/(\rho c_p)$, the faster will a solid body take up heat due to a change of temperature surrounding the body. Usually we have two basic geometries: Either a solid body is perforated by pipes where the heat transfer fluid flows, or the heat transfer fluid is passing through a heap of smaller solids bodies (pebbles, rocks) and flows around the bodies.

A natural solid storage option is the storage of heat or cold in the ground. The seasonal and daily ambient temperature changes lead to changing temperature distribution in the surface layers of the ground. While daily fluctuations with a period of 24 h can only impact the temperature in the ground up to a depth of typically 0.3–0.4 m, the annual variations influence the ground temperature up to

Table 8.5 Thermophysical data of some potential massive solid storage media as well as their volume-specific heat capacity, weight-specific costs and costs per thermally generated thermal kilowatt hour (kWh_{th}), weight-specific costs and costs per thermally generated kWh, partially extracted from (Libby, 2009), calculations for hot media operating temperature

Fluid	T_{min} [°C]	T_{max} [°C]	ρ [kg/ m^3]	k [W/ (mK)]	$(c_p \cdot \rho)$ [kWh/ (m^{3}K)]	Cost/kg $(US\$ /kg)$	Cost/kWh$_{th}$ $(US\$ /kWh_{th})$
Water	50	140	1000	0.67	1.161	10^{-4}	
Sandstone	200	300	1700	1.0	0.614	0.15	4.2
Fibre-reinforced concrete	200	400	2200	1.5	0.519	0.05	1.0
Table salt (solid)	200	500	2160	7.0	0.510	0.15	1.5
Cast Iron	200	400	7200	37.0	1.12	1.0	32.0
Cast steel	200	700	7800	40.0	1.30	5.0	60.0
Silicon bricks	200	700	1820	1.5	0.505	1.0	7.0
Magnesium bricks	200	1200	3000	5.0	0.958	2.0	6.0

a depth of 6–8 m. This means that below this depth the ground temperature is nearly stationary. Heat or cold can be transported to this layer and charge and discharge the ground using vertical pipes or thermal probes circulating the heat transfer fluid. As it is difficult to insulate the ground in that depth thermally, only large storage volumes are useful as then the thermal losses at the edges are small compared to the overall One limiting factor in charge time is then the property of the soil. If aquifer water passes the ground, then the heat can be carried away by convection. Then, a storage of heat in the ground is not effective.

In a building the structure of the building itself may be a heat storage. If the thermal mass of the building structure is large (say from using high mass construction materials like concrete) and the thermal insulation value of the building surface is high, the thermal capacity of the building structure, resulting in relatively small temperature changes compared to light-weight building when no heating is available.

8.5 Latent Heat Storage (LHS)

Latent storage is characterized by the ideally isothermal release or uptake of thermal energy in a material which undergoing a phase-change. The phase-change might be between the solid, liquid, and gaseous state of a material; however, it is also possible that different structural phases of a solid material (an internal reordering of atoms in a crystal structure) might be utilized.

8.5.1 Integration of LHS

Integration of latent heat storage in an application is favoured when large temperature changes should be avoided. For instance, if a building shall be kept during summer daytime at low temperatures a constant comfort temperature is preferable even when solar gains and internal loads add heat to the building.

8.5.1.1 Use of Latent Heat Storage

LHS is favourably used in two basic situations depicted by Fig. 8.13:

- storing heat isothermally from a two-phase heat transfer fluid (HTF),
- keeping a temperature constant without active cooling/heating.

The first situation depicted in Fig. 8.13a schematically indicates the isothermal charging and de-charging of a latent heat storage over a large amount of stored heat. This can be accomplished in two ways. Either the medium transferring the heat to and from the storage is also performing a phase-change, or a very high mass-flow of a sensible heat transfer medium prevents a substantial change in temperature. For example, steam (indicated by the upper dashed horizontal curve in the graph) can charge a cold solid LHS where the condensation heat is transferred to the LHS, hence melting the solid storage material. A certain temperature (10–20 °K) difference is needed in the heat exchange. As long as the LHS is not completely molten up, the temperatures stay constant. For discharging the LHS and taking out the heat water at a slightly lower temperature (the lower dashed horizontal curve in the left picture) has to be put into contact to the LHS. If the water is at the same pressure as the steam in the charging process, the evaporation temperature is above the melting point of the LHS. Then, the storage may only transfer little heat to the water, until the temperature of the melting point is reached by the water. On the other hand, when the pressure is lowered such that

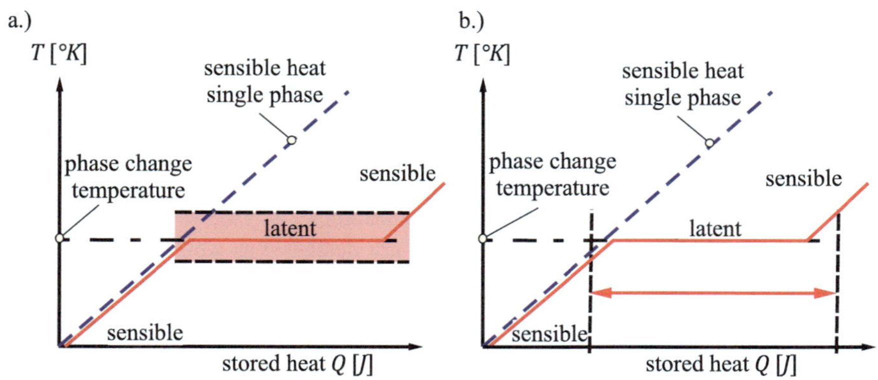

Fig. 8.13 Schemes of latent heat storage adapted from (Mehling & Cabeza, 2008) **a** use as temperature control; **b** storage and supply of heat or cold with small temperature change

the evaporation temperature is below the melting point of the LHS, then heat again can be transferred from one isothermal phase-change to the other. Of course, the lower pressure of the generated steam has a lower exergy content of the steam as consequence. The exergy loss for a charging/de-charging cycle of a LHS using sensible heat transfer fluids would be much large, however.

A second situation is depicted Fig. 8.13b. Here, the temperature of the storage stays constant even if a large amount of heat is added to the store. Imagine the walls of a room incorporating PCM material with a melting point at a comfort temperature of say 23 °C. If heat is added the room surface temperature stay the same and the heat is used for melting up the PCM. Only after the full melting the sensible heat of the building walls lead to noticeable temperature rise above the comfort temperature. This may lead to an avoidance of active cooling energy if the de-charging of the room walls (by cool air during night-time) restores the capacity to keep the temperature constant. This situation is similar to the effect of ice cubes in a drink. Although in summer the drink would soon become warm due to heat transfer from the ambient, the melting of ice cubes keeps it a 0 °C until the ice is molten.

The term "latent heat" describes the heat of all possible phase changes, i.e. solid–solid (change in lattice structure), solid–liquid and liquid–vapor. However, the terms "latent heat storage" and "phase change material" (PCM) are usually used in the technical literature only for the first two types of phase change and not for liquid–vapor phase transition. The reason is that in the case of a liquid–vapor phase change, the temperature of the phase change depends essentially on the boundary conditions, so that it is associated with a pressure and temperature difference between charging and discharging. In this context, the terms "latent heat" and "phase change material" (PCM) in thermal storage also refer only to phase changes between solids and liquids.

8.5.1.2 Charging and Discharging Process

As described in the previous paragraphs the charging and discharging process of a latent heat storage is isothermal. Ideally the phase-change occurs at the same temperature in both directions. This is essentially given in the case of supplying heat. However, while cooling below the melting temperature often no solidification is observed. Only, if some crystallization starts the solidification sets in. This effect is called sub-cooling and is illustrated in Fig. 8.14. Only, the onset- of crystallization enables the release of the latent heat stored in the phase change material (PCM). Thus, for technical applications sub-cooling behaviour is an essential issue. A slow nucleation process means a slow growth of solid nuclei as vice versa a fast one an instantaneous release of heat, which may be not desirable as well.

Physically, solidification of a liquid means that the amount of solid phase increases. At the beginning of solidification, there are almost no solidified particles. These smallest solidified particles, mostly called nucleus, grow by solidification of the liquid phase at its surface. For this to occur, they must release the stored latent heat to reach an energetic minimum. However, there is competition between the surface energy, which is proportional to the surface area ($\sim r^2$), and the latent heat

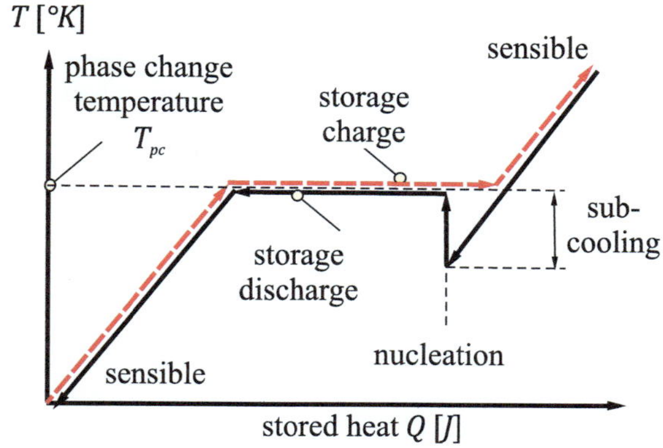

Fig. 8.14 Effect of subcooling of a PCM material when discharging a thermal accumulator modified from (Mehling & Cabeza, 2008)

released during phase change, which is proportional to the volume (~r^3). If the nucleus is small, the heat released by crystallization may be less than the surface energy gained. This means that there is an energetic barrier. Thus, only if there is a nucleus with a sufficiently large radius does solidification begin. In some cases, this requires temperatures well below the necessary melting temperature. This means that the sub-cooling behaviour can be manipulated via the number of particles at the beginning of solidification near the melting temperature. Sub-cooling may be reduced or avoided by adding appropriate nucleation seeds to the material, by surface treatments or by stirring the PCM with a mechanical device with a temperature below the sub-cooling point (cold finger technique).

8.5.2 Materials

8.5.2.1 Properties

When sensitive heat is used, the storage tank temperature increases with the stored energy, whereas latent heat storage tanks can absorb larger amounts of heat at a constant temperature. Since the efficiency of solar collectors decreases with increasing temperature, latent heat storage tanks offer advantages here. The amount of latent heat (or enthalpy) Δh_m that storage materials can absorb during a phase change (solid–liquid under relatively small volume and pressure changes or from liquid vapor) is relatively large. The thermal energy storage capacity Q in (J) of these materials can be calculated in the following way:

$$Q = m \cdot \eta_C \cdot \left(c_{p,s} \cdot \left(T_{min} - T_{pc}\right) + \Delta h_m + c_{p,l} \cdot \left(T_{max} - T_{pc}\right)\right), \qquad (8.22)$$

where m denotes the mass, η_C the storage conversion efficiency, $c_{p,s}$ and $c_{p,l}$ the specific heat capacity in the solid and the liquid state and T_{pc} the phase change temperature. T_{min} and T_{max} are the minimum and the maximum temperatures of the charge/discharge process.

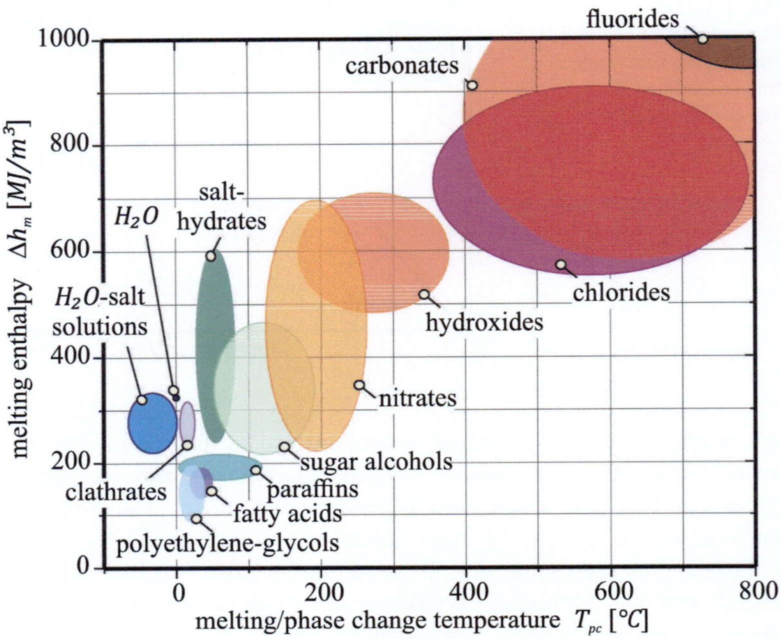

Fig. 8.15 Studied classes of phase change materials (PCMs) with ranges of their typical melting temperature or volumetric enthalpy of fusion modified from (Mehling & Cabeza, 2008)

Phase change materials exist with a wide spread of temperatures where the phase-change occurs, and also with a wide range of latent heats (melting or evaporation enthalpies). Figure 8.15 gives an overview over different material classes which are used in solar energy. PCM with melting points below 0 °C can be used for cold storages, and on the other side metals are also phase-change materials although melting often at temperature above several hundred to thousand degrees. Tables 8.6 to 8.9 provide some specific data of some so-called "Phase *Change* Materials" PCMs. PCM materials are characterized by the fact that their latent heat of fusion, heat of solution or heat of absorption is considerably greater than the heat they can store due to their normal specific heat capacity (without the phase change effect). The exploitation of the solid–liquid phase transition is the most frequently used principle. When charging the contents of commercial latent heat storage tanks, special salts (especially nitrate or chlorine salts) or paraffins are usually melted as a storage medium, which absorb a great deal of heat energy (heat of fusion) for this purpose. Since this process is reversible, the storage medium releases exactly this amount of heat when it solidifies. A comparison of phase change materials that can be used at higher temperatures with liquid thermal storage media is shown in Table 8.3 which also shows the thermal stability limits of the phase change materials, more details on candidate materials and the further literature may be taken from (Gomez, 2011). The example of water-steam shows

Table 8.6 Selected PCM materials in from of eutectic water salt solutions and salt hydrides modified from (Mehling and Cabeza 2008)

Material	T_{pc} [°C]	Melting enthalpy Δh_m [kJ/kg]	k [W/(mK)]	ρ liquid $\left[\text{kg/m}^3\right]$
$NaCl(22.4wt\%) - H_2O$	−21.2	222	–	1165
$KCl(19.5wt\%) - H_2O$	−10.7	283	–	1165
H_2O	0	333	0.6	998
$Na_2HPO_4 - 12H_2O$	35 – 44	280	0.476	1442
$Ba(OH)_2 - 8H_2O$	78	265 – 280	0.653	1937
$Mg(NO_3)_2 - 6H_2O$	90	149 – 163	0.49	1550
$MgCl_2 - 6H_2O$	117	165 – 169	0.57	1450

Table 8.7 Selected paraffine type and fatty acids PCM materials (Mehling & Cabeza, 2008; Ravotti et al., 2020)

Material	T_{pc} [°C]	Melting enthalpy Δh_m [kJ/kg]	k [W/(mk)]	ρ liquid $\left[\text{kg/m}^3\right]$
n-Octadecane $C_{18}H_{38}$	28	200 – 245	0.148	774
n-Octacosane $CH_3(CH_2)_{26}CH_3$	61	202	–	800
Polyethylene	110 – 135	200	0.33 – 0.4	998
Acetic acid	16.7	184	0.153	1005
Tristearin	56	191	0.17	862
Acetamide	81	241	–	1160
Methyl fumarate	102	242	0.31	1370

that the temperature is no fixed value, but dependent on physical boundary conditions. This is especially true for the phase-change from gaseous to fluid phase. The boiling point where the solid phase completely disappears changes dramatically depending on the atmospheric pressure—whereas for 1 bar the boiling point is about 100 °C, for 0.1 bar it is 47 °C, for 10 bar 180 °C and for 100 bar 311 °C.

8.5.2.2 Advantages

The advantage of this heat storage technology is based on storing as much heat energy as possible isothermally at the melting or boiling temperature of the storage material used. A further advantage is high energy density which is available in the phase change. The isothermal charging and de-charging are especially advantageous in steam production for power plant applications. The use of a phase transition is much more effective than simply heating the medium. For technical applications of latent heat storage with fluid and solid phases, recrystallization just below the melting temperature is generally desirable. Modern latent heat storage

Table 8.8 Selected sugar alcohols PCM materials and non-paraffins from (Sharma et al., 2009)

Material	T_{pc} [°C]	Melting enthalpy Δh_m $\left[\text{kJ/kg}\right]$	k [W/(mK)]	ρ Liquid $\left[\text{kg/m}^3\right]$
Xylitol	94	263		1500
Erythritol	120	340	0.326	1300
Galactitol	188	351	0.33 – 0.4	1520
Glyolic acid	63	109	0.276	1490
Alpha napthol	96	163	0.44	1100
Stibene	124	167	0.31	971

Table 8.9 Selected solid–liquid high-temperature PCM materials

Material	T_{pc} [°C]	Melting enthalpy Δh_m $[kJ/kg]$	k [W/(mK)]	ρ liquid$\left[\text{kg/m}^3\right]$
Sodium nitrate $NaNO_3$	306	182.3	0.67	2260
Potassium nitrate KNO_3	334	267	0.50	2110
Potassium hydroxide KOH	360	150	0.50	2040
Salt ceramics $(NaCO_3 - BaCO_3/MgO)$	500 – 850	420	0.50	2600
Table salt $NaCl$	801	7035	5.00	2170
Sodium carbonate (Na_2CO_3)	851	520	5.00	2260

materials based on salt or paraffin have physical properties developed for various applications and are available for almost all temperature ranges (Fig. 8.15). Figure 8.16 shows the storable energy as a function of the temperature of a phase change material from the low temperature range in comparison with water and the solid graphite.

8.5.2.3 Examples

When using latent heat storage for solar collector systems providing low temperature space heat to a building in winter, the investment is higher than just using a water tank, but the energy density of the storage is much higher and hence the technology is more space-saving due to the exploitation of latent heat. Before using a latent heat storage system in a classically heated house, the building insulation should be improved as a priority to reduce the heat demand.

The phase change liquid-gaseous is used on a technical scale in pressurized steam-water storage tanks. Although water has a relatively large weight-specific storage capacity nevertheless it is expensive to build large storage volumes suitable

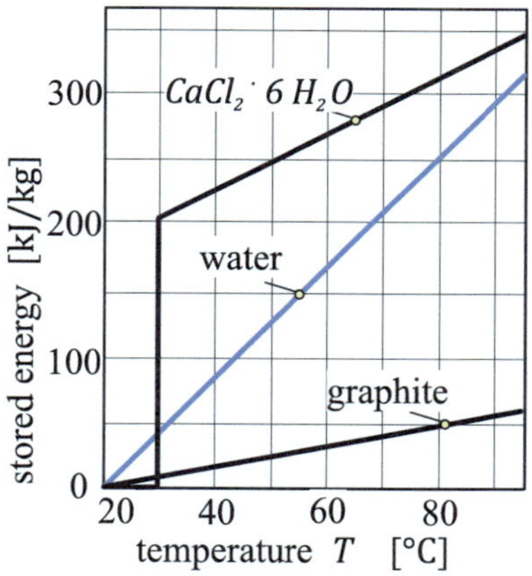

Fig. 8.16 Storable energy contents of a phase change material from the low temperature range compared with water and graphite

for high pressures. The principle of these so-called steam accumulators is described and discussed in detail in the Sect. 8.5.3 on gas–liquid phase-change.

8.5.2.4 Criteria

As there is a nearly infinite number of possibilities for phase-change materials to be selected for thermal energy storage, it is necessary to establish criteria for the selection. The following list give important criteria for the assessment:

- melting/boiling point,
- high phase change enthalpy,
- cycling stability,
- chemical stability (no degradation),
- high thermal conductivity (high thermal powers),
- low volume change,
- high density and
- economically viable

The first two criteria are obvious from the view of suitability to a specific application. The phase-change temperature has to be adapted to process conditions, as already described in the previous Sect. 8.5.1.1. A high phase-change enthalpy is always useful to have as compact storage as possible. The cycling stability, however, is very much depending on the use of the material. It means that the phase-change for thermally cycling the material in charging and de-charging processes is stable. There should be no reduction in enthalpy nor a change of melting temperature. To be distinguished should be chemical stability which means that the material is not reacting with container materials or the surrounding atmosphere. Oxidation or corrosion processes would change the chemistry and hence also the parameters for the thermal storage system. A high thermal conductivity (in solid PCM) is required to enable a high de-charging thermal power. If solid material with low conductivity forms around a heat exchanger surface the heat transfer rate will decrease substantially. In the liquid phase low thermal conductivity can be offset by convection of the surrounding phase, which indicates that here the viscosity should be low. During a phase change occurs a reordering of molecules. They may form a crystal structure, change the geometric structure in a solid phase, or even form a compressible (gas) or incompressible fluid (liquid) with freely moving molecules. As the storage materials usually are kept in a container, a corresponding change in volume might lead to problems. Very well known in everyday life is the freezing of water in bottles, which lead to a cracking of the glass due to the volume increase. Generally during freezing the volume decreases (for instance in hydrates). But even that may lead to problems. When a liquid freeze to a crystalline structure having a smaller volume then the effective contact area to the heat exchanger might be reduced which can be avoided by careful design. A high mass density of the material is helpful as very often excess heat added to the storage even when the phase-change is complete will lead to sensible temperature change. Low mass density is generally coupled to low heat capacity. Hence excess heat might lead to unwanted temperature rise which could be a problem for chemical stability. And last not least it is clearly a target to find cost-effective thermal energy storage solutions with economically viable materials.

8.5.3 Gas–Liquid Phase Change

The most prominent example of a gas–liquid phase change to be used in thermal energy storage is the change from water to steam. Technically this physical principle is used in so-called steam accumulators in power plants or industrial steam networks to avoid steam loss from intermittency of generation (Sun et al., 2017; Tamme, 2010). There are mainly two options to add heat to the water/steam storage:

- sliding pressure steam accumulator,
- constant pressure steam accumulator.

These two options are used in commercial systems and described in Sect. 8.9.4 Although other gas–liquid phase changes might be utilized, for example refrigeration media which can be used also in so-called organic Rankine cycles, the water-steam pair is the only phase-change storage of this type realized in commercial applications by now.

8.5.4 Liquid–Solid Phase Change

Nearly all latent heat storage systems which are investigated and developed in R&D are based on the liquid–solid phase-change. The process can be handled easier as there is no high pressure necessarily included. However, the solid phase also poses technical challenges. We discuss therefore, the limiting issue of low thermal conductivity (k_{PCM}) in the solid phase in detail, which leads to a design problem for the heat exchanger power.

8.5.4.1 Challenge of Low Heat Conductivity in the Solid Phase

A phase change from the liquid to the solid phase in the discharging mode of a latent heat storage is facing the problem of low heat conductivity (k_{PCM}) in the solid phase. A heat exchanger used for adding or extracting the heat from the storage is cooling down the PCM. The material close to the HX surface solidifies and forms a kind of thermal insulation barrier. This situation is sketched in Fig. 8.17.

Most PCM exhibit small specific heat conductivities of the order of $k_{PCM} = 0.2 - 2\text{W}/(\text{mK})$, which is low compared to steel $k_{SS} = 15 - 30\text{W}/(\text{mK})$ mostly used in heat exchangers. Therefore, in order to keep the $(U \cdot A)$ value of the heat exchanger on a reasonable level, the thickness of the growing solid PCM layer has to be limited. The U-value in $(\text{W}/(\text{m}^2\text{K}))$ is given for the configuration illustrated Fig. 8.17 by

$$U = \frac{1}{\frac{1}{h_{HTF}} + \frac{t_{SS}}{k_{steel}} + \frac{t_{PCM,solid}}{k_{PCM,solid}} + \frac{1}{h_{PCM,fluid}}} \tag{8.23}$$

As the effective $(U \cdot A)$ increases during solidification of the PCM material, the thermal power transferred during discharge is decreasing over time as well. This can be achieved either by having small volumes between heat exchanger surfaces, or by dynamically removing the solid layer from the heat exchanger surface. These two approaches characterize and distinguish passive latent heat storage from active latent heat storage.

Figure 8.18 illustrates an example for a discharge operation of a PCM storage in sliding pressure operation with a steam production of 0.3 kg/s in the forced convective mode. The hot temperature (T_{hot}) and the cold temperature (T_{cold}) represent the temperatures of the HTF fluid recorded before entering the tube collectors at the hot and cold end of the PCM storage modules, while the lines T_{MQ1} to T_{MQ3} denote the temperatures measured in the phase change material at the three different axial positions $MQ1$ (close to the hot end), $MQ2$ (middle) and $MQ3$ (close

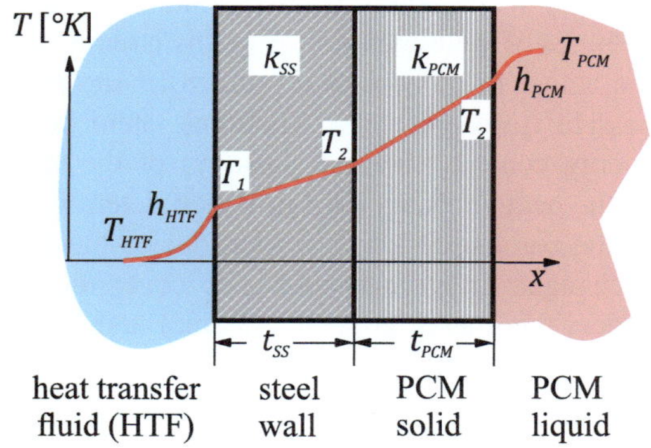

Fig. 8.17 Heat transfer through a heat exchanger wall covered with a layer of solidified PCM

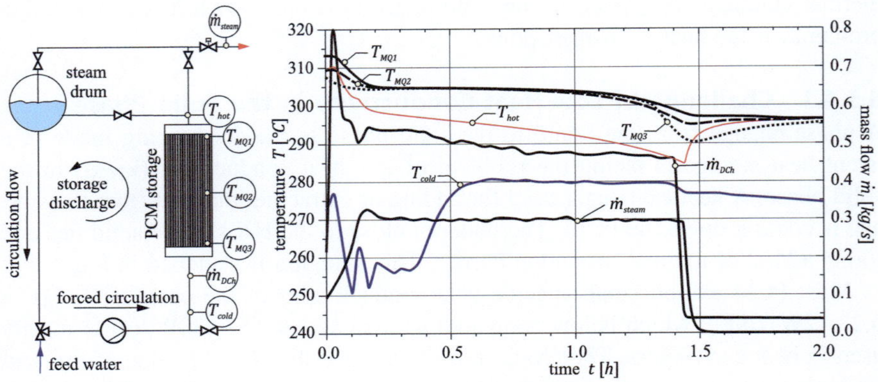

Fig. 8.18 Example for discharge operation of the PCM storage in sliding pressure operation with a steam production of 0.3 kg/s from (Laing et al. 2013)

to the cold end) of the PCM storage module. $\dot{m}_{DCh}$ is the total water/steam mass flow rate through the storage module, $\dot{m}_{steam}$ is the dry steam produced by the storage module.

8.5.5 Active and Passive Storage Designs

The conventional passive LHS design is characterized by a container of PCM with an internal heat exchanger. As the PCM solidifies, it cannot be used as a heat transfer fluid as such. However, it is also possible to actively transport solid material. A pre-requisite is that the material is put into a shape which can be transported easily. This means that size is limited and, hence, smaller volumes of solid material will be transported, either as pellets, granules, solid balls, or plates. This solid material can be stored easily in a container, and when it shall be charged it will be transported to a heat exchanger facility for melting up the material. Then we have a liquid which has to be transported to a liquid container tank using a different transport principle. The real problem which has to be solved is the reverse process. When liquid material solidifies continuously it is difficult to shape small volumes of solid material. And this issue is one crucial points which distinguishes the active LHS systems. Figure 8.19 shows schematically the two principles of passive (top) and active (bottom) latent heat storage. In the active system the energy content (thermal capacity) of the storage is decoupled and independent of the heat exchanger rating which determines the charging and discharging rate. In the passive design, however, due to the solidification of the material, the heat exchanger structure itself has to cover the whole volume of the container, and hence, the energy content (volume) and heat transfer rate (heat exchanger area) are not independent.

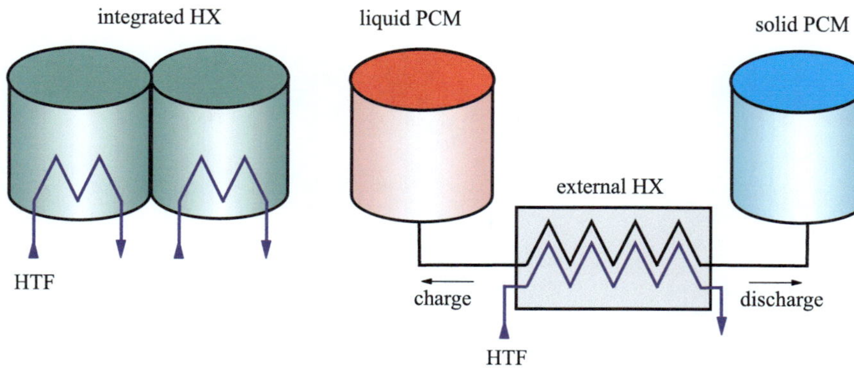

Fig. 8.19 Principle set-up of a passive latent heat storage (left) with embedded HX in the storage tank versus active latent heat storage (right) where PCM material is transported to an external HX

8.5.5.1 Passive Latent Heat Storages

Embedded tube-fin heat exchanger

The most common design for a passive LHS is a container which incorporates an internal heat exchanger structure. These structures may be simple finned tube designs or sophisticated alumina extruded structures, as shown in Fig. 8.20a and b. Those structures are then integrated in a larger container and Fig. 8.20c gives an impression on the arrangement and Fig. 8.20d shows a complete structure. In both fin concepts the heat transported in the piping is conducted to the PCM by fin plate vertical to the pipes. The fins can be fabricated of aluminium or graphite, depending on the salt mixture used. Both materials have a very high thermal conductivity, however, they react different with Nitrate salt materials. Depending on the salt mixtures, which are needed for specific melting temperatures, one or the other material has problems with reacting with the salt. For stability reasons graphite has been used for applications below 250 °C, such as salt mixtures as $LiNO_3 - NaNO_3$ mixtures ($T_{pc} = 194\ °C$) or $NaNO_3 - KNO_3$ ($T_{pc} = 224\ °C$), exceeding this temperature limit aluminium fins showed stability up to 350 °C. The incorporation of a heat exchanger in a solid PCM storage is expensive, and due to the limitations of the heat conductivity of Nitrate salts the distances between the fins were quite small. This led to a high-volume fraction of metal (or graphite) in the storage, which reduces the latent storage density on one hand side, and which increases cost on the other side.

Macroscopic encapsulation

An alternative design is to reverse the engineering: Instead of containing the HTF in pipes within the PCM, the PCM is contained and the HTF flows freely around the PCM containers. This leads to encapsulated PCM designs. PCM material is filled into rods, spheres or other stable and thermally conducting containers. Figure 8.21 shows some macroscopic encapsulation examples. The dimension of the rod or sphere diameters is a few centimetres. However, this encapsulation has

Fig. 8.20 Heat exchanger structures **a** and **b** embedded in PCM storage volume, **c** set-up of a modules and **d** assembled module; all developed at DLR for a molten salt $NaNO_3$ with melting point of $T_{pc} = 304\ °C$, modified from (Johnson, 2016)

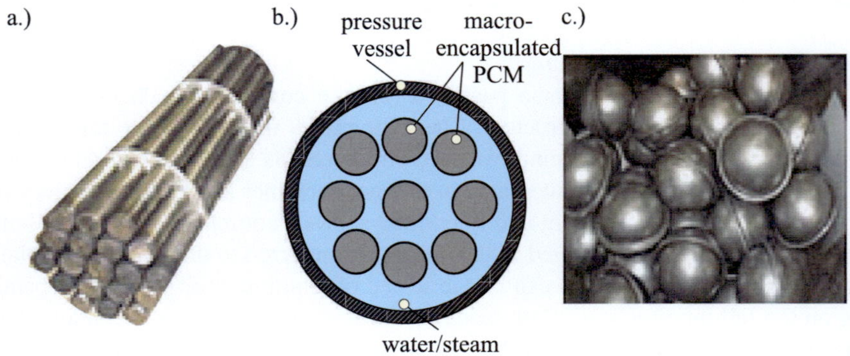

Fig. 8.21 Example of macroscopic encapsulation using rods (**a**, **b**) or spheres (**c**) modified from (Steinmann, 2021)

to consider the thermal expansion of the material, and the filling process is an effort which is reflected in the overall costs.

Microscopic encapsulation
Stimulated by the difficulties of the macroscopic encapsulation microscopic encapsulation has been investigated too. The advantage could be that microencapsulated PCM could be an industrial mass product and raw material which can be used and made-up for different applications. Microencapsulation is a technology which is used already in many parts of industry, they may contain dyes, drugs, herbicides, or biological elements, to give some examples. The reasons for microencapsulation can be differentiated depending on how and when the capsules shall release the material inside:

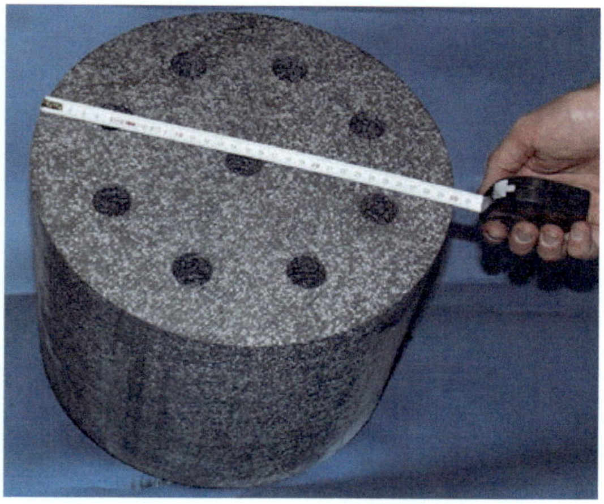

Fig. 8.22 PCM/expanded graphite composite segment manufactured by isostatic cold compression from (Steinmann, 2021)

- no release,
- release at a specific time,
- controlled release over a period of time.

Microcapsules with permanent containment of a liquid shall immobilize that material, and possibly protect it from reaction or external forces. The chemical company BASF has developed some years ago encapsulated paraffins in polymeric microspheres with a diameter of about 20 μm. These can be put inside a mortar or gypsum to enhance the heat capacity of light-weight building material, or they can be put as suspension in a carrier liquid.

Another way of embedding paraffin in material is by pressing liquid paraffin in hollow pores of porous concrete or graphite powder. In such a way an embedding of PCM material can be achieved, which—contrary to the microencapsulation—is not completely contained. During a melting process some leakage from the matrix material may occur. Figure 8.22 shows an example of a graphite matrix block with embedded PCM material, which exhibits a high heat conductivity of the order of k = 15 W/(mK). The larger holes are intended for pipes containing the heat transfer fluid for charging and discharging the storage material.

8.5.5.2 Active Latent Heat Storage

Screw-Heat exchanger (SHE)

An interesting option of an active latent heat storage was discussed by (Zipf et al., 2016) where the transport of PCM material is actively done by a transport screw system which simultaneously is also a heat exchanger. The transport screws transport solid granular material from one side of the HX to the other. The screw is transpired by a hot fluid which provides the heat to heat the screw surfaces and hence is used to melt up the PCM granules. The liquified PCM then may be drained into a container, as indicated in Fig. 8.23.

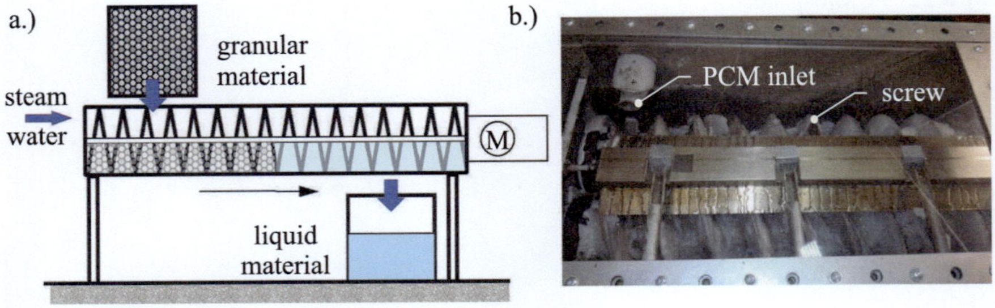

Fig. 8.23 **a** Principle of the storage charging process in a screw heat exchanger (SHE) and **b** photograph of the inlet section of an experimental device modified

In the discharging process the hot liquid phase is sprayed onto a SHE with cold HTF. The PCM material crystallizes on the screw surface but may not form a thick layer as two screws rotating in the opposite way may scratch away the solid layer and breaks it up to flakes which can be transported further. The flakes in a second step may be further reduced in size by a mechanical crusher before stored in granular form in a simple container.

The advantage of this active LHS is that the thermal power (determined by the HX area) is decoupled from the storage capacity (determined by the material choice and the tank volume). In this sense the principle might be beneficial especially when large storage capacity is needed. From the permanent removal of solid material from the rotating screw heat exchanger, there is no formation of a thick solid layer, and hence the rate of heat transfer is rather constant and not decreasing over time. In a laboratory prototype it could be shown that the principle works even with horizontal SHE devices. The Fig. 8.23 shows a small prototype where the feasibility of the concept had been proven. A mass flow of about 150 kg/h could be achieved with this small device. With a crusher the granular material of frozen nitrate salts reached an apparent density of 950 kg/m^3. Although SHE is used for recovering the heat of hot ashes in industrial applications using this principle economical storage could not yet be achieved for the temperature range of 400 °C. Here, steam with high-pressure has to be safely contained inside the moving screws, which means that bearings have to be tight even for typical steam pressures. Whereas in a second prototype heating with steam was demonstrated, the pressures in a practical power plant application in the range of 7–12 MPa would have been too high for cost-effective solutions.

Other types

A different active LHS approach has been tested with the principle of conveyor belts. Container with liquid PCM are transported to a heat exchanger section where during a suitable passage time the material is cooled down enough to solidify. The process can be reversed easily by running the conveyor belt with a hot heat exchanger pipe. Figure 8.24 depicts the set-up of such a PCM concept. Although in principle both active LHS described using nitrate salts with melting temperatures

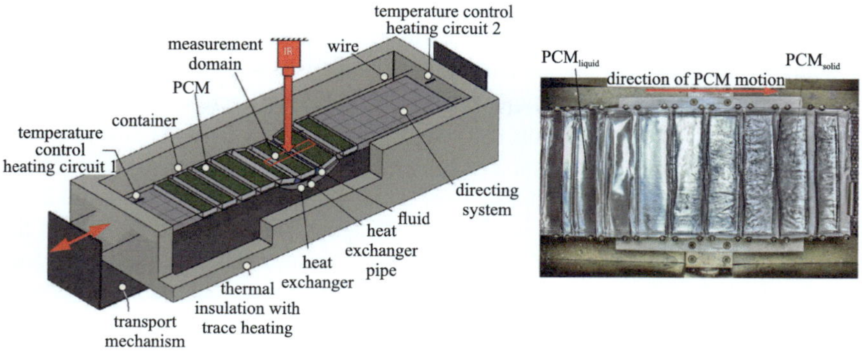

Fig. 8.24 Experimental facility to demonstrate active PCM storage using a movable transport band and top view of the demonstrator modified from (Pointner & Steinmann, 2016)

around 200 °C would be ideal for industrial applications with steam networks of saturation temperatures 160–190 °C (with pressures of 0.4–2 MPa), no commercial product has been developed yet.

8.6 Chemical Storage

Reaction storage materials are substance pairs in which a reversible chemical reaction or physical surface reaction occurs. The central concept of a reaction store is the reversibility of the process. By supplying thermal energy from solar radiation at high temperature levels, a dissociation or conversion of the educts can be achieved, breaking up chemical bonds or with lower temperature overcome surface-physical forces. Chemical process engineering separation methods then can be used to separate the products obtained and store them individually in different containers. If energy is required, the two products can be brought into contact again and the amount of energy resulting from the reaction can be thermally utilized—similar as heat from a thermal oxidation like burning gas or coal may produce high temperature heat. The class of reaction storages can be divided into four different groups:

- catalytic reactions,
- thermal dissociation reactions,
- adsorption processes,
- absorption processes.

whereby the first two classes are chemical reaction in the strict sense, while the latter ones describe a surface related process, which can be physical or chemical.

Based on this rather fundamental consideration thermochemical energy storage (TCES) can be conceived third fundamental principle of heat storage, along with sensible and latent heat storage. In contrasts to the first two principles TCES systems do not store heat but the potential for heat release, which is essential since because the effort for thermal insulation can be substantially reduced. Other advantages are potentially high energy storage densities of up to 500 kWh/m^3 and dispatchable power production, which makes them attractive for medium- and long-term storage. Depending on the concept selected, TCES also enable decoupling from the point of generation and consumer, for example, transfer of chemical products by pipeline or rail/road transport, which is only possible with large losses in the case of thermal storage. Another advantage of TECS-based storage is that chemical reactions can be used to achieve temperatures that exceed the temperature level of the original generation source. In thermochemical storage systems, heat (ΔH_r) is applied to trigger a reaction of an educt A and B and into products B and C. Thus, the reaction of A and B with the addition of heat to C and D is an endothermic reaction that characterizes the loading of the storage tank. Conversely, heat is released when products C and C are mixed together and react exothermically to form the initial reactants A and B representing the discharge of the storage.

$$A + B + \Delta H_R \underset{discharge}{\overset{charge}{\rightleftarrows}} C + D. \tag{8.24}$$

The energetic storage capacity (Q) depends on the mass of the storage material (m), the molar mass $\mathcal{M}$ and the molar reaction enthalpy (ΔH_R),

$$Q = m \cdot \frac{\Delta H_R}{\mathcal{M}}. \tag{8.25}$$

The turning point temperature T_t allows to evaluate the suitability of a chemical reaction for the temperature range defined by an application. At T_t, the reaction rate constant is equal to unity (1), so that for temperatures $T > T_t$, the forward reaction is favoured, and heat is stored in chemical form, while for temperatures $T < T_t$, heat is thus released during discharge. Thus, T_t provides a guide for classifying reactions into low, intermediate, and high temperature ranges. Unfortunately, this estimate introduced by (Wentworth & Chen, 1975) is only valid at one bar. The magnitude of T_t can be estimated from the standard formation enthalpies (ΔH_R^0) and the formation entropy (ΔS_R^0) using the equation

$$T_t = \frac{\Delta H_R^0}{\Delta S_R^0}. \tag{8.26}$$

In addition to the general requirements for storage, chemical reactants must meet some specific requirements:

- in addition to reversibility, no side reactions should occur,
- the temperature level should match the available heat source and the desired use,

- both the charging and discharging reactions should be fast enough to achieve sufficient power densities,
- the specific volume of the products should be small, and,
- the products should be easily separable and storable.

This last requirement also refers to the state form of reactants such as products, which can be in gaseous, liquid but also solid form. This directly results in a form of classification of thermochemical energy storage (TCES), which is used in (Steinmann, 2021). There is a myriad of reversible chemical equilibrium reactions that can serve as the basis for a TCES and Table 8.10 provides an overview of the classes, some typical reactions, the turning point temperature and the energy density.

8.6.1 Catalytic Reactions

Catalytic reaction storages are chemical equilibrium reactions that are completely reversible. Two products are formed by the chemical reaction of two educts under heat supply (for example produced from solar radiation). This takes place during the charging process of the storage tank. When discharging, the substances are put into a reactor with lower temperature and the energy is released in the reverse reaction. The basic chemical reaction formula

$$A + B + \Delta H_R \underset{discharge}{\overset{charge}{\rightleftarrows}} C + D. \tag{8.27}$$

An example is the reaction of methane (CH_4) and carbon dioxide (CO_2) to carbon monoxide and hydrogen with the addition of heat, see equation. Both medias can be stored and thus stored separately after a process engineering separation. The heat released needed for the forward reaction is $247 kJ/mol$ methane, and the same amount is released in the backward reaction.

$$CH_4 + CO_2 + 247kJ \underset{discharge}{\overset{charge}{\rightleftarrows}} 2CO + 2H_2. \tag{8.28}$$

Dry methane reforming with CO_2 by solar energy has been demonstrated in several experiments to the $100kW$ range, see e.g. (Buck et al., 1991) or (Wörner & Tamme, 1998), conversion rates of methane in the range of 70–80% using Rhodium as catalyst.

8.6.2 Thermal Dissociation Reactions

In thermal dissociation reactions from one educt two products are produced in a mixture. AB is split into two substances A and B under heat input at a high

Table 8.10 Examples of proposed chemical reactions for thermochemical energy storage (TECS) in the medium and high temperature domain from, (Steinmann, 2021; Tamme et al., 2012)

Phase state	Class	Reactions	T_t [°C]	p [MPa]	ΔH_R[kJ/mol]	q [kWh/kg]
Gas- gas	Reforming	$CH_4 + H_2O \leftrightarrow CO + 3H_2$ $CH_4 + CO_2 \leftrightarrow 2CO + 2H_2$	687 687	$2 - 15$ 0.35	250 247	4.34
	Dissociation (catalytic)	$NH_3 \leftrightarrow 1/2N_2 + 3/2H_2$ $SO_3 \leftrightarrow SO_2 + 1/2O_2$	195 767	$10 - 30$ $0.1 - 0.5$	59 109	0.51 0.38
	Sulfuric acid decomp	$H_2SO_4 \leftrightarrow SO_3 + H_2O$	–	$0.1 - 5$	297	0.84
Liquid- gas	Concentrat./ dilution	$NaOH(m+n)H_2O \leftrightarrow$ $(NaOHmH_2O_l) + (nH_2O)_g$	–	–	44.5	–
Solid–gas	Hydroxides	$Mg(OH)_2 \leftrightarrow MgO + H_2O$ $Ca(OH)_2 \leftrightarrow CaO + H_2O$	531 752	$0.01 - 1$ $0.01 - 1$	84 109	0.40 0.39
	Hydrates	$MgH_2 \leftrightarrow Mg + H_2$ $Mg_2NiH_2 \leftrightarrow Mg_2Ni + 2H_2$	293 253	$0.1 - 10$ *n.a.*	75 128	0.79 *N.a.*
	Peroxide	$BaO_2 \leftrightarrow BaO + 1/2O_2$ $KO_2 \leftrightarrow 1/2K_2O + 3/4O_2$	782 668	*n.a.* *n.a.*	75 101	0.66 0.64
	Carbonates	$CaCO_3 \leftrightarrow CaO + CO_2$ $BaCO_3 \leftrightarrow BaO + CO_2$	896 1497	1 1	167 212	0.46 *N.a.*
	Redox	$2Co_3O_4 \leftrightarrow 6CoO + O_2$ $6Mn_2O_3 \leftrightarrow 4Mn_3O_4 + O$	1208 1188	0.1 0.1	205 90	0.12 0.05

pressure p_1 in a charging process, which have to be stored separately in two containers. In many cases the products evolve in two different phases, which makes them easy to separate. This is a chemical reaction with the following formula type:

$$AB_{gas/liquid/solid} + \Delta H_R \underset{discharge\ p_2,\ T_2}{\overset{charge\ p_1,\ T_1}{\rightleftarrows}} A_{gas/liquid/solid} + B_{gaseous/liquid}. \quad (8.29)$$

With respect to solar utilization as well the ammonia as the sulfur-based dissociation/synthesis have been studied, since its temperature level of 400–700 °C suits well with the operational temperature range of concentrating solar power systems, however, the ammonia systems require high pressures as indicated in Table 8.10. Although the ammonia reaction has compared to methane reformation the advantage of almost no side reactions the major drawback it is low reaction enthalpy. Here, sulphur-based systems exhibit a much large potential, since they allow the combination e.g. with the sulfuric acid decomposition and the elemental sulphur combustion an extremely high storage density of (2.6 kWh/kg) at low raw product cost. Also, the simplicity of product separation, since the exhibit gaseous, liquid, and solid state seems at a first glance attractive, however, the harmonization of numerous reaction steps and the material challenges to be solved especially for sulfuric acid are still critical.

Here, TCES storage concepts based on simple solid/gas pairs as the hydration/dehydration of metal hydroxides like the conversion of calcium hydroxide $Ca(OH)_2$ under heat to calcium oxide CaO and water vapor shown in

$$Ca(OH)_{2_{solid}} + 109.2\text{kJ} \rightleftarrows CaO_{solid} + 2H_2O_{\text{gas}}. \quad (8.30)$$

are much simpler, and moreover, the reactants are non-toxic, abundant, and cheap. Product separation here is no major challenge since the water vapor can be easily extracted from the solid calcium oxide (CaO). Also, a catalyst is not required, and the educts/products can be stored at ambient pressure. Similar as with the phase change materials (PCM) the low thermal conductivity of the solid reactants poses a challenge especially to the heat exchangers. The technology and equipment of a pilot system is described e.g. in (Criado et al., 2017).

The process of dissociation of a gas B from a liquid solution AB is shown in the van't Hoff diagram, in which the pressure p is plotted over the inverse of the temperature T, as sketched in Fig. 8.25. The van't Hoff diagram shows the equilibrium constant K of a chemical reversible reaction versus temperature. It is given by the ratio of the molar concentrations of the product species over those of the educt species, taking the stoichiometric coefficients of the reaction into account as exponents of the concentrations. Endothermic reactions have a negative slope, while exothermic reactions have a positive one.

In Fig. 8.25 when a liquid AB is heated up from point U to point 1, then pressure increases until it exceeds the equilibrium pressure, and then evaporation

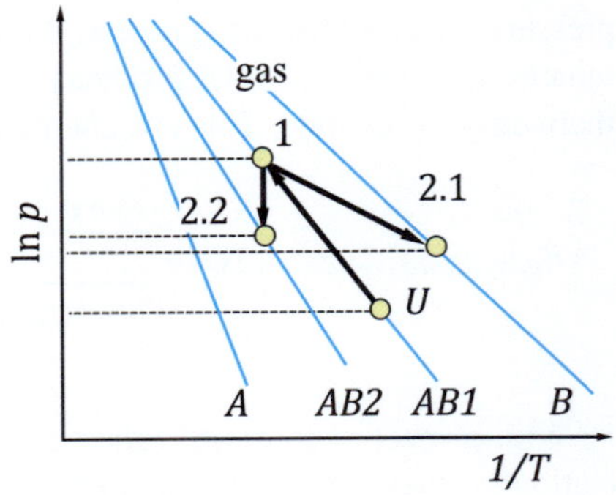

Fig. 8.25 Van't Hoff diagram for the thermal dissociation of a gas B from the liquid solution AB

produces gas B and the liquid $AB2$ with a lower concentration of dissolved gas at a lower temperature (reaction enthalpy is extracted). When the gas from the storage material containers is fed into the reactor again and put into contact with $AB2$, then in the reverse reaction, due to the reaction of gas with the liquid the pressure and heat is released to raise the temperature. A plant scheme of for the operation of a chemical storage using dissociation reaction is shown in Fig. 8.26. The heat exchanger takes up the reaction heat from the reaction of A and B brought into contact and is stored as the compound AB again in a storage. If surplus heat at sufficient temperature is available, AB can be dissociated again into the two products A and B.

8.6.3 Metal Hydrides

Metals consist of regular atomic structures with the conduction band electrons moving freely between the atomic nuclei. As a proton as the positive part of the hydrogen atom is very small, it is possible that protons penetrate into the metallic structure and move to interstitial sites between the ions of the metal grid. The associated electron is added to the free electrons of the metal, which rearrange to find the best charge distribution to electrically screen the proton. Hydrogen gas can dissociate at metal surfaces to from protons and electrons which then are dissolved in the metal structure. This is an exothermic process, because the metal grid exerts an attractive force on the protons, energy is released as the proton is incorporated into the metal. Hydrogen in the reverse reaction can be produced by heating up the metal grid, from which it is released as a gas. The reaction reads generally to:

$$MH_n + \Delta H_R \underset{discharge}{\overset{charge}{\rightleftarrows}} M + n/2\, H_2. \tag{8.31}$$

There is a large number of metals which can react reversibly with hydrogen in the temperature range between –100 °C and 1100 °C; an overview is provided e.g. in Möller. This type of storage is primarily of interest for solid state storage of

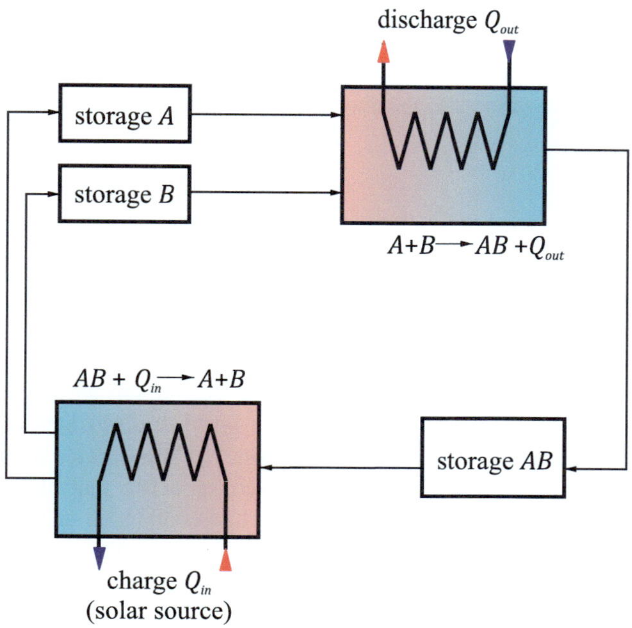

Fig. 8.26 System diagram of the operation of a TCES based on thermal dissociation reactions

hydrogen but can be similarly considered for heat storage. An example of such a relatively cheap metal hydride is Lithium, Calcium or Magnesium hydride LiH, CaH_2 and MgH_2. For other than thermal purposes such as the use of hydrogen storage also other metals like Niob or Palladium are used. However, for thermal storage the following reaction could be utilized:

$$MgH_2 + 75\text{kJ} \underset{discharge}{\overset{charge}{\rightleftarrows}} Mg + H_2. \tag{8.32}$$

From Table 8.10 we have seen that for the reaction of Mg in Eq. (8.32) has a turning point temperature of 293 °C at one bar. When MgH_2 is heated, MgH_2 dissociates in a closed system, the hydrogen pressure increases and so does the dissociation temperature; this is then 500 °C at 10 MPa. The hydrogen released during charging can be stored either as compressed gas or in a second low temperature metal hydride system. For CSP applications, storage systems are required that allow temperatures above 600 °C, but promising hydride systems are also possible here, see for example, (Corgnale et al., 2014). As with all hydrogen systems, it is primarily the integrity of the metal that is compromised during cyclic intercalation and the metal structure is degraded.

8.6.4 Adsorption Processes

Sorption in physics is describing the uptake of a substance into a substance, where the nature of the substance and its shape is open. Sorption can be distinguished as

- absorption and
- adsorption.

Absorption refers to a process where a substance completely penetrates a liquid or solid and both substances together form a homogeneous mixed phase. Adsorption is describing the accumulation of gases or vapor or even dissolved substances (the adsorptive) on a solid surface (the adsorbent) or on the interface of a liquid.

The understanding of why fluids accumulate near a surface developed at the beginning of the twentieth century. Due to their peripheral position, the outermost atoms of a solid have different properties—caused by unsaturated atomic bonding forces—than the atoms inside. This has as a consequence that they can interact with gas and even liquid molecules when these approaches the surface. According to the resulting relatively weak bonding strength between the adsorbed molecule and the solid, a distinction is made between physisorption and chemisorption.

In chemisorption, substances are bound by chemical bonds on the surface of a solid. Chemisorption can have considerable consequences for the adsorbed substance. Other substances can be formed by breaking and forming chemical bonds and a desorbed substance (desorbate) can be a product of a chemical reaction. In chemisorption, high reaction temperatures may be needed with reaction enthalpies of up to 200 kJ/mol. This is a clear criterion for differentiating between physisorption and chemisorption.

Physisorption is based on electrostatic forces of attraction that hold the polar molecules on the surface of the solid. Neither surface nor adsorbed molecule are chemically changed. The energy released during physisorption has the same order of magnitude as the enthalpy of condensation and is between 10 and 70 kJ/mol and. This energy is not sufficient to break chemical bonds, the molecules largely retain their properties. Also, for physical adsorption there is an equilibrium concentration based on the reverse process, desorption. The system always strives for a balance between adsorbing and desorbing a substance. The surface on which adsorption takes place can be the surface of a solid or the surface of a liquid. The adsorbing molecules reach the interface through their non-directional, thermally or concentration driven molecular movement. In the following sections the adsorption of gases to solid interfaces is described. Porous materials are very good for adsorption processes because within a small volume they contain a huge surface area.

The reaction enthalpy or temperatures needed for separation or desorption may be used as criterion to distinguish physi- and chemisorption. Table 8.11 gives an overview of the characteristic properties of physisorption and chemisorption.

In the *physical adsorption* on a surface, the adsorbed molecule (adsorbate) is attracted to the surface by van der Waal forces. Due to the attraction force

Table 8.11 Comparison of the characteristic properties of physisorption and chemisorption (Glaser, 2005)

Fluid	Physisorption	Chemisorption
Adsorbents	Solids	Specific solids
Adsorptives	Gases/liquids	Chemically reactive gases/liquids
Bonding conditions	Van der Waals forces	Chemical bonds
Adsorption heat	Low (heat of condensation) Approx. 10–70 kJ/mol	High (heat of reaction) Approx. 200 kJ/mol
Activation Energy	Low	High
Temperature range	Low temperatures <100 °C	High temperatures >100 °C
Sorption arrangement	Multi-layers possible	Monolayer

adsorption is an exothermic process. The enthalpies released in the process are the condensation enthalpy (Δh_{cond}) and the binding enthalpy (Δh_{bind}).

$$\Delta h_{Ad} = \Delta h_{cond} + \Delta h_{bind}. \tag{8.33}$$

The condensation enthalpy (Δh_{cond}) is the larger part with 50–100% contribution depending on the load. This indicates that substances with high evaporation or condensation enthalpy should be favorably chosen as adsorptives.

The adsorption energy of physisorption is not strong enough to change the chemical structure of an adsorbed particle, but rather polarizes it. A kind of reaction equilibrium can be formulated between the adsorbate or adsorptive, the free surface S of the solid (adsorbent) and the surface with adsorbate S_{ads}:

$$A_{gas} + S \underset{\leftarrow}{\overset{\rightarrow}{}} S_{ads}. \tag{8.34}$$

where in A_{gas} are the particles in the gas phase or in solution (adsorptive), S_{ads} is surface with adsorbate and S *is* the free surface sites of the adsorbent. When particles with kinetic energy hit the surface, as illustrated in Fig. 8.27 they are either reflected or adsorbed. Adsorption is possible when the particles can release their kinetic energy E_{kin} in an inelastic collision. Then, it is released as adsorption energy E_{ad} in the form of heat (see Fig. 8.27b). The reversible process is, therefore, exothermic, and particles can leave the surface when the adsorbent is heated up. For this process the activation or desorption energy E_{des} corresponds to the adsorption energy E_{ad}. Adsorbed particles condense on a surface as liquid and as such can move freely along the surface in a kind of two-dimensional liquid. The equilibrium concentration depends on the properties and size of the surface, the properties and pressure of the adsorbent, and the temperature. At room temperatures and atmospheric pressure, the equilibrium very often is shifted very much towards adsorption, with ideally a monomolecular layer on the surface, whereas for higher temperatures >100 °C the desorbed gas phase is favored.

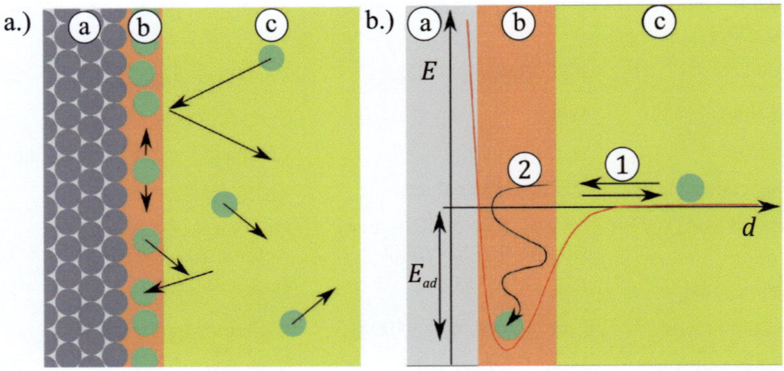

Fig. 8.27 **a** Sketch illustrating adsorption on surfaces adsorbent (**a**), adsorbate at the interface **b** gas phase or solution with adsorptive (**c**). **b** Sketch of attractive potential depending on distance from surface for adsorption (d = *distance*, E = *energy*, E_{ad} = *adsorption energy*, (1) reflection and (2) adsorption

In practice, further layers can form after the formation of the monolayer, but strictly speaking this is no adsorption but an interaction between the adsorbate particles. The multilayer arrangement is physically a condensation of the adsorptive. As the heat of condensation (binding energy of the adsorbates to each other) is only a part of the adsorption energy, the released energy is smaller than for the monolayer adsorption.

Generally, the surfaces of the adsorbents do not have an ideally flat structure but have numerous defects or lattice faults. These defects sites are particularly attractive for adsorbate particles, as shown in Fig. 8.28. The mobility of the adsorbate allows these sites to be occupied regardless of where the adsorbent was adsorbed. The capacity of an adsorbent, therefore, increases with the number of defects.

For this reason, the largely amorphous activated carbon is a very powerful adsorbent compared graphite which has a large smooth surface area. Other strongly adsorbing substances such as zeolites are not amorphous, but on the other hand have an open porous structure with a huge internal surface. The adsorption effect is not only used for thermal storage but also plays an important role in air conditioning and in catalysis. Some literature on this wider fields of applications are

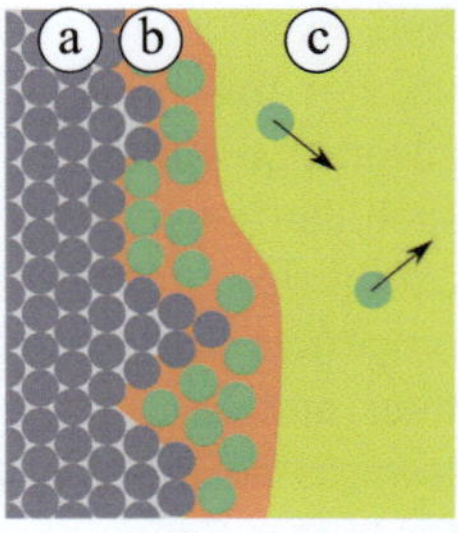

Fig. 8.28 Sketch showing adsorption on rough surfaces; adsorbent (**a**), adsorbate at the interface (**b**) gas phase or solution with adsorptive (**c**)

(Bathen & Breitbach, 2001; Schwieger & Toufar, 2007; Bruch et al., 2007). In everyday practice dry adsorbents for water vapor are used for storing and conserving materials which are easily corroded. Silica gel is used in the package box when electronic equipment is transported over large distances. If humidity of air is high then inside the box the water vapor is adsorbed in the silica gel as a sink, and therefore, the sensitive parts are not impacted.

In the following Tables 8.12 and 8.13 some properties of adsorptives and adsorbents which are used in demonstration or practical equipment using adsorption or absorption processes are provided. The applications relevant to solar energy are storage and solar cooling devices, where the sorption effect is use in a thermodynamic closed cycle.

All the adsorptives can be used with solid adsorbents, however, ammonia is also used in absorption with water as an absorbent.

- Silica gels can adsorptively bind large amounts of water/water vapor. Silica gels are highly porous, glass-like substances which are produced in a thermal process from a colloidal silicate suspension ($SiO_2 + H_2O$), where by activation takes place by removing the water at high temperatures. The material consists mainly of SiO_2 and has an inner surface of 500 to 800 m^2/g due to the manufacturing process. The usable adsorption volume is between 25 and 30% of the total silica gel volume.

Table 8.12 Thermodynamic properties of selected adsorptives

Adsorptive	Molecular chem. formula	Boiling temperature at p = 1013 mbar [°C]	Dipole moment m = e.l$m = e \cdot l$ $[3.3 \cdot 10^{-30}\ Cbm]$	Evaporation enthalpy [kJ/kg]
Ammonia	NH_3	–34	1.476	1368
Methanol	CH_3OH	65	1.67	1102
Ethanol	C_2H_5OH	79	1.70	842
Water	H_2O	100	1.87	2258

Table 8.13 Chemical composition and specific surface of some adsorbents

Adsorbent	Molecular formula	Surface area $[m^2/g]$
Alumina	Al_2O_3	$100 - 400$
Adsorber resins	e.g. styrene	$100 - 800$
Silica gel	SiO_2	$500 - 800$
Carbon molecular sieve	C(as anthracite)	$500 - 1000$
Zeolites	$Al_2O_3,\ SiO_2$	$500 - 1200$
Activated carbon	C	$800 - 2000$

- Zeolites is the name of the group of hydrous metal aluminosilicates. In nature, there are about 40 different types of zeolites, mostly formed by volcanic activity and showing strong impurities. They are therefore mostly unsuitable for refrigeration applications. Synthetically produced zeolites consist mainly of aluminium oxide (Al_2O_3) and have an inner surface of approx. 800 to 1200 m²/g. They are modified by ion exchange or by doping with hygroscopic salts (hydrophilic surface) and thus their energy storage capacity is considerably improved. Zeolites require higher temperatures (>100 °C) for desorption than silica gels and therefore provide higher temperatures in the discharge process. They can therefore also be used as thermochemical storage media in heating systems with higher flow temperatures.
- In the chemical industry, hydrophobic activated carbon or activated coke are often used as adsorbents, which are produced from wood, coal or even petroleum coke. After shaping the individual parts, the material is first carbonized at low temperatures and then activated at high temperatures. This removes all moisture from the carbon, resulting in a highly porous substance with a wide pore spectrum.

Table 8.13 shows the chemical composition and specific surface area of some important adsorbents.

Adsorption processes are an interesting alternative in solar chillers and air conditioning. Important for the technical use of an adsorption process is the knowledge of the time course of the adsorption up to the equilibrium position. In the case of adsorption on porous materials, this process is determined on the one hand by the diffusion rate of the adsorbent and on the other hand by the heat dissipation of the sorption heat released by the adsorbent (heat transfer coefficient). At the beginning of adsorption, when the adsorbent is in the unloaded state, the adsorption process is fast. With increasing loading, this process slows down and finally comes to a standstill when a state of equilibrium has been reached, a so-called adsorption equilibrium.

- Pressure p (for gases),
- temperature T and
- adsorbed mass m_{ads} or relative load x_i,

which depend on the adsorbed components and the adsorbent used. Adsorption is favored at high partial pressure and low temperature, whereas desorption requires an increase in temperature by heat input and/or a reduction of the partial pressure. Due to the strong molecular binding forces, polar gases are particularly suitable as adsorptives. The greater the dipole moment of the gas molecules, the higher the energy density achieved in the sorption process. Table 8.12 shows the thermophysical properties of selected adsorptives.

For zeolites and silica gels the Fig. 8.29 shows the equilibrium curves for a certain loading x_i for the vapor partial pressure versus temperature of the absorbent. The higher the temperature the lower can be the loading for a certain vapor

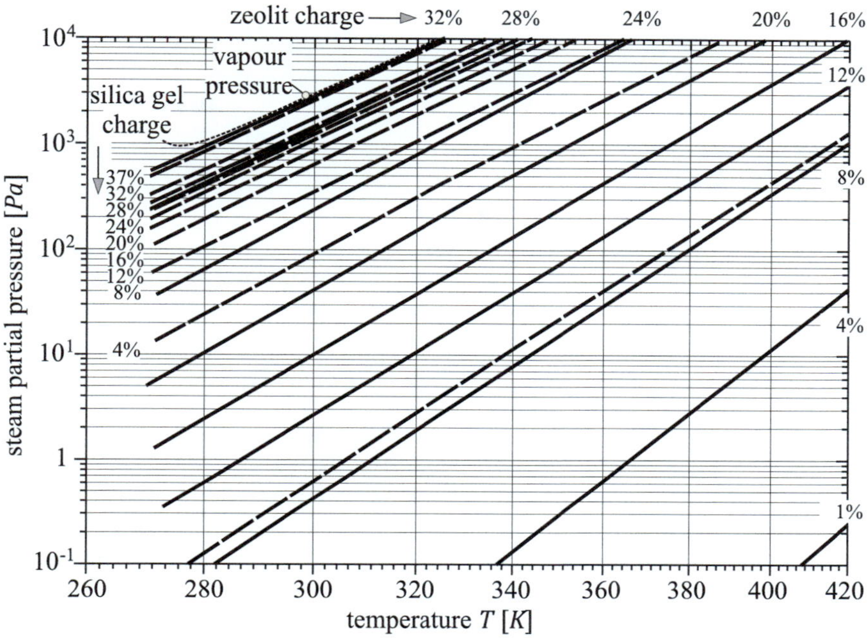

Fig. 8.29 Comparison of selected isosters of silica gel (dashed) and zeolite (solid) curves for various loading x_i in a temperature range up to 150 °C modified from (Hauer, 2002)

pressure, which implies that the storage capacity is dependent on the charging temperature.

8.6.5 Absorption

Similar to the adsorption of substances to a surface, the absorption of substances in a homogenous absorbent liquid is based on weak attractive forces, and only relatively low temperature are needed to reverse the process to separate again the substances. The most common absorbent is liquid water. Salts, bases, acids, and gases can be dissolved in water, the solution process is temperature dependent—the reaction kinetics is reversible.

In Table 8.14 some examples for absorbates which dissolve in water are listed. The reactions may be used for thermal storage purposes, however the main use of them is the working fluid the cooling cycle of an absorption chiller. The absorbate is expelled from the solution by thermal heat using a heat exchanger. Because the evaporation and condensation of a refrigeration medium like $LiBr$ may occur at very different temperatures, depending on the pressure. The absorbent water is used then to extract the cooling medium from the evaporator. The solution has to be regenerated at high temperature and higher pressure using external heat.

Table 8.14 Examples of absorbent-absorbate couples

Absorbent	Adsorbate	Absorption enthalpy $[kj/mol]_{absarbat}$	Molar mass $[g/mol]_{absorbat}$	Storage density $[kWh/kg]_{absorbat}$
$H_2O(l)$	$LiBr(s)$	49.04	86.85	0.157
$H_2O(l)$	$NH_3(g)$	34.18	17.03	0.558
$H_2O(l)$	$H_2SO_4(l)$	95.28	98.08	0.270

The exact principle is described in Chap. 12. Due to the low bonding energy of a medium in the water solution, which is based on polarities, therefore low discharge temperature can be utilized for the whole absorption process.

8.6.6 Thermochemical Energy Storage Systems (TCES)

We have seen that a number of reversible reactions with different reaction enthalpies, depending which forces are responsible for the attractive potential between the different substances, exist which can be in principle used for storing thermal energy. Heat is used normally to dissociate the substances and overcome the attractive forces, and the heat in turn can be released again when the substances will react again at lower temperatures. Storage systems designed on these principles are called thermo-chemical energy storage (TCES) systems.

The advantage of thermochemical heat accumulators over conventional heat accumulators in the form of a water tank is their higher storage density. Theoretically, values of 200–300 kWh/m^3 are achievable compared to only about 60 kWh/m^3 for water. In practice, however, only about 130 kWh/m^3 have been achieved in the best pilot plants to date. A major advantage is that the stored energy can be stored over long periods of time without loss. Storage tanks based on sorption can be trimmed to an appropriate working temperature. Due to the low binding energy here, the required temperatures are low. For silica gel charging temperatures of about 40–100 °C and for zeolites temperatures in the range of about 130–300 °C are typical. For metal hydrides already, typical temperatures of about 280–500 °C are necessary. Storage based on real chemical dissociation reactions is rather used for high-temperature processes. Reaction temperatures of 500–2500 °C are needed depending on the substances. For all storage systems we need.

- a reactor volume where the substances can react with each other;
- a method to transfer heat into the reacting volume;
- a method to retrieve the released heat in the discharging process.

In literature, the storage tanks controlled by adsorption processes are often referred to as thermochemical heat storage tanks. They store heat by endothermic reactions and release it again by exothermic reactions. The best known example of a thermochemical heat storage is the so-called sorption storage: A tank contains granules of silica gel (zeolite), which is hygroscopic, highly porous and with a large inner surface (one gram may have an inner surface of several hundred square meters). Silica gels have the property of attracting water vapor and adsorbing it on their surface (adsorption), releasing heat. Conversely, heat energy must be used to dry silica gels (desorption). Technically, the silica gel is stored in granular form in a boiler with a heat exchanger inside. The thermal conductivity of granular silica gel or zeolite is very low, so we have a similar problem as in latent heat storage with solidification. The heat exchanger surface needs to be in close contact to the adsorbent releasing the heat, therefore the design of a highly conducting manifold or matrix of the heat exchanger is key to power rate for charging and discharging via a heat transfer fluid.

Due to the sorption equilibrium depending on temperature and vapor pressure as shown in Fig. 8.30 (left) the storage capacity of an adsorption storage depends very much on the temperature provided by the solar system to dry the material. Similarly, for a specific desorption temperature, i.e. a dry sorption material in a specific state, the temperature of the discharging air saturated with water–vapor determines in turn the outlet temperature of the dry air after the sorption process. This is shown for a sample on the right-hand side of the Fig. 8.30. The warmer the air, the higher is the water content of the saturated air, and the larger the sorption energy leading to a heating of the carrier air.

Storage systems using sorption technology can be built in different ways. But contrary to sensible or latent heat storage, the temperature of the medium which carries the discharged heat is not determined by the charge state of the TES fully, as explained above. So, a storage system usually is more complex than an ordinary sensible storage. Figure 8.31 shows the hydraulic scheme of a compact thermal energy storage developed in the European COMTES (Combined Development of Compact Thermal Energy Storage Technologies) project (van Helden et al., 2016;

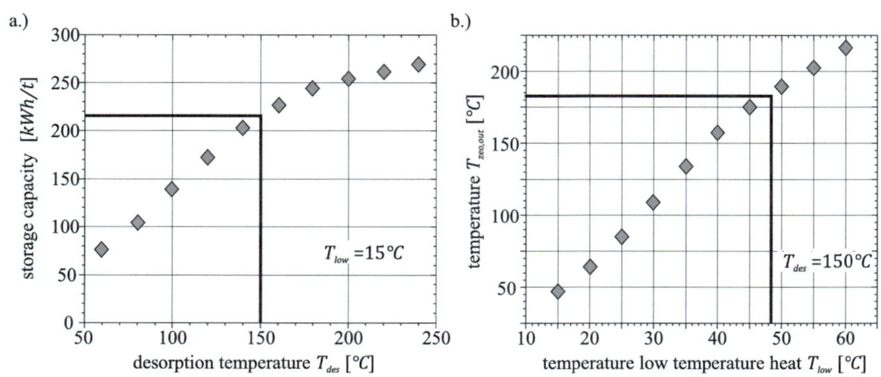

Fig. 8.30 **a** Storage capacity as a function of the desorption temperature (T_{des}) indicating that storage capacity increases with increasing temperature of waste heat; **b** dependence of the discharge temperature ($T_{zeo,out}$) on the temperature of the low temperature heat (T_{low}). Both graphs for zeolite 13*X* from (Hauer & Gläser, 2011)

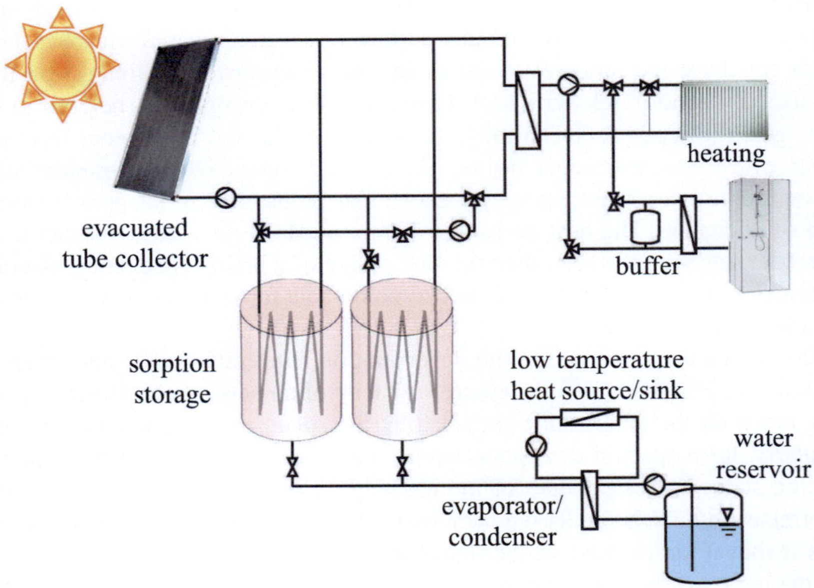

Fig. 8.31 Possible configuration of an adsorption storage coupled to a solar thermal system from (van Helden et al., 2016)

van Helden and Rommel). A storage volume of two times $1000l$ filled with binder less $13X$ zeolite was realized in stainless steel vessels. The adsorption material is characterized by high adsorption capacity and fast adsorption kinetics. The idea behind is to develop a long-term seasonal storage for low-energy houses. To get a high-capacity high-performance vacuum flat plate solar thermal collectors were considered as the high-temperature heat source for drying of the material during summer. For low temperature heating the heat for evaporation is taken from the environment, using an air heat exchanger unit. In addition, the solar thermal collectors deliver low-temperature heat for evaporation in case of insufficient ambient conditions. The storage here is *a packed-bed storage*.

In packed bed systems, the packed bed of sorbent functions both as storage and as reactor vessel, without transport of the sorbent. Another variant is the *separate reactor system*, where—for charging—the sorbent is taken from a tank with the humid material, transported to a separate reactor where it is dried and charged, and then finally it is stored in a second container for charged dry material. For discharging the process is reversed. In this variant the storage volume is decoupled from the reaction.

Systems using solid particles for adsorption tend to be of the packed bed type because the transport problem of solid material, while systems using liquid sorption have less problems with the separate reactor type. However, a few solid adsorption reactors separated from the storage tanks can also be found.

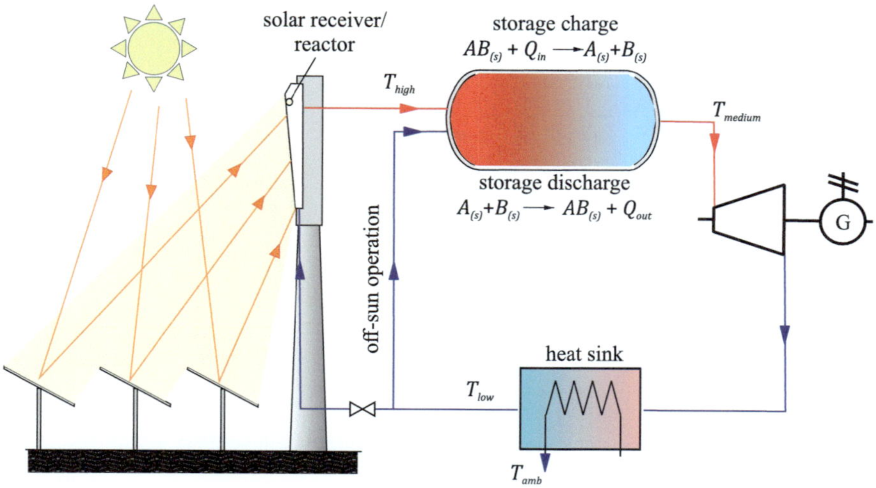

Fig. 8.32 Flow diagram and operating principle of thermochemical energy storage system integrated with solar thermal power plant for continuous power production modified from (André & Abanades, 2020)

A second classification is between *open systems* taking the sorbate from the ambient (e.g., water vapor from the air) and *closed systems* in which the sorbate remains within the system, and only the evaporation heat is taken from the ambient. Figure 8.32 is an example of a closed system which is less dependent on ambient temperatures. Open systems work at ambient pressure and use sorbate components already present in the air, such as water vapor. The water vapor is transported as moist air by a fan.

On the other hand, closed systems are more flexible in the choice of the sorbate, and also other liquids such as ammonia or methanol can be used. Inert gases are removed as much as possible from the system to improve heat and mass transfer between absorbate and absorbent. The system is evacuated, and the system pressure depends on the equilibrium pressure of the sorbate under operation. If water vapor no ventilator is necessary, but the vapor pressure difference between sorbent and evaporator is sufficient for a passive transport.

Similar to sorption storage system also chemical reaction storage systems may work with separate reactor or integrated reactor within the storage, as illustrated in Fig. 8.33a, b. The advantage of separated storages is certainly that reaction materials can be stored in individual containers for a very long time without any possible reverse reaction degrading the storage capacity.

$Ca(OH)_2/CaO$ has been demonstrated to date to be the most interesting studied thermochemical energy storage system based on metal hydroxides for high temperature applications and has prompted tests in lab-scale reactors and thermogravimetry analysis (TGA). An experimental proof of the reversibility of the chemical reactions is shown in Fig. 8.34 for reaction of calcium oxide (CaO) to

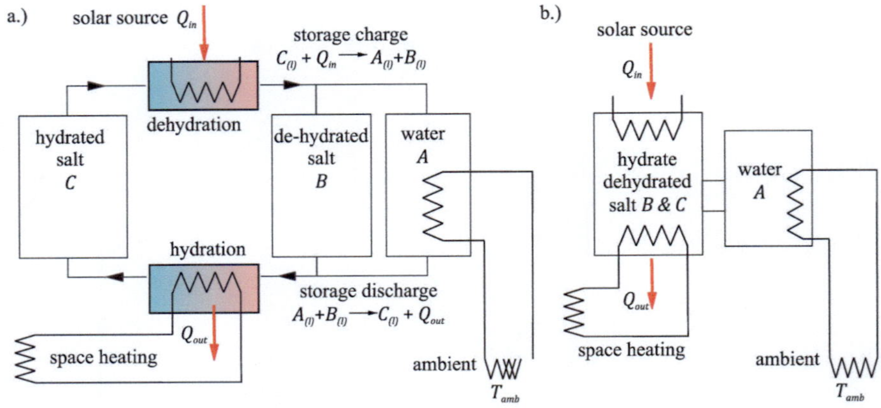

Fig. 8.33 Closed TCES concept options for domestic utilization; **a** separate reactor vesus integrated reactor

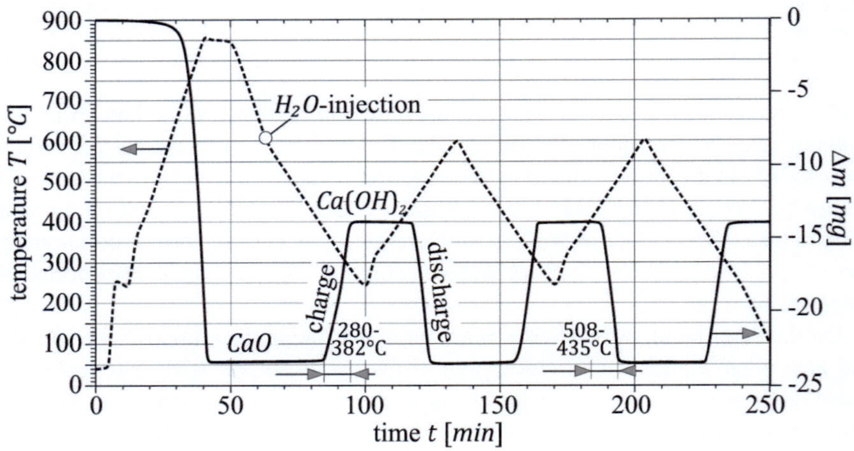

Fig. 8.34 Thermogravimentric analysis (TGA) of the cycling stability of TCES based on $Ca(OH)_2/CaO$ dissociation reaction for high temperatur applications from (André & Abanades, 2020). Charge/discharge cycles under 21 mol% H_2O

hydroxide (see Eq. (8.30)). Despite this successful demonstration still the enhancement of material stability is required to reduce the sintering effect. $Ca(OH)_2$ particles, which show to agglomerate faster than CaO particles in the presence of H_2O, and the presence of H_2O also accelerate the agglomeration of CaO particles. Also, the reaction kinetics and the active surface poses challenges. Hereby, the rehydration step is slower than the dehydration step (Criado et al., 2017) and the hydration of CaO agglomerated lump is more challenging than that of fresh particles.

Although storage systems based on sorption or chemical reactions have very high storage capacities, they are not yet in a commercial stage. This is due to some inherent limitations with respect to a successful implementation in practical systems:

- The requirement of low temperature heat being available at all times constitutes a barrier for the heat storage option because at least 5 to 10 °C are needed for evaporation to take place, independent of ambient air temperatures.
- Specialized designs are required for effective storage reactor beds, particularly for the solid sorbents, where an adequate rate of heat and mass transfer is needed to produce sufficient power.
- Solidification/freezing of the working pair materials containing water may pose a risk, especially while flowing through the hydronic piping network.
- Thermal losses during the starting and closing operational stages of the storage may be high, due to a high discharge temperature delivered from the storage. The sensible heat fraction is lost after a short storage time.
- In closed loop systems, the vacuum has to be maintained, which needs extra effort and maintenance.
- The dense materials used in the thermochemical or sorption storage are most suitable for heat pump or cooling applications due to their storage density and power density.
- Thermochemical storage systems are still very expensive in terms of both system and materials cost.

In Table 8.15 the three principles of thermal energy storage described by sensible, latent, and thermochemical energy storage, are characterized focussing on the high-temperature range and the main features are outlined.

8.7 Commercial Water Storage Systems

For most solar thermal systems worldwide pressure less water tanks storing hot water are the most common TES options. As the water boils close to 100 °C at standard atmospheric pressure, the maximum operating temperature for safety reasons is chosen to be typically between 85 and 90 °C. Water is a very cheap and abundant material with a very high specific thermal heat capacity (c_p). The tanks with appropriate heat-exchangers, therefore, determine the main investment costs.

There are water tanks for drinking water, which have to fulfil some hygienic requirements, buffer tanks for non-potable water from the heating or solar loop, and combinations of those. Buffer tanks are usually made out of ordinary carbon steel or polymer material.

Pressurized water storages at reasonable low pressures shift the boiling point a few degrees higher and thus may be used with a maximum temperature of 110–130 °C. This enlarges the overall specific capacity of water storage due to large temperature spread, but at higher cost and complexity. Material and construction

Table 8.15 Comparison of main characteristics of thermal energy storage concepts for the use in CSP, adapted from (Pardo et al., 2014; Sunku Prasad et al., 2019)

Storage concept	Sensible heat storage (SHS)	Latent heat storage (LHS)	Thermochemical energy storage (TCES)
Gravimetric energy density $[kWh/kg]$	0.02 – 0.03	0.05 – 0.1	0.5 – 1
Volumetric energy density $[kWh/m^3]$	~50	~100	~500
Storage temperature	Charging temp	Charging temp	Room temp
Development status	Commercial	Demonstration	Pilot
Energy storage period	Limited	Limited	Unlimited
Energy transport	Short distance	Short distance	Long distance (>100 km)
Technology complexity	Simple	Medium	Complex
Drawbacks	High thermal losses, large quantity of storage material	Thermal losses over time, corrosive materials, low heat conductivity	Expensive investment, complex technology

have to withstand pressure of a few bars and have to be air-tight. Usually, steel of glass-fibre enforced tanks are used for that.

The sizes of commercial tank water storages for solar systems range from about 100 l to 200 m^3. Larger seasonal storages in the ground will be discussed also in the chapter.

8.7.1 Requirements

Low-temperature storage tanks are mostly used to heat domestic hot water, to support the heating system and for low temperature process heat. The thermal energy required to charge the storage tank (Q_{total}) is made up of the direct heating of the useful fluid (Q_{use}), the storage losses (Q_{loss}), and the circulation losses (Q_{circ}):

$$Q_{total} = Q_{use} + Q_{loss} + Q_{circ}. \tag{8.35}$$

The storage losses depend both on the temperature in the storage tank and on the ambient temperature, which is determined by the location of the storage tank. The storage losses, which are based on the European standard (EN 12,977 -3 2018) are calculated as follows:

$$Q_{\text{loss}} = U \cdot A \cdot \Delta T \cdot t, \tag{8.36}$$

in which U refers to the product-dependent heat transfer coefficient of the storage tank [W/(m^{2}K)], A denotes its surface area [m^2], ΔT *is* the average temperature difference between the storage tank water and the ambient [°K] and t the period of storage tank losses under consideration.

To ensure in larger buildings that hot water is available for every user within a short time, a circulation pipe is usually installed when the hot water heated in a central boiler room, because long pipes lead to unwanted dead times drafting. In this pipe the hot water is circulated at certain hours or all day long, which leads to an increased heat loss. This additional loss Q_{circ} is often underestimated. The power loss can be estimated with a good approximation when the length of the pipe is known. For well-insulated circulation pipes as a rule of thumb is approximately 8 W/m for pipes inside the building or in well-insulated outside walls. For less well insulated pipes, at least 10 W/m pipe should be assumed. Wall and ceiling openings are considered by adding 10% as there only half the insulation thickness is needed according to the regulations, i.e. for well-insulated pipes in a new building.

$$Q_{\text{circ}} = 1.1 \cdot l_{circ} \cdot 8[W/m], \tag{8.37}$$

where l_{circ} is the length of the circulation line in $[m]$. Table 8.16 lists the minimum thickness of insulation required for different pipe diameters to comply with the German Building Energy Act (Gebäude Energie Gesetz—Building Energy Act (in German) 2020).

The thickness of the storage insulation depends individual design. For the period considered, it must reduce the heat loss and maintain the stored heat as well as possible. The thickness is limited by technical necessity and economic considerations. The requirements for storage tank insulation are.

Table 8.16 Required insulation thicknesses for hot water pipes with different pipe diameters from (Gebäude Energie Gesetz—Building Energy Act (in German) 2020)

Line	Type of pipes/fittings inner diameter $[mm]$	Minimum insulation layer thickness related to a thermal conductivity of $k = 0.035$ W/(mK)
1	$\rightarrow 22$	20
2	$22 - 35$	30
3	$40 - 100$	Same as inner diameter
4	> 100	100
5	Pipes/fittings according to lines 1- 4 in wall, ceiling penetrations, in crossing area of lines, at line connection points, at central line network distributors	1/2 of the requirements of lines $1 - 4$
	Refrigeration distribution, chilled water piping, fittings of room air conditioning, air conditioning refrigeration systems	6

Table 8.17 Material choice and recommended siphon depth of thermosiphon connection to domestic water storage from (Streich, 2010)

Heat conduction	Recommended material	Siphon depth t [cm]
Very low	Plastic or stainless steel corrugated pipe	4
Low	Stainless steel pipe 1 mm thick	7
Moderate	Stainless steel pipe >>1 mm thick	10
Large	Copper or steel pipe	Not recommended

- non-flammability,
- the lowest possible thermal conductivity,
- only small or preferably no air gaps between insulation and storage wall.

For an energy-efficient solar storage tank a good thermal insulation is essential. The preferred thermal conductivity k is below 0.035 W/(mK). The bottom and top of the storage tank as well as the support-contact to the ground should be insulated, it should fit well tightly, should cover flanges, plugs, etc. and be connected to the pipe connections without gaps. Domestic hot water storage tanks are often equipped with a foamed insulating jacket at the factory. As a guideline a value of 0.5 W/K relative to one cubic metre should not be exceeded. As a rule, 15–20 cm thick insulation is sufficient. The connections to the storage tank must be designed in such a way that the storage tank cannot be discharged overnight due to natural circulation. This can be counteracted by appropriate measures such as a thermosiphon as illustrated in Fig. 8.35 or a check valve.

Line openings for cold water are to be provided in the lower, cold storage area; connections for warm water should be installed in the upper, warm storage area. Since the storage tank insulation should not be interrupted, especially in the uppermost part of the storage tank, it is advisable to route the hot water connection internally (in the storage tank itself) or externally (between the storage tank wall and the outer skin of the insulation) through the floor insulation. In the internal version, the storage tank stratification is disrupted and the hot water in the pipe is cooled down.

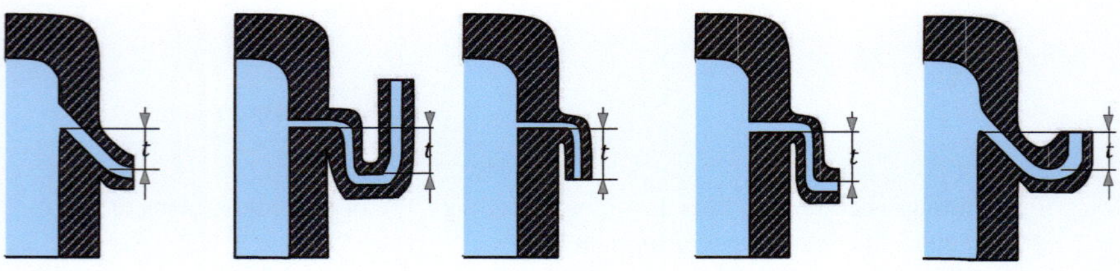

Fig. 8.35 Recommendations for pipe connection to thermal storage tanks for different variants of the thermosiphon as protection against unwanted cooling losses, the required siphon depth t and material choice may be taken from Table 8.17

8.7.2 Water Tank Storage Types

Water storage tanks are being produced in many slightly different design options. Starting from the application profile, the basic type of storage tank is selected first. An unpressurized storage tank is limited in temperature to 90 °C, because above under atmospheric pressure the evaporation of the hot water is appreciable. For domestic hot water and heating this is sufficient, however, in some process heat applications with pressurized storage (up to 3–4 bar) upper storage temperatures of 120–130 °C can be reached. A pressurized storage needs to be stable under internal pressure, and more over shall be airtight. This leads to higher cost, and therefore is avoided where possible.

A second options is the feasibility of storing potable water or not. Potable water needs generally high-quality stainless-steel tank walls or tanks with an enamel coating inside, as otherwise hygienic requirements could not be met. The fluid in a potable water tank shall not be mixed with heat transfer fluid from the solar loop. If less stringent specifications have to be met, so-called buffer storage tanks at lower costs can be used. Combination of both types in one tank is given by the so-called combi-stores.

This is followed by a further differentiation into an internal or external type of heat transfer to the storage tank. Finally, if additional heating is desired, which is called the bivalent option, it must be decided whether this should be direct or indirect. Without claiming completeness Fig. 8.36 shows the different options for unpressurized storage water tanks. The central element of a solar heating system for domestic hot water (DHW) preparation is the domestic hot water storage tank, and in the case of simultaneous heating support, it is the buffer storage tank. There are countless versions of both basic types. Due to its high availability, low cost and non-toxicity, water is the preferred storage medium in residential areas.

There are a variety of options for placing the heat exchanger inside the water tank, depending on both the size of the solar heat source and the system concept. The system concept includes central information such as the number of water tanks, pipe routing (single or dual pipe network), centralized or decentralized domestic water heating, etc. Figure 8.37 gives an overview of possible heat exchanger arrangements in a water storage tank.

Water characterized by a particularly high specific weight and volume related heat. In liquid form, it can also be used as a heat transfer medium for loading and unloading the storage tank.

8.7.2.1 Hot Water Storage (Potable Water)

In systems with direct heating of the drinking water by the collector circuit, drinking water storage tanks are used for the entire volume. In buffer storage tank concepts, they only serve as standby storage tanks for the hot water supply. Domestic hot water (DHW) storage tanks must be protected against corrosion caused by the oxygen that continuously enters with the DHW. For this reason, stainless steel tanks, steel tanks with internal coatings of enamel or plastic tanks are used (Stryi-Hipp, 2011). Although stainless steel tanks are more corrosion-resistant and more

application	basic storage type	heat transfer storage design	bivalent option (re-heating)
potable water 	service or potable water storage	internal / external: o flow through principle; o pipe coil in storage; o heat exchanger and pump	direct: o electr. heating in tank; indirect: o external heating flow-through-principle; o electring heating in tank
potable water + heating  & 	buffer storage	internal: o tank-in-tank system; external: o connection to buffer storage and pump; o separate tank and pump	direct: o electr. heating in tank; o firing with gas /oil; indirect: o internal via HX in tank (fed by other heat source); o external re-heating in flow-through-principle
heating 	buffer storage	internal / external: flow-trough-principle; o pipe coil in tank; o HX and pump	direkt: o electr. heating in tank; o firing with gas /oil; indirekt: o internal via HX in tank (fed by other heat source)); o external re-heating in flow-through-principle

Fig. 8.36 Classification of water heat storage systems for residential use

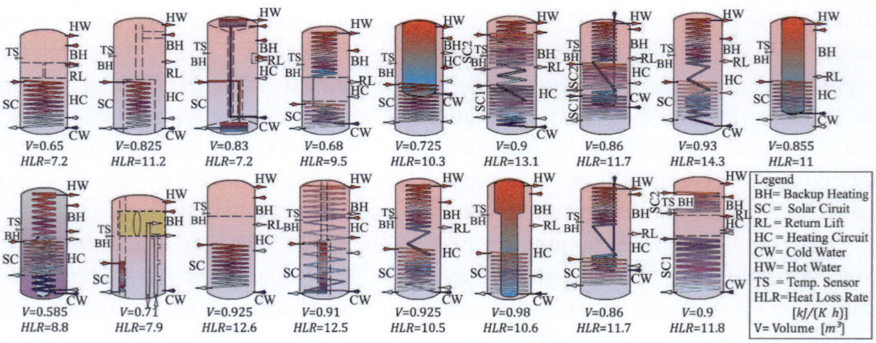

Fig. 8.37 Options to arrange heat exchanger in water storage tanks for residential use (Stryi-Hipp, 2011)

durable, they are considerably more expensive than the other variants. Enameled storage tanks must be equipped with either a magnesium or an external current anode (see for example Fig. 8.38) for corrosion protection reasons.

Depending on the design of the overall system, there are different types of drinking water storage tanks. Especially for solar systems that are only used for hot water preparation, a storage tank with two internal heat exchangers is very common. Here, the storage tank is supplied by the solar circuit via the lower heat exchanger, while the upper heat exchanger, which is supplied by a conventional heat generator, takes over the additional heating of the upper storage tank area if required, from which hot water is also drawn (see Fig. 8.38). In some storage tanks

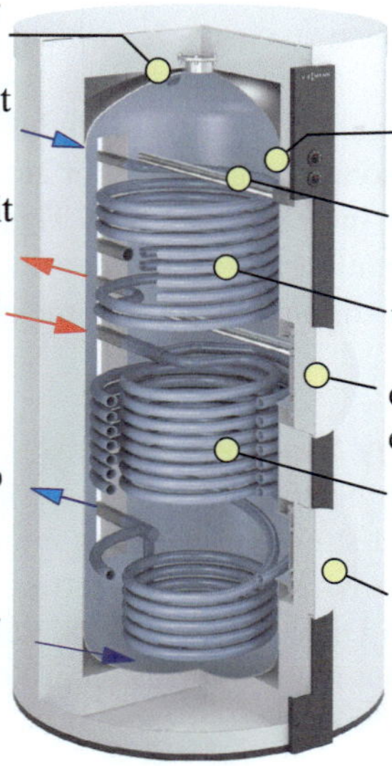

Fig. 8.38 Version of a domestic hot water cylinder with solar connection Viessmann—Vitocell-100-B (Viessmann Climate Solutions SE, 2023)

the upper heat exchanger is replaced by an electric heating cartridge. The division of the heat-up volumes should be as follows: upper heat-up volume 40% and lower heat-up volume 60% of the storage volume. In one and two-family houses, storage tanks with a capacity of 300–500 l are common.

The solar circuit heat exchanger should reach as far down in the tank as possible so that the tank contents can be heated up to the tank bottom as depicted in Fig. 8.38. A connection of the post-heating heat exchanger in the upper part of the storage tank guarantees a rapid heating of the standby volume without depriving the solar circuit of the possibility to effectively store even small amounts of solar energy in the colder part of the storage tank. The technical implementation in Fig. 8.38, showing a storage tank that has two heat source inputs (solar supply and conventional fossil fuel firing/electric heating cartridges). This version is called also a bivalent storage tank.

For domestic hot water storage tanks with a volume of more than 400 l, it is required to heat the tank up to 60 °C once a day to prevent the formation of legionella. As the energy required for this type of heating is relatively high, downstream heat exchangers, which operate on the flow principle, are increasingly being used. In large solar systems, drinking water storage tanks are being used less and less often due to the difficult drinking water hygiene. In this case, systems with a two-storage tank system are mainly used, in which the drinking water storage tank is heated by the buffer tank as a boiler.

The energy is introduced into the service water storage tank via pre-installed pipe registers or via finned tube heat exchangers, which can be retrofitted via flanges as required.

The storage tank temperature should be limited to slightly more than 60 °C for domestic hot water tanks, both for solar charging and for auxiliary heating, as lime precipitates at higher temperatures increases strongly. This has the consequence that the dissolved calcium compounds form solid deposits at the hottest points, which hinder the heat transfer. In addition, crumbling lime gradually settles on the floor of the storage tank (silting up).

8.7.2.2 Buffer Storage (Energy Storage)

Buffer storage tanks are filled with heating water and serve the purpose of long-term storage of solar heat. The heat stored in them can either be fed directly into the heating system (central heating backup) or transferred to the drinking water via a heat exchanger. Compared to DHW storage tanks, they offer the advantage of being cheaper because they do not require corrosion protection. In addition, the stored heating water can be heated up to 90 °C—there is no danger of calcification with treated heating water—which means that more energy can be stored in the storage tank for the same volume. In closed systems with overpressure in the energy storage circuit, steel storage tanks made of $St37$ steel are usually used. These can be used without any problems and without further corrosion protection, as the system is only filled with water once and the overpressure prevents the penetration of fresh oxygen. The Fig. 8.39 shows a buffer storage tank for heating hot water supply with a potential external auxiliary fossil fuel heating system. In principle, this combination version can also be operated without solar radiation.

Efficient buffer or combi tanks have a riser pipe for the solar fluid with outlet nozzles at different heights. The heated water rises in the pipe until it has the same temperature as the water in the storage tank. In this way, the thermal stratification in the tank is maintained and convection losses within the tank are reduced.

Figure 8.40 shows a purely solar-operated stratified charge storage tank, which is more suitable for low flow systems than long-term storage. In the stratified storage tank, the heated fluid, which has a lower density than the bottom of the tank, rises in the riser pipe up to the point in the tank where a temperature congruence occurs. This works only if the impulse and velocity of the solar fluid is small, hence it is restricted to low-flow systems.

In standard solar thermal systems, the water storage is used to cover the demand for $1 - 2$ days without large solar gains. However, there are concepts, even for individual houses, which require a much larger storage volume and storage period. If not only a small fraction of space heating load shall be covered by solar, a medium to long-term seasonal storage is required. A first requirement is certainly a very low heating demand due to extremely good insulation level of the building. But, then in principle large storages of several m^3 volume incorporated in houses can store the heat from autumn into the winter month. The Swiss solar pioneer Jenni Energietechnik has designed such buildings, typically with storage fabricated inside the house or put into the house with a crane during construction. A volume

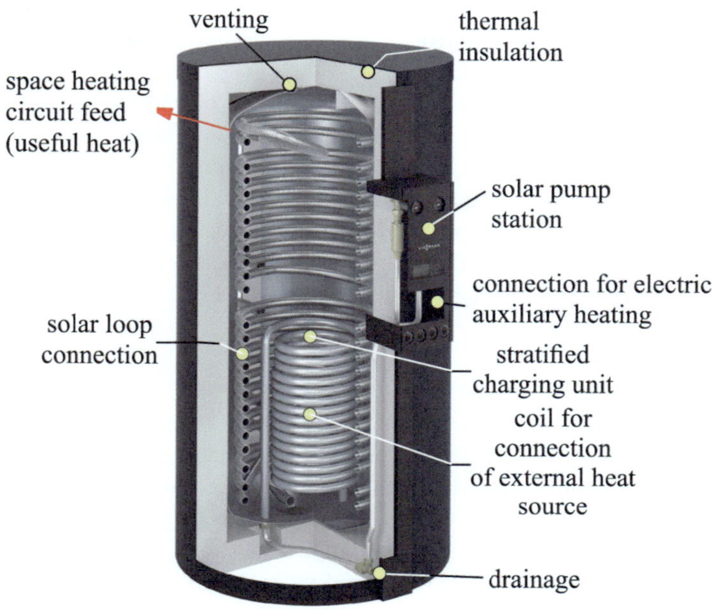

Fig. 8.39 Buffer storage tank for heating hot water with heating option—Vitocell—340 M (Viessmann Climate Solutions SE, 2023)

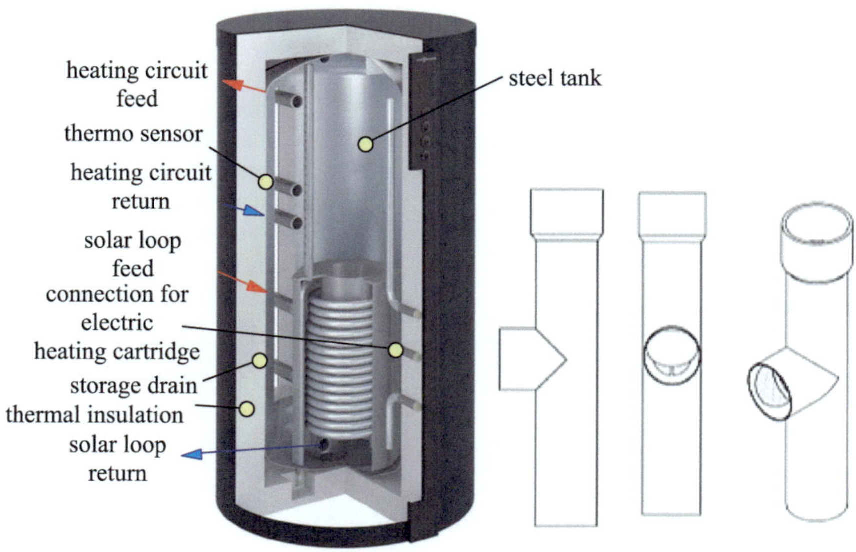

Fig. 8.40 Heating water buffer tank with stratified charging device Vitocell—160E (Viessmann Climate Solutions SE, 2023) left; on the right hand side outlet openings in the riser pipe are shown. Flaps in the outlet may support the stratification by reducing the outlet velocity

of 20 m^3 or more is needed for a well-insulated single-family house and a sketch of such a system can be taken from the Fig. 8.42. Such a tank may be welded on-site or shipped as a completed unit.

In open/pressure less systems, storage tanks made of various plastics can also be used besides steel tanks to reduce the risk of corrosion. However, the maximum temperature in the plastic tanks there must be safely limited due to the load limits of the plastic. Buffer storage tanks are often supplied with foam insulation tailored by the manufacturer or with hard shell thermal insulation made of polystyrene, PUR foam or melamine. Larger storage tanks are thermally insulated in several layers with mineral wool, which is sealed with a sheet metal jacket, or with foam insulation materials.

8.7.2.3 Combined Storage

The combination cylinder is a combination of buffer and drinking water storage. In their basic design, combi cylinders are buffer cylinders for the heating system, with a second closed cylinder or a spiral heat exchanger, both holding the volume of the standby requirement for domestic hot water, is located in the upper area. Systems with this storage tank are also known as "Tank-in-Tank systems" (see Fig. 8.41). The solar storage tank is charged via an external heat exchanger and is suitable for medium to large collector surfaces (from approximately 20 m^2).

In addition to the compact design of the combi-reservoir, a major advantage of this type of storage tank is that the desired useful temperature can be provided quickly if there is sufficient irradiation, as there is always a stand-by volume with a high temperature in the upper part of the combi-reservoir. With combi-reservoirs,

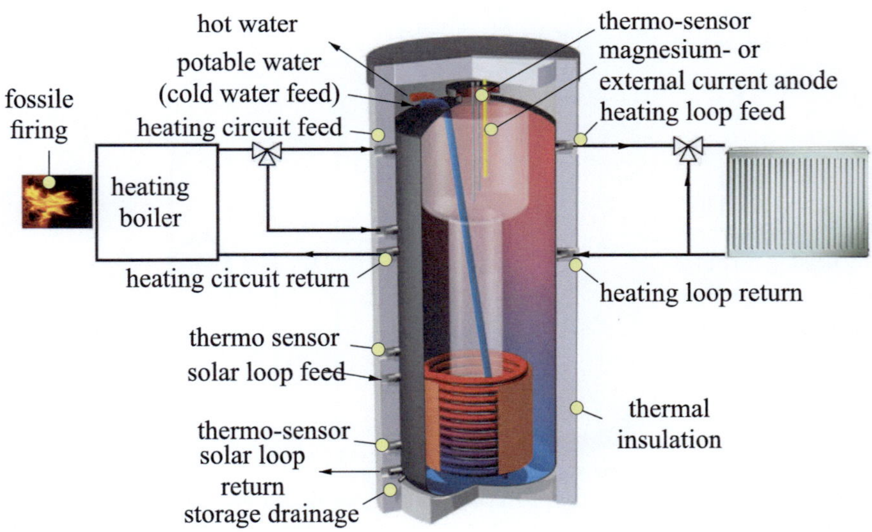

Fig. 8.41 Schematic diagram of a tank-in-tank combination storage tank with auxiliary fossil heating for the provision of hot water and heating water

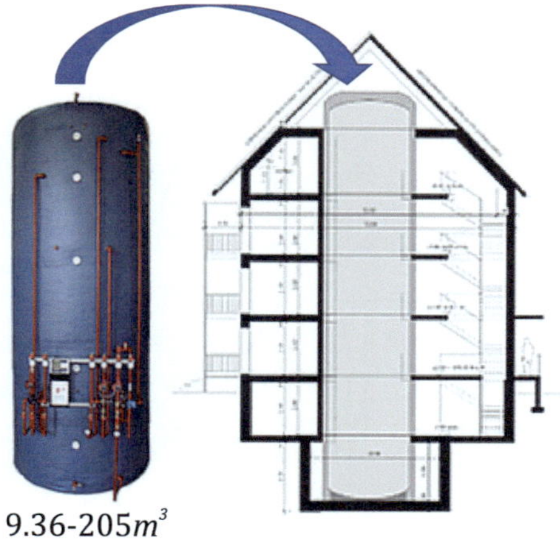

Fig. 8.42 Large seasonal storage incorporated in a building (Dasch, Hilz et al., 2013), © Jenni Energietechnik AG www.jenni.ch

the solar heating system can be operated in all modes (low—flow, matched flow and high—flow, see Sect. 8.4.2 Forced circulation systems). A disadvantage of the combination storage tanks is the hydraulic separation of the solar circuit from the storage circuit by an external plate heat exchanger, which requires a second pump circuit.

These large storages can also be used for hot process water in industry. Commercially on may order 10–200 m^3 sized tanks produced in the factory and transported to a site by rail and by trucks.

8.7.3 Seasonal Storages in Heating Grids

Just as in the domestic sector, energy storage systems can also be implemented with large storage volumes; the principle of stable stratification also applies there. Seasonal heat accumulators are provided in the network of solar-supported local heating supply systems (Mangold et al.). The aim is to provide both heating energy and domestic hot water for a large number, say 100 to 500, residential units with a solar coverage of more than 50%. District heating solar fields have been installed in many countries in the Northern hemisphere, for example in Denmark, Sweden, France, Austria, Canada, and Germany. Denmark is particularly successful in economic terms. Sweden has pioneered work based on rock cavern storages which is available there in many places. Generally solar fields from several 10^3–$10^4$$m^2$ $10^3 - 10^4 m^2$ are connected to a central heating system. Due to scaling laws the specific cost is decreasing with size, as depicted in Fig. 8.43. The reason is that the surface of a storage is costly due to wall construction and thermal insulation whereas the volume is cheap (essentially consisting of water). A second viewpoint is that thermal losses are proportional to the surface area of a storage and the storage capacity to the volume. With larger volumes the surface-to-volume ratio drops and hence the losses become comparably smaller. Less insulation may be used and long-term storage is economically feasible.

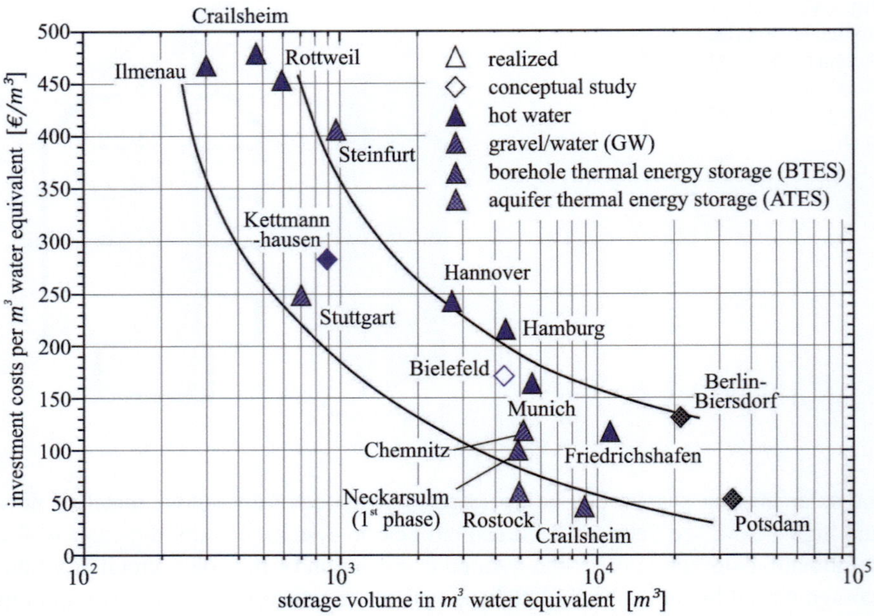

Fig. 8.43 Specific cost reduction of seasonal storage in Germany depending on volume from (Schmidt et al. 2004)

Four different basic thermal storage systems have been used in different locations, which are shown in Fig. 8.44.

The first one is the tank thermal energy storage (TTES) based on a concrete storage tank built into the ground, see Fig. 8.44a. This type can be used everywhere as long as area able to bear large weights is available (Sibbitt et al., 2012; Reuss et al., 2020; Battaglia and Haller op., 2018). A subtype of this type of storage is cavern storage (CTES), which is characterized by the storage volume being located at greater depths, for example, natural or mined caverns. The pit thermal storage (PTES), shown in Fig. 8.44b is similar, except that the walls have not static function and consist of polymer foil for holding the hot water inside and thermal insulation against the ground. Earthworks require a ground which is not consisting of rock or similar rigid material. Borehole thermal energy storage (BTES) needs relatively soft ground material like clay which should also be able to store and conduct heat. Here the storage material is the ground itself, with deep loops of piping introduced into the boreholes to exchange heat to a fluid as illustrated in Fig. 8.44c (Reuss et al., 2020). Finally, aquifer thermal energy storage (ATES) uses natural water reservoirs partially embedded in the ground and filled with water or water/gravel mixtures as sketched in Fig. 8.44d. For these kinds of TES, it is important that the water reservoir is localized and not connected to an underground river which carries away the heat. The production costs of these plants scale with the area or the required storage volumes. Nevertheless, more and more combined

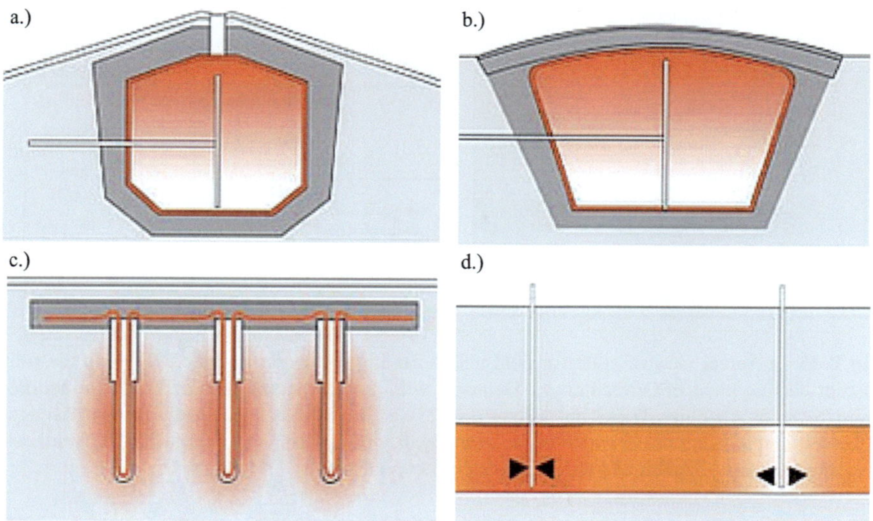

Fig. 8.44 Classification of large seasonal heat storage technologies from (Pauschinger 2015); **a** tank thermal energy storage (TTES, 60–80 kWh/m^3), **b** pit thermal energy storage (PTES, 15–30 kWh/m^3), **c** borehole thermal energy storage (BTES, 60–80 kWh/m^3 15–30 kWh/m^3) and **d** aquifer thermal energy storage (ATES, 30–40 kWh/m^3)

systems are also becoming established at the municipal level, with the help of which solar thermal energy can achieve a considerable degree of coverage in some cases.

Denmark for the moment is the most advanced market for solar district heating (SDH). The cost of heat in these systems is commercially attractive and competitive. Depending on location and system design the levelized cost of heating (LCOH) for realized projects in the last 10 years is between 2 and 6 ct/kWh heat. Currently the largest storage is installed in Denmark scoping a storage volume of $2 \cdot 10^5$ m^3 in 13 m depth, which is fed by a solar collector area of $7 \cdot 10^4$ m^2 with an operation temperature of the stored water of 80 °C (max. 95 °C) and provides a yearly yield of 28 GWh. Another large-scale plant is shown in Fig. 8.45a; the plant in Dronninglund, Denmark, which gives an idea of the size of the plant (peak load 12 MW$_{th}$ output, approx. $3.76 \cdot 10^4$m^2 collector area).

Also, in Germany SDH is considered an interesting option with much better economy than small solar thermal system for single houses, however, in the past only in big cities district heating grids had been installed, whereas in Denmark there is a much larger tradition in these technologies and hence a better chance for introducing solar. Therefore, in Germany up to now only smaller projects with public funding have explored the opportunities and demonstrated the potential, especially in conditions with large concentration of heat consumers. Figure 8.45b shows such a combined system, which has been installed in the municipality of Eggenstein-Leopoldshafen and which links many consumers with

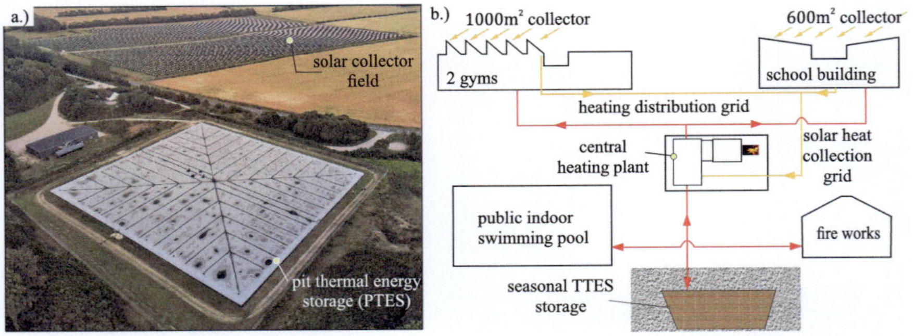

Fig. 8.45 **a** Aereal view of collector field with a Pit thermal energy storage (PTES) in the solar district heating plant in Dronninglund, Denmark with a peak thermal load of $12MW_{th}$ modified from (Sifnaios, Gauthier, Trier, Fan, & Jensen, 2023); **b** Sketch of the local heating network and the solar settlement using the example of the municipality of Eggenstein-Leopoldshafen (Gemeinde Eggenstein Leopoldshafen and Gde. Eggenstein-Leopoldshafen)

Table 8.18 Key data of the local heating network and the solar settlement using the example of the municipality of Eggenstein-Leopoldshafen (Gemeinde Eggenstein Leopoldshafen, 2010)

Gross floor area	$1.2 \cdot 10^4 m^2$
Solar collector area	$1.6 \cdot 10^3 m^2$
Local heating network	300m
Long-term heat storage	$4.5 \cdot 10^3 m^3$ gravel
Top gas-fired boilers	$2 \times 600kW$
Heat pump	60kW
Buffer storage	$30m^3$
Solar coverage ratio	35 – 40%

different requirement profiles. In Table 8.18 the key data of the combination concept are listed. The systems comprise $6.7{\cdot}10^3\,m^2$ collector surface and storage volumes of $1.1{\cdot}10^4\,m^3$ and $1.3{\cdot}10^4\,m^3$, respectively. For a location in northern Germany, the design calculations resulted in a solar energy yield per year of $260 - 280\,kWh/(m^2a)$ and a solar coverage ratio of 53%, while in southern Germany $380\,kWh/(m^2a)$ and 70% are achievable.

8.8 Medium Temperature Storage for Industrial Process Heat Applications

In the medium temperature range above 120 °C to about 250 °C no commercial storage option is available. Very often industrial process heat is supplied to the processes in the form of steam. Ideally one would consider a latent heat storage using a phase-change material suitable for these temperatures.

Possible candidates PCM materials which have melting temperatures in this range are many organic materials. Interesting examples are Phenols (e.g.

Hydoquinone,$T_m = 168-172$ °C), sugar alcohols (e.g. D-Mannitol,$T_m = 166$ °C) and Polymers (e.g. HDPE, $T_m = 130-140$ °C).

Sugar alcohols have been investigated because they exhibit a high energy density, as illustrated for three types in Fig. 8.46, low volumetric change, are inexpensive and not toxic. Moreover, they have a good compatibility with container materials, and are recyclable. Negative issues up to now are that the candidate materials like D-Mannitol have a large subcooling behaviour, which is shown in Fig. 8.46b, low cyclic stability and—as organic materials—also low thermal conductivity, which leads to difficulties in the heat transfer to a cooling medium when part of the sugar alcohol is already solidified (Solé et al., 2014).

a.)

substance	melting point [°C]	specific enthalpy [J/g]	chemical structure formula	visual photo
galaktitol	187.3	357.69	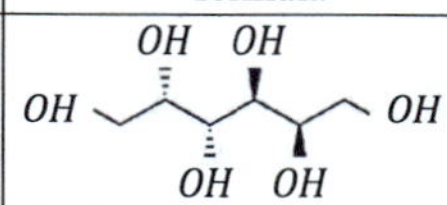	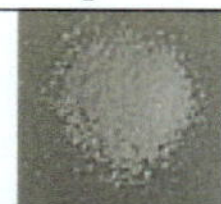
D-mannitol	166.1	308.24	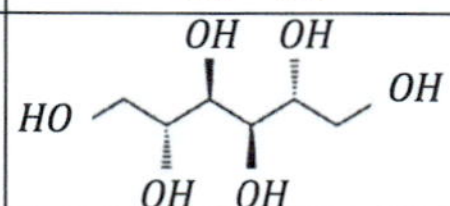	
Myo-inositol	230.0	260.00	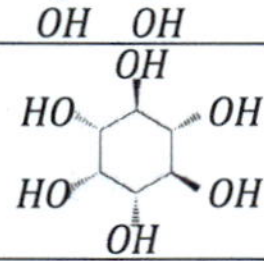	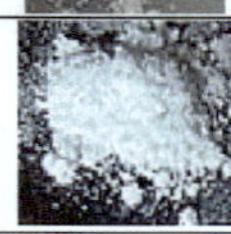

b.)

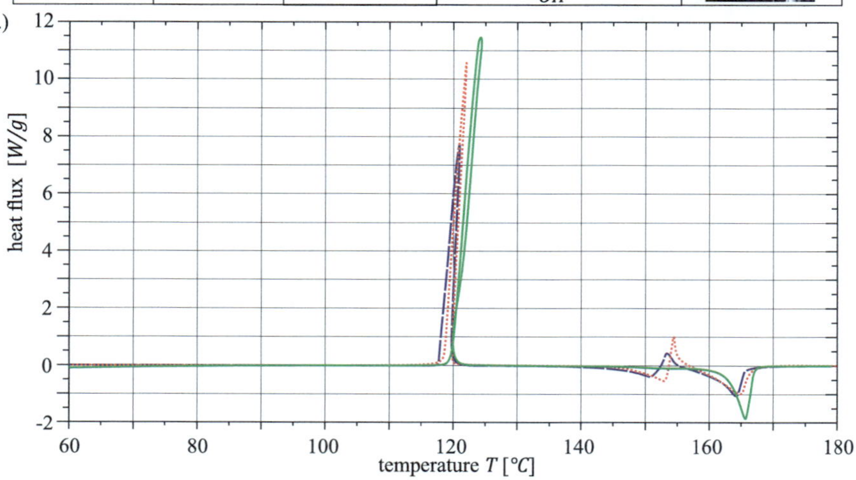

Fig. 8.46 **a** Properties of some of sugar alcohols (Alditols); **b** Sub-cooling of D-Mannitol measured with a differential scanning calorimeter (DSC) where uptaken or released heat is determined versus temperature change (Solé et al., 2014)

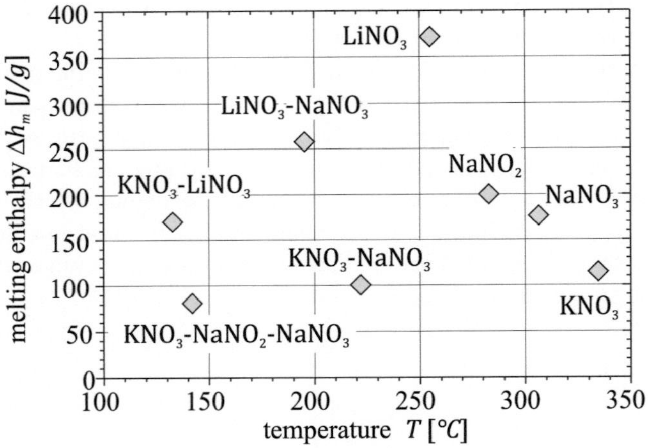

Fig. 8.47 Melting enthalpy and melting temperature for some mixtures of Nitrate and Nitrite salts. from (Zhou & Eames, 2016; Gil et al., 2010)

Latent heat materials with slightly higher melting temperatures can be found in the Nitrates. Here, inorganic salt mixtures are found with melting temperatures in the range of 130 to 350 °C, as shown in Fig. 8.47. The abundant sodium nitrate salts are a good basis for using these mixtures in the solar process heat systems mentioned above (Zhou & Eames, 2016).

In R&D projects possible candidates for this temperature range have been investigated and storage systems based on finned heat exchangers (Bonk & Bauer, 2021) have been built (see also Fig. 8.48). Using these developments already commercial LHS systems in three variants with different PCM materials have been brought to the market for process steam applications by the company SGL Carbon (Schmitt et al., 2010), see Fig. 8.22. The bases for these development was sodium nitrate $NaNO_3$ (melting temperature $T_m = 308$ °C), a binary mixture $NaNO_3/KNO_3$ ($T_m \approx 246$ °C, 60/40 wt%, (Bonk & Bauer, 2021) and a ternary mixture $KNO_3/NaNO_2/NaNO_3$ ($T_m \approx 142$ °C, 53/40/7 wt%, (Bauer et al., 2013). The heat exchanger fins were made-depending on the material—of aluminium or graphite plates. The heat exchanger material had been optimized in order to avoid corrosion of the fins. However, due to little demand from the market, the commercialization was stopped.

8.9 High Temperature Storage

In the high temperature field of solar thermal power (CSP) storage temperature above 350 °C are desired. The operational temperature of a typical parabolic trough field in a solar power station reaches 390 °C, so a storage material is needed which is stable at these temperatures and is available in large quantities at a low price. Also, here Nitrate salts are already being used in many commercial

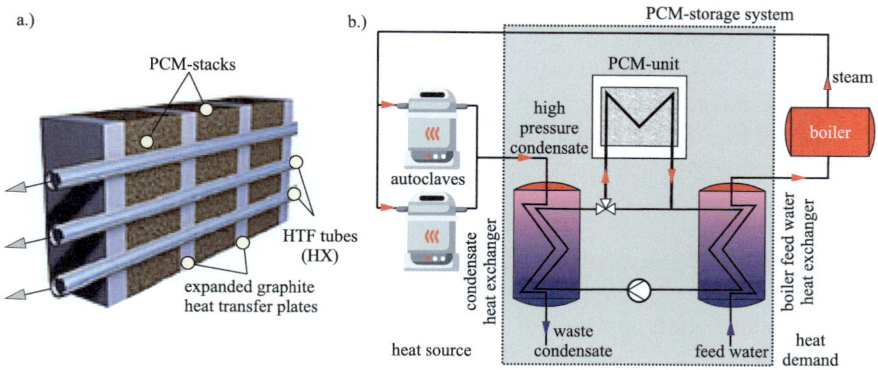

Fig. 8.48 Principle of passive latent heat storage unit (**a**) and system integration with embedded heat exchanger system (**b**) from (Schmitt et al., 2020; Steinmann, 2020)

installations. However, here the sensible heat of specific salt mixtures in the liquid state is utilized. The melting point in this case shall be as low as possible because this restricts the temperature span and hence the specific thermal capacity of the TES. The inventory of a liquid sensible storage shall not freeze. The attractivity of Nitrate salts and their binary, tertiary or even quaternary mixtures lies in the abundance of most salts, and in the possibility to use these liquid molten salts also as heat transfer materials in the solar loops. Then no costly heat exchangers between the solar loop and the storage volume have to be installed. The liquid material is directly charged into the adequate tank. This in a one-tank option is similar to the buffer storage concept for water storages. Two-tank options however have a cold tank and a hot tank, where the material is pumped from one tank to the other, depending on whether charging or discharging occurs. There is R&D looking into materials with much higher temperature stability, such as chlorides or carbonates. The corrosiveness of these materials up to now is a major problem. Therefore, also liquid metal combinations like NaK—known from nuclear reactor technology—are investigated.

Several thermal energy storage systems in the high temperature field are already proven and can be used in commercial projects. They can be divided into four groups (Steinmann, 2020; Medrano et al., 2010):

- two tank direct storage,
- two tank indirect storage,
- concrete storage,
- steam accumulators.

Steam accumulators have been used already in solar power plants with direct steam generation (DSG) in the collectors, however, mainly for short time periods 0.25 – 1.0 h storage capacity. The one tank stratified storage has been used already (with oil and molten salt as storage medium) in small projects. In the past also, filler materials have been added to the oil storage to reduce costs. However, as oil incorporates a fire risk, the molten salt option with different compatible filler materials is still under investigation. Other storage options, for example high-temperature latent heat storage or packed bed storage using small solid particles or rocks are in a demonstration state. Chemical heat storage for high temperatures is an active field of international research.

8.9.1 High-Temperature Storage Materials

For high-temperature applications storage materials are used in a liquid version only, which we have learned already in the latent heat section, where the phase-change properties of these materials are utilized at a lower temperature level. A wide range of inorganic salts can be used for sensible high-temperature storage, for temperatures above their melting point. Some of them might also be used as heat transfer fluids or as phase-change materials for the medium temperature range.

In Table 8.19 some typical candidates of nitrates, chlorides, carbonates and even fluorides are listed as well as examples of binary and ternary mixtures. On the positive side of salt and salt mixtures there is the high availability, low costs for most salts and good cyclic stability. The melting temperatures of the pure salts are in the range of $250 - 850°C$ and too high for such an application. However binary, ternary or even quaternary mixtures of salts have much lower melting points and can be used as storage fluids, as indicated already in Tables 8.2 and 8.3. However, it is important that the stability of the mixture can be guaranteed over a long time. For instance, the nitrates usually are in a dynamic equilibrium with nitrites. At higher temperature the reaction equilibrium shifts towards the reduce nitrites and oxygen is released, requiring measures in the piping grid such as autocatalytic recombiners to mitigate unintended ignition.

The thermofluiddynamic and material compatibility engineering aspects and constraints of liquid salts as well as those of the liquid metals as far as there is operational experience available may be taken from Sect. 5.3.2.4 (salts) and 5.32.5 (liquid metals) and the literature cited therein.

Liquid metals itself despite their excellent large specific heat conductivity are primarily for cost reasons not suited for large scale thermal storage purposes. But even if their specific price could be substantially reduced, they pose significant engineering challenges in the context of a large-scale storage. The class of the heavy liquid metals such as gallium or lead and its eutectic alloys place extreme demands on the seismic integrity of the storage foundations, as well as the bearing and suspension, due to their very high specific density, not to mention very challenging demands on the structural material and coolant conditioning. These requirements are largely foreign to the other technically used liquid metal class such as the alkali metals and their alloys. If there were a corresponding demand, a low price would also be conceivable due to their frequent occurrence in the earth's crust. But their density is lower than that of water, so that the storage density is rather small. In addition, their high reactivity with air or water necessitates

Table 8.19 Thermophysical properties of some inorganic salts, binary and ternary mixtures. (Gil et al., 2010; Steinmann, 2020)

Salt	Melting temp. T_m [°C]	Melting enthalpy $\Delta h_{melt}\left[\text{kJ/kg}\right]$	Density ρ_l $\left[\text{kg/m}^3\right]$
$LiNO_3$	254	360	1780
$NaNO_3$	307	172	1850
KNO_3	333	266	1890
$MgCl_2$	714	452	2140
$NaCl$	800	492	2160
Na_2CO_3	854	276	2533
KF	857	452	2370
K_2CO_3	897	236	2290
$LiNO_3NaNO_3(49/51)$	194	265	1770
$KClLiCl(55/45)$	355	236	1650
$KFZrF_4(32/68)$	390	N.a.	2800
$KNO_3LiNO_3NaNO_3(52/29/18)$	120	N.a.	1780
$KNO_3NaNO_2NaNO_3(53/40/7)$	120	60	1790

special safety requirements, the feasibility of which has been demonstrated over many years in compact plants such as nuclear reactors, but the level of experience in extensive open-space plants such as solar thermal systems eludes operational experience.

For the high-temperature range also solid materials can be used which are stable and do not exhibit a phase-change below the maximum operation temperature. Natural rocks, ceramics, slag waste material or even high-temperature resistant concrete are examples. (Ioan Sarbu & Calin Sebarchievici, 2018; Steinmann, 2020). The application of solid sensible storage is mainly motivated by the low cost of the materials and their abundancy. Solid materials have the disadvantage that heat can be stored only by conduction processes, which are locally. Therefore, the uptake and release of heat always takes more time when compared to convection in liquid materials.

8.9.2 Indirect 2-Tank Storage

Historically this concept, in which both the molten salt and mineral oil fluid options were realized, was successfully operated at Barstow California Power Plant (SEGS-1) with an electrical output of $14MW_{el}$ with a mineral oil based version at a T_{min}/T_{max} = 240/370 °C with a storage capacity of 115 MW_{th}. When the thermal energy was removed via an external steam generator, the turbine could thus be operated for approximately $3h$ at nominal load.

Fig. 8.49 Realized thermal molten salt storage for solar thermal power plants. **a** SEGS-1 solar tower power (National Renewable Energy Laboratory and U.S. Department of Energy (DOE) 2000); **b** Andasol parabolic trough power plant (Solar Millennium AG, 2008)

Some decades ago, a concept of an indirect two-tank storage system has been developed in which molten salt acts as a storage medium and is heated by the heat transfer fluid (oil or water) circulating in the circuit of a solar power plant via a heat exchanger. The thermal energy storage system is loaded by transferring the heat transfer fluid from the solar field or tower to the salt via a heat exchanger. For this purpose, the cold liquid salt is conveyed from the cold storage tank and transported in countercurrent through the heat exchanger, where it heats up. The heated salt is finally stored in the hot tank. When the storage energy is required, the system is operated in the opposite direction. The salt heats the heat transfer fluid that feeds the steam generator, with whose steam a turbine generates the electricity. This concept is called indirect because it uses two different fluids for storage and direct system operation in the solar field, and is shown in Fig. 8.49a.

Several parabolic trough power plants in Spain use this storage concept, as it combines several advantages. On the one hand, there is sufficient experience with thermal oil-operated plants, good availability of the components and the required salt storage facilities do not require any special construction effort with regard to pressure resistance. On the other hand, there is the additional expenditure for heat exchangers and pumps with the associated control technology. The cost of storage in this concept is based on the use of heat exchangers and the relatively small temperature differences between the cold and hot fluid in the storage system. Therefore, development is currently concentrating mainly on reducing the number of components with the aim of lowering the costs of energy storage, see e.g. (Kelly & Kearney, 2006). This indirect concept is shown in Fig. 8.50.

However, when thermal energy from a direct steam generating collector (DSG) shall be used and be charged into a storage system, a sensible storage is not optimal as discussed already. A latent heat storage (LHS) system would be more adequate (Vogel, 2016). However, the exergy loss can be reduced if several temperature levels are being used in a solar plant. Figure 8.51 shows such an indirect sensible heat storage concept using 3 tanks with different optimized sizes. The overheated steam can be produced from the smallest high-temperature storage.

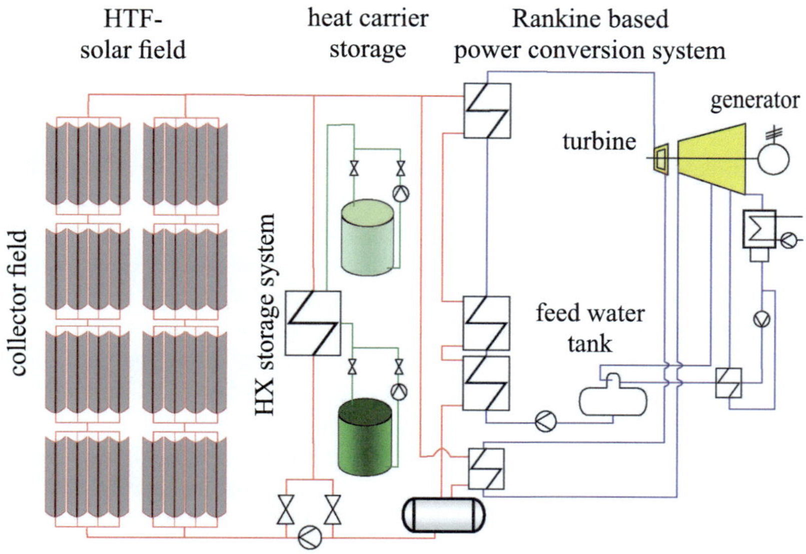

Fig. 8.50 System diagram of a concept with indirect thermal two-tank liquid storage

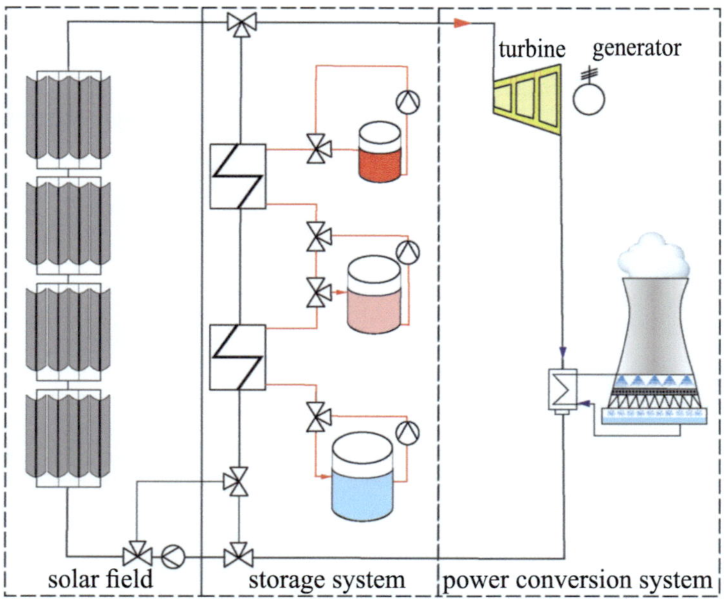

Fig. 8.51 Variation of indirect 3-tank system (for improved coupling of direct steam collectors) from (Vogel, 2016)

8.9.3 Direct 2-Tank Storage

Figure 8.52 shows a plant concept where HTF and storage liquid are identical. Then no heat exchanger is needed, and the temperature loss is not existent with charging and discharging. This is called a direct thermal storage according to the two-tank direct storage principle. Historically this concept has been implemented the first time in the second expansion stage of the solar tower in Barstow, California, (SEGS 2-Solar Two) where a mixture of $NaNO_3$ (60%) and KNO_3 (40%) was used as heat transfer fluid, which has a melting point of 221 °C and a stability temperature of 565 °C. The tank volume of the thermal storage was twice 875 m^3 and corresponded to an energy content of 115 MWh_{th}. The full temperature range of the molten salt between melting stability temperature cannot be used, in practice it allows a safe operating temperature range of 290–565 °C. Compared to the indirect two-tank storage with thermo-oil as HTF where only 290 to 390 °C (100 °K) can be used, the energy content of the same volume nearly triples. The material cost due to the huge quantities needed is about 50% of the storage cost, which indicates that the direct storage in this case is a much cost-effective option. Heat exchangers are not needed. However other cost (prevention of freezing in the solar field and in the piping, corrosion prevention) also need some attention. In total the systems have a much lower cost per stored kWh than the indirect storage.

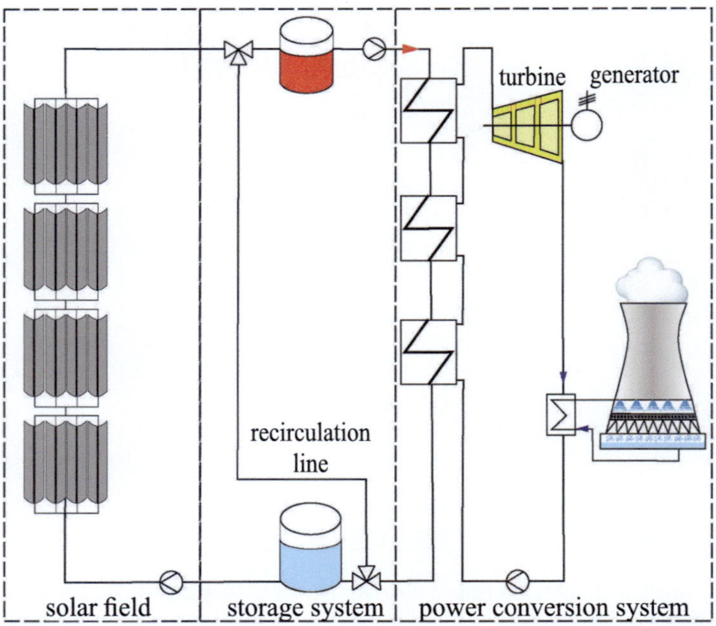

Fig. 8.52 System diagram of a solar power plant with a direct thermal two-tank liquid storage in storage operation

8.9.4 Steam Accumulators

In a sliding pressure steam accumulator, often called Ruths storage as depicted in Fig. 8.53a steam is used for charging. Here, superheated, or saturated wet steam generated by a heat source is pressed into a pressure vessel filled with pressurized water to a very large fraction (typically in the range of 85–95%) at a temperature just below saturation temperature. The life or waste steam injected via a charge manifold, a nozzle into the convection pipe condenses due to the high pressure and requires less volume. The condensation causes a temperature and pressure rise. In case steam is needed, a valve is opened and temperature as well as pressure decrease maintaining stable saturation conditions. Due to the pressure reduction steam is flashed into the piping. Of course, the amount of steam is strongly dependent on the saturation pressure at the specific water temperature, as illustrated in Fig. 8.53b. During discharge the steam pressure is be reduced as the temperature is decreasing also. Ruths storages or steam are used in process heat applications as well as in solar thermal power plants, for more details see Sect. 10.5.3.1 or the technical book by (Steinmann, 2021).

Another possibility is a constant pressure steam accumulator, often termed as expansion type storage (Steinmann & Eck, 2006). Hereby, heat is added to the storage using saturated pressured liquid to be filled into the pressure vessel. In some applications where heat will be transferred from a different medium, also heat can be added using an internal heat exchanger. The discharging of steam from the pressure vessel is conducted by opening a valve at the bottom of the vessel. Although there is also a pressure drop in the accumulator, this is lower compared to the sliding pressure accumulator and almost all water can be discharged. The hot water leaving the pressure vessel through the pipe is flash evaporated in an external evaporator, allowing to separate steam from water, which is illustrated in Fig. 8.54a. The water is stored at low temperature for later refilling. The almost isentropic thermodynamic change of state from a condition A to B in the

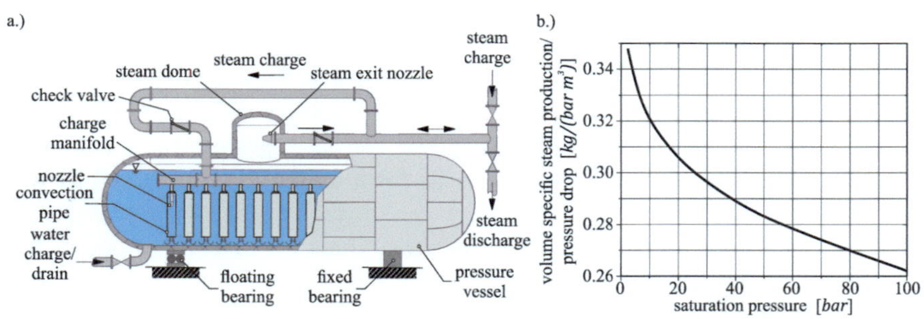

Fig. 8.53 **a** Schematic of a sliding pressure steam accumulator; **b** Mass of produced saturated steam as a function of pressure reduction of 1*bar* in a steam accumulator dependent on saturation pressure in vessel; mass of produced steam refers to volume of liquid water from (Steinmann & Eck, 2008)

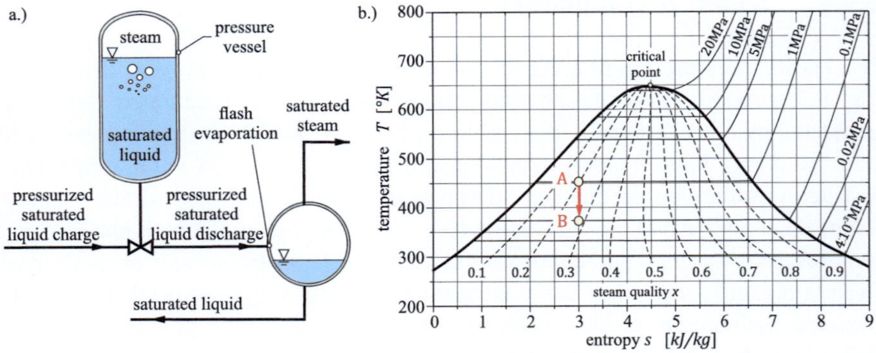

Fig. 8.54 **a** pipe routing in expansion type steam accumulator **b** Visualization of the discharging process using the temperature-entropy ($T - s$) diagram of waterT − s

temperature-enthalpy ($T - s$) diagram is depicted in Fig. 8.54b. When the pressure is lowered from A to B, the equilibrium quality x of the two-phase mixture in the vessel is increased which indicates that a large mass of the steam phase is available, and the wetness is reduced.

8.9.5 Stratified Storage Tanks and Filling Materials

Another option is storage in a single tank thermocline, where both the hot and cold fluids are stored in a single tank. In a thermocline, the hot storage fluid is located at the top and the cold fluid at the bottom. In the zone in between, there is an area where the temperature changes with a steep gradient from a high to a low temperature level. This temperature jump is called a thermocline. The term thermocline is usually used in connection with temperature stratification in oceans.

Figure 8.55 shows the schematic integration of a storage thermocline into a solar thermal power plant. The main advantage of a thermocline storage system is that the largest part of the storage fluid can be replaced by a less expensive filling material with a high specific heat capacity (e.g. sand or rock, cf. Figure 8.56a), as was demonstrated in a test storage system with a capacity of $25 MWh$ (Brosseau et al., 2004). The authors were able to prove the basic performance, but found that with increasing operating time, deposits form on the filling material as well as precipitations in the salt.

In a thermocline storage tank of diameter D, which is filled with a porous bed of a quartz rock of height h, an isothermal layer is formed at the bottom and top of the tank in which no filling material is used, as indicated in Fig. 8.56b. These distribution areas are therefore necessary to ensure even inflow and outflow conditions at the inlet and outlet of the storage facility in order to achieve good thermal stratification in the storage facility. The thermal fluid (mostly molten salt) flows through the filler bed from top to bottom in the loading half-cycle and heats the filler material. In the discharge cycle, the discharge pump transports the fluid

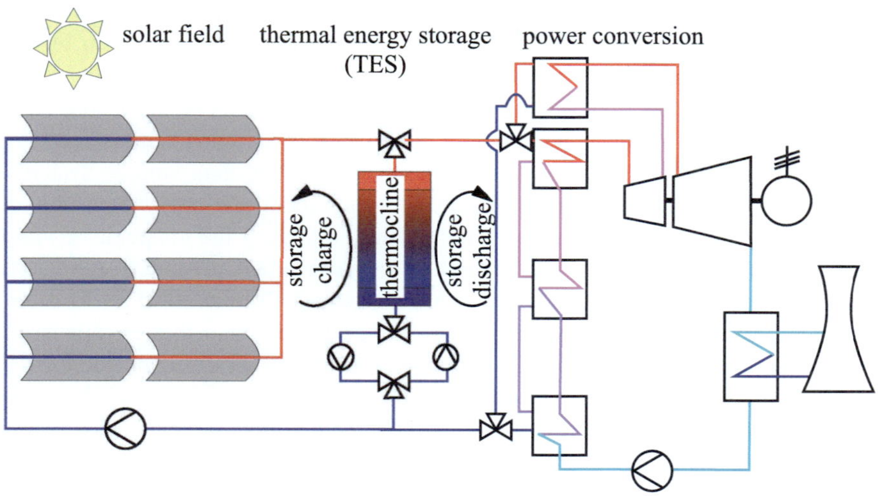

Fig. 8.55 Integration of a single-tank thermclinestorage system into the energy generation process of a solar thermal power plant with the storage and removal of thermal energy

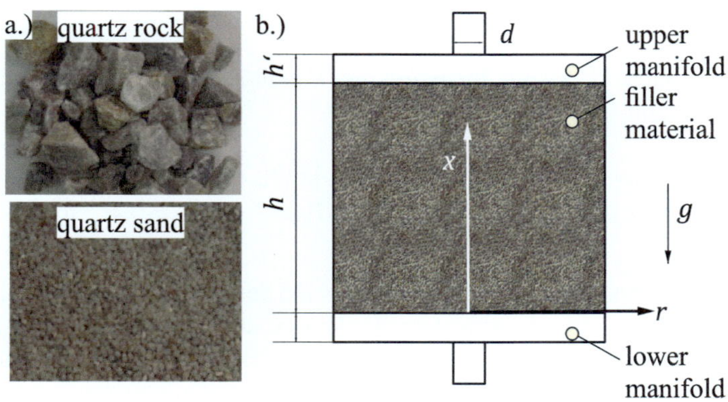

Fig. 8.56 **a** quartz rock or sand as thermal storage material of a single-tank thermocline storage (Brosseau et al., 2004); **b** Schematic diagram for the calculation of the required storage volume and dimensions of a thermocline storage tank

in the opposite direction through the storage bed and the heat is fed into the power plant cycle. With continuous discharge, the height of the hot section of the storage tank decreases continuously.

Only a part of the thermal energy at high temperatures stored in the storage can be recovered as useful energy in the sense that it can be used for electrical power generation. The ratio of high-temperature energy storage to the amount stored is called *efficiency metric*. A high discharge efficiency results from a low discharge power and at the same time high thermal energy. Since the discharge power is composed of the product of the flow rate, which is proportional to the Reynolds number Re, and the convective conductive heat transfer, which is given by the molecular Prandtl number Pr, the discharge efficiency decreases rapidly with a large product $(Re \cdot Pr)$, see (Yang & Garimella, 2010). Depending on the costs of the thermofluid, the costs per kWh of stored energy are considerably lower in the case of thermocline storage systems compared to two-tank storage systems. However, to maintain the storage capacity of a thermocline, it must be ensured at all times that there is a temperature shift within the tank. If the temperature jump spreads over the entire tank, the efficiency of the storage tank is more than modest.

In the following, simple analytical considerations shall be used to try to determine the required dimensions of a single-storage thermocline that is capable of determining a certain amount of high-temperature energy at different withdrawal rates and times. First of all, one estimates the storage diameter D and the average size of the thermal filling material d_s. The mean cross-sectional area of the accumulator A is given by $A = \pi \cdot D^2$, with which the power density (P/A) and the energy density (Q/A) of the accumulator can be calculated directly. The dimensionless discharge power $(Re \cdot Pr)$ and the usable thermal energy $(H \cdot \eta)$ are given by the Eqs. (8.38) and (8.39):

$$\frac{P}{A} = u_m \cdot \rho_{l,\text{cold}} \cdot c_{p,l,\text{cold}} \cdot (T_\text{hot} - T_\text{cold}) = k_l \cdot \frac{(T_\text{hot} - T_\text{cold})}{d_s} \cdot Re \cdot Pr, \quad (8.38)$$

in which the index "l" stands for liquid and k is the specific thermal conductivity of the thermofluid.

$$\begin{aligned} \tfrac{Q}{A} &= \left[\varepsilon \cdot \rho_l \cdot c_{p,l} + (1-\varepsilon) \cdot \rho_s \cdot c_{p,s}\right] \cdot (T_\text{hot} - T_\text{cold}) \cdot h \cdot \eta, \\ &= \left[\varepsilon \cdot \rho_l \cdot c_{p,l} + (1-\varepsilon) \cdot \rho_s \cdot c_{p,s}\right] \cdot (T_\text{hot} - T_\text{cold}) \cdot d_s \cdot H \cdot \eta, \end{aligned} \quad (8.39)$$

where ε is the porosity of the bed. H indicates the desired height of the thermocline reservoir and η the discharge efficiency. In the next step take a value $H = H \cdot \eta$ and calculate a Reynolds number Re from the product $(Re \cdot Pr)$. With the values of the Reynolds number and the height H, a storage efficiency η can be determined with the following equation (Yang & Garimella, 2010):

$$\begin{aligned} &\eta = 1 - 0.1807 Re^{0.1801} \cdot (H/100)^m \quad \text{with} \\ &m = 2.34 \cdot 10^{-3} \cdot Re^{-0.6151} + 5.5 \cdot 10^{-4} Re - 0.485. \end{aligned} \quad (8.40)$$

This results in a new storage height H by division with $(H\eta)$ for efficiency. The last steps have to be repeated until the relative change of the new storage height and the previous one falls below a certain limit value which has been determined before. The dimensional height of the thermocline tank h finally results from $h = d_s \cdot H$.

In the literature the modelling of heat and mass transfer with the filler material in place has been described e.g. in (Votyakov & Bonanos, 2015; Cascetta et al., 2016). The stress exerted by the ratcheting of the filler material due to the cyclic

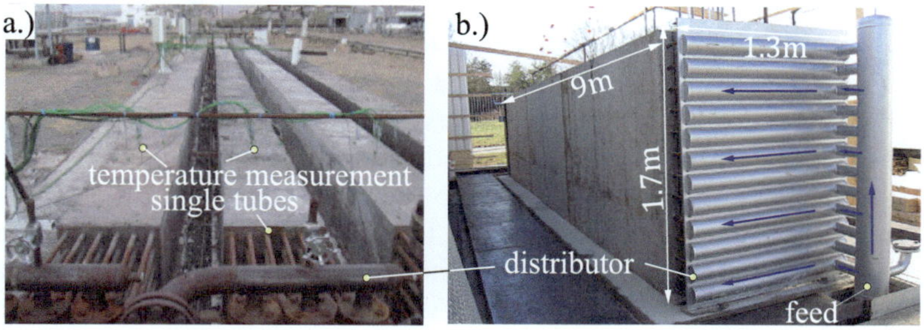

Fig. 8.57 Concrete-solid storage **a** Test storage PSA, Almeria 4 modules with 25 kW each, 80 kWh test operation 2004/2005 **b** Test storage DLR, Stuttgart 100 kW, 400 kWh test operation 2006 (Laing & Tamme, 2008)

temperature changes in the thermocline tank is critical for the tank shell, and countermeasures have to be taken (Sassine et al., 2018). The exergy loss of such a single thermocline tank is small although a two-tank system here has superior behaviour (Odenthal et al., 2018).

8.9.6 Solid Storage Options

Rock or rock beds are and have often been used as solid-state reservoirs, provided that favorable development is possible due to the type of rock. In some cases, air can also be used as a heat transfer fluid. Table 8.5 outlines the capacity of some massive storage types. The table shows that solid heat transfer media have a similar heat capacity to liquid transfer media. However, if there are no local storage possibilities in the form of "almost free" resources, solid storage tanks also represent an infrastructure for the transport of media to and from the storage tank. Under no circumstances should environmental aspects be ignored. The storage of considerable amounts of energy in direct terrestrial storage facilities can have significant effects on the local soil structure as well as its water balance. The advantages of solid state storage are:

- high long-term thermal stability over a wide temperature range,
- thermal expansions similar to those of pipe registers,
- good handling during production (modularity, series production),
- low mobility,
- low costs and a
- good availability of the components for media transport.

Figure 8.57a and b show concrete solids storage tanks as they were successfully used for test purposes for solar thermal systems. Based on these experiments, basic storage modules with a thermal storage volume of 4 MWH_{th} were developed.

The corresponding storage blocks, which are provided with boreholes for media transport, have a weight of approximately 400 t (Laing et al., 2012).

For the successful implementation as a storage material the development of high-temperature concrete had to be addressed. Water inside of normal concrete would evaporate at high temperatures and due to the closed material would crack the concrete blocks. Another issue to be solved is the permanent contact between pipes and surrounding concrete during the thermal expansion of the two different materials. Additional interface materials with good thermal conductivity are added in spite the extra cost to guarantee a good heat transfer into and from the concrete (Laing et al., 2012).

High-temperature rock bed storage using air as a heat transfer fluid is an interesting option to store solar heat for a combined-cycle or gas turbine power plant. A model for the heat transfer between packed bed and air has been developed and experimentally validated (Hänchen et al., 2011). It has been used for a parametric study of the packed bed dimension, fluid mass flow rate, particle diameter, and solid phase material to characterize charging/discharging characteristics, daily cyclic operation, overall thermal efficiency, and capacity ratio. For a continuous operation, the overall efficiency exceeds 90% under optimal condition. Local material may be used, however the rock properties should guarantee a stable longtime operation with no degradation of the material (Allen et al., 2014). The modelling of a complete power plant has been endeavored for example in (Heller & Gauché, 2013). For air as a heat transfer fluid, other solid concepts are available. The so-called Cowper storage, used in industry for example for regenerative heat in glass melting processes of float glass production, where ceramic bricks store the heat, has been used also in the demonstration solar tower Jülich, see e.g. (Zunft et al., 2010).

8.9.7 Particle Storage

Solid particles can be used for receivers with direct absorption of concentrated solar radiation. The same particles with typical size less than 1 mm diameter then can be stored in a simple storage without the need of a heat exchanger. This idea has the charming aspect of possibly using even very cheap particles like natural sands, however, the concept has some complexities. The long-term stability of the particles must be ensured. The transport of the particles using for example conveyor belts needs parasitic energy. The main challenge probably is the heat transfer of the stored heat from the particle to the working fluid of the power cycle, where usually air for a Brayton cycle is considered. A blocking of the heat exchanger by the particles has to be avoided, while a good heat transfer from particle to air is needed (Baumann et al., 2011) have investigated early the problems of such a heat exchanger using natural sands (A1-Hansari, 2013) looked into the suitability of different desert sands. Silica-based particles (i.e., sand) are inexpensive and abundant, but they lack high solar absorptance, and some sands sinter to larger conglomerates. Commercially available ceramic particles can be produced with high absorptivity and have a high durability. They can be easily transported relatively using conveyor-belts or screw transport systems. An overvisew over particle properties and appropriate solar tower receiver designs is given by (Ho, 2016), mainly focusing on the main commercially available ceramic particles. A comprehensive overview of the state of the art of the potential and research gaps of solid particles in concentrated solar thermal power is given by (Calderón et al., 2019). Apart from the thermophysical properties of ceramic particles a variety of natural alternatives are also listed. The options for the particle-air heat exchangers like fluidized bed are shortly described as well.

References

Agyenim, F., Hewitt, N., Eames, P., & Smyth, M. (2010). A review of materials, heat transfer and phase change problem formulation for latent heat thermal energy storage systems (LHTESS). *In Renewable and Sustainable Energy Reviews, 14*(2), 615–628. https://doi.org/10.1016/j.rser.2009.10.015

Al-Hansari, H. et.al. (2013, September 17-20). Characterization and sintering potential of solid particles for use in high temperature thermal energy storage system. In : Proc. SolarPACES 2013. Proc. 19th SolarPACES International Symposium. Las Vegas, NV.

Allen, K. G., Backström, T. W. von, Kröger, D. G., & Kisters, A. F. M. (2014). Rock bed storage for solar thermal power plants: Rock characteristics, suitability, and availability. *In Solar Energy Materials and Solar Cells 126*, 170–183. https://doi.org/10.1016/j.solmat.2014.03.030.

Barlev, D., Vidu, R., & Stroeve, P. (2011). Innovation in concentrated solar power. *In Solar Energy Materials and Solar Cells, 95*(10), 2703–2725.

Bathen, Dieter, & Breitbach, Marc. (2001). Grundlagen der Adsorption. In Dieter Bathen, Marc Breitbach (Eds.): Adsorptionstechnik. Berlin: Springer (VDI-Buch), pp. 49–101.

Battaglia, Mattia, & Haller, Michel (op. 2018): Stratification in Large Thermal Storage Tanks. In Andreas Häberle (Ed.), EuroSun2018. Proceedings of the ISES EuroSun 2018 Conference. ISES EuroSun 2018 Conference—12th International Conference on Solar Energy for Buildings and Industry. Rapperswil, CH, 10.09.2018—13.09.2018. Freiburg: International Solar Energy Society, pp. 1–11, checked on 6/29/2021.

Bauer, Th., Pfelger, N., Laing, D., Steinmann, W. -D., Eck, M., & Kaesche, S. (2013). High temperature molten salts for solar power application. In F. Lantelme, H. Groult (Eds.), Molten Salts Chemistry: From Lab To Applications: Elsevier, p. 592.

Baumann, Torsten, Teixeira Boura, Cristiano José, Göttsche, Joachim, Hoffschmidt, Bernhard, Schmitz, Stefan, & Zunft, Stefan. (2011, September 20-23). Air-sand heat exchangers: Materials and flow properties. In: Proc. SolarPACES 2011. Proc. 17th SolarPACES International Symposium. Granada, Spain.

Bonk, A., & Bauer, T. (2021). Solar salt —Thermal property analysis-report on thermo-physical properties of binary NaNO3-KNO3 mixtures in a range of 59–61 wt% NaNO3. DLR-FB-2021–19. Scientific Report. DLR. Stuttgart.

Brosseau, Doug A., Hlava, Paul F., & Kelly, Michael J. (2004). Testing thermocline filler materials and molten-salt heat transfer fluids for thermal energy storage systems used in parabolic trough solar power plants. SAND2004–3207. Sandia National Laboratories. Albuquerque, New Mexico, United States.

Bruch, L. W., Cole, M. W., & Zaremba, E. (2007). *Physical adsorption: Forces and phenomena.* Dover Publications.

Buck, R., Muir, J. F., & Hogan, R. E. (1991). Carbon dioxide reforming of methane in a solar volumetric receiver/reactor: The CAESAR project. *In Solar Energy Materials, 24*(1–4), 449–463.

Calderón, Alejandro, Barreneche, Camila, Palacios, Anabel, Segarra, Mercè, Prieto, Cristina, Rodriguez Sanchez, Alfonso, & Fernández, Ana Inés. (2019). Review of solid particle materials for heat transfer fluid and thermal energy storage in solar thermal power plants. *In Energy Storage, 1*(4). https://doi.org/10.1002/est2.63.

Cascetta, Mario, Cau, Giorgio, Puddu, Pierpaolo, & Serra, Fabio. (2016). A comparison between CFD simulation and experimental investigation of a packed-bed thermal energy storage system. In Combined Special Issues: ECP 2011 and IMPRES 2010 98, pp. 1263–1272. https://doi.org/10.1016/j.applthermaleng.2016.01.019.

Corgnale, C., Hardy, B., Motyka, T., Zidan, R., Teprovich, J., & Peters, B. (2014). Screening analysis of metal hydride based thermal energy. *In Renewable & Sustainable Energy Reviews, 38*, 821–833.

Criado, Y., Huille, A., & Rougé, S. (2017). Experimental investigation and model validation of a CaO/Ca(OH)2 fluidized bed reactor for thermochemical energy storage applications. *In Industrial and Engineering Chemistry Research, 313*, 844–852.

EN 12977 -3. (2018). Thermal solar systems and components—Custom built systems—Part 3: Performance test methods for solar water heater stores. Beuth Verlag GmbH.

Ernström, Tova Crosser. (2020). Thermal energy storage. In Celsius initiative, 8/17/2020. Retrieved 24 June, 2021, from https://celsiuscity.eu/thermal-energy-storage/.

Fernández, F., Díaz, J., Folgueras, M., & Suárez, I. (2021). Stratification analysis and behaviour of a real industrial thermocline thermal energy storage tank for cogeneration purposes. *In Processes, 9*(1), 120. https://doi.org/10.3390/pr9010120

Fertahi, Saïf ed-Dîn, Jamil, A., & Benbassou, A. (2018). Review on solar thermal stratified storage tanks (STSST): Insight on stratification studies and efficiency indicators. *In Solar Energy, 176*, 126–145. https://doi.org/10.1016/j.solener.2018.10.028.

Gebäude Energie Gesetz -Building Energy Act (in German). (2020, November). Requirements for thermal insulation of pipelines and fittings. § 69 Heat distribution and hot water pipes and fittings Subsection 8.3: Thermal insulation of pipes and fittings.

Gemeinde Eggenstein Leopoldshafen. (2010). Solares Nahwärmekonzept. Retrieved from https://www.egg-leo.de/de/Umwelt-Wirtschaft/Umwelt/Energieprojekte/Nahwaermekonzept-Eggenstein.

Gil, Antoni, Medrano, Marc, Martorell, Ingrid, Lázaro, Ana, Dolado, Pablo, Zalba, Belén, & Cabeza, Luisa, F. (2010). State of the art on high temperature thermal energy storage for power generation. Part 1—Concepts, materials and modellization. *In Renewable and Sustainable Energy Reviews, 14*(1), 31–55. https://doi.org/10.1016/j.rser.2009.07.035.

Glaser, H. (2005). *Beiträge zur Betriebsoptimierung solarthermisch betriebener Adsorptionskälteanlagen*. Fakultät III—Prozesswissenschaften; Technische Universität Berlin.

Gomez, J. C. (2011): High-temperature phase change materials (PCM) candidates for thermal Energy Storage (TES) Applications. NREL, U.S. Department of Energy, Office of Energy. 1617 Cole Boulevard, Golden, Colorado 80401.

Gutierrez, Andrea, Miró, Laia, Gil, Antoni, Rodríguez-Aseguinolaza, Javier, Barreneche, Camila, & Calvet, Nicolas et al. (2016). Industrial waste materials and by-products as thermal energy storage (TES) materials: A review. In. SOLARPACES 2015: International Conference on Concentrating Solar Power and Chemical Energy Systems. Cape Town, South Africa, 13–16 October 2015: Author(s) (AIP Conference Proceedings), p. 50019, checked on 6/30/2021.

Hadorn, J. -C. (2005a). Thermal energy storage for solar and low energy buildings-Statew of the art. Edited by Programme, Internat. Energy Agency -Solar Heating & cooling: IEA Solar Heating and Cooling Task, p. 32.

Hadorn, Jean-Christophe (Ed.). (2005b). Thermal energy storage for solar and low energy buildings. State of the art. International Energy Agency. IEA: IEA.

Haller, M. Y., Cruickshank, C. A., Streicher, W., Harrison, S. J., Andersen, E., & Furbo, S. (2009a). Methods to determine stratification efficiency of thermal energy storage processes—Review and theoretical comparison. *In Solar Energy 83*, 1847–1860.

Haller, M. Y., Cruickshank, C. A., Streicher, W., Harrison, S. J., Andersen, E., & Furbo, S. (2009a). Methods to determine stratification efficiency of thermal energy storage processes—Review and theoretical comparison. *In Solar Energy, 83*(10), 1847–1860. https://doi.org/10.1016/j.solener.2009.06.019

Haller, Michel Y., Streicher, Wolfgang, Andersen, Elsa, & Furbo, Simon. (2009c). Compararative analysis of thermal energy storage stratification efficiency–A new method combines advantages of previous approaches. In : EffStock 2009—The 11th Intl Conference on Thermal Energy Storage for Efficiency and Sustainability. With assistance of H. Severinson, S. Gehlin. Stockholm, pp. 1–9.

Hänchen, M., Brückner, S., & Steinfeld, A. (2011). High-temperature thermal storage using a packed bed of rocks—Heat transfer analysis and experimental validation. *In Applied Thermal Engineering, 31*, 1798–1806.

Heller, L., & Gauché, P. (2013). Modeling of the rock bed thermal energy storage system of a combined cycle solar thermal power plant in South Africa. *In Solar Energy, 93*, 345–356. https://doi.org/10.1016/j.solener.2013.04.018

Ho, C. K. (2016). A review of high-temperature particle receivers for concentrating solar power. *In Applied Thermal Engineering, 109*, 958–969. https://doi.org/10.1016/j.applthermaleng.2016.04.103

Kelly, Bruce, & Kearney, D. (2006). Thermal storage commercial plant design study for a 2-tank indirect molten salt system. NREL/SR-550–40166. National Renewable Energy Laboratory. Golden, Colorado, USA.

Kleinbach, E. M., Beckman, W. A., & Klein, S. A. (1993). Performance study of one-dimensional models for stratified thermal storage tanks. In Solar Energy, pp. 155–166.

Laing, D., Bahl, C., Bauer, T., Fiss, M., Breidenbach, N., & Hempel, M. (2012). High-temperature solid-media thermal energy storage for solar thermal power plants. *In Proc IEEE, 100*(2), 516–524. https://doi.org/10.1109/JPROC.2011.2154290

Libby, C. (2009). Program on technology innovation: Evaluation of concentrating solar thermal energy storage systems. Technical Report 1018464. Electric Power Research Institute-EPRI. 3420 Hillview Avenue,Palo Alto, CA 94304.

Mangold, D., Riegger, M., Schmidt, T. (2021). Solare Nahwärme- und Lnagzeitspeicher. FORSCHUNGSBERICHT ZUM BMU-VORHABEN 0329607L. Solites. Stuttgart, checked on 7/15/2021.

Marshall, R., Barragan, F. A. Simulation study of the IEA system test procedure. University of Wales. Cardiff.

Medrano, Marc, Gil, Antoni, Martorell, Ingrid, Potau, Xavi, & Cabeza, Luisa F. (2010). State of the art on high-temperature thermal energy storage for power generation. Part 2—Case studies. *In Renewable and Sustainable Energy Reviews, 14*(1), 56–72. https://doi.org/10.1016/j.rser.2009.07.036.

Mehling, H., & Cabeza, L. F. (2008). *Heat and cold storage with PCM- An up to date introduction into basics and applications*. Springer Verlag.

Odenthal, C., Klasing, F., & Bauer, T. (2018). Parametric study of the thermocline filler concept based on exergy. *In Journal of Energy Storage, 17*, 56–62. https://doi.org/10.1016/j.est.2018.01.009

Ortega, I., Faik, A., Gil, A., Rodríguez-Aseguinolaza, J., & D'Aguanno, B. (2015). Thermophysical Properties of a Steel-making by-product to be used as thermal energy storage material in a packed-bed system. *In Energy Procedia, 69*, 968–977. https://doi.org/10.1016/j.egypro.2015.03.180

Pardo, P., Deydier, A., Anxionnaz-Minvielle, Z., Rougé, S., Cabassud, M., & Cognet, P. (2014). A review on high temperature thermochemical heat energy storage. *In Renewable Sustainable Energy Reviews, 32*, 591–610.

Ravotti, R., Worlitschek, J., Pulham, C. R., & Stamatiou, A. (2020). Triglycerides as novel phase-change materials: a review and assessment of their thermal properties. *In Molecules, 25*(23), 5572ff.

Reuss, Manfred et al. (2020). Quality management in design, construction and operation of borehole systems. IEA-ECES-ANNEX-27-Final-Report-20201118. Retrieved 19 May, 2021, from https://iea-eces.org/wp-content/uploads/public/IEA-ECES-ANNEX-27-Final-Report-20201118.pdf.

Sassine, Nahia, Donzé, Frédéric-Victor, Harthong, Barthélémy, & Bruch, Arnaud. (2018). Thermal stress numerical study in granular packed bed storage tank. *In Granular Matter 20*(3). https://doi.org/10.1007/s10035-018-0817-y.

Schmitt, Rainer, Oettinger, Oswin, Steinmann, Wolf-Dieter, & Johnson, Maike. (2010). PCM-graphite latent heat storage systems for industrial process heat recovery. *In Advances in Science and Technology, 74*. https://doi.org/10.4028/www.scientific.net/AST.74.259.

Schwieger, W., & Toufar, H. (2007). *Anorganisch-poröse Materialien: Präparation, Verarbeitung und Industrielle Herstellung*. Springer Verlag.

Sharma, A., Tyagi, V. V., Chen, C. R., & Buddhi, D. (2009). Review on thermal energy storage with phase change materials and applications. *In Renewable and Sustainable Energy Reviews, 13*, 318–345.

Sibbitt, B., McClenahan, D., Djebbar, R., Thornton, J., Wong, B., Carriere, J., & Kokko, J. (2012). The performance of a high solar fraction seasonal storage district heating system—Five years of operation. *In Energy Procedia, 30*, 856–865. https://doi.org/10.1016/j.egypro.2012.11.097

Solé, A., Neumann, H., Niedermaier, S., Martorell, I., Schossig, P., & Cabeza, L. F. (2014). Stability of sugar alcohols as PCM for thermal energy storage. *In Solar Energy Materials and Solar Cells, 126*, 125–134. https://doi.org/10.1016/j.solmat.2014.03.020

Steinmann, W.-D. (2021). *Thermal energy storage for medium and high temperatures-concepts and applications*. Springer Fachmedien.

Steinmann, Wolf-Dieter. (2020). Thermal energy storage systems for concentrating solar power plants. Chapter 11. In: Concentrating Solar Power Technology Principles, Developments and Applications. With assistance of Keith Lovegrove, Wes Stein. 2nd ed. Duxford, UK: Elsevier Woodhead Publishing, pp. 399–442.

Steinmann, W.-D., & Eck, M. (2006). Buffer storage for direct steam generation. *In Solar Energy, 80*(10), 1277–1282. https://doi.org/10.1016/j.solener.2005.05.013

Streich, M. (2010). … ein Thermosiphon ? -Gegen Wärmewanderschaft. In SBZ-Monteur, pp. 10–13.

Stryi-Hipp, Gerhard. (2011). Ansätze zur Bewertung und Optimierung von solarthermischen Anlagen. Erkenntnisse aus dem Ökotest-Vergleichstest. In : Proc. OTTI-Solarthermie Symposium. Solarthermie Symposium. Staffelstein, pp. 11-12. Mai 2011. OTTI. Regensburg, checked on 5/19/2021.

Sun, W., Hong, Y., & Wang, Y. (2017). Operation optimization of steam accumulators as thermal energy storage and buffer units. *In Energies, 10*(1), 17. https://doi.org/10.3390/en10010017

Sunku Prasad, J., Muthukumar, P., Desai, F., Basu, D. N., & Rahman, M. M. (2019). A critical review of high-temperature reversible thermochemical energy storage systems. *In Applied Energy, 254*, 113733.

Tamme, R., Bauer, Th., & Hahne, E. (2012). Heat storage media. *In Ullmann's Encyclopedia of Industrial Chemistry, 17,* 421–438.

Tamme, Rainer. (2010). Optimised industrial process heat and power generation with thermal energy storage. IEA ECES Annex 19—Final Report. Retrieved 19 May, 2021, from https://iea-eces.org/wp-content/uploads/public/Annex-19-Final-Report-07-2010.pdf.

van Helden, & Wim; Rommel (2021) Matthias: Compact-thermal-storage-position-paper. IEA SHC Task 42—Report 2015. Retrrieved 19 May, 2021, from https://task42.iea-shc.org/Data/Sites/1/publications/IEA-SHC-Compact-Thermal-Storage-Position-Paper.pdf.

van Helden, Wim, Yamaha, Motoi, Rathgeber, Christoph, Hauer, Andreas, Huaylla, Fredy, Le Pierrès, Nolwenn et al. (2016). IEA SHC Task 42 / ECES Annex 29—working group B: Applications of compact thermal energy storage. *In Energy Procedia 91*, 231–245. https://doi.org/10.1016/j.egypro.2016.06.210.

Vogel, Alexander. (2016). Konzepte für Speicher in thermischen Solarkraftwerken. Dissertation. Technische Universität Braunschweig; Fraunhofer-Institut für Solare Energiesysteme.

Votyakov, E. V., Bonanos, A. M. (2015). Algebraic model for thermocline thermal storage tank with filler material. *In Solar Energy 122,* 1154–1157. https://doi.org/10.1016/j.solener.2015.10.047.

Wentworth, W. E., & Chen, E. (1975). Simple thermal decomposition reactions for storage of solar thermal energy. *In Solar Energy, 18*(3), 205–214.

Wörner, A., & Tamme, R. (1998). CO2 reforming of methane in a solar driven volumetric. *In Catalyst Today, 46*, 165–174.

Yang, Z., & Garimella, S. V. (2010). Thermal analysis of solar thermal energy storage in a molten-salt thermocline. *In Solar Energy, 84*, 974–985.

Zhou, D., & Eames, P. (2016). Thermal characterisation of binary sodium/lithium nitrate salts for latent heat storage at medium temperatures. *In Solar Energy Materials and Solar Cells, 157*, 1019–1025. https://doi.org/10.1016/j.solmat.2016.08.017

Zipf, V., Willert, D., & Neuhäuser, A. (2016). Active latent heat storage with a screw heat exchanger—experimental results for heat transfer and concept for high pressure steam. *In AIP Conference Proceedings, 1734*, 050044–050110.

Zunft, Stefan, Wahl, Eberhard, Krüger, Michael, Dreißigacker, Volker; Göhring, Felix. (2010). Jülich solar power tower—experimental evaluation of the storage subsystem and performance calculations.

Low Temperature Systems for Buildings

9

Abstract

Although the solar collector is the key component in a solar thermal system, depending on the temperature range needed and the heat transfer fluid found useful for this, many other system components like controls, pumps, heat exchangers, valves etc. are needed. When we investigate the application in buildings, room heating and hot water are the dominant applications, both restricted to a low temperature range up to 60 °C. The solar cooling and climatization of buildings are a rather special application which needs higher temperatures in the collector circuit, therefore it is treated in a separate Chap. 13. A large variety of system concepts are found nowadays, with passive and active heat transport to the consumer, which are described after a classification in this chapter. The components of a collector circuit are described and specified. The planning and dimensioning of a solar thermal collector systems has many aspects and can use simple rule-of-thumb design, but generally utilizes numerical simulation tools, where the time series of temperatures and heat flows in a solar thermal system are modelled for all individual components. These design rules and the use of simulation tools are detailed in the last sections of the chapter.

Keywords

Heating and hot water demand • Liquid-based solar thermal systems • Air-collector systems • Active and passive system • Solar district heating • System components and piping • System design • System simulation

9.1 Introduction

In the low temperature application, i.e. for demand temperatures below 100 °C, thermal solar systems have a large variety of different system layouts and optimized designs. The applications which are most common in this temperature range

R. Stieglitz and W. Platzer, *Solar Thermal Energy Systems*,
https://doi.org/10.1007/978-3-031-43173-9_9

are the production of hot water for domestic and non-domestic use, and the provision of room heating to buildings. Both can be supplied by active systems, which operate based on a forced circulation of a heat transfer fluid (usually a water–glycol mixture or air), or passive solar systems where the heat transport is based only on temperature differences (natural convection), conduction and thermal radiation.

As active solar systems with a forced circulation of a fluid in the loop a variety of specific system components must be discussed in their functionality, dimensioning and interaction. Since the total system of collector, pump, pipe network and many other components must be adapted to managing the time-variable resource of solar irradiation, time-dependent system operation and control issues have to be considered. Approaches and methods for calculating the overall system performance are presented in the chapter using the example of solar thermal low temperature systems.

9.1.1 Application and Demand

9.1.1.1 Heat Demand in a Building

In a heating dominated climate the supply of heat for compensating thermal losses of the building in the winter is a seasonal demand, which is fundamentally opposite to the seasonal supply of solar irradiation. Therefore, it is generally a very difficult task to provide solar heating to a building. The demand of hot water for showering, household purposes on the other hand is rather constant, and varies mainly only due to small changes in the seasonal groundwater temperature, and of course due to absences of inhabitants (e.g. due to holidays). Both, seasonal demand curves usable solar heat and incident solar irradiation are shown in Fig. 9.1 exemplary for a typical single family house situated in Central Europe on the Northern hemisphere.

The fraction of the heat demand of a building, which is covered by a solar thermal system is called *solar fraction* (S_F). This concept is explained in more

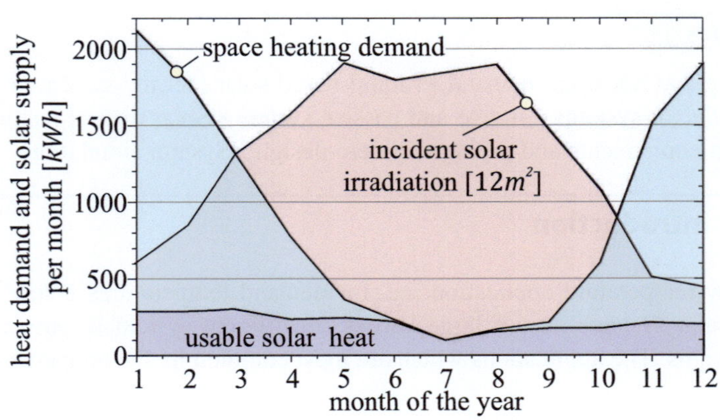

Fig. 9.1 Seasonal heat demand of a typical single family house in Central Europe

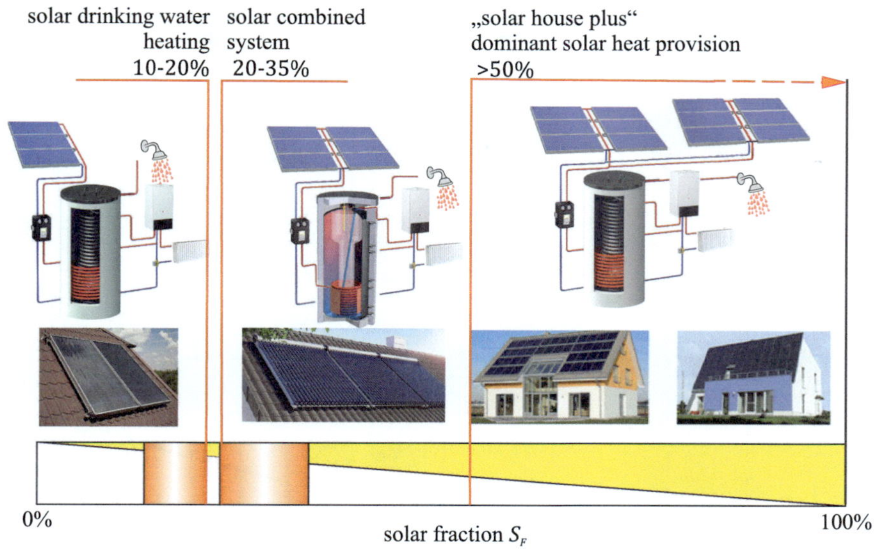

Fig. 9.2 Strategies to cover a larger share of the heating demand for a buildings by solar. From left: Cover hot water demand for the summer half-year; solar-assisted heating system for covering also heating demand in the transition periods; high solar fraction for a low energy building with optimized solar thermal system ("solar house"); full coverage of minimum heating demand by an optimized solar thermal system with seasonal storage

detail below in Sect. 9.1.3.2. Typical values for heating dominated climates are in the range of 10–15% for systems which are dimensioned to cover mainly the domestic hot water demand and are designed to cover on a sunny design day the hot water demand of a building (see Fig. 9.2) on the left-hand side. If also a part of the heating demand in the transition periods autumn and spring should be covered by a solar-assisted heating, then 15–25% of the total heat demand are typically covered. Larger solar fractions require substantial measures as well on the solar thermal system design (with large solar collector fields, good orientation for winter sun, large buffer storage and low heating demand due to well-insulated building shells. In the extreme of 100% solar share in Central Europe houses have been built without a second heat source, but they required a huge tank for seasonal storage. The solar autonomy is possible but is very difficult to attain economically (Jenni, 2010).

9.1.1.2 Maximum Solar Fraction and Bivalent System Approach

The previous paragraph indicates that the building performance as such is a paramount feature influencing the possibility to supply the heating by solar. High heat demand in winter leads to a situation where even with all envelope surfaces oriented towards the sun (roof, facades oriented towards the South (Northern hemisphere)) the collector area is not large enough to produce enough solar gains. Of course, this is also influenced by the thermal losses of a collector type. But let

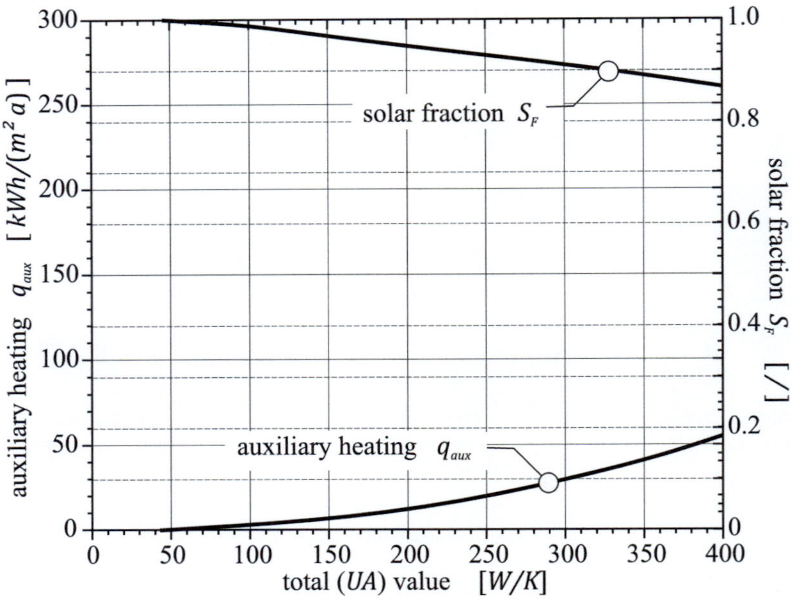

Fig. 9.3 Computed hypothetical maximum solar fraction (S_F) for a solar heated building of living area 165 m^2 and 45° tilted roof oriented South/North at location Würzburg/Germany $(U \cdot A) = 150$ W/K ventilation plus transmission losses is equivalent to an average U-value 0.3 $W/(m^2K)$

us assume a solar heated building in Germany (average climate Würzburg). What would happen with a building equipped on all roof and façade area with collectors of negligible loss which is the upper limit of physics? The results of such a solar heated house in terms of auxiliary power required and achievable S_F is illustrated in Fig. 9.3.

It can be derived from Fig. 9.3 that even for the unrealistic case that all solar radiation on a building surface is converted to heat the house with 100% efficiency, there would be a necessity of auxiliary heating from about 150 W/K upwards which is below the current requirement for new buildings in Germany. Older buildings exhibit often considerably higher thermal losses. The calculation has been executed with a simplified model, assuming that the building has a thermal capacity to store heat for a day, i.e. solar gains cannot be transferred to the next day.

In a next step we consider typical passive solar gains and internal gains in the building resulting in a heating demand. How much of that demand could then be covered by a collector system on the South oriented roof (area 136 m^2)? Figure 9.4 shows that even with a fully covered roof and an idealized collector without thermal losses an auxiliary heating is needed for heating demand above 50 $kWh/(m^2a)$. The reason is that on some days the incoming radiation is not large enough to cover the heating demand. For a solar collector field with thermal losses considered the required auxiliary heating is even larger, but still the

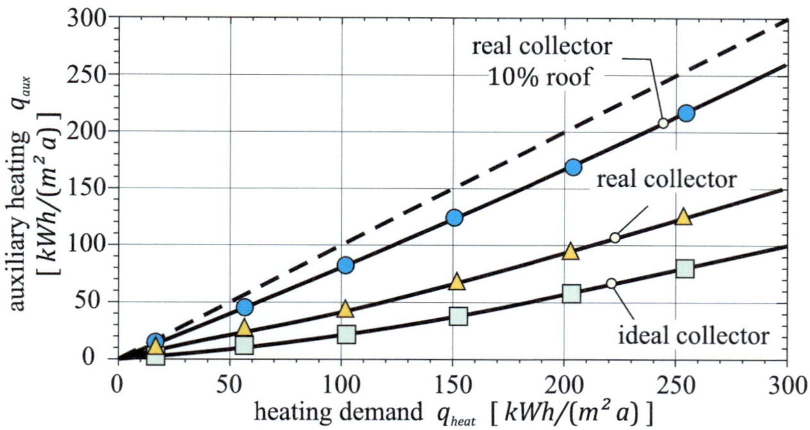

Fig. 9.4 Auxiliary heating needed in the building of Fig. 9.3 when a solar collector systems assists the heating system (dashed line: auxiliary heating = heat demand, i.e. no solar contribution; south roof)

solar fraction can be considered substantial (up to 50%). For a small collector area of 13.6 m^2—a size which is typical for today's solar-assisted systems for single-family houses—the auxiliary energy becomes the dominant part of the heating and solar fractions are around 10–20%. Consequently, we may conclude that except in extreme cases solar thermal collector systems cannot provide 100% of the heating energy in a new building according to the building standard or even in a house of the building stock. This means that a bivalent system is required with two different heat sources. Traditionally, gas and oil boilers supplied the heating in the past when non-renewable source were considered. As a renewable option wood stoves and pellet boilers were available. And, if renewable electricity is available, a fully renewable heating system can be designed using a heat-pump as counterpart for solar thermal. Both systems have their specific advantages and disadvantages which is subsequently.

9.1.2 Interaction with Building Performance

We have seen that the building performance has an effect on the performance of the solar thermal heating systems. The larger the demand, the higher can be solar utilization which implies that the economics of a solar thermal system will be generally better with a badly insulated building, as more energy can be supplied by the solar system. However, also auxiliary heating demand is increasing, therefore, an optimum balance between reduction of thermal losses and the utilization of solar gains is an economically driven design task, and the optimum depends very much on the specific costs of the technologies for reducing the energy demand of a building and for supplying it with sufficient heat. On the other hand, this is not the only kind of interaction between built environment and solar systems.

- Thermal loss

The main parameter influencing the performance is the thermal loss of a building which has two parts: transmission losses through the solid walls and building elements, and ventilation and infiltration losses due to the exchange of air having different temperatures between the ambient and the indoor environment.

- Thermal capacity

The thermal capacity of a building is important as it governs the heat storage effects and the temperature swings in a building. Buildings with high thermal mass and low thermal loss keep the heat over a long time. Low thermal mass (lightweight) and high thermal losses on the contrary react very fast to changes of the ambient temperature or solar radiation. If the thermal capacity allows to store heat over 24 h or more then the utilization of solar gains is better. This aspect has a main impact on the utilization of passive solar gains through facades, but it also affects the utilization of the active solar heating systems. For the latter one, however, the size of the water storage coupled to the solar field is more important.

- Solar gains direct and heating demand solar system

Buildings with a high share of direct solar gains through windows and glazed facades do not require a lot of heating on sunny winter days, i.e. the day where the active solar system also operates best. Therefore, the dimensioning of an active solar heating system does also depend on the extent of passive solar use in the building.

- Low energy housing seasonally better adapted

Although the utilization of an active solar heating system increases with heating demand, low energy buildings have also advantages for solar thermal systems. The seasonal imbalance of hot water demand in summer and heating plus hot water demand in winter becomes smaller. Therefore, the oversizing of a system for summer operation to achieve a larger solar share also in winter is less pronounced. The problem of long stagnation periods in summer is becomes smaller for low energy buildings.

- Building integration

Building integration has many aspects from building physics to architecture and aesthetics. The main idea is just not to add a solar collector to a building without looking for synergies. One may use the functionalities provided by a solar collector also for other purposes in the building. The thermal insulation of a building element can be improved by a collector. Similarly, the sound protection might be improved by a proper construction. The collector as such can provide the rain-tightness needed for the building elements, e.g. when completely integrated in a roof construction. Transparently glazed collectors may also provide daylight. A review of functionalities and different technologies for *BIST* (building integrated solar thermal) is given by Aelenei et al. (2016).

9.1.3 Characterization of Systems

9.1.3.1 Components and Functionalities

The main components and functions of a solar thermal system for buildings are shortly introduced in the following Table 9.1. As a solar thermal system for buildings might have very different characteristics, these components can be very different in construction and operation.

For instance, passive solar thermal systems reducing the heat demand of a building (like windows or solar walls) have usually no active medium transporting the heat from the collector to the demand site—the heat is transported by passive means (conduction in a wall, natural convection of air or thermal radiation). Therefore, there is no control needed for the heat transport—however, there are possibilities to control the gain, e.g. by using shading devices in windows and collectors). Active solar thermal systems use a liquid or air as a heat transport medium. Here, a pump or ventilator is used for driving the transport system, being controlled by a control unit. The different forms and functionalities are therefore, discussed in detail in the Sect. 9.2 for active systems based on liquid, in Sect. 9.3 for active air-collector systems, and in Sect. 9.6 for passive solar systems.

If one focuses first on active solar thermal systems, the components of Table 9.1 are connected to a system which schematically is shown in Fig. 9.5. The solar

Table 9.1 Main components and functions in a solar thermal system for hot water and heating

Component	Function
Solar collector	Conversion of solar radiation to heat
Thermal storage	Storage of generated heat to cover the heat demand
Auxiliary heating system/backup system	Supply of heat in times where no solar heat is available, and the storage is empty
Heat exchangers for separating the different hydraulic circuits: potable water, solar collector, and heating circuit	Separation of the different hydraulic circuits with different liquids and transfer of heat from one to the other
Control system with solar loop pump	Optimal operation of the solar collector field to maximize the heat generated from the sun considering the temperature levels

hydraulic circuit or solar loop transports the heat generated in the solar collectors to a storage. The transfer to the storage is facilitated by internal or external heat exchangers. If the solar heat provided is not sufficient the back heating system which in the simplest case may be an electrical heating rod complements the solar heat such that the required heat can be provided at a suitable temperature level. The control unit operates the solar collector loop only if the irradiation is large enough to add heat to the storage. Also, it switches of the solar collectors if the storage is fully charged, and no heat is needed.

As discussed in the section on heating demand, solar thermal systems for heat demand in buildings are generally bivalent systems, which implies that a backup or auxiliary heating system is required to provide heat at times where no solar heat is available. Monovalent systems may only be practical if demand and solar supply is in parallel (e.g. for supplying hot water for showering in a public outdoor swimming pool) or if the requirement on reliable supply is low. The backup system might be also based on renewable energy (e.g. PV-heat pump system, biomass boiler) or a conventional energy source (gas or oil boiler).

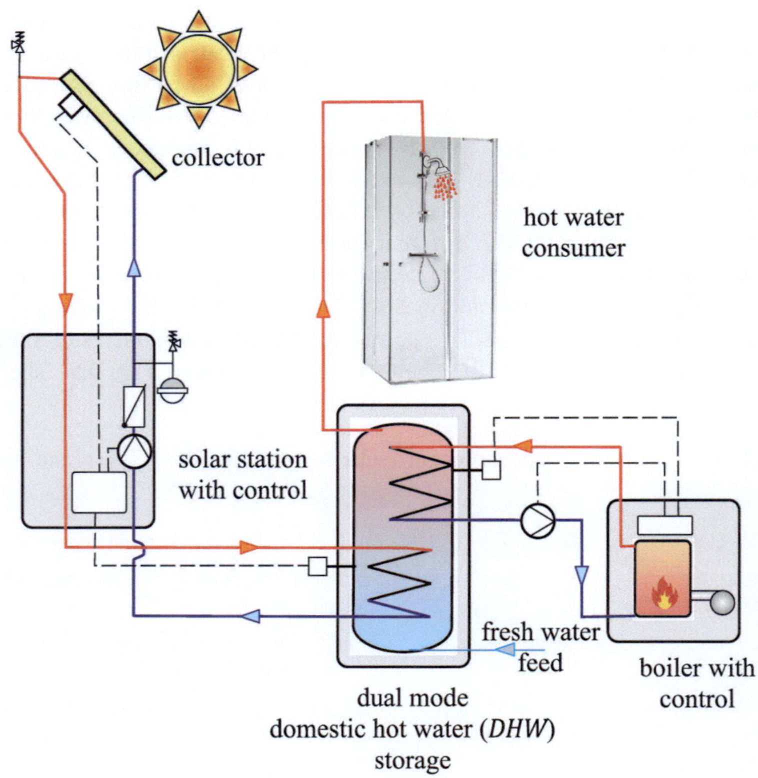

Fig. 9.5 Main functionalities and components of a solar system (dashed signal lines)

The heat transfer media in the hydraulic circuits for low temperature solar thermal systems are either based on water or air, because these media have a lot of advantages, especially the low cost. We will first discuss the liquid-based systems. Water with or without anti-freezing additive (typical 30–40 wt% glycol) has very high specific thermal capacity and therefore, needs small piping diameters for transporting the heat to the consumer. Air on the contrary has a very low heat capacity and therefore, requires large volume flows using air channels with large diameters to transport the heat. The specific features, advantages, and challenges of such a system will be discussed in Sect. 9.3.

9.1.3.2 Key Performance Indicators: Solar Fraction and Annual Utilization

The solar fraction is usually defined as the part of the heat load supplied by the solar thermal system. However, this solar system gain cannot be defined unanimously in all cases. Let us consider the solar thermal system as depicted in Fig. 9.6. Here we measure a *solar collector gain* Q_{coll} using the mass flow through the collector and the temperature rise between inlet and outlet of the collector field. This quantity should be rather close to the *solar collector loop gain* Q_{loop}. Some small heat loss through piping and other loop components reduces the amount of heat which is transferred to the storage via the heat exchanger. The *solar system gain* Q_{sys} includes the thermal energy storage and designates therefore, the heat transferred via second heat exchanger to the demand side. In this example the solar system gain is well-defined and can be measured. The system gain is smaller than the loop gain, because the storage tank also has heat loss:

$$Q_{coll} > Q_{loop} > Q_{sys}. \tag{9.1}$$

For this case the solar fraction (S_F) is well-defined by

$$S_F = \frac{\text{solar system gain}}{\text{delivered heat}}. \tag{9.2}$$

The heat demand of hot water or the delivered heat can be measured by the mass flow of the heated water, using the water source temperature before the solar heat exchanger and the outlet temperature at the tap.

However, how can the system gain be measured in a storage with integrated auxiliary heating by an electric rod or an internal heat exchanger fed by a gas boiler?

Then the storage heat losses cannot be attributed to the solar or the auxiliary heat source. If the solar system gain is defined as identical to the solar loop gain, we get an upper limit for the solar fraction S_F' (cf. Eq. 9.3), if we take the heat output of the storage and subtract the amount of heat delivered by the auxiliary system, we get a lower limit for solar fraction S_F'' (see Eq. 9.4). In the first case all heat losses of the storage and piping are attributed to the conventional heating

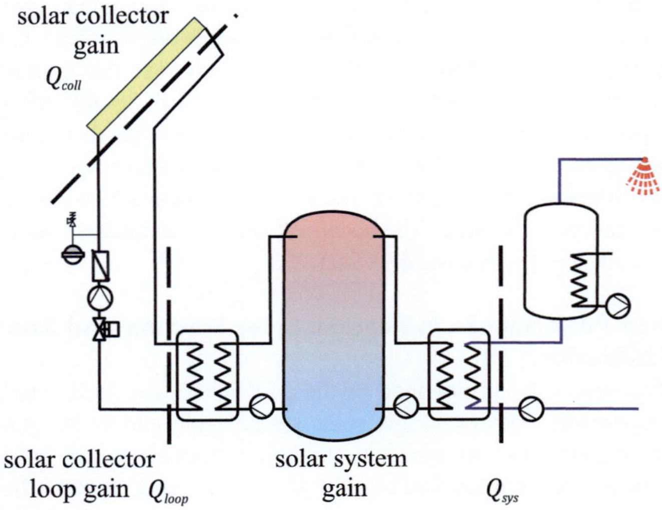

Fig. 9.6 Solar thermal system gain and system boundaries

system, in the latter case all to the solar thermal system. Both definitions are not fair.

$$S_F' = \frac{\text{solar loop gain}}{\text{delivered heat}}, \tag{9.3}$$

$$S_F'' = 1 - \frac{\text{auxiliary heating}}{\text{delivered heat}}. \tag{9.4}$$

A compromise could be that the solar fraction is defined by the individual fractional contributions to the system without considering the distribution and storage losses:

$$S_F = \frac{\text{solar loop gain}}{\text{solar loop gain + auxiliary heat}}. \tag{9.5}$$

Similarly, we have difficulties to define the annual solar utilization factor ($\eta_{ann,sys}$) where the relevant system gains are divided by the solar irradiance on the complete solar collector field over a year:

$$\eta_{ann,sys} = \frac{\text{solar system gain}}{\text{solar irradiance on collectors}}. \tag{9.6}$$

The annual solar collector or loop utilization factor is well-defined. The difference to the system utilization is that the (larger) collector or loop gain is used in the nominator.

9.1.4 Fully Renewable Systems

Up to today in most solar thermal domestic hot water and heating systems, as indicated in Fig. 9.5, the auxiliary heating which is necessary to obtain a functioning system providing heat at all times is realized using a gas or oil boiler, and for hot water alone often by direct electric heating. A 100% renewable system, which should be the requirement of the future and is increasingly asked for, needs an auxiliary renewable source. Two main options will be described shortly here: using boilers with wood (wood chips or pellets) as renewable fuel is a common approach or using heat pumps with fully renewable electricity. Synthetic renewable fuel produced with green hydrogen seems to be not a feasible cost-efficient route at least for the next decades, as these fuels are likely needed for other purposes (e.g. traffic and industry).

9.1.4.1 Solar Thermal Systems Combined with Wood Boilers

Wood as fuel is used in two different forms as wood pellets and log wood. For the first option mechanical transport and feed systems have been developed, which implies that continuous automatic firing is possible. Therefore, an integration of a pellet boiler is not very different from the use of a gas boiler. However, for log wood as illustrated in Fig. 9.7 often discontinuous operation is used. Water storage is used to match the solar heat production with the demand.

Looking into the biomass production rates it is not sensible to use the limited resource wood, which can be used for many more purposes sustainably like construction and production purposes, for paper and packaging, for replacing plastics and conventional building materials, just for heating. Other renewable sources have to be used to save wood for these purposes. A conversion of all households

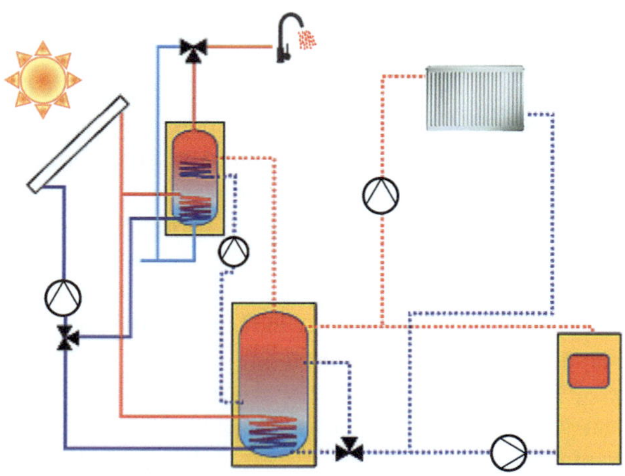

Fig. 9.7 Simplified scheme of a solar thermal hot water and heating system with a log wood boiler as auxiliary source to heat the buffer storage

in most countries from gas and oil to wood is not a sensible path for the energy transformation in general.

9.1.4.2 Solar Thermal Heat Pump Systems

Heat pump systems are often qualified and described as renewable heating systems as they take the majority of their heat from the environment as low temperature ambient heat. The three different sources of ambient heat are ambient air, ground water and ground soil. Ambient air is the most common source as it is available everywhere, and the extraction of heat via air to the brine loop of air source heat pump (ASHP). They can take heat from outside with temperatures down to −15 °C and convert it into usable heat for low temperature heating systems like underfloor heating or large area radiators. For maximising efficiency, a fan is used to draw air across a heat exchanger usually installed outside which creates some noise. The heat in this air is transferred to a refrigerant which evaporates. A compressor, using electricity, then compresses this vapour causing it to rise in temperature. At this upper temperature, the heat can be transferred to the heating and hot water systems in the building.

For extraction of the geothermal heat by a ground source heat pump (GSHP), a pipe system is laid horizontally and in serpentine form below the frost line, in Central Europe around 1–2 m below the surface. A brine medium made of frost-proof liquid circulates in the pipe system, which absorbs the thermal energy and transfers it to the heat exchanger. The size of the collector area required depends of course on the heat demand of the building. In practice, it is often 1–2 times the heated area. Geothermal collectors absorb thermal energy from near the surface, provided by solar radiation, ambient air, and rainwater. Consequently, it is preferable that the area above the pipe system is not asphalted or built upon. For larger buildings also borehole heat exchangers with several meters to several hundred meters depth may be used. The borehole loop technology depends on the ground characteristics. The geothermal temperature gradient from the lower levels of the ground transports heat into the borehole environment.

Water source heat pumps (WSHP) are working identically to GSHP, however the ambient heat is extracted by open or closed loops (e.g. in boreholes or pond mats) from water in lakes, rivers, or ground water reservoirs. GSHP and WSHP need an environmental permit for extracting the heat.

Another heat source for a heat pump could be waste heat, which is quite common in the industrial sector, however for buildings this in not so common. Office buildings with large-server rooms could use the waste heat of the computers. If a controlled ventilation system is installed in the building the heat of the exhaust air could be recovered and used by the heat pump.

It is also possible to use uncovered solar thermal collectors as heat source for the ambient heat which is transformed to higher temperatures by using the heat pump. The advantage of uncovered collectors is that they can be also used without solar irradiation. In this case they serve as heat exchangers to ambient air from which the heat is extracted.

A true renewable heating system needs more than a heat pump. The electricity for the heat pump itself should be also renewable. Therefore, monovalent renewable heating systems are possible with a heat pump providing the required temperature level for the heating and for hot water with ambient heat transformed and driven by renewable electricity from solar panels for example. Heat pumps are characterized with the coefficient of performance COP, where the heat taken from a specified temperature level (e.g. air at 2 °C) and transformed to a design temperature (e.g. 35 °C) is divided by the required electric power:

$$COP_h = \frac{\overline{\dot{Q}}_{HP,h}}{\overline{P}_{el,HP}}. \tag{9.7}$$

The heat pump may also be used for cooling, where at the lower temperature heat is extracted of a medium and more heat rejected at a higher temperature. The connection between these quantities is given in

$$COP_c = \frac{\overline{\dot{Q}}_{HP,c}}{\overline{P}_{el,HP}} = \frac{\overline{\dot{Q}}_{HP,h} + \overline{P}_{el,HP}}{\overline{P}_{el,HP}} = COP_h + 1. \tag{9.8}$$

Standard values in Europe are for air–water heat pumps $A2/W35$ (i.e. air at 2 °C as source and water at 35 °C as output). For ground source brine heat pumps the conditions are $B0/W35$. The letters designate the respective medium. If the heating system is not a low temperature heating system but needs higher forward flow temperatures such as 55 °C, the COP_h for e.g. $A2W55$ can be given, which according to thermodynamics (Carnot) is much smaller than for 35 °C. The temperature dependence with respect to source temperature can be often approximated by a linear curve in the source temperature T_s

$$COP_h = a + b \cdot T_s. \tag{9.9}$$

However, in real heating systems, depending on the outdoor temperature, the heat demand changes and hence the forward flow temperature for the floor heating or the radiators. Changing outdoor temperature, source temperature and heating temperature over a year therefore lead to a complex averaging of conditions for the heat pump. If hot water at a specified temperature (say 60 °C) shall also be included, that has also an impact on the annually averaged performance, which is given by the seasonal performance factor SPF. There are definitions of different SPF possible depending on the system boundaries considered. The simplest one is using only taking into account the electricity used for compressor and control system of the heat pump:

$$SPF_0 = \frac{Q_h}{E_{el,HP}}. \tag{9.10}$$

SPF_1 and SPF_2 also include the electricity consumption of the brine or water pumping and the backup heater consumption, respectively. The later one is used only for extreme low temperatures when the heat pump heating power would be too small. There are different options: To stop the compressor and heat only with an electrical heating rod, or to use both in parallel. SPF_2 is in practice about 0.2–0.3 lower than SPF_0, cf. Miara et al. (2014). Generally, it can be said, that the SPF is smaller the higher the required temperatures are, and the lower the source temperatures are. Therefore, an interesting idea is to couple solar thermal systems with heat pump systems in the so-called solar thermal and heat pump systems (SAHPS), as described for example in Sparber et al. (2011). The idea is to use solar thermal and heat pumps in order to increase the overall system seasonal performance factor by adequately integrating solar thermal collectors.

The hydraulic schemes of the single systems, especially the degree of integration of the solar thermal and electrical driven heat pump unit vary significantly. The same is true for the used low temperature heat source for the heat pump where ground heat exchangers (horizontal and vertical), ambient air, specific solar collectors, phase change material storages (e.g. ice/water) and exhaust air are implemented. To classify these combined heating systems using heat pumps and solar collectors a graphical scheme has been developed by Frank et al. (2010) including all options.

In the template of Fig. 9.8 possible heat sources, heat sinks or applications and energy sources are available with the basic system components solar collector, heat pump, storages, and a backup system. Energy flows between these sources/sinks and the system components are represented by various arrows, depending on the type of flow.

To show the main subclasses of SAHPS system the scheme is based primarily on the heat source used by the heat pump as one characteristic aspect. In the following examples we use the most common heat source air or solar collectors (which might be also PVT-collectors). Two main types are:

- "*parallel systems*" characterized by the fact that the solar thermal collector field and the heat pump are delivering heat in parallel to the heat sinks, often via a central heat storage, and
- "*serial systems*" characterized by the fact that the heat delivered by the solar thermal collector is used as low temperature heat source directly or indirectly by the heat pump.

It should be mentioned that these integration classes are not exclusive, i.e. a system can be serially and parallelly integrated. Also, a third type "regeneration" has been mentioned where solar collectors are used to raise the ground temperature for GSHPS.

In the serial integration of Fig. 9.9 an uncovered solar collector serving also as a heat exchanger to ambient air is a good solution, because this may also provide ambient heat at night time.

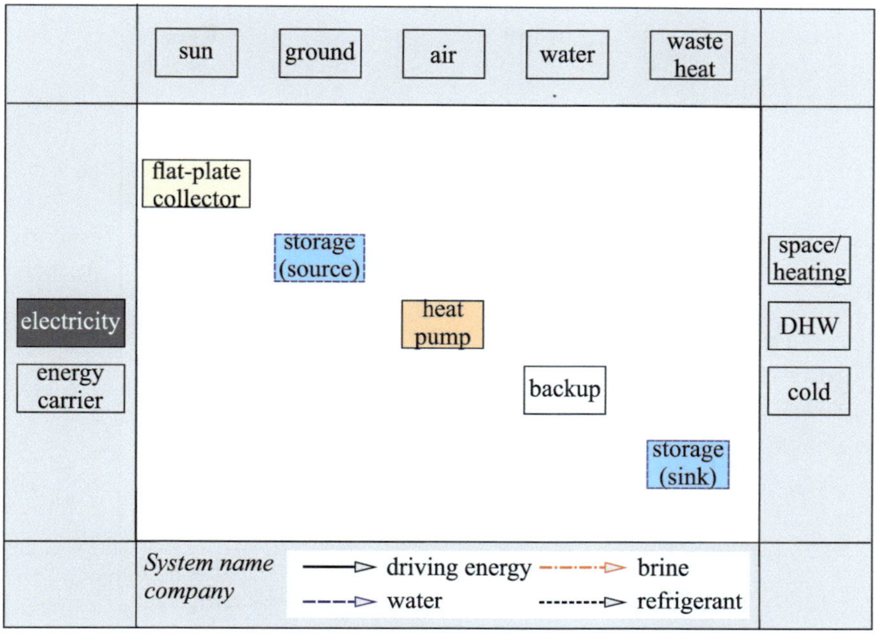

Fig. 9.8 Graphical template for the classification of different SAHPS (after IEA Task 44/Annex 38)

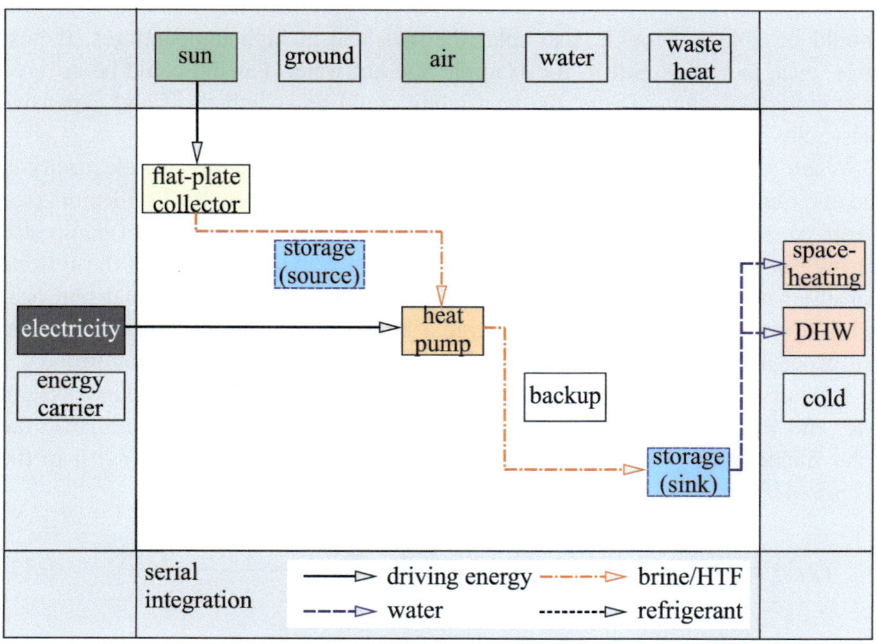

Fig. 9.9 Example of serial integration of solar thermal and heat pump

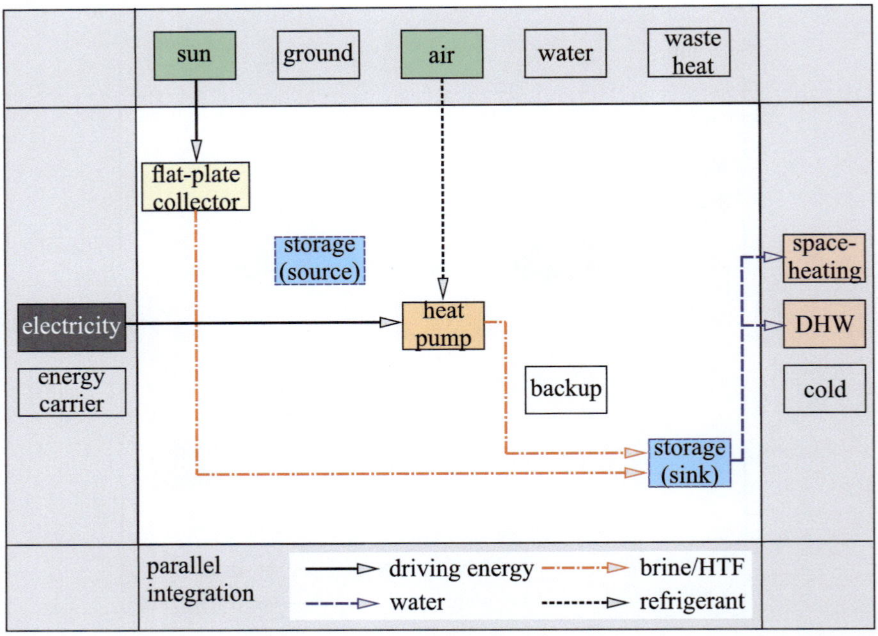

Fig. 9.10 Example of parallel integration of solar thermal and heat pump

On the contrary, the solar collector for the parallel integration in Fig. 9.10 should be able to provide also solar thermal heat at high temperatures, if possible, even in winter. But if, for example, the hot water demand could be covered in summertime, then at least the heat pump could be switched off and expensive electricity could be saved.

When one thinks about stand-alone 100% renewable systems, the electricity in the diagram could be provided by PV-panels on the roof, or by PVT-collectors (see Chap. 6). A battery part may be needed then to be able to run the heat pump also during low insolation time or at night. This certainly adds to the cost of the system. An alternative operation would be that the PV runs the solar thermal and heat pump system through daytime to charge a large enough thermal storage to cover hot water demand and heating demand for 24 h every day, even in wintertime.

If no or only partially renewable electricity is being used for a heat pump system with and without solar thermal providing 100% of the heat needed, similar to the solar fraction of a solar thermal system a renewable heating fraction f_{REH} of the HPS/SAHPS can be defined.

$$f_{REH} = \frac{E_{el,HPS} \cdot \left(SPF - 1 + f_{el,RE}\right) + Q_{SOLS} - Q_{loss,storage}}{Q_{DHW} + Q_{Heating}}, \tag{9.11}$$

where $E_{el,HPS}$ is the electricity consumption of the heat pump system with a renewable fraction $f_{el,RE}$, Q_{SOLS} is the solar thermal heat provided *directly* to the combi storage (but not the heat from the collector used as heat source for the heat pump in a serial connection) and $Q_{loss,storage}$ the storage heat losses (for both heat from heat pump and solar collectors). If $f_{el,RE} = 1$ the renewable heating fraction should also be 100%: $f_{REH} = 1$. This definition includes the ambient heat from the heat source as renewable and subtracts from the total heat provided by the heat pump the non-renewable electricity. For a heat pump system without solar collectors or only serial integration $Q_{SOLS} = 0$.

A fully renewable heating system using wood boiler, or a heat pump can be seen similarly treated as a bivalent system using a gas or oil boiler or direct electrical heating. However, the use of ambient heat with heat pumps has a big advantage as it saves wood for other purposes, and due to the electrification a fully renewable system is achieved when the national electricity grid system achieves 100% renewables. Therefore, it seems a very cost-effective way to achieve this target. Individual solutions have the advantage of higher degree of autonomy (say in cases of a blackout or a national energy crisis) but the investment usually is higher. For large heating grids a compromise with respect to autonomy might be achieved, but the integration of a large heat pump and a large (also distributed) solar thermal collector system promises a much better economy than the small-scale individual solutions for single and double family houses. Personal preferences might guide the selection of the approach for the own household or building.

Not all the solar thermal systems described in the next subchapters have sufficient specifications for the integration in an entirely renewable auxiliary heating. For example, passive solar the auxiliary system for example must provide not only the remaining heating load, but also the hot water demand. Solar air heating the storage would be a useful component improving efficiency also in the auxiliar system, even if the solar heated air is provided directly to the building. The best variants, integration variants and the optimal dimensioning have to be worked out still.

9.2 Solar Thermal Liquid-Based Systems

There is a wide range of suggested and implemented solar thermal low temperature systems. Reasons for this are manifold: different boundary conditions existing at the location, such as space requirements, building architecture, building traditions and legal regulations, different approaches to overcome challenges of solar thermal systems operation, specific technical design optimizations and marketing strategies of companies combining their individual components or even socio-political conditions such as market prizes for electricity and fuels or even subsidy programs.

Starting from some basic conceptual differences we then discuss a variety of different hydraulic layouts and their favourite applications. Here, the basic differences relate to the size of the solar thermal system, the number of circuits and heat exchangers between them, and the types and number of storages. Also, the integration of the solar thermal system into the conventional supply system (or the backup systems for the solar system) plays a role.

There are many possible variations of system components in a low temperature system. Depending on the choice of individual components consequential selection of further components is necessary. In the following part, however, we try to describe different fundamental principles of implementing the solar circuit, which have consequences in the component selection, but the choices also lead to completely different topology and operation of the complete systems.

A basic decision is whether the heat transport in the solar collector circuit is active or passive. The active transport by forced convection needs one or more pumps controlled by an active system controller. The flow conditions may be exactly controlled, and mass flows can be assigned to different operational states. The passive transport on the other hand relies on the buoyancy in the gravitational field, i.e. the density difference of hot and cold liquid. These density differences lead to smaller pressure differences in the loop than those achievable with an electric driven mechanical pump, therefore, the movement of the liquid is rather slow. Furthermore, natural convection systems rely very much on the placement of the components as the hottest liquid always will move to the top of the system. Due to its lower density, the water heated in the collector rises to the top and stays in the storage tank located above the collector. Cold water moves to the bottom of the storage tank and flows back to the collector through the return pipe. This implies that the storage of heat (the hot water tank) will be above the collector field, whereas in forced convection systems the storage can be placed at any place relative to the collector. In fact, very often the storage due to its heavy weight is placed in the basements of buildings.

9.2.1 Natural Circulation Systems

Advantages of natural circulation systems (also called "thermosyphon systems") are that they do not need electricity for pumping and for a control unit. The investment costs are much lower than for forced circulation systems. The thermosyphon systems widely used in the Mediterranean region often has an open circuit. This means that potable water flows directly through the collectors and is discharged from the storage. It is often referred to as the single-circuit concept having no heat exchanger between collector and potable water circuit. The pressure inside the system is depending on the mains water pressure. The liquid is constantly replaced by fresh water when hot water is drafted. Alternatively, two-circuit systems are also possible, where the collector circuit transfers heat to the consumer circuit via a heat exchanger located in the storage tank.

Figure 9.11 on the left side shows a the simple one-circuit scheme, where the hot water from the solar collector directly feeds a potable hot water tank. An electric heater is implemented there as a backup system. In the two-circuit scheme solar circuit and hot water circuit may be also separated as shown in Fig. 9.11 on the right, where for example inside the insulated buffer storage connected to the solar collector there is a separated potable water tank. Other options are an internal coil heat exchanger holding the potable water, which has sufficient internal volume.

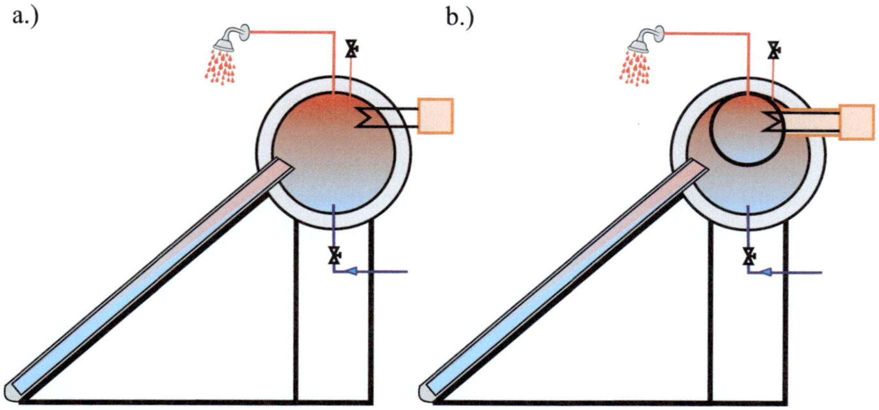

Fig. 9.11 Thermosyphon systems for potable water storage with direct connection to solar collectors (left) and potable water storage inside buffer storage with separated circuits (right)

More often, the storage is a potable water tank, and the solar collector transfers heat to it via an internal heat exchanger placed in the bottom.

Very often the complete thermosyphon system is installed on top of the roof—well-known are the many systems in the Mediterranean where flat roofs are very common. It is nevertheless also possible to integrate these types of systems into a tilted roof. Then, the storage might by inside the roof structure just in the attics if static considerations allow this. In practice, the systems are used for producing hot water mainly in climatic regions with little heating demand. Dimensioning and operation would be a major challenge if solar house heating in winter would be the target.

The mass flow $\dot{m}$ through the collector is determined by the temperature-dependent density change of the heat transfer fluid generated by the absorbed solar radiation in the collector. It is influenced by the following parameters:

- height difference H of the storage tank above the collector outlet,
- temperature difference $\Delta T = (T_o - T_i)$ between solar collector outlet and inlet,
- hydrodynamic resistance Δp.

If one excludes multi-phase flow and assumes finite height (H) and temperature (ΔT) differences, the mass flow of a thermosyphon system is limited by the pressure losses in the pipes. It follows that large pipe diameters, short pipe lengths and few fittings with low pressure loss should be used for thermosiphon structures. Thermosyphon systems are thus systems with a self-regulating mass-flow: when the solar irradiation is high, for a certain flow the temperature rise will be large, hence the mass flow will increase. This in turn reduces the temperature lift between outlet and inlet and in consequence the buoyancy driving the flow.

A qualitative estimation of the mass-flow can be given using the following approach. The collector equation based on inlet temperature shows the useful heat

($\dot{Q}_u$) for a collector:

$$\dot{Q}_u = A_c \cdot F_R \cdot [S - U_L \cdot (T_{abs} - T_{amb})], \tag{9.12}$$

in which A_c denotes the collector area, F_R the heat removal factor, S the solar power converted into heat and U_L the collector heat loss coefficient. The temperatures are the absorber temperatures and that of the ambient. But the useful power can be also defined by the temperature lift between inlet and outlet of the collector (ΔT_f) taken up by the heat transfer fluid given by:

$$\dot{Q}_u = \dot{m} \cdot c_p \cdot (T_o - T_i) = \dot{m} \cdot c_p \cdot \Delta T_f, \tag{9.13}$$

in which c_p describes the heat capacity of the fluid. This leads to the equation for the mass flow

$$\dot{m} = \frac{A_c \cdot F_R \cdot [S - U_L \cdot (T_i - T_a)]}{c_p \cdot \Delta T_f}. \tag{9.14}$$

Now, assuming that for a small temperature rise, the heat removal factor (F_R) and the collector heat loss coefficient (U_L) are over the collector length, the heat removal factor can be written as:

$$F_R = F' \cdot \frac{\dot{m} \cdot c_p}{A_c \cdot F' \cdot U_L} \cdot \left[1 - exp\left(-\frac{A_c \cdot F' \cdot U_L}{\dot{m} \cdot c_p}\right)\right], \tag{9.15}$$

where F' is the collector efficiency factor. Equations (9.3) and (9.4) may be combined and solved for the mass flow, resulting in:

$$\dot{m} = -\frac{F' \cdot A_c \cdot U_L}{c_p \cdot \ln\left[1 - \frac{U_L \cdot \Delta T_f}{S - U_L \cdot (T_i - T_a)}\right]}. \tag{9.16}$$

Using this formula, the mass flow for a certain measured temperature lift can be estimated. Alternatively, assuming a temperature lift the buoyancy forces can be calculated which have to be in balance with hydraulic friction forces of the system. This would imply a certain mass flow. If the result does not comply with Eq. (9.5) an iterative procedure can lead to correct value. In practice temperature lifts for small thermosyphon systems have been observed in the range of 10 K.

For the simple calculation of the buoyancy of a thermosiphon system, the temperature dependence of the density of water $\rho(T)$ is assumed in the following way:

$$\rho(T) = \rho_0 \cdot [1 - \gamma(T - T_0)] = \rho_0 \cdot (1 - \gamma \Delta T), \tag{9.17}$$

where γ is the thermal expansion coefficient and ρ_0 represents the specific density of water at a reference temperature T_0. At large temperature differences a linear

Table 9.2 Temperature dependence of the volumetric expansion coefficient (γ) of water at standard pressure ($p = 1013$ mbar)

Temperature [°C]	20	40	60	80
$\gamma\ [10^{-3}/\mathrm{K}]$	0.209	0.386	0.522	0.640

relationship of the thermal expansion coefficient is no longer given, so that for this case the density is calculated in the following way:

$$\rho(T) = \rho_0 \cdot \exp\left(- \int_{T=T_0}^{T=T} \gamma(T) dT \right) \tag{9.18}$$

Since thermosyphon systems are essentially operated at low pressures and temperatures (<100 °C), the values for water can be taken from Table 9.2.

The mass flow in a designed system regulates itself automatically. It is determined by the equilibrium of the buoyancy pressure difference (Δp_{buoy}) and the hydraulic pressure loss (Δp_{loss}), which depends on the flow in the collector loop, i.e. $\Delta p_{buoy} = \Delta p_{loss}$. The pressure difference resulting from buoyancy (Δp_{buoy}) can be calculated by

$$\Delta p_{buoy} = g \cdot H \cdot (\rho_{in} - \rho_{out}) = g \cdot H \cdot \rho_{in} \cdot \overline{\gamma} \cdot (T_{out} - T_{in}), \tag{9.19}$$

where g is the acceleration due to gravity with $g = 9.81\ \mathrm{m/s^2}$.

The hydraulic pressure losses of the circuit results from the sum of the pressure losses of the connecting pipes, the elbows and collector structures according to Chap. 5.

Although the construction of a gravity system requires relatively little installation work because no pumps are needed, the outdoor placement of storage results in poorer efficiencies compared to systems with forced circulation. In Central Europe, therefore, systems based almost exclusively on the forced circulation principle are used. Nevertheless, the advantages of a thermosyphon system should be clearly noted. In addition to the low acquisition costs, they allow easy installation even in self-assembly (modular system), operation without electricity. They have low or no maintenance costs and an unproblematic emptying of the collector is easily possible.

9.2.2 Forced Circulation Systems

For the forced convection or forced circulation systems we have more degrees of freedom to design the whole system. The storage may be placed at an arbitrary location, and the heat transfer can be facilitated by external or internal heat exchangers with a very good heat exchanger efficiency. Due to the forced mass

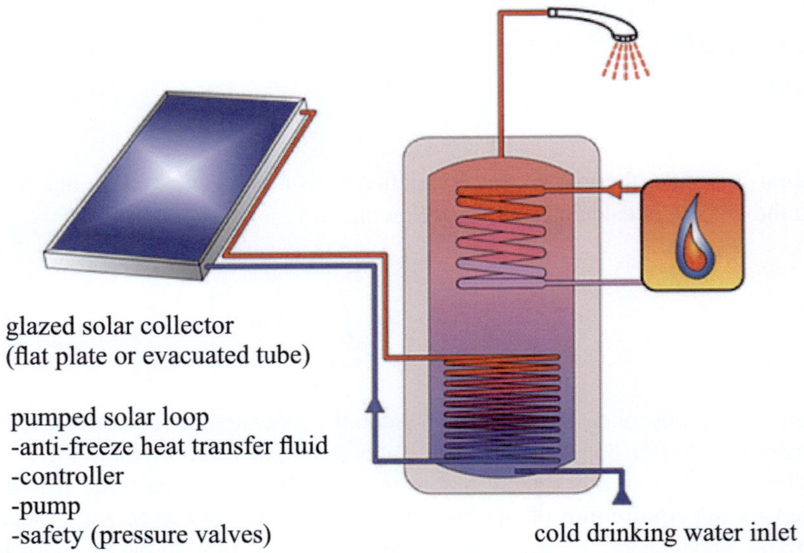

Fig. 9.12 Closed forced circulation pure domestic hot water system ($DHWS$)

flow the heat transfer conditions within the heat exchanger can be adapted. A schematic example of a solar domestic hot water system is shown in Fig. 9.12. The advantage of the system is the controlled operation and the optimized heat transfer between the solar collector loop and the hot water circuit, as the flow conditions in the heat exchangers can be adapted to very high heat transfer rates. On the other hand, pump and controller plus other loop components like temperature sensors are necessary and especially for small systems represent a considerable addition to the investment. For the forced circulation we have two additional subvariants. One is the closed solar circuit and the other the open drainback circuit.

Thermosyphon systems are dominant in Southern Europe while forced circulation systems are dominant in Central and Northern Europe, as can be seen in Fig. 9.13.

9.2.3 Pressurized and Non-pressurized Systems

Solar collector systems have to operate and survive in a wide variety of meteorological conditions—in hot summer days and cold winter nights. Therefore, we have to consider the two extreme cases of freezing and collector stagnation. The heat transfer fluid in the collector circuit is based on water which freezes at 0 °C.

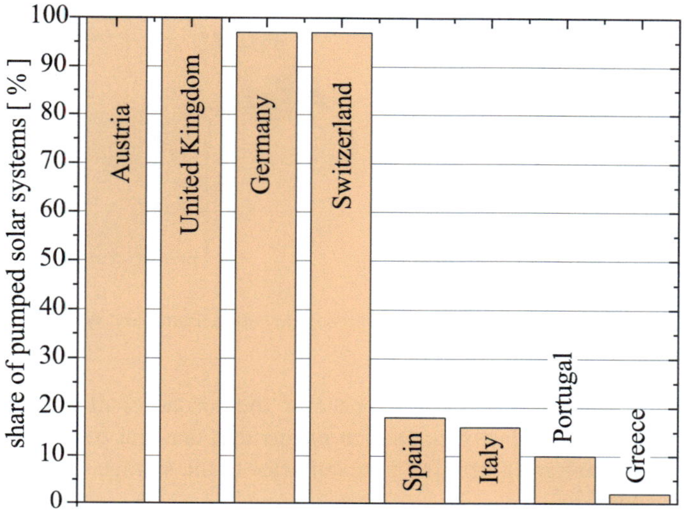

Fig. 9.13 Percentage of forced circulation systems in European countries from (Vogelsanger, 2005)

In order to prevent freezing even in cold winter nights in the outdoor connections to the collector field there are two possibilities available. The first one is the addition of a liquid which lowers the freezing point of water. The mixture widely used in a water–glycol mixture where the percentage of ethylene or propylene glycol determines the freezing point. For 40 vol% of glycol the freezing point is lowered to −24 °C which is sufficient for most locations. In more extreme climates the glycol fraction should be increased (50 vol% lowers to −37 °C). A second option to prevent freezing is surprisingly simple in principle: just emptying the solar circuit which may be filled by pure water. This leads to the so-called drain-back system, illustrated in Fig. 9.14. Whenever the collector system is not operating, the water is drained and collected in a collection tank at the bottom of the system. When starting up, a solar circuit pump has to lift again the water against gravity and press it through the complete circuit, which means that additionally to the friction forces the gravitational forces have to be overcome by the pump. As a consequence, the pump power has to be larger having a higher head. The principle of the drain-back system is an *open or closed unpressurized* system, while the standard circulation uses a *closed pressurized* system option. Pressurized circuits are widely used; they are mainly made of metal and are therefore, suitable for pressures above 3 bar. They have to be closed, which has a positive effect on the corrosion potential: Since after some time the oxygen originally present in the circuit is used up, no further corrosion occurs afterwards. In unpressurized systems oxygen can be avoided by replacing the air with nitrogen (implying a closed circuit as in the pressurized case), or the piping is made of non-corrosive materials like polymeric tubes (making open circuits possible). Open systems require the use of stainless materials (mostly stainless steels) in pumps.

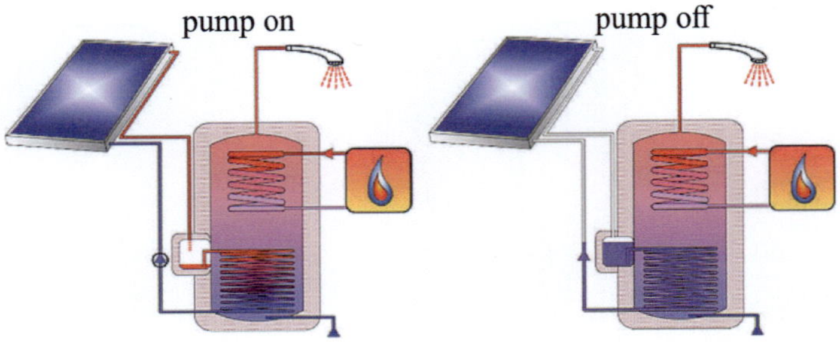

Fig. 9.14 Two operating modes of the open forced circulation drain-back system

Pressurizing a system is a consequence of the second challenge, namely the collector stagnation. Collector stagnation means that the heat transfer fluid is not circulated and transferring heat from the collector to the storage. There are several possible reasons:

- the storage is full, and no heat can be released any more to it,
- the pump or the control unit is defect and/or
- the electricity supply for control unit and pump is interrupted.

We will deal with the different phases of stagnation and the means to prevent it or at least reduce the stagnation hours in Sect. 9.7.5. Here, only the different ways how closed and open circuit systems deal with stagnation are sketched. In an open drain-back system again the liquid water is drained back, and the collector field should be emptied. Then no steam can be formed from water staying in the collectors, leading to overpressure and steam/liquid loss in the system. In a closed system the heat transfer fluid is pressurized above atmospheric pressure, to a level between 4 and 6 bar. Then the evaporation temperature is rising dramatically (to 143 °C for 4 bar and 159 °C for 6 bar), and the probability of steam formation is very much reduced. Flat-plate collectors for example have high heat losses at these temperature levels and due to the minimum efficiency even at high irradiation values it takes long to evaporate the collector water content completely.

In the following section the operation of a drain-back system and its requirements are discussed in more detail.

9.2.4 Drain-Back Systems

The drain-back concept, illustrated in Fig. 9.14, requires that each pipe in the collector circuit is laid in a descending direction from the highest point of the plant towards a storage volume. This allows the entire fluid volume to be completely emptied into the container volume by gravity.

In return, a gas—usually air or nitrogen—flows from the container volume into the collector. Particular points to be observed when designing drain-back systems are:

- Pipe gradient of 1.5–3 cm/m (minimum 1 cm/m). If the pipes are not fastened properly, thermal expansions over time can lead to fluid sacks. However, a continuous gradient is also required for pipe fittings and connecting pieces between the collectors. The collector piping itself should be inclined is such a way that emptying is possible. Particular attention must be paid to this when collectors are connected in series one above the other. In the case of collectors connected in parallel, the collector connections between the supply and return lines of the individual collectors are critical.
- Special attention must be paid to the cross-section of the supply line. On the one hand side, the collector must be ventilated via this pipe, and on the other hand, the air must be entrained by the water when the system starts up via this pipe. Generally, this requires a smaller cross-section than the return line.
- The solar circuit pump must not only be able to overcome the flow resistance during normal operation, but it must also be able fill the collector and lift the liquid over the difference in height between drainage vessel and collector. In order to be able to fill collectors connected in parallel without problems, a volume flow of about 40–80 l/(m^2h) is necessary (high flow system). The high mass flow during start-up can be reduced after a few minutes.

This list of requirements for the drain-back systems indicates that assembly and design require specially trained personnel (Hausner et al., 2003).

Originally, drain-back systems were developed to provide a frost-proof system concept without special heat transfer media, but only by using water as a fluid. The solar circuit can thus be connected to the water supply like a conventional heating system. With the increasingly observed occurrence of stagnation problems in solar systems due to evaporating medium, drain-back systems are gaining renewed interest, as the collector is filled with air/nitrogen during stagnation and no evaporation processes can occur.

There are variants of drain back arrangements related to the placement of the drainage vessel. During operation of the pump an air/nitrogen volume is located in an empty container connected to the collector circuit. When the pump is switched off, this gas cushion rises to the location of the container volume and replaces the liquid there. Some systems have a drain-back vessel integrated into the hot water storage, some provide a separate drainage container, and other just use empty space in a buffer tank.

Drain-back systems have proven their functionality on several thousand installations in the Netherlands. There, 80–90% of the plants have been built according to this concept. In most cases, however, the collector areas are small with only several square meters in size and the systems just provide hot water. Large plants with a collector surface of approx. 1000 m^2 were also put into operation based on this functional principle—however, it has been shown that the piping and installation costs are considerably higher than for plants of conventional design, see e.g. Botpaev et al. (2016).

9.2.5 Low, Matched and High-Flow Solar Loop

Another classification of low temperature solar systems is based on the specific volume flow per square meter of collector surface and accordingly there are so-called high flow ($\geq$ 30 l/(m^{2}h)) and low flow systems ($\approx$5–25 l/(m^{2}h)). Matched flow systems vary the flow according to the temperature requirements of the system. By adjusting the flow, even with variable solar irradiation levels a target output temperature can be reached. The choice of system flow has paramount impact on the hydraulic operation.

High-flow systems target at a maximum collector efficiency in two ways: first of all, the temperature rise in the collector and hence the average fluid temperature is low. This leads to low thermal losses and high collector field efficiency. In addition, due to the turbulent flow within the collector internal piping caused by high mass flow, the heat transfer from absorber to the fluid is very good. As a consequence, to reach a certain temperature within a high-flow collector system, even with high irradiation levels, multiple passages of the fluid during a day between storage and collector are necessary.

Low-flow systems on the other hand focus on a minimization of hydraulic losses. This relates to the mechanical pumping energy needed for circulating the liquid through the collector field. And secondly the low velocities at the entry point of the thermal energy storage improves in many cases the stratification of the storage water (Hollands & Lightstone, 1989), whereas the high impulse of high-flow circuits may destroy an existing stratification by adding too much kinetic energy to a storage volume. The operation of low-flow systems leads to higher temperature differences between the collector and the storage tank, i.e. high exergy creation, which remain in the storage even during operation. Due to the higher average temperatures the collector efficiency is reduced slightly, but even with medium irradiation, high temperatures without need of multiple passages of the fluid during a day are possible. A further disadvantage of the low flow may be that the heat transfer between fluid and absorber pipes is reduced because the flow type often is laminar.

Matched flow systems target very specifically at a certain temperature level in operation. The flow is adapted to the solar irradiation and the thermal losses of the collector field in such a way that ideally the outlet temperature of the solar field is constant, at the cost of a variable mass flow. Also, in this case once-through operation is the standard case. Collector efficiency is lower than in high-flow concepts.

High flow systems are particularly suitable for small collector areas up to 25 m^2. Above that range the diameters of the pipes leading to and from the solar field increases very much. Pumping power is becoming an appreciable part in the overall energy balance. Low flow systems, on the other hand, are generally used for collector surfaces of more than 10–15 m^2. Compared to high flow systems of the same surface area, low flow collector fields also allow more cost-effective piping, with smaller diameters as the hydraulic resistance is dramatically reduced. More compact heat exchangers (high temperature differences!) and the use of less powerful pumps is possible. Because of these advantages, large plants are usually operated in the form of low flow systems.

Systems with very small header pipe cross-sections can only be operated as low-flow systems, as otherwise the flow resistance increases too much. Within the absorber, small pipe cross-sections are desired so that the collector itself has a low heat capacity and heats up quickly, making heat energy available quickly. Also, serpentine collectors are preferred due to the longer passage length for a water volume compared to harp collectors.

Matched flow is also used in very large collector fields (Shrestha et al., 2018). When the system requires a certain temperature level and is operating very often without backup system, then matched flow could be a good option. However, the integration of storage also has a certain influence on the choice. If the matched flow system targets at a demand temperature of 60 °C, then a storage would not be charged to higher temperatures. In a low flow system, the 60 °C storage water still can be raised to 90 °C, thus increasing the storage capacity of the same water tank. This indicates that the decision on a flow system is depending not only on collector efficiency and hydraulic parameters, but also on heat demand, system, and storage design.

9.2.6 Classification of Hydraulic System Concepts

Hydraulic concepts of liquid solar thermal systems for buildings provide a large number of different successfully implemented variants. They can be classified according to the following parameters:

- *Solar hot water (SHW), solar heating (SH) or combined solar hot water and heating system*: In hot water systems providing potable water, the solar energy is transferred directly from the solar collector loop with a heat exchanger to the potable water in suitable potable hot water tank. The potable water is the storage medium. In systems with heating water, the solar energy is transferred to heating water in a buffer or combi tank. From the heating water then the transfer is via a secondary heat exchanger to the potable water. Pure solar heating systems rarely are used, but they would only need a buffer storage tank, not potable water storage.
- *Single-storage systems–multi-storage systems*: In the selection of storage several aspects have to be considered. Firstly, compact storage is advantageous, because smaller area for the same volume results in lower heat losses (the volume/surface ratio is decisive here). On the other hand, tall storage cylinders are

favourable for stratification (hot water is on the top and cold water in the bottom of the tank). Larger volumes provide a better volume/surface ration, but they require high room heights which often is not available in normal buildings. The required storage volume must be divided into several tanks.

- *Type of thermal energy storage*: Heating water can be stored in a simple buffer storage using ordinary steel or polymers. The heating water is circulated in the systems. Potable water needs high quality drinking water storage using steel coated with enamel or stainless steel as tank walls. There is a continuous exchange of the energy storage medium, which is the potable water.
- *Integration of auxiliary heating*: Auxiliary heating (the second heat source of a bivalent system) can be integrated at several points in the hydraulic system, in the storage tank itself, using an internal or external heat exchanger to the storage tank or directly in the flow of the heating water.

9.2.6.1 Solar Domestic Hot Water Hydraulic Concepts (SDHW)

The differences in the hydraulic configurations of $SDHW$ systems mainly concern the separation of hydraulic circuits and the use of heat exchangers. These decisions have consequences on the storage types. The circuit on the demand side is the potable water circuit with fresh water at low temperature coming from the water supply of the building, which should be heated up. The circuit on the solar supply side—except for very small simple thermosyphon systems—contains a non-potable liquid (with or without freeze protection). Fresh water would stimulate corrosion in the collector circuit and could be contaminated in a metal piping network. Therefore, these two circuits are separated by a heat exchanger. This also allows the pressurization of the collector circuit. For auxiliary backup heating we have more than one option. Either it can be done by an electric heater in the storage container or the potable water loop, or we have a third circuit connecting a conventional boiler with the system. For the storage we can have potable water storage or buffer storage for non-potable water (Fig. 9.15). The heat between the circuits can be transferred using internal or external heat exchangers related to the storage tanks.

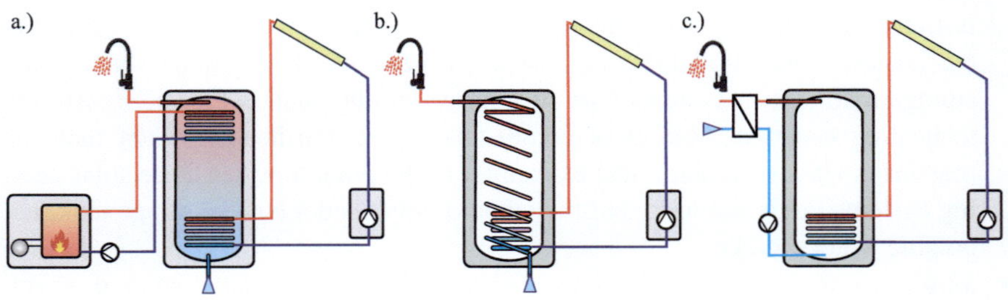

Fig. 9.15 Different storage concepts for drinking water **a** potable water storage; **b** combi storage; **c** buffer storage

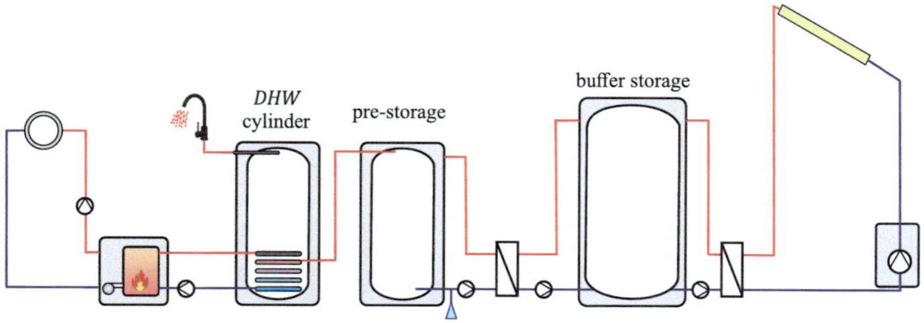

Fig. 9.16 Large *SDHW* system with several storages and external and internal heat exchangers

While smaller *SDHW* systems often use a storage vessel made from stainless steel or carbon steel with internal enamel coating which can be directly used for drinking water (Fig. 9.15a), other approaches use a combi storage where the hot water is kept in an internal heat exchanger coil or an internal smaller tank (Fig. 9.15b). For larger systems this would be too expensive therefore, buffer storages which contain hot technical water of the heating system are used, and the hot drinking water is produced by an external heat exchanger, a so-called fresh-water station (Fig. 9.15c).

As for large solar thermal collector systems a larger total storage volume is needed, multi-tank concepts are implemented. The different tanks might have different specifications—some are used for technical heating water, others for potable water (see for example Fig. 9.16). The use of external heat exchangers allows a better control of the heat flows within the storage system, but on the other hand additional pumps and sensors are needed.

For freeze protection in larger systems with a buffer storage, extra considerations have to be taken into account. The solution of Fig. 9.17 is required for long piping on roof to protect the heat exchanger: In wintertime, when a water–glycol mixture is cooled down to temperatures below 0 °C in the collector field, operation may start early when the collector sensor due to irradiation senses temperatures above possibly low buffer storage temperatures. If the pump of the collector loop then starts immediately circulating the liquid, out of the long collecting pipes of the solar field liquid below 0 °C may be circulated to the external plate-heat exchanger charging the storage. This situation would lead to a freezing problem inside the heat exchanger because the secondary circuit contains pure water. Therefore, before circulating the collector loop liquid through the heat exchanger, a short-circuit is created until the liquid in the primary circuit exceeds the freezing point.

The high temperatures which may be achieved in a buffer storage, on the other side, may lead to calcification problems, if the heating water inside the buffer directly heats potable water in a heat exchanger (see Fig. 9.18). As the calcium carbonate precipitation potential increases very much with temperature, and specifically above 40–50 °C, potable hard water should be heated up only to service

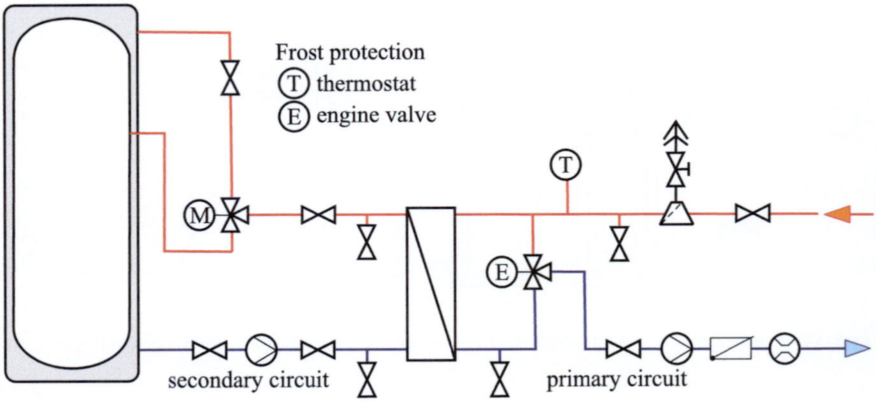

Fig. 9.17 Freeze protection layout for a large system with buffer storage

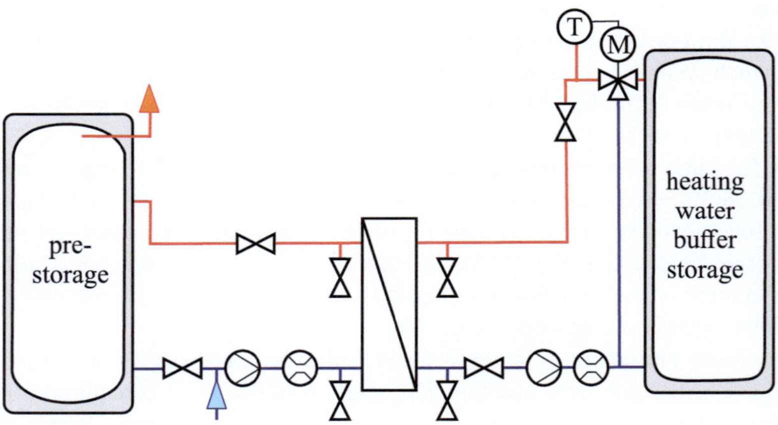

Fig. 9.18 Option to avoid excessive calcification in a hot water storage

temperatures. Therefore, in Fig. 9.18 higher buffer storage temperatures are mixed with return water from the heat exchanger in order to supply the lowest possible temperature being sufficient to meet the demand. With the controlled mixing valve calcification in the potable water storage can be largely avoided.

9.2.6.2 Solar Domestic Hot Water and Solar Heating Hydraulic Concepts

When we treat solar domestic hot water systems ($SDHWS$) a rather constant demand over the year at a temperature level of 40–50 °C has to be covered. Now, when solar heating of a building shall be assisted by the solar thermal system, a seasonally strongly dependent heat demand should be met additionally. The temperature level of this demand is significantly impacted by the building insulation level and the conventional existing heating system. Large-area low temperature heating systems (wall or floor heating) need lower feed temperatures than traditional radiator or convector systems with small areas. Required feed and return temperatures in the heating system have effects on the storage capacity and the solar thermal system efficiency. For example, a condensing gas boiler needs low return temperatures for the condensation of the flue gas humidity. Ideally that implies that the solar system may only add heat to the heating water after the flue gas condenser. On the other hand, the solar system efficiency is higher for low temperatures—a condensation of the boiler flue gas after the solar system has added heat however reduces the condensing efficiency. There is an optimization required. Therefore, many different hydraulic options are possible for the combination of $SDHW$ and solar heating (SH).

Generally, in a combined system may have a potable water storage and another a buffer storage for heating water. The direct storage of potable water has advantages, like the low return temperatures achieved for the solar thermal system leading to high efficiency. In addition, one heat exchanger less is required to transfer heat from the solar circuit to the potable water, as the heating water circuit is avoided. On the other hand, water storage tanks require considerably higher investments than buffer tanks made of conventional, uncoated steel due to the internal coating or the use of stainless steel. The addition of solar heat is limited to a temperature around 60–70 °C, because otherwise the calcification from heated fresh water would be too excessive, leading to a shortened service life or higher maintenance costs. Experience with decalcification plants at high temperatures is rare. In addition, large drinking water reservoirs bring with them special requirements for water hygiene—see the Sect. 9.7.5 on Legionellae prevention.

Water storages may be avoided in designs, where from the heating circuit heat is transferred via a suitable heat exchanger to the potable water. This may be an internal heat exchanger in a buffer tank (for example a tank-in-tank system, a combi storage, a spiral HX in tank) or an external heat exchanger (usually a plate HX) operated with an additional pump. The first options are more economical for small systems, the latter ones for larger systems.

In reality, especially in multi-family dwellings, the single buffer or hot water tanks can be split up in two or more individual tanks due to space requirements. Small room height leads rather to an installation of four 1500 l tanks with 2 m height each instead of the installation of a single 6 m^3 tank with 3–4 m height.

There are two different concepts of using the solar heat and the thermal energy storage for the heating of a building.

In the first concept shown in Fig. 9.19a both solar circuit and backup heating circuit heat the buffer storage to the required temperature level. The hot water in both concepts is prepared by an internal spiral heat exchanger starting at the colder bottom and exiting at the hot top of the buffer tank. The heating water for the radiator heating is directly taken from the buffer tank at an intermediate height. The outlet to the radiator is placed well above the solar heat exchanger. Solar heat can be used if the temperatures are meeting the demand supply flow temperature. If the solar system is not able to provide the correct temperature, then the boiler

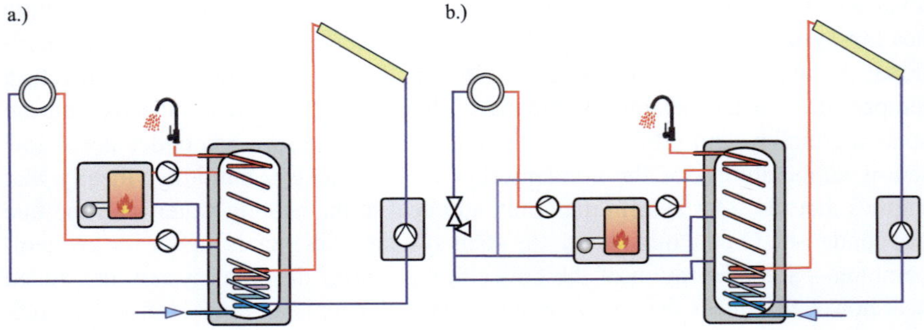

Fig. 9.19 Different concepts of using solar heat for heating the building in a combi system; **a** heating from the buffer storage; **b** preheating the heating return (return-lifting)

charges the buffer storage in a direct flow loop. The return pipe to the boiler is below the supply pipe to the radiators. Of course, temperature sensors have to be installed inside the buffer storage to control hot water supply and heating water supply temperature at the relevant exit. An integrated control unit has to control solar loop and boiler circuit.

In the second concept, shown in Fig. 9.19b, the boiler only heats the stand-by section of the buffer storage for the potable water (the upper part of the storage), not the lower section dedicated to the heating. This lower part is heated by the solar loop. When heating is needed and the return temperature from the radiators is lower than that of the water in the heating section of the buffer, the cold water is directed to the buffer by a three-way valve. At an intermediate level the violet line in Fig. 9.19b indicates water preheated by solar which then is directed to the boiler where it is raised to the required temperature for the radiator heating. In this way the solar preheating and the boiler post-heating is done sequentially. The advantages of this latter concept are that the average storage temperature of the buffer can be lower in the lower section, resulting in lower heat losses, and the control of the heating temperature can be done by the boiler control, i.e. this is a good concept for adding a solar thermal system to an existing boiler heating system.

A variant of the buffer storage with an internal spiral heat exchanger for potable water is the so-called tank-in tank storage, where a stainless-steel hot water tank is contained in a larger buffer storage as shown in Fig. 9.20. The whole potable water storage container serves as heat exchanger. It is important the fresh water supply enters from the bottom of the buffer. In a slim first section of the water storage the fresh water will be preheated. In the upper wider stand-by volume the surrounding buffer volume can be heated up not only by solar but also by backup heat from the boiler.

Of course, the single buffer storage volume can be adequately split-up in a hot water storage and a separate buffer as well, as indicated in Fig. 9.21. This may be used for intermediate system sizes.

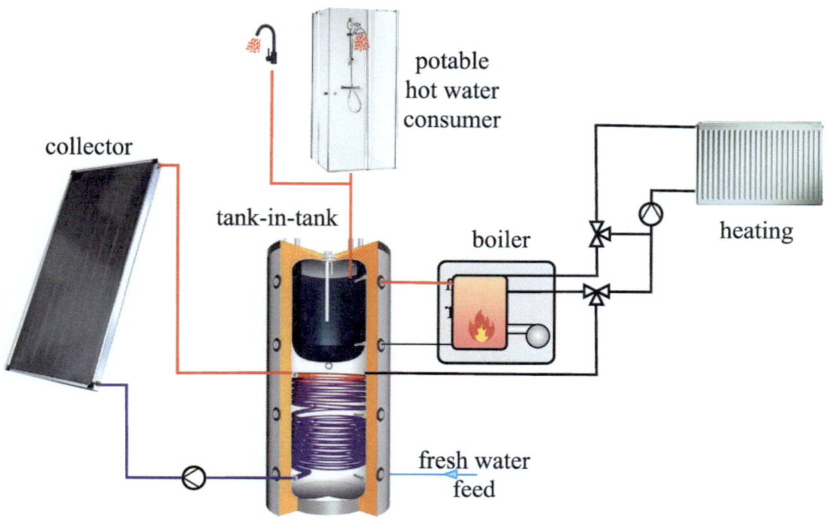

Fig. 9.20 Return-lifting with a tank-in-tank storage system

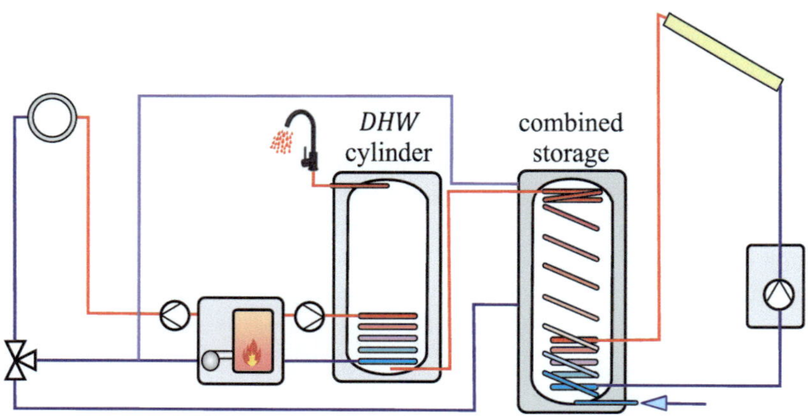

Fig. 9.21 Return-lifting in a two-storage system

When the solar systems become larges (>25 m^2) then external heat exchangers with pumps become more economic. A schematic hydraulic design for such a system is shown in Fig. 9.22. Of course, the large buffer storage needed here could in fact be realized physically as a series connection of several smaller buffers.

When more storages are used the hydraulic possibilities are becoming more and mor complex. Serial or parallel or mixed serial-parallel connections of storage arrays are possible. On the other hand, more heat exchangers, more sensors, and a more complex control logic has to be implemented. Therefore, details in the hydraulic circuits are decisive in providing an energy-efficient and economic

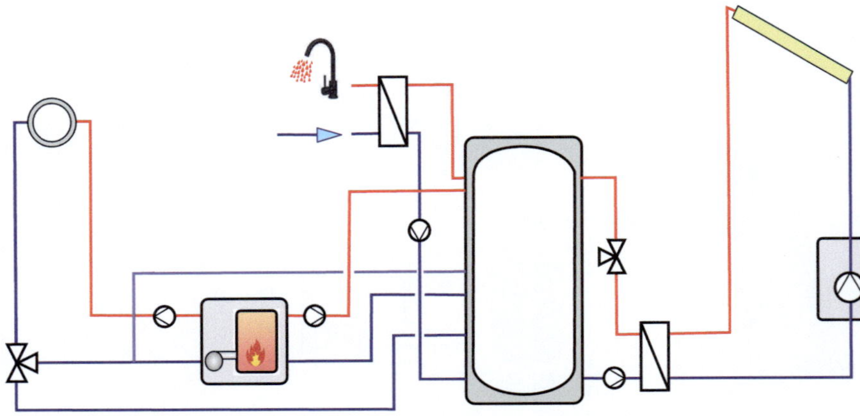

Fig. 9.22 Return-lifting in a large solar system using external heat exchangers

solutions. As an example, in Fig. 9.23 one can see two alternative solutions for charging two storage tanks in parallel—one using two pumps, and the other one using one pump and a controlled three-way valve.

The external heat exchanger or fresh-water station providing the hot potable water has to be able to heat up a considerable amount of water flow in very short time. This needs a high mass flow and pumping power in the storage loop, and a large heat exchanger area. Especially for large buildings with many flats, this might be problematic. Therefore, an alternative hydraulic scheme is shown in Fig. 9.24, which stores a amount of potable hot water in a central tank, circulating that in a classical two-pipe system to individual flats, where small hot water storages hold the standby volume for an individual family. The comfort provided by this solution has to be evaluated against the high heat losses of the two-pipe network due to the permanent circulation and to the small, distributed water storages. An

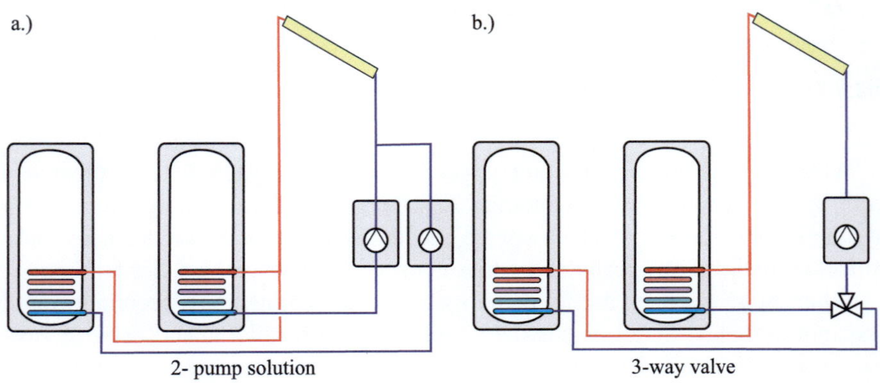

Fig. 9.23 Two options to operate the parallel charging of two storage tanks

idea to reduce these losses is the use of decentral water storages which only are preheated by a lower temperature network. The final temperature is then provided by electrical heaters, boilers, or heat-pumps individually for each flat, as discussed e.g. in Arnold et al. (2016).

9.2.6.3 General Consideration Related to Hydraulic Options

The separation of hydraulic circuits by heat exchangers is necessary but creates costs and on the other hand also has an impact of required temperature level. In each heat exchange process a temperature difference is needed between cold and hot flow. If for example hot water at 45 °C is needed, then the source for the heat (often a storage) has to provide 50–55 °C depending on the heat exchanger size and mass flows. This implies that the solar thermal system has to provide heat at a higher temperature level, which reduces its efficiency being depended on the average fluid temperature in the collector field. Therefore, heat exchangers should be avoided when not necessary because of hygienic, pressure or liquid type restrictions. On the other hand, heat exchangers allow a targeted and controlled heat transfer at certain times and can add to the overall management of a complex system e.g. when being able to charge a storage battery of several tanks depending on the individual charging status and the supply temperature. The latter effect may rise system efficiency when a good control logic is implemented.

The cost argument is more easily evaluated. The cost of additional heat exchangers and charging pumps often is counterbalanced by lower storage costs for lower quality buffer storage compared to potable water storage. A buffer storage under atmospheric pressure is much cheaper than a pressure vessel being able to withstand say a 6 bar pressure of a collector loop directly connected without a heat exchanger.

Similarly, the kind and number of storage tanks is an important design question. Potable water tanks are more expensive because of drinking water requirements (stainless steel, enamel steel). It is difficult to heat them over 60 °C because of strong calcification risk above that temperature due to fresh water with dissolved calcium carbonate. The buffer storages for heating water are less expansive, can be heated up to more than 90 °C because the calcium carbonate in the heating water is removed after some time.

When considering the number of storages, one may say that for single storages the volume to surface ratio is better compared to the case when the same volume is split up in a storage battery. Less piping is required, and a simpler control system can be used. This leads to the following advantages

- lower heat losses,
- low space requirement due to compactness,
- lower investments in control system, number of components, piping, insulation etc.

The advantages of multi-storage systems are

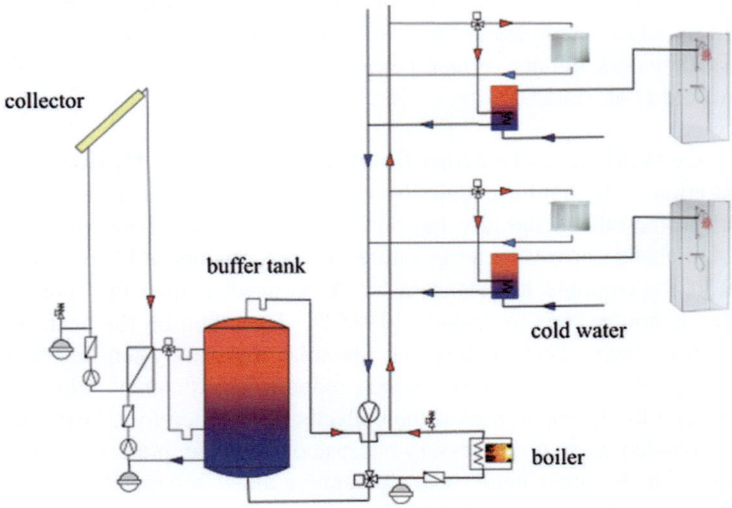

Fig. 9.24 Hydraulic flow diagram of a two-pipe network with central hot water tank and decentralised domestic hot water heating

- Easy separation of potable and drinking water circuits
- Better control of charging and discharging of individual tanks
- Often lower investments in storage tanks

Very often the piping network and storage integration are not independent. In a building with existing two-line network or in a new building decentralized small water storages can be charged in flats of multi-family houses. Then hot water circulation is not necessary because of the proximity of hot water storage and consumer. If the hot water supply is using only a single line from a central heating room to supply all flats, then circulation which implies also additional heat losses is needed to guarantee short response times for hot water draft. The central storage version, therefore, is more suitable for compact, smaller buildings than for large buildings with many flats. A further viewpoint in the choice of hydraulic design is the issue of Legionellae prevention. Legionella Pneumophila are aqueous bacillus which cause the so-called Legionnaire's disease which is potentially lethal. The growth of these Legionellae has to be avoided in hot water in buildings for domestic use. The growth is promoted by medium water temperatures of 30–42 °C, especially in non-flowing quiet water. By raising the temperatures and by avoiding "dead corners" in the hot water hydraulic systems as prescribed by the Legionellae regulations (see e.g. Bundesministerium für Gesundheit, 2018) this risk can be minimized. However, higher temperature levels and higher circulation flows have a negative impact on system efficiency. Therefore, the choice of the best system and operation is important. Details on the Legionellae prevention will be discussed in Sect. 9.7.4.

9.3 Solar Thermal Air-Collector Systems

Air collectors use the freely available heat transfer medium air in the collector loop, which has many practical advantages. Within the temperature limits used for heating systems it may not freeze, is non-toxic, does not degrade and ventilators and ducts for moving the air are well established in building technology. However, the low density of a gas leads to the necessity of moving large volumes of air in order to transport the heat within the collector loop. That again leads to large cross-sections of ducts and a high electricity demand for the ventilators.

The classical air collectors are flat-plate collectors and very similar in design to a liquid flat-plate collector. One difference is that there are a number of absorber geometries and designs. There is no direct need for closed pipes or channels connected to the absorber. The air flow may be passing the absorber sheet at the front between absorber and front glazing, behind the absorber in the gap between insulation and absorber sheet. A hybrid operation often starts with the air starting above the absorber, flowing through a matrix-type or perforated absorber, and continuing then behind. In order to increase the heat transfer from the absorber to the air, fins or corrugated sheets are used to increase the heat exchange area. Depending on the temperature level required even unglazed transpired collectors are used. The collector types are briefly described in Chap. 6.

9.3.1 Closed Loop Systems

In buildings with a standard temperature requirement it is more common to use closed-loop air heating systems. There are two possibilities with air as working fluid of the solar collector loop and the heating medium of the indoor space heating loop:

- air–air systems and
- air–water systems.

In the first category air is also used as the medium for heat supply to the building. In countries like United States or in Scandinavia air-heating systems are very common even in dwellings. In Europe air-heating nowadays is becoming more popular in Passive Houses because the required low heat demand can be guaranteed without excessive air volume flows and without the need for high temperatures. In most countries many office buildings do have air-heating, because very often air-conditioning is required due to high internal and solar gains and humidity control. Then a distribution system using large air ducts, filters, heat exchangers, sensors, sound dampers and ventilators is already installed. It is straightforward to mix solar heated air with air from another source (electrical or gas heater). The existing infrastructure needed for a comfortable air-heating system can be used also by the solar installation. As storage a sensible heat storage is a good option, where

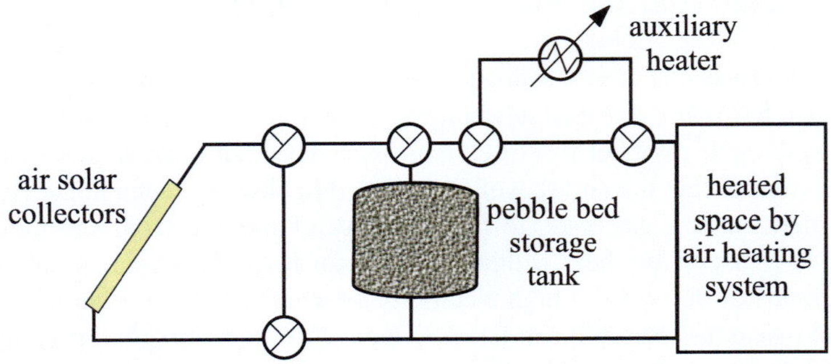

Fig. 9.25 Simplified active solar air-heating system with a pebble-bed storage

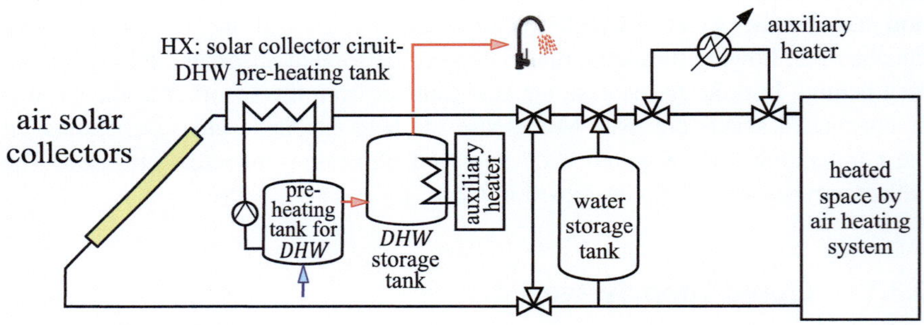

Fig. 9.26 Simplified active solar air heating system with hot water preparation

heat from the air can be stored in solid media like pebbles, hollow bricks or similar. In a simplified form the system is shown in Fig. 9.25.

It is possible to use air as the heat transfer medium in a solar collector loop and water as the heat transfer medium in indoor space heating systems. However, the thermal efficiency of such a combination of heat transfer media is low. A large heat-exchange surface area is required for air to exchange heat to water and a high flow of air is required as a result the energy needed for fans is high and noise problems can arise. On the other hand, the preparation of hot-water in any case needs an air-to-water heat exchanger. Therefore, this system may be the right choice for an active solar domestic hot water and heating system. A simplified system is sketched in Fig. 9.26.

9.3.2 Open Loop Systems

A very cost-effective way to use solar air collectors is the pre-heating of ventilation air needed in buildings with a large demand of fresh air, e.g. in schools, industrial and commercial buildings. The ambient air is preheated by a glazed or unglazed solar air-collector and then introduced into the ventilation system of the building. If possible, a heat recovery from the exhaust air will increase the energy performance. For buildings with low requirements on temperature and/or large internal heat sources like industrial productions sites, bus terminals or similar there is no additional heat source required. However, in commercial or residential buildings the ventilation air should be heated additionally by duct heaters.

Solar air collectors are often installed as façade collectors using the low altitude sun in winter-time more efficiently than roof collectors, for which one configuration is illustrated in Fig. 9.27. Costs are reduced by the building integrations as the collectors have a multiple function and protect the building against rain, wind and cold.

Air is drawn through the solar collector and is warmed in in a cavity between the solar collector and building wall. High efficiencies are possible because the solar collector absorber is only a few degrees warmer than the outdoor air. Therefore, there is little heat loss and most of the solar radiation is transferred to heat the

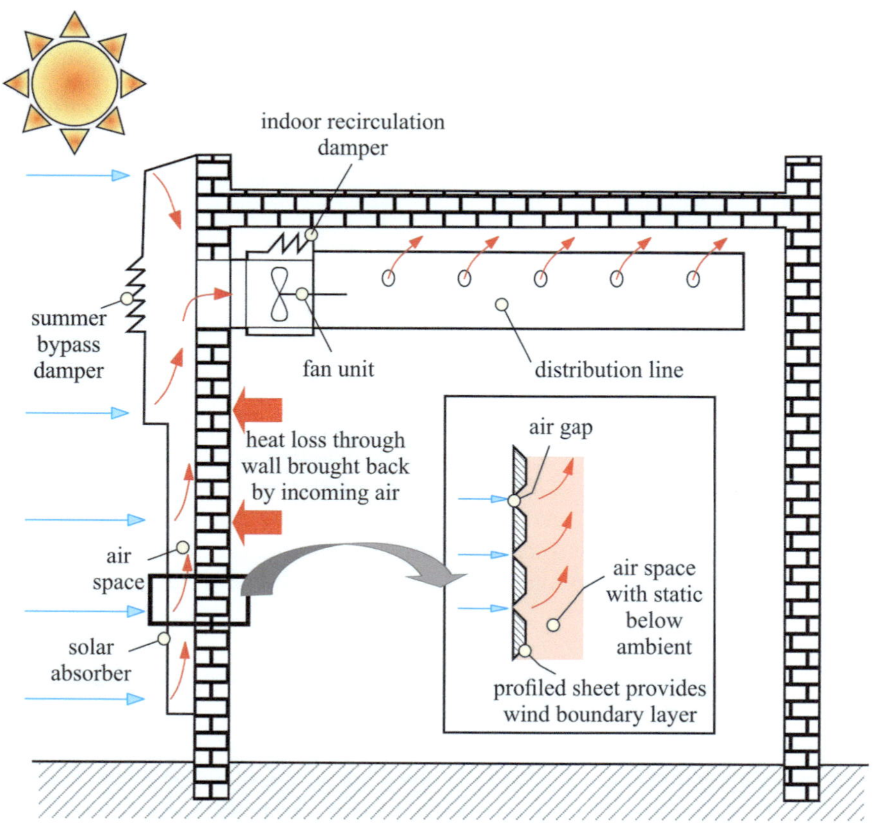

Fig. 9.27 Schematic view of solar preheating of ventilation air (modified from Conserval Engineering)

Fig. 9.28 Solar air collector wall integrated into the façade of the The Fire and Emergency Services Training Institute (FESTI) at Toronto airport Ontario (Canada), modified from Conserval Engineering Inc.—SolarWall (2011)

air. Depending on the project and climate often unglazed transpired air-collectors can be used (Wang et al., 2016). As indicated in Fig. 9.27 the air is then circulated through a distribution duct by a ventilator. In summertime a bypass damper allows ambient air to be fed directly into the building or process when no heating is required.

Open loop solar air heating systems can be building integrated and typically reduce between 20 and 30% of the conventional energy used to heat the buildings, for which one example is given by Fig. 9.28. The main markets worldwide are Canada, Australia, Japan, and the United States with actually about 1.3 Mio m^2 air collector area installed in total, see Weiss and Spörk-Dürr (2022).

9.3.3 Performance and Market of Solar Air Collector Fields

The installation of an air-duct network and the operation of a solar air-collector field has some complexities which are not existing in liquid systems. First of all, the airtightness of the duct network is usually not perfect. Air Leakage has to be considered and taken into account in performance. Also, the efficiency of air collectors is much more dependent on air velocities, see Welz et al. (2014). This means that collector efficiency testing is more complex and has to consider air flows (ISO 9806, 2017). To optimize the real performance a tool has been developed by Fraunhofer ISE, cf. Welz et al. (2015).

9.4 Solar District Heating (SDH)

In SDH a large field of solar thermal collectors supplies solar energy to the heat network of a neighbourhood, a village, or a town. This field is supplemented by a heating center, which provides additional energy to meet all the heating needs of connected residential, public or office buildings. The heat network can likewise be supplied with collected energy of distributed collectors installed on the roofs of the individual buildings. SDH is the renewable extension of district heating grids which provide heat to large neighbourhoods from sources like central heating power plants, combined heat and power, biomass heating or waste heat from industry.

Around 400 SDH systems are up and running around the world and 10–20 more are added each year. In many towns and cities, district energy plays a key role in supporting climate action and cutting emissions. In most cases, solar energy contributes up to 20% to annual heat demand. Using seasonal storage can increase this solar fraction to 60% or more, see Nielsen and Sørensen (2016). By optimizing the large installations and taking advantage of the statistical distribution of demand patterns in communities, the cost per supplied heat is reaching very low and competitive values.

When designing a SDH an extension of an existing heating grid certainly saves costs due to the existing piping network. However, typical supply/return temperatures are often in the range of 75–95/55–65 °C and can reach 90–145 °C in the transport pipes during the heating season (Sayegh et al., 2018). The high supply temperatures in winter restricts direct solar thermal supply to the summer months and sets limits to the solar fraction at which a solar thermal installation can be operated economically. More modern district heating (DH) aims at lower temperatures in the range of 75 °C/40 °C feeding low temperature heating systems. Improved insulation of the piping (also vacuum insulation is used) can reduce the heat losses of the distribution and facilitates smaller demand density. Heat pumps can be used to raise the temperatures reached by solar thermal collectors. In some cases, even concentrating collectors are being used to raise the supply temperatures, cf. Tschopp et al. (2020).

9.4.1 Concepts

Depending on the availability and cost of ground area SDH uses distributed solar fields integrated on the building envelopes of the buildings and central solar fields in a free field installation. The latter one is less costly in the specific installation cost. Moreover, the collection of solar heat from the individual roofs needs a second grid of transport pipes which raises cost and thermal losses. Whenever it is possible, one or few central solar fields should be preferable.

The heat distribution losses depend on the size of the piping network, on its thermal insulation and on the supply and return temperature levels. The installation shall be sufficiently deep buried in the ground to reduce the thermal losses to the environment, as the ground temperature in winter is increasing with depth.

The collector network transfers the solar heat to central heat transfer stations, where it can be stored in buffer storage which can also be heated by an auxiliary heat source (e.g. gas boiler, biomass boiler or heat pump unit). The distribution grid then is fed via a heat exchanger from the buffer storage. In some cases (if space is available and the ground conditions are good) a seasonal storage can be added to the network, for which the concept is depicted in Fig. 9.29 and a realisation in Denmark in Fig. 9.30.

The central heating station can be equipped with high-efficiency equipment, saving installation costs for individual buildings. This has to counterbalance the additional costs and heat losses of the distribution grid. The higher the demand density is, the lower is the specific cost of the distribution. In traditional high-temperature district heating (sometimes even working with steam) a critical demand density of more than 100 TJ/km^2 was needed. With today's 4th Generation district heating (Lund et al., 2014) densities of 30–100 TJ/km^2 are also possible.

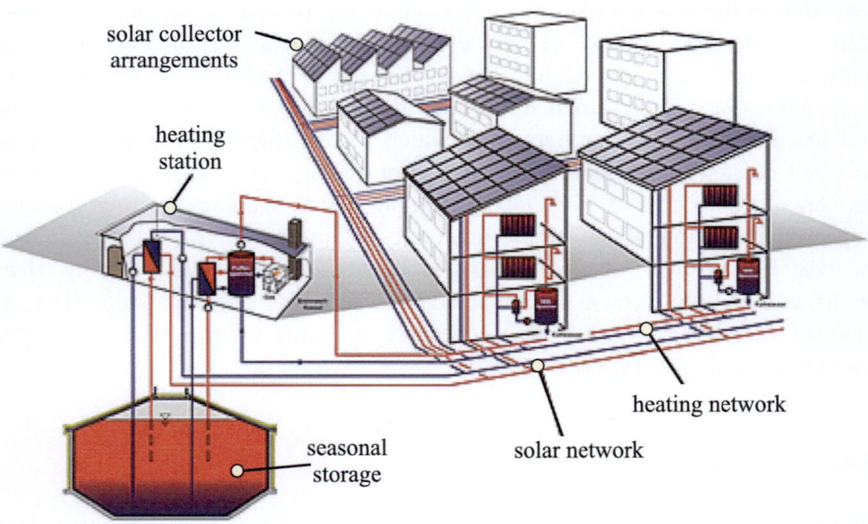

Fig. 9.29 Schematic view of heating grid with central storage and heating station and distributed solar field (modified from Mangold Solites-Steinbeis Forschungsinstitut für solare und zukunftsfähige thermische Energiesysteme, 2022)

Fig. 9.30 Solar district heating plant Marstal in Denmark (33,000 m^2 solar collectors and 75,000 m^3 pit heat storage), (Marstal Fjernvarme a.m.b.a., 2023)

9.4.2 Storage

Large storage volumes are needed for district heating grids, which leads to an adaptation of the tank concept which we have encountered in building systems. The thermal losses and the mechanical forces on the tank walls lead to a preferred integration into the ground. Such installations are less costly than free standing tanks, and moreover, they are visually more acceptable as they are integrated into the landscape.

Several concepts sketched in Fig. 9.31 for large storages have been developed. Some of them are only possible in few locations. For example, the aquifer thermal energy storage (ATES) cannot be used if the aquifers underground are carrying away the energy (underground rivers). Only localized aquifers with still water can be used. The borehole thermal energy storage concepts (BTES) necessitate a borehole with surrounding ground material which is able to conduct and store the heat during a charging process efficiently. Huge concrete tanks in the tank thermal energy storage (TTES) are expensive and up to now sizes up to $1.2 \cdot 10^4$ m^3 have been realized. The largest and also very cost-effective storages can be built as covered closed pits filled with water as in the pit thermal energy storage (PTES) concept. The charging and de-charging pipe system is similar to the systems used in tanks. The pit ground is covered with insulation material and the water tightness is assured by a polymer liner foil. The lid needs to be stable as persons should be able to walk on it. As a covered pit it can be integrated well in the landscape and even be covered with plants.

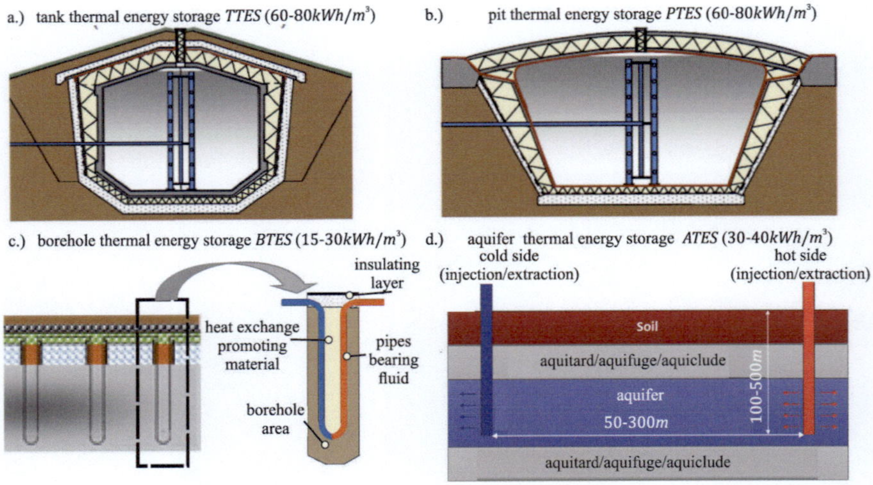

Fig. 9.31 Four different concepts for large thermal energy storage in heating grids

Seasonal storage needs extremely low heat losses, which is possible when the storage is very large and compact. The reason is that the heat losses are proportional to the surface of a tank volume: for a cubic storage the surface is proportional to L^2 when L is the side length. The volume and energy content, however, is proportional to L^3. This means that the losses become less and less important, the larger the dimension L of a storage volume is. A second positive aspect of large storage volumes is that the specific costs drop substantially. Figure 9.32 shows an overview of realized storages in solar district heating grids with specific costs and sizes. The PTES systems in Denmark reach up to more than $2 \cdot 10^5$ m^3 volume at costs around 25 €/m^3.

9.4.3 Economics

The economics of SDH can be very positive. Very important for a low heat cost are the market conditions and the existing infrastructure. The Danish example shows that solar heat can be provided for very competitive prices at about 30 €/MWh. One of the reasons is that district heating with low temperatures in the network (forward 70–80 °C and return 35–45 °C) is very common. Small communities have available cheap ground around the networks. Due to low installation costs for the solar fields of around 200–250 €/m^2 and a low-cost seasonal storage option (PTES) even high solar fractions larger than 25% are feasible. There are benefits from combining the technologies of combined heat and power units (CHP using biomass), wind electricity generation coupled with heat pumps and solar thermal heat, cf. Fig. 9.33.

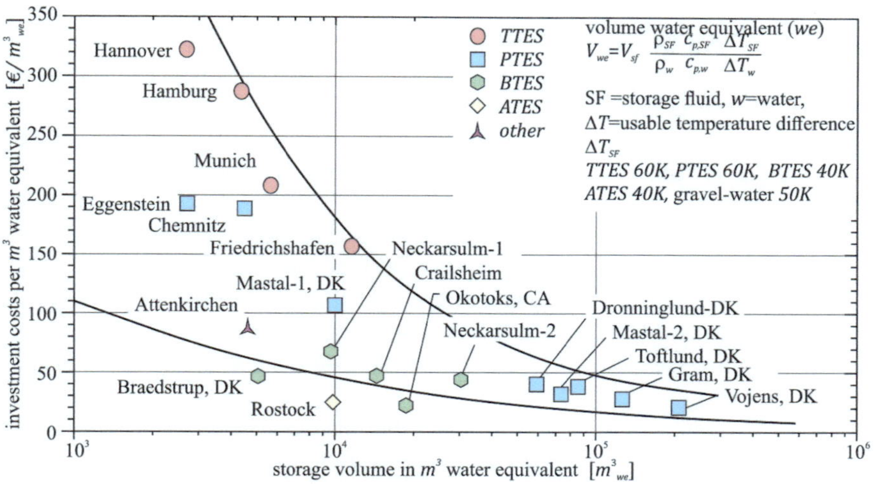

Fig. 9.32 Specific investment cost for large water storages realized in Central Europe based on currency value 2017 (modified from IKZ HAUSTECHNIK, 2022); (HW/TTES: Tank thermal energy storage, GW/PTES: Pit thermal energy storage, ATES: Aquifer Thermal Energy Storage; BTES: Borehole Thermal Energy Storage)

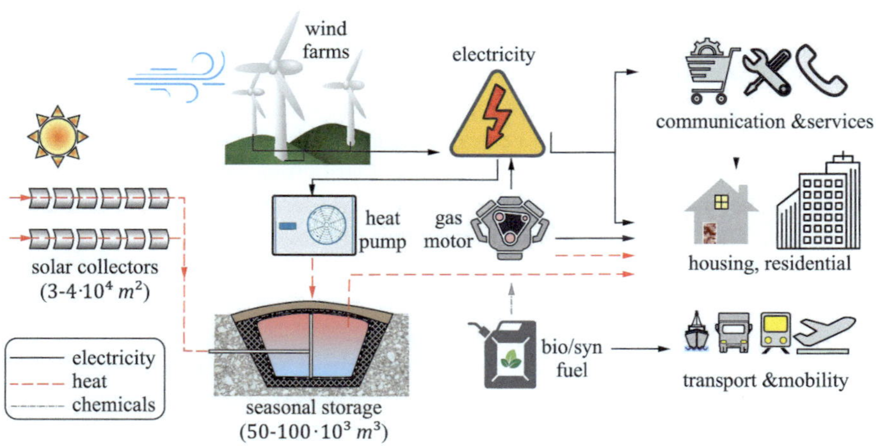

Fig. 9.33 Combination of technologies leads to optimized and 100% renewable systems, modified from Nielsen (2014)

The solar field produces free heat (after having done the investment) preferably in the summer half year. The heat pumps may produce cheap heat and raise temperature levels whenever there is a surplus of wind in the electricity grid. The heat pumps act as a fast capacity regulation for the volatile wind electricity, which produces revenues in the electricity market. On the other hand, when there is little wind available, the CHP gas motor can produce high-value electricity and additionally charges waste heat into the grid and the storage. The large seasonal storage is used to store the heat whenever it is available and discharges it to the grid whenever the demand is there. The heat pumps also have a benefit in the interaction with the storage: low temperature heat might be transformed to higher temperatures, which in turn enhances the specific storage capacity, leading to reduced storage sizes.

So, in this example the storage acts as convenor to give flexibility to the whole system and allows a good integration of the combined technologies. When in future more wind and photovoltaic electricity are available in the grid, the demand for CHP electricity is decreasing, and hence the demand for solar thermal energy increasing. This approach is a possibility of achieving a 100% renewable heating supply.

Another example of a successful renewable district heating grid is the Drake Landing Solar Community (DSLC) project (Mesquita et al., 2017). It is master planned neighbourhood in Okotoks, Alberta. The planners have integrated energy efficient technologies with solar thermal. As the first of its kind in North America, DLSC is heated by a district system with a large seasonal underground storage based on borehole heat exchangers (BTES). The system meets 90% of the space heating requirements from solar energy, resulting in less dependency on fossil fuels.

Despite the higher heat demand, the climatic conditions are favourable for solar heating in combination with seasonal storage due to the high solar radiation of 1750 kWh/(m^2a) on the tilted collector plane. A distributed solar collector field of approximately 2300 m^2 delivers the heat to two water tanks short-term storage of 240 m^3 located in the heating central. A general layout of the system is shown in Fig. 9.34. These tanks serve as the central distribution unit delivering heat via the district heating pipes. From here also the long-term storage is charged and discharged. The low permeability of the upper geological layers at the location qualifies the site for BTES for long-term storage. The BTES consists of 144 boreholes with depth 37 m. The borehole field is covered with an insulation layer and soil on top of which a landscaped park is built on.

A more recent urban project is the district heating system in Freiburg for the new building area Gutleutmatten, see e.g. Elci (2018). Space is limited in an urban environment, and property costs are high. This leads to a disregard of solutions with free field installations. Distributed solar collectors on individual houses are the preferred approach. Here decentralized solar thermal systems with a total collector area of 2200 m^2 and decentralized heat storage with a total volume of 160 m^3 in the buildings are connected to an existing district heating grid, where the heating station is equipped with bio-methane CHP and natural gas boilers. The grid is operating with 75 °C forward and 40 °C return and covers a heat demand of 2600 MWh/a—about 30% will be supplied by solar. The operation of the grid has some specifics: First of all, during summer where only hot water has to be supplied, the grid is turned off in favor of local solar thermal supply. Secondly, in wintertime the heat is distributed in impulse packages to individual buildings.

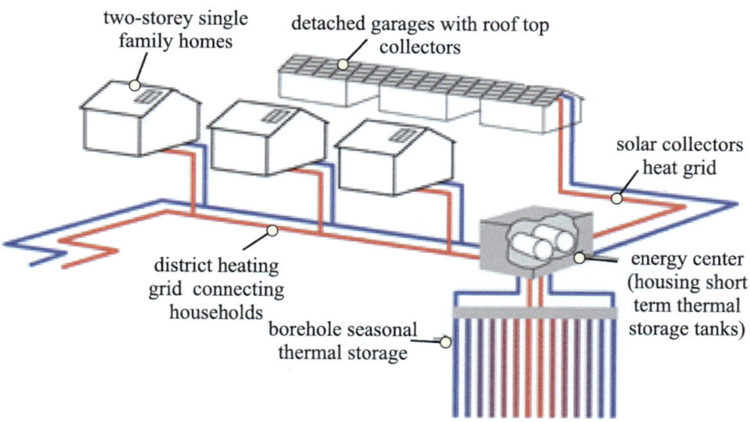

Fig. 9.34 Concept drake landing (Canada) modified from Mesquita et al. (2017)

Thirdly, the individual storages are operated in a cooperative way. These features minimize the distribution losses up to 30% as demonstrated by Oliva et al. (2020).

The market of solar district heating is rapidly expanding, and district heating in general is an important option for the future energy transition. In a district heating grid a multi-source approach can be realized much easier and more economical than in an individual building. High-quality components can be integrated in the grid, meeting the demands of different seasons, of environmental protection measures (e.g. particulate matter prevention in biomass boilers) and utilizing the local or regional sources (wind, solar, ambient heat from lakes or seaside, waste heat from industry or heat from waste incineration). The flexibility to design a 100% renewable system is much higher than for individual houses. And the economies of scale are in favor of larger capacity components, lading to much lower heat costs. Many new projects and descriptions of knowledge, designs options and operational experience can be found, for example in Solar District Heating (2023) and IEA Solar Heating and Cooling—Task 68 (2020). An article specifically dealing with the integration of heat pumps into solar district heating systems is Abokersh et al. (2021). Marx et al. investigated the energy efficient integration in systems with seasonal storage (Marx et al., 2014).

9.5 Components of the Solar Circuit

For systems intended for year-round operation in Central and Northern Europe, due to the risk of frost in the winter months, since winter operation requires an anti-freeze additive in the collector circuit, closed systems are required. In forced circulation systems, unlike natural circulation systems, a circulation pump is integrated into the circuit to circulate the heat transfer medium. The pump is necessary because the collector is usually located at the highest point of the system and the storage tank in the heated room system is often positioned at the lowest point of the building. Circulation due to difference in density is, therefore, not possible. In practice, forced circulation systems are usually installed in Central Europe for the following reasons:

- Higher efficiency than natural circulation systems at lower ambient temperatures.
- Little space availability for the storage tanks at locations above the collector.
- Usually integration of a supplementary heating system into the storage tank or use of the storage tank as a central element of the entire building heating system. Both aspects require the whole system to be housed in one room or nearby rooms.

Even a simple forced circulation system consists of a large number of pipes, components and sensors that must be installed to reach a functioning system. Figure 9.35 shows schematically the structure of a typical low temperature solar system for the production of domestic hot water and heating water with the associated components. As shown in the diagram, the storage tank is one of the central elements linking the heat generation process (solar with or without auxiliary fossil fuel firing) to the consumer. At the same time, it compensates for fluctuations in the heat supply. However apart from that, a large number of small components is necessary, which are explained in the following sections. A part of these components can be grouped together as a predesigned unit, called a solar pipe group, which includes at least temperature measurements for the heat transfer fluid (HTF), pump, controller unit, but other components may be added. The structure of a simple solar group and its components is illustrated in Fig. 9.36.

9.5.1 Solar Pump (in Collector Circuit)

In solar systems, the heat transfer medium must be pumped through the hydraulic piping system to transport the energy absorbed in the collector. Since it is a closed system, no geodetic heights are overcome. The circulating pump only has to overcome the flow resistance of the pipes, which is given by the hydraulic resistances of the internals (pipes, branches, elbows, etc.), their length and the viscosity of the fluid by the application of pump power. For solar thermal applications the typical water–glycol mixing ratio is 3:2 (60%:40%). For the basic design of the pump, the pressure losses must first be determined. In the solar circuit, the pressure losses consist of the following components:

- collector field,
- pipelines for flow and return, fittings, and screw connections and
- heat exchanger

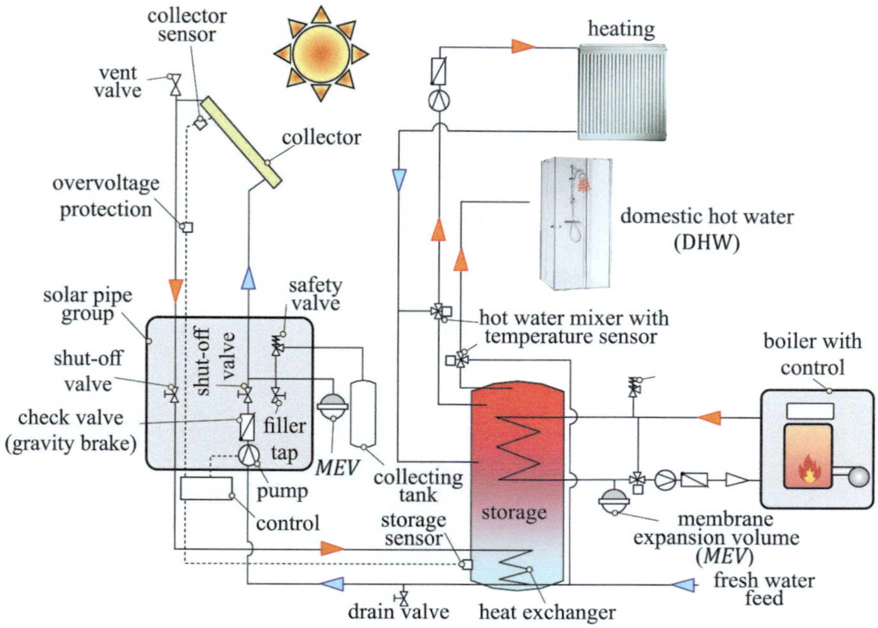

Fig. 9.35 Schematic sketch of a forced circulation solar circuit with the associated components for domestic hot water heating with auxiliary fossil fuel firing

together. The simplest target value can be assumed to be a distribution of the pressure losses in a ratio of 1:1:1. For primary design, the maximum pressure drop in collector arrays should not exceed the values specified in the Table 9.3.

The mixture used as antifreeze has a considerably higher kinematic viscosity compared to pure water. If water at 20 °C has a kinematic viscosity of $\nu = 10^{-6}$ m^2/s, the mixture has a considerably larger value of $\nu = 1.19 \cdot 10^{-5}$ m^2/s. However, since it is a Newtonian medium just like water, the pressure loss increases considerably due to the higher viscosity. The pipeline network should, therefore, have the lowest possible flow resistance and the flow velocity of the medium should be kept low. As a rule, values between 0.5 m/s and a maximum of 1 m/s are aimed for in technical systems. The higher pressure losses also cause a higher flow noise. In addition, the viscosity of the heat transfer medium mixture changes depending on the antifreeze and the temperature. As the proportion of antifreeze increases and the temperature decreases, the viscosity increases rapidly. The resulting change in the pressure loss factor in round pipelines is shown in Fig. 9.37 for various propylene-water mixtures as a function of temperature compared to water. The design of the required pump capacity must, therefore, be based on the lowest operating temperature of the heat transfer medium. As a target value for the pressure drop in the pipelines plus all fittings with the exception of the heat exchangers), values of a maximum of 100 mbar (corresponding to 1 m of water column) should be aimed for. Very small pipe cross-sections result in low

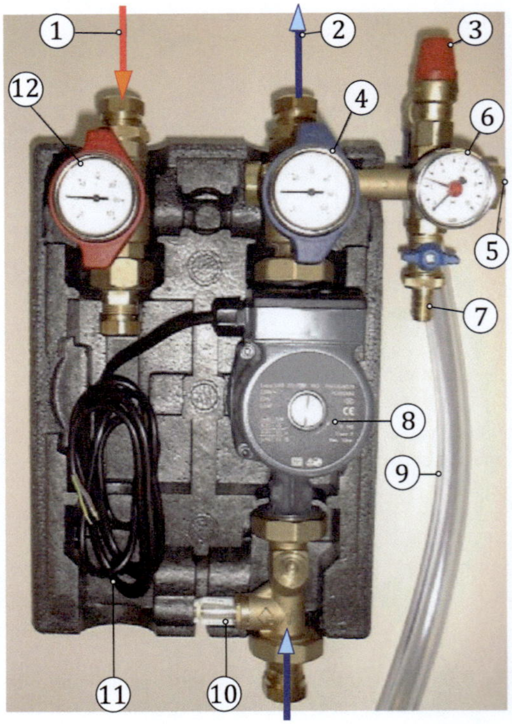

Fig. 9.36 Components of a simple solar pipe group connecting the collector to the storage tank. Simplest case of a forced circulation system without external heat exchanger. 1 Flow pipe, 2 Return pipe, 3 Safety valve, 4 Thermometer with shut-off valve, 5 Connection diaphragm expansion vessel (MEV), 6 Pressure gauge, 7 Filling valve, 8 Pump, 9 Blow-off hose, 10 Flow limiter, 11 Electrical pump connection, 12 Shut-off valve and thermometer flow pipe

Table 9.3 Reference values for the permissible pressure drop in solar collector arrays

Collector field size [m^2]	–6	6–30	30–100
Permissible pressure drop [mbar = 10^2; Pa]	50–100	100–200	200–400

heat losses and material costs, but high pump capacities and vice versa. Losses resulting from screw connections and fittings can be converted to an equivalent line length with the aid of relative pressure loss factors depending on nomograms.

The pressure drop of a heat exchanger can usually be found in the data sheets of the manufacturer. In contrast to ready-installed coil heat exchangers, the pressure drop, and capacity of plate heat exchangers can be individually adjusted, but the pressure drop should never exceed 100 mbar.

By adding the pressure losses of the components mentioned above, the total pressure loss of the system is obtained. The curve of the total pressure loss as a function of the heat transfer volume flow is called the system characteristic curve. An appropriate pump must be selected to match the system curve, always aiming for a pump that can be operated at the highest speed and that requires as little throttling as possible. The point of intersection between a system line and a pump line at a given speed is the duty point. Before the pump design, the duty points of the system, i.e. the volume flow in the system and the corresponding pressure

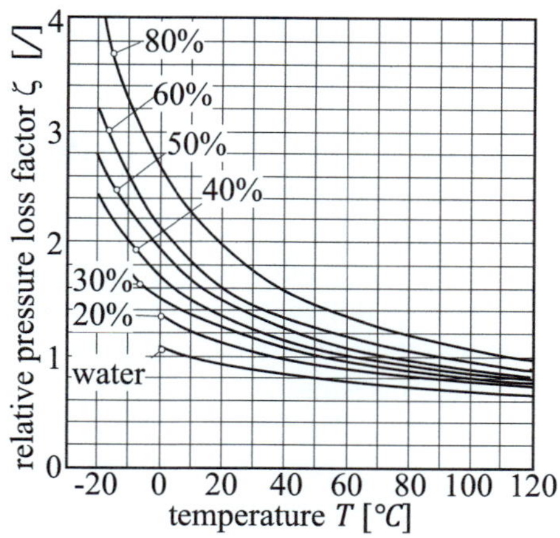

Fig. 9.37 Relative pressure loss factor ς as a function of temperature and concentration (in % volume of antifreeze) of a water-propylene glycol mixture (data sheets e.g. from Tyforop Chemie GmbH, 2011)

drop at the design point must be known. So, if the system's duty point is fixed, the pump line should ideally intersect this point. This means that the pump delivers exactly the required flow rate and pressure head at its best efficiency, as shown in Fig. 9.38 in a general manner; the pumps should be designed so that the duty point is in the upper third of the flow rate and as close as possible to the maximum speed, as centrifugal pumps have their best efficiency in this range. For more details on pumps and their design fundamentals see in the Appendix C.

In practice, however, the duty point very rarely lies on the pump line. It is usually not possible to hit it exactly—except with frequency speed-controlled pumps.

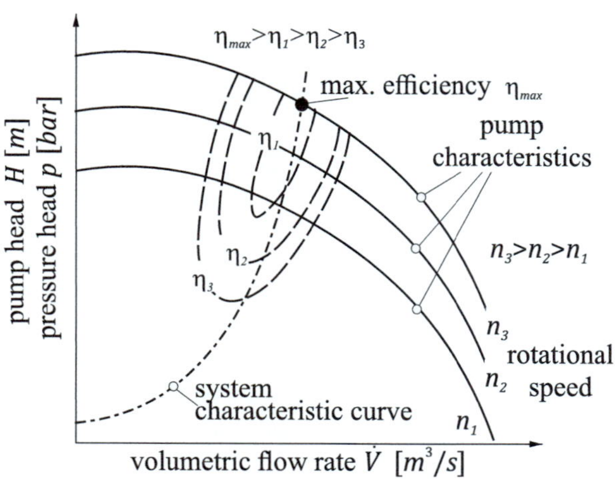

Fig. 9.38 Schematic representation of the pump characteristic curve (so-called shell diagram) of a centrifugal pump as a function of the flow rate. In the ideal design case, the system characteristic curve intersects with the optimum pump operating point

Therefore, the flow rate is often throttled to the correct operating point with a throttle valve. However, this type of pump design has the following disadvantages:

- increased power consumption (more than the operating point requires).
- additional built-in throttle valve.

In frequency-controlled pumps, on the other hand, an electronic system measures the flow rate, differential pressure or differential temperature before and after the pump and thus adjusts the pump performance, which also shifts the characteristic curve. This adjustment is infinitely variable between the highest and the lowest pump capacity. The difference to not speed-controlled pumps is that the power consumption of the pump corresponds to the power actually required, which allows additional energy savings of up to 40% in full load operation and up to 90% in partial load operation. The use of a frequency-controlled pump is definitely recommended, because despite the somewhat higher initial costs, the pump requires considerably less energy during its entire service life.

Often the solar radiation is not sufficient to keep the system at the optimal operating point. This is where control engineering intervention is required. There are several options:

- *Two-point control*. It is equivalent to switching the pump on and off. The advantages of this system are that there are almost no additional investment costs, and the pump is always operated at its design point. The disadvantage of this is that the pump, motor, and power supply system are loaded by the switching operations and that the system is always operated at full volume flow and thus also with high friction losses. In solar systems it means a permanent thermal alternating load of the collector.
- With *throttle control*, the flow is throttled by a control valve in the pipeline. This type of control is very common for small and medium size centrifugal pumps. It is very easy to implement. As the piping resistance increases when the valve is throttled, the back pressure at the pump increases and a new flow rate is set according to the pump characteristic curve. In the characteristic curve diagram of the pump in Fig. 9.38 is shown in such a way that the pipeline characteristic curve becomes steeper and steeper with increasing throttling and the point of intersection of both characteristic curves shifts to the left in the direction of smaller flow rates. The throttle control causes additional energy losses for two reasons. Firstly, hydraulic energy is converted into heat at the control valve. On the other hand, the operating point at throttling moves away from the design point of the pump and the motor and thus generally leads to a deterioration in efficiency.
- A *speed control* influences the hydraulic performance of the pump and thus the pump characteristic curve. In a positive displacement pump, the flow rate is directly proportional to the speed or stroke frequency. In a centrifugal pump, the hydraulic power output depends on the peripheral speed of the impeller. Depending on the resistance of the pipeline, a flow rate is then set. For example,

in centrifugal pumps, pressure and volume flow can be controlled by varying the speed, whereas in positive displacement pumps only the volume flow can be controlled. The pump characteristic curve shifts in parallel to the lower left in the $\dot{V}$–H diagram. In thermal solar systems, which only have friction losses and do not have to overcome a static pressure difference, the point of intersection of the characteristic curves and thus the operating point moves along a line of constant efficiency when the speed is varied. If the pump is designed to operate at full load at its best, it can be run at optimum efficiency even at partial load by speed control. In these cases, speed control in partial load operation can save a considerable amount of energy costs. This adaptability of the volume flow allows an optimal utilization of the collector, which means that even with low irradiation and a very small volume flow, energy can be introduced into the storage tank, or the storage tank can be charged at the optimum temperature. The shape of the pump characteristic curve also has an influence. With steep characteristic curves, as typically found in axial and semi-axial pumps, the savings are greater than with flat (e.g. radial pumps).

The properties of the heat transfer medium in particular make it clear that solar pumps differ from conventional heating pumps. While heating circulation pumps are designed for higher volume flows and low delivery heads, solar systems require small volume flows at higher pressure losses caused by the pumping medium of the glycol–water mixture. Heating pumps, therefore, only operate with low efficiencies in solar circuits.

For solar applications, mostly glanded pumps (inline pumps) or centrifugal pumps designed as glandless pumps are used. Glanded pumps have a higher temperature resistance and a higher delivery head, but they are more expensive than glandless pumps. For the latter design, both impeller and casing must be glycol and temperature resistant (impeller, casing).

In addition to centrifugal pumps, gear pumps are also used especially for low-flow systems. Such pumps allow very high heads at a very low mass flow. This pump is particularly suitable for a collector array with many collectors connected in series, as the ratio of pressure head to volume flow is many times greater than with centrifugal pumps.

A new generation of centrifugal pumps are the so-called *EC* pumps (brushless electronically *commutated* synchronous motor, *EC* = Electronic *Commutated*). They differ from conventional asynchronous motors (AC drive) only in the electrical drive mode of the motor, while the impellers for conveying are identical. In an *EC* drive, the brushless electronically commutated synchronous motor has a permanent magnet rotor and therefore, requires no energy for its magnetization. The torque is generated by the alternating current supply of the pairs of coils in the stator. The controlling electronics determine the rotor position via sensors or by analyzing the induced voltage. This electronic commutation and thus the rotor speed are infinitely variable. This can save up to 70% of the operating energy. At the same time, this type of drive enables improved partial load behaviour. In contrast to AC motors, motor efficiencies of up to 80% can be achieved even at low power levels up to 100 W. In comparison to EC systems, conventional drives have an efficiency of only 50% at low electrical power consumption in the same power class, see e.g. Meier (2009).

For air as a heat transfer fluid, ventilators are needed instead of pumps. As large volumes of air have to be transported, large channels and ventilators with sufficient power rating have to be installed. For both system designs—open or closed circuit—the gravitational force is not significant, but the pressure loss in the air channels is decisive.

9.5.2 Heat Exchangers

Forced circulation systems in particular require heat exchangers, since the thermal energy from the collector must be efficiently transferred between the primary heat transfer medium (water–glycol mixture or air) and the secondary medium, the service water. In solar systems, the function of the heat exchanger is of essential importance, as the temperatures at which the heat is transferred have a considerable influence on the operating condition and efficiency of the collector.

The design objective of the heat exchanger is to transfer the power supplied by the collector to the liquid circuit of the heat storage tank with the lowest possible temperature differences. The basic procedure for dimensioning heat exchangers is outlined in detail in Appendix C. In low temperature solar systems essentially two types of heat exchangers are widespread in application, the plate heat exchanger (PHE) and the coil heat exchanger.

- Plate type heat exchanger (PHE)

In plate heat exchangers, a package of thin, profiled stainless steel plates is pressed into position between two support plates by means of clamping bolts. The mechanical structure of a plate heat exchanger is illustrated in Fig. 9.39a. In almost all available technical PHE's the heat transfer takes place in counter current flow mode of the two fluids as indicated in Fig. 9.39b. The sealing of the plates against each other is realized either by elastomeric seals or by means of soldering. The mechanical design of the sealed PHE allows easy disassembly for easy cleaning and, by adding or removing plates, easy subsequent power adjustment if necessary. Demands of this kind can, if at all, only be fulfilled with tube bundle heat exchangers in a complicated way or not at all. The inlet and outlet connections are normally arranged on the frame plate, in one plane, which makes installation much easier than with shell-and-tube heat exchangers. The combination of mechanically high-strength stainless steel sheets and elastic as well as thermally resilient gaskets results in a plate package that is extremely resistant to pulsating vibrations and thermal cycling. Plate heat exchangers cannot be integrated into a storage tank.

Due to their technical features, plate heat exchangers are the right choice if the design requires rapid water heating or if the plant is subject to strong cyclical thermal and mechanical loads. The installation of plate heat exchangers is essential

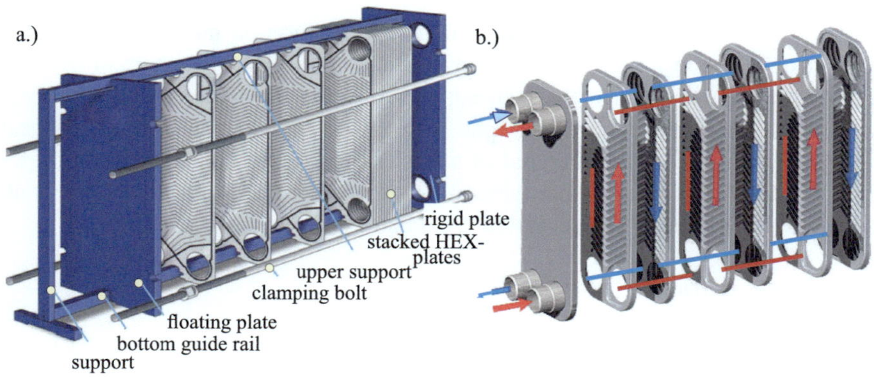

Fig. 9.39 **a** Mechanical design of a plate heat exchanger. **b** Flow into a plate heat exchanger according to the counterflow principle

for low-flow systems with a hot water storage stratification by means of a riser pipe.

Like all heat exchangers, solar heat exchangers are also exposed to a certain amount of contamination during operation. Liquids, such as hard tap water (service water of the secondary side) and the heat transfer fluid often show undesirable deposits, especially on the heating surfaces of the secondary side (side to be heated). The deposits depend primarily on the degree of contamination of the water and the heating surface temperature. Deposits and dirt on heating surfaces are undesirable, as they drastically reduce the thermal efficiency of the heat exchangers. Here too, plate heat exchangers have advantages over tube bundles.

With PHE, the forced, turbulent flow in the plate channels effectively retards fouling and deposits. The consequence of the lower heating surface temperatures is the reduction of deposits on the secondary side of the heating surfaces. This results in longer service lives and thus lower maintenance and operating costs than with tube bundles. If the heating surfaces are nevertheless dirty, they can be maintained and cleaned with considerably less effort in PHE's. The main arguments in favour of a plate heat exchanger are its high efficiency and thus its compact design compared to conventional coils and bundles. Typical ratios of plate heat exchangers to tube bundles are 1:7, weight 1:6, heat transfer medium content 1:10 for the same thermal performance.

Basically, the design is carried out with regard to the required heat output. This is determined via the collector surface and the flow through the collectors. The heat exchanger should be able to deliver the full capacity of the collector to the storage tank at the strongest solar radiation.

External plate heat exchangers are usually used for rapid heating of domestic water, provided that solar radiation permits. The integration into the solar circuit often takes place in a bypass to the conventional control with the help of suitable temperature sensors. Since the heat supply and thus the loading of the storage tank often takes place over larger time scales, the time characteristic of the heat exchanger, which is more in favour of a PHE, is of secondary importance. Therefore, in many conventional solar thermal low temperature systems the storage tank is charged by means of a pipe coil heat exchanger.

- Coil heat exchanger

Coil heat exchangers can be designed as plain tube, cf. Fig. 9.40a, or finned tube heat exchangers as depicted in Fig. 9.40b. They typically exhibit heat transfer coefficient values in the range of 200–500 W/(m^{2}K).

The smooth tube heat exchanger has a greater heat exchanger capacity per square meter of exchanger surface, but compared to the finned tube heat exchanger, this requires a multiple of the tube length. Copper is usually used as a material because of its very good specific thermal conductivity. The advantages of this type of heat exchanger are its relatively simple design and its lower pressure drop compared to plate heat exchangers. This means that the demands on the solar pump are considerably lower. They can be integrated into the storage with little effort, as the Fig. 9.41 illustrates. This eliminates the need for additional space and installation. However, internal heat exchangers are usually only available as accessories for the respective storage tank and are therefore, bound by the geometry to its dimensions, whereas external heat exchangers of any size and capacity can be selected independently of the storage tank. For the design of a heat exchanger integrated

a.) b.)

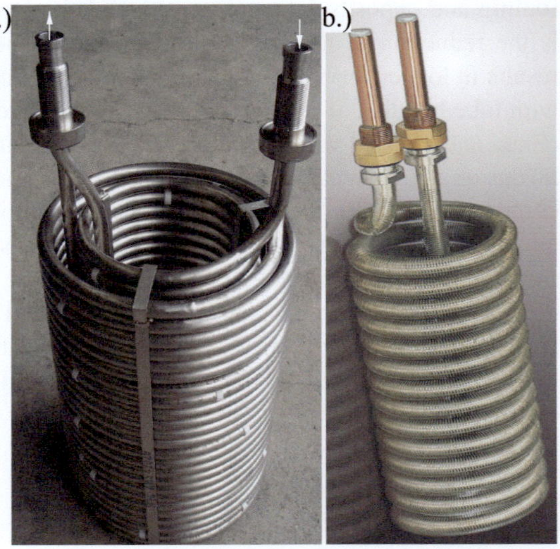

Fig. 9.40 **a** Double loop stainless steel smooth tube heat exchanger from a process plant; **b** copper finned tube heat exchanger (example from Wieland-Werke AG, 2023)

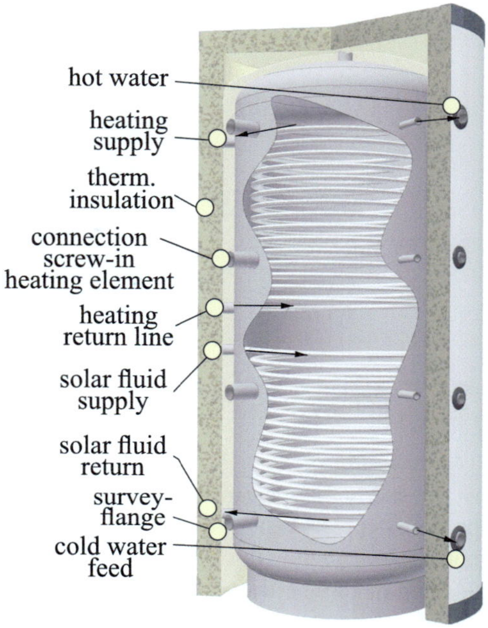

Fig. 9.41 Integration of coil heat exchangers in solar hot water low temperature storage

into the solar water storage tank, the following rule of thumb can be applied for an average logarithmic temperature difference of 10 °K.

- smooth-tube HX: approx. 0.2 m^2 heat exchanger surface per m^2 collector surface.
- finned tube-HX: approx. 0.3–0.4 m^2 heat exchanger surface per m^2 collector surface.

The basics for the design of the heat transfer performance are described in calculation leaflets of the manufacturers, in the VDI heat atlas (Stephan et al., 2019) or may be taken from Appendix C. The output dimensioning itself is a question of the collector surface and the heat demand of the consumer and is discussed in connection with Sect. 9.7.

Lime deposits can significantly reduce the effective transfer surface of heat exchangers. A layer thickness of only 2 mm reduces the heat transfer performance of a heat exchanger by 20%, with 5 mm thickness already by more than 40%. During continuous operation with service water above 60 °C, massive calcification phenomena must be expected, especially with high water hardness levels, which in extreme cases can clog the entire pipe or the heat transfer area. Depending on the lime-bearing fluid side, calcification can occur on the inside as well as on the outside of the pipe, as shown in Fig. 9.42a. For continuous operation above 60 °C, the use of finned tube heat exchangers is problematic because lime accumulates between the fins. In smooth tube heat exchangers, the lime is largely blasted off

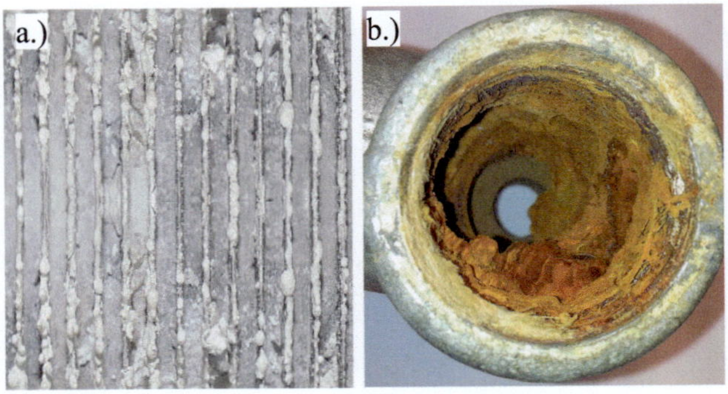

Fig. 9.42 **a** Close-up view of calcification on the outside of a finned tube heat exchanger. Lime scale in a steel service water pipe (**b**)

by the thermal expansion of the tube. If drinking water flows through the coil, the lime deposits can, in the worst case, narrow the cross-section so much that no more water can flow through (see Fig. 9.42b). Regular inspection and, if necessary, cleaning of service water pipes through which water has flowed must therefore, be provided for.

For air collector systems, air–water heat exchangers are needed, when also potable water shall be heated, or a water storage is to be used. For heating of rooms one option is also to direct the warm air directly into the rooms (air heating system) instead of indirectly heating the heating liquid.

9.5.3 Heat Transfer Fluid

The task of the heat transfer fluid is to absorb the heat captured in the solar absorber and transport it to the heat exchanger, where the energy is transferred to the end user. The thermophysical properties of the fluid play an excellent role in achieving high thermodynamic efficiency, but technoeconomic factors such as cycle stability, aging, material and environmental compatibility, safety, operating behaviour, price, and maintenance costs are also particularly relevant for the economics.

9.5.3.1 Water–Glycol

Pure water is the most suitable heat transfer medium, as it is available in almost unlimited quantities, is environmentally friendly, non-toxic, and cheap, and has one of the highest specific heat capacities ($c_p(T = 50°\mathrm{C}) = 4170\,\mathrm{J/(kgK)}$. However, year-round operation of a solar thermal system in Central Europe or North America requires the addition of an antifreeze to prevent frost damage. Most plants use water–glycol mixtures using either propylene glycol or ethylene glycol. The addition of these organic substances reduces the heat capacity of the heat transfer medium. A mixture of 60% water and 40% propylene glycol has a specific heat capacity of only ($c_p(T = 50°C) = 3810$ J/(kgK). Figure 9.43 shows the effect of the antifreeze agent on the thermophysical properties of specific density ρ, heat capacity c_p, heat conductivity k, and dynamic viscosity μ at different volume fractions in percent of the antifreeze of a commercial manufacturer. Although the density increases with increasing volume fraction, the heat capacity decreases. More important for the heat transfer, however, is that in particular the thermal conductivity decreases significantly with higher antifreeze fraction. More relevant as the logarithmic scale of the dynamic viscosity indicates is that the behaviour of the specific dynamic viscosity of water glycol mixtures is considerably higher than that of pure water especially at lower temperatures, necessitating the use of special pumps in addition to increased pressure losses.

The central positive aspect of adding glycol is that, in addition to lowering the freezing point—(when a mixture of 60% water and 40% 1,2 propylene glycol is added, approx. $T = -28°C$) no large ice crystals are formed even during potential freezing. Rather, a tough ice slurry is created, which no longer has a blast effect even at even lower temperatures, as the crystals can shift against each other. Frost

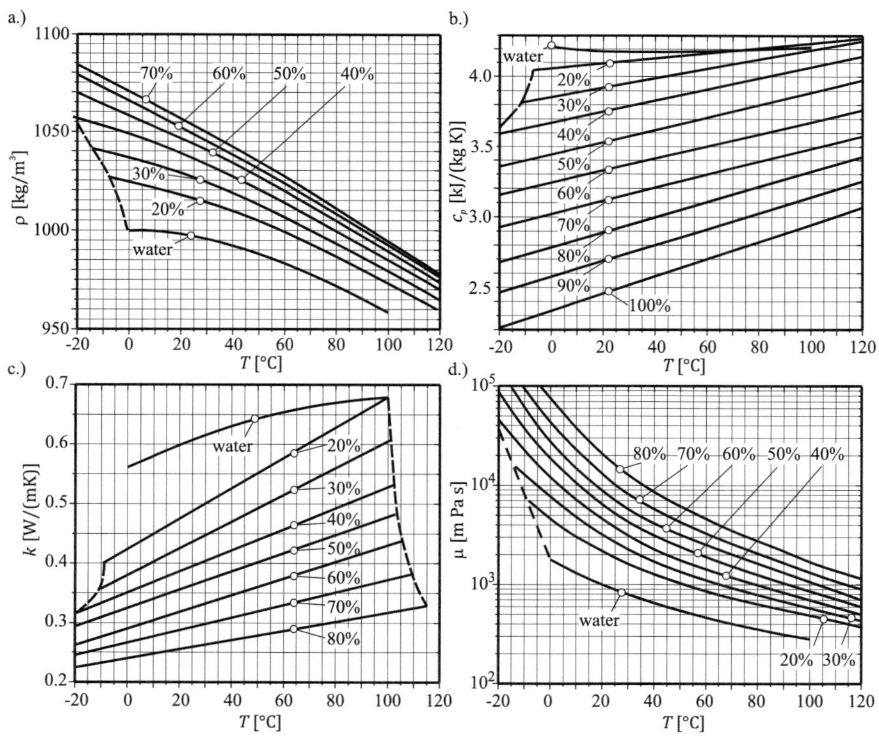

Fig. 9.43 Effect of volume fraction of a propylene glycol based antifreeze from a commercial manufacturer (Tyforop Chemie GmbH, 2011) at different temperatures compared to water. **a** Specific density ρ, **b** specific heat capacity c_p, **c** specific heat conductivity k and **d** dynamic viscosity μ

resistance can be achieved, for example, by using an antifreeze meter (hydrometer for density measurement) or a refractometer (optical concentration measurement of dissolved solids).

Propylene glycol-based antifreeze agents used today also fulfil the task of corrosion protection through added corrosion-inhibiting substances, so-called corrosion inhibitors. The heat transfer medium mixture is therefore an active ingredient combination of inhibitors, stabilizers, and alkalizing components, which is essentially intended to counteract a drop in the pH-value and thus corrosion of the solar circuit components. In addition, degradation is slowed down when the heat transfer medium is subjected to thermal stress, especially at temperatures above 130 °C. The degradation of glycol is mainly based on the mechanisms of oxidation and thermal decomposition.

The oxidation process of the glycols leads to the formation of organic acids, which cause a drop in the pH-value and thus have a corrosive effect. The alkalizing components added to the mixture of active ingredients neutralize the acids. The neutralization thus leads to a consumption of the alkaline bases. The amount of base remaining in the mixture is described by the so-called reserve alkalinity and is thus a measure of the corrosion protection remaining in the heat transfer medium. The pH-value can, therefore, be used to check whether the heat transfer medium is corrosive or can still be used. The pH-value should exceed 7.5. The pH-value can be measured, for example, chemically (with indicator paper) or by a potentiometric method (pH-meter—a device with an electrode held in the liquid).

Thermal decomposition of the heat transfer medium is caused by high temperatures, such as those that occur when the collector is at a stagnation. Above about 160 °C many of the alkalizing components decompose and the decomposition process increases rapidly with fluid temperature. As a result, tar-like, polymeric decomposition products are formed in the heat transfer medium, which are no longer soluble in the mixture and can thus lead to sticking within the solar circuit or to sedimentation, as depicted in Fig. 9.44a.

The worldwide antifreeze market consists of various ethylene glycol (EG) and propylene glycol (PG) products mostly delivered as concentrates and prefabricated ready-to-use products, in which the main difference between EG and PG is the level of toxicity. PG has very low toxicity, whereas ethylene glycol is toxic and requires measures to avoid accidental ingestion or exposure to humans. Pure glycol mixture are corrosive so commonly available products additives (so-called inhibitors) and buffer substances to protect metallic parts and counteract chemical reactions in the fluid (especially acid formation). Usually, the manufacturer supplies its products with a dedicated solid/liquid inhibitor package being adapted to the operating conditions envisaged, either max. or min. temperatures, metal materials used, phase transitions. The composition of commercially available concentrates and the functionality of the individual additive components are outlined in Table 9.4.

The combination of high temperatures and oxygen access to the solar system also accelerates the aging or degradation process of the heat transfer medium

Fig. 9.44 **a** Cracked solar fluid with solid formation (Cosmo GmbH); **b** visible aging phenomenon of solar fluids after appropriate temperature exposure (https://www.inowatec.de); **c** coaxial tube of a vacuum tube collector clogged with polymeric decomposition products of the heat transfer medium from (Scheuren, 2008)

Table 9.4 Composition of commercially available glycol-based concentrates with the functionality of additive components for solar thermal applications (modified from Baack, 2020)

Base concentrate	Weight fraction [%]	Functionality
1,2-Propylene glycol $C_3H_6(OH)_2$	92–94	Antifreeze
Organic acids salts	3–4	Corrosion protection for steels
Silicates	<1	Aluminium protection
Triazole	<0.2	Protection for non-ferrous metals
Borax	1.0–1.5	Alkali-reserve
Potassium hydroxide	<1	Neutralizer
Water	3–4	Solvent for additives
Stabilizers, dyes	0.1–0.3	Hardness stabilisers/labelling

many times over. The photograph of Fig. 9.44b illustrates how the liquid decomposes over time in the plant, changing colour. First it becomes yellowish- then yellow-orange, orange-brown to deep black. This degradation process can be very quick, as the Fig. 9.44c illustrates, which shows internal coaxial tubes clogged with polymeric decomposition products, which took place within a few days. The starting point of the scenario was a modification of a collector circuit in which the heat transfer medium was drained from the pipes but not completely removed from the vacuum tubes. In addition, the collectors remained uncovered. The presence of oxygen, in conjunction with the high temperatures resulting from lack of cooling, completely decomposed the heat transfer medium in the vacuum tubes. Replacement of the vacuum tubes remained unavoidable.

A current research effort is to use nanoparticles (including metal oxides, SiC, Al_2O_3) to increase the thermal conductivity and heat capacity of water–glycol mixtures and thus increase their heat transfer capacity. One relevant drawback which cannot overcome by water-anti freezing mixtures is that similar to water the change of the Prandtl-number Pr is almost one order of magnitude already in the temperature window up to 100 °C, which must be adequately considered in thermal–hydraulic computations.

Detailed descriptions of the chemical and physical processes occurring during the degradation of the heat transfer medium can be found in Henderson et al. (1981), Rossiter (1985) and Wedel and Bezzel (2000), while the latter also considers different water–glycol mixtures with and without inhibitors.

9.5.3.2 Air

Technically, there are many advantages to using air as a heat transfer medium. First of all, the price, the availability, the absence of toxic hazards speaks in favour of air. If properly designed, there is no danger of frost and air can be used directly in the heating process, e.g. for ventilation and drying, without a heat exchanger. A potential system stagnation, for example if no heat can be extracted from the collector because the heat accumulator has reached its maximum temperature, can be easily controlled. No dangerous system conditions such as pressure surges due to overheating can occur in the collector supply lines or in the heat network. Leakages in the ventilation pipe system also do not cause any structural damage as is possible with water or other pressurized systems. Especially this advantage is decisive for the application in the facade. Even with low solar radiation as in winter and with sub-zero temperatures, the solar air collectors can be operated in the air preheating as a support for the building heating. Due to the low heat capacity, air collectors reach the desired operating temperature very quickly. Large solar systems can be easily and inexpensively implemented, and no cost-intensive safety systems or control strategies are required, such as emergency coolers. The lightweight design of solar air collectors is ideal for lightweight buildings. Air as a heat transfer medium itself does not require maintenance, aging is not an issue, only collector circuit components such as air filters need to be checked or cleaned at regular intervals. With an air–water heat exchanger, a simple and safe system integration into existing hot water systems can be realized—and with all the advantages of air. And not to forget—a large number of collector circuit components like safety valves or expansion vessel are not needed at all.

9.5.4 Membrane Expansion Vessel (MEV)

Since every fluid has a density dependent on the temperature, every closed, liquid-filled pipe and container system requires an expansion vessel that compensates for this change in density in a suitable manner. The dimensioning of the expansion vessel for a collector circuit with selective absorbers differs from the dimensioning for conventional hydraulic circuits in which the vaporous state of the heat transfer medium does not occur. The correct dimensioning is discussed at the end of this section. Today, mainly membrane expansion vessels (MEV) are in use, which comply with certain standards, e.g. in Europe on DIN EN 13831 (2007). The MEV is a closed metal tank with an expandable membrane in the middle (for which another standard has to be met, e.g. in Europe (DIN 4807-2, 1991) must be fulfilled), separates two media, gas and solar fluid. In most cases, nitrogen under admission pressure is used for the gas component. Different expansion vessels are used for heating applications and solar systems, since in solar systems the membrane must also be resistant to glycol. The Fig. 9.45 shows a sectional view of a membrane expansion vessel typical in solar technology in its correct installation form with the gas space facing downwards. The membrane expansion vessel (MEV) has three main functions:

- Absorption of the volume of liquid that is created by volume expansion of the fluid, as a result of solar heat input.
- Absorption of the liquid volume by possible vapor formation in the collector during stagnation processes.
- Absorption of the fluid reservoir.

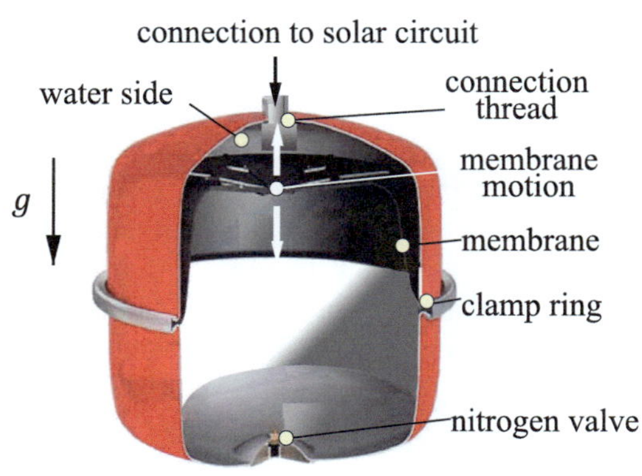

Fig. 9.45 Cross-sectional cut through a membrane expansion vessel (MEV) of a typical solar thermal system

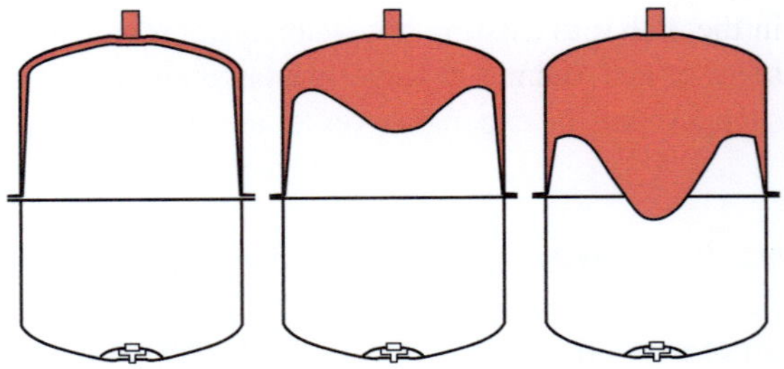

Fig. 9.46 Position of the diaphragm in a membrane expansion vessel (MEV) vessel at cold system (left), at operating pressure (middle) and at maximum pressure (right)

In the MEV there should always be a minimum amount of liquid content so that the membrane does not touch the upper edge and is therefore, able to maintain pressure. If only this liquid supply is left in the vessel, the expansion vessel must generate such a preliminary pressure that no negative pressure is created at the top of the system. When the collectors evaporate at stagnation, the liquid receiver must also buffer the hot liquid displaced from the collector (typically up to 130 °C) to the maximum permissible temperature of the membrane (usually 90 °C). The different positions of the membrane at different loads are shown in the Fig. 9.46.

The gas pre-pressure to be specified on the underside of the MEV has the task of pressing the expansion volume of the heat transfer fluid back into the system when it cools down and also of protecting the expansion vessel against overfilling. If there is no pressure or if it is too small, the expansion tank absorbs so much water even at low system temperatures that there is no room for the volume expansion of the solar fluid during the heat-up phase. In a closed system, the collector content evaporates when the system is shut down and the MEV must also be able to absorb the entire volume of steam. Otherwise, the system pressure would rise above the set pressure of the safety relief valve and the system would lose a lot of fluid.

Calculation methods for the volume of the membrane expansion vessel often show a too optimistic because too small volume, since a possible collector stagnation is only partially or not at all taken into account. For correct dimensioning, the liquid absorption of the MEV must first be determined by the volumetric expansion of the heat transfer fluid, which can be roughly estimated according to

$$V_e = V_{total} \cdot \beta \cdot (T_{max} - T_{min}) \tag{9.20}$$

where V_e is the expansion volume, V_{total} the total volume of heat transfer fluid in the primary solar circuit and β the volume expansion coefficient of the heat transfer fluid (typically V_e is about 9% of V_{total} for a temperature difference of 120 K for water-glycol mixtures). For the volume to be taken up due to steam formation in the collector during stagnation, a surcharge of 10% is generally assumed, so that

$$V_{steam} = 1.1 \cdot V_{coll}, \tag{9.21}$$

applies, in which V_{steam} represents the vapor volume and V_{coll} the fluid volume of all absorber tubes/surfaces in the collector field. The liquid storage volume V_{liquid}

in the MEV is assumed to be approx. 1% of the total fluid volume of the system, see e.g. Braun and Marko (1996), i.e.

$$V_{liquid} = 0.01 \cdot V_{\text{total}}. \tag{9.22}$$

This results in the minimum required tank volume $V_{n,min}$:

$$V_{n,min} > \frac{V_{liquid} + V_e + V_{steam}}{N} = \frac{0.1 \cdot V_{\text{total}} + 1.1 \cdot V_{coll}}{N}, \tag{9.23}$$

where N is the efficiency specified by the manufacturer. The efficiency indicates the volume of the entire expansion vessel that can actually be used without "overstretching" the membrane, causing damage, and reducing its service life. As a rule, the useful effect should not exceed the value $N = 0.5$, otherwise the vessel membrane may be overstretched. The efficiency can be estimated in the following way:

$$N = \frac{p_e - p_0}{p_e + 1}. \tag{9.24}$$

Here, the overpressure in the unit of measurement *bar* is to be used for the pressure. p_e is the final system pressure, which is also the safety valve pressure minus the closing pressure (= response pressure—20%), and p_0 is the vessel admission pressure. The inlet pressure in the expansion tank should be at least equal to the static system height. It ensures that the system pressure is still sufficient even in the upper area of the collector and that no air is sucked into the system when the heat transfer medium cools down. With a system height of $H = 20$ m this means a vessel pressure p_0 of $p_0 = 2$ bar. Calculation example: If the safety valve responds at 4 bar and the inlet pressure is 1.3 bar, the efficiency is $N = 0.45$.

The diaphragm expansion vessel should always be mounted suspended with the gas cushion facing down. The following effect would occur with standing installation: If hot water flows past the expansion tank, this hot water also rises into the expansion tank, as it is lighter than the cold water in the tank. Since the MEV is not thermally insulated, it looses a lot of heat on the one hand, and on the other hand in collector systems a hot heat transfer medium plug can enter the MEV during start-up (collector heated up at stagnation), which can lead to the destruction of the membrane. The MEV must be connected to the collector in such a way that it cannot be shut off (DIN EN 12828, 2014). However, a shut-off valve for inspection purposes that is secured against unintentional closing is permitted.

9.5.5 Air Bleed Valves

Air or gases dissolved in the HTF are the most common cause of malfunctions in cooling, heating, and solar systems. In most cases, improperly installed or incorrectly operated installations such as air separators, sludge traps or similar are the cause of the malfunction, which must be identified. From a technical point of view, there are no sealed systems, as gases diffuse or are sucked in via sealing elements (such as O-rings, stuffing boxes). The reasons are manifold and range from aging of the sealing elements, unfavourable system pressure conditions or incorrect treatment of the system fluid.

The most common occurrence is air in the system, where pressure and temperature determine the amount of air that may be dissolved in the water. Henry's law describes the solubility of volatile gases in liquids and states that the concentration of the gas (c_g) dissolved in the fluid is directly proportional to the partial pressure of the gas (p) over the liquid. The constant of proportionality is the Henry's constant $k_{H,pC}$.

$$p = k_{H,\mathrm{pC}} \cdot c_g. \tag{9.25}$$

Figure 9.47 shows the consequences of the Henry's law. For example, if water saturated with air is heated from 20 to 80 °C at a constant pressure of 2 bar, an air volume of 18 l per m^3 of water is released into the system. According to Henry's law, the amount of air released increases with rising temperature and decreasing pressure.

Gases can be dissolved in the fluid by various causes. The following processes can be considered as development processes:

- Microbubble formation on thermally highly loaded surfaces of heat generators (heating boilers, collector absorbers) or by cavitation (in pipe constrictions, fittings, branches).

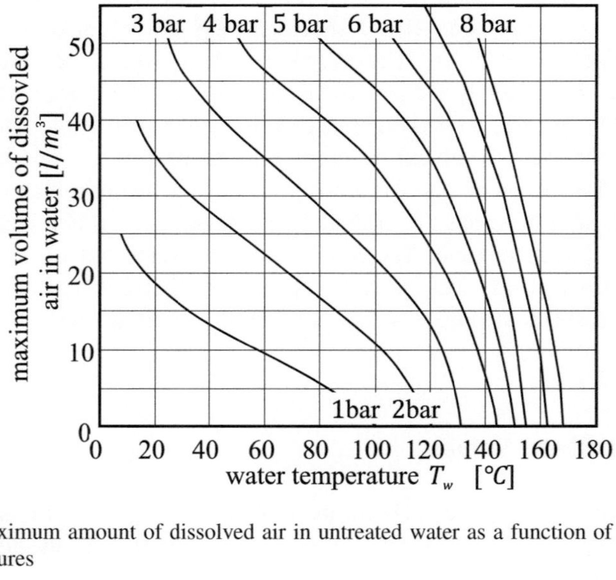

Fig. 9.47 Maximum amount of dissolved air in untreated water as a function of temperature at different pressures

- Hydrogen and methane gas formation. Hydrogen formation occurs in steel pipe networks after the so-called Schikorr reaction, which describes the conversion of iron hydroxide into iron (II, III) oxide in an anaerobic corrosion of iron. It can also occur in connection with hydrogen sulphide formation in pipes with copper materials (e.g. pipe bundles of heat exchangers, copper-brazed plate heat exchangers). Methane gases are produced by biological decomposition of organic material (fats, oils, dirt particles, ...) under exclusion of air in untreated filling water.
- Gas formation by electrolysis is inherent in every system in which different metals are present. As a result of the differences between the metals in the electrochemical voltage series, a current flow occurs which can lead to the production of hydrogen and oxygen.

The air bleed valves must therefore, always be installed in such a way that the gases can be removed as completely as possible, since each gas cushion acts like a small expansion vessel and can thus shift the zero point of the system (pressure level). Air also leads to corrosion of components. Therefore, there has to be an air bleed valve at the highest point of a solar circuit. There are several technical options for venting the pipelines.

- In baffle plate separators (compare Fig. 9.48a) there is a perforated baffle plate. As the flow velocity at the separator inlet decreases and the entering medium flows against the sheet metal, the entrained air bubbles rise on the sheet metal and collect at the topmost point where the vent valve is located.
- Automatic vent pots (compare Fig. 9.48b) have a float which actuates the vent valve. When filled, the pot is filled with liquid and the float is in the highest position so that the vent valve remains closed. If the float tank fills with air—(as the water level drops)—, the vent valve opens, and the air can escape. As a result, the water level rises, and the vent valve closes. When used in a solar circuit, the following problem arises: When steam is formed in the collector, the steam rises and escapes via the automatic venting plug. In case of strong vapor formation, a two-phase mixture may escape. Antifreeze is lost in the process. After cooling, the system pressure may fall below the specified minimum value. For this reason, these air vents can only be used with a ball valve in front of the valve, which has to be shut off, but this means that they lose their automatic venting function.
- Cyclone air separators are used for ventilation systems installed centrally in the boiler room, see Fig. 9.48c. In them, the flowing fluid is set into a rotary motion by the pipe connections tangentially attached to the separator wall. As a result of the rotation of the fluid, the heavier medium is pressed against the wall, while the lighter medium air (contained in the fluid) collects in the center of the vessel. The vent can be removed from the system manually or with the help of a screwed-on automatic floating vent (Fig. 9.48c). A further advantage of a cyclone separator is that no air vents need to be installed at the top point of the system, for example in converted attics.

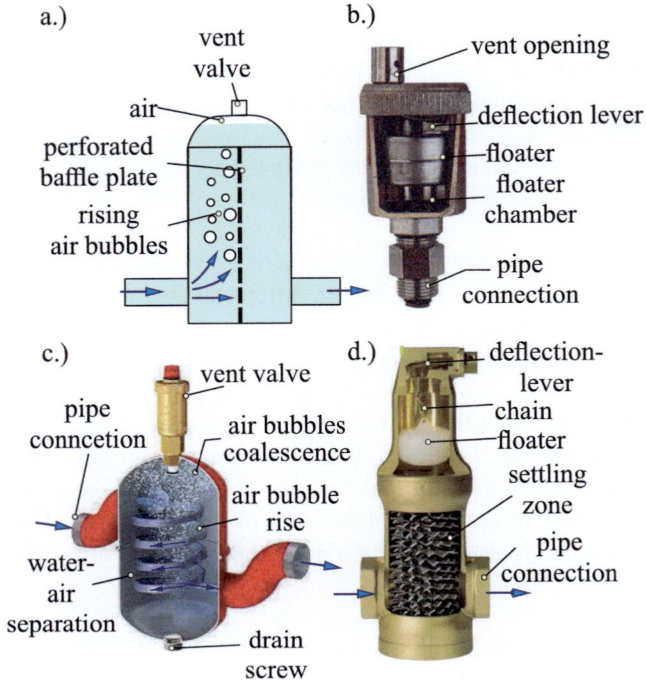

Fig. 9.48 Venting options of a solar circuit using a baffle plate separator (**a**), an automatic float vent plug (**b**), a cyclone separator (**c**) and an absorption air separator (**d**)

- Alternatively, an absorption air separator (cf. Fig. 9.48d) can be installed as a continuous air separator. In its interior, the micro bubbles contained in the water lean against a filling of wire mesh, metal granules or other solids. The pressure difference applied from the outside is not sufficient to transport the bubbles through the narrow meshes. The filtered microbubbles rise upwards in this calmed zone into the air collection chamber and can also be removed from the system via a floating vent.

9.5.6 Safety Valves

Pressurized solar systems must be equipped with a safety valve according to codes and standards similar to all other heat generating circulation systems. These safety valves basically serve to protect closed heating systems against excess pressure and are also subject to standardization. The safety valve provides additional protection in the event of a fault in the collector circuit leading to a pressure rise in the system despite the expansion tank. The safety valve responds as soon as a threshold pressure is exceeded. As a result, the system blows off steam or liquid from the

Table 9.5 Recommended inlet cross section of the safety valve depending on the collector area, see Brandstetter (2008)

Entry cross-section	$DN16$	$DN20$	$DN25$	$DN32$	$DN40$
Collector area [m^2]	50	100	200	350	600

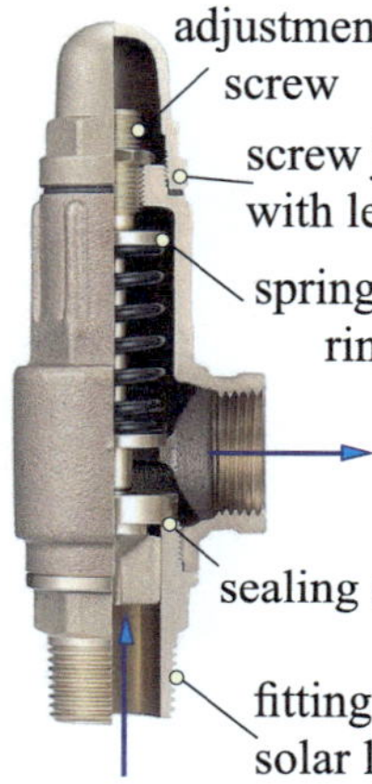

Fig. 9.49 Technical design of a angle safety valve

safety valve. An associated loss of antifreeze must be compensated in any case. Similar to the expansion tank, the safety valve must never be located between two shut-off valves or non-return valves. The set pressure of the safety relief valve depends on the highest permissible pressure of the circuit and ranges normally between 3 and 6 bar. The safety valve is connected to a blow-off pipe that releases steam into a temperature and glycol resistant collecting tank that can hold the entire volume of the solar circuit. The values given in Table 9.5 can be used as a guideline for dimensioning the inlet cross section of a safety valve in relation to the collector area; a technical design is shown in Fig. 9.49.

9.5.7 Check Valves

So-called check or non-return valves are installed in a solar circuit to prohibit flow in a return direction. The reason can be seen in the following problem of self-discharge of the storage tank via the collector circuit. This may happen for example at night, when the collector is colder than the storage, through natural circulation. For this reason, a check valve is installed as a "gravity brake", which is either installed as a flange component (Fig. 9.50a). or inserted in a pipe within the network (Fig. 9.50b). The check valve is positioned in the flow direction upstream of the membrane expansion vessel (MEV), as shown in Fig. 9.35. This guarantees that the MEV can receive the liquid heat transfer medium expelled by evaporation in the collector, from both sides of the collector. If the non-return valve were mounted above the expansion tank, it could only take up heat transfer medium from the collector outlet side. As a rule, a non-return valve (non-return valve or non-return flap) is installed in the solar station, which restricts flow to one direction. The element blocks the flow in the return direction.

Depending on the design, the lock is provided by a valve or by a flap. Various types of constructions have established themselves on the market, such as systems

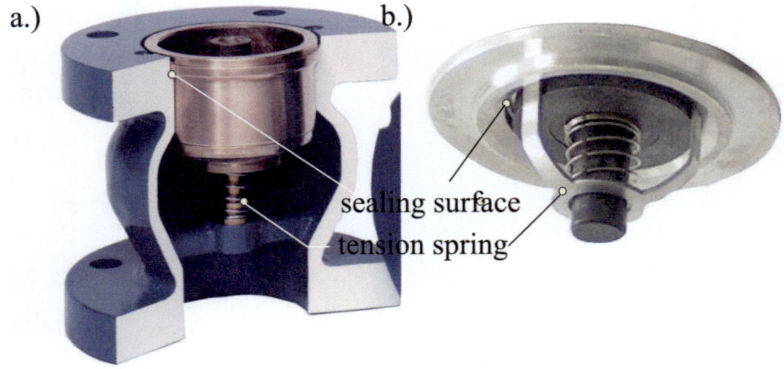

Fig. 9.50 Check valve as backflow preventer for solar thermal systems; **a** flange design (Val-Matic Valve & Manufacturing Corporation, 2020) and **b** disc design as pipe insert

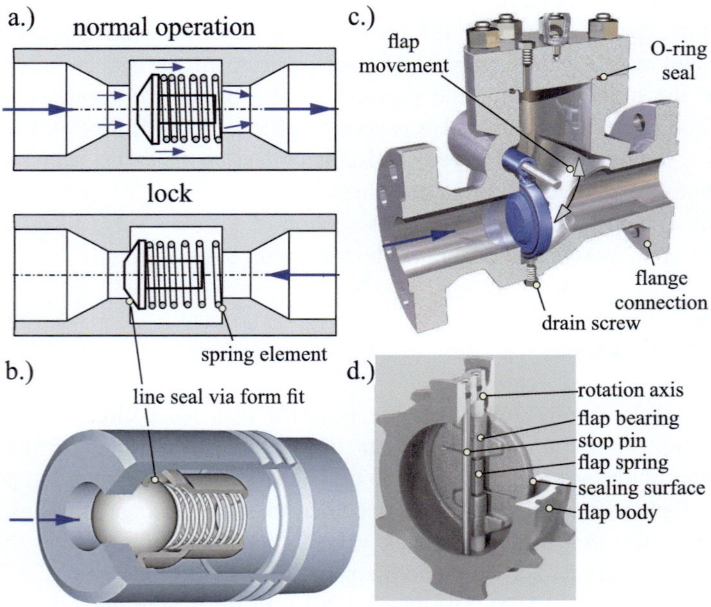

Fig. 9.51 **a** Principle function of a non-return valve. **b** Non-return valve for horizontal and vertical installation with a ball element, **c** non-return valve for horizontal installation with flange connection and **d** schematic application design with surface seal

with a form-fitting seal (line pressing, see Fig. 9.51a, b) or those with surface pressing (Fig. 9.51c, d).

The valve type to be selected depends on the installation position (horizontal, vertical) and whether gravity plays a role in the functionality. Essential in solar circuits is the glycol and temperature resistance of the sealing surface and internals.

9.5.8 Piping

The connection between collectors, solar heat exchangers and components is almost exclusively made with pipes. The pressure losses in the pipe network, together with the characteristics of the collectors and their interconnection, determine the pump characteristics. In addition, the pipework and pipe connections must meet the following requirements:

- Temperature resistance up to the frost range (usually in the order of −20 to 35 °C) and at least 250 °C as the highest potential stagnation temperature for non-concentrating collectors in summer.
- Pressure resistance up to the maximum operating pressure of the system (up to 6 bar).
- Long-term compatibility with the heat transfer fluid used (antifreeze).
- Simple, reliable, and flexible connection.

In the collector circuit, pipes made of copper, steel, stainless steel and corrugated stainless steel pipes are used. Plastic pipes are less suitable due to their lack of temperature resistance and are used mainly for uncovered absorber mats. In small-scale plant construction, copper pipes in combination with soft and hard soldering has established itself as the common joining technique. Flexible corrugated stainless-steel pipes are often used to connect the collectors on the roof because of the relatively large temperature fluctuations and the associated thermal expansion. Steel pipes are relatively complex to process (welding, bending, cutting, threading, sealing) and are used in larger solar systems due to the higher costs of copper pipes from a dimension of $DN28$ upwards, increasing dramatically with diameter. When connecting different metal pipelines, the corrosion aspect must be considered because of the electrochemical potential differences of the metals.

The solar circuit piping and the fittings must be insulated throughout. The material used in the solar circuit must be temperature-resistant insulation. Mainly foams are used that are resistant to temperatures of up to 150–170 °C. Outdoor pipes insulation should also be UV-resistant, weatherproof, and watertight. In addition to the classic separately laid combination of pipe, thermal insulation and electrical cable, there are also pre-fabricated pipe units, so-called twin pipes or "twin tubes" (Fig. 9.52). They consist of a pair of copper pipes or corrugated stainless steel hoses for solar supply and return including thermal insulation and collector sensor cable. The advantages of flexible corrugated pipes are as follows:

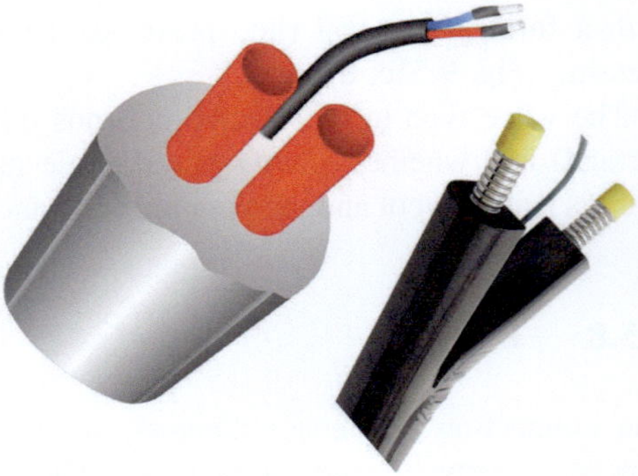

Fig. 9.52 Schematic diagram of a flexible twin pipe with insulation and integrated instrumentation cable and technical design (right)

- easy installation of the finished pipe in one piece and therefore, no soldered joints (fewer sources of leakage),
- small bending radii without kinks,
- integrated insulation and collector sensor cable,
- corrosion resistance.

Disadvantages of this construction method are:

- a greater pressure loss for corrugated hoses and
- a higher specific material price per meter.

As a result of the heat supply varying over time, the relative elongation of the pipes due to the different temperatures for installation and operation plays a role. The linear thermal expansion depends largely on the thermal expansion coefficient α of the material, which for example for steel is $\alpha \approx 1.1 \cdot 10^{-5}\mathrm{K}^{-1}$ and for copper already $\alpha \approx 1.6 \cdot 10^{-5}\mathrm{K}^{-1}$. The stress caused by the expansion must be compensated by the pipe without exceeding the permissible material strength

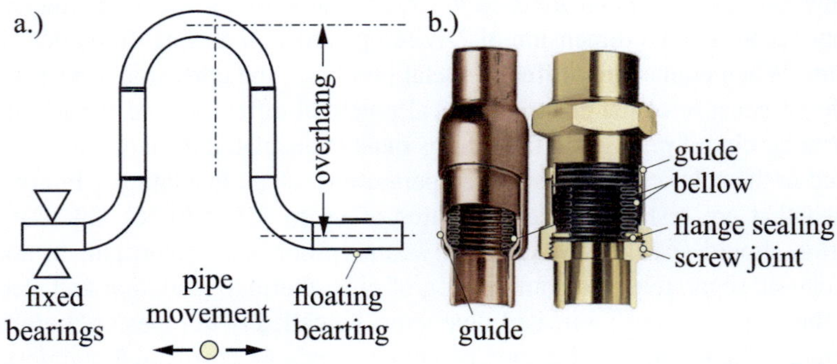

Fig. 9.53 **a** Principle of a pipe loop compensator. **b** Photograph of axial compensators with internal compensator and protective pipe, (left) soldered version made of copper and (right) screwed version made of brass/steel

parameters. Technically, stress compensation is achieved by means of expansion compensators or so-called expansion joints. Clamps that fix the pipe are selected as fixed and sliding points so that the compensation elements can fulfil their task. Pipe loop compensators (Fig. 9.53a) are often used because of their simple production. In cases where the installation of loop compensators is impossible due to space limitations, expansion compensators with integrated expansion bellows are used. Figure 9.53b shows such axial expansion joints with integrated bellows in a screwed or soldered version.

9.6 Passive Solar Thermal Systems

Passive solar thermal systems work without the need of a mechanical fluid transport using pumps of ventilators. They use passive heat transport principles instead. Therefore, the architectural design, integration and arrangement is a key factor. Already in the ancient cultures passive solar utilization and solar protection was known and utilized. Form there we know already the principles of passive solar energy utilization. Today's technologies, however, can make the principles much more powerful and efficient. Depending on how the heat is transferred from the solar absorber to the inhabited room we distinguish:

- Direct gain systems,
- Indirect gain systems,
- Convective systems and
- Storage wall systems.

In the direct gain system, the solar radiation is entering the room or building through a window or transparent glazing element, whereas in an indirect gain system the heat is first generated in one zone of the building, and the transferred by natural convection of heated air or through a wall to a heated zone (cf. Fig. 9.54a, b). A conservatory, a wintergarden, or a double-façade construction forms such an unheated zone providing at certain times heat to the heated zone of a building or at least reducing the heat losses of these parts.

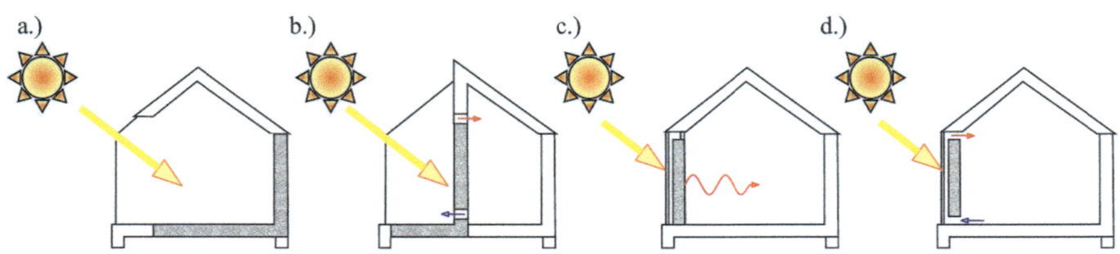

Fig. 9.54 Principles of passive solar energy utilization: **a** direct gain system **b** indirect gain system **c** storage wall system **d** convective system

When the space acting as absorber for the solar radiation becomes more a technical façade component than an unheated space we speak about solar wall systems, see Fig. 9.54c, d. When the wall accumulates the solar heat and only gradually releases the heat by conduction and radiation into the interior, we call that a storage wall. If, however, hot air shall be produced through contact to an absorber to be convected into the interior room, we speak of a convective wall or Trombe wall after the French architect Felix Trombe (see for example in Stazi, 2017). In this case the heat capacity of the absorber wall should be preferably low in order to produce high temperatures leading to a convective force.

9.6.1 Windows

Windows transmit solar radiation into a building and hence provide light and heat. On the other hand, they provide protection against rain and wind, and have a thermal insulation value, described by the U-value of the window. The measure for the solar radiation contribution as internal heat source is described by the total energy transmittance or *g-value* (according to DIN-EN-410, 2011). It describes for glazing the sum of the energy that is transported from outside to inside as direct solar radiation and as secondary heat gain (due to the superposition principle a part of the solar heat absorbed in the individual glazing components is guided into the interior, thus, reducing a net heat loss). The Fig. 9.55 graphically sketches the definition of the g-value. The g-value is usually determined by a combination of measurement (transmittance and reflectance) and calculation (U-value). The processes in the window (heat conduction, convection etc.) are not subject of the definition, but only the external heat flows towards the interior. The g-value is thus defined as follows

$$g = \frac{\dot{q}_{transmitted} + \dot{q}_{absorbed,rad} + \dot{q}_{absorbed,conv}}{E_w} = \tau_e + q_i. \tag{9.26}$$

where $\dot{q}_{transmitted}$, $\dot{q}_{absorbed,rad}$, $\dot{q}_{absorbed,conv}$ are the heat fluxes arising from the solar radiation transmission, and the heat flows caused by absorbed radiation in the glazing which are partly guided to the interior and released by convection and thermal radiation from the inner surface. E_w is the solar irradiation on the glazing. The secondary heat gain q_i is of course dependent on the thermal resistances from the absorption location to the interior and the exterior, i.e. on the U-value of the glazing.

The classical determination of the *g-value* according to standard procedures such as DIN-EN-410 (2011) can only be carried out exactly if clear glass is used. Diffusing surfaces or even structures (e.g. printed or enamelled patterns) can be only approximated. The calculation assumes a vertical incidence of radiation with a homogeneous structure of the glass layers. Therefore, calorimetric measurement methods have been developed to measure directly the g-value of complex glazings (Kuhn et al., 2000) or even shading devices (Kuhn et al., 2001). A high

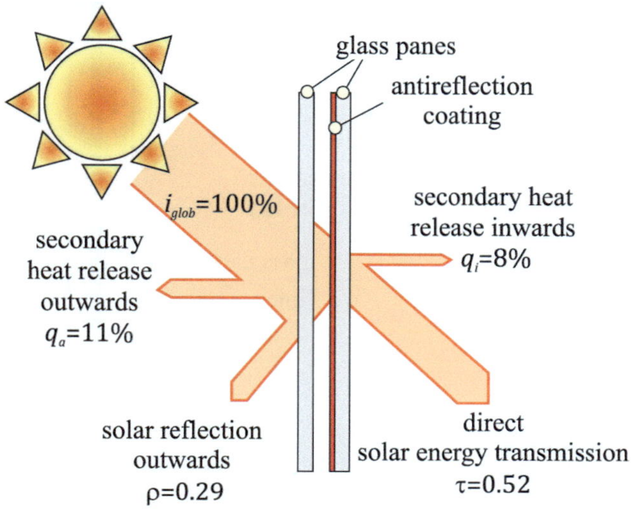

Fig. 9.55 Sketch for the calculation g-value for a window glazing (here for $g = q_i + \tau_e = 0.6$)

total energy transmittance means a high level of solar gains which is useful during the heating season. A low g-value is on the contrary extremely useful in hot summertime. Therefore, temporary solar protection devices like blinds or lamellae are used. They are more effective when the solar protection layer is on the outside of a glazing. The total energy transmittance g differs significantly from the heat transfer coefficient U_g describing the heat loss proportional to the temperature difference between interior and exterior. Coatings based on metallic thin films are needed to provide the low emissivity of modern insulating glazing. They are usually applied to the inner of two panes towards the interspace (position 3 as counted from the outside), reduce the long-wave infrared radiation exchange, but let short-wave visible light through so they are optically very transparent, which is illustrated in Fig. 9.56. The U_g-values of the multi-pane insulating glazing can be further improved with a noble gas filling with Argon or Krypton.

In the past decades the field of windows and glazing has changed rapidly, resulting in a low cost highly reliable technology. Due to the different requirements on a window in summer or winter time, daylighting and solar protection, there was a strong attempt to develop switchable "intelligent" solar control glazings. Several physical principles have been investigated. These so-called smart glazings are regarded as an attractive alternative to conventional sun protection systems such as external or internal blinds, slats, where the glass panes perform the tasks

- sun protection,
- glare protection,
- thermal insulation and
- providing a dynamic energy balance of the window.

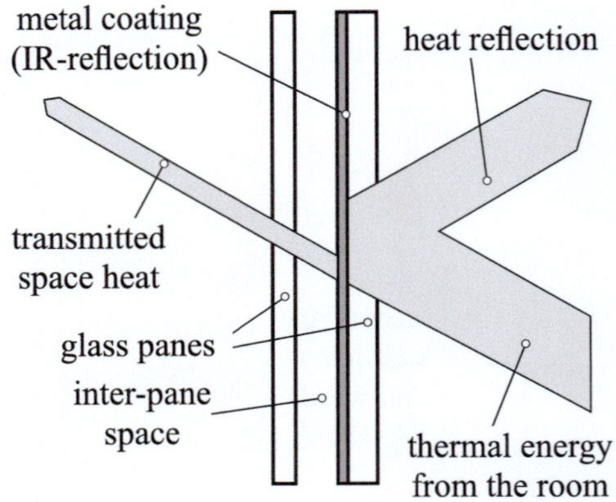

Fig. 9.56 Heat balance at the window panes

These types of switchable and controllable sun protection glasses should not be confused with the static sun protection glasses, which cannot change their properties. The change of state of dynamic smart glazings are capable of changing their optical properties under specific boundary or system conditions, for example by a temperature change, an irradiation change or by an electric voltage change which can be easily applied and controlled. The optical properties may switch between low and high absorption, or clear and diffusing/scattering states. A number of technology principles have been analysed, for more details see e.g. Platzer (2003), such as:

- electrochromic glazing: change from a clear view to dark blue colour as a function of an applied electrical voltage;
- gasochromic glazing: variation of transparency and color, regulated by a small change in hydrogen concentration in the interspace;
- photochromic glazing: switching from clear view to grey-brown colour;
- photoelectrochromic glazing: combination of electrochromic layers and an electrochemical solar cell, change from clear transparency to blue colour;
- thermochromic glazing; change in absorption due to a temperature dependent absorption of metal oxide film;
- thermotropic glazing: change from a clear state to a scattering state resulting in a white diffusing appearance;
- PDLC Polymer-Dispersed-Liquid-Crystal, limited transparency, product is used in the market as a switchable privacy screen in the interior;
- SPD suspended particle devices, comparable to PDLC (prototype stage).

Table 9.6 provides data of some typical glazings in terms of design, light transmittance (τ_v), energy transmittance (g) and U_g.

Coloured glasses with low transmission are also used for sun protection. Green glasses in particular take on the task of reducing thermal radiation, as in motor vehicles, for example, in order to reduce the heating of the interior. However, these glasses lead to a limited color rendering due to their tinting. Furthermore,

Table 9.6 Characteristic values of glazings: light transmittance (τ_v), energy transmittance (g) and thermal transmittance (U_g). DGU ≡ double glazed unit; TGU ≡ triple glazed unit; *≡ indicates coating position

Glazing type	Gas filling	Layers [mm]	τ_v [/]	g [/]	U_g $[\mathrm{W}/(\mathrm{m}^2\mathrm{K})]$
DGU, no coating	Air	$6-16-4$	81	76	2.9
DGU, hard, heat mirror	Argon	$6-14-*4$	75	72	1.9
DGU, soft, heat mirror	Argon	$6-16-*4$	79	60	1.2
DGU, soft, heat mirror	Krypton	$6-10-*4$	76	58	0.9
DGU, soft, solar control	Argon	$6*-16-4$	66	34	1.2
DGU, soft, solar control	Argon	$6*-16-4$	20	21	1.2
TGU, soft, heat mirror	Argon	$4*-10-4-10-*4$	71	50	0.8
TGU, soft, heat mirror	Krypton	$4*-10-4-10-*4$	71	50	0.6

sun protection glasses can be installed. These are intended to reduce the transmitted solar radiation and at the same time does not restrict the light transmission. These are glasses with a low total energy transmittance (g-value) and a high light transmittance (τ_v-value).

The structure of an insulating glass unit is shown in the Fig. 9.57. The usual U_g-value of a double-glazed insulating glass unit with simple thermal function layers is approx. 1.1 $\mathrm{W}/(\mathrm{m}^2\mathrm{K})$. The glass panes are separated at their outer edge by spacers made of aluminum, stainless steel, or plastic. The sealant used in an edge seal structurally bonds the glass panes and spacer bar together while providing a high level of moisture vapor and gas diffusion resistance and allowing flexion to accommodate glass movement. In dual-sealed insulating glazing units (IGU) (cf. Fig. 9.57b) these functions are performed by separate primary and secondary sealants whereas single-sealed IG units (Fig. 9.57a) have only one (secondary) sealant. As primary thermoplastic sealant polyisobutylene PIB is used. The secondary sealant provides the adhesive strength between the glass panes. Polyurethane (PU), silicone (Si) and polysulphide (PS) are widely used, but also hot-melt butyl- or epoxy-based sealants may also be used. Today, the majority of IG units manufactured globally are dual sealed. The spacer bars, usually 10–20 mm thick aluminum or steel profiles, are filled with a drying agent to dry the air trapped in the space between the panes. This consists of substances that fall into the material family of silica gels or molecular sieves (zeolites).

The metallic profiles constitute a thermal bridge between the panes. Low temperatures in the edge-of-glass region increase the potential for condensation, which, in turn, can lead to mold growth and deterioration of window frames. Warm edge technology (WET) reduces this problem for example by using a thermoplastic material (Thermo Plastic Spacer TPS) which has a lower thermal conductivity, and already contains the drying agent, so that subsequent filling is no longer necessary. This elastic spacer is glued on the glass. This now also allows smaller workshops to economically produce tailor-made multi-pane insulating glass elements. After

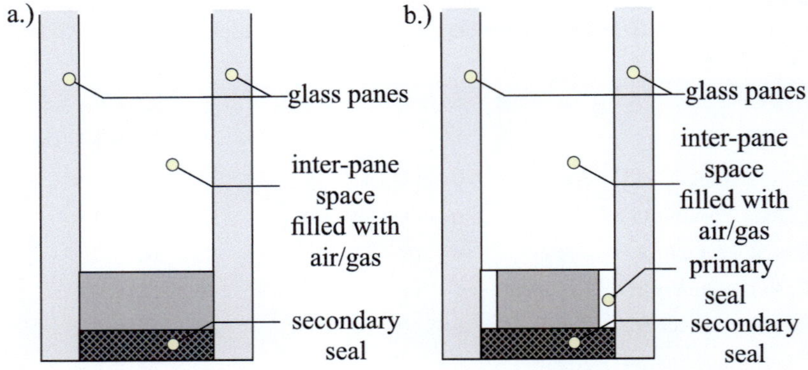

Fig. 9.57 Construction of an insulating glazing unit (IGU) with single (**a**) and dual seal (**b**)

filling the space between the panes with inert gas (air, Argon or in some cases the more expensive Krypton), the edge joint formed by the spacer and the two glass edges is then sealed with a secondary sealant only.

Within the group of building components, windows show the lowest insulation level of all exterior components. However, considerable solar gains can be achieved, so that with a well-chosen window arrangement and orientation, the passive solar gains can largely compensate for the heat losses. The heat transfer coefficient U-value in building physics describes the heat flow through 1 m^2 of a building component proportional to a given temperature difference between indoor and outdoor air. In the SI system it has the unit $W/(m^2K)$. For instance, in the German law on building energy requirements (German Building Energy Act, 2020), there is some flexibility on the heat transfer coefficient U of different building elements as long as they comply with prescribed standards—here, (DIN EN 4108-2, 2013), however, the building is assessed in the so-called reference building approach.

The energy consumption of the reference building of the same size with specified maximum values for the heat transfer coefficient shall not be exceeded. In previous ordinances, tables with maximum U-values for the first installation, for replacement and refurbishment of building components were prescribed. Based on European standards like (DIN EN ISO 10077-1, 2020), the insulation values of a complete window are calculated according to three components:

- frame U_f,
- glazing U_g and
- glass edge psi.

The values U_f and U_g are calculated according to the respective area. The heat transfer at the glass edge psi is treated as a specific thermal conductivity, i.e. related to the length, so that it is given in $W/(mK)$. Thus, the calculation formula for the total thermal insulation value (U_w) of a window is given by:

$$U_W = \frac{U_f \cdot A_f + U_g \cdot A_g + l \cdot psi}{A_W}. \tag{9.27}$$

in which A_i represents the area of the respective segment (f or g) and l is the length of the glazing edge. The U_W-value is usually larger than the glazing U-value (U_g).

The design of the window and the frame plays an outstanding role, especially with low U_g-values. Since the choice of materials is limited in terms of service life, cost, availability and processability, a great deal of attention is paid to the technical design of the entire window frame. Thermal bridges shall be minimized, large variations in temperature shall be taken, and rain penetration be avoided. Therefore, design varies from material to material and requires material-specific solutions, as the designs for plastic and wood, which is illustrated in Fig. 9.58.

The heat transfer coefficients U_W of windows with different glazings are calculated and listed in Table 9.7 for window frame shares of 30%. For special glazing spacers made of aluminium or steel for glazing bars etc., corrections ΔU_W of the total heat transfer coefficient may be necessary. These shall be specified by the manufacturer. The table clearly shows that triple glazing has a much lower heat transfer coefficient than double or single glazing. As the thermal transmittance of the glass U_g decreases, the importance of the thermal transmittance of the frame U_f increases significantly. At low U_g values the technical design of the frame plays a dominant role for example in passive house windows.

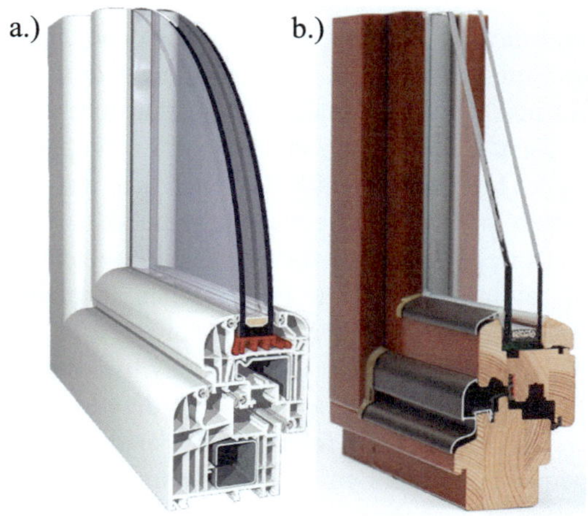

Fig. 9.58 Triple insulating glazing in plastic (**a**) (WELL spol. s.r.o., 2011) and wooden window profile (**b**) (Hofegger, 2011)

Table 9.7 Extract from the total heat transfer coefficients U_W of windows of different types and French windows as a function of the heat transfer coefficient of the glazing U_g and frame U_f assuming a window frame proportion of 30%

Glazing type	U_g	U_f [W/(m^{2}K)]								
	[W/(m^{2}K)]	1.0	1.4	1.8	2.2	2.6	3.0	3.4	3.8	7.0
Single glazing	5.7	4.3	4.4	4.5	4.6	4.8	4.9	5.0	5.1	6.1
Double glazing	3.3	2.7	2.8	2.9	3.1	3.2	3.4	3.5	3.6	4.4
	2.3	2.1	2.2	2.3	2.4	2.6	2.7	2.8	2.9	3.8
	1.7	1.6	1.8	1.9	2.0	2.2	2.3	2.4	2.5	3.3
	1.1	1.2	1.3	1.5	1.6	1.7	1.9	2.0	2.1	2.9
Triple glazing	2.3	2.0	2.1	2.2	2.4	2.5	2.7	2.8	2.9	3.7
	1.7	1.6	1.7	1.8	1.9	2.1	2.2	2.4	2.5	3.3
	1.1	1.2	1.3	1.5	1.6	1.7	1.9	2.0	2.1	2.9
	0.5	0.8	0.9	1.0	1.2	1.3	1.4	1.6	1.7	2.5

9.6.2 Transparent Thermal Insulation

9.6.2.1 Materials

Transparent insulation materials (TIM) belong to a material class which provides thermal insulation similar to conventional insulation materials but have a high solar transmission and g-value. In most cases, however, these materials are not *image-* or *sight-permeable*. These insulation materials are often also referred to as *translucent*. In distinction to transparency, translucency can be described as light transmission and transparency as image or view transmission. Therefore, "transparent insulation" is usually rather "translucent insulation". The functional principle of TIM is also found in nature, for example in the fur of polar bears.

As Fig. 9.59a, b shows, polar bear fur has a hollow chamber structure that largely prevents convection near the skin. The fur consists of hollow lightguides that direct light onto the black skin, which acts as an absorber layer. The layer of fat under the skin is a thermal insulator. The heat radiated by the body (infrared wavelength range) is reflected by the fibre, further reducing heat loss. An analogous process is used today by technical textiles, as the Fig. 9.59c shows.

TIM have been classified according to the internal structure. In principle, there are three structural forms in the macroscopic area: absorber-parallel, absorber-perpendicular and chamber structures, as shown by Fig. 9.60. A further category is called quasi-homogeneous, that includes materials whose appearance is homogeneous on a macroscopic scale (like a porous foam, a fibre wool or similar) but has irregular microscopic structure. The classification is indicative and not strict. For example, tilted honeycomb structures which open channels tilted some degree against the surface normal are also counted as absorber-perpendicular, as long as

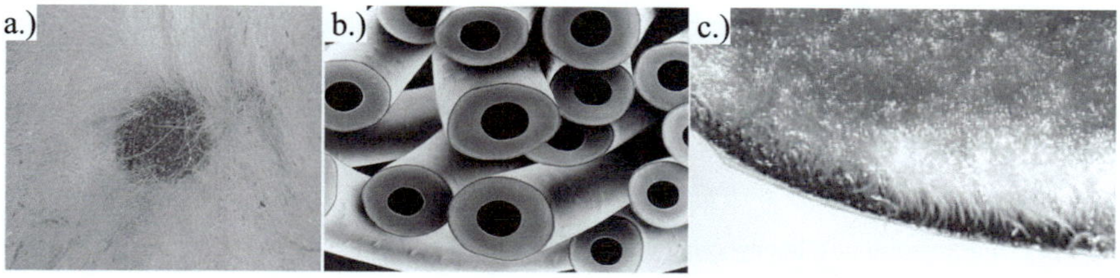

Fig. 9.59 Polar bear fur with underlying epidermis (**b**). SEM image of the fur structure of the polar bear, which are formed as hollow tubes, and textile replica of the polar bear fur (**c**) from Stegmeier et al. (2009)

the solar radiation transport is not characterized by reflection losses to the outside but by forward reflection gains along the tilt axis.

A type A TI insulation illustrated in consists of glass panes or plastic foils that are parallel to the absorber surface. Classical insulating glazings belong to this category. Here, a small distance between the panes or foils significantly reduces convective heat transfer to the outside. The convection can be even more reduced by choosing an appropriate noble gas as filler or even by evacuating the gap between the glasses. In this case of a vacuum glazing the external atmospheric pressure has to be countered by using micro pillars between the glasses in order to avoid breakage. The radiative heat transfer between the foils or glasses can be substantially reduced in all cases. According to Fresnel's formula, however, the transmission of solar radiation is also reduced due to the many surface reflections on the panes and intermediate layers.

More common and more robust in terms of wind loads or strength are absorber-perpendicular type B TIs in which so-called honeycomb or capillary structures

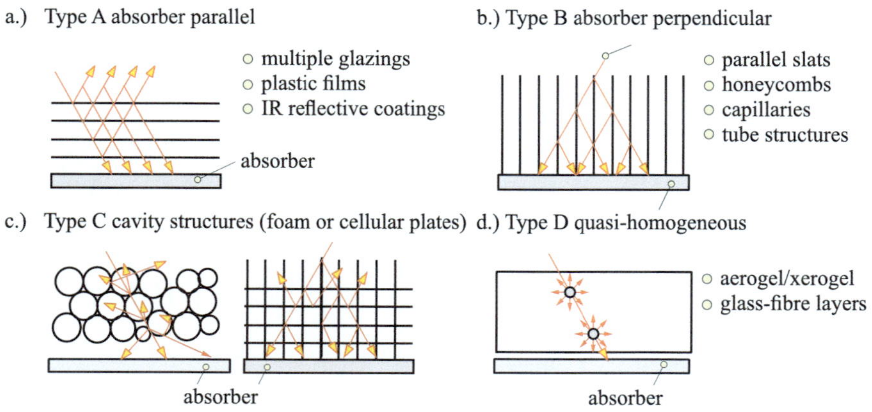

Fig. 9.60 Classification of transparent thermal insulation (TI) materials modified according (Platzer, 1988); **a** Type A: absorber-parallel, **b** Type B: absorber-perpendicular, **c** Type C: chamber/cavity structure and **d** Type D quasi-homogenous

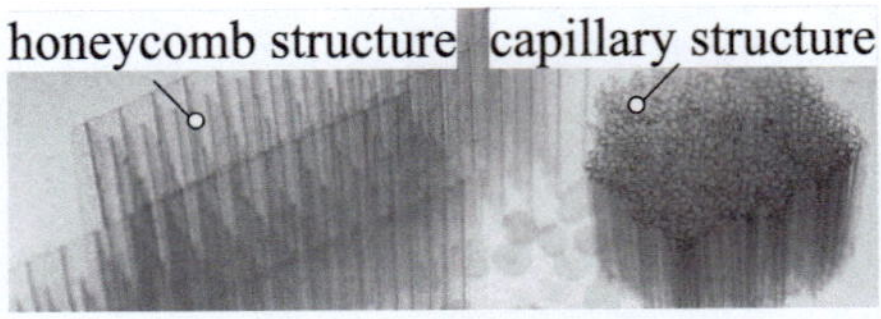

Fig. 9.61 Photograph of a honeycomb and a capillary structure of type B TI

are arranged perpendicular to the absorber surface, as shown in the Fig. 9.61. They offer depending on the aspect ratio of the structure a significantly improved reduction of convective heat transfer compared to type A. The main characteristic is that solar reflections on the cell walls are directed towards the absorber. This means that light that is not normally incident is also transmitted and reflected to the absorber. The result is a high solar transmission.

The Fig. 9.62 shows for capillary and honeycomb structures the ideal shape and typical imperfections found in real products.

Type C TI systems have chamber structures consisting of transparent foams such as acrylic foam, corrugated and embossed foils or hollow glass spheres, as depicted in Fig. 9.63a. This type is in a first approximation a superposition of absorber-parallel and absorber-perpendicular cell walls leading to closed cells. The advantage is certainly a better thermal resistance as convection can be reduced more effectively in the closed cells. On the other side the optical properties are not as favourable: high reflection and in case of more irregular cells scattering leads to large optical losses.

Finally, type D illustrated in Fig. 9.63b exhibits some resemblance to type C structures and in fact the border is not well-defined. It is rather the dimension of the cells or pores and the regularity of the material which is different. Ideally in type D the structure has a sub-wavelength dimension and is not visible for the eye. This is the case for transparent silica aerogels with a pore size between 10

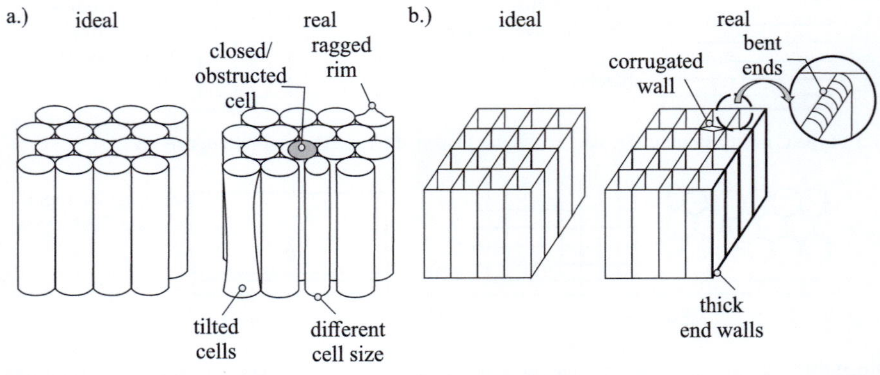

Fig. 9.62 Sketch of two versions of absorber-normal TI insulation capillary (**a**) and honeycomb-type (rectangular) structures of type B with ideal shape and typical imperfections

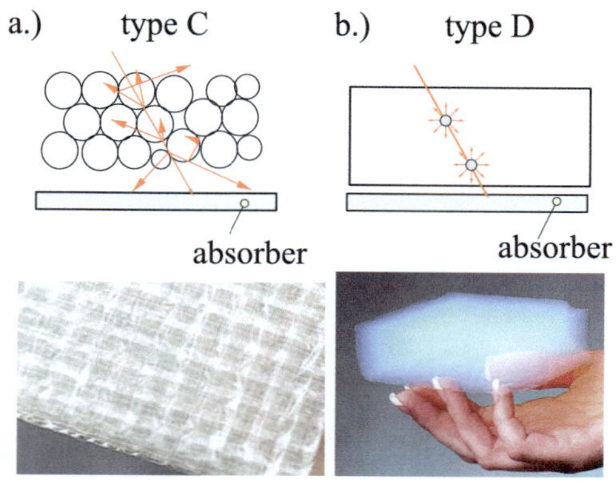

Fig. 9.63 **a** Sketch of the radiation path of a type C TI insulation and example of a technical implementation. **b** Type D thermal insulation in a high porosity aerogel

and 20 nm. This material has a very high porosity, and the refractive index of the apparently homogeneous material is close to unity, which means that there is hardly any refraction and reflection of the light. The optical transmission is high. Due some scattering of light at the nanoporous structure (similar to Rayleigh scattering), the light reflected by a slab of transparent silica aerogel shows a shift into the blue, as Fig. 9.63 exhibits.

It is important to note that TI materials can be characterized by a heat conductivity, but as the convective and radiative heat transport changes with layer thickness not in a linear law, the effective conductivity is a function of thickness, which highlighted in Fig. 9.64. The conductivity value also changes with temperature—as is the case for mineral wool or foam insulation—but due to the larger conductivity value the absolute change is again larger.

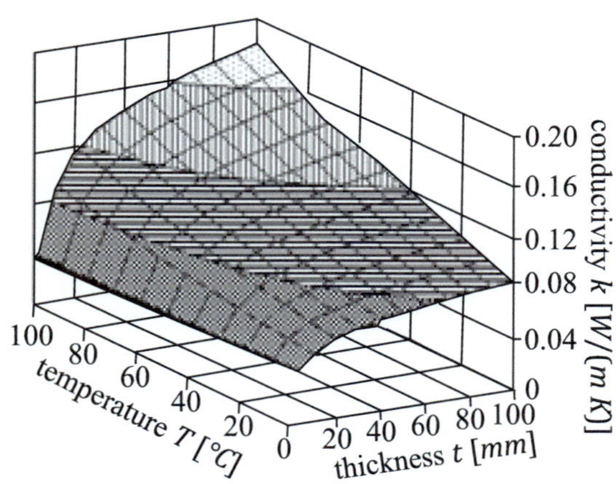

Fig. 9.64 Effective heat conductivity k in W/(mK) of a TI material as a function of layer thickness and mean temperature

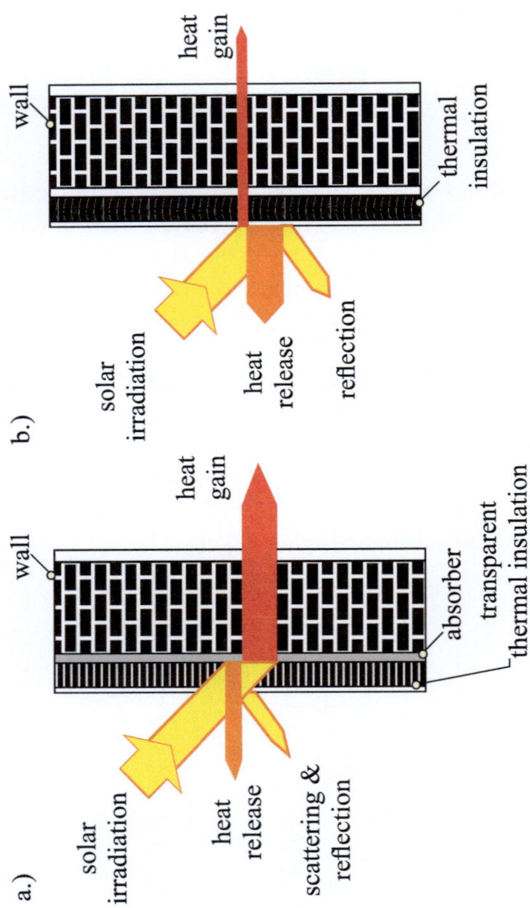

Fig. 9.65 Schematic comparison of the energy flow of a transparent insulation (TI) **(a)** compared to conventional non-transparent thermal insulation **(b)**

9.6.2.2 Functional Principle

The TI storage wall constructions are based on the principle that solar gains are created under irradiation on the outer surface of an absorbing wall, while the TI materials provide a high thermal resistance to the outside. In this way the wall is converted into a solar collector. If the absorber wall is built from a heavy weight building material like concrete, clay, limestone or heavy bricks, the wall acts as a sensible storage. If light-weight absorbers are used, they should be able to withstand high temperatures—these variants are found in constructions of TI ventilated facades, see e.g. Platzer and Goerdt (2000). Further details on the definition, materials and building projects can be found in Kerschberger et al. (1998).

Figure 9.65a and b show the principle of the energy flow within a transparently insulated wall (b) in comparison with an opaque wall (a). The solar radiation coming from the sun penetrates through the TI material of the building envelope, reaches a dark absorber surface and is converted into heat radiation there. This surface thus acts as a collector absorber. The TI material reduces drastically the convective and radiative heat transport to the outside air and causes the heat to be conducted through a solid massive wall layer into the interior. The TI material itself must be closed on at least one side to prevent penetration of rain, but also to reduce the risk of convection.

There are façade constructions with an air gap between TIM and absorber and without air gap. This has some impact on the thermal resistance, on condensation prevention on the front glass. An air gap is needed when a selective absorber is positioned on the solar wall, as direct contact of absorber and TI material destroys the low emissivity by conductive coupling. The thermal conductivity and heat storage capacity of a massive storage wall usually is sufficient to restrict the maximum temperatures to a level where the plastic materials still are stable. However, light-weight low density products like porous concrete are not useful in this context. Often a shading device is placed in front of the TI construction to regulate the heat flows. Passive shading (overhangs, static lamellae) can be used for certain orientations.

Due to the transient heat conduction, depending on the thickness, heat capacity, density, and thermal conductivity of the wall behind the TI construction, there is a time lag t_ϕ of several hours between the generation of solar heat at the absorber and the release of heat into the interior, as depicted in Fig. 9.66. This means that a TI wall may be designed to deliver heat to the room in the evening or night hours. A further feature of the transient heat conduction is the temperature-amplitude variation ΔT_a: although the absorber temperature between night and day may cycle between typically 10 and 70 °C on a sunny winter day, the resulting temperature variation indoors in only a few degrees (typically 20–35 °C). The time-lag leads to a very positive utilization of the solar gains. A combination of direct gain windows and TI wall may supply the room heating over 24 h during a sunny day.

The passive heating principle, therefore, has its advantages, however, the operation cannot be switched off when it is desirable, e.g. on hot summer days. Here, an effective solar protection using external blinds or louvers is an important feature for guaranteeing the thermal comfort inside the room. Similar to windows a

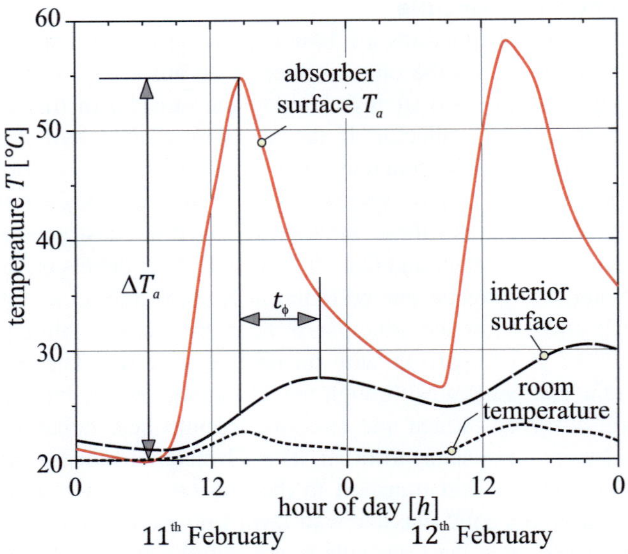

Fig. 9.66 Typical absorber and inner wall surface temperature variation over a day for a South oriented TI façade (ΔT_a = temperature swing on absorber; t_ϕ = time lag between interior and exterior temperature maximum)

small solar gain remains even with the best solar protection device. Nevertheless, also here the time-lag of the solar gain helps at least to some extent: unwanted remaining solar gains can be ventilated away in the cooler night-time much better than during the hot day.

TI systems can also be used to support the house heating system by installing a latent heat storage tank on the rear side facing the building. A technical design of such a system is shown in Fig. 9.67.

Another variation is the ventilated TI-façade shown schematically in Fig. 9.68. The idea here is that in summertime the generated solar heat is not transferred from the absorber to the wall behind which is separated by an air gap. But ventilated away by opening the lower and upper part of the façade construction. When the heating is needed, insulated flaps can be closed to prevent the ventilation loss. The principle has been proven and applied to a building working well, however the cost of well-designed thermally insulated and air-tight flaps is quite high.

9.6.2.3 Calculation of the Energy Transport in a TI Facade

The calculation of the solar gains of a solar wall is based on the thermal network describing the construction. A nodal or one-dimensional treatment of a simple TI storage wall without ventilation is shown in Fig. 9.69. For the evaluation the thermal resistances R_i are required, which are:

- thermal resistance R_{TI} of the TI from absorber to ambient air (including surface heat transfer coefficients in an air gap and to ambient air) and the

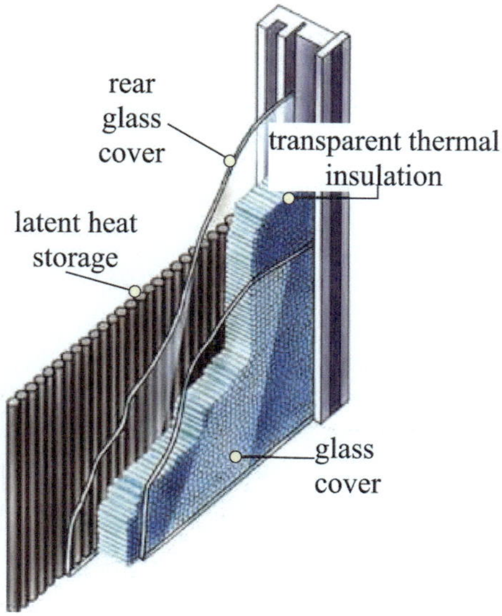

Fig. 9.67 Technical design of a TI insulation with coupled latent heat storage according to Hausner (2005)

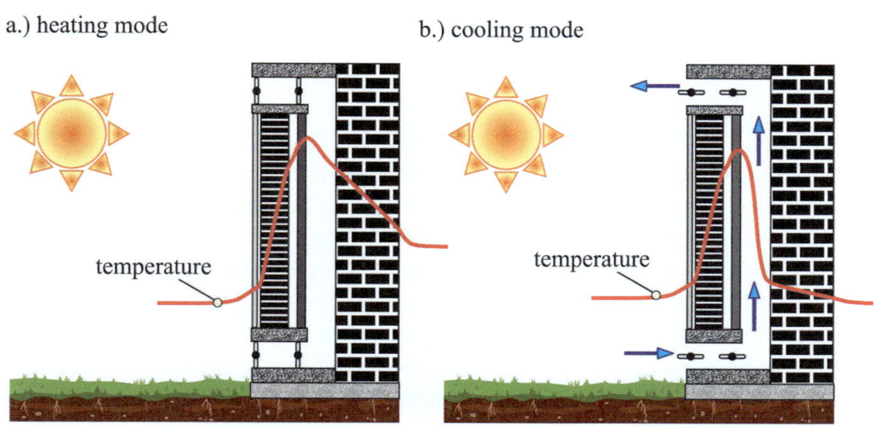

Fig. 9.68 Principle of a ventilated TI-wall in heating (**a**) and cooling mode (**b**) with indication of temperature profile

Fig. 9.69 Thermal network for a solar storage wall with TI

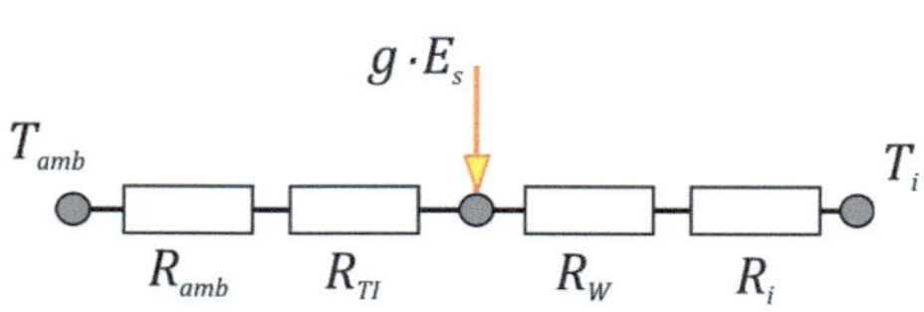

- thermal resistance R_W of the massive wall from the absorber to the interior air (including the surface heat transfer coefficients to indoor air).

Both, have the physical unit $\left(\mathrm{m}^2\mathrm{K}\right)/\mathrm{W}$ are computed as:

$$R_W = \frac{d_W \cdot}{\lambda_W} + R_i, \tag{9.28}$$

$$R_{TI} = R_{gap} + \frac{d_{TI} \cdot}{\lambda_{TI}} + R_a, \tag{9.29}$$

in which d_i and λ_i denote the thickness and the specific heat conductivity of the wall (index "W") and the thermal insulation (index "TI"), and R_i is the thermal resisitance of the gap and the ambient air. The solar gain at the absorber is determined by the g-value of the TI-construction and the solar irradiation E_S in $(\mathrm{W/m^2})$:

$$q_{abs} = g \cdot E_S. \tag{9.30}$$

Of course, this formula can be developed in a more detailed way when the split between direct sun and diffuse irradiation from sky and ground are separated. From a consideration of the resulting heat flux to the interior and exterior one calculates the stationary net solar heat gain to the interior room for indoor temperature (T_i) and exterior ambient temperature (T_{amb}).

$$q_{net} = \frac{R_W}{R_W + R_{TI}} \cdot g \cdot E_S. \tag{9.31}$$

A dynamic treatment needs a numerical simulation of the network, including the heat capacities of the thermal nodes. The conductive heat transport in the wall has to be discretized with a suitable number of nodes then. A stationary treatment often is sufficient for an approximative estimation. The physical transport equations can be linearized in the small temperature range, which is relevant, and then due to the superposition principle the average heat gain or loss can be calculated using the average indoor and outdoor temperatures (for a week or a month). In fact, the standards using monthly calculation of heat demand in buildings, e.g. DIN EN ISO 13789 (2018), are all based on the stationary Eq. (9.32).

They additionally include factors describing frame fractions, shading by obstacles, orientation of the façade. The utilizability of the solar gains in the standards is given by a simple function which is identical for windows and solar TI walls neglecting the positive storage and time shift effect of the wall and reads to

$$Q_{\mathrm{TWD}} = F_F \cdot F_S \cdot g \cdot U \cdot R_{\mathrm{TI}} \cdot E_s \cdot A \cdot t, \tag{9.32}$$

in which F_F describes the frame factor, F_S the shading factor, R_{TI} the thermal resistance of TI component, U the total U-value of opaque wall plus TI component, E_s the global irradiance on TI component, A area of TI-wall and the time, respectively.

From the considerations presented we can conclude some important features of TI wall systems:

- As the thermal resistance R_W of the wall behind the TI material increases, the solar efficiency of the façade decreases. Solid bricks, sand-lime bricks and concrete are more suitable than aerated concrete or other "light" walls. For a specific building material, the thickness is a parameter to be optimized: thinner walls have higher efficiency, but less storage capacity and a smaller time shift and a larger temperature variation indoors.
- TI walls achieve the best results in cold and sunny regions. The length of the heating season increases the utilizable or useful monthly solar gains.
- South-facing TI applications have the best specific energy values especially in winter, but unshaded east and west facades may also have positive energy balances.
- For a parameter analysis on TI-insulated residential buildings a lot of parameters can be varied, and reductions in heating energy calculated per square meter of the TI-insulated exterior wall. The calculation is based on the physical solar gains described by Eq. (9.32) and the utilizability of the solar gains in the building, which is dependent on other factors like solar gains through windows, internal gains, length of heating season and actual heat demand. The physical solar gain is proportional to a number of parameters, having the largest impact on the energy savings.

The physical solar gain is proportional to the following factors which have a high impact:

- transmission through the TI material (*g-value*).
- frame factor F_F of the TI system (100% minus frame fraction);
- shading factor F_S describing the reduction of irradiation through plants, neighbouring buildings,
- area share of the TI and
- irradiation sum on the TI-system which depends on orientation—here South facing vertical walls are best (in the Northern hemisphere) as they have a relatively high solar irradiation and good solar incidence on the TI due to the low sun in wintertime and are less efficient in summertime due to high sun incident on the TI.

There are some other parameters which have also medium effect on the solar gain described by Eq. (9.32)

- *R*-value of the TI-system—the higher R_{TI} the larger the gains;
- *R*-value of the massive wall behind the TI-system (depending on external wall thickness and heat conductivity of the wall material)—the lower R_W the higher the gains.

Other parameters are affecting the utilizability of the solar gains during the heating season:

- thermal insulation standard of the building,
- location of building—with climate affecting the length of heating season and the irradiation levels, the latitude the incidence angles of the sun;
- building orientation—affecting irradiation and incidence angles,
- internal heat sources and solar gains through windows,
- indoor temperatures and
- interior segmentation—in open designs the heat can be better distributed into other zones.

In a well-designed building the TI-system will not be over-dimensioned and limited to around 0.1–0.15 m^2 TI per m^2 living area. Older buildings with a long heating season obviously have the best utilization, whereas far a low energy building with a short heating season of 3–4 months the utilization is reduced accordingly to these months. Generally, the solar gains during the heating season can be largely utilized. Factors influencing the dynamic response of the TI wall system as there are:

- bulk density of the outer wall and,
- thickness of the wall.

Both affect the time needed for the solar gains to reach the interior room. They have only a marginal effect on the overall utilization but may be important for specific purposes (i.e. when a room heating is only needed in the evening hours). Unwanted seasonal solar gains outside the heating season may sometimes be reduced to a large extent by passive solar shading, for example at the South oriented façade using an overhang or balcony or a prismatic glazing in the system, as illustrated in Fig. 9.70.

The different classification of *g-value* and *U-value* correlates with an essential property of TI constructions: The savings potential through radiation gains is larger by a factor of 4 than that through reduction of heat losses. In this sense, a transparent insulation (TI) façade construction is more of a solar collector than a thermal insulation. The bulk density of the exterior wall has a significant influence on the TI yield. The reason is the increased heat penetration and the better storage capacity of heavy wall constructions. Therefore, when using a TI, heavy materials must be provided, and gross densities of over 1200–1400 kg/m^3 should be aimed for. Opaque building envelopes, on the other hand, should be built light because this improves thermal insulation. A big influence on the benefits of a TI is naturally the location and orientation of the façade, which is illustrated in Fig. 9.71. At locations with high radiation gains and long heating periods at the same time, the highest TI utilization values result because some summer heat contributions can also be used. The TI simulations largely coincide with the experiences made in recent years on houses with passive or hybrid solar energy use (windows and

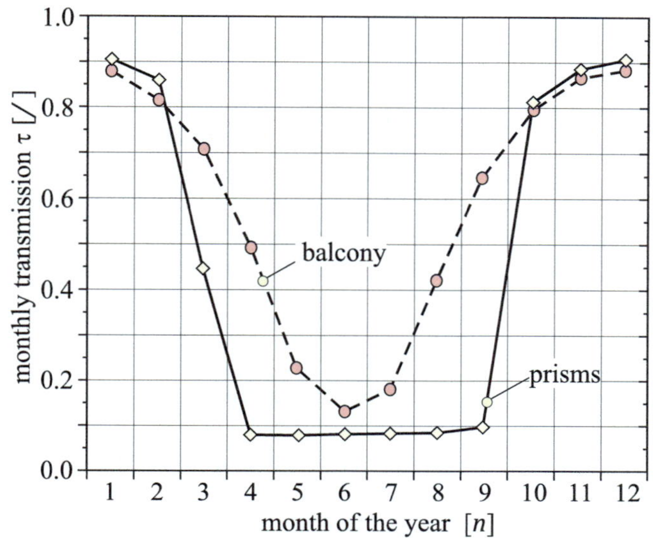

Fig. 9.70 Reduction of monthly direct solar irradiance by static shading devices on a south oriented façade (latitude $\Phi = 48° N$)

window collectors) and agree quite well with the simplified monthly calculation of the standards, see e.g. Platzer (1993).

Fig. 9.71 Reduction of specific annual heating demand for a row house with windows and TI-wall/opaque standard insulation at a South facing façade for different German locations. The TI-area is varied between 0 and the maximum opaque wall available; the opaque wall area is thermally insulated by either TI or opaque insulation; from Platzer (2005)

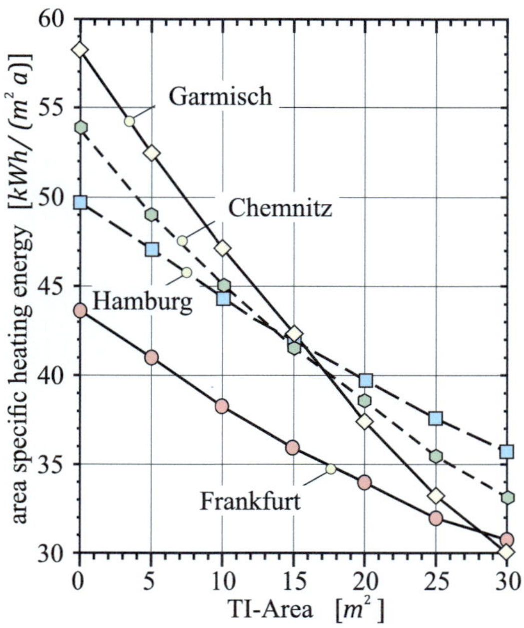

9.6.2.4 Architectural Integration

In this book we will only mention that architectural integration has to consider many factors when designing a house with TI-façade. In a solar building naturally the time of solar gains and the use of the rooms in the building should be correlated. The TI-façade specifically should be integrated in the appearance and functionality of the building and not be attached as a foreign matter.

Depending on the product the TI facades have different appearance. The photographs in Fig. 9.72 shall present a small number of buildings with different products and approaches, where the architectural integration was considered in a good way.

9.6.3 Conservatories and Winter Gardens

A well-known and popular variant of an indirect passive solar system is an unheated wintergarden or conservatory. Solar radiation heats a transparent building extension, mainly in dwellings. The construction is based on window technology, where framing elements are combined with glazings. In some cases, multiple ribbon polymer sheets from polyacrylic are also used instead of conventional glazings (transparent insulation type C). Usually, vertical walls and roofs are combined as a transparent shell in front of a heated zone. The rain tightness of such a construction has to be guaranteed. If internal shading devices are used the thermal stress on the roof glazings is considerable. The quality and the components of the construction may resemble a conservatory for plants or the windows of a living room, from single glazing with simple clamping profile to thermal insulation glazing in a complete frame with lip seals. There exist wooden, plastic and metal framing versions similar to the window technology.

Fig. 9.72 Multi-family house in Stutensee (D) using a transparent exterior insulation finish systems (TEIFS) without external shading; Project row house in Donaueschingen (D) using capillary glazings in a transom-mullion façade and external venetian blinds as shading device; both left (Kerschberger et al., 1998), Project Senior residence in Domat/Ems (CH) using a combined glazing element with $U = 0.5\ \mathrm{W/(m^2K)}$ including prismatic glazing as seasonal shading element and a translucent PCM-storage layer equivalent to a 10 cm concrete wall in thermal mass (© Dietrich Schwarz Architekten AG ETH/SIA-Projekt Domat/Ems) (right)

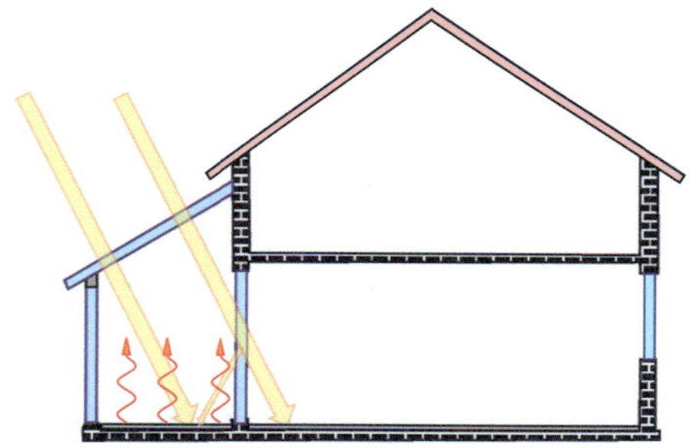

Fig. 9.73 Sketch of a wintergarden

The function of an unheated wintergarden—only then a reduction of heating energy can be achieved—consist in the thermal buffer effect with respect to transmission and ventilation heat losses. The latter ones can be reduced when fresh air for the building is preheated by the solar wintergarden, as indicated in Fig. 9.73. However, as the buffer space may reach rather low temperatures during night or during days with little irradiation, the control of air exchange between wintergarden and living space is critical. Very often it is achieved by opening a door between the zones. This may lead to an unwanted heating of the wintergarden leading to additional heating demand. On the other hand, the wintergarden certainly can be an extension of the living area especially in intermediate seasons. Especially with large roof glazing area the protection against overheating in summertime should be paid attention, as sketched in Fig. 9.73. Ventilation and shading are the means to prevent excessive temperatures.

9.6.4 Double-Skin Facades

While a wintergarden serves as extra living space for certain times of the year, the idea of a double-skin façade has been developed for commercial and high-rise buildings. The principle of converting the building shell into a solar collector is the same, but the floor area is very small, as the distance between outer transparent layer and inner wall or window layer is often just sufficient to incorporate solar protection devices and allowing maintenance access to them. Therefore, double skin facades usually have horizontal floors, e.g. using walkable steel grids or similar. One complexity of the use in multi-storey buildings is the transmission of sound, when internal windows of the facades are opened. The sound can be reflected by the external glazing to other rooms. Figure 9.74 illustrate design options and indicate the flow path.

The energetic effect depends very much on the management and control of the air flows and the solar protection devices. Is there an individual control for each room, for each orientation or just for the whole building? As commercial buildings are large and compact, and host a number of technical equipment, heating is in many cases less important than cooling. Therefore, the reduction of heating due

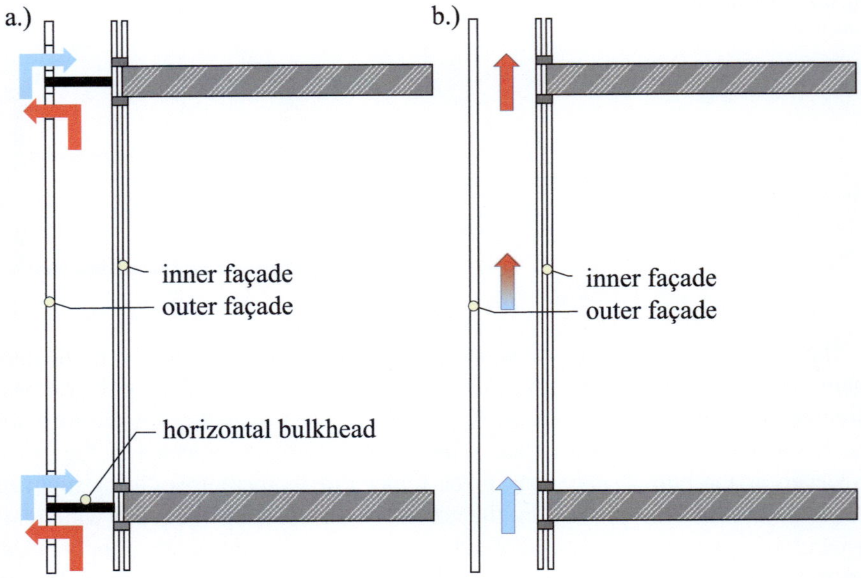

Fig. 9.74 Sketch of a double-skin façade cross section: **a** thermal separation of storeys, **b** multi-storey façade

to a double-skin façade has a lesser importance than that of a wintergarden in a private dwelling.

9.7 Planning and Dimensioning

9.7.1 Assessment Approach

When designing solar low-temperature systems, the first thing to check the demand and define the intended application. Should the solar thermal system just be used to prepare hot water or should it support the heating demand of a building or both. The dimensioning of the solar collector area and of the thermal energy storage are the key factors of the planning. The key figures characterizing the design are the annual solar fraction S_F and the annual utilization of the system $\eta_{ann,sys}$ as discussed in Sect. 9.1.3.

When we dimension a solar thermal system, we may increase the solar fraction by increasing the area of the solar collector field. But on the other hand, an oversized collector field leads to excess temperatures and to unusable heat, for example in summer, when the heat demand is low, and the solar potential is large. This would mean that the annual utilization factor is decreasing. We, therefore, have an optimization problem here.

- Solar domestic hot water preparation ($SDHW$)

In this concept, the area of the solar collectors and the volume of hot water storage is based on the consumption profiles during the year. A typical design rule is that the solar heat produced on one sunny day should be sufficient for two days of average demand. In summer in Central Europe then the need of an auxiliary heating of the hot water is marginal. In the transitional period and in the winter months, however, the system may only contribute a fraction of the demand. The overall yearly solar fraction S_F then is in the range of 50–70%. Large differences may occur if the auxiliary heating system (a boiler, a heat pump, a pellet stove) cannot be optimally controlled and is not matching the solar system. A good interaction of the complete bivalent system is necessary. Users may raise the solar fraction to some extent, if the adapt their hot water usage to the solar supply, e.g. by showering in the afternoon.

- Solar domestic hot water preparation and solar heating support (SDHW + SH)

In this system concept, a larger collector system is needed than in pure SDHW systems, because the solar system should even support a low temperature heating system during the heating season. Even when the solar thermal system in winter is not able to fully cover the hot water demand (at 55–60 °C) it may contribute to the heating system with supply temperatures well below the hot water temperature. The following additional requirements must be met:

- A suitable good thermal insulation standard of the building to limit heating demand.
- Integration of passive solar technologies (high-quality windows, conservatories, and atriums).
- Heating with a large area heating system such as a panel heating system (floor or wall heating) or generously designed radiators to supply the heat to the room with low temperatures.

The realisation of a solar thermal housing project should pass through three stages. Before the actual detailed planning can take place, especially in the case of existing residential properties, a detailed recording of the given conditions should be carried out. Within the three levels, there are several sub-items to be considered:

- Planning
  - general requirements,
  - determination of heat demand profile over a typical year,
  - inventory of the hot water and heating system,
  - consideration of cable and piping placement, location of the storage tanks,
  - selection of roof or façade area for collectors,
  - conceptual integration of the conventional system with the solar system.
- Dimensioning

  - system concept development based on the defined boundary conditions,
  - basic dimensioning of collector area to reach a reasonable solar fraction,
  - basic dimensioning of storage size to cover the hot water and heating demand for 1–2 days,
  - optimization of solar fraction and economics by simultaneously varying collector area and storage volume.
- Approval and technical implementation
  - installation of a solar system does not usually require permission,
  - in the case of historic buildings an approval is needed by the heritage conservation office,
  - for large weights the statics of the installation area should be checked,
  - safety aspects like wind load and snow load resistance are covered by standards and have to be considered by the installer,
  - the installer should prove a functioning system, where malfunctioning can be detected by the end-user; a simple monitoring routine in the controller is preferable.

9.7.2 Planning Stage

In the planning stage essentially the actual state and conceptional options are considered to develop further specification.

General Requirements
Since collector arrays are usually designed for a life span of about 20 years, the status of the available roofs (or facades in case of vertical installation) plays an important role. This includes whether statics allows the installation of collectors, whether the roof cover or façade is worn out and badly insulated or renovated. Is there a sufficiently large roof or façade surface in a suitable orientation that can accommodate the collector array without the collectors shading each other? Is the dedicated area for the collectors sufficiently large and not shaded by obstacles (other buildings, large trees, or parts of the own building)? Is the roof surface divided, so that the collectors must be arranged in different orientations? The available roof surface does not have to be aligned exactly to the south, deviations of 45° show only marginal losses. The pitch of the steep roofs is between 25° and 55°. If the roof has an inclination of less than 25°, an elevation of the collectors can be considered, especially for heating applications. Larger tilt angles favour the energy yield in winter (for systems for heating support) whereas small tilt angles increase the energy yield in summer (for systems for hot water preparation).

Assessment of hot water and heating demand
While the hot water demand is relatively constant of the year and is dependent on service temperature, mains water temperature, number of persons in the building and their average daily hot water consumption. Holiday seasons or weekends

change the user pattern in dwelling and also commercial buildings. Standardized hot water demand profiles on hourly, daily, and monthly basis have been developed to guide the planer when no measured demand profiles are available; an example is provided by VDI 4655 (2021). Cold water supply is usually quite close the ground temperature which has small variations over the year, but in average is close to mean annual ambient temperature.

As a rule of thumb, a consumption of 40 l of hot water per capita and day should be sufficient. In a single-family house of a family of four, this corresponds to an annual energy consumption of 2725 kWh/a, assuming a heating range of 40 K (domestic hot water withdrawal $T = 50$ °C and mains water 10 °C).

A rough estimation for Germany with global irradiation on the roof of 1000–1200 $\mathrm{kWh}/(\mathrm{m}^2\mathrm{a})$ exhibits that solar domestic hot water heating requires only a small amount of area for the collector surface. Table 9.8 shows typical values of the hot water demand of a single-family house and the corresponding collector area and storage volume assuming a 60% solar fraction and 50% annual efficiency. If one assumes an optimal south orientation of the collectors under an angle of inclination of $\beta = 45°$, only gross collector surfaces in the order of a few square meters are required. If the collectors are not aligned properly, the reduced solar yield can be compensated by increasing the collector surface area.

The heating demand of a building, however, is very much dependent on ambient temperature and season, as indicated in Fig. 9.75a. The solar radiation is contrary to the energy demand of a house, so that with regard to the climatic conditions Central or Northern Europe offers relatively unfavourable boundary conditions. In the season in which the maximum heat demand exists, the solar radiation supply is lowest and vice versa. In a commercial sense, supply and demand are therefore, countercyclical, as the Fig. 9.75b illustrates.

Therefore, especially at the beginning and end of the heating period (September, October and March, April, May) a low heating energy demand can be covered by solar radiation. This has long been exploited in residential construction through passive use of solar energy. The use of solar collectors can further increase the solar heating share. But, in the coldest months, using classical flat-plate collectors, even with huge areas the demand cannot be matched. A strategy to come to solar houses, therefore, has to exploit two options:

Table 9.8 Typical service water demand values of a single-family house and correlating service water storage volumes and gross collector area (standard flat-plate) assuming a 60% coverage by solar thermal energy, an optimal south orientation of the collector at an angle $\beta = 45°$

Daily domestic hot water requirement at approx. 50°C [l/day]	Volume of the domestic hot water storage tank [l]	Gross collector area of a flat plate collector [m^2]
100	200	3–4
200	400	5–6
300	500–750	8–9
400	750–1000	10–12

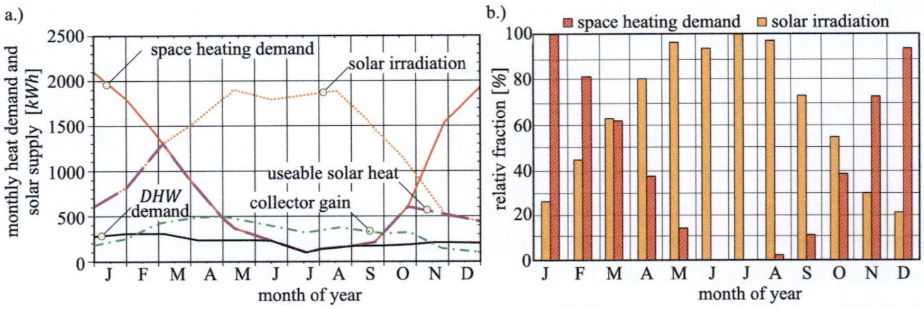

Fig. 9.75 **a** Typical heat demand for heating (modern single-family house, central European climate) and hot water demand compared to the solar irradiation sum on a 12 m^2 solar roof collector and the solar collector gain; **b** Seasonal development of the relative solar radiation energy supply and the relative space heating demand in Germany

- reducing the heating demand by energy efficiency measures in the building and/ or
- using seasonal storage for covering the winter gap.

The heating requirements of buildings are subject to a multitude of external factors:

- climatic conditions;
  - seasonal course of the outdoor temperature,
  - wind exposure,
  - solar radiation.
- Building concept;
  - thermal insulation,
  - window area percentage,
  - building surface to volume fraction,
  - integration (single family house to row house to large dwelling).
- User behaviour and requirements
  - Desired room temperature,
  - Intensity of use (offices, dwellings, schools differ),
  - Heated area percentage (rooms not heated permanently),
  - Number of users and electrical equipment (internal gains),
  - ventilation (natural, mechanical, heat recovery, controlled or free).

Although the real heating demand of a building is depending on those factors, for planning purposes a standardized estimation of the heating demand is important. This standardization has to neglect some individual aspects, but it should consider important factors like the regional climate. In this sense, the standardization has improved a lot during the last decades. Today's standards like (DIN EN ISO 13789, 2018) calculate on a monthly basis, thus also giving a realistic estimation of passive solar gains. The calculation of a standard heat demand Q_H of a building has to consider transmission heat demand Q_T and the ventilation heat demand Q_V.

The standard transmission heat requirement is the sum of the heat flows which the room releases through heat conduction through walls, windows, doors, ceiling, and floor. The proportion of ventilation heat requirement is calculated from the energy required to heat the incoming outside air to the room temperature level with or without heat recovery Internal and solar gains through solar active facades reduce the heating demand, as well as unheated buffer spaces. The remaining monthly heat demand then has to be covered by a heating system, which can be a bivalent solar thermal system (e.g. consisting of a solar thermal collector system in combination with a heat-pump or a biomass boiler).

Inventory of the hot water and heating system

This aspect is particularly important for existing systems. For the boiler plant the remaining operating time has to be clarified, possibly a new boiler is cheaper than a cost-intensive adaptation. The condition of the other components potentially to be integrated into the solar system should also be checked for their remaining service life (degradation due to corrosion, ageing of the insulation, etc.). Old potable water storage tanks can often be used as buffer storage. The power loss can be determined experimentally by means of a night measurement.

Pipe and cable installation and location of the storage tanks

Can short and direct piping/sensor cabling from roof to tank be installed without expensive construction measures? Where can the storage tank be installed and/or are constructional measures required to absorb the mechanical loads? Does the tank have to be divided into several freestanding tanks because the room height does not allow one single tank? In the case of new buildings, the location of the pipe channels, the dimensions of the technical rooms and the installation routes should be considered in the planning.

Selection of roof or façade area for collectors

For the required collector area rule-of-the-thumb dimensioning is possible when looking at hot water and heating demand. Since collector arrays are usually designed for a life span of about 20 years, the status of the available roofs (or facades in case of vertical installation) plays an important role. This includes whether statics allows the installation of collectors, whether the roof cover or façade is worn out and badly insulated or renovated. Is there a sufficiently large roof or façade surface in a suitable orientation that can accommodate the collector array without the collectors shading each other? Is the dedicated area for the collectors sufficiently large and not shaded by obstacles (other buildings, large trees, or parts of the own building)? Is the roof surface divided, so that the collectors must be arranged in different orientations? The available roof surface does not have to be aligned exactly to the south, deviations of 45° show only marginal losses. The pitch of the steep roofs is between 25° and 55°. If the roof has an inclination of less than 25°, an elevation of the collectors can be considered, especially for heating applications. Larger tilt angles favour the energy yield in winter (for systems for heating support) whereas small tilt angles increase the energy yield in summer (for systems for hot water preparation). If possible, the collector field should be compact and not distributed too much. This may be opposed by the available installation situation, the roof statics, or the requirements of the protection of historical monuments.

Connection and coordination of the conventional system with the solar system
For the further procedure information on the conventional hot water system of the object is essential. Already available components and performance of the conventional system should be considered in the dimensioning of the system. Also, a conceptual integration of the two sources in important. Operation in part load conditions, efficiencies of the non-solar system have to be respected. Which standby tank is required for the hot water and is there sufficient space available for this tank volume? What peak loads are expected (hourly peaks, tap peaks) and how could they be met? It must be checked whether these statements are sufficiently backed up by empirical values from the regulations/technical literature or experience with comparable objects. For heating systems, the heating areas and supply and return temperatures influence the overall utilization of the solar thermal system. Different integration schemes are available for heating especially (see Sect. 9.2.2).

Following the collection of the above data, an initial pre-selection of system variants can be made which are technically reasonable in view of the demand situation, the size of the system, peak loads, the existing hot water system, and the available non-solar supply capacity. Apart from this aspect, efficiency and technical risk play a major role. Thus, after this stage it is possible to have a simple conceptual design of the required solar thermal system including a rough estimation of collector surface and storage tank volume.

9.7.3 Dimensioning Stage

The dimensioning of the *conventional part* of the system, i.e. the conventional auxiliary heating and the standby storage tank, is completely independent of the dimensioning of the solar collector array. The auxiliary capacity and the size of the standby storage tank form a unit and must be precisely matched to each other. For the dimensioning of the standby storage tank, the occurring daily consumption profile is decisive and therefore, has to be dimensioned as for conventional system without collector system. In combination with the auxiliary heating, the standby storage tank guarantees the supply security of the object and must be designed in such a way that the maximum hourly peak (maximum peak tapping) can be covered in any case. One of the basic rules of efficient storage design is that the standby storage should be as small as possible but as large as necessary. Generally accepted methods for solar system dimensioning are available for calculating the required effective storage volume and the required post-heating capacity and mainly two types find application:

- demand indicator method according to (DIN4708, 1994),
- simultaneity factor method (Dell et al., 2010).

9.7.3.1 Simplified Dimensioning Tools

Simplified dimension tools can be design charts based on a large number of detailed simulations. They summarize the results based on the main parameters and are often provided by collector manufacturers for their own products. In such a way a quick assessment of the dimensioning can be done. The disadvantage of these methods is, of course, that the charts are developed often for specific products, for specific climates and use assumptions. With the help of dimensioning diagrams, such as the one shown in Fig. 9.76 for multi-party apartment buildings in Germany, the collector and storage tank dimensions can be designed for different solar fractions of a SDHW system. The collector area designed according to the diagram corresponds to the required collector area when the collector is oriented south at an angle of inclination to the horizontal of $\beta = 40°$. If the orientation and inclination of the area planned for the collector array deviates from those in the dimensioning diagram, the size of the collector array can be corrected by a factor.

Example Given is a compact new building with 50 residential units for 125 persons and an average hot water consumption of 30 l per person and day. At the location the average annual irradiation is 1020 kWh/m^2 and the roof orientation is 30° south-west. The roof pitch is $\beta = 40°$ and a roof integration of the collectors is possible. A central building services system is planned with a new technical room and a circulation pipe of approx. 250 m length. The available roof area is 25×6 m^2 and the highest mounting point is 25 m. A system with a cost–benefit optimum is required.

Solution From the number of persons and the consumption profile are known, we find point 1 in the Fig. 9.76. On the right-hand scale, the storage volume 1, the so-called standby storage, can be found with approximately 1100 l. By drawing a parallel to the abscissa (*x-axis*) you get point 2 for a climatic region in Germany. With the help of a parallel to the ordinate through point 2, one obtains point 3 for the desired solar fraction (S_F) of 45% (which is roughly a cost–benefit optimum, whereas 35% has lower specific energy cost and 55% stands for higher benefits). If one now draws another horizontal line through the marked intersection to the right edge of the diagram, the required collector area 105 m^2 and the corresponding buffer storage volume (storage tank 2) of 5800 l can be read off. The collector area read from the diagram refers to the active collector aperture area (net, irradiation, or aperture area), which is specified by the collector manufacturer and is required to indicate the performance of the solar system. The aperture area does not correspond to the gross area occupied by the collector. As the collector brand and its dimensions are usually not known in the preliminary planning phase, the following formula can be roughly

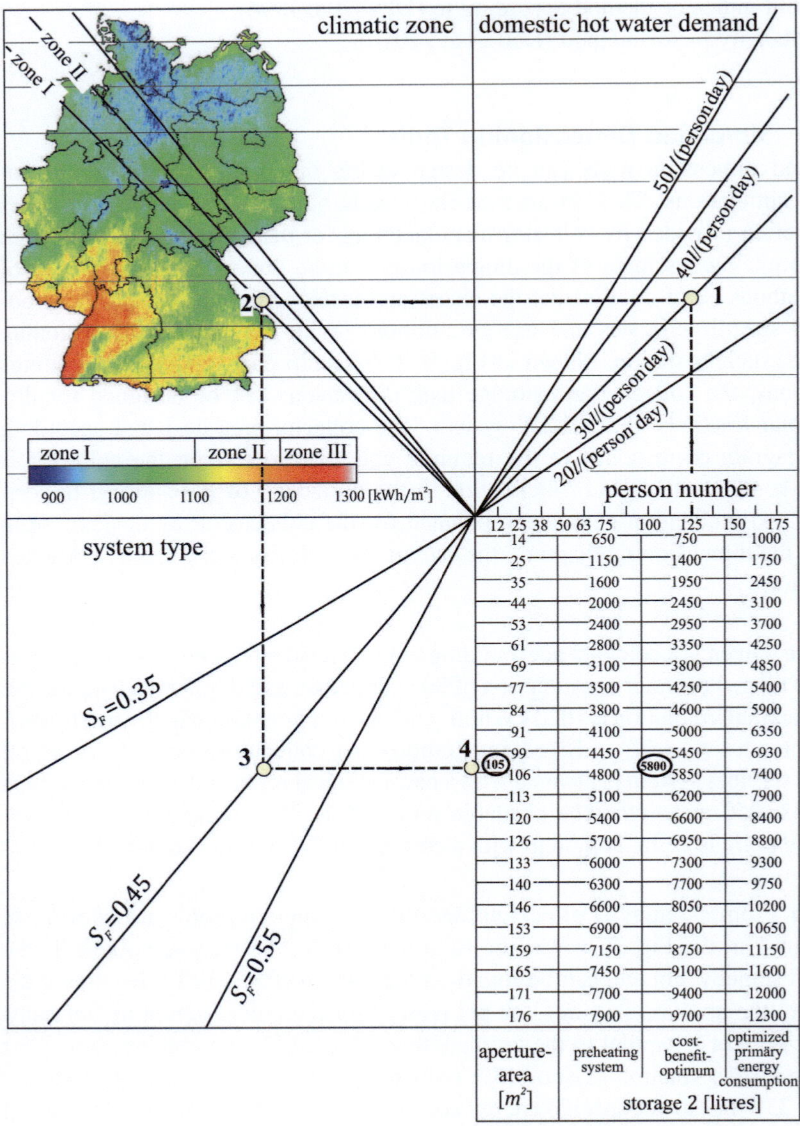

Fig. 9.76 Dimensioning diagram for multi-family houses to graphically determine the volumes of the standby storage tank (storage 1) and the buffer storage tank volume (storage 2) modified according to Dell et al. (2010)

Table 9.9 Correction factors (f_{corr}) to determine the corrected aperture area (A_{corr}) for different roof pitch β and azimuth angles γ if a roof is not exactly south facing

Roof pitch	Deviation from the south direction γ [°]				
	45	30	0	−30	−30
20	1.05	1.05	1.05	1.05	1.05
25	1.05	1.05	1.05	1.05	1.05
30	1.05	1.05	1.1	1.05	1.05
35	1.05	1.05	1.1	1.05	1.1
40	1.05	1.05	1.1	1.05	1.1
45	1.05	1.05	1.1	1.05	1.1
50	1.1	1.05	1.1	1.1	1.15
55	1.1	1.1	1.05	1.1	1.15
60	1.15	1.1	1.1	1.15	1.2

applied:

$$A_{\text{gross}} = 1.15 \cdot A_{ap}. \tag{9.33}$$

If the roof area available deviates from the south orientation and the pitch angle $\beta = 40°$ assumed in the dimensioning diagram, the resulting collector area must be corrected. The aperture area A_{ap} is then multiplied by f_{corr}, so that the corrected aperture area (A_{corr}) is $A_{corr} = \left(f_{corr} \cdot A_{ap}\right)$. Typical correction factors f_{corr} may be taken from Table 9.9.

In literature a number of other simplified dimensioning tools are described. The textbook of Duffie and Beckman (2020), for example, describes the f-chart method, which in essence is a correlation method for a specific system design. The key trick which makes it useful for a wider range of dimensions is the use of dimensionless groups X and Y, one which describes the monthly average collector field losses divided by the monthly heat load, and the other one the corresponding solar gains divided by the heating load. For a certain range of parameter variations and climates using the simulation program TRNSYS (TRNSYS Coordinator, 2017) a correlation curve f for the solar fraction as a polynomial of X and Y. This curve then can be used to estimate the monthly solar contribution to the load for any parameter within the validity range of the method. Correction factor for storage volume and heat exchanger efficiency have been developed. As the dimensionless numbers X and Y are easily calculated in a table using monthly meteorological data and tilt corrections, the calculation was very fast even with slow computers some decades ago. The result has reasonable accuracy on the monthly and annual level. Therefore, in a comparative technology tool like RETSCREEN (NRCAN, 2022) this method is still used.

A second approach is the calculation of solar utilizability Φ. Here the critical radiation level I_{Tc} of a collector, which gives the radiation where just the heat losses were compensated, is compared to the solar radiation during the day. Only for radiation above the critical level the solar collector produces useful heat. The critical level depends of course on the collector temperature. Combining the statistics of daily and monthly global radiation with this idea, estimations for monthly net collector gains can be calculated. In a combination with the f-chart-Method, the so-called $\overline{\Phi}$, f-chart method is more flexible as it allows to treat system with different load demand temperatures and also variable storage temperatures during the month. It is implemented in the F-CHART Software (F-CHART Software, 2022a, 2022b).

Nowadays, modern computers are so much faster than decades ago, that the calculation speed of the methods does not require to use them any more with all their limitations and inaccuracies.

9.7.3.2 System Concept Development and Integration into Circuits

Once the most important component sizes (collector and storage tank) have been determined in an approximate and simplified way, it is important to detail further the hydraulic concept and the integration into a multivalent system, as the solar thermal collector field in the overwhelming number of cases cannot be a heating system covering 100% of the demand. The simplest bivalent system uses as auxiliary source an electrical heating rod in the relevant storage section. Also, here the standby volume being permanently heated by the auxiliary and the buffer volume (which is only heated by the solar field) have to be defined by the positioning of the rod.

In general, a multivalent system consists of several heat sources, the solar collector field, a gas boiler, an electric heating, a heat pump system, a biomass boiler, or other heat generating devices. It is important to find a suitable hydraulic integration and a control strategy fitting to the characteristic operational modes of the individual heat sources. A bivalent system is required, but sometimes additional sources might be connected in favourable configurations, e.g. when some biomass is available at low cost or waste heat from a process can be integrated at some times.

The system concept has to specified as far as the simulation tool allows to model the details. Commercial tools use a library with pre-configured hydraulic concepts, and often allow to specify the performance characteristics of the components or even the commercial product to be used. Using this type of simulation, the operation of a system can be analyzed in detail and be optimized.

For the selection of the respective system concept, the type of building is important. Since the building structure and the building services system are different for each installation, the proposed system concepts must also be adapted to the respective situation. The descriptions concern the structure, the pipeline network, the location of the components and the building services system. As a selection criterion for the most suitable system concept, a distinction is made between a central building services system and a decentralized building services system.

- Elongated and large buildings require a different profile of hydraulic supply compared to compact buildings as pipe lengths are much large influencing heat losses, delay times, circulation needs and hydraulic resistances.
- In a compact building, the building services of the solar system should be centrally organized; this applies especially to smaller buildings.
- In the case of large and elongated buildings, the use of decentralized solar systems for individual residential units may be an alternative to a centralized two-pipe network with central supply.
- The placement and sizing of storage facilities is complex, because on the one hand side large storage are favourable, and on the other side room size restrictions apply and the distance to the consumers should be small to have small delivery delay times, which favours small storages.

9.7.3.3 Dimensioning Using Energy System Simulation Tools

When the system concept is specified and components are selected, an improved detailed dimensioning is carried out nowadays with the help of commercially available dimensioning programs T*SOL (Valentin Software, 2022a, 2022b), POLYSUN (Velasolaris, 2020) or others. The starting point is the first selection of collector and storage dimensions. Then optimization runs varying the two main parameters collector area and storage volume were executed with the aim to determine a cost curve for the delivered heat (hot water and heating) depending on a targeted solar fraction. When simulations indicate an operational problem or show a high loss of performance in a specific component, a redesign of the system is advisable (e.g. by choosing a different heat exchanger area or by adding more insulation to the buffer storage). We describe some basics of a numerical energy simulation in the Sect. 9.8. In this section we rather describe the application of such tools, which require a certain strategy and discipline in order to reach reasonable and useful results. An optimization based on simulation needs a strategy.

- First of all, the optimization target has to be selected. Low energy cost per kWh is often a target but may sometimes lead to unwanted results. The performance of a solar thermal system is not only based on economics, but often environmental and energy saving aspects shall also be respected. In buildings also thermal comfort and visual acceptance plays a role. The autonomy and resilience of a solution becomes increasingly important. For a combination of several objectives in one simulation strategy a combined performance parameter is a useful possibility, where weighting factors determine the relative importance of a performance indicator (Platzer, 2021).
- Secondly it is important to decide which simulation parameters are kept fixed and which ones are used as variables. Fixed parameters not necessarily mean fixed inputs to simulation runs. For example, a fixed specific mass flow could imply that for every run with a different collector area also the mass flow in the solar loop has to be changed. Collector area and storage volume are in most cases the main variables in such a simulation, allowing to display the results

also graphically in a 3D-surface. However, other parameters might be added to extend the parameter space, for example mass flow, heat exchanger area, in a systematic way

- Thirdly when an optimum of the basic parameters is found, one might vary individual components (e.g. collector products) in order to check commercial solutions around the optimum.
- It is advisable to start first with a coarse and not too fine discretization of parameter variations. When the approximate area of a global optimum has been found, fine tuning can be done in a second step.
- Documentation of the optimization strategy, the fixed and variable parameters, and the boundary conditions (demand profile, location, climate etc.) in combination with the results and findings should complement good simulation work.

9.7.4 Avoidance of Legionellae

Legionella Pneumophila are aqueous bacillus with a size from 0.3–0.9 mm width and 1.5–5 mm length. Legionella reproduces only inside other cells. Bacterial-feeding amoebae also live in the wet environment where Legionella is found. After the Legionella is eaten by the amoeba, it is encapsulated inside the amoeba, where it continues to grow and multiply. By releasing Legionella bacteria from the amoeba, they can disperse into the environment again and form a biofilm with other bacteria. They can be released to water used in human environment, into aerosols or droplets so that humans can inhale them, as illustrated in Fig. 9.77. In humans, this cycle is repeated, with the resulting effect that the human lung cells are infected. Legionella are cause for the so-called Legionnaire's disease which can be potentially lethal. The infection is possible due to inhalation of aerosols containing the bacillus. It has been observed for example with showering using infected water. The bacillus affects the aveoles of the lung. The formation of Legionellae is promoted by:

- moderately warm water temperature (30–42 °C),
- quiet water without movement,
- pipe parts of no frequent use,
- decrease of water flow,
- certain pipe materials.

For this reason, legislation requires prevention against the formation of the bacillus in hot water systems. The prevention measures have to be reasonable and feasible for different project sizes. In Germany, small systems in single or double family houses with a total water volume of less than 400 l and a pipe volume of less than 3 l between heater outlet and draw off point, no measures are required. For

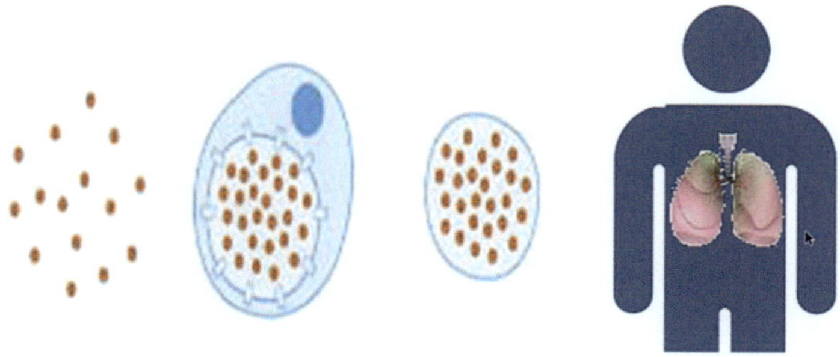

Fig. 9.77 Scheme showing the growth of Legionellae and inhalation of human lungs. From the left to the right: individual Legionella are uptaken by an amoeba, where the reproduce and can be released in larger number to an aerosol droplet, which can be inhaled by humans

larger systems in multi-family houses specific legionella prevention measures are required in order to guarantee legionellae prevention:

- heat water storage to 60 °C once a day,
- regular microbiological tests.

The first solution is based on the circulation of hot water through a two-line piping system using a separate heat exchanger due to exergy reasons, as sketched in Fig. 9.78. The circulation mass flow is continuous at a low constant level, transporting hot water above the critical limit to all pipes where water otherwise would be stationary. Of course, this implies a higher heat loss of the system than without circulation.

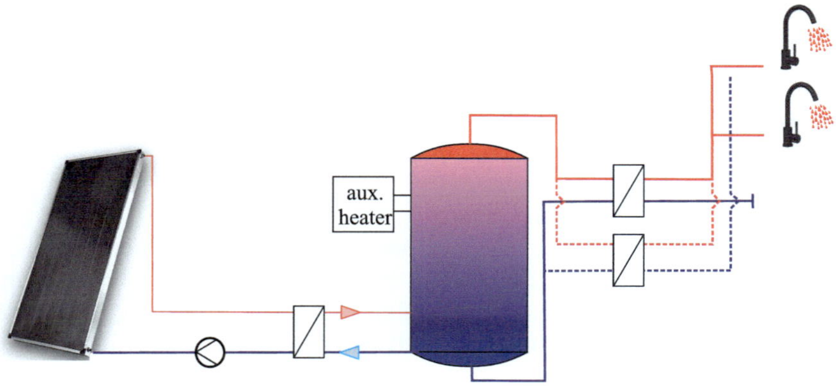

Fig. 9.78 Legionella prevention using circulation of hot water (dotted line circulation with separate heat exchanger)

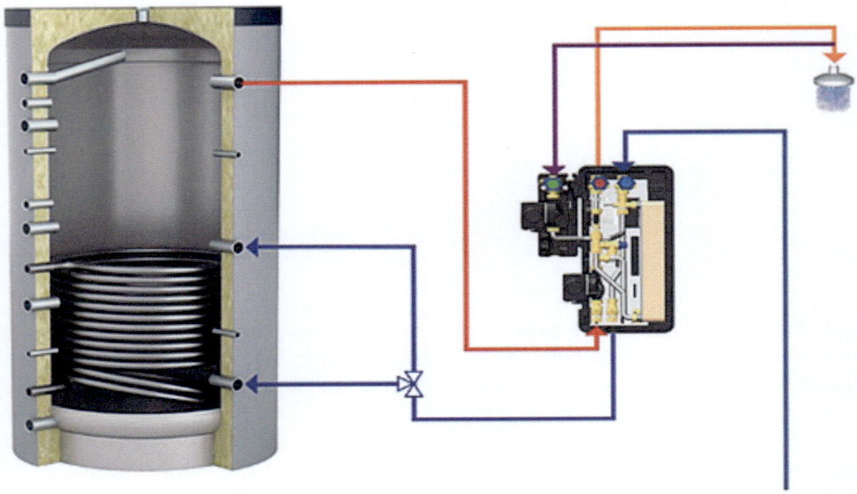

Fig. 9.79 Legionella prevention using decentral fresh water stations

A second solution which does not have this disadvantage uses so-called fresh-water stations. Decentral heat exchanger unit in each apartment of the multi-family building provide a sufficient reduction of heated volume of potable water to prevent legionellae. The fresh water is heated up by water from a buffer storage when needed, and only a small fraction is left warm between heat exchanger and tap line. Such a concept is illustrated in Fig. 9.79.

9.7.5 Reduction and Prevention of Stagnation Times

Stagnation is defined as an operational state of the collector where heat losses exactly compensate the solar gains, so the net useful heat is zero, because the circulation of HTF in the collector loop in such a case is stopped. There can be several reasons in the collector loop for such a situation:

- a defect of the collector loop pump;
- some blocking of valves or other components in the collector loop;
- due to strong gas intrusion/dissolution of gas in the collector loop and a malfunction or non-existence of a deaerator a gas bubble in the collector field obstructs the circulation (pump too weak to move against gravitational forces);
- a failure of the control system leads to a non-operating pump.

Apart from these failures of the system stagnation may also be the result of ordinary operational control:

- due to a fully charged thermal storage and no demand the circulation stops in order to avoid overheating of the thermal storage and/or
- the user manually switches off the system (e.g. for maintenance).

A stagnation scenario is only very rarely caused by a failure of the control system or the pump. The higher the plant's share of coverage, the greater the excess energy supply, which leads to stagnation due to full storages in the summer months. In this case, the collector surface is oversized with respect to the summer radiation. The downtime in such systems can be several 100 h per year, see e.g. Lustig (2002). The consequence of stagnation is that the collectors assume the highest possible absorber temperature for the prevalent irradiation. The stagnation behaviour of a solar system is becoming more and more important due to the ever more powerful absorbers and efficient system components. As a consequence, at high solar irradiation levels temperatures way above the evaporation temperature of water are reached, leading to steam formation. The five subsequent phases of such a stagnation process have been described already in Sect. 6.7 in detail. These are:

- Phase 1: Liquid expansion of the heat transfer fluid.
- Phase 2: Steam formation with fluid displacement and rapid pressure increase.
- Phase 3: Fluid boiling in the collector—saturated steam only.
- Phase 4: Boiling in the collector—saturated and superheated steam coexistent.
- Phase 5: Refilling of the collector.

In addition to the collector circuitry, the pipe routing and the arrangement of the fittings play an important role. It is important to design the system in a way to reduce the problems occurring with stagnation. In order to reduce undesirable effects such as excessive temperature and pressure loads on the components, the quantity of residual liquid in the stagnating collectors must be reduced. There are two different concepts to reduce stagnation problems in collectors. These are covering the problems in the circuit, but of course not the problem temperature stress and durability of absorbers and coatings. When temperatures well above 200°C are reached for many hours a year, the lifetime of selective absorber coatings is reduced, c.f. Köhl et al. (2004). The first concept are so-called intrinsically safe systems in which a stagnation does not place any additional load on the system components. The following systems are intrinsically safe thermal solar systems:

- operation as drain back system,
- use of heat pipe collectors,

The following measures are leading to intrinsically safe thermal solar systems:

- optimization of collector piping and array interconnection,
- heat dissipation through targeted condensation,
- increase of system pressure.

The second concept for avoiding collector stagnation assumes a complete exclusion of potential evaporation situations in the collector. It is based on sophisticated control strategies of the solar loop. The strategies can be simulated numerically for the respective plant in detail (even before their technical implementation). The idea is that potential stagnation processes are specifically excluded or minimized by means of an active control intervention in the pump, heat exchanger, etc.

In the following, the different measures of both concepts to reduce collector stagnation problems as far as possible are presented and the basic ideas of control strategies are explained.

9.7.5.1 Drain Back Systems

As already discussed in Sect. 9.2.4 a drain back system empties automatically when the pumping stops. This prevents overheating and steam formation for the heat transfer fluid. The thermal stress on the overheated absorbers and the other collector materials, on the other side, cannot be avoided.

9.7.5.2 Heat-Pipe Collector Systems

Heat-pipe collectors, the technical design of which was dealt with in Chap. 6, consist of closed heat pipes, inside which there is a slightly boiling liquid and an absorber fin. Under solar radiation, the liquid in the pipe of the heat pipe evaporates; the steam flows upwards and condenses on the header pipe, through which the heat transfer fluid of the plant circuit flows. The condensate flows down again and evaporates there again. In case of stagnation, the collector is no longer flowed through, which means that the liquid there also evaporates: the heat transport from the tube to the condenser collapses, so that the vapor volume in the collector circuit is limited to the collector area.

9.7.5.3 Optimized Collector Array Circuit

Another possibility to reduce the problems is the optimization of the collector piping and the collector field connection. The goal is to achieve the best possible emptying behaviour of the collectors. A collector field shows minor stagnation problems if no stagnating liquid volumes (so-called "bags" or "syphons") occur, which increase the steam volume in the system when boiling dry. This can be seen in Fig. 9.80a: The steam is formed in the top hottest part of the collector and presses out most of the liquid still in the collector. On the contrary on the right side, the liquid remains in in the collector which increases the total steam volume created eventually. Similarly, to the individual collector it should be emphasized that interconnections of collectors should consider the fast expulsion of liquid from the field. Ideally, a collector field should not have coolant volumes in which the fluid can collect and stagnate.

Thus, for maximal draining in case of overload both the collector-internal piping as well as the connections collectors should be taken into account.

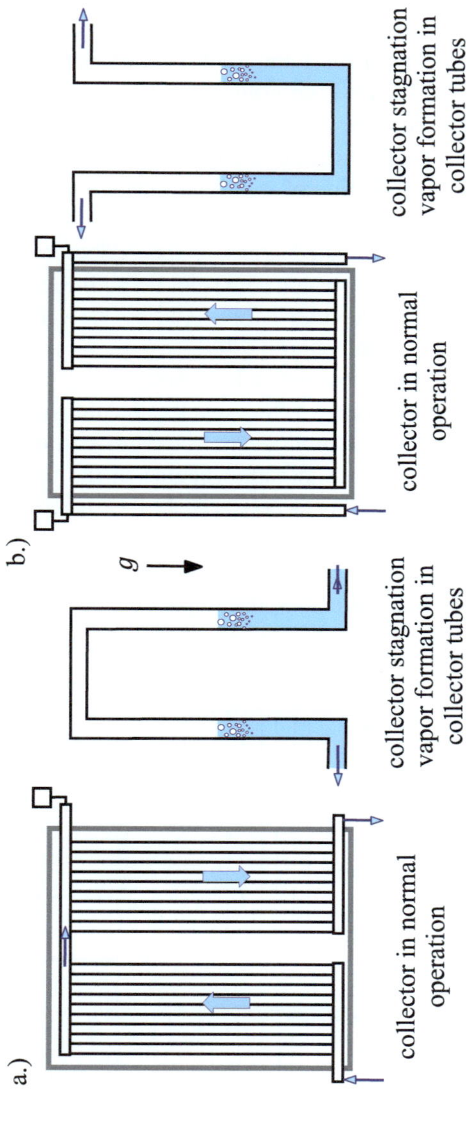

Fig. 9.80 Collector piping and connections with favorable drain behaviour (**a**); Collector piping and connections with unfavorable drain behaviour (**b**)

Collector internal piping
A prerequisite for optimum draining behaviour is that at least one of the two collector connections is led out on the underside of the collector. Piping systems such as the serpentine type, Fig. 9.81a, the harp type, cf. Fig. 9.81b show in experimental measurements a draining factor of almost 100%, whereby a slightly poorer draining behaviour is achieved with absorber tubes that are designed exactly horizontally as in Fig. 9.81c, since there a layered flow occurs in the parallel tubes, which slows down the removal of the fluid. With connections typical for vacuum tube collectors (see Fig. 9.81d, e), both collector connections are located at the upper end of the collector. As a result, the emptying factor is much smaller for these variants; it is about 30% for version in Fig. 9.81e and about 50% for option in Fig. 9.81d. Here, a deep penetration of steam into the downstream pipeline was observed. In addition, condensate flows directly back from the pipes through the connections at the top into superheated areas of the collector. As a result, pulsating evaporation processes occur, which represent an additional alternating load on the collector. An improved behaviour is observed by the variant Fig. 9.81f, as experimentally demonstrated by Scheuren (2008).

A horizontal alignment of the absorber tubes shows no (gravitationally based) preferred direction during emptying. This results in unfavourable stagnation behaviour in a field connection, as high steam velocities can occur in the upper absorber tubes. This occurs when evaporation begins in the uppermost absorber tube and a stratified or annular flow is established. At the end of the absorber pipe the steam flow is mixed with the liquid in the pipe segments below. During this process, considerable noise can be generated by cavitation or water hammer. For long rows of serially connected collectors, the steam velocity can increase to such

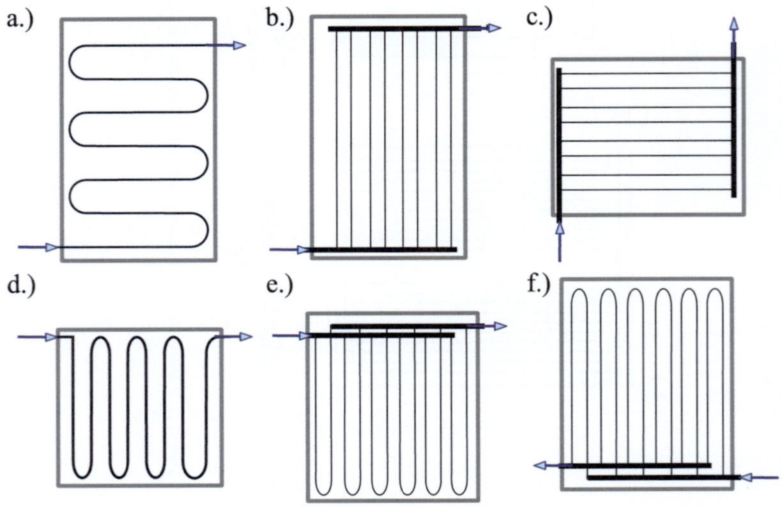

Fig. 9.81 Sketches of different types of internal collector piping

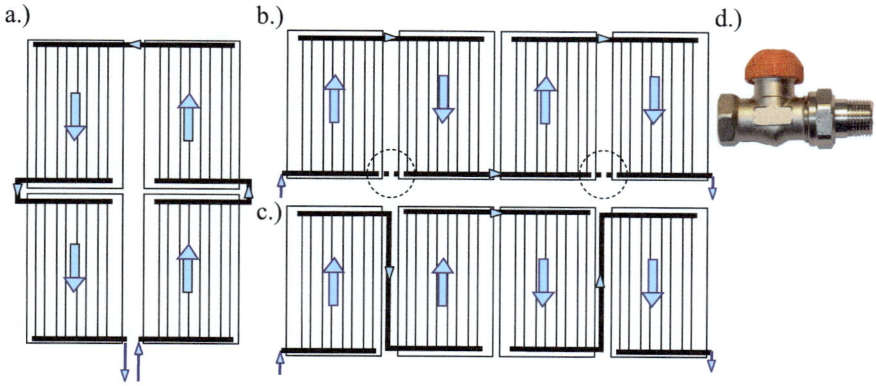

Fig. 9.82 Examples of serially connected collectors with Tichelmann flow control with favorable draining behaviour (**a**), unfavorable draining behaviour (**b**) and improved draining behaviour (**c**); (dashed circle) installation location of thermostatic valves (**d**) to improve draining behaviour

an extent that these noises and the associated pressure peaks can reach down to the cellar, see Kimmel et al. (2001).

Collector field connections

Collector arrays consist of an interconnection of several individual collectors. Parallel connection of collectors does not pose a problem, however, for serial connections unfavourable positioning has to be avoided, (see Fig. 9.82). A collector that is as such well self-draining can have a considerably lower draining factor if the field connection is unfavourable (Fig. 9.82b). At least one of the two connections to the collector array has to lead downwards in order to enable favourable drainage.

In Fig. 9.82 four collectors connected in series are shown in three configurations. Although a single collector of this type has a drain factor of nearly 100%, the overall drain behavior can vary significantly.

In Fig. 9.82a, the field is optimally drained since the liquid can be displaced through both connection lines. The entire field is emptied during the first evaporation phase in the upper collector area. Figure 9.82b shows an unfavourable case since the central collectors can only displace the liquid upward through the two collectors at sides. Therefore, more liquid stay in the array and a longer empty boiling phase of the field is expected. With regard to the emptying behaviour, the interconnection option in Fig. 9.82c represents an improvement. All collectors can internally displace remaining liquid downwards during the first evaporation. However, the necessary risers (vertical pipes) required for this interconnection can lead to pressure rise in the system if the thermal insulation is insufficient, as condensate can form in them, which can flow back into overheated areas of the collectors. In addition, liquid plugs in the pipe can cause water hammer due to strong acceleration during evaporation when hitting elbows or tees.

An intrinsically safe stagnation behaviour of any arrangement can be achieved by means of thermostatic valves which open at a certain temperature and allows emptying via a common collecting pipe. The Fig. 9.82d shows such a thermostatic valve and in the Fig. 9.82b the corresponding installation locations are marked with dashes. The thermostat option allows greater flexibility of the collector field connection and is an important aid in the event of structural restrictions (roof shape, free roof area, piping, etc.).

9.7.5.4 Heat Dissipation Through Targeted Condensation

In a conventional solar thermal system, there is no component in the circuit which can be used as a significant heat sink in the event of a stagnation. At night it is possible to discharge the storage tank via the collector field due to the temperature reduction outside, but this procedure is not very efficient. The integration of a defined heat sink is a measure to condense the steam generated during stagnation to prevent overheating of the storage tank.

Heat dissipation in pipelines

There is always an inherent heat loss through the pipes of the solar circuit, however, since most of the pipes are thermally insulated, their cooling capacity is limited. Therefore, if no stagnation cooler is integrated into the solar circuit, the insulated pipes act as a natural single "heat sink". The specific heat loss current per meter of cooling pipe length depends on insulation thickness and pipe diameter and can be calculated with standard heat transfer methods (see Chap. 4). In Table 9.10 specific heat losses for steam filled pipes were calculated for two different insulation levels (thickness of insulation 50% or 100% of pipe diameter). The results are much lower than the values for uninsulated pipes, which are outlined in Table 9.11.

Preconnection vessels

A further possibility for limiting the system pressure is the use of a vessel upstream of the membrane expansion vessel (MEV). From a technical point of view, an arrangement as shown in Fig. 9.83b is preferable, as steam bubbles can penetrate the cold upstream vessel from below and condense immediately. In this way the MEV remains protected from steam, but pressure oscillations occur due to direct condensation in the upstream vessel.

The heat loss of the vessel by radiation and convection has to consider that the temperature is below condensation temperature. Table 9.12 shows specific maximum cooling capacities of intermediate vessels of different volumes at condensation temperature. It can be seen from the table that upstream vessels are not very suitable for heat dissipation since their specific cooling capacity is lower than for insulated pipes. The task of the upstream vessel is to protect the MEV by storing cold solar fluid, which is pressed into the MEV when stagnation occurs. This capacitive buffer thus ensures that the temperature in the MEV remains comparatively low, but it increases the amount of solar fluid required in the collector circuit.

Table 9.10 Length-specific heat losses of steam-filled pipes at a temperature difference $\Delta T = (T_i - T_{amb}) = 115K$ with two insulation thicknesses $t_{ins} = 0.5d_a$ and $t_{ins} = d_a$ with $\lambda_{ins} = 0.05W/(mK)$

Pipe dimension $d_a \times t_w$ [mm × mm]	18 × 1	22 × 1	28 × 1.5	35 × 1.5	42 × 1.5	54 × 2	76.1 × 2	88.9 × 2	108 × 2.5
$Q_R(t_{ins} = 0.5d_a)$ [W/m]	36.5	38.4	40.5	42.3	43.6	45.1	46.9	47.5	48.3
$Q_R(t_{ins} = d_a)$ [W/m]	27.6	28.3	29.2	29.8	30.3	30.8	31.4	31.6	31.8

Table 9.11 Length-specific heat losses of steam-filled uninsulated pipes at a temperature difference $\Delta T = (T_i - T_{amb}) = 115$K and a vertical or horizontal pipe arrangement

Pipe dimension d_a [mm]	10	20	30	50	70	100
Q_R vertical [W/m]	51	103	154	257	360	515
Q_R horizontal [W/m]	74	133	189	295	397	546

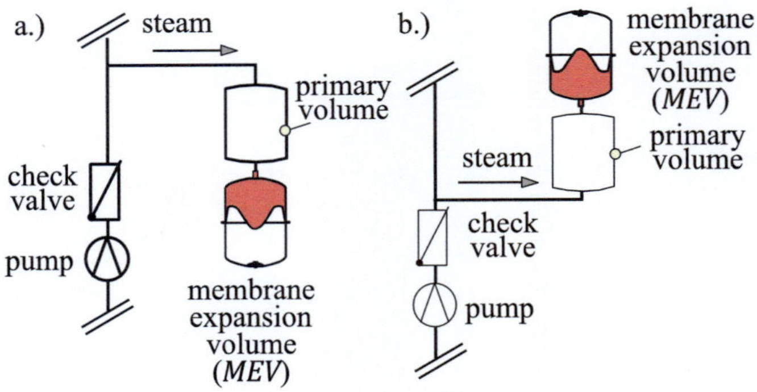

Fig. 9.83 Potential installation configurations of a ballast vessel to protect the membrane expansion vessel (MEV) during stagnation processes

Table 9.12 Specific cooling capacity of steam-filled intermediate vessels of different volumes at a temperature difference $\Delta T = (T_i - T_{amb}) = 115K$; geometry assumption: height = diameter ($T_{amb} = 25°$C)

Volume of primary vessel [l]	5	10	15	20	50	80
Cooling capacity [W/l]	51	40	35	32	23	20

Finned Tube Cooler

Finally, the most effective option to dissipate heat to the environment is via finned tube coolers, which consist of a tube with pressed-on aluminium fins. The heat exchange efficiency of such finned tube coolers can be calculated using the relations of engineering textbooks as outlined in Chap. 5, such as Kreith and Goswami (2005) or Stephan et al. (2019).

Active Cooling

A large number of other active cooling methods can be considered, such as additional water coolers, delivery via the heating system, table coolers, cooling towers etc. In addition to the achievable cooling capacity of the individual methods, the location for integrating the cooler into the primary circuit plays a central role. The aim is to reduce potential downtime. The Fig. 9.84 shows three different variants.

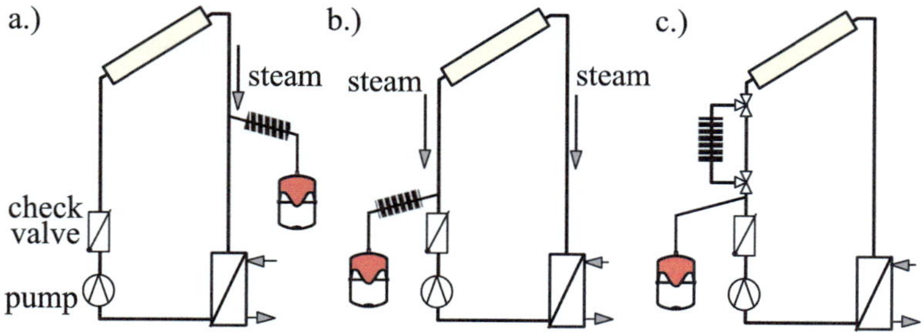

Fig. 9.84 Possibilities to avoid stagnation by integrating an active or passive cooler into the primary circuit of a solar thermal system

In variant (a), the cooler is integrated in the branch to the MEV which is not transpired during normal operation. Due to the MEV positioning in the forward flow, in case of stagnation steam propagation is only possible via this line. The steam is given a clear direction and the pump is effectively protected against high temperatures. If only one line is available for steam propagation, the draining behaviour may be impaired under certain circumstances. The formation of liquid sacks is possible, which would not occur in variant (b).

The integration according to Fig. 9.84b shows a classic component arrangement with an additional stagnation cooler in the MEV branch. In this configuration the steam has no preferred direction. Either the steam condenses there (internal heat exchanger, danger of steam formation in the storage tank) or it spreads further to the pump. In common solar stations, the branch to the MEV, the non-return valve and the pump are located close together. When steam reaches the branch, pump temperatures of over 110 °C are possible solely through conductive heat conduction.

The option illustrated in Fig. 9.84c outlines an operational cooling, which should prevent the formation of steam. If the storage tank temperature or the collector outlet temperature rises above a threshold value when the solar circuit pump is switched on, a stagnation cooler, which is located in a branch parallel to the solar circuit, is switched on and flowed through. In the case of operational cooling, the required cooling capacity, and the amount of heat to be dissipated is by far the highest and at the same time the fluid and all other solar circuit components are subjected to the lowest load.

9.7.5.5 System Pressure Increase

A strategy for load reduction is the choice of a high system pressure. This measure increases the boiling temperature within the collector field and the efficiency, while at the same time the steam production capacity decreases. Due to the higher boiling temperature, the heat load increases and, if hydrocarbons (seals, …) are used, also their ageing. The conflict of objectives is thus: do the solar circuit components or the solar fluid require increased protection?

9.7.5.6 Control Strategies to Reduce Stagnation Times

Control strategies are an attractive method of reducing downtimes, as they do not require any structural changes to the plant. By programming a suitable control algorithm, which is adapted to the specifications of the solar system, the solar pump is switched on also for cooling purposes, which results in an increased pump power requirement. Two strategies are pursued, usually used in combination.

- Collector cooling function—reduction of solar gains during daytime.
- Storage night cooling function—disposal of stored heat during nighttime.

The collector cooling function is integrated in most common solar controllers. It is activated when the storage temperature is between a first limit value of the storage temperature (T_{max1}) and the maximum permissible storage temperature (T_{max2}). If nevertheless the storage tank temperature exceeds the maximum permissible storage tank temperature (T_{max2}) then the solar pump is completely switched off.

Triggered by storage temperature, during daytime in *collector cooling mode,* the solar pump is switched on when the collector temperatures are between a first temperature limit value of the fluid in the collector ($T_{K,max1}$) and the maximum permissible collector cooling fluid temperature ($T_{K,max2}$). The pump is switched off if the collector temperature is lower than the first temperature limit minus a safety temperature interval (usually $5-10°K$). This type of control ensures that the solar circuit pump is run in pulsed operation as soon as an initial limit value of the storage tank temperature is exceeded. By starting the pump operation close to the first evaporation in the collector field, which is given by the temperature ($T_{K,max2}$), the collectors worked at a low efficiency. The reaching of the maximum storage tank temperature (T_{max2}) and thus the occurrence of stagnation is delayed by a certain time. This strategy is combined with a recooling function of the storage tank, which ensures that the tank cools below the temperature (T_{max1}) at night.

The use of the collector cooling function is *only useful if the excess storage heat can also be dissipated*. For this purpose, the collector array is used as a heat sink during nighttime, which is called *storage night cooling.* If the collector temperature drops significantly and measurably below the temperature of the storage tank, which is usually the case during the night, the solar circuit pump is operated, pumping hot water to the collector array. Storage night cooling works best with collectors having large heat losses.

9.8 System Simulation of Solar Thermal Systems

To calculate the system behaviour of complex solar thermal and other thermal systems, various system simulation programs have been developed in recent decades, particularly in chemical and process engineering as well as in power plant engineering. But even with modern computers, a complete three-dimensional implementation of a pipeline network is hardly possible even for a stationary operating condition. Transient effects, such as those additionally caused by temporal changes in the heat sources (e.g. position of the sun, clouds or even failure of active components such as pumps, valves etc.), lead to a further significantly increased complexity, since in such cases, for example, the direction of flow within the circuit with branches is no longer predetermined.

In a system simulation program, therefore, a simplified view of the overall system is given, primarily by using scale separation methods. The easiest way to explain the basic idea of scale separation is to take a pipe as an example. Pipelines usually have a length to diameter ratio L/D that is significantly greater than unity. Starting from the inlet, the flow can be divided in the isothermal case into different sub-segments, such as the segment of the incoming flow and the outgoing flow. Elbows or headers and manifolds, for example, appear at the ends of the pipeline. Once the individual segments have been identified, the respective pressure loss or heat transfer relationships can be applied to the individual sections from standard engineering handbooks, e.g. Kreith and Goswami (2005) or Stephan et al. (2019). Because $L/D \gg 1$ applies, the calculation of these segments can be reduced to a one-dimensional approximation preserving momentum and energy conservation, since the influence of multidimensionality is of secondary importance. For this purpose, the system is divided into so-called nodes.

Thus, the entire pipeline network can either be divided into one-dimensional sections or, in the case of very complex sections where a three-dimensional calculation is absolutely necessary, the result of the one-dimensional calculation is transferred as a boundary condition to a detailed three-dimensional fluid dynamic simulation.

In the following parts, the basic ideas of a system simulation are presented, which mostly use finite difference methods to be able to represent the system dynamics in a time-resolved manner.

9.8.1 Energy Flow with Plug-Flow Modelling

Current system programs often use the finite difference technique for the calculation of the models with the help of Euler methods implemented within the components. In order to achieve a high degree of flexibility, system programs use a distributed independent calculation of individual components ($UNITS$) and do not set up large equation matrices like, for example, computational fluid dynamic (CFD) programs. The advantage of the distributed calculation is that the calculation sequence of the $UNITS$ *is* individually related to the system diagram. This enables, for example, the modelling of the dead time behaviour of pipeline networks as well as an efficient mass flow balancing, since these are non-linear systems.

Especially, in branched pipeline systems conventional solvers offer little efficient possibilities for error control. The plug-flow technique is suitable for modelling stratified storage tanks in order to map the temperature stratification as accurately as possible. Here the storage is divided into homogeneous isothermal zones depending on the input and shifted in time. The segments can be of different sizes and may be enlarged by mixing. In order to map the fluid flow through the line elements, a plug is defined as follows on the basis of a high time resolution:

$$\Delta Q_{Plug} = \dot{M} \cdot c_p \left(T_{Plug} - T_{ref} \right) \cdot h \text{ using } T_{ref} = 0\ ^\circ\text{C}. \tag{9.34}$$

The reference temperature T_{ref} is defined as 0 °C in order to avoid triggering unnecessary calculation operations. The calculation step h therefore, determines the amount of energy of the plug that is exchanged between the hydraulic $UNITS$. A hydraulic $UNIT$, therefore, requires at least two input variables (*INPUTS*) to describe the heat flow, i.e. ΔQ_{Plug} and $\dot{M}$

$$\text{inp}[i_{unit}][1] = \Delta Q_{\text{Plug}}; \text{inp}[i_{unit}][2] = \dot{M}. \tag{9.35}$$

The index i_{unit} identifies the unique $UNIT$ in the simulation script, the second index indicates the channel name of the input. Usually only temperature and mass flow are used in the interface description to recalculate the inlet heat flow of a $UNIT$ (TRNSYS Coordinator, 2017). This calculation step is omitted in the description by the plug flow modelling. There, are other significant advantages of the method:

- The energy can be stored as a discrete profile in the dead time register.
- Branch heat flows in nodes are balanced directly by the *plugs*.

While the discrete temperature profile technique is often used in the piping module to map the dead time effects, the plug is also used directly in the discrete node equation:

$$\begin{aligned} M \cdot c_p \cdot \frac{dT_{node}}{dt} &= \sum_{i=1}^{\bar{n}} \dot{Q}_i; \\ M \cdot c_p \cdot \left(T_{node,i} - T_{node,k-1} \right) &\approx \Delta Q_{plug,in} - \Delta Q_{plug,out} \\ &\quad - U \cdot A \cdot \left(T_{node,k} - T_U \right) \cdot h \end{aligned} \tag{9.36}$$

In the simulation program an integration or a summation over the entire simulation time is carried out to calculate the current node temperature.

$$\begin{aligned} Q_{\text{node}} &= Q_{node,intial} + \sum_{i=1}^{n} \left(\sum_{t=0}^{k} \dot{Q}_{i,t} \cdot h \right) + \sum_{j=1}^{2} \left(\sum_{t=0}^{k} \dot{Q}_{j,t} \cdot h \right); \\ \sum_{t=0}^{k} \dot{Q}_{i,t} \cdot h &= \sum_{t=0}^{k-1} \dot{Q}_{i,t} + \dot{Q}_{i,k} \cdot h; \\ \sum_{t=0}^{k} \dot{Q}_{j,t} \cdot h &= Q_{\text{Plug},j,t} + \Delta Q_{\text{Plug},j,k}; \\ &\text{with } i : \text{branch index and } j = 1, 2 : \text{plugindex}. \end{aligned} \tag{9.37}$$

The Eq. (9.37)a represents the energy conservation of a node, in that the state of charge results from the initial energy and the energy supplied or removed. The balancing is carried out over the entire simulation period to minimize the numerical approximation error. The relation (9.37)b shows the way of calculating these branch energies. In the current time step $t = k$, only the calculation of $\left(\dot{Q}_{i,k} \cdot h\right)$ and addition of a sum value (a field size) stored in the storage is performed. In the case of a branch heat flow (Eq. 9.37c), the *plug* is directly passed and summed via the *plug flow* mechanism without using a multiplication operation. Furthermore, this means that the branch energy, which is emitted from one $UNIT$ as *Plug Flow*, is available as branch entry energy to exactly the following $UNIT$. This keeps the numerical errors (which occur when multiplying and dividing a discretized floating-point number) small. Finally, the node temperature is calculated directly from the charge state Q_{node} if the heat capacity is known, provided there is a linear relationship between enthalpy and temperature, as given in Eq. (9.38):

$$T_{node} = \frac{Q_{node}}{M \cdot c_p}. \tag{9.38}$$

While the load state in a system program is usually marked with a variable, the field of the node temperatures $T_{node}[i_{node}]$ is managed in a field.

9.8.2 Error Analysis in System Simulations

The complexity of solar systems makes it necessary, especially with modular simulation tools, to use precautions for automatic error detection. Here the software concepts often differ. The greatest time expenditure when carrying out a system simulation of a solar system based on modular concepts is usually not spent on the design of the system, as there is a large number of graphical user interfaces for this purpose, which already contain many validated segments and components. A significantly higher effort is required for the error analysis and its localization, especially if complex structures with potential two-phase flows occur. In most cases two fundamental methods are used, which allow a fast error detection (e.g. by simulation abort):

- Mass flow balancing
- System Energy Balancing

Mass flow balancing checks the correct return of the mass flow to the pump in branched hydraulic networks. The basis for this is the restriction to closed systems. System energy balancing involves the evaluation of the reserved output variables of each $UNIT$, where total system losses, gains and internal energy are transferred to the capacities. The background of this concept is a fixed subsystem boundary, which distinguishes between "heat flows in the system" and "heat flows to the system environment".

9.8.2.1 Mass Flow Balancing

In a complete solar system calculation, each closed hydraulic circuit is calculated separately, starting from the pump, by calculating each subsequent component with an already updated $OUTPUT$ *vector.* Convergence problems caused by the arrangement of $UNITS$ are thus avoided in principle. The hydraulic sequence is determined with the aid of a subprogram. It guarantees the actuality of the input variables (*INPUT*) especially in branched piping systems. If, for example, the mass flow is divided at a T-piece (so-called '*flow_diverter*'), it must be recombined in the associated *mixer.* The paired appearance of distributors and mixers, which occurs in every domestic water or heating system, are essential components of system programs.

The mass flow is divided by a *control* variable ('*control*') at the distributor, which is either fixed or supplied by the *controller.* Within a branch, branching may be repeated. The Fig. 9.85 schematically shows a sequence (circled numbers), which is determined by the sequence subroutine. At the branching element, the procedure initially selects the upper connection (4) *at* random until the evaluation of the mixer (5) shows that the second input still has to be updated for the calculation of the outlet mass flow. Consequently, the work will continue at point (6) to complete the second branch. Similar to the parenthesis calculation in mathematics, the "content" of each branch must first be fulfilled before the outlet size of the mixer can be determined. Ultimately, the mass flow which was imposed by the pump must also appear again at the inlet of the pump in the same time step t_k, otherwise the mass flow balance will terminate the simulation at this point.

$$\left|\left(\dot{M}_{out,k} - \dot{M}_{in,k}\right)\right| \geq \varepsilon_{\dot{M}} \Rightarrow \text{termination mass flow balance,} \tag{9.39}$$

where $\varepsilon_{\dot{M}}$ is the maximum permissible parameterizable error of the mass flow.

Dead times in the mass flow of a component or a branch, e.g. the filling of a container, can also lead to abort in the system simulation. Such behaviour can be avoided if an overall mass balance of the system is carried out.

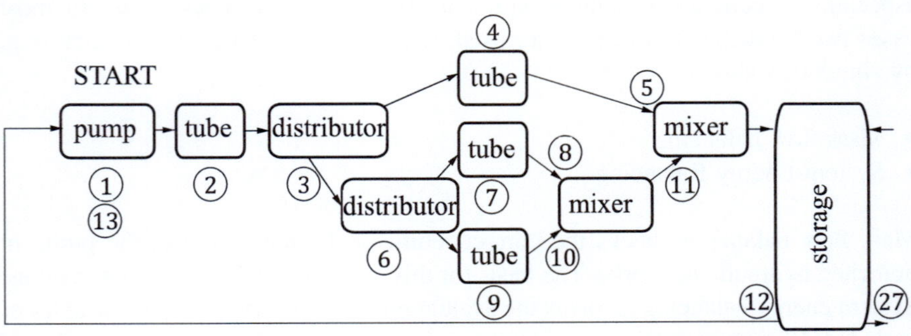

Fig. 9.85 Branching conventions: The hydraulic network is executed during the simulation step, starting from the pump, according to the numbers entered. A branching at the distributor is followed until an outlet mass flow of a mixer can no longer be clearly determined modified from (Lustig, 2002)

9.8.2.2 System Energy Balancing

For further error control, which is indispensable for the simulation of complex networks, the integration of system energy balancing is an important function: The internal and external system energy flows are separated, and the external energy flows are transmitted to the outside via an output vector ($OUTPUT$ vector-"*out*"). For this purpose, special areas are reserved with the help of reserved fields, which indicate the system gain, loss, and stored energy, allowing a global balancing to be carried out.

$$out_{auctual}[i_{unit}][l] \equiv \Delta Q_{gain};\ out_{auctual}[i_{unit}][l+1] \equiv \Delta Q_{loss};$$
$$out_{auctual}[i_{unit}][l+2] \equiv Q_{capacity}. \tag{9.40}$$

The stored energy $Q_{capacity}$ corresponds to the heat content (charge) of all thermal capacities that are in the component (e.g. buffer water) and refers to the already defined reference temperature of $T_{ref} = 0°\text{C}$.

System gains ΔQ_{gain} and losses ΔQ_{loss} are also expressed as energy in the current time step and refer to those heat flows that do not flow within the $UNIT$ network. For example, the heat flow of a collector to the following pipe element is a "system-internal" heat flow, while the heat loss of the collector to the environment must be declared as "system-external". A superordinate procedure checks the system energy at the end of each simulation time step h by balancing profits, losses, and stored energy.

$$\left| Q_{sys,init} + \sum_{i=1}^{unit-max} \left(\sum_{t=0}^{k} \left(\Delta Q_{gain,i,t} - \Delta Q_{loss,i,t} \right) \right) - \sum_{i=1}^{unit-max} Q_{capacity,i} \right| \geq \varepsilon_{Q_{sys}},$$
$$\Rightarrow \text{termination system energy balance.} \tag{9.41}$$

Equation (9.41) the energy balance equation of the thermal system with *unit-max* components. During initialization at the time step $t = 0$, each $UNIT$ outputs the initialization energy contained in the thermal capacities. At the end of the simulation time step $t = 0$, the system initialization energy $Q_{sys,init}$ is formed from the sum of all components. This energy remains constant over the entire simulation period and is always compared with the current system state $\sum Q_{capacity,i}$ and with the cumulated gain and loss energy Q_{gain} and $Q_{loss,}$. If this results in $\varepsilon_{Q_{sys}}$ an error greater than the parameterized error, the simulation is aborted at this point.

If, for example, a pipeline component is incorrectly linked at the output, the system energy balance responds, because the *plug* $\Delta Q_{plug,out}$ that is issued is not declared as a system loss ΔQ_{loss}. A prerequisite of this convention is that only closed hydraulic circuits with a clearly defined system boundary are considered. Therefore, for example, the tap circuit of the domestic water is closed by integrating a return thermostat and a pump. With this approach, the load behaviour can be balanced at the thermostat.

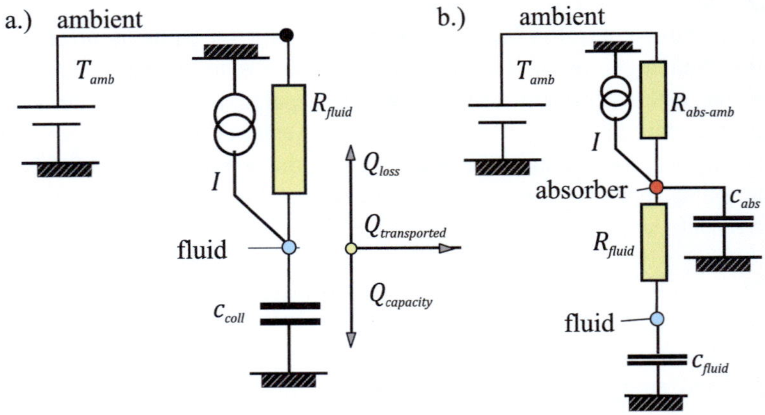

Fig. 9.86 Schematic diagram of an equivalent electrical circuit in a single node model (**a**) and a double node model (**b**). I = solar irradiance

9.8.3 Solar Collector Modelling

Many collector models used in simulation programs are based on the single node model. Electrical equivalent circuit diagrams are used for this purpose, as shown schematically in the Fig. 9.86a. The solar radiation is shown in this equivalent circuit diagram as a current source. Voltages or voltage drops correspond to temperatures or temperature differences. Flows correspond to heat flows. The electrical capacity corresponds to the heat capacity. In this diagram, the latter is a combination of absorber and fluid capacity. This also applies to the loss resistance R_{fluid}, which represents an effective value from the heat transfer between absorber and environment and from the absorber to the fluid.

The Fig. 9.86b shows a more detailed model that divides the collector into two nodes. The absorber node is coupled to the environment via the thermal resistance of the absorber to the ambient ($R_{abs-amb}$). There is also a separate thermal absorber capacity. The second resistance represents the heat transport from the absorber to the fluid. This is largely dependent on the absorber design and the flow conditions in the absorber tube. The ratio of the two resistors leads to the absorber efficiency factor F' with $F' = R_{fluid}/\left(R_{fluid} + R_{abs-amb}\right)$, which has already been introduced in Chap. 6.

The thermal capacity of the collector is considered in the dynamic description of the collector. A distinction is also made here between the 1, 2 and 3-node models, which divide the collector into vertical planes. The 1-node model (cf. Fig. 9.86a) assumes a homogeneous temperature distribution in absorber and fluid:

$$M \cdot c_{coll} \cdot \frac{d\overline{T}_{fluid}}{dt} = A \cdot \left(\eta_0 \cdot i_{glob} - k\left(\overline{T}_{fluid} - T_{amb}\right)\right). \tag{9.42}$$

The Eq. (9.42) describes the thermal behaviour of the collector above the average $\overline{T}_{fluid}$ fluid temperature, considering the inertia effects due to the delayed heating and cooling.

A more detailed description can be found, for example, in the subdivision according to Fig. 9.86b. In such a description, the system is described by two state equations, which are represented by the Eqs. (9.43):

$$\begin{aligned} M \cdot c_{abs} \cdot \frac{dT_{abs}}{dt} &= A \cdot \left(\eta_0 \cdot i_{glob} - \frac{1}{R_{abs-amb}}(T_{abs} - T_{amb})\right) \\ M \cdot c_{fluid} \cdot \frac{dT_{fluid}}{dt} &= A \cdot \frac{1}{R_{fluid}}\left(T_{abs} - T_{fluid}\right) + \dot{M} \cdot c_p \cdot \left(T_{in} - T_{fluid}\right) \end{aligned} \tag{9.43}$$

The discretization of the collector in the direction of fluid flow allows the description of the temperature behaviour even at low flow velocities because a nonlinear rise in temperature can be mapped. On the left side, the Fig. 9.87 shows a schematic diagram of a simple test system of a solar thermal circuit and its essential components. The Fig. 9.87 depicts the transfer of the schematic system into a system simulation program. With the help of system simulation, in which the components and the associated physical models are embedded, it is now possible to determine the time behaviour of the system as a function of the input variables.

In practical systems, the solar thermal circuit is much more complex and includes a storage tank, auxiliary heaters, heat exchangers and many other components and control equipment. The Fig. 9.88 shows a block diagram of a conventional system, as it is used for solar domestic hot water heating in the low temperature range. It consists of a closed solar circuit, usually operated with a water–glycol mixture, in which the collector is located as a heat source, and which feeds the heat via a heat exchanger into the domestic hot water tank of the house. If necessary, this storage water can be heated to a set value specified by the user with the aid of a fossil or electric auxiliary heater and thus be used directly for space heating. Domestic hot water is heated by a heat exchanger which is integrated into the domestic hot water circuit. For adequate control and regulation, a certain number of level, pressure and temperature measuring points are required to ensure the operational safety and functionality of the system under nominal conditions as well as in case of failure of central components.

Especially near the maximum load of the collector, a simple one-dimensional view of the collector and the circulation network is no longer appropriate and a finer local view, especially in the area of the absorber, is required to detect the formation of two-phase fluid states and their propagation in the pipeline network, which ultimately lead to a stagnation.

In such a case, either a CFD calculation must be carried out within a segment (unit), which often exceeds the available computer resources and time, or multi-node models are introduced in the unit, which can describe the internal heat transfer processes in the collector locally. The Fig. 9.89 shows such a multi-node model which, in addition to the axial heat conduction in the fluid and the structures, captures the processes in the circumferential direction with the associated thermal masses (which in the electrical sense are represented by capacitances and mean a time delay) with the corresponding interactions.

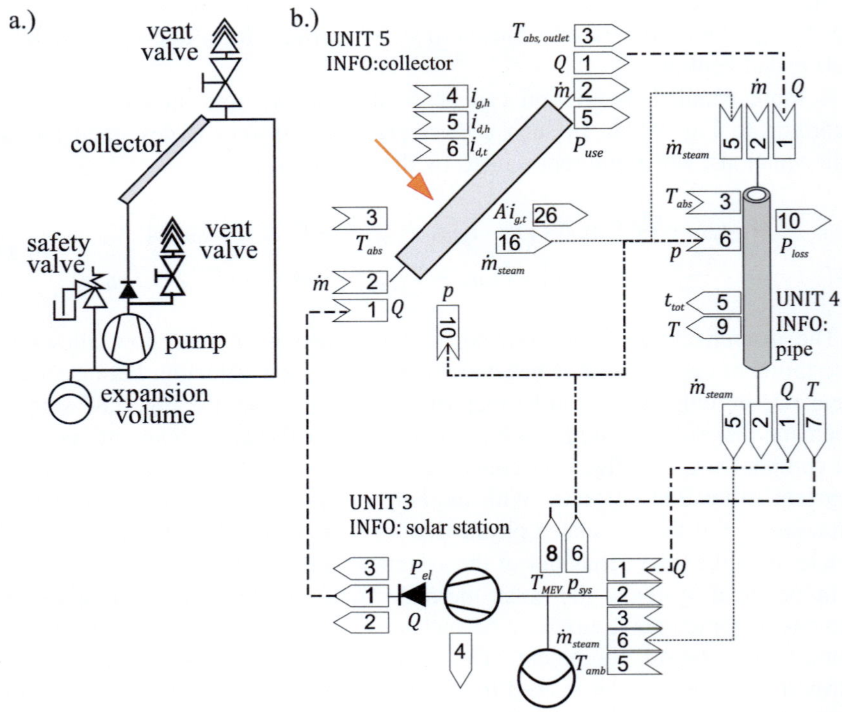

Fig. 9.87 **a** Simplified flow chart of a solar thermal circuit; **b** Figure of the flow chart from **a** in a system simulation program with determination of the sub-areas and the transfer parameters

It must be emphasized that multi-node models require a thorough knowledge of the physics models and correlations to be used in the code or linked via external subroutine. This holds especially for the validity ranges of the correlations implemented. In return, however, the integral modelling allows to quickly execute a large number of parameter variations even when using a conventional personal computer and thereby is the method of choice to identify potential weaknesses in the construction and placement of the components as well as possible weaknesses in the construction of the collector field and much more. With the help of multi-node models, local quantities or temporal gradients can also be resolved. Although the local resolution may be not comparable to the capability of current computational fluid dynamics and thermo-mechanical 3D simulations, the results of an integral modelling can be used as boundary condition for a 3D detailed simulation in case of engineering challenges.

Figure 9.90 illustrates a result of a multi-node calculation in a comparison of measured and calculated temperatures of the low-temperature circuit of a house system (from Fig. 9.88) at two discrete points for several consecutive summer days. The graphs clearly show that during the summer shown here, there were several cases of collector stagnation characterized by fast gradients. The results

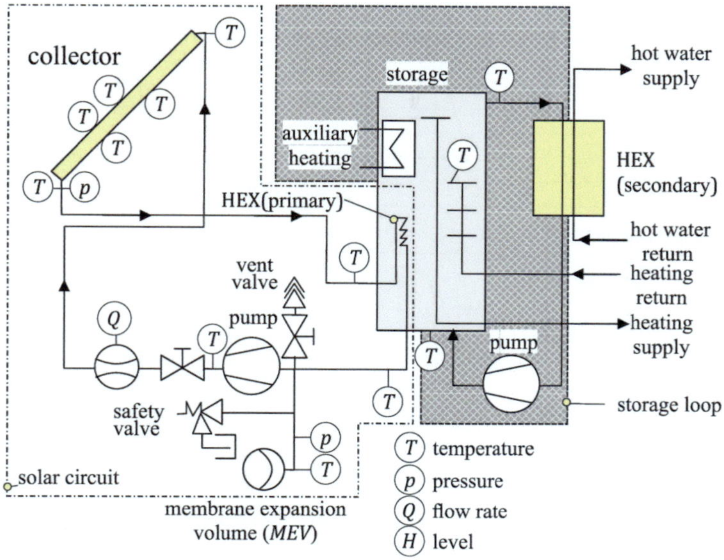

Fig. 9.88 Block diagram of a conventional flat plate collector system in two-circuit design

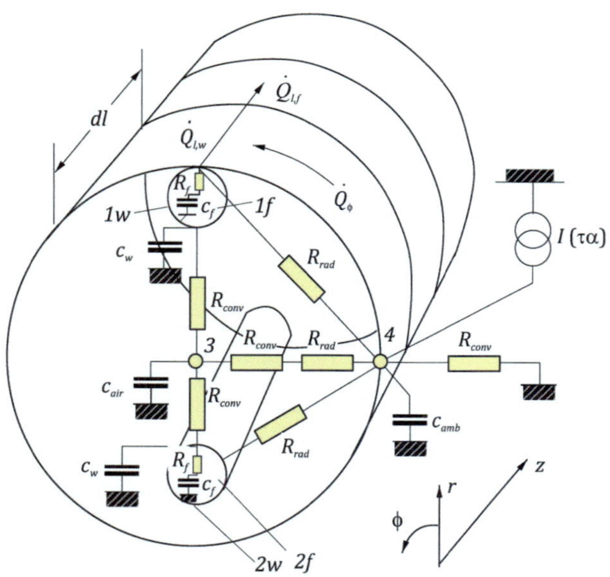

Fig. 9.89 Multi-node model to describe the internal heat transfer process in an evacuated tube collector segment, modified from Lustig (2002)

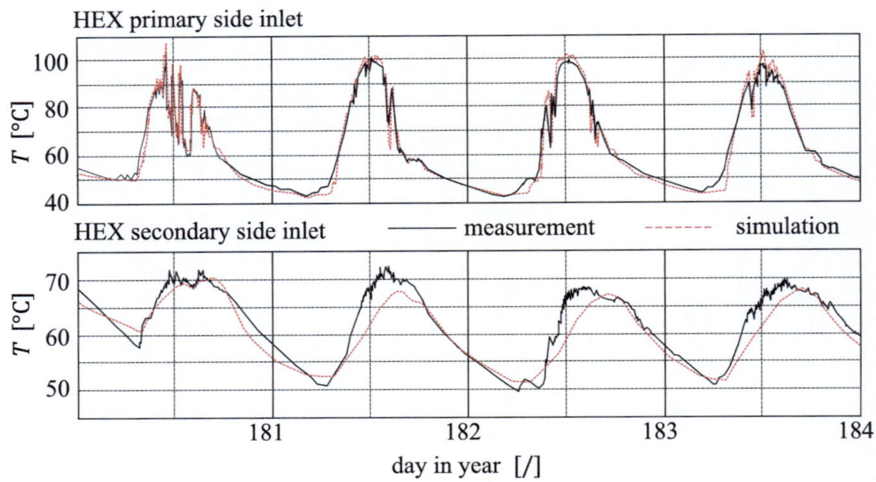

Fig. 9.90 Comparison of measurement and system simulation of temperatures at the domestic hot water storage tank inlet and outlet of the system in Fig. 9.88 on several consecutive summer days; modified from (Lustig, 2002)

of the system simulation obtained with the help of the multi-node model exhibit a good temporal agreement with the measurement results on the primary side. It is clear to see that the primary side of the system is well reproduced, while the secondary side correctly describes the magnitude of the temperature change but the time response with some offset. It is shown here that, even with node models, the user must have very good knowledge of the flow in the heat exchanger in order to be able to predict the time behaviour with high accuracy.

References

Abokersh, M., Vallès, M., Kangkana, S., Cabeza, L., & Boer, D. (2021). Techno-economic analysis of control strategies for heat pumps integrated into solar district heating systems. *Journal of Energy Storage, 42*, 12.

Aelenei, L., Smyth, M., Platzer, W., Norton, B., Kennedy, D., Kalogirou, S., & Maurer, C. (2016). Solar thermal systems—Towards a systematic characterization of building integration. *Energy Procedia, 91*, 897–906. https://doi.org/10.1016/j.egypro.2016.06.256

Arnold, O., Mercker, O., Steinweg, J., & Rockendorf, G. (2016). Efficiency analysis of solar assisted heat supply systems in multi-family houses. In *Proceedings of EuroSun*. Retrieved July 1, 2022, from https://proceedings.ises.org/

Baack, S. (2020). *Heat metering with glycol-water-mixtures*. Technischen Universität Berlin.

Botpaev, R., Louvet, Y., Perers, B., Furbo, S., & Vajen, K. (2016). Drainback solar thermal systems: A review. *Solar Energy, 128*, 41–60. https://doi.org/10.1016/j.solener.2015.10.050

Brandstetter, F. (2008). *Ausbildungsskriptum Solarwärme*. Österreichisches Forschungs- und Prüfzentrum Arsenal G.m.b.H.

Braun, P., & Marko, A. (1996). *Thermische Solarenergienutzung an Gebäuden*. Springer.

Bundesministerium für Gesundheit. (2018). *Trinkwasserverordnung und Legionellen (in German)*. German Government. Retrieved July 4, 2022, from https://www.bundesgesundheitsministerium.de/fileadmin/Dateien/3_Downloads/T/Trinkwasserverordnung/Stammtext_TrinkwV_und_Legionellen_250418.pdf

Conserval Engineering Inc.—SolarWall. (2011). *Solar wall at the Fire and Emergency Services Training Institute (FESTI), ON—Canada—Greater Toronto Airport Authority*. (2. W. Conserval Engineering Inc. - SolarWall, Editor) Retrieved May 3, 2023, from https://commons.wikimedia.org/wiki/File:GTAA_-_SolarWall.jpg

Dell, G., Nagl, M., & Egger, C. (2010). *Solarguide—Manual for the planning of thermal solar systems for multi-family houses (in German)*. O.Ö. Energiesparverband. Austrian Solar Innovation Center, ASiC. Retrieved July 8, 2022, from https://forschung.fh-ooe.at/fileadmin/user_upload/forschung/landingpages/asic/SolarGuide.pdf

DIN 4708. (1994). *Central heat-water-installations; terms and calculation-basis*. Beuth Verlag GmbH. https://doi.org/10.31030/2480131

DIN 4807-2. (1991). *Expansion vessels; open and closed expansion vessels used in thermic systems; calculation, requirements and test*. Beuth Verlag.

DIN EN 12828. (2014). Heating systems in buildings—Design for water-based heating systems.

DIN EN 13831. (2007). *Closed expansion vessels with built in diaphragm for installation in water.* Deutsches Institut für Normung. Beuth Verlag. https://doi.org/10.31030/1381685

DIN EN 4108-2. (2013). *Thermal protection and energy economy in buildings—Part 2: Minimum requirements to thermal insulation*. Beuth Verlag GmbH.

DIN EN ISO 10077-1. (2020). *Thermal performance of windows, doors and shutters—Calculation of thermal transmittance—Part 1: General*. Beuth Verlag GmBH.

DIN EN ISO 13789. (2018). *Thermal performance of buildings—Transmission and ventilation heat transfer coefficients—Calculation method*. Beuth Verlag GmbH. https://doi.org/10.31030/2560190

DIN-EN-410. (2011). *Glass in building—Determination of luminous and solar characteristics of glazing*. Beuth Verlag GmbH.

Duffie, J. A., Beckmann, W. A., & Blair, N. (2020). *Solar engineering of thermal process*. Wiley.

Elci, M. (2018). *Smarte und Dezentrale Solare Fernwärme-Solare Energie- und Systemforschung (in German)*. Fraunhofer ISE, Fraunhofer Verlag. https://publica.fraunhofer.de/handle/publica/282312

F-CHART Software. (2022). F-CHART—solar system analysis and design program. Retrieved July 8, 2022, from https://fchartsoftware.com/

Frank, E., Haller, M., Herkel, S., & Ruschenburg, J. (2010). Systematic classification of combined solar thermal and heat pump systems. In *Proceedings of the EUROSUN international conference on solar heating, cooling and buildings*.

German Building Energy Act. (2020). Law on saving energy and using renewable energies for heating and cooling in buildings (translated from German). (G. M. Justice, Ed.) *Federal Law Gazette* (BGBl. I S. 1728).

Hausner, R. (2005). *Measurement of transparent thermal insulation at the demonstration object Graz Maria Trost (in German)*. Final Report, Institut für Nachhaltige Technologien, AEE-INTEC Arbeitsgemeinschaft erneuerbare Energie.

Hausner, R., Fink, C., Wagner, W., Riva, R., & Hillerns, F. (2003). *Entwicklung von thermischen Solarsystemen mit unproblematischem Stagantionsverhalten*. Berichte aus Energie- und Umweltforschung 9/2003, Impulsprogramm Nachhaltig Wirtschaften. Bundesministerium für Verkehr, Innovation und Technologie Österreich.

Henderson, T. R., Clark, C. R., Marshall, T. C., Hanson, R. L., & Hobbs, C. H. (1981). Heat degradation studies of solar heat transfer fluids. *Solar Energy, 27*(2), 121–128.

Hofegger, C. (2011). Windows doors sun protection plastic windows Ideal 6000 (in German).

Hollands, K., & Lightstone, M. F. (1989). A review of low-flow, stratified-tank solar water heating systems. *Solar Energy, 43*, 97–105. https://doi.org/10.1016/0038-092X(89)90151-5

IEA SHC Task 68 Efficient Solar District Heating Systems. (2020). *IEA solar heating and cooling—Task 68*. https://task68.iea-shc.org/

IKZ HAUSTECHNIK. (2022). Seasonal heat storage systems. In S. V. KG, (Ed.), *The course of the energy turnaround, they are gaining new importance*. Retrieved February 27, 2023, from https://www.ikz.de/detail/news/detail/saisonale-waermespeicher/

ISO 9806. (2017). *ISO 9806:2017 Solar energy—Solar thermal collectors—Test methods*. ISO TC 180.

Jenni, J. (2010). *Das Sonnehaus-Grundlagenund Rüstzeug zum Bauen von Solaranlagen mit hohem Deckungsgrad aus Sicht des Praktikers (in German)*. Jenni Energietechnik AG. Retrieved July 1, 2022, from www.jenni.ch

Kerschberger, A., Platzer, W., & Weidlich, B. (1998). *Transparent thermal insulation: Products, projects, planning information (translated, in German)*. Vieweg+Teubner Verlag.

Kimmel, O., Streicher, W., & Heimrath, R. (2001). *Messung und Analyse von Ursachen und Auswirkungen von Kondensationsschlägen beim Ausdampfen von großen thermischen Kollektorfeldern*. Jubiläumsfonds der Österreichischen Nationalbank.

Köhl, M., Heck, M., Brunold, S., Frei, U., Carlsson, B., & Möller, K. (2004). Advanced procedure for the assessment of the lifetime of solar absorber coatings. *Solar Energy Materials and Solar Cells, 81*(1–4), 275–289. https://doi.org/10.1016/j.solmat.2004.01.041

Kreith, F., & Goswami, D. Y. (2005). *The CRC handbook of mechanical engineering* (2nd ed.). CRC-Press LLC.

Kuhn, T. E., Bühler, C., & Platzer, W. J. (2001). Evaluation of overheating protection with sunshading systems. *Solar Energy, 69*(Supplement 6), 59–74. https://doi.org/10.1016/S0038-092X(01)00017-2

Kuhn, T., Goerdt, W., Wienold, J., & Platzer, W. J. (2000). Complete characterization of complex glazings and shading devices.

Lund, H., Werner, S., Wiltshire, R., Svendsen, S., Thorsen, J. E., Hvelplund, F., & Mathiesen, B. V. (2014). 4th Generation District Heating (4GDH). *Energy, 68*, 1–11. https://doi.org/10.1016/j.energy.2014.02.089

Lustig, K. (2002). *Experimentelle Untersuchungen zum Stillstandsverhalten thermischer Solaranlagen*. Dissertation, Universität Karlsruhe, Fak. Maschinenbau, Karlsruhe.

Marstal Fjernvarme a.m.b.a. (2023). Marstal district heating. Retrieved February 27, 2023, from https://www.solarmarstal.dk/

Marx, R., Bauer, D., & Drueck, H. (2014). Energy efficient integration of heat pumps into solar district heating systems with seasonal thermal energy storage. *Energy Procedia, 57*, 2706–2715.

Meier, F. (2009). Stromsparende Pumpen für Heizungen und Solaranlagen. BINE Projektinfo.

Mesquita, L. C., McClenahan, D., Thornton, J. W., Carriere, J., & Wong, B. (2017). Drake landing solar community: 10 Years of operation. In *ISES solar world congress* (p. 12). https://doi.org/10.18086/swc.2017.06.09

Miara, M., Günther, D., Langner, R., & Helmling, S. (2014). Efficiency of heat pumps in real operating conditions—Results of three monitoring campaigns in Germany. *REHVA Journal* 7–12.

Nielsen, J. E. (2014). The booming market for solar district heating—Opportunities and challenges. In *Proc. SHC conference*.

Nielsen, J. E., & Sørensen, P. A. (2016). Renewable district heating and cooling technologies with and without seasonal storage. In S.-H. G., *Renewable heating and cooling* (pp. 197–220). Woodhead Publishing. https://doi.org/10.1016/B978-1-78242-213-6.00009-6

NRCAN. (2022). *RETScreen—The RETScreen® Clean energy management software*. Retrieved October 24, 2022, from https://www.nrcan.gc.ca/maps-tools-and-publications/tools/modelling-tools/retscreen/7465

Oliva, A., Kim, Y., Lämmle, M. H., Ripka, A., & Gebele, C. (2020). Measurement results of a district heating system with decentralized incorporated solar thermal energy for an energy, cost effective and electricity grid favorable intermitting operation. In I. S. Society (Ed.), *Proceedings of EuroSun 2020* (p. 11). https://doi.org/10.18086/eurosun.2020.02.06

Platzer, W. J. (1988). Solare Transmission und Wärmetransportmechanismen bei Transparente Wärmedämmmaterialien. Fakultät für Physik. Albert-Ludwigs-Universität.

Platzer, W. J. (1993). A simple monthly calculation procedure for transparently insulated houses. In *Proceedings of the 6th international meeting TI6. Transparent insulation technology, 3–5 June 1993*, (pp. 62–65). Franklin Company Consultants Ltd.

Platzer, W. J. (2003). Properties and application of switchable facade technology. In *Proceedings of the 26th ISES solar world congress, 16–19 July 2003.*

Platzer, W. J. (2005). Die Solare Umweltwand mit Transparenter Wärmedämmung (TWD). In E. E.-u. GmbH (Ed.), Internationale Kongressmesse RENEXPO.

Platzer, W. J. (2021). Total performance assessment method for industrial process heat systems. In *Proceedings of the ISES solar world congress 2021* (pp. 1–9). International Solar Energy Society.

Platzer, W., & Goerdt, W. (2000). Solar walls with ventilated transparent insulation. Performance properties, problems and materials. In *2000, Proc. Eurosun, international conference on solar heating, cooling and buildings, 19–22 June 2000.*

Rossiter, W. J. (1985). An investigation of the degradation of aqueous ethylene. Glycol and propylene glycol solutions using ion chromatography. *Solar Energy Materials, 11*, 455–467.

Sayegh, M. A., Jadwiszczak, P., Axcell, B. P., Niemierka, E., Bryś, K., & Jouhara, H. (2018). Heat pump placement, connection and operational modes in European district heating. *Energy and Buildings, 166*, 122–144. https://doi.org/10.1016/j.enbuild.2018.02.006

Scheuren, J. (2008). *Untersuchungen zum Stagnationsverhalten solarthermischer Kollektorfelder*. Universität Kassel.

Shrestha, N. L., Frotscher, O., Urbaneck, T., Oppelt, T., Göschel, T., Uhlig, U., & Frey, H. (2018). Hermal and hydraulic investigation of large-scale solar collector field. *Energy Procedia, 149*, 605–614. https://doi.org/10.1016/j.egypro.2018.08.225

Solites—Steinbeis Forschungsinstitut für solare und zukunftsfähige thermische Energiesysteme. (2022). Retrieved February 27, 2023, from https://www.solites.de/projekte/

Sparber, W., Vajen, K., Herkel, S., Ruschenburg, J., Thür, A., Fedrizzi, R., & D'Antoni, M. (2011). *Overview on solar thermal plus heat pump systems and review of monitoring*. IEA Task 44/ Annex 38 publication/outcome, International Energy Agency.

Stazi, F. (2017). Chapter Five—Passive envelopes. In F. Stazi (Eds.), *Thermal inertia in energy efficient building envelopes* (pp. 175–255). Butterworth-Heinemann.

Stegmeier, T., Linke, M., & Planck, H. (2009). Bionics in textiles: Flexible and translucent thermal insulations for solar thermal applications. Philosophical Transactions of the Royal Society A. 1749–1758.

Stephan, P., Kabelac, S., & Kind, M., Mewes, D. (2019). *VDI-Wärmeatlas* (12th ed.). Springer Vieweg.

TRNSYS Coordinator. (2017). TRNSYS-A TRaNsient SYstems Simulation Program. (U. o. Wisonsin, Ed.). Retrieved July 8, 2022, from https://sel.me.wisc.edu/trnsys/

Tschopp, D., Tian, Z., Berberich, M., Fan, J., Perers, B., & Furbo, S. (2020). Large-scale solar thermal systems in leading countries: A review and comparative study of Denmark, China, Germany and Austria. *Applied Energy, 270*, 114997. https://doi.org/10.1016/j.apenergy.2020.114997

Tyforop Chemie GmbH. (2011). *Datenblätter gebrauchsfertiger, rückstandsfrei verdampfender Spezial-Wärmeträgerflüssigkeit für Solaranlagen mit hoher und höherer thermischer Belastung.*

Valentin Software. (2022). *T*SOL—Software*. Retrieved July 4, 2022, from https://valentin-software.com/produkte/tsol/

Val-Matic Valve & Manufacturing Corporation. (2020). *Silent check valves—Product information.* Retrieved February 27, 2023, from https://www.valmatic.com/

VDI 4655. (2021). *Reference load profiles of residential buildings for power, heat and domestic hot water as well as reference generation profiles for photovoltaic plants.* Beuth Verlag GmbH.

Velasolaris. (2020). *Polysun – Energiesysteme präzise simulieren und effizient Planen.* Retrieved July 4, 2022, from https://www.velasolaris.com/software/

Vogelsanger, P. (2005). Neue Systemideen aus Europa. *NEGST WP1: Next generation of solar systems* (p. 36p). AEE-Intec. Retrieved July 1, 2022, from https://www.aee-intec.at/0uploads/dateien213.pdf

Wang, Y., Shukla, A., & Liu, S. (2016). A state of art review on methodologies for heat transfer and energy flow characteristics of the active building envelopes. *Renewable and Sustainable Energy Reviews, 78*, 1102–1116. https://doi.org/10.1016/j.rser.2017.05.015

Wedel, S., & Bezzel, E. (2000). *Heat transfer fluids for solar DHW systems* (2. Auflage ed.). (D. T. Institute, Ed.) Solar Energy Centre Denmark.

Weiss, W., & Spörk-Dürr, M. (2022). Solar heat worldwide. Global market development and trends 2021. Detailed market figures 2020. In I.-S. Programme (Ed.). *Edition 2022.* AEE—Institute for Sustainable Technologies. Retrieved July 8, 2022, from https://www.iea-shc.org/Data/Sites/1/publications/Solar-Heat-Worldwide-2022.pdf

WELL spol. s.r.o. (2011). Product catalogue.

Welz, C., Hermann, M., Kramer, K., & Thoma, C. (2015). *AIRCOW program.* Fraunhofer ISE.

Welz, C., Maurer, C., Di Lauro, P., Stryi-Hipp, G., & Hermann, M. (2014). Mass flow, pressure drop, and leakage dependent modeling and characterization of solar air collectors. *Energy Procedia, 48*, 250–263.

Wieland-Werke, A. G. (2023). *Heat exchanger—Product information.*

Solar Fields of Concentrating Collectors

10

Abstract

Based on the characteristics of individual collector units, a solar field has to be assembled which is sufficiently large to generate the required thermal power at a sufficiently high temperature. The heat transfer fluid is pumped through a solar field, transporting the heat from the collectors to the demand side heat exchanger or directly to the consumer. Additional optical losses may occur in a solar field due to mutual shading and blocking. Thermal losses, especially at nighttime, and hydraulic pressure losses to be overcome by pumping power, shall be minimized. Changes in the thermophysical properties of single-phase fluids as well as phase change of two-phase fluids have to be considered. The basics of solar thermal loop control are presented, boundary conditions for the operation and maintenance of the components are illustrated, and finally some line- and point-focusing systems are compared with respect to their operating characteristics.

Keywords

Solar field characteristics • Mirror and heliostat field design • Operational strategies • Monitoring • Thermal and hydraulic losses in the connecting piping • Comparison of line and point focusing systems

10.1 Solar Field Characteristics

In this chapter not only single concentrating collectors are being treated, but specific questions related to the performance of a complete large field of collectors are dealt with. There are optical effects, for example from mutual shading of collector field elements. The optical efficiency of a complete solar field can be defined in different ways, depending on the view-point of interest. The thermal field characteristics includes additional thermal losses of connecting piping, of the headers which distribute the heat transfer fluid to collector loops and collect the hot fluid from the collectors. Hydraulic losses due to friction have to be considered in the

R. Stieglitz and W. Platzer, *Solar Thermal Energy Systems*,
https://doi.org/10.1007/978-3-031-43173-9_10

overall design. Large piping diameters lead to small velocities and hence to small pressure losses, but the costs are increasing. On optimization is necessary.

The field of collectors can be arranged in different way, taking possibly given area limitations (the shape of a specific property or restrictions due to topography) into account. The piping structure and especially the field size is a dominating characteristic. A compact size may need less area and hence reduce costs. On the other hand, mutual shading could be the consequence. Also, service roads in large fields are needed operational purposes. For example, washing vehicles have to pass the mirrors of concentrating collectors and therefore the distances between lines or rings of collector elements should allow maintenance operations.

The ground area for installing concentrating collectors preferentially would be a plane, and usually it is horizontal. Then the installation of the concentrating and tracking elements is easier than an installation with variable tilt and height. However, the heliostat field with individual heliostat height is conceivable in a slightly hilly area. More computing power is needed then because the positioning and tracking is not as regular as for a plane. For line concentrating collectors it would be preferential to have a plane tilted towards the midday elevated sun (South in the Northern hemisphere) to reduce cosine losses, but tilted planes on roofs or on a regular hill slope are quite limited in size. Nevertheless, a very small tilt of a collector line could be useful in some cases as it allows a passive draining of the collector line.

The most important area for a solar collector plant is the gross installation area A_{gross}. This includes the solar field itself with the gross solar field area A_{sf} including the necessary area between the solar collectors and the area required for non-solar parts like the storage area, the control center, cooling towers, heat exchangers to the load and—for a power plant—turbine house and electrical installations for grid connection.

The solar collector area characterized by the aperture area A_{ap} and the mirror area A_{mirror}. Generally, the area sizes are different:

$$f A_{gross} > A_{sf} > A_{mirror} > A_{ap}. \tag{10.1}$$

For characterizing the solar field A_{sf} is the best choice, of course. The optical active aperture should be as large as sensible for a given solar field area, but mutual shading and blocking prohibit a too dense spacing of the collector lines or the individual heliostats or dish collectors. The different collector types have a different ability to cover the solar field area.

For parabolic trough collectors PTC, the mutual shading of the different collector lines is the critical restriction which leads roughly to a ratio of aperture to field size of 1:3.

For the linear Fresnel collectors LFC the most compact packing is possible. Mutual shading of lines is less critical as only the shade of a receiver line could have an impact of a neighboring mirror field. The typical ratio between aperture and field is 2:3.

For heliostat field of a solar tower the situation is more complicated. The heliostat field shape and density are dependent on the receiver view factor (cavity receiver versus 360° receiver), and the size of the field, as the radial distance between heliostats grows with the radius due to blocking and shading considerations. For a 360° receiver field a typical ratio between aperture and field is around 1:5 up to 1:6 for a tower of 200 m height and a solar power plant of 100 MW_{el}.

10.2 Line Focussing Systems

Line-focusing collector systems have the largest market penetration of all high-temperature collector systems, such that more than 80% of the world's operating CSP power plants are based on line-focusing collector technologies, of which more than 90% are parabolic trough power plant types according to SolarPaces-Solar Power &Chemical Energy Systems (2022). In the following sections, we focus first on a full description of the PTC variant for high-temperature application, with the aim of calculating the collector efficiency. The optical collector efficiency of an individual collector is essentially calculated as a factorisation of the different effects by introducing collector specific functions (incident angle modifies) in an analytical expression depending essentially on the direction of the incoming radiation. Thermal losses are subtracted. Although both, Parabolic Trough, and linear Fresnel, are line concentrating technologies, the optical paths in the concentrators and also the spatial distribution of optical conversion and heat transfer in the receivers differ. It should be mentioned, however, that the energy performance is not mainly determined by the collector type, but to a high extent by the quality of the individual commercial product or prototype.

10.2.1 Optical Performance of Solar Collector Fields

The optical performance of solar collector fields has to consider several aspects. In line focussing systems the individual collectors are arranged in a long row connected through the heat transfer fluid piping. First of all, there is an interaction of individual collector rows by mutual shading. Also, depending on the length of the rows, the end effects which can be considerable of an individual collector in a collector test loop may become negligible. Mutual shading depends on the collector type, because for linear Fresnel the large reflecting area is mounted in all practical designs on a single height level. The high linearly extended receiver towers may shade possibly some neighbouring rows; however, the mutual interaction of rows otherwise is negligible.

On the other hand, linear Fresnel rows can be arranged in the compact linear Fresnel reflector concepts (CLFR) (Mills & Morrison, 2000) in a way, that primary mirrors might send the concentrated radiation to receivers in neighbouring rows.

10.2.1.1 Mutual Shading of PTC Collector Rows

In a solar field several parabolic trough collectors (PTC) are arranged in series forming a row, and several rows are arranged in parallel to constitute an array.

We´ve seen that a part of the physically available collector aperture area cannot be used due to end losses occurring at high incidence angles, which act similar as cosine losses and cannot be avoided in a single collector row. However, the arrangement of parallel rows of collectors also causes a shadow to fall from one row to the next adjacent row when the sun is low, which reduces the yield. In the following, the size of the shaded area will be determined using trigonometry.

The relevant quantity for the determination of the portion of the shading requires the knowledge of the sun position and contains thus apart from the day of the year (δ), the time (ω), the geographical latitude of the place (Φ), the length (L_{rec}) and width (W) of the aperture of the collector row, the distance between neighboring collector rows (S), the orientation of the collector (γ; $\gamma = 0°$ corresponding to south), the focal length of the collector (f), the dimensions of the receiver (d_{rec}). The optical conversion of the effective solar irradiance to the aperture area also requires knowledge of the material properties (absorptivity-α, transmissivity-τ, reflectivity-ρ), which are also dependent on the incidence angle, but usually this angular dependence is neglected for yield calculations.

Let us consider a PTC field consisting of parallel collector rows, where each row consists of an equal number of identical collectors connected in series. The dimensions of aperture area of each row as well as the centre-to-centre distance between rows referred to as pitch corresponds to that depicted in in Fig. 10.1a. When a collector row is shaded by an adjacent row, let L_S and H_S be the corresponding length and height of the shadow cast on the aperture plane of the shaded row. The shadow area of a rectangular aperture caused by one row on the other always has a rectangular shape, because the aperture planes of both rows are always parallel to each other and rectangular. The orientation of collector field plays a significant role in its performance. It is defined by angle α, the angle made in the horizontal plane between horizontal line due south and projection of trough's focal line on horizontal plane and measured in anticlockwise direction. Figure 10.1b shows the aperture plane of two adjacent rows arranged in a pitch S as seen along their axis of rotation. The height of the shadow cast by one row on the other can be calculated at any time in the following way:

$$H_S = W - S \cos\beta, \tag{10.2}$$

for which only positive values of H_S have to be taken as negative ones do not provide any contribution to shading.

The difference between the tracking angle β, given as the angle between the aperture normal and the local surface normal, and the incident angle should be kept as small as possible. The general expression for β is obtained by minimizing the incidence angle on an inclined surface with any orientation, at any location and at any time of day to be

$$\beta = \tan^{-1}\left(\frac{\sin\Phi\cos\delta\cos\gamma\cos\omega + \cos\delta\ \sin\gamma\sin\omega - \cos\Phi\sin\delta\ \cos\gamma}{\cos\Phi\cos\delta\cos\omega + \sin\Phi\sin\delta}\right), \tag{10.3}$$

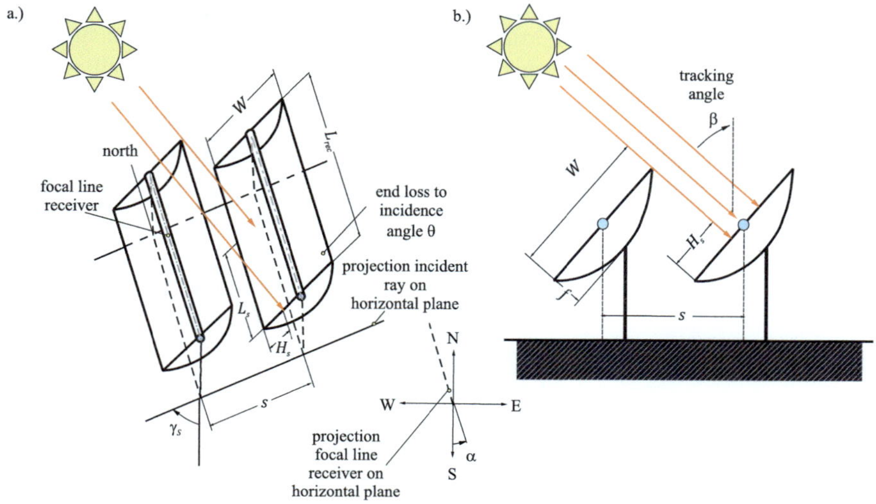

Fig. 10.1 **a** 3D shading in PTC arrays; **b** sketch of PTC illustrating shaded height impact

where γ denotes the surface azimuth angle, which depends on the solar azimuth angle γ_s as well as on the field orientation angle α due to the permanent tracking of the troughs. In the next step, correlations must be determined for the azimuth angle γ and the solar azimuth angle γ_S, which are obtained by using the geometry shown in Fig. 10.1.

$$\gamma = \begin{cases} 90^\circ + \alpha & \text{if } \alpha \leq \gamma_S \leq 180^\circ + \alpha \\ 270^\circ + \alpha \text{ if } & 180 + \alpha < \gamma_S < 360 + \alpha \end{cases}, \tag{10.4}$$

$$\gamma_S = \begin{cases} -\cos^{-1}\left(\frac{\cos\theta_Z \sin\Phi - \sin\delta}{\sin\theta_Z \cos\Phi}\right) & \textit{if } -180^\circ \leq \omega \leq 0^\circ \\ \cos^{-1}\left(\frac{\cos\theta_Z \sin\Phi - \sin\delta}{\sin\theta_Z \cos\Phi}\right) & \textit{if } 0^\circ < \omega < 180^\circ \end{cases}, \tag{10.5}$$

in which θ_Z represents the zenith angle. The shadow height H_S and the shadow length L_S cast by one row on the other is thus given by:

$$H_S = W - S\cos\beta; \; L_S = (L - |S\tan(\gamma - \gamma_S)|). \tag{10.6}$$

In Eq. (10.6) only positive values of the shadow length are relevant. The product of H_S and L_S yields the shaded area A_S. To quantify the shading over a time interval $[t_1, t_2]$ a shading factor (F_{shade}) is introduced, which is given as the ratio of the sum of the shaded areas to the total collector aperture areas available over the same time period.

$$F_{shade} = \frac{\int_{t_1}^{t_2} A_S \, dt}{\int_{t_1}^{t_2} W \, L_{rec} \, dt}. \tag{10.7}$$

The determination of the energy lost due to shading in the collector field requires the inclusion of solar irradiation. The amount of energy incident on an unit aperture area E_{inc} of the trough over a given period of time without shading is

$$E_{inc} = \int_{t_1}^{t_2} DNI \ \cos\theta_i \ dt, \tag{10.8}$$

while the energy available considering shading E_{avl} per unit aperture area is

$$E_{avl} = \int_{t_1}^{t_2} \left(\left(1 - \frac{A_S}{W L_{rec}} \right) DNI \ \cos\theta_i \right) dt. \tag{10.9}$$

The fraction of normal radiation available for optical conversion after accounting for shadowing is defined as the energy availability factor (*EAF*) given by

$$EAF = \frac{E_{avl}}{\int_{t_1}^{t_2} DNI\, dt}. \tag{10.10}$$

We have now developed relationships for all collector rows that experience shading. However, there is now one row that remains unshaded at any time during the day regardless of the pitch, while all behave like the intermediate rows. Mathematically, then, the shading behaviour of these two types of rows over a day can be interpreted as one row remaining completely unshaded and one row behaving like an intermediate row. The expression for the effective shading factor $F_{shade,eff}$ or effective energy availability factor EAF_e after taking these extreme row types into account can thus be written as

$$F_{shade,eff} = \frac{(n-1)}{n} F_{shade}; \quad EAF_e = \frac{((n-1)E_{avl} + E_{inc})}{n \cdot \int_{t_1}^{t_2} DNI \ dt}, \tag{10.11}$$

where n specifies the number of rows in the collector field.

By this assessment different field orientations of the collector field, the pitch between the rows and their daily as well as annual yield can be assessed. Using real weather data for locations a true separation of latitude effects which are created by the geometrical relation between solar orbit over time and orientation of the collector, and climatic effects is difficult. Therefore, a possibility has been proposed to separate these effects by creating artificial weather data from extra-terrestrial data using a specific clearness index for all latitudes, see e.g. Platzer (2019). This work shows that line-focussing collectors generally have a more pronounced seasonal optical performance than solar tower systems. Orientation is very important: For example, the seasonal variation may be reduced appreciably for LFC by rotating it from *NS* to *EW* orientation, but at the cost of higher aperture area.

For a complete power plant performance not only the optical characteristics, but storage size and dumping changes the seasonal variations appreciably for high latitudes, where a high capacity factor shall be achieved in an area with lower DNI. A comparison between technologies and plant alternatives is a complex task which exceeds a purely optical treatment. Nonetheless, some general statements can be made for the PTC considered here:

- The percentage change in the value for the annual EAF_e remains below 1% if $S/W > 2.5$ and $n > 25$.
- The annual effective shading factor $F_{shade,eff}$ decreases for increasing S/W and increases with latitude Φ. The latter because of the relatively lower position of sun in the sky at high Φ.
- The annual $F_{shade,eff}$ for North–South oriented collector arrays ($\alpha = 0°$) is highest at all latitudes, because the magnitude of collector tilt along the day in North–South orientation is larger than for East–West orientation and a greater tilt means a greater shaded height.
- However, the annual EAF_e for North–South oriented field is higher than for East–West mounted collector arrays even if those experience lesser shading. This originates from the fact that the incident angles are lower for North–South oriented fields resulting in a higher annual incident energy per unit aperture area (E_{inc}).

10.2.1.2 Optical Effects in LFC Field Performance

Due to the robust, simple, and cost-effective structure and the lower requirements for maintenance and operation, an LFC collector is also more attractive for mixed applications. Two aspects may speak less in favor of an LFC than a PTC collector.

- optical efficiency at high transversal incidence angles, and
- inhomogeneous heat generation over a day due to the different incidence angle modifiers (as discussed in Sect. 7.2.2.1).

The question of length-specific heat losses does not differentiate the two collector types, as for both types unevacuated receiver tubes with high thermal loss and vacuum tube receiver tubes with low losses are being used in different products. The PTC shows higher optical efficiency than the LFC because the segmented primary reflectors result in significant optical losses due to blocking and shadowing. For short collector lines, the row end optical losses of the LFC are more important than those of the PTC because the ratio of receiver height (h_{rec}) to collector length (L_{col}) in linear Fresnel receivers (except for power plant applications) is larger than for PTC. For a short 100 m collector and receiver height 6–8 m the ratio for LFC would be in the order of 0.1, which increases the shaded portion of the receiver, especially at low sun angles (see Sect. 7.2.2.3, Eq. 7.39). Because there is a longitudinal and a transversal $IAM(\theta_T, \theta_L)$ in the LFC, which are both smaller than one (in contrast to the PTC), there is intrinsically a smaller optical efficiency than in the PTC, which is especially noticeable at higher geographical latitudes. On the other hand, the row-shading effect of PTC rows also has a strong impact on the optical efficiency for low sun altitude in the mornings and evenings.

With respect to the material specific properties of the collector system, expressed by the reflectivity of the mirrors (ρ_{pr}), the transmissivity of the mostly borosilicate used glass (τ_{rec}) as cover of the secondary reflector or in form of a hull enveloping the selective absorber tube and finally the absorptivity of the receiver the (α_{rec}), the data of a LFC receiver correspond almost identically to that of a parabolic trough collector (PTC), so that any improvement of an LFC collector is associated to the design of the mirror-receiver configuration. In principle there are four measures to increase LFC collector efficiency:

- increasing absorber thermal performance,
- design of the primary reflector shape and arrangement,
- design of the secondary reflector, and finally
- reduction of optical losses.

The individual measures will be briefly outlined below, details of which can be found in the literature listed.

Compact Linear Fresnel Reflector Concept

The primary reflector is a key component of the LFC system because it redirects the solar rays to the absorber or secondary reflector, and therefore its structure determines the final optical efficiency. Innovative primary reflector designs aim at minimizing shadowing and blocking effects, such as the Compact Linear Fresnel Reflector (CLFR) concept, where the solar rays are directed to two linear receivers, without compromising the ground use efficiency as illustrated in Fig. 10.2. Depending on whether the receiver is two-sided or one-sided, this elimination of shading and blocking effects of the CLFR may lead to losses in the concentration ratio.

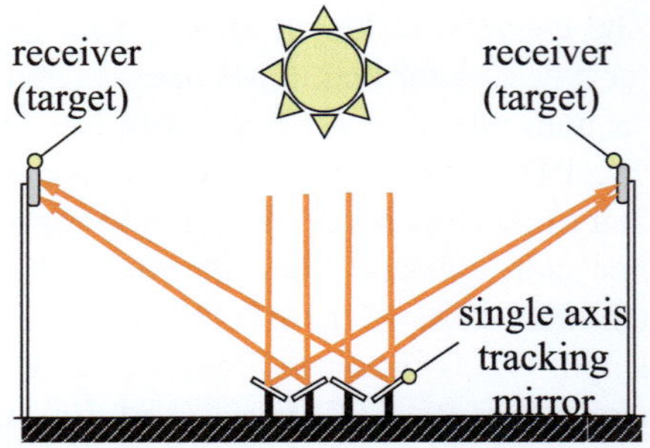

Fig. 10.2 Schematic diagram of the CLFR concept after Mills and Morrison (2000) consisting of interleaving of mirrors without shading between mirrors

10.2.2 Heat Transfer Fluid System

Heat transfer fluid selection can involve complicated, multi-dimensional decisions where factors such as thermal stability, pumpability, pressure requirements, and more must be weighted in an effort to achieve the optimum balance of performance and economy in a particular system.

Relevant quantities for the operation of a solar thermal based system are

- Hot/cold cycle operation temperature limit. The usable temperature difference essentially determines the efficiency of thermodynamic power conversion processes in electrical power generation as well as the associated economical costs (system pressure, pumps/compressors, heat exchangers, integration of thermal storage). While the upper limit is usually determined by the thermal stability of the fluid (either by dissociation, the lower operating limit is given either by the solidification temperature or by the pumpability).
- Kinematic viscosity ν (m^2/s), specific heat capacity c_p(J/(kgK)), density (kg/m^3), specific heat conductivity k (W/(mL)) and thermal (volume) expansion coefficient β (1/K) define the design of the thermal transfer units, where the first three parameters essentially determine the dimensioning of the active components such as pumps, valves, compressors, etc. in the plant, and the thermal conductivity determines the design of the heat transfer interfaces (heat exchangers). The volumetric expansion coefficient determines the pressure range of the plant.
- Other thermophysical quantities like vapor pressure are not only of operational interest (formation of two-phase flows, cavitation, etc.), but should be considered for environmental or health aspects (as oral/skin or water toxicity).
- Interface compatibility with structural materials and chemical potential. Except for pure heat transfer fluids many used solar fluids are binary/ternary mixtures with additives preventing either unprotected freezing or corrosion of the structure materials. Scaling (oxidation) and cracking (overheating) can very quickly consume the protective components of the solar fluid cracks out. In case of aquatic solution, the excess acidification of the fluid is corrosive and leads to the formation of tarlike decomposition products, which are no longer soluble and can lead to adhesions within the solar circuit up to the destruction of the solar system. But also, pure liquids may exhibit corrosion potential in combination with structure materials to their electrochemical potential or dissolved impurities and gases. With regard to chemical potential, flammability, miscibility with water/air and reactions with other operating media, such as lubricants, are relevant.

All these factors must be considered when designing the system in terms of performance, service life, availability, and reliability in order to develop an economical and sustainable system. Naturally, the choice of the heat transfer medium is one of the fundamental decisions in the design of a solar thermal functional unit, and therefore the aspects mentioned above, related to the thermophysical limitations, play a prominent role, since the properties can only be changed marginally and with considerable development effort. For example, a high specific heat capacity in combination with a suitable temperature span allows small apparatus dimensions with a simultaneously high transferred energy quantity as well as a high energy storage density (kJ/m^3) and, with suitable thermal insulation, low heat losses and thus long bridging times. A high specific thermal conductivity ensures not only a fast transfer of heat at small temperature differences, which enables rapid loading and unloading of a sensible heat storage, but also compact component size and hence a smaller apparatus. Since the transfer of fluid from the place of optical conversion to the demand side usually requires a pumping device, the kinematic viscosity plays a role, since this significantly determines the pressure loss in the pipework.

Due to the prominent role of the heat transfer fluid, the applicable temperature range of the different heat transfer fluid types is first highlighted, which also has a significant influence on the choice of operational instrumentation (Sect. 10.4) and central components like pumps or sensors. In the following subsections, a discussion of the individual heat transfer fluid takes place with regard to thermophysical properties, specific required design as well as operational measures.

In principle all fluids allowing for a heat transfer from a hot surface can be used as heat carriers in solar thermal applications., however, the type of applications limits the utilization of specific classes as the Fig. 10.3 show for the operation temperature range of different fluids. In case of low temperature applications air, water/water glycol mixtures as well as silicone or synthetic organic fluids are favoured, mainly because of their flexibility with respect to various applications ranging from simple provision of space heating to low-temperature process heat, or even electrical energy generation using ORC or Kalina processes.

When a certain temperature limit is exceeded, either an evaporation occurs associated with significant specific volume change or pressure rise, or the thermal stability limit is surpassed, which leads to an irreversible change in the fluid properties or even to a decomposition of the heat transfer fluid. By adding additives or nanoparticles, this limit can be shifted upwards or downwards by several tens of degrees, but fundamental limitations of the base fluid cannot be removed. At temperatures higher than 400 °C, as targeted to achieve high thermodynamic efficiencies, either liquid salts, liquid metals, or compressed gases such as helium, supercritical CO_2 are the means of choice, whereby liquid metals such as salts require, with few exceptions, trace heating of all pipework as well as specific operating strategies to prevent freezing of the fluid at ambient temperatures. On the other hand, gases require correspondingly high pressures (>7.5 MPa) and correspondingly high specific compression work. In addition, liquid salts are subject to an upper temperature limit given by thermal dissociation, while for liquid metals the upper limit is given by the liquid-gaseous phase transition.

10.2.2.1 Methods for Comparison of Heat Transfer Fluid Performance

The choice of the heat transfer fluid is a-priori not prescribed and most commercial system providers rely on traditional technologies available on the market along

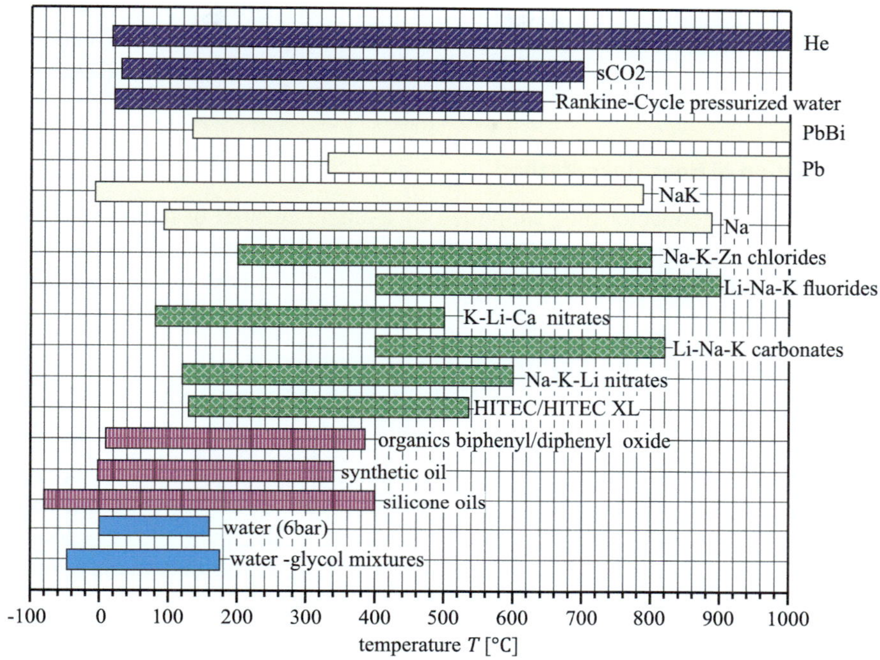

Fig. 10.3 Typical operational ranges of various heat transfer fluids in solar thermal applications

the entire heat transfer chain including the power conversion system and thermal storage technology. Although the performance of heat transfer fluids is determined to a large extend by the thermo-physical properties, not only a single property of the fluid, but a combination of the properties is of relevance. To evaluate the effects of combined thermophysical properties, researchers have suggested various figures of merit (FOMs), where the earliest date back to Mouromtseff (1942) or Bonilla (1957) combining the thermo-physical quantities ρ, c_p, μ and λ to a single figure of merit (FOM) either focussing on the heat transfer or on the pumping requirements. Since the performance of solar heat transfer fluids is determined by the combined effects of the heat capacity of the fluid, the convective heat transfer from wall to fluid, and the hydraulic pressure loss characterized by pumping power, an approach is required incorporating all three aspects. In the last decade essentially two methods have been developed:

- Optimization of heat sinks based on minimization of pumping power introduced by Lenert et al. (2012).
- Minimum system entropy production by Zhang and Li (2017).

By both methods a figure of merit (FOM) argument in technical terms is formulated allowing to evaluate the performance of different solar thermal heat transfer fluids at the same boundary conditions. Both approaches are based solely on engineering and thermal–hydraulic considerations and do not incorporate any operational limitations, material-specific or economic considerations into the outcome. For the sake of simplicity both methods are subsequently sketched by the example of a single-phase flow.

Optimization of Heat Sinks Method

Consider a solar collector tube simplified by a single channel of diameter D and length L, exposed to a uniform heat flux $\dot{q}$ throughout the entire length and neglecting temperature distribution inside the solid wall, then the heat stored in the fluid is given by.

$$\rho\, u_o A\, c_p\left(T_{l,o} - T_{l,i}\right) = \dot{q}\ P\ L, \tag{10.12}$$

in which A and P are the cross-section and wetted perimeter, ρ and c_p the density and specific heat capacity, u_o the mean velocity and $T_{l,o}$ and $T_{l,i}$ the mean liquid temperatures at tube in- and outlet. The limitation of a solar collector, however, is either given by the thermal stability limit of the fluid or the selective absorber coating on the outside of the tube wall$T_{w,out}$. The maximum temperature difference between the tube wall and fluid inlet $\Delta T_{max} = \left(T_{w,o} - T_{l,i}\right)$ can be related to the heat flux simply by rewriting in the way $\left(T_{l,o} - T_{l,i}\right) = \left(T_{w,o} - T_{l,i}\right) - \left(T_{w,o} - T_{l,o}\right) = \Delta T_{max} - \dot{q}/h$. The pumping power to overcome viscous friction P_v is given by the product of the pressure loss Δp along the duct and the volumetric flow rate through it ($u_o A$). Normalizing the pumping power with the thermal power and introducing dimensionless pumping (Π) and thermal loads (γ) yields:

$$\overline{P}_v = \frac{\Delta p u_o A}{\dot{q}\,P\,L} = \frac{1}{8}\frac{\Pi}{\Theta}\frac{f\,L^3}{D^3}\mathrm{Re}^3 \text{ with } \Pi = \frac{\rho \nu^3}{kL^2\Delta T_{\max}};\ \gamma = \frac{\dot{q}\,L}{k\Delta T_{\max}}, \tag{10.13}$$

in which f is the Darcy friction factor, k the heat conductivity of the fluid and ν its kinematic viscosity. Inserting Eq. (10.13) in relation (10.12) and introducing a normalized temperature rise $\Theta = \left(T_{l,out} - T_{l,in}\right)/\Delta T_{max}$, yields:

$$1 = \frac{D}{L}\frac{\gamma}{Nu} + \Theta. \tag{10.14}$$

The normalized temperature rise can be expressed as function of the Reynolds number and Prandtl number in the way $\Theta = 4\Upsilon/(RePr)$. Assuming a turbulent pipe flow engineering relation for the friction factor f and the Nusselt number Nu can be estimated, e.g. $f = 0.184Re^{0.2}$, $Nu = 0.023Re^{0.8}Pr^{0.4}$. Substituting all relationships into the equation for dimensionless pumping power (Eq. (10.13)), it modifies in the following way.

$$\overline{P}_v \sim \left(\frac{\Pi\gamma^{2.4}}{\mathrm{Pr}^{1.6}}\right)\left(\frac{1}{\Theta^{0.4}(1-\Theta)^3}\right). \tag{10.15}$$

Since applied heat flux $\dot{q}$ and the maximum temperature difference ΔT_{max} have been assumed fixed the normalized pumping power is determined by the thermo-physical properties of the fluid and the Reynolds number. Optimizing the normalized temperature rise Θ by means of a search of the extreme value from the derivative of the second term of Eq. (10.15) to Θ, yields for this simple case of tube $\Theta = 2/17$. Inserting this value in normalized pumping power yields

$$\overline{P}_v \sim \left(\frac{\mu^{1.4}}{\rho^2 c_p^{1.6} \lambda^{1.8}}\right)\left(\frac{\dot{q}^{2.4}}{\Delta T_{max}^{3.4}}\right) L^{0.4}. \tag{10.16}$$

Thus, the normalized pumping power and thereby the performance of heat transfer fluids is determined by the combination of fluid properties expressed by the first term of Eq. (10.16). Since the normalized pumping power needs to be minimized, the figure of merit (FOM) of solar thermal heat transfer fluids becomes.

$$FOM = \left(\frac{\rho^2 c_p^{1.6} k^{1.8}}{\mu^{1.4}}\right). \tag{10.17}$$

Of course, this is a rather simplified example based on one single engineering-based design limitation, but it can be extended to entire piping networks with several constraints requiring then numerical solution.

Minimum System Entropy Production

The minimum system entropy production ansatz aims to find a relationship of the system entropy production to all thermophysical fluid properties at given conditions, which jointly determine the heat transfer, pressure loss and thus the entropy production. In contrast to the previous method the entropy is chosen as quantity to be minimized. Consider a solar circuit as illustrated in Fig. 10.4, where the fluid enters a collector row (col) with the inlet temperature $T_{l,i}$ heated up by the solar irradition $\dot{q}$ to $T_{l,o}$ and entering a heat exchanger (hex) to cool down the fluid again to $T_{l,i}$, in which the heat $\dot{Q}_o$ is extracted. An optimal heat transfer fluid (HTF) should design the heat transfer so that the outer wall temperature, $T_{w,o}$ of the solar collector does not exceed the allowable temperature at any operating point and at the same time wall temperature, $T_{hex,i}$ in the delivering heat exchanger to the consumer side should not be too low, since $T_{hex,i}$ represents the maximum temperature of the high-temperature heat storage side of the useful thermal cycle and thus significantly affects the thermodynamic efficiency. For an optimal design, the required pumping power should be as small as possible in order to transfer the thermal energy collected by the solar collector to the working fluid of solar circuit. The method assumes that the dimensioning of the piping and heat exchanger with the thermophysical properties of the heat transfer fluid are sufficient to ensure the heat transfer rate $\dot{Q}$ per unit time for the solar thermal system.

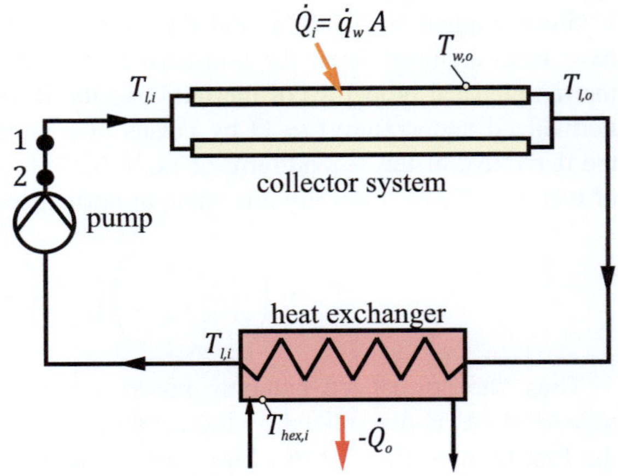

Fig. 10.4 Sketch of a solar thermal circuit with a heat transfer fluid to illustrate the minimum system entropy production ansatz

Using the second law of thermodynamics, the entropy production ($\dot{m}\ s_2$) rate for a steady state operation from point 1 to point 2 along the entire circuit as a function of the respective heat transfer fluid can be determined as follows:

$$\dot{m}\ s_2 = \dot{m}\ s_1 + \int\limits_{col} \frac{d\dot{Q}_i}{T_{w,o}} + \int\limits_{hex} \frac{d\dot{Q}_o}{T_{hex,i}} - \frac{\dot{m}}{\rho}\frac{1}{T_m}\Delta p + \dot{m}\ s_{gen}, \tag{10.18}$$

in which T_m denotes the average temperature of the liquid $T_m = \left(T_{l,o} + T_{l,i}\right)/2$, ρ the specific mean density, Δp the pressure loss along the loop and s_{gen} the generated entropy. Assuming a counter-current heat exchanger and constant heat flux $\dot{Q}_i$ as input source the integrals can be simplified to ratios of the total heat rate and the average temperatures of the walls being $\dot{Q}_i/T_w$ and $\dot{Q}_o/T_{hex}$, in which T_w and T_{hex} are the average values for the collector and the heat exchanger walls. In a closed loop from 1 to 2, the entropy at the beginning and at the end point, i.e. $\dot{m}s_2 = \dot{m}s_1$, is equal, so that the equation for the generated entropy $\dot{S}_{gen}$ has the following form:

$$\dot{S}_{gen} = \dot{m}\ s_{gen} = \frac{\dot{Q}_o}{T_{hex}} - \frac{\dot{Q}_i}{T_w} + \frac{\dot{m}}{\rho}\frac{1}{T_m}\Delta p = \frac{\dot{Q}_o}{T_{hex}} - \frac{\dot{Q}_i}{T_w} + \frac{\dot{V}}{T_m}\Delta p, \tag{10.19}$$

in which $\dot{V}$ is the volumetric flow rate in the loop. The Eq. (10.19) illustrates that the entropy production is composed of the two heat transfer units (heat exchanger and collector) and the viscous dissipation in loop piping network, which must be provided by the pump. Simplifying the analysis by assuming the absence of heat losses yields that $\dot{Q}_o = \dot{Q}_i = \dot{Q}$. Considering the heat transfer processes in the solar collector field and in the heat exchanger, there are:

$$\dot{Q} = A_w\ h_w\left(T_w - \left(T_{l,o} + T_{l,i}\right)/2\right) \text{ and } \dot{Q} = A_{hex}\ h_{hex}\left(\left(T_{l,o} + T_{l,i}\right)/2 - T_{hex}\right), \tag{10.20}$$

where A_w and A_{hex} denote the heat transfer surface areas of both unit and h_w and h_{hex} the corresponding heat transfer coefficients. This allows the temperatures T_w and T_{hex} to be expressed as a function of the heating power $\dot{Q}$. Finally, the volume flow $\dot{V}$ as a function of the heating power must be determined, which results from the relationship $\dot{Q} = \rho \dot{V} c_p (T_{l,o} - T_{l,i})$. If these relations are inserted into the Eq. (10.19), the result is:

$$\dot{S}_{gen} = \frac{\dot{Q}}{(T_{l,o} + T_{l,i})/2 - \dot{Q}/(A_{hex}h_{hex})} - \frac{\dot{Q}}{(T_{l,o} + T_{l,i})/2 + \dot{Q}/(A_w h_w)} + \frac{\dot{Q}\Delta p}{0.5\rho c_p (T_{l,o} - T_{l,i})(T_{l,o} + T_{l,i})}. \quad (10.21)$$

In addition to the three components of entropy production, Eq. (10.21) clearly shows that large values of the heat transfer coefficient h_{hex} and h_w lead to lower entropy production, while the pressure drop Δp always contributes to higher entropy production. It shows that improvement of heat transfer either by surface area increase or better properties of heat transfer fluid leads temperature T_{hex} closer to T_w, which increases thermodynamic efficiency. The pressure drop in the total system includes the losses of the heat exchanger and piping, which can be additively cumulated along the pipe circuit. With the volumetric flow rate of the heat transfer fluid known, the pressure losses in the piping segments can be determined depending on the flow type (laminar or turbulent). The next step of the procedure is to represent the heat transfer coefficients and the pressure drop coefficient as a function of the Reynolds number and the Prandtl number with the help of pressure drop and heat transfer correlations recommended for the different fluid types, as it was also used in the first described procedure. Substituting the relationships into the equation for the entropy generation rate then provides a relationship between the heat output, temperatures, and a combination of the thermophysical parameters of each heat transfer fluid, which has a similar composition as the FOM shown in Eq. (10.17). This relation is then analysed to search for a local minimum of entropy generation defining the most suitable operation temperature window for the liquids considered. By this approach the system can be analysed in much more detail, however, the effort is considerably greater especially since in the example presented here a set of simplifications were used to represent the logics. Thus, this methodology requires numerical procedures for the investigation of realistic complex solar thermal circuit networks.

10.2.2.2 Properties of Individual Heat Transport Fluids (HTF)

The task of the heat transfer fluid is to absorb the radiation converted into heat in the solar absorber and to transport it with the lowest possible pressure and heat loss to the consumer, which generates heat, cold or exergy in the form of electricity. As described above, the thermophysical properties of the fluid play an excellent role in achieving high thermodynamic efficiency, but technoeconomic factors such as cycle stability, aging, material and environmental compatibility, safety, operating behaviour, price, and maintenance costs are also particularly relevant for economic efficiency and are essentially determined by the purpose or type of application, and in some cases differ fundamentally from one fluid to another.

Water and Steam (DSG)

Pure water is the most suitable heat transfer medium, as it is available in almost unlimited quantities, is environmentally friendly, non-toxic, and cheap, and has one of the highest specific heat capacities ($c_P(T = 50\ °C) = 4170$ J/(kgK)). While in low temperature collectors liquid water is a possible HTF, for higher temperatures even with increasing pressures the two-phase region of the fluid water is reached. Then a phase change occurs in a two-phase region with phase coexistence. Above the critical pressure of $p_{crit} = 22.1$ MPa the supercritical region is entered, where the distinction between liquid and gas is vanishing.

Due to its single-phase nature, pressurized water requires little compression work and at the same time exhibits high enthalpies of vaporization, so that with suitable steam superheating, large turbine work is possible and thus, by means of a Clausius Rankine process, also high thermodynamic efficiencies. Unfortunately, however, water has only a limited temperature range due to its critical point ($T_{crit} = 374.12$ °C) at $p_{crit} = 22.1$ MPa. Therefore, a direct utilization in a boiling water collector requires a priori high pressures and a direct storage of sensible heat, e.g. in the form of steam drums at Fresnel or Parabolic trough power plants to bridge solar irradiation fluctuations occurring for instance during the passage of cloud fields, is only possible on a small scale at simultaneous high pressure. Larger power plant concepts that are to operate economically thus require an indirect thermal storage solution.

For use in boiling water collectors, water-steam applications, and in all Clausius Rankine-based power plant cycles, the water-steam relationships as a function of pressure and temperature are relevant. The corresponding state data are freely available in the central database of the IAPWS (2021) and almost all commercial program packages use the correlations proposed there.

The water-chemical operation of pressurized water-based power plant processes requires not only a tight, closed system that is degassed, but also cooling water conditioning, because both temperature levels and transients have an influence on precipitation, which leads to fouling (worsened heat transfer), but also on corrosion phenomena in the steam- and water-bearing plant components. Acidic fluids are highly conducive to corrosion, so that only alkaline or neutral operation is possible. In the alkaline mode, the condensate and feed water are inoculated with an alkalizing agent (usually ammonia in the case of once-through boilers). A pH-Value pH > 9 reduces the occurrence of large-scale corrosion. Oxygen dissolved in the fluid leads to the formation of magnetite, which forms an effective protective layer on steels against further corrosion. This protective layer is only damaged by high mechanical and thermal alternating stresses (frequent start-up and shutdown with large temperature gradients). At these points of damage, the protective layer regenerates, resulting in material erosion due to corrosion. If the feed water is almost free of salt, the operating fluid can be conditioned with oxidizing agents (so-called neutral operation). With the addition of oxygen or hydrogen peroxide, a protective hematite top layer is formed over the magnetite layer by oxidation. Hematite is more resistant compared to magnetite. However, these top layers have the negative effect that they inevitably lead to the occurrence of stress corrosion cracking. The pre-requisite for this is the presence of critical conditions in terms of mechanical stress, fluid properties and the materials used. Stress corrosion cracking is not limited to the water- and steam-carrying boiler components but is a phenomenon that occurs in the entire water-steam cycle and is often underestimated.

Air and Air/Particle Mixtures

Technically, there are many advantages to using air as a heat transfer medium. First of all, the price, the availability, the absence of toxic hazards speaks in favour of air. If properly designed, there is no danger of frost and air can be used directly in the heating process, e.g. for ventilation and drying, without a heat exchanger. A potential system stagnation, for example if no heat can be extracted from the collector because the heat accumulator has reached its maximum temperature, can be easily controlled. No dangerous system conditions such as pressure surges due to overheating can occur in the collector supply lines or in the heat network. Leakages in the ventilation pipe system also do not cause any structural damage as is possible with water or other pressurized systems. Especially this advantage is decisive for the application in the facade. Even with low solar radiation as in winter and with sub-zero temperatures, the solar air collectors can be operated in the air preheating as a support for the building heating. Due to the low heat capacity, air collectors reach the desired operating temperature very quickly. Large solar systems can be easily and inexpensively implemented, and no cost-intensive safety systems or control strategies are required, such as emergency coolers. The lightweight design of solar air collectors is ideal for lightweight buildings. Air as a heat transfer medium itself does not require maintenance, aging is not an issue, only collector circuit components such as air filters need to be checked or cleaned at regular intervals. With an air–water heat exchanger, a simple and safe system integration into existing hot water systems can be realized—and with all the advantages of air.

Another positive aspect is that the thermophysical properties hardly change over a wide temperature range, so that the change in the Prandtl number hardly plays a role over a wide temperature and pressure range. Figure 10.5 shows the most important thermophysical properties of dry air up to 900 °C for various pressures, from which it can be seen that a temperature dependence of the properties except for pressure dependent density is only visible at very low temperatures, which are hardly relevant for solar thermal energy. Conversely, this also means that a compression of the air at the same flow velocity only increases the mass transport through the flow channel but has only a marginal influence on the heat transfer itself. Thus, the heat surface power density that can be dissipated from a heated surface using air alone as a heat transfer medium is limited to relatively low values.

Nevertheless, compressed air is again finding its way into some newer concepts, including solar towers with so-called particle receivers, in recent years. In these concepts, air serves primarily as a transport medium and for fluidizing particles.

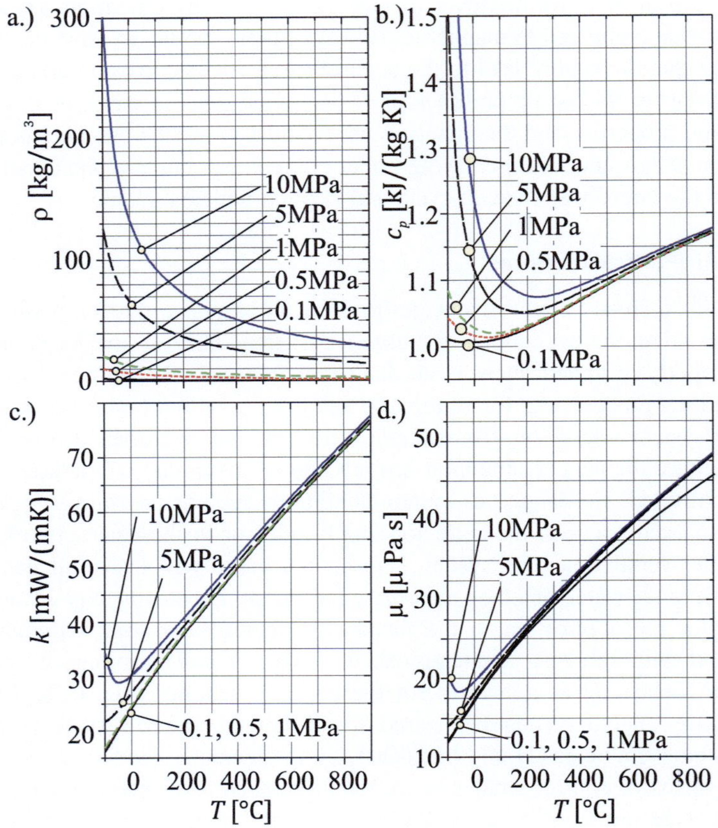

Fig. 10.5 Thermophysical properties of dry air at various pressures as a function of the temperature. **a** specific density ρ, **b** specific heat capacity c_p, **c** specific heat conductivity k and **d** dynamic viscosity μ

There are two basic concepts here, that of gas/particle flow in the direction of gravity, and the so-called fluidized bed concept. By specific selection of the thermophysical properties of the particles (mostly ceramics, sand, or similar, which have a higher thermal conductivity than air) and a suitable size distribution, sufficient heat removal from the thermally highly loaded surface can be achieved, more on this in the section on high temperature receivers.

In the fluidized bed, a bulk of solid particles is loosened and carried by an upward flowing gas or liquid (generally: fluid) to such an extent that the solid layer as a whole exhibit a liquid-like behaviour. For this purpose, the gas volume flow must exceed a certain minimum velocity, the so-called loosening velocity, because otherwise it moves through their cavities of the particle bed without changing the packing structure. Above the minimum fluidization, the gas fraction exceeding the minimum fluidization permeates the layer in the form of practically solid-free bubbles, and the gas bubbles grow mainly by coalescence during ascent unless built-in

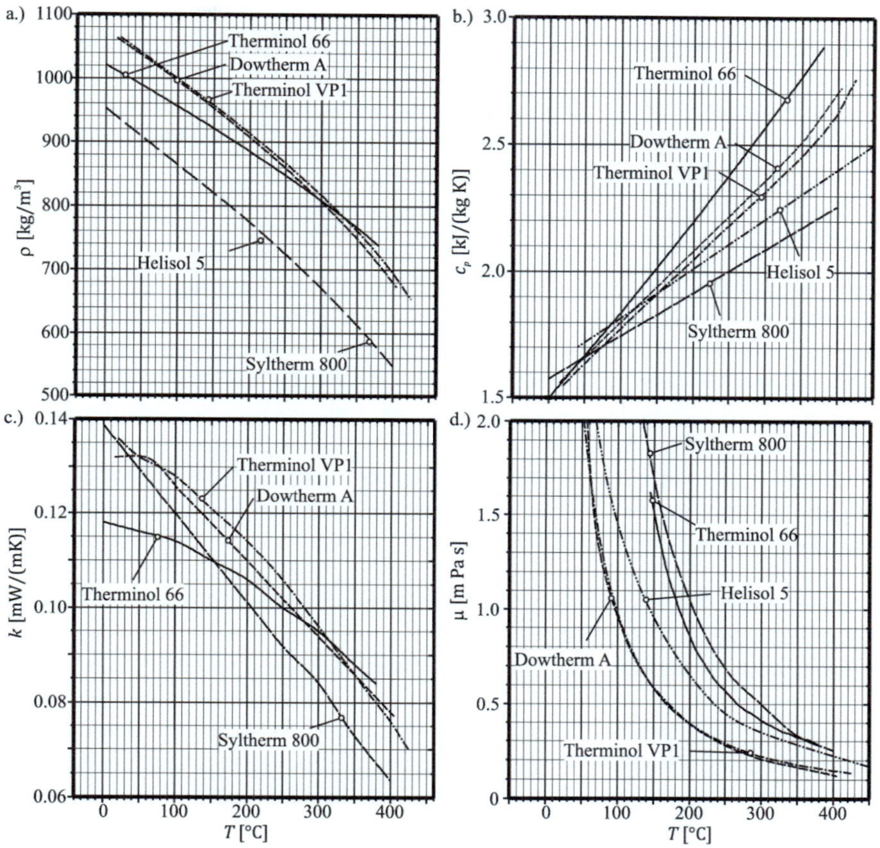

Fig. 10.6 Thermophysical properties of different heat transfer oils as a function of the temperature. **a** specific density ρ, **b** specific heat capacity c_p, **c** specific heat conductivity k and **d** dynamic viscosityμ. Therminol-66 (Hydrogenated terphenyl) (Solutia Europe S.A./N.V.-Applied Chemistry), Therminol-VP1 (Diphenyl oxide/diphenyl, Solutia (2022), Dowtherm A (Diphenyl Oxide/Biphenyl Blend, Dow Chemicals (2017), Syltherm 800 (Dimethyl Polysiloxane, Dow Inc. (1997) Helisol 5A (Polydimethylsiloxane, (Wacker Chemie AG, 2021)

"chicanes" (chevrons) are provided to break them up (e.g., by installing horizontal heat exchanger tubes). As long as the layer has the delineable surface, fluidization in dense (solid) phase prevails. If the fluid velocity reaches the sinking velocity of the solid particles, the entire layer is discharged. In the gas–solid fluidized bed, bubble formation and defined layer surface disappear at high porosities. However, because of the formation of solid strands, such an expanded fluidized bed can be maintained at a gas velocity that is several times the sinking velocity of the solid particles. Because of the high solids discharge, these fluidized beds can only be maintained by continuous circulation of the solids with the aid of a recirculating cyclone, which is why they are referred to as "circulating fluidized beds".

Fluidized beds or gas-particle flows exhibit similar behaviour to two-phase flows and flow maps exist as a function of the relationships the properties of gas and particles and their relative velocity, which can be used to determine the pressure drop and heat transfer. For details on the modelling of gas-particle flows, see Gibilaro (2015), and for engineering correlations and design approaches, please refer to the work by Kunii and Levenspiel (2013).

Thermal Oils

Thermal oils are used for oil cooling and for heating industrial equipment and processes in general, mostly in closed circuits. They can have different properties depending on their chemical composition. They can be divided into low-viscosity, easily flammable oils with low solidification and boiling points and higher-viscosity, hardly flammable oils with higher solidification and boiling points. Organic heat transfer fluids are also used in power engineering, as they enable heat transfer up to 400 °C with comparatively low demands on the apparatus technology (pumps, heat exchangers, instrumentation). According to their origin, they can be divided into three classes as follows:

- mineral oils (for example diesel oils) and white/paraffinic oils,
- synthetic oils (e.g. silicone oils),
- aromatic hydrocarbons (for example Diphenyl-Diphenyl oxide-DP/DPO).

Mineral oils, or better heat transfer oils, are hydrocarbons and are derived from petroleum. They are relatively inexpensive but are usually flammable and chemically aggressive to many sealing materials. But the statement about flammability must be put into perspective. Heat transfer oils based on mineral oil are difficult to ignite at room temperature when delivered; only after thermal stress do low-boiling decomposition products form, which lower the oil's flash point. Suitable maintenance measures can be taken to eliminate the low-boiling products; alternatively, the addition of additives allows a high flash point to be maintained. Mineral heat transfer oils do not attack common industrially used seals. However, there are synthetic oils that require special sealing systems.

Silicone oils are polymeric methyl siloxanes or polymeric phenyl siloxanes. Silicone oils are non-corrosive and non-toxic. Although silicone oils are relatively new to the heat transfer market; although costly, they do exhibit extreme resistance to thermal and oxidative degradation. However, the specific heat capacities are relatively small, so that high volume flows with a correspondingly high-pressure drop are required.

Aromatic hydrocarbons have particularly high thermal stability due to their molecular structure and can be used for applications up to 400 °C (some even claiming up to 450 °C). In solar thermal parabolic trough power plants, a eutectic mixture of biphenyl (BP) and diphenyl ether (DPO) is therefore predominantly used (trade names include Dowtherm A, Diphyl or Therminol VP-1). While they offer good thermal characteristics, they tend to be costly and less friendly to both the environment and worker health and safety, moreover, they are not recommended for use in open systems. Silicone fluids (e.g. Syltherm 800), on the other hand, have so far only been used in test facilities.

A variety of "made to order" organic fluids have been developed; a eutectic mixture of diphenyl and diphenyl oxide is one of the most common and is due to the temperature limits also suited for linear concentrating solar systems as parabolic trough or Fresnel plants. One of the leading arguments is that even "so-called" high temperature oils are liquid at room temperature not requiring any trace heating of piping. Additionally, they show a low viscosity requiring much less system pressure for the envisaged process temperature. But organic fluids exhibit additional advantages, as they do not expand when they solidify and are not corrosive to most materials. However, any liquid oil operation requires closed sealed systems. Moreover, the heat transfer oil of choice must be carefully matched to the temperature operational window of the application. A relatively minor change in process conditions may require a different fluid, and changing fluids may require changes in piping sizes, flow rates and other system parameters, which are oil specific. Even if it sounds strange some of the heat exchanger oils are poor lubricants. The fact that organic fluids burn is most significant for loss prevention. The most common heat transfer fluids are Class III B combustible liquids meaning a flash point of larger 93 °C (200 °F).

Oxidative Degradation

Scientifically speaking, oxidative degradation is the reaction of oxygen (in air) with the fluid by a free radical mechanism. This process forms larger molecules that end up as polymers or solids. These elements can then thicken the fluid, thereby increasing its viscosity. The more viscous a fluid becomes, the more difficult it will be to pump. Its heat transfer characteristics will also be compromised, and the acidity or Total Acid Number (*TAN*) of the fluid will increase and there will be a greater risk of coke formation within the system. As with a lot of chemical reactions, oxidation occurs more rapidly as temperatures increase. The reaction rate is hardly measurable at room temperature but as the temperature rises, the risk of oxidative degradation increases exponentially in the absence of special measures such as inert blanketing of expansion tanks. Put simply, oxidation happens when hot fluid comes in contact with air. Signs of fluid oxidation become evident with the formation of sludge within the system—especially in low flow areas such as reservoirs or expansion tanks.

Thermal Degradation

Thermal degradation or thermal cracking is the breaking of the fluid's carbon–carbon bonds at the molecular level through temperatures beyond the stability limit. This form smaller fragments called "free radicals" and in some cases, this is as far as the reaction goes. In others, the fragments may react with each other to form larger polymeric molecules. In heat transfer terminology, these outcomes are known as "low boilers" and "high boilers". Low boilers presence is associated

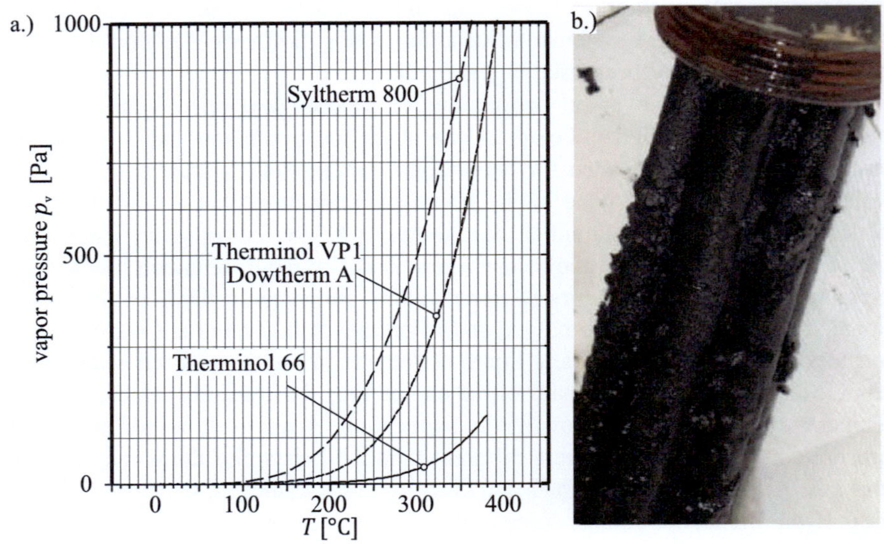

Fig. 10.7 **a** temperature dependent vapor pressure of various heat transfer oils **b** fouling/carbonization on a heating cartridge after using an organic HTF

with a reduction of the flash point and viscosity of the thermal fluid as well as an increase in vapor pressure. The increased vapor pressure can affect overall system efficiency and can cause pump cavitation. The reduction in the flash point is relevant with respect to safety and operating concerns. Thermal degradation occurring at temperatures exceeding 400 °C does not only lead to a break-up of carbon–carbon but also hydrogen atoms separate from their carbon atoms—leading to the formation of coke (see e.g. Fig. 10.7b). This yields an increase in the viscosity of the fluid for as long as they remain in solution. However, once their solubility limit is exceeded solids are formed that can foul the heat transfer surfaces. The fouling of the heat transfer surfaces is very rapid, and the system will soon cease to operate. Thus, thermal degradation is the result of overheating the oil past its boiling point. As the fluid boils, it produces a lighter component usually in the form of vapours. Continual overheating or cracking can cause reduced viscosity. It can also pose safety concerns with the creation of the lighter components. This in turn reduces the overall flash point, fire point and auto-ignition temperatures of the fluid—which can pose a serious safety concern.

Fig. 10.6a–d show the temperature-dependent thermophysical data of commercially available thermal oils. While the density of the oils is similar, with the exception of silicone oils, the heat capacity, thermal conductivity and, above all, viscosity depend essentially on the product, and must be considered in the design of pumps and heat exchangers. The central key is to a long operation period is maintaining the oil-based heat-transfer-fluid (HTF) quality by:

- match heat transfer fluid to the process temperature demands,
- avoid problem-causing contaminants,
- protect against heat-transfer-fluid oxidation and excessive purging and venting,
- identify and correct causes of high-temperature excursions,
- establish and maintain appropriate administrative and engineering controls.

As part of the risk assessment of technical circuits, possible fire and explosion hazards must be investigated in the area of hazardous substances. For all heat transfer fluid, the flash point and the lower explosion limit in particular are used as key material parameters.

An important substance parameter that contributes to the formation of explosive atmospheres is the vapor pressure, but its importance is often underestimated. This can have fatal consequences and must not be ignored in the risk assessment. The vapor pressure is a measure of the number of molecules that leave a liquid phase and build up a pressure in a closed system, which depends only on the temperature and the respective liquid and is therefore a measure of the "mobility" of solvents. The vapor pressure, for its part, determines the boiling point and flash point of the solvents. A liquid boil exactly when the vapor pressure in the liquid is equal to the atmospheric pressure. Heat transfer oils with a high vapor pressure or high temperature sensitivity therefore reach this point much faster when heated than those with a low vapor pressure and therefore also have a correspondingly low boiling point.

The flash point is reached when thermally induced evaporation directly above the liquid releases vapor in such quantities that a flame appears immediately on contact with an ignition source. Accordingly, the flash point for oils with high vapor pressures is much lower than for those with low vapor pressures. Figure 10.7a show the vapor pressures of some heat transfer oils, indicating that the heat transfer oils develop significant vapor pressures at high temperatures. Vapor pressure should also be considered in other issues such as inhalation exposure; in particular, the safety data sheets of the respective manufacturer should be consulted. Another important issue with the use of oils is the unintended hydrogen generation near the maximum temperature of the respective oil, which is a function of time, necessitating hydrogen getters to preserve vacuum in absorber tubes of parabolic troughs. Here, silicone-based heat transfer oils show smaller rates. Also, compatibility with potential thermal storage fluids such as solar salts speak in favour of silicone-based carriers. What cannot be prevented is the aging of the organic fluid, which is countered primarily by the addition of additives.

Operation with thermal oils requires regular monitoring of the heat transfer fluid integrity during operation through appropriate measures, such as sampling, determination of the condition and possible aging, during which appropriate procedures and occupational safety measures must be taken based on the safety regulations for the respective thermal oil.

Molten Salts

Solar salts play a central role in the consideration of solar thermal high temperature systems (>400 °C), as they allow both as heat transfer fluid and direct storage of sensible heat. One reason for this is the liquid state of solar salts over a wide temperature range, which in combination with the low viscosity allows good circulation. In addition, many of the salt compounds are non-toxic, non-flammable or even explosive and exhibit a low vapor pressure in contrast to thermal oils. A decisive argument, especially for the storage of sensitive heat, is the high volumetric heat capacity combined with high availability and low prices. In addition, salts have a low enthalpy of reaction with water, air, or other compounds (e.g. lubricants, oils, etc.) used in power plant construction and show high solubility to impurities. The latter is disadvantageous at high temperatures, as salts thus attack the structural material, especially corrosively. The temperature stability of salt compounds depends essentially on the thermal decomposition reactions (primary and secondary chemical equilibrium reactions) of the respective salt mixtures, which are not only a function of the temperature but also of the heating rate.

A crucial aspect of the use of salts is their high melting point, usually above 200 °C, which requires the use of pipeline heat tracing. Therefore, a development focus is on using mixtures to both lower the melting point towards 150 °C and to drive the dissociation limit to 700°C. Since salts are ionic compounds, the molten salts interact with metals, fillers and also gases and air.

Most of the commercially available solar salts are binary or ternary mixtures of sodium nitrate ($NaNO_3$) or calcium nitrate ($Ca(NO_3)_2$) and potassium nitrate (KNO_3) with admixtures of sodium nitrite or sodium carbonate (Na_2CO_3), respectively, whereby the composition determines the melting point or the thermal decomposition temperature. The fact that molten salts are mixtures requires that their constituents are in complex chemical equilibrium reactions relative to one another and are chemically active. The composition is thus dependent on the temperature and the composition of the gas phase, as well as the impurities in the salt. Thus, the thermophysical properties in operation, similar to the oils, do not remain constant, but initially change significantly during commissioning and only after a longer operating phase is a state reached in which the molten salt only decomposes at a low, almost constant rate. However, this also means that the thermophysical material properties are not as exact as, for example, in the case of pure substances such as water, air, oils or liquid metals. The following chemical reactions are relevant for the behaviour of liquid salts in thermal plants.

$$\begin{gathered} NaNO_3 \rightleftharpoons NaNO_2 + 1/2O_2, \\ 2NaNO_2 \rightleftharpoons Na_2O + 2NO + 1/2O_2, \\ NO_2 \rightleftharpoons NO + 1/2O_2, \\ H_2O + NO_3^- \rightleftharpoons HNO_3 + OH^-. \end{gathered} \tag{10.22}$$

The first reaction describes the nitrate/nitrite equilibrium, in which there is a release of oxygen from 400 °C, which oxidizes iron atoms, which are then dissolved by the wall in the salt. Thus, a high oxygen partial pressure protects against the formation of a high nitrite concentration, while a low oxygen partial pressure leads to the increased formation of nitrite in the melt and thus to a lower density decrease with increasing temperature. Starting from approx. 330 °C (and stronger with increasing temperature) sodium nitrite decomposes to sodium oxide, toxic nitric oxide (NO) and oxygen. The mixture of nitric oxide with oxygen is in turn in balance with the formation of nitrogen dioxide (NO_2). The concentration of oxide ions may influence the corrosive effect of melts hydrolysed by contact with humidity. It can be concluded that the formation of alkaline oxides and their inhibitory effect is not effective in nitrate reduction in aqueous salts. The corrosion reaction is primarily due to oxidation, followed by the dissolution of the metal oxides in the melt. Fused salts often act as a flux and remove oxide layers such as magnetite from the container materials, which can usually have a protective effect. Nitrate melts emit nitrogen oxides at high heat rates, so that in open systems both nitrogen oxides and oxygen are permanently removed from the equilibrium system and in the long term a higher alkalinity of the salt melt is obtained, which directly causes a higher corrosivity of the salt. All salt mixtures relevant to solar engineering are hygroscopic. The contact of nitrate salts with carbon dioxide leads to the formation of carbonates, which have a positive effect on the thermal stability of the salt and increase the decomposition temperature. Due to the complexity of the anion and cation equilibrium, which are temperature dependent, the operation of liquid salt plants requires instrumentation and control logic that reliably prevents both freezing and exceeding of maximum temperatures. According to the state of the art, methods for monitoring the anion distribution (Na^+, K^+, Ca^{2+}, . . .) are available to predictively calculate their distribution. Likewise, nitrate-nitrite equilibria can be reliably evaluated as a function of partial pressure of oxygen (p_{O_2})and temperature, which determines the reaction kinetics. It is more difficult to assess the carbonate content in molten salt with possible CO_2 gas contact or the oxide concentration in molten salt without CO_2 contact; the same applies to the alkalinity. Similar to thermal oils, liquid salt operated plants require regular monitoring of the heat transfer fluid chemistry during operation in order to detect performance degradation at an early stage and to initiate suitable countermeasures. The choice of structural material plays an excellent role in liquid salts due to the complex heat transfer fluid chemistry. Although austenitic stainless steels exhibit very good corrosion behaviour and form a smaller oxide layer than ferritic martensitic steels, their usability is difficult at temperatures above 500 °C and they are less efficient as structural material e.g. for heat exchangers or boilers (Fernández et al., 2015). Although corrosion products are identified in ferritic materials, the corrosion rate can be kept negligible by means of an appropriate heat transfer fluid conditioning. All solar salts show a high surface tension, which makes it difficult to fill narrow cavities and bores, especially at low temperatures, for example for measuring instruments. Figure 10.8 illustrates the temperature dependence of the thermo-physical properties of some solar salts. In addition to the different melting points, which can drop from over 200 to 140 °C for ternary mixtures compared with binary ones, different temperature dependencies are shown for density, heat capacity and heat conductivity, which depend on the respective composition. A common

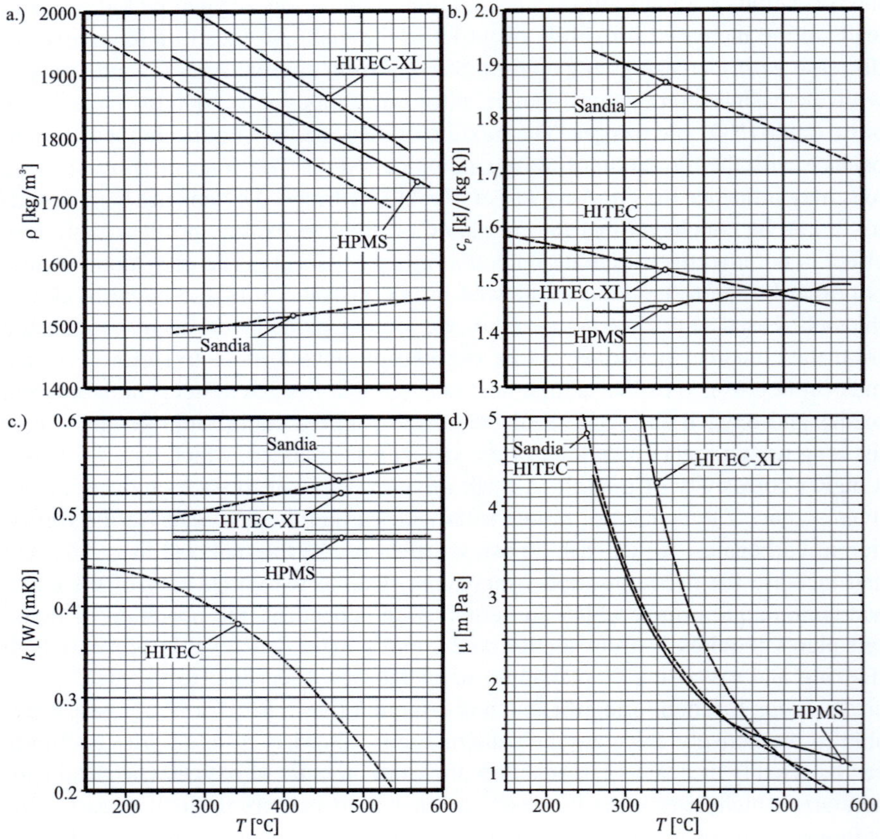

Fig. 10.8 Temperature dependent thermophysical properties of various molten salts: **a** specific density ρ, **b** specific heat capacity c_p, **c** specific heat conductivity k and **d** dynamic viscosity μ. HPMS ($NaNO_3$—60 wt.%—KNO_3—40 wt.%, (Puppe et al. nd)), Sandia (Zavoico, 2001), HITEC (KNO_3—53 wt.%—$NaNO_2$—40 wt.%—$NaNO_3$—7 wt.%) and HITEC-XL (KNO_3—43 wt.%—$Ca(NO_3)_2$—42 wt.%—$NaNO_3$—15 wt.%) both from Siegel et al. (2011)

behaviour of all salts is the low viscosity at high temperatures, which allows good pumpability; conversely, viscosity increases significantly with decreasing temperature. As already mentioned, the thermophysical properties, in contrast to the other heat transfer fluids, are substantially dependent on the adjusted heat transfer fluid chemistry, so that they are subject to a fluctuation range of up to 10–20%.

Liquid Metals

Liquid metals are intrinsically good electrical conductors and thus have high thermal conductivity according to the Wiedemann–Franz law. They are all characterized by a Prandtl number smaller than $Pr < 0.1$. Together with the low viscosity, liquid metals thus exhibit good heat transfer properties, which allow large surface power densities and thus extremely compact dimensions for heat exchangers and receivers. The heat transfer coefficient being substantially larger than that of the solar salts together with the high thermal conductivity reduces the risk of hot spots and thus reduce pipe stresses. This can also mean dependent on the design slower flow rates for the same level of heat transfer. Additionally, liquid metals exhibit high thermal stability, high boiling points that allows for efficient thermodynamic cycles in power plant. Moreover, almost all liquid metals show low vapor pressures, lowering operational constraints by unintended ingestions.

The use of liquid metals is limited mainly due to two aspects, the weight specific price and the low specific heat capacity, which discourages their use as direct storage medium for sensible heat compared to solar salts, as the volume would increase considerably. Various authors mentioned in Pacio (2013) have analysed the different heat transfer fluid for solar thermal applications and found that liquid metals offer great potential for cost reduction, especially where efficient heat transfer at high temperatures is relevant, for example in solar towers. High removable surface thermal power densities, allow the use of larger concentration factors and thus greater receiver efficiency than other heat transfer fluids. Higher receiver outlet temperatures not only result in higher thermodynamic efficiency, but topping processes are also conceivable in principle (e.g. Joule Brayton or thermoelectric conversion with the aid of alkali-metal thermoelectric conversion AMTEC), although these require increased plant engineering effort and are currently still in the infancy of technical development.

There are a multitude of metals and a myriad of eutectic metal alloys, so that similar to liquid salts, a natural restriction of a technically useful melting point of below 300 °C is first considered, which limits the number of possible metals and their alloys to the alkali metals (Li, Na, K) and the metals of Group IIIA (Ga, In), IVA (Sn, Pb) and VA (Bi) of the Periodic Table of the Elements. Due to the thermophysical properties of liquid metals, there are a number of studies that have considered and applied liquid metals as a cooling medium in nuclear physics experiments or in nuclear reactors, from which a wealth of operational experience has accumulated over many decades. These experiments exhibited that lithium requires a complex heat transfer fluid control not only with respect to the oxygen and nitrogen but also the carbon content, which makes in-situ purification extremely complex, so that application of pure liquid lithium is limited to rather confined small-scaled solutions that require specific lithium properties.

Gallium initially appears attractive because of its low melting point (~30 °C), and liquid gallium is remarkable in several respects. It tends to form a supercooled melt very easily and if the melt is placed in the refrigerator, for example, it will remain liquid down to −20 °C. Only when seed crystals are added does crystallization occur. The second special feature of the liquid is that its volume increases during solidification similar as with ice; its density consequently decreases. Gallium thus shows a melting anomaly analogous to water, most semimetals, and some minor other metals (bismuth, plutonium). The increase in volume during melting is 3.23%. Apart from the extremely high price, which makes large volume use almost prohibitive, pure gallium dissolves almost any pure metal at high rates at about 400 °C, and the effect tends to increase with metal alloys such as steels. The corrosion rate increases with increasing temperature (Yatsenko et al., 2008), so that gallium in pure form appears technically unsuitable as a cooling m medium. Similar arguments as for gallium hold for other low melting point liquid metals as indium ($T_m \approx 157$ °C), bismuth ($T_m \approx 271$ °C) and tin H($T_m \approx 232$ °C) in terms of corrosion, while the first two exhibit quite high prices as well.

This limits the selection of technically promising heat transfer fluid to sodium Na ($T_m \approx 98$ °C), the eutectic sodium–potassium alloy $Na^{22}K^{78}$ ($T_m \approx -11$ °C), and lead Pb ($T_m \approx 327$ °C) and lead–bismuth $Pb^{45}Bi^{55}$ ($T_m \approx 127$ °C), each of which pose their own individual challenges for safety, operation, and material selection, respectively. Many articles also mention mercury (Hg) as a potentially useful liquid metal, the only pure metal that is liquid at room temperature ($T_m \approx -38$ °C). However, mercury releases atomic mercury vapor into the environment even at room temperature, something that only noble gases can do apart from mercury. Mercury is one of the toxic heavy metals; although absorption via the digestive tract is comparatively harmless, inhaled vapours have a strong toxic effect, since it destroys proteins by interacting with their sulphur atoms or breaks down sulphur bridges, which are indispensable for the complex structure and thus the function of proteins, especially in the central nervous system. Mercury therefore requires special authorization regulations, so that it is not considered here.

The thermal-physical data of the pure metals as well as of the two alloys are shown in Fig. 10.9, which all have been evaluated with high precision in the context of nuclear research programs with a high accuracy. Figure 10.9a clearly shows that lead and its alloy exhibit an extreme density, which technically poses a particular challenge with regard to foundations and supports in piping networks, or when installed in towers, to the structural design, while the alkali metals have smaller densities than water. At first glance, the specific heat capacities of the alkali metals are considerably bigger, but this is put into perspective, since the volumetric thermal storage density is composed of the product of density and heat capacity, so that this is below the solar salts for the same temperature difference for all liquid metals. The dynamic viscosity is similar in magnitude to those of the solar salts and thermal oils, however, over a temperature range of several hundred degrees Kelvin with significantly smaller gradients, which facilitates circulation in the circuit. The central advantage lies in particular in the extremely high specific thermal conductivity, which enables high heat transfer at small temperature differences (and thereby low thermal material stresses) at comparatively low flow velocities (i.e. low hydraulic pressure losses).

All liquid metals are opaque and are characterized by high electrical and thermal conductivity, which requires specially adapted instrumentation, especially at the high temperatures in which they can show their advantages. For this purpose, reliable measuring instruments for flow, pressure, local velocity, pore fraction, level and temperature have been developed in the last decades, covering almost the whole range of possible operating conditions, details can be found for example in (OECD-NEA, 2015). Nevertheless, liquid metals require specific safety measures,

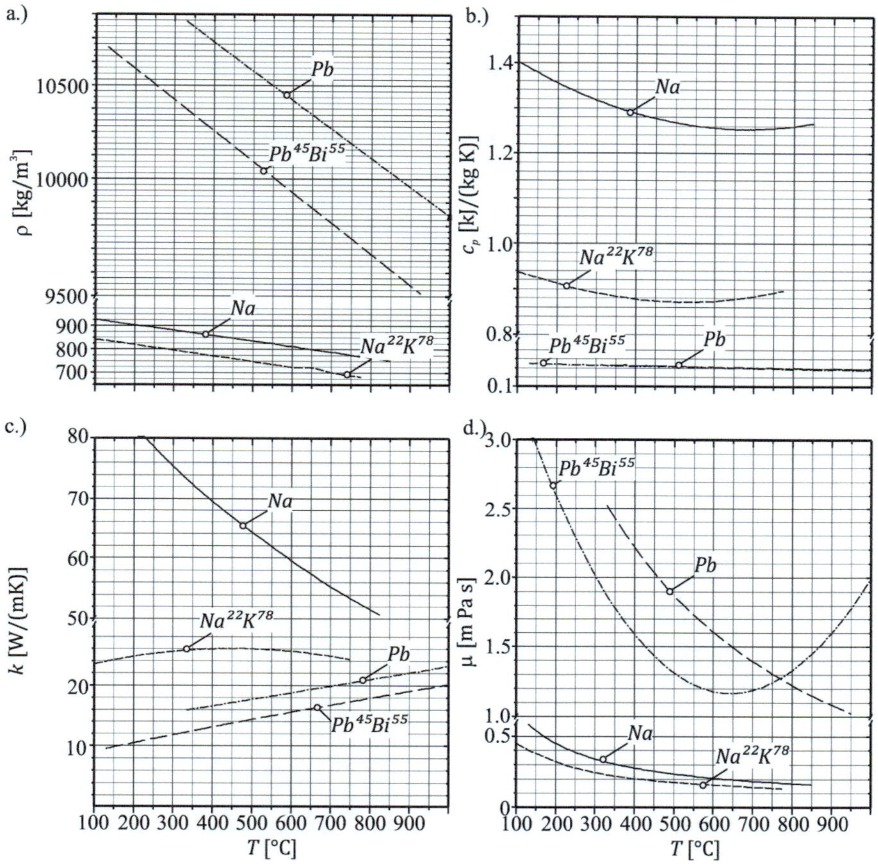

Fig. 10.9 Temperature dependent thermophysical properties of sodium (Na), eutectic sodium–potassium ($Na^{22}K^{78}$), lead (Pb) and lead–bismuth-alloy $Pb^{45}Bi^{55}$ from OECD-NEA (2015): **a** specific density ρ, **b** specific heat capacity c_p, **c** specific heat conductivity k and **d** dynamic viscosity μ

sophisticated operating strategies, and just like all other fluids, monitoring of heat transfer fluid chemistry during operation.

Safety Considerations

While lead and its alloys are largely chemically inert in contact with air and water, vapours and dusts of the metal and oxides are toxic if inhaled or ingested, requiring ventilation measures, appropriate filtration equipment, and hygiene concepts during operation. Similar to thermal oils, concepts to prevent release into the soil or groundwater must be provided.

The last two aspects do not play a role for alkali metals, but they show a high reactivity towards water and air. Although the combustion of sodium and air has relatively low burning rates and energy releases (about four times less than diesel fuel), spray fires in particular pose a challenge, which can occur when released from a pressurized circuit in case of a pipe rupture. In the spray, such as occurred in the case of a leak in the sodium-cooled SPSS receiver in the Almeria solar tower, the auto-ignition temperature drops to 120 °C. Common safety concepts for overhead lines include the fulfilment of the following concepts:

- prevention of leaks through appropriate design measures,
- consistent application of quality assurance concepts,
- leak detection systems,
- leak-before-break concepts, and
- industrially proven fire protection systems.

The sodium-water interaction represents the second safety risk. This situation is of great importance for the design of steam generators, as water/steam can escape into the sodium circuit in case of a pipe rupture. If adjacent pipes are affected, significant damage can occur before the circuit is drained. Based on evolutionary developments in the context of sodium-cooled reactors, the following safety measures have proven effective in preventing sodium-water interactions:

- periodic monitoring of the interfaces,
- measurement of the hydrogen content in the sodium,
- pressure relief systems and rupture discs.

Heat Transfer Fluid Chemistry and Structural Material Choice

Sodium and its alloys do not place any special requirements on the structural materials with regard to corrosion, so that all steel types, refractory metals, and nearly other metals except for aluminium can be used up their structural operation temperature limits. The purification of the fluid can be carried out in-situ by means of a cold trap connected in the bypass of the main loop system, in which the fluid is cooled below the solubility temperature of oxides dissolved in the fluid. The impurities are deposited on a steel sieve or steel wire mesh. The heat transfer fluid quality established can be simply and accurate determined, for example, by means of a plugging meter which detects the melting point.

Lead and lead alloys place considerably higher demands on material selection, detailed information can be found in OECD-NEA (2015). Ferritic martensitic steels can be used up to 550 °C, while austenitic steels require special treatment or the application of protective layers. With the exception of refractory metals, lead and alloys dissolve all metals, so that suitable design measures must be provided. Despite the chemical inertness of lead and lead alloys, the circuits must be kept tight from the atmosphere and corresponding oxygen control systems must be installed which maintain the oxygen partial pressure within a specific concentration window. These prevent lead oxide from forming on the one hand, which has a melting point of more than 1600 °C, but on the other hand, serve an oxygen partial pressure in the lead which ensures establishing and maintaining of a stable magnetite layer on the steels, since otherwise a dissolution attack takes place.

Gas-Hydrogen, Helium and Carbon Dioxide CO_2

Air at various pressure has been has considered up to pressures of 10 MPa, however, its heat transfer behaviour is mainly due to its small heat conductivity rather poor. Although various heat transfer enhancement approaches have been developed to increase its heat transfer coefficient as surface structuring, micro-channels, air-based systems seem to be suitable only in conjunction with particles or in volumetric receivers made of foams (ceramic or metallic).

Compressed helium (He), and carbon dioxide (CO_2) are used in nuclear facilities (fusion experiments, nuclear reactors) to cool thermally highly loaded surfaces. Due to their largely chemical inertness, there is great freedom with respect to structural material selection and heat transfer fluid integrity control is almost completely eliminated. However, due to their low density (see Fig. 10.10a), large flow velocities are required for heat transfer, requiring large specific pumping efforts, especially for gases monatomic gases due to the large adiabatic coefficient. Large flow velocities also mean high wall shear stresses, which can lead to erosion of the structural material.

Therefore, filter systems must be provided for cleaning. Helium and CO_2, in particular, allow temperatures in excess of 1000 °C, which enables high thermodynamic efficiencies for power plant applications, and specific turbines have already been developed for both fluids. For CO_2, operation in a supercritical state called $s - CO_2$- is also possible, since the critical point is at relatively low temperatures ($p_{crit} = 7.375$ MPa, $T_{crit} = 304.5$ °K), with a significant increase in thermal conductivity. A review of $s - CO_2$ applications in nuclear power plants (Dostal, 2004) illustrated its potential for advanced power cycles and also for concentrated solar energy studies analysed its potential (Turchi et al., 2013) although an experimental validation is still lacking.

Potentially, compressed hydrogen (H_2) can also be used as a cooling fluid, and it is already used for cooling steam turbine bearings due to its high heat capacity and good thermal conductivity. In thermal power plants, the use of compressed hydrogen is still in its infancy, mainly for safety reasons but also because of the high diffusion through structural materials and the associated hydrogen embrittlement.

10.2.3 Thermal Field Losses

In the solar field of line-concentrating collectors the HTF is usually heated from a minimum temperature determined by storage temperatures, operational constraints of the HTF (e.g. freezing point, viscosity of the fluid) to a maximum temperature determined by solar collector performance, application requirements and HTF stability limits. The temperatures are in most cases and times considerably above the ambient temperature. Piping is needed to collect all output from the collector rows and return it to the demand side. Therefore, not only the collector thermal

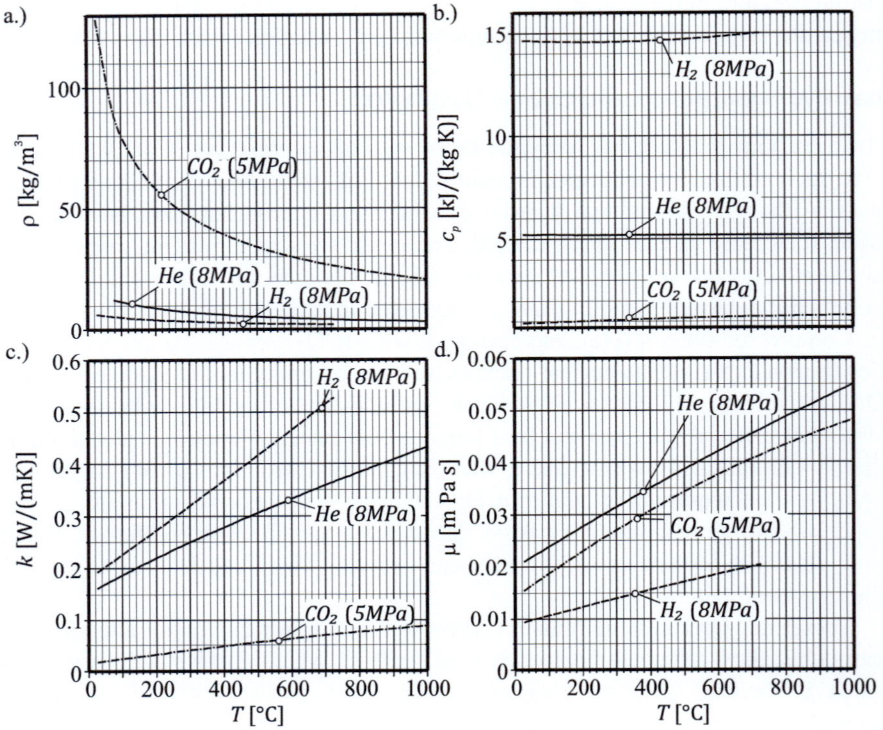

Fig. 10.10 Temperature dependent thermophysical properties of helium (He) at 8 MPa, hydrogen (H_2) at 8 MPa and Carbon dioxide (CO_2) from Bell et al. (2016): **a** specific density ρ, **b** specific heat capacity c_p, **c** specific heat conductivity k and **d** dynamic viscosity μ

losses are important but also the thermal losses of connecting pipework. In order to minimize losses thermal insulation is utilized. The required insulation material has to cope with the high outlet temperatures and should be installed to be durable in the rough outdoor environment. Often a metallic casing is chosen to protect the low-density insulation material, let it be a high-temperature foam, mineral wool, or powder insulation. The necessary thickness of the cylindrical shell around the pipes depends on the temperature level, the pipe length, and the insulation material itself. The basics of the calculation of heat loss of a pipe with insulation in a cylindrical geometry is described in Chap. 4 and detailed in Sect. 10.2.5.1. Of course, this is still an idealization, as additional elements are included in the solar loop such as flanges, valves, sensors. However, the additional heat loss due to the thermal bridges for the specific complex geometries may be negligible considering the total loop. More important is the insulation of these elements for other reasons. The temperature field should not be disturbed at the sensor positions resulting in false sensor readings. Especially for molten salt as a liquid a lower temperature at a specific point might be problematic regarding the freezing risk. Freezing prevention is required for low ambient temperatures and high solidification temperatures.

It can be achieved with different measures: one possibility is nighttime circulation of stored hot liquid within the loop, so that thermal energy from the storage can be used to keep the temperature high enough. A second possibility of Joule heating is especially used with molten salts for the piping and the receivers. Here the electrical resistance of the metallic steel pipes is used in combination with applied voltage to create sufficient heat to avoid freezing. For pumps, flanges, and valves it is easier to wrap the elements with electrical heating mats. Freezing of molten salt in a loop is to be avoided at any time, because the solidified material stops any circulation and therefore melting up is not possible anymore with solar heat from the receivers.

In a solar field a target temperature has to be reached according to the requirements of the process behind. Therefore, the mass flow of the loop is adapted to reach along the full loop the required temperature. For Direct Steam Generation (DSG) in the collectors the absorbed solar heat has to be sufficiently high to be able to evaporate the correct amount of water as described also in the Sect. 10.4. This means a minimum length is required. Increasing the length, however, also increases the thermal losses of the complete field. These considerations limit the lower and upper limits of the collector loops. Many loops then are fed by large header pipes. The thermal losses of these headers have to be included in the calculation. Optimizing for system costs and operational costs a compact configuration with many loops started from each header line is useful. The geometrical layout of the header pipes often follows a linear line (I-configuration) or for more loops a central header is split up into two branches at a 90° angle on each side of the I (H-configuration). Reducing thermal and hydraulic field losses as well as incorporating as many loops as necessary with a reasonable loop length on a prescribed area is the design task. With typical insulation limits and associated cost, the size of individual solar fields using line-focussing collectors is limited to about a thermal power at design conditions of about 500 MW_{th}. Therefore one of the largest projects up to now, the Noor Energy 1/DEWA IV project in Dubai uses three separate 200 MW parabolic trough segments to feed a 600 MW turbine in one CSP Project (https://solarpaces.nrel.gov/project/noor-energy-1-dewa-iv-3x-200mw-trough-segment). When thermal and hydraulic field losses can be limited with new materials and HTF at reasonable costs, this limit may move to an even larger value.

10.2.4 Solar Tracking System

10.2.4.1 PTC Tracking System

Parabolic troughs as every concentrating system have to track the sun in order to reach a continuous concentration of the direct solar radiation in order to provide maximum efficiency at the given site and environmental conditions. In contrast to linear Fresnel in a parabolic trough the whole assembly (mirror and absorber) is simultaneously single axis tracking the sun's position.

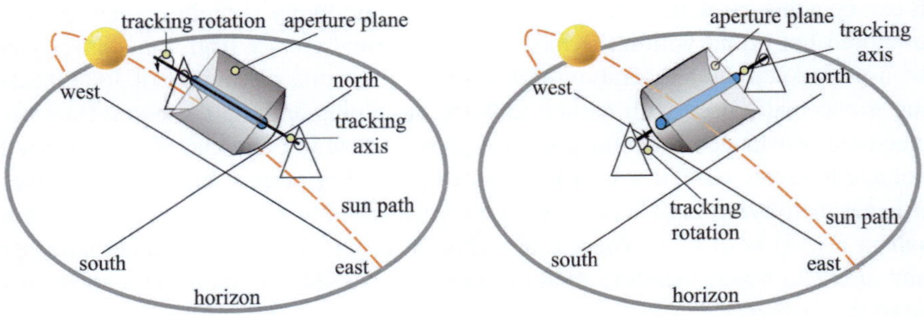

Fig. 10.11 Definition of north–south tracking (left) and east–west tracking (right)

Two types of collector tracking directions are used in PTCs, namely north–south ($N - S$) and east–west ($E - W$) tracking, according to the direction of rotation of the collector aperture plane as sketched in Fig. 10.11. In north–south tracking, the receiver axis and the rotation axis are aligned in the east–west direction, which means that the collector aperture is rotated from north to south. Conversely, in east–west tracking, the aperture plane and the rotation axis are aligned in the north–south direction and the aperture plane rotates from east to west. The north–south tracking has the advantage of lower energy consumption during tracking, but is associated with higher end losses, while with east–west tracking systems it is the other way around: lower end effect but higher energy consumption. Without prejudice to solar field design, north–south tracking is almost exclusively used to achieve maximum thermal yield.

In order to realize a tracking of the receiver, a solar field sensor, which detects the irradiation, an actuator consisting of a motor and a gearbox for the rotation of the collector, and a position sensor, which detects the angular position of the collector, are required. All these elements are part of a regulation and control chain and are themselves subject to manufacturing tolerances or inaccuracies and errors that must be taken into account in the regulation, regardless of whether they are sensors or drives or gears.

Active tracking systems are standard in concentrating solar power plants, with a further subdivision into closed-loop and open-loop systems, more information on details may be taken from Moga et al. (2015) or Lee et al. (2009). The main difference between these types of active tracking systems is the way of signal conversion and feedback with the actuator.

Closed-loop active tracking systems consist of three components, a light sensor, a control system, and a drive system. The sensor sends a signal to the control system when the collector is not properly aligned with the sun, prompting the control system, where a comparison is made. If it is misaligned, the controller sends a signal to the actuator, which applies torque to the structure until the vector value falls below a threshold. Closed-loop tracking systems are therefore pure feedback controls. Light receiving diodes or PV cells are the most commonly used sensors for closed loop tracking. The main advantage of this tracking method is the high precision tracking, but it is susceptible to the sensitivity of the light sensor and its positioning/s in the solar field or due to wanted or environmental shading of the sensor. A key disadvantage is the inability of the control system to recover or find the direction of the sun during long periods of cloud cover.

Open-loop tracking systems consist of a controller and a drive. During operation, the algorithm programmed in the control system, based for example on a clock, signals the actuator to change the collector position. Open-loop tracking systems can be further subdivided into pure time-controlled tracking systems or altitude/azimuth tracking systems. Timed tracking systems use algorithms based only on time, so that the collector is controlled by incremental movements. Altitude/azimuth trackers use algorithms based on astronomical data as a function of location and time.

Hybrid tracking has been developed to overcome the disadvantages of closed-loop and open-loop tracking. The basic strategy of hybrid tracking is to use an algorithm (open-loop) to track the sun and reduce misalignment by sensors (closed-loop). A typical hybrid loop tracking consists of the algorithm calculating the position of the sun and activating the drive, and later correcting the misalignment by the light sensor if it occurs.

Increased computer capabilities such as wireless measurement signal transmission and the use of data buses have led to the widespread adoption of hybrid control systems, which not only include the position of the collector in relation to the sun, but also model-based representations of the individual receiver circuit strings and the reference to the respective overall power plant operation strategy. Such systems not only consider the plant itself and its different operating modes, such as start-up or shutdown, standby, control mode, but automatic switch to safety positions in case of extreme weather situations e.g. provided by external information provided by weather forecast systems.

Tracking, especially with parabolic receivers, places extreme demands on the drives and gearboxes, not only in terms of precision but also because of the enormous masses that have to be operated without backlash under highly variable environmental conditions over long periods of time with as little maintenance as possible. Both hydraulic drives, which the figure outlines, and either hydraulic or worm gear units driven directly by electric motors have become established on the market. Figure 10.12 illustrates some options of tracking systems and drives being installed in solar fields.

10.2.4.2 Tracking LFC

Tracking a LFC is very similar to the PTC tracking described above. However, there are a few important differences. First of all, one has to consider the angular speed for rotation of the mirror rows. Consider a North–South oriented collector with East–West tracking. For a PTC in a place without horizon the minimum and maximum tilt angles are $-90°$ and $+90°$ during the course of a day. For an LFC with a fixed receiver, a mirror row with a start tilt of β_0, where the sun at the Eastern horizon is reflected to the receiver above, has to move 90° during the whole day, until at $\beta_0 = +90°$ the mirror reflects the sun at sunset in the West

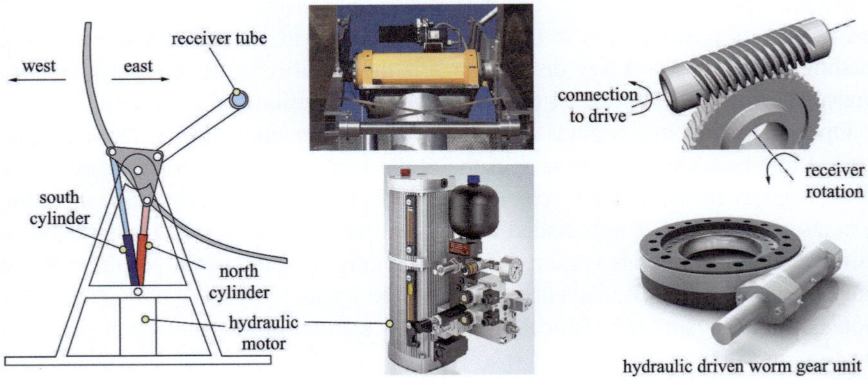

Fig. 10.12 Hydraulic drive operated receiver tracking principle (left), hydraulic motor gear box unit (top, middle); functional principle of backlash-free actuators based on a screw drive with a constructive design (right)

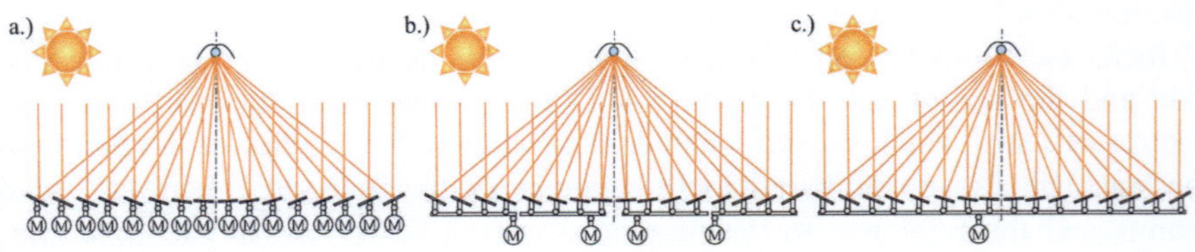

Fig. 10.13 LFC primary mirror row rotation with **a** individual tracking, **b** group tracking and **c** common tracking of all rows

onto the receiver. The rotational speed of the mirror row is 50% of the rotational speed of a PTC. If the sun is in the transversal plane of a PTC, in one hour the PTC has to track the full hour angle of 15°, whereas the reflecting mirror of a LFC has to rotate only 7.5°.

A second difference is that the individual mirror rows can be tracked individually or grouped together. The reason is that each mirror row, although being tilted at a different angle at any time, the rotational speed of the mirror rows is always the same (Fig. 10.13).

10.2.5 Collector Field Design of Line Focussing Systems

Linear focusing collectors, regardless of whether they are parabolic troughs or linear Fresnel systems, use the same method when designing the mirror field structure. However, since they have different optical and thermal characteristics, the design methodology results in different, cost-effective mirror field arrangements for the different technologies. Simply spoken, a cost-optimal mirror field design of a parabolic trough power plant does not mean a cost-optimal LFC power plant with the same dimensions. The type of use of the collector system is primarily considered in the design of the collector field, i.e. whether it has a singular use or should have a multifunctional character. Singular uses include operation as a power plant for electricity generation, seawater desalination, provision of process heat, etc. while multi-functional modes of operation can be the provision of heat for cooling/heating purposes plus process heat or combined process heat and electricity generation. The temperature level to be generated results directly from the envisaged application profile. The temperature level, which is usually given by the collector outlet temperature of the heat transfer fluid used as a set-value for the operation, also determines the choice of the heat transfer medium.

If no TES is available, the orientation of the collector field is very important for the hourly generation during a day. Due to the different incidence angle modifiers the parabolic trough field and the linear Fresnel fields produce different profiles for East–West and North–South orientation. For the latter one the PTC fields produce a "hat profile" with a relatively constant thermal power output for a clear sky when the sun is high enough to avoid the mutual shading of the rows. The LFC field, however, continuously increases the power output until noon. When fields are designed to produce the same thermal energy for a design day with clear sky, the LFC therefore has a higher momentary power output around noon. When TES is used to manage these fluctuations, the hourly profile is not relevant anymore.

The HTF selected certainly has an important impact on the field design. On the one hand side the temperature-dependent viscosity of the HTF determines the pressure losses in the field. The larger the field and the row lengths, the higher the mass transport inside the header pipes and the collector rows, as a specific maximum outlet temperature below the stability limit of the HTF shall not be exceeded. In combination with the selected piping diameter this determines the mass velocities and hence the pressure losses in the piping. Also, the thermal losses over large distances in header piping may lead to slightly different hot temperatures when the mass flows of different field parts are mixed together, leading to small exergy losses. A local shading by clouds of a field part might also lead to different temperatures. For direct steam generation DSG it has to be guaranteed that the row length of a collector is long enough to evaporate the targeted amount of steam. The control of the evaporation process is complex and dealt with in Sect. 10.4.

In order to formulate an appropriate heat quantity for the various systems in the solar field and thus the required solar field size, it is necessary to specify a design point in which the solar field power is nominal. This point is commonly fixed on the 21st June at solar noon (12 : 00 solar time) in the northern hemisphere. The location of the solar system must also be specified (geographical longitude and latitude) using a typical meteorological year (*TMY*), as this represents the basis for the yield calculations of the solar system and allows to compare different design options with respect to their daily, seasonal, and overall annual performance.

In order to summarize, the mirror field design targets at maximizing the thermal annual energy yield for a given site for a given pre-scribed temperature or power level to be achieved by the mirror field. The achievement of a maximum attainable thermal yield is only apparently a singular optimization parameter since it must be achieved under the fulfillment of various constraints.

- On the one hand, the mirror rows must be arranged in such a way that the optical efficiency is as high as possible. In other words, the thermal yield per square meter of mirror surface must be as high as possible.
- At the same time, the maximum permissible receiver temperature occurring at the collector row outlet must not be exceeded. This usually requires a division into several collector rows connected in parallel. In this case, the collector outlet temperatures of the individual collector rows should not be too different, even if the irradiation values vary over time, such as the passage of a cloud field, so that no excessive thermal stresses occur in the collecting pipes.
- Furthermore, the collector rows must be connected in such a way that the permissible hydraulic pressure losses at a given heat transfer fluid velocity in a row do not become too large or exceed that of the pump, since otherwise too strong pumps associated with too great wall thicknesses of the receiver pipes are required.

In addition to fulfilling these central main criteria, the geometric arrangement of the mirror field must also be designed in such a way that maintenance and regular cleaning of the mirror field is possible and that the control and regulation of the thermal heat generation is robust and reliable. This shows that despite a singular primary objective, the real mirror field optimization process is a multi-objective optimization problem.

10.2.5.1 Collector Field Configuration and Layout

When modelling the solar thermal performance of a solar field, it is usually assumed that the entire solar field equipment single collector as well as collector circuits behave similarly, no matter if it is a parabolic trough technology or a linear Fresnel system.

The basic element of a solar field is the linear collector, consisting of the reflector, and the receiver, where the optical conversion to heat occurs and is transferred to the heat transfer fluid. In the context of the solar field, these elements are called Heat Collector Elements (HCEs). The series connection of HCEs forms what is known as a Solar Collector Assemblies (SCAs), which is the smallest independent structural unit equipped with solar tracking capability; each SCA can therefore track or not track (defocus) the sun as desired by the operator.

The SCAs are arranged in circuits, where a circuit is an independent hydraulic segment consisting of several SCAs that receive the heat transfer fluid (HTF) from a manifold (called the cold header pipe) and return it to another pipe (called the hot header pipe) at higher temperatures. Such a circuit can be several hundred meters long. One of the most important considerations in the design of a solar field is the choice of the number of SCAs connected in series in each loop, which is based on the level of temperature to be reached at the outlet from the loop and the performance of the individual HCE. A typical solar field layout consists

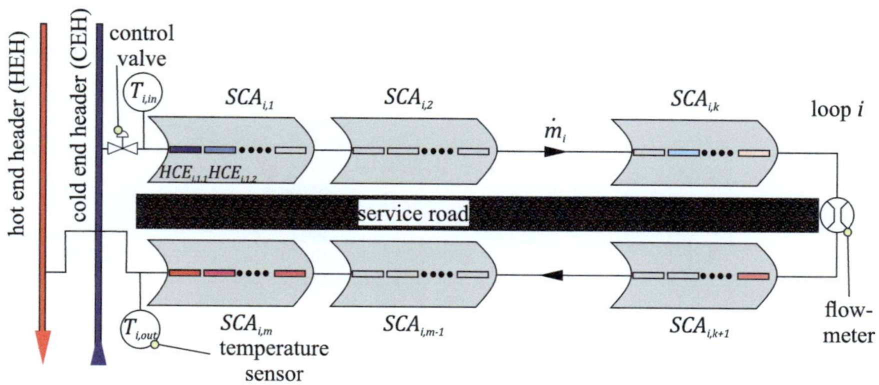

Fig. 10.14 Sketch of a solar loop of a linear focussing collector system comprised of a set of heat collecting elements groups to solar collector assembly (SCA) connected in series to form a loop the SCA-loop i, requiring a flow meter, temperature sensors at in-/outlet and a control valve to regulate flow-rate/mass flow $\dot{m}$

of identical circuits operating under the same conditions; although some authors have proposed the subdivision of the solar field into specialized sectors operating under different conditions (Rovira et al., 2011). Figure 10.14 shows schematic representation of a circuit (loop) composed of SCAs connected in series, as well as the designation of the sub elements and the general notation used in the following.

Increasing the number of SCAs in the loop results in a higher outlet temperature of the heat transfer fluid under the same operating conditions. But due to the series connection of the SCAs in a circuit, the circulating mass flow of the HTF in each SCA is identical. Control valves are located after the branch from the cold manifold (often termed as cold end header-CEH) of a subfield to one loop, so that the flow of the heat transfer fluid (HTF) in each loop can be controlled independently. Adjustment of the required mass flow requires temperature signals, so at least one temperature sensor is installed at the inlet and outlet of the loop and a flow meter, which is usually U-shaped, as outlined in Fig. 10.14.

Since the solar field consists of a large number of loops, one pump is not sufficient to supply all circuits in series with the required heat transfer fluid flow, since the pressure loss would be too large or, conversely, the pump would have to generate a pressure exceeding the allowable pressure of a collector receiver tube. Therefore, several loops are connected in parallel. With this parallel connection, the loops form a separate subfield, as indicated in Fig. 10.15a. If several subfields are supplied by one pump, each subfield also requires a control valve and temperature sensors at the respective inlet and outlet to ensure that the set temperature at each subfield loop outlet is reached in the hot collector pipes.

If the consumer requires a large thermal output—for example, a turbine of a power block—this results in several subfields that are supplied by one or several pumps; more than a hundred loops are not uncommon in large plants. Figure 10.15b. sketches a solar field piping system comprised of four subfields.

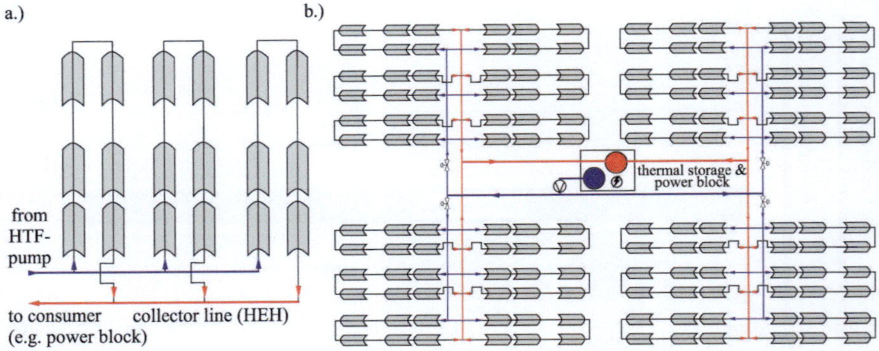

Fig. 10.15 **a** Sketch of the piping scheme of a linear focussing collector loop in one subfield; **b** example of piping scheme of four subfields to a thermal storage/power block (H-field)

The simplest configuration would be an I-field, where a long HEH and CEH line coming from the power block central services many collector loops. In order to restrict the length two I's may be addressed in the so-called H-field configuration as shown in Fig. 10.15b. Larger fields split up the headers into more sections (combinations of I- and H-field), sometimes with the additional complexity of non-rectangular total field area.

The operator must adjust the total flow of the heat transfer fluid so that it is delivered at the optimum operating temperature to the consumer (for example a power unit) or storage system. In addition, the operator must ensure that the heat transfer fluid temperature remains below the maximum permissible values for the receiver or the heat transfer fluid everywhere in the pipe grid, as the latter could decompose (for example, oils or salts). The basics of control and regulation of thermal loops are addressed in Sect. 10.4.

The thermal insulation of the header pipes has to be considered in the thermal field losses. The thermal loss modelling of a complete field also considering the hydraulic pressure loss and the required pumping power is addressed in the next section.

10.2.5.2 Modelling of the Thermal Loss of the Mirror Field

The determination of the daily, monthly or annual yield of a linear focussing solar thermal system is carried out with the help of system simulation programs, the best-known representatives of which are, for example, System Advisor Model (SAM), NREL-National Renewable Energy Laboratory, 2022), ColSim (Fraunhofer Institute for Solar Energy System- ISE, 2022) or TRNSYS (Thermal Energy System Specialists, LLC, 2019). The entire model to obtain the thermal yield necessitates information on:

- hydraulic circuit model,
- a thermal pipe model and

- a plant model connecting the pipes with a pump and a heat sink (either a power conversion system or a thermal storage or both).

To operate a loop also a control scheme is required, which indicates in what time instance useful thermal power is generated. Here, it is for simplicity considered, that useful thermal energy is generated once a certain temperature threshold is exceeded.

Hydraulic Circuit Model

When calculating a solar circuit, each closed hydraulic circuit is calculated separately based on the hydraulic components connected in series. For this purpose, conservation of mass is used as the conservation equation by starting from the pump and calculating each subsequent component with an already updated value, starting from the pump. The calculation requires:

- inner diameter and length of each element,
- the number of pipe fittings (90° bends, T-junctions, reducers, mixer, divertor),
- valve characteristics and position and
- pressure head and flow rate of the pump

If, for example, the mass flow splits at a branch, it must be recombined at the associated mixer. The pairing of diverters and mixers, which occur in every solar circuit, are essential components of system programs.

The division of the mass flow is done by a control variable at the manifold, which is either fixed or supplied by the controller. Within a branch, re-branching is allowed. Figure 10.16 shows schematically a sequence (circled digits) which is determined by a sequence subroutine.

At the branching element, the procedure first arbitrarily selects the upper connection (4) until the evaluation of the mixer (5) shows that the second input still

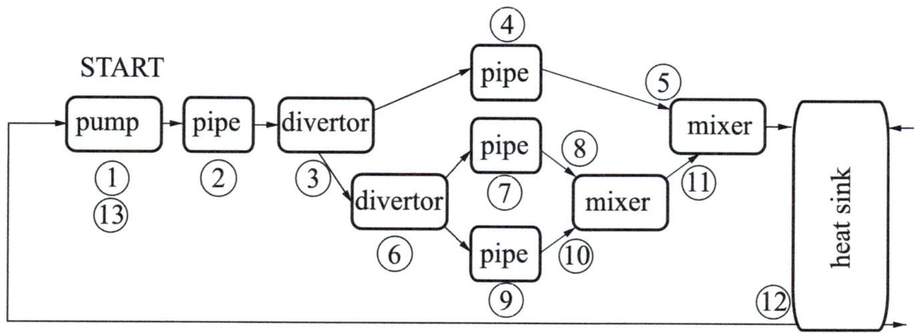

Fig. 10.16 Branching conventions: The hydraulic network is executed during the simulation step, starting from the pump, according to the entered digits. Thereby, a branching at the distributor is followed until an outlet mass flow of a mixer can no longer be determined unambiguously

has to be updated for the calculation of the outlet mass flow. Consequently, processing continues at digit (6) to complete the second branch. Comparable to the bracket calculation in mathematics, the "content" of each branch must first be satisfied before the outlet mass flow of the mixer can be determined. Finally, the mass flow imposed by the pump must appear again at the inlet of the pump in the same time step $t^{(k)}$, otherwise the mass flow balance will abort the simulation at this point.

$$\left|\left(\dot{m}_{out}^{(k)} - \dot{m}_{in}^{(k)}\right)\right| \geq \varepsilon_{\dot{m}} \Rightarrow \textit{termin ation mass flow balance}, \tag{10.23}$$

where $\varepsilon_{\dot{m}}$ is the maximum permissible parameterizable error of the mass flow rate. By that the flow rates in each branch can be determined and the pressure level is obtained at any location of the piping grid according to the explicit formulation of the friction factors of each loop element, for more details see Sect. 5.5.

Thermal Pipe Model

The core of all system simulation programs are the pipe segment modules since the entire solar thermal loop systems is composed of pipe elements. The pipe elements are decomposed into small volumetric pipe sections, which are connected to each other to form a loop.

These finite volume segments, assumed to be axisymmetric, consist of the fluid volume, the pipe material and, if applicable, the insulation material enclosing the pipe segment. Within such a pipe control volume$n$, shown in Fig. 10.17a and b, respectively, the conservation of energy between the inlet and outlet is formulated, taking into account all heat fluxes across the control space boundaries. These heat fluxes include heat conduction across the structures between the upstream downstream $(n + 1)$ as well as the upstream nodes $(n - 1)$, heat input via the solar radiation to the structure as well as to the fluid within the pipe segment n as well as heat loss from the pipe segment via convective heat transfer or thermal radiation. To represent the thermal pipe segment, an electrical equivalent circuit is used as an analogon. In it, solar radiation incident on an absorber tube denotes a current source, while voltages or voltage gradients correspond to temperatures or temperature differences. Likewise, electric currents are equivalent to thermal heat fluxes and electric capacities to thermal heat capacity. This modelling approach involves two key assumptions. First, it assumes a scale separation approach, i.e., the pipe diameter and the considered control volume length of a node is much smaller than the total length of the pipeline network. The second assumes a perfectly mixed fluid within a control volume with homogeneous internal energy, which is justified when the energy contained in the control volume is relatively small compared to the energy transfer across its boundaries, for example, at small control volumes or high flow velocities.

The temporal change of the internal energy $(d(mu)_{n,1}/dt)$ of the fluid volume within a node n is balanced by the fluid enthalpy change, the heat losses and gains across the boundaries and the heat fluxes up and downstream within the fluid at the boundaries in and out of the fluid domain as illustrated in Eq. (10.24).

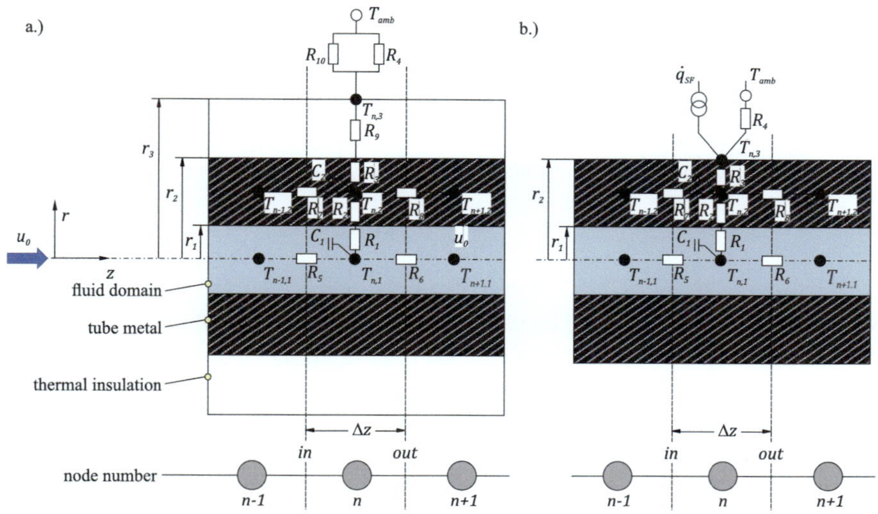

Fig. 10.17 Electrical equivalent circuit for describing heat transfer in a nodal piping element in a system simulation program; **a** insulated pipe; **b** typical absorber pipe element

The enthalpy change is expressed the product of the heat transfer fluid mass flow ($\dot{m}$) and the enthalpy difference ($h_{in} - h_{out}$) and the others as the heat fluxes multiplied, specific thermal resistances and characteristic temperature differences in the following way:

$$\frac{d(mu)_{n,1}}{dt} = \dot{m}\left(h_{n,in} - h_{n,out}\right) + \frac{\left(T_{n,2} - T_{n,1}\right)}{R_1 + R_2}\Delta z + \left(\frac{T_{n-1,1} - T_{n,1}}{R_5} + \frac{T_{n+1,1} - T_{n,1}}{R_6}\right)A_1, \tag{10.24}$$

in which $T_{n,i}$ represent the volume averaged temperatures of node n and the neighboring nodes $n-1$ and $n+1$ in in the fluid (1), the metal wall (2) and the insulation (3). A_1 is the cross section of the fluid containing pipe domain and Δz is the axial length of the node. Equation (10.24) neglects the heat generation by viscous dissipation, which is small compared to the heat fluxes across the boundaries. R_i with $i \in [1, 10]$ denotes the specific thermal resistances in (mK)/W or $\left(m^2K\right)/W$ in the individual domains. Considering an incompressible fluid, we get

$$\begin{aligned} \frac{d(mu)_{n,1}}{dt} &= \rho_{n,1}A_1\Delta c_{pn,1}\frac{\left(T_{n,1}-T_{n,1}^{(0)}\right)}{\Delta t}, \\ \dot{m}\left(h_{n,in} - h_{n,out}\right) &= \rho_{n,1}A_1u_0\left(c_{pn,1}\left(T_{n,1in} - T_{n,1out}\right) - \frac{\left(p_{n,in}-p_{n,out}\right)}{\rho_{n,1}}\right), \end{aligned} \tag{10.25}$$

in which the superscript refers to the data value of the temperature in the preceding time step, and $\rho_{n,1}$ is the specific fluid density, $c_{pn,1}$ its heat capacity, $T_{n,1in}$ and

$T_{n,1out}$ denote the temperatures at in- outlet of the node and $p_{n,in}$, $p_{n,out}$ the pressures at both boundaries of the control volume.

The balance equations require then to be linked to the fluid dynamic transport via the Peclet number Pe which can be perceived as the ratio of the convective heat transport to the heat transport by heat conduction. The nodal Peclet number following the formulation by Patankar (1980) is then given by:

$$Pe = \rho_{n,1} c_{pn,1} \frac{u_0 \Delta z}{\lambda_{n,1}}, \tag{10.26}$$

where $\lambda_{n,1}$ is the specific heat conductivity of the heat transfer fluid in (W/(mK)). If $Pe > 10$, the axial conduction within the fluid can be neglected, which expresses that the convective transport dominates compared to heat conduction, substantially simplifying the problem in the way that it gets mathematically hyperbolic. This means that nodal value of the temperature $T_{n,i}$ only depends on the value of the previous node and can be explicitly calculated. This approach is often termed as "donor-cell" approach. If $Pe < 10$, a power law formulation is required for simultaneous solving the equation systems of all calculation nodes of the considered pipe section, which is called in literature finite volume method "FVM" and constitutes an unsteady advection–diffusion heat transfer problem.

Neglecting axial heat conduction ("donor cell approach") allows to assume a linear profile between inlet and outlet of the nodal volume (i.e. $T_{n,1} = \left(T_{n,1in} + T_{n,1out}\right)/2$) leading to:

$$\begin{aligned} &T_{n,1out}\left(\frac{\rho_{n,1}}{2}\frac{\Delta z}{\Delta t} + \rho_{n,1} u_0 + \frac{\Delta z}{2A_1 c_{pn,1}(R_1 + R_2)}\right) - T_{n,2}\left(\frac{\Delta z}{A_1 c_{pn,1}(R_1 + R_2)}\right) \\ &= T^{(0)}_{n,1}\left(\frac{\rho_{n,1}}{2}\frac{\Delta z}{\Delta t}\right) + T_{n,1in}\left(-\frac{\rho_{n,1}}{2}\frac{\Delta z}{\Delta t} + \rho_{n,1} u_0 - \frac{\Delta z}{2A_1 c_{pn,1}(R_1 + R_2)}\right) \\ &\quad + \frac{u_0}{c_{pn,1}}\left(p_{n,in} - p_{n,out}\right). \end{aligned} \tag{10.27}$$

In case of the FVM approach the equation to solve reads to:

$$\begin{aligned} &T_{n,1out}\left(\frac{\rho_{n,1}}{2}\frac{\Delta z}{\Delta t} + \rho_{n,1} u_0 + \frac{\Delta z}{2A_1 c_{pn,1}(R_1 + R_2)} + a_{n-1,1} - a_{n+1,1}\right) \\ &\quad + T_{n,2}\left(-\frac{\Delta z}{A_1 c_{pn,1}(R_1 + R_2)}\right) + T_{n-1,1}\left(-a_{n-1,1}\right) + T_{n+1,1}\left(-a_{n+1,1}\right) \\ &= T^{(0)}_{n,1}\left(\frac{\rho_{n,1}}{2}\frac{\Delta z}{\Delta t}\right) + T_{n,1in}\left(-\frac{\rho_{n,1}}{2}\frac{\Delta z}{\Delta t} + \rho_{n,1} u_0 - \frac{\Delta z}{2A_1 c_{pn,1}(R_1 + R_2)}\right) \\ &\quad + \frac{u_0}{c_{pn,1}}\left(p_{n,in} - p_{n,out}\right), \end{aligned} \tag{10.28}$$

in which the coefficients $a_{i,1}$ depend on the FVM solution scheme. In Patankar (1980) the ansatz:

$$
\begin{aligned}
a_{n-1,1} &= \frac{f(Pe)}{c_{pn,1}R_5} + \max\left(0,\, \rho_{n-1,1}u_0\right); \\
a_{n+1,1} &= \frac{f(Pe)}{c_{pn,1}R_6} + \max\left(0,\, \rho_{n+1,1}u_0\right),
\end{aligned}
\tag{10.29}
$$

is used, which depends on a power law formulation in the way

$$
f(Pe) = \max\left(0,\, (1 - 0.1|Pe|)^5\right).
\tag{10.30}
$$

The fluid confining wall temperature is then given by:

$$
\begin{aligned}
\rho_{n,2}A_2\Delta z c_{pn,2}\frac{\left(T_{n,2} - T_{n,2}^{(0)}\right)}{\Delta t} &= \left(\frac{T_{n,1} - T_{n,2}}{(R_1 + R_2)} + \frac{T_{n,3} - T_{n,2}}{(R_3 + R_9)}\right)\Delta z \\
&\quad + \left(\frac{T_{n1,2} - T_{n,2}}{R_7} + \frac{T_{n+1,2} - T_{n,2}}{R_8}\right)A_2,
\end{aligned}
\tag{10.31}
$$

in which A_2 denotes the pipe wall cross section and $\rho_{n,2}$, $c_{pn,2}$ are the specific density and heat capacity of the pipe wall metal and $T_{n,3}$ the mean outer wall surface temperature. In the absence of an insulation material, e.g. for an absorber tube, R_9 vanishes since no insulation is present.

Considering a donor cell approach, we get for the wall temperature the following equation for a node:

$$
\begin{aligned}
&T_{n,2}\left(\frac{\rho_{n,2}A_2c_{pn,2}}{\Delta t} + \frac{1}{(R_1 + R_2)} + \frac{1}{(R_3 + R_9)}\right) + T_{n,1out}\left(-\frac{1}{2(R_1 + R_2)}\right) \\
&\quad + T_{n,3}\left(-\frac{1}{(R_3 + R_9)}\right) = T_{n,2}^{(0)}\left(\frac{\rho_{n,2}A_2c_{pn,2}}{\Delta t}\right) + T_{n,1in}\left(\frac{1}{2(R_1 + R_2)}\right),
\end{aligned}
\tag{10.32}
$$

while for the FVM condition, we obtain:

$$
\begin{aligned}
&T_{n,2}\left(\frac{\rho_{n,2}A_2c_{pn,2}}{\Delta t} + \frac{1}{(R_1 + R_2)} + \frac{1}{(R_3 + R_9)} + \frac{A_2}{R_7\Delta z} + \frac{A_2}{R_8\Delta z}\right) \\
&\quad + T_{n,1}\left(-\frac{1}{(R_1 + R_2)}\right) + T_{n,3}\left(-\frac{1}{(R_3 + R_9)}\right) + T_{n-1,2}\left(-\frac{A_2}{R_7\Delta z}\right) \\
&\quad + T_{n+1,2}\left(-\frac{A_2}{R_8\Delta z}\right) = T_{n,2}^{(0)}\left(\frac{\rho_{n,2}A_2c_{pn,2}}{t}\right).
\end{aligned}
\tag{10.33}
$$

Solving for the tube wall outer surface temperature now, there are two further options, depending if an absorber tube or an insulated pipe is considered.

For an insulated pipe, we get for the insulation surface temperature $T_{n,3}$:

$$T_{n,3}\left(\frac{1}{R_4}+\frac{1}{R_{10}}+\frac{1}{(R_3+R_9)}\right)+T_{n,2}\left(-\frac{1}{(R_3+R_9)}\right)=T_{amb}\left(\frac{1}{R_4}+\frac{1}{R_{10}}\right), \tag{10.34}$$

in which T_{amb} is the ambient air dry-bulb temperature, R_4 and R_{10} are the convective and radiative heat transfer resistance from the outside of the insulation to the ambient.

For an uninsulated and uncovered absorber tube one obtains for the tube surface temperature $T_{n,3}$:

$$T_{n,3}\left(\frac{1}{R_4}+\frac{1}{R_{10}}+\frac{1}{R_3}\right)+T_{n,2}\left(-\frac{1}{R_3}\right)=T_{amb}\left(\frac{1}{R_4}+\frac{1}{R_{10}}\right)+\dot{q}_{abs}, \tag{10.35}$$

where $\dot{q}_{abs}$ is the linear power absorbed by the collector in (W/m). The convective and radiative heat transfer resistance R_4 and R_{10} from the pipe surface to the ambient depends on the receiver construction. The loss to T_{amb} is only valid for uncovered absorber tubes exposed to ambient air. For absorbers surrounded by a glass tube Eq. (10.34) can be modified. For vacuum receivers only, the radiative thermal heat transfer to the surrounding glass tube has to be considered as R_9, for air-filled receivers R_9 has to represent the combined radiative and convective heat transfer to the cover glass. The convective part here is dependent on the natural convection stimulated by the temperature difference between glass cover and absorber surface.

What we need still to determine are the specific thermal resistances R_i with $i \in [1, 10]$ in order to solve the previous equations. The convective heat transfer from the heat transfer fluid towards the wall specifies R_1 reading to

$$R_1=\frac{1}{2\pi r_1}\frac{1}{h(Nu, Pr)}, \tag{10.36}$$

in which $h(Nu, Pr)$ is the convective heat transfer coefficient in $(\mathrm{W}/(\mathrm{m}^2\mathrm{K}))$ depending on Nusselt number (Nu) specifying the thermal boundary conditions and the type of heat transfer fluid (Pr) as elaborated in Chap. 5. The thermal resistance R_2 of the radial conductive heat transfer through the duct wall is calculated by:

$$R_2=\frac{\ln((r_1+r_2)/2r_1)}{2\pi k_w}, \tag{10.37}$$

where k_w is the thermal conductivity of the wall. The thermal resistance of radial conductive heat transfer through the steel tube second half R_3 shares a similar formulation to R_2 but with adapted geometrical parameters. In the case of an insulated pipe segment, the thermal resistance is determined accordingly where k_{ins} is the thermal conductivity of the insulation material.

$$R_9 = \frac{\ln(r_3/r_2)}{2\pi k_{ins}}. \quad (10.38)$$

For the heat transfer to the ambient the convective and radiative heat transfer coefficients must be determined. To evaluate the convective heat transfer coefficient h from the outer surface to the environment the heat transfer correlations for free convection in air are required to obtain R_4. Depending on shape and conditions the corresponding Nusselt number relations are formulated in Sect. 5.1.15 External flows while as a simplified first guess of the convective heat transfer coefficient by free convection in air the formulation of Kakac et al. (1985)

$$h_{conv} = 1.6\left(T_{n,3} - T_{amb}\right)^{1/4}, \quad (10.39)$$

is used in many codes.

$$R_4 = \frac{1}{2\pi r_3} \frac{1}{h_{conv}}. \quad (10.40)$$

The radiative heat transfer from a cylinder surface to the environment should take into account the different radiative temperatures of the sky and the ground. For the radiative heat loss form the pipe surface towards the ambient expressed by the thermal resistance R_{10}

$$R_{10} = \frac{1}{2\pi r_3} \frac{1}{h_{rad}}, \quad (10.41)$$

a heat transfer coefficient h is required. The view factor F of a cylinder to a hemisphere (e.g. the sky) or an infinite plane (ground) is given by 1/2. A weighted ambient temperature T_{amb} for the radiative heat transfer could be defined depending on what is seen be the tube. In a LFC secondary the tube mainly sees the secondary reflector.

The specific heat losses of a covered air-filled or evacuated absorber tube to the ambient should be better provided by corresponding standard procedures. Here not only the ideal freshly made tubes but also broken tubes or tubes with hydrogen contamination could be characterized.

The conductive heat transfers up- and downstream in the fluid R_5, R_6 and the metal tube R_7, R_8 to the neighboring nodes $n-1$ and $n+1$ is given by:

$$R_5 = R_6 = \frac{\Delta z}{k_f} \quad \text{and} \quad R_7 = R_8 = \frac{\Delta z}{k_w}, \quad (10.42)$$

where k_f is the specific heat conductivity of the fluid.

The only parameter now missing is the solar collector optical model for which the integral formulations of the parabolic trough collector (PTC, cf. Chap. 7) or the linear Fresnel collector (LFC, cf. Chap. 7) have to be introduced in order to specify the local and time dependent linear absorber heat flux $\dot{q}_{abs}$ (Eq. (10.36)). This linear heat flux is obtained in a general manner as:

$$\dot{q}_{abs} = \eta_0 \cdot IAM(\theta_T, \theta_L) \cdot DNI \cdot W \cdot f_{clean} \cdot f_{end}, \tag{10.43}$$

where η_0 is the nominal optical efficiency a normal incidence, $IAM(\theta_T, \theta_L)$ describes the incidence angle modifier of the corresponding collector system, DNI the direct normal intensity, W the solar aperture width, f_{clean} the solar collector cleanliness correction factor and f_{end} the row end losses correction factor.

Regarding the LFC collector systems the *IAM* is factorized multiplicatively in a transversal and longitudinal part, in which the *IAM* -factors are calculated by step-wise linear interpolation of data tables provided by the original equipment manufacturer (OEM), see also Chap. 7.

For PTC collectors the approach for the *IAM*-calculation is similar; following here the ansatz used in the SAM code as specified in Wagner and Gilman (2011),

$$\begin{aligned} IAM(\theta_T) &= \frac{(W_{row}+W)}{W} \cdot \max(0, \min(1, \cos\theta_T)); \\ IAM(\theta_i) &= \max\left(0, \min\left(1, a_0 + a_1 \cdot \frac{\theta_i}{\cos\theta_i} + a_2 \cdot \frac{\theta_i^2}{\cos\theta_i}\right)\right), \end{aligned} \tag{10.44}$$

in which W_{row} is the width of the gap between two solar collector rows and a_i denote the collector specific coefficients. θ_i has to be used in radian (rad).

In order to consider the impact of the solar collector inclination in direction of the absorber tube axis (longitudinal inclination) the row end losses f_{end} read to

$$f_{end} = 1 - \frac{h_{rec}}{L_{row}} \cdot (\tan\theta_i + \tan\theta_{incl}), \tag{10.45}$$

which is an extension compared to a single collector (cf. Eq. (10.8)) taking into account the solar collector inclination θ_{incl} in the longitudinal direction in (°) and L_{row} specifies the length of a row. This in turn requires the definition of the transversal and incident angles to be modified in order to take the longitudinal inclination into account, which yields:

$$\begin{aligned} \tan\theta_T &= \frac{|\sin(\theta_a - \theta_{ori})| \cdot \sin\theta_z}{\cos(\theta_a - \theta_{ori}) \cdot \sin\theta_z \cdot \sin\theta_{incl} + \cos\theta_z \cos\theta_{incl}}, \\ \sin\theta_i &= \cos(\theta_a - \theta_{ori}) \cdot \sin\theta_z \cdot \cos\theta_{incl} - \cos\theta_z \sin\theta_{incl}. \end{aligned} \tag{10.46}$$

In Eq. (10.46) θ_z specifies the zenith angle, $\gamma = (\theta_a - \theta_{ori})$ the sun azimuth angle corrected from the solar collector orientation (θ_{ori}), θ_a the sun azimuth angle, θ_T the transversal angle and θ_i the incident angle.

Plant Model

Regarding the plant model all loops are modelled in a segmented manner according to their physical presence. For an exact representation, each solar collector loop must be simulated separately, because its input flow parameters vary slightly from other loops due to its individual temperature and pressure losses along the piping path at the cold end header (CEH). In most simulation programs only, a single solar collector loop is calculated completely, namely the farthest one from the solar field terminal points, because it exhibits the highest pressure drop and CEH losses. Thereby, a conservativeness is achieved, because for modelling of the hot end header (HEH), it is then assumed that all loop outlets have the same fluid properties (pressure, internal energy, temperature), while for a given mass flow, fluid temperature would in practice be higher for loops closer to the solar field terminal point.

Once the required mass flow of the HTF is calculated for given temperatures at the entrance and exit of each loop, it is then possible to calculate the thermal power delivered by the solar field as the sum of the contributions of all loops. A priori, the layout of the header pipes, which can be diverse, is not known. Thus, the calculation of the losses in the headers is an estimation. However, since the aim is to evaluate different configurations of the same solar field, in the process of comparing a field to another, assuming similar piping layouts, piping losses are of equal magnitude and therefore they cancel each other in the comparison process. So, the thermal power delivered by the solar field to the power block, neglecting energy losses in the hot header pipe, is:

$$P_{th,SF} = \sum_{i=1}^{N_{loop}} \dot{m}_i \, \Delta h_{i,j}, \tag{10.47}$$

where $\Delta h_{i,j}$ is the increase in specific enthalpy in each loop i composed of j solar collector assemblies (*SCA*). In principle it is possible to analyse in terms of statistics different loop performances by means of scattering performance parameters, which allows to analyse the robustness of a design solution or vice versa maximum acceptable limit of collector parameter scattering. Especially, the latter is important in considerations to reduce manufacturing and maintenance costs, details may be taken from Colmenar-Santos et al. (2014).

10.2.5.3 Results of Collector Field Optimization

There is a myriad of publications on the optimization of linear concentrator systems, whether parabolic trough or linear Fresnel systems, as well as many comparisons of linear Fresnel collectors with parabolic troughs, see e.g. Morin et al. (2012), Abbas et al. (2016) or Sait et al. (2015). Crucial for the optimization of each system is the overall objective of the system analysis, which includes the thermal yield calculation of the respective solar field. Target values can be, for example, the plant utilization factor of the power plant unit, the absolute output of the power plant unit, the absolute cost minimum of the generated thermal output or the minimum of the electricity production costs (*LCoE*) and many other aspects. Depending on the optimization point of view, Fresnel or parabolic trough plants can be optimal for the particular application.

However, parabolic trough and linear Fresnel systems show different characteristics and sensitivities which will be discussed in this context.

The design and dimensioning of the mirror field is strongly related to the other main plant components such as the existence and dimensions of a thermal energy storage (*TES*) and the power cycle, which are strongly correlated through several interdependent parameters. Therefore, the optimisation of the plants in terms of design and component sizing with respect to target indicators poses a non-linear problem (e.g. change in a given design parameter may provide benefit in terms of performances while inducing drawbacks in terms of costs or availability, or the opposite, while also influencing the impact of other parameters).

LFC/PTC Performance Dependent on Collector Orientation

Figure 10.18 shows the flux intensity curve along one day of a parabolic trough (PTC) and a linear Fresnel collector (LFC) for the same location on two characteristic days in summer and winter for north–south oriented (a) and east–west oriented collectors (b). It can be seen that when there is no shading and the longitudinal component of the incident light does not matter, the flux intensity in the central part of an LFC receiver is the same as in a PTC, e.g., on June 21st for an approximately long period for LFC designs and at noon for $E - W$-oriented fields. Of course, this is not the case when the position of the sun in the transverse plane is low, as this results in shading losses in the LFCs that reduce the amount of energy reflected on the mirrors, e.g., in the morning and afternoon on any given day for LFC designs and in the winter for $E - W$ orientations. It is noteworthy that the yield of both collector types is more uniform over the year for $E - W$ oriented collectors, although for both types the year-integral yield is larger in north–south orientation.

LFCs with secondary reflectors achieve very high flux intensities equivalent to those of a PTC. Only when shading effects are very important, such as in winter, the radiation intensity of an LFC drops. By appropriate measures (high fill factors,

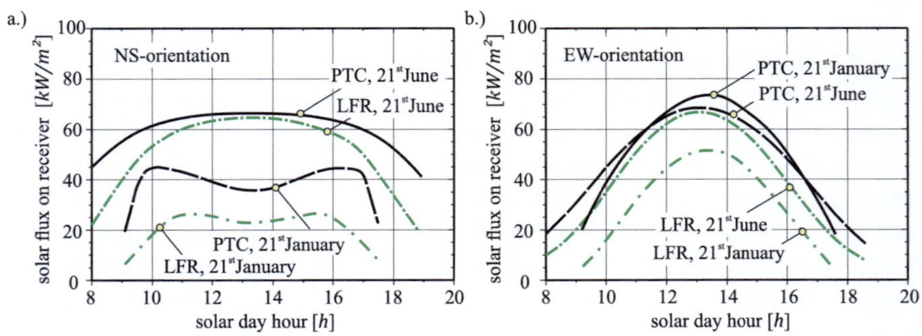

Fig. 10.18 Flux intensity (kW/m^2) on the receiver of a Fresnel collector (LFR-Fresdemo design) with secondary reflector and a parabolic trough collector (PTC), both with NS and EW orientation and for June and January calculated for Almeria; modified from Abbas et al. (2016)

multi-tube receivers,) LFC can even achieve higher flux intensities on the receiver than a PTC. It is important to increase the mass flow accordingly.

The previous consideration of the thermal collector yield calculation is reasonable, as long as one considers the solar mirror field alone, since it allows to compare different design options (field layout, collector options) and technologies (LFC, PTC) basically with respect to their performance.

If one takes the simplified approach that a minimum receiver solar flux density threshold (I_{min}) is sufficient to generate useful power, one obtains for the solar field as comparative quantities the annual energy efficiency $\eta_{en,ann}$ and the annual exergy efficiency $\eta_{ex,ann}$, as defined in Eqs. (10.48) and (10.49) and shown in Table 10.1 for the Almeria site as an example for two collector orientations. In the table the yields attainable without any storage for the specific PTC are by about 20% superior to those of the FRESDEMO LFC collector (DSG) and $N-S$ orientations are attaining a higher yield. It is surprising that the difference in both observed efficiencies between $N-S$ and $E-W$ alignment is much smaller for LFCs than for PTCs. This also depends on the geographical latitude, and it should be noted that the differences between LFC and PTC decrease with decreasing latitude, so that close to the equator the difference reduce to a few percent only. The result shown may not be generalized, because the choice of HTF, the change of the old-fashioned receiver technology of FRESDEMO to a vacuum receiver, and the installation of a thermal energy storage, which allows to compensate diurnal fluctuations, may change the picture completely.

$$\eta_{en,ann} = \frac{\int_0^{365} \dot{Q}_{u,sf}\, dt}{\int_0^{365} G_{b,n} \cdot \cos\theta\, dt}. \tag{10.48}$$

$$\eta_{ex,ann} = \frac{\int_0^{365} \dot{Q}_{u,sf} \cdot \left(1 - \frac{T_{amb}}{T_{out,sf}}\right) dt}{\int_0^{365} G_{b,n} \cdot \cos\theta\, dt}; \tag{10.49}$$

where the useful thermal heat from the solar field $\dot{Q}_{u,sf}$ is

$$\dot{Q}_{u,sf} = \dot{m} \cdot c_p \cdot \left(T_{out,sf} - T_{in,sf}\right). \tag{10.50}$$

Using $\dot{m}$ the total mass flow in the solar field, c_p the specific heat capacity of the HTF and $T_{out,sf}$ and $T_{in,sf}$ being the outlet and inlet temperatures of the solar field at the heat exchanger.

Comparison of Annual Performance of LFC and PTC Technology on a Power Plant Level

For a realistic consideration of the design of a mirror field, however, and its optimization with regard to the respective target value, however, an overall plant consideration is required, which also includes the consumer side (for example the power plant unit) but also all transfer losses as well as the own demand required

Table 10.1 Computed annual efficiencies and optical losses for selected PTC and LFC collector technologies assuming a receiver flux intensity threshold $I_{min} = 25$ kW/m^2 for Almeria from Abbas et al. (2016)

Technology orientation (%)	LFC N/S	EW	PTC NS	EW
Useful energy efficiency η_{use}	58.7	57.5	69.9	64.4
Exergy efficiency η_{ex}	28.2	27.4	31.8	29.4
Cosine losses	23.3	24.3	22.1	21.4
Receiver shading	1.2	1.3	0.0	0.0
Shading	3.9	2.7	0.1	0.0
Absorbed by mirrors	7.6	7.4	5.3	4.9
Blocking	0.2	0.2	0.0	0.0
Lateral losses	4.8	3.9	2.1	2.0

for operation. This is typically done by balancing the energy input to the overall system and the energy losses from the entire system. The energy input to the system is constituted by:

- Thermal energy generation by optical conversion in the solar field (product of aperture area and *DNI*).
- Energy supply for:
  - freeze protection of heat transfer fluid (e.g. molten salts, or liquid metals)
  - pump operation, e.g. to circulate the HTF in the solar field, to operate the thermal energy storage, to run the thermodynamic cycle in case of solar-electric power conversion.
  - plant internal consumption for all auxiliary systems required to operate the plant such as instrumentation and control (I&C), tracking system motor operation, building, heating, ventilation air conditioning (HVAC), ...

The energy losses are comprised by all losses to the ambient environment consisting of:

- optical losses by the solar collector given by material intrinsic properties of the absorber and by the respective collector geometry,
- heat losses within the solar field for the absorber and the connecting the piping grid,
- heat losses by the thermal energy storage,
- consumer based losses:
  - transfer heat losses such as steam generator heat losses in Rankine-cycle or heat exchangers in Joule-Brayton cycle for heat transfer from the solar field to the consumer grid,
  - thermodynamic conversion cycle losses,
  - power-cycle start-up losses prescribed machine inertia for switching from idle mode to power operation,

- defocusing losses, if the generated solar thermal power can no longer be absorbed by the thermal storage or consumer,
- other losses: such as transformer losses in the generator, ohmic electrical losses in the lines to the grid operator, etc.

A central element for quality assurance of the overall plant optimization, which is executed during the planning phase almost exclusively by means of numerical simulations, is the validation of the tools by means of so-called benchmark solutions or code-to-code comparisons. For the depiction of the plant, it is important to ensure that all relevant components are modelled adequately. Moreover, a correct spatial nodalization of solar field and equipment as well as adequate time-stepping procedure has to demonstrated during the validation & verification (V&V) in order to provide physically correct results with respect to the target parameter. A typical target comparison parameter is, for example, the annual thermal yield or the levelized cost of electricity (LCoE) in case of power stations.

The individual simulation programs differ essentially in the level of detail of the representation of the thermal losses (in the solar field, the thermal storage, etc.) or the modelling of the consumer side (integral representation vs. component-wise representation), which allow different solution procedures. Integral models like SAM (NREL-National Renewable Energy Laboratory, 2022) allows implicit solution procedures and thus time step widths of one hour as well as nodal widths in mirror filed of the order of 100 m, while in the other extremely detailed representations e.g. Bachelier and Stieglitz (2017) mostly rely on explicit procedures and nodalizations of a few meters only requiring time step widths of 60 s. The latter allows very detailed parameter sensitivity studies solutions, but conversely require correspondingly higher computation times.

Using a simple example of a direct evaporation solar thermal power plant of the 50 MW_{el} class without thermal storage, the capabilities of such models for optimizing the mirror field design but also the fundamentally different characteristics of Fresnel and parabolic trough collectors are to be analysed. After determining the location and the targeted power class of the power plant, the optimization target value, either the Levelized Electricity Costs (*LCoE*) or the thermal energy of the field to be generated by the mirror field, results directly.

In order to achieve cost parity of Fresnel (type FRESDEMO) and parabolic trough (type EUROTROUGH), the mirror field of the Fresnel collector field requires an aperture area about 40% larger than the parabolic trough, since its optical efficiency is smaller mainly by the fact that the collector aperture is always horizontal (cosine losses), as extensively discussed already in Sect. 7.4, but still LFC exhibits lower solar field costs. Since the distance of the road between two rows of LFC-collectors can be kept smaller than that of the parabolic trough, the absolute land consumption is about 20% smaller with the selected collector technology. The greater optical efficiency of the parabolic trough is additionally associated with smaller heat losses by the evacuated receiver tubes which combined yield a greater thermal efficiency of the PTC field compared to the LFC technology. However, this has little influence on the annual efficiency of the power

Table 10.2 Computed dimensions, power levels and efficiencies for a direct steam generation power plant of the 50 MW class at cost parity of linear Fresnel (LFC) and parabolic trough (PTC) technology for the site Daggett, CA (USA, $\Phi = 34°51'$) from Morin et al. (2012)

Technology	Unit	PTC	LFC
Solar field cost	%	100	53
Solar field size(= aperture area)	$10^3 \cdot m^2$	265	369
Land use	$10^3 \cdot m^2$	927	738
Net electric power output/year	$GW \cdot h$	107	90
Annual optical efficiency $\eta_{opt,a}$	%	57	–
Annual thermal efficiency η_{use}	%	84	74
Annual power block efficiency η_{th} ($\approx$ thermodynamic efficiency)	%	33	32
Total annual solar to electric efficiency	%	14.5	8.7

conversion system since the mirror field has been optimized with regard to the electrical power to be achieved at the same cost. Overall, both designs have a modest solar to electrical efficiency in the order of 10%, with that of the parabolic trough clearly exceeding that of the LFC (Table 10.2).

Comparison of Temporal and Seasonal LFC/PTC Performance for Power Generation

However, not only the year-integral magnitudes of the two technologies are different but also their diurnal dynamics, which Fig. 10.19 illustrates for the similar example above on one day near the equinox (March 21st) and another in the summer solstice (June 15th) for both technologies. The figure illustrates the causes for the different annual efficiency levels of the respective technologies when assuming cost parity. The vertical bars show the power actually used to generate electricity, while the dotted bars indicate the period of time in which the solar field is not able to provide the minimum thermal power for generating electricity from the turbine (25% of the turbine nominal power was assumed as lowest power threshold here). This is termed as minimal power dumping. Conversely, there is also maximum power dumping on sunny days, i.e. the solar field provides thermal energy that exceeds the maximum rated turbine power (100% of the rated power). In such cases, the mirrors are defocused, which is why these losses are also referred to as defocus losses. Due to the design of the Fresnel collector, it delivers significantly less thermal energy to the power plant block, especially in the early morning and late afternoon. This leads to a reduction in the operating time of the turbine and thus to an increase in the costs per electric kWh generated. But, since the solar field costs of Fresnel technology are significantly lower, significantly larger optimal solar field is obtained. This leads to an excess of thermal energy around noon on summer days, which can no longer be absorbed by the turbine and either has to be stored in a TES (which is difficult for a DSG collector but no problem for s molten salt DMS collector) or requires defocusing the mirrors. This proportion is significantly greater in Fresnel mirror fields than in the parabolic trough considered. The consequence of these optical and thermal disadvantages of DSG Fresnel

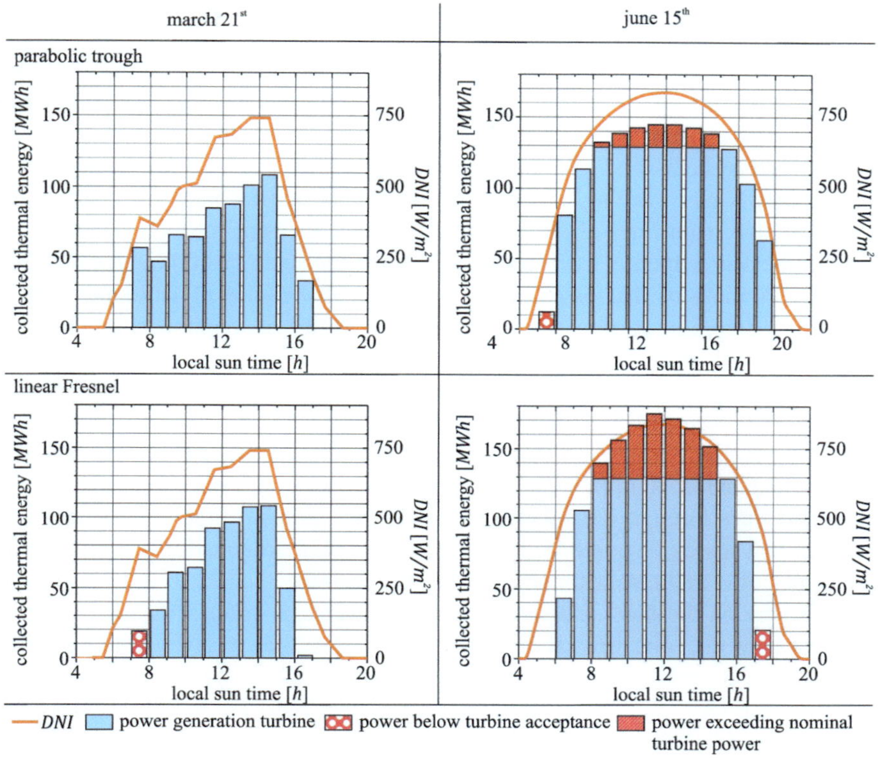

Fig. 10.19 Simulated daily cycle at two distinct days of Direct Normal Intensity (*DNI*) and its conversion into power generation/withdrawal of a direct steam generation (DSG) plant for two different technologies: parabolic trough (PTC) and linear Fresnel (LFR) assuming cost parity of *LCoE* at the site Dagget, CA (USA, $\Phi = 34°51'$) from Morin et al. (2012)

technology compared to parabolic troughs is shown here for the specific Fresdemo collector type and the absence of a thermal energy storage (*TES*) on which this study is based. However, the lower optical efficiency is independent of the selected heat transfer fluid of the solar field. More recent studies (Bachelier, 2021) show that solar power plants based on Fresnel technology with thermal energy storage, even when operated with direct molten salt technology, require approx. 40% larger aperture areas than optimized PTC-based systems as already discussed in 10.1. The size of the optimal LFC mirror field decreases slightly with decreasing geographical latitude, but not substantially. On the other hand, the same study claims a 13% lower LCOE compared to the PTC power plant due to the much lower area-specific collector costs. The integration of a thermal energy storage does not change the picture completely, although the defocusing losses decrease significantly. For N-S orientation of both collector type, the seasonal performance change of a LFC is more pronounced than the one of a PTC (cf. Platzer 2019).

10.3 Point Focussing Systems

In point-focused solar receiver concepts, the solar radiation is concentrated onto a small area by means of mirrors, the so-called heliostats. The achieved concentration ratios (C) of the solar radiation are considerably higher than those of line-focused systems and allow correspondingly high receiver temperatures but require a two-axis tracking of the mirror elements due to the solar beam path. There are two main lines of development for point-focusing systems. Despite the biaxial tracking, which is used to precisely track the sun's position, optical losses account for most of the losses, and the optical efficiencies of large-area point-focusing systems in particular are of the same order of magnitude as those of line-focussing systems. A key advantage of point-focusing systems is the compactness of the receiver volume. A higher solar beam concentration translates to higher heat transfer fluid temperatures and thus higher thermodynamic efficiencies of a downstream power conversion system. Simultaneously, the central receiver concept realized in point focusing concepts eliminates the need for circulating the heat fluid lines through the solar field, where heat is lost to the environment, so that the heat losses of point-focusing systems in total are smaller. There are two main lines of development for point-focusing receivers.

These are on the one hand so-called solar towers, in which with the help of a heliostat field the solar irradiation is focused on a central receiver installed close to the top in a tower. Each of the individual biaxial tracking mirrors is therefore similar to a Fresnel mirror, which redirects the solar radiation to the receiver with different focal lengths. The basic beam path of a solar tower concept is shown in Fig. 10.20a. With the help of large heliostat arrays, solar towers with more than 100 MW nominal readings have already been put into operation.

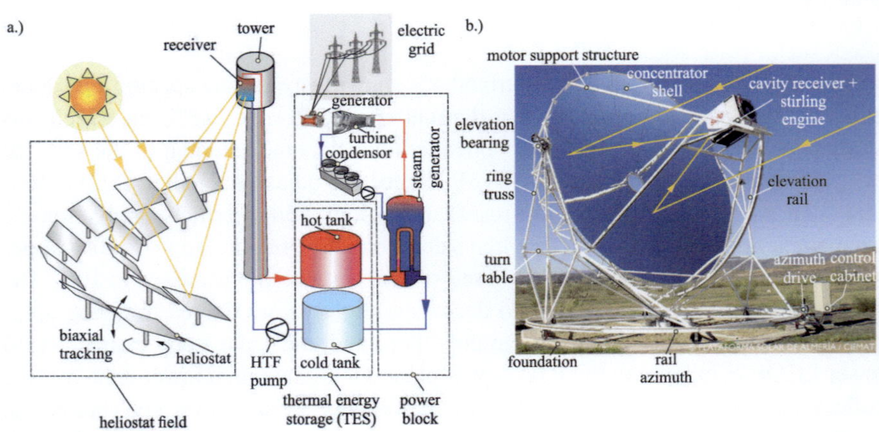

Fig. 10.20 Components and radiation solar beam path of point-focusing solar thermal systems; **a** solar tower concept **b** solar dish concept (© CIEMAT, Plataforma Solar de Almería)

The second line of development are so called parabolic dishes, where the solar radiation is directed either by a monolithic mirror comparable to a satellite dish or by parabolic segmented mirror facets to a focal point, where the receiver coupled to a heat engine (Stirling engine or micro gas turbine) is located. By the fixed focal distance between the mirror and the receiver, the entire setup composed of mirror and receiver plus heat engine unit has to track the position of the sun biaxially, as illustrated in Fig. 10.20b. As the entire system is designed as full functional unit the dish systems perform power scaling by replicating the base unit. An intrinsic disadvantage is that the storage of thermally generated excess power is technically only possible to a limited extent. Therefore, dish systems are mostly used where either high temperatures or electricity is to be generated to a limited extent independently of the grid.

Without going into the different technology options for point-focussing systems, it is useful to recall the basic idea for the development of high-concentration collectors as presented in Sect. 7.1.1, showing that point concentration with geometrical concentration factors C_g in the range of 400–2500 is more efficient in achieving high temperatures exceeding 1000 °C. One application is the combination with a heat engine, as the efficiency of these engines increases with temperature (Carnot limit). The other application requiring very high temperatures are receiver reactors where high temperature chemical reactions have to be driven by the solar heat. Therefore, the development of central receivers is essentially targeting at high collector outlet temperatures.

10.3.1 Solar Tower and Heliostat field Concepts

The design of solar towers requires a complex determination of the optical path via a heliostat to the receiver in order to achieve an efficient, highly concentrating optical system. At the same time, the concentrating solar radiation hitting the receiver during optical conversion generates temperatures that reach the limits of the currently available structural and functional materials. In order to ensure efficient and, above all, sufficiently quantitative predictive control of the heliostats and thus power generation in the receiver at all times, both the control and regulation of the thermal and the widely spatially distributed optical system (heliostats) with the sensors must be optimal and be coordinated at any time. This requires a large number of calculations that are difficult to access analytically, which is why the development of solar towers in the past was essentially limited to experimental demonstrators. Only the massive increase in the computing power available, supplemented by innovative, high-precision production processes, advances in structural and functional material development, wireless signal transmission and refined modeling approaches make it possible today to predict and design efficient solar towers and their operational behavior. This computational design architecture has made the development of large solar towers so attractive that the growth rate in the construction of new solar tower systems, often referred to as solar power towers (SPT), has exceeded that of linear concentrator systems over the past decade. A concentrating solar tower concept essentially consists of three core components.

- A solar tower with a receiver installed at a certain height. If the conversion of the thermal receiver power does not take place in the tower itself, it contains the lines for the heat transfer fluid supply and removal from the receiver.
- The receiver itself, in which the optical conversion of the solar radiation into heat takes place and which is dissipated with the help of the heat transfer fluid. The concept and design of the receiver essentially determine the arrangement of the heliostat field and the height of the tower.
- The mirror field, which is decoupled from the heat transfer fluid system, consists of heliostats that track the position of the sun on two axes, in which the solar radiation is focused on the receiver with the help of target point strategies based on measurement data from sensors in the solar field.

All three components must be precisely coordinated to enable a cost-optimal, competitive solar tower system and are discussed in detail in the following sections. The core component, which determines the arrangement of the heliostats around the tower, is the receiver mounted on the tower, which can be up to 250 m in height, and there are several designs. At present, it is clear that no receiver technology is economically superior to any other.

10.3.1.1 Impact of Receiver Concept on Heliostat Field Design

As we have seen from the receiver design types, a typical design limitation or constraint for high temperature receivers is to define a limited receiver aperture that is the basic design at a desired area power density inMW/m^2, e.g. a billboard or cavity receiver rather than a cylindrical external receiver. While a cylinder has a 360° field of view to the environment, a billboard type is limited to a 180° field of view, often even less, although a downward sloping front aperture can bring a larger ground area into the field of view. Of course, the field of view determines the area which can be used for heliostats. The approximately circular heliostat fields are implemented for 360° receivers, especially when the solar tower plant is at a low latitude. At high latitudes the field becomes more asymmetric. And cavity receivers or billboard receivers utilize a field opposite to the sun (North of receiver in Northern hemisphere), reflecting the radiation back to the receiver. Using a polar aligned billboard design with a 120° field of view to the heliostats, high area performance can be achieved around noon, as most heliostats have an angle of incidence of sunlight close to 0° ("cosine effect" is marginal). However, if the azimuth of the sun changes (at other times of the day), the optical power on the target aperture decreases rapidly. Conversely, with an external cylindrical receiver, only a moderate power is obtained around noon, but when the azimuth of the sun changes from its noon value, which can be more than 100° in summer, it results in only small changes in power density. This is because the nearly circular nature of the surrounding heliostat field, approximately maintains the average value of the cosine over the entire field. Therefore, for similar dimensions, the annual power for the ambient field of a cylindrical external receiver concept is more balanced and larger over the year than that for the polar field of the billboard or cavity concept. However, for small systems, with intelligent design, a billboard design can be much cheaper than a cylindrical external receiver.

If, however, the selected receiver concept provides for the use of working fluids allowing for higher receiver outlet temperatures and simultaneously low heat transfer coefficients (e.g. gases), a cavity concept can hardly be avoided, since the heat radiation losses with cavity concepts are considerably lower than those of an external receiver directly facing the environment. The effects of determining the receiver concept therefore directly involve a restriction in the design and shape of the heliostat array.

One of the most important advantages of central receivers is the relatively short pipeline lengths for storing excess power from the solar field, both in comparison with line-concentrating concepts, but decentralized tower concepts, which are currently also considered, partially eliminate this advantage.

Effect of the permissible heat flux density on the receiver design

Most commercial receivers only allow surface power densities in the order of up to 10 MW/m^2 (e.g. for molten salt) with a certain safety margin from a heat transfer point of view, so that larger receivers are used than the size of the minimum optically achievable focal spot power density. The ratio of height to diameter of an external receiver, often referred to as an aspect ratio, should be larger than unity, so that heliostats can be effective in the near range, since they see optically only a substantially shortened receiver. Since the target variable of a receiver is the heat transfer fluid outlet temperature, a minimum pipe length is required for a given flux density distribution, or the flow guidance can be designed in such a way that the permissible flux density can be large, which usually leads to an aspect ratio of between 1 and 2. The permissible solar flux is then achieved by defining several Heliostat target planes (two to six) on the receiver, to each of which specific portions of the heliostats are assigned. An intelligent strategy that estimates the radius of the beam and shifts the image of the heliostats associated with the outer target planes so that their beam remains just on the receiver results in a very low spillage, while the peak flux is significantly reduced (by a factor of 2–4, depending on the optimization strategy). A second requirement may be that the ratio between the north and south side of the flux density be limited in order to ensure a solar flux which is as uniform as possible on the receiver surfaces and, above all, to minimize circumferential flux gradients and thus thermal stresses. This can be achieved, inter alia, by reinforcing the field border on the "weak" side at the expense of the other side.

Heat Transfer Fluid Flow Configuration

Another option to distinguish receiver types is the type of connection to the environment; there are both open and closed receiver concepts.

- Open concepts imply that the heat transfer fluid is air. In these open concepts, the optical conversion either takes place at a fixed structure where the generated heat is directly transferred to the heat transfer fluid, e.g. air or a gravity driven fluid film. The other option is direct optic conversion on particles which are carried by an air flow or are exposed to the concentrating radiation in the form of a free-falling particle curtain in the gravitational field. Open concepts have the advantage that there are no heat transfer losses due to heat conduction via walls or other structures. Conversely, they are susceptible to convective/radiative heat losses through wind.
- In closed receiver concepts, the optical conversion is realized at the receiver panels. The generated heat flux is transported conductively to the interface with the heat transfer fluid (*HTF*), where it is by forced convection transferred to a liquid, gas, or fluidized bed. Closed concepts inevitably imply a pressure-resistant pipe network as well as corresponding conveying machines (e.g. pumps, compressors, ….). In contrast to open concepts, closed arrangements allow both large heat transfer fluid mass flow densities and, via a temperature-adapted choice of heat transfer fluid, significantly larger heat transfer coefficients to be achieved than open arrangements. This enables very compact receiver dimensions, which also mean low area-specific heat losses to the ambient.

There are also several experimentally validated concepts with regard to flow guidance and routing. The flow guidance within the receiver plays an important role in the receiver development, as it determines both the heat dissipation quantity and the temperature distribution in the structural material of the receiver, because the thermomechanical stress distribution results from the temperature distribution, especially during transients. All receiver concepts presented in the remainder of this chapter aim at a homogeneous temperature distribution in order to keep the thermo-mechanical stress levels low. The optimisation parameter of the high-temperature receiver with regard to the application is the provision of the highest possible heat transfer fluid outlet temperature from the receiver over the longest possible period of the day. For this purpose, there are several options for flow control. The simplest solution is the simple upward flow through a tube or a plate and the adjustment of the flow in such a way that the required heat transfer fluid outlet temperature is reached at the upper end of the tube, as illustrated in Fig. 10.21a. Often, however, the vertical extension of the receiver is not sufficient to achieve the desired temperature lift along its length, which is either due to the thermophysical properties of the cooling medium or the limitation of the heat transfer. This then requires increasing the residence time of the cooling fluid in the irradiated surface by extending the flow path length.

An increase in the residence time can be achieved, among other things, with a meandering flow path, as shown in Fig. 10.21b by splitting a tube array into segments (top) through which flow passes in succession or a plate design with an integrated meander flow channel (bottom). If even longer residence times are required, the concept of a quasi-volumetric receiver can be used, in which cooling

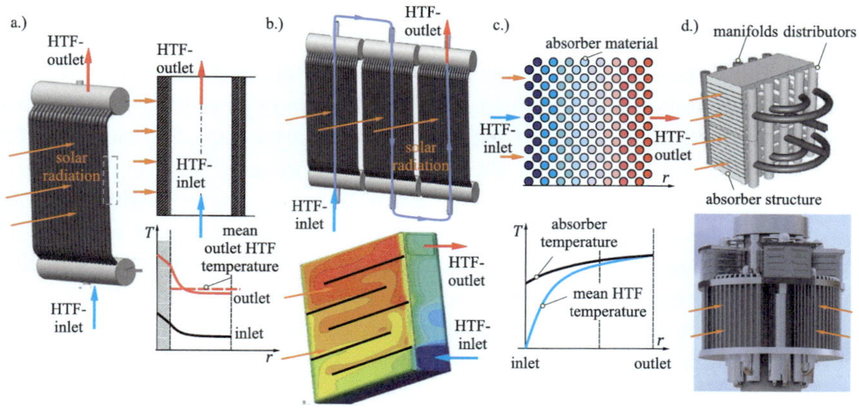

Fig. 10.21 **a** Billboard receiver and its typical HTF temperature evolution along a single tube (right); **b** Meander flow receiver to enlarge HTF residence type (top) subsequent bundle array type (bottom) monolithic billboard type receiver (for sodium/salt as HTF); **c** volumetric open air receiver; **d** closed gas/direct steam generator volumetric receiver with indirect cooling modified from Kalbhor (2014)

channels are positioned normal to the solar irradiation and flow is alternated several times in order to achieve the desired target outlet temperature. The concept of an open volumetric air receiver and a quasi-volumetric closed receiver is shown in Fig. 10.21c and d.

The basic form of the flow channels is either tubes or tube bundle receivers or plate elements. Both designs allow a modular construction and are therefore suitable for both external and cavity receiver designs. The general structure of the tube receiver consists of parallel tubes arranged in rows, whose outer surfaces are heated by the concentrated solar rays. A tubular receiver is simple, easy to build, inexpensive and easy to monitor and maintain. The tubular design allows for parallel but also staggered tube arrangements. Another key advantage is that heat transfer relationships for tubes are reliably known and available with high accuracy for both single-phase and two-phase flows at many Prandtl numbers (i.e. quasi for arbitrary fluids) regardless of their orientation, which substantially facilitates the predictive performance determination of the receiver.

Monolithic plate concepts are becoming increasingly important, as they allow a high degree of receiver coverage and thus have large heat flux densities and at the same time smaller temperature gradients than tube concepts. Due to their construction, they have a lower compressive strength than tube concepts; but additive manufacturing processes, especially in low-pressure receiver concepts with liquid salts or liquid metals as heat carriers, allow fluid-adapted high surface utilisation rates at extremely high surface power densities.

10.3.1.2 Heliostats and Its Field Arrangement

The design of a solar tower system, i.e. the determination of the number and position of the heliostats and the size and position of the receiver on the tower, is a problem with high number of degrees of freedom and can only be solved with computational help, the results of which allow to analyse the efficiency and the yield of the heliostat field. This is always a two-step process.

First, a large heliostat field is generated as input and after positioning the heliostats, for this initial field the annual energy that each heliostat provides to the receiver is evaluated. In order to calculate this energy, yield an optical model is required to determine the optical efficiency not only of each individual mirror, but also its interaction with the neighbouring mirrors, which scopes numerous loss effects occurring in the heliostat field, namely:

- cosine loss, i.e., the reduction of effective mirror size due to off-axis reflection,
- shadowing of specular surfaces by neighbouring heliostats,
- blocking of reflected light by adjacent heliostats,
- attenuation of reflected light between heliostat and receiver due to absorption and scattering processes in the atmosphere,
- reflected light reaching the receiver plane but not entering the aperture ("spillage") due to incomplete heliostat alignment and/or an oversized reflection image.

If the annual yield of the initial heliostat field is known as well as its annual integral optical efficiency; however, one is still far away from the optimal heliostat arrangement for the tower and the question arises, how can an optimal arrangement look like?

An ideal heliostat array shows the highest annual energy yield and simultaneously the highest optical efficiency. For the customer of a solar tower, the heliostat field with the highest annual energy yield is naturally the most interesting, since it can produce the maximum output. However, if only this target value is considered, a heliostat field is obtained which contains a very large number of heliostats with only low efficiency and which is also very expensive to build and operate. Thus, the objective function of a high annual yield conflicts with the objective function of an optically efficient heliostat array. Therefore, a multi-objective optimization is performed, which is always a compromise between the two objectives.

The heliostat function is to reflect the solar image and focus it to a fixed position on the receiver installed at the tower. To do this, the heliostat must track the position of the sun throughout the day. Because many hundreds or thousands of heliostats with a lot of mirror surface reflect the solar image onto a comparatively small area of the receiver, high concentrations of sunlight are achieved. The largest distance a heliostat can be away from the tower is about 1 km. Figure 10.22 shows the different basic positions of the field as a function of the location of the power plant on Earth. In the northern hemisphere of the Earth the Sun is south of the power plant, so the field must be placed north of the tower. The reverse is true in the southern hemisphere. Heliostat fields located only on one side of the tower

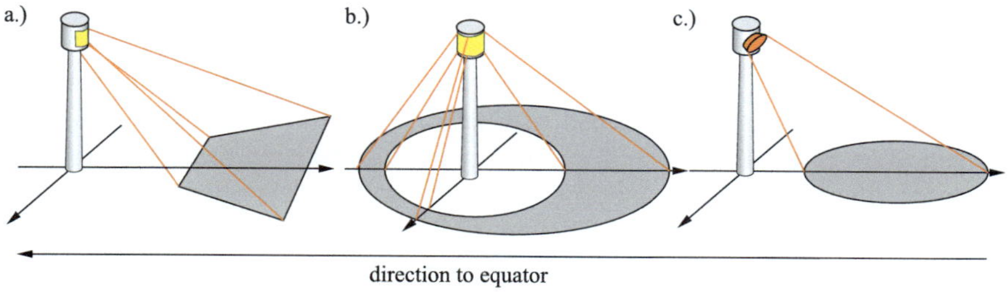

Fig. 10.22 Basic shapes of a heliostat field depending on the receiver type and the distance to the equator; **a** north-field for rectangular cavity receiver; **b** surround field for external receiver and **c** elliptic field for receiver with secondary concentrator

use a cavity receiver. Near the equator, a surround field can be used in combination with an external circumferential receiver because the sun is directly over the tower. Receiver designs that use secondary concentrators are characterized by an elliptically placed field, since light reflected from more angular positions in the field would not reach the secondary concentrators.

The design process of the heliostat field is an iterative process, which takes place in different phases.

(1) At the beginning there is a given receiver concept with an assumed thermal power, as well as a basic arrangement for a mirror field with defined mirror dimensions at a given location (geographical longitude and latitude (Φ)).
(2) For this basic configuration the instantaneous optical efficiency as well as the fraction of the different loss mechanisms (atmospheric attenuation, collection factor, shadowing, and blockage) as a function of time is determined and at the end the annual optical efficiency.
(3) Subsequently, an optimization of the arrangement of the mirrors (considering various limiting constraints, including available area, heliostat spacing, maximum area power density on the target, ...) is determined using different methodological optimization procedures.
(4) Next, an analysis of the robustness against aim point strategies of the heliostat field is performed for the then determined intermediate optimal configuration.
(5) In the last step of the optical procedure, an analysis is performed with respect to the sensitivity for the selected mirror dimensions, receiver dimensions and the optimal tower height. If larger deviations are found, the found compromise solution is the starting point for a further optimization starting with step (2).
(6) If this is not the case, the last step is a thermal analysis of all radiation loss mechanisms in different operating modes, the result of which can also include a new re-initialization with step (2).

10.3.2 Heliostat Fields

The optical efficiency of a solar tower plant is not only determined by the individual heliostat design as described in Chap. 7, but also by the arrangement of the heliostats in a complete field which usually is considered to be on a plane—although slightly hilly area could be also utilized if the elevations do not block the radiation to and from the heliostats. If the heliostats are widely dispersed, they do not interact optically with each other easily. But this would require a huge area compared to actual heliostat fields, implying high area costs and additional deteriorating the optical efficiency due to the very large distances of the outer heliostat which should focus on the receiver. For this reason, it is important that heliostats are grouped together as closely as possible—without interfering with each other mechanically in rotation and minimizing the optical interaction by blocking reflected radiation and shading of adjacent heliostats from the sun. It is as crucial to find a suitable field layout as it is to find suitable heliostats for a selected receiver technology and capacity which has an impact on receiver aperture area.

10.3.2.1 Heliostat Field Design Methods

Because of the relative position between heliostats, between heliostats and solar tower and the moving sun optical losses occur related to cosine effect, shadowing, blocking, atmospheric attenuation, and spillage. Several algorithms and methodologies are found in literature to minimize these losses. The current state-of the-art uses radially staggered layouts, where heliostats are arranged in circular segments as arrays where the gaps between heliostats of the inner radius are used to position in a second-row heliostats. The arrangement is similar to singers in a choir looking at the conductor.

To optimize such a layout and to design tower height and radially staggered heliostat fields as indicated in Fig. 10.23 with high optical efficiency, computer programs have been developed already in the eighties of last century. The RCELL positions heliostats in rings around towers (Lipps, 1978). The CAMPO algorithm considers several zones and starts with a high density radially staggered field which exhibits highest blocking and shading losses. Then the radii of the rings and the azimuthal distances are expanded systematically based on two parameters until an optimum between distance and blocking/shading losses is reached (Collado & Guallar, Campo: Generation of regular heliostat fields, 2012). Also, the DELSOL program splits up the radial rings into several sectors and divides the field into several zones distributed in radial and azimuthal direction. For each zone a different heliostat density is calculated (Kistler, 1986). These codes use simplifications and calculate the optical efficiencies for certain design days. In more modern times as computers were becoming faster and faster, the possibility arose to generate yearly efficiency maps based on the cosine factor, the spillage factor and the atmospheric attenuation coefficient of the site using real Direct Normal Irradiance (DNI) data, within a reasonable computing time. Sánchez and Romero (2006) propose a method where the yearly energy available at any point in a site for a given tower height, the so-called yearly normalized energy surface (YNES) is calculated. Then

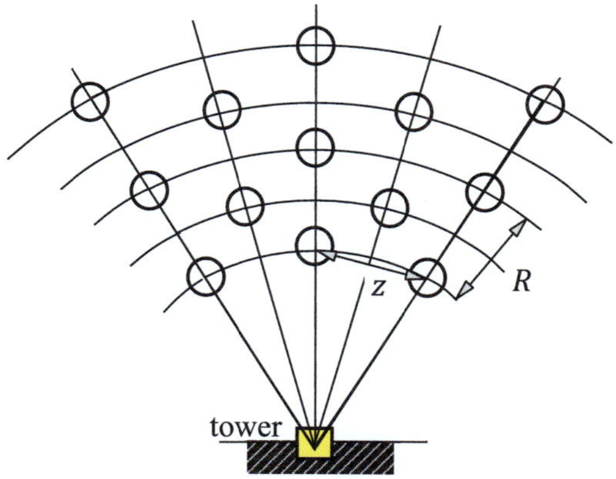

Fig. 10.23 Principle of radially staggered layout after Sánchez and Romero (2006)

the sequential positioning of the heliostats at points with high potential can be found. Blocking and shading is included by recalculation of the YNES. A completely different method is the Fermat spiral proposed by Nonne (2012) where a biometric approach is used. The density of heliostats here is homogeneous than in radially staggered configurations.

To incorporate even specific restraints like available area shape or height above a reference plane more specific algorithms are presented. An example method to create more sophisticated layouts can be found in Schöttl et al. (2019) where an optimization method is proposed for arbitrary polygon areas around a solar tower. In the study of Mutuberria et al. (2015) the authors try to determine the benefits and drawback of the different layout methods. They maximised the annual field efficiency for a defined location Seville in Spain for solar towers using a cavity receiver with different algorithms. The Fermal spiral and the Campo algorithm for radially staggered fields showed the best performance for very large solar tower projects of 150 MW_{th}. For smaller projects of 100 and 120 MW_{th}. the DELSOL program reached the same optical efficiency. But although the optical efficiencies in Fig. 10.24b–e are nearly identical the heliostat arrangement is very different. For determining the heliostat field optical performance for a certain time and location the following section indicates the different steps of modelling. An annual efficiency then has to consider the hourly irradiation levels of DNI.

10.3.2.2 Modelling Heliostat Field Performance

The first step is to determine the position of the sun in relation to a target point where the heliostat is located, which is given by the geographic latitude (Φ). The position is determined by the solar hour (ω) and the declination angle (δ), see also Chap. 2.

$$\sin\alpha_s = \sin\Phi\sin\delta + \cos\Phi\cos\delta\cos\omega;\ \sin\gamma_s = \frac{\cos\delta\sin\omega}{\cos\alpha_s}, \tag{10.51}$$

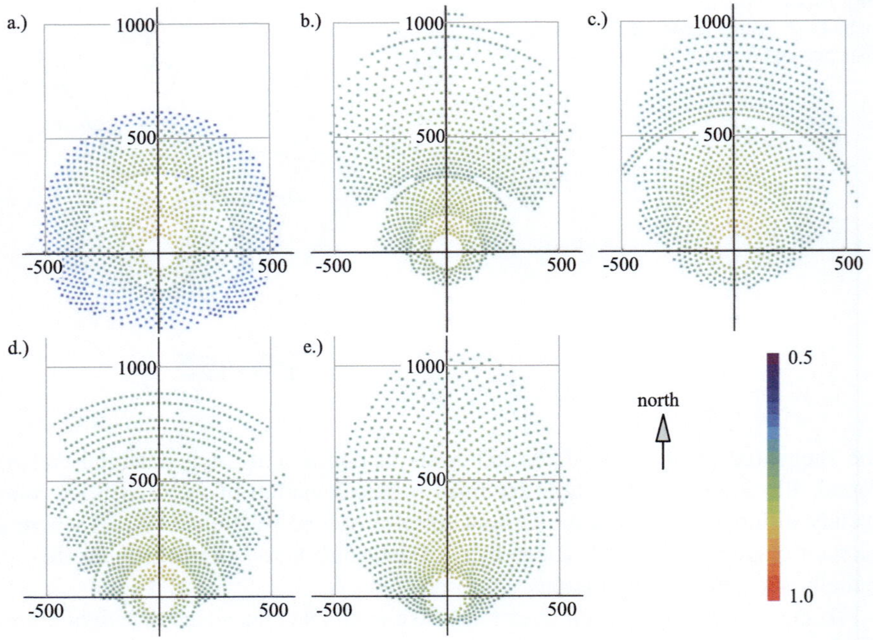

Fig. 10.24 Heliostat field layouts obtained by optimization of a 120 MW_{th} solar tower scenario **a** Dense radially staggered **b** Campo **c** graphical method **d** DELSOL **e** Fermat spiral (Mutuberria et al. 2015)—Colour of the heliostat points (blue = low and red = high efficiency)

in which denote α_s the sun's altitude and γ_s the azimuth angle. Then, the unit vector from the center of the heliostat to the center of the sun can be expressed in terms of a ground coordinate system (x, y, z) as shown in Fig. 10.25.

$$\overrightarrow{e_s} = (-\sin\gamma_s\cos\alpha_s, -\cos\gamma_s\cos\alpha_s, \sin\alpha_s). \tag{10.52}$$

The unit normal vector of the heliostat surface $(\overrightarrow{n_h})$ at its central point is given by

$$\overrightarrow{n_s} = \frac{\overrightarrow{e_s} + \overrightarrow{e_t}}{\left|\overrightarrow{e_s} + \overrightarrow{e_t}\right|}, \tag{10.53}$$

in which $\overrightarrow{e_t}$ is the unit vector of the individual heliostat center to the target point on the receiver. This vector is obtained once the coordinates of the heliostat centre and the receiver centre in terms of the ground coordinate system are known. Regarding the altitude-azimuth tracking strategy, the corresponding altitude and azimuth angles (γ) are described by

$$\sin\ \alpha_{trac} = \overrightarrow{n_{h,z}} \text{ and } \tan\ \gamma_{trac} = \frac{\overrightarrow{n_{h,x}}}{\overrightarrow{n_{h,y}}}, \tag{10.54}$$

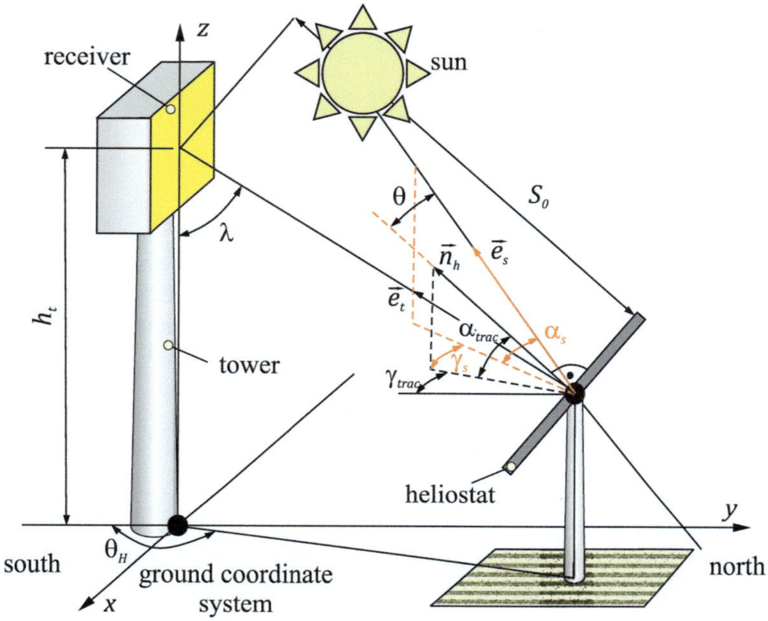

Fig. 10.25 Schematic sketch for defining angles and position of the heliostat with respect to the receiver plane

in which $\overrightarrow{n_{h,x}}$, $\overrightarrow{n_{h,y}}$ and $\overrightarrow{n_{h,z}}$ are the components of the corrdinates of the unit vector $\overrightarrow{n_h}$ with respect to the ground coordinate system.

The energy arriving at the receiver is equal to the energy of direct normal irradiance *(DNI)* available at each instant multiplied by the effective area (A_{eff}) of the heliostats. Thus, this energy ($I(t)$) at any time can be expressed as follows:

$$I(t) = DNI \cdot A_{eff}. \tag{10.55}$$

The effective area of each heliostat $A_{eff,i}$ is smaller than the real area (A_i) because of optical losses, so the effective area of each heliostat can be described in the following way:

$$A_{eff,i}(t) = A_i - A_{loss,i}(t). \tag{10.56}$$

Heliostat optical losses include cosine, atmospheric attenuation, spillage, blocking, and shadowing, similar to the optical losses of a linear Fresnel reflector array. In the heliostat case the movement of the reflectors has two degrees of freedom whereas in the linear Fresnel case only one. Since the orientation of each heliostat changes permanently as a function of time of day and season, the losses also change as a function of time, and so does the available *DNI*. The annual yield intercepted at the receiver is thus an integral of all m heliostats in the field of an entire year ($8760h$):

$$I = \int_{t=0}^{8760} I(t) \cdot \left(\sum_{i=1}^{m} A_{eff,i}(t) \right). \tag{10.57}$$

It is more convenient to express the optical losses of heliostats in terms of efficiencies, where the efficiency term is the ratio of the effective heliostat area due to optical losses to the total heliostat area. Thus, the effective area of a heliostat can be represented in Eq. (10.57) as the area of the heliostat multiplied by the product of individual optical efficiencies.

$$A_{eff,i}(t) = A_i \cdot \eta_i(t) = A_i \underbrace{\left(\rho_i \cdot \eta_{cos,i} \cdot \eta_{at,i} \cdot \eta_{sb,i} \cdot \eta_{sp,i}\right)}_{\eta_{opt,i}}, \tag{10.58}$$

where ρ_i is the mean reflectance of each mirror i, and η_{cos}, η_{at}, η_{sb} and $\eta_{sp,i}$ denote the time dependent cosine factor, the atmospheric attenuation efficiency, the shadowing and blocking efficiency and the intercept or spillage factor of each mirror each being a function of time (t). The product in the bracket can be considered as an optical error of the heliostat i, i.e $\eta_{opt,i}$.

If we now assume in a simplified way that the atmospheric damping and spillage of a heliostat i depends only on the location of the heliostat within the field and that all m heliostats in the field are of identical size, we can determine the collected energy by substituting (10.58) into (10.57):

$$I = A \int_{t=0}^{8760} I(t) \cdot \left(\sum_{i=1}^{m} \rho_i \cdot \eta_{cos,i} \cdot \eta_{at,i} \cdot \eta_{sb,i} \cdot \eta_{sp,i} \right). \tag{10.59}$$

Since each of these efficiencies depends on the position of heliostat i within the field with respect to the distance to the receiver on the tower, the annual intercepted energy is a function of all coordinates of the heliostats. The next step is to find the expressions for these efficiencies for the heliostat coordinates of each individual heliostat i in relation to the receiver. At this point at the latest, system characteristics such as the receiver type and the heliostat shape and, if necessary, the target point strategy are to be determined.

Cosine Loss

The Fig. 10.26 shows the effects of cosine loss, which describes the radiation intensity reflected from a heliostat. The instantaneous cosine efficiency η_{cos} is given by the vector $\overrightarrow{e_s}$ and $\overrightarrow{n_h}$ given by

$$\eta_{cos} = \cos\theta = \overrightarrow{e_s} \cdot \overrightarrow{n_h}. \tag{10.60}$$

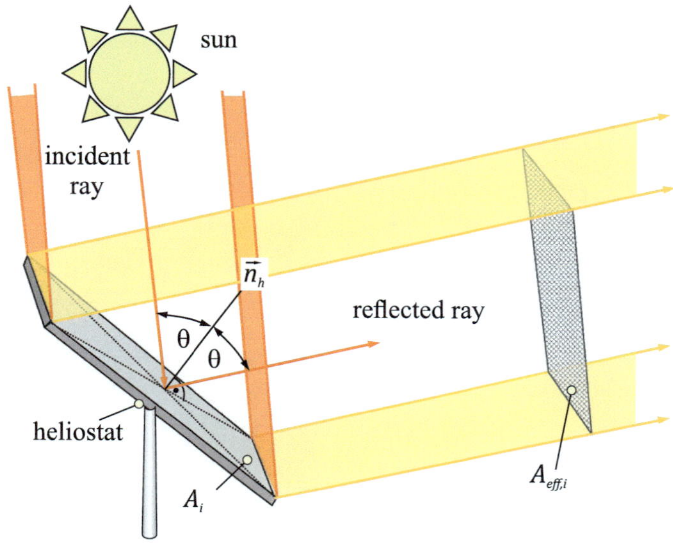

Fig. 10.26 Schematic representation of the cosine effect in a heliostat

The deficit of (10.60) is the missing correlation with the tower. With the help of trigonometry, it is possible to produce these quickly and we obtain,

$$\eta_{cos} = \frac{\sqrt{2}}{2} \sqrt{\sin \alpha_s \cos \lambda - \cos(\theta_H - \gamma_s) \cos \alpha_s \sin \lambda + 1}, \tag{10.61}$$

where α_s represents the solar altitude elevation angle, γ_s the solar azimuth angle, λ represents the angle between the reflected beam incident on the heliostat center and the vertical direction. θ_H is the azimuth angle of the heliostat relative to the tower base (see Fig. 10.25). By integrating η_{cos} over the year, one obtains the annual cosine efficiency of the heliostat, which represents the greatest source of losses in a heliostat field.

Atmospheric Attenuation

The atmospheric transmission efficiency strongly depends on the weather condition and the distance between the heliostat and the receiver. The reflected solar beam is weakened in the atmosphere before arriving at the receiver due to absorption and scattering. The atmospheric attenuation can be simplified as a function of the distance S_0 between the heliostat and the receiver, see cf. Fig. 10.25. The equations of η_{at} can be calculated in the visual range of about 40 km as a function of the distance between the heliostat and the receiver in meters according to Schmitz et al. (2006):

$$\eta_{at} = \begin{cases} 0.99321 - 1.176 \cdot 10^{-4} \cdot S_0 + 1.97 \cdot 10^{8} \cdot S_0^2 & \text{for } S_0 \leq 1000 \text{ m} \\ \exp\left(1.106 \cdot 10^{-4} \cdot S_0\right) & \text{for } S_0 > 1000 \text{ m} \end{cases}. \tag{10.62}$$

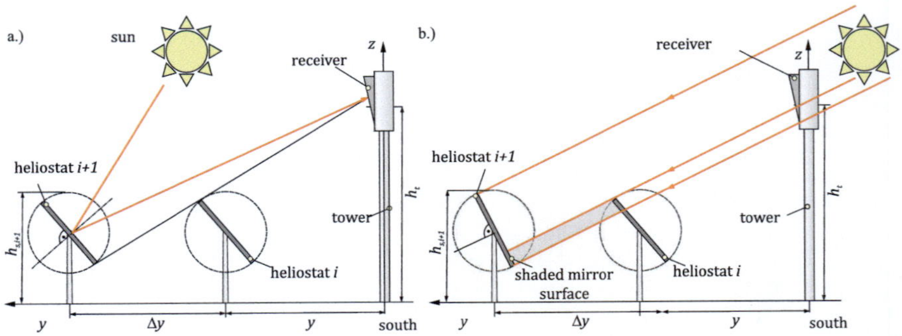

Fig. 10.27 Principle sketch for shadowing and blocking of neighbouring heliostats. **a** Minimum distance Δy to avoid blocking **b** Shading of mirror surface of heliostat $i+1$ by heliostat i

Shading and Blocking

The Fig. 10.27a illustrates that a certain minimum distance Δy between two heliostats is required that the radiation reflected from heliostat $i+1$ is not blocked from heliostat i. To completely eliminate blocking of the heliostats, the distance Δy must be large enough so that the line can be drawn from the center of the receiver to the upper tangent of the mirror swivel circle of heliostat i and this can be drawn further to the lower tangent of the mirror swivel circle of heliostat $i+1$. Approximately; the minimum distance can be calculated as follows

$$\Delta y = y \cdot \frac{h_{s,i+1}}{h_t}, \tag{10.63}$$

where y is the horizontal distance of the heliostat pivot point of heliostat i to the receiver center. As the distance y from the tower increases and the ratio of $h_{s,i+1}/h_t$ grows, the distance between each heliostat increases. However, the distance between the heliostats must not be too large, otherwise the tracking of the sun will become too complex, and the area required for the heliostats will increase rapidly.

The shading is similar to the blocking addressed before, with the only difference that here the light rays hitting the mirror surface of heliostat i cast a shadow on the heliostat behind it, $i+1$ or laterally behind it. Figure 10.27b shows the partial shadowing of heliostat mirror $i+1$ by heliostat i.

The shading and blocking effect between neighbouring heliostats are the most complex factor affecting optical efficiency and is always the most difficult to calculate. With the exception of the overall convolution around the target heliostat, the most accurate method to date is the Monte-Carlo ray tracing method (MCRT) based on statistical theory, similarly as already explained for line-focussing collectors.

The most common simple mapping methods involve some simplifying assumptions, such as.

- the surface of all heliostats is flat and has no optical defects,
- the image projected from another heliostat onto the objective heliostat is a matching rectangle,
- four edges of the rectangle are parallel to the corresponding edges of the objective heliostat.

Under these assumptions, the shape of the projection by shadowing or blocking heliostat onto the objective heliostat is identical and parallel to the surface of the objective heliostat. Then the shadowing or blocking surface can be calculated accordingly. In practice, however, the shape of the projection is not the same rectangle as that of the target heliostat. It is not even a rectangle, but rather a rotating parallelogram. The reason for this is that contrary to the assumptions of the simple mapping method, in the real heliostat field the four edges of the neighbouring heliostats are not perfectly parallel to those of the target heliostat. Therefore, the shapes of the neighbouring heliostat faces can be rotated, reduced, or enlarged after projection. During the field optimization, the shape of the heliostat array and the location of each heliostat are constantly changing, so the shading and blocking must be determined again for each mirror of the heliostat array. There are also some methods that before calculating the shadowing and blocking factor of the $i - th$ heliostat, first identify the heliostats that can potentially shadow or block the target heliostat. Then, the arrangement of the original neighbour group is changed. After that, the shadowing and blocking factor can be calculated by the accurate ray tracing method with the updated neighbour group. If the neighbouring heliostats actually shadow or block, the number of heliostats around the target heliostat can be determined. Regardless of the predicted or actual number, all heliostats are determined during the calculation process.

The conflict of goals that arises when optimizing the heliostat field has already been explained at the beginning of this chapter. It is possible either to aim for the highest possible annual yield or to search for the optimum of the highest optical efficiency of the heliostat field. But it is not possible to get an optimal solution for both goals. There are now several methods to deal with this conflict of objectives by means of so-called multi-objective optimization methods, where ultimately a compromise of annual yield and high efficiency is achieved. Mathematically, the multi-objective problem is transformed into a problem with an objective function, e.g. the annual yield, where the other objective function, assumed to be the efficiency, is defined as a constraint of the problem (i.e. the annual yield). The idea of the constrained objective function is pursued by all optimization tools, they differ only in the way of implementation of the logic, ranging from analytical geometric arrangement functions, permutation algorithms to the use of genetic methods. At the end of the optimization process, there is a set of solutions (i.e. geometric arrangements of the heliostat array) that represents a good compromise between the two objectives and is subject to analysis in terms of the desired properties by the customer. Literature on the design of heliostat mirror fields, the optimization methods used, and the software can be found in Collado and Guallar (2013), Cruz et al. (2017) and Pfahl et al. (2017) the sources cited therein.

Intercept Factor or Spillage

The heliostat field reflects the direct normal irradiation (*DNI*) and directs it to the receiver surface. Some of the radiation is lost and misses the receiver surface due to the optical error of the overall system—this is called spillage. Thus, the intercept or spillage factor ($\eta_{sp,i}$) describes the fraction of the total power transmitted by a heliostat that actually hits the receiver, that is, it is equal to one minus the spillage factor $(1-f_{int})$.

Its mathematical expression is the integration of the solar flux function on the image plane (receiver), where the boundaries are defined by projecting the contours on the image plane in the direction of the central reflected beam $\overrightarrow{e_t}$ of the heliostat. This unit vector $\overrightarrow{e_t}$ depends on the location of the heliostat and its actual target point. Several factors, briefly outlined below, contribute to a reflected image at the receiver surface that deviates from its intended path to the receiver and is distorted in shape as it originates from a single heliostat. Figure 10.28 illustrates the scattered radiation loss for a solar tower with north field only, where the spillage is the radiation that does not hit the (rectangular yellow) receiver surface.

The standard intercept model assumes that all heliostats have well-canted concentrating facets with spherical curvature, so that the expression for the flux density received by the image plane has a circular normal distribution, and the intercepted fraction of power is given by (Schwarzbözl et al., 2009):

$$f_{int} = \frac{1}{2\pi\sigma_{tot}^2} \int_{x'} \int_{y'} \exp\left(-\frac{x'^2 + y'^2}{2\sigma_{tot}^2}\right) dy' dx', \tag{10.64}$$

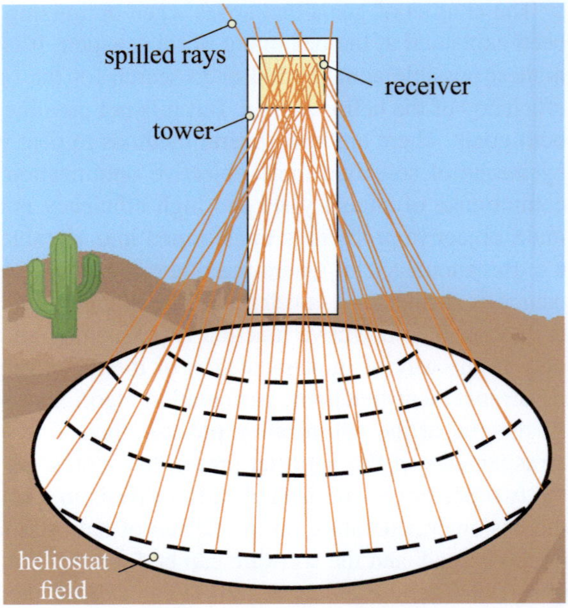

Fig. 10.28 Sketch of radiation loss due to spillage

in which σ_{tot} is the result of the total standard deviation measured on the image plane, i.e. a plane perpendicular to the unit vector $\overrightarrow{e_t}$ and with coordinates $(x\prime, y\prime)$. The effective deviation σ_{tot} is the convolution of four Gaussian error functions that contribute to the image formed on the receiver. These are the sun shape error σ_{sun} (typically assumed to $\sigma_{sun} \approx 2.5$ mrad), the beam quality σ_{bq}, the astigmatic effect σ_{ast}, which is related to the image dimensions of the heliostat reflected onto the image plane, and finally the tracking error σ_{trac}, which takes into account the deviation of the central reflected beam from the aiming point.

$$\sigma_{tot} = \sqrt{S_0^2 \cdot \left(\sigma_{sun}^2 + \sigma_{bq}^2 + \sigma_{ast}^2 + \left(2 \cdot \sigma_{trac}^2\right)\right)}, \qquad (10.65)$$

where S_0 is the actual distance from the heliostat centre to the actual aim point.

In the beam quality factor (σ_{bq}), deviations of the mirror curvature from the ideal shape and imperfections of the reflecting surface due to waviness and roughness from manufacturing and assembly are considered. These are called slope errors (σ_{slope}), compare Chap. 3. Statistically, they are assumed to be uniformly distributed and described by a circular normal distribution. The slope error σ_{slope} is defined with respect to the surface normal vector; therefore, its effect is doubled in the reflected beam.

$$\sigma_{bq}^2 = \left(2 \cdot \sigma_{slope}\right)^2. \qquad (10.66)$$

Typical values reported for σ_{bq} for central towers are of the order of $\sigma_{bq} \approx$ 1 mrad, see Collado and Guallar (2013).

In the case of off-centre reflection ($\theta \neq 0°$) with a spherical or parabolic concentrator, parallel light rays are focused in two focal lines instead of a single focal point, as outlined in Fig. 10.29. When the ratio of focal length to target diameter is large, i.e., $f/d \gg 1$, rays in the frontal (or tangential) plane (defined by the incident and reflected central rays) are concentrated in a line focus at a distance $f \cdot \cos\theta$ from the mirror, while rays in the perpendicular sagittal or median plane are concentrated in a line focus $f/\cos\theta$ from the mirror. In between, at distance f, the reflected ray has minimal dimensions $d \cdot (1 - \cos\theta)$. At an arbitrary distance D from the mirror, the image dimensions in the tangential and sagittal (or median) planes (H_t, W_s) are given by

$$H_t = d \cdot \left|\frac{D}{f} - \cos\theta\right| \; and \; W_s = d \cdot \left|\frac{D}{f}\cos\theta - 1\right|. \qquad (10.67)$$

A heliostat usually consists of several facets, and astigmatism occurs in the facet and in the whole heliostat. Assuming that each heliostat is correctly canted for$\theta = 0°$, the image dimensions are described as a superposition of astigmatism in the heliostat and in the facet due to off-axis reflection. The root mean square of the image expansion in the tangential and sagittal planes can then be treated as the

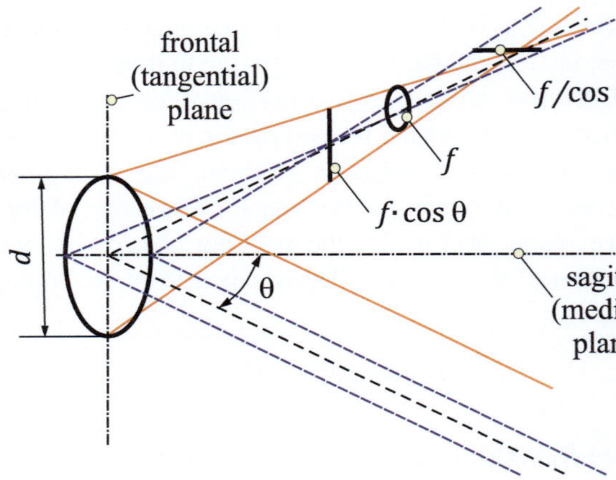

Fig. 10.29 Astigmatism in the case of off-center reflection of light from a spherical concentrator (θ incidence angle, f focal length of the concentrator, d diameter of the concentrator)

diameter of a pillbox distribution. Thus, the additional widening of the reflected beam as a result of astigmatism can be represented as an optical error as follows:

$$\sigma_{ast} = \frac{\sqrt{0.5 \cdot \left(H_t^2 + W_s^2\right)}}{D}. \tag{10.68}$$

10.3.2.3 Entire Field Performance and Heliostat Aiming Point Strategy

In order to determine the instant overall efficiency of the heliostat field $\eta_{opt}(t)$, we´ve to sum up all weighted optical efficiencies by

$$\eta_{opt}(t) = \frac{\sum_{i=1}^{m} \eta_{opt,i}(t) \cdot A_i}{A}, \tag{10.69}$$

in which m is the number of heliostats, A_i the area of a single heliostat, and A the area of the entire heliostat field.

Aiming strategies are used to distribute the radiation provided by the heliostat array as evenly as possible over the active receiver surface, since at the receiver the coupling of the optics with the thermal-hydraulics is accomplished. This is for several reasons:

- The radiation must be converted into heat in the receiver by means of optical conversion and dissipated by the heat carrier.
- The maximum permissible absorber temperature must not be exceeded in order to avoid degradation of the absorber and to keep the thermo-mechanical stresses low.
- At the same time, the receiver outlet temperature should be as high as possible to maintain a high efficiency of the downstream power conversion system and this at simultaneously possible small heat transfer fluid mass flows to keep the pumping work small.

- An inhomogeneous heat distribution on the receiver, e.g. due to the choice of a (central) target point, leads to locally high temperatures on the absorber, which cannot be absorbed by the heat carrier and are lost as heat radiation.

Heliostats offer by their biaxial tracking to align the target point of the heliostat to different points in the image plane (on the receiver) and thus to control the flux density distribution on the receiver. The flux density on the receiver depends on the design of the heliostat field considering: Position of the sun, shape of the sun (time of day and season), number, size and optical quality of the heliostats, and the focal length and canting of the heliostats. However, the most important factor in operation is the distribution of the individual target points of the heliostats on the receiver. The following qualitatively presented aiming point strategies are valid for external as well as cavity receivers of different design, which are fluid cooled (single or two phase). The basic idea refers to a limitation of the solar flux in the image plane, more details can be found in the original report (Kistler, 1986). First of all, it should be noted that the target point strategy changes permanently during operation when the projected image on the receiver changes, so that the target point strategy represents a regulation and control variable of the entire system and is therefore not a static variable. The following basic options as aiming point strategies illustrated in Fig. 10.30 should also be understood in this sense.

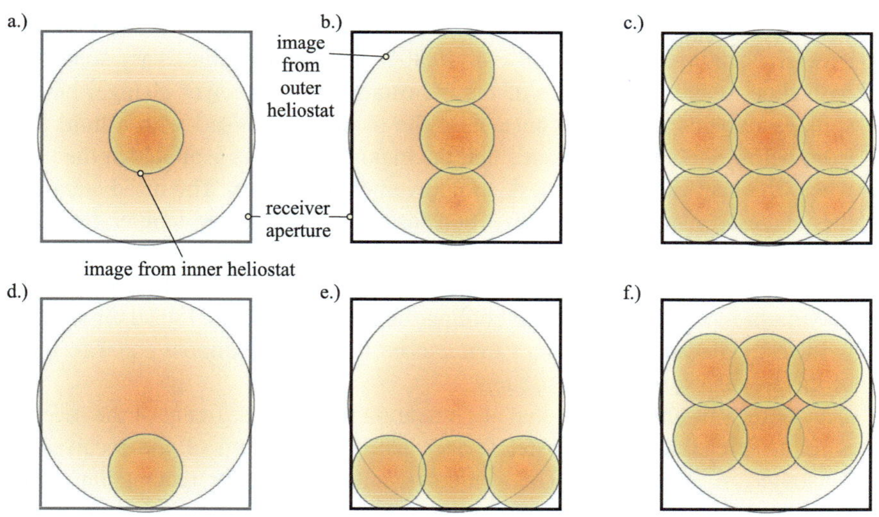

Fig. 10.30 Different options of heliostat aiming point strategies when operating a central tower receiver: **a** Central point method; **b** One-dimensional alignment; **c** Two-dimensional aiming point method; **d** Single aiming point at receiver foot; **e** one-dimensional lower section targeting and **f** user defined aiming point strategy

- In the single aiming point strategy as the simplest of all strategies all heliostats are aligned to the centre of the receiver as seen from the heliostat surface. This option illustrated in Fig. 10.30a produces the maximum flux on the receiver.
- One-dimensional alignment: The images of the heliostats are stacked over the "height" of the receiver aperture until the scattered light component (spillage) starts to increase. As shown in Fig. 10.30b, the smaller images of the inner (or best focused) heliostats can be distributed to more target points than the larger images of the outer (or less focused) heliostats. This strategy reduces not only the maximum flux but also the flux gradients on the receiver. This targeting option is particularly well suited for external receivers, as well as for cavity or flat plate receivers with elliptical shape. However, as the size of the images from the heliostats on the receiver changes with time, the one-dimensional alignment also changes with time, so both the number and position of the target points change.
- Two-dimensional target acquisition is similar to the previous option, except that the images are distributed in two dimensions, as shown in Fig. 10.30c. This results in even smaller peak fluxes. However, this strategy is only suitable for rectangular cavity or rectangular flat plate receivers. When used with elliptical receivers, the spillage will be larger. It is not very suitable for use in external receivers, since much of the reflected radiation flux is incident on the receiver only at a very small angle of incidence, where absorption is small.
- Method with a singular aiming point at the lower receiver section. Here, the heliostats are aligned as close as possible to the lower receiver section without excessive spillage, as shown in Fig. 10.30d. The aiming points vary somewhat with time as the image sizes of the heliostats on the receiver change, but the spillage remains relatively constant. If the receiver is cooled with a fluid flow from below, this method ensures that the maximum flux is available at the point where the heat carrier is coldest, so no overheating occurs. The disadvantage is the increased radiation and convection losses, as the average temperature over the receiver surface becomes greater. If the heat transfer medium flows into the receiver from above, the radiation and convection losses are minimized, but the maximum flux occurs where the cooling tubes are hottest and thus their strength is lowest. The maximum fluxes that can be realized are almost equal to those of the singular target point strategy.
- The strategy of one-dimensional alignment on the lower segment of the receiver aims at spreading the heliostat images along the bottom of the receiver, as illustrated in Fig. 10.30e. This option is suitable only for rectangular cavity openings or rectangular flat plates.
- User-defined target strategy (Fig. 10.30e): A user-defined target strategy can be defined for each zone, in which a rectangular dot grid is assigned to each receiver zone. This target strategy is not time-dependent. Here, the target points are fixed no matter what time of day or year, even if the image size and thus the spillage change over time.

It is also possible to combine the aiming strategies of the 2D aiming point method (Fig. 10.30c) and the 1D lower receiver section (Fig. 10.30e). In this case, the inner heliostats close to the tower are aligned to the bottom of the aperture or the flat receiver, while the heliostats further out distribute their focal target points over the entire aperture. This reduces the inner spillage, i.e. a radiation flux onto uncooled segments of the receiver or the roof. This method is suitable for cavity receivers.

10.3.2.4 Typical Result of Heliostat Field Simulation

The value of the optimal heliostat field efficiency for existing solar tower systems is mostly proprietary information and it is often not possible to make a direct comparison, but comparative calculations can be made. Once a compromise solution has been found for an optimal heliostat field from an optical point of view, a check is made to analyse how the annual collector efficiency or field efficiency varies with the receiver radius and the optical height of the tower. Figure 10.31a show the simulated annual efficiency of an external receiver field (perimeter field) as a function of receiver radius, while Fig. 10.31b depicts field efficiency as a function of tower height for the same field. The absolute numbers of the simulations should always be taken with caution and considered relative because the optimal field efficiencies change when other input parameters are assumed, especially tower height, which is closely related to blockage. In addition, both the inclination and the tracking error of the heliostats are of course key parameters and information about them is proprietary in nature by the manufacturer and therefore rather limited in access.

Taking these caveats into account, it can be seen from the comparison of Fig. 10.31a and b that the sensitivity of receiver size is higher than that of tower

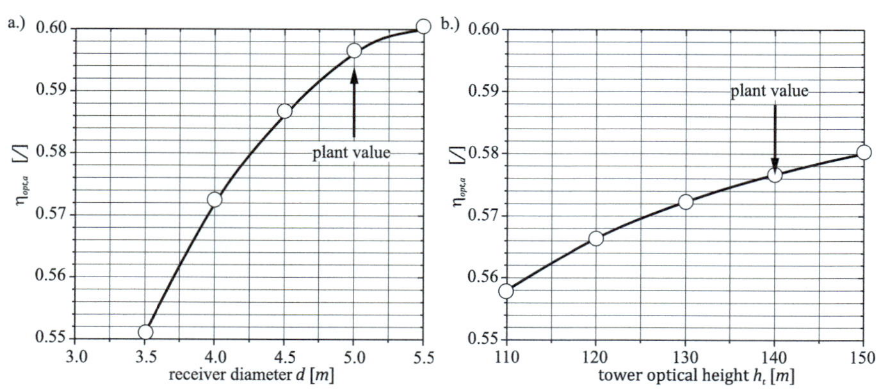

Fig. 10.31 Calculated annual optical efficiency of a heliostat array for an external liquid salt receiver (Gemasolar, Spain, $\Phi = 37°33'N$); **a** sensitivity as a function of receiver diameter d; **b** dependence as a function of optical tower height h_t (m); modified from Collado and Guallar (2013)

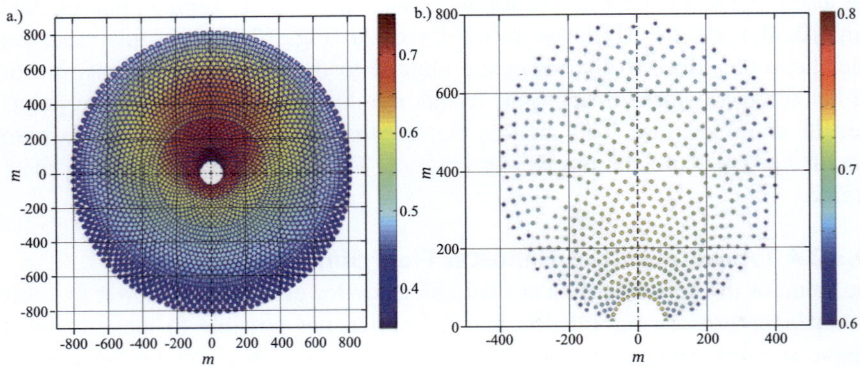

Fig. 10.32 Simulated annual optical efficiency of an optimized heliostat field **a** external receiver at $(\Phi = 26°14'N)$ rom Atif and Al-Sulaiman (2015); **b** cavity receiver at PS10 ($\Phi = 37.4°N$) from Zhang et al. (2016)

height (greater pitch), although again it should be noted that the optimized layouts and their efficiencies could be altered at other heights.

The optical efficiencies in themselves are not yet fully meaningful quantities representing performance of the composite of heliostat array and tower, because the dimensions of the receiver are closely related to thermal losses, which are not considered in an optical calculation.

This optical efficiency consists of a wide range of values of each heliostat, which ranges from $\eta_{opt,a} = 0.3 - 0.8$ depending on the size of the heliostat field. Figure 10.32a show the calculated annual optical efficiency of each individual heliostat in the mirror field for the Dhahran location ($\Phi = 26°14'$N, Saudi Arabia) for a cylindrical external receiver. It is easy to see that the efficiency of the north-facing mirrors is considerably higher than that of the south. Likewise, the near-tower heliostats have a value of about 0.8, while the values at the extreme edge of the mirror field hardly exceed 0.35. Typical annual efficiencies are barely above 60% as shown in Fig. 10.31. A similar distribution of the annual efficiencies is also found in the case of cavity receiver. Also, the tower-close mirrors exhibit a higher efficiency than the external mirrors, as shown in Fig. 10.32b. However, the deviations of the annual optical efficiencies of the individual mirrors are significantly smaller than in peripheral fields, although the annual efficiency of the entire field is of similar magnitude.

10.4 Control and Regulation of Thermal Loops in Concentrating Systems

The main objective of the control system of a solar thermal plant is to maintain the thermal energy supply at the receiver outlet at a given set point even in the presence of disturbances such as solar radiation (e.g. due to clouds), fluid inlet conditions, deterioration of reflectivity (e.g. due to pollution) and changes in demand of the consumer (process heat, power conversion system). This results in a control problem that can initially be described as nonlinear, time-varying with pure time delays.

Compared to a conventional fossil fuel heat source, which can be easily controlled by regulating the fuel supply and eventually reach a steady state, a solar system can never reach a steady state a-priori, since the position of the sun and the irradiance are constantly changing and, in addition, other disturbances affect the system. Therefore, control stability is of crucial importance in the control of a solar thermal system. Abrupt disturbances (e.g. due to clouds) can substantially change the power input and lead to undesired or even structurally dangerous component or even system conditions, where the system would have to be shut down, wasting valuable energy. This may also require additional components to increase the thermal inertia (steam drums or buffer tanks, heat tracing, etc.) in order to meet the operator's requirements, which conversely means increased investment costs.

A stable control system is therefore the basic pre-requisite for reducing downtime and thus increasing the availability of the plant. This is particularly relevant in an industrial environment, where a stable, highly available heat source at specific setpoints is not only a standard requirement, but also taken for granted.

10.4.1 Considerations to Control Solar Thermal Loop Systems

Let us first consider a simple solar circuit consisting of a solar absorber of a pump and a heat load, as shown in Fig. 10.33a. The absorber tube represents the heat source ($\dot{Q}_{net}$) into which the heat transfer fluid enters at a temperature of T_i. When flowing through the absorber, the fluid is heated to the outlet temperature T_o. The mass flow through the circuit $\dot{m}$ is adjusted by means of a pump in such a way that a constant outlet temperature T_o is maintained. In the case of the concentrated solar thermal systems, the energy input to the absorber tubes can also be controlled by manipulating the mirror setting.

Since the dynamics of the pipeline network are determined from the point of view of the control of the absorber outlet temperature, it is expedient to set up the energy balance by a control volume of the absorber pipe, as shown in Fig. 10.33b leading to

$$\dot{Q}_{solar} = \left(\dot{H}_{out} - \dot{H}_{in}\right) + \dot{Q}_{loss} + \frac{dQ_{abs}}{dt}, \qquad (10.70)$$

in which $\dot{Q}_{solar}$ is solar heat flux absorbed, $\dot{Q}_{loss}$ represents the the heat loss in the control volume, $\dot{H}_{out}$ and $\dot{H}_{in}$ denote the enthalpy fluxes at out- and inlet and dQ_{abs}/dt describes the temporal heat change along the absorber tube. The change of the enthalpy between in- and outlet can be given as

$$\left(\dot{H}_{out} - \dot{H}_{in}\right) = \dot{m}c_p(T_o - T_i), \qquad (10.71)$$

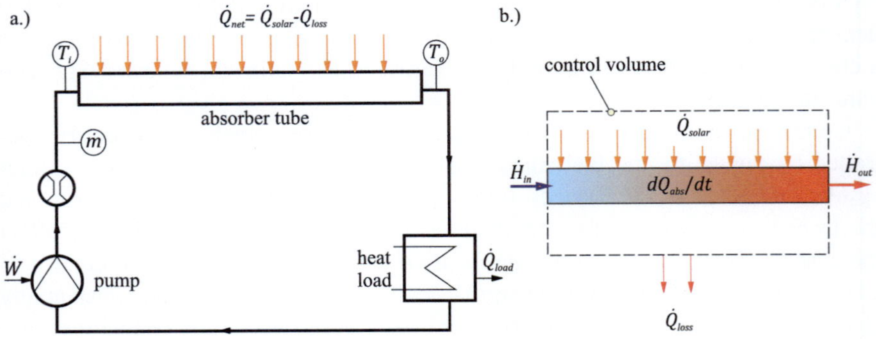

Fig. 10.33 **a** schematic structure of a single-phase solar thermal circuit; **b** control volume of the absorber tube as the starting point of the outlet temperature control

where $\dot{m}$ is the mass flow, c_p the heat capacity of the heat transfer fluid, T_o and T_i denote the mean out- and inlet temperatures, respectively.

During the actual control process, the change in the absorber temperature is close to the outlet temperature, so that the temporal change of the thermal absorber energy is relatively small (<5% during operation of the collector and zero in the steady state). This allows to reformulate the temporal change of the thermal absorber energy approximately as:

$$\frac{dQ_{abs}}{dt} = C_{abs}\frac{dT_o}{dt}, \tag{10.72}$$

in which C_{abs} describes the integral thermal mass of fluid and structure of the absorber assembly in (J/K). Rearranging Eq. (10.71) and assuming $\dot{Q}_{net} = \dot{Q}_{solar} - \dot{Q}_{loss}$ we get:

$$\dot{Q}_{net} = \dot{m}c_p(T_o - T_i) + C_{abs}\frac{dT_o}{dt}. \tag{10.73}$$

Resorting this with respect to the set-point temperature T_o this results in:

$$T_o + \underbrace{\frac{C_{abs}}{\dot{m}c_p}}_{\tau}\frac{dT_o}{dt} = T_i + \frac{\dot{Q}_{net}}{c_p}\underbrace{\frac{1}{\dot{m}}}_{F}. \tag{10.74}$$

With the help of a Laplace transformation, the preceding differential equation can be transposed into frequency space, and we obtain the following result using $\tau = C_{abs}/\dot{m}c_p$ and $F = 1/\dot{m}$ after some rearrangements:

$$T_o(s) = \left(\frac{\dot{Q}_{net}}{c_p}F(s) + T_i(s)\right)\left(\frac{1}{1+\tau s}\right). \tag{10.75}$$

Since the heat transfer fluid requires a finite time to flow through the absorber tube from the inlet to the outlet, a dead time can be introduced to account for this fact. This time delay is a function of the heat capacity of the absorber volume and is represented in the Laplacian space as exponential functions in the way $\exp(-s\kappa)$, where κ represents the dead time. Thus, Eq. (10.75) can be modified to account for the response of the system caused by the dead time.

$$\underbrace{T_o(s)}_{\text{output}} = \left(\underbrace{\frac{\dot{Q}_{net}}{c_p} F(s)}_{\text{input}} + \underbrace{T_i(s)}_{\text{disturbance}} \right) \underbrace{\left(\frac{1}{1+\tau s} \right)}_{\text{absorber/plant}} \underbrace{\exp(-s\kappa)}_{\text{dead time}}. \tag{10.76}$$

This simplified model considers the collector inlet temperature treated as a disturbance variable, since it depends mainly on the load, which cannot be controlled by the solar thermal circuit. The input variable to achieve the control variable is the reciprocal of the mass flow rate (i.e. F), which is obtained by means of a change in the pump mass flow rate ($\dot{m}$). The shortcomings of the model described using Eq. (10.76) include three aspects:

- The relationship between system input and output is non-linear since the output variable depends on F and not on $\dot{m}$.
- The system time constant depends on the system input through relationship $\tau = C_{abs}/\dot{m}c_p$.
- The absorber/system parameter itself is a time-dependent function.

The three identified challenges make it difficult to uniquely control such a thermal piping network using linear time invariant (LTI) control and therefore requires more advanced methods, which are essentially based on a distributed element model. Thereby, the solar thermal loop system is still described in a simplified manner, but all the entire piping network is depicted in form of nodes.

Let us first consider the solar thermal problem as a simple first-order control problem, as suggested by the left-hand side of differential Eq. (10.74). If we assume that τ would be a constant and the right-hand side of the equation is zero or a constant, then we would have a system that is negatively exponential tending towards a limit and does not consider inertia. If we complement this system by considering a thermal inertia, assuming that all the thermal inertia can be approximated by a single pure time delay, we obtain the following transfer function in Laplacian space

$$G(s) = \frac{b}{as+1} \exp(-s\kappa), \tag{10.77}$$

where a describes the system time constant, b is the static gain, and κ is the dead time. This model approach is called a "*lumped element*" model. The main simplifying assumption for such a model is that the absorber tube is considered in its entirety as a single element. It neglects the fact that there is a temperature distribution along the tube. Due to its simplicity, it is well suited to experimentally identify individual parameters representing, for example, inertia, which is a great advantage over higher order models.

In the first step, we now want to analyse the extent to which a simplified "lumped parameter" model can represent a solar thermal circuit. To determine the static gain (b) and the dead time (κ), we consider the response of the system to an abrupt change of the input variable, here the mass flow rate ($\dot{m}$) to the controlled variable, i.e. the absorber outlet temperature T_o. The basic idea of the method is to apply an input of a certain magnitude and frequency and record how the process responds to the step input. A step function is ideal because it includes all frequencies and can be implemented experimentally. Figure 10.34 show the response of the system to a step change in mass flow as a function of time of a Fresnel test system operated with water at real conditions. The increase in mass flow leads to a reduction in absorber exit temperature at nearly constant irradiance. If one now evaluates the system parameters in terms of a lumped-parameter approach, two methods are offered to analyse the response function:

- In the first method, as depicted in Fig. 10.34a shows straight line is drawn through the inflection point of the response curve, since the slope of the inflection point line in a first order system determines the maximum rate of change of the output. This can be interpreted to mean that if the output of the system changes at a constant rate, the steady-state end state is reached after a time constant. Evaluating the system according to this criterion, we obtain for the static gain $b = K_o/K_i = -8.94\text{Ks/kg}$, the time constant $\alpha = 61.7$ s and the dead time $\kappa = 12.7$ s.
- The second option for identifying the system parameters of the same experiment identifies two time variables t_1 and t_2, which indicate the times at which the

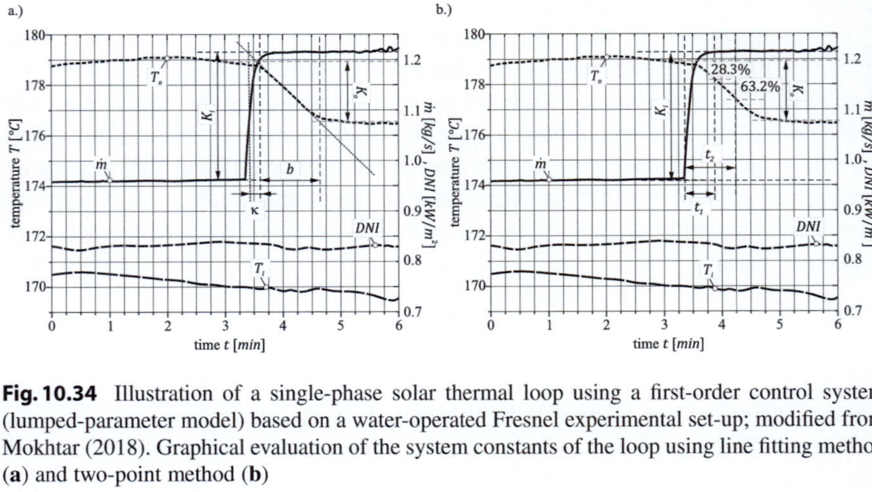

Fig. 10.34 Illustration of a single-phase solar thermal loop using a first-order control system (lumped-parameter model) based on a water-operated Fresnel experimental set-up; modified from Mokhtar (2018). Graphical evaluation of the system constants of the loop using line fitting method (**a**) and two-point method (**b**)

system reaches 28.3 and 63.2%, respectively, of the steady-state equilibrium condition of the control variable T_o, see Fig. 10.34b. This leads to the following result for the static gain $b = K_o/K_i = -8.94\text{Ks/kg}$, the time constant $a = 1.5 \cdot (t_2 - t_1) = 31.9$ s and the dead time $\kappa = 1.5 \cdot (t_1 - 1/3 \cdot t_2) = 19$ s.

Thus, we see that both methods yield substantially different results for the same system, the cause of which is an inadequate representation of the system response using a first-order system. The second method gives a smaller time constant, which is compensated with a larger dead time, while the method is more error-prone due to the line fitting. Hence, we require a more sophisticated control system.

10.4.2 PID Based Control Strategies and Options for Solar Loop Systems

Most thermal cycle networks can be approximated by a linearized high-order system, since they are usually composed of a large number of smaller first-order dynamical elements, the synthesis of which constitutes a linearizable high-order model, see Camacho and Bordons (2007). In the case of the absorber tubes of the receiver tube(s), these small elements correspond to mass volume elements distributed along the length of the absorber. We've used this type of modelling already in the design and dimensioning of the mirror field to also assess the thermal yield, see Sect. 10.2.5 Collector field design of line focussing systems.

The only difference between the control model and the design model of the mirror field is that a control variable, usually the collector outlet temperatureT_o, is specified. In control engineering, this method, which is ultimately based on the solution of a set of equations of linear algebraic differential equations, is called distributed element method. This method is very time-consuming and, unfortunately, as a control model, does not protect against oscillations, possible overshooting of the control variable (such as the collector outlet temperature). For the operation of a power plant and its control, however, the operator needs a real-time capable control and regulation system, which handles possible changes of input variables, e.g. increased heat load demand, changes of boundary conditions, such as irradiation fluctuations, under consideration of operational limitations (maximum pump throughput, permissible maximum temperatures, etc.) in a suitable manner quickly, efficiently and, if possible, predictively, in order to achieve the desired controlled variable (set-value).

Unlike fossil fuel power generation processes, where the output of the heat source is predetermined from the product of the fuel feed (manipulated variable—$\dot{m}$) and a constant specific lower calorific value of the fuel, a solar thermal power plant is an infinite large outdoor heat exchanger whose heat source cannot be manipulated. Rather, the heat source meaning the solar radiation must be treated as a disturbance variable from the control point of view.

As we have seen in the introductory considerations, simple linear time invariant (LTI) controls such as *PID* control is not able to deal with changing dynamics, due to non-linearities and disturbances. However, if conventional *PID* controllers are used, they must be tuned with low gain, resulting in slow response time, or if narrowly tuned, they can result in significant overshoot of the controlled variable when the dynamics of the process change due to changes in environmental and/or operating conditions, or the controlled variable, such as the pump, is ramped up or down any number of times. Both of these have an impact on the number of hours the field can effectively operate and the life of the components. In addition, if the control specifications are very tight and the control system has the process operating at high frequencies where the uncertainties are greater, or plants with complex dynamics that cannot be approximated by simple low-order linear models, modified control strategies are inevitable.

To summarize, solar thermal loop networks are characterized by nonlinearity, complexity, and changing dynamics under simultaneously changing environmental conditions, which make control difficult, while simultaneously fast and efficient control requires a simplification of the model description. The main characteristics are:

- solar radiation acts as a fast perturbation with respect to the dominant system time constant of the circulatory network;
- time-varying transport delay between input and output variables, due to the inertia (delay) of the manipulated variable (heat transfer fluid flow) in all pipelines;
- the use of linearized dynamics leads to the fact that the dynamics at each operating point is different.

This results for control modelling that it must reproduce the physical model in simplified form so that a system excitation by a signal excitation signal (heat transfer fluid flow or solar radiation) the changes at the system output are small. A detailed presentation of different control logics and a classification can be found in the review articles, Camacho et al. (2007a, 2007b) as well as the text book (Camacho & Bordons, 2007); in the following, some of the most commonly used ones are shown.

PID-Control

A *PID* controller contains three control terms Proportional (P), Integral (I) and Derivative (D), which are applied to the controller output to achieve the setpoint accurately and in an optimal manner. The block diagram in Fig. 10.35a shows basic functionality of a *PID* controller that continuously determines a deviation (error) $e(t)$, which is the difference between the set-point $r(t)$ and a measured process variable $y(t)$, and applies a correction based on proportional, integral, and derivative terms. The controller attempts to minimize the error over time by setting a control variable $u(t)$, for example valve position or flow rate change, to a new value determined by a weighted sum of all control terms. Defining $u(t)$ as the

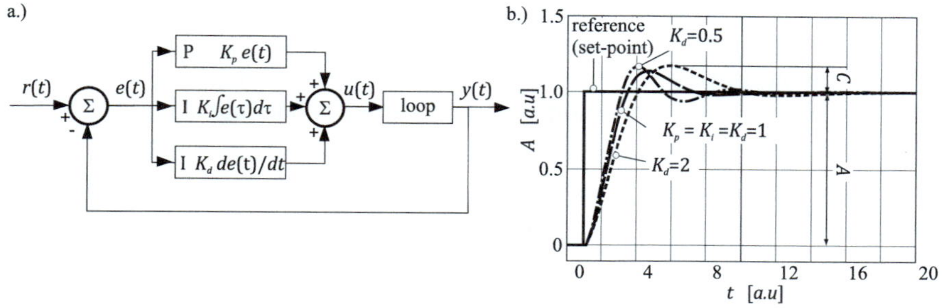

Fig. 10.35 **a** Schematic structure of the information flow in a *PID* controller, **b** Qualitative response function of a *PID* controller at different gain factors K_d

controller output, the final form of the *PID* algorithm is:

$$u(t) = K_p e(t) + K_i \int_0^t (\tau)d\tau + K_d \frac{de(t)}{dt}, \qquad (10.78)$$

in which K_p is the proportional gain, K_i the integral gain and K_d the derivative gain, which are all tuning parameters. t denotes the instantaneous time and τ is the integration variable ($0 \le \tau \le t$). Thus, the Laplace transfer form of the *PID* controller is

$$G(s) = K_p + K_i/s + K_d s. \qquad (10.79)$$

Tuning a control loop is the adjustment of its control parameters (proportional band/gain, integral gain/reset, derivative gain/rate) to the optimal values for the desired control behaviour. Here, stability (no unlimited oscillation) is a basic requirement, but beyond that, different systems have different behaviour, different applications have different requirements, and the requirements may conflict with each other. *PID* tuning is not trivial, even if there are only three constant parameters. Accordingly, there are different methods for control loop tuning, such as Ziegler-Nichols, Tyreus Luyben, Cohen-Coon amongst others. Designing and tuning a *PID* controller seems conceptually intuitive but can prove difficult in practice when multiple (and often conflicting) goals such as short settling times, which are intrinsically present in *PID* controllers (see cf. Fig. 10.35b), and high stability are to be achieved. Therefore, designs will be repeatedly adjusted by computation until the closed-loop system performs as desired or compromises are made. Thermal loop systems exhibit some degree of non-linearity as discussed. At full load where the deviations are small, *PID* controllers may operate, but fail to operate during start-up/shutdown; using an operating condition-based parameter set, this can be compensated for or corrected by gain scheduling. If the *PID* controller parameters (the gains of the proportional, integral, and differential elements) are incorrectly chosen, the controlled process input can be unstable, i.e. its output diverges, with or without oscillation, and is limited only by the system bounds. The instability is caused by excessive gain or resonance, especially when a significant delay (dead time) is present. In general, stabilization of the response is required, and the process may not oscillate at any combination of process conditions and set points, although sometimes marginal stability (limited oscillation) is acceptable or desirable. Mathematically, the origin of instability can be seen in the Laplace domain. The total loop transfer function is:

$$H(s) = \frac{G(s)P(s)}{1 + G(s)P(s)}, \tag{10.80}$$

in which $G(s)$ is the *PID* transfer function and $P(s)$ that of the total system. The system is called unstable if the transfer function of the closed loop diverges for some$s$, which occurs if$G(s) \cdot P(s) = -1$. Stability is assured provided that match $G(s) \cdot P(s) < 1$ for frequencies that have high phase shifts. A more general formalism for this effect is known as the Nyquist stability criterion.

The fundamental difficulty with *PID* control is that it is a feedback control system with constant parameters and no direct mapping of the physical process, so the effect is a response and therefore a trade-off. *PID* controllers generally have difficulty with nonlinearities and usually pass them on to the response time. They therefore do not respond to changing process behaviour (for example, the process changes after it heats up) and have delays in responding to large disturbances.

For example, *PID* s are problematic in a flow control application that we used to control the absorber outlet temperature. It results in variable loop sensitivity depending on operating point, which requires damping action to avoid instability. To evaluate the control result, the magnitude of the difference between the set point and the process variable is used without considering time. This method of determining controller performance is called the peak overshoot ratio (*POR*) criterion, see Fig. 10.35b, and is defined as follows:

$$POR = \frac{C}{A}. \tag{10.81}$$

The *POR* should be as small as possible. In the case of flow control of temperature, this can be compensated, for example, by using the nonlinear characteristic of the valve or the pump power in the control algorithm.

Noise in Derivative Term of the Controller

One problem with the derivative term is that the measurement data is available faster than the piping network can react due to its inertia; in other words, high-frequency measurements, or a noisy measurement signal (process noise) can cause large changes in the output. In such a case, the measurement data can be filtered using a low-pass filter to eliminate the noise components; low-noise instrumentation is therefore important. The other option is to turn off differential band so that the *PID* controller works like a *PI* controller.

As we have seen in the considerations at the beginning of the section, a *PID* control even with the inclusion of a dead time is not an optimal solution, not only because it is error-prone, but also because of the variability of the dead times. The main improvement of a *PID* control is to provide forward control using model-based prior knowledge about the system behaviour and to use the *PID* control only for verification. Alternatively, *PID*s can be modified in a more minor way by changing the parameters (either operating point dependent gain scheduling) or improving the measurement (higher accuracy and/or low-pass filtering) or cascading multiple *PID* controllers. We will consider these in the following.

PID with Feedforward Control

The deficits of a PID controller can be substantially reduced with the help of feedforward control by integrating model representations of the process to be controlled into the feedforward control; feedforward control falls into the category of model-based control methods. The basic idea of the models is to take into account the effects of measured disturbances on the control variable. The feedforward controllers are usually combined with feedback controllers to capture all the deviations that the feedforward controller does not consider. Figure 10.36a shows the schematic diagram of the control loop of a feed-forward control with parallel *PID* controllers. The feedforward control is intended to serve two purposes; one is to provide measurable disturbances, which in the case of the simple single-phase heat carrier loop consist mainly of changes in inlet temperature and/or solar irradiation. The other purpose is to provide linearization of the system close to the operating point. Without feedforward control, the feedback controller would have to operate under highly fluctuating conditions, so by incorporating a feedforward controller, the feedback controller is only needed to correct for changes that the feedforward model did not account for. In the control chain, the outputs of the *PID* and the feedforward controller are added, so that the control signal is generated in the following way:

$$\underbrace{\dot{m}_{cont}}_{\text{output}} = \underbrace{\dot{m}_{FFW}}_{\text{feed forward term}} - \underbrace{\dot{m}_{PID}}_{\text{PID term}}, \tag{10.82}$$

in which $\dot{m}_{cont}$ is the mass flow rate transferred from the controller to the process, $\dot{m}_{PID}$ the corrective term produced by the *PID* controller and $\dot{m}_{FFW}$ the feedforward control term.

In the example considered, the feedforward controller is to calculate the required mass flow rate on the basis of a system model. Therefore, we have to transform Eq. (10.73) and obtain the following for the required mass flow of the feedforward controller

$$\dot{m}_{FFW} = \frac{\dot{Q}_{net} - C_{abs}(dT_r/dt)}{c_p(T_r - T_i)}, \tag{10.83}$$

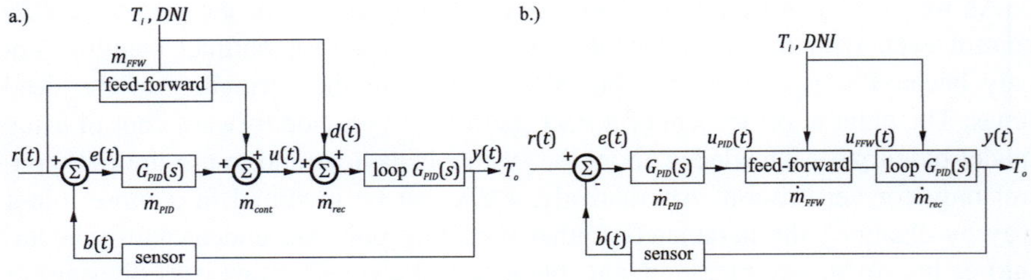

Fig. 10.36 Schematic design of a forward control of a thermal circuit in a parallel arrangement **a** and a series connection **b** with typical solar field quantities

in which T_r denotes the set-point temperature. Thus, inclusion of the model requires knowledge of the inlet temperature (T_i), the change in absorber internal energy, and the net power transferred to the fluid in the absorber $\dot{Q}_{net}$, which in turn is a function of direct solar irradiance (*DNI*), specular reflectance, total optical efficiency, solar altitude, and heat losses.

This determination of the feedforward term according to Eq. (10.83) only is not sufficient since it generates undesired reactions and leads to instability. This is because when the set point temperature T_r approaches the inlet temperature T_i, an infinitely large value results, leading to a very large flow rate. Conversely, when T_i is greater than T_r, the required mass flow is negative, causing the pump to shut down and, as a result, overheating of the system occurs. This situation is beyond the control of the mass flow controller. The correct action of the mass flow controller in the latter case would be to increase the flow to the maximum so that the temperature rise of the inlet temperature is reduced to a minimum. This unphysical behaviour of the pure feedforward controller is thus a consequence of the non-linearity of the system. To eliminate such behaviour, the feedforward term can be linearized or limited to a suitable safe range. This can be achieved by relaxing the feedforward control term with a factor ($C_0 < 1$), adding a small base mass flow ($C_1 \cdot \dot{m}_{rec}$) and a constant (C_2) in the following way.

$$\dot{m}_{FFW} = C_0\left(\frac{\dot{Q}_{net} - C_{abs}(dT_r/dt)}{c_p(T_r - T_i)}\right) + C_1 \cdot \dot{m}_{rec} + C_2. \tag{10.84}$$

By the constant C_2, the flow is shifted to larger values, so that overheating is counteracted. If the mass flow gets negative, overheating is detected, and the mass flow is increased to a high value accordingly. Alternatively, feedforward control can be accomplished by connecting the controllers in series as shown in Fig. 10.36b.

The result of such a feedforward control by means of a parallel configuration is shown by the example of a thermo-oil cooled solar thermal loop in Fig. 10.37. It can be seen that there is no excessive overshooting of the collector outlet temperature T_o and the set point temperature is quickly reached.

Nevertheless, this feedforward control model requires the knowledge of the current *DNI* value, which is not only subject to considerable temporal fluctuations

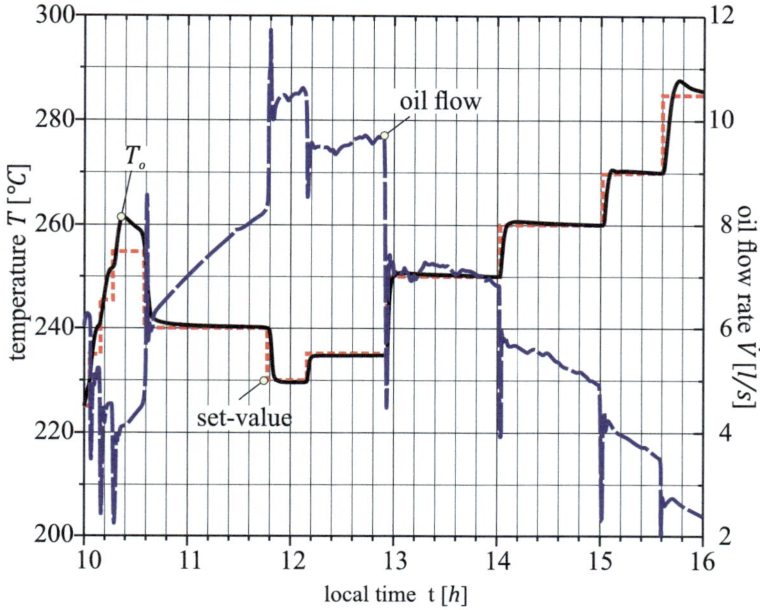

Fig. 10.37 Measured collector outlet temperatures of an oil-cooled solar thermal loop using a parallel feedforward control system modified from Camacho and Bordons (2007)

and thus represents a permanent disturbance of the control, but it is also costly and requires constant maintenance of the measuring device. An alternative is to estimate the net power of the collector $\dot{Q}_{net}$ from previous data on inlet and outlet temperatures (T_i, T_o) and mass flow rate. The result is then used for the current time step to calculate the feedforward term. Due to the thermal inertia of the circuit, it is useful to use a weighted moving average of the past data so that any overshoot of the target setpoint remains small; this leads to the following approach.

$$\overline{\dot{Q}} = \sum_{i=0}^{k} w_i \cdot \left(\dot{Q}_{net} - C_{abs}(dT_r/dt)\right)(t - i\Delta t), \tag{10.85}$$

in which k is the number of time steps, Δt is the time step length and w_i is the weight of each time step, which can be experimentally adapted. Thus, the resulting feedforward term reads to:

$$\dot{m}_{rec} = \frac{\overline{\dot{Q}}}{c_p \cdot (T_r - T_i)}. \tag{10.86}$$

It must be mentioned that this methodology for determining the feedforward term carries stability risks, since the feedforward term depends on both the controlled variable (T_o) and the manipulated variable ($\dot{m}_{rec}$), of which the temperature changes only slowly due to inertia while the mass flow leads to an immediate change of the calculated $\overline{\dot{Q}}$ and provokes extreme values (maximum or minimum) in the mass flow.

This problem is mitigated by longer time windows with moving average and some smoothing of the calculated $\overline{\dot{Q}}$. Furthermore, the calculation of the feedforward term can also induce undesirable feedback paths since it uses the controlled parameter as part of the calculation.

10.4.3 Alternative Control Strategies for Distributed Collector Systems

So far, it has only been considered that the flow is the manipulated variable to achieve the desired output. The background of the measure is that the solar yield as far as possible is maximized. However, all point- or line-focusing solar thermal systems also offer the possibility to control the solar power in addition to the mass flow. In the simplest case of using feedforward control, the mass flux is replaced by the solar thermal power on the receiver in Eq. (10.83). This assumes that the mirror power of each mirror segment or heliostat is a-priori known. The controller calculates the relative contribution of each mirror at each time point based on the current position of the sun. It then sorts the mirrors based on their relative power using a previously selected sorting algorithm. Once the mirrors are sorted, the controller selects the mirror combinations that most closely match the required power. This technique for selecting mirror combinations reduces the impact of the discrete nature of the manipulated parameter, but intrinsically results in dumping a fraction of the solar irradiance output.

There are a number of other control options that are beyond the scope of this paper; a detailed description can be found in Camacho and Bordons (2007). Here, we will only briefly present the basic ideas of the different control logics.

Adaptive Control (AC)

The basic idea of adaptive control is to adapt the controller to dynamic changes in the process. It can be said that adaptive control is a special type of nonlinear control in which the state variables can be divided into two groups that move on two different time scales. The state variables that change faster correspond to the process (internal loop) variables, while the state variables that change slower correspond to the estimated process (or controller) parameters. Adaptive controllers are traditionally classified into one of the following families: model reference adaptive controllers (MRAC) and self-optimizing controllers (STC). Distributed collector system (DCS) control for regulating the flow rate in the collector array usually uses a combination of feedback and feedforward control, which incorporates a nonlinear mechanism to work effectively. However, many adaptive control schemes exhibit oscillatory behaviour when rapid responses to setpoint changes and disturbances are required, the cause of which is the existence of resonant dynamics at a low frequency. This is remedied by the use of frequency-based internal model control (IMC), in which a frequency-based interpolation model is integrated in both the model of the plant and the controller in such a way that a specific frequency response of the controlled system is specified at the selected interpolation points. Frequency-based adaptive control requires knowledge of the frequency response of the plant at all interpolation points; this can be obtained by using banks of bandpass filters for the input and output of the plant.

Gain Scheduling (GS)

Some controllers have the ability to adapt to changes in process dynamics but are not considered true adaptive controllers. This is the case with gain scheduling (GS) controllers, where the process dynamics can be related to the value of some process variables that can be measured with respect to the operating point or environmental conditions. If the dynamic characteristics of the process can be derived from measurable variables, the controller parameters can be calculated from these variables.

In this type of control, only the inner loop emerges, and the parameter update can be considered like a feedforward that changes the controller gains. Explicitly detecting nonlinearities of the path and exploiting them could lead to improvements in performance and stability, but at the cost of increasing the size of the controller feedforward. This means that the output of the controllers used is not the flow rate but the temperature setpoint for this series compensator, which already dampens the nonlinear characteristics of the plant to some extent.

Time Delay Compensation (TDC)

Time delay compensation (TDC) aims to design controllers without considering the pure delay of the plant and expects the closed-loop response after the delay to be the one expected by the design without considering the delay in the dynamics of the process. Several TDC schemes have been developed for controlling collector arrays using a simplified transfer function model including resonance characteristics. Mostly, these controllers are based on a parallel control structure since it effectively combats the resonant dynamics of the system. The implementation of the control law is very simple, since the controller has only three adjustment parameters that can be set with a classical step identification test, and a filter does not need to be adjusted.

Optimal Control (LQG)

This method is based on the hypothesis that the system to be controlled is linear, the criterion is quadratic, and the random perturbations are Gaussian. The problem is solved separately: first, a state estimate is obtained by an observer (deterministic consideration) or by a Kalman filter (stochastic case). A linear control law is applied to this estimate, which is feedback, and thus the full controller is obtained. This method is extremely sensitive to inaccuracies in the parameters and to structural changes. To solve the problems arising from the deterioration of robustness due to the introduction of the observer, the method known as LQG with Loop Transfer Recovery Mechanism (LQG/LTR) can be used.

Nonlinear Model Predictive Control (NMPC) techniques.

Since most controlled processes exhibit inherently nonlinear behaviour, there are incentives to develop MPC control strategies based on nonlinear process models obtained from both physical principles and data. In these cases, a nonlinear programming problem must be solved in real time at each sampling period, rather than the quadratic linear problem typical of standard MPC. The main difficulties of these methods are that the theoretical analysis of the characteristics of the control loop (stability and robustness) is very complicated because of the nonlinearity, and that when solving a nonlinear programming problem in each sampling period, it is difficult to guarantee the convergence of the algorithm in a reasonable time.

In addition, there are a number of other control options such as nonlinear control, robust control, fuzzy logic control, or neural network controllers.

The control systems considered up to now essentially dealt with one controlled variable, the absorber outlet temperature or alternatively the thermal power, which is sufficient as an individual control even for distributed collector arrays. Since they essentially deal with only one input and one output variable, the control mode is called *SISO* (single input–single output). In the case of solar towers or even widely distributed direct steam systems, control via one input or output variable is no longer possible, so that a supervisory controller is required to monitor the operation of the various controllers in the solar field and to handle the system transitions between different operating modes. This controller is called a Balance of Plant (*BOP*) monitoring controller and has the following tasks:

- Power staging
  To avoid abrupt changes in solar input power, the *BOP* controller regulates the power input from the solar field. This is achieved by determining the number of mirrors focused on the absorber at a given time. For example, in a direct evaporation system, when the vapor pressure exceeds a certain limit, the supervisory controller gradually reduces the power supplied by the solar field. Conversely, when more power is needed, it gradually increases it. Similarly, when a cloud field passes, the mirrors are taken out of focus and refocused.
- Cloud detection (weather control)
  During operation, it is important to detect whether a cloud is over the solar field, or the atmosphere is hazy. This also requires control via the *BOP* controller, which can adjust system parameters in anticipation of the dynamics. For example, if a cloud field crosses the solar field, a power staging or flow reduction must be applied to maintain the outlet temperature. Cloud detection is either based on a direct *DNI* measurement or combined with a weather forecast. For example, two system flags (binary quantity) can be used to describe the cloud situation in the *BOP* controller. When the cloud detection flag is set, the controller defocuses mirrors away from the focal point and when the cloud

has passed, the flag is reset, and power is incremented. The power ramping continues until the cloud detection flag is completely reset.

- Scheduling the recirculation flow setpoint.
  The mass flow in the solar field is controlled by the recirculation controller. The mass flow setpoint is specified by the *BOP* controller and adjusted according to the operating conditions. When the solar array is running normally, a reduced flow rate is used to conserve electrical energy; for example, this is required for single-phase heat transfer media such as salts or liquid metals to prevent freezing of the heat transfer fluid during night-time operation. In direct evaporating systems, under certain operating situations, such control may be required to actually increase the flow rate to improve absorber cooling and avoid unfavourable two-phase flow patterns. In the latter application, this type of control becomes necessary in situations such as low solar irradiance (low *DNI*), cloud penetration, overheating, at the end of operation, and during the flooding process.
- Control of exceptions and emergency treatment and detection
  The *BOP* control system ensures the operation of the solar thermal plant within the previously defined safe operating limits. Therefore, one of its most important tasks is to identify emergency situations at an early stage, initiate countermeasures or repair the damage and, if all measures fail, transfer the plant into a "safe" idle mode. This requires the control system to issue control exceptions when necessary. Operating situations where the supervisory controller will intervene, and issue control exceptions include for example:
  - Loss of flow due to pump failure, clogging, leakage, and cavitation. Detection by the *BOP* controller requires instantaneous backlash defocusing to prevent receiver overheating.
  - Overpressure and/or overheating: If the *BOP* controller detects the exceeding of a predefined limit for pressure, temperature, or other operating variable, it first defocuses the mirrors to reduce the respective operating variable. It thus represents the first line of defence against overpressure or overtemperature.

As we have seen, every control system is based on the measurement of numerous operating parameters in the solar field. In the previous chapters, we have already shown methods to measure these parameters, such as irradiance (Chap. 3), temperature and heat flux (Chap. 4), pressure and mass flow (Chap. 5). But some of the operating variables change during operation and require continuous maintenance of the system. Among the most important time-dependent variables is soiling and aging, especially of the mirrors/heliostats, which is discussed below.

10.5 Solar Field Operation and Maintenance

The instantaneous optical efficiency of a concentrating solar thermal system (CST) according to Eq. (10.59) results from the product individual efficiencies of the respective mirror i:

$$\eta_{opt,i} = \rho_i \cdot \eta_{at,i} \cdot \eta_{cos,i} \cdot \eta_{sb,i} \cdot \eta_{sp,i}, \tag{10.87}$$

in which $\eta_{cos,i}$ represents the cosine factor, $\eta_{at,i}$ the absorption and scattering of the reflected radiation in the air mass between the collector and the receiver, $\eta_{sp,i}$ is the spillage factor, which is the fraction of the reflected radiation, which falls outside the receiver opening; $\eta_{sb,i}$ is the shading and blocking effect, and ρ_i is the ratio between the incident and reflected radiation at the mirror surface. While the last three terms are determined exclusively by the relative position of the collector and receiver at a specific time and interactions of a particular collector with its neighbours, represent the first two terms of the right side describing the reflection at the interface and the attenuation of the light beam in the atmosphere from the mirror to the receiver, quantities substantially influenced by local meteorological and environmental conditions. In contrast to their efficiencies, they cannot be determined exactly by means of Monte-Carlo methods but are subject to stochastic meteorologically determined fluctuations at the respective location.

The weather dependence and thus the attenuation of the reflected light by atmospheric changes, i.e. the term $\eta_{at,i}$, evades manipulability by the system operator.

But, the mirror reflectivity, which can be conceived as an efficiency quantity, offers the possibility of technical manipulation. So far, we have only considered the absolute magnitude of the reflectivity ρ_i and found that the spectral reflection is essentially defined by the reflective material (silver or aluminium) and the optical properties of the surface (transmissivity of the glass and its refractory index). In the case of well-sealed mirrors, the reflectivity changes only marginally during the service life. In the literature, reflectivity changes in the size of 0.2% per year for thick glass mirrors are mentioned here, see Kennedy and Terwilliger (2005). This degradation originates mainly from corrosion and abrasion of the various reflective and protective layers under environmental impact.

Much more significant is an effect that occurs on a considerably shorter time scale, namely mirror pollution or soiling, which involves several physical mechanisms, such as dust separation, air dust concentration, mirror orientation and wind speed. In contrast to slow degradation, dust separation can reduce the transmissivity of the glass cover by up to 20% in extreme cases, but also rain and other dirt can reduce spectral reflection by up to 10% during a few days only.

The contamination and the resulting reduction of the reflectivity enter directly and moreover linearly proportional the solar yields and thus the economic efficiency of a solar thermal system. Subsequently, the physical effects contributing to pollution are analysed, and options to mitigate them either by means of cleaning or anti-pollution coatings. Finally, a glance is taken on inevitable cleaning strategies required to be scheduled, what their cost and environmental impact especially with respect to water consumption.

10.5.1 Soiling

Soiling is the deposition of impurities on glasses, mirrors, and pipes of concentrating solar thermal systems (CST). They lead to significant reduction of thermal power generation in the receiver, which may make a plant unprofitable and therefore appropriate countermeasures must be taken. On the other hand, the corresponding costs for cleaning can significantly increase the price of the generated energy. Soiling can easily cause more than 1% power loss per day and is a site-specific phenomenon significantly affected by local atmospheric conditions. The most relevant form of soiling depends on the location of the plant and there are different causes such as:

- mineral dust deposits,
- bird droppings,
- biofilms of bacteria, algae, lichens, mosses, or fungi,
- plant residues or pollen,
- engine exhaust or industrial emissions or agricultural emissions (feed dusts, crop dusts).

Soiling on glass mainly leads to optical losses by light absorption or backward scattering, depending on the area shaded by the soiling particles as well as the dust composition and particle size distribution. Figure 10.38a shows the contamination of a linear Fresnel mirror after a spray rain event, resulting in large area dirt deposition. Prior to and during the event, local weather conditions were recorded, including the concentration of particulate matter in the atmosphere, as shown in Fig. 10.38b. The graph shows that the distribution of both large (mean diameter 10 μm) and smaller particles (mean diameter 2.5 μm) behaved the same in terms of trend but differed in magnitude over the days.

The two grey shaded bars in Fig. 10.38b indicate two rain events, the first being a steady rain that barely affects the mean mirror reflectivity in a negative way, so that visually all that could be seen on the mirrors were streaks caused by natural drying of raindrops on a surface. Even after this event, the reflectivity does not decrease abruptly but slowly deteriorates gradually but monotonously due to particles settling on the mirror. This is contrasted by the spray rain event on May 22nd, which lowers the mean reflectivity by more than 15%. At the same time, the fine dust load was high, resulting in a mottled mirror surface on which muddy raindrops fall and dry.

Compared to photovoltaics, the soiling-related losses are 8–14 times larger for concentrating solar systems, since most of the forward-scattered light no longer

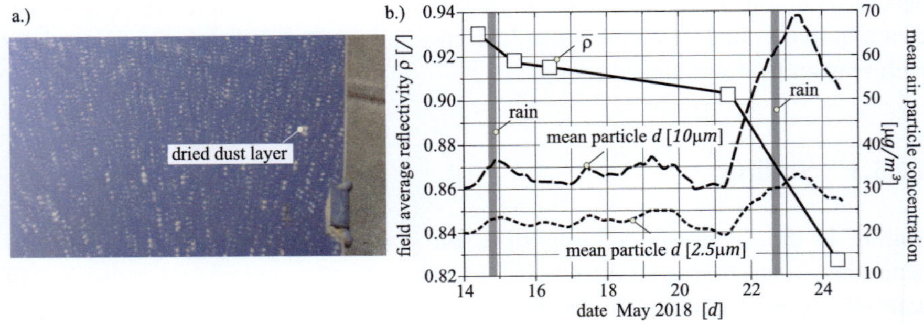

Fig. 10.38 **a** Spotty mirror surface after a spray on a linear Fresnel mirror at an atmospheric high dust loading (22nd May of the right image). **b** Recordings of the mean mirror reflectivity and the atmospheric dust loading of a linear Fresnel mirror field over several days after mirror cleaning; both figures modified from Bonanos et al. (2020)

reaches the receiver due to the limited acceptance angle of the collector, see Ilse et al. (2019).

The physics of dust deposition and attachment is complex due to the many influencing factors, ranging from weather, site, and system specifications to surface nano-properties, as well as their temporal variability (e.g., diurnal, or seasonal weather changes). Airborne dust concentration, along with rain frequency, is considered the most important determinant of pollution because rain, when it occurs in sufficient quantity, can clean polluted surfaces quite effectively. Conversely, rain can also have negative effects, as shown in the spray rain event in Fig. 10.38b, in which wet deposits of aerosol particles washed out of the atmosphere and dry. Wind speed is also an important parameter because it affects particle deposition mechanisms and determines the balance between deposition and resuspension. Relative humidity and dew enhance dust adhesion to surfaces by capillary forces, particle caking, and cementation to a not insignificant degree. These moisture-induced adhesion mechanisms are also important in arid zones such as deserts because the night-time temperature drop cools glass surfaces to temperatures lower than that of the ambient air. If the temperature falls below the dew point temperature, water droplets form on the collector surfaces or mirrors.

In addition to reversible optical losses, soiling can lead to permanent deterioration of glass and mirror materials. If cleaning is neglected, layers of dust can virtually cement, or lichen and/or fungi can become practically indelible. Mechanical or chemical cleaning then leads to scratches or abrasion of typical *a*nti-*r*eflective *c*oatings (ARC) or even to glass corrosion. In addition, mechanical stresses during cleaning or thermal shocks when a hot component is cleaned with cold water can lead to glass breakage or the propagation of microcracks.

Partial shading due to uneven fouling can lead to the formation of local hot spots, especially in line-focusing collector systems, as the heat transfer fluid circulates throughout the solar field. Both with Fresnel but especially with Parabolic Trough, increased dust loading leads to accelerated degradation of the receivers by melting of particles on the glass envelope of the absorber tube, which is harder to clean.

In general, however, dust deposits in all concentrating systems that require single- or dual-axis tracking contribute to increased stress on bearings, ball joints and all mechanically moving elements of the solar field, so that permanent monitoring and maintenance is required.

10.5.1.1 Prediction and Forecast of Mirror Soiling

From the preceding consideration of the effect of mirror contamination, immediate conclusions can be drawn for the operation of a concentrating solar thermal system.

- The effective mirror reflectivity, often specified as the cleanliness factor, must be determined regularly by measurement in order to record the degradation over time. This also means recording the environmental conditions.
- To maintain efficiency, models must be developed that can predictively forecast the degradation state as a function of time after cleaning in order to plan the timing of inevitable cleaning actions, including cost considerations.
- A regular visual inspection of the solar field is indispensable since extreme events are hardly accessible predictively.

To measure the reflectivity in a solar field, portable multiple wavelength specular reflectometers are usually used, which can be used to determine the local reflectivity over several grid points of the solar spectrum from the visible to the near infrared range. However, since the measurement of reflectivity at a discrete location on a mirror is a random variable; it could be a particularly dirty position or, conversely, a particularly clean spot, questions immediately arise as to how many such measurements at how many locations are required to obtain a reliable predictive value.

For this purpose, different methods of statistical mathematics can be used. One option is based on a time series analysis using dynamic linear Gaussian state space time series models relying on reflectance measurements at fixed positions, see Bouaddi et al. (2018).

As shown by Bonanos et al. (2020), the specular reflectivities of both line-concentrating and point-concentrating mirrors are subject to a Gaussian normal distribution if enough statistically representative locations on a mirror/heliostat are measured. To check whether a normal distribution of the residuals exists, the Shapiro-Wilks test can be executed. To determine the spatial distribution of the points, we assume that the variation of the reflectance from one point to another within the facet can be understood as a linear combination of two effects. One is the local position within the facet and the other is random effects. To determine the effect of the specific points used to estimate the mean reflectance, Bonano performed an analysis of variance (ANOVA, (ASTM-International, 2017)) for groups of 3 values (3 groups consisting of 9 points on heliostats and 4 groups consisting of 12 points on Fresnel mirrors). The result of the ANOVA is a p value. If it is greater than the p value for the significance level (e.g., $p = 0.05$ for 95% confidence), then the hypothesis that the means of the different groups are not equal can be rejected. Their measurements and analysis show that the mean of any 3 points within a mirror facet/heliostat is a good estimate of the mean reflectance of a linear module or a single heliostat. However, they also note that "user experience" is required to select appropriate points, i.e., those without local contamination, and spaced appropriately to cover a large area of the mirror.

Now, not all heliostats or linear focusing modules do not behave identically and the number of zones with different behaviour in a distributed mirror field still needs to be determined, for which cluster analysis is performed. This is done using the k-means algorithm, where the data is divided into k mutually exclusive clusters so that the objects within each cluster are as close to each other as possible. The "distance" of the points in the clusters is evaluated as the Euclidean distance between the average reflectance of the cluster and the reflectance of the individual measurements. The number of clusters to be identified is determined by the user. Once clusters with only one point are identified, the solution is assumed to be converged. Cluster behaviour in a collector field can result from several factors, for example, if different mirror zones are washed at different times, these mirrors belong to different clusters. Similarly, local conditions can lead to the generation of multiple clusters, such as proximity to fields, roads, or other wind conditions. Cluster analysis is used to identify mirrors that are similar in reflectivity. Using ANOVA again, the number of heliostats to be measured in real terms can be determined in order to determine the average reflectance of the cluster error and the confidence interval. It can be seen that only very few heliostats/modules mostly in the order of 3 heliostats per cluster are required to obtain a deviation (error) of $\delta = 2.5\%$ at 95% confidence.

To summarize, per cluster a small number of mirrors is required to capture the average reflectance of a cluster, with for each mirror module also only a few measurement points (usually 3) being sufficient to obtain an accurate estimate of its reflectance average. This procedure is an iterative process, which also has to be adjusted again and again during operation based on operational experience. For each of these clusters, we then obtain a predictive value of the temporal evolution of reflectivity as a function of time after the last cleaning operation, which allows the planning of cleaning intervals.

10.5.1.2 Mitigation of Mirror Soiling

If one analyses the pollution mechanisms phenomenologically, they can be categorized into two classes, the dry processes and the processes caused with wetness by rain or dew, as Fig. 10.39 illustrates. This gives access to which methods can be used to reduce soiling.

In the dry processes, particles can either adhere to the surface or bounce off due to a suitable coating. Related to this is the second dry process of resuspension, in which even minor environmental influences such as wind lead to delayed removal of particles from the surface. In the wet process, wind action from rain or dew can cause either clumping (rearrangement and compaction of particles) or cementation

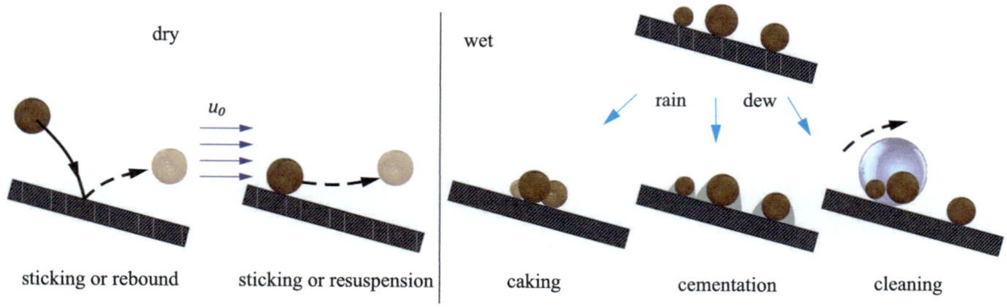

Fig. 10.39 Sketch of the effect of dry and wet mechanisms in the contamination of mirrors

(formation of chemical and/or solid bridges between particles and surfaces due to moisture). In the best case, this can be removed by aquatic cleaning.

Anti-soiling Coatings (ASC)

Thus, to reduce soiling of mirrors, rebound, resuspension, and water purification must be increased while cementation and clumping must be minimized. Anti-soiling coatings (*ASC*) under development are based on three main physical mechanisms:

- Hydrophilic coatings often based on silica have a high surface energy that leads to the formation of small contact angles between the coating and water droplets so that the surface is wetted. This allows very thin water films to form during washing or rain, which facilitates dirt removal.
- Hydrophobic coatings have a low surface energy, which results in a large contact angle, there is no wetting, so that small water droplets form on the surface, which roll easily over the surface, entraining existing dirt particles. These coatings consist of either silicon dioxide or titanium dioxide nanoparticles.
- Coatings based on titanium dioxide use the photocatalytic effect, which utilizes the decomposition of organic matter in the presence of UV radiation.

An overview of anti-soiling coatings (*ASC*), their operating characteristics and performance can be found in Wette et al. (2019) and the literature listed therein. According to the operating results listed therein, the main advantage of the coatings is the reduction of the soiling effect and improvement of the washability of the reflectors and a reduction in the number of cleaning cycles, which however reduces during lifetime. By now, the degree of maturity is not yet so great that the anti-soiling effect is maintained over the entire service life and its degradation is difficult to predict. However, even anti-soiling coatings are subject to physical limitations:

- Since the rate of contamination is essentially a function of location, coatings must be tailored to specific site conditions.

- Particles smaller than the thickness of the viscous boundary layer of the flow around the mirror experience little force effect from the flow around them and adhere. If dew forms, the particles adhere even more firmly to the surface. However, it is precisely these particle sizes that make up the majority of all particles and, even with a coating, an increasingly thick layer of dirt is formed.
- Due to unavoidable cleaning but also abrasive effects of the environment, the performance of the coating is reduced.

ASC's has so far been used only to a limited extent on the market, since they do not eliminate the need for cleaning, but allow longer periods between cleanings.

Tilt Angle and Solar Tracker Positioning

Virtually all reflectometer measurements in mirror fields show that the soiling rate decreases significantly at steeper tilt angles of the mirrors. In addition, depending on the location, several studies show greater soiling during the night. A reduction of soiling is observed when the mirrors are positioned vertically or inverted overnight, so that the cleaning intervals could be extended and thus the water consumption could be reduced.

10.5.1.3 Mirror Cleaning Approaches

Electrodynamic Screens

Electrostatic or electrodynamic dust shields are often used in laboratories as cleaning systems for loaded surfaces. Here, a time-varying electric field is generated over a surface in such a way that dust particles are repelled from the surface by Coulomb interaction. These fields can be generated by metallic thin films applied to the supporting structure or to the back of the mirror surface by means of electrodes, which are pulsed with high alternating voltages. The transfer to real mirror fields has not been successful so far, because environmental influences (humidity, dust, robustness of the electronics) have affected the performance of the systems; also, the effectiveness of dust cemented on the surface is limited.

Aquatic Cleaning Processes

At present, aquatic cleaning is the state of the art in the fight against pollution, either manually, but mostly using automats or even robots (especially Fresnel system or heliostats). The different cleaning methods can be divided into four categories,

- natural cleaning by rain,
- truck or hose systems (brush cleaning, Fig. 10.40a),
- sprinkler-like systems with fixed nozzles (pressure nozzle cleaning or spray cleaning, Fig. 10.40b), and
- robotic cleaning systems with water recovery (Fig. 10.40c).

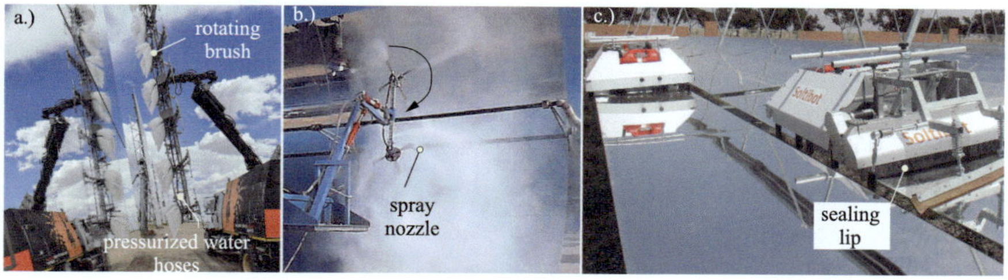

Fig. 10.40 **a** mobile cleaning vehicle with pressurized water and rotating brushes for heliostat cleaning modified from National Renewable Energy Laboratory-NREL (2017) **b** reddening spray water system, **c** self-propelled cleaning system on a Fresnel mirror with water recovery from Solituga SRL P.I. (2022)

The economics of cleaning also determine the viability of other contamination reduction technologies, which require evaluation of their effectiveness in reducing contamination losses and their potential costs. The immediate costs of cleaning include not only the capital costs of the mobile carrier transport systems, but also the ongoing costs of maintaining the cleaning equipment, its consumables, staff working hours, and, most importantly, water costs. The latter are not negligible for concentrating solar thermal systems, which are preferably installed in arid regions, since water is a scarce commodity there and represents a considerable cost factor. Therefore, the focus of a number of studies is on the development of processes that reduce specific water consumption, i.e. water use per square meter, as shown in Fig. 10.40c.

Water is the method of choice for cleaning heliostats/mirrors, but the question is what type of water is best, tap water, demineralized water, with or without detergent. The results of the studies reported in the literature show that the use of demineralized water is the best to reduce the deposition of other foreign substances carried by the solution, and cleaning in the early morning, when the mirrors still have a residual moisture from the dew, is the best time, since the contaminants can be easily rinsed off the surfaces without damaging them. Depending on the location, the use of cleaning solutions in combination with water cleaning in a multi-step process can also be advantageous. Cleaning technologies are now capable of restoring almost 100% of the original reflectance with one cleaning. The state of the art is high pressure water cleaning with demineralized water and scrubbing, with a water consumption of 0.3–0.7 l/m^2, see Fernández-García et al. (2014).

Mechanical Methods

Direct mechanical cleaning of surfaces is the simplest method of dry cleaning, but it damages the integrity of the surface by irreversible scratching. An alternative method is compressed air cleaning, where the cleaning performance can be specifically adapted by nozzle configuration, multi-nozzle arrangement as well as setting angle and translatory/rotatory movement of the nozzle heads. These processes can also be coupled with aquatic cleaning processes. However, various field measurements have shown that water-based systems are far superior to air-only systems.

10.5.1.4 Methods to Plan and Schedule Mirror Cleaning

As we have seen, pollution is subject to a certain stochasticity, in which diurnal, seasonal characteristics of the weather at a defined location are the main influencing variable. From the predictive predictions of the pollution level, which we obtain by permanently measuring individual selected mirror positions within a segment/module and extrapolating over all clusters, we obtain a prediction of the current optical performance of all mirrors. This average reflectivity of the mirror field naturally decreases monotonically from the original value ρ_0 as a function of time and thus also the thermal output of the receivers, which we provide either directly to the customer as heat or also as electricity. This reduces the yield of the solar thermal system. We now have three options to schedule the cleaning cycles.

- The simplest method sets a mean wear of the mirrors as a function of time after the last cleaning. This gives a fixed time interval from which each mirror in the mirror field is cleaned. This means, however, that eventually cleaning will be done even if the solar irradiance is not very high and thus the yield of the field is low. This procedure also does not consider whether there is rain or not, seasonal differences, and whether it makes sense to clean some clusters of the field, while for others that contribute little to the yield, cleaning can be postponed for days. This method results in the highest frequency of cleaning operations and thus the highest water consumption and maintenance requirements, and ultimately the highest costs. In NREL's (SAM), 50–70 cleaning intervals are stored as the default value, depending on the technology.
- The second variant consists of a condition-based maintenance cleaning cycle (CBM&C), in which the mirrors in a cluster are cleaned only when the average reflectivity falls below a certain threshold. This method approach considers the actual optical condition of the mirror field, as well as weather events and seasonal differences, so that the cleaning frequency can be reduced, as illustrated by Fig. 10.41a.

The intrinsic shortcoming of both methods mentioned above is that they rely only on the optical state of the field via the average reflectivity in the clusters. However, the yield of the solar thermal system depends not only on the reflectivity but also on the thermodynamic efficiency of the system, which is high only when the direct normal irradiance (DNI) is high.

This leads to the third method of optimization strategies of cost optimization. The cost optimization is composed of two components, the productivity decreases due to specular reflectivity decline and the direct maintenance costs due to cleaning operations. The area specific costs $L(t_k)$ due to reflectivity losses within a time

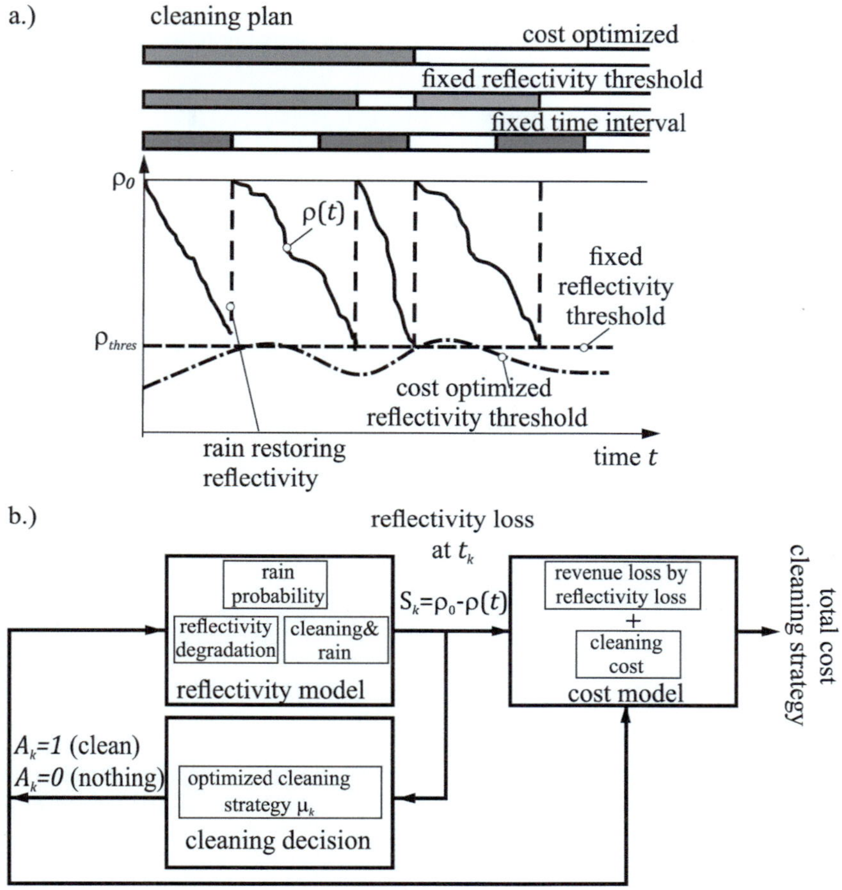

Fig. 10.41 **a** Time spans between cleanings (intervals) using different criteria as triggers for a cleaning, fixed time intervals, attaining a minimum reflectivity, cost-optimized reflectivity; **b** Calculation procedure for determining cost-optimized cleaning intervals modified from Ba et al. (2017)

interval $\left[t_k, t_{k+1}\right]$ are calculated as follows

$$L(t_k) = \int_{t_k}^{t_{k+1}} p(\tau) \cdot DNI(\tau) \cdot \eta(\tau) \cdot (\rho_0 - \rho(\tau))\, \mathrm{d}\tau, \tag{10.88}$$

in which $\rho(t)$ and ρ_0 describe the measured and fully-clean mirror reflectivities, $p(t)$ denotes market price of our goods (process heat, electricity) at time t, $DNI(t)$ the actual solar direct normal irradiance and $\eta(\tau)$ is the total plant efficiency containing the optical and thermal losses except for the reflectivity. Since the temporal variation of the relative change of the reflectivity $(\rho_0 - \rho(\tau))$ being of the order of days compared to the other quantitites changing on minutes/hour basis is slow it can be considered constant within this time interval allowing to rewrite Eq. (10.88)

$$L(t_k) = \underbrace{(\rho_0 - \rho(t_k))}_{S_k} \underbrace{\int_{t_k}^{t_{k+1}} p(\tau) \cdot DNI(\tau) \cdot \eta(\tau) \cdot \mathrm{d}\tau}_{H_k} = S_k \cdot H_k. \quad (10.89)$$

Thus, we mapped the area-specific cost loss due to degradation as the product of the reflectivity loss and the potential area-specific economic return. We are now missing the cost of a cleaning operationC_{cl}, which consists of area-specific costs for water, labor, and other overhead (depreciation, wear, and tear on materials, etc.) given by

$$C_{cl} = C_{water} + C_{labor} + C_{overhead}. \quad (10.90)$$

Thus, the total area-specific costs at time t_k result from the sum of the degradation costs ($L(t_k)$) and the cleaning costs (C_{cl}), the latter depending on whether or not a cleaning action takes place. This results in:

$$C_k(S_k, A_k) = S_k \cdot H_k + C_{cl} A_k, \quad (10.91)$$

where at a value $A_k = 0$ no cleaning takes place and at $A_k = 1$ cleaning is executed. The optimization now consists of developing a cleaning strategy that minimizes $C_k(S_k, A_k)$. These costs, however, are subject to a variety of stochastic events; these include.

- soiling rates as a function of weather conditions,
- the occurrence of rain,
- seasonally varying statistical characteristics, and also
- effectiveness of cleaning measures (especially rainfall).

Therefore, we must instead minimize the expected total maintenance cost, denoted as follows:

$$\min_{\pi} \left\{ \sum_{k=0}^{K} C_k(S_k, \mu_{k-1} S_{k-1}) \right\}, \quad (10.92)$$

where the cleaning strategy is denoted by $\pi = (\mu_0, \mu_1, ..., \mu_k, ..., \mu_K)$, which is a set of functions μ_K. Here, the reflectivity change is mapped into an action at each time interval$k$. The objective is to develop a cleaning strategy that minimizes the expected total cleaning cost. The details of this methodology can be found, among others, in the work of Ba et al. (2017) and the calculation path is illustrated in Fig. 10.41b. With the help of these calculations, which are based on a large number of Monte Carlo calculations, the number of cleaning cycles required can be reduced by about one third and the total costs by 10–15%.

A complete comprehensive treatment of the water management of a solar thermal plant and its optimization has been treated by Rohani (2020). In the thesis suitable dynamic simulation models have been developed that can replicate the water use behaviour and mechanism of a CSP plant and were validated against real plant data. This includes integrated water and wastewater streams and treatment processes throughout the plant. For example, washing cycles can be optimized with respect to many parameters: cleaning costs, cleaning quality, labour availability, recycling of water, optical efficiency, thermal output of the solar field to a storage, electricity generation to the grid, revenues in the grid, market prices and guaranteed power generation as well as imposed load-shedding. All these factors may affect the cleaning strategy. A dynamical and real-time optimization tool therefore can be of high economic value. A result of Rohani (2020) therefore was, that an optimized cleaning schedule based on continuous monitoring of spatial distribution of soiling rate and cleanliness could decrease the water consumption by up to 19% and the cost of cleaning by 25% without any negative effect on the plant performance.

10.5.2 Sensors in the Solar Field

10.5.2.1 Optical and Mechanical Measurement Charges

During the assembly of the mirror facets, a correct alignment of the facet structure on the support structure has to be established. In addition, the support structure must meet geometric requirements. Photogrammetric measurement methods are often used to measure the shape and also the deformation under the influence of gravity and loads. Deflectometric measurement methods are utilized to sense the deviation of the mirror pitch or the mirror shape from its nominal value. For this purpose, a series of fringe patterns reflected on the mirror surface is recorded by a camera and evaluated. In the case of a solar tower system, the image can be taken from a central point at tower height, while in the case of a parabolic trough solar field, the camera is located on the ground or suspended below a flying platform. Photogrammetric and deflectometric techniques overlap here in interaction with automatic image processing and evaluation.

The concentrators of solar thermal power plants generally have to track the sun's path ("*tracking*"). Therefore, either the control via a solar algorithm or the tracking via a controlled variable is suitable. For example, inclinometers, encoders, and sun sensors are used. Sun sensors in the area of parabolic trough systems work, for example, with a shading structure that partially shades two photovoltaic cells (PV) mounted at a distance and connected in series. If the concentrator is not oriented perpendicular to the sun, the amount of shading is greater and the current from the PV cells is not maximized. This type of tracking system is needed in large numbers when operating solar thermal power plants. The engineering challenges here are the correct orientation of the sensor on a structure that is not completely free of deformation, the initial orientation of the concentrators, contamination of the sensor, drift, long-term stability, accuracy, and sensor price. Intelligent sensors or algorithms then allow e.g. for the compensation of a deformation of the heliostats or a torsion of a parabolic trough by gravitational forces, friction, or wind. Individual parabolic trough collectors can be partially defocused in case of excess temperatures. Heliostats of solar tower power plants are assigned to individual target points in the area of the receiver in order to enable a temperature distribution of the receiver that is as homogeneous as possible to achieve minimum losses. This opens up the field of an optimized dynamic target point strategy which reacts specifically to disturbances. A pre-requisite is that the individual target points of the heliostats are known with high accuracy. The target points are determined by focusing a heliostat on a white screen, determining its focal spot in a camera image and thus its target point at the time of exposure. More advanced methods also allow the determination of the heliostat orientation by means of cameras focused on the heliostat field. The need in this area is seen in sensors for intelligent measurement and control of the drive of parabolic troughs or heliostats, either in the form of autonomous units, or connected to a central control unit.

An important parameter for the yield and efficiency of solar tower power plants is the solar flux density distribution in the aperture area of receivers, which, integrated over the area, corresponds to the amount of energy irradiated into the receiver. This information is also used to optimize target point strategy to be applied to the heliostats. Besides computer simulations and other measurement techniques, a flux density measurement with a moving Lambertian target is often performed. The brightness reflected into a CCD camera is calibrated via radiometers. Here, radiometers or PV cells are required as sensors, which can measure flux density values in the order of over a thousand suns as independently of the spectrum as possible in a stable manner.

10.5.2.2 Thermal Sensors

Typical thermal measurement tasks in solar thermal power plants are the measurement of the temperature distribution on an external receiver or in a cavity receiver under concentrated solar radiation. In some cases, receivers are covered by a quartz glass window to reduce convective and radiative losses or to close off a receiver reactor from the environment. For the spatial acquisition of the temperature distributions, infrared measurement technology with specific filters is usually used. In addition to the non-contact infrared technology, commercially available resistance thermometers and thermocouples are used.

10.5.2.3 Fluidic Measurement Tasks and Fluids

Open volumetric air receivers of solar tower power plants use ambient air as the heat transfer medium. The air heated in the receiver transfers its heat to a secondary circuit and is returned to the receiver due to its residual heat content, where a large part of it is recirculated. Tasks in R&D are the determination of the share of recirculated air in the total air flow and its influence by wind at various operating conditions as well as design measures to increase the recirculation rate. This requires measurement techniques that resolve the flow conditions in a larger volume of air in front of the receiver non-intrusively. Here, currently LIDAR-based scanner systems are under development. Other measurement methods use the receiver and the air circulating in it by obtaining information about the recirculation rate via mass balances (tracer gas) or heat balances or transient temperature excitations.

Conventional solar thermal parabolic trough power plants mostly use thermal oils as heat transfer medium. Efficiency measurements are made for acceptance and optimization of plants, and their significance depends on the accuracy of the specific heat capacity value of the thermal oil present. Since these oils can age over time and thereby the thermo-physical properties alter, a flow-through calorimeter is used to determine the specific heat capacity at operating conditions. Typical parabolic trough receivers consist of a selectively coated steel absorber tube and a surrounding glass cladding tube. The space between the tubes is evacuated to minimize heat loss. Decomposition reactions of the thermal oil release minute amounts of hydrogen, which can diffuse into the evacuated space and destroy the thermal insulation effect. Here, sensors at dedicated positions track the hydrogen content in the annular gap.

10.5.2.4 Recording of Environmental Conditions

Examples for the measurement of environmental conditions as the evaluation of the solar direct radiation (*DNI*), the measurement of the contamination of mirrors and absorbers or the extinction of light in a solar tower power plant on the way from the heliostat to the receiver have been already addressed. Commercial solar trackers with pyrheliometers, two pyranometers and shadow ball are used for high precision *DNI* measurements. For the automatic measurement of the soiling of mirrors, measuring equipment is used in which one pyrheliometer is pointed directly into the sun, and the second looks into the sun via a (soiled) mirror. To measure the extinction of light from the heliostat to the receiver, e.g. LIDAR systems, transmissiometers, MOR scattered light sensors, aerosol particle counters and simulations with radiative transfer models, or the use of pyrheliometers at different altitudes are common practise.

The situation is different when it comes to siting solar thermal power plants. Here, it is often a question of remote desert locations and several different location options must be compared with each other. Rotating Shadowband Irradiometer (RSI) is typically used to measure the direct solar radiation for the site search. Here, the signal dip of a photovoltaic cell with diffuser disk is evaluated while a shadow band rotates over the cell and briefly shades the cell. Besides its advantages of a fast response time and a lower susceptibility to contamination, the photovoltaic cell with diffuser disk has the disadvantage of a pronounced spectral acceptance behaviour compared to the thermal sensors of the pyrheliometers.

In the area of the measurement of the extinction between heliostat and receiver, it is analysed during the site search to what extent correlations can be derived, which correlate the atmospheric extinction with easily measured quantities and other available site information. It is also relevant for the site search whether the same pollution rates are available for different site options. During the operating phase of a solar thermal power plant, sensors must be used that can be maintained and operated by the operating personnel. For optimal power plant operation, the question arises, for example, as to which clusters of the collector fields need to be cleaned and how often. This is where the use of low-cost contamination measuring devices comes into question. The basis for an optimal operating strategy of a solar thermal power plant is, in addition to longer-term weather forecasts, a solar direct radiation measurement and a short-term forecast (now-casting), e.g. with cloud cameras and intelligent evaluation methodology.

10.5.3 HTF Ullage System

The stability of the HTF is not only a problem for the hydraulics. The thermal dissociation of thermo-oils does produce hydrogen gas in minute quantities. The hydrogen should not enter the vacuum receivers by diffusion through the steel tube, as then the vacuum is destroyed. Hydrogen has a high gas conductivity and would increase the thermal collector losses. Therefore, several measures are used to prevent an accumulation of hydrogen gas in the receiver. First of all, diffusion barriers are coated onto the steel tube below the absorber coating. Secondly getter material is included in the vacuum receivers which adsorbs hydrogen gas. And very important, the ullage system is used for controlling the status of the HTF. Analysis of the fluid at regular times can give an indication whether the fluid should be replaced or refilled partly. Hydrogen gas can by extracted from the fluid in the ullage system by venting through valves. In this way the concentration of hydrogen molecules should be kept low in the fluid.

10.6 Comparison of Total Collector Performance of Parabolic Trough, Linear Fresnel, and Solar Towers

It is not easy to conclude from the optical or thermal performance of the individual line- or point-focusing systems, which of the concepts is better or worse, because the suitability for the respective application, for example electrical power generation, process heat use, chemical reactors, has to be considered. A choice maximizing energy or exergy efficiency not necessarily is the best economical choice. Temperature requirements as well as temporal demand plays a decisive role in the optimization of a system. For example, if solar thermal heat and electricity shall be used to generate green hydrogen in an electrolyser or even liquid synthetic fuels (synfuels), the required operational hours for the equipment has a crucial impact on the project economics. In this example, storage seems to be a key element for successful operation. Other aspects that can shift the optimum are financing options, water availability and cost, local labour, spatial limitations, and many other parameters more. Project planning issues will be dealt with in Chap. 11.

Here only a simple comparison of the technologies PTC, LFC and CRS with purely solar thermal electricity generation will be given. The projects have been simulated using the software Solar Advisor Model SAM (https://sam.nrel.gov/). According to the commercial availability three different solar power plants technologies with 100 MW_{el} net generation capacity have been simulated:

- PTC plant with thermal oil as heat transfer fluid;
- LFC with molten salt as heat transfer fluid;
- CRS with molten salt as heat transfer fluid.

Each of the technologies has been simulated with the parameters

- solar multiple SM = 1.4 and no thermal energy storage (TES capacity 0 h);
- solar multiple SM = 3.2 and thermal energy storage (TES capacity 10 h).

The cases have been simulated exemplary for the location Hurghada (Egypt). The results in Table 10.3 show that as well with and without thermal energy storage TES the required optimized aperture area is smallest for the solar towers (CRS) and largest for the linear Fresnel system (LFC). Reflector areas are very similar for the two line concentrating technologies. On the other hand, the annual gross electric generation depends on the TES—without TES the generation is largest for the parabolic trough PTC. With TES the solar tower has the largest generation. Although the generation here is lowest for the LFC, the area specific value here is the highest of all technologies. This is due to the very compact configuration using a linear Fresnel collector. The mutual shading of lines is negligible and therefor the different lines in a field can be very closely spaced. The required total solar field per MW electrical capacity is minimal for the Fresnel. For other locations due to other values of irradiation DNI, atmospheric attenuation and ambient temperature and humidity (having an impact on the cooling technology efficiency other values might be the result. So, it is certainly not immediately clear which optical concentration, which technology is best suitable for a concrete solar power plant. In feasibility studies the different options have to be evaluated. In the Chap. 11 we will discuss the different solar power plant concepts. Of, course the economic evaluation may change, very much depending on the individual technology costs. The solar field here has a very large share in the cost break-down.

Table 10.3 Comparison of solar power plants of different optical concentration (parabolic trough PTC, linear Fresnel LFC, and solar tower with central receiver CRS) using identical solar multiple SM (location Hurghada, latitude $\Phi = 27.15°N$, Egypt)

	TES 0h $SM = 1.4$			TES 10h $SM = 3.2$		
	PTC	LFC	CRS	PTC	LFC	CRS
Solar field area (ha)	156	106	376	357	225	1010
Aperture area (ha)	60	64	57	137	147	132
Reflector area (ha)	66	66	69	151	151	167
Thermal output (GWh_{th})	758	535	630	1733	1214	1552
Electric generation (GWh_{et})	233	209	211	540	491	564
Specific area (ha/MW_{el})	1.56	1.06	3.76	3.57	2.25	10.10
Specific production (kWh/m^2)	149	197	56	151	218	56

References

Abbas, R., Montes, M., Rovira, A., & Martinez-Val, J. (2016). Parabolic trough collector or linear Fresnel collector? A comparisonof optical features including thermal quality based on commercial solutions. *Solar Energy, 124*, 198–215. https://doi.org/10.1016/j.solener.2015.11.039

ASTM-International. (2017). *Standard practice for calculating sample size to estimate, with specified precision, the average for a characteristic of a lot or process.* 100 Barr Harbor Drive, PO Box C700, West Conshohocken, PA, 19428-2959. ASTM Committee E11.

Atif, M., & Al-Sulaiman, F. (2015). Optimization of heliostat field layout in solar central receiver systems on annual basis using differential evolution algorithm. *Energy Conversion and Management, 95*, 1–9. https://doi.org/10.1016/j.enconman.2015.01.089

Ba, H., Cholette, M., Wang, R., Borghesani, P. ..., Ma, L., & Steinberg, T. (2017). Optimal condition-based cleaning of solar power collectors. *Solar Energy, 157*, 762–777. https://doi.org/10.1016/j.solener.2017.08.076

Bachelier, C. (2021). *Techno-economical analyses of linear solar thermal electric plants for dispatchable power generation.* KIT Faculty for Mechanical Engineering. Karlsruher Institut für Technologie KIT.

Bachelier, C., & Stieglitz, R. (2017). Design and optimisation of linear Fresnel power plants based on the direct molten salt concept. *Solar Energy, 152*, 171–192. https://doi.org/10.1016/j.solener.2017.01.060

Bell, I. H., Laesecke, A., Lemmon, E., Meyer, C., Muzny, C., & Perkins, R. (2016). *Thermophysical properties of carbon dioxide and CO_2-rich mixtures.* 100 Bureau Drive, Gaithersburg, MD 20899, NIST-National Institute of Standards and Technology.

Bonanos, A., Montenon, A., & Blanco, M. (2020). Estimation of mean field reflectance in CST applications. *Solar Energy, 208*, 1031–1038. https://doi.org/10.1016/j.solener.2020.08.073

Bonilla, C. (1957). *Nuclear engineering*. McGraw-Hill Book Company.

Bouaddi, S., Fernández-García, A., Sansom, C., Sarasua, J. A., Wolfertstetter, F., Bouzekri, H., ... Azpitarte, I. (2018). A review of conventional and innovative-sustainable methods for cleaning reflectors in concentrating solar power plants. *Sustainability, 10*, 3937, 25p. https://doi.org/10.3390/su10113937

Camacho, E., & Bordons, C. (2007). *Model predictive control* (2nd ed.). Springer. https://doi.org/10.1007/978-0-85729-398-5

Camacho, E., Rubio, F., Berenguel, M., & Valenzuela, L. (2007a). A survey on control schemes for distributed solar collector fields. Part II: Advanced control approaches. *Solar Energy, 81*, 1252–1272. https://doi.org/10.1016/j.solener.2007.01.001

Camacho, E., Rubio, F., Berenguel, M., & Valenzuela, L. (2007b). A survey on control schemes for distributed solar collector fields. Part I: Modeling and basic control approaches. *Solar Energy, 81*, 1250–1251. https://doi.org/10.1016/j.solener.2007.01.002

Collado, F., & Guallar, J. (2012). Campo: Generation of regular heliostat fields. *Renewable Energy, 46*, 49–59.

Collado, F., & Guallar, J. (2013). A review of optimized design layouts for solar power tower plants with campo code. *Renewable and Sustainable Energy Reviews, 20*, 142–154. https://doi.org/10.1016/j.rser.2012.11.076

Colmenar-Santos, A., Munuera-Perez, F., Tawfik, M., & Castro-Gil, M. (2014). A simple method for studying the effect of scattering of the performance parameters of parabolic trough collectors on the control of a solar field. *Solar Energy, 99*, 215–230. https://doi.org/10.1016/j.solener.2013.11.004

Cruz, N., Redondo, J., Berenguel, M., Alvarez, J., & Ortigosa, P. (2017). Review of software for optical analyzing and optimizing heliostat fields. *Renewable and Sustainable Energy Reviews, 72*, 1001–1018. https://doi.org/10.1016/j.rser.2017.01.032

Dostal, V. (2004). *A super-critical carbon dioxide cycle for next generation nuclear reactors.* Ph.D. thesis, MIT, Department of Nuclear Engineering.

Dow Chemicals. (2017). DOWTHERMTM A. Midland, MI 48674, USA. Retrieved January 17, 2023, from https://www.dow.com/documents/en-us/productdatasheet/176/176-01463-01-dowtherm-a-tds.pdf

Dow Inc. (1997). *SYLTHERM™ 800 Heat transfer Fluid—Product technical data.* Retrieved January 17, 2023, from https://www.dow.com/content/dam/dcc/documents/en-us/app-tech-guide/176/176-01435-01-syltherm-800-heat-transfer-fluid.pdf

Fernández, A., Galleguillos, H., Fuentealba, E., & Pérez, F. (2015). Corrosion of stainless steels and low-Cr steel in molten $Ca(NO_3)_2$–$NaNO_3$–KNO_3 eutectic salt for direct energy storage in CSP plants. *Solar Energy Materials & Solar Cells, 141*, 7–13. https://doi.org/10.1016/j.solmat.2015.05.004

Fernández-Garcíaa, A., Álvarez-Rodrigo, L., Martínez-Arcos, L., Aguiard, R. .., & Márquez-Payés, J. (2014). Study of different cleaning methods for solar reflectors used in CSP plants. *Energy Procedia, 49*, 80–89. https://doi.org/10.1016/j.egypro.2014.03.009

Fraunhofer Institute for Solar Energy System—ISE. (2022). ColSim—Collector Simulation Environment (C. Wittwer, Compiler) Heidenhofstr. 2, 79110 Freiburg, Germany. Retrieved January 17, 2023, from https://www.colsim.org/

Gibilaro, L. (2015). *Fluidization dynamics.* (Butterworth-Heinemann, Ed.)

IAPWS. (2021). IAPWS releases, supplementary releases, guidelines, and advisory notes. (I. A. Steam, Ed.). Retrieved April 27, 2021, from http://iapws.org/release.html

Ilse, K., Micheli, L., Figgis, B., Lange, K., Daßler, D., Hanifi, H., … Bagdahn, J. (2019). Techno-economic assessment of soiling losses and mitigation strategies for solar power generation. *Joule, 3*, 2303–2321. https://doi.org/10.1016/j.joule.2019.08.019

Kakac, S., Aung, W., & Viskanta, R. (1985). *Natural convection: Fundamentals and applications.* Hemisphere Publishing.

Kalbhor, A. (2014). Modified solar central receiver in concentrated solar pwer systems. *International Journal of Engineering and Technical Research (IJETR), 2*(12), 257–264.

Kennedy, C., & Terwilliger, K. (2005). Optical durability of candidate solar reflectors. *Journal of Solar Energy Engineering, 127*(2), 262–269. https://doi.org/10.1115/1.1861926

Kistler, B. (1986). *A user's manual for DELSOL3: A computer code for calculating the optical performance and optimal system design for solar thermal central receiver plants.*

Kunii, D., & Levenspiel. (2013). *Fluidization engineering* (Brenner, Ed.) (2nd ed.). Butterworth-Heinemann-Elsevier.

Lee, C.-Y., Chou, P.-C., Chiang, C.-M., & Lin, C.-F. (2009). Sun tracking systems: A review. *Sensors, 9*, 3875–3890. https://doi.org/10.3390/s90503875

Lenert, A., Nam, Y., & Wang, E. (2012). Heat transfer fluids. *Annual Review of Heat Transfer, 15*, 93–129. https://doi.org/10.1615/AnnualRevHeatTransfer.2012004122
Lipps, F.V.-H. (1978). A cell wise method for the optimization of large central receiver systems. *Solar Energy, 20*, 505–516.
Mills, D. R., & Morrison, G. (2000). Compact Linear Fresnel Reflector solar thermal powerplants. *Solar Energy, 68*(3), 263–283. https://doi.org/10.1016/S0038-092X(99)00068-7
Moga, D., Sita, I. V., Stroia, N., Dobra, P., Moga, R., & Petreus, D. (2015, May 25). Sensing and control strategies in tracking solar systems. In *Proceddings of International Conference on Control Systems and Science* (pp. 989–995). IEEE. https://doi.org/10.1109/CSCS.2015.143
Mokhtar, M. M. (2018). *Control of solar thermal linear fresnel collector plants in single phase and direct steam generation modes.* Dissertation, Karlsruher Institut f¨ur Technologie (KIT), KIT Faculty of Mechanical Engineering, Karlsruhe.
Morin, G. D., Platzer, W. E., & Häberle, A. (2012). Comparison of linear fresnel and parabolic trough collector power plants. *Solar Energy, 86*, 1–12. https://doi.org/10.1016/j.solener.2011.06.020
Mouromtseff, I. (1942). Water and forced-air cooling of vacuum tubes. *Proceedings of the Institute of Radio Engineers, 30*, 1–15.
Mutuberria, A., Pascual, J., Guisado, M. V., & Mallor, F. (2015). Comparison of heliostat field layout design methodologies and impact on power plant efficiency. *Energy Procedia, 69*, 1360–1370.
National Renewable Energy Laboratory—NREL. (2017, August 4). *Mechanized washing truck for cleaning tracking heliostats.* Retrieved May 3rd, 2023, from Crescent Dunes Solar Thermal Facility: https://www.flickr.com/photos/nrel/36362104745/in/photostream/
Nonne, C. T. (2012). Heliostat filed optimization. A new computationally efficient model and biomimietic layout. *Solar Energy, 86*, 792–803.
NREL-National Renewable Energy Laboratory. (2022). System Advisor Model (SAM) (N. R. (NREL), Ed.), USA. Retrieved January 17, 2023, from https://sam.nrel.gov/
OECD-NEA. (2015). *Handbook on lead-bismuth eutectic alloy and lead properties, materials compatibility, thermal-hydraulics and technologies.* NEA. No. 7268, OECD, Nuclear Science.
Pacio, J. W. (2013). Assessment of liquid metal technology status and research paths for their use as efficient heat transfer fluids in solar central receiver systems. *Solar Energy, 93*, 11–22. https://doi.org/10.1016/j.solener.2013.03.025
Patankar, S. V. (1980). *Numerical heat transfer and fluid flow.* Hemisphere Publishing Corporation.
Pfahl, A., Coventry, J., Röger, M., Wolfertstetter, F., Vásquez-Arango., J. G., Arjomandi, M., … Liedke, P. (2017). Progress in heliostat development. *Solar Energy, 152*, 3–37. https://doi.org/10.1016/j.solener.2017.03.029
Platzer, W. (2019). Separation of latitude-dependent and climate-dependent effects of CSP power plant performance. In *International Conference SOLARPACES 2018, Casablanca, Morocco*, 2–5 October 2018. AIP Publishing.
Puppe, M., Giuliano, S., Frantz, C., Uhlig, R., Flesch, R., Schumacher, R., … Hattendorf. (n.d.). Techno-economic optimization of molten salt solar tower plants In *AIP Conference Proceedings. SolarPACES 2017* (A. Publishing, Ed., pp. 040033-1–040033-13). https://doi.org/10.1063/1.5067069
Rohani, S. (2020). *Reducing the water footprint of concentrating solar power plants.* Ph.D. thesis. Universität Freiburg, Freiburg.
Rovira, A., Montes, M. J., Valdes, M., & Martínez-Val, J. (2011). Energy management in solar thermal power plants with double thermal storage system and subdivided solar field. *Applied Energy, 88*, 4055–4066. https://doi.org/10.1016/j.apenergy.2011.04.036
Sait, H., Martinez-Val, J.-M., Abbas, R., & Munoz-Anton, J. (2015). Fresnel-based modular solar fields for performance/cost optimization in solar thermal power plants: A comparison with parabolic trough collectors. *Applied Energy, 141*, 175–189. https://doi.org/10.1016/j.apenergy.2014.11.074
Sánchez, M., & Romero, M. (2006). Methodology for generation of heliostat field layout in central receiver systems based on yearly normalized energy surfaces. *Solar Energy, 80*(7), 861–874.

Schmitz, M., Schwarzbözl, P., Buck, R., & Pitz.-Paal, R. (2006). Assessment of the potential improvement due to multiple apertures in central receiver systems with secondary concentrators. *Solar Energy, 80*, 111–120. https://doi.org/10.1016/j.solener.2005.02.012

Schöttl, P., Rohani, S., Leonardi, E., Pisani, L., Les, I., Mutuberria, M., & Nitz, P. (2019). Solar field heliostat selection based on polygon optimization and boundaries. *AIP Conference Proceedings, 2126*, 8. https://doi.org/10.1063/1.5117565

Schwarzbözl, P., Pitz-Paal, R., & Schmitz, M. (2009, September 15–18). Visual HFLCAL—A software tool for layout and optimisation of heliostat fields. In *Proceedings of SolarPACES* (p. 8).

Siegel, N., Bradshaw, R., Cordaro, J., & Kruizenga, A. (2011, August 7–10). Thermophysical property measurement of nitrate salt heat transfer fluids. In *Proceedings of ASME 2011 5th International Conference on Energy Sustainability* (pp. 439–446). https://doi.org/10.1115/ES2011-54058

SolarPaces-Solar Power &Chemical Energy Systems. (2022, December). *CSP projects around the world*. Apartado 39,04200 Tabernas, Almería, Spain. Retrieved from www.solarpaces.org

Solituga SRL P.I. (2022). Cleaning robot Soltibot®. Via Roma, 54 47035 Gambettola (FC), Italy. https://doi.org/10.3030/654443

Solutia Europe S.A./N.V. Applied Chemistry. (n.d.). THERMINOL®66. Rue Laid Burniat 3—Parc Scientifique—Fleming, B-1348 Louvain-la-Neuve (Sud), Belgium. Retrieved January 17, 2023, from http://twt.mpei.ac.ru/tthb/hedh/htf-66.pdf

Solutia. (2022). Therminol VP1. P.O. Box 66760, St. Louis, MO 63166-6760, United States. Retrieved January 17, 2023, from https://www.sintelub.com/wp-content/uploads/PDS/37.therminol_vp1.pdf

Thermal Energy System Specialists, LLC. (2019). TYNSYS-Transient System Simulation Tool. 3 North Pinckney Street, Suite 202, Madison, WI 53703, USA. Retrieved January 17, 2023, from www.tess-inc.com

Turchi, G., Z., M., N. T., & Wagner, M. J. (2013). Thermodynamic study of advanced supercritical carbon dioxide cycles for concentrated solar power. *ASME Journal of Solar Energy Engineering, 135*, 041007-1/7.

Wacker Chemie AG. (2021). HELISOL® 5A. Hanns-Seidel-Platz 4, 81737 Munich, Germany. Retrieved January 17, 2023, from https://www.wacker.com/h/en-us/medias/HELISOL-5A-en-2021.10.13.pdf

Wagner, M. J., & Gilman, P. (2011). *Technical manual for the SAM physical trough model*. 1617 Cole Boulevard,Golden, Colorado 80401, National Renewable Energy Laboratory. Retrieved from www.nrel.gov

Wette, F., Sutter, F., & Fernández-García, A. (2019). Evaluation of anti-soiling coatings for CSP reflectors under realistic outdoor conditions. *Solar Energy, 191*, 574–584. https://doi.org/10.1016/j.solener.2019.09.031

Yatsenko, S. P., Sabirzyanov, N. A., & Yatsenko, A. S. (2008). Dissolution rates and solubility of some metals in liquid gallium and aluminum (I. Publishing, Ed.) *Journal of Physics—Conference Series, 98*, 062032. https://doi.org/10.1088/1742-6596/98/6/062032

Zavoico, A. B. (2001). *Solar power tower design basis document, Revision 0*. Sandia National Lab. (SNL-NM), Albuquerque, NM (USA). https://doi.org/10.2172/786629

Zhang, Y., & Li, P. (2017). March). Minimum system entropy production as the FOM of high temperature heat transfer fluids for CSP systems. *Solar Energy, 152*, 80–90. https://doi.org/10.1016/j.solener.2017.03.021

Zhang, M., Yang, L., Xu, C., & Du, X. (2016). An efficient code to optimize the heliostat field and comparisons between the biomimetic spiral and staggered layout. *Renewable Energy, 87*, 720–730. https://doi.org/10.1016/j.renene.2015.11.015

Solar Thermal Power Plants

11

Abstract

Solar heat can be used for electricity production by means of coupling a collector technology with an energy conversion cycle. In the chapter, at first the principles of the thermal cyclic process are elaborated, and then different ideal single and two-phase cyclic processes are discussed in terms of thermodynamics, plant arrangement as well as optimization options and limitations. In order to categorize and analyse power plants technically and economically rating parameters are used, which are introduced in a next subchapter. Economic performance and competitiveness are essential for all electricity generation systems; therefore, cost structures of thermal solar power plants are elaborated and options for optimization are indicated. Complete solar thermal power station concepts, divided into plants based on non-concentrating and concentrating fields, are presented further with respect to their architecture, operational aspects as well as their prospects and limitations. Also, the aspects for hybridisation and cogeneration are addressed.

Keywords

Thermal energy conversion processes • Technical and economic rating parameters • Solar thermal power plant concepts • Hybridization of solar power plants

11.1 Types of Solar Thermal Electric Power Plants

The use of solar thermal energy to generate electricity primarily involves coupling the heat generated by the solar collectors into a thermodynamic cycle to produce mechanical work. Here, it is desirable to keep the heat losses in the hydraulic piping network low. At the same time, an economic analysis of the overall power plant

R. Stieglitz and W. Platzer, *Solar Thermal Energy Systems*,
https://doi.org/10.1007/978-3-031-43173-9_11

concept is required to ensure that high availability can be achieved at a competitive cost, measured for example in a price per kWh_{el} generated. If the power plant concept permits, this also requires the consideration of the integration of thermal energy storage, which decouples the power generation from the immediate availability of solar radiation, but at the same time also increases the dimensions of the solar field and thus their production costs.

The quality of the generated heat and thus the efficiency of the downstream conversion processes in a thermal power plant, generally increases with increasing collector outlet temperature. Conversely, however, the efficiency of a solar thermal collector system decreases monotonically with increasing temperature difference between collector outlet temperature and ambient temperature due to heat losses. The multiplicative link between collector efficiency and thermodynamic efficiency thus results in an optimal collector outlet temperature, so that the power conversion system must be optimally adapted to the collector characteristics. At collector outlet temperatures of more than 350–400 °C, there are a number of conventional mature power engines for converting the heat into work, which require only an adaptation of the machine to the conditions of the collector system, while at lower collector outlet temperatures special thermodynamic cycle processes or working machines must be used to achieve a high system efficiency and thus competitive power generation costs.

If one does not consider the solar field as the only solitary heat source for driving a power engine, but rather considers the use of fossil or renewable fuels as a supplementary heat source for power generation, one enters the field of hybrid power plants. The use of chemical energy sources as a supplementary heat source allows power plant operation to be decoupled from solar irradiation, thus increasing the availability and utilization of the engines, and making them easier to plan. Conversely, the use of fossil fuels such as gas and oil eliminate the argument of CO_2-free power generation.

There is a range of different technical concepts for converting solar thermal heat into electricity, of which Table 11.1 shows some types that have been technically demonstrated or successfully implemented on the market to date. Roughly, a classification can be made into three technical areas, the optical system, the heat transfer medium, and the power conversion system. However, the respective sectors are linked via the quality of the solar thermal heat generated and thus via the heat transfer fluid (HTF), so that there are essentially three families of solar thermal power plant types.

- Non-concentrating solar power plant concepts

In non-concentrating power plant concepts the entire global horizontal irradiation can be converted, which is realized in a volumetric receiver as e.g. in a solar pond or a surface such as in an upwind-draft. Due to the limited power density the systems need not be pressurized. The heat is either directly converted into kinetic energy (chimney) or indirectly via a heat exchanger transferred to a thermodynamic circuit suitable for low temperatures to drive a power engine. This type

Table 11.1 Characteristics of some solar thermal power plant concepts with indications of technical key data

Characteristics	Non-concentrating		Line-concentrating		Point-concentrating	
Type of power plant	Solar pond	Solar chimney	Parabolic trough	Linear Fresnel	Solar tower	Solar dish
*Optical system*x						
Imaging system	None	None	Mirror	Mirror	Mirror	Mirror
Mirror shape	–	–	Parabolic	Slightly bent	Flat/slightly bent	Paraboloid
Mirror tracking	–	–	1-Axis	1-Axis	2-Axis	2-Axis
Mean concentration factor C[/]	1	1	$\mathcal{O}$ (< 100)	$\mathcal{O}$ (< 100)	$\mathcal{O}$ (1000)	$\mathcal{O}$(>1000)
Mirror-receiver distance (focus)	None	None	$\mathcal{O}$ (1 m)	$\mathcal{O}$ (1–10 m)	$\mathcal{O}$ (100–1000 m)	$\mathcal{O}$ (1–10 m)
Receiver design	Volumetric	Volumetric/ surface	Surface	Surface	Volumetric or surface	Surface
*HTF in solar field and thermal storage*x						
Type(s) of heat transfer fluid (HTF)	Brine	Air	Oil, salt, water/steam	Oil, salt, water/steam	Oil, salt, water/steam, (gas)	Helium, s–CO_2
HTF integration in solar field	Collector integrated	Open loop in collector	Collector attached	Collector attached	None	None
HTF coupling to power cycle	Indirect	Direct	Indirect/direct	Indirect/direct	Indirect/direct	Direct
HTF temp. realized (feasible) (°C)	90	60–70	350–420 (565)	270–540 (565)	250–565 (850)	750–800 (1000)
Thermal storage integration	None	Yes indirect	Yes dir./indir.	Yes dir./indir.	Yes dir./indir.	None
*Power conversion*x						

(continued)

Table 11.1 (continued)

Characteristics	Non-concentrating		Line-concentrating		Point-concentrating	
Type of power plant	Solar pond	Solar chimney	Parabolic trough	Linear Fresnel	Solar tower	Solar dish
Power conversion systems	ORC/Kalina	Wind turbine	Rankine	Rankine	Rankine, gas turbine, combined cycle	Rankine/stirling
Hybridisation with other heat sources	None	None	Yes	Yes	Yes	Yes
Heat rejection/ cooling	Water	Air	Water/air	Water/air	Water/air	Air

of power plant concept is highly integrated and can only be expanded to a very restricted extent in terms of output. The integration of storage is also only possible to a very limited extent, so that full-load operation is not possible in the absence of solar irradiation.

- Line-concentrating solar power plants

Line-concentrating systems are built in long parallel rows characterized by single-axis tracking mirror systems and tubular receivers. Due to this row arrangement, line-concentrating systems including all the connecting pipework are often referred to as solar fields or solar farms. In line-concentrating types of power plants, the heat transfer fluid (HTF) circulates through the entire solar field before it is then fed to the power conversion system either directly or by means of a heat exchanger. In contrast to non-concentrating systems line focussing systems can use essentially only the direct normal irradiation (DNI) of the solar irradiation, and diffuse radiation is not used, for more details see Chap. 10. The parabolic trough concept (PTC), which has the highest market penetration of all solar thermal power plants, has been particularly successful, and the linear Fresnel collector type (LFR) has become established as well. One of the main advantages of linear concentrating systems is the modularity in solar field design.

- Point-focussing solar power plants

Point-concentrating systems require a two-axis tracking of the mirrors to achieve high concentration ratios C. There are two main types; the first one uses large solar towers, on which the solar radiation is focused by large planar mirrors onto a central receiver, with several receiver technologies. The second one is dish systems, in which paraboloids are used to direct the solar radiation onto a small receiver mounted in the focus. In the receiver of the latter usually a direct conversion engine is positioned, for example, a Stirling engine. Dish systems of this type only allow an increase in power by installing a large number of individual systems and thus linking them at grid level. A variant of the parabolic dish is the Scheffler dish—an off-axis paraboloid with fixed focus mounted in polar orientation. Here the receiver needs not to be moved, which is the reason that these kinds of dishes are used not only for power generation but also for solar cooking with a fixed pot or a steam generator.

In Sect. 11.2, different thermal conversion processes are first presented with regard to their mode of operation, their efficiency and the essential mechanical equipment, and then technical and economic evaluation parameters for characterizing solar thermal power plants are introduced. Section 11.4 provides some insight in the conceptual planning methodology of solar thermal power plants. Since non-concentrating power plant concepts represent intrinsically integrated concepts the optical conversion, operating principle and realisation is illustrated as a whole in

Sect. 11.5. The following section is dedicated to the characterization of larger scale solar power plants and presents different concepts and discusses their performance.

11.2 Thermal Energy Conversion Cycles

In technical systems, the conversion of thermal energy via mechanical into electrical energy takes place in thermodynamic cyclic processes. The choice of the cycle process depends essentially on the process temperature generated in the receiver and the heat transfer fluid, which transports the heat from the receiver either to a thermal storage or to a thermal power engine. In thermal power plants steam turbines based on the two-phase Clausius-Rankine process have a long tradition and work very efficient in the temperature range which can be provided by linear focusing systems and current solar tower systems. Although turbines used in solar projects have a lower complexity (two pressure stages) and specific cost than for example the ones used in coal-fired stations, the availability of the components or the operating experience in power generation is an asset. For future projects achieving temperatures much higher than $T > 600$ °C single-phase cycles, such as Joule-Brayton process or Stirling process, or supercritical processes can exhibit superior thermodynamic efficiencies.

In principle, high-temperature systems permit also combined single- and two-phase processes that are familiar from conventional power plant technology, such as combined cycle power plants. If the focus is not exclusively on exploiting the solar source, there are a number of hybrid solar-fossil concepts such as integrated solar combined cycles with integrated recuperative gas turbine, to which other cycle processes such as double recuperative and double expanding propane cycles can also be connected.

Both, cyclic power conversion processes and thermal storage system issues can only be considered separately to a limited extent since the associated heat transfer processes in the heat exchanger or the direct use in the power engine can often only run efficiently within a limited temperature window. The next sections elaborate the basics of the thermal based power generation processes and to link them with the requirements for thermal energy storage systems.

11.2.1 Principles of Energy-Converting Thermal Cyclic Processes

Circular processes are an essential core of technical thermodynamics. They describe a process consisting of several process steps in which the final state corresponds to the initial state. In this cyclic process, the working fluid periodically passes through the various thermodynamic states. In principle, there are two different ways of looking at the process, a fluid-related view (e.g. steam or gas turbine process) or periodic changes of state in a control room (e.g. reciprocating or piston engine such as diesel, or Stirling engine).

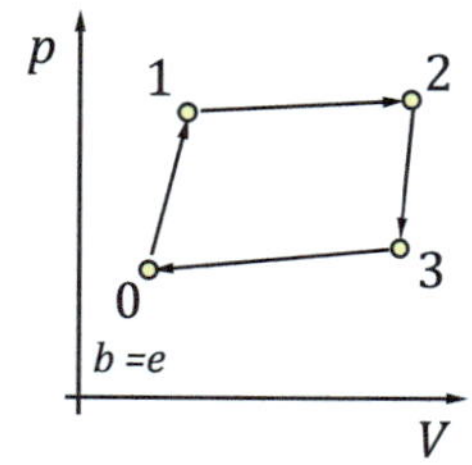

Fig. 11.1 Schematic sketch of a cycle process in the $p - V$ diagram

Regardless of the point of view, the central requirement of any thermodynamic cycle is the identity of the thermodynamic state variables after the passage of a cycle, so that all state variables (temperature—T, pressure—p, internal energy—U, entropy—S) correspond at the beginning (b) and at the end (e) of a cycle, so that $T_b = T_e$, $p_b = p_e$, $U_b = U_e$ and $S_b = S_e$. However, the process variables change after one cycle and so does the heat and as a consequence the work. Figure 11.1 shows an arbitrary thermodynamic cycle in the $p - V$ diagram.

If we now consider a closed overall system, the 1st law of thermo-dynamics reads:

$$dU = dQ + dW, \tag{11.1}$$

in which Q represents the heat energy and W the work done. In a closed system, the mass does not change, so the mass change is $dm = 0$. Now, if we go through the circular process from the initial state (0) and return to it clockwise, at the starting point the internal energy U of the circular integral does not change; which yields:

$$\oint dU = 0. \tag{11.2}$$

This results in a fundamental result of the thermo-dynamic cycle processes, which states: In the cycle process, as much heat is released as work is done, and thus the following applies:

$$\oint dQ + \oint dW = 0 \quad \text{or} \quad \sum dQ + \sum dW = 0. \tag{11.3}$$

From this, the nature of the thermodynamic cycle can be analysed. If net heat energy is supplied to the system, i.e. $\left(\oint dQ > 0\right)$, Eq. (10.3) states that the net work must be negative. Thus, work is supplied out of the system. This is a right-handed cycle (clockwise, cf. Fig. 11.2 on the left). All heat engine processes belong to the right-handed, i.e. work-delivering, cycles. Correspondingly, all work consuming processes $\left(\oint dQ < 0\right)$ are called left-handed processes (or counter-clockwise). In contrast to the clockwise processes, their benefit can lie in a heat gain, for example in a heat pump, or in a heat extraction, such as in a refrigerating machine.

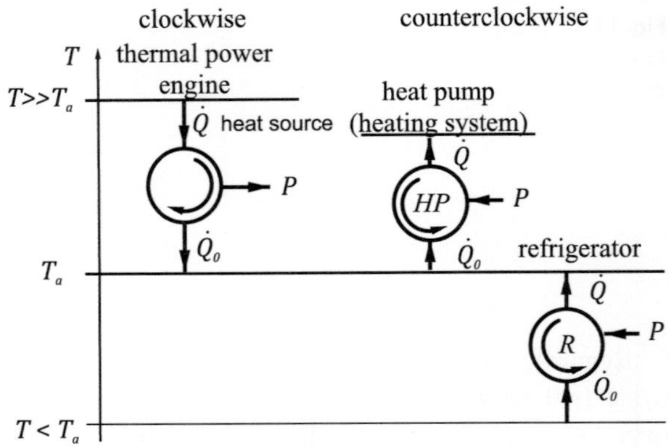

Fig. 11.2 Distinction between right-handed circular processes (heat engines, left) and left-handed processes using work to raise (heat pump) or lower (refrigeration machine) the temperature level

The ratio of benefit to effort is called efficiency. In clockwise systems, the efficiency is uniformly given as the thermal efficiency η_{th}, which is defined as follows:

$$\eta_{th} = \frac{|P|}{\dot{Q}}. \tag{11.4}$$

In contrast, in left-handed processes, only the effort, i.e. the supplied mechanical power P, is unambiguously given, whereas the benefit in the case of the heat pump is $\dot{Q}$ in the case of the refrigerating machine is $\dot{Q}_0$. If one restricts oneself to clockwise thermal power processes, the work and thus the benefit is often added or removed in several partial steps in large-scale industrial processes in order to achieve an efficiency increase. The same applies to the effort. In general, the sign rule must be observed, i.e. energy supplied to the system is considered positive while energy dissipated is assigned a negative sign.

For circular processes with a continuous mass flow, the thermal efficiency is defined as follows:

$$\eta_{th} = \frac{|P|}{\dot{Q}} = \frac{-\sum_{process,\ ij} \dot{m} \cdot w_{t,ij}}{\sum_{process,\ ij} \dot{m} \cdot q_{t,ij}} = \frac{-\sum_{process,\ ij} w_{t,ij}}{\sum_{process,\ ij} q_{t,ij}}, \tag{11.5}$$

where $w_{t,ij}$ is the specific work output at the transition from state i to j and $q_{t,ij}$. denotes correspondingly the specific heat input at the transition of process step i to j.

11.2.2 Single Phase Conversion Cycles

11.2.2.1 Carnot Cycle

The significance of the purely theoretical Carnot cycle is that it indicates the highest thermal efficiency and thus the optimum that cannot be exceeded by any other thermo-dynamic cycle in which the working medium moves between

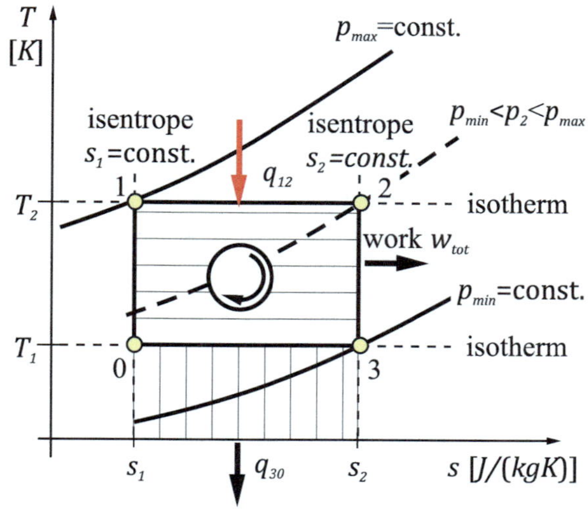

Fig. 11.3 Illustration of the Carnot cycle in the $T - s$ diagram

the same two temperatures. Thus, the Carnot efficiency η_C, determined from the working temperatures, indicates the maximum proportion of the supplied heat that can be converted into mechanical work. The Carnot process consists of two isothermal and two isentropic changes of state and forms a rectangle in the temperature-entropy ($T - s$) diagram shown in Fig. 11.3. The process sequence is as follows:

- From state 0 to 1, isentropic compression occurs, and the substance is raised from temperature level T_1 to level T_2 without any loss of thermal energy; it is $\Delta q_{01} = 0$.
- In contact with the hot temperature reservoir T_2, the medium expands from state 1 to 2 at constant temperature T_2 (isothermal). The amount of heat q_{12} supplied in this process is: $q_{12} = \int_1^2 T ds = T_2(s_2 - s_1) > 0$
- At the transition from 2 to 3, the fluid expands isentropically while performing mechanical work; thus, $\Delta q_{23} = 0$. The temperature drops to the temperature level T_1.
- In contact with the cold reservoir, heat q_{30} is extracted from the working fluid. The heat extracted is given by the integral $q_{30} = \int_3^0 T ds = T_1(s_1 - s_2) < 0$ and is thus less than zero.

Since all state variables in the system, i.e. also the internal energy, return to their initial values after passing through the full cycle, i.e. $\Delta u = 0$. The useful work is thus calculated from $|w| = q_{12} + q_{30} = (T_2 - T_1)(s_2 - s_1)$. The Carnot efficiency

η_C is thus obtained from:

$$\eta_c = \frac{|w|}{\dot{q}_{12}} = \frac{(T_2 - T_1)}{T_2} = 1 - \frac{T_1}{T_2}. \tag{11.6}$$

Mechanical energy is converted in all four phases of the Carnot process. However, the total mechanical energy gained after passing through a cycle depends only on the amount of heat supplied and removed. The mechanical work gained corresponds to the area enclosed by points 0123. The Carnot efficiency is always less than 1. Physically, this means; there is no real machine that can only extract heat from a reservoir and is able to convert it completely into work.

Figure 11.4 illustrates that mechanical work is converted in all four phases of the Carnot process. In the graph, an ideal gas is assumed to represent the specific volume change work required for the changes of state. The following equations apply to an ideal gas:

$$p \cdot V = n \cdot R \cdot T = N \cdot k_B \cdot T, \tag{11.7}$$

where p is the pressure, V is the volume, n is the amount of substance, R is the universal gas constant ($R = 8.314$ J/(mol K)), T is the temperature, N is the number of particles, and k_B is Boltzmann's constant ($k_B = 1.38 \cdot 10^{-23}$ J/K).

The specific heat capacities of an ideal gas at constant volume c_V and at constant pressure c_p and thus the adiabatic index (isentropic coefficient) κ are given by the following relations:

$$c_V = \frac{f}{2} N \cdot k_B; \; c_p = \left(\frac{f}{2} + 1\right) \cdot N \cdot k_B; \; \kappa = \frac{c_p}{c_V} = \frac{f+2}{f}, \tag{11.8}$$

where f is the degree of freedom of the gas. The degree of freedom of a monoatomic gas such as helium is $f = 5$, while diatomic gases such as air have a degree of freedom of $f = 7$. The isentropic change of state of an ideal gas can be calculated by

$$p \cdot V^k = const., \tag{11.9}$$

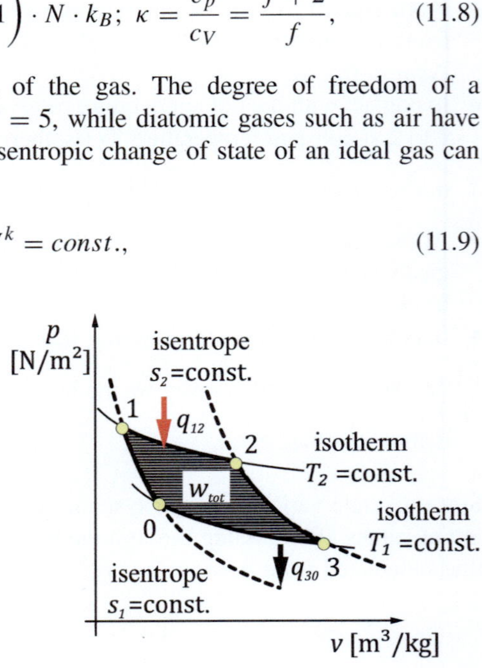

Fig. 11.4 Carnot cycle of an ideal gas in the $p - v$ diagram

and the entropy change Δs of an ideal gas is determined by the expression:

$$\Delta s = (s_2 - s_1) = c_V \ln\left(\frac{T_2}{T_1}\right) + N \cdot k_B \cdot \ln\left(\frac{v_2}{v_1}\right). \tag{11.10}$$

This allows one to determine the individual work steps w_{ij} within the Carnot cycle:

- From 0 to 1 an isentropic compression occurs. The specific work w_{01} required is $w_{01} = c_V(T_2 - T_1)$.
- For the isothermal expansion from 1 to 2, $w_{12} = RT_2 \ln(p_1/p_2)$ is obtained.
- The isentropic expansion from 2 to 3 yields $w_{23} = c_V(T_2 - T_1)$.
- Finally, isothermal compression from 3 to 0 requires the following work $w_{30} = RT_1 \ln(p_0/p_3)$

The total work w_{tot} that can be discharged from the cycle thus results in:

$$w_{tot} = w_{01} + w_{30} - w_{12} - w_{23}. \tag{11.11}$$

To resolve the pressure ratio (p_0/p_3), the isentropic equation (Eq. 11.9) is modified in the following way:

$$\left(\frac{T_1}{T_2}\right) = \left(\frac{p_1}{p_2}\right)^{\frac{\kappa-1}{\kappa}}; \tag{11.12}$$

and we get for w_{tot}:

$$w_{tot} = R \cdot (T_2 - T_1) \cdot \ln\left(\frac{p_2}{p_1}\right) < 0; \tag{11.13}$$

a result that is obviously less than zero and thus useful work is dissipated from the system. Thus, again the Carnot efficiency η_C can be determined as follows:

$$\eta_C = \frac{|w|}{\dot{q}_{12}} = \frac{(T_2 - T_1)}{T_2} = 1 - \frac{T_1}{T_2}; \tag{11.14}$$

which of course must correspond to that of Eq. (11.6). Some fundamental results of the idealized Carnot process should be noted again:

- The efficiency depends only on two temperature levels T_2 and T_1.
- The work done will be all the greater if the maximum temperature T_2. is as large as possible and the minimum temperature T_1 (which is mostly limited by the ambient temperature T_a) is as small as possible.
- Only the reversibility of the process ensures the maximum achievable Carnot efficiency η_C.

In reality, the Carnot efficiency is not achieved for essentially four reasons.

- Heat transfer is only possible with the help of a driving temperature difference (see Chap. 4). This means that the coldest temperature in the cycle is higher than that of the heat sink and that the highest temperature of the cycle is lower than that of the heat source. As a result, the thermal efficiency is immediately lower than that of the Carnot process.
- Real technical compression and expansion processes are irreversible. In the volume change operations, gap losses as well as friction occur, both of which are dissipative in nature, as discussed in Chap. 5. This reduces the useful work (cf. Figs. 11.3 and 11.4) and consequently the efficiency.
- The process steps require a transfer of the fluid from one component to the other. During this transfer, losses occur such as pressure and heat losses, which are purely dissipative in nature, and thus also reduce the efficiency.
- Isothermal heat transfer (i.e. $T = const.$) is difficult to achieve because a temperature gradient occurs within the transport medium on the transfer side.

In principle, as a result of compressions and decompressions, the Carnot-cycle could be realized approximately in the following forms

- in a piston machine—as a closed system
  - The compression in step $3 \rightarrow 0$ should take place with cooling, which would be possible by choosing the coolant and its flow intensity.
  - Further compression $0 \rightarrow 1$ should take place without heat exchange.
  - Relief $1 \rightarrow 2$ should take place when heat is supplied via the cylinder; on the approximately same stroke distance $0 \rightarrow 1 = 1 \rightarrow 2$, this can be achieved by switching on of a heat flow around the piston skirt.
- in a turbo-machine with mass flow through a concatenation of two compressors and two turbines (or compressor and turbine stages).
  - $3 \rightarrow 0$—compressor or compressor stage with strong cooling,
  - $0 \rightarrow 1$—heat-insulated compressor (or compressor stage),
  - $1 \rightarrow 2$—turbine or turbine stage with strong heat supply, e.g. by a combustion chamber,
  - $2 \rightarrow 3$—heat-insulated turbine or turbine stage.

For example, if we consider the Carnot process in a turbine arrangement and limit the maximum pressure ratio p_2/p_1, which is usually given by the number of compressor stages arranged in series, the final temperature T_2 (cf. Eq. 11.12) is automatically fixed. It leads to relatively small values. Compared to the Joule-Brayton gas turbine process discussed later, this results in small specific cycle work values (w_{tot}) for the Carnot cycle process, which are reduced even further at partial load. Conversely, if the temperature T_2 is set to a high value in order to attain achieve a high efficiency, this automatically leads to very high pressure ratios, which are technically hardly feasible.

Fig. 11.5 Schematic sketch for the calculation of the effective mean temperature T_m

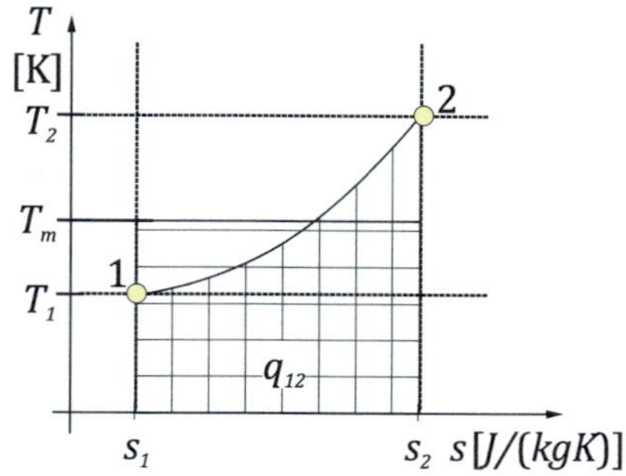

At this point it is useful to introduce a thermodynamic or effective mean temperature (T_m), which we can later use as a quantity to compare thermodynamic cycle processes. This quantity is defined so that the transferred heats q_{12m} and q_{12}, which are outlined in Fig. 11.5, are identical, which yields:

$$q_{12m} = T_m \cdot (s_2 - s_1) = q_{12} = \int_1^2 T(s)ds. \tag{11.15}$$

For an isobaric change of state ($p = const.$), the entropy changes from 1 to 2 yields the following result:

$$(s_2 - s_1) = \int_1^2 ds = \int_1^2 \frac{c_p}{T}\, dT = c_p \ \ln\left(\frac{T_2}{T_1}\right). \tag{11.16}$$

Accordingly, the integral in Eq. (11.15) results in:

$$\int_1^2 T(s)\, ds = \int_1^2 T \cdot \frac{c_p}{T}\, dT = c_p(T_2 - T_1), \tag{11.17}$$

and thus, solving Eq. (11.15) according to the thermodynamic mean temperature T_m, we obtain the relation:

$$T_m = \frac{(T_2 - T_1)}{\ln\left(\frac{T_2}{T_1}\right)}. \tag{11.18}$$

Another key finding when considering the calculation of the thermodynamic mean temperature is that to achieve high thermodynamic efficiency, it is advantageous if the temperature changes only slightly, which is technically extremely complex.

11.2.2.2 Stirling Cycle

The Stirling process consists of two isothermal and two isochoric changes of state and forms an albeit curved quadrilateral in both the $p - v$ diagram and the $T - s$ diagram, as shown in Fig. 11.6.

Technically realized is the Stirling engine by a cylinder in which two pistons are coupled via a gear box. One is a displacement piston and the other a working piston. The task of the displacer piston is to move the working gas back and forth between a space with a constant low temperature (expansion space: $T_{\min} = T_1 = const.$) and a space with a constant high temperature (compression space: $T_{\max} = T_2 = const.$). The heat exchanger or regenerator located between the working and displacer pistons, or between the compression and expansion pistons, has the function of a thermodynamic reservoir that alternately absorbs and releases heat, which is indicated in Fig. 11.6. This is technically realized in the form of metallic sponges consisting of matrix structures of fine wires or strips. The individual work cycles of the Stirling process are outlined in Fig. 11.7:

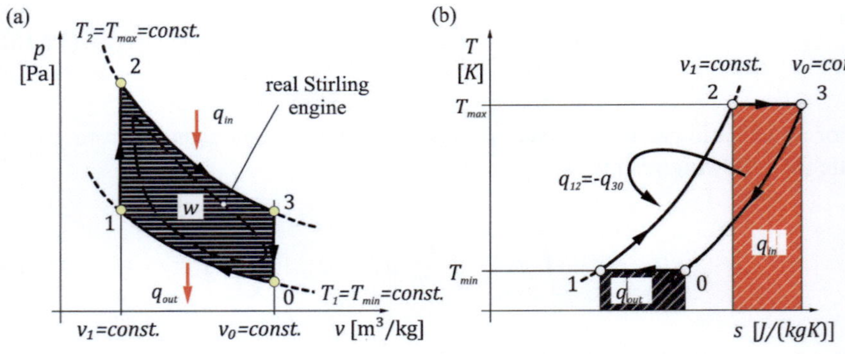

Fig. 11.6 a Stirling cycle states in the $p - v$ diagram (**a**) and $T - s$ diagram (**b**)

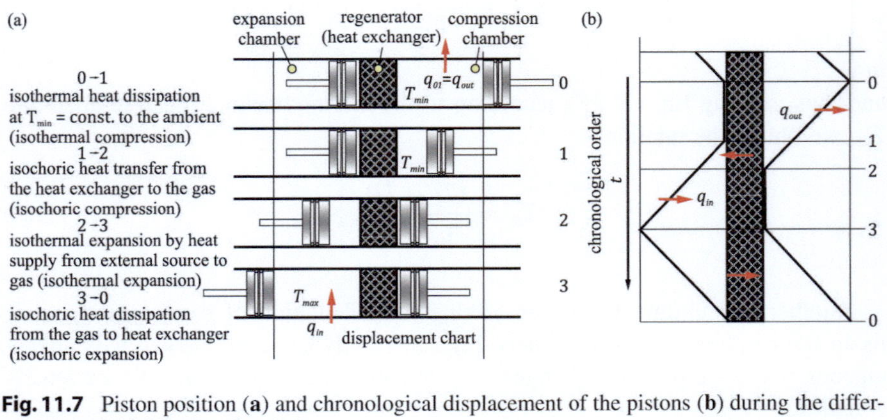

Fig. 11.7 Piston position (**a**) and chronological displacement of the pistons (**b**) during the different steps of a Stirling engine cycle

- A at the beginning of the cycle, the compression piston is at the outer dead centre and the expansion piston at the inner dead centre. The entire working medium is in the cold compression chamber. Since the volume is maximum, the pressure and temperature take minimum values (0).
- The compression piston moves in the direction of its inner dead centre while the expansion piston still remains at its inner dead centre, i.e. isothermal compression from 0 to 1. The pressure increases to the value (1) while the temperature remains constant at $T_{\min} = T_1$, since heat is dissipated to the environment during compression.
- In isochoric compression from 1 to 2, both pistons move simultaneously; the compressor piston moves to its inner dead centre and the expansion piston to its outer dead centre. The volume between the two pistons therefore remains constant (isochoric, $v = const.$). The working medium is pushed through the porous metal matrix of the regenerator and thereby heated from $T_{\min}$ to $T_{\max}$ and leaves the regenerator at $T_{\max}$. The temperature increases from $T_{\min}$ to $T_{\max}$ at constant volume causes the pressure to rise (2).
- During isothermal expansion with $T_{\max}$, the expansion piston moves further to its outer dead center, while the compression piston remains stationary at its inner dead centre. The expansion due to the increase in volume at $T_{\max}$ causes the pressure to drop to the value (3).
- To complete the cycle, the expansion and compression pistons move synchronously, i.e. at constant volume (isochoric, $v = const.$); the expansion piston to its inner dead center and the compression piston to its outer dead centre. The working medium is thereby pushed back through the regenerator and releases its heat to the regenerator.

Thus, the specific useful work of the Stirling cycle (w_{tot}) can be determined from the heat balance:

$$-w_{tot} = q_{in} - |q_{out}| = \underbrace{q_{12} + q_{23}}_{q_{in}} + \underbrace{q_{30} + q_{01}}_{q_{out}}. \quad (11.19)$$

Assuming an ideal recuperator (heat exchanger), we get $q_{12} = -q_{30}$ as indicated in Fig. 11.7c and thus Eq. (11.19) alters to

$$-w_{tot} = q_{23} + q_{12}. \quad (11.20)$$

In the isothermal compression from 0 to 1, in which the pressure increases from p_0 to p_1, heat is released to the environment, and one obviously obtains a negative result:

$$q_{01} = RT_{\min} \ln\left(\frac{p_0}{p_1}\right) = p_0 v_0 \ln\left(\frac{v_1}{v_0}\right) < 0. \quad (11.21)$$

During isochoric compression from 1 to 2, the pressure increases from p_1 to p_2 as the working gas is pushed through the regenerator and ideally heated from

$T_{\min}$ to $T_{\max}$. The transferred heat quantity q_{12} is:

$$q_{12} = c_v(T_{\max} - T_{\min}) > 0. \tag{11.22}$$

During the isothermal expansion from 2 to 3, the pressure decreases but heat is supplied to the working gas by an external heat source, e.g. concentrating solar radiation from a parabolic dish. This results in a positive heat quantity q_{23}:

$$q_{23} = RT_{\max} \ln\left(\frac{p_2}{p_3}\right) = p_2 v_2 \ln\left(\frac{v_3}{v_2}\right) > 0. \tag{11.23}$$

For isochoric heat removal from 3 to 0, the heat transfer from the working gas to the recuperator results in a negative value:

$$q_{30} = c_v(T_{\min} - T_{\max}) < 0. \tag{11.24}$$

This allows us to determine the specific useful work we are looking for:

$$-w_{\text{tot}} = RT_{\min} \ln\left(\frac{p_0}{p_1}\right) + RT_{\max} \ln\left(\frac{p_2}{p_3}\right) = RT_{\max} \ln\left(\frac{v_1}{v_0}\right)\left(\left(\frac{T_{\min}}{T_{\max}}\right) - 1\right). \tag{11.25}$$

For the thermal efficiency of the Stirling cycle η_{th} then, using the relations for the isothermal changes of state ($p_0 v_0 = p_1 v_1$) and because of the isochoric changes of state 3 to 0 or from 1 to 2, the following result is obtained after some transformations:

$$\eta_{th} = 1 - \frac{|q_{out}|}{q_{in}} = 1 - \frac{|q_{01}|}{q_{23}} = 1 - \frac{T_{\min} \ln\left(\frac{v_0}{v_1}\right)}{T_{\max} \ln\left(\frac{v_3}{v_2}\right)} = 1 - \frac{T_{\min}}{T_{\max}} = \eta_C. \tag{11.26}$$

The amazing thing about the solution is that the maximum thermal efficiency, assuming an ideal heat exchanger (or regenerator), depends only on the temperature difference between hot and cold space and is fully equivalent to the Carnot efficiency, although none of the state changes are isentropic. In the case of the Stirling cycle, however, this is subject to the condition that the heat dissipated to isochore 3–0 must be completely returned to the process as heat recuperation during isochore 1–2, which remains an idealized notion in view of a real irreversible heat transfer. In terms of magnitude, the specific work achievable with the ideal Stirling process corresponds exactly to the area enclosed by the two isothermal and isochoric curves in the $p - v$ diagram (compare Fig. 11.6a). In general, the following applies to thermodynamic cyclic processes:

> *The area enclosed in the $p - v$ diagram corresponds to the converted net work w_{tot}.*

The idealized Stirling cycle shown cannot be realized with an actually functioning machine. The piston does not stand still during the isochoric process steps 1–2 and 3–0, but performs a harmonic motion in time. If the engine runs at f revolutions per second, the volume changes periodically between the extreme values volume values (V_0, V_m) a:

$$V(t) = \frac{1}{2}(V_0 + V_3) + \frac{1}{2}(V_0 + V_3)\cos(\omega t), \tag{11.27}$$

where $\omega = 2\pi f$ is the angular frequency. The motion of the displacer is also harmonic. It can be shown that the circular process is most favourable when the position of the compression piston precedes that of the expansion piston by 90 °C.

If the warm gas flows from the expansion chamber (often called displacer) into the compression chamber in step $3 \rightarrow 0$, it gives off heat to the regenerator. In step $1 \rightarrow 2$, the cold gas flows back through the recuperator and absorbs the heat stored there. However, complete recovery of the heat energy cannot be achieved in this way either. The $p - v$ diagram of a real Stirling machine then looks as indicated in Fig. 11.6a (dashed line).

Advantages of the Stirling engine are obvious; it does not require valves, valve actuation and control and is thus theoretically largely maintenance-free. The choice of working fluid is also largely free, as the type of external heat source is irrelevant to the operation of the engine itself. A Stirling engine does not build up pressure peaks and pressure changes occur smoothly. This results in almost noiseless and vibration-free operation. Ideally no refilling of gear oil is necessary, and some Stirling engines even operate oil-free and are considered as well suited for the energy-saving coupled generation of electricity and heat (cogeneration) in the cycle. However, long term operation experience is still lacking. With high power density—i.e. compact design—Stirling engines require high operating pressures and speeds. In principle, if the process is reversed, the Stirling engine can also be used as a refrigerating machine without any design modifications, which can be exploited in the field of air conditioning.

The problem areas of the Stirling engine include sealing problems between the expansion and compression pistons and, above all, the technical realization of an ideal heat exchanger.

Stirling engines are generally externally heated piston engines and are thus subject to limitations in terms of maximum absolute engine power. External heating allows the use of most heat sources to drive them. This includes combustion but also prominently solar energy. High-performance Stirling engines operate at the thermal limits of the materials of construction used, so that conversion efficiencies of up to 40% are possible at temperatures of 800 °C, with typical engine speeds of 30–70 revolutions per second.

The Stirling cycle is not initially tied to any particular machine design type, but technically only reciprocating engines have been implemented to date, the three main types of which are alpha, beta and gamma arrangements as shown in Fig. 11.8. These differ in piston arrangement, stroke, and arrangement of heat source and sink, respectively.

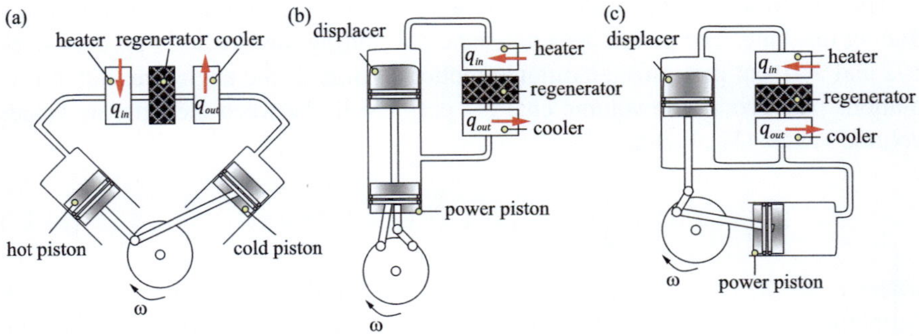

Fig. 11.8 Schematic design of different types of Stirling engine: alpha (**a**), beta (**b**) and gamma (**c**) type

- The alpha type consists of two separate cylinders: a hot working cylinder and a cold compression cylinder. Both sides of the cylinder head are connected by a channel, which generally contains the regenerator. Instead of the displacer piston, there is a second working piston (see Fig. 11.8a).
- In the beta type, the displacer and working pistons move in a common cylinder, which is terminated by the working piston. The regenerator can be arranged in a bypass but can also be integrated directly into the displacer piston for small outputs (see Fig. 11.8b).
- In the gamma type, the displacer and working pistons move in two different cylinders that are connected to each other. The connection point can be at the cold or hot end (see Fig. 11.8c).

A special form of reciprocating piston machine is the free-piston machine. In the free-piston Stirling engine, the displacer and working pistons are mechanically decoupled and there are no crank drives. Therefore, the "free" movements of displacer and working pistons hardly lead to mechanical friction, and the forces are much easier to control.

In the Stirling cycle, the working gas is alternately heated and cooled at a constant temperature and volume. Early Stirling engines have used air at atmospheric conditions because it has a relatively high density for a gas and can therefore be used directly in the cycle, with gas loss through seals also a minor problem. However, high efficiency requires high temperatures, which quickly lead to oxidation. Due to their high thermal conductivity, hydrogen and helium are suitable for high-speed, high-performance Stirling engines, their density deficit being compensated by high pressures, thus allowing a high mean working gas pressure. Above and below this mean pressure (boost pressure), compression and expansion vary. Thermodynamically, hydrogen allows higher efficiencies than helium, but at high temperatures hydrogen embrittlement of materials occurs, so helium is usually used. To maximize performance, high performance engines operate at pressures up to 20 MPa, which places increased demands on the seals between the high-pressure area of the engine and the parts that are under ambient pressure.

11.2.2.3 Ericsson Cycle

The Ericsson process composed of two isothermal and two isobaric process state changes which are illustrated in the $p - V$ and $T - s$ diagram in Fig. 11.9a, b. The state changes comprise:

- $0 \rightarrow 1$: Isothermal compression ($T_1 = T_{\min} = const.$) from a low (p_1) to a high pressure (p_2) with simultaneous gas cooling (q_{out}).
- $1 \rightarrow 2$: Isobaric ($p_2 = const.$) heat addition (q_{12}) at high pressure, where heat is added by means of a regenerator (heat exchanger) located between the compression chamber and the expansion chamber.
- $2 \rightarrow 3$: Isothermal expansion ($T_2 = T_{max} = const.$) from a high to a low pressure, with heat being supplied externally (q_{in}) to the expansion space.
- $3 \rightarrow 0$: Isobaric ($p_1 = const.$) heat removal at a low pressure, where the heat (q_{30}) is removed into the regenerator and can be used for the next cycle.

In the ideal Ericsson cycle, the regenerator (heat exchanger) is designed in such a way that the same amount of heat is exchanged between the working machine, which can be a piston or turbine, and the compression chamber being a piston or a compressor, i.e. the following applies $q_{12} = -q_{30}$. The specific useful work w results from the heat balance to:

$$-w_{tot} = q_{in} - |q_{out}| = q_{23} + q_{30} + q_{01} + q_{12} = q_{23} + q_{01}. \tag{11.28}$$

The isothermal compression at temperature T_1 from p_1 to p_2 leads to:

$$q_{01} = p_1 v_0 \ln\left(\frac{p_1}{p_2}\right) = -RT_{\min} \ln\left(\frac{p_2}{p_1}\right) < 0. \tag{11.29}$$

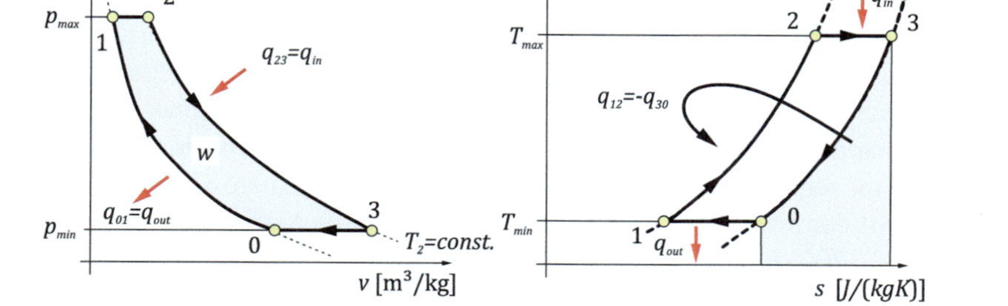

Fig. 11.9 Thermodynamic depiction of an Ericsson cycle in the $p - v$ diagram (**a**) and $T - s$ diagram (**b**)

In the isothermal expansion from 2 to 3, in which the pressure decreases from p_2 to p_1, we obtain

$$q_{23} = p_2 v_2 \ln\left(\frac{p_2}{p_1}\right) = RT_{\max} \ln\left(\frac{p_2}{p_1}\right) > 0. \tag{11.30}$$

If one introduces a compression ratio in the form $\pi = p_2/p_1$, the specific useful work (w_{tot}) is given by:

$$-w_{tot} = RT_{\min}\left(\frac{T_{\max}}{T_{\min}} - 1\right)\ln \pi. \tag{11.31}$$

Analogously, one can determine the thermal efficiency (η_{th}), which leads to the following result:

$$\eta_{th} = 1 - \frac{|q_{out}|}{q_{in}} = 1 - \frac{T_{\min}}{T_{\max}} = \eta_C \tag{11.32}$$

The thermal efficiency of the Ericson process is thus identical to that of the Carnot cycle and would therefore be an ideal process. However, the practical execution fails because of the technical realization of an isothermal compression (requires cooling) and an isothermal expansion (requires heat supply). Approximately, this can be achieved by a stepwise compression with intercooling during compression and a stepwise expansion with interheating.

In contrast to the Stirling hot gas engine, an Ericsson engine, often referred to in the literature as the Joule-Ericsson or Ackeret process, can be designed not only as a piston engine but also as a turbo-machine as indicated in Fig. 11.10c. The latter permits high mass throughputs and thus, in principle, also high outputs. The Ericsson engine as a hot gas piston engine can be designed both as an open process and as a closed process, as depicted in Fig. 11.10a, b, respectively, without going here into the details of the connecting rod and crankshaft design. The Ericsson engine can be operated either with a closed circuit with a cooler (in which case the system can be operated at high pressure and allows the use of working fluids such as helium or hydrogen), or with an open circuit with or without regenerator. In this case, the working fluid is air, which can be expanded to atmospheric pressure. When Ericsson engines are used, an open circuit is usually considered, since it meets both the requirements of cogeneration at low power (micro-CHP) or to allow a low cost of conversion of solar energy into electricity (Alaphilippe et al., 2007; Bonnet et al., 2005). In these engines, the compressor and the expansion cylinder run at a low speed to limit the pressure losses caused by valves, as well as to reduce mechanical losses. This solution ensures the best performance.

When comparing the Ericsson and Stirling cycles for a given temperature ratio ($T_{\max}/T_{\min}$) and a given volume, the Stirling cycle generates more specific net work output above a pressure ratio threshold (π), and when comparing both cycles for a given pressure ratio and maximum volume, the Ericsson cycle generates more

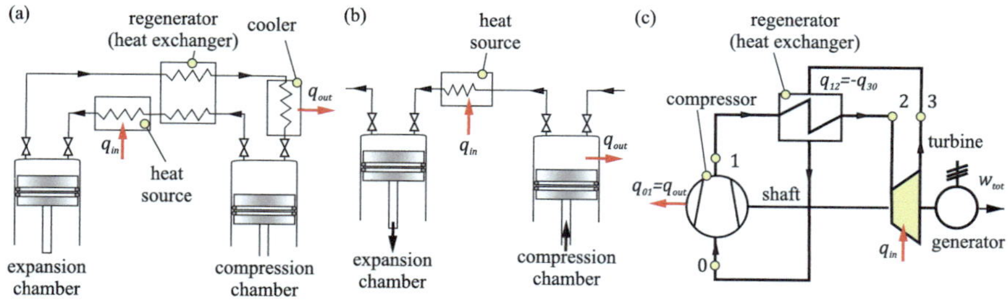

Fig. 11.10 Different designs of the Ericsson cycle in a closed process with recuperator (**a**) and an open process (**b**) in the design of a reciprocating engine as well as a closed turbine cycle (**c**)

network output above a temperature ratio threshold, studies of which are analysed in detail in (Costa & MacDonald, 2018).

A central disadvantage of the Ericsson cycle is the intensive heat exchange required and the resulting thermal load for a relatively compact machine for closed cycles. Even in large power plants, implementation is challenging, because the continuous cooling during the targeted isothermal compression as well as gradual heat input during the targeted isothermal expansion has not been technically solved by now. Nevertheless, the principle remains of interest because its partial implementation through suitable heat exchangers, using highly efficient heat recuperators can noticeably increase the efficiency of the machine and thus, at a small cost, provide an alternative to gasoline and diesel engines that require internal combustion of fossil carriers. An important advantage of the Ericsson cycle over the widely known Stirling engine often goes unnoticed. The volume of the heat exchanger does not affect the thermal efficiency.

11.2.2.4 Joule-Brayton Process (Gas Turbine Process)

The Joule-Brayton cycle is a right-handed cycle that describes a process occurring in gas turbines and jet engines. Figure 11.11a schematically shows the structure and components of a closed, single-shaft Joule-Brayton cycles. Ideally, the four work cycles consist of:

- an isentropic compression ($dp > 0$) with a reduction in volume ($dv < 0$) and, if possible, without heat input ($dq = 0$) in a compressor from 0 to 1;
- an isobaric ($dp = 0$) heat supply ($dq > 0$) in the combustion chamber accompanied by an increase in volume ($dv > 0$) from 1 to 2;
- a heat loss-free ($dq = 0$), isentropic expansion ($dp < 0$) accompanied by a further volume increase ($dv > 0$) in the turbine from 2 to 3 driving the generator, and finally
- an isobaric ($dp = 0$) heat dissipation ($dq < 0$) in a cooler from 3 to 0 leading to a volume reduction ($dv < 0$).

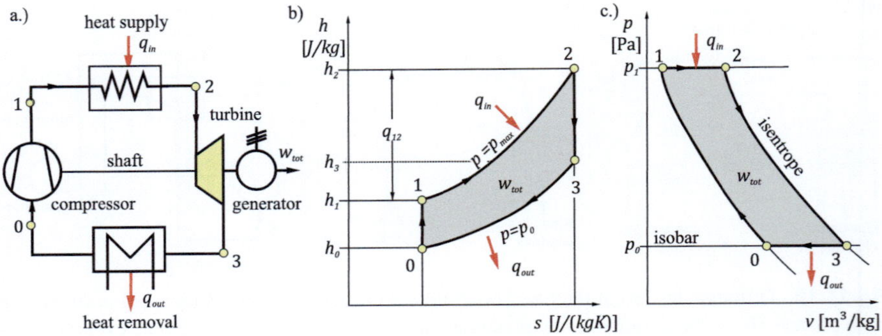

Fig. 11.11 **a** Mechanical implementation of a simple Joule-Brayton process in a single-shaft engine; representation of the Joule-Brayton process in $h - s$ (**b**) and $p - V$ diagram (**c**)

The thermodynamic illustration of the ideal Joule-Brayton process in the $h-s$ and $p-v$ diagrams is shown in Fig. 11.11b, c. The useful work w_{tot}, which can be supplied to the generator, results from the area enclosed by the points 0–1–2–3, which is highlighted in gray. The area below the curve $p = p_0$ is lost uselessly. It is immediately apparent from the diagram that the highest possible efficiencies can only be achieved with high pressures and high temperatures.

Assuming constant specific heat capacities (c_p), the specific useful work ($-w_{tot}$) from the heat balance is given by

$$-w_{tot} = q_{in} + q_{out} = \sum_{ij} q_{ij} = q_{12} + q_{30} = c_p(T_2 - T_1 + T_0 - T_3). \quad (11.33)$$

If one introduces a compression ratio $\pi = p_1/p_0$ and uses the isentropic relation for ideal gases, which gives the following result for the temperatures

$$T_1 = T_0 \pi^{\frac{\kappa-1}{\kappa}}, T_3 = T_2 \pi^{\frac{1-\kappa}{\kappa}}, \quad (11.34)$$

the equation for the specific useful work is obtained,

$$-w_{tot} = c_p T_1 \left(1 - \pi^{\frac{\kappa-1}{\kappa}} + \frac{T_3}{T_1} - \frac{T_3}{T_1}\left(\frac{1}{\pi}\right)^{\frac{\kappa-1}{\kappa}}\right), \quad (11.35)$$

or, using $c_p/R = \kappa/(\kappa - 1)$, the following is obtained:

$$-w_{tot} = \frac{\kappa}{\kappa - 1} R T_1 \left(\left(\pi^{\frac{\kappa-1}{\kappa}} - 1\right)\left(\frac{T_3}{T_1}\left(\frac{1}{\pi}\right)^{\frac{\kappa-1}{\kappa}} - 1\right)\right). \quad (11.36)$$

The thermal efficiency η_{th}, which is defined as the ratio of benefit to expense for a heat engine, is given by:

$$\eta_{th} = \frac{w_{tot}}{q_{in}} = 1 - \frac{q_{out}}{q_{in}} = 1 - \frac{q_{30}}{q_{12}} = 1 - \frac{(T_3 - T_0)}{(T_2 - T_1)} = 1 - \left(\frac{1}{\pi}\right)^{\frac{\kappa-1}{\kappa}} \quad (11.37)$$

Two conclusions follow immediately from the preceding equation:

- The greater the adiabatic coefficient κ, the greater the achievable thermal efficiency for a given pressure ratio π. Therefore, the Joule-Brayton process can be operated most effectively with helium ($\kappa = 5/3$) as gas.
- The thermal efficiency of the Joule-Brayton cycle is a pure function of the pressure ratio π. It increases degressively with the pressure ratio. That is, the higher the pressure ratio π, the greater the efficiencies that can be achieved. With a simple Joule-Brayton process, thermodynamic efficiencies $\eta_{th} \approx 45\%$ are achievable with temperatures of 1000 °C and pressure ratios $\pi \approx 30$.

For a maximum turbine inlet temperature T_2 given by the structural material used, an optimum temperature $T_{1,opt}$ after compression can be determined at which the cycle yields the greatest possible useful work (w_{tot}). This is:

$$T_{1,opt} = \sqrt{T_0 \cdot T_2}. \quad (11.38)$$

One development goal of the Joule-Brayton gas turbine process is to increase the maximum turbine inlet temperature T_2. The Fig. 11.12 illustrates the thermal efficiency as a function of the pressure ratio at different turbine inlet temperatures T_2. From the graph, it can be seen that for a given temperature T_2, there is an optimum pressure ratio at which maximum thermal efficiency can be achieved. By using the waste heat from the expansion machine, the thermal efficiency can be further improved with the aid of a recuperator to preheat the compressed gas. This is possible if the turbine outlet temperature T_3 is above the temperature at the compressor outlet (T_1). This amount of heat then no longer has to be supplied externally. However, this measure only makes sense for moderate pressure ratios. Figure 11.13 show the corresponding block diagram and the representation in the $T - s$ diagram. The efficiency is then calculated as follows:

$$\eta_{th} = 1 - \left(\frac{T_{\min}}{T_{\max}}\right)\pi^{\frac{\kappa-1}{\kappa}}. \quad (11.39)$$

With the aid of a two-stage gas turbine process, the efficiency can be increased even further. Figure 11.14 depicts the two-stage cycle in the $T - s$ diagram and the associated block diagram. The two-stage process increases the equipment requirements not inconsiderably. In the two-stage process, the gas is first compressed from the ground state 0 to the pressure p_1 after 1*. At this mean pressure ($p_1 > p_0$), the gas is cooled to the state 1**, which reduces the compressor work; this is followed

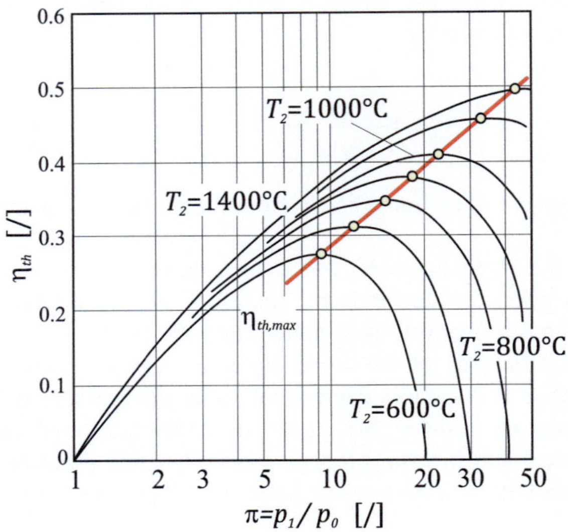

Fig. 11.12 Thermal efficiency η_{th} of a simple Joule-Brayton process as a function of pressure ratio $\pi = p_1/p_0$ at different turbine inlet temperatures

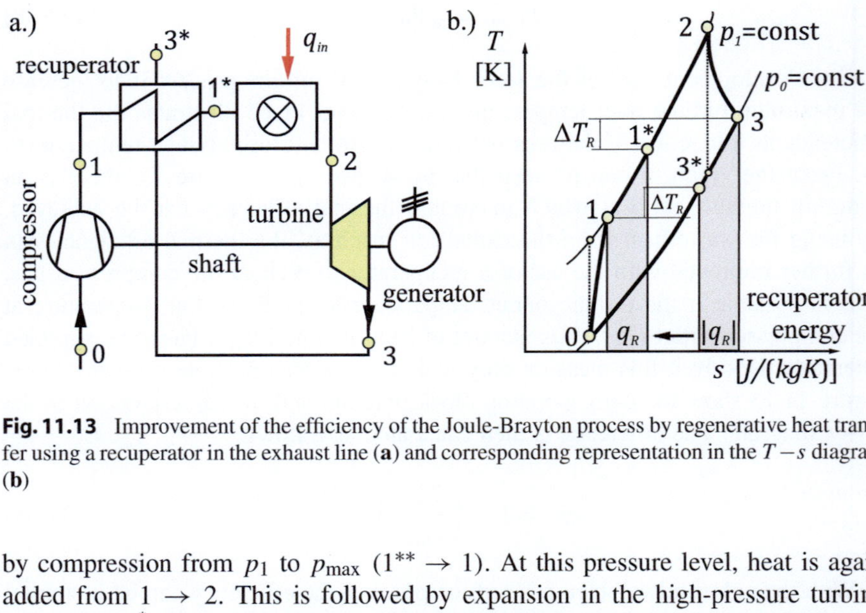

Fig. 11.13 Improvement of the efficiency of the Joule-Brayton process by regenerative heat transfer using a recuperator in the exhaust line (**a**) and corresponding representation in the $T-s$ diagram (**b**)

by compression from p_1 to p_{max} ($1^{**} \rightarrow 1$). At this pressure level, heat is again added from $1 \rightarrow 2$. This is followed by expansion in the high-pressure turbine from $2 \rightarrow 3^*$. The exhaust gas from the high-pressure turbine is again heated in the "*combustion chamber*" ($3^* \rightarrow 3^{**}$), where it is fed to the low-pressure turbine and the expansion of $3^{**} \rightarrow 3$ takes place with energy being released in the turbine. This two-stage process ultimately results in the generation of a higher mean

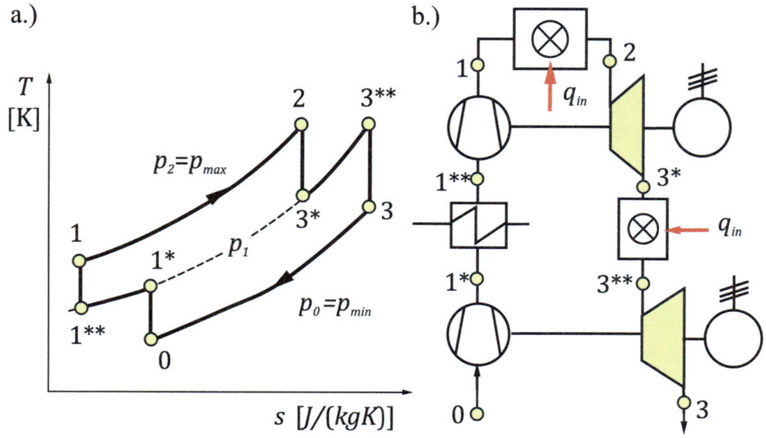

Fig. 11.14 Schematic representation of a two-stage Joule-Brayton process in the $T-s$ diagram (**a**) and the corresponding plant flow diagram (**b**)

temperature and thus a higher thermodynamic efficiency η_{th}. It should not be forgotten that in multi-stage arrangements, a very large amount of enthalpy circulates in the cycle compared to the turbine work. Changes in intermediate cooling and heating thus have a more lasting effect on the thermodynamic efficiency than in the simple Joule-Brayton process.

By introducing further staged compression processes with respective heat removal and a corresponding multistage expansion with respective heat supply and regeneration, the thermodynamic efficiency is

- increased through the regeneration of temperature ranges to be bridged and,
- the compression work can be reduced by the heat dissipation, and
- the expansion work can be increased by the heat supply in the expansion stages.

With an infinite number of stages of compression and heat dissipation, the process transitions to isothermal compression. The process is then described by the Ericsson or Ackeret cycle, whose efficiency corresponds to that of the Carnot process.

Regarding the application in solar thermal power plants, with their relatively strong fluctuating power supply given by the solar irradiation, some aspects should not be forgotten. With flexible combustion chamber temperatures and more or less fixed pressure ratio (given by the inertia of the compressor shaft), the efficiency of the thermodynamic cycle can change not insignificantly. Possible target point optimizations with regard to thermodynamic efficiency simply by selecting multistage compressor and turbine arrangements can be counterproductive, since the efficiency gains possible at an operating point are marginal compared with the additional equipment required for this purpose, and part-load operation in particular is problematic. Under fluctuating conditions, it is relatively easy for the efficiency to even deteriorate.

So far, solar thermal power plants based purely on the Joule-Brayton process for energy generation exist only on a laboratory scale. In open receiver concepts, the energy would be supplied at approximately atmospheric pressure, which limits the attainable efficiency of a Joule-Brayton process. By using pressure-charged closed receivers concepts, on the other hand, Joule-Brayton concepts become attractive. For solar gas heating, the receiver must operate under operating conditions defined by the gas turbine. For simple air-driven turbines, tests have been successfully performed at air inlet temperatures of 250–400 °C and a pressure of 0.17–1.5 MPa. However, the thermal efficiency is comparatively low at about 20% (cf. Fig. 11.12). If volumetric pressure receivers with metal mesh absorbers, which have already been successfully tested at temperatures up to 845 °C (Pitz-Paal et al., 2002), prove themselves in long-term tests, it would be possible to achieve thermal efficiencies of the order of 40% when using helium-driven turbines. Further development of this type of receiver makes it possible in principle to increase the temperature up to 1100 °C, provided those ceramic materials can be used as absorber materials. Such temperature levels then allow the Joule-Brayton cycle to be used as a combined cycle gas-steam process (gas and steam cycle (CCGT), see section combined cycles 11.2.4) or even for the direct solar generation of hydrogen.

11.2.2.5 Alternative Gas Cycle Processes with Supercritical Fluids

Even though single-phase thermal cycles allow for very large thermal efficiencies, these are usually tied to large pressure ratios (π) and very high temperatures that are difficult to achieve using solar thermal collectors. At the same time, their partial load behaviour is not easy to handle because their thermal inertia is also very low, resulting in rapid changes in temperature in time, which in turn lead to high thermomechanical stresses and early fatigue as well as creep. Although the specific useful power of the turbine requires an adiabatic coefficient as large as possible, one would rather have a value as small as possible for compression in order to keep the specific compressor work and, above all, the irreversible entropy generation in the compressor low.

One way to reduce the compression work of a working circuit is to use supercritical fluids, since their behaviour halfway from liquid to gas is largely incompressible, which reduces the work required to overcome a given pressure difference and thus the amount of energy circulating in the circuit.

The beneficial effect of mechanical compression near the critical point is illustrated in Fig. 11.15a by the CO_2 pressure versus specific volume plot. Assuming ideal isothermal compression of a uniform mass of CO_2 between two given pressure levels (from 7 to 16 MPa), the gray shaded area shows the work required when the temperature of the fluid at the compressor inlet is approximately the critical temperature, while the shaded area shows the additional work required when the temperature is increased to 1.5 °C. On the other hand, the thermophysical properties of a fluid change more with temperature and pressure as the critical point

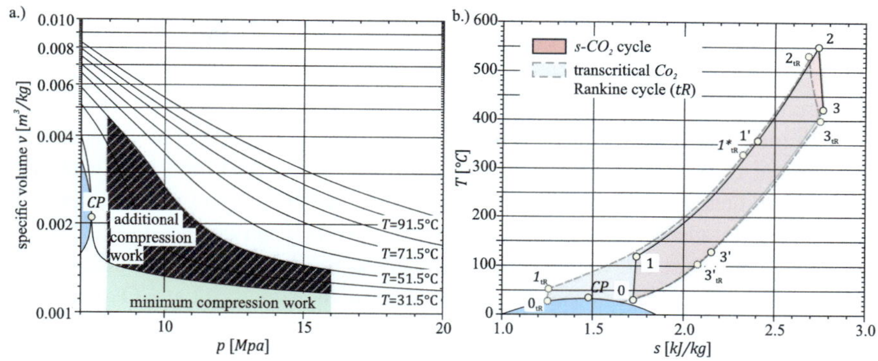

Fig. 11.15 **a** Specific volume v of CO_2 as function of pressure; **b** $T - s$ diagram of the s–CO_2 and transcritical CO_2 cycle both with recuperation

is approached. The critical CO_2 point also limits the maximum pressure ratio that can be achieved in the cycle. Since the minimum pressure of the cycle is fixed by the critical point of CO_2 (7.375 MPa, 30.98 °C) and the maximum pressure is fixed by technological constraints (typically about 25–30 MPa), only a maximum pressure ratio of $\pi = 4$ can be achieved (which is extremely low compared to the value of 200 of the two-phase Rankine steam technology).

In recent years, supercritical carbon dioxide s–CO_2 systems in particular have been considered (Dostal et al., 2004) and also in the context of solar application (Yamaguchi et al., 2006). Unlike air, carbon dioxide's critical point is close to ambient conditions, at 7.38 MPa and 1.1 °C, making it easily affordable by conventional cooling systems. This makes CO_2 an excellent choice for use in supercritical cycles. One disadvantage of supercritical gas cycle processes is their real gas behaviour. In the case of a real gas, such as CO_2, thermal efficiency is a function of pressure, therefore cycle efficiency is negatively affected by pressure and pressure drop.

Supercritical cycle processes are closed-loop single phase fluid cycles similar as the Joule-Brayton concept, using only higher pressures. The thermodynamic representation of the gas processes itself corresponds to the Joule-Brayton cycle, as indicated in the Fig. 11.15b.

In the transcritical Brayton cycle, the state of the working fluid is also always above the critical point and merely moves closer to the critical point. Since the point 0 then still represents a gaseous medium, the compression of the working fluid can still be brought to the upper pressure level with the aid of a compressor, just as in the s–CO_2 cycle. The rest of the cycle corresponds to the s–CO_2 cycle. If the higher pressure level of the transcritical CO_2-Brayton cycle p_1 corresponds to that of the s–CO_2 cycle, a higher amount of energy can be generated at the lower pressure level p_0 than in the pure s–CO_2 cycle due to a lower compressibility factor. This means that the power consumption of the compressor is lower.

However, the transcritical CO_2 Brayton cycle can only be realized in a narrow pressure range.

The main difference between supercritical carbon dioxide (s–CO_2) operation compared to helium-based Joule-Brayton cycles is in the operating conditions. The helium cycle operates at high temperatures and medium pressures, while the supercritical CO_2 cycle operates at medium temperatures and high pressures. The low specific heat in the case of s–CO_2 and the smaller temperature difference in the heat input causes, that the mass flow required to generate the same power is much larger than in the helium cycle. However, the high pressure and thus the high liquid density reduces the mass flow rate for the supercritical CO_2 cycle (by a factor of about 5). This leads to more compact plant components, especially for turbomachinery, and gives more flexibility in designing for low pressure drop, since high density reduces velocity, which exists as a quadratic term in the pressure drop equations and thus becomes more important than density. In addition, the efficiency of the s–CO_2 cycle is not as sensitive to the pressure drop as the helium cycle due to the high pressure (the penalty for efficiency for the same total pressure drop would be less in the case of s–CO_2). This is very important for the design of recuperators. Thus, the heat to be regenerated that is not lost can be more than twice as high in the case of the supercritical s–CO_2 cycle as in the helium cycle. This presents a challenge in the design of the recuperators. If one wishes to maintain the same magnitude as in the helium cycle, it is necessary to increase the pressure drop in the recuperators. However, this can be done without compromising efficiency or achieving reasonable recuperator volumes.

The reason why the supercritical CO_2 cycle achieves a higher efficiency than the helium cycle at the same temperature is temperature-entropy behaviour. Looking at the helium and s–CO_2 cycles operating between the same maximum and minimum temperatures, the average temperature at which heat removal occurs in the s–CO_2 cycle is similar to that of the helium. But the mean temperature at which heat input occurs is much higher in the s–CO_2 cycle. The higher temperature for heat input brings the s–CO_2 cycle closer to the Carnot cycle and results in an efficiency improvement over helium. This is mainly caused by the flatter isobaric curve of the s–CO_2 due to its proximity to the critical point. This is the main reason why the s–CO_2 cycle is very efficient even at intermediate temperatures. The low compressor work (~30% of turbine power, compared to ~44% for helium) makes intercooling unattractive. This greatly simplifies the cycle design by eliminating the need to introduce intercoolers. In the helium cycle, each intercooler, although they can be very compact, requires separate cooling.

Comparing the Joule-Brayton based 880 °C helium cycle with the s–CO_2 cycle, it can be seen that the average heat removal temperature is significantly higher for the helium cycle. This is an advantage when designing the pre-cooler for the helium cycle since the higher temperature difference between the working fluid and the cooling water reduces the required size of the pre-cooler and its pressure drop. On the other hand, this disadvantage is reduced by the lower sensitivity of the s–CO_2 cycle to the pressure drop. In general, the design of the pre-cooler is not a significant problem for the s–CO_2 cycle since the improvement of the heat transfer coefficient near the critical point is another advantageous factor.

From a purely thermodynamic point of view, it can be stated that the s–CO_2 cycle always performs better than the helium cycle at the same turbine inlet temperature, although the pressures are somewhat higher than for helium (~20 MPa). However, as far as the temperature range of 500–650 °C is concerned, which is relevant for thermal engines and is limited primarily by the limitations of the structural materials, the s–CO_2-cycle performs better than both the supercritical steam and the superheated steam Rankine cycle.

The reason why cycle concepts other than just the simple Brayton cycles with intercooling and reheating are studied in s–CO_2 plants is the existence of the pinch point problem in the recuperator of the cycle. The pinch point is the location in the recuperator with the lowest—in the limit zero—temperature difference. Due to the radical temperature and pressure dependence of the specific heat, the temperature difference between the hot and cold fluid of the s–CO_2 inside the recuperator varies considerably. Thus, even for the single-phase condition of the CO_2 working fluid, the minimum value of the temperature difference is not always reached at the recuperator inlet or outlet, but sometimes somewhere along the recuperator. This effect significantly reduces the effectiveness of the recuperator and immediately lowers the thermal efficiency of the cycle. To avoid the pinch point problem, measures must be taken to balance the flow heat capacities on the hot and cold sides of the recuperator. However, when a larger total volume of heat exchangers is provided, the efficiency of the cycle improves, but the rate of efficiency improvement decreases as the volume increases. Thus, at some point, the efficiency improvement is equalized by the additional cost of heat exchangers.

The cycle efficiency of an s–CO_2 cycle decreases linearly with increasing compressor inlet temperature. Up to an inlet temperature of about 45 °C, the optimum pressure ratio is significantly affected. From a steady state perspective, compressor inlet temperature does not significantly affect cycle optimization, so cycles with compressor inlet temperatures further away from the critical temperature still achieve significantly better efficiency than an ideal gas cycles under the same conditions. But, for example, increasing the compressor inlet temperature to 50 °C results in a drop in efficiency of about 5% caused by additional compressor work as indicated in Fig. 11.13b. Optimal compressor inlet temperature and low compression ratios therefore require operation close to the critical temperature, which makes the design of the pre-cooler and the surrounding heat sink particularly challenging.

In terms of turbomachinery technology, s–CO_2 as operation fluid does not pose any technological challenges for both compressors and turbines at all power levels, and practical operating experience is available both on a small scale up to CO_2-cooled British nuclear gas cooled reactors in the MW range. CO_2 turbines exhibit a higher efficiency (about 2%), because they have a fewer number of stages and are shorter (by a factor of five compared to helium and by a factor of 25 compared to a steam turbine). In contrast to helium and Rankine cycles, only one stage is required, whereas a Rankine cycle requires high, medium, and low pressure stages for the same power.

Therefore, special attention must be paid to the design of the pre-cooler and the heat sink system. The s–CO_2-cycle also optimizes itself around a relatively small temperature difference between heat source inlet and outlet, which is why it is also suitable for combined heat-work machine utilization at low solar thermal receiver temperatures. However, the narrow optimal operating range requires highly efficient recuperators. An efficient s–CO_2 cycle requires recuperation of approximately twice (!!!) as much heat as is supplied by the heat source. Only with the availability of PCHE devices (Printed Circuit Heat Exchangers), which have been commercially available since about the year 2000, this can be realized with sufficiently high efficiency in correspondingly compact devices; otherwise, the advantage of extremely compact turbines is nullified by excessively large heat exchangers.

Another shortcoming that should not be underestimated is also the current lack of simple effective procedures to operate highly efficient part-load operation of the s–CO_2 recompression cycle. This need not be a flaw if sufficiently large thermal storage facilities enable economically competitive in a base-load operation mode. Technically, efficient part-load operation would only be possible via turbine optimization if, for example, these allow for a blade adjustment.

An alternative to the s–CO_2, which is not very easy to control, is a transcritical CO_2 Rankine Cycle ($t_{c,R}$-Rankine Cycle). Like the Clausius-Rankine cycle, the transcritical CO_2 Rankine cycle involves a condensation process and is thus a two-phase process. The CO_2 passes through the two-phase region during the cooling process and is completely liquefied. This is one main difference from the (s–CO_2 or transcritical) Brayton cycles. Another difference is that instead of a compressor, a pump is used to pressurize the working fluid. Figure 11.15b shows a $T-s$-diagram (dashed line) of a simple transcritical CO_2 Rankine cycle. The inlet to the pump (point 0) is located on the saturated liquid line below the critical point (CP). Again, the schematic representation of the rest of the cycle is the same as that of the supercritical s–CO_2 cycle with recuperator. The specific working power of the pump is significantly lower than that of the compressor of the two Brayton cycles. This is due to the high density (liquid state) of the CO_2 and the resulting small compressibility factor. With the same boundary conditions for the transcritical Rankine Cycle as for the s–CO_2 or the transcritical Brayton Cycles (p_2, T_2), significantly more electrical energy can be obtained. A special feature of isothermal condensation is the pinch point problem in the cooling section. If the temperature of the cooling water is given, the water mass flow has to be increased to liquefy the CO_2. Depending on the ambient conditions (e.g. water sources), this can lead to technical and economic problems.

11.2.3 Two-Phase Thermal Cyclic Processes

The vast majority of thermal energy conversion processes are not based on single-phase cycles but on two-phase cycles. The essential basis for understanding why this is so requires an analysis of work done/discharged and heat added/discharged during changes of state in a thermal work cycle, which also considers the equations of state and specifics of the respective fluid in the two-phase region. Two-phase cycles are used in a number of conversion processes, ranging from the classic Clausius-Rankine cycle (water-steam) to the Organic Rankine Cycle (ORC), Kalina processes and others. The choice of working fluids is equally diverse, ranging from pure substances (water, CO_2, organic fluids) to ideal mixtures and zeotropic mixtures.

For a singlehase ideal gas, we know that the curve of the pressure (p) and specific volume (v) is constant, i.e. $pv = const$. For a real gas where a phase-transition to a liquid happens, there is a one-to-one correspondence between the condensation (or evaporation) temperature and the pressure. These values are called saturation pressure and saturation temperature. In the two-phase region where gas and liquid coexist, we only need to specify one variable to determine the state of the system. For a saturation temperature T_S the saturation pressure p_S is set and vice versa, if p_S is given then T_S is fixed. In summary, for two phases in equilibrium, $p_S = p(T_S)$. This implies a fundamental difference from the single-phase cycle: we can add or remove heat isothermally.

Let us consider the pressure–volume behaviour of a liquid–vapour system at constant temperature. There is a domain where liquid (l) and vapor (g) coexist, the quality region, which is bounded by the liquid saturation curve on the left and the vapour saturation curve on the right. This domain is roughly dome-shaped and is therefore often referred to as *vapor dome*. Outside this regime, the equilibrium state is a single phase. Left to the critical point the slope of the isotherms, which scale almost hyperbolically with T, is very large, which is due to the low compressibility of most liquids. This, however, means that the work of changing the volume to generate a large pressure is small. Further, the behaviour of the isotherms at temperatures below the *critical point* (CP) in the region to the right of the vapor dome approaches those of an ideal gas as the pressure decreases, and the ideal gas relation is a good approximation in this region. This behaviour shown in Fig. 11.16a for water is found for all isotherms passing through the vapor dome. At a sufficiently high temperature, especially at a temperature corresponding to the pressure at the top of the vapor dome, there is no transition from liquid to vapour at all and the fluid continuously changes from a liquid-like behaviour to a gas-like behaviour. This region is called the super-critical region; for water, the critical temperature is 74 °C and the corresponding critical pressure is 22.1 MPa.

When we now consider working machine processes, we need to describe changes of state that occur when we reduce the pressure or equivalently increase the specific volume, starting, for example, from a state of high pressure and low specific volume (the upper left side of the isotherm in Fig. 11.16a). The

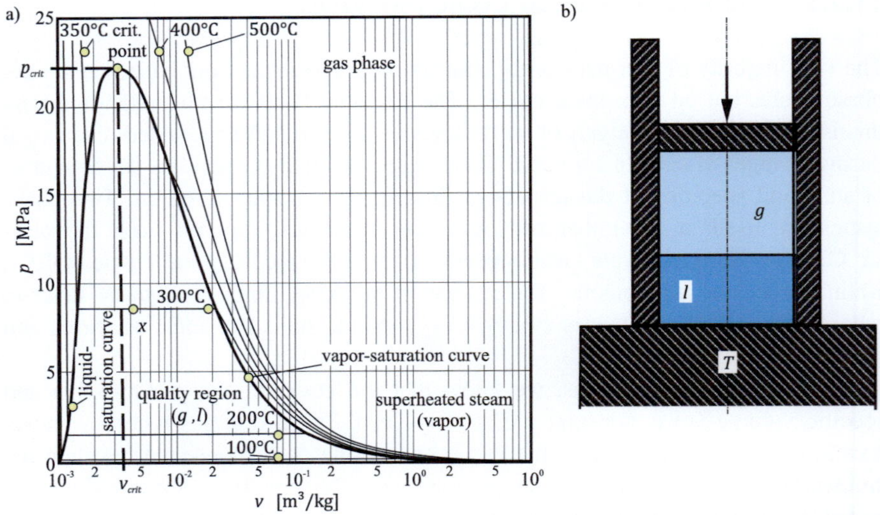

Fig. 11.16 **a** $p - v$-Diagram of water with the changes of state at different temperatures, **b** liquid vapor equilibrium in a two-phase medium

behaviour in this region is liquid-like with very low compressibility. As the pressure decreases, the volume changes little until the limit of the vapour dome is reached. When this point is reached, the pressure is fixed because the temperature is constant. When a piston is retracted, the specific volume increases as more liquid evaporates and more vapour is produced. During this process, as the expansion is isothermal (we have set it that way), heat is transferred to the system. The specific volume increases at constant pressure until the right-hand boundary of the vapour dome is reached. At this point, all the liquid has turned to vapour, and the system behaves like a single-phase fluid again. For water at temperatures close to room temperature, the behaviour in this region would essentially be that of a perfect gas. To the right of the vapour dome, the fluid behaviour again qualitatively corresponds to that of a perfect gas.

We investigate the work and heat transfer in quasi-static processes with two-phase systems. For clarification, we consider the system as a liquid–vapour mixture in a container whose volume can be varied by the movement of a piston, as shown in Fig. 11.16b. The system is kept at a constant temperature by contact with a heat reservoir of temperature T. The pressure is therefore also constant, but the volume V can change. For a given mass, the volume is proportional to the specific volume, so that the point x in Fig. 11.16a must move to the left or right when v changes. This implies that the amount of mass in each of the two phases also changes, and hence the quality, as mass is transferred from one phase to the other. We want to find the heat and work transfer associated with the mass change in each phase. The volume change can be related to the mass changes in the two phases.

$$dV = v_l dm_l + v_g dm_g. \tag{11.40}$$

Since the system mass is constant $\left(m = m_l + m_g = \text{ const.}\right)$ the differential change dm vanishes, i.e. $dm = 0 = dm_l + dm_g$. If we define the quantity dm_{lg} as the mass transferred from the liquid to the vapor state, we get $dm_{lg} = dm_g = -dm_l$. In terms of dm_{lg} the volume change of the system is the described by:

$$dV = \left(v_g - v_l\right)dm_{lg}. \tag{11.41}$$

Thereby we can evaluate the work done,

$$dW = pdV = p\left(v_g - v_l\right)dm_{lg}. \tag{11.42}$$

The internal energy U of the system can be expressed as a sum the mass in each phase (m_i) and the specific internal energy (u_i) in the way $U = u_l m_l + u_g m_g$, so that the differential change dU is given by $dU = \left(u_g - u_l\right)dm_{lg}$. Note that the specific internal energy of the two-phase system can be expressed in a similar way as the specific volume in terms of the vapour quality (α) and the specific internal energy of each phase

$$u = \alpha u_g + (1 - \alpha)u_l. \tag{11.43}$$

Using the 1st law of thermodynamics $dQ = dU + dW$, we obtain

$$dQ = \left(u_g - u_l\right)dm_{lg} + p\left(v_g - v_l\right)dm_{lg} = dm_{lg}\left(\left(u_g + pv_g\right) - (u_l + pv_l)\right), \tag{11.44}$$

$$dQ = \left(h_g - h_l\right)dm_{lg}. \tag{11.45}$$

We get as a result that the heat required for the phase transition is proportional to the difference in specific enthalpy between vapour and liquid. Since the pressure and temperature are constant during the phase transition, the specific internal energy and the specific enthalpy are also constant for the liquid phase and the gas phase. The heat needed per unit mass (q) for transformation between the two phases is $q = h_{lg}$. The notation h_{lg} refers to the specific enthalpy change between the liquid state and the vapor state. The expression for the amount of heat needed (q) is a particular case of the general result that in any reversible process at constant pressure, the heat flowing into, or out of, the system is equal to the enthalpy change. Heat is absorbed if the change is from solid to liquid (heat of fusion), liquid to vapor (heat of vaporization), or solid to vapor (heat of sublimation).

Since the specific phase change enthalpies can take on very large values, for water with a pressure of 0.1 MPa $h_g = 2.676$ kJ/kg and $h_l = 419$ kJ/kg, the heat added q_{in} complete phase change is $q_{in} = h_{fg} = 2.257\text{kJ/kg}$. However, if this is compared with the specific work done (w, Eq. 11.42), this is only $w = p\left(v_g - v_l\right) = 10^5\left(1.63 - 1.044 \cdot 10^{-3}\right)\,\text{kJ/kg} \approx 163\,\text{kJ/kg}$ (from Fig. 11.16a) and thus only about 7.5%. Likewise, the change in the specific internal energy

($\Delta u = q - w$) is significantly greater than the work done. This has lasting consequences for thermodynamic cycle. For example, condensation processes, which are a reversal of the evaporation process, involve a large heat rejection out of the thermal cycle and thus an unrecoverable loss. This negatively influences the achievable thermal efficiency η_{th}.

11.2.3.1 Carnot Cycle as a Two-Phase Power Cycle

An idealized Carnot Cycle using a two-phase fluid as the working medium is depicted in Fig. 11.17 in the $p - v$ and $T - s$ diagram, which indicate the two-phase domain characterized by the vapor dome. The individual cycle steps of this ideal work cycle, which is close to the flash cycle used in geothermal energy, are:

- Isothermal heat supply at temperature T_2 in state 0 with saturated liquid (all mass in liquid state) until complete evaporation of the liquid at 1. The heat supply by the source q_{in} for to an isothermal expansion.
- Reversible adiabatic (i.e. isentropic) expansion from 1 to 2, which lowers the temperature to T_1. For conventional with a wet expansion (non-retrograde fluids), a mixture of liquid and vapour is present at 2.
- Isothermal compression from 2 at T_1 to state 3, where heat q_{out} is transferred to the heat sink.
- Reversible adiabatic (i.e. isentropic) compression from 3 to 0, where the vapour condenses to liquid and the initial state 0 is reached.

The thermal efficiency of the two-phase cycle is thus given by

$$\eta_{th} = \frac{q_{in} - q_{out}}{q_{in}} = \frac{T_2(s_2 - s_1) - T_1(s_2 - s_1)}{T_2(s_2 - s_1)} = 1 - \frac{T_1}{T_2}, \tag{11.46}$$

and thus, entirely corresponding to that of the single-phase cycle.

The question arises as to whether the two-phase Carnot cycle can also be used in real terms as a power engine process for generating electricity. Both the heat

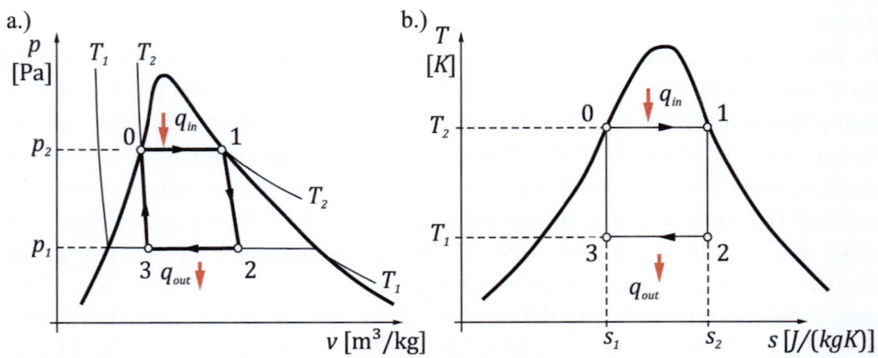

Fig. 11.17 Carnot cycle with two-phase medium in $p - v$ (**a**) and $T - s$ (**b**) representation

absorption and the heat dissipation take place at constant temperature and constant pressure in the two-phase range. These can be well approximated by a boiler for heat input and a condenser for heat removal. Furthermore, with an efficient turbine, a reasonable approximation to reversible adiabatic expansion can be achieved, since the steam is expanded with only small losses. The key technical challenge is the compression part of the cycle. If compression is performed slowly, there is equilibrium between the liquid and the vapour, but the pressure increase can be lower than desired and is associated with considerable heat loss to the environment. Conversely, rapid compression causes the two phases to assume very different temperatures (the temperature of the liquid increases only slightly during compression, while the temperature of the vapour phase increases greatly). The equilibrium between the two phases cannot be maintained and approximation of reversibility is no longer given.

Another disadvantage of this idealised process is that all the heat is supplied at the same temperature. However, in order to achieve high efficiency, the heat supply should be at a higher temperature than that of the critical point, so that the heat supply no longer takes place in the two-phase range. An isothermal heat supply is difficult to achieve under these circumstances. Furthermore, if the heat source and the circuit are considered together, the heat source can only be cooled to the highest temperature of the circuit. Thus, if the heat source has a varying temperature while the system requires a constant temperature heat supply, irreversible heat loss and thus entropy generation is inevitable.

In summary, the practical application of the two-phase Carnot cycle is limited due

- to the inefficient compression process,
- the low specific work per cycle,
- the upper temperature limit for operation in the two-phase flow regime and
- the irreversibility in heat transfer from the heat source,

and thus, of academic nature only. In the next section, we examine the Rankine cycle, which is much more compatible with the characteristics of two-phase media and available machinery for carrying out the processes.

11.2.3.2 Clausius-Rankine Process (Steam Turbine Process)

The Clausius-Rankine cycle is used in more than 95% of all solar thermal power plants built to date to convert thermal energy into electrical energy. It is a closed process (steam turbine process), which in its basic design consists of four components, the feedwater pump, the steam generator, the turbine and finally the condenser, as shown in Fig. 11.18a.

The working medium is water, which flows through the system in a closed circuit and periodically changes its state of aggregation. The feedwater pump delivers liquid water from the condenser to the steam generator ($0 \rightarrow 1$) and increases the pressure to the selected saturation pressure. The feedwater pump is driven either by an electric motor or by the turbine. The steam generator is usually divided into

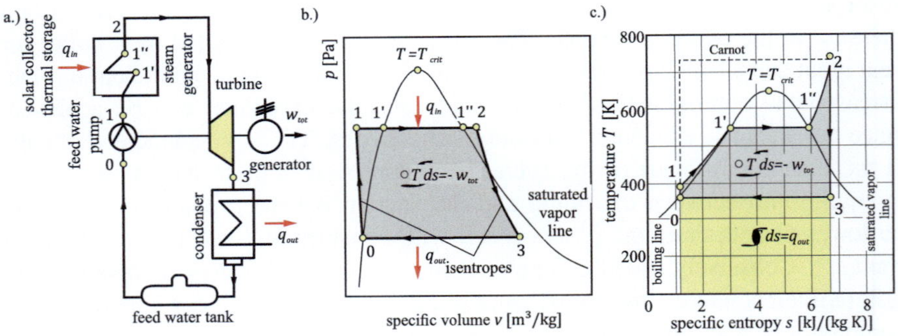

Fig. 11.18 Structure and components of a closed steam turbine circuit (**a**) and cycle mapping in the $p - v$ (**b**) and $T - s$ diagram (**c**)

three parts; in the first segment, the fluid is heated up to the boiling line $(1')$, followed by the evaporator by generating saturated steam by isobaric heat supply ($1' \rightarrow 1''$). Subsequently, the saturated steam is converted into superheated steam by adding further isobaric heat in the superheater as the temperature rises $(1'' \rightarrow 2)$. In an ideal case, the superheated steam expands isentropically in the turbine from a high pressure (p_2) to a low pressure with work release back to the wet steam region ($2 \rightarrow 3$). From the turbine, the wet steam flows to the condenser where it is liquefied again under isobaric heat release to a heat sink ($3 \rightarrow 0$). An idealised Clausius-Rankine cycle, as shown in Fig. 11.18b, c in the $p-v$ and $T-s$ diagrams, thus involves the following changes of state:

- $0 \rightarrow 1$ isentropic pressure increase,
- $1 \rightarrow 2$ isobaric heat input (preheating, evaporation, and steam superheating),
- $2 \rightarrow 3$ isentropic steam expansion,
- $3 \rightarrow 0$ isobaric heat removal.

The calculation of the technical work results from the heat balance as the difference between the heat supplied and the heat dissipated.

$$-w_{tot} = q_{in} - |q_{out}| = (q_{11'} + q_{1'1''} + q_{1''2'}) - |q_{30}|. \tag{11.47}$$

If we represent the specific heats (q_{ij}) in the form of specific enthalpies (h_i), we get:

$$-w_{tot} = (h_2 - h_1) - |(h_0 - h_3)| = h_0 - h_1 + h_2 - h_3. \tag{11.48}$$

Since the working fluid changes its state of aggregation during the cycle the water vapour tables must be used to calculate the enthalpy differences. As with the other cycles, the thermal efficiency η_{th} is calculated as follows

$$\eta_{th} = \frac{q_{in} - q_{out}}{q_{in}} = 1 - \frac{h_3 - h_0}{h_2 - h_1}. \tag{11.49}$$

In the Rankine cycle, the average temperature at which heat is supplied is less than the maximum temperature T_2, which can be easily recognized in Fig. 11.18c, so the thermal efficiency is less than that of a Carnot cycle, which operates between the same maximum and minimum temperatures. Heat absorption takes place at constant pressure via $1 \rightarrow 2$, but only fraction from $1 \rightarrow 2$ is isothermal; in contrast heat release takes place via $3 \rightarrow 0$ at constant temperature and pressure. To analyse in the efficiency of the Rankine cycle, we define a mean effective temperature T_{mi}, in terms of the heat exchanged and the entropy differences in the way:

$$q_{12} = q_{in} = T_{m2}(s_2 - s_1); q_{30} = q_{out} = T_{m1}(s_3 - s_0). \tag{11.50}$$

By this the thermal efficiency η_{th} of the cycle is

$$\eta_{th} = \frac{T_{m2}(s_2 - s_1) - T_{m1}(s_3 - s_0)}{T_{m2}(s_2 - s_1)} = 1 - \frac{h_3 - h_0}{h_2 - h_1}, \tag{11.51}$$

and because the compression and expansion processes are isentropic, so the entropy differences are related by

$$(s_2 - s_1) = (s_3 - s_0), \tag{11.52}$$

the thermal efficiency η_{th} can be written in terms of the mean effective temperatures as

$$\eta_{th} = 1 - \frac{T_{m1}}{T_{m2}}. \tag{11.53}$$

For the Rankine cycle, $T_{m1} \approx T_1$ but $T_{m2} < T_2$. From this equation we see not only the reason why the efficiency of the cycle is lower than that of a Carnot cycle, but also the direction in which we must move in the design of the cycle if we want to increase the thermal efficiency; we need an increase in T_{m2}. From the quantitative representation in the $T - s$ diagram (Fig. 11.18c), it can be seen that most of the heat input is used for evaporation. The advantage of the steam power process compared to the processes with inert gases is the large specific cycle work with a comparatively small work for compression by the feed water pump (small specific volume of the liquid) to achieve the pressure increase. The heat supplied to or removed from the process can be read off as an area in the $T - s$ diagram, since $ds = dq/T$. The difference between supplied q_{in} and removed heat q_{out} can be converted into useful work w_{tot}. From the $T - s$ diagram it can be seen that the thermal efficiency increases with increasing temperatures of the heat supply and decreasing temperatures of the heat removal. High process temperatures thus enable high efficiencies.

To further increase the efficiency of real processes, the average temperature of the heat supply is increased by regenerative feedwater pre-heating and reheating. The feedwater pre-heating in an economizer is realised by tapping steam from the turbine stages and is therefore called regenerative because the heat is served from the thermal cycle itself and is not supplied by external sources. During reheating, the steam from the high-pressure stage of the turbine is reheated. This also raises the average temperature of the heat supply T_{m2}, provided the extraction is chosen correctly. In addition, the intermediate superheating during the expansion of the steam in the turbine means that the dew point line (or saturated vapor line) is undershot at lower temperatures. The combination of a higher mean temperature (T_{m2}) and a lower final temperature (T_{m1}) ultimately provides a higher thermal efficiency. The schematic structure of the cycle including reheating and intermediate superheating is sketched in Fig. 11.19a. The intermediate superheating results in a changed sequence of states in the $T-s$ diagram, whose typical temperatures, steam contents and pressures are shown in Fig. 11.19b. Apart from efficiency considerations intermediate heating has practical advantages: When the steam in the turbine relaxes below the dew point line, the size of water droplets increase when the end point is deep in the two-phase region. To maintain the integrity of the turbine blades in this so-called end wetting, a maximum mass content of the droplets of about 13% should be guaranteed (steam quality $<$ 13%). This limitation is shifted towards lower temperatures by the intermediate superheating.

The design of the steam circuit for a thermal-electrical conversion has the highest temperature of the heat source as a central input variable. For a solar-only thermal power plants without hybridization, this corresponds to the highest collector outlet temperature, irrespective which collector technology is chosen (Morin et al., 2004). It should be noted that higher live steam temperatures lead to better thermal efficiencies of the cycle, but at the same time the heat losses of the collector increase, too. For linear concentration collectors (linear Fresnel, parabolic trough) the heat losses increase significantly with rising collector outlet temperature, however, the limiting factor in most commercial plants today is not the heat loss but the temperature stability limit of the synthetic oil used in the collector loop.

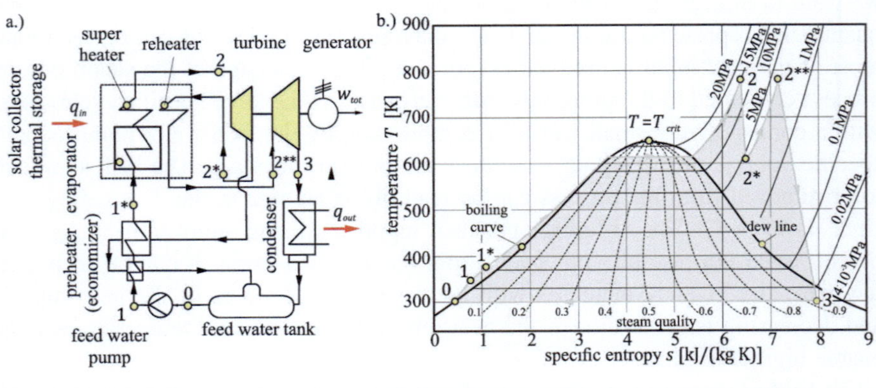

Fig. 11.19 Schematic sketch of a Clausius Rankine process with reheating and feedwater preheating in the plant flow diagram (**a**); $T-s$ diagram of a Clausius Rankine process with reheating and feedwater preheating with typical temperatures, steam contents and pressures (**b**)

11.2.3.3 Alternative Two-Phase Conversion Cycles

Alternative cycle processes primarily address the low-temperature conversion of thermal energy into electricity. Almost all of these low-temperature cycles are based on two-phase cycles, which benefit from the low compression work of a heat transfer medium that is liquid at low temperatures and thus largely incompressible. During heat absorption, a phase transition from liquid to gas takes place, which then allows the extraction of useful work with the aid of a turbine for expansion to the initial pressure. The heat transfer medium is then liquefied again in a condenser, with the heat of condensation leaving the cycle as heat loss.

Due to the two-phase nature of the cycles, the closer the heat input from the solar heat source follows the boiling line of the heat transfer fluid of the thermodynamic cycle, the more efficient the process, as the area within the cycle is maximized, considering the boiling properties of the fluid. Ideally, the process flow of isentropic compression and isobaric heat supply includes the dome of the two-phase region of the heat transfer fluid of the thermodynamic cycle. After the most isentropic expansion possible, condensation of the heat transfer medium occurs due to the external heat sink. The then supercritical process can then ideally be represented approximately as a triangular process as in the $T-s$ diagram of Fig. 11.20a. The triangular process, which is mainly used in refrigeration, is composed of the process steps:

- isentropic compression,
- the isobaric supply of heat from the solar source to the cycle,
- an isentropic expansion in a working machine,
- and an isothermal heat removal through condensation of the cycle fluid by an external heat sink.

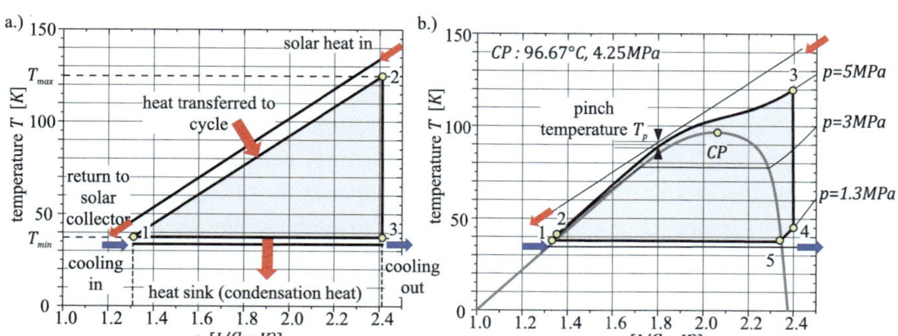

Fig. 11.20 **a** Matching to a heat transfer medium that cools down during heat release leads to a triangle-shaped circular process in the $T-s$ diagram. Ideal triangular matching process adapted from Förster and Schmidt (2006). **b** Approximation of a super-critical propane cycle ($R290$) by a triangular process

The thermal efficiency ($\eta_{th,\Delta}$) of this process can be calculated in a similar manner as the Carnot efficiency with the help of the highest ($T_{\max}$) and the lower cycle ($T_{\min}$) temperature:

$$\eta_{th,\Delta} = \frac{T_{\max} - T_{\min}}{T_{\max} + T_{\min}}. \tag{11.54}$$

Assuming an upper process temperature of 125.5 °C and a lower temperature of 7.5 °C, the efficiency of the triangular process yields 12.34%, which is lower than the Carnot efficiency of 21.976%. However, it provides a more realistic upper limit for estimating the efficiency of two-phase cycle processes at a comparatively low temperature level of the thermal heat source than a Carnot assessment. In a real heat transfer fluid, the idealized triangular process cannot be realized exactly due to the shape of the boiling line. Figure 11.20b shows, however, using the example of the refrigerant ($R290$) (propane), that when a supercritical process is assumed, the quality of a triangular process is approximately achieved, which represents the optimum. A triangular process can also be approximated with a super-critical Clausius-Rankine process at high pressures and temperatures (close to the critical state with 74 °C and 22.06 MPa. In the classical water/steam cycle, the following measures are used to increase the efficiency:

- raising the live steam temperature by increasing the evaporating pressure,
- increasing the live steam temperature by superheating,
- reheating,
- lowering of the condenser pressure and
- regenerative feedwater preheating.

These options elaborated for the Clausius-Rankine cycle to enhance efficiency apply identically also to triangle processes. All of the above measures are subject to certain limitations when used in solar thermal applications. For example, increasing the maximum steam temperature by increasing the evaporating pressure is linked to the maximum collector outlet temperature or the capacity limits of the cycle compressor. The same also applies to reheating, because for this it must be ensured that the thermally available collector capacity is sufficient. Regenerative fluid heating (recuperation) using unused heat to preheat the feed fluid is similarly feasible, if the temperature difference provided is sufficiently far away from the pinch point temperature. The point at which the curves of heat input (through the collector system) and that of the thermodynamic cycle converge most closely is the so-called pinch temperature (pinch) indicated in Fig. 11.20b. It is at this point that most design constraints exist. By identifying this point and starting to design the heat exchanger network around it, the energetically optimal heat transfer between hot and cold flows, can be achieved (more is explained later in detail in Sect. 11.2.6). Condenser pressure is usually dictated by the heat released to the environment by the conditions prevailing at a site, especially if the system is operated at a sunny site with high temperatures and evaporative cooling. When optimization of solar and process efficiencies results in lower collector or heat transfer temperatures, other cycle processes such as the Organic Rankine Cycle (ORC) or Kalina process are advantageous. They are used in geothermal plants and the first one also in solar ponds.

11.2.3.4 Organic Rankine Cycle (ORC)

Thermodynamic organic Rankine cycles (ORC) are particularly suitable for power plant development of low temperature heat sources using low boiling point organic working fluids and/or for plants that require miniaturization due to their simple configuration. Geothermal heat and solar heat are usually converted by ORC systems through direct heat transfer. Although low-temperature heat can have different sources, there are common parameters that allow analysis and categorization. The optimization objective of an ORC cycle for a given closed heat source is to maximize the power output and efficiency,

$$\dot{W}_{net} = \eta_{th} \cdot \dot{Q}, \tag{11.55}$$

where $\dot{W}_{net}$ represents the extractable power and $\dot{Q}$ represents the heat supplied by the heat source. For closed heat source cycles such as solar collectors, the heat output $\dot{Q}$ supplied to the ORC system is fixed. Therefore, optimizing the cycle efficiency (η_{th}) is equivalent to optimizing the total power output $\dot{W}_{net}$. The temperatures provided by the collectors range from approx. 80–350 °C. Based on the temperature of the heat source and the critical temperature of the working fluid, the cycles can be divided into transcritical and subcritical cycles. This distinction has a significant impact on the efficiency of energy recovery and the design of the heat exchanger.

The capacity of the heat source ($\dot{Q}_{hs}$) has a direct influence on the system capacity and results for solar systems from

$$\dot{Q}_{hs} = A \cdot I, \tag{11.56}$$

where A is the cross-sectional area in (m^2) of the collector and I is the solar irradiation in (W/m^2).

The design of an ORC circuit based on the heat source requires the analysis of thermodynamics with regard to the choice of the circuit (transcritical or subcritical) and the working medium. Only the temperature of the heat source and the temperature difference of the pinch in a $T-s$ chart reflect the choice of the cycle and the working agent. The heat source temperature is the parameter that most influences the efficiency of the cycle and the choice of the working fluid. The inlet and outlet temperatures determine the optimal cycle design, e.g. whether the circuit is operated with a recuperator or not. The heat source determines the smallest pinch temperature difference in the evaporator. The heat sources can be divided into two types according to their inlet and outlet temperatures. Figure 11.21a show the $T-s$ diagram of a heat source with limited capacity that does not have a limit on the outlet temperature, while Fig. 11.21b characterises a heat source that has

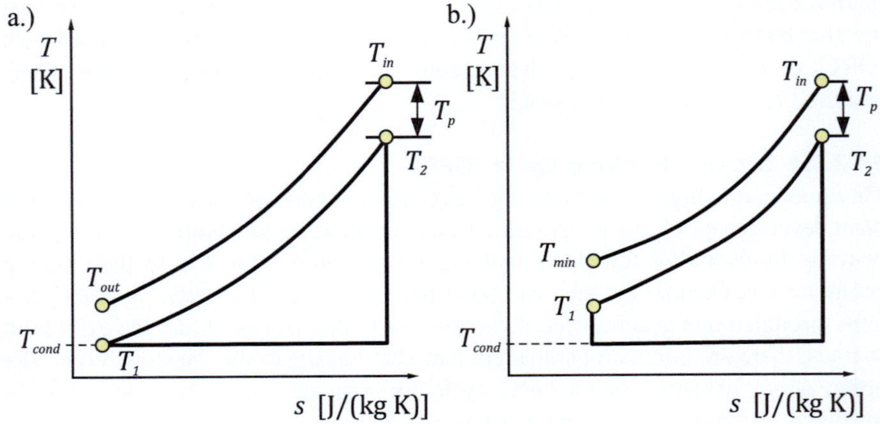

Fig. 11.21 **a** Thermodynamic categorization of heat sources; **a** finite capacity heat source without an outlet temperature limitation; **b** finite capacity heat source with outlet temperature limitation ($T_{\min}$)

a minimum exit temperature $T_{\min}$. In the case of closed heat sources, $T_{\min}$ is the outlet temperature. With a closed circuit, it is useful to select the minimum outlet temperature of the heat source ($T_{\min}$) as large as possible in order to achieve a large area in thee cycle and thus a high efficiency.

For heat sources without temperature limitation, we get the triangular efficiency from Eq. (11.54). But if the heat source changes isobaric state, the temperature from T_1 to T_2 does not follow a straight line, because the differential enthalpy change (dh) is given by the following relation:

$$dh = Tds + vdp \rightarrow \left(\frac{\partial h}{\partial T}\right)_p = T\left(\frac{\partial s}{\partial T}\right)_p \text{ with } \left(\frac{\partial h}{\partial T}\right)_p = c_{p,ORC}. \tag{11.57}$$

If we consider a real fluid and apply a fixed pinch point, then the real evaporation temperature is lower than the one at a constant T/c_p and thus the real efficiency is also lower. For the work output of the cycle per unit mass flow we obtain

$$w = \int_{T_1}^{T_1} \left(1 - \frac{T_{cond}}{T}\right) Tds \text{ with } Tds = c_{p,ORC} dT. \tag{11.58}$$

The heat absorbed per unit mass by the ORC cycle is given by $q = c_{p,ORC}(T_2 - T_1)$ and we get then for the efficiency.

$$\eta_{th} = \frac{w}{q} = 1 - \frac{T_{cond}}{(T_2 - T_1)} \ln \frac{T_2}{T_1}. \tag{11.59}$$

Assuming an ideal heat exchanger (i.e. pinch temperature difference $T_p = 0$), the net power for heat source without an outlet temperature limitation results in:

$$\dot{W}_{net,ideal} = \eta_{th} \cdot \dot{Q} = \dot{m}_{hs} \cdot c_{p,hs} \cdot \left(T_{in} - T_{cond} \left(1 - \ln \frac{T_{in}}{T_{cond}} \right) \right). \quad (11.60)$$

For finite capacity heat sources such as solar collectors, the calculation procedure is similar, but the outlet temperature limit ($T_{\min}$) is now the lowest temperature instead of the condenser temperature, so that for an ideal cycle for a real fluid, the following relationship is obtained for the net power:

$$\dot{W}_{net,ideal} = \eta_{th} \cdot \dot{Q} = \dot{m}_{hs} \cdot c_{p,hs} \cdot \left(T_{in} - T_{\min} - T_{cond} \ln \frac{T_2}{T_1} \right). \quad (11.61)$$

Cycle optimization according to the first law of thermodynamics and selecting the working fluid based on it has some weaknesses and essentially two arguments against it reflecting the optimal system performance.

- For example, high efficiency can be achieved by increasing the evaporating temperature, but this may result in lower utilization of the heat source and thus lower net output power.
- Optimizing efficiency alone can lead to unrealistic operating conditions, for example, if a high critical temperature ORC working fluid is selected for a low temperature heat source. This then leads to small vapor densities and requires large heat exchanger areas and thus large components in the cycle.

An alternative and most widely used performance indicator is exergy efficiency (η_{ex}), which is defined as

$$\eta_{ex} = \frac{\dot{W}_{net}}{\dot{W}_{net,\,ideal}} = \frac{\dot{W}_{net}}{\dot{Q}_{hs}\left(1 - T_{ref}/T_{hs}\right)}, \quad (11.62)$$

and thereby describes the ratio of the achieved power output to the power output of the ideal cycle or in engineering terms how good the heat source is used. Here, $\dot{W}_{net}$ denotes the actual cycle output, and $\dot{W}_{net,ideal}$ is the ideal one for a given reference temperature T_{ref}.

The Rankine cycle with organic working fluid (ORC = Organic Rankine Cycle) is similar to the classic Rankine cycle in terms of components. The main difference lies in thermodynamic properties of the organic working fluid, which impact the process parameters of pressure and temperature, being considerably lower than those of a water-steam power plant. The working fluid of an ORC is selected in such a way that its thermodynamic properties optimally match the available heat source; in particular, the temperature level of the available heat source determines the fluid under consideration. The fluid should evaporate at low temperatures, be non-toxic and have no harmful effect on the climate. Among others, fluids from the groups of kerosene, aromatics, heterocyclic compounds and alcohols are used. Zeotropic mixtures such as ammonia-water or propane-isobutane mixtures also promise better adaptability to the heat source due to non-isothermal phase transitions. The use of organic working fluids requires separate engineering solutions of the system technology, such as

- the turbines used. They are usually special turbines since the working fluid differs significantly from water in terms of molecular weight and lower specific heat capacity.
- the chemistry of the working fluid. Organic working fluids are often corrosive and require surface coating of the components.
- the sealing of the piping system, which is more complex than for water.

Some typical working fluids of ORC processes are summarized in Table 11.2. Many of them are also used in compression chillers. Therefore, the discussion of climate friendly selection and avoidance of hydrogen-containing hydrofluorocarbons (HFCs), perfluorocarbons (PFCs) or other greenhouse gases with a high GHG potential applies also to ORC systems.

Table 11.2 Critical pressure and critical temperature of some fluids considered for ORC processes from modified from (NIST, 2018)

Fluid (ASHRAE 15 & 34 designation)	Chemical formula	Critical pressure p_{crit} [MPa]	Critical temperature T_{crit} [°C]
Carbon dioxide (*R*744)	CO_2	7.39	30.98
Propane (*R*290)	C_3H_8	4.25	96.74
Fluoromethane (*R*41)	CH_3F	5.90	44.13
Trifluoroethane (*R*143*a*)	$C_2H_3F_3$	3.76	72.71
Tetrafluoroethane (*R*134*a*)	$C_2H_2F_4$	4.06	101.03
Octafluoroethane (*RC*318)	C_4F_8	2.78	115.23
n-Butane (*R*600)	C_4H_{10}	3.80	151.98
Isobutane (*R*600*a*)	C_4H_{10}	3.64	134.70
n-Pentane (*R*601)	C_5H_{12}	3.37	196.60
Isopentane (*R*601*a*)	C_5H_{12}	3.38	187.20
Ammonia (*R*717)	NH_3	11.33	132.25
Water (*R*718)	H_2O	22.06	373.95

As shown in Table 11.2, the critical points of organic fluids are at lower temperatures and pressures than those of water. In addition, these fluids also have a lower heat of vaporization allowing for faster evaporation and thus more efficient use of limited capacity heat sources. Due to the shape of the two-phase diagram, ORC fluid-based cycles allow better adaptation to a triangular cycle scheme. The optimum represents a large dimensionless evaporation temperature P, which is formed in the following way:

$$P = \frac{c_{p,ORC} \cdot T_S}{\Delta h_{V,ORC}}, \tag{11.63}$$

where $c_{p,ORC}$ is the specific heat capacity of the ORC fluid, T_s its vaporization temperature and $\Delta h_{V,ORC}$ the heat of vaporization. Figure 11.22a show the $T - s$ diagrams of some of the ORC fluids mentioned in Table 11.2 compared with water (Fig. 11.22b). Some of the organic working fluids exhibit retrograde behaviour, such as isobutane ($R600a$). For these working fluids, the expansion does not proceed along the condensation curve, but it always remains outside the wet vapor region. Superheating of the fluid increases as a result of expansion in the turbine. Consequently, the exhaust steam at the turbine outlet is always superheated and this heat must first be dissipated along the isobars before the working fluid condenses or depending on the heat content can be used for recuperation.

Figure 11.23 shows a schematic diagram of a simple ORC cycle without recuperation. Here, the heat q_{in} absorbed in the receiver is conveyed to the evaporator by means of a primary fluid. In order to use the heat optimally, the ORC fluid is preheated in a recuperator (economizer). In the secondary circuit—the actual ORC cycle—the working fluid passes through the economizer ($2 \rightarrow 3$) after the pressure increase by the feed pump ($1 \rightarrow 2$) and then through the evaporator ($3 \rightarrow 4$). The steam is expanded in the turbine ($4 \rightarrow 5$). At the outlet of the turbine, superheated steam (5) is usually present. Consequently, part of the waste heat must be

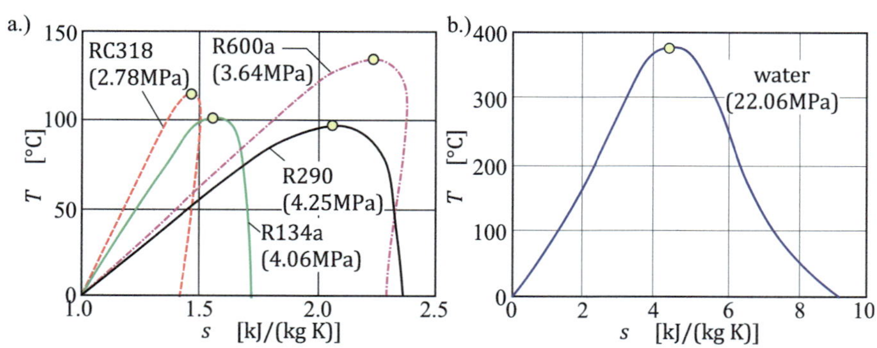

Fig. 11.22 Comparison of the $T - s$ diagrams with wet steam regions of four organic ORC fluids (**a**) from Table 10.1 with the data of water (**b**)

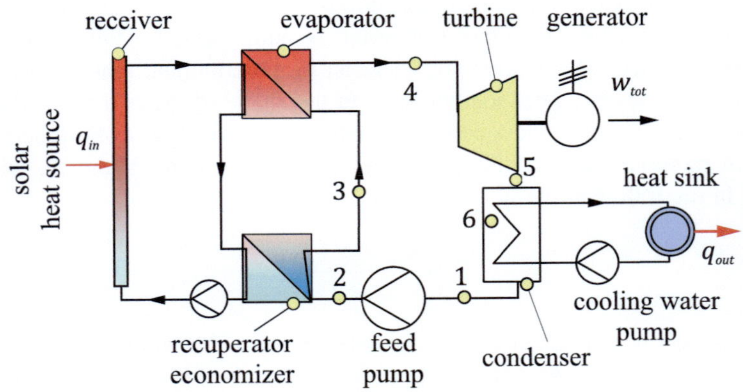

Fig. 11.23 Block diagram of the component linkage of a simple Clausius-Rankine process with organic working fluid (ORC)

removed at a temperature above the condensing temperature. The steam is cooled down in the condenser along an isobar ($5 \rightarrow 6$) and subsequently liquefied again ($6 \rightarrow 1$).

The corresponding $T-s$ diagram of the simple ORC process using a retrograde fluid is illustrated in Fig. 11.24. The thermal efficiency (η_{th}) of the ORC cycle is thus given by:

$$\eta_{th} = \frac{q_{in} - q_{out}}{q_{in}} = \frac{(h_4 - h_1) - (h_5 - h_1)}{(h_4 - h_1)}, \tag{11.64}$$

if the pumping power is added to the heat input. q_{in} and q_{out} denote the heat input and heat output, and h_i the specific enthalpy of the respective state "i". If we express the enthalpies by:

$$(h_4 - h_1) = T_{m,\,in}(s_4 - s_1) \text{ and } (h_5 - h_1) = T_{m,\,out}(s_5 - s_1), \tag{11.65}$$

where $T_{m,in}$. represents the mean temperature of the energy input and $T_{m,out}$ is that of the heat removal, and also assume that the relaxation in the turbine is isentropic ($\Delta s_{4-5} = 0$), we obtain the following relationship for the thermal efficiency η_{th}:

$$\eta = \frac{T_{m,\,in} - T_{m,\,out}}{T_{m,\,in}} = 1 - \frac{T_{m,\,out}}{T_{m,\,in}}. \tag{11.66}$$

This corresponds to the formulation of the Carnot efficiency as for the water/steam cycle, with the difference that a mean temperature must also be used for the waste heat. This can be determined in the following way:

$$T_{m,\,out} = \frac{(T_5 s_5 - T_6 s_6) + (T_6 s_6 - T_1 s_1)}{s_5 - s_1} = \frac{T_5 s_5 - T_1 s_1}{s_5 - s_1}. \tag{11.67}$$

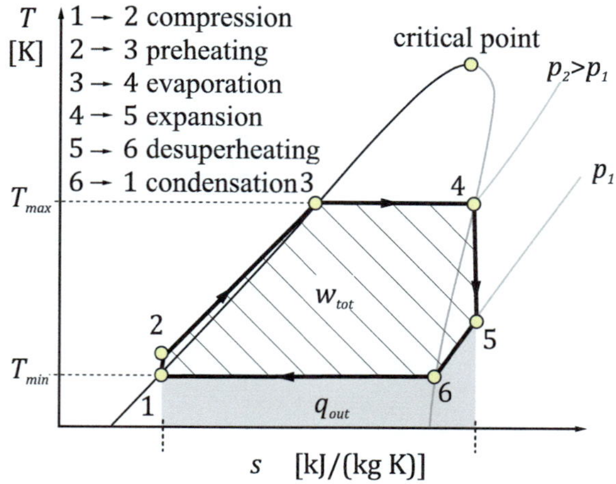

Fig. 11.24 $T - s$ diagram of a simple ORC process with retrograde fluid

Thus also $T_{m,in}$ can be found with $q_{in} = \left(T_{m,in}\Delta s_{4-1}\right)$ and $q_{\text{out}} = \left(T_{m,\text{ out}}\,\Delta s_{5-1}\right)$. As a result, one obtains:

$$T_{m,\,in} = \frac{q_{in}}{q_{out}} T_{m,\,out}. \tag{11.68}$$

For the working fluids used in ORC processes, a distinction is made between dry (retrograde) and moist fluids based on the slope of the condensation curve (dew curve) in the $T - s$ diagram. For dry working fluids, superheat increases during relaxation, as shown in the $T - s$ diagram in Fig. 11.25a. The thermal efficiency of a reversible cycle can be read from the $T - s$ diagram as the area ratio of the area $(w_{\text{tot}} + q_{\text{out}})$ to the area (q_{out}). The ratio of the area $4 - 4^* - 5^* - 5$ added by superheating to the area $B - 4 - 4^* - B^*$ is worse than that for the superheating of the water/steam cycle, which aimed at reducing the final wetness at the end of the expansion in the turbine. Thus, the heat added during superheating is paid for with a lower thermal efficiency. As a consequence, the superheated steam ORC process with retrograde fluids does more work but delivers lower thermal efficiencies than that of the saturated steam process. Consequently, for retrograde fluids, superheating of the live steam does not increase efficiency or effectiveness.

The deterioration in thermal efficiency due to the desuperheating from $5 \rightarrow 6$ required for retrograde fluids makes regenerative recuperation of the working fluid particularly attractive. This is because some of the heat generated during desuperheating of the vaporous heat transfer fluid can be used to preheat the liquid working fluid. Figure 11.25b illustrates the regenerative preheating of the working fluid in a recuperator in the $T - s$ diagram and Fig. 11.26 illustrates the corresponding ORC cycle block diagram.

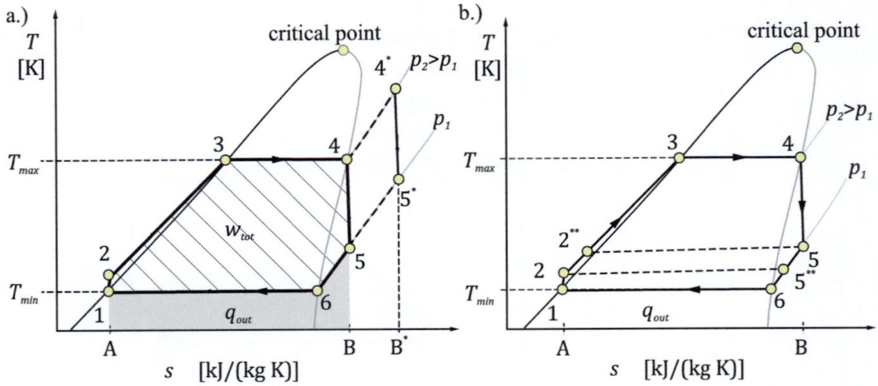

Fig. 11.25 $T-s$ diagram of an ORC process with superheating with a retrograde fluid (**a**) and a configuration with regenerative preheating (recuperation) (**b**)

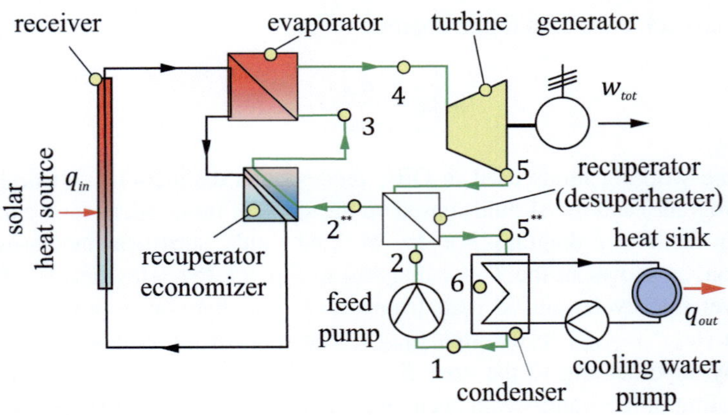

Fig. 11.26 Block diagram of the component linkage of a ORC cycle with recuperation (desuperheater)

Which fluid proves to be optimal?

It has already been mentioned that the ratio $c_{p,ORC}$ and enthalpy of vaporization $\Delta h_{V,ORC}$ should be as large as possible. As the upper cycle pressure p_2 approaches the critical pressure p_{crit}, the enthalpy of evaporation decreases and the specific heat capacity of the liquid working fluid increases. Simultaneously, the cycle process approaches a triangular shape in the $T-s$ diagram. Thus, for an ORC process, it is favorable if the critical temperature of the working fluid to be selected is slightly above the upper process temperature. The requirement for the highest possible enthalpy of evaporation alone is not useful, since the rectangular Carnot process is not very suitable for retrograde fluids with the optimal triangular characteristics of heat source and heat sink. The working fluid should allow to realize

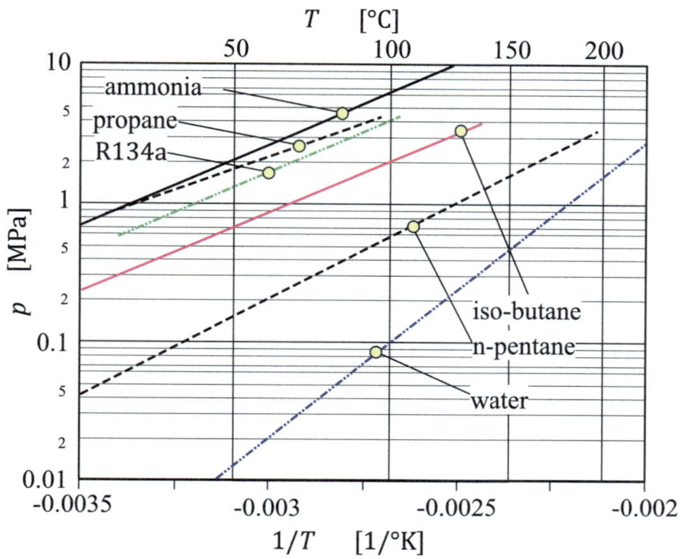

Fig. 11.27 Vapor pressure curves as a function of negative reciprocal absolute temperature (Raoult diagram) for some fluids from Köhler (2005)

the largest possible pressure ratio $\pi = p_2/p_1$ at the given temperature boundary conditions. According to the Clausius–Clapeyron equation, the slope of the vapor pressure curve of the Raoult diagram shown in Fig. 11.27 is proportional to the enthalpy of vaporization $\Delta h_{V,ORC}$ (Traupel, 1971). With the aid of the diagram, the temperature can be directly assigned to the pressure position of the respective working fluid. Thus, working fluids with large evaporation enthalpy promise advantages, the cause of which is a larger expected pressure ratio of the turbine and not a possible Carnotization of the cyclic process. Table 11.3 provides an approximate overview of the relationship between solar collector types, characteristic collector outlet temperature levels, and potential conversion technologies.

11.2.3.5 Kalina Cycle

In the Kalina process (USA Patentnr. U.S. Patent No. 4,346,561, 1982), as in the ORC cycles, there is no direct use of the warm fluid, but a heat transfer by means of a recuperator to a secondary working medium. The Kalina process uses an ammonia/water mixture for this purpose. The zeotropic mixture does not evaporate at a constant temperature like a pure fluid, but over a wide temperature range, thus meeting the requirement for a triangular cycle. The desorber with the release of a gas rich in ammonia takes the place of an evaporator, and the absorption of this gas, expanded in the turbine, in a solution poor in ammonia corresponds to the condenser. Depending on the available heat source/sink, it is thus possible to build up a tailor-made working machine cycle by selecting the operating pressure and the concentration difference. Different cycle designs exist; known are the KSC5

Table 11.3 Solar collector types with typical outlet temperatures and corresponding conversion technologies

Collector	Collector design	Typical outlet temperature *T* [°C]	Conversion technologies
Solar pond	Volumetric receiver	75–90	ORC
Static collectors	Flat plate receiver	<150	ORC (or heat)
	Vacuum tube collector	<220	ORC (or heat)
Line-concentrating	Linear Fresnel collector	150–450	ORC, steam cycle
	Parabolic trough	200–500	ORC, steam cycle
Point-concentrating	Solar tower	500 (−750)	Steam cycle
	Parabolic dish	750–1000	Stirling process

suitable for separate firing or waste heat from industrial plants as heat source, the KSC6 as combination process with an upstream water/steam process or the KSC11 for operation with geothermal heat input. Wagar et al. (2010) describe processes with six flashes for combination with solar plants. The schematic diagram of a Kalina plant is shown in Fig. 11.28.

Figure 11.29a shows the boiling and dew lines of the ammonia-water mixture for a high and a low pressure (*p*1 and *p*2), which is essential for understanding the changes of state, especially those with concentration and associated temperature changes. At the absorber outlet (1), the base solution of the zeotropic mixture is directly on the boiling line of the lower process pressure. The feed pump compresses the liquid base solution from p_1 to p_2 (1 → 2), where it is subcooled. With the absorption heat released in the low-temperature recuperator (10 → 11),

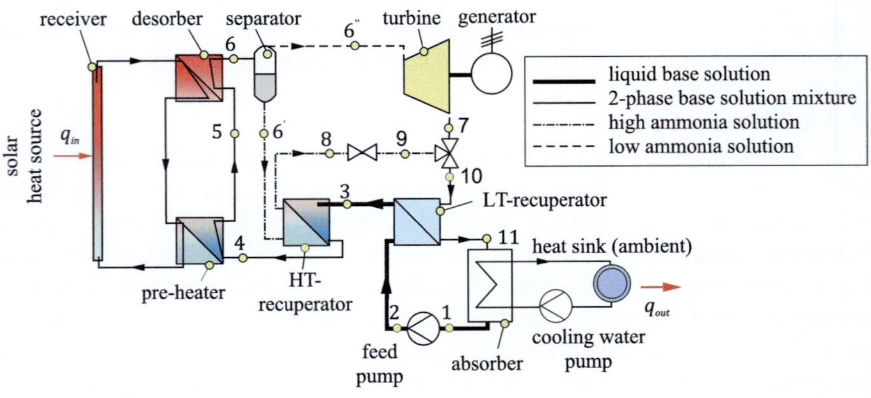

Fig. 11.28 Components of a basic KALINA cycle. HT = High temperature, LT = Low temperature

the solution is preheated ($2 \rightarrow 3$) and further heated in a high-temperature recuperator (HT) ($2 \rightarrow 3$). The base solution now has vapor components. The HT recuperator extracts heat from the warm, low-ammonia solution moving from $6'$ to 8. Only in the pre-heater ($4 \rightarrow 5$) and the desorber ($5 \rightarrow 6$) is external heat supplied by a primary fluid, such as a solar collector. In the preheater, the base solution is heated to the boiling point (5) and then partially evaporated in the desorber (6). Two phases are present at the desorber outlet (6), which are separated in the separator into a low-ammonia $(6')$ and an ammonia-rich vapor $(6'')$. The ammonia-rich vapor is expanded in the turbine to p_1 down to the maximum permissible final wetness at the turbine outlet (dashed line below the dew line) $(6'' \rightarrow 7)$. The separated low-ammonia liquid preheats the base solution in the HT recuperator and is expanded in a throttle $(6'' \rightarrow 7'' \rightarrow 8)$. The throttling leads across the dew line into the two-phase region (equilibrium points $9'$ and $9''$). Both the ammonia-poor and ammonia-rich fluid streams are now at the same pressure. Both are mixed (10, equilibrium points $10'$ and $10''$). The absorption heat is partly used within the process for regenerative pre-heating of the basic solution ($10 \rightarrow 11$) and the non-usable part is dissipated to the environment in the absorber ($11 \rightarrow 1$).

Figure 11.29b shows a comparison of a classical Rankine cycle with a Kalina cycle in the $T-s$ diagram. The advantage of the Kalina process lies in the more favourable heat transfer ratios during evaporation and condensation, since the changes of state do not take place isothermally, in contrast to the Rankine process. Instead, the properties of mixtures are specifically used to bring about

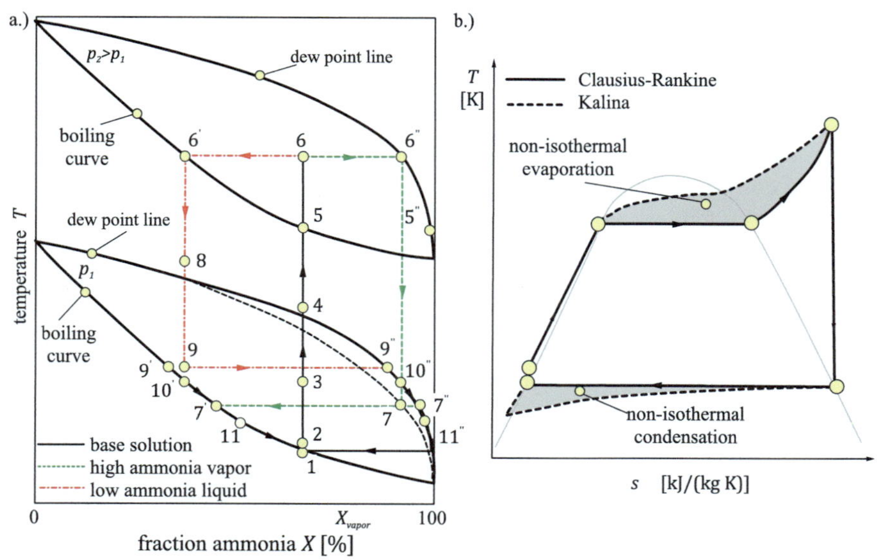

Fig. 11.29 **a** Representation of the Kalina cycle in the boiling diagram of the ammonia-water mixture according to Köhler (2005); **b** Comparison of Rankine and Kalina cycle in the $T-s$ diagram

temperature changes through concentration changes. This is done by changing the concentration of the individual phases of water and ammonia at constant total concentration and constant pressure. The mixture evaporates under continuously rising temperatures or condenses under continuously falling temperatures. Essentially, two effects lead to an improved thermal efficiency compared to the Rankine process:

- Due to the non-isothermal evaporation or condensation, the temperature differences in the heat transfers and thus the heat transfer losses are lower.
- The mean temperature of the heat transfer (upper process temperature) is raised, and the mean temperature of the condensation (lower process temperature) is lowered.

But the advantage of small temperature differences in heat transfer (high Nusselt numbers) means inherently large heat transfer areas (large heat exchangers). In addition, heat transfer is small in diffusion and absorption processes. Technical challenges are also posed by the chemical stability of the ammonia or other solvents, which means considerable additional technical and monetary expense in the design and construction of the components.

11.2.3.6 Non-nominal Operation Correction Factors in Two-Phase Cyclic Processes

The basis for the design of a thermal power plant are nominal operating conditions, i.e. the achievement of the nominal thermal load and also the assumption of a given temperature of the heat sink, i.e. the ambient temperature. However, at many times during the course of the year these are not achieved, and non-nominal operating conditions exist; for example,

- non-nominal ambient air temperature,
- production of an insufficient amount of steam, or
- pressures or temperatures that deviate from nominal values,

all of which affect component efficiency and ultimately the achievable electrical yield. In single-phase heat transfer fluids of power conversion systems, these deviations from the nominal point are rather small due to the low heat capacity of the fluid, but in two-phase conversion cycles they have to be taken into account in an appropriate way.

The simplest and usually practiced method to consider non-nominal operating conditions is the introduction of so-called efficiency correction factors. With the help of these, characteristic curves can be created which show the dependence of the power generation (gross and net) on the thermal power, the air temperature, the steam pressure, and the steam temperature.

In the first step, the changes in steam pressures and temperatures over time are calculated using the thermal power provided by the solar field or thermal storage tank as input parameters, assuming that the specifications of the individual components (e.g. isentropic efficiency of the turbine, pump efficiencies, etc.) remain unchanged. After determining time-dependent variables (thermal power, ambient air temperature), an additional correction factor is calculated for each value duo. A typical power plant output (50 MW or 100 MW) at nominal operating conditions (usually 25 °C ambient air with steam parameters at the turbine inlet of, for example, 40 °C, 10 MPa) is taken as the reference value, which represents the 100% nominal value. With the help of commercial but also freely available computational programs for steam cycles, the correction factors for the thermal power can then be normalized in relation to this reference value, resulting in a dimensionless scale. This also allows the consideration of other power plant configurations or an upscaling or downscaling of the power of a certain power plant type. Figure 11.30a shows the calculated gross efficiency (f_{gross}) as a function of thermal load (ratio of thermal power $\dot{Q}$ to nominal thermal power $\dot{Q}_0$) for different ambient temperatures using a linear Fresnel power plant as an example. For example, if the current thermal load of the cycle is 75% of its rated value at an ambient temperature of 45 °C, the electrical efficiency of the cycle is 90% of its rated value. Figure 11.30b illustrates, for the same power plant, the gross-net efficiency (net power is gross power minus internal power input, $f_{gross-net}$) as a function of thermal load for various ambient temperatures.

Equations (11.69)–(11.71) describe the consideration of the correction factors with the respective efficiencies for all combinations of thermal load and ambient temperature as determined by the steam cycle calculation program.

$$\eta_{gross} = \eta_{gross,0} \cdot f_{gross}\left(\dot{Q}/\dot{Q}_0, T_a\right), \tag{11.69}$$

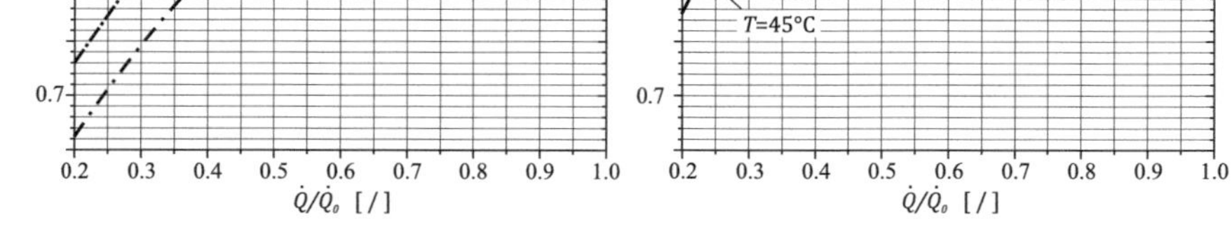

Fig. 11.30 **a** Gross thermal-electric efficiency factor $\left(f_{gross}\right)$ as a function of the normalized thermal load $\left(\dot{Q}/\dot{Q}_0\right)$ and **b** gross thermal to net efficiency factor $\left(f_{net}\right)$ as a function of the normalized thermal load $\left(\dot{Q}/\dot{Q}_0\right)$ assuming a reference steam cycle with an electric power of 100 MW and reference conditions of 10 MPA, 540 °C steam temperature at 25 °C ambient air temperature T_a (modified from Bachelier, 2021)

$$\eta_{gross-net} = \eta_{gross-net,0} \cdot f_{net}\left(\dot{Q}/\dot{Q}_0, T_a\right), \tag{11.70}$$

$$\eta_{net} = \eta_{gross} \cdot \eta_{gross-net}, \tag{11.71}$$

where η_{gross} is the gross efficiency at non-rated conditions and $\eta_{gross,0}$ is the same efficiency at nominal conditions, f_{gross} is the efficiency correction factor at the considered thermal power $\dot{Q}$ in $[w]$ and ambient air temperature T_a in [°C]. $\eta_{gross-net}$ describes the gross-net efficiency at non-rated conditions, while $\eta_{gross-net,0}$ is the same efficiency under nominal conditions, f_{net} denotes the efficiency correction factor, and η_{net} is the net efficiency of the power plant cycle.

11.2.4 Combined Cycle Gas Turbine Processes (CCGT)

The gas turbine cycle can be coupled with a steam process if the amount of energy available is sufficiently high, so that the Joule-Brayton process is used in the high-temperature section, while the residual heat drives a steam process (Clausius-Rankine). This coupled process is called a gas and steam process or combined cycle (CC). In principle, efficiencies of up to 58% can be achieved with the CCGT process. Figure 11.31 shows a $T - s$ diagram of a CCGT process with different recuperative process steps. Even if the waste heat (q_{R1}) from the gas turbine process is used to preheat the compressed gas, the exhaust gas (state 3*) still contains a usable amount of energy. With the help of a heat exchanger, the amount of energy q_{R2} can be transferred to a steam process in which it raises the compressed feedwater from state 1 to 1*. However, the enthalpy increases from 1* to 2, i.e. the remaining preheating, evaporation and superheating of the water, usually requires an additional heat source. After expansion of the steam in the turbine, part of the residual steam heat (q_{R3}) can be used again for feedwater preheating. The losses of such a CCGT process are thus minimised and limited to the hatched areas illustrated in Fig. 11.31a.

CCGT power plants are a more efficient and thus more environmentally friendly way of generating electricity than simple pure fossil-fuelled thermal power plants. By the utilization of fossil sources, unless being CO_2 free produced, however, they are not CO_2 free. Furthermore, the performance of CCGT power plants is closely related to the atmospheric conditions at an individual location and to the time. As a result of the volumetric machine behaviour of gas turbines, a significant yield loss occurs at high air temperatures (first-order factor) and in conditions with low relative humidity and low pressure (second-order factors), which ultimately lead to yield reductions even in a conventional "solar-free" CCGT power plant. But the times with high air temperatures correspond to the times with more direct normal solar irradiation; thus, it is possible to integrate the heat generated by solar thermal collectors into the CCGT cycle in order to mitigate the reduced output of the CCGT power plants. This integration of solar heat into a CCGT cycle is called an integrated solar combined cycle (ISCC). The objective of ISCC

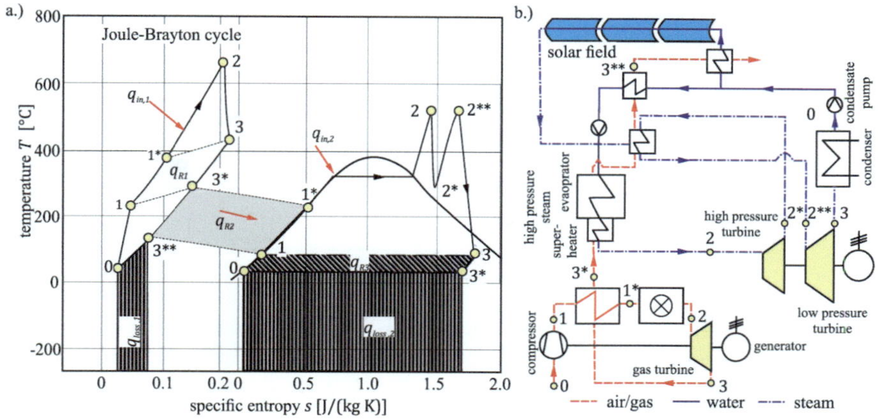

Fig. 11.31 **a** Representation of the principle of a CCGT cycle and the heat flows in a $T - s$ diagram. **b** Example of a ISCC plant concept with low pressure steam line solar feed

is mainly to improve cycle performance, even if environmental conditions vary. Since these ISCC's open up the option of adding heat at various points in the topping cycle as well as in the bottom cycle and, if necessary, also to support recuperation processes, it makes sense first to identify potential figures of merit/evaluation criteria to rate the quality of solar integration in a CCGT.

The simplest way to calculate the annual yield is to calculate it as the sum of the hours $(n \cdot \Delta t)$ of an operating state multiplied by the power in this state produced by gas turbine $(\dot{P}_{gt})$ and steam turbine $(\dot{P}_{st})$, which yields the thermal efficiency (η_{th}) as the ratio of the total power production in relation to the total thermal energy supplied, which is given by the mass flux of fossil fuel $(\dot{m}_f)$ times the specific calorific value of the fuel (h_f) and the total solar thermal power provided by the solar field $(\dot{Q}_{sol})$:

$$\eta_{th} = \frac{\sum n \cdot \Delta t \left(\dot{P}_{gt} + \dot{P}_{st}\right)}{\sum n \cdot \Delta t \left(\dot{m}_f h_f + \dot{Q}_{sol}\right)}. \tag{11.72}$$

The thermal efficiency is useful for evaluating the thermodynamic efficiency of the entire thermal process, but it is not helpful as an evaluation criterion when using the integral formulation as in the preceding. Especially in the case of ISCC, the integral formulation of the thermal efficiency shows two central weaknesses. It does not consider the contribution of the individual heat source and overestimates the contribution of the solar source when it is added to the topping cycle or underestimates the solar contribution when it is introduced to the bottom cycle, as this has a lower thermal efficiency.

An alternative approach would be to use the heat rate (HR), which is the reciprocal of the thermal efficiency, i.e. $HR = 1/\eta_{th}$, but only considers the consumption of fossil fuels. The corresponding evaluation criterion is then the fuel saving. However, this does not consider the fact that the solar contribution also changes the capacity of an ISCC power plant.

Another option of evaluation would be to use the incremental solar-to-electricity efficiency, η_{inc}, (Dersch et al., 2004), which quantifies the ratio of incremental production by solar contribution to total solar-generated power and results as follows:

$$\eta_{inc} = \frac{\sum n \cdot \Delta t\left(\left(\dot{P}_{gt} + \dot{P}_{st}\right)_{ISCC} - \left(\dot{P}_{gt} + \dot{P}_{st}\right)_{CCGT}\right)}{\sum n \cdot \Delta t \cdot \dot{Q}_{sol}}. \tag{11.73}$$

However, this type of efficiency approach is only suitable for comparing ISCC with a pure CSP power plant. Furthermore, the incremental efficiency only makes sense in cases where the plants are operated with a boosting strategy but are not operated in fuel saving mode, because in such a case the numerator of Eq. (11.73) becomes zero.

The methodology of attributing the contribution of each heat source to the overall process leads to the expression of the solar-electric efficiency, $\eta_{sol-elec}$, which is given by:

$$\eta_{sol-elec} = \frac{\sum n \cdot \Delta t\left(\Delta P - \eta_{CCGT}\,\Delta \dot{m}_f \cdot h_f\right)}{\sum n \cdot \Delta t \cdot \dot{Q}_{sol}}, \tag{11.74}$$

where ΔP is the incremental power of the ISCC configuration over the CCGT one and η_{CCGT} denotes the thermal efficiency at the time increment considered. Unfortunately, this type of efficiency definition as proposed by Li and Yang (2014) depends on the choice of a reference configuration without solar contribution, which leads to an uncertainty that can be influenced by the analysis expert.

Performance metric that does not depend on a reference is the weighting of the exergy flux contribution of each heat source to the CCGT cycle compared to the total exergy flux in the cycle, formulated as an internal solar-electric power factor (E_{ise}) by Rovira et al. (2020).

$$E_{ise} = \sum \left(\frac{n \cdot \Delta t\left(\dot{P}_{gt} + \dot{P}_{st}\right) \cdot \dot{E}_{\mathrm{sol} \rightarrow \mathrm{cycle}}}{\dot{E}_{f \rightarrow cycle} + \dot{E}_{\mathrm{sol} \rightarrow \mathrm{cycle}}}\right), \tag{11.75}$$

in which $\dot{E}_{\mathrm{sol} \rightarrow \mathrm{cycle}}$ describes the exergy flux provided by the solar field to the entire cycle and $\dot{E}_{f \rightarrow \mathrm{cycle}}$ that of the fossile fuel. From this power ratio the internal solar-to-electricity efficiency depicting the ratio of internal solar to electricity production to the total energy supply is obtained which reads to,

$$\eta_{ise} = \frac{E_{ise}}{\sum n \cdot \Delta t \cdot \dot{Q}_{sol}}, \tag{11.76}$$

which is independent of any reference, but only describing a technical ratio.

For the construction and technical evaluation, all previous considerations for the comparative selection of concepts and their evaluation are useful, but under economic considerations, only the Levelized Cost of Electricity ($LCOE$) is used, which is introduced in the next sub-chapter.

State of the art in integration of solar heat (mostly of PTC type) into a combined gas turbine-steam cycle and the most common choice is the coupling into the steam cycle, where plants of up to a solar thermal field power of about 80 MW have been realized. There are also various options for solar heat injection into steam cycles, both for simple and advanced steam turbine cycles, as illustrated in Fig. 11.31b. Analyses there are both the solar heat input to the steam cycle for high pressure evaporation (Behar et al., 2014) and in some cases to a small degree of superheating (Abdel Dayem et al., 2014), pre-heating (Griffin et al., 2009), where mainly the size of the solar feed temperature and thus the collector technology used and its yield just as the degree of hybridization, which still allow efficient power plant operation.

There are also studies that propose solar integration at the low or medium pressure level of the heat recovery steam generator ($HRSG$). The reasons for integrating at this level are better heat exchanger sizing and better integration with the steam turbine (Calise et al., 2018).

In principle, solar heat can also be applied to preheat the air of the gas turbine, but then central receiver tower designs (Okoroigwe & Madhlopa, 2016) are required due to the high temperatures involved. Rovira et al. (2018) compare the annual performance of different ISCC and hybrid systems using different solar concentration technologies: parabolic trough collectors (PTC), linear Fresnel collectors, and central tower receivers (CTR) that integrate solar energy into either the bottoming cycle or the gas turbine air. This work shows that configurations with solar integration on the steam cycle (i.e., conventional ISCCs) result in a higher annual yield and use more conventional and reliable solar concentration technologies (e.g. PTC). However, plant configurations with solar integration at the gas turbine usually result in significantly higher solar-to-electricity efficiencies (up to even $\eta_{\text{ise}} > 50\%$).

Partial recuperation can lead to a reduction in fuel consumption without changing the operating conditions of the gas turbine and thus without reducing efficiency. The introduction of partial recuperation in ISCC allows the use of conventional solar concentrating technologies (PTC with heat transfer fluid), where solar injection is introduced in the lower cycle (as in conventional ISCC), and at the same time fuel savings equivalent to non-conventional technologies with solar integration in the gas turbine.

11.2.5 Approaches for Improving Thermal Cycle Efficiency

The focus in a thermal plant design must consider all heat sources, heat storage and exchange options as well as heat losses and is of primary importance to elaborate the most effective system solution from a cost point of view, since different power generating systems are in cost competition in the energy market. Before cost optimization can be performed, a thermodynamically near-optimal design must first be identified to act as the starting point for cost optimization. One of the largest loss fractions of any thermal based power generation plant is the thermodynamic cycle. Here, some guidelines for minimizing the sources of thermodynamic inefficiency are be pointed out, which in principle can also be applied to two-phase energy conversion cyclic processes. To improve thermodynamic efficiency, it is necessary to deal directly with inefficiencies that contribute to exergy destruction and loss, such as heat transfer, mixing, and friction, including unconstrained expansion of gases and liquids. To analyse these, they must be quantified, at least approximately.

Since any efficiency improvement measure is associated with a design change, it is always necessary to consider whether the yield improvement achieved over the operating lifetime by eliminating the source of inefficiency is higher than the cost of external provision of additional electrical or mechanical power to provide the destroyed exergy. Thus, the focus of thermodynamic process design optimization is primarily directed on low-cost conventional measures to improve component thermodynamic performance. Another major source of inefficiency is a non-ideal heat transfer process, and thus must be avoided. The following general guidelines shall be considered in the optimization process.

Heat transfer and loss optimization of process and components

- The higher the temperature T at which heat transfer occurs, if $T > T_a$, where T_a denotes ambient temperature, the more valuable the heat transfer and the greater the need to avoid any heat loss to the ambient and cooling water. Heat losses above T_a should be reduced as much as possible.
- The lower the temperature at which heat transfer occurs, the more important efficient heat transfer is and the greater the need to avoid direct heat loss to the environment.
- Since the exergy dissipation associated with heat transfer between fluid streams varies inversely with temperature level, the lower the temperature level, the more important it is to minimize the temperature difference between streams.
- Avoid the use of intermediate heat transfer fluids/loops when energy is exchanged by heat transfer between two streams. This not only increases the thermal inertia and the partial load behaviour and leads to additional heat losses.

Although gas-driven energy conversion cycles can generate energy at much more compact dimensions than two-phase cycles, they have some intrinsic fluid-related deficiencies. First, the regenerative heat transfer is much less efficient compared to fluid media—reasons are both the smaller specific density and the low heat capacity—, but most importantly the magnitude of the exit temperature from the working machine. From an efficiency optimization point of view, an approach to the ambient temperature T_a is desirable in order to achieve maximum utilization of the specific working content of the thermodynamic cycle, but this in turn requires large component dimensions, so that in principle usable exergy is lost by releasing the heat to the environment. However, since the ambient temperature fluctuates, especially in dry regions, the efficiency is also subject to the fluctuations of the ambient temperature as the ultimate heat sink.

Flow optimization in pipes and components

Although irreversibilities associated with viscous friction, unrestrained expansion and mixing are mostly smaller than those of heat transfer, they require some attention, especially in gas-driven thermal cycles due to the high flow velocities. The following basic rules should be taken into account:

- The design of the lower temperature stages of turbines and compressors (the last turbine stages and the first compressor stages) requires a higher priority than the other turbomachinery stages.
- Minimize the use of throttling; consider whether regenerative expansions are a cost-effective alternative to pressure reduction.
- Avoid processes with excessive thermodynamic driving forces (differences in temperature and pressure). In particular, minimize mixing of flows that differ greatly in temperature or pressure, as these only create entropy.
- The greater the mass flow, the greater the need to effectively use the exergy of the flow.
- The lower the temperature level, the greater the need to minimize friction.

Design tools

With the help of process simulation software, thermodynamic efficiency can be increased. Many of the programs, most of which have a nodal structure, allow the behaviour of an entire system or of system components under certain conditions to be easily mapped and corresponding thermal analyses and dimensioning to be carried out. Cost sensitivities and optimization can then be performed on this basis. Many of the most widely used commercially available process simulation programs, such as ASPEN PLUS, RELAP, TRACE, EBSILON or APROS, are of the sequential-modular type and assemble components nodally one by one, so that both steady-state and transient calculations and analyses can be performed. For the solar thermal field, there are also a number of process simulation programs such as COLSIM, TRYNSYS or SAM, which already include solar radiation or collector models.

11.2.6 Operation and Components of Steam (Two-Phase) Thermal Cycle Processes

Especially at high power levels, two-phase conversion processes play a central role, but at the same time, power plant-based solar thermal heat generation processes offer a variety of options, ranging from direct steam generation in the collector itself, in which the collector field itself is a free-space heat exchanger, to conventional solutions in which solar heat merely replaces a fossil heat source. In addition, the temporal volatility of the source also places special demands on the dimensioning and, above all, the mode of operation of the two-phase cycles and therefore requires adaptation. Finally, solar thermal power plants are often located in arid regions, where the availability of water as a heat sink in the condenser is not given.

11.2.6.1 Operation Mode and Cycle Options of the Steam Rankine Cycles

With regard to the power control of steam power plants, a distinction is made between a sliding and fixed pressure mode of operation. At partial load, the pressure drop in the turbine is lower than at nominal power due to the lower mass flow according to Stodola's cone law.

In the sliding pressure mode, the pressure in the evaporator adapts to the mass flow. In the fixed-pressure mode, the pressure in the evaporator remains constant regardless of the mass flow. The reduced steam pressure at the turbine inlet during partial load is realised by means of throttling. Due to the exergy losses during throttling, the sliding pressure mode is used to the greatest extent in modern fossil-fired steam power plants. Morin et al. (2004), however, show that there is hardly any difference in efficiency between a sliding and a fixed pressure mode for the operating parameters considered for solar collectors.

With varying pressure ratios of a sliding pressure mode in the collector, there are varying enthalpy differences between collector inlet and collector outlet of the individual sections. As a result, even with a fixed division of the different collector segments, the achievable mass flows of the individual sections change. However, since the lowest mass flow determines the total mass flow, part of the mirrors must be defocused for adjustment. This increases the adjustment losses due to defocusing in the case of sliding pressure operation. Therefore, a fixed-pressure method is usually used, which is also easier to realise by means of a central feed pump. When cooling the condenser, a distinction is made between air and flow cooling.

The type of steam process to be selected depends on the size and type of plant. Simple steam processes are often used in smaller linear-focusing power plants and solar towers, while complex ones are used in large scaled solar field power plants.

- Typical setup of a simple solar steam process and process parameters

In the simple steam process, a relatively large feedwater tank is used. The gases dissolved in the recycled condensate and in the make-up feedwater would lead to undesired corrosion in the steam circuit. By heating the feedwater tank to boiling temperature by means of tapped steam from the turbine, the feedwater is thermally degassed. The feedwater tank is thus in principle a mixing preheater. If no steam storage tank is available, it is the only tank in the plant (Fig. 11.32a). Due to the fluctuating radiation supply and the optical efficiency of a collector, which varies

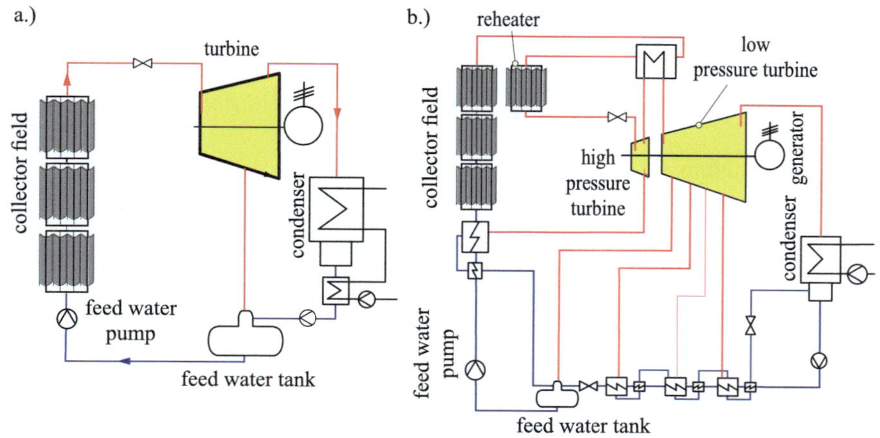

Fig. 11.32 **a** Simple direct steam process with preheating by the feedwater tank; **b** Advanced direct steam cycle with reheating and five pre-heaters (economizers)

over the course of the day, partial utilisation of the turbine can occur. In the partial load, however, the thermal efficiency is reduced. The minimum permissible mass flow of the turbine is usually limited to approx. 20% of the nominal mass flow. If the solar field can deliver a higher mass flow with the required live steam parameters than the nominal mass flow, there are two possibilities. Firstly, a thermal buffer tank can be fed or the power coupling to the fluid must be adjusted by turning out (defocusing) the mirrors. The amount of energy discarded by defocusing is referred to by the technical term "dumping".

- Typical set-up of an advanced solar steam process and process parameters

The advanced steam process shown in Fig. 11.32b is equipped with several preheaters and reheaters. The intermediate superheating of the steam expanded in the high-pressure turbine stage is realised indirectly via a heat exchanger with the steam from the collector. This means that the steam, which is already somewhat expanded and thus has a lower density, does not have to be heated in the collector, where it would generate relatively high pressure losses due to the high velocities. Due to the larger number of pre-heaters, a higher collector inlet temperature is achieved. This and the intermediate superheating increase the average temperature of the heat supply. This higher average temperature not only increases the thermal efficiency of the cycle, but also the heat losses of the collector. The thermal efficiency of the cycle can be improved to approximately 37%. Whether the increased equipment expenditure for the measures increasing the thermal efficiency is still worthwhile in the case of a slight deterioration of the collector is usually the subject of a cost–benefit analysis, especially because the complex steam process requires a significantly more complex control and exhibits a more lethargic system dynamic behaviour.

11.2.6.2 Components of a Two-Phase Thermal Conversion Cycle

As we have seen in the preceding sections, a two-phase thermodynamic cycle process consists of more components than its single-phase counterpart. It comprises the following components in the order of the cycle process steps

- feedwater pump,
- pre-heater (economizer),
- steam generator and steam super-heater,
- steam turbine (one or more stages),
- condenser with condensate pump,
- feed water tank with deaerator assembly.

Specific two-phase components in water-steam systems are, in particular, the steam generator, the condenser and the feedwater tank with the deaerator, the function and design of which will be briefly explained below.

Steam generator

We have already implicitly become acquainted with a special design of steam generator, that of direct steam generation in the solar receiver (either in a central tower or line-focusing power plant). With this, power plant-relevant live steam temperatures (>400 °C) can be robustly achieved at high pressures (>5 MPa) depending on the available irradiation. However, the direct coupling of the power conversion cycle with the solar field has a central deficit; the degree of utilisation of the plant. This is only limited to the duration of the solar irradiation, which is subject to daily fluctuations as well as a diurnal and seasonal cycle. A thermal surplus power of the solar field can therefore only be stored to a limited extent in direct evaporation systems, for example in steam drums, the size of which is determined by the strength of the container material, depending on the pressures and temperatures of the selected power conversion. The "range" of the steam storage thereby reaches the order of magnitude of an hour at best.

Conversely, high power plant utilisation rates can be achieved by using thermal storage tanks, which allow the power plant cycle to operate in the event of solar irradiation fluctuations or the absence of irradiation. These cycle concepts operated with thermal oils, molten salts or metals then require a steam generator.

In the indirect solar thermal power plant types, in which the heat transfer medium is not the same as that of the power conversion system, the steam generator is no longer a compact unit as known from fossil-fired power plants but consists of at least three heat exchangers that are interconnected. These are:

- Pre-heater or economiser, which preheats the supercooled liquid water (feedwater) to its saturation temperature. It is designed as a counterflow heat exchanger and by design steam is not generated inside the preheater tubes.
- Evaporator, in which the phase transition from water to steam takes place, and a

- Superheater, in which the saturated steam is superheated above its saturation temperature. This type of steam is called live steam. The pressure in the superheater corresponds to that in the evaporator. The essential parameter of the superheater is the maximum steam temperature, which determines not only the material choice for the superheaters but has also to match the highest admissible operating temperature of the steam turbine.

Depending on the design of the power plant circuit, a reheater can also be integrated. The fluidic connection of the heat transfer fluid of the solar circuit and the water/steam system of the steam generator in a simple steam generator and a cycle with reheating is shown in Fig. 11.33. The corresponding heat exchange behaviour, which is also referred to in the literature as the $\dot{Q}_{th} - T$ diagram, is shown qualitatively in for both concepts, where the numbers in the $\dot{Q}_{th} - T$ diagram correspond to the individual positions of the flow diagrams depicted in Figs. 11.18 and 11.19. From the $\dot{Q}_{th} - T$ diagram it can be read how the vapour temperature/pressure determines the liquid temperatures. Under non-nominal operating conditions (e.g. non-nominal fluid or water flow, varying steam pressure, ...) the relative contributions of the different heat exchange stages (economiser, evaporator, superheater) to the total heat exchange in the steam generator vary. To design a steam generator, such operating conditions must be calculated individually for each heat exchanger stage by iterative calculation of the heat exchange area. However, if only the operating characteristics are to be determined, a simplified model assuming constant pinch-point temperature differences (say a few °K) between the two streams and a constant specific heat capacity of the fluid during heat transfer is sufficient. A steam generator has two such pinch points characterized by ΔT_{Pi}, where the temperature differences between water/steam and the heat transfer fluid are minimal; this is:

- the inlet to the evaporator (ΔT_{P1} referring to the points $7/1^*$ in Fig. 11.33) and
- at the exit of the superheater/reheater (ΔT_{P2} referring to the points $5/2$ in Fig. 11.33).

Analysing Fig. 11.33, most of the thermal power $\dot{Q}_{th}$ is transferred in the evaporator. In the superheater, a large amount of power is also added to the steam, but this essentially consists of generating a sufficiently high temperature difference of the steam to make it available for turbine inlet. Figure 11.33b also shows that the thermal power transferred in the reheater is rather comparatively low.

Steam generators can be divided into two basic types:

- firetube boiler and,
- watertube boiler.

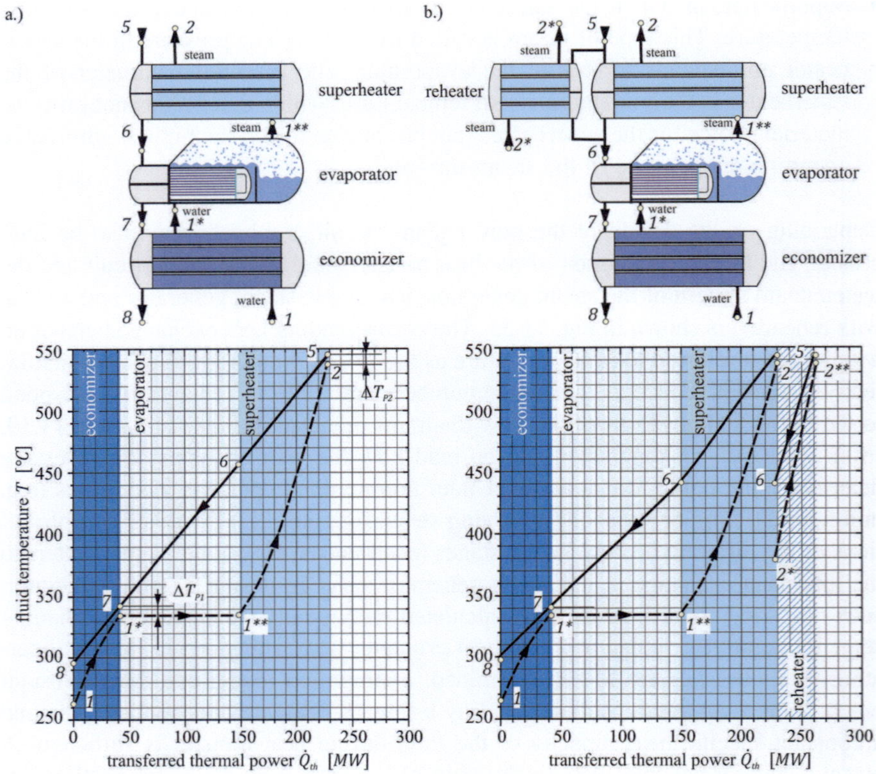

Fig. 11.33 Evolution of the fluid temperatures of water/steam (dashed line) and heat transfer fluid (here salt, solid line) as a function of the thermal power in the individual components of a steam generator for a simple steam cycle (**a**) and a cycle with intermediate superheating for a 100 MW_{el} electric solar power plant modified from Bachelier (2021)

Although named firetube boilers, these types of boilers use heat from a gas, e.g. the exhaust gas from a gas turbine or that of an air-cooled receiver to evaporate water. Therefore, this boiler design is a type of heat recovery steam generator (HRSG), which requires an external furnace because steam is generated solely by passing hot gases through its tubes. Heat recovery steam generators are used in (conventional) combined-cycle plants, but they can also be used in solar tower power plants with air receiver technology and Integrated Solar Combined Cycle Systems (ISCCS) using solar thermal heat generation to reduce fossil fuel consumption. Firetube boilers can typically not be used in large-scale power stations as their design limits them to lower capacities, pressures, and temperatures. Typical operating pressures are of the order of 2–4 MPa, while watertube designs of nowadays fossil stations allow for more than 23 MPa (super-critical water). In watertube boilers, the water and evaporated steam flows through a great number of small diameter pipes housed in a pressure resistant vessel. The smaller dimensions allow for much higher inner pressure loads. Water tube boilers are predominantly used in large-scale power plants and are highly scalable. Within this class of boilers there are three subfamilies,straight tube boilers, bent tube/U-tube (or header type) boilers and helical coil tube boilers. More details on boiler design, operation and control and the corresponding literature can be found in Hesleton (2005).

Condenser

The condensation of the steam leaving the turbine represents a huge amount of heat and is the main limiting factor in the efficiency of a Clausius-Rankine process. The reason is the high enthalpy of condensation of the water. How does this happen? The live steam leaving the turbine with a low water content is fed to the condenser. In the condenser, this live steam flows around the tubes through which a coolant flows and condenses on them. This liquid, then called condensate, collects due to gravity in the lower part of the condenser in the so-called hot well, as illustrated in Fig. 11.34a, from where it can be fed back into the steam cycle. In order to keep the efficiency as high as possible, the condensation temperature must be as low as possible, ideally at the level of the ambient temperature. Thus, the cooling capacity is directly dependent on the coolant temperature provided by the environment. The most stable heat sink with the least daily and seasonal fluctuations is an easily accessible water source in the environment with a sufficiently large water volume (capacity), which is the reason why almost all thermoelectric power plants are built as close as possible to rivers, lakes, or the sea, regardless of the primary heat source (fossil, nuclear or even solar). Evaporative cooling with water is more efficient than dry cooling not only because of the high evaporation and condensation enthalpy of water combined with small temperature differences, but also because the equipment required, and the associated investment costs are lower. Unfortunately, solar thermal power plants are often located in arid regions (deserts) with limited access to large amounts of water as required by evaporative cooling. Transporting water from distant sources or treating water of inferior quality can make this unattractive for steam cycle cost reasons and make dry cooling more attractive.

Condenser cooling types and their evaluation

Continuous or once-through cooling: In this case, fresh water is constantly taken from a body of water (river or lake), passed through the condenser of the power plant and finally, warmed by about 10 °C, returned. The body of water cools down again after a few kilometers, passing the heat on to the ambient air. Once-through cooling is possible on rivers with high volume flows or on the coast. Water is taken from the receiving water and then cleaned via coarse and fine screens and passed to the condenser. Without effluent cooling, the heated water is returned directly to the river or sea, as depicted in Fig. 11.34a, without the addition of chemicals. More often, additional effluent cooling is realized by means of a cooling tower. In this tower, the water is additionally cooled and enriched with oxygen before it is returned to the river. Often, effluent cooling is necessary because the discharge conditions into open waters are strictly regulated by law. The water requirement for once-through cooling is large. For example, with a heating margin of about

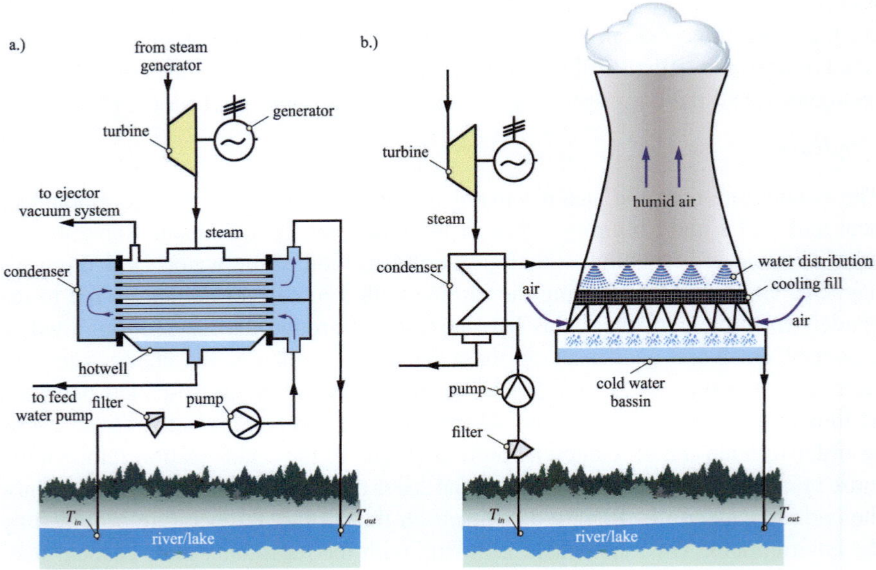

Fig. 11.34 Schematic set-up of a steam condenser (**a**) and of flow routing for once-through (**a**) and effluent cooling by a natural draft cooling tower (**b**)

10 °C, a cooling water requirement of about 3.5 m^3/s per 100 MW of installed electrical power must be expected, which is barely available at most sites. In this cooling type, the entire waste heat is discharged to the body of water.

Effluent cooling differs from once-through cooling that the cooling water heated in the condenser passes through a cooling tower to be cooled before it is returned to the water reservoir. Most of the waste heat is transferred to the ambient air by evaporative cooling through the cooling tower. Both processes are also referred to as freshwater cooling. The air volume flow can be generated by natural draught or forced draught by fans. Advantages of the natural draft cooling tower are the lower operating costs, but the disadvantage is the large space requirement. The advantage of the forced draft design, which has become increasingly important in recent years, is again a lower overall height and smaller footprint and thus a higher specific cooling capacity. This is countered by comparatively high operating costs due to the additional consumption of electrical energy by the fans. In addition, the investment costs are higher due to the fans.

In recirculating cooling the water heated in the condenser is re-cooled in a cooling tower, as in the case of effluent cooling, but then fed back to the condenser. Open-circuit cooling is the most commonly used type of recirculating cooling. The water is circulated, cooled down in the cooling tower, and recirculated again and again, as shown in Fig. 11.35a. In open re-cooling, some of the water evaporates. This has the effect of concentrating the content of dissolved and undissolved substances. This effect is called thickening of the cooling water. To avoid this, part of the water is drained off and make-up water (fresh water) is added. Furthermore, chemical water treatment is necessary to minimize lime and algae formation. Losses due to evaporation and spraying must also be compensated. As in the case of once-through cooling with drain cooling, the air volume flow can be achieved via fans or via the natural draft effect or a combination of both (fan assisted natural draft cooling tower). The waste heat load for the water body can be reduced if effluent cooling is operated, i.e. part of the absorbed heat is released to the atmosphere via a cooling tower. With closed-circuit cooling, the water requirement and the waste heat discharge into the water body can be further limited, whereby the majority of the heat is discharged to the atmosphere and the cooling water is used several times. Table 11.4 provides indicative values for the amounts of heat discharged to the water and to the atmosphere.

The hybrid cooling tower is a mixture of wet and dry cooling (closed circuit) and is often referred in literature as plume abated cooling. It is classified as recirculation cooling. There is a wide range of applications and designs, whereby the terms are sometimes used contradictorily. Generally, in hybrid cooling, part of the hot water is passed through a heat exchanger before trickling into the cooling

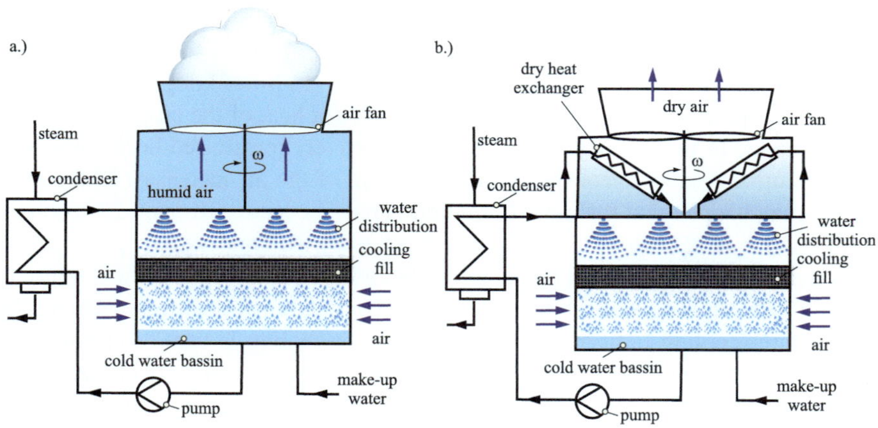

Fig. 11.35 Flow scheme of a recirculation-based condenser cooling by means of an induced mechanical draft cooling tower (**a**) and a plume abated hybrid cooling tower (**b**)

Table 11.4 Share of heat dissipation to water or atmosphere depending on the selected condenser cooling process

Condenser cooling principle	Fraction of heat transferred to water (%)	Fraction of heat transferred to atmosphere (%)
Once through	100	0
Effluent cooling	30–90	70–10
Recirculation cooling	10	90
Air condensation	0	100

tower, as shown in Fig. 11.35b, and this pre-cools the water with air. This is the dry cooling part and ensures that the air is preheated before exiting the cooling tower. There is also mixing of saturated air from the wet section and the non-saturated air flow from the dry section. Due to the increased air outlet temperature, it is possible to reduce cloud formation. This leads to reduced buoyancy, higher pressure losses and the fans represent an additional electrical consumer, so that this concept is actually not profitable, but it plays an important role for solar thermal power plants, especially because of the saving of evaporation losses of approx. $10 - 15\%$ compared to the recirculation concept, especially in arid regions. Another advantage is that at high outdoor temperatures, especially in deserts, the additional evaporative cooling can achieve a significantly better cooling performance compared to pure dry cooling.

The last category are pure air-cooled condensers as sketched in Fig. 11.36. The hot coolant flow from the condenser is split in the air-cooled heat exchanger in a large number of small finned tubes. The finned tubes serve to increase the heat transfer surface area. At the top of the heat exchanger fans are located providing a forced air flow so that the air flows around the coolant tubes and the heat is removed from the coolant by forced convection.

Cooling processes, like other thermodynamic processes, represented in state diagrams. The corresponding state diagram of humid air is called psychrometric chart or $h - x$-diagram, in which h denotes the enthalpy and x the humidity ratio. The humidity ratio x expresses the vapour content in kg per kilogram dry air. In the classical process of evaporative cooling, the latent heat of the evaporated water mass is added to the air in addition to the convectively transferred heat. Thus, the air heats up and simultaneously the water content of the air increases. In the process, it may happen that the air is humidified up to the saturated range. Recooling is targeting to reduce the water temperature, whereby the temperature difference of the water cooling between the condenser outlet and the cold water basin is referred to as the cooling zone width $z = T_{out} - T_{in}$. The theoretical maximum achievable temperature (cooling limit temperature) for open recooling is thus determined by the wet bulb temperature T_{wb} of the ambient air. Thus, it is theoretically possible to cool the water below the ambient air temperature T_{amb}. The efficiency of open

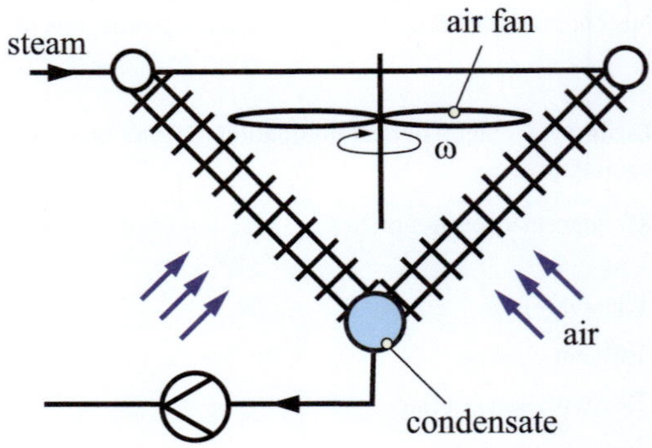

Fig. 11.36 Principle of an air-cooled condenser

and closed cooling tower variants differs. In natural draft cooling towers, the cooling effect depends on the temperature difference between the two media, water, and air, as well as on the influence of wind, while in forced draft cooling towers the cooling effect can be influenced by the volume flow generated by the fans. Cooling towers are evaluated in terms of approach and range, where

- the approach denotes the temperature difference between the temperature of the cooled water and the wet bulb temperature T_{wb} of the air entering the tower,
- the range is the temperature difference between the water entering and leaving the tower.

For a cooling tower based purely on evaporative cooling, the maximum efficiency of the cooling tower η_{CT} is limited by the wet bulb temperature T_{wb} of the cooling air, and it is given by

$$\eta_{CT} = \frac{(T_{in} - T_{out})}{(T_{in} - T_{wb})}, \tag{11.77}$$

where T_{in} describes the water inlet temperature to the tower, T_{out} the water outlet temperature from the tower, and the wet bulb temperature of the air. Typical cooling tower efficiency are in the range of about 0.7 and typical temperature difference $(T_{in} - T_{out})$ range from 5–15 °C. More details on fundamentals, computational assessment, design, and best practices of cooling towers may be found in Hill et al. (1990) or Hensley (2009).

11.3 Characterisation of Solar Power Plants

The characterization of solar power plants can be done according to both technical and economic parameters, although both are often closely intertwined. In principle, a solar thermal power plant contains only four central functional systems,

- a solar field/collector system, in which solar short-wave radiation (visible light and near-infrared radiation) is converted into heat (long-wave radiation) either directly or by means of concentrator technologies (mirrors/lenses);
- a heat-to-electricity conversion system implemented using a thermodynamic cycle; summarized as power block,
- and a thermal storage device that allows heat-to-power conversion even during periods of absence of solar irradiation.
- Optionally an additional heat source (hybridization).

The coupling of the solar field with the power conversion including a storage of the thermal energy, is as the Fig. 11.37 shows, not linearly subsequent linked. As a result, an immediate comparison of different mainly concentrating technologies (line or point concentrating), especially using an economic parameter, depends on the type of market model, the use of the solar park (e.g. combined heat-power use) or the electrical grid architecture (peak load operation, load follow operation).

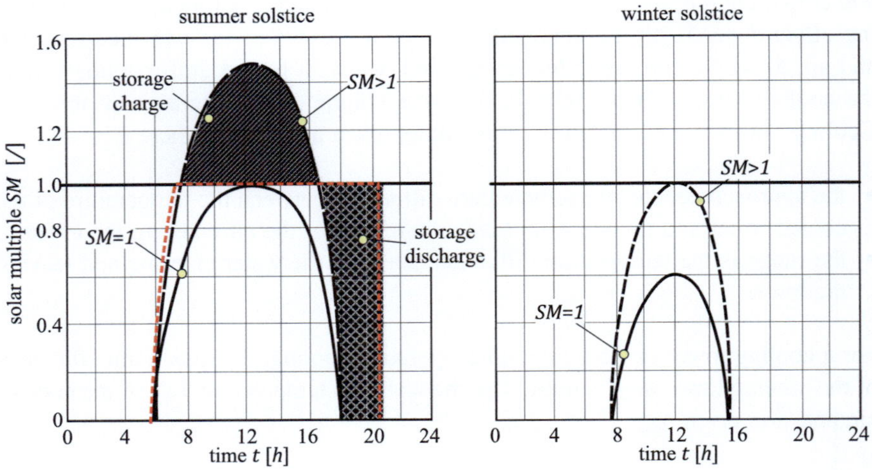

Fig. 11.37 Solar multiple as a function of time of day at summer and winter solstice for $SM = 1$ and $SM > 1$, respectively, for ideal solar conditions. Dimensioning of the solar field with $SM = 1.5$ at summer solstice allows thermal storage charge/discharge during summer, but yields reaching rated power in winter (right) only at noon

11.3.1 Technical and Economic Rating Parameters

11.3.1.1 Technical Rating Parameters

The specification of a solar thermal electric plant (STE) requires not only the definition of the concentrator technology (point or line-concentrating) and the corresponding power conversion cycle. Usually, the design specifies a certain number of operating hours per year during which electrical power is generated. In order to describe how well the power plant park is utilized during a year in the operating mode, an integral is formed from electrically generated power in relation to the nominal power, which is referred to as the capacity factor (CF).

$$CF = \frac{\int_{t=0h}^{t=8760h} P_{el}(t)dt}{P_{el,des}a}, \tag{11.78}$$

in which $P_{el}(t)$ describes the generated electric power delivered by the plant and $P_{el,des}$ the rated electric power and a the duration of a calender year ($a = 8760$ h). The capacity factor thus does not indicate the number of hours during the year when the rated power is reached; this quantity is given by the number of full load hours (FLH). From this consideration it is immediately obvious that the provision of an additional heat source either in the form of a thermal storage system (TES) or with the help of fossil/regenerative chemical heat carriers, i.e. a hybridization, increases the capacity factor (CF) and thus the plant utilization. Considering the power plant's internal provision of heat for electricity generation, the TES should be large enough to allow operation not only during short-term irradiation variations, but also after sunset; more details on the sizing and options of a TES are given in Chap. 8.

This requirement, in turn, implies that the solar field must be sized to not only provide the thermal energy needed to directly operate the genset/consumer, but also to generate "excess" thermal energy to "charge" the thermal energy storage device, thus temporally decoupling the thermal energy conversion from the consumption. The TES system availability allows timely decoupling from solar irradiation and power generation and enables a dispatchable power generation, which is particularly relevant in electrical grids with time-dependent electricity prices. Thus, *STE* plants including a large-scale thermal storage provide a unique feature in the context of renewable energy. To enable charging/discharging of a TES the thermal power provided by the solar field expressed by the solar multiple SM needs to be larger than that of the rated design power of the conversion cycle. Thus, the relevant rating parameters are the solar multiple (SM) and the capacity factor (CF), which are defined as follows:

$$SM = \frac{\dot{Q}_{SF,\,\mathrm{des}}}{P_{th,\,\mathrm{des}}}, \tag{11.79}$$

in which $\dot{Q}_{SF}$ is the thermal power of the entire solar field considered under design conditions, $P_{th,des}$ the design thermal power of the consumer.

The solar multiple (SM) describes the ratio of the maximum thermal output of the thermal solar field at the design point to the nominal output of the turbine. The capacity factor (CF) weights annual electricity generation in relation to the maximum possible annual electric power output at nominal power. The capacity of the thermal energy storage is defined in terms of hours of equivalent full load operation, so that, for example, 12 h of storage means that the thermal energy storage (TES, see Chap. 8) is capable of providing thermal energy for 12 h of operation at rated power. The choice of the thermal storage fluid and its maximum targeted temperature level enables the sizing of the thermal storage volume. Another relevant parameter for the sizing of the solar field is the solar field design ratio (SFR). It is defined as the ratio of the HTF mass flow actually used for the design and sizing of the piping circuit and pumps of the collector field to the required mass flow $\dot{m}_{th}$ denoting the required theoretical HTF mass flow at the design point of the solar field (summer solstice for a given DNI and zero incidence angle), i.e.

$$SFR = \frac{\dot{m}_{real}}{\dot{m}_{th}}. \tag{11.80}$$

Usually $SFR < 1$ because, due to meteorological variations, the nominal operating conditions used to define the effective aperture area of the field may occur only a few hours per year. Therefore, it is not practical to design the entire piping loop and pumps exactly to this nominal point, as this will result in higher cost, needed only a few hours of operation per year. The choice of a SFR value smaller than unity lowers the costs on the one hand but reduces the actual energy yield from the mirror field to some extent. Depending on the linear concentrator technology used, situations inevitably arise in which thermal power generated by the solar field must be discarded by defocusing the mirrors, which is referred to as dumping.

The optimization of a particular system configuration therefore requires a suitable combination of all three design parameters (SM, CF and SFR), especially when searching for optimal sizing for the selected target parameter, for example, the minimum levelized cost of electricity ($LCoE$).

Figure 11.37 shows qualitatively the solar multiple as a function of time of day at the summer solstice (June 21st) and at the winter solstice (December 21st) at any point on the northern hemisphere. A value of SM of one implies equivalence of solar field thermal output and power plant output. The technical term SM does not appear in conventional power plants. There, a similar term, called design ratio DR is used, where a DR-value slightly above one is aimed for—usually 1.05–1.1, meaning that the turbine may run 5–10% above rated power in order to be able to meet the consumer's demand in case of overload. A value of $DR \sim \mathcal{O}(1)$ ensures optimum efficient operation of all power plant components.

For an electricity-generating solar thermal power plant, $SM = 1$ at the time of maximum solar irradiation means that ideally once a year is the solar field able to provide the nominal thermal power that can be processed by the thermodynamic power plant cycle. The solar field is not able to charge any type of thermal storage. As a consequence, the capacity factor (CF) of the installed power plant is low. Thus, only $SM > 1$ allows a loading of the thermal storage with simultaneous nominal load operation of the working machine. Even for solar thermal electric plants (STE) without any thermal energy storage, the solar multiple should be large enough to ensure a sufficiently high power block utilization. Typical values for parabolic trough and solar tower plants are in the range of $SM \approx 1.2 - 1.4$, see e.g. Montes et al. (2009), while linear Fresnel plants are designed with a SM being approximately 20% larger due to their lower optical efficiency.

Considering a STE with a thermal energy storage (TES) the thermal power can be provided to the power conversion system of the power block either by the solar field or the TES. So, if the thermal energy provided by the solar field decreases in the evening hours, the discharge of the thermal storage allows the continued operation of the power block decoupled from the solar irradiation, in extreme cases theoretically a 24/7 operation could be imagined. The power plant operation can be continued as long as the thermal energy content of the storage tank or the minimum temperature of the thermodynamic power plant process allows it.

Here it makes sense to introduce a parameter for the thermal energy storage capacity. The number of (electrical) full load hours FLH which can be generated by a fully charged TES in the power plant unit is often used here, which is defined as follows;

$$FLH = \eta_{RT} \frac{m \cdot c_p \cdot (T_{hot} - T_{cold})}{3600\,\text{s/h} \cdot P_{el,0}}, \tag{11.81}$$

in which η_{RT} represents the round-trip efficiency being close to 100% for the large TES, m the active mass of the storage liquid, c_p its specific heat capacity, and T_{hot} and T_{cold} the hot charge and the cold return temperature during charging. $P_{el,0}$ denotes the nominal power of the power plant unit. The amount of thermal energy used for power generation during the discharge process is slightly smaller than the amount charged from the solar field into the storage tank.

Heat losses occur overnight even with good thermal insulation (usually $<1\%$ per day). These will be compensated during the next morning, when the sun is rising with low power, before power operation can be started. Often the thermal storage is used to preheat the solar field in the morning, in order to enable early power operation and/or to save external electrical energy required to keep the HTF liquid if using electric trace heating. So far, $SM = 4 - 7$ have been realized for solar towers, corresponding to a thermal storage capacity of 10 h (Crescent Dunes, $SM = 4$) or 17.5 h (Atacama I, $SM = 7$) electrical full load hours ($EFLH$) of the power plant unit. Another reason to choose SM sufficiently large is seasonal variations that increase with increasing geographic latitude.

The optimum SM for a certain storage configuration is defined by cost optimization. Although the integration of even larger thermal storage would allow the power plant to operate at full load for more than a few hours (typically 3–12 h), it makes little sense for several reasons. First, the size of the solar field increases, which in line-concentrating power plants causes heat losses in the field and pressure losses in the lines to increase. The net solar power generation stagnates, while the investment costs increase. Solar towers show a similar effect, which is caused by the decreasing optical efficiency of the heliostats with increasing distance from the central receiver. A sensible heat based TES on the order of 6 h already allows annual capacity factors on the order of 45% or more (González-Roubaud et al., 2017), while 12 h storage volume may reach up to 60%. Nevertheless, an arbitrarily large SM does not result in a capacity factor of unity; rather, studies show that regardless of the particular power plant design (line-concentrating or point-concentrating), the relationship of SM and CF can be divided into three domains. Figure 11.38 outlines this by a simulation considering a direct molten salt linear Fresnel power plant with a direct two-tank molten storage. Therein, the capacity factor is drawn as a function of SM for different thermal storage (TES) sizes.

- Domain I, characterized by a linear relation of SM and CF. e.g. an $SM = 1.3 - 2.5$ for a 9 h TES in Fig. 11.38. Within this region the solar field is not large enough to fully charge the TES at any time during a year fully. Conversely, at no point in time a mirror defocussing is required.
- Domain II, asymptotic range (here for $SM \sim 2.6 - 5.5$ for a TES capacity of 9 h): At the beginning of this domain, the solar field is sufficiently large to fully charge the TES at least once per year. However, a further increase of the SM leads to an increase of the number of hours with fully charged TES but also to an increase of the defocusing losses. The growth rate of the capacity factor with the SM therefore decreases as the SM increases, resulting in an asymptotic value of the capacity factor (CF), here e.g. 51% for a TES with

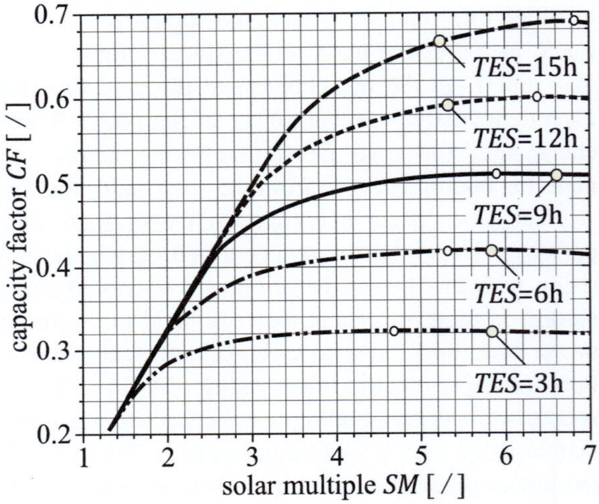

Fig. 11.38 Simulated plant capacity factor (CF) as a function of the solar multiple (SM), for various TES sizes in hours [h] of a direct molten salt linear Fresnel plant assuming a nominal power of 100 MW_{el} at Riyadh (KSA) $(\Phi = 46°48')$ modified from Bachelier (2021)

9 h. Both, the magnitude of CF and the limiting SM essentially depend on the power plant concept, the storage concept but also the control logic in operating thermal loop together with the TES. Asymptotic capacity factors of more than 75% are conceivable.

- Domain III, decreasing range (in the example here for $SM > 6$ with a 9 h TES). In this domain, the energy yield has already reached a maximum and the electrical power that can be generated with the solar field increases only imperceptibly. At the same time, however, for line-concentrating systems the heat losses in the solar field and the pressure losses, and thus the pumping power, increase. For solar towers, with increasing SM, the mirror (or optic) efficiency decreases due to the increasing distance from heliostat to tower, and simultaneously the time in which the heliostats have to be defocused increases, so that the thermal energy generated by the solar field hardly increases, while the losses increase disproportionately. As a result, the annual net yield of electrical energy decreases.

It has to be mentioned, of course, that the optimum size of the TES and the corresponding SM depends very much on the local climate and the investment costs of the components. For climates with less DNI and more overcast or cloudy days, the TES serves as backup for the hours where otherwise no full load operation would be possible. For climates with very high DNI and mainly clear days TES serves mainly to continue operation in the evening and night hours.

In principle, TES capacities of more than 24 h are also conceivable, allowing energy for power generation to be stored for several days. The situation may be interesting, if solar power plants are included in a transregional electrical grid as a base load reserve or electricity market in which the power plant is located is based on a time-updated pricing model. There, the predictability of power generation is of excellent importance for the economic operation of a power plant. Such configurations are rather questionable, however, for sunny locations where every day the storage can be recharged again. In a purely capacity-driven electricity price model, on the other hand, the optimization of the power plant focuses only on one target variable, the minimization of the price per generated kilowatt-hour (or levelized cost of electricity, $LCoE$). In this case, excessively large thermal storage units make no economic sense, since they only provide an insignificant additional energy yield over the year.

The capacity factor (CF) is closely related to the solar multiple (SM). Considering a $SM = 1$ without thermal energy storage (TES), the rated power output of the plant is only reached at one point in time, which is not economically viable. However, since steam-cycled turbines, can produce electric power already for nominal thermal solar field outputs of about 20%, the capacity factor (CF) achieved in in STE is around 20% for line-concentrating systems and for optical reasons slightly more for point-concentrating systems. Power plant types with an already low thermal storage capacity, such as point-concentrating dish-Stirling systems or concentrating power plant types relying direct-steam generation receiver technologies, are designed with an SM above unity, but as soon as the value of unity is exceeded during operation, they require that either a mirror defocussing or to discharge the generated heat into the environment without being used. This amount of thermal energy discarded in the process is referred to as "dumping".

It is useful to introduce a term that also characterizes a hybrid system, where not only solar heat but also other heat sources can be used to drive the heat engine. The fraction of the solar energy source (SF) in the heat input to the power unit, which is defined as follows:

$$SF = \frac{Q_{SF}}{Q_{tot}} = \frac{Q_{SF}}{Q_{SF} + Q_{com}}, \qquad (11.82)$$

where Q_{SF} denotes the fraction of heat energy provided by the solar field, Q_{tot} is the total heat production of the hybrid power plant, being the sum of the solar field input and Q_{com} describing the amount of heat provided by any complementary heat carrier (natural gas, biomass, geothermal). It is important for the correct representation of SF that it is based on the annual integral values, which results from the time series of a typical meteorological year. According to the solar supply fraction (SF), hybrid power plants can be divided into three categories, cf. Nathan et al. (2018):

- Solar Aided Combustion Generation ($SF < 0.33$), which is derived from using solar feed water for preheating in a Rankine cycle, with 30% being approximately the upper limit of heat provided.
- Balanced Solar-Combustion Assisted Generation ($0.33 < SF < 0.66$);
- Combustion-supported solar electric generation ($SF > 0.66$).

Considering now at the total cost of the power plant, the cost of the solar field is proportional to its aperture area while the cost of the thermal energy storage

(TES) is proportional to its volume (and thus its energy capacity), so for both the cost increases approximately linearly. However, since the electrical energy generated no longer increases linearly with the solar multiple of the solar field and the storage hours, the cost per kilowatt-hour generated ($LCoE$) of the system inevitably reaches a minimum value, and as solar field sizes continue to increase, the specific cost ($LCoE$) increases again. Without going into detail here about the economics of a particular type of power plant, there is therefore a cost-optimal power size for a particular defined STE configuration for technical reasons. This means, however, that an arbitrary upscaling of a power plant capacity does not make economic sense; it makes much more sense to build two instead of one power plant at one site.

11.3.1.2 Economic Assessment Indicators

Static or dynamic calculation methods?

Solar thermal electric power plants as other power plant investment projects consider a financial accounting period over two to three decades so that a comparative economic analysis of different technology options by means of static methods, such as cost comparison calculations, profit comparison calculations, or return on investment (ROI) are not suitable. The static methods are characterized by calculating with costs and revenues and usually relate to them single-period investment decisions, i.e. the figures of one year are representative for the entire lifetime and the time-dependent value of the money is not considered. Thereby, expenses and income are not depreciated and discounted to reflect the changing present value of money spent or earned in the future. For this reason, static investment calculation methods are unsuitable both for the purpose of profitability calculation and for the comparison of different technologies with high up-front investment cost and long return time on investment.

Therefore, dynamic methods are used almost exclusively, which consider the time value of money as well as the interest rate risk. These dynamic methods include the cash flows and do not use accounting-based target/actual quantities, so that they ensure a higher degree of planning reliability by taking an overall period view. The disadvantages of the dynamic methods are, apart from the higher computational effort, above all the collection of extensive planning data, which also include secured scaling laws (e.g. upscaling or downscaling of variable parameters) for technical components as well as for overhead costs, just as sensitivity studies for price development of the product factor (electricity, heat) as well as the interest rate development of the cost side during the project life.

Dynamic cost modelling approaches

The most common method for comparing power plants with different generation and cost structures is the levelized cost of energy ($LCoE$), i.e. a price in any monetary unit per amount of energy generated in kWh, for example €/kWh. Ecological and environmental performance may be included in the holistic assessment of a project. Multi-criteria evaluation tools are discussed in the multi-criteria decision tools. Conflicting criteria need an interaction prior or interactively in the decision-making process. A possibility is a combined weighted performance indicator reflecting the preferences of an investor or a regulating body (Platzer, 2021).

The comparator $LCoE$ has become established as a relatively simple method of comparing different power plant concepts and technology options. In real projects, on the other hand, an estimate of costs and revenues over the entire lifetime of the project is made, which also considers the degradation of efficiencies, the replacement of components, maintenance intervals and failures, financing instruments and their changes in terms and conditions in an appropriate manner. The basic idea behind the $LCoE$ method is to calculate the sum of all cumulative costs for the construction and operation of a plant and to normalize this with the sum of the annual electricity generation volume. The $LCoE$ can be calculated either on the basis of the net present value method (NPV) or the so-called annuity method, for details see Kost et al. (2018).

When applying the net present value method, the expenses for the investment as well as the cash flows of income and expenses during the plant's lifetime are calculated by discounting in context for a common reference date. For this purpose, the expenses of all expenditures are divided by the cash flow received (present value) of the electricity generation. Discounting electricity generation seems incomprehensible at first from a physical point of view, but discounting is used to determine the net present value (NPV) of a future payment. The basic idea is that the electricity generated implicitly corresponds to the income from the sale of this energy. Thus, the further in the future this income lies, the smaller the present value. The total annual expenditure during the entire operating period consists of the capital expenditure and the operating costs incurred during the entire operating period. This results in the following relationship for the calculation of the $LCoE$ of a new plant:

$$LCoE = \frac{I_0 + \sum_{t=1}^{n} \frac{A_t}{(1+i)^t}}{\sum_{t=1}^{n} \frac{E_{el,t}(1-d)^t}{(1+i)^t}}, \tag{11.83}$$

where I_0 denotes the absolute capital expenditure to build the plant, often referred to as $CAPEX$. A_t indicates the annual costs within year t and $E_{el,t}$ is the amount of energy produced in year t (output power times operating time in year t), i is the effective interest rate in [%], n the operational lifetime of the power plant in years and t is the year of operation within the lifetime from $1, \ldots, n$. The factor $1/(1+i)^t$ is called the discount factor of the period or the present value factor (NPV). Finally, d denotes the degradation factor of the plant, e.g. declining performance of mirrors or absorbers. The investment costs I_0 include all costs for the construction of the plant; in addition to the direct costs for the acquisition of land, this also includes the associated additional overhead costs such as general project expenses, e.g. company mark-up, project development fee, construction risk, grid connection costs, project permission fees and the like plus some contingency. These are referred to as engineering procurement construction (EPC).

The total annual costs A_t, often referred to as $OPEX$ (operational expenditure), are composed of fixed and variable costs for operating the power plant, its maintenance, repairs, and expenses for insurance:

$$A_t = O_{fixed} + O_{var} + O_{res}, \tag{11.84}$$

where O_{fixed} is the fixed running costs consisting of personnel, annual insurance, and replacement of equipment, which can be assumed as fraction of the $CAPEX$. Correspondingly, O_{var} is the variable costs expenses for water, legal costs referring to electricity delivered, freeze protection of storage and piping, which are all depending on the size of the plant. These variable costs not only refer to aperture area of the solar field but also to the sizing the Finally, O_{res} is the residual value or the expenses for plant disposal. The proportion of debt and equity can be explicitly included in the analysis through the weighted average cost of capital ($WACC$) over the discount factor (interest rate—i). The $WACC$ is strongly country dependent and has to be handled with care for any renewable technology, see Ondraczek et al. (2015). The discount factor depends on the amount of the investor's equity, the demanded or expected return on equity over the life of the asset, the cost of financing using preferential loans or bank credits, and the proportion of debt contributed. By discounting all expenses and the amount of electricity delivered over the entire lifetime, comparability of the $LCoE$ value is ensured. Table 11.5 shows typical values used in previous calculations to evaluate the $LCoE$ of tentative projects.

The calculation of the $LCoE$ according to the annuity method can be understood as a simplification of the net present value method. The basic idea of the annuity method is to determine an annuity, i.e. the average, constant period surplus of the investment. The actual payment flows are transformed into an equivalent (same cash value), equidistant (same payment intervals) and uniform (same payment amounts) payment series. In contrast to the net present value, it is not the total target value that is determined, but the target value per period. It is thus also a dynamic procedure. In full formulation, the $LCoE$ is defined as the quotient of the annualised investment and operating costs and the average electricity yield, using the following relationship (Augsburger, 2013):

$$LCoE = \frac{I_0 \cdot ANF + A_t}{E_{el,t}}, \tag{11.85}$$

where ANF is the annuity or capital recovery factor calculated in the following manner:

$$ANF = \frac{WACC \cdot (1 + WACC)^n}{(1 + WACC)^n - 1}, \tag{11.86}$$

Table 11.5 Financing parameters required for a $LCoE$ calculation of solar thermal electricity plants (STE); * OEM scaling dependent on quotation and market situation containing economy of scale factors (Augsburger, 2013; Collado & Guallar, 2016; Mihoub et al., 2017; Morin et al., 2012; Seitz et al., 2017)

Description	Parameter	Phys. unit	Typ. value
Lifetime	n	Years	25–30
Interest rate	i	%	8–12
EPC(as overhead to $CAPEX$)		%	5–8
Degradation factor	d	%	0.1–0.2
Fixed operational costs O_{fixed}			
• Management maintenance and insurance (% of direct EPC)/year		%	5–12
• Staff (fix-labour cost per year)		\$/year	Country dep.
• Ground tax/lease		$\$/(\text{year m}^2)$	Country dep.
• Replacement of equipment (% of $CAPEX$)		%	1–2
Variable cost per year O_{var}			
• Labour staff (1/1000 m^2 aperture area)			0.03
• Water (depending on cleaning strategy)		$\$/(MWh_{el})$	1–2
• Maintenance of (in % of its individual $CAPEX$)			
• Solar field (SF)		\$/year	0.5–0.7
• Thermal energy storage (TES)		\$/year	0.2–0.3
• Power block (PB)		\$/year	1
Simplified variable costs O_{var}		$\$/(\text{year m}^2)$	OEM\confidential
Residual value O_{res} (in % $CAPEX$)		%	1–2
$CAPEX$ related quantities			
Solar field (currency per m^2 aperture area)		$\$/(m^2)$	Site depen.
Site preparation (currency per m^2 total area)		$\$/(m^2)$	Site depen.
Power block (scaling law of type $a \cdot P^x$)		$\$/(MW_{el})$	OEM*
Erection costs (scaling law of type $a \cdot P^x$)		$\$/(MW_{el})$	OEM*
Profit (% of direct EPC costs)		%	8–15

in which $WACC$ describes the *w*eighted *a*verage *c*ost of *c*apital, which consists of equity (E), debt (D), return rate on equity (k_E), return rate on debt (k_D) as follows:

$$WACC = \frac{E}{E+D}k_E + \frac{D}{E+D}k_D. \tag{11.87}$$

The selection problem between competing investment alternatives always arises when the payment series have different structural characteristics resulting from different investment periods and different temporal distribution of cash flows (from income and expenditure) within the lifetime. The differences between the net present value method and the annuity method arise particularly in the case of large useful lives, because the more periods are available for the annuity distribution of the net present value, the smaller the annuity turns out to be, which is a simple distribution effect. The more realistic method is thus the NPV method.

Future cost projection—Learning curve models

In addition to analysing the $LCoE$, it is possible to calculate the future development of plant prices and the corresponding $LCoE$ using learning curves generated from market projections.

The concept of learning curves describes the empirical finding that the cost of an industrial product decreases by a more or less constant percentage with each cumulative doubling of its produced volume. This percentage is generally referred to as the learning rate (LR) and ranges from 5 to 20%, depending on the product or technology. Learning curves thus establish a relationship between the cost of a product and the experience in the cumulative production of that product, so that the product cost can be represented as a function of cumulative production.

$$I_t = I_{t=0} \cdot \left(\frac{n_t}{n_{t=0}} \right)^{-l}. \tag{11.88}$$

Here, I_t are the costs of the cumulative quantity n_t of the product produced at a certain time t, while $n_{t=0}$ represents the output quantity at the reference time $t = 0$ and the corresponding costs. l describes the (positive) learning parameter. When cumulative production doubles, the costs fall to 2^{-l} of the original costs. This figure, the so-called progress ratio (PR), is often used to compare different learning curves and is calculated as $PR = 2^{-l}$. From the progress ratio, the learning rate can be extracted, which describes the cost reduction when cumulative production or capacity is doubled, and is calculated as:

$$LR = 1 - PR = 1 - 2^{-l}. \tag{11.89}$$

With the help of the learning curves, the plant price I_t can be predicted assuming a certain value for LR (or PR) independent of the time. However, this assumes that the learning effect for technologies is continuous. Assuming product market scenarios in the future, annual figures can then be assigned to the cumulative product quantities in each case and a temporal development of the price is obtained, for example for electricity generation a temporal development of the $LCoE$ (Platzer & Dinter, 2015).

Significance of the $LCoE$

The use of the $LCoE$ as a methodology allows the comparison of different electricity generation technologies to be transferred to a cost level. Provided that all data are available, the $LCoE$ enables comprehensible transparency and has therefore become internationally accepted as a fundamental cost metric. With the help of the dynamic methodology, which takes into account both static factors and dynamic variables such as $OPEX$ with its monetary value over the entire life cycle, the result can be represented in just one figure.

However, the simplified representation in only one figure also contains risks, because this methodology is also fraught with uncertainties. The methodological deficits are mainly due to the fact that forecast values (reliable prognoses) are required for the calculation of all values over the entire lifetime of the power plant; this includes not only the development of interest rates over the lifetime but also the development of expenditures. In addition, many projects are based on a number of preconditions whose projection into the future is difficult to predict. These range from tax relief on depreciation and/or construction, guaranteed invariant purchase prices for the electricity produced, remuneration for CO_2-free electricity generation and many more. Although revenues are not reflected in the $LCoE$, but only expenses, the absolute values can change substantially in a market dominated by renewable energies and other factors, such as guaranteed available electricity generation for grid stabilisation purposes, can play a significant role. These must then be used as additional criteria. Nevertheless, the $LCoE$ will not lose importance as a technology comparison criterion as long as all boundary conditions are imposed on the individual technology in a similar justified manner.

11.3.2 Cost and Efficiency of Solar Thermal Electric Plants (STE)

Since solar thermal power plant projects can have different market objectives, e.g. pure solar power generation, cogeneration, or combined seawater desalination/ power generation, it is difficult to directly compare different projects at different sites and their performance, although an economic comparison of different technologies at one site is always feasible. This is due to different climatic conditions in different locations, and to the respective market and its boundary conditions, like consumer profiles. However, it is of vital importance to compare different technology options for an intended new project. Here the cost structure of the power plant project as well as the various revenue options should be considered. The question of technology selection for the solar field, heat transfer medium, storage type as well as the dimensioning of a solar thermal electricity (STE) plant should not be prejudiced by individual interest but selected in a technology open way. Here the size effects play a role not only in the case of the investment costs, such as the dimensioning of the solar field, of the thermal energy storage or the power conversion system, in which linear economy of scale as well as exponential scaling laws are present, but also in the variable costs of operation. A cost-optimal design of the entire STE system (from the components to the system) must take both aspects into account. This results in a multitude of possible configurations, each of which has an individual optimum creating a Pareto front.

At present, about 180 solar thermal power plants are operated with a capacity of approximately $7.6\,GW_{el}$, of which the majority (70%) are parabolic troughs present in all electric power output ranges, as shown Fig. 11.39a. Solar towers

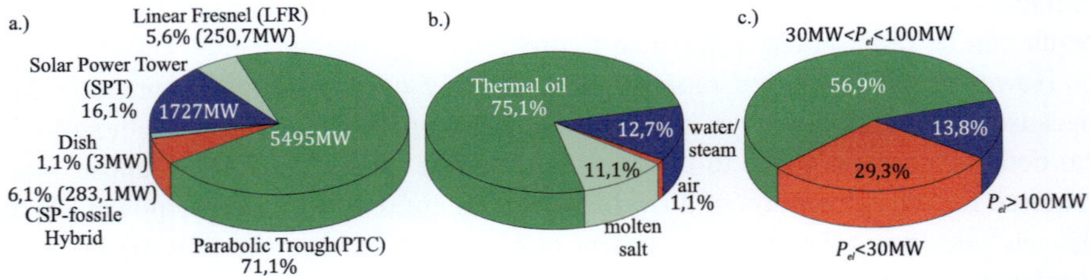

Fig. 11.39 Current status of operable solar thermal electric (STE) plants according to NREL (2022); **a** share of plants with respect to technology and installed power, **b** share of solar field HTF used in STE, **c** Power output classes of operating STE

gained importance in recent years, especially for large scaled plants, and similar to linear Fresnel shift from direct steam to molten salt as HTF. Dish is negligible in the market, and hybrid solar-fossil power plants occupy a small market niche considered as transition technology with limited renewable fraction. In more than 130 power plants, thermal oil is used as the heat transfer fluid while only about 20 power plants use molten salt or steam in the solar field, see Fig. 11.39b. The two-phase Clausius-Rankine cycle almost exclusively dominates at present as power conversion system and less than half of all power plants use thermal energy storage (TES), most of which rely on an an indirect two-tank salt storage while about 20 plants use a direct two-tank thermal energy storage. Thermal energy storage systems based on thermoclines and vapor storage are rather the exception. However, in future, plants without TES will be rather the exception, because photovoltaic plants with a similar capacity factor as STE plants without TES have lower costs. Approximately 60% of the power plants are in the power class 30 to 100 MW_{el}, while only slightly more than 20 have a capacity exceeding 100 MW_{el}, see Fig. 11.39c.

11.3.2.1 Line-Concentrating STE

Starting from parabolic trough power plants with a thermal oil as fluid in the solar field as the most technically mature option as reference, one can analyse the influence of other potential heat carriers on the yield and on the normalised $LCoE$ assuming a reference plant with a nominal power of the order of 50 MW_{el}. In order to separate the individual effects from each other, we discuss first STEs without thermal storage using different heat transfer fluids (molten salts, gases (Nitrogen or CO_2) and direct steam generation (DSG) as alternatives to thermal oil in the solar field.

PTC without TES

All studies show that the $LCoE$ can be reduced by more than 10% in configurations without the integration of a TES if the live steam temperatures can be increased significantly above 400 °C, which is a natural limitation of thermal oils as a heat transfer fluid. Although the heat losses in the solar field increase, the efficiency of power conversion in the Rankine cycle increases at the same time. When considering direct steam-generating PTC power plants, the improvements due to the higher temperature increases even more (approx. $2 - 3\%$) the yield if regenerative feedwater preheating is used (Desai et al., 2014; Feldhoff et al., 2010). This effect is more pronounced at sites with high annual DNI values than at sites exhibiting medium annual DNI values. Increasing the turbine inlet pressure also has a positive effect on reducing the $LCoE$ in DSG systems, which is due in particular to the improved part-load behaviour of the steam turbine (lower outlet wetness) at lower irradiation values. An optimum seems to be at approx. 480 °C at 10 MPa.

The most attractive aspect of DSG operating variants is the reduction of investment, operating and maintenance costs, which are significantly reduced by more than 10%, since the solar field is smaller and the expenditure for the oil systems as well as the heat exchangers are omitted for steam generation. At the same time, however, the capital expenditure for the significantly more pressure-resistant piping network and the absorber increases by about 3% (Feldhoff et al., 2012). The total annual costs of the DSG system without storage are therefore about 7–8% lower than the annual costs of a thermal oil system. The annual net electricity yield of the DSG system is slightly higher than the yield of currently available synthetic oil HTF system (T_{max} < 410 °C). This difference is between 0.1 and 4%, depending on the difference in the yield results of the power models used.

In principle, even higher temperatures can be generated in the solar field with gases and higher live steam temperatures, but higher heat losses and the high compressor work in particular lead to no advantage over thermal oil as shown in studies so far (Binecinto et al., 2014). Molten salts allow significantly higher solar field outlet temperatures than thermal oils, so that the optimal parameters of a Rankine cycle can be achieved, with costs for the salt being significantly lower than for synthetic oil low. The advantage of saving oil treatment systems (e.g. for hydrogen removal) must be seen against the additional costs for a freezing protection for molten salt. This freeze protection (at approx. 230 °C) requires an additional expenditure for trace heating, so that at the end the achievable $LCoE$ values are on the level of the reference version with thermal oil (Mihoub et al., 2017).

PTC with TES

The integration of a thermal energy storage system initially increases not only the investment costs ($CAPEX$) due to the construction of one or two tanks filled with the storage liquid (molten salt), which in a first approximation increase quadratically with its external diameter (compare e.g. Bachelier, 2021), but also due to additional pumps, pipes and instrumentation. The $O\&M$ costs also increase slightly depending on the storage liquid used. In almost all cases analysed in the literature, however, with current costs the $LCoE$ decrease with integration of TES due to the enhanced electricity generation. It increases its capacity factor (CF) resulting in revenues exceeding the additional expenditure for $CAPEX$ and $OPEX$.

For each choice of heat transfer medium in the solar field and the TES, there is a cost-optimal storage size given by the limitations of the components, but the main advantage of TES is the significant increase in the capacity factor. Of course, a larger thermal energy storage requires a larger aperture area of the solar field, here a cost-optimal solar multiple can be observed in the order of about $SM \approx 3$. Possible cost advantages of a direct steam generation power plant disappear as soon as thermal energy storage over several hours becomes necessary for reasons to achieve a minimum capacity factor. If the TES capacity exceeds 2–3 h, the $LCoE$ of a DSG power plant is greater than that of a plant with oil or molten salt as heat transfer medium. Although the DSG solar field is about 5% smaller for the same power output. The high pressures in the steam receiver increase not only the receiver costs but also those of the entire piping network due to the required pressure resistance. With increasing power plant capacity and growing TES capacity also the difference increases to the disadvantage of the DSG system because of the complexity of designing a suitable TES system for steam. The second main cost driver of the DSG system is the cost of the thermal energy storage system, whose share of the total costs for the DSG system increases to a quarter of the total costs of the power plant. Figure 11.40 shows the relative cost structure of a 100 MW_{el} power plant with 9 h TES storage with HTF thermal oil compared to an optimised DSG with a 9 h TES consisting of a 3-tank molten salt system with phase change material. Feldhoff et al. (2012) analyzed a variety of TES options and configurations (thermal storage with phase change material, indirect oil storage, mixtures of both, optimization of the steam cycle at different storage temperature levels, etc.) and were able to determine $LCoE$ reductions. But nevertheless, the analysis results show no improvement compared to the reference system, with the cost of the thermal energy storage system being identified as the biggest hurdle.

A comparison of different line-concentrating collector technologies such as parabolic trough (PTC) and linear Fresnel collectors (LFC) is complex, as they not only differ significantly in terms of optical efficiency and collector length-specific heat losses, but also in specific installation costs, see in Chap. 10. For comparison of both technologies, we assume a large power plant with an electrical capacity of 100 MW_{el} at the same site (Riyadh, KSA, $\Phi = 24.4°$ N, $DNI = 2266$ kWh/(m^{2}a)), whose concept is based on a direct 2-tank molten salt storage with a TES capacity of 15 h, in order to maximise the differences. In a cost-optimal configuration, a concept based on PTC collector technology requires both a one-third smaller solar multiple (SM) and thus a smaller aperture area than the linear Fresnel collector LFC system. For a cost-optimal STE, a PTC system requires $SM \approx 3$ while an LFC requires about $SM \approx 4$. The annual energy fluxes and utilisation of solar irradiance also differ substantially, as shown in Fig. 11.41a. The main reasons are the considerably smaller optical losses and defocusing losses in the mirror field of the PTC system, in addition to shorter lines that require less heating to avoid salt freezing, although this hardly matters as the size of the power plant increases, since these are essentially contested by the TES. On the other hand, since the shading losses of an LFC collector are much smaller than those

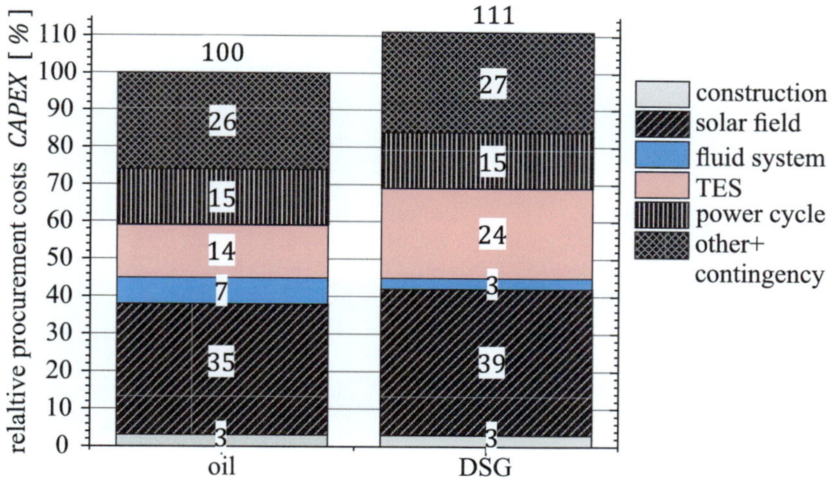

Fig. 11.40 Comparison of the procurement cost structure of 100 MW_{el} STE plant with a 9 h TES for an plant using oil as HTF with a direct steam concept using advance TES concepts (modified from Feldhoff et al., 2012)

of the large PTC mirror system, the LFC rows can be placed closer together. As a result, there is only a marginal difference between the two collector types in terms of land consumption. There are also no major differences in the power conversion and the control system. Nevertheless, the net yield in relation to the solar irradiation is about 50% higher in a PTC collector system than in an LFC collector with the same electrical power (23.3% vs. 15.3%), as illustrated by the annual integral energy fluxes in Fig. 11.41a. Despite the smaller aperture area and the higher optical efficiency, there is no cost advantage of the PTC collector type for the considered configuration in the $CAPEX$ for the construction of the system, since the expenses for the solar field in particular are considerably higher than those of the LFC systems, which can be constructed cost-effectively in lightweight construction, as the relative cost structure in Fig. 11.41b illustrates. In both linear concentrating technologies, the main $CAPEX$ cost drivers are the costs for the construction of the solar field, with a clear disadvantage for the PTC collectors.

With regard to $OPEX$, one can assume similar cost structures for both types of power plants; the studies analysed in the literature show deviations in the order of approx. 10–20%. In the case of LFR collectors, the larger mirror surface is noticeable, but the simpler mirror geometry and the use of automatic cleaning robots with water recovery facilitate the cleaning effort (Askari & Ameri, 2017; Giostri et al., 2012; Morin et al., 2012). In total, there seems to be a slight advantage for PTC collectors in the market, because for LFR systems the data basis is not as robust as for PTC collectors and is subject to greater uncertainty.

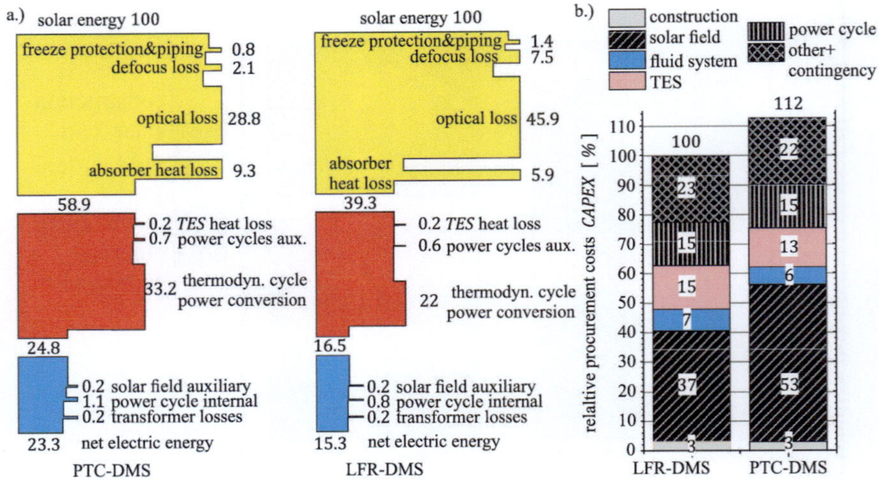

Fig. 11.41 Comparison of the annual integrated energy conversion chain (**a**) and $CAPEX$ structure (**b**) of a PTC and LFR based 100 MW_{el} STE plant with 15 h TES storage in a direct molten salt configuration using a steam Rankine cycle (both configuration are cost optimized in terms of aperture dimension; modified from Bachelier, 2021)

11.3.2.2 Point-Concentrating STE—Solar Tower

Solar towers allow additional degrees of freedom in optimising the mirror field, the different options of which have already been discussed in Chap. 10 and depending on the location specific costs per generated kilowatt hour can be lower for large power plant outputs ($\geq$50 MW_{el}) than for parabolic troughs (PTC). This applies in particular to solar towers that achieve receiver outlet temperatures exceeding 500 °C. For these temperatures possible heat transfer fluids are water-steam, molten salts, liquid metals (Na or NaK) or compressed gases (He or s–CO_2). Gas requires too much specific pumping (ventilator) power. Liquid metal sodium-cooled receivers with very compact receiver dimensions, however, require a molten salt storage due to high specific fluid costs (and thus an indirect TES storage concept associated with additional costs). All studies show a cost-optimal design for external receiver designs on solar power towers.

Solar power towers without TES

The data base for commercial solar towers is sparse due to the few projects realised so far, but from an economic point of view ($LCoE$), direct steam driven systems (DSG) are competitive for STE concepts without significant thermal energy storage. Often a Ruth storage of about 0.25–0.5 h capacity is used to compensate temporary shading by clouds, but larger capacities would be too expensive for these pressurized vessels. For a stable operation of a Rankine steam cycle over a longer time, DSG towers need a fossil auxiliary firing, which has to be provided by means of a natural gas or biogas burner and causes additional fuel costs and

CO_2 emissions. The achievable capacity factors without any fossil firing and TES and reach approx. CF ~ 20–25%.

Solar power towers with TES

For solar tower power plant concepts containing a thermal energy storage large-scale TES of more than 6–18 h FLH are pursued depending on market requirements and climate, in which a direct 2-tank molten salt storage concept represents the reference standard. Cost-optimal solar field dimensions for charging the TES with FLH in this range need heliostat fields—depending also on the DNI and latitude of the location, with SM between 2.8 and 4.2. For a 50 MW_{el} electric power plant with a 15 h TES, $SM = 3.5$ results in a capacity factor of up to $CF \approx 0.60$ close to 5000 full load hours (FLH). In power plant configurations with a large TES (>9 h), solar towers show an approx. 20% ($rel.$) higher capacity factor than PTC.

Figure 11.42 depicts the cost $CAPEX$ structures that arise for such a configuration, and it can be seen that the main cost factor are the heliostats. This is also one reason why solar towers have a cost-optimal electrical output capacity being in the range of 100 MW_{el}. Larger net outputs would be technically possible, but the high heliostat costs and the decreasing optical efficiency with increasing distance from the receiver tower lead to an increasing $LCoE$, therefore it makes more economic sense to build two towers with corresponding solar fields for large outputs rather than one large one, see Augsburger (2013). If one compares the cost structure of a solar tower using molten salt as HTF with that of a PTC power plant using thermal oil as HTF (cf. Fig. 11.40), it can be seen that the share of expenses for the TES and the fluid system of a solar tower is lower due to the more compact construction and the higher temperature level; at the same time, the solar field-related costs (heliostats, receiver and tower) amount to more than 50%. If that would be compared to a Linear Fresnel power plant using molten salt, the expenses for TES would be similar to the tower plant. In terms of the absolute size of $CAPEX$ expenditures, the investment share in a solar tower project is larger than in a power plant based on PTC technology, although the difference becomes smaller as the net electrical output increases.

11.3.2.3 Metrics for Characterizing Fossil Hybridization of Solar Power Plants

Solar thermal power plants without storage can be operated more efficiently and, above all, with a large capacity factor and high dispatchability, if fossil-fuel backup options support or supplement solar thermal power generation. Also, with still low fossil fuel prizes in many countries, the combined $LCoE$ of such a plant would be lower. Depending on carbon taxes and fuel shortages this might change in future. A technical reason is that the efficiency of the power unit decreases during part load operation, which can be avoided with auxiliary firing. Even daily and seasonal fluctuations may be compensated with auxiliary firing with fossil carriers (regardless of the fossil carrier whether biomass, syngas, or natural gas). To an even greater extent, however, auxiliary firing increases the capacity factor of the power

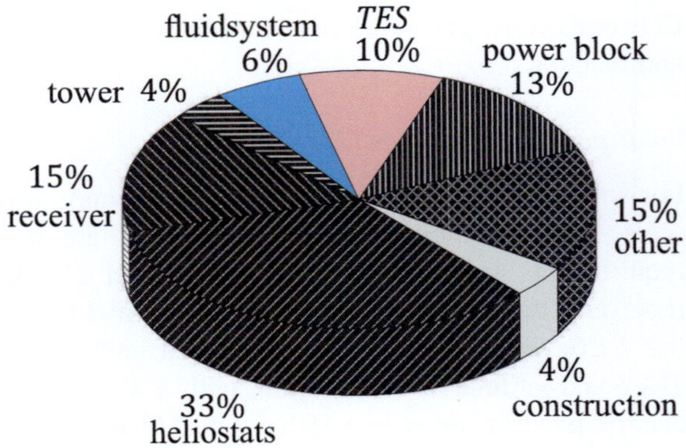

Fig. 11.42 Cost shares of the individual $CAPEX$ relevant trades of the 50 MW_{el} solar tower STE with 15 h TES based on a direct molten salt concept with 2-tank storage from Fichtner (2010), similar in Collado and Guallar (2016)

plant and thus the amount of electricity effectively produced. Even if an expensive biomass burner is used (in a factor of 5 larger $CAPEX$ than a comparably powerful natural gas burner in currency units per $kWth$), a hybridized power plant shows significantly lower electricity production costs than an exclusively solar thermal based power plant.

A second type of hybrid CSP power plants are conventional fossil power plants, where the CSP solar field is used to substitute partially heat from the boiler by solar heat. In this sense the solar field is a used as a fuel saver. The heat can be integrated in a conventional power plant, e.g. a coal-fired power plant, at different integration points in the whole process. It might be used for preheating the water or it might even add solar steam to the steam produced by the boiler.

Any thermal storage support system (TES) increases the solar fraction (SF) of the hybrid system, especially with combustion-intensive operating strategies such as load-following operation or even base-load generation. All studies to date show that for hybrid power plants, increasing the fuel fraction leads to lower $LCoE$ compared to purely solar-based STE. In addition to the relatively low $CAPEX$ of combustion systems per unit of energy generated, especially at high power, the operating costs of the combustion system are also lower compared to the sum of TES and solar field, even if fuel prices double. With increasing carbon tax and even further increasing fuel costs (e.g. natural gas) this effect may become smaller. On the other hand, other alternatives exist which make solar hybrid plants more competitive, which is the hybridisation with photovoltaics (PV) discussed further down.

Hybridization and thermal energy storage TES as strategies to increase the capacity factor of a solar power plant are alternative approaches and not complementary. A combination of both will mainly increase the investment sum. Remarkably, the analyses also show that a solar share of $SF = 30 - 50\%$ of the nominal electric power generation allows both a high operational suitability and great flexibility of the power plant with a simultaneously high solar share in the power generation. In addition, such concepts have the advantage of requiring a significantly smaller solar field (almost by a factor of 2) and thus smaller investment costs, while at the same time ensuring a significantly larger capacity factor.

From these considerations, it becomes clear that an exclusive evaluation of solar thermal power plants (STE) in terms of the capacity factor and the electricity production costs ($LCoE$) is not sufficient. Neither aspects of the reduction of CO_2 emissions by the solar source nor the operation of the power plant with regard to the stability of the power grid by means of a dispatchable power supply are included in these evaluation parameters, so that essential strengths of a STE with regard to resource consumption or dispatchable power generation are omitted. In connection with the consideration of hybrid power plants, it makes sense to introduce further parameters that characterize these hybrids.

A qualitative way to characterize hybrid power plant types is by sharing components. Here, a distinction can be made between indirect hybrids, in which the working fluids of conventional and solar power generation are separated, and directly coupled hybrids, in which both systems share components such as the turbine and condenser. An example of the latter is hybrid DSG power plants, where a steam generator can be driven by either a fossil burner or the solar field.

The annual consideration is important because the net gain of a hybrid depends on the temporal variability of the solar resources. Thus, the system efficiency of a hybrid in steady-state operation can be worse than that of the respective individual technology but still achieve a net gain because advantages such as avoided (or reduced) start-up and shutdown losses and a smaller and thus less expensive overall infrastructure effort more than compensate for this disadvantage in terms of $LCoE$. The net energy utilization factor (η_{net}) is given by the following expression

$$\eta_{net} = \frac{E_{el}}{Q_{SF} + Q_{com}}, \tag{11.90}$$

where E_{el} represents the electrically generated energy of the overall system. By using the solar field, the CO_2 emission is reduced compared to a pure fossil power plant and the annual CO_2 abatement results from

$$\Delta CO_2 = \left(\frac{E_{el}}{\eta_{net}} - E_{fossil} \right) \cdot f_{CO_2}, \tag{11.91}$$

in which f_{CO_2} denote the specific CO_2 emissions resulting from the fuel and E_{fossil} is the thermal energy supplied from a fossil carrier. The specific carbon emission factor (f_{CO_2}) can be deduced from the carbon content of the fuel (c_c), m_f is the annual consumed mass of the fuel, and the conversion factor $44/12$, which reflects the relative molar mass of CO_2 versus pure carbon:

$$f_{CO_2} = \frac{44}{12} \frac{m_f \cdot c_c}{E_{fossil}}. \tag{11.92}$$

Table 11.6 Typical exergy values, LHV's, and f_{CO_2} values of fuels most commonly used in hybrid cycles from Sheu et al. (2012)

Fuel	Exergy value [MJ/kg]	LHV [MJ/kg]	f_{CO_2} [kg CO_2/MJ]
Natural gas (CH_4)	52	50	0.055
Coal (C)	26	23	0.159
Mineral oil (C_8H_{18})	45.5	43	0.072

The total thermal energy supplied from a fossil carrier (E_{fossil}) is given by

$$E_{fossil} = m_f \cdot LHV \tag{11.93}$$

where LHV indicates the lower calorific value in [MJ/kg]; typical values are given in Table 11.6.

Another performance criterion for hybrid power plant concepts can be specific water consumption targeting at a reduction of the specific water consumption per generated electric kilowatt hour (f_{H_2O}), which is of great relevance especially in arid locations. It is defined as the ratio of the volume of annual water consumption (V_{H_2O}) to annual electricity production (E_{el}) in the following way:

$$f_{H_2O} = \frac{V_{H_2O}}{E_{el}}. \tag{11.94}$$

11.3.2.4 Solar Hybridization of Solar Thermal Power Plants

Apart from fossil or biomass hybridization of STE power plant also a combination with a photovoltaic (PV) solar field is an interesting option for electricity production. In this case for both technologies the primary resource is the sun. The main reason for this development is the cost advantage of solar PV over CSP for instantaneous power production, which is fluctuating, however. The advantage of cost-effective thermal energy storage integrated in a CSP power plant in this way can be combined with the advantage of low-cost non-dispatchable PV power generation. In principle this may happen on a national scale with locally distant power plants, but for a single power producer it can make sense to collocate both power plants in one project and also exploit some synergies (e.g. control-room, grid connection, operation, and maintenance staff etc.). Generally, we have three different integration approaches for combining CSP and PV:

- PV power can be used in order to operate the CSP plant (pumps, controls etc.) during daytime. The logic behind this integration is that the cheap PV power is used when the CSP solar field operates during daytime, in order to reduce the difference between gross and net electricity generation from the CSP plant. This approach is only possible in collocation and PV power is not fed into the grid but used within the power plant.

- The generation of electricity is clearly divided between PV generation during daytime and CSP generation using TES for generation during night-time or times when the PV power is not sufficient. In this approach the CSP part of the power plant is designed to complement the PV generation, mainly by taking heat from the large TES to run the turbine. Solar field sizes may be appreciably decreased as their main purpose is to charge the TES and not to directly deliver heat to the power cycle. As a consequence of this operation mode is that the operation hours of the turbine are reduced. Economically this makes sense only as long the installation cost of the PV field is lower than the cost of the solar thermal collector field. The objective could be to be able to run a combined project as a baseline power plant. The strict target to complement PV leads to many part-load operation hours of the turbine. In total this approach might be an idealized case.
- The third approach is to run PV generation during daytime and CSP with TES for supplementing the generation at all times during day and night. In contrast to the previous case, however, CSP and PV may generate electricity in parallel when abundant solar radiation is available, and the TES storage is fully charged. The daily generation profile of such a plant is not constant but especially in summertime is clearly peaked around sunshine hours. Solar field sizes similarly to the previous case may be decreased but depending on the possible revenues of day production they have to be optimized economically. Operation hours of the turbine are high and even the part-load operation hours compared to the previous approach might be reduced. At least the load would be often high during part-load operation when complementing PV.

An example of such a hybrid PV/CSP power plant is the Atacama I/Cerro Dominador project in Chile halfway between Antofagasta and Calama.[1] It is operational since 2021 and consists of a solar tower of 250 m height with molten salt receiver and a circular field of 10, 600 heliostats. The steam turbine has a rated power of 110 MW_{el}. In order to reach a high capacity factor, the CSP part uses a 17.5 h large direct two-tank molten salt storage. The CSP part is complemented with a 110 MW_{el} flat-plate photovoltaic field on two side of the heliostats. The power plant is connected to the Northern grid in Chile which is powering many mining projects which need 24/7 power. The targeted annual electricity production should be around 950 GWh, which related to the 110 MW_{el} of the CSP turbine would give a capacity factor of 98%. The fraction of PV generation in this would be about 30%. Another CSP-PV-hybrid project is the huge Noor I DEWA project in Abu Dhabi with 600 MW of parabolic trough, 100 MW of solar tower and 250 MW of PV, which is under construction.[2]

[1] https://solarpaces.nrel.gov/project/atacama-I-cerro-dominador-110mw-csp-100-mw-pv.

[2] https://www.solarpaces.org/shanghai-electric-announces-thermal-storage-testing-milestone-at-the-700-mw-dewa-csp/.

Generally it has been shown that CSP power plants with TES and hybridized with PV may reach between 80 and 90% CF at much lower $LCoE$ compared with CSP only plants even with linearly concentrating collectors (Giaconia & Grena, 2021; Platzer, 2014, 2015).

The advantage of smaller volume TES for the same storage capacity, of course, currently can only be utilized with molten salt as HTF (DMS = direct molten salt), for example using the DMS Linear Fresnel system or in a DMS solar tower.

Apart from hybridizing CSP with an additional PV power plant, also more integrated approaches are being discussed due to the spectrally selective absorption of PV cells which can be placed in CSP optical components, and thus convert solar radiation only partially direct into electricity by the photovoltaic effect, using the remaining part of the spectrum for producing heat. These concepts are certainly not yet close to economic feasibility (Ju et al., 2017).

11.3.3 Multi-objective Optimization Approaches

As we have seen, planning power generation systems means considering multiple often conflicting goals. This is especially true for hybrid solar power systems, where many different performance indicators must be considered. It is desirable for hybrid power plants to produce electricity at the most competitive prices, with as little CO_2 emissions as possible. Also, the reduction of water consumption is important in arid areas. These different goals are in conflict with each other. With the currently decreasing but still high costs for solar components and the still relatively cheap fossil fuel available depending on the location, a purely cost-oriented design leads to a low degree of solarization of hybrid power plants and induces high operational CO_2 emissions as a consequence. Conversely, a pure CO_2 emission minimization-based may lead to concepts that may not be competitive. Thus, no optimum can be found for simultaneously achieving all goals.

To investigate the trade-offs that have to be made between the reduced fuel consumption (resulting from the solar heat input) and the additional investment costs for the solar collector equipment, it is therefore, useful to perform a multi-objective optimization. In this way, the various trade-offs between objectives can be explored, and decision makers are presented with a set of solutions from which they can then choose the desired trade-off between the objectives. The optimal trade-off curve is given by the designs that lie on the Pareto-optimal front. A Pareto-optimal design is characterized by the fact that there is no other power plant design that is simultaneously better with respect to all the objectives defined for it, i.e., moving from a Pareto-optimal point to a feasible point away from the Pareto curve leads to a deterioration of at least one objective without an improvement in the others. The Pareto optimal logic of a multi-criteria evaluation scheme is outlined in Fig. 11.43. Details on the evaluation for solar thermal power plants can be found in Spelling et al. (2014) or Peterseim et al. (2013), among others, but are beyond the context of a textbook.

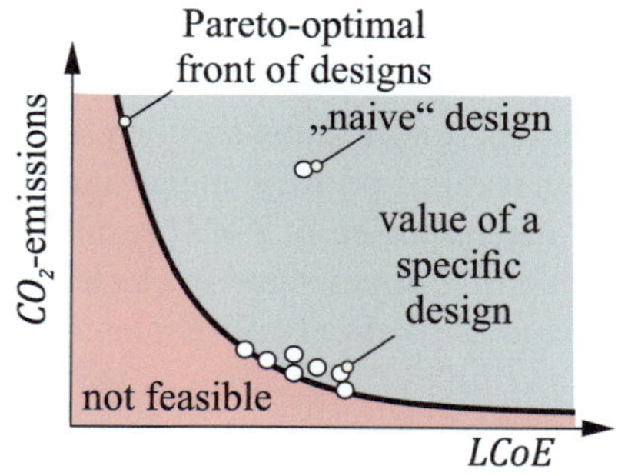

Fig. 11.43 Basic idea of the multicriteria optimization of a hybrid solar thermal power plant based on Pareto logic of two properties CO_2-emission and $LCoE$

As an example of Pareto logic, we assume that a power plant should have both low CO_2 and small $LCoE$. So, we have two properties CO_2 emission and $LCoE$. The larger the CO_2 emission and the larger the $LCoE$, the worse the power plant is. If we draw the pairs of technically feasible design values in the diagram, we get the gray shaded area. For the same CO_2 emissions, the power plant that has a smaller $LCoE$ is better. When the improvement of one target meets the deterioration of the other, the components are no longer Pareto-comparable. With respect to Fig. 11.43 values further to the left and below are Pareto superior to values further to the right or above. All components on the continuous curve thus represent an optimum ("the best") and are called Pareto-optimal. All points below the curve are technically not feasible with the models. A decrease in one value is then only possible if the other increases. On the Pareto curve, "further left" forces "further up"; conversely, "further down" forces "further right".

In addition to the parameters discussed so far, such as CO_2 emissions, $LCoE$, water consumption or land use, other measurable but also socio-commercial factors, which can be weighted, can play a role in an extended target optimization. The cumulative energy demand associated with the manufacturing, construction, and decommissioning (CED_C) phase of the power plant, as well as the cumulative energy demand for operation and maintenance (CED_O), are among the factors that can be weighted when comparing several technologies with each other. It can be seen that the CED_C is somewhat higher in the hybridization scenarios (mostly only a few %) than in the solar-only scenarios, which is mostly due to the purchase of additional combustion equipment to burn the fuel. However, the CED_O results vary significantly depending on the degree of hydridization (SF), since they consider the calorific value of the fired fuel. The parameter that captures the resource consumption of raw materials and the consumption during operation is the so-called Energy Payback Time (EPT) in years, which can be determined as follows:

$$EPT = \frac{CED_C}{\left(\frac{E_{el}}{\eta} - CED_O\right)}, \tag{11.95}$$

in which E_{el} represents the annual net electricity generation of the plant (in $MJ/year$), CED_C the cumulative primary energy demand associated with the extraction, manufacturing, construction, and decommissioning of the CSP plant (in MJ), and CED_O denotes the cumulative primary energy demand associated with operation and maintenance (in $MJ/year$). The average utilization factor of the conversion of primary energy into electricity is represented by η. In this context, the determination of CED results consistently show that hybridization scenarios with a high fossil share in electricity generation show values far more than twice as high as solar-only STE-plants, compare e.g. Corona and San Miguel (2015).

11.4 Conceptual Planning of Solar Power Generation

It is obvious that the objective of building a solar thermal power plant is to generate income with electricity provided to the market. The cost structures discussed above reflect average values in the world market, which might nevertheless vary in regional markets. Local market conditions, import and export taxes, transport costs, competitive options and other factors have an impact on the local pricing. The electricity market is organized in a different way in countries, resulting in very different revenues for a unit, a kWh_{el}. But not only economical conditions change, also the physical, social, political, and financial conditions change drastically. Amongst all the solar resource is very different depending on latitude and climate. Planning a new solar power station in a project development phase needs to answer the questions, where, how, and when to build that plant. The characterizing parameters of the previous chapter are useful for a comparison of alternative options, but a lot of detailed work is behind such an analysis before the integrated key performance numbers can be specified.

11.4.1 Project Phases in the Construction of a Power Plant

When we talk about conceptual planning, this happens in a project development phase where the project is initiated until solar power plant concept, the location, the legal framework, the partners in the project and the project financing scheme has been developed. Only then the work of engineering the details, of procurement and construction can begin. This second phase then is usually followed by a phase of testing the power plant performance in order to check whether guaranteed performance is achieved by the different components and the whole system. This phase is called the warranty period. Figure 11.44 shows the three main project phases, project development, engineering, procurement and construction phase and warranty period for an idealized project.

Within the three different phases many individual tasks are executed, which will be only mentioned here not claiming completeness, as any solar project has to encounter different challenges.

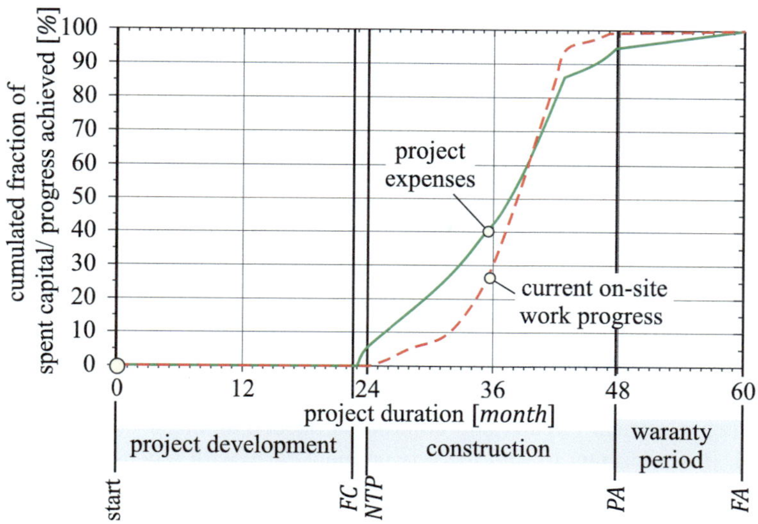

Fig. 11.44 Idealized time development of project expenses and actual work progress on site for a typical solar thermal power plant project (FC = financial close, NTP = note to proceed, PA = provisional acceptance, FA = final acceptance)

Project development phase

After finding a suitable location for a project a prefeasibility study of 1 − 2 years duration has to prepare the basis for a contractual agreement between the project developer, the intended project owner, and a financing party. This agreement, the financial close FC is only reached, when all technical, financial, and legal risks have been satisfactorily answered by the project development in the prefeasibility. Usually the financing institutions ask for a so-called due diligence (DD), which is executed with the help of experts on both sides. The results of the prefeasibility study will be critically reviewed in this process. The topical items dealt with here are

- National and regional authorization proceedings (specific to the site, from construction authorization to environmental impact assessment).
- Bankable solar resource (local solar data to predict power generation of project lifetime).
- Basic conceptual design feasibility study.
- First cost evaluation through tendering.
- Contracts
  - LLA—land lease agreement.
  - PPA—power purchase agreement.
  - FSA—fuel supply agreement (e.g. gas for hybrid operation).
  - WSA—water supply agreement (for cleaning, cooling mainly).
- Special conditions to FC of renewable energy (RE) projects.
- Others, ….

With the financial close the roles within the project consortium are also defined. The project company (the so-called "special vehicle" SPV) is formed with possible participation of project developers, energy providers, technology companies and financing institutions. The idea of the SPV is to limit the risk of such a project for the mother companies. For the next construction phase a suitable company for the realization of the project is needed. This often is the so-called EPC company, an abbreviation for Engineering, Procurement and Construction. Very often the final operation is outsourced to the O&M provider (operation and maintenance).

Project construction phase

The project construction starts with a preparation of the building by groundbreaking, setting up a construction site with infrastructure and earthworks. In parallel the review of the basic engineering and the detailed engineering of heat exchanger and piping, electrical circuits and controls is needed amongst other details to be clarified. It is important to start the procurement early with requests for quotations (RFQ), negotiation of guarantees for components and purchase orders, especially for the key components of the plant (collector assembly, mirrors, receivers, turbine etc.). Delivery times otherwise may delay a project which would make the project costlier. The local manufacturing and assembly processes have to be defined and the quality of delivered components be assessed. The different sectors of constructions from civil to mechanical to electrical works has to be coordinated, and after an installation of sensors, instrumentation and controls the operation of the individual parts of the plant has to be verified after finalization of each part. This is the pre-commissioning of components and subsystems. When all subsystems are working according to specification, the commissioning of the whole plant with start-up procedures and control procedures defined can be initiated. After a successful commissioning the EPC is ready to transfer the plant to the owner and O&M company after a provisional acceptance (PA) of the owner.

Warranty phase

Due to the specific complexity of solar projects where the sun and ambient conditions naturally have a large impact on the performance, it is not possibly to determine in the commissioning phase whether the targeted and agreed performance can be fulfilled. Seasonal variations lead to an uncertainty in the prediction of annual generation which is not acceptable by the owner. Also, smaller problems with can be easily remedied may turn up during different operation states. Therefore, a warranty period of usually a complete year is defined where malfunctioning in certain contexts can be remedied, and where the seasonal effects are becoming unimportant due to the performance monitoring of a full year. With the available measured DNI and the annual performance—using accepted performance models—the correct functionality of the complete power plant may be stated then, leading to a final acceptance. After that only guarantee cases defined in the purchase contracts will be accepted by the technology provider companies and the EPC company.

The acceptance testing is a costly and long-lasting effort. The models and procedures have to be agreed and described in contracts between the parties. The international standards organization therefore tries to harmonize these approaches for a faster negotiation and execution. Standards relating to linear solar collectors exist, and standards for field acceptance testing and plant acceptance are in preparation (see the website of the International Electric Commission IEC https://www.iec.ch and check for the technical committee TC 117 on Solar thermal electric plants to see the different publications and projects).

11.4.2 Location Analysis

The very first step in a planning phase within the project development is the identification and subsequent analysis of potential locations for a solar power plant. Here, most projects make use of a top-down approach, as depicted in Fig. 11.45

With very general information an investor or project developer will select a suitable macro region (for example Southern Spain, Australia, or South Africa) based on information on the general country conditions, such as:

- Is there a need for additional power and, what are the regulations of the energy market and access?
- Is the political and economic situation stable and profitable?
- Are climate and solar irradiation levels suitable for a solar power plant?

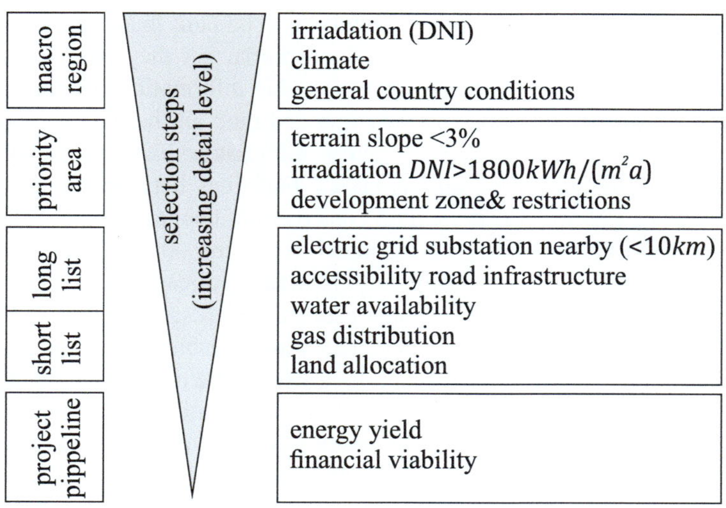

Fig. 11.45 Top-down approach for site selection

Within the macro region geographically the values of parameters influencing the decision will vary. Ideally rather flat terrain eases the construction process of industrial building in general, and of solar power plants especially. Mountainous areas therefore often will be excluded. Similarly, other geographical factors lead to an exclusion of areas. Territories as urban development zones, valuable agricultural land, lakes, rivers, and forests are too valuable in general to accommodate large solar fields with dimensions of several square kilometres. Natural protection parks, water areas of lakes and rivers or wildlife parks similarly have to be excluded like on the other hand archeologically sites or military zones. Deserts with wandering dunes pose a risk for an existing solar field, and the ground should be stable which leads to an exclusion of volcanic activities or swamps. Apart from the exclusions a positive solar yield should be possible. Generally, a DNI level of 1800 $\text{kWh}/\left(\text{m}^2\text{a}\right)$ is often considered as a limit for the feasibility of a CSP project. However, this number is only an indication, and it is dependent on the developing cost reductions of the technology, on demands and revenues to be found on the country's electricity market. With all these consideration and exclusion areas the project development will arrive hopefully at certain priority regions, which may be sufficiently large spots to place a large power plant on them.

The first longlist of possible project location then develops considering preferential conditions of geography, which influence the project cost wise and thus economically in a positive way. No specific design is needed yet at this point. Any design of a power station to be built would need low property prices, access to water, electricity for consumption and a substation to feed in the generated power afterwards. When hybrid co-firing is intended access to natural gas or other fuel is required. As well for the construction process as for the operation later transport is needed, therefore roads in generally have to be built to access the location. The criteria evaluated here are no exclusion criteria, but they pose costs which are dependent on area and distances. A geographic information system (GIS, see also Chap. 2) is a suitable tool for evaluating and rank all the priority regions. A GIS tool uses a program where geographical information like topography, roads, power lines can be defined in various layers. Functions between the geographic variables can be defined. Costs of road construction, connecting electricity lines or water and gas pipelines may be evaluated depending on the exact location of the project and the existing infrastructure. Nowadays already first estimations of locally resolved DNI irradiation is also available from satellite data evaluations (see Chap. 2). By map overlaying areas which are suitable for the installation of large concentrating solar power plants a long list can be established, as indicated in Fig. 11.46. Combined with a simple conceptual evaluation already revenues can be attributed to the different sites. The ranking of costs and benefits then results in an ordered list of project locations, a long list (hopefully).

To reduce the effort of adapting the conceptual design of the power plant to all these locations a short list of the top projects in the longlist is taken. Designs are adapted to DNI conditions and other restrictions like area shape available. Using yield simulation, the possible energy yield of a power plant at these locations is determined. Costs are estimated depending on the optimized design (in

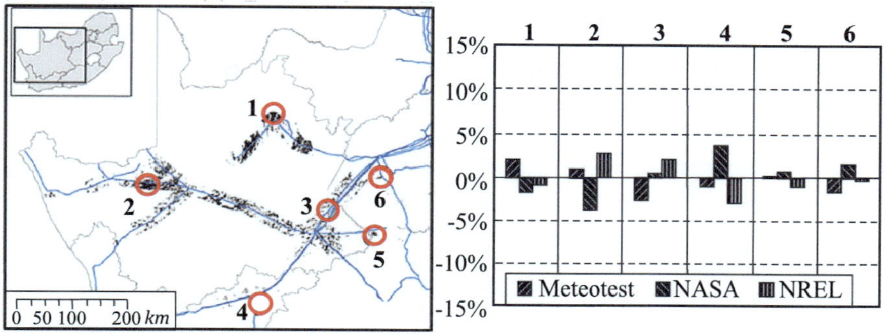

Fig. 11.46 Using a GIS-tool to evaluate six different locations in the Western Cape province in South Africa with respect to solar resource (using 3 different data bases Meteotest, NASA NREL) and access to water for cooling purposes after Fluri (2009)

terms of SM of the solar field and FLH capacity of the thermal storage). And with a first financial analysis using the conceptual design the cost–benefit analysis for each location can be made. The ranking which not necessarily is the same as on the short list then defines a project pipeline for this macro region. Of course, the ordering might change in the further development, as possibly additional view-points or information turns up, changing the overall performance of the project at the location. Several projects had to be postponed or even cancelled because of environmental impacts which could not have been foreseen in the first place.

11.4.3 Conceptual Design

A conceptual design is based on a decision which type of technology may be used in the project. First a power plant scheme has to be selected where collector technology with optics and heat transfer fluid HTF, storage type and integration, power cycle and capacity, hybridization options have to be defined. The next step is a dimensioning of the various components in order to reach a minimum cost of electricity competitive in the market. Due to the special requirements of solar projects not only an assessment at design point is sufficient, but a full annual yield simulation on an hourly basis is required to optimize the design. The solar field size (SM) and the thermal storage capacity in FLH can be varied in such an optimization phase, but also temperature levels and mass flows of heat transfer media. A techno-economic optimization requires in this stage proper cost estimates of the different components. For the simulation sometimes, technology companies develop their own programs, other rely on public available or commercial tools in the market. As public domain tools, two can be mentioned which have already a lot of systems predefined and components parameterized in a component data base: The GREENIUS tool from DLR (Germany) to be found on https://www.dlr.de/ and the System Advisor Model SAM from NREL (United States) to be found on https://sam.nrel.gov/.

Conceptual design is not a detailed plan how to build a plant. It does not consider important engineering tasks necessary for an adequate functioning of the plant. Piping dimensions, thermal insulation levels, heat exchanger dimensioning and heat transfer analysis, selection of flow regimes, selection of hydraulic circuits topography are examples where the detailed engineering still has to give answers. Also, the controls are idealized and simplified in a conceptual design. The real operation of a controller, the placement of sensors and the interaction with real materials is not included. Although the programs mentioned above include some variants for example of collectors, many specifications of individual components like drives or pumps have to be developed in a later phase, and moreover, suitable components be sourced in the world market.

11.4.4 Economic Performance Simulation

The economic performance indicator which is usually taken as a target for optimizing a plant design with respect to the component dimensions, is the $LCoE$ which incorporates investment $CAPEX$ and operation cost $OPEX$. Also financing, assurance and engineering costs can be included. This means that all these costs have to be estimated in a rational and up-to-date way to achieve designs adequate for the project, and which are accepted by the lender's engineers, the experts on the financing side, as sound and realistic. The resulting $LCoE$ then can be compared with possible competitive technologies or market prices. When extra revenues have to be included in the analysis, say public support measures (e.g. feed-in tariffs) or partial funding due to a riskier technology to be tested, a complete cash-flow analysis should be set up. It is important to understand that the value of money may be different depending on the year of the project operation. If a revenue or a cost occurs now, the value is much higher than when it occurs after 10 or 20 years. For an investor a Euro or Dollar spent now is lost for other profitable business. So, a Euro now has more value than a Euro in 10 years, because in these 10 years it alternatively could have been invested with a certain profit. This expected minimum profit is described by the assumed discount rate d. The money is discounted by a factor by

$$\frac{1}{(1+i)^t}, \tag{11.96}$$

where i is the discount rate and t the year to be considered in future. For the present year $t = 0$. For a complete analysis of all expenses and revenues the summation of the total financial project period (for power plants 25–30 years) is being done. The resulting sum is the net present value NPV of the project:

$$NPV = \frac{\sum_{t=0}^{t=N} E_t - I_t - A_t}{(1+i)^t} + \frac{\mathrm{R}}{(1+i)^N}, \tag{11.97}$$

where E_t are the revenues, I_t the investment cost, A_t the O&M costs for year t and R the residual value at the end of the analysis period after N years. In this version of the formula inflation is not included. The present value therefore should not be confused with the decreasing value of money due to inflation. Taxes as well as tax reductions also can be included in the simulation of the financial flows. The mixture of own capital, preferential loans (e.g. from development banks) and actual bank loans is only useful, when the exact conditions and the specific project are known.

The different business cases for financial models vary and may give rather different results. A commercial power purchase agreement (PPA) can be concluded between a plant owner and a single large consumer. In a utility contract different prices for certain time periods (time-of-delivery TOD schedules) can be defined, reflecting the variable demand in the grid. Many parameters of these financial modelling are found and described in the SAM environment.

11.4.5 Feasibility and Performance Assessment

Feasibility of a project is considered when based on technical, legal, financial, and economic reasons the benefit for the project owner seems small enough to justify the risks of the projects. Risks are often judged in a very subjective way. Unknown technologies are obviously considered quite risky, considering that the promised revenues and operational costs might be estimated too optimistic. The existence of reference projects for solar power plant technologies is therefore very important. Preferential funding or entrepreneurship of the technology companies is needed to build the first kind of a power plant design or to use a new component. Such projects are not feasible and rather called pilot projects (in a very early phase of development with operation only proven in a laboratory environment) or demonstration project (when the demonstration should show the scalability in a real environment). For large projects the requirement of feasibility is crucial. Very often banks are part of the financing. Therefore, a feasible project in this sense is also called bankable.

The performance assessment in terms of technical and economic performance is done in a techno-economical assessment or optimization. The future performance of a solar project especially has therefore been based on sound predictions of the solar resource based on a long-term experience. This question is being dealt with in Chap. 2. On the other hand, also, durability and performance degradation have to be assessed. Operational procedures have to be defined so that environmental deterioration for example be soiling of the transparent and reflecting parts of the optical concentrator is kept at a minimum. So, washing procedures leading to operational costs have to be established and considered in a project.

11.5 Non-concentrating Power Plants

Compared to concentrating power plants, non-concentrating power plants, such as solar chimney power plants or solar ponds, are characterized by a small volumetric power density in the collector area, so that their total output is directly coupled to the solar irradiation and the absolute output of a power plant is thus rather low. This makes them suitable for stand-alone operation under given site-specific conditions, and they require only a small amount of maintenance and operation compared to other STE, which, however, has to be compensated by a correspondingly high land consumption per kWh_{el} generated.

11.5.1 Updraft Power Plants

11.5.1.1 Functional Principle

The solar updraft power plant is not a solar-concentrating high-temperature system, but it is an option to convert low temperature solar heat into electrical energy. The atmospheric thermal updraft power plant, also called solar chimney, combines three physical principles:

- greenhouse effect,
- chimney draft and
- conversion of air motion into a mechanical rotation within a turbine to drive a generator.

It essentially consists of three components, a glass roof collector, a chimney, and a wind turbine. Figure 11.47 illustrates the principle of operation. Updraft power plants for electrical power generation were already considered by Günther (1931). The proposal of the French meteorologist Bernard Dubos from 1929, illustrated in Fig. 11.47 dates even earlier (Dubos, 1954). He proposed to install 1.000–1.500 m high risers with a diameter of 10 m on the steep slopes of the Atlas chain or the Hoggar mountains in North Africa. The tubes, made of heat-insulating lightweight concrete, were to allow the air heated under the glass roofs to rise upwards and drive the blades of wind turbines. The plan was to feed the electricity generated into the local infrastructure and the European power grid.

In the updraft power plant, the incident solar radiation heats the air under a collector roof covered with glass or foil, which spreads out in a circle around the chimney standing in the middle. The air, heated only slightly by approx. 15–20 °C, flows to the chimney due to its lower density and rises upwards. The actual warm air collector is formed by the open glass roof together with the floor underneath, which should have a high absorptivity to increase efficiency. The glass roof is

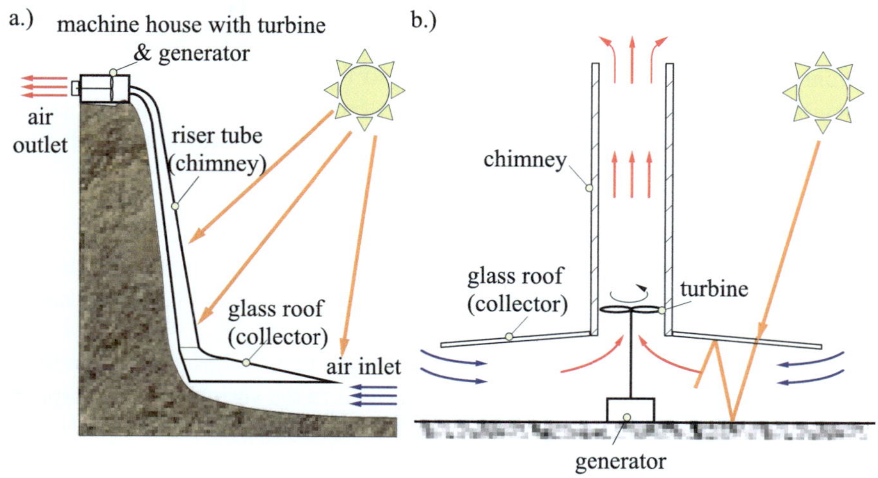

Fig. 11.47 Functional principle of the updraft power plant according to Dubos 1929 (Dubos, 1954) and **b** classical updraft power plant concept modified from Weinrebe (2004)

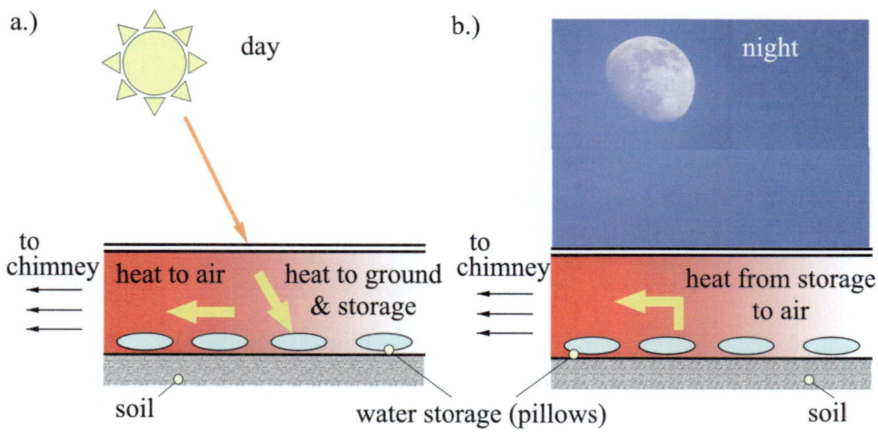

Fig. 11.48 Principle of heat storage in an updraft power plant with schematic representation of the heat flows in daytime operation (**a**) and at night (**b**) modified from Kreetz (1997)

connected to the base of the chimney in an airtight manner and the buoyancy creates a chimney suction that draws warm air out of the collector and allows cold air to flow in from outside. This ensures a continuous updraft in the chimney. The kinetic energy contained in the air flow is converted into mechanical energy with the help of pressure-staged wind turbines and into electrical energy via generators.

Continuous 24 h operation can be realised with updraft power plants (Schlaich & Schiel, 2001) by water-filled hoses or bags installed at the bottom of the warm air collector, which heat up during the day and transfer the stored thermal energy back to the air at night, as shown in Fig. 11.48. The night operation is enhanced by cooler ambient temperatures. The water containers used for this only need to be filled once and there is no further need for water, which is a significant advantage over concentrating systems that use conventional power plant technology. This method is particularly attractive in arid areas with little water, such as deserts.

11.5.1.2 Power and Efficiency of a Chimney Power Plant

The essential relationships and scaling laws of updraft power plants are derived in the following; a more detailed analysis can be found in dos Santos Bernardes et al. (2003).

The air heated in the collector has a lower density than that of the surroundings. The lighter column of air in the tower is connected to the surrounding atmosphere at the base of the tower (collector exit) and at the top of the tower. Thus, the density difference creates a buoyancy flow, as a result of which air is sucked in again from the surroundings, which is called the stack effect. The total pressure difference Δp_{tot} between the tower vessel and the environment is calculated according to:

$$\Delta p_{tot} = g \cdot \int_{z=0}^{H_t} [\rho_{amb} - \rho_t(z)] \cdot dz, \tag{11.98}$$

where ρ_{amb} and $\rho_t(z)$ are the density of the environment and the air in the tower at height z, respectively. H_t is the height of the tower and g the acceleration due to gravity. The total pressure difference in the tower is composed additively of the static pressure Δp_{stat}, the dynamic pressure Δp_{dyn} and the viscous friction pressure drop in the stack Δp_{visc}:

$$\Delta p_{tot} = \Delta p_{stat} + \Delta p_{dyn} + \Delta p_{visc}. \tag{11.99}$$

Due to the relatively low flow velocities (<20 m/s) and the large tower diameters, the viscous losses of the flow through the tower walls are negligible, it is $\Delta p_{visc} \approx 0$. The actual division into the static and dynamic components of the pressure ultimately depends on how much energy the turbine extracts from the flow. Without a turbine, the maximum flow velocity $v_{t,max}$ is reached and the entire pressure difference is converted into kinetic energy, i.e. the flow is accelerated. Assuming that the temperature profiles in the tower and the surroundings are parallel (this corresponds to a constant heat loss along the stack height), the lift flow velocity can be expressed with the help of a temperature-modified Torricelli equation.

$$v_{t,max} = \sqrt{2 \cdot g \cdot H_t \cdot \frac{\Delta T}{T_{amb}(z=0)}}, \tag{11.100}$$

where ΔT is the temperature difference between the temperature at the collector outlet (mean tower inlet temperature) and the ambient temperature. $T_{amb}(z=0)$

is the ambient temperature at ground level. The power P_{tot} contained in the stack updraft results from:

$$P_{tot} = \dot{V} \cdot \Delta p_{tot} = \Delta p_{tot} \cdot A_t \cdot v_t, \tag{11.101}$$

with A_t the cross-sectional area of the tower and v_t the mean buoyancy velocity of the air in the tower (chimney). If the total pressure difference is converted into kinetic energy, the result is:

$$P_{tot} = \frac{1}{2}\dot{m} \cdot v_{t,max}^2. \tag{11.102}$$

However, since the static pressure is reduced in the pressure difference provided by the updraft from Eq. (11.98) at a pressure-stepped turbine and this has a technical efficiency η_{turb}, the power that can be converted in real terms is significantly lower.

To derive the scaling, an efficiency analysis is first carried out. The net power P_{el} of the solar chimney power plant supplied to the grid is composed of the solar irradiation power provided at the collector surface and the efficiencies of the collector, the tower, the turbine, and the generator in the following way:

$$P_{el} = \dot{Q}_{Solar} \cdot \eta_{tot} = \dot{Q}_{Solar} \cdot \eta_{col} \cdot \eta_t \cdot \eta_{turb} \cdot \eta_{gen}, \tag{11.103}$$

where the solar irradiation power $\dot{Q}_{Solar}$ results from the product of the global irradiation density DNI and the collector area A_{col}. The efficiencies of the collector, the tower, the turbine, and the generator are given by η_{col}, η_t, η_{turb} and η_{gen}. The tower efficiency η_t is a new parameter and results from the quotient of the updraft power P_{tot} and the thermal power of the tower:

$$\eta_t = \frac{P_{ges}}{\dot{Q}}. \tag{11.104}$$

The heat output of the tower is given by $\dot{Q} = \dot{m} c_p \Delta T$, where ΔT is the temperature difference between tower inlet and outlet. The maximum tower efficiency is obtained by substituting the maximum flow velocity (Eq. 11.100) into P_{tot} (Eq. 11.102) and yields for the maximum tower efficiency η_t.

$$\eta_t = \frac{\dot{m} \cdot v_{t,\max}^2}{2 \cdot \dot{m} \cdot c_p \cdot \Delta T} = \frac{g \cdot H_t}{c_p \cdot T_{amb}(z = 0)}. \tag{11.105}$$

Thus, the tower efficiency increases linearly with the height of the tower. For example, a tower height H_t of 1000 m and an ambient temperature of 20 °C ($T = 293.15$ K) results in a tower efficiency of $\eta_t = 3.35\% (c_{p,air} = 1$ kJ/(kgK)). Assuming a collector efficiency $\eta_{col} = 0.7$, a turbine efficiency $\eta_{turb} = 0.8$ and an ideal generator ($\eta_{gen} = 1$), the total efficiency η_{tot} is in the order of approx.

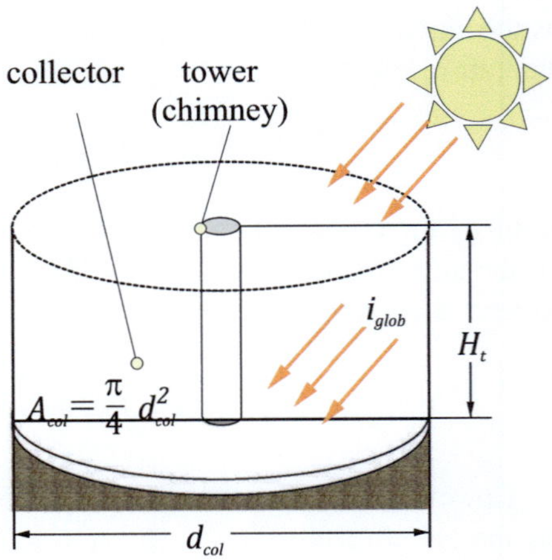

Fig. 11.49 Schematic sketch of relation of output power P_{el} of an updraft power plant to the collector area A_{col} and the tower height H_t

1.88%. For a 100 MW power plant, an area of 5.5 km^2 would be required. Equations (11.103) and (11.105) show that the output of the solar chimney power plant is proportional to the collector area and the tower height. Thus, the electrical output power P_{el} is proportional to the volume of the cylinder shown in Fig. 11.49 and the tower height H_t.

11.5.1.3 Components of the Updraft Power Plant

Collector and storage

The warm air of an updraft power plant is generated in a conventional air collector whose sun-facing side consists of a translucent glass or plastic roof located 2–6 m above the ground, depending on the power plant's output. In addition to translucency, the collector roof must be durable and cost-effective. In the case of the prototype power plant in Manzanares, Spain, glass (cf. Fig. 11.50) proved to be the most resistant material, withstanding even heavy storms, and also proving to be self-cleaning due to occasional rain showers. The translucent roof is permeable to solar radiation, but impermeable to the long-wave thermal radiation emitted by the heated collector floor. As a result, the floor under the roof heats up considerably and releases this heat to the air flowing radially from the outside towards the tower.

Although plastic membranes are cheaper, they become brittle over time and lose their transparency. In addition, plastics are transparent in the infrared wavelength range. However, they allow for easy drainage in case of rain or condensation and were also used for the prototype in Manzanares. For this purpose, the plastic membranes were clamped into profiles and braced in the middle to the ground with a plastic plate with a drainage opening (cf. Fig. 11.51). The height of the air collector increases towards the tower. This ensures that the flow velocity does not increase too much and thus friction losses remain low. Low flow velocities also mean low deflection losses into the tower. The growth under the transparent cover shown

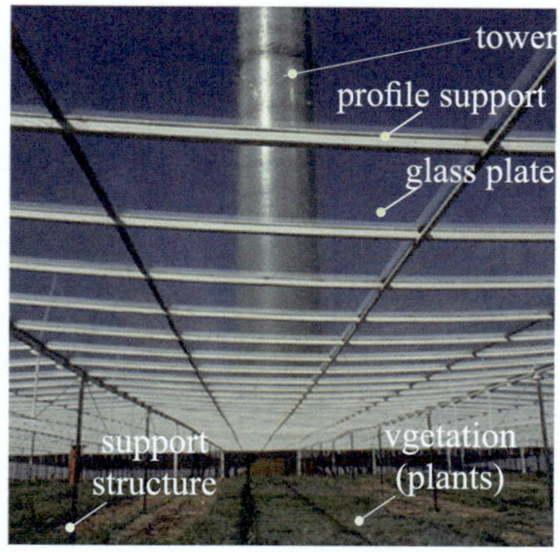

Fig. 11.50 Prototype glass air collector in Manzanares, Spain (Weinrebe, 2004)

in Fig. 11.50 was not pursued further because the plants release moisture. Water vapour is lighter than air at the same pressure and temperature and would therefore generate more buoyancy. However, more heat must be used for evaporation, so the buoyancy of the heated dry air is greater for the same heat input.

With the help of water-filled storage tanks, solar irradiation fluctuations can be compensated, and, in addition, partial night operation can be realised. These storage tanks can be water hoses or cushions that are laid out on the collector floor. The performance characteristics of the heat accumulator can be controlled via the average height of the water depth in the storage reservoirs, as shown in Fig. 11.52. Due to their large heat capacity, large water depths mean a slow temperature increase during the heating phase (Chap. 5) but long storage runtimes during de-storage. Shallow water depths, on the other hand, provide a high relative useful capacity for a short time, but only have a short storage time. In most

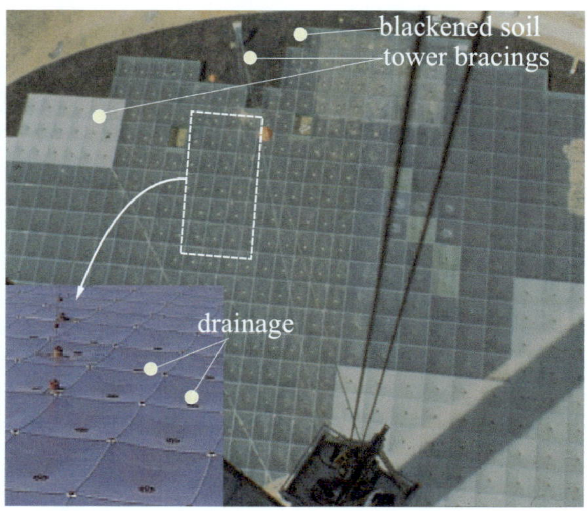

Fig. 11.51 Design of the collector roof of an updraft power plant with a plastic membrane

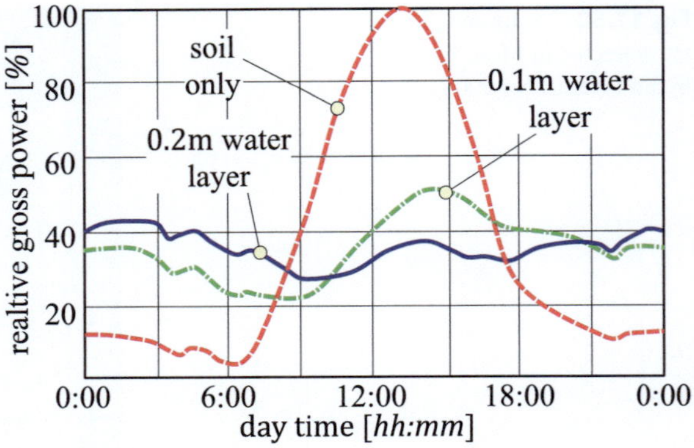

Fig. 11.52 Influence of heat-storing water tubes under the collector cover on the diurnal variation of the power supply of an updraft power plant (Weinrebe, 2004)

cases, the amount of water in the hoses is chosen to correspond to an average water depth of 5–20 cm. Since the heat transfer between the tubes and the water by natural convection is much greater than between the radiation-absorbing surface of the ground and the underlying soil layers, even at minimum water flow rates, and since the heat capacity of the water is about five times higher than that of the ground, the water in the tubes stores part of the solar heat and only releases it again at night when the air temperatures in the collector are lower. In principle, this allows the system to operate continuously.

Tower (chimney) of the updraft power plant

The tower stack constitutes the heat engine of the updraft power plant. The lift of the air heated in the collector is, in a first approximation, proportional to the temperature increase ΔT of the air achieved in the collector (cf. Eq. 11.98) and the tower volume, which is the product of the height and the diameter. In a large updraft power plant, the collector raises the temperature of the outside air by up to 30 K and, as a result, updraft velocities of the order of 15 m/s occur in the stack, which also allow maintenance during operation. There are two main construction methods for high towers or chimneys: Steel truss or cable net tower—clad with trapezoidal sheeting or membranes. Several analyses showed that for potentially interesting sites, the reinforced concrete tube is the most durable and cost-effective solution.

In principle, the tower of an updraft power plant is nothing other than a large natural draught cooling tower. Large towers are complex structures in which a multitude of loads must be considered. The main static loads are the stress distributions caused by the dead weight and the thermal loads on the inside and caused by solar radiation. In operation, for example, a rotationally symmetrical temperature difference $\Delta T = -45$ K must be assumed at the base in winter when designing natural draught chimneys. Due to the solar radiation, a non-rotationally symmetrical effective temperature difference between the shell surfaces of $\Delta T = +25$ K must be assumed when the tower is at rest (Lenz & Meiswinkel, 2005).

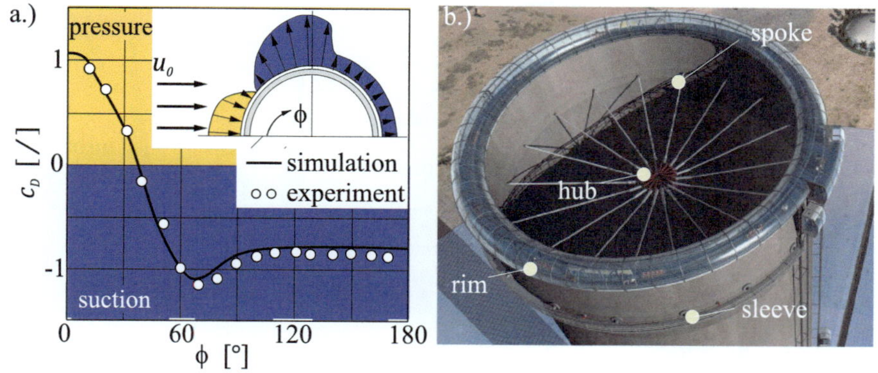

Fig. 11.53 **a** Calculated distribution of the drag coefficient in the flow around a circular cylinder (Ma et al., 2000); **b** Spoked wheel with spokes made of strand bundles at the top of the tower of a 5 MW solar chimney power plant (Weinrebe, 2004)

Thin tubes deform under wind load to an oval cross-section, especially at the suction flanks (cf. Fig. 11.53a). As a result, the meridian stresses become very high, and the tower can be deformed oval. The ovalisation can be effectively counteracted similar to a bicycle rim by inserting spokes (cf. Fig. 11.53b). Spokes have a similar effect to bulkheads but offer less resistance to the updraft.

An elastic vertical support of the tower on the ground leads to a significant reduction in the frequency of the vibrational movement of a high tower and there in particular of the circumferential waves most damaging to the structure. Seismic excitation or even wind loads can generate vibrations that lead to deformations of the tower.

Wind turbine

Turbines are used to extract mechanical energy from the air flow. The turbines of an updraft power plant do not operate in a speed-stepped manner but mostly as a jacketed, pressure-stepped wind turbine generator set, in which the static pressure is reduced in a similar way to a hydroelectric power plant. In principle, two construction methods of an updraft turbine are conceivable, a completely axial construction, which is more conceivable for turbines with a large tower height, or an axial radial construction. Figure 11.54a and b illustrate both designs.

In order to calculate the flow in turbomachines, it is necessary to consider the relative system rotating with the impeller in addition to the consideration from the space-fixed absolute system. The absolute velocity c is composed vector-additively of the circumferential velocity u and the relative velocity w measured in the rotating reference system.

$$\vec{c} = \vec{u} + \vec{w} \text{ and without vector indication } c = u + w. \tag{11.106}$$

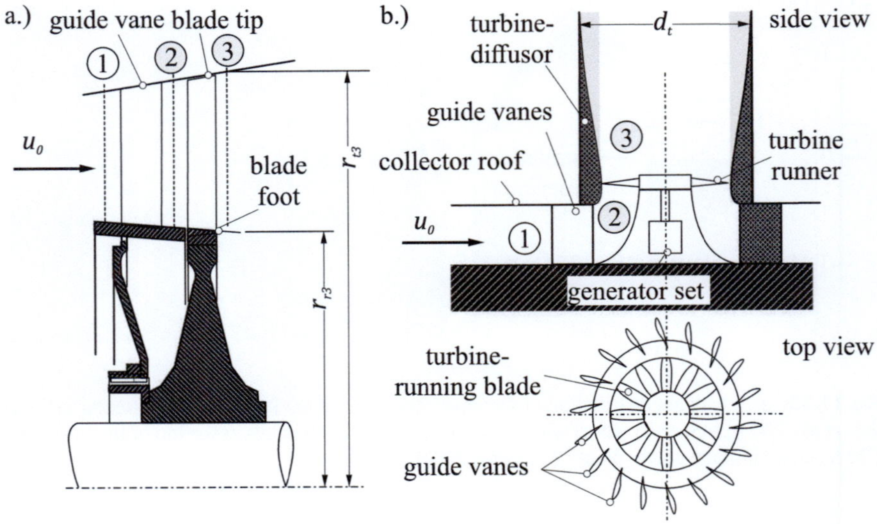

Fig. 11.54 Guide vane and rotor design of turbines for power generation in updraft power plants. **a** Pure axial turbine and **b** radial-axial arrangement of the turbine

The turbomachine main equation is derived for stationary flows, which does not quite do justice to the fact that the rotating blades brush past stationary components such as the stator blades or the housing tongue but is confirmed by practical experience. The derivation implies that the flow should follow the blade contour exactly, assuming the blades to be infinitely thin and infinite in number (so-called blade congruence). All moments are referred to the axis of rotation and, in addition, the moment at the end surfaces is supposed to be zero. Losses due to gap currents are not considered, nor are frictional influences.

The derivation is carried out under the nomenclature of the velocity triangles of the different turbine designs as shown in Fig. 11.55. As a result of the simplifications, the angular momentum theorem yields the following result with regard to the torque M:

$$M = \dot{m} \cdot (r_3 \cdot c_{3u} - r_2 \cdot c_{2u}), \tag{11.107}$$

where the index "u" indicates the circumferential component. The power P at the shaft is the product of torque and angular velocity Ω:

$$P = M \cdot \Omega = \dot{m} \cdot (r_3 \cdot c_{3u} - r_2 \cdot c_{2u}). \tag{11.108}$$

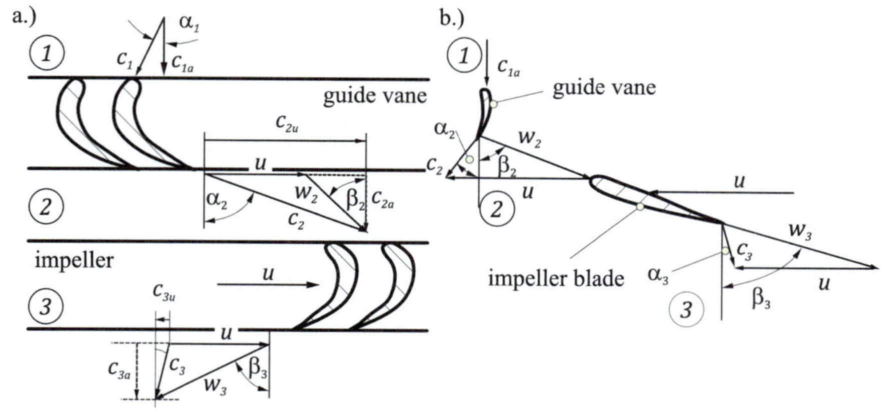

Fig. 11.55 Velocity triangles on turbine guide vanes and impellers. **a** Axial turbine; **b** radial-axial turbine. Index "*a*" denotes the axial component

Often only the specific stage work y is given, which results from the quotient of shaft power and mass flow, and the result is Euler's equation for fluid machines with:

$$y = \frac{P}{\dot{m}} = (u_2 \cdot c_{2u} - u_3 \cdot c_{3u}). \tag{11.109}$$

This Eq. (11.109) is valid for

- compressible and incompressible flows,
- with and without friction of the flow in the blade grid,

and thus, also for hydraulic and thermal fluid flow machines. By transforming it, an illustrative interpretation is possible; variously, the technical literature also speaks of the application of the cosine theorem.

$$y = \underbrace{\frac{c_2^2 - c_3^2}{2}}_{1} + \underbrace{\frac{u_2^2 - u_3^2}{2}}_{2} - \underbrace{\frac{w_2^2 - w_3^2}{2}}_{3}. \tag{11.110}$$

The individual fraction can be interpreted in the following way:

(1) change in the specific kinetic energy in the absolute system, which occurs in the turbine due to the deceleration of the flow.
(2) difference in specific kinetic energy resulting from the change in centrifugal forces at the inlet and outlet.
(3) change of the specific kinetic energy in the relative system, which indicates the specific acceleration work of the flow in a turbine.

For the specific stage work y in the case of incompressible flow, it follows from the Bernoulli equation with energy input and losses assuming the same geodetic height:

$$y = \frac{\Delta p}{\rho} = \underbrace{\frac{c_3^2 - c_2^2}{2} + \underbrace{\frac{p_3 - p_2}{\rho}}_{\text{static pressure change}}}_{\text{total pressure change}} + \frac{\Delta p_{\text{loss}}}{\rho}, \qquad (11.111)$$

in which ρ denotes the density. For compressible flows and assuming equal geodetic heights, the specific step work y corresponds to the total enthalpy change Δh_t in the following way

$$y = \Delta h_t = \underbrace{\frac{c_3^2 - c_2^2}{2} + \underbrace{h_3 - h_2}_{\text{static enthalpy change}}}_{\text{change of total enthalpy}} . \qquad (11.112)$$

In turbomachinery design, characterisation is often carried out with the aid of dimensionless numbers. The dimensional analysis yields the flow coefficient φ for a dimensionless volume flow and the load coefficient ψ for the dimensionless pressure.

$$\varphi = \frac{c_a}{u}, \quad \psi = \frac{\Delta p/\rho}{u^2/2} = \frac{2 \cdot y}{\pi^2 \cdot n^2 \cdot D^2}, \qquad (11.113)$$

where ρ is the density of the fluid, n. the speed of the turbine and D the mean diameter.

The derived dimensionless numbers can now be used to design a turbine. The gas flows into the turbine with the absolute velocity c_1, the static pressure p_1 and the absolute temperature T_1. in the guide vane ring. In the guide vane, expansion of the gas to the state p_2. and T_2 usually occurs, with the gas leaving the guide vane channels at the velocity $c_2 > c_1$. at the angle α_2.. The rotor blade should—in order to avoid shock losses—correspond to the inlet angle β_2. After deflection and usually subjected to further expansion, the gas leaves the rotor blade channels with p_3 and T_3 at the absolute velocity c_3 at angle β_3 relative to the blade, in absolute terms at angle α_3. The angle α_3 is called the swi angle. In a single-stage turbine, as in updraft power plants, an attempt is made to allow the gas to flow into the stage without swirl, $\alpha_1 \approx 0°$, as well as to allow the gas to flow out of the stage without swirl, i.e. $\alpha_3 \approx 0°$. The expression $c_{2u} + c_{3u}$ represents the specific swirl change of the gas flowing through the stage. This gives the effective output torque of the stage itself. The specific power is obtained from the main turbine equation:

$$y = \vec{u} \cdot \vec{c}_2 - \vec{u} \cdot \vec{c}_3 = u \cdot (c_{2u} + c_{3u}). \qquad (11.114)$$

The change in the axial velocity component $c_{2a} - c_{3a}$ causes an axial thrust load on the rotor, which is partially compensated by the pressure difference in front of and behind the rotor.

In the further considerations, the axial velocity component in the stage is to be regarded as constant, i.e. $c_{1a} = c_{2a} = c_{3a} = c_a$. Thus, the specific stage work (y) can also be given in the form of the blade geometry and as a result one obtains:

$$y = u \cdot c_a \cdot (\tan \beta_2 + \tan \beta_3). \tag{11.115}$$

The efficiency of the turbine η_{turb} results from the real enthalpy difference across the turbine compared to the ideal enthalpy difference. If a loss enthalpy ΔH_{loss} is introduced, the turbine efficiency η_{turb} results in:

$$\eta_{turb} = \frac{\Delta H}{\Delta H + \Delta H_{loss}} = \left(1 + \frac{\Delta H_{loss}}{\Delta H}\right)^{-1}. \tag{11.116}$$

In the case of low enthalpy losses ($\Delta H_{loss}/\Delta H \ll 1$), the previous equation can be represented in the form of a total pressure loss and an average density and results in Lewis et al. (1996):

$$\eta_{turb} = \left(1 + \frac{\Delta p_{turb,loss}/\rho}{\Delta H}\right)^{-1}. \tag{11.117}$$

The total pressure loss in the turbine $\Delta p_{turb,loss}$ is composed of the losses at the stator (index "s"), the rotor ("r") and the diffuser ("d"), whose loss coefficients (ς_i) result from the relative velocity components and the outlet density as follows:

$$\zeta_s = \frac{\Delta p_s}{(1/2) \cdot \rho \cdot c_2^2}; \ \zeta_r = \frac{\Delta p_r}{(1/2) \cdot \rho \cdot w_3^2} \text{ and } \zeta_d = \frac{\Delta p_d}{(1/2) \cdot \rho \cdot c_3^2}. \tag{11.118}$$

Replacing the enthalpy reduction by the pressure number ψ yields the turbine efficiency:

$$\eta_{tur} = \left[1 + \underbrace{\left(\zeta_s \cdot \frac{c_2^2}{u^2} + \zeta_r \cdot \frac{w_2^2}{u^2} + \zeta_d \frac{c_2^2}{u^2}\right)/\psi}_{L}\right]^{-1} = (1 + L)^{-1}, \tag{11.119}$$

in which L represents the mass flow-specific proportion of the total energy that is no longer available at the turbine diffuser outlet.

Another important parameter in the design of turbines is the degree of reaction r. The kinematic degree of reaction describes the ratio of the enthalpy converted in the rotor blades of a turbine stage (into flow energy and mechanical work) to the total (isentropic) enthalpy gradient of a turbine stage. In an ideal axial turbine blade, the relative total enthalpy over the rotor is maintained. It is given by $H_r = h_r + w^2/2$. Thus, the static enthalpy difference over the rotor blades is given by:

$$h_r = \frac{(w_3^2 - w_2^2)}{2} = \frac{(c_a^2 + w_{u3}^2 - c_a^2 - w_{u2}^2)}{2} = \frac{(w_{u3} - w_{u2})(w_{u3} + w_{u2})}{2}. \tag{11.120}$$

According to Euler's main turbine Eq. (11.109), the total enthalpy difference across the turbine is:

$$\Delta H = u \cdot (c_{u2} - c_{u3}) = u \cdot (w_{u2} - w_{u3}). \tag{11.121}$$

If one now formulates the degree of reaction r in the form of relative or absolute velocity components, we obtain:

$$r = -\frac{(w_{u2} + w_{u3})}{2 \cdot u} = 1 - \frac{(c_{u3} + c_{u2})}{2 \cdot u}. \tag{11.122}$$

With the help of the pressure number ψ, the delivery number φ and the degree of reaction r, the loss fraction L can be represented as a function of dimensionless quantities and we get for L:

$$L = \left(\frac{1}{\psi}\right)\left[\varphi^2 + \frac{\psi^2}{16} - \left(\frac{\psi}{2}\right)(1-r) + (1-r)^2\right] \cdot \sum \zeta + (1-r)\zeta_s + \left[\frac{1}{2} + \frac{(2r-1)}{\psi}\right] \cdot \zeta_r, \tag{11.123}$$

where $\sum \zeta = (\zeta_s + \zeta_r + \zeta_d)$ is the sum of the stator, rotor, and diffuser losses. The choice of the degree of reaction for a given combination of the pressure number and the delivery number determines the shape of the velocity triangles and the relative magnitude of the stator and rotor exit velocities. In this context, turbine losses can be reduced if

- reducing the individual losses in the stator, rotor, and diffuser, i.e. ζ_s, ζ_r and ζ_d, or
- minimising their influence coefficients, which are functions of the pressure, delivery, and reaction coefficients.

In solar chimneys there is no pressure recovery in the tower but an acceleration pressure loss. Therefore, the static pressure at the turbine diffuser outlet is crucial. The flow cross-section of the turbine is smaller than that of the tower because of the hub, and the outer diameter must be smaller than that of the tower in order to obtain a technically reasonable radius of curvature that ensures a favourable flow inlet to the turbine inlet for all boundary conditions. According to Eq. (11.23), minimum losses at a given pressure and delivery number result at a turbine reaction degree of $r > 0.5$. By definition, the reaction degree r cannot exceed 1, since otherwise a compressor and no turbine would exist anymore. In any case, this means a pressure-staged turbine. When designing a wind turbine, it makes sense to look for an optimal pressure number for a given reaction degree and a delivery number. If the loss coefficients ζ_s, ζ_r and ζ_d are independent of the pressure number, Eq. (11.23) can be differentiated with respect to ψ and the result set to zero and thereby the optimum pressure number is obtained. This shows that the optimal reaction degree of the turbine is approx. $r \approx 0.8$ and thus turbine efficiencies with $\eta_{turb} > 0.9$ can be realised. With the determination of r, the blade geometries are also fixed.

However, the choice of a pressure-staged turbine also has disadvantages. Overpressure stages require a seal between the rotor blade and the casing, as they have a more or less high pressure difference upstream and downstream of the rotor blade. Another aspect is at least as important. Due to the solar irradiation, the updraft power cannot be planned, and overpressure turbines behave much more problematically than constant-pressure turbines with regard to partial load behaviour. This disadvantage can be counteracted by a Curtis stage or an adjustable guide vane apparatus in the inlet. Further details on design and efficient control can be found in the articles Fluri and von Backström (2008) and dos Santos Bernardes and von Backström (2010). Either one turbine can be installed centrally in the tower or several axial rotors arranged around the circumference of the tower base, as illustrated in Fig. 11.56.

In the power plants realised so far, such as the one in Manzanares, air velocity upstream and downstream of the turbine is approximately the same. The power extracted is proportional to the product of the volume flow and the pressure drop at the turbine. With regard to maximum energy yield, the aim of turbine control is to maximise this product in all operating conditions. Via the blade pitch, the turbine controls the pressure drop and thus the air velocity and the air flow in the system. If the blades are perpendicular to the inflow, the turbine does not rotate. If the blades are parallel to the flow and allow the airflow to pass through

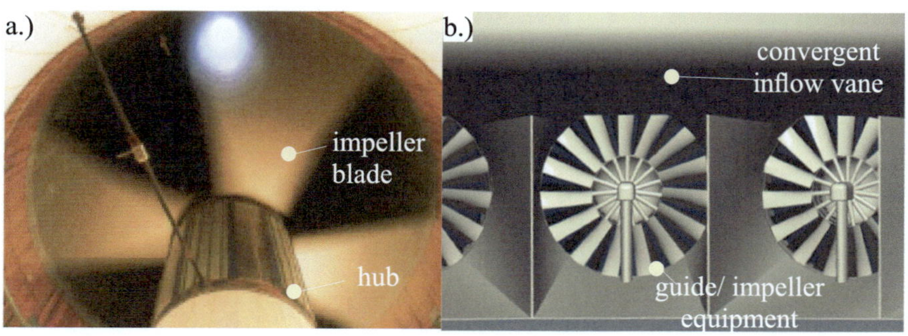

Fig. 11.56 **a** Turbine of the updraft prototype plant in Manzanares from Weinrebe (2004); **b** CAD drawing of circumferentially arranged axial turbines positioned side by side from Fluri and von Backström (2008)

undisturbed, the pressure drop at the turbine is zero and no electrical power is generated. Between these two positions, there is an optimal blade position. The power yield is at its maximum when the pressure drop at the turbine is around 80% of the total available pressure difference.

Prototypes and plans

As early as 1980, an experimental plant with a tower height of 200 m, a tower diameter of slightly more than 10 m and a 45,000 m^2 glass roof ($d_{col} \approx 250$ m) was built in Manzanares, Spain, which started operation in 1982 (cf. Fig. 11.57a). The solar to electric efficiency was reported to be $0.1 - 0.2\%$ and the heated air reached velocities of about 12 m/s. The typical mean temperature rise in the collector was about 20 °K and the nominal power was designed for 50 kW. In 1987, the system was in operation for a total of 3197 h, which corresponded to an average daily operating time of almost 9 h. As soon as the flow velocity reached a certain value (≥ 2.5 m/s), the plant started automatically, and the power was fed into the grid. The ground had become so light relatively quickly that a layer of bitumen had to be sprayed on. In 1987, a black foil was applied, which raised the temperature by about 7 °C and thus increased the electricity yield by about 30%. Overall, both components and the system as a whole proved to be highly reliable before a hurricane destroyed the chimney in February 1989. Ned's Corner Station in the north of Australia's state of Victoria, where a first large-scale solar chimney power plant with a capacity of approx. 200 MW generated by 32 turbines is considered to have chances of realisation. The planned station with reinforced concrete tower tube and steel/glass air solar collector (alternatively with plastic covering) would then be the world's largest civil engineering structure (cf. Fig. 11.57b). About 30 km^2 of glass or transparent plastic and $4 \cdot 10^5$ m^3 of concrete must be used to build the chimney, which is about 1 km high and 130 m in diameter, and the associated roof structure (diameter 6–7 km). By laying closed water hoses under the collector roof, a continuous 24 h operation can be guaranteed. The air under the greenhouse

Fig. 11.57 **a** Photograph of the 50 KW updraft power plant prototype in Manzanares, Spain (Weinrebe, 2004); **b** Computer animation of the 200 MW updraft power plant planned in Mildura, Australia (Kammas, 2011)

roof, which rises slightly towards the centre, is heated by the sun to up to 35 °C above ambient temperature and flows upwards in the chimney at up to 15 m/s. When the air enters the chimney, it is heated by the sun. When the air enters there, it is expected to reach 70 °C.

11.5.2 Solar Pond

11.5.2.1 Physical Functionality

Solar ponds also termed as salinity gradient ponds a very special variety of a non-concentrating solar thermal collector system. They occupy a special position among solar thermal systems, as the pond acts simultaneously as volumetric collector and as heat reservoir.

A solar pond is a shallow salt lake in which layers of fluid with different salt content are stably stacked on top of each other. The salt content of the fluid in the bottom layer, and thus its specific density, is significantly larger than that of the surface fluid layer. If solar radiation penetrates the solar pond, a density-dependent absorption of solar radiation takes place in the individual fluid layers, which leads to a volumetric heating, especially of the deeper layers within the salt pond. Temperatures of more than 80–90 °C can be reached there. The density layer due to the salt content counteracts the density reduction caused by the heating, so that the onset of a density-dependent buoyant convective flow is prevented. This means that the heat energy remains stored in the layer of the pond near the bottom and can be converted into electricity either by means of a pump or a heat exchanger integrated directly into the bottom of the solar pond via a turbine generator block.

With sufficient size, 24/7 operation is possible in principle. Since the available temperature differences only reach between 30 and 60 °K, the thermodynamic efficiency of such systems is comparatively modest. In addition, the small temperature differences available require alternative thermodynamic cycle processes for power generation that use working media that already evaporate at low temperatures. The Organic Rankine Cycle (ORC) and the Kalina process are technical options here.

11.5.2.2 Mathematical Description

A solar pond consists mainly of three distinctly different zones, a so-called upper convective zone (UCZ), a non-convective zone (NCZ) and a lower convective zone (LCZ), as illustrated in Fig. 11.58. The upper relatively thin convective zone has an almost constant density, close to that of the sea water, while in the non-convective zone the density increases almost linearly with the depth z. The lower convective zone is called the halocline. This transition layer is called the halocline or salinity spring layer. In the lower convective zone, which according to measurements takes up the greatest thickness, the density is again almost constant and almost corresponds to that of a saturated brine. The solar radiation incident on the solar pond at the angle of incidence θ_i is absorbed in the salt solution and stored for the most part in the LCZ layer, heating it up. In the non-convective zone, heat transport occurs primarily through a relatively slow heat conduction process and

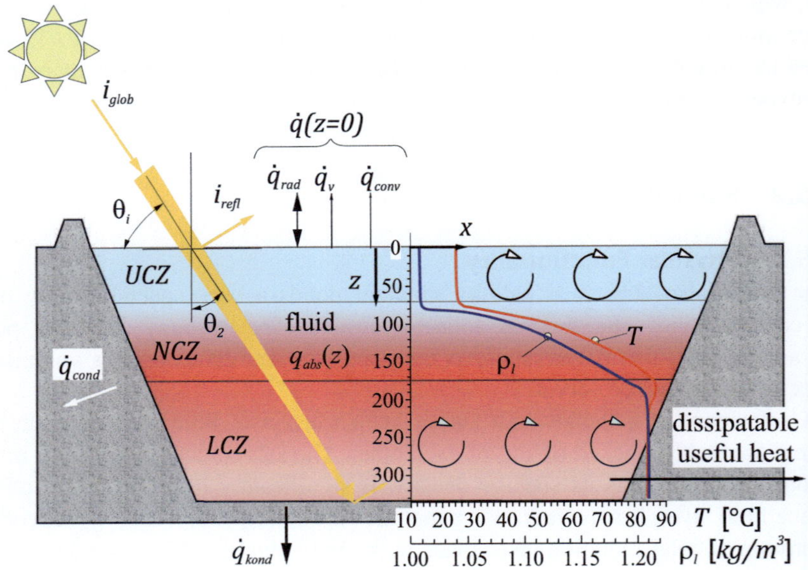

Fig. 11.58 Schematic sketch for stratification, heat losses and heat balance in a solar pond and introduced coordinate system. UCZ = upper convective zone, NCZ = stagnant zone and LCZ = lower convective zone. The temperature and density profiles are from measurements of Leblanc et al. (2011)

thus this layer acts like a thermal insulation. The benefit of a solar pond is thus limited to the amount of energy stored. This is offset by the necessary constructional expenditure as well as the maintenance and operating costs. The thermal performance is ultimately a function of

- the solar irradiation,
- the heat losses over all edges to the surroundings (ground as well as atmosphere),
- the internal heat losses of the lower convective zone to the layers above,
- the volume of the pond and thus the storage capacity and finally
- the efficiency of the heat exchanger complex to the power conversion system.

All layers of a solar pond are subject to temporal dynamics and have their own specific time constants, some of which differ by orders of magnitude. A complete correct analysis therefore requires a simultaneous solution of mass, momentum, and energy conservation. The big challenge is the coupling of the simultaneous heat and salt transport. Especially the mixing is a coupled process. In the phenomenon known as double-diffusive convection, convective motion is driven by a density gradient in which two components with different diffusivities coexist. Both provide an opposite directional contribution to the vertical density gradient. Due to the complexity of the problem, only a few simulations with consideration of the double-diffusive character of the salt stratification exist, such as Suarez et al. (2010).

To describe the problem, the conservation equations of mass for water and the species (salts), momentum and energy conservation are coupled with the equation for fluid density, where a linear dependence on temperature and salinity is assumed. The result is:

$$\frac{\partial \rho}{\partial t} + \nabla \cdot (\rho\, \vec{u}) = 0; \tag{11.124}$$

$$\frac{\partial}{\partial t}(\rho\, c) + \nabla \cdot (\rho\, \vec{u}\, c) = \nabla \cdot (\rho\, \kappa_c\, \nabla c); \tag{11.125}$$

$$\frac{\partial}{\partial t}(\rho\, \vec{u}) + \nabla \cdot (\rho\, \vec{u}\, \vec{u}) = -\nabla p + \mu\, \Delta \vec{u} + \rho\, \vec{g}; \tag{11.126}$$

$$\rho\, c_p \frac{dT}{dt} = \nabla \cdot (\kappa_T\, \nabla T) + \beta_T\, T\, \frac{dp}{dt} + \varphi_h; \tag{11.127}$$

$$\rho(c, T) = \rho_0 \cdot [1 - \beta_T (T - T_0) + \beta_c (c - c_0)]. \tag{11.128}$$

In the Eqs. (11.124)–(11.128) c denotes the salt concentration, T the temperature, $\vec{u}$ the velocity and $\vec{g}$ the gravity vector and φ_h the internal heat source due to solar radiation. p describes the pressure, β_T the thermal expansion coefficient and β_c that due to the salinity change. The thermal and concentration-related diffusivities are given by κ_T and κ_c, respectively, and the initial state is described by the index "0". To solve the problem, in addition to the initial data, the thermo-physical data of the substances involved must be known and detailed boundary conditions must be specified. The formulation of the boundary conditions includes a large number of effects to be considered and ultimately determines the quality of the prediction.

The thermal boundary condition at the upper border of the fluid domain to the surroundings at $z = 0$ is given by a heat flux condition. The net heat flux $\dot{q}(z = 0)$ is composed additively of the radiation from the pond $\dot{q}_{rad}$, the evaporation $\dot{q}_v$ of the fluid and the conductive-convective heat losses $(\dot{q}_{cond} + \dot{q}_{conv})$:

$$\dot{q}(z = 0) = \dot{q}_{rad} + \dot{q}_v + (\dot{q}_{cond} + \dot{q}_{conv}), \tag{11.129}$$

where the radiation exchange $\dot{q}_{rad}$ is composed of a short-wave component, which occurs due to the reflection of sunlight in the water, and a long-wave component, which occurs due to the heat radiation into the environment due to the increased temperature. The component of the solar radiation reflected by the fluid results from

$$i_{refl} = (1 - \rho) \cdot i_{glob}, \tag{11.130}$$

where the reflection coefficient ρ of water can be calculated with the help of Fresnel's formulae (see Chap. 3 Optics) and by means of the refractive indices for air and water. When calculating, it should be noted that the reflection depends on the diurnal variation of the angle of incidence θ_i. A simplified calculation scheme is given by Wang and Akbarzadeh (1983). As a rule, the short-wave component can be neglected, as it only contributes to a small extent to the radiated power density.

The long-wave thermal radiation across the air–liquid interface is composed of two radiation components:

- thermal radiation from the atmosphere/environment onto the water $\dot{q}_{amb\to l}$ and
- the heat radiation from the water of the pond into the atmosphere $\dot{q}_{l\to amb}$,

which can be evaluated using the correlation proposed by Hendersson-Sellers (1986):

$$\dot{q}_{amb\to l} = 0.97 \cdot \varepsilon_{amb} \cdot \sigma \cdot T_{amb}^4, \tag{11.131}$$

where T_{amb} is the ambient temperature in Kelvin and ε_{amb} is the atmospheric emissivity, which is essentially dependent on the cloud cover fraction and can be approximated in the following way:

$$\varepsilon_{amb} = \begin{cases} 0.84 - \left(0.100 - 9.973 \cdot 10^{-6} p_v\right)(1-C) + 3.491 \cdot 10^{-5} p_v 0 \le C \le 0.6 \\ 0.87 - \left(0.175 - 29.92 \cdot 10^{-6} p_v\right)(1-C) + 2.693 \cdot 10^{-5} p_v 0.6 < C \le 1 \end{cases}. \tag{11.132}$$

C indicates the cloud fraction in the sky in the preceding equations. The vapour pressure p_v is used in Pa. and also mostly estimated, whereby the following relationship is often used:

$$p_v = x \cdot \underbrace{2.1718 \cdot 10^{10} \cdot \exp\left(\frac{4157}{T_{amb} - 33.91}\right)}_{p_{v,sat}}, \tag{11.133}$$

where x is the relative humidity in percent and the remaining expression represents the saturation vapour pressure $p_{v,sat}$. For the heat radiation of the pond to the surrounding air, the water surface can be approximately understood as a black radiator and $\dot{q}_{l\to amb}$ is calculated according to Rosenberry et al. (2007):

$$\dot{q}_{l\to amb} = 0.972 \cdot \sigma \cdot T^4(z=0), \tag{11.134}$$

The evaporation or latent heat flux across the air–liquid interface includes both a forced convective component resulting from the air flow over the water surface and a free convective component resulting from unstable atmospheric conditions.

The latter is generated by density differences between the water surface and the air. Both effects are geometrically averaged in the following way

$$\dot{q}_v = -\sqrt{\left(\dot{q}^2_{conv,eva} + \dot{q}^2_{buoy,eva}\right)}. \tag{11.135}$$

The convective evaporation heat flux $\dot{q}_{conv,v}$ at the interface due to wind is approximated with the following relationship

$$\dot{q}_{\text{conv },v} = 3.1 \cdot 10^{-2} \cdot u(z = -2m) \cdot \left(p_{v,\text{ sat}}(z = 0) - p_v\right), \tag{11.136}$$

where $u(z = -2m)$ is the mean wind speed at a height of two metres above the salt water, pd from Eq. (11.133) is used and the saturation vapour pressure at the air–water interface is used for $p_{v,sat}(z = 0)$, where $T_l(z = 0) = T_{amb}$ at the interface. The subsequent correlation is used for the evaporation heat flow by free convection

$$q_{buoy,v} = \begin{cases} 2.7 \cdot 10^2 \cdot \left(T_{l,v} - T_{amb,v}\right)^{\frac{1}{3}} \cdot \left(p_{v,sat} - p_v\right) & \text{for } T_{l,v} > T_{U,v} \\ 0 & \text{for } T_{l,v} \leq T_{amb,v} \end{cases}, \tag{11.137}$$

in which virtual ambient and water temperatures $T_{amb,v}$ and $T_{l,v}$ are used, estimated using:

$$T_{l,v} = \frac{T_l(z = 0)}{\left(1 - 0.378 \cdot \frac{p_{v,sat}(z=0)}{p_{amb}}\right)} \text{ and } T_{amb,v} = \frac{T_{amb}}{\left(1 - 078 \cdot \frac{p_v}{p_{amb}}\right)}, \tag{11.138}$$

where p_{amb} is the ambient pressure in Pa.

Convection leads not only leads to an evaporation but also to convective heat flux $\dot{q}_{conv}$, which is driven by the temperature difference between the water surface and the air as well as the wind speed. There are a multitude of correlations; many publications resort to the following relationship validated by experiments mentioned in Losordo and Piedrahita (1991):

$$\dot{q}_{conv} = -1.5701 \cdot u(z = -2m) \cdot (T_l(z = 0) - T_{amb}). \tag{11.139}$$

The heat losses at the side and at the bottom depend on how much of the solar radiation reaches the bottom and the edges, since the largest part is to be used as a heat source for heating the brine and another part is lost through shading due to the respective solar altitude. The heat losses at the boundaries to the soil must be supplemented by the heat losses through heat conduction to the surrounding sediment, which are given by Fourier's law. This results in for the conductive heat flux $\dot{q}_{cond}$ in:

$$\dot{q}_{cond} = \dot{q}_{rad,R} - (-\lambda \nabla T_R), \tag{11.140}$$

Table 11.7 Wavelength-dependent contributions to the determination of the transmission coefficient τ in brine from Jaefarzadeh (2004)

Wavelength λ [nm]	η	Absorption coefficient solar $\mu[1/m]$
200–600	0.237	0.032
600–750	0.193	0.450
750–900	0.167	3.000
900–1200	0.179	35.000

where the index "R" denotes the boundary coordinates.

The short-wave solar radiation that penetrating the water body causes the saline fluid to heat up. According to the Lambert Beer law, the attenuation of the radiation occurs negatively exponentially with the penetration depth z into the fluid (cf. Chap. 3 Optics). The transmission coefficient τ is a function of the light wavelength λ, the penetration depth z, the refraction angle θ_2 and the absorption coefficient μ. Since all coefficients depend on each other, a section-wise approximation of the transmission coefficient according to Rabl and Nielsen (1975) has become established in the literature:

$$\tau = \sum_{j=1}^{4} \eta_j \cdot \exp\left(\frac{\mu \cdot z}{\cos\theta_2}\right), \tag{11.141}$$

where the values listed in Table 11.7 are used. The salt concentration can thus be used to control the absorption and power storage in the fluid. A good overview is given in the article by Wang and Seyed-Yagoobi (1994).

For the heat flux introduced into the fluid per specific surface area, we then obtain:

$$i_{dir} = (1-\rho) \cdot \tau(z) \cdot i_{glob}(z=0), \tag{11.142}$$

where i_{dir} (or DNI) is the direct solar irradiance flux in $[W/m^2]$. Differentiation of equation gives the volumetric heat input in the brine to be

$$\varphi_h = -\frac{\partial i_{dir}}{\partial z} = (1-\rho) \cdot i_{dir}(z=0) \cdot \frac{\partial \tau(z)}{\partial z}. \tag{11.143}$$

In addition to the thermal boundary conditions, kinematic boundary conditions must also be formulated to solve the problem. At the air–liquid interface, the wind velocity vector $\vec{u}$ generates a shear stress at the liquid interface due to its speed, which acts on the water as a force and can be formulated as follows

$$\vec{\tau} = \rho_U(z=0) \cdot c_W \cdot |\vec{u}| \cdot \vec{u}, \tag{11.144}$$

where c_W is the drag coefficient, which depends on the surface shape (for example waves) and $\vec{\tau}$ is the surface tension tensor. Although no salt leaves the water-filled control volume, evaporation as well as salt precipitation on the floor and walls lead to an apparent salt flux F_v, which can usually be estimated in the following way, see cf. Delmore (1980):

$$F_v = c \cdot [a \cdot \rho \cdot -\dot{q}_v / \Delta h_v] \rho_U (z = 0) \cdot c_W \cdot |\vec{u}| \cdot \vec{u}. \tag{11.145}$$

In it, a indicates the deposition rate in [m/s], $\dot{q}_{eva}$ the heat flux of evaporation and Δh_v the enthalpy of evaporation.

Further assumptions concern the permeability of the soil, whereby an impermeable soil is usually assumed, which is technically realised by designing the solar pond with foils or liners. This results in the adhesion condition of the fluid to the wall at all fixed boundaries of the solar pond.

11.5.2.3 Stability Transfer and Vertical Interfacial Movement in Solar Ponds

Solar ponds are not functional without a stable non-convective stratification. With the help of a simplified two-dimensional consideration, a stationary stability limit can be determined. The condition for density in the NCZ thus requires at least a constant value to prevent flipping due to gravity. However, the condition for the stationary case is not sufficient for dynamic stability of the solar pond, since in the transient case additional potential energy occurs in the form of a destabilising temperature distribution caused by an unbalanced internal heat release, which originates from the volumetric input due to the absorption of solar irradiation, see cf. Rebai et al. (2006).

If a large difference occurs between the diffusion of the thermal energy and the salinity (i.e. if a double-diffusive system exists), then the gradient of the salinity in the non-convective zone (NCZ) must be greater than the gradient required to fulfil the steady-state stability condition. To fulfil the dynamic stability, the following critical Rayleigh number results:

$$Ra_{T,crit} = \left(\frac{\nu + \kappa_c}{\nu + \kappa_T}\right) \cdot Ra_c + \left(1 + \frac{\kappa_S}{\kappa_T}\right) \cdot \left(1 + \frac{\kappa_S}{\nu}\right) \cdot \frac{27}{4} \cdot \pi^4, \tag{11.146}$$

where ν is the kinematic viscosity and the Rayleigh numbers Ra_T and Ra_c occurring in it give the ratio of the buoyancy forces (either due to thermal "T" or solubility-induced expansion or contraction "c") to the viscous drag forces, defined as follows:

$$Ra_T = \frac{g \beta_T \Delta T L_c^3}{\nu \kappa_T} \text{ and } Ra_c = \frac{g \beta_c \Delta c L_c^3}{\nu \kappa_T}, \tag{11.147}$$

where L_c is the characteristic length of the layer in question. Strictly speaking, condition (11.146) only applies to the limiting case of constant gradients of temperature and salinity in the layer, which is rarely given in a real solar pond. With the help of empirical data, it could be shown for simplified cases that the edges of the inner convective layers can be reproduced well with the help of a power law:

$$G_c = 28 \cdot G_T^{0.63}, \tag{11.148}$$

where G_c and G_T are the salinity and temperature gradients near the layer boundaries of each interface. This empirical relationship is not a stability limit, rather it describes possible temperature and salinity gradients near the interfaces for which the existence of stationary interfaces is given. This condition, called boundary layer equilibrium, allows a prediction of the behaviour of the solar pond when the temperature or salinity gradient changes. For example, if the salinity gradient is greater than the value predicted in Eq. (11.148), the thickness of the non-convective layer will increase. As a result, the adjacent convective layers will shrink to compensate for this growth. If the salinity gradient is lower than the predicted one, a reduction of the gradient occurs, i.e. a reduction of the non-convective layer. Then the adjacent convective liquid layers increase.

It should be noted in this context that even the steady-state operation of a "low tech" technology such as solar ponds requires detailed knowledge of interacting phenomena in order to ensure at least steady-state operation.

11.5.2.4 Heat Extraction

To convert the heat stored in the solar pond into energy, the heat must be extracted from the pond, whereby in most cases the heat extraction takes place near the upper border of the lower convective zone (LCZ). Heat extraction can either be done directly by pumping out the saline solution, which is called an external process and has been tested in the El Paso Solar Pond (Lu et al., 2004). The other way is an internal process in which the heat transfer fluid directly transfers the heat through heat exchangers integrated into the pond, which has been realized in the demonstration plant at Pyramid Hill in Australia (Andrews & Akbarzadeh, 2002). Both processes are shown schematically in Fig. 11.60 and have proven their functionality in experimental operation.

Fig. 11.59 **a** Construction and dimensions of the experimental solar pond in El Paso **b** Instrumentation tower of the plant for monitoring the altitude, temperature, salinity and turbidity

The approximately 3000 m^2, 3.3 m deep solar pond in El Paso shown in Fig. 11.59 which operated from 1985 to 2003, used only an external heat extraction method. In this method, hot saturated brine is extracted from the solar pond at the top of the LCZ (heat storage zone) and fed to an external heat exchanger.

Special collectors, so-called extraction diffusers, are used for heat storage, which may only slightly disturb the salt content stratification. These diffusers are installed in a height-adjustable manner so that the extraction can take place at the point with the highest temperature of the storage zone. The corresponding reflux diffuser is installed below the extraction point. This method allows flexible placement of the heat extraction and return with the highest efficiencies currently available in the Pond. In addition, the cold brine is reintroduced directly at the bottom of the Pond, resulting in a reduction of heat losses to the bottom. An inherent disadvantage of the external process is the corrosion of the piping and heat exchanger when using metallic components. The iron ions dissolved in the brine led to the formation of algae and ultimately to an undesirable turbidity of the pond. As a consequence, all components within the pond were covered with a plastic layer or made of plastic right away. In Pyramid Hill, Australia, a solar pond, also of about 3000 m^2 surface area and 2.2 m depth was commissioned in 2000 and used for inland salt production (cf. Fig. 11.61).

Another special feature is the use of floating rings, which prevents the formation of larger waves at high wind speeds. In the Australian pond, the heat transfer takes place in the pond itself. For this purpose, fresh water is pumped through pipes embedded near the bottom of the pond, which act as heat exchangers. In a second external heat exchanger, this heated fresh water transfers the heat to the heat transfer fluid of the useful circuit, which feeds it to the respective application.

Here, too, corrosion plays an excellent role. The use of plastics is possible, but they only have a low thermal conductivity and therefore require large heat transfer surfaces. Another disadvantage is their low density. Since both water and

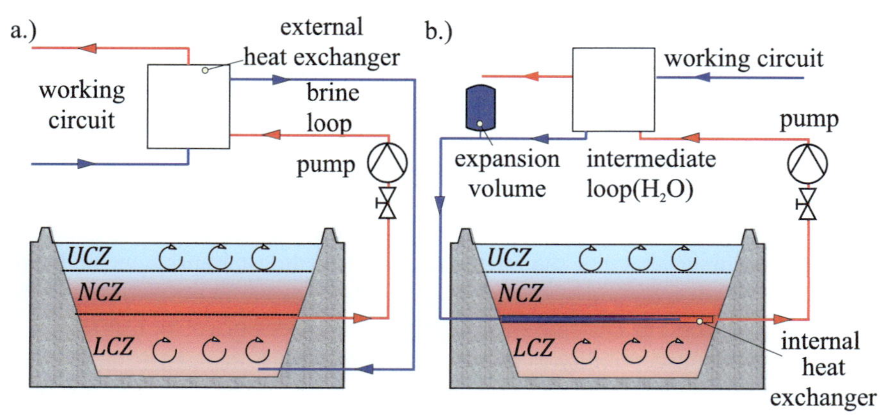

Fig. 11.60 Extraction process to extract the thermal energy stored in the solar pond. **a** External process and **b** internal process

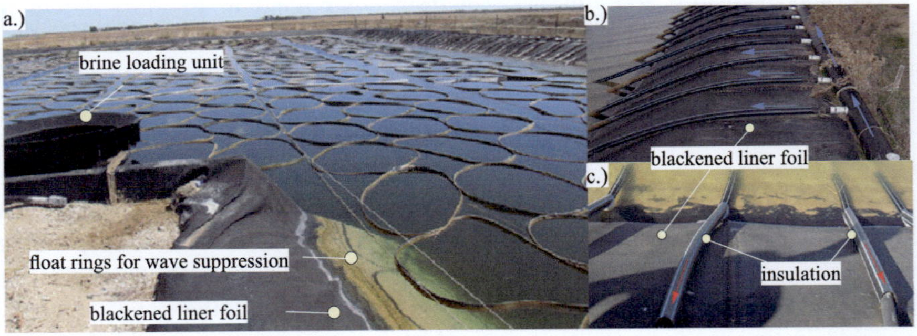

Fig. 11.61 **a** Solar pond in Pyramid Hills, Australia; pipe routing for internal heat extraction from solar ponds in Pyramid Hills for inlet manifold of the cold water (**b**) and return of the hot water (**c**)

the plastics in question have a lower density than the hot brine ($\rho \approx 1.2-1.3$ kg/l), a support structure is required inside the pond to keep the heat transfer tubes in the desired position. In addition to the second heat exchanger required in this concept, the level at which heat is extracted from the solar pond cannot be controlled, so flexibility is limited compared to the external concept.

11.5.2.5 Operational and Technical Measures in Solar Pond Operation

Regarding the choice of salt, there are several options, but since the functionality of a solar pond is based on setting a salt gradient, the salt(s) must meet several criteria:

- Sufficiently high solubility with water. Only a high solubility allows the setting of a large gradient and a stratification profile.
- Transparency to solar radiation. Turbid mixtures lead to a large amount of moisture evaporating from the solar pond and most of the heat being released at the surface.
- Low cost and locally high availability to ensure competitiveness.
- Low impact on the environment (groundwater, …).

In most cases, saline brines (NaCl) or magnesium chloride brines ($MgCl_2$) or mixtures of both components are used in the solar ponds built so far. The latter not least because seawater inherently contains both components. Saline solutions have the advantage that the saturated brine is still almost transparent and thus allows deep ponds with a high storage capacity. Magnesium chloride only has the advantage over sodium chloride of a higher specific density ($\rho \approx 1300$ kg/m^3) and thus a lower susceptibility to temperature fluctuations.

The salinity gradient is one of the core elements of a solar pond and therefore the central task for the operation of such a system is the efficient and effective construction of a given salinity profile along the pond elevation.

One of the most common methods of constructing a salinity gradient is by injecting water. Here, the solar pond is partially filled with a saturated brine and then fresh as well as brackish water is injected into the brine successively at increasing heights in such a way that the concentrated brine is gradually diluted, thus creating a gradient. The critical parameter here is the densimetric Froude number, which is a dimensionless quantity and, in this context, represents the ratio of the kinetic energy and the buoyancy energy of the injected fluid. It is calculated from:

$$Fr_{\Delta\rho} = \sqrt{\frac{\rho_l \cdot u_0^2}{g \cdot \Delta\rho \cdot d}}, \tag{11.149}$$

where ρ_l is the density of the surrounding brine, u_0 is the mean exit velocity from the diffuser, g is the gravitational constant and l is the hydraulic height of the exit diffuser. $\Delta\rho$ is the density difference between the injected and the surrounding fluid. A critical Froude number of $Fr_{\Delta\rho,crit} = 18$ was determined from a large number of experiments. If the Froude number is less than this value, the fluid rises and mixes above the diffuser. For Froude numbers significantly greater than $Fr_{\Delta\rho,crit}$, the injected fluid entrains significant amounts of fluid below the diffuser layer. Either case results in a less than favourable salinity gradient distribution. In the successive redistribution method, a layer of fresh water is injected last at the top, which then fills the pond.

Another simple method is that of natural diffusion, where the pond is first half filled with salt and water to create a concentrated brine. The rest of the pond is then filled with fresh water. Diffusion then creates the required density gradient, which is practical for small ponds but leads to unacceptably long times for larger systems.

An alternative approach is a stacking process, in which the pond is filled with several layers of brines of different salinity one after the other. If charging from the top, the bottom layer is set first. If the layers are stacked from below, the almost salt-free top layer is filled in first and then the layers of higher density are let in.

During operation of the solar pond, the salinity gradient must be permanently maintained and thus the density and temperature distribution must be constantly recorded in order to correct instabilities that occur within the density gradient zone. Any disintegration of this gradient zone affects the efficiency of the pond and is inherently a dynamic equilibrium between convective motion and diffusive transport processes. In the context of capturing the natural movement of the two internal interfaces, the storage zone is constantly set in convective motion by solar irradiation, while the upper convective zone is set in motion both by wind but also by water evaporation increasing salinity. Thus, there is hardly a stationary operating state, even if the daily changes and thus the gradients are small.

In principle, there are two operating modes, the rising water level, and the falling level.

- In the rising mode, saturated brine/brine is injected so that the gradient experiences an upward velocity. The injection causes the salinity gradient to steepen near the upper boundary layer to the upper convective zone, thus minimising gradient erosion. For this purpose, brine is injected at the bottom of the solar pond, but the upward velocity that sets in contributes to the natural concentration equilibrium created by diffusion.
- The falling pond process involves a withdrawal (power extraction) and a reconcentration of the brine near the lower boundary layer or in the storage zone near the bottom. There is no surface flushing, but rather a salt reservoir inserted into the pond ensures that the diffused amount of salt is replaced.

For long-term operation of a solar pond, the brine must be clear so that the solar radiation can penetrate to the bottom. Algae and bacteria can substantially change the turbidity of the brine. Chemical monitoring of the brine is therefore essential to ensure bacteria, algae and salt and acidity levels to achieve a high transmission coefficient. For this purpose, the acid content in the layers, as well as the temperature and transparency, are permanently monitored. In addition to biological activity, however, wind can also lead to the entry of particles into the uppermost layers of a solar pond and thus to turbidity of the brine. For this reason, the edge of the pond is usually raised. This also helps to reduce increased convection in the top layer caused by the wind. Wind-generated waves also increase the reflection of light at the surface. Wave suppression rings are successfully used to reduce wave amplitudes (see Fig. 11.61a).

11.6 Concentrating Solar Thermal Electric Power Plant Concepts (STE)

A concentrating solar thermal power plant consists of three connected functional units in the basic configuration. The first unit is constituted by the solar thermal heat generation, which is the subject of Chaps. 7 (high temperature concentrating collectors) and 10 (high temperature collector fields). The second unit although not mandatory for the operation of the solar power plant but enhancing its availability is built by the thermal energy storage system (TES), being discussed in Chap. 8. Finally, in the third unit, the thermal energy is converted into mechanical energy for electricity generation via a turbine with the aid of a generator or any other expansion machine through a thermodynamic cycle as elaborated in Sect. 11.2. For each of the units and their plant integration different technical options have been developed allowing a large number of different power plant concepts, as indicated in Fig. 11.62. However, techno-economic analyses show that market competitiveness currently exists for only a limited number of the possible permutations described in the following sections. Therefore, here are some remarks why not all technically possible permutations are described below.

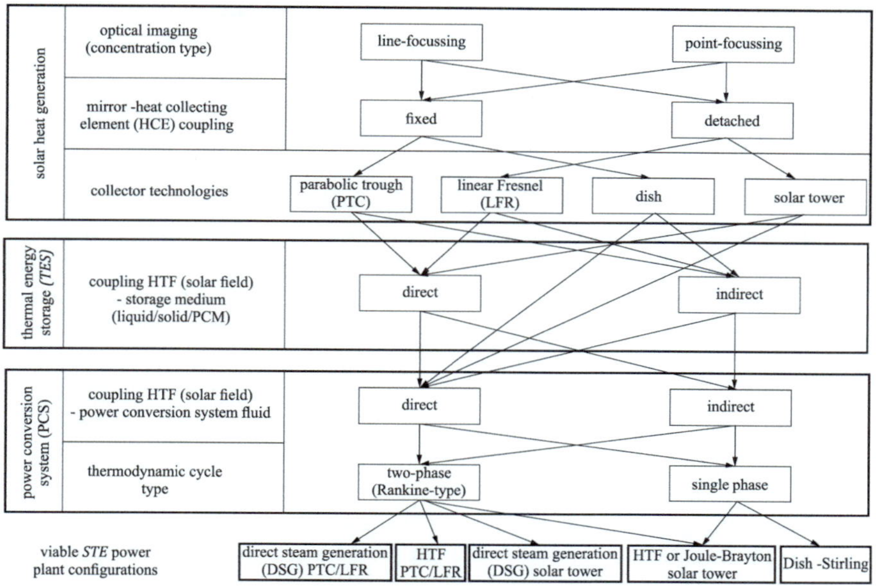

Fig. 11.62 Available technical options for the realization of solar thermal power plant (STE) and derived technical power plant families

- Direct steam generating power plant types (DSG no matter if PTC/LFR or solar towers) are only interesting on markets where combined electricity and process heat are demanded on the market because of the complex thermal energy storage.
- In linear CSP systems, indirect heat transfer fluid systems (with thermal oil, salt) are attractive because of the comparatively low-cost storage possibility and also allow heat supply in addition to dispatch.
- For solar towers, design with salt as heat transfer fluid seems to be the most feasible option. Looking into the future, more attractive concepts with higher temperatures, e.g. with particle receivers (see Chap. 7) or pressure-charged Joule-Brayton designs are conceivable as competitive.
- Dish-Stirling have little or no thermal storage capability and, at the same time, the cost price of photovoltaics (PV) has decreased substantially.

11.6.1 Direct Steam Generation with Linear Focussing STE Concepts

Direct steam-generating solar thermal linear collectors represent an interesting technological option for solar thermal power generation for several reasons. In contrast to power plant concepts using synthetic oil as the heat transfer medium, steam is not subject to temperature limitations in the order of 400 °C, but allows temperatures of approximately 480 °C, for which commercially available power plant components exist and which also increase the efficiency of the thermodynamic cycle. If water/steam can be used for both the solar field and the power plant unit, the otherwise required heat exchanger system for steam generation is not necessary, since the solar field is virtually a large evaporator and superheater. This results in a simpler overall design of the power plant with fewer components. Finally, the use of demineralized water is ecologically safe. Water is cheap, non-toxic and non-flammable, and manageably corrosive if an appropriate water chemistry is applied, which facilitates the choice of materials for piping as well as instrumentation. Although direct steam generation appears attractive at first glance due to its simplicity, it is subject to certain challenges and limitations. These include, but are not limited to:

- stability of the water-steam two-phase flow in the absorber tubes of the evaporator section of the solar field. Here, uniform cooling of the absorber tubes to avoid local hot spots must be ensured under all operating conditions;
- structural integrity of the absorber tubes and the connections between the collectors and the absorber tubes at high pressure and high temperatures;
- system for the control of the plant, especially during transient periods of radiation;
- integration of a thermal energy storage (TES) system adapted to the special properties of the two-phase fluid (water/steam).

For direct steam generation in the solar field collectors, there are three principal options, once through mode, the injection principle, and the recirculation concept, which are outlined in Fig. 11.63. Although once through mode is the simplest concept, it poses the greatest challenges in terms of operation because, especially in the area of the evaporator, which makes up most of the collector length, the two-phase flow patterns change permanently along the flow path depending on the solar irradiance. This makes the controllability of superheated steam parameters at the collector exit extremely difficult. In addition, this mode of operation also requires permanent regulation of the coolant flow rate. In the injection mode, part of the water flow is extracted from the feed water and injected into the absorbers along the evaporator path. The basic idea is that sufficient cooling of the absorber tubes is ensured at all times and no hot spots occur. The shortcoming of the concept, besides the considerable effort of injection at many points, is that the heat transfer behaviour of the two-phase flow changes depending on flow pattern and irradiation, requiring an injection flow control dependent on the vapor content. Therefore, the recirculation concept has been the most conservative concept so far. In this concept, a separation of water and steam takes place at the end of the evaporation section with the aid of a separator, and the water fraction is pumped back into the preheater and mixed with the feed water. To ensure that no dry-out occurs in the evaporator end section, the amount of water supplied to the evaporator inlet is always set higher than the amount that can be evaporated. The excess of water in the evaporation zone guarantees good wetting of the absorber tubes and

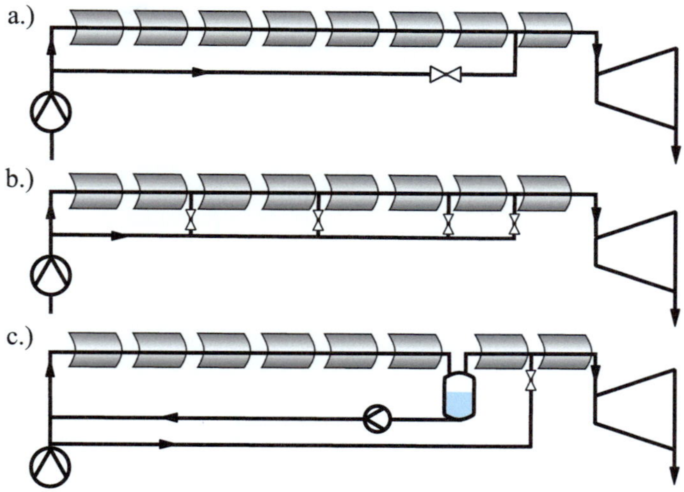

Fig. 11.63 Basic concepts of solar steam generation in linear concentrating collectors; **a** once through concept, **b** injection mode and **c** recirculation concept

almost eliminates layer formation of the two phases. The steam separated in the separator is transferred to the inlet of the superheater section. This type of DSG system is very well controllable by means of modern control algorithms, compare (Mokhtar, 2018) but requires higher equipment effort (steam separator, pressure, and temperature resistant recirculation pump).

A simple plant of a direct steam generation (DSG) STE concept is shown in Fig. 11.64. The feedwater pump delivers the water, depending on the allowable pressure of the receiver tubes, into the preheater and evaporator receiver tubes of the solar field, operational circuits reach pressures up to 10 MPa. At the end of the evaporator section, there is a separator that returns excess (non-evaporated water) from the separator sump collector to the feed line and preheats the feed water. From the separator, steam is fed into the superheater receiver tubes. To keep defocusing losses small, a small amount of feed water can be mixed into the superheater section. At the end of the superheater section, the superheated steam (at temperatures of 450 °C) enters the turbine separator, from where it is discharged into the high-pressure turbine via a pressure control valve that controls the turbine inlet pressure. If the radiation is not sufficient, the steam can be brought to the desired temperature by means of a fossil-fired backup system, e.g. with natural gas. Part of the high-pressure steam feeds a steam drum (Ruth storage) that can be used to bridge short-term irradiation fluctuations. Most of the steam from the high-pressure turbine is fed to the low-pressure turbine, where it is expanded to condenser pressure. Part of the steam from the high-pressure turbine can be fed to the feedwater tank via the deaerator for regenerative feedwater preheating. In the condenser, the steam is condensed by releasing heat to the environment, for example with the aid of cooling water from the cooling tower, and the water

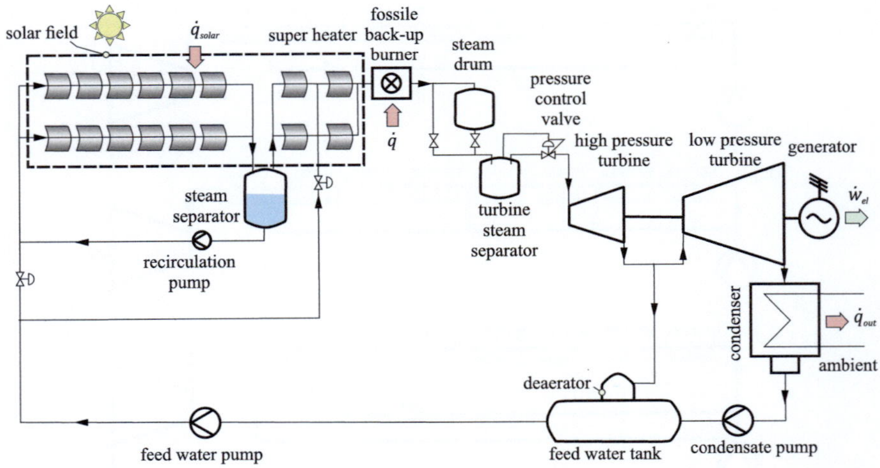

Fig. 11.64 Fluid flow scheme of *DSG* linear concentrating *STE* without large scale thermal energy storage

accumulating in the condenser sump is conveyed by means of the condensate pump to the feedwater tank, from where it is returned to the feedwater pump. Depending on the concept, the pressure in the feedwater tank is up to 0.5 MPa.

The design of a *DSG* power plant is extremely compact and based on easily available industrial components. In addition, *DSG*-based concepts allow simultaneous provision of process steam and power generation and thus represent attractive power plant concepts for combined power generation and process steam utilization for agricultural products and industrial processes, respectively. Linear Fresnel collector technologies in particular are more widespread than PTC plants due to their low collector costs, modular, compact design. Figure 11.65 shows such a *DSG*-LFR plant concept.

One of the disadvantages of direct steam generation is that the steam cycle is run at a constant pressure in the evaporator, i.e. at or near the fixed pressure mode, irrespective of the mass flow. At part load, however, a reduced steam pressure at the turbine inlet is required, which is realized either by throttling the inlet to the turbine, which means exergy losses, or by a sliding pressure control, which, however, has narrow operating limits due to the fixed condenser pressure. Without *TES*, *DSG* linear-concentrating power plants show annual efficiencies on the order of about 17–19, with the modified sliding-pressure mode tending to outperform the fixed-pressure mode absolute by approximately 1%, cf. Ravelli et al. (2016). The main shortcoming of a direct steam-generating power plant is the low capacity factor and thus the question of integrating a thermal energy storage into the water/steam system.

In order to benefit maximally from high vapor temperatures at the collector outlet and thus high thermodynamic efficiencies of the Rankine cycle of a *DSG*-*STE*, different thermal energy storage options are available, whereby the basic structure

Fig. 11.65 Direct steam generation LFR plant with two steam accumulators (Photo © Frenell GmbH, 2022)

always includes an indirect molten salt storage tank for the high temperature part, allowing storage fluid temperatures exceeding 500 °C. The storage configuration can be either an indirect 2-tank storage tank consisting of a hot and a cold tank or a modified configuration where a third salt tank is located at an intermediate temperature between the hot and cold tank allowing steam superheating (Vogel, 2016). At lower temperature, there is the option of using thermal storage tanks using of phase change materials (PCM) to preheat water. Technically promising PCM storages—but not yet commercial—are for example storages with salt (e.g. $NaNO_3$) with a melting point of 306 °C, compare Seitz et al. (2017). Each of these two power plant configuration options with a TES includes a second steam generator train in parallel with the solar field steam generator train. In the TES version 1, the optimum temperature of the intermediate sensible heat storage tank must be selected to allow evaporation of the water in the evaporator, typically around 400 °C, while for a DSG-TES concept with a phase change material storage tank, temperatures of 250–310 °C are favourable to allow it to act as an efficient economizer. The hot molten salt tanks are used to superheat the steam. Figure 11.66 illustrate the integration of both options as alternatives in a DSG power plant. Analyses of DSG with molten salt 2-tank storage show comparable characteristics and $LCoE$ to direct power plant designs using molten salt in both solar field and TES. The performance of a DSG-type storage is better than that of the HTF oil-based power plant configuration with indirect molten salt TES, is the reason being mainly the higher efficiency of power conversion at higher steam temperatures. The use of PCM in the thermal energy storage chain leads to an increase in TES specific storage capacity (in kWh/kg) of up to 30–40%, cf. Bachelier et al. (2015). Since the specific cost of a PCM material such as $NaNO_3$ is comparatively lower than that of most solar salts, the integration of such approaches offers

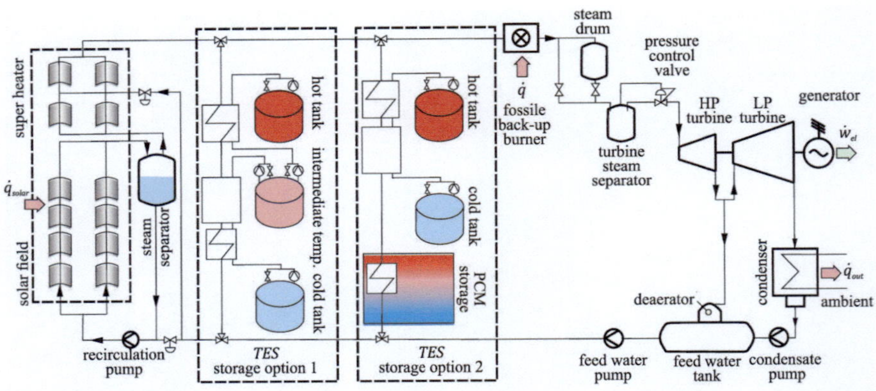

Fig. 11.66 Direct steam generation (DSG) power plant designs with different TES integration options

a cost optimization potential to achieve competitive $LCoE$. Currently, the maturity of the technology and operational experience is not advanced enough to allow commercialization.

11.6.2 Heat Transfer Fluid (HTF) Based Linear Focussing STE

Most solar thermal power plants (STE) are based on linear concentrating heat transfer fluid (HTF) technologies, mostly parabolic troughs, with installed capacities up to 250 MW_{el}. HTF-based power plant configurations are divided into two major subfamilies: direct HTF power plant types, where the term direct means that the heat transfer fluid in the solar field is identical to that in the thermal energy storage system. In contrast, indirect HTF power plants refer to configurations where the HTF in the solar field and thermal energy storage are different (Vogel, 2016).

11.6.2.1 Indirect HTF Based Linear Focussing STE

The majority of all $STEs$ with thermal energy storage capacities of up to 11 h (FLH) are based on an indirect HTF concept, in which the heat transfer medium in the solar field is a heat transfer oil, while in the TES, molten salts are used almost exclusively in a two-tank arrangement consisting of a hot and cold tank. This basic arrangement is outlined in Fig. 11.67.

Due to the temperature limitation of thermal oils in the order of 400 °C, power conversion, especially in large plants $\geq$50 MW_{el}, is done in Rankine cycles, so that thermodynamic efficiencies of 38–40% are achieved in practice. The majority of power plants use molten salts mainly consisting of a mixture of $NaNO_3$ and KNO_3 (60:40, see Chap. 5), which have a melting point of about 240 °C, so that the temperature of the cold tank is usually around 280–290 °C to provide a sufficient temperature margin against freezing of the salts for safety reasons. This results in large thermal energy storage volumes for the targeted thermal storage

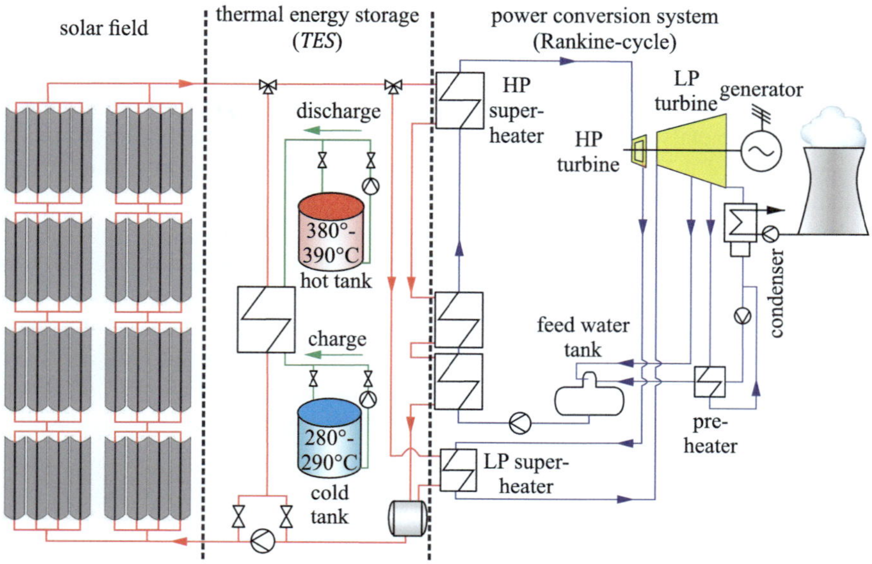

Fig. 11.67 Loop and component arrangement of indirect linear-focusing STE

capacities due to the small temperature difference between hot and cold tank. In addition, these indirect concepts require additional oil-salt heat exchangers, pumps, and piping valves to circulate the salt during storage charging and discharging, which leads to further degradation of thermal efficiency and associated parasitic losses. Although the components are industrially available and easily scalable, they represent additional expenses in both $CAPEX$ and $OPEX$. In addition, a thermooil-based indirect power plant concept enables HTF oil from the solar field circuit to be used as a preheating stage of the Rankine cycle up to the temperature level of the capacitor. This means that part of the regenerative feed water preheating of the Rankine cycle (by tapping the turbine) can be eliminated, since the HTF oil can handle all preheating processes, which partly compensates for the disadvantages compared to the liquid salt concept. Nevertheless, the higher realizable temperatures and thus efficiencies at comparatively low costs speak for the salt concept. It is true that advanced salt mixtures such as HITEC or HITEC-XL, allow melting points in the order of 130–140 °C (compare Bauer et al., 2013), so that the disadvantage of required trace heating can be to a large extend avoided. Nevertheless, these salt mixtures are by more than 30% more costly than conventional solar salt (Caraballo et al., 2021).

11.6.2.2 Direct HTF Linear Focussing STE

Direct storage concepts with heat transfer fluids, which allow temperatures of 550 °C (and potentially more), enable thermodynamic efficiencies of the Rankine cycle well above 40% and seem at first glance predestined to enable smaller

LCoE. Here the molten nitrate salts are used, often the so-called Solar Salt, a mixture of $NaNO_3$ and KNO_3 (60:40). Other mixtures for example with $LiNO_3$ are feasible but costlier. Other liquids, such as molten chlorides or carbonates, for even higher temperatures are being researched and corrosion issues have to be solved there. Liquid metals as sodium or lead alloys, allow temperature differences, for the operation of a Rankine cycle comparable to fossil power plants at relatively low melting temperatures, such plant concepts are not competitive for safety and operational considerations as well as for cost reasons. Despite the obvious advantages of direct molten salt power plant designs, even the use of advanced molten salts with low melting temperatures down to 140 °C requires appropriate heat-tracing systems to prevent the salt from freezing. Although the operation of a direct molten salt (DMS) power plant looks simple in terms of the assemblies involved, it requires dedicated operating modes to ensure safe operation. Figure 11.68 shows the structure of such a power plant concept. The individual operating modes of a DMS plant include.

- Start-up /shut down of the solar field

Salt is circulated from the cold tank through the solar field where the collectors continuously raise the salt temperature from the inlet temperature of approximately 290 °C. This circulation occurs through a bypass line back to the cold tank and is maintained until a threshold temperature is reached at the exit of the solar field and the bypass valve is closed. When the threshold temperature is reached, the solar field switches to "normal operation".

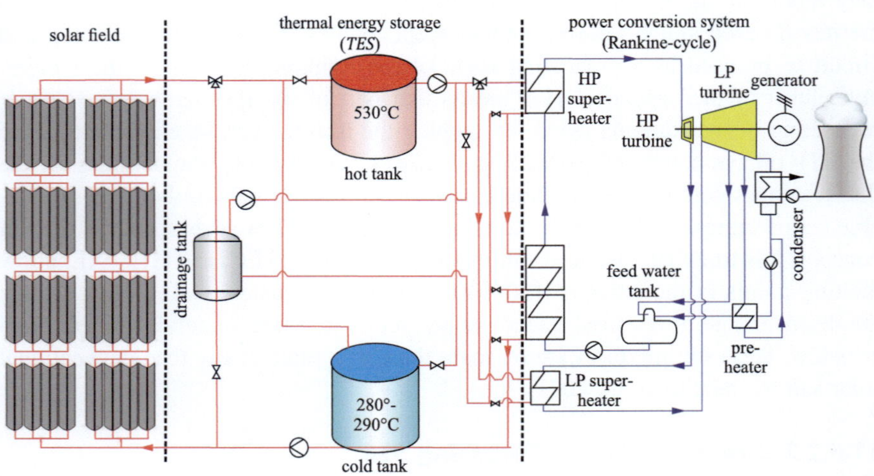

Fig. 11.68 Simplified plant schematic with component interconnection of a direct linear focusing molten salt operated power plant (DMS)

- Normal operation of the solar field

In this mode, the salt mass flow is transported from the cold tank through the collectors, heated in the solar field, and transported into the hot tank the solar. The temperature control is established by changing the mass flow and focusing the primary mirrors. If the hot tank is fully charged, defocusing in the solar field is required.

- Solar field night and frost protection operation

To prevent freezing of the solar salt during night operation, a pump transports heat from the cold tank into the solar field. After passing through the solar field, the cooled salt is returned via the bypass line to the cold tank, which gradually cools down. For large TES, this cooling is very small because the volume of the TES is very large compared to the solar field. However, if the cooling of the cold tank becomes too great, its temperature can be stabilized by mixing hot salt from the hot tank into the cold tank by means of a pump. During the operation of the steam generator (and thus a power generation at night) the cooling of the cold tank slows down significantly because the outlet temperature of the steam generation system continuously supplies the cold tank with fluid at the nominal temperatures of the cold tank.

- Steam generator operation/-shut down

The steam generating system is started up by gradually increasing the salt mass flow and temperature by mixing the mass flows from the hot tank and the drainage tank. The steam generator can only start up when sufficient thermal energy is stored in the hot tank. The volume of the drainage tank is kept constant by a regulator that balances the outlet mass flow with the mass flow coming from the cold tank. The shutdown of the steam generator is basically the reverse of the start-up.

- Steam generator normal operation (power generation)

During power generation, the steam generator is supplied with constant temperatures from the hot tank. Power changes of the Rankine cycle are adjusted by varying the mass flow of salt. The salt leaving the steam generator is discharged to the cold tank, from where it is again conveyed to the solar field.

- Drainage of the solar field:

The solar field can be emptied independently of the steam generating system, which makes sense for two reasons. Firstly, if maintenance work is required on the solar field, it can be emptied independently of the steam generating system, and secondly, if maintenance of the steam generator is required, salt melt can remain in the pipes in the field. If the steam generator is switched off, salt melt can continue to be pumped through the steam generator channels via the drainage tank to prevent it from freezing. The salt flows back into the cold tank. Since the heat losses in the steam generator part are very small compared to those of the solar field due to its compact design, no stabilization is required by mixing from the hot tank.

The drainage tank is installed at the lowest point of the system. All the salt contained in the pipes of the solar field and the steam generator system is stored in the drainage tank. The atmospheric drainage tank is also used to decouple from the solar field in case of leakage of the steam generation system operating at high pressures. Drainage tank is always heated so that in case of emergency drainage, damage due to thermal shock can be prevented. Large-scale solar fields often have multiple drainage tanks, which require adaptation of individual operating modes or plant-specific control strategies.

Direct HTF power plant concepts conceptually allow a further simplification of the integration of a thermal energy sparger via a thermocline. This thermocline can consist either exclusively of the liquid heat transfer fluid or some fluid is replaced by a low-cost filler material with a high heat capacity, e.g. sand, rock, slag, or concrete material (see Chap. 8), into which the heat transfer fluid stores thermal energy during the charging process or retrieves it during the discharging process. The charging and discharging of the accumulator take place by reversing the direction of the flow in the accumulator, which is realized by means of suitable pumps and valves, as shown in the simplified operating principle of this concept in Fig. 11.69. The main advantage of this concept, which has so far only been tested

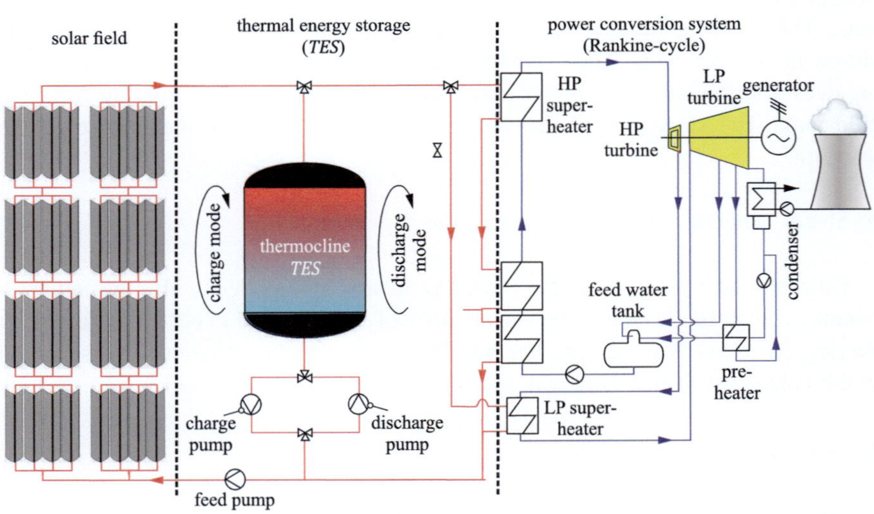

Fig. 11.69 Schematic diagram of the integration of a single-tank thermocline energy storage system into a linear focusing solar thermal power plant, indicating the flow direction for storage and discharge of thermal energy

on an experimental scale, is the elimination of a second tank and the associated heat losses, as well as the reduced volumes and thus the fast response time. On the other hand, this advantage is bought by the fact that start-up and shutdown processes or standby operations, which require mixing processes, are much more difficult to map operationally, so that this system has not yet reached the maturity to be used in large power plants.

11.6.3 Point-Concentrating Solar STE Plants

Depending on the characteristics of each plant component, there exists a big variety of solar power tower plants both at a commercial and at a research stage, which originates from the fact that point-focussing optics allows for high power densities and thus high heat transfer fluid or fluidized solid particles mixture temperatures not only enabling a high thermodynamic efficiency but also allowing for combined cycle operation of the power conversion systems.

11.6.3.1 DSG Based Solar Tower Concepts

Analogous to line-concentrating power plants, solar towers also offer the option of generating steam directly in the receiver and converting it directly into electricity in a Rankine cycle with the use of a turbine. To achieve high thermodynamic efficiencies, modern DSG-based solar towers are operated at live steam pressures and temperatures of 10–16 MPa and about 520 °C, respectively, such as Ivanpah or Khi Solar One, and thus at significantly higher values than the first DSG tower power plants PS10/PS20 with 4 MPA at about 250–300 °C. From an operational point of view, the biggest challenge is dealing with irradiance variations with respect to steam generation. The simplest compensation option is fossil-fired burners downstream of the steam generator. Already in PS10/PS20, expensive steam accumulators with a capacity of 1 h FLH were used to compensate for irradiation variations. An efficient type of steam accumulator is to divide the accumulators into a high-pressure and a low-pressure battery, as implemented for example in the Khi Solar One power plant for a 2 h TES. Figure 11.70 shows the simplified plant diagram.

A major disadvantage of steam accumulation is its relatively low temperature of the outlet saturated steam, compared with normal DSG operating temperatures, which is constrained by the wall size and the material of the pressure vessels of the accumulator. The low steam temperature is not desired since it decreases the thermal cycle efficiency, and it increases the risk of damaging the steam turbines at part load operations. Therefore, steam-accumulator type TES systems in DSG based plants are usually coupled with a superheater. The superheater can be in the simplest case a fossile natural gas burner or being established by the interaction of low and high-pressure accumulator via a heat exchanger. In the charging mode, the feed water flows into the evaporator and absorbs heat until it becomes saturated steam at a pressure above 10 MPa. Some of the saturated steam is directed to the two groups of steam accumulators for storage, while the remainder flows to

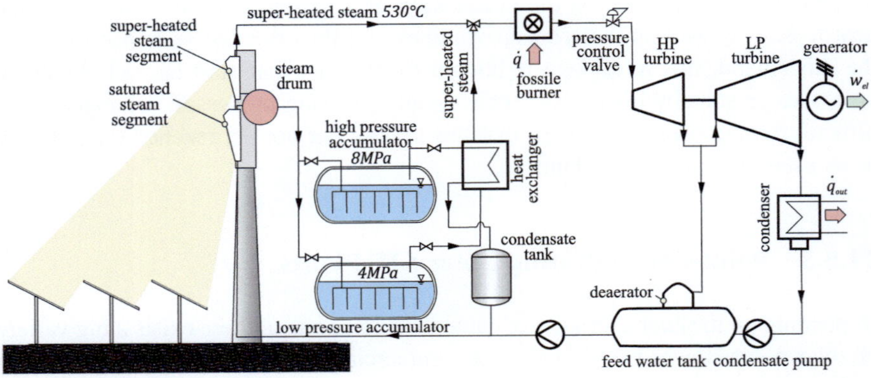

Fig. 11.70 Design of a direct steam generating power plant with integrated two-stage steam accumulator as thermal energy storage

the superheater segment of the receiver where it is superheated to about 530 °C at 12 MPa. The superheated steam is then used to drive the steam turbine to generate electricity. In convective steam storage systems such as Khi Solar One, the superheat steam accumulator (SSA—charge pressure max. 8 MPA) is filled first, followed by the base steam accumulators (BSA, max. charge pressure 4 MPa). The charging process is finished when the pressure in the steam accumulators has reached the maximum permissible value.

The superheat group is always charged to a higher pressure and temperature level than the base group. This is due to the fact that steam with a higher temperature is required for the superheating process in discharge mode. In unloading mode, the base steam accumulator discharges saturated steam at the stored pressure. This steam flows through the storage superheater, is superheated with higher temperature steam from the superheat storage tanks, and then flows into the steam turbine to generate electricity. Superheating is important to avoid the formation of water droplets in the steam turbine and to increase the thermal efficiency of the cycle. Despite the great operational experience with steam accumulators, their efficiency is limited because their design only allows steam temperatures of $290°-300°$, which lowers the thermodynamic efficiency. Figure 11.71 illustrates the steam accumulator array of a *DSG* operated solar tower.

Although the solar electric efficiency and the cycle efficiency of *DSG* operated solar towers are $3-5\%$ higher than those of linear-focused power plant types, their capacity factor is limited to about $23-27\%$ despite the use of steam accumulators. Due to the small storage capacity and even with the use of a high fraction of fossil steam firing, only an order of $CF \approx 30\%$ is achievable. Nevertheless, such power plants can be profitable for a local market under certain conditions, e.g. priority feed-in of electricity, guaranteed purchase prices, corresponding tax depreciation or linkage via process heat with industrial production. However, a dispatchable electric power generation with different functionalities in the electric grid as base

Fig. 11.71 Steam accumulator banks of the 50 MW_{el} DSG power plant Khi Solar One/South Africa (LEMCO Group of Companies, 2023)

load or load following/balancing load power plant is difficult or even impossible to achieve due to the storage problem.

11.6.3.2 HTF-Rankine Based Solar Tower STE Concepts

The disadvantage of DSG power plant types with respect to a dispatchable power generation is eliminated by power plant concepts operated directly with molten salts, in which molten salts are used both in the solar field and in a construction mostly designed as 2-tank storage (TES). These concepts allow in principle a $24/7$ operation mode, but at the cost of a correspondingly large solar field ($SM > 2.8$). This requirement, which is noticeable in the high $CAPEX$ costs of the solar field of solar towers in the $LCoE$, can be reduced by fossil fuel auxiliary firing, but this then translates into a correspondingly higher CO_2 footprint. DMS solar tower systems such as Gemasolar, which has a TES of 15 h FLH, have operatively demonstrated $24/7$ operational capability. The plant demonstrated a capacity of $CF \approx 63\%$ in operation. Larger plants such as Crescent Dunes in the 100 MW_{el} have a projected capacity factor of well over 50%, which they have never achieved operationally (Boretti et al., 2019). Some of the concerns often mentioned in the media are mainly related to large plants above 50 MW_{el} and here there are two main reasons why the projected capacity factor in the solar tower plants is not achieved, which are not related to the highest thermally loaded component the receiver:

- The leakage of the molten salt tank. Three CSP solar tower projects, such as Crescent Dunes (USA), Gemasolar (Spain) and NOOR3 (Morocco), experienced leaks in the molten salt tank. Until the leak was stopped, power generation also had to be suspended.
- Problems of the steam turbine at partial load operation, especially in plants without back-up systems.

Both failure scenarios are based on engineering and materials causes, because the handling and operation of molten salt is anything but trivial and standard industrial technology, even if it is improving rapidly. Operation with liquid salts places high demands on materials and systems, both in terms of design, operation, and operating transients, and in terms of operating experience, which to date is only available over a period of a decade. With consistent consideration of best practice guidelines and quality assurance in project planning, construction and operation, cf. Mehos et al. (2020), almost all salt tower power plants show a good track record and, above all, that these difficulties are manageable, as recent performance records of the SUPCON Delingha 50 MW_{el} tower (China) proof. The plant schematic is largely the same as the DMS linear-focusing power plant schematic shown in Fig. 11.68

Also based on the direct storage power plant concept is the 50 MW_{el} beam down power plant Yumen Xinneng (China), whose optics differ from a conventional solar tower. Here, the heliostats first focus the irradiation on a secondary concentrator located on a tower, which projects it onto a molten salt receiver located on the ground. A two-tank molten salt system with a capacity of 9 h FLH also serves as thermal storage. Apart from the beam guidance, this system concept shown in Fig. 11.72a does not differ from the other DMS concepts. This also applies to the liquid sodium operated direct liquid metal power plant concept illustrated in Fig. 11.72b. Due to the specific thermophysical properties of the liquid metal, power densities of more than 2 MW/m^2 can be realized on the receiver (about twice that of a molten salt receiver). This allows compact receivers on relatively low towers, which are illuminated by near-tower heliostats with high optical efficiency. Hot sodium from multiple towers is collected and stored in this demonstrator facility in a 2-tank sodium storage tank with a capacity of $3FLH$. In principle, sodium allows thermal storage temperatures more than 200 °C higher than the molten salts, which dissociate exceedingly approx. 550–570 °C. However, these cannot be technically implemented so far due to the available power engineering equipment such as material technology constraints.

11.6.3.3 Joule-Brayton Based Solar Tower STE Plant Concepts

Unlike liquid-cooled receivers, gas-cooled receivers are not subject to temperature limitations of coolant integrity with respect to the heat transfer medium. The receiver absorbs the solar radiation reflected from the heliostat and converts it into heat at a high temperature level using optical conversion. Depending on the design of the receiver and the type of heat transfer fluid, the upper heat carrier temperatures can range from 250 °C to theoretically 1400 °C, so that due to the

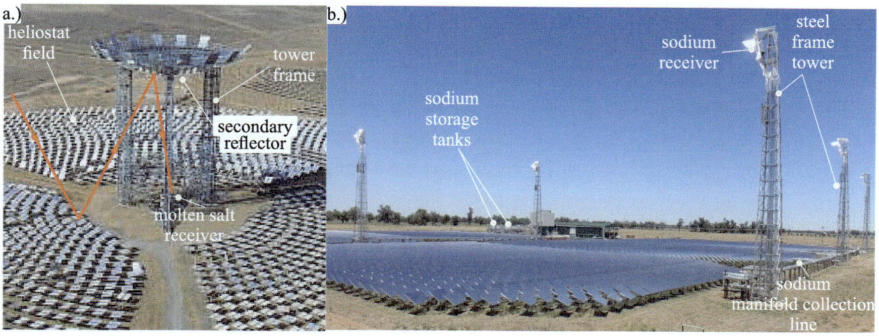

Fig. 11.72 **a** Arrangement of heliostat field, and solar beam path line in the 50 MW_{el} direct molten salt beam-down tower power plant Yumen Xinneng (China) (BCP Solar Technology Co., Ltd, 2023); **b** Arrangement of heliostats in the direct liquid sodium-operated multi-tower 1.1 MW_{el} demonstration power plant in Jemalong (Australia) from VastSolar (2023)

principally high possible concentration ratios, Brayton-based solar towers allow a greater variability in power generation in contrast to linear-focusing solar field concepts:

- Very high overall efficiency due to the high thermodynamic efficiency of the power conversion system at higher temperatures (well above 1000 °C) and a
- Comparatively easy integration of fossil carriers for hybridization with a variety of options, and thus a potential for dispatchable power generation with high annual capacity factors ($CF = 0.4–0.8$) with or without the use of thermal heat storage.

For the exploration of different central receiver systems, there are a number of publications looking at the analysis of different heat transfer gases such as compressed air, $s - CO_2$, or helium, cf. Behar et al. (2013). In particular, hybridization allows for different options of electricity generation, either primarily based on a Brayton cycle and using a downstream ORC or Rankine process as bottoming process, or a fossil-fired gas turbine process whose exhaust heat supports the Rankine cycle, cf. Blanco and Santigosa (2017). Focusing on the Brayton cycle as the core element, the gas turbine cycle benefits from its higher efficiency compared to Rankine-based cycles at temperatures above 550–600 °C, as enabled by solar towers in particular. In addition, the operating characteristics of the gas turbines are well known, at least for the air-cooled systems, even at part-load operation. This is a-priori an advantage, since part of a power plant concept is already commercially available, but also represents a disadvantage, since the efficiency is already limited by the fluctuating heat source, and thus the collector and thermal energy storage integration is of greater importance than in Rankine-based power plant concepts due to the smaller heat storage capacity of the gases as heat carriers and the associated larger temporal gradients. This is all the truer as the cost of heat generation using optical conversion is relatively high compared to chemical heat carriers, and thus the importance of efficiency plays a greater role in the calculation of electricity costs.

The best conversion efficiency of about 50% higher than pure Rankine-based power units is achieved by gas turbine systems in a combined cycle configuration (with a steam cycle as bottoming cycle), which makes the use of solar gas turbine systems very attractive because a high thermal conversion efficiency simultaneously reduces the size of the solar field for a given power output. Since the solar field represents the largest item of the $CAPEX$ in a central tower system, any decrease translates into an immediate reduction in the cost of electricity ($LCoE$).

The heat integration of the solar heat into the gas turbine process takes place in the receiver segment. Since the heat capacity and heat transfer coefficients of even pressurized gas systems are small and limited, the achievable peak heat fluxes of gas-cooled receivers themselves are below the values of molten salt receivers (which achieve approx. 1 MW/m^2). As a result, for example, compressed air-charged experimental solar tower receivers currently reach about 1000 °C, which is considerably lower than modern turbine inlet temperatures of fossil gas turbines of about 1400–1550 °C. Solar receiver adapted gas turbine systems with turbine inlet temperatures (TIT) around 1000 °C have been developed for years but have not yet reached market maturity, cf. Blanco and Santigosa (2017), so in most power plant scaled demonstrator experiments solar heat serves as a support to the conventional gas turbine combustor to achieve high TIT. For this, the conventional gas turbine must be appropriately "solar" adapted, which includes the following:

- Modification of the air flow, which includes preheating of the air by means of decoupling heat from the exhaust gas flow or intercooling and a staged compression process.
- Solar-adapted combustor, with parallel or series connection of the receiver air stream with a conventional combustor.
- Customized gas turbine control and regulation with respect to the receiver.

Figure 11.73 shows an example of the plant schematic of a compressed air driven solar gas turbine cycle with integrated intercooling and thermal heat storage in a parallel combustor with typical pressures and temperatures. In this process, called BCSS (Bottoming Combined Cycle Process), the gas turbine is operated either solar only or hybrid with a fuel and the warm exhaust gas serves as a heat source for the bottoming cycle, which can be either a Rankine or an ORC cycle.

As a result, operation using this plant concept results in high power output during sunny hours (by gas turbine plus bottoming cycle) and reduced power output during periods of low irradiance (mainly by the bottoming cycle). Brayton concepts also require thermal energy storage (regenerators) to achieve a sufficiently high capacity factor, but this necessitate both high pressures but especially temperatures beyond 700 °C compared to line-concentrating plants. Commercially available regenerators for high temperatures are, for example, so-called Cowper storages, which are used in blast furnaces in metal melting. These storage tanks

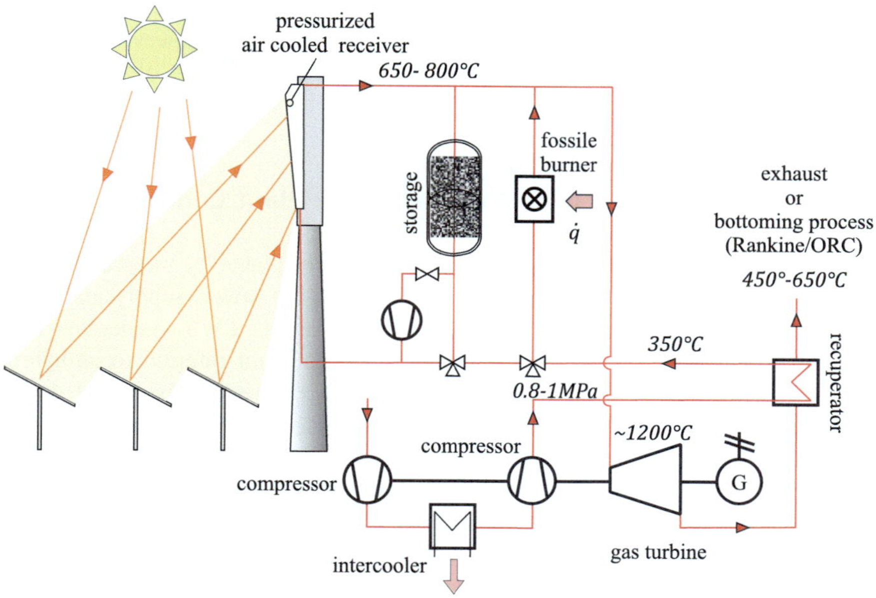

Fig. 11.73 Air flow and system diagram of a solar gas turbine process with typical temperatures and pressures

are designed either as fixed beds consisting of ceramic lattice bricks with a high heat capacity, or as packed beds in the form of ceramic spherical beds, whereby the gas heats the solid storage tank during charging and, conversely, extracts heat from the solid storage material during discharging.

The main limitations of all gas turbine processes are, on the one hand, the availability of power plant certified materials at high pressures for temperatures above 700 °C with sufficient cycle stability. In contrast to the relatively compact design of a gas turbine, the spatial separation of the solar combusto from the turbine inlet poses a design challenge to be solved technically. Another option to increase the thermodynamic efficiency of the gas turbine process is to reduce the compressor work of the circulator. For several reasons, supercritical carbon dioxide (s–CO_2) can be used as a heat transfer medium. (s–CO_2) has a lower adiabatic coefficient than compressed air and, above all, helium, so that the specific compressor work is reduced compared with air and, above all, helium. Other advantages of (s–CO_2) are the low critical temperature at high pressures, which makes heat transfer more difficult, especially in recuperative plant concepts, and above all the extremely high power density in the turbine section, which allows considerably smaller turbines than for air or helium. The challenges and options of (s–CO_2) is certainly warrant long-term research and development, discussed in more depth in Stein and Buck (2017) or White et al. (2021), but the technology is currently still far from being ready for the market.

The step towards higher process temperatures beyond 1000 °C is opened up by particle receivers without requiring excessive pressures; however, they are currently at a comparable level to the *s*–CO_2 systems in terms of maturity of component developments.

11.6.4 Fossil Hybrid Solar Thermal Power Plant Concepts

Improving the thermal efficiency of fossil-fired power plants by integrating solar technologies and vice versa enables a more predictable power supply and security of supply while producing lower CO_2 emissions per kWh_{el} generated than a purely fossil-fired power plant. The type of power plant referred to as hybrid allows capacity factors of significantly more than $CF = 80\%$. due to the use of fuels that can be planned at any time and, in addition to the provision of electricity, also heat, i.e. polygeneration. With the hybridization of the power plant by fossil carriers, however, the goal of CO_2-free power generation is abandoned, so that these concepts represent a technically proven option for power generation with reduced CO_2 emissions, but the goal of CO_2-free power generation is given in the long term only with the use of renewable fuels. Hybridization, however, should not be just a pretext to avoid the necessary fundamental change of our power sector towards renewables. Within this change, solar thermal heat generation can make an important contribution to clean power generation with higher efficiency, double use of conventional power plant technologies and possibly avoidance of large additional costs and material resources.

Nevertheless, it should be pointed out in this context that the avoidance of CO_2 emissions must not be considered singularly. The reduction of CO_2 in ISCCS plants has to be measured on the one hand in comparison to competing technologies such as carbon capture and storage (CCS) and on the other hand also in terms of the resource vector required by the power plant for construction and during operation, which is documented in the parameter of the energy payback time (EPT), cf. Eq. (11.95). Only a weighted consideration of efficiency, consideration of comparative technologies and ecological footprint allows a differentiated assessment in the economic and social context.

An integrated solar combined cycle power plant (ISCCS) essentially consists of three key assemblies: a combined cycle gas turbine (CCGT), a solar steam generator (SSG), and a solar field configuration including the receiver. Most of the ISCCS power plants in operation today use PTC technology, although a solar tower-based concept (ST-ISCCS) is one of the potential improvement options, as it can provide heat transfer fluid temperatures well above 450 °C, and thus allows a more versatile integration in the combined cycle (CC)-plant architecture.

Depending on the temperature level generated with concentrating solar thermal (CSP), the heat can be used either in the Rankine cycle, for example, to generate additional steam that is used in the steam turbine portion of the power plant to generate electricity, or to preheat the compressed air in the gas turbine prior to entering the combustor. Therefore, ISCCS power plants are conceptually able to effectively embed solar heat into the power plant configuration and achieve high capacity factors while maintaining significantly lower operating and capital costs, especially of a solar only plant, and also offer the possibility of higher operating flexibility compared to a purely fossil-fuelled power plant. Fossil hybridization is distinguished between two fundamentally different types of power plants:

- Solar-assisted conventional power plants or
- fossil support of solar power plants.

Figure 11.74 outlines in a simplified system diagram where heat from solar concentrator technologies could be integrated un a combined cycle plant as an example.

Based on some examples shown in the subsequent section, the influence of hybridization on the possible cost reduction in hybrid solar power plants with and without thermal storage is demonstrated with special consideration of the influence of combustion processes. However, all analyses also show that a substantial cost reduction can only be achieved with sufficiently low gas prices.

Comparing a solar-only PTC power plant with a direct molten salt storage of different thermal energy storage capacity (in FLH) with a gas turbine-only option and two hybrid options with a solar-combined-cycle technology, the results shown in Table 11.8 are obtained. Compared to the solar-only system, the hybrid system shows lower installation costs ($CAPEX$) and significantly smaller electricity production costs both in absolute terms and, to a lesser extent, solar electricity production costs for comparable annual energy production. In addition, if the solar fraction is to be of the order of 50% and more with simultaneously large annual

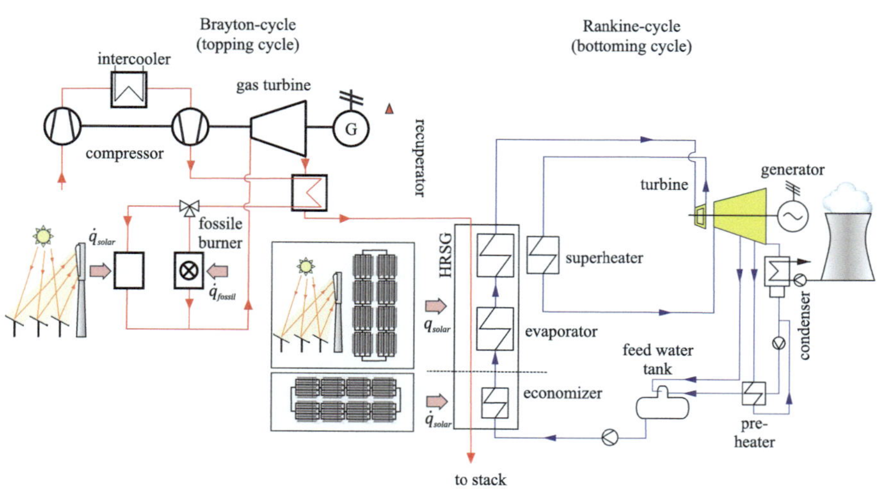

Fig. 11.74 Options for integrating heat generated with concentrating collector technology into a combined cycle power plant for electricity generation

solar efficiency, one must aim for a more even system load, which means reducing the number of start-up/shut-down ramps per year. But again, the cost result of hybridization is significantly dependent on the level of the specific gas price.

Two options are available here. The first involves keeping the 100 MW_{el} steam power unit as in the Solar-only plant and firing the gas turbine only during the period of solar irradiation. This operating strategy aims to maximize generation during daytime hours. Conversely, a steam power unit half the size can be selected if the gas turbine is operated only when solar irradiation is not present. With this option, a uniform output is achieved throughout the day, with a capacity factor $CF \approx 77\%$ (based on the 50 MW_{el} rated capacity) then being achieved (however, the solar fraction drops below 50%). Both hybrid power plant configurations show a significant advantage for the gas turbine heat rate due to the integration through

Table 11.8 Comparison of solar only, gas turbine only, and hybrid configurations for a PTC plant based hybrid using molten-HTF with TES-sizes for Dagget, California ($\Phi = 34°51'$); (Turchi & Ma, 2014)

Parameters	Unit	Solar only + TES ($FLH = 6$) ($SM = 2$)	Gas turbine only	Hybrid-solar TES ($FLH = 3$) ($SM = 1.5$)	Hybrid-solar + TES ($FLH = 4$) ($SM = 1.5$)
Steam turbine	MW	100	–	100	50
Gas turbine	MW	–	39.8	37.8	37.8
Annual steam turbine energy	GWh	379	–	315	173
Annual gas turbine energy			140	136	165
Total electric energy generated	GWh	379	140	375	338
Solar fraction SF	/	1.0	0	0.64	0.42
Annual solar utilization factor	MWh_{el}/MWh	0.16	–	0.162	0.16
Effective heat rate efficiency (gas)	/	–	0.414	0.483	0.474
Relative total $CAPEX$	/	1.0	0.074	0.845	0.473
Relative total $LCoE$		1.0	0.5	0.81	0.679
Absolute gas $LCoE$	\$/MWh_{el}		95	98	90
Absolute solar $LCoE$	\$/MWh_{el}	190	–	187	182

the waste heat recovery and the subsequent use in the steam cycle, so that the effective gas utilization rate increases from 41% for the gas turbine alone to about 48% in the hybrid plant. This type of detailed time-dependent calculations for integrated combined cycle power plants considers the heat transport chain from the solar field via the gas turbine linked to the Heat Recovery Steam Generator (HRSG) of the steam power plant section, taking all components into account. A large number of limitations occur in the part-load operation of the gas turbine as well as in the sub-segments of the HRSG and the steam turbine part, which lead to the fact that simple operating strategies for the solar field, such as constant heat transfer fluid outlet temperature, or constant electric power generation are either not feasible at all or with a loss of efficiency. Details of the methodology can be found in Broderick et al. (2017).

Hybridization by no means allows arbitrary integration of the heat source from a collector technology into a gas or steam turbine cycle or even combined power plant processes. The decisive factor is the outlet temperature level of the thermal heat generated by the solar field. Line-concentrating collectors such as parabolic troughs or linear Fresnel reach collector outlet temperatures in the order of 550 °C, which can be most efficiently integrated into a two-phase steam process. Conversely, solar towers would allow heat transfer fluid temperatures on the order of 1000 °C when a suitable heat transfer fluid is selected and a suitable receiver taking the high temperature stresses would be available. Then the heat above 1000 °C could be efficiently coupled into a gas turbine cycle as a preheater upstream of the combustor, for example. The way in which collector technology can be integrated into a combined cycle power plant process is illustrated in Fig. 11.75 by comparing the heat input to a conventional combined cycle power plant process with the ones using linear concentrating collectors or solar towers.

11.6.4.1 Solar-Assisted Conventional Power Plants

Solar-assisted power plant types feed the heat generated with the collectors, for example, into the feedwater preheating or into the steam section of a combined cycle power plant. CSP-based feedwater preheating reduces then the amount of bleed steam from the turbine stages required for regenerative preheating, allowing more steam to be used for power generation. CSP-based heat generation from solar towers can also be used to preheat the air of a gas turbine, reducing the amount of fuel used by the gas turbine.

For the solar assisted power plant types, the solar fraction (SF) generated with CSP in the total power generation is comparatively low and ranks between $SF = 0.6–14\%$, cf. Behar et al. (2014). However, since the fossil efficiency of the power plant of the mostly natural gas-fuelled power plants is not affected by the solar feed-in and hardly any technical problems are to be expected, this variant of hybridization is the most widespread technical option, although from an CO_2-emission point of view their future is debateable. Due to the omission of the additional investment for solar integration in the power plant section, the financial risk is considerably reduced compared to solar-only variants. These combined cycle power plants use natural gas in the gas turbine and utilize the waste heat

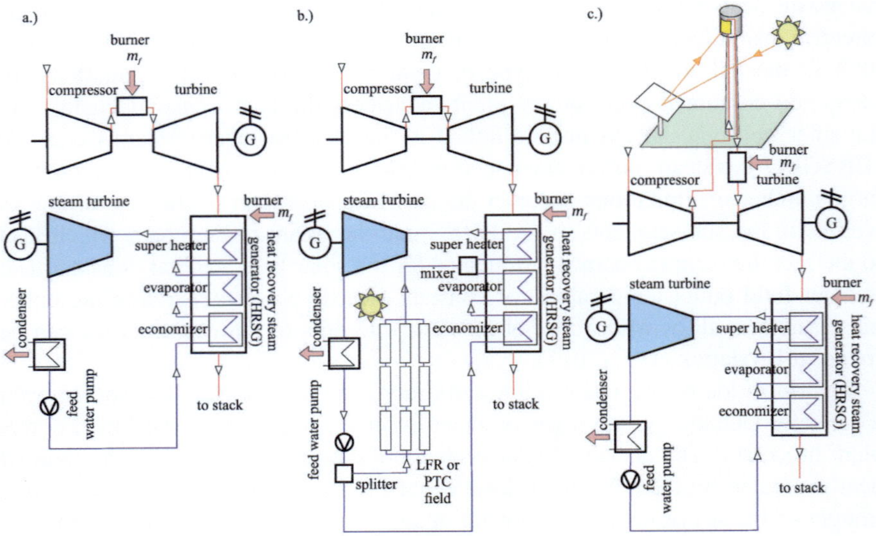

Fig. 11.75 Schematic diagram of a combined cycle (gas-steam) power plant concept with details of the main components; **a** conventional fossil combined cycle (CC) power plant; **b** integration of a linear concentrating collector technology in the CC; **c** integration of a solar tower in the CC

from the combustion gases to heat a steam cycle with the aid of a heat recovery steam generator (HRSG).

However, if the solar share is to be increased, the amount of steam generated with CSP must be increased. In a CCGT power plant, this requires either the gas turbine to go to partial load or the steam turbine to be oversized. The former is associated with considerable losses in efficiency of the gas turbine, so that in the case of additional solar power the gas turbine converts the fuel used into electricity less efficiently, and as a result the additional solar benefit is significantly reduced. If, on the other hand, the steam turbine is over dimensioned, it also goes into partial load operation when there is insufficient solar irradiation or at night, resulting in lower efficiency than in a plant optimized purely for fossil operation. The additional gas burned, especially at night, must therefore first be compensated during the day when there is solar irradiation, until the solar field actually contributes. The greater the oversizing of the steam turbine, the higher the solar contribution can be, but at the expense of reduced efficiency due to poorer partial load behaviour. Thus, an incorrect design of an ISCCS can result in more fuel consumption, and thus CO_2-emissions, being emitted per electrical kilowatt hour generated than in a reference power plant without solar support, when viewed over the course of a year. One of the future technical challenges is to efficiently coordinate power plant technology, storage and fluctuating solar field supply. Therefore, the solar fraction (SF) of most ISCCS power plants installed up to the GW_{el} scale is low for technical reasons. Relative to the overall economic risk of CSP integration in ISCCS power plants, feedwater preheating in existing gas or coal-fired power plants is therefore more sensible.

All studies show that pure fossil-fuelled CCGT power plants reveal a higher energetic efficiency than ISCCS power plants with integrated line-concentrating collectors (PTC or LFR). However, if the solar field is adequately tailed to the power conversion system ISCCS also exhibit significantly lower CO_2 emissions per produced kWh_{el}, so that the sole energetic efficiency analysis cannot be the only target-oriented argument. In order to clearly identify the weak points of these combined systems, the exergy analysis can be used as a complement to the energy/efficiency analysis. The combined energy and exergy analyses show that in a combined cycle process, the gas turbine combustor accounts for the largest share of exergy losses, followed by the solar field, while the largest energy losses occur in the condenser of the steam section and in the stack of the gas turbine (via the exhaust gases), cf., Reddy et al. (2012) and Dersch et al. (2004).

11.6.4.2 Fossil Support of Solar Thermal Power Plants

The situation is different if it is possible to feed the high temperatures of solar tower power plants directly into the gas turbine cycle (topping cycle), or to keep the partial load times of the steam turbine as low as possible using appropriate storage concepts. One possibility to achieve hybridization with a high solar fraction is a gas-fired evaporator connected in parallel to the solar field. Since the gas can be converted at best with the thermal efficiency of the steam process downstream of the collector field, and this is almost 50% lower compared to the values of a combined cycle power plant, the burner should only be used as a support. A reasonable operating strategy is to use the parallel fossil-fired evaporator only, when the solar field produces steam, but not sufficient to run the turbine at full load. Then co-firing compensates for small fluctuations in irradiance and allows the turbine's rated mass flow to be reached more often. Fossil co-firing increases solar efficiency because the number of partial load hours is reduced. Depending on the dimensioning of the burner and the solar field, this can also significantly reduce the $LCoE$. The real added value of solar tower combined cycle concepts is to utilize high temperature heat through the topping cycle down to the bottoming Rankine cycle, thus interfering little with the optimized architecture of the combined cycle, which increases plant utilization. This is analysed in detail in Okoroigwe and Madhlope (2016), although they also conclude that technical maturity and upscaling to large power plant capacities will require some R&D effort.

Turchi and Ma (2014) analyse both gas turbine/solar hybrids and gas turbine combined cycle (CCGT) processes. Considering a coupling of gas turbines and a thermal oil operated parabolic trough STE without thermal energy storage (TES), they find that for solar CCGT-hybrids, a cost-optimal mode of operation is for a given low gas price to start the gas turbine before sunrise and stop after sunset (each with a one-hour offset). Starting the gas turbine before sunrise allows the exhaust heat to be used to start the steam turbine. When the gas turbine is operating at full load, all of the exhaust heat is fed into the steam cycle if sufficient capacity

Table 11.9 Comparison of two different solar only with a hybrid STE (gasturbine CC + solar) assuming thermal oil PTC for without TES modified from Turchi and Ma (2014)

Parameters	Unit	Solar only ($SM = 1.1$)	Solar only ($SM = 1.4$)	Hybrid-solar ($SM = 1.1$)
Steam turbine	MW	100	100	100
Gas turbine	MW			37.8
Steam turbine energy	GWh	210	246	231
Total electric energy generated	GWh	210	246	375
Solar fraction SF	/	1.0	1.0	0.57
Annual solar utilization factor	MWh_{el}/MWh	0.159	0.145	0.161
Relative total $CAPEX$	/	1	1.18	1.12
Relative total $LCoE$		1	0.978	0.756
Absolute gas $LCoE$	\$/MWh_{el}			95
Absolute solar $LCoE$	\$/MWh_{el}	184	180	168

is available. Shutting down the gas turbine during periods of high solar radiation can improve annual efficiency, but these periods often coincide with high demand, so it makes more sense to run the turbine all day. A low solar multiple of $SM \approx 1.1$ allows the gas turbine's thermal contribution to be used in the steam cycle during most of the year, with solar energy rarely having to be discarded. Table 11.9 shows, for a 100 MW_{el}, a comparison of a solar-only power plant at two different solar multiples SM compared to a CCGT-hybrid. One should highlight, that no storage option is included here for solar-only plants, which is considered suboptimal today. While the capital expenditure for the hybrid option increases by only 12% for the same electrical output (and thus less than the increase in solar field size), the energy fed into the electrical grid increases by 78% and the electricity production costs are reduced by about 25%. The annual solar efficiency of the hybrid is also slightly higher than that of the solar-only type. However, the solar fraction is then only about $SF \approx 57\%$.

11.6.4.3 Concluding Remarks on Solar Thermal Hybrid Plants

In the long run, it does not make sense to burn fossil fuel with a comparatively poor efficiency in order to reduce the cost of electricity ($LCoE$) of solar power plants. However, if gas can be used in small quantities to compensate power fluctuations of the solar field collectors, with predominantly solar-powered power plants, this variant is very interesting. The extent to which solar thermal energy can be generated at "competitive" prices will not be discussed in this context, as it depends on dramatically changing boundary conditions e.g. the fuel prices such as natural gas.

One important aspect should not be forgotten. The energy demand of a society is characterized by its need of electricity at any time, as determined by the branches of industry, services as well as the private households. The temporal match of generation and demand has a crucial effect on the hourly price of energy. With a foreseeable reduction of fossil resources to eventually carbon neutrality, electrification of mobility and industrial production, the match between fluctuating renewable generation and the demand profile will be more difficult. The security of supply requires either the development and improvement of large-volume energy storage, the embedding of renewable energies in an energy mix as well as resource-saving technologies that are able to flexibly integrate or exchange different forms of energy and energy technologies at plant and grid level.

Hybridization, which combines two energy sources in a single plant, offers a number of advantages over power plant technologies that use only a single heat source. The advantages are not only variability and increased flexibility, but also include the ability to use a common infrastructure that allows for either greater efficiency or increased capacity. In the short term, hybrid systems between solar energy and fossil fuels offer the potential to reduce the use of fossil fuels, while in the longer term they offer the entry point of carbon neutral or even carbon negative power generation. In some cases, even, a fuelling by renewable green fuels (e.g. produced from hydrogen generated using PV and wind electricity), could be a better option in the total energy m than excessive storage investments in spite of efficiency considerations.

In the context of this chapter, the aspect of direct solar thermal heat utilization in power plant processes was considered. However, high-temperature heat in particular broadens the spectrum of solar thermal integration into, for example, CO_2. capture technologies such as oxy-fuel combustion or chemical looping combustion, since these use the same components for both CO_2 capture and solar energy storage. Just using the same overall infrastructure is likely to reduce the cost of such systems. This extremely broad field is still in the early stages of development and further elaboration is beyond the scope of this book, so the reader is referred to related review articles analysing and discussing energy and material cogeneration aspects, such as Powell et al. (2017) or Nathan et al. (2018) and the literature cited therein.

References

(NREL), N. R. (2022). Concentrating solar power projects. (N. R. Laboratory, Ed.). Retrieved January 19, 2022, from https://solarpaces.nrel.gov/

Abdel Dayem, A., Nabil Metwally, M., Alghamdi, A., & Marzouk, E. M. (2014). Numerical simulation and experimental validation of integrated solar combined power plant. *Energy Procedia, 50*, 290–305. https://doi.org/10.1016/j.egypro.2014.06.036

Alaphilippe, M., Bonnet, S., & Stouffs, P. (2007). Low power thermodynamic solar energy conversion: Coupling of a parabolic trough concentrator and an Ericsson engine. *International Journal of Thermodynamics, 10*, 37–45.

Andrews, J., & Akbarzadeh, A. (2002). *Solar pond project: Stage 1—Solar ponds for industrial process heating.* Renewable Energy Commercialisation program. RMIT, Australian Greenhouse Office.

Askari, I. B., & Ameri, M. (2017). The application of Linear Fresnel and Parabolic Trough solar fields as thermal source to produce electricity and fresh water. *Desalination, 415*, 90–103. https://doi.org/10.1016/j.desal.2017.04.005

Augsburger, G. (2013). *Thermo-economic optimisation of large solar tower power plants.* École Polytechnique Fédérale de Lausanne-EPFL.

Bachelier, C. (2021). *Techno-economical analyses of linear solar thermal electric plants for dispatchable power generation.* KIT Faculty for Mechanical Engineering, Karlsruher Institut für Technologie KIT.

Bachelier, C., Selig, M., Mertins, M., Stieglitz, R., Zipf, V., Neuhäuser, A., & Steinmann, W. D. (2015). Systematic analysis of Fresnel CSP plants with energy storage. (S. 2. International Conference on Concentrating Solar Power and Chemical Energy Systems, Ed.) *Energy Procedia, 69*, 1201–1210. https://doi.org/10.1016/j.egypro.2015.03.207

Bauer, T., Pfelger, N., Laing, D., Steinmann, W.-D., Eck, M., & Kaesche, S. (2013). High temperature molten salts for solar power application. In F. Lantelme & H. Groult (Eds.), *Molten salts chemistry: From lab to applications* (p. 592). Elsevier.

BCP Solar Technology Co., Ltd. (2023). Tower CSP systems. Retrieved Febraury 17, 2023, from http://www.xinchen-csp.com/

Behar, O., Khellaf, A., & Mohammedi, K. (2013). A review of studies on central receiver solar thermal power plants. *Renewable and Sustainable Energy Reviews, 23*, 12–39. https://doi.org/10.1016/j.rser.2013.02.017

Behar, O., Khellaf, A., Mohammedi, K., & Ait-Kaci, S. (2014). A review of integrated solar combined cycle system (ISCCS) with a parabolic trough technology. *Renewable and Sustainable Energy Reviews, 39*, 223–250. https://doi.org/10.1016/j.rser.2014.07.066

Binecinto, M., Gozalez, L., Zarza, E., Diez, L., & Munoz-Anton, J. (2014). Performance model and annual yield comparison of parabolic-trough solar thermal power plants with either nitrogen or synthetic oil as heat transfer fluid. *Energy Conversion and Management, 87*, 238–249. https://doi.org/10.1016/j.enconman.2014.07.017

Blanco, M., & Santigosa, L. R. (2017). *Advances in concentrating solar thermal research and technology.* Woodhead Publishing, Elsevier Ltd. https://doi.org/10.1016/C2014-0-04054-3

Bonnet, S., Alaphilippe, M., & Stouffs, P. (2005). Energy, exergy and cost analysis of a micro-cogeneration based on an Ericsson engine. *International Journal of Thermal Sciences, 44*, 1161–1168.

Boretti, A., Castelletto, S., & Al-Zubaidy, S. (2019). Concentrating solar power tower technology: Present status and outlook. *Nonlinear Engineering, 1*, 10–31.

Broderick, P. G., Brandt, A. R., & Durlofsky, L. (2017). Operational optimization of an integrated solar combined cycle under practical time-dependent constraints. *Energy, 141*, 1569–1584. https://doi.org/10.1016/j.energy.2017.11.059

Calise, F., d'Accadia, M., Libertini, L., & Vicidomini, M. (2018). Thermoeconomic analysis of an integrated solar combined cycle power plant. *Energy Conversion and Management, 171*, 1038–1051. https://doi.org/10.1016/j.enconman.2018.06.005

Caraballo, A., Galan-Casado, S., Caballero, A., & Serena, S. (2021). Molten salts for sensible thermal energy storage: A review and an energy performance analysis. *Energies, 14*, 1197–1212. https://doi.org/10.3390/en14041197

Collado, F., & Guallar, J. (2016). Two-stages optimised design of the collector field of solar power tower plants. *Solar Energy, 135*, 884–896. https://doi.org/10.1016/j.solener.2016.06.065

Corona, B., & San Miguel, G. (2015). Environmental analysis of a Concentrated Solar Power (CSP) plant hybridised with different fossil and renewable fuels. *Fuel, 145*, 63–69. https://doi.org/10.1016/j.fuel.2014.12.068

Costa, R. F., & MacDonald, B. D. (2018). Comparison of the net work output between stirling and ericsson cycles. *Energies, 11*, 670. https://doi.org/10.3390/en11030670

Delmore, V. (1980). Numerical simulation of thermohaline convection in the upper ocean. (C. U. Press, Ed.) *Journal of Fluid Mechanics, 96*(4), 803–826. https://doi.org/10.1017/S0022112080002376

Dersch et al., 2004 Dersch, J., Geyer, M., Herrmann, U., Jones, S. A., Kelly, B., Kistner, R., et al. (2004). Trough integration into power plants—A study on the performance and economy of integrated solar combined cycle systems. *Energy, 29*, 947–959. https://doi.org/10.1016/S0360-5442(03)00199-3

Desai, N. B., Kedarfe, S. B., & Bandyopadhyay, S. (2014). Optimization of design radiation for concentrating solar thermal power plants without storage. *Solar Energy, 107*, 98–112. https://doi.org/10.1016/j.solener.2014.05.046

dos Santos Bernardes, M. A., Voß, A., & Weinrebe, G. (2003). Thermal and technical analyses of solar chimneys. *Solar Energy, 75*, 511–524.

Dostal, V., Driscoll., M., & Hejzlar, P. A. (2004). *A supercritical carbon dioxide cycle for next generation nuclear reactors.* MIT topical report.

dos Santos Bernardes, M. A., & von Backström, T. W. (2010). Evaluation of operational control strategies applicable to solar chimney power plants. *Solar Energy, 84*(2), pp. 277–288.

Dubos, F. (1954). Harnessing the winds. In W. Ley (Ed.), *Engineers dream.* Viking Press.

Feldhoff, J. F., Benitez, D., Eck, M., & Riffelmann, K. (2010). Economic potential of solar thermal power plants with direct steam generation compared with HTF plants. *Journal of Solar Energy Engineering, 132*, 041001-1–041001-9.

Feldhoff, J., Schmitz, K., Eck, M., Schnatbaum-Laumann, L., Laing, D., Ortiz-Vives, F., & Schulte-Fischedick, J. (2012). Comparative system analysis of direct steam generation and synthetic oil parabolic trough power plants with integrated thermal storage. *Solar Energy, 86*, 520–530. https://doi.org/10.1016/j.solener.2011.10.026

Fichtner. (2010). *Technology assessment of concentrated solar thermal power technologies for a site specific project in South Africa.* Final Report 5442P10/Ficht-6156498-v1, Worlbank.

Fluri, T. (2009). The potential of concentrating solar power in South Africa. *Energy Policy, 12*, 5075–5080.

Fluri, T. P., & von Backström, T. W. (2008). Performance analysis of the power conversion unit of a solar chimney power plant. *Solar Energy, 82*(11), 999–1008.

Förster, H., & Schmidt, K.-P. (2006). Kreisprozesse für die maximale Nutzung von Wärmeangeboten im Temperaturbereich von 80 bis 250°C in Industrie. *Geothermie und Solarwirtschaft, 1924*, 289–303.

Frenell GmbH. (2022). Concentrated Solar Power—sustainable, efficient and proven. Karlsruhe, Germany. https://www.frenell.de/

Giaconia, A., & Grena, R. (2021). A model of integration between PV and thermal CSP technologies. *Solar Energy*, 149–159.

Giostri, A., Binotti, M., Silva, P., Macchi, E., & Manzolini, G. (2012). Comparison of two linear collectors in solar thermal plants: Parabolic trough versus fresnel. *Journal of Solar Energy Engineering, 135*, 011001-1–011001-9.

González-Roubaud, E., Pérez-Osorio, D., & Prieto, C. (2017). Review of commercial thermal energy storage in concentrated solar power plants: Steam vs. molten salts. *Renewable and Sustainable Energy Reviews, 80*, 133–148. https://doi.org/10.1016/j.rser.2017.05.084

Griffin, P., Huschka, K., & Morin, G. (2009). Software for design, simulation, and cost estimation of solar thermal power and heat cycles. In *Proceedings of the SolarPACES.*

Günther, H. (1931). *In hundert Jahren – Die künftige Energieversorgung der Welt.* (F. V. Gesellschaft der Naturfreunde, Ed.) Kosmos.

Hendersson-Sellers, B. (1986). Calculating the surface energy balance for lake and reservoir modeling: A review. *Reviews of Geophysics, 24*, 625–649.

Hensley, J. (2009). *Cooling tower fundamentals* (2nd ed.). SPX Cooling Technologies Inc.

Hesleton, K. (2005). *Boiler operators handbook.* Thew Fairmont Press Inc.

Hill, G., Pring, E., & Osborn, P. (1990). *Cooling towers—Principles and practice* (2nd ed.). Butterowrth-Heinemann.

Jaefarzadeh, M. (2004). Thermal behavior of a small salinity-gradient solar pond with wall shading effect. *Solar Energy, 77*(3), 281–290. https://doi.org/10.1016/j.solener.2004.05.013

Ju, X., Xu, C., Hu, Y., Han, X., Wei, G., & Du, X. (2017). A review on the development of photovoltaic/concentrated solar power (PV-CSP). *Solar Energy Materials and Solar Cells*, 305–327.

Kalina, A. I. (1982). *U.S. Patent No. 4,346,561.*

Kammas, A. (2011). *The book of synergy.* Frank Media.

Köhler, S. (2005). *Geothermisch angetriebene Dampfkraftprozesse Analyse und Prozessvergleich binärer Kraftwerke.* Dissertation, Berlin.

Kost, C. S., Jülch, V., Nguyen, H.-T., & Schlegl, T. (2018). *Levelized cost of electricity renewable energy technologies.* Fraunhofer Institute for Solar Energy Systems-ISE.

Kreetz, H. (1997). *Theoretische Untersuchungen und Auslegung eines temporären Wasserspeichers für das Aufwindkraftwerk.* TU Berlin.

Leblanc, J., Akbarzadeh, A., Andrews, J., Lu, H., & Golding, P. (2011). Heat extraction methods from salinity-gradient solar ponds and introduction of a novel system of heat extraction for improved efficiency. *Solar Energy*, 3103–3142.

LEMCO Group of Companies. (2023). KHI solar plant. Retrieved February 17, 2023, from https://lemco.co.za/our-projects/steel-structures-khi-solar-one-plant-renewable-energy/

Lenz, J., & Meiswinkel, R. (2005). Die VGB-Richtlinie "Bautechnik bei Kühltürmen" und ihre Anpassung an dien eue DIN 1045 / DIN 1055. *Beton- Und Stahlbeton, 100*(1), 12–19.

Lewis, R. I., Lewin, D. R., Seider, W. D., & Seader, J. D. (1996). Turbomachinery performance analysis. John Wiley & Sons Inc.

Li, Y., & Yang, Y. (2014). Thermodynamic analysis of a novel integrated solar combined cycle. *Applied Energy, 122*, 133–142. https://doi.org/10.1016/j.apenergy.2014.02.017

Losordo, T., & Piedrahita, R. (1991). Modeling temperature variation and thermal stratification in shallow water aquaculture ponds. *Ecological Modelling, 54*(3–4), 189–226.

Lu, H., Swift, A. H., Hein, H. D., & Walton, J. C. (2004). Advancements in salinity gradient solar pond technology based on sixteen years of operational experience. *ASME Journal of Solar Energy Engineering, 126*(2), 759–767. https://doi.org/10.1115/1.1667977

Ma, X., Karamos, G. S., & Karniadakis, G. E. (2000). Dynamics and low-dimensionality of a turbulent near wake. *Journal of Fluid Mechanics, 410*, 29–65.

Mehos, M., Price, H., Cable, R., Kearney, D., Kelly, B., Kolb, G., & Morse, F. (2020). *Concentrating solar power best practise study.* National Renewable Energy Laboratory (NREL).

Mihoub, S., Chermiti, A., & Beltagy, H. (2017a). Methodology of determining the optimum performances of future concentrating solar thermal power plants in Algeria. *Energy, 122*, 801–810. https://doi.org/10.1016/j.energy.2016.12.056

Mokhtar, M. M. (2018). *Control of solar thermal linear fresnel collector plants in single phase and direct steam generation modes.* Dissertation, Karlsruher Institut f¨ur Technologie (KIT), KIT Faculty of Mechanical Engineering, Karlsruhe.

Montes, M. J., Abanades, A., Martinez-Val, J., & Valdes, M. (2009). Solar multiple optimization for a solar-only thermal power plant, using oil as heat transfer fluid in the parabolic trough collectors. *Solar Energy, 83*, 2165–2176. https://doi.org/10.1016/j.solener.2009.08.010

Morin, G., Dersch, J., Platzer, W., Eck, W., & Häberle, A. (2012). Comparison of linear fresnel and parabolic trough collector plants. *Solar Energy, 86*, 1–12. https://doi.org/10.1016/j.solener.2011.06.020

Morin, G., Lerchenmüller, H., Mertins, M., Ewert, M., Fruth, M., Bockamp, S., et al. (2004). Plug-in strategy for market introduction of fresnel-collectors. In *Proceedings of the 12th international symposium SolarPaces.*

Nathan, G., Jafarian, M., Dally, B., Saw, W., Ashman, P., Hu, E., & Steinfeld, A. (2018). Solar thermal hybrids for combustion power plant: A growing opportunity. *Progress in Energy and Combustion Science, 64*, 4–28. https://doi.org/10.1016/j.pecs.2017.08.002

NIST. (2018). NIST chemistry WebBook. (N. I. Technology, Ed.) U.S. Secretary of Commerce on behalf of the United States of America. Retrieved May 1, 2021, from https://webbook.nist.gov/

Okoroigwe, E., & Madhlope, A. (2016). An integrated combined cycle system driven by a solar tower: A review. *Renewable and Sustainable Energy Reviews, 57*, 337–350. https://doi.org/10.1016/j.rser.2015.12.092

Ondraczek, J., Komendantova, N., & Patt, A. (2015). WACC the dog: The effect of financing costs on the levelized cost of solar PV power. *Renenwable Energy, 75*, 888–898. https://doi.org/10.1016/j.renene.2014.10.053

Peterseim, J. H., White, S., Tadros, A., & Hellwig, U. (2013). Concentrated solar power hybrid plants, which technologies are best suited for hybridisation? *Renewable Energy, 57*, 520–532. https://doi.org/10.1016/j.renene.2013.02.014

Pitz-Paal, R., Buck, R., & Hoffschmidt, B. (2002). *Solarturmkraftwerkssysteme.* FVS Themen 2002.

Platzer, W. (2015). Combined solar thermal and photovoltaic power plants—An approach to 24h solar electricity? In *Proceedings of the 21st SolarPACES international symposium, 15–16 Oct, 2015, Capetown.*

Platzer, W. (2021). Total performance assessment method for industrial process heat systems. In *ISES solar world congress 2021, 25–29 October, 2021.*

Platzer, W. J. (2014). PV–enhanced solar thermal power. *Energy Procedia, 57*, 477–486.

Platzer, W., & Dinter, F. (2015). A learning curve for solar thermal power. In *Proceedings of the 21st SolarPACES international symposium, 15–16 Oct, 2015, Capetown.*

Powell, K., Rashid, K., Ellingwood, K., Tuttle, J., & Iverson, B. (2017). Hybrid concentrated solar thermal power systems: A review. *Renewable and Sustainable Energy Reviews, 80*, 215–237. https://doi.org/10.1016/j.rser.2017.05.067

Rabl, A., & Nielsen, C. (1975). Solar ponds for space heating. *Solar Energy, 17*(1), 1–12.

Ravelli, S., Franchini, G., Perdichizzi, A., Rinaldi, S., & Valcarenghi, V. (2016). Modeling of direct steam generation in concentrating solar power plants (A. 1.-1. 71st Conference of the Italian Thermal Machines Engineering Association, Ed.) *Energy Procedia, 101*, 464–471.

Rebai, L. K., Mojtabi, A. K., Safi, M. J., & Mohamad, A. A. (2006). A linear stability study of the gradient zone of a solar pond. *Transactions of the ASME Journal of Solar Energy Engineering, 128*(3), 383–393. https://doi.org/10.1115/1.2210498

Reddy, V., Kaushik, S., & Tyagi, S. (2012). Exergetic analysis of solar concentrator aided natural gas fired combined cycle power plant. *Renewable Energy, 39*(1), 114–125. https://doi.org/10.1016/j.renene.2011.07.031

Rosenberry, D. O., Winter, T. C., Buso, D. C., & Likens, G. E. (2007). Comparison of 15 evaporation methods applied to a small mountain lake in the northeastern USA. *Journal of Hydrology, 340*(3–4), 149–166.

Rovira, A., Abbas, R., Sanchez, C., & Munoz, M. (2020). Proposal and analysis of an integrated solar combined cycle with partial recuperation. *Energy, 198*, 117379. https://doi.org/10.1016/j.energy.2020.117379

Rovira, A., Sánchez, C., Valdés, M., Abbas, R., Barbero, R., José Montes, M., et al. (2018). Comparison of different technologies for integrated solar combined cycles: Analysis of concentrating technology and solar integration. *Energies, 11*(5), 1064ff. https://doi.org/10.3390/en11051064

Schlaich, J., & Schiel, W. (2001). Solar chimneys. In R. A. Meyers (Ed.), *Encyclopedia of physical science and technology* (3rd ed.). Academic Press.

Seitz, M., Johnson, M., & Hübner, S. (2017). Economic impact of latent heat thermal energy storage systems within direct steam generating solar thermal power plants with parabolic troughs. *Energy Conversion and Management, 143*, 286–294. https://doi.org/10.1016/j.enconman.2017.03.084

Sheu, E. J., Mitsos, A., Eter, A., Mokheimer, E., Habib, M., & Al-Qutub, A. (2012). A review of hybrid solar–fossil fuel power generation systems and performance metrics. *Journal of Solar Energy Engineering, 134*, 041006-1–041006-17.

Spelling, J., Laumert, B., & Fransson, T. (2014). A comparative thermoeconomic study of hybrid solar gas-turbine power plants. *Journal of Engineering for Gas Turbines and Power, 136*, 011801-1–011801-9.

Stein, H., & Buck, R. (2017). Advanced power cycles for concentrated solar power. *Solar Energy, 152*, 91–105. https://doi.org/10.1016/j.solener.2017.04.054

Suarez, F., Tyler, S., & Childress, A. E. (2010). A fully coupled, transient double-diffusive convective model for salt-gradient solar ponds. *International Journal of Heat and Mass Transfer, 53*, 1718–1730.

Traupel, W. (1971). *Die Grundlagen der Thermodynamik*. G. Braun Verlag.

Turchi, C. S., & Ma, Z. (2014). Co-located gas turbine/solar thermal hybrid designs for power production. *Renewable Energy, 64*, 172–179. https://doi.org/10.1016/j.renene.2013.11.005

VastSolar. (2023). *Scalable, cost-effective concetrated solar thermal power system*. Retrieved February 17, 2023, from https://www.vastsolar.com/

Vogel, A. (2016). *Konzepte für Speicher in thermischen Solarkraftwerken*. Dissertation, Technische Universität Braunschweig; Fraunhofer-Institut für Solare Energiesysteme.

Wagar, W. R., Zamfirescu, C., & Dincer, I. (2010). Thermodynamic performance assessment of an ammonia–water Rankine cycle for power and heat production. *Energy Conversion and Management, 52*(12), 2501–2509.

Wang, J., & Seyed-Yagoobi, J. (1994). Effects of water turbidity and salt concentration levels on penetration of solar radiation under water. *Solar Energy, 52*, 429–438.

Wang, Y. F., & Akbarzadeh, A. A. (1983). A parametric study on solar ponds. *Solar Energy, 30*, 555–562.

Weinrebe, G. (2004). Das Aufwindkraftwerk-Wasserkraftwerk der Küste. *Nova Acta Leopoldina NF, 91*(339), 117–141.

White, M. T., Bianchi, G., Chai, L., Tassou, S., & Sayma, A. (2021). Review of supercritical CO_2 technologies and systems for power generation. *Applied Thermal Engineering, 185*, 116447. https://doi.org/10.1016/j.applthermaleng.2020.116447

Yamaguchi, H., Zhang, X., Fujima, K., Enomoto, M., & Sawada, N. (2006). Solar energy powered Rankine cycle using supercritical CO_2. *Applied Thermal Engineering, 25*, 2345–2354. https://doi.org/10.1016/j.applthermaleng.2006.02.029

Solar Process Heat

12

Abstract

The application of solar thermal systems to cover the heat demand of industries is aa relatively new and developing field. The heat demand in industry worldwide is huge and is needed to drive many different processes. Temperature requirements, demand profiles and integration complexities shown in the chapter vary from case to case. For temperatures above the typical 50–80 °C of a solar thermal hot water system, improved medium temperature collectors have been developed. For these collectors the thermal efficiency at higher temperatures between 100 °C and 400 °C is achieved by vacuum technologies, improved transparent covers or by optical concentration. A direct and strong economic competition to renewable supply of heat is the optimization of heat recovery potentials using a well-known tool like Pinch technology. Energy efficiency and renewable energies have to be considered in parallel to arrive at the best solution. Up to now the standardization level is low, leading to individual engineering solutions. The philosophies cover direct supply of solar thermal heat to individual processes looking for the best match of technology and process, to the integration of solar thermal heat into existing supply and distribution systems. A classification and characterization of the solar integration using different types of heat exchangers is shown. This is also demonstrated and described in some detail for real installations.

Keywords

Heat demand • Process heat • Temperature level • Demand profile • Heat recovery • Integration of solar heat • Heat exchanger networks • Pinch technology

R. Stieglitz and W. Platzer, *Solar Thermal Energy Systems*,
https://doi.org/10.1007/978-3-031-43173-9_12

12.1 Introduction

The application of solar thermal heat for industrial use is not completely new, its role in the worldwide markets until today is small. Solar domestic hot water systems for small and larger buildings are on the contrary widespread in most countries. One of the reasons is certainly that the preparation of hot water for domestic use is a relatively standardized task: One needs to supply hot water at a service temperature of 45–65 °C, the upper temperature mainly being required in order to avoid legionellae problems. A hot water storage for a demand of 1–2 days often is sufficient, and although the individual demand of hot water is varying, the dimensioning of a solar thermal system supplying the water needed for a single family is straightforward. Pre-dimensioned systems are available on the market and usually there is no need for a more sophisticated engineering. In industry, however, the situation is completely different. Starting with very low temperatures of 25 °C for washing or curing materials to extremely high temperatures with many hundred degrees e.g. for melting processes or chemical reactions the temperature levels in industrial processes have a wide span. Also, the size of the industrial plant and its energy or heat demand varies substantially. The quality and the production output of industrial goods can be influenced easily by details in the processing parameters such as heating and cooling rates, pressure, temperature stability in certain periods. The media to be heated is not just water but can vary from sensitive food ingredients to aggressive acids or bases which even might react at some point. Heat transfer to and from the media depends on physical phases, different shape, and mass relations. Therefore, somebody designing a solar thermal system for supplying heat to industrial process needs not only knowledge about a broad range of solar components and their properties, using that for producing a specified daily heat amount at a specified temperature level. It is also important to know the requirements and operation of the individual industrial heat supply system, which should be supported by solar. Technical solutions to supply heat at suitable points of a complex production system have to be found and checked. And a communication with the process engineers for understanding the impacts of any change in the existing system on the process output, in quantity and quality, is a key issue.

For this reason, we do not start immediately with describing solar system designs but try to give a short introduction into the variety of conventional industrial process heat systems. Also, methods to improve energy efficiency will be shown. A kilowatt-hour (kWh) of thermal energy which has not to be supplied by improving a system is as good as one supplied by a renewable source. As we will see the potential market for solar energy in industrial applications is huge, despite the small share such systems have at the moment. A better conceptualization and standardization of solar heat for industrial processes ($SHIP$) is a pathway to larger and more implementation which is currently being developed.

12.2 Heat Demand in Industry

In order to show the size of the potential market for the industrial application as an example the final energy consumption of the European Union is given. In Fig. 12.1a the yearly final energy consumption within the countries of the European Union (EU-28) is shown for the year 2020. The total final energy consumption has been 37086 PJ, and in the sectors of households (domestic use), transport, industry an almost equal share of more than a quarter in each is allocated, while the rest is devoted to the tertiary sector (services) about 13% and a small fraction to agriculture and others. In industry about one third of the final energy demand is required for electricity and about two thirds for heat. There are several studies trying to derive a heat demand in different temperature intervals and different industry sectors which is depicted in Fig. 12.1b.

Analysing the heat demand according to temperature levels in industry it is clear that is very much country dependent. Integrating e.g. across Europe within the EU-28, however, it becomes visible that a large fraction is consumed for high temperature heat, mainly steel and metal smelting/processing, while food and paper industries require much lower temperature levels, as illustrated in Fig. 12.2a. If we look at the sources of heat and cold generation at the different temperature levels, as shown in Fig. 12.3b, it becomes apparent that renewable energies have so far only played a minor role; electricity is also only used to a small extent in the majority of cases to provide cooling.

When we look into a comprehensive study (Lauterbach et al., 2011) in process heat demand per temperature range in Germany (see Fig. 12.3), we have to note that in this study concentrating technologies are not considered for the determination of the potential solar contribution, as it was considered that with the relatively

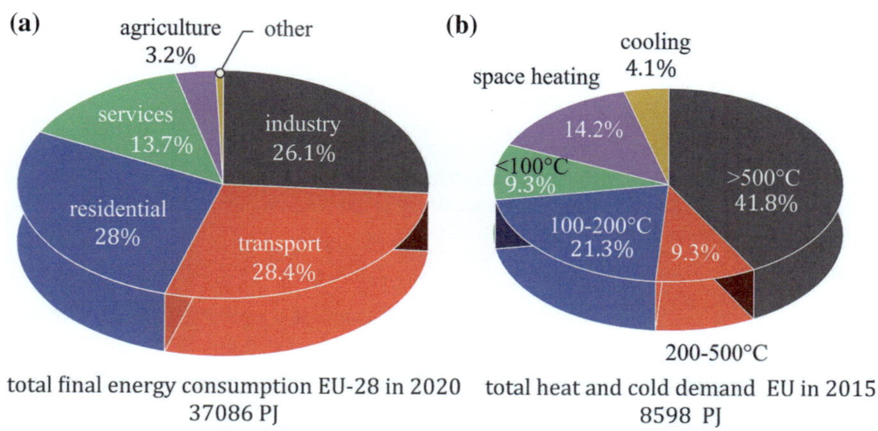

Fig. 12.1 **a** EU-28 final energy consumption by sector (Eurostat-Statistics Explained, 2022); **b** heat consumption in industry broken down by temperature level (Heat Roadmap Europe, 2017)

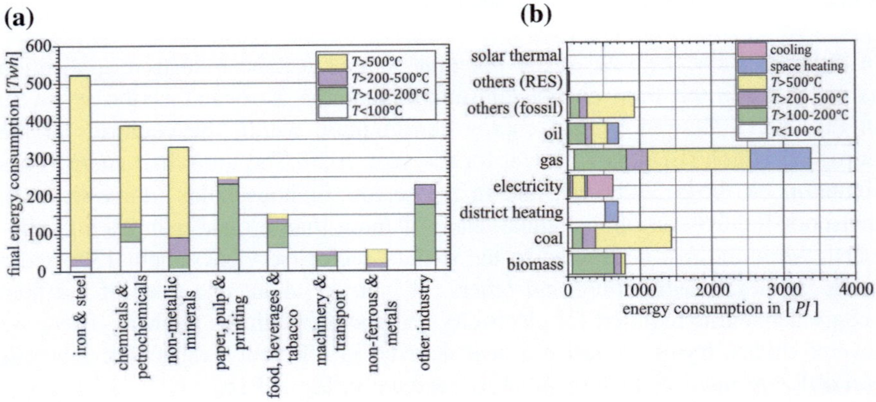

Fig. 12.2 **a** Breakdown of industrial heating and cooling demand in the EU-28 to different kinds of industry from (Rehfeldt et al., 2018), **b** energy sources used for heating and cooling demand in EU-28 from (Heat Roadmap Europe, 2017); 1 TWh = 3.6 PJ

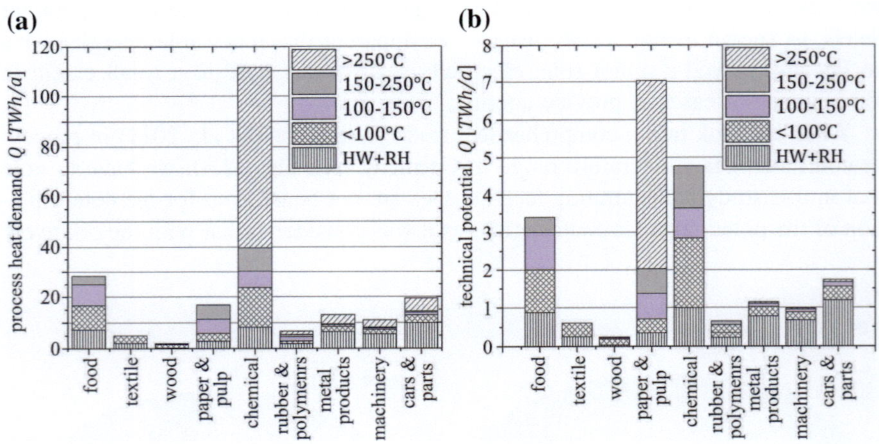

Fig. 12.3 Heat demand (**a**) and technical potential (**b**) at different temperature ranges below 250 °C for selected industry branches in German according to data from (Lauterbach et al., 2011). *HW*: hot water, *RH*: room heating

high latitude of Germany and high diffuse fraction the performance of concentrating collectors is too low. However, more recently (Krüger, 2021) it has been shown that parabolic trough and Linear Fresnel can contribute as much in Germany as flat-plate or vacuum tube collectors, given that the thermal losses at high temperatures are much smaller.

To sum up, about 16% of the final energy consumption in Europe is industrial process heat which is 32% of the total heat demand. A significant share can be provided at temperatures below 250 °C (e.g. 31% in Germany). Already with this restriction to low temperatures and other restrictions like solar radiation potential and area limitations, suitability of roofs and buildings, a huge potential for solar collector installations is calculated in the studies—for Germany about 16 TWh/a can be covered with solar thermal (3.1% of industrial heat demand in Germany) and in Europe about 72 TWh/a (about 3.7% of industrial heat demand, see Vannoni et al., 2008). The latter number corresponds to about 140–180 Mio.m^2 installed collector area for this new application. As a comparison the total area of solar collectors installed in Europe 2019 (all other applications) are 84 Mio.m^2, cf. (Weiss & Spörk-Dürr, 2021).

Main factors contribute to a reduction of total heat demand to the technical potential. The most prominent factors are:

- area availability of suitable roofs and building areas or low-cost ground installation,
- mismatch of temporal availability of solar heat with heat demand schedules (determined by day/night, workhours, working days and holidays).

In the industry heat is needed for many application areas, in which many individual unit operations have to be performed. Some of these in the temperature range below 200 °C are listed here:

- drying and dehydration,
- preheating (input or raw material),
- concentration,
- pasteurization and sterilization,
- washing and cleaning,
- chemical reactions,
- supply of hot water or steam,
- space heating of production halls.

Others in much higher temperature ranges are, e.g.

- ore melting,
- metal casting
- reduction of oxides,
- calcination.

We will look into the industrial process in more detail within the next chapter.

12.3 The Industrial Process

Industrial plants aim to deliver products which are produced over a sequence of operations in which they are formed and changed—which in total constitutes an industrial process. The different steps making up the complete process may require a treatment which need different forms of energy. According to the energy form the operations can be largely classified as mechanical, electrical, and thermal operations. The latter ones are all operations that require or produce heat (or extracting heat for cooling), such as drying, distillation, evaporation etc.

The different sequences of operations, their operating schedules and their necessary energy intensity determine the energy demand profile of the industrial production site. There may be a significant difference between this energy demand profile based on the process requirements and the energy supply profile, which describes the supply of energy as heat or electricity: Especially the heat supply profile will be at higher temperatures to ensure a reasonable temperature gradient for distribution of heat supply. The energy supply profile can differ largely if storage as temporal buffer between supply and demand can be utilized. The heat supply system often considers the highest temperature, at least in a section of the production. Lower temperature demands often then can be covered partially or completely by waste heat of the high-temperature processes. It is an important task of energy efficiency studies to optimize the heat-recovery potential in the company. In order to determine the potential of energy efficiency using heat recovery measures the pinch point analysis described in more detail in Sect. 11.5.1 is a valuable tool.

12.3.1 Conventional Heat Generation Systems in Industry

The design of conventional fuel-based heat supply systems in industry is mainly influenced by two factors: the fuel and the heat transfer fluid (HTF) (e.g. thermal oil, superheated steam, saturated steam, pressurized hot water, air) utilized. In some parts of the world coal is still the primary energy source, but in Europe and North America most industries cover the heat demand mainly with gas, LPG, and oil. Traditionally, biomass is still used widely in countries such as India, where access to oil and gas is costlier, but modern biomass equipment (e.g. pellet boilers) is increasingly raising interest in industrial countries using a renewable energy source. As replacement to natural gas also biogas as a renewable source plays a role, especially where a potential to produce biogas from agricultural waste products exists. Several thermal processes are based on electrical heating, and most cooling is based on electricity. However, especially in East Asia absorption cooling is playing rather a major role and solar thermal cooling is investigated. Finally, the increasing networks of district heat allow industries to directly heat their processes via externally produced heat.

While electrical processes and district grids provide already the energy as used in the process, the use of fuels requires an energy conversion plant on site to produce the thermal heat. We list in Table 12.1 typical conversion technologies for the most common HTF in use. Obviously, the selection of fuel largely has an impact on the design of boilers, e.g. a gas boiler differs largely from a solid fuel (coal/biomass) boiler. Important performance characteristics of the different types are their start-up and shut-down times as well as time needed to adjust the heating

Table 12.1 Thermal energy conversion technologies (LPG ≡ Liquified Pressurized Gas)

Conversion technology	Fuel	Heat transfer fluid (HTF)
Boiler	Gas, LPG, oil, coal, biomass, biogas (also in combination with natural gas)	Steam, hot water, thermal oil
Cogeneration systems	Gas, LPG, oil, coal, biomass, biogas (also in combination with natural gas)	Steam, hot water, thermal oil
Burner	Gas, LPG, oil, coal, biomass, biogas (also in combination with natural gas)	Hot air
Heat pump	Electricity	Hot water, hot air, thermal oil

power supply to demand variations. Solid fuel boilers need more time to reach a full combustion and thus require much longer response times for starting-up, shutting-down and for load changes. A possible consequence is that heat storages are implemented in such systems to compensate the slow response. Liquid and gas fuel boilers and burners on the other hand react much faster on load changes and can be controlled easily within a large range of load variations. Their part-load efficiencies are higher than for solid equipment. Heat pump systems also for higher temperatures in industry are raising increasingly interest. Limitations due to the physical principle of the thermodynamical reverse compressor cycle lead, however, to a lower performance the higher the temperature to be reached. Also, the availability of sufficient environmental heat to be transformed into a heat at higher temperatures by the heat pump has to be considered in the planning process.

The choice of the conversion equipment is based on the choice of the heat distribution medium. In most industries the distribution medium is steam or (pressurized) hot water, therefore, mainly steam boilers are in use. In larger plants often, cogenerations systems are used for an efficient production of electricity and heat. Steam boilers are more complex than hot water boilers in many aspects, as the steam-condensate cycle requires more care and additional handling steps than hot water systems (water treatment, condensate recovery and degasification). Safety aspects due to the high pressures and risks of leakages require special attention.

Large plants need specifically trained personnel. Open condensate systems may lead to water losses and significant energy losses if not handled with care. The reason that many plant operators choose steam is its advantageous high energy density, enabling small distribution tube diameters and high heat transfer rates, ensuring a constant temperature due to condensation. Thus, steam systems are often the traditional way of heat supply. Hot water systems are easier to operate and maintain, but they have the disadvantage of larger distribution pipe diameters and heat transfer areas (in comparison to steam). In order to stay in the liquid phase, for temperatures above 100 °C water systems need pressurization. The maximum pressure is a design parameter of the distribution system also defining the maximum temperature of the saturated steam. Typically, we find plants using 0.8 to 2 MPa leading to a maximum temperature of 170–212 °C. Synthetic or silicon oils are alternative liquids which can be operated at lower pressures without phase-change for a given temperature when compared to water. These liquids can be used

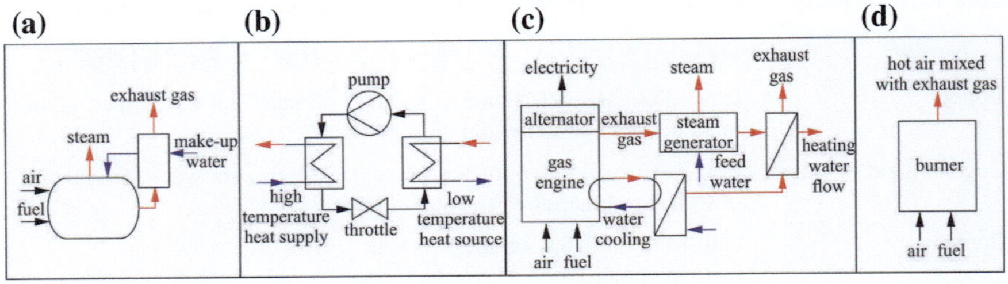

Fig. 12.4 Schemes of conversion technologies with basic components; **a** boiler; **b** heat pump; **c**cogeneration systems and **d** direct heating (e.g. burner)

up to 400 °C at reasonably low pressures but with a little lower specific heat capacities and higher material costs. Silicone oil is often used in cogeneration systems such as Organic Rankine Cycles (ORC) operated at 200–300 °C. Finally, air can be used as heat transfer fluid (HTF), however, due to its very low heat capacity per volume hot air is often not used exclusively as heat distribution medium but rather heated locally at the processes requiring hot air (such as drying). Advantages are the possibility to use it in very high temperature processes even above 1000 °C, the non-corrosiveness and the low cost. In plants where the only significant heat demand is for convective drying (as e.g. in tea factories, malting plants) a hot air burner might be the main heat supply device. Figure 12.4 illustrates the different schemes of conversion technologies and their basic components.

Thermal storages are useful components in many thermal industrial heating systems. Most storages are being integrated in the distribution and supply system, but some storages are already integrated on the process level, especially when the process media themselves can be stored. When used within the supply and distribution system the main purpose of thermal energy storages is to keep the boiler/burner load at a rather constant level and to reduce start/stop operations. Therefore, storages are placed mainly in slowly reacting supply systems, such as biomass or coal boilers.

Hot water storages (pressurized or non-pressurized) are the most cost-efficient and popular versions, while more expensive steam storages are rather the exception. Obviously, this stems from the fact that most steam boilers in industrialized countries are operated on natural gas/LPG or oil which can follow load changes more easily. However, when solar heat shall be integrated storages play a more important role. In slowly reacting steam systems operated with coal boilers steam storages are more common to level off fluctuating demands. Pressurized water storages (Ruth storage) providing flashed steam are used then. Other storage concepts, such as latent and thermo-chemical heat storage are not yet commercially available and play an insignificant role in the existing heat systems in industry although they have a high potential for some specific applications.

It is possible to integrate solar heat in all types of supply and distribution networks mentioned above. However, for each type of network very specific integration concepts and solar thermal system designs are required (see Sects. 12.5.2 and 12.8.2).

12.3.2 Frequent Thermal Unit Operations

In industry a large variety of processes exists depending on products and requirements. However, they can be designed from relatively few single process steps that repeat themselves. In order to classify these different steps, the concept of *unit operations* was introduced already 1905 by Arthur D. Little. The basic idea of unit operations is to classify operation steps based on similar physical processes even if some parameters may be varying. The unit operations can be grouped into categories, such as heat transfer, mass transfer, fluid flow and mechanical operations based on the basic operating principle, see for example (McCabe et al., 2004). In heat transfer operations the basic principle is transferring heat (based on a temperature gradient), while mass transfer operations rely on the mechanisms of molecular diffusion. For describing processes in industry with thermal energy demand, mainly heat transfer operations (e.g. heating, cooling, evaporation, drying, or distillation) and mass transfer operations (e.g. extraction, absorption, distillation, drying, or crystallization), are relevant classes of unit operations. Obviously, in most cases mass transfer operations include also heat transfer, because the different mass flows carry different internal heat. Examples are distillation or drying. So, there is not always a clear distinction into which category one operation falls. This is the reason why categorization of unit operations used by different authors is not always completely consistent.

More important than the classification is the finding that one unit operation, such as drying, may be performed with very similar process equipment, working with the same principle, no matter in which industrial sector it occurs. Of course, materials might vary—e.g. in food industry stainless steel heat exchangers may be used avoiding contaminations while in the petrochemical industry this is not necessary. A consequence is that the integration possibilities of different energy sources, such as solar process heat, could be similar for the same unit operation in different production processes. The concept of unit operations is thus useful to classify the industrial process step and shows amongst other characteristics the implementation possibilities for solar heat, which has been realised in the Wiki-Web “Efficiency Finder” hosted by the Technical University of Graz (AEE—Institut für Nachhaltige Technologien, 2013).

However, even if in two industries the basic process layout of a unit operation is similar and using similar equipment, there can be of course many specific aspects to each unit operation in each sector or subsector. For example, drying in the food industry and in the textile industry is similar, but energy intensity, process temperature and operating schedule may differ significantly. For evaluating the potential for solar thermal energy, one should certainly know all these characteristics describing energy demand and temperature requirement of the processes. A compilation of typical heat demand and temperature ranges for the most important processes in different industry sectors is basic knowledge needed in the design of a solar process heat plant.

Food industry is an important sector in most countries worldwide, and moreover, due to the low temperatures needed, it seems to be especially suited for solar energy integration. Many studies already have investigated especially solar integration in this sector. Therefore, as an example we present specific data on primary energy demand and temperature levels compiled for all processes having relevant thermal energy demand in the dairy industry as one subsector. The Figs. 12.5 and 12.6 provide an overview on how much energy is required at which temperature level, which is a basic information requirement also for planning the solar thermal system design.

In Fig. 12.7 an overview is given over several industry sectors using low and medium temperature heat. When comparing this to the previous tables on may see that temperature ranges are of course wider, when the subsector is not specified, the more details one knows about the products and production processes e.g. in food products, the smaller becomes the uncertainty of the targeted temperature needed.

When solar thermal systems shall cover at least a part of the thermal loads of industry, the operational schedules leading to a time-profile of the demand is crucial. A good match between solar heat generation and thermal load profile is benefitting the utilisation and economics of a solar system. Usually, load profiles are determined via the production cycle of the company, where start-up, closing, process duration, shift hours, breaks, logistics and raw material availability are

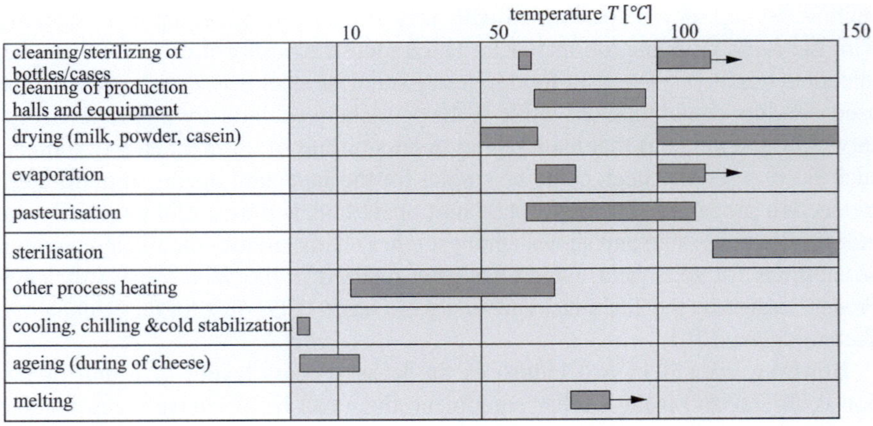

Fig. 12.5 Temperature levels of unit operations in the milk processing industry (Brunner et al., 2012). Two bars signify that depending on the technology the process may run between these two different temperature levels

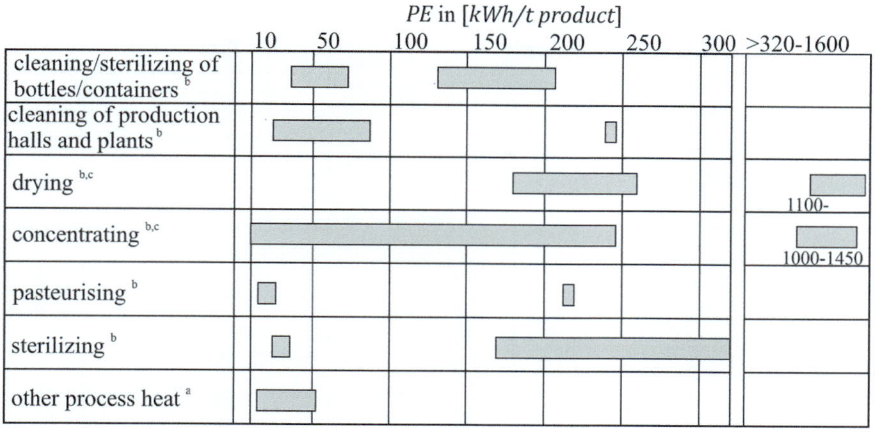

Fig. 12.6 Primary energy demand of unit operations in the milk processing industry (Brunner et al., 2012); Two bars signify that depending on the technology the process may run between these two different temperature levels **a** initial data in useful energy, conversion to primary energy PE: 80% boiler efficiency and *PE*-factor for natural gas and heating oil 1.17 **b** initial data in final energy, conversion using PE-factor for natural gas and heating oil 1.17, **c** data from Dutch Dairy Statistics (http://www.prodzuivel.nl/): data in specific energy consumption (*SEC*) of fuel and heat—estimation of *EE*

amongst the influencing factors. The overall thermal load profile of an industrial process is the sum of the load profiles of often many operating steps that require heat at different times with varying intensity at various temperature levels. A good load management may level out the individual energy peaks created by the single process steps into a smoother overall demand.

In general, we distinguish between daily, weekly, and seasonal profiles. Daily profiles occur over 24 h, with typical profiles shown in Fig. 12.8.

The *weekly profile* as illustrated in Fig. 12.9. This profile is important for the dimensioning of the equipment and especially the heat storage. Most relevant/important examples of typical weekly profiles are constant operation and 5-day peak.

Finally, the *annual load* profile depicted in Fig. 12.10 is especially important for solar process heat plants. The most relevant three load profiles include constant, summer peak, and space heating. The lower level of space heating profiles depends of course also on many factors like the building quality, internal heat sources and suitable comfort levels.

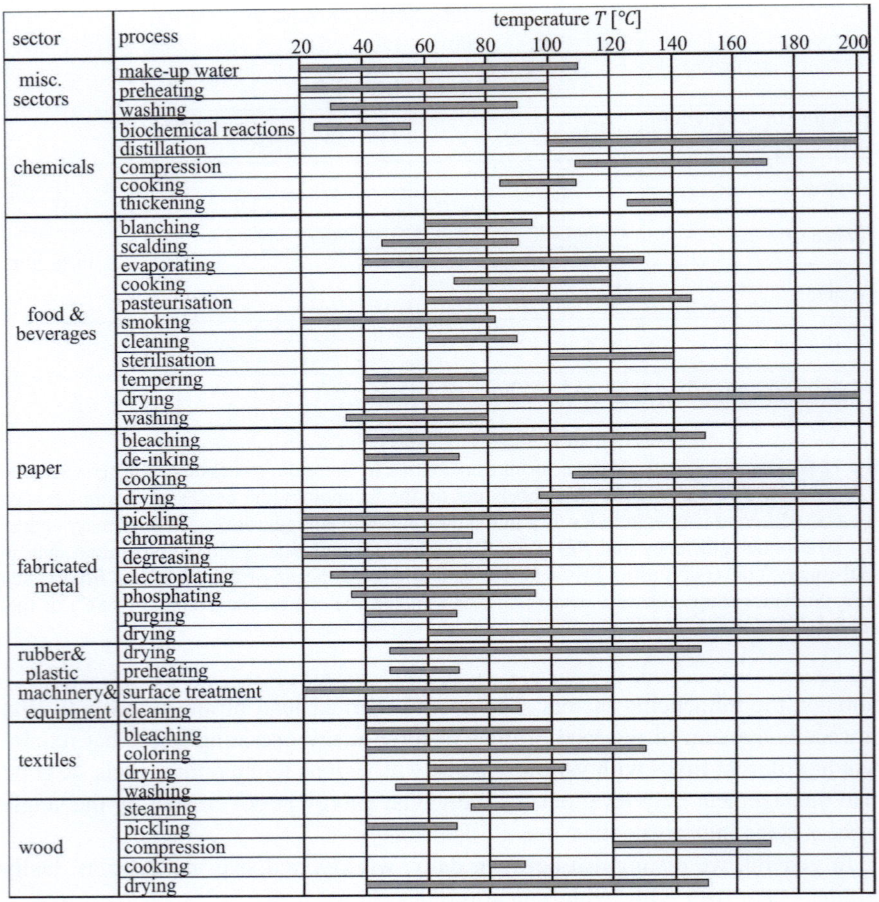

Fig. 12.7 Typical temperature ranges in industrial processes from Lauterbach et al. (2011)

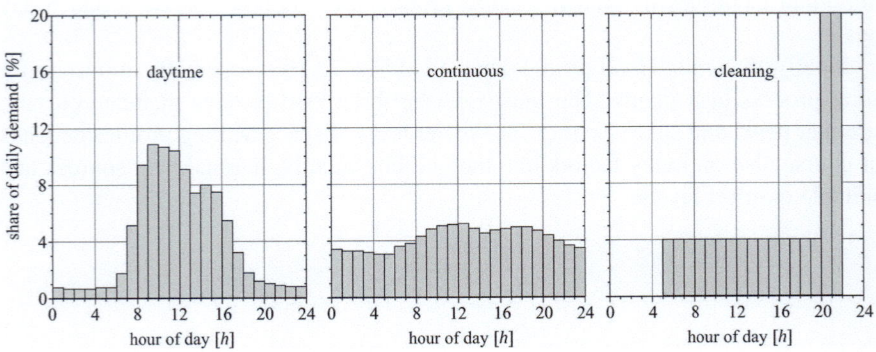

Fig. 12.8 Different typical daily profiles like Daytime production (8–18 h operation), continuous (24/7 h operation), and cleaning (peak evening demand)

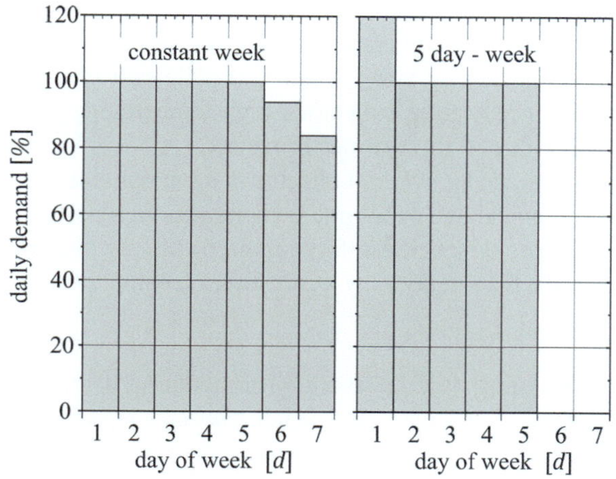

Fig. 12.9 Options of typical weekly operation profiles of process plants

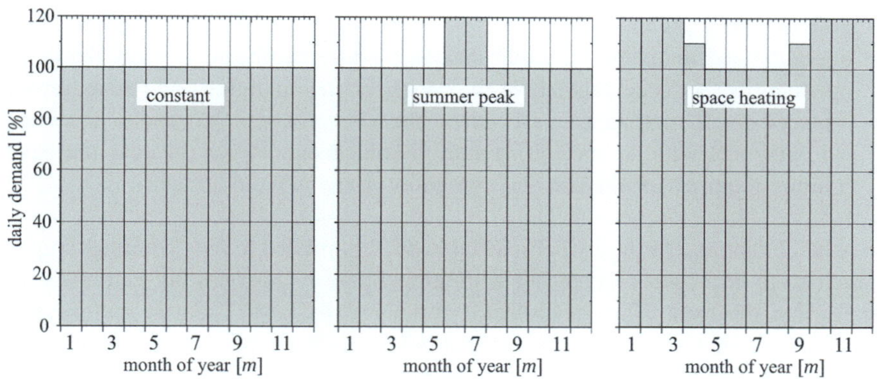

Fig. 12.10 Different generic annual operation profiles of a process plant

12.4 Solar Thermal Process Heat Collectors

In general, all existing solar thermal collectors can be used to deliver heat to industrial processes since process heat applications are manifold, ranging from low ($T < 100$ ºC) to medium ($100\,°C < T < 250\,°C$) to high ($T > 250\,°C$) operating temperatures, see e.g. (Frank, Hess, & Zahler, 2012). Such applications are already suited by either well-established stationary collector technologies (usually, $T < 150\,°C$) or by a growing number of tracked collector technologies in pre-commercial or commercial stages (usually, $T > 80\,°C$, especially for higher concentration ratios). The selection depends not only on the temperature level but also on other factors such as climate and stagnation characteristics. Whereas for temperatures below 120 °C many products are commercially available since many years or even decades, the adaption and further development of collectors for large collector fields and/or higher temperatures has been carried out in the past 10 years. New concepts were introduced like vacuum flat plate collectors, easy connection systems, improved concentrating collectors, high-temperature insulation materials to reduce costs and enhance thermal performance.

The main criteria to decide which collector is more suitable for a certain application will be the technical and economic performance in most cases (but certainly space/area requirements or specific safety requirements are also important). Other selection aspects (of which many may affect either technical or cost performance) are:

- modularity: collectors should be suitable for medium ($10^2 m^2$) or large ($10^3 m^2$) solar fields. Easy and fast collector installation and repair/replacement procedures as well as the potential for hydraulic field layouts with low costs for connecting parts and low-pressure drop are needed;
- robustness and safety: process heat collectors must present material properties and design features suitable for a reliable and safe operation under the conditions of an industrial environment, i.e. also for higher temperatures and pressures if required;
- operation and maintenance requirements: in case that the solar plant should be operated by the technical staff of the company, the operation and (basic, repair/replacement) maintenance procedures must be accessible to end-user technical personnel without special training in solar technologies (besides the basic training required for the use and operation of any new equipment);

In solar collectors, the thermal efficiency is directly related to the operating temperature due to heat losses increasing with temperatures. This does not automatically imply that the most efficient collector system is the best. Thermal performance must be related to the final cost of heat being produced at a given temperature. Looking at specific performance, this might render a cheaper collector with low efficiency more suitable to a given process than a more expensive collector with high efficiency. However, to produce the same amount of energy, a low efficiency collector requires a larger area. Considering that available area might also be a constraint to industrial end-users, this shows that the selection of the optimum collector is a quite complex task with many consequences to be considered.

Different solar collector technologies emerge from adoption of different thermal performance strategies and optical designs, showing their best performances in rather different temperature ranges. Not neglecting concepts such as evacuated flat-plate collectors or fixed mirror moving absorber type concentrators, the most common technologies currently available as marketed products are:

- *flat-plate collectors (FPC)*: (selective) flat absorber with back and side thermal insulation and single flat glazing cover; hydraulic circuit attached to the back of the absorber surface; stationary collector suitable to the low temperature range ($T < 100°C$);

- *evacuated tube collectors (*ETC*)*: selective absorber layer coating the outer surface of the inner glass wall of a Dewar evacuated tube; hydraulic circuit based on a U-pipe or on heat pipes, mounted inside the evacuated tube sleeve; stationary collector suitable to the low and lower boundary of medium temperature ranges ($T < 120 - 150°C$);
- *Compound parabolic concentrator (CPC) collectors*: stationary line-focus concentrator (with low concentration factor) designed after non-imaging optics concepts for ideal concentrators; might be combined with evacuated tubes (with external concentrator reflectors) or with flat (or flat-type) absorbers with external glazing; depending on the absorber and on the effective concentration factor is suitable to the low and medium temperature ranges ($T < 100 - 120\,°C$);
- *Parabolic trough collector (PTC)*: tracking line-focus concentrator designed after the parabola geometrical feature of reflecting any ray incident on its aperture parallel to its axis to the parabola focus; one-axis tracking around the longitudinal (absorber) axis; coupled with evacuated or non-evacuated (single-pass) absorber tubes; depending on the absorber and on the effective concentration factor is suitable to the medium temperature ranges ($80\,°C < T < 400\,°C$);
- *Linear Fresnel collector (LFC)*: tracking line-focus concentrator designed after the Fresnel principle of dividing a parabola into segments displaced in (or close to) a horizontal plane; individual mirror one-axis tracking around the longitudinal axis; coupled with evacuated or non-evacuated (single-pass) absorber tubes located at a vertical displacement related to its focal length; used with a secondary concentrator located around the absorber to enhance its optical behaviour; depending on the absorber and on the effective concentration factor is suitable to the medium temperature ranges ($80\,°C < T < 400\,°C$);

Especially for the concentrating collectors the upper temperature limit is very much dependent on the specific product, and moreover, on the HTF being used. The same optical collector design can be used successfully in a collector using pressurized hot water up to 200 °C, whereas the operation with thermal oil may allow 400 °C. All these products have certain features e.g. how to avoid losses, different concepts for modularity and the connection between collectors, characteristics such as the specific weight and the suitability for certain climates etc. Also, the specific costs of different collector types and concepts may have a significantly different progression over the collector field size. One of the most important features decisive for the choice of a certain collector is the overheating/stagnation handling—regarding this, solutions can be either on the collector side or the system side, e.g. using additional components or certain system integration concepts (see Chap. 8). Summarizing, for each solar process heat project the choice may have to be made out of several suitable collectors. To give a basic orientation, a general distribution of technologies according to their suitability for different temperature ranges and to their classification in terms of positioning is sketched in Fig. 12.11.

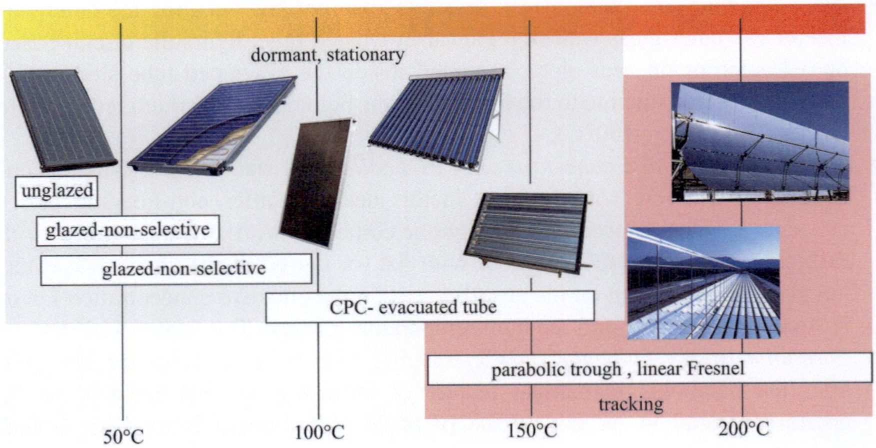

Fig. 12.11 Distribution of technologies according to their suitability for different temperature ranges and to their classification in terms of positioning

12.4.1 Solar Collector Performance

The performance of a solar collector depends not only on its thermal properties, which determine how much heat is lost to the surroundings when its temperature raises, but also on its optical properties, which determine the amount of irradiation which effectively hits the absorber and is transformed into heat. As so, solar collector efficiency is not a fixed value, as it depends on the collector operating temperature, but is rather represented by an efficiency curve.

The measurement of solar collector optical and thermal performance parameters is experimentally achieved by means of standardized procedures, based on the measurement of instantaneous power values. The international standard available for this purpose is (ISO9806, 2017), which defines testing procedures and collector characterization parameters following one of two collector models: steady-state or quasi-dynamic. Collector testing can be done indoors and outdoors—for concentrating collectors the indoor version is not practical due to the required homogeneous and directional beam irradiation simulating direct sun. Solar process heat collectors typically are assembled in a factory and only installed at a site based on delivered parts. Therefore, a testing and certification of a standard collector is possible (see Chap. 6). This is contrary to the very large solar fields for solar thermal power plants, where really the assembly even of optical concentrators is done on site—and as the collector quality thus depends on local assembly procedures, no factory certification scheme is useful but only a testing on site. As reference area the collector gross area is used in all newer standards instead of the aperture area and for the planning of a suitable solar field size. This should be checked when dealing with a specific collector.

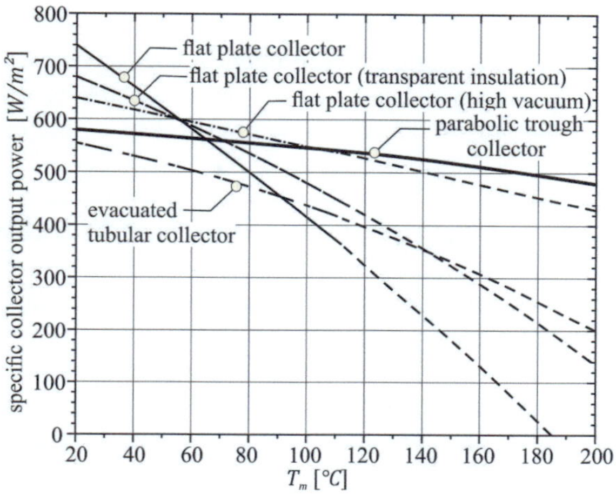

Fig. 12.12 Examples of representative power curves of different solar collector technologies from (Frank et al., 2015), specific output power related to gross collector area. Boundary conditions: hemisph.l intensity $G_h = 1000$ W/m^2, 15% diffuse, ambient temp. $T_a = 20\,°\text{C}$

Especially for stationary collector rows the mutual shading forbids a too narrow arrangement. Due to the higher temperatures in some collectors the existing standards have to be enhanced by additional guidelines which have been developed within IEA-SHC Task 49, see (Fahr & Kramer, 2015).

Figure 12.12 shows examples of representative power curves for different solar collector technologies. Optical and thermal characterization parameters obtained using the standardized test procedures are available at the Solar Keymark database, which is a voluntary third-party certification mark for solar thermal products, demonstrating to end-users that a product conforms to the relevant European standards and fulfils additional requirements. The Solar Keymark is used in Europe and increasingly recognized worldwide. Yield calculations follow a time integration of instantaneous efficiency values. Simplified or more accurate yield calculation models (and system design tools) can hence be implemented following simplified or more detailed profiles for load (temperatures and power) and climate (irradiation and ambient temperature) data series.

12.4.2 Selection Criteria

The selection of the most suitable solar collector for a given application depends on technological and non-technological criteria. The technological criteria regard the collector efficiency when operating at a prescribed temperature. Such criteria determine the type(s) of solar collector suiting the application in study, as depicted in Fig. 12.11.

After this first selection, a comparison of collector technologies follows a comparison of collector efficiency within the desired operating temperature range (see Fig. 12.12). Of course, such criteria are directly linked to a non-technological criterion: collector price. From the economical point of view, efficiency translates, for a given collector price, in an overall energy production cost €/kWh.

Apart from collector cost, among the non-technological criteria one can also include other aspects which might render technically unsuitable a given collector choice, such as: installation ground requirements (covers, façades, land use, etc.); occurrence of stagnation conditions; O&M requirements, etc.).

12.5 Integration of Solar Heat in Industrial Heat Supply

The integration of solar heat into an industrial process heat supply network is a complex planning task. Not only the design and selection of a solar thermal system has to be done. Considering the investment sums for the purpose of reducing the fossil fuel and fossil electricity consumption not only a pure addition of a solar thermal system has to be optimized, but a close look also should be given to the energy efficiency measures accessible in the thermal network. Of course, this also includes reduction of heat demand by classical measures like better insulation of piping and components. One main aspect is the use of waste heat from high-temperature processes within the plant in order to supply other processes at a lower temperature. For this task the technical term is *process integration*, which should not be confused with *solar heat integration*.

Industrial energy efficiency has many aspects which shall be addressed shortly here. Generally, this task needs a strong commitment to energy management by the management. The role of employees should not be underestimated as they can give realistic hints and show practical limits to measures discussed. A sound energy analysis based on an energy audit is a key factor in improving the energy efficiency, to know which fields should be prioritized and which possible measures are available.

Of course, the analysis and consequentially the proposed measures shall be based on state-of-the-art technology options, which have to be described in a transparent and reproducible way in order to be able to cross-check the consistency of measures.

A priorization of implementation of measures certainly follows the economical assessment, how much energy and hence cost could be reduced with as little as possible investment. But also, other aspects play a role. When looking into the overall energy consumption a company might choose first to work on efficiency in the thermal area, and then in the electrical area, because resources have to be used in a focused way, and related professions and problems should be dealt with in one period.

Last not least the question of financing the required investment has to consider standards and specifications, cost offers by different manufactures, and the quality of their services and products.

When we turn now to our special area, the thermal energy management, energy efficiency is mainly related to two categories:

- improvement of equipment (efficiency),
- waste heat management (heat recovery).

Table 12.2 Typical energy efficiency in installed industrial equipment

Equipment	Typical yearly efficiency [%]
Steam boilers (NG)	80
Steam boilers (oil, coal)	80–85
Waste-heat boilers	60–70
Paper drying	50
distilation column	25–40
Cement calciner	30–70
Pulping	60–70

For the improvement of efficiency, it is important to know which typical performance figures can be reached by state-of-the-art installations, and which performance can be assessed in the existing plant. For that an energy audit should determine demand profiles and relate that to the output of the plant and of the production. From these findings the necessity and the potential of an energy efficiency improvement can be deduced. For comparison, Table 12.2 shows typical efficiency ranges of industrial equipment.

12.5.1 Process Integration and Waste Heat Recovery

Many industrial processes require that heat be added to and removed from the process by heat flows, the so-called *streams*. This is especially frequent in the chemical process industry, but also in other industry sector like food and beverages. Materials are heated or cooled to an appropriate process temperature. A specified condition of temperature and pressure has to be reached. The product of the process may be heated or cooled to stop reactions, to preserve the quality and reach the correct temperature for storage and transportation. Thus, at all times heat is either added or removed to a variety of process streams by *utilities* (heat sources like a boiler or heat sinks like a chiller) or by heat exchange between process streams. In an integrated plant therefore, a *heat exchanger network (HEN)* exists for this purpose.

An economic combination has to be found with a minimum number of heat exchanger units and to the recovery of the maximum possible heat with them. Heat can be recovered by exchanging it between *hot process streams* that need to be cooled and *cold process streams* that need to be heated, in addition to the heating and cooling utilities. The task is to synthesize a good heat exchanger network. However, a very large number of alternatives exist. A quick and systematic assessment is possible with the aid of "pinch technology". *Process Integration (PI)* is a general approach for the design of energy efficient process systems and Pinch Technology is a tool to achieve it. *PI* should not be confused with integration of solar heat—it considers heat recovery between processes, see e.g. (Linnhoff et al., 1994).

In the late 1970s pinch technology emerged as a tool for the design of heat exchanger networks providing the engineers with simple concepts of heat, power, and thermodynamics, which can be used interactively in each stage of design.

On may say the pinch technology is a tool for analysis and decisions. The experiences in the past decades were extremely good, as energy savings of 25–50% of energy could be realized in large companies like (ICI, BASF, UC, ...), and the payback times were typically in the order of a year (Linnhoff & Eastwod, 1987).

In practice, there are a number of situations where heat integration is not desirable. Examples include long distances (costly piping), safety (heat exchange between hydrocarbon streams and oxygen rich streams), product purity (potential leakage in heat exchangers), operability (start-up and shut-down), controllability and flexibility.

Pinch technology is based on thermodynamic principles. In the following paragraphs the basic elements of pinch technology are described:

- streams,
- utilities,
- composite and grand composite curves,
- pinch point.

More details can be found in the books of (Kemp, 2007) or (Trivedi et al., 2007).

12.5.1.1 Streams

Stream can be shown in a temperature-enthalpy diagram. "hot streams" are the streams that need cooling (i.e. heat sinks) while "cold streams" are the streams that need heating (i.e. heat sources). The supply temperature of the stream is denoted as T_s and target temperature as T_t. The heat capacity flow rate (CP) is given by the product of the mass flowrate ($\dot{m}$) times the specific heat capacity (c_p), in which c_p is the specific heat capacity of the stream in (kJ/(K \cdot kg)) and $\dot{m}$ the mass flowrate in (kg/s).

$$Q = \Delta H = \begin{cases} \dot{m} \cdot c_p \cdot \Delta T = \mathrm{CP} \cdot \Delta T & \text{sensible heat} \\ \dot{m} \cdot h_{\mathrm{lat}} = \mathrm{CP} \cdot 1 & \text{latent heat} \end{cases}, \tag{12.1}$$

In denotes the latent heat or heat of vaporization for streams with a phase change given in (kJ/kg). The cold and hot streams can be pictured as arrows in a diagram temperature versus enthalpy change. The slope of this arrow line is inversely proportional to the heat capacity flow CP for the sensible heat streams, as illustrated in Fig. 12.13.

Like indicated in in Eq. (12.1) latent heat streams do not change the temperature in the phase transition. In the representation in a similar temperature-enthalpy diagram this means that condensation in a hot or evaporation in a cold stream is depicted by a horizontal line as illustrated in Fig. 12.14.

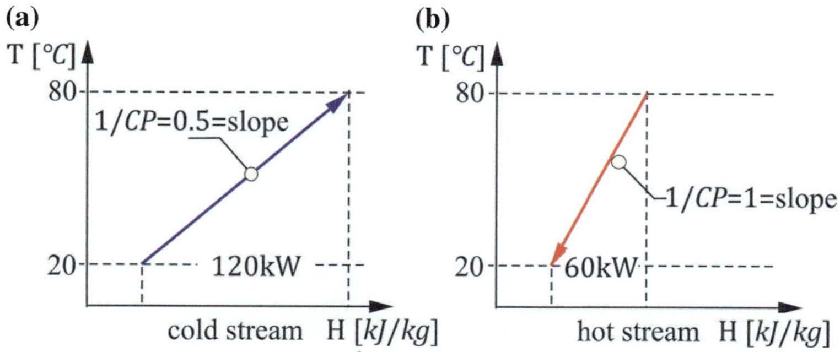

Fig. 12.13 Representation of a cold stream heated from 20 °C to 80 °C (**a**), and a hot stream (**b**) cooled from 80 °C down to 20 °C

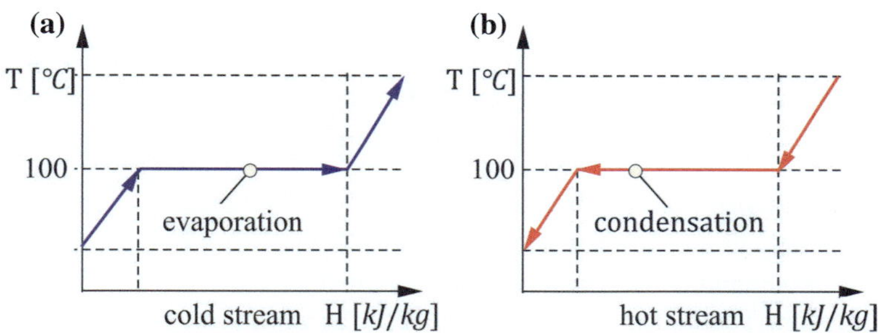

Fig. 12.14 Example of a temperature-enthalpy representation of a cold (**a**) and a hot stream (**b**) associated with a phase change

Essentially, the question is which hot stream can provide heat they need to get rid of, to cold streams which have to be heated. And here the optimal heat exchange between hot and cold streams is in question. Temperature differences need to exist in order to allow a transfer of heat—as temperature gradients are required to achieve an efficient heat transfer. And also, the suitable mass or better capacity flow rates have to be identified minimizing pumping/compression power. To illustrate the situation also a network presentation can be given, where the heat exchange between horizontal streams is shown as a vertical line connecting the relevant streams. Such a simple example with one cold and two hot streams in depicted Fig. 12.15. Here, the heat Q_1 is transferred from the hot stream H_1 to the cold stream C, tand the heat Q_2 from H_2 to C, thus providing the complete heat demand of the cold stream by heat recovery. The situation in the temperature-enthalpy diagram is depicted on the right-hand side of Fig. 12.15.

In a case were such a complete coverage of the heat demand by heat recovery is not possible, either because the demand is larger than the available heat, or because the temperatures of the available waste heat from hot streams is not sufficiently high, a certain part of the heat, in the high temperature section usually, has to be

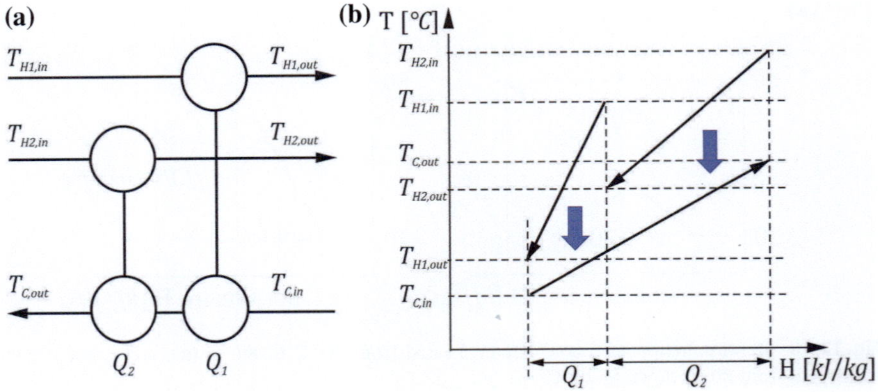

Fig. 12.15 Two hot streams providing heat to one cold stream: **a** network representation and **b** temperature-enthalpy diagram

provided by a heating system, a so-called hot utility. Technically, this is realized often by a combustion boiler or a heat pump or even an electric heater.

12.5.1.2 Utilities

A simple situation where a hot utility is needed is shown in Fig. 12.16. Here, the overall heat demand of the cold stream is larger than the waste heat provided by the hot stream. The hot utility in the network diagram is depicted as a circle with an "*H*" inside. It provides the amount of heat Q_H which is needed still even when all the heat provided by the hot stream Q has been transferred to the cold stream.

Figure 12.17 shows a similar situation where a hot stream has to be cooled down further by a cold utility *C*, e.g. a compression chiller. The additional cold needed is Q_C and indicated in the temperature-enthalpy diagram on the right-hand side. We note that if a hot stream is extending on the left-hand side by an amount Q_C the cold stream, and similarly on the right-hand side the cold stream exceeds

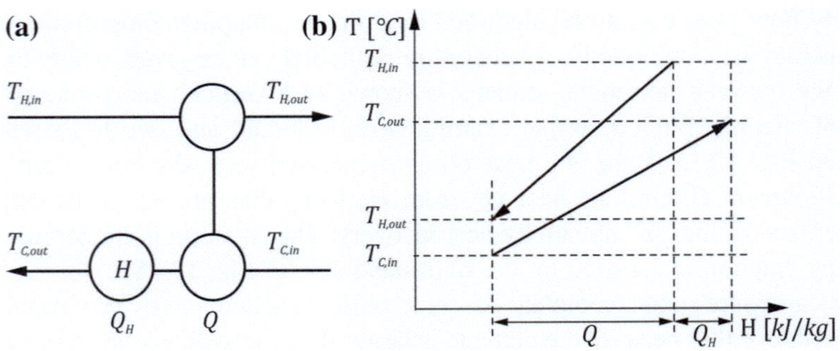

Fig. 12.16 Heat exchange plus hot utility in network representation (**a**) and (**b**) temperature-enthalpy diagram

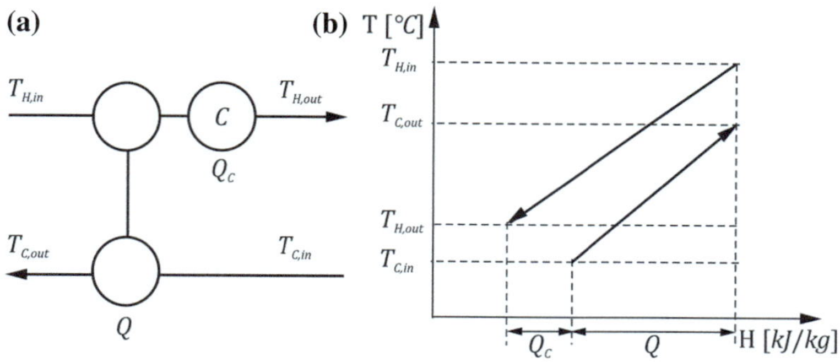

Fig. 12.17 Heat exchange plus cold utility in network representation (**a**) and (**b**) temperature-enthalpy diagram

by Q_H the hot stream, utilities are needed to provide this energy. In principle in such a network, therefore a cold and a hot utility may be in principle be needed.

The heat exchange between two streams, depicted by the vertical lines in the network diagrams, is dependent on several parameters. First of all, of course on the temperature differences during heat exchange between the streams. The heat exchanger area (m^2) and the heat transfer coefficient for cold and hot sides of heat exchangers ($W/(m^2K)$) then determine the rate of transfer. For utilities used in the network and available in the process additional information needed is for example water temperature, or steam pressure level and temperature, etc.

For an economic treatment of course also heat exchanger costs and marginal utility costs, as opposed to average utility costs, are needed for optimization.

12.5.1.3 Composite Curve

All hot streams and all cold streams can be added similar to vectors to form a composite hot or cold stream. The resulting curve lumps together all heat available in one temperature interval. However, in this process some information is lost on in the vertical arrangement. Figure 12.18 shows a simple example where two hot streams are added to a composite curve. The resulting curve however could also be the composite of three or more streams. Insofar the individual stream information is lost.

When now taking the example of Fig. 12.19 and adding a cold stream in the same enthalpy interval the left hand-side of Fig. 12.20 seen. The smallest temperature between these streams is on the left-hand side $(T_{H,out} - T_{C,in})$. This temperature difference determines the necessary size of a heat exchanger which should transfer heat from the hot stream to the cold stream. If it is too small for a cost-efficient design because the area (UA) is too large, then the cold stream cold be moved horizontally to the right in order to increase the minimum temperature difference. How is that be done? An additional heat input Q_H at the higher temperature level is provided by a hot utility, by heating the cold stream. The hot stream has only to provide heat $\Delta H - Q_H$ to the cold stream and, therefore, is

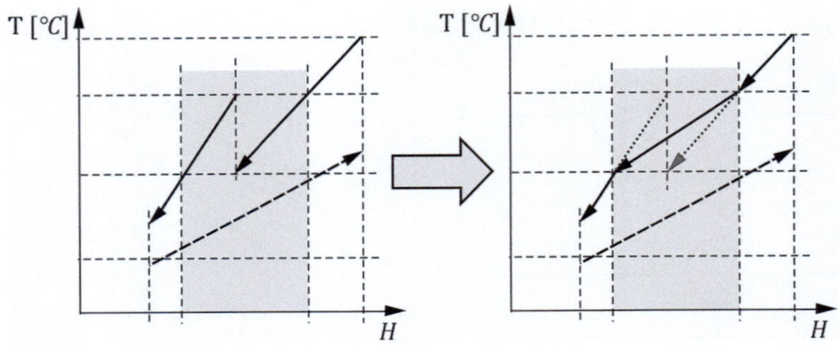

Fig. 12.18 Construction of a composite curve in the temperature-enthalpy diagram

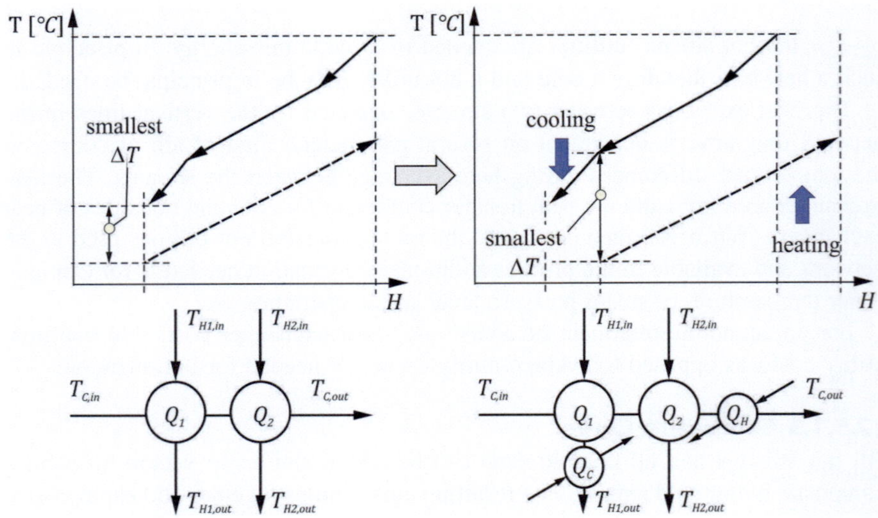

Fig. 12.19 Moving curves horizontally, in network representation (bottom) and temperature-enthalpy diagram (top)

not yet cooled down to the final lowest temperature. If that is required by some reason (environmental or process), a cold utility has to be used to cool the hot stream down to the final temperature level $T_{H,out}$.

The network diagram of the situations just described shows two heat exchange processes for Q_1 and Q_2 are shown here as the composite curve originally stemmed from two hot streams. On the right-hand side of Fig. 12.19 then the hot and cold utility is added.

12.5.1.4 Pinch Point

The minimum temperature difference between more complex composite curves may be positioned at any enthalpy level. The position is called the pinch point. As

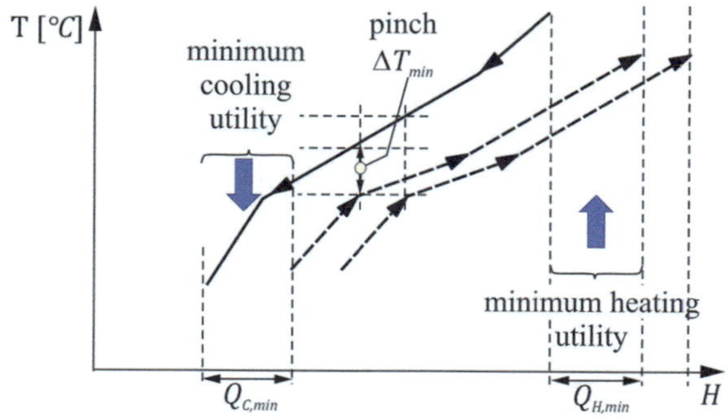

Fig. 12.20 Pinch point defining minimum hot and cold utilities

pointed out the minimum temperature ΔT_{min} at then pinch point determines the area A for heat exchanger ($A = Q/(U \cdot \Delta T_{\mathrm{min}})$).

From the energy point of view moving the cold composite curve left reduces the required energy for heating and cooling, moving right it increases the required energies. For a given minimum temperature difference ΔT_{min}, we find the minimum required cold and hot utilities by comparing the enthalpy differences between the end points of the cold and hot composite curves, as depicted in Fig. 12.20.

The pinch point divides the problem in two parts: above the pinch heat is required and the system is a net heat sink. Below the pinch the system rejects heat is so a net heat source. Thus, violating the three "golden rules":

- heat must not be transferred across the pinch;
- there must be no external cooling above the pinch;
- there must be no external heating below the pinch,

will increase the required hot and cold utility by the so-called cross-pinch heat transfer and the amounts of unnecessary external cooling and heating.

12.5.1.5 Example

In our example we want to use the waste heat of a thermooil stream to produce steam with the following conditions:

- hot stream: Thermooil
  - $h(100\,°\mathrm{C}) = 145.2\,\mathrm{k}J/kg$
  - $h(200\,°\mathrm{C}) = 336.5\,\mathrm{k}J/kg$

  1050 kW with a mass flow $\dot{m} = 5.5\,\mathrm{kg/s}$
- cold stream: Heating up water from 20–120 °C
  - evaporation: $h_{lat} = 2258\,\mathrm{kJ/kg}$
  - $h(20\,°\mathrm{C}) = 83.9\,\mathrm{kJ/kg}$ and $h(120\,°\mathrm{C}) = 2716.3\,\mathrm{kJ/kg}$

  1050 kW with a mass flow $\dot{m} = 0.4\,\mathrm{kg/s}$

Such a process is reflected in the temperature enthalpy diagram as illustrated in Fig. 12.21.

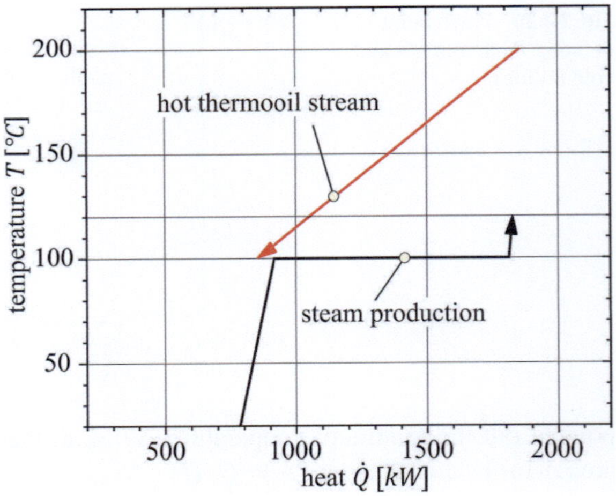

Fig. 12.21 Example of producing steam in a heat exchanger (HX) driven by a hot thermooil stream

12.5.1.6 Multiple Targeting

It is also possible to supply energy via several utility levels e.g. steam pressure levels, refrigeration levels, hot pressurized water circuit or furnace flue gas and other options. The general objective is to maximise the use of the cheaper utilities and minimise the expensive utility levels. So low pressure steam (*LP*) is preferred over high pressure steam (*HP*) and natural cooling water is preferred to refrigeration circuits. The composite curves however do provide overall energy targets but do not clearly indicate at which different utility levels the energy should be provided.

The composite curves in Fig. 12.22a provide targets for the extreme utility levels *HP* steam and cooling water. Figure 12.22b shows the construction of the composite curves if *LP* steam consumption replaces part of the HP steam consumption. The *LP* steam load is added to the hot composite curve as shown in Fig. 12.22b. As the *LP* steam consumption increases a ΔT_{min} temperature difference is reached between the composite curves. This is the maximum *LP* consumption that can replace the *HP* steam consumption.

The composite curves are difficult to use for setting loads with several utility levels. A clear visual representation of the selected utilities and the associated enthalpy change can be provided by the so-called grand composite curve (*GCC*).

12.5.1.7 Grand Composite Curve

The construction of the *GCC* is illustrated in Fig. 12.23. The first step is to shift the temperature of the cold composite curve by $(1/2 \cdot \Delta T_{\text{min}})$ and decreasing the hot composite temperature by $(1/2 \cdot \Delta T_{\text{min}})$. This temperature shifting ensures that even when the utility levels touch the grand composite curve, the minimum temperature difference is maintained between the utility levels and the process streams. This makes it easier to target for multiple utilities. As a result, the composite curves touch each other at the pinch (b). The grand composite curve (c) is then constructed from the enthalpy (horizontal) differences between the shifted composite curves at different temperatures (shown by distance α in Fig. 12.23).

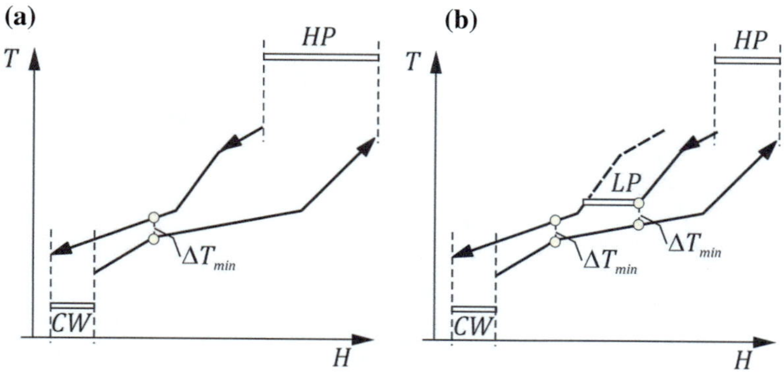

Fig. 12.22 Using composite curves for multiple utilities targeting (cold water = *CW*)

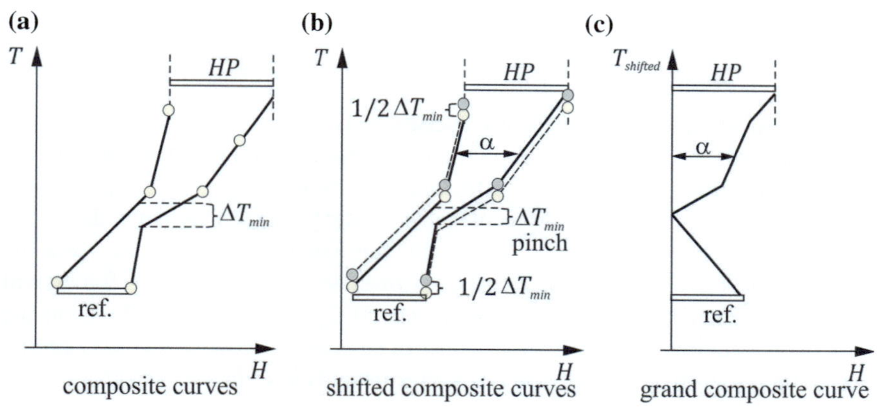

Fig. 12.23 Construction of grand composite curve (*GCC*) in the temperature enthalpy diagram

An example of multiple targeting is shown in Fig. 12.24. The situation with two utilities *HP* steam and refrigeration is modified to adding medium pressure (*MP*) steam and cold water (*CW*) at the respective temperature levels (b). A violation of autility pinch (cross utility pinch heat flow) results in shifting heat load from a cheaper utility to a more expensive one.

Figure 12.24c shows a different possibility of utility levels where furnace heating is used instead of HP steam. Considering that furnace heating is more expensive than *MP* steam, the use of *MP* steam is maximised. In the temperature range above the *MP* steam level, the heating duty has to be supplied by the furnace flue gas. The flue gas flowrate is set as shown in Fig. 12.24c by drawing a sloping line starting from the *MP* steam temperature to the theoretical flame temperature (T_{TFT}).

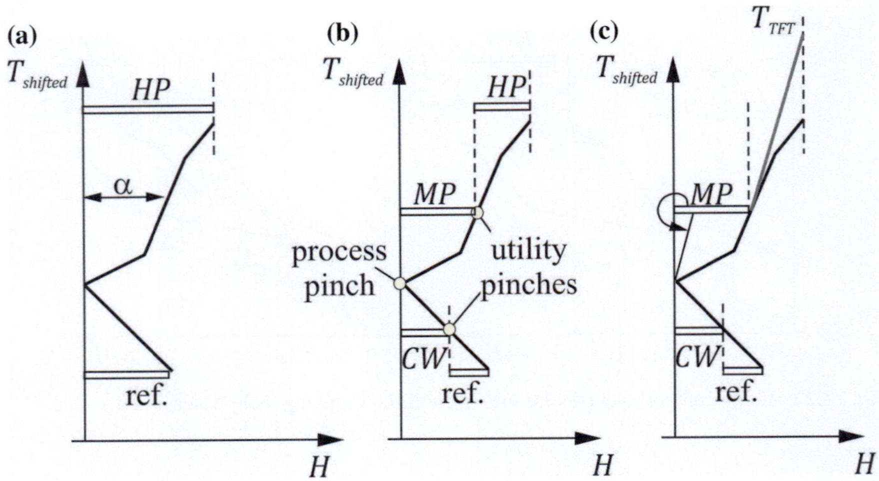

Fig. 12.24 Using the *GCC* for multiple utility

12.5.1.8 Advanced Use of Pinch Technology

Using the tools of pinch technology also cost optimizations can be performed. The optimum design of an energy efficient heat exchange network often results in a trade-off between the equipment and operating costs. This depends on the chosen $\Delta T_{\min}$. The lower $\Delta T_{\min}$ is chosen, the lower the energy costs and the higher the heat exchanger capital costs, as lower temperature driving forces will result in the need for greater heat exchanger (HX) area. A large $\Delta T_{\min}$, on the other hand, requires more energy because of less heat recovery and less capital costs. The trade-off is further complicated in a retrofit situation, where a capital investment has already been made.

The theoretically minimum area of HX for a given temperature difference (ΔT) is calculated by assuming always vertical heat transfer between the hot and cold composite curves. The network minimum area ($A_{\min}$) of HX is then given by

$$A_{\min} = \sum_i \left[\frac{1}{\Delta T_{LM}} \sum_j \frac{q_j}{h_j} \right], \tag{12.2}$$

in which i denotes the i-th enthalpy interval, j the j-th-stream, ΔT_{LM} the logarithmic mean temperature difference in the interval and q_j the enthalpy change of the j-th stream.

Calculating $A_{\min}$ automatically implies assuming ideal counter-current heat exchange in all streams. Combining that with the minimum number of HX units $N_{\min} = N - 1$, where N is the number of process and utility streams in the network, and with the cost information on heat exchanger area, cost functions for operating and investment cost can be constructed versus the given $\Delta T_{\min}$. The result of the optimization process is graphically depicted in Fig. 12.25.

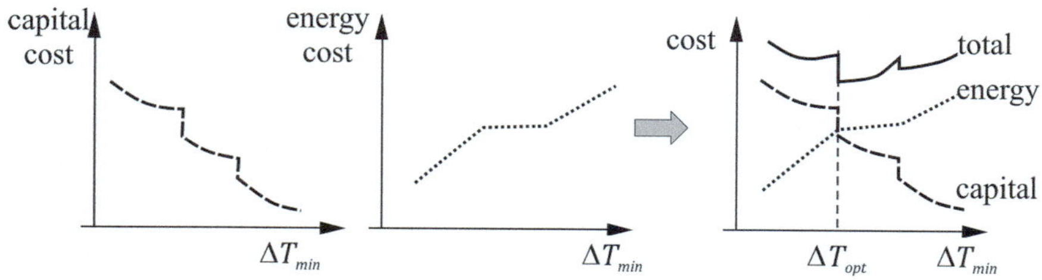

Fig. 12.25 Trade-off between energy and capital costs giving the cost-optimum for $\Delta T_{\min}$

Retrofit within existing heat exchange networks (HENs), exchange or modifications of processes in a network, modifications of chemical distillation columns can all be treated with pinch technology, for more details the textbook e.g. of (Kemp, 2007) is suggested. The placement of heat engines and heat pumps is relatively simple when considering the *GCC*. Heat engines should either be integrated with both temperature levels above the pinch (effectively using 100% of heat in order to provide work) or below the pinch (free waste heat use), but not across the pinch. On the contrary, a heat pump is appropriately placed across the pinch where it can save both in hot and cold utilities. Here, not only process pinch but also utility pinches may be considered.

12.5.2 Integration of Solar Thermal Energy into Supply Networks

When process integration is trying to achieve an energy efficient heat-exchanger network with high percentage of heat recovery, the integration of solar thermal energy into a process heat system is a completely different topic. Solar thermal energy in this context shall be used to partially provide heat to an existing or planned supply system of process heat within an industrial site, and in the next paragraphs we will deal with the different options, how the solar energy may be coupled into the system. Although there are simple cases where a heat source is only needed for one single heat sink defining the full demand, in general there will be several heat consumers which receive the heat from a heat generator such as a steam or hot water boiler or a district heating grid.

This means that we generally find a heat supply system, consisting of one or more heat generators and a heat distribution system delivering the heat to specific processes. Therefore, we have the basic distinction whether solar heat will be integrated at the supply level in the heat supply system, or at the process level where solar heat is directly delivered to individual processes using a separate distribution network. In Fig. 12.26 schematically the situation is illustrated for a steam supply system where the heat is provided by a steam boiler and then distributed to three different processes. In this case two low temperature processes will be supplied by external heat exchangers in which the steam condenses, and the condensate is return back to the steam boiler. A third open process needs steam supply in the

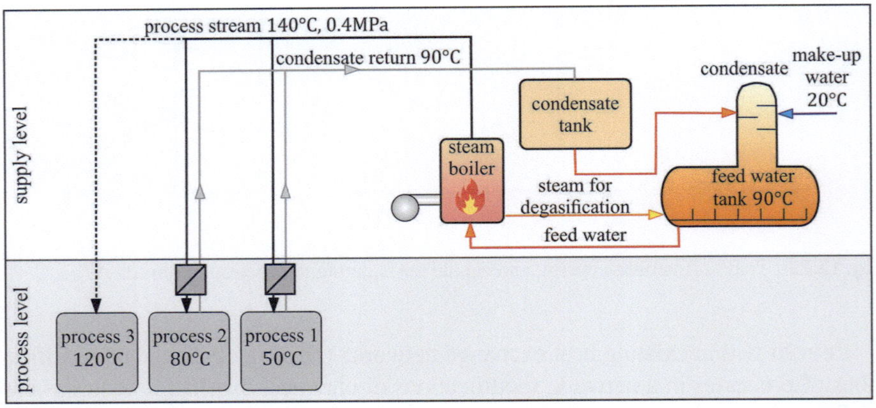

Fig. 12.26 Process heat network with supply and process level

temperature range 120 °C which is escaping to the environment. Therefore, the water within the steam supply system has to be replenished, and cold make-up water is being fed into the feed-water tank to compensate the steam losses.

Looking at Fig. 12.27 we may identify four different positions for a solar heat integration, which have different specifications and technical solutions. If a lot of make-up water is being used (e.g. in a laundry where a lot of steam is being lost in the open ironing process) the preheating of make-up water at a low temperature level can easily be realized using a flat-plate or vacuum tube collector system. Integration point 2 however needs a direct feed of steam of pressure 0.4 MPa and with at least 140 °C steam temperature. That could be provided for example by a direct steam generating concentrating collector, a Linear Fresnel, or a Parabolic Trough collector where water is evaporated directly within the absorber section. Other options are collectors with sensible heat transfer fluid at sufficiently high temperature where steam then can be generated in an evaporator. Integration points 3 and 4 are both in the temperature range 70–90 °C where the pressurized water shall be lifted in temperature, however without exceeding the evaporation temperature. Enthalpy coming from a solar loop heat exchanger can be transferred to the liquid water. Typical temperature lifts are about 20 K. When a modern gas boiler with economizer (heat recovery from the exhaust gas as shown in Fig. 12.27) is used, then the integration point 3 is rather useless, as the required condensate return temperature lift can be provided by the heat recovery process.

The other main option of process level integration is shown schematically in Fig. 12.28. Here, the example of providing solar heat to the process with the lowest temperature level is shown, because the efficiency of a solar collector system decreases with working temperatures. In the example an internal heat exchanger for the process vessel is indicated, but other options are possible and will be described later.

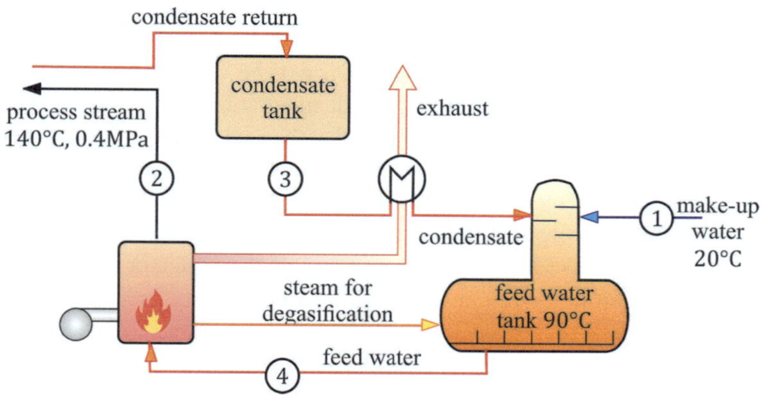

Fig. 12.27 Possible integration points for solar heat in the supply level within a process steam network

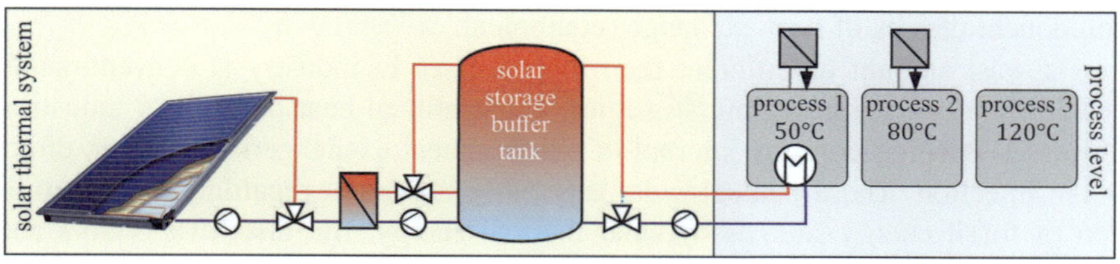

Fig. 12.28 Example of process level integration of solar heat

Depending on the specific situation several advantages and disadvantages should be considered when discussing solar thermal energy integration on either supply or process level.

Obviously, in many cases with several processes one may identify a process with lower temperature requirement than the supply temperature of the distribution network. In these cases, a solar thermal system with non-concentrating collectors and high solar efficiency may be used. Similar efficiencies can be found for make-up water heating although the target temperature 90 °C is already quite high. If the solar thermal system may be installed in close proximity of the process, then small heat distribution losses favour the direct integration. On the other hand, supply level integration in any case has a higher flexibility. Even if a process at some time will not be needed and dismantled in the production, solar heat can be supplied to all other remaining and to new processes installed with the need of a re-installation. A solar system however designed and optimized for a specific process only would need a complete redesign, and possibly a re-installation in a different configuration at a different place. Also, the flexibility with respect to demand profiles is better with the supply level integration: Individual processes may need the heat a very specific times in a day or week, and solar heat has to be stored in order to cover these demands in the process level installation. In supply level, the likelihood of other off-takers of the solar heat for any time is much higher. Storage dimensioning and losses often will be smaller for the same solar field size. On the other hand, especially with integration at the highest temperature level in the distribution network, the solar thermal system efficiency will be lower with supply level integration. In most cases here concentrating collectors are needed to provide temperatures above 100 °C with good efficiency. The selection of concentrating collectors is also much more site-dependent, because a high *DNI* is necessary to have many operation hours of the collector and thus a good economic performance.

12.6 Heat Exchangers and Integration Points

Every integration point is related to a unique way how solar heat is transferred or other energy is converted into heat to the system. The component responsible for that is the heat exchanger. Figure 12.29 gives an impression of the different component classes of heat exchanger equipment.

The vast amount of different thermal processes in industry is conventionally heated by a few basic heating strategies. Conventional heat delivery to industrial processes is being done by internal or external heat exchangers as well as direct steam injection. These strategies are especially important regarding the substitution of fossil energy sources by solar thermal energy. But also, evaporators and dryers are of high relevance. These component classes can be found in almost

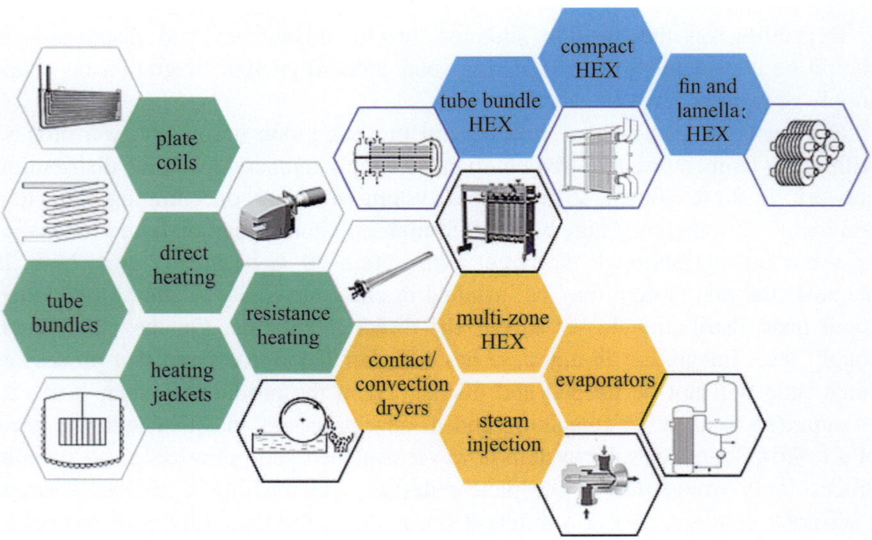

Fig. 12.29 Different heat exchanger equipment adapted from (Muster-Slawitsch et al., 2015)

every industry sector, of course fulfilling different requirements with respect to dimensions, materials being used and safety aspects.

12.6.1 Heat Exchanger Types in Industry

Below, some general information on different heat exchanger types is given. The aim of this paragraph is not complete information on heat exchanger (HEX) types and their description, but rather to show how heat exchangers can be classified. Their dimension and type classification have been introduced already in Sect. 5.5.3, for more information on heat exchanger types and details the reader is referred to (VDI, 2013) or (Holman, 2009). Basically, we can distinguish between external heat exchangers placed outside a process bath/vessel/machinery or internal heat exchangers placed inside a process bath/vessel/machinery.

Different types of *external heat exchangers* are used to heat industrial processes. A suitable HEX type is chosen according to applied temperature and pressure, phase change, available space, or specific properties of the heated media such as viscosity and corrosiveness addressed in Chap. 5. According to (Shah & Sekulic, 2003) the most important types are tubular-, compact-, and extended surface HEX.

Tubular HEX are the most common heat exchanger types in industry and cover tube bundle- and double pipe HEX. Tubular HEX is made of a closed tubular jacket that encloses a bundle of tubes. This creates a shell space which is encircled by a medium. The other medium flows through the tube bundle, which consists of plain or ribbed tubes (Vauck & Müller, 1999).

Compact HEX cover plate- and spiral HEX that are suitable for media with different viscosity. The most common compact HEX type is the plate heat exchanger. Within the food industry often sealed plate heat exchangers are used. The plates typically have strong profiles and are held together by clamping screws and frames. Spiral HEX consists of two equal-width steel strips that are spirally wound at an equal distance to an axis. The resulting channels are closed at both ends by removable covers with gaskets. These HEX types are mainly used for liquid–liquid applications, condensation, or gas cooling (Shah & Sekulic, 2003). For applications where one of the mediums has a much lower heat capacity (e.g. water–air), extended surface heat exchangers are used. The extension of the HEX surface is realized by lamella or fins. Some examples of external heat exchangers are shown in Fig. 12.30. These examples show that an external HEX can be part of a unit operation or the unit operation itself. Based on the specific application the heat exchanger is used to heat a product or process medium.

Another possibility is the supply of heat to baths, tanks, or equipment *internal heat exchangers*. The most important types of internal HX are heating jackets, heating coils, and tube bundles. Also plate coil heat exchanger and direct heating elements such as resistance heaters are used. Some examples of these internal HX types are illustrated in Fig. 12.31.

Heating jackets can be found in different variations such as full jackets, welded half pipes or profiles and so-called dimple jackets. The utilization of full jackets is

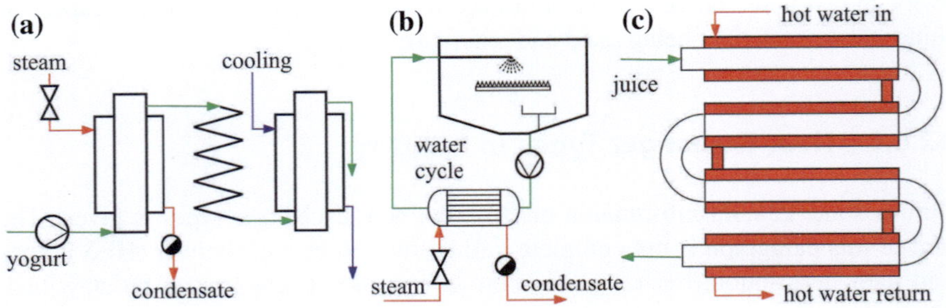

Fig. 12.30 Examples of processes that are heated by external heat exchangers. **a** Pasteurisation of yogurt with steam heated scraped surface heat exchangers; **b** tunnel pasteurizer heated by external tube bundle; **c** pasteurization of fruit juice via double pipe HEX modified from (Schmitt, 2014)

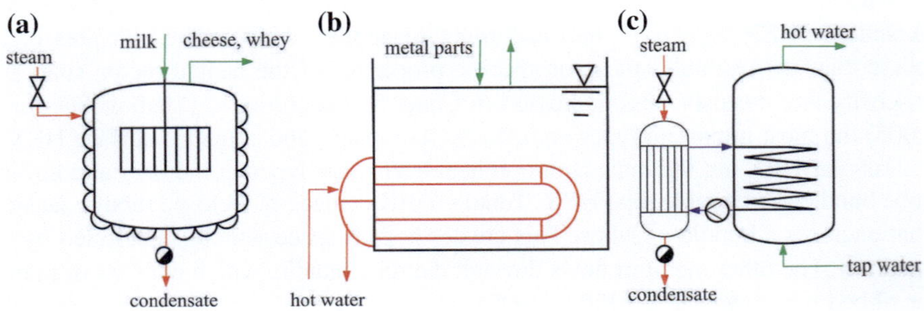

Fig. 12.31 Process heating via internal heat exchangers; **a** cheese production—a vessel with heating jacket is heated via steam; **b** heating of degreasing bath via internal tube bundle without shell and **c** hot water supply with internal heating coil that is heated indirectly via steam by an intermediate hot water circuit modified from (Schmitt, 2014)

limited to a pressure of the heat transfer medium up to 0.6 MPa and vessel sizes up to 10 m^3. For larger volumes or higher pressures, welded coils, profiles, or half pipes are used, see (Vauck & Müller, 1999).

Heating coils are the most simple and cheapest heat exchanger types. Therefore, a tube is helically wound around a cylinder and fitted with ports for inlet and outlet of the heating medium. Although the capacity of heating coils can be affected by pitch and diameter of the tube, the specific heat transfer rate is rather low due to the low flow rate on the outside of the tube.

Although so-called panel- or plate coils are classified as compact heat exchangers by (Shah & Sekulic, 2003), their main field of application corresponds rather to internal heat exchangers. Their construction equals roll bond absorbers. They are often used as internal heating element or heating jacket to heat vessels or baths. Based on the material used, operating pressures of 0.7 to 1.8 MPa are possible. Due to the simple production process, this heat exchanger type is very cheap (Shah & Sekulic, 2003).

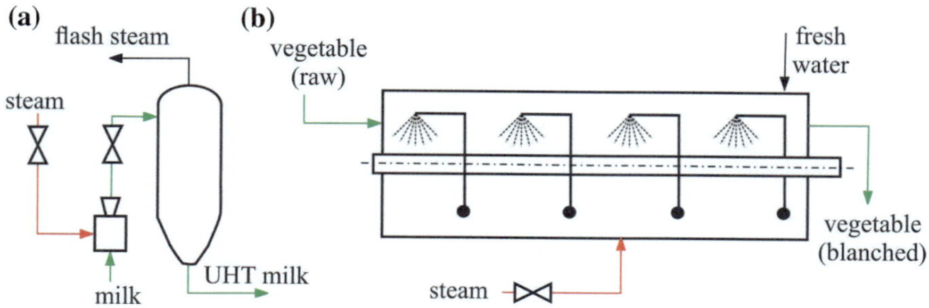

Fig. 12.32 Process heating with direct steam injection. Left: Ultra high temperature sterilization of milk via direct steam injection and flash evaporation. Right: Belt blancher for vegetables heated by direct steam injection (Schmitt, 2014)

In principle, electric heating elements are able to provide very high temperatures. However, they are also used to provide lower temperature ranges because of procedural reasons. So-called rod-type or pipe heating elements can be placed in fluids to heat them (Vauck & Müller, 1999).

12.6.2 Direct Heating Systems

There are two important types of direct heating systems: direct combustion and steam injection. Using direct combustion systems, the firing flame room and flue gas pipes of an oil or gas burner are directly disposed in the facility to be heated (e.g. water bath for cleaning) and transfer the heat to the product or process medium.

Direct steam injection can be used to heat baths, tanks, and fluid streams or to provide an air–steam atmosphere. The direct heating of baths or vessels via steam can be realized by steam injection at the bottom of the vessel/bath using perforated tubes and coils or nozzles. By using nozzles, the product or process medium within the vessel/bath is sucked in, mixed with steam, and returned into the vessel/bath. Based on the specific properties of the heated product or process medium, the injected steam has to meet high standards. Two examples are illustrated in Fig. 12.32.

12.6.3 Evaporators and Dryers

Also, evaporators and dryers can be equipped with internal or external heat exchangers. They can be categorized within an own group. Since they are often heated by special heating mechanisms, they can be considered separate and not based on the placement of heat exchanger.

Evaporator types can be classified according to the evaporation mechanism. Important evaporator types are kettle reboilers, natural and forced circulation evaporators, climbing and falling-film evaporators. Mainly, continuous steam heated evaporators are used, which are often designed as multiple effect evaporation systems, for details see e.g. (Goedecke, 2011).

For drying, a multitude of different dryers is used within industry that can differ significantly in construction and mode of operation. Dryers are classified into convective-, contact-, radiation-, dielectric-, and freeze dryers. Out of these types, convective- and contact dryers can be supplied with solar thermal energy. Convective dryers mostly use air that is heated to dry products. Contact dryers use heated surfaces that are in direct contact with the product. These surfaces such as cylinders, plates, or screws, are mainly heated by steam.

12.7 Classification of Solar Integration

In Fig. 12.33 a classification of solar integration possibilities is given, split up into integration on process and on supply level. The various concepts depending on heat transfer fluid and other parameters are labelled as PL_X for concepts on the process level and SL_X as concepts for the supply level. The integration concepts on supply level are grouped based on the heat transfer medium. The integration concepts on process level are grouped based on the additional technology that is required for the integration of solar heat (external or internal heat exchanger, solar steam supply). The classification has been developed in the IEA Task 49 of the IEA Solar Heating and Cooling Program.

Benefits of classification:

- simplifies overview of industrial heat consumers,
- highlights importance of current designs of heating equipment,
- assistance for the identification of relevant heat sinks,
- shows possibilities for the integration of solar heat,
- indicates the effort for the integration of solar heat.

In the following subchapters, all integration concepts are briefly introduced and illustrated by a process flow sheet. The integration concepts on supply level are grouped based on the heat transfer medium. The integration concepts on process level are grouped based on the additional technology that is required for the integration of solar heat (external or internal heat exchanger, solar steam supply).

12.7.1 Supply Level Integration Using Steam

If high temperature levels are needed for the parallel integration of solar heat (solar steam generation), suitable collector concepts have to be applied. Due to the fact that the required temperature level for parallel integration of solar heat

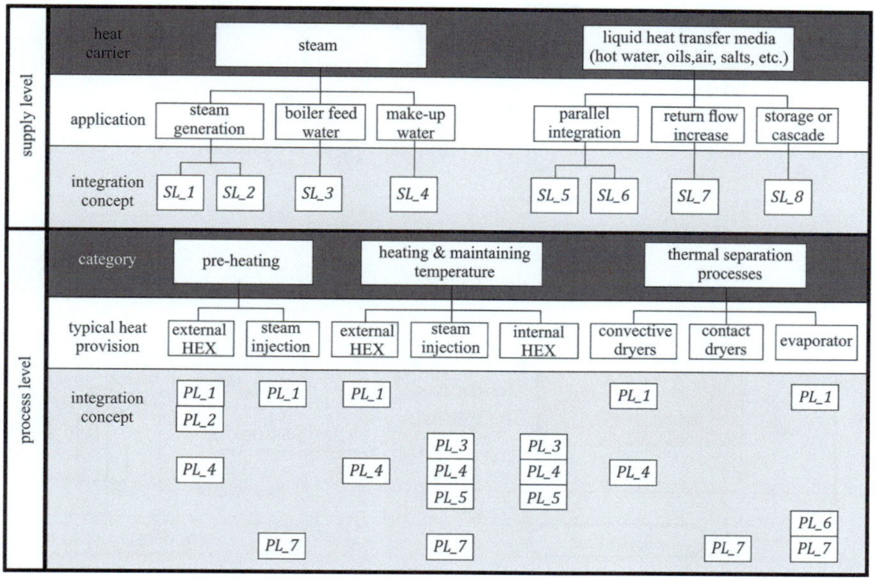

Fig. 12.33 Classification of integration points according to IEA modified from (Muster-Slawitsch et al., 2015)

in conventional steam systems is usually above 150 °C, concentrating collectors might be applied. Therefore, irradiation data of the location have to be considered for the choice of the appropriate collector concept.

SL_S_PD—direct solar steam generation

For direct solar steam generation, a steam drum is required that is fed by concentrating collectors with direct steam generation (DSG). Boiler feed water is fed to the solar collectors and partially evaporated. The water-steam mixture is fed to the steam drum where it is separated. The accumulated water is fed back to the collector loop. In case of sufficient pressure in the steam drum, the steam is fed into the conventional steam circuit (cf. Fig. 12.34).

While solar steam is fed into the existing circuit the conventional steam boiler reduces its steam production similar to operation periods with reduced load. The actual loss of efficiency of the steam boiler caused by the solar heating system is influenced by the ratio of installed solar power to conventional capacity, the typical load, and steam boiler performance (scope of modulation). This integration concept was realized with in several demonstration plants.

SL_S_PI—indirect solar steam generation

Regarding the interaction of the solar heating system with the conventional steam supply system this integration concept is similar to the prior one (SL_1/SL_S_PD). However, in this case concentrating collectors with pressurized water or thermal oil as heat transfer medium are used to feed a special heat exchanger for

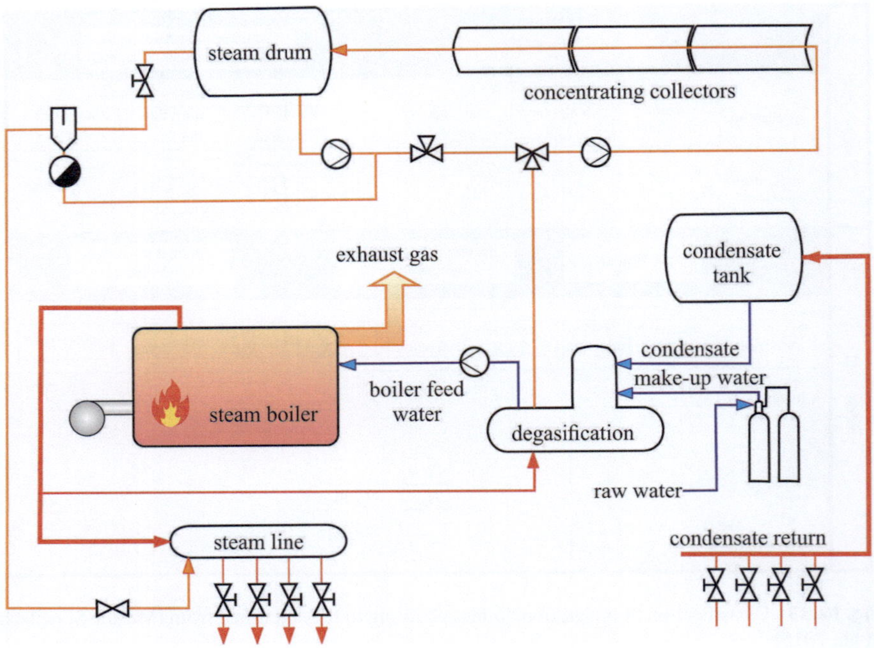

Fig. 12.34 Integration concept for direct solar steam generation (SL_S_PD)

evaporation (typically a kettle type reboiler). The heat exchanger is operated at the same pressure as the conventional steam system and the configuration is illustrated in Fig. 12.35.

SL_S_FW—solar heating of boiler feed water

The serial integration of solar heat for heating of boiler feed water takes place after degasification. Usually the degasification is operated around 105 °C. The set temperature of boiler feed water depends on the supply pressure of the steam boiler and is usually in the range of 160–220 °C (equals 0.6 to 2.3 MPa). Typically, boiler feed water is (or should be) preheated by an economizer before entering the steam boiler (by approx. 30 °K). The remaining temperature difference can be supplied by solar heat. Based on the dimensioning of economizers, a solar preheating of boiler feed water prior to the economizer is not always possible. Such a set-up is shown in Fig. 12.36.

SL_S_MW– solar heating of make-up water

The last integration concept for steam systems is preheating of make-up water which is required to compensate losses of condensate caused by flash evaporation, blow down and direct steam consumption. In this configuration, illustrated in Fig. 12.37 make-up water is fed into the degasification after desalting. In principle,

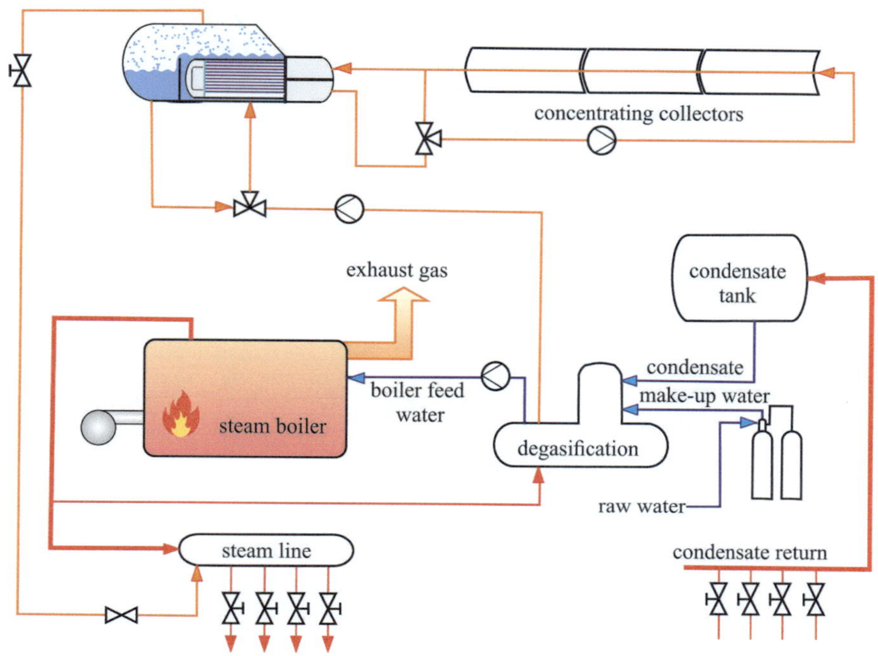

Fig. 12.35 Integration concept for indirect solar steam generation (SL_S_PI)

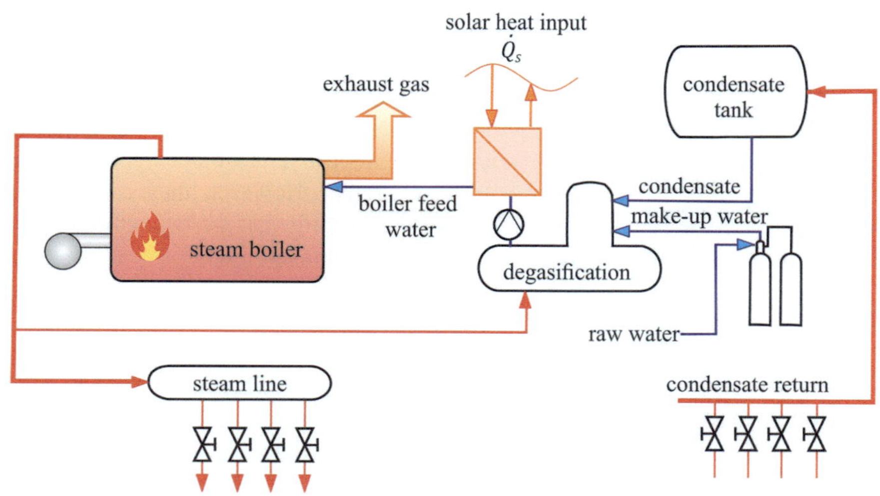

Fig. 12.36 Integration concept for preheating of boiler feed water (SL_S_FW)

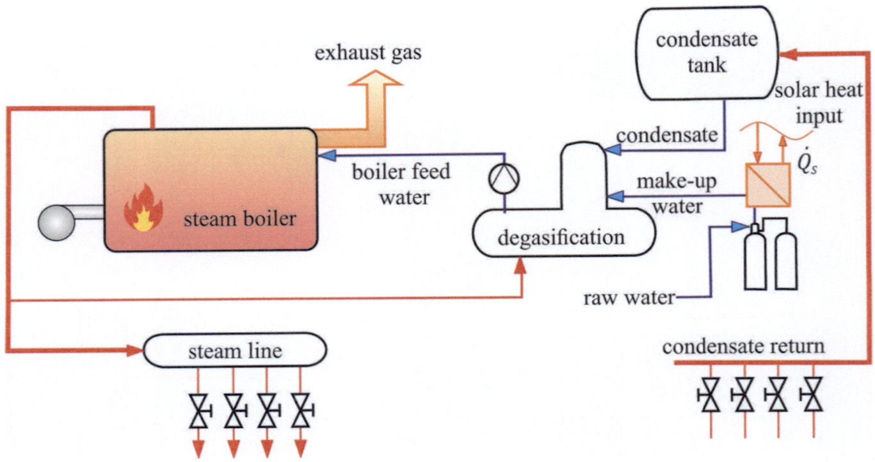

Fig. 12.37 Integration concept for preheating of make-up water (SL_S_MW)

the amount of make-up water can be heated by solar energy prior to degasification. However, several waste heat sources within the boiler house can significantly decrease the potential for implementation of solar heat.

12.7.2 Supply Level with Liquid Heat Transfer Media (SL_L)

For liquid heat transfer media different integration concepts can be applied. Apart from parallel and serial integration of solar heat it is also possible to heat storages or cascades which are integrated into or even existing already in the conventional heat supply system. Due to the relevance in industry, the main application of these integration concepts is for hot water circuits. However, they can also be applied to synthetic heat transfer media and thermal oils that are typically used for a temperature range of 250–400 °C or even to molten salt loops.

SL_L_P—parallel integration

For the parallel integration of solar heat into hot water circuits, the return is divided prior to the conventional boiler and partially directed through the solar heating system and fed back into the flow. Therefore, the control of the solar collector system has to ensure that the return is heated to the required flow temperature. The integration can be realized directly (SL_L_PD) or indirectly (SL_L_PI). For the direct integration the water of the heating circuit flows directly through the collectors (cf. Fig. 12.38), while a heat exchanger is used for the indirect integration.

Sometimes also combined heating circuits can be found in Industry that use steam as primary heat transfer medium to generate hot water as secondary heat transfer medium via heat exchangers or hot water cascades. This could be turned

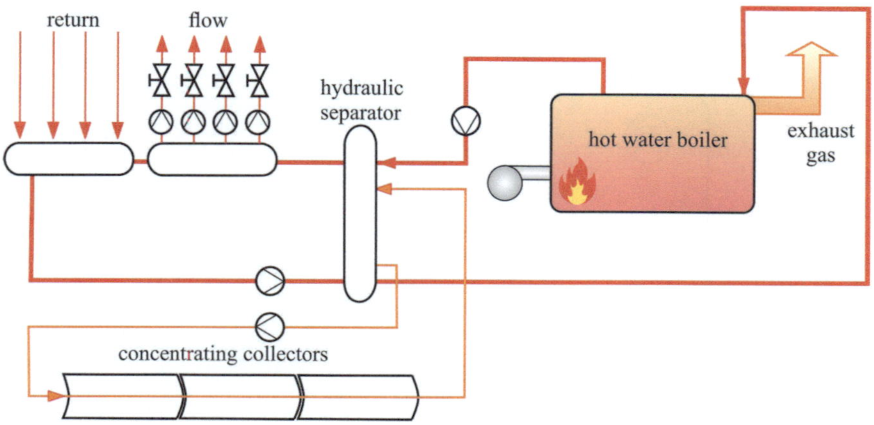

Fig. 12.38 Integration concept for direct parallel integration at supply level with liquid heat transfer media (SL_L_PD)

around in companies with little low-pressure steam demand. Here, hot water is used as primary heat transfer medium and steam produced from hot water in a small steam drum.

SL_L_RF—solar return flow boost

The serial integration of solar heat is similar to integration concept SL_L_PI, with the difference that the return is preheated, as illustrated in Fig. 12.39. Therefore, no fixed set temperature has to be supplied by the solar heating system. Return line heating might not be desired by the industrial plant operator if the efficiency of the conventional heating system is affected in a negative way (e.g. condensing boiler, combined heat and power plant (*CHP* or cogeneration) or district heating).

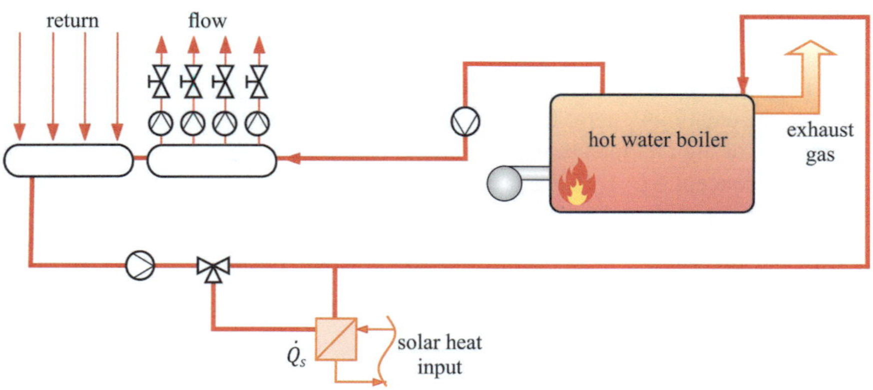

Fig. 12.39 Integration concept for serial integration at supply level with liquid heat transfer media (SL_L_RF)

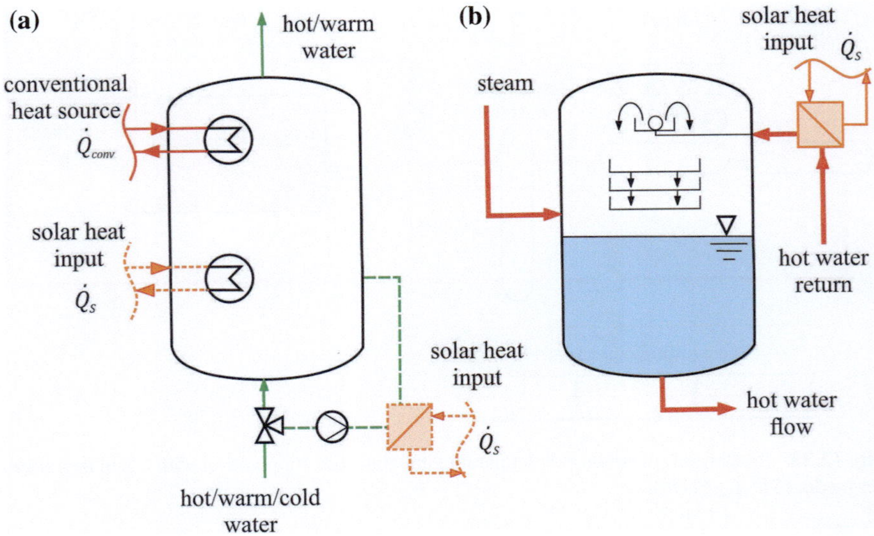

Fig. 12.40 Integration concept for solar heating of hot water storages (**a**) or cascades (SL_L_SC, **b**)

SL_L_SC—solar heating of storages or cascades

There are lots of different types and connection possibilities for hot water storages in industry. Realized are storages with fixed or variable volume fed by hot water or buffer storages that are energetically charged and discharged. Based on the actual number of storages and their configuration solar heat can be used to preheat input streams or to maintain the temperature of the storage, as indicated in Fig. 12.40. Additionally, hot water cascades can be supplied with solar heat. Usually the hot water return is fed into a cascade and heated up by injection of steam. The integration concept for hot water cascades is similar to SL_L_RF, which is shown in Fig. 12.40b since the return flow is preheated before entering the cascade.

The integration concepts presented are not limited to the integration of solar heat. They solely recommend a suitable interface to couple an additional heat source to the existing system. In principle, other heat sources such as cogeneration, heat pumps, or high temperature waste heat can be used similarly. For example, the integration concept SL_L_RF can also be applied to heat pumps. In this case the return of the conventional heating circuit is heated by a heat pump to the required flow temperature. Then a heat source at a lower temperature level (e.g. ambient or waste heat) is required, see e.g. Fig. 12.41.

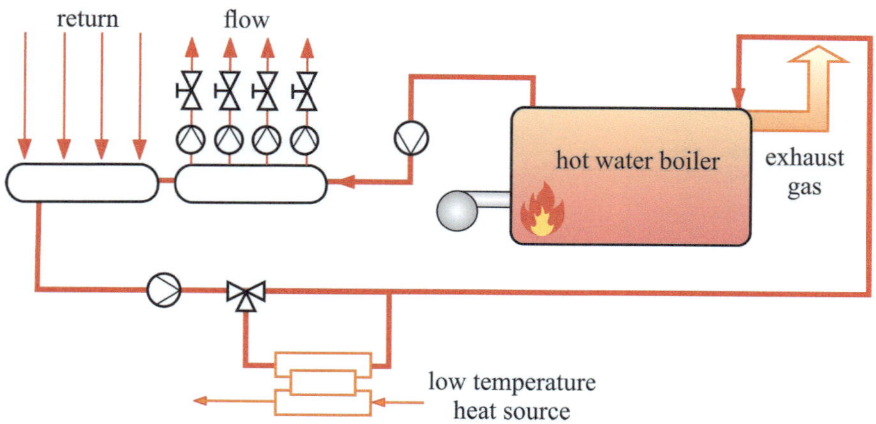

Fig. 12.41 Integration concept SL_L_RF adapted for a heat pump

12.7.3 Process Level External Heat Exchanger (PL_E)

At process level three general classes for integration of solar heat are distinguished. The first class is the integration of an external heat exchanger supplied with solar energy. For this class, four different concepts can be applied: serial connection of external HEX for heating of product or process media (PL_E_PM), serial connection of external HEX for heating of intermediate hot water circuit (PL_E_IC), integration of external HEX for heating of bath, machinery or tank (PL_E_HB) and integration of external heat exchanger for heating of input streams (PL_E_IS). In the following, these concepts are introduced.

PL_E_PM—external HEX for heating of product or process medium

This integration concept is basically the easiest since an additional solar driven heat exchanger is installed in serial to the conventional heat exchanger to heat a product or process medium (e.g. water, lye, fruit juice, etc.). It can be applied to all three categories of heat consumers "(pre-) heating of fluid streams", "heating and maintaining temperature of baths, machineries, or tanks" and "thermal separation processes". The serial integration of a solar driven heat exchanger can be realized in a way that only part of the entire product or process medium stream is heated up and hence heat exchanger surface, pump and pipe dimensions can be kept small (Fig. 12.42).

PL_E_IC—external HEX for heating of intermediate hot water circuit

Within the food and beverage sector intermediate hot water circuits can be used to heat sensitive products. Therefore, the conventional heat transfer medium is used to heat this hot water circuit to a temperature slightly above the required set temperature of the product (e.g. pasteurization of milk). In these cases, integration concept PL_E_IC illustrated in Fig. 12.47 can be applied. This concept equals

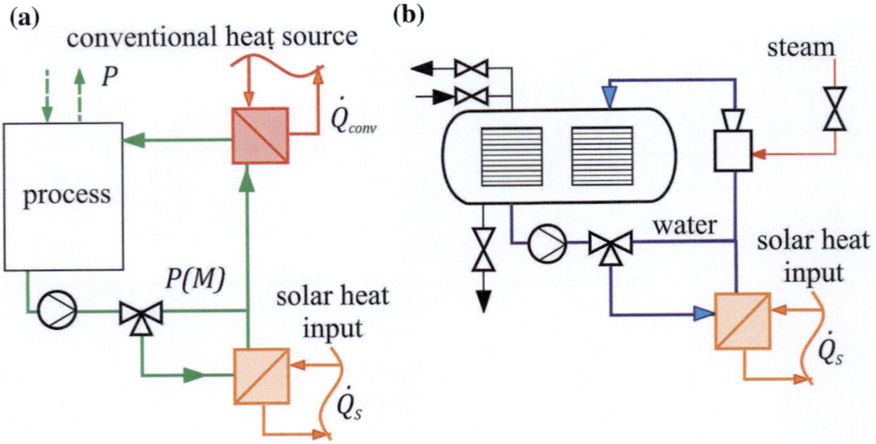

Fig. 12.42 Integration concept for solar heating of product or process media with external heat exchanger (PL_E_PM). General integration flow sheet (**a**) and illustrated by sterilization process with autoclave (**b**)

more or less PL_E_PM. Due to the fact that solar heat is used to heat water, the integration is usually easier in case of the required heat exchanger or safety installations compared to the direct solar heating of products. If a process is directly heated by a conventional hot water circuit, the integration concepts PL_E_IC and SL_L_RF can be combined. Therefore, the return is preheated with solar energy directly behind the process since the average temperature of the central return line is usually higher (Fig. 12.43).

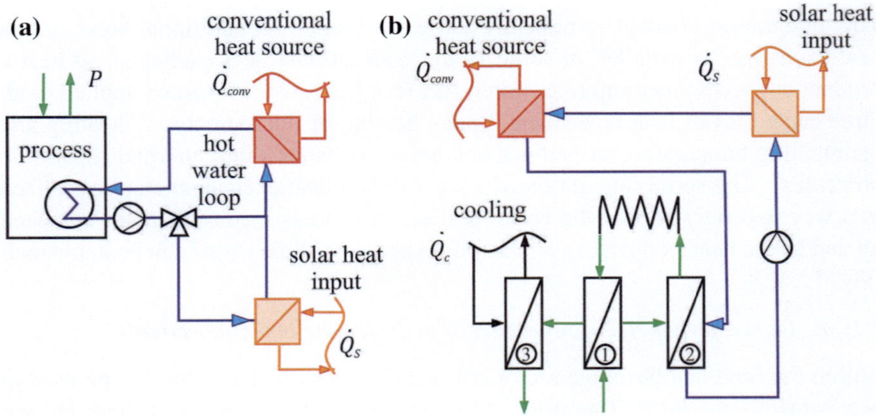

Fig. 12.43 Integration concept for solar heating of intermediate hot water circuits with external heat exchanger (PL_E_IC). General integration flow sheet (**a**) and illustrated by pasteurization process with multi-zone plate heat exchanger and external heating zone (**b**)

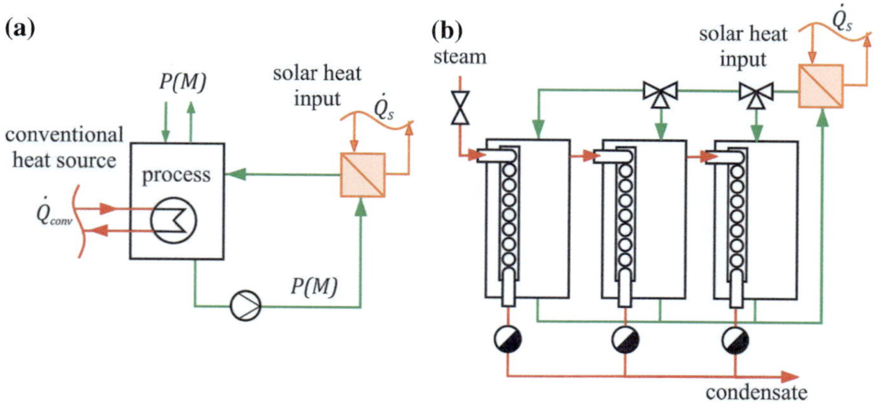

Fig. 12.44 Integration concept for solar heating of bath, machinery, or tank with external heat exchanger (PL_E_HB). General integration flow sheet (**a**) and illustrated by electro plating process heating with internal plate coils (**b**)

PL_E_HB—external HEX for heating of bath, machinery, or tank

This integration concept is used to heat baths, machineries, or tanks via an additional external heat exchanger. Therefore, the product or process medium is drawn from the bath, machinery, or tank, heated by solar energy, and fed back as shown in Fig. 12.44. It's crucial for this integration concept to identify a suitable position or point in time within the process where the temperature of the product or process medium is relatively low. Else, this concept would lead to a parallel integration of solar heat to the conventional heat supply. If the product or process medium must be kept in a very narrow temperature range (due to operational issues), the resulting temperature difference for solar heating might be very small.

PL_E_IS—external HEX for heating of input streams

The fourth possibility for integration of external heat exchangers is solar heating of input streams. Lots of processes require continuous or batch feed-in of product or process media streams. E.g. feed-in of fresh water for compensation of evaporation losses or fresh air for drying applications. Based on the specific process it can be reasonable to preheat these input streams. The process flow sheet for this integration concept is shown Fig. 12.45a. The dashed lines indicate possibilities for the conventional heating of the existing process. Additionally, the integration concept is illustrated with an example, see Fig. 12.45b. The displayed scalding bath for chicken slaughtering is conventionally heated by direct steam injection. Based on water discharge by carcasses and evaporation losses a significant amount of fresh water is fed to the scalding bath, which can be preheated by solar energy. Possibly, this integration concept can be combined with PL_E_PM and PL_E_HB.

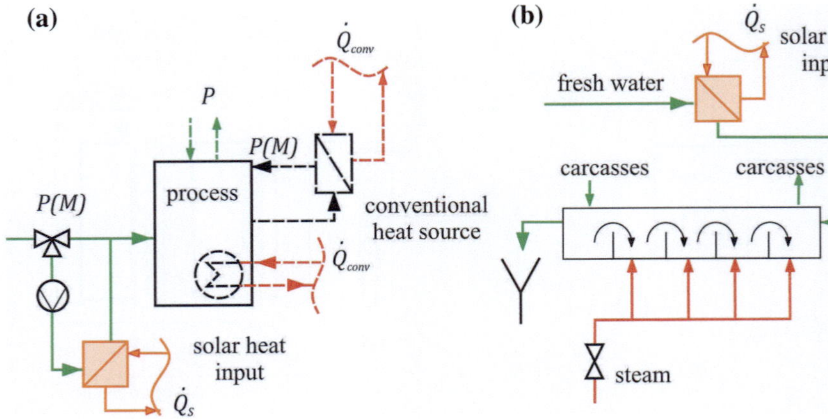

Fig. 12.45 Integration concept for solar heating of input streams with external heat exchanger (PL_E_IS). General integration flow sheet (**a**) and illustrated by scalding process for chicken slaughtering heated via direct steam injection (**b**)

12.7.4 Process Level Internal Heat Exchanger (PL_I)

Besides the integration of external heat exchangers, processes can be supplied with thermal energy by internal heat exchangers. Therefore, the second class for the integration of solar heat on process level is the integration of an additional heat exchanger surface within baths, machineries, or tanks (PL_I). Such an integration concept is used for the category "heating and maintaining temperature of baths, machineries, or tanks". Usually, the additional solar driven internal heating surface is installed in parallel to the conventional heat source. For this concept it is important to consider the available space in the process equipment since it is usually limited due to the treated product (e.g. metal parts) or components (e.g. stirrers, strainers, etc.) (Fig. 12.46).

12.7.5 Process Level Steam Supply with Reduced Pressure (PL_S)

The third class for the integration of solar heat covers solar steam generation with reduced pressure. Therefore, two concepts can be applied to supply steam with vacuum (PL_S_V) or with low pressure at a temperature range of 110–135 °C (PL_S_LP).

PL_S_V solar steam generation at vacuum

Solar steam generation with vacuum can be applied for numerous evaporation processes e.g. in food and beverage industry, since typical process temperatures are in the range of 40–80 °C. Therefore, an evacuated storage with a conventional heat exchanger is required. The condensate from evaporation process is preheated

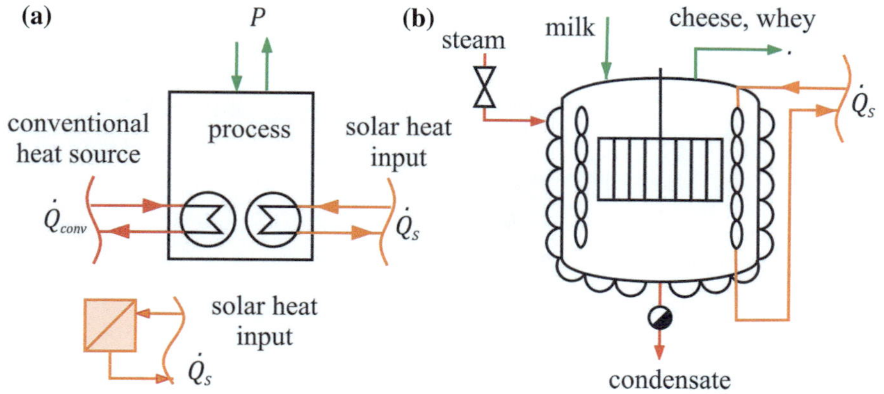

Fig. 12.46 Integration concept for solar heating of bath, machinery or tank with internal heat exchanger (PL_I). General integration flow sheet (**a**) and illustrated by curd vessel for cheese production heated with conventional heating jacket and additional solar driven dimple plates (**b**)

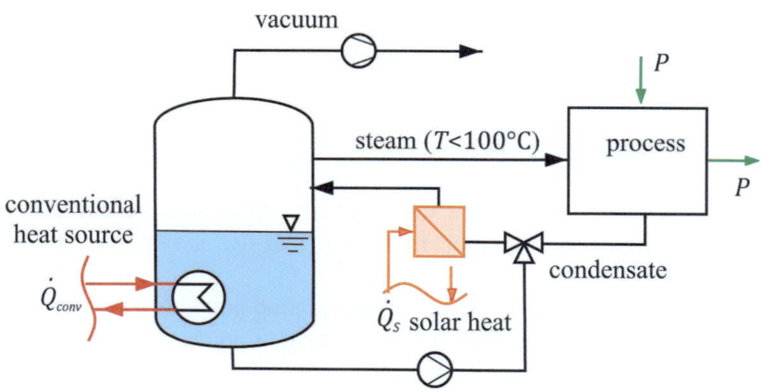

Fig. 12.47 Integration concept for solar steam generation with vacuum for heating of evaporators (PL_S_V)

by solar energy and fed into the evacuated storage. The evaporated condensate is fed back into the evaporation process. Such an integration is shown in Fig. 12.47.

PL_S_LP solar steam generation at low pressure

For solar steam supply with reduced pressure of 0.15 – 0.3 MPa (according to 110–135 °C) a kettle reboiler is required that is supplied with solar heat. Feed water or condensate is fed into the reboiler, evaporated, and used to heat the process. Therefore, the conventional steam demand for injection heating is reduced. Based on the required installations this integration concept is similar to SL_S_PI on supply level. The only difference is the reduced temperature level compared to

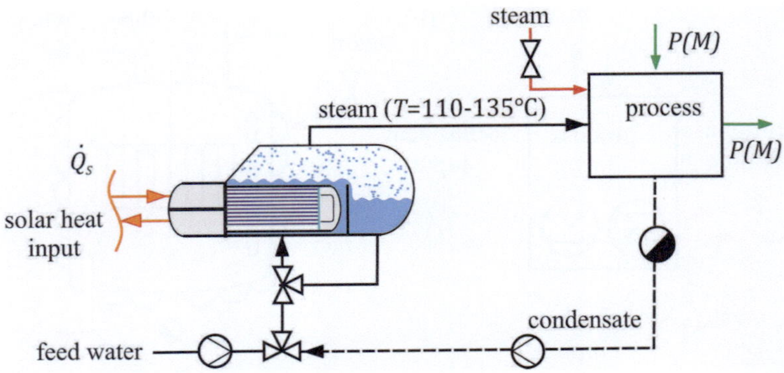

Fig. 12.48 Integration concept for solar steam generation with reduced pressure for direct steam injection (PL_S_LP)

conventional steam systems in Industry. Therefore, this integration concept can also be applied under moderate climatic conditions.

12.8 Planning Guide

12.8.1 Stepwise Approach

The design of a solar heat-based or solar heat-only integrated process heat system typically follows a stepwise approach consisting of five major elements outlined in Fig. 12.49.

The effort for identification of suitable integration points for solar heat as well as the number and complexity of possible integration points can vary significantly between industrial sectors and individual factories. To assist the necessary steps for a feasibility assessment, the methodology illustrated in Fig. 12.50 can be used. This methodology can be divided into three main parts: pre-feasibility assessment (steps 1 and 2), feasibility study (steps 3 to 7) and decision/further activities (steps 8 and 9). In the following, the single steps are briefly described.

The purpose of the pre-feasibility assessment is to quickly find out if solar heat can be used in a company. Therefore, basic information should be gathered after the initial contact. This can be done by phone or by a questionnaire that is sent to the company. Most important information is temperature level of consumed process heat, available roof- or ground areas for solar heating systems,

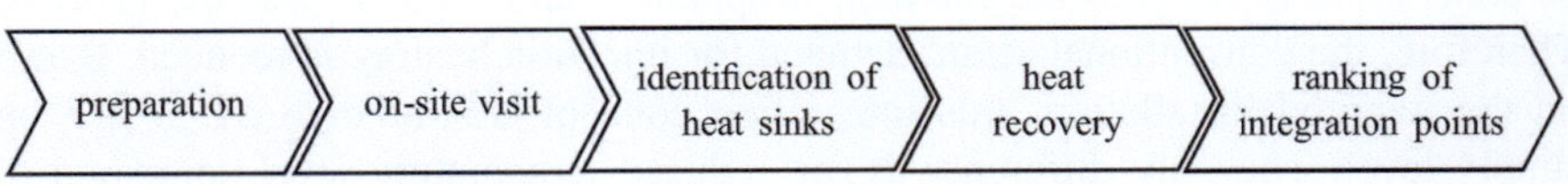

Fig. 12.49 Stepwise approach to planning a solar process heat system

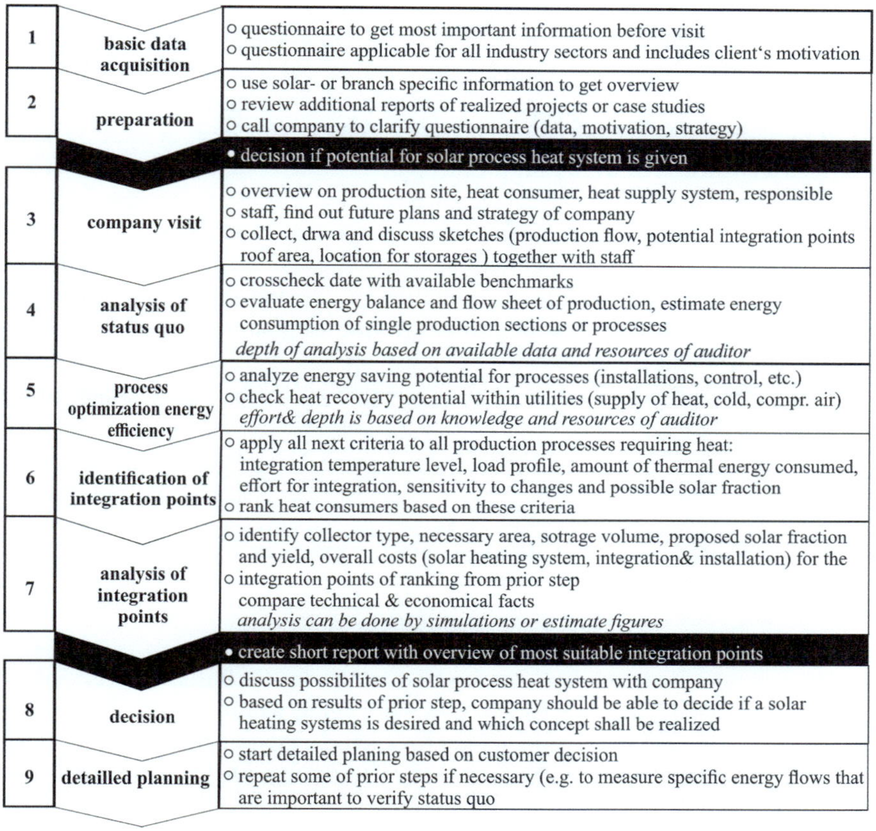

Fig. 12.50 Assessment methodology for solar heat integration, modified from (Muster-Slawitsch et al., 2015)

production times and investment policy of the company. Additional, branch specific information and reports of realized projects or case studies can help to assess the general possibilities for integration solar heat within a specific sector. Based on this information a decision should be made if solar heat can be implemented or not.

If the potential for a solar process heat system is given, a feasibility study is carried out. Starting with a company visit it is intended to get an overview of the production site, heat consumers, and heat supply system as well as to find out about future plans and the strategy of the company. Therefore, it is helpful to collect, draw, and discuss sketches (production flow, possible integration points, roof area, location for storages, etc.) with the technical staff of the company. Based on these information and relevant data provided by the company the status quo is analysed by crosschecking gathered data with available benchmarks, drawing energy balances and flow sheets of production processes, and by trying to estimate the energy consumption of single production sections or processes. The actual depth of this analysis is based on available data and resources of the auditor.

The next step would be to estimate the potential for heat recovery and energy efficiency measures. Similar to step 4, the effort and depth of this step is influenced by the knowledge and resources of the auditor. However, at least common heat recovery measures within utilities (supply of heat, cold, and compressed air) should be considered. The advantage of a detailed process integration study is that besides the identification of heat recovery options the study helps to identify which processes at which temperature level are sensible candidates for solar heat integration. Based on the energy consumption of different production sections or processes and the consideration of energy savings, integration points for solar process heat can be identified (step 6). For these pre-selected integration points, the following criteria can be applied to all production processes with thermal energy consumption:

- integration temperature level,
- load profile (daily, weekly, annually),
- amount of thermal energy consumed annually effort for integration,
- sensitivity to changes,
- achievable solar fraction.

Based on these criteria the heat consumers can be ranked. Based on this ranking the best integration points are analysed in more detail within step 7. Therefore, a suitable collector type must be identified for each integration point and the necessary area and storage volume, the proposed solar fraction, yield, and overall costs must be estimated. The costs should cover the solar heating system, integration, and installation. This analysis can be done by simulations or estimative figures. Considering the technical and economic facts of the considered integration points, the ranking of step 6 might be rearranged.

Based on the results of the feasibility study, the different possibilities for integrating solar process heat should be discussed with the company. With the information of effort for implementation, estimated costs and expected savings the company should be able to decide if a solar heating system is desired and which concept shall be realized. Based on the decision the detailed planning is carried out. Therefore, it might be necessary to repeat some of the prior steps again (e.g. to measure specific energy flows that are important to verify status quo).

12.8.2 Identification of Integration Points

Generally, more than heat sink (on process and/or supply level) can be assisted with solar heat within an industrial plant. While the main differences of process and supply level integration may be summarized within a few aspects (see Table 12.3), the correct choice of the most suitable integration point is a complex task.

Table 12.3 Basic aspects of solar process heat integration on process- and supply level

Criteria	Process level	Supply level
Detailed process data	Required (at least for target processes)	Steam, hot water, thermal oil
Preliminary process integration analysis	Essential for appropriate solar system design	Steam, hot water, thermal oil
Flexibility to adapt to later changes in processes	Low	Hot air
Collector efficiency	Potentially higher (thanks to lower operating T)	
Solar heat contribution potential	Restricted (to share of the supported process(es) on the overall heat demand)	
Heat storage to cope with variability of radiation and out-of-phase of source and sink	Depends on the profile of the selected process stream(s) with respect to solar availability, and sizing of solar contribution	Hot water, hot air, thermal oil

This subchapter introduces a method to rank different integration points by analysing their suitability for solar heat supply. The term 'point' is used instead of unit operation or sink to highlight the fact that some sinks could be assisted at different levels (e.g. heat exchangers) and temperatures.

The method implies two steps. The first step, also called the pre-integration step, mainly focuses on the sink's heat consumption before the solar thermal system is integrated. By means of general suitability indicators, heat sink candidates can be identified. The second step considers the possible integration scheme/concept adopted for each of the candidates and applies technical criteria for ranking them. A matrix using the suitability indicators and integration criteria is then presented as a ranking tool. Limits of the method are mentioned in the last chapter.

The indicators and criteria described below are collected from different case studies and are generally independent of the industrial sector. Planners should take them into account to ensure an objective and thorough assessment of all integration possibilities. Due to the large number of suitability indicators, it is suggested to use the ranking matrix as a tool to facilitate the evaluation of different integration schemes. Of course, other tools like the cost–benefit analysis may be applied. The proposed ranking method has to be seen as a filtering tool of most suitable integration candidates. It should enhance the planning process by reducing the number of relevant integration points. A more detailed techno-economic evaluation of the candidates is envisaged in the assessment methodology within "analysis of integration points" (step 7).

The data collected in the process integration step provide a good basis for the method introduced. Technical information on the unit operations and the heat supply and distribution network as well as the operational schedules and the required temperatures are used to identify suitable integration points within the industrial plant. The identification method also bases upon details like the size of existing heat exchangers or the operation strategy that should be basically prepared for process integration.

12.8.3 Evaluation of Heat Sinks and Integration Concepts

The first step considers a set of indicators that are relevant for solar heat integration. The indicators are grouped into three sections: demand, schedule, and technology. For each section, the indicators are listed in Table 12.4 according to their impact on future solar integration (the indicator in the first row is more significant than the one below etc.).

The suitability indicators have been chosen based on a set of considerations related to energy consumption and operation characteristic. Every indicator can be quantified through a variable. The suitability range of each variable will not be limited in this document; however general rules can be applied to evaluate them as described in the following paragraphs.

The operational temperature T_{op} describes the typical temperature of the process medium during operation. For operations without an intermediate medium, the temperature of the heat carrier being in contact with the treated product is taken as indicator. ΔT is the temperature range by which the process medium needs to be heated up. Generally, the set point temperature of the planned solar thermal system has to be beyond T_{op} where ΔT determines the collector mean temperature. Therefore, higher solar thermal system efficiency and gains are expected

Table 12.4 List of suitability indicators in pre-integration phase modified from (Muster-Slawitsch et al., 2015)

Sector	Indicator name	Indicator variable	Unit
Demand	Operational temperature	T_{op}	°C
	Temperature range	ΔT	°C
	Annual heat demand	Q_a	kWh/a
	Storage capacity	Q_{st}	kWh
	Storage charging	$\dot{Q}_{st}$	kW
	Utilization	t_{op}	h/a
	Mean load	$P_{av} = Q_a / t_{op}$	kW
	Recirculation	rec	%
Schedule	Daily demand coincidence	$C_d = Q_{mid} / Q_{day}$	%
	Demand seasonality	$C_a = Q_{sum} / Q_a$	%
	Demand uniformity	$uni = P_{av,day} / P_{may,day}$	%
	Rescheduling	t_i	h
Technology	Equipment supply quality	$\pm T_e$	°C
	Product supply quality	$\pm T_p$	°C

with lower T_{op} and larger ΔT. In other words, sinks with low operational temperature and larger ΔT are more suitable than those requiring higher temperatures and having hot process medium circuits.

The process utilisation t_{op} accounts for the annual number of hours during which the operation is running. Especially for tropical climate zones the longer the process time, the higher the potential for solar integration. Depending on the industrial segment, the range of high suitability differs. For example, in building material industry most of the operations meet the criteria of moderate to high operation times (about 4000 h/year), so these values are considered as indicative of a moderate suitability. In the building material industry operation times above 6000 h correspond to high potential and for those operations with utilizations less than 3000 h low potential is assigned.

The annual heat demand Q_a is one of the main criteria to be considered before integrating solar energy. It strongly determines the solar yield that can be achieved for the corresponding sink. This indicator has to be evaluated in a comparison between all sinks within the production site. A milk drying process with a high demand share (e.g. 1/3 of the annual consumption in a dairy plant) is more interesting than the space heating system of the reception office.

The mean load P_{av} is very useful for pre-dimensioning of the solar heat exchanger and storage tank. High mean capacities along with constant loads allow large solar installations with low specific costs and generally make the sink more attractive for solar feed-in; however, the potential of system capacity has to be seen within the climatic and spatial boundary conditions of the plant. If for example the available surface for collector installation allows an aperture of 1000 m^2, large sink capacities over 1 MW are as well significant as moderate capacities around 600 kW.

The recirculation ratio *rec* provides an indication if the unit operation is running in open, closed, or semi-closed mode. Since open operations without mass or heat recovery are very suitable for solar heat integration, the indicator potential is the opposite of the variable itself: the higher the ratio, the lower the potential.

The available storage Q_{str} and charging $\dot{Q}_{str}$ capacity indicate how flexible the operation is in regard to fluctuating supply and demand profiles. Generally, large capacities facilitate the integration of supplementary heat sources. Short-term (diurnal) to medium-term (weekly) storages can be seen as of high potential whereas sinks without own heat accumulators are of lower significance.

The coincidence of solar radiation with heat demand is the most important indicator in the schedule section. Especially for sinks with low storage capacities and loads the achievable solar gains can strongly depend on this indicator. The coincidence is evaluated on a yearly basis through the thermal demand seasonality C_a which characterises the ratio of the summer (heat demanded from April to September for sites in the northern hemisphere) to the annual consumptions. A daily evaluation C_d may also be useful for sinks with low seasonal dependency. In this case, C_d is calculated as the ratio of the average thermal demand between 11 AM and 3 PM to the daily one. Once both indicators are calculated, heat sinks with values above 0.6 have to be grouped as most interesting ones where coincidences around 0.3 show low suitability for solar feed-in.

The thermal demand uniformity *unif* characterises the ratio between average and peak load during one operation cycle. A low uniformity below 0.3 indicates the need for additional heat utilities (e.g. the main boiler) to handle high peak loads. A high uniformity over 0.7 makes the solar plant much easier to design for planners. Sinks with higher uniformities *unif* have to be privileged in the evaluation.

If rescheduling of a unit operation is allowed and technically feasible, its coincidence potential can be enhanced adequately, and better conditions for solar integration can be therefore obtained. The indicator suitability is measured by evaluating the operation interval t_i that can be shifted to the time between 11 AM and 3 PM. If many sinks can be rescheduled, the highest suitability factor will be assigned to the sink with the lowest impact on the whole production time plan.

The last two indicators describe the requirements of the unit equipment and of the product based on their sensitivity to temperature fluctuations in the supply line. The sink with the largest allowed fluctuation range ($\pm\Delta T$) has the highest potential for solar heat integration and so on. Both indicators are of the same significance for solar heat integration.

Once the heat sinks with higher suitability for solar feed-in are identified, the second evaluation step can be started under consideration of technical criteria/ requirements. Here, the integration schemes are evaluated as a whole. The set of criteria defined in Table 12.5 is based on constraints that apply if solar thermal systems are integrated into the existing heat supply system. Some criteria are hydraulic or control specific, others are generally related to the spatial location of the sink and to its existing supply utility. Like in Table 12.4, the criteria are listed top-down according to their significance for solar heat integration. The impact of some criteria is explained in the paragraph below.

From an operational point of view, the supply reliability and safety are much more important than the solar gains which can be achieved in an industrial plant. Therefore, the first criterion to be considered when feeding-in solar energy is the production safety. An indication for a non-suitable integration scheme could be the need for production interruption during maintenance of the solar installation. A design of a solar thermal system without solar heat storage and backup is not the best choice in combination with a heat sink requiring high supply quality in terms of feed temperature and pressure stability. The intermittent solar radiation may lead to low solar gains or unnecessary process shut-downs.

Heat exchangers and storages are important cost drivers in solar process heat installations. Therefore, integration schemes allowing -if any- small sized components are favourable. Several changes in the control hard- and software of the main supply utility or of the process machinery itself may put the production continuity at risk and therefore have to be avoided. Heat recovery and CHP are competitive technologies which also have to be taken into account by the planner.

The criteria load balancing and modulation bring the interaction between the solar thermal system as heat source and the existing heat utility into play. In some

Table 12.5 Basic criteria for solar process heat integration on process- and supply level

Criteria	Description
Process continuity	No sink interruption needed during maintenance or repair of solar thermal system
HX sizing	Existing heat exchanger has to be used for solar feed-in
Storage sizing	Addition of storage capacity has to be avoided
Control Hardware	No extension in supply utility control hardware is required
Control Software	No changes in supply utility software are required
Estimated solar yield	High solar yield estimated
Replacement of CHP	The utilization of (planned) CHP utility will not be reduced
Replacement of WH/HR	The use of waste heat / heat recovery (compressors) will not be unnecessary reduced
Load balancing	Utilization of excessively operating supply utilities will be reduced
Distance to solar	The solar plant has to be close ($< 50\,\text{m}$) to the supply line of the sink
Multi-supply	Other heat sinks can be easily co-assisted
Modulation	The supplying or back-up heat utility can be modulated
Fouling risk	No fouling risk for the added HEX—if any
Auxiliary energy	No significant pressure differential to be overcome, only pump element for collector circulation is needed
Dependency on radiation course	The estimated solar yield is slightly dependent on the direct fraction of solar radiation

cases, a boiler is operated at nearly full capacity, so that solar integration at the appropriate supply level would balance heat loads and enhance the scheduling flexibility within the site. Generally, higher solar gains are expected if the back-up system can be modulated down to low capacities and if more heat sinks are technically easy to assist. The integration concept of direct solar steam generation (*SL_S_PD*) is not suitable for regions with low or particularly intermittent beam irradiation due to the discontinuity of supply. That´s why schemes with low dependency on the radiation course are advantageous. For simplicity, the criteria of Table 12.5 are reorganised in four sections: reliability, costs, benefit, and efficiency (Table 12.6).

Table 12.6 Basic aspects of solar process heat integration on process- and supply level modified from (Muster-Slawitsch et al., 2015)

Sector	Indicator name
Reliability	Process continuity
	Load balancing
	Control hardware
	Control software
	Fouling risk
Cost	HX sizing
	Storage sizing
	Distance to solar
	Auxiliary energy
Benefit	Estimated solar yield
	Multi-supply
	Modulation
	Dependency on radiation
Efficiency	Replacement of CHP
	Replacement of WH/HR

12.9 Example Installations

12.9.1 Brewery Göss/Austria

Breweries have a high low temperature demand and often a high potential for process optimization. Manufacturing of malt and beer requires large amounts of electrical and thermal energy. According to Mauthner et al. (2014), the entire heat demand of breweries and malting plants can be supplied at temperature levels between 25 °C and 105 °C, with processes like malt drying, bottle cleaning, pasteurization or mashing being very promising for solar process heat (*SPH*).

Sector specific *SPH* concepts for this sector have been developed and a number of successful *SPH* projects have been carried out recently. Figure 12.51: shows a flat-plate collector field at the Goess Brewery in Leoben, Austria, which was installed in June 2013.

Two steam supplied vessels (mash tuns) were retrofitted with internal plate heat exchangers to allow the mash tuns to be hated with hot water instead of steam. This hot water supply is fed by waste district heat from a biomass plant as well as from the large-scale SPH installation (100 collectors with 1500 m^2 gross area in total and about 1 MW_{th} peak power). The collector field feeds a 200 m^3 pressurized water buffer storage, which was installed outside the brewery (see yellow cylinder in the background of Fig. 12.51). Figure 12.52 shows the system concept of the *SPH* plant.

Fig. 12.51 Flat-plate collector field of 1375 m^2 at the Goess Brewery (AEE INTEC, 2022)

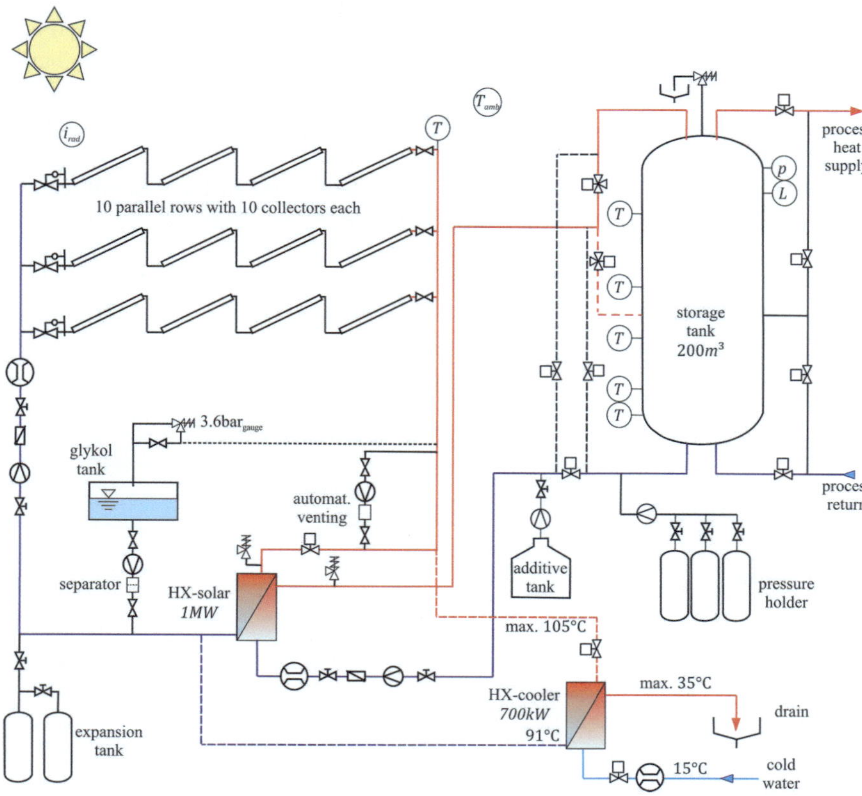

Fig. 12.52 System concept of the SPH installation at Goess Brewery (AEE INTEC, 2022)

The collector loop contains a water/glycol mixture and charges the solar buffer storage via a charging heat exchanger. External stratified charging is done depending on the charging flow temperature level and the temperature within the storage. Non-return valves eliminate losses due to natural convection through the collectors and the charging loop at night. To prevent the charging loop from freezing on cold winter mornings, the charging heat exchanger is automatically bypassed until the whole collector loop has been heated up. Stagnation (i.e. steam production in the field when the solar supply exceeds the demand) is prevented in two ways: First, night cooling is implemented, i.e. hot water from the storage top can be pumped through the collectors at night to cool the storage. If necessary, additionally a water/water heat exchanger can cool down the collector field.

Figure 12.53 shows the solar heat integration into the retrofitted mash tuns. The mashing process requires consecutive temperature increase from initially 58 °C to finally 78 °C in three steps. If solar heat above these temperatures is available, heat from the buffer is pumped through the retrofitted plate heat exchangers. The return flow from the process back to the storage is stratified according to the temperature level. If the temperature in the solar energy storage is not high enough to supply the entire mashing process, but still higher than the process return temperature, the solar heat is further heated in-line by district heating from a biomass combined heat and power plant. If the temperature within the solar buffer is too low, the buffer is bypassed and the mash tuns are supplied by waste heat only. If both systems cannot supply either the temperature or the energy quantity needed, the existing steam supply system acts as parallel backup.

Simulations indicate that almost 30% of the heat demand for mashing can be supplied by SPH and the remaining process heat demand can be covered by waste

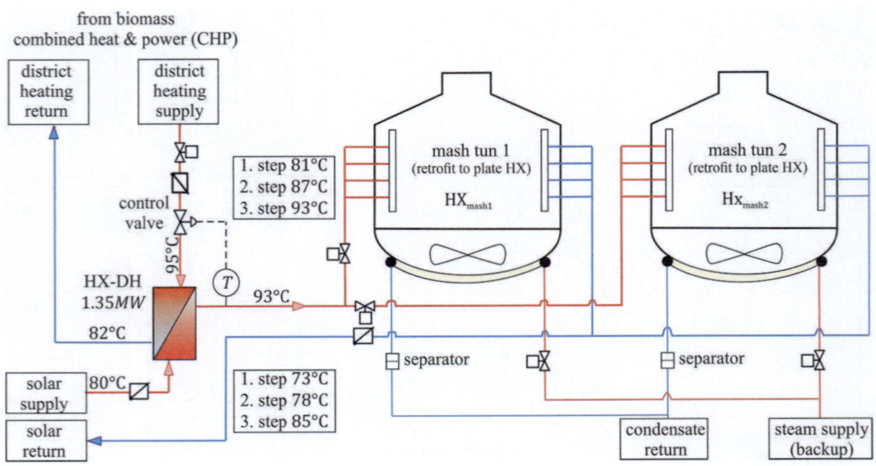

Fig. 12.53 Solar heat and biomass district heat integration into the mash tuns of Goess brewery (Mauthner et al., 2014)

heat from biomass. In sum, round 1570 MWh of natural gas per year, corresponding to round $3.8 \cdot 10^4$ tons of CO_2 equivalents can be saved in future by this hybrid system, see (Mauthner et al., 2014).

The system is expected to have relatively low annual specific solar gains of $280\,\text{kWh}/(\text{m}^2 a)$ due to the for standard flat-plates comparably high supply temperatures. The overall invest was 420 €/m^2, not including subsidies. About 50% of these costs have been funded by the EU.

12.9.2 Copper Mine Gabriela Mistral, Chile

In the mining sector we find a number of interesting applications for *SPH*. The selection of required processes is of course very much dependent on the materials and their form to be treated. Mining generally needs very high investments, which makes the sector very risk sensitive and conservative about applying new technologies. However, there is a need for reduction of CO_2 emissions, since customers increasingly care about the carbon footprint caused by their supply chain. Adding to that, the sector sees an important advantage of *SPH* in the fact that solar heating costs are highly predictable over the lifetime of a *SPH* plant, which highly reduces investment risks.

During the last years, mining industries have installed several very large-scale *SPH* plants, and more projects are being planned. The currently largest *SPH* system worldwide is the Minera Gabriela Mistral, pictured from bird view in Fig. 12.54. The mine Gabriela Mistral in Chile, situated in the Atacama Desert, is producing copper ore. There, energy is mainly needed for breaking the ore, for transporting it out of the mine, for pumps irrigating the outdoor heap leaching process, as well as for electrolytic copper winning from the sulphide solution. Copper mines are usually very remote, so that energy supply has to be provided by tank trucks delivering diesel or oil. In the electrolysis process of copper winning, copper of high purity is deposited on the cathode plates. The process requires a stable temperature in order to produce high quality. For this electrolytic purification, a copper solution is being heated in a bath to a temperature between 46 °C and 51 °C. Prior to the installation of the *SPH* plant, the mine used $8 \cdot 10^3 \text{m}^3$ of oil per year to produce $1.2 \cdot 10^5 t$ of copper for this process.

The *SPH* plant at Minera Gabriela Mistral as seen in an aerial picture from above in Fig. 12.54 has a flat-plate collector area of $3.93 \cdot 10^4 \text{m}^2$ and a heat storage of $4 \cdot 10^3 \text{m}^3$. It was commissioned in October 2013 and has a thermal peak power of $27.5\,\text{MW}_{\text{th}}$. The plant is located 100 km south of the city of Calama, in a region with nearly no rainfall and regular sandstorms. Regardless of the harsh environment, the Atacama Desert is pre-destined for *SPH*, since due to the high altitude of 2000–4000 m above sea level the dry and thin atmosphere leads to very clear days during 90% of the year. Additionally, the Atacama Desert is located near the equator between $\Phi = 18°$ and 25° southern latitude, so that there is only little seasonal variation in irradiance.

Fig. 12.54 Flat-plate collector field of $3.93 \cdot 10^4 m^2$ at Codelco's "Minera Gabriela Mistral" in the Atacama Desert, Chile modified from (Behar et al., 2021)

The system delivers heat for electrolytic copper winning to the *Gabriela Mistral* copper mine, which belongs to the state-owned mining company Codelco. The schematic plant layout is illustrated in Fig. 12.55. The primary (collector) loop of the plant works at 85 °C flow and 55 °C return flow, the secondary loop at $80\,°C/60C$. The $80\,C$ flow temperature is needed to heat the baths. The secondary solar loop is directly connected to the boiler heating network. Due to the large storage, the *SPH* system supplies heat for 24 h. The collectors are cleaned by an automated dry-cleaning device and their tilt is seasonally adjusted to ensure maximum solar gains. The system annually delivers 50 GWh of heat, with an extremely high specific solar gain of $1,272\,\text{kWh}/\left(\text{m}^2 a\right)$. It covers 85% of the energy needed for electro winning, equal to savings of $2 \cdot 10^4 l$ oil per day or 250 road tankers annually. The remaining heat demand is delivered by the existing oil burner.

In Fig. 12.55 one can also identify the high storage vessel located next to the collector field composed of flat plate collectors as illustrated can be seen in Fig. 12.56. Sunmark Solutions (Sunmark-Arcon, 2014) installed Minera Gabriela Mistral as a turn-key project. The project was realized by the project developer Energía Llaima SpA, who organized financing of the plant. Energía Llaima is contracted by Codelco, who only buy the solar heat actually produced. This way, Codelco did not have to take any risks, neither technically nor economically, nor did they have to put any effort into realization of the project (Nasirov & Agostini, 2018).

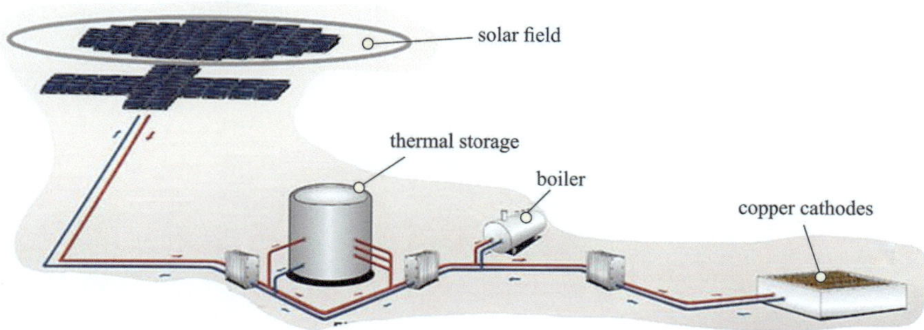

Fig. 12.55 Schematical view of solar field integration in electrowinning plant of copper mine Gabriela Mistral from (Sunmark-Arcon, 2014)

Fig. 12.56 View of installation with solar field from (Behar et al., 2021)

12.9.3 Food Industry (NEP Parabolic Trough)

Dairy industry requires heat for a number of processes in a wider range of temperatures. Low temperature is needed for washing purposes, but pasteurization, cheese production needs higher temperature. The company Emmi in Sairlegnier, Switzerland, produces cheese and has a conventional gas boiler with a pressurized water heat distribution system. A parabolic trough collector field of 627 m^2 aperture was installed by NEP Solar at the site of the dairy to reduce the energy bill, shown in Fig. 12.57. The field has an installed power of 360 kW_{th} and operates at 140–180 °C in a water–glycol loop, which has been chosen due to the cold winter conditions locally.

An existing heat storage of 15 m^3 integrated into the supply side has been used to store the solar heat to cover the demand. Solar heat is transferred via a 360 kW plate heat exchanger to either of two integration points, see Fig. 12.58. Firstly, solar heat is provided to the supply side, into the existing 15 m^3 supply heat storage. The option is chosen as long as the boiler is switched off because the storage

Fig. 12.57 Parabolic trough collector field by NEP Solar, installed on the roof of the company Emmi in Switzerland (Minder, 2013)

remains above 25% charging state. When storage state drops below 25% then the boiler is switched on, and the solar heat is provided to the return line of the hot water boiler for pre-heating.

The system is designed to deliver 255 MWh heat per year. The specific installation cost was around 480 €/m^2, with planning costs and integrations costs coming on top of that (AEE INTEC, 2022). The solar fraction of the whole system is 15%. Due to the demonstration character of this innovative system, subsidies were used to support the economy of such an installation.

12.9.4 Pharmaceutical Industry (Fresnel)

In 2015 the company Industrial Solar GmbH, which had developed the high quality Linear Fresnel collector LF-11 for different heat transfer fluids (pressurized hot water, thermo oil and direct steam), has installed a Fresnel collector field on the factory roof of RAM Pharma (RAM) in Sahab, Jordan to provide steam for the pharmaceutical production process and cut energy costs and CO_2 emissions, see Fig. 12.60.

The system is comprised of 18 Industrial Solar LF-11 Fresnel collectors with a total aperture area of 396 m^2 and a peak capacity of 223 kWh$_{th}$. The integration of the direct steam production is done via a steam drum as shown in Fig. 12.61. The plant is a hybrid system of a Fresnel collector and a diesel boiler, which successfully provides saturated steam at 4.5 bar for a pharmaceutical factory in Amman, Jordan. Here, a 2 m^3 steam drum serves simultaneously to separate water

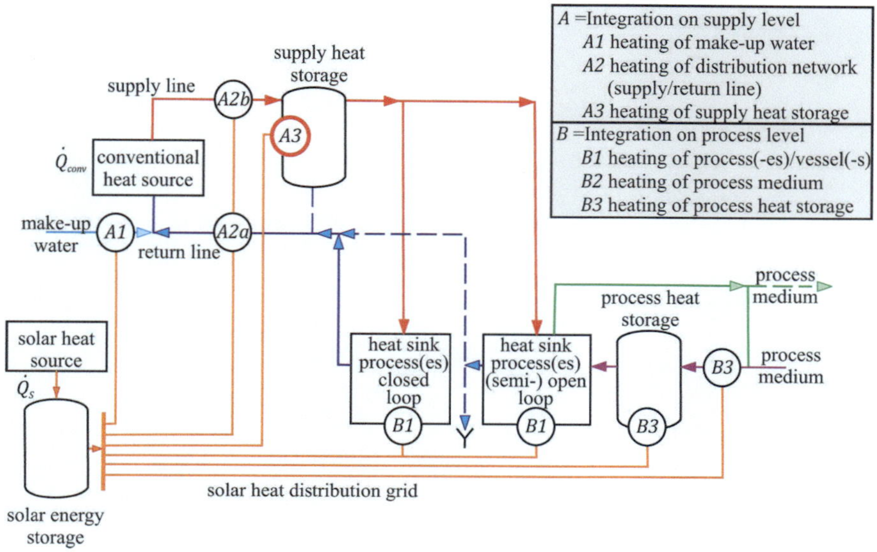

Fig. 12.58 Hydraulic integration of the PTC-field at Emmi into the existing supply system modified from (AEE INTEC, 2022); process medium options: product, (fresh) water, (drying) air

and steam, as a balancing vessel for evaporation volumes, as a short-term storage tank according to the Ruths storage principle and as a pressure maintenance vessel for the solar system (Fig. 12.59).

In a remodelling, an alternative circuit was realised where water and saturated steam are separated in a cyclone. The steam drum now is only used as a start-up

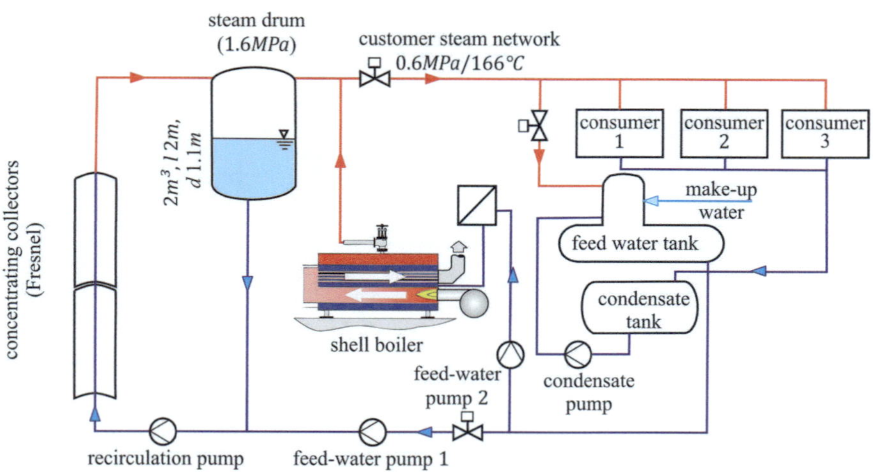

Fig. 12.59 Integration scheme of the DSG Fresnel collector field into the RAM Pharma plant

Fig. 12.60 Photograph of the RAM Pharma roof installation (Berger et al., 2016)

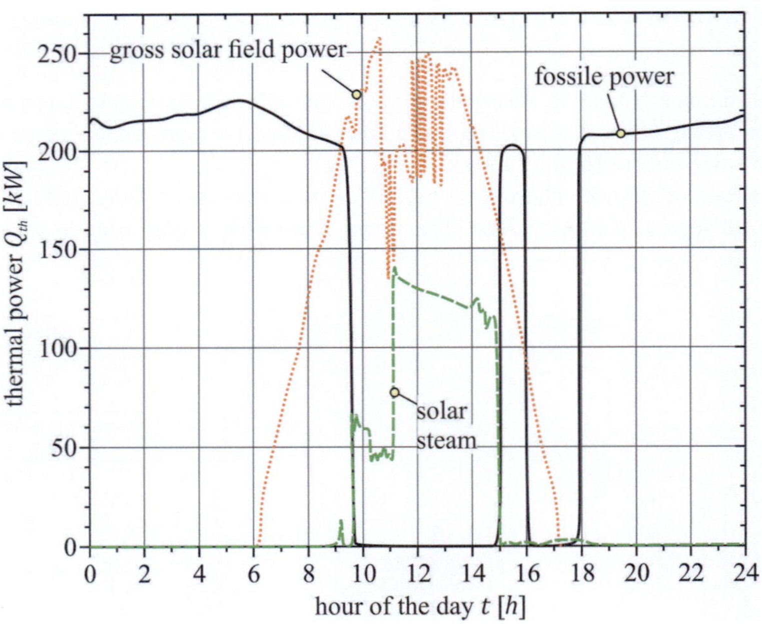

Fig. 12.61 Steam production by fossil and solar boilers during a day from (Krüger et al., 2017)

tank. The pressure maintenance of the boiler is also realised by the solar field, so that no intermediate firing with the boiler is necessary during the solar supply, as illustrated in Fig. 12.61.

12.9.5 Surface Treatment—Zehnder

The company Zehnder is producing radiators, heat pumps, and other heating components in Gränichen, Switzerland with about 250 employees at five to six working days per week. The company has a heating demand of approx. $1.67 \cdot 10^3 \mathrm{MWh/a}$, which was previously covered with heating oil and liquefied petroleum gas (LPG).

The radiators are pre-treated in baths at a temperature of ca. 60C before they are coated. About 40% of the heating demand of Zehnder is caused by these baths. Figure 12.62 shows the solar collector field of the company, which was commissioned in July 2012. The field is orientated along the roof edge with about 25° turned towards the south-east. The collectors are sloped 45° and unshaded. Since the production hall was newly constructed before the collectors were installed, the necessary sub-structure for the collectors had already been considered beforehand.

Figure 12.63 shows the integration of the plant into the hydraulic system. Solar heat is directly integrated via a storage into the heating network of the new production hall; thus, the collector field works like an additional boiler. The installation is a so-called Aqua System, operating with water only as a heat carrier fluid. Therefore, no heat exchanger between collector loop and heating network is necessary. Depending on the available power of the field, the solar loop is either used as return flow booster for the boilers or to directly deliver the required flow temperature of 85 °C to heat the baths.

Because the baths themselves can be used as buffers, only a very small additional storage tank of $5\,\mathrm{m}^3$ is installed. This storage acts as a hydraulic separator,

Fig. 12.62 Field of 80 high performance evacuated tube collectors with CPC reflectors on the roof of Zehnder in Gränichen, Switzerland, with a gross area of $394\,\mathrm{m}^2$ from Schmitt and Hess (2016)

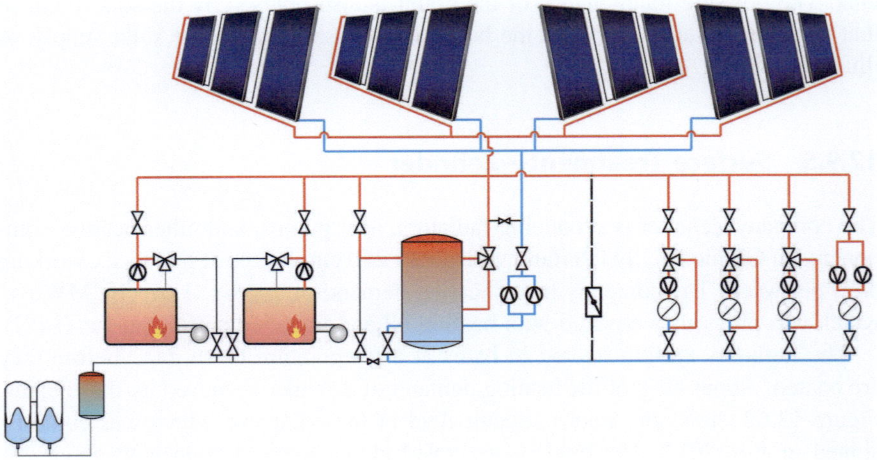

Fig. 12.63 Integration of the evacuated tube collector field into the heating network parallel to the existing boilers, (Schmitt & Hess, "Best practice" Series of Case Study Reports from Demonstration Projects, 2016)

compensating the different mass flows through the solar loop, the boilers, and the baths.

At very good summer days the plant covers the whole heating demand of the baths. Annually, the solar fraction at this process is 30%. Alternatively, also the heating of the production hall can be supplied by solar. Considering all material and installation costs of the system including the steel sub-structure for mounting, the solar plant had specific investment costs of about 1000 €/m^2, not including subsidies. Thanks to a grant from the Swiss Federal Office of Energy, the plant is expected to have a payback period slightly above five years.

According to the planner, the simulated annual gain of the system is $400\,\text{kWh}/\left(\text{m}^2 a\right)$, summing up to 158MWh/a, which means annual savings in LPG of about 200 MWh. The system has been provided as a turnkey system and was installed under Guaranteed Solar Results contracting. In this case, this means if the system delivers less than 90% of the simulated annual energy gain as a mean value over the first three years, the solar company is paying the costs of the fuel not saved. In the first two years of operation, the measured solar gains matched the simulated system gains very well.

12.9.6 Drying—Alimentos Campestres

Our last example is a relatively simple installation typical for the drying process in an agro-food application. Air collectors often are used to heat fresh air and thus reduce the relative humidity in the air. This air can be used to extract humidity from fruits, vegetables, wood chips or other materials when the hot air is put into

Fig. 12.64 Air collector field for indirect drying of fruits with aperture area of 400 m^2 (CONA, 2022)

contact with the product. This may be a possibility to increase value and/or shelf life and handling of the product. A temperature control is necessary in order to prevent excessive temperatures harming the product.

The company Alimentos Campestres S.A. in Ciudad de Guatemala uses solar air collectors in combination with rock bed storage for continuous drying. The plant shown in Fig. 12.64 can dehydrate up to 3000 kg of fresh fruits in one batch. The collectors consist of wooden frames, white solar security glazing and an absorber made of special aluminium profiles to ensure high heat conductivity. Ventilators suck the air through the collectors and into the horizontal drying chambers. Indirect drying (no direct sunlight) allows the product to preserve colour, taste, and vitamins. High hygienic standards are met since filters are applied to prevent the product from getting in contact with dust or insects.

References

AEE—Institut für Nachhaltige Technologien. (2013). Matrix of industrial process indicators. AEE INTEC. Retrieved from http://wiki.zero-emissions.at/index.php?title=Main_Page

AEE INTEC. (2022). Database for applications of solar heat integration in industrial processes-SHIP. (F. o. 49/IV, Ed.) Retrieved April 11, 2022, from http://ship-plants.info/solar-thermal-plants/125-goess-brewery-austria

Behar, O., Pena, R., Kouro, S., Kracht, W., Fuentealba, E., Moran, L., & Sbarbaro, D. (2021). The use of solar energy in the copper mining processes: A comprehensive review. *Cleaner Engineering and Technology, 4*, 100259, 13. https://doi.org/10.1016/j.clet.2021.100259

Berger, M., Meyer-Grünefeldt, M., Küger, D., Hennecke, K., Mokhtar, M., & Zahler, C. (2016). First year of operational experience with a solar process steam system for a pharmaceutical company in Jordan. *Energy Procedia, 91*, 591–600. https://doi.org/10.1016/j.egypro.2016.06.209

Brunner, C., Herzog, U., & Muster-Slawitsch, B. (2012). *Potential for solar thermal integration via new technologies.* AEE-Intec.

CONA. (2022). Retrieved from CONA Solare Lufttrocknung. https://www.cona.at/index.php?id=solare_trocknung&L=302

Eurostat-Statistics Explained. (2022). Energy statistics—an overview. European Commission. Retrieved June 27, 2022, from https://ec.europa.eu/eurostat/statistics-explained/index.php?title=Energy_statistics_-_an_overview#Final_energy_consumption

Fahr, S., & Kramer, K. (2015). *Guideline on testing procedures for collectors used in solar process heat.* Technical Report A.3.1, V3.0, SolarIEA SHC Task 49 Solar Process Heat for Production and Advanced Applications.

Frank, E., Hess, S., & Zahler, C. (2012). *General requirements and relevant parameters for process heat collectors and specific collector loop components.* IEA SHC Task 49 Deliverable A 1.1., SPF Institut für Solartechnik.

Frank, E., Mauthner, F., & Fischer, S. (2015). *Overheating prevention and stagnation handling in solar process heat applications.* Technical Report A.1.2, IEA SHC Task 49-Solar Process Heat for Production and Advanced Applications.

Goedecke, R. (2011). *Fluidverfahrenstechnik: Grundlagen, Methodik, Technik, Praxis.* Weinheim: Wiley-VCH Verlag GmbH & Co. KGaA.

Heat Roadmap Europe. (2017). Heating and cooling -fact and figures. In B. V. Mathiesen (Ed.) *The transformation towards a low-carbon heating & cooling sector*. A. U. Department of Planning, & Compiler). Retrieved from https://heatroadmap.eu/wp-content/uploads/2019/03/Brochure_Heating-and-Cooling_web.pdf

Holman, J. (2009). *Heat transfer* (10th ed.). McGraw-Hill Education.

ISO 9806. (2017). *Solar energy—solar thermal collectors—test methods.* ICS 27.160 Solar energy enigneering. ISO/TC 180 Solar Energy.

Kemp, I. C. (2007). *Pinch analysis and procress integratgion* (2nd ed.). Butterworth-Heinemann.

Krüger, D. R. (2021). Chancen für den Einsatz konzentrierender Kollektoren in Mitteleuropa. *31. Symposium Solarthermie und innovative Wärmesysteme 2021, 27–30.04.2021, Online Konferenz.*

Krüger, D., Berger, M., Mokhtar, M., Willwerth, L., & Zahler, C. (2017). Experiences with industrial solar steam generation in Jordan. *AIP Conference Proceedings, 1850*, 180003–180009. https://doi.org/10.1063/1.4984570

Lauterbach, C., Schmitt, B., & Vajen, K. (2011). *The potential of solar process heat in Germany -Part 1 of the final report on the research project "SOPREN—solar process heat and energy efficiency" (in German).* Universität Kassel.

Linnhoff, B., & Eastwod, A. (1987). Experience from the application of pinch technology to ethylene plants in a given site context. *AIChE Annual Meeting* (Paper No 70a).

Linnhoff, B., Townsend, D. W., Boland, D., Hewitt, G. F., Guy, A. R., & Marsland, R. H. (1994). *A user guide on process integration for the efficient use of energy* (Rev. 1 ed.).

Mauthner, F., Hubmann, M., Brunner, C., & Fink, C. (2014). Manufacture of malt and beer with low temperature solar process heat. *Energy Procedia, 48*, 1188–1193. https://doi.org/10.1016/j.egypro.2014.02.134

McCabe, W., Smith, J., & Harriott, P. (2004). *Unit operations of chemical engineering* (7th ed.). London: McGraw Hill Higher Education.

Minder, S. (2013). Example of concentrated solar systems (PTC) in the dairy industry in Switzerland. *SHC Workshop on solar process heat for industry* (p. 33). NEP Solar AG. Retrieved from www.nep-solar.com

Muster-Slawitsch, B., Ben Hassine, I., Helmke, A., Hess, S., Krummenacher, P., Schmitt, B., & Schnitzer, H. (2015). *Integration guideline.* International Energy Agency-IEA. Retrieved July 1, 2022, from http://task49.iea-shc.org/publications

Nasirov, S., & Agostini, C. (2018). Mining experts' perspectives on the determinants of solar technologies. *Renewable and Sustainable Energy Reviews, 95*, 194–202. https://doi.org/10.1016/j.rser.2018.07.038

Rehfeldt, M., Fleiter, T., & Toro, F. (2018). A bottom-up estimation of the heating and cooling demand in European industry. *Energy Efficiency, 11*, 1057–1082. https://doi.org/10.1007/s12053-017-9571-y

Schmitt, B. (2014). *Integration of solar heating plants for supply of process heat in industrial companies.* Aachen: Shaker Verlag.

Schmitt, B., & Hess, S. (2016). *"Best practice" series of case study reports from demonstration projects. Solar process heat for production and advanced applications.* IEA SHC Task 49.

Shah, R., & Sekulic, D. P. (2003). *Fundamentals of heat exchanger design.* Wiley.

Sunmark-Arcon. (2014). Pampa Elvira Project. Energieplatz 1,9300 St. Veit/Glan, Austria. Retrieved April 11, 2022, from info@greenonetec.com

Trivedi, K., Fouche, E., & Parmenter, K. (2007). Process energy efficiency: Pinch technology. In F. Kreith, & D. Godwami (Eds.), *Handbook of energy efficiency and renewable energy* (p. 1560). CRC Press LLC.

Vannoni, C., Battisti, R., & Drigo, S. (2008). *Potential for solar heat in industrial processes.* Madrid: CIEMAT.

Vauck, W. R., & Müller, H. A. (1999). *Grundoperationen chemischer verfahrenstechnik* (11th ed.). Wiley-VCH.

VDI. (2013). *VDI-Wärmeatlas-Berechnungsblätter für den Wärmeübergang* (11. Auflage ed.). (D. D. V., Ed.) Springer-Vieweg Verlag GmbH.

Weiss, W., & Spörk-Dürr, M. (2021). *Solar heat worldwide- global market development and trends 2020.* AEE—Institute for Sustainable Technologies.

Solar Climatization and Refrigeration 13

Abstract

Solar heat can also be used as a thermal drive to operate refrigeration and air conditioning systems. Starting from the definition of refrigeration and air conditioning, a quantification of the power demand and a description of different methods for refrigeration and cooling is given. In particular, thermally driven refrigeration processes such as closed-loop absorption and adsorption processes are presented, their operation and plant design are explained, and their limitations are discussed. A further class of air-conditioning is formed by open sorption-based systems, the basic ideas and performance of which are discussed. Finally, a comparison of the technical options is given.

Keywords

Cooling/refrigeration definition • Coefficient of performance (COP) • Compression and thermal driven chillers • Adsorption/absorption chillers • Air conditioning in open & closed systems

13.1 Cooling and Air Conditioning

Air conditioning to create comfortable indoor air conditions is a central element of building services engineering in office and administrative buildings. Modern architecture, with its use of large glass surfaces as facade elements, and the increased use of electronic equipment producing waste heat, play a significant role in the increased energy requirements for air-conditioning. The use of building materials with low heat storage capacity also has a negative effect on the required cooling load. Since in many applications the removal of heat via natural ventilation for example at night is no longer possible, it is carried out in buildings with the aid of air-conditioning systems designed as air–water or air-only systems (DIN EN 13,779, 2009). Climate change with hotter summers and a larger number of tropical nights limit the use of natural ventilation with night-cooling even in moderate climates even more. The designation of a system as an air-conditioning system

R. Stieglitz and W. Platzer, *Solar Thermal Energy Systems*,
https://doi.org/10.1007/978-3-031-43173-9_13

is only given if the air treatment functions of heating, cooling, humidifying and dehumidifying can be performed in the system. Systems that do not meet this criterion are referred to as partial air-conditioning systems or ventilation systems. In addition to the user's demand for comfort in the air-conditioned building, the economic operation and the ecological assessment of the building services system are becoming more important. However, a building service system can only meet expectations if all relevant subsystems, including building physics, are taken into account during the planning phase.

A distinction should be made between refrigeration, cooling and air conditioning systems. Although refrigeration and cooling is sometimes used synonymously, we use refrigeration systems when the temperature of a medium is reduced and maintained below a desired room temperature or ambient, and cooling system when we talk about the temperature reduction of a room, whereas air conditioning systems control both temperature and humidity. A refrigeration system is a general process, where heat is transferred from one medium to another. Often it is used for food preservation or for cooling processes in industry. But a refrigeration system might be a part of an air-conditioning system, e.g. the system cooling a liquid in a distribution system which drives dehumidifiers and air coolers.

For an evaluation and comparison of the different solar air-conditioning systems, it is crucial to consider the complete air-conditioning process—consisting of conditioning the air temperature and humidity. The complex planning process for solar thermal air conditioning is well described in (Henning H.-M., 2004). Both cooling and air conditioning of a room or building cause a change in the energy content of the air, which is expressed in terms of the thermodynamic potential of the enthalpy H or the specific enthalpy h. In the case of cooling only, the absolute energy content of the air remains constant. For cooling alone, the absolute humidity of the air x, which is measured in kilograms of water per kilogram of dry air, remains constant. For cooling, therefore holds

$$\Delta H_{air} = \Delta H_{air}(T,\ x = const.) \tag{13.1}$$

where T is the temperature in $[°C]$. In the case of air conditioning,the specific humidity x changes in addition to the temperature, so that the following applies in this case:

$$\Delta H_{air} = \Delta H_{air}(T,\ x = const.) \tag{13.2}$$

This difference can be clearly seen in theh − x psychrometric chart ($h - x$ diagram), which is used to represent the thermodynamic state of the air. In the Mollier-diagram depicted in Fig. 13.1, a cooling process is characterized by a vertical path in the diagram ($x = const$), in which a decrease in enthalpy occurs due to a decrease in temperature at constant humidity. The course of the climatization

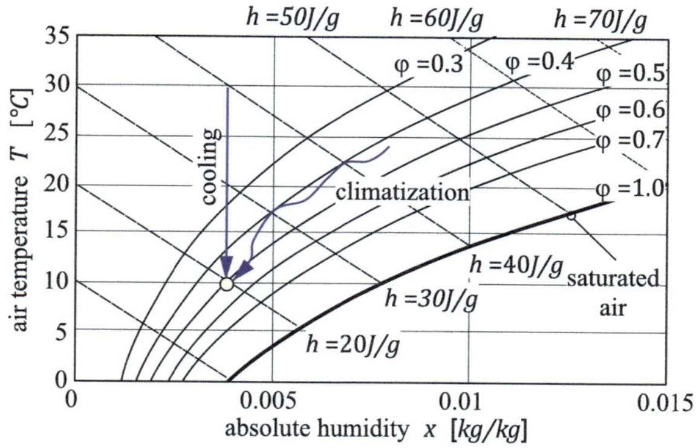

Fig. 13.1 Schematic evolution of cooling and climatization of air in the psychrometric chart (Mollier or $h - x$ diagram); $\varphi =$ relative humidity

process, on the other hand, can be any path in two-dimensional $T - x$ space that is not parallel to the temperature axis.

The absolute humidity x and relative humidity φ are linked via Dalton's law. In an ideal gas, the mass of water vapor m_w in the air is determined by the partial pressure p_v, the volume V, the gas constant R_v and the temperature of the vapor T by the following equation:

$$m_v = \frac{p_v \cdot V}{R_v \cdot T}. \tag{13.3}$$

From Dalton's law it follows that the sum of all partial pressures p_i for ideal gases is equal to the total pressure of the mixture p_{tot}. In the case of moist air, the water vapor pressure is subtracted from the air pressure and the residual pressure is divided among the gases in proportion to their shares in the dry air. Thus, with the help of Eq. (13.3), the mass ratio x can be rewritten into a ratio of the partial pressures of water and air and it is obtained (using additionally $p \cdot V = m \cdot R_s \cdot T$ with $R_{air/s}$ as specific gas constant of air and steam, respectively):

$$x = \frac{m_s}{m_{air}} = \frac{R_a \cdot p_s}{R_s \cdot p_{air}} = \frac{R_{air} \cdot p_s}{R_s \cdot (p_{atm} - p_s)} = \frac{R_{air}}{R_s \cdot \left(\frac{p_{atm}}{p_s} - 1\right)}. \tag{13.4}$$

By defining relative humidity φ as the ratio of water vapor partial pressure and saturation pressure p_{sat} in the form $\varphi = p_s/p_{sat}(T)$, the vapor pressure in Eq. (13.4) can be eliminated by the relative humidity, leading to

$$x = \frac{R_{air}}{R_s \cdot \left(\frac{p_{atm}}{\varphi \cdot p_{sat}(T)} - 1\right)}. \tag{13.5}$$

In addition to the temperature, the saturation pressure $p_{sat}(T)$ at the given temperature is required to calculate x from φ via Eq. (13.5). The saturation vapor pressure for water vapor can be calculated using the Magnus formula (Grigull, Staub, & Schiebener, 1990):

$$p_{sat}(T) = \exp\left(-\frac{6094.46}{T} + 21.125 - 2.7245 \cdot 10^{-2} T + 1.689 T^2 + 2.4576 \cdot \ln T\right) [Pas]. \quad (13.6)$$

in which the temperature T is to be specified in Kelvin and which is valid for a temperature range from –100 °C to +100 °C.

The arbitrariness of the process flow in the air conditioning path allows a variety of possible process designs. All sub-processes can be divided into two general processes, humidity and temperature conditioning. This will be discussed exemplarily on the basis of two different air conditioning processes in the $h-x$ diagram, in which ambient air at (T = 32 °C, $x = 12g/kg$, $\varphi = 0.4$) is to be adjusted to a room air condition of (T = 26 °C, $x = 10g/kg$, $\varphi = 0.47$).

In the first example, cooling and dehumidification are only realized via cold, as shown in the $h-x$ chart in Fig. 13.2. For de-humidification via cold, the air is first cooled from its initial state to the corresponding saturation temperature (or dew point temperature T = 17 °C). Subsequently, dehumidification and further cooling along the saturation line (φ = 1.0) is performed by adding cold until the desired humidity is reached. The required room temperature of T = 26 °C is achieved in the last step by heating with the aid of a reheater.

In sorption-assisted air conditioning, the process of which is depicted in the $h-x$ diagram in Fig. 13.3, the air is first sorptively dehumidified to the desired humidity$x$. In this process, the air is first heated to approx. 39.5 °C by sorption.

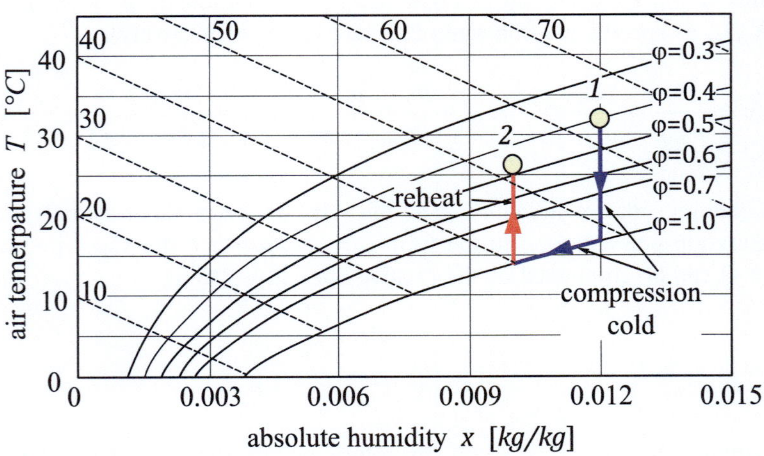

Fig. 13.2 Process control of a purely cold-driven air conditioning system from state 1 to state 2 in the psychrometric chart ($h-x$ diagram)

In the second process step, a cooling process is used to cool the air to the desired final temperature.

Both example processes require cooling energy. The required energy results from the sum of the enthalpy differences of the individual process steps from state 1 to state 2. However, a comparison of the two cases shows that a considerably lower enthalpy difference of the cooling process is required in the sorption-supported process (cold enthalpy difference in example 1 of $24kJ/kg$ and in example 2 only $14kJ/kg$). In this example, it proves unfavorable to dehumidify with the aid of refrigeration. This also formulates the goal of efficient air conditioning. For given utilization and climatic boundary conditions, process control must be designed in such a way that a minimum energy requirement can be determined.

13.2 Evaluation Quantities of Refrigeration and Air Conditioning Processes

The evaluation of technical processes results from the ratio of the usable power to the power input. This efficiency is referred to as the "coefficient of performance" (COP) and is defined in the following way.

$$COP = \frac{\text{useful power}}{\text{power consumed}}. \tag{13.7}$$

In the case of chillers, the useful power is the cooling power generated in a refrigeration process; accordingly, the COP value of a chiller is often referred to

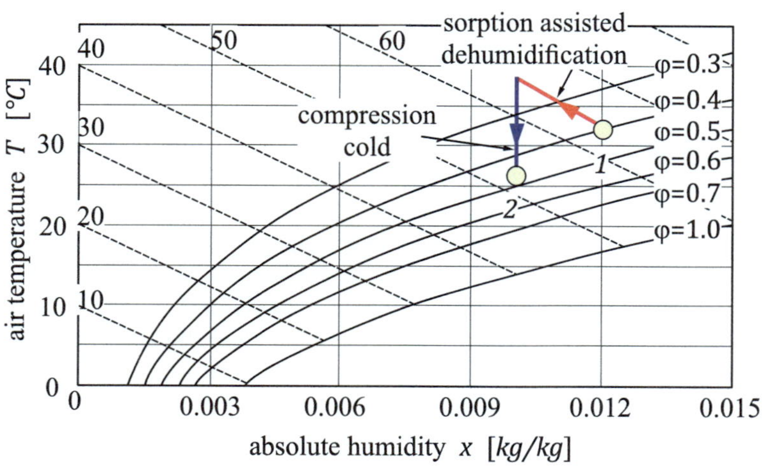

Fig. 13.3 Air conditioning from state 1 to state 2 using a mixed cooling process in the $h - x$ diagram

as the cooling coefficient. Either a thermal or an electrical process can be used to generate the cooling power, so that the useful power (cooling) is related to the expenditure process; thus:

$$COP_{cold,elec} = \frac{\dot{Q}_{cold}}{\dot{Q}_{elec}} \text{ or } COP_{cold,therm} = \frac{\dot{Q}_{cold}}{\dot{Q}_{therm}}. \tag{13.8}$$

In contrast, in the case of a thermally driven refrigeration process, the total thermal power required $\dot{Q}_{therm}$ is composed of the auxiliary ($\dot{Q}_{aux}$) and solar thermal ($\dot{Q}_{solar}$) power.

$$\dot{Q}_{therm} = \dot{Q}_{aux} + \dot{Q}_{solar}. \tag{13.9}$$

The different reference quantities make a direct comparison between thermally and electrically driven processes difficult, which is why a COP_{prim} related to the primary energy input is often used. Here, all energy sources required to provide the drive energy and other auxiliary energies such as electrical energy for fans, pumps and auxiliary heat are included with their specific conversion efficiencies η_i, so that the refrigeration COP_{prim} related to the primary energy input can be represented in the following way

$$COP_{prim} = \frac{\dot{Q}_{cold}}{\frac{P_{elec}}{\eta_{elec}} + \frac{\dot{Q}_{aux}}{\eta_{therm}} + \dot{Q}_{solar}}. \tag{13.10}$$

The ideal reversible refrigeration process can be described as a reversal of the Carnot process. In this ideal limiting case, the η_{cold} efficiency of the refrigeration process η_{cold} can be specified via the temperature levels of the process steps and is in this case write to:

$$\eta_{cold} = \frac{T_{cold}}{T_{condenser} - T_{cold}} = \frac{1}{\frac{T_{condenser}}{T_{cold}} - 1}, \tag{13.11}$$

where T_{cold} is the temperature of the generated cold and $T_{condenser}$ is the temperature of the waste heat transferred to the ambient and the temperature in degrees Kelvin is to be used for both quantities.

In air-conditioning, the exergy refers to the energy portion of the drive energy that is converted into air-conditioning power in the process and can ultimately be used to dissipate the cooling load. Figure 13.4 shows the energy flow of an air conditioning process in comparison between an ideal (reversible) and an irreversible cooling process. Figure 13.4 illustrates that the total cooling load ($\dot{Q}$) dissipated is delivered to the environment as anergy and that the exergy flow ($\dot{E}_Q$) is required for this purpose. In the case of irreversibly operating air-conditioning machines, there is an exergy loss rate ($\dot{E}_{loss}$). This means that the anergy flux to be dissipated and the drive power (P) increase by the exergy loss rate ($\dot{E}_{loss}$). Real air-conditioning processes are irreversible and inherently result in a higher

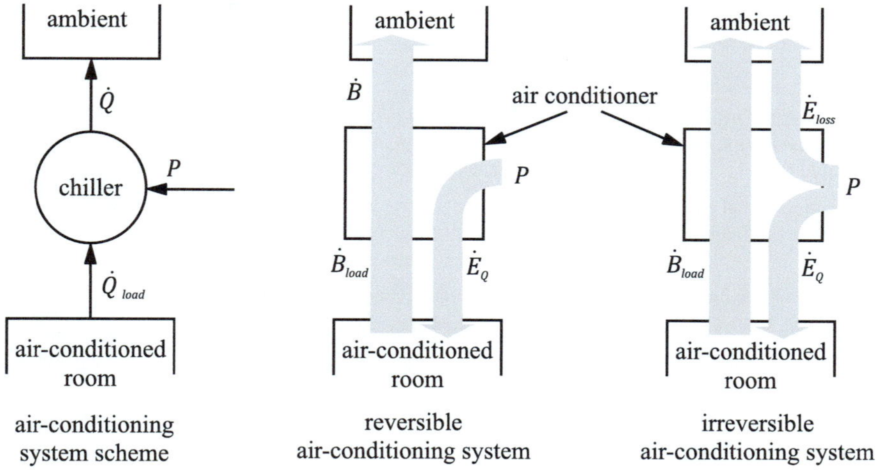

Fig. 13.4 Energy flow diagram of different air conditioning systems. Energy flows in reversible and irreversible air conditioning machines; ($\dot{E}$) denotes the exergy flow and ($\dot{B}$) the anergy flow

power consumption compared to the ideal process. The parameter resulting from this type of consideration is the exergetic efficiency ς:

$$\varsigma = \frac{\text{usable exergy flow}}{\text{exergy flow applied}} = \frac{\dot{E}_Q}{\dot{E}_Q + \dot{E}_V}, \tag{13.12}$$

13.3 Indoor Climate, Comfort and Cooling Load

The condition of indoor air determines the well-being of the people who spend time in it. With the participation of physicians, biologists, social scientists and technicians, national regulations or even international standards (ANSI/ASHRAE Standard 62.1, 2007) have been developed that include two aspects:

- Compliance with a functional room comfort.
- Determining the load parameters and thus the required cooling load.

Room comfort (or comfort perception) is determined by three classes of influencing parameters, shown in Table 13.1. The regulations specify guide values or calculation methods for the individual influencing parameters, which are mainly based on the following criteria.

- air humidity,
- air temperature,
- temperature of the enclosing surfaces,
- air movement.

Table 13.1 Influence classes and parameters for determining room comfort for the design of an air conditioning system

Persons	Room (enclosure)	Ventilation system
Type of activity (administrative work, physical production activity,)	Spatial constraints (Window area and orientation, room geometry, wall colour,..)	Air exchange (circulation volume,..)
Clothing (normal clothing, protective clothing, …)	Heat sources (communication electronics, production machines, ….)	Air purity requirements (humidity, particle content,..)
Duration of use (stay, frequency, …)	Pollutant sources (vapours, dust generation,..)	Air flow (supply and exhaust, available space for conditioning)
Number of persons	Relation between surface and air temperature	

The humidity in a room designated as "comfortable" must not exceed a maximum humidity of $x = 11.5g/kg$, which corresponds to a relative humidity of $\varphi = 0.65$, and the lower limit of relative humidity is $\varphi = 0.3$. Thermal comfort is given by the operative room temperature T_{op} (or "perceived temperature"), which is composed of equal parts of the room air temperature (T_r) and the radiation temperature (T_{rad}).

$$T_{op} = \frac{1}{2} \cdot (T_r + T_{rad}). \tag{13.13}$$

The radiant temperature is the sum of the local radiant temperatures and includes not only the thermal radiation of the walls but also that of all other heat sources within the room. In addition, the set point of the operative room temperature ($T_{op,set}$) depends on the outdoor temperature. At present, the following values are defined for the comfortable operative room temperature $T_{op,set}$ (cf. Fig. 13.5).

$$T_{op,set} = \begin{cases} 22^\circ C \leq T_{op} \leq 25^\circ C & T_{amb} \leq 26^\circ C \\ \frac{T_{amb}+40^\circ C}{3} \leq T_{op} \leq \frac{T_{amb}+49^\circ C}{3} & 26 < T_{amb} \leq 32^\circ C \end{cases}. \tag{13.14}$$

In addition to humidity and temperature, air circulation also influences the comfort level. The various standards specify guideline values for air exchange depending on the room type and use and are based on a hygienic consideration. The recommended values of the minimum required outdoor air flow shown in Table 13.2 are given either in person-related or area-related form.

The limitations of the comfort range by indoor humidity and operative temperature are illustrated by the gray shaded area in Fig. 13.6 in the psychrometric chart ($h - x$ diagram).

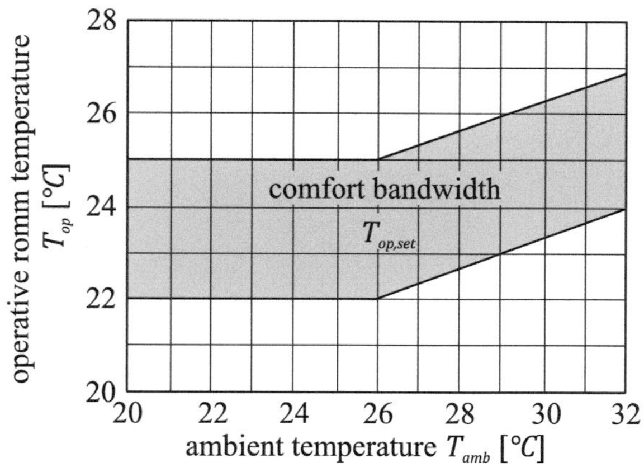

Fig. 13.5 Comfort range of operative temperature ($T_{op,set}$) as a function of the ambient temperature (T_{amb})

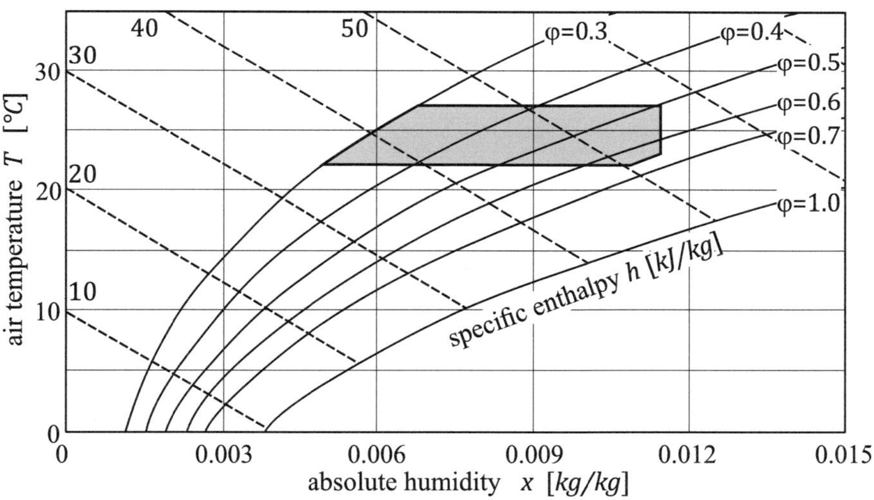

Fig. 13.6 Comfort range of humidity and temperature in the psychrometric chart ($h - x$ diagram)

13.4 Determination of the Cooling Capacity

The comfort requirements of the individual object determine the required air conditioning capacity, which must be supplied to the room as heating capacity (or heating load) in winter and as cooling capacity (or cooling load) in summer. On the one hand, the required cooling capacity is determined by environmental variables, such as the air temperature of the environment T_{amb}, the air humidity x and the solar irradiation i_{glob}, and on the other hand by internal sources. This results

Table 13.2 Required minimum person- and area-related external air flow of air circulation for different room use types

			External air flow
Room type	Application example	Person related air flow$\left(m^3/(h \cdot person)\right)$	Area-related volumetric air flow $\left(m^3/(h \cdot m^2)\right)$
Work rooms	Single office, open plan office	40 60	4 6
Meeting rooms	Concert hall, theatre, conference room		$10 - 20$
Class rooms	Reading room, classroom, seminar room, lecture hall	$20 - 30$	$12 - 15$
Rooms with public traffic	Sales room, restaurant	$20 - 30$	$3 - 12$ 8

in the required cooling or heating capacity ($\dot{Q}_{load}$):

$$\dot{Q}_{load} = \rho_{air} \cdot \dot{V} \cdot (h_{amb} - h_r) + \dot{Q}_{solar} + \dot{Q}_{trans} + \sum\nolimits_i \dot{Q}_{sens,i} + \sum\nolimits_i \dot{Q}_{latent,i}. \tag{13.15}$$

where ρ_{air} denotes the density of the air, $\dot{V}$ is the volumetric flow rate of the air exchange between indoor and outdoor environment (infiltration plus ventilation), and h_{amb} and h_r are the specific enthalpies of the environment and the room, respectively. A value of $\dot{Q}_{load} > 0$ indicates a cooling load, while $\dot{Q}_{load} < 0$ corresponds to a heating load. $\dot{Q}_{solar}$ reflects the solar input through the windows and $\dot{Q}_{trans}$ the energy input or loss by transmission through the exterior walls. $\sum_i^n \dot{Q}_{sens,i}$ represents the sum of sensible heat inputs from, for example, electrical equipment, lighting, and/or people, and $\sum_i^n \dot{Q}_{latent,i}$ comprises the sum of internal power due to evaporation of water for example from people, plants, cooking and showering. The cooling load correlates more strongly with the solar irradiation the larger contribution from outside is. If there is a strong temporal correlation, only small cold storage volumes might be needed. A consequence of Eq. (13.15) is that the easiest way to reduce the active cooling power requirement by reducing the individual contributions. This can be achieved for example by reducing the air exchange rate to the minimum level specified in Table 13.2 or by reducing the solar energy input through the targeted use of shading elements and good thermal insulation. The use of energy-saving equipment and optimal use of daylight also reduces the load due to internal sources. These aspects are therefore important, since active air conditioning involves not only investment costs but also operating costs, which are mainly determined by the total cooling capacity and the ventilation rates.

13.5 Heating Ventilation Air Cooling (HVAC) and Refrigeration Technologies

Solar cooling offers the possibility to integrate regenerative energy sources energetically and into the building technology. Depending on how the solar energy is converted and how the refrigeration process operates, solar cooling can in principle be produced in different ways. Solar radiation can be converted directly into electrical energy using photovoltaic cells or into heat by solar collectors. Based on these principles, several forms of air conditioning can be realized by using different physical principles. For the transformation of electricity or heat into corresponding useful cooling electrical, electrical–mechanical as well as thermal driven systems can be used, whereby the dividing line between the mixed processes/systems is difficult to draw, since active components such as fans and ventilators require electricity as the driving force even when using thermally driven chillers. An overview of the basic functionality of HVAC systems, their performance capabilities and limitations can be found, for example, in the reports by (Goetzler, Zogg, Young, & Johnson, 2014) or (Lewis, Chaer, & Tassou, 2007) and the corresponding literature cited therein.

Purely electrical or electro-caloric cooling processes are based on solid state technology, as illustrated by their key features in Fig. 13.7. These systems are barely able to integrate efficiently any external heat sources to realize the cooling effect.

With the exception of evaporative systems, combined electro-mechanical cooling methods, such as those shown in Fig. 13.8 in their essential characteristics, also exhibit only a limited ability to incorporate heat sources for cooling purposes. However, vapor compression cycles (VCC) and their technology based on a combined electro-mechanical concept are the reference technology in heating, ventilation and air conditioning (HVAC) since more than a century. These conventional VCC systems use still refrigerants contributing not insignificantly to global climate change, especially when released into the atmosphere. Although numerous alternative refrigerants with lower Greenhouse Warming potential (GWP) such as hydrocarbons, ammonia, carbon dioxide, and hydrofluorocarbons (HFOs) have been analysed, many of them trade off the greenhouse advantage for disadvantages in terms of toxicity, flammability, lower efficiency, and/or higher equipment costs. In order to provide a benchmark for chiller-based generation and to introduce performance parameters, it is useful to first consider an ideal (Carnot) or real (Rankine) vapor compression cycle (VCC), even though this hardly provides any option for integrating solar thermal heat.

As shown in Fig. 13.8, most of the refrigeration processes are similar to the single-phase and two-phase work generation cycles already discussed in Chap. 10. In contrast to the working machine cycles, however, the useful quantity (here the "useful cold") as well as the direction of rotation in the $T - s$ diagram (, which is now counter clockwise,) changes.

Using heat as a driving process allows also to provide cooling via various combined mechanical or electrical system technologies. As shown in Fig. 13.9, there

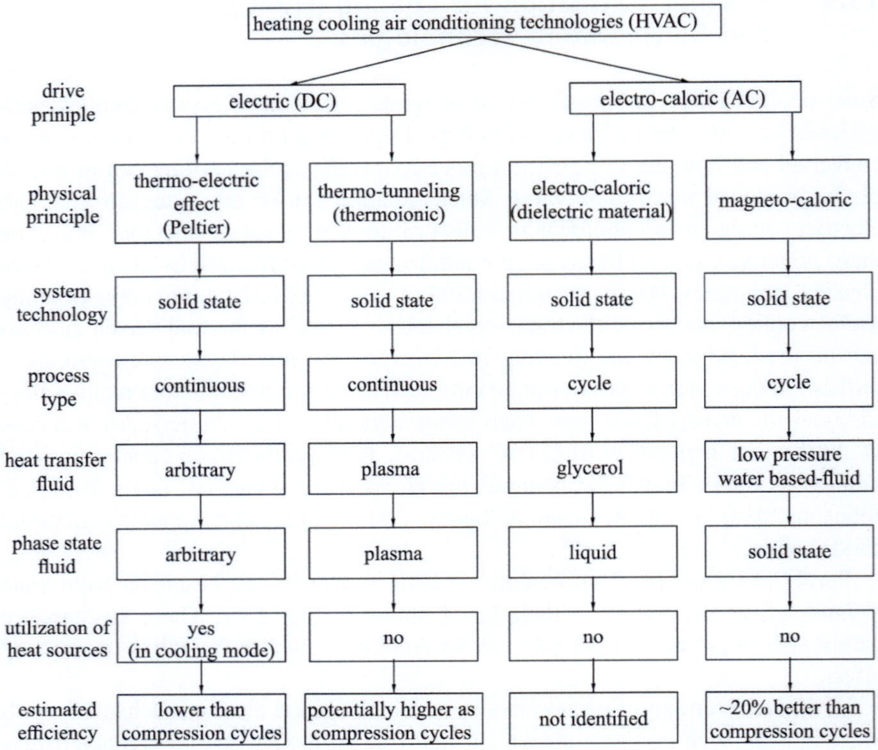

Fig. 13.7 Key features of potential electric and electro-caloric heating ventilation and cooling technologies (HVAC)

are essentially two main families of heat pump systems and sorption-supported processes. The various types differ not only in the mode of operation, which can be cyclic or continuous, but also in the type of system, which can be either open, closed or a hybrid between both. The choice of the operating fluid as well as the fluid state (liquid, gaseous, two-phase) during operation determines the temperature level at which the thermal source is an efficient drive.

In the following, we will mainly consider thermally driven systems that have already been developed. Another distinguishing criterion in Fig. 13.9 is the type of dehumidification. If a chiller as a closed system is used, the air dehumidification is realized by a dew point undershoot, which is achieved by condensing water vapor. This undershooting of the dew point thus determines the lowest temperature needed. This is similar to the use of electrically driven compression chillers. In the case of open systems, sorption-assisted air-conditioning systems (SAC systems) or solar desiccant cooling (SDC) uses direct dehumidification of the air by adsorption in rotating drying wheels. Therefore, in SAC systems not only refrigeration is generated, but air conditioning is also provided. The cooling process takes

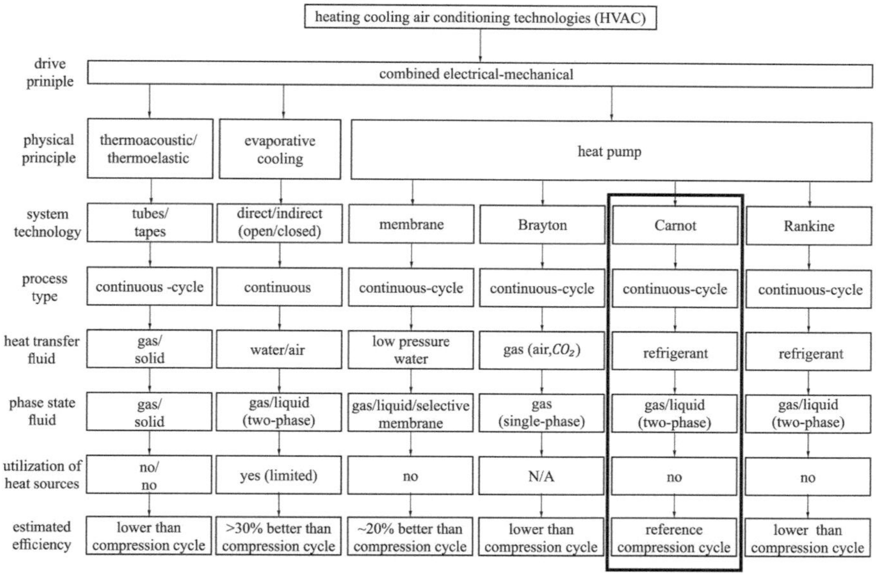

Fig. 13.8 Key features of electro-mechanic HVAC technologies

place isobarically (at ambient pressure), which is an advantage over closed system concepts in terms of technical effort.

Exclusively cooling systems, thermally driven absorption and adsorption chillers, are characterized by a closed refrigerant circuit. This includes all electrical systems and closed thermal processes. They provide their cooling capacity in the form of a heat sink for a heat transfer medium, which can be used via a cooling ceiling or as a cooling circuit.

Due to the large number of possible cooling systems, but also the different options to incorporate both environmental heat and collector-generated heat into cooling processes, there are a number of hybrid cooling processes in which one or more technologies are combined in such a way that they efficiently provide the required cooling load for the respective application and its boundary conditions. Here we limit ourselves to the presentation of the main principles and methods in order to elaborate the fundamentals and modes of operation, otherwise reference is made to the relevant technical literature.

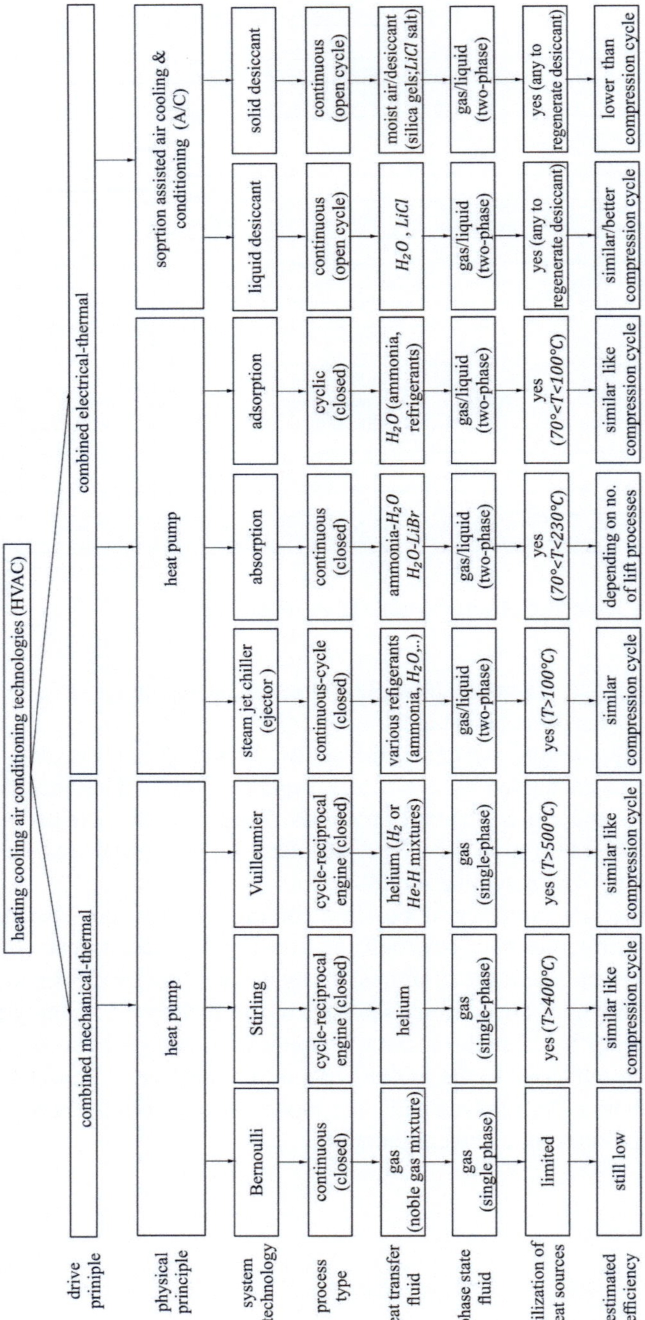

Fig. 13.9 Characteristics of different thermal driven HVACS technologies

13.6 Refrigeration and Air Conditioning Process Control

13.6.1 Ideal Vapor Compression Refrigerator, Carnot's Comparative Process

The two-phase counterclockwise Carnot process is shown in the $T - s$ diagram as in Fig. 13.10 composed of two isotherms and two isentropes. It is best approximated when the isentropes run in the wet steam region during evaporation and during condensation, respectively.

Therefore, the method mainly used in heat pump and air conditioning refrigeration is the cold vapor compression method. A refrigerant evaporates on the cold side, absorbing as much heat of evaporation as possible from the room to be cooled, for example the refrigerator. After compression in a compressor, it is liquefied again on the warm side. Liquefaction takes place with the heat of condensation being released to the environment, the heating system or to another recooling medium. It is then returned to the evaporating pressure in a throttle. In all plants operating according to this principle, the dependence of the evaporation and condensation temperature on the pressure is exploited. Figure 13.10 shows the simplified process flow in a refrigerating machine.

The individual components listed with numbers in Fig. 13.10 are:

- Evaporator (1): In the evaporator, the refrigerant evaporates at the temperature T_0 and the pressure p_0 and absorbs the heat q_0 (heat of evaporation) from the space to be cooled, whose temperature can subsequently decrease to the value T_0 at maximum.
- Compressor (2): In the compressor, the refrigerant vapor is increased from pressure p_0 to pressurep_c. Thereby, the temperature increases to the valueT_c.

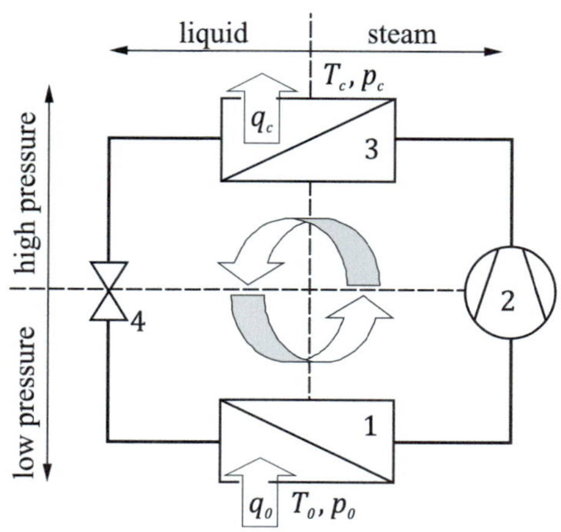

Fig. 13.10 Schematic representation of the process flow in a refrigeration machine, where ne process flow in a refrigeration machine, where numbers represent the components. 1=evaporator, 2=compressor, 3=condenser and 4=expansion valve

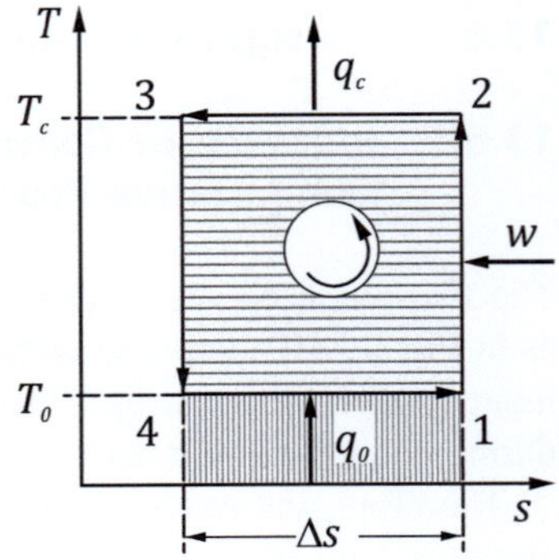

Fig. 13.11 Representation of an ideal left-hand Carnot cycle (ideal refrigeration compression engine) in the $T - s$ diagram

- Condenser (3): In the condenser, the refrigerant is liquefied at the temperature T_c and the pressurep_c, releasing the condensation heat (or heat of condensation)q_c.
- Expansion valve (4): The refrigerant is expanded from the high pressure p_c to the lower pressure p_0.

In the ideal left-hand cycle, the temperature entropy diagram ($T - s$) shows the process as shown in Fig. 13.11. It consists of two isentropes and two isotherms. The individual steps here are:

$1 \to 2$: isentropic compression of the gaseous refrigerant (in the real case $\Delta s > 0$).

$2 \to 3$: Isobaric and isothermal condensation at temperature T_c, in which heat q_c is transferred to a colder environment; (in reality often an isobaric supercooling).

$3 \to 4$ Isentropic expansion of the refrigerant from pressure p_c to pressure p_0, which in practice is often a narrow cross-section. In the real case, this is adiabatic throttling with approximately constant enthalpy ($h \approx const.$), in which evaporation often already occurs in the two-phase region.

$4 \to 1$: Isobaric and isothermal evaporation at temperature T_0 and pressure p_0, in which heat q_0 is transferred from the space to be cooled to the refrigerant.

In the cyclic process, the energy supplied is always equal to the energy dissipated, and the following applies to the Carnot process

$$q_0 + w = q_c. \tag{13.16}$$

The difference between a heat pump and a chiller is only the use of the condensation heat (q_c) or the evaporation heat (q_0). In practice, left-hand Carnot processes are represented in the $\log p - h$ chart (cf. Fig. 13.12). In this, q_0, w and q_c can be read directly as distances (corresponding to enthalpy differences). In it takes place from:

- $1 \to 2$ isentropic compression with superheating. The superheat temperature is far above the condensation temperature.

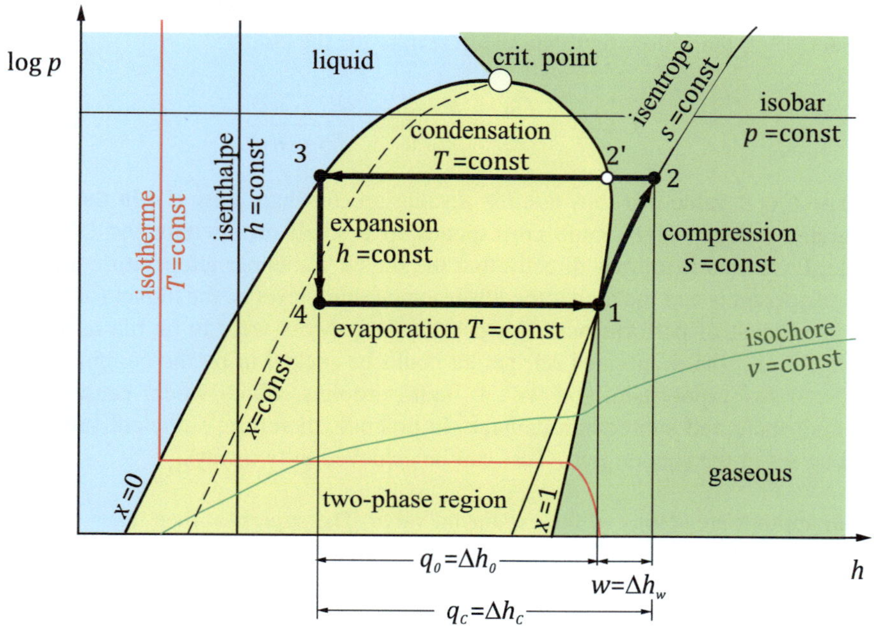

Fig. 13.12 Process flow of an ideal left-hand Carnot refrigeration process in the $\log p - h$ chart

- $2 \rightarrow 2\prime$ a heat release of the superheated refrigerant. In this case, the superheat is transferred to the medium to be heated, with the refrigerant still remaining gaseous.
- $2\prime \rightarrow 3$ isobaric liquefaction, in which the heat of liquefaction is transferred to the medium to be heated, and
- $3 \rightarrow 4$ expansion at constant enthalpy. The refrigerant is expanded to the evaporating pressure. (Isentropic expansion would require a working machine, such as a turbine. The expansion would arrive at $4\prime$ slightly to the left of 4 in the $\log p - h$ chart. The work gained would then be returned to the compressor. However, the work is small, so a turbine is not needed. The left-hand process with isenthalpic expansion is, therefore, called "ideal").
- $4 \rightarrow 1$ an isobaric evaporation. The evaporation energy is extracted from the medium to be cooled.s

The capacities of the evaporator ($\dot{Q}_0$) and the condenser ($\dot{Q}_c$), which are thermal in nature, and the mechanical compressor drive capacity (P) to be provided are obtained as:

$$\dot{Q}_0 = \dot{m} \cdot (h_1 - h_4);\ \dot{Q}_C = \dot{m} \cdot (h_2 - h_3) \text{ and } P = \dot{m} \cdot (h_2 - h_1), \qquad (13.17)$$

in which $\dot{m}$ denotes the mass flow of the cooling fluid through the chiller.

Based on the definition of the Coefficient of Performance (COP) introduced in Sect. 13.2, the following relationship results for the chiller:

$$COP = \frac{\dot{Q}_0}{P} = \frac{\dot{m} \cdot q_0}{\dot{m} \cdot w} = \frac{(h_1 - h_4)}{(h_2 - h_1)} \tag{13.18}$$

The COP value can thus assume a value greater than unity (!). In the $T - s$ diagram, a large COP value corresponds to a small area within the left-hand cycle. From this it follows directly that the higher the upper temperature level of the refrigerant is and the higher the lower temperature level is, the higher (or better) the coefficient of performance. However, if the process were to be run as shown in Fig. 13.12, the evaporated refrigerant could be sucked in by the compressor in the saturated vapor state, and with it liquid droplets, which would cause liquid hammer there and subsequent damage. In practice, therefore, sufficient superheat will be set in the suction port. This can be achieved in two ways:

- an appropriate setting of the expansion valve. The superheat then takes place in the evaporator and can be used as cooling capacity;
- by installing a heat exchanger. In this case, the refrigerant is additionally subcooled on the condensing side and the heat is transferred to the evaporator for superheating. Since the drive work remains approximately the same, this improves the coefficient of performance (COP, Eq. (13.7)), since the useful cooling capacity is increased.

By using the heat exchanger, both the condenser capacity and the evaporator capacity can be increased for the theoretically same compressor capacity, as Fig. 13.13 outlines. But the heat exchanger capacity cannot be chosen arbitrarily, because a compressor does not allow arbitrarily high suction as well as compressor discharge temperatures due to lubrication and mechanical load.

Real vapor compression chiller (VCC) and refrigerant selection.

Heat absorption in the cold suction line, as well as the heat loss from the compressor and, depending on the design, also from the compressor motor, results in additional overheating, which is noticeable as a loss. Technical compression is not isentropic, but polytropic. In addition, the lines, the fittings, the evaporator and the condenser all exhibit a hydraulic pressure loss, which is introduced into the fluid as heat through internal friction and reduces the efficiency. In the condenser, there are also heat losses to the environment. This shifts the process flow and Fig. 13.14 shows schematically the flow of a real process without heat exchanger.

Since heat losses occur in a compressor, not all the energy supplied by the compressor motor contributes to the condenser capacity. The heat loss is mainly dependent on the compressor design and the operating temperature. Encapsulated compressors with motor cooling by the refrigerant circuit achieve efficiencies of about 90%, while compressors with separate motor cooling achieve efficiencies on the order of 80%.

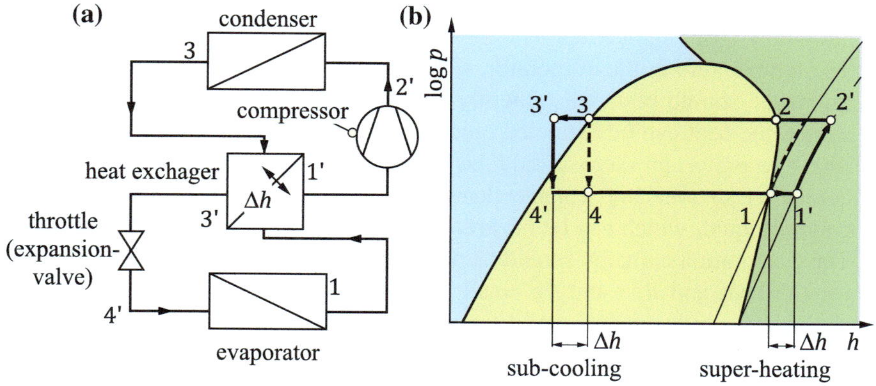

Fig. 13.13 Increase of evaporator and condenser capacity of a chiller with the same compressor capacity by using a heat exchanger. **a** System sketch **b** Representation in $\log p - h$ chart

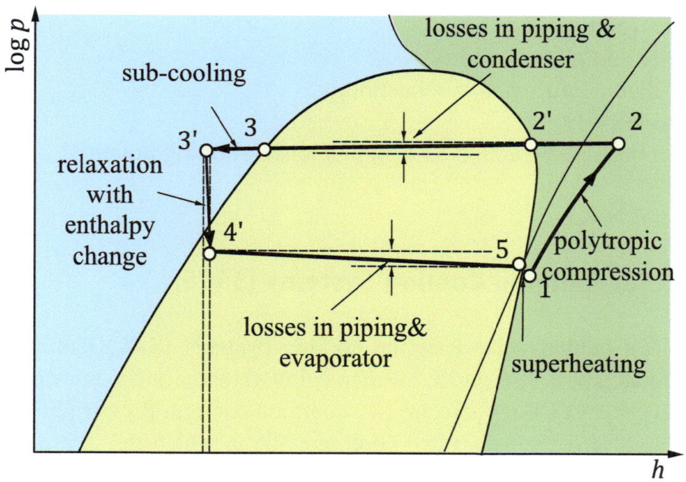

Fig. 13.14 Real chiller process in $\log p - h$ chart

The real energy balance must be corrected by the real process. But even neglecting losses and auxiliary drives, Eq. (13.18) still holds for the cold vapor compression principle. The refrigerant of a compression chiller must satisfy some boundary conditions, of which the choice of a suitable vapor pressure curve is the most important.

- The temperature in the condenser, which transfers the heat (q_c) to the outside, should be slightly above the ambient temperature at the condenser pressure (p_c) to ensure sufficient heat transfer.

- Condenser pressure (p_c) with the condenser should be less than $2MPa$ bar to allow low-cost engineering.
- The temperature in the evaporator, which extracts heat from the cold room at a low level, should be at least slightly below the desired cold room temperature so that the heat can be extracted quickly and efficiently.
- The evaporator pressure should be slightly below atmospheric pressure ($< 0.1MPa$) so that any leakage does not result in immediate leakage of the cooling liquid, which can be hazardous to health.
- The evaporation enthalpy should be high, as it ensures a large transport capacity for the heat and thus only a small coolant mass flow is required. This also significantly reduces the cost of the compressor.

In addition, further operational, environmental, safety and health requirements must be fulfilled by the cooling medium such as:

- chemical stability (aging, degradation, oxidation),
- compatibility with structural (pipes, valves, etc.) and functional materials (seals) and media (pump oil),
- availability and price,
- minimal flammability and combustibility,
- low or no toxicity,
- low or no ozone destruction as well as global warming potential.

13.6.2 Solar Rankine Cooling Systems (SRCS)

The main idea behind solar Rankine cooling cystems (SRCS) is to use solar-thermal generated heat to produce mechanical work and drive a conventional vapor compression cycle (VCC). There are two common configurations of SRCS designs, the first using a separate power and cooling cycle in which the turbine (expander) of the work generation process, usually based on an organic Rankine cycle (ORC, see Chap. 11) and the compressor of the VCC are mechanically coupled via a shaft. The other configuration is an integrated cycle, using a combined turbine-compressor unit and a common condenser for both cycles. In the latter design, the same working fluid (refrigerant) is used in both circuits, avoiding leakage and mixing problems. The integrated design is also simpler, but at the same time reduces the flexibility of the system.

Figure 13.15 shows a simple schematic of the two common SRCS configurations. In the Organic Rankine Cycle (ORC), the high-pressure fluid in the generator (1) absorbs heat from the solar collector and vaporizes it. The steam (2) enters the turbine and generates mechanical energy as it expands. At the outlet of the turbine, the pressure of the working fluid is equal to the pressure of the condenser (3). In the condenser, the working fluid becomes a saturated liquid by releasing heat to a heat sink (4). With the help of a pump, the pressure of the liquid

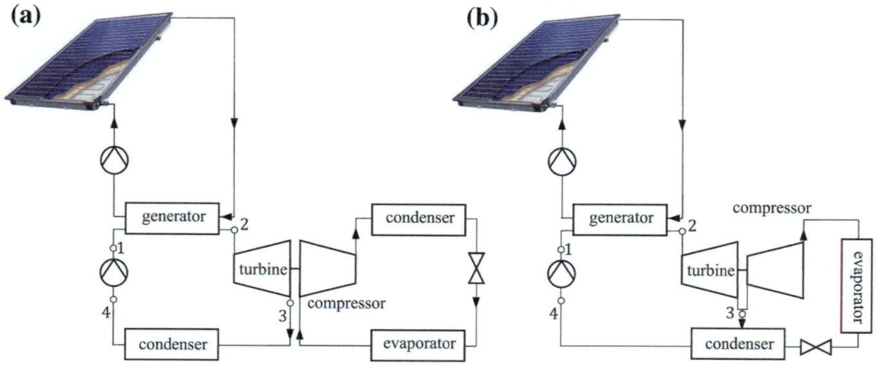

Fig. 13.15 Block diagram of Rankine solar cooling system with separate power and refrigeration cycles (**a**) and integrated power/refrigeration cycle (**b**) modified from (Zeyghami et al., 2015)

is increased almost isentropically, and the working fluid is again fed to the generator (boiler) as a subcooled liquid (1). The mechanical energy generated by the turbine is used as drive energy to operate the compressor of the vapor compression cycle (VCC).

The choice of working fluid is one of the most important design parameters for an SRCS. Water is the most widely used refrigerant for the Rankine cycle for power generation. However, the use of water requires high temperatures and pressures means expensive equipment with low overall performance. Therefore, dry or isentropic fluids, as used in ORC cycles, are preferred working fluids. The molecular weight of the refrigerant is another property that affects system performance; fluids with higher molecular weight are in principle more favourable in this respect. All other requirements for the refrigerant's properties, such as the critical point of the fluid, stability and compatibility with structural and sealing materials, environmental aspects, safety, availability, and cost, are the same as for other chillers.

Regarding the effects of operating conditions on SRCS performance, lower condenser temperatures (T_c) and higher evaporator temperatures (T_e) are favorable for higher cooling efficiency. SRCS systems are able to take advantage of high generator temperatures whenever they are available. The thermodynamic efficiency of ORC working processes increases with increasing inlet temperatures; unfortunately, however, the efficiency of solar collectors decreases with increasing temperature, showing the opposite trend. Due to this opposing behaviour of both systems involved, a refrigerant-dependent optimal generator temperature results, which provides the highest total COP.

ORC turbine selection is another important other design parameter that should be considered depending on the working fluid and size of the system. For fluids with a high boiling point, a turbine should be used, and for fluids with lower boiling points, a positive displacement expander would be more suitable. Turbines are also more suitable for large systems with high flow rates.

13.6.3 Vapor Jet Chillers (Ejector Chillers)

The ejector chiller is a conventional cold vapor process, as also occurs in a compression chiller. Instead of a mechanical compressor, compression takes place with the aid of thermal compression. In the ejector jet compressor, refrigerant vapor is transported from the evaporator pressure to the condenser pressure level with the aid of a suitable motive medium. In this refrigeration process, the choice of refrigerant plays a decisive role; in particular, the dynamic behaviour in the partial load range and a pressure-independent COP characterize the ejector jet refrigeration machine.

13.6.3.1 Operational Principle of an Ejector Pump

The core element of this type of chiller is the jet compressor shown in Fig. 13.16a, which consists of a driving nozzle, a suction nozzle, the mixing chamber, and the diffuser.

In the jet compressor, the motive fluid flows under high pressure (3) into the motive nozzle where it is accelerated due to the geometry of the nozzle (4). The nozzle can be shaped as a subsonic or also as a supersonic nozzle (Laval nozzle). The potential energy of the drive (pressure) is converted along the nozzle into kinetic flow energy, which increases continuously up to the area of the narrowest cross-section. Depending on the type of motive nozzle, the motive gas in a

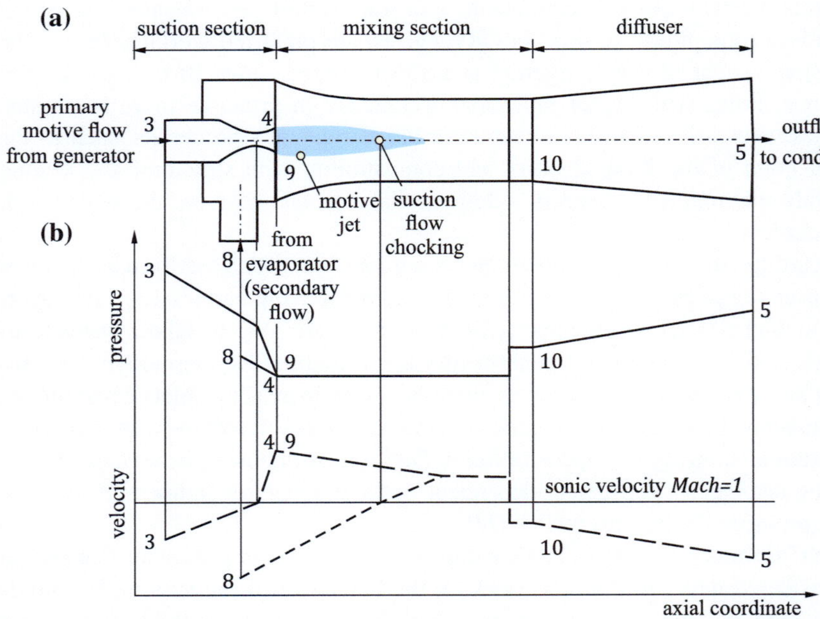

Fig. 13.16 Schematic evolution of pressure and velocity along of ejector axis modified from (Little & Garimella, 2016)

Laval nozzle is accelerated to supersonic speed during expansion. This causes the static pressure to drop below the evaporator pressure, so that refrigerant vapor is permanently sucked out of the evaporator (8) via the suction nozzle (9) and can flow in. The two streams from the motive nozzle and the suction nozzle mix first in the mixing chamber (4/9) and then in the mixing tube. The velocities of both streams adjust in the course of mixing, and there are mixing tube designs with constant axial area or constant axial pressure. In these, the primary motive stream is decelerated and, conversely, the secondary (suction) stream is accelerated. The resulting velocity of the mixture is still very high after the mixing process (10). In the supersonic diffuser, the high flow velocity is used by deceleration to increase the pressure. At this deceleration, compression of the mixture occurs at the smallest cross-section due to a compression shock.

Pressure is recovered in the diffuser from 10 to 5 as the total flow at the ejector outlet is decelerated to a low velocity. The desired compression effect produced by the ejector is the pressure increase from suction inlet 3 to ejector outlet 5. The two critical sections of the ejector (jet compressor) are the driving nozzle and the mixing section. The conditions in the motive nozzle determine the flow geometry of the motive jet and are critical in determining whether or not two-phase flow is present in refrigerant applications. The interactions between the motive flow and the suction flow in the mixing section determine the overall performance of the ejector. The pressure increase must be large enough to allow the vapor to be subsequently re-liquefied in the condenser. The evolution of the pressure and the flow speed along the jet compressor is illustrated in Fig. 13.16b.

The nozzle geometry influences the operation of the injector. The nozzle shape can be convergent, i.e. the injector operates in the subsonic range and can reach sonic velocity at the intake outlet. In most cases, however, the nozzles are shaped like a Laval nozzle convergent-divergent so that the flow through the injector can reach supersonic velocities. The choice between the two types of ejectors depends largely on the specific characteristics of the end application. Subsonic injectors are not designed to produce a significant pressure increase (fluid compression) but require a low pressure drop.

Regardless of whether the injector is subsonic or supersonic, there are three modes of operation, shown in Fig. 13.17a at a constant primary pressure (p_3) and Fig. 13.17b and c at a back pressure (p_5).

In the subsonic critical mode, the primary flow is throttled and the secondary mass flow is constant, while in the subcritical mode, the primary flow is not throttled, resulting in a strong dependence of the secondary mass flow on the back pressure value (p_5). In the operating condition (backflow), the secondary flow reverses so that the ejector is inoperative.

In the critical mode, the supersonic ejector shows a constant suction of the secondary flow (entrainment ratio) due to the compression of the primary and secondary flows, while in the subcritical mode, only the primary flow is simply compressed. At a constant back pressure (p_5), the ratio between primary and secondary flow changes only slightly. In the malfunction mode (backflow), the secondary flow is reversed, causing the ejector to malfunction.

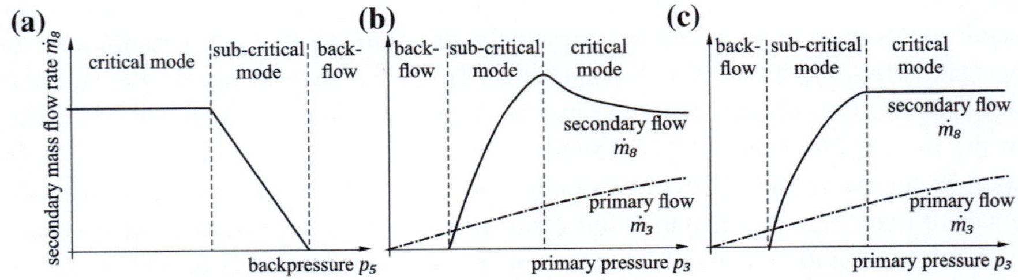

Fig. 13.17 Ejector operational mode dependence on pressure and secondary mass flow rate ($\dot{m}_8$) o; **a** sub- and supersonic ejectors at fixed primary pressure (p_3), **b** sub-sonic ejector at fixed backpressure (p_5) and **c** supersonic ejector at fixed back pressure (p_5)

An important phenomenon related to secondary flow is the throttling phenomenon, which limits the maximum flow through the injector in critical mode. This limitation causes the cooling capacity (CC) and the coefficient of performance (COP) to remain constant, the definition of which will be done at a later time. The propagating density waves of the compression shock from the primary nozzle create an area of the propellant gas as it leaves the primary nozzle where there is no mixing with the secondary flow. This creates a converging channel for the secondary flow and the flow is continuously accelerated. The acceleration depends on the operating conditions. In contrast, in subcritical operation, the ejectors are affected by the backpressure (p_5); when the backpressure is increased, the shock wave moves upstream into the mixing chamber and acts on the mixing of both streams. A further increase in back pressure eventually causes the primary flow to return to the suction chamber. In addition to the complex geometry, the flow in the ejector also depends on whether oblique or vertical compression shocks occur and how they interact with the boundary and shear layers. Ultimately, the fluid dynamics layer determines the performance of the ejector and numerous studies are reported in literature. In this context here, only the dissipative effect of the shock waves is important, since it causes compression and a transition from supersonic to subsonic flow.

13.6.3.2 Operational Key Figures and Characteristics

The properties of the flow in the motive nozzle and in the mixing, section have a strong influence on the operation of the ejector. Firstly, the choking of the motive flow in the motive nozzle defines the maximum possible motive flow. Second, the conditions at the outlet of the ejector determine whether the suction flow is choked or not. Several parameters are used to describe the performance of ejectors in cooling circuits.

First, the so-called characteristic curve shown in Fig. 12.16 shows the change in mass entrainment ratio (MER) as a function of ejector outlet pressure (p_5), where MER is defined in the following way:

$$MER = \frac{\dot{m}_8}{\dot{m}_3}. \tag{13.19}$$

The suction mass flow rate that can be entrained (capacity) and the pressure rise from suction conditions to ejector discharge conditions (lift) are of great importance in quantifying ejector performance. While MER defines the capacity of the ejector, the suction pressure ratio (SPR) determines the lift.

$$SPR = \frac{p_5}{p_8}, \tag{13.20}$$

where the static pressure at the ejector outlet and the static pressure at the suction inlet are used at positions 5 and 8, respectively. The entrainment ratio evaluates the efficiency of the refrigeration circuit, and the compression ratio (lift) is a measure of the operating range of the circuit.

The cooling capacity (CC) is given by,

$$CC = \dot{m}_8 \cdot (h_8 - h_7), \tag{13.21}$$

and describes the amount of useful cold.

The ejector efficiency (η_e), as used by ASHRAE (ASHRAE, 2012), is defined as the ratio of actual compression energy recovered to the theoretical energy available from the motive stream

$$\eta_e = \frac{(\dot{m}_8 + \dot{m}_3) \cdot (h_5 - h_8)}{\dot{m}_3 \cdot (h_3 - h_8)}. \tag{13.22}$$

The coefficient of performance (COP) is the ratio between the evaporation heat power $\dot{Q}_{eva}$ (cooling effect or useful) and the total energy supplied to the circuit, which is composed of the heat supplied to the generator ($\dot{Q}_{gen}$) and the pumping power (P_{pump}) required to lift the pressure from the condenser pressure (p_5) to the generator level (p_3).

$$COP = \frac{\dot{Q}_{eva}}{\dot{Q}_{gen} + P_{pump}} = \frac{\dot{m}_8 \cdot (h_8 - h_7)}{\dot{m}_3 \cdot (h_2 - h_6)} = MER \cdot \frac{(h_8 - h_7)}{(h_2 - h_6)}. \tag{13.23}$$

The COP can also be represented by enthalpies, as shown in the right-hand side of Eq. (13.23), so it is evident that it is related to the mass entrainment ratio (MER).

13.6.3.3 Phase State and Refrigerant Selection

Depending on the primary and secondary flow conditions, the flow in the ejector can be either single-phase (gas–gas or liquid–liquid) or two-phase. A two-phase ejector can be classified by the type of two-phase flow into:

- a condensing ejector (the primary flow condenses in the ejector) and
- a two-phase ejector (where the flow at the outlet is two-phase).

When the fluid exiting the ejector is two-phase, both a liquid and vapor state are present, where either.

- the primary fluid is a liquid entraining a gas, or
- the primary fluid is a high-pressure vapor entraining a liquid secondary stream.

Detailed modelling of such a hydrodynamic process is challenging. As can be seen from these considerations, thermodynamic properties of the particular refrigerant play a central role in the selection. Here, the shape of the temperature entropy condensation or saturation vapor line of the fluid is crucial. Depending on the slope of the saturation vapor line in the $T - s$ diagram, fluids are designated as wet, isentropic, and dry fluids, as shown in Fig. 13.18 using three example fluids. A wet fluid such as water has a negative slope, so small droplets can form at the nozzle outlet if during expansion the saturation line is undershoot, which then block the effective area. This reduces the efficiency and can mechanically damage the component. Such effects are avoided if the motive fluid in the evaporator is superheated, but this is at the expense of efficiency. Therefore, isentropic fluids and dry fluids seem more favourable for ejector cooling circuits at first glance. Nevertheless, isentropic and dry fluids can also enter the two-phase zone during isentropic expansion. If the saturation temperature is below the critical value, the expansion can lead to the same problems as with wet fluids. Safe operation with dry and isentropic fluids is ensured when the temperature of the ejector cooling system does not approach the critical value. In practical applications, this problem is reduced by the dynamic loss of the fluid, since the condition at the nozzle is very close to the saturation line of the vapor.

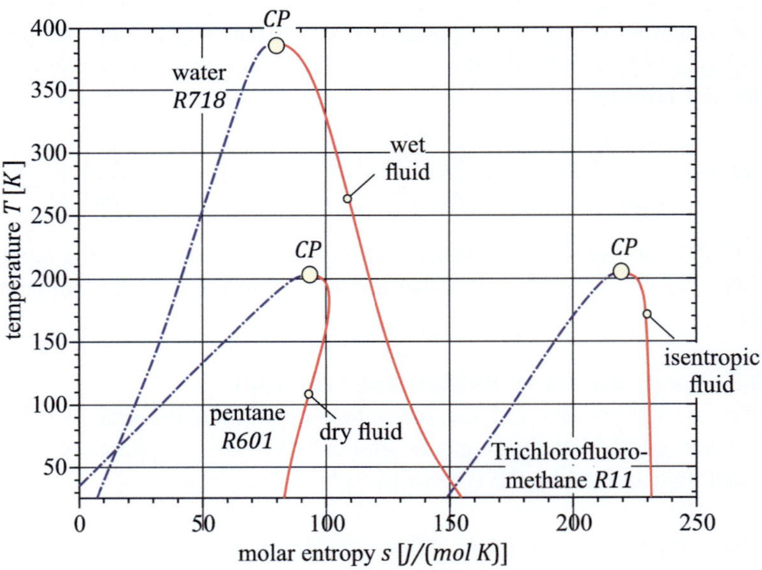

Fig. 13.18 Classification of refrigerants by shape of the saturation line in wet (water), dry (pentane) and isentropic ($R11$)

But other fluid properties affect coolant selection. A large latent heat of vaporization is desirable to minimize the required circulating mass flow rate for a given cooling capacity. A low evaporating pressure and temperature allows for safer operation and less expensive circuit design, as well as the use of lower temperature waste heat sources, such as even non-concentrating solar thermal collectors to evaporate the drive stream. A large molecular mass aids momentum transfer between the drive and suction streams, resulting in higher entrainment of the suction stream. And, as with any refrigerant, it is desirable to simultaneously have chemically stable, non-toxic, non-corrosive, environmentally friendly and cost-effective behaviour with low viscosity and high thermal conductivity; more detailed information on refrigerant selection for ejector systems may be taken from (Zeyghami et al., 2015) or (Besagni, Mereu, & Inzoli, 2016).

For this reason, many realized systems use air and water as working fluids. However, especially for cooling applications, other fluids such as ammonia, $R134a$, $R141b$, $R142b$, $R245fa$ and $R744$ (CO_2) are more suitable. $R245fa$ is of particular interest because its low global warming potential (GWP) and low boiling temperatures make it a good candidate for refrigeration cycles. R744 is also attractive, in addition to its low GWP, because of the large difference between typical high and low operating pressures and the use of lower mean drive temperature differentials required for supercritical operation. Table 13.3 shows some typical refrigerants for ejector-based chillers at typical operational parameters.

13.6.3.4 Design of a Standard Ejector Chiller

The components of the standard cycle single ejector refrigeration chiller are shown in Fig. 13.19, also depicts a qualitative representation in the pressure-enthalpy chart for both a wet (cf. Fig. 13.19c) a dry (Fig. 13.19b) motive jet. In the generator (boiler, solar collector) the heat energy is supplied to evaporate the refrigerant at the generator pressure (p_3) from $1 \to 2$. When the saturation conditions (2) are reached, the flow at high pressure (primary flow, $\dot{m}_3$) is passed through the ejector nozzle, which sucks the secondary flow ($\dot{m}_8$) from the evaporator, which consists of steam at low pressure (p_8). In the mixing zone of the ejector, mixing of the two streams takes place, and the resulting mixed stream leaves the ejector (5) and is directed to the condenser. There, condensation takes place and a heat flow is released to a heat sink mostly the environment, so that liquid refrigerant is present at the condenser outlet (6). Then a division of the liquid flow takes place:

One partial flow is expanded via a valve to the evaporator pressure and thus to a lower temperature (7) and fed into the evaporator to achieve the desired cooling effect; which is either used directly as useful cold or stored in a cold storage, from where it is then available again to the ejector (8). The other partial flow is brought up to the generator pressure (p_3) by means of a pump ($6 \to 1$) and again fed to the generator. The tasks of the ejector are to "entrain" the steam from the evaporator (8) and to recompress it before it leaves the evaporator and is sent to the condenser.

Table 13.3 Typical refrigerants for ejector based chillers with data of at typical operation temperature, global warming potential (GWP) and fluid

Refrigerant code	$R717$	$R718$	$R718$	$R134a$	$R141b$	$R245fa$
Name	Ammonia	Water	CO_2	Tetrafluorethan	Dichlorofluoroethane	Pentafluoropropane
Boiling point [°C] @0.1MPa	−33.5 °	99.9 °	–	−26.1 °	32.0 °	15.1 °
$p[kPa]$@100 °C	6255	101	–	3975	677	1269
mol. Mass [g/mol]	17	18	44	102	117	134
latent heat [kJ/kg] @10 °C	200	2476.9	197.1	190.8	233.3	212.8
GWP	0	0	1	1300	630	950
wet/dry	Wet	Wet	Wet	Wet	Dry	Dry

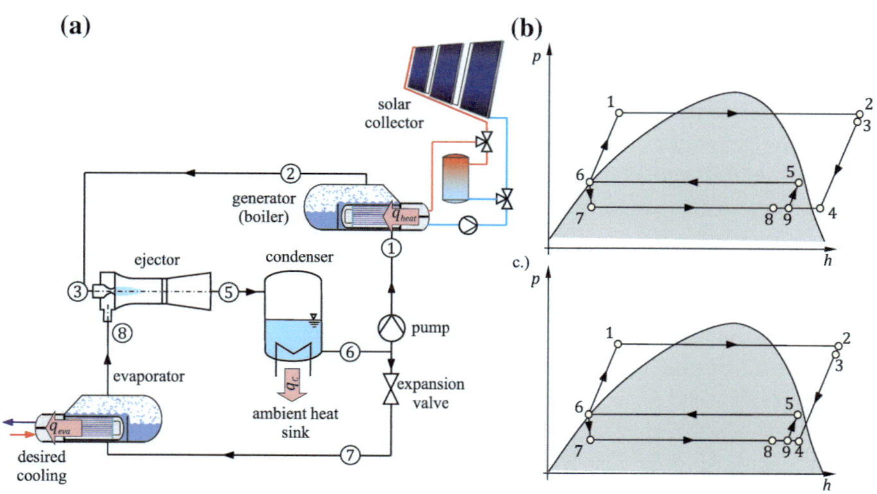

Fig. 13.19 **a** Schematic arrangement of components in a single ejector refrigeration cycle fed by solar collector system; qualitative depcition of the cycle in a pressure enthalpy chart $(p - h)$ with indication of cycle points for a dry **b** and a wet **c** jet operation

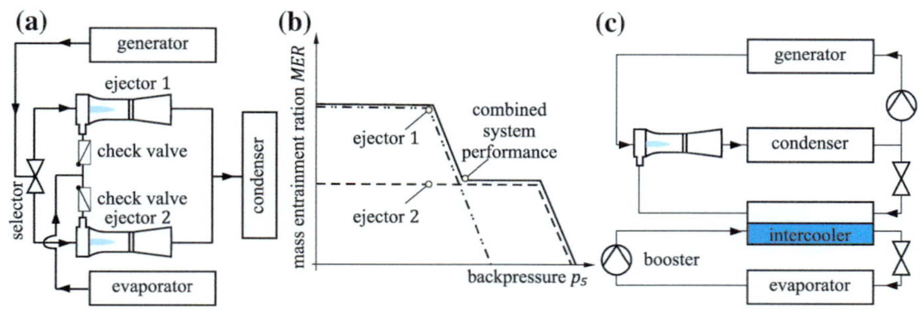

Fig. 13.20 **a** Schematic component arrangement of a double cascade ejector chiller and **b** the resulting composite operating curve as a superposition of the MER curve over the output back-pressure for each ejector; **c** Solar ejector cooling systems with integrated booster

The main features of the standard cycle are:

- Adjusting the generator and evaporator operating conditions, i.e., the ejector operates under critical conditions and provides constant COP and CC (when the critical pressure is exceeded, the secondary flow is reduced, resulting in a significant decrease in MER and COP).
- An increase in generator pressure (p_3) results in a decrease in MER but an increase in critical COP. This is related to the increase in primary mass flow and the associated increase in expansion angle, which leads to a decrease in the effective area within the mixing section of the ejector; thus, a lower secondary

flow is entrained. However, the momentum of both the primary current and the mixture increases, so that the position of the shock wave moves downstream within the ejector domain. This causes the critical pressure to increase and CC and COP to be reduced.

- Holding generator conditions constant, an increase in pressure in the evaporator causes MER and condenser critical pressure to increase. This is due to the reduction in the angle of the shock wave of the underexpanded jet, resulting in a larger effective area, which leads in turn to an increase in secondary flow entrainment. The momentum of the motive jet core decreases, but the total momentum with respect to the mixed flow is higher due to the large secondary pressure. As a result, the shock position shifts further downstream, allowing the ejector to operate against a high back pressure. This leads to an increase in CC and COP.

For an ejector cooling system driven by solar heat, the COP of the ejector sub-cycle is between 0.1 and 0.5, while the generator outlet temperature (T_2) and the overall COP also depend on the collector used. The total COP_{tot} is composed of the product of the collector efficiency (η_c) and the ejector cycle COP.

$$COP_{tot} = \eta_c \cdot COP. \tag{13.24}$$

Typical generator temperatures realized in practice with different refrigerants range up to about 150 °C, while condenser temperatures range between 20 °C – 40 °C depending on the refrigerant. The useful cold that can be extracted from the evaporator is usually around 10 °C, further details can be found in (Besagni, Mereu, & Inzoli, 2016).

13.6.3.5 Performance Improvement Options for a Ejector Chiller

The ejector, which is the main element of the ejector refrigeration cycle, is a static component and shows the best performance close its design point while under varying operating conditions the performance is rather poor. Any increase in condenser temperature and pressure reduces the COP and the required condenser heat exchanger area but requires an increase in the mechanical work required by the compressor.

In order to provide a high COP under different operating conditions, several ejectors with different dimensions can be arranged in parallel between a common inlet and the condenser. This arrangement is called cascaded configuration. In the cascading arrangement shown in Fig. 13.20a, a valve selects the ejector to be used at the given operating conditions depending on the condenser pressure, so that the selected ejector operates at or near design conditions. The improvement of the double-cascading system results from the superposition of the MER as a function of the respective ejector outlet pressure, as qualitatively shown in Fig. 13.20b.

An alternative option to cascading is a continuous adjustment of the ejector geometry. It allows by means of an axially sliding adjustable ejector nozzle a continuous cross-sectional area change, which enables area ratio adaption depending on the condenser conditions.

To improve the performance of the ejector cycle, a compressor can also be used to increase the pressure of the refrigerant leaving the evaporator. The vapor from the evaporator is compressed to an intermediate pressure and then fed to the ejector. The increased suction pressure from the intercooler to the ejector results in a higher mass flow in the ejector circuit, a higher total COP, and a lower possible evaporator temperature. To maintain an intermediate pressure, an intercooler can be used to act as a heat and mass exchanger. A simple diagram of a solar ejector cooling system with a booster and an intercooler is shown in Fig. 13.20c. In this configuration, some of the generator heat in the loop is replaced by the work of the booster. With the technical complex booster solution, the required collector area can be reduced by about 60%.

In some system concepts, the influence of preheaters (regenerators) and the integration of a precooler, as schematically shown in Fig. 13.21, has been analysed to increase efficiency. In this, the liquid refrigerant returning to the generator is preheated in the regenerator with the hot refrigerant from the ejector's exhaust jet, while the liquid refrigerant is cooled by the pre-cooler with the cold refrigerant from the evaporator before reaching the evaporator. The refrigerant coming from the condenser is heated and cooled before passing through the boiler (generator) and the evaporator, reducing the heat input required for the generator, but also reducing the cooling load for the evaporator of the system.

Due to the lower required heat input but also the reduced cooling load, the COP can be improved by about 20%, but this requires two additional heat exchangers with the associated investment costs, and above all complicates the control and regulation. The result is that the introduction of a recooler can increase the COP of the system considerably, but the integration of a recooler cannot.

Other optimization aspirations concern the ejector refrigeration system without a pump. The pump causes only low costs (—or power consumption; approx. $\sim 0.2 - 1\%$), but requires more maintenance than the other system components, as it is the only moving part in the ejector chiller. Therefore, to replace the pump, several solutions were found, such as.

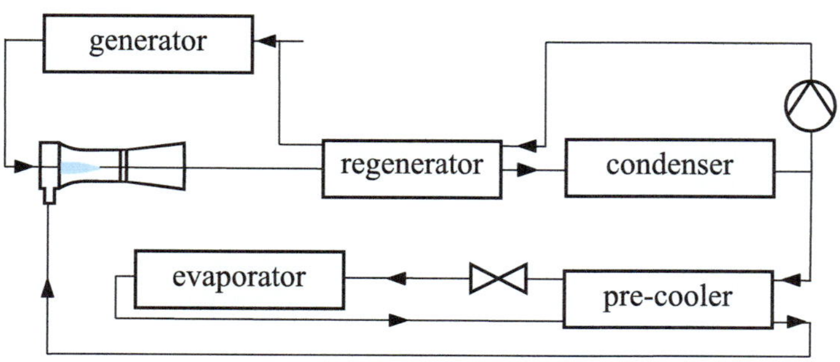

Figure. 13.21 System lay out of a single ejector refrigeration system with pre-heater and pre-cooler

- gravity/rotary ejector refrigeration systems;
- bi-ejector refrigeration systems;
- thermopump effect ejector systems; and
- heat pipe/ejector cooling systems;

for more details see (Besagni, Mereu, & Inzoli, 2016). However, such pumpless designs are not very suitable for solar thermal systems, especially because of the large thermal inertia of the solar collector, which affects the performance during start-up and shut-down. Furthermore, the use of an additional burner is required during non-design operation for additional heating and to avoid thermal transients.

In addition to the already mentioned option of using several ejectors to increase the operating range, there is also the possibility of either using a two-stage ejector or of integrating a second ejector into the suction line of the main injector, which uses a partial flow of the compressed refrigerant from the condenser as motive fluid before it enters the generator. Both options aim at an increased pressure rise and thus an increase of the COP, which are in the order of $10 - 30\%$, but are extremely sensitive to variations especially of the generator temperature, so that they are less suitable for solar thermal systems without appropriate thermal energy storage. As an alternative to multi-ejector systems, several evaporators operating at different pressure levels can be used to feed one or more ejectors. Theoretically, this results in similar increases in efficiency as multi-ejector systems, but also in the corresponding complexity of control and regulation.

Most ejector chillers operate below the critical point (CP) of a fluid, as do almost all refrigerants or water/vapor or water/ammonia mixtures. However, a transcritical ejector refrigeration system (TERS) uses a refrigerant that operates above the critical conditions. In TERS systems, the propellant generation process occurs at supercritical pressure, and the density of the primary working fluid decreases until the vapor condition is reached. The supercritical vapor expands through the ejector nozzle and entrains the flow from the evaporator. To maintain the required output, operation of the transcritical process requires control of the high pressure. In these cycles, both pump outlet pressure and generator outlet temperature are critical operating parameters. In addition, the ejector may exhibit two-phase flows that depend on operating conditions, particularly primary flow pressure and generator temperature. Compared to a conventional vapor compression cycle, an ejector replaces the throttle valve in this configuration. The ejector increases the compressor suction pressure compared to a standard cycle, resulting in higher system efficiency since less compression work is required. However, this arrangement presents some difficulties in controlling operating conditions because of the close relationship between ejector output quality and mass entrainment ratio (MER). Significant COP improvements of CO_2-TERS systems in the range of $10 - 20\%$ have been observed compared to conventional cycles, but experimental validation is pending; techno-economic analyses for large-scale systems are also still in the early stages. The role of the ejector in modelling these cycles is particularly critical.

13.6.3.6 Performance Adjustment and Solar Thermal Ejector-Based Chillers

In the case of strongly fluctuating refrigeration loads, economical operation of a single-stage steam jet refrigeration system becomes unlikely. Therefore, a modular capacity adjustment with controllable steam jet compressors is required. In continuous operation, both the capacity and the motive steam pressure must be controlled. Generally, four concepts are used for capacity adjustment:

- Adding and disconnecting additional jet compressors (ejectors),
- bypass control,
- capacity adjustment via changing evaporator temperature, self-regulation/ suction steam throttling, and/or
- nozzle control.

Steam jet-based ejector chillers with up to one MW cooling capacity have been realized so far mainly with water and ammonia-water mixtures, see e.g. (Jähnig & Pollerberg, 2010) or (Joemann, et al., 2016). Especially purely water-based ejector chillers are used where process cooling is required in addition to process steam heat, e.g. in the food industry, e.g. dairies, breweries, etc. The advantages of steam jet-based chillers are obvious.

- very robust and reliable equipment with simple design low cost materials with low maintenance and low cost refrigerant (water);
- no need for hydraulic decoupling (water);
- operating parameters can be freely selected over a wide range on the cold water, cooling water and drive heat sides;
- Low process and control engineering effort with good dynamic operating behaviour (fast load changes),
- no aging of the refrigerant (crystallization, decomposition).

In addition to the manageable COP, the disadvantages of water-based ejector cooling systems include high motive steam consumption and correspondingly high cooling water consumption, since the cooling water may only be heated slightly, as well as a narrow temperature range for use as a chiller, which cannot be less than 0 °C for water.

The influence and role of the solar collectors is critical because, as indicated in Eq. (13.24), the total COP_{tot} depends on those of the collector and the ejector nozzle. Therefore, small differences in solar collector efficiency can result in a proportionally larger difference in COP_{tot}. If one analyzes the exergy losses of a solar-based ejection chiller, the solar collector accounts for the two largest contributions and the ejector accounts for a much smaller portion, with all others even smaller.

Therefore, fluctuations in solar irradiance are a critical issue in purely solar ejector chillers, as they do not allow for a constant generator temperature. If a fixed ejector geometry is used, the refrigeration cycle would not consistently deliver the intended COP. At low ambient temperatures, the circuit is limited by throttling, and at high ambient temperatures, the ejector requires more power than can be supplied by the collector. A larger throat cross-section can accommodate a larger solar collector area and a wider range of generator temperatures, but the component may be oversized (especially under non-design conditions) and increase cost. In contrast, a nozzle diameter that is too small limits the range of generator temperatures. For all of the above reasons, a variable area ejector is attractive.

In recent years, coupled Rankine and solar ejector refrigeration cycles have been operated experimentally and shown to be energy efficient, reliable, and flexible. However, efforts are still needed to optimize these cycles and to develop, in particular, time-dependent models that can account for transient phenomena in all components of the cycle.

The collector efficiency of the operated systems varies between 0.1 and 0.65 depending on the technology (mostly parabolic troughs), ambient and operating conditions, while the COP of the ejector subcycle is between 0.1 and 0.5. Thus, the maximum overall COP achieved so far is in the range of $COP_{tot} \approx 0.3$, indicating that there is still quite a large potential for increase.

Steam ejector systems are also suitable per se for cogeneration, in which the motive steam generated in the generator is first expanded in a turbine, which drives a generator, and is only injected into the ejector after the turbine outlet. Current studies essentially consider fluids such as those used in organic Rankine cycles (ORC), which are discussed in Chap. 11.

Similarly, steam ejector systems can also be combined with purely thermally driven sorption-based processes for cooling, as discussed in the following section. Although all studies here show that significantly higher COP values can be achieved than with pure ejector cycles, these have so far been limited to theoretical studies, see e.g. (Besagni, Mereu, & Inzoli, 2016).

13.6.4 Thermally Driven Chillers (Sorption Chillers)

In sorption-driven chillers, a mechanical compressor is not needed. The sorption system replaces the compressor by absorbing or adsorbing the refrigerant vapor at low evaporator pressure and feeding it to the condenser at a higher condensing pressure by desorption (expulsion of the sorbate refrigerant). Desorption is achieved by adding heat to the sorbent (mostly salt solutions). This effect is the actual thermal drive of the system. A prerequisite for the process is that the two substances are soluble in each other in any ratio in the temperature interval used. The sorption system, which is the driving part, is also called a thermal compressor. Figure 13.22 shows the schematic diagram of a closed absorption chiller. In the $p-T$ diagram, the process operation is as sketched in Fig. 13.23.

There are two types of closed sorption chillers.

- absorption units with liquid working materials
- adsorption systems with solid sorbents.

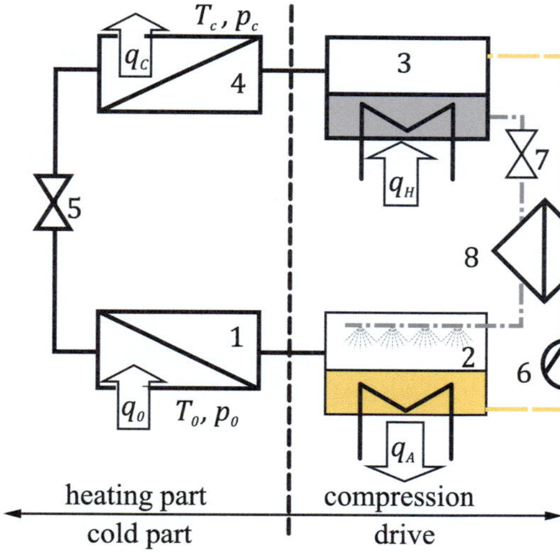

Fig. 13.22 Component arrangement of a closed absorption chiller. Evaporator (1), absorber (2), expander or generator (3), condenser or condenser (4), expansion valve (5), pump (6), throttle (7) and heat exchanger (8). (---) refrigerant rich solution and (-·-·-) refrigerant depleted solution, respectively

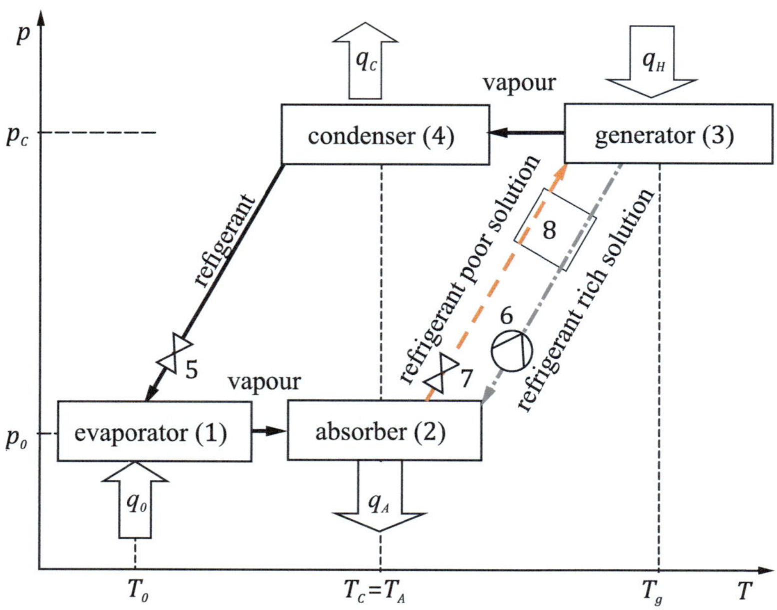

Fig. 13.23 Process and media flow of a closed absorption chiller in the $p-T$ diagram with indication of the aggregate states. (---) solution rich in refrigerant or (-·-·-) solution poor in refrigerant. Components: Evaporator (1), absorber (2), discharger or generator (3), condenser or liquefier (4), expansion valve (5), pump (6), throttle (7), and heat exchanger (8)

13.6.4.1 Closed Cycle Absorption Chillers Using Liquids

The absorption chiller combines the following basic ideas: The temperature required to evaporate a liquid decreases with decreasing pressure and as a result the refrigerant evaporates already at low temperatures due to the low absolute pressure in the evaporator (1). The heat of evaporation (q_0) required for evaporation is taken from the space to be cooled and represents the usable cold. The evaporation process would stop if the saturation pressure of the refrigerant in the evaporator was reached. Therefore, the refrigerant vapor is continuously withdrawn in a second step. Here, the properties of particular solutions are utilized to bind (absorb) refrigerant vapor. This step takes place in the so-called absorber (2). The heat of absorption (q_a) released during absorption is transferred to a cooling medium. The absorption process would also stop if the solution were saturated with the refrigerant. The solution enriched with refrigerant is called a rich solution. The rich solution is continuously compressed to the condensing pressure (p_c) by the absorber (2) by means of a pump (6) and delivered to the generator or discharger (3). In the generator (3), the refrigerant is evaporated from the solution by supplying heat (q_h) generated by solar collectors or occurring as waste heat from thermal processes. This process takes place at a significantly increased pressure (p_c) and temperature level (T_g). In the condenser (4), the refrigerant is liquefied again while releasing heat (q_c) to a coolant. This requires heat dissipation by providing cooling power from an external cooling circuit, for example a cooling tower. From the condenser (4, condenser), the refrigerant is then expanded in the expansion valve (5) to the evaporating pressure (p_0). In the evaporator (1), it again absorbs heat from the space to be cooled. The regenerated solution in the generator, often referred to as the poor solution, is returned to the absorber (2) in a closed circuit via a throttle (7) and a heat exchanger (8), which preheats the rich solution. There it is trickled to maximize the surface area for absorption of the refrigerant. The connection (phase contact) of the refrigerant with the sorption material shifts the vapor pressure curve of the refrigerant. This effect enables a (thermal) compression of the refrigerant vapor by adding heat. The drawn solvent circuit with concentrated and diluted (poor) solution is necessary to ensure continuous operation. The process control in the $p - T$ chart and the media flow between the components is illustrated in Fig. 13.23.

The challenge in the design of an absorption chiller lies on the one hand in the adaptation to the externally provided temperature levels and on the other hand in the coupled sequence of heat and mass transfer processes. Cooling water, heating medium and required refrigerant temperature determine the parameters condensation and evaporation pressure as well as the degassing width. The degassing width is the concentration difference between refrigerant-rich (ξ_r) and refrigerant-poor solution (ξ_a). Thermodynamic operability requires a positive degassing width, i.e.$\Delta\xi = (\xi_r - \xi_a) > 0$. However, a small degassing width requires a high specific solvent circulation, which has a negative effect on the heat ratio due to the irreversibility of the individual heat transfers. Therefore, for practical plant operation, approximating the solution concentrations to the thermodynamic limit case (infinite solution circulation) is not practical. The specific solution circulation (or relative mass flow) of refrigerant-rich (μ_r) or—poor solution (μ_a) related to the refrigerant flow ($\dot{m}_{ref}$) can be determined graphically with the $h - \xi$ diagram for the respective refrigerant-solvent mixture. It is determined mathematically with the aid of

$$\mu_r = \frac{\dot{m}_r}{\dot{m}_{ref}} = \frac{\xi_d - \xi_a}{\xi_r - \xi_a} \text{ and } \mu_a = \frac{\dot{m}_a}{\dot{m}_{ref}} = \mu_r - 1 = \frac{\xi_d - \xi_r}{\xi_r - \xi_a}. \tag{13.25}$$

To achieve a high heat ratio, a high absorber efficiency (η_{ab}) is required:

$$\eta_{ab} = \frac{\xi_r - \xi_a}{\xi_{r,theo} - \xi_a}, \tag{13.26}$$

where $\xi_{r,theo}$ describes the theoretical saturation concentration. For this purpose, sufficiently large phase boundary surfaces must be provided. However, since these surfaces are also heat transfer surfaces, unintentional, excessive subcooling of the solution can occur (a certain amount of subcooling is required as a driving force for the absorption process). As a result, more heat is required to reheat the solution in the heat exchanger and in the generator (expulsor). A detailed discussion of the individual sub-aspects of the solution design and the design of the technical components is described in (Herold et al., 1996) or (Jungnickel, Agsten, & Kraus, 2001), among others.

Absorption chillers are used where larger amounts of waste heat energy are available for the discharge process. Therefore, before the rise of thermal solar energy, they were usually found only in large-scale plants. However, there are also small devices that operate as absorbers, such as gas-powered camping refrigerators. Since liquid absorption chillers require media in both circuits that are matched to each other, so-called substance pairs are needed; the most common substance pairs are listed in Table 13.4.

The power balance of the refrigeration plant results from the sum of the incoming and outgoing power quantities:

$$\dot{Q}_0 + \dot{Q}_h + P = \dot{Q}_c + \dot{Q}_a + \dot{Q}_{loss}, \tag{13.27}$$

where $\dot{Q}_0$ is the evaporator capacity (cooling capacity), $\dot{Q}_h$ the thermal driving capacity, P the pump drive capacity, $\dot{Q}_c$ the condenser capacity, $\dot{Q}_a$ the absorber capacity and $\dot{Q}_{loss}$ the sum of all heat losses. When an absorption chiller is operated in cooling mode the measure for the quality is often called EER(energy

Table 13.4 Typical substance pairs in commercially available absorption chillers

Refrigerant	Solvent
Ammonia ($R717$)	Water
Water ($R718$)	Lithium bromide
$R22$ (banned in the EU as of 2010)	Tetraethylene glycol dimethyl ether

efficiency ratio) instead of COP (coefficient of performance for the heating mode However, in literature also COP is used for cooling. For the real chiller this is:

$$EER = COP_C = \frac{\dot{Q}_0}{\dot{Q}_h + P}. \tag{13.28}$$

The thermal EER of an absorption refrigeration system is calculated in the ideal limiting case, for which the pump power is negligible ($P = 0$) and for temperatures $T_c = T_a$ (no heat losses), solely from the heat of evaporation Δh_{eva} of the refrigerant and the heat of solution or binding Δh_{latent} in the way:

$$EER = COP_C = \frac{\Delta h_{eva}}{\Delta h_{eva} + \Delta h_{latent}}. \tag{13.29}$$

In real single-stage plants, values around $COP_C/EER \approx 0.7$ can be realized.

In the absorption refrigeration process, there are thermodynamically two interconnected circuits, one running on the clockwise side (heating process) and one running on the counterclockwise side (cooling process), as shown in the $T - s$ diagram in Fig. 13.24. The energy supplied by the clockwise heating process and the energy supplied by the counter clockwise cooling process, which represents the useful part, are the same. The areas enclosed by the clockwise heating process, which supplies the drive energy, and those enclosed by the counter clockwise cooling process, which represents the useful part, are equal in the ideal comparative process, so that the following applies:

$$\Delta s_1 \cdot (T_h - T_a) = \Delta s_2 \cdot (T_c - T_0). \tag{13.30}$$

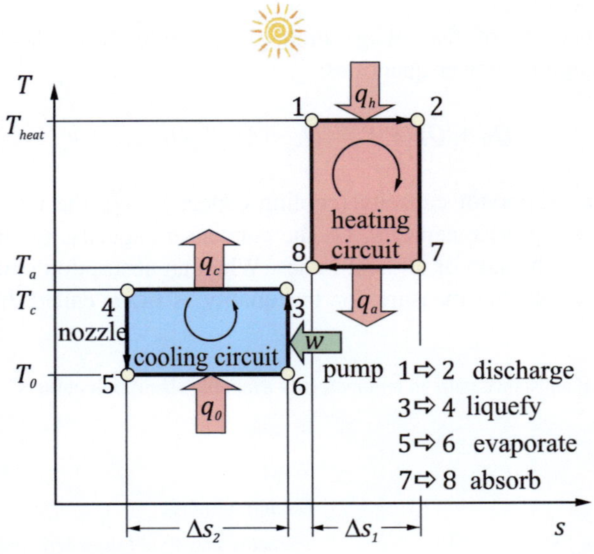

Fig. 13.24 Absorption chiller process in $T - s$ chart

If T_a and $T_c$are equal, the Carnot coefficient of performance COP_{CC} can be calculated as follows:

$$COP_{CC} = \frac{(T_h - T_c)}{T_h} \cdot \frac{T_0}{(T_c - T_0)}. \tag{13.31}$$

In practice, however, $T_a > T_c$and Carnot's heat ratio ζ_{KC} is given by

$$COP_{CC} = \frac{(T_h - T_a)}{T_h} \cdot \frac{T_0}{(T_c - T_0)}. \tag{13.32}$$

Figure 13.25 shows the ideal Carnot COP_{CC} of an absorption machine. The higher the temperature difference between the heating source (T_h) and the condenser (T_c), the higher the heat ratio. It also increases with decreasing difference of $(T_c - T_0)$. The quality ratio of the absorption machines η_{CC} can then be determined as follows,

$$\eta_{CC} = \frac{COP_C}{COP_{CC}}, \tag{13.33}$$

and it is suitable for actual operation because the real heat ratio is always smaller than that of the Carnot process.

The system diagram of a closed absorption chiller (cf. Fig. 13.26) clearly shows that such an apparatus contains a generation and an absorption process. For this

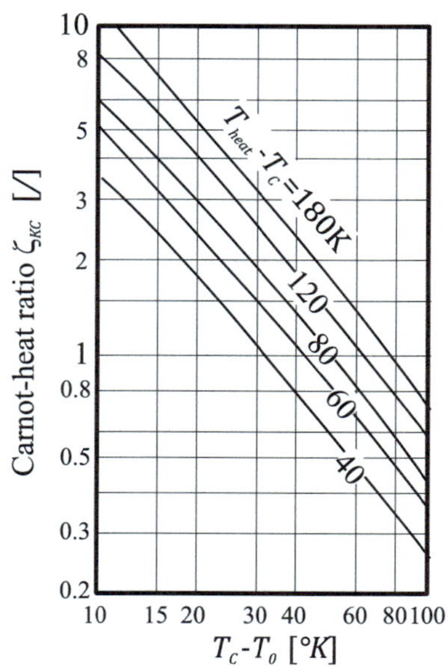

Fig. 13.25 Carnot COP_{CC} of an absorption chiller as a function of the temperature difference of the condenser and evaporator $(T_c - T_0)$

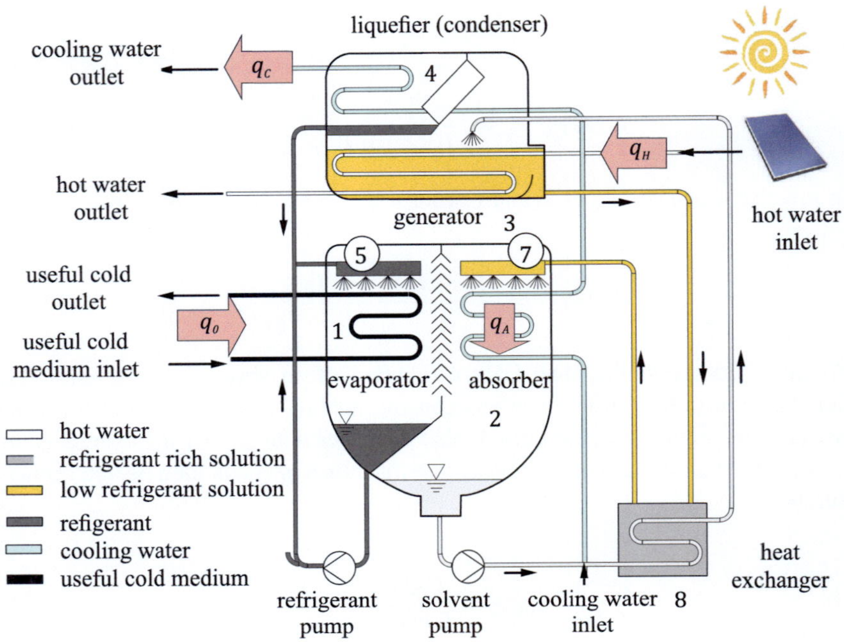

Fig. 13.26 System diagram of a closed absorption chiller with liquid working fluid

reason, this connection is also referred to as a "single-effect" variant. In order to use the drive heat used in the generator as effectively as possible, a number of variants have been developed for this process control. In the so-called double-effect process, a second generator is interposed to support the regeneration of the diluted solution, and a second condenser is interposed at an intermediate temperature and pressure level. The condensation heat of the condenser at high temperature is used as driving heat in the second generator. In the ideal limit case, the evaporation heat Δh_{eva} can be used once at the first generator and a second time as condensation heat coming from the condenser in the second generator, so that the limit value for the thermal $COP_C = 2$ results. Real machines reach values around$COP_C \approx 1.2$.

Another circuit for efficient utilization of the regeneration heat used is the dual-loop process described in (Reimann, 1984). This variant consists of two separate single-effect machines, which exchange heat with each other but not the refrigerant. Rather, both material loops are completely separated from each other. The sorption and condensation heat of sorption system 1 is used as driving heat for generator 2 of the second sorption system, while the condensation heat of system 2 is used for evaporation of the refrigerant in system 1.

A good suitability for the use of solar energy is also shown by the process called Single-Effect Double-Lift circuit developed by ZAE Bayern (Lamp & Ziegler,

1998; Schweigler, Hellmann, Preissner, Demmel, & Ziegler, 1997). In it, an additional single-effect process via a second sorption system ensures that the heat used for propulsion can be used at a total of three generators at different temperature levels. If a heat transfer medium flows through the three generators one after the other in a sequence that decreases according to their temperature level, the heat transfer medium cools down significantly more than in a single-effect system. This characteristic is significant for the integration of solar collectors, as it allows high efficiency to be achieved even at low temperatures. The COP_C of this type of system lies between that of a single-effect system (often around $COP_C \approx 0.7$) and that of a double-lift system (around $COP_C \approx 0.4$) and is determined by the distribution of the heat supplied to the three generators.

13.6.4.2 Closed Adsorption Chillers with Solid Sorbents

In contrast to absorption chillers, the sorption process in adsorption chillers takes place on a solid, so that no cycle is possible between the enriched and the dried sorbent. For continuous operation of the sorption process, therefore, at least two sorption chambers must be operated, which alternately adsorb the refrigerant and are themselves regenerated. In the case of adsorption machines, a distinction is made between pure adsorption and dry adsorption processes.

Adsorption is based on the accumulation of refrigerant molecules on the inner surfaces of a highly porous sorbent. Since adsorption of the refrigerant involves condensation, it is favoured by low temperature and high pressure, reduces the volume of the refrigerant, and releases energy in the form of heat. Desorption involves evaporation and thus has the opposite effect. At constant pressure, this results in the temperature always being equal to the boiling point of the refrigerant, provided that refrigerant is still present in both states of aggregation. This is the basic prerequisite for a refrigerating machine and for a heat accumulator. Since the refrigerant is to adhere to the sorbent, materials that are very finely porous, i.e. have a very large inner surface area, are particularly suitable, such as.

- activated carbon,
- zeolite,
- silica gel.

In dry absorption, on the other hand, the working substance enters into a reversible chemical bond with the solid, for example the reaction of ammonia with salts such as calcium chloride.

The simplest form of a closed adsorption-based system is the adsorption dryer, in which moisture removal takes place by adsorption on hygroscopic materials. The advantages of this system are a modular design, self-cleaning and the avoidance of bacterial build-up, because no condensates are produced. However, this closed process only includes a drying effect without cooling. Nevertheless, drying is possible even at temperatures below 0 °C. The mode of operation of an adsorption dryer can be understood most easily by looking at the plant layout and the corresponding $T - x$ diagram, as illustrated in Fig. 13.27.

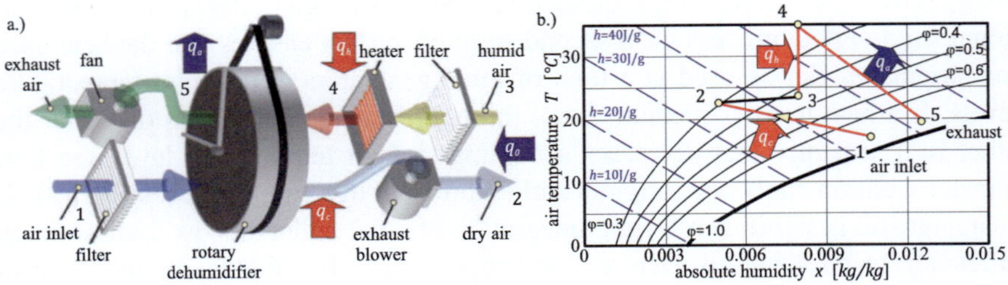

Fig. 13.27 Heat-assisted adsorption dryer; system design (**a**), process execution in $T - x$ chart (**b**)

- First, condensation takes place at rotary dehumidifier quasi-adiabatic, and the condensation heat is transferred to air, resulting in a temperature increase (2).
- The dry air absorbs the room humidity as a load (3).
- The humid air is heated by a heat source (4).
- Then desorption takes place on the rotary dehumidifier wheel (5).

Refrigerating machines operating on the adsorption principle were already being manufactured at the beginning of the 20th century. However, they have been almost completely superseded by compression systems. Since the beginning of the 1990s, however, plants with water as the working fluid and silica gel as the sorbent have again been manufactured in East Asia.

A closed pure adsorption chiller consists of two working chambers filled with sorbent, a condenser, and an evaporator (cf. Fig. 13.28). The process itself is discontinuous and closed. Thus, at least two sorbent vessels are required for quasi-continuous refrigeration, so that refrigerant is adsorbed in one while the regeneration cycle takes place in the other. One of the most common material pairs is silica gel as sorbent and water as refrigerant. A cycle in an adsorption refrigeration machine, runs in the following stages:

- First, the refrigerant attached to the sorbent silica gel is expelled in the heat exchanger 2 supplying heat (q_{heat}) in the generator.
- The refrigerant (water) is liquefied in the condenser and the heat of condensation (q_c) is dissipated to the cooling water.
- The condensate is collected, sprayed into the evaporator and evaporated at under pressure. In the evaporator, the heat (q_0) is extracted from the chilled water and is available as useful cooling.
- In heat exchanger 1, the refrigerant vapor (water vapor) is adsorbed and the heat (q_a) generated during adsorption is dissipated to the cooling water.

By simply diverting the heating and cooling water circuits between heat exchangers 1 and2, the generator (expulsion) and adsorber functions are reversed at the end of a cycle and the process starts again. Typical cycles last around 400s, with a 20s switch-over phase between two duty cycles, so that the total cycle lasts around

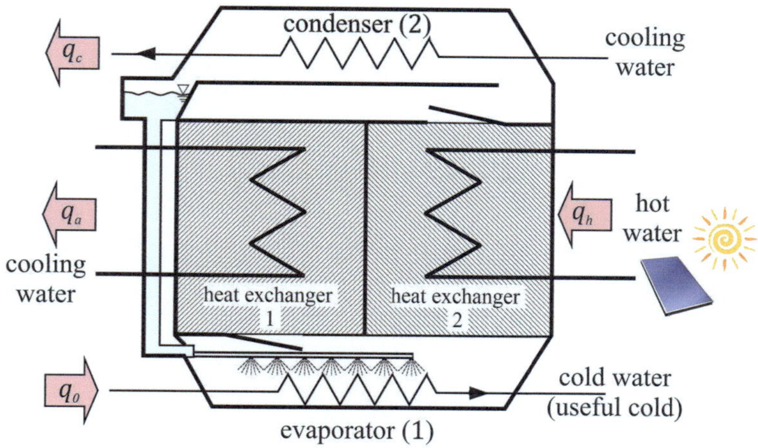

Fig. 13.28 Schematic structure of an adsorption chiller

seven minutes. During the changeover phase, both chambers are flown through in series so that heat is recovered. In the ideal case, only the sensible heat for the chamber heating is recovered. However, if only a small temperature swing is required, adsorption already starts in the switch-over phase, so that not only sensible heat but also adsorption enthalpy is recovered. Such an operating condition can occur when supplying chilled ceiling systems. The adsorption chiller can then achieve a cooling ratio COP, as defined in Eq. (13.8), of above $COP_C = 0.7$.

Figure 13.29 illustrates the idealized internal operation of the chiller in the $p - T$ diagram. The isosters (state in which the degree of coverage of the adsorbent remains constant) illustrate increasing degrees of loading from right to left. The left boundary line defines the fully charged, i.e. saturated, sorbent and corresponds to the vapor pressure curve of water in a substance pairing with water as refrigerant. The condenser (2) and the evaporator (1) lie on this curve. Adsorption is terminated at evaporator pressure and condenser temperature, which defines point (3). After adsorption, both valves are closed and the sorbent is heated at constant concentration until it reaches the condenser pressure (4). Now the valve to the condenser is opened and under constant pressure and increasing temperature water is expelled until the desorption end temperature is reached, which marks point (5). Then the valve closes and at constant concentration cooling down to evaporator pressure (6) takes place. At this point, the valve to the evaporator is opened and under constant pressure and decreasing temperature, adsorption occurs to the adsorption end temperature (3). However, the process illustrated in Fig. 13.29 is of rather theoretical nature and applies only to the reversible process. In reality, the following aspects have to be considered:

- The external material flows (heating water, cold water, cooling water) have cycle-variable spreads.

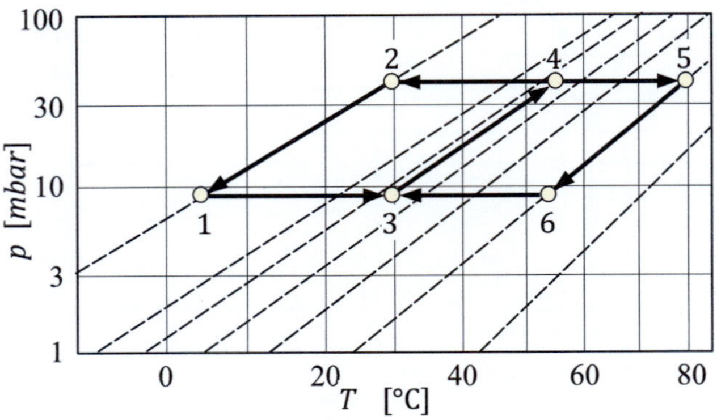

Fig. 13.29 Ideal absorption refrigeration machine process in the isostere diagram (isostere = constant degree of coverage of the adsorbent)

- The heat-transferring surfaces are subject to gradients that depend on the output.
- Mass transfer between steam and sorbent is a function of concentration difference (and inert gas fraction).
- Pressure losses occur at chamber dampers, which depend on the volume flow.

The major technical problems are the freezing of the evaporator, which causes a considerable repair effort, and the difficulties regarding the uncertainties in the specific cyclic plant behaviour, especially in the partial load range, and the resulting demands on the control system. Therefore, when planning adsorption refrigeration plants, the fluctuating refrigeration temperature has to be considered above all. At times the temperature is too high and at times too low. If the refrigeration network itself does not have a very large capacity, a special cold storage system must be installed. It should be noted that not only stored, but also actually mixed. Another challenge is the close approach to the freezing point required in classical air-conditioning refrigeration systems. If an average cooling temperature of 6 °C is required, a temperature of 4 °C may occur at times. If the volume flow breaks off at this point, freezing damage can occur. The refrigeration machine is only reliably protected against the consumers being switched off while the refrigeration circuit continues to run. It is best to design the air conditioning system for a slightly higher chilled water temperature. This is also more favourable in the interest of a large cooling capacity.

An advantage of the adsorption chillers is not only the low heating temperature but the wide range of application. The adsorption chiller works reliably even under parameters where absorption chillers have already shut down due to real danger of crystallization or due to control problems. It is therefore particularly suitable for systems with low flow temperatures (<90 °C) and thus especially for solar systems. Furthermore, adsorption machines are characterized by simple design, ease of operation and high reliability, since they contain only a small number of active electrical components. In particular, processes using silica gel are well

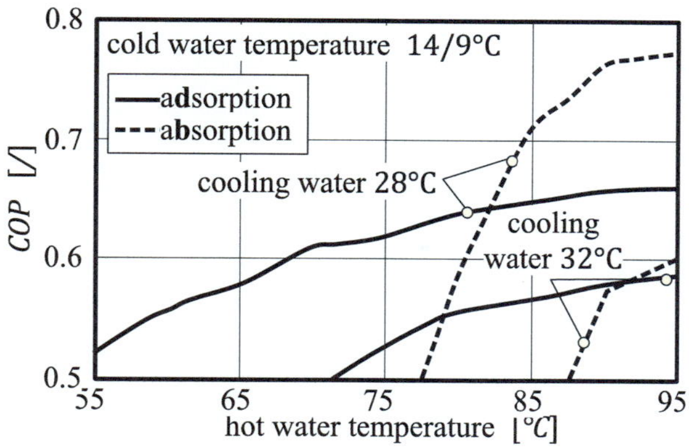

Fig. 13.30 Comparison of achievable cooling ratios (COP_C) of adsorption and absorption chillers as a function of heating water temperature at two different cooling water temperatures

suited for the integration of solar heat due to the relatively low regeneration temperatures. The comparison of the characteristic curves of an adsorption chiller and an absorption chiller in Fig. 13.30 illustrates this. The generated chilled water with temperatures of about 12 °C is suitable for use in static cooling elements such as chilled ceilings. Disadvantages of adsorption chillers are the high price and especially the size and mass of the plants. Nevertheless, some advantages such as relatively simple operation and low maintenance speak for an increased use of such devices.

13.6.4.3 Air Conditioning in Open Systems

Open systems are characterized by the fact that they operate at ambient pressure. Here, solar heat can be used in the simplest case for heating and ventilation, for example, for dehumidifying rooms, drying materials, etc. The simplest method of humidity regulation using solar heat is illustrated in Fig. 13.31. Ambient air is drawn in by a fan and heated by solar heat, decreasing the humidity level (1 → 2). The dry air absorbs the heat of evaporation of water (q_e) in the room as a load and is discharged to the outside (2 → 3). This simplest form of ventilation is, of course, highly dependent on outdoor temperature and is most efficient in winter, and less good in spring or autumn. The lowest efficiency is in summer, when humidity and outdoor temperatures are high, but then the demand for space heating is also the lowest.

Another method of using solar heat is condensation drying. Here, the air is cooled to below the dew point and moisture condenses. The individual process steps outlined in Fig. 13.32 are as follows.

- Suction of supply air (1) by exhaust fan.
- Evaporator temperature < dew point temperature of air leading to condensation on evaporator walls (2).

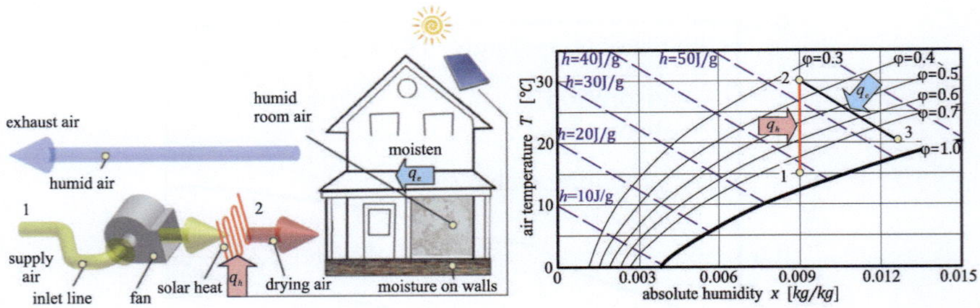

Fig. 13.31 Operation of a simple solar thermal based heating and drying system (**a**); system design and representation in $T - x$ diagram (**b**)

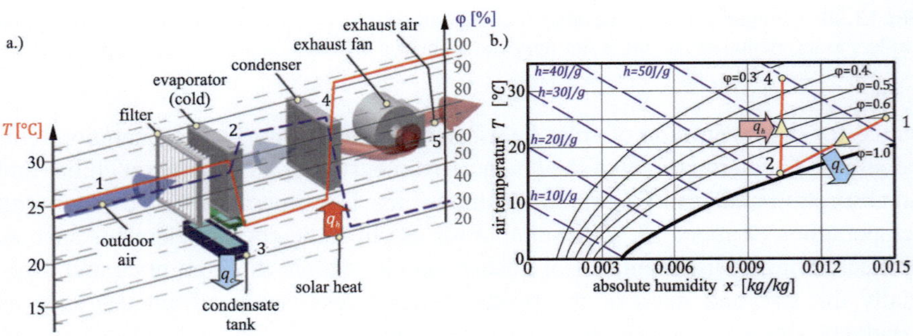

Fig. 13.32 Illustration of the system design of a condensation dryer (**a**) with corresponding $T - x$ diagram (**b**)

- Precipitation of water (3) and discharge of water.
- Cooled & dry air is heated in evaporator (4) by solar heat.
- Return of dry and hot air to exit point.

13.6.4.4 Sorption-Assisted Air Conditioning in Open Systems

Sorption-assisted open-air conditioning systems have become increasingly prominent in the air conditioning and refrigeration industry in recent years. Systems of this type are referred to as Desiccant Cooling Systems (DCS) in the refrigeration sector and as Desiccative and Evaporative Cooling (DEC) in the air conditioning sector. In Europe, the term sorption assisted cooling (SAC) is also used. SAC systems differ from thermally driven absorption and adsorption chillers in that they represent an open system. Both the refrigerant water and the medium air to be cooled are in contact and mutual exchange with the environment on the one hand and with each other on the other hand. Figure 13.33 shows the basic structure of a SAC system with its most important components. The system structure corresponds to the Pennington cycle (Pennington, 1955), which is referred to in the literature as the ventilation cycle.

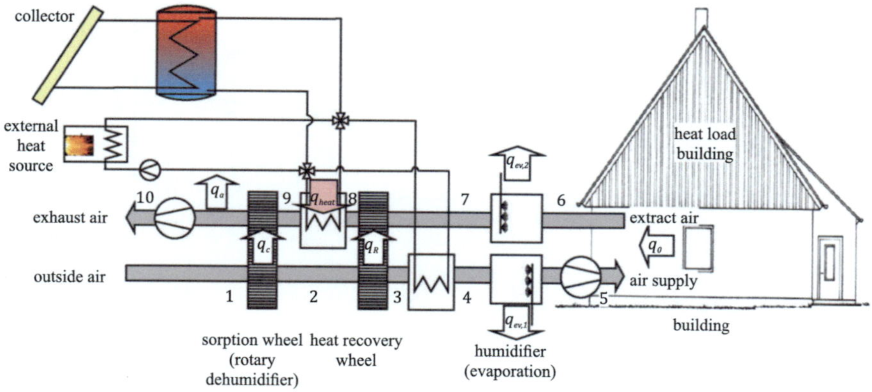

Fig. 13.33 Schematic of a solar assisted drying and evaporative cooling process (SAC—ventilation cycle) indicating components and heat fluxes

There are two further differences between SAC units and pure ab—or adsorption chillers. SAC units operate at a constant pressure level at ambient pressure and thus no pressure change is required between the evaporator and condenser. The second difference is in the useful power generated. In closed systems, a chiller is provided that can be used for further application. A SAC system directly conditions the temperature and humidity of the air flowing through it, so that the entire system must be regarded as a complete air-conditioning unit. In contrast, absorption and adsorption chillers must additionally drive an air conditioning unit with their produced cold in order to fully condition the room air.

Thermodynamically, air conditioning in a SAC system is represented in the $T - x$ diagram as shown in Fig. 13.34, with the numbering corresponding to that in Fig. 13.33. The process flow (cf. Table 13.5) illustrates the characteristics of a SAC system:

- the cooling load of the room is removed by ventilation with conditioned ambient air;
- sensible cooling is provided exclusively by evaporative cooling and heat recovery;
- sorption increases evaporative potential and serves to control humidity;
- refrigerant water is in direct contact with room air;
- direct influence on the condition of the room air;
- supply air temperature is limited by the dew point of the humidified supply air.

As with other air-conditioning systems, the components of a SAC system can be combined and arranged in any other way.

One variant is the recirculation cycle, which represents a closed circuit for the cooling medium. Here, the room supply and exhaust air is routed in a closed circuit and the regeneration of the dehumidification rotor is carried out with ambient air.

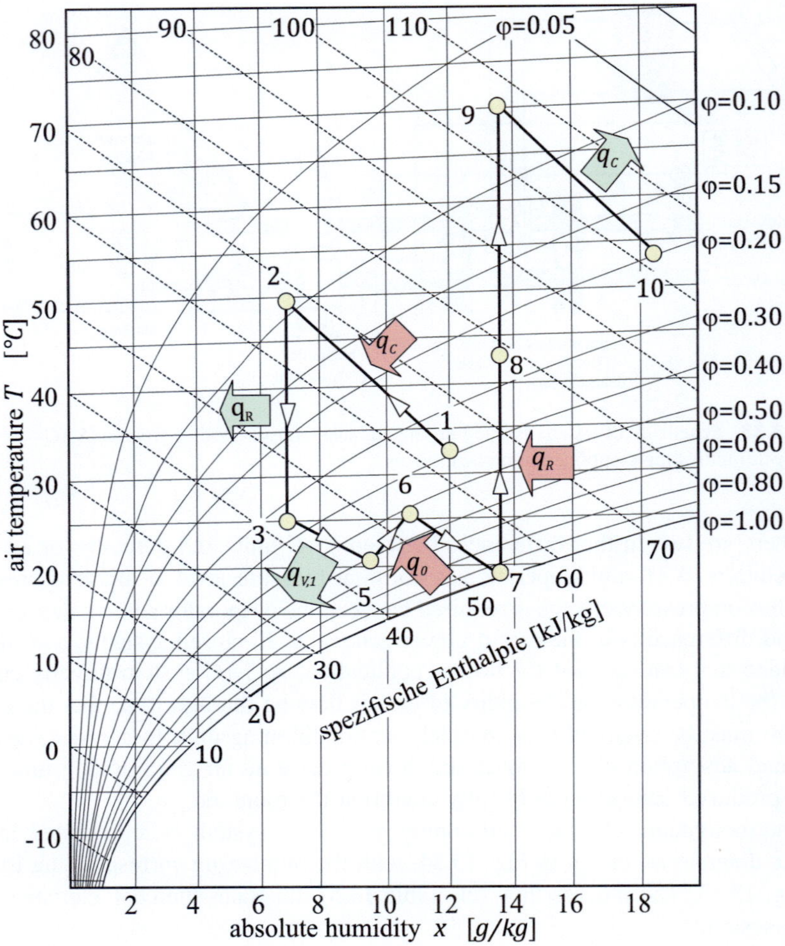

Fig. 13.34 Change of state of the air during a SAC air-conditioning process (DEC, in the ventilation cycle) in the $T - x$ diagram according to the numbering in Fig. 13.33

The structure of the recirculation cycle is illustrated in Fig. 13.35a. The advantage over the ventilation cycle is that the room exhaust air, which is cooler and often drier than the ambient air, contains a greater cooling potential than the ambient air. As can be seen from the $T - x$ diagram of a typical operating condition of Fig. 13.35b, the lower cooling potential of the ambient air is used for sensible cooling of the supply air at the heat recovery unit in this system setup. This can be seen as a disadvantage compared to the ventilation cycle.

Table 13.5 Sequence steps of the process control in a *SAC* system indicating the steps shown in Fig. 13.33 and the respective heat source and sink

State change	Process sequence	Heat source/ -sink
$1 \rightarrow 2$	Outside air is drawn in and dried in the rotary dehumidifier. Since this takes place almost adiabatically, the air is heated by the released condensation heat of the water vapor and the bond heat at the sorbent	q_c
$2 \rightarrow 3$	The dehumidified supply air is cooled in counterflow to the exhaust air from the building using the heat recovery wheel	q_r
$3 \rightarrow 4$	The supply air is heated in an air heater, which is only used for heating. (It is not active in cooling mode)	
$4 \rightarrow 5$	Cooling of the supply air using evaporative cooling by a humidifier. Adjustable humidifiers provide additional control	$q_{e,1}$
$5 \rightarrow 6$	Absorption of the cooling load of the room in which heat caused by internal and external heat sources leads to an increase in air temperature in the building and the air is additionally enriched with water vapor	Q_0
$6 \rightarrow 7$	Removal of exhaust air from the building and humidification as close as possible to the saturation point in order to make maximum use of the cooling potential (evaporative cooling) of the exhaust air	$q_{e,2}$
$7 \rightarrow 8$	Heat absorption of the exhaust air (from the dried supply air) via the heat recovery wheel without any change in humidity. Thereby cold transfer to the supply air	q_r
$8 \rightarrow 9$	In the air heater, the actual input heat is supplied to the exhaust air. Heating to the required regeneration temperature reduces the relative humidity of the exhaust air	q_h
$9 \rightarrow 10$	The heated exhaust air flows through the drying wheel and absorbs the water bound in the sorbent material in the form of steam (regeneration of the sorbent). The warm regeneration air desorbs the wheel by adiabatic moisture absorption. The water vapor is discharged to the environment with the exhaust air	q_a

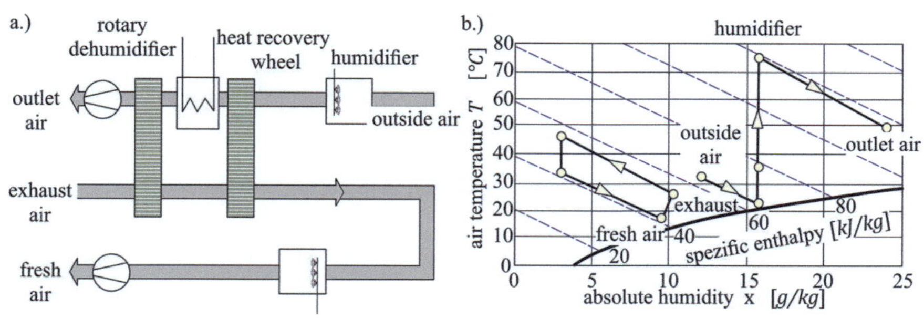

Fig. 13.35 Sorption assisted air conditioning (*SAC*) in the recirculation cycle; **a** arrangement of system components and **b** representation in the $T - x$ diagram

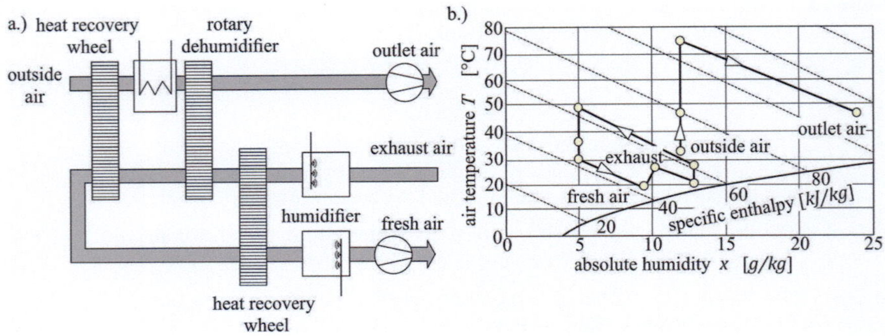

Fig. 13.36 Sorption-assisted air conditioning (SAC) in the Dunkle cycle; **a** arrangement of system components and **b** representation in the $T-x$ diagram

The respective advantages of the ventilation cycle and the recirculation cycle are combined in the Dunkle cycle concept, the structure of which is shown in Fig. 13.36a. Here, both the cooling potential of the exhaust air for the supply air cooling and the lower cooling temperature at the heat exchanger are utilized by humidifying the exhaust air, as can be seen from the corresponding $T-x$ diagram in Fig. 13.36b. However, a more complex system design results from an additional heat exchanger, which requires a higher capital expenditure.

A fundamental disadvantage of the two recirculation modes compared to the ventilation cycle is the limited supply of fresh air. In the recirculation and Dunkle cycle, fresh air enters the room to be conditioned only proportionally. Due to high comfort requirements, this can necessitate additional, usually costly, air treatment.

At this point it becomes clear that the design of suitable air conditioning systems must take into account the different possible ambient air conditions and the expected room cooling loads. In open processes, the key change of state to reduce supply air enthalpy is heat exchange via the heat recovery wheel. This is because in the ideal case the sum of the enthalpy changes of sorption and evaporation is zero ($\Delta h = 0$). Only at the wheel a significant enthalpy change occurs (cf. Figure 13.34). Assuming this ideal case, the cooling capacity ($\dot{Q}_{cold}$) is defined by the enthalpy change:

$$\dot{Q}_{cold} = \rho_{air} \cdot \dot{V} \cdot (h_1 - h_5) = \dot{m} \cdot (h_2 - h_3), \tag{13.34}$$

where ρ_{air} is the density of the air and $\dot{V}$ is the volume flow passing through the system. To increase the cooling capacity, isothermal sorption would be considerably more favorable than isenthalpy, since it already causes a significant enthalpy jump during sorption ($1 \rightarrow 2$) (compare Fig. 13.34).

To evaluate the performance of the SAC system, one can use a COP value with respect to the refrigeration—(COP_{cold}) as well as the air conditioning performance (COP_{ac}) as follows:

$$
\begin{aligned}
COP_{cold} &= \frac{\dot{m}_{freshair}}{\dot{m}_{reg}} \cdot \frac{\left(h_{amb} - h_{freshair}\right)}{\Delta h_{reg}} \\
&= \frac{\dot{V}_{freshair} \cdot \rho_{air}}{\dot{V}_{reg} \cdot \rho_{air}} \cdot \frac{(h_1 - h_5)}{(h_9 - h_8)} COP_{ac} \\
&= \frac{\dot{m}_{freshair}}{\dot{m}_{reg}} \cdot \frac{\left(h_{exhaust} - h_{freshair}\right)}{\Delta h_{reg}} \\
&= \frac{\dot{V}_{freshair} \cdot \rho_{air}}{\dot{V}_{reg} \cdot \rho_{air}} \cdot \frac{(h_6 - h_5)}{(h_9 - h_8)},
\end{aligned}
\quad (13.35)
$$

where the index "reg" denotes the regeneration air and amb stands for ambient. (Lavan et al., 1982) discuss the basics for the calculation of COP refrigeration in open air conditioning systems and trace them back to the calculation form for reversible refrigeration machines with the help of the introduction of Carnot temperaturesT^*. For this purpose, characteristic temperatures for evaporator, condenser and generator are determined and the SAC process is transferred to a chiller. The determination of the Carnot temperatures forms the transformation rule. According to this procedure, the following relationship results:

$$
COP_{cold} = \frac{1 - \frac{T_{condenser}}{T_{generator}}}{\frac{T_{condenser}}{T_{evaporator}} - 1} \text{ with } T_{condenser} = \frac{(h_{outlet} - h_{exhaust})}{s_{outlet} - s_{exhaust}}
$$

$$
\text{and } T_{generator} = \frac{\left(h_{reg,outlet} - h_{reg,inlet}\right)}{\left(s_{reg,outlet} - s_{reg,inlet}\right)}, \; T_{evaporator} = \frac{\left(h_{freshair} - h_{amb}\right)}{\left(s_{freshair} - s_{amb}\right)},
\quad (13.36)
$$

in which s denotes the specific entropy.

The COP_{ac} values determined in real installations are usually rather low, with values around 0.5. This is due to the high enthalpy difference in the regeneration air flow. In addition to the required drive heat, however, the value of COP_{AC} depends on other parameters. Studies on sorption behavior show maximum values of $0.85 - 1.05$, which can already be achieved with silica gel as sorbent (Collier, Cale, & Lavan, 1986). With an ideal sorbent material, values up to 1.4 can be realized. In addition to the sorbent materials and the system components, the achievable cooling performance is determined by a combination of passive adiabatic cooling (via the humidifiers) and active thermally driven cooling (sorptive dehumidification). Since the passive portion is not accounted for as an expense when determining the performance figures, these performance figures are not directly comparable with those of other systems. The consequence of this is that for combined cooling, the COP_{cold} and COP_{ac} values fall as the cooling capacity increases, since more heat must be invested as the capacity increases, and at

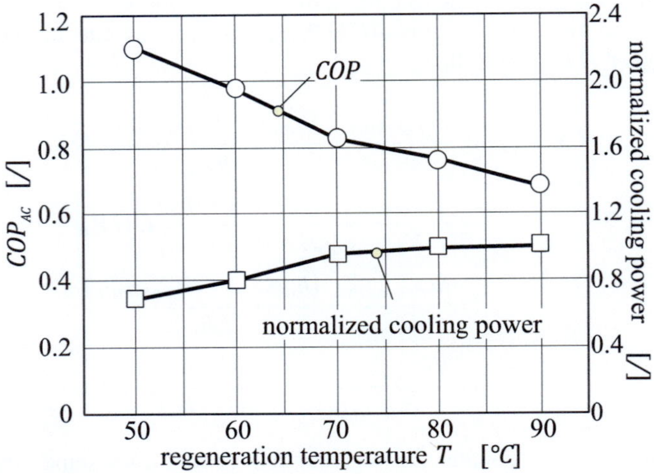

Fig. 13.37 Calculated coefficient of performance COP_{AC} and cooling capacity as a function of regeneration temperature at an ambient temperature $T_{amb} = 32\,°C$ and a humidity of $x = 12g/kg$ (Erpenbeck, 1999)

the same time the proportion of passive cooling falls. Figure 13.37 illustrates this with a calculation example.

13.7 Assessment of Solar Air Conditioning

The difference between solar generated cooling and thermal use of solar irradiation is that in cooling the solar generated energy is not used directly but rather serves as a drive for energy conversion. If the efficiency of the cooling or air-conditioning system is not to be determined by means of the Coefficient of Performance (COP), but rather the quality of the conversion of the heat used by solar thermal energy, then another variable must be defined. Based on the energy supplied to the system, the solar fraction is evaluated as Fraction of Non-Purchased Energy (FNP) (Ileri, 1997):

$$FNP = \frac{q_{Solar}}{q_{total}} \text{ with } q_{total} = q_{drive} + w_{elec} \tag{13.37}$$

It is important to note in Eq. (13.37) that q_{Solar} is the gross heat gain of the collector and not the solar heat needed to drive the system.

Thus, FNP is an obvious quantity to measure solar benefit for solar heating applications, since solar heat is used directly in the heating process. The savings on the primary energy source can thus be easily grasped. For a thermally driven cooling process, a modification of the FNP is required because the solar contributed heat is not directly determined by its contribution to the total energy input,

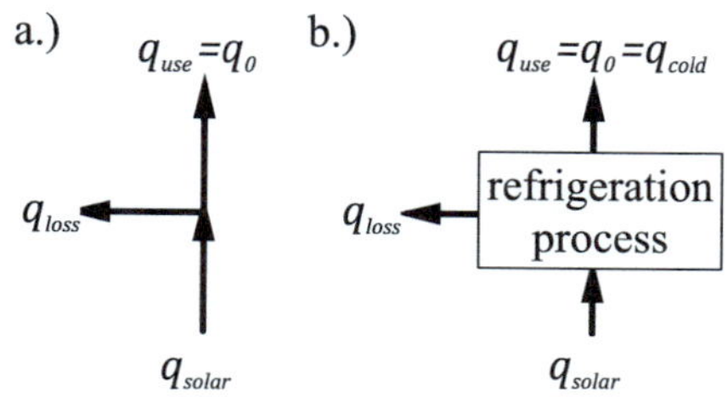

Fig. 13.38 Schematic representation of the energy flow with solar utilization in a heating system (**a**) and a thermally driven refrigeration system (**b**)

but by its contribution to the operation of the thermodynamic cooling process, as illustrated by Fig. 13.38.

If the solar energy would be accounted only as waste heat due to its thermal level in the cooling process, the FNP value would be constant and independent of whether the solar heat is used completely, partially or not at all. Therefore, a second quantity is introduced in a cooling process with the Solar Performance Coefficient (SPC), in which the efficiency of the solar used energy is defined as follows:

$$SPC = 1 - \frac{q_{aux}}{q_0}. \tag{13.38}$$

The quotient in Eq. (13.38) indicates how much specific auxiliary heat (q_{aux}) must be provided in relation to the cooling load (q_0). Thus, the SPC is a quantity independent of the amount of available solar heat. The solar benefit thus occurs only indirectly and is documented by the fact that with increasing efficiency of solar use q_{aux} decreases and as a consequence the SPC value increases. In the ideal case of heating (loss-free use of heat and neglect of electrical auxiliary energy), FNP and SPC are identical.

$$FNP = \frac{q_{solar}}{q_{total}} = \frac{q_0 - q_{aux}}{q_0} = 1 - \frac{q_{aux}}{q_0} = SPC. \tag{13.39}$$

The SPC value corresponds to another widely used metric to describe solar-powered systems, the solar fraction (η_{SF})ηSF, which is given in the following way:

$$\eta_{SF} = \frac{solar\ covered\ load}{total\ load\ to\ be\ covered} = 1 - \frac{load\ covered\ by\ auxiliary\ power}{total\ load\ to\ be\ covered}. \tag{13.40}$$

Equation (13.40) shows the identity of SPC and η_{SF}. Only their determination is different. The calculation of the solar fraction η_{SF} requires the knowledge of the used fraction of the solar heat, while the determination of the SPC value requires the magnitude of the auxiliary heat.

To save energy in building air conditioning, two basic requirements must be met. These are, on the one hand, the separation of cooling and dehumidification and, on the other hand, the limitation of air exchange due to hygienic requirements. Both requirements can only be fulfilled by combining previously known processes, which in many applications results in hardly justifiable equipment and thus investment costs.

Open thermal systems (SAC) offer the advantages of an energetically favorable separation of dehumidification and cooling at comparatively low drive temperatures. But the entire cooling load must be dissipated via air exchange (recirculating air cooling), since cooling of liquid media (chilled water network) is not possible (Schwaerzler & Rudischer, 1996). With high cooling loads, this restriction leads to air exchange rates that are far above the hygienically necessary level and significantly increase the energy consumption for ventilation. Static or "silent cooling", which is increasingly preferred for comfort air conditioning, can only be realized in combination with other processes. Another disadvantage is that the water used for humidification and thus cooling must meet air hygiene requirements, the cost-effective provision of which can be problematic, especially in southern countries. At locations with high air humidity, the effectiveness of SAC is thus limited.

The use of liquid sorbents (e.g. $LiCl$) in open sorption systems is the only process that offers the advantage that the (solar) energy can be stored loss-free and with high energy density by storing concentrated solution (Kessling, Laevemann, & Peltzer, 1998). In addition, in contrast to DEC with solid sorbents, carryover of odors from the exhaust to the supply air is excluded. Due to thermodynamic laws, the use of open systems is basically only possible for air conditioning applications, but not for the generation of low temperatures. The closed heat transformation processes require higher drive temperatures to achieve the same COP s, which necessitates the use of more efficient and thus more expensive collectors, depending on the location. However, they allow simultaneous use of solar energy for domestic hot water preparation or heating support, thus achieving higher efficiency of the overall system. On the other hand, they do not allow separation of cooling and dehumidification. Dehumidification can only be achieved by falling below the dew point, which requires correspondingly low temperatures. On the other hand, the chilled water provided by these systems is suitable for use in static cooling elements, allowing both "silent" cooling, which is perceived as more pleasant, and minimization of supply air flows to meet hygienic requirements. Absorption chillers are also capable of providing storage and freezing refrigeration, with higher drive temperatures requiring the use of concentrating collectors.

Another advantage of all thermal processes is that, in addition to solar-generated heat, they can also be driven by waste heat from technological processes as well as heat from combined heat and power plants. Especially in air-conditioning applications, the use of heat from cogeneration for cooling makes sense, since there is no need for heating in summer and the heat generated during power generation is available as surplus.

All thermal processes have a comparatively high water consumption for recooling, which is a significant application limitation in regions with water scarcity. In tropical regions, the effectiveness of evaporative cooling is limited by the high humidity.

In particular, the electrical energy requirements of the currently dominant compression refrigeration technology for cold generation/air conditioning and the associated costs to be incurred will increasingly establish alternative refrigeration processes in the future, as they enable increased integration of solar energy. In this context, absorption and adsorption refrigeration technology as well as the SAC process are suitable options. At present, it is not possible to predict which technology will become the most widespread in which area of application, since all processes have specific advantages and disadvantages that must be weighed against each other. Another decisive factor will be the extent to which the costs of the systems can be reduced and their reliability increased. The underlying physical-technological principles have been known for a long time and have been applied for decades. They are mostly used in plants with large outputs.

For coupling with solar energy, the systems still can be improved. Work is needed on component development as well as system integration (e.g. coupling with storage and back-up systems). In the area of SAC technology, the objectives are to increase the dehumidification performance and the lifetime of the sorption wheels, to use improved sorbents, to improve heat recovery and absorbers when using liquid sorbents, and to thermodynamically optimize the process control.

The work in the field of absorption refrigeration is aimed at the development of machines with low refrigerating capacity. The focus is on realizing the smallest possible temperature differences between internal and external media, improving the heat ratio and partial load behaviour, dispensing with (wet) cooling towers, and reducing size and costs. In addition, individual components, such as pumps, are not yet available for the low power range, or are only available to a limited extent. In the case of diffusion-absorption chillers, work is concentrating on (gas bubble) pump technology and on increasing the cooling capacity and heat ratio. Table 13.6 shows a spectrum of requirements and operating conditions of solar thermal driven processes for air conditioning.

Table 13.6 Requirements and operating conditions of some thermal processes for solar cooling or air conditioning: SE = single effect; DE = double effect; SE-DL = single effect/double lift; FP = flat plate collector; VTC = vacuum tube collector; PTC/LFC = parabolic trough collector/linear Fresnel collector; AC = air collector according to data from (Lävemann, 2002; Henning H.-M., 1996, Sumathy, Huang, & Li, 2002)

Process	Adsorption chillers	Absorption chillers	Sorption assisted air conditioning (SAC)
Refrigerant circuit	Closed	Closed	Open
Process principle	Cold water generation (dehumidification via dew point undershoot possible)	Cold water generation (dehumidification via dew point undershoot possible)	Dehumidification (sorptive) and evaporative cooling
Sorbent type	Solid	Liquid	Solid/liquid
Typical substance systems	Water silica gel, ammonia salt	Water-$LiBr$, ammonia-water	Solid: H_2O-silica gel, H_2O $LiCl$ cellulose Liquid: H_2O $LiCl$, H_2O -$CaCl_2$
Market available sizes	$50-430kW$	$35kW-5MW$	$20-350kW$
Heat ratio COP	$0.3-0.7$	$0.5-0.75$(SE) $1.1-1.2$ (DE)	Solid: $0.5-1$ Liquid:>1
Typical drive temperatures [$°C$]	$60-95$	$65-95$(SE-DL) $80-110$(SE) $130-160$ (DE)	Solid: $45-95$ Liquid:$45-70$
Solar collector technology	FP, VTC, (PTC/LFC)	FP, VTC, PTC/LFC	FP, AC
Remarks	Quasi-stationary operation due to 2 chambers, large mass and volume of equipment		Not suitable for cold water grid, good storage possibility with liquid version (due to loss-free chem. storage)

References

ANSI/ASHRAE Standard 62.1. (2007). *Ventilation for Acceptable Indoor Air Quality.* American Society of Heating,Refrigerating and Air-Conditioning Engineers, Inc., Atlanta, GA, USA.

ASHRAE. (2012). *Handbook—HVAC Systems and Equipment.* (ASHRAE, Ed.) Atlanta Georgia 30329.

Besagni, G., Mereu, R., & Inzoli, F. (2016). Ejector refrigeration: A comprehensive review. *RenewableandSustainableEnergyReviews, 52,* 373–407. doi:https://doi.org/10.1016/j.rser.2015.08.059.

Collier, R. K., Cale, T. S., & Lavan, Z. (1986). *Advanced Desiccant Materials.* Chikago, Il, USA: Gas Research Institute.

DIN EN 13779. (2009). *Lüftung von Nichtwohngebäuden—Allgemeine Grundlagen und Anforderungen für Lüftungs- und Klimaanlagen und Raumkühlsysteme.* Berlin: Beuth Verlag GmbH.

Erpenbeck, T. (1999). *Sorptionsgestützte solare Klimatisierung—Experimentelle und theoretische Analyse von Potential und Systemverhalten.* Universität Karlsruhe.

Goetzler, W., Zogg, R., Young, J., & Johnson, C. (2014). *Energy Savings Potential and RD&D Opportunities for Non-Vapor-Compression HVAC Technologies.* Office of Energy Efficiency and Renewable Energy, Building Technologies Office. Burlington, MA 01803: U.S. Department of Energy. Retrieved 10 4, 2022, from http://www.buildings.energy.gov

Grigull, U., Staub, J., & Schiebener, P. (1990). *Steam Tables in SI-Units—Wasserdampftafeln.* Springer-Verlagdima gmbh.

Henning, H. -M. (1996). Aktive solarthermische Systeme für die Gebäudeklimatisierung. In A. Marko, P. Braun, & W. Roth (Eds.), *Thermische Solarenergienutzung an Gebäuden.* Berlin: Springer Verlag.

Henning, H.-M. (Ed.). (2004). *Solar-assisted air-conditioning in buildings—A handbook fpr planners.* Springer.

Herold, K. E., Radermacher, R., & Klein, S. A. (1996). *Absorption Chillers and Heat Pumps.* CRC Press.

Ileri, A. (1997, April). A discussion on performance parameters for solar-aided absorption cooling systems. *Renewable Energy, 10*(4), 617–624.

Jähnig, D., & Pollerberg, C. (2010). *Solarthermische Kälteerzeugung mit Parabolrinnenkollektorsystem und Dampfstrahlkältemaschine (in German).* Division of Energy and Environmental Technologies. Radetzkystraße 2, 1030 Wien, Austria: Bundesministerium für Verkehr, Innovation und Technologie.

Joemann, M., Pollerberg, C., Kauffeld, M., Oezcan, T., Bauer, I., Grave, H., & Dietzmann, T. (2016). *Process steam and cooling generation with solar collectors, steam jet chiller and latent heat storages (in German).* Osterfelderstraße 3, D46047 Oberhausen, Germany: Fraunhofer-Institut für Umwelt, Sicherheits- und energietechnik (UMSICHT).

Jungnickel, H., Agsten, R., & Kraus, W. E. (2001). *Grundlagen der Kältetechnik* (3. stark bearb. Auflage ed.). Verlag Technik /Huss Medi.

Kessling, W., Laevemann, E., & Peltzer, M. (1998, March). Energy storage in open cycle liquid desiccant cooling systems. *International Journal of Refrigeration, 21*(2), 150–156.

Lamp, P., & Ziegler, F. (1998, March). European research on solar-assisted air conditioning. *International Journal of Refrigeration, 21*(2), 89–99.

Lavan, Z., Monnier, J. B., & Worek, W. M. (1982). Second law analysis of desiccant cooling systems. *ASME Journal Solar Energy Engineering, 104*, 229–236.

Lävemann, E. (2002). Betriebserfahrungen mit einer solar gestützten Flüssigsorptionsanlage in einem Bürogebäude in Amberg. *Tagungsband 2.Symposium Solares Kühlen in der Praxis.* Stuttgart: Hochschule für Technik.

Lewis, J., Chaer, I., & Tassou, S. (2007). *Reviews of Alternative Refrigeration Technologies-Fostering the development of technologies &practises to reduce the nergy inputs to the refrigeration of food.* Centre for Energy and Built Environment Research, School of Engineering and Design. London, United Kingdom: Brunel University.

Little, A., & Garimella, S. (2016). A critical review linking ejector flow phenomena with component- and system-level performance. *International Journal of Refrigeration, 70*, 243–268. https://doi.org/10.1016/j.ijrefrig.2016.05.015

Reimann, R. C. (1984, December 10). Advanced absorption heat pump cycles. *Proc. DOE/ORNL Heat Pump Conf.*

Schwaerzler, C., & Rudischer, R. (1996). Solarthermische Pilotanlage zur Klimatisierung. *Tagungsband Sechstes Symposium Thermische Solarenergie*, pp. 240–244.

Schweigler, C., Hellmann, H. -M., Preissner, M., Demmel, S., & Ziegler, F. (1997). Operation and Performance of a 350 kW (100 RT) Single-Effect/Double Lift Absorption Chiller. In P. Treffinger & H. Hertlein (Eds.), *Solar Sorptive Cooling* (pp. 93–106). Köln,16./17.10: Workshop des Forschungsverbund Sonnenenergie.

Sumathy, K., Huang, Z. C., & Li, Z. (2002, February). Solar absorption cooling with low grade heat source — a strategy of development in South China. *Solar Energy, 72*(2), 155–165.

Zeyghami, M., Goswami, D. Y., & Stefanakos, E. (2015). A review of solar thermo-mechanical refrigeration and cooling methods. *Renewable and Sustainable Energy Reviews, 51*, 1428–1445. https://doi.org/10.1016/j.rser.2015.07.011

International Systems of Units, Fundamental Constants and Conversion Factors

A

In this appendix a number of units, symbols, conversion factors and definitions of quantities used in the previous chapters are listed. However, symbols are often ambiguous because the number of Latin and Greek letters is limited, so in different scientific, engineering, and cultural context the same letters may be used for different physical quantities. The Greek letter λ—amongst many other usage—can represent the optical wavelength, but also the heat conductivity of a material. On the other hand, different letters are sometimes used by different authors for the same quantity. Therefore, readers should be aware that they should always try to understand the context when using a certain formula or a symbol for a quantity (Tables A.1, A.2, A.3, A.4, A.5, A.6, A.7 and A.8).

A more comprehensive list of definitions and solar terminology can be found in the relevant standards of the technical committees ISO TC180 (ISO 9488, 2021) and IEC TC 117 (IEC TS 62862-1-1, 2018).

Table A.1 SI units

Quantity	Unit	Symbol
Amount of a substance	mol	n
Electric current	A	I
Force	N	F
Length	m	L, $\lceil \vec{x} \rceil$
Radiation intensity	W/sr	I_λ
Magnetic field strength	T	$\vec{B}$
Mass	kg	m
Molar mass	g/mol	$\mathcal{M}$
Temperature	°C, K	T
Time	s, h, a	t

R. Stieglitz and W. Platzer, *Solar Thermal Energy Systems*,
https://doi.org/10.1007/978-3-031-43173-9

Table A.2 Prefixes of engineering physical quantities

Prefix	Symbol	Multiplier
Exa	P	10^{+18}
Peta	E	10^{+15}
Tera	T	10^{+12}
Giga	G	10^{+9}
Mega	M	10^{+6}
Kilo	k	10^{+3}
Milli	m	10^{-3}
Micro	μ	10^{-6}
Nano	n	10^{-9}

Table A.3 Physical constants

Quantity	Symbol	Value
1st Planck radiation constant	h	$6.62607015 \times 10^{-34}$ (J s)
2nd Planck radiation constant	$c_2 = (h/k_B)c$	1.438775×10^{-2} (m K)
Avogadro number	A_V	$\left(6.022147 \times 10^{23}/\text{mol}\right)$
Boltzmann constant	k_B	1.380649×10^{-23} J/K
Elementary charge	e	1.602175×10^{-19} A s
Gas constant	R	8.314462618 J/(mol K)
Loschmidt number	N_0	2.68678×10^{25} m^{-3}
Magnetic permittivity	μ_0	$4\pi \times 10^{-7}$ H/m
Speed of light	c	2.99792×10^{8} m/s
Stefan-Boltzmann constant	σ	$5.670374419 \times 10^{-8}\ \text{W}/\left(\text{m}^2\text{K}^4\right)$
Wien's displacement constant	λ_{max}	2897.8 (µm K)
Vacuum permittivity	ε_0	$8.854 \times 10^{-12}\ \left(\text{A}^2\,\text{s}^4\right)/\left(\text{kg}\,\text{m}^3\right)$
View half-angle sun earth	θ_s	$\pm 0.267° = \pm 4.65$ mrad

Table A.4 Energy conversion factors

Start unit	Target units				
	(GJ)	(TWh)	(toe)	(tce)	(MMBTU)
1 Gigajoule GJ	1	0.278	0.0246	0.034	0.947
1 Megawatthour MWh	3.6	1	0.0861	0.123	3.410
1 toe (tons of oil equivalent)	41.869	11.63	1	1.429	39.659
1 tce (tons of coal equivalent)	29.308	8.14	0.7	1	27.761
1 Mio. Brit. therm. unit MMBTU	1.056	0.293	0.0252	0.036	1

Table A.5 Dimensionless physical parameters

Parameter	Abbreviation	Context
Biot number	Bi	Heat transfer
Boltzmann number	Bo	Heat transfer
Cavitation number	Ca	Two-phase flow
Convection-conduction number	Cn	Heat transfer
Courant number	C	Momentum transfer
Damköhler number	Da	Heat transfer
Eckert number	Ec	Heat transfer
Euler number	Eu	Momentum transfer
Fourier number	Fo	Heat transfer
Froude number	Fr	Momentum transfer
Galilean number	Ga	Two-phase flow
Grashof number	Gr	Heat transfer
Knudsen number	Kn	Vacuum
Linear scattering albedo	ω	Heat transfer
Martinelli parameter	X	Two-phase flow
Nusselt number	Nu	Heat transfer
Peclet number	Pe	Heat transfer
Prandtl number	Pr	Heat transfer
Rayleigh number	Ra	Heat transfer
Reynolds number	Re	Momentum transfer
Richardson number	Ri	Heat transfer
Stark number	N	Heat transfer
Stokes number	Sto	Momentum transfer
Strouhal number	Str	Momentum transfer
Thermal activity ratio	TAR	Heat transfer
Weber number	Wb	Momentum transfer

Table A.6 Specific optical quantities

Quantity	Symbol	Unit
Absorptance/absorptivity	α	–
Geometric concentration ratio	C_g	–
Global horizontal irradiance	G_h	W/m^2
Diffuse irradiance	G_{dh}	W/m^2
Direct/beam normal irradiance	G_{bn}	W/m^2
Direct/beam horizontal irradiance	G_{bh}	W/m^2
Emissivity/emittance	ε	–
Incidence angle	θ_i	rad, °
Intercept factor	γ	–
Optical concentration ratio	C, C_{opt}	–
Optical error	σ_{opt}	mrad
Plane angle	θ, φ	rad, °
Radiance	L_e	$W/(sr\,m^2)$
Irradiance	E_e	W/m^2
Radiant intensity	$I_{e,\Omega}$	$W/(sr)$
Radiant power/flux	Φ	W
Reflectance	ρ	–
Reflectivity	R, ρ	–
Reflection coefficient	r	–
Refractory index	$n\ (real)$	–
	$k\ (imaginary)$	–
Solar absorptivity	α_s	–
Solar multiple	SM	–
Solid angle	Ω	sr
Thermodynamic temperature	T	K
Transmittance	τ	–
Transmission coefficient	t	–
Transmissivity	T, τ	
Total extinction coefficient	μ	1/m
Spectral atmospheric scattering coefficient	β_λ	1/m
Wavelength	λ	m
Wave number	κ	–

Table A.7 Physical and solar quantities: symbols and units

Quantity	Symbol	Unit
Acceleration/gravity	a, g	m^2/s
Accommodation coefficient	α	–
Adiabatic index. Laplace coefficient	κ	–
Air-mass factor	AM	–
Anergy	B	J
Angular frequency	ω	$1/s$
Area	A	m^2
Azimuth angle of surface normal of A	γ	rad,°
Declination angle earth-sun	δ	rad,°
Density	ρ	kg/m^3
Dipole moment	m	Cb m
Dynamic viscosity	μ	kg/(m s), $(N\ s)/m^2$, (Pa s)
Electric field strengths	E	V/m
Electric potential	U, ϕ	V
Enthalpy	H	J
Entropy	S	J/K
Exergy	Ex	J
Frequency	f	Hz, (1/s)
Friction factor	f_D	–
Gebhart factor	B_{ij}	–
Gibbs free energy	G	J/mol
Geographical latitude	Φ	rad,°
Spectral solar irradiance	E_λ	$W/(\mu m\ m^2)$
Global irradiance (direct beam, diffuse)	G, (G_b, G_d)	W/m^2
Global irradiation (beam, diffuse)	H, (H_b, H_d)	kWh/m^2, J/m^2
Heat capacity ($p = const$, $v = const$)	c_p, c_v	J/(kg K)
Heat flux density	$\dot{q}$	W/m^2
Heat transfer coefficient	h, U	$W/(m^2\ K)$
Hour angle (solar position)	ω	rad,°
Humidity (absolute)	x	kg/kg
Incidence angle	θ_i	rad,°
Internal energy	U	J
Kinematic viscosity	ν	m^2/s, (N m s)/kg, (J s)/kg
Load coefficient	ψ	–
Lower calorific value	LHV	J/kg

(continued)

Table A.7 (continued)

Quantity	Symbol	Unit
Magnetic field strength	B	T
Mean free path	ℓ	m
Optical depth	δ_λ	m
Plane angle	θ, φ	rad,°
Poisson ratio	ν	–
Power	P	W, J/s
Pressure	p	Pa, N/m^2, $kg/(m^2s)$
Shear rate	$\dot{\gamma}$	1/s
Shear stress	τ	Pa
Solid angle	Ω	sr
Specific heat	c_p	J/(kg K)
Stream function	Ψ	m^2/s
Sun elevation angle/solar altitude	α_s	rad,°
Sun solar surface temperature	T_S	°C, K
Thermal conductance	G	W/K
Thermal conductance per unit area	Λ	$W/(m^2K)$
Thermal conductivity	k	W/(m K)
Thermal diffusivity	a	m^2/s
Thermal effusivities	b	$J/(m^2K\sqrt{s})$
Thermal power	$\dot{Q}$	W
Thermal resistance	R_{th}	K/W
Thermal resistance per unit area	R_{th}	$(m^2K)/W$
Temperature	T	°C, K
Torque	M	N m
Turbulence intensity	ϕ	m/s
Vapour quality	α, x	–
View factor	F_{ij}	–
Void fraction	μ, α	–
Volume	V	m^3
Volumetric expansion coefficient	β	–
Vorticity	$\vec{\omega}$	1/s
Wavelength	λ	m
Wave number	κ	–
Work, energy, heat	W, E, Q	J
Young modulus	E	Pa
Zenith angle	θ_z	rad,°

Table A.8 Solar quantities: symbols and definitions

Quantity	Symbol	Unit	Definition
Air-mass factor	AM	–	Measure of the length of the path through the atmosphere to sea level traversed by direct sun rays with reference to the path length along the vertical
Direct normal irradiation	DNI	kWh/m^2a	Annual sum of direct solar irradiation on the surface always facing the sun, used for a characterization of a location
Global irradiance (direct beam, diffuse)	G, (G_b, G_d)	W/m^2	Incoming radiant power on a horizontal plane
Solar azimuth angle	γ_s	rad,°	Solar position angle w.r.t. meridian plane
Solar altitude	α_s	rad,°	Solar position angle w.r.t. horizontal plane
Solar zenith angle	θ_z	rad,°	Solar position angle w.r.t. zenith, zenith angle equal to 90° minus altitude angle
Sky temperature	T_{sky}	°C, K	Equivalent blackbody radiation temperature of the atmospheric longwave radiation received at a horizontal surface
Collector capacitance rate or dimensionless mass flow rate	γ	–	Collector mass flow rate related to thermal loss (see Chap. 6 Eq. (6.25))
Collector efficiency	η	–	Ratio of the thermal power removed by the heat transfer fluid, to the product of a defined collector area (gross, absorber or aperture) and the solar irradiance on the collector
Collector efficiency factor	F'	–	Fraction of thermal resistance from absorber to ambient of resistance fluid to ambient
Collector flow factor	F''	–	Factor dependent on the mass flow through the collector which defines the temperature distribution within the collector
Collector heat removal factor	F_R	–	$F_R = F' \cdot F''$
Collector tilt angle	β	rad,°	Tilt angle between collector and horizontal plane
Effective collector transmittance-absorptance product	$(\tau\alpha)_e$	–	Optical collector efficiency including secondary heat gains by radiation absorbed e.g. in the cover glass and secondary concentrators

(continued)

Table A.8 (continued)

Quantity	Symbol	Unit	Definition
Collector peak optical efficiency	$\eta_{opt,0}$	–	Collector optical efficiency when direct solar radiation is perpendicular to the collector aperture plane
Intercept factor	γ	–	The ratio of solar radiation reflected or refracted towards the absorber to geometrically captured by it, considering the complete receiver aperture area (point-focus solar systems) or length (line-focus systems)
Acceptance angle of a concentrating collector	θc	rad,°	It is the angular range (2 * θc) over which all parallel rays intercepted by the collector hit the absorber without moving all or part of the collector
Transversal angle of incidence	θ_T	rad,°	Angle between the collector aperture normal and the projection of the sun beam into the transversal plane (plane perpendicular to the collector axis)
Longitudinal angle of incidence	θ_L	rad,°	Angle between the aperture normal and the projection of the sun beam into the longitudinal plane (plane defined by the collector axis and the collector aperture normal
Longitudinal solar angle	θ_{LS}	rad,°	Angle between the sun beam vector and the projection of the sun beam into the transversal plane
Collector axis		–	This term is defined for line-focus solar thermal collectors. Straight line resulting from the intersection of the collector aperture plane and a plane containing the linear receiver and perpendicular to the collector aperture plane
Collector loop		–	Set of line-focus solar thermal collectors assembled in series, in such a way that the same heat transfer fluid mass flow is circulating through the absorber tube of each. A loop is composed of one or more collector rows connected in series
Heat transfer fluid	HTF	–	Fluid used to carry heat from the collector/receiver ta heat exchanger or storage

(continued)

Table A.8 (continued)

Quantity	Symbol	Unit	Definition
Solar field/collector array		–	Group of solar collectors that are closely connected in series, in parallel or in combination of both modes, with one hydraulic input and one hydraulic output
Tracking collector		–	Solar collector that moves to follow the apparent motion of the sun during the day, rotating about one or two axes
Photovoltaic thermal solar collector	PVT	–	A hybrid device used to convert solar radiation to thermal and electrical energy
Collector aperture area	A_a	m^2	Maximum projected area which accepts the solar radiation
Collector gross area	A_g	m^2	The area of the flat surface defined by the outer perimeter of the collector, including framing and gaps
Gross plant electricity production	$E_{el,gross}$	J, MWh	Electricity produced by the plant generating equipment
Net plant electricity production	$E_{el,net}$	J, MWh	Electricity delivered to the grid
Storage hours		h	Ratio between storage capacity, in MWh_{th}, and maximum demand thermal power, in MW
Equivalent full load hours	$EFLH$	h	Annual energy generation divided by rated power as a measure of load matching
Capacity factor	CF	–	The ratio of gross plant heat or electricity production in a year divided by the product of nominal power of the plant and the number of hours in a year
Solar fraction/solar share	SF	–	Ratio of the net plant heat or electricity production from solar radiation to the total net plant heat or electricity production over a given period of time
Solar multiple	SM	–	Ratio of the net thermal power of the solar field at design point, to the rated thermal power required for maximum demand (e.g. of a turbine)
Auxiliary energy	E_{aux}	J, MWh	Electricity consumed by pumps, fans and controls in a solar thermal system
Supplementary energy	E_{sup}	J, MWh	Energy, other than solar, used to supplement the output provided by the solar system

References

IEC TS 62862-1-1. (2018). *Solar thermal electric plants—Part 1–1: Terminology.* IEC.
IEC TS 62862-1-1. (2018). *Solar thermal electric plants—Part 1–1: Terminology.* IEC.

Measurements of Thermal Quantities

B

The permanent determination of thermal state variables, such as temperature, mass flow/flow rate, heat flux, pressure or, if applicable, the two-phase fraction of the heat carrier fluid circulating in the various sections (for example, collectors, storage tanks, heat engines, ...) of a solar thermal plant form the backbone for controlling and regulating the system. The operation and control of the systems, in turn, requires a high degree of automation and thus process control technology, with the help of which operating procedures are documented and operational reliability is ensured. The central process data are recorded by measuring devices. Since the recorded measured values are the basis for the execution of actions, such as the change of the pump capacity or the adjustment of a valve position, the following requirements are placed on the measurement devices of the process data:

- high accuracy, speed and time constant, stability, drift, if necessary self-calibration,
- robustness, simple function/operation, maintenance requirements,
- verifiability even without test bench, reproducibility,
- self-monitoring with error message if necessary self-diagnosis,
- communication capability.

Before going into the individual physical principles of recording thermohydraulic state variables such as volume flow, pressure or phase fraction, it is useful to recall the basic features of a measurement.

B.1 Measurement Requirements, Measurement Chain and Errors

A measurement of any physical quantity consists of three elements, the measurement set-up, the measurement request and the quantification of the measurement error.

R. Stieglitz and W. Platzer, *Solar Thermal Energy Systems*,
https://doi.org/10.1007/978-3-031-43173-9

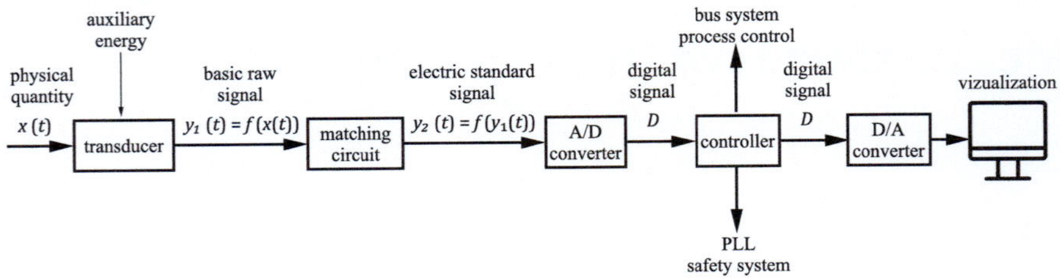

Fig. B.1 Schematic structure of a measuring chain for recording a state variable

Measurement Set-up

In most cases, not a single device is used to record a measured value, but several elements. The elements can either be structurally combined in one unit or arranged separately (e.g. temperature sensor stationary installed in the pipeline and measuring device mobile). The path of the measurement signal to be acquired from the recording of the measured variable to the provision to the output device or the process control system is illustrated in Fig. B.1.

Basically, every measurement setup—regardless of the type of measurement—consists of the following components:

- Transducer (sensor, probe): The quantity to be measured (measurand) is detected by the transducer and converted into an electrical signal, or a property of the transducer that can be used in an electrical circuit (e.g. resistance, capacitance) changes due to the measurand. Sensors make use of a special physical operating principle which must meet certain requirements and is therefore subject to limitations. They are the main distinguishing feature of the measuring devices for the respective state variable and will be discussed in the following sections. In the case of passive sensors, an external auxiliary energy is also required, while active sensors provide an electrical signal by themselves.
- Matching circuit (amplification): The matching circuit has the task of matching the electrical output signal of the sensor in terms of voltage/resistance to the downstream measuring element (often an A/D converter) and is usually designed as an amplifier providing standard signals in the form of electric voltages/currents.
- A/D converter: In order to be able to further process or store the signals recorded by the sensor and amplified by the matching circuit, these are converted into digital signals by the A/D converter. The main parameters of an A/D converter are sampling rate and resolution, for example 8, 10, 12, 16, 20 or 24 bits. Another parameter is the input voltage range, i.e. the limits within which the signal to be converted can be processed (e.g. 0–10 V).
- Data processing: Before signals are stored or forwarded to a controller, the data must be processed in a suitable manner; for example, the digital signals must be translated into the bus protocol for coupling to a bus system or a programmable logic controller.

- D/A converter: With the help of the D/A converter, the digital signals are converted back into analogue signals (reverse process to A/D conversion). The analogue signals then obtained can in turn be displayed on the screen or output to another display type.

Measurement Requirements
The requirements for the measurement of a process variable contains:

- Accuracy: The accuracy of a measurement indicates how far the measured value x_m deviates from the true value x_t. It is specified in the form of a measurement uncertainty (absolute or relative error), whereby a confidence level should be specified for the measurement uncertainty.
- Speed: The speed describes how quickly the measured value follows the signal to be measured. The speed is often specified by the time constant of the measuring device. The most common is the use of the time constant T, which indicates after which time 63% of the final value of the output signal is reached after a step change of the input signal.
- Reproducibility: Reproducibility of a measurement is a fundamental requirement that guarantees traceability and thus requires complete documentation of the measurement set-up and performance.
- Simplicity: The simplicity of a measurement is usually in direct contradiction to the first three basic requirements. The simplicity of a measurement setup usually also reflects the investment costs. Therefore, especially from an economic point of view, the simplicity of a measurement must also be considered.

Measurement Uncertainty
The aim of a measurement is to determine the true value x_w. In every measurement, the measured value obtained always corresponds to the true value only up to a finite difference, this is referred to as measurement deviation or error. The reasons for measurement uncertainties are multi-fold, they are partly due to the measuring device or to environmental influences such as ambient temperature, air pressure or humidity. Also, the correct application as well as the ability to read a measured value correctly are involved. Every measurement is subject to uncertainties (error-prone). Therefore, when specifying the measured value, a statement about the error must be given in addition to the numerical value and the unit. Furthermore, a confidence interval is assigned to the respective uncertainty:

$$\underbrace{x(t)}_{\substack{\text{measurement}\\\text{result}}} = \underbrace{x_m(t)}_{\substack{\text{measurement}\\\text{value}}} \pm \underbrace{\Delta x(t)}_{\substack{\text{measurement}\\\text{uncertainty}\\\text{(confidence interval)}}} \quad . \tag{B.1}$$

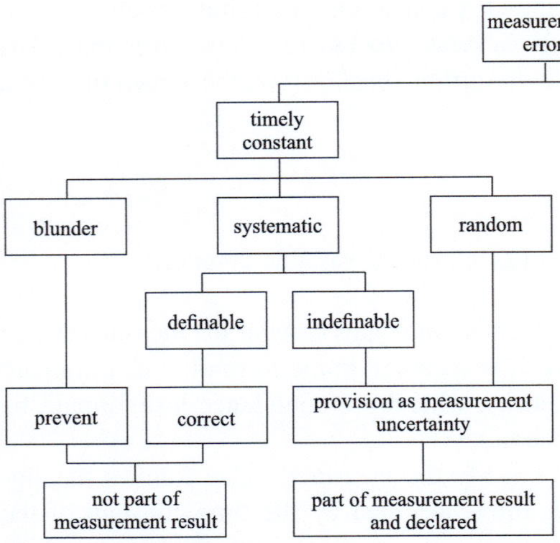

Fig. B.2 Categorization of different measurement error types

Basically, a distinction is made between time-constant and time-dependent errors. In the case of time-constant errors, a distinction is made between blunder measurement errors, systematic errors and random errors. Time-dependent errors, on the other hand, are divided into dynamic errors and drift. However, only random errors and systematic errors, which cannot be determined, are indicated as measurement uncertainty in the result. Figure B.2 outlines the categorization of different measurement error types.

Time-dependent measurement errors are:

- Dynamic errors: A dynamic error arises from the time behaviour of the measuring device in which the output variable follows the input variable with a time delay. The dynamic error is the difference between the measured quantity and the measured value resulting from this time delay.
- Drift: Drift is the maximum change of a metrological characteristic, e.g. the zero point under reference conditions within a defined time interval and is therefore also referred to as the long-term stability of a measuring instrument.

Time constant measurement errors can be divided into three classes:

- *Blunder errors* are caused by improper or incorrect use of a measuring device, such as the measured variable not being within the measuring range or the measuring device being incorrectly installed/connected.
- *Systematic errors* have a certain magnitude and sign, i.e. they are either positive or negative. In repeat measurements, they always occur with the same magnitude and sign. The repeat measurement does not provide any information about systematic deviations. Only a comparison measurement with another measuring

instrument of higher accuracy reveals systematic deviations. The comparative measurement makes systematic deviations detectable and correctable. Examples of such measurement errors are aging (long-term stability), offset errors of an amplifier or non-calibrated measuring instruments.

- *Random errors* are unpredictable in size and amount. A correction is not possible. However, frequent repetition of the individual measurement or manufacturer's specifications allow an assessment of the measured value. They can be described with statistical methods. Examples of random errors are imperfections of the measuring device (deformation, wear, adjustments), or indeterminable environmental influences such as temperatures, vibrations or contamination.

Specifications of the Measurement Uncertainty (Confidence Level)

The measurement uncertainty Δx is a measure of how "good" a measurement result is. It is composed by geometric addition of the (unknown) systematic contribution Δx_s and the random contribution Δx_r. Known systematic deviations (errors) should be considered and compensated by correction. Unknown systematic deviations us can be taken from manufacturer's documentation or must be estimated. Random deviations Δx_r can be determined by the standard deviation of a repeat measurement series or taken from data sheets. The measurement uncertainty is added to the measurement result x as a deviation range with the associated confidence level $\Delta x(t)$.

$$\underbrace{x}_{\substack{\text{measurement}\\\text{result}}} = \underbrace{\overline{x}}_{\substack{\text{mean value of}\\\text{measurements}}} \pm \underbrace{\Delta x}_{\substack{\text{measurement}\\\text{uncertainty}\\\text{(confidence interval)}}} \quad \text{with } \Delta x = \sqrt{\Delta x_s^2 + \Delta x_r^2} \tag{B.2}$$

For a single measurement, the measurement uncertainty can be taken from data sheets or from already existing measurement series. In some cases, a confidence level is already specified in the data sheet.

B.2 Temperature Measurements

B.2.1 Classification of Temperature Measurement Techniques

Temperature is one of seven basic values in the current SI-System of Units and at the same time, probably the most important parameter in measurement technology; it accounts for 75–80% of the worldwide sensor market (Childs et al., 2000). The range of techniques for the measurement of temperature is extensive, utilizing such diverse phenomena as thermoelectricity, temperature dependent variation of the resistance of electrical conductors, fluorescence and spectral characteristics. Additionally, the requirements for a temperature measurement may allow direct contact with the medium. Alternatively, this may not be possible or desirable and a non-invasive method may be used.

For convenience, the various measurement techniques can be classified into three categories depending on the nature of contact which exists between the measuring device and the solid, liquid or gaseous medium of interest.

- *Methods with direct contact sensor-medium (invasive)*: in this type of method, the sensor tends to locally equilibrate itself with the medium. If there is perfect adiabaticity of the sensor with the environment, its temperature is equal to that of the medium. However, in thermometric devices, this adiabaticity is usually not perfect.
- *Methods with contact without physical connection with the environment (semi-invasive)*: in some cases, the temperature readings are carried out using an optical mean therefore no physical connection exists between the sensor and the environment. In this category, we can find surface temperature measurement with deposited thermosensitive material such as liquid crystals or photoluminescent salts as e.g. thermographic phosphors or heat sensitive paints.
- *Methods without contact (non-invasive)*: in this method sensors are far from the medium. Despite there are still interactions between them, the sensor is no more in equilibrium with the medium. Such methods are essentially based on radiative heat transfer.

Before discussing the individual techniques, some general preparation aspects should be answered. The process of measuring temperature—inserting the sensor and taking a reading—could not in principle be simpler, but in practice there are many reasons why the reading may not be correct. This is a list of important points to consider when measuring temperature and the subsequent fundamental questions should be answered before selecting a technology:

- Does the probe make proper contact? How the ambient looks like (presence of electric/magnetic/high frequency fields, heat transfer or radiation transfer)?
- Can the sensor withstand the environment? Maximum temperature or pressure range and appearing gradients which cause likely drift or failure because of stresses due to expansion and contraction, contamination, oxidation or corrosion.
- Can we use a remote sensing technique? And how is the sensor calibrated?

Also, the application temperature measurements can be roughly divided in three categories:

- precision temperature measurements for scientific and basic research,
- technical temperature measurements for measurement and control technology,
- temperature monitoring using temperature indicators.

B.2.2 Temperature Measurement by Thermal Expansion

The expansion of materials of gases, liquids and solids depend on temperature and thus can be utilized to measure temperature. Fluid thermometry covers a range of devices from the constant volume and constant pressure gas thermometers used in calibration to liquid-in-glass thermometers. The expansion of solids is used in devices such as bimetallic strips. To use this effect for temperature measurement in practice, the specific properties of the material have to be taken into account. Considering a solid body, the length change (ΔL) of a bar exposed to a temperature change (ΔT) as a first approximation is proportional to the bar length (L):

$$(T_{meas} - T_0) = \Delta T = \frac{\Delta L}{\alpha L} + C_1 \cdot (T_{meas} - T_0) + C_2 \cdot (T_{meas} - T_0)^2, \quad \text{(B.3)}$$

in which T_0 denotes the reference temperature and C_1 and C_2 the calibration constants accounting for non-linear material behaviour.

The proportionality factor α (linear thermal expansion coefficient) relates to a material specific property. The integration of this equation along the length at a given temperature, yields the length of the bar at the temperature T. Since the proportionality factor a can only be considered as linear over small temperature ranges, higher order terms must be included in the calculation for larger temperature differences. The technical application of this sensor principle leads to bar and bimetal thermometers, being preferential installed in industrial applications as local indicators or emergency switches.

The dependence of a liquid volume on temperature can be utilized analogously, while α is replaced by a cubic expansion coefficient β, which is also a fluid/gas property. Liquid filled thermometers are encountered as glass thermometers (clinical thermometers, filament thermometers) and used for local temperature monitoring of liquids, gases and steam in pipelines and tanks.

A variant is the liquid/gas filled spring-loaded thermometer, consisting of a capillary tube filled with liquid and placed in a metal housing. Temperature changes produce a pressure change transmitted via a membrane to an elastic, deformable spring. In case of gas pressure thermometers, the ideal gas equation is used to evaluate the temperature relationships of the gas. The pressure thermometers are used for local measurements and as temperature indicators, e.g., in machines, however, they that the measurement body is completely surrounded/immersed in the medium whose temperature is being measured.

The accuracy of gas thermometry measurements depends on the care taken and the temperature range. For example, Pavese and Steur (1987) reported an accuracy of 0.5 mK for temperatures between 0.5 and 30 °K. The method is generally expensive and restricted to detailed scientific experiments, calibration and metronomy standards laboratories.

B.2.3 Resistance Temperature Detectors (RTD)

Resistance thermometer sensors use a wire, film, chip or bead whose electrical resistance changes with temperature. The sensor is located near the tip of a closed protective tube, to make a probe which can be inserted into the measurement environment.

B.2.3.1 Metallic Probes

As we've seen in in Sect. 4.2.4 the thermal conductivity of metals is related to their electrical conductivity, which is ideally described by the Wiedemann–Franz law and thus can be used to determine the temperature. To the leading order the electrical conductivity in metals is determined by the electrons. Due to temperature the lattice atoms oscillate, and a temperature rise increases the oscillation amplitude leading to a temperature dependent increase of the electrical resistance. This effect is described as a positive temperature coefficient (T_C) of the electrical resistance and is utilized as the measurement effect. Additionally, flaws in the crystalline structure of the metal interfere with the electron flow. These flaws include foreign or missing lattice electrons, lattice faults at the grain boundaries and atoms in the lattice interstices. Since these interference effects are temperature independent, they result in an additional constant resistance value. By the latter, the relationship between temperature and electrical resistance is no longer linear but can be approximated by a polynomial. Metals, which are suitable for use as resistance thermometers, should have a high T_C, so that the temperature dependent resistance changes are pronounced.

Platinum, in spite of its high price, has become dominant as the resistance material for industrial applications the its resistance $R(T)$ to temperature relationships are especially easy to describe by:

$$R(T) = R_0 \cdot \left(1 + A \cdot T + B \cdot T^2 + C \cdot (T - 100) \cdot T^3\right), \tag{B.4}$$

in which R_0 is the resistance of the thermometer at 0 °C and the coefficients A, B, and C for platinum resistance thermometers must satisfy Standard IEC 60751 (2008). The constant C is only necessary for $T < 0$ °C. For other materials than platinum T has to be replaced by $(T - T_0)$ with T_0 as the reference temperature to which the element is calibrated.

Their respective sensitivities, A, are rather weak and of the order of 10^{-3}/°K as indicated in Table B.1, but their accuracy is large and higher than that of the thermocouples and also in the specified temperature range, their stability is excellent. The resistor probes have an almost linear answer. A resistance measurement device or a power supply with a low voltage voltmeter has to be used inducing a current about a few mA through the thermo-resistive probes. One has to take care about self-heating or Joule effect in order to limit temperature bias by the applied current source. Of importance is especially the stability of the current source.

Various options are available for measuring the electrical resistance of an RTD depending on the desired accuracy and some of these are illustrated in Fig. B.3. In

Table B.1 Electric resistance versus temperature for various metals

	Relative resistance $(R(T)/R_0)$ at 0 °C										
Metal	Resistivity (μΩ/cm)	0	100	200	300	400	500	600	700	800	900
Alumel	28.1	1	1.239	1.428	1.537	1.637	1.726	1.814	1.899	1.982	2.066
Copper	0.13156	1	1.431	1.862	2.299	2.747	3.210	3.695	4.208	4.752	5.334
Iron	0.97857	1	1.650	2.464	3.485	4.716	6.162	7.839	9.790	12.01	12.79
Manganin	0.18482	1	1.002	0.996	0.991	0.983	–	–	–	–	–
Nickel	6.38	1	1.663	2.501	3.611	4.847	5.398	5.882	6.327	6.751	7.156
Platinum	9.83	1	1.392	1.773	2.142	2.499	2.844	3.178	3.500	3.810	4.109
$90Pt-10Rh$	18.4	1	1.166	1.330	1.490	1.646	1.798	1.947	2.093	2.234	2.370
Silver	1.5	1	1.408	1.827	2.256	2.698	3.150	3.616	4.094	4.586	5.091

the fixed bridge circuit, the value of R_1 is adjusted until the current flow through the bridge output, if a galvanometer is used or the voltage read, is zero. Under these conditions the value of R_1 equals the unknown resistance and the temperature can be determined from a calibration, such as equation above. This circuit suffers from inaccuracies in accounting for lead wire resistance and is not used for accurate measurements. In the three-leg RTD circuit (Fig. B.3b), the lead resistance of the middle leg is common to both halves of the bridge. Its resistance cancels out when the bridge is balanced and reduces lead wire error considerably. The four-leg configuration exhibits the best performance, eliminating lead wire resistance uncertainties. With a four-wire RTD, in which two leads are connected to each end of the sensor, one can connect a constant current supply to two of the leads and use a digital voltmeter (DVW) to measure the voltage from the sensor's other two leads.

If the input impedance of the DVM is tens of MΩ, there is a negligible loading error from even a 1 kΩ RTD and negligible lead-wire resistance errors. It is possible to use ac supplies with bridges, potentiometers, and four-wire current/voltage systems and this has the advantages of easier amplification and avoidance of spurious dc signals. The inductive ratio bridge, for example, is an ac method incorporating precision wound resistors for the ratio arms. It is capable of high accuracy and high stability. Methods using both dc and ac bridges were reviewed by Connolly (1998).

For practical applications, the thermo-resistive probes are composed of a metallic layer deposited on a flat electrical insulating substrate (epoxy resin, ceramic, mica…) or cylindrical (glass, pyrex….). The size and the shape of these thermo-resistive probes make them useful for average temperature measurement. In addition, their time constant is much larger than that of thermocouples due to

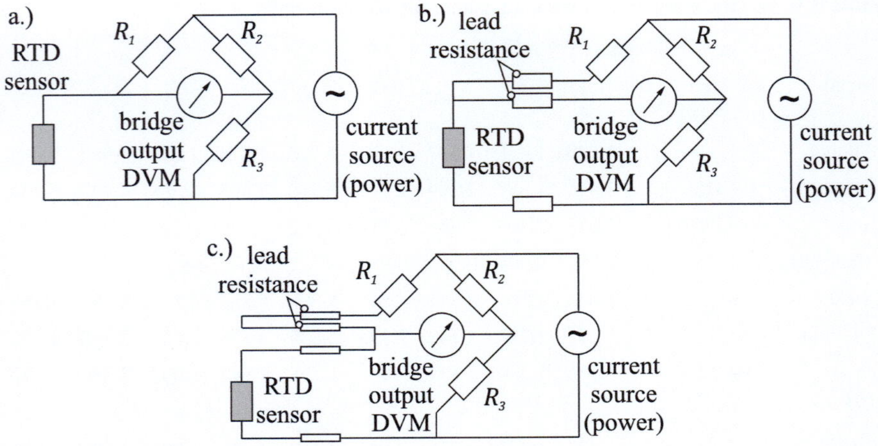

Fig. B.3 Resistance temperature device (RTD) electric acquisition circuits. **a** Fixed bridge; **b** 3-leg and **c** four leg

their insulating substrate. Therefore, they will be used preferentially for temperature measurement in stationary mode. Commercially RTD are available in many shapes and for a large bandwidth of applications, as shown in Fig. B.4a.

B.2.3.2 Thermistors

If accuracy is less critical, then a cheaper form of resistance temperature device than platinum resistance thermometers is the thermistor. Thermistors consist of a semiconductor whose resistance is sensitive to temperature. Modern thermistors are usually mixtures of oxides of nickel, manganese, iron, copper, cobalt, magnesium, titanium, other metals, and doped ceramics. They are manufactured by sintering particles in controlled atmospheres. Thermistors can have both positive or negative temperature coefficients, where for negative temperature coefficient (NTC) types, the resistance increases very strongly as the temperature falls. A typical resistance temperature characteristic is $1\,\Omega/0.01\,°C$. Thermistors are commercially available in bead, disk, rod, washer, and flake form. Beads as small as 0.07 mm in diameter on leads of 0.01 mm diameter are possible. These devices can be encapsulated in epoxy resin, glass, or be painted. Stable thermistors are available using mixed metal oxides for temperatures less than approximately 250 °C. For temperatures above 300 °C the refractory metal oxide devices are appropriate, while for temperatures of the order of 700 °C, devices utilizing zirconia doped with rare earth oxides can be used. They are well suited for use in small probes with fast response due to their small heat capacity. Since the resistances are large, generally several kiloohms, two-wire connections can usually be used without significant error. Thermistors are not standardised, and the manufacturer's specification must be referred to. To a first approximation the temperature can be determined from

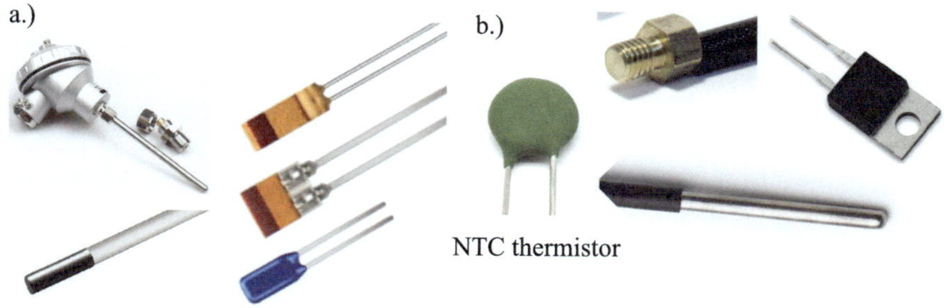

Fig. B.4 Some examples of RTD temperature sensors (**a**) and thermistors (**b**) commercially available

the relationship

$$R(T) = R_0 \cdot \exp\left(1 - B\left(\frac{1}{T} - \frac{1}{T_0}\right)\right), \tag{B.5}$$

where R_0 is the resistance at T_0 and B is a constant for the particular thermistor material. The high resistivity of thermistors of up to $1\,\mathrm{k}\Omega$ negates the need for a four-wire bridge circuit. The accuracy of these devices can be as good as $\pm 0.05\,^\circ\mathrm{C}$, however depending on the calibration effort. The disadvantage of thermistors is their susceptibility to de-calibration and drift due to changes in the semi-conductor materials. Nonetheless, they are widely used in industrial and commercial applications, cheap and available in many forms as illustrated in Fig. B.4b.

B.2.3.3 Thermocouples

Thermocouples are the workhorse of temperature measurement with different devices capable of monitoring temperatures between -270 and $3000\,^\circ\mathrm{C}$. Thermocouples are commonly in almost all thermal piping networks because of their low cost, simplicity, very rugged in protective metal sheathes, variable size, and their relatively wide temperature range. Their sensitivity is adequate for many applications as is their speed of response, but they are less accurate than resistance temperature devices (RTD). The basis of these devices is the Seebeck effect. This is the production of an electromotive force (emf) in a circuit of two dissimilar conductors experiencing a thermal gradient. Thereby, a potential difference (voltage) is generated in in an electrical conductor when there is a temperature difference along its length.

Assume first two dissimilar metal conductors A and B connected to a loop being at the temperature T_1 and T_2 ($T_1 \neq T_2$). At each of the joints a Peltier electromotive force $\Delta\phi$ is created, depending on the material and the temperature T, and the electromotive force (emf) is proportional to the Peltier coefficient Π_{AB}, i.e. $\Pi_{AB} = \Delta\phi$, as depicted in Fig. B.5a. In an isothermal circuit composed by different materials, the sum of the Peltier emf vanishes. The introduction of a third metal in the circuit, as depicted in Fig. B.5b, does not alter the thermoelectric electro-motive force (emf) produced as long as not net thermal gradient is present across its junctions. This circuit can form the basis of a practical thermoelectric measuring circuit as shown Fig. B.5c where in the loop a measurement device recording the potential difference is integrated. For this circuit to provide meaningful results the temperature T_2 must be known and is often called the reference junction temperature. In most applications, however, a reference junction voltage corresponding to T_2 is provided by an electronic bridge circuit incorporating a resistance temperature device. The circuit is mounted at the location of the measuring instruments input terminals where the resistance temperature device can sense that temperature. The measuring instrument can be a digital voltmeter (DVM) or the amplifier/analogue digital converter combination of a direct reading in temperature controller or readout. The input impedance of the voltage measuring instrument has to be large compared to the resistance of the thermocouple and its extension wires. This condition is easily met in practice and will prevent a significant current from flowing through the circuit, which could cause other thermoelectric effects.

From solid state physics aspect, the Seebeck effect relevant for establishment of a potential difference originates from temperature dependence of the electron potentials, which cause a charge shift in an electrical conductor when it is placed in a non-homogeneous temperature field with a temperature gradient $\nabla T = (T_1 - T_2)$. In a simplified explanation the free charge carrier in a metal—the electrons—in a one side warmed conductor distribute themselves in a non-homogeneous manner so that a potential difference (thermal voltage) is generated along the path dx in a metal A from T_1 to T_2. Thus, a Seebeck potential voltage E_S is generated in a homogeneous wire according to:

$$E_S = S_A \cdot \frac{dT}{dx} \cdot dx, \tag{B.6}$$

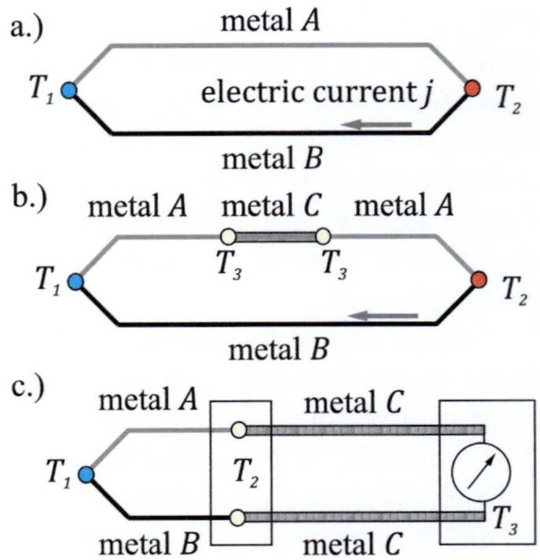

Fig. B.5 Thermocouple circuit formation

where S_A is the absolute Seebeck coefficient of the material A with the physical unit (μV/°K). In a physical sense the Seebeck coefficient can be thought of as the heat per carrier per degree K to transport a specific heat. Thus, S_A can be for a homogeneous material approximately computed to $S_A \approx c/e$, where c is the heat capacity and e the elementary charge with $e = 1.602 \times 10^{-19}$ A s. In normal metals only, small fraction around the Fermi energy (E_F) contribute, so the thermopower is very small and one gets:

$$S_A = \left(\frac{k_B T}{E_F}\right) \cdot \frac{k_B}{e}, \tag{B.7}$$

with k_B the Boltzmann constant and $E_F = \left(\left(h^2/\left(8\pi^2 m\right)\right)\left(3\pi^2 n\right)^{2/3}\right)$, in which m ist he molar mass, n the particle density and h Planck's constant.

At the cold end more electrons accumulate while at the hot end, the electron quantity is decreasing. Therefore, it is plausible that even in a single electrical conductor in a temperature field a thermal potential difference is generated. This thermal voltage can be measured if a second dissimilar metal B with a different temperature dependence of the thermal electric effect is added. By the two materials A and B a net Seebeck potential ΔE_S is generated

$$\Delta E_S = \int_{T=T_1}^{T=T_2} S_A dT + \int_{T=T_2}^{T=T_1} S_B dT = \int_{T=T_1}^{T=T_2} (S_A - S_B) dT. \tag{B.8}$$

It is important that this thermal potential difference effect is the result of a volume diffusion effect of the charge carriers (electrons) and not a contact potential phenomenon between the two materials. Therefore, it is understandable that the thermal voltage is produced along the entire length of the thermocouple to the junction and not only at the connection between the two legs.

In thermoelectric practise the absolute Seebeck coefficients are not used. Instead Seebeck coefficients are determined to a reference material, mostly platinum. From those the relative Seebeck coefficients S_{AB} are determined fort he individual material combinations of conventional thermocouple, for more details see Benedickt (1984). These relative Seebeck coefficients S_{AB} represent for a given material combination the net charge in thermal electromotive force caused by a unit temperature difference.

The number of thermocouple material combinations available is extensive and over 150 types are tabulated. The criteria for thermocouple selection include cost, maximum and minimum operating temperatures, chemical stability, material compatibility, atmospheric protection, mechanical limitations, duration of exposure, sensor lifetime, sensitivity, and output emf. Ten thermocouple combinations have been standardised in IEC 60584 (2013) for industrial use:

- Five base-metal (e.g. copper-nickel, nickel–chromium) types J, T, E, K and N. These are relatively inexpensive and can be used (variously) down to −270 °C and up to about 1200 °C,

- high temperature types, designated types *R*, *S* and *B*, use wires of platinum and platinum–rhodium alloys and applied for up to about 1600 °C and
- two thermocouples of tungsten-rhenium alloys, types *C* and *A*, which can be used even beyond 2000 °C;

and some of their features are listed in Table B.2.

All the thermometers considered so far must contact the process being measured, and they must be calibrated at temperature fixed points or by comparison with standard thermometers. Fixed-point cells are used for the calibration of platinum resistance thermometers and thermocouples at the highest level. However, only a few thermometers can be measured while the phase change—melting or freezing—lasts, and several fixed points are needed to cover a given range. The method is time-consuming and hence expensive, and it is only suitable for a limited number of high specification thermometers, which are designated as 'standards'. The alternative, for larger numbers of thermometers of lower specification, is to use a temperature-controlled environment (bath, furnace or 'dry block calibrator'), into which the thermometers to be calibrated are inserted with one or two standard thermometers or thermocouples as reference instruments.

The bath or furnace is set to control at a series of steady temperatures covering the range of the calibration, which can be chosen at will. At each stable point, measurements are made of all the thermometers in a suitable scanning procedure, and the temperature is determined from the standards. The calibration is thereby achieved. The thermal environment must be uniform in temperature over the critical volume, and stable (or only slowly drifting) during the measurement period. The uniformity and stability must be checked before use, and the conditions are also tested during the calibration if more than one standard thermometer is used.

Thermocouples and wires come in a variety of packages and insulations to handle a wide variety of applications. The actual thermocouple is no more than a weld bead on the end of the two material wires. These can be extremely small, with the smallest thermocouple wire being around 0.001 mm in diameter. This can create a micro thermocouple with a response time under 0.05 s. The response time of a thermocouple is defined as the time it takes to reach 62.3% of an instantaneous temperature change. On the market there is a variety of probes and sheath materials. Probes are typically thermocouples placed inside a stainless steel, or other material tube. This tube can be open on the end exposing the junction, or closed, encasing the junction, as illustrated in Fig. B.6, for which each type requires specific electric cabling to measure a correct temperature, as discussed in principles in Benedickt (1984). In addition, this junction can be either isolated from the sheath material, or welded to it. The sheath material depends on the temperature range and ambient of the application. Grounded tip thermocouples reveal fast response times and higher sensitivity than ungrounded ones but are vulnerable to ground loops, circuitous paths or electric currents between conductive sheath of the thermocouple and some other point in the thermocouple circuit.

Table B.2 Standardized thermocouples according to Childs et al. (2000)

Material pairing	Designation letter	Stability within temp. range	Cost	Output (μV/°C)	Temp. range (°C)
Iron/constantan (CuNi-alloy)	*J*	Low	Low	26 @ −190 °C 63 @ 800 °C	−196 – 700
Copper/constantan (CuNi-alloy)	*T*	Low	Low	15 @ −200 °C 60 @ 350 °C	−262 – 850
Nickel chromium alloy/copper nickel alloy	*E*	Low	Low to mid	68 @ 100 °C 81 @ 500 °C 77 @ 900 °C	−268 – 800
Nickel chromium alloy (chromel)/ nickel aluminium (alumel)	*K*	Low	Low	40 (from 250–1000 °C) 35 @ 1300 °C	−250 – 1100
Nickel chromium silicon (nicrosil)/ nisilicon magneisium (nisil)	*N*	Low	Low to mid	37 @ 1000 °C	0–1250
Platin 10%—rhodium/platin	*S*	High	High	6 (from 0–100 °C)	0–1500
Platin 30%—rhodium/platin 6%—rhodium	*B*	High	High	5 (from 0–100 °C)	0–1750
Platin 13%—rhodium/platin	*R*	High	High	10 @ 1000 °C	0–1600

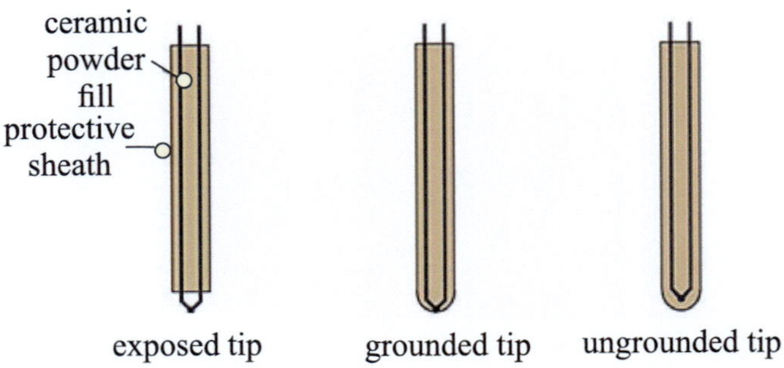

Fig. B.6 Types of thermocouple designs commercially available

B.2.4 Non-invasive Temperature Measurement Techniques

In high temperature receivers the temperature distribution on the surface is of importance for the mirror control strategy but the embedding a thermocouple there may be prohibitive due to the oxidizing ambient or for heat transfer reasons. Also, regular monitoring the heat losses of large scaled objects as solar tanks storing sensitive heat may require hundreds of thermocouples, which is impractical. Moreover, invasive instrumentation can degrade with time; exceeding only temporarily material limiting thresholds, it can disintegrate completely. Non-invasive methods are not bound by this constraint. Most non-invasive techniques measure temperature from the electromagnetic thermal radiation spectrum, especially Infrared devices are sensitive to that part of the spectrum. Optical techniques such as absorption and emission spectroscopy, scattering and luminescence, are sensitive in the visible region mostly because lasers are used as part of the system. One notable exception is acoustic temperature measurement, which relies on the measurement of the speed of sound.

B.2.4.1 Infrared Thermometers

The operation principle of radiation thermometers irrespective if they are infrared thermometers or radiation pyrometers is similar to a camera. It consists of an optical system, using lenses or mirrors and a filter which selects the wavelength range (waveband) over which the downstream arranged detector is sensitive, as illustrated in Fig. B.7. The radiation is focused onto a detector whose output indicates the intensity of the radiation and hence the temperature.

Dependent on the wavelength of the emitted radiation of the target surface to be monitored the detector may be a ‘photo-detector’, in which the incident photons give rise to an electric current, or it may be ‘thermal’, in which case it senses the temperature rise produced by the energy absorbed. The detector is generally temperature-controlled to make its response more consistent. For low temperature applications, it must be cooled to improve the signal-to-noise ratio and to reduce the intensity of its own thermal radiation. Thermal radiation is mostly in the infrared and is not visible, but as the temperature increases beyond about 700 °C

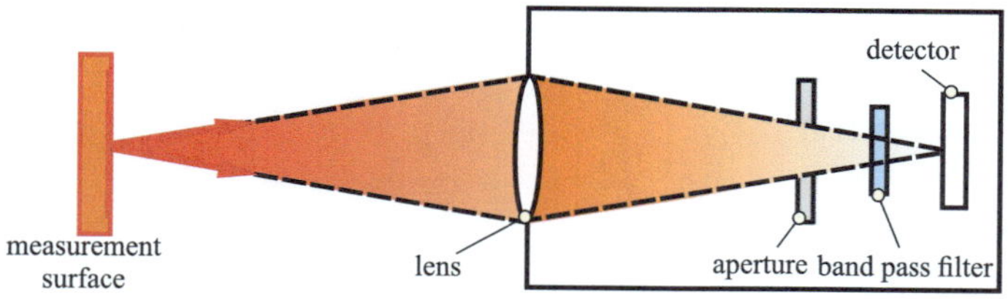

Fig. B.7 Schematic arrangement of a radiation thermometer measuring the temperature of a hot sample

a dull 'red heat' can be seen, which gradually brightens to orange, yellow and finally a brilliant white heat according to Planck's law. The intensity is strongly dependent on temperature, so radiation thermometry is a sensitive method of temperature measurement, even low temperatures down to—50 °C are commercially feasible with devices.

The fraction of blackbody emission $e_{b,meas}(\lambda, T)$ in a spectral band between λ_1 and λ_2 is given by:

$$e_{b,meas}(\lambda, T) = \frac{\int_{\lambda=0}^{\lambda=\lambda_2} e_{\lambda b}(\lambda, T)d\lambda - \int_{\lambda=0}^{\lambda=\lambda_1} e_{\lambda b}(\lambda, T)d\lambda}{\sigma T^4}, \tag{B.9}$$

and is required to determine the proportion of blackbody emission between the two wavelengths. This information can then be used to determine whether a particular detector has an appropriate spectral sensitivity and whether the optical path transmits this radiation bandwidth.

Radiation thermometers require a calibration to a known source of thermal radiation, e.g. a blackbody cavity, so that the measured intensity of the radiation can be accurately translated into a temperature. According to Wiens displacement law blackbodies radiate at close to the maximum possible intensity at any given wavelength and temperature. They are so called because they are also perfect absorbers of all radiation incident on them, and hence appear perfectly black when cold. One challenge for infrared thermography is that the sample surfaces always emit less radiation than a blackbody at the same temperature and the measurements must be corrected for the '*emissivity*' of this surface.

It is generally advantageous to operate at as short a wavelength as possible, because the sensitivity is greater (the curves are more widely separated). A wavelength in the range 0.65–1 μm, using a silicon (Si) photodiode detector, is commonly used at high temperatures, $T > 700\,°C$. At low-to-moderate temperatures, where the intensity is low, it is necessary to operate some way into the infrared and over a substantial band of wavelengths, to obtain a useful signal. Some commonly used infrared detector materials and their wave band is given in Table B.3.

Although radiation-based thermography allows for remote sensing some of the advantages offset by some significant disadvantages and sources of error, which must be accounted for. These are the:

Table B.3 Some commonly used infrared detector materials and their wave band modified from Childs et al. (2000)

Material	Wave band λ (μm)	Material	Wave band λ (μm)
Al_2O_3	0.2–11	KBr	0.2–32
CaF	0.13–10	PbSe	2–5
Ge	1–12	SiO_2	0.2–4
InAs	1–3.8	TiO_2	0.4–5.6
InSb	1–6	ZnS	8–12

- *Emissivity*: The radiation emitted from an object depends not only on its temperature but also on the surface properties, and thus on its *emissivity*, which must be known or estimated if an accurate temperature is to be obtained otherwise significantly large errors (tens of degrees is quite common) arise because of an incorrect emissivity estimation.
- *Reflections*: Thermal radiation emitted by heaters or lighting is partially reflected by the sample and add to the radiation which is detected. Large errors can be caused by reflections from furnace heaters at high temperatures, unless pre-cautions are taken. When measuring low temperatures, the artificial heaters include lights, human beings or even ambient environment.
- *Environment*: The thermometer must have a clear line of sight and dust, water vapour or carbon dioxide in the atmosphere can scatter or absorb the radiation before it reaches the thermometer, and hence cause an error in the measurement. Some application may allow the use a fibre-optic light-pipe to transmit the radiation to the remote detector, but require also calibration since absorption also occurs within the fibre.
- *Optical imperfections*: Finally, imperfections in the optics lead to imperfections in the field of view, so that the 'target size', or 'spot size', of the instrument may be significantly larger than is intended or claimed.

A comprehensive overview on infrared technique, sources of uncertainty, including how to quantify these sources is given in Minkina and Dudzik (2009). Figure B.8 shall illustrate the radiation exchange between the sample of the temperature T_S to be measured, the IR-thermometer and the surrounding background at the temperature T_B. In this configuration the radiation received by a radiation thermometer focused on the sample surface can be written as the sum of two parts, the emitted and the reflected radiance and reads to:

$$J_{total} = \varepsilon \cdot J(T_S) + (1 - \varepsilon) \cdot J(T_B), \tag{B.10}$$

where is the J_{total} radiance of the sample surface, ε the sample emissivity, $J(T_S)$ is the radiance of a blackbody at temperature T_S and $J(T_B)$ is the radiance at the background temperature. Only the first term corresponds to the radiation emitted by the sample surface, while the second one originates from the radiation reflected by the background onto the sample and back to the thermometer. Assuming an emitting surface obeying Kirchhoff's law ($\alpha = \varepsilon$), the reflected fraction of the radiation is given by product of the radiance at the background temperature (T_B) and the reflectivity of the sample surface ($1 - \varepsilon$).

There are three distinct cases of interaction between target, thermometer and surroundings:

- $T_S \gg T_B$, then the first term dominates and the temperature can be calculated for a given estimated value for the emissivity. It is beneficial if the operating wavelength of the thermometer is chosen short ($<1\,\mu$m), because the sensitivity is higher as the Planck curves are more widely separated and an error in estimating the emissivity produces less error in the derived value of temperature.
- For $T_S \approx T_B$ the radiation reflected from the background is significant and the background temperature must be taken into account. To an extent, the background compensates for the effect of the surface emissivity: in the specific case where $T_S = T_B$, $J_{total} = J(T_S)$ and the whole system of surface and surroundings behaves as a blackbody cavity at temperature T_S.
- $T_B > T_S$ implies that the second term will again be significant and may even dominate. It is now advantageous to use a thermometer which operates at a longer wavelength, because the difference in the radiances $J(T_S)$ and $J(T_B)$ is then smaller—i.e. the Planck curves are less widely separated—and the effect of errors in estimating the emissivity or the background temperature are less serious.

Radiation detectors can be broadly grouped into three categories depending on the temperature range intended to be measured, which are:

Fig. B.8 Sketch of radiation exchange between a surface, its background and an IR (infrared) thermometer illustrating that the detected radiation is the sum of the emitted radiation and the reflected radiation

- disappearing filament optical pyrometers, which is similar to a refracting telescope. Here, an electrically heated tungsten filament is placed in the focal plane of the objective lens and a red filter is located between the lamp and the detector. The pyrometer is sighted on the target and the image is formed in the same plane as the lamp filament. The magnified image of the lamp filament is superimposed on the target. By adjusting the current through the filament, its luminance or brightness can be matched to that it disappears on the detector.
- Thermal detectors convert the absorbed electromagnetic radiation into heat, which is sensed electrical resistance used by the bolometers, thermoelectric emf by means of thermocouples or pyroelectric detectors using crystal wafers, which produce surface electric charges when heated.
- Quantum, photon, or photoelectric detectors measure the direct excitation of electrons to conduction states by incident photons. Types of quantum detector include photo emissive, photoconductive, and photovoltaic. Photon detectors respond to individual photons by releasing or displacing electrical charge carriers by the photoelectric effect, photovoltaic effect or photoelectromagnetic effect.

B.2.4.2 Thermography or Thermal Imaging

Thermography or thermal imaging involves determining the spatial distribution of thermal energy emitted from the surface of an object to build up a two-dimensional image of a substantial area or even a complete scene. This technique is an extension of the thermometer to a sensor plane and thus allows mapping of the temperature distribution and can be used in conjunction with software analysis to provide quantitative information. Thermographic techniques are used in the detection of body temperatures, satellite imaging, and in automatic guidance systems.

Noise or random fluctuations present in detectors place limitations on the lower limit of the temperature range and affect the overall precision. The noise can be considered to arise from two sources: detector noise originating in the detector itself and radiation noise due to the random variations in the rate at which a heated surface emits photons. The noise equivalent power gives an indication of the smallest amount of radiation that can be detected and is equal to the amount of incoming radiation required to produce a signal equal to the noise. The accuracy of a radiation thermometer depends on attenuation of the radiation between the target and the thermometer, background radiation reflected from the target and detected by the thermometer, uncertainty of the target emissivity, and errors caused by the thermometer itself. In order to reduce the effect of uncertainty in the emissivity on the temperature measurement, a narrow bandwidth is desirable. Additionally, the size of the measuring object must be equal to or bigger than the viewing field of the measurement device in order to reach accurate results. The spot diameter changes accordingly to the distance of the sensor. Compared to temperature sensors thermal imagers have a so-called field of view (FOV), not just a point; therefore, many suppliers provide exchangeable lenses for various measuring distances between camera and object.

B.3 Heat Flux Measurement

A heat flux sensor is a transducer that generates an electrical signal proportional to the total heat rate ($\dot{Q}$) applied to the surface A of the sensor, it thus measures an area specific heat flux ($\dot{q}$) with units $\left(\mathrm{W/m^2}\right)$. The heat flux to be measured can have different origins; but originates in general from convective, radiative as well as conductive sources. Heat flux sensors are sold commercially under various names, such as heat flux transducers, heat flux gauges, heat flux plates, some of which are single-purpose as e.g. heat flux sensors, like pyranometers for solar radiation measurement. Other heat flux sensors include Gardon gauges (also known as a circular-foil gauge), thin-film thermopiles, and Schmidt-Boelter gauges.

B.3.1 Sensors Based on Spatial Temperature Gradient

According to Fourier's first law the heat flux ($\dot{q}$) across a material surface can be determined at a location if the temperature gradient can be measured at that position, however, the positioning of a temperature sensor with the requisite accuracy inside the material is often challenging. Thus, sensors to measure heat flux are either applied on the surface or mounted in a hole in the material. From the simplest arrangement to determine a heat flux as indicated in Fig. B.9a, one can derive the heat flux perpendicular to the surface directly in steady state conditions by acquiring the temperature difference ($T_1 - T_2$) and the knowledge of the sensor thickness (δ).

$$\dot{q} = \frac{k}{\delta} \cdot (T_1 - T_2). \tag{B.11}$$

But, the thickness of the sensor and thermal conductivity k of the sensor layer not known a priori with sufficient accuracy for any particular sensor to preclude direct calibrations of each sensor. An adhesive layer may also be required between the sensor and surface to attach the sensor, which adds an additional thermal resistance and increases the thermal disruption. Thus, complementing temperature measurements on the sensor and on the surrounding undisturbed material are recommended to quantify this disruption.

As we've seen in in the temperature measurement devices thermocouples are capable to generate their own voltage output (expressed by an emf) corresponding to the temperature difference between two junctions. Connecting in series n_A and n_B thermoelectric elements of the material A and B in a thermopile arrangement connected together at both sides of the sensing surface as indicated in Fig. B.9b, we create an even number N of thermocouple pairs, as the number of thermocouple junctions is $N = ((n_A - 1) + n_B)/2$. Applying a heat flux to the sensor yields to the establishment of a temperature difference on both sides of the planar arrangement, which in turn generates a thermoelectric voltage ΔE_S, which is

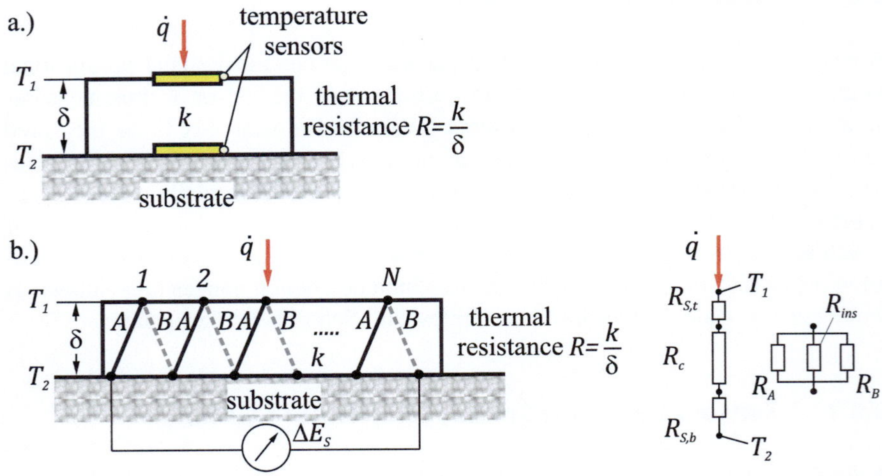

Fig. B.9 **a** Planar heat flux sensor concept; **b** thermopile concept to measure the heat flux by electromotive force

given by:

$$\Delta E_S = N \cdot S_{AB} \cdot (T_1 - T_2), \tag{B.12}$$

in which S_{AB} is the Seebeck coefficient for the thermoelectric material, mostly metallic alloys. Any pair of conductors (e.g., copper–constantan) can be used for the legs of the thermopile, but the output leads should be of the same material so that additional thermocouple junctions are not created. The heat flux through the sensor is then obtained as $\dot{q} = (T_1 - T_2)/R$, where R is the total thermal resistance in $(\mathrm{W}/(\mathrm{m}^2\mathrm{K}))$ of the thermopile, which is composed of the thermal resistance of the surface layers on both sides of the plate ($R_{S,t}$, $R_{S,b}$) as e.g. by welds or glues and ($R_{S,c}$) the thermal resistance of core of the plate consisting of material A and B as well as the electrical insulation material in which the wires are embedded. While the thermal resistances $R_{S,t}$, $R_{S,b}$ and $R_{S,c}$ can be considered as to be connected in series along the heat flux, the thermal resistance of the core can be treated as parallel connection of the thermal resistance of the volume fractions of material of the insulation and the materials A and B, respectively, as indicated in Fig. B.9b and in detail described by Gifford et al. (2010). The corresponding theoretical sensitivity of the sensor (S) is then given by:

$$S = \frac{\Delta E_S}{\dot{q}} = N \cdot S_{AB} \cdot R. \tag{B.13}$$

Hence, by piling up a large number of thermocouple pairs, where in practise up to 10^4 pairs have been realized, sensitivities are sufficient to measure heat fluxes as low as $0.1\,\mathrm{Wm}^{-2}$ are feasible. Regarding the temporal resolution we require the solutions of transient heat transfer for a sheet exposed to a sudden temperature jump an analytic solution can be found e.g. in Carslaw and Jaeger (1986). Assuming that the heat flux facing surface approaches temperature of the ambient T_∞, the response time t is obtained to

$$t = 1.5 \cdot \frac{\delta^2}{a}\,(\mathrm{s}), \tag{B.14}$$

where a denotes the effective thermal diffusivity of the sensor. Thus, thin films with a high effective diffusivity can monitor transient phenomena, and micro-foil sensors of a thickness of 75–400 μm are commercially available to allow for response times of 20 ms and for specific applications also microfabrication techniques attained sensors of thickness of 2 μm.

The measurement results of heat flux sensor rely on the sensor calibration. The reference standard is assumed to accurately measure the reference heat flux. The sensitivity for the test sensor is obtained by dividing its voltage output by the heat flux measured by the reference sensor. The heat flux through the test and reference gauges may be due to radiation, convection, conduction, or mixed modes of heat transfer. It has been shown that the most useful calibration results are obtained when calibration is performed using the mode(s) of heat transfer that the sensor will experience in actual use (Gifford et al., 2010).

For the heat conduction calibration of the heat flux sensor mainly two calibration methods are commonly used. The first involves the use of a one-sided guarded hot plate (GHP), where a uniform flux is generated between two plates held at different temperatures. The sensor to be calibrated is positioned in the intermediate zone. The applied heat flux is calculated from the temperature drop across a reference specimen of known thickness and thermal conductivity. The second method used is to compute the heat flux by means of a null detection technique using a two-sided GHP apparatus. The GHP intermediate heater is adjusted until the heat flux through sensor vanishes. Thereby, the heat flux to sensor can be directly calculated from the electric power applied to the heater and its active cross-sectional area.

Regarding the sketch outline in Fig. B.10b illustrates only the principle; however, many geometrical forms have been realized. One option is to wrap thin pieces of two types of metal foil alternately around a thin plastic (Kapton) sheet and butt-welded on either side to form thermocouple junction. Another option is to use thin-film sputtering techniques. Similar sensor design has been realized with the help of screen-printing techniques of conductive inks, in which a copper/nickel thermocouple pair are taken with a dielectric ink for the thermal resistance layer. The inks have then been printed onto anodized aluminium shim stock for the substrate. In figure some commercially available sensors are depicted in Fig. B.10a.

A specific type of thermopiles is the so-called the Schmidt-Boelter gauge (SB gauge) also known as Wire-Wound gauge, which also operates on the basis of an axially heat flow, i.e. perpendicular to the receiving surface. The principal set-up of an SB-gauge is depicted in Fig. B.10b. The sensing element is again a thermopile

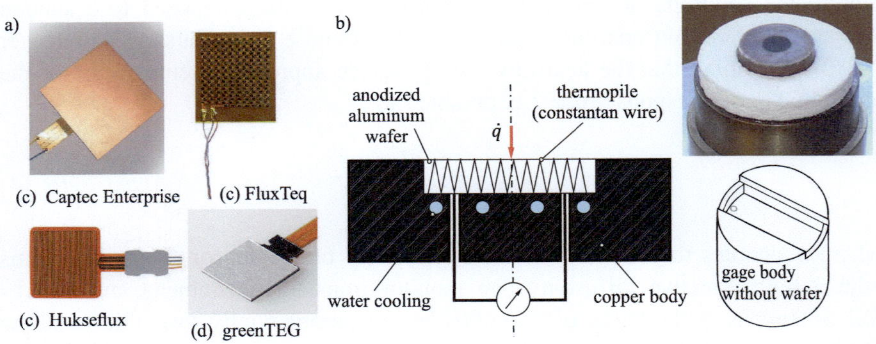

Fig. B.10 **a** Commercially available thermopiles based on the emf principle. **b** Schmidt-Boelter gauge

made of constantan wires, with a varying number of junctions, and the wafer is usually consisting of anodized aluminium. The number of junctions depends on the range of heat flux which the gauge is designed to measure. The body consists of copper containing dependent on the type water cooling channels, to ensure that the receiving disk is cooled both radially and axially. Regarding the calibration a similar effort has to be spent as for all thermopiles accounting for conduction, convection and radiation.

B.3.2 Gardon-Gauge

The circular foil or Gardon gauge consists of a hollow cylinder of one thermocouple material with a thin foil of a second thermocouple material attached to one end. The foil typically made of constantan is coated with a diffuse, absorbing coating attached to a copper heat sink A wire of the first material is attached to the centre of the foil to complete a differential thermocouple pair between the centre and edge of the foil as illustrated in Fig. B.11. The body and wire are commonly made of copper with the foil made of constantan. If a Gardon gauge is used to measure radiation flux and assuming the thermal radiation is uniformly incident onto the gauge foil surface a temperature gradient in the radial direction develops such that temperature difference is proportional to the heat flux.

Thus, is type of sensor can be conceived diffusion type of heat flux gauge, because the thermal energy is collected by the disk and then transported to the heat sink, connected to the edge of the disk. As a result, the actual measured temperature difference is a function of both the total heat transfer to the disk, and the distribution of heat flux over the disk surface. The direct integration of a sensing element in form of an integral radial acting thermocouple producing an emf makes it quite robust.

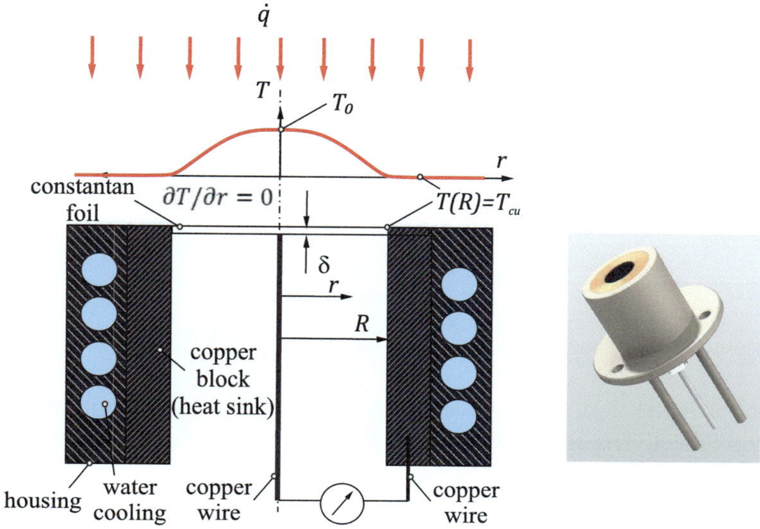

Fig. B.11 Cut through a Gardon gauge (left) and commercial version (right)

The calculation of the heat flux corresponds entirely to the problem considered in for steady state heat conduction, with a boundary condition at $r = 0$ of $\partial T/\partial r = 0$ and for $r = R$ we get $T(R) = T_{cu}$. Solving the the steady state heat conduction equation for a constant imposed heat flux ($\dot{q}$) in a cylindrical coordinate system, reading to:

$$\frac{\dot{q}}{\delta k} + \frac{1}{r}\frac{\partial T}{\partial r} + \frac{\partial^2 T}{\partial r^2} = 0, \tag{B.15}$$

where k is the specific heat conductivity of the constantan foil and allowing for the variation in thermal conductivity with temperature of the thin foil by means of using a Kirchhoff-transformation in the form $k = k_0(1 + c \cdot T)$, in which c accounts for the fractional variation in thermal conductivity per unit temperature increase, we obtain an analytical solution for the heat flux:

$$\dot{q} = \frac{4k_0\delta}{R^2}(T_0 - T_{Cu}) \cdot \left(1 + \frac{c}{2} \cdot (T_0 - T_{Cu})\right). \tag{B.16}$$

Also, the thermal response time t of a Gardon Gauge can be approximately assessed by means of a one-dimensional analysis. Without referring to this in detail, here only the result is provided:

$$t = \frac{R}{4a}, \tag{B.17}$$

in which a again denotes the thermal diffusivity.

Calibration of Gardon gauges has often been undertaken using a radiation heat source. Gardon gauges provide accurate results for systems where radiation is the dominant mode of heat transfer. When convection is also significant, large errors can result owing to non-uniformities in the foil temperature. For the case where these sensors are used to determine the heat flux when convection is more significant than radiation in the contribution to the total heat flux correction ratios ($\dot{q}_{total}/\dot{q}_{radiation}$) have been successfully introduced, see Kuo and Kulkarni (1991), reading to:

$$\frac{\dot{q}_{total}}{\dot{q}_{radiation}} = \frac{1 + \left(R\sqrt{h/(\delta k)}/2\right)^2}{1 + \left(R\sqrt{h/(\delta k)}/4\right)^2}, \tag{B.18}$$

in which h is the heat transfer coefficient at the constantan foil surface towards the environment. The error decreases as $R\sqrt{h/(\delta k)}$ decreases and thus small foil radii are preferable, however, reducing sensitivity.

B.3.3 Calorimetric Methods

This type of methods relies on the rate of change in temperature with time at a location near to or on the surface of interest. Analysis of these results with an appropriate form of the conduction equation or heat balance and accurate knowledge of the material properties allows a quantification of the heat flux. There exists a wide verity of this heat flux techniques such as slug calorimeters, plug gauges, null point calorimeters, coaxial thermocouple gauges, thin-skin and thin-film sensors allowing for monitoring of entire surfaces.

Slug Calorimeters
A typical slug calorimeter consists of a volume of material thermally insulated from its surroundings is illustrated in Fig. B.12a. Typically, a single temperature measurement at the base of the slug is made and this is assumed to represent the temperature of the entire mass of the slug. This assumption is valid within a few per cent if the condition for the Biot number $Bi = (h \cdot d)/\lambda < 0.1$ is met, where d denotes the depth of the slugs and λ its specific heat conductivity. For the slug a material with high thermal conductivity is required. The measurement principle is based on a time dependent energy conservation consideration, which reads for this simple example to:

$$\dot{q} \cdot A = \left(m \cdot \frac{\partial}{\partial t}(cT)\right) + \dot{q}_{loss} + \dot{q}_{out}, \tag{B.19}$$

in which c ist he heat capacity of the slug, A the surface area facing the heat flux. Thus, a calorimeter balances heat stored to heat dissipated.

Neglecting losses through the thermal insulation, and assuming uniform temperature and constant properties the heat transferred to the calorimeter is equal to the energy stored and the heat flux is given by

$$\dot{q} = \frac{mc}{A}\frac{dT}{dt}, \tag{B.20}$$

assuming also a constant specific heat capacity c. Depending on the nature of the thermal boundary conditions, several solutions of the equation can be obtained as shown in the chapter transient heat transfer. If the heat flux ($\dot{q}$) is constant we get in the simplest form:

$$\dot{q} = \frac{mc}{A}\frac{\left(T_f - T_0\right)}{t_f - t_0}, \tag{B.21}$$

with the subscript f denoting the time at which the measurement is taken compared to reference time indexed by 0. Therefore, the heat flux can be determined from measurement of the transient temperature if the thermal properties are known. The technique is usually expanded to consider the thermal losses to the bulk material through the thermal insulation, which modifies Eq. (B.20) to

$$\dot{q} = \frac{mc}{A}\frac{dT}{dt} + U_{ins}\Delta T, \tag{B.22}$$

in which ΔT denotes temperature difference between the slug and the bulk material. The constant U_{ins} for a given sensor must be determined by calibration. If the sensor is exposed to a convective flow with constant heat transfer coefficient, we obtain the solution of got in Sect. 4.2.3.5 reading to:

$$\frac{(T - T_\infty)}{(T_0 - T_\infty)} = \exp\left(-\frac{h \cdot A}{m \cdot c} \cdot t\right). \tag{B.23}$$

The heat transfer coefficient h appears in the exponential term and in practice can be evaluated from the gradient of the natural logarithm of the temperature ratio against time. It is recommended for all of these slug calorimeters to take care of the specific application type they are designed for and to compare the calibration data sheets of the suppliers with respect to the specific characteristic numbers of the own application. Even that does not prevent to recalibrate on-site.

One-Dimensional Transient Conduction Analysis

The measurement devices mentioned herein, can be purchased as coaxial thermocouples or surface junction thermocouples. Although the name suggests that the measurement principles is based on the Seebeck effect this is not correct. The basic principle of this method is the penetration of the thermal pulse into the surface is small in comparison with the relevant surrounding dimensions, so it relies on a small Fourier number approach ($Fo = (at)/L^2 \ll 1$). Thus, this technique is applied for short term test e.g. to evaluate the irradiation at a solar thermal receiver panel or its fluctuations.

The coaxial thermocouples are generally fabricated by one metallic element swaged over the second element with insulating material in between them as indicated in Fig. B.12b. A surface junction is formed by grinding the front surface that provides micro scratches leading to small scale plastic deformation. Thus, it offers distinct advantages over the conventional heat transfer gauges and the fast-response characteristic makes it suitable for the short-duration measurement studies and for fast responses.

In general, these thermal sensors measure transient temperatures and surface heating rates are then predicted by using suitable one-dimensional heat conduction modelling. Since the thermal penetration distance during experimental run times is small compared to the linear dimension of the gauge, the system can be modelled by considering unsteady, linear conduction of heat in a one-dimensional semi-infinite solid

$$\frac{\partial^2 T_1}{dx^2} = \frac{\rho_1 c_1}{k_1}\frac{\partial T_1}{\partial t} \text{ and } \frac{\partial^2 T_2}{dx^2} = \frac{\rho_2 c_2}{k_2}\frac{\partial T_2}{\partial t}. \tag{B.24}$$

For uniform initial conditions, if a heat flux ($\dot{q}$) is applied instantaneously, the transient surface temperature for the thermocouple at the junction is $T_S(t)$. When substrate thermal properties are treated as constant, the heat flux passing through surface l_1 is calculated by using Duhamel's superposition integral as given below, for more details see Taler (1996):

$$\dot{q}(t) = \frac{\sqrt{\rho c k}}{\sqrt{\pi}} \int_{t=0}^{t} \frac{dT}{dt}\frac{1}{\sqrt{t-\tau}} d\tau. \tag{B.25}$$

Quite recent experiments complemented by numerical analysis have shown that the 1D semi-infinite heat conduction model assumed is satisfied when the Fourier number is smaller than 1/16; in such cases the error is less than 1%, see Zhang et al. (2020).

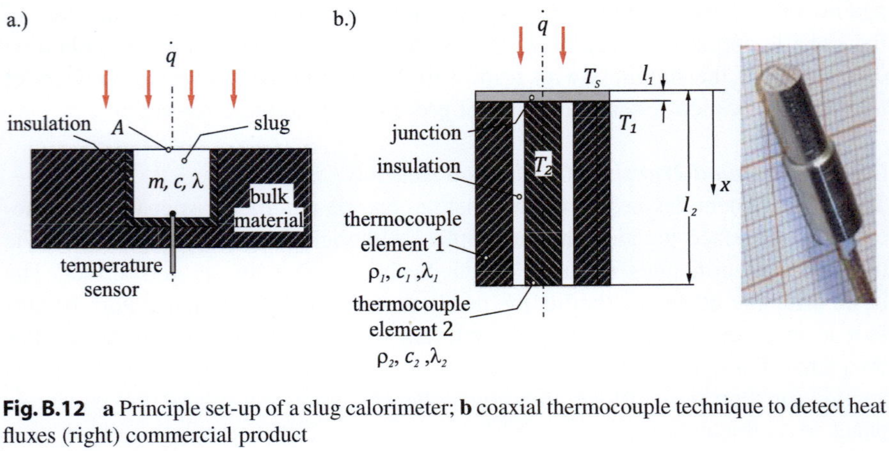

Fig. B.12 **a** Principle set-up of a slug calorimeter; **b** coaxial thermocouple technique to detect heat fluxes (right) commercial product

B.4 Flow Rate and Mass Flow Rate Devices

The flow rate is the ratio of the quantity of the flowing medium and the time in which this quantity flows through a pipe cross-section. The flow rate is specified in volume units or in mass units, depending on the definition of the quantity. Mass flow and volume flow are linked via density. Since the viscosity and density of many media also depend on other boundary conditions such as temperature and pressure, these variables may have to be measured additionally for correct volume flow determination. A large number of methods exist for determining the flow rate, so that the methods presented here are limited to those in closed pipelines, since these are of central importance in thermal measurement technology. For flow rate measurement in closed pipelines, a distinction is made between flow volume meters and flow rate meters. Flow volume meters, also called totalizers, are devices that delimit and record partial volumes of the flow to be measured and integrate them into a total volume. Totalizers can also work with the indirect method of measured data acquisition, which is realized by recording the flow velocity or by determining the kinetic energy, so that the mean volumetric flow rate is obtained.

Central to any measurement is the selection of the appropriate measurement method to be used. For this purpose, the manufacturers of measuring instruments provide structured decision support matrices that enable a staggered and targeted selection.

The starting point of these matrices is always the characterization of the fluid according to the state of aggregation (gaseous, liquid, vaporous, solid) and the essential properties of the fluid (clean/dirty, chemical aggressiveness, solid content, dissolved gases in the liquid, electrical conductivity, viscosity, two-phase mixture, abrasion behaviour).

The second step is the determination of the measuring section, which includes the following aspects:

- pipe diameter,
- permissible pressure loss through the device,
- maximum operating pressure,
- operational temperature range,
- flow pattern (laminar, turbulent, or flow pattern types for two-phase flows),
- structural inlet and outlet sections,
- flow direction (uni-directional or bi-directional),
- requirement for explosion protection,
- pipe and gasket material,
- installation location and position (horizontal, vertical, in-line, offset).

The third level is then formed by the characteristics of the measuring device, which includes the ordered measurement accuracy, measurement dynamics, repeatability, self-calibration capability, stability and drift.

Based on the assignment of points to the criteria, possible measurement principles can then be identified, with the help of which the measurement objective can

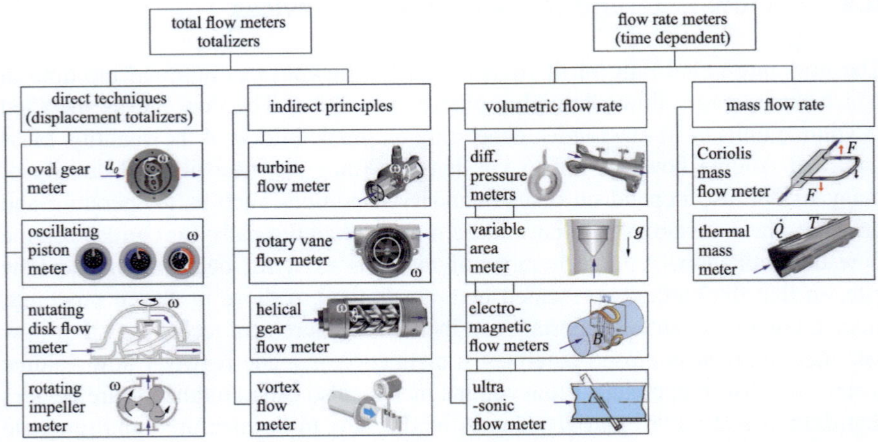

Fig. B.13 Commercially available measurement principles for total flow meters (totalizers) and flow rate meters

be realized. In the case of several possible measuring principles, the aspect of economic efficiency decides which measuring principle is finally used. Figure B.13 outlines the various physical principles available to measure total flow meters and flow rate meters in closed pipelines.

B.4.1 Total Flow or Volume Flow Meters (Totalizers)

Total flow meters are not only important elements to record the customer volumetric flow rate but also represent an indicator for maintenance intervals for pipeline networks. Commercially available total flow meters are divided into indirect and direct methods. Direct volume meters are essentially based on the positive displacement principle and thus have movable measuring chambers with clearly denoted volumes that are driven by the fluid. To this class belong oval gear, oscillating piston, nutating disk and rotating impeller flow meters. Indirect volumetric meters include also positive displacement meters but also flow meters relying on momentum change such as measuring vanes (also known as turbine meters) and flow meters, which by pre-conditioning of the flow use hydraulic phenomena such as oscillatory flow meters. Irrespective of the measurement principle of the indirect class, also counting pulses are generated by each of the methods that represent a certain partial volume, which in this case, however, is not delimited.

Positive Displacement Meters

A major advantage of displacement counters is the ability to accurately measure pulsating currents. Measurement errors occur mainly due to gap losses, i.e. leaks between fixed and moving measuring chamber walls. The viscosity of the fluid plays a significant role here. Due to wear of the material, the gap losses become larger, thus underestimating the flow rates over time. All displacement meters are susceptible to contamination due to the small clearance between the moving and fixed walls of the measuring chamber. and fixed measuring chamber walls. Careful filtering of the fluid is therefore mandatory. Gas inclusions must be eliminated by gas separators.

- *Oval gear meter*: Oval gear meters are displacement-type volume meters that transport defined incremental volumes in individual measuring chambers. The measuring element consists of two high precision toothed oval gears, which are driven by the flow of the medium and mesh with each other. In this way, a defined volume is transported for each rotation of the pair of oval gears. The number of rotations is a measure of the amount of fluid that has passed through the meter. The rotations are detected by a sensor element. Oval gear meters are applicable for high and low viscosity liquids and allow in principle for bi-directional flow. Furthermore, they require no dedicated in- or outlet flow conditioning and in case of high-quality construction allow for a long service life and high reliability. They exhibit long-term stability, a high measurement accuracy and repeatability and are easy to install.
- *Oscillating Piston Flow Meters*: Liquid enters a precision-machined chamber containing an oscillating (rotating) piston. The position of the piston divides the chamber into compartments containing an exact volume. Liquid pressure drives the piston to oscillate and rotate on its centre hub. The movements of the hub are sensed through the flowmeter wall by a follower magnet. Each revolution of the piston hub is equivalent to a fixed volume of fluid, which is indicated as flow by an indicator/totalizer. Close clearances between the piston and the chamber ensure minimum liquid slip for highly accurate and repeatable flow measurement of each volume cycle, while the maximum fluid viscosity is limited mostly to 4 Pa s.
- *Nutating Disc Flow Meters*: Here the fluid enters a precision-machined chamber containing a disc which nutates (wobbles). The position of the disc divides the chamber into compartments containing an exact volume. Liquid pressure drives the disc to wobble and a roller cam causes the nutating disc to make a complete cycle. This motion is translated into rotary motion by means of a ball and shaft, which is attached to the disc. The movements of the disc are transmitted by gear train to an indicator/totalizer or pulse transmitter. There are inherently more leakage paths in this design and it tends to be used where longer flow meter life is required rather than high performance; however, close clearances between the disc and chamber ensure minimum leakage for accurate and repeatable measure of each volume cycle. Typically, the industrial products allow for viscosities up to 11 Pa s.
- *Rotating (Lobed) Impeller Flow Meters*: A gear drive external to the measuring chamber synchronizes the impellers. During each rotation four crescent shaped volumes are moved through the measuring chamber. The number of rotations is proportional to the flow. The rotation is coupled using an adjustable fine-tooth gear train to the totalizer. An unmeasured flow, which is a function of the

pressure drop, flows through the gaps. The negative error is compensated by an adjustment. The viscosity of flow increases at high pressures and reduces the losses in the gaps which compensates for the higher losses which would otherwise exist due to the higher-pressure difference. The pressure drop results from the mechanical and dynamic resistances in the meter. The dynamic portion of the pressure losses increases appreciably with increasing load. Lobed impeller meters are very susceptible to contamination. Since contamination affects the pressure drop it must be monitored and the meter cleaned when required.

- *Helical (gear) Flow Meter*: This flow meter type has rotor scrolls which accept and release continuous flow. While the flow is being transported through the centre portion of the meter, it is trapped by the flights of the scrolls, these form positive displacement pockets. Smooth flow is assured and no pulsation is created by the meter itself. A sensor detects the motion of the rotor and produces a high-resolution pulse output. Helical flow meters get their name from the shape of their gears or rotors. These rotors resemble the shape of a helix, which is a spiral-shaped structure. As the fluid flows through the meter, it enters the compartments in the rotors, causing the rotors to rotate. Flow rate is calculated from the speed of rotation. Helical flow meters are also Positive Displacement PD Meters with a very favourable pressure drop behaviour. They are therefore perfectly suited for the flow measurement of very high viscosity fluids. Thanks to the minimal pressure loss and low shearing, helical flow meters are especially well suited for measuring high-viscosity media. Additionally, they exhibit a large measuring range, a very high resolution, outstanding reproducibility, wide measuring span and in combination with a robust construction they reveal a long service life.

Inferential Meters

In them, a pressure difference drives the medium through the flow channel and, through its momentum, moves an impeller or propeller whose number of revolutions is directly proportional to the volume flow. However, unlike the positive displacement meter, the enclosed volume is not geometrically defined. The most prominent representative is turbine volumetric flow meters, but this type also includes Woltmann meters, propeller or impeller designs, which do not necessarily have to occupy the entire pipe cross-section to measure the volumetric flow.

- *Turbine meter*: A turbine meter usually comprises an axially mounted bladed rotor assembly (the turbine) running on bearings and mounted concentrically within the fluid flow. To ensure a reliable reading a support assembly containing an upstream and downstream flow straightening section is installed. The rotor is driven by the medium (gas or liquid) impinging on the blades. Consider that the blades make an angle α with the body, then the angular speed of the blade is given by

$$\tan \alpha = \frac{\omega R}{u_0}, \quad \text{(B.26)}$$

in which u_0 denotes the mean velocity of the fluid ($u_0 = \dot{V}/A$ with the volumetric flow rate and the effective flow area of the turbine), R the radius of the blade and ω the angular speed of the blade. Each revolution is recorded as a count, which is realized by means of a magnet fitted within the rotor assembly, that induces a single pulse in an externally mounted pick-up coil. From the above expression, the volumetric flow rate $\dot{V}$ can be related with the angular speed, as

$$\omega = k\dot{V} \text{ with } k = \frac{\tan \alpha}{RA}. \quad \text{(B.27)}$$

To improve the resolution, the externally mounted pick-up coil is integrated with a permanent magnet and the rotor blades are made of a magnetically permeable ferrous material. As each blade passes the pick-up coil, it cuts the magnetic field produced by the magnet and induces a voltage pulse in the coil. To improve the resolution even further, especially in large turbine meters (200 mm and beyond) where the rotor operates at much lower angular velocities, small magnetic bars are inserted in a non-magnetic rim that is fitted around the blades. As the turbine rotation signal is acquired non-invasively, no tapping required in the pipe and hence the meter can be applied in high pressure environment at a high accuracy (±0.5% of flow) over a wide measurement range with excellent repeatability (±0.05%). Temperature limitations refer mainly to the structural and bearing materials used, so that high temperature applications are accessible. The main limitation of a turbine meter is related to the wear of the rotor and its bearing, which requires regular maintenance and a clean fluid. Another disadvantage is that the flow rate measured depends on the knowledge of the fluid viscosity and thus needs a calibration for each fluid, which is especially important at low flow rates. Further disadvantages are a certain minimum installation length (min. 10 diameter upstream and 5 diameters downstream) and a non-swirling flow.

Oscillatory Flow Meters

Oscillatory flow measurement systems involve three primary metering principles: vortex street formation, vortex swirl (precession) and Coanda effect, while the latter applies mainly to open channel flows. In all three, by means of a flow conditioner, called primary device, generates an oscillatory motion of the fluid whose frequency is detected by a secondary acquisition or measuring device to generate an output signal proportional to fluid velocity.

Vortex Flow Meters

Vortex meters use a dimensioned bluff or shedder bar, generating downstream a regular oscillating pattern of counter-rotating vortices known as Kármán vortex street, if a certain Reynolds number is exceeded. The differential pressure changes that occur as the vortices are formed and shed downstream the bluff body are used to actuate a sealed sensor at a frequency proportional to the vortex shedding.

The vortex shedding frequency (f) is directly proportional to flow velocity, since the Strouhal number (St), defined as

$$f = \frac{St u_0}{d_h}, \tag{B.28}$$

in which u_0 is the mean flow velocity and d_h the hydraulic diameter of the bluff body, is a constant. According to this relation, the drift in a vortex flow meter is not a problem as long as the system does not leave its operating range. It is also independent of the density, viscosity, temperature, pressure and thermal as well as electric fluid conductivity, as long as the Reynolds number (Re) stays within defined limits. Consequently, irrespective of whether the meter is used for measuring steam, gas or liquids, it will have virtually the same calibration characteristic and the same meter proportionality factor. This is the major advantage of the vortex meter which translates into a constant, long-term calibration not involving any in-service adjustment or tuning. The constancy of the Strouhal number over a wide Reynolds number domain is, of course, an idealized assumption that is not given, especially at smaller Reynolds numbers. In particular, optimization of the bluff body shape has increased the measurement accuracy, so that most current design show delta shaped bodies, or are T-shaped bars. Since the vortices are generated by a flow detachment at the shedder bar, the amplitude of the vortex signal depends on the dynamic pressure of the fluid it is proportional to the fluid density and the square of the velocity. As a consequence, the dynamic sensitivity range of the vortex sensor needs to be quite large. Regarding the flow meter size. the vortex shedding frequency decreases as the size of the bluff body or meter increases, while the signal strength falls off as the size decreases. Thus, vortex meters are available within a limited dimensional size ranging from 15 to 200 mm (Crabtree, 2009). Many vortex meters use non-wetted, external sensors connected to internal parts that move or twist due to vortex shedding, thus the technology applied for the sensors depends on the manufacturer and a variety of sensor techniques such as thermal sensors, mechanical sensors, capacitive either diaphragm or piezoelectric based pressure sensors, but also strain gauges or ultrasonic devices are used, each exhibiting a different turn-down ratios, sensitivities or temperature limitations. Independent of the sensor technology used, the natural vortex frequency is converted into a digital signal which is then processed through the meter's electronics to calculate flow.

Vortex shedding flowmeter require clean low viscosity liquids (<1 Pa s), gases and steam to obtain specified accuracy and pipe Reynolds numbers exceeding $Re = 3 \times 10^4$. Due to its operation principles a vortex meter can only measure flow in one direction and requires full flooding of the ducts cross-section and shows a low pressure drop. An essential drawback is that a vortex meter vortex accuracy is based on a well hydraulically developed and symmetrical fluid velocity profile, free from distortion or swirl, existing in the pipe similar as turbine meters, ultra-sonic flow meters or orifice plates, for which the manufacturer provided in the data sheet corresponding recommendations.

B.4.2 Flow Rate Meters (Time Dependent)

Flow rate meters can be meters can be roughly divided into two families, the actual volumetric flow meters and the mass flow meters. The category of volumetric flow meters is further subdivided according to the physical measuring principle into the differential pressure or variable area methods, the electromagnetic flow meters (DC and AC) and the ultrasonic methods, while the mass flow methods include the Coriolis meter and the thermal mass flow meter.

Differential Pressure Meters
Despite many new developments, the differential pressure method is still the most widely used measuring method. The measuring principle of differential-pressure flowmeters is based on a pressure difference caused by constriction of the flow cross-section, which is directly dependent on the respective flow rate. A basic distinction is made between four different types of orifice plates used for the differential pressure measuring method:

- *Orifice plates.* The orifice plate generates a pressure difference by means of a circular constriction in the pipe cross-section. The associated increase in flow velocity reduces the static pressure. The pressure difference upstream and downstream of the orifice plate is a measure of the flow rate.
- *Nozzles.* The use of a nozzle as a throttling element results in lower pressure losses as with an orifice plate. The disadvantage of nozzles is that they are more difficult to manufacture than orifices and are therefore considerably more expensive.
- *Venturi tubes.* When using Venturi tubes or Venturi nozzles, only very small pressure losses have to be accepted compared to orifices. Due to the manufacturing costs, the price is much higher than that of orifices.
- *Prandtl probe (pitot tube).* The Prandtl probe is a measuring device for mobile measurement of the flow velocity of a medium. With the help of the pitot tube the static pressure and the total pressure in e.g. the air flow are measured. The difference between the total pressure and the static pressure results in the dynamic pressure, which is a measure of the flow velocity at this position.

If we consider the flow of a fluid through a pipe segment and reduce the cross-section, as shown in Fig. B.14a, the velocity of the medium increases at this point. According to Bernoulli's equation, however, the energy content of a flowing substance, which is made up of the sum of static, dynamic and height energy, is constant along a flow path.

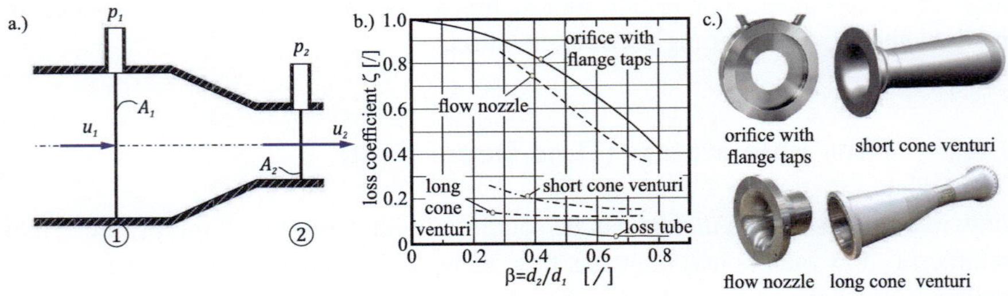

Fig. B.14 **a** Principle set-up of a difference pressure based flow meter; **b** permanent pressure loss as a function of the constriction ratio β for different orifice type of differential pressure meters; **c** design of commercially available pressure meters

Thus, the increase in velocity causes a decrease in static pressure.

$$\frac{p_1}{\rho g} + \frac{u_1^2}{2g} + h_1 = \frac{p_2}{\rho g} + \frac{u_2^2}{2g} + h_2, \tag{B.29}$$

where g represents the gravity vector, ρ the specific density and p_i, u_{i1} and h_i denote the static pressure, mean velocity and heights at the positions 1 and 2. If the heights are the same and the fluid is assumed to be incompressible ($\rho = const.$), so that the continuity equation simplifies to $(u_1 A_1) = (u_2 A_2)$, the volumetric flow rate ($\dot{V}$) can be determined to:

$$\dot{V} = u_1 A_1 = \frac{A_2}{\sqrt{1 - (A_2/A_1)^2}} \sqrt{\frac{2(p_2 - p_1)}{\rho}}. \tag{B.30}$$

Equation (B.30) applies exactly only to laminar, inviscid flows. To take the effects related to turbulence and viscosity into account a discharge coefficient (C_D) is introduced multiplicatively. C_D is affected by changes in the duct diameter ratio (or constriction ratio) β defined as the ratio of the smaller to the larger diameter of the pipe, i.e. $\beta = d_2/d_1$, the Reynolds number Re, pipe roughness, the sharpness of the leading edge of the orifice, and the positions at which the differential pressure across the plate are recorded. Generally, for most orifices the discharge coefficient ranges from 0.6 to 0.9. Reference texts and standards are available that list typical values and tolerances for C_D in standard installations. Compressibility effects are compensated by the introduction of an expansibility factor ε. Both, C_D and ε are determined from equations and tables documented in international standards, in Europe ISO 5167-1 (2004) and in the United states API 2530 (1985). These standards not only specify C_D and ε, but also the geometry and installation conditions for the orifice plate, Venturi tube, and nozzle, and are essentially a design guide for the correct use of the most available types of differential pressure flowmeter. The installation recommendations are intended to ensure that fully developed turbulent flow conditions exist within the measurement section of the flowmeter.

On the one hand Eq. (B.30) illustrates the greatest strength of the orifice plate, which is that measurement performance can be confidently predicted without the need for calibration if the device is manufactured, installed, and operated matching international standards. Additionally, the device is cheap to manufacture, has no moving parts, is reliable, and can be used for metering most clean gases, liquids, and steam even at high pressures and temperatures. But, Eq. (B.30) shows also two intrinsic limitations common to all differential pressure systems, which are the square root relationship of differential pressure and flow rate ($\dot{V} \sim \sqrt{\Delta p}$) severely limits the turn-down ratio (typical maximum of 5 : 1 and less); and if density (ρ) is not constant, it must be known or measured. The latter is in practice mostly irrelevant in incompressible media, since the density changes are marginal but are of relevance in gas flows. The most relevant limitation, however is that orifice meters create a permanent pressure loss, which depends on the constriction ratio β and the type of meters used. Figure B.14b shows the permanent pressure losses of the different types of flow meters and Fig. B.14c some commercially available flow meters.

Variable Area Flow Meters (Rotameter)

In order to be able to determine the flow of gases and liquids simply and at the same time relatively accurately, the float flow measurement principle can be applied, which is sketched in Fig. B.15. In this case, a vertically positioned tube, which widens conically towards the top, is flowed through from bottom to top. The upward flowing measuring medium lifts the float (often called bob) located in the tube until the annular gap between the float and the tube wall is so large that the forces acting on the float are in equilibrium and a floating state is thus achieved. Transparent measuring tubes are most commonly used, where the height of the float can be read directly as flow information; for higher temperatures or pressures, designs are available that realize a direct conversion via magnets, for example.

The evaluation of the equilibrium position defines the flow rate. From Eq. (B.30) the pressure difference of an incompressible fluid is given. This, however, contains the pressure difference not known. But, what can be easily conducted is to formulate the force equilibrium, in which the weight of the float (F_g) must be balanced by the difference of the downward and upward forces (F_d, F_u)

$$F_g = F_u - F_d = gV_f\left(\rho_f - \rho\right) = p_1 A_f - p_2 A_f \Rightarrow (p_1 - p_2) = \frac{V_f}{A_f}\left(\rho_f - \rho\right), \tag{B.31}$$

in which V_f denotes the volume of the float, ρ and ρ_f the density of the fluid and the float and A_f is the area of the float. Substituting the above expression in Eq. (B.30) one obtains for the volume flow rate:

$$\dot{V} = \frac{C_D\left(A_t - A_f\right)}{\sqrt{1 - \left(\left(A_t - A_f\right)/A_t\right)^2}} \sqrt{\frac{2g}{\rho}\frac{V_f}{A_f}\left(\rho_f - \rho\right)}, \tag{B.32}$$

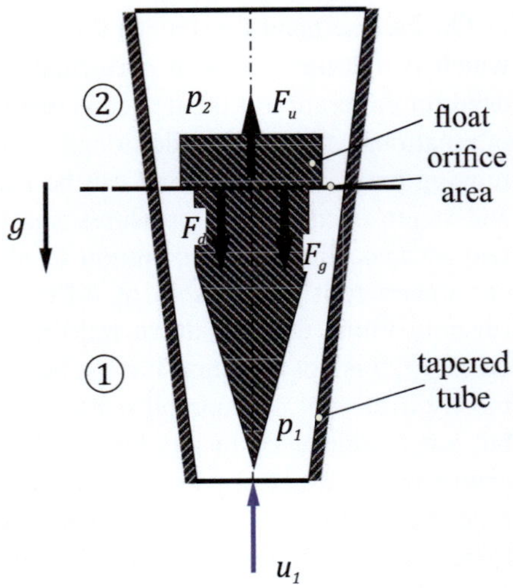

Fig. B.15 Basic construction of a variable area flow meter (Rotameter)

where A_t denotes area of the tube at equilibrium (corresponding to the dotted line) in Fig. B.15. If the distance between float and tapered tube is small, i.e. $\left((A_t - A_f)/A_t\right)^2 \rightarrow 0$, one gets a linear relationship of flow rate and area difference, $\dot{V} = K(A_t - A_f)$, with K being a lumped parameter. Designing the tube in a way that its cross-section increases linearly with the vertical displacement x the volumetric flow rate is proportional to the displacement. Additionally, the construction of the float determines the performance of the rotameter. Aside from a vertical placement it shall create a uniform turbulence to ensure insensitivity to viscosity but maintain sensitive to variations of the fluid density. According to these requirements, the shape the floats exhibits commonly a sharp edge and several angular grooves at the top, so that the fluid passing through these grooves, causes the rotation of the float and the turbulence inducing a flow separation at the top reduces the viscous forces. The simple and practical design allows it to be used primarily as a low-cost device for monitoring cooling circuits, as it can withstand high pressures and temperatures, allows opaque measurement media and two-phase flows, and is also suitable for high flow rates. Other advantages include the absence of an external power supply, independence from electrical conductivity, the ability to measure the smallest gas and liquid volume flows, and high repeatability.

Electromagnetic Flowmeter

Permanent magnet flow meters (PMF) are mostly used if the installation volume is rather small or where also very low flow rates have to be resolved. According to the Faraday law an electrically conducting fluid flowing perpendicular to a magnetic field ($\vec{B}$) induces an electric field ($\vec{E}$). The strength of this electric field is proportional to the mean flow velocity (u) and can be measured with diametrically opposed electrodes on the pipe walls perpendicular to flow direction and magnetic field, as illustrated in Fig. B.16a. From the Maxwell equations the Poisson equation for an electromagnetic flow meter can be derived,

$$\nabla^2 \phi = \nabla \cdot \left(\vec{u} \times \vec{B} \right), \tag{B.33}$$

in which ϕ is the electric potential and $\vec{u}$, $\vec{B}$ are the vector quantities for velocity and magnetic field strength. The partial differential equation (B.33) cannot be solved in general, but with simplified boundary conditions and with the aid of the Green function a transformation can be made, and a solution can be given for a finite region of a flow meter with point electrodes. For the voltage at the electrodes, derived from the electric potential difference $\Delta\phi$, one obtains with a weighting vector W, which indicates the contribution of the flow in a single volumetric element to the effective measurable voltage,

$$\Delta\phi = \int_{x,y,z} W\left(\vec{u} \times \vec{B} \right) dxdydz. \tag{B.34}$$

For an axially symmetric flow profile and an infinite, homogenous magnetic field follows the simple dependency,

$$\Delta\phi = k \cdot u_0 \cdot \overline{B} \cdot d, \tag{B.35}$$

exhibiting a linear proportionality between the measured electrode voltage $\Delta\phi$ and the mean flow velocity u_0 depending on the mean magnetic field strength $\overline{B}$ and the pipe diameter d. The parameter k is device specific constant, usually provided by the manufacturer. The weighting distribution W accounts for the influence of an asymmetric flow profile on the measurement accuracy, for more details see OECD-NEA (2015). Since higher gradients in the weighting field cause large measurement uncertainties, the use of point electrodes should introduce higher uncertainties in the signal to be measured than using line electrodes. In reality this can be observed, but the deviations are always small compared to the signal amplitude, normally lower than 1%. A different situation occurs, when impurities accumulate in the at or in the vicinity of the electrodes. In this case a massive change of the signal voltage arises and destroys the linearity of the flow meter system. Hence, Eq. (B.35) no longer holds and the flow meter becomes useless.

The magnetic field, which should be as large as possible to generate a high useful signal, can be generated either by a permanent magnet or by means of excitation coils which have a ferromagnetic iron core. The externally excited designs can be operated with AC mains voltage or pulsating DC voltage. The only requirement for the piping material is that it is made of non-magnetizable pipe materials, such as austenitic steel or other stainless steels, so that the externally applied magnetic field is not weakened.

Contrary to what is claimed in some metrology books, it does not matter if the electrodes have only electrical contact with the fluid and the rest of the flow channel is covered with an insulating liner. The reason is that the flowmeter itself is a low impedance voltage source and by connecting it to a high impedance measuring device, the pipe wall is only an electrical resistance in the measuring circuit. Thus, for example, volume flows of liquid salts or metals can be measured reliably and precisely even at extremely high temperatures.

This flowmeter principle does not make any extreme demands on the specific electrical conductivity of the fluid either, as long as the internal electrical resistance of the fluid in the measuring tube is considerably lower than the ohmic input resistance of the measuring device. Even small impurities in the tap water are sufficient to ensure the functionality of the flowmeter, which is often called magnetic inductive. It should be noted that the conductivity of liquids can vary with temperature and care should be taken to ensure the performance of the liquid in marginal conductivity applications is not affected by the operating temperatures. Most liquids have a positive temperature coefficient of conductivity. However negative coefficients are possible in a few liquids.

The conventional designs do not solve the problem of 'electrode coating' in which an insulating deposit effectively isolates the electrodes. However, especially for AC operated divides a capacitive coupled flowmeter has been developed in which the electrodes are replaced by capacitive plates buried in the liner as one of the plates as indicated in Fig. B.16b; the meter lining acts as the dielectric; and the second capacitive plate is formed by the metallic electrode that is embedded in the tube liner. In other designs, the electrodes take the form of two large plates bonded to the outside of a ceramic flow tube and a preamplifier mounted directly on the flow tube. The latter design allows to measure liquids with specific electric conductivity levels down to $0.05\,\mu S/cm$. All non-electric contact based inductive flow meters have as positive feature, that they exhibit no gaps or crevices, no risk of electrode damage due to abrasion; no leakage; and no electro-chemical degradation.

With the help of combined sensor arrangements, AC based systems also allow the detection of volume flows in partially filled pipelines. Figure B.16c shows the magnetic field distribution at two different values of excitation. If the pipe is full, the maximum signal between the electrodes E_1 is obtained, while at the reversed excitation the signal annihilates. If the level in the tube decreases, a smaller signal is obtained in the upper part, while the signal in the lower part remains identical. With the aid of calculation algorithms, the wetted pipe surface and ultimately the volume flow can be determined.

An alternative method using electromagnetic effects to measure the flow rate, which provides two net signals is the AC electromagnetic frequency flow meter (EMFM). The general measurement principle of such an induction flow meter is that the motion of an electrically conducting fluid in an imposed field B produces an induced field B' which is proportional to the flow rate in the first order. This method involves an effect which is proportional to the specific electric fluid conductivity σ. The advantage of the AC electromagnetic flow meter is that its output

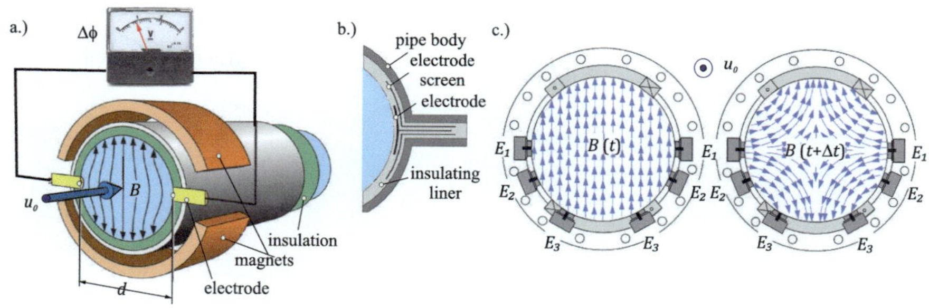

Fig. B.16 **a** Functional principle of a magnetic inductive flow meters; **b** non-invasive capacitive acquisition of a flow meter signal for an AC based magnetic inductive flow meter; **c** electrode arrangement of an AC based inductive flow meter to measure flow rate in partially filled tube

is directly electrical and no transducer is required which allows a large temporal resolution. Moreover, no direct contact of the sensor with the operation fluid is necessary for data acquisition, avoiding material compatibility issues. The set-up consists of two coils A are supplied with alternating electric currents in opposite direction, producing an AC magnetic field $B(t)$ as illustrated in Fig. B.17. In the absence of a fluid motion, the resulting magnetic field recorded at the centre coil B is symmetrical and no signal is induced in the sensing coil B according to the induction equation. As soon as flow occurs, the magnetic field lines are dragged downstream and a signal appears in coil B, which is proportional to the flow rate to the first order. The flow direction is obtained by the sign of the RMS value of the sensing coil B. The magnitude of the RMS value in the sensing coil is proportional to the magnetic Reynolds number of the fluid flow Re_m which is calculated according to:

$$Re_m = \mu_0 \sigma(T) u_0 d, \tag{B.36}$$

where μ_0 is the magnetic permeability of vacuum, given as $4\pi \cdot 10^{-7}$A s/(Vm), σ the specific electric conductivity of the fluid as a function of the temperature T, u_0 the mean flow velocity within the duct and d its diameter. Because the transport and the diffusion of the magnetic field by the electric conducting liquid expressed by magnetic Reynolds number Re_m as well as the dimensions of the duct (d) play a role, they contribute multiplicatively to acquired signal, so that this type of flow meters requires a certain size (typically >20 mm in d). What makes the device appreciable is that a second net signal given by the phase shift $\Delta\varphi$ between applied and induced field is obtained for free, for more details see Schulenberg and Stieglitz (2010).

Ultrasonic Flowmeter

Flow rates can be determined non-intrusively with the aid of ultrasonic waves. Two main methods have been established for this purpose; the ultrasonic transit time

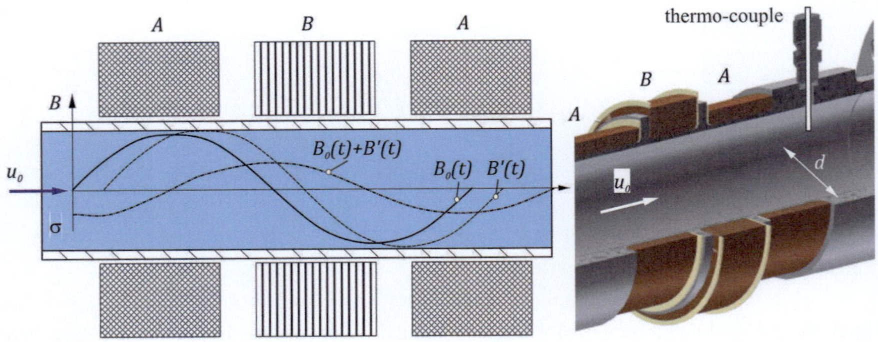

Fig. B.17 Set-up of an electro-magnetic frequency flow meter (EMFM), left and practical realization (right)

method (UTT) and the ultra-sound Doppler velocimetry (UDV) method, which differ in principle with regard to the processing of the measurement signals.

Ultra-sound Transit time (UTT) flow meter

The UTT meters measure the difference in time from when an ultrasonic signal is transmitted from the first transducer A until it crosses the pipe through the fluid and is received by the second transducer B. Both transducers are bi-functional acting as sound emitter and receiver. By means of a comparison of upstream and downstream measurements a time difference is evaluated. The principal set-up is illustrated in Fig. B.18a. If there is no flow, the travel time will be the same in both directions and the transit time t_0 is given by the distance l from A to B and the sound speed of the liquid (c) in a simple manner by $t_0 = l/c$. In case of flow with the mean velocity u_0 the sound is transported faster if traveling in the same direction as the flow and slower if moving against it. The transit times from A to B (t_{AB}) and vice versa (t_{AB}) of the signals can then be calculated by:

$$t_{AB} = \frac{l}{c + u_0 \cos\theta} \text{ and } t_{BA} = \frac{l}{c - u_0 \cos\theta}, \tag{B.37}$$

in which θ denotes the angle of inclinitation with respect to the tube axis. Transformation yields to an expression for the mean flow velocity u_0:

$$u_0 = \frac{L}{2\cos\theta}\left(\frac{1}{t_{AB}} - \frac{1}{t_{BA}}\right). \tag{B.38}$$

In practice the determination of the transit time difference from the phase shift between the two signals $\Delta t = (t_{AB} - t_{BA})$ is common and with the postulate $(u_0/c)^2 \ll 1$ one obtains

$$u_0 = \frac{c^2}{2l\cos\theta}\Delta t. \tag{B.39}$$

Compared to Eq. (B.38) in expression (B.39) the mean velocity depends on the sound speed only. This disadvantage is accepted because the determination of the usually very short transit times appearing in Eq. (B.37) is technically difficult and less accurate than the identification of its difference. Due to the fact that beside the sound speed only geometric measures occur in the equation, the ultrasound transit time method can be used for calibration of other measurement methods if the sound speed in the fluid is known, and a fully developed flow profile is present.

As the cross-sectional area of the pipe section is known, the product of area and mean velocity results in the volumetric flow rate. The accuracy of measurement is determined by the ability of the instrument to measure accurately the transit time, which results in dense liquids exhibiting a high sound speed in small tubes to nanoseconds and even smaller values. Obviously, with enlarging longer path lengths, this stringent time measurement requirement becomes easier to meet, so that many manufacturers of commercial equipment use one or more reflections of the pipe walls to enlarge the effective path length, as indicated in Fig. B.18b.

Since the measured time of the sound transport is made up of the sum of the instantaneous velocities at each point across the diameter of the pipe and thus represents an integral value, the transit time meter only provides a picture of the total flow profile along the path of the beam. Having that in mind, the validity of the measurement can only be relied on if the flow distribution across the cross-section is not affected by asymmetries or swirl. Moreover, an a priori knowledge of the flow profile within the intended measurement range is required. If e.g. a laminar to turbulent transition occurs, the velocity profile alters as indicated in Fig. B.18a and the laminar-to-turbulent transition error can be up to 33%. Hence, any UTT based flow rate measurement requires a preconditioning of the flow in the tube, for which the manufacturer guidelines are recommended to follow.

Since the ultrasonic signal must traverse the pipe to be received by the sensor, the liquid cannot be comprised of a significant number of solids or bubbles (changing the sound-speed), or the high frequency sound will be abated and too weak to travel across the pipe. Hence, the UTT methods is in principle a time of flight (TOF) measurement technique, in which two ultrasonic sensors are either in direct contact with the fluid or contacted via a pipe to the fluid. The TOF is linear to the average flowrate in the pipe. Thus, the flowrate can be calculated from the given TOF.

To generate ultrasound mainly piezoelectric materials are used (quartz, lead zirconate, barium titanate) whose Curie-temperature, above which they get instable and irreversibly destroyed, limits the applicable temperature range to a maximum of about 150 °C. For measurements in high temperature fluids like liquid metals or solar salts tow options are available; either to use cooled clamp-on systems or to decouple the emitter/receiver system by means of wave guides, which ensure the transport of the sound pressure signal as loss-free as possible Such wave guides are currently commercially available up to temperatures of up to 450 °C.

In order to transport a significant energy through the structure into the fluid the material thickness of the emitter has to be adapted to the fluid. Here, the impedance Z which is the product of the sound speed c and the density ρ has to be considered. The permittance D of an acoustic wave through a solid–fluid interface can be expressed as a function of the thickness of the material t, the wavelength λ of the sound wave and the impedance of the fluid and the liquid Z_i according to

$$D = \frac{1}{\sqrt{1 + \frac{1}{4}\left(m - \frac{1}{m}\right)^2 \sin^2\left(\frac{2\pi t}{\lambda}\right)}} \text{ and } m = \frac{Z_1}{Z_2}. \tag{B.40}$$

This permittance is only achieved if the interface is perfectly physically wetted and no mode transitions of the wave at the interface appear. Other sources of uncertainties using the ultra-sound are the absorption and the sound scattering. Finally, the anisotropy of the sound speed has to be considered. In case of measuring flows containing a temperature gradient, the sound speed varies with the square root of the temperature and thus the temperature has to be known and adequate countermeasures to be foreseen. The mode transition of the sound wave at any interface leads not only to a deflection of the sound wave, as sketched in Fig. B.18c but also causes a significant decrease of the signal magnitude is solved by a specific sensor design. The mode transitions can in principle occur due to:

- longitudinal to transverse wave transitions,
- surface waves like Rayleigh-waves,
- creep waves and
- ring sound.

Despite all the aspects care that has to be taken of for a UTT flow rate measurement and the compared to other methods relatively high price, UTT has some advantages. Among them one is important so that transducers can be clamped externally to the walls of the pipe to provide portable non-intrusive flow measurement systems that can be installed within a few minutes to virtually any pipe and by choosing an adequate frequency measurement are accessible to various structural materials encompassing metal, plastic, ceramic, asbestos cement and internally and externally coated pipes. Clamp-on transducers are also often used in permanent installations that cannot justify a permanent in-line meter but nonetheless require periodic metering.

Ultrasound Doppler Velocimetry (UDV)
The Doppler ultrasonic flow meter operates on effect that the frequencies of the sound waves received by an observer are dependent upon the motion of the source or observer in relation to the source of the sound. A Doppler ultrasonic flow meter uses a transducer to emit an ultrasonic beam into the stream flowing through the pipe. For the flow meter to operate, there must be obstacles (solid particles, air bubbles) in the stream to reflect the ultrasonic beam. The motion of particles shifts the frequency of the beam, which is received by a second transducer. In principle also UDV is a non-intrusive method to measure instantaneously entire velocity profiles, which makes it largely independent from a sophisticated flow conditioning, except

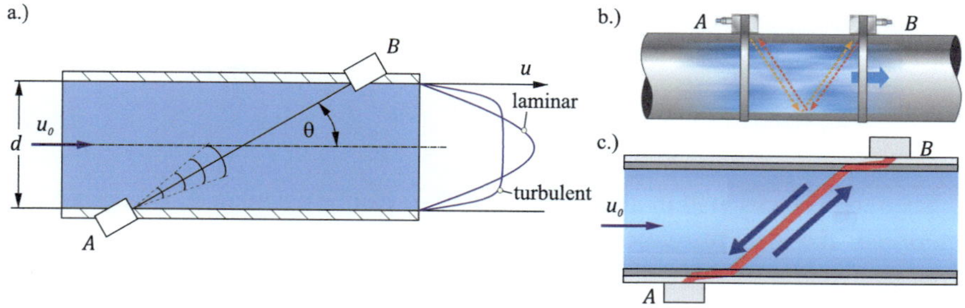

Fig. B.18 **a** Principle of a UTT flow rate measurement; **b** sound pass enlargement by reflection; **c** refraction of a sound-wave at interfaces

for swirl, which must be prevented. The related velocity information in a UDV is obtained from the shift in position of scatters between ultrasound pulses and not exactly from the doppler frequency shift of the echoes. Hence, the velocity information is calculated from a correlation function. The advantage of the UDV compared to the UTT approach is that the flow rate is not obtained as an integral value but scanned along the entire sound path at several positions.

The location P of the measurement volume is given by the product of the sound speed (c) and the actual traveling time of the pulse (t_d). The spatial shift of the position ΔP from position ① to positions ② as indicated in Fig. B.19, between two pulses applying a pulse repetition time t_{prf} is given by

$$P = \frac{c\, t_d}{2} \text{ and } \Delta P = (P_2 - P_1) = u(y) t_{prf} \cos\theta = \frac{c\,(t_2 - t_1)}{2}. \quad \text{(B.41)}$$

The time difference $(t_2 - t_1)$ is recorded by means of the phase shift of the echoed signals and finally, the velocity $u(y)$ at a discrete coordinate z can be evaluated by:

$$\Delta d = 2\pi f_e (t_2 - t_1) \text{ and } u(z) = \frac{c \Delta d}{2 f_e t_{prf} \cos\theta} = \frac{c f_d}{2 f_e \cos\theta}, \quad \text{(B.42)}$$

where f_e is the echo velocity and f_d representing the Doppler shift.

Of course, an upper limit of the maximum measurable velocity and the maximum measurement depth exists. This constraint is expressed by

$$u_{max} = \frac{c}{4 t_{prf} f_e \cos\theta}; \; y_{max} = \frac{c\, t_{prf}}{2}. \quad \text{(B.43)}$$

The UDV technique allows the instantaneous acquisition of whole velocity profiles. This is performed by a reconstruction of the velocity information in the following way. First, the length of the acquired echoes reflected from the particles within in fluid is divided into intervals of the temporal length $\Delta\tau_i$. The run time τ_i is associated with a measurement position in the duct y_i via the relation

$$y_i = \frac{c\ \tau_i}{2}. \tag{B.44}$$

Then the velocity within a discrete interval u_i can be calculated. This procedure determines the local resolution and the size of the measurement volume, which is given by

$$\Delta y_i = \frac{c\ \Delta\tau_i}{2} \text{ and } \Delta y_e = \frac{c\ \Delta\tau_e}{2}. \tag{B.45}$$

From the procedure explained above it is clear that numerous effects contribute to the acquisition of a reliable signal and in fact about 100 parameters are necessary to define the measurement process. The enormous sound speed of dense media requires extremely fast acquisition and data processing systems to obtain reliable velocity information. Besides these quite sophisticated demands on the data transmission recording and processing units several constraints regarding the environmental conditions have to be fulfilled. One of the most crucial problems is the elevated temperature of more than 200 °C in typical liquid metal applications, which cannot be sustained by commercially available Barium titanate sensors ($BaTiO_3$). Thus, wave guides have been developed to decouple the temperature from the sensor, where the technologic limit currently is currently about 600 °C.

Mass Flow Meters

Mass flow is a primary unit of flow measurement and thereby unaffected by biases arising from viscosity, density, conductivity, pressure and temperature. As a result, it is inherently more accurate and meaningful for measuring material transfer. In most applications, however, mass flow is measured inferentially, by volumetric flow metres by multiplication with the density.

Coriolis Mass Flowmeter (CMF)

The measurement principle of a gyrostatic flow meter is taking advantage of acquiring the Coriolis force, which appears in rotating and oscillating (vibrating) systems. The setup consists of a curved tube clamped between two fixed points and vibrated by an exciter. When a medium flow through the tube, it causes a change in frequency, a phase shift and a change in amplitude that can be detected.

Figure B.20a shows such a set-up, in which the tube is excited by an external force $\overrightarrow{F}$, where the excitation frequency is kept at the natural frequency of the tube, which minimizes the energy needed for vibration. The general expression for the Coriolis force is $\overrightarrow{F}_C = 2 \cdot m \cdot \overrightarrow{u} \times \overrightarrow{\omega}$, where the mass flow $\overrightarrow{q}$ is given by $\overrightarrow{q} = m \cdot \overrightarrow{u}$ and $\overrightarrow{\omega}$ is the rotation vector. In case the fluid is at rest the Coriolis force in the vibrating tube vanishes. In case of a fluid flow at the inlet section, the Coriolis force tends to decelerate the movement of the oscillating tube, whereas, for the outlet section, the Coriolis force tends to accelerate the movement. In the

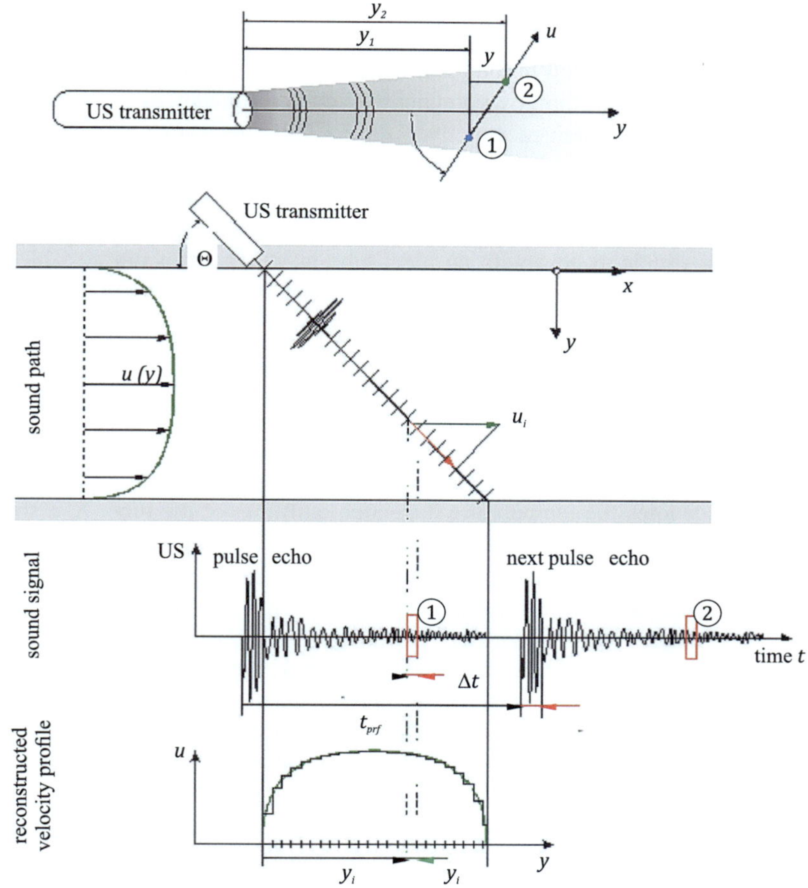

Fig. B.19 Operation principle of the ultra-sound Doppler velocimeter (UDV) to evaluate the flow rate in a tube; bottom right industrial UDV-Sensor with high temperature wave guide for solar applications

middle of the tube, the Coriolis force is always zero, since $\vec{q}$ is parallel to $\vec{\omega}$, so that the cross product $\vec{q} \times \vec{\omega}$ gets zero.

As soon as the fluid begins to flow, the Coriolis force induces a phase shift along the tube. In the arm through which fluid flows away from the axis of rotation must exert a force on the fluid to increase its angular momentum, so it is lagging behind the overall vibration. The arm through which fluid is pushed back towards the axis of rotation must exert a force on the fluid to decrease the fluid's angular momentum again; hence that arm leads the overall vibration. The inlet arm and the outlet arm vibrate with the same frequency as the overall vibration, but when there is mass flow the two vibrations are out of synchronization: the inlet arm is behind, while the outlet arm is ahead. The two vibrations are shifted in phase with respect to each other, and the degree of phase-shift measured between two sensor positions,

S_1 and S_2 is a measure for the amount of mass flowing through the tubes. Since the oscillation is kept at the natural frequency of the system, the frequency alters with changing density of the fluid in the tube; i.e., the natural frequency increases with decreasing density. Therefore, by knowing the actual frequency of the system, the density of the fluid is obtained directly. Another direct measurement, in addition to mass flow and density, is the fluid temperature, which is measured by the CMF. The general problem to derive an exact solution for the measurement signal as a function of the excitation is very complex, and an analytical solution can only be obtained for a simple system with an ideal tube in which an incompressible and inviscid fluid flows; fortunately, the fundamental dependencies do not change and thus a general relationship can be provided so that e.g. the mass flow of a u-shaped Coriolis flow meter is given as:

$$\dot{V} = \frac{K_u - I_u \omega^2}{2Kd^2} \tau \,, \tag{B.46}$$

in which K_u denotes the temperature dependent stiffness of the tube, K a shape-dependent factor, d the width, τ the time lag, ω the vibration frequency and I_u the inertia of the tube. As the inertia of the tube depend on its contents, knowledge of the fluid density is needed for the calculation of an accurate mass flow rate. If the density changes too often for manual calibration to be sufficient, the Coriolis flow meter can be adapted to measure the density as well. The natural vibration frequency of the flow tubes depends on the combined mass of the tube and the fluid contained in it. By setting the tube in motion and measuring the natural frequency, the mass of the fluid contained in the tube is immediately obtained and dividing the mass by the known volume of the tube yields the density of the fluid.

The calibration of a CMF is affected by stiffness changes of the flow tubes, caused by temperature and pressure. But these can be compensated for through pressure and temperature zero and span compensation factors. Additional effects such as pitting, cracking, coating, erosion or corrosion will alter the tube rigidity and thereby cause shifts in the calibration factor over time due, however, these can be handled by regular meter calibration or verification checks.

There are many different designs of Coriolis mass flowmeters in which the primary sensor comprises an array of one or two bent tubes through which the fluid to be measured flows. This most common type exhibits a high pressure drop. However, there are also designs based on straight tubes that do not have this disadvantage. Most of the mass flow meters can be only operated in one flow direction.

Even if the Coriolis measuring principle suggests the suitability for the measurement of two-phase flows, this is not true, because with a high two-phase fraction two problems arise. An inhomogeneous mixture leads to an irregularly fluctuating density and thus to a constantly fluctuating resonant frequency, which can bring the measuring instrument out of phase. A second problem is that the Coriolis method assumes that all particles in the medium are accelerated along paths that follow the motion of the tube. At high gas fractions, the particles in the centre of the

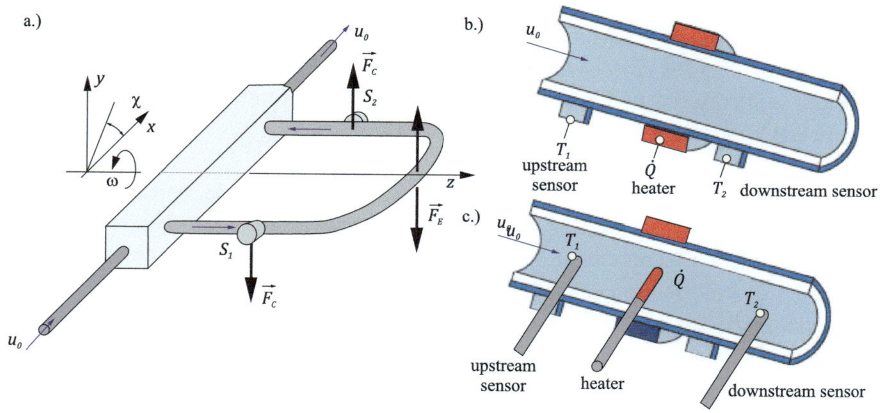

Fig. B.20 **a** Operation principle of a Coriolis mass flow meter (CMF); set-up of a thermal mass flow meter in an externally heated version (**b**) and an immersion type device (**c**)

tube will no longer participate the motion of the tube. Conversely, the Coriolis forces of the mass particles in the centre of the tube will no longer act on the tube walls. The consequence is that the measured value systematically becomes too small. Most manufacturer therefore limit the void fraction to be acceptable for an accurate measurement of less than 10%.

Thus, the major advantage of a CMF scopes the capability of a direct, in-line and accurate mass flow measurement of both liquids and gases over a large range with no dependence on temperature, pressure, viscosity, conductivity and density of and with a simultaneous acquisition not only of mass flow but also density and temperature from only one sensor. The drawbacks are aside from the price the limitations in the tube diameter barely exceeding 150 mm in diameter.

Thermal Mass Flow Meter

The measuring principle of thermal mass flow meters is based on the thermal properties of the flowing medium and is thus intrinsically correlated with the specific heat capacity and the thermal conductivity. Since mass flow $\dot{m}$ is defined as the differential of the change in mass over time, i.e. dM/dt, the measurement signal of a thermal mass flow measurement is related to the average volume flow rate.

$$\dot{m} = \rho \cdot \dot{V} + \left(V \frac{d\rho}{dt} \right), \tag{B.47}$$

in which ρ is the specific density of the fluid. For an incompressible fluid this relation simplifies to the first term only. If a channel of constant cross-section A is considered, the measurement of the mass flow is reduced to the product of the mean flow velocity (u_0) and mean density (ρ), and thus a mechanical quantity such as the mass flow is converted into an electrical signal via a thermal measured quantity. Since the specific heat capacity c_p of gases and fluids is largely independent of pressure and changes only slightly with small temperature differences, it is proportional to the density and thus to the mass. Assuming two different types of heater controls and two evaluation modes, three different types of thermal mass flowmeters can be realized:

- Thermal mass flowmeters measuring the heat transfer at a defined heated object (Modes: detection of heating at constant heater temperature, change of heater temperature at constant heating power). This class is given by hot-wire, hot-film sensors or hot-element sensors.
- Calorimetric sensors measuring the spatial distortion of temperature profile around the heater caused by the flowing medium.
- Sensors measuring the transport time of a heat pulse over a pre-scribed distance along the flow path, which belong to the thermal time of flight (TOF) sensor class.

According to the fundamental operation principle either by introducing a known amount of heating power ($\dot{Q}$) into the flowing liquid and measuring an associated temperature change or by maintaining a probe at a constant temperature (or temperature difference) and measuring the electric heating power to do so, they are mainly used for gases since liquids associated with their high heat capacity requires large heating powers. The components of a basic thermal mass flow meter include two temperature sensors and an electric heater between them, but design dependent the heater can in the fluid flow or externally attached to the pipe, see Fig. B.20b. According to the operation principle the thermal mass flow meters function only in one flow direction, however, the feature only marginal pressure losses due to the minor or even absent disturbance of the flow. Nonetheless, they require a flow conditioning as the signal is assumed to be the integral of the behaviour of the entire cross-section.

In the direct-heat version, a fixed amount of heat ($\dot{Q}$) is added by an electric heater, as shown in Fig. B.20c. As the process fluid flows through the pipe, a temperature detector (RTDs or thermocouple) measure the temperature rise from upstream temperature T_1 to the downstream temperature T_2, while the amount of electric heat introduced is held constant. The mass flow ($\dot{m}$) is calculated on the basis of the measured temperature difference ($T_2 - T_1$), a manufacturer and design dependent device coefficient (K), the electric heat rate ($\dot{Q}$) according to,

$$\dot{m} = \frac{k \cdot \dot{Q}}{\left(c_p \cdot (T_2 - T_1)\right)}, \tag{B.48}$$

in which c_p is the specific heat capacity of the fluid.

B.5 Pressure Measurements

In contrast to the volume/mass flow, the pressure itself is initially a scalar quantity and thus independent of direction; the pressure units, on the other hand, are a measure of the force (F) acting on a given area (A). The unit of pressure is Pascal $Pa = N/m^2$. The pressure is divided into absolute pressure (p_{abs}), gauge pressure (p_g) and differential pressure. Absolute pressure is the sum of gauge pressure and atmospheric pressure (p_{atm}), i.e. $p_{abs} = p_g + p_{atm}$.

Pressure gauges form the main group of pressure sensors in piping instrumentation, measuring pressure mostly in relation to atmospheric pressure, i.e. gauge pressure. Pressure gauge sensors are components that change their shape when exposed to pressure. These devices include diaphragms, capsules, bellows, and Bourdon tubes, as illustrated in Fig. B.21a–e.

Diaphragms consist of a thin layer or film of material supported on a rigid frame. The pressure to be measured can be applied to one side of the film for gauge sensing or to both sides of the film for differential or absolute pressure sensing. The material of the measuring foil depends on the application; for low pressure or low temperature systems, the spectrum ranges from rubber to plastic, while silicone or metal foils are used for higher pressures. At very high pressures, stainless steel or other metals are used. When pressure is applied to the membrane, the foil deforms. This movement can be detected with the help of a sensor. Strain gauges, but also piezoelectric, capacitive, magnetic or inductive sensors are used in commercial products. The sensor elements either transfer the deformation directly into an electrical signal or are integrated a fast Wheatstone bridge circuit to provide an electrical output signal proportional to the pressure.

In most pipeline pressure sensors, the diaphragm is made of a semiconductor material (usually silicon) with a semiconductor strain gauge embedded in the diaphragm. In addition to being inexpensive to manufacture, semiconductor strain gauges have high gauge factors, which increases the sensitivity of the sensors. Typically, the Wheatstone bridge signal conditioner is built into the transducer (all branches of the bridge are active strain gages), and the strain gages are connected for temperature compensation.

Pressures can also be sensed with LVDT (linear variable differential transformer) devices. The arrangement consists of a flexible chamber called a capsule, connected to the pressure measurement point, and an LVDT to detect displacement. This design is more expensive than those using strain gauges (SGs), but can be more durable in an application that requires long life.

Capacitive sensors are occasionally used in pressure transmitters. Here, one or more solid metal plates are placed directly above or below a metal diaphragm, creating a capacitor that is an effective secondary element. The displacement of the diaphragm due to the pressure to be measured changes the average gap distance, which changes the capacitance between the two plates. Capacitive pressure transducers are mainly used when measuring very small pressures or pressure differences (down to 0.1 Pa), as they are capable of resolving extremely small displacements.

Transducers used for high frequency pressure measurements usually use a piezoelectric sensing element directly connected to the membrane. The piezoelectric material is very stiff, and the transducers have a high natural frequency in many applications. If the diaphragm (or other displacing element) is very flexible, the natural frequency will be low and the transducer output signal will be misleading for high frequency pressure measurements. Piezoelectric pressure transducers can have natural frequencies up to 150 kHz and can be used up to about 30 kHz.

Often the fluid temperature exceeds the maximum temperature of the sensor, then a separation of the sensor from the membrane occurs. The membrane deformation is then transferred to the sensor via an incompressible coupling medium (oil, liquid metal), whereby the temporal resolution of the sensor is reduced.

To ensure correct pressure measurement, the diaphragm must be in "flush" contact with the fluid. This is necessary for two reasons. If a cavity were present, it would be very slow to equalize to the real pressure. Secondly, the natural frequency would be reduced and thus the ability of the sensor to react to transients would be impaired.

In high resolution measurements it may appear that misreading of the wall static pressure occurs, which are related due to the finite size of the pressure tapping. Experimental results indicate that the correction term as a fraction of the wall stress continues to increase as the hole Reynolds number $Re_d = (u_\tau d)/\nu$ increases, in which d is the dimensionless tap diameter, u_τ the wall shear stress velocity ($u_\tau = \sqrt{\tau_w/\rho}$) and ν the kinematic viscosity. For small holes relative to the pipe diameter the results follow a single curve, but for larger holes the data diverge from this universal behaviour at a point that depends on the ratio of the holes diameter to the pipe diameter. This effect becomes pronounced especially for large Reynolds numbers. If $d/D > 0.1$ the global flow pattern of the pipe flow above the measurement tap is altered leading to a miscellaneous pressure reading.

Also, the depth to diameter ratio of the pressure tap l/d plays a role and can cause a pressure error, since it dictates the eddy system establishing within the cavity, as Fig. B.21f sketches. If $l/d > 2$, the error for all Reynolds numbers remains constant and small. Consequently, a wide cavity behind the tapping reduces the error, while a contraction in diameter from the tapping to the pressure gauge increase the error.

References

API-2530. (1985). *Manual of petroleum measurement standards Chapter 14—Natural gas fluids measurement Section 3—Orifice metering of natural gas and other related hydrocarbon fluids*. American Petroleum Institute.

Benedickt, R. P. (1984). *Fundamentals of temperature, pressure, and flow measurements*. Wiley.

Carslaw, H., & Jaeger, J. C. (1986). *Conduction of heat in solids* (2nd ed.). Oxford University Press.

Childs, P. R., Greenwood, J. R., & Long, C. A. (2000). Review of temperature measurement. *Review of Scientific Instruments, 71*(8), 2959–2978. https://doi.org/10.1063/1.1305516

Connolly, J. (1998). *Industrial resistance thermometers—Handbook of temperature measurement—Vol. 2, Resistance and liquid in glass thermometry* (B. R. R., Ed.). Springer Science & Business Media.

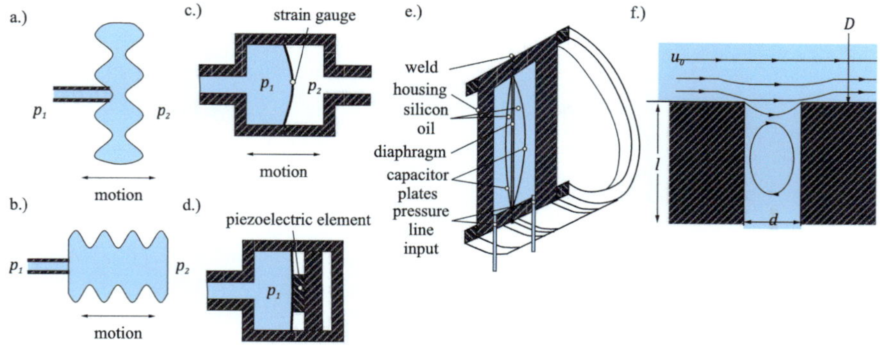

Fig. B.21 Functional principle of different pressure gauge types: **a** capsule, **b** bellow, **c** diaphragm, **d** piezoelectric, **e** capacitive pressure difference

Crabtree, M. A. (2009). *Industrial flow measurement.* University of Huddersfield. Retrieved May 8th, 2021, from http://eprints.hud.ac.uk/

Gifford, A. D., Huble, D. O., Pullins, C. A., & Diller, T. E. (2010). Durable heat flux sensor for extreme temperature. *Journal of Thermophysics and Heat Transfer, 24*(1), 69–76. https://doi.org/10.2514/1.42298

IEC-60584. (2013). *Thermocouples—Part 1: EMF specifications and tolerances.* International Electrotechnical Commission.

IEC-60751. (2008). *Industrial platinum resistance thermometers and platinum temperature sensors.* International Electrotechnical Commission (TC 65/SC 65B—Measurement and control devices).

ISO-5167-1. (2004). *Measurement of fluid flow by means of pressure differential devices—Part 1 Orifice plates, nozzles and Venturi tubes inserted in circular cross-section conduits running full.* International Organization for Standardization.

Kuo, C. H., & Kulkarni, A. K. (1991). Analysis of heat flux measurement by circular foil gages in a mixed convection/radiation environment. *Journal of Heat Transfer, 113*, 1037–1040.

Minkina, W., & Dudzik, S. (2009). *Infrared Thermography: Errors and Uncertainties.* Wiley.

OECD-NEA. (2015). *Handbook on lead-bismuth eutectic alloy and lead properties, materials compatibility, thermal-hydraulics and technologies.* NEA. No. 7268, OECD, Nuclear Science.

Pavese, F., & Steur, P. M. (1987). 3He constant-volume gas thermometry: Calculations for a temperature scale between 0.8 K and 25 K. *Journal of Low Temperature Physics, 69*, 91–117. https://doi.org/10.1007/BF00681625

Schulenberg, T., & Stieglitz, R. (2010). Flow measurement techniques in heavy liquid metals. *Flow Measurement Techniques in Heavy Liquid Metals, 240*, 2077–2087.

Taler, J. (1996). Theory of transient experimental techniques for surface. *International Journal of Heat and Mass Transfer, 39*(17), 3733–3748.

Zhang, S., Wang, Q., Li, J., & Zhang, X. C. (2020). Coaxial thermocouples for heat transfer measurements in long-duration high enthalpy flows. *Sensors, 20*(18), 5254ff. https://doi.org/10.3390/s20185254

Hydraulic Components

C

C.1 Pumps and Their Description

By far the majority of pump installations refer to the *velocity head* type, in which the pressure rise is achieved by giving the fluid movement and at the exit of the machine, this movement is translated into a pressure increase by slowing down the fluid. There are more than 20 differentiating features to characterize a pump, among these e.g. the number of stages, the position of the pump shaft, single- or multi-flow impellers, design of the pump casing, so that there exist a vast variety of different pump types and pumping principles. However, centrifugal pumps are mostly used for closed line networks, and we will limit ourselves to their description here.

C.1.1 Basics of Incompressible Turbomachinery

The starting point for calculating the characteristic curve of a pump is the Bernoulli equation (cf. Eq. (C.1)), which represents the constancy of the energy of a fluid element along a current path. For the non-stationary flow of a Newtonian fluid with friction between points 1 and 2 of a current filament, it reads as follows:

$$p_1 + \frac{\rho}{2}v_1^2 + \rho g h_1 = p_2 + \frac{\rho}{2}v_2^2 + \rho g h_2 + \underbrace{\sum_{i=1}^{n} f_D \frac{\rho}{2} v_i^2}_{\text{friction losses}} + \underbrace{\rho L \frac{dv}{dt}}_{\text{acceleration}}, \tag{C.1}$$

in which p, v, ρ and g denote pressure, density, velocity and geodetic height f_D is the Darcy friction factor of the pipe element i and L the entire length of the piping network. The movement of the pump rotor performs work and generates heat and at the same time changes the enthalpy. In a general form, the energy equation can be written in the following differential form.

$$dY + dq = d\underbrace{\left(h + \frac{v^2}{2} + gz\right)}_{e}, \tag{C.2}$$

R. Stieglitz and W. Platzer, *Solar Thermal Energy Systems*,
https://doi.org/10.1007/978-3-031-43173-9

where Y and q are the work and heat transferred to the fluid, e represents the total energy, while h is the enthalpy. Note that the enthalpy is related to the inner energy u via

$$h = u + \frac{p}{\rho} = c_p T, \tag{C.3}$$

and thus, expresses the sum of the internal energy u and the ability to perform hydrodynamic work via the pressure p. Therein, c_p is the specific heat capacity at constant pressure and T is the temperature. Note, that the inner energy u is given by $u = c_v T$, in which c_v is the heat capacity at constant volume. Enthalpy is directly linked to the second law of thermodynamics and thus to specific entropy (s)

$$dh = d\left(u + \frac{p}{\rho}\right) = dq + \frac{1}{\rho}dp = Tds + \frac{1}{\rho}dp. \tag{C.4}$$

Inserting these relations in the in the second law of thermodynamics ($ds = dq/T + ds_{irrev}$), we get for the mechanical work dY:

$$dY = \underbrace{\frac{1}{\rho}dp + d\left(\frac{v^2}{2} + gz\right)}_{\text{useful=reversible work}} + \underbrace{Tds_{irrev}}_{\text{losses=heat}}. \tag{C.5}$$

When considering compressors, the change in density plays a role, so that the specific work Y is the relevant parameter, while for fans the change in total pressure (Δp_t) is a characterization feature. For pumps, where the fluid can well be considered as incompressible in a first approximation, the head H is used as a parameter instead of the work. The formulations are:

$$\begin{aligned} &\text{Compressors: } Y = c_p(T_{2,s} - T_1) + \tfrac{v_2^2 - v_1^2}{2} = h_{2,s} - h_1 \quad \left(\tfrac{\text{m}^2}{\text{s}^2}\right]; \\ &\text{Fans: } p_t = Y\rho = p_2 - p_1 + \rho\tfrac{v_2^2 - v_1^2}{2} = p_{2,t} - p_{1,t} \quad (\text{Pa}); \\ &\text{Pumps: } H = \tfrac{Y}{g} = \tfrac{p_2 - p_1}{\rho g} + \tfrac{v_2^2 - v_1^2}{2g} + z_2 - z_1 \quad (\text{m}). \end{aligned} \tag{C.6}$$

Turbomachinery is the term used to describe any device in which energy is either transferred to or extracted from a continuously flowing fluid by the dynamic action of one or more moving rows of blades. Essentially, a rotating row of blades, a rotor or impeller, changes the stagnant enthalpy of the fluid flowing through it. These enthalpy changes are closely related to the pressure changes in the fluid and, to a first approximation, ventilators, compressors, turbines and also pumps are the same machines.

Euler's turbine equation is used to establish a relationship between the specific work Y and the pump geometry and speeds in the impeller. For this purpose, let us consider the flow in a pump from a position 1 (inlet, suction side) to the location 2 (outlet, discharge side) as depicted in the Fig. C.1a. Through the inlet a fluid with density ρ, absolute velocity $\vec{c}$ and volumetric flow rate $\dot{V}$ flows and we determine the moment $\vec{M}$ of the force $\vec{F}$ acting at the inlet and outlet of a generalized turbomachine.

$$\vec{M} = \vec{F} \times \vec{r} = \frac{d}{dt}(m\vec{c}) \times \vec{r} = \dot{m}(\vec{r} \times \vec{c}), \tag{C.7}$$

in which $\dot{m}$ denotes the mass flow and $\vec{c}$ the absolute velocity, which is the vector sum of the relative velocity $\vec{w}$ and the circumferential velocity $\vec{u}$, i.e. $\vec{c} = \vec{u} + \vec{w}$. When the difference of moments at the inlet and outlet are determined, the power required to rotate the pump is obtained:

$$\begin{aligned} P = \dot{m}Y = \omega M &= \dot{m}\omega((r_2 \times c_2) - (r_1 \times c_1)), \\ &= \dot{m}(c_{2u}u_2 - c_{1u}u_1). \end{aligned} \tag{C.8}$$

In this equation called the Euler turbine equation, the subscript "u" indicates the circumferential portion of the absolute velocity and the specific work Y is directly obtained as $Y = (c_{2u}u_2 - c_{1u}u_1)$. From the Euler turbine equation, we get immediately the head of the pump H via

$$\Delta p = \rho g H = \rho(c_{2u}u_2 - c_{1u}u_1) \Rightarrow H = \frac{(c_{2u}u_2 - c_{1u}u_1)}{g}, \tag{C.9}$$

and find that the pump head is independent of the fluid density ρ. This allows you to directly determine the volume flow rate $\dot{V}$ through the pump, which results in

$$\dot{V} = c_{2m}A_2 = \pi D_2 b_2 c_{2m}, \tag{C.10}$$

where b_2 is the depth of the flow channel at the outlet and D_2 is the diameter there (see Fig. C.1). From Eqs. (C.9) and (C.10) it is obvious that an increase of the pressure Δp occurs as soon as the circumferential speed c_{2u} (i.e. the speed of the shaft n) is increased but at the same time the volume flow $\dot{V}$ decreases (because c_{2m} decreases). Conversely, if the pressure difference Δp is reduced (i.e. c_{2u} decreases), this leads to an increase in volume flow $\dot{V}$ since c_{2m} increases.

From these ideal considerations, one can also determine the hydraulic efficiency η_h of a pump as pressure difference times volume flow rate divided by pump power P,

$$\eta_h = \frac{gH}{(c_{2u}u_2 - c_{1u}u_1)}. \tag{C.11}$$

The part of the vane work corresponding to the vane losses leads to an increase of the internal energy of the fluid, while the rest causes a useful change of the fluid state (gH).

Thus, to determine the flow rate $\dot{V}$ through an ideal pump, the velocity triangles play a central role. We know that the absolute velocity $\vec{c}$ is the vector sum of the

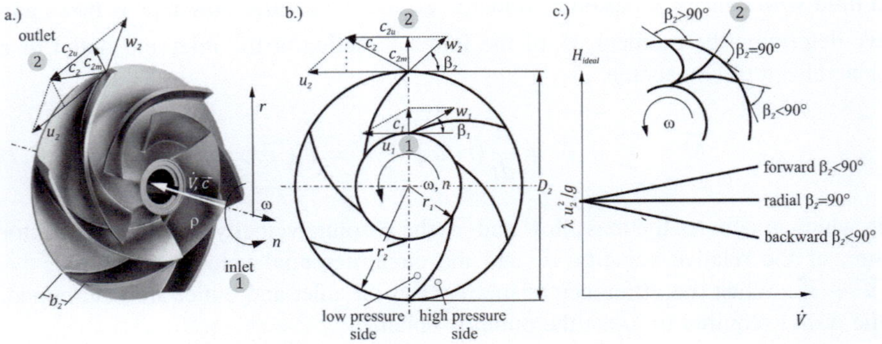

Fig. C.1 **a** Impellor of a radial centrifugal pump and **b** the associated velocity triangles at in- and outlet; **c** ideal pump head H_{ideal} as a function of the flow rate $\dot{V}$ for different outlet bent angles β_2

circumferential $\vec{u}$ and relative velocities $\vec{w}$ and that the circumferential velocity is given directly by the speed n of the pump shaft, i.e. $u = \pi D_i n$. Furthermore, we know the direction of the relative velocity $\vec{w}$, since this is tangential to the angle β_i at the inlet and outlet of the rotor blades. This allows you to find the velocity c_{2u} using trigonometry to $c_{2u} = u_2 - c_{2m}/\tan\beta_2$, as illustrated in Fig. C.1b.

In reality, the outlet angle of the flow does not correspond exactly to β_2, so that the head is reduced, which is usually taken into account by the introduction of a slip factor λ. Since the rotor blades at the outlet block a part of the outlet due to their spatial dimensions, a so-called blockage factor Ψ is introduced, which is constant and always smaller than one ($\Psi < 1$). Assuming a swirl-free inlet flow ($c_{1u} = 0$), we obtain an equation for the theoretical ideal head of a radial pump H_{ideal} which has the following appearance:

$$H_{ideal} = \lambda \frac{c_{2u} u_2}{g} = \lambda \left(\frac{u_2^2}{g} - \frac{u_2}{g} \frac{c_{2m}}{\tan\beta_2} \right) = \frac{\lambda u_2}{g} \left(u_2 - \frac{\dot{V}}{\pi D_2 b_2 \Psi \tan\beta_2} \right). \tag{C.12}$$

Analysing Eq. (C.12) the performance curve H_{ideal} as a function of the flow rate $\dot{V}$ is a horizontal line for a radial blade ($\beta_2 = 90°$) and not changing with the flowrate $\dot{V}$, while the flow rate is increasing for forward curved blades ($\beta_2 > 90°$) and correspondingly decreased for backward bent blades ($\beta_2 < 90°$), as illustrated in Fig. C.1c.

C.1.2 Pump Characteristic Curve, Losses and Dimensionless Numbers

In the real case, the flow in an impellor is not ideal; for example, the flow only follows the rotor blade curvature at the design point, while deviations occur at other numbers of revolutions, and secondly, the flow within the pump is by no means loss-free. The direction of the inlet flow at point 1 of the pump changes as function of $\dot{V}$. Only at the design point with the design speed do the smallest losses occur. Any upward or downward deviation design flow rate $\dot{V}_d$ leads to the formation of detachment areas on the upper or lower side of the blade, respectively, which are associated with a hydraulic pressure loss. Although these losses vary from pump to pump and depend on many parameters such as blade profile, angles, number, but also inlet/outlet area ratios, diffuser types, they can be expressed in a general form of a head loss factor H_g by a proportionality of the type $H_g \sim (\dot{V} - \dot{V}_d)^2$.

Another significant loss factor is viscous friction losses in the impeller itself. The rotation of the impeller results in shear in the gaps of the casing, and the flow along the blades also leads to the formation of viscous boundary layers. Like all viscous or friction losses, the losses scale in a first approximation with the square of the velocity and thus of the flow rate leading to a reduction of the head, i.e. $H_f \sim \dot{V}^2$. If both types of loss are subtracted from the straight line of the ideal characteristic curve $H_{ideal} - \dot{V}$, the real head characteristic curve H_{real} of the pump is obtained, which is also termed throttle curve as illustrated in Fig. C.2.

Different fluid properties change the fraction of power flow within the pump, for example, a greater viscosity leads to increased impeller friction and blade losses. However, as we calculated the hydraulic pump efficiency η_h is calculated, which is the ratio of the useful power (P_u) expressed as the geodetic height to the power expended, which results from the product of the torque balance with the angular velocity at the shaft coupling of the pump, we found that in the ideal case (Eq. (C.11)), that it is independent of the thermo-physical fluid properties. But, in reality the effective gross pumping efficiency η_{gr} is composed of a product of mechanical efficiency η_m, the impeller efficiency η_i, the casing efficiency η_c and the volumetric efficiency η_v in the way:

$$\eta_{gr} = \frac{P_u}{P} = \eta_m \eta_i \eta_c \eta_v, \tag{C.13}$$

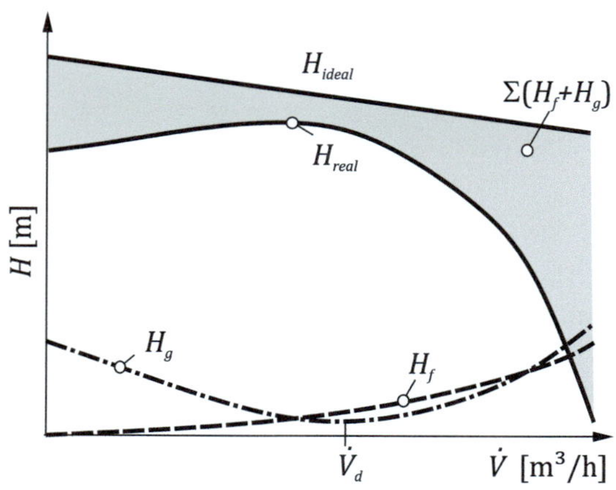

Fig. C.2 Schematical operational characteristics ($H - \dot{V}$) of a centrifugal pump where the useful power (P_u) represents the shaft power that remains after overcoming all losses from the supplied power (P) to pump the fluid and produce the change of state. Mechanical efficiency, typically $\eta_m = 95–99.5\%$, is the portion of shaft power that must be expended to overcome frictional resistance in bearings and seals. Impeller efficiency, η_i, whose magnitude spans a wide range of 60–95% depending on the pump type, is a measure of how much of the available shaft power ($\eta_m P$) is converted into kinetic energy in the fluid and how much power is lost due to blade losses and disk friction. Casing efficiency, η_c, describes how much kinetic energy generated by the impeller is recovered as static pressure rise compared to losses that occur as increased internal energy in the form of irreversible temperature rise. Kinetic energy can be recovered as static pressure rise in two types of diffusers; vane and vaneless. A vaned diffuser will suffer the same velocity triangle problems as the impeller with off-design flows. The vaneless diffuser or volute does not have blade angles to cause extra losses but does operate with flow paths away from the optimum. Volumetric efficiency, $\eta_v = 93–99\%$), weights the flow at the pump outlet to the flow through the impeller. Internal leakage, slip, allows a small amount of fluid to circulate around the impeller, reducing the effective flow rate of the pump.

The above individual losses are not constant but vary depending on which operating point of the pump curve the pump is operating at. At low flow rates (high heads), bearing friction increases due to the greater axial thrust and radial loads. As already mentioned in the case of the ideal pump, the power conversion into kinetic energy in the impeller depends on the velocity triangles, so that the lowest losses are only achieved at the design point, often also referred to as the Best Efficiency Point (BEP). At other flows, mismatches of blade and liquid angles occur, causing additional losses.

The dimensionless pressure number ψ describes the normalized pressure rise and is often used to describe single-stage fluid machinery. In connection with the flow coefficient φ, the pressure coefficient is also described as a parameter of the power conversion of the stage. The pressure number relates the specific nozzle work Y to the square of the circumferential speed u of the impeller of the machine. In this definition, the nozzle work can be replaced by the mass-specific technical work, neglecting friction forces, and further described by the main Eulerian turbine equation with velocity components.

$$\psi = \frac{2Y}{u_2^2} = 2\left(\frac{c_{2u}}{u_2} - \frac{u_1 c_{1u}}{u_2^2}\right) = \frac{2gH}{u_2^2} = \frac{2\Delta p}{\rho\pi^2 D_2^2 n^2}. \tag{C.14}$$

The dimensionless flow coefficient φ, also called delivery coefficient or volume coefficient, describes the throughput in a pump. It can be understood as the ratio of the volumetric flow $\dot{V}$ at the outlet of the impeller to an imaginary volumetric flow formed from the circular area of the outer diameter D_2 of the impeller and its circumferential velocity u, which can be replaced by the rotational speed n given by $n = u_2/(\pi D_2)$.

$$\varphi = \frac{4\dot{V}}{\pi D_2^2 u_2} = \frac{4\dot{V}}{\pi^2 D_2^3 n}. \tag{C.15}$$

Fluid machinery can be described in a uniform manner even when the size and speed are changed by means of the similarity laws. The prerequisites for applying the similarity laws are as follows:

- The compared pumps should be identical or geometrically similar.
- The compared pumps should operate at similar operating points, i.e. with geometrically similar speed diagrams
- Any difference in efficiency should be negligible.

When comparing two pumps 1 and 2 that meet these conditions, the following relationships apply between their respective speeds n_1 and n_2, expressed as impeller diameters D_1 and D_2, and their performance.

$$\begin{aligned}
&\text{Pump head:} && \tfrac{H_2}{H_1} = \left(\tfrac{n_2}{n_1}\right)^2 \left(\tfrac{D_2}{D_1}\right)^2;\\
&\text{flow rate:} && \tfrac{Q_2}{Q_1} = \left(\tfrac{n_2}{n_1}\right) \left(\tfrac{D_2}{D_1}\right)^3;\\
&\text{power:} && \tfrac{P_2}{P_1} = \left(\tfrac{\rho_2}{\rho_1}\right) \left(\tfrac{n_2}{n_1}\right)^2 \left(\tfrac{D_2}{D_1}\right)^5;\\
&\text{ray theorem for velocity triangles:} && \tfrac{c_2}{c_1} = \tfrac{u_2}{u_1} = \left(\tfrac{n_2}{n_1}\right) \left(\tfrac{D_2}{D_1}\right);\\
&\text{mass flow:} && \tfrac{\dot{m}_2}{\dot{m}_1} = \left(\tfrac{\rho_2}{\rho_1}\right) \left(\tfrac{c_2}{c_1}\right) \left(\tfrac{D_2}{D_1}\right)^2.
\end{aligned} \tag{C.16}$$

Moreover, also the operating behaviour at different rotation speeds of the same pump is also largely fixed by the similarity laws; for example, if the speed is changed, the speed triangles remain similar and the shift of the pump operation point can be easily calculated.

Comparing the equations for the dimensionless pressure number ψ (Eq. (C.14)) and the flow coefficient φ (Eq. (C.15)) and applying similarity laws, one is able reduce either one of the two parameters u_2 or D_2, which yields a dimensionless speed quantity σ:

$$\sigma = \frac{\sqrt{\varphi}}{\psi^{3/4}} = \frac{\sqrt{\pi}}{4\sqrt{2}\,g^{3/4}} \underbrace{n\frac{\sqrt{\dot{V}}}{H^{3/4}}}_{n_s}. \tag{C.17}$$

Note, that σ depends only on the revolution number n, for which the data of the best-efficiency point (BEP) is to be taken. Specific speed n_s (sometimes called pump specific speed) is a dimensionless rating for centrifugal pump performance; although it is not actually a speed but depends on shaft speed and thus a function of the flow rate and differential head at the pump's Best Efficiency Point. Specific

Speed is an important factor in the design and selection of pumps and especially pump impellers. When selecting a pump for a particular application, and knowing the shaft speed, flow and differential head, a calculation of specific speed allows to identify the impeller shape (radial, mixed, or axial) most suited to the task. If the specific speed is less than 40, radial impellers are suitable for the application, as they produce high head through centrifugal action as the fluid is expelled perpendicular to the pump shaft. Compared with other impeller designs, however, flow rates are low. Efficiencies are also lower because of high friction and mechanical losses. At the other end of the scale, axial flow impellers are preferred if the specific speed is greater than 175. With this design, fluid enters and is discharged parallel to the rotating shaft, which gives a high flow rate but at a low head.

The specific speed defines not only the shape of the impeller but describes in particular the characteristics of the pump head, efficiency, and power consumption of a pump as a function of the volume flow delivered, which is qualitatively displayed in Fig. C.3.

C.1.3 Cavitation—Universal Pump Characteristics

Complementing to the dimensionless speed σ, one can also define a dimensionless diameter δ in the form $\delta = \left[\left(\sqrt{\pi}D/2\right)\left(\sqrt[4]{2Y/\dot{V}^2}\right)\right]$, where $Y = \Delta p_{BEP}/\rho$ represents the operating point with the optimal efficiency. Using the Cordier diagram, which represents the dimensionless speed σ as a function of the dimensionless diameter δ, one can determine the type of machine (axial, diagonal or radial) and the diameter of the impeller that reaches this operating point with the best efficiency. On the other hand, if a certain duty point (i.e. flow rate and head) is to be reached with an impeller of a certain diameter, the Cordier diagram provides the speed at which this duty point is reached with the best efficiency. For this reason, the Cordier diagram, illustrated in Fig. C.4a is widely used today in the design of pumps (or in turbomachinery in general).

So far, we have only considered the output side of the pump, but to fully represent a pump we must also consider the input side limitations. If the liquid pressure in a pump suction is reduced to the vapour pressure of the liquid, the liquid evaporates. So-called cavitation occurs when the vapour bubbles return to liquid form in areas of higher pressure, this phenomenon causes intense local pressure spikes associated with noise and vibrations. Cavitation is not desirable since it reduces drastically the hydraulic pump performance ($H - \dot{V}$) and via impellor damage the operating life of the pump.

The risk of cavitation is greatest where the pressure within the pump is lowest, which in a centrifugal pump occurs on the suction side of the impeller blades, slightly downstream of the front edges where the liquid has started to accelerate. At the pump's inlet connection at the same elevation as the shaft the pressure is somewhat greater, by Δp. At this point, which lies at the static suction head (h_S) above the liquid free surface, the flow velocity is (c_S). By applying Bernoulli's

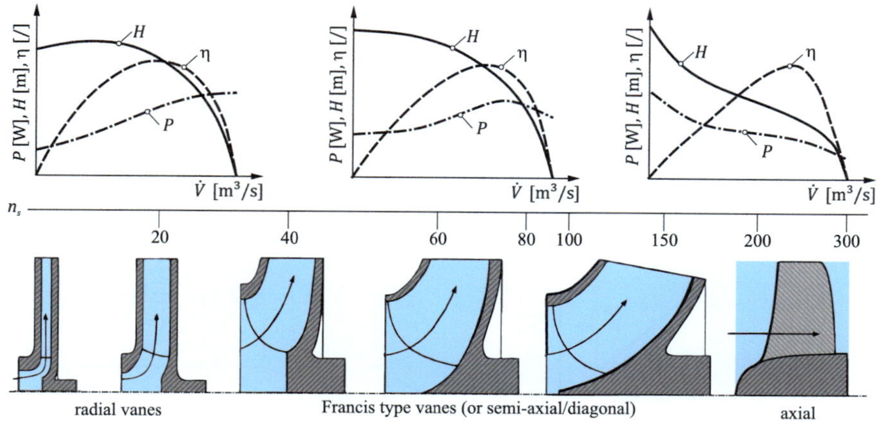

Fig. C.3 Dependency of pump head (H), efficiency (η) and power consumption (P) as a function of the volumetric flow rate ($\dot{V}$) for different dimensionless pumping speeds (n_s)

equation, the flow through the suction pipe yields the suction pressure (p_S):

$$p_S = p_a - \rho h_S - \rho h_{fS} - \rho \frac{c_S^2}{2} \text{ and } p_{min} = p_s - \Delta p, \tag{C.18}$$

in which ρ is the fluids density, p_{min} the lowest pressure in the pump, p_a the ambient pressure, Δp the local pressure reduction and h_{fS} the friction losses in the suction pipe as indicted in Fig. C.4b. Δp covers any losses within the pump, after the suction connection, such as velocity increases or friction losses. To avoid cavitation, which occurs mainly at large flow rates through the pump, it is not sufficient that the pressure on the suction side is greater than the vapor pressure ($p_S > p_v$). Since there is complex flow in the pump, it is possible that $p < p_S$ exists locally if the velocity is large enough. To ensure cavitation-free operation, the *net positive suction head* ($NPSH$) must be defined. It is convenient to divide the absolute pressure on the pressure side (p_S) into two parts, the vapor pressure (p_S) plus the part above it, denoted by ($\rho g NPSH$):

$$p_S = p_v + \rho g NPSH. \tag{C.19}$$

In this way, the $NPSH$ value indicates the net "standby" pressure above the vapor pressure that is available before cavitation occurs. There are two different $NPSH$ values: the available ($NPSH_a$) and the required ($NPSH_r$).

- The available $NPSH_a$ is a property of the hydraulic circuit network to which the pump is connected, i.e. geometry, loss coefficients, etc. of the piping and reservoirs, and can be expressed as

$$NPSH_a = \frac{p_a - p_v}{\rho g} - h_S - h_{fS}, \tag{C.20}$$

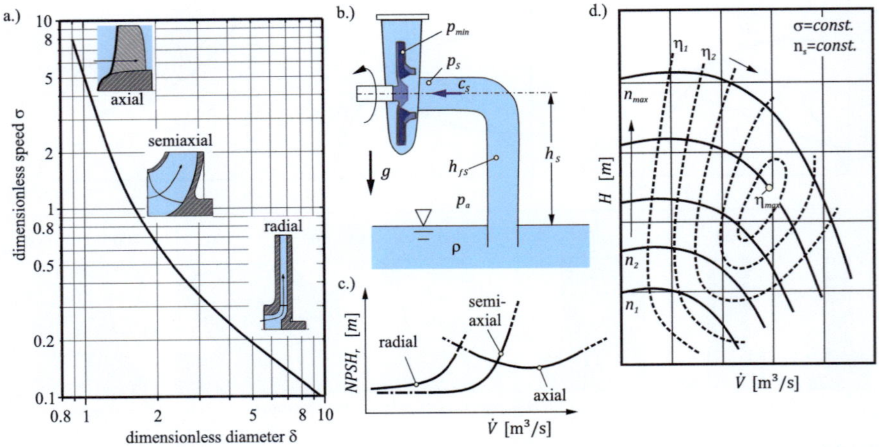

Fig. C.4 **a** Cordier diagram for selecting the optimum impeller design for a pumping application; **b** schematic sketch for deriving the net positive suction head ($NPSH$); **c** qualitative curve of the dependence of the required $NPSH_r$ as a function of the volume flow ($\dot{V}$) for different impeller types; **d** universal pump characteristic curve of a pump family with shell diagram of the efficiency

where $h_{fS}(\dot{V})$ represents the friction losses on the suction side piping.

- The required $NPSH_r$ value can be found in the manufacturer pump catalogue. It is usually dependent on the flow rate in a similar way to the head.

The condition for avoiding cavitation is, that the available $NPSH$ value must be greater than the required $NPSH$ value, i.e. $NPSH_a < NPSH_r$ and a typical $NPSH$-curve of a manufacturer. Depending on the impeller design, cavitation can occur not only on the pressure side, but also on the suction side, but mainly in the case of semi-axial or axial impellers, as shown qualitatively in Fig. C.4c.

Thus, the pumps are completely described and for a certain pump type the pump characteristic curve can be represented dimensionless, in which the efficiencies are entered as shell curves in the characteristic diagram of the delivery head, as depicted in Fig. C.4d. The fact that the efficiency curves also close again at the top can be explained by the cavitation influence, which increases with increasing speed. as the speed increases.

C.1.4 Effect of Viscous Liquids on the Performance of a Centrifugal Pump

Most pumps are designed for water at 20 °C; however, many solar thermal applications starting from domestic use (mostly water–glycol mixtures) to solar thermal power plants circulate other fluids. As long as Newtonian fluids are involved, the universal pumping diagrams can be converted to other fluids and the performance characteristics estimated for the modified application. Fluids with higher viscosity and/or higher density than water, affect centrifugal pump performance in different ways:

- Power consumption increases, meaning a larger motor may be required to meet flow requirements.
- Head, flow rate and pump efficiency decrease.

The similarity laws can be used to determine corrections to the pump curves for pumps for highly viscous liquids. This requires knowledge of the operating point, i.e. the required volume flow and the pumping head; the kinematic viscosity of the pumped liquid (say "*s*") at the delivery temperature should also be known. The correction factors for the pumping height (k_H) and the shaft power (k_P) can be determined on the basis of the data sheets of the pump manufacturers. With the data k_H and k_P, the equivalent head for clean water (H_w) and the corrected actual shaft power (P_S) can be calculated according to the following formula:

$$H_w = k_H H_s;\ P_s = k_P P_w \left(\frac{\rho_w}{\rho_s} \right), \tag{C.21}$$

in which H_s is the required height of the viscous fluid "*s*", P_S the shaft power of the clean water case, and ρ_w, ρ_s the densities of water and the viscous fluid, respectively. Pump selection is made according to the normal data sheets/characteristics for water. The calculation procedure for the pump and motor selection process includes the following steps:

- Calculation of the correct head H_w (based on H_s and k_H).
- Selection of a pump that will provide power according to the corrected power point ($\dot{V}_s$, H_w).
- Reading the power consumption P_w at the duty point ($\dot{V}_s$, H_w),
- Based on P_w, k_P, ρ_w and ρ_s, calculate the corrected required shaft power P_s.
- Check if $P_s < P_{w,max}$ of the motor. If it is, the motor can be used. Otherwise, a stronger motor must be selected.

C.2 Hydraulics Circuits

C.2.1 Pump Circuits

Solar thermal systems are mostly closed systems in which large amounts of heat circulate. What these types of closed systems have in common is that they circulate a fluid that acts as a carrier of heat energy. Closed systems are characterized as systems with pumps that must overcome only the sum of frictional losses generated by all components. Together with the pipes and fittings, all these components make up a system characteristic curve.

Since the dependence of the coefficients h_f and h_m on the Reynolds number is small and the velocity is proportional to the flow rate, the equation of the pipe characteristic with $h_l = const.$ can be decomposed into a constant and a quadratically flow rate dependent part in the following way, $H = H_{stat} + h_l\dot{V}^2$, where the static height (H_{stat}) vanishes if the circuit is not open to the ambient pressure, e.g. via thermal storage tanks. The set pressure in a closed system, which is represented by the system characteristic curve, is simply a parabola with the starting point $(\dot{V}, H) = (0, 0)$.

The operating point of a hydraulic circuit is the intersection of the throttle curve with the pipe characteristic. There, the head generated by the pump and the head required by the pipeline coincide. It should, if the pump is properly matched to the pipeline, be near the point of best pump efficiency. The equilibrium between the pump and the pipeline is stable if, at the operating point, the pipe curve rises more steeply above the volume flow than the throttle curve, because then a disturbance is compensated for by itself and the operating values always return to the point of intersection. During start-up, a pump must cooperate, starting from zero volume flow, with all points of the pipe characteristic between its vertex and the steady-state operating point. Because of the horizontal tangent of the pipe parabola at the vertex, stable operation at all times is only possible if the throttle curve is everywhere negatively inclined to the $\dot{V}$-axis. Such a pump characteristic is called stable, while otherwise it is divided into a stable and an unstable range. In the unstable range, a centrifugal pump cannot work in continuous operation.

Change of the operating point.

The equilibrium between the pump and the pipeline can be used to adjust the operating points by specifically changing either the pipeline characteristic or the throttling curve of the pump. Most often this involves adjusting for changes in flow rate, less often head or drive power.

- *Throttling*. The simplest way to influence the system curve is to adjust a throttling device (valve or slide valve) in the pipeline. Because of its simplicity, the so-called throttling control is preferred, where the pressure-side shut-off device serves as a throttle. Throttling in the suction line would reduce the holding pressure head of the system and potentially lead to cavitation and should therefore be avoided. Energy losses are unfavourable with throttling because the friction component of the pipe characteristic is increased. In addition, there are pressure fluctuations caused by periodic vortex shedding at the partially opened throttle.
- *Bypass*. Instead of allowing the pump to operate at a lower flow rate, a bypass can be used. flow rate, an adjustable portion of the pumped fluid can be bypassed through a shunt line, a so-called bypass, back into the suction tank. This type of control is advantageous when the performance curve drops above the flow rate, i.e. for high-speed machines.
- *Speed adjustment*. The most economical mode of operation and at the same time a gentle treatment of the pump is to adjust the speed to the volume flow to be pumped. to be pumped. Here, the operating points are located as intersections with the various throttle curves on the unchanged system characteristic curve. Particularly favourable conditions exist when the pipe parabola has a low

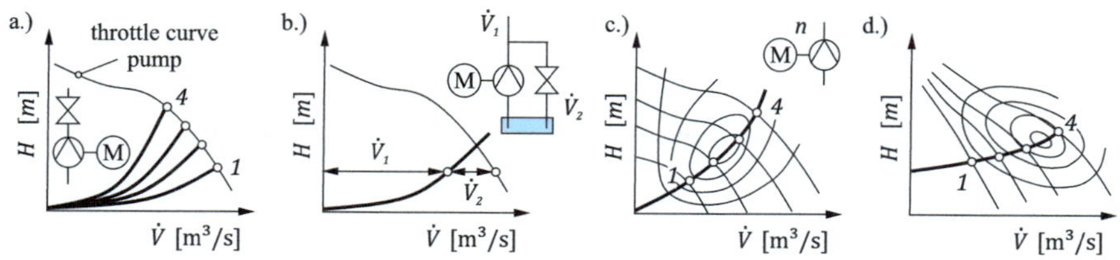

Fig. C.5 Methods of changing the operating point in a hydraulic piping network in the $H - \dot{V}$ diagram: **a** throttling; **b** bypass; **c** speed adjustment; **d** blade adjustment

static component and is influenced almost exclusively by the frictional losses. Its vertex then lies near the origin of the coordinates, and it coincides approximately with the parabola of the same impact state, which connects all optimum operating points. With larger static system head, the pump does enter the area of lower efficiencies at smaller volume flows, but compared to throttling, the advantage remains that only the required head is ever generated and the power requirement remains correspondingly small. The disadvantage is that a variable speed drive generally requires more effort.

- *Blade adjustment*. Adjustable guide vanes, are less likely to be considered for pumps. The $H - \dot{V}$ point curve of the pump is only slightly changed by the change of the guide vane position, and above all the guide vanes are excited by the unsteady flow with its periodic changes to vibrations, which can lead to alarming amplitudes with adjustable design. On the other hand, pumps with semi-axial impellers (less frequently those with axial impellers) use guide vanes to generate a pre-swirl in the same direction as the impeller rotation at low flow rates and in the opposite direction at high flow rates. The operating points are the points of intersection of the modified pump characteristic curve with the pipe characteristic curve. Finally, for pumps with axial and semi-axial impellers, it is also possible to adjust the impeller blades. The performance map of such a pump has efficiency vanes whose largest diameter is located approximately parallel to the $\dot{V}$-axis. Variable vane pumps therefore operate particularly economically with a flat piping characteristic, as is the case, for example, with line-concentrating solar thermal power plants (Fresnel or parabolic trough).

Figure C.5a–d illustrates the progression of operating points from 1 to 4 for the four different pump control types.

C.2.2 Multiple Pumps in the Pipeline Network

There are several reasons why more than one pump may be required to operate a hydraulic pipe spreading network. For example, some heating or drinking water concepts require more than one pump due to their control strategies or the integration of conventional heating systems, as well as large absorber line lengths of solar farm concepts can lead to pressure losses that can no longer be realized with one pump; possible options are then a series connection of pumps, or a parallel connection of pumps combined with a segmentation of the solar field.

Kirchhoff's laws apply in a modified form to the interconnection of pumps.

Branched Pipes in Series Connection

Series connection of pumps is used where large delivery heads (large pressure losses) occur. It is therefore suitable for large delivery heads in combination with comparatively small flow rates, because by dividing the delivery head among several pumps, a higher specific speed can be maintained for each individual pump, thus increasing efficiency. Multistage pumps can be thought of as a compact, integral form of series connection. Systems with multiple pumps, where the pumps are located in close proximity to each other, are treated simply by calculating a common characteristic curve for the entire "pump ensemble." According to Kirchhoff's law, for series operation the flow rate does not change, but the resulting $H - \dot{V}$ characteristic curve is obtained by adding the actual head differences for each value of flow, which is called vertical summation in literature, compare Fig. C.6a. The common pump curve is then compared with the system curve of the piping losses. The resulting efficiency at different flows is calculated from the relation:

$$\eta = \frac{\rho g \dot{V}(H_1 + H_2)}{P_1 + P_2}, \tag{C.22}$$

in which P_1, P_2 are the shaft powers of the pumps 1 and 2.

When two similar pumps are connected in series, the same efficiency curve applies to the "package" as to each individual pump (except for compressible liquids). Pumps 1 and 2 may not be able to operate individually in the system if one of the pumps does not develop sufficient head to overcome the static pressure (H_{stat}). Depending on the pump type and size, precautions may need to be taken to prevent one pump from running without the other. The second pump may cavitate if the first pump is not running, and both pumps may overheat, which is prevented by appropriate metering devices and control logic for pump protection.

For most fluids, the same flow will pass through both pumps. The inlet pressure for pump 2 is the pump outlet pressure of pump 1, so the pressure rating of the pump and the shaft seals may be affected. The axial thrusts, which depend on the differences between the internal and atmospheric pressures, will also be different for the two pumps. For low differential pressures, identical pumps can be used. As the differential pressure increases, the second pump may need to be significantly modified to accommodate the increased inlet pressure.

Often a series operation is used to overcome the problem of low $NPSH_a$. To solve the problem and to avoid very costly solutions, a low capacity booster pump is operated in series with the main pump. The booster pump's only function is to deliver $NPSH_a$, so its differential head is very small compared to the main pump and so is the power consumed; efficiency is not really important. Usually this is a single-stage pump, possibly two-stage, running slowly to match the $NPSH_a$.

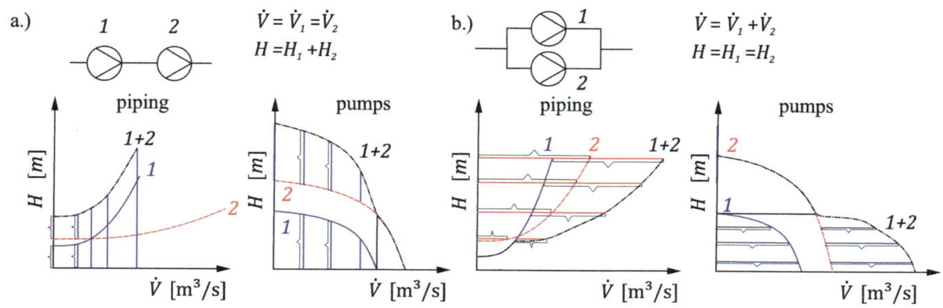

Fig. C.6 Schematical impact of operation of two pumps 1 and 2 on system and pump head curves: **a** series pump connection; **b** parallel pump connection

In this type of series pump installation, the pumps are completely different and reflect their respective functions. Pumps that handle more viscous fluids use a low pressure positive displacement pump.

Branched Pipes in Parallel Connection

According to Kirchhoff's law, when pumps are operated in parallel, the resulting $H - \dot{V}$ curve is obtained by adding the actual flows for each head value, which is called horizontal summation, compare Fig. C.6b. Parallel pump operation is used to achieve large flows and to cope with highly fluctuating flows by operating a different number of pumps. The effect of parallel operation depends largely on the shape of the system curve. For steep plant curves dominated by friction, parallel operation achieves only a small increase in flow. A much better effect is achieved when the plant curve is dominated by static head. The resulting efficiency curve for a pump package connected in parallel is calculated from the following relationship.

$$\eta = \frac{\rho g H\left(\dot{V}_1 + \dot{V}_2\right)}{P_1 + P_2}. \tag{C.23}$$

For two similar pumps, the efficiency values for each pump are the same except that the Q values are shifted by a factor of two are shifted by a factor of two. Unstable pumps and flat pump curves should be avoided when pumping in parallel. In any case, the intended operating range should not fall within the unstable range without taking special precautions. Special minimum flow valves or orifices can be installed to prevent the pump from reaching the apex of the curve. Typical problems with parallel pump operation include irregular flow due to different operating points and the inability of a pump to open its check valve at start-up. Check valves are installed to prevent backflow from stationary or standby pumps. Different types of pumps can be operated in parallel to achieve best operating efficiency at all flows.

Pressure Boosting

Booster pumps are mainly used in long pipelines and large circulation systems. A common problem is that the pressure in a distribution network is not sufficient for certain higher points, especially in heating systems; or when an additional tapping point (collector or heat consumer) is connected to an existing network. Pressure boosting is a form of series connection in which pumps are spaced farther apart. The advantages of pressure boosting are:

- A better distribution of the given pressure in the pumping system is achieved
- Pump and energy costs can be reduced, (applies to branched piping systems).

In order to ensure a certain flow/pressure at a distant tapping point, it is possible either to install a pump with a large head, or to install a second pump at a downstream section, which has the advantage of reducing the average pipeline pressure, as illustrate in Fig. C.7a. For high-temperature systems, this is less of an option, as they must be resistant to high temperatures.

Branched piping systems are usually sized according to the pressure requirements of the most distant tapping point. The required pump specification is therefore based on this pressure demand and the total flow consumption. The use of booster principles can therefore be an alternative, as shown in the example shown in Fig. C.7b. The pump data for the single pump option yields for flow rate $\dot{V} = 1$, required pumping height $H = 0.5 + 0.5 = 1$ and pumping power $P = 1$, while for the two pump option we get $\dot{V}_1 = 1$, $H_1 = 0.5 + 0.1 = 0.6$, $P_1 = 0.6$ and $\dot{V}_2 = 0.5$, $H_2 = 0.5 - 0.1 = 0.4$, $P_1 = 0.2$. Thus, the total installed pump power of the booster option is $P = P_1 + P_2 = 0.8$ is 20% smaller. The lower power requirement of the option with a booster pump is explained by the fact that the pressure in line BD, which is not required, is eliminated. In the version with one pump, this overpressure must be throttled so that the tapping point D does not overflow and reduce the flow at C.

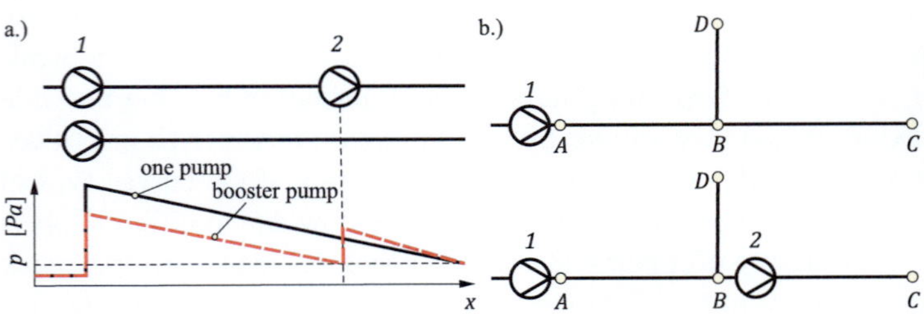

Data supplied		
length	$\dot{V}$ [/]	H_{sys} [/]
AB	1	0.5
BC	0.5	0.5
BD	0.5	0.1

Fig. C.7 **a** Schematic pressure distribution in a pipeline with and without boosting option; **b** example of conditions for pump sizing in a branched pipe system

C.3 Heat Exchangers

Heat exchangers are essential components of every thermal system. It is difficult to define the term heat exchanger exactly, since heat is transported during the generation, transport and storage of heat. In this sense, a boiler, a solar absorber, a pipe carrying hot water as well as a thermal storage tank is also a heat exchanger. In this context, heat exchangers are compact technical components that are used to efficiently exchange heat between two circuits. For this purpose, a classification of heat exchanger types is given first. Subsequently, general laws of heat exchangers are derived, before the operating characteristics are discussed and demonstrated using the example of co-current and counter-current heat exchangers.

C.3.1 Heat Exchanger Types and Their Classification

There are a large number of different types of heat exchangers, the listing of which in terms of their characteristics and special properties is beyond the scope of this paper. Roughly, they can be divided into three classes, the tube heat exchanger, the plate heat exchanger, and the regenerative heat exchanger, as shown in Fig. C.8.

By specifically designing the surface structure of the heat transfer surface, the heat transfer between the individual media can often be significantly improved. Although the structuring of the surface leads to increased pressure losses, the improvement in heat transfer can be greater than the additional pressure loss due

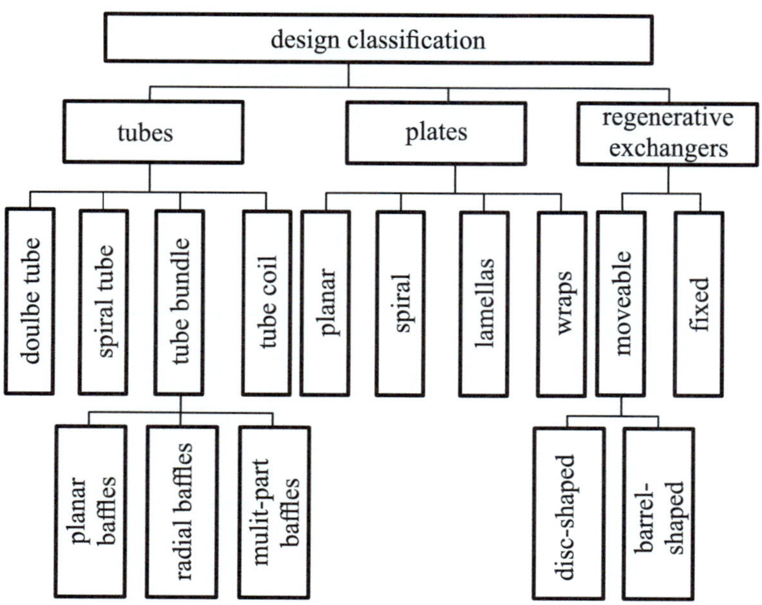

Fig. C.8 Design classification of heat exchanger types

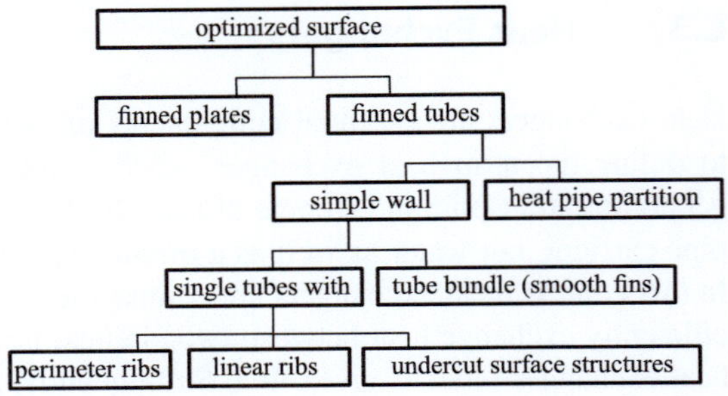

Fig. C.9 Classification of heat exchangers with optimized heat transfer behaviour through improvement of surface structures

to the nonlinearity of the turbulent heat transfer. Figure C.9 shows a spectrum of possible surface modifications in heat exchangers to increase heat transfer.

C.3.2 General Laws of Heat Exchangers

The general case of a technical heat exchanger can be represented in a simplified way by the picture shown in Fig. C.10. At the inlet of one side, medium 1 with temperature T_1' and mass flow $\dot{m}_1$ enters the heat exchanger. On the way of the medium 1 to the outlet, the temperature of the medium 1 to the outlet changes to the value T_1'' due to the heat dissipation to the medium 2. The heat transfer takes place by a transfer of energy from the fluid 1 across the wall separating the media to the fluid 2.

The magnitude of the process depends on the heat transfer coefficient h at the available surface area A. Heat exchangers can be designed according to different principles. The best-known designs are co-current, counter-current and cross-current heat exchangers. Depending on the design, different local heat transfer coefficients and temperature curves result. By introducing a design-dependent mean heat transfer coefficient h and a mean temperature difference ΔT_m, one can specify the heat flux:

$$\dot{Q} = h \cdot A \cdot \Delta T_m, \tag{C.24}$$

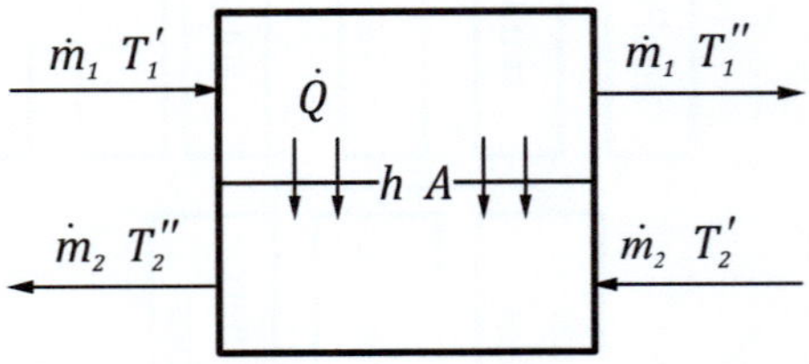

Fig. C.10 Abstracted model of a counter current heat exchanger in which ΔT_m is defined by

$$\Delta T_m = \frac{1}{A}\int_A (T_1 - T_2)dA, \tag{C.25}$$

where $T_1 - T_2 = f(A)$ is a local temperature difference between both fluids. Assuming that the heat transfer coefficient h is known and independent of the local temperature difference, one can integrate over the heat exchanger length and the solution of the integral yields the following relationships of the mean logarithmic temperature difference ΔT_m for a co-current heat exchanger or counter-current heat exchanger:

$$\begin{aligned}\text{co-current}: \quad &\Delta T_m = \frac{(T_1''-T_2'')-(T_1'-T_2')}{\ln\left[\frac{(T_1''-T_2'')}{(T_1'-T_2')}\right]};\\ \text{counter-current}: \; &\Delta T_m = \frac{(T_1''-T_2')-(T_1'-T_2'')}{\ln\left[\frac{(T_1''-T_2')}{(T_1'-T_2')}\right]}.\end{aligned} \tag{C.26}$$

Simplified, the following notation is also used in the literature

$$\Delta T_m = \frac{\Delta T_o - \Delta T_i}{\ln\left(\frac{\Delta T_o}{\Delta T_i}\right)}, \tag{C.27}$$

where ΔT_i is the temperature difference of the inlet side and ΔT_o that of the outlet side. ΔT_i is therefore $\Delta T_i = \left(T_1' - T_2''\right)$ for the counter-current heat exchanger and $\left(T_1'' - T_2'\right)$ for the co-current heat exchanger, and ΔT_o is $\Delta T_o = \left(T_1'' - T_2'\right)$ for the counter-current heat exchanger and $\Delta T_o = \left(T_1'' - T_2''\right)$ for the cocurrent heat exchanger. Thus, the transferred heat flow $\dot{Q}$ can be recorded at three positions (cf. Fig. C.10), as heat flow,

- released by the medium 1 by $\dot{Q} = \dot{m}_1 c_{p1}\left(T_1' - T_1''\right)$,
- which passes through the heat exchanger surface A given by $\dot{Q} = hA\Delta T_m$ and is
- finally transferred to the medium 2 reading to $\dot{Q} = \dot{m}_2 c_{p2}\left(T_2'' - T_2'\right)$.

A central variable here is the heat transfer coefficient h. It is coupled by the term of thermal resistance R with the heat transfer coefficients h_1 and h_2 as well as the specific molecular thermal conductivity λ:

$$R = \frac{1}{h \cdot A} = \frac{1}{h_1 \cdot A_1} + \frac{l}{k \cdot A} + \frac{1}{h_2 \cdot A_2}. \tag{C.28}$$

In Eq. (C.28), the largest unknowns are the determination of the heat transfer coefficients h_i, since these depend on the flow velocities of the fluids. They must be assumed first. After the heat exchanger has been calculated, a correction calculation is necessary. Since the heat flow in steady state is the same at all three positions, the following couplings can be derived:

$$\dot{m}_1 \cdot c_{p1} \cdot \left(T_1' - T_1''\right) = h \cdot A \cdot \Delta T_m = \dot{m}_2 \cdot c_{p2} \cdot \left(T_2'' - T_2'\right). \tag{C.29}$$

Equation (C.29) is valid only if there are no heat losses, the heat generation in the fluid due to viscous dissipation is neglected, and no work is done on the fluid. It can be transformed, and the quotients expressed as NTU (Number of Transfer Unit, *heat capacity flow* or *dimensionless transfer area* or *dimensionless residence time*) in the following way:

$$\begin{aligned} NTU_1 &= \frac{h \cdot A}{\dot{m}_1 \cdot c_{p1}} = \frac{\left(T_1' - T_1''\right)}{\Delta T_m} = \frac{\Delta T_1}{\Delta T_m}; \\ NTU_2 &= \frac{h \cdot A}{\dot{m}_2 \cdot c_{p2}} = \frac{\left(T_2'' - T_2'\right)}{\Delta T_m} = \frac{\Delta T_2}{\Delta T_m}. \end{aligned} \tag{C.30}$$

In this, the product $(h \cdot A)$ can be understood as heat transfer capacity and energy storage capacity. The interpretation of the numerator is easy to understand, because a large transfer area A and a high heat transfer coefficient h result in a high heat transfer capacity. This is simultaneously associated with a small residence time of the heat-transporting fluid in the component. The quotient NTU_2/NTU_1 is introduced as the heat capacity flow ratio ($R_2 = NTU_1/NTU_2$). On the other hand, a high mass flow rate for a given heat exchanger area will allow only a part of the heat flux to pass. The interpretation as residence time is supported by the argument that the mass flow can also be represented by $\dot{m}_i = m_i/t$, and the time t thus enters the numerator. Furthermore, the representation in the form of the NTU specification can be regarded as a specification for the technical realization, while the quotient of $\Delta T/\Delta T_m$ characterizes the thermal task. Further it is usual to introduce the temperature change of the heat flow by normalization to the largest temperature difference (at the inlet) a value P_1 or P_2 as efficiency, which are defined in the following way:

$$P_1 = \frac{\left(T_1' - T_1''\right)}{\left(T_1' - T_2'\right)} \text{ and } P_2 = \frac{\left(T_2'' - T_2'\right)}{\left(T_1' - T_2'\right)}. \tag{C.31}$$

The quantity P_1 is called the operating characteristic. In the same sense, the mean temperature difference ΔT_m can be normalized to ΔT_m^+:

$$\Delta T_m^+ = \frac{\Delta T_m}{\left(T_1' - T_2'\right)}. \tag{C.32}$$

When determining the maximum transferable heat quantities, the flow routing must be considered. This script is limited to co-current or counter-current heat exchangers. For the temperature difference at the outlet of the heat exchanger results for both designs:

$$\begin{aligned} &\text{co-current}: && \Delta T_{\min} = \left(T_1'' - T_2''\right) \geq 0; \\ &\text{counter-current}: && \Delta T_{\min} = \begin{cases} \left(T_1'' - T_2''\right) \geq 0 \text{ for } & \dot{m}_1 c_{p1} < \dot{m}_2 c_{p2} \\ \left(T_1' - T_2''\right) \geq 0 \text{ for } & \dot{m}_2 c_{p2} < \dot{m}_1 c_{p1} \end{cases}. \end{aligned} \tag{C.33}$$

Thus, from the energy balance, the temperature at the outlet of a co-current heat exchanger reads to:

$$\text{co-current: } T_2'' \leq \frac{\frac{\dot{m}_1 c_{p1}}{\dot{m}_2 c_{p2}} T_1' + T_2'}{\frac{\dot{m}_1 c_{p1}}{\dot{m}_2 c_{p2}} + 1}, \tag{C.34}$$

and for the special case $\dot{m}_1 c_{p1} = \dot{m}_2 c_{p2}$:

$$\text{co-current: } T_2'' \leq \frac{T_1' + T_2'}{2}. \tag{C.35}$$

The corresponding course of the temperature in fluid 1 and 2 along the run length of a co-current heat exchanger is sketched qualitatively in Fig. C.11.

For counter-current flow, the corresponding boundary condition for the fluid outlet temperature depends on the ratio of the heat capacity flows, cf. Fig. C.12. If one compares the maximum exchangeable heat quantities for co-current and counter current flow;

$$\begin{aligned} &\text{co-current}: && \dot{Q} \leq \dot{m}_1 \cdot c_{p1} \cdot \left(T_1' - T_2''\right); \\ &\text{counter-current}: && \begin{aligned} &\dot{Q} \leq \dot{m}_1 \cdot c_{p1} \cdot \left(T_1' - T_2''\right) \text{ for } && \dot{m}_1 c_{p1} < \dot{m}_2 c_{p2} \\ &\dot{Q} \leq \dot{m}_2 \cdot c_{p2} \cdot \left(T_1' - T_2'\right) \text{ for } && \dot{m}_2 c_{p2} < \dot{m}_1 c_{p1} \end{aligned}, \end{aligned} \tag{C.36}$$

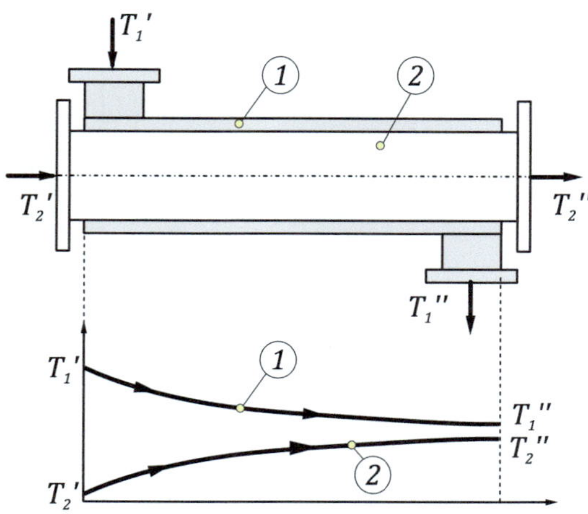

Fig. C.11 Qualitative axial temperature profile in a co-current heat exchanger it becomes clear that a heat exchanger with counter current flow of the fluids can transfer more heat than one in a co-current configuration.

The total amount of heat transferred in the heat exchanger from fluid 1 to fluid 2 results from the integration of the local values. For this purpose, the planar system shown as illustrated in Fig. C.13 is to be considered. For a planar system, assuming an average temperature in each slice of fluids 1 and 2, the simplified result for the differential heat flow is as follows

$$d\dot{Q} = \frac{1}{\frac{1}{h_1} + \frac{t_W}{k} + \frac{1}{h_2}} \cdot \left(\overline{T}_1 - \overline{T}_2\right) 2\pi \cdot r_o\, dz = \frac{1}{\frac{1}{h_1} + \frac{t_W}{k} + \frac{1}{h_2}} \cdot (T_1 - T_2)\, dA_o, \tag{C.37}$$

with $2\pi r_o dz = 2\pi d_o dz = dA_o$. This can be summarized as follows:

$$d\dot{Q} = h \cdot (T_1 - T_2)\, dA_o. \tag{C.38}$$

For a thick-walled tube ($t_W/d_o \sim O(1)$) the following result is obtained:

$$d\dot{Q} = \frac{1}{\frac{1}{h_1} \cdot \frac{d_o}{d_i} + \frac{d_o}{2k} \ln\left(\frac{d_o}{d_i}\right) + \frac{1}{h_2}} \cdot (T_1 - T_2)\, dA_o = k \cdot (T_1 - T_2)\, dA_o. \tag{C.39}$$

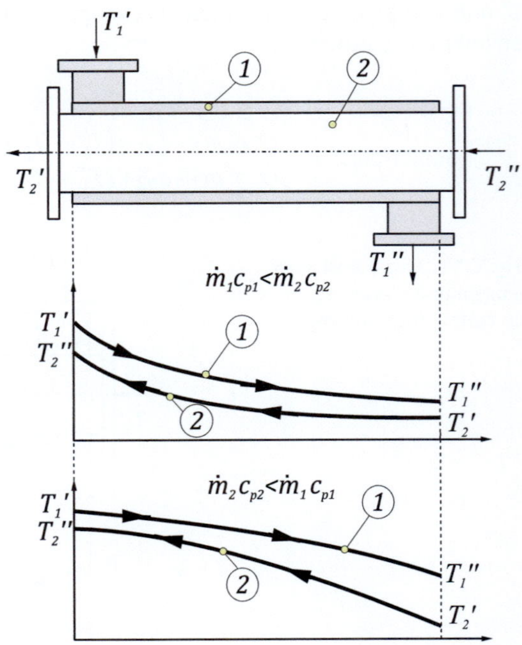

Fig. C.12 Qualitative axial temperature profile in a counterflow heat exchanger

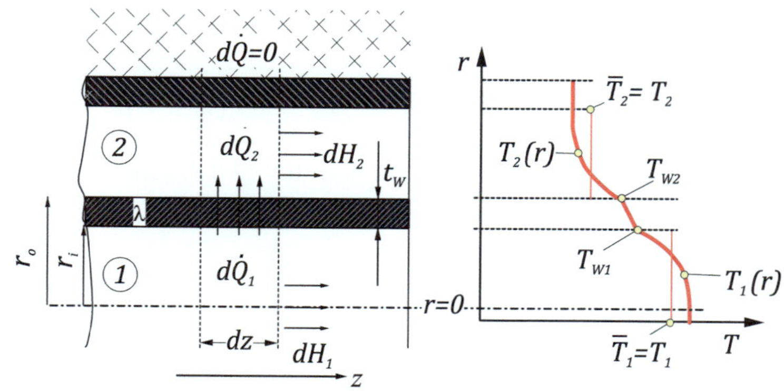

Fig. C.13 Sketch for calculating local temperatures in a heat exchanger

The respective heat transfer coefficients h_i are obtained from characteristic functions of the type $Nu = C Re^m Pr^n$, see Chap. 5. The balance of the heat flux at a control volume of fluid 1 gives:

$$d\dot{Q}_1 = \dot{q}_1 \cdot \pi \cdot d_i \cdot dz = d\dot{H}_1 = \dot{m}_1 \cdot c_{p1} \cdot dT_1. \tag{C.40}$$

and respectively applies to the control volume of the fluid 2:

$$d\dot{Q}_2 = \dot{q}_2 \cdot \pi \cdot d_i \cdot dz = d\dot{H}_2 = \dot{m}_2 \cdot c_{p2} \cdot dT_2. \tag{C.41}$$

In the steady state, the heat leaving fluid 1 is equal to that entering fluid 2, so that $d\dot{Q}_1 = -d\dot{Q}_2 = -d\dot{Q}$; as a consequence, we obtain:

$$d\dot{Q} = \dot{m}_2 \cdot c_{p2} \cdot dT_2 = -\dot{m}_1 \cdot c_{p1} \cdot dT_1 = h \cdot \pi \cdot d_o \cdot (T_1 - T_2) \cdot dz. \tag{C.42}$$

This results in two differential equations:

$$\frac{dT_1}{dz} = -\frac{h \cdot \pi \cdot d_o}{\dot{m}_1 \cdot c_{p1}} \cdot (T_1 - T_2); \quad \frac{dT_2}{dz} = \frac{h \cdot \pi \cdot d_o}{\dot{m}_2 \cdot c_{p2}} \cdot (T_1 - T_2). \tag{C.43}$$

The subtraction of the two equations leads to:

$$\frac{d(T_1 - T_2)}{dz} = -h \cdot \pi \cdot d_o \cdot \left(\frac{1}{\dot{m}_1 \cdot c_{p1}} + \frac{1}{\dot{m}_2 \cdot c_{p2}} \right) \cdot (T_1 - T_2). \tag{C.44}$$

Assuming that the heat transfer coefficient h is known and, in addition, independent of the local temperature difference, it is possible to integrate over the length L of the heat exchanger for a co-current heat exchanger, for example, and obtain:

$$\ln \left(\frac{(T_1'' - T_2'')}{(T_1' - T_2')} \right) = -h \cdot \pi \cdot d_o \cdot \left(\frac{1}{\dot{m}_1 \cdot c_{p1}} + \frac{1}{\dot{m}_2 \cdot c_{p2}} \right) \cdot L. \tag{C.45}$$

This relation can be transformed by introducing the total balance for the two fluids:

$$\dot{Q} = \dot{m}_2 \cdot c_{p2} \cdot \left(T_2'' - T_2'\right) = -\dot{m}_1 \cdot c_{p1} \cdot \left(T_1'' - T_1'\right), \tag{C.46}$$

and the following result is obtained for the heat flux $\dot{Q}$ transferred:

$$\dot{Q} = -k \cdot A_o \cdot \left[\frac{\left(T_1'' - T_2''\right) - \left(T_1' - T_2'\right)}{\ln\left(\frac{\left(T_1'' - T_2''\right)}{\left(T_1' - T_2'\right)}\right)} \right]. \tag{C.47}$$

If we compare this equation with the approach valid for a constant temperature difference between the fluids,

$$\dot{Q} = h \cdot A_o \cdot \Delta T \text{and} \Delta T = (T_1 - T_2), \tag{C.48}$$

so, can be written in general

$$\dot{Q} = h \cdot A_o \cdot \Delta T_m. \tag{C.49}$$

Here, $\cdot \Delta T_m$ is the average driving temperature difference termed in process engineering as log-mean temperature difference or $LMTD$. In the special case of nearly constant fluid temperatures is:

$$\Delta T_m = (T_1 - T_2) \text{for } T_1, T_2 = const., \tag{C.50}$$

or as derived above in the case of a co-current heat exchanger:

$$\Delta T_m = \frac{\Delta T_o - \Delta T_i}{\ln\left(\frac{\Delta T_o}{\Delta T_{oi}}\right)}. \tag{C.51}$$

The result corresponds to that of Eq. (C.27).

C.3.3 Dimensionless Numbers and Operating Characteristics of Heat Exchangers

Heat exchangers can be described similarly as fluid flows with the aid of characteristic dimensionless numbers. If the amount of heat exchanged in an apparatus is related to the maximum amount of heat that can be exchanged under the given inlet conditions of the fluids, the so-called operating characteristic F is obtained, which in some plants is also referred to as the exchanger efficiency, degree of heating or degree of cooling.

$$F = \frac{\dot{Q}}{\dot{Q}_{max}} = \frac{\left(T_1' - T_1''\right)}{\left(T_1' - T_2'\right)}. \tag{C.52}$$

The operating characteristic F corresponds to the value P_1 as derived in Eq. (C.31). The maximum amount of heat can be transferred by an infinitely large counter-current heat exchanger, for the determination of which it is necessary to distinguish two cases:

- Case 1: Counter-current heat exchanger with $\dot{m}_1 \cdot c_{p1} < \dot{m}_2 \cdot c_{p2}$

  For the case of an infinitely long counter-current exchanger, the outlet temperature of fluid 1 $T''_{1\infty}$ is almost equal to the inlet temperature of fluid 2 T'_2. This results in a maximum transmittable heat flow of:

$$\dot{Q}_{max} = \dot{m}_1 \cdot c_{p1} \cdot \left(T'_1 - T'_2\right). \tag{C.53}$$

  In reality, a heat exchanger has a finite length and thus the transferred heat flow is lower; one then obtains temperature profile as qualitatively shown in Fig. C.14a.

- Case 2: Counter-current heat exchanger with $\dot{m}_1 \cdot c_{p1} > \dot{m}_2 \cdot c_{p2}$

  For a large heat capacity flow ratio of fluid 1 with respect to fluid 2, the outlet temperature of fluid 2 $T''_{2\infty}$ tends toward the inlet temperature T'_1 for very long exchangers. The maximum heat that can be transferred is thus:

$$\dot{Q}_{max} = \dot{m}_2 \cdot c_{p2} \cdot \left(T'_1 - T'_2\right). \tag{C.54}$$

  Qualitatively, the temperature curve for this case is as sketched in Fig. C.14b.

This gives the following general values for the maximum transmissible heat flux:

$$\dot{Q}_{max} = \left(\dot{m} \cdot c_p\right)_{min} \cdot \left(T'_1 - T'_2\right), \tag{C.55}$$

and as a consequence, follows for the operating characteristic F the relation:

$$F = \frac{\dot{Q}}{\dot{Q}_{max}} = \frac{\dot{m}_1 \cdot c_{p1} \cdot \left(T'_1 - T''_1\right)}{\left(\dot{m} \cdot c_p\right)_{min} \cdot \left(T'_1 - T'_2\right)} = \frac{\dot{m}_2 \cdot c_{p2} \cdot \left(T''_2 - T'_2\right)}{\left(\dot{m} \cdot c_p\right)_{min} \cdot \left(T'_1 - T'_2\right)}. \tag{C.56}$$

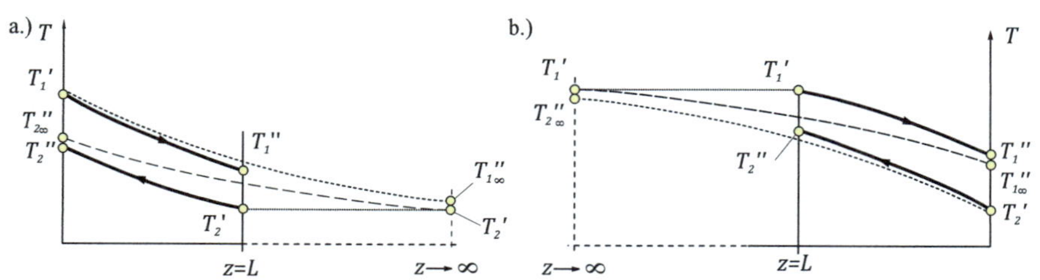

Fig. C.14 Qualitative temperature profile of a counter-current heat exchanger with **a** $\dot{m}_1 \cdot c_{p1} < \dot{m}_2 \cdot c_{p2}$; **b** $\dot{m}_1 \cdot c_{p1} > \dot{m}_2 \cdot c_{p2}$

If we now agree that 1 denotes the fluid with the smaller heat capacity flow, i.e. that the ratio of the heat capacity flows is in the range:

$$0 < \frac{\dot{m}_1 \cdot c_{p1}}{\dot{m}_2 \cdot c_{p2}} < 1, \tag{C.57}$$

the following results for the operating characteristic:

$$F = \frac{\left(T_1' - T_1''\right)}{\left(T_1' - T_2'\right)} = P_1. \tag{C.58}$$

The operating characteristic F depends on the heat transfer conditions, the flow configuration and the heat capacity currents. These influencing variables can be summarized in the form of two characteristics, so that $F = f(\kappa, C^*)$. The first one characterizes the ratio of heat transfer to heat capacity current and is called heat transfer characteristic κ:

$$\kappa = \frac{h \cdot A}{\dot{m}_i \cdot c_{pi}} = NTU_i. \tag{C.59}$$

and corresponds to the definition of the Number of Transfer Units (NTU). The second parameter is the ratio of heat capacity currents C^*, which is defined as follows:

$$C^* = \frac{\dot{m}_1 \cdot c_{p1}}{\dot{m}_2 \cdot c_{p2}} \text{ with } 0 \leq C^* \leq 1. \tag{C.60}$$

This allows us to analyse different heat exchangers

- Small heat exchangers ($\kappa \rightarrow 0$). If small heat exchangers are considered, the heat transfer coefficient κ tends towards 0 due to the small surface area. This means that the fluid temperatures change only slightly and thus the heat flow is given by:

$$\dot{Q} = F \cdot \dot{m}_1 \cdot c_{p1} \cdot \left(T_1' - T_2'\right), \tag{C.61}$$

the result is immediate:

$$F = \frac{h \cdot A}{\dot{m}_1 \cdot c_{p1}} \cdot \frac{\left(T_1' - T_2'\right)}{\left(T_1' - T_2'\right)} = \kappa. \tag{C.62}$$

This result is independent of the fluid flow configuration (i.e. it is valid for both co-current and counter-current heat exchangers). The derivation of the operating characteristic according to the heat transfer coefficient κ provides:

$$\left(\frac{dF}{d\kappa}\right)_{\kappa \rightarrow 0} = 1 = \frac{\Delta T_m}{\left(T_1' - T_2'\right)}. \tag{C.63}$$

If, in this case, the values of the operating characteristic F are plotted against the heat transfer coefficient κ in the operating characteristic diagram, an original straight line with a slope of 1 is obtained.

- Heat capacity current of one stream much larger than that of the other ($C^* \to 0$). If $\dot{m}_1 \cdot c_{p1} \ll \dot{m}_2 \cdot c_{p2}$ is assumed, then Eq. (C.44) reduces to:

$$\frac{d(T_1 - T_2)}{(T_1 - T_2)} = \cdot - \frac{h}{\dot{m}_1 \cdot c_{p1}} \cdot dA, \tag{C.64}$$

and at the same time the temperature of the fluid 2 also changes only slightly, so that $T_2 \approx T_2' \approx T_2''$. If, in addition, the heat transfer coefficient h is constant, the integration of Eq. (C.64) yields:

$$\ln\left(\frac{T_1'' - T_2}{T_1' - T_2}\right) = -\kappa \text{or} \left(\frac{T_1'' - T_2}{T_1' - T_2}\right) = \exp(-\kappa). \tag{C.65}$$

If we introduce the relationship derived above for the operating characteristic F (Eq. (C.58)), we obtain the following result:

$$F = 1 - \exp(-\kappa) \text{ for } C^* \to 0, \tag{C.66}$$

which is valid for all flow configurations. For very large heat exchangers, where the heat transfer characteristic κ becomes very large ($\kappa \gg 1$), $F = 1$ results immediately.

In cases where, in addition, the heat transfer coefficient is determined only by a transition thermal resistance, as is often the case with condenser and evaporator tubes or also with gas/liquid heat exchangers, further simplification can be made and it follows:

$$\kappa = \frac{4L}{d} \cdot \frac{Nu}{Re \cdot Pr} \frac{h \cdot A}{\dot{m}_1 \cdot c_{p1}} \approx \frac{h_i \cdot \pi \cdot d \cdot L}{(\rho \cdot u_m \cdot c_p) \cdot \frac{\pi}{4} \cdot d^2} = \frac{4 \cdot h_i}{(\rho \cdot u_m \cdot c_p)} \cdot \frac{L}{d}, \tag{C.67}$$

By introducing the Nusselt number Nu ($Nu = h_i d/\lambda$), the hydraulic Reynolds number Re ($Re = u_m d/\nu$) and the molecular Prandtl number Pr ($Pr = (\nu \rho c_p)/\lambda$), the expression for the heat transfer parameter can be further simplified. The result is:

$$\kappa = \frac{4 \cdot L}{d} \cdot \frac{Nu}{Re \cdot Pr} = \frac{4 \cdot L}{d} \cdot St, \tag{C.68}$$

in which St denotes the Stanton-number for the heat transfer. Applying this relationship to the operating characteristic, we obtain:

$$F = 1 - \exp\left(-\frac{4 \cdot L}{d} \cdot St\right) \text{ for } C^* \to 0 \text{ and } h \approx h_i. \tag{C.69}$$

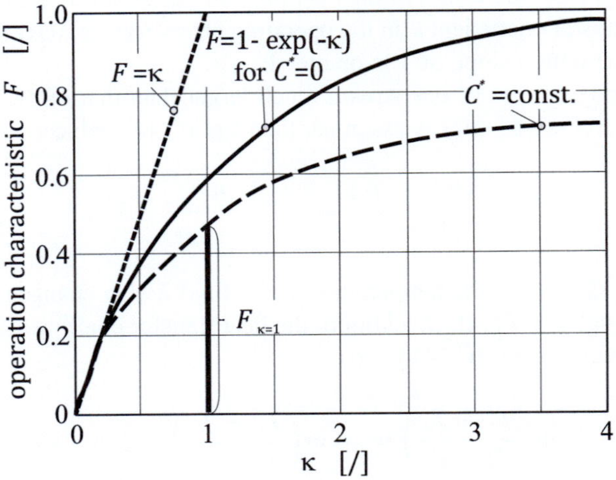

Fig. C.15 Typical operating characteristics diagram as a function of the heat transfer parameter for a given heat capacity flux ratio C^*

If general heat exchangers are considered, the values of the operating characteristics found experimentally or numerically are shown in so-called operating characteristic diagrams $F = F(\kappa, C^*)$ for a given fluid flow path. Such an operating characteristic diagram is shown in Fig. C.15, in which the two special cases for $\kappa \to 0$ and $C^* = 0$ valid for all fluid flow paths are drawn.

The course of the lines of constant values for the ratio of the heat capacity fluxes $0 \leq C^* \leq 1$ depends on the fluid flow configuration. For simple fluid flow conduction, e.g. co-current, counter-current or cross-current set-ups, the operating characteristic curve can be derived theoretically or analytically. For more complicated flow configurations, this curve must be determined experimentally or numerically. If the operating characteristic diagram is available for a heat exchanger, the operating point of the apparatus can be entered and the value of the operating characteristic determined. This makes it very easy to determine the amount of heat exchanged via

$$\dot{Q} = F \cdot \dot{Q}_{max} = F \cdot \dot{m}_1 \cdot c_{p1} \cdot \left(T_1' - T_2'\right), \tag{C.70}$$

or the mean driving temperature difference (from Eq. (C.49)) can be extracted:

$$\Delta T_m = F_{\kappa=1} \cdot \left(T_1' - T_2'\right). \tag{C.71}$$

C.3.4 Operating Characteristics of Co-current and Counter-Current Heat Exchangers

After formulating the differential balance equations of the heat flows in fluid 1 and 2, an integration of the local heat transfer from the inlet to element dz assuming a constant heat transfer coefficient h leads to the following relationship (compare Fig. C.16):

$$\ln\left(\frac{T_1(z) - T_2(z)}{(T_1' - T_2')}\right) = -h \cdot A_z \cdot \left(\frac{1}{\dot{m}_1 \cdot c_{p1}} + \frac{1}{\dot{m}_2 \cdot c_{p2}}\right). \tag{C.72}$$

This equation can be further transformed and we obtain:

$$\frac{(T_1(z) - T_2(z))}{(T_1' - T_2')} = -\exp(-\kappa_z \cdot (1 + C^*)) \text{ with } \kappa_z = \frac{h \cdot A_z}{\dot{m}_1 \cdot c_{p1}}. \tag{C.73}$$

If we introduce into this relation the operating characteristic at position z,

$$F(z) = \frac{(T_1' - T_1(z))}{(T_1' - T_2')}, \tag{C.74}$$

and we note that the heat flux $\dot{Q}_z$ exchanged up to the point z is,

$$\dot{Q}_z = \dot{m}_1 \cdot c_{p1} \cdot (T_1' - T_1(z)) = -\dot{m}_2 \cdot c_{p2} \cdot (T_2' - T_2(z)), \tag{C.75}$$

then, after some transformations, one obtains the operating characteristic of the section of the apparatus with:

$$F(z) = \frac{1 - \exp(-\kappa_z \cdot (1 + C^*))}{(1 + C^*)}. \tag{C.76}$$

For the operating characteristics of the overall apparatus of a co-current heat exchanger, the following results:

$$F = \frac{1 - \exp(-\kappa \cdot (1 + C^*))}{(1 + C^*)}, \tag{C.77}$$

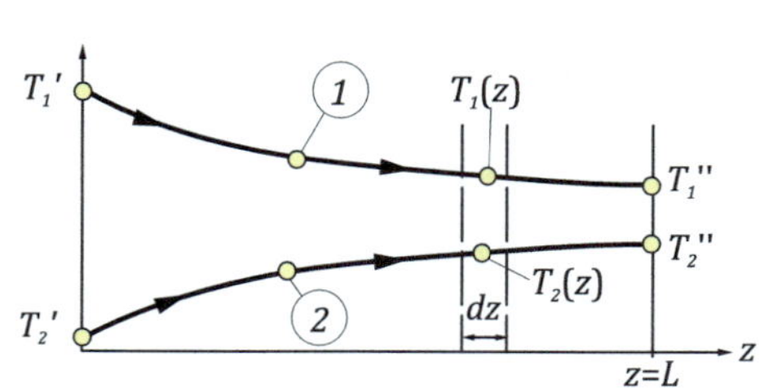

Fig. C.16 Schematic progression of temperatures in a co-current heat exchanger

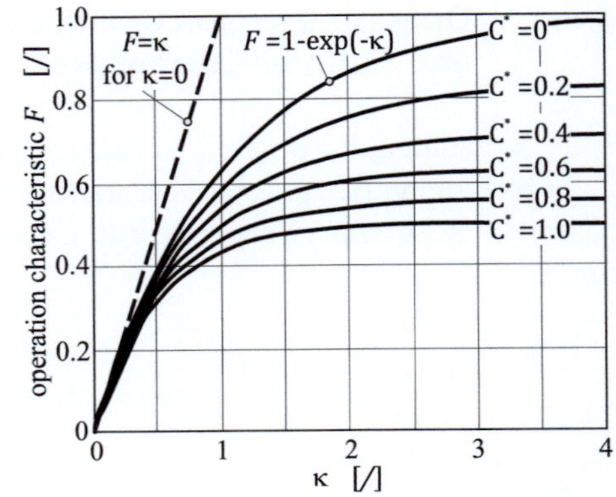

Fig. C.17 Operating characteristic F of a co-current heat exchanger as a function of κ at various C^*

Also, the operating characteristic F of the co-current heat exchanger is a function of the heat transfer characteristic κ and that of the heat capacity current ratio C^*, as depicted in Fig. C.17.

Again, several special cases may be discussed:

- $C^* \to 0$. When the heat capacity flow ratio disappears, the operating characteristic quickly tends towards the value 1, i.e. fluid flow 1 has almost the temperature of fluid 2 at the end of the heat exchanger, especially when the heat transfer is very good. It is then:

$$F = 1 - \exp(-\kappa). \tag{C.78}$$

- $C^* = 1$. For co-current heat exchangers with $C^* = 1$, which are technically more common in solar thermal applications, the result is

$$F = \frac{1 - \exp(-2\kappa)}{2}. \tag{C.79}$$

  If the heat transfer in these heat exchangers is very good, i.e. $\kappa \to \infty$, we obtain $F = 0.5$. In other words, at the outlet of the heat exchanger the initial temperature difference has halved.
- $\kappa \to \infty$. If the heat transfer between the two media across the interface is high, Eq. (C.77) reduces to:

$$F = \frac{1}{(1 + C^*)}. \tag{C.80}$$

  This means that with a high heat transfer (large Nusselt numbers), only the heat capacity flow ratio plays an excellent role.

An analogous procedure yields the following result for the counter-current heat exchanger:

$$\ln\left(\frac{T_1(z) - T_2(z)}{(T_1' - T_2')}\right) = -h \cdot A_z \cdot \left(\frac{1}{\dot{m}_1 \cdot c_{p1}} + \frac{1}{\dot{m}_2 \cdot c_{p2}}\right), \tag{C.81}$$

and after several transformations, the following expression is obtained for the temperature of the counterflow of the partial heat exchanger at the point z:

$$\frac{(T_1(z) - T_2(z))}{(T_1' - T_2'')} = -\exp(-\kappa_z \cdot (1 - C^*)) \text{ with } \kappa_z = \frac{h \cdot A_z}{\dot{m}_1 \cdot c_{p1}}, \tag{C.82}$$

where a constant heat transfer coefficient h along z was also assumed here. The temperature difference of the entire heat exchanger reads then to:

$$\frac{(T_1'' - T_2')}{(T_1' - T_2'')} = -\exp(-\kappa_z \cdot (1 - C^*)). \tag{C.83}$$

If we introduce into this relation the operating characteristic F of the total apparatus and note that $(T_2'' - T_2') = C^* \cdot (T_1' - T_1'')$, the final result for the countercurrent heat exchanger is:

$$F = \frac{1 - \exp(-\kappa_z \cdot (1 - C^*))}{1 - C^* \cdot \exp(-\kappa_z \cdot (1 - C^*))}. \tag{C.84}$$

Again, several limiting cases can be discussed:

- $C^* \to 0$. When the heat capacity current ratio disappears, the functional relationship of the operating characteristic reduces to the following relation:

$$F = 1 - \exp(-\kappa). \tag{C.85}$$

- $\kappa \to \infty$ (for all C^*). If the heat transfer is excellent, the heat capacity ratio is no longer important and the operating characteristic has the value $F = 1$. This means the temperature at the outlet of fluid 1 has assumed the value of the inlet temperature of fluid 2.
- $C^* \to 1$. If both heat capacity currents correspond to each other, Eq. (C.84) reduces to:

$$F = \frac{\kappa}{(\kappa + 1)}. \tag{C.86}$$

If the relationship $F = \kappa_z/(\kappa + 1)$ also applies to the local variables, a linear temperature curve over the heat exchanger surface results.

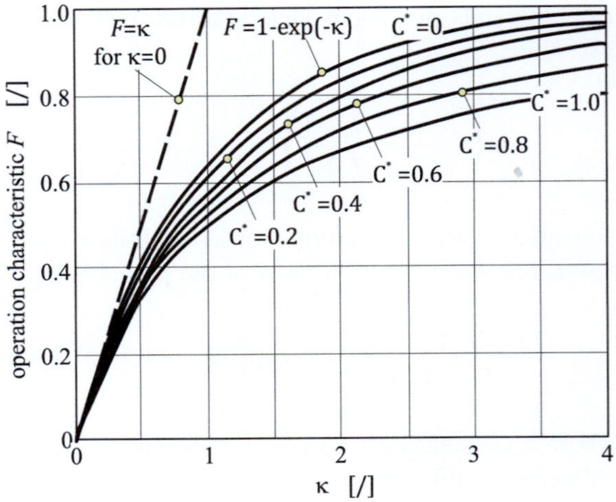

Fig. C.18 Operating characteristic F of a counter-current heat exchanger as a function of κ at different C^*

Figure C.18 shows the operating characteristic diagram of a counter-current heat exchanger as a function of the heat transfer parameter at different heat capacity currents. A comparison of the operating characteristics of the counter-current heat exchanger with that of a co-current heat exchanger (cf. Fig. C.17) shows that the heat transfer in the counter-current principle is considerably more efficient.

C.3.5 Alternative Design Approach of Heat Exchangers

The starting point of an alternative method for the design of heat exchangers, as shown among others in Kreith and Goswami (2005) and which, depending on the task, may have advantages or disadvantages compared to the traditional method presented in the previous section, are the two energy balance equations for the fluids and the approach for the heat transfer.

$$\dot{Q} = \dot{m}_1 \cdot c_{p1} \cdot \left(T_1'' - T_1'\right) = \dot{m}_2 \cdot c_{p2} \cdot \left(T_2' - T_2''\right) = h \cdot A \cdot \Delta T_m. \quad \text{(C.87)}$$

Following this, dimensionless temperature changes of the material flows P_i are defined in the following way:

$$P_1 = \frac{\left(T_1' - T_1''\right)}{\left(T_1' - T_2'\right)} \quad \text{and} \quad P_2 = \frac{\left(T_2'' - T_2'\right)}{\left(T_1' - T_2'\right)}. \quad \text{(C.88)}$$

A comparison of Eq. (C.88) with the definition (Eq. (C.58)) shows that the quantity P_1 corresponds to the operating characteristic F. Furthermore, this method uses the Number of Transfer Units (NTU) as input variable, which are defined as follows,

$$NTU_1 = \frac{h \cdot A}{\dot{m}_1 \cdot c_{p1}} \text{ and } NTU_2 = \frac{h \cdot A}{\dot{m}_2 \cdot c_{p2}}, \quad \text{(C.89)}$$

and the ratio of the heat capacity currents R_i in the form:

$$R_1 = \frac{\dot{m}_1 \cdot c_{p1}}{\dot{m}_2 \cdot c_{p2}} \quad \text{and} \quad R_2 = \frac{\dot{m}_2 \cdot c_{p2}}{\dot{m}_1 \cdot c_{p1}}. \tag{C.90}$$

Obviously, the quantity R_1 corresponds to the heat capacity current ratio C^* described in the other method. Furthermore, one introduces a dimensionless mean temperature difference Θ:

$$\Theta = \frac{\Delta T_m}{(T_1' - T_2')}, \tag{C.91}$$

and additionally, a correction factor f for the determination of ΔT_m, which is defined as follows:

$$\Delta T_m = f \cdot \Delta T_{ln,counter-current} = \frac{P_1}{NTU_1} \cdot (T_1' - T_2'). \tag{C.92}$$

These parameters result in the following relationships:

$$\frac{P_1}{P_2} = \frac{NTU_1}{NTU_2} = \frac{1}{R_1} = R_2 \quad \text{and} \quad \Theta = \frac{P_1}{NTU_1} = \frac{P_2}{NTU_2}. \tag{C.93}$$

For the example of a simple co-current heat exchanger, Fig. C.19a, b shows a comparison of an operating characteristic diagram (F, κ diagram) with that of the alternative method (P_1, P_2 diagram) with an entered operating point.

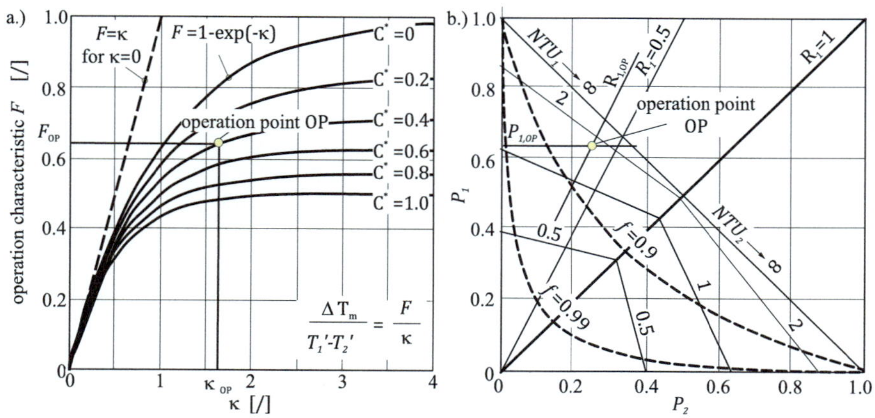

Fig. C.19 **a** Operating characteristic diagram of a DC heat exchanger and corresponding representation in the P_1, P_2 diagram (**b**)

Reference

Kreith, F., & Goswami, D. Y. (2005). *The CRC handbook of mechanical engineering* (2nd ed.). CRC-Press LLC.

Index

R. Stieglitz and W. Platzer, *Solar Thermal Energy Systems*,
https://doi.org/10.1007/978-3-031-43173-9

C

D

G

Q

R

S

U

MIX
Papier aus verantwortungsvollen Quellen
Paper from responsible sources
FSC® C105338

If you have any concerns about our products,
you can contact us on
ProductSafety@springernature.com

In case Publisher is established outside the EU,
the EU authorized representative is:
Springer Nature Customer Service Center GmbH
Europaplatz 3, 69115 Heidelberg, Germany

Printed by Libri Plureos GmbH
in Hamburg, Germany